About Pearson

Pearson is the world's learning company, with presence across 70 countries worldwide. Our unique insights and world-class expertise comes from a long history of working closely with renowned teachers, authors and thought leaders, as a result of which, we have emerged as the preferred choice for millions of teachers and learners across the world.

We believe learning opens up opportunities, creates fulfilling careers and hence better lives. We hence collaborate with the best of minds to deliver you class-leading products, spread across the Higher Education and Test Preparation spectrum.

Superior learning experience and improved outcomes are at the heart of everything we do. This product is the result of one such effort.

Your feedback plays a critical role in the evolution of our products and you can reachus@pearson.com. We look forward to it.

THOMAS'

CALCULUS

FIFTEENTH EDITION

Based on the original work by

GEORGE B. THOMAS, JR.
Massachusetts Institute of Technology

as revised by

JOEL HASS
University of California, Davis

CHRISTOPHER HEIL
Georgia Institute of Technology

PRZEMYSLAW BOGACKI
Old Dominion University

MAURICE D. WEIR
Naval Postgraduate School

Pearson

Acknowledgments of third-party content appear on page C-1, which constitutes an extension of this copyright page.

Authorized adaptation from the United States edition, entitled *Thomas' Calculus,* 15th Edition by George B. Thomas, Jr., published by Pearson Education Inc. © 2023.

Indian Subcontinent Adaptation

ISBN 978-81-198-9660-8

First Impression, 2024
Fourth Impression, 2026
Fifth Impression

Published by Pearson India Education Services Pvt. Ltd, CIN: U72200TN2005PTC057128.

Head Office: 1st Floor, Berger Tower, Plot No. C-001A/2, Sector 16B, Noida - 201 301, Uttar Pradesh, India.
Registered Office: 6th Floor, Tower A, Unit A and B, International Tech Park, CapitaLand, Chennai, 200 Feet Radial Road, Zamin Pallavaram, Old Pallavaram, Chennai – 600117, Tamil Nadu India

Website: in.pearson.com, Email: companysecretary.india@pearson.com

Printed in India at: Rahul Print O Pack

Contents

Preface ix

1

Functions 1

1.1 Functions and Their Graphs 1
1.2 Combining Functions; Shifting and Scaling Graphs 14
1.3 Trigonometric Functions 21
1.4 Graphing with Software 29
Questions to Guide Your Review 33
Practice Exercises 34
Additional and Advanced Exercises 35
Technology Application Projects 37

2

Limits and Continuity 38

2.1 Rates of Change and Tangent Lines to Curves 38
2.2 Limit of a Function and Limit Laws 45
2.3 The Precise Definition of a Limit 56
2.4 One-Sided Limits 65
2.5 Continuity 72
2.6 Limits Involving Infinity; Asymptotes of Graphs 83
Questions to Guide Your Review 96
Practice Exercises 97
Additional and Advanced Exercises 99
Technology Application Projects 101

3

Derivatives 102

3.1 Tangent Lines and the Derivative at a Point 102
3.2 The Derivative as a Function 106
3.3 Differentiation Rules 115
3.4 The Derivative as a Rate of Change 124
3.5 Derivatives of Trigonometric Functions 133
3.6 The Chain Rule 139
3.7 Implicit Differentiation 147
3.8 Related Rates 152
3.9 Linearization and Differentials 160
Questions to Guide Your Review 172
Practice Exercises 173
Additional and Advanced Exercises 176
Technology Application Projects 179

4

Applications of Derivatives 180

4.1 Extreme Values of Functions on Closed Intervals 180
4.2 The Mean Value Theorem 188

4.3 Monotonic Functions and the First Derivative Test 195
4.4 Concavity and Curve Sketching 200
4.5 Applied Optimization 213
4.6 Newton's Method 225
4.7 Antiderivatives 230
Questions to Guide Your Review 239
Practice Exercises 240
Additional and Advanced Exercises 242
Technology Application Projects 245

5 Integrals 246

5.1 Area and Estimating with Finite Sums 246
5.2 Sigma Notation and Limits of Finite Sums 256
5.3 The Definite Integral 263
5.4 The Fundamental Theorem of Calculus 276
5.5 Indefinite Integrals and the Substitution Method 287
5.6 Definite Integral Substitutions and the Area Between Curves 294
Questions to Guide Your Review 304
Practice Exercises 305
Additional and Advanced Exercises 307
Technology Application Projects 310

6 Applications of Definite Integrals 311

6.1 Volumes Using Cross-Sections 311
6.2 Volumes Using Cylindrical Shells 322
6.3 Arc Length 329
6.4 Areas of Surfaces of Revolution 335
6.5 Work and Fluid Forces 340
6.6 Moments and Centers of Mass 350
Questions to Guide Your Review 361
Practice Exercises 362
Additional and Advanced Exercises 364
Technology Application Projects 365

7 Transcendental Functions 366

7.1 Inverse Functions and Their Derivatives 366
7.2 Natural Logarithms 377
7.3 Exponential Functions 386
7.4 Exponential Change and Separable Differential Equations 399
7.5 Indeterminate Forms and L'Hôpital's Rule 411
7.6 Inverse Trigonometric Functions 420
7.7 Hyperbolic Functions 435
7.8 Relative Rates of Growth 445
Questions to Guide Your Review 451
Practice Exercises 451
Additional and Advanced Exercises 455

8 Techniques of Integration 457

8.1 Using Basic Integration Formulas 457
8.2 Integration by Parts 462
8.3 Trigonometric Integrals 470
8.4 Trigonometric Substitutions 476
8.5 Integration of Rational Functions by Partial Fractions 481
8.6 Integral Tables and Computer Algebra Systems 488
8.7 Numerical Integration 494
8.8 Improper Integrals 504
8.9 Probability 515
Questions to Guide Your Review 528
Practice Exercises 529
Additional and Advanced Exercises 531
Technology Application Projects 534

9 First-Order Differential Equations 535

9.1 Solutions, Slope Fields, and Euler's Method 535
9.2 First-Order Linear Equations 543
9.3 Applications 549
9.4 Graphical Solutions of Autonomous Equations 555
9.5 Systems of Equations and Phase Planes 564
Questions to Guide Your Review 570
Practice Exercises 570
Additional and Advanced Exercises 572
Technology Application Projects 572

10 Infinite Sequences and Series 573

10.1 Sequences 573
10.2 Infinite Series 586
10.3 The Integral Test 596
10.4 Comparison Tests 602
10.5 Absolute Convergence; The Ratio and Root Tests 607
10.6 Alternating Series and Conditional Convergence 614
10.7 Power Series 621
10.8 Taylor and Maclaurin Series 632
10.9 Convergence of Taylor Series 637
10.10 Applications of Taylor Series 644
Questions to Guide Your Review 653
Practice Exercises 654
Additional and Advanced Exercises 656
Technology Application Projects 658

11 Parametric Equations and Polar Coordinates 659

11.1 Parametrizations of Plane Curves 659
11.2 Calculus with Parametric Curves 668
11.3 Polar Coordinates 677

11.4 Graphing Polar Coordinate Equations 681
11.5 Areas and Lengths in Polar Coordinates 685
11.6 Conic Sections 690
11.7 Conics in Polar Coordinates 698
Questions to Guide Your Review 704
Practice Exercises 705
Additional and Advanced Exercises 707
Technology Application Projects 709

12 Vectors and the Geometry of Space 710

12.1 Three-Dimensional Coordinate Systems 710
12.2 Vectors 715
12.3 The Dot Product 726
12.4 The Cross Product 734
12.5 Lines and Planes in Space 740
12.6 Cylinders and Quadric Surfaces 749
Questions to Guide Your Review 755
Practice Exercises 755
Additional and Advanced Exercises 757
Technology Application Projects 759

13 Vector-Valued Functions and Motion in Space 760

13.1 Curves in Space and Their Tangents 760
13.2 Integrals of Vector Functions; Projectile Motion 769
13.3 Arc Length in Space 778
13.4 Curvature and Normal Vectors of a Curve 782
13.5 Tangential and Normal Components of Acceleration 788
13.6 Velocity and Acceleration in Polar Coordinates 794
Questions to Guide Your Review 797
Practice Exercises 798
Additional and Advanced Exercises 800
Technology Application Projects 801

14 Partial Derivatives 802

14.1 Functions of Several Variables 802
14.2 Limits and Continuity in Higher Dimensions 810
14.3 Partial Derivatives 819
14.4 The Chain Rule 831
14.5 Directional Derivatives and Gradient Vectors 841
14.6 Tangent Planes and Differentials 850
14.7 Extreme Values and Saddle Points 860
14.8 Lagrange Multipliers 869
14.9 Taylor's Formula for Two Variables 879
14.10 Partial Derivatives with Constrained Variables 883
Questions to Guide Your Review 887
Practice Exercises 888
Additional and Advanced Exercises 891
Technology Application Projects 893

15 Multiple Integrals 894

15.1 Double and Iterated Integrals over Rectangles 894
15.2 Double Integrals over General Regions 899
15.3 Area by Double Integration 908
15.4 Double Integrals in Polar Form 911
15.5 Triple Integrals in Rectangular Coordinates 918
15.6 Applications 928
15.7 Triple Integrals in Cylindrical and Spherical Coordinates 938
15.8 Substitutions in Multiple Integrals 950
Questions to Guide Your Review 959
Practice Exercises 960
Additional and Advanced Exercises 962
Technology Application Projects 964

16 Integrals and Vector Fields 965

16.1 Line Integrals of Scalar Functions 965
16.2 Vector Fields and Line Integrals: Work, Circulation, and Flux 972
16.3 Path Independence, Conservative Fields, and Potential Functions 985
16.4 Green's Theorem in the Plane 996
16.5 Surfaces and Area 1008
16.6 Surface Integrals 1018
16.7 Stokes' Theorem 1028
16.8 The Divergence Theorem and a Unified Theory 1041
Questions to Guide Your Review 1053
Practice Exercises 1054
Additional and Advanced Exercises 1056
Technology Application Projects 1058

17 Second-Order Differential Equations (Online - Refer to inside front cover for steps to access)

17.1 Second-Order Linear Equations
17.2 Nonhomogeneous Linear Equations
17.3 Applications
17.4 Euler Equations
17.5 Power-Series Solutions

18 Complex Functions (Online - Refer to inside front cover for steps to access)

18.1 Complex Numbers
18.2 Functions of a Complex Variable
18.3 Derivatives
18.4 The Cauchy-Riemann Equations
18.5 Complex Power Series
18.6 Some Complex Functions
18.7 Conformal Maps
Questions to Guide Your Review
Additional and Advanced Exercises

19 Fourier Series and Wavelets (Online - Refer to inside front cover for steps to access)

19.1 Periodic Functions
19.2 Summing Sines and Cosines
19.3 Vectors and Approximation in Three and More Dimensions
19.4 Approximation of Functions
19.5 Advanced Topic: The Haar System and Wavelets
Questions to Guide Your Review
Additional and Advanced Exercises

Appendix A AP-1

A.1 Real Numbers and the Real Line AP-1
A.2 Mathematical Induction AP-6
A.3 Lines, Circles, and Parabolas AP-9
A.4 Proofs of Limit Theorems AP-19
A.5 Commonly Occurring Limits AP-22
A.6 Theory of the Real Numbers AP-23
A.7 The Distributive Law for Vector Cross Products AP-26
A.8 The Mixed Derivative Theorem and the Increment Theorem AP-27

Appendix B (Online - Refer to inside front cover for steps to access)

B.1 Determinants
B.2 Extreme Values and Saddle Points for Functions of More than Two Variables
B.3 The Method of Gradient Descent

Answers to Odd-Numbered Exercises AN-1

Applications Index AI-1

Subject Index I-1

A Brief Table of Integrals T-1

Credits · C-1

Preface

Thomas' Calculus, Fifteenth Edition, continues its tradition of clarity and precision in calculus with a modern update to the popular text. The authors have worked diligently to add exercises, revise figures and narrative for clarity, and update many applications to modern topics. *Thomas' Calculus* remains a modern and robust introduction to calculus, focusing on developing conceptual understanding of the underlying mathematical ideas. This text supports a calculus sequence typically taken by students in STEM fields over several semesters. Intuitive and precise explanations, thoughtfully chosen examples, superior figures, and time-tested exercise sets are the foundation of this text. We continue to improve this text in keeping with shifts in both the preparation and the goals of today's students, and in the applications of calculus to a changing world.

As Advanced Placement Calculus continues to grow in popularity for high school students, many instructors have communicated mixed reviews of the benefit for today's university and community college students. Some instructors report receiving students with an overconfidence in their computational abilities coupled with underlying gaps in algebra and trigonometry mastery, as well as poor conceptual understanding. In this text, we seek to meet the needs of the increasingly varied population in the calculus sequence. We have taken care to provide enough review material (in the text and appendices), detailed solutions, and a variety of examples and exercises, to support a complete understanding of calculus for students at varying levels. Within the text, we present the material in a way that supports the development of mathematical maturity, going beyond memorizing formulas and routine procedures, and we show students how to generalize key concepts once they are introduced. References are made throughout, tying new concepts to related ones that were studied earlier. After studying calculus from *Thomas,* students will have developed problem-solving and reasoning abilities that will serve them well in many important aspects of their lives. Mastering this beautiful and creative subject, with its many practical applications across so many fields, is its own reward. But the real gifts of studying calculus are acquiring the ability to think logically and precisely; understanding what is defined, what is assumed, and what is deduced; and learning how to generalize conceptually. We intend this book to encourage and support those goals.

New to This Edition

We welcome to this edition a new coauthor, Przemyslaw Bogacki from Old Dominion University. Przemek joined the team for the 4th edition of University Calculus and now joins the *Thomas' Calculus* team. Przemek has also written the new appendix on Optimization covering determinants, extreme values, and gradient descent.

The most significant update to this 15th edition includes new online chapters on Complex Functions, Fourier Series and Wavelets, and the new appendix on Optimization. These chapters can provide material for students interested in more advanced topics. The details are outlined below in the chapter descriptions.

We have also made the following updates:

- Many updated graphics and figures to bring out clear visualization and mathematical correctness.
- Many wording clarifications and revisions.
- Many instruction clarifications for exercises, such as suggesting where the use of a calculator may be needed.
- Notation of inverse trig functions favoring arcsin notation over $\sin^{-1}$, etc.

Content Enhancements

Chapter 1

- Section 1.2. Revised Example 4 to clarify the distinction between vertical and horizontal scaling of a graph.
- Section 1.3. Added new Figure 1.46, illustrating a geometric proof of the angle sum identities.

Chapter 2

- Section 2.2. New Example 11, illustrating the use of the Sandwich Theorem, with corresponding new Figure 2.14.
- Section 2.4. New subsection on "Limits at Endpoints of an Interval" added. New Example 2 added, illustrating limits at a boundary point of an interval.
- Section 2.5. Exercises 37–40 on limits involving trigonometric functions moved from Chapter 3.
- Additional and Advanced Exercises. Exercises 29–38 on limits involving trigonometric functions moved from Chapter 3.

Chapter 3

- Section 3.8. Updated and improved related rates problem strategies, and correspondingly revised Examples 2–6.

Chapter 4

- Section 4.3. Added new Exercises 63–64.
- Section 4.4. Added new Exercises 103–104.
- Section 4.5. Updated and improved strategies for solving applied optimization problems. Added new Exercises 33–34.
- Section 4.7. Added Table 4.4 of integration formulas.

Chapter 5

- Section 5.1. The Midpoint Rule and the associated formula for calculating an integral numerically were given a more central role and used to introduce a numerical method.
- Section 5.3. New basic theory Exercise 89. Integrals of functions that differ at one point.
- Section 5.6. New Exercises 83–84. Compare areas using graphics and computation.

Chapter 6

Section 6.2. Discussion of cylinders in Example 1 clarified.

Chapter 7

- Clarified derivative formulas involving *x* versus those involving a differentiable function *u*.
- Section 7.1. Revised Figure 7.5 illustrating the relationship between slopes of graphs of inverse functions. Added new Exercises 45–48 involving the derivatives of inverse functions.
- Section 7.2. Expanded Example 1.
- Section 7.3. Expanded Example 6 to clarify the computation of the derivative of x^x.
- Section 7.5. Improved the discussion of indeterminate forms. Expanded Example 1. Added Exercises 19–20.
- Section 7.6. Added new Example 5 to illustrate differentiating a composition involving the arctangent function. Rewrote the introduction to the subsection on the derivative of arcsec *x*.

Chapter 8

- Section 8.3. Clarified computing integrals involving powers of sines and cosines. Exercise 42 replaced. Exercises 51 and 52 added.
- Section 8.4. Ordering of exercises was updated.
- Section 8.5. Discussion of the method of partial fractions rewritten and clarified.
- Section 8.7. New subsection on the Midpoint Rule added. Discussion of Error Analysis expanded to include the Midpoint Rule. Exercises 1–10 expanded to include the Midpoint Rule.
- Section 8.8. Discussion of infinite limits of integration clarified. Material on Tests for Convergence and Divergence, including the Direct Comparison Test and the Limit Comparison Test, their proofs, and associated examples, all revised. New Exercises 69–80 added.

Chapter 9

- Section 9.2. Added Figure 9.9.
- Section 9.4. Added a new application of the logistic function showing its connection to Machine Learning and Neural Networks. Added New Exercises 21–22 on the Logistic Equation.

Chapter 10

- Section 10.2. Solution to Example 2 replaced. Solution to Example 8 replaced.
- Section 10.3. Solution to Example 5 revised.
- Section 10.5. Exercise 71 added.
- Section 10.6. Proof of Theorem 15 replaced. Discussion of Theorem 16 revised.

- Section 10.7. Discussion of absolute convergence added to the solution of Example 3. Figure 10.21 revised. New Exercises 40–41 added. Exercise 66 entirely rewritten.
- Section 10.8. Ordering of Exercises was revised. New Exercises 47 and 52 added.
- Section 10.9. Discussion of Taylor series between Examples 4 and 5 rewritten.
- Section 10.10. Exercise 9 replaced.
- Practice Exercises. New Exercises 45–46 added.
- Additional and Advanced Exercises. New Exercises 30–31 added.

Chapter 12

- Section 12.2. New subsection on Vectors in n Dimensions added, with corresponding new Figure 12.19, and new Exercises 60–65.
- Section 12.3. New subsection on The Dot Product of Two n-Dimensional Vectors added, with new Example 9, and new Exercises 53–56.
- Section 12.6. Discussion of cylinders revised.

Chapter 13

Section 13.5. New Exercises 1–2 and 5–6 added.

Chapter 14

- Section 14.2. Added a Composition Rule to Theorem 1 and expanded Example 1.
- Section 14.3. Rewrote the concept of differentiability for functions of several variables to improve clarity.
- Expanded Example 8.
- Section 14.4. Added new Exercises 62–63 on the chain rule with multiple variables.
- Section 14.5. Added a new subsection on gradients for Functions of More Than Three Variables.
- A new Example 7 illustrates a gradient of a 3-variable function.
- New Exercises 45–52 involve gradients of functions with several variables.
- Section 14.7. Added a definition of the Hessian matrix.
- Clarified Example 6.
- Section 14.8. Clarified the use of Lagrange Multipliers throughout, with a more explicit discussion of how to use them for finding maxima and minima.

Chapter 15

- Section 15.2. Added discussion of the properties of limits of iterated double integrals.
- Rewrote Exercises 1–8. Added new Exercises 19–26.
- Section 15.5. Added discussion of the properties of limits of iterated triple integrals. Revised and expanded Example 2.
- Section 15.7. Revised Figure 15.55 to clarify the shape of a spherical wedge involved in triple integration.

Appendices

Rewrote Appendix A.7 to replace the prime notation with the subscript notation.

New Online Appendix B

B.1 Determinants
B.2 Extreme Values and Saddle Points for Functions of More than Two Variables
B.3 The Method of Gradient Descent

This new appendix covers many topics relevant to students interested in Machine Learning and Neural Networks.

New Online Chapter 18—Complex Functions

This new online chapter gives an introduction to complex functions. Section 1 is an introduction to complex numbers and their operations. It replaces Appendix A.7. Section 2 covers limits and continuity for complex functions. Section 3 introduces complex derivatives and Section 4 the Cauchy-Riemann Equations. Section 5 develops the theory of complex series. Section 6 studies the standard functions such as $\sin z$ and $\text{Log } z$, and Section 7 ends the chapter by introducing the theory of conformal maps.

New Online Chapter 19—Fourier Series and Wavelets

This new online chapter introduces Fourier series, and then treats wavelets as a more advanced topic.

It has sections on

19.1 Periodic Functions
19.2 Summing Sines and Cosines
19.3 Vectors and Approximation in Three and More Dimensions
19.4 Approximation of Functions
19.5 Advanced Topic: The Haar System and Wavelets

Continuing Features

Rigor The level of rigor is consistent with that of earlier editions. We continue to distinguish between formal and informal discussions and to point out their differences. Starting with a more intuitive, less formal approach helps students understand a new or difficult concept so they can then appreciate its full mathematical precision and outcomes. We pay attention to defining ideas carefully and to proving theorems appropriate for calculus students, while mentioning deeper or subtler issues they would study in a more advanced course. Our organization and distinctions between informal and formal discussions give the instructor a degree of flexibility in the amount and depth of coverage of the various topics. For example, while we do not prove the Intermediate Value Theorem or the Extreme Value Theorem for continuous functions on a closed finite interval, we do state these theorems precisely, illustrate their meanings in numerous examples, and use them to prove other important results. Furthermore, for those instructors who desire greater depth of coverage, in Appendix A.6 we discuss the reliance of these theorems on the completeness of the real numbers.

Writing Exercises Writing exercises placed throughout the text ask students to explore and explain a variety of calculus concepts and applications. In addition, the end of each chapter contains a list of questions for students to review and summarize what they have learned. Many of these exercises make good writing assignments.

End-of-Chapter Reviews and Projects In addition to problems appearing after each section, each chapter culminates with review questions, practice exercises covering the entire chapter, and a series of Additional and Advanced Exercises with more challenging or synthesizing problems. Most chapters also include descriptions of several **Technology Application Projects** that can be worked by individual students or groups of students over a longer period of time. These projects require the use of *Mathematica* or *Maple*.

Writing and Applications This text continues to be easy to read, conversational, and mathematically rich. Each new topic is motivated by clear, easy-to-understand examples and is then reinforced by its application to real-world problems of immediate interest to students. A hallmark of this book has been the application of calculus to science and engineering. These applied problems have been updated, improved, and extended continually over the last several editions.

Technology In a course using the text, technology can be incorporated according to the taste of the instructor. Each section contains exercises requiring the use of technology; these are marked with a T if suitable for calculator or computer use, or they are labeled **Computer Explorations** if a computer algebra system (CAS, such as *Maple* or *Mathematica*) is required.

Acknowledgments

We would like to express our thanks to the people who made many valuable contributions to this edition as it developed through its various stages.

Accuracy Checkers

Jennifer Blue
Roger Lipsett
Patricia Nelson
Thomas Wegleitner

Reviewers for the Fifteenth Edition

Philip Veer	*Johnson County Community College*
Kent Kast	*Campion Academy*
Richard Rupp	*Del Mar College*
Brian Albright	*Concordia University*
Sara Weiss	*Dallas College*
Shawn Chiappetta	*University of Sioux Falls*
Marshall Ransom	*Georgia Southern University*

We wish to thank the many people at Pearson who have contributed to the success of this book. We appreciate the efforts of the production, design, manufacturing, marketing, and sales departments.

If you have any comments or suggestions, we would like to hear from you.

Joel Hass

Christopher Heil

Przemyslaw Bogacki

The publishers would like to thank Professor P. Sivaramakrishna Das and C. Vijayakumari for their valuable inputs and enhancing the content of this book to suit the requirements of Indian Universities. They are highly accomplished educators in the realm of mathematics, possessing vast experience of 52 and 40 years, instructing students at both the undergraduate and postgraduate levels. They have shared their profound expertise in mathematics across a spectrum of institutions, including Arts, Science, and Engineering Colleges. Their dedicated contributions have played a pivotal role in shaping the academic landscape.

1 Functions

OVERVIEW In this chapter we review what functions are and how they are visualized as graphs, how they are combined and transformed, and ways they can be classified.

1.1 Functions and Their Graphs

Functions are a tool for describing the real world in mathematical terms. A function can be represented by an equation, a graph, a numerical table, or a verbal description; we will use all four representations throughout this text. This section reviews these ideas.

Functions; Domain and Range

The temperature at which water boils depends on the elevation above sea level. The interest paid on a cash investment depends on the length of time the investment is held. The area of a circle depends on the radius of the circle. The distance an object travels depends on the elapsed time.

In each case, the value of one variable quantity, say y, depends on the value of another variable quantity, which we often call x. We say that "y is a function of x" and write this symbolically as

$$y = f(x) \qquad (\text{"}y \text{ equals } f \text{ of } x\text{"}).$$

The symbol f represents the function, the letter x is the **independent variable** representing the input value to f, and y is the **dependent variable** or output value of f at x.

> DEFINITION A **function** f from a set D to a set Y is a rule that assigns a single value $f(x)$ in Y to each x in D.

A rule that assigns more than one value to an input x, such as the rule that assigns to a positive number both the positive and negative square roots of the number, does not describe a function.

The set D of all possible input values is called the **domain** of the function. The domain of f will sometimes be denoted by $D(f)$. The set of all output values $f(x)$ as x varies throughout D is called the **range** of the function. The range might not include every element in the set Y. The domain and range of a function can be any sets of objects, but often in calculus they are sets of real numbers interpreted as points of a coordinate line. (In Chapters 13–16, we will encounter functions for which the elements of the sets are points in the plane, or in space.)

Often a function is given by a formula that describes how to calculate the output value from the input variable. For instance, the equation $A = \pi r^2$ is a rule that calculates the area A of a circle from its radius r. When we define a function f with a formula $y = f(x)$ and the domain is not stated explicitly or restricted by context, the domain is assumed to be the largest set of real x-values for which the formula gives real y-values. This is called the **natural domain** of f. If we want to restrict the domain in some way, we must say so. The domain of $y = x^2$ is the entire set of real numbers. To restrict the domain of the function to, say, positive values of x, we would write "$y = x^2, x > 0$."

Changing the domain to which we apply a formula usually changes the range as well. The range of $y = x^2$ is $[0, \infty)$. The range of $y = x^2, x \geq 2$, is the set of all numbers obtained by squaring numbers greater than or equal to 2. In set notation (see Appendix A.1), the range is $\{x^2 | x \geq 2\}$ or $\{y | y \geq 4\}$ or $[4, \infty)$.

When the range of a function is a set of real numbers, the function is said to be **real-valued**. The domains and ranges of most real-valued functions we consider are intervals or combinations of intervals. Sometimes the range of a function is not easy to find.

FIGURE 1.1 A diagram showing a function as a kind of machine.

A function f is like a machine that produces an output value $f(x)$ in its range whenever we feed it an input value x from its domain (Figure 1.1). The function keys on a calculator give an example of a function as a machine. For instance, whenever you enter a nonnegative number x and press the $\sqrt{x}$ key, the calculator gives an output value (the square root of x).

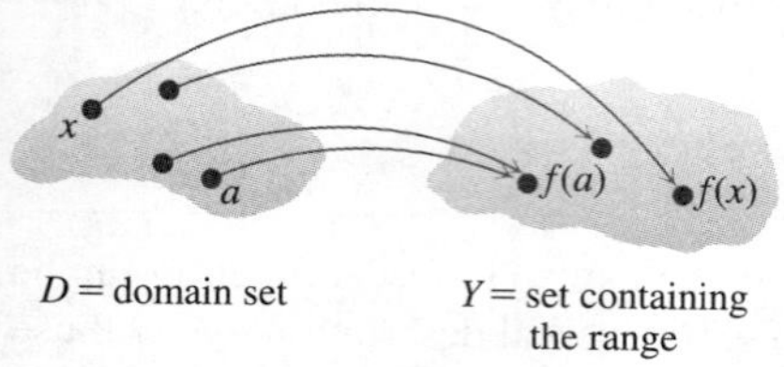

FIGURE 1.2 A function from a set D to a set Y assigns a unique element of Y to each element in D.

A function can also be pictured as an **arrow diagram** (Figure 1.2). Each arrow associates to an element of the domain D a single element in the set Y. In Figure 1.2, the arrows indicate that $f(a)$ is associated with a, $f(x)$ is associated with x, and so on. Notice that a function can have the same *output value* for two different input elements in the domain (as occurs with $f(a)$ in Figure 1.2), but each input element x is assigned a *single* output value $f(x)$.

EXAMPLE 1 Verify the natural domains and associated ranges of some simple functions. The domains in each case are the values of x for which the formula makes sense.

Function	Domain (x)	Range (y)
$y = x^2$	$(-\infty, \infty)$	$[0, \infty)$
$y = 1/x$	$(-\infty, 0) \cup (0, \infty)$	$(-\infty, 0) \cup (0, \infty)$
$y = \sqrt{x}$	$[0, \infty)$	$[0, \infty)$
$y = \sqrt{4 - x}$	$(-\infty, 4]$	$[0, \infty)$
$y = \sqrt{1 - x^2}$	$[-1, 1]$	$[0, 1]$

Solution The formula $y = x^2$ gives a real y-value for any real number x, so the domain is $(-\infty, \infty)$. The range of $y = x^2$ is $[0, \infty)$ because the square of any real number is nonnegative and every nonnegative number y is the square of its own square root: $y = \left(\sqrt{y}\right)^2$.

The formula $y = 1/x$ gives a real y-value for every x except $x = 0$. For consistency in the rules of arithmetic, *we cannot divide any number by zero*. The range of $y = 1/x$, the set of reciprocals of all nonzero real numbers, is the set of all nonzero real numbers, since $y = 1/(1/y)$. That is, for $y \neq 0$ the number $x = 1/y$ is the input that is assigned to the output value y.

The formula $y = \sqrt{x}$ gives a real y-value only if $x \geq 0$. The range of $y = \sqrt{x}$ is $[0, \infty)$ because every nonnegative number is some number's square root (namely, it is the square root of its own square).

In $y = \sqrt{4 - x}$, the quantity $4 - x$ cannot be negative. That is, $4 - x \geq 0$, or $x \leq 4$. The formula gives nonnegative real y-values for all $x \leq 4$. The range of $\sqrt{4 - x}$ is $[0, \infty)$, the set of all nonnegative numbers.

The formula $y = \sqrt{1 - x^2}$ gives a real y-value for every x in the closed interval from -1 to 1. Outside this domain, $1 - x^2$ is negative and its square root is not a real number. The values of $1 - x^2$ vary from 0 to 1 on the given domain, and the square roots of these values do the same. The range of $\sqrt{1 - x^2}$ is $[0, 1]$. ■

Graphs of Functions

If f is a function with domain D, its **graph** consists of the points in the Cartesian plane whose coordinates are the input-output pairs for f. In set notation, the graph is

$$\{(x, f(x)) \mid x \in D\}.$$

The graph of the function $f(x) = x + 2$ is the set of points with coordinates (x, y) for which $y = x + 2$. Its graph is the straight line sketched in Figure 1.3.

The graph of a function f is a useful picture of its behavior. If (x, y) is a point on the graph, then $y = f(x)$ is the height of the graph above (or below) the point x. The height may be positive or negative, depending on the sign of $f(x)$ (Figure 1.4).

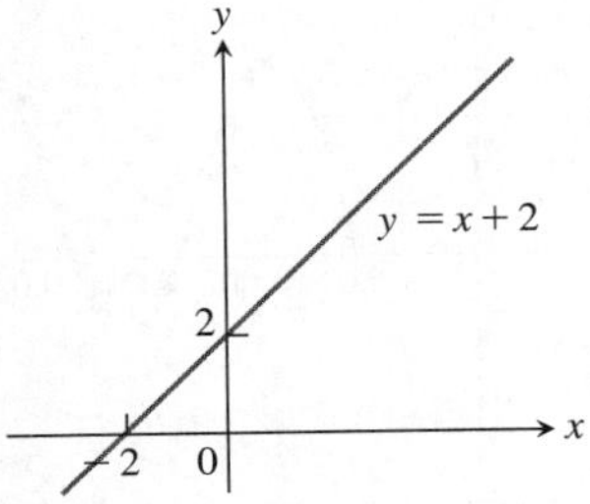

FIGURE 1.3 The graph of $f(x) = x + 2$ is the set of points (x, y) for which y has the value $x + 2$.

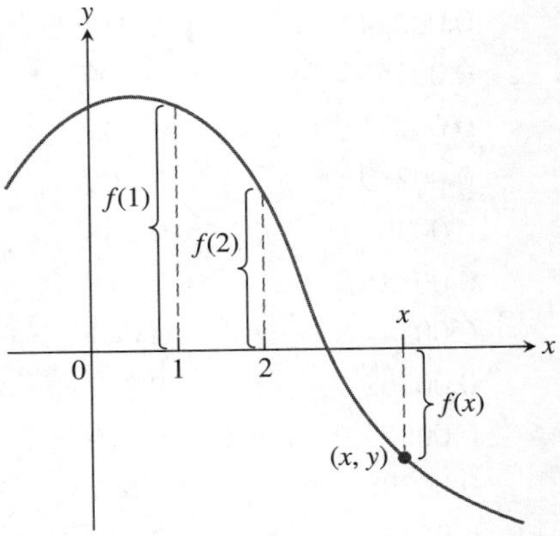

FIGURE 1.4 If (x, y) lies on the graph of f, then the value $y = f(x)$ is the height of the graph above the point x (or below x if $f(x)$ is negative).

x	$y = x^2$
−2	4
−1	1
0	0
1	1
$\frac{3}{2}$	$\frac{9}{4}$
2	4

EXAMPLE 2 Graph the function $y = x^2$ over the interval $[-2, 2]$.

Solution Make a table of xy-pairs that satisfy the equation $y = x^2$. Plot the points (x, y) whose coordinates appear in the table, and draw a *smooth* curve (labeled with its equation) through the plotted points (see Figure 1.5). ■

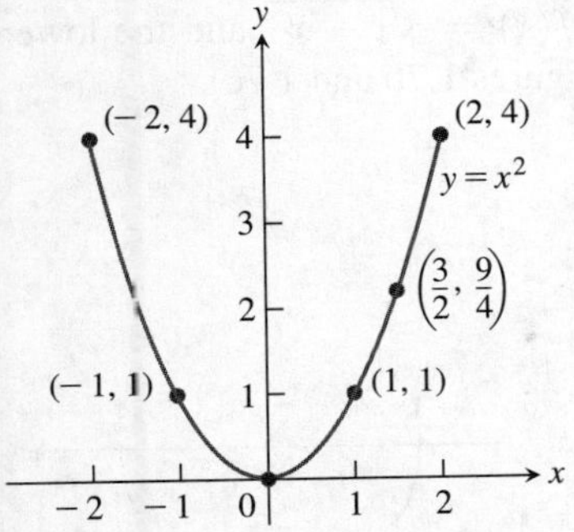

FIGURE 1.5 Graph of the function in Example 2.

How do we know that the graph of $y = x^2$ doesn't look like one of these curves?

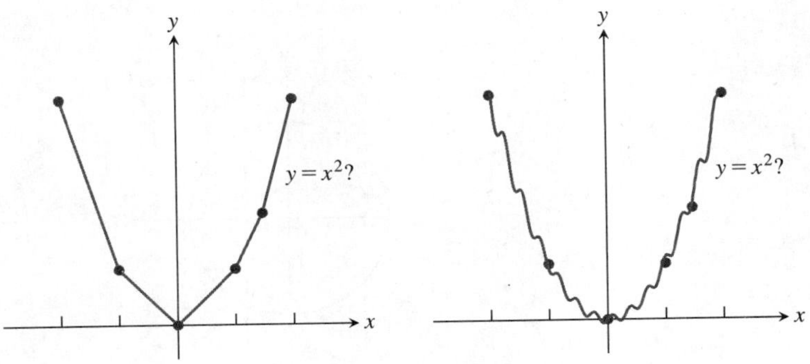

To find out, we could plot more points. But how would we then connect *them*? The basic question still remains: How do we know for sure what the graph looks like between the points we plot? Calculus answers this question, as we will see in Chapter 4. Meanwhile, we will have to settle for plotting points and connecting them as best we can.

Time	Pressure
0.00091	−0.080
0.00108	0.200
0.00125	0.480
0.00144	0.693
0.00162	0.816
0.00180	0.844
0.00198	0.771
0.00216	0.603
0.00234	0.368
0.00253	0.099
0.00271	−0.141
0.00289	−0.309
0.00307	−0.348
0.00325	−0.248
0.00344	−0.041
0.00362	0.217
0.00379	0.480
0.00398	0.681
0.00416	0.810
0.00435	0.827
0.00453	0.749
0.00471	0.581
0.00489	0.346
0.00507	0.077
0.00525	−0.164
0.00543	−0.320
0.00562	−0.354
0.00579	−0.248
0.00598	−0.035

Representing a Function Numerically

A function may be represented algebraically by a formula and visually by a graph (Example 2). Another way to represent a function is **numerically**, through a table of values. From an appropriate table of values, a graph of the function can be obtained using the method illustrated in Example 2, possibly with the aid of a computer. The graph consisting of only the points in the table is called a **scatterplot**.

EXAMPLE 3 Musical notes are pressure waves in the air. The data associated with Figure 1.6 give recorded pressure displacement versus time in seconds of a musical note produced by a tuning fork. The table provides a representation of the pressure function (in micropascals) over time. If we first make a scatterplot and then draw a smooth curve that approximates the data points (t, p) from the table, we obtain the graph shown in the figure.

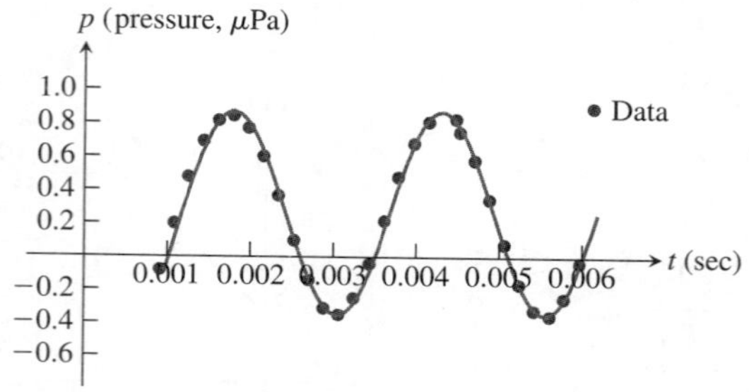

FIGURE 1.6 A smooth curve approximating the plotted points gives a graph of the pressure function represented by the accompanying tabled data (Example 3). ■

The Vertical Line Test for a Function

Not every curve in the coordinate plane can be the graph of a function. A function f can have only one value $f(x)$ for each x in its domain, so *no vertical line* can intersect the graph of a function at more than one point. If a is in the domain of the function f, then the vertical line $x = a$ will intersect the graph of f at the single point $(a, f(a))$.

A circle cannot be the graph of a function, since some vertical lines intersect the circle twice. The circle graphed in Figure 1.7a, however, contains the graphs of two functions of x, namely the upper semicircle defined by the function $f(x) = \sqrt{1 - x^2}$ and the lower semicircle defined by the function $g(x) = -\sqrt{1 - x^2}$ (Figures 1.7b and 1.7c).

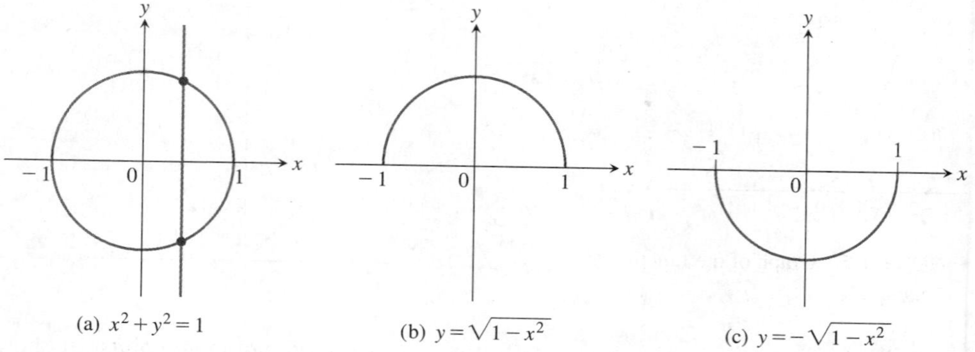

FIGURE 1.7 (a) The circle is not the graph of a function; it fails the vertical line test. (b) The upper semicircle is the graph of the function $f(x) = \sqrt{1 - x^2}$. (c) The lower semicircle is the graph of the function $g(x) = -\sqrt{1 - x^2}$.

Piecewise-Defined Functions

Sometimes a function is described in pieces by using different formulas on different parts of its domain. One example is the **absolute value function**

$$|x| = \begin{cases} x, & x \geq 0 \quad \text{First formula} \\ -x, & x < 0 \quad \text{Second formula} \end{cases}$$

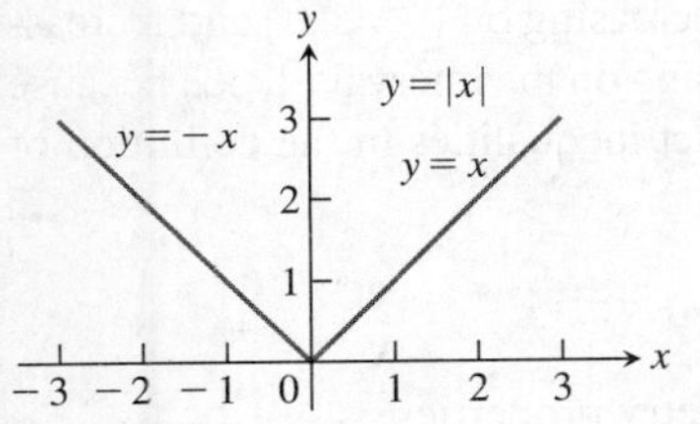

FIGURE 1.8 The absolute value function has domain $(-\infty, \infty)$ and range $[0, \infty)$.

whose graph is given in Figure 1.8. The right-hand side of the equation means that the function equals x if $x \geq 0$, and equals $-x$ if $x < 0$. Piecewise-defined functions often arise when real-world data are modeled. Here are some other examples.

EXAMPLE 4 The function

$$f(x) = \begin{cases} -x, & x < 0 \quad \text{First formula} \\ x^2, & 0 \leq x \leq 1 \quad \text{Second formula} \\ 1, & x > 1 \quad \text{Third formula} \end{cases}$$

is defined on the entire real line but has values given by different formulas, depending on the position of x. The values of f are given by $y = -x$ when $x < 0$, $y = x^2$ when $0 \leq x \leq 1$, and $y = 1$ when $x > 1$. The function, however, is *just one function* whose domain is the entire set of real numbers (Figure 1.9). ■

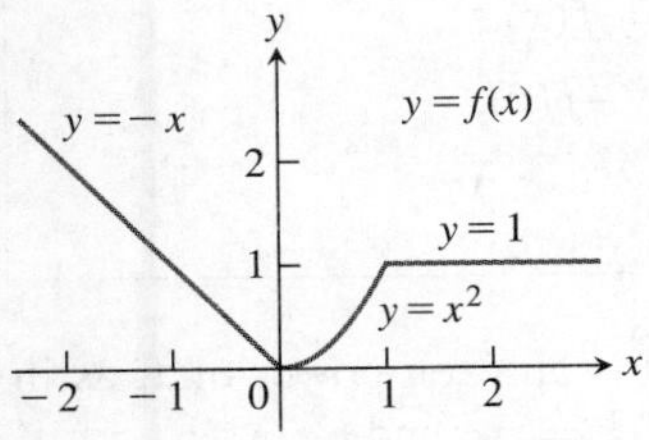

FIGURE 1.9 To graph the function $y = f(x)$ shown here, we apply different formulas to different parts of its domain (Example 4).

EXAMPLE 5 The function whose value at any number x is the *greatest integer less than or equal to* x is called the **greatest integer function** or the **integer floor function**. It is denoted $\lfloor x \rfloor$. Figure 1.10 shows the graph. Observe that

$$\lfloor 2.4 \rfloor = 2, \quad \lfloor 1.9 \rfloor = 1, \quad \lfloor 0 \rfloor = 0, \quad \lfloor -1.2 \rfloor = -2,$$
$$\lfloor 2 \rfloor = 2, \quad \lfloor 0.2 \rfloor = 0, \quad \lfloor -0.3 \rfloor = -1, \quad \lfloor -2 \rfloor = -2.$$ ■

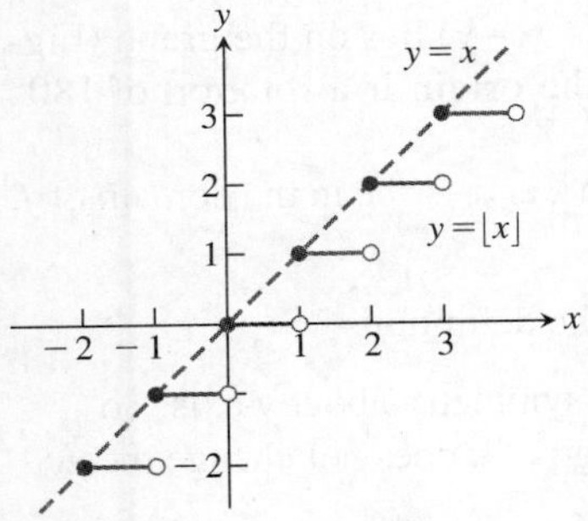

FIGURE 1.10 The graph of the greatest integer function $y = \lfloor x \rfloor$ lies on or below the line $y = x$, so it provides an integer floor for x (Example 5).

EXAMPLE 6 The function whose value at any number x is the *smallest integer greater than or equal to* x is called the **least integer function** or the **integer ceiling function**. It is denoted $\lceil x \rceil$. Figure 1.11 shows the graph. For positive values of x, this function might represent, for example, the cost of parking x hours in a parking lot that charges \$1 for each hour or part of an hour. ■

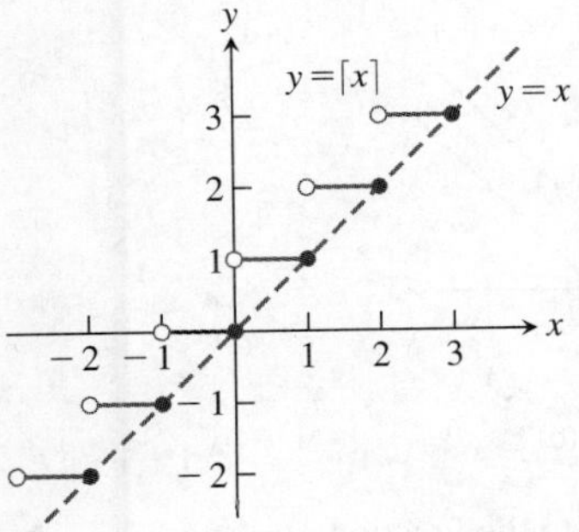

FIGURE 1.11 The graph of the least integer function $y = \lceil x \rceil$ lies on or above the line $y = x$, so it provides an integer ceiling for x (Example 6).

Increasing and Decreasing Functions

If the graph of a function climbs or rises as you move from left to right, we say that the function is *increasing*. If the graph descends or falls as you move from left to right, the function is *decreasing*.

> **DEFINITIONS** Let f be a function defined on an interval I and let x_1 and x_2 be two distinct points in I.
>
> 1. If $f(x_2) > f(x_1)$ whenever $x_1 < x_2$, then f is said to be **increasing** on I.
> 2. If $f(x_2) < f(x_1)$ whenever $x_1 < x_2$, then f is said to be **decreasing** on I.

It is important to realize that the definitions of increasing and decreasing functions must be satisfied for *every* pair of points x_1 and x_2 in I with $x_1 < x_2$. Because we use the inequality $<$ to compare the function values, instead of $\leq$, it is sometimes said that f is *strictly* increasing or decreasing on I. The interval I may be finite (also called bounded) or infinite (unbounded).

EXAMPLE 7 The function graphed in Figure 1.9 is decreasing on $(-\infty, 0)$ and increasing on $(0, 1)$. The function is neither increasing nor decreasing on the interval $(1, \infty)$ because the function is constant on that interval, and hence the strict inequalities in the definition of increasing or decreasing are not satisfied on $(1, \infty)$. ■

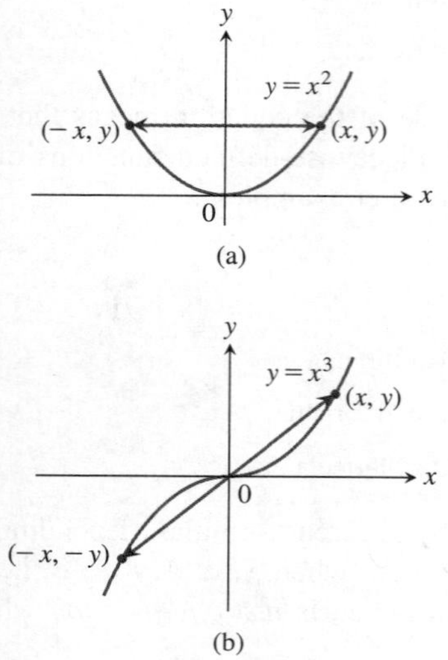

FIGURE 1.12 (a) The graph of $y = x^2$ (an even function) is symmetric about the y-axis. (b) The graph of $y = x^3$ (an odd function) is symmetric about the origin.

Even Functions and Odd Functions: Symmetry

The graphs of *even* and *odd* functions have special symmetry properties.

DEFINITIONS A function $y = f(x)$ is an

even function of x if $f(-x) = f(x)$,
odd function of x if $f(-x) = -f(x)$,

for every x in the function's domain.

The names *even* and *odd* come from powers of x. If y is an even power of x, as in $y = x^2$ or $y = x^4$, it is an even function of x because $(-x)^2 = x^2$ and $(-x)^4 = x^4$. If y is an odd power of x, as in $y = x$ or $y = x^3$, it is an odd function of x because $(-x)^1 = -x$ and $(-x)^3 = -x^3$.

The graph of an even function is **symmetric about the y-axis**. Since $f(-x) = f(x)$, a point (x, y) lies on the graph if and only if the point $(-x, y)$ lies on the graph (Figure 1.12a). A reflection across the y-axis leaves the graph unchanged.

The graph of an odd function is **symmetric about the origin**. Since $f(-x) = -f(x)$, a point (x, y) lies on the graph if and only if the point $(-x, -y)$ lies on the graph (Figure 1.12b). Equivalently, a graph is symmetric about the origin if a rotation of 180° about the origin leaves the graph unchanged.

Notice that each of these definitions requires that both x and $-x$ be in the domain of f.

EXAMPLE 8 Here are several functions illustrating the definitions.

$f(x) = x^2$ Even function: $(-x)^2 = x^2$ for all x; symmetry about y-axis. So $f(-3) = 9 = f(3)$. Changing the sign of x does not change the value of an even function.

$f(x) = x^2 + 1$ Even function: $(-x)^2 + 1 = x^2 + 1$ for all x; symmetry about y-axis (Figure 1.13a).

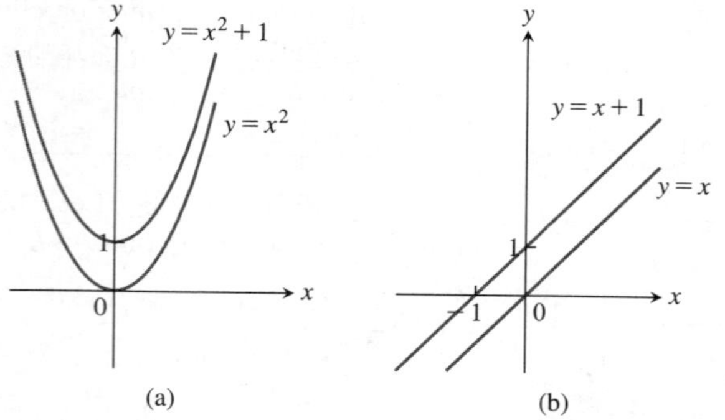

FIGURE 1.13 (a) When we add the constant term 1 to the function $y = x^2$, the resulting function $y = x^2 + 1$ is still even and its graph is still symmetric about the y-axis. (b) When we add the constant term 1 to the function $y = x$, the resulting function $y = x + 1$ is no longer odd, since the symmetry about the origin is lost. The function $y = x + 1$ is also not even (Example 8).

$f(x) = x$ Odd function: $(-x) = -x$ for all x; symmetry about the origin. So $f(-3) = -3$ while $f(3) = 3$. Changing the sign of x changes the sign of the value of an odd function.

$f(x) = x + 1$ Not odd: $f(-x) = -x + 1$, but $-f(x) = -x - 1$. The two are not equal.

Not even: $(-x) + 1 \neq x + 1$ for all $x \neq 0$ (Figure 1.13b). ■

Common Functions

A variety of important types of functions are frequently encountered in calculus.

Linear Functions A function of the form $f(x) = mx + b$, where m and b are fixed constants, is called a **linear function**. Figure 1.14a shows an array of lines $f(x) = mx$. Each of these has $b = 0$, so these lines pass through the origin. The function $f(x) = x$, where $m = 1$ and $b = 0$, is called the **identity function**. Constant functions result when the slope is $m = 0$ (Figure 1.14b).

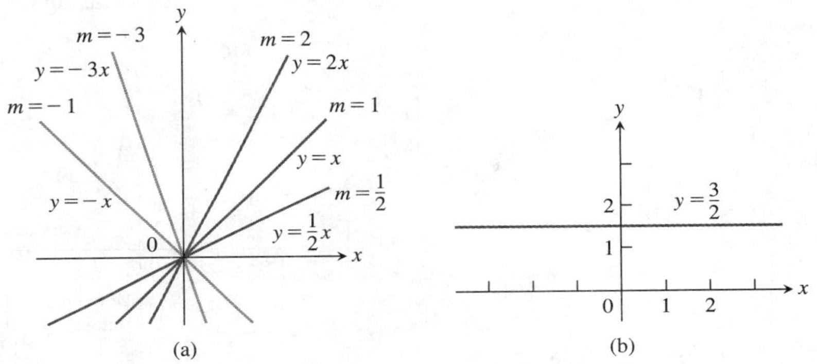

FIGURE 1.14 (a) Lines through the origin with slope m. (b) A constant function with slope $m = 0$.

> **DEFINITION** Two variables y and x are **proportional** (to one another) if one is always a constant multiple of the other—that is, if $y = kx$ for some nonzero constant k.

If the variable y is proportional to the reciprocal $1/x$, then sometimes it is said that y is **inversely proportional** to x (because $1/x$ is the multiplicative inverse of x).

Power Functions A function $f(x) = x^a$, where a is a constant, is called a **power function**. There are several important cases to consider.

(a) $f(x) = x^a$ with $a = n$, a positive integer.

The graphs of $f(x) = x^n$, for $n = 1, 2, 3, 4, 5$, are displayed in Figure 1.15. These functions are defined for all real values of x. Notice that as the power n gets larger, the curves tend to flatten toward the x-axis on the interval $(-1, 1)$ and to rise more steeply for $|x| > 1$. Each curve passes through the point $(1, 1)$ and through the origin. The graphs of functions with even powers are symmetric about the y-axis; those with odd powers are symmetric about the origin. The even-powered functions are decreasing on the interval $(-\infty, 0]$ and increasing on $[0, \infty)$; the odd-powered functions are increasing over the entire real line $(-\infty, \infty)$.

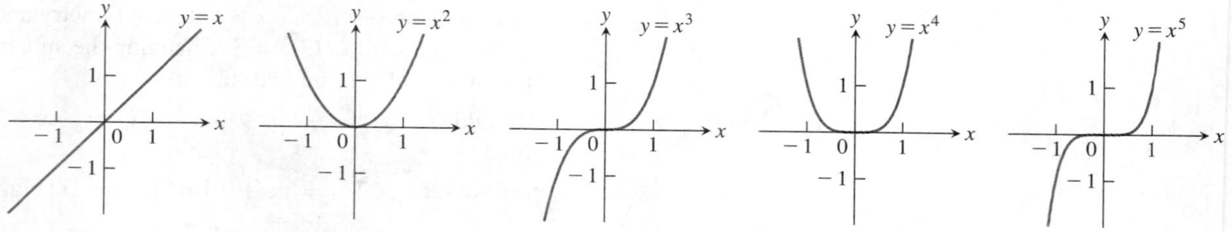

FIGURE 1.15 Graphs of $f(x) = x^n, n = 1, 2, 3, 4, 5$, defined for $-\infty < x < \infty$.

(b) $f(x) = x^a$ with $a = -1$ or $a = -2$.

The graphs of the functions $f(x) = x^{-1} = 1/x$ and $f(x) = x^{-2} = 1/x^2$ are shown in Figure 1.16. Both functions are defined for all $x \neq 0$ (you can never divide by zero). The graph of $y = 1/x$ is the hyperbola $xy = 1$, which approaches the coordinate axes far from the origin. The graph of $y = 1/x^2$ also approaches the coordinate axes. The graph of the function $f(x) = 1/x$ is symmetric about the origin; this function is decreasing on the intervals $(-\infty, 0)$ and $(0, \infty)$. The graph of the function $f(x) = 1/x^2$ is symmetric about the y-axis; this function is increasing on $(-\infty, 0)$ and decreasing on $(0, \infty)$.

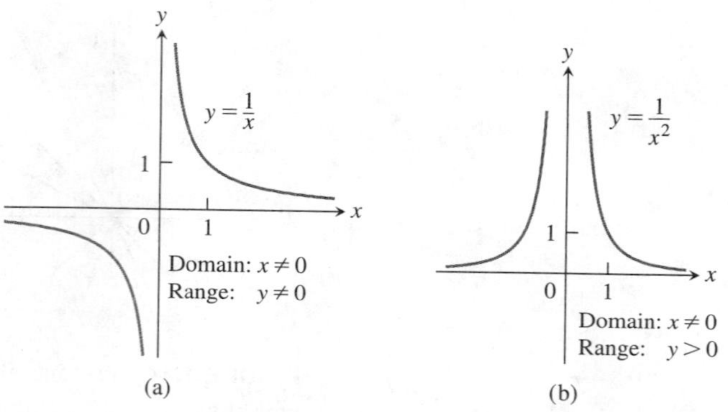

FIGURE 1.16 Graphs of the power functions $f(x) = x^a$. (a) $a = -1$. (b) $a = -2$.

(c) $f(x) = x^a$ with $a = \frac{1}{2}, \frac{1}{3}, \frac{3}{2}$, or $\frac{2}{3}$.

The functions $f(x) = x^{1/2} = \sqrt{x}$ and $f(x) = x^{1/3} = \sqrt[3]{x}$ are the **square root** and **cube root** functions, respectively. The domain of the square root function is $[0, \infty)$, but the cube root function is defined for all real x. Their graphs are displayed in Figure 1.17, along with the graphs of $y = x^{3/2}$ and $y = x^{2/3}$. (Recall that $x^{3/2} = (x^{1/2})^3$ and $x^{2/3} = (x^{1/3})^2$.)

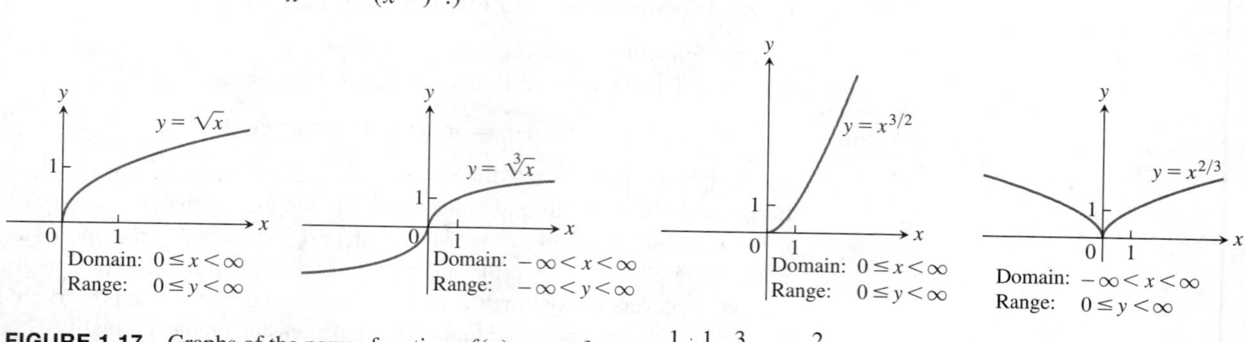

FIGURE 1.17 Graphs of the power functions $f(x) = x^a$ for $a = \frac{1}{2}, \frac{1}{3}, \frac{3}{2}$, and $\frac{2}{3}$.

Polynomials A function p is a **polynomial** if

$$p(x) = a_n x^n + a_{n-1}x^{n-1} + \cdots + a_1 x + a_0,$$

where n is a nonnegative integer and the numbers $a_0, a_1, a_2, \ldots, a_n$ are real constants (called the **coefficients** of the polynomial). All polynomials have domain $(-\infty, \infty)$. If the leading coefficient $a_n \neq 0$, then n is called the **degree** of the polynomial. Linear functions with $m \neq 0$ are polynomials of degree 1. Polynomials of degree 2, usually written as $p(x) = ax^2 + bx + c$, are called **quadratic functions**. Likewise, **cubic functions** are polynomials $p(x) = ax^3 + bx^2 + cx + d$ of degree 3. Figure 1.18 shows the graphs of three polynomials. Techniques to graph polynomials are studied in Chapter 4.

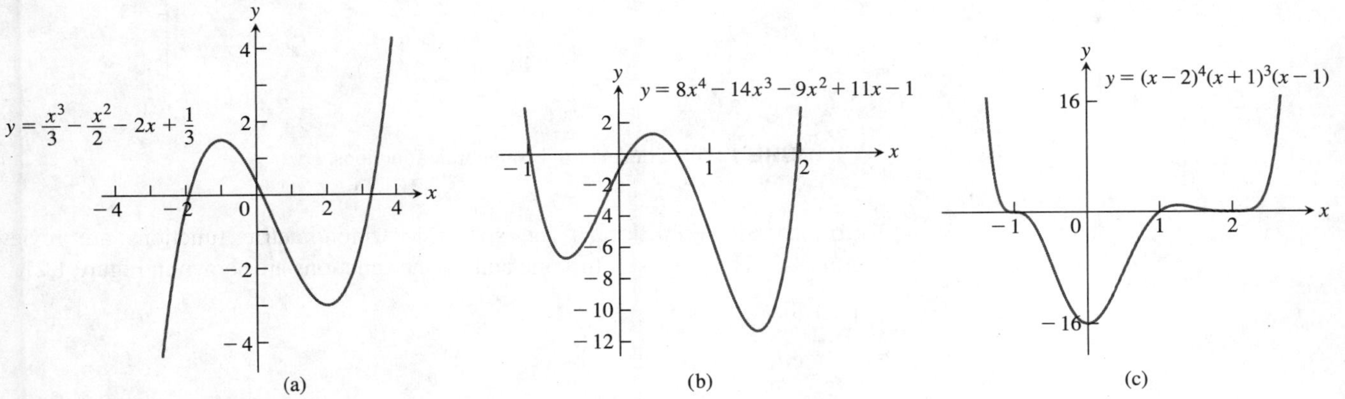

FIGURE 1.18 Graphs of three polynomial functions.

Rational Functions A **rational function** is a quotient or ratio $f(x) = p(x)/q(x)$, where p and q are polynomials. The domain of a rational function is the set of all real x for which $q(x) \neq 0$. The graphs of three rational functions are shown in Figure 1.19.

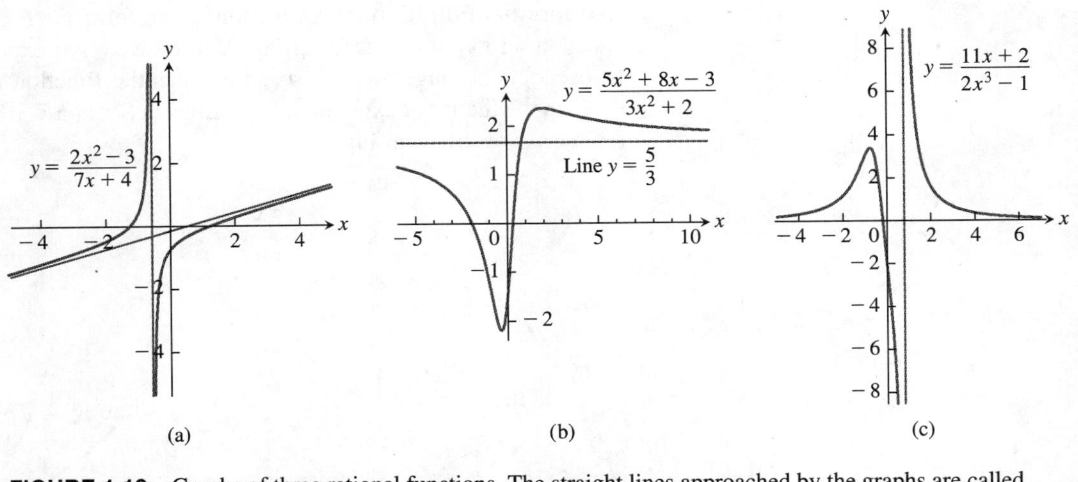

FIGURE 1.19 Graphs of three rational functions. The straight lines approached by the graphs are called *asymptotes* and are not part of the graphs. We discuss asymptotes in Section 2.6.

Algebraic Functions Any function constructed from polynomials using algebraic operations (addition, subtraction, multiplication, division, and taking roots) lies within the class of **algebraic functions**. All rational functions are algebraic, but also included are more complicated functions (such as those satisfying an equation like $y^3 - 9xy + x^3 = 0$, studied in Section 3.7). Figure 1.20 displays the graphs of three algebraic functions.

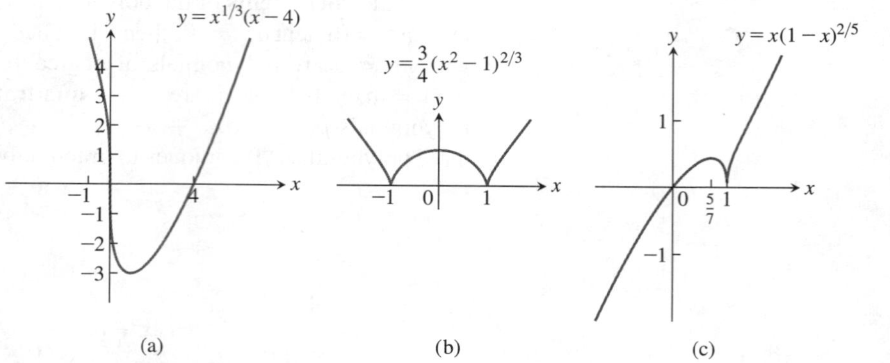

FIGURE 1.20 Graphs of three algebraic functions.

Trigonometric Functions The six basic trigonometric functions are reviewed in Section 1.3. The graphs of the sine and cosine functions are shown in Figure 1.21.

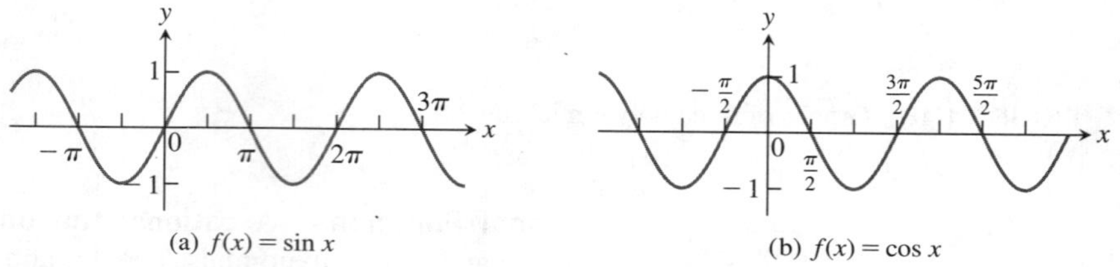

FIGURE 1.21 Graphs of the sine and cosine functions.

Exponential Functions A function of the form $f(x) = a^x$, where $a > 0$ and $a \neq 1$, is called an **exponential function** (with base a). All exponential functions have domain $(-\infty, \infty)$ and range $(0, \infty)$, so an exponential function never assumes the value 0. We develop the theory of exponential functions in Section 7.3. The graphs of some exponential functions are shown in Figure 1.22.

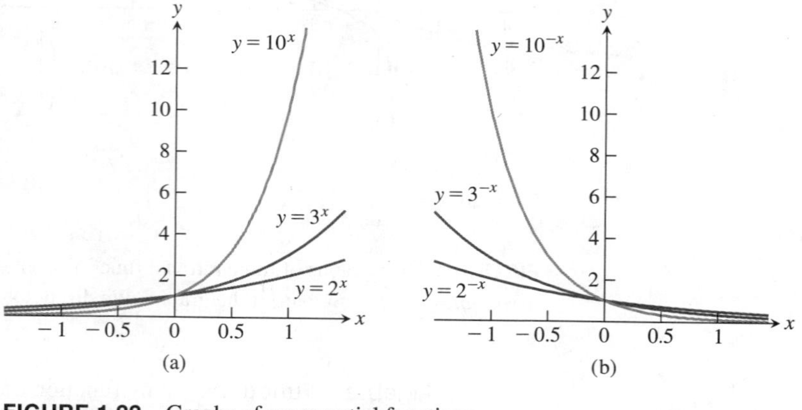

FIGURE 1.22 Graphs of exponential functions.

Logarithmic Functions These are the functions $f(x) = \log_a x$, where the base $a \neq 1$ is a positive constant. They are the *inverse functions* of the exponential functions, and we define and develop the theory of these functions in Section 7.2. Figure 1.23 shows the graphs of four logarithmic functions with various bases. In each case the domain is $(0, \infty)$ and the range is $(-\infty, \infty)$.

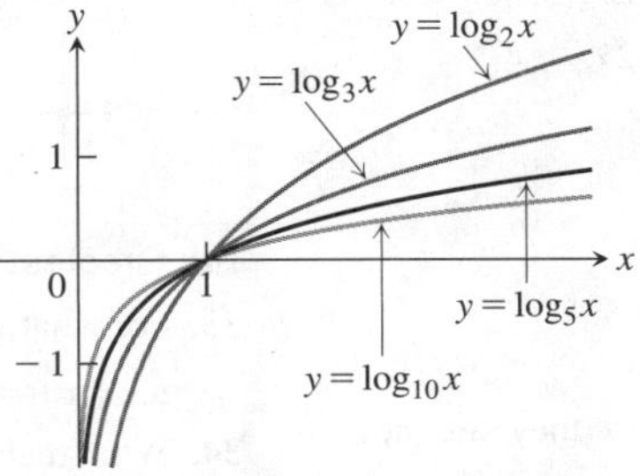

FIGURE 1.23 Graphs of four logarithmic functions.

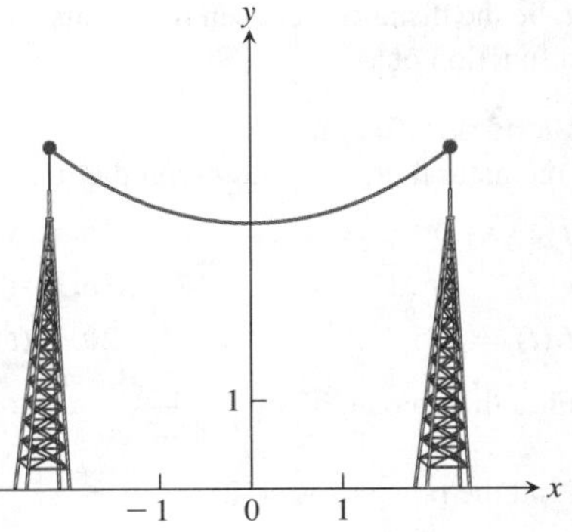

FIGURE 1.24 Graph of a catenary or hanging cable. (The Latin word *catena* means "chain.")

Transcendental Functions These are functions that are not algebraic. They include the trigonometric, inverse trigonometric, exponential, and logarithmic functions, and many other functions as well. The **catenary** is one example of a transcendental function. Its graph has the shape of a cable, like a telephone line or electric cable, strung from one support to another and hanging freely under its own weight (Figure 1.24). The function defining the graph is discussed in Section 7.7.

EXERCISES 1.1

Functions

In Exercises 1–6, find the domain and range of each function.

1. $f(x) = 1 + x^2$

2. $f(x) = \dfrac{1}{x+3} - 5$

3. $F(x) = \sqrt{5x + 10}$

4. $g(x) = \sqrt{x^2 - 3x}$

5. $f(x) = \dfrac{x^2 - 3x - 4}{x+1}$

6. $G(t) = \dfrac{2}{t^2 - 16}$

In Exercises 7 and 8, which of the graphs are graphs of functions of x, and which are not? Give reasons for your answers.

7. a.

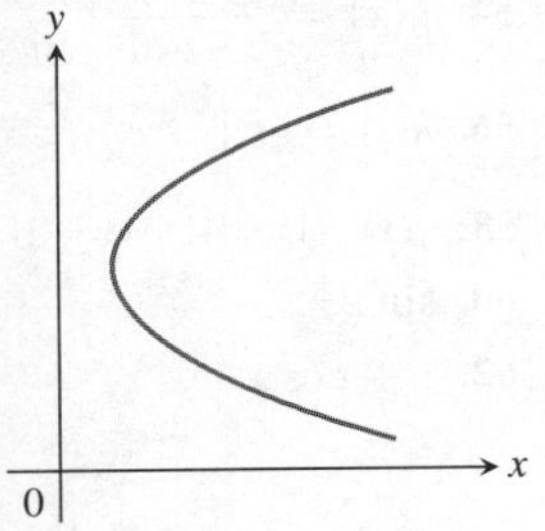

b.

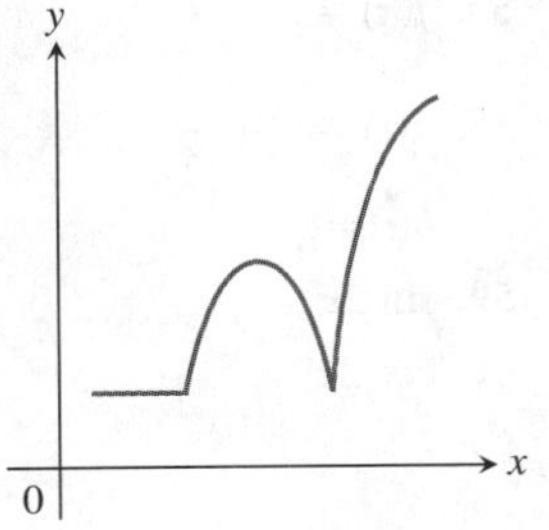

8. a.

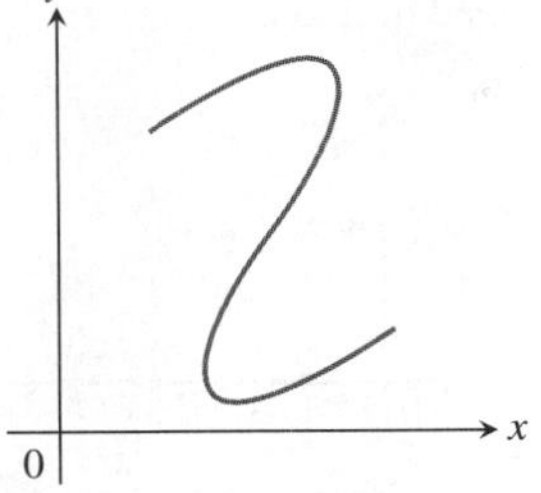

b.

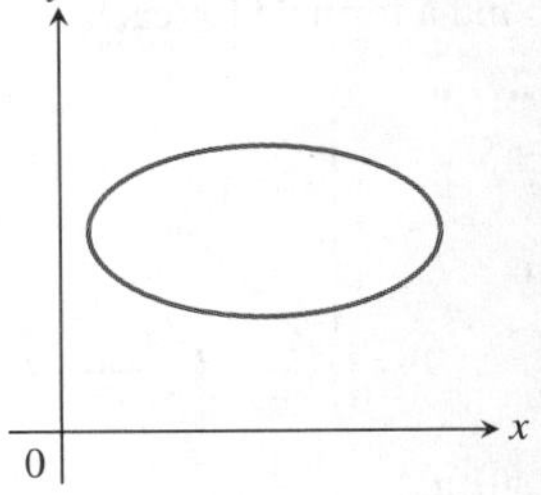

Finding Formulas for Functions

9. Express the area and perimeter of an equilateral triangle as a function of the triangle's side length x.

10. Express the side length of a square as a function of the length d of the square's diagonal. Then express the area as a function of the diagonal length.

11. Express the edge length of a cube as a function of the cube's diagonal length d. Then express the surface area and volume of the cube as a function of the diagonal length.

12. A point P in the first quadrant lies on the graph of the function $f(x) = \sqrt{x}$. Express the coordinates of P as functions of the slope of the line joining P to the origin.

13. Consider the point (x, y) lying on the graph of the line $y = x + 2$. Let L be the distance from the point (x, y) to the origin $(0, 0)$. Write L as a function of x.

14. Consider the point (x, y) lying on the graph of $y = \sqrt{x - 3}$. Let L be the distance between the points (x, y) and $(4, 0)$. Write L as a function of y.

Functions and Graphs

Find the natural domain and graph the functions in Exercises 15–20.

15. $f(x) = 5 - 2x$
16. $f(x) = 1 - 2x - x^2$
17. $g(x) = \sqrt{|x|}$
18. $y = \sqrt{x + 4}$
19. $F(t) = t/|t|$
20. $G(t) = 1/|t|$

21. Find the domain of $y = \dfrac{x + 3}{4 - \sqrt{x^2 - 9}}$.

22. Find the range of $y = 2 + \sqrt{9 + x^2}$.

23. Graph the following equations and explain why they are not graphs of functions of x.

a. $|y| = x$ b. $y^2 = x^2$

24. Graph the following equations and explain why they are not graphs of functions of x.

a. $|x| + |y| = 1$ b. $|x + y| = 1$

Piecewise-Defined Functions

Graph the functions in Exercises 25–28.

25. $f(x) = \begin{cases} x, & 0 \le x \le 1 \\ 2 - x, & 1 < x \le 2 \end{cases}$

26. $f(x) = \begin{cases} 1 - x^2, & x \ge 0 \\ 1 - x, & x < 0 \end{cases}$

27. $F(x) = \begin{cases} 4 - x^2, & x \le 1 \\ x^2 + 2x, & x > 1 \end{cases}$

28. $f(x) = \begin{cases} x^2 + 1, & x \ge 0 \\ 2, & x < 0 \end{cases}$

Find a formula for each function graphed in Exercises 29–32.

29. a.

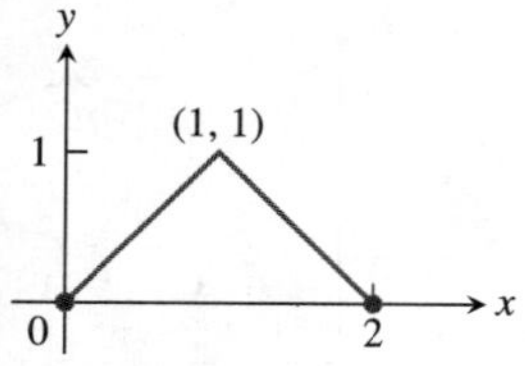

b.

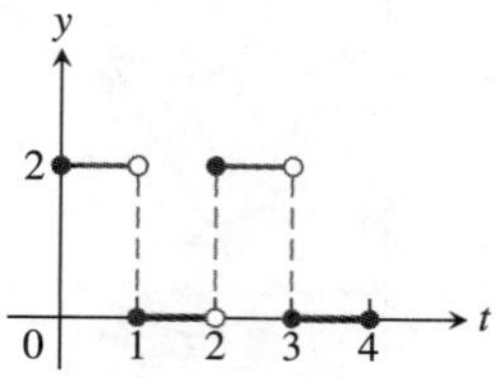

30. a.

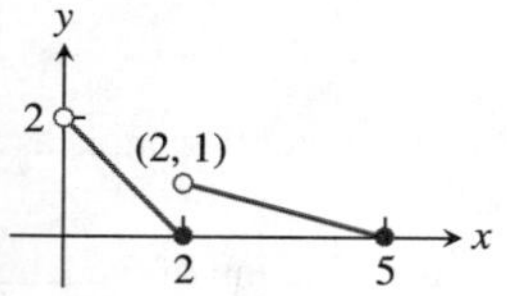

b.

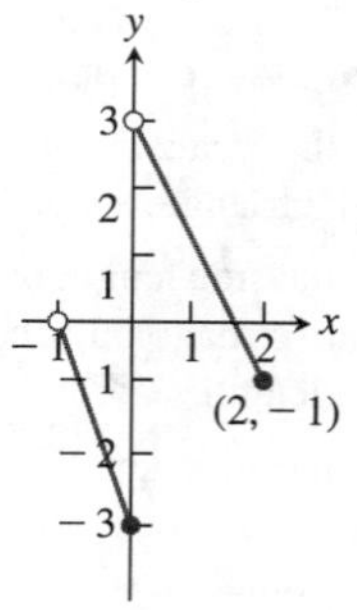

31. a.

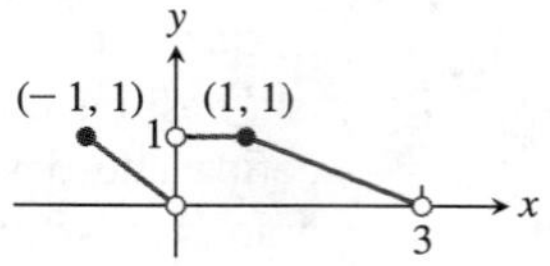

b.

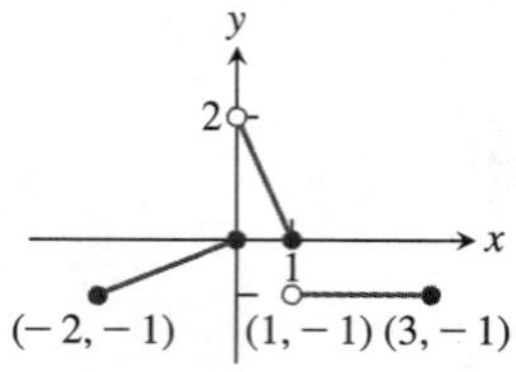

32. a.

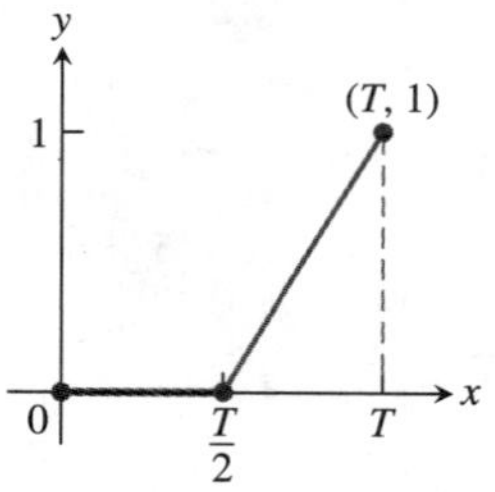

b.

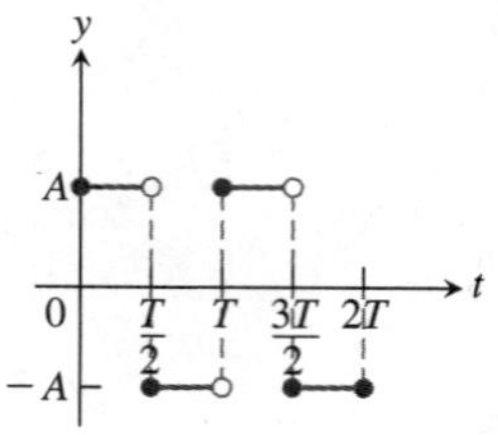

The Greatest and Least Integer Functions

33. For what values of x is

a. $\lfloor x \rfloor = 5$ b. $\lceil x \rceil = 5$

34. What real numbers x satisfy the equation $\lfloor x \rfloor = \lceil x \rceil$?

35. Does $\lceil -x \rceil = -\lfloor x \rfloor$ for all real x? Give reasons for your answer.

36. Graph the function

$$f(x) = \begin{cases} \lfloor x \rfloor, & x \ge 0 \\ \lceil x \rceil, & x < 0. \end{cases}$$

Why is $f(x)$ called the *integer part* of x?

Increasing and Decreasing Functions

Graph the functions in Exercises 37–46. What symmetries, if any, do the graphs have? Specify the intervals over which the function is increasing and the intervals where it is decreasing.

37. $y = -x^3$
38. $y = -\dfrac{1}{x^2}$
39. $y = -\dfrac{1}{x}$
40. $y = \dfrac{1}{|x|}$
41. $y = \sqrt{|x|}$
42. $y = \sqrt{-x}$
43. $y = x^3/8$
44. $y = -4\sqrt{x}$
45. $y = -x^{3/2}$
46. $y = (-x)^{2/3}$

Even and Odd Functions

In Exercises 47–62, say whether the function is even, odd, or neither. Give reasons for your answer.

47. $f(x) = 3$
48. $f(x) = x^{-5}$
49. $f(x) = 3x^2 - 4$
50. $f(x) = x^2 + x$
51. $g(x) = x^3 + x$
52. $g(x) = x^4 + 3x^2 - 1$
53. $f(x) = \dfrac{x}{x^3 + 3}$
54. $g(x) = \dfrac{x}{x^2 - 1}$
55. $h(t) = \dfrac{1}{t - 1}$
56. $h(t) = |t^3|$
57. $h(t) = t^3 - t^2$
58. $f(x) = |x - 1| + |x + 1|$
59. $\sin 2x$
60. $\sin x^2$
61. $\cos 3x$
62. $1 + \cos x$

Theory and Examples

63. The variable x is proportional to y, and $x = 30$ when $y = 120$. Determine y when $x = 55$.

64. Kinetic energy The kinetic energy K of a mass is proportional to the square of its velocity v. If $K = 12{,}960$ joules when $v = 18$ m/sec, what is K when $v = 10$ m/sec?

65. The variables r and s are inversely proportional, and $r = 6$ when $s = 4$. Determine s when $r = 10$.

66. According to Boyle's law, at the constant temperature, the volume of the gas (V) and pressure (P) are inversely proportional to each other. If $P = 25.5$ lb/in^2 when $V = 2000$ in^3, then find V when $P = 42$ lb/in^2.

67. A box with an open top is to be constructed from a rectangular piece of cardboard with dimensions 14 in. by 22 in. by cutting out equal squares of side x at each corner and then folding up the sides as in the figure. Express the volume V of the box as a function of x.

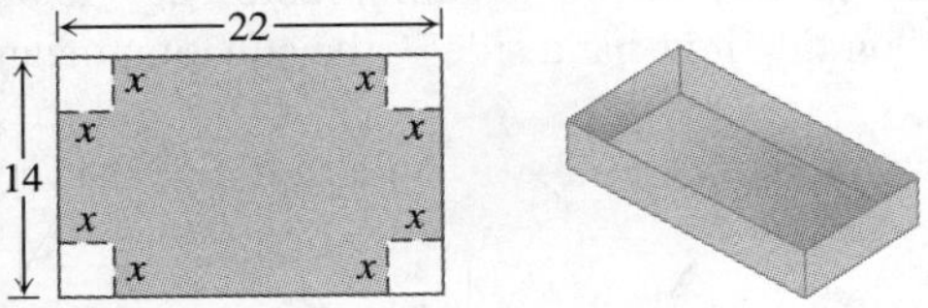

68. The accompanying figure shows a rectangle inscribed in an isosceles right triangle whose hypotenuse is 2 units long.

a. Express the y-coordinate of P in terms of x. (You might start by writing an equation for the line AB.)

b. Express the area of the rectangle in terms of x.

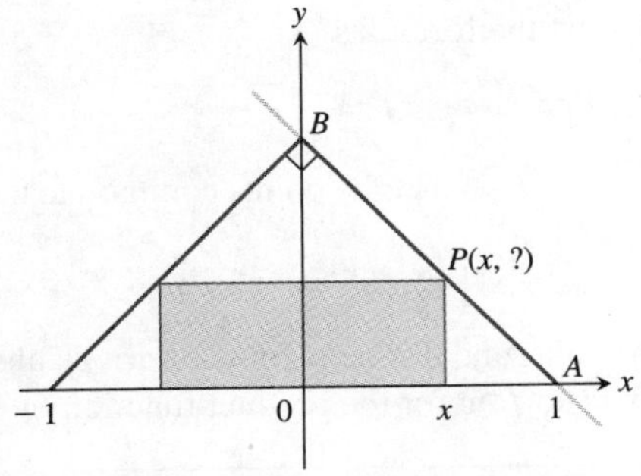

In Exercises 69 and 70, match each equation with its graph. Do not use a graphing device, and give reasons for your answer.

69. a. $y = x^4$ **b.** $y = x^7$ **c.** $y = x^{10}$

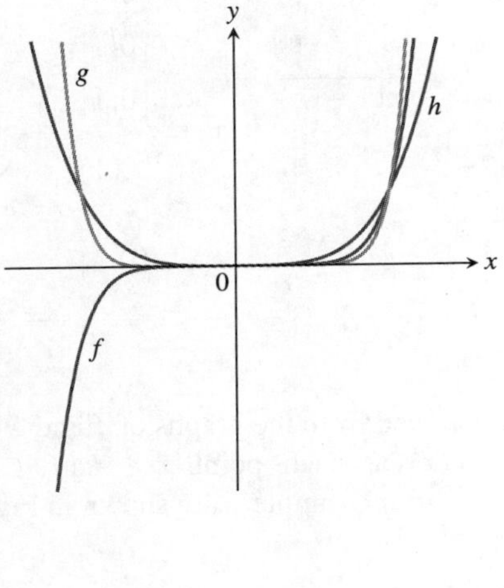

70. a. $y = 5x$ **b.** $y = 5^x$ **c.** $y = x^5$

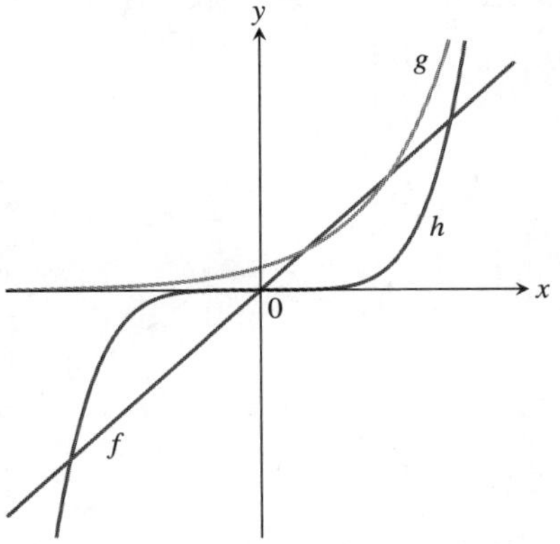

T 71. a. Graph the functions $f(x) = x/2$ and $g(x) = 1 + (4/x)$ together to identify the values of x for which

$$\frac{x}{2} > 1 + \frac{4}{x}.$$

b. Confirm your findings in part (a) algebraically.

T 72. a. Graph the functions $f(x) = 3/(x - 1)$ and $g(x) = 2/(x + 1)$ together to identify the values of x for which

$$\frac{3}{x-1} < \frac{2}{x+1}.$$

b. Confirm your findings in part (a) algebraically.

73. For a curve to be *symmetric about the x-axis*, the point (x, y) must lie on the curve if and only if the point $(x, -y)$ lies on the curve. Explain why a curve that is symmetric about the x-axis is not the graph of a function, unless the function is $y = 0$.

74. Five hundred books sell for \$ 50 each, resulting in a revenue of \$ 25,000. For each \$ 10 increase in the price, 35 fewer books are sold. Write the revenue R as a function of the number x of \$10 increase.

75. Fencing needs to be done around a pen in the shape of an isosceles right triangle with legs of length x cm and hypotenuse of length h cm. If fencing costs \$8/cm for the legs and \$12/cm for the hypotenuse, write the total cost C of construction as a function of h.

76. Industrial costs A power plant sits next to a river where the river is 800 ft wide. Laying a new cable from the plant to a location in the city 2 mi downstream on the opposite side costs \$180 per foot across the river and \$100 per foot along the land.

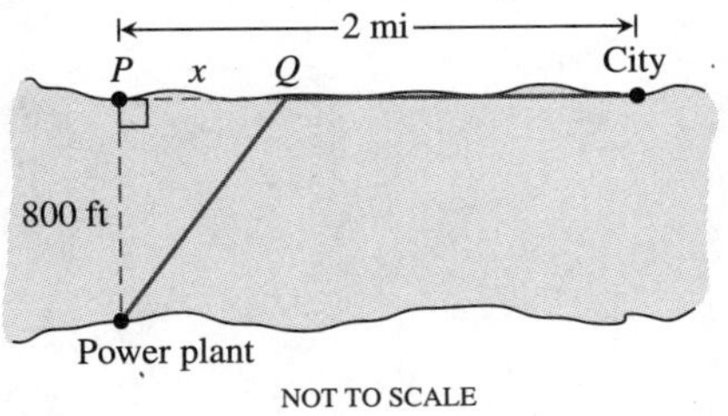

a. Suppose that the cable goes from the plant to a point Q on the opposite side that is x ft from the point P directly opposite the plant. Write a function $C(x)$ that gives the cost of laying the cable in terms of the distance x.

b. Generate a table of values to determine whether the least expensive location for point Q is less than 2000 ft or greater than 2000 ft from point P.

1.2 Combining Functions; Shifting and Scaling Graphs

In this section we look at the main ways functions are combined or transformed to form new functions.

Sums, Differences, Products, and Quotients

Like numbers, functions can be added, subtracted, multiplied, and divided (except where the denominator is zero) to produce new functions. If f and g are functions, then for every x that belongs to the domains of both f and g (that is, for $x \in D(f) \cap D(g)$), we define functions $f + g$, $f - g$, and fg by the formulas

$$\begin{aligned}(f + g)(x) &= f(x) + g(x)\\ (f - g)(x) &= f(x) - g(x)\\ (fg)(x) &= f(x)g(x).\end{aligned}$$

Notice that the + sign on the left-hand side of the first equation represents the operation of addition of *functions*, whereas the + on the right-hand side of the equation means addition of the real numbers $f(x)$ and $g(x)$.

At any point of $D(f) \cap D(g)$ at which $g(x) \neq 0$, we can also define the function f/g by the formula

$$\left(\frac{f}{g}\right)(x) = \frac{f(x)}{g(x)} \qquad (\text{where } g(x) \neq 0).$$

Functions can also be multiplied by constants: If c is a real number, then the function cf is defined for all x in the domain of f by

$$(cf)(x) = cf(x).$$

EXAMPLE 1 The functions defined by the formulas

$$f(x) = \sqrt{x} \quad \text{and} \quad g(x) = \sqrt{1 - x}$$

have domains $D(f) = [0, \infty)$ and $D(g) = (-\infty, 1]$. The points common to these domains are the points in

$$[0, \infty) \cap (-\infty, 1] = [0, 1].$$

The following table summarizes the formulas and domains for the various algebraic combinations of the two functions. We also write $f \cdot g$ for the product function fg.

Function	Formula	Domain
$f + g$	$(f + g)(x) = \sqrt{x} + \sqrt{1 - x}$	$[0, 1] = D(f) \cap D(g)$
$f - g$	$(f - g)(x) = \sqrt{x} - \sqrt{1 - x}$	$[0, 1]$
$g - f$	$(g - f)(x) = \sqrt{1 - x} - \sqrt{x}$	$[0, 1]$
$f \cdot g$	$(f \cdot g)(x) = f(x)g(x) = \sqrt{x(1 - x)}$	$[0, 1]$
f/g	$\frac{f}{g}(x) = \frac{f(x)}{g(x)} = \sqrt{\frac{x}{1 - x}}$	$[0, 1)$ $(x = 1$ excluded$)$
g/f	$\frac{g}{f}(x) = \frac{g(x)}{f(x)} = \sqrt{\frac{1 - x}{x}}$	$(0, 1]$ $(x = 0$ excluded$)$

■

The graph of the function $f + g$ is obtained from the graphs of f and g by adding the corresponding y-coordinates $f(x)$ and $g(x)$ at each point $x \in D(f) \cap D(g)$, as in Figure 1.25. The graphs of $f + g$ and $f \cdot g$ from Example 1 are shown in Figure 1.26.

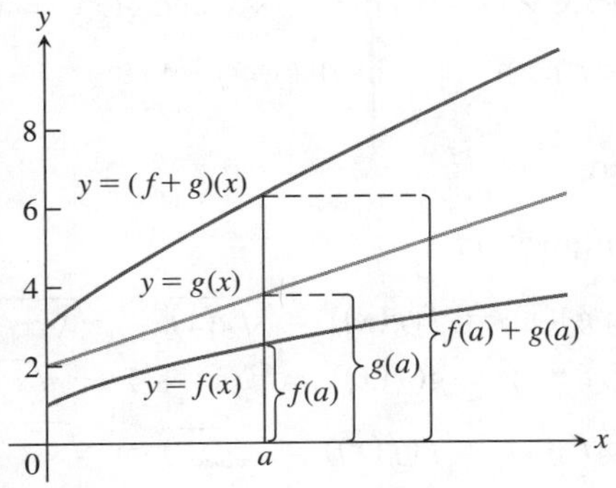

FIGURE 1.25 Graphical addition of two functions.

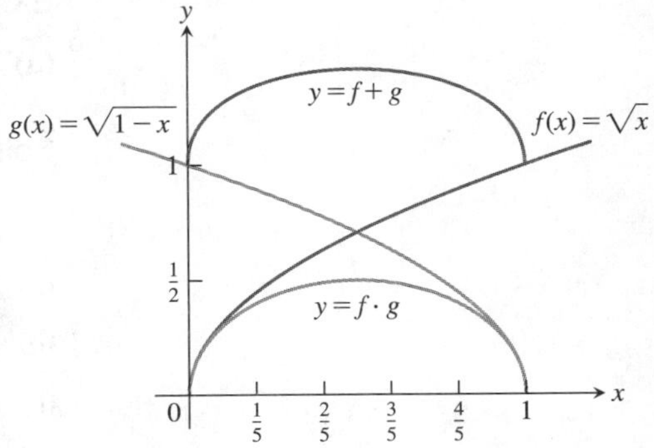

FIGURE 1.26 The domain of the function $f + g$ is the intersection of the domains of f and g, the interval $[0, 1]$ on the x-axis where these domains overlap. This interval is also the domain of the function $f \cdot g$ (Example 1).

Composing Functions

Composition is another method for combining functions. In this operation the output from one function becomes the input to a second function.

> **DEFINITION** If f and g are functions, the function $f \circ g$ ("f composed with g") is defined by
>
> $$(f \circ g)(x) = f(g(x))$$
>
> and called the **composition** of f and g. The domain of $f \circ g$ consists of the numbers x in the domain of g for which $g(x)$ lies in the domain of f.

To find $(f \circ g)(x)$, *first* find $g(x)$ and *second* find $f(g(x))$. Figure 1.27 pictures $f \circ g$ as a machine diagram, and Figure 1.28 shows the composition as an arrow diagram.

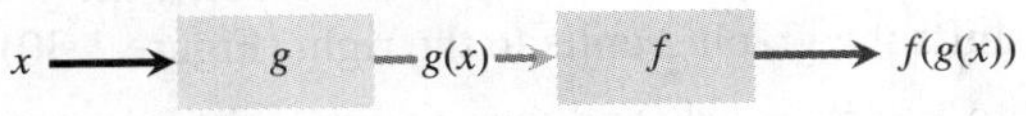

FIGURE 1.27 The composition $f \circ g$ uses the output $g(x)$ of the first function g as the input for the second function f.

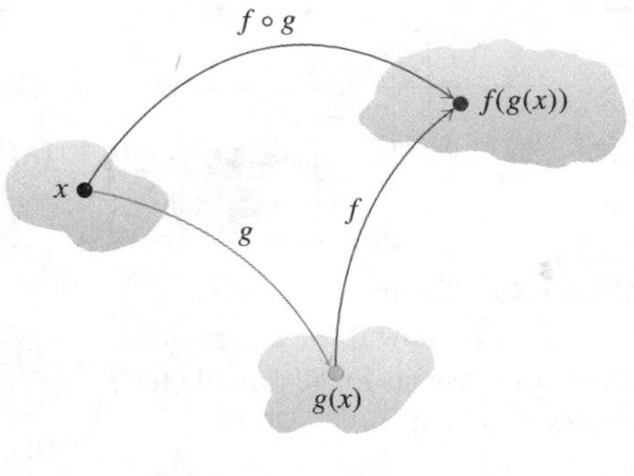

FIGURE 1.28 Arrow diagram for $f \circ g$. If x lies in the domain of g and $g(x)$ lies in the domain of f, then the functions f and g can be composed to form $(f \circ g)(x)$.

To evaluate the composition $g \circ f$ (when defined), we find $f(x)$ first and then find $g(f(x))$. The domain of $g \circ f$ is the set of numbers x in the domain of f such that $f(x)$ lies in the domain of g.

The functions $f \circ g$ and $g \circ f$ are usually quite different.

EXAMPLE 2 If $f(x) = \sqrt{x}$ and $g(x) = x + 1$, find

(a) $(f \circ g)(x)$ **(b)** $(g \circ f)(x)$ **(c)** $(f \circ f)(x)$ **(d)** $(g \circ g)(x)$.

Solution

Composition	Domain
(a) $(f \circ g)(x) = f(g(x)) = \sqrt{g(x)} = \sqrt{x+1}$	$[-1, \infty)$
(b) $(g \circ f)(x) = g(f(x)) = f(x) + 1 = \sqrt{x} + 1$	$[0, \infty)$
(c) $(f \circ f)(x) = f(f(x)) = \sqrt{f(x)} = \sqrt{\sqrt{x}} = x^{1/4}$	$[0, \infty)$
(d) $(g \circ g)(x) = g(g(x)) = g(x) + 1 = (x+1) + 1 = x + 2$	$(-\infty, \infty)$

To see why the domain of $f \circ g$ is $[-1, \infty)$, notice that $g(x) = x + 1$ is defined for all real x but $g(x)$ belongs to the domain of f only if $x + 1 \geq 0$, that is to say, when $x \geq -1$. ■

Notice that if $f(x) = x^2$ and $g(x) = \sqrt{x}$, then $(f \circ g)(x) = \left(\sqrt{x}\right)^2 = x$. However, the domain of $f \circ g$ is $[0, \infty)$, not $(-\infty, \infty)$, since $\sqrt{x}$ requires $x \geq 0$.

Shifting a Graph of a Function

A common way to obtain a new function from an existing one is by adding a constant to each output of the existing function, or to its input variable. The graph of the new function is the graph of the original function shifted vertically or horizontally, as follows.

Shift Formulas

Vertical Shifts

$y = f(x) + k$	Shifts the graph of f *up* k units if $k > 0$ Shifts it *down* $\lvert k \rvert$ units if $k < 0$

Horizontal Shifts

$y = f(x + h)$	Shifts the graph of f *left* h units if $h > 0$ Shifts it *right* $\lvert h \rvert$ units if $h < 0$

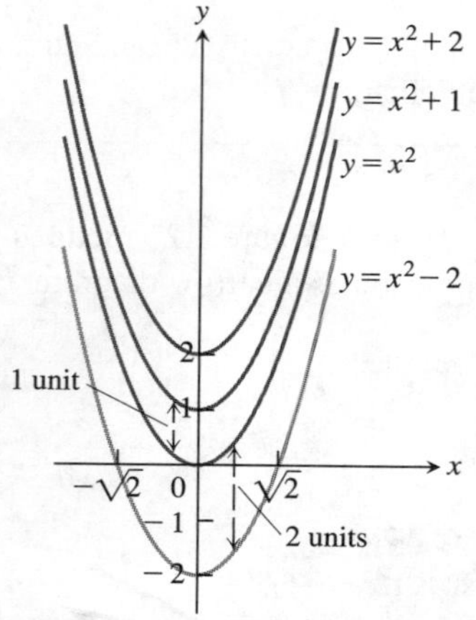

FIGURE 1.29 To shift the graph of $f(x) = x^2$ up (or down), we add positive (or negative) constants to the formula for f (Examples 3a and b).

EXAMPLE 3

(a) Adding 1 to the right-hand side of the formula $y = x^2$ to get $y = x^2 + 1$ shifts the graph up 1 unit (Figure 1.29).

(b) Adding -2 to the right-hand side of the formula $y = x^2$ to get $y = x^2 - 2$ shifts the graph down 2 units (Figure 1.29).

(c) Adding 3 to x in $y = x^2$ to get $y = (x + 3)^2$ shifts the graph 3 units to the left, while adding -2 shifts the graph 2 units to the right (Figure 1.30).

(d) Adding -2 to x in $y = |x|$, and then adding -1 to the result, gives $y = |x - 2| - 1$ and shifts the graph 2 units to the right and 1 unit down (Figure 1.31). ■

Scaling and Reflecting a Graph of a Function

To scale the graph of a function $y = f(x)$ is to stretch or compress it, vertically or horizontally. This is accomplished by multiplying the function f, or the independent variable x, by an appropriate constant c. Reflections across the coordinate axes are special cases where $c = -1$.

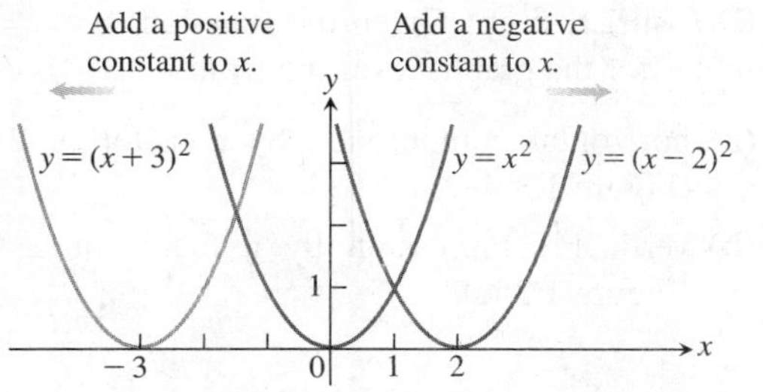

FIGURE 1.30 To shift the graph of $y = x^2$ to the left, we add a positive constant to x (Example 3c). To shift the graph to the right, we add a negative constant to x.

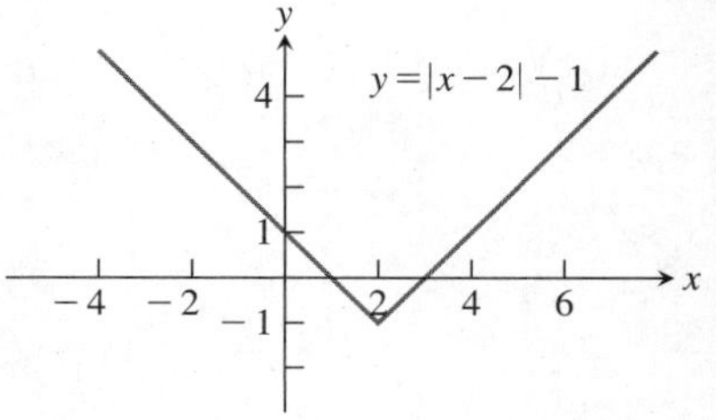

FIGURE 1.31 The graph of $y = |x|$ shifted 2 units to the right and 1 unit down (Example 3d).

Vertical and Horizontal Scaling and Reflecting Formulas

For $c > 1$, the graph is scaled:

$y = cf(x)$	Stretches the graph of f vertically by a factor of c.
$y = \frac{1}{c} f(x)$	Compresses the graph of f vertically by a factor of c.
$y = f(cx)$	Compresses the graph of f horizontally by a factor of c.
$y = f(x/c)$	Stretches the graph of f horizontally by a factor of c.

For $c = -1$, the graph is reflected:

$y = -f(x)$	Reflects the graph of f across the x-axis.
$y = f(-x)$	Reflects the graph of f across the y-axis.

EXAMPLE 4 Here we scale and reflect the graph of $y = \sqrt{x} + 1$.

(a) Vertical: Multiplying the right-hand side of $y = \sqrt{x} + 1$ by 3 to get $y = 3(\sqrt{x} + 1)$ stretches the graph vertically by a factor of 3, whereas multiplying by 1/3 compresses the graph vertically by a factor of 3 (Figure 1.32).

(b) Horizontal: The graph of $y = \sqrt{3x} + 1$ is a horizontal compression of the graph of $y = \sqrt{x} + 1$ by a factor of 3, and $y = \sqrt{x/3} + 1$ is a horizontal stretching by a factor of 3 (Figure 1.33).

(c) Reflection: The graph of $y = -(\sqrt{x} + 1)$ is a reflection of $y = \sqrt{x} + 1$ across the x-axis, and $y = \sqrt{-x} + 1$ is a reflection across the y-axis (Figure 1.34). ■

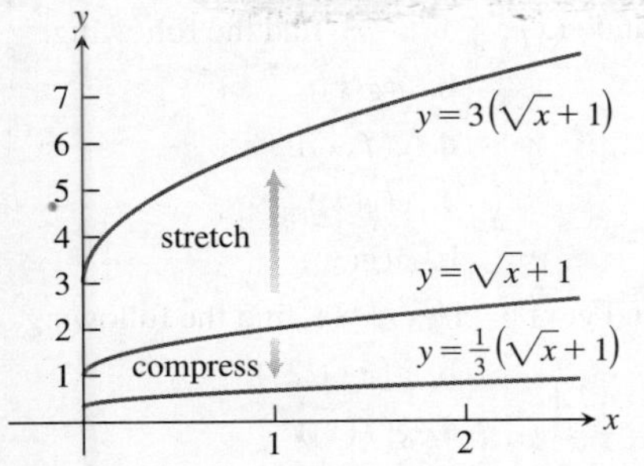

FIGURE 1.32 Vertically stretching and compressing the graph of $y = \sqrt{x} + 1$ by a factor of 3 (Example 4a).

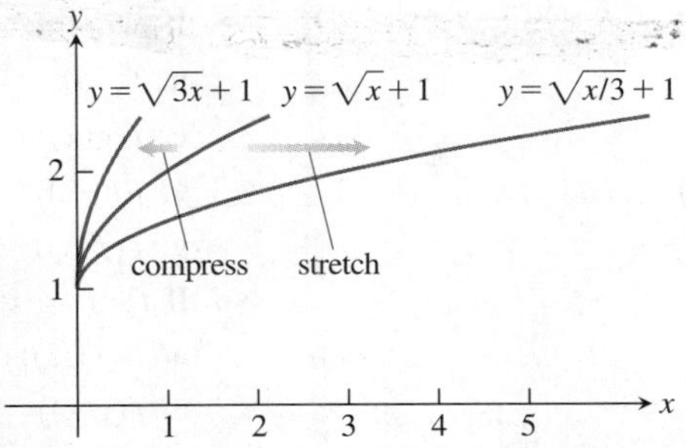

FIGURE 1.33 Horizontally stretching and compressing the graph of $y = \sqrt{x} + 1$ by a factor of 3 (Example 4b).

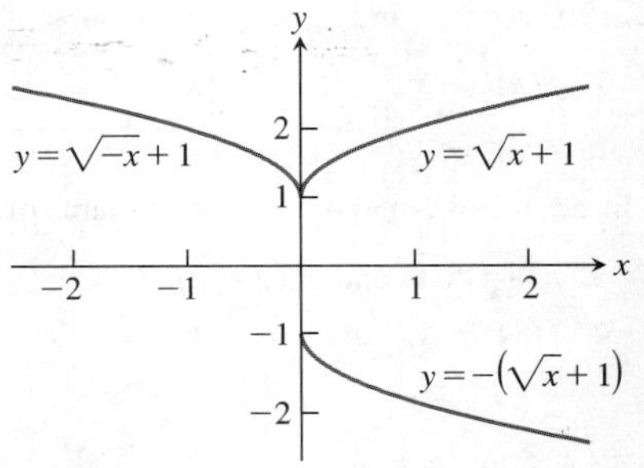

FIGURE 1.34 Reflections of the graph of $y = \sqrt{x} + 1$ across the coordinate axes (Example 4c).

EXAMPLE 5 Given the function $f(x) = x^4 - 4x^3 + 10$ (Figure 1.35a), find formulas for the graphs resulting from

(a) horizontal compression by a factor of 2 followed by reflection across the y-axis (Figure 1.35b).

(b) vertical compression by a factor of 2 followed by reflection across the x-axis (Figure 1.35c).

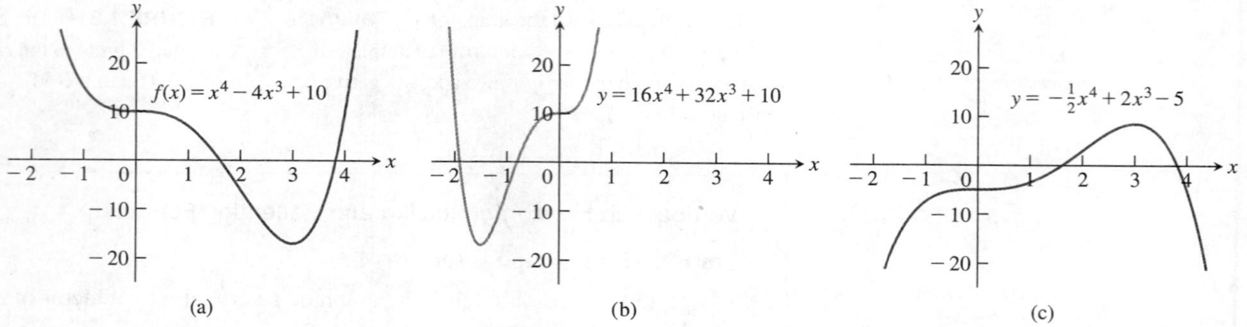

FIGURE 1.35 (a) The original graph of f. (b) The horizontal compression of $y = f(x)$ in part (a) by a factor of 2, followed by a reflection across the y-axis. (c) The vertical compression of $y = f(x)$ in part (a) by a factor of 2, followed by a reflection across the x-axis (Example 5).

Solution

(a) We multiply x by 2 to get the horizontal compression, and by -1 to give reflection across the y-axis. The formula is obtained by substituting $-2x$ for x in the right-hand side of the equation for f:

$$\begin{aligned} y = f(-2x) &= (-2x)^4 - 4(-2x)^3 + 10 \\ &= 16x^4 + 32x^3 + 10. \end{aligned}$$

(b) The formula is

$$y = -\frac{1}{2}f(x) = -\frac{1}{2}x^4 + 2x^3 - 5. \quad \blacksquare$$

EXERCISES 1.2

Algebraic Combinations

In Exercises 1 and 2, find the domains of $f, g, f + g$, and $f \cdot g$.

1. $f(x) = x, \quad g(x) = \sqrt{x - 1}$

2. $f(x) = \sqrt{x + 1}, \quad g(x) = \sqrt{x - 1}$

In Exercises 3 and 4, find the domains of $f, g, f/g$, and g/f.

3. $f(x) = \sqrt{x}$ and $g(x) = |x - 3|$

4. $f(x) = 1, \quad g(x) = 1 + \sqrt{x}$

Compositions of Functions

5. If $f(x) = x + 5$ and $g(x) = x^2 - 3$, find the following.

a. $f(g(0))$ **b.** $f(g(1))$

c. $f(g(x))$ **d.** $g(f(x))$

e. $f(f(-2))$ **f.** $g(g(4))$

g. $f(f(x))$ **h.** $g(g(x))$

6. If $f(x) = x - 1$ and $g(x) = 1/(x + 1)$, find the following.

a. $f(g(1/3))$ **b.** $g(f(1/2))$

c. $f(g(x))$ **d.** $g(f(x))$

e. $f(f(-2))$ **f.** $g(g(5))$

g. $f(f(x))$ **h.** $g(g(x))$

In Exercises 7–10, write a formula for $f \circ g \circ h$.

7. $f(x) = x + 1, \quad g(x) = 3x, \quad h(x) = 4 - x$

8. $f(x) = 3x + 4, \quad g(x) = 2x - 1, \quad h(x) = x^2$

9. $f(x) = \sqrt{x+3}, \; g(x) = \dfrac{1}{x-2}, \; h(x) = \dfrac{1}{x}$

10. $f(x) = \dfrac{x+2}{3-x}, \quad g(x) = \dfrac{x^2}{x^2+1}, \quad h(x) = \sqrt{2-x}$

Let $f(x) = x - 3, \; g(x) = \sqrt{x}, \; h(x) = x^3,$ and $j(x) = 2x.$ Express each of the functions in Exercises 11 and 12 as a composition involving one or more of f, g, h, and j.

11. a. $y = \sqrt{x} - 3$ **b.** $y = 2\sqrt{x}$

c. $y = x^{1/4}$ **d.** $y = 4x$

e. $y = \sqrt{x^3} - 3$ **f.** $y = 8x^3 - 3$

12. a. $y = 2x - 3$ **b.** $y = x^{3/2}$

c. $y = x^9$ **d.** $y = 2x - 6$

e. $y = \sqrt{x-3}$ **f.** $y = \sqrt{x^3 - 3}$

13. Copy and complete the following table.

	$g(x)$	$f(x)$	$(f \circ g)(x)$
a.	$x - 7$	$\sqrt{x}$	?
b.	$x + 2$	$3x$	?
c.	$g(x) = ?$	$f(x) = \sqrt{x^2 - 6}$	$(f \circ g)(x) = \sqrt{x^4 - 6}$
d.	$\dfrac{x}{x-1}$	$\dfrac{x}{x-1}$	?
e.	?	$1 + \dfrac{1}{x}$	x
f.	$g(x) = \dfrac{5}{x}$	$f(x) = ?$	$(f \circ g)(x) = 2x$

14. Copy and complete the following table.

	$g(x)$	$f(x)$	$(f \circ g)(x)$
a.	$\dfrac{1}{x-1}$	$\lvert x \rvert$	?
b.	$g(x) = ?$	$f(x) = \dfrac{2x-10}{x}$	$(f \circ g)(x) = \dfrac{2x - 10x^2}{x}$
c.	?	$\sqrt{x}$	$\lvert x \rvert$
d.	$\sqrt{x}$	?	$\lvert x \rvert$

15. Evaluate each expression using the given table of values:

x	−2	−1	0	1	2
$f(x)$	1	0	−2	1	2
$g(x)$	2	1	0	−1	0

a. $f(g(-1))$ **b.** $g(f(0))$

c. $f(f(-1))$ **d.** $g(g(2))$

e. $g(f(-2))$ **f.** $f(g(1))$

16. Evaluate each expression using the functions

$$f(x) = 2 - x, \quad g(x) = \begin{cases} -x, & -2 \le x < 0 \\ x - 1, & 0 \le x \le 2. \end{cases}$$

a. $f(g(0))$ **b.** $g(f(3))$ **c.** $g(g(-1))$

d. $f(f(2))$ **e.** $g(f(0))$ **f.** $f(g(1/2))$

In Exercises 17 and 18, **(a)** write formulas for $f \circ g$ and $g \circ f$ and **(b)** find the domain of each.

17. $f(x) = \sqrt{x+1}, \; g(x) = \dfrac{1}{x}$

18. $f(x) = x^2, \; g(x) = 1 - \sqrt{x}$

19. Let $f(x) = \dfrac{x-8}{2}$. Find the function $y = g(x)$ such that $(f \circ g)(x) = 2x$.

20. Let $f(x) = x^2 + 8$. Find the function $y = g(x)$ such that $(f \circ g)(x) = 2x + 5$.

21. A balloon's volume V is given by $V = 2s^2 + 7s + 10 \text{ cm}^3$, where s is the ambient temperature in °C. The ambient temperature s at time t minutes is given by $s = 2t - 3$ °C. Write the balloon's volume V as a function of time t.

22. Use the graphs of f and g to sketch the graph of $y = f(g(x))$.

a.

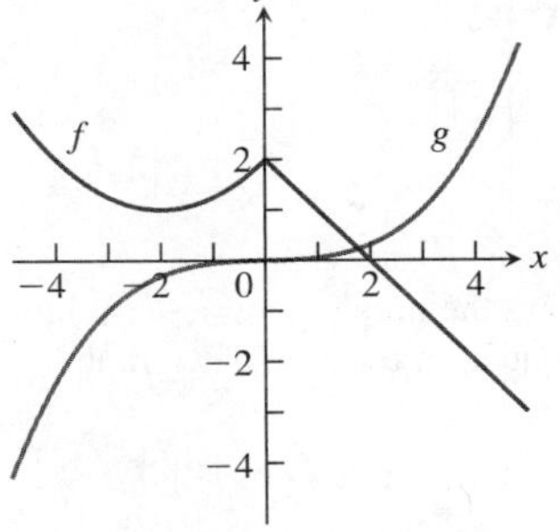

b.

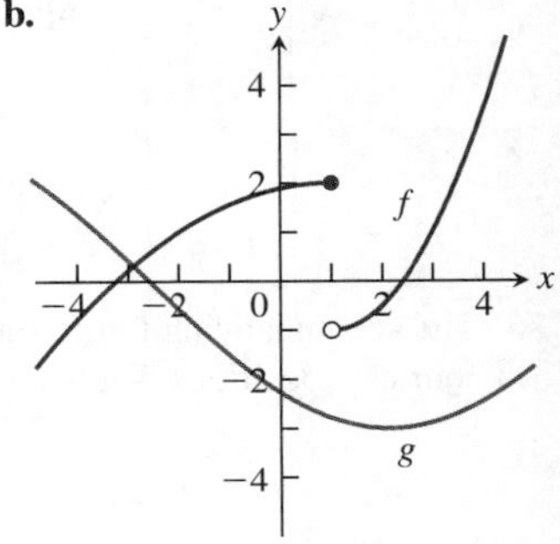

Shifting Graphs

23. The accompanying figure shows the graph of $y = -x^2$ shifted to two new positions. Write equations for the new graphs.

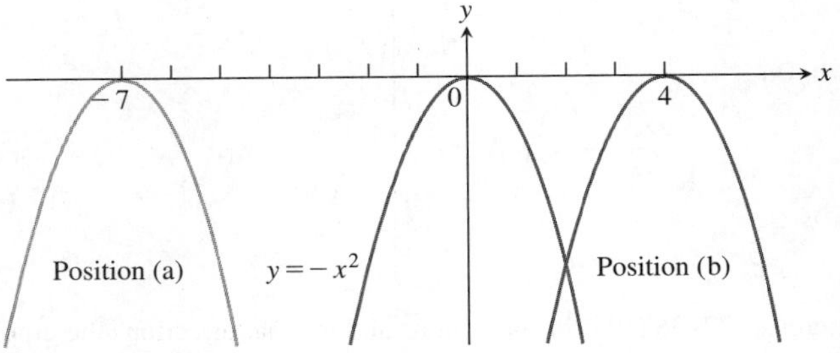

24. The accompanying figure shows the graph of $y = x^2$ shifted to two new positions. Write equations for the new graphs.

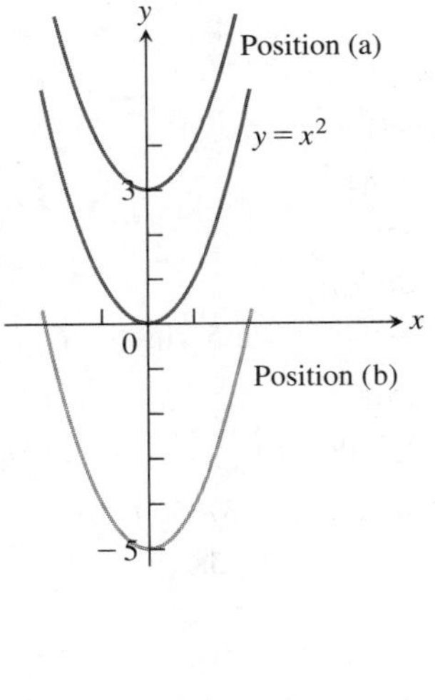

25. Match the equations listed in parts (a)–(d) to the graphs in the accompanying figure.

a. $y = (x - 1)^2 - 4$ **b.** $y = (x - 2)^2 + 2$

c. $y = (x + 2)^2 + 2$ **d.** $y = (x + 3)^2 - 2$

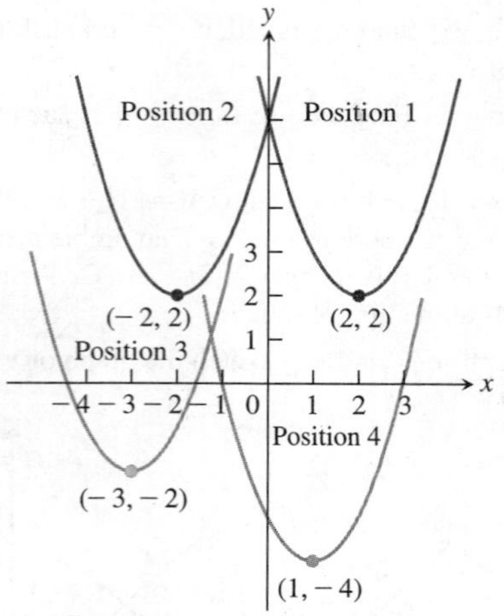

26. The accompanying figure shows the graph of $y = -x^2$ shifted to four new positions. Write an equation for each new graph.

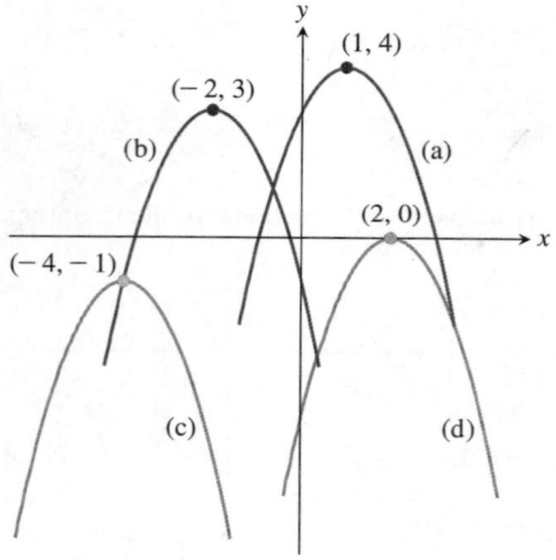

Exercises 27–36 tell how many units and in what directions the graphs of the given equations are to be shifted. Give an equation for the shifted graph. Then sketch the original and shifted graphs together, labeling each graph with its equation.

27. $x^2 + y^2 = 49$ Down 3, left 2

28. $x^2 + y^2 = 25$ Up 3, left 4

29. $y = x^3$ Left 1, down 1

30. $y = x^{2/3}$ Right 1, down 1

31. $y = \sqrt{x}$ Left 0.81

32. $y = -\sqrt{x}$ Right 3

33. $y = 2x - 7$ Up 7

34. $y = \frac{1}{2}(x + 1) + 5$ Down 5, right 1

35. $y = 1/x$ Up 1, right 1

36. $y = 1/x^2$ Left 2, down 1

Graph the functions in Exercises 37–56.

37. $\sqrt{x-2}$ **38.** $y = \sqrt{9 - x}$

39. $|x| + 1$ **40.** $y = |1 - x| - 1$

41. $\sqrt{x-1} + 2$ **42.** $y = 1 - \sqrt{x}$

43. $y = (x + 1)^{2/3}$ **44.** $y = (x - 8)^{2/3}$

45. $y = 1 - x^{2/3}$ **46.** $y + 4 = x^{2/3}$

47. $y = \sqrt[3]{x - 1} - 1$ **48.** $y = (x + 2)^{3/2} + 1$

49. $y = \dfrac{1}{x - 2}$ **50.** $y = \dfrac{1}{x} - 2$

51. $y = \dfrac{1}{x} + 2$ **52.** $y = \dfrac{1}{x + 2}$

53. $y = \dfrac{1}{(x - 1)^2}$ **54.** $y = \dfrac{1}{x^2} - 1$

55. $y = \dfrac{1}{x^2} + 1$ **56.** $y = \dfrac{1}{(x + 1)^2}$

57. The accompanying figure shows the graph of a function $f(x)$ with domain $[0, 2]$ and range $[0, 1]$. Find the domains and ranges of the following functions, and sketch their graphs.

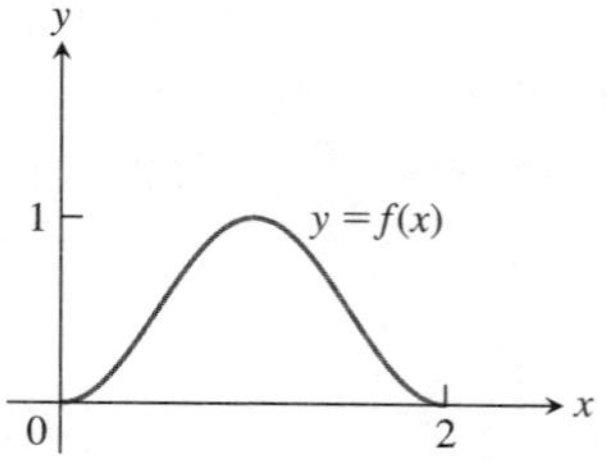

a. $f(x) + 2$ **b.** $f(x) - 1$

c. $2f(x)$ **d.** $-f(x)$

e. $f(x + 2)$ **f.** $f(x - 1)$

g. $f(-x)$ **h.** $-f(x + 1) + 1$

58. The accompanying figure shows the graph of a function $g(t)$ with domain $[-4, 0]$ and range $[-3, 0]$. Find the domains and ranges of the following functions, and sketch their graphs.

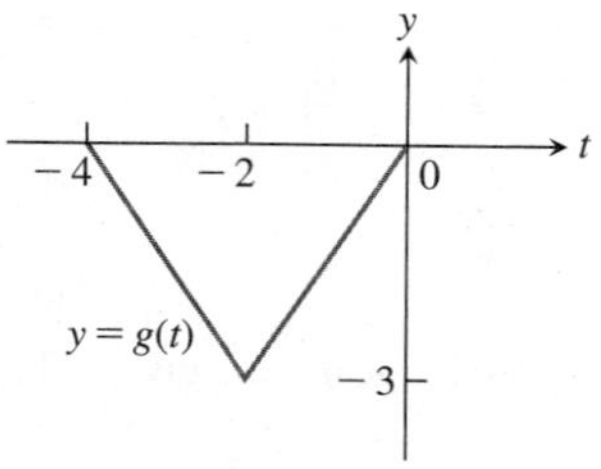

a. $g(-t)$ **b.** $-g(t)$

c. $g(t) + 3$ **d.** $1 - g(t)$

e. $g(-t + 2)$ **f.** $g(t - 2)$

g. $g(1 - t)$ **h.** $-g(t - 4)$

Vertical and Horizontal Scaling

Exercises 59–68 tell in what direction and by what factor the graphs of the given functions are to be stretched or compressed. Give an equation for the stretched or compressed graph.

59. $y = x^2$, stretched vertically by a factor of 3

60. $y = x^2 - 1$, compressed horizontally by a factor of 2

61. $y = x + \dfrac{1}{x}$, compressed vertically by a factor of 2

62. $y = 1 + \dfrac{1}{x^2}$, stretched horizontally by a factor of 3

63. $y = \sqrt{x+4}$, compressed horizontally by a factor of 2

64. $y = \sqrt{4x+3}$, stretched vertically by a factor of 4

65. $y = \sqrt{4 - x^2}$, stretched horizontally by a factor of 2

66. $y = \sqrt{3 - x^2}$, compressed vertically by a factor of 2

67. $y = 1 - x^3$, compressed horizontally by a factor of 3

68. $y = 2 - x^2$, stretched horizontally by a factor of 2

Graphing

In Exercises 69–76, graph each function not by plotting points, but by starting with the graph of one of the standard functions presented in Figures 1.14–1.17 and applying an appropriate transformation.

69. $y = -\sqrt{2x+1}$ 70. $y = \sqrt{1 - \dfrac{x}{2}}$

71. $y = (x-1)^3 + 2$ 72. $y = (1-x)^3 + 2$

73. $y = \dfrac{1}{2x} - 1$ 74. $y = \dfrac{2}{x^2} + 1$

75. $y = -\sqrt[3]{x}$ 76. $y = (-2x)^{2/3}$

77. Graph the function $y = |x^2 - 1|$.

78. Graph the function $y = \sqrt{|x|}$

Combining Functions

79. Assume that f is an even function, g is an odd function, and both f and g are defined on the entire real line $(-\infty, \infty)$. Which of the following (where defined) are even? odd?

a. gf b. f/g c. g/f
d. $f^2 = ff$ e. $g^2 = gg$ f. $f \circ g$
g. $g \circ f$ h. $f \circ f$ i. $g \circ g$

80. Can a function be both even and odd? Give reasons for your answer.

T 81. (*Continuation of Example 1.*) Graph the functions $f(x) = \sqrt{x}$ and $g(x) = \sqrt{1-x}$ together with their (a) sum, (b) product, (c) two differences, (d) two quotients.

T 82. Let $f(x) = x - 7$ and $g(x) = x^2$. Graph f and g together with $f \circ g$ and $g \circ f$.

1.3 Trigonometric Functions

This section reviews radian measure and the basic trigonometric functions.

Angles

Angles are measured in degrees or radians. The number of **radians** in the central angle $A'CB'$ within a circle of radius r is defined as the number of "radius units" contained in the arc s subtended by that central angle. If we denote this central angle by θ when measured in radians, this means that $\theta = s/r$ (Figure 1.36), or

$$s = r\theta \qquad (\theta \text{ in radians}). \tag{1}$$

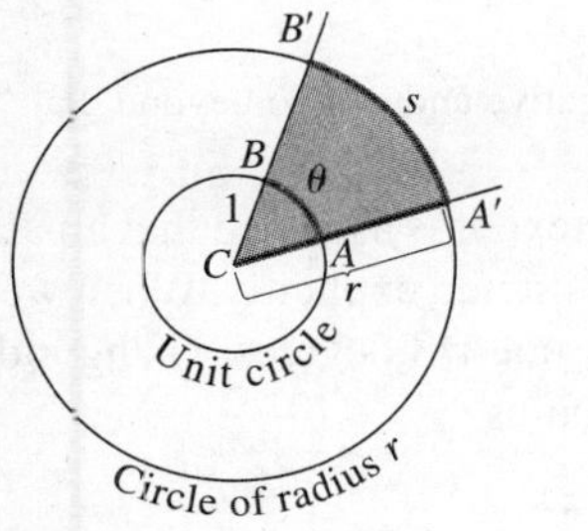

FIGURE 1.36 The radian measure of the central angle $A'CB'$ is the number $\theta = s/r$. For a unit circle of radius $r = 1$, θ is the length of arc AB that central angle ACB cuts from the unit circle.

If the circle is a unit circle having radius $r = 1$, then from Figure 1.36 and Equation (1), we see that the central angle θ measured in radians is just the length of the arc that the angle cuts from the unit circle. Since one complete revolution of the unit circle is 360° or 2π radians, we have

$$\pi \text{ radians} = 180° \tag{2}$$

and

$$1 \text{ radian} = \frac{180}{\pi} (\approx 57.3) \text{ degrees} \quad \text{or} \quad 1 \text{ degree} = \frac{\pi}{180} (\approx 0.017) \text{ radians.}$$

Table 1.1 shows the equivalence between degree and radian measures for some basic angles.

TABLE 1.1 Angles measured in degrees and radians

Degrees	−180	−135	−90	−45	0	30	45	60	90	120	135	150	180	270	360
θ (radians)	$-\pi$	$\frac{-3\pi}{4}$	$\frac{-\pi}{2}$	$\frac{-\pi}{4}$	0	$\frac{\pi}{6}$	$\frac{\pi}{4}$	$\frac{\pi}{3}$	$\frac{\pi}{2}$	$\frac{2\pi}{3}$	$\frac{3\pi}{4}$	$\frac{5\pi}{6}$	π	$\frac{3\pi}{2}$	2π

An angle in the xy-plane is said to be in **standard position** if its vertex lies at the origin and its initial ray lies along the positive x-axis (Figure 1.37). Angles measured counterclockwise from the positive x-axis are assigned positive measures; angles measured clockwise are assigned negative measures.

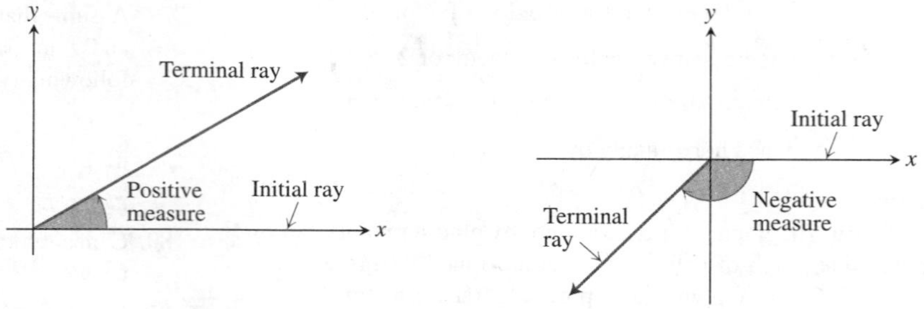

FIGURE 1.37 Angles in standard position in the xy-plane.

Angles describing counterclockwise rotations can go arbitrarily far beyond 2π radians or 360°. Similarly, angles describing clockwise rotations can have negative measures of all sizes (Figure 1.38).

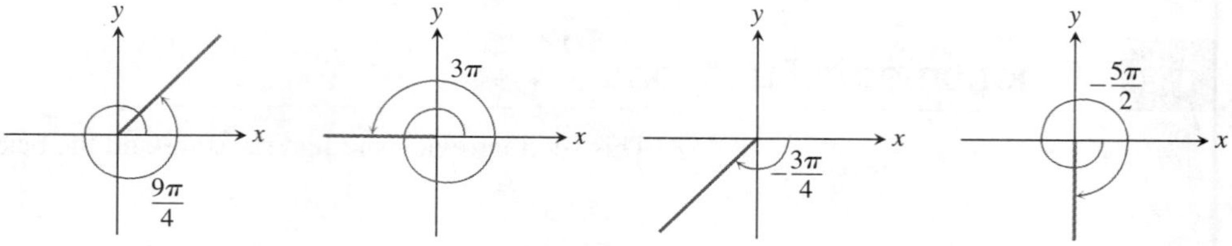

FIGURE 1.38 Nonzero radian measures can be positive or negative and can go beyond 2π.

Angle Convention: Use Radians From now on in this text, it is assumed that all angles are measured in radians unless degrees or some other unit is stated explicitly. When we talk about the angle $\pi/3$, we mean $\pi/3$ radians (which is 60°), not $\pi/3$ degrees. Using radians simplifies many of the operations and computations in calculus.

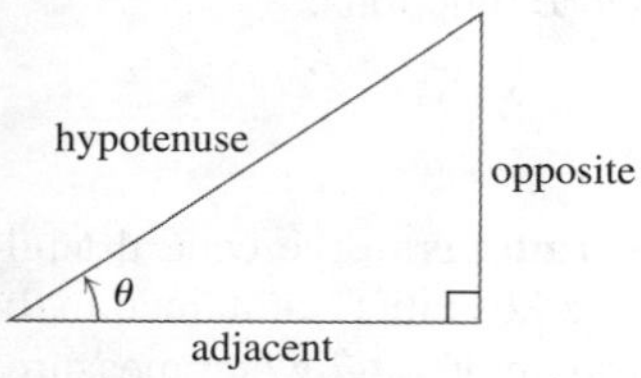

$$\sin\theta = \frac{\text{opp}}{\text{hyp}} \qquad \csc\theta = \frac{\text{hyp}}{\text{opp}}$$

$$\cos\theta = \frac{\text{adj}}{\text{hyp}} \qquad \sec\theta = \frac{\text{hyp}}{\text{adj}}$$

$$\tan\theta = \frac{\text{opp}}{\text{adj}} \qquad \cot\theta = \frac{\text{adj}}{\text{opp}}$$

FIGURE 1.39 Trigonometric ratios of an acute angle.

The Six Basic Trigonometric Functions

The trigonometric functions of an acute angle are given in terms of the sides of a right triangle (Figure 1.39). We extend this definition to obtuse and negative angles by first placing the angle in standard position in a circle of radius r. We then define the trigonometric functions in terms of the coordinates of the point $P(x, y)$ where the angle's terminal ray intersects the circle (Figure 1.40).

$$\textbf{sine:} \quad \sin\theta = \frac{y}{r} \qquad \textbf{cosecant:} \quad \csc\theta = \frac{r}{y}$$

$$\textbf{cosine:} \quad \cos\theta = \frac{x}{r} \qquad \textbf{secant:} \quad \sec\theta = \frac{r}{x}$$

$$\textbf{tangent:} \quad \tan\theta = \frac{y}{x} \qquad \textbf{cotangent:} \quad \cot\theta = \frac{x}{y}$$

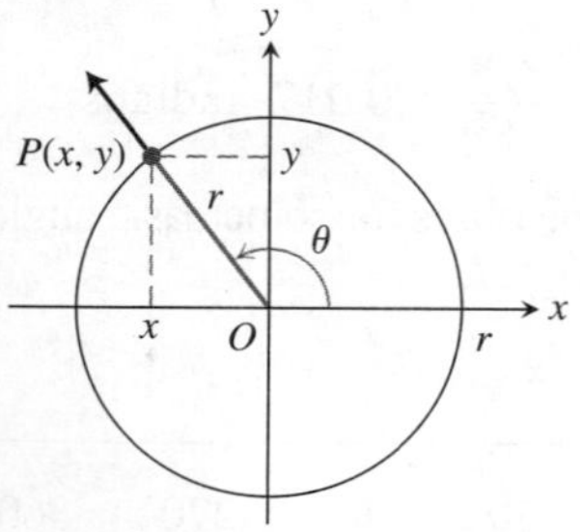

FIGURE 1.40 The trigonometric functions of a general angle θ are defined in terms of x, y, and r.

These extended definitions agree with the right-triangle definitions when the angle is acute.

Notice also that whenever the quotients are defined,

$$\tan\theta = \frac{\sin\theta}{\cos\theta} \qquad \cot\theta = \frac{1}{\tan\theta}$$

$$\sec\theta = \frac{1}{\cos\theta} \qquad \csc\theta = \frac{1}{\sin\theta}$$

As you can see, $\tan\theta$ and $\sec\theta$ are not defined if $x = \cos\theta = 0$. This means they are not defined if θ is $\pm\pi/2, \pm 3\pi/2, \ldots$. Similarly, $\cot\theta$ and $\csc\theta$ are not defined for values of θ for which $y = 0$, namely $\theta = 0, \pm\pi, \pm 2\pi, \ldots$.

The exact values of these trigonometric ratios for some angles can be read from the triangles in Figure 1.41. For instance,

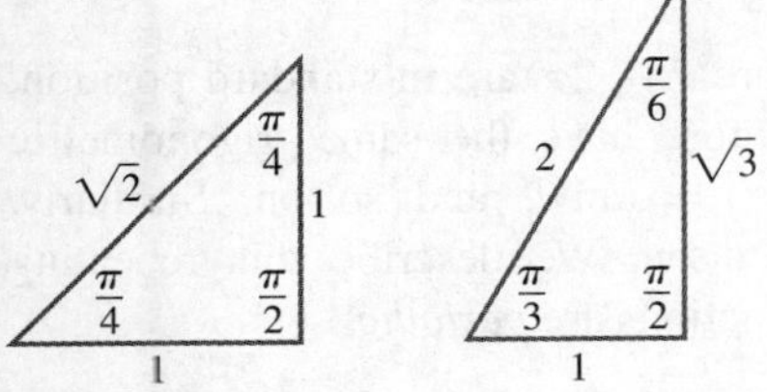

FIGURE 1.41 Radian angles and side lengths of two common triangles.

$$\sin\frac{\pi}{4} = \frac{1}{\sqrt{2}} \qquad \sin\frac{\pi}{6} = \frac{1}{2} \qquad \sin\frac{\pi}{3} = \frac{\sqrt{3}}{2}$$

$$\cos\frac{\pi}{4} = \frac{1}{\sqrt{2}} \qquad \cos\frac{\pi}{6} = \frac{\sqrt{3}}{2} \qquad \cos\frac{\pi}{3} = \frac{1}{2}$$

$$\tan\frac{\pi}{4} = 1 \qquad \tan\frac{\pi}{6} = \frac{1}{\sqrt{3}} \qquad \tan\frac{\pi}{3} = \sqrt{3}$$

The ASTC rule (Figure 1.42) is useful for remembering when the basic trigonometric functions are positive or negative. For instance, from the triangle in Figure 1.43, we see that

$$\sin\frac{2\pi}{3} = \frac{\sqrt{3}}{2}, \qquad \cos\frac{2\pi}{3} = -\frac{1}{2}, \qquad \tan\frac{2\pi}{3} = -\sqrt{3}.$$

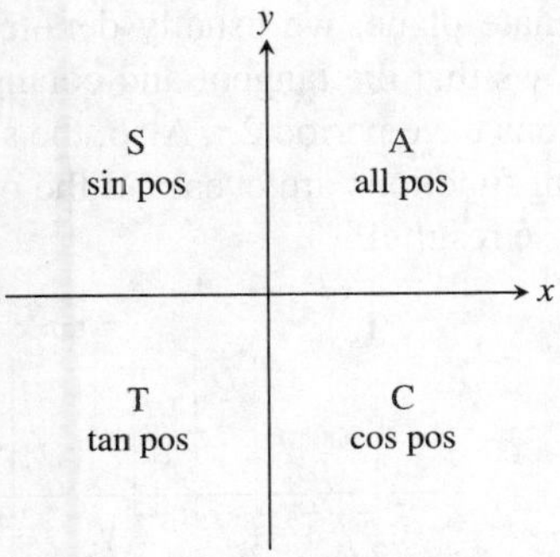

FIGURE 1.42 The ASTC rule, remembered by the statement "All Students Take Calculus," tells which trigonometric functions are positive in each quadrant.

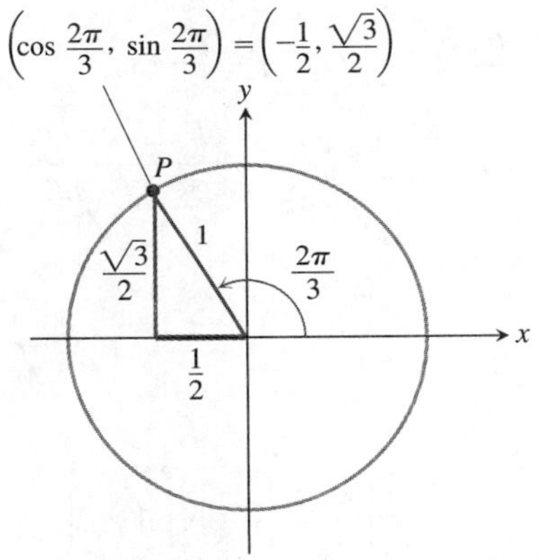

FIGURE 1.43 The triangle for calculating the sine and cosine of $2\pi/3$ radians. The side lengths come from the geometry of right triangles.

Using a similar method we obtain the values of $\sin\theta$, $\cos\theta$, and $\tan\theta$ shown in Table 1.2.

TABLE 1.2 Values of $\sin\theta$, $\cos\theta$, and $\tan\theta$ for selected values of θ

Degrees	−180	−135	−90	−45	0	30	45	60	90	120	135	150	180	270	360
θ (radians)	$-\pi$	$\frac{-3\pi}{4}$	$\frac{-\pi}{2}$	$\frac{-\pi}{4}$	0	$\frac{\pi}{6}$	$\frac{\pi}{4}$	$\frac{\pi}{3}$	$\frac{\pi}{2}$	$\frac{2\pi}{3}$	$\frac{3\pi}{4}$	$\frac{5\pi}{6}$	π	$\frac{3\pi}{2}$	2π
$\sin\theta$	0	$\frac{-\sqrt{2}}{2}$	−1	$\frac{-\sqrt{2}}{2}$	0	$\frac{1}{2}$	$\frac{\sqrt{2}}{2}$	$\frac{\sqrt{3}}{2}$	1	$\frac{\sqrt{3}}{2}$	$\frac{\sqrt{2}}{2}$	$\frac{1}{2}$	0	−1	0
$\cos\theta$	−1	$\frac{-\sqrt{2}}{2}$	0	$\frac{\sqrt{2}}{2}$	1	$\frac{\sqrt{3}}{2}$	$\frac{\sqrt{2}}{2}$	$\frac{1}{2}$	0	$-\frac{1}{2}$	$\frac{-\sqrt{2}}{2}$	$\frac{-\sqrt{3}}{2}$	−1	0	1
$\tan\theta$	0	1		−1	0	$\frac{\sqrt{3}}{3}$	1	$\sqrt{3}$		$-\sqrt{3}$	−1	$\frac{-\sqrt{3}}{3}$	0		0

Periodicity and Graphs of the Trigonometric Functions

When an angle of measure θ and an angle of measure $\theta + 2\pi$ are in standard position, their terminal rays coincide. The two angles therefore have the same trigonometric function values: $\sin(\theta + 2\pi) = \sin\theta$, $\tan(\theta + 2\pi) = \tan\theta$, and so on. Similarly, $\cos(\theta - 2\pi) = \cos\theta$, $\sin(\theta - 2\pi) = \sin\theta$, and so on. We describe this repeating behavior by saying that the six basic trigonometric functions are *periodic*.

Periods of Trigonometric Functions

Period π: $\tan(x + \pi) = \tan x$
$\cot(x + \pi) = \cot x$

Period 2π: $\sin(x + 2\pi) = \sin x$
$\cos(x + 2\pi) = \cos x$
$\sec(x + 2\pi) = \sec x$
$\csc(x + 2\pi) = \csc x$

DEFINITION A function $f(x)$ is **periodic** if there is a positive number p such that $f(x + p) = f(x)$ for every value of x. The smallest such value of p is the **period** of f.

When we graph trigonometric functions in the coordinate plane, we usually denote the independent variable by x instead of θ. Figure 1.44 shows that the tangent and cotangent functions have period $p = \pi$, and the other four functions have period 2π. Also, the symmetries in these graphs reveal that the cosine and secant functions are even and the other four functions are odd (although this does not prove those results).

Even
$\cos(-x) = \cos x$
$\sec(-x) = \sec x$

Odd
$\sin(-x) = -\sin x$
$\tan(-x) = -\tan x$
$\csc(-x) = -\csc x$
$\cot(-x) = -\cot x$

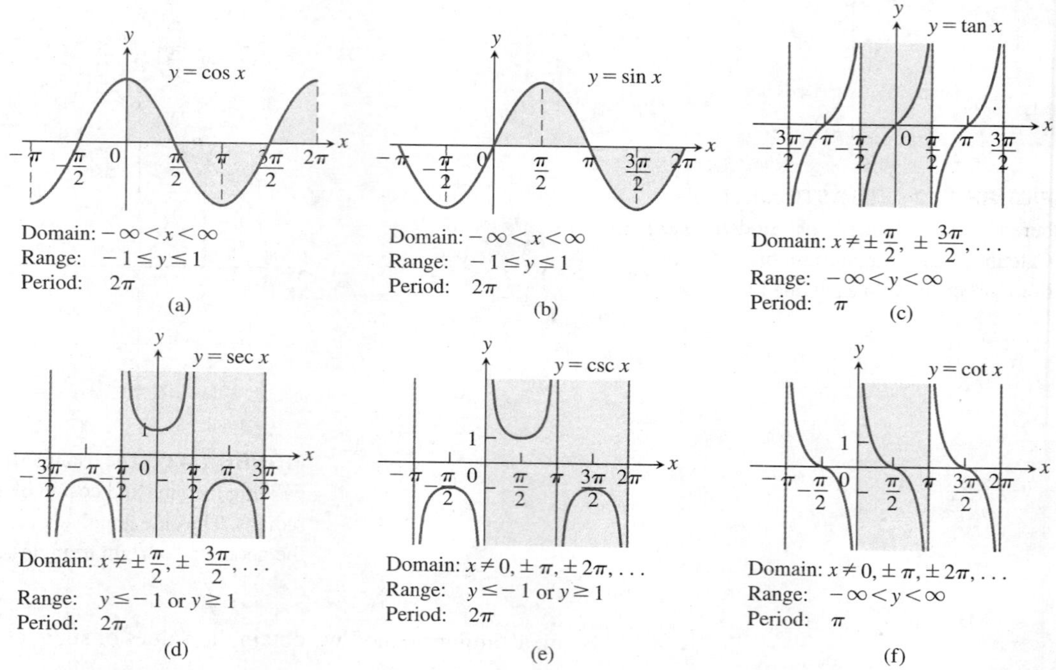

FIGURE 1.44 Graphs of the six basic trigonometric functions using radian measure. The shading for each trigonometric function indicates its periodicity.

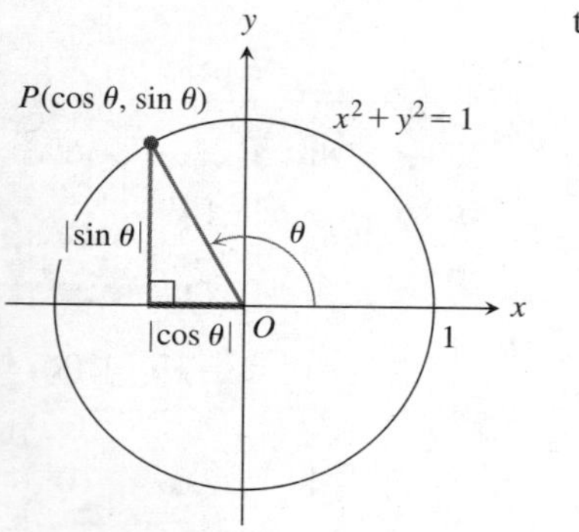

FIGURE 1.45 The reference triangle for a general angle θ.

Trigonometric Identities

The coordinates of any point $P(x, y)$ in the plane can be expressed in terms of the point's distance r from the origin and the angle θ that ray OP makes with the positive x-axis (Figure 1.40). Since $x/r = \cos\theta$ and $y/r = \sin\theta$, we have

$$x = r\cos\theta, \quad y = r\sin\theta.$$

When $r = 1$ we can apply the Pythagorean theorem to the reference right triangle in Figure 1.45 and obtain the equation

$$\cos^2\theta + \sin^2\theta = 1. \tag{3}$$

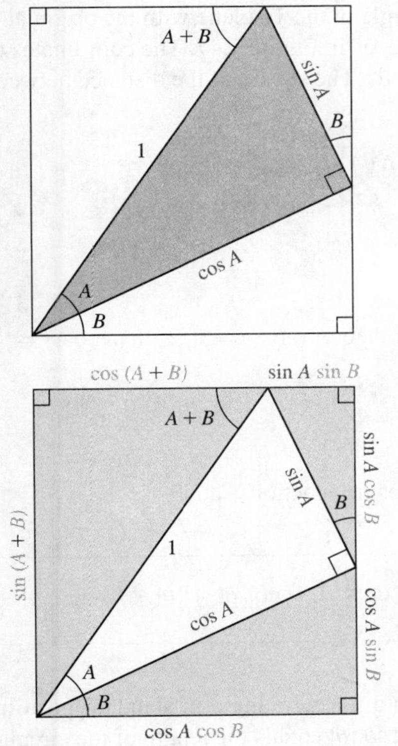

FIGURE 1.46 In a geometric proof of the angle sum identities we compare the opposite sides of the rectangle, which are equal. This assumes that A, B, and $A + B$ are acute, but the identities hold for all values of A and B.

This equation, true for all values of θ, is the most frequently used identity in trigonometry. Dividing this identity in turn by $\cos^2\theta$ and $\sin^2\theta$ gives

$$1 + \tan^2\theta = \sec^2\theta$$
$$1 + \cot^2\theta = \csc^2\theta$$

The following formulas hold for all angles A and B (see Figure 1.46 and Exercise 58).

Addition Formulas

$$\cos(A + B) = \cos A \cos B - \sin A \sin B$$
$$\sin(A + B) = \sin A \cos B + \cos A \sin B \tag{4}$$

There are similar formulas for $\cos(A - B)$ and $\sin(A - B)$ (Exercises 35 and 36). All the trigonometric identities needed in this text derive from Equations (3) and (4). For example, substituting θ for both A and B in the addition formulas gives

Double-Angle Formulas

$$\cos 2\theta = \cos^2\theta - \sin^2\theta$$
$$\sin 2\theta = 2\sin\theta\cos\theta \tag{5}$$

Additional formulas come from combining the equations

$$\cos^2\theta + \sin^2\theta = 1, \qquad \cos^2\theta - \sin^2\theta = \cos 2\theta.$$

We add the two equations to get $2\cos^2\theta = 1 + \cos 2\theta$ and subtract the second from the first to get $2\sin^2\theta = 1 - \cos 2\theta$. This results in the following identities, which are useful in integral calculus.

Half-Angle Formulas

$$\cos^2\theta = \frac{1 + \cos 2\theta}{2} \tag{6}$$

$$\sin^2\theta = \frac{1 - \cos 2\theta}{2} \tag{7}$$

The Law of Cosines

If a, b, and c are sides of a triangle ABC and if θ is the angle opposite c, then

$$c^2 = a^2 + b^2 - 2ab\cos\theta. \tag{8}$$

This equation is called the **law of cosines**.

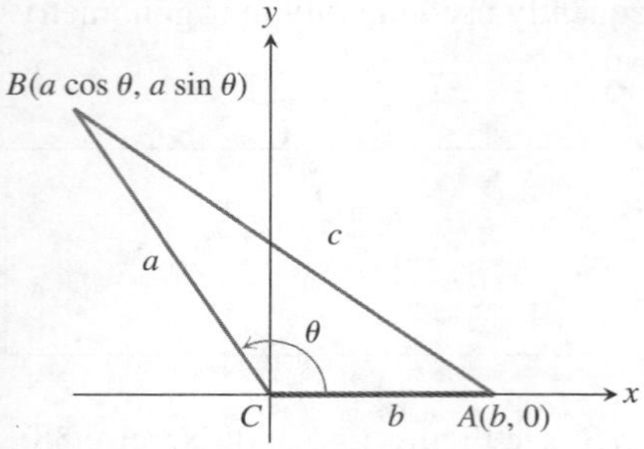

FIGURE 1.47 The square of the distance between A and B gives the law of cosines.

To see why the law holds, we position the triangle in the xy-plane with the origin at C and the positive x-axis along one side of the triangle, as in Figure 1.47. The coordinates of A are $(b, 0)$; the coordinates of B are $(a\cos\theta, a\sin\theta)$. The square of the distance between A and B is therefore

$$\begin{aligned} c^2 &= (a\cos\theta - b)^2 + (a\sin\theta)^2 \\ &= a^2\underbrace{(\cos^2\theta + \sin^2\theta)}_{1} + b^2 - 2ab\cos\theta \\ &= a^2 + b^2 - 2ab\cos\theta. \end{aligned}$$

The law of cosines generalizes the Pythagorean theorem. If $\theta = \pi/2$, then $\cos\theta = 0$ and $c^2 = a^2 + b^2$.

Two Special Inequalities

For any angle θ measured in radians, the sine and cosine functions satisfy

$$-|\theta| \le \sin\theta \le |\theta| \qquad \text{and} \qquad -|\theta| \le 1 - \cos\theta \le |\theta|.$$

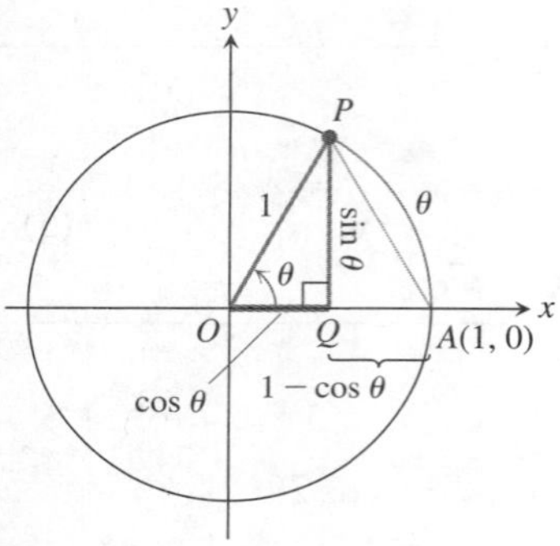

FIGURE 1.48 From the geometry of this figure, drawn for $\theta > 0$, we get the inequality $\sin^2\theta + (1 - \cos\theta)^2 \le \theta^2$.

To establish these inequalities, we picture θ as a nonzero angle in standard position (Figure 1.48). The circle in the figure is a unit circle, so $|\theta|$ equals the length of the circular arc AP. The length of line segment AP is therefore less than $|\theta|$.

Triangle APQ is a right triangle with sides of length

$$QP = |\sin\theta|, \qquad AQ = 1 - \cos\theta.$$

From the Pythagorean theorem and the fact that $AP < |\theta|$, we get

$$\sin^2\theta + (1 - \cos\theta)^2 = (AP)^2 \le \theta^2. \tag{9}$$

The terms on the left-hand side of Equation (9) are both positive, so each is smaller than their sum and hence is less than or equal to θ^2:

$$\sin^2\theta \le \theta^2 \qquad \text{and} \qquad (1 - \cos\theta)^2 \le \theta^2.$$

By taking square roots, this is equivalent to saying that

$$\sin\theta| \le |\theta| \qquad \text{and} \qquad |1 - \cos\theta| \le |\theta|,$$

so

$$-|\theta| \le \sin\theta \le |\theta| \qquad \text{and} \qquad -|\theta| \le 1 - \cos\theta \le |\theta|.$$

These inequalities will be useful in the next chapter.

Transformations of Trigonometric Graphs

The rules for shifting, stretching, compressing, and reflecting the graph of a function summarized in the following diagram apply to the trigonometric functions we have discussed in this section.

Vertical stretch or compression; reflection about $y = d$ if negative

Vertical shift

$$y = af(b(x + c)) + d$$

Horizontal stretch or compression; reflection about $x = -c$ if negative

Horizontal shift

The transformation rules applied to the sine function give the **general sine function** or **sinusoid** formula

$$f(x) = A\sin\left(\frac{2\pi}{B}(x - C)\right) + D,$$

where $|A|$ is the *amplitude*, $|B|$ is the *period*, C is the *horizontal shift*, and D is the *vertical shift*. A graphical interpretation of the various terms is given below.

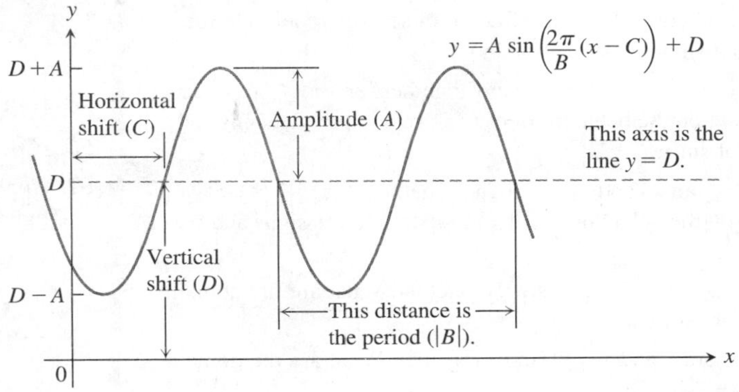

EXERCISES 1.3

Radians and Degrees

1. On a circle of radius 12 m, how long is an arc that subtends a central angle of **(a)** $3\pi/4$ radians? **(b)** 120°?

2. A central angle in a circle of radius 8 is subtended by an arc of length 10π. Find the angle's radian and degree measures.

3. You want to make an angle of 90° by marking an arc on the perimeter of 20 in. diameter disk and drawing lines from the ends of the arc to the disk's centre. To the nearest tenth of an inch, how long should the arc be?

4. If you roll a 1 m diameter wheel forward 35 cm over level ground, through what angle will the wheel turn? Answer in radians (to the nearest tenth) and degrees (to the nearest degree).

Evaluating Trigonometric Functions

5. Copy and complete the following table of function values. If the function is undefined at a given angle, enter "UND." Do not use a calculator or tables.

θ	$-\pi$	$-2\pi/3$	0	$\pi/2$	$3\pi/4$
$\sin\theta$					
$\cos\theta$					
$\tan\theta$					
$\cot\theta$					
$\sec\theta$					
$\csc\theta$					

6. Copy and complete the following table of function values. If the function is undefined at a given angle, enter "UND." Do not use a calculator or tables.

θ	$-3\pi/2$	$-\pi/3$	$-\pi/6$	$\pi/4$	$5\pi/6$
$\sin\theta$					
$\cos\theta$					
$\tan\theta$					
$\cot\theta$					
$\sec\theta$					
$\csc\theta$					

In Exercises 7–12, one of $\sin x$, $\cos x$, and $\tan x$ is given. Find the other two if x lies in the specified interval.

7. $\sin x = \frac{5}{13},\ x \in \left[\frac{\pi}{2}, \pi\right]$

8. $\tan x = 2,\quad x \in \left[0, \frac{\pi}{2}\right]$

9. $\cos x = \frac{1}{3},\quad x \in \left[-\frac{\pi}{2}, 0\right]$

10. $\cos x = -\frac{5}{13},\quad x \in \left[\frac{\pi}{2}, \pi\right]$

11. $\tan x = \frac{3}{4},\ x \in \left[\pi, \frac{3\pi}{2}\right]$

12. $\sin x = -\frac{1}{2},\quad x \in \left[\pi, \frac{3\pi}{2}\right]$

Graphing Trigonometric Functions

Graph the functions in Exercises 13–22. What is the period of each function?

13. $\sin 3x$

14. $\sin(x/2)$

15. $\cos 2x$

16. $\cos\frac{\pi x}{2}$

17. $-\sin\frac{\pi x}{3}$

18. $-\cos 2\pi x$

19. $\sin\left(x - \frac{\pi}{2}\right)$

20. $\cos\left(x + \frac{\pi}{2}\right)$

21. $\sin\left(x - \frac{\pi}{4}\right) + 1$ **22.** $\cos\left(x + \frac{2\pi}{3}\right) - 2$

Graph the functions in Exercises 23–26 in the *ts*-plane (*t*-axis horizontal, *s*-axis vertical). What is the period of each function? What symmetries do the graphs have?

23. $s = \cot 2t$ **24.** $s = -\tan \pi t$

25. $s = \sec\left(\frac{\pi t}{2}\right)$ **26.** $s = \csc\left(\frac{t}{2}\right)$

T **27. a.** Graph $y = \cos x$ and $y = \sec x$ together for $-3\pi/2 \le x \le 3\pi/2$. Comment on the behavior of $\sec x$ in relation to the signs and values of $\cos x$.

b. Graph $y = \sin x$ and $y = \csc x$ together for $-\pi \le x \le 2\pi$. Comment on the behavior of $\csc x$ in relation to the signs and values of $\sin x$.

T **28.** Graph $y = \tan x$ and $y = \cot x$ together for $-7 \le x \le 7$. Comment on the behavior of $\cot x$ in relation to the signs and values of $\tan x$.

29. Graph $y = \sin x$ and $y = \lfloor \sin x \rfloor$ together. What are the domain and range of $\lfloor \sin x \rfloor$?

30. Graph $y = \sin x$ and $y = \lceil \sin x \rceil$ together. What are the domain and range of $\lceil \sin x \rceil$?

Using the Addition Formulas

Use the addition formulas to derive the identities in Exercises 31–36.

31. $\cos\left(2x - \frac{\pi}{2}\right) = \sin 2x$ **32.** $\cos\left(x + \frac{3\pi}{2}\right) = \sin x$

33. $\sin\left(2x - \frac{\pi}{2}\right) = -\cos 2x$ **34.** $\sin\left(x + \frac{3\pi}{2}\right) = \cos x$

35. $\sin(B - A) = \sin B \cos A - \cos B \sin A$

36. $\cos(B - A) = \cos B \cos A + \sin B \sin A$

37. What happens if you take $B = A$ in the trigonometric identity $\cos(A - B) = \cos A \cos B + \sin A \sin B$? Does the result agree with something you already know?

38. What happens if you take $B = 2\pi$ in the addition formulas? Do the results agree with something you already know?

In Exercises 39–42, express the given quantity in terms of $\sin x$ and $\cos x$.

39. $\cos(2\pi + x)$ **40.** $\sin\left(\frac{\pi}{2} - x\right)$

41. $\sin\left(\frac{3\pi}{2} - x\right)$ **42.** $\cos\left(\frac{3\pi}{2} + x\right)$

43. Evaluate $\sin \frac{11\pi}{12}$ as $\sin\left(\frac{\pi}{4} + \frac{2\pi}{3}\right)$.

44. Evaluate $\cos \frac{7\pi}{12}$ as $\cos\left(\frac{\pi}{4} + \frac{\pi}{3}\right)$.

45. Evaluate $\cos \frac{\pi}{12}$. **46.** Evaluate $\sin \frac{5\pi}{12}$.

Using the Half-Angle Formulas

Find the function values in Exercises 47–50.

47. $\cos^2 \frac{\pi}{8}$ **48.** $\cos^2 \frac{5\pi}{12}$

49. $\cos^2 \frac{\pi}{12}$ **50.** $\sin^2 \frac{3\pi}{8}$

Solving Trigonometric Equations

For Exercises 51–54, solve for the angle θ, where $0 \le \theta \le 2\pi$.

51. $\sin^2 \theta = \frac{3}{4}$ **52.** $\sin^2 \theta = \cos^2 \theta$

53. $\sin 2\theta - \cos \theta = 0$ **54.** $\cos 2\theta + \cos \theta = 0$

Theory and Examples

55. The tangent sum formula The standard formula for the tangent of the sum of two angles is

$$\tan(A + B) = \frac{\tan A + \tan B}{1 - \tan A \tan B}.$$

Derive the formula.

56. (*Continuation of Exercise 55.*) Derive a formula for $\tan(A - B)$.

57. Apply the law of cosines to the triangle in the accompanying figure to derive the formula for $\cos(A - B)$.

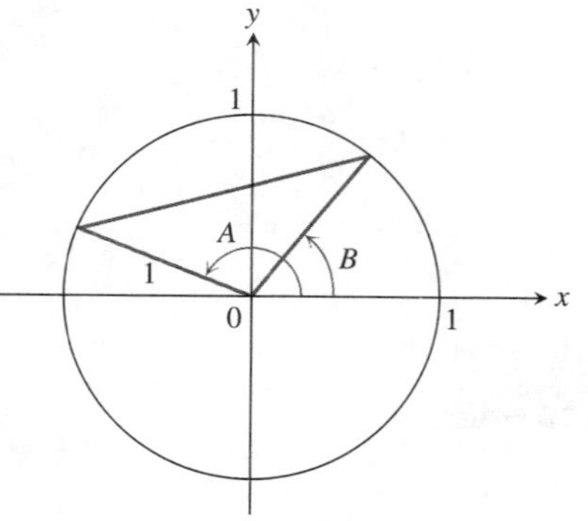

58. a. Apply the formula for $\cos(A - B)$ to the identity $\sin \theta = \cos\left(\frac{\pi}{2} - \theta\right)$ to obtain the addition formula for $\sin(A + B)$.

b. Derive the formula for $\cos(A + B)$ by substituting $-B$ for B in the formula for $\cos(A - B)$ from Exercise 35.

59. A triangle has sides $a = 3$ and $b = 4$ and angle $C = 90°$. Find the length of side c.

60. A triangle has sides $a = 2$ and $b = 3$ and angle $C = 40°$. Find the length of side c.

61. The law of sines The law of sines says that if a, b, and c are the sides opposite the angles A, B, and C in a triangle, then

$$\frac{\sin A}{a} = \frac{\sin B}{b} = \frac{\sin C}{c}.$$

Use the accompanying figures and the identity $\sin(\pi - \theta) = \sin \theta$, if required, to derive the law.

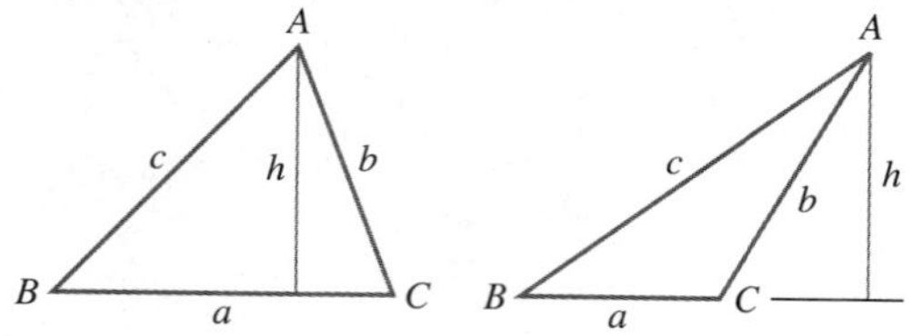

62. A triangle has sides $a = 2$ and $b = 3$ and angle $C = 60°$ (as in Exercise 59). Find the sine of angle B using the law of sines.

T 63. A triangle has side $c = 2$ and angles $A = \pi/4$ and $B = \pi/3$. Find the length a of the side opposite A.

64. Consider the length h of the perpendicular from point B to side b in the given triangle. Show that

$$h = \frac{b \tan \alpha \tan \gamma}{\tan \alpha + \tan \gamma}$$

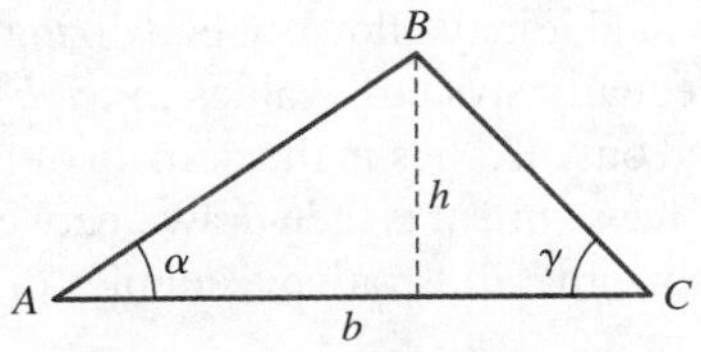

65. Refer to the given figure. Write the radius r of the circle in terms of α and θ.

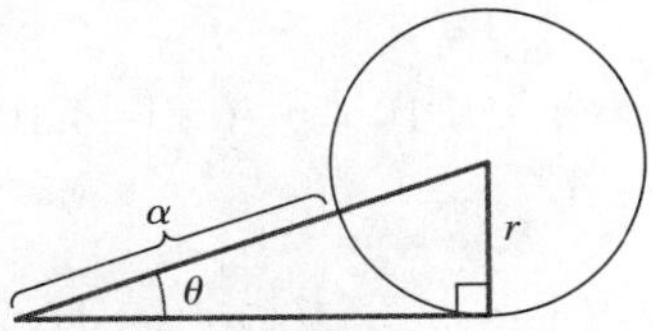

T 66. **The approximation $\sin x \approx x$** It is often useful to know that, when x is measured in radians, $\sin x \approx x$ for numerically small values of x. In Section 3.11, we will see why the approximation holds. The approximation error is less than 1 in 5000 if $|x| < 0.1$.

a. With your grapher in radian mode, graph $y = \sin x$ and $y = x$ together in a viewing window about the origin. What do you see happening as x nears the origin?

b. With your grapher in degree mode, graph $y = \sin x$ and $y = x$ together about the origin again. How is the picture different from the one obtained with radian mode?

General Sine Curves

For

$$f(x) = A \sin\left(\frac{2\pi}{B}(x - C)\right) + D,$$

identify A, B, C, and D for the sine functions in Exercises 67–70 and sketch their graphs.

67. $y = 2\sin(x + \pi) - 1$

68. $y = \frac{1}{2}\sin(\pi x - \pi) + \frac{1}{2}$

69. $y = -\frac{2}{\pi}\sin\left(\frac{\pi}{2}t\right) + \frac{1}{\pi}$

70. $y = \frac{L}{2\pi}\sin\frac{2\pi t}{L}, \quad L > 0$

COMPUTER EXPLORATIONS

In Exercises 71–74, you will explore graphically the general sine function

$$f(x) = A \sin\left(\frac{2\pi}{B}(x - C)\right) + D$$

as you change the values of the constants A, B, C, and D. Use a CAS or computer grapher to perform the steps in the exercises.

71. **The period B** Set the constants $A = 3$ and $C = D = 0$.

a. Plot $f(x)$ for the values $B = 1, 3, 2\pi, 5\pi$ over the interval $-4\pi \le x \le 4\pi$. Describe what happens to the graph of the general sine function as the period increases.

b. What happens to the graph for negative values of B? Try it with $B = -3$ and $B = -2\pi$.

72. **The horizontal shift C** Set the constants $A = 3, B = 6, D = 0$.

a. Plot $f(x)$ for the values $C = 0, 1$, and 2 over the interval $-4\pi \le x \le 4\pi$. Describe what happens to the graph of the general sine function as C increases through positive values.

b. What happens to the graph for negative values of C?

c. What smallest positive value should be assigned to C so the graph exhibits no horizontal shift? Confirm your answer with a plot.

73. **The vertical shift D** Set the constants $A = 3$, $B = 6$, $C = 0$.

a. Plot $f(x)$ for the values $D = 0, 1$, and 3 over the interval $-4\pi \le x \le 4\pi$. Describe what happens to the graph of the general sine function as D increases through positive values.

b. What happens to the graph for negative values of D?

74. **The amplitude A** Set the constants $B = 6$ and $C = D = 0$.

a. Describe what happens to the graph of the general sine function as A increases through positive values. Confirm your answer by plotting $f(x)$ for the values $A = 1, 5$, and 9.

b. What happens to the graph for negative values of A?

1.4 Graphing with Software

Many computers, calculators, and smartphones have graphing applications that enable us to graph very complicated functions with high precision. Many of these functions could not otherwise be easily graphed. However, some care must be taken when using such graphing software, and in this section we address some of the issues that can arise. In Chapter 4 we will see how calculus helps us determine that we are accurately viewing the important features of a function's graph.

Graphing Windows

When software is used for graphing, a portion of the graph is visible in a **display** or **viewing window**. Depending on the software, the default window may give an incomplete or misleading picture of the graph. We use the term *square window* when the units or scales used on both axes are the same. This term does not mean that the display window itself is square (usually it is rectangular), but instead it means that the x-unit is the same length as the y-unit.

When a graph is displayed in the default mode, the x-unit may differ from the y-unit of scaling in order to capture essential features of the graph. This difference in scaling can cause visual distortions that may lead to erroneous interpretations of the function's behavior. Some graphing software enables us to set the viewing window by specifying one or both of the intervals, $a \leq x \leq b$ and $c \leq y \leq d$, and it may allow for equalizing the scales used for the axes as well. The software selects equally spaced x-values in $[a, b]$ and then plots the points $(x, f(x))$. A point is plotted if and only if x lies in the domain of the function and $f(x)$ lies within the interval $[c, d]$. A short line segment is then drawn between each plotted point and its next neighboring point. We now give illustrative examples of some common problems that may occur with this procedure.

EXAMPLE 1 Graph the function $f(x) = x^3 - 7x^2 + 28$ in each of the following display or viewing windows:

(a) $[-10, 10]$ by $[-10, 10]$ **(b)** $[-4, 4]$ by $[-50, 10]$ **(c)** $[-4, 10]$ by $[-60, 60]$

Solution

(a) We select $a = -10$, $b = 10$, $c = -10$, and $d = 10$ to specify the interval of x-values and the range of y-values for the window. The resulting graph is shown in Figure 1.49a. It appears that the window is cutting off the bottom and top parts of the graph and that the interval of x-values is too large. Let's try the next window.

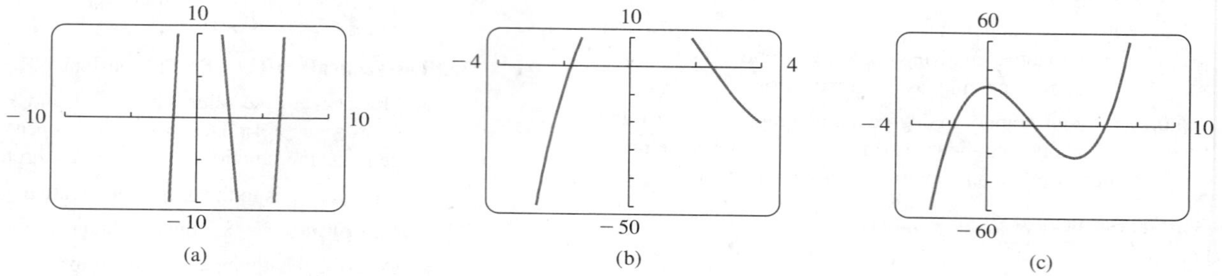

FIGURE 1.49 The graph of $f(x) = x^3 - 7x^2 + 28$ in different viewing windows. Selecting a window that gives a clear picture of a graph is often a trial-and-error process (Example 1). The default window used by the software may automatically display the graph in (c).

(b) We see some new features of the graph (Figure 1.49b), but the top is still missing and we need to view more to the right of $x = 4$ as well. The next window should help.

(c) Figure 1.49c shows the graph in this new viewing window. Observe that we get a more complete picture of the graph in this window, and it is a reasonable graph of a third-degree polynomial. ■

EXAMPLE 2 When a graph is displayed, the x-unit may differ from the y-unit, as in the graphs shown in Figure 1.49. The result is distortion in the picture, which may be misleading. The display window can be made square by compressing or stretching the units on one axis to match the scale on the other, giving the true graph. Many software systems have built-in options to make the window "square." If yours does not, you may have to bring to your viewing some foreknowledge of the true picture.

Figure 1.50a shows the graphs of the perpendicular lines $y = x$ and $y = -x + 3\sqrt{2}$, together with the semicircle $y = \sqrt{9 - x^2}$, in a nonsquare $[-4, 4]$ by $[-6, 8]$ display window. Notice the distortion. The lines do not appear to be perpendicular, and the semicircle appears to be elliptical in shape.

Figure 1.50b shows the graphs of the same functions in a square window in which the x-units are scaled to be the same as the y-units. Notice that the scaling on the x-axis for Figure 1.50a has been compressed in Figure 1.50b to make the window square. Figure 1.50c gives an enlarged view of Figure 1.50b with a square $[-3, 3]$ by $[0, 4]$ window. ■

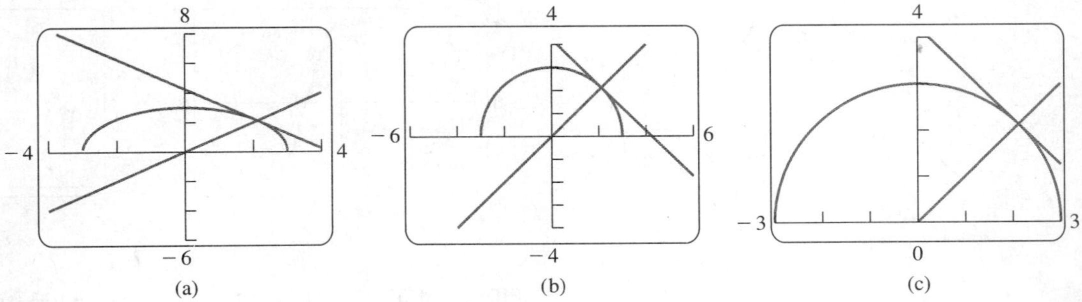

FIGURE 1.50 Graphs of the perpendicular lines $y = x$ and $y = -x + 3\sqrt{2}$ and of the semicircle $y = \sqrt{9 - x^2}$ appear distorted (a) in a nonsquare window, but clear (b) and (c) in square windows (Example 2). Some software may not provide options for the views in (b) or (c).

If the denominator of a rational function is zero at some x-value within the viewing window, graphing software may produce a steep, near-vertical line segment from the top to the bottom of the window. Example 3 illustrates steep line segments.

Sometimes the graph of a trigonometric function oscillates very rapidly. When graphing software plots the points of the graph and connects them, many of the maximum and minimum points are actually missed. The resulting graph is then very misleading.

EXAMPLE 3 Graph the function $f(x) = \sin 100x$.

Solution Figure 1.51a shows the graph of f in the viewing window $[-12, 12]$ by $[-1, 1]$. We see that the graph looks very strange because the sine curve should oscillate periodically between -1 and 1. This behavior is not exhibited in Figure 1.51a. We might experiment with a smaller viewing window, say $[-6, 6]$ by $[-1, 1]$, but the graph is not better (Figure 1.51b). The difficulty is that the period of the trigonometric function $y = \sin 100x$ is very small $(2\pi/100 \approx 0.063)$. If we choose the much smaller viewing window $[-0.1, 0.1]$ by $[-1, 1]$ we get the graph shown in Figure 1.51c. This graph reveals the expected oscillations of a sine curve. ■

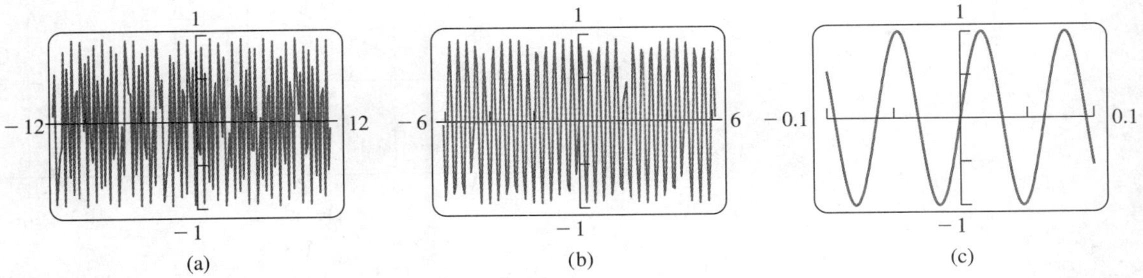

FIGURE 1.51 Graphs of the function $y = \sin 100x$ in three viewing windows. Because the period is $2\pi/100 \approx 0.063$, the smaller window in (c) best displays the true aspects of this rapidly oscillating function (Example 3).

EXAMPLE 4 Graph the function $y = \cos x + \dfrac{1}{200}\sin 200x$.

Solution In the viewing window $[-6, 6]$ by $[-1, 1]$ the graph appears much like the cosine function with some very small sharp wiggles on it (Figure 1.52a). We get a better look when we significantly reduce the window to $[-0.2, 0.2]$ by $[0.97, 1.01]$, obtaining the graph in Figure 1.52b. We now see the small but rapid oscillations of the second term, $(1/200)\sin 200x$, added to the comparatively larger values of the cosine curve. ■

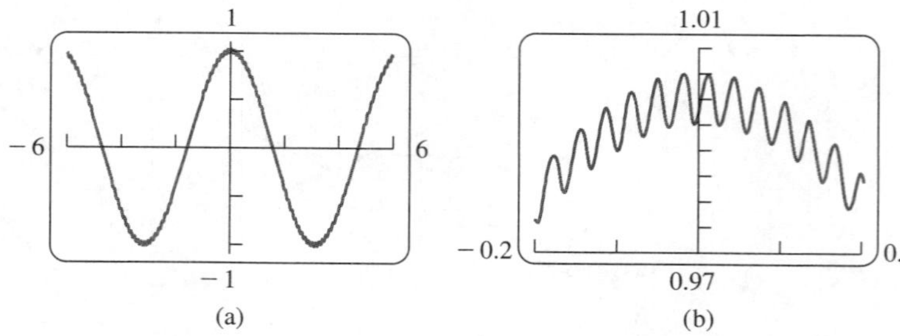

FIGURE 1.52 (a) The function $y = \cos x + \frac{1}{200}\sin 200x$. (b) A close-up view, blown up near the y-axis. The term $\cos x$ clearly dominates the second term, $\frac{1}{200}\sin 200x$, which produces the rapid oscillations along the cosine curve. Both views are needed for a clear idea of the graph (Example 4).

Obtaining a Complete Graph

Some graphing software will not display the portion of a graph for $f(x)$ when $x < 0$. Usually that happens because of the algorithm the software is using to calculate the function values. Sometimes we can obtain the complete graph by defining the formula for the function in a different way, as illustrated in the next example.

EXAMPLE 5 Graph the function $y = x^{1/3}$.

Solution Some graphing software displays the graph shown in Figure 1.53a. When we compare it with the graph of $y = x^{1/3} = \sqrt[3]{x}$ in Figure 1.17, we see that the left branch for $x < 0$ is missing. The reason the graphs differ is that the software algorithm calculates $x^{1/3}$ as $e^{(1/3)\ln x}$. Since the logarithmic function is not defined for negative values of x, the software can produce only the right branch, where $x > 0$. (Logarithmic and exponential functions are discussed in detail in Chapter 7.)

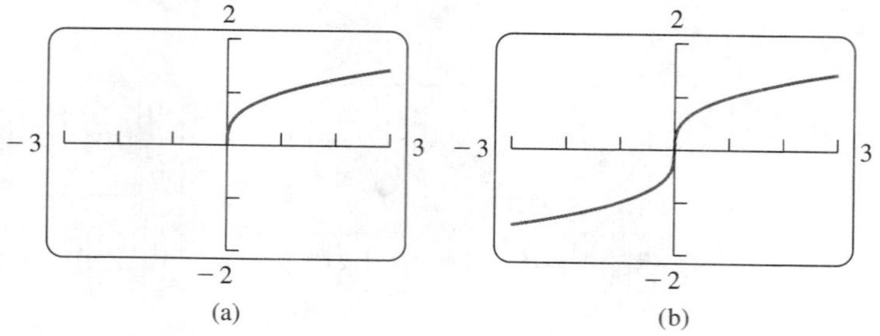

FIGURE 1.53 The graph of $y = x^{1/3}$ is missing the left branch in (a). In (b) we graph the function $f(x) = \frac{x}{|x|} \cdot |x|^{1/3}$, obtaining both branches. (See Example 5.)

To obtain the full picture showing both branches, we can graph the function

$$f(x) = \frac{x}{|x|} \cdot |x|^{1/3}.$$

This function equals $x^{1/3}$ except at $x = 0$ (where f is undefined, although $0^{1/3} = 0$). A graph of f is displayed in Figure 1.53b. ■

EXERCISES 1.4

Choosing a Viewing Window

T In Exercises 1–4, use graphing software to determine which of the given viewing windows displays the most appropriate graph of the specified function.

1. $f(x) = x^4 - 7x^2 + 6x$
   a. $[-1, 1]$ by $[-1, 1]$ b. $[-2, 2]$ by $[-5, 5]$
   c. $[-10, 10]$ by $[-10, 10]$ d. $[-5, 5]$ by $[-25, 15]$
2. $f(x) = x^3 - 2x^2 - 3x$
   a. $[-1, 1]$ by $[-5, 5]$ b. $[-3, 3]$ by $[-10, 10]$
   c. $[-5, 5]$ by $[-20, 20]$ d. $[-20, 20]$ by $[-100, 100]$
3. $f(x) = x^5 - 8x^3 + 10x + 6$
   a. $[-1, 1]$ by $[-1, 1]$ b. $[-5, 5]$ by $[-10, 10]$
   c. $[-3, 3]$ by $[-10, 20]$ d. $[-4, 4]$ by $[-20, 20]$
4. $f(x) = \sqrt{5 + 4x - x^2}$
   a. $[-2, 2]$ by $[-2, 2]$ b. $[-2, 6]$ by $[-1, 4]$
   c. $[-3, 7]$ by $[0, 10]$ d. $[-10, 10]$ by $[-10, 10]$

Finding a Viewing Window

T In Exercises 5–30, find an appropriate graphing software viewing window for the given function and use it to display that function's graph. The window should give a picture of the overall behavior of the function. There is more than one choice, but incorrect choices can miss important aspects of the function.

5. $f(x) = x^4 - x^3 - 6x^2$
6. $f(x) = \frac{x^3}{3} - \frac{x^2}{2} - 2x + 1$
7. $f(x) = x^5 - 5x^4 + 10$
8. $f(x) = x^4 - 2x^2$
9. $f(x) = x\sqrt{9 - x^2}$
10. $f(x) = x^3(4 - x^2)$
11. $y = 2x - 3x^{2/3}$
12. $y = x^{1/3}(x^2 - 8)$
13. $y = 5x^{2/5} - 2x$
14. $y = x^{2/3}(5 - x)$
15. $y = |x^2 - 1|$
16. $y = |x^2 - x|$
17. $y = \frac{x + 5}{x + 3}$
18. $y = 1 - \frac{1}{x + 3}$
19. $f(x) = \frac{x^2 + 2}{x^2 + 1}$
20. $y = \frac{x^2 - 2}{x^2 + 2}$
21. $f(x) = \frac{x - 1}{x^2 - x - 6}$
22. $y = \frac{10}{x^2 - 16}$
23. $f(x) = \frac{6x^2 - 15x + 6}{4x^2 - 10x}$
24. $f(x) = \frac{x^2 - 3}{x - 2}$
25. $y = \sin 150x$
26. $y = 3\cos 60x$
27. $y = \cos\left(\frac{x}{20}\right)$
28. $y = \frac{1}{10}\sin\left(\frac{x}{10}\right)$
29. $y = x + \frac{1}{10}\sin 30x$
30. $y = x^2 + \frac{1}{50}\cos 100x$

Use graphing software to graph the functions specified in Exercises 31–36. Select a viewing window that reveals the key features of the function.

31. Graph the lower half of the circle defined by the equation $x^2 + 2x = 4 + 4y - y^2$.
32. Graph the upper branch of the hyperbola $y^2 - 25x^2 = 8$.
33. Graph four periods of the function $f(x) = -\tan 2x$.
34. Graph two periods of the function $f(x) = 3\cot\frac{x}{2} + 1$.
35. Graph the function $f(x) = \sin 3x + \cos 2x$.
36. Graph the function $f(x) = \sin^3 x$.

CHAPTER 1 Questions to Guide Your Review

1. What is a function? What is its domain? Its range? What is an arrow diagram for a function? Give examples.
2. What is the graph of a real-valued function of a real variable? What is the vertical line test?
3. What is a piecewise-defined function? Give examples.
4. What are the important types of functions frequently encountered in calculus? Give an example of each type.
5. What is meant by an increasing function? A decreasing function? Give an example of each.
6. What is an even function? An odd function? What symmetry properties do the graphs of such functions have? What advantage can we take of this? Give an example of a function that is neither even nor odd.
7. If f and g are real-valued functions, how are the domains of $f - g$, $f - g$, fg, and f/g related to the domains of f and g? Give examples.
8. When is it possible to compose one function with another? Give examples of compositions and their values at various points. Does the order in which functions are composed ever matter?
9. How do you change the equation $y = f(x)$ to shift its graph vertically up or down by $|k|$ units? Horizontally to the left or right? Give examples.
10. How do you change the equation $y = f(x)$ to compress or stretch the graph by a factor $c > 1$? Reflect the graph across a coordinate axis? Give examples.
11. What is radian measure? How do you convert from radians to degrees? Degrees to radians?
12. Graph the six basic trigonometric functions. What symmetries do the graphs have?
13. What is a periodic function? Give examples. What are the periods of the six basic trigonometric functions?
14. Starting with the identity $\sin^2\theta + \cos^2\theta = 1$ and the formulas for $\cos(A + B)$ and $\sin(A + B)$, show how a variety of other trigonometric identities may be derived.
15. How does the formula for the general sine function $f(x) = A\sin((2\pi/B)(x - C)) + D$ relate to the shifting, stretching, compressing, and reflection of its graph? Give examples. Graph the general sine curve and identify the constants A, B, C, and D.
16. Name three issues that arise when functions are graphed using a calculator or computer with graphing software. Give examples.

CHAPTER 1 Practice Exercises

Functions and Graphs

1. Express the area and circumference of a circle as functions of the circle's radius. Then express the area as a function of the circumference.
2. Express the radius of a sphere as a function of the sphere's surface area. Then express the surface area as a function of the volume.
3. A point P in the first quadrant lies on the parabola $y = x^2$. Express the coordinates of P as functions of the angle of inclination of the line joining P to the origin.
4. A hot-air balloon rising straight up from a level field is tracked by a range finder located 500 ft from the point of liftoff. Express the balloon's height as a function of the angle the line from the range finder to the balloon makes with the ground.

In Exercises 5–8, determine whether the graph of the function is symmetric about the y-axis, the origin, or neither.

5. $y = x^{1/5}$
6. $y = x^{2/5}$
7. $y = x^2 - 2x - 1$
8. $y = e^{-x^2}$

In Exercises 9–16, determine whether the function is even, odd, or neither.

9. $y = x^2 + 1$
10. $y = x^5 - x^3 - x$
11. $y = 1 - \cos x$
12. $y = \sec x \tan x$
13. $y = \dfrac{x^4 + 1}{x^3 - 2x}$
14. $y = x - \sin x$
15. $y = x + \cos x$
16. $y = x \cos x$
17. Suppose that f and g are both odd functions defined on the entire real line. Which of the following (where defined) are even? odd?
    a. fg **b.** f^3 **c.** $f(\sin x)$ **d.** $g(\sec x)$ **e.** $|g|$
18. If $f(a - x) = f(a + x)$, show that $g(x) = f(x + a)$ is an even function.

In Exercises 19–32, find the **(a)** domain and **(b)** range.

19. $y = |x| - 2$
20. $y = -2 + \sqrt{1 - x}$
21. $y = \sqrt{16 - x^2}$
22. $y = 3^{2-x} + 1$
23. $y = 2e^{-x} - 3$
24. $y = \tan(2x - \pi)$
25. $y = 2\sin(3x + \pi) - 1$
26. $y = x^{2/5}$
27. $y = \cos(x - 3) + 1$
28. $y = -1 + \sqrt[3]{2 - x}$
29. $y = 5 - \sqrt{x^2 - 2x - 3}$
30. $y = 2 + \dfrac{3x^2}{x^2 + 4}$
31. $y = 4\sin\left(\dfrac{1}{x}\right)$
32. $y = 3\cos x + 4\sin x$ (*Hint:* A trig identity is required.)
33. State whether each function is increasing, decreasing, or neither.
    a. Volume of a sphere as a function of its radius
    b. Greatest integer function
    c. Height above Earth's sea level as a function of atmospheric pressure (assumed nonzero)
    d. Kinetic energy as a function of a particle's velocity
34. Find the largest interval on which the given function is increasing.
    a. $f(x) = |x - 2| + 1$ **b.** $f(x) = (x + 1)^4$
    c. $g(x) = (3x - 1)^{1/3}$ **d.** $R(x) = \sqrt{2x - 1}$

Piecewise-Defined Functions

In Exercises 35 and 36, find the **(a)** domain and **(b)** range.

35. $y = \begin{cases} \sqrt{-x}, & -4 \le x \le 0 \\ \sqrt{x}, & 0 < x \le 4 \end{cases}$
36. $y = \begin{cases} -x - 2, & -2 \le x \le -1 \\ x, & -1 < x \le 1 \\ -x + 2, & 1 < x \le 2 \end{cases}$

In Exercises 37 and 38, write a piecewise formula for the function.

37.
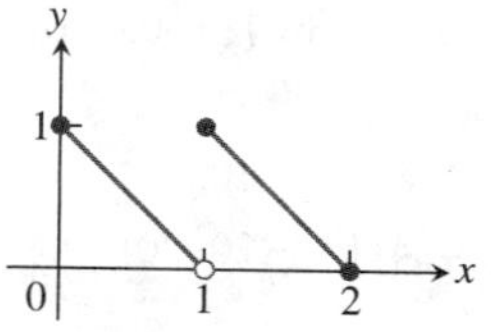

38.
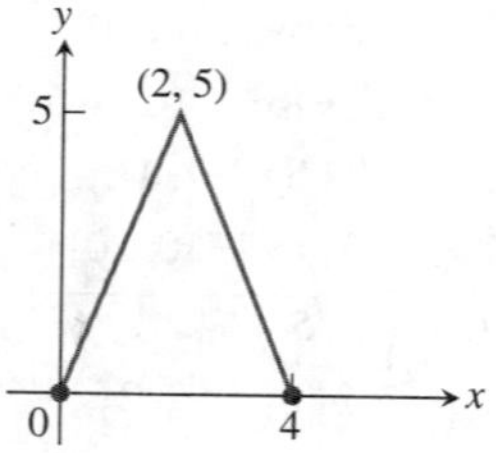

Composition of Functions

In Exercises 39 and 40, find

a. $(f \circ g)(-1)$. **b.** $(g \circ f)(2)$.
c. $(f \circ f)(x)$. **d.** $(g \circ g)(x)$.

39. $f(x) = \dfrac{1}{x}, \quad g(x) = \dfrac{1}{\sqrt{x + 2}}$
40. $f(x) = 2 - x, \quad g(x) = \sqrt[3]{x + 1}$

In Exercises 41 and 42, **(a)** write formulas for $f \circ g$ and $g \circ f$ and find the **(b)** domain and **(c)** range of each.

41. $f(x) = 2 - x^2, \quad g(x) = \sqrt{x + 2}$
42. $f(x) = \sqrt{x}, \quad g(x) = \sqrt{1 - x}$

For Exercises 43 and 44, sketch the graphs of f and $f \circ f$.

43. $f(x) = \begin{cases} -x - 2, & -4 \le x \le -1 \\ -1, & -1 < x \le 1 \\ x - 2, & 1 < x \le 2 \end{cases}$
44. $f(x) = \begin{cases} x + 1, & -2 \le x < 0 \\ x - 1, & 0 \le x \le 2 \end{cases}$

Composition with absolute values In Exercises 45–52, graph f_1 and f_2 together. Then describe how applying the absolute value function in f_2 affects the graph of f_1.

	$f_1(x)$	$f_2(x)$
45.	x	$\lvert x\rvert$
46.	x^2	$\lvert x\rvert^2$
47.	x^3	$\lvert x^3\rvert$
48.	$x^2 + x$	$\lvert x^2 + x\rvert$
49.	$4 - x^2$	$\lvert 4 - x^2\rvert$
50.	$\dfrac{1}{x}$	$\dfrac{1}{\lvert x\rvert}$
51.	$\sqrt{x}$	$\sqrt{\lvert x\rvert}$
52.	$\sin x$	$\sin\lvert x\rvert$

Shifting and Scaling Graphs

53. Suppose the graph of g is given. Write equations for the graphs that are obtained from the graph of g by shifting, scaling, or reflecting, as indicated.

a. Up $\frac{1}{2}$ unit, right 3

b. Down 2 units, left $\frac{2}{3}$

c. Reflect about the y-axis

d. Reflect about the x-axis

e. Stretch vertically by a factor of 5

f. Compress horizontally by a factor of 5

54. Describe how each graph is obtained from the graph of $y = f(x)$.

a. $y = f(x - 5)$ **b.** $y = f(4x)$

c. $y = f(-3x)$ **d.** $y = f(2x + 1)$

e. $y = f\left(\frac{x}{3}\right) - 4$ **f.** $y = -3f(x) + \frac{1}{4}$

In Exercises 55–58, graph each function, not by plotting points, but by starting with the graph of one of the standard functions presented in Figures 1.15–1.17, and applying an appropriate transformation.

55. $y = -\sqrt{1 + \frac{x}{2}}$ **56.** $y = 1 - \frac{x}{3}$

57. $y = \frac{1}{2x^2} + 1$ **58.** $y = (-5x)^{1/3}$

Trigonometry

In Exercises 59–62, sketch the graph of the given function. What is the period of the function?

59. $y = \cos 2x$ **60.** $y = \sin \frac{x}{2}$

61. $y = \sin \pi x$ **62.** $y = \cos \frac{\pi x}{2}$

63. Sketch the graph $y = 2\cos\left(x - \frac{\pi}{3}\right)$.

64. Sketch the graph $y = 1 + \sin\left(x + \frac{\pi}{4}\right)$.

In Exercises 65–68, ABC is a right triangle with the right angle at C. The sides opposite angles A, B, and C are a, b, and c, respectively.

65. a. Find a and b if $c = 2$, $B = \pi/3$.

b. Find a and c if $b = 2$, $B = \pi/3$.

66. a. Express a in terms of A and c.

b. Express a in terms of A and b.

67. a. Express a in terms of B and b.

b. Express c in terms of A and a.

68. a. Express $\sin A$ in terms of a and c.

b. Express $\sin A$ in terms of b and c.

69. Height of a pole Two wires stretch from the top T of a vertical pole to points B and C on the ground, where C is 10 m closer to the base of the pole than is B. If wire BT makes an angle of $35°$ with the horizontal and wire CT makes an angle of $50°$ with the horizontal, how high is the pole?

70. Height of a weather balloon Observers at positions A and B 2 km apart simultaneously measure the angle of elevation of a weather balloon to be $40°$ and $70°$, respectively. If the balloon is directly above a point on the line segment between A and B, find the height of the balloon.

T **71. a.** Graph the function $f(x) = \sin x + \cos(x/2)$.

b. What appears to be the period of this function?

c. Confirm your finding in part (b) algebraically.

T **72. a.** Graph $f(x) = \sin(1/x)$.

b. What are the domain and range of f?

c. Is f periodic? Give reasons for your answer.

CHAPTER 1 Additional and Advanced Exercises

Functions and Graphs

1. Are there two functions f and g such that $f \circ g = g \circ f$? Give reasons for your answer.

2. Are there two functions f and g with the following property? The graphs of f and g are not straight lines but the graph of $f \circ g$ is a straight line. Give reasons for your answer.

3. If $f(x)$ is odd, can anything be said of $g(x) = f(x) - 2$? What if f is even instead? Give reasons for your answer.

4. If $g(x)$ is an odd function defined for all values of x, can anything be said about $g(0)$? Give reasons for your answer.

5. Graph the equation $|x| + |y| = 1 + x$.

6. Graph the equation $y + |y| = x + |x|$.

Derivations and Proofs

7. Prove the following identities.

a. $\frac{1 - \cos x}{\sin x} = \frac{\sin x}{1 + \cos x}$

b. $\frac{1 - \cos x}{1 + \cos x} = \tan^2 \frac{x}{2}$

8. Explain the following "proof without words" of the law of cosines. (*Source:* Kung, Sidney H., "Proof Without Words: The Law of Cosines," *Mathematics Magazine*, Vol. 63, no. 5, Dec. 1990, p. 342.)

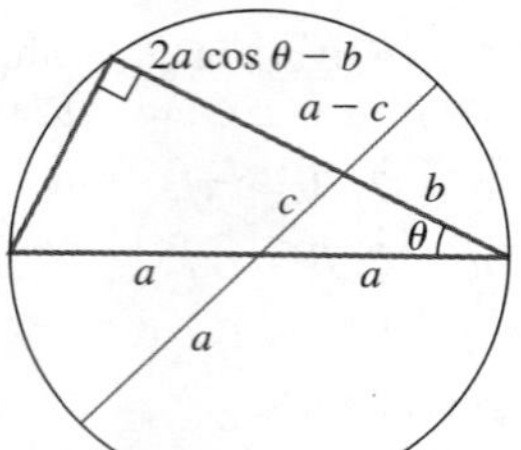

9. Show that the area of triangle ABC is given by $(1/2)ab\sin C = (1/2)bc\sin A = (1/2)ca\sin B$.

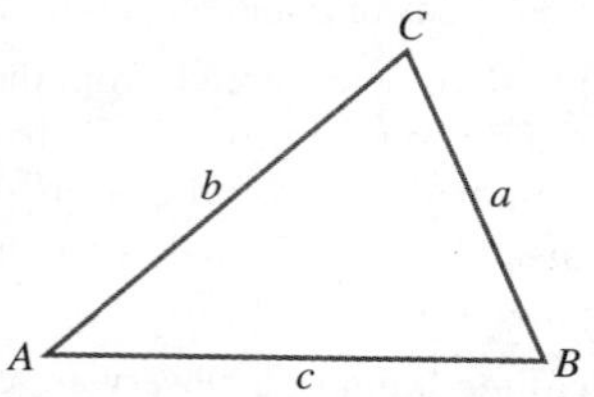

10. Show that the area of triangle ABC is given by $\sqrt{s(s-a)(s-b)(s-c)}$ where $s = (a+b+c)/2$ is the semiperimeter of the triangle.

11. Show that if f is both even and odd, then $f(x) = 0$ for every x in the domain of f.

12. a. Even-odd decompositions Let f be a function whose domain is symmetric about the origin, that is, $-x$ belongs to the domain whenever x does. Show that f is the sum of an even function and an odd function:

$$f(x) = E(x) + O(x),$$

where E is an even function and O is an odd function. (*Hint:* Let $E(x) = (f(x) + f(-x))/2$. Show that $E(-x) = E(x)$, so that E is even. Then show that $O(x) = f(x) - E(x)$ is odd.)

b. Uniqueness Show that there is only one way to write f as the sum of an even and an odd function. (*Hint:* One way is given in part (a). If also $f(x) = E_1(x) + O_1(x)$ where E_1 is even and O_1 is odd, show that $E - E_1 = O_1 - O$. Then use Exercise 11 to show that $E = E_1$ and $O = O_1$.)

Effects of Parameters on Graphs

T **13.** What happens to the graph of $y = ax^2 + bx + c$ as

a. a changes while b and c remain fixed?

b. b changes (a and c fixed, $a \neq 0$)?

c. c changes (a and b fixed, $a \neq 0$)?

T **14.** What happens to the graph of $y = a(x+b)^3 + c$ as

a. a changes while b and c remain fixed?

b. b changes (a and c fixed, $a \neq 0$)?

c. c changes (a and b fixed, $a \neq 0$)?

Geometry

15. An object's center of mass moves at a constant velocity v along a straight line past the origin. The accompanying figure shows the coordinate system and the line of motion. The dots show positions that are 1 sec apart. Why are the areas $A_1, A_2, \ldots, A_5$ in the figure all equal? As in Kepler's equal area law (see Section 13.6), the line that joins the object's center of mass to the origin sweeps out equal areas in equal times.

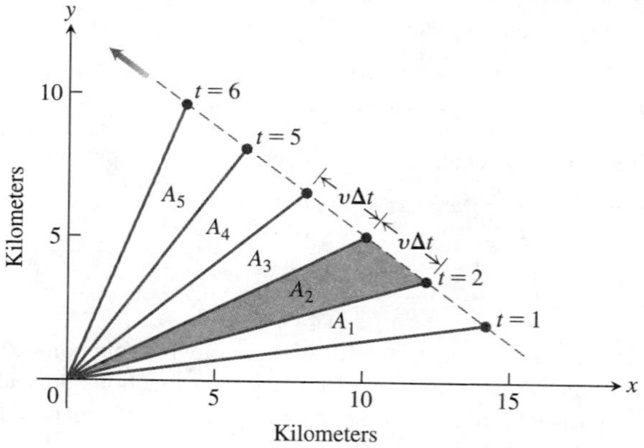

16. a. Find the slope of the line from the origin to the midpoint P of side AB in the triangle in the accompanying figure $(a, b > 0)$.

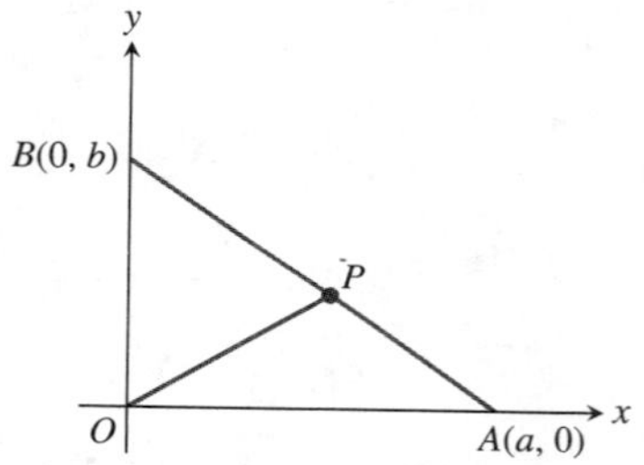

b. When is OP perpendicular to AB?

17. Consider the quarter-circle of radius 1 and right triangles ABE and ACD given in the accompanying figure. Use standard area formulas to conclude that

$$\frac{1}{2}\sin\theta\cos\theta < \frac{\theta}{2} < \frac{1}{2}\frac{\sin\theta}{\cos\theta}.$$

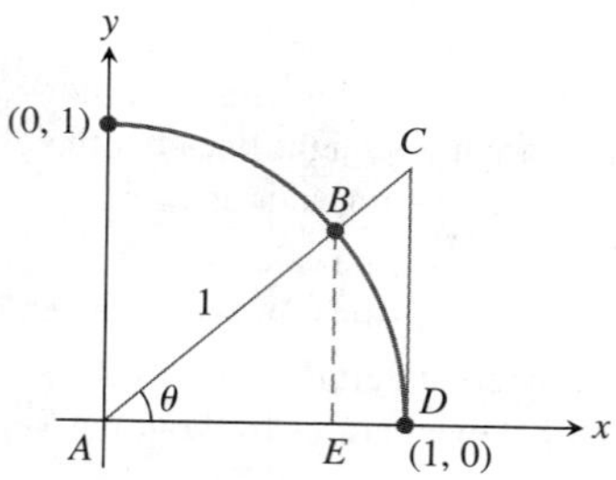

18. Let $f(x) = ax + b$ and $g(x) = cx + d$. What condition must be satisfied by the constants a, b, c, d in order that $(f \circ g)(x) = (g \circ f)(x)$ for every value of x?

CHAPTER 1 Technology Application Projects

Mathematica/Maple Projects

Projects can be found within MyLab Math.

- ***An Overview of Mathematica***
  An overview of *Mathematica* sufficient to complete the *Mathematica* modules appearing on the Web site.
- ***Modeling Change: Springs, Driving Safety, Radioactivity, Trees, Fish, and Mammals***
  Construct and interpret mathematical models, analyze and improve them, and make predictions using them.

2

Limits and Continuity

OVERVIEW In this chapter we develop the concept of a limit, first intuitively and then formally. We use limits to describe the way a function varies. Some functions vary *continuously*; small changes in x produce only small changes in $f(x)$. Other functions can have values that jump, vary erratically, or tend to increase or decrease without bound. The notion of limit gives a precise way to distinguish among these behaviors.

2.1 Rates of Change and Tangent Lines to Curves

Average and Instantaneous Speed

HISTORICAL BIOGRAPHY
Galileo Galilei
(1564–1642)
www.bit.ly/2OpdNBs

In the late sixteenth century, Galileo discovered that a solid object dropped from rest (initially not moving) near the surface of the earth and allowed to fall freely, will fall a distance proportional to the square of the time it has been falling. This type of motion is called **free fall**. It assumes negligible air resistance to slow the object down, and it assumes that gravity is the only force acting on the falling object. If y denotes the distance fallen in feet after t seconds, then Galileo's law is

$$y = 16t^2 \text{ ft},$$

where 16 is the (approximate) constant of proportionality. (If y is measured in meters instead, then the constant is close to 4.9.)

More generally, suppose that a moving object has traveled distance $f(t)$ at time t. The object's **average speed** during an interval of time $[t_1, t_2]$ is found by dividing the distance traveled $f(t_2) - f(t_1)$ by the time elapsed $t_2 - t_1$. The unit of measure is length per unit time: kilometers per hour, feet (or meters) per second, or whatever is appropriate to the problem at hand.

Average Speed
When $f(t)$ measures the distance traveled at time t,

$$\text{Average speed over } [t_1, t_2] = \frac{\text{distance traveled}}{\text{elapsed time}} = \frac{f(t_2) - f(t_1)}{t_2 - t_1}.$$

EXAMPLE 1 A rock breaks loose from the top of a tall cliff. What is its average speed

(a) during the first 2 sec of fall?

(b) during the 1-sec interval between second 1 and second 2?

Δ is the capital Greek letter Delta.

Solution The average speed of the rock during a given time interval is the change in distance, Δy, divided by the length of the time interval, Δt. (The capital Greek letter Delta, written Δ, is traditionally used to indicate the increment, or change, in a variable. Increments like Δy and Δt are reviewed in Appendix A.3 and pronounced "delta y" and "delta t.") Measuring distance in feet and time in seconds, we have the following calculations:

(a) For the first 2 sec: $\dfrac{\Delta y}{\Delta t} = \dfrac{16(2)^2 - 16(0)^2}{2 - 0} = 32 \dfrac{\text{ft}}{\text{sec}}$

(b) From sec 1 to sec 2: $\dfrac{\Delta y}{\Delta t} = \dfrac{16(2)^2 - 16(1)^2}{2 - 1} = 48 \dfrac{\text{ft}}{\text{sec}}$ ■

We want a way to determine the speed of a falling object at a single instant t_0, instead of using its average speed over an interval of time. To do this, we examine what happens when we calculate the average speed over shorter and shorter time intervals starting at t_0. The next example illustrates this process. Our discussion is informal here but will be made precise in Chapter 3.

EXAMPLE 2 Find the speed of the falling rock in Example 1 at $t = 1$ and $t = 2$ sec.

Solution We can calculate the average speed of the rock over a time interval $[t_0, t_0 + h]$, having length $\Delta t = (t_0 + h) - (t_0) = h$, as

$$\frac{\Delta y}{\Delta t} = \frac{16(t_0 + h)^2 - 16t_0^2}{h} \frac{\text{ft}}{\text{sec}}. \tag{1}$$

We cannot use this formula to calculate the "instantaneous" speed at the exact moment t_0 by simply substituting $h = 0$, because we cannot divide by zero. But we *can* use it to calculate average speeds over shorter and shorter time intervals starting at either $t_0 = 1$ or $t_0 = 2$. When we do so, by taking smaller and smaller values of h, we see a pattern (Table 2.1).

TABLE 2.1 Average speeds over short time intervals $[t_0, t_0 + h]$

$$\text{Average speed: } \frac{\Delta y}{\Delta t} = \frac{16(t_0 + h)^2 - 16t_0^2}{h}$$

Length of time interval h	Average speed over interval of length h starting at $t_0 = 1$	Average speed over interval of length h starting at $t_0 = 2$
1	48	80
0.1	33.6	65.6
0.01	32.16	64.16
0.001	32.016	64.016
0.0001	32.0016	64.0016

The average speed on intervals starting at $t_0 = 1$ seems to approach a limiting value of 32 as the length of the interval decreases. This suggests that the rock is falling at a speed of 32 ft/sec at $t_0 = 1$ sec. Let's confirm this algebraically.

If we set $t_0 = 1$ and then expand the numerator in Equation (1) and simplify, we find that

$$\frac{\Delta y}{\Delta t} = \frac{16(1+h)^2 - 16(1)^2}{h} = \frac{16(1+2h+h^2) - 16}{h}$$
$$= \frac{32h + 16h^2}{h} = 32 + 16h. \qquad \text{Can cancel } h \text{ when } h \neq 0$$

For values of h different from 0, the expressions on the right and left are equivalent and the average speed is $32 + 16h$ ft/sec. We can now see why the average speed has the limiting value $32 + 16(0) = 32$ ft/sec as h approaches 0.

Similarly, setting $t_0 = 2$ in Equation (1), for values of h different from 0 the procedure yields

$$\frac{\Delta y}{\Delta t} = 64 + 16h.$$

As h gets closer and closer to 0, the average speed has the limiting value 64 ft/sec when $t_0 = 2$ sec, as suggested by Table 2.1. ■

The average speed of a falling object is an example of a more general idea, an average rate of change.

Average Rates of Change and Secant Lines

Given any function $y = f(x)$, we calculate the average rate of change of y with respect to x over the interval $[x_1, x_2]$ by dividing the change in the value of y, $\Delta y = f(x_2) - f(x_1)$, by the length $\Delta x = x_2 - x_1 = h$ of the interval over which the change occurs. (We use the symbol h for Δx to simplify the notation here and later on.)

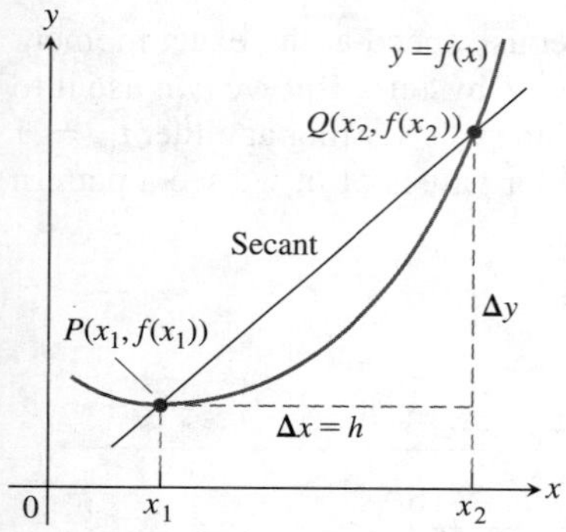

FIGURE 2.1 A secant to the graph $y = f(x)$. Its slope is $\Delta y/\Delta x$, the average rate of change of f over the interval $[x_1, x_2]$.

> **DEFINITION** The **average rate of change** of $y = f(x)$ with respect to x over the interval $[x_1, x_2]$ is
>
> $$\frac{\Delta y}{\Delta x} = \frac{f(x_2) - f(x_1)}{x_2 - x_1} = \frac{f(x_1 + h) - f(x_1)}{h}, \qquad h \neq 0.$$

Geometrically, the rate of change of f over $[x_1, x_2]$ is the slope of the line through the points $P(x_1, f(x_1))$ and $Q(x_2, f(x_2))$ (Figure 2.1). In geometry, a line joining two points of a curve is called a **secant line**. Thus, the average rate of change of f from x_1 to x_2 is identical to the slope of secant line PQ. As the point Q approaches the point P along the curve, the length h of the interval over which the change occurs approaches zero. We will see that this procedure leads to the definition of the slope of a curve at a point.

Defining the Slope of a Curve

We know what is meant by the slope of a straight line, which tells us the rate at which it rises or falls—its rate of change as a linear function. But what is meant by the *slope of a curve* at a point P on the curve? If there were a *tangent line* to the curve at P—a line that grazes the curve like the tangent line to a circle—it would be reasonable to identify the slope of the tangent line as the slope of the curve at P. We will see that, among all the lines that pass through the point P, the tangent line is the one that gives the best approximation to the curve at P. We need a precise way to specify the tangent line at a point on a curve.

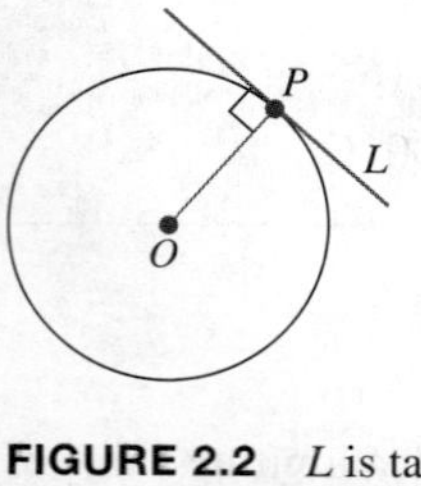

FIGURE 2.2 L is tangent to the circle at P if it passes through P perpendicular to radius OP.

Specifying a tangent line to a circle is straightforward. A line L is tangent to a circle at a point P if L passes through P and is perpendicular to the radius at P (Figure 2.2). But what does it mean to say that a line L is tangent to a more general curve at a point P?

HISTORICAL BIOGRAPHY
Pierre de Fermat
(1601–1665)
www.bit.ly/2NRJEeC

To define tangency for general curves, we use an approach that analyzes the behavior of the secant lines that pass through P and nearby points Q as Q moves toward P along the curve (Figure 2.3). We start with what we *can* calculate, namely the slope of the secant line PQ. We then compute the limiting value of the secant line's slope as Q approaches P along the curve. (We clarify the limit idea in the next section.) If the limit exists, we take it to be the slope of the curve at P and *define* the tangent line to the curve at P to be the line through P with this slope.

The next example illustrates the geometric idea for finding the tangent line to a curve.

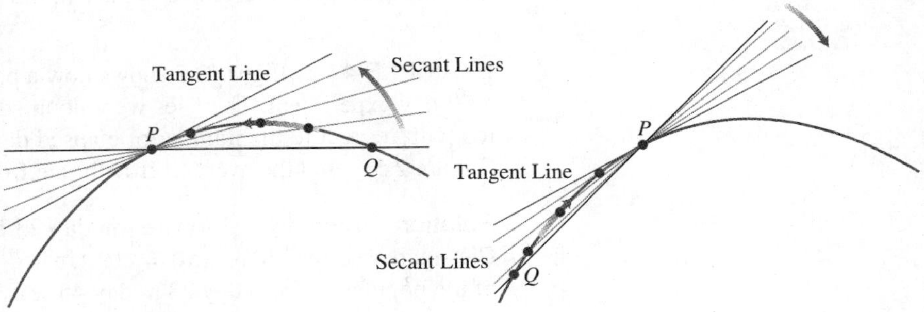

FIGURE 2.3 The tangent line to the curve at P is the line through P whose slope is the limit of the secant line slopes as $Q \to P$ from either side.

EXAMPLE 3 Find the slope of the tangent line to the parabola $y = x^2$ at the point $(2, 4)$ by analyzing the slopes of secant lines through $(2, 4)$. Write an equation for the tangent line to the parabola at this point.

Solution We begin with a secant line through $P(2, 4)$ and a nearby point $Q(2 + h, (2 + h)^2)$, as shown in Figure 2.4. We then write an expression for the slope of the secant line PQ and investigate what happens to the slope as Q approaches P along the curve:

$$\text{Secant line slope} = \frac{\Delta y}{\Delta x} = \frac{(2 + h)^2 - 2^2}{h} = \frac{h^2 + 4h + 4 - 4}{h}$$

$$= \frac{h^2 + 4h}{h} = h + 4.$$

If $h > 0$, then Q lies above and to the right of P, as in Figure 2.4. If $h < 0$, then Q lies to the left of P (not shown). In either case, as Q approaches P along the curve, h approaches zero and the secant line slope $h + 4$ approaches 4. We take 4 to be the parabola's slope at P.

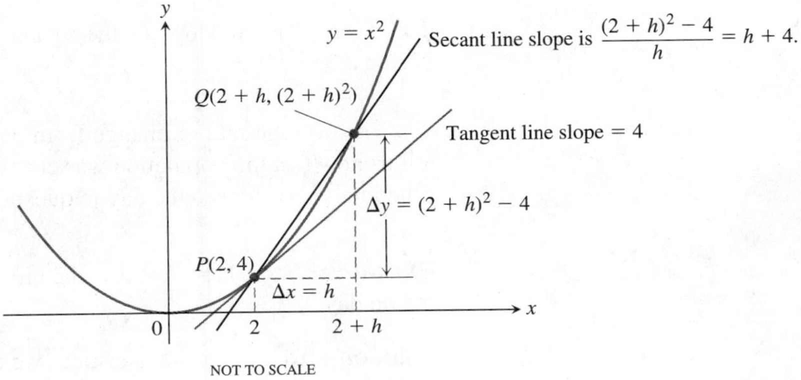

FIGURE 2.4 Finding the slope of the parabola $y = x^2$ at the point $P(2, 4)$ as the limit of secant line slopes (Example 3).

The tangent line to the parabola at P is the line through P with slope 4:

$$y = 4 + 4(x - 2) \quad \text{Point-slope equation}$$
$$y = 4x - 4. \quad \text{Simplify.}$$

■

Rates of Change and Tangent Lines

The rates at which the rock in Example 2 was falling at the instants $t = 1$ and $t = 2$ are called *instantaneous rates of change*. Instantaneous rates of change and slopes of tangent lines are closely connected, as we see in the following examples.

EXAMPLE 4 Figure 2.5 shows how a population p of fruit flies (*Drosophila*) grew in a 50-day experiment. The flies were counted at regular intervals, the counted values plotted with respect to the number of elapsed days t, and the points joined by a smooth curve (Figure 2.5). Find the average growth rate from day 23 to day 45.

Solution There were 150 flies on day 23 and 340 flies on day 45. Thus the number of flies increased by $340 - 150 = 190$ in $45 - 23 = 22$ days. The average rate of change of the population from day 23 to day 45 was

$$\frac{\Delta p}{\Delta t} = \frac{340 - 150}{45 - 23} = \frac{190}{22} \approx 8.6 \text{ flies/day.}$$

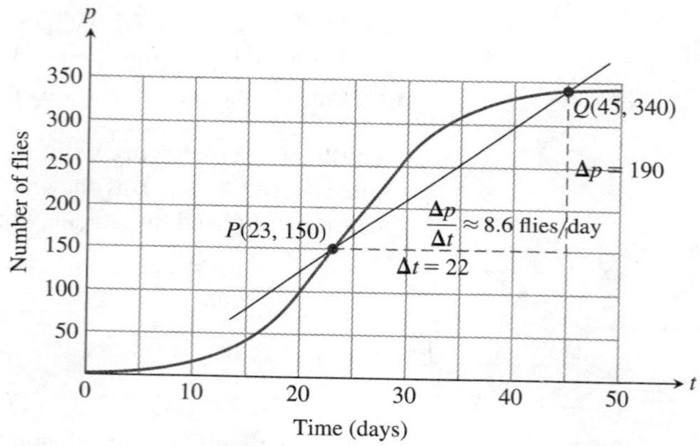

FIGURE 2.5 Growth of a fruit fly population in a controlled experiment. The average rate of change over 22 days is the slope $\Delta p/\Delta t$ of the secant line (Example 4).

This average is the slope of the secant line through the points P and Q on the graph in Figure 2.5. ■

The average rate of change from day 23 to day 45 calculated in Example 4 does not tell us how fast the population was changing on day 23 itself. For that we need to examine time intervals closer to the day in question.

EXAMPLE 5 How fast was the number of flies in the population of Example 4 growing on day 23?

Solution To answer this question, we examine the average rates of change over shorter and shorter time intervals starting at day 23. In geometric terms, we find these rates by calculating the slopes of secant lines from P to Q, for a sequence of points Q approaching P along the curve (Figure 2.6).

Q	Slope of $PQ = \Delta p/\Delta t$ (flies/day)
$(45, 340)$	$\dfrac{340 - 150}{45 - 23} \approx 8.6$
$(40, 330)$	$\dfrac{330 - 150}{40 - 23} \approx 10.6$
$(35, 310)$	$\dfrac{310 - 150}{35 - 23} \approx 13.3$
$(30, 265)$	$\dfrac{265 - 150}{30 - 23} \approx 16.4$

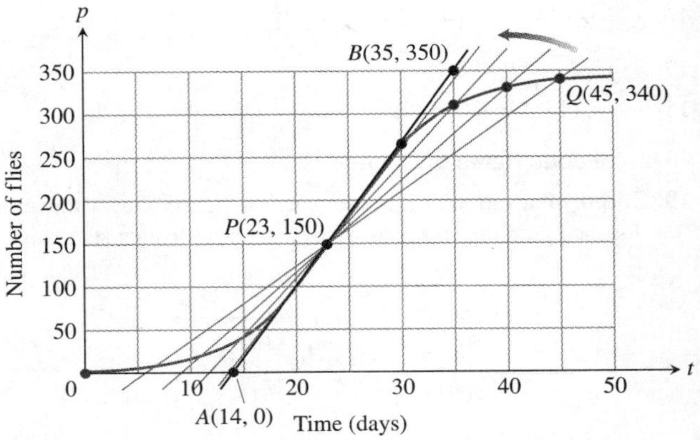

FIGURE 2.6 The positions and slopes of four secant lines through the point P on the fruit fly graph (Example 5).

The values in the table show that the secant line slopes rise from 8.6 to 16.4 as the t-coordinate of Q decreases from 45 to 30, and we would expect the slopes to rise slightly higher as t continued decreasing toward 23. Geometrically, the secant lines rotate counterclockwise about P and seem to approach the black tangent line in the figure. Since the line appears to pass through the points $(14, 0)$ and $(35, 350)$, its slope is approximately

$$\frac{350 - 0}{35 - 14} = 16.7 \text{ flies/day.}$$

On day 23 the population was increasing at a rate of about 16.7 flies/day. ■

The instantaneous rate of change is the value the average rate of change approaches as the length h of the interval over which the change occurs approaches zero. The average rate of change corresponds to the slope of a secant line; the instantaneous rate corresponds to the slope of the tangent line at a fixed value. So instantaneous rates and slopes of tangent lines are closely connected. We give a precise definition for these terms in the next chapter, but to do so we first need to develop the concept of a *limit*.

EXERCISES 2.1

Average Rates of Change

In Exercises 1–6, find the average rate of change of the function over the given interval or intervals.

1. $f(x) = x^3 + 1$
  a. $[2, 3]$ **b.** $[-1, 1]$

2. $g(x) = x^2 - 2x$
  a. $[1, 3]$ **b.** $[-2, 4]$

3. $h(t) = \cot t$
  a. $[\pi/4, 3\pi/4]$ **b.** $[\pi/6, \pi/2]$

4. $g(t) = 2 + \cos t$
  a. $[0, \pi]$ **b.** $[-\pi, \pi]$

5. $R(\theta) = \sqrt{4\theta + 1};\ [0, 2]$

6. $P(\theta) = \theta^3 - 4\theta^2 + 5\theta;\ [1, 2]$

Slope of a Curve at a Point

In Exercises 7–18, use the method in Example 3 to find **(a)** the slope of the curve at the given point P, and **(b)** an equation of the tangent line at P.

7. $y = x^2 - 5,\quad P(2, -1)$

8. $y = 7 - x^2,\quad P(2, 3)$

9. $y = x^2 - 2x - 3,\quad P(2, -3)$

10. $y = x^2 - 4x,\quad P(1, -3)$

11. $y = x^3,\quad P(2, 8)$

12. $y = 2 - x^3,\quad P(1, 1)$

13. $y = x^3 - 12x,\quad P(1, -11)$

14. $y = x^3 - 3x^2 + 4,\quad P(2, 0)$

15. $y = \dfrac{1}{x},\quad P(-2, -1/2)$

16. $y = \dfrac{x}{2-x}, \quad P(4, -2)$

17. $y = \sqrt{x}, \quad P(4, 2)$

18. $y = \sqrt{11 - 2x} \quad P(1, 3)$

Instantaneous Rates of Change

T **19. Speed of a car** The accompanying figure shows the time-to-distance graph for a sports car accelerating from a standstill.

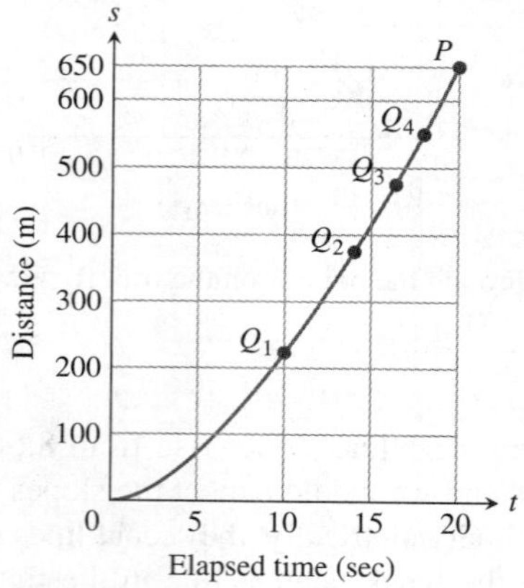

a. Estimate the slopes of secant lines PQ_1, PQ_2, PQ_3, and PQ_4, arranging them in order in a table like the one in Figure 2.6. What are the appropriate units for these slopes?

b. Then estimate the car's speed at time $t = 20$ sec.

T **20.** The accompanying figure shows the plot of distance fallen versus time for an object that fell from the lunar landing module a distance 80 m to the surface of the moon.

a. Estimate the slopes of the secant lines PQ_1, PQ_2, PQ_3, and PQ_4, arranging them in a table like the one in Figure 2.6.

b. About how fast was the object going when it hit the surface?

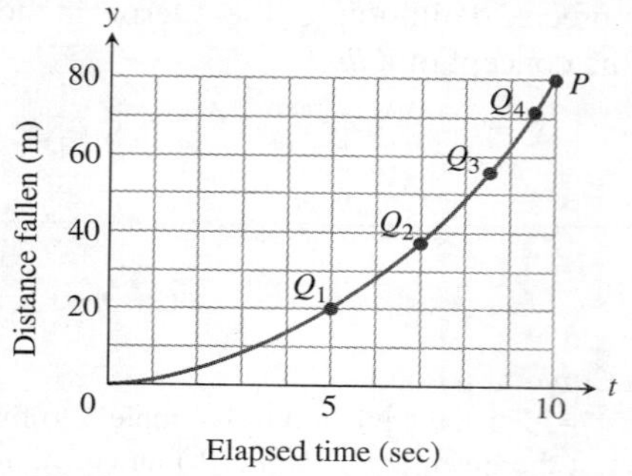

T **21.** The profits of a small company for each of the first five years of its operation are given in the following table:

Year	Profit in $1000s
2017	6
2018	27
2019	62
2020	111
2021	174

a. Plot points representing the profit as a function of year, and join them by as smooth a curve as you can.

b. What is the average rate of increase of the profits between 2019 and 2021?

c. Use your graph to estimate the rate at which the profits were changing in 2019.

T **22.** Make a table of values for the function $F(x) = \dfrac{(x+3)}{(x-3)}$ at the points $x = 1.5, x = 1.4, x = 1, x = 0.5$.

a. Find the average rate of change of $F(x)$ over the intervals $[0.5, x]$ for each $x \neq 0.5$ in your table.

b. Extending the table if necessary, try to determine the rate of change of $F(x)$ at $x = 2$.

T **23.** Let $g(x) = \sqrt{x}$ for $x \geq 0$.

a. Find the average rate of change of $g(x)$ with respect to x over the intervals $[1, 2], [1, 1.5]$ and $[1, 1 + h]$.

b. Make a table of values of the average rate of change of g with respect to x over the interval $[1, 1 + h]$ for some values of h approaching zero, say $h = 0.1, 0.01, 0.001, 0.0001, 0.00001$, and 0.000001.

c. What does your table indicate is the rate of change of $g(x)$ with respect to x at $x = 1$?

T **24.** Let $f(t) = \dfrac{5}{t}$ for $t \neq 0$.

a. Find the average rate of change of f with respect to t over the intervals (i) from $t = 2$ to $t = 3$, and (ii) from $t = 2$ to $t = T$.

b. Make a table of values of the average rate of change of f with respect to t over the interval $[2, T]$, for some values of T approaching 2, say $T = 2.1, 2.01, 2.001$ and 2.0001.

c. What does your table indicate is the rate of change of f with respect to t at $t = 2$?

d. Calculate the limit as T approaches 2 of the average rate of change of f with respect to t over the interval from 2 to T. You will have to do some algebra before you can substitute $T = 2$.

25. The accompanying graph shows the total distance s traveled by a bicyclist after t hours.

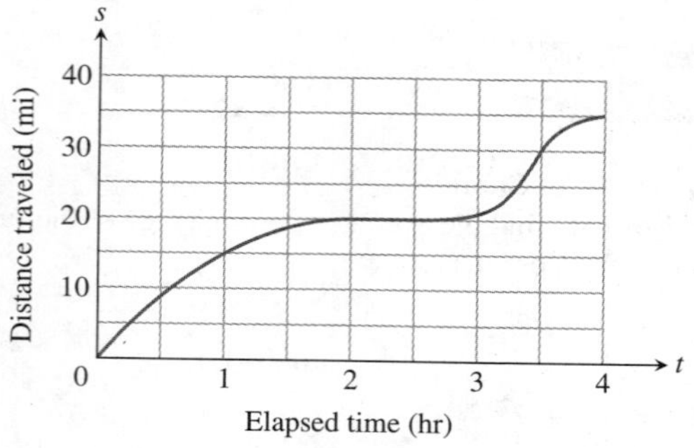

a. Estimate the bicyclist's average speed over the time intervals $[0, 1]$, $[1, 2.5]$, and $[2.5, 3.5]$.

b. Estimate the bicyclist's instantaneous speed at the times $t = \frac{1}{2}$, $t = 2$, and $t = 3$.

c. Estimate the bicyclist's maximum speed and the specific time at which it occurs.

26. The accompanying graph shows the total amount of gasoline A in the gas tank of an automobile after it has been driven for t days.

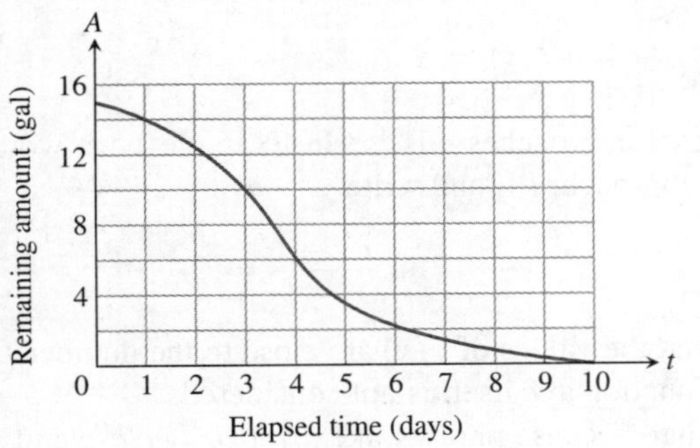

a. Estimate the average rate of gasoline consumption over the time intervals $[0, 3]$, $[0, 5]$, and $[7, 10]$.

b. Estimate the instantaneous rate of gasoline consumption at the times $t = 1$, $t = 4$, and $t = 8$.

c. Estimate the maximum rate of gasoline consumption and the specific time at which it occurs.

2.2 Limit of a Function and Limit Laws

HISTORICAL ESSAY

Limits

`www.bit.ly/2P04ZyV`

In Section 2.1 we saw how limits arise when finding the instantaneous rate of change of a function or the tangent line to a curve. We begin this section by presenting an informal definition of the limit of a function. We then describe laws that capture the behavior of limits. These laws enable us to quickly compute limits for a variety of functions, including polynomials and rational functions. We will present the precise definition of a limit in Section 2.3.

Limits of Function Values

Frequently, when studying a function $y = f(x)$, we find ourselves interested in the function's behavior *near* a particular point c, but not *at* c itself. An important example occurs when the process of trying to evaluate a function at c leads to division by zero, which is undefined. We encountered this when seeking the instantaneous rate of change in y by considering the quotient function $\Delta y/h$ for h closer and closer to zero. In the next example we explore numerically how a function behaves near a particular point at which we cannot directly evaluate the function.

EXAMPLE 1 How does the function

$$f(x) = \frac{x^2 - 1}{x - 1}$$

behave near $x = 1$?

Solution The given formula defines f for all real numbers x except $x = 1$ (since we cannot divide by zero). For any $x \neq 1$, we can simplify the formula by factoring the numerator and canceling common factors:

$$f(x) = \frac{(x - 1)(x + 1)}{x - 1} = x + 1 \quad \text{for} \quad x \neq 1.$$

The graph of f is the line $y = x + 1$ with the point $(1, 2)$ *removed*. This removed point is shown as a "hole" in Figure 2.7. Even though $f(1)$ is not defined, it is clear that we can make the value of $f(x)$ as close as we want to 2 by choosing x close enough to 1 (Table 2.2). ■

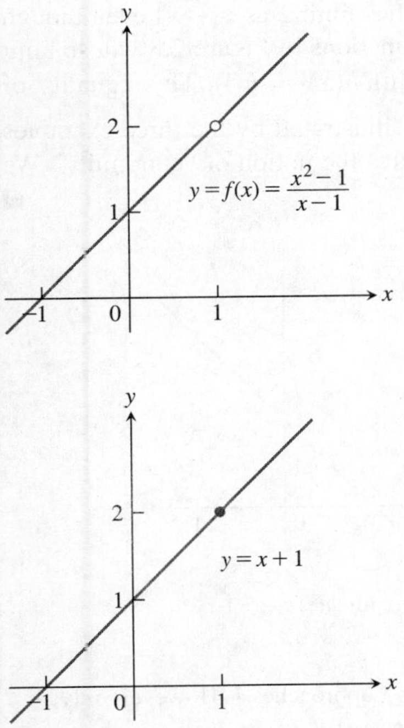

FIGURE 2.7 The graph of f is identical to the line $y = x + 1$ except at $x = 1$, where f is not defined (Example 1).

We will illustrate some other types of behavior near a point in Example 3.

An Informal Description of the Limit of a Function

We now give an informal definition of the limit of a function f at an interior point of the domain of f. Suppose that $f(x)$ is defined on an open interval about c, *except possibly at* c

TABLE 2.2 As x gets closer to 1, $f(x)$ gets closer to 2.

x	$f(x) = \dfrac{x^2 - 1}{x - 1}$
0.9	1.9
1.1	2.1
0.99	1.99
1.01	2.01
0.999	1.999
1.001	2.001
0.999999	1.999999
1.000001	2.000001

itself. If $f(x)$ is arbitrarily close to the number L (that is, as close to L as we like) for all x sufficiently close to c, other than c itself, then we say that f approaches the **limit** L as x approaches c, and write

$$\lim_{x \to c} f(x) = L.$$

This is read "the limit of $f(x)$ as x approaches c is L." In Example 1 we would say that $f(x)$ approaches the *limit* 2 as x approaches 1, and write

$$\lim_{x \to 1} f(x) = 2, \qquad \text{or} \qquad \lim_{x \to 1} \frac{x^2 - 1}{x - 1} = 2.$$

Essentially, the definition says that the values of $f(x)$ are close to the number L whenever x is close to c. The value of the function at c itself is not considered.

Our definition here is informal, because phrases like *arbitrarily close* and *sufficiently close* are imprecise; their meaning depends on the context. (To a machinist manufacturing a piston, *close* may mean *within a few thousandths of an inch*. To an astronomer studying distant galaxies, *close* may mean *within a few thousand light-years*.) Nevertheless, the definition is clear enough to enable us to recognize and evaluate limits of many specific functions. We will need the precise definition given in Section 2.3 when we set out to prove theorems about limits or study complicated functions. Here are several more examples exploring the idea of limits.

EXAMPLE 2 The limit of a function does not depend on how the function is defined at the point being approached. It does not even matter whether the function is defined at that point. Consider the three functions in Figure 2.8. The function f has limit 2 as $x \to 1$ even though f is not defined at $x = 1$. The function g has limit 2 as $x \to 1$ even though $2 \neq g(1)$. The function h is the only one of the three functions in Figure 2.8 whose limit as $x \to 1$ equals its value at $x = 1$. For h, we have $\lim_{x \to 1} h(x) = h(1)$. This equality of limit and function value has an important meaning. As illustrated by the three examples in Figure 2.8, equality of limit and function value captures the notion of "continuity." We study this in detail in Section 2.5. ■

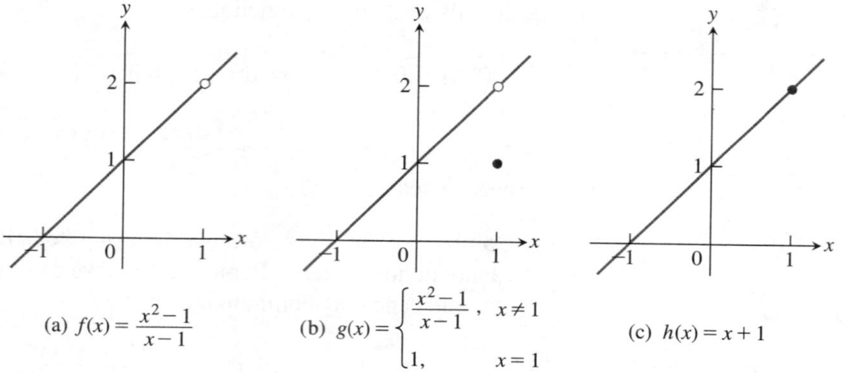

FIGURE 2.8 The limits of $f(x)$, $g(x)$, and $h(x)$ all equal 2 as x approaches 1. However, only $h(x)$ has the same function value as its limit at $x = 1$ (Example 2).

The process of finding a limit can often be broken up into a series of steps involving limits of basic functions, which are combined using a sequence of simple operations that we will develop. We start with two basic functions.

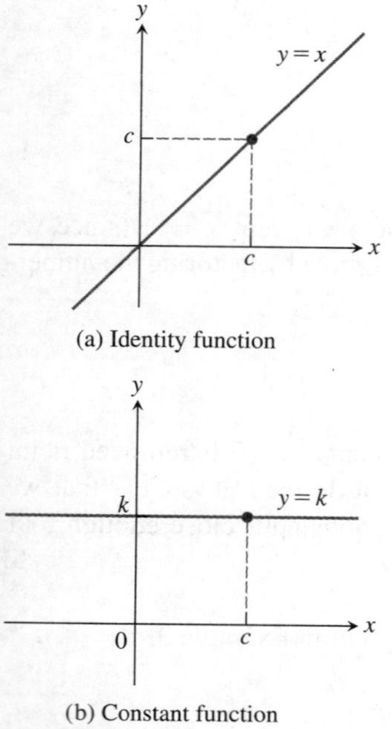

FIGURE 2.9 The functions in Example 3 have limits at all points c.

EXAMPLE 3 We find the limits of the identity function and of a constant function as x approaches $x = c$.

(a) If f is the **identity function** $f(x) = x$, then for any value of c (Figure 2.9a),

$$\lim_{x \to c} f(x) = \lim_{x \to c} x = c.$$

(b) If f is the **constant function** $f(x) = k$ (function with the constant value k), then for any value of c (Figure 2.9b),

$$\lim_{x \to c} f(x) = \lim_{x \to c} k = k.$$

For instances of each of these rules we have

$$\lim_{x \to 3} x = 3 \qquad \text{Limit of identity function at } x = 3$$

and

$$\lim_{x \to -7} 4 = \lim_{x \to 2} 4 = 4. \qquad \text{Limit of constant function } f(x) = 4 \text{ at } x = -7 \text{ or at } x = 2$$

We prove these rules in Example 3 in Section 2.3. ■

A function may not have a limit at a particular point. Some ways that limits can fail to exist are illustrated in Figure 2.10 and described in the next example.

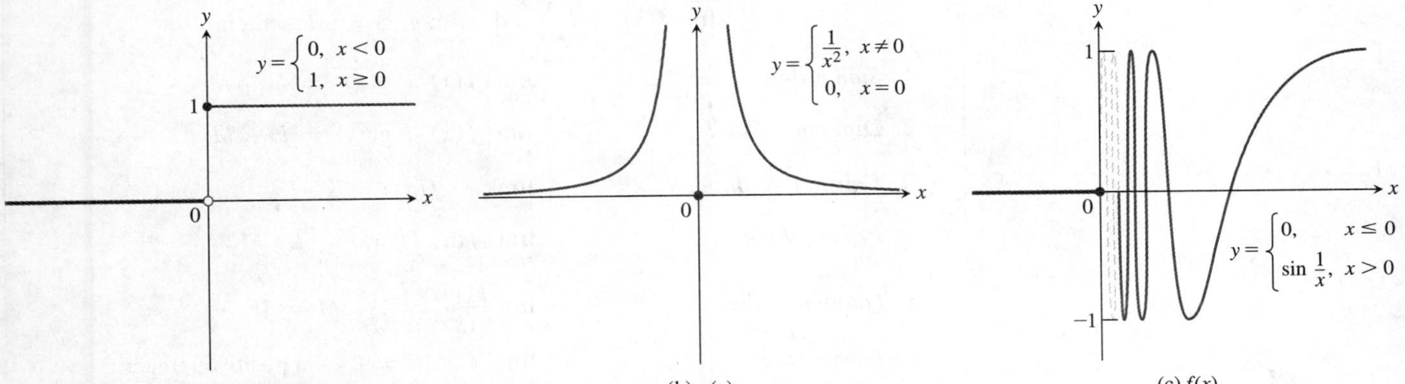

FIGURE 2.10 None of these functions has a limit as x approaches 0 (Example 4).

EXAMPLE 4 Discuss the behavior of the following functions, explaining why they have no limit as $x \to 0$.

(a) $U(x) = \begin{cases} 0, & x < 0 \\ 1, & x \geq 0 \end{cases}$

(b) $g(x) = \begin{cases} \dfrac{1}{x^2}, & x \neq 0 \\ 0, & x = 0 \end{cases}$

(c) $f(x) = \begin{cases} 0, & x \leq 0 \\ \sin \dfrac{1}{x}, & x > 0 \end{cases}$

Solution

(a) This function *jumps*: The **unit step function** $U(x)$ has no limit as $x \to 0$ because its values jump at $x = 0$. For negative values of x arbitrarily close to zero, $U(x) = 0$. For positive values of x arbitrarily close to zero, $U(x) = 1$. There is no *single* value L approached by $U(x)$ as $x \to 0$ (Figure 2.10a).

(b) This function *grows too "large" to have a limit*: $g(x)$ has no limit as $x \to 0$ because the values of g grow arbitrarily large as $x \to 0$ and therefore do not stay close to *any* fixed real number (Figure 2.10b). We say that the function g is *not bounded*.

The function U in Example 4 is known as the *Heaviside function*, after Oliver Heaviside (1850–1925).

(c) This function *oscillates too much to have a limit*: $f(x)$ has no limit as $x \to 0$ because the function's values oscillate between $+1$ and -1 in every open interval containing 0. The values do not stay close to any single number as $x \to 0$ (Figure 2.10c). ■

A function that oscillates may or may not have a limit. In Example 11 we will see a function that oscillates wildly, but nevertheless does have a limit. The problem with the function f discussed in Example 4 is not that it oscillates, but that it oscillates *too much* for a limit to exist.

The Limit Laws

A few basic rules allow us to break down complicated functions into simple ones when calculating limits. By using these laws, we can greatly simplify many limit computations.

THEOREM 1—Limit Laws
If L, M, c, and k are real numbers and

$$\lim_{x\to c} f(x) = L \quad \text{and} \quad \lim_{x\to c} g(x) = M, \quad \text{then}$$

1. *Sum Rule:* $\lim_{x\to c} (f(x) + g(x)) = L + M$
2. *Difference Rule:* $\lim_{x\to c} (f(x) - g(x)) = L - M$
3. *Constant Multiple Rule:* $\lim_{x\to c} (k \cdot f(x)) = k \cdot L$
4. *Product Rule:* $\lim_{x\to c} (f(x) \cdot g(x)) = L \cdot M$
5. *Quotient Rule:* $\lim_{x\to c} \dfrac{f(x)}{g(x)} = \dfrac{L}{M}, \quad M \neq 0$
6. *Power Rule:* $\lim_{x\to c} [f(x)]^n = L^n$, n a positive integer
7. *Root Rule:* $\lim_{x\to c} \sqrt[n]{f(x)} = \sqrt[n]{L} = L^{1/n}$, n a positive integer

(If n is even, we assume that $f(x) \geq 0$ for x in an interval containing c.)

The Sum Rule says that the limit of a sum is the sum of the limits. Similarly, the next rules say that the limit of a difference is the difference of the limits; the limit of a constant times a function is the constant times the limit of the function; the limit of a product is the product of the limits; the limit of a quotient is the quotient of the limits (provided that the limit of the denominator is not 0); the limit of a positive integer power (or root) of a function is the integer power (or root) of the limit (provided that the root of the limit is a real number).

There are simple intuitive arguments for why the properties in Theorem 1 are true (although these do not constitute proofs). If x is sufficiently close to c, then $f(x)$ is close to L and $g(x)$ is close to M, from our informal definition of a limit. It is then reasonable that $f(x) + g(x)$ is close to $L + M$; $f(x) - g(x)$ is close to $L - M$; $kf(x)$ is close to kL; $f(x)g(x)$ is close to LM; and $f(x)/g(x)$ is close to L/M if M is not zero. We prove the Sum Rule in Section 2.3, based on a rigorous definition of the limit. Rules 2–5 are proved in Appendix A.5. Rule 6 is obtained by applying Rule 4 repeatedly. Rule 7 is proved in more advanced texts. The Sum, Difference, and Product Rules can be extended to any number of functions, not just two.

EXAMPLE 5 Use the observations $\lim_{x\to c} k = k$ and $\lim_{x\to c} x = c$ (Example 3) and the limit laws in Theorem 1 to find the following limits.

(a) $\lim_{x\to c} (x^3 + 4x^2 - 3)$ **(b)** $\lim_{x\to c} \dfrac{x^4 + x^2 - 1}{x^2 + 5}$ **(c)** $\lim_{x\to -2} \sqrt{4x^2 + 3}$

Solution

(a) $\lim_{x \to c} (x^3 + 4x^2 - 3) = \lim_{x \to c} x^3 + \lim_{x \to c} 4x^2 - \lim_{x \to c} 3$ Sum and Difference Rules

$= c^3 + 4c^2 - 3$ Power and Multiple Rules and limit of a constant function

(b) $\lim_{x \to c} \dfrac{x^4 + x^2 - 1}{x^2 + 5} = \dfrac{\lim_{x \to c} (x^4 + x^2 - 1)}{\lim_{x \to c} (x^2 + 5)}$ Quotient Rule: Note that $(x^2 + 5) > 0$ for all x.

$= \dfrac{\lim_{x \to c} x^4 + \lim_{x \to c} x^2 - \lim_{x \to c} 1}{\lim_{x \to c} x^2 + \lim_{x \to c} 5}$ Sum and Difference Rules

$= \dfrac{c^4 + c^2 - 1}{c^2 + 5}$ Power Rule and limit of a constant function

(c) $\lim_{x \to -2} \sqrt{4x^2 + 3} = \sqrt{\lim_{x \to -2} (4x^2 + 3)}$ Root Rule with $n = 2(4x^2 + 3 \geq 0)$

$= \sqrt{\lim_{x \to -2} 4x^2 + \lim_{x \to -2} 3}$ Difference Rule

$= \sqrt{4(-2)^2 + 3}$ Power and Multiple Rules and limit of a constant function

$= \sqrt{16 + 3}$

$= \sqrt{19}$ ■

Evaluating Limits of Polynomials and Rational Functions

Theorem 1 simplifies the task of calculating limits of polynomials and rational functions. To evaluate the limit of a polynomial function as x approaches c, just substitute c for x in the formula for the function. To evaluate the limit of a rational function as x approaches a point c *at which the denominator is not zero*, substitute c for x in the formula for the function. (See Examples 5a and 5b.) We state these results formally as theorems.

THEOREM 2—Limits of Polynomials
If $P(x) = a_n x^n + a_{n-1}x^{n-1} + \cdots + a_0$, then

$$\lim_{x \to c} P(x) = P(c) = a_n c^n + a_{n-1}c^{n-1} + \cdots + a_0.$$

THEOREM 3—Limits of Rational Functions
If $P(x)$ and $Q(x)$ are polynomials and $Q(c) \neq 0$, then

$$\lim_{x \to c} \frac{P(x)}{Q(x)} = \frac{P(c)}{Q(c)}.$$

EXAMPLE 6 The following calculation illustrates Theorems 2 and 3:

$$\lim_{x \to -1} \frac{x^3 + 4x^2 - 3}{x^2 + 5} = \frac{(-1)^3 + 4(-1)^2 - 3}{(-1)^2 + 5} = \frac{0}{6} = 0$$

Since the denominator of this rational expression does not equal 0 when we substitute -1 for x, we can just compute the value of the expression at $x = -1$ to evaluate the limit. ■

Eliminating Common Factors from Zero Denominators

Theorem 3 applies only if the denominator of the rational function is not zero at the limit point c. If the denominator is zero, canceling common factors in the numerator and

Identifying Common Factors
If $Q(x)$ is a polynomial and $Q(c) = 0$, then $(x - c)$ is a factor of $Q(x)$. Thus, if the numerator and denominator of a rational function of x are both zero at $x = c$, they have $(x - c)$ as a common factor.

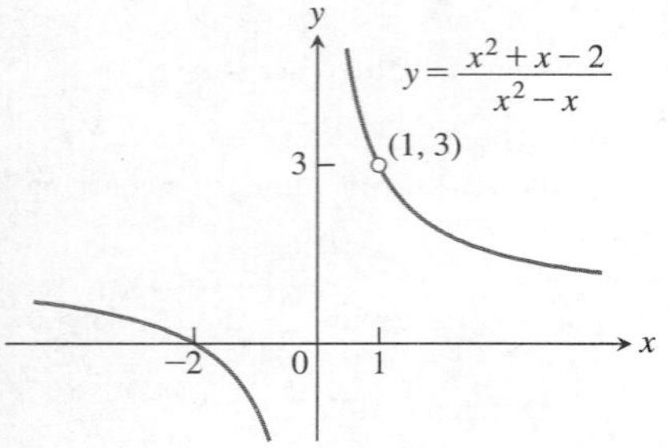

(a)

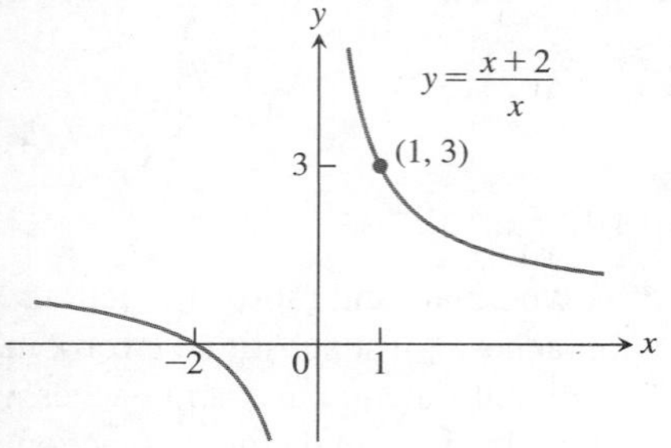

(b)

FIGURE 2.11 The graph of $f(x) = (x^2 + x - 2)/(x^2 - x)$ in part (a) is the same as the graph of $g(x) = (x + 2)/x$ in part (b) except at $x = 1$, where f is undefined. The functions have the same limit as $x \to 1$ (Example 7).

denominator may reduce the fraction to one whose denominator is no longer zero at c. If this happens, we can find the limit by substitution in the simplified fraction.

EXAMPLE 7 Evaluate

$$\lim_{x \to 1} \frac{x^2 + x - 2}{x^2 - x}.$$

Solution We cannot substitute $x = 1$ because it makes the denominator zero. We test the numerator to see if it, too, is zero at $x = 1$. It is, so it has a factor of $(x - 1)$ in common with the denominator. Canceling this common factor gives a simpler fraction with the same values as the original for $x \neq 1$:

$$\frac{x^2 + x - 2}{x^2 - x} = \frac{(x - 1)(x + 2)}{x(x - 1)} = \frac{x + 2}{x}, \quad \text{if } x \neq 1.$$

Using the simpler fraction, we find the limit of these values as $x \to 1$ by evaluating the function at $x = 1$, as in Theorem 3:

$$\lim_{x \to 1} \frac{x^2 + x - 2}{x^2 - x} = \lim_{x \to 1} \frac{x + 2}{x} = \frac{1 + 2}{1} = 3.$$

See Figure 2.11. ■

Using Calculators and Computers to Estimate Limits

We can try using a calculator or computer to guess a limit numerically. However, calculators and computers can sometimes give false values and misleading evidence about limits. Usually the problem is associated with rounding errors, as we now illustrate.

EXAMPLE 8 Estimate the value of $\lim_{x \to 0} \dfrac{\sqrt{x^2 + 100} - 10}{x^2}$.

Solution Table 2.3 lists values of the function obtained on a calculator for several points approaching $x = 0$. As x approaches 0 through the points $\pm 1, \pm 0.5, \pm 0.1$, and ± 0.01, the function seems to approach the number 0.05.

As we take even smaller values of x, $\pm 0.0005, \pm 0.0001, \pm 0.00001$, and ± 0.000001, the function appears to approach the number 0.

Is the answer 0.05 or 0, or some other value? We resolve this question in the next example. ■

Using a computer or calculator may give ambiguous results, as in Example 8. A computer cannot always keep track of enough digits to avoid rounding errors in computing the values of $f(x)$ when x is very small. We cannot substitute $x = 0$ in the problem, and the

TABLE 2.3 Computed values of $f(x) = \dfrac{\sqrt{x^2 + 100} - 10}{x^2}$ near $x = 0$

x	$f(x)$	
± 1	0.049876	approaches 0.05?
± 0.5	0.049969	
± 0.1	0.049999	
± 0.01	0.050000	
± 0.0005	0.050000	approaches 0?
± 0.0001	0.000000	
± 0.00001	0.000000	
± 0.000001	0.000000	

numerator and denominator have no obvious common factors (as they did in Example 7). Sometimes, however, we can create a common factor by using algebra.

EXAMPLE 9 Evaluate

$$\lim_{x\to 0}\frac{\sqrt{x^2+100}-10}{x^2}.$$

Solution This is the limit we considered in Example 8. We can create a common factor by multiplying both numerator and denominator by the conjugate radical expression $\sqrt{x^2+100}+10$ (obtained by changing the sign after the square root). The preliminary algebra rationalizes the numerator:

$$\begin{aligned}\frac{\sqrt{x^2+100}-10}{x^2} &= \frac{\sqrt{x^2+100}-10}{x^2}\cdot\frac{\sqrt{x^2+100}+10}{\sqrt{x^2+100}+10} && \text{Multiply and divide by the conjugate.}\\ &= \frac{x^2+100-100}{x^2\left(\sqrt{x^2+100}+10\right)} && \text{Simplify.}\\ &= \frac{x^2}{x^2\left(\sqrt{x^2+100}+10\right)} && \text{Common factor } x^2\\ &= \frac{1}{\sqrt{x^2+100}+10}. && \text{Cancel } x^2 \text{ for } x \neq 0.\end{aligned}$$

Therefore,

$$\begin{aligned}\lim_{x\to 0}\frac{\sqrt{x^2+100}-10}{x^2} &= \lim_{x\to 0}\frac{1}{\sqrt{x^2+100}+10}\\ &= \frac{1}{\sqrt{0^2+100}+10} && \text{Use limit laws: Sum Rule, Power Rule, Root Rule, and Quotient Rule (denominator not 0).}\\ &= \frac{1}{20} = 0.05.\end{aligned}$$

This calculation provides the correct answer, resolving the ambiguous computer results in Example 8. ■

We cannot always manipulate the terms in an expression to find the limit of a quotient where the denominator becomes zero. In some cases the limit might then be found with geometric arguments (see the proof of Theorem 6 in Section 2.4), or through methods of calculus (developed in Section 7.5). The next theorem shows how to evaluate difficult limits by comparing them with functions having known limits.

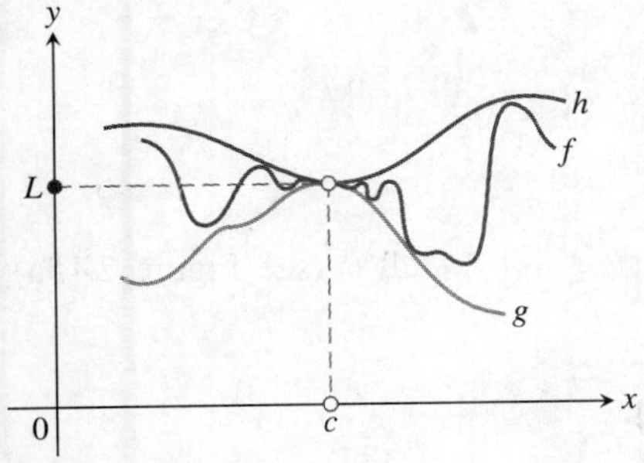

FIGURE 2.12 The graph of f is sandwiched between the graphs of g and h.

The Sandwich Theorem

The following theorem enables us to calculate a variety of limits. It is called the Sandwich Theorem because it refers to a function f whose values are sandwiched between the values of two other functions g and h that have the same limit L at a point c. Being trapped between the values of two functions that approach L, the values of f must also approach L (Figure 2.12). A proof is given in Appendix A.5.

THEOREM 4—The Sandwich Theorem
Suppose that $g(x) \leq f(x) \leq h(x)$ for all x in some open interval containing c, except possibly at $x = c$ itself. Suppose also that

$$\lim_{x\to c} g(x) = \lim_{x\to c} h(x) = L.$$

Then $\lim_{x\to c} f(x) = L$.

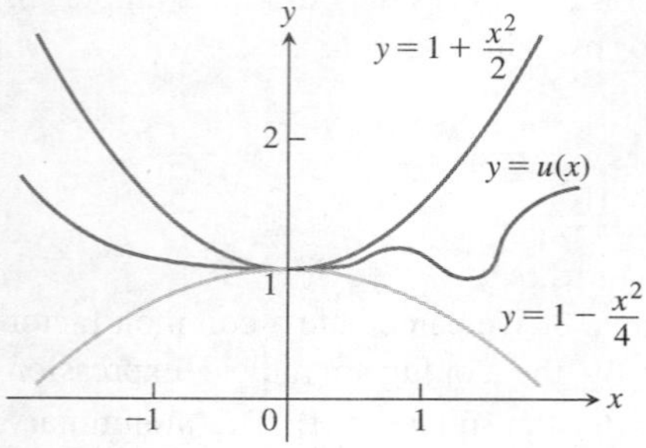

FIGURE 2.13 Any function $u(x)$ whose graph lies in the region between $y = 1 + (x^2/2)$ and $y = 1 - (x^2/4)$ has limit 1 as $x \to 0$ (Example 10).

The Sandwich Theorem is also called the Squeeze Theorem or the Pinching Theorem.

EXAMPLE 10 Given a function u that satisfies

$$1 - \frac{x^2}{4} \le u(x) \le 1 + \frac{x^2}{2} \quad \text{for all } x \ne 0,$$

find $\lim_{x \to 0} u(x)$, no matter how complicated u is.

Solution Since

$$\lim_{x \to 0} (1 - (x^2/4)) = 1 \quad \text{and} \quad \lim_{x \to 0} (1 + (x^2/2)) = 1,$$

the Sandwich Theorem implies that $\lim_{x \to 0} u(x) = 1$ (Figure 2.13). ■

We use the Sandwich Theorem to show that it is possible for a function that oscillates to have a limit.

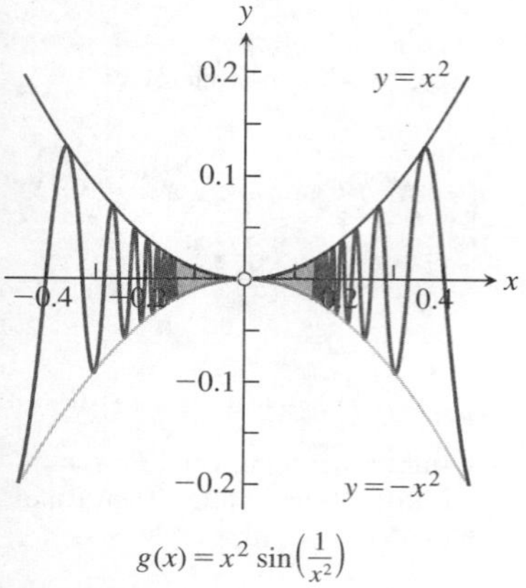

FIGURE 2.14 The graph of the function g (Example 11). It is not defined at $x = 0$. Even though the function oscillates, it has a limit as $x \to 0$. The value of $g(x)$ always lies between x^2 and $-x^2$.

EXAMPLE 11 How does the function $g(x) = x^2 \sin(1/x^2)$ behave near $x = 0$?

Solution The formula defines $g(x)$ for all real numbers x except $x = 0$. The graph of g is shown in Figure 2.14. We can see that the graph oscillates, but we can use the Sandwich Theorem to find the limit of $g(x)$ as x approaches 0. If $x \ne 0$, then $1/x^2$ is a positive real number. Since the range of the sine function is the interval $[-1, 1]$, it follows that $-1 \le \sin(1/x^2) \le 1$ for all $x \ne 0$. Even though we may not know the exact value of $g(x) = x^2 \sin(1/x^2)$, we do know that it lies between $-x^2$ and x^2. Since

$$\lim_{x \to 0} x^2 = 0 \quad \text{and} \quad \lim_{x \to 0} (-x^2) = 0,$$

the Sandwich Theorem implies that $\lim_{x \to 0} g(x) = 0$. ■

EXAMPLE 12 The Sandwich Theorem helps us establish several important limit rules:

(a) $\lim_{\theta \to 0} \sin \theta = 0$

(b) $\lim_{\theta \to 0} \cos \theta = 1$

(c) For any function f, $\lim_{x \to c} |f(x)| = 0$ implies $\lim_{x \to c} f(x) = 0$.

Solution

(a) In Section 1.3 we established that $-|\theta| \le \sin \theta \le |\theta|$ for all θ (see Figure 2.15a). Since $\lim_{\theta \to 0} (-|\theta|) = \lim_{\theta \to 0} |\theta| = 0$, we have

$$\lim_{\theta \to 0} \sin \theta = 0.$$

(b) From Section 1.3, $0 \le 1 - \cos \theta \le |\theta|$ for all θ (see Figure 2.15b). Since $\lim_{\theta \to 0} |\theta| = 0$ and $\lim_{\theta \to 0} 0 = 0$, we have $\lim_{\theta \to 0} (1 - \cos \theta) = 0$ so

$$\lim_{\theta \to 0} \cos \theta = \lim_{\theta \to 0} (1 - (1 - \cos \theta)) = 1 - \lim_{\theta \to 0} (1 - \cos \theta) = 1 - 0 = 1.$$

(c) Since $-|f(x)| \le f(x) \le |f(x)|$ and $-|f(x)|$ and $|f(x)|$ have limit 0 as $x \to c$, it follows that $\lim_{x \to c} f(x) = 0$. ■

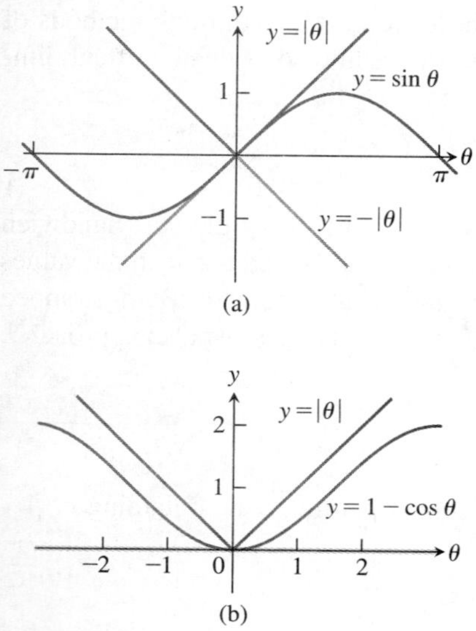

FIGURE 2.15 The Sandwich Theorem confirms the limits in Example 12.

Example 12 shows that the sine and cosine functions are equal to their limits at $\theta = 0$. We have not yet established that for any c, $\lim_{\theta \to c} \sin \theta = \sin c$, and $\lim_{\theta \to c} \cos \theta = \cos c$. These limit formulas do hold, as will be shown in Section 2.5.

EXERCISES 2.2

Limits from Graphs

1. For the function $g(x)$ graphed here, find the following limits or explain why they do not exist.

   a. $\lim_{x\to 0} g(x)$ b. $\lim_{x\to 1} g(x)$ c. $\lim_{x\to 2} g(x)$ d. $\lim_{x\to 3} g(x)$

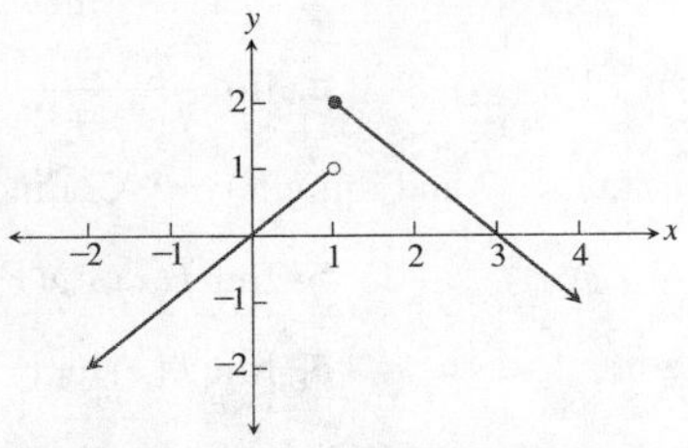

2. For the function $f(t)$ graphed here, find the following limits or explain why they do not exist.

   a. $\lim_{t\to -2} f(t)$ b. $\lim_{t\to -1} f(t)$ c. $\lim_{t\to 0} f(t)$ d. $\lim_{t\to -0.5} f(t)$

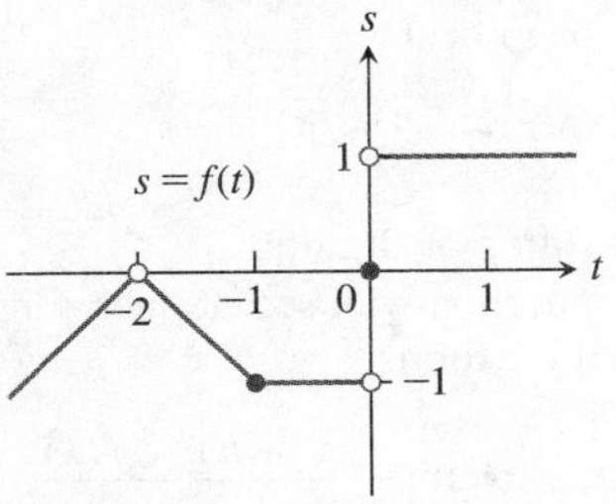

3. Which of the following statements about the function $y = f(x)$ graphed here are true, and which are false?

   a. $\lim_{x\to 0} f(x)$ exists.

   b. $\lim_{x\to 0} f(x) = 0$

   c. $\lim_{x\to 0} f(x) = 1$

   d. $\lim_{x\to 1} f(x) = 1$

   e. $\lim_{x\to 1} f(x) = 0$

   f. $\lim_{x\to c} f(x)$ exists at every point c in $(-1, 1)$.

   g. $\lim_{x\to 1} f(x)$ does not exist.

   h. $f(0) = 0$

   i. $f(0) = 1$

   j. $f(1) = 0$

   k. $f(1) = -1$

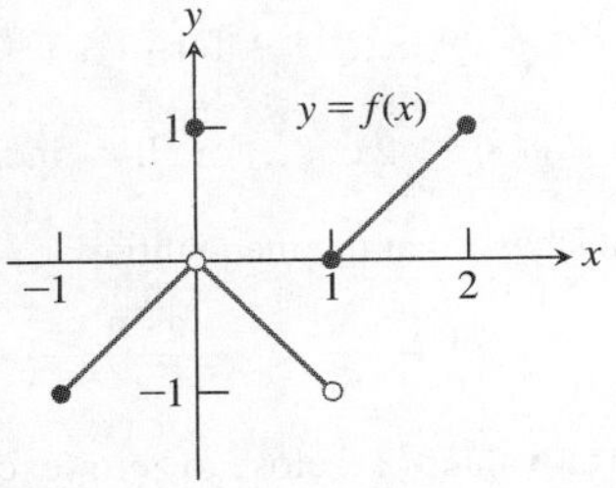

4. Which of the following statements about the function $y = f(x)$ graphed here are true, and which are false?

   a. $\lim_{x\to 2} f(x)$ does not exist.

   b. $\lim_{x\to 2} f(x) = 2$

   c. $\lim_{x\to 1} f(x)$ does not exist.

   d. $\lim_{x\to c} f(x)$ exists at every point c in $(-1, 1)$.

   e. $\lim_{x\to c} f(x)$ exists at every point c in $(1, 3)$.

   f. $f(1) = 0$

   g. $f(1) = -2$

   h. $f(2) = 0$

   i. $f(2) = 1$

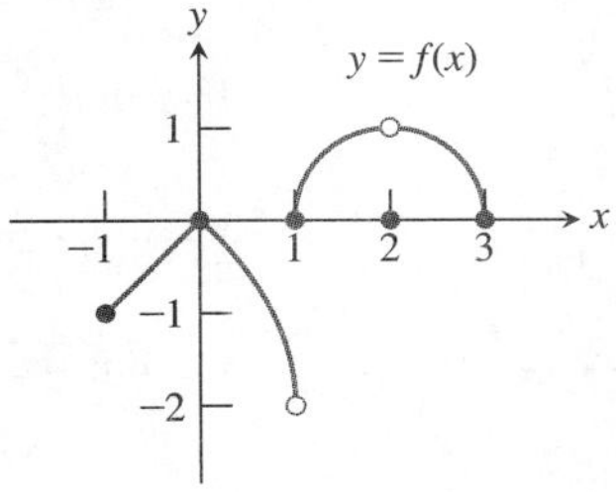

Existence of Limits

In Exercises 5 and 6, explain why the limits do not exist.

5. $\lim_{x\to 0} \dfrac{2x}{|2x|}$

6. $\lim_{x\to 1} \dfrac{1}{x-1}$

7. Suppose that a function $f(x)$ is defined for all real values of x except $x = c$. Can anything be said about the existence of $\lim_{x\to c} f(x)$? Give reasons for your answer.

8. Suppose that a function $f(x)$ is defined for all x in $[-1, 1]$. Can anything be said about the existence of $\lim_{x\to 0} f(x)$? Give reasons for your answer.

9. If $\lim_{x\to 2} f(x) = 10$, must f be defined at $x = 2$? If it is, must $f(2) = 10$? Can we conclude anything about the values of f at $x = 2$? Explain.

10. If $f(1) = 5$, must $\lim_{x\to 1} f(x)$ exist? If it does, then must $\lim_{x\to 1} f(x) = 5$? Can we conclude *anything* about $\lim_{x\to 1} f(x)$? Explain.

Calculating Limits

Find the limits in Exercises 11–22.

11. $\lim_{x\to -2} (x^3 - 15)$

12. $\lim_{x\to 2} (-x^2 + 5x - 2)$

13. $\lim_{x\to 5} 10(t-8)(t-12)$

14. $\lim_{x\to -2} (x^3 - 2x^2 + 4x + 8)$

15. $\lim_{x\to 2} \dfrac{2x+5}{11-x^3}$

16. $\lim_{s\to 2/3} (8-3s)(2s-1)$

17. $\lim_{x\to -1/2} 4x(3x+4)^2$

18. $\lim_{y\to -3} \dfrac{y+6}{y^2+4y+12}$

19. $\lim_{y\to -3} (5-y)^{4/3}$

20. $\lim_{z\to 4} \sqrt{z^2-10}$

21. $\lim_{h\to 0} \dfrac{3}{\sqrt{3h+1}+1}$

22. $\lim_{h\to 1} \dfrac{\sqrt{6h+10}-5}{h^2}$

Limits of quotients Find the limits in Exercises 23–42.

23. $\lim_{x\to 5} \dfrac{x-5}{x^2-25}$

24. $\lim_{x\to 3} \dfrac{x+4}{x^2+7x+12}$

25. $\lim_{x\to -5} \dfrac{x^2+3x-10}{x+5}$

26. $\lim_{x\to 2} \dfrac{x^2-7x+10}{x-2}$

27. $\lim_{t\to 2} \dfrac{t^2+4t+3}{t^2-9}$

28. $\lim_{t\to -1} \dfrac{t^2+3t+2}{t^2-t-2}$

29. $\lim_{t\to -2} \dfrac{-2x-4}{x^3+2x^2}$

30. $\lim_{y\to 0} \dfrac{10y^4-40}{5y^3+10y^2}$

31. $\lim_{x\to 1} \dfrac{x^{-1} - 1}{x - 1}$

32. $\lim_{x\to 0} \dfrac{\dfrac{1}{x-1} + \dfrac{1}{x+1}}{x}$

33. $\lim_{u\to -1} \dfrac{u^4 - 81}{u^3 - 27}$

34. $\lim_{v\to 2} \dfrac{v^3 - 8}{v^4 - 16}$

35. $\lim_{x\to 9} \dfrac{\sqrt{x} - 3}{x - 9}$

36. $\lim_{x\to 9} \dfrac{9x - x^2}{3 - \sqrt{x}}$

37. $\lim_{x\to 1} \dfrac{x - 1}{\sqrt{x+3} - 2}$

38. $\lim_{x\to -1} \dfrac{\sqrt{x^2+8} - 3}{x + 1}$

39. $\lim_{x\to 3} \dfrac{\sqrt{2x^2+7} - 5}{5x - 2}$

40. $\lim_{x\to -2} \dfrac{x + 2}{\sqrt{x^2+5} - 3}$

41. $\lim_{x\to -3} \dfrac{2 - \sqrt{x^2 - 5}}{x + 3}$

42. $\lim_{x\to 7} \dfrac{10 - x}{11 - \sqrt{x^2 + 15}}$

Limits with trigonometric functions Find the limits in Exercises 43–50.

43. $\lim_{x\to 0} (2\sin x - 1)$

44. $\lim_{x\to 0} \cos^2 x$

45. $\lim_{x\to 0} \sec x$

46. $\lim_{x\to \pi} \tan x$

47. $\lim_{x\to 0} \dfrac{1 + x + \sin x}{3\cos x}$

48. $\lim_{x\to 0} (x^2 - 1)(2 - \cos x)$

49. $\lim_{x\to \pi} \sqrt{x+2}\, \sin\left(\dfrac{x}{2}\right)$

50. $\lim_{x\to 0} \sqrt{7 + \sec^2 x}$

Using Limit Rules

51. Suppose $\lim_{x\to 0} f(x) = 1$ and $\lim_{x\to 0} g(x) = -5$. Name the rules in Theorem 1 that are used to accomplish steps (a), (b), and (c) of the following calculation.

$$\lim_{x\to 0} \frac{2f(x) - g(x)}{(f(x)+7)^2} = \frac{\lim_{x\to 0}(2f(x) - g(x))}{\lim_{x\to 0}(f(x)+7)^2} \qquad \text{(a)}$$

(We assume the denominator is nonzero.)

$$= \frac{\lim_{x\to 0} 2f(x) - \lim_{x\to 0} g(x)}{\left(\lim_{x\to 0}(f(x)+7)\right)^2} \qquad \text{(b)}$$

$$= \frac{2\lim_{x\to 0} f(x) - \lim_{x\to 0} g(x)}{\left(\lim_{x\to 0} f(x) + \lim_{x\to 0} 7\right)^2} \qquad \text{(c)}$$

$$= \frac{(2)(1) - (-5)}{(1+7)^2} = \frac{7}{64}$$

52. Let $\lim_{x\to 1} h(x) = 5$, $\lim_{x\to 1} p(x) = 1$, and $\lim_{x\to 1} r(x) = 2$. Name the rules in Theorem 1 that are used to accomplish steps (a), (b), and (c) of the following calculation.

$$\lim_{x\to 1} \frac{\sqrt{5h(x)}}{p(x)(4 - r(x))} = \frac{\lim_{x\to 1}\sqrt{5h(x)}}{\lim_{x\to 1}(p(x)(4 - r(x)))} \qquad \text{(a)}$$

(We assume the denominator is nonzero.)

$$= \frac{\sqrt{\lim_{x\to 1} 5h(x)}}{\left(\lim_{x\to 1} p(x)\right)\left(\lim_{x\to 1}(4 - r(x))\right)} \qquad \text{(b)}$$

$$= \frac{\sqrt{5\lim_{x\to 1} h(x)}}{\left(\lim_{x\to 1} p(x)\right)\left(\lim_{x\to 1} 4 - \lim_{x\to 1} r(x)\right)} \qquad \text{(c)}$$

$$= \frac{\sqrt{(5)(5)}}{(1)(4 - 2)} = \frac{5}{2}$$

53. Suppose $\lim_{x\to c} f(x) = 10$ and $\lim_{x\to c} g(x) = -5$. Find

a. $\lim_{x\to c}(f(x) + g(x))$

b. $\lim_{x\to c} 5f(x)g(x)$

c. $\lim_{x\to c}(2f(x) + 3g(x))$

d. $\lim_{x\to c} \dfrac{g(x)}{f(x) + g(x)}$

54. Suppose $\lim_{x\to 4} f(x) = 0$ and $\lim_{x\to 4} g(x) = -3$. Find

a. $\lim_{x\to 4} (g(x) + 3)$

b. $\lim_{x\to 4} xf(x)$

c. $\lim_{x\to 4} (g(x))^2$

d. $\lim_{x\to 4} \dfrac{g(x)}{f(x) - 1}$

55. Suppose $\lim_{x\to b} f(x) = 7$ and $\lim_{x\to b} g(x) = -3$. Find

a. $\lim_{x\to b} (f(x) + g(x))$

b. $\lim_{x\to b} f(x)\cdot g(x)$

c. $\lim_{x\to b} 4g(x)$

d. $\lim_{x\to b} f(x)/g(x)$

56. Suppose that $\lim_{x\to -2} p(x) = 4$, $\lim_{x\to -2} r(x) = 0$, and $\lim_{x\to -2} s(x) = -3$. Find

a. $\lim_{x\to -2} (p(x) + r(x) + s(x))$

b. $\lim_{x\to -2} (p(x)\cdot r(x)\cdot s(x))$

c. $\lim_{x\to -2} (-4p(x) + 5r(x))/s(x)$

Limits of Average Rates of Change

Because of their connection with secant lines, tangents, and instantaneous rates, limits of the form

$$\lim_{h\to 0} \frac{f(x+h) - f(x)}{h}$$

occur frequently in calculus. In Exercises 57–62, evaluate this limit for the given value of x and function f.

57. $f(x) = 5x^2,\quad x = 1$

58. $f(x) = x^2,\quad x = -2$

59. $f(x) = 3x - 4,\quad x = 2$

60. $f(x) = \dfrac{5}{x},\quad x = -1$

61. $f(x) = \sqrt{x},\quad x = 7$

62. $f(x) = \sqrt{3x + 1},\quad x = 0$

Using the Sandwich Theorem

63. If $\sqrt{9 - 5x^2} \le f(x) \le \sqrt{x^2 + 9}$ for $-1 \le x \le 1$, find $\lim_{x\to 0} f(x)$.

64. If $2 - x^2 \le g(x) \le 2\cos x$ for all x, find $\lim_{x\to 0} g(x)$.

65. **a.** It can be shown that the inequalities

$$1 - \frac{x^2}{6} < \frac{x\sin x}{2 - 2\cos x} < 1$$

hold for all values of x close to zero (except for $x = 0$). What, if anything, does this tell you about

$$\lim_{x\to 0} \frac{x\sin x}{2 - 2\cos x}?$$

Give reasons for your answer.

T **b.** Graph $y = 1 - (x^2/6)$, $y = (x\sin x)/(2 - 2\cos x)$, and $y = 1$ together for $-2 \le x \le 2$. Comment on the behavior of the graphs as $x \to 0$.

66. **a.** Suppose that the inequalities

$$\frac{1}{2} - \frac{x^2}{24} < \frac{1 - \cos x}{x^2} < \frac{1}{2}$$

hold for values of x close to zero, except for $x = 0$ itself. (They do, as you will see in Section 9.9.) What, if anything, does this tell you about

$$\lim_{x \to 0} \frac{1 - \cos x}{x^2}?$$

Give reasons for your answer.

T **b.** Graph the equations $y = (1/2) - (x^2/24)$, $y = (1 - \cos x)/x^2$, and $y = 1/2$ together for $-2 \le x \le 2$. Comment on the behavior of the graphs as $x \to 0$.

Estimating Limits

T You will find a graphing calculator useful for Exercises 67–74.

67. Let $f(x) = (x^2 - 9)/(x + 3)$.

a. Make a table of the values of f at the points $x = -3.1, -3.01, -3.001$, and so on as far as your calculator can go. Then estimate $\lim_{x \to -3} f(x)$. What estimate do you arrive at if you evaluate f at $x = -2.9, -2.99, -2.999, \ldots$ instead?

b. Support your conclusions in part (a) by graphing f near $c = -3$ and using Zoom and Trace to estimate y-values on the graph as $x \to -3$.

c. Find $\lim_{x \to -3} f(x)$ algebraically, as in Example 7.

68. Let $g(x) = \dfrac{x^2 - 4}{x - 2}$.

a. Make a table of the values of g at the points $x = 2.1, 2.01, 2.001$, and so on through successive approximations of $\sqrt{4}$. Estimate $\lim_{x \to 2} g(x)$.

b. Support your conclusion in part (a) by graphing g near $c = 2$ and using Zoom and Trace to estimate y – values on the graph as x approaches 2.

c. Find $\lim_{x \to 2} g(x)$ algebraically.

69. Let $G(x) = (x + 6)/(x^2 + 4x - 12)$.

a. Make a table of the values of G at $x = -5.9, -5.99, -5.999$, and so on. Then estimate $\lim_{x \to -6} G(x)$. What estimate do you arrive at if you evaluate G at $x = -6.1, -6.01, -6.001, \ldots$ instead?

b. Support your conclusions in part (a) by graphing G and using Zoom and Trace to estimate y-values on the graph as $x \to -6$.

c. Find $\lim_{x \to -6} G(x)$ algebraically.

70. Let $h(x) = \dfrac{x^2 - 6x + 5}{x^2 - 7x + 10}$.

a. Make a table of the values of h at $x = 2.9, 2.99, 2.999$, and so on. Then estimate $\lim_{x \to 3} h(x)$. What estimate do you arrive at if you evaluate h at $x = 3.1, 3.01, 3.001, \ldots$ instead?

b. Support your conclusion in part (a) by graphing h near $c = 3$ and using Zoom and Trace to estimate y – values on the graph as x approaches 3.

c. Find $\lim_{x \to 3} h(x)$ algebraically.

71. Let $f(x) = (x^2 - 1)/(|x| - 1)$.

a. Make tables of the values of f at values of x that approach $c = -1$ from above and below. Then estimate $\lim_{x \to -1} f(x)$.

b. Support your conclusion in part (a) by graphing f near $c = -1$ and using Zoom and Trace to estimate y-values on the graph as $x \to -1$.

c. Find $\lim_{x \to -1} f(x)$ algebraically.

72. Let $F(x) = (x^2 + 3x + 2)/(2 - |x|)$.

a. Make tables of values of F at values of x that approach $c = -2$ from above and below. Then estimate $\lim_{x \to -2} F(x)$.

b. Support your conclusion in part (a) by graphing F near $c = -2$ and using Zoom and Trace to estimate y-values on the graph as $x \to -2$.

c. Find $\lim_{x \to -2} F(x)$ algebraically.

73. Let $g(\theta) = (\sin \theta)/\theta$.

a. Make a table of the values of g at values of θ that approach $\theta_0 = 0$ from above and below. Then estimate $\lim_{\theta \to 0} g(\theta)$.

b. Support your conclusion in part (a) by graphing g near $\theta_0 = 0$.

74. Let $G(t) = (1 - \cos t)/t^2$.

a. Make tables of values of G at values of t that approach $t_0 = 0$ from above and below. Then estimate $\lim_{t \to 0} G(t)$.

b. Support your conclusion in part (a) by graphing G near $t_0 = 0$.

Theory and Examples

75. If $x^4 \le f(x) \le x^2$ for x in $[-1, 1]$ and $x^2 \le f(x) \le x^4$ for $x < -1$ and $x > 1$, at what points c do you automatically know $\lim_{x \to c} f(x)$? What can you say about the value of the limit at these points?

76. Suppose that $g(x) \le f(x) \le h(x)$ for all $x \ne 5$ and suppose that

$$\lim_{x \to 5} g(x) = \lim_{x \to 5} h(x) = 10$$

Can we conclude anything about the values of f, g and h at $x = 5$? Could $f(5) = 0$? Could $\lim_{x \to 5} f(x) = 0$? Give reasons for your answers.

77. If $\lim_{x \to 10} \dfrac{f(x) - 10}{x - 5} = 12$, find $\lim_{x \to 10} f(x)$.

78. If $\lim_{x \to -2} \dfrac{f(x)}{x^2} = 1$, find

a. $\lim_{x \to -2} f(x)$, **b.** $\lim_{x \to -2} \dfrac{f(x)}{x}$.

79. **a.** If $\lim_{x \to 2} \dfrac{f(x) - 5}{x - 2} = 3$, find $\lim_{x \to 2} f(x)$.

b. If $\lim_{x \to 2} \dfrac{f(x) - 5}{x - 2} = 4$, find $\lim_{x \to 2} f(x)$.

80. If $\lim_{x \to 1} \dfrac{f(x)}{x^2} = 5$, find

a. $\lim_{x \to 1} f(x)$ **b.** $\lim_{x \to 1} \dfrac{f(x)}{x}$

T **81. a.** Graph $g(x) = x\sin(1/x)$ to estimate $\lim_{x\to 0} g(x)$, zooming in on the origin as necessary.

b. Confirm your estimate in part (a) with a proof.

T **82. a.** Graph $h(x) = x^2\cos(1/x^3)$ to estimate $\lim_{x\to 0} h(x)$, zooming in on the origin as necessary.

b. Confirm your estimate in part (a) with a proof.

COMPUTER EXPLORATIONS

Graphical Estimates of Limits

In Exercises 83–88, use a CAS to perform the following steps:

a. Plot the function near the point c being approached.

b. From your plot, guess the value of the limit.

83. $\lim_{x\to 2} \dfrac{x^4 - 16}{x - 2}$

84. $\lim_{x\to -1} \dfrac{x^3 - x^2 - 5x - 3}{(x+1)^2}$

85. $\lim_{x\to 0} \dfrac{\sqrt[3]{1+x} - 1}{x}$

86. $\lim_{x\to 3} \dfrac{x^2 - 9}{\sqrt{x^2+7} - 4}$

87. $\lim_{x\to 0} \dfrac{1 - \cos x}{x \sin x}$

88. $\lim_{x\to 0} \dfrac{2x^2}{3 - 3\cos x}$

2.3 The Precise Definition of a Limit

We now turn our attention to the precise definition of a limit. The early history of calculus saw controversy about the validity of the basic concepts underlying the theory. Apparent contradictions were argued over by both mathematicians and philosophers. These controversies were resolved by the precise definition, which allows us to replace vague phrases like "gets arbitrarily close to" in the informal definition with specific conditions that can be applied to any particular example. With a rigorous definition, we can avoid misunderstandings, prove the limit properties given in the preceding section, and establish many important limits.

To show that the limit of $f(x)$ as $x \to c$ equals the number L, we need to show that the gap between $f(x)$ and L can be made "as small as we choose" if x is kept "close enough" to c. Let us see what this requires if we specify the size of the gap between $f(x)$ and L.

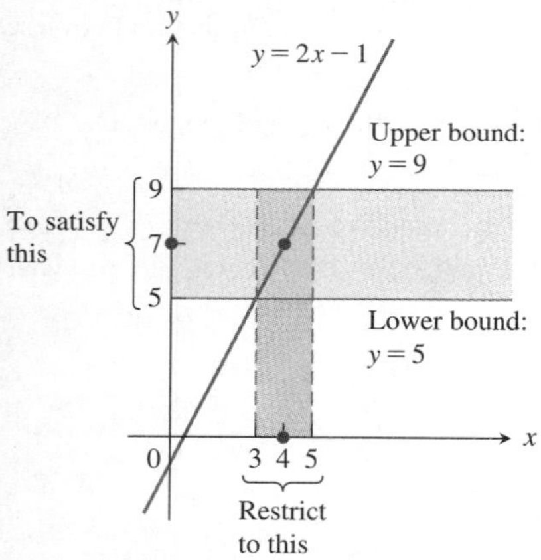

FIGURE 2.16 Keeping x within 1 unit of $x = 4$ will keep y within 2 units of $y = 7$ (Example 1).

EXAMPLE 1 Consider the function $y = 2x - 1$ near $x = 4$. Intuitively it seems clear that y is close to 7 when x is close to 4, so $\lim_{x\to 4}(2x - 1) = 7$. However, how close to $x = 4$ does x have to be so that $y = 2x - 1$ differs from 7 by, say, less than 2 units?

Solution We are asked: For what values of x is $|y - 7| < 2$? To find the answer we first express $|y - 7|$ in terms of x:

$$|y - 7| = |(2x - 1) - 7| = |2x - 8|.$$

The question then becomes: What values of x satisfy the inequality $|2x - 8| < 2$? To find out, we solve the inequality:

$$|2x - 8| < 2$$
$$-2 < 2x - 8 < 2 \quad \text{Removing absolute value gives two inequalities.}$$
$$6 < 2x < 10 \quad \text{Add 8 to each term.}$$
$$3 < x < 5 \quad \text{Solve for } x.$$
$$-1 < x - 4 < 1. \quad \text{Solve for } x - 4.$$

Keeping x within 1 unit of $x = 4$ will keep y within 2 units of $y = 7$ (Figure 2.16). ■

In the previous example we determined how close x must be to a particular value c to ensure that the outputs $f(x)$ of some function lie within a prescribed interval about a limit value L. To show that the limit of $f(x)$ as $x \to c$ actually equals L, we must be able to show that the gap between $f(x)$ and L can be made less than *any prescribed error*, no matter how small, by holding x close enough to c. To describe arbitrary prescribed errors, we introduce two constants, δ (delta) and ε (epsilon). These Greek letters are traditionally used to represent small changes in a variable or a function.

δ is the Greek letter delta.
ε is the Greek letter epsilon.

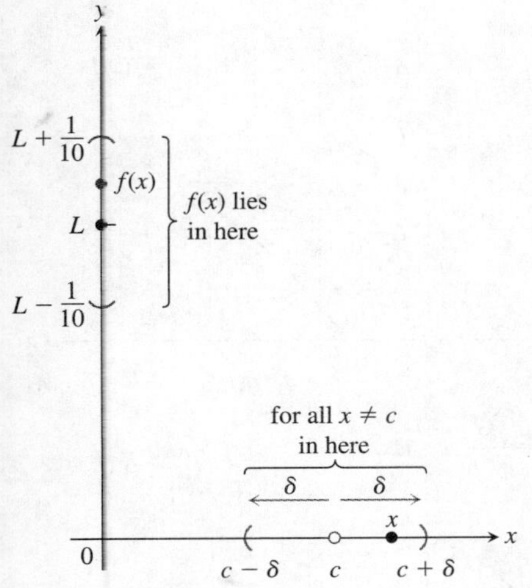

FIGURE 2.17 How should we define $\delta > 0$ so that keeping x within the interval $(c - \delta, c + \delta)$ will keep $f(x)$ within the interval $\left(L - \frac{1}{10}, L + \frac{1}{10}\right)$?

Definition of Limit

Suppose we are watching the values of a function $f(x)$ as x approaches c (without taking on the value c itself). Certainly we want to be able to say that $f(x)$ stays within one-tenth of a unit from L as soon as x stays within some distance δ of c (Figure 2.17). But that in itself is not enough, because as x continues on its course toward c, what is to prevent $f(x)$ from jumping around within the interval from $L - (1/10)$ to $L + (1/10)$ without tending toward L? We can be told that the error can be no more than 1/100 or 1/1000 or 1/100,000. Each time, we find a new δ-interval about c so that keeping x within that interval satisfies the new error tolerance. And each time the possibility exists that $f(x)$ might jump away from L at some later stage.

The figures on the next page illustrate the problem. You can think of this as a quarrel between a skeptic and a scholar. The skeptic presents ε-challenges to show there is room for doubt that the limit exists. The scholar counters every challenge with a δ-interval around c which ensures that the function takes values within ε of L.

How do we stop this seemingly endless series of challenges and responses? We can do so by proving that for *every* error tolerance ε that the challenger can produce, we can present a matching distance δ that keeps x "close enough" to c to keep $f(x)$ within that ε-tolerance of L (Figure 2.18). This leads us to the precise definition of a limit.

DEFINITION Let $f(x)$ be defined on an open interval about c, except possibly at c itself. We say that the **limit of $f(x)$ as x approaches c is the number L**, and write

$$\lim_{x \to c} f(x) = L,$$

if, for every number $\varepsilon > 0$, there exists a corresponding number $\delta > 0$ such that

$$|f(x) - L| < \varepsilon \quad \text{whenever} \quad 0 < |x - c| < \delta.$$

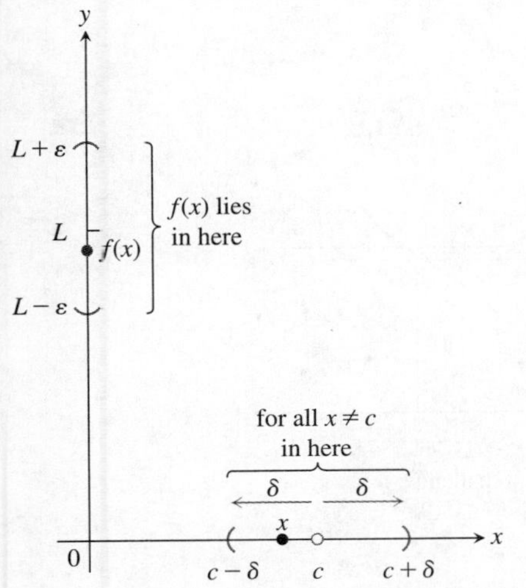

FIGURE 2.18 The relation of δ and ε in the definition of limit.

To visualize the definition, imagine machining a cylindrical shaft to a close tolerance. The diameter of the shaft is determined by turning a dial to a setting measured by a variable x. We try for diameter L, but since nothing is perfect we must be satisfied with a diameter $f(x)$ somewhere between $L - \varepsilon$ and $L + \varepsilon$. The number δ is our control tolerance for the dial; it tells us how close our dial setting must be to the setting $x = c$ in order to guarantee that the diameter $f(x)$ of the shaft will be accurate to within ε of L. As the tolerance for error becomes stricter, we may have to adjust δ. The value of δ, how tight our control setting must be, depends on the value of ε, the error tolerance.

The definition of limit extends to functions on more general domains. It is only required that each open interval around c contain points in the domain of the function other than c. See Additional and Advanced Exercises 47–50 for examples of limits for functions with complicated domains. In the next section we will see how the definition of limit applies at points lying on the boundary of an interval.

Examples: Testing the Definition

The formal definition of limit does not tell how to find the limit of a function, but it does enable us to verify that a conjectured limit value is correct. The following examples show how the definition can be used to verify limit statements for specific functions. However, the real purpose of the definition is not to do calculations like this, but rather to prove general theorems so that the calculation of specific limits can be simplified, such as the theorems stated in the previous section.

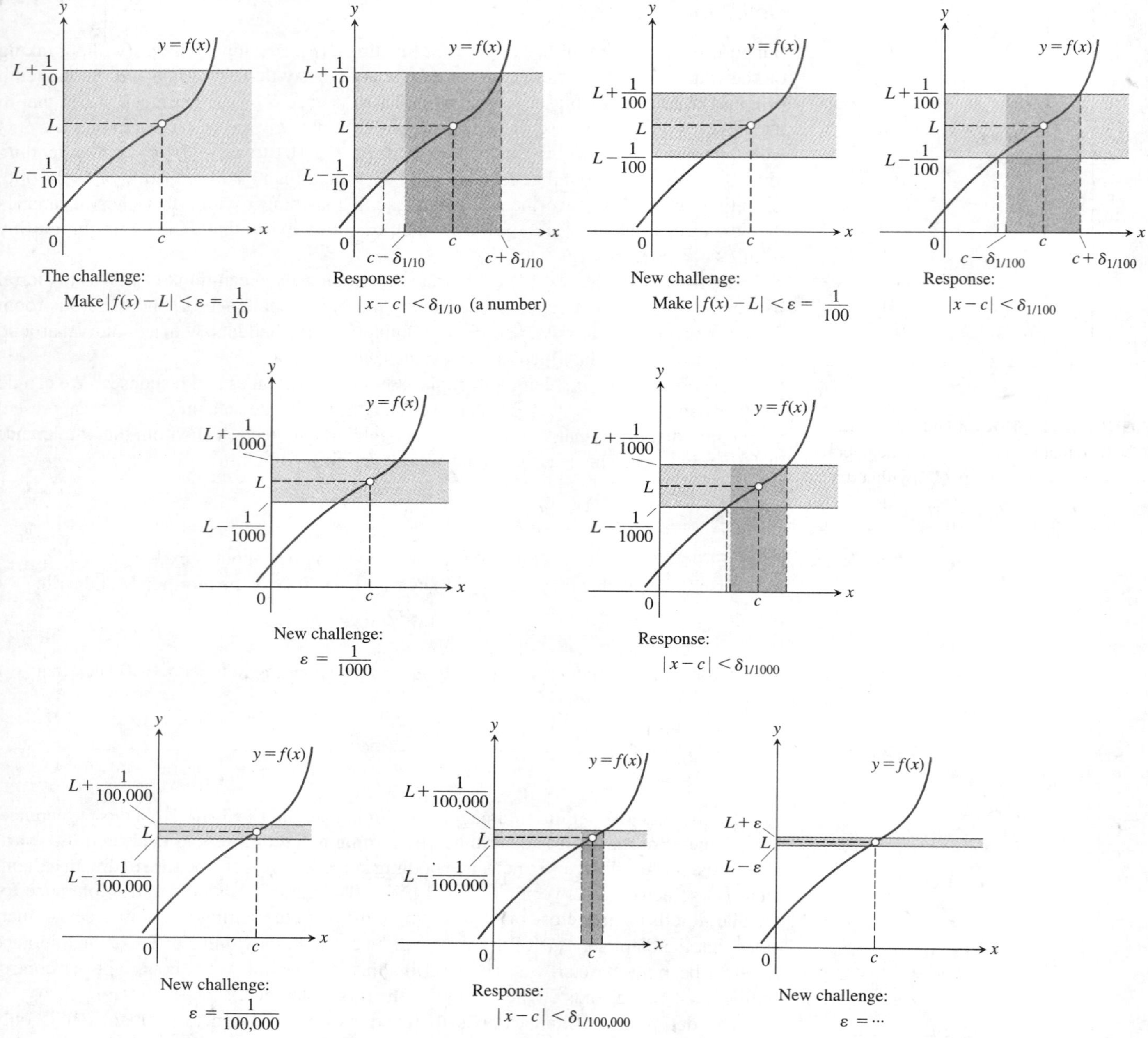

EXAMPLE 2 Show that

$$\lim_{x \to 1} (5x - 3) = 2.$$

Solution Set $c = 1$, $f(x) = 5x - 3$, and $L = 2$ in the definition of limit. For any given $\varepsilon > 0$, we have to find a suitable $\delta > 0$ so that if $x \neq 1$ and x is within distance δ of $c = 1$, then it is true that $f(x)$ is within distance ε of $L = 2$. That is, we must show that if x satisfies

$$0 < |x - 1| < \delta,$$

then $f(x)$ will satisfy

$$|f(x) - 2| < \varepsilon.$$

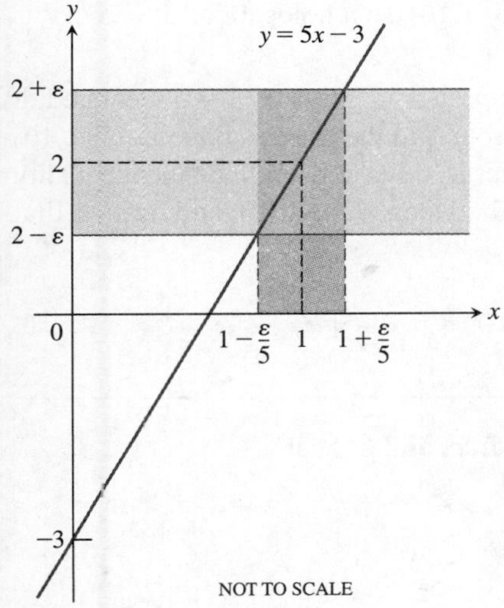

FIGURE 2.19 If $f(x) = 5x - 3$, then $0 < |x - 1| < \varepsilon/5$ guarantees that $|f(x) - 2| < \varepsilon$ (Example 2).

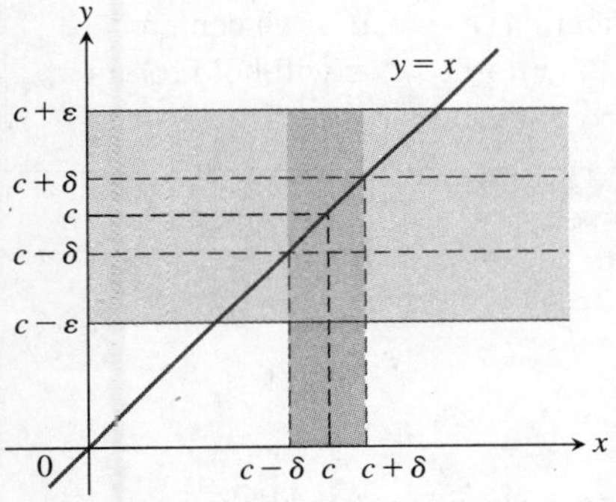

FIGURE 2.20 For the function $f(x) = x$, we find that $0 < |x - c| < \delta$ will guarantee $|f(x) - c| < \varepsilon$ whenever $\delta \leq \varepsilon$ (Example 3a).

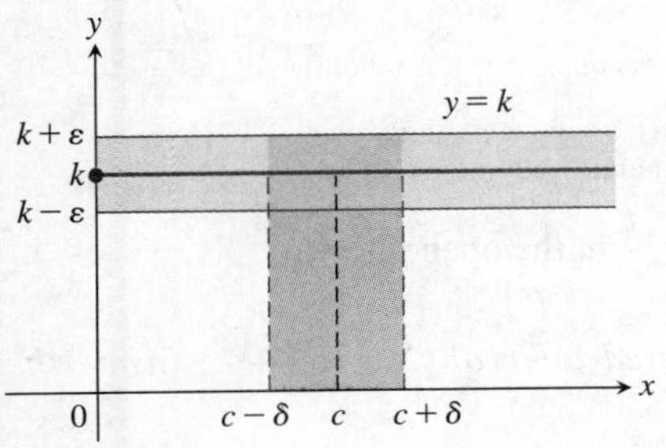

FIGURE 2.21 For the function $f(x) = k$, we find that $|f(x) - k| < \varepsilon$ for any positive δ (Example 3b).

We find δ by working backward from the ε-inequality, $|f(x) - 2| < \varepsilon$:

$$|(5x - 3) - 2| = |5x - 5| < \varepsilon \qquad \text{Substitute.}$$
$$5|x - 1| < \varepsilon \qquad \text{Factor.}$$
$$|x - 1| < \varepsilon/5. \qquad \text{Simplify.}$$

Thus, we can take $\delta = \varepsilon/5$ (Figure 2.19). If $0 < |x - 1| < \delta = \varepsilon/5$, then

$$|(5x - 3) - 2| = |5x - 5| = 5|x - 1| < 5(\varepsilon/5) = \varepsilon,$$

which proves that $\lim_{x \to 1} (5x - 3) = 2$.

The value of $\delta = \varepsilon/5$ is not the only value that will make $0 < |x - 1| < \delta$ imply $|5x - 5| < \varepsilon$. Any smaller positive δ will do as well. The definition does not ask for the "best" positive δ, just one that will work. ■

EXAMPLE 3 Prove the following results, which were presented graphically in Section 2.2.

(a) $\lim_{x \to c} x = c$

(b) $\lim_{x \to c} k = k$ $\quad k$ constant

Solution

(a) Let $\varepsilon > 0$ be given. We must find $\delta > 0$ such that

$$|x - c| < \varepsilon \qquad \text{whenever} \qquad 0 < |x - c| < \delta.$$

The implication will hold if δ equals ε or any smaller positive number (Figure 2.20). This proves that $\lim_{x \to c} x = c$.

(b) Let $\varepsilon > 0$ be given. We must find $\delta > 0$ such that

$$|k - k| < \varepsilon \qquad \text{whenever} \qquad 0 < |x - c| < \delta.$$

Since $k - k = 0$, we will always have $|k - k| < \varepsilon$. Therefore we can use any positive number for δ, and the implication will hold (Figure 2.21). This proves that $\lim_{x \to c} k = k$. ■

Finding Deltas Algebraically for Given Epsilons

In Examples 2 and 3, the interval of values about c for which $|f(x) - L|$ was less than ε was symmetric about c and we could take δ to be half the length of that interval. When the interval around c on which we have $|f(x) - L| < \varepsilon$ is not symmetric about c, we can take δ to be the distance from c to the interval's *nearer* endpoint.

EXAMPLE 4 For the limit $\lim_{x \to 5} \sqrt{x - 1} = 2$, find a $\delta > 0$ that works for $\varepsilon = 1$. That is, find a $\delta > 0$ such that

$$\left|\sqrt{x - 1} - 2\right| < 1 \quad \text{whenever} \quad 0 < |x - 5| < \delta$$

Solution We organize the search into two steps.

1. *Solve the inequality $\left|\sqrt{x - 1} - 2\right| < 1$ to find an interval containing $x = 5$ on which the inequality holds for all $x \neq 5$.*

$$\left|\sqrt{x - 1} - 2\right| < 1$$
$$-1 < \sqrt{x - 1} - 2 < 1$$
$$1 < \sqrt{x - 1} < 3$$
$$1 < x - 1 < 9$$
$$2 < x < 10$$

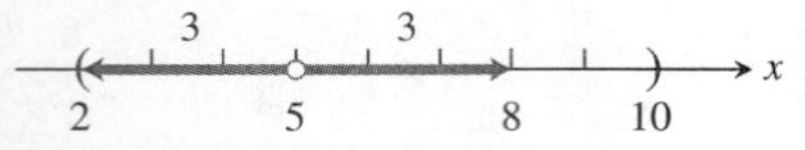

FIGURE 2.22 An open interval of radius 3 about $x = 5$ will lie inside the open interval $(2, 10)$.

The inequality holds for all x in the open interval $(2, 10)$, so it holds for all $x \neq 5$ in this interval as well.

2. *Find a value of $\delta > 0$ to place the centered interval $5 - \delta < x < 5 + \delta$* (centered at $x = 5$) *inside the interval* $(2, 10)$. The distance from 5 to the nearer endpoint of $(2, 10)$ is 3 (Figure 2.22). If we take $\delta = 3$ or any smaller positive number, then the inequality $0 < |x - 5| < \delta$ will automatically place x between 2 and 10 and imply that $|\sqrt{x-1} - 2| < 1$ (Figure 2.23):

$$\left|\sqrt{x-1} - 2\right| < 1 \quad \text{whenever} \quad 0 < |x - 5| < 3. \qquad \blacksquare$$

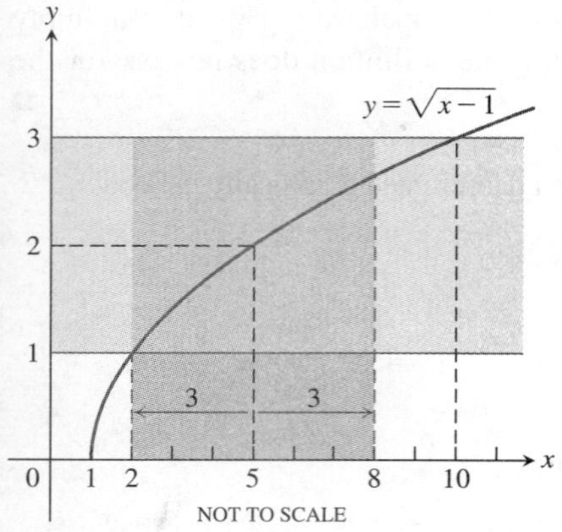

FIGURE 2.23 The function and intervals in Example 4.

How to Find Algebraically a δ for a Given f, L, c, and $\varepsilon > 0$

The process of finding a $\delta > 0$ such that

$$|f(x) - L| < \varepsilon \quad \text{whenever} \quad 0 < |x - c| < \delta$$

can be accomplished in two steps.

1. *Solve the inequality* $|f(x) - L| < \varepsilon$ to find an open interval (a, b), containing c on which the inequality holds for all $x \neq c$. Note that we do not require the inequality to hold at $x = c$. It may hold there or it may not, but the value of f at $x = c$ does not influence the existence of a limit.
2. *Find a value of $\delta > 0$* that places the open interval $(c - \delta, c + \delta)$ centered at c inside the interval (a, b). The inequality $|f(x) - L| < \varepsilon$ will hold for all $x \neq c$ in this δ-interval.

EXAMPLE 5 Prove that $\lim_{x \to 2} f(x) = 4$ if

$$f(x) = \begin{cases} x^2, & x \neq 2 \\ 1, & x = 2. \end{cases}$$

Solution Our task is to show that given $\varepsilon > 0$, there exists a $\delta > 0$ such that

$$|f(x) - 4| < \varepsilon \quad \text{whenever} \quad 0 < |x - 2| < \delta.$$

1. *Solve the inequality $|f(x) - 4| < \varepsilon$ to find an open interval containing $x = 2$ on which the inequality holds for all $x \neq 2$.*

   For $x \neq c = 2$, we have $f(x) = x^2$, and the inequality to solve is $|x^2 - 4| < \varepsilon$:

$$|x^2 - 4| < \varepsilon$$
$$-\varepsilon < x^2 - 4 < \varepsilon$$
$$4 - \varepsilon < x^2 < 4 + \varepsilon$$
$$\sqrt{4 - \varepsilon} < |x| < \sqrt{4 + \varepsilon} \qquad \text{Assumes } \varepsilon < 4\text{; see below.}$$
$$\sqrt{4 - \varepsilon} < x < \sqrt{4 + \varepsilon}. \qquad \text{An open interval about } x = 2 \text{ that solves the inequality}$$

   The inequality $|f(x) - 4| < \varepsilon$ holds for all $x \neq 2$ in the open interval $(\sqrt{4 - \varepsilon}, \sqrt{4 + \varepsilon})$ (Figure 2.24).

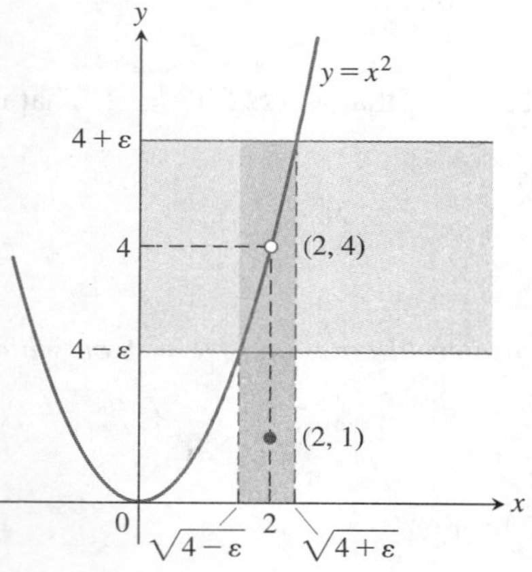

FIGURE 2.24 An interval containing $x = 2$ so that the function in Example 5 satisfies $|f(x) - 4| < \varepsilon$.

2. *Find a value of $\delta > 0$ that places the centered interval $(2 - \delta, 2 + \delta)$ inside the interval $(\sqrt{4 - \varepsilon}, \sqrt{4 + \varepsilon})$.*

   Take δ to be the distance from $x = 2$ to the nearer endpoint of $(\sqrt{4 - \varepsilon}, \sqrt{4 + \varepsilon})$. In other words, take $\delta = \min\{2 - \sqrt{4 - \varepsilon}, \sqrt{4 + \varepsilon} - 2\}$, the *minimum* (the smaller) of the two numbers $2 - \sqrt{4 - \varepsilon}$ and $\sqrt{4 + \varepsilon} - 2$. If δ has this or any smaller positive value, the inequality $0 < |x - 2| < \delta$ will automatically place x between $\sqrt{4 - \varepsilon}$ and $\sqrt{4 + \varepsilon}$ to make $|f(x) - 4| < \varepsilon$. For all x,

$$|f(x) - 4| < \varepsilon \quad \text{whenever} \quad 0 < |x - 2| < \delta.$$

This completes the proof for $\varepsilon < 4$.

If $\varepsilon \geq 4$, then we take δ to be the distance from $x = 2$ to the nearer endpoint of the interval $(0, \sqrt{4 + \varepsilon})$. In other words, take $\delta = \min\{2, \sqrt{4 + \varepsilon} - 2\}$. (See Figure 2.24.) ■

Using the Definition to Prove Theorems

We do not usually rely on the formal definition of limit to verify specific limits such as those in the preceding examples. Rather, we appeal to general theorems about limits, in particular the theorems of Section 2.2. The definition is used to prove these theorems (Appendix A.5). As an example, we prove part 1 of Theorem 1, the Sum Rule.

EXAMPLE 6 Given that $\lim_{x \to c} f(x) = L$ and $\lim_{x \to c} g(x) = M$, prove that

$$\lim_{x \to c} (f(x) + g(x)) = L + M.$$

Solution Let $\varepsilon > 0$ be given. We want to find a positive number δ such that

$$|f(x) + g(x) - (L + M)| < \varepsilon \quad \text{whenever} \quad 0 < |x - c| < \delta.$$

Regrouping terms, we get

$$\begin{aligned} |f(x) + g(x) - (L + M)| &= |(f(x) - L) + (g(x) - M)| \\ &\leq |f(x) - L| + |g(x) - M|. \end{aligned}$$

Triangle Inequality: $|a+b| \leq |a| + |b|$

Since $\lim_{x \to c} f(x) = L$, there exists a number $\delta_1 > 0$ such that

$$|f(x) - L| < \varepsilon/2 \quad \text{whenever} \quad 0 < |x - c| < \delta_1.$$

Can find δ_1 since $\lim_{x \to c} f(x) = L$

Similarly, since $\lim_{x \to c} g(x) = M$, there exists a number $\delta_2 > 0$ such that

$$|g(x) - M| < \varepsilon/2 \quad \text{whenever} \quad 0 < |x - c| < \delta_2.$$

Can find δ_2 since $\lim_{x \to c} g(x) = M$

Let $\delta = \min\{\delta_1, \delta_2\}$, the smaller of δ_1 and δ_2. If $0 < |x - c| < \delta$ then $|x - c| < \delta_1$, so $|f(x) - L| < \varepsilon/2$, and $|x - c| < \delta_2$, so $|g(x) - M| < \varepsilon/2$. Therefore,

$$|f(x) + g(x) - (L + M)| < \frac{\varepsilon}{2} + \frac{\varepsilon}{2} = \varepsilon.$$

This shows that $\lim_{x \to c} (f(x) + g(x)) = L + M$. ■

EXERCISES 2.3

Centering Intervals About a Point

In Exercises 1–6, sketch the interval (a, b), on the x-axis with the point c inside. Then find a value of $\delta > 0$ such that $a < x < b$ whenever $0 < |x - c| < \delta$.

1. $a = 2, \quad b = 8, \quad c = 4$

2. $a = 1, \quad b = 7, \quad c = 2$

3. $a = -5/2, \quad b = -1/2, \quad c = -2$

4. $a = -7/2, \quad b = -1/2, \quad c = -3/2$

5. $a = 2/7, \quad b = 5/7, \quad c = 1/2$

6. $a = 2.7591, \quad b = 3.2391, \quad c = 3$

Finding Deltas Graphically

In Exercises 7–14, use the graphs to find a $\delta > 0$ such that

$$|f(x) - L| < \varepsilon \quad \text{whenever} \quad 0 < |x - c| < \delta.$$

7.

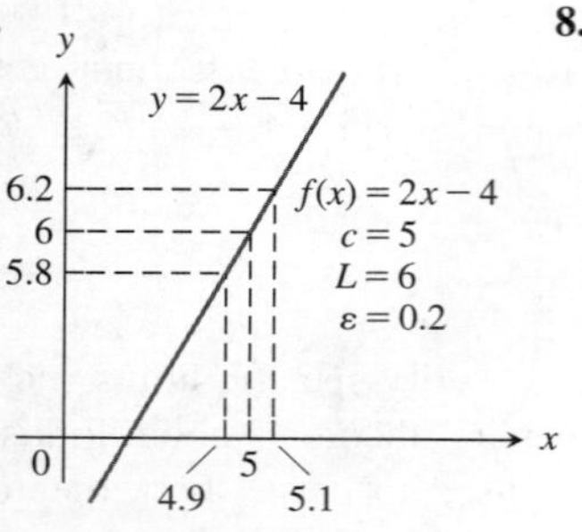

NOT TO SCALE

8.

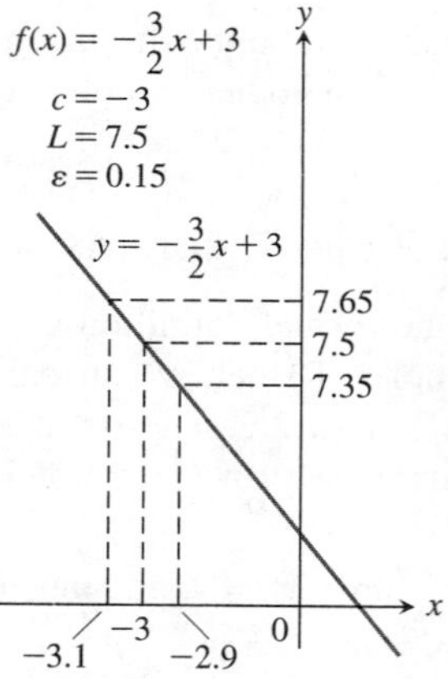

NOT TO SCALE

9.

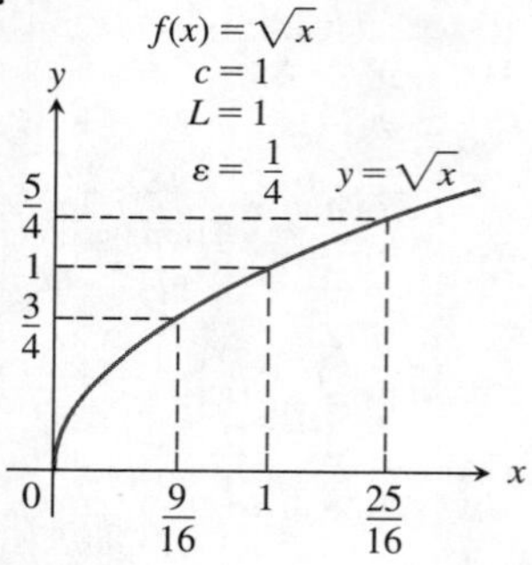

10.

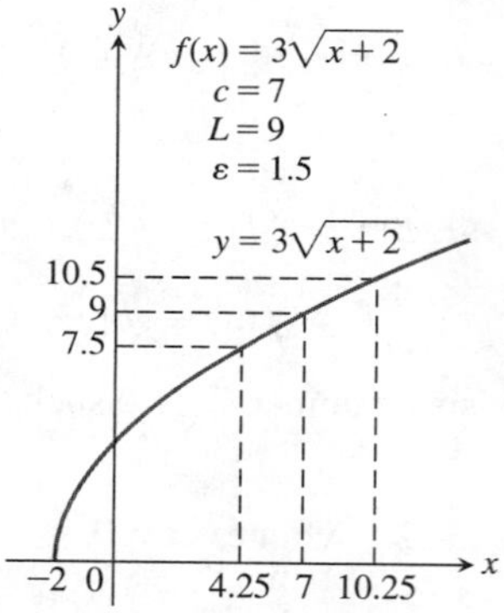

11.

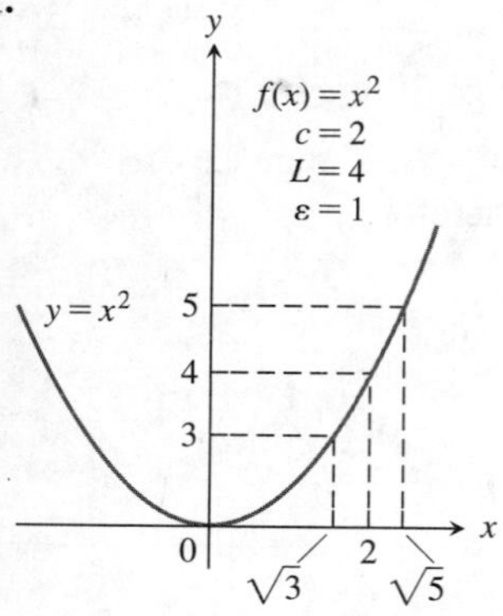

NOT TO SCALE

12.

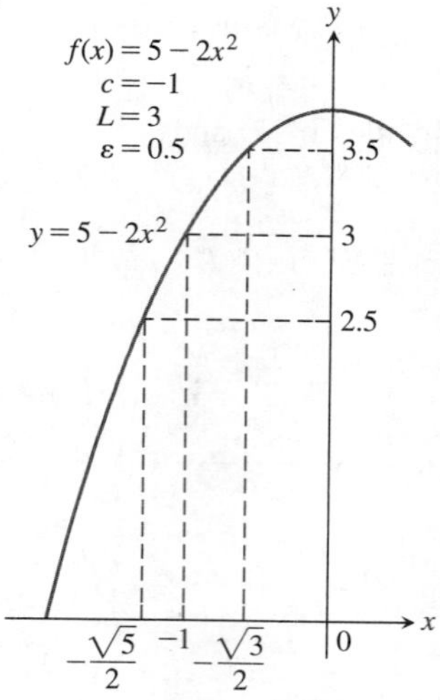

13.

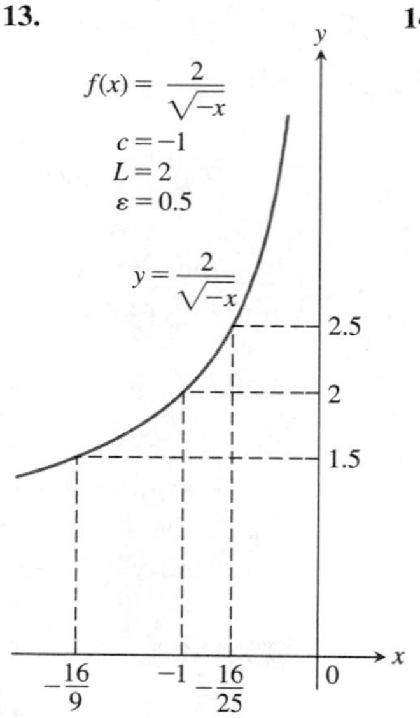

14.

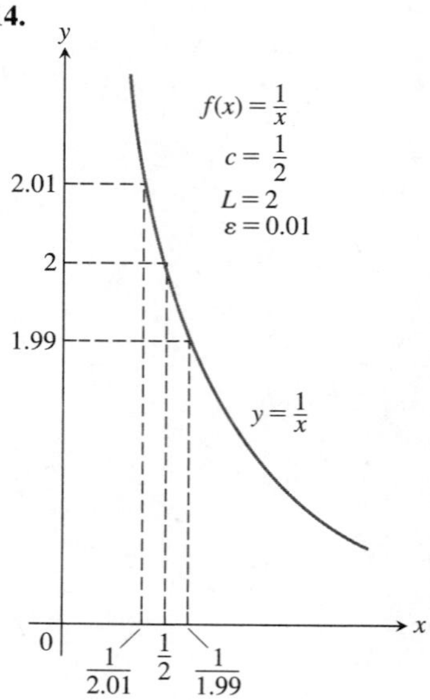

NOT TO SCALE

Finding Deltas Algebraically

Each of Exercises 15–30 gives a function $f(x)$ and numbers L, c, and $\varepsilon > 0$. In each case, find the largest open interval about c on which the inequality $|f(x) - L| < \varepsilon$ holds. Then give a value for $\delta > 0$ such that for all x satisfying $0 < |x - c| < \delta$, the inequality $|f(x) - L| < \varepsilon$ holds.

15. $f(x) = x + 5, \quad L = 8, \quad c = 3, \quad \varepsilon = 0.01$

16. $f(x) = 2x - 2, \quad L = -6, \quad c = -2, \quad \varepsilon = 0.02$

17. $f(x) = \sqrt{x + 1}, \quad L = 1, \quad c = 0, \quad \varepsilon = 0.1$

18. $\sqrt{2x} \quad L = 2, \quad c = 2, \quad \varepsilon = 0.1$

19. $f(x) = \sqrt{19 - x}, \quad L = 3, \quad c = 10, \quad \varepsilon = 1$

20. $\sqrt{x - 1} \quad L = 2, \quad c = 5, \quad \varepsilon = 1$

21. $f(x) = 1/x, \quad L = 1/4, \quad c = 4, \quad \varepsilon = 0.05$

22. $f(x) = 4 + x - 3x^2 \quad L = 2, \quad c = 1, \quad \varepsilon = 0.01$

23. $f(x) = x^2, \quad L = 4, \quad c = -2, \quad \varepsilon = 0.5$

24. $\dfrac{1}{x} \quad L = 1, \quad c = 1, \quad \varepsilon = 0.07$

25. $f(x) = x^2 + 4, \quad L = 29, \quad c = 5, \quad \varepsilon = 0.01$

26. $f(x) = 120/x, \quad L = 5, \quad c = 24, \quad \varepsilon = 1$

27. $f(x) = mx, \quad m > 0, \quad L = 2m, \quad c = 2, \quad \varepsilon = 0.03$

28. $f(x) = mx, \quad m > 0, \quad L = 3m, \quad c = 3, \quad \varepsilon = c > 0$

29. $f(x) = mx + b, \quad m > 0, \quad L = (m/2) + b,$
$c = 1/2, \quad \varepsilon = c > 0$

30. $f(x) = mx + b, \quad m > 0, \quad L = m + b, \quad c = 1, \quad \varepsilon = 0.05$

Using the Formal Definition

Each of Exercises 31–36 gives a function $f(x)$, a point c, and a positive number ε. Find $L = \lim_{x \to c} f(x)$. Then find a number $\delta > 0$ such that

$$|f(x) - L| < \varepsilon \quad \text{whenever} \quad 0 < |x - c| < \delta.$$

31. $f(x) = 5 - 3x \quad c = 4, \quad \varepsilon = 0.001$

32. $f(x) = -3x - 2, \quad c = -1, \quad \varepsilon = 0.03$

33. $f(x) = x^2 - 1 \quad c = 1, \quad \varepsilon = 0.005$

34. $f(x) = \dfrac{x^2 + 6x + 5}{x + 5}, \quad c = -5, \quad \varepsilon = 0.05$

35. $\sqrt{x+1} \quad c = 24, \quad \varepsilon = 0.1$

36. $f(x) = 4/x, \quad c = 2, \quad \varepsilon = 0.4$

Prove the limit statements in Exercises 37–50.

37. $\lim_{x \to 4} (9 - x) = 5$

38. $\lim_{x \to 2} 7x + 4 = 18$

39. $\lim_{x \to 9} \sqrt{x - 5} = 2$

40. $\lim_{x \to 2} x^2 + 3 = 7$

41. $\lim_{x \to 1} f(x) = 1 \quad \text{if} \quad f(x) = \begin{cases} x^2, & x \neq 1 \\ 2, & x = 1 \end{cases}$

42. $\lim_{x \to -2} f(x) = 4 \quad \text{if} \quad f(x) = \begin{cases} x^2, & x \neq -2 \\ 1, & x = -2 \end{cases}$

43. $\lim_{x \to 1} \dfrac{1}{x} = 1$

44. $\lim_{x \to \sqrt{4}} \dfrac{1}{x^2} = \dfrac{1}{4}$

45. $\lim_{x \to -3} \dfrac{x^2 - 9}{x + 3} = -6$

46. $\lim_{x \to 1} \dfrac{x^2 - 1}{x - 1} = 2$

47. $\lim_{x \to 1} f(x) = 2 \quad \text{if} \quad f(x) = \begin{cases} 4 - 2x, & x < 1 \\ 6x - 4, & x \geq 1 \end{cases}$

48. $\lim_{x \to 0} f(x) = 0 \quad \text{if} \quad f(x) = \begin{cases} 2x, & x < 0 \\ x/2, & x \geq 0 \end{cases}$

49. $\lim_{x \to \frac{\pi}{4}} \tan x = 1$

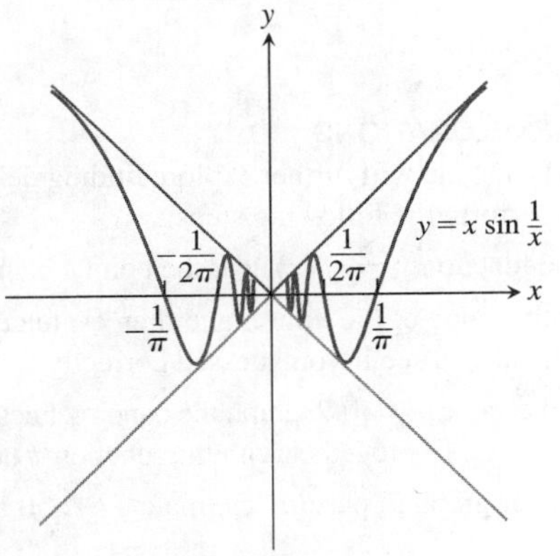

50. $\lim_{x \to 0} x^2 \sin \dfrac{1}{x} = 0$

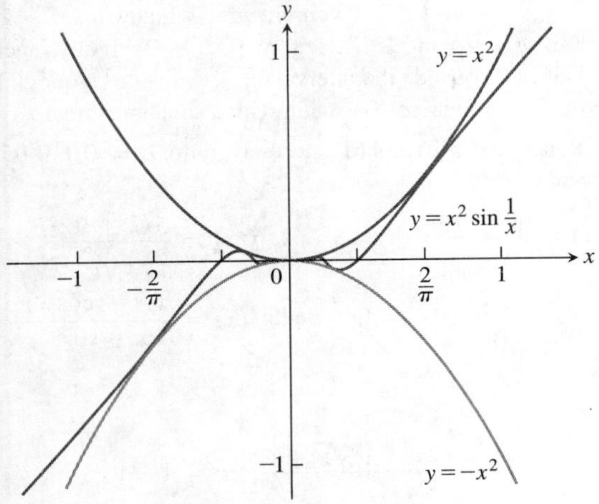

Theory and Examples

51. Define what it means to say that $\lim_{x \to 0} g(x) = k$.

52. Prove that $\lim_{x \to c} f(x) = L$ if and only if $\lim_{h \to 0} f(h + c) = L$.

53. A wrong statement about limits Show by example that the following statement is wrong.

> The number L is the limit of $f(x)$ as x approaches c if $f(x)$ gets closer to L as x approaches c.

Explain why the function in your example does not have the given value of L as a limit as $x \to c$.

54. Another wrong statement about limits Show by example that the following statement is wrong.

> The number L is the limit of $f(x)$ as x approaches c if, given any $\varepsilon > 0$, there exists a value of x for which $|f(x) - L| < \varepsilon$.

Explain why the function in your example does not have the given value of L as a limit as $x \to c$.

T **55. Grinding engine cylinders** Before contracting to grind engine cylinders to a cross-sectional area of $9\,\text{in}^2$, you need to know how much deviation from the ideal cylinder diameter of $c = 3.385\,\text{in}$. you can allow and still have the area come within $0.01\,\text{in}^2$ of the required $9\,\text{in}^2$. To find out, you let $A = \pi(x/2)^2$ and look for the largest interval in which you must hold x to make $|A - 9| \leq 0.01$. What interval do you find?

56. Manufacturing electrical resistors Ohm's law for electrical circuits like the one shown in the accompanying figure states that $V = RI$. In this equation, V is a constant voltage, I is the current in amperes, and R is the resistance in ohms. Your firm has been asked to supply the resistors for a circuit in which V will be 120 volts and I is to be 5 ± 0.1 amp. In what interval does R have to lie for I to be within 0.1 amp of the value $I_0 = 5$?

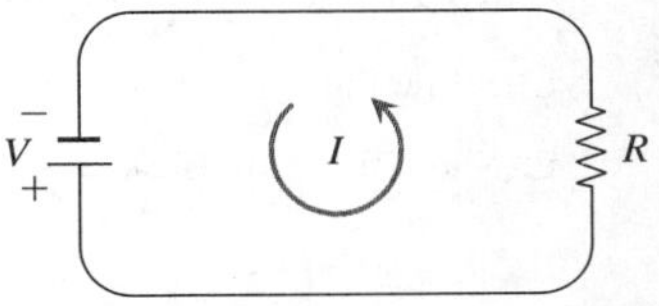

When Is a Number L Not the Limit of $f(x)$ As $x \to c$?

In Exercises 57–60 we will consider what it means to *not* have a limit.

Showing L is not a limit We can prove that $\lim_{x \to c} f(x) \neq L$ by providing an $\varepsilon > 0$ such that no possible $\delta > 0$ satisfies the condition

$$|f(x) - L| < \varepsilon \quad \text{whenever} \quad 0 < |x - c| < \delta.$$

We accomplish this for our candidate ε by showing that for each $\delta > 0$ there exists a value of x such that

$$0 < |x - c| < \delta \quad \text{and} \quad |f(x) - L| \geq \varepsilon.$$

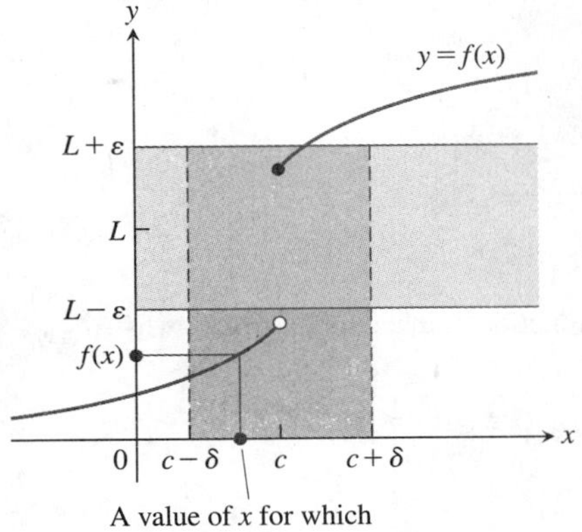

A value of x for which
$0 < |x - c| < \delta$ and $|f(x) - L| \geq \varepsilon$

57. Let $f(x) = \begin{cases} x, & x < 1 \\ x + 1, & x > 1. \end{cases}$

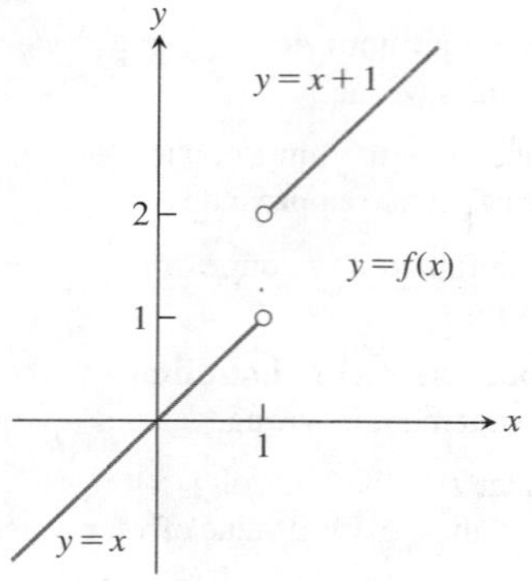

a. Let $\varepsilon = 1/2$. Show that no possible $\delta > 0$ satisfies the following condition:

$$|f(x) - 2| < 1/2 \quad \text{whenever} \quad 0 < |x - 1| < \delta.$$

That is, show that for each $\delta > 0$, there is a value of x such that

$$0 < |x - 1| < \delta \quad \text{and} \quad |f(x) - 2| \geq 1/2.$$

This will show that $\lim_{x \to 1} f(x) \neq 2$.

b. Show that $\lim_{x \to 1} f(x) \neq 1$.

c. Show that $\lim_{x \to 1} f(x) \neq 1.5$.

58. Let $h(x) = \begin{cases} x^2, & x < 2 \\ 3, & x = 3 \\ 2, & x > 2. \end{cases}$

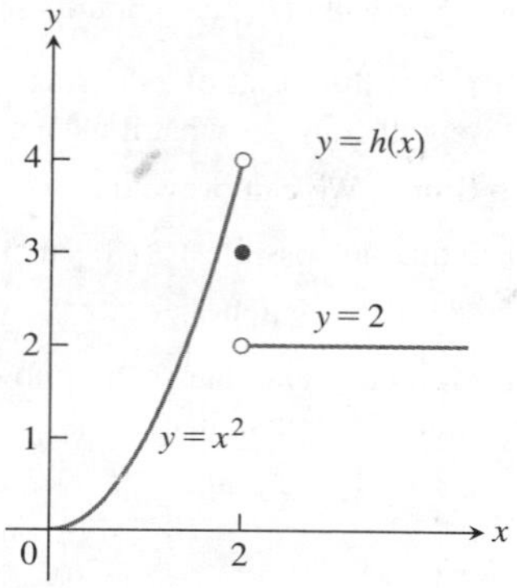

Show that

a. $\lim_{x \to 2} h(x) \neq 4$

b. $\lim_{x \to 2} h(x) \neq 3$

c. $\lim_{x \to 2} h(x) \neq 2$

59. For the function graphed here, explain why

a. $\lim_{x \to 4} f(x) \neq 5$

b. $\lim_{x \to 4} f(x) \neq 5.8$

c. $\lim_{x \to 4} f(x) \neq 4$

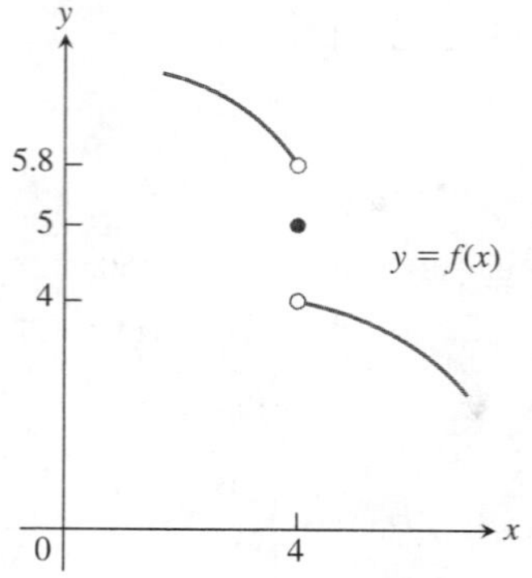

60. a. For the function graphed here, show that $\lim_{x \to -2} g(x) \neq 4$.

b. Does $\lim_{x \to -2} g(x)$ appear to exist? If so, what is the value of the limit? If not, why not?

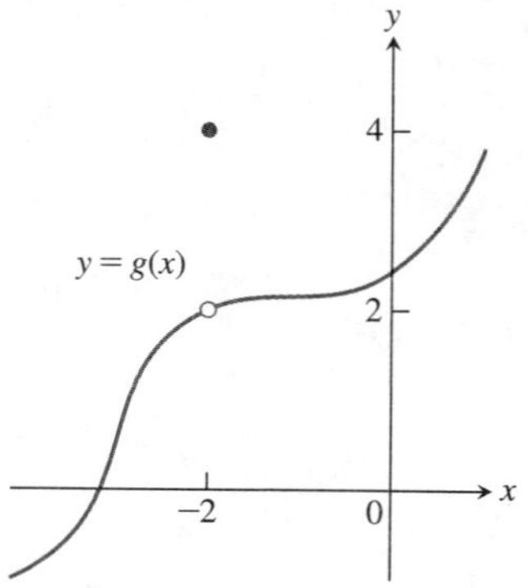

COMPUTER EXPLORATIONS

In Exercises 61–66, you will further explore finding deltas graphically. Use a CAS to perform the following steps:

a. Plot the function $y = f(x)$ near the point c being approached.

b. Guess the value of the limit L and then evaluate the limit symbolically to see if you guessed correctly.

c. Using the value $\varepsilon = 0.2$, graph the banding lines $y_1 = L - \varepsilon$ and $y_2 = L + \varepsilon$ together with the function f near c.

d. From your graph in part (c), estimate a $\delta > 0$ such that

$$|f(x) - L| < \varepsilon \quad \text{whenever} \quad 0 < |x - c| < \delta.$$

Test your estimate by plotting f, y_1, and y_2 over the interval $0 < |x - c| < \delta$. For your viewing window use $c - 2\delta \leq x \leq c + 2\delta$ and $L - 2\varepsilon \leq y \leq L + 2\varepsilon$. If any function values lie outside the interval $[L - \varepsilon, L + \varepsilon]$, your choice of δ was too large. Try again with a smaller estimate.

e. Repeat parts (c) and (d) successively for $\varepsilon = 0.1, 0.05$, and 0.001.

61. $f(x) = \dfrac{x^4 - 81}{x - 3}, \quad c = 3$

62. $f(x) = \dfrac{5x^3 + 9x^2}{2x^5 + 3x^2}, \quad c = 0$

63. $f(x) = \dfrac{\sin 2x}{3x}, \quad c = 0$

64. $f(x) = \dfrac{x(1 - \cos x)}{x - \sin x}, \quad c = 0$

65. $f(x) = \dfrac{\sqrt[3]{x} - 1}{x - 1}, \quad c = 1$

66. $f(x) = \dfrac{3x^2 - (7x + 1)\sqrt{x} + 5}{x - 1}, \quad c = 1$

2.4 One-Sided Limits

In this section we extend the limit concept to *one-sided limits*, which are limits as x approaches the number c from the left-hand side (where $x < c$) or the right-hand side (where $x > c$) only. These allow us to describe functions that have different limits at a point, depending on whether we approach the point from the left or from the right. One-sided limits also allow us to say what it means for a function to have a limit at an endpoint of an interval.

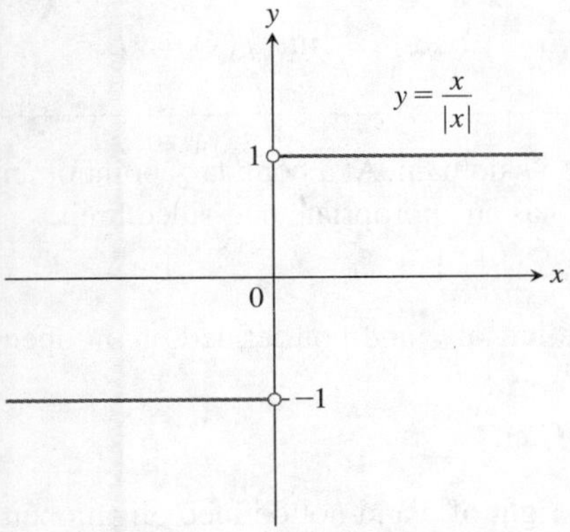

FIGURE 2.25 Different right-hand and left-hand limits at the origin.

Approaching a Limit from One Side

Suppose a function f is defined on an interval that extends to both sides of a number c. In order for f to have a limit L as x approaches c, the values of $f(x)$ must approach the value L as x approaches c from either side. Because of this, we sometimes say that the limit is **two-sided**.

If f fails to have a two-sided limit at c, it may still have a one-sided limit, that is, a limit if the approach is only from one side. If the approach is from the right, the limit is a **right-hand limit** or **limit from the right**. Similarly, a **left-hand limit** is also called a **limit from the left**.

The function $f(x) = x/|x|$ (Figure 2.25) has limit 1 as x approaches 0 from the right, and limit -1 as x approaches 0 from the left. Since these one-sided limit values are not the same, there is no single number that $f(x)$ approaches as x approaches 0. So $f(x)$ does not have a (two-sided) limit at 0.

Intuitively, if we consider only the values of $f(x)$ on an interval (c, b), where $c < b$, and the values of $f(x)$ become arbitrarily close to L as x approaches c from within that interval, then f has **right-hand limit** L at c. In this case we write

$$\lim_{x \to c^+} f(x) = L.$$

The notation "$x \to c^+$" means that we consider only values of $f(x)$ for x greater than c. We don't consider values of $f(x)$ for $x \leq c$.

Similarly, if $f(x)$ is defined on an interval (a, c), where $a < c$, and $f(x)$ approaches arbitrarily close to M as x approaches c from within that interval, then f has **left-hand limit** M at c. We write

$$\lim_{x \to c^-} f(x) = M.$$

The symbol "$x \to c^-$" means that we consider the values of f only at x-values less than c.

These informal definitions of one-sided limits are illustrated in Figure 2.26. For the function $f(x) = x/|x|$ in Figure 2.25 we have

$$\lim_{x \to 0^+} f(x) = 1 \qquad \text{and} \qquad \lim_{x \to 0^-} f(x) = -1.$$

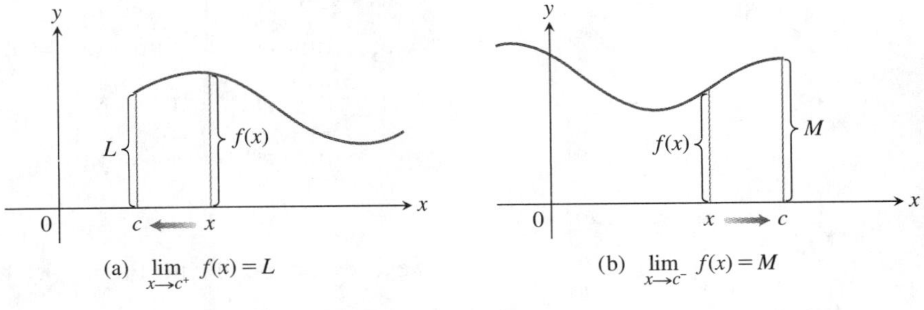

FIGURE 2.26 (a) Right-hand limit as x approaches c. (b) Left-hand limit as x approaches c.

One-sided limits have all the properties listed in Theorem 1 in Section 2.2. The right-hand limit of the sum of two functions is the sum of their right-hand limits, and so on. The theorems for limits of polynomials and rational functions hold with one-sided limits, as does the Sandwich Theorem. One-sided limits are related to limits at interior points in the following way.

THEOREM 5
Suppose that a function f is defined on an open interval containing c, except perhaps at c itself. Then $f(x)$ has a limit as x approaches c if and only if it has both a limit from the left at c and a limit from the right at c, and these one-sided limits are equal:

$$\lim_{x\to c} f(x) = L \quad \Leftrightarrow \quad \lim_{x\to c^-} f(x) = L \quad \text{and} \quad \lim_{x\to c^+} f(x) = L.$$

Theorem 5 applies at interior points of a function's domain. At a boundary point of an interval in its domain, a function has a limit when it has an appropriate one-sided limit.

Limits at Endpoints of an Interval

- If f is defined on an open interval (b, c) to the left of c and not defined on an open interval (c, d) to the right of c, then
$$\lim_{x\to c} f(x) = \lim_{x\to c^-} f(x).$$
- If f is defined on an open interval (c, d) to the right of c and not defined on an open interval (b, c) to the left of c, then
$$\lim_{x\to c} f(x) = \lim_{x\to c^+} f(x).$$

(The definition of a limit on an arbitrary domain is discussed in Additional and Advanced Exercises 39–42.)

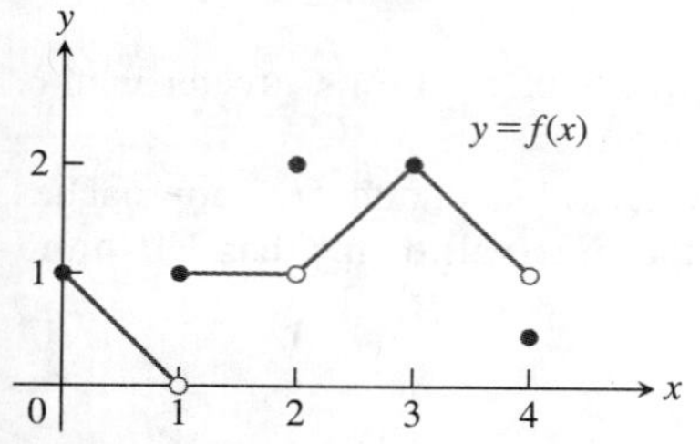

FIGURE 2.27 Graph of the function in Example 1.

EXAMPLE 1 For the function graphed in Figure 2.27,

At $x = 0$: $\lim_{x\to 0^-} f(x)$ does not exist, — f is not defined to the left of $x = 0$.

$\lim_{x\to 0^+} f(x) = 1$, — f has a right-hand limit at $x = 0$.

$\lim_{x\to 0} f(x) = 1$. — f has a limit at domain endpoint $x = 0$.

At $x = 1$: $\lim_{x\to 1^-} f(x) = 0$, — Even though $f(1) = 1$.

$\lim_{x\to 1^+} f(x) = 1$,

$\lim_{x\to 1} f(x)$ does not exist. — Right- and left-hand limits are not equal.

At $x = 2$: $\lim_{x\to 2^-} f(x) = 1$,

$\lim_{x\to 2^+} f(x) = 1$,

$\lim_{x\to 2} f(x) = 1$. — Even though $f(2) = 2$.

At $x = 3$: $\lim_{x\to 3^-} f(x) = \lim_{x\to 3^+} f(x) = \lim_{x\to 3} f(x) = f(3) = 2$.

At $x = 4$: $\lim_{x\to 4^-} f(x) = 1$, — Even though $f(4) \neq 1$.

$\lim_{x\to 4^+} f(x)$ does not exist, — f is not defined to the right of $x = 4$.

$\lim_{x\to 4} f(x) = 1$. — f has a limit at domain endpoint $x = 4$.

At every other point c in $[0, 4]$, $f(x)$, has limit $f(c)$. ■

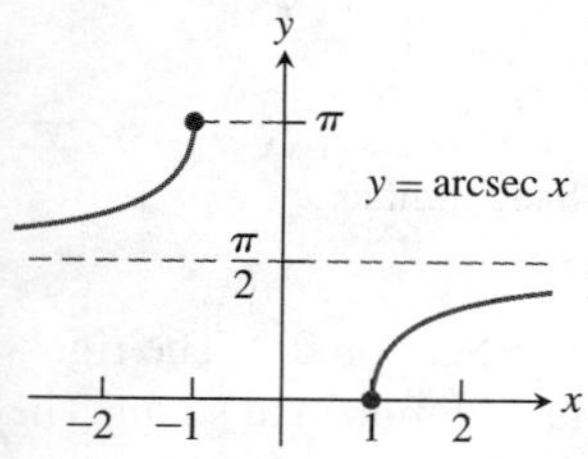

FIGURE 2.28 The arcsec function has limits at $x = \pm 1$.

EXAMPLE 2 The graph of the function arcsec x, which we will discuss in detail in Chapter 7, is shown in Figure 2.28. The domain of arcsec x is a union of the intervals $(-\infty, -1]$ and $[1, \infty)$. For the boundary points of these intervals,

At $x = -1$: $\lim_{x\to -1^-} \operatorname{arcsec} x = \pi$, — $\operatorname{arcsec} x$ has a limit from the left at $x = -1$.

$\lim_{x\to -1^+} \operatorname{arcsec} x$ does not exist, — $\operatorname{arcsec} x$ is not defined on $(-1, 1)$.

$\lim_{x\to -1} \operatorname{arcsec} x = \pi$. — $\operatorname{arcsec} x$ has a limit at $x = -1$.

At $x = 1$: $\lim_{x \to 1^-} \operatorname{arcsec} x$ does not exist, $\quad$ arcsec x is not defined on $(-1, 1)$.

$\lim_{x \to 1^+} \operatorname{arcsec} x = 0,$ $\quad$ arcsec x has a limit from the right at $x = 1$.

$\lim_{x \to 1} \operatorname{arcsec} x = 0.$ $\quad$ arcsec x has a limit at $x = 1$.

At every c in $(-\infty, -1)$ and every c in $(1, \infty)$, the limit of arcsec x as $x \to c$ is equal to arcsec c. However, for each c in the interval $(-1, 1)$, $\lim_{x \to c} \operatorname{arcsec} x$ does not exist. ■

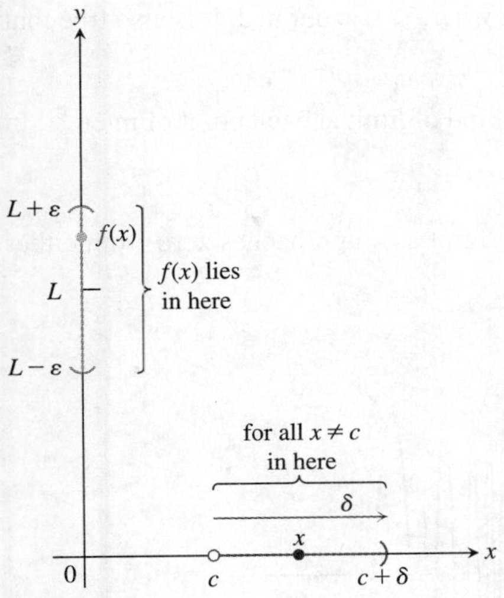

FIGURE 2.29 Intervals associated with the definition of right-hand limit.

Precise Definitions of One-Sided Limits

The formal definition of the limit in Section 2.3 is readily modified for one-sided limits.

> **DEFINITIONS** (a) Assume the domain of f contains an interval (c, d) to the right of c. We say that $f(x)$ has **right-hand limit** L **at** c, and write
>
> $$\lim_{x \to c^+} f(x) = L$$
>
> if for every number $\varepsilon > 0$ there exists a corresponding number $\delta > 0$ such that
>
> $$|f(x) - L| < \varepsilon \quad \text{whenever} \quad c < x < c + \delta.$$
>
> (b) Assume the domain of f contains an interval (b, c) to the left of c. We say that f has **left-hand limit** L **at** c, and write
>
> $$\lim_{x \to c^-} f(x) = L$$
>
> if for every number $\varepsilon > 0$ there exists a corresponding number $\delta > 0$ such that
>
> $$|f(x) - L| < \varepsilon \quad \text{whenever} \quad c - \delta < x < c.$$

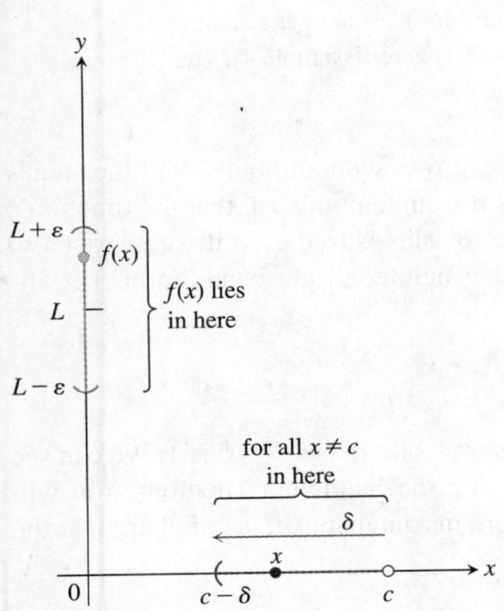

FIGURE 2.30 Intervals associated with the definition of left-hand limit.

The definitions are illustrated in Figures 2.29 and 2.30.

EXAMPLE 3 Prove that

$$\lim_{x \to 0^+} \sqrt{x} = 0.$$

Solution Let $\varepsilon > 0$ be given. Here $c = 0$ and $L = 0$, so we want to find a $\delta > 0$ such that

$$\left|\sqrt{x} - 0\right| < \varepsilon \quad \text{whenever} \quad 0 < x < \delta,$$

or

$$\sqrt{x} < \varepsilon \quad \text{whenever} \quad 0 < x < \delta. \qquad \sqrt{x} \geq 0 \text{ so } |\sqrt{x}| = \sqrt{x}$$

Squaring both sides of this last inequality gives

$$x < \varepsilon^2 \quad \text{if} \quad 0 < x < \delta.$$

If we choose $\delta = \varepsilon^2$ we have

$$\sqrt{x} < \varepsilon \quad \text{whenever} \quad 0 < x < \delta = \varepsilon^2,$$

or

$$\left|\sqrt{x} - 0\right| < \varepsilon \quad \text{whenever} \quad 0 < x < \varepsilon^2.$$

According to the definition, this shows that $\lim_{x \to 0^+} \sqrt{x} = 0$ (Figure 2.31).

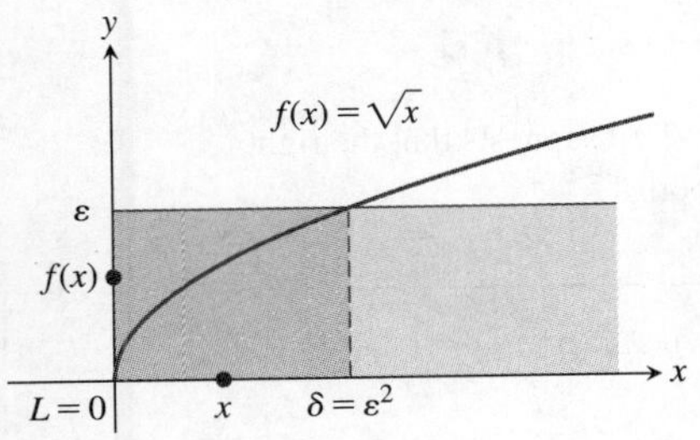

FIGURE 2.31 $\lim_{x \to 0^+} \sqrt{x} = 0$ in Example 3.

Note that since 0 is an endpoint of the domain where $\sqrt{x}$ is defined, it is also true that $\lim_{x \to 0} \sqrt{x} = 0$. ■

The functions examined so far have had some kind of limit at each point of interest. In general, that need not be the case.

EXAMPLE 4 Show that $y = \sin(1/x)$ has no limit as x approaches zero from either side (Figure 2.32).

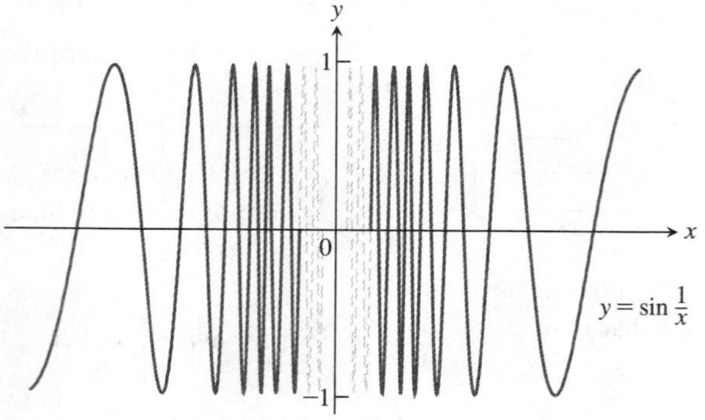

FIGURE 2.32 The function $y = \sin(1/x)$ has neither a right-hand nor a left-hand limit as x approaches zero (Example 4). The graph here omits values very near the y-axis.

Solution As x approaches zero, its reciprocal, $1/x$, grows without bound, and the values of $\sin(1/x)$ cycle repeatedly from –1 to 1. There is no single number L that the function's values stay increasingly close to as x approaches zero. This is true even if we restrict x to positive values or to negative values. The function has neither a right-hand limit nor a left-hand limit at $x = 0$. ■

Limits Involving $(\sin \theta)/\theta$

A central fact about $(\sin \theta)/\theta$ is that in radian measure its limit as $\theta \to 0$ is 1. We can see this in Figure 2.33 and confirm it algebraically using the Sandwich Theorem. You will see the importance of this limit in Section 3.5, where instantaneous rates of change of the trigonometric functions are studied.

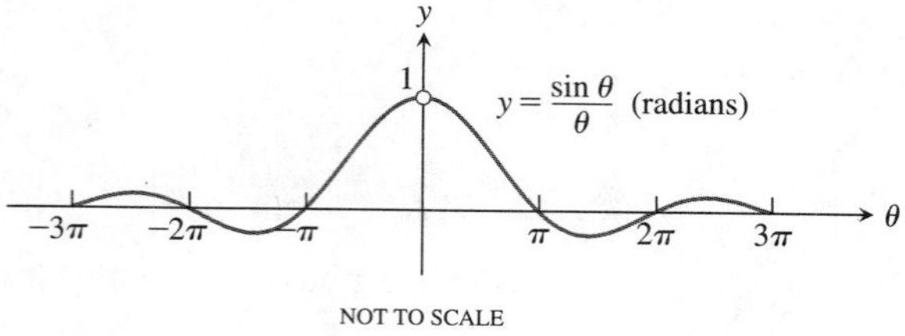

FIGURE 2.33 The graph of $f(\theta) = (\sin \theta)/\theta$ suggests that the right- and left-hand limits as θ approaches 0 are both 1.

THEOREM 6—Limit of the Ratio $\sin \theta/\theta$ as $\theta \to 0$

$$\lim_{\theta \to 0} \frac{\sin \theta}{\theta} = 1 \quad (\theta \text{ in radians}) \tag{1}$$

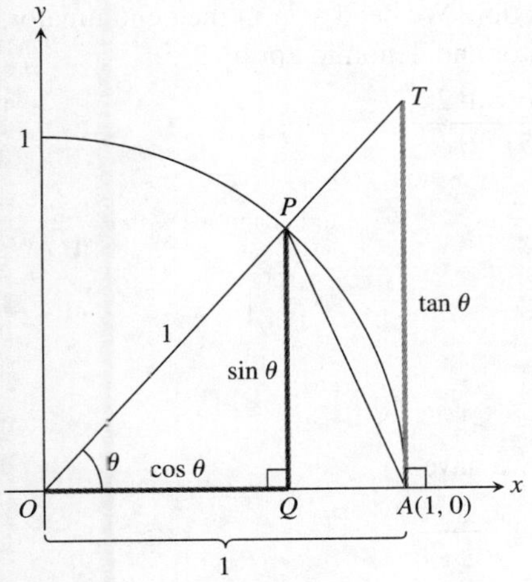

FIGURE 2.34 The ratio $TA/OA = \tan\theta$, and $OA = 1$, so $TA = \tan\theta$.

The use of radians to measure angles is essential in Equation (2): The area of sector OAP is $\theta/2$ only if θ is measured in radians.

Proof The plan is to show that the right-hand and left-hand limits are both 1. Then we will know that the two-sided limit is 1 as well.

To show that the right-hand limit is 1, we begin with positive values of θ less than $\pi/2$ (Figure 2.34). Notice that

$$\text{Area } \Delta OAP < \text{area sector } OAP < \text{area } \Delta OAT.$$

We can express these areas in terms of θ as follows:

$$\text{Area } \Delta OAP = \frac{1}{2}\text{base} \times \text{height} = \frac{1}{2}(1)(\sin\theta) = \frac{1}{2}\sin\theta$$

$$\text{Area sector } OAP = \frac{1}{2}r^2\theta = \frac{1}{2}(1)^2\theta = \frac{\theta}{2} \qquad (2)$$

$$\text{Area } \Delta OAT = \frac{1}{2}\text{base} \times \text{height} = \frac{1}{2}(1)(\tan\theta) = \frac{1}{2}\sin\theta.$$

Thus,

$$\frac{1}{2}\sin\theta < \frac{1}{2}\theta < \frac{1}{2}\tan\theta.$$

This last inequality goes the same way if we divide all three terms by the number $(1/2)\sin\theta$, which is positive, since $0 < \theta < \pi/2$:

$$1 < \frac{\theta}{\sin\theta} < \frac{1}{\cos\theta}.$$

Taking reciprocals reverses the inequalities:

$$1 > \frac{\sin\theta}{\theta} > \cos\theta.$$

Since $\lim_{\theta\to 0^+}\cos\theta = 1$ (Example 12b, Section 2.2), the Sandwich Theorem gives

$$\lim_{\theta\to 0^+}\frac{\sin\theta}{\theta} = 1.$$

To consider the left-hand limit, we recall that $\sin\theta$ and θ are both *odd functions* (Section 1.1). Therefore, $f(\theta) = (\sin\theta)/\theta$ is an *even function*, with a graph symmetric about the y-axis (see Figure 2.33). This symmetry implies that the left-hand limit at 0 exists and has the same value as the right-hand limit:

$$\lim_{\theta\to 0^-}\frac{\sin\theta}{\theta} = 1 = \lim_{\theta\to 0^+}\frac{\sin\theta}{\theta},$$

so $\lim_{\theta\to 0}(\sin\theta)/\theta = 1$ by Theorem 5. ■

EXAMPLE 5 Show that **(a)** $\lim_{y\to 0}\frac{\cos y - 1}{y} = 0$ and **(b)** $\lim_{x\to 0}\frac{\sin 2x}{5x} = \frac{2}{5}.$

Solution

(a) Using the half-angle formula $\cos y = 1 - 2\sin^2(y/2)$, we calculate

$$\lim_{y\to 0}\frac{\cos y - 1}{y} = \lim_{y\to 0} -\frac{2\sin^2(y/2)}{y}$$

$$= -\lim_{\theta\to 0}\frac{\sin\theta}{\theta}\sin\theta \qquad \text{Let } \theta = y/2.$$

$$= -(1)(0) = 0. \qquad \text{Eq. (1) and Example 12a in Secion 2.2}$$

(b) Equation (1) does not apply to the original fraction. We need a $2x$ in the denominator, not a $5x$. We produce it by multiplying numerator and denominator by $2/5$:

$$\begin{aligned}\lim_{x\to 0}\frac{\sin 2x}{5x} &= \lim_{x\to 0}\frac{(2/5)\cdot\sin 2x}{(2/5)\cdot 5x}\\ &= \frac{2}{5}\lim_{x\to 0}\frac{\sin 2x}{2x} && \text{Eq. (1) applies with } \theta = 2x.\\ &= \frac{2}{5}(1) = \frac{2}{5}.\end{aligned}$$

■

EXAMPLE 6 Find $\displaystyle\lim_{t\to 0}\frac{\tan t \sec 2t}{3t}$.

Solution From the definition of $\tan t$ and $\sec 2t$, we have

$$\begin{aligned}\lim_{t\to 0}\frac{\tan t\sec 2t}{3t} &= \lim_{t\to 0}\frac{1}{3}\cdot\frac{1}{t}\cdot\frac{\sin t}{\cos t}\cdot\frac{1}{\cos 2t}\\ &= \frac{1}{3}\lim_{t\to 0}\frac{\sin t}{t}\cdot\frac{1}{\cos t}\cdot\frac{1}{\cos 2t}\\ &= \frac{1}{3}(1)(1)(1) = \frac{1}{3}. && \text{Eq. (1) and Example 12b in Section 2.2}\end{aligned}$$

■

EXAMPLE 7 Show that for nonzero constants A and B.

$$\lim_{\theta\to 0}\frac{\sin A\theta}{\sin B\theta} = \frac{A}{B}.$$

Solution

$$\begin{aligned}\lim_{\theta\to 0}\frac{\sin A\theta}{\sin B\theta} &= \lim_{\theta\to 0}\frac{\sin A\theta}{A\theta}A\theta\frac{B\theta}{\sin B\theta}\frac{1}{B\theta} && \text{Multiply and divide by } A\theta \text{ and } B\theta.\\ &= \lim_{\theta\to 0}\frac{\sin A\theta}{A\theta}\frac{B\theta}{\sin B\theta}\frac{A}{B} && \lim_{u\to 0}\frac{\sin u}{u} = 1, \text{ with } u = A\theta\\ &= \lim_{\theta\to 0}(1)(1)\frac{A}{B} && \lim_{v\to 0}\frac{v}{\sin v} = 1, \text{ with } v = B\theta\\ &= \frac{A}{B}.\end{aligned}$$

■

EXERCISES 2.4

Finding Limits Graphically

1. Which of the following statements about the function $y = f(x)$ graphed here are true, and which are false?

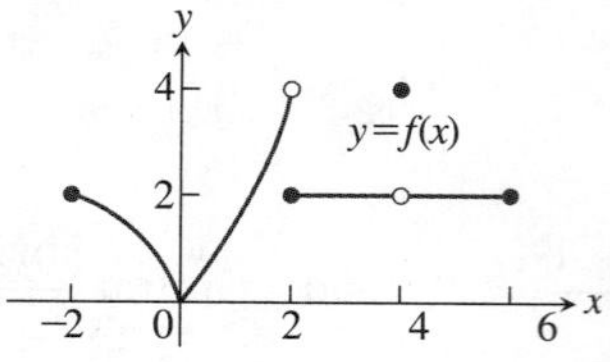

a. $\lim_{x\to -2^+} f(x) = 2$

b. $\lim_{x\to 4} f(x)$ does not exist

c. $\lim_{x\to 4} f(x) = 4$

d. $\lim_{x\to 2^-} f(x) = 4$

e. $\lim_{x\to 2^+} f(x) = 2$

f. $\lim_{x\to -2} f(x)$ does not exist

g. $\lim_{x\to 0^+} f(x) = \lim_{x\to 0^-} f(x)$

h. $\lim_{x\to c} f(x)$ exists at every c in the open interval $(-2, 2)$

i. $\lim_{x\to c} f(x)$ exists at every c in the open interval $(1, 3)$

j. $\lim_{x\to -2^-} f(x) = 0$

k. $\lim_{x\to 6^+} f(x)$ does not exist

2. Which of the following statements about the function $y = f(x)$ graphed here are true, and which are false?

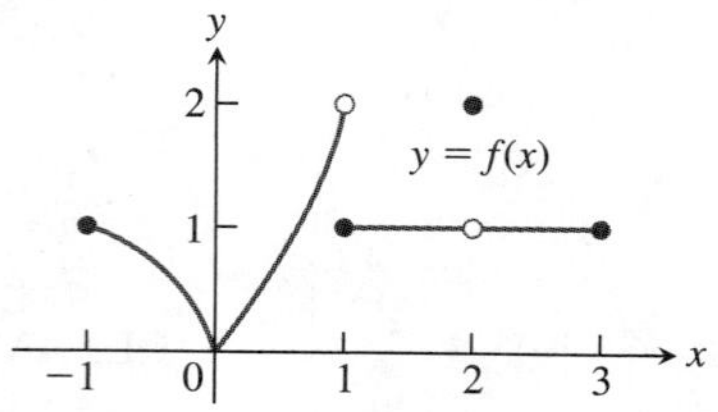

a. $\lim_{x\to -1^+} f(x) = 1$

b. $\lim_{x\to 2} f(x)$ does not exist.

c. $\lim_{x\to 2} f(x) = 2$

d. $\lim_{x\to 1^-} f(x) = 2$

e. $\lim_{x\to 1^+} f(x) = 1$

f. $\lim_{x\to 1} f(x)$ does not exist.

g. $\lim_{x\to 0^+} f(x) = \lim_{x\to 0^-} f(x)$

h. $\lim_{x\to c} f(x)$ exists at every c in the open interval $(-1, 1)$.

i. $\lim_{x\to c} f(x)$ exists at every c in the open interval $(1, 3)$.

j. $\lim_{x\to -1^-} f(x) = 0$

k. $\lim_{x\to 3^+} f(x)$ does not exist.

3. Let $f(x)=\begin{cases}3-x, & x<2\\ \dfrac{x}{2}+1, & x>2\end{cases}$.

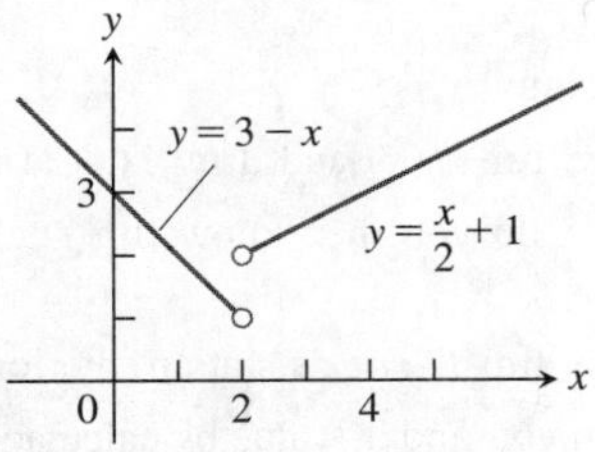

a. Find $\lim_{x\to 2^+} f(x)$ and $\lim_{x\to 2^-} f(x)$.

b. Does $\lim_{x\to 2} f(x)$ exist? If so, what is it? If not, why not?

c. Find $\lim_{x\to 3^+} f(x)$ and $\lim_{x\to 3^-} f(x)$.

d. Does $\lim_{x\to 3} f(x)$ exist? If so, what is it? If not, why not?

4. Let $f(x)=\begin{cases}3-x, & x<2\\ 2, & x=2\\ \dfrac{x}{2}, & x>2.\end{cases}$

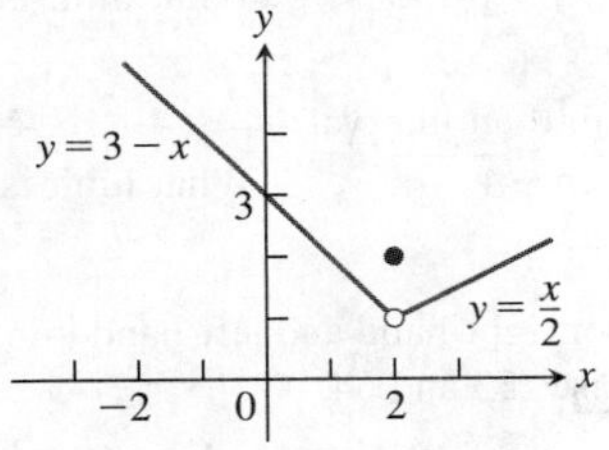

a. Find $\lim_{x\to 2^+} f(x)$, $\lim_{x\to 2^-} f(x)$, and $f(2)$.

b. Does $\lim_{x\to 2} f(x)$ exist? If so, what is it? If not, why not?

c. Find $\lim_{x\to -1^-} f(x)$ and $\lim_{x\to -1^+} f(x)$.

d. Does $\lim_{x\to -1} f(x)$ exist? If so, what is it? If not, why not?

5. Let $f(x)=\begin{cases}1, & x\le 0\\ \sin x, & x>0\end{cases}$

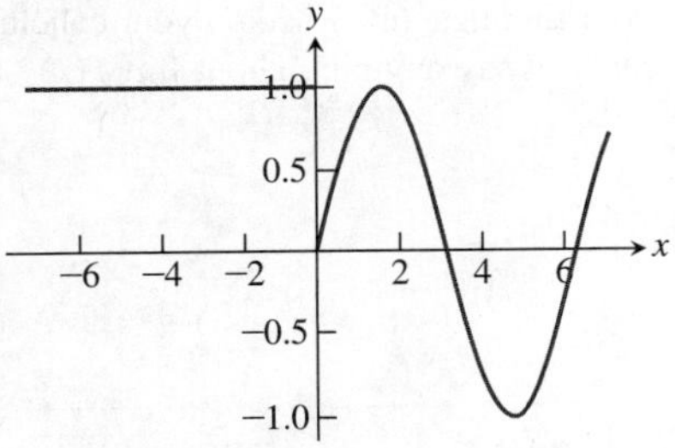

a. Does $\lim_{x\to 0^+} f(x)$ exist? If so, what is it? If not, why not?

b. Does $\lim_{x\to 0^-} f(x)$ exist? If so, what is it? If not, why not?

c. Does $\lim_{x\to 0} f(x)$ exist? If so, what is it? If not, why not?

6. Let $g(x)=\sqrt{x}\sin(1/x)$.

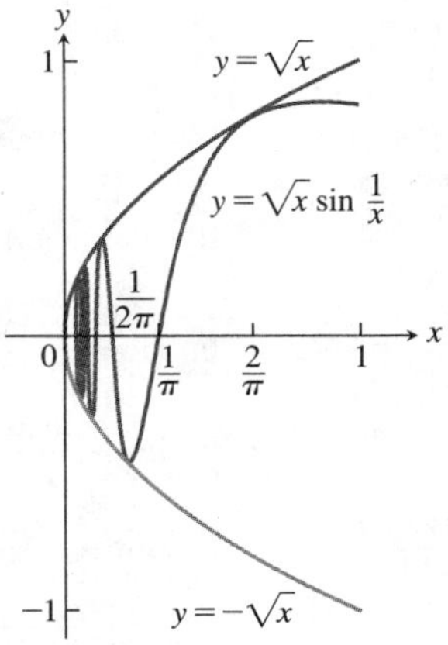

a. Does $\lim_{x\to 0^+} g(x)$ exist? If so, what is it? If not, why not?

b. Does $\lim_{x\to 0^-} g(x)$ exist? If so, what is it? If not, why not?

c. Does $\lim_{x\to 0} g(x)$ exist? If so, what is it? If not, why not?

7. a. Graph $f(x)=\begin{cases}x^3, & x\neq 1\\ 0, & x=1.\end{cases}$

b. Find $\lim_{x\to 1^-} f(x)$ and $\lim_{x\to 1^+} f(x)$.

c. Does $\lim_{x\to 1} f(x)$ exist? If so, what is it? If not, why not?

8. a. Graph $f(x)=\begin{cases}1-x^2, & x\neq 1\\ 2, & x=1.\end{cases}$

b. Find $\lim_{x\to 1^+} f(x)$ and $\lim_{x\to 1^-} f(x)$.

c. Does $\lim_{x\to 1} f(x)$ exist? If so, what is it? If not, why not?

Graph the functions in Exercises 9 and 10. Then answer these questions.

a. What are the domain and range of f?

b. At what points c, if any, does $\lim_{x\to c} f(x)$ exist?

c. At what points does the left-hand limit exist but not the right-hand limit?

d. At what points does the right-hand limit exist but not the left-hand limit?

9. $f(x)=\begin{cases}\sqrt{1-x^2}, & 0\le x<1\\ 1, & 1\le x<2\\ 2, & x=2\end{cases}$

10. $f(x)=\begin{cases}x, & -1\le x<0, \text{ or } 0<x\le 1\\ 1, & x=0\\ 0, & x<-1 \text{ or } x>1\end{cases}$

Finding One-Sided Limits Algebraically

Find the limits in Exercises 11–20.

11. $\lim_{x\to -0.5^-}\sqrt{\dfrac{x+5}{x+13}}$

12. $\lim_{x\to 1^+}\sqrt{\dfrac{x-1}{x+2}}$

13. $\lim_{x\to -3^+}\left(\dfrac{2x}{2x+3}\right)\left(\dfrac{x+2}{x^2-7}\right)$

14. $\lim\limits_{x\to -1^-}\left(\dfrac{1}{x+1}\right)\left(\dfrac{x+6}{x}\right)\left(\dfrac{3-x}{7}\right)$

15. $\lim\limits_{h\to 1^+}\dfrac{\sqrt{h^2+5h+10}-\sqrt{h}}{h}$

16. $\lim\limits_{h\to 0^-}\dfrac{\sqrt{6}-\sqrt{5h^2+11h+6}}{h}$

17. a. $\lim\limits_{x\to -3^+}(x+4)\dfrac{|x+3|}{x+3}$ **b.** $\lim\limits_{x\to -3^-}(x+4)\dfrac{|x+3|}{x+3}$

18. a. $\lim\limits_{x\to 2^+}\dfrac{\sqrt{2x}(x-2)}{|x-2|}$ **b.** $\lim\limits_{x\to 2^-}\dfrac{\sqrt{2x}(x-2)}{|x-2|}$

19. a. $\lim\limits_{x\to 0^+}\dfrac{|\sin x|}{x}$ **b.** $\lim\limits_{x\to 0^-}\dfrac{|\sin x|}{x}$

20. a. $\lim\limits_{x\to 0^+}\dfrac{1-\cos x}{|\cos x-1|}$ **b.** $\lim\limits_{x\to 0^-}\dfrac{\cos x-1}{|\cos x-1|}$

Use the graph of the greatest integer function $y=\lfloor x\rfloor$, Figure 1.10 in Section 1.1, to help you find the limits in Exercises 21 and 22.

21. a. $\lim\limits_{\theta\to 3^+}\dfrac{\lfloor\theta\rfloor}{\theta}$ **b.** $\lim\limits_{\theta\to 3^-}\dfrac{\lfloor\theta\rfloor}{\theta}$

22. a. $\lim\limits_{t\to 4^+}(t-\lfloor t\rfloor)$ **b.** $\lim\limits_{t\to 4^-}(t-\lfloor t\rfloor)$

Using $\lim\limits_{\theta\to 0}\dfrac{\sin\theta}{\theta}=1$

Find the limits in Exercises 23–46.

23. $\lim\limits_{\theta\to 0}\dfrac{\sin\sqrt{5\theta}}{\sqrt{5\theta}}$

24. $\lim\limits_{t\to 0}\dfrac{\sin kt}{t}$ (k constant)

25. $\lim\limits_{x\to 0}\dfrac{\sin 5x}{9x}$

26. $\lim\limits_{h\to 0^-}\dfrac{h}{\sin 3h}$

27. $\lim\limits_{x\to 0}\dfrac{\tan 2x}{x}$

28. $\lim\limits_{h\to 0}\dfrac{4h}{\tan h}$

29. $\lim\limits_{x\to 0}\dfrac{x\csc 2x}{\cos 5x}$

30. $\lim\limits_{x\to 0}6x^2(\cot x)(\csc 2x)$

31. $\lim\limits_{h\to 0}\dfrac{2h+2h\cos h}{\cos h\sin h}$

32. $\lim\limits_{x\to 0}\dfrac{x^2-x+\sin x}{2x}$

33. $\lim\limits_{\theta\to 0}\dfrac{1-\cos\theta}{\sin 2\theta}$

34. $\lim\limits_{x\to 0}\dfrac{3x-3x\cos x}{\sin^2 3x}$

35. $\lim\limits_{t\to 0}\dfrac{\sin(1-\cos t)}{1-\cos t}$

36. $\lim\limits_{h\to 0}\dfrac{\sin(\sin h)}{\sin h}$

37. $\lim\limits_{\theta\to 0}\dfrac{\sin\theta}{\sin 3\theta}$

38. $\lim\limits_{x\to 0}\dfrac{\sin 5x}{\sin 4x}$

39. $\lim\limits_{\theta\to 0}3\theta\cos\theta$

40. $\lim\limits_{\theta\to 0}\sin\theta\cot 2\theta$

41. $\lim\limits_{x\to 0}\dfrac{\tan 3x}{\sin 8x}$

42. $\lim\limits_{x\to 0}\dfrac{\sin 2x\tan 3x}{x\tan 4x}$

43. $\lim\limits_{\theta\to 0}\dfrac{\tan\theta}{\theta^2\cot 3\theta}$

44. $\lim\limits_{y\to 0}\dfrac{y\tan 4y}{\cos^2 y\tan^2 2y}$

45. $\lim\limits_{x\to 0}\dfrac{1-\cos 3x}{2x}$

46. $\lim\limits_{x\to 0}\dfrac{\cos^2 2x-\cos 2x}{x^2}$

Theory and Examples

47. Once you know $\lim_{x\to a^+}f(x)$ and $\lim_{x\to a^-}f(x)$ at an interior point of the domain of f, do you then know $\lim_{x\to a}f(x)$? Give reasons for your answer.

48. If you know that $\lim_{x\to c}f(x)$ exists at an interior point of a domain interval of f, can you find its value by calculating $\lim_{x\to c^+}f(x)$? Give reasons for your answer.

49. Suppose that f is an odd function of x. Does knowing that $\lim_{x\to 0^+}f(x)=3$ tell you anything about $\lim_{x\to 0^-}f(x)$? Give reasons for your answer.

50. Suppose that f is an even function of x. Does knowing that $\lim_{x\to 2^-}f(x)=7$ tell you anything about either $\lim_{x\to -2^-}f(x)$ or $\lim_{x\to -2^+}f(x)$? Give reasons for your answer.

Formal Definitions of One-Sided Limits

51. Given $\varepsilon>0$, find an interval $I=(5,5+\delta)$, $\delta>0$, such that if x lies in I, then $\sqrt{x-5}<\varepsilon$. What limit is being verified and what is its value?

52. Given $\varepsilon>0$, find an interval $I=(4-\delta,4)$, $\delta>0$, such that if x lies in I, then $\sqrt{4-x}<\varepsilon$. What limit is being verified and what is its value?

Use the definitions of right-hand and left-hand limits to prove the limit statements in Exercises 53 and 54.

53. $\lim\limits_{x\to 0^-}\dfrac{x}{|x|}=-1$

54. $\lim\limits_{x\to 2^+}\dfrac{x-2}{|x-2|}=1$

55. Greatest integer function Find **(a)** $\lim_{x\to 400^+}\lfloor x\rfloor$ and **(b)** $\lim_{x\to 400^-}\lfloor x\rfloor$; then use limit definitions to verify your findings. **(c)** Based on your conclusions in parts (a) and (b), can you say anything about $\lim_{x\to 400}\lfloor x\rfloor$? Give reasons for your answer.

56. One-sided limits Let $f(x)=\begin{cases}x^2\sin(1/x), & x<0\\ \sqrt{x}, & x>0.\end{cases}$

Find **(a)** $\lim_{x\to 0^+}f(x)$ and **(b)** $\lim_{x\to 0^-}f(x)$; then use limit definitions to verify your findings. **(c)** Based on your conclusions in parts (a) and (b), can you say anything about $\lim_{x\to 0}f(x)$? Give reasons for your answer.

2.5 Continuity

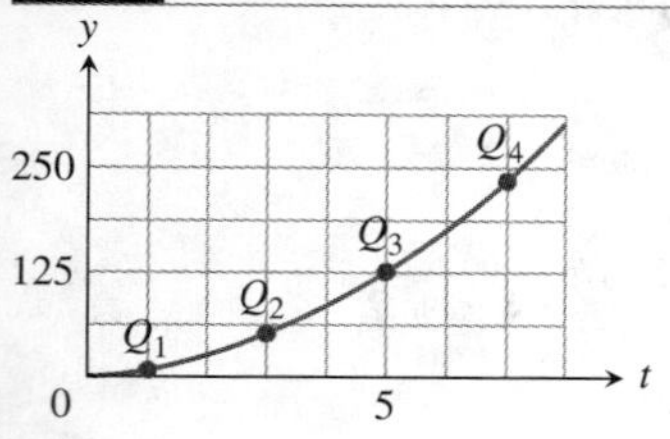

FIGURE 2.35 Connecting plotted points.

When we plot function values generated in a laboratory or collected in the field, we often connect the plotted points with an unbroken curve to show what the function's values are likely to have been at the points we did not measure (Figure 2.35). In doing so, we are assuming that we are working with a *continuous function*, so its outputs vary regularly and consistently with the inputs, and do not jump abruptly from one value to another without taking on the values in between. Intuitively, any function $y=f(x)$ whose graph can be sketched over its domain in one unbroken motion is an example of a continuous function. Such functions play an important role in the study of calculus and its applications.

Continuity at a Point

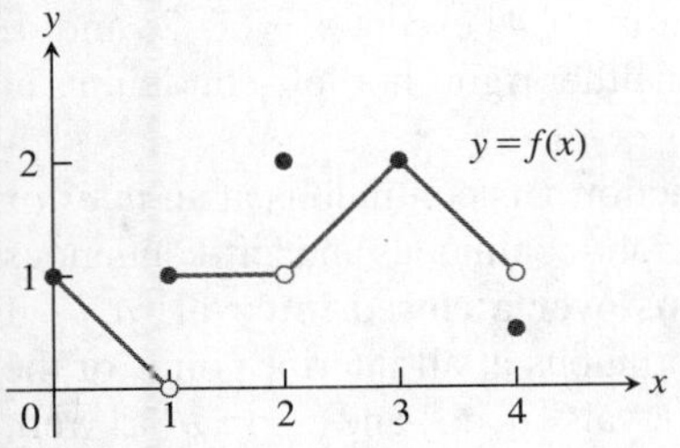

FIGURE 2.36 The function is not continuous at $x = 1$, $x = 2$, and $x = 4$ (Example 1).

To understand continuity, it helps to consider a function like that in Figure 2.36, whose limits we investigated in Example 1 in the last section.

EXAMPLE 1 At which numbers does the function f in Figure 2.36 appear to be not continuous? Explain why. What occurs at other numbers in the domain?

Solution First we observe that the domain of the function is the closed interval $[0, 4]$, so we will be considering the numbers x within that interval. From the figure, we notice right away that there are breaks in the graph at the numbers $x = 1$, $x = 2$, and $x = 4$. The break at $x = 1$ appears as a jump, which we identify later as a "jump discontinuity." The break at $x = 2$ is called a "removable discontinuity" since by changing the function definition at that one point, we can create a new function that is continuous at $x = 2$. Similarly, $x = 4$ is a removable discontinuity.

Numbers at which the graph of *f* has breaks:

At the interior point $x = 1$, the function fails to have a limit. It does have both a left-hand limit, $\lim_{x\to 1^-} f(x) = 0$, as well as a right-hand limit, $\lim_{x\to 1^+} f(x) = 1$, but the limit values are different, resulting in a jump in the graph. The function is not continuous at $x = 1$. However, the function value $f(1) = 1$ is equal to the limit from the right, so the function *is* continuous from the right at $x = 1$.

At $x = 2$, the function does have a limit, $\lim_{x\to 2} f(x) = 1$, but the value of the function is $f(2) = 2$. The limit and function values are not the same, so there is a break in the graph and f is not continuous at $x = 2$.

At $x = 4$, the function does have a left-hand limit at this right endpoint, $\lim_{x\to 4^-} f(x) = 1$, but again the value of the function $f(4) = \frac{1}{2}$ differs from the value of the limit. We see again a break in the graph of the function at this endpoint and the function is not continuous from the left.

Numbers at which the graph of *f* has no breaks:

At $x = 3$, the function has a limit, $\lim_{x\to 3} f(x) = 2$. Moreover, the limit is the same value as the function there, $f(3) = 2$. The function is continuous at $x = 3$.

At $x = 0$, the function has a right-hand limit at this left endpoint, $\lim_{x\to 0^+} f(x) = 1$, and the value of the function is the same, $f(0) = 1$. The function is continuous from the right at $x = 0$. Because $x = 0$ is a left endpoint of the function's domain, we have that $\lim_{x\to 0} f(x) = 1$ and so f is continuous at $x = 0$.

At all other numbers $x = c$ in the domain, the function has a limit equal to the value of the function, so $\lim_{x\to c} f(x) = f(c)$. For example, $\lim_{x\to 5/2} f(x) = f(5/2) = 3/2$. No breaks appear in the graph of the function at any of these numbers and the function is continuous at each of them. ■

The following definitions capture the continuity ideas we observed in Example 1.

DEFINITIONS Let c be a real number that is either an interior point or an endpoint of an interval in the domain of f.

The function f is **continuous at *c*** if

$$\lim_{x\to c} f(x) = f(c).$$

The function f is **right-continuous at *c*** (or **continuous from the right**) if

$$\lim_{x\to c^+} f(x) = f(c).$$

The function f is **left-continuous at *c*** (or **continuous from the left**) if

$$\lim_{x\to c^-} f(x) = f(c).$$

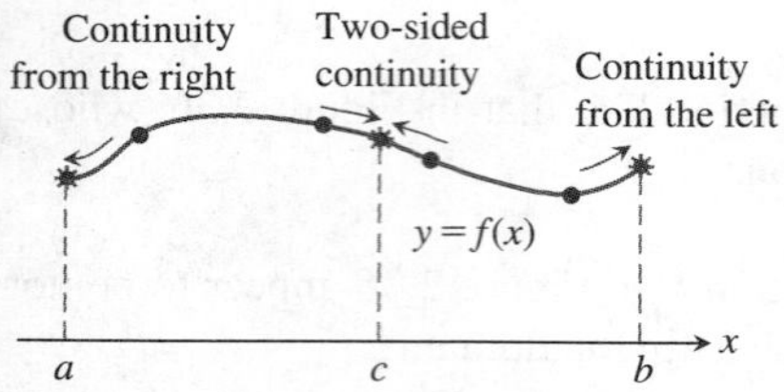

FIGURE 2.37 Continuity at points a, b, and c.

The function f in Example 1 is continuous at every x in $[0, 4]$ except $x = 1$, 2, and 4. It is right-continuous but not left-continuous at $x = 1$, neither right- nor left-continuous at $x = 2$, and not left-continuous at $x = 4$.

From Theorem 5, it follows immediately that a function f is continuous at an *interior* point c of an interval in its domain if and only if it is both right-continuous and left-continuous at c (Figure 2.37). We say that a function is **continuous over a closed interval** $[a, b]$ if it is right-continuous at a, left-continuous at b, and continuous at all interior points of the interval. This definition applies to the infinite closed intervals $[a, \infty)$ and $(-\infty, b]$ as well, but only one endpoint is involved. If a function is not continuous at point c of its domain, we say that f is **discontinuous at** c and that f has a discontinuity at c. Note that a function f can be continuous, right-continuous, or left-continuous only at a point c for which $f(c)$ is defined.

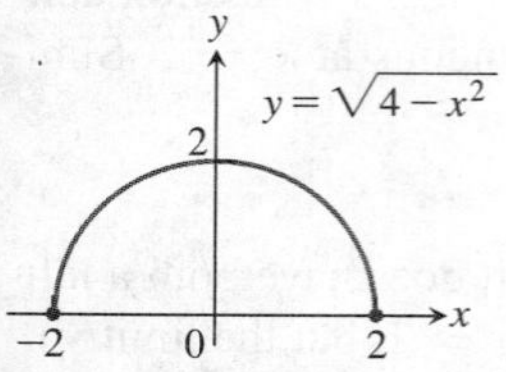

FIGURE 2.38 A function that is continuous over its domain (Example 2).

EXAMPLE 2 The function $f(x) = \sqrt{4 - x^2}$ is continuous over its domain $[-2, 2]$ (Figure 2.38). It is continuous at all points of this interval, including the endpoints $x = -2$ and $x = 2$. ■

EXAMPLE 3 The unit step function $U(x)$, graphed in Figure 2.39, is right-continuous at $x = 0$, but is neither left-continuous nor continuous there. It has a jump discontinuity at $x = 0$. ■

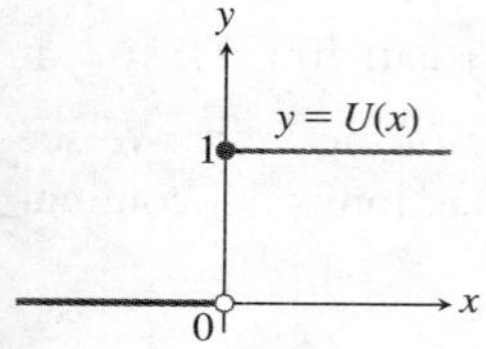

FIGURE 2.39 A function that has a jump discontinuity at the origin (Example 3).

At an interior point or an endpoint of an interval in its domain, a function is continuous at points where it passes the following test.

> **Continuity Test**
> A function $f(x)$ is continuous at a point $x = c$ if and only if it meets the following three conditions.
> **1.** $f(c)$ exists (c lies in the domain of f).
> **2.** $\lim_{x\to c} f(x)$ exists (f has a limit as $x \to c$).
> **3.** $\lim_{x\to c} f(x) = f(c)$ (the limit equals the function value).

For one-sided continuity, the limits in parts 2 and 3 of the test should be replaced by the appropriate one-sided limits.

EXAMPLE 4 The function $y = \lfloor x \rfloor$ introduced in Section 1.1 is graphed in Figure 2.40. It is discontinuous at every integer n, because the left-hand and right-hand limits are not equal as $x \to n$:

$$\lim_{x\to n^-} \lfloor x \rfloor = n - 1 \quad \text{and} \quad \lim_{x\to n^+} \lfloor x \rfloor = n.$$

Since $\lfloor n \rfloor = n$, the greatest integer function is right-continuous at every integer n (but not left-continuous).

The greatest integer function is continuous at every real number other than the integers. For example,

$$\lim_{x\to 1.5} \lfloor x \rfloor = 1 = \lfloor 1.5 \rfloor.$$

In general, if $n - 1 < c < n$, n an integer, then

$$\lim_{x\to c} \lfloor x \rfloor = n - 1 = \lfloor c \rfloor.$$ ■

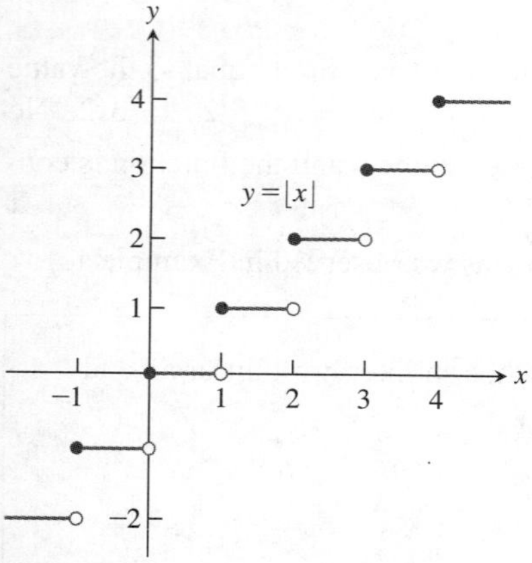

FIGURE 2.40 The greatest integer function is continuous at every noninteger point. It is right-continuous, but not left-continuous, at every integer point (Example 4).

Figure 2.41 displays several common ways in which a function can fail to be continuous. The function in Figure 2.41a is continuous at $x = 0$. The function in Figure 2.41b does not contain $x = 0$ in its domain. It would be continuous if its domain were extended so that $f(0) = 1$. The function in Figure 2.41c would be continuous if $f(0)$ were 1 instead of 2. The discontinuity in Figure 2.41c is **removable**. The function has a limit as $x \to 0$, and we can remove the discontinuity by setting $f(0)$ equal to this limit.

The discontinuities in Figure 2.41d through f are more serious: $\lim_{x \to 0} f(x)$ does not exist, and there is no way to improve the situation by appropriately defining f at 0. The step function in Figure 2.41d has a **jump discontinuity**: The one-sided limits exist but have different values. The function $f(x) = 1/x^2$ in Figure 2.41e has an **infinite discontinuity**. The function in Figure 2.41f has an **oscillating discontinuity**: It oscillates so much that its values approach each number in $[-1, 1]$ as $x \to 0$. Since it does not approach a single number, it does not have a limit as x approaches 0.

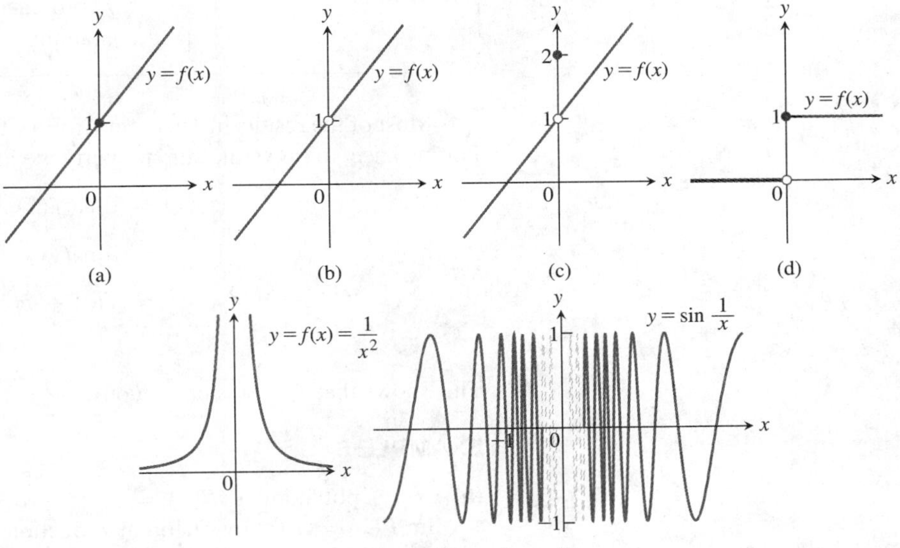

FIGURE 2.41 The function in (a) is continuous at $x = 0$; the functions in (b) through (f) are not.

Continuous Functions

We now describe the continuity behavior of a function throughout its entire domain, not only at a single point. We define a **continuous function** to be one that is continuous at every point in its domain. This is a property of the *function*. A function always has a specified domain, so if we change the domain, then we change the function, and this may change its continuity property as well. If a function is discontinuous at one or more points of its domain, we say it is a **discontinuous** function.

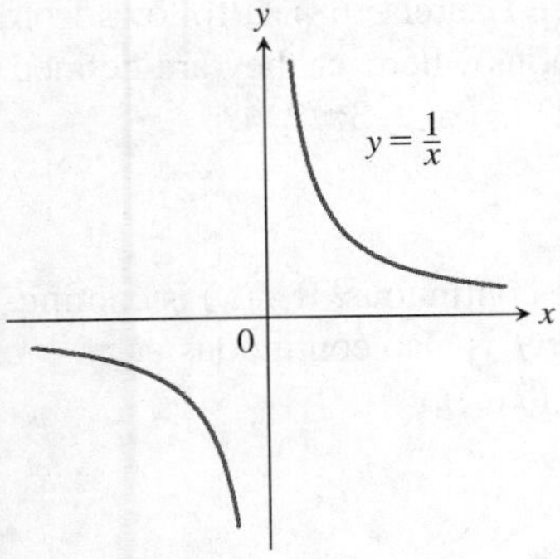

FIGURE 2.42 The function $f(x) = 1/x$ is continuous over its natural domain. It is not defined at the origin, so it is not continuous on any interval containing $x = 0$ (Example 5).

EXAMPLE 5

(a) The function $f(x) = 1/x$ (Figure 2.42) is a continuous function because it is continuous at every point of its domain. The point $x = 0$ is not in the domain of the function f, so f is not continuous on any interval containing $x = 0$. Moreover, there is no way to extend f to a new function that is defined and continuous at $x = 0$. The function f does not have a removable discontinuity at $x = 0$.

(b) The identity function $f(x) = x$ and constant functions are continuous everywhere by Example 3, Section 2.3. ■

Algebraic combinations of continuous functions are continuous wherever they are defined.

THEOREM 7—Properties of Continuous Functions
If the functions f and g are continuous at $x = c$, then the following algebraic combinations are continuous at $x = c$.

1. *Sums:* $f + g$
2. *Differences:* $f - g$
3. *Constant multiples:* $k \cdot f$, for any number k
4. *Products:* $f \cdot g$
5. *Quotients:* f/g, provided $g(c) \neq 0$
6. *Powers:* f^n, n a positive integer
7. *Roots:* $\sqrt[n]{f}$, provided it is defined on an interval containing c, where n is a positive integer

Most of the results in Theorem 7 follow from the limit rules in Theorem 1, Section 2.2. For instance, to prove the sum property we have

$$\begin{aligned}
\lim_{x \to c} (f + g)(x) &= \lim_{x \to c} (f(x) + g(x)) \\
&= \lim_{x \to c} f(x) + \lim_{x \to c} g(x) && \text{Sum Rule, Theorem 1} \\
&= f(c) + g(c) && \text{Continuity of } f, g \text{ at } c \\
&= (f + g)(c).
\end{aligned}$$

This shows that $f + g$ is continuous.

EXAMPLE 6

(a) Every polynomial $P(x) = a_n x^n + a_{n-1} x^{n-1} + \cdots + a_0$ is continuous because $\lim_{x \to c} P(x) = P(c)$ by Theorem 2, Section 2.2.

(b) If $P(x)$ and $Q(x)$ are polynomials, then the rational function $P(x)/Q(x)$ is continuous wherever it is defined ($Q(c) \neq 0$) by Theorem 3, Section 2.2. ■

EXAMPLE 7 The function $f(x) = |x|$ is continuous. If $x > 0$, we have $f(x) = x$, a polynomial. If $x < 0$, we have $f(x) = -x$, another polynomial. Finally, at the origin, $\lim_{x \to 0} |x| = 0 = |0|$. ■

The functions $y = \sin x$ and $y = \cos x$ are continuous at $x = 0$ by Example 12 of Section 2.2. Both functions are continuous everywhere (see Exercise 68). It follows from Theorem 7 that all six trigonometric functions are continuous wherever they are defined. For example, $y = \tan x$ is continuous on $\cdots \cup (-\pi/2, \pi/2) \cup (\pi/2, 3\pi/2) \cup \cdots$.

Continuity of Compositions of Functions

Functions obtained by composing continuous functions are continuous. If $f(x)$ is continuous at $x = c$ and $g(x)$ is continuous at $x = f(c)$, then $g \circ f$ is also continuous at $x = c$ (Figure 2.43). In this case, the limit of $g \circ f$ as $x \to c$ is $g(f(c))$.

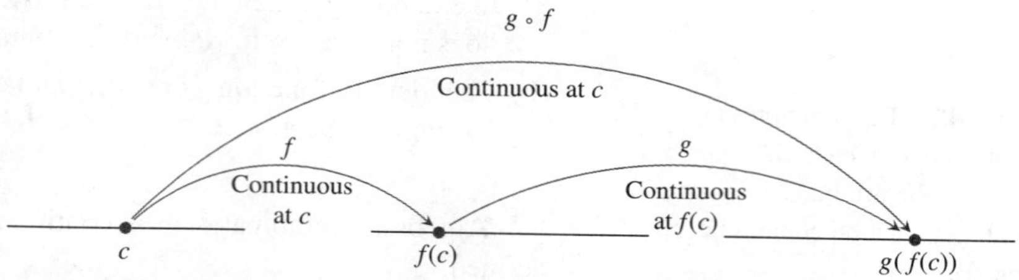

FIGURE 2.43 Compositions of continuous functions are continuous.

> **THEOREM 8—Compositions of Continuous Functions**
> If f is continuous at c, and g is continuous at $f(c)$, then the composition $g \circ f$ is continuous at c.

Intuitively, Theorem 8 is reasonable because if x is close to c, then $f(x)$ is close to $f(c)$, and since g is continuous at $f(c)$, it follows that $g(f(x))$ is close to $g(f(c))$.

The continuity of compositions holds for any finite number of compositions of functions. The only requirement is that each function be continuous where it is applied. An outline of a proof of Theorem 8 is given in Exercise 6 in Appendix A.5.

EXAMPLE 8 Show that the following functions are continuous on their natural domains.

(a) $y = \sqrt{x^2 - 2x - 5}$ **(b)** $y = \dfrac{x^{2/3}}{1 + x^4}$

(c) $y = \left|\dfrac{x - 2}{x^2 - 2}\right|$ **(d)** $y = \left|\dfrac{x \sin x}{x^2 + 2}\right|$

Solution

(a) The square root function is continuous on $[0, \infty)$ because it is a root of the continuous identity function $f(x) = x$ (Part 7, Theorem 7). The given function is then the composition of the polynomial $f(x) = x^2 - 2x - 5$ with the square root function $g(t) = \sqrt{t}$, and is continuous on its natural domain.

(b) The numerator is the cube root of the identity function squared; the denominator is an everywhere-positive polynomial. Therefore, the quotient is continuous.

(c) The quotient $(x - 2)/(x^2 - 2)$ is continuous for all $x \neq \pm\sqrt{2}$, and the function is the composition of this quotient with the continuous absolute value function (Example 7).

(d) Because the sine function is everywhere-continuous (Exercise 68), the numerator term $x \sin x$ is the product of continuous functions, and the denominator term $x^2 + 2$ is an everywhere-positive polynomial. The given function is the composition of a quotient of continuous functions with the continuous absolute value function (Figure 2.44). ■

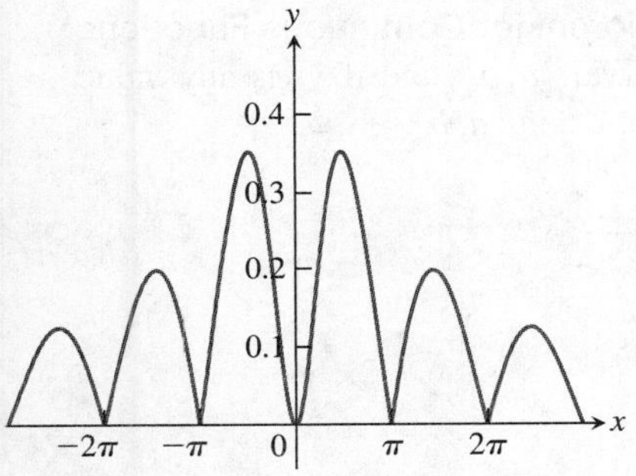

FIGURE 2.44 The graph suggests that $y = |(x \sin x)/(x^2 + 2)|$ is continuous (Example 8d).

Theorem 8 is actually a consequence of a more general result, which we now prove. It states that if the limit of $f(x)$ as x approaches c is equal to b, then the limit of the composition function $g \circ f$ as x approaches c is equal to $g(b)$.

> **THEOREM 9—Limits of Continuous Functions**
> If $\lim_{x \to c} f(x) = b$ and g is continuous at the point b, then
> $$\lim_{x \to c} g(f(x)) = g(b).$$

Proof Let $\varepsilon > 0$ be given. Since g is continuous at b, there exists a number $\delta_1 > 0$ such that

$$|g(y) - g(b)| < \varepsilon \quad \text{whenever} \quad 0 < |y - b| < \delta_1.$$

$\lim_{y \to b} g(y) = g(b)$ since g is continuous at $y = b$.

Note that if $|y - b| = 0$, so that $y = b$, then the inequality $|g(y) - g(b)| < \varepsilon$ holds for any positive ε, and therefore we have

$$|g(y) - g(b)| < \varepsilon \quad \text{whenever} \quad |y - b| < \delta_1. \tag{1}$$

Since $\lim_{x \to c} f(x) = b$, there exists a $\delta > 0$ such that

$$|f(x) - b| < \delta_1 \quad \text{whenever} \quad 0 < |x - c| < \delta.$$

Definition of $\lim_{x \to c} f(x) = b$

If we let $y = f(x)$, we then have that

$$|y - b| < \delta_1 \quad \text{whenever} \quad 0 < |x - c| < \delta,$$

which implies from Equation (1) that $|g(y) - g(b)| = |g(f(x)) - g(b)| < \varepsilon$ whenever $0 < |x - c| < \delta$. From the definition of limit, it follows that $\lim_{x \to c} g(f(x)) = g(b)$. This gives the proof for the case where c is an interior point of the domain of f. The case where c is an endpoint of the domain is entirely similar, using an appropriate one-sided limit in place of a two-sided limit. ■

EXAMPLE 9 Applying Theorem 9, we have

$$\begin{aligned}\lim_{x \to \pi/2} \cos\left(2x + \sin\left(\frac{3\pi}{2} + x\right)\right) &= \cos\left(\lim_{x \to \pi/2} 2x + \lim_{x \to \pi/2} \sin\left(\frac{3\pi}{2} + x\right)\right) \\ &= \cos(\pi + \sin 2\pi) = \cos \pi = -1.\end{aligned}$$ ■

Intermediate Value Theorem for Continuous Functions

A function is said to have the **Intermediate Value Property** if whenever it takes on two values, it also takes on all the values in between.

THEOREM 10—The Intermediate Value Theorem for Continuous Functions
If f is a continuous function on a closed interval $[a, b]$, and if y_0 is any value between $f(a)$ and $f(b)$, then $y_0 = f(c)$ for some c in $[a, b]$.

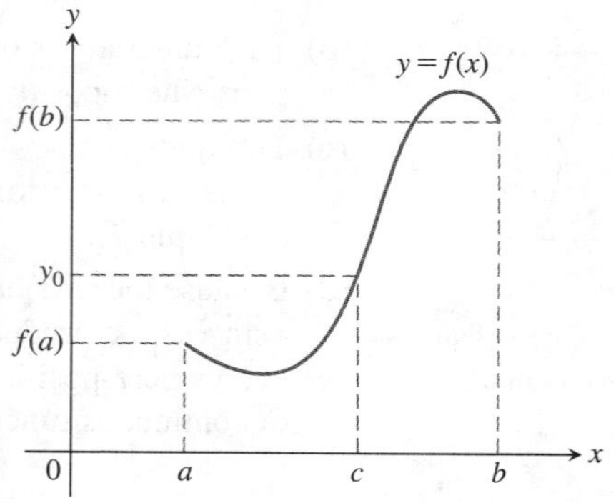

Theorem 10 says that continuous functions over *finite closed* intervals have the Intermediate Value Property. Geometrically, the Intermediate Value Theorem says that any horizontal line $y = y_0$ crossing the y-axis between the numbers $f(a)$ and $f(b)$ will cross the curve $y = f(x)$ at least once over the interval $[a, b]$.

The proof of the Intermediate Value Theorem depends on the completeness property of the real number system. The completeness property implies that the real numbers have no holes or gaps. In contrast, the rational numbers do not satisfy the completeness property, and a function defined only on the rationals would not satisfy the Intermediate Value Theorem. See Appendix A.7 for a discussion and examples.

The continuity of f on the interval is essential to Theorem 10. If f fails to be continuous at even one point of the interval, the theorem's conclusion may fail, as it does for the function graphed in Figure 2.45 (choose y_0 as any number between 2 and 3).

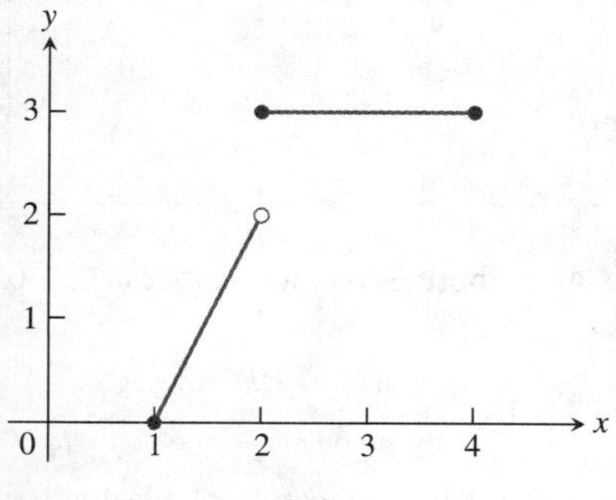

FIGURE 2.45 The function

$$f(x) = \begin{cases} 2x - 2, & 1 \le x < 2 \\ 3, & 2 \le x \le 4 \end{cases}$$

does not take on all values between $f(1) = 0$ and $f(4) = 3$; it misses all the values between 2 and 3.

A Consequence for Graphing: Connectedness Theorem 10 implies that the graph of a function that is continuous on an interval cannot have any breaks over the interval. It will be **connected**—a single, unbroken curve. It will not have jumps such as the ones found in the graph of the greatest integer function (Figure 2.40), or separate branches as found in the graph of $1/x$ (Figure 2.42).

A Consequence for Root Finding We call a solution of the equation $f(x) = 0$ a **root** of the equation or a **zero** of the function f. The Intermediate Value Theorem tells us that if

f is continuous, then any interval on which f changes sign contains a zero of the function. Somewhere between a point where a continuous function is positive and a second point where it is negative, the function must be equal to zero.

In practical terms, when we see the graph of a continuous function cross the horizontal axis on a computer screen, we know it is not stepping across. There really is a point where the function's value is zero.

EXAMPLE 10 Show that there is a root of the equation $x^3 - x - 1 = 0$ between 1 and 2.

Solution Let $f(x) = x^3 - x - 1$. Since $f(1) = 1 - 1 - 1 = -1 < 0$ and $f(2) = 2^3 - 2 - 1 = 5 > 0$, we see that $y_0 = 0$ is a value between $f(1)$ and $f(2)$. Since f is a polynomial, it is continuous, and the Intermediate Value Theorem says there is a zero of f between 1 and 2. Figure 2.46 shows the result of zooming in to locate a root near $x = 1.32$. ■

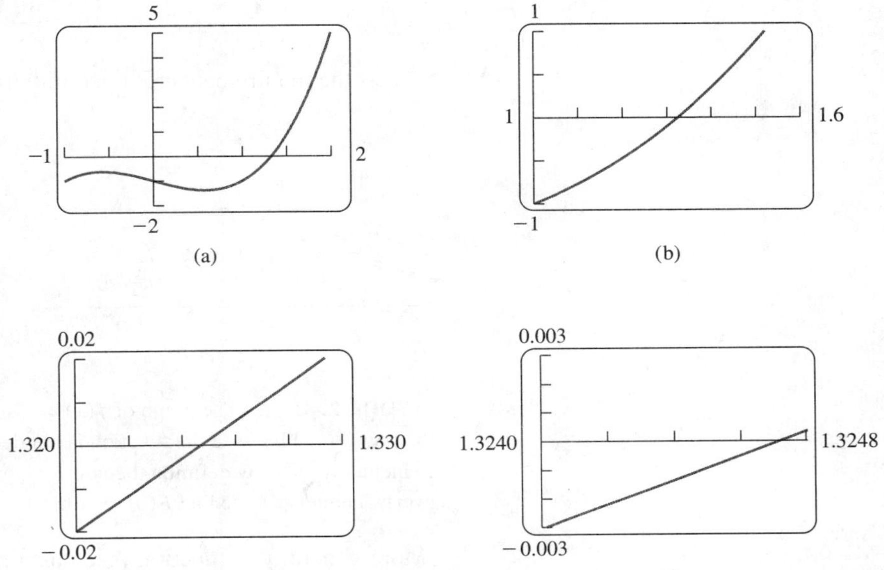

FIGURE 2.46 Zooming in on a zero of the function $f(x) = x^3 - x - 1$. The zero is near $x = 1.3247$ (Example 10).

EXAMPLE 11 Use the Intermediate Value Theorem to prove that the equation

$$\sqrt{2x + 5} = 4 - x^2$$

has a solution (Figure 2.47).

Solution We rewrite the equation as

$$\sqrt{2x + 5} + x^2 - 4 = 0,$$

and set $f(x) = \sqrt{2x + 5} + x^2 - 4$. Now $g(x) = \sqrt{2x + 5}$ is continuous on the interval $[-5/2, \infty)$ since it is formed as the composition of two continuous functions, the square root function with the nonnegative linear function $y = 2x + 5$. Then f is the sum of the function g and the quadratic function $y = x^2 - 4$, and the quadratic function is continuous for all values of x. It follows that $f(x) = \sqrt{2x + 5} + x^2 - 4$ is continuous on the interval $[-5/2, \infty)$. By trial and error, we find the function values $f(0) = \sqrt{5} - 4 \approx -1.76$ and $f(2) = \sqrt{9} = 3$. Note that f is continuous on the finite closed interval $[0, 2]$, which is a subset of the domain $[-5/2, \infty)$. Since the value $y_0 = 0$ is between the numbers $f(0) = -1.76$ and $f(2) = 3$, by the Intermediate Value Theorem there is a number $c \in [0, 2]$ such that $f(c) = 0$. We have found a number c that solves the original equation. ■

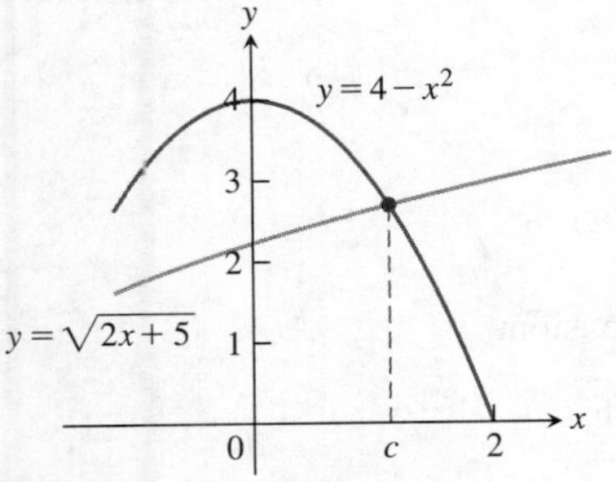

FIGURE 2.47 The curves $y = \sqrt{2x + 5}$ and $y = 4 - x^2$ have the same value at the number $x = c$ where $\sqrt{2x + 5} + x^2 - 4 = 0$ (Example 11).

Continuous Extension to a Point

Sometimes the formula that describes a function f does not make sense at a point $x = c$. It might nevertheless be possible to extend the domain of f to include $x = c$, creating a new function that is continuous at $x = c$. For example, the function $y = f(x) = (\sin x)/x$ is continuous at every point except $x = 0$, since $x = 0$ is not in its domain. Since $y = (\sin x)/x$ has a finite limit as $x \to 0$ (Theorem 6), we can extend the function's domain to include the point $x = 0$ in such a way that the extended function is continuous at $x = 0$. We define the new function

$$F(x) = \begin{cases} \dfrac{\sin x}{x}, & x \neq 0 \quad \text{Same as original function for } x \neq 0 \\ 1, & x = 0. \quad \text{Value at domain point } x = 0 \end{cases}$$

The new function $F(x)$ is continuous at $x = 0$ because

$$\lim_{x \to 0} \frac{\sin x}{x} = F(0),$$

so it meets the requirements for continuity (Figure 2.48).

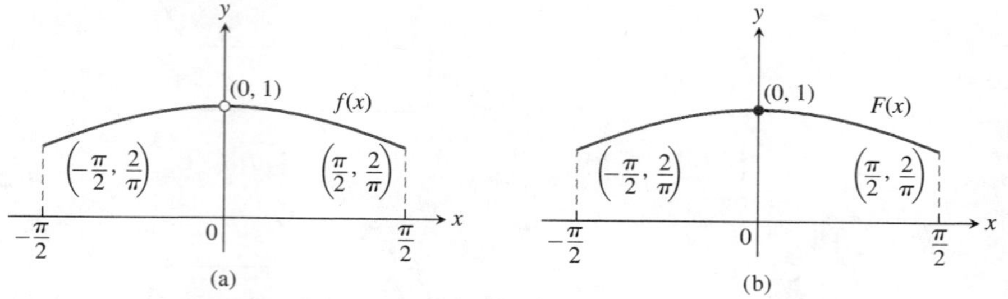

FIGURE 2.48 (a) The graph of $f(x) = (\sin x)/x$ for $-\pi/2 \leq x \leq \pi/2$ does not include the point $(0, 1)$ because the function is not defined at $x = 0$. (b) We can extend the domain to include $x = 0$ by defining the new function $F(x)$ with $F(0) = 1$ and $F(x) = f(x)$ everywhere else. Note that $F(0) = \lim_{x \to 0} f(x)$ and $F(x)$ is a continuous function at $x = 0$.

More generally, a function (such as a rational function) may have a limit at a point where it is not defined. If $f(c)$ is not defined, but $\lim_{x \to c} f(x) = L$ exists, we can define a new function $F(x)$ by the rule

$$F(x) = \begin{cases} f(x), & \text{if } x \text{ is in the domain of } f \\ L, & \text{if } x = c. \end{cases}$$

The function F is continuous at $x = c$. It is called the **continuous extension of** f to $x = c$. For rational functions f, continuous extensions are often found by canceling common factors in the numerator and denominator.

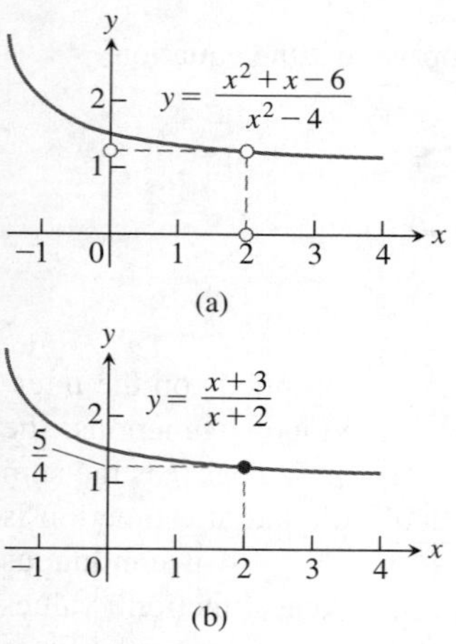

FIGURE 2.49 (a) The graph of $f(x)$ and (b) the graph of its continuous extension $F(x)$ (Example 12).

EXAMPLE 12 Show that

$$f(x) = \frac{x^2 + x - 6}{x^2 - 4}, \quad x \neq 2$$

has a continuous extension to $x = 2$, and find that extension.

Solution Although $f(2)$ is not defined, if $x \neq 2$ we have

$$f(x) = \frac{x^2 + x - 6}{x^2 - 4} = \frac{(x - 2)(x + 3)}{(x - 2)(x + 2)} = \frac{x + 3}{x + 2}.$$

The new function

$$F(x) = \frac{x + 3}{x + 2}$$

is equal to $f(x)$ for $x \neq 2$, but is continuous at $x = 2$, having there the value of 5/4. Thus F is the continuous extension of f to $x = 2$, and

$$\lim_{x \to 2} \frac{x^2 + x - 6}{x^2 - 4} = \lim_{x \to 2} f(x) = \frac{5}{4}.$$

The graph of f is shown in Figure 2.49. The continuous extension F has the same graph except with no hole at $(2, 5/4)$. Effectively, F is the function f extended across the missing domain point at $x = 2$ so as to give a continuous function over the larger domain. ■

EXERCISES 2.5

Continuity from Graphs

In Exercises 1–4, say whether the function graphed is continuous on $[-1, 3]$. If not, where does it fail to be continuous and why?

1.

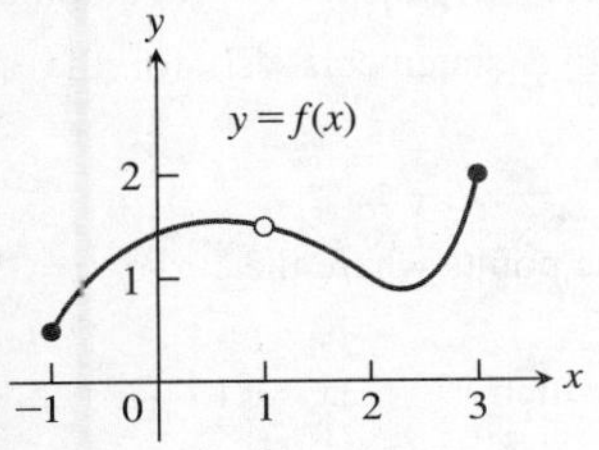

2.

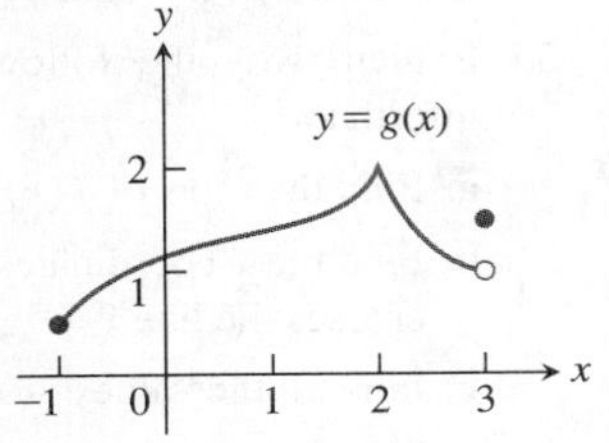

3.

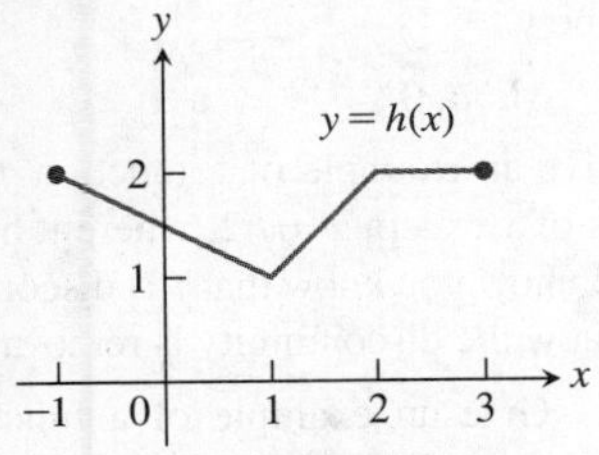

4.

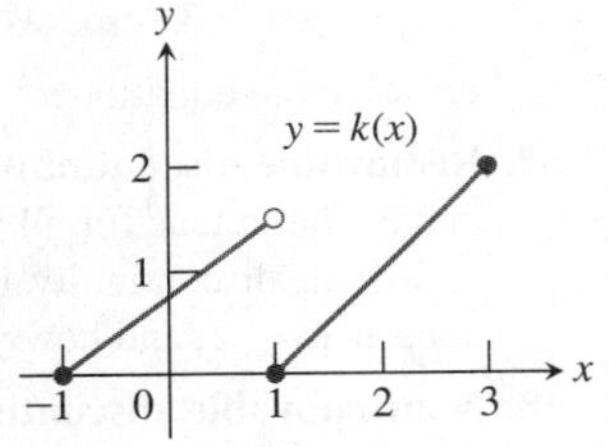

Exercises 5–10 refer to the function

$$f(x) = \begin{cases} x^2 - 1, & -1 \le x < 0 \\ 2x, & 0 < x < 1 \\ 1, & x = 1 \\ -2x + 4, & 1 < x < 2 \\ 0, & 2 < x < 3 \end{cases}$$

graphed in the accompanying figure.

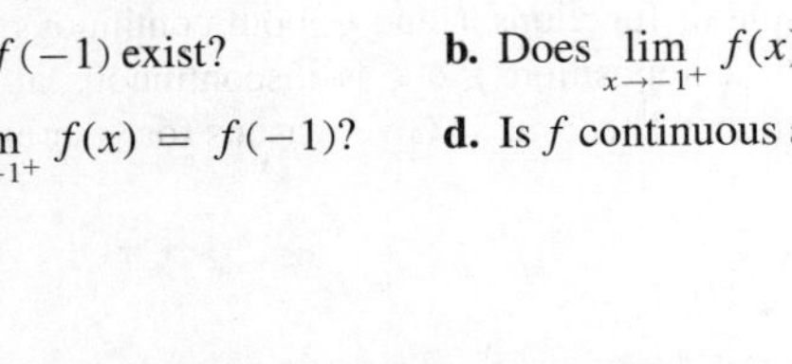

5. a. Does $f(-1)$ exist?

b. Does $\lim_{x \to -1^+} f(x)$ exist?

c. Does $\lim_{x \to -1^+} f(x) = f(-1)$?

d. Is f continuous at $x = -1$?

6. a. Does $f(1)$ exist?

b. Does $\lim_{x \to 1} f(x)$ exist?

c. Does $\lim_{x \to 1} f(x) = f(1)$?

d. Is f continuous at $x = 1$?

7. a. Is f defined at $x = 3$? (Look at the definition of f.)

b. Is f continuous at $x = 3$?

8. At what values of x is f continuous?

9. What value should be assigned to $f(2)$ to make the extended function continuous at $x = 2$?

10. To what new value should $f(1)$ be changed to remove the discontinuity?

Applying the Continuity Test

At which points do the functions in Exercises 11 and 12 fail to be continuous? At which points, if any, are the discontinuities removable? Not removable? Give reasons for your answers.

11. The function defined in Exercise 1, Section 2.4

12. The function defined in Exercise 2, Section 2.4

At what points are the functions in Exercises 13–32 continuous?

13. $y = \dfrac{1}{x-5} - 5x$

14. $y = \dfrac{1}{(x+2)^2} + 4$

15. $y = \dfrac{x+2}{x^2 - 7x + 10}$

16. $y = \dfrac{x+3}{x^2 - 3x - 10}$

17. $y = |2x - 1| + 2\sin x$

18. $y = \dfrac{1}{|x| + 1} - \dfrac{x^2}{2}$

19. $y = \dfrac{\cos x}{5x}$

20. $y = \dfrac{x+2}{\cos x}$

21. $y = \csc 2x$

22. $y = \cot \dfrac{\pi x}{2}$

23. $y = \dfrac{x \tan x}{x^2 + 1}$

24. $y = \dfrac{\sqrt{x^4 + 1}}{1 + \sin^2 x}$

25. $y = \sqrt{5x + 10}$

26. $y = \sqrt[4]{3x - 1}$

27. $y = (7x - 5)^{1/3}$

28. $y = (2 - x)^{1/5}$

29. $g(x) = \begin{cases} \dfrac{x^2 + x - 30}{x - 5}, & x \neq 5 \\ 11, & x = 5 \end{cases}$

30. $f(x) = \begin{cases} \dfrac{x^3 - 8}{x^2 - 4}, & x \neq 2, x \neq -2 \\ 3, & x = 2 \\ 4, & x = -2 \end{cases}$

Limits Involving Trigonometric Functions

Find the limits in Exercises 31–40. Are the functions continuous at the point being approached?

31. $\lim_{x\to 0} 4\sin(x-\sin x)$

32. $\lim_{t\to 0} \sin\left(\frac{\pi}{2}\cos(\tan t)\right)$

33. $\lim_{x\to 0} \frac{\sec x - 1}{x}$

34. $\lim_{x\to 0} \tan\left(\frac{\pi}{4}\cos(\sin x^{1/3})\right)$

35. $\lim_{t\to 0} \frac{\tan x - \sin x}{x^3}$

36. $\lim_{x\to\pi/6} \sqrt{\csc^2 x + 5\sqrt{3}\tan x}$

37. $\lim_{x\to 0} \sec\left[\cos x + \pi\tan\left(\frac{\pi}{4\sec x}\right) - 1\right]$

38. $\lim_{x\to 0} \sin\left(\frac{\pi + \tan x}{\tan x - 2\sec x}\right)$

39. $\lim_{t\to 0} \tan\left(1 - \frac{\sin t}{t}\right)$

40. $\lim_{\theta\to 0} \cos\left(\frac{\pi\theta}{\sin\theta}\right)$

Continuous Extensions

41. Define $g(4)$ in a way that extends $g(x) = (x^2-16)/x-4$ to be continuous at $x = 4$.

42. Define $h(2)$ in a way that extends $h(t) = (t^2+3t-10)/(t-2)$ to be continuous at $t = 2$.

43. Define $f(3)$ in a way that extends $f(s) = s^3 - 27/(s^2-9)$ to be continuous at $s = 3$.

44. Define $g(4)$ in a way that extends

$$g(x) = (x^2-16)/(x^2-3x-4)$$

to be continuous at $x = 4$.

45. For what values of a is

$$f(x) = \begin{cases} 3x^2+5, & x<5 \\ 5ax, & x\ge 5 \end{cases}$$

continuous at every x?

46. For what value of b is

$$g(x) = \begin{cases} x, & x<-2 \\ bx^2, & x\ge -2 \end{cases}$$

continuous at every x?

47. For what values of a is

$$f(x) = \begin{cases} a^2x-2a, & x\ge 2 \\ 12, & x<2 \end{cases}$$

continuous at every x?

48. For what values of m is

$$g(m) = \begin{cases} \frac{2x+m}{m-5}, & x\le 0 \\ 2x^2-5 & x>0 \end{cases}$$

continuous at every x?

49. For what values of a and b is

$$f(x) = \begin{cases} -2, & x\le -1 \\ ax-b, & -1<x<1 \\ 3, & x\ge 1 \end{cases}$$

continuous at every x?

50. For what values of a and b is

$$g(x) = \begin{cases} ax+2b, & x\le 0 \\ x^2+3a-b, & 0<x\le 2 \\ 3x-5, & x>2 \end{cases}$$

continuous at every x?

Theory and Examples

51. A continuous function $y = f(x)$ is known to be negative at $x = 0$ and positive at $x = 1$. Why does the equation $f(x) = 0$ have at least one solution between $x = 0$ and $x = 1$? Illustrate with a sketch.

52. Explain why the equation $\cos x = x$ has at least one solution.

53. Roots of a cubic Show that the equation $x^3 - 15x + 1 = 0$ has three solutions in the interval $[-4, 4]$.

54. A function value Show that the function $F(x) = (x-a)^2 \cdot (x-b)^2 + x$ takes on the value $(a+b)/2$ for some value of x.

55. Solving an equation If $f(x) = x^3 - 8x + 10$, show that there are values c for which $f(c)$ equals **(a)** π; **(b)** $-\sqrt{3}$; **(c)** 5,000,000.

56. Explain why the following five statements ask for the same information.

a. Find the roots of $f(x) = x^3 - 3x - 1$.

b. Find the x-coordinates of the points where the curve $y = x^3$ crosses the line $y = 3x + 1$.

c. Find all the values of x for which $x^3 - 3x = 1$.

d. Find the x-coordinates of the points where the cubic curve $y = x^3 - 3x$ crosses the line $y = 1$.

e. Solve the equation $x^3 - 3x - 1 = 0$.

57. Removable discontinuity Give an example of a function $f(x)$ that is continuous for all values of x except $x = 2$, where it has a removable discontinuity. Explain how you know that f is discontinuous at $x = 2$, and how you know the discontinuity is removable.

58. Nonremovable discontinuity Give an example of a function $g(x)$ that is continuous for all values of x except $x = -1$, where it has a nonremovable discontinuity. Explain how you know that g is discontinuous there and why the discontinuity is not removable.

59. A function discontinuous at every point

a. Use the fact that every nonempty interval of real numbers contains both rational and irrational numbers to show that the function

$$f(x) = \begin{cases} 1, & \text{if } x \text{ is rational} \\ 0, & \text{if } x \text{ is irrational} \end{cases}$$

is discontinuous at every point.

b. Is f right-continuous or left-continuous at any point?

60. If functions $f(x)$ and $g(x)$ are continuous for $0 \le x \le 1$, could $f(x)/g(x)$ possibly be discontinuous at a point of $[0, 1]$? Give reasons for your answer.

61. If the product function $h(x) = f(x)\cdot g(x)$ is continuous at $x = 0$, must $f(x)$ and $g(x)$ be continuous at $x = 0$? Give reasons for your answer.

62. Discontinuous compositions of continuous functions Give an example of functions f and g, both continuous at $x = 0$, for which the composition $f \circ g$ is discontinuous at $x = 0$. Does this contradict Theorem 8? Give reasons for your answer.

63. Never-zero continuous functions Is it true that a continuous function that is never zero on an interval never changes sign on that interval? Give reasons for your answer.

64. Stretching a rubber band Is it true that if you stretch a rubber band by moving one end to the right and the other to the left, some point of the band will end up in its original position? That is, if x is a position on the rubber band before stretching and $f(x)$ is the position of that point after stretching, must there be some x such that $f(x) = x$? Give reasons for your answer.

65. A fixed point theorem Suppose that a function f is continuous on the closed interval $[0, 1]$ and that $0 \le f(x) \le 1$ for every x in $[0, 1]$. Show that there must exist a number c in $[0, 1]$ such that $f(c) = c$ (c is called a **fixed point** of f).

66. The sign-preserving property of continuous functions Let f be defined on an interval (a, b) and suppose that $f(c) \neq 0$ at some c where f is continuous. Show that there is an interval $(c - \delta, c + \delta)$ about c where f has the same sign as $f(c)$.

67. Prove that f is continuous at c if and only if

$$\lim_{h \to 0} f(c + h) = f(c).$$

68. Use Exercise 67 together with the identities

$$\sin(h + c) = \sin h \cos c + \cos h \sin c,$$
$$\cos(h + c) = \cos h \cos c - \sin h \sin c$$

to prove that both $f(x) = \sin x$ and $g(x) = \cos x$ are continuous at every point $x = c$.

Solving Equations Graphically

T Use the Intermediate Value Theorem in Exercises 69–74 to prove that each equation has a solution. Then use a graphing calculator or computer grapher to solve the equations.

69. $x^3 - 3x - 1 = 0$

70. $2x^3 - 2x^2 - 2x + 1 = 0$

71. $x(x - 1)^2 = 1$ (one root)

72. $x^3 - 15x + 1 = 0$ (three roots)

73. $\cos x = x$ (one root). Make sure you are using radian mode.

74. $2 \sin x = x$ (three roots). Make sure you are using radian mode.

2.6 Limits Involving Infinity; Asymptotes of Graphs

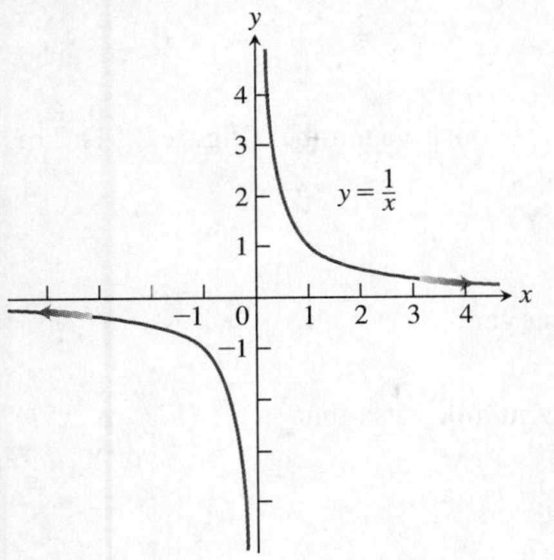

FIGURE 2.50 The graph of $y = 1/x$ approaches 0 as $x \to \infty$ or $x \to -\infty$.

In this section we investigate the behavior of a function when the magnitude of the independent variable x becomes increasingly large, or $x \to \pm\infty$. We further extend the concept of limit to *infinite limits*. Infinite limits provide useful symbols and language for describing the behavior of functions whose values become arbitrarily large in magnitude. We use these ideas to analyze the graphs of functions having *horizontal* or *vertical asymptotes*.

Finite Limits as $x \to \pm\infty$

The symbol for infinity (∞) does not represent a real number. We use ∞ to describe the behavior of a function when the values in its domain or range outgrow all finite bounds. For example, the function $f(x) = 1/x$ is defined for all $x \neq 0$ (Figure 2.50). When x is positive and becomes increasingly large, $1/x$ becomes increasingly small. When x is negative and its magnitude becomes increasingly large, $1/x$ again becomes small. We summarize these observations by saying that $f(x) = 1/x$ has limit 0 as $x \to \infty$ or $x \to -\infty$, or that 0 is a *limit of* $f(x) = 1/x$ *at infinity and at negative infinity*. Here are precise definitions for the limit of a function whose domain contains positive or negative numbers of unbounded magnitude.

DEFINITIONS

1. We say that $f(x)$ has the **limit L as x approaches infinity** and write

$$\lim_{x \to \infty} f(x) = L$$

if, for every number $\varepsilon > 0$, there exists a corresponding number M such that for all x in the domain of f

$$|f(x) - L| < \varepsilon \quad \text{whenever} \quad x > M.$$

2. We say that $f(x)$ has the **limit L as x approaches negative infinity** and write

$$\lim_{x \to -\infty} f(x) = L$$

if, for every number $\varepsilon > 0$, there exists a corresponding number N such that for all x in the domain of f

$$|f(x) - L| < \varepsilon \quad \text{whenever} \quad x < N.$$

Intuitively, $\lim_{x\to\infty} f(x) = L$ if, as x moves increasingly far from the origin in the positive direction, $f(x)$ gets arbitrarily close to L. Similarly, $\lim_{x\to-\infty} f(x) = L$ if, as x moves increasingly far from the origin in the negative direction, $f(x)$ gets arbitrarily close to L.

The strategy for calculating limits of functions as $x \to +\infty$ or as $x \to -\infty$ is similar to the one for finite limits in Section 2.2. There we first found the limits of the constant and identity functions $y = k$ and $y = x$. We then extended these results to other functions by applying Theorem 1 on limits of algebraic combinations. Here we do the same thing, except that the starting functions are $y = k$ and $y = 1/x$ instead of $y = k$ and $y = x$.

The basic facts to be verified by applying the formal definition are

$$\lim_{x\to\pm\infty} k = k \qquad \text{and} \qquad \lim_{x\to\pm\infty} \frac{1}{x} = 0. \tag{1}$$

We prove the second result in Example 1, and leave the first to Exercises 91 and 92.

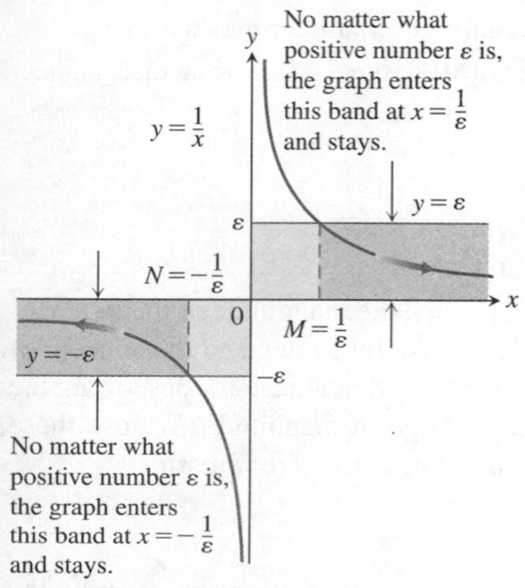

FIGURE 2.51 The geometry behind the argument in Example 1.

EXAMPLE 1 Show that

(a) $\lim_{x\to\infty} \frac{1}{x} = 0$ **(b)** $\lim_{x\to-\infty} \frac{1}{x} = 0.$

Solution

(a) Let $\varepsilon > 0$ be given. We must find a number M such that

$$\left|\frac{1}{x} - 0\right| = \left|\frac{1}{x}\right| < \varepsilon \qquad \text{whenever} \qquad x > M.$$

The implication will hold if $M = 1/\varepsilon$ or any larger positive number (Figure 2.51). This proves $\lim_{x\to\infty} (1/x) = 0$.

(b) Let $\varepsilon > 0$ be given. We must find a number N such that

$$\left|\frac{1}{x} - 0\right| = \left|\frac{1}{x}\right| < \varepsilon \qquad \text{whenever} \qquad x < N.$$

The implication will hold if $N = -1/\varepsilon$ or any number less than $-1/\varepsilon$ (Figure 2.51). This proves $\lim_{x\to-\infty} (1/x) = 0$. ■

Limits at infinity have properties similar to those of finite limits.

> **THEOREM 11**
>
> All the Limit Laws in Theorem 1 are true when we replace $\lim_{x\to c}$ by $\lim_{x\to\infty}$ or $\lim_{x\to-\infty}$.
>
> That is, the variable x may approach a finite number c or $\pm\infty$.

EXAMPLE 2 The properties in Theorem 11 are used to calculate limits in the same way as when x approaches a finite number c.

(a)
$$\begin{aligned}\lim_{x\to\infty}\left(5 + \frac{1}{x}\right) &= \lim_{x\to\infty} 5 + \lim_{x\to\infty}\frac{1}{x} && \text{Sum Rule}\\ &= 5 + 0 = 5 && \text{Known limits}\end{aligned}$$

(b)
$$\begin{aligned}\lim_{x\to-\infty}\frac{\pi\sqrt{3}}{x^2} &= \lim_{x\to-\infty} \pi\sqrt{3}\cdot\frac{1}{x}\cdot\frac{1}{x}\\ &= \lim_{x\to-\infty}\pi\sqrt{3}\cdot\lim_{x\to-\infty}\frac{1}{x}\cdot\lim_{x\to-\infty}\frac{1}{x} && \text{Product Rule}\\ &= \pi\sqrt{3}\cdot 0\cdot 0 = 0 && \text{Known limits}\end{aligned}$$

■

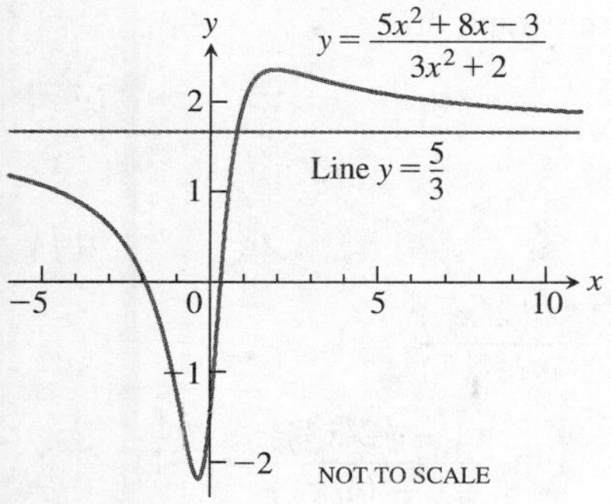

FIGURE 2.52 The graph of the function in Example 3a. The graph approaches the line $y = 5/3$ as $|x|$ increases.

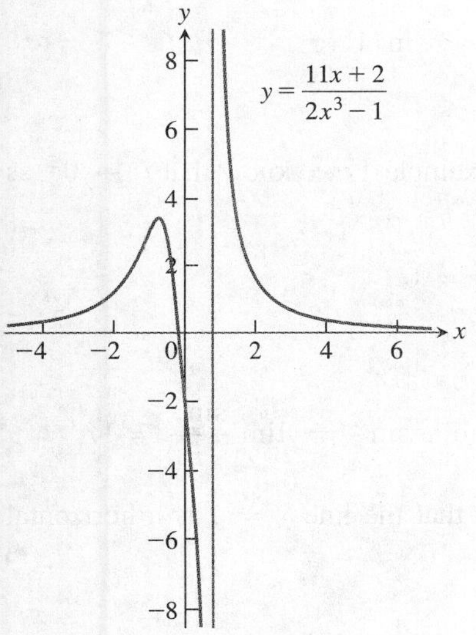

FIGURE 2.53 The graph of the function in Example 3b. The graph approaches the x-axis as $|x|$ increases.

Limits at Infinity of Rational Functions

To determine the limit of a rational function as $x \to \pm\infty$, we first divide the numerator and denominator by the highest power of x in the denominator. The result then depends on the degrees of the polynomials involved.

EXAMPLE 3 These examples illustrate what happens when the degree of the numerator is less than or equal to the degree of the denominator.

(a) $$\lim_{x\to\infty} \frac{5x^2 + 8x - 3}{3x^2 + 2} = \lim_{x\to\infty} \frac{5 + (8/x) - (3/x^2)}{3 + (2/x^2)}$$ Divide numerator and denominator by x^2.

$$= \frac{5 + 0 - 0}{3 + 0} = \frac{5}{3}$$ See Fig. 2.52.

(b) $$\lim_{x\to-\infty} \frac{11x + 2}{2x^3 - 1} = \lim_{x\to-\infty} \frac{(11/x^2) + (2/x^3)}{2 - (1/x^3)}$$ Divide numerator and denominator by x^3.

$$= \frac{0 + 0}{2 - 0} = 0$$ See Fig. 2.53. ■

Cases for which the degree of the numerator is greater than the degree of the denominator are illustrated in Examples 9 and 13.

Horizontal Asymptotes

If the distance between the graph of a function and some fixed line approaches zero as a point on the graph moves increasingly far from the origin, we say that the graph approaches the line asymptotically and that the line is an *asymptote* of the graph.

Looking at $f(x) = 1/x$ (see Figure 2.50), we observe that the x-axis is an asymptote of the curve on the right because

$$\lim_{x\to\infty} \frac{1}{x} = 0$$

and on the left because

$$\lim_{x\to-\infty} \frac{1}{x} = 0.$$

We say that the x-axis is a *horizontal asymptote* of the graph of $f(x) = 1/x$.

> **DEFINITION** A line $y = b$ is a **horizontal asymptote** of the graph of a function $y = f(x)$ if either
>
> $$\lim_{x\to\infty} f(x) = b \quad \text{or} \quad \lim_{x\to-\infty} f(x) = b.$$

The graph of a function can have zero, one, or two horizontal asymptotes, depending on whether the function has limits as $x \to \infty$ and as $x \to -\infty$.

The graph of the function

$$f(x) = \frac{5x^2 + 8x - 3}{3x^2 + 2}$$

sketched in Figure 2.52 (Example 3a) has the line $y = 5/3$ as a horizontal asymptote on both the right and the left because

$$\lim_{x\to\infty} f(x) = \frac{5}{3} \quad \text{and} \quad \lim_{x\to-\infty} f(x) = \frac{5}{3}.$$

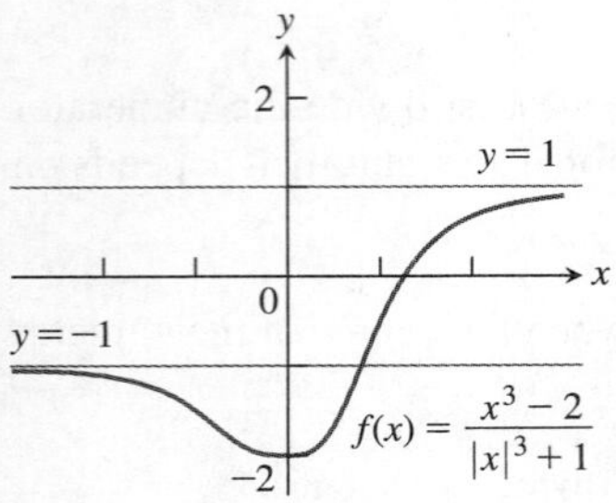

FIGURE 2.54 The graph of the function in Example 4 has two horizontal asymptotes.

EXAMPLE 4 Find the horizontal asymptotes of the graph of

$$f(x) = \frac{x^3 - 2}{|x|^3 + 1}.$$

Solution We calculate the limits as $x \to \pm\infty$.

$$\text{For } x > 0: \quad \lim_{x\to\infty} \frac{x^3-2}{|x|^3+1} = \lim_{x\to\infty} \frac{x^3-2}{x^3+1} = \lim_{x\to\infty} \frac{1-(2/x^3)}{1+(1/x^3)} = 1.$$

$$\text{For } x < 0: \quad \lim_{x\to-\infty} \frac{x^3-2}{|x|^3+1} = \lim_{x\to-\infty} \frac{x^3-2}{(-x)^3+1} = \lim_{x\to-\infty} \frac{1-(2/x^3)}{-1+(1/x^3)} = -1.$$

The horizontal asymptotes are $y = -1$ and $y = 1$. The graph is displayed in Figure 2.54. Notice that the graph crosses the horizontal asymptote $y = -1$ for a positive value of x. ■

Sometimes it is helpful to transform a limit in which x approaches ∞ to a new limit by setting $t = 1/x$ and seeing what happens as t approaches 0.

EXAMPLE 5 Find **(a)** $\lim_{x\to\infty} \sin(1/x)$ and **(b)** $\lim_{x\to\pm\infty} x\sin(1/x)$.

Solution

(a) We introduce the new variable $t = 1/x$. From Example 1, we know that $t \to 0^+$ as $x \to \infty$ (see Figure 2.50). Therefore,

$$\lim_{x\to\infty} \sin\frac{1}{x} = \lim_{t\to0^+} \sin t = 0.$$

(b) We calculate the limits as $x \to \infty$ and $x \to -\infty$:

$$\lim_{x\to\infty} x\sin\frac{1}{x} = \lim_{t\to0^+} \frac{\sin t}{t} = 1 \quad \text{and} \quad \lim_{x\to-\infty} x\sin\frac{1}{x} = \lim_{t\to0^-} \frac{\sin t}{t} = 1.$$

The graph is shown in Figure 2.55, and we see that the line $y = 1$ is a horizontal asymptote. ■

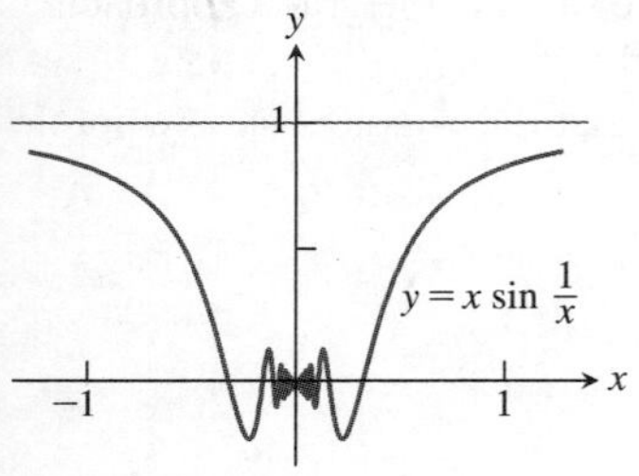

FIGURE 2.55 The line $y = 1$ is a horizontal asymptote of the function graphed here (Example 5b).

Similarly, we can investigate the behavior of $y = f(1/x)$ as $x \to 0$ by investigating $y = f(t)$ as $t \to \pm\infty$, where $t = 1/x$.

EXAMPLE 6 Find $\lim_{x\to0^+} x\left\lfloor \frac{1}{x} \right\rfloor$.

Solution We let $t = 1/x$ so that

$$\lim_{x\to0^+} x\left\lfloor \frac{1}{x} \right\rfloor = \lim_{t\to\infty} \frac{1}{t}\lfloor t \rfloor$$

From the graph in Figure 2.56, we see that $t - 1 \le \lfloor t \rfloor \le t$, which gives

$$1 - \frac{1}{t} \le \frac{1}{t}\lfloor t \rfloor \le 1 \qquad \text{Multiply inequalities by } \tfrac{1}{t} > 0.$$

It follows from the Sandwich Theorem that

$$\lim_{t\to\infty} \frac{1}{t}\lfloor t \rfloor = 1.$$

so 1 is the value of the limit we seek. ■

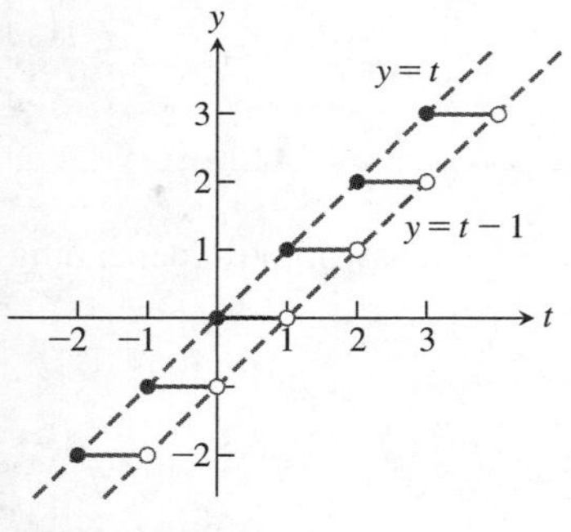

FIGURE 2.56 The graph of the greatest integer function $y = \lfloor t \rfloor$ is sandwiched between $y = t - 1$ and $y = t$.

The Sandwich Theorem also holds for limits as $x \to \pm\infty$. You must be sure, though, that the function whose limit you are trying to find stays between the bounding functions for all very large x in the positive direction (if $x \to \infty$) or all very large x in the negative direction (if $x \to -\infty$).

EXAMPLE 7 Using the Sandwich Theorem, find the horizontal asymptote of the curve

$$y = 2 + \frac{\sin x}{x}.$$

Solution We are interested in the behavior as $x \to \pm\infty$. Since

$$0 \le \left|\frac{\sin x}{x}\right| \le \left|\frac{1}{x}\right|$$

and $\lim_{x\to\pm\infty} |1/x| = 0$, we have $\lim_{x\to\pm\infty} (\sin x)/x = 0$ by the Sandwich Theorem. Hence,

$$\lim_{x\to\pm\infty}\left(2 + \frac{\sin x}{x}\right) = 2 + 0 = 2,$$

and the line $y = 2$ is a horizontal asymptote of the curve on both left and right (Figure 2.57).

This example illustrates that a curve may cross one of its horizontal asymptotes many times. ■

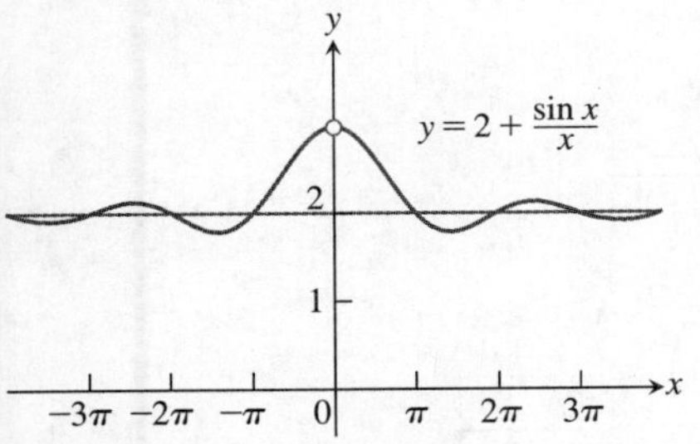

FIGURE 2.57 A curve may cross one of its asymptotes infinitely often (Example 7).

EXAMPLE 8 Find $\lim_{x\to\infty}\left(x - \sqrt{x^2+16}\right)$.

Solution Both of the terms x and $\sqrt{x^2+16}$ approach infinity as $x \to \infty$, so what happens to the difference in the limit is unclear (we cannot subtract ∞ from ∞ because the symbol does not represent a real number). In this situation we can multiply the numerator and the denominator by the conjugate radical expression to obtain an equivalent algebraic expression:

$$\lim_{x\to\infty}\left(x - \sqrt{x^2+16}\right) = \lim_{x\to\infty}\left(x - \sqrt{x^2+16}\right)\frac{x+\sqrt{x^2+16}}{x+\sqrt{x^2+16}}$$

Multiply and divide by the conjugate.

$$= \lim_{x\to\infty}\frac{x^2-(x^2+16)}{x+\sqrt{x^2+16}} = \lim_{x\to\infty}\frac{-16}{x+\sqrt{x^2+16}}.$$

As $x \to \infty$, the denominator in this last expression becomes arbitrarily large, while the numerator remains constant, so we see that the limit is 0. We can also obtain this result by a direct calculation using the Limit Laws:

$$\lim_{x\to\infty}\frac{-16}{x+\sqrt{x^2+16}} = \lim_{x\to\infty}\frac{-\dfrac{16}{x}}{1+\sqrt{\dfrac{x^2}{x^2}+\dfrac{16}{x^2}}} = \frac{0}{1+\sqrt{1+0}} = 0.$$ ■

Oblique Asymptotes

If the degree of the numerator of a rational function is 1 greater than the degree of the denominator, the graph has an **oblique** or **slant line asymptote**. We find an equation for the asymptote by dividing numerator by denominator to express f as a linear function plus a remainder that goes to zero as $x \to \pm\infty$.

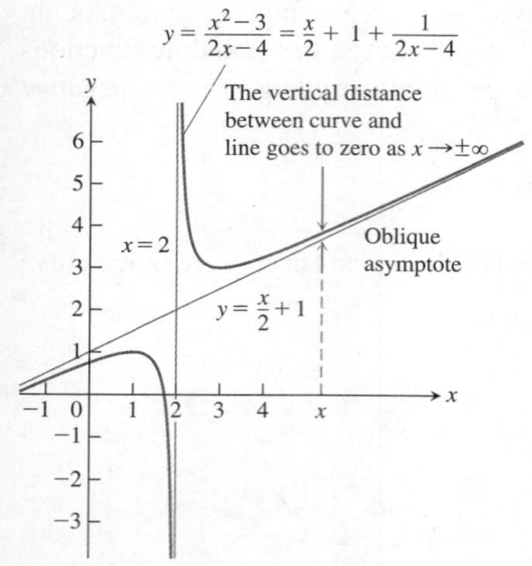

FIGURE 2.58 The graph of the function in Example 9 has an oblique asymptote.

EXAMPLE 9 Find the oblique asymptote of the graph of

$$f(x) = \frac{x^2 - 3}{2x - 4}$$

in Figure 2.58.

Solution We are interested in the behavior as $x \to \pm\infty$. We divide $(2x - 4)$ into $(x^2 - 3)$:

$$\begin{array}{r} \frac{x}{2} + 1 \\ 2x - 4 \overline{\big)\, x^2 + 0x - 3} \\ \underline{x^2 - 2x} \\ 2x - 3 \\ \underline{2x - 4} \\ 1 \end{array}$$

This tells us that

$$f(x) = \frac{x^2 - 3}{2x - 4} = \underbrace{\left(\frac{x}{2} + 1\right)}_{\text{linear } g(x)} + \underbrace{\left(\frac{1}{2x - 4}\right)}_{\text{remainder}}$$

As $x \to \pm\infty$, the remainder, whose magnitude gives the vertical distance between the graphs of f and g, goes to zero, making the slanted line

$$g(x) = \frac{x}{2} + 1$$

an asymptote of the graph of f (Figure 2.58). The line $y = g(x)$ is an asymptote both to the right and to the left. ■

Notice in Example 9 that if the degree of the numerator in a rational function is greater than the degree of the denominator, then the limit as $|x|$ becomes large is $+\infty$ or $-\infty$, depending on the signs assumed by the numerator and denominator.

Infinite Limits

Let us look again at the function $f(x) = 1/x$. As $x \to 0^+$, the values of f grow without bound, eventually reaching and surpassing every positive real number. That is, given any positive real number B, however large, the values of f become larger still (Figure 2.59).

Thus, f has no limit as $x \to 0^+$. It is nevertheless convenient to describe the behavior of f by saying that $f(x)$ approaches ∞ as $x \to 0^+$. We write

$$\lim_{x \to 0^+} f(x) = \lim_{x \to 0^+} \frac{1}{x} = \infty.$$

In writing this equation, we are *not* saying that the limit exists. Nor are we saying that there is a real number ∞, for there is no such number. Rather, this expression is just a concise way of saying that $\lim_{x \to 0^+} (1/x)$ *does not exist because* $1/x$ *becomes arbitrarily large and positive as* $x \to 0^+$.

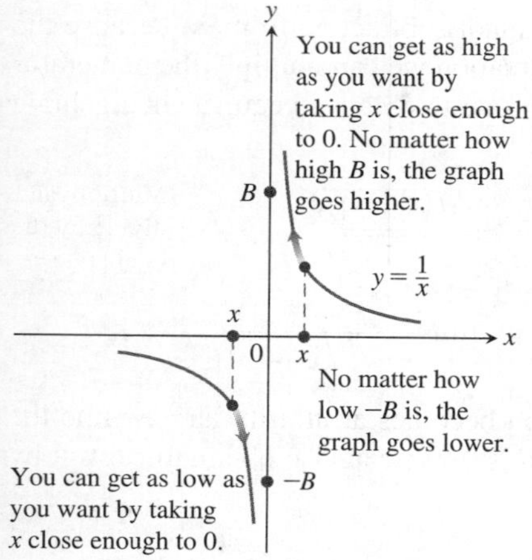

FIGURE 2.59 One-sided infinite limits:

$$\lim_{x \to 0^+} \frac{1}{x} = \infty \quad \text{and} \quad \lim_{x \to 0^-} \frac{1}{x} = -\infty.$$

As $x \to 0^-$, the values of $f(x) = 1/x$ become arbitrarily large and negative. Given any negative real number $-B$, the values of f eventually lie below $-B$. (See Figure 2.59.) We write

$$\lim_{x \to 0^-} f(x) = \lim_{x \to 0^-} \frac{1}{x} = -\infty.$$

Again, we are not saying that the limit exists and equals the number $-\infty$. There *is* no real number $-\infty$. We are describing the behavior of a function whose limit as $x \to 0^-$ *does not exist because its values become arbitrarily large and negative.*

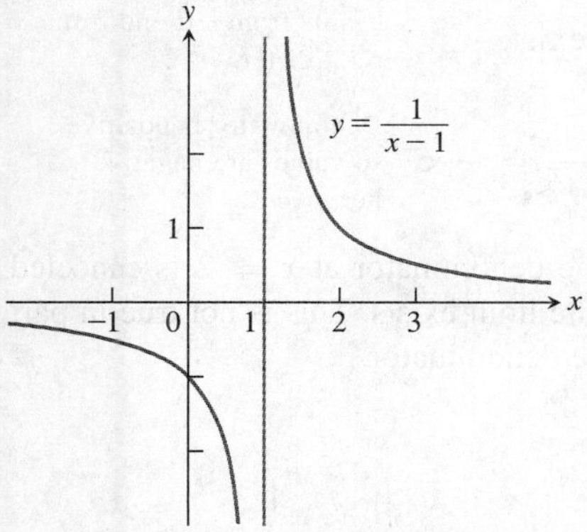

FIGURE 2.60 Near $x = 1$, the function $y = 1/(x-1)$ behaves the way the function $y = 1/x$ behaves near $x = 0$. Its graph is the graph of $y = 1/x$ shifted 1 unit to the right (Example 10).

EXAMPLE 10 Find $\lim_{x \to 1^+} \frac{1}{x-1}$ and $\lim_{x \to 1^-} \frac{1}{x-1}$.

Geometric Solution The graph of $y = 1/(x-1)$ is the graph of $y = 1/x$ shifted 1 unit to the right (Figure 2.60). Therefore, $y = 1/(x-1)$ behaves near 1 exactly the way $y = 1/x$ behaves near 0:

$$\lim_{x \to 1^+} \frac{1}{x-1} = \infty \quad \text{and} \quad \lim_{x \to 1^-} \frac{1}{x-1} = -\infty.$$

Analytic Solution Think about the number $x - 1$ and its reciprocal. As $x \to 1^+$, we have $(x-1) \to 0^+$ and $1/(x-1) \to \infty$. As $x \to 1^-$, we have $(x-1) \to 0^-$ and $1/(x-1) \to -\infty$. ■

EXAMPLE 11 Discuss the behavior of

$$f(x) = \frac{1}{x^2} \quad \text{as} \quad x \to 0.$$

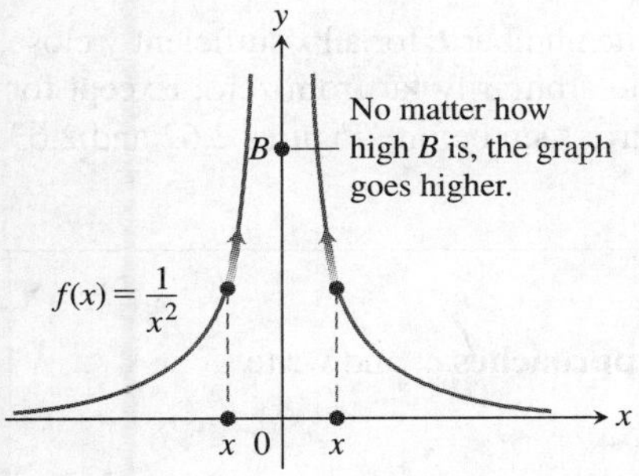

FIGURE 2.61 The graph of $f(x)$ in Example 11 approaches infinity as $x \to 0$.

Solution As x approaches zero from either side, the values of $1/x^2$ are positive and become arbitrarily large (Figure 2.61). This means that

$$\lim_{x \to 0} f(x) = \lim_{x \to 0} \frac{1}{x^2} = \infty.$$

The function $y = 1/x$ shows no consistent behavior as $x \to 0$. We have $1/x \to \infty$ if $x \to 0^+$, but $1/x \to -\infty$ if $x \to 0^-$. All we can say about $\lim_{x \to 0} (1/x)$ is that it does not exist. The function $y = 1/x^2$ is different. Its values approach infinity as x approaches zero from either side, so we can say that $\lim_{x \to 0} (1/x^2) = \infty$. ■

EXAMPLE 12 These examples illustrate that rational functions can behave in various ways near zeros of the denominator.

(a) $\lim_{x \to 2} \frac{(x-2)^2}{x^2-4} = \lim_{x \to 2} \frac{(x-2)^2}{(x-2)(x+2)} = \lim_{x \to 2} \frac{x-2}{x+2} = 0$ Can substitute 2 for x after algebraic manipulation eliminates division by 0.

(b) $\lim_{x \to 2} \frac{x-2}{x^2-4} = \lim_{x \to 2} \frac{x-2}{(x-2)(x+2)} = \lim_{x \to 2} \frac{1}{x+2} = \frac{1}{4}$ Again substitute 2 for x after algebraic manipulation eliminates division by 0.

(c) $\lim_{x \to 2^+} \frac{x-3}{x^2-4} = \lim_{x \to 2^+} \frac{x-3}{(x-2)(x+2)} = -\infty$ The values are negative for $x > 2$, x near 2.

(d) $\lim_{x \to 2^-} \frac{x-3}{x^2-4} = \lim_{x \to 2^-} \frac{x-3}{(x-2)(x+2)} = \infty$ The values are positive for $x < 2$, x near 2.

(e) $\lim_{x\to 2}\frac{x-3}{x^2-4}=\lim_{x\to 2}\frac{x-3}{(x-2)(x+2)}$ does not exist. — Limits from left and from right differ.

(f) $\lim_{x\to 2}\frac{2-x}{(x-2)^3}=\lim_{x\to 2}\frac{-(x-2)}{(x-2)^3}=\lim_{x\to 2}\frac{-1}{(x-2)^2}=-\infty$ — Denominator is positive, so values are negative near $x=2$.

In parts (a) and (b), the effect of the zero in the denominator at $x = 2$ is canceled because the numerator is zero there also. Thus a finite limit exists. This is not true in part (f), where cancellation still leaves a zero factor in the denominator. ■

EXAMPLE 13 Find $\lim_{x\to-\infty}\frac{2x^5-6x^4+1}{3x^2+x-7}$.

Solution We are asked to find the limit of a rational function as $x \to -\infty$, so we divide the numerator and denominator by x^2, the highest power of x in the denominator:

$$\begin{aligned}\lim_{x\to-\infty}\frac{2x^5-6x^4+1}{3x^2+x-7} &= \lim_{x\to-\infty}\frac{2x^3-6x^2+x^{-2}}{3+x^{-1}-7x^{-2}}\\ &= \lim_{x\to-\infty}\frac{2x^2(x-3)+x^{-2}}{3+x^{-1}-7x^{-2}}\\ &= -\infty, \qquad x^{-n}\to 0,\ x-3\to-\infty\end{aligned}$$

because the numerator tends to $-\infty$ while the denominator approaches 3 as $x \to -\infty$. ■

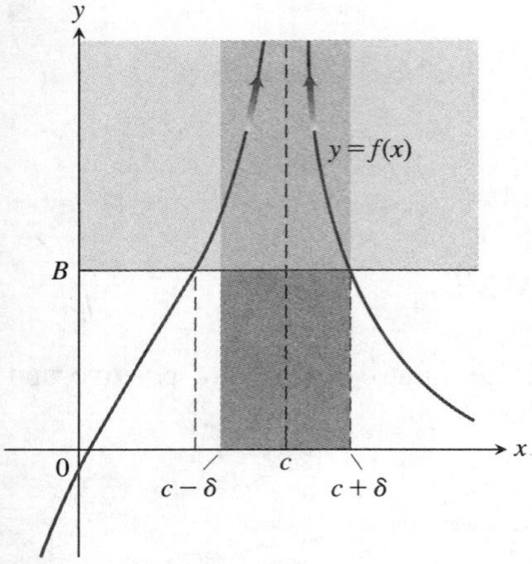

FIGURE 2.62 For $c-\delta<x<c+\delta$, the graph of $f(x)$ lies above the line $y = B$.

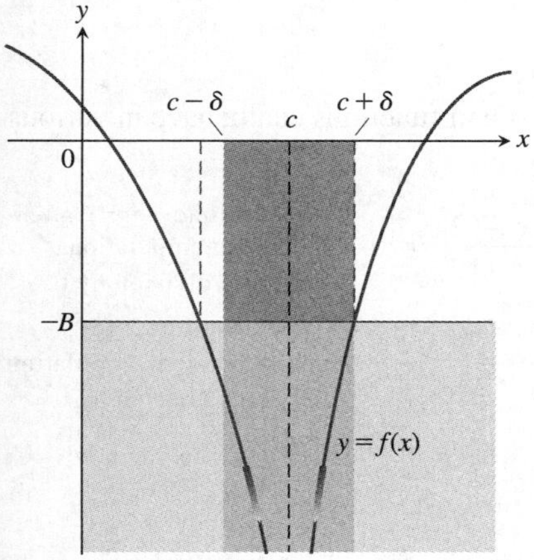

FIGURE 2.63 For $c-\delta<x<c+\delta$, the graph of $f(x)$ lies below the line $y = -B$.

Precise Definitions of Infinite Limits

Instead of requiring $f(x)$ to lie arbitrarily close to a finite number L for all x sufficiently close to c, the definitions of infinite limits require $f(x)$ to lie arbitrarily far from zero. Except for this change, the language is very similar to what we have seen before. Figures 2.62 and 2.63 accompany these definitions.

DEFINITIONS

1. We say that **$f(x)$ approaches infinity as x approaches c**, and write
$$\lim_{x\to c} f(x) = \infty,$$
if for every positive real number B there exists a corresponding $\delta > 0$ such that
$$f(x) > B \quad \text{whenever} \quad 0 < |x-c| < \delta.$$

2. We say that **$f(x)$ approaches negative infinity as x approaches c**, and write
$$\lim_{x\to c} f(x) = -\infty,$$
if for every negative real number $-B$ there exists a corresponding $\delta > 0$ such that
$$f(x) < -B \quad \text{whenever} \quad 0 < |x-c| < \delta.$$

The precise definitions of one-sided infinite limits at c are similar and are stated in the exercises.

EXAMPLE 14 Prove that $\lim_{x\to 0}\frac{1}{x^2}=\infty$.

Solution Given $B > 0$, we want to find $\delta > 0$ such that
$$\frac{1}{x^2} > B \quad \text{whenever} \quad 0 < |x-0| < \delta.$$

Now,

$$\frac{1}{x^2} > B \quad \text{if and only if} \quad x^2 < \frac{1}{B}$$

or, equivalently,

$$|x| < \frac{1}{\sqrt{B}}.$$

Thus, choosing $\delta = 1/\sqrt{B}$ (or any smaller positive number), we see that

$$\text{if } 0 < |x| < \delta \quad \text{then} \quad \frac{1}{x^2} > \frac{1}{\delta^2} \geq B.$$

Therefore, by definition,

$$\lim_{x\to 0} \frac{1}{x^2} = \infty. \quad \blacksquare$$

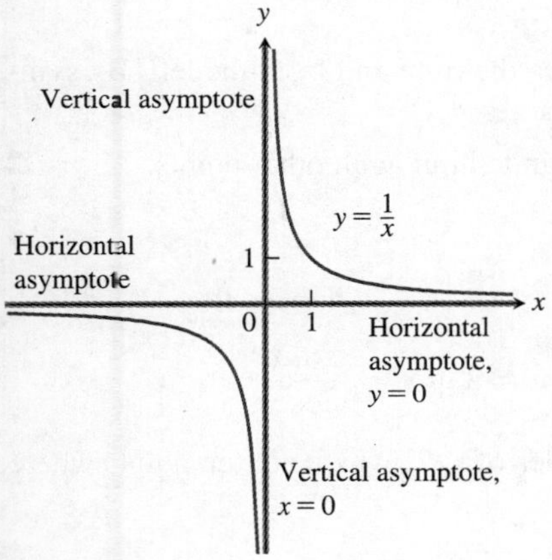

FIGURE 2.64 The coordinate axes are asymptotes of both branches of the hyperbola $y = 1/x$.

Vertical Asymptotes

Notice that the distance between a point on the graph of $f(x) = 1/x$ and the y-axis approaches zero as the point moves nearly vertically along the graph and away from the origin (Figure 2.64). The function $f(x) = 1/x$ is unbounded as x approaches 0 because

$$\lim_{x\to 0^+} \frac{1}{x} = \infty \quad \text{and} \quad \lim_{x\to 0^-} \frac{1}{x} = -\infty.$$

We say that the line $x = 0$ (the y-axis) is a *vertical asymptote* of the graph of $f(x) = 1/x$. Observe that the denominator is zero at $x = 0$ and the function is undefined there.

> **DEFINITION** A line $x = a$ is a **vertical asymptote** of the graph of a function $y = f(x)$ if either
>
> $$\lim_{x\to a^+} f(x) = \pm\infty \quad \text{or} \quad \lim_{x\to a^-} f(x) = \pm\infty.$$

EXAMPLE 15 Find the horizontal and vertical asymptotes of the curve

$$y = \frac{x+3}{x+2}.$$

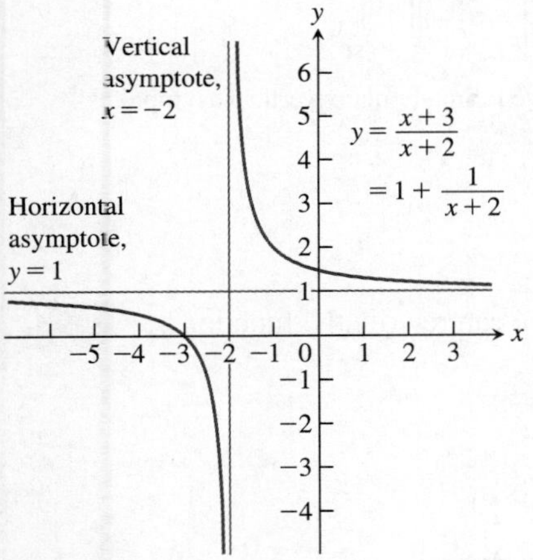

FIGURE 2.65 The lines $y = 1$ and $x = -2$ are asymptotes of the curve in Example 15.

Solution We are interested in the behavior as $x \to \pm\infty$ and the behavior as $x \to -2$, where the denominator is zero.

The asymptotes are revealed if we recast the rational function as a polynomial with a remainder, by dividing $(x + 2)$ into $(x + 3)$:

$$\begin{array}{r} 1 \\ x+2\overline{)\,x+3} \\ \underline{x+2} \\ 1 \end{array}$$

This result enables us to rewrite y as

$$y = 1 + \frac{1}{x+2}.$$

We see that the curve in question is the graph of $f(x) = 1/x$ shifted 1 unit up and 2 units left (Figure 2.65). The asymptotes, instead of being the coordinate axes, are now the lines $y = 1$ and $x = -2$. As $x \to \pm\infty$, the curve approaches the horizontal asymptote $y = 1$; as $x \to -2$, the curve approaches the vertical asymptote $x = -2$. $\blacksquare$

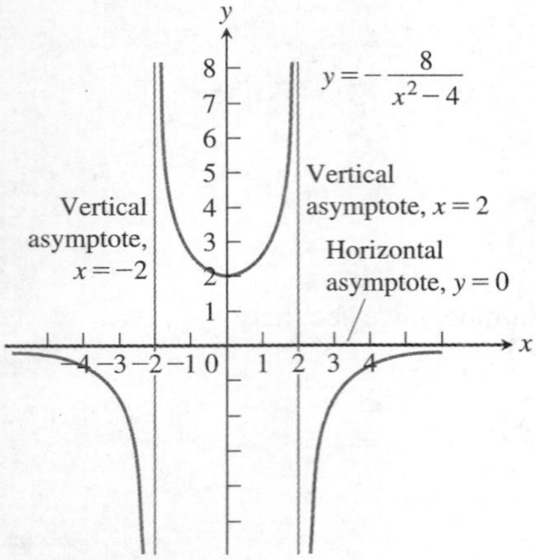

FIGURE 2.66 Graph of the function in Example 16. Notice that the curve approaches the x-axis from only one side. Asymptotes do not have to be two-sided.

EXAMPLE 16 Find the horizontal and vertical asymptotes of the graph of

$$f(x) = -\frac{8}{x^2 - 4}.$$

Solution We are interested in the behavior as $x \to \pm\infty$ and as $x \to \pm 2$, where the denominator is zero. Notice that f is an even function of x, so its graph is symmetric with respect to the y-axis.

(a) *The behavior as $x \to \pm\infty$.* Since $\lim_{x\to\infty} f(x) = 0$, the line $y = 0$ is a horizontal asymptote of the graph to the right. By symmetry it is an asymptote to the left as well (Figure 2.66). Notice that the curve approaches the x-axis from only the negative side (or from below). Also, $f(0) = 2$.

(b) *The behavior as $x \to \pm 2$.* Since

$$\lim_{x\to 2^+} f(x) = -\infty \quad \text{and} \quad \lim_{x\to 2^-} f(x) = \infty,$$

the line $x = 2$ is a vertical asymptote both from the right and from the left. By symmetry, the line $x = -2$ is also a vertical asymptote.

There are no other asymptotes because f has a finite limit at all other points. ■

EXAMPLE 17 The curves

$$y = \sec x = \frac{1}{\cos x} \quad \text{and} \quad y = \tan x = \frac{\sin x}{\cos x}$$

both have vertical asymptotes at odd-integer multiples of $\pi/2$, which are the points where $\cos x = 0$ (Figure 2.67).

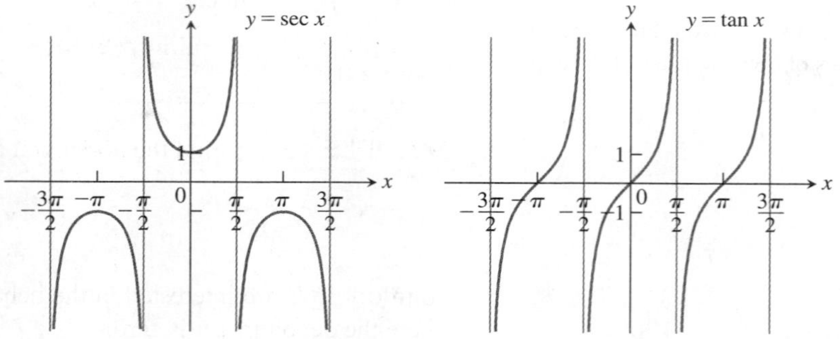

FIGURE 2.67 The graphs of $\sec x$ and $\tan x$ have infinitely many vertical asymptotes (Example 17). ■

Dominant Terms

In Example 9 we saw that by using long division, we can rewrite the function

$$f(x) = \frac{x^2 - 3}{2x - 4}$$

as a linear function plus a remainder term:

$$f(x) = \left(\frac{x}{2} + 1\right) + \left(\frac{1}{2x - 4}\right).$$

This tells us immediately that

For x large: $\quad f(x) \approx \dfrac{x}{2} + 1 \qquad \dfrac{1}{2x-4}$ is near 0.

For x near 2: $\quad f(x) \approx \dfrac{1}{2x-4} \qquad$ This term is very large in absolute value.

If we want to know how f behaves, this is one possible way to find out. It behaves like $y = (x/2) + 1$ when $|x|$ is large and the contribution of $1/(2x - 4)$ to the total value of f is insignificant. It behaves like $1/(2x - 4)$ when x is so close to 2 that $1/(2x - 4)$ makes the dominant contribution.

We say that $(x/2) + 1$ **dominates** when x approaches ∞ or $-\infty$, and we say that $1/(2x - 4)$ dominates when x approaches 2. **Dominant terms** like these help us predict a function's behavior.

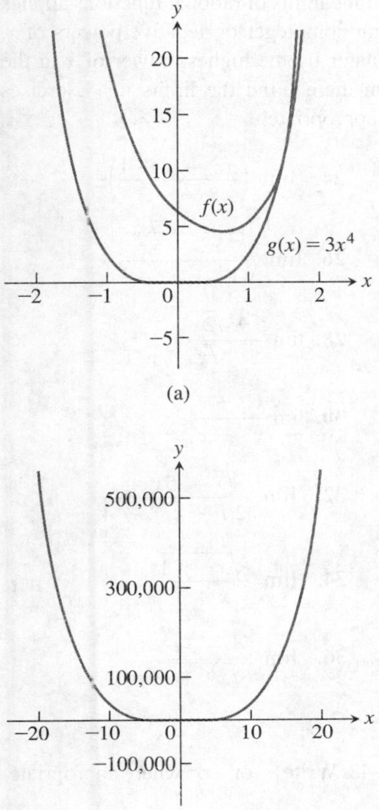

FIGURE 2.68 The graphs of f and g are (a) distinct for $|x|$ small, and (b) nearly identical for $|x|$ large (Example 18).

EXAMPLE 18 Let $f(x) = 3x^4 - 2x^3 + 3x^2 - 5x + 6$ and $g(x) = 3x^4$. Show that although f and g are quite different for numerically small values of x, they behave similarly for very large $|x|$, in the sense that their ratios approach 1 as $x \to \infty$ or $x \to -\infty$.

Solution The graphs of f and g behave quite differently near the origin (Figure 2.68a), but appear as virtually identical on a larger scale (Figure 2.68b).

We can test that the term $3x^4$ in f, represented graphically by g, dominates the polynomial f for numerically large values of x by examining the ratio of the two functions as $x \to \pm\infty$. We find that

$$\begin{aligned}\lim_{x\to\pm\infty} \frac{f(x)}{g(x)} &= \lim_{x\to\pm\infty} \frac{3x^4 - 2x^3 + 3x^2 - 5x + 6}{3x^4} \\ &= \lim_{x\to\pm\infty} \left(1 - \frac{2}{3x} + \frac{1}{x^2} - \frac{5}{3x^3} + \frac{2}{x^4}\right) \\ &= 1,\end{aligned}$$

which means that f and g appear nearly identical when $|x|$ is large. ■

EXERCISES 2.6

Finding Limits

1. For the function f whose graph is given, determine the following limits. Write ∞ or $-\infty$ where appropriate.

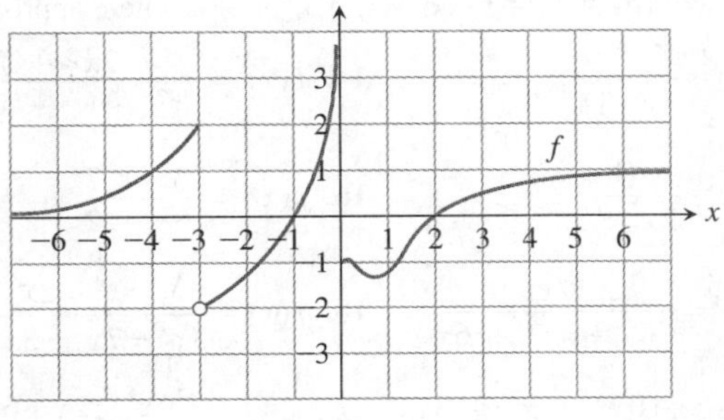

a. $\lim_{x\to 2} f(x)$
b. $\lim_{x\to 2^+} f(x)$
c. $\lim_{x\to -3^-} f(x)$
d. $\lim_{x\to -3} f(x)$
e. $\lim_{x\to 0^+} f(x)$
f. $\lim_{x\to 0^-} f(x)$
g. $\lim_{x\to 0} f(x)$
h. $\lim_{x\to \infty} f(x)$
i. $\lim_{x\to -\infty} f(x)$

2. For the function f whose graph is given, determine the following limits. Write ∞ or $-\infty$ where appropriate.

a. $\lim_{x\to 4} f(x)$ **b.** $\lim_{x\to 2^+} f(x)$ **c.** $\lim_{x\to 2^-} f(x)$

d. $\lim_{x\to 2} f(x)$ **e.** $\lim_{x\to -3^+} f(x)$ **f.** $\lim_{x\to -3^-} f(x)$

g. $\lim_{x\to -3} f(x)$ **h.** $\lim_{x\to 0^+} f(x)$ **i.** $\lim_{x\to 0^-} f(x)$

j. $\lim_{x\to 0} f(x)$ **k.** $\lim_{x\to \infty} f(x)$ **l.** $\lim_{x\to -\infty} f(x)$

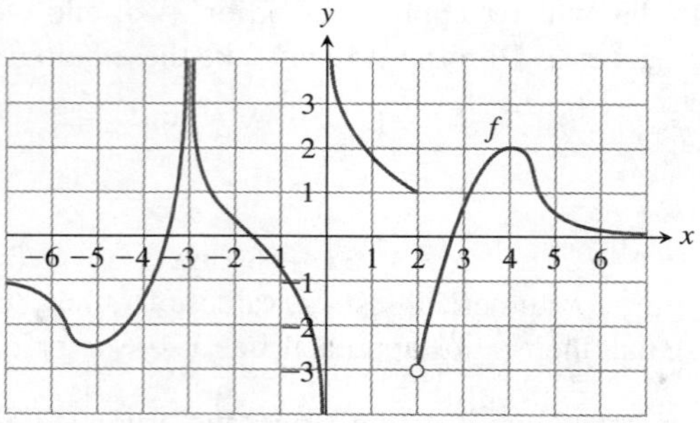

In Exercises 3–8, find the limit of each function **(a)** as $x \to \infty$ and **(b)** as $x \to -\infty$. (You may wish to visualize your answer with a graphing calculator or computer.)

3. $f(x) = \dfrac{2}{x} - 3$ **4.** $f(x) = 2\pi - \dfrac{5}{x^2}$

5. $g(x) = \dfrac{1}{2 + (1/x)}$ **6.** $f(x) = \dfrac{5}{10 - (7/x^2)}$

7. $h(x) = \dfrac{-5 + (7/x)}{3 - (1/x^2)}$ **8.** $f(x) = \dfrac{8 - (1/x^2)}{9 + (\sqrt{5}/x)}$

Find the limits in Exercises 9–12.

9. $\lim_{x\to\infty} \dfrac{\sin 4x}{2x}$ **10.** $\lim_{\theta\to-\infty} \dfrac{\cos\theta}{3\theta}$

11. $\lim_{t\to-\infty} \dfrac{5 - 2t + \cos t}{2t - \sin t}$ **12.** $\lim_{r\to\infty} \dfrac{r + \sin r}{2r + 7 - 5\sin r}$

Limits of Rational Functions

In Exercises 13–22, find the limit of each rational function **(a)** as $x \to \infty$ and **(b)** as $x \to -\infty$. Write ∞ or $-\infty$ where appropriate.

13. $f(x) = \dfrac{7x + 9}{8x + 15}$ **14.** $f(x) = \dfrac{2x^3 + 7}{x^3 - x^2 + x + 7}$

15. $f(x) = \dfrac{5x + 2}{2x^2 + 5}$ **16.** $f(x) = \dfrac{3x + 7}{x^2 - 2}$

17. $h(x) = \dfrac{7x^3}{x^3 - 3x^2 + 6x}$ **18.** $f(x) = \dfrac{5x^4 + 2x}{3x^4 + 7x^2 - 3x + 5}$

19. $g(x) = \dfrac{10x^5 + x^4 + 31}{x^6}$ **20.** $f(x) = \dfrac{9x^2 - 2x + 6}{2x^3 + x^2 - 5x + 7}$

21. $f(x) = \dfrac{3x^7 + 5x^2 - 1}{6x^3 - 7x + 3}$ **22.** $h(x) = \dfrac{5x^8 - 2x^3 + 9}{3 + x - 4x^5}$

Limits as $x \to \infty$ or $x \to -\infty$

The process by which we determine limits of rational functions applies equally well to ratios containing noninteger or negative powers of x. Divide numerator and denominator by the highest power of x in the denominator and proceed from there. Find the limits in Exercises 23–36. Write ∞ or $-\infty$ where appropriate.

23. $\lim_{x\to\infty} \sqrt{\dfrac{9x^2 - 5}{3x^2 + 5x}}$ **24.** $\lim_{x\to-\infty} \left(\dfrac{x^2 + x - 1}{8x^2 - 3}\right)^{1/3}$

25. $\lim_{x\to-\infty} \left(\dfrac{4 + 3x^3}{4x^2 + 9x}\right)^3$ **26.** $\lim_{x\to\infty} \sqrt{\dfrac{x^2 - 5x}{x^3 + x - 2}}$

27. $\lim_{x\to\infty} \dfrac{2\sqrt{x} + x^{-1}}{3x - 7}$ **28.** $\lim_{x\to\infty} \dfrac{5 + \sqrt{2x}}{3 - \sqrt{x}}$

29. $\lim_{x\to-\infty} \dfrac{\sqrt[3]{x} - \sqrt[5]{x}}{\sqrt[3]{x} + \sqrt[5]{x}}$ **30.** $\lim_{x\to\infty} \dfrac{x^{-2} + x^{-1}}{x^{-4} + x^{-2}}$

31. $\lim_{x\to\infty} \dfrac{2x^{5/3} - x^{1/3} + 7}{x^{8/5} + 3x + \sqrt{x}}$ **32.** $\lim_{x\to-\infty} \dfrac{\sqrt[3]{x} - 5x + 3}{2x + x^{2/3} - 4}$

33. $\lim_{x\to\infty} \dfrac{\sqrt{4x^2 + 1}}{2x + 1}$ **34.** $\lim_{x\to-\infty} \dfrac{\sqrt{x^2 + 1}}{x + 1}$

35. $\lim_{x\to\infty} \dfrac{x - 3}{\sqrt{4x^2 + 25}}$ **36.** $\lim_{x\to-\infty} \dfrac{4 - 3x^3}{\sqrt{x^6 + 9}}$

Infinite Limits

Find the limits in Exercises 37–48. Write ∞ or $-\infty$ where appropriate.

37. $\lim_{x\to 0^+} \dfrac{2}{5x}$ **38.** $\lim_{x\to 0^-} \dfrac{5}{2x}$

39. $\lim_{x\to 3^-} \dfrac{5}{2x - 6}$ **40.** $\lim_{x\to 3^+} \dfrac{1}{x - 3}$

41. $\lim_{x\to -8^+} \dfrac{2x}{x + 8}$ **42.** $\lim_{x\to -2^+} \dfrac{7x}{5x + 10}$

43. $\lim_{x\to 7} \dfrac{4}{(x - 7)^2}$ **44.** $\lim_{x\to 0} \dfrac{-1}{x^2(x + 1)}$

45. a. $\lim_{x\to 0^+} \dfrac{5}{x^{1/2}}$ **b.** $\lim_{x\to 0^-} \dfrac{5}{x^{1/2}}$

46. a. $\lim_{x\to 0^+} \dfrac{2}{x^{1/5}}$ **b.** $\lim_{x\to 0^-} \dfrac{2}{x^{1/5}}$

47. $\lim_{x\to 0} \dfrac{4}{x^{2/5}}$ **48.** $\lim_{x\to 0} \dfrac{1}{x^{2/3}}$

Find the limits in Exercises 49–52. Write ∞ or $-\infty$ where appropriate.

49. $\lim_{x\to(\pi/2)^-} \cot x$ **50.** $\lim_{x\to(-\pi/2)^+} \sec x$

51. $\lim_{\theta\to 0^-} (1 + \tan\theta)$ **52.** $\lim_{\theta\to 0} (2 - \cot\theta)$

Find the limits in Exercises 53–58. Write ∞ or $-\infty$ where appropriate.

53. $\lim \dfrac{1}{x^2 - 4}$ as

a. $x \to 2^+$ **b.** $x \to 2^-$

c. $x \to -2^+$ **d.** $x \to -2^-$

54. $\lim \dfrac{x}{x^2-1}$ as

a. $x \to 1^+$ **b.** $x \to 1^-$

c. $x \to -1^+$ **d.** $x \to -1^-$

55. $\lim \left(\dfrac{x^2}{2} - \dfrac{1}{x}\right)$ as

a. $x \to 0^+$ **b.** $x \to 0^-$

c. $x \to \sqrt[3]{2}$ **d.** $x \to -1$

56. $\lim \dfrac{x-4}{3x+9}$ as

a. $x \to -3^+$ **b.** $x \to -3^-$

c. $x \to 2^+$ **d.** $x \to 0^-$

57. $\lim \dfrac{x^2-3x+2}{x^3-2x^2}$ as

a. $x \to 0^+$ **b.** $x \to 2^+$

c. $x \to 2^-$ **d.** $x \to 2$

e. What, if anything, can be said about the limit as $x \to 0$?

58. $\lim \dfrac{x^2-8x+15}{x^2-9x}$ as

a. $x \to 3^+$ **b.** $x \to -3^+$

c. $x \to 0^-$ **d.** $x \to 5^+$

e. What if anything, can be said about the limit as $x \to 0$?

Find the limits in Exercises 59–62. Write ∞ or $-\infty$ where appropriate.

59. $\lim \left(2 - \dfrac{3}{t^{1/3}}\right)$ as

a. $t \to 0^+$ **b.** $t \to 0^-$

60. $\lim \left(\dfrac{2}{t^{1/3}} - 5\right)$ as

a. $t \to 0^+$ **b.** $t \to 0^-$

61. $\lim \left(\dfrac{1}{x^{2/3}} + \dfrac{2}{(x-1)^{2/3}}\right)$ as

a. $x \to 0^+$ **b.** $x \to 0^-$

c. $x \to 1^+$ **d.** $x \to 1^-$

62. $\lim \left(\dfrac{1}{x^{1/3}} - \dfrac{1}{(x-1)^{4/3}}\right)$ as

a. $x \to 0^+$ **b.** $x \to 0^-$

c. $x \to 1^+$ **d.** $x \to 1^-$

Graphing Simple Rational Functions

Graph the rational functions in Exercises 63–68. Include the graphs and equations of the asymptotes and dominant terms.

63. $y = \dfrac{1}{3x-1}$ **64.** $y = \dfrac{1}{x+1}$

65. $y = \dfrac{1}{2x+4}$ **66.** $y = \dfrac{-3}{x-3}$

67. $y = \dfrac{x+3}{x+2}$ **68.** $y = \dfrac{2x}{x+1}$

Domains and Asymptotes

Determine the domain of each function in Exercises 69–72. Then use various limits to find the asymptotes.

69. $y = 4 + \dfrac{3x^2}{x^2+1}$ **70.** $y = \dfrac{2x}{x^2-1}$

71. $y = \dfrac{\sqrt{x^2+4}}{x}$ **72.** $y = \dfrac{x^3}{x^3-8}$

Inventing Graphs and Functions

In Exercises 73–76, sketch the graph of a function $y = f(x)$ that satisfies the given conditions. No formulas are required—just label the coordinate axes and sketch an appropriate graph. (The answers are not unique, so your graphs may not be exactly like those in the answer section.)

73. $f(0) = 0, f(1) = 2, f(-1) = -2, \lim_{x\to-\infty} f(x) = -1$, and $\lim_{x\to\infty} f(x) = 1$

74. $f(0) = 0, \lim_{x\to\pm\infty} f(x) = 0, \lim_{x\to0^+} f(x) = 2$, and $\lim_{x\to0^-} f(x) = -2$

75. $f(0) = 0, \lim_{x\to\pm\infty} f(x) = 0, \lim_{x\to1^-} f(x) = \lim_{x\to-1^+} f(x) = \infty$, $\lim_{x\to1^+} f(x) = -\infty$, and $\lim_{x\to-1^-} f(x) = -\infty$

76. $f(2) = 1, f(-1) = 0, \lim_{x\to\infty} f(x) = 0, \lim_{x\to0^+} f(x) = \infty$, $\lim_{x\to0^-} f(x) = -\infty$, and $\lim_{x\to-\infty} f(x) = 1$

In Exercises 77–80, find a function that satisfies the given conditions and sketch its graph. (The answers here are not unique. Any function that satisfies the conditions is acceptable. Feel free to use formulas defined in pieces if that will help.)

77. $\lim_{x\to\pm\infty} f(x) = 0, \lim_{x\to2^-} f(x) = \infty$, and $\lim_{x\to2^+} f(x) = \infty$

78. $\lim_{x\to\pm\infty} g(x) = 0, \lim_{x\to3^-} g(x) = -\infty$, and $\lim_{x\to3^+} g(x) = \infty$

79. $\lim_{x\to-\infty} h(x) = -1, \lim_{x\to\infty} h(x) = 1, \lim_{x\to0^-} h(x) = -1$, and $\lim_{x\to0^+} h(x) = 1$

80. $\lim_{x\to\pm\infty} k(x) = 1, \lim_{x\to1^-} k(x) = \infty$, and $\lim_{x\to1^+} k(x) = -\infty$

81. Suppose that $f(x)$ and $g(x)$ are polynomials in x and that $\lim_{x\to\infty}\left(\dfrac{f(x)}{g(x)}\right) = 5$. Can you conclude anything about $\lim_{x\to-\infty}\left(\dfrac{f(x)}{g(x)}\right)$? Give reasons for your answer.

82. Suppose that $f(x)$ and $g(x)$ are polynomials in x. Can the graph of $f(x)/g(x)$ have an asymptote if $g(x)$ is never zero? Give reasons for your answer.

83. How many horizontal asymptotes can the graph of a given rational function have? Give reasons for your answer.

Finding Limits of Differences When $x \to \pm\infty$

Find the limits in Exercises 84–90. (*Hint*: Try multiplying and dividing by the conjugate.)

84. $\lim_{x\to\infty}\left(\sqrt{4x+3} - \sqrt{4x+2}\right)$

85. $\lim_{x\to\infty}\left(\sqrt{x^2+25} - \sqrt{x^2-1}\right)$

86. $\lim_{x\to-\infty}\left(\sqrt{x^2+3} + x\right)$

87. $\lim_{x\to\infty}\left(3x+\sqrt{9x^2+5x-2}\right)$

88. $\lim_{x\to\infty}\left(\sqrt{9x^2-x}-3x\right)$

89. $\lim_{x\to\infty}\left(\sqrt{x^2+3x}-\sqrt{x^2-2x}\right)$

90. $\lim_{x\to\infty}\left(\sqrt{4x^2+2x}-\sqrt{4x^2-2x}\right)$

Using the Formal Definitions

Use the formal definitions of limits as $x \to \pm\infty$ to establish the limits in Exercises 91 and 92.

91. If f has the constant value $f(x) = k$, then $\lim_{x\to\infty} f(x) = k$.

92. If f has the constant value $f(x) = k$, then $\lim_{x\to-\infty} f(x) = k$.

Use formal definitions to prove the limit statements in Exercises 93–96.

93. $\lim_{x\to 0}\dfrac{-1}{x^2} = -\infty$

94. $\lim_{x\to\infty}\dfrac{1}{|2x|} = \infty$

95. $\lim_{x\to 3}\dfrac{-2}{(x-3)^2} = -\infty$

96. $\lim_{x\to -4}\dfrac{1}{(x+4)^2} = \infty$

97. Here is the definition of **infinite right-hand limit**.

> Suppose that an interval (c, d) lies in the domain of f. We say that $f(x)$ approaches infinity as x approaches c from the right, and write
>
> $$\lim_{x\to c^+} f(x) = \infty,$$
>
> if, for every positive real number B, there exists a corresponding number $\delta > 0$ such that
>
> $$f(x) > B \quad \text{whenever} \quad c < x < c + \delta.$$

Modify the definition to cover the following cases.

a. $\lim_{x\to c^-} f(x) = \infty$

b. $\lim_{x\to c^+} f(x) = -\infty$

c. $\lim_{x\to c^-} f(x) = -\infty$

Use the formal definitions from Exercise 97 to prove the limit statements in Exercises 98–102.

98. $\lim_{x\to 0^+}\dfrac{1}{x} = \infty$

99. $\lim_{x\to 0^-}\dfrac{1}{x} = -\infty$

100. $\lim_{x\to 3^-}\dfrac{1}{x-3} = -\infty$

101. $\lim_{x\to 3^+}\dfrac{1}{x-3} = \infty$

102. $\lim_{x\to 1^-}\dfrac{1}{1-x^2} = \infty$

Oblique Asymptotes

Graph the rational functions in Exercises 103–108. Include the graphs and equations of the asymptotes.

103. $y = \dfrac{x^2}{x-1}$

104. $y = \dfrac{x^2+1}{x-1}$

105. $y = \dfrac{x^2-4}{x-1}$

106. $y = \dfrac{x^2-1}{2x+4}$

107. $y = \dfrac{x^2-1}{x}$

108. $y = \dfrac{x^3+1}{x^2}$

Additional Graphing Exercises

T Graph the curves in Exercises 109–112. Explain the relationship between the curve's formula and what you see.

109. $y = \dfrac{x}{\sqrt{4-x^2}}$

110. $y = \dfrac{-1}{\sqrt{4-x^2}}$

111. $y = x^{2/3} + \dfrac{1}{x^{1/3}}$

112. $y = \sin\left(\dfrac{\pi}{x^2+1}\right)$

T Graph the functions in Exercises 113 and 114. Then answer the following questions.

a. How does the graph behave as $x \to 0^+$?

b. How does the graph behave as $x \to \pm\infty$?

c. How does the graph behave near $x = 1$ and $x = -1$?

Give reasons for your answers.

113. $y = \dfrac{3}{2}\left(x - \dfrac{1}{x}\right)^{2/3}$

114. $y = \dfrac{3}{2}\left(\dfrac{x}{x-1}\right)^{2/3}$

CHAPTER 2 Questions to Guide Your Review

1. What is the average rate of change of the function $g(t)$ over the interval from $t = a$ to $t = b$? How is it related to a secant line?

2. What limit must be calculated to find the rate of change of a function $g(t)$ at $t = t_0$?

3. Give an informal or intuitive definition of the limit

$$\lim_{x\to c} f(x) = L.$$

Why is the definition "informal"? Give examples.

4. Does the existence and value of the limit of a function $f(x)$ as x approaches c ever depend on what happens at $x = c$? Explain and give examples.

5. What function behaviors might occur for which the limit may fail to exist? Give examples.

6. What theorems are available for calculating limits? Give examples of how the theorems are used.

7. How are one-sided limits related to limits? How can this relationship sometimes be used to calculate a limit or prove it does not exist? Give examples.

8. What is the value of $\lim_{\theta\to 0}((\sin\theta)/\theta)$? Does it matter whether θ is measured in degrees or radians? Explain.

9. What exactly does $\lim_{x\to c} f(x) = L$ mean? Give an example in which you find $\delta > 0$ for a given f, L, c, and $\varepsilon > 0$ in the precise definition of limit.

10. Give precise definitions of the following statements.

a. $\lim_{x\to 2^-} f(x) = 5$ **b.** $\lim_{x\to 2^+} f(x) = 5$

c. $\lim_{x\to 2} f(x) = \infty$ **d.** $\lim_{x\to 2} f(x) = -\infty$

11. What conditions must be satisfied by a function if it is to be continuous at an interior point of its domain? At an endpoint?

12. How can looking at the graph of a function help you tell where the function is continuous?

13. What does it mean for a function to be right-continuous at a point? Left-continuous? How are continuity and one-sided continuity related?

14. What does it mean for a function to be continuous on an interval? Give examples to illustrate the fact that a function that is not continuous on its entire domain may still be continuous on selected intervals within the domain.

15. What are the basic types of discontinuity? Give an example of each. What is a removable discontinuity? Give an example.

16. What does it mean for a function to have the Intermediate Value Property? What conditions guarantee that a function has this property over an interval? What are the consequences for graphing and solving the equation $f(x) = 0$?

17. Under what circumstances can you extend a function $f(x)$ to be continuous at a point $x = c$? Give an example.

18. What exactly do $\lim_{x\to\infty} f(x) = L$ and $\lim_{x\to-\infty} f(x) = L$ mean? Give examples.

19. What are $\lim_{x\to\pm\infty} k$ (k a constant) and $\lim_{x\to\pm\infty} (1/x)$? How do you extend these results to other functions? Give examples.

20. How do you find the limit of a rational function as $x \to \pm\infty$? Give examples.

21. What are horizontal and vertical asymptotes? Give examples.

CHAPTER 2 Practice Exercises

Limits and Continuity

1. Graph the function

$$f(x) = \begin{cases} 1, & x \le -1 \\ -x, & -1 < x < 0 \\ 1, & x = 0 \\ -x, & 0 < x < 1 \\ 1, & x \ge 1. \end{cases}$$

Then discuss, in detail, limits, one-sided limits, continuity, and one-sided continuity of f at $x = -1, 0,$ and 1. Are any of the discontinuities removable? Explain.

2. Repeat the instructions of Exercise 1 for

$$f(x) = \begin{cases} 0, & x \le -1 \\ 1/x, & 0 < |x| < 1 \\ 0, & x = 1 \\ 1, & x > 1. \end{cases}$$

3. Suppose that $f(t)$ and $f(t)$ are defined for all t and that $\lim_{t\to t_0} f(t) = -7$ and $\lim_{t\to t_0} g(t) = 0$. Find the limit as $t \to t_0$ of the following functions.

a. $3f(t)$ **b.** $(f(t))^2$

c. $f(t) \cdot g(t)$ **d.** $\dfrac{f(t)}{g(t) - 7}$

e. $\cos(g(t))$ **f.** $|f(t)|$

g. $f(t) + g(t)$ **h.** $1/f(t)$

4. Suppose the functions $f(x)$ and $g(x)$ are defined for all x and that $\lim_{x\to 0} f(x) = 1/2$ and $\lim_{x\to 0} g(x) = \sqrt{2}$. Find the limits as $x \to 0$ of the following functions.

a. $-g(x)$ **b.** $g(x) \cdot f(x)$

c. $f(x) + g(x)$ **d.** $1/f(x)$

e. $x + f(x)$ **f.** $\dfrac{f(x) \cdot \cos x}{x - 1}$

In Exercises 5 and 6, find the value that $\lim_{x\to 0} g(x)$ must have if the given limit statements hold.

5. $\lim_{x\to 0} \left(\dfrac{4 - g(x)}{x}\right) = 1$ **6.** $\lim_{x\to -4} \left(x \lim_{x\to 0} g(x)\right) = 2$

7. On what intervals are the following functions continuous?

a. $f(x) = x^{1/3}$ **b.** $g(x) = x^{3/4}$

c. $h(x) = x^{-2/3}$ **d.** $k(x) = x^{-1/6}$

8. On what intervals are the following functions continuous?

a. $f(x) = \tan x$

b. $g(x) = \csc x$

c. $h(x) = \dfrac{\cos x}{x - \pi}$

d. $k(x) = \dfrac{\sin x}{x}$

Finding Limits

In Exercises 9–28, find the limit or explain why it does not exist.

9. $\lim \dfrac{x^2 - 4x + 4}{x^3 + 5x^2 - 14x}$

a. as $x \to 0$ **b.** as $x \to 2$

10. $\lim \dfrac{x^2 + x}{x^5 + 2x^4 + x^3}$

a. as $x \to 0$ **b.** as $x \to -1$

11. $\lim_{x\to 1} \dfrac{1 - \sqrt{x}}{1 - x}$

12. $\lim_{x\to a} \dfrac{x^2 - a^2}{x^4 - a^4}$

13. $\lim_{h\to 0} \dfrac{(x + h)^2 - x^2}{h}$

14. $\lim_{x\to 0} \dfrac{(x + h)^2 - x^2}{h}$

15. $\lim_{x\to 0} \dfrac{\frac{1}{2+x} - \frac{1}{2}}{x}$

16. $\lim_{x\to 0} \dfrac{(2+x)^3 - 8}{x}$

17. $\lim_{x\to 1} \dfrac{x^{1/3} - 1}{\sqrt{x} - 1}$

18. $\lim_{x\to 64} \dfrac{x^{2/3} - 16}{\sqrt{x} - 8}$

19. $\lim_{x\to 0} \dfrac{\tan(2x)}{\tan(\pi x)}$

20. $\lim_{x\to \pi^-} \csc x$

21. $\lim_{x\to \pi} \sin\left(\dfrac{x}{2} + \sin x\right)$

22. $\lim_{x\to \pi} \cos^2(x - \tan x)$

23. $\lim_{x\to 0} \dfrac{8x}{3\sin x - x}$

24. $\lim_{x\to 0} \dfrac{\cos 2x - 1}{\sin x}$

In Exercises 25–28, find the limit of $g(x)$ as x approaches the indicated value.

25. $\lim_{x\to 0^+} (4g(x))^{1/3} = 2$

26. $\lim_{x\to \sqrt{5}} \dfrac{1}{x + g(x)} = 2$

27. $\lim_{x\to 1} \dfrac{3x^2 + 1}{g(x)} = \infty$

28. $\lim_{x\to -2} \dfrac{5 - x^2}{\sqrt{g(x)}} = 0$

T Roots

29. Let $f(x) = x^3 - x - 1$.

a. Use the Intermediate Value Theorem to show that f has a zero between -1 and 2.

b. Solve the equation $f(x) = 0$ graphically with an error of magnitude at most 10^{-8}.

c. It can be shown that the exact value of the solution in part **(b)** is

$$\left(\frac{1}{2} + \frac{\sqrt{69}}{18}\right)^{1/3} + \left(\frac{1}{2} - \frac{\sqrt{69}}{18}\right)^{1/3}.$$

Evaluate this exact answer and compare it with the value you found in part (b).

T 30. Let $f(\theta) = \theta^3 - 2\theta + 2$.

a. Use the Intermediate Value Theorem to show that f has a zero between -2 and 0.

b. Solve the equation $f(\theta) = 0$ graphically with an error of magnitude at most 10^{-4}.

c. It can be shown that the exact value of the solution in part (b) is

$$\left(\sqrt{\frac{19}{27}} - 1\right)^{1/3} - \left(\sqrt{\frac{19}{27}} + 1\right)^{1/3}.$$

Evaluate this exact answer and compare it with the value you found in part (b).

Continuous Extension

31. Can $f(x) = x(x^2 - 1)/|x^2 - 1|$ be extended to be continuous at $x = 1$ or -1? Give reasons for your answers. (Graph the function—you will find the graph interesting.)

32. Explain why the function $f(x) = \sin(1/x)$ has no continuous extension to $x = 0$.

T In Exercises 33–36, graph the function to see whether it appears to have a continuous extension to the given point a. If it does, use Trace and Zoom to find a good candidate for the extended function's value at a. If the function does not appear to have a continuous extension, can it be extended to be continuous from the right or left? If so, what do you think the extended function's value should be?

33. $f(x) = \dfrac{x - 1}{x - \sqrt[4]{x}}, \quad a = 1$

34. $g(\theta) = \dfrac{5\cos\theta}{4\theta - 2\pi}, \quad a = \pi/2$

35. $h(t) = (1 + |t|)^{1/t}, \quad a = 0$

36. $k(x) = \dfrac{x}{1 - 2^{|x|}}, \quad a = 0$

Limits at Infinity

Find the limits in Exercises 37–46.

37. $\lim_{x\to \infty} \dfrac{2x + 3}{5x + 7}$

38. $\lim_{x\to -\infty} \dfrac{2x^2 + 3}{5x^2 + 7}$

39. $\lim_{x\to -\infty} \dfrac{x^2 - 4x + 8}{3x^3}$

40. $\lim_{x\to \infty} \dfrac{1}{x^2 - 7x + 1}$

41. $\lim_{x\to -\infty} \dfrac{x^2 - 7x}{x + 1}$

42. $\lim_{x\to \infty} \dfrac{x^4 + x^3}{12x^3 + 128}$

43. $\lim_{x\to \infty} \dfrac{\sin x}{|x|}$ (If you have a grapher, try graphing the function for $-5 \le x \le 5$.)

44. $\lim_{\theta\to \infty} \dfrac{\cos\theta - 1}{\theta}$ (If you have a grapher, try graphing $f(x) = x(\cos(1/x) - 1)$ near the origin to "see" the limit at infinity.)

45. $\lim_{x\to \infty} \dfrac{x + \sin x + 2\sqrt{x}}{x + \sin x}$

46. $\lim_{x\to \infty} \dfrac{x^{2/3} + x^{-1}}{x^{2/3} + \cos^2 x}$

Horizontal and Vertical Asymptotes

47. Use limits to determine the equations for all vertical asymptotes.

a. $y = \dfrac{x^2 + 4}{x - 3}$

b. $f(x) = \dfrac{x^2 - x - 2}{x^2 - 2x + 1}$

c. $y = \dfrac{x^2 + x - 6}{x^2 + 2x - 8}$

48. Use limits to determine the equations for all horizontal asymptotes.

a. $y = \dfrac{1 - x^2}{x^2 + 1}$

b. $f(x) = \dfrac{\sqrt{x} + 4}{\sqrt{x + 4}}$

c. $g(x) = \dfrac{\sqrt{x^2 + 4}}{x}$

d. $y = \sqrt{\dfrac{x^2 + 9}{9x^2 + 1}}$

49. Determine the domain and range of $y = \dfrac{\sqrt{16 - x^2}}{x - 2}$.

50. Assume that constants a and b are positive. Find equations for all horizontal and vertical asymptotes for the graph of $y = \dfrac{\sqrt{ax^2 + 4}}{x - b}$.

CHAPTER 2 Additional and Advanced Exercises

1. **Lorentz contraction** In relativity theory, the length of an object, say a rocket, appears to an observer to depend on the speed at which the object is traveling with respect to the observer. If the observer measures the rocket's length as L_0 at rest, then at speed v the length will appear to be

$$L = L_0\sqrt{1 - \frac{v^2}{c^2}}.$$

This equation is the Lorentz contraction formula. Here, c is the speed of light in a vacuum, about 3×10^8 m/sec. What happens to L as v increases? Find $\lim_{v\to c^-} L$. Why was the left-hand limit needed?

2. **Controlling the flow from a draining tank** Torricelli's law says that if you drain a tank like the one in the figure shown, the rate y at which water runs out is a constant times the square root of the water's depth x. The constant depends on the size and shape of the exit valve.

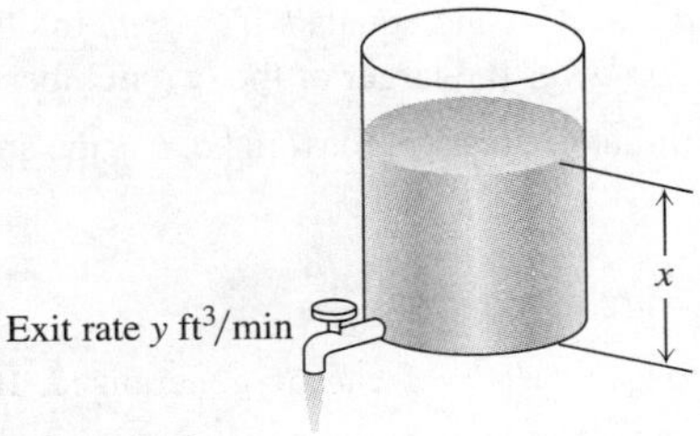

Suppose that $y = \sqrt{x}/2$ for a certain tank. You are trying to maintain a fairly constant exit rate by adding water to the tank with a hose from time to time. How deep must you keep the water if you want to maintain the exit rate

a. within $0.2\ \text{ft}^3/\text{min}$ of the rate $y_0 = 1\ \text{ft}^3/\text{min}$?

b. within $0.1\ \text{ft}^3/\text{min}$ of the rate $y_0 = 1\ \text{ft}^3/\text{min}$?

3. **Thermal expansion in precise equipment** As you may know, most metals expand when heated and contract when cooled. The dimensions of a piece of laboratory equipment are sometimes so critical that the shop where the equipment is made must be held at the same temperature as the laboratory where the equipment is to be used. A typical aluminum bar that is 10 cm wide at 70°F will be

$$y = 10 + (t - 70) \times 10^{-4}$$

centimeters wide at a nearby temperature t. Suppose that you are using a bar like this in a gravity wave detector, where its width must stay within 0.0005 cm of the ideal 10 cm. How close to $t_0 = 70$°F must you maintain the temperature to ensure that this tolerance is not exceeded?

4. **Stripes on a measuring cup** The interior of a typical 1-L measuring cup is a right circular cylinder of radius 6 cm (see accompanying figure). The volume of water we put in the cup is therefore a function of the level h to which the cup is filled, the formula being

$$V = \pi 6^2 h = 36\pi h.$$

How closely must we measure h to measure out 1 L of water ($1000\ \text{cm}^3$) with an error of no more than 1% ($10\ \text{cm}^3$)?

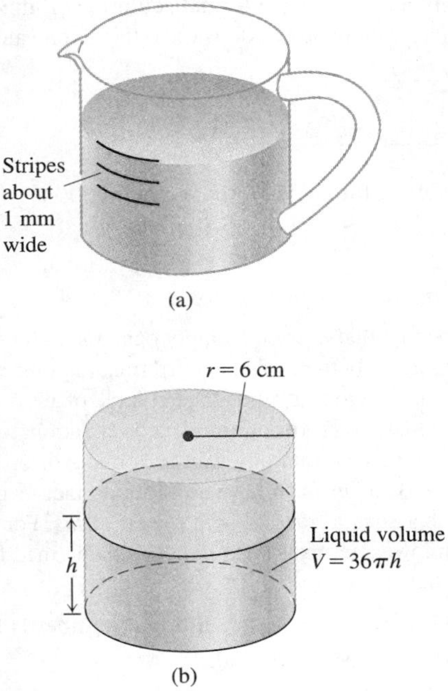

A 1-L measuring cup (a), modeled as a right circular cylinder (b) of radius $r = 6$ cm

Precise Definition of Limit

In Exercises 5–8, use the formal definition of limit to prove that the function is continuous at c.

5. $f(x) = x^2 - 7,\quad c = 1$

6. $g(x) = 1/(2x),\quad c = 1/4$

7. $h(x) = \sqrt{2x - 3},\quad c = 2$

8. $F(x) = \sqrt{9 - x},\quad c = 5$

9. **Uniqueness of limits** Show that a function cannot have two different limits at the same point. That is, if $\lim_{x\to c} f(x) = L_1$ and $\lim_{x\to c} f(x) = L_2$, then $L_1 = L_2$.

10. Prove the limit Constant Multiple Rule:

$$\lim_{x\to c} kf(x) = k\lim_{x\to c} f(x) \quad \text{for any constant } k.$$

11. **One-sided limits** If $\lim_{x\to 0^+} f(x) = A$ and $\lim_{x\to 0^-} f(x) = B$, find

a. $\lim_{x\to 0^+} f(x^3 - x)$

b. $\lim_{x\to 0^-} f(x^3 - x)$

c. $\lim_{x\to 0^+} f(x^2 - x^4)$

d. $\lim_{x\to 0^-} f(x^2 - x^4)$

12. **Limits and continuity** Which of the following statements are true, and which are false? If true, say why; if false, give a counterexample (that is, an example confirming the falsehood).

a. If $\lim_{x\to c} f(x)$ exists but $\lim_{x\to c} g(x)$ does not exist, then $\lim_{x\to c}(f(x) + g(x))$ does not exist.

b. If neither $\lim_{x\to c} f(x)$ nor $\lim_{x\to c} g(x)$ exists, then $\lim_{x\to c} (f(x)+g(x))$ does not exist.

c. If f is continuous at x, then so is $|f|$.

d. If $|f|$ is continuous at c, then so is f.

In Exercises 13 and 14, use the formal definition of limit to prove that the function has a continuous extension to the given value of x.

13. $f(x) = \dfrac{x^2-1}{x+1}, \quad x = -1$

14. $g(x) = \dfrac{x^2-2x-3}{2x-6}, \quad x = 3$

15. A function continuous at only one point Let

$$f(x) = \begin{cases} x & \text{if } x \text{ is rational} \\ 0 & \text{if } x \text{ is irrational.} \end{cases}$$

a. Show that f is continuous at $x = 0$.

b. Use the fact that every nonempty open interval of real numbers contains both rational and irrational numbers to show that f is not continuous at any nonzero value of x.

16. The Dirichlet ruler function If x is a rational number, then x can be written in a unique way as a quotient of integers m/n, where $n > 0$ and m and n have no common factors greater than 1. (We say that such a fraction is in *lowest terms*. For example, 6/4 written in lowest terms is 3/2.) Let $f(x)$ be defined for all x in the interval $[0, 1]$ by

$$f(x) = \begin{cases} 1/n & \text{if } x = m/n \text{ is a rational number is lowest terms} \\ 0 & \text{if } x \text{ is irrational.} \end{cases}$$

For instance, $f(0) = f(1) = 1, f(1/2) = 1/2, f(1/3) = f(2/3) = 1/3, f(1/4) = f(3/4) = 1/4$, and so on.

a. Show that f is discontinuous at every rational number in $[0, 1]$.

b. Show that f is continuous at every irrational number in $[0, 1]$. (*Hint*: If ε is a given positive number, show that there are only finitely many rational numbers r in $[0, 1]$ such that $f(r) \geq \varepsilon$.)

c. Sketch the graph of f. Why do you think f is called the "ruler function"?

17. Antipodal points Is there any reason to believe that there is always a pair of antipodal (diametrically opposite) points on Earth's equator where the temperatures are the same? Explain.

18. If $\lim_{x\to c}(f(x)+g(x)) = 3$ and $\lim_{x\to c}(f(x)-g(x)) = -1$, find $\lim_{x\to c} f(x)g(x)$.

19. Roots of a quadratic equation that is almost linear The equation $ax^2 + 2x - 1 = 0$, where a is a constant, has two roots if $a > -1$ and $a \neq 0$, one positive and one negative:

$$r_+(a) = \frac{-1+\sqrt{1+a}}{a}, \qquad r_-(a) = \frac{-1-\sqrt{1+a}}{a}.$$

a. What happens to $r_+(a)$ as $a \to 0$? As $a \to -1^+$?

b. What happens to $r_-(a)$ as $a \to 0$? As $a \to -1^+$?

c. Support your conclusions by graphing $r_+(a)$ and $r_-(a)$ as functions of a. Describe what you see.

d. For added support, graph $f(x) = ax^2 + 2x - 1$ simultaneously for $a = 1, 0.5, 0.2, 0.1$, and 0.05.

20. Root of an equation Show that the equation $x + 2\cos x = 0$ has at least one solution.

21. Bounded functions A real-valued function f is **bounded from above** on a set D if there exists a number N such that $f(x) \leq N$ for all x in D. We call N, when it exists, an **upper bound** for f on D and say that f is bounded from above by N. In a similar manner, we say that f is **bounded from below** on D if there exists a number M such that $f(x) \geq M$ for all x in D. We call M, when it exists, a **lower bound** for f on D and say that f is bounded from below by M. We say that f is **bounded** on D if it is bounded from both above and below.

a. Show that f is bounded on D if and only if there exists a number B such that $|f(x)| \leq B$ for all x in D.

b. Suppose that f is bounded from above by N. Show that if $\lim_{x\to c} f(x) = L$, then $L \leq N$.

c. Suppose that f is bounded from below by M. Show that if $\lim_{x\to c} f(x) = L$, then $L \geq M$.

22. Max $\{a, b\}$ and min $\{a, b\}$

a. Show that the expression

$$\max\{a, b\} = \frac{a+b}{2} + \frac{|a-b|}{2}$$

equals a if $a \geq b$ and equals b if $b \geq a$. In other words, $\max\{a, b\}$ gives the larger of the two numbers a and b.

b. Find a similar expression for $\min\{a, b\}$, the smaller of a and b.

Generalized Limits Involving $\dfrac{\sin\theta}{\theta}$

The formula $\lim_{\theta\to 0}(\sin\theta)/\theta = 1$ can be generalized. If $\lim_{x\to c} f(x) = 0$ and $f(x)$ is never zero in an open interval containing the point $x = c$, except possibly at c itself, then

$$\lim_{x\to c} \frac{\sin f(x)}{f(x)} = 1.$$

Here are several examples.

a. $\displaystyle\lim_{x\to 0} \frac{\sin x^2}{x^2} = 1$

b. $\displaystyle\lim_{x\to 0} \frac{\sin x^2}{x} = \lim_{x\to 0} \frac{\sin x^2}{x^2} \lim_{x\to 0} \frac{x^2}{x} = 1 \cdot 0 = 0$

c. $\displaystyle\lim_{x\to -1} \frac{\sin(x^2-x-2)}{x+1}$

$$= \lim_{x\to -1} \frac{\sin(x^2-x-2)}{(x^2-x-2)} \cdot \lim_{x\to -1} \frac{(x^2-x-2)}{x+1}$$

$$= 1 \cdot \lim_{x\to -1} \frac{(x+1)(x-2)}{x+1} = -3$$

d. $\displaystyle\lim_{x\to 1} \frac{\sin(1-\sqrt{x})}{x-1} = \lim_{x\to 1} \frac{\sin(1-\sqrt{x})}{1-\sqrt{x}} \frac{1-\sqrt{x}}{x-1}$

$$= \lim_{x\to 1} \frac{(1-\sqrt{x})(1+\sqrt{x})}{(x-1)(1+\sqrt{x})}$$

$$= \lim_{x\to 1} \frac{1-x}{(x-1)(1+\sqrt{x})} = -\frac{1}{2}$$

Find the limits in Exercises 23–28.

23. $\displaystyle\lim_{x\to 0} \frac{\sin(1-\cos x)}{x}$

24. $\displaystyle\lim_{x\to 0^+} \frac{\sin x}{\sin\sqrt{x}}$

25. $\displaystyle\lim_{x\to 0} \frac{\sin(\sin x)}{x}$

26. $\displaystyle\lim_{x\to 0} \frac{\sin(x^2+x)}{x}$

27. $\displaystyle\lim_{x\to 2} \frac{\sin(x^2-4)}{x-2}$

28. $\displaystyle\lim_{x\to 9} \frac{\sin(\sqrt{x}-3)}{x-9}$

Trigonometric Limits

Find the limits in Exercises 29–36.

29. $\lim_{x\to 0} \dfrac{\sin x}{2x^2 - x}$

30. $\lim_{x\to 0} \dfrac{3x - \tan 7x}{2x}$

31. $\lim_{r\to 0} \dfrac{\sin r}{\tan 2r}$

32. $\lim_{\theta\to 0} \dfrac{\sin(\sin\theta)}{\theta}$

33. $\lim_{\theta\to(\pi/2)^-} \dfrac{4\tan^2\theta + \tan\theta + 1}{\tan^2\theta + 5}$

34. $\lim_{\theta\to 0^+} \dfrac{1 - 2\cot^2\theta}{5\cot^2\theta - 7\cot\theta - 8}$

35. $\lim_{x\to 0} \dfrac{x\sin x}{2 - 2\cos x}$

36. $\lim_{\theta\to 0} \dfrac{1 - \cos\theta}{\theta^2}$

Show how to extend the functions in Exercises 37 and 38 to be continuous at the origin.

37 $g(x) = \dfrac{\tan(\tan x)}{\tan x}$

38. $f(x) = \dfrac{\tan(\tan x)}{\sin(\sin x)}$

Oblique Asymptotes

Find all possible oblique asymptotes in Exercises 39–42.

39. $y = \dfrac{2x^{3/2} + 2x - 3}{\sqrt{x} + 1}$

40. $y = x + x\sin\dfrac{1}{x}$

41. $y = \sqrt{x^2 + 1}$

42. $y = \sqrt{x^2 + 2x}$

Showing an Equation Is Solvable

43. Assume that $1 < a < b$ and $\dfrac{a}{x} + x = \dfrac{1}{x - b}$. Show that this equation is solvable for x.

More Limits

44. Find constants a and b so that each of the following limits is true.

a. $\lim_{x\to 0} \dfrac{\sqrt{a + bx} - 1}{x} = 2$ **b.** $\lim_{x\to 1} \dfrac{\tan(ax - a) + b - 2}{x - 1} = 3$

45. Evaluate $\lim_{x\to 1} \dfrac{x^{2/3} - 1}{1 - \sqrt{x}}$.

46. Evaluate $\lim_{x\to 0} \dfrac{|3x + 4| - |x| - 4}{x}$.

Limits on Arbitrary Domains

The definition of the limit of a function at $x = c$ extends to functions whose domains near c are more complicated than intervals.

> **General Definition of Limit**
>
> Suppose every open interval containing c contains a point other than c in the domain of f. We say that $\lim_{x\to c} f(x) = L$ if, for every number $\varepsilon > 0$, there exists a corresponding number $\delta > 0$ such that for all x in the domain of f, $|f(x) - L| < \varepsilon$ whenever $0 < |x - c| < \delta$.

For the functions in Exercises 47–49,

a. Find the domain.

b. Show that at $c = 0$ the domain has the property described above.

c. Evaluate $\lim_{x\to 0} f(x)$.

47. The function f is defined as follows: $f(x) = x$ if $x = 1/n$ where n is a positive integer, and $f(0) = 1$.

48. The function f is defined as follows: $f(x) = 1 - x$ if $x = 1/n$ where n is a positive integer, and $f(0) = 1$.

49. $f(x) = \sqrt{x\sin(1/x)}$

50. Let g be a function with domain the rational numbers, defined by

$$g(x) = \frac{2}{x - \sqrt{2}} \text{ for rational } x.$$

a. Sketch the graph of g as well as you can, keeping in mind that g is defined only at rational points.

b. Use the general definition of a limit to prove that $\lim_{x\to 0} g(x) = -\sqrt{2}$.

c. Prove that g is continuous at the point $x = 0$ by showing that the limit in part (b) equals $g(0)$.

d. Is g continuous at other points of its domain?

CHAPTER 2 Technology Application Projects

Mathematica/Maple Projects

Projects can be found within MyLab Math.

- ***Take It to the Limit***
  Part I
  Part II (Zero Raised to the Power Zero: What Does It Mean?)
  Part III (One-Sided Limits)
  Visualize and interpret the limit concept through graphical and numerical explorations.
  Part IV (What a Difference a Power Makes)
  See how sensitive limits can be with various powers of x.

- ***Going to Infinity***
  Part I (Exploring Function Behavior as $x \to \infty$ or $x \to -\infty$)
  This module provides four examples to explore the behavior of a function as $x \to \infty$ or $x \to -\infty$.
  Part II (Rates of Growth)
  Observe graphs that *appear* to be continuous, yet the function is not continuous. Several issues of continuity are explored to obtain results that you may find surprising.

3

Derivatives

OVERVIEW In Chapter 2 we discussed how to determine the slope of a curve at a point and how to measure the rate at which a function changes. Now that we have studied limits, we can make these notions precise and see that both are interpretations of the *derivative* of a function at a point. We then extend this concept from a single point to the *derivative function*, and we develop rules for finding this derivative function easily, without having to calculate limits directly. These rules are used to find derivatives of most of the common functions reviewed in Chapter 1, as well as combinations of them.

3.1 Tangent Lines and the Derivative at a Point

In this section we define the slope and tangent line to a curve at a point, and the derivative of a function at a point. The derivative gives a way to find both the slope of a graph and the instantaneous rate of change of a function.

Finding a Tangent Line to the Graph of a Function

To find a tangent line to an arbitrary curve $y = f(x)$ at a point $P(x_0, f(x_0))$, we use the procedure introduced in Section 2.1. We calculate the slope of the secant line through P and a nearby point $Q(x_0 + h, f(x_0 + h))$. We then investigate the limit of the slope as $h \to 0$ (Figure 3.1). If the limit exists, we call it the slope of the curve at P and define the tangent line at P to be the line through P having this slope.

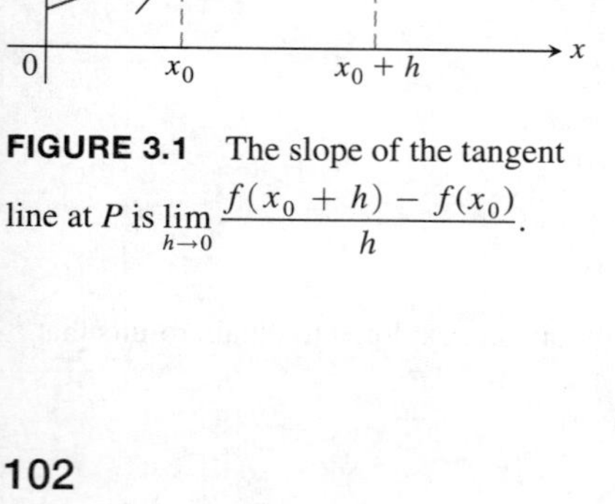

FIGURE 3.1 The slope of the tangent line at P is $\lim_{h\to 0} \dfrac{f(x_0 + h) - f(x_0)}{h}$.

> **DEFINITIONS** The **slope of the curve** $y = f(x)$ at the point $P(x_0, f(x_0))$ is the number
>
> $$\lim_{h\to 0} \frac{f(x_0 + h) - f(x_0)}{h} \qquad \text{(provided the limit exists).}$$
>
> The **tangent line** to the curve at P is the line through P with this slope.

In Section 2.1, Example 3, we applied these definitions to find the slope of the parabola $f(x) = x^2$ at the point $P(2, 4)$ and the tangent line to the parabola at P. Let's look at another example.

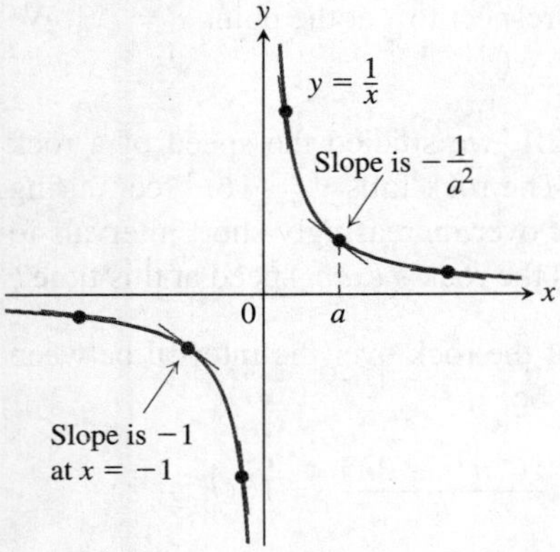

FIGURE 3.2 The tangent lines are steep when x is close to 0, and they become less steep as the point of tangency moves away (Example 1).

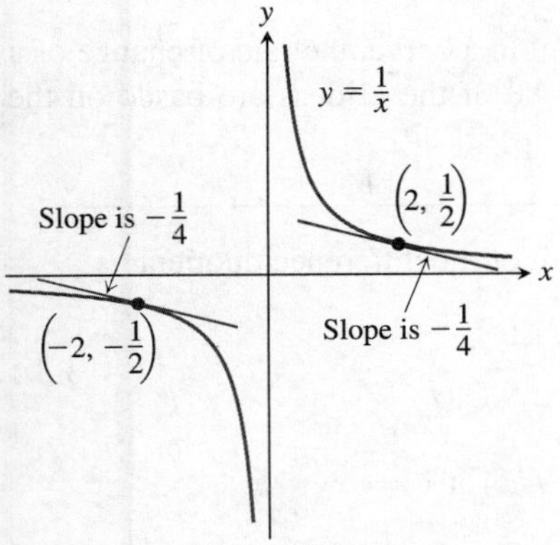

FIGURE 3.3 The two tangent lines to $y = 1/x$ having slope $-1/4$ (Example 1).

The notation $f'(x_0)$ is read "f prime of x_0."

EXAMPLE 1

(a) Find the slope of the curve $y = 1/x$ at any point $x = a \neq 0$. What is the slope at the point $x = -1$?

(b) Where does the slope equal $-1/4$?

(c) What happens to the tangent line to the curve at the point $(a, 1/a)$ as a changes?

Solution

(a) Here $f(x) = 1/x$. The slope at $(a, 1/a)$ is

$$\lim_{h\to 0}\frac{f(a+h)-f(a)}{h} = \lim_{h\to 0}\frac{\dfrac{1}{a+h}-\dfrac{1}{a}}{h} = \lim_{h\to 0}\frac{1}{h}\frac{a-(a+h)}{a(a+h)}$$

$$= \lim_{h\to 0}\frac{-h}{ha(a+h)} = \lim_{h\to 0}\frac{-1}{a(a+h)} = -\frac{1}{a^2}.$$

Notice how we had to keep writing "$\lim_{h\to 0}$" before each fraction until the stage at which we could evaluate the limit by substituting $h = 0$. The number a may be positive or negative, but not 0. When $a = -1$, the slope is $-1/(-1)^2 = -1$ (Figure 3.2).

(b) The slope of $y = 1/x$ at the point where $x = a$ is $-1/a^2$. It will be $-1/4$, provided that

$$-\frac{1}{a^2} = -\frac{1}{4}.$$

This equation is equivalent to $a^2 = 4$, so $a = 2$ or $a = -2$. The curve has slope $-1/4$ at the two points $(2, 1/2)$ and $(-2, -1/2)$ (Figure 3.3).

(c) The slope $-1/a^2$ is always negative if $a \neq 0$. As $a \to 0^+$, the slope approaches $-\infty$ and the tangent line becomes increasingly steep (Figure 3.2). We see this situation again as $a \to 0^-$. As a moves away from $x = 0$ in either direction, the slope approaches 0 and the tangent line levels off, becoming closer and closer to a horizontal line. ■

Rates of Change: Derivative at a Point

The expression

$$\frac{f(x_0+h)-f(x_0)}{h}, \quad h \neq 0$$

is called the **difference quotient of f at x_0 with increment h.** If the difference quotient has a limit as h approaches zero, that limit is given a special name and notation.

DEFINITION The **derivative of a function f at a point x_0**, denoted $f'(x_0)$, is

$$f'(x_0) = \lim_{h\to 0}\frac{f(x_0+h)-f(x_0)}{h},$$

provided this limit exists.

The derivative has more than one meaning, depending on what problem we are considering. The formula for the derivative is the same as the formula for the slope of the curve $y = f(x)$ at a point. If we interpret the difference quotient as the slope of a secant line, then the derivative gives the slope of the curve $y = f(x)$ at the point $P(x_0, f(x_0))$. If we interpret the difference quotient as an average rate of change (Section 2.1), then the derivative gives the function's instantaneous rate of change with respect to x at the point $x = x_0$. We study this interpretation in Section 3.4.

EXAMPLE 2 In Examples 1 and 2 in Section 2.1, we studied the speed of a rock falling freely from rest near the surface of the earth. The rock falls $y = 16t^2$ feet during the first t sec, and we use a sequence of average rates over increasingly short intervals to estimate the rock's speed at the instant $t = 1$. What is the rock's *exact* speed at this time?

Solution We let $f(t) = 16t^2$. The average speed of the rock over the interval between $t = 1$ and $t = 1 + h$ seconds, for $h > 0$, is found to be

$$\frac{f(1+h) - f(1)}{h} = \frac{16(1+h)^2 - 16(1)^2}{h} = \frac{16(h^2 + 2h)}{h} = 16(h + 2).$$

The rock's speed at the instant $t = 1$ is then

$$f'(1) = \lim_{h \to 0} \frac{f(1+h) - f(1)}{h} = \lim_{h \to 0} 16(h + 2) = 16(0 + 2) = 32 \text{ ft/sec.}$$ ■

Summary

We have been discussing slopes of curves, lines tangent to a curve, the rate of change of a function, and the derivative of a function at a point. All of these ideas are based on the same limit.

All of the following are interpretations for the limit of the difference quotient

$$\lim_{h \to 0} \frac{f(x_0 + h) - f(x_0)}{h}.$$

1. The slope of the graph of $y = f(x)$ at $x = x_0$
2. The slope of the tangent line to the curve $y = f(x)$ at $x = x_0$
3. The rate of change of $f(x)$ with respect to x at $x = x_0$
4. The derivative $f'(x_0)$ at $x = x_0$

In the next sections, we allow the point x_0 to vary across the domain of the function f.

EXERCISES 3.1

Slopes and Tangent Lines

In Exercises 1–4, use the grid and a straight edge to make a rough estimate of the slope of the curve (in y-units per x-unit) at the points P_1 and P_2.

1.

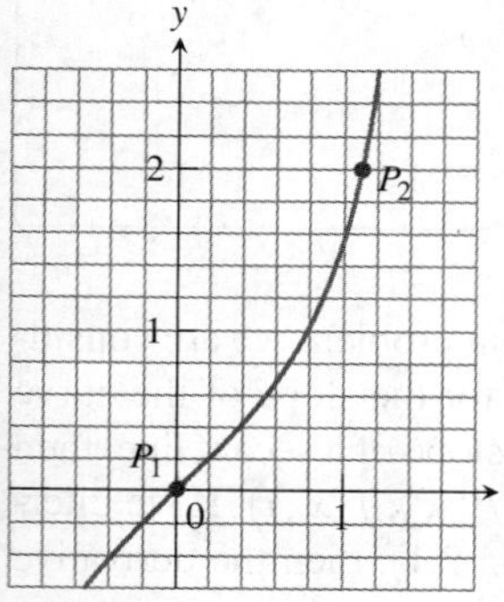

2.

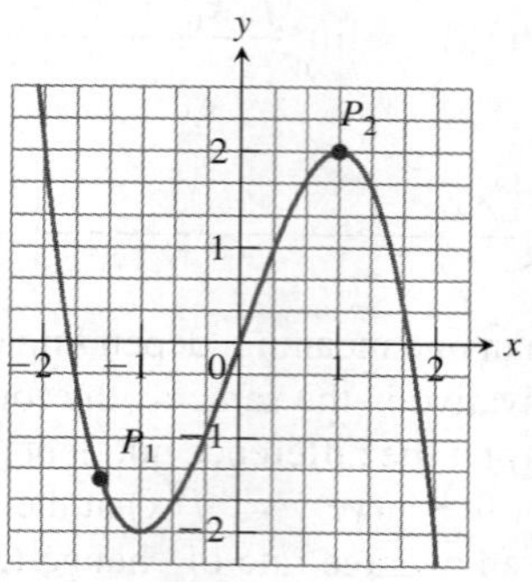

3.

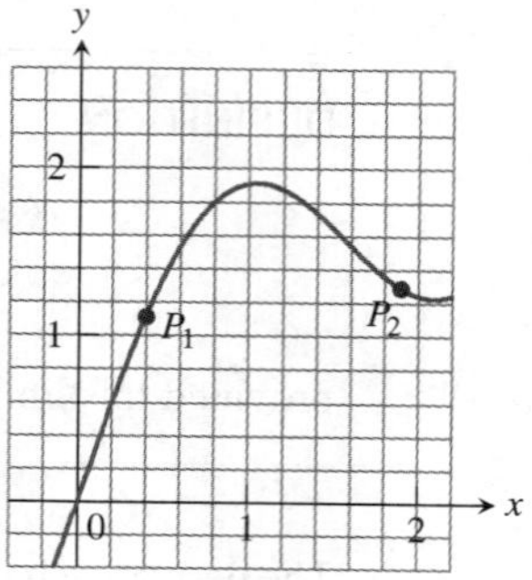

4.

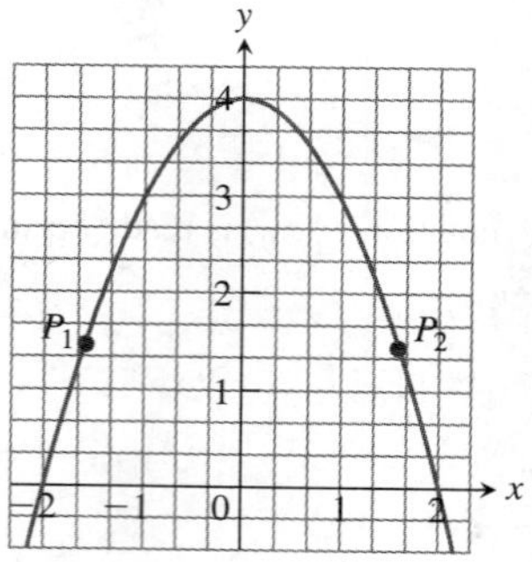

In Exercises 5–10, find an equation for the tangent line to the curve at the given point. Then sketch the curve and tangent line together.

5. $y = 7 - x^2, \quad (1, 6)$ **6.** $y = (x - 1)^2 + 1, \quad (1, 1)$

7. $y = 4\sqrt{x}, \quad (4, 8)$ **8.** $y = \frac{1}{x^2}, \quad (-1, 1)$

9. $y = 2x^3, \quad (-1, -2)$ **10.** $y = \frac{1}{x^3}, \quad \left(-2, -\frac{1}{8}\right)$

In Exercises 11–18, find the slope of the function's graph at the given point. Then find an equation for the line tangent to the graph there.

11. $f(x) = 2x^2 + 5, \quad (1, 7)$ **12.** $f(x) = x - 2x^2, \quad (1, -1)$

13. $g(x) = \frac{x}{x - 2}, \quad (3, 3)$ **14.** $g(x) = -\frac{4}{x^2}, \quad (2, -1)$

15. $h(t) = t^3, \quad (2, 8)$ **16.** $h(t) = t^3 + 3t, \quad (1, 4)$

17. $f(x) = \sqrt{2x}, \quad (2, 2)$ **18.** $f(x) = \sqrt{x + 1}, \quad (8, 3)$

In Exercises 19–22, find the slope of the curve at the point indicated.

19. $y = 3x - x^2, \quad x = 2$ **20.** $y = x^3 - 2x + 7, \quad x = -2$

21. $y = \frac{2}{2x - 5}, \quad x = 3$ **22.** $y = \frac{x - 1}{x + 1}, \quad x = 0$

Interpreting Derivative Values

23. Growth of yeast cells In a controlled laboratory experiment, yeast cells are grown in an automated cell culture system that counts the number P of cells present at hourly intervals. The number after t hours is shown in the accompanying figure.

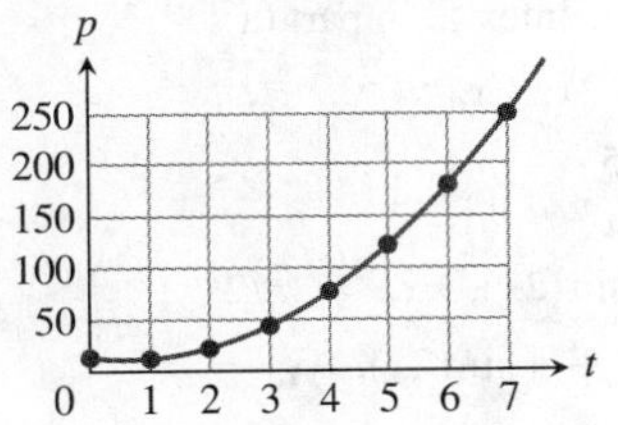

a. Explain what is meant by the derivative $P'(5)$. What are its units?

b. Which is larger, $P'(2)$ or $P'(3)$? Give a reason for your answer.

c. The quadratic curve capturing the trend of the data points is given by $P(t) = 6.10t^2 - 9.28t + 16.43$. Find the instantaneous rate of growth when $t = 5$ hours.

24. Effectiveness of a drug On a scale from 0 to 1, the effectiveness E of a pain-killing drug t hours after entering the bloodstream is displayed in the accompanying figure.

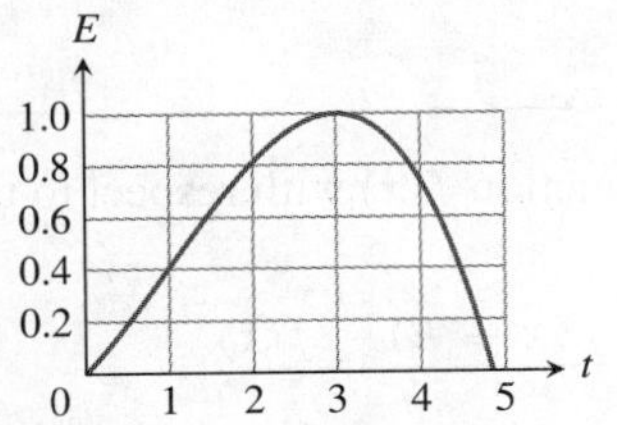

a. At what times does the effectiveness appear to be increasing? What is true about the derivative at those times?

b. At what time would you estimate that the drug reaches its maximum effectiveness? What is true about the derivative at that time? What is true about the derivative as time increases in the 1 hour *before* your estimated time?

At what points do the graphs of the functions in Exercises 25 and 26 have horizontal tangent lines?

25. $f(x) = 4x^2 + 2x - 5$ **26.** $g(x) = x^3 - 3x$

27. Find equations of all lines having slope -1 that are tangent to the curve $y = 1/(x - 1)$.

28. Find an equation of the straight line having slope $1/3$ that is tangent to the curve $y = \sqrt{2x}$.

Rates of Change

29. Object dropped from a tower An object is dropped from the top of a 200 m high tower. Its height above ground after t sec is $200 - 5t^2$ m. How fast is it falling 5 sec after it is dropped?

30. Speed of a rocket At t sec after liftoff, the height of a rocket is $3t^2$ ft. How fast is the rocket climbing 10 sec after liftoff?

31. Circle's changing circumference What is the rate of change of the circumference of a circle ($C = 2\pi r$) with respect to the radius when the radius is $r = 5$?

32. Ball's changing volume What is the rate of change of the volume of a ball ($V = (4/3)\pi r^3$) with respect to the radius when the radius is $r = 2$?

33. Show that the line $y = mx + b$ is its own tangent line at any point $(x_0, mx_0 + b)$.

34. Find the slope of the tangent line to the curve $y = \frac{1}{\sqrt{2x}}$ at the point where $x = 4$.

Testing for Tangent Lines

35. Does the graph of

$$f(x) = \begin{cases} x^2 \cos(1/x), & x \neq 0 \\ 0, & x = 0 \end{cases}$$

have a tangent line at the origin? Give reasons for your answer.

36. Does the graph of

$$g(x) = \begin{cases} x \sin(1/x), & x \neq 0 \\ 0, & x = 0 \end{cases}$$

have a tangent line at the origin? Give reasons for your answer.

Vertical Tangent Lines

We say that a continuous curve $y = f(x)$ has a **vertical tangent line** at the point where $x = x_0$ if the limit of the difference quotient is ∞ or $-\infty$. For example, $y = x^{1/3}$ has a vertical tangent line at $x = 0$ (see accompanying figure):

$$\lim_{h \to 0} \frac{f(0 + h) - f(0)}{h} = \lim_{h \to 0} \frac{h^{1/3} - 0}{h}$$

$$= \lim_{h \to 0} \frac{1}{h^{2/3}} = \infty.$$

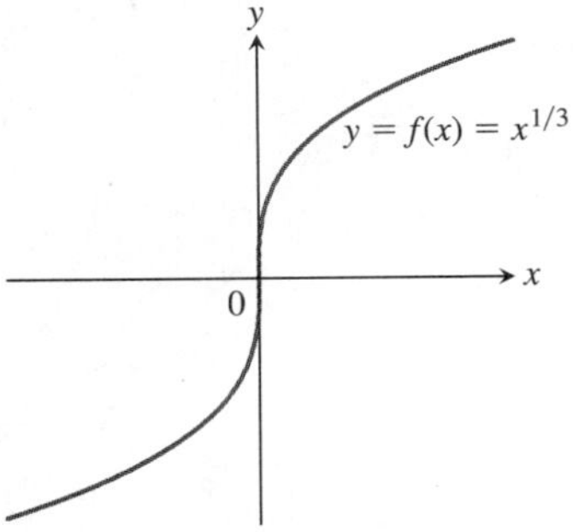

VERTICAL TANGENT LINE AT ORIGIN

However, $y = x^{2/3}$ has *no* vertical tangent line at $x = 0$ (see next figure):

$$\lim_{h\to 0}\frac{g(0+h)-g(0)}{h} = \lim_{h\to 0}\frac{h^{2/3}-0}{h}$$
$$= \lim_{h\to 0}\frac{1}{h^{1/3}}$$

does not exist, because the limit is ∞ from the right and $-\infty$ from the left.

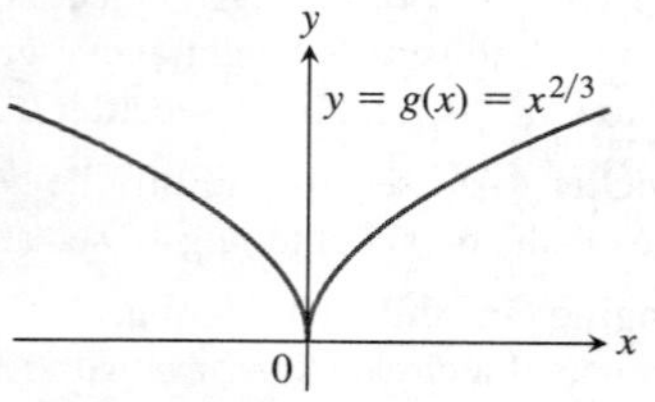

NO VERTICAL TANGENT LINE AT ORIGIN

37. Does the graph of

$$f(x) = \begin{cases} -2, & x < 0 \\ 0, & x = 0 \\ 2, & x > 0 \end{cases}$$

have a vertical tangent line at the origin? Give reasons for your answer.

38. Does the graph of

$$U(x) = \begin{cases} 0, & x < 0 \\ 1, & x \geq 0 \end{cases}$$

have a vertical tangent line at the point $(0,1)$? Give reasons for your answer.

T Graph the curves in Exercises 39–48.

a. Where do the graphs appear to have vertical tangent lines?

b. Confirm your findings in part (a) with limit calculations. But before you do, read the introduction to Exercises 37 and 38.

39. $y = x^{2/7}$

40. $y = x^{4/5}$

41. $y = x^{1/5}$

42. $y = x^{3/5}$

43. $y = 2x^{1/5} - x$

44. $y = x^{5/3} - 5x^{2/3}$

45. $y = x^{2/3} - (x-1)^{1/3}$

46. $y = x^{1/3} + (x-1)^{1/3}$

47. $y = \begin{cases} -\sqrt{|x|}, & x \leq 0 \\ \sqrt{x}, & x > 0 \end{cases}$

48. $y = \sqrt{|9 - x|}$

COMPUTER EXPLORATIONS

Use a CAS to perform the following steps for the functions in Exercises 49–52:

a. Plot $y = f(x)$ over the interval $(x_0 - 1/2) \leq x \leq (x_0 + 3)$.

b. Holding x_0 fixed, the difference quotient

$$q(h) = \frac{f(x_0+h) - f(x_0)}{h}$$

at x_0 becomes a function of the step size h. Enter this function into your CAS workspace.

c. Find the limit of q as $h \to 0$.

d. Define the secant lines $y = f(x_0) + q\cdot(x - x_0)$ for $h = 3, 2$, and 1. Graph them, together with f and the tangent line, over the interval in part (a).

49. $f(x) = x^3 + 2x, \quad x_0 = 0$

50. $f(x) = x + \dfrac{5}{x}, \quad x_0 = 1$

51. $f(x) = x + \sin(2x), \quad x_0 = \pi/2$

52. $f(x) = \cos x + 4\sin(2x), \quad x_0 = \pi$

3.2 The Derivative as a Function

In the last section we defined the derivative of $y = f(x)$ at the point $x = x_0$ to be the limit

$$f'(x_0) = \lim_{h\to 0}\frac{f(x_0+h) - f(x_0)}{h}.$$

HISTORICAL ESSAY
The Derivative
www.bit.ly/2xLhYy3

We now investigate the derivative as a *function* derived from f by considering the limit at each point x in the domain of f.

DEFINITION The **derivative** of the function $f(x)$ with respect to the variable x is the function f' whose value at x is

$$f'(x) = \lim_{h\to 0}\frac{f(x+h) - f(x)}{h},$$

provided the limit exists.

We use the notation $f'(x)$ in the definition, rather than $f'(x_0)$ as before, to emphasize that f' is a function of the independent variable x with respect to which the derivative function $f'(x)$ is being defined. The domain of f' is the set of points in the domain of f for

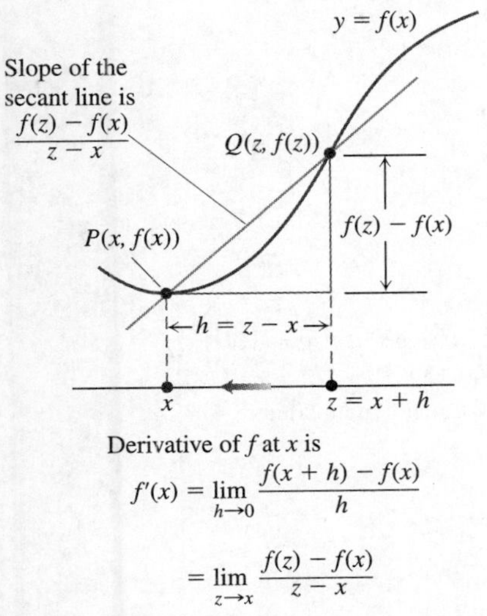

FIGURE 3.4 Two forms for the difference quotient.

which the limit exists, which means that the domain may be the same as or smaller than the domain of f. If f' exists at a particular x, we say that f is **differentiable (has a derivative) at x**. If f' exists at every point in the domain of f, we call f **differentiable**.

If we write $z = x + h$, then $h = z - x$ and h approaches 0 if and only if z approaches x. Therefore, an equivalent definition of the derivative is as follows (see Figure 3.4). This formula is sometimes more convenient to use when finding a derivative function, and it focuses on the point z that approaches x.

Alternative Formula for the Derivative

$$f'(x) = \lim_{z \to x} \frac{f(z) - f(x)}{z - x}$$

Calculating Derivatives from the Definition

The process of calculating a derivative is called **differentiation**. To emphasize the idea that differentiation is an operation performed on a function $y = f(x)$, we use the notation

$$\frac{d}{dx} f(x)$$

as another way to denote the derivative $f'(x)$. Example 1 of Section 3.1 illustrated the differentiation process for the function $y = 1/x$ when $x = a$. For x representing any point in the domain, we get the formula

$$\frac{d}{dx}\left(\frac{1}{x}\right) = -\frac{1}{x^2}.$$

Derivative of the Reciprocal Function

$$\frac{d}{dx}\left(\frac{1}{x}\right) = -\frac{1}{x^2}, \quad x \neq 0$$

Here are two more examples in which we allow x to be any point in the domain of f.

EXAMPLE 1 Differentiate $f(x) = \dfrac{x}{x - 1}$.

Solution We use the definition of derivative, which requires us to calculate $f(x + h)$ and then subtract $f(x)$ to obtain the numerator in the difference quotient. We have

$$f(x) = \frac{x}{x - 1} \quad \text{and} \quad f(x + h) = \frac{(x + h)}{(x + h) - 1}, \text{ so}$$

$$f'(x) = \lim_{h \to 0} \frac{f(x + h) - f(x)}{h} \qquad \text{Definition}$$

$$= \lim_{h \to 0} \frac{\dfrac{x + h}{x + h - 1} - \dfrac{x}{x - 1}}{h} \qquad \text{Substitute.}$$

$$= \lim_{h \to 0} \frac{1}{h} \cdot \frac{(x + h)(x - 1) - x(x + h - 1)}{(x + h - 1)(x - 1)} \qquad \frac{a}{b} - \frac{c}{d} = \frac{ad - cb}{bd}$$

$$= \lim_{h \to 0} \frac{1}{h} \cdot \frac{-h}{(x + h - 1)(x - 1)} \qquad \text{Simplify.}$$

$$= \lim_{h \to 0} \frac{-1}{(x + h - 1)(x - 1)} = \frac{-1}{(x - 1)^2}. \qquad \text{Cancel } h \neq 0 \text{ and evaluate.}$$

■

EXAMPLE 2

(a) Find the derivative of $f(x) = \sqrt{x}$ for $x > 0$.

(b) Find the tangent line to the curve $y = \sqrt{x}$ at $x = 4$.

Derivative of the Square Root Function

$$\frac{d}{dx}\sqrt{x} = \frac{1}{2\sqrt{x}}, \quad x > 0$$

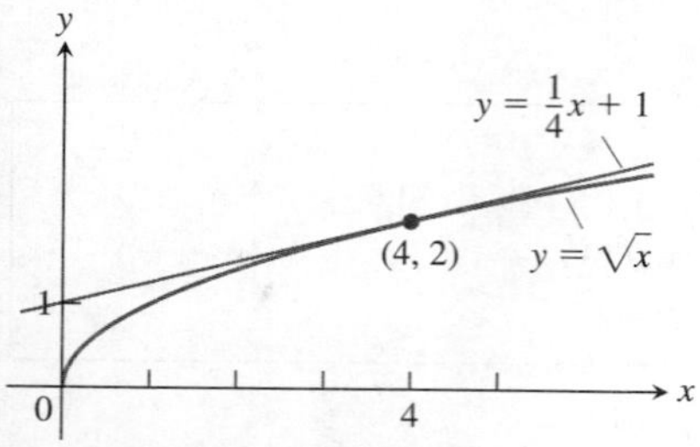

FIGURE 3.5 The curve $y = \sqrt{x}$ and its tangent line at $(4, 2)$. The tangent line's slope is found by evaluating the derivative at $x = 4$ (Example 2).

Solution

(a) We use the alternative formula to calculate f':

$$\begin{aligned} f'(x) &= \lim_{z \to x} \frac{f(z) - f(x)}{z - x} \\ &= \lim_{z \to x} \frac{\sqrt{z} - \sqrt{x}}{z - x} \\ &= \lim_{z \to x} \frac{\sqrt{z} - \sqrt{x}}{(\sqrt{z} - \sqrt{x})(\sqrt{z} + \sqrt{x})} && \frac{1}{a^2 - b^2} = \frac{1}{(a-b)(a+b)} \\ &= \lim_{z \to x} \frac{1}{\sqrt{z} + \sqrt{x}} = \frac{1}{2\sqrt{x}}. && \text{Cancel and evaluate.} \end{aligned}$$

(b) The slope of the curve at $x = 4$ is

$$f'(4) = \frac{1}{2\sqrt{4}} = \frac{1}{4}.$$

The tangent line is the line through the point $(4, 2)$ with slope $1/4$ (Figure 3.5):

$$y = 2 + \frac{1}{4}(x - 4)$$

$$y = \frac{1}{4}x + 1.$$

■

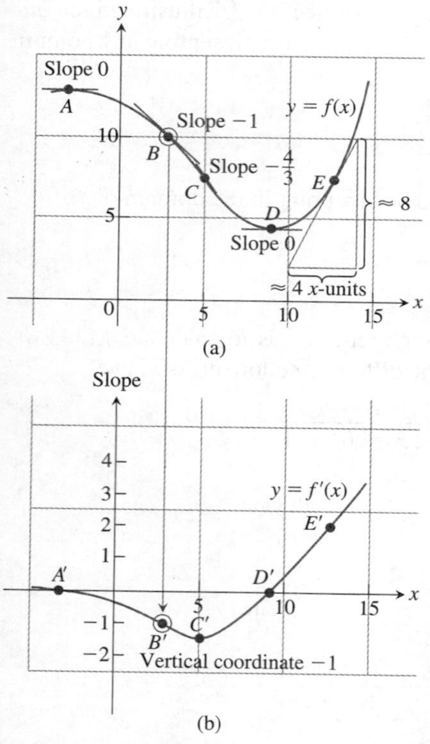

FIGURE 3.6 We made the graph of $y = f'(x)$ in (b) by plotting slopes from the graph of $y = f(x)$ in (a). The vertical coordinate of B' is the slope at B, and so on. The slope at E is approximately $8/4 = 2$. In (b) we see that the rate of change of f is negative for x between A' and D'; the rate of change is positive for x to the right of D'.

Notation

There are many ways to denote the derivative of a function $y = f(x)$, where the independent variable is x and the dependent variable is y. Some common alternative notations for the derivative are

$$f'(x) = y' = \frac{dy}{dx} = \frac{df}{dx} = \frac{d}{dx}f(x) = D(f)(x) = D_x f(x).$$

The symbols d/dx and D indicate the operation of differentiation. We read dy/dx as "the derivative of y with respect to x," and df/dx and $(d/dx)\, f(x)$ as "the derivative of f with respect to x." The "prime" notations y' and f' originate with Newton. The d/dx notations are similar to those used by Leibniz. The symbol dy/dx should not be regarded as a ratio; it simply denotes a derivative.

To indicate the value of a derivative at a specified number $x = a$, we use the notation

$$f'(a) = \left.\frac{dy}{dx}\right|_{x=a} = \left.\frac{df}{dx}\right|_{x=a} = \left.\frac{d}{dx}f(x)\right|_{x=a}.$$

For instance, in Example 2,

$$f'(4) = \left.\frac{d}{dx}\sqrt{x}\,\right|_{x=4} = \left.\frac{1}{2\sqrt{x}}\right|_{x=4} = \frac{1}{2\sqrt{4}} = \frac{1}{4}.$$

Graphing the Derivative

We can often make an approximate plot of the derivative of $y = f(x)$ by estimating the slopes on the graph of f. That is, we plot the points $(x, f'(x))$ in the xy-plane and connect them with a curve that represents $y = f'(x)$.

EXAMPLE 3 Graph the derivative of the function $y = f(x)$ in Figure 3.6a.

Solution We sketch the tangent lines to the graph of f at frequent intervals and use their slopes to estimate the values of $f'(x)$ at these points. We plot the corresponding $(x, f'(x))$ pairs and connect them with a curve as sketched in Figure 3.6b.

■

What can we learn from the graph of $y = f'(x)$? At a glance we can see

1. where the rate of change of f is positive, negative, or zero;
2. the rough size of the growth rate at any x;
3. where the rate of change itself is increasing or decreasing.

Differentiability on an Interval; One-Sided Derivatives

A function $y = f(x)$ is **differentiable on an open interval** (finite or infinite) if it has a derivative at each point of the interval. It is **differentiable on a closed interval** $[a, b]$ if it is differentiable on the interior (a, b) and if the limits

$$\lim_{h\to 0^+} \frac{f(a+h) - f(a)}{h} \qquad \textbf{Right-hand derivative at } a$$

$$\lim_{h\to 0^-} \frac{f(b+h) - f(b)}{h} \qquad \textbf{Left-hand derivative at } b$$

exist at the endpoints (Figure 3.7).

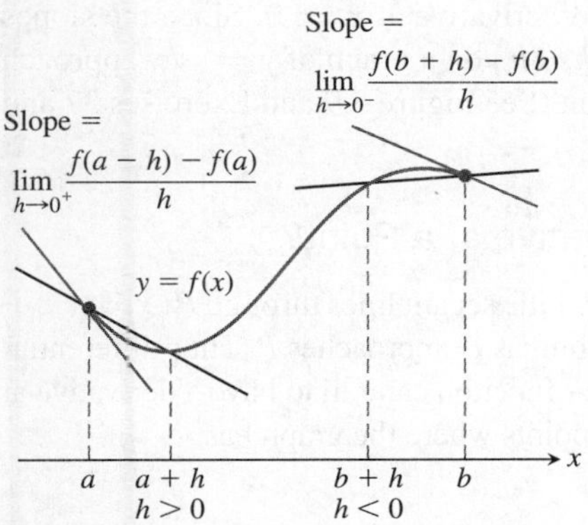

FIGURE 3.7 Derivatives at endpoints of a closed interval are one-sided limits.

Right-hand and left-hand derivatives may or may not be defined at any point of a function's domain. Because of Theorem 5, Section 2.4, a function has a derivative at an interior point if and only if it has left-hand and right-hand derivatives there, and these one-sided derivatives are equal.

EXAMPLE 4 Show that the function $y = |x|$ is differentiable on $(-\infty, 0)$ and on $(0, \infty)$ but has no derivative at $x = 0$.

Solution The graph of the function $y = mx + b$ is a straight line with slope m. Thus, to the right of the origin, when $x > 0$,

$$\frac{d}{dx}(|x|) = \frac{d}{dx}(x) = \frac{d}{dx}(1 \cdot x) = 1. \qquad |x| = x \text{ since } x > 0,\ \frac{d}{dx}(mx+b) = m$$

To the left, when $x < 0$,

$$\frac{d}{dx}(|x|) = \frac{d}{dx}(-x) = \frac{d}{dx}(-1 \cdot x) = -1 \qquad |x| = -x \text{ since } x < 0$$

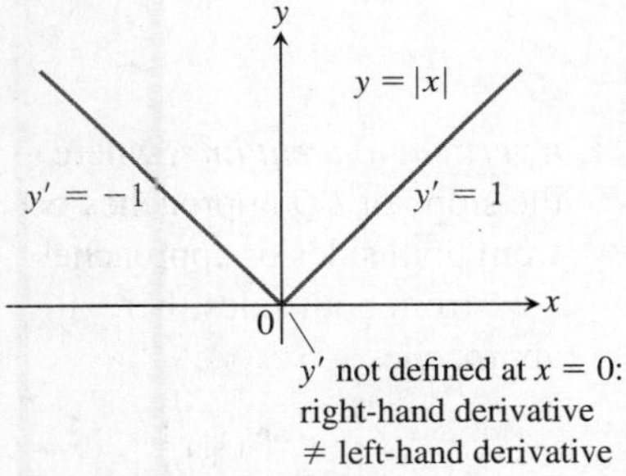

FIGURE 3.8 The function $y = |x|$ is not differentiable at the origin where the graph has a "corner" (Example 4).

(Figure 3.8). The two branches of the graph come together at an angle at the origin, forming a non-smooth corner. There is no derivative at the origin because the one-sided derivatives differ there:

$$\begin{aligned}\text{Right-hand derivative of } |x| \text{ at zero} &= \lim_{h\to 0^+} \frac{|0+h| - |0|}{h} = \lim_{h\to 0^+} \frac{|h|}{h} \\ &= \lim_{h\to 0^+} \frac{h}{h} \qquad |h| = h \text{ when } h > 0 \\ &= \lim_{h\to 0^+} 1 = 1\end{aligned}$$

$$\begin{aligned}\text{Left-hand derivative of } |x| \text{ at zero} &= \lim_{h\to 0^-} \frac{|0+h| - |0|}{h} = \lim_{h\to 0^-} \frac{|h|}{h} \\ &= \lim_{h\to 0^-} \frac{-h}{h} \qquad |h| = -h \text{ when } h < 0 \\ &= \lim_{h\to 0^-} -1 = -1.\end{aligned}$$

■

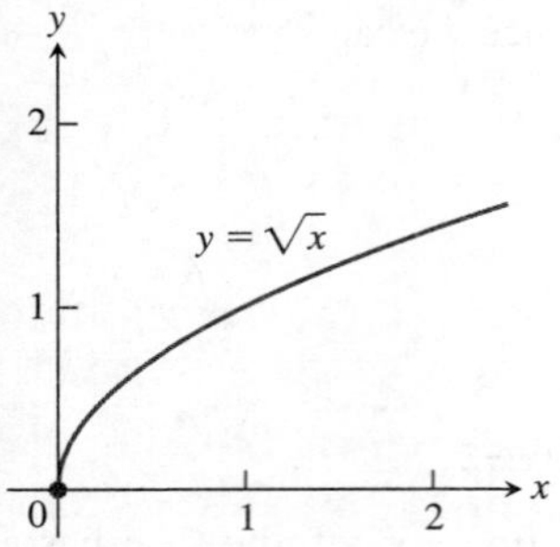

FIGURE 3.9 The square root function is not differentiable at $x = 0$, where the graph of the function has a vertical tangent line.

EXAMPLE 5 In Example 2 we found that for $x > 0$,

$$\frac{d}{dx}\sqrt{x} = \frac{1}{2\sqrt{x}}.$$

We apply the definition to examine whether the derivative exists at $x = 0$:

$$\lim_{h \to 0^+} \frac{\sqrt{0 + h} - \sqrt{0}}{h} = \lim_{h \to 0^+} \frac{1}{\sqrt{h}} = \infty.$$

Since the (right-hand) limit is not finite, there is no derivative at $x = 0$. Since the slopes of the secant lines joining the origin to the points $(h, \sqrt{h})$ on a graph of $y = \sqrt{x}$ approach ∞, the graph has a *vertical tangent line* at the origin. (See Figure 3.9 and Exercises 37 and 38 in Section 3.1.) ■

When Does a Function *Not* Have a Derivative at a Point?

A function has a derivative at a point x_0 if the slopes of the secant lines through $P(x_0, f(x_0))$ and a nearby point Q on the graph approach a finite limit as Q approaches P. Thus differentiability is a "smoothness" condition on the graph of f. A function can fail to have a derivative at a point for many reasons, including the existence of points where the graph has

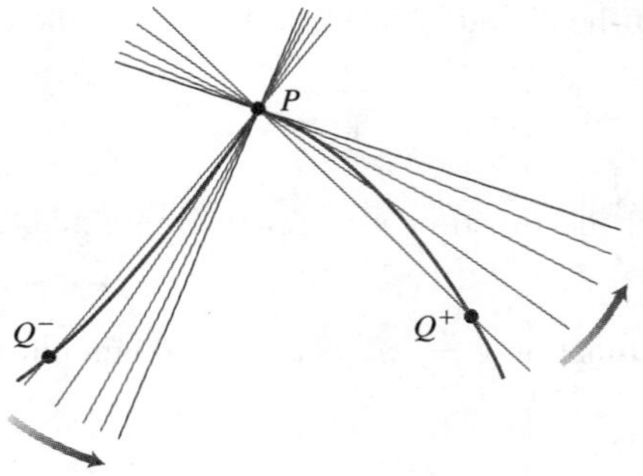

1. a *corner*, where the one-sided derivatives differ

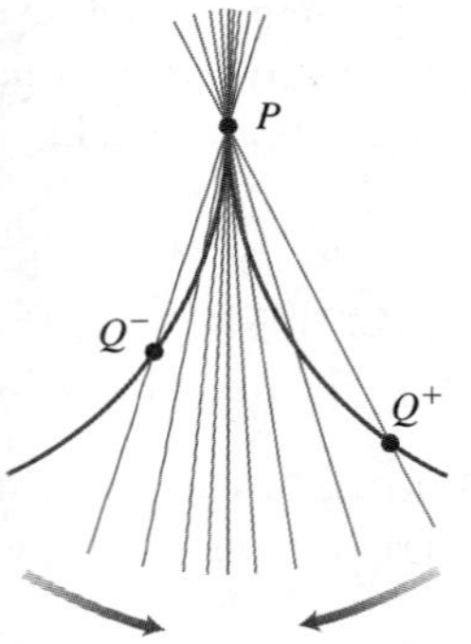

2. a *cusp*, where the slope of PQ approaches ∞ from one side and $-\infty$ from the other

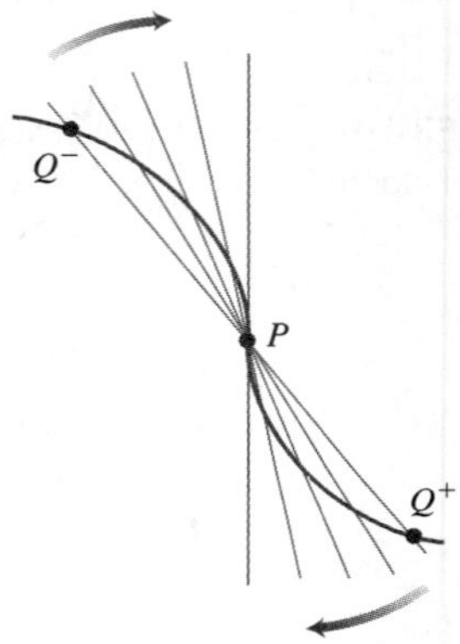

3. a *vertical tangent line*, where the slope of PQ approaches ∞ from both sides or approaches $-\infty$ from both sides (here, it approaches $-\infty$)

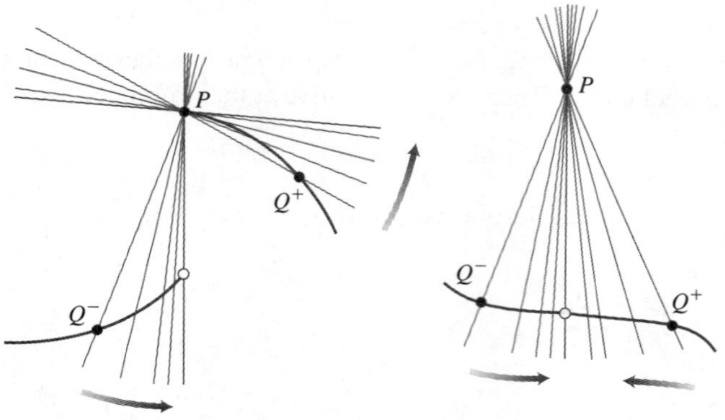

4. a *discontinuity* (two examples shown)

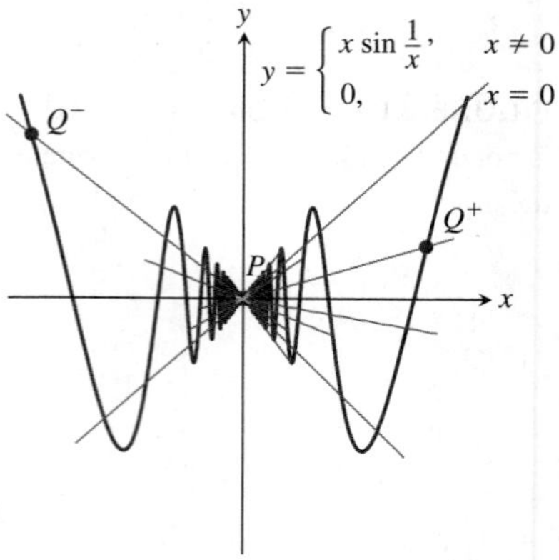

5. wild oscillation

The last example shows a function that is continuous at $x = 0$, but whose graph oscillates wildly up and down as it approaches $x = 0$. The slopes of the secant lines through 0 oscillate between -1 and 1 as x approaches 0, and do not have a limit at $x = 0$.

Differentiable Functions Are Continuous

A function is continuous at every point where it has a derivative.

> **THEOREM 1—Differentiable Implies Continuous** If f has a derivative at $x = c$, then f is continuous at $x = c$.

Proof Given that $f'(c)$ exists, we must show that $\lim_{x\to c} f(x) = f(c)$, or, equivalently, that $\lim_{h\to 0} f(c+h) = f(c)$. If $h \neq 0$, then

$$f(c+h) = f(c) + (f(c+h) - f(c)) \qquad \text{Add and subtract } f(c).$$

$$= f(c) + \frac{f(c+h)-f(c)}{h}\cdot h. \qquad \text{Divide and multiply by } h.$$

Now take limits as $h \to 0$. By Theorem 1 of Section 2.2,

$$\lim_{h\to 0} f(c+h) = \lim_{h\to 0} f(c) + \lim_{h\to 0}\frac{f(c+h)-f(c)}{h}\cdot \lim_{h\to 0} h$$
$$= f(c) + f'(c)\cdot 0$$
$$= f(c) + 0$$
$$= f(c).$$

■

Similar arguments with one-sided limits show that if f has a derivative from one side (right or left) at $x = c$, then f is continuous from that side at $x = c$.

Theorem 1 says that if a function has a discontinuity at a point (for instance, a jump discontinuity), then it cannot be differentiable there. The greatest integer function $y = \lfloor x \rfloor$ fails to be differentiable at every integer $x = n$ (Example 4, Section 2.5).

Caution The converse of Theorem 1 is false. A function need not have a derivative at a point where it is continuous, as we saw with the absolute value function in Example 4. ●

EXERCISES 3.2

Finding Derivative Functions and Values

Using the definition, calculate the derivatives of the functions in Exercises 1–6. Then find the values of the derivatives as specified.

1. $f(x) = 5 - 2x^2;\ f'(-2), f'(0), f'(-1)$
2. $F(x) = (x-1)^2 + 1;\quad F'(-1), F'(0), F'(2)$
3. $g(x) = \dfrac{1}{2t^2};\ g'\left(\dfrac{1}{2}\right), g'(-1), g'\left(\sqrt{2}\right)$
4. $k(z) = \dfrac{1-z}{2z};\quad k'(-1), k'(1), k'\left(\sqrt{2}\right)$
5. $p(\theta) = \sqrt{2\theta}\ ;\ p'(1), p'(2), p'\left(\dfrac{8}{9}\right)$
6. $r(s) = \sqrt{2s+1};\quad r'(0), r'(1), r'(1/2)$

In Exercises 7–12, find the indicated derivatives.

7. $\dfrac{dy}{dx}$ if $y = 5x^2$
8. $\dfrac{dr}{ds}$ if $r = s^3 - 2s^2 + 3$
9. $\dfrac{ds}{dt}$ if $s = \dfrac{2t}{3t+5}$
10. $\dfrac{dv}{dt}$ if $v = t - \dfrac{1}{t}$
11. $\dfrac{dp}{dt}$ if $p = 2q^{1/3}$
12. $\dfrac{dz}{dw}$ if $z = \dfrac{1}{\sqrt{w^2-1}}$

Slopes and Tangent Lines

In Exercises 13–16, differentiate the functions and find the slope of the tangent line at the given value of the independent variable.

13. $f(x) = 3x + \dfrac{5}{2x},\quad x = 2$
14. $k(x) = \dfrac{1}{2+x},\quad x = 2$
15. $s = t^3 - t^2,\quad t = -1$
16. $y = \dfrac{2x+5}{2-x},\quad x = -2$

In Exercises 17–18, differentiate the functions. Then find an equation of the tangent line at the indicated point on the graph of the function.

17. $y = f(x) = \dfrac{8}{\sqrt{x-2}},\quad (x, y) = (6, 4)$
18. $w = g(z) = 2 + \sqrt{9-z},\quad (z, w) = (5, 4)$

In Exercises 19–22, find the values of the derivatives.

19. $\left.\dfrac{ds}{dt}\right|_{t=-2}$ if $s = 5 - 2t^2$
20. $\left.\dfrac{dy}{dx}\right|_{x=\sqrt{3}}$ if $y = 1 - \dfrac{1}{x}$
21. $\left.\dfrac{dr}{d\theta}\right|_{\theta=0}$ if $r = \dfrac{5}{\sqrt{9-\theta}}$
22. $\left.\dfrac{dw}{dz}\right|_{z=4}$ if $w = z + \sqrt{z}$

Using the Alternative Formula for Derivatives

Use the formula

$$f'(x) = \lim_{z \to x} \frac{f(z) - f(x)}{z - x}$$

to find the derivative of the functions in Exercises 23–26.

23. $f(x) = \dfrac{5}{2x+3}$

24. $f(x) = x^2 - 3x + 4$

25. $g(x) = \dfrac{x}{x-1}$

26. $g(x) = 2 + 2\sqrt{x}$

Graphs

Match the functions graphed in Exercises 27–30 with the derivatives graphed in the accompanying figures (a)–(d).

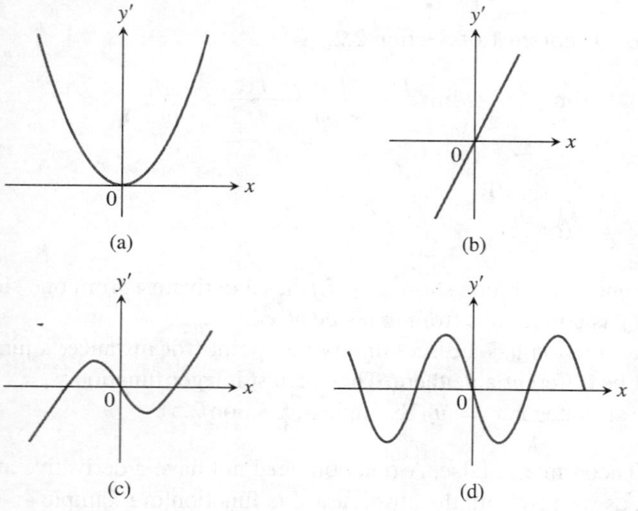

27.

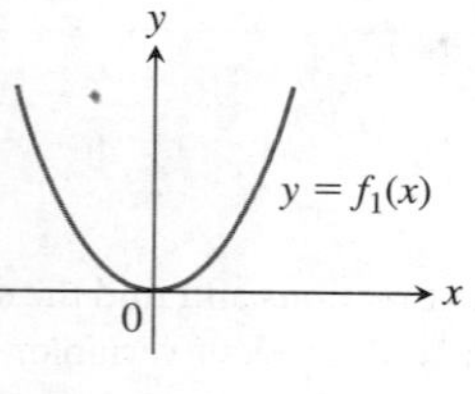

28.

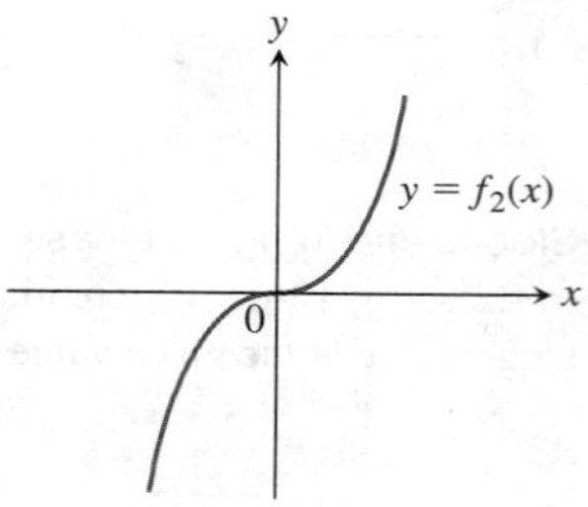

29.

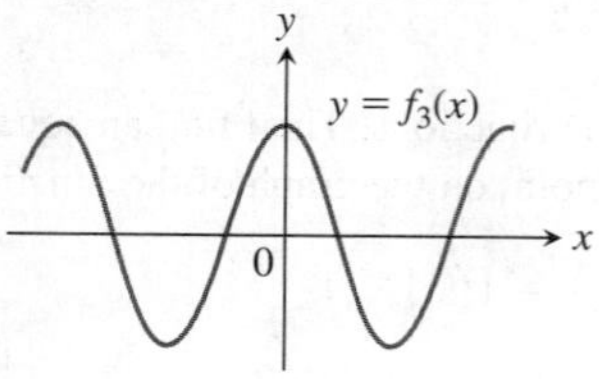

30.

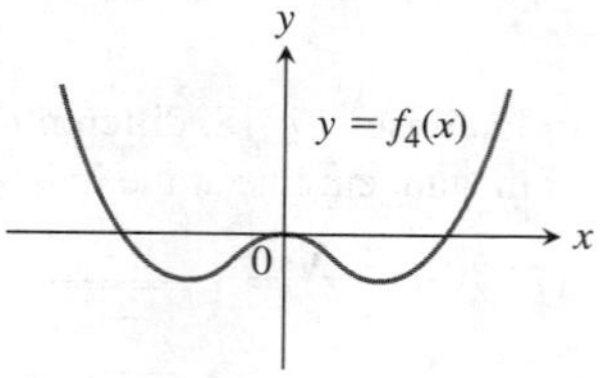

31. Consider the function f graphed here. The domain of f is the interval $[-4, 6]$ and its graph is made of line segments joined end to end.

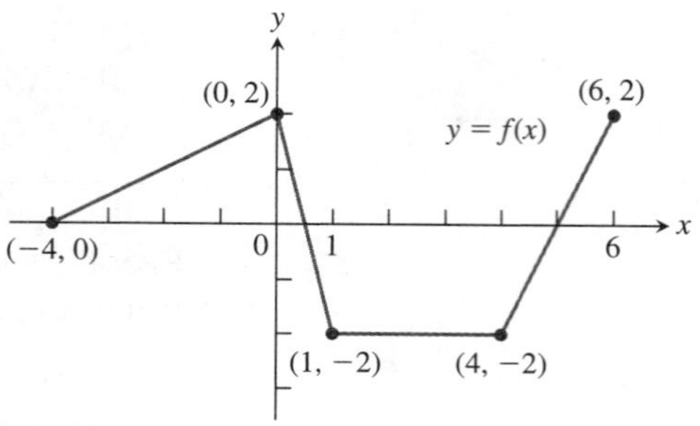

a. At which points of the domain interval is f' not defined? Give reasons for your answer.

b. Graph the derivative of f.
The graph should show a step function.

32. Recovering a function from its derivative

a. Use the following information to graph the function f over the closed interval $[-2, 3]$.

i) The graph of f is made of closed line segments joined end to end.

ii) The graph starts at the point $(-2, 3)$.

iii) The derivative of f is the step function in the figure shown here.

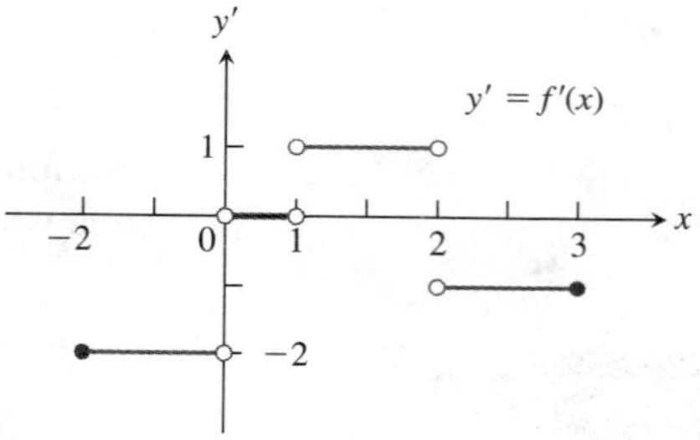

b. Repeat part (a), assuming that the graph starts at $(-2, 0)$ instead of $(-2, 3)$.

33. Growth in the economy The graph in the accompanying figure shows the average annual percentage change $y = f(t)$ in the U.S. gross national product (GNP) for the years 2005–2011. Graph dy/dt (where defined).

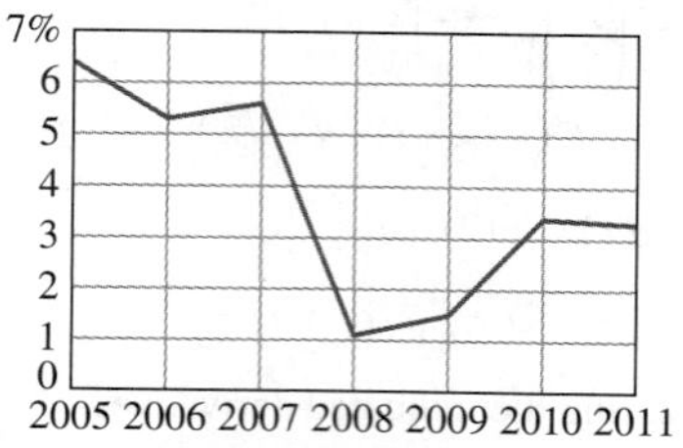

34. Fruit flies (*Continuation of Example 4, Section 2.1.*) Populations starting out in closed environments grow slowly at first, when there are relatively few members, then more rapidly as the number of reproducing individuals increases and resources are still abundant, then slowly again as the population reaches the carrying capacity of the environment.

a. Use the graphical technique of Example 3 to graph the derivative of the fruit fly population as a function of time (in days). The graph of the population is reproduced here.

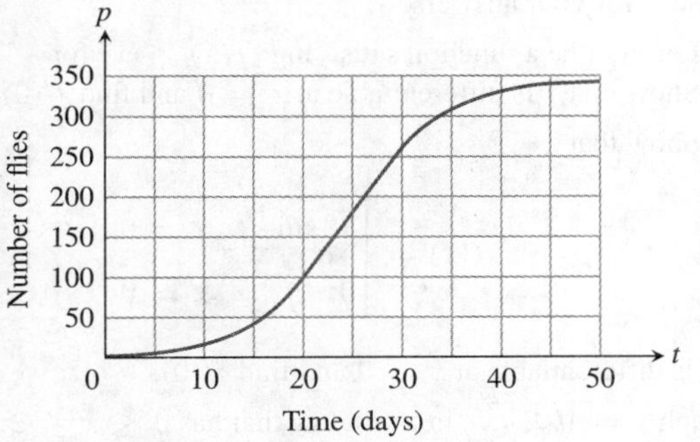

b. During what days does the population seem to be increasing fastest? Slowest?

35. Temperature The given graph shows the outside temperature T in °F, between 5 A.M. and 5 P.M.

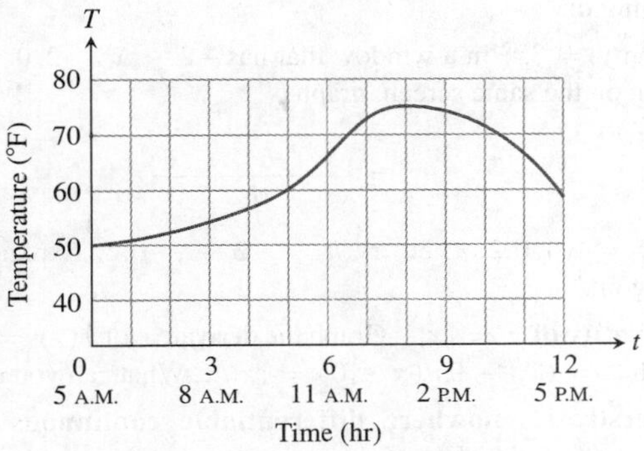

a. Estimate the rate of temperature change at the times

i) 6 A.M. ii) 8 A.M. iii) 1 P.M. iv) 3 P.M.

b. At what time does the temperature increase most rapidly? Decrease most rapidly? What is the rate for each of those times?

c. Use the graphical technique of Example 3 to graph the derivative of temperature T versus time t.

36. Average single-family home prices P (in thousands of dollars) in Sacramento, California, are shown in the accompanying figure from the beginning of 2006 through the end of 2015.

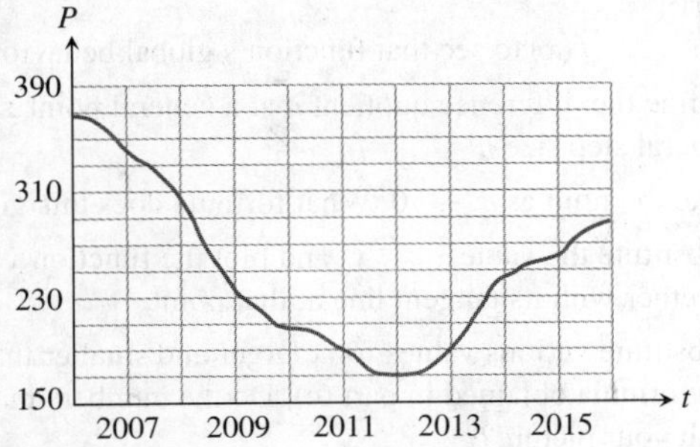

a. During what years did home prices decrease? increase?

b. Estimate home prices at the end of

i) 2007 ii) 2012 iii) 2015

c. Estimate the rate of change of home prices at the beginning of

i) 2007 ii) 2010 iii) 2014

d. During what year did home prices drop most rapidly and what is an estimate of this rate?

e. During what year did home prices rise most rapidly and what is an estimate of this rate?

f. Use the graphical technique of Example 3 to graph the derivative of home price P versus time t.

One-Sided Derivatives

Compute the right-hand and left-hand derivatives as limits to show that the functions in Exercises 37–40 are not differentiable at the point P.

37.

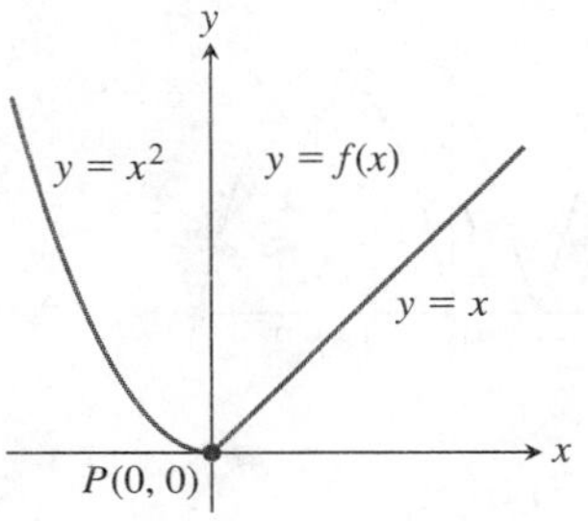

38.

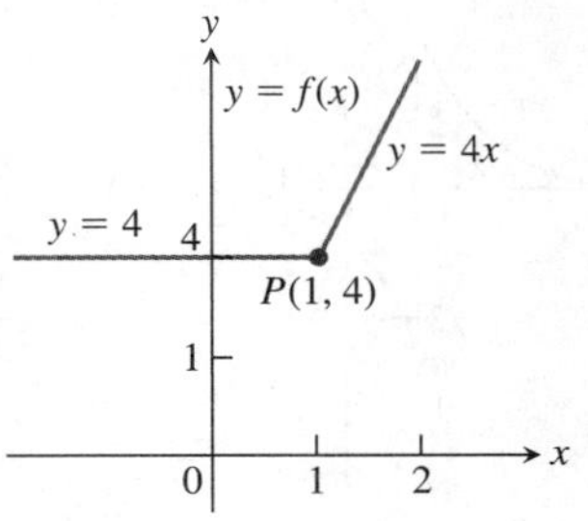

39.

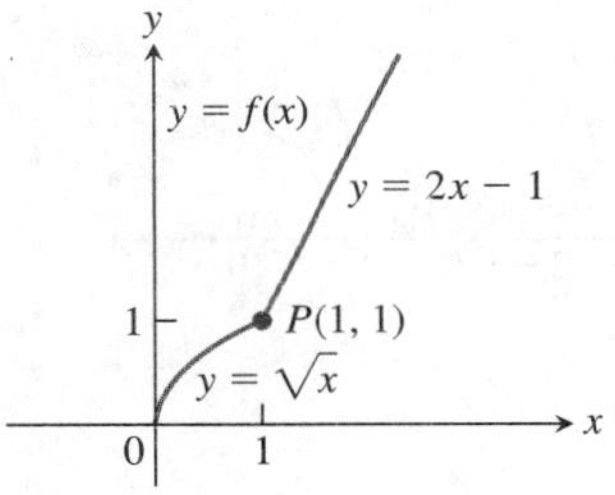

40.

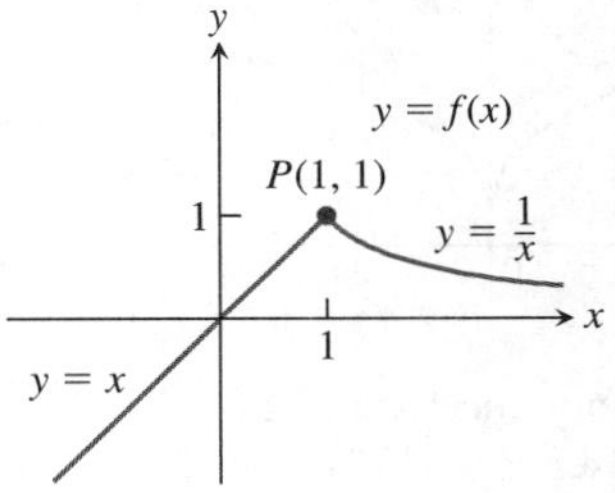

In Exercises 41–44, determine whether the piecewise-defined function is differentiable at $x = 0$.

41. $f(x) = \begin{cases} 2x - 1, & x \geq 0 \\ x^2 + 2x + 7, & x < 0 \end{cases}$

42. $g(x) = \begin{cases} \dfrac{x}{1+x}, & x \geq 0 \\ x^2, & x < 0 \end{cases}$

43. $f(x) = \begin{cases} 2x + \tan x, & x \geq 0 \\ x^2, & x < 0 \end{cases}$

44. $g(x) = \begin{cases} 2x - x^3 - 1, & x \geq 0 \\ x - \dfrac{1}{x + 1}, & x < 0 \end{cases}$

Differentiability and Continuity on an Interval

Each figure in Exercises 45–50 shows the graph of a function over a closed interval D. At what domain points does the function appear to be

a. differentiable?

b. continuous but not differentiable?

c. neither continuous nor differentiable?

Give reasons for your answers.

45.

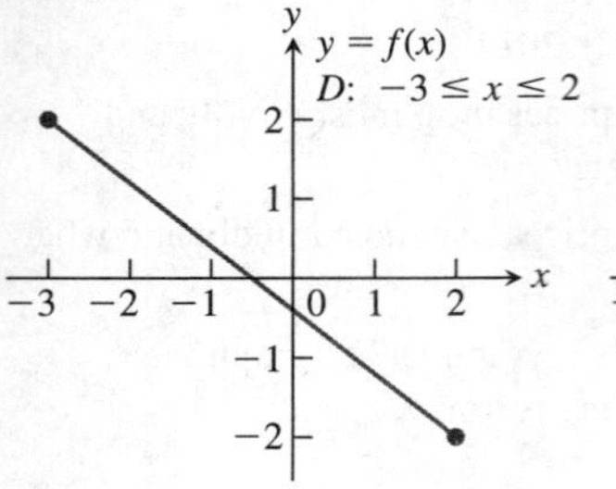

46.

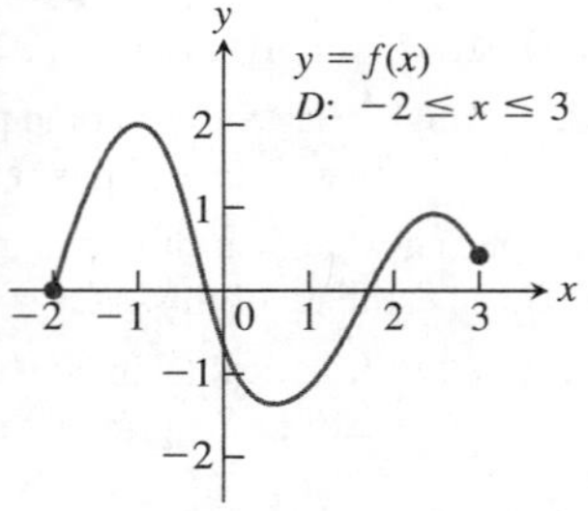

47.

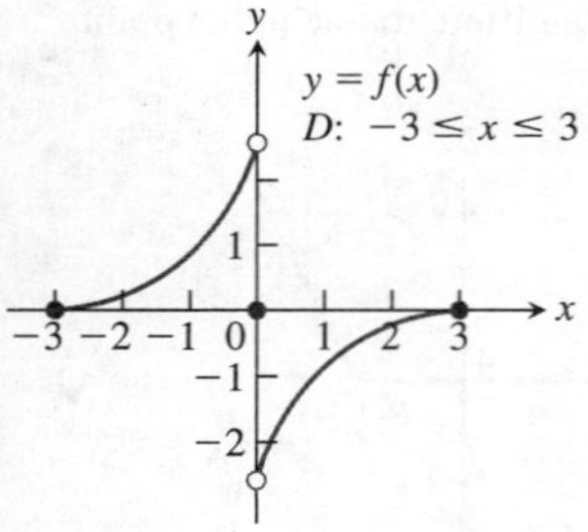

48.

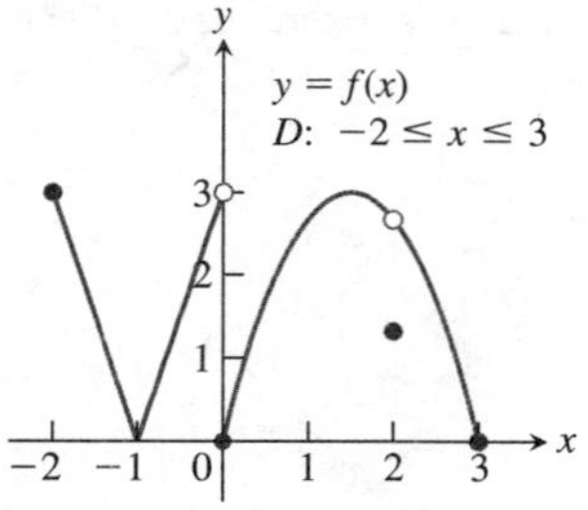

49.

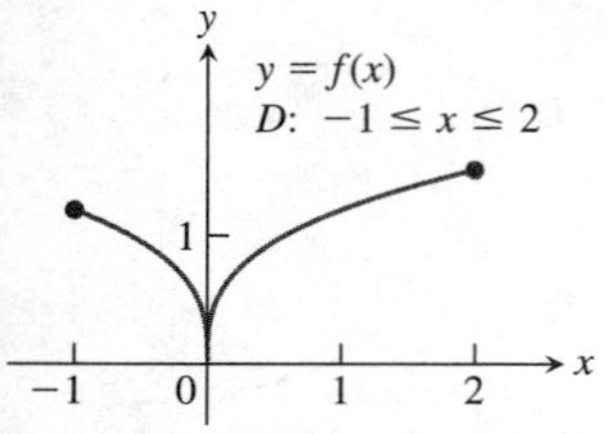

50.

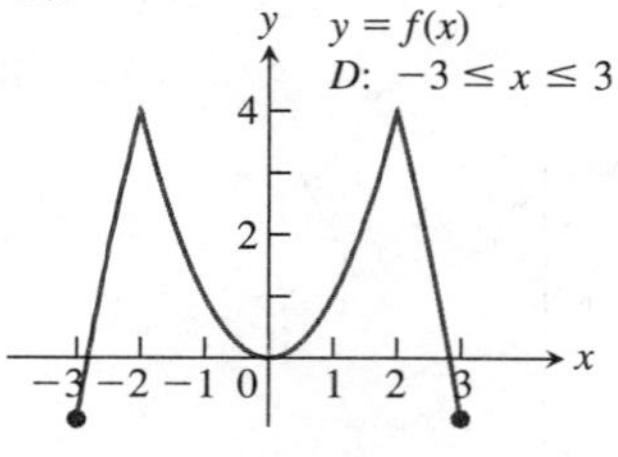

Theory and Examples

In Exercises 51–54,

a. Find the derivative $f'(x)$ of the given function $y = f(x)$.

b. Graph $y = f(x)$ and $y = f'(x)$ side by side using separate sets of coordinate axes, and answer the following questions.

c. For what values of x, if any, is f' positive? Zero? Negative?

d. Over what intervals of x-values, if any, does the function $y = f(x)$ increase as x increases? Decrease as x increases? How is this related to what you found in part (c)? (We will say more about this relationship in Section 4.3.)

51. $y = -x^2$ **52.** $y = -1/x$

53. $y = x^3/3$ **54.** $y = x^4/4$

55. Tangent to a parabola Does the parabola $y = 4x^2 - 12x + 3$ have a tangent line whose slope is 4. If so, find an equation for the line and the point of tangency. If not, why not?

56. Tangent line to $y = \sqrt{x}$ Does any tangent line to the curve $y = \sqrt{x}$ cross the x-axis at $x = -1$? If so, find an equation for the line and the point of tangency. If not, why not?

57. Derivative of $-f$ Does knowing that a function $f(x)$ is differentiable at $x = x_0$ tell you anything about the differentiability of the function $-f$ at $x = x_0$? Give reasons for your answer.

58. Derivative of multiples Does knowing that a function $f(x)$ is differentiable at $x = 5$ tell you anything about the differentiability of the function $2f$ at $x = 5$? Give reasons for your answer.

59. Limit of a quotient Suppose that functions $g(t)$ and $h(t)$ are defined for all values of t and $g(0) = h(0) = 0$. Can $\lim_{t\to 0} g(t)/h(t)$ exist? If it does exist, must it equal zero? Give reasons for your answers.

60. a. Let $f(x)$ be a function satisfying $|f(x)| \le x^2$ for $-1 \le x \le 1$. Show that f is differentiable at $x = 0$ and find $f'(0)$.

b. Show that

$$f(x) = \begin{cases} x^2 \sin\dfrac{1}{x}, & x \neq 0 \\ 0, & x = 0 \end{cases}$$

is differentiable at $x = 0$ and find $f'(0)$.

T **61.** Graph $y = 1/(2\sqrt{x})$ in a window that has $0 \le x \le 2$. Then, on the same screen, graph

$$y = \frac{\sqrt{x+h} - \sqrt{x}}{h}$$

for $h = 1, 0.5, 0.1$. Then try $h = -1, -0.5, -0.1$. Explain what is going on.

T **62.** Graph $y = 3x^2$ in a window that has $-2 \le x \le 2, 0 \le y \le 3$. Then, on the same screen, graph

$$y = \frac{(x+h)^3 - x^3}{h}$$

for $h = 2, 1, 0.2$. Then try $h = -2, -1, -0.2$. Explain what is going on.

63. Derivative of $y = |x|$ Graph the derivative of $f(x) = |x|$. Then graph $y = (|x| - 0)/(x - 0) = |x|/x$. What can you conclude?

T **64. Weierstrass's nowhere differentiable continuous function** The sum of the first eight terms of the Weierstrass function $f(x) = \sum_{n=0}^{\infty} (2/3)^n \cos(9^n \pi x)$ is

$$g(x) = \cos(\pi x) + (2/3)^1 \cos(9\pi x) + (2/3)^2 \cos(9^2\pi x) + (2/3)^3 \cos(9^3\pi x) + \cdots + (2/3)^7 \cos(9^7\pi x).$$

Graph this sum. Zoom in several times. How wiggly and bumpy is this graph? Specify a viewing window in which the displayed portion of the graph is smooth.

COMPUTER EXPLORATIONS

Use a CAS to perform the following steps for the functions in Exercises 65–70.

a. Plot $y = f(x)$ to see that function's global behavior.

b. Define the difference quotient q at a general point x, with general step size h.

c. Take the limit as $h \to 0$. What formula does this give?

d. Substitute the value $x = x_0$ and plot the function $y = f(x)$ together with its tangent line at that point.

e. Substitute various values for x larger and smaller than x_0 into the formula obtained in part (c). Do the numbers make sense with your picture?

f. Graph the formula obtained in part (c). What does it mean when its values are negative? Zero? Positive? Does this make sense with your plot from part (a)? Give reasons for your answer.

65. $f(x) = x^3 + x^2 - x, \quad x_0 = 1$

66. $f(x) = x^{1/3} + x^{2/3}, \quad x_0 = 1$

67. $f(x) = \dfrac{4x}{x^2 + 1}, \quad x_0 = 2$

68. $f(x) = \dfrac{x - 1}{3x^2 + 1}, \quad x_0 = -1$

69. $f(x) = \sin 2x, \quad x_0 = \pi/2$

70. $f(x) = x^2 \cos x, \quad x_0 = \pi/4$

3.3 Differentiation Rules

This section introduces several rules that allow us to differentiate constant functions, power functions, polynomials, rational functions, and certain combinations of them, simply and directly, without having to take limits each time.

Powers, Multiples, Sums, and Differences

A basic rule of differentiation is that the derivative of every constant function is zero.

Derivative of a Constant Function
If f has the constant value $f(x) = c$, then

$$\frac{df}{dx} = \frac{d}{dx}(c) = 0.$$

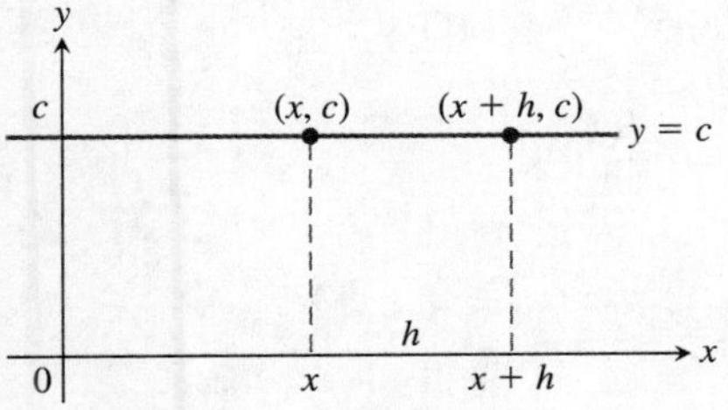

FIGURE 3.10 The rule $(d/dx)(c) = 0$ is another way to say that the values of constant functions never change and that the slope of a horizontal line is zero at every point.

Proof We apply the definition of the derivative to $f(x) = c$, the function whose outputs have the constant value c (Figure 3.10). At every value of x, we find that

$$f'(x) = \lim_{h\to 0} \frac{f(x+h) - f(x)}{h} = \lim_{h\to 0} \frac{c - c}{h} = \lim_{h\to 0} 0 = 0. \quad \blacksquare$$

We now consider powers of x. From Section 3.1, we know that

$$\frac{d}{dx}\left(\frac{1}{x}\right) = -\frac{1}{x^2}, \quad \text{or} \quad \frac{d}{dx}(x^{-1}) = -x^{-2}.$$

From Example 2 of the last section we also know that

$$\frac{d}{dx}\left(\sqrt{x}\right) = \frac{1}{2\sqrt{x}}, \quad \text{or} \quad \frac{d}{dx}(x^{1/2}) = \frac{1}{2}x^{-1/2}.$$

These two examples illustrate a general rule for differentiating a power x^n. We first prove the rule when n is a positive integer.

Derivative of a Positive Integer Power
If n is a positive integer, then

$$\frac{d}{dx}x^n = nx^{n-1}.$$

HISTORICAL BIOGRAPHY
Richard Courant
(1888–1972)
www.bit.ly/2QonUE5

Proof of the Positive Integer Power Rule The formula

$$z^n - x^n = (z - x)(z^{n-1} + z^{n-2}x + \cdots + zx^{n-2} + x^{n-1})$$

can be verified by multiplying out the right-hand side. Then, from the alternative formula for the definition of the derivative,

$$\begin{aligned} f'(x) &= \lim_{z\to x} \frac{f(z) - f(x)}{z - x} = \lim_{z\to x} \frac{z^n - x^n}{z - x} \\ &= \lim_{z\to x} (z^{n-1} + z^{n-2}x + \cdots + zx^{n-2} + x^{n-1}) \qquad n \text{ terms} \\ &= nx^{n-1}. \end{aligned} \quad \blacksquare$$

The Power Rule is actually valid for all real numbers n, not just for positive integers. We have seen examples for a negative integer and fractional power, but n could be an irrational number as well. Here we state the general version of the rule, but postpone its proof until Chapter 7.

Power Rule (General Version)
If n is any real number, then

$$\frac{d}{dx}x^n = nx^{n-1},$$

for all x where the powers x^n and x^{n-1} are defined.

EXAMPLE 1 Differentiate the following powers of x.

(a) x^3 **(b)** $x^{2/3}$ **(c)** $x^{\sqrt{2}}$ **(d)** $\frac{1}{x^4}$ **(e)** $x^{-4/3}$ **(f)** $\sqrt{x^{2+\pi}}$

Applying the Power Rule
Subtract 1 from the exponent and multiply the result by the original exponent.

Solution

(a) $\frac{d}{dx}(x^3) = 3x^{3-1} = 3x^2$

(b) $\frac{d}{dx}(x^{2/3}) = \frac{2}{3}x^{(2/3)-1} = \frac{2}{3}x^{-1/3}$

(c) $\frac{d}{dx}(x^{\sqrt{2}}) = \sqrt{2}x^{\sqrt{2}-1}$

(d) $\frac{d}{dx}\left(\frac{1}{x^4}\right) = \frac{d}{dx}(x^{-4}) = -4x^{-4-1} = -4x^{-5} = -\frac{4}{x^5}$

(e) $\frac{d}{dx}(x^{-4/3}) = -\frac{4}{3}x^{-(4/3)-1} = -\frac{4}{3}x^{-7/3}$

(f) $\frac{d}{dx}\left(\sqrt{x^{2+\pi}}\right) = \frac{d}{dx}\left(x^{1+(\pi/2)}\right) = \left(1+\frac{\pi}{2}\right)x^{1+(\pi/2)-1} = \frac{1}{2}(2+\pi)\sqrt{x^{\pi}}$ ■

The next rule says that when a differentiable function is multiplied by a constant, its derivative is multiplied by the same constant.

Derivative Constant Multiple Rule
If u is a differentiable function of x, and c is a constant, then

$$\frac{d}{dx}(cu) = c\frac{du}{dx}.$$

Proof

$$\frac{d}{dx}cu = \lim_{h\to 0}\frac{cu(x+h)-cu(x)}{h} \qquad \text{Derivative definition with } f(x) = cu(x)$$

$$= c\lim_{h\to 0}\frac{u(x+h)-u(x)}{h} \qquad \text{Constant Multiple Rule for Limits}$$

$$= c\frac{du}{dx} \qquad u \text{ is differentiable.}$$ ■

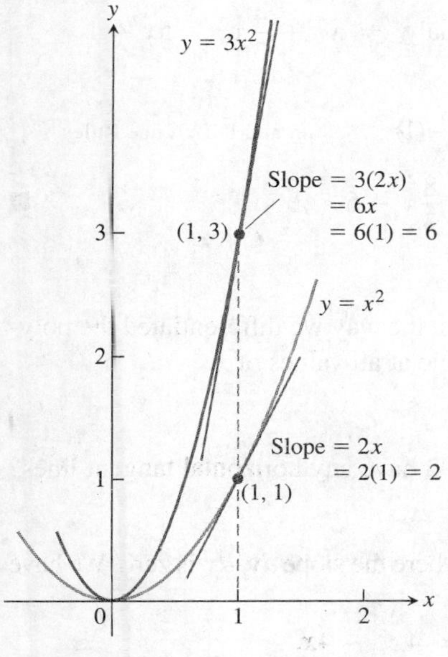

FIGURE 3.11 The graphs of $y = x^2$ and $y = 3x^2$. Tripling the y-coordinate triples the slope (Example 2).

Denoting Functions by u and v
The functions we are working with when we need a differentiation formula are likely to be denoted by letters like f and g. We do not want to use these same letters when stating general differentiation rules, so instead we use letters like u and v that are not likely to be already in use.

EXAMPLE 2

(a) The derivative formula

$$\frac{d}{dx}(3x^2) = 3 \cdot 2x = 6x$$

says that if we rescale the graph of $y = x^2$ by multiplying each y-coordinate by 3, then we multiply the slope at each point by 3 (Figure 3.11).

(b) Negative of a function

The derivative of the negative of a differentiable function u is the negative of the function's derivative. The Constant Multiple Rule with $c = -1$ gives

$$\frac{d}{dx}(-u) = \frac{d}{dx}(-1 \cdot u) = -1 \cdot \frac{d}{dx}(u) = -\frac{du}{dx}. \quad \blacksquare$$

The next rule says that the derivative of the sum of two differentiable functions is the sum of their derivatives.

Derivative Sum Rule

If u and v are differentiable functions of x, then their sum $u + v$ is differentiable at every point where u and v are both differentiable. At such points,

$$\frac{d}{dx}(u + v) = \frac{du}{dx} + \frac{dv}{dx}.$$

Proof We apply the definition of the derivative to $f(x) = u(x) + v(x)$:

$$\begin{aligned}\frac{d}{dx}[u(x) + v(x)] &= \lim_{h\to 0}\frac{[u(x+h) + v(x+h)] - [u(x) + v(x)]}{h}\\ &= \lim_{h\to 0}\left[\frac{u(x+h) - u(x)}{h} + \frac{v(x+h) - v(x)}{h}\right]\\ &= \lim_{h\to 0}\frac{u(x+h) - u(x)}{h} + \lim_{h\to 0}\frac{v(x+h) - v(x)}{h} = \frac{du}{dx} + \frac{dv}{dx}. \quad \blacksquare\end{aligned}$$

Combining the Sum Rule with the Constant Multiple Rule gives the **Difference Rule**, which says that the derivative of a *difference* of differentiable functions is the difference of their derivatives:

$$\frac{d}{dx}(u - v) = \frac{d}{dx}[u + (-1)v] = \frac{du}{dx} + (-1)\frac{dv}{dx} = \frac{du}{dx} - \frac{dv}{dx}.$$

The Sum Rule also extends to finite sums of more than two functions. If $u_1, u_2, \ldots, u_n$ are differentiable at x, then so is $u_1 + u_2 + \cdots + u_n$, and

$$\frac{d}{dx}(u_1 + u_2 + \cdots + u_n) = \frac{du_1}{dx} + \frac{du_2}{dx} + \cdots + \frac{du_n}{dx}.$$

A proof by mathematical induction for any finite number of terms is given in Appendix A.2.

EXAMPLE 3 Find the derivative of the polynomial $y = x^3 + \frac{4}{3}x^2 - 5x + 1$.

Solution
$$\begin{aligned}\frac{dy}{dx} &= \frac{d}{dx}x^3 + \frac{d}{dx}\left(\frac{4}{3}x^2\right) - \frac{d}{dx}(5x) + \frac{d}{dx}(1) \qquad \text{Sum and Difference Rules}\\ &= 3x^2 + \frac{4}{3}\cdot 2x - 5 + 0 = 3x^2 + \frac{8}{3}x - 5\end{aligned}$$
■

We can differentiate any polynomial term by term, the way we differentiated the polynomial in Example 3. All polynomials are differentiable at all values of x.

EXAMPLE 4 Does the curve $y = x^4 - 2x^2 + 2$ have any horizontal tangent lines? If so, where?

Solution The horizontal tangent lines, if any, occur where the slope dy/dx is zero. We have

$$\frac{dy}{dx} = \frac{d}{dx}(x^4 - 2x^2 + 2) = 4x^3 - 4x.$$

Now solve the equation $\frac{dy}{dx} = 0$ for x:

$$\begin{aligned}4x^3 - 4x &= 0\\ 4x(x^2 - 1) &= 0\\ x &= 0, 1, -1.\end{aligned}$$

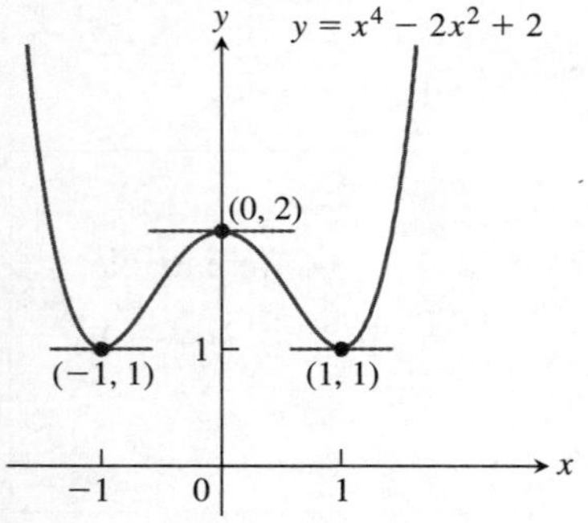

FIGURE 3.12 The curve in Example 4 and its horizontal tangent lines.

The curve $y = x^4 - 2x^2 + 2$ has horizontal tangent lines at $x = 0, 1$, and -1. The corresponding points on the curve are $(0, 2)$, $(1, 1)$, and $(-1, 1)$. See Figure 3.12. ■

Products and Quotients

While the derivative of the sum of two functions is the sum of their derivatives, the derivative of the product of two functions is *not* the product of their derivatives. For instance,

$$\frac{d}{dx}(x \cdot x) = \frac{d}{dx}(x^2) = 2x, \qquad \text{while} \qquad \frac{d}{dx}(x)\cdot\frac{d}{dx}(x) = 1 \cdot 1 = 1.$$

The derivative of a product of two functions is the sum of *two* products, as we now explain.

Derivative Product Rule
If u and v are differentiable at x, then so is their product uv, and

$$\frac{d}{dx}(uv) = u\frac{dv}{dx} + \frac{du}{dx}v.$$

The derivative of the product uv is u times the derivative of v plus the derivative of u times v. In prime notation, $(uv)' = uv' + u'v$. In function notation,

$$\frac{d}{dx}[f(x)g(x)] = f(x)g'(x) + f'(x)g(x), \quad \text{or} \quad (fg)' = fg' + f'g. \qquad (1)$$

EXAMPLE 5 Find the derivative of $y = (x^2 + 1)(x^3 + 3)$.

Solution

(a) From the Product Rule with $u = x^2 + 1$ and $v = x^3 + 3$, we find

$$\frac{d}{dx}[(x^2+1)(x^3+3)] = (x^2+1)(3x^2) + (2x)(x^3+3) \qquad \frac{d}{dx}(uv) = u\frac{dv}{dx} + \frac{du}{dx}v$$

$$= 3x^4 + 3x^2 + 2x^4 + 6x$$

$$= 5x^4 + 3x^2 + 6x.$$

(b) This particular product can be differentiated as well (perhaps better) by multiplying out the original expression for y and differentiating the resulting polynomial:

$$y = (x^2+1)(x^3+3) = x^5 + x^3 + 3x^2 + 3$$

$$\frac{dy}{dx} = 5x^4 + 3x^2 + 6x.$$

This is in agreement with our first calculation. ■

Picturing the Product Rule

Suppose $u(x)$ and $v(x)$ are positive and increase when x increases, and $h > 0$.

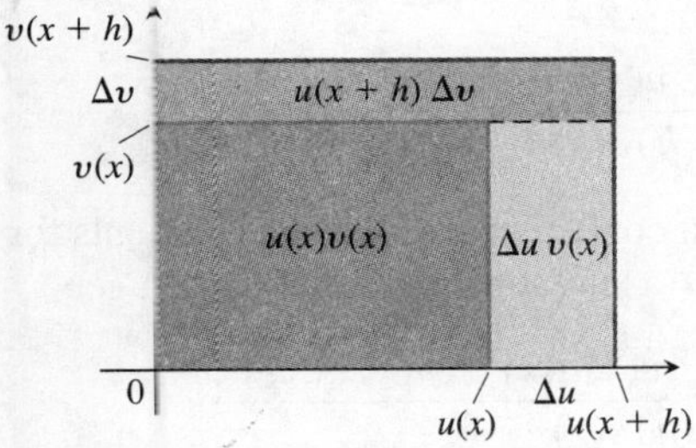

Then the change in the product uv is the difference in areas of the larger and smaller "boxes," which is the sum of the areas of the upper and right-hand. That is,

$$\Delta(uv) = u(x+h)v(x+h) - u(x)v(x)$$
$$= u(x+h)\Delta v + \Delta u v(x).$$

Division by h gives

$$\frac{\Delta(uv)}{h} = u(x+h)\frac{\Delta v}{h} + \frac{\Delta u}{h}v(x).$$

The limit as $h \to 0^+$ yields the Product Rule.

Proof of the Derivative Product Rule

$$\frac{d}{dx}(uv) = \lim_{h\to 0}\frac{u(x+h)v(x+h) - u(x)v(x)}{h}$$

To change this fraction into an equivalent one that contains difference quotients for the derivatives of u and v, we subtract and add $u(x+h)v(x)$ in the numerator:

$$\frac{d}{dx}(uv) = \lim_{h\to 0}\frac{u(x+h)v(x+h) - u(x+h)v(x) + u(x+h)v(x) - u(x)v(x)}{h}$$

$$= \lim_{h\to 0}\left[u(x+h)\frac{v(x+h)-v(x)}{h} + \frac{u(x+h)-u(x)}{h}v(x)\right]$$

$$= \lim_{h\to 0}u(x+h)\cdot\lim_{h\to 0}\frac{v(x+h)-v(x)}{h} + \lim_{h\to 0}\frac{u(x+h)-u(x)}{h}\cdot v(x).$$

As h approaches zero, $u(x+h)$ approaches $u(x)$ because u, being differentiable at x, is continuous at x. The two fractions approach the values of dv/dx at x and du/dx at x. Therefore,

$$\frac{d}{dx}(uv) = u\frac{dv}{dx} + \frac{du}{dx}v.$$ ■

The derivative of the quotient of two functions is given by the Quotient Rule.

> **Derivative Quotient Rule**
> If u and v are differentiable at x and if $v(x) \neq 0$, then the quotient u/v is differentiable at x, and
>
> $$\frac{d}{dx}\left(\frac{u}{v}\right) = \frac{v\dfrac{du}{dx} - u\dfrac{dv}{dx}}{v^2}.$$

In function notation,

$$\frac{d}{dx}\left[\frac{f(x)}{g(x)}\right] = \frac{g(x)f'(x) - f(x)g'(x)}{[g(x)]^2}.$$

EXAMPLE 6 Find the derivative of $y = \dfrac{t^2 - 1}{t^3 + 1}$.

Solution

We apply the Quotient Rule with $u = t^2 - 1$ and $v = t^3 + 1$:

$$\frac{dy}{dt} = \frac{(t^3 + 1) \cdot 2t - (t^2 - 1) \cdot 3t^2}{(t^3 + 1)^2} \qquad \frac{d}{dt}\left(\frac{u}{v}\right) = \frac{v(du/dt) - u(dv/dt)}{v^2}$$

$$= \frac{2t^4 + 2t - 3t^4 + 3t^2}{(t^3 + 1)^2}$$

$$= \frac{-t^4 + 3t^2 + 2t}{(t^3 + 1)^2}.$$

■

Proof of the Derivative Quotient Rule

$$\frac{d}{dx}\left(\frac{u}{v}\right) = \lim_{h \to 0} \frac{\dfrac{u(x + h)}{v(x + h)} - \dfrac{u(x)}{v(x)}}{h}$$

$$= \lim_{h \to 0} \frac{v(x)u(x + h) - u(x)v(x + h)}{hv(x + h)v(x)}$$

To change the last fraction into an equivalent one that contains the difference quotients for the derivatives of u and v, we subtract and add $v(x)\,u(x)$ in the numerator. We then get

$$\frac{d}{dx}\left(\frac{u}{v}\right) = \lim_{h \to 0} \frac{v(x)u(x + h) - v(x)u(x) + v(x)u(x) - u(x)v(x + h)}{hv(x + h)v(x)}$$

$$= \lim_{h \to 0} \frac{v(x)\dfrac{u(x + h) - u(x)}{h} - u(x)\dfrac{v(x + h) - v(x)}{h}}{v(x + h)v(x)}.$$

Taking the limits in the numerator and denominator now gives the Quotient Rule. Exercise 62 outlines another proof. ■

The choice of which rules to use in solving a differentiation problem can make a difference in how much work you have to do. Here is an example.

EXAMPLE 7 Find the derivative of

$$y = \frac{(x - 1)(x^2 - 2x)}{x^4}.$$

Solution Using the Quotient Rule here will result in a complicated expression with many terms. Instead, use some algebra to simplify the expression. First expand the numerator and divide by x^4:

$$y = \frac{(x - 1)(x^2 - 2x)}{x^4} = \frac{x^3 - 3x^2 + 2x}{x^4} = x^{-1} - 3x^{-2} + 2x^{-3}.$$

Then use the Sum, Constant Multiple, and Power Rules:

$$\frac{dy}{dx} = -x^{-2} - 3(-2)x^{-3} + 2(-3)x^{-4}$$

$$= -\frac{1}{x^2} + \frac{6}{x^3} - \frac{6}{x^4}.$$

■

Second- and Higher-Order Derivatives

If $y = f(x)$ is a differentiable function, then its derivative $f'(x)$ is also a function. If f' is also differentiable, then we can differentiate f' to get a new function of x denoted by f''. So $f'' = (f')'$. The function f'' is called the **second derivative** of f because it is the derivative of the first derivative. It is written in several ways:

$$f''(x) = \frac{d^2y}{dx^2} = \frac{d}{dx}\left(\frac{dy}{dx}\right) = \frac{dy'}{dx} = y'' = D^2(f)(x) = D_x^2 f(x).$$

The symbol D^2 means that the operation of differentiation is performed twice.

If $y = x^6$, then $y' = 6x^5$ and we have

$$y'' = \frac{dy'}{dx} = \frac{d}{dx}(6x^5) = 30x^4.$$

Thus $D^2(x^6) = 30x^4$.

If y'' is differentiable, its derivative, $y''' = dy''/dx = d^3y/dx^3$, is the **third derivative** of y with respect to x. The names continue as you imagine, with

$$y^{(n)} = \frac{d}{dx}y^{(n-1)} = \frac{d^ny}{dx^n} = D^ny$$

denoting the ***n*th derivative** of y with respect to x for any positive integer n.

We can interpret the second derivative as the rate of change of the slope of the tangent line to the graph of $y = f(x)$ at each point. You will see in the next chapter that the second derivative reveals whether the graph bends upward or downward from the tangent line as we move off the point of tangency. In the next section, we interpret both the second and third derivatives in terms of motion along a straight line.

How to Read the Symbols for Derivatives

y'	"y prime"
y''	"y double prime"
$\frac{d^2y}{dx^2}$	"d squared y by dx squared"
y'''	"y triple prime"
$y^{(n)}$	"y super n"
$\frac{d^ny}{dx^n}$	"d to the n of y by dx to the n"
D^n	"d to the n"

EXAMPLE 8 The first four derivatives of $y = x^3 - 3x^2 + 2$ are

First derivative:	$y' = 3x^2 - 6x$
Second derivative:	$y'' = 6x - 6$
Third derivative:	$y''' = 6$
Fourth derivative:	$y^{(4)} = 0.$

All polynomial functions have derivatives of all orders. In this example, the fifth and later derivatives are all zero. ■

EXERCISES 3.3

Derivative Calculations

In Exercises 1–12, find the first and second derivatives.

1. $y = -x^2 + 5$ **2.** $y = x^2 + x + 8$

3. $s = 7t^3 - 2t^5$ **4.** $w = 3z^7 - 7z^3 + 21z^2$

5. $y = \frac{4x^3}{3} - x$ **6.** $y = \frac{2x^3}{5} - \frac{5x^2}{7} + \frac{3x}{4}$

7. $w = 3z^{-2} - \frac{1}{z}$ **8.** $s = -2t^{-1} + \frac{4}{t^2}$

9. $y = 5x^3 + 7x^2 - 10x - 3x^{-2} + 6x^{-3}$ **10.** $y = 4 - 2x - x^{-3}$

11. $r = \frac{1}{3s^2} - \frac{5}{2s}$ **12.** $r = \frac{12}{\theta} - \frac{4}{\theta^3} + \frac{1}{\theta^4}$

In Exercises 13–16, find y' **(a)** by applying the Product Rule and **(b)** by multiplying the factors to produce a sum of simpler terms to differentiate.

13. $y = (5 - 2x^2)(2x^3 - 3x + 4)$ **14.** $y = (2x + 3)(5x^2 - 4x)$

15. $y = (x^2 + 1)\left(x + 6 + \frac{2}{x}\right)$ **16.** $y = (1 + x^2)(x^{3/4} - x^{-3})$

Find the derivatives of the functions in Exercises 17–28.

17. $y = \frac{x+2}{x-2}$ **18.** $z = \frac{4 - 3x}{3x^2 + x}$

19. $g(x) = \frac{x^2 - 9}{x + 5}$ **20.** $f(t) = \frac{t^2 - 1}{t^2 + t - 2}$

21. $v = (2 - t)(7 + 2t^2)^{-2}$

22. $w = (2x - 7)^{-1}(x + 5)$

23. $f(s) = \dfrac{\sqrt{4s} - 9}{\sqrt{4s} + 9}$

24. $u = \dfrac{5x + 1}{2\sqrt{x}}$

25. $v = \dfrac{1 + x - 4\sqrt{x}}{x}$

26. $r = 5\left(\dfrac{5}{\sqrt{\theta}} + 5\sqrt{\theta}\right)$

27. $y = \dfrac{1}{(x^2 - 1)(x^2 + x + 1)}$

28. $r = \dfrac{(x+5)(x+2)}{(x-5)(x-2)}$

Find the derivatives of all orders of the functions in Exercises 29–32.

29. $y = \dfrac{x^4}{2} - \dfrac{3}{2}x^2 - x$

30. $y = \dfrac{x^5}{120}$

31. $y = (x - 5)(x + 7)(x + 9)$

32. $y = (4x^2 + 3)(2 - x)x$

Find the first and second derivatives of the functions in Exercises 33–38.

33. $y = \dfrac{x + 9}{2x}$

34. $s = \dfrac{t^2 + 5t - 1}{t^2}$

35. $r = \dfrac{(\theta - 1)(\theta^2 + \theta + 1)}{\theta^3}$

36. $u = \dfrac{(2x^2 + 5x)(2x^2 - 3x + 5)}{x^4}$

37. $w = \left(\dfrac{1 + 3z}{3z}\right)(3 - z)$

38. $p = \dfrac{(2q^2 + 9)}{(q-2)^3 + (q+2)^3}$

39. Suppose u and v are functions of x that are differentiable at $x = 0$ and that

$$u(0) = 7, \quad u'(0) = -5, \quad v(0) = -2, \quad v'(0) = 4$$

Find the values of the following derivatives at $x = 0$.

a. $\dfrac{d}{dx}(uv)$ b. $\dfrac{d}{dx}\left(\dfrac{u}{v}\right)$ c. $\dfrac{d}{dx}\left(\dfrac{v}{u}\right)$ d. $\dfrac{d}{dx}(5v - 3u)$

40. Suppose u and v are differentiable functions of x and that

$$u(1) = 2, \quad u'(1) = 0, \quad v(1) = 5, \quad v'(1) = -1.$$

Find the values of the following derivatives at $x = 1$.

a. $\dfrac{d}{dx}(uv)$ b. $\dfrac{d}{dx}\left(\dfrac{u}{v}\right)$ c. $\dfrac{d}{dx}\left(\dfrac{v}{u}\right)$ d. $\dfrac{d}{dx}(7v - 2u)$

Slopes and Tangent Lines

41. a. **Normal line to a curve** Find an equation for the line perpendicular to the tangent line to the curve $y = x^3 - 4x + 1$ at the point $(2, 1)$.

b. **Smallest slope** What is the smallest slope on the curve? At what point on the curve does the curve have this slope?

c. **Tangent lines having specified slope** Find equations for the tangent lines to the curve at the points where the slope of the curve is 8.

42. a. **Horizontal tangent lines** Find equations for the horizontal tangent lines to the curve $y = x^3 - 3x - 2$. Also find equations for the lines that are perpendicular to these tangent lines at the points of tangency.

b. **Smallest slope** What is the smallest slope on the curve? At what point on the curve does the curve have this slope? Find an equation for the line that is perpendicular to the curve's tangent line at this point.

43. Find the tangent lines to *Newton's serpentine* (graphed here) at the origin and the point $(1, 2)$.

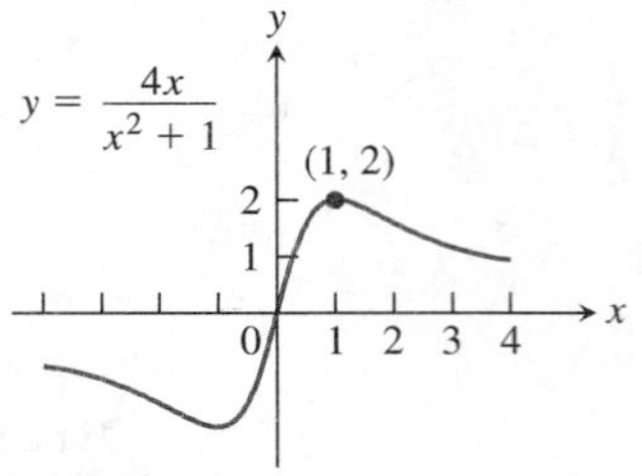

44. Find the tangent line to the *Witch of Agnesi* (graphed here) at the point $(2, 1)$.

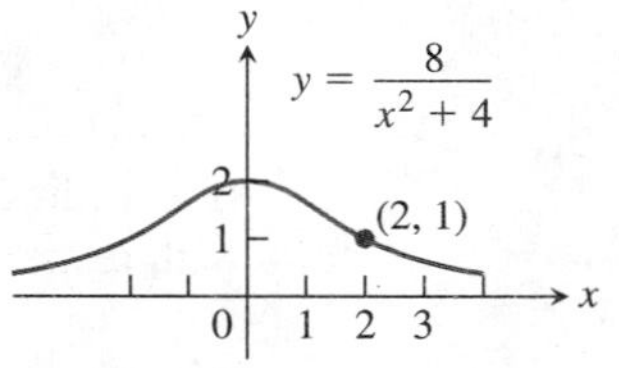

HISTORICAL BIOGRAPHY
Maria Gaetana Agnesi
(1718–1799)

45. **Quadratic tangent to identity function** The curve $y = ax^2 + bx + c$ passes through the point $(1, 2)$ and is tangent to the line $y = x$ at the origin. Find a, b, and c.

46. **Quadratics having a common tangent line** The curves $y = x^2 + ax + b$ and $y = cx - x^2$ have a common tangent line at the point $(1, 0)$. Find a, b, and c.

47. Find all points (x, y) on the graph of $f(x) = 5x^2 - 2x$ with tangent lines parallel to the line $y = 8x - 5$.

48. Find all points (x, y) on the graph of $g(x) = \frac{1}{3}x^3 - \frac{3}{2}x^2 + 1$ with tangent lines parallel to the line $8x - 2y = 1$.

49. Find all points (x, y) on the graph of $y = x/(x - 2)$ with tangent lines perpendicular to the line $y = 2x + 3$.

50. Find all points (x, y) on the graph of $f(x) = x^2$ with tangent lines passing through the point $(3, 8)$.

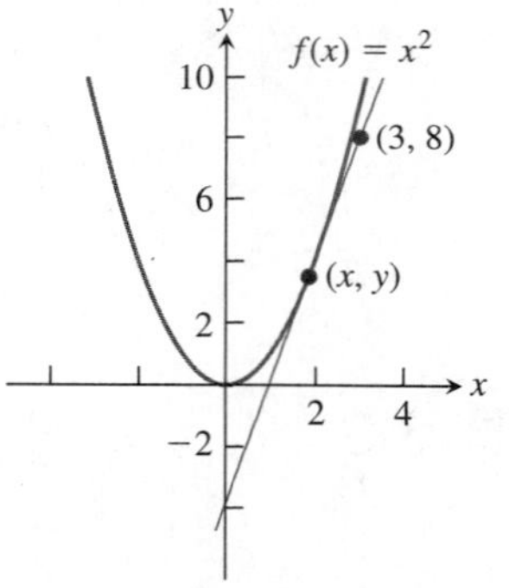

51. Assume that functions f and g are differentiable with $f(1) = 2$, $f'(1) = -3$, $g(1) = 4$, and $g'(1) = -2$. Find the equation of the line tangent to the graph of $F(x) = f(x)g(x)$ at $x = 1$.

52. Assume that functions f and g are differentiable with $f(2) = 3$, $f'(2) = -1$, $g(2) = -4$, and $g'(2) = 1$. Find an equation of the line perpendicular to the line tangent to the graph of $F(x) = \dfrac{f(x) + 3}{x - g(x)}$ at $x = 2$.

53. **a.** Find an equation for the line that is tangent to the curve $y = 2x^3 - x$ at the point $(-1, -1)$.

T **b.** Graph the curve and tangent line together. The tangent intersects the curve at another point. Use Zoom and trace to estimate the point's coordinates.

T **c.** Confirm your estimates of the coordinates of the second intersection point by solving the equations for the curve and tangent line simultaneously.

54. **a.** Find an equation for the line that is tangent to the curve $y = x^3 - 6x^2 + 5x$ at the origin.

T **b.** Graph the curve and tangent line together. The tangent line intersects the curve at another point. Use Zoom and Trace to estimate the point's coordinates.

T **c.** Confirm your estimates of the coordinates of the second intersection point by solving the equations for the curve and tangent line simultaneously.

Theory and Examples

For Exercises 55 and 56, evaluate each limit by first converting each to a derivative at a particular x-value.

55. $\lim_{x \to 1} \frac{x^{50} - 1}{x - 1}$ 56. $\lim_{x \to -1} \frac{x^{2/9} - 1}{x + 1}$

57. Find the value of a that makes the following function differentiable for all x-values.

$$g(x) = \begin{cases} ax, & x \geq 0 \\ 2x^2 - x, & x < 0 \end{cases}$$

58. Find the values of a and b that make the following function differentiable for all x-values.

$$f(x) = \begin{cases} ax + b, & x > -1 \\ bx^2 - 3, & x \leq -1 \end{cases}$$

59. The general polynomial of degree n has the form

$$P(x) = a_n x^n + a_{n-1} x^{n-1} + \cdots + a_2 x^2 + a_1 x + a_0,$$

where $a_n \neq 0$. Find $P'(x)$.

60. **The body's reaction to medicine** The reaction of the body to a dose of medicine can sometimes be represented by an equation of the form

$$R = M^2\left(\frac{C}{2} - \frac{M}{3}\right),$$

where C is a positive constant and M is the amount of medicine absorbed in the blood. If the reaction is a change in blood pressure, R is measured in millimeters of mercury. If the reaction is a change in temperature, R is measured in degrees, and so on.

Find dR/dM. This derivative, as a function of M, is called the sensitivity of the body to the medicine. In Section 4.5, we will see how to find the amount of medicine to which the body is most sensitive.

61. Suppose that the function v in the Derivative Product Rule has a constant value c. What does the Derivative Product Rule then say? What does this say about the Derivative Constant Multiple Rule?

62. **The Reciprocal Rule**

a. The *Reciprocal Rule* says that at any point where the function $v(x)$ is differentiable and different from zero,

$$\frac{d}{dx}\left(\frac{1}{v}\right) = -\frac{1}{v^2}\frac{dv}{dx}.$$

Show that the Reciprocal Rule is a special case of the Derivative Quotient Rule.

b. Show that the Reciprocal Rule and the Derivative Product Rule together imply the Derivative Quotient Rule.

63. **Generalizing the Product Rule** The Derivative Product Rule gives the formula

$$\frac{d}{dx}(uv) = u\frac{dv}{dx} + \frac{du}{dx}v$$

for the derivative of the product uv of two differentiable functions of x.

a. What is the analogous formula for the derivative of the product uvw of *three* differentiable functions of x?

b. What is the formula for the derivative of the product $u_1u_2u_3u_4$ of *four* differentiable functions of x?

c. What is the formula for the derivative of a product $u_1u_2u_3 \cdots u_n$ of a finite number n of differentiable functions of x?

64. **Power Rule for negative integers** Use the Derivative Quotient Rule to prove the Power Rule for negative integers, that is,

$$\frac{d}{dx}(x^{-m}) = -mx^{-m-1}$$

where m is a positive integer.

65. **Cylinder pressure** If gas in a cylinder is maintained at a constant temperature T, the pressure P is related to the volume V by a formula of the form

$$P = \frac{nRT}{V - nb} - \frac{an^2}{V^2},$$

in which a, b, n, and R are constants. Find dP/dV. (See accompanying figure.)

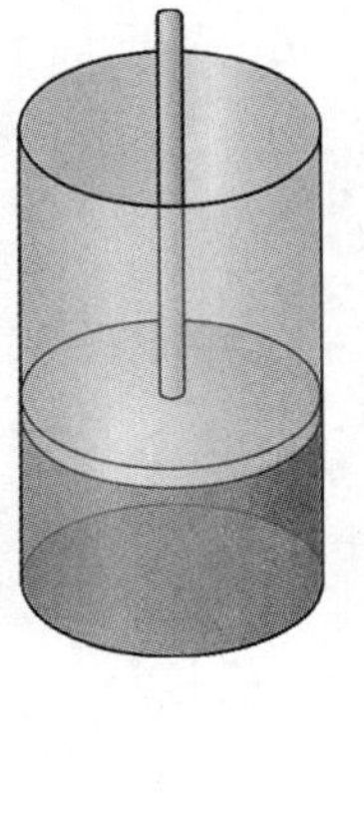

66. The best quantity to order One of the formulas for inventory management says that the average weekly cost of ordering, paying for, and holding merchandise is

$$A(q) = \frac{km}{q} + cm + \frac{hq}{2},$$

where q is the quantity you order when things run low (shoes, TVs, brooms, or whatever the item might be); k is the cost of placing an order (the same, no matter how often you order); c is the cost of one item (a constant); m is the number of items sold each week (a constant); and h is the weekly holding cost per item (a constant that takes into account things such as space, utilities, insurance, and security). Find dA/dq and d^2A/dq^2.

3.4 The Derivative as a Rate of Change

In this section we study applications where derivatives model the rates at which things change. It is natural to think of a quantity changing with respect to time, but other variables can be treated in the same way. For example, an economist may want to study how the cost of producing steel varies with the number of tons produced, or an engineer may want to know how the power output of a generator varies with its temperature.

Instantaneous Rates of Change

If we interpret the difference quotient $(f(x + h) - f(x))/h$ as the average rate of change in f over the interval from x to $x + h$, we can interpret its limit as $h \to 0$ as the instantaneous rate at which f is changing at the point x. This gives an important interpretation of the derivative.

DEFINITION The **instantaneous rate of change** of f with respect to x at x_0 is the derivative

$$f'(x_0) = \lim_{h \to 0} \frac{f(x_0 + h) - f(x_0)}{h},$$

provided the limit exists.

Thus, instantaneous rates are limits of average rates.

It is conventional to use the word *instantaneous* even when x does not represent time. The word is, however, frequently omitted. When we say *rate of change*, we mean *instantaneous rate of change*.

EXAMPLE 1 The area A of a circle is related to its diameter by the equation

$$A = \frac{\pi}{4} D^2.$$

How fast does the area change with respect to the diameter when the diameter is 10 m?

Solution The rate of change of the area with respect to the diameter is

$$\frac{dA}{dD} = \frac{\pi}{4} \cdot 2D = \frac{\pi D}{2}.$$

When $D = 10$ m, the area is changing with respect to the diameter at the rate of $(\pi/2)10 = 5\pi \text{ m}^2/\text{m} \approx 15.71 \text{ m}^2/\text{m}$. ■

Motion Along a Line: Displacement, Velocity, Speed, Acceleration, and Jerk

Suppose that an object (or body, considered as a whole mass) is moving along a coordinate line (an s-axis), usually horizontal or vertical, so that we know its position s on that line as a function of time t:

$$s = f(t).$$

The **displacement** of the object over the time interval from t to $t + \Delta t$ (Figure 3.13) is

$$\Delta s = f(t + \Delta t) - f(t),$$

and the **average velocity** of the object over that time interval is

$$v_{av} = \frac{\text{displacement}}{\text{travel time}} = \frac{\Delta s}{\Delta t} = \frac{f(t + \Delta t) - f(t)}{\Delta t}.$$

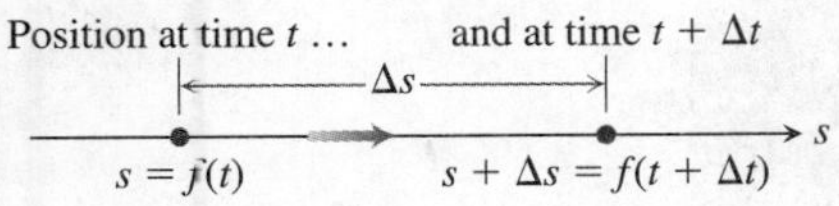

FIGURE 3.13 The positions of a body moving along a coordinate line at time t and shortly later at time $t + \Delta t$. Here the coordinate line is horizontal.

To find the body's velocity at the exact instant t, we take the limit of the average velocity over the interval from t to $t + \Delta t$ as Δt approaches zero. This limit is the derivative of f with respect to t.

DEFINITION **Velocity (instantaneous velocity)** is the derivative of position with respect to time. If a body's position at time t is $s = f(t)$, then the body's velocity at time t is

$$v(t) = \frac{ds}{dt} = \lim_{\Delta t \to 0} \frac{f(t + \Delta t) - f(t)}{\Delta t}.$$

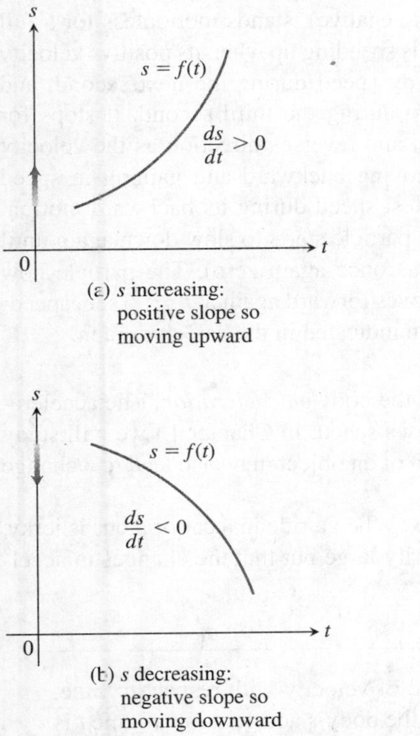

FIGURE 3.14 For motion $s = f(t)$ along a straight line (the vertical axis), $v = ds/dt$ is (a) positive when s increases and (b) negative when s decreases.

Besides telling how fast an object is moving along the horizontal line in Figure 3.13, its velocity tells the direction of motion. When the object is moving forward (s increasing), the velocity is positive; when the object is moving backward (s decreasing), the velocity is negative. If the coordinate line is vertical, the object moves upward for positive velocity and downward for negative velocity. The curves in Figure 3.14 represent position along the line over time; they do not portray the path of motion, which lies along the vertical s-axis.

If we drive to a friend's house and back at 30 mph, say, the speedometer will show 30 on the way over but it will not show -30 on the way back, even though our distance from home is decreasing. The speedometer always shows *speed*, which is the absolute value of velocity. Speed measures the rate of progress regardless of direction.

DEFINITION **Speed** is the absolute value of velocity.

$$\text{Speed} = |v(t)| = \left|\frac{ds}{dt}\right|$$

EXAMPLE 2 Figure 3.15 shows the graph of the velocity $v = f'(t)$ of a particle moving along a horizontal line as in Figure 3.13 (as opposed to the graph of a position function $s = f(t)$, such as in Figure 3.14). In the graph of the velocity function, it's not the slope of the curve that tells us whether the particle is moving forward or backward along the line (which is not shown in the figure), but rather the sign of the velocity. Figure 3.15

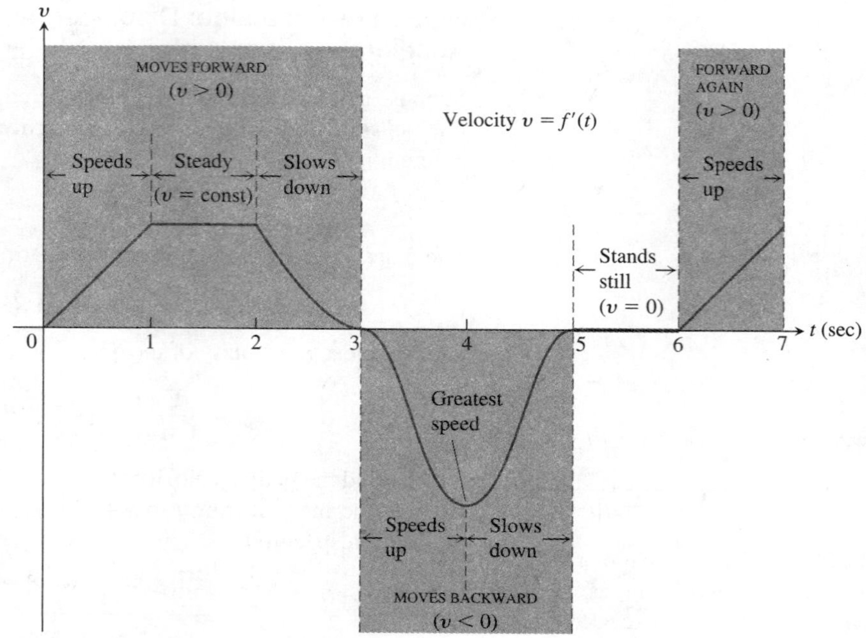

FIGURE 3.15 The velocity graph of a particle moving along a horizontal line, discussed in Example 2.

shows that the particle moves forward for the first 3 sec (when the velocity is positive), moves backward for the next 2 sec (the velocity is negative), stands motionless for a full second, and then moves forward again. The particle is speeding up when its positive velocity increases during the first second, moves at a steady speed during the next second, and then slows down as the velocity decreases to zero during the third second. It stops for an instant at $t = 3$ sec (when the velocity is zero) and reverses direction as the velocity starts to become negative. The particle is now moving backward and gaining in speed until $t = 4$ sec, at which time it achieves its greatest speed during its backward motion. Continuing its backward motion at time $t = 4$, the particle starts to slow down again until it finally stops at time $t = 5$ (when the velocity is once again zero). The particle now remains motionless for one full second, and then moves forward again at $t = 6$ sec, speeding up during the final second of the forward motion indicated in the velocity graph. ■

HISTORICAL BIOGRAPHY
Bernard Bolzano
(1781–1848)
www.bit.ly/2zICCA0

The rate at which a body's velocity changes is the body's *acceleration*. The acceleration measures how quickly the body picks up or loses speed. In Chapter 13 we will study motion in the plane and in space, where acceleration of an object may also lead to a change in direction.

A sudden change in acceleration is called a *jerk*. When a ride in a car or a bus is jerky, it is not that the accelerations involved are necessarily large but that the changes in acceleration are abrupt.

DEFINITIONS **Acceleration** is the derivative of velocity with respect to time. If a body's position at time t is $s = f(t)$, then the body's acceleration at time t is

$$a(t) = \frac{dv}{dt} = \frac{d^2s}{dt^2}.$$

Jerk is the derivative of acceleration with respect to time:

$$j(t) = \frac{da}{dt} = \frac{d^3s}{dt^3}.$$

Near the surface of Earth, all bodies fall with the same constant acceleration. Galileo's experiments with free fall (see Section 2.1) lead to the equation

$$s = \frac{1}{2}gt^2,$$

where s is the distance fallen and g is the acceleration due to Earth's gravity. This equation holds in a vacuum, where there is no air resistance, and closely models the fall of dense, heavy objects, such as rocks or steel tools, for the first few seconds of their fall, before the effects of air resistance are significant.

The value of g in the equation $s = (1/2)gt^2$ depends on the units used to measure t and s. With t in seconds (the usual unit), the value of g determined by measurement at sea level is approximately 32 ft/sec^2 (feet per second squared) in English units, and $g = 9.8\ \text{m/sec}^2$ (meters per second squared) in metric units. (These gravitational constants depend on the distance from Earth's center of mass, and are slightly lower on top of Mt. Everest, for example.)

The jerk associated with the constant acceleration of gravity ($g = 32\ \text{ft/sec}^2$) is zero:

$$j = \frac{d}{dt}(g) = 0.$$

An object does not exhibit jerkiness during free fall.

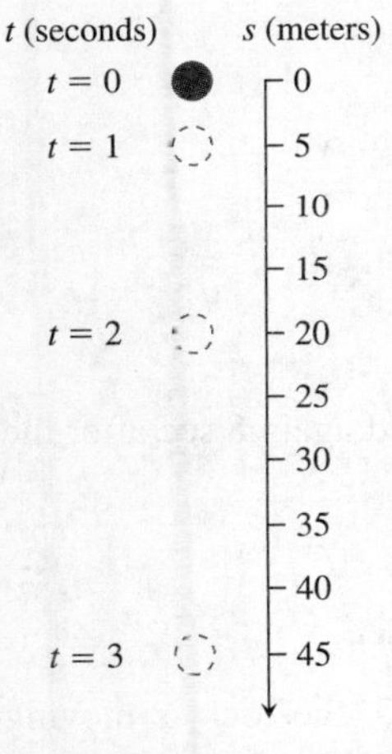

FIGURE 3.16 A ball bearing falling from rest (Example 3).

EXAMPLE 3 Figure 3.16 shows the free fall of a heavy ball bearing released from rest at time $t = 0$ sec.

(a) How many meters does the ball fall in the first 3 sec?

(b) What are its velocity, speed, and acceleration when $t = 3$?

Solution

(a) The metric free-fall equation is $s = 4.9t^2$. During the first 3 sec, the ball falls

$$s(3) = 4.9(3)^2 = 44.1\ \text{m}.$$

(b) At any time t, *velocity* is the derivative of position:

$$v(t) = s'(t) = \frac{d}{dt}(4.9t^2) = 9.8t.$$

At $t = 3$, the velocity is

$$v(3) = 29.4\ \text{m/sec}$$

in the downward (increasing s) direction. The *speed* at $t = 3$ is

$$\text{speed} = |v(3)| = 29.4\ \text{m/sec}.$$

The *acceleration* at any time t is

$$a(t) = v'(t) = s''(t) = 9.8\ \text{m/sec}^2.$$

At $t = 3$, the acceleration is $9.8\ \text{m/sec}^2$. ■

EXAMPLE 4 A dynamite blast blows a heavy rock straight up with a launch velocity of 160 ft/sec (about 109 mph) (Figure 3.17a). It reaches a height of $s = 160t - 16t^2$ ft after t sec.

(a) How high does the rock go?

(b) What are the velocity and speed of the rock when it is 256 ft above the ground on the way up? On the way down?

(c) What is the acceleration of the rock at any time t during its flight (after the blast)?

(d) When does the rock hit the ground again?

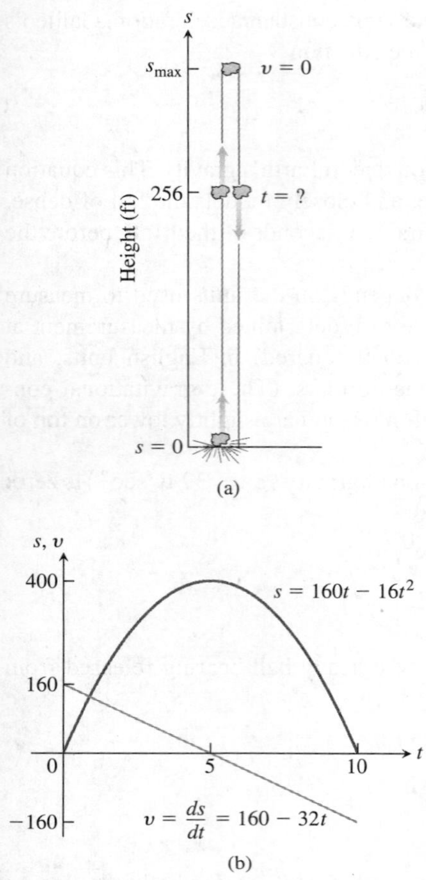

FIGURE 3.17 (a) The rock in Example 4. (b) The graphs of s and v as functions of time; s is largest when $v = ds/dt = 0$. The graph of s is *not* the path of the rock: It is a plot of height versus time. The slope of the plot is the rock's velocity, graphed here as a straight line.

Solution

(a) In the coordinate system we have chosen, s measures height from the ground up, so the velocity is positive on the way up and negative on the way down. The instant the rock is at its highest point is the one instant during the flight when the velocity is 0. To find the maximum height, all we need to do is to find when $v = 0$ and evaluate s at this time.

At any time t during the rock's motion, its velocity is

$$v = \frac{ds}{dt} = \frac{d}{dt}(160t - 16t^2) = 160 - 32t \text{ ft/sec.}$$

The velocity is zero when

$$160 - 32t = 0 \quad \text{or} \quad t = 5 \text{ sec.}$$

The rock's height at $t = 5$ sec is

$$s_{max} = s(5) = 160(5) - 16(5)^2 = 800 - 400 = 400 \text{ ft.}$$

See Figure 3.17b.

(b) To find the rock's velocity at 256 ft on the way up and again on the way down, we first find the two values of t for which

$$s(t) = 160t - 16t^2 = 256.$$

To solve this equation, we write

$$\begin{aligned} 16t^2 - 160t + 256 &= 0 \\ 16(t^2 - 10t + 16) &= 0 \\ (t - 2)(t - 8) &= 0 \\ t &= 2 \text{ sec}, t = 8 \text{ sec.} \end{aligned}$$

The rock is 256 ft above the ground 2 sec after the explosion and again 8 sec after the explosion. The rock's velocities at these times are

$$\begin{aligned} v(2) &= 160 - 32(2) = 160 - 64 = 96 \text{ ft/sec.} \\ v(8) &= 160 - 32(8) = 160 - 256 = -96 \text{ ft/sec.} \end{aligned}$$

At both instants, the rock's speed is 96 ft/sec. Since $v(2) > 0$, the rock is moving upward (s is increasing) at $t = 2$ sec; it is moving downward (s is decreasing) at $t = 8$ because $v(8) < 0$.

(c) At any time during its flight following the explosion, the rock's acceleration is a constant

$$a = \frac{dv}{dt} = \frac{d}{dt}(160 - 32t) = -32 \text{ ft/sec}^2.$$

The acceleration is always downward and is the effect of gravity on the rock. As the rock rises, it slows down; as it falls, it speeds up.

(d) The rock hits the ground at the positive time t for which $s = 0$. The equation $160t - 16t^2 = 0$ factors to give $16t(10 - t) = 0$, so it has solutions $t = 0$ and $t = 10$. At $t = 0$, the blast occurred and the rock was thrown upward. It returns to the ground 10 sec later. ■

Derivatives in Economics and Biology

Economists have a specialized vocabulary for rates of change and derivatives. They call them *marginals*. In a manufacturing operation, the *cost of production* $c(x)$ is a function of x, the number of units produced. The **marginal cost of production** is the rate of change of cost with respect to level of production, so it is dc/dx.

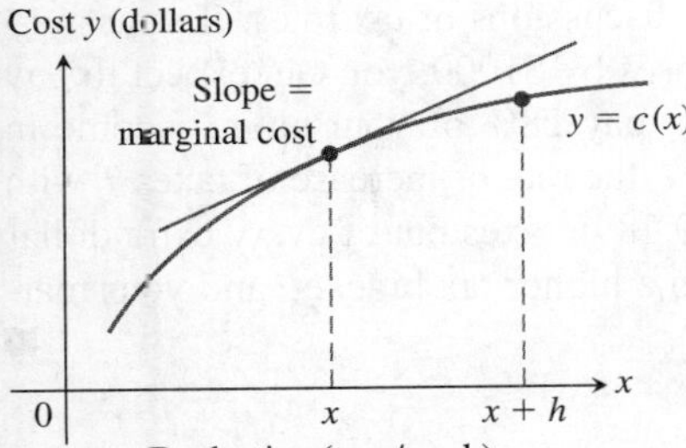

FIGURE 3.18 Weekly steel production: $c(x)$ is the cost of producing x tons per week. The cost of producing an additional h tons is $c(x+h) - c(x)$.

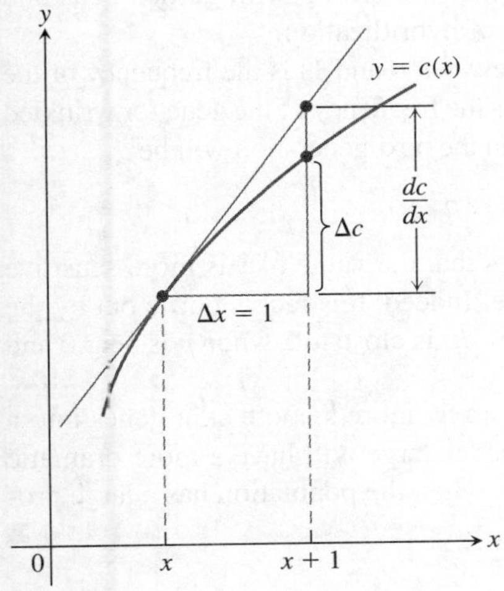

FIGURE 3.19 The marginal cost dc/dx is approximately the extra cost Δc of producing $\Delta x = 1$ more unit.

Suppose that $c(x)$ represents the dollars needed to produce x tons of steel in one week. It costs more to produce $x + h$ tons per week, and the cost difference, divided by h, is the average cost of producing each additional ton:

$$\frac{c(x+h) - c(x)}{h} = \begin{array}{l}\text{average cost of each of the additional}\\ h \text{ tons of steel produced.}\end{array}$$

The limit of this ratio as $h \to 0$ is the *marginal cost* of producing more steel per week when the current weekly production is x tons (Figure 3.18):

$$\frac{dc}{dx} = \lim_{h \to 0} \frac{c(x+h) - c(x)}{h} = \text{marginal cost of production.}$$

Sometimes the marginal cost of production is loosely defined to be the extra cost of producing one additional unit:

$$\frac{\Delta c}{\Delta x} = \frac{c(x+1) - c(x)}{1},$$

which is approximated by the value of dc/dx at x. This approximation is acceptable if the slope of the graph of c does not change quickly near x. Then the difference quotient will be close to its limit dc/dx, which is the rise in the tangent line if $\Delta x = 1$ (Figure 3.19). The approximation often works well for large values of x.

Economists often represent a total cost function by a cubic polynomial

$$c(x) = \alpha x^3 + \beta x^2 + \gamma x + \delta,$$

where δ represents *fixed costs*, such as rent, heat, equipment capitalization, and management costs. The other terms represent *variable costs*, such as the costs of raw materials, taxes, and labor. Fixed costs are independent of the number of units produced, whereas variable costs depend on the quantity produced. A cubic polynomial is usually adequate to capture the cost behavior on a realistic quantity interval.

EXAMPLE 5 Suppose that it costs

$$c(x) = x^3 - 6x^2 + 15x$$

dollars to produce x radiators when 8 to 30 radiators are produced and that

$$r(x) = x^3 - 3x^2 + 12x$$

gives the dollar revenue from selling x radiators. Your shop currently produces 10 radiators a day. About how much extra will it cost to produce one more radiator a day, and what is your estimated increase in revenue and increase in profit for selling 11 radiators a day?

Solution The cost of producing one more radiator a day when 10 are produced is about $c'(10)$:

$$c'(x) = \frac{d}{dx}\left(x^3 - 6x^2 + 15x\right) = 3x^2 - 12x + 15$$

$$c'(10) = 3(100) - 12(10) + 15 = 195.$$

The additional cost will be about \$195. The marginal revenue is

$$r'(x) = \frac{d}{dx}\left(x^3 - 3x^2 + 12x\right) = 3x^2 - 6x + 12.$$

The marginal revenue function estimates the increase in revenue that will result from selling one additional unit. If you currently sell 10 radiators a day, you can expect your revenue to increase by about

$$r'(10) = 3(100) - 6(10) + 12 = \$252$$

if you increase sales to 11 radiators a day. The estimated increase in profit is obtained by subtracting the increased cost of \$195 from the increased revenue, leading to an estimated profit increase of \$252 − \$195 = \$57. ■

EXAMPLE 6 Marginal rates frequently arise in discussions of tax rates. If your marginal income tax rate is 28% and your income increases by \$1000, you can expect to pay an extra \$280 in taxes. This does not mean that you pay 28% of your entire income in taxes. It just means that at your current income level I, the rate of increase of taxes T with respect to income is $dT/dI = 0.28$. You will pay \$0.28 in taxes out of every extra dollar you earn. As your income increases, you may land in a higher tax bracket, and your marginal rate will increase. ■

Sensitivity to Change

When a small change in x produces a large change in the value of a function $f(x)$, we say that the function is sensitive to changes in x. The derivative $f'(x)$ is a measure of this sensitivity. The function is more sensitive when $|f'(x)|$ is larger (when the slope of the graph of f is steeper).

EXAMPLE 7 Genetic Data and Sensitivity to Change

The Austrian monk Gregor Johann Mendel (1822–1884), working with garden peas and other plants, provided the first scientific explanation of hybridization.

His careful records showed that if p (a number between 0 and 1) is the frequency of the gene for smooth skin in peas (dominant) and $(1 - p)$ is the frequency of the gene for wrinkled skin in peas, then the proportion of smooth-skin peas in the next generation will be

$$y = 2p(1 - p) + p^2 = 2p - p^2.$$

The graph of y versus p in Figure 3.20a suggests that the value of y is more sensitive to a change in p when p is small than when p is large. Indeed, this fact is borne out by the derivative graph in Figure 3.20b, which shows that dy/dp is close to 2 when p is near 0 and close to 0 when p is near 1.

The implication for genetics is that introducing a few more smooth skin genes into a population where the frequency of wrinkled-skin peas is large will have a more dramatic effect on later generations than will a similar increase when the population has a large proportion of smooth-skin peas. ■

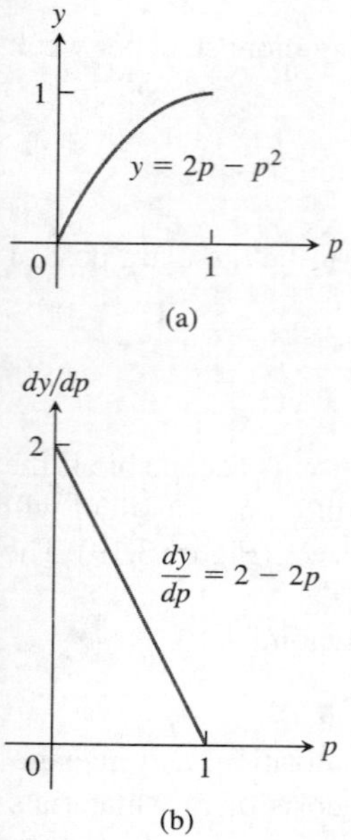

FIGURE 3.20 (a) The graph of $y = 2p - p^2$, describing the proportion of smooth-skin peas in the next generation. (b) The graph of dy/dp (Example 7).

EXERCISES 3.4

Motion Along a Coordinate Line

Exercises 1–6 give the positions $s = f(t)$ of a body moving on a coordinate line, with s in meters and t in seconds.

a. Find the body's displacement and average velocity for the given time interval.

b. Find the body's speed and acceleration at the endpoints of the interval.

c. When, if ever, during the interval does the body change direction?

1. $s = 2t^2 - 5t + 2, \quad 0 \le t \le 3$

2. $s = 6t - t^2, \quad 0 \le t \le 6$

3. $s = -t^3 + 3t^2 - 3t, \quad 0 \le t \le 3$

4. $s = (t^4/4) - t^3 + t^2, \quad 0 \le t \le 3$

5. $s = \dfrac{9}{t^2} - \dfrac{3}{t}, \quad 1 \le t \le 3$

6. $s = \dfrac{25}{t+5}, \quad -4 \le t \le 0$

7. Particle motion At time t, the position of a body moving along the s-axis is $s = t^3 - 15t^2 + 48t$ m.

a. Find the body's acceleration each time the velocity is zero.

b. Find the body's speed each time the acceleration is zero.

c. Find the total distance travelled by the body from $t = 0$ to $t = 5$.

8. Particle motion At time $t \ge 0$, the velocity of a body moving along the horizontal s-axis is $v = t^2 - 4t + 3$.

a. Find the body's acceleration each time the velocity is zero.

b. When is the body moving forward? Backward?

c. When is the body's velocity increasing? Decreasing?

Free-Fall Applications

9. Free fall on Mars and Jupiter The equations for free fall at the surfaces of Mars and Jupiter (s in metres, t in seconds) are $s = 2.56t^2$ on Mars and $s = 12.45t^2$ on Jupiter. How long does it take a rock falling from rest to reach a velocity of 27.8 m/sec (about 100 km/h) on each planet?

10. Lunar projectile motion A rock thrown vertically upward from the surface of the moon at a velocity of 24 m/sec (about 86 km/h) reaches a height of $s = 24t - 0.8t^2$ m in t sec.

a. Find the rock's velocity and acceleration at time t. (The acceleration in this case is the acceleration of gravity on the moon.)

b. How long does it take the rock to reach its highest point?

c. How high does the rock go?

d. How long does it take the rock to reach half its maximum height?

e. How long is the rock aloft?

11. Finding g on a small airless planet Explorers on a small airless planet used a spring gun to launch a ball bearing vertically upward from the surface at a launch velocity of 24 m/sec. Because the acceleration of gravity at the planet's surface was g_s m/sec^2, the explorers expected the ball bearing to reach a height of $s = 24t - \frac{3}{2}g_s t^2$ m t sec later. The ball bearing reached its maximum height 8 sec after being launched. What was the value of g_s.

12. Speeding bullet A 45-caliber bullet shot straight up from the surface of the moon would reach a height of $s = 832t - 2.6t^2$ ft after t sec. On Earth, in the absence of air, its height would be $s = 832t - 16t^2$ ft after t sec. How long will the bullet be aloft in each case? How high will the bullet go?

13. Free fall from the Tower of Pisa Had Galileo dropped a cannonball from the Tower of Pisa, 196 ft above the ground, the ball's height above the ground t sec into the fall would have been $s = 196 - 25t^2$

a. What would have been the ball's velocity, speed, and acceleration at time t?

b. About how long would it have taken the ball to hit the ground?

c. What would have been the ball's velocity at the moment of impact?

14. Galileo's free-fall formula Galileo developed a formula for a body's velocity during free fall by rolling balls from rest down increasingly steep inclined planks and looking for a limiting formula that would predict a ball's behavior when the plank was vertical and the ball fell freely; see part (a) of the accompanying figure. He found that, for any given angle of the plank, the ball's velocity t sec into motion was a constant multiple of t. That is, the velocity was given by a formula of the form $v = kt$. The value of the constant k depended on the inclination of the plank.

In modern notation—part (b) of the figure—with distance in meters and time in seconds, what Galileo determined by experiment was that, for any given angle θ, the ball's velocity t sec into the roll was $v = 9.8(\sin\theta)t$ m/sec.

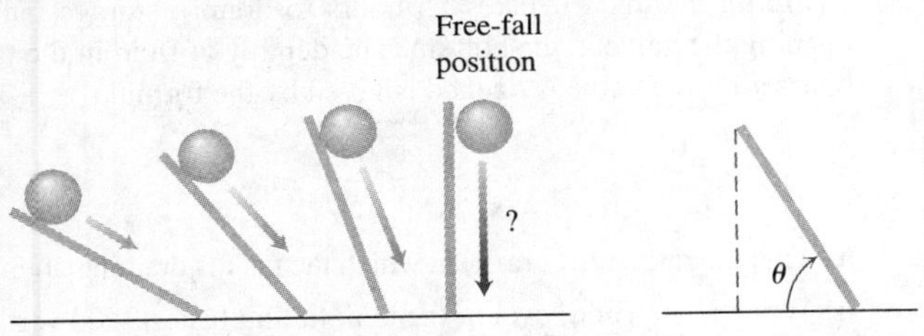

(a) (b)

a. What is the equation for the ball's velocity during free fall?

b. Building on your work in part (a), what constant acceleration does a freely falling body experience near the surface of Earth?

Understanding Motion from Graphs

15. The accompanying figure shows the velocity $v = ds/dt = f(t)$ (m/sec) of a body moving along a coordinate line.

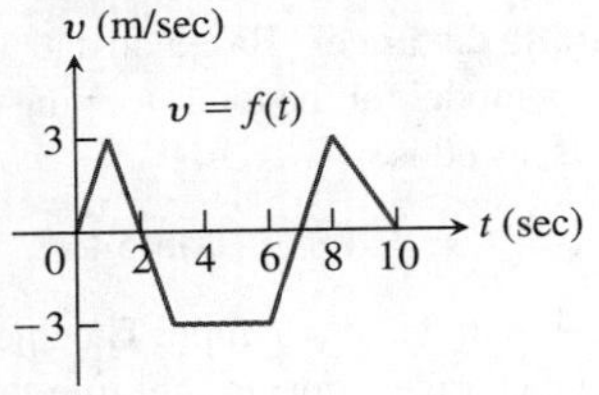

a. When does the body reverse direction?

b. When (approximately) is the body moving at a constant speed?

c. Graph the body's speed for $0 \leq t \leq 10$.

d. Graph the acceleration, where defined.

16. A particle P moves on the number line shown in part (a) of the accompanying figure. Part (b) shows the position of P as a function of time t.

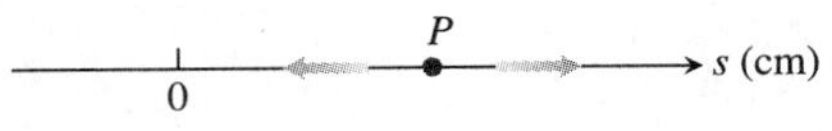

(a)

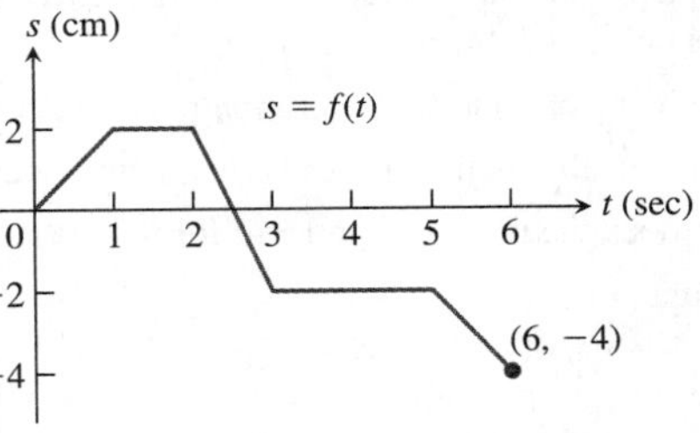

(b)

a. When is P moving to the left? Moving to the right? Standing still?

b. Graph the particle's velocity and speed (where defined).

17. Launching a rocket When a model rocket is launched, the propellant burns for a few seconds, accelerating the rocket upward. After burnout, the rocket coasts upward for a while and then begins to fall. A small explosive charge pops out a parachute shortly after the rocket starts down. The parachute slows the rocket to keep it from breaking when it lands. The figure here shows velocity data from the flight of the model rocket. Use the data to answer the following.

a. How fast was the rocket climbing when the engine stopped?

b. For how many seconds did the engine burn?

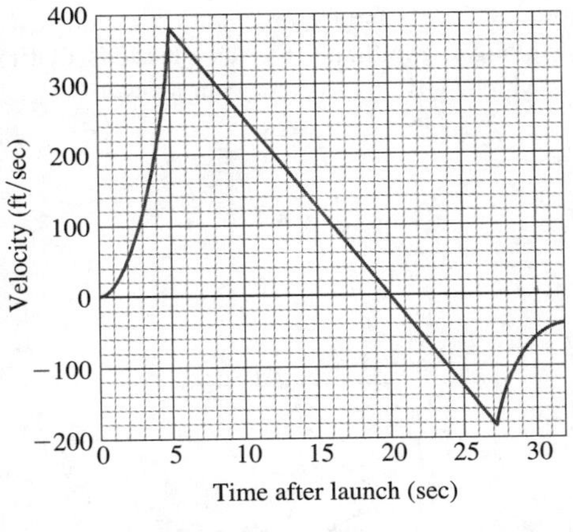

c. When did the rocket reach its highest point? What was its velocity then?

d. When did the parachute pop out? How fast was the rocket falling then?

e. How long did the rocket fall before the parachute opened?

f. When was the rocket's acceleration greatest?

g. When was the acceleration constant? What was its value then (to the nearest integer)?

18. The accompanying figure shows the velocity $v = f(t)$ of a particle moving on a horizontal coordinate line.

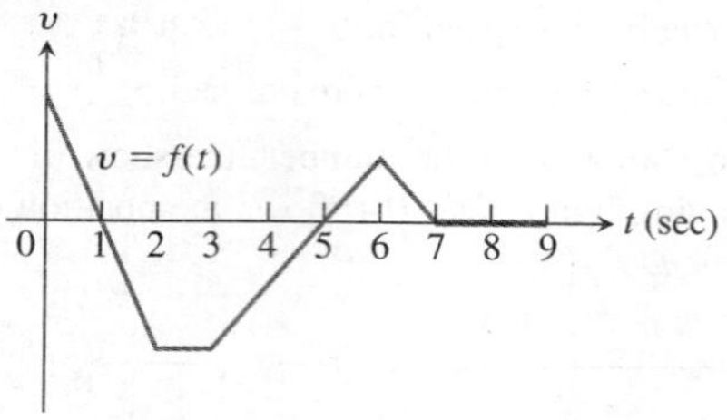

a. When does the particle move forward? Move backward? Speed up? Slow down?

b. When is the particle's acceleration positive? Negative? Zero?

c. When does the particle move at its greatest speed?

d. When does the particle stand still for more than an instant?

19. The graphs in the accompanying figure show the position s, velocity $v = ds/dt$, and acceleration $a = d^2s/dt^2$ of a body moving along a coordinate line as functions of time t. Which graph is which? Give reasons for your answers.

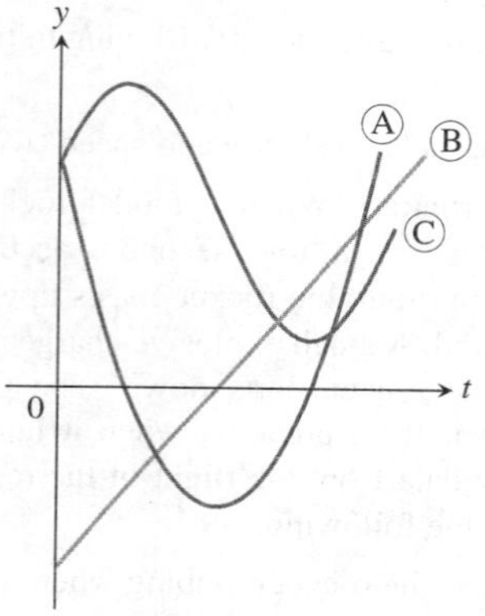

20. The graphs in the accompanying figure show the position s, the velocity $v = ds/dt$, and the acceleration $a = d^2s/dt^2$ of a body moving along a coordinate line as functions of time t. Which graph is which? Give reasons for your answers.

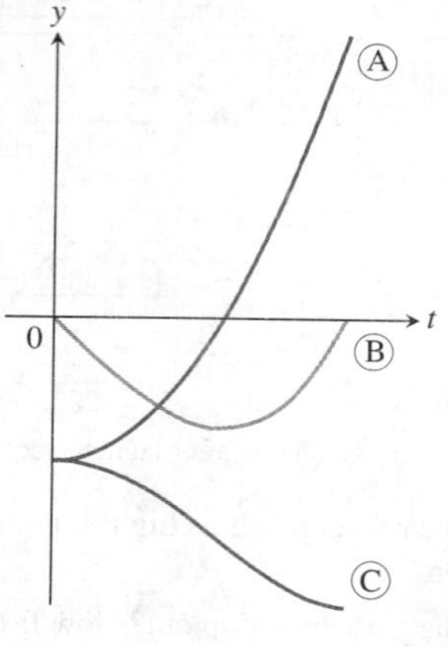

Economics

21. Marginal cost Suppose that the dollar cost of producing x washing machines is $c(x) = 1000 + 50x - 0.2x^2$

a. Find the average cost per machine of producing the first 50 washing machines.

b. Find the marginal cost when 100 washing machines are produced.

c. Show that the marginal cost when 100 washing machines are produced is approximately the cost of producing one more washing machine after the first 100 have been made, by calculating the latter cost directly.

22. Marginal revenue Suppose that the revenue from selling x washing machines is

$$r(x) = 20{,}000\left(1 - \frac{1}{x}\right)$$

dollars.

a. Find the marginal revenue when 100 machines are produced.

b. Use the function $r'(x)$ to estimate the increase in revenue that will result from increasing production from 100 machines a week to 101 machines a week.

c. Find the limit of $r'(x)$ as $x \to \infty$. How would you interpret this number?

Additional Applications

23. Bacterium population When a bactericide was added to a nutrient broth in which bacteria were growing, the bacterium population continued to grow for a while, but then stopped growing and began to decline. The size of the population at time t (hours) was $b = 10^6 + 10^4t - 10^3t^2$. Find the growth rates at

a. $t = 0$ hours.

b. $t = 5$ hours.

c. $t = 10$ hours.

24. Body surface area A typical human's body surface area S in square meters can be modeled by the formula $S = \frac{1}{60}\sqrt{wh}$, where h is the height in centimeters, and w the weight in kilograms, of the person. Find the rate of change of body surface area with respect to weight for humans of constant height $h = 180$ cm (roughly 5′9″). Does S increase more rapidly with respect to weight at lower or higher body weights? Explain.

T **25. Draining a tank** It takes 15 hours to drain a storage tank by opening the valve at the bottom. The depth y of fluid in the tank t hours after the valve is opened is given by the formula

$$y = 5\left(2 - \frac{t}{15}\right)^2 \text{ m.}$$

a. Find the rate dy/dt (m/h) at which the tank is draining at time t.

b. When is the fluid level in the tank falling fastest? Slowest? What are the values of dy/dt at these times?

c. Graph y and dy/dt together and discuss the behavior of y in relation to the signs and values of dy/dt.

26. Draining a tank The number of gallons of water in a tank t minutes after the tank has started to drain is $Q(t) = 200(30 - t)^2$. How fast is the water running out at the end of 10 min? What is the average rate at which the water flows out during the first 10 min?

27. Vehicular stopping distance Based on data from the U.S. Bureau of Public Roads, a model for the total stopping distance of a moving car in terms of its speed is

$$s = 1.1v + 0.054v^2,$$

where s is measured in ft and v in mph. The linear term $1.1v$ models the distance the car travels during the time the driver perceives a need to stop until the brakes are applied, and the quadratic term $0.054v^2$ models the additional braking distance once they are applied. Find ds/dv at $v = 35$ and $v = 70$ mph, and interpret the meaning of the derivative.

28. Inflating a balloon The volume $V = (4/3)\pi r^3$ of a spherical balloon changes with the radius.

a. At what rate (ft^3/ft)does the volume change with respect to the radius when $r = 5$ ft?

b. By approximately how much does the volume increase when the radius changes from 5 to 5.5 ft?

29. Airplane takeoff Suppose that the distance an aircraft travels along a runway before takeoff is given by $D = (10/9)t^2$, where D is measured in meters from the starting point and t is measured in seconds from the time the brakes are released. The aircraft will become airborne when its speed reaches 200 km/h. How long will it take to become airborne, and what distance will it travel in that time?

30. Volcanic lava fountains Although the November 1959 Kilauea Iki eruption on the island of Hawaii began with a line of fountains along the wall of the crater, activity was later confined to a single vent in the crater's floor, which at one point shot lava 1900 ft straight into the air (a Hawaiian record). What was the lava's exit velocity in feet per second? In miles per hour? (*Hint*: If v_0 is the exit velocity of a particle of lava, its height t sec later will be $s = v_0 t - 16t^2$ ft. Begin by finding the time at which $ds/dt = 0$. Neglect air resistance.)

Analyzing Motion Using Graphs

T Exercises 31–34 give the position function $s = f(t)$ of an object moving along the s-axis as a function of time t. Graph f together with the velocity function $v(t) = ds/dt = f'(t)$ and the acceleration function $a(t) = d^2s/dt^2 = f''(t)$. Comment on the object's behavior in relation to the signs and values of v and a. Include in your commentary such topics as the following:

a. When is the object momentarily at rest?

b. When does it move to the left (down) or to the right (up)?

c. When does it change direction?

d. When does it speed up and slow down?

e. When is it moving fastest (highest speed)? Slowest?

f. When is it farthest from the axis origin?

31. $s = 200t - 16t^2, \quad 0 \le t \le 12.5$ (a heavy object fired straight up from Earth's surface at 200 ft/sec)

32. $s = t^2 - 3t + 2, \quad 0 \le t \le 5$

33. $s = t^3 - 6t^2 + 7t, \quad 0 \le t \le 4$

34. $s = 4 - 7t + 6t^2 - t^3, \quad 0 \le t \le 4$

3.5 Derivatives of Trigonometric Functions

Many phenomena of nature are approximately periodic (electromagnetic fields, heart rhythms, tides, weather). The derivatives of sines and cosines play a key role in describing periodic changes. This section shows how to differentiate the six basic trigonometric functions.

Derivative of the Sine Function

To calculate the derivative of $f(x) = \sin x$, for x measured in radians, we combine the limits in Example 5a and Theorem 6 in Section 2.4 with the angle sum identity for the sine function (see Figure 1.46):

$$\sin(x + h) = \sin x \cos h + \cos x \sin h.$$

If $f(x) = \sin x$, then

$$f'(x) = \lim_{h\to 0} \frac{f(x+h) - f(x)}{h} = \lim_{h\to 0} \frac{\sin(x+h) - \sin x}{h} \qquad \text{Derivative definition}$$

$$= \lim_{h\to 0} \frac{(\sin x \cos h + \cos x \sin h) - \sin x}{h} \qquad \text{Identity for } \sin(x+h)$$

$$= \lim_{h\to 0} \frac{\sin x(\cos h - 1) + \cos x \sin h}{h}$$

$$= \lim_{h\to 0}\left(\sin x \cdot \frac{\cos h - 1}{h}\right) + \lim_{h\to 0}\left(\cos x \cdot \frac{\sin h}{h}\right)$$

$$= \sin x \cdot \underbrace{\lim_{h\to 0} \frac{\cos h - 1}{h}}_{\text{limit } 0} + \cos x \cdot \underbrace{\lim_{h\to 0} \frac{\sin h}{h}}_{\text{limit } 1}$$

$$= \sin x \cdot 0 + \cos x \cdot 1 = \cos x. \qquad \text{Example 5a and Theorem 6, Section 2.4}$$

The derivative of the sine function is the cosine function:

$$\frac{d}{dx}(\sin x) = \cos x.$$

EXAMPLE 1 We find derivatives of a difference, a product, and a quotient, each of which involves the sine function.

(a) $y = x^2 - \sin x$: $\quad \dfrac{dy}{dx} = 2x - \dfrac{d}{dx}(\sin x)$ $\qquad$ Difference Rule

$\qquad = 2x - \cos x$

(b) $y = x^2 \sin x$: $\quad \dfrac{dy}{dx} = x^2 \dfrac{d}{dx}(\sin x) + 2x \sin x$ $\qquad$ Product Rule

$\qquad = x^2 \cos x + 2x \sin x$

(c) $y = \dfrac{\sin x}{x}$: $\quad \dfrac{dy}{dx} = \dfrac{x \cdot \dfrac{d}{dx}(\sin x) - \sin x \cdot 1}{x^2}$ $\qquad$ Quotient Rule

$\qquad = \dfrac{x \cos x - \sin x}{x^2}$ ■

Derivative of the Cosine Function

With the help of the angle sum formula for the cosine function (see Figure 1.46),

$$\cos(x + h) = \cos x \cos h - \sin x \sin h,$$

we can compute the limit of the difference quotient:

$$\begin{aligned}
\frac{d}{dx}(\cos x) &= \lim_{h \to 0} \frac{\cos(x + h) - \cos x}{h} && \text{Derivative definition} \\
&= \lim_{h \to 0} \frac{(\cos x \cos h - \sin x \sin h) - \cos x}{h} && \text{Cosine angle sum identity} \\
&= \lim_{h \to 0} \frac{\cos x (\cos h - 1) - \sin x \sin h}{h} \\
&= \lim_{h \to 0} \left(\cos x \cdot \frac{\cos h - 1}{h}\right) - \lim_{h \to 0} \left(\sin x \cdot \frac{\sin h}{h}\right) \\
&= \cos x \cdot \lim_{h \to 0} \frac{\cos h - 1}{h} - \sin x \cdot \lim_{h \to 0} \frac{\sin h}{h} \\
&= \cos x \cdot 0 - \sin x \cdot 1 && \text{Example 5a and Theorem 6, Section 2.4} \\
&= -\sin x.
\end{aligned}$$

> **The derivative of the cosine function is the negative of the sine function:**
>
> $$\frac{d}{dx}(\cos x) = -\sin x.$$

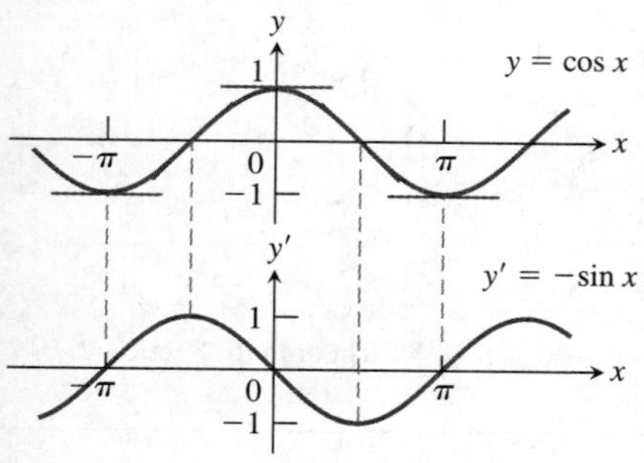

FIGURE 3.21 The curve $y' = -\sin x$ as the graph of the slopes of the tangent lines to the curve $y = \cos x$.

Figure 3.21 shows a way to visualize this result by graphing the slopes of the tangent lines to the curve $y = \cos x$.

EXAMPLE 2 We find derivatives of the cosine function in combinations with other functions.

(a) $y = 5x + \cos x$:

$$\frac{dy}{dx} = \frac{d}{dx}(5x) + \frac{d}{dx}(\cos x) \qquad \text{Sum Rule}$$

$$= 5 - \sin x$$

(b) $y = \sin x \cos x$:

$$\begin{aligned}\frac{dy}{dx} &= \sin x \frac{d}{dx}(\cos x) + \left(\frac{d}{dx}(\sin x)\right)\cos x && \text{Product Rule}\\ &= (\sin x)(-\sin x) + (\cos x)(\cos x)\\ &= \cos^2 x - \sin^2 x\end{aligned}$$

(c) $y = \dfrac{\cos x}{1 - \sin x}$:

$$\begin{aligned}\frac{dy}{dx} &= \frac{(1-\sin x)\frac{d}{dx}(\cos x) - \cos x \frac{d}{dx}(1-\sin x)}{(1-\sin x)^2} && \text{Quotient Rule}\\ &= \frac{(1-\sin x)(-\sin x) - (\cos x)(0 - \cos x)}{(1-\sin x)^2}\\ &= \frac{1-\sin x}{(1-\sin x)^2} && \sin^2 x + \cos^2 x = 1\\ &= \frac{1}{1-\sin x}\end{aligned}$$

■

Simple Harmonic Motion

Simple harmonic motion models the motion of an object or weight bobbing freely up and down on the end of a spring, with no resistance. The motion is periodic and repeats indefinitely, so we represent it using trigonometric functions. The next example models motion with no opposing forces (such as friction).

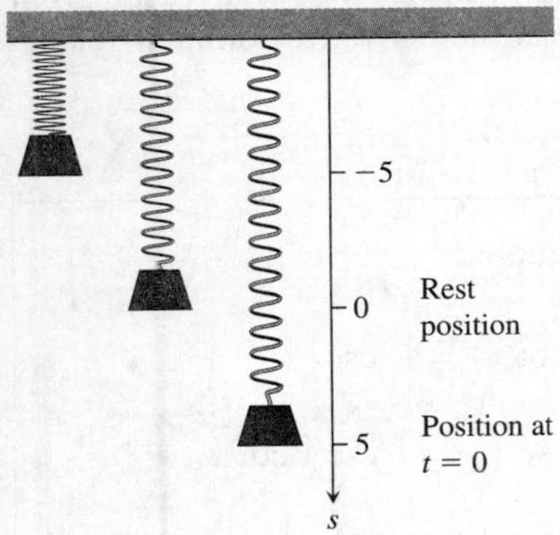

FIGURE 3.22 A weight hanging from a vertical spring and then displaced oscillates above and below its rest position (Example 3).

EXAMPLE 3 A weight hanging from a spring (Figure 3.22) is stretched down 5 units beyond its rest position and released at time $t = 0$ to bob up and down. Its position at any later time t is

$$s = 5 \cos t.$$

What are its velocity and acceleration at time t?

Solution We have

Position: $s = 5\cos t$

Velocity: $v = \dfrac{ds}{dt} = \dfrac{d}{dt}(5\cos t) = -5\sin t$

Acceleration: $a = \dfrac{dv}{dt} = \dfrac{d}{dt}(-5\sin t) = -5\cos t.$

Notice how much we can learn from these equations:

1. As time passes, the weight moves down and up between $s = -5$ and $s = 5$ on the s-axis. The amplitude of the motion is 5. The period of the motion is 2π, the period of the cosine function.
2. The velocity $v = -5\sin t$ attains its greatest magnitude, 5, when $\cos t = 0$, as the graphs in Figure 3.23 show. Hence, the speed of the weight, $|v| = 5|\sin t|$, is greatest when $\cos t = 0$, that is, when $s = 0$ (the rest position). The speed of the weight is zero when $\sin t = 0$. This occurs when $s = 5\cos t = \pm 5$, at the endpoints of the interval of motion. At these points the weight reverses direction.
3. The weight is acted on by the spring and by gravity. When the weight is below the rest position, the combined forces pull it up, and when it is above the rest position, they pull it down. The weight's acceleration is always proportional to the negative of its displacement. This property of springs is called *Hooke's Law*, and is studied further in Section 6.5.

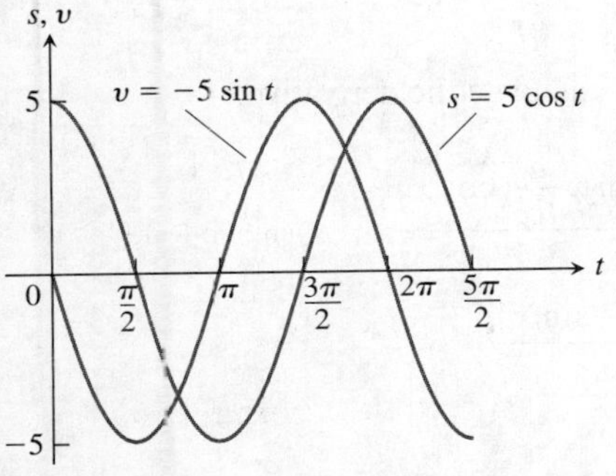

FIGURE 3.23 The graphs of the position and velocity of the weight in Example 3.

4. The acceleration, $a = -5\cos t$, is zero only at the rest position, where $\cos t = 0$ and the force of gravity and the force from the spring balance each other. When the weight is anywhere else, the two forces are unequal, and acceleration is nonzero. The acceleration is greatest in magnitude at the points farthest from the rest position, where $\cos t = \pm 1$. ■

EXAMPLE 4 The jerk associated with the simple harmonic motion in Example 3 is

$$j = \frac{da}{dt} = \frac{d}{dt}(-5\cos t) = 5\sin t.$$

It has its greatest magnitude when $\sin t = \pm 1$, not at the extremes of the displacement but at the rest position, where the acceleration changes direction and sign. ■

Derivatives of the Other Basic Trigonometric Functions

Because $\sin x$ and $\cos x$ are differentiable functions of x, it follows from the Quotient Rule that the related functions

$$\tan x = \frac{\sin x}{\cos x}, \qquad \cot x = \frac{\cos x}{\sin x}, \qquad \sec x = \frac{1}{\cos x}, \qquad \text{and} \qquad \csc x = \frac{1}{\sin x}$$

are differentiable at every value of x at which they are defined. Their derivatives are given by the following formulas. Notice the negative signs in the derivative formulas for the cofunctions.

The derivatives of the other trigonometric functions:

$$\frac{d}{dx}(\tan x) = \sec^2 x \qquad \frac{d}{dx}(\cot x) = -\csc^2 x$$

$$\frac{d}{dx}(\sec x) = \sec x \tan x \qquad \frac{d}{dx}(\csc x) = -\csc x \cot x$$

To show a typical calculation, we find the derivative of the tangent function. The other derivations are left for Exercise 54.

EXAMPLE 5 Find $d(\tan x)/dx$.

Solution We use the Derivative Quotient Rule to calculate the derivative:

$$\frac{d}{dx}(\tan x) = \frac{d}{dx}\left(\frac{\sin x}{\cos x}\right) = \frac{\cos x \frac{d}{dx}(\sin x) - \sin x \frac{d}{dx}(\cos x)}{\cos^2 x} \qquad \text{Quotient Rule}$$

$$= \frac{\cos x \cos x - \sin x(-\sin x)}{\cos^2 x}$$

$$= \frac{\cos^2 x + \sin^2 x}{\cos^2 x}$$

$$= \frac{1}{\cos^2 x} = \sec^2 x.$$ ■

EXAMPLE 6 Find y'' if $y = \sec x$.

Solution Finding the second derivative involves a combination of trigonometric derivatives.

$$
\begin{aligned}
y &= \sec x \\
y' &= \sec x \tan x && \text{Derivative rule for secant function} \\
y'' &= \frac{d}{dx}(\sec x \tan x) \\
&= \sec x \frac{d}{dx}(\tan x) + \tan x \frac{d}{dx}(\sec x) && \text{Derivative Product Rule} \\
&= (\sec x)(\sec^2 x) + (\tan x)(\sec x \tan x) && \text{Derivative rules} \\
&= \sec^3 x + \sec x \tan^2 x
\end{aligned}
$$

■

EXERCISES 3.5

Derivatives

In Exercises 1–18, find dy/dx.

1. $y = -10x + 3\cos x$
2. $y = \dfrac{3}{x} + 5\sin x$
3. $y = x^2 \cos x$
4. $y = \sqrt{x}\sec x + 3$
5. $y = \csc x - 4\sqrt{x} + 7$
6. $y = x^2 \cot x - \dfrac{1}{x^2}$
7. $f(x) = \sin x \tan x$
8. $g(x) = \dfrac{\cos x}{\sin^2 x}$
9. $y = x\sec x + \dfrac{1}{x}$
10. $y = (\sin x + \cos x)\sec x$
11. $y = \dfrac{\cot x}{1 + \cot x}$
12. $y = \dfrac{\cos x}{1 + \sin x}$
13. $y = \dfrac{4}{\cos x} + \dfrac{1}{\tan x}$
14. $y = \dfrac{\cos x}{x} + \dfrac{x}{\cos x}$
15. $y = (\sec x + \tan x)(\sec x - \tan x)$
16. $y = \dfrac{x\sin x + \cos x}{(x\sin x - \cos x)^2}$
17. $y = \dfrac{\sin x}{1 + \tan x}$
18. $g(x) = (2 - x)\tan^2 x$

In Exercises 19–22, find ds/dt.

19. $s = \tan t - t$
20. $s = t^2 - \sec t + 1$
21. $s = \dfrac{1 + \csc t}{1 - \csc t}$
22. $s = \dfrac{\sin t}{1 - \cos t}$

In Exercises 23–26, find $dr/d\theta$.

23. $r = 4 - \theta^2 \sin\theta$
24. $r = \theta\sin\theta + \cos\theta$
25. $r = \sec\theta\csc\theta$
26. $r = (1 + \sec\theta)\sin\theta$

In Exercises 27–32, find dp/dq.

27. $p = 5 + \dfrac{1}{\cot q}$
28. $p = (1 + \csc q)\cos q$
29. $p = \dfrac{\sin q + \cos q}{\cos q}$
30. $p = \dfrac{\tan q}{1 + \tan q}$
31. $p = \dfrac{q\sin q}{q^2 - 1}$
32. $p = \dfrac{3q + \tan q}{q\sec q}$

33. Find y'' if
    a. $y = \csc x$.
    b. $y = \sec x$.
34. Find $y^{(4)} = d^4 y/dx^4$ if
    a. $y = -2\sin x$.
    b. $y = 9\cos x$.

Tangent Lines

In Exercises 35–38, graph the curves over the given intervals, together with their tangent lines at the given values of x. Label each curve and tangent line with its equation.

35. $y = \sin x, \quad -3\pi/2 \le x \le 2\pi$
    $x = -\pi, 0, 3\pi/2$
36. $y = \tan x, \quad -\pi/2 < x < \pi/2$
    $x = -\pi/3, 0, \pi/3$
37. $y = \sec x, \quad -\pi/2 < x < \pi/2$
    $x = -\pi/3, \pi/4$
38. $y = 1 + \cos x, \quad -3\pi/2 \le x \le 2\pi$
    $x = -\pi/3, 3\pi/2$

T Do the graphs of the functions in Exercises 39–44 have any horizontal tangent lines in the interval $0 \le x \le 2\pi$? If so, where? If not, why not? Visualize your findings by graphing the functions with a grapher.

39. $y = x + \sin x$
40. $y = 2x + \sin x$
41. $y = x - \cot x$
42. $y = x + 2\cos x$
43. $y = \dfrac{\sec x}{3 + \sec x}$
44. $y = \dfrac{\cos x}{3 - 4\sin x}$

45. Find all points on the curve $y = \tan x, -\pi/2 < x < \pi/2$, where the tangent line is parallel to the line $y = 2x$. Sketch the curve and tangent lines together, labeling each with its equation.

46. Find all points on the curve $y = \cot x, 0 < x < \pi$, where the tangent line is parallel to the line $y = -x$. Sketch the curve and tangent lines together, labeling each with its equation.

In Exercises 47 and 48, find an equation for **(a)** the tangent line to the curve at P and **(b)** the horizontal tangent line to the curve at Q.

47.

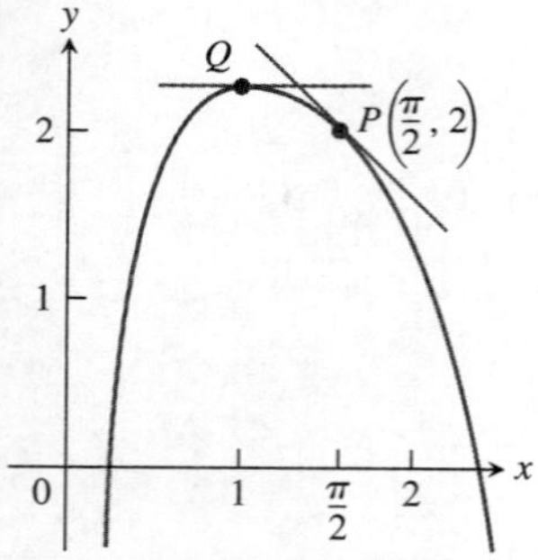

$y = 4 + \cot x - 2\csc x$

48.

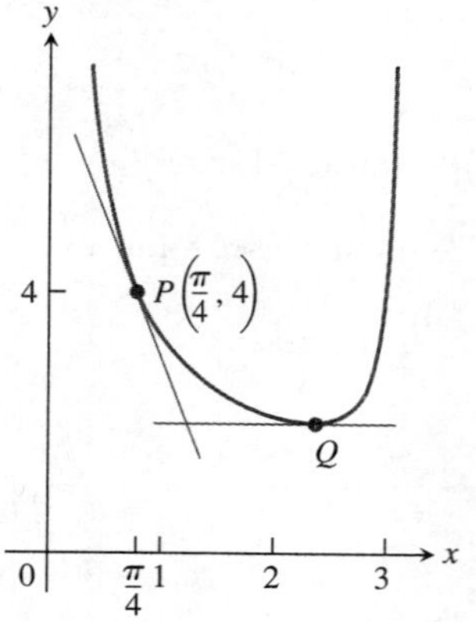

$y = 1 + \sqrt{2}\csc x + \cot x$

Theory and Examples

The equations in Exercises 49 and 50 give the position $s = f(t)$ of a body moving on a coordinate line (s in meters, t in seconds). Find the body's velocity, speed, acceleration, and jerk at time $t = \pi/4$ sec.

49. $s = 2 - 2\sin t$

50. $s = \sin t + \cos t$

51. Is there a value of c that will make

$$f(x) = \begin{cases} \dfrac{\sin^2 3x}{x^2}, & x \neq 0 \\ c, & x = 0 \end{cases}$$

continuous at $x = 0$? Give reasons for your answer.

52. Is there a value of b that will make

$$g(x) = \begin{cases} x + b, & x < 0 \\ \cos x, & x \geq 0 \end{cases}$$

continuous at $x = 0$? Differentiable at $x = 0$? Give reasons for your answers.

53. By computing the first few derivatives and looking for a pattern, find the following derivatives.

a. $\dfrac{d^{999}}{dx^{999}}(\cos x)$ **b.** $\dfrac{d^{110}}{dx^{110}}(\sin x - 3\cos x)$

c. $\dfrac{d^{73}}{dx^{73}}(x \sin x)$

54. Derive the formula for the derivative with respect to x of

a. $\sec x$. **b.** $\csc x$. **c.** $\cot x$.

55. A weight is attached to a spring and reaches its equilibrium position ($x = 0$). It is then set in motion resulting in a displacement of

$$x = 10 \cos t,$$

where x is measured in centimeters and t is measured in seconds. See the accompanying figure.

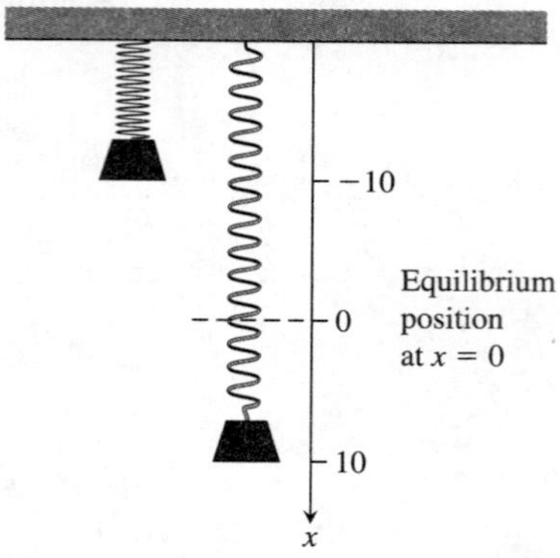

a. Find the spring's displacement when $t = 0$, $t = \pi/3$, and $t = 3\pi/4$.

b. Find the spring's velocity when $t = 0$, $t = \pi/3$, and $t = 3\pi/4$.

56. Assume that a particle's position on the x-axis is given by

$$x = 3\cos t + 4\sin t,$$

where x is measured in feet and t is measured in seconds.

a. Find the particle's position when $t = 0$, $t = \pi/2$, and $t = \pi$.

b. Find the particle's velocity when $t = 0$, $t = \pi/2$, and $t = \pi$.

T **57.** Graph $y = \cos x$ for $-\pi \leq x \leq 2\pi$. On the same screen, graph

$$y = \frac{\sin(x + h) - \sin x}{h}$$

for $h = 1, 0.5, 0.3$, and 0.1. Then, in a new window, try $h = -1, -0.5$, and -0.3. What happens as $h \to 0^+$? As $h \to 0^-$? What phenomenon is being illustrated here?

T **58.** Graph $y = -\sin x$ for $-\pi \leq x \leq 2\pi$. On the same screen, graph

$$y = \frac{\cos(x + h) - \cos x}{h}$$

for $h = 1, 0.5, 0.3$, and 0.1. Then, in a new window, try $h = -1, -0.5$, and -0.3. What happens as $h \to 0^+$? As $h \to 0^-$? What phenomenon is being illustrated here?

T **59. Centered difference quotients** The *centered difference quotient*

$$\frac{f(x + h) - f(x - h)}{2h}$$

is used to approximate $f'(x)$ in numerical work because (1) its limit as $h \to 0$ equals $f'(x)$ when $f'(x)$ exists, and (2) it usually gives a better approximation of $f'(x)$ for a given value of h than the difference quotient

$$\frac{f(x + h) - f(x)}{h}.$$

See the accompanying figure.

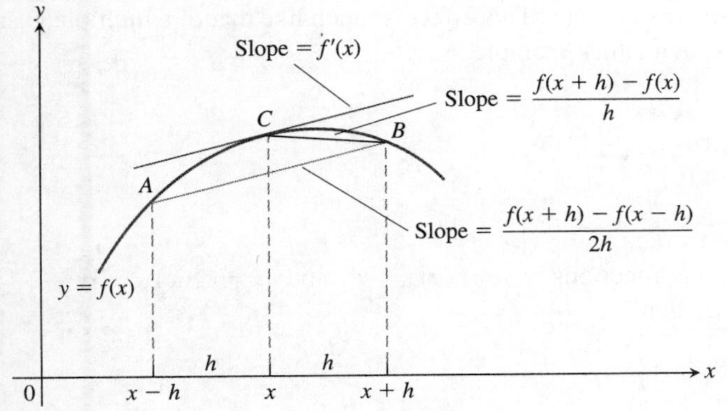

a. To see how rapidly the centered difference quotient for $f(x) = \sin x$ converges to $f'(x) = \cos x$, graph $y = \cos x$ together with

$$y = \frac{\sin(x + h) - \sin(x - h)}{2h}$$

over the interval $[-\pi, 2\pi]$ for $h = 1, 0.5$, and 0.3. Compare the results with those obtained in Exercise 57 for the same values of h.

b. To see how rapidly the centered difference quotient for $f(x) = \cos x$ converges to $f'(x) = -\sin x$, graph $y = -\sin x$ together with

$$y = \frac{\cos(x + h) - \cos(x - h)}{2h}$$

over the interval $[-\pi, 2\pi]$ for $h = 1, 0.5$, and 0.3. Compare the results with those obtained in Exercise 58 for the same values of h.

60. A caution about centered difference quotients (*Continuation of Exercise 59.*) The quotient

$$\frac{f(x + h) - f(x - h)}{2h}$$

may have a limit as $h \to 0$ when f has no derivative at x. As a case in point, take $f(x) = |x|$ and calculate

$$\lim_{h \to 0} \frac{|0 + h| - |0 - h|}{2h}.$$

As you will see, the limit exists even though $f(x) = |x|$ has no derivative at $x = 0$. *Moral:* Before using a centered difference quotient, be sure the derivative exists.

61. Slopes on the graph of the tangent function Graph $y = \tan x$ and its derivative together on $(-\pi/2, \pi/2)$. Does the graph of the tangent function appear to have a smallest slope? A largest slope? Is the slope ever negative? Give reasons for your answers.

62. Exploring $(\sin kx)/x$ Graph $y = (\sin x)/x$, $y = (\sin 2x)/x$, and $y = (\sin 4x)/x$ together over the interval $-2 \le x \le 2$. Where does each graph appear to cross the y-axis? Do the graphs really intersect the axis? What would you expect the graphs of $y = (\sin 5x)/x$ and $y = (\sin(-3x))/x$ to do as $x \to 0$? Why? What about the graph of $y = (\sin kx)/x$ for other values of k? Give reasons for your answers.

3.6 The Chain Rule

How do we differentiate $F(x) = \sin(x^2 - 4)$? This function is the composition $f \circ g$ of two functions $y = f(u) = \sin u$ and $u = g(x) = x^2 - 4$ that we know how to differentiate. The answer, given by the *Chain Rule*, says that the derivative is the product of the derivatives of f and g. We develop the rule in this section.

C: y turns B: u turns A: x turns

FIGURE 3.24 When gear A makes x turns, gear B makes u turns and gear C makes y turns. By comparing circumferences or counting teeth, we see that $y = u/2$ (C turns one-half turn for each B turn) and $u = 3x$ (B turns three times for A's one), so $y = 3x/2$. Thus, $dy/dx = 3/2 = (1/2)(3) = (dy/du)(du/dx)$.

Derivative of a Composite Function

The function $y = \frac{3}{2}x = \frac{1}{2}(3x)$ is the composition of the functions $y = \frac{1}{2}u$ and $u = 3x$. We have

$$\frac{dy}{dx} = \frac{3}{2}, \quad \frac{dy}{du} = \frac{1}{2}, \quad \text{and} \quad \frac{du}{dx} = 3.$$

Since $\frac{3}{2} = \frac{1}{2} \cdot 3$, we see in this case that

$$\frac{dy}{dx} = \frac{dy}{du} \cdot \frac{du}{dx}.$$

If we think of the derivative as a rate of change, this relationship is intuitively reasonable. If $y = f(u)$ changes half as fast as u, and $u = g(x)$ changes three times as fast as x, then we expect y to change $3/2$ times as fast as x. This effect is much like that of a multiple gear train (Figure 3.24). Let's look at another example.

EXAMPLE 1 The function

$$y = (3x^2 + 1)^2$$

is obtained by composing the functions $y = f(u) = u^2$ and $u = g(x) = 3x^2 + 1$. Calculating derivatives, we see that

$$\begin{aligned} \frac{dy}{du} \cdot \frac{du}{dx} &= 2u \cdot 6x \\ &= 2(3x^2 + 1) \cdot 6x \qquad \text{Substitute for } u. \\ &= 36x^3 + 12x. \end{aligned}$$

Calculating the derivative from the expanded formula $(3x^2 + 1)^2 = 9x^4 + 6x^2 + 1$ gives the same result:

$$\frac{dy}{dx} = \frac{d}{dx}(9x^4 + 6x^2 + 1) = 36x^3 + 12x.$$ ■

The derivative of the composite function $f(g(x))$ at x is the derivative of f at $g(x)$ times the derivative of g at x. This is known as the Chain Rule (Figure 3.25).

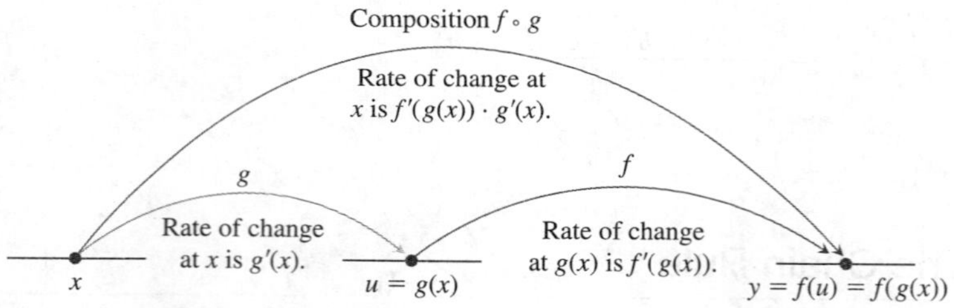

FIGURE 3.25 Rates of change multiply: The derivative of $f \circ g$ at x is the derivative of f at $g(x)$ times the derivative of g at x.

THEOREM 2—The Chain Rule If $f(u)$ is differentiable at the point $u = g(x)$ and $g(x)$ is differentiable at x, then the composite function $(f \circ g)(x) = f(g(x))$ is differentiable at x, and

$$(f \circ g)'(x) = f'(g(x)) \cdot g'(x).$$

In Leibniz's notation, if $y = f(u)$ and $u = g(x)$, then

$$\frac{dy}{dx} = \frac{dy}{du} \cdot \frac{du}{dx},$$

where dy/du is evaluated at $u = g(x)$.

A Proof of One Case of the Chain Rule: Let Δu be the change in u when x changes by Δx, so that

$$\Delta u = g(x + \Delta x) - g(x).$$

Then the corresponding change in y is

$$\Delta y = f(u + \Delta u) - f(u).$$

If $\Delta u \neq 0$, we can write the fraction $\Delta y/\Delta x$ as the product

$$\frac{\Delta y}{\Delta x} = \frac{\Delta y}{\Delta u} \cdot \frac{\Delta u}{\Delta x} \tag{1}$$

and take the limit as $\Delta x \to 0$:

$$\begin{aligned} \frac{dy}{dx} &= \lim_{\Delta x \to 0} \frac{\Delta y}{\Delta x} \\ &= \lim_{\Delta x \to 0} \frac{\Delta y}{\Delta u} \frac{\Delta u}{\Delta x} \\ &= \lim_{\Delta x \to 0} \frac{\Delta y}{\Delta u} \cdot \lim_{\Delta x \to 0} \frac{\Delta u}{\Delta x} \qquad \text{(Note that } \Delta u \to 0 \text{ as } \Delta x \to 0 \text{ since } g \text{ is continuous.)} \\ &= \lim_{\Delta u \to 0} \frac{\Delta y}{\Delta u} \cdot \lim_{\Delta x \to 0} \frac{\Delta u}{\Delta x} \\ &= \frac{dy}{du} \cdot \frac{du}{dx}. \end{aligned}$$

The problem with this argument is that if the function $g(x)$ oscillates rapidly near x, then Δu can be zero even when $\Delta x \neq 0$, so the cancelation of Δu in Equation (1) would be invalid. A complete proof requires a different approach that avoids this problem, and we give one such proof in Section 3.9. ■

EXAMPLE 2 An object moves along the x-axis so that its position at any time $t \geq 0$ is given by $x(t) = \cos(t^2 + 1)$. Find the velocity of the object as a function of t.

Solution We know that the velocity is dx/dt. In this instance, x is a composition of two functions: $x = \cos(u)$ and $u = t^2 + 1$. We have

$$\frac{dx}{du} = -\sin(u) \qquad x = \cos(u)$$

$$\frac{du}{dt} = 2t. \qquad u = t^2 + 1$$

By the Chain Rule,

$$\begin{aligned} \frac{dx}{dt} &= \frac{dx}{du} \cdot \frac{du}{dt} \\ &= -\sin(u) \cdot 2t \\ &= -\sin(t^2 + 1) \cdot 2t \\ &= -2t \sin(t^2 + 1). \end{aligned}$$

■

Ways to Write the Chain Rule

$$(f \circ g)'(x) = f'(g(x)) \cdot g'(x)$$

$$\frac{dy}{dx} = \frac{dy}{du} \cdot \frac{du}{dx}$$

$$\frac{dy}{dx} = f'(g(x)) \cdot g'(x)$$

$$\frac{d}{dx} f(u) = f'(u) \frac{du}{dx}$$

"Outside-Inside" Rule

A difficulty with the Leibniz notation is that it doesn't state specifically where the derivatives in the Chain Rule are supposed to be evaluated. So it sometimes helps to write the Chain Rule using functional notation. If $y = f(g(x))$, then

$$\frac{dy}{dx} = f'(g(x)) \cdot g'(x).$$

In words, differentiate the "outside" function f and evaluate this derivative at the "inside" function $g(x)$ left alone; then multiply by the derivative of the "inside function."

EXAMPLE 3 Differentiate $\sin(x^2 + x)$ with respect to x.

Solution We apply the Chain Rule directly and find

$$\frac{d}{dx}\underbrace{\sin(x^2 + x)}_{\text{inside}} = \underbrace{\cos(x^2 + x)}_{\substack{\text{inside}\\ \text{left alone}}} \cdot \underbrace{(2x + 1)}_{\substack{\text{derivative of}\\ \text{the inside}}}.$$

■

Repeated Use of the Chain Rule

We sometimes have to use the Chain Rule two or more times to find a derivative.

HISTORICAL BIOGRAPHY
Johann Bernoulli
(1667–1748)
www.bit.ly/2OWKvXL

EXAMPLE 4 Find the derivative of $g(t) = \tan(5 - \sin 2t)$.

Solution Notice here that the tangent is a function of $5 - \sin 2t$, whereas the sine is a function of $2t$, which is itself a function of t. Therefore, by the Chain Rule,

$$\begin{aligned} g'(t) &= \frac{d}{dt}\tan(5 - \sin 2t) \\ &= \sec^2(5 - \sin 2t) \cdot \frac{d}{dt}(5 - \sin 2t) && \text{Derivative of } \tan u \text{ with } u = 5 - \sin 2t \\ &= \sec^2(5 - \sin 2t) \cdot \left(0 - \cos 2t \cdot \frac{d}{dt}(2t)\right) && \text{Derivative of } -\sin u \text{ with } u = 2t \\ &= \sec^2(5 - \sin 2t) \cdot (-\cos 2t) \cdot 2 \\ &= -2(\cos 2t)\sec^2(5 - \sin 2t). \end{aligned}$$

■

The Chain Rule with Powers of a Function

If n is any real number and f is a power function, $f(u) = u^n$, the Power Rule tells us that $f'(u) = nu^{n-1}$. If u is a differentiable function of x, then we can use the Chain Rule to extend this to the **Power Chain Rule**:

$$\frac{d}{dx}(u^n) = nu^{n-1}\frac{du}{dx}. \qquad \frac{d}{du}(u^n) = nu^{n-1}$$

EXAMPLE 5 The Power Chain Rule simplifies computing the derivative of a power of an expression.

(a) $$\begin{aligned} \frac{d}{dx}(5x^3 - x^4)^7 &= 7(5x^3 - x^4)^6\frac{d}{dx}(5x^3 - x^4) && \text{Power Chain Rule with } u = 5x^3 - x^4,\ n = 7 \\ &= 7(5x^3 - x^4)^6(15x^2 - 4x^3) \end{aligned}$$

(b) $$\begin{aligned} \frac{d}{dx}\left(\frac{1}{3x - 2}\right) &= \frac{d}{dx}(3x - 2)^{-1} \\ &= -1(3x - 2)^{-2}\frac{d}{dx}(3x - 2) && \text{Power Chain Rule with } u = 3x - 2,\ n = -1 \\ &= -1(3x - 2)^{-2}(3) \\ &= -\frac{3}{(3x - 2)^2} \end{aligned}$$

In part (b) we could also find the derivative with the Quotient Rule.

(c) $$\begin{aligned} \frac{d}{dx}(\sin^5 x) &= 5\sin^4 x \cdot \frac{d}{dx}\sin x && \text{Power Chain Rule with } u = \sin x,\ n = 5, \text{ because } \sin^n x \text{ means } (\sin x)^n,\ n \neq -1 \\ &= 5\sin^4 x\cos x \end{aligned}$$

■

EXAMPLE 6 In Example 4 of Section 3.2 we saw that the absolute value function $y = |x|$ is not differentiable at $x = 0$. However, the function is differentiable at all other real numbers, as we now show. Since $|x| = \sqrt{x^2}$, we can derive the following formula, which gives an alternative to the more direct analysis seen before.

$$\frac{d}{dx}(|x|) = \frac{d}{dx}\sqrt{x^2}$$

$$= \frac{1}{2\sqrt{x^2}} \cdot \frac{d}{dx}(x^2) \qquad \text{Power Chain Rule with } u = x^2,\ n = 1/2,\ x \neq 0$$

$$= \frac{1}{2|x|} \cdot 2x \qquad \sqrt{x^2} = |x|$$

$$= \frac{x}{|x|}, \quad x \neq 0.$$

■

Derivative of the Absolute Value Function

$$\frac{d}{dx}(|x|) = \frac{x}{|x|}, \quad x \neq 0$$

$$= \begin{cases} 1, & x > 0 \\ -1, & x < 0 \end{cases}$$

EXAMPLE 7 Show that the slope of every line tangent to the curve $y = 1/(1 - 2x)^3$ is positive.

Solution We find the derivative:

$$\frac{dy}{dx} = \frac{d}{dx}(1 - 2x)^{-3}$$

$$= -3(1 - 2x)^{-4} \cdot \frac{d}{dx}(1 - 2x) \qquad \text{Power Chain Rule with } u = (1 - 2x), n = -3$$

$$= -3(1 - 2x)^{-4} \cdot (-2)$$

$$= \frac{6}{(1 - 2x)^4}.$$

At any point (x, y) on the curve, the denominator is nonzero, and the slope of the tangent line is

$$\frac{dy}{dx} = \frac{6}{(1 - 2x)^4},$$

which is the quotient of two positive numbers. ■

EXAMPLE 8 The formulas for the derivatives of both $\sin x$ and $\cos x$ were obtained under the assumption that x is measured in radians, *not* degrees. The Chain Rule gives us new insight into the difference between the two. Since $180° = \pi$ radians, $x° = \pi x/180$ radians, where $x°$ is the size of the angle measured in degrees.

By the Chain Rule,

$$\frac{d}{dx}\sin(x°) = \frac{d}{dx}\sin\left(\frac{\pi x}{180}\right) = \frac{\pi}{180}\cos\left(\frac{\pi x}{180}\right) = \frac{\pi}{180}\cos(x°).$$

See Figure 3.26. Similarly, the derivative of $\cos(x°)$ is $-(\pi/180)\sin(x°)$.

The factor $\pi/180$ would propagate with repeated differentiation, showing an advantage for the use of radian measure in computations. ■

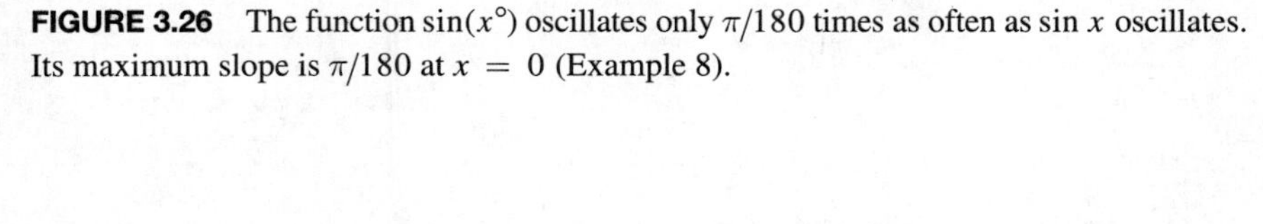

FIGURE 3.26 The function $\sin(x°)$ oscillates only $\pi/180$ times as often as $\sin x$ oscillates. Its maximum slope is $\pi/180$ at $x = 0$ (Example 8).

EXERCISES 3.6

Derivative Calculations

In Exercises 1–8, given $y = f(u)$ and $u = g(x)$, find $dy/dx = f'(g(x))g'(x)$.

1. $y = 4u - 5, \quad u = (1/3)x^3$
2. $y = 2u^3, \quad u = 8x - 1$
3. $y = \sin^2 u, \quad u = 2x + 5$
4. $y = \cos u, \quad u = -x/3$
5. $y = 2\sqrt{u}, \quad u = \cos x$
6. $y = \sin u, \quad u = x - \cos x$
7. $y = \tan u, \quad u = \pi x^2$
8. $y = -\operatorname{cosec} u, \quad u = x^2 + 5x$

In Exercises 9–18, write the function in the form $y = f(u)$ and $u = g(x)$. Then find dy/dx as a function of x.

9. $y = (2x + 1)^5$
10. $y = (2 - 5x)^8$
11. $y = \left(1 - \frac{x}{7}\right)^{-7}$
12. $y = \left(\frac{2\sqrt{x}}{4} - 5\right)^{-5}$
13. $y = \left(\frac{x^2}{8} + x - \frac{1}{x}\right)^4$
14. $y = \sqrt{3x^2 - 4x + 6}$
15. $y = \sec(\tan x)$
16. $y = \cot\left(\pi - \frac{1}{x}\right)$
17. $y = \tan^3 x$
18. $y = 5\cos^{-4} x$

Find the derivatives of the functions in Exercises 19–40.

19. $\sqrt{5 - 2t}$
20. $q = \sqrt[3]{2r - r^2}$
21. $s = \frac{2}{5\pi}\sin 2t + \frac{1}{4\pi}\cos 3t$
22. $s = \sin\left(\frac{3\pi t}{2}\right) + \cos\left(\frac{3\pi t}{2}\right)$
23. $r = (\sec\theta + \tan\theta)^{-1}$
24. $r = 6(\sec\theta - \tan\theta)^{3/2}$
25. $y = x^2\sin^4 x + x\cos^{-2} x$
26. $y = \frac{2}{x}\cos^{-5} x - \frac{x}{5}\sin^3 x$
27. $y = \frac{1}{18}(3x - 2)^6 + \left(4 - \frac{1}{2x^2}\right)^{-1}$
28. $y = (2 - 5x)^{-2}\left(\frac{1}{x} + 2\right)^4$
29. $y = (4x + 3)^4(x + 1)^{-3}$
30. $y = (3x - 5)^{-2}(x^2 - 4x)^4$
31. $h(x) = x\tan(2\sqrt{x}) + 7$
32. $k(x) = x^2\sec\left(\frac{1}{x}\right)$
33. $f(x) = \sqrt{5 + 2x\tan x}$
34. $g(x) = \frac{\tan 3x}{(x + 7)^4}$
35. $f(\theta) = \frac{\cos\theta}{1 + \sin\theta}$
36. $g(t) = \left(\frac{1 + \sin 3t}{3 - 2t}\right)^{-1}$
37. $r = \sin(\theta^2)\cos(2\theta)$
38. $r = \sec\sqrt{\theta}\tan\left(\frac{1}{\theta}\right)$
39. $q = \sin\left(\frac{t}{\sqrt{t + 1}}\right)$
40. $q = \cot\left(\frac{\sin t}{t}\right)$

In Exercises 41–58, find dy/dt.

41. $y = \sin^2(\pi t - 2)$
42. $y = \sec^2 \pi t$
43. $y = (1 + \cos 2t)^{-4}$
44. $y = (1 + \cot(t/2))^{-2}$
45. $y = (t\tan t)^{10}$
46. $y = (t^{-3/4}\sin t)^{4/3}$
47. $y = \left(\frac{t^2}{t^3 - 4t}\right)^3$
48. $y = \left(\frac{2t - 5}{3t + 5}\right)^{-2}$
49. $y = \cos(\sin(t - 4))$
50. $y = \cos\left(5\sin\left(\frac{t}{3}\right)\right)$
51. $y = \left(1 + \tan^4\left(\frac{t}{12}\right)\right)^3$
52. $y = \frac{1}{6}(1 + \cos^2(7t))^3$
53. $y = \cot^3(\sin^2 t)$
54. $y = 4\sin\left(\sqrt{1 + \sqrt{t}}\right)$
55. $y = \tan^2(\sin^3 t)$
56. $y = \sin^4(\operatorname{cosec}^2 2t)$
57. $y = 3t(2t^2 - 5)^4$
58. $y = \sqrt{3t + \sqrt{2 + \sqrt{1 - t}}}$

Second Derivatives

Find y'' in Exercises 59–64.

59. $y = \left(2 + \frac{5}{x}\right)^4$
60. $y = (1 - \sqrt{x})^{-1}$
61. $y = \frac{1}{9}\cot(3x - 1)$
62. $y = 5\tan\left(\frac{x}{5}\right)^4$
63. $y = x(2x + 1)^4$
64. $y = 2x^2(3x^3 - 5)^5$

For each of the following functions, solve both $f'(x) = 0$ and $f''(x) = 0$ for x.

65. $f(x) = x(x - 4)^3$
66. $f(x) = \sec^2 x - 2\tan x$ for $0 \le x \le 2\pi$

Finding Derivative Values

In Exercises 67–72, find the value of $(f \circ g)'$ at the given value of x.

67. $f(u) = 2u^4 + 5, \quad u = g(x) = 2\sqrt{x}, \quad x = 1$
68. $f(u) = 1 - \frac{1}{u}, \quad u = g(x) = \frac{1}{1 - x}, \quad x = -1$
69. $f(u) = \cot\frac{\pi u}{10}, \quad u = g(x) = 5\sqrt{x}, \quad x = 1$
70. $f(u) = u + \frac{1}{\cos^2 u}, \quad u = g(x) = \pi x, \quad x = 1/4$
71. $f(u) = \frac{5u}{u^3 + 2}, \quad u = g(x) = 5x^2 + x + 1, \quad x = 0$
72. $f(u) = \left(\frac{u - 1}{u + 1}\right)^2, \quad u = g(x) = \frac{1}{x^2} - 1, \quad x = -1$
73. Assume that $f'(5) = -2$, $g'(3) = 3$, $g(3) = 5$, and $y = f(g(x))$. What is y' at $x = 3$?
74. If $r = \sin(f(t))$, $f(0) = \pi/3$, and $f'(0) = 4$, then what is dr/dt at $t = 0$?
75. Suppose that functions f and g and their derivatives with respect to x have the following values at $x = 2$ and $x = 3$.

x	$f(x)$	$g(x)$	$f'(x)$	$g'(x)$
2	8	2	1/3	−3
3	3	−4	2π	5

Find the derivatives with respect to x of the following combinations at the given value of x.

a. $2f(x), \quad x = 2$
b. $f(x) + g(x), \quad x = 3$
c. $f(x) \cdot g(x), \quad x = 3$
d. $f(x)/g(x), \quad x = 2$
e. $f(g(x)), \quad x = 2$
f. $\sqrt{f(x)}, \quad x = 2$
g. $1/g^2(x), \quad x = 3$
h. $\sqrt{f^2(x) + g^2(x)}, \quad x = 2$

76. Suppose that the functions f and g and their derivatives with respect to x have the following values at $x = 0$ and $x = 1$.

x	$f(x)$	$g(x)$	$f'(x)$	$g'(x)$
0	1	1	5	1/3
1	3	-4	$-1/3$	$-8/3$

Find the derivatives with respect to x of the following combinations at the given value of x.

a. $5f(x) - g(x),\quad x = 1$ **b.** $f(x)g^3(x),\quad x = 0$

c. $\dfrac{f(x)}{g(x)+1},\quad x = 1$ **d.** $f(g(x)),\quad x = 0$

e. $g(f(x)),\quad x = 0$ **f.** $(x^{11} + f(x))^{-2},\quad x = 1$

g. $f(x + g(x)),\quad x = 0$

77. Find ds/dt when $\theta = 3\pi/2$ if $s = \cos\theta$ and $d\theta/dt = 5$.

78. Find dy/dt when $x = 1$ if $y = x^2 + 7x - 5$ and $dx/dt = 1/3$.

Theory and Examples

What happens if you can write a function as a composition in different ways? Do you get the same derivative each time? The Chain Rule says you should. Try it with the functions in Exercises 79 and 80.

79. Find dy/dx if $y = x$ by using the Chain Rule with y as a composition of

a. $y = (u/5) + 7$ and $u = 5x - 35$

b. $y = 1 + (1/u)$ and $u = 1/(x - 1)$.

80. Find dy/dx if $y = x^{3/2}$ by using the Chain Rule with y as a composition of

a. $y = u^3$ and $u = \sqrt{x}$

b. $y = \sqrt{u}$ and $u = x^3$.

81. Find the tangent to $y = \left(\dfrac{2x-1}{2x+1}\right)^2$ at $x = 0$.

82. Find the tangent line to $y = \sqrt{x^2 - x + 7}$ at $x = 2$.

83. a. Find the tangent line to the curve $y = 2\tan(\pi x/4)$ at $x = 1$.

b. Slopes on a tangent curve What is the smallest value the slope of the curve can ever have on the interval $-2 < x < 2$? Give reasons for your answer.

84. Slopes on sine curves

a. Find equations for the tangent lines to the curves $y = \sin 2x$ and $y = -\sin(x/2)$ at the origin. Is there anything special about how the tangent lines are related? Give reasons for your answer.

b. Can anything be said about the tangent lines to the curves $y = \sin mx$ and $y = -\sin(x/m)$ at the origin (m a constant $\neq 0$)? Give reasons for your answer.

c. For a given m, what are the largest values the slopes of the curves $y = \sin mx$ and $y = -\sin(x/m)$ can ever have? Give reasons for your answer.

d. The function $y = \sin x$ completes one period on the interval $[0, 2\pi]$, the function $y = \sin 2x$ completes two periods, the function $y = \sin(x/2)$ completes half a period, and so on. Is there any relation between the number of periods $y = \sin mx$ completes on $[0, 2\pi]$ and the slope of the curve $y = \sin mx$ at the origin? Give reasons for your answer.

85. Running machinery too fast Suppose that a piston is moving straight up and down and that its position at time t sec is

$$s = A\sin(2\pi bt)$$

with A and b positive. The value of A is the amplitude of the motion, and b is the frequency (number of times the piston moves up and down each second). What effect does doubling the frequency have on the piston's velocity, acceleration, and jerk? (Once you find out, you will know why some machinery breaks when you run it too fast.)

86. Temperatures in Fairbanks, Alaska The graph in the accompanying figure shows the average Fahrenheit temperature in Fairbanks, Alaska, during a typical 365-day year. The equation that approximates the temperature on day x is

$$y = 37\sin\left[\frac{2\pi}{365}(x - 101)\right] + 25$$

and is graphed in the accompanying figure.

a. On what day is the temperature increasing the fastest?

b. About how many degrees per day is the temperature increasing when it is increasing at its fastest?

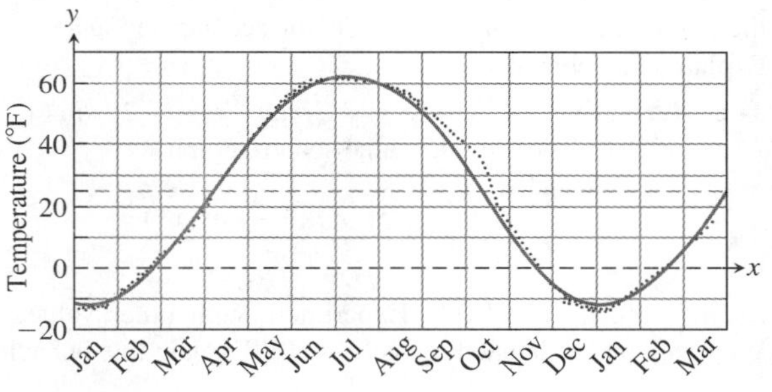

87. Particle motion The position of a particle moving along a coordinate line is $s = \sqrt{6+2t}$ with s in meters and t in seconds. Find the particle's velocity and acceleration at $t = 5$ sec.

88. Constant acceleration Suppose that the velocity of a falling body is $v = k\sqrt{s}$ m/sec (k a constant) at the instant the body has fallen s m from its starting point. Show that the body's acceleration is constant.

89. Falling meteorite The velocity of a heavy meteorite entering Earth's atmosphere is inversely proportional to $\sqrt{s}$ when it is s km from Earth's center. Show that the meteorite's acceleration is inversely proportional to s^2.

90. Particle acceleration A particle moves along the x-axis with velocity $dx/dt = f(x)$. Show that the particle's acceleration is $f(x)f'(x)$.

91. Temperature and the period of a pendulum For oscillations of small amplitude (short swings), we may safely model the relationship between the period T and the length L of a simple pendulum with the equation

$$T = 2\pi\sqrt{\frac{L}{g}},$$

where g is the constant acceleration of gravity at the pendulum's location. If we measure g in centimeters per second squared, we measure L in centimeters and T in seconds. If the pendulum is made of metal, its length will vary with temperature, either increasing or decreasing at a rate that is roughly proportional to L. In symbols, with u being temperature and k the proportionality constant,

$$\frac{dL}{du} = kL.$$

Assuming this to be the case, show that the rate at which the period changes with respect to temperature is $kT/2$.

92. Chain Rule Suppose that $f(x) = x^2$ and $g(x) = |x|$. Then the compositions

$$(f \circ g)(x) = |x|^2 = x^2 \quad \text{and} \quad (g \circ f)(x) = |x^2| = x^2$$

are both differentiable at $x = 0$ even though g itself is not differentiable at $x = 0$. Does this contradict the Chain Rule? Explain.

T **93. The derivative of sin 2x** Graph the function $y = 2\cos 2x$ for $-2 \le x \le 3.5$. Then, on the same screen, graph

$$y = \frac{\sin 2(x+h) - \sin 2x}{h}$$

for $h = 1.0, 0.5$, and 0.2. Experiment with other values of h, including negative values. What do you see happening as $h \to 0$? Explain this behavior.

T **94. The derivative of cos (x^2)** Graph $y = -2x\sin(x^2)$ for $-2 \le x \le 3$. Then, on the same screen, graph

$$y = \frac{\cos((x+h)^2) - \cos(x^2)}{h}$$

for $h = 1.0, 0.7$, and 0.3. Experiment with other values of h. What do you see happening as $h \to 0$? Explain this behavior.

Using the Chain Rule, show that the Power Rule $(d/dx)x^n = nx^{n-1}$ holds for the functions x^n in Exercises 95 and 96.

95. $x^{1/4} = \sqrt{\sqrt{x}}$

96. $x^{3/4} = \sqrt{x\sqrt{x}}$

97. Consider the function

$$f(x) = \begin{cases} x\sin\left(\frac{1}{x}\right), & x > 0 \\ 0, & x \le 0 \end{cases}$$

a. Show that f is continuous at $x = 0$.

b. Determine f' for $x \ne 0$.

c. Show that f is not differentiable at $x = 0$.

98. Consider the function

$$f(x) = \begin{cases} x^2\cos\left(\frac{2}{x}\right), & x \ne 0 \\ 0, & x = 0 \end{cases}$$

a. Show that f is continuous at $x = 0$.

b. Determine f' for $x \ne 0$.

c. Show that f is differentiable at $x = 0$.

d. Show that f' is not continuous at $x = 0$.

99. Verify each of the following statements.

a. If f is even, then f' is odd.

b. If f is odd, then f' is even.

COMPUTER EXPLORATIONS

Trigonometric Polynomials

100. As the accompanying figure shows, the trigonometric "polynomial"

$$s = f(t) = 0.78540 - 0.63662\cos 2t - 0.07074\cos 6t$$
$$-0.02546\cos 10t - 0.01299\cos 14t$$

gives a good approximation of the sawtooth function $s = g(t)$ on the interval $[-\pi, \pi]$. How well does the derivative of f approximate the derivative of g at the points where dg/dt is defined? To find out, carry out the following steps.

a. Graph dg/dt (where defined) over $[-\pi, \pi]$.

b. Find df/dt.

c. Graph df/dt. Where does the approximation of dg/dt by df/dt seem to be best? Least good? Approximations by trigonometric polynomials are important in the theories of heat and oscillation, but we must not expect too much of them, as we see in the next exercise.

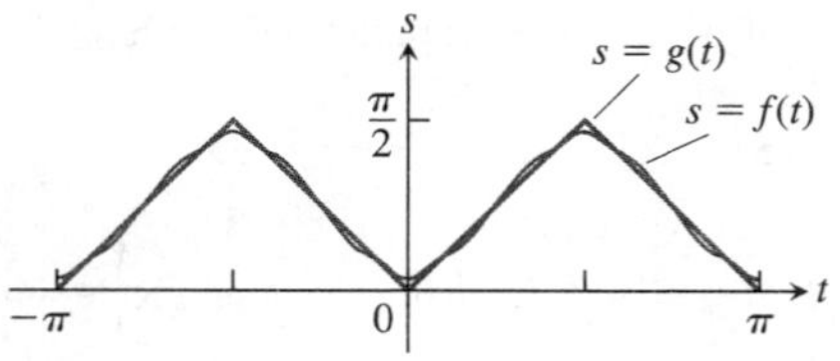

101. (*Continuation of Exercise 100.*) In Exercise 100, the trigonometric polynomial $f(t)$ that approximated the sawtooth function $g(t)$ on $[-\pi, \pi]$ had a derivative that approximated the derivative of the sawtooth function. It is possible, however, for a trigonometric polynomial to approximate a function in a reasonable way without its derivative approximating the function's derivative at all well. As a case in point, the trigonometric "polynomial"

$$s = h(t) = 1.2732\sin 2t + 0.4244\sin 6t + 0.25465\sin 10t$$
$$+ 0.18189\sin 14t + 0.14147\sin 18t$$

graphed in the accompanying figure approximates the step function $s = k(t)$ shown there. Yet the derivative of h is nothing like the derivative of k.

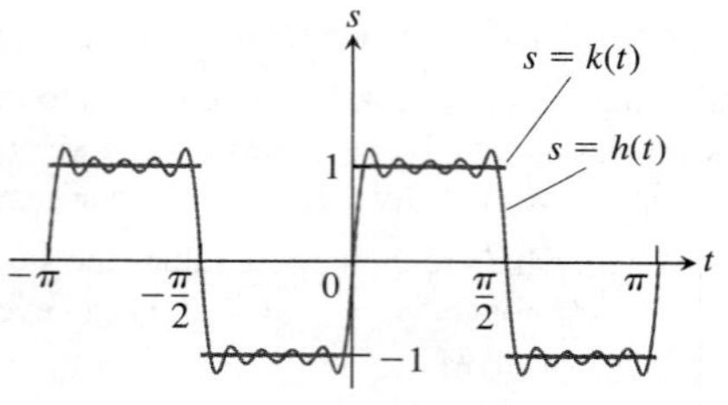

a. Graph dk/dt (where defined) over $[-\pi, \pi]$.

b. Find dh/dt.

c. Graph dh/dt to see how badly the graph fits the graph of dk/dt. Comment on what you see.

3.7 Implicit Differentiation

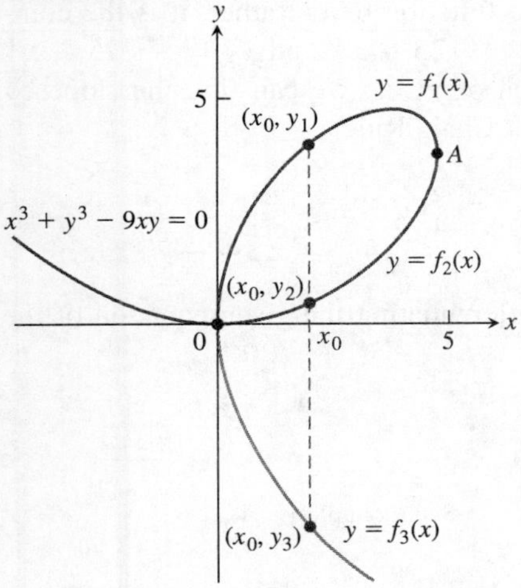

FIGURE 3.27 The curve $x^3 + y^3 - 9xy = 0$ is not the graph of any one function of x. The curve can, however, be divided into separate arcs that *are* the graphs of functions of x. This particular curve, called a *folium*, dates to Descartes in 1638.

Most of the functions we have dealt with so far have been described by an equation of the form $y = f(x)$ that expresses y explicitly in terms of the variable x. We have learned rules for differentiating functions defined in this way. A different situation occurs when we encounter equations like

$$x^3 + y^3 - 9xy = 0, \qquad y^2 - x = 0, \qquad \text{or} \qquad x^2 + y^2 - 25 = 0.$$

(See Figures 3.27, 3.28, and 3.29.) Each of these equations defines an *implicit* relation between the variables x and y, meaning that a value of x may determine more than one value of y, even though we do not have a simple formula for the y-values. In some cases we may be able to solve such an equation for y as an explicit function (or even several functions) of x. When we cannot put an equation $F(x, y) = 0$ in the form $y = f(x)$ to differentiate it in the usual way, we may still be able to find dy/dx by *implicit differentiation*. This section describes the technique.

Implicitly Defined Functions

We begin with examples involving familiar equations that we can solve for y as a function of x and then calculate dy/dx in the usual way. Then we differentiate the equations implicitly, and find the derivative. We will see that the two methods give the same answer. Following the examples, we summarize the steps involved in the new method. In the examples and exercises, it is always assumed that the given equation determines y implicitly as a differentiable function of x so that dy/dx exists.

EXAMPLE 1 Find dy/dx if $y^2 = x$.

Solution The equation $y^2 = x$ defines two differentiable functions of x that we can actually find, namely $y_1 = \sqrt{x}$ and $y_2 = -\sqrt{x}$ (Figure 3.28). We know how to calculate the derivative of each of these for $x > 0$:

$$\frac{dy_1}{dx} = \frac{1}{2\sqrt{x}} \quad \text{and} \quad \frac{dy_2}{dx} = -\frac{1}{2\sqrt{x}}.$$

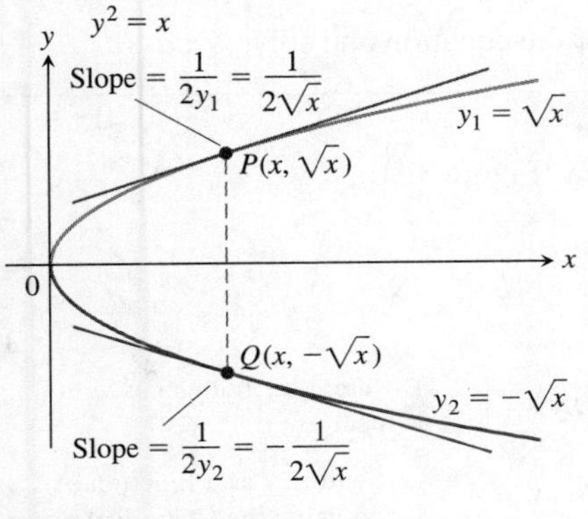

FIGURE 3.28 The equation $y^2 - x = 0$, or $y^2 = x$ as it is usually written, defines two differentiable functions of x on the interval $x > 0$. Example 1 shows how to find the derivatives of these functions without solving the equation $y^2 = x$ for y.

But suppose that we knew only that the equation $y^2 = x$ defined y as one or more differentiable functions of x for $x > 0$ without knowing exactly what these functions were. Can we still find dy/dx?

The answer is yes. To find dy/dx, we simply differentiate both sides of the equation $y^2 = x$ with respect to x, treating $y = f(x)$ as a differentiable function of x:

$$y^2 = x$$

$$2y\frac{dy}{dx} = 1 \qquad \text{The Chain Rule gives } \frac{d}{dx}(y^2) = \frac{d}{dx}[f(x)]^2 = 2f(x)f'(x) = 2y\frac{dy}{dx}.$$

$$\frac{dy}{dx} = \frac{1}{2y}.$$

This one formula gives the derivatives we calculated for *both* explicit solutions $y_1 = \sqrt{x}$ and $y_2 = -\sqrt{x}$:

$$\frac{dy_1}{dx} = \frac{1}{2y_1} = \frac{1}{2\sqrt{x}} \quad \text{and} \quad \frac{dy_2}{dx} = \frac{1}{2y_2} = \frac{1}{2(-\sqrt{x})} = -\frac{1}{2\sqrt{x}}. \qquad \blacksquare$$

EXAMPLE 2 Find the slope of the circle $x^2 + y^2 = 25$ at the point $(3, -4)$.

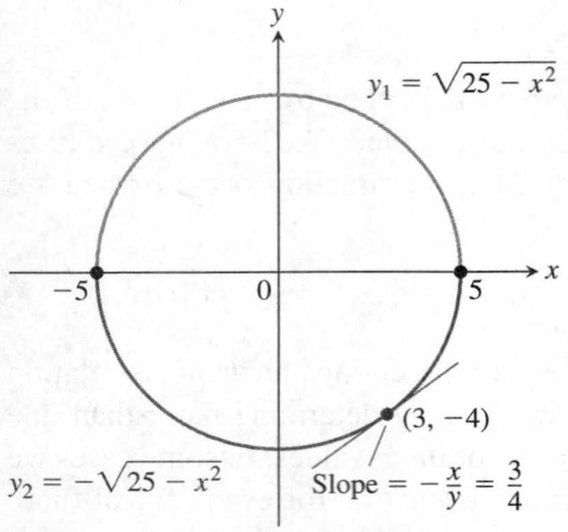

FIGURE 3.29 The circle combines the graphs of two functions. The graph of y_2 is the lower semicircle and passes through $(3, -4)$.

Solution The circle is not the graph of a single function of x. Rather, it is the combined graphs of two differentiable functions, $y_1 = \sqrt{25 - x^2}$ and $y_2 = -\sqrt{25 - x^2}$ (Figure 3.29). The point $(3, -4)$ lies on the graph of y_2, so we can find the slope by calculating the derivative directly, using the Power Chain Rule:

$$\left.\frac{dy_2}{dx}\right|_{x=3} = -\left.\frac{-2x}{2\sqrt{25 - x^2}}\right|_{x=3} = -\frac{-6}{2\sqrt{25 - 9}} = \frac{3}{4}. \qquad \frac{d}{dx}\left(-(25 - x^2)^{1/2}\right) = -\frac{1}{2}(25 - x^2)^{-1/2}(-2x)$$

We can solve this problem more easily by differentiating the given equation of the circle implicitly with respect to x:

$$\frac{d}{dx}(x^2) + \frac{d}{dx}(y^2) = \frac{d}{dx}(25)$$

$$2x + 2y\frac{dy}{dx} = 0 \qquad \text{See Example 1.}$$

$$\frac{dy}{dx} = -\frac{x}{y}.$$

The slope at $(3, -4)$ is $-\left.\frac{x}{y}\right|_{(3,-4)} = -\frac{3}{-4} = \frac{3}{4}.$

Notice that unlike the slope formula for dy_2/dx, which applies only to points below the x-axis, the formula $dy/dx = -x/y$ applies everywhere the circle has a slope—that is, at all circle points (x, y) where $y \neq 0$. Notice also that the derivative involves *both* variables x and y, not just the independent variable x. ■

To calculate the derivatives of other implicitly defined functions, we proceed as in Examples 1 and 2: We treat y as a differentiable implicit function of x and apply the usual rules to differentiate both sides of the defining equation.

Implicit Differentiation

1. Differentiate both sides of the equation with respect to x, treating y as a differentiable function of x.
2. Collect the terms with dy/dx on one side of the equation and solve for dy/dx.

EXAMPLE 3 Find dy/dx if $y^2 = x^2 + \sin xy$ (Figure 3.30).

Solution We differentiate the equation implicitly.

$$y^2 = x^2 + \sin xy$$

$$\frac{d}{dx}(y^2) = \frac{d}{dx}(x^2) + \frac{d}{dx}(\sin xy) \qquad \text{Differentiate both sides with respect to } x\ldots$$

$$2y\frac{dy}{dx} = 2x + (\cos xy)\frac{d}{dx}(xy) \qquad \ldots\text{ treating } y \text{ as a function of } x \text{ and using the Chain Rule.}$$

$$2y\frac{dy}{dx} = 2x + (\cos xy)\left(y + x\frac{dy}{dx}\right) \qquad \text{Treat } xy \text{ as a product.}$$

$$2y\frac{dy}{dx} - (\cos xy)\left(x\frac{dy}{dx}\right) = 2x + (\cos xy)y \qquad \text{Collect terms with } dy/dx.$$

$$(2y - x\cos xy)\frac{dy}{dx} = 2x + y\cos xy$$

$$\frac{dy}{dx} = \frac{2x + y\cos xy}{2y - x\cos xy} \qquad \text{Solve for } dy/dx.$$

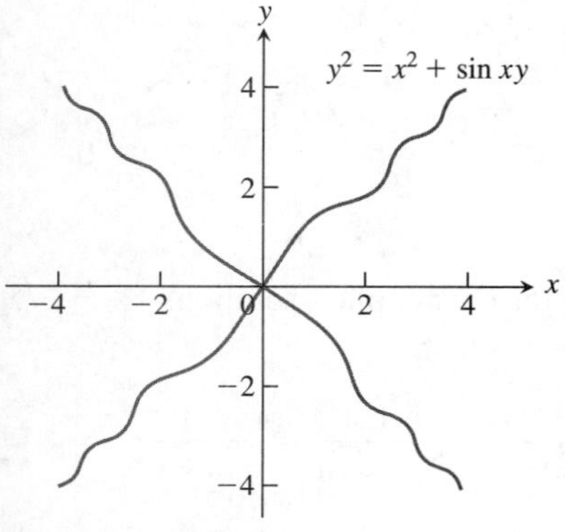

FIGURE 3.30 The graph of the equation in Example 3.

Notice that the formula for dy/dx applies everywhere that the implicitly defined curve has a slope. Notice again that the derivative involves *both* variables x and y, not just the independent variable x. ■

Derivatives of Higher Order

Implicit differentiation can also be used to find higher derivatives.

EXAMPLE 4 Find d^2y/dx^2 if $2x^3 - 3y^2 = 8$.

Solution To start, we differentiate both sides of the equation with respect to x in order to find $y' = dy/dx$.

$$\frac{d}{dx}(2x^3 - 3y^2) = \frac{d}{dx}(8) \qquad \text{Treat } y \text{ as a function of } x.$$

$$6x^2 - 6yy' = 0$$

$$y' = \frac{x^2}{y}, \quad \text{when } y \neq 0 \qquad \text{Solve for } y'.$$

We now apply the Quotient Rule to find y''.

$$y'' = \frac{d}{dx}\left(\frac{x^2}{y}\right) = \frac{2xy - x^2y'}{y^2} = \frac{2x}{y} - \frac{x^2}{y^2} \cdot y'$$

Finally, we substitute $y' = x^2/y$ to express y'' in terms of x and y.

$$y'' = \frac{2x}{y} - \frac{x^2}{y^2}\left(\frac{x^2}{y}\right) = \frac{2x}{y} - \frac{x^4}{y^3}, \quad \text{when } y \neq 0$$ ■

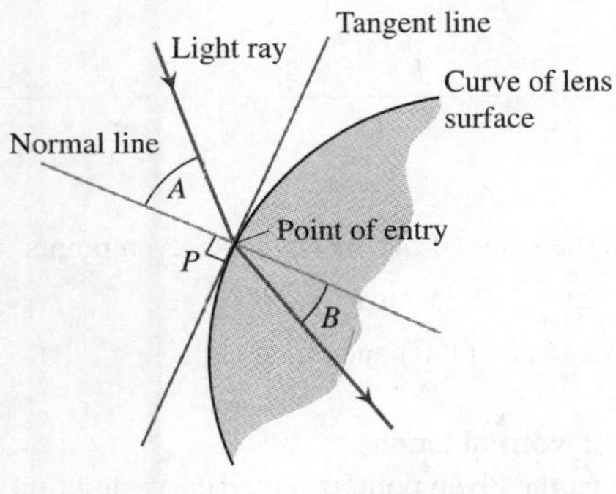

FIGURE 3.31 The profile of a lens, showing the bending (refraction) of a ray of light as it passes through the lens surface.

Lenses, Tangent Lines, and Normal Lines

In the law that describes how light changes direction as it enters a lens, the important angles are the angles the light makes with the line perpendicular to the surface of the lens at the point of entry (angles A and B in Figure 3.31). This line is called the *normal line* to the surface at the point of entry. In a profile view of a lens like the one in Figure 3.31, the **normal line** is the line perpendicular (also said to be *orthogonal*) to the tangent line of the profile curve at the point of entry.

EXAMPLE 5 Show that the point $(2, 4)$ lies on the curve $x^3 + y^3 - 9xy = 0$. Then find the tangent line and normal line to the curve there (Figure 3.32).

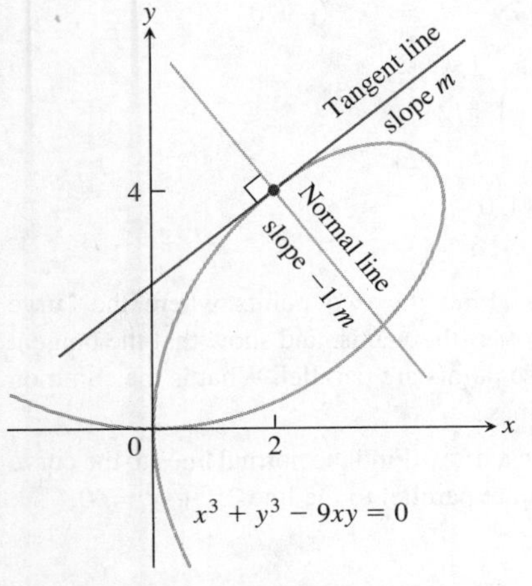

FIGURE 3.32 Example 5 shows how to find equations for the tangent line and normal line to the folium of Descartes at $(2, 4)$.

Solution The point $(2, 4)$ lies on the curve because its coordinates satisfy the equation given for the curve: $2^3 + 4^3 - 9(2)(4) = 8 + 64 - 72 = 0$.

To find the slope of the curve at $(2, 4)$, we first use implicit differentiation to find a formula for dy/dx:

$$x^3 + y^3 - 9xy = 0$$

$$\frac{d}{dx}(x^3) + \frac{d}{dx}(y^3) - \frac{d}{dx}(9xy) = \frac{d}{dx}(0) \qquad \text{Differentiate both sides with respect to } x.$$

$$3x^2 + 3y^2\frac{dy}{dx} - 9\left(x\frac{dy}{dx} + \frac{dx}{dx}y\right) = 0 \qquad \text{Treat } xy \text{ as a product and } y \text{ as a function of } x.$$

$$(3y^2 - 9x)\frac{dy}{dx} + 3x^2 - 9y = 0$$

$$3(y^2 - 3x)\frac{dy}{dx} = 9y - 3x^2$$

$$\frac{dy}{dx} = \frac{3y - x^2}{y^2 - 3x}. \qquad \text{Solve for } dy/dx.$$

We then evaluate the derivative at $(x, y) = (2, 4)$:

$$\left.\frac{dy}{dx}\right|_{(2,4)} = \left.\frac{3y - x^2}{y^2 - 3x}\right|_{(2,4)} = \frac{3(4) - 2^2}{4^2 - 3(2)} = \frac{8}{10} = \frac{4}{5}.$$

The tangent line at $(2, 4)$ is the line through $(2, 4)$ with slope $4/5$:

$$y = 4 + \frac{4}{5}(x - 2)$$
$$y = \frac{4}{5}x + \frac{12}{5}.$$

The normal line to the curve at $(2, 4)$ is the line perpendicular to the tangent line there, the line through $(2, 4)$ with slope $-5/4$:

$$y = 4 - \frac{5}{4}(x - 2)$$
$$y = -\frac{5}{4}x + \frac{13}{2}.$$

Slopes of two nonvertical perpendicular lines are negative reciprocals of each other (see Appendix A.3). ■

EXERCISES 3.7

Differentiating Implicitly

Use implicit differentiation to find dy/dx in Exercises 1–14.

1. $x^2y + xy^2 = 6$

2. $x^3 + y^3 = 18xy$

3. $2xy + y^2 = x + y$

4. $x^3 - xy + y^3 = 1$

5. $x^2(x - y)^2 = x^2 - y^2$

6. $(3xy + 7)^2 = 6y$

7. $y^2 = \dfrac{x - 1}{x + 1}$

8. $x^3 = \dfrac{2x - y}{x + 3y}$

9. $x\cot^2 y + y = 0$

10. $xy = \cot(xy)$

11. $x^m y^n = (x + y)^{m+n}$

12. $x\sqrt{1+y} + y\sqrt{1+x} = 0$

13. $y\sin\left(\dfrac{1}{y}\right) = 1 - xy$

14. $x\cos(2x + 3y) = y\sin x$

Find $dr/d\theta$ in Exercises 15–18.

15. $\theta^{1/2} + r^{1/2} = 1$

16. $r - 2\sqrt{\theta} = \dfrac{3}{2}\theta^{2/3} + \dfrac{4}{3}\theta^{3/4}$

17. $\sin(r\theta) = \dfrac{1}{2}$

18. $\cos r + \cot\theta = r$

Second Derivatives

In Exercises 19–26, use implicit differentiation to find dy/dx and then d^2y/dx^2. Write the solutions in terms of x and y only.

19. $x^2 + y^2 = 1$

20. $x^{2/3} + y^{2/3} = 1$

21. $y^2 = x^2 + 2x$

22. $y^2 - 2x = 1 - 2y$

23. $2\sqrt{y} = x - y$

24. $y^3 + 3x^2 + x^3 = 0$

25. $3 + \sin y = y - x^3$

26. $\sin y = x\cos y - 2$

27. If $x^3 + y^3 = 16$, find the value of d^2y/dx^2 at the point (2, 2).

28. If $xy + y^2 = 1$, find the value of d^2y/dx^2 at the point $(0, -1)$.

In Exercises 29 and 30, find the slope of the curve at the given points.

29. $y^2 + x^2 = y^4 - 2x$ at $(-2, 1)$ and $(-2, -1)$

30. $(x^2 + y^2)^2 = (x - y)^2$ at $(1, 0)$ and $(1, -1)$

Slopes, Tangent Lines, and Normal Lines

In Exercises 31–40, verify that the given point is on the curve and find the lines that are **(a)** tangent and **(b)** normal to the curve at the given point.

31. $x^3 + y^3 = 9x^2, (1, 2)$

32. $y(x-2)(x-3) - x + 7 = 0, (7, 0)$

33. $x^2y^2 = 9, \quad (-1, 3)$

34. $y\,e^{\frac{x}{a}} = b, (0, b)$

35. $6x^2 + 3xy + 2y^2 + 17y - 6 = 0, \quad (-1, 0)$

36. $x^2 - \sqrt{3}xy + 2y^2 = 5, \quad (\sqrt{3}, 2)$

37. $2xy + \pi\sin y = 2\pi, \quad (1, \pi/2)$

38. $x\sin 2y = y\cos 2x, \quad (\pi/4, \pi/2)$

39. $y = 2\sin(\pi x - y), \quad (1, 0)$

40. $x^2\cos^2 y - \sin y = 0, \quad (0, \pi)$

41. Parallel tangent lines Find the two points where the curve $x^2 + xy + y^2 = 7$ crosses the x-axis, and show that the tangent lines to the curve at these points are parallel. What is the common slope of these tangent lines?

42. Normal lines parallel to a line Find the normal lines to the curve $xy + 2x - y = 0$ that are parallel to the line $2x + y = 0$.

43. The eight curve Find the slopes of the curve $y^4 = y^2 - x^2$ at the two points shown here.

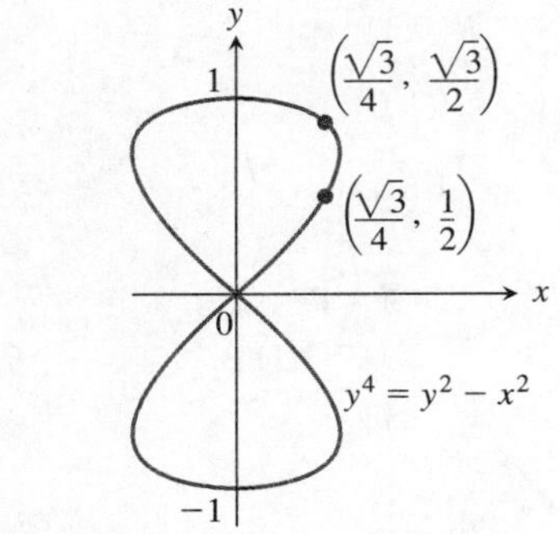

44. The cissoid of Diocles (from about 200 B.C.) Find equations for the tangent line and normal line to the cissoid of Diocles $y^2(2 - x) = x^3$ at $(1, 1)$.

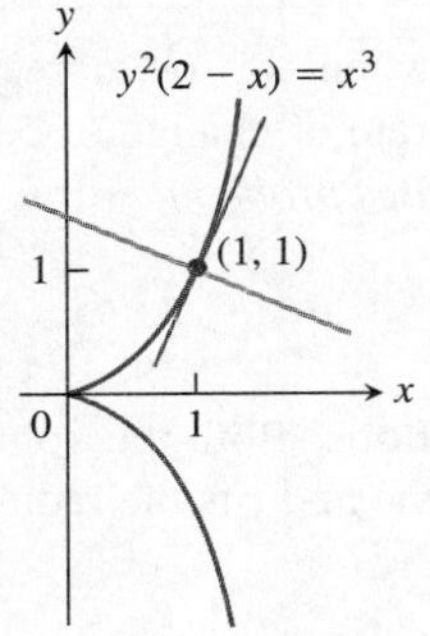

45. The devil's curve (Gabriel Cramer, 1750) Find the slopes of the devil's curve $y^4 - 4y^2 = x^4 - 9x^2$ at the four indicated points.

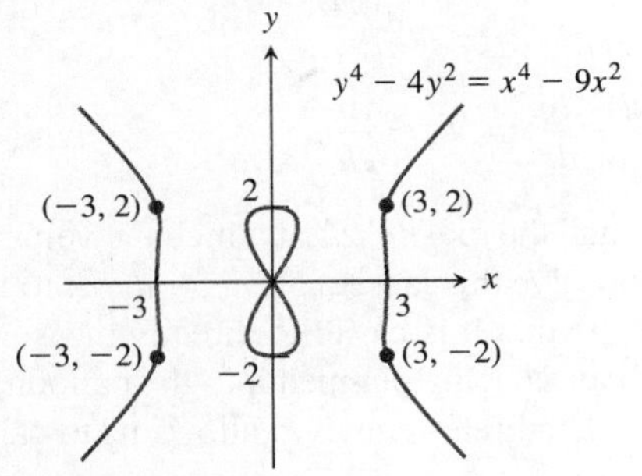

46. The folium of Descartes (See Figure 3.27)

a. Find the slope of the folium of Descartes $x^3 + y^3 - 9xy = 0$ at the points $(4, 2)$ and $(2, 4)$.

b. At what point other than the origin does the folium have a horizontal tangent line?

c. Find the coordinates of the point A in Figure 3.27 where the folium has a vertical tangent line.

Theory and Examples

47. Intersecting normal line The line that is normal to the curve $x^2 + 2xy - 3y^2 = 0$ at $(1, 1)$ intersects the curve at what other point?

48. Power rule for rational exponents Let p and q be integers with $q > 0$. If $y = x^{p/q}$, differentiate the equivalent equation $y^q = x^p$ implicitly and show that, for $y \neq 0$,

$$\frac{d}{dx}x^{p/q} = \frac{p}{q}x^{(p/q)-1}.$$

49. Normal lines to a parabola Show that if it is possible to draw three normal lines from the point $(a, 0)$ to the parabola $x = y^2$ shown in the accompanying diagram, then a must be greater than $1/2$. One of the normal lines is the x-axis. For what value of a are the other two normal lines perpendicular?

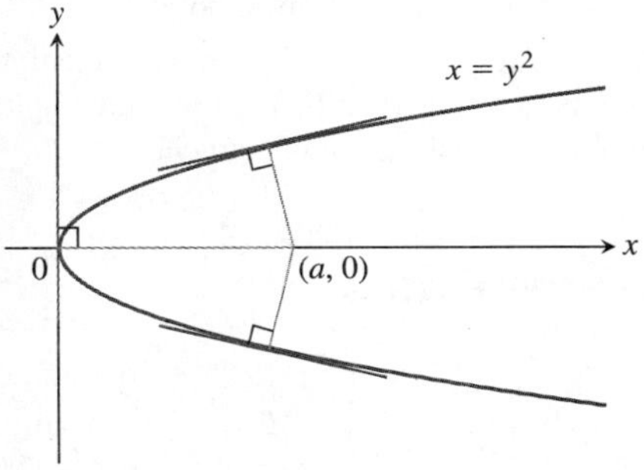

50. Is there anything special about the tangent lines to the curves $y^2 = x^3$ and $2x^2 + 3y^2 = 5$ at the points $(1, \pm 1)$? Give reasons for your answer.

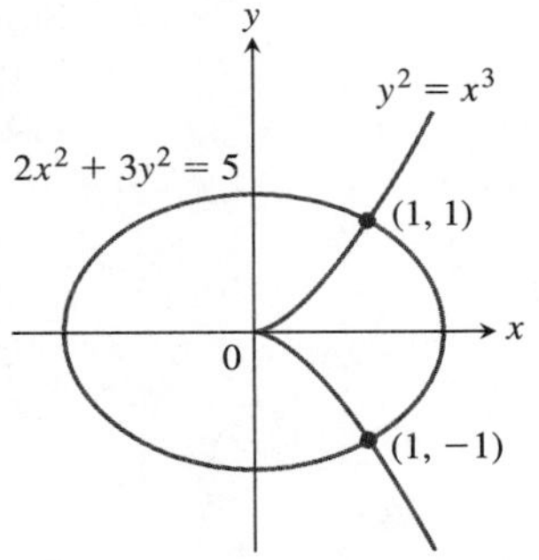

51. Verify that the following pairs of curves meet orthogonally.

a. $x^2 + y^2 = 4, \quad x^2 = 3y^2$

b. $x = 1 - y^2, \quad x = \frac{1}{3}y^2$

52. The graph of $y^2 = x^3$ is called a **semicubical parabola** and is shown in the accompanying figure. Determine the constant b so that the line $y = -\frac{1}{3}x + b$ meets this graph orthogonally.

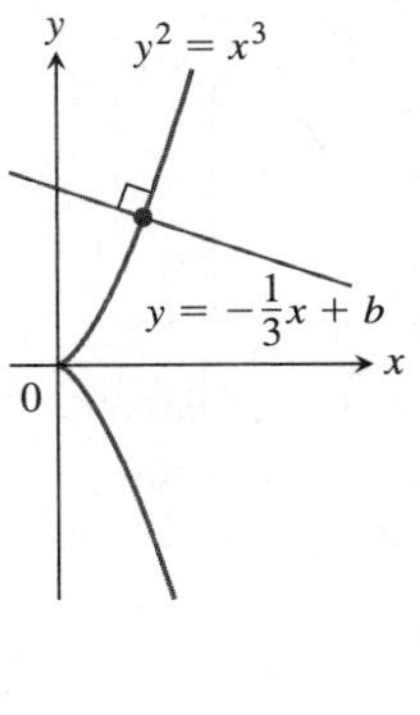

In Exercises 53 and 54, find both dy/dx (treating y as a differentiable function of x) and dx/dy (treating x as a differentiable function of y). How do dy/dx and dx/dy seem to be related?

53. $xy^3 + x^2y = 6$

54. $x^3 + y^2 = \sin^2 y$

COMPUTER EXPLORATIONS

Use a CAS to perform the following steps in Exercises 55–62.

a. Plot the equation with the implicit plotter of a CAS. Check to see that the given point P satisfies the equation.

b. Using implicit differentiation, find a formula for the derivative dy/dx and evaluate it at the given point P.

c. Use the slope found in part (b) to find an equation for the tangent line to the curve at P. Then plot the implicit curve and tangent line together on a single graph.

55. $x^3 - xy + y^3 = 7, \quad P(2,1)$

56. $x^5 + y^3x + yx^2 + y^4 = 4, \quad P(1,1)$

57. $y^2 + y = \dfrac{2+x}{1-x}, \quad P(0,1)$

58. $y^3 + \cos xy = x^2, \quad P(1,0)$

59. $x + \tan\left(\dfrac{y}{x}\right) = 2, \quad P\left(1, \dfrac{\pi}{4}\right)$

60. $xy^3 + \tan(x+y) = 1, \quad P\left(\dfrac{\pi}{4}, 0\right)$

61. $2y^2 + (xy)^{1/3} = x^2 + 2, \quad P(1,1)$

62. $x\sqrt{1+2y} + y = x^2, \quad P(1,0)$

3.8 Related Rates

In this section we look at questions that arise when two or more related quantities are changing. The problem of determining how the rate of change of one of them affects the rates of change of the others is called a *related rates problem.*

Related Rates Equations

Suppose we are pumping air into a spherical balloon. Both the volume and radius of the balloon are increasing over time. If V is the volume and r is the radius of the balloon at an instant of time, then

$$V = \frac{4}{3}\pi r^3.$$

Using the Chain Rule, we differentiate both sides with respect to t to find an equation relating the rates of change of V and r,

$$\frac{dV}{dt} = \frac{dV}{dr}\frac{dr}{dt} = 4\pi r^2 \frac{dr}{dt}.$$

So if we know the radius r of the balloon and the rate dV/dt at which the volume is increasing at a given instant of time, then we can solve this last equation for dr/dt to find how fast the radius is increasing at that instant. Note that it is easier to directly measure the rate of increase of the volume (the rate at which air is being pumped into the balloon) than it is to measure the increase in the radius. The related rates equation allows us to calculate dr/dt from dV/dt.

Very often the key to relating the variables in a related rates problem is drawing a picture that shows the geometric relations between them, as illustrated in the following example.

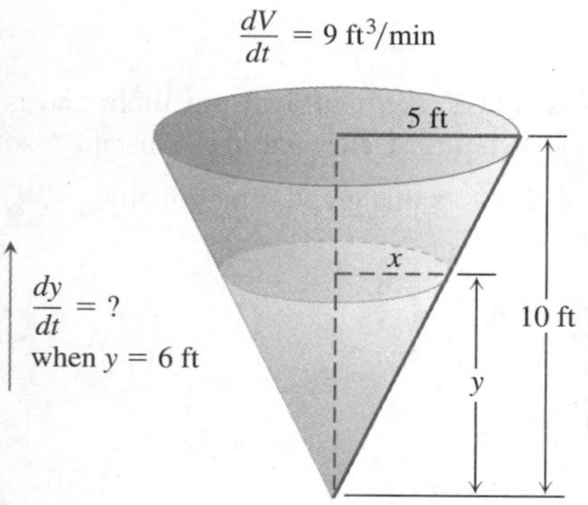

FIGURE 3.33 The geometry of the conical tank and the rate at which water fills the tank determine how fast the water level rises (Example 1).

EXAMPLE 1 Water runs into a conical tank at the rate of 9 ft^3/min. The tank stands point down and has a height of 10 ft and a base radius of 5 ft. How fast is the water level rising when the water is 6 ft deep?

Solution Figure 3.33 shows a partially filled conical tank. The variables in the problem are

V = volume (ft^3) of the water in the tank at time t (min)

x = radius (ft) of the surface of the water at time t

y = depth (ft) of the water in the tank at time t.

We assume that V, x, and y are differentiable functions of t. The constants are the dimensions of the tank. We are asked for dy/dt when

$$y = 6 \text{ ft} \quad \text{and} \quad \frac{dV}{dt} = 9 \text{ ft}^3/\text{min}.$$

The water forms a cone with volume

$$V = \frac{1}{3}\pi x^2 y.$$

This equation involves x as well as V and y. Because no information is given about x and dx/dt at the time in question, we need to eliminate x. The similar triangles in Figure 3.33 give us a way to express x in terms of y:

$$\frac{x}{y} = \frac{5}{10} \quad \text{or} \quad x = \frac{y}{2}.$$

Therefore, we find

$$V = \frac{1}{3}\pi\left(\frac{y}{2}\right)^2 y = \frac{\pi}{12}y^3$$

to give the derivative

$$\frac{dV}{dt} = \frac{\pi}{12}\cdot 3y^2\frac{dy}{dt} = \frac{\pi}{4}y^2\frac{dy}{dt}.$$

Finally, use $y = 6$ and $dV/dt = 9$ to solve for dy/dt.

$$9 = \frac{\pi}{4}(6)^2\frac{dy}{dt}$$

$$\frac{dy}{dt} = \frac{1}{\pi} \approx 0.32$$

At the moment in question, the water level is rising at about 0.32 ft/min. ■

Related Rates Problem Strategy

1. Let t denote time, and **choose names** for all of the variables that change over time (we will assume that those variables are differentiable functions of t). Identify any quantities that remain constant (these do not need to be given names). In most problems it will be very helpful to **draw a picture** that depicts the setup of the problem.
2. Write an **equation that relates the variables** (and any constants that are present). You may have to combine two or more equations to get a single equation that relates the variable whose rate you want to the variables whose rates or values you know.
3. Differentiate with respect to t to obtain a **related rates equation**.
4. Substitute all of the numerical values provided in the problem into the related rates equation. You may need to use the equation(s) relating the variables (which you obtained in Step 2), or use other relationships (such as trigonometric identities), until you reach the point at which *the only remaining unknown quantity is the rate of change that you are asked to find.* Solve for this unknown.

EXAMPLE 2 A hot air balloon rising straight up from a level field is tracked by a range finder 150 m from the liftoff point. At the moment the range finder's elevation angle is $\pi/4$, the angle is increasing at the rate of 0.14 rad/min. How fast is the balloon rising at that moment?

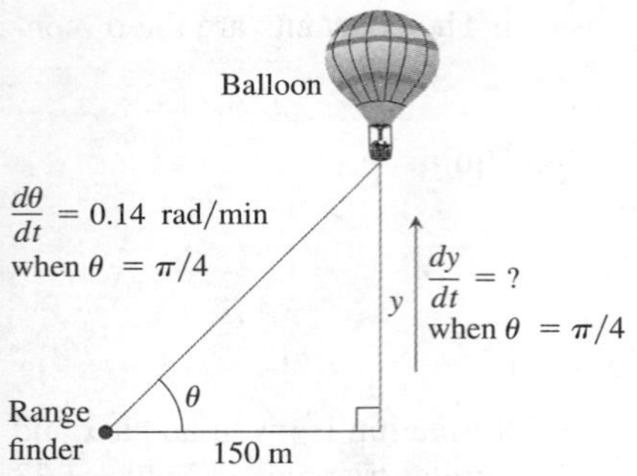

FIGURE 3.34 The rate of change of the balloon's height is related to the rate of change of the angle the range finder makes with the ground (Example 2).

Solution We draw a picture (Figure 3.34) and name the variables that appear in the problem (which we assume are differentiable functions of t, where time is measured in minutes):

$\theta =$ the angle in radians the range finder makes with the ground, and

$y =$ the height in meters of the balloon above the ground.

One constant in the picture is the distance from the range finder to the liftoff point (150 m). There is no need to give this distance a special symbol.

Trigonometry (Section 1.3) yields

$$\frac{y}{150} = \tan\theta \quad \text{or} \quad y = 150\tan\theta. \qquad \text{Equation relating the variables}$$

By differentiating with respect to t using the Chain Rule, we obtain

$$\frac{dy}{dt} = 150(\sec^2\theta)\frac{d\theta}{dt}. \qquad \text{Related rates equation}$$

Substituting the known values $\dfrac{d\theta}{dt} = 0.14$ rad/min and $\theta = \dfrac{\pi}{4}$ gives

$$\frac{dy}{dt} = 150\left(\sec^2\frac{\pi}{4}\right)(0.14) = 150\left(\sqrt{2}\right)^2(0.14) = 42. \qquad \sec\frac{\pi}{4} = \sqrt{2}$$

Thus, at the moment in question, the balloon is rising at the rate of 42 m/min. ■

EXAMPLE 3 A police cruiser, approaching a right-angled intersection from the north, is chasing a speeding car that has turned the corner and is now moving straight east. When the cruiser is 0.6 mi north of the intersection and the car is 0.8 mi to the east, the police determine with radar that the distance between them and the car is increasing at 20 mph. If the cruiser is moving at 60 mph at the instant of measurement, what is the speed of the car?

Solution We picture the car and cruiser in the coordinate plane, using the positive x-axis as the eastbound highway and the positive y-axis as the northbound highway (Figure 3.35). Note that the police cruiser is traveling south on this highway. We let t represent time in hours and set

$x =$ position in miles of car at time t

$y =$ position in miles of cruiser at time t

$s =$ distance in miles between car and cruiser at time t

where we assume that x, y, and s are differentiable functions of t.

We differentiate the distance equation between the car and the cruiser,

$$s^2 = x^2 + y^2 \qquad \text{Equation relating the variables}$$

(we could instead have used the formula $s = \sqrt{x^2 + y^2}$), and obtain

$$2s\frac{ds}{dt} = 2x\frac{dx}{dt} + 2y\frac{dy}{dt} \qquad \text{Related rates equation}$$

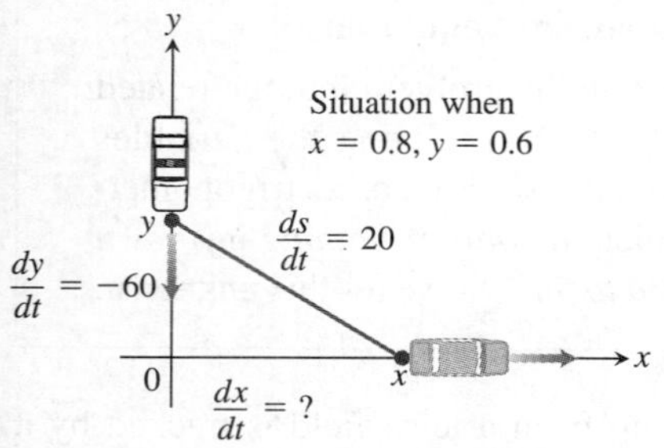

FIGURE 3.35 The speed of the car is related to the speed of the police cruiser and the rate of change of the distance s between them (Example 3).

We want to find dx/dt given that $x = 0.8$, $y = 0.6$, $dy/dt = -60$, and $ds/dt = 20$. Note that dy/dt is negative since the y-coordinate is decreasing. Using the equation $s^2 = x^2 + y^2$, we obtain $s^2 = 0.8^2 + 0.6^2 = 0.64 + 0.36 = 1$, so $s = 1$. Substituting into the related rates equation yields

$$2(1)(20) = 2(0.8)\frac{dx}{dt} + 2(0.6)(-60)$$

$$\frac{dx}{dt} = \frac{40 + 72}{1.6} = 70.$$

At the moment in question, the car's speed is 70 mph. ■

EXAMPLE 4 A particle P moves clockwise at a constant rate along a circle of radius 10 m centered at the origin. The particle's initial position is $(0, 10)$ on the y-axis, and its final destination is the point $(10, 0)$ on the x-axis. Once the particle is in motion, the tangent line at P intersects the x-axis at a point Q (which moves over time). If it takes the particle 30 sec to travel from start to finish, how fast is the point Q moving along the x-axis when it is 20 m from the center of the circle?

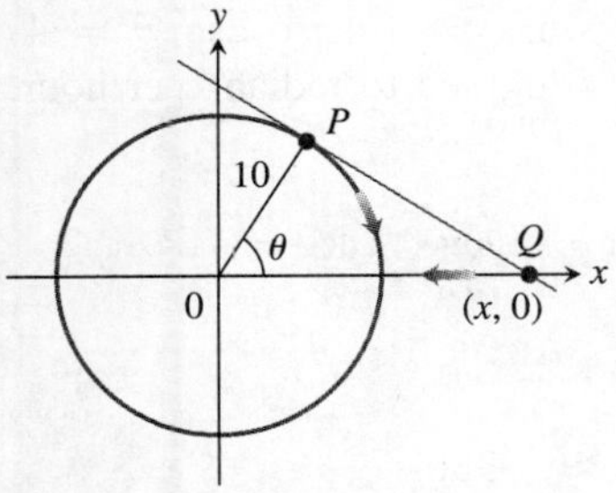

FIGURE 3.36 The particle P travels clockwise along the circle (Example 4).

Solution We picture the situation in the coordinate plane with the circle centered at the origin (see Figure 3.36). We let t represent time and let θ denote the angle in radians from the x-axis to the radial line joining the origin to P. Since the particle travels from start to finish in 30 sec, it is traveling along the circle at a constant rate of $\pi/2$ radians in 1/2 min, or π rad/min. In other words, $d\theta/dt = -\pi$, with t being measured in minutes. The negative sign appears because θ is decreasing over time.

Setting $x(t)$ to be the distance in meters at time t from the point Q to the origin, we see from Figure 3.36 that

$$x \cos \theta = 10. \qquad \text{Equation relating the variables}$$

Differentiation with respect to t gives

$$x(-\sin \theta)\frac{d\theta}{dt} + \frac{dx}{dt}\cos \theta = 0. \qquad \text{Related rates equation}$$

We want to find dx/dt given that $x = 20$ and $d\theta/dt = -\pi$. The equation $x \cos \theta = 10$ implies $\cos \theta = 10/20 = 1/2$. Furthermore, for angles θ in the first quadrant, the identity $\sin^2 \theta + \cos^2 \theta = 1$ yields $\sin \theta = \sqrt{1 - (1/2)^2} = \sqrt{3}/2$. Substituting into the related rates equation, we obtain

$$(20)\left(-\frac{\sqrt{3}}{2}\right)(-\pi) + \frac{dx}{dt}\cdot\frac{1}{2} = 0$$

$$\frac{dx}{dt} = -20\sqrt{3}\pi.$$

Note that x is decreasing because dx/dt is negative. At the moment in question, the point Q is moving toward the origin at the speed of $20\sqrt{3}\pi \approx 109$ m/min. ■

EXAMPLE 5 A jet airliner is flying at a constant altitude of 12,000 ft above sea level as it approaches a Pacific island. The aircraft comes within the direct line of sight of a radar station located on the island (at sea level), and the radar indicates that the initial angle between sea level and its line of sight to the aircraft is 30°. How fast (in miles per hour) is the aircraft approaching the island when first detected by the radar instrument if it is turning upward at the rate of 2/3 deg/sec in order to keep the aircraft within its direct line of sight?

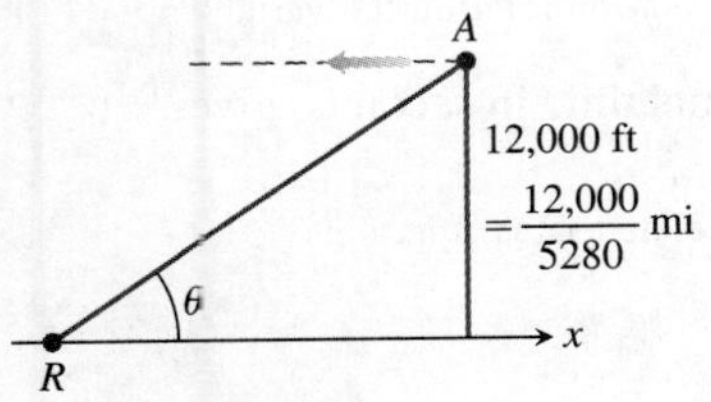

FIGURE 3.37 Jet airliner A traveling at constant altitude toward radar station R (Example 5).

Solution The aircraft A and the radar station R are pictured in the coordinate plane, using the positive x-axis as the horizontal distance in miles at sea level from R to A, and the positive y-axis as the vertical altitude in miles above sea level. We let t represent time in hours and observe that the altitude $y = 12{,}000/5280$ mi is a constant (converted from feet).

We will let θ denote the line-of-sight angle in radians. From Figure 3.37, we see that

$$\frac{12{,}000/5280}{x} = \tan \theta \quad \text{or} \quad x = \frac{12{,}000}{5280}\cot \theta. \qquad \text{Equation relating the variables}$$

Differentiation with respect to t gives

$$\frac{dx}{dt} = -\frac{1200}{528}\csc^2 \theta \frac{d\theta}{dt}. \qquad \text{Related rates equation}$$

When $\theta = \pi/6$ (converted to radians from 30 degrees), $\sin^2\theta = 1/4$, so $\csc^2\theta = 4$. Converting the given rate of change of the angle from 2/3 deg/sec to radians per hour, we find

$$\frac{d\theta}{dt} = \frac{2}{3}\left(\frac{\pi}{180}\right)(3600) \text{ rad/hr.} \qquad 1 \text{ hr} = 3600 \text{ sec}, 1 \text{ deg} = \pi/180 \text{ rad}$$

Substitution into the equation for dx/dt then gives

$$\frac{dx}{dt} = \left(-\frac{1200}{528}\right)(4)\left(\frac{2}{3}\right)\left(\frac{\pi}{180}\right)(3600) \approx -380.$$

The negative sign appears because the distance x is decreasing, so the aircraft is approaching the island at a speed of approximately 380 mi/hr when first detected by the radar. ■

Note that the solution of Example 5 involved several unit conversions: from feet to miles, from seconds to hours, and from degrees to radians. When solving related rates problems, we should check that consistent units are used.

EXAMPLE 6 Figure 3.38a shows a rope running through a pulley at P and bearing a weight W at one end. The other end is held 5 ft above the ground in the hand M of a worker. Suppose the pulley is 25 ft above ground, the rope is 45 ft long, and the worker is walking rapidly away from the vertical line PW at the rate of 4 ft/sec. How fast is the weight being raised when the worker's hand is 21 ft away from PW?

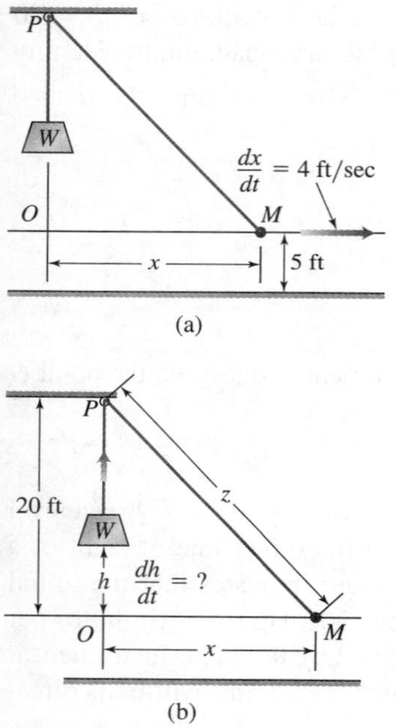

FIGURE 3.38 A worker at M walks to the right, pulling the weight W upward as the rope moves through the pulley P (Example 6).

Solution We let OM be the horizontal line of length x ft from a point O directly below the pulley to the worker's hand M at any instant of time (Figure 3.38). Let h be the height in feet of the weight W above O, and let z denote the length in feet of rope from the pulley P to the worker's hand. Note that the height of P above O is 20 ft because O is 5 ft above the ground, and also note that the angle at O is a right angle.

We have the following relationships (see Figure 3.38b):

$$20 - h + z = 45 \qquad \text{Total length of rope is 45 ft.}$$

$$20^2 + x^2 = z^2. \qquad \text{Angle at } O \text{ is a right angle.}$$

If we solve for $z = 25 + h$ in the first equation, and substitute into the second equation, we have

$$20^2 + x^2 = (25 + h)^2. \qquad \text{Equation relating the variables} \quad (1)$$

Differentiating both sides with respect to t, which represents time in seconds, gives

$$2x\frac{dx}{dt} = 2(25 + h)\frac{dh}{dt}. \qquad \text{Related rates equation}$$

From Equation (1),

$$20^2 + 21^2 = (25 + h)^2$$

so that

$$(25 + h)^2 = 841, \qquad \text{or} \qquad 25 + h = 29.$$

Substituting this value along with the given values $x = 21$ and $\frac{dx}{dt} = 4$ into the related rates equation yields

$$2(21)(4) = 2(29)\frac{dh}{dt}.$$

This gives

$$\frac{dh}{dt} = \frac{21}{29} \cdot 4 = \frac{84}{29} \approx 2.9 \text{ ft/sec}$$

as the rate at which the weight is being raised when $x = 21$ ft. ■

EXERCISES 3.8

1. **Area** Suppose that the radius r and area $A = \pi r^2$ of a circle are differentiable functions of t. Write an equation that relates dA/dt to dr/dt.

2. **Surface area** Suppose that the radius r and surface area $S = 4\pi r^2$ of a sphere are differentiable functions of t. Write an equation that relates dS/dt to dr/dt.

3. Assume that $y = 10x$ and $dx/dt = 5$. Find dy/dt.

4. Assume that $2x + 3y = 12$ and $dy/dt = -2$. Find dx/dt.

5. If $y = 2x^2$ and $dx/dt = 4$, then what is dy/dt when $x = -2$?

6. If $x = y^3 - y$ and $dy/dt = 5$, then what is dx/dt when $y = 2$?

7. If $x^2 + y^2 = 16$ and $dx/dt = -5$, then what is dy/dt when $x = 2$ and $y = -2$?

8. If $x^2y^3 = 4/27$ and $dy/dt = 1/2$, then what is dx/dt when $x = 2$?

9. If $L = \sqrt{x^2 + y^2}$, $dx/dt = -2$ and $dy/dt = 5$, find dL/dt when $x = 3$ and $y = 4$?

10. If $r + s^2 + v^3 = 12$, $dr/dt = 4$, and $ds/dt = -3$, find dv/dt when $r = 3$ and $s = 1$.

11. If the original 12 m edge length x of a cube decreases at the rate of 2 m/min, when $x = 5$ m at what rate does the cube's
    a. surface area change? **b.** volume change?

12. A cube's surface area increases at the rate of 96 in²/sec. At what rate is the cube's volume changing when the edge length is $x = 4$ in?

13. **Volume** The radius r and height h of a right circular cylinder are related to the cylinder's volume V by the formula $V = \pi r^2 h$.
    a. How is dV/dt related to dh/dt if r is constant?
    b. How is dV/dt related to dr/dt if h is constant?
    c. How is dV/dt related to dr/dt and dh/dt if neither r nor h is constant?

14. **Volume** The radius r and height h of a right circular cone are related to the cone's volume V by the equation $V = (1/3)\pi r^2 h$.
    a. How is dV/dt related to dh/dt if r is constant?
    b. How is dV/dt related to dr/dt if h is constant?
    c. How is dV/dt related to dr/dt and dh/dt if neither r nor h is constant?

15. **Changing voltage** The voltage V (volts), current I (amperes), and resistance R (ohms) of an electric circuit like the one shown here are related by the equation $V = IR$. Suppose that V is increasing at the rate of 1 volt/sec while I is decreasing at the rate of 1/3 amp/sec. Let t denote time in seconds.

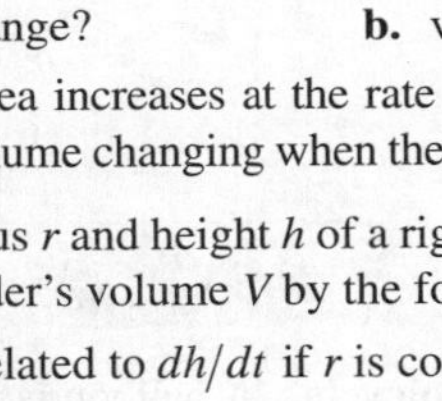

    a. What is the value of dV/dt?
    b. What is the value of dI/dt?
    c. What equation relates dR/dt to dV/dt and dI/dt?
    d. Find the rate at which R is changing when $V = 12$ volts and $I = 2$ amps. Is R increasing, or decreasing?

16. **Electrical power** The power P (watts) of an electric circuit is related to the circuit's resistance R (ohms) and current I (amperes) by the equation $P = RI^2$.
    a. How are dP/dt, dR/dt, and dI/dt related if none of P, R, and I are constant?
    b. How is dR/dt related to dI/dt if P is constant?

17. **Distance** Let x and y be differentiable functions of t, and let $s = \sqrt{x^2 + y^2}$ be the distance between the points $(x, 0)$ and $(0, y)$ in the xy-plane.
    a. How is ds/dt related to dx/dt if y is constant?
    b. How is ds/dt related to dx/dt and dy/dt if neither x nor y is constant?
    c. How is dx/dt related to dy/dt if s is constant?

18. **Diagonals** If x, y, and z are lengths of the edges of a rectangular box, then the common length of the box's diagonals is $s = \sqrt{x^2 + y^2 + z^2}$.
    a. Assuming that x, y, and z are differentiable functions of t, how is ds/dt related to dx/dt, dy/dt, and dz/dt?
    b. How is ds/dt related to dy/dt and dz/dt if x is constant?
    c. How are dx/dt, dy/dt, and dz/dt related if s is constant?

19. **Area** The area A of a triangle with sides of lengths a and b enclosing an angle of measure θ is

$$A = \frac{1}{2}ab \sin \theta.$$

    a. How is dA/dt related to $d\theta/dt$ if a and b are constant?
    b. How is dA/dt related to $d\theta/dt$ and da/dt if only b is constant?
    c. How is dA/dt related to $d\theta/dt$, da/dt, and db/dt if none of a, b, and θ are constant?

20. **Heating a plate** When a circular plate of metal is heated in an oven, its radius increases at the rate of 0.02 cm/min. At what rate is the plate's area increasing when the radius is 60 cm?

21. **Changing dimensions in a rectangle** The length l of a rectangle is decreasing at the rate of 2 cm/sec while the width w is increasing at the rate of 2 cm/sec. When $l = 12$ cm and $w = 5$ cm, find the rates of change of **(a)** the area, **(b)** the perimeter, and **(c)** the lengths of the diagonals of the rectangle. Which of these quantities are decreasing, and which are increasing?

22. **Changing dimensions in a rectangular box** Suppose that the edge lengths x, y, and z of a closed rectangular box are changing at the following rates:

$$\frac{dx}{dt} = 2 \text{ m/sec}, \quad \frac{dy}{dt} = -3 \text{ m/sec}, \quad \frac{dz}{dt} = 2 \text{ m/sec}$$

    Find the rates at which the box's **(a)** volume, **(b)** surface area, and **(c)** diagonal length $s = \sqrt{x^2 + y^2 + z^2}$ are changing at the instant when $x = 5$, $y = 4$ and $z = 3$.

23. **A sliding ladder** A 13-ft ladder is leaning against a house when its base starts to slide away (see accompanying figure). By the time the base is 12 ft from the house, the base is moving at the rate of 5 ft/sec.
    a. How fast is the top of the ladder sliding down the wall then?

b. At what rate is the area of the triangle formed by the ladder, wall, and ground changing then?

c. At what rate is the angle θ between the ladder and the ground changing then?

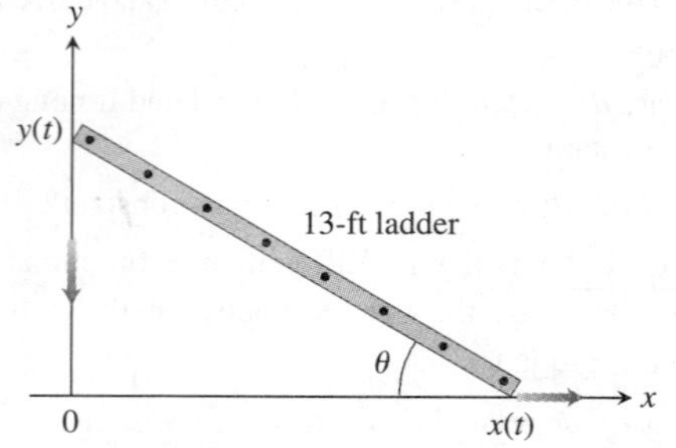

24. Commercial air traffic Two commercial airplanes are flying at an altitude of 40,000 ft along straight-line courses that intersect at right angles. Plane A is approaching the intersection point at a speed of 450 knots (nautical miles per hour; a nautical mile is 2000 yd). Plane B is approaching the intersection at 480 knots. At what rate is the distance between the planes changing when A is 8 nautical miles from the intersection point and B is 15 nautical miles from the intersection point?

25. Flying a kite A girl flies a kite at a height of 300 ft, the wind carrying the kite horizontally away from her at a rate of 25 ft/sec. How fast must she let out the string when the kite is 500 ft away from her?

26. Boring a cylinder The mechanics at Lincoln Automotive are reboring a 6-in.-deep cylinder to fit a new piston. The machine they are using increases the cylinder's radius one-thousandth of an inch every 3 min. How rapidly is the cylinder volume increasing when the bore (diameter) is 3.800 in.?

27. A growing sand pile Sand falls from a conveyor belt at the rate of 15 m^3/min onto the top of a conical pile. The height of the pile is always three-eighths of the base diameter. How fast are the **(a)** height and **(b)** radius changing when the pile is 4 m high? Answer in centimeters per minute.

28. A draining conical reservoir Water is flowing at the rate of 50 m^3/min from a shallow concrete conical reservoir (vertex down) of base radius 45 m and height 6 m.

a. How fast (in centimeters per minute) is the water level falling when the water is 5 m deep?

b. How fast is the radius of the water's surface changing then? Answer in centimeters per minute.

29. A draining hemispherical reservoir Water is flowing at the rate of 6 m^3/min from a reservoir shaped like a hemispherical bowl of radius 13 m, shown here in profile. Answer the following questions, given that the volume of water in a hemispherical bowl of radius R is $V = (\pi/3)y^2(3R - y)$ when the water is y meters deep.

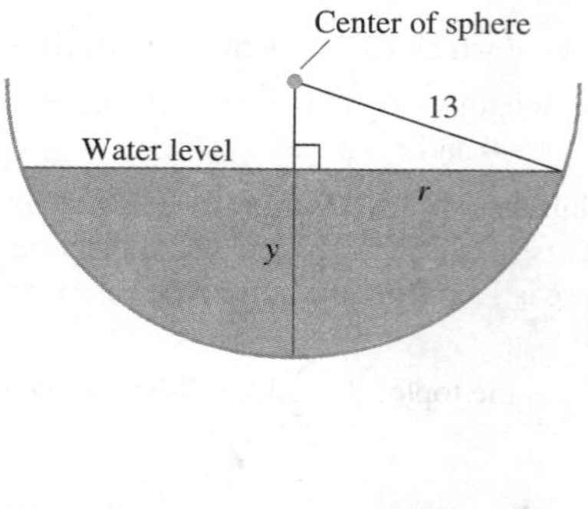

a. At what rate is the water level changing when the water is 8 m deep?

b. What is the radius r of the water's surface when the water is y m deep?

c. At what rate is the radius r changing when the water is 8 m deep?

30. A growing raindrop Suppose that a drop of mist is a perfect sphere and that, through condensation, the drop picks up moisture at a rate proportional to its surface area. Show that under these circumstances the drop's radius increases at a constant rate.

31. The radius of an inflating balloon A spherical balloon is inflated with helium at the rate of 64π ft^3/min. How fast is the balloon's radius increasing at the instant the radius is 4 ft? How fast is the surface area increasing?

32. Hauling in a dinghy A dinghy is pulled toward a dock by a rope from the bow through a ring on the dock 6 ft above the bow. The rope is hauled in at the rate of 2 ft/sec.

a. How fast is the boat approaching the dock when 10 ft of rope are out?

b. At what rate is the angle θ changing at this instant (see the figure)?

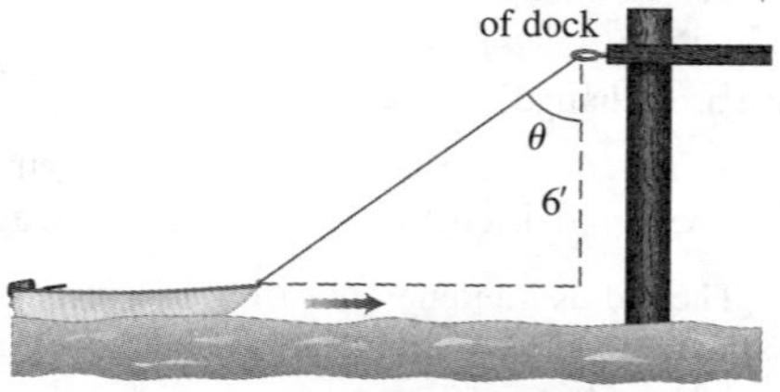

33. A balloon and a bicycle A balloon is rising vertically above a level, straight road at a constant rate of 1 ft/sec. Just when the balloon is 65 ft above the ground, a bicycle moving at a constant rate of 17 ft/sec passes under it. How fast is the distance $s(t)$ between the bicycle and the balloon increasing 3 sec later?

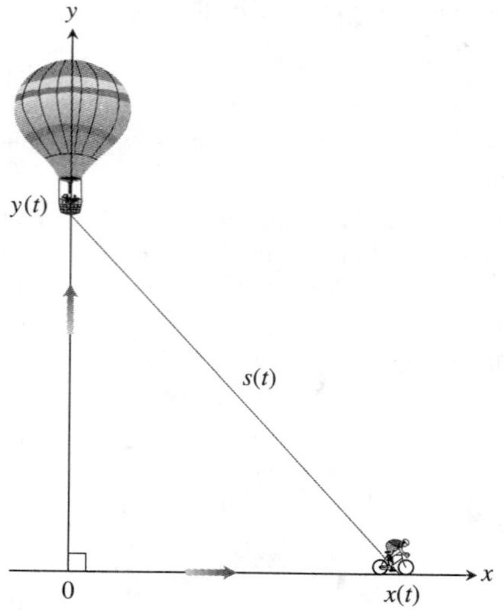

34. Making coffee Coffee is draining from a conical filter into a cylindrical coffeepot at the rate of 10 in^3/min.

a. How fast is the level in the pot rising when the coffee in the cone is 5 in. deep?

b. How fast is the level in the cone falling then?

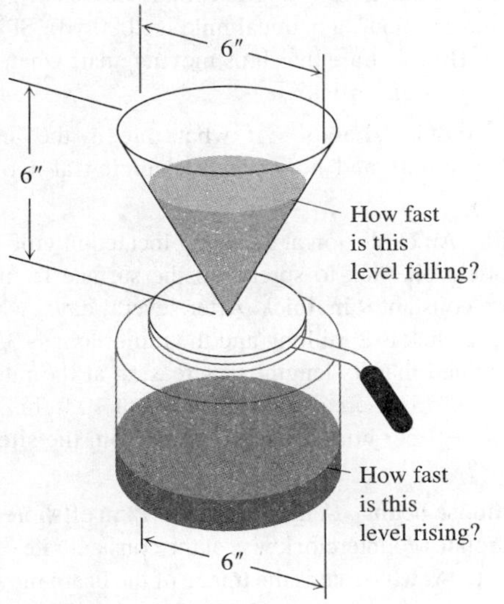

35. Cardiac output In the late 1860s, Adolf Fick, a professor of physiology in the Faculty of Medicine in Würzberg, Germany, developed one of the methods we use today for measuring how much blood your heart pumps in a minute. Your cardiac output as you read this sentence is probably about 7 L/min. At rest it is likely to be a bit under 6 L/min. If you are a trained marathon runner running a marathon, your cardiac output can be as high as 30 L/min.

Your cardiac output can be calculated with the formula

$$y = \frac{Q}{D},$$

where Q is the number of milliliters of CO_2 you exhale in a minute and D is the difference between the CO_2 concentration (ml/L) in the blood pumped to the lungs and the CO_2 concentration in the blood returning from the lungs. With $Q = 233$ ml/min and $D = 97 - 56 = 41$ ml/L,

$$y = \frac{233\ \text{ml/min}}{41\ \text{ml/L}} \approx 5.68\ \text{L/min},$$

fairly close to the 6 L/min that most people have at basal (resting) conditions. (Data courtesy of J. Kenneth Herd, M.D., Quillan College of Medicine, East Tennessee State University.)

Suppose that when $Q = 233$ and $D = 41$, we also know that D is decreasing at the rate of 2 units a minute but that Q remains unchanged. What is happening to the cardiac output?

36. Moving along a parabola A particle moves along the parabola $y = x^2$ in the first quadrant in such a way that its x-coordinate (measured in meters) increases at a steady 10 m/sec. How fast is the angle of inclination θ of the line joining the particle to the origin changing when $x = 3$ m?

37. Motion in the plane The coordinates of a particle in the metric xy-plane are differentiable functions of time t with $dx/dt = -1$ m/sec and $dy/dt = -5$ m/sec. How fast is the particle's distance from the origin changing as it passes through the point $(5, 12)$?

38. Videotaping a moving car You are videotaping a race from a stand 132 ft from the track, following a car that is moving at 180 mi/h (264 ft/sec), as shown in the accompanying figure. How fast will your camera angle θ be changing when the car is right in front of you? A half second later?

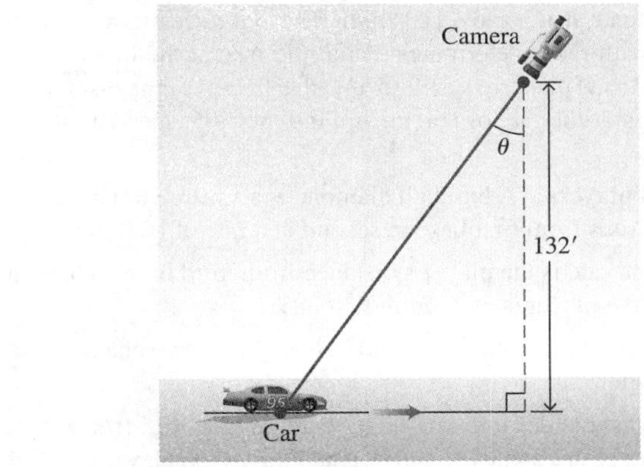

39. A moving shadow A light shines from the top of a pole 50 ft high. A ball is dropped from the same height from a point 30 ft away from the light. (See accompanying figure.) How fast is the shadow of the ball moving along the ground 1/2 sec later? (Assume the ball falls a distance $s = 16t^2$ ft in t sec.)

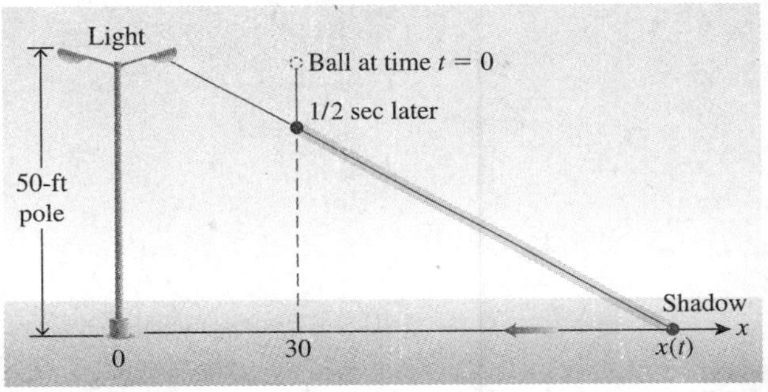

40. A building's shadow On a morning of a day when the sun will pass directly overhead, the shadow of an 80-ft building on level ground is 60 ft long. At the moment in question, the angle θ the sun makes with the ground is increasing at the rate of 0.27°/min. At what rate is the shadow decreasing? (Remember to use radians. Express your answer in inches per minute, to the nearest tenth.)

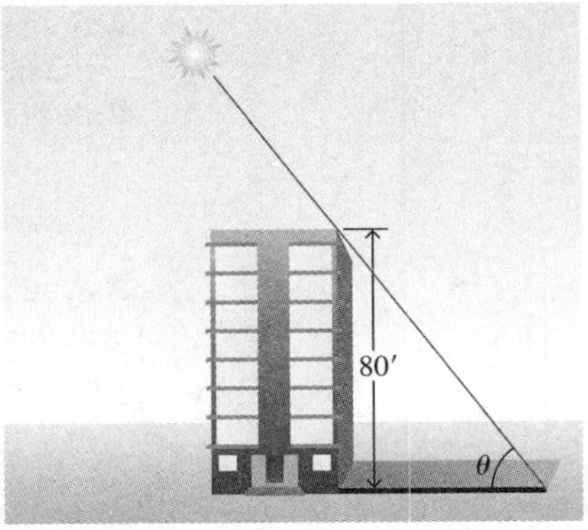

41. A melting ice layer A spherical iron ball 9 in. in diameter is coated with a layer of ice of uniform thickness. If the ice melts at the rate of 5 in^3/min, how fast is the thickness of the ice decreasing when is 4 in. thick? How fast is the outer surface area of ice decreasing?

42. Highway patrol A highway patrol plane flies 3 mi above a level, straight road at a steady 120 mi/h. The pilot sees an oncoming car and with radar determines that at the instant the line-of-sight distance from plane to car is 5 mi, the line-of-sight distance is decreasing at the rate of 160 mi/h. Find the car's speed along the highway.

43. Baseball players A baseball diamond is a square 90 ft on a side. A player runs from first base to second at a rate of 16 ft/sec.

a. At what rate is the player's distance from third base changing when the player is 30 ft from first base?

b. At what rates are angles θ_1 and θ_2 (see the figure) changing at that time?

c. The player slides into second base at the rate of 15 ft/sec. At what rates are angles θ_1 and θ_2 changing as the player touches base?

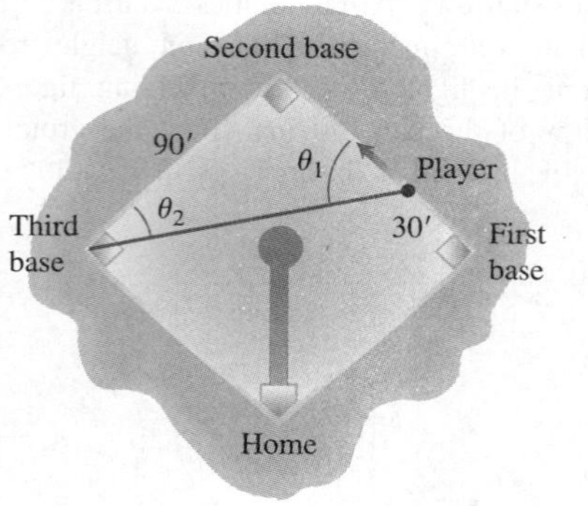

44. Ships Two ships are steaming straight away from a point O along routes that make a 120° angle. Ship A moves at 14 knots (nautical miles per hour; a nautical mile is 2000 yd). Ship B moves at 21 knots. How fast are the ships moving apart when $OA = 5$ and $OB = 3$ nautical miles?

45. Clock's moving hands At what rate is the angle between a clock's minute and hour hands changing at 4 o'clock in the afternoon?

46. Oil spill An explosion at an oil rig located in gulf waters causes an elliptical oil slick to spread on the surface from the rig. The slick is a constant 9 in. thick. After several days, when the major axis of the slick is 2 mi long and the minor axis is 3/4 mi wide, it is determined that its length is increasing at the rate of 30 ft/hr, and its width is increasing at the rate of 10 ft/hr. At what rate (in cubic feet per hour) is oil flowing from the site of the rig at that time?

47. A lighthouse beam A lighthouse sits 1 km offshore, and its beam of light rotates counterclockwise at the constant rate of 3 full circles per minute. At what rate is the image of the beam moving down the shoreline when the image is 1 km from the spot on the shoreline nearest the lighthouse?

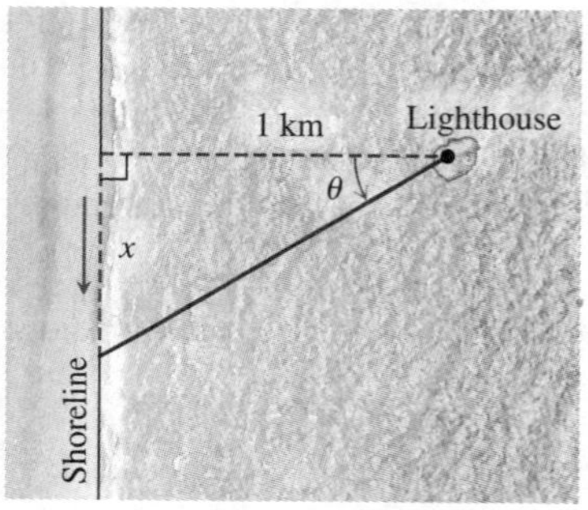

3.9 Linearization and Differentials

It is often useful to approximate complicated functions with simpler ones that give the accuracy we want for specific applications and, at the same time, are easier to work with than the original functions. The approximating functions discussed in this section are called *linearizations*, and they are based on tangent lines. Other approximating functions, such as polynomials, are discussed in Chapter 10.

We introduce new variables dx and dy, called *differentials*, and define them in a way that makes Leibniz's notation for the derivative dy/dx a true ratio. We use dy to estimate error in measurement, which then provides for a precise proof of the Chain Rule (Section 3.6).

Linearization

As you can see in Figure 3.39, the tangent line to the curve $y = x^2$ lies close to the curve near the point of tangency. For a brief interval to either side, the y-values along the tangent line give good approximations to the y-values on the curve. We observe this phenomenon by zooming in on the two graphs at the point of tangency, or by looking at tables of values for the difference between $f(x)$ and its tangent line near the x-coordinate of the point of tangency. The phenomenon is true not just for parabolas; every differentiable curve behaves locally like its tangent line.

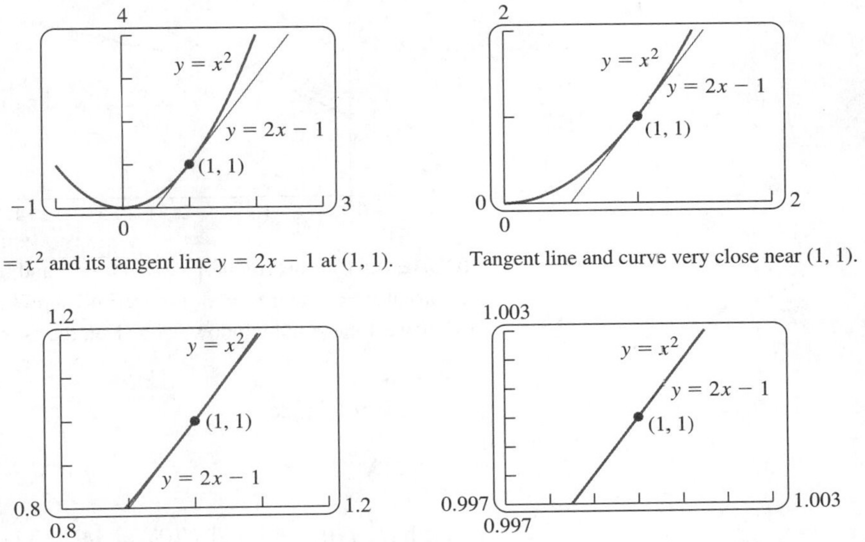

$y = x^2$ and its tangent line $y = 2x - 1$ at (1, 1).

Tangent line and curve very close near (1, 1).

Tangent line and curve very close throughout entire x-interval shown.

Tangent line and curve closer still. Computer screen cannot distinguish tangent line from curve on this x-interval.

FIGURE 3.39 The more we magnify the graph of a function near a point where the function is differentiable, the flatter the graph becomes and the more it resembles its tangent line.

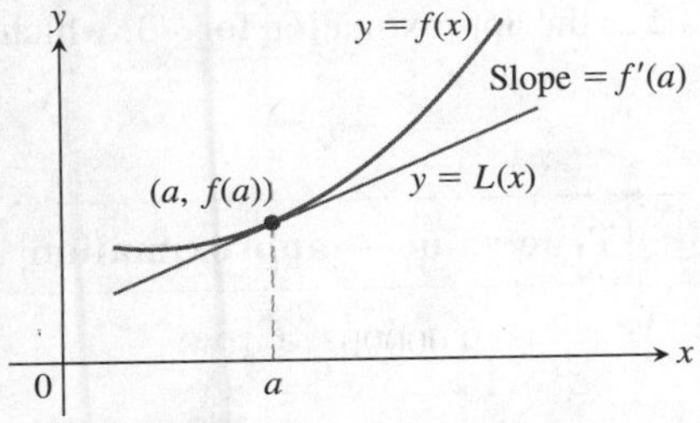

FIGURE 3.40 The tangent line to the curve $y = f(x)$ at $x = a$ is the line $L(x) = f(a) + f'(a)(x - a)$.

In general, the tangent line to $y = f(x)$ at a point $x = a$, where f is differentiable (Figure 3.40), passes through the point $(a, f(a))$, so its point-slope equation is

$$y = f(a) + f'(a)(x - a).$$

Thus, this tangent line is the graph of the linear function

$$L(x) = f(a) + f'(a)(x - a).$$

As long as this line remains close to the graph of f as we move off the point of tangency, $L(x)$ gives a good approximation to $f(x)$.

DEFINITIONS If f is differentiable at $x = a$, then the approximating function

$$L(x) = f(a) + f'(a)(x - a)$$

is the **linearization** of f at a. The approximation

$$f(x) \approx L(x)$$

of f by L is the **standard linear approximation** of f at a. The point $x = a$ is the **center** of the approximation.

EXAMPLE 1 Find the linearization of $f(x) = \sqrt{1 + x}$ at $x = 0$ (Figure 3.41).

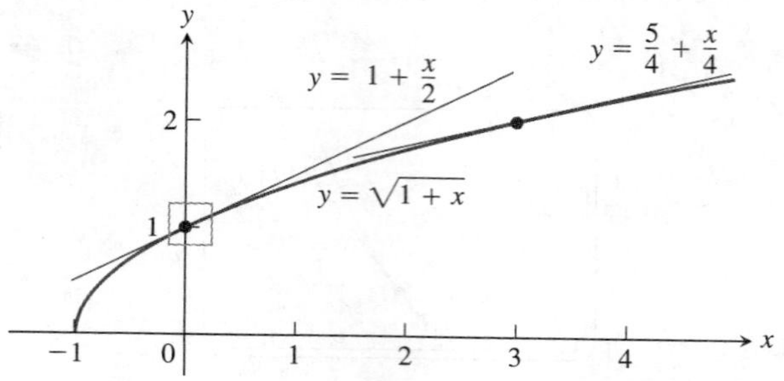

FIGURE 3.41 The graph of $y = \sqrt{1 + x}$ and its linearizations at $x = 0$ and $x = 3$. Figure 3.42 shows a magnified view of the small window about 1 on the y-axis.

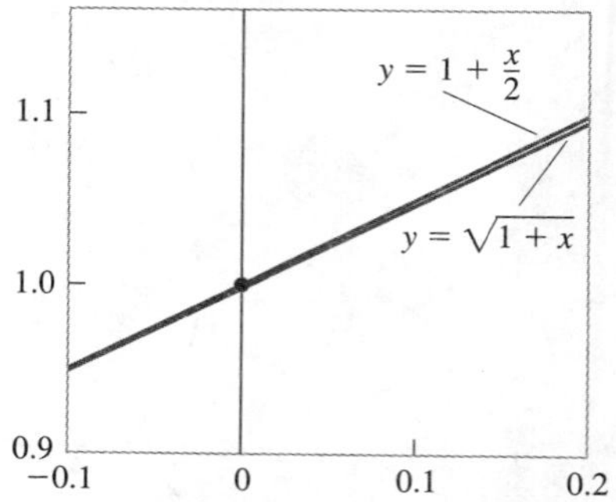

FIGURE 3.42 Magnified view of the window in Figure 3.41.

Solution Since

$$f'(x) = \frac{1}{2}(1 + x)^{-1/2},$$

we have $f(0) = 1$ and $f'(0) = 1/2$, giving the linearization

$$L(x) = f(a) + f'(a)(x - a) = 1 + \frac{1}{2}(x - 0) = 1 + \frac{x}{2}.$$

See Figure 3.42. ■

The following table shows how accurate the approximation $\sqrt{1 + x} \approx 1 + (x/2)$ from Example 1 is for some values of x near 0. As we move away from zero, we lose accuracy. For example, for $x = 2$, the linearization gives 2 as the approximation for $\sqrt{3}$, which is not even accurate to one decimal place.

Approximation	True value	|True value − approximation|
$\sqrt{1.005} \approx 1 + \dfrac{0.005}{2} = 1.00250$	1.002497	$0.000003 < 10^{-5}$
$\sqrt{1.05} \approx 1 + \dfrac{0.05}{2} = 1.025$	1.024695	$0.000305 < 10^{-3}$
$\sqrt{1.2} \approx 1 + \dfrac{0.2}{2} = 1.10$	1.095445	$0.004555 < 10^{-2}$

Do not be misled by the preceding calculations into thinking that whatever we do with a linearization is better done with a calculator. In practice, we would never use a linearization to find a particular square root. The utility of a linearization is its ability to replace a complicated formula by a simpler one over an *entire interval of values*. If we have to work with $\sqrt{1 + x}$ for x in an interval close to 0 and can tolerate the small amount of error involved over that interval, we can work with $1 + (x/2)$ instead. Of course, we then need to know how much error there is. We further examine the estimation of error in Chapter 10.

A linear approximation normally loses accuracy away from its center. As Figure 3.41 suggests, the approximation $\sqrt{1+x} \approx 1 + (x/2)$ is too crude to be useful near $x = 3$. There, we need the linearization at $x = 3$.

EXAMPLE 2 Find the linearization of $f(x) = \sqrt{1+x}$ at $x = 3$. (See Figure 3.41.)

Solution We evaluate the equation defining $L(x)$ at $a = 3$. With

$$f(3) = 2, \qquad f'(3) = \frac{1}{2}(1+x)^{-1/2}\Big|_{x=3} = \frac{1}{4},$$

we have

$$L(x) = 2 + \frac{1}{4}(x - 3) = \frac{5}{4} + \frac{x}{4}.$$

At $x = 3.2$, the linearization in Example 2 gives

$$\sqrt{1+x} = \sqrt{1+3.2} \approx \frac{5}{4} + \frac{3.2}{4} = 1.250 + 0.800 = 2.050,$$

which differs from the true value $\sqrt{4.2} \approx 2.04939$ by less than one one-thousandth. The linearization in Example 1 gives

$$\sqrt{1+x} = \sqrt{1+3.2} \approx 1 + \frac{3.2}{2} = 1 + 1.6 = 2.6,$$

a result that is off by more than 25%. ■

EXAMPLE 3 Find the linearization of $f(x) = \cos x$ at $x = \pi/2$ (Figure 3.43).

Solution Since $f(\pi/2) = \cos(\pi/2) = 0$, $f'(x) = -\sin x$, and $f'(\pi/2) = -\sin(\pi/2) = -1$, we find the linearization at $a = \pi/2$ to be

$$\begin{aligned} L(x) &= f(a) + f'(a)(x - a) \\ &= 0 + (-1)\left(x - \frac{\pi}{2}\right) \\ &= -x + \frac{\pi}{2}. \end{aligned}$$ ■

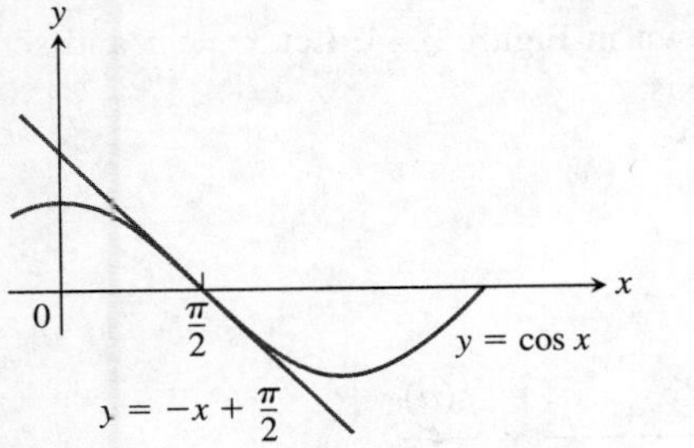

FIGURE 3.43 The graph of $f(x) = \cos x$ and its linearization at $x = \pi/2$. Near $x = \pi/2$, $\cos x \approx 2x + (\pi/2)$ (Example 3).

An important linear approximation for roots and powers is

$$(1+x)^k \approx 1 + kx \qquad (x \text{ near } 0;\ \text{any number } k)$$

(Exercise 15). This approximation, which is good for values of x sufficiently close to zero, has broad application. For example, when x is small,

$$\sqrt{1+x} \approx 1 + \frac{1}{2}x \qquad k = 1/2$$

$$\frac{1}{1-x} = (1-x)^{-1} \approx 1 + (-1)(-x) = 1 + x \qquad k = -1;\ \text{replace } x \text{ by } -x.$$

$$\sqrt[3]{1+5x^4} = (1+5x^4)^{1/3} \approx 1 + \frac{1}{3}(5x^4) = 1 + \frac{5}{3}x^4 \qquad k = 1/3;\ \text{replace } x \text{ by } 5x^4.$$

$$\frac{1}{\sqrt{1-x^2}} = (1-x^2)^{-1/2} \approx 1 + \left(-\frac{1}{2}\right)(-x^2) = 1 + \frac{1}{2}x^2 \qquad k = -1/2;\ \text{replace } x \text{ by } -x^2.$$

Approximations Near $x = 0$

$$\sqrt{1+x} \approx 1 + \frac{x}{2}$$

$$\frac{1}{1-x} \approx 1 + x$$

$$\frac{1}{\sqrt{1-x^2}} \approx 1 + \frac{x^2}{2}$$

Differentials

We sometimes use the Leibniz notation dy/dx to represent the derivative of y with respect to x. Contrary to its appearance, it is not a ratio. We now introduce two new variables dx and dy with the property that when their ratio exists, it is equal to the derivative.

DEFINITION Let $y = f(x)$ be a differentiable function. The **differential dx** is an independent variable. The **differential dy** is

$$dy = f'(x)\,dx.$$

Unlike the independent variable dx, the variable dy is always a dependent variable. It depends on both x and dx. If dx is given a specific value and x is a particular number in the domain of the function f, then these values determine the numerical value of dy. Often the variable dx is chosen to be Δx, the change in x.

EXAMPLE 4

(a) Find dy if $y = x^5 + 37x$.

(b) Find the value of dy when $x = 1$ and $dx = 0.2$.

Solution

(a) $dy = (5x^4 + 37)\,dx$

(b) Substituting $x = 1$ and $dx = 0.2$ in the expression for dy, we have

$$dy = (5 \cdot 1^4 + 37)\,0.2 = 8.4. \qquad \blacksquare$$

The geometric meaning of differentials is shown in Figure 3.44. Let $x = a$ and set $dx = \Delta x$. The corresponding change in $y = f(x)$ is

$$\Delta y = f(a + dx) - f(a).$$

The corresponding change in the tangent line L is

$$\begin{aligned}\Delta L &= L(a + dx) - L(a)\\ &= \underbrace{f(a) + f'(a)[(a + dx) - a]}_{L(a+dx)} - \underbrace{f(a)}_{L(a)}\\ &= f'(a)\,dx.\end{aligned}$$

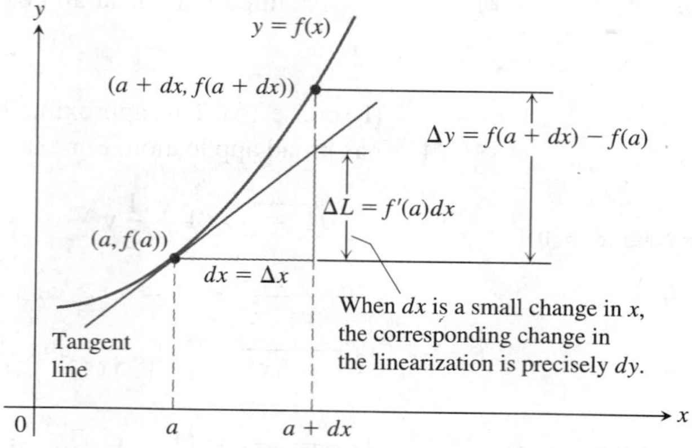

FIGURE 3.44 Geometrically, the differential dy is the change ΔL in the linearization of f when $x = a$ changes by an amount $dx = \Delta x$.

That is, the change in the linearization of f is precisely the value of the differential dy when $x = a$ and $dx = \Delta x$. Therefore, dy represents the amount the tangent line rises or falls when x changes by an amount $dx = \Delta x$.

If $dx \neq 0$, then the quotient of the differential dy by the differential dx is equal to the derivative $f'(x)$ because

$$dy \div dx = \frac{f'(x)\,dx}{dx} = f'(x) = \frac{dy}{dx}.$$

We sometimes write

$$df = f'(x)\,dx$$

in place of $dy = f'(x)\,dx$, calling df the **differential of f**. For instance, if $f(x) = 3x^2 - 6$, then

$$df = d(3x^2 - 6) = 6x\,dx.$$

Every differentiation formula like

$$\frac{d(u + v)}{dx} = \frac{du}{dx} + \frac{dv}{dx} \quad \text{or} \quad \frac{d(\sin u)}{dx} = \cos u \frac{du}{dx}$$

has a corresponding differential form like

$$d(u + v) = du + dv \quad \text{or} \quad d(\sin u) = \cos u\,du.$$

EXAMPLE 5 We can use the Chain Rule and other differentiation rules to find differentials of functions.

(a) $d(\tan 2x) = \sec^2(2x)\,d(2x) = 2\sec^2 2x\,dx$

(b) $d\left(\dfrac{x}{x + 1}\right) = \dfrac{(x + 1)\,dx - x\,d(x + 1)}{(x + 1)^2} = \dfrac{x\,dx + dx - x\,dx}{(x + 1)^2} = \dfrac{dx}{(x + 1)^2}$ ■

Estimating with Differentials

Suppose we know the value of a differentiable function $f(x)$ at a point a and want to estimate how much this value will change if we move to a nearby point $a + dx$. If $dx = \Delta x$ is small, then we can see from Figure 3.44 that Δy is approximately equal to the differential dy. Since

$$f(a + dx) = f(a) + \Delta y, \quad \Delta x = dx$$

the differential approximation gives

$$f(a + dx) \approx f(a) + dy$$

when $dx = \Delta x$. Thus the approximation $\Delta y \approx dy$ can be used to estimate $f(a + dx)$ when $f(a)$ is known, dx is small, and $dy = f'(a)\,dx$.

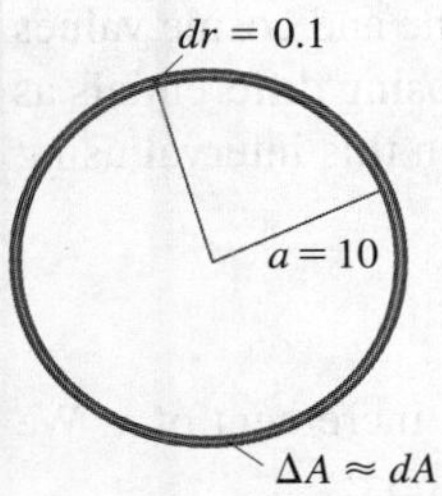

FIGURE 3.45 When dr is small compared with a, the differential dA gives the estimate $A(a + dr) = \pi a^2 + dA$ (Example 6).

EXAMPLE 6 The radius r of a circle increases from $a = 10$ m to 10.1 m (Figure 3.45). Use dA to estimate the increase in the circle's area A. Estimate the area of the enlarged circle and compare your estimate to the true area found by direct calculation.

Solution Since $A = \pi r^2$, the estimated increase is

$$dA = A'(a)\,dr = 2\pi a\,dr = 2\pi(10)(0.1) = 2\pi \text{ m}^2.$$

Thus, since $A(r + \Delta r) \approx A(r) + dA$, we have

$$\begin{aligned} A(10 + 0.1) &\approx A(10) + 2\pi \\ &= \pi(10)^2 + 2\pi = 102\pi. \end{aligned}$$

The area of a circle of radius 10.1 m is approximately 102π m^2.

The true area is

$$\begin{aligned} A(10.1) &= \pi(10.1)^2 \\ &= 102.01\pi \text{ m}^2. \end{aligned}$$

The error in our estimate is $0.01\pi \text{ m}^2$, which is the difference $\Delta A - dA$. ■

When using differentials to estimate functions, our goal is to choose a nearby point $x = a$ where both $f(a)$ and the derivative $f'(a)$ are easy to evaluate.

EXAMPLE 7 Use differentials to estimate

(a) $7.97^{1/3}$ **(b)** $\sin(\pi/6 + 0.01)$.

Solution

(a) The differential associated with the cube root function $y = x^{1/3}$ is

$$dy = \frac{1}{3x^{2/3}}\,dx.$$

We set $a = 8$, the closest number near 7.97 where we can easily compute $f(a)$ and $f'(a)$. To arrange that $a + dx = 7.97$, we choose $dx = -0.03$. Approximating with the differential gives

$$\begin{aligned} f(7.97) = f(a + dx) &\approx f(a) + dy \\ &= 8^{1/3} + \frac{1}{3(8)^{2/3}}(-0.03) \\ &= 2 + \frac{1}{12}(-0.03) = 1.9975. \end{aligned}$$

This gives an approximation to the true value of $7.97^{1/3}$, which is 1.997497 to 6 decimal places.

$\sin(a + dx) \approx \sin a + (\cos a)\,dx$

(b) The differential associated with $y = \sin x$ is

$$dy = \cos x\,dx.$$

To estimate $\sin(\pi/6 + 0.01)$, we set $a = \pi/6$ and $dx = 0.01$. Then

$$\begin{aligned} f(\pi/6 + 0.01) = f(a + dx) &\approx f(a) + dy \\ &= \sin\frac{\pi}{6} + \left(\cos\frac{\pi}{6}\right)(0.01) \\ &= \frac{1}{2} + \frac{\sqrt{3}}{2}(0.01) \approx 0.5087. \end{aligned}$$

For comparison, the true value of $\sin(\pi/6 + 0.01)$ to 6 decimal places is 0.508635. ■

The method in part (b) of Example 7 can be used in computer algorithms to give values of trigonometric functions. The algorithms store a large table of sine and cosine values between 0 and $\pi/4$. Values between these stored values are computed using differentials as in Example 7b. Values outside of $[0, \pi/4]$ are computed from values in this interval using trigonometric identities.

Error in Differential Approximation

Let $f(x)$ be differentiable at $x = a$ and suppose that $dx = \Delta x$ is an increment of x. We have two ways to describe the change in f as x changes from a to $a + \Delta x$:

The true change:	$\Delta f = f(a + \Delta x) - f(a)$
The differential estimate:	$df = f'(a)\,\Delta x.$

How well does df approximate Δf?

We measure the approximation error by subtracting df from Δf:

$$
\begin{aligned}
\text{Approximation error} &= \Delta f - df \\
&= \Delta f - f'(a)\Delta x \\
&= \underbrace{f(a + \Delta x) - f(a)}_{\Delta f} - f'(a)\Delta x \\
&= \underbrace{\left(\frac{f(a + \Delta x) - f(a)}{\Delta x} - f'(a)\right)}_{\text{Call this part } \varepsilon.} \cdot \Delta x \\
&= \varepsilon \cdot \Delta x.
\end{aligned}
$$

As $\Delta x \to 0$, the difference quotient

$$\frac{f(a + \Delta x) - f(a)}{\Delta x}$$

approaches $f'(a)$ (remember the definition of $f'(a)$), so the quantity in parentheses becomes a very small number (which is why we called it ε). In fact, $\varepsilon \to 0$ as $\Delta x \to 0$. When Δx is small, the approximation error $\varepsilon\, \Delta x$ is smaller still.

$$\underbrace{\Delta f}_{\substack{\text{true}\\ \text{change}}} = \underbrace{f'(a)\Delta x}_{\substack{\text{estimated}\\ \text{change}}} + \underbrace{\varepsilon\, \Delta x}_{\text{error}}$$

Although we do not know the exact size of the error, it is the product $\varepsilon \cdot \Delta x$ of two small quantities that both approach zero as $\Delta x \to 0$. For many common functions, whenever Δx is small, the error is still smaller.

Change in $y = f(x)$ near $x = a$

If $y = f(x)$ is differentiable at $x = a$ and x changes from a to $a + \Delta x$, the change Δy in f is given by

$$\Delta y = f'(a)\, \Delta x + \varepsilon\, \Delta x, \tag{1}$$

in which $\varepsilon \to 0$ as $\Delta x \to 0$.

In Example 6 we found that

$$\Delta A = \pi(10.1)^2 - \pi(10)^2 = (102.01 - 100)\pi = (\underbrace{2\pi}_{dA} + \underbrace{0.01\pi}_{\text{error}})\,\text{m}^2$$

so the approximation error is $\Delta A - dA = \varepsilon\Delta r = 0.01\pi$ and $\varepsilon = 0.01\pi/\Delta r = 0.01\pi/0.1 = 0.1\pi$ m.

Proof of the Chain Rule

Equation (1) enables us to give a complete proof of the Chain Rule. Our goal is to show that if $f(u)$ is a differentiable function of u and $u = g(x)$ is a differentiable function of x, then the composition $y = f(g(x))$ is a differentiable function of x. Since a function is differentiable if and only if it has a derivative at each point in its domain, we must show that whenever g is differentiable at x_0 and f is differentiable at $g(x_0)$, then the composition is differentiable at x_0 and the derivative of the composition satisfies the equation

$$\left.\frac{dy}{dx}\right|_{x=x_0} = f'(g(x_0)) \cdot g'(x_0).$$

Let Δx be an increment in x and let Δu and Δy be the corresponding increments in u and y. Applying Equation (1), we have

$$\Delta u = g'(x_0)\,\Delta x + \varepsilon_1\,\Delta x = (g'(x_0) + \varepsilon_1)\,\Delta x,$$

where $\varepsilon_1 \to 0$ as $\Delta x \to 0$. Similarly,

$$\Delta y = f'(u_0)\,\Delta u + \varepsilon_2\,\Delta u = (f'(u_0) + \varepsilon_2)\,\Delta u,$$

where $\varepsilon_2 \to 0$ as $\Delta u \to 0$. Notice also that $\Delta u \to 0$ as $\Delta x \to 0$. Combining the equations for Δu and Δy gives

$$\Delta y = (f'(u_0) + \varepsilon_2)(g'(x_0) + \varepsilon_1)\,\Delta x,$$

so

$$\frac{\Delta y}{\Delta x} = f'(u_0)g'(x_0) + \varepsilon_2 g'(x_0) + f'(u_0)\varepsilon_1 + \varepsilon_2\varepsilon_1.$$

Since ε_1 and ε_2 go to zero as Δx goes to zero, the last three terms on the right vanish in the limit, leaving

$$\left.\frac{dy}{dx}\right|_{x=x_0} = \lim_{\Delta x \to 0} \frac{\Delta y}{\Delta x} = f'(u_0)g'(x_0) = f'(g(x_0)) \cdot g'(x_0). \qquad \blacksquare$$

Sensitivity to Change

The equation $df = f'(x)\,dx$ tells how *sensitive* the output of f is to a change in input at different values of x. The larger the value of f' at x, the greater the effect of a given change dx. As we move from a to a nearby point $a + dx$, we can describe the change in f in three ways: absolute, relative, and percentage.

	True	**Estimated**
Absolute change	$\Delta f = f(a + dx) - f(a)$	$df = f'(a)\,dx$
Relative change	$\dfrac{\Delta f}{f(a)}$	$\dfrac{df}{f(a)}$
Percentage change	$\dfrac{\Delta f}{f(a)} \times 100$	$\dfrac{df}{f(a)} \times 100$

EXAMPLE 8 You want to calculate the depth of a well from the equation $s = 16t^2$ by timing how long it takes a heavy stone you drop to splash into the water below. How sensitive will your calculations be to a 0.1-sec error in measuring the time?

Solution The size of ds in the equation

$$ds = 32t\,dt$$

depends on how big t is. If $t = 2$ sec, the change caused by $dt = 0.1$ is about

$$ds = 32(2)(0.1) = 6.4 \text{ ft}.$$

Three seconds later at $t = 5$ sec, the change caused by the same dt is

$$ds = 32(5)(0.1) = 16 \text{ ft}.$$

For a fixed error in the time measurement, the error in using ds to estimate the depth is larger when it takes a longer time before the stone splashes into the water. That is, the estimate is more sensitive to the effect of the error for larger values of t. $\blacksquare$

EXAMPLE 9 Newton's second law,

$$F = \frac{d}{dt}(mv) = m\frac{dv}{dt} = ma,$$

is stated with the assumption that mass is constant, but we know this is not strictly true because the mass of an object increases with velocity. In Einstein's corrected formula, mass has the value

$$m = \frac{m_0}{\sqrt{1 - v^2/c^2}},$$

where the "rest mass" m_0 represents the mass of an object that is not moving and c is the speed of light, which is about 300,000 km/sec. Use the approximation

$$\frac{1}{\sqrt{1 - x^2}} \approx 1 + \frac{1}{2}x^2 \tag{2}$$

to estimate the increase Δm in mass resulting from the added velocity v.

Solution When v is very small compared with c, v^2/c^2 is close to zero and it is safe to use the approximation

$$\frac{1}{\sqrt{1 - v^2/c^2}} \approx 1 + \frac{1}{2}\left(\frac{v^2}{c^2}\right) \qquad \text{Eq. (2) with } x = \frac{v}{c}$$

to obtain

$$m = \frac{m_0}{\sqrt{1 - v^2/c^2}} \approx m_0\left[1 + \frac{1}{2}\left(\frac{v^2}{c^2}\right)\right] = m_0 + \frac{1}{2}m_0v^2\left(\frac{1}{c^2}\right),$$

or

$$m \approx m_0 + \frac{1}{2}m_0v^2\left(\frac{1}{c^2}\right). \tag{3}$$

Equation (3) expresses the increase in mass that results from the added velocity v. ■

Converting Mass to Energy

Equation (3) derived in Example 9 has an important interpretation. In Newtonian physics, $(1/2)\,m_0v^2$ is the kinetic energy (KE) of the object, and if we rewrite Equation (3) in the form

$$(m - m_0)\,c^2 \approx \frac{1}{2}m_0v^2,$$

we see that

$$(m - m_0)\,c^2 \approx \frac{1}{2}m_0v^2 = \frac{1}{2}m_0v^2 - \frac{1}{2}m_0\,(0)^2 = \Delta(\text{KE}),$$

or

$$(\Delta m)c^2 \approx \Delta(\text{KE}).$$

So the change in kinetic energy $\Delta(\text{KE})$ in going from velocity 0 to velocity v is approximately equal to $(\Delta m)c^2$, the change in mass times the square of the speed of light. Using $c \approx 3 \times 10^8$ m/sec, we see that a small change in mass can create a large change in energy.

EXERCISES 3.9

Finding Linearizations

In Exercises 1–5, find the linearization $L(x)$ of $f(x)$ at $x = a$.

1. $f(x) = 2x^3 - 5x + 2, \quad a = 2$

2. $f(x) = \sqrt{x^2 + 9}, \quad a = -4$

3. $f(x) = 2x + \dfrac{5}{x}, \quad a = 1$

4. $f(x) = \sqrt[3]{x}, \quad a = -8$

5. $f(x) = \tan x, \quad a = \pi$

6. Common linear approximations at $x = 0$ Find the linearizations of the following functions at $x = 0$.

a. $\sin x$ **b.** $\cos x$ **c.** $\tan x$

Linearization for Approximation

In Exercises 7–12, find a linearization at a suitably chosen integer near a at which the given function and its derivative are easy to evaluate.

7. $f(x) = x^2 + 2x, \quad a = 0.1$

8. $f(x) = x^{-1}, \quad a = 0.9$

9. $f(x) = 4x^2 + 2x - 5, \quad a = 0.9$

10. $f(x) = 1 + x, \quad a = 8.1$

11. $f(x) = \sqrt{x}, \quad a = 8.5$

12. $f(x) = \dfrac{x}{x+1}, \quad a = 1.3$

13. Show that the linearization of $f(x) = (1 + x)^k$ at $x = 0$ is $L(x) = 1 + kx$.

14. Use the linear approximation $(1 + x)^k \approx 1 + kx$ to find an approximation for the function $f(x)$ for values of x near zero.

a. $f(x) = (1 - x)^5$ **b.** $f(x) = \dfrac{2}{1 - x}$

c. $f(x) = \dfrac{1}{\sqrt{1 + x}}$ **d.** $f(x) = \sqrt{2 + x^2}$

e. $f(x) = (5 + 2x)^{1/3}$ **f.** $f(x) = \sqrt[3]{\left(1 - \dfrac{x}{2 + x}\right)^2}$

15. Faster than a calculator Use the approximation $(1 + x)^k \approx 1 + kx$ to estimate the following.

a. $(1.0002)^{50}$ **b.** $\sqrt[3]{1.009}$

16. Find the linearization of $f(x) = \sqrt{x + 1} + \sin x$ at $x = 0$. How is it related to the individual linearizations of $\sqrt{x + 1}$ and $\sin x$ at $x = 0$?

Derivatives in Differential Form

In Exercises 17–28, find dy.

17. $y = x^3 - 3\sqrt{x}$ **18.** $y = x\sqrt{1 - x^2}$

19. $y = \dfrac{2x}{1 + x^2}$ **20.** $y = \dfrac{2\sqrt{x}}{3(1 + \sqrt{x})}$

21. $2y^{3/2} + xy - x = 0$ **22.** $xy^2 - 4x^{3/2} - y = 0$

23. $y = \sin(5\sqrt{x})$ **24.** $y = \cos(x^2)$

25. $y = 4\tan(x^3/3)$ **26.** $y = \sec(x^2 - 1)$

27. $y = 3\csc(1 - 2\sqrt{x})$ **28.** $y = 2\cot\left(\dfrac{1}{\sqrt{x}}\right)$

Approximation Error

In Exercises 29–34, each function $f(x)$ changes value when x changes from x_0 to $x_0 + dx$. Find

a. the change $\Delta f = f(x_0 + dx) - f(x_0)$;

b. the value of the estimate $df = f'(x_0)\,dx$; and

c. the approximation error $|\Delta f - df|$.

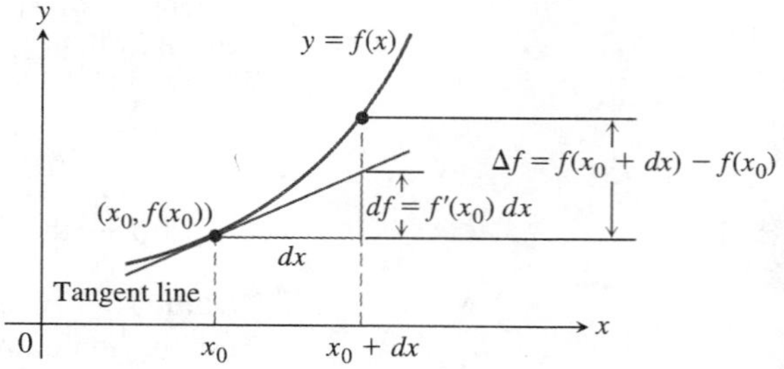

29. $f(x) = x^2 + 2x, \quad x_0 = 1, \quad dx = 0.1$

30. $f(x) = 2x^2 + 4x - 3, \quad x_0 = -1, \quad dx = 0.1$

31. $f(x) = x^3 - x, \quad x_0 = 1, \quad dx = 0.1$

32. $f(x) = x^4, \quad x_0 = 1, \quad dx = 0.1$

33. $f(x) = x^{-1}, \quad x_0 = 0.5, \quad dx = 0.1$

34. $f(x) = x^3 - 2x + 3, \quad x_0 = 2, \quad dx = 0.1$

Differential Estimates of Change

In Exercises 35–40, write a differential formula that estimates the given change in volume or surface area.

35. The change in the volume $V = (4/3)\pi r^3$ of a sphere when the radius changes from r_0 to $r_0 + dr$

36. The change in the volume $V = x^3$ of a cube when the edge lengths change from x_0 to $x_0 + dx$

37. The change in the surface area $S = 6x^2$ of a cube when the edge lengths change from x_0 to $x_0 + dx$

38. The change in the lateral surface area $S = \pi r\sqrt{r^2 + h^2}$ of a right circular cone when the radius changes from r_0 to $r_0 + dr$ and the height does not change

39. The change in the volume $V = \pi r^2 h$ of a right circular cylinder when the radius changes from r_0 to $r_0 + dr$ and the height does not change

40. The change in the lateral surface area $S = 2\pi rh$ of a right circular cylinder when the height changes from h_0 to $h_0 + dh$ and the radius does not change

Applications

41. The radius of a circle is increased from 2.00 to 2.02 m.

a. Estimate the resulting change in area.

b. Express the estimate as a percentage of the circle's original area.

42. The diameter of a tree was 10 in. During the following year, the circumference increased 2 in. About how much did the tree's diameter increase? The tree's cross-sectional area?

43. Estimating volume Estimate the volume of material in a cylindrical shell with length 30 in., radius 6 in., and shell thickness 0.5 in.

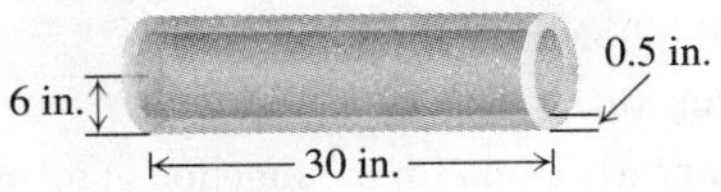

44. Estimating height of a building A surveyor, standing 30 ft from the base of a building, measures the angle of elevation to the top of the building to be 75°. How accurately must the angle be measured for the percentage error in estimating the height of the building to be less than 4%?

45. The radius r of a circle is measured with an error of at most 2%. What is the maximum corresponding percentage error in computing the circle's

a. circumference? **b.** area?

46. The edge x of a cube is measured with an error of at most 0.5%. What is the maximum corresponding percentage error in computing the cube's

a. surface area? **b.** volume?

47. Tolerance The height and radius of a right circular cylinder are equal, so the cylinder's volume is $V = \pi h^3$. The volume is to be calculated with an error of no more than 1% of the true value. Find approximately the greatest error that can be tolerated in the measurement of h, expressed as a percentage of h.

48. Tolerance

a. About how accurately must the interior diameter of a 10-m-high cylindrical storage tank be measured to calculate the tank's volume to within 1% of its true value?

b. About how accurately must the tank's exterior diameter be measured to calculate the amount of paint it will take to paint the side of the tank to within 5% of the true amount?

49. The diameter of a sphere is measured as 100 ± 1 cm and the volume is calculated from this measurement. Estimate the percentage error in the volume calculation.

50. Estimate the allowable percentage error in measuring the diameter D of a sphere if the volume is to be calculated correctly to within 3%.

T **51. The effect of flight maneuvers on the heart** The amount of work done by the heart's main pumping chamber, the left ventricle, is given by the equation

$$W = PV + \frac{V\delta v^2}{2g},$$

where W is the work per unit time, P is the average blood pressure, V is the volume of blood pumped out during the unit of time, δ ("delta") is the weight density of the blood, v is the average velocity of the exiting blood, and g is the acceleration of gravity.

When P, V, δ, and v remain constant, W becomes a function of g, and the equation takes the simplified form

$$W = a + \frac{b}{g} (a, b \text{ constant}).$$

As a member of NASA's medical team, you want to know how sensitive W is to apparent changes in g caused by flight maneuvers, and this depends on the initial value of g. As part of your investigation, you decide to compare the effect on W of a given change dg on the moon, where $g = 5.2 \text{ ft/sec}^2$, with the effect the same change dg would have on Earth, where $g = 32 \text{ ft/sec}^2$. Use the simplified equation above to find the ratio of dW_{moon} to dW_{Earth}.

T **52. Drug concentration** The concentration C in milligrams per milliliter (mg/ml) of a certain drug in a person's bloodstream t hr after a pill is swallowed is modeled by

$$C(t) = \frac{4t}{1 + t^3} + 0.06t.$$

Estimate the change in concentration when t changes from 20 to 30 min.

53. Unclogging arteries The formula $V = kr^4$, discovered by the physiologist Jean Poiseuille (1797–1869), allows us to predict how much the radius of a partially clogged artery has to be expanded in order to restore normal blood flow. The formula says that the volume V of blood flowing through the artery in a unit of time at a fixed pressure is a constant k times the radius of the artery to the fourth power. How will a 10% increase in r affect V?

54. Measuring acceleration of gravity When the length L of a clock pendulum is held constant by controlling its temperature, the pendulum's period T depends on the acceleration of gravity g. The period will therefore vary slightly as the clock is moved from place to place on Earth's surface, depending on the change in g. By keeping track of ΔT, we can estimate the variation in g from the equation $T = 2\pi(L/g)^{1/2}$ that relates T, g, and L.

a. With L held constant and g as the independent variable, calculate dT and use it to answer parts (b) and (c).

b. If g increases, will T increase or decrease? Will a pendulum clock speed up or slow down? Explain.

T **c.** A clock with a 100-cm pendulum is moved from a location where $g = 980 \text{ cm/sec}^2$ to a new location. This increases the period by $dT = 0.001$ sec. Find dg and estimate the value of g at the new location.

55. Quadratic approximations

a. Let $Q(x) = b_0 + b_1(x - a) + b_2(x - a)^2$ be a quadratic approximation to $f(x)$ at $x = a$ with these properties:

i. $Q(a) = f(a)$

ii. $Q'(a) = f'(a)$

iii. $Q''(a) = f''(a)$.

Determine the coefficients b_0, b_1, and b_2.

b. Find the quadratic approximation to $f(x) = 1/(1 - x)$ at $x = 0$.

T **c.** Graph $f(x) = 1/(1 - x)$ and its quadratic approximation at $x = 0$. Then zoom in on the two graphs at the point $(0, 1)$. Comment on what you see.

d. Find the quadratic approximation to $g(x) = 1/x$ at $x = 1$. Graph g and its quadratic approximation together. Comment on what you see.

T **e.** Find the quadratic approximation to $h(x) = \sqrt{1 + x}$ at $x = 0$. Graph h and its quadratic approximation together. Comment on what you see.

f. What are the linearizations of f, g, and h at the respective points in parts (b), (d), and (e)?

56. The linearization is the best linear approximation Suppose that $y = f(x)$ is differentiable at $x = a$ and that $g(x) = m(x - a) + c$ is a linear function in which m and c are constants. If the error $E(x) = f(x) - g(x)$ were small enough near $x = a$, we might think of using g as a linear approximation of f instead of the linearization $L(x) = f(a) + f'(a)(x - a)$. Show that if we impose on g the conditions

1. $E(a) = 0$ — The approximation error is zero at $x = a$.
2. $\lim_{x \to a} \frac{E(x)}{x - a} = 0$ — The error is negligible when compared with $x - a$.

then $g(x) = f(a) + f'(a)(x - a)$. Thus, the linearization $L(x)$ gives the only linear approximation whose error is both zero at $x = a$ and negligible in comparison with $x - a$.

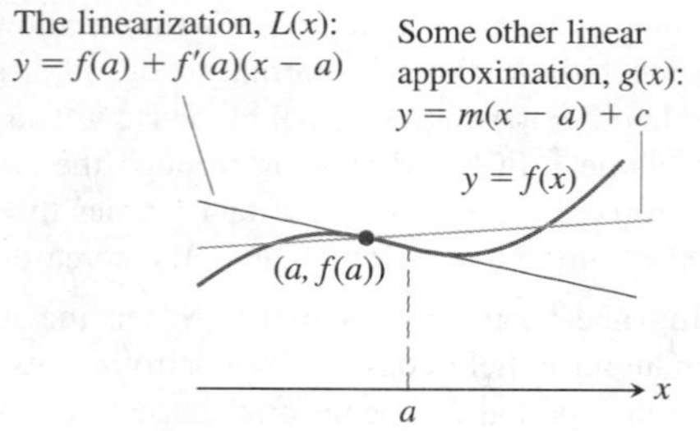

COMPUTER EXPLORATIONS

In Exercises 57–60, use a CAS to estimate the magnitude of the error in using the linearization in place of the function over a specified interval I. Perform the following steps:

a. Plot the function f over I.

b. Find the linearization L of the function at the point a.

c. Plot f and L together on a single graph.

d. Plot the absolute error $|f(x) - L(x)|$ over I and find its maximum value.

e. From your graph in part (d), estimate as large a $\delta > 0$ as you can that satisfies

$$|x - a| < \delta \quad \Rightarrow \quad |f(x) - L(x)| < \varepsilon$$

for $\varepsilon = 0.5, 0.1$, and 0.01. Then check graphically to see whether your δ-estimate holds true.

57. $f(x) = x^3 + x^2 - 2x, \quad [-1, 2], \quad a = 1$

58. $f(x) = \dfrac{x - 1}{4x^2 + 1}, \quad \left[-\dfrac{3}{4}, 1\right], \quad a = \dfrac{1}{2}$

59. $f(x) = x^{2/3}(x - 2), \quad [-2, 3], \quad a = 2$

60. $f(x) = \sqrt{x} - \sin x, \quad [0, 2\pi], \quad a = 2$

CHAPTER 3 Questions to Guide Your Review

1. What is the derivative of a function f? How is its domain related to the domain of f? Give examples.
2. What role does the derivative play in defining slopes, tangent lines, and rates of change?
3. How can you sometimes graph the derivative of a function when all you have is a table of the function's values?
4. What does it mean for a function to be differentiable on an open interval? On a closed interval?
5. How are derivatives and one-sided derivatives related?
6. Describe geometrically when a function typically does *not* have a derivative at a point.
7. How is a function's differentiability at a point related to its continuity there, if at all?
8. What rules do you know for calculating derivatives? Give some examples.
9. Explain how the three formulas

   a. $\dfrac{d}{dx}(x^n) = nx^{n-1}$ **b.** $\dfrac{d}{dx}(cu) = c\dfrac{du}{dx}$

   c. $\dfrac{d}{dx}(u_1 + u_2 + \cdots + u_n) = \dfrac{du_1}{dx} + \dfrac{du_2}{dx} + \cdots + \dfrac{du_n}{dx}$

   enable us to differentiate any polynomial.
10. What formula do we need, in addition to the three listed in Question 9, to differentiate rational functions?
11. What is a second derivative? A third derivative? How many derivatives do the functions you know have? Give examples.
12. What is the relationship between a function's average and instantaneous rates of change? Give an example.
13. How do derivatives arise in the study of motion? What can you learn about an object's motion along a line by examining the derivatives of the object's position function? Give examples.
14. How can derivatives arise in economics?
15. Give examples of still other applications of derivatives.
16. What do the limits $\lim_{h \to 0} ((\sin h)/h)$ and $\lim_{h \to 0} ((\cos h - 1)/h)$ have to do with the derivatives of the sine and cosine functions? What *are* the derivatives of these functions?
17. Once you know the derivatives of $\sin x$ and $\cos x$, how can you find the derivatives of $\tan x$, $\cot x$, $\sec x$, and $\csc x$? What *are* the derivatives of these functions?
18. What is the rule for calculating the derivative of a composition of two differentiable functions? How is such a derivative evaluated? Give examples.
19. If u is a differentiable function of x, how do you find $(d/dx)(u^n)$ if n is an integer? If n is a real number? Give examples.
20. What is implicit differentiation? When do you need it? Give examples.
21. How do related rates problems arise? Give examples.
22. Outline a strategy for solving related rates problems. Illustrate with an example.
23. What is the linearization $L(x)$ of a function $f(x)$ at a point $x = a$? What is required of f at a for the linearization to exist? How are linearizations used? Give examples.
24. If x moves from a to a nearby value $a + dx$, how do you estimate the corresponding change in the value of a differentiable function $f(x)$? How do you estimate the relative change? The percentage change? Give an example.

CHAPTER 3 Practice Exercises

Derivatives of Functions

Find the derivatives of the functions in Exercises 1–42.

1. $y = x^5 - 0.125x^2 + 0.25x$
2. $y = 3 - 0.7x^3 + 0.3x^7$
3. $y = x^3 - 3(x^2 + \pi^2)$
4. $y = x^7 + \sqrt{7}x - \dfrac{1}{\pi + 1}$
5. $y = (x + 1)^2(x^2 + 2x)$
6. $y = (2x - 5)(4 - x)^{-1}$
7. $y = (\theta^2 + \sec\theta + 1)^3$
8. $y = \left(-1 - \dfrac{\csc\theta}{2} - \dfrac{\theta^2}{4}\right)^2$
9. $s = \dfrac{\sqrt{t}}{1 + \sqrt{t}}$
10. $s = \dfrac{1}{\sqrt{t} - 1}$
11. $y = 2\tan^2 x - \sec^2 x$
12. $y = \dfrac{1}{\sin^2 x} - \dfrac{2}{\sin x}$
13. $s = \cos^4(1 - 2t)$
14. $s = \cot^3\left(\dfrac{2}{t}\right)$
15. $s = (\sec t + \tan t)^5$
16. $s = \csc^5(1 - t + 3t^2)$
17. $r = \sqrt{2\theta\sin\theta}$
18. $r = 2\theta\sqrt{\cos\theta}$
19. $r = \sin\sqrt{2\theta}$
20. $r = \sin(\theta + \sqrt{\theta + 1})$
21. $y = \dfrac{1}{2}x^2\csc\dfrac{2}{x}$
22. $y = 2\sqrt{x}\sin\sqrt{x}$
23. $y = x^{-1/2}\sec(2x)^2$
24. $y = \sqrt{x}\csc(x + 1)^3$
25. $y = 5\cot x^2$
26. $y = x^2\cot 5x$
27. $y = x^2\sin^2(2x^2)$
28. $y = x^{-2}\sin^2(x^3)$
29. $s = \left(\dfrac{4t}{t + 1}\right)^{-2}$
30. $s = \dfrac{-1}{15(15t - 1)^3}$
31. $y = \left(\dfrac{\sqrt{x}}{1 + x}\right)^2$
32. $y = \left(\dfrac{2\sqrt{x}}{2\sqrt{x} + 1}\right)^2$
33. $y = \sqrt{\dfrac{x^2 + x}{x^2}}$
34. $y = 4x\sqrt{x + \sqrt{x}}$
35. $r = \left(\dfrac{\sin\theta}{\cos\theta - 1}\right)^2$
36. $r = \left(\dfrac{1 + \sin\theta}{1 - \cos\theta}\right)^2$
37. $y = (2x + 1)\sqrt{2x + 1}$
38. $y = 20(3x - 4)^{1/4}(3x - 4)^{-1/5}$
39. $y = \dfrac{3}{(5x^2 + \sin 2x)^{3/2}}$
40. $y = (3 + \cos^3 3x)^{-1/3}$
41. $y = 5x^{3.6}$
42. $y = \sqrt{2}x^{-\sqrt{2}}$

Implicit Differentiation

In Exercises 43–50, find dy/dx by implicit differentiation.

43. $xy + 2x + 3y = 1$
44. $x^2 + xy + y^2 - 5x = 2$
45. $x^3 + 4xy - 3y^{4/3} = 2x$
46. $5x^{4/5} + 10y^{6/5} = 15$
47. $\sqrt{xy} = 1$
48. $x^2y^2 = 1$
49. $y^2 = \dfrac{x}{x + 1}$
50. $y^2 = \sqrt{\dfrac{1 + x}{1 - x}}$

In Exercises 51 and 52, find dp/dq.

51. $p^3 + 4pq - 3q^2 = 2$
52. $q = (5p^2 + 2p)^{-3/2}$

In Exercises 53 and 54, find dr/ds.

53. $r\cos 2s + \sin^2 s = \pi$
54. $2rs - r - s + s^2 = -3$

55. Find d^2y/dx^2 by implicit differentiation:
    a. $x^3 + y^3 = 1$ **b.** $y^2 = 1 - \dfrac{2}{x}$
56. **a.** By differentiating $x^2 - y^2 = 1$ implicitly, show that $dy/dx = x/y$.
    b. Then show that $d^2y/dx^2 = -1/y^3$.

Numerical Values of Derivatives

57. Suppose that functions $f(x)$ and $g(x)$ and their first derivatives have the following values at $x = 0$ and $x = 1$.

x	$f(x)$	$g(x)$	$f'(x)$	$g'(x)$
0	1	1	−3	1/2
1	3	5	1/2	−4

Find the first derivatives of the following combinations at the given value of x.

a. $6f(x) - g(x),\ x = 1$ **b.** $f(x)g^2(x),\ x = 0$
c. $\dfrac{f(x)}{g(x) + 1},\ x = 1$ **d.** $f(g(x)),\ x = 0$
e. $g(f(x)),\ x = 0$
f. $(x + f(x))^{3/2},\ x = 1$
g. $f(x + g(x)),\ x = 0$

58. Suppose that the function $f(x)$ and its first derivative have the following values at $x = 0$ and $x = 1$.

x	$f(x)$	$f'(x)$
0	9	−2
1	−3	1/5

Find the first derivatives of the following combinations at the given value of x.

a. $\sqrt{x}f(x),\ x = 1$ **b.** $\sqrt{f(x)},\ x = 0$
c. $f(\sqrt{x}),\ x = 1$ **d.** $f(1 - 5\tan x),\ x = 0$
e. $\dfrac{f(x)}{2 + \cos x},\ x = 0$ **f.** $10\sin\left(\dfrac{\pi x}{2}\right)f^2(x),\ x = 1$

59. Find the value of dy/dt at $t = 0$ if $y = 3\sin 2x$ and $x = t^2 + \pi$.
60. Find the value of ds/du at $u = 2$ if $s = t^2 + 5t$ and $t = (u^2 + 2u)^{1/3}$.
61. Find the value of dw/ds at $s = 0$ if $w = \sin(\sqrt{r} - 2)$ and $r = 8\sin(s + \pi/6)$.
62. Find the value of dr/dt at $t = 0$ if $r = (\theta^2 + 7)^{1/3}$ and $\theta^2 t + \theta = 1$.
63. If $y^3 + y = 2\cos x$, find the value of d^2y/dx^2 at the point $(0, 1)$.
64. If $x^{1/3} + y^{1/3} = 4$, find d^2y/dx^2 at the point $(8, 8)$.

Applying the Derivative Definition

In Exercises 65 and 66, find the derivative using the definition.

65. $f(t) = \dfrac{1}{2t + 1}$
66. $g(x) = 2x^2 + 1$

67. a. Graph the function

$$f(x) = \begin{cases} x^2, & -1 \le x < 0 \\ -x^2, & 0 \le x \le 1. \end{cases}$$

b. Is f continuous at $x = 0$?

c. Is f differentiable at $x = 0$?

Give reasons for your answers.

68. a. Graph the function

$$f(x) = \begin{cases} x, & -1 \le x < 0 \\ \tan x, & 0 \le x \le \pi/4. \end{cases}$$

b. Is f continuous at $x = 0$?

c. Is f differentiable at $x = 0$?

Give reasons for your answers.

69. a. Graph the function

$$f(x) = \begin{cases} x, & 0 \le x \le 1 \\ 2 - x, & 1 < x \le 2. \end{cases}$$

b. Is f continuous at $x = 1$?

c. Is f differentiable at $x = 1$?

Give reasons for your answers.

70. For what value or values of the constant m, if any, is

$$f(x) = \begin{cases} \sin 2x, & x \le 0 \\ mx, & x > 0 \end{cases}$$

a. continuous at $x = 0$?

b. differentiable at $x = 0$?

Give reasons for your answers.

Slopes, Tangent Lines, and Normal Lines

71. Tangent lines with specified slope Are there any points on the curve $y = (x/2) + 1/(2x - 4)$ where the slope is $-3/2$? If so, find them.

72. Tangent lines with specified slope Are there any points on the curve $y = x - 1/(2x)$ where the slope is 2? If so, find them.

73. Horizontal tangent lines Find the points on the curve $y = 2x^3 - 3x^2 - 12x + 20$ where the tangent line is parallel to the x-axis.

74. Tangent intercepts Find the x- and y-intercepts of the line that is tangent to the curve $y = x^3$ at the point $(-2, -8)$.

75. Tangent lines perpendicular or parallel to lines Find the points on the curve $y = 2x^3 - 3x^2 - 12x + 20$ where the tangent line is

a. perpendicular to the line $y = 1 - (x/24)$.

b. parallel to the line $y = \sqrt{2} - 12x$.

76. Intersecting tangent lines Show that the tangent lines to the curve $y = (\pi \sin x)/x$ at $x = \pi$ and $x = -\pi$ intersect at right angles.

77. Normal lines parallel to a line Find the points on the curve $y = \tan x, -\pi/2 < x < \pi/2$, where the normal line is parallel to the line $y = -x/2$. Sketch the curve and normal lines together, labeling each with its equation.

78. Tangent lines and normal lines Find equations for the tangent and normal lines to the curve $y = 1 + \cos x$ at the point $(\pi/2, 1)$. Sketch the curve, tangent line, and normal line together, labeling each with its equation.

79. Tangent parabola The parabola $y = x^2 + C$ is to be tangent to the line $y = x$. Find C.

80. Slope of a tangent line Show that the tangent line to the curve $y = x^3$ at any point (a, a^3) meets the curve again at a point where the slope is four times the slope at (a, a^3).

81. Tangent curve For what value of c is the curve $y = c/(x + 1)$ tangent to the line through the points $(0, 3)$ and $(5, -2)$?

82. Normal lines to a circle Show that the normal line at any point of the circle $x^2 + y^2 = a^2$ passes through the origin.

In Exercises 83–88, find equations for the lines that are tangent, and the lines that are normal, to the curve at the given point.

83. $x^2 + 2y^2 = 9, \quad (1, 2)$

84. $(x + 1)^3 + y^2 = 2, \quad (0, 1)$

85. $xy + 2x - 5y = 2, \quad (3, 2)$

86. $(y - x)^2 = 2x + 4, \quad (6, 2)$

87. $x + \sqrt{xy} = 6, \quad (4, 1)$

88. $x^{3/2} + 2y^{3/2} = 17, \quad (1, 4)$

89. Find the slope of the curve $x^3y^3 + y^2 = x + y$ at the points $(1, 1)$ and $(1, -1)$.

90. The graph shown suggests that the curve $y = \sin(x - \sin x)$ might have horizontal tangent lines at the x-axis. Does it? Give reasons for your answer.

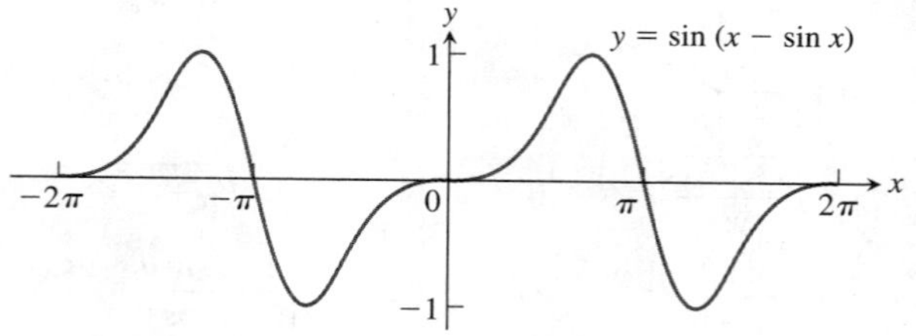

Analyzing Graphs

Each of the figures in Exercises 91 and 92 shows two graphs, the graph of a function $y = f(x)$ together with the graph of its derivative $f'(x)$. Which graph is which? How do you know?

91.

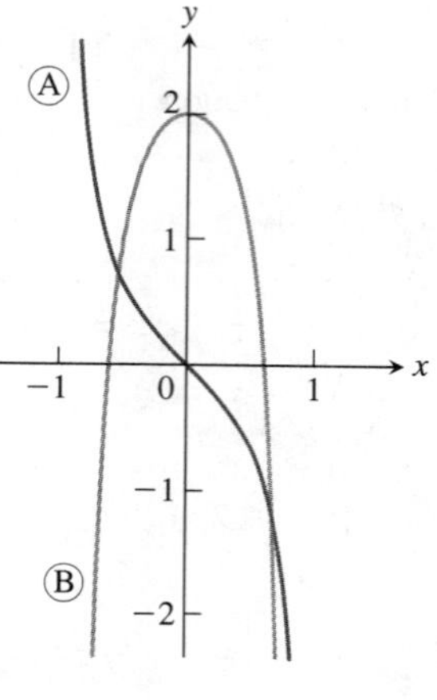

92.

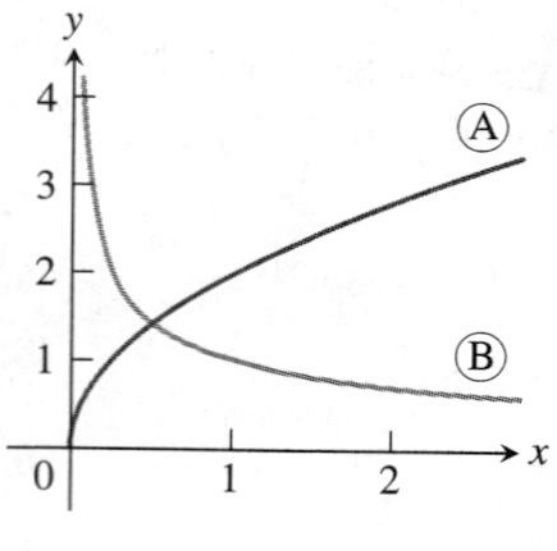

93. Use the following information to graph the function $y = f(x)$ for $-1 \le x \le 6$.

i) The graph of f is made of line segments joined end to end.

ii) The graph starts at the point $(-1, 2)$.

iii) The derivative of f, where defined, agrees with the step function shown here.

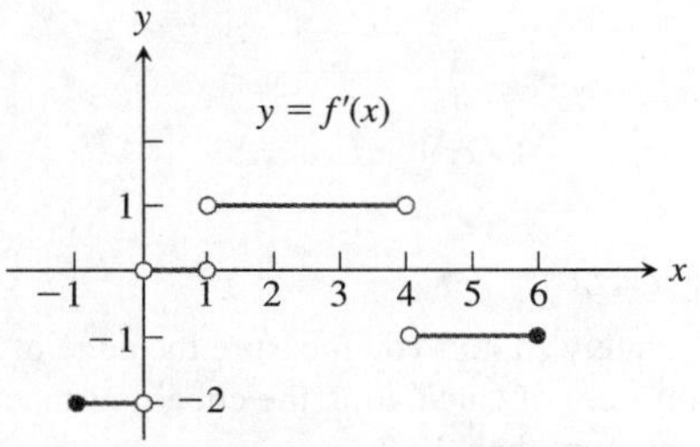

94. Repeat Exercise 93, supposing that the graph starts at $(-1, 0)$ instead of $(-1, 2)$.

Related Rates

95. Right circular cylinder The total surface area S of a right circular cylinder is related to the base radius r and height h by the equation $S = 2\pi r^2 + 2\pi rh$.

a. How is dS/dt related to dr/dt if h is constant?

b. How is dS/dt related to dh/dt if r is constant?

c. How is dS/dt related to dr/dt and dh/dt if neither r nor h is constant?

d. How is dr/dt related to dh/dt if S is constant?

96. Right circular cone The lateral surface area S of a right circular cone is related to the base radius r and height h by the equation $S = \pi r\sqrt{r^2 + h^2}$.

a. How is dS/dt related to dr/dt if h is constant?

b. How is dS/dt related to dh/dt if r is constant?

c. How is dS/dt related to dr/dt and dh/dt if neither r nor h is constant?

97. Circle's changing area The radius r of a circle is changing at the rate of $-2/\pi$ m/sec. At what rate is the circle's area changing when $r = 10$ m?

98. Cube's changing edges The volume of a cube is increasing at the rate of 1200 cm^3/min at the instant its edges are 20 cm long. At what rate are the lengths of the edges changing at that instant?

99. Resistors connected in parallel If two resistors of R_1 and R_2 ohms are connected in parallel in an electric circuit to make an R-ohm resistor, the value of R can be found from the equation

$$\frac{1}{R} = \frac{1}{R_1} + \frac{1}{R_2}.$$

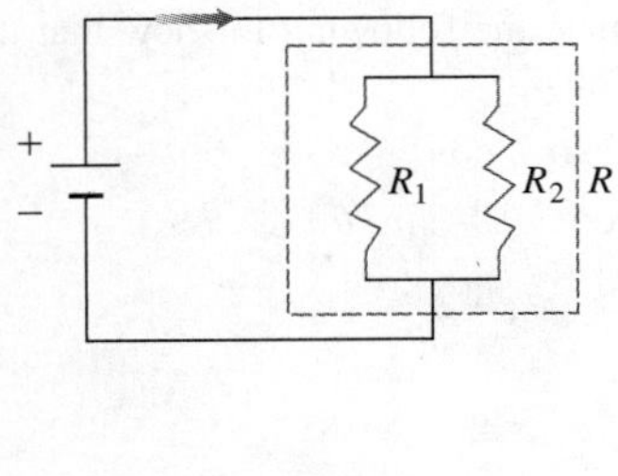

If R_1 is decreasing at the rate of 1 ohm/sec and R_2 is increasing at the rate of 0.5 ohm/sec, at what rate is R changing when $R_1 = 75$ ohms and $R_2 = 50$ ohms?

100. Impedance in a series circuit The impedance Z (ohms) in a series circuit is related to the resistance R (ohms) and reactance X (ohms) by the equation $Z = \sqrt{R^2 + X^2}$. If R is increasing at 3 ohms/sec and X is decreasing at 2 ohms/sec, at what rate is Z changing when $R = 10$ ohms and $X = 20$ ohms?

101. Speed of moving particle The coordinates of a particle moving in the metric xy-plane are differentiable functions of time t with $dx/dt = 10$ m/sec and $dy/dt = 5$ m/sec. How fast is the particle moving away from the origin as it passes through the point $(3, -4)$?

102. Motion of a particle A particle moves along the curve $y = x^{3/2}$ in the first quadrant in such a way that its distance from the origin increases at the rate of 11 units per second. Find dx/dt when $x = 3$.

103. Draining a tank Water drains from the conical tank shown in the accompanying figure at the rate of 5 ft^3/min.

a. What is the relation between the variables h and r in the figure?

b. How fast is the water level dropping when $h = 6$ ft?

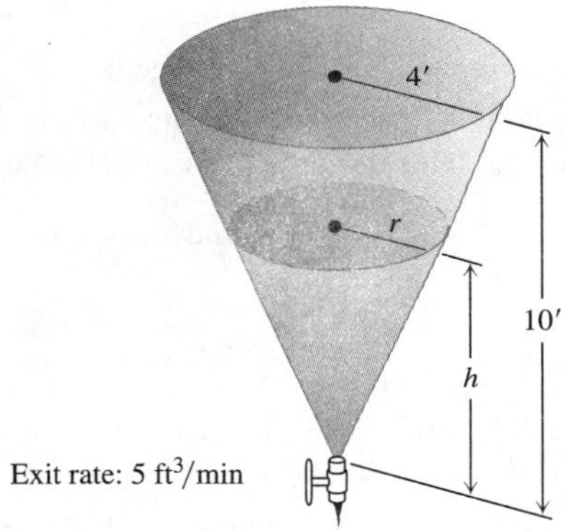

104. Rotating spool As television cable is pulled from a large spool to be strung from the telephone poles along a street, it unwinds from the spool in layers of constant radius (see accompanying figure). If the truck pulling the cable moves at a steady 6 ft/sec (a touch over 4 mph), use the equation $s = r\,\theta$ to find how fast (radians per second) the spool is turning when the layer of radius 1.2 ft is being unwound.

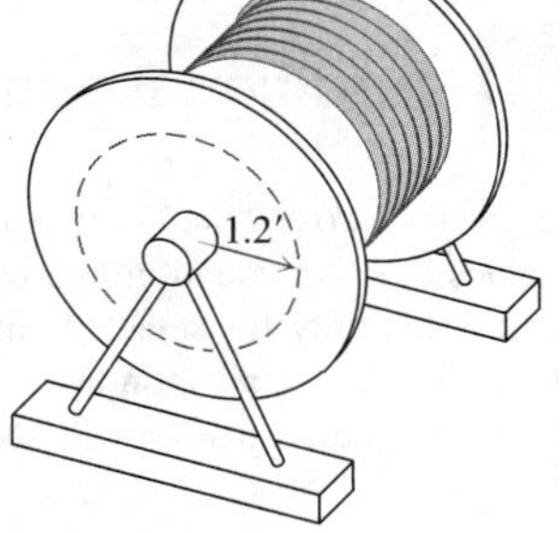

105. Moving searchlight beam The figure shows a boat 1 km offshore, sweeping the shore with a searchlight. The light turns at a constant rate, $d\theta/dt = -0.6$ rad/sec.

a. How fast is the light moving along the shore when it reaches point A?

b. How many revolutions per minute is 0.6 rad/sec?

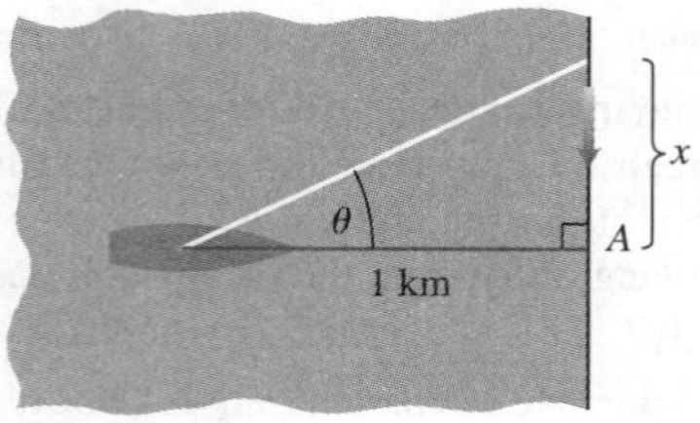

106. Points moving on coordinate axes Points A and B move along the x- and y-axes, respectively, in such a way that the distance r (meters) along the perpendicular from the origin to the line AB remains constant. How fast is OA changing, and is it increasing or decreasing, when $OB = 2r$ and B is moving toward O at the rate of $0.3r$ m/sec?

Linearization

107. Find the linearizations of

a. $\tan x$ at $x = -\pi/4$ **b.** $\sec x$ at $x = -\pi/4$.

Graph the curves and linearizations together.

108. We can obtain a useful linear approximation of the function $f(x) = 1/(1 + \tan x)$ at $x = 0$ by combining the approximations

$$\frac{1}{1+x} \approx 1 - x \quad \text{and} \quad \tan x \approx x$$

to get

$$\frac{1}{1 + \tan x} \approx 1 - x.$$

Show that this result is the standard linear approximation of $1/(1 + \tan x)$ at $x = 0$.

109. Find the linearization of $f(x) = \sqrt{1 + x} + \sin x - 0.5$ at $x = 0$.

110. Find the linearization of $f(x) = 2/(1 - x) + \sqrt{1 + x} - 3.1$ at $x = 0$.

Differential Estimates of Change

111. Surface area of a cone Write a formula that estimates the change that occurs in the lateral surface area of a right circular cone when the height changes from h_0 to $h_0 + dh$ and the radius does not change.

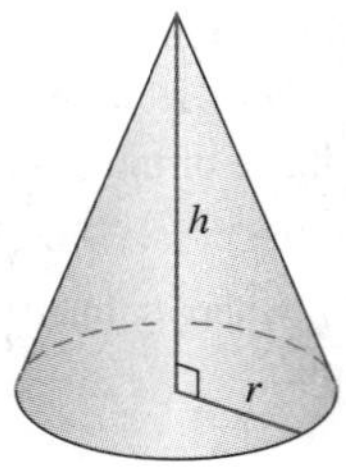

$V = \frac{1}{3}\pi r^2 h$

$S = \pi r\sqrt{r^2 + h^2}$

(Lateral surface area)

112. Controlling error

a. How accurately should you measure the edge of a cube to be reasonably sure of calculating the cube's surface area with an error of no more than 2%?

b. Suppose that the edge is measured with the accuracy required in part (a). About how accurately can the cube's volume be calculated from the edge measurement? To find out, estimate the percentage error in the volume calculation that might result from using the edge measurement.

113. Compounding error The circumference of the equator of a sphere is measured as 10 cm with a possible error of 0.4 cm. This measurement is used to calculate the radius. The radius is then used to calculate the surface area and volume of the sphere. Estimate the percentage errors in the calculated values of

a. the radius. **b.** the surface area. **c.** the volume.

114. Finding height To find the height of a lamppost (see accompanying figure), you stand a 6-ft pole 20 ft from the lamp and measure the length a of its shadow, finding it to be 15 ft, give or take an inch. Calculate the height of the lamppost using the value $a = 15$ and estimate the possible error in the result.

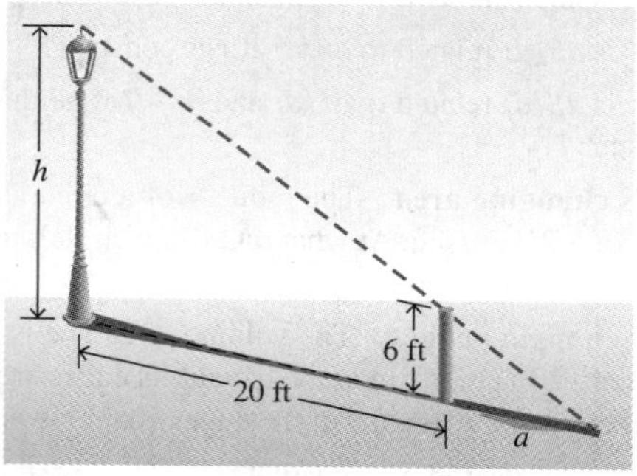

CHAPTER 3 Additional and Advanced Exercises

1. An equation like $\sin^2\theta + \cos^2\theta = 1$ is called an **identity** because it holds for all values of θ. An equation like $\sin\theta = 0.5$ is not an identity because it holds only for selected values of θ, not all. If you differentiate both sides of a trigonometric identity in θ with respect to θ, the resulting new equation will also be an identity.

Differentiate the following to show that the resulting equations hold for all θ.

a. $\sin 2\theta = 2\sin\theta\cos\theta$

b. $\cos 2\theta = \cos^2\theta - \sin^2\theta$

2. If the identity $\sin(x + a) = \sin x \cos a + \cos x \sin a$ is differentiated with respect to x (with a assumed to be a constant), is the resulting equation also an identity? Does this principle apply to the equation $x^2 - 2x - 8 = 0$? Explain.

3. **a.** Find values for the constants a, b, and c that will make

$$f(x) = \cos x \quad \text{and} \quad g(x) = a + bx + cx^2$$

satisfy the conditions

$$f(0) = g(0), \quad f'(0) = g'(0), \quad \text{and} \quad f''(0) = g''(0).$$

b. Find values for b and c that will make

$$f(x) = \sin(x + a) \quad \text{and} \quad g(x) = b \sin x + c \cos x$$

satisfy the conditions

$$f(0) = g(0) \quad \text{and} \quad f'(0) = g'(0).$$

c. For the determined values of a, b, and c, what happens for the third and fourth derivatives of f and g in each of parts (a) and (b)?

4. **Solutions to differential equations**

a. Show that $y = \sin x$, $y = \cos x$, and $y = a \cos x + b \sin x$ (a and b constants) all satisfy the equation

$$y'' + y = 0.$$

b. How would you modify the functions in part (a) to satisfy the equation

$$y'' + 4y = 0?$$

Generalize this result.

5. **An osculating circle** Find the values of h, k, and a that make the circle $(x - h)^2 + (y - k)^2 = a^2$ tangent to the parabola $y = x^2 + 1$ at the point $(1, 2)$ and that also make the second derivatives d^2y/dx^2 have the same value on both curves there. Circles like this one that are tangent to a curve and have the same second derivative as the curve at the point of tangency are called *osculating circles* (from the Latin *osculari*, meaning "to kiss"). We will encounter them again in Chapter 13.

6. **Marginal revenue** A bus will hold 60 people. The number x of people per trip who use the bus is related to the fare charged (p dollars) by the law $p = [3 - (x/40)]^2$. Write an expression for the total revenue $r(x)$ per trip received by the bus company. What number of people per trip will make the marginal revenue dr/dx equal to zero? What is the corresponding fare? (This fare is the one that maximizes the revenue.)

7. **Industrial production**

a. Economists often use the expression "rate of growth" in relative rather than absolute terms. For example, let $u = f(t)$ be the number of people in the labor force at time t in a given industry. (We treat this function as though it were differentiable even though it is an integer-valued step function.)

Let $v = g(t)$ be the average production per person in the labor force at time t. The total production is then $y = uv$. If the labor force is growing at the rate of 4% per year $(du/dt = 0.04u)$ and the production per worker is growing at the rate of 5% per year $(dv/dt = 0.05v)$, find the rate of growth of the total production, y.

b. Suppose that the labor force in part (a) is decreasing at the rate of 2% per year while the production per person is increasing at the rate of 3% per year. Is the total production increasing, or is it decreasing, and at what rate?

8. **Designing a gondola** The designer of a 30-ft-diameter spherical hot air balloon wants to suspend the gondola 8 ft below the bottom of the balloon with cables tangent to the surface of the balloon, as shown. Two of the cables are shown running from the top edges of the gondola to their points of tangency, $(-12, -9)$ and $(12, -9)$. How wide should the gondola be?

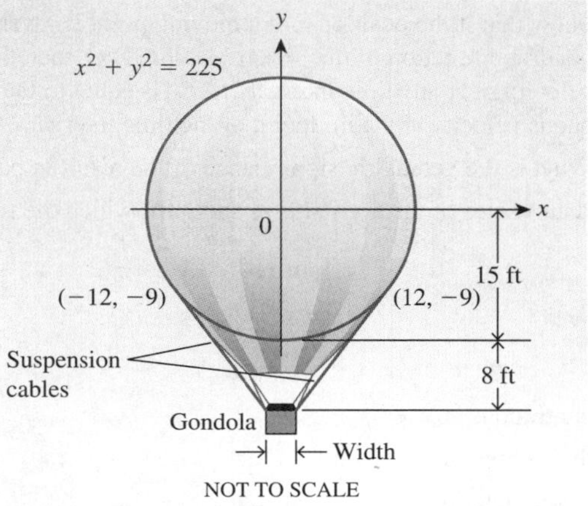

9. **Pisa by parachute** On August 5, 1988, Mike McCarthy of London jumped from the top of the Tower of Pisa. He then opened his parachute in what he said was a world record low-level parachute jump of 179 ft. Make a rough sketch to show the shape of the graph of his speed during the jump. (*Data from: Boston Globe*, Aug. 6, 1988.)

10. **Motion of a particle** The position at time $t \geq 0$ of a particle moving along a coordinate line is

$$s = 10 \cos(t + \pi/4).$$

a. What is the particle's starting position $(t = 0)$?

b. What are the points farthest to the left and right of the origin reached by the particle?

c. Find the particle's velocity and acceleration at the points in part (b).

d. When does the particle first reach the origin? What are its velocity, speed, and acceleration then?

11. **Shooting a paper clip** On Earth, you can easily shoot a paper clip 64 ft straight up into the air with a rubber band. In t sec after firing, the paper clip is $s = 64t - 16t^2$ ft above your hand.

a. How long does it take the paper clip to reach its maximum height? With what velocity does it leave your hand?

b. On the moon, the same acceleration will send the paper clip to a height of $s = 64t - 2.6t^2$ ft in t sec. About how long will it take the paper clip to reach its maximum height, and how high will it go?

12. **Velocities of two particles** At time t sec, the positions of two particles on a coordinate line are $s_1 = 3t^3 - 12t^2 + 18t + 5$ m and $s_2 = -t^3 + 9t^2 - 12t$ m. When do the particles have the same velocities?

13. Velocity of a particle A particle of constant mass m moves along the x-axis. Its velocity v and position x satisfy the equation

$$\frac{1}{2}m(v^2 - v_0^2) = \frac{1}{2}k(x_0^2 - x^2),$$

where k, v_0, and x_0 are constants. Show that whenever $v \neq 0$,

$$m\frac{dv}{dt} = -kx.$$

14. Average and instantaneous velocity

a. Show that if the position x of a moving point is given by a quadratic function of t, $x = At^2 + Bt + C$, then the average velocity over any time interval $[t_1, t_2]$ is equal to the instantaneous velocity at the midpoint of the time interval.

b. What is the geometric significance of the result in part (a)?

15. Find all values of the constants m and b for which the function

$$y = \begin{cases} \sin x, & x < \pi \\ mx + b, & x \geq \pi \end{cases}$$

is

a. continuous at $x = \pi$.

b. differentiable at $x = \pi$.

16. Does the function

$$f(x) = \begin{cases} \dfrac{1 - \cos x}{x}, & x \neq 0 \\ 0, & x = 0 \end{cases}$$

have a derivative at $x = 0$? Explain.

17. a. For what values of a and b will

$$f(x) = \begin{cases} ax, & x < 2 \\ ax^2 - bx + 3, & x \geq 2 \end{cases}$$

be differentiable for all values of x?

b. Discuss the geometry of the resulting graph of f.

18. a. For what values of a and b will

$$g(x) = \begin{cases} ax + b, & x \leq -1 \\ ax^3 + x + 2b, & x > -1 \end{cases}$$

be differentiable for all values of x?

b. Discuss the geometry of the resulting graph of g.

19. Odd differentiable functions Is there anything special about the derivative of an odd differentiable function of x? Give reasons for your answer.

20. Even differentiable functions Is there anything special about the derivative of an even differentiable function of x? Give reasons for your answer.

21. Suppose that the functions f and g are defined throughout an open interval containing the point x_0, that f is differentiable at x_0, that $f(x_0) = 0$, and that g is continuous at x_0. Show that the product fg is differentiable at x_0. This process shows, for example, that although $|x|$ is not differentiable at $x = 0$, the product $x|x|$ *is* differentiable at $x = 0$.

22. (*Continuation of Exercise 21.*) Use the result of Exercise 21 to show that the following functions are differentiable at $x = 0$.

a. $|x| \sin x$

b. $x^{2/3} \sin x$

c. $\sqrt[3]{x}(1 - \cos x)$

d. $h(x) = \begin{cases} x^2 \sin(1/x), & x \neq 0 \\ 0, & x = 0 \end{cases}$

23. Is the derivative of

$$h(x) = \begin{cases} x^2 \sin(1/x), & x \neq 0 \\ 0, & x = 0 \end{cases}$$

continuous at $x = 0$? How about the derivative of $k(x) = xh(x)$? Give reasons for your answers.

24. Let $f(x) = \begin{cases} x^2, & x \text{ is rational} \\ 0, & x \text{ is irrational.} \end{cases}$

Show that f is differentiable at $x = 0$.

25. Point B moves from point A to point C at 2 cm/sec in the accompanying diagram. At what rate is θ changing when $x = 4$ cm?

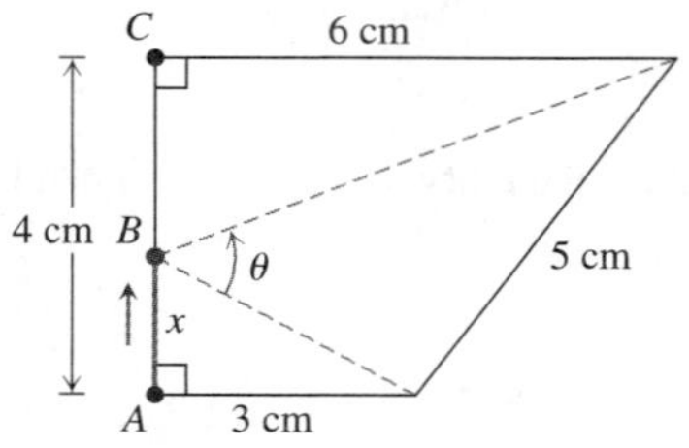

26. Suppose that a function f satisfies the following two conditions for all real values of x and y:

i) $f(x + y) = f(x) \cdot f(y)$.

ii) $f(x) = 1 + xg(x)$, where $\lim_{x \to 0} g(x) = 1$.

Show that the derivative $f'(x)$ exists at every value of x and that $f'(x) = f(x)$.

27. The generalized product rule Use mathematical induction to prove that if $y = u_1 u_2 \cdots u_n$ is a finite product of differentiable functions, then y is differentiable on their common domain, and

$$\frac{dy}{dx} = \frac{du_1}{dx}u_2 \cdots u_n + u_1\frac{du_2}{dx} \cdots u_n + \cdots + u_1 u_2 \cdots u_{n-1}\frac{du_n}{dx}.$$

28. Leibniz's rule for higher-order derivatives of products Leibniz's rule for higher-order derivatives of products of differentiable functions says that

a. $\dfrac{d^2(uv)}{dx^2} = \dfrac{d^2u}{dx^2}v + 2\dfrac{du}{dx}\dfrac{dv}{dx} + u\dfrac{d^2v}{dx^2}.$

b. $\dfrac{d^3(uv)}{dx^3} = \dfrac{d^3u}{dx^3}v + 3\dfrac{d^2u}{dx^2}\dfrac{dv}{dx} + 3\dfrac{du}{dx}\dfrac{d^2v}{dx^2} + u\dfrac{d^3v}{dx^3}.$

c. $\dfrac{d^n(uv)}{dx^n} = \dfrac{d^nu}{dx^n}v + n\dfrac{d^{n-1}u}{dx^{n-1}}\dfrac{dv}{dx} + \cdots$

$$+ \frac{n(n-1)\cdots(n-k+1)}{k!}\frac{d^{n-k}u}{dx^{n-k}}\frac{d^kv}{dx^k}$$

$$+ \cdots + u\frac{d^nv}{dx^n}.$$

The equations in parts (a) and (b) are special cases of the equation in part (c). Derive the equation in part (c) by mathematical induction, using

$$\binom{m}{k} + \binom{m}{k+1} = \frac{m!}{k!(m-k)!} + \frac{m!}{(k+1)!(m-k-1)!}.$$

29. The period of a clock pendulum The period T of a clock pendulum (time for one full swing and back) is given by the formula $T^2 = 4\pi^2 L/g$, where T is measured in seconds, $g = 32.2 \text{ ft/sec}^2$, and L, the length of the pendulum, is measured in feet. Find approximately

a. the length of a clock pendulum whose period is $T = 1$ sec.

b. the change dT in T if the pendulum in part (a) is lengthened 0.01 ft.

c. the amount the clock gains or loses in a day as a result of the period's changing by the amount dT found in part (b).

30. The melting ice cube Assume that an ice cube retains its cubical shape as it melts. If we call its edge length s, its volume is $V = s^3$ and its surface area is $6s^2$. We assume that V and s are differentiable functions of time t. We assume also that the cube's volume decreases at a rate that is proportional to its surface area. (This latter assumption seems reasonable enough when we think that the melting takes place at the surface: Changing the amount of surface changes the amount of ice exposed to melt.) In mathematical terms,

$$\frac{dV}{dt} = -k(6s^2), \qquad k > 0.$$

The minus sign indicates that the volume is decreasing. We assume that the proportionality factor k is constant. (It probably depends on many things, such as the relative humidity of the surrounding air, the air temperature, and the incidence or absence of sunlight, to name only a few.) Assume a particular set of conditions in which the cube lost 1/4 of its volume during the first hour, and assume that the volume is V_0 when $t = 0$. How long will it take the ice cube to melt?

CHAPTER 3 Technology Application Projects

Mathematica/Maple Projects

Projects can be found at pearsonhighered.com/Thomas or within MyLab Math.

- ***Convergence of Secant Slopes to the Derivative Function***
  You will visualize the secant line between successive points on a curve and observe what happens as the distance between them becomes small. The function, sample points, and secant lines are plotted on a single graph, while a second graph compares the slopes of the secant lines with the derivative function.

- ***Derivatives, Slopes, Tangent Lines, and Making Movies***
  Parts I–III. You will visualize the derivative at a point, the linearization of a function, and the derivative of a function. You will learn how to plot the function and selected tangent lines on the same graph.
  Part IV (Plotting Many Tangent Lines)
  Part V (Making Movies). Parts IV and V of the module can be used to animate tangent lines as one moves along the graph of a function.

- ***Convergence of Secant Slopes to the Derivative Function***
  You will visualize right-hand and left-hand derivatives.

- ***Motion Along a Straight Line: Position → Velocity → Acceleration***
  Observe dramatic animated visualizations of the derivative relations among the position, velocity, and acceleration functions. Figures in the text can be animated.

4 Applications of Derivatives

OVERVIEW One of the most important applications of the derivative is its use as a tool for finding the optimal (best) solutions to problems. For example, what are the height and diameter of the cylinder of largest volume that can be inscribed in a given sphere? What are the dimensions of the strongest rectangular wooden beam that can be cut from a cylindrical log of given diameter? How many items should a manufacturer produce to maximize profit?

In this chapter we apply derivatives to find extreme values of functions, to determine and analyze the shapes of graphs, and to solve equations numerically. We also investigate how to recover a function from its derivative. The key to many of these applications is the Mean Value Theorem, which connects the derivative and the average change of a function.

4.1 Extreme Values of Functions on Closed Intervals

This section shows how to locate and identify extreme (maximum or minimum) values of a function from its derivative. Once we can do this, we can solve a variety of optimization problems (see Section 4.5). The domains of the functions we consider are intervals or unions of separate intervals.

> **DEFINITIONS** Let f be a function with domain D. Then f has an **absolute maximum** value on D at a point c if
>
> $$f(x) \leq f(c) \qquad \text{for all } x \text{ in } D$$
>
> and an **absolute minimum** value on D at c if
>
> $$f(x) \geq f(c) \qquad \text{for all } x \text{ in } D.$$

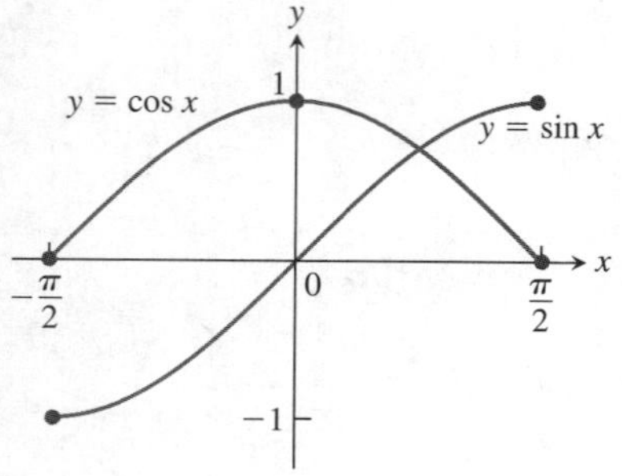

FIGURE 4.1 Absolute extrema for the sine and cosine functions on $[-\pi/2, \pi/2]$. These values can depend on the domain of a function.

Maximum and minimum values are called **extreme values** of the function f. Absolute maxima or minima are also referred to as **global** maxima or minima.

For example, on the closed interval $[-\pi/2, \pi/2]$ the function $f(x) = \cos x$ takes on an absolute maximum value of 1 (once) and an absolute minimum value of 0 (twice). On the same interval, the function $g(x) = \sin x$ takes on a maximum value of 1 and a minimum value of -1 (Figure 4.1).

Functions defined by the same equation or formula can have different extrema (maximum or minimum values), depending on the domain. A function might not have a maximum or minimum if the domain is unbounded or fails to contain an endpoint. We see this in the following example.

EXAMPLE 1 The absolute extrema of the following functions on their domains can be seen in Figure 4.2. Each function has the same defining equation, $y = x^2$, but the domains vary.

Function rule	Domain D	Absolute extrema on D
(a) $y = x^2$	$(-\infty, \infty)$	No absolute maximum Absolute minimum of 0 at $x = 0$
(b) $y = x^2$	$[0, 2]$	Absolute maximum of 4 at $x = 2$ Absolute minimum of 0 at $x = 0$
(c) $y = x^2$	$(0, 2]$	Absolute maximum of 4 at $x = 2$ No absolute minimum
(d) $y = x^2$	$(0, 2)$	No absolute extrema

■

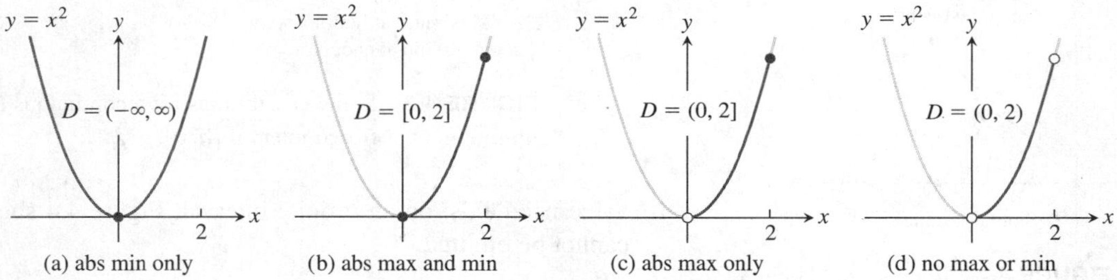

FIGURE 4.2 Graphs for Example 1.

HISTORICAL BIOGRAPHY
Daniel Bernoulli
(1700–1789)
www.bit.ly/2xPvS1Z

Some of the functions in Example 1 do not have a maximum or a minimum value. The following theorem asserts that a function which is *continuous* over (or on) a finite *closed* interval $[a, b]$ has an absolute maximum and an absolute minimum value on the interval. We look for these extreme values when we graph a function.

THEOREM 1—The Extreme Value Theorem
If f is continuous on a closed interval $[a, b]$, then f attains both an absolute maximum value M and an absolute minimum value m in $[a, b]$. That is, there are numbers x_1 and x_2 in $[a, b]$ with $f(x_1) = m$, $f(x_2) = M$, and $m \le f(x) \le M$ for all x in $[a, b]$.

The proof of the Extreme Value Theorem requires a detailed knowledge of the real number system (see Appendix A.7) and we will not give it here. Figure 4.3 illustrates possible locations for the absolute extrema of a continuous function on a closed interval $[a, b]$. As we observed for the function $y = \cos x$, it is possible that an absolute minimum (or absolute maximum) may occur at two or more different points of the interval.

The requirements in Theorem 1 that the interval be closed and finite, and that the function be continuous, are essential. Without them, the conclusion of the theorem need not hold. Example 1 shows that an absolute extreme value may not exist if the interval fails to be both closed and finite. The function $y = x$ over $(-\infty, \infty)$ shows that neither extreme

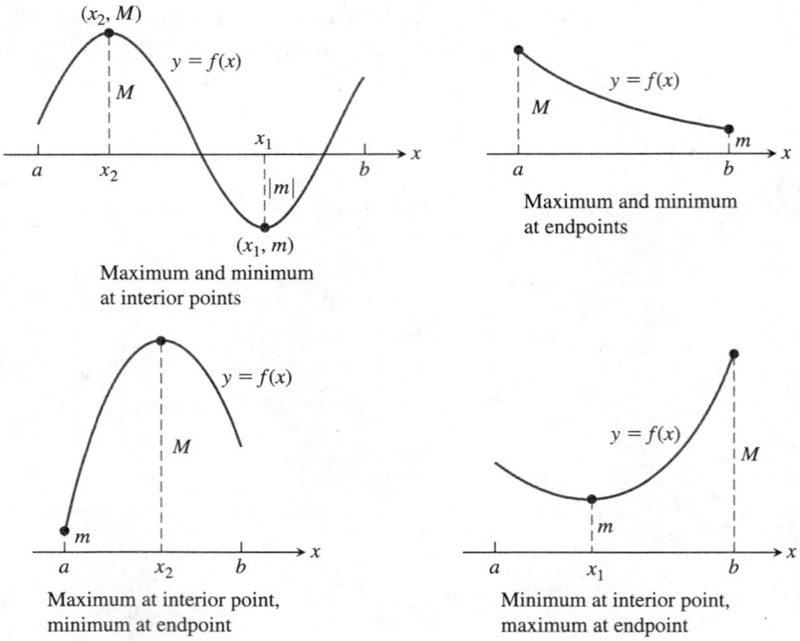

FIGURE 4.3 Some possibilities for a continuous function's maximum and minimum on a closed interval $[a, b]$.

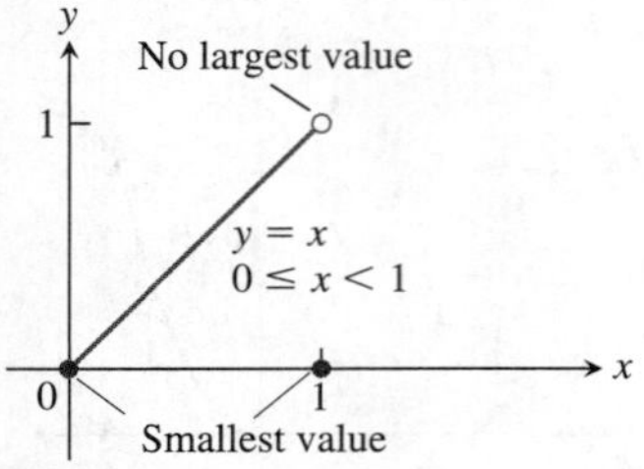

FIGURE 4.4 Even a single point of discontinuity can keep a function from having either a maximum or a minimum value on a closed interval. The function

$$y = \begin{cases} x, & 0 \le x < 1 \\ 0, & x = 1 \end{cases}$$

is continuous at every point of $[0, 1]$ except $x = 1$, yet its graph over $[0, 1]$ does not have a highest point.

value need exist on an infinite interval. Figure 4.4 shows that the continuity requirement cannot be omitted.

Local (Relative) Extreme Values

Figure 4.5 shows a graph with five points where a function has extreme values on its domain $[a, b]$. The function's absolute minimum occurs at a even though at e the function's value is smaller than at any other point *nearby*. The curve rises as x approaches c from the left, then falls to the right of c, making $f(c)$ a maximum locally. The function attains its absolute maximum at d. We now define what we mean by local extrema.

> **DEFINITIONS** A function f has a **local maximum** value at a point c within its domain D if $f(x) \le f(c)$ for all $x \in D$ lying in some open interval containing c.
>
> A function f has a **local minimum** value at a point c within its domain D if $f(x) \ge f(c)$ for all $x \in D$ lying in some open interval containing c.

If the domain of f is the closed interval $[a, b]$, then f has a local maximum at the endpoint $x = a$ if $f(x) \le f(a)$ for all x in some half-open interval $[a, a + \delta)$, $\delta > 0$. Likewise, f has a local maximum at an interior point $x = c$ if $f(x) \le f(c)$ for all x in some open interval $(c - \delta, c + \delta)$, $\delta > 0$, and a local maximum at the endpoint $x = b$ if $f(x) \le f(b)$ for all x in some half-open interval $(b - \delta, b]$, $\delta > 0$. The inequalities are reversed for local minimum values. In Figure 4.5, the function f has local maxima at c and d and local minima at a, e, and b. Local extrema are also called **relative extrema**. Some functions can have infinitely many local extrema, even over a finite interval. One example is the function $f(x) = \sin(1/x)$ on the interval $(0, 1]$. (We graphed this function in Figure 2.41.)

An absolute maximum is also a local maximum. Being the largest value overall, it is also the largest value in its immediate neighborhood. Hence, *a list of all local maxima will*

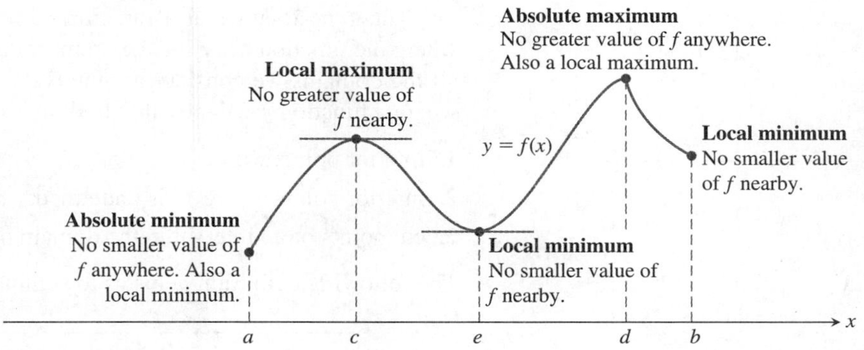

FIGURE 4.5 How to identify types of maxima and minima for a function with domain $a \le x \le b$.

automatically include the absolute maximum if there is one. Similarly, *a list of all local minima will include the absolute minimum if there is one*.

Finding Extrema

The next theorem explains why we usually need to investigate only a few values to find a function's extrema.

> **THEOREM 2—The First Derivative Theorem for Local Extreme Values**
> If f has a local maximum or minimum value at an interior point c of its domain, and if f' is defined at c, then
> $$f'(c) = 0.$$

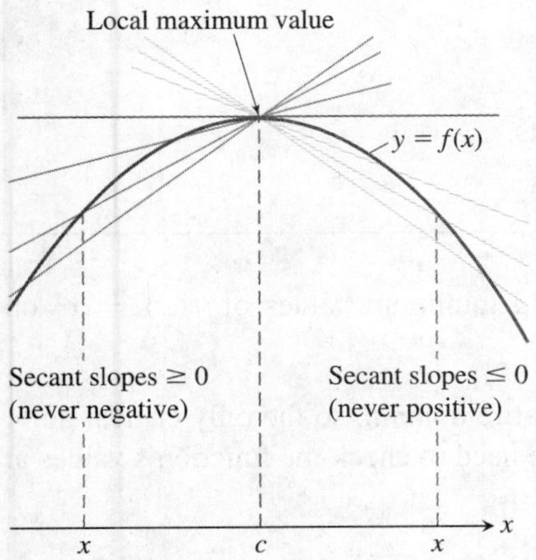

FIGURE 4.6 A curve with a local maximum value. The slope at c, simultaneously the limit of nonpositive numbers and nonnegative numbers, is zero.

Proof To prove that $f'(c)$ is zero at a local extremum, we show first that $f'(c)$ cannot be positive and second that $f'(c)$ cannot be negative. The only number that is neither positive nor negative is zero, so that is what $f'(c)$ must be.

To begin, suppose that f has a local maximum value at $x = c$ (Figure 4.6) so that $f(x) - f(c) \le 0$ for all values of x near enough to c. Since c is an interior point of f's domain, $f'(c)$ is defined by the two-sided limit

$$\lim_{x \to c} \frac{f(x) - f(c)}{x - c}.$$

This means that the right-hand and left-hand limits both exist at $x = c$ and equal $f'(c)$. When we examine these limits separately, we find that

$$f'(c) = \lim_{x \to c^+} \frac{f(x) - f(c)}{x - c} \le 0. \quad \text{Because } x - c > 0 \text{ and } f(x) \le f(c) \tag{1}$$

Similarly,

$$f'(c) = \lim_{x \to c^-} \frac{f(x) - f(c)}{x - c} \ge 0. \quad \text{Because } x - c < 0 \text{ and } f(x) \le f(c) \tag{2}$$

Together, Equations (1) and (2) imply $f'(c) = 0$.

This proves the theorem for local maximum values. To prove it for local minimum values, we simply use $f(x) \ge f(c)$, which reverses the inequalities in Formulas (1) and (2). ■

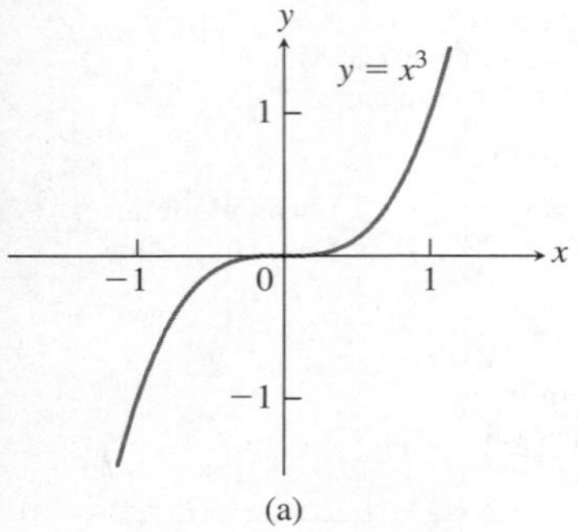

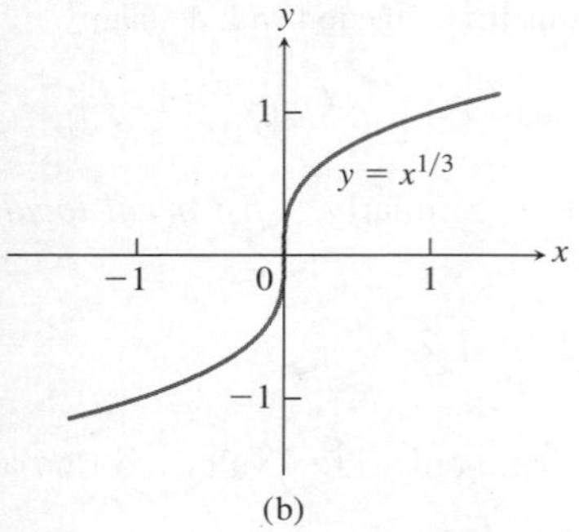

FIGURE 4.7 Critical points without extreme values. (a) $y' = 3x^2$ is 0 at $x = 0$, but $y = x^3$ has no extremum there. (b) $y' = (1/3)x^{-2/3}$ is undefined at $x = 0$, but $y = x^{1/3}$ has no extremum there.

Theorem 2 says that a function's first derivative is always zero at an interior point where the function has a local extreme value and the derivative is defined. If we recall that all the domains we consider are intervals or unions of separate intervals, the only places where a function f can possibly have an extreme value (local or global) are

1. interior points where $f' = 0$, At $x = c$ and $x = e$ in Fig. 4.5
2. interior points where f' is undefined, At $x = d$ in Fig. 4.5
3. endpoints of an interval in the domain of f. At $x = a$ and $x = b$ in Fig. 4.5

The following definition helps us to summarize these results.

DEFINITION An interior point of the domain of a function f where f' is zero or undefined is a **critical point** of f.

Thus the only domain points where a continuous function on a closed and finite interval can assume extreme values are critical points and endpoints. However, be careful not to misinterpret what is being said here. A function may have a critical point at $x = c$ without having a local extreme value there. For instance, both of the functions $y = x^3$ and $y = x^{1/3}$ have critical points at the origin, but neither function has a local extreme value at the origin. Instead, each function has a *point of inflection* there (see Figure 4.7). We define and explore inflection points in Section 4.4.

If the interval is not closed or not finite (such as $a < x < b$ or $a < x < \infty$), we have seen that absolute extrema need not exist. However, many problems that ask for extreme values call for finding the absolute extrema of a continuous function on a closed and finite interval. Theorem 1 assures us that such values exist; Theorem 2 tells us that they are taken on only at critical points and endpoints. Often we can simply list these points and calculate the corresponding function values to find what the largest and smallest values are, and where they are located.

Finding the Absolute Extrema of a Continuous Function f on a Finite Closed Interval

1. Find all critical points of f on the interval.
2. Evaluate f at all critical points and endpoints.
3. Take the largest and smallest of these values.

EXAMPLE 2 Find the absolute maximum and minimum values of $f(x) = x^2$ on $[-2, 1]$.

Solution The function is differentiable over its entire domain, so the only critical point occurs where $f'(x) = 2x = 0$, namely $x = 0$. We need to check the function's values at $x = 0$ and at the endpoints $x = -2$ and $x = 1$:

Critical point value: $f(0) = 0$

Endpoint values: $f(-2) = 4$

$f(1) = 1.$

The function has an absolute maximum value of 4 at $x = -2$ and an absolute minimum value of 0 at $x = 0$. ■

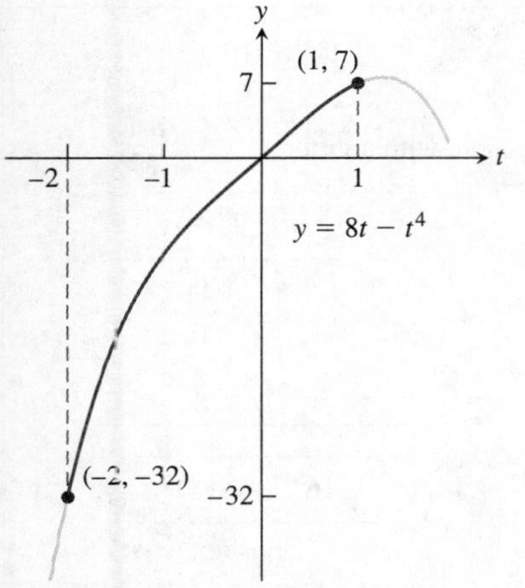

FIGURE 4.8 The extreme values of $g(t) = 8t - t^4$ on $[-2, 1]$ (Example 3).

EXAMPLE 3 Find the absolute maximum and minimum values of $g(t) = 8t - t^4$ on $[-2, 1]$.

Solution The function is differentiable on its entire domain, so the only critical points occur where $g'(t) = 0$. Solving this equation gives

$$8 - 4t^3 = 0 \quad \text{or} \quad t = \sqrt[3]{2} > 1,$$

a point not in the given domain. The function's absolute extrema therefore occur at the endpoints, $g(-2) = -32$ (absolute minimum), and $g(1) = 7$ (absolute maximum). See Figure 4.8 ■

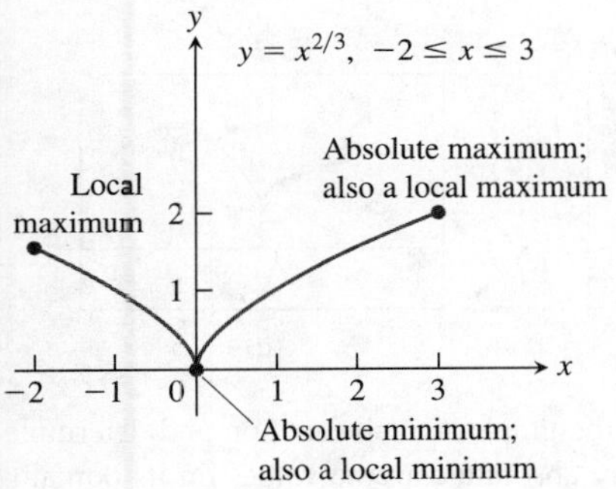

FIGURE 4.9 The extreme values of $f(x) = x^{2/3}$ on $[-2, 3]$ occur at $x = 0$ and $x = 3$ (Example 4).

EXAMPLE 4 Find the absolute maximum and minimum values of $f(x) = x^{2/3}$ on the interval $[-2, 3]$.

Solution We evaluate the function at the critical points and endpoints and take the largest and smallest of the resulting values.

The first derivative,

$$f'(x) = \frac{2}{3}x^{-1/3} = \frac{2}{3\sqrt[3]{x}},$$

has no zeros but is undefined at the interior point $x = 0$. The values of f at this one critical point and at the endpoints are

$$\begin{aligned} &\text{Critical point value:} && f(0) = 0 \\ &\text{Endpoint values:} && f(-2) = (-2)^{2/3} = \sqrt[3]{4} \\ & && f(3) = (3)^{2/3} = \sqrt[3]{9}. \end{aligned}$$

We can see from this list that the function's absolute maximum value is $\sqrt[3]{9} \approx 2.08$, and it occurs at the right endpoint $x = 3$. The absolute minimum value is 0, and it occurs at the interior point $x = 0$ where the graph has a cusp (Figure 4.9). ■

Theorem 1 leads to a method for finding the absolute maxima and absolute minima of a differentiable function on a finite closed interval. On more general domains, such as $(0, 1)$, $[2, 5)$, $[1, \infty)$, and $(-\infty, \infty)$, absolute maxima and minima may or may not exist. To determine if they exist, and to locate them when they do, we will develop methods to sketch the graph of a differentiable function. With knowledge of the asymptotes of the function, as well as the local maxima and minima, we can deduce the locations of the absolute maxima and minima, if any. For now we can find the absolute maxima and the absolute minima of a function on a finite closed interval by comparing the values of the function at its critical points and at the endpoints of the interval. For a differentiable function on a closed and finite interval $[a, b]$, these are the only points where the extrema have the potential to occur.

EXERCISES 4.1

Finding Extrema from Graphs

In Exercises 1–6, determine from the graph whether the function has any absolute extreme values on $[a, b]$. Then explain how your answer is consistent with Theorem 1.

1.

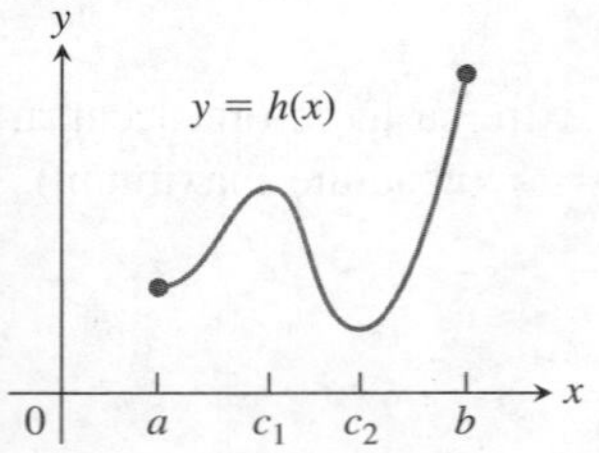

2.

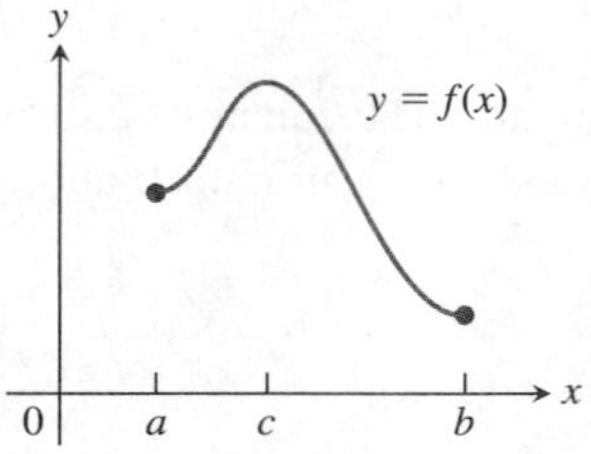

3.

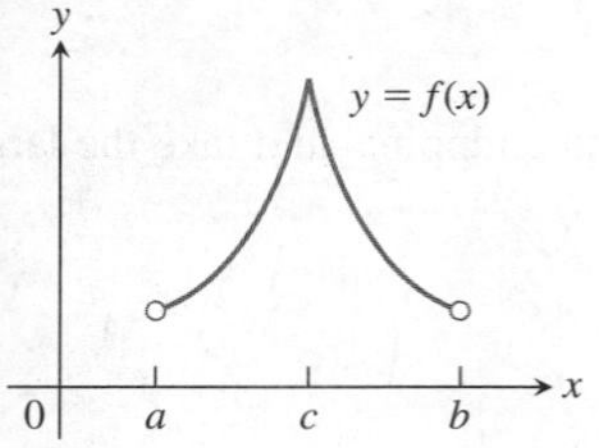

4.

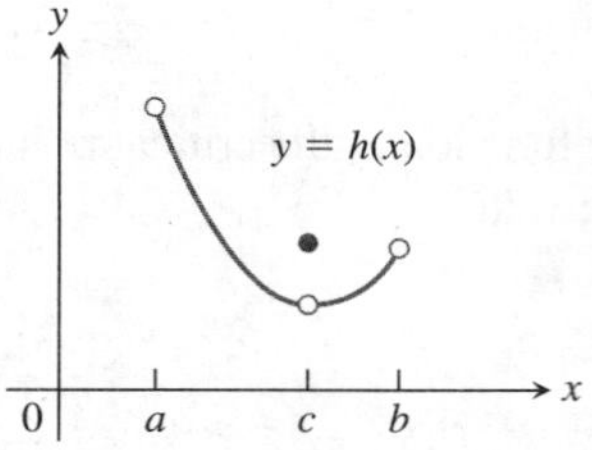

5.

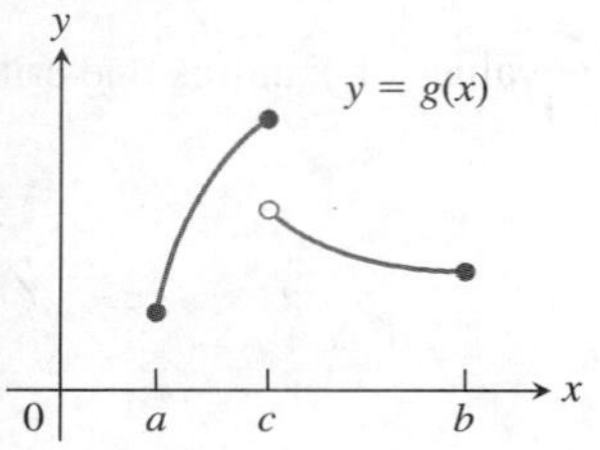

6.

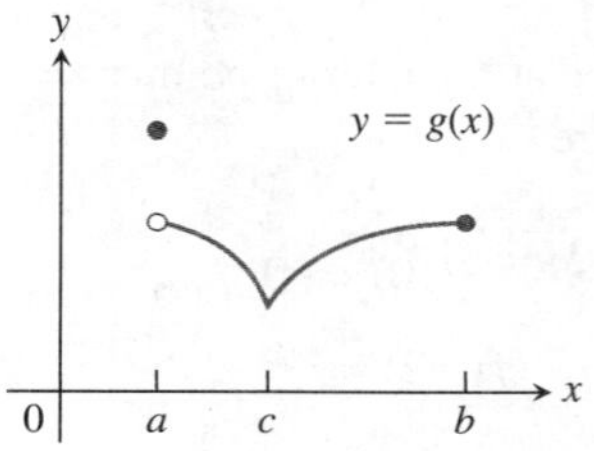

In Exercises 7–10, find the absolute extreme values and where they occur.

7.

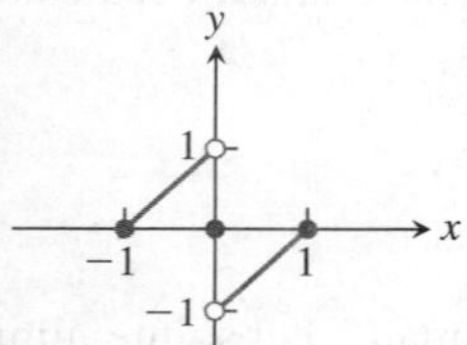

8.

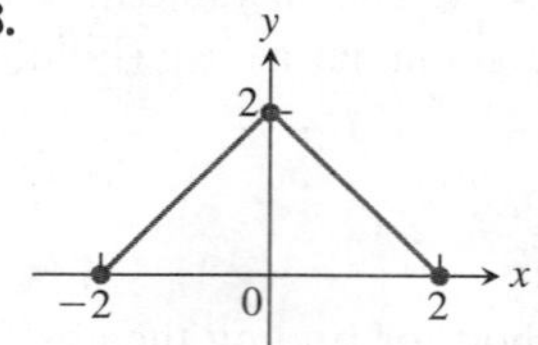

9.

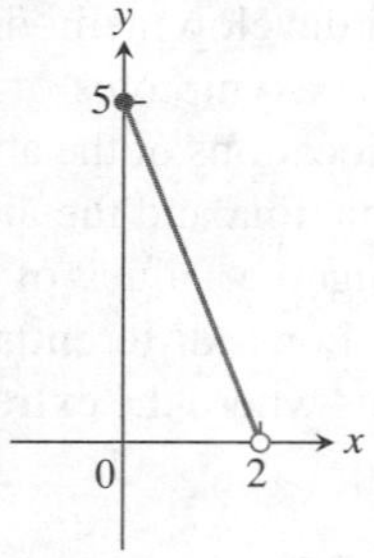

10.

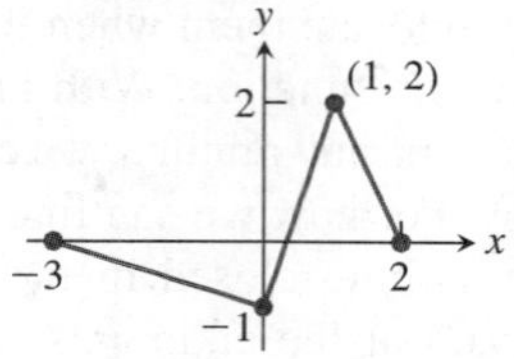

In Exercises 11–14, match the table with a graph.

11.

x	$f'(x)$
a	0
b	0
c	5

12.

x	$f'(x)$
a	0
b	0
c	−5

13.

x	$f'(x)$
a	does not exist
b	0
c	−2

14.

x	$f'(x)$
a	does not exist
b	does not exist
c	−1.7

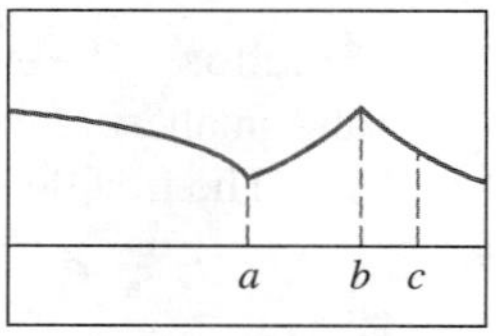

(a)

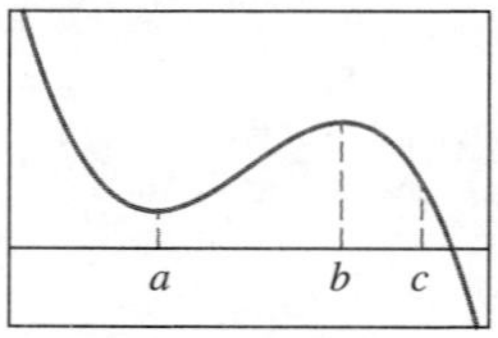

(b)

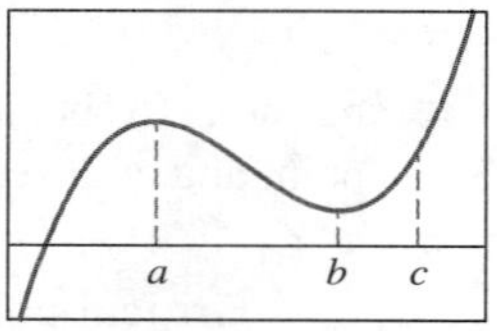

(c)

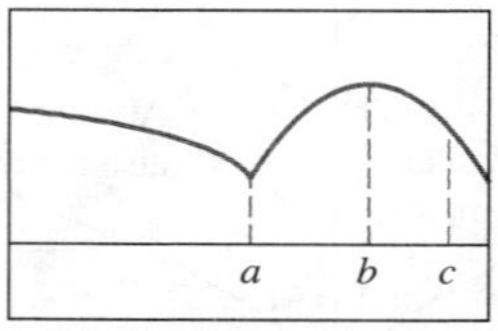

(d)

In Exercises 15–20, sketch the graph of each function and determine whether the function has any absolute extreme values on its domain. Explain how your answer is consistent with Theorem 1.

15. $f(x) = |x|, \quad -1 < x < 2$

16. $y = 2 - x^2, \quad -1 < x < 1$

17. $g(x) = \begin{cases} -x, & 0 \le x < 1 \\ x - 1, & 1 \le x \le 2 \end{cases}$

18. $h(x) = \begin{cases} \dfrac{1}{x}, & -1 \le x < 0 \\ \sqrt{x}, & 0 \le x \le 4 \end{cases}$

19. $y = 3 \sin x, \quad 0 < x < 2\pi$

20. $f(x) = \begin{cases} x + 1, & -1 \le x < 0 \\ \cos x, & 0 < x \le \dfrac{\pi}{2} \end{cases}$

Absolute Extrema on Finite Closed Intervals

In Exercises 21–36, find the absolute maximum and minimum values of each function on the given interval. Then graph the function. Identify the points on the graph where the absolute extrema occur, and include their coordinates.

21. $f(x) = \frac{2}{3}x - 5, \quad -2 \le x \le 3$

22. $f(x) = -x - 4, \quad -4 \le x \le 1$

23. $f(x) = x^2 - 1, \quad -1 \le x \le 2$

24. $f(x) = 4 - x^3, \quad -2 \le x \le 1$

25. $F(x) = -\frac{1}{x^2}, \quad 0.5 \le x \le 2$

26. $F(x) = -\frac{1}{x}, \quad -2 \le x \le -1$

27. $h(x) = \sqrt[3]{x}, \quad -1 \le x \le 8$

28. $h(x) = -3x^{2/3}, \quad -1 \le x \le 1$

29. $g(x) = \sqrt{4 - x^2}, \quad -2 \le x \le 1$

30. $g(x) = -\sqrt{5 - x^2}, \quad -\sqrt{5} \le x \le 0$

31. $f(\theta) = \sin\theta, \quad -\frac{\pi}{2} \le \theta \le \frac{5\pi}{6}$

32. $f(\theta) = \tan\theta, \quad -\frac{\pi}{3} \le \theta \le \frac{\pi}{4}$

33. $g(x) = \csc x, \quad \frac{\pi}{3} \le x \le \frac{2\pi}{3}$

34. $g(x) = \sec x, \quad -\frac{\pi}{3} \le x \le \frac{\pi}{6}$

35. $f(t) = 2 - |t|, \quad -1 \le t \le 3$

36. $f(t) = |t - 5|, \quad 4 \le t \le 7$

In Exercises 37–40, find the function's absolute maximum and minimum values and say where they occur.

37. $f(x) = x^{4/3}, \quad -1 \le x \le 8$

38. $f(x) = x^{5/3}, \quad -1 \le x \le 8$

39. $g(\theta) = \theta^{3/5}, \quad -32 \le \theta \le 1$

40. $h(\theta) = 3\theta^{2/3}, \quad -27 \le \theta \le 8$

Finding Critical Points

In Exercises 41–50, determine all critical points and all domain endpoints for each function.

41. $y = \sin x\,(1 + \cos x)$

42. $f(x) = 6x^2 - x^3$

43. $f(x) = x(4 - x)^3$

44. $g(x) = (x - 1)^2(x - 3)^2$

45. $y = x^2 + \frac{2}{x}$

46. $f(x) = \frac{x^2}{x - 2}$

47. $y = x^2 - 32\sqrt{x}$

48. $g(x) = \sqrt{2x - x^2}$

49. $y = x^3 + 3x^2 - 24x + 7$

50. $y = x - 3x^{2/3}$

Theory and Examples

In Exercises 51 and 52, give reasons for your answers.

51. Let $f(x) = (x - 2)^{2/3}$.

a. Does $f'(2)$ exist?

b. Show that the only local extreme value of f occurs at $x = 2$.

c. Does the result in part (b) contradict the Extreme Value Theorem?

d. Repeat parts (a) and (b) for $f(x) = (x - a)^{2/3}$, replacing 2 by a.

52. Let $f(x) = |x^3 - 9x|$.

a. Does $f'(0)$ exist?

b. Does $f'(3)$ exist?

c. Does $f'(-3)$ exist?

d. Determine all extrema of f.

In Exercises 53 and 54, show that the function has neither an absolute minimum nor an absolute maximum on its natural domain.

53. $y = x^{11} + x^3 + x - 5$

54. $y = 3x + \tan x$

55. A minimum with no derivative The function $f(x) = |x|$ has an absolute minimum value at $x = 0$ even though f is not differentiable at $x = 0$. Is this consistent with Theorem 2? Give reasons for your answer.

56. Even functions If an even function $f(x)$ has a local maximum value at $x = c$, can anything be said about the value of f at $x = -c$? Give reasons for your answer.

57. Odd functions If an odd function $g(x)$ has a local minimum value at $x = c$, can anything be said about the value of g at $x = -c$? Give reasons for your answer.

58. No critical points or endpoints exist We know how to find the extreme values of a continuous function $f(x)$ by investigating its values at critical points and endpoints. But what if there *are* no critical points or endpoints? What happens then? Do such functions really exist? Give reasons for your answers.

59. Find the maximum possible slope for a tangent line to the curve

$$y = \frac{8}{1 + 3e^{-x}}$$

60. Cubic functions Consider the cubic function

$$f(x) = ax^3 + bx^2 + cx + d.$$

a. Show that f can have 0, 1, or 2 critical points. Give examples and graphs to support your argument.

b. How many local extreme values can f have?

61. Maximum height of a vertically moving body The height of a body moving vertically is given by

$$s = -\frac{1}{2}gt^2 + v_0 t + s_0, \qquad g > 0,$$

with s in meters and t in seconds. Find the body's maximum height.

62. Let $f(x) = \begin{cases} |x-2| + a^2 - 2a - 9 & \text{if } x > 2 \\ 2x - 3 & \text{if } x \ge 2 \end{cases}$

Find the values of a if $f(x)$ has a local minimum at $x = 2$

T Graph the functions in Exercises 63–66. Then find the extreme values of the function on the interval and say where they occur.

63. $f(x) = |x - 2| + |x + 3|, \quad -5 \le x \le 5$

64. $g(x) = |x - 1| - |x - 5|, \quad -2 \le x \le 7$

65. $h(x) = |x + 2| - |x - 3|, \quad -\infty < x < \infty$

66. $k(x) = |x + 1| + |x - 3|, \quad -\infty < x < \infty$

COMPUTER EXPLORATIONS

In Exercises 67–72, you will use a CAS to help find the absolute extrema of the given function over the specified closed interval. Perform the following steps.

a. Plot the function over the interval to see its general behavior there.

b. Find the interior points where $f' = 0$. (In some exercises, you may have to use the numerical equation solver to approximate a solution.) You may want to plot f' as well.

c. Find the interior points where f' does not exist.

d. Evaluate the function at all points found in parts (b) and (c) and at the endpoints of the interval.

e. Find the function's absolute extreme values on the interval and identify where they occur.

67. $f(x) = x^4 - 8x^2 + 4x + 2, \quad [-20/25, 64/25]$

68. $f(x) = -x^4 + 4x^3 - 4x + 1, \quad [-3/4, 3]$

69. $f(x) = x^{2/3}(3 - x), \quad [-2, 2]$

70. $f(x) = 2 + 2x - 3x^{2/3}, \quad [-1, 10/3]$

71. $f(x) = \sqrt{x} + \cos x, \quad [0, 2\pi]$

72. $f(x) = x^{3/4} - \sin x + \frac{1}{2}, \quad [0, 2\pi]$

4.2 The Mean Value Theorem

We know that constant functions have zero derivatives, but could there be a more complicated function whose derivative is always zero? If two functions have identical derivatives over an interval, how are the functions related? We answer these and other questions in this chapter by applying the Mean Value Theorem. First we introduce a special case, known as Rolle's Theorem, which is used to prove the Mean Value Theorem.

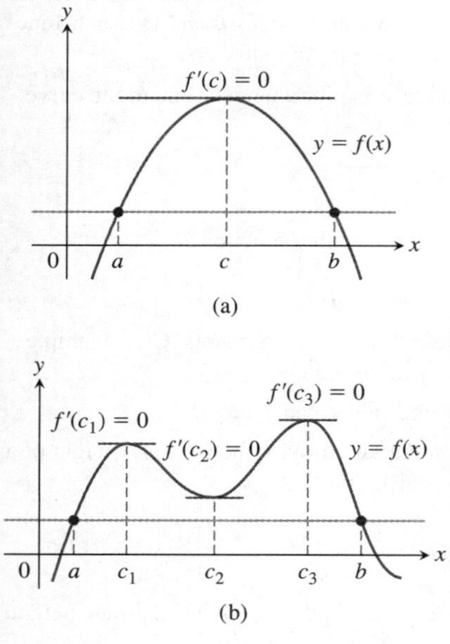

FIGURE 4.10 Rolle's Theorem says that a differentiable curve has at least one horizontal tangent between any two points where it crosses a horizontal line. It may have just one (a), or it may have more (b).

Rolle's Theorem

As suggested by its graph, if a differentiable function crosses a horizontal line at two different points, there is at least one point between them where the tangent to the graph is horizontal and the derivative is zero (Figure 4.10). We now state and prove this result.

> **THEOREM 3—Rolle's Theorem**
> Suppose that $y = f(x)$ is continuous over the closed interval $[a, b]$ and differentiable at every point of its interior (a, b). If $f(a) = f(b)$, then there is at least one number c in (a, b) at which $f'(c) = 0$.

Proof Being continuous, f assumes absolute maximum and minimum values on $[a, b]$ by Theorem 1. These can occur only

1. at interior points where f' is zero,
2. at interior points where f' does not exist,
3. at endpoints of the interval, in this case a and b.

By hypothesis, f has a derivative at every interior point. That rules out possibility (2), leaving us with interior points where $f' = 0$ and with the two endpoints a and b.

HISTORICAL BIOGRAPHY
Michel Rolle
(1652–1719)
www.bit.ly/2P2MFW7

If either the maximum or the minimum occurs at a point c between a and b, then $f'(c) = 0$ by Theorem 2 in Section 4.1, and we have found a point for Rolle's Theorem.

If both the absolute maximum and the absolute minimum occur at the endpoints, then because $f(a) = f(b)$ it must be the case that f is a constant function with $f(x) = f(a) = f(b)$ for every $x \in [a, b]$. Therefore, $f'(x) = 0$ and the point c can be taken anywhere in the interior (a, b). ■

The hypotheses of Theorem 3 are essential. If they fail at even one point, the graph may not have a horizontal tangent (Figure 4.11).

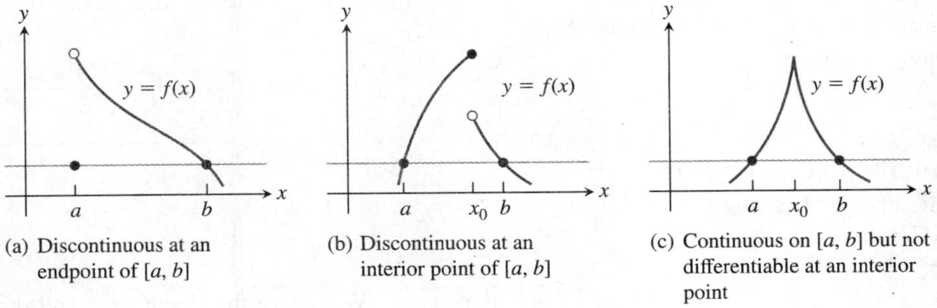

FIGURE 4.11 There may be no horizontal tangent line if the hypotheses of Rolle's Theorem do not hold.

Rolle's Theorem may be combined with the Intermediate Value Theorem to show when there is only one real solution of an equation $f(x) = 0$, as we illustrate in the next example.

EXAMPLE 1 Show that the equation

$$x^3 + 3x + 1 = 0$$

has exactly one real solution.

Solution We define the continuous function

$$f(x) = x^3 + 3x + 1.$$

Since $f(-1) = -3$ and $f(0) = 1$, the Intermediate Value Theorem tells us that the graph of f crosses the x-axis somewhere in the open interval $(-1, 0)$. (See Figure 4.12.) Now, if there were even two points $x = a$ and $x = b$ where $f(x)$ was zero, Rolle's Theorem would guarantee the existence of a point $x = c$ between them where f' was zero. However, the derivative

$$f'(x) = 3x^2 + 3$$

is never zero (because it is always positive). Therefore, f has no more than one zero. ■

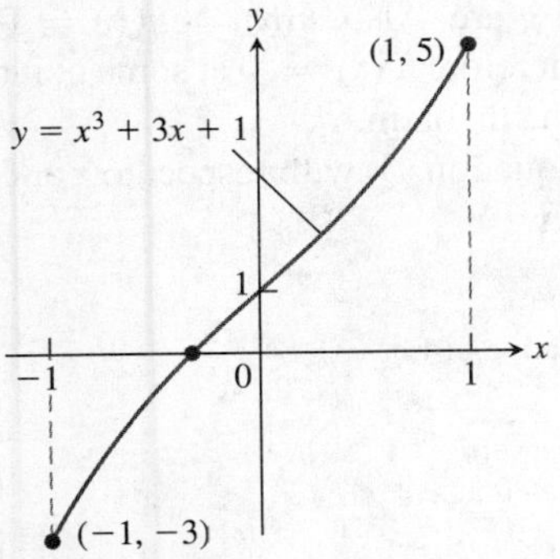

FIGURE 4.12 The only real zero of the polynomial $y = x^3 + 3x + 1$ is the one shown here where the curve crosses the x-axis between −1 and 0 (Example 1).

We will use Rolle's Theorem to prove the Mean Value Theorem.

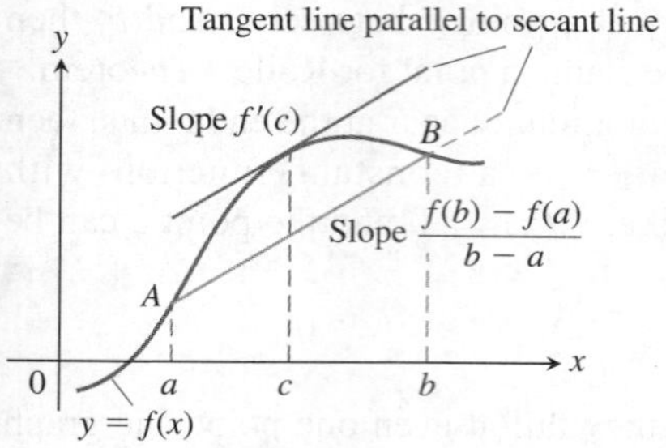

FIGURE 4.13 Geometrically, the Mean Value Theorem says that somewhere between a and b the curve has at least one tangent line parallel to the secant line that joins A and B.

The Mean Value Theorem

The Mean Value Theorem, which was first stated by Joseph-Louis Lagrange, is a slanted version of Rolle's Theorem (Figure 4.13). The Mean Value Theorem guarantees that there is a point where the tangent line is parallel to the secant line that joins A and B.

THEOREM 4—The Mean Value Theorem
Suppose $y = f(x)$ is continuous over a closed interval $[a, b]$ and differentiable on the interval's interior (a, b). Then there is at least one point c in (a, b) at which

$$\frac{f(b) - f(a)}{b - a} = f'(c). \tag{1}$$

HISTORICAL BIOGRAPHY
Joseph-Louis Lagrange
(1736–1813)
www.bit.ly/2IsuzKu

Proof We picture the graph of f and draw a line through the points $A(a, f(a))$ and $B(b, f(b))$. (See Figure 4.14.) The secant line is the graph of the function

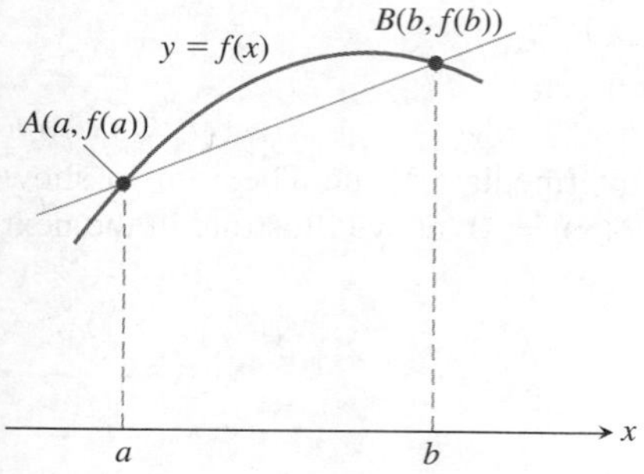

FIGURE 4.14 The graph of f and the secant line AB over the interval $[a, b]$.

$$g(x) = f(a) + \frac{f(b) - f(a)}{b - a}(x - a) \tag{2}$$

(point-slope equation). The vertical difference between the graphs of f and g at x is

$$\begin{aligned} h(x) &= f(x) - g(x) \\ &= f(x) - f(a) - \frac{f(b) - f(a)}{b - a}(x - a). \end{aligned} \tag{3}$$

Figure 4.15 shows the graphs of f, g, and h together.

The function h satisfies the hypotheses of Rolle's Theorem on $[a, b]$. It is continuous on $[a, b]$ and differentiable on (a, b) because both f and g are. Also, $h(a) = h(b) = 0$ because the graphs of f and g both pass through A and B. Therefore $h'(c) = 0$ at some point $c \in (a, b)$. This is the point we want for Equation (1) in the theorem.

To verify Equation (1), we differentiate both sides of Equation (3) with respect to x and then set $x = c$:

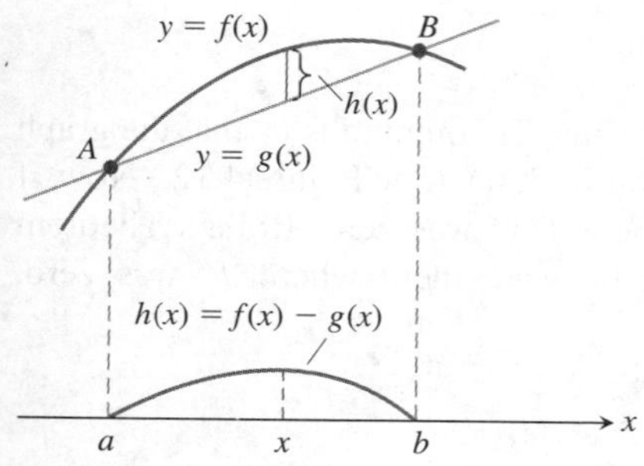

FIGURE 4.15 The secant line AB is the graph of the function $g(x)$. The function $h(x) = f(x) - g(x)$ gives the vertical distance between the graphs of f and g at x.

$$h'(x) = f'(x) - \frac{f(b) - f(a)}{b - a} \qquad \text{Derivative of Eq. (3)}$$

$$h'(c) = f'(c) - \frac{f(b) - f(a)}{b - a} \qquad \text{Evaluated at } x = c$$

$$0 = f'(c) - \frac{f(b) - f(a)}{b - a} \qquad h'(c) = 0$$

$$f'(c) = \frac{f(b) - f(a)}{b - a}, \qquad \text{Rearranged}$$

which is what we set out to prove. ■

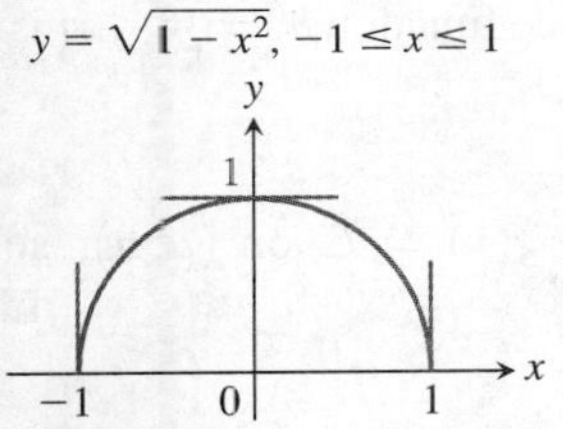

FIGURE 4.16 The function $f(x) = \sqrt{1 - x^2}$ satisfies the hypotheses (and conclusion) of the Mean Value Theorem on $[-1, 1]$ even though f is not differentiable at -1 and 1.

The hypotheses of the Mean Value Theorem do not require f to be differentiable at either a or b. One-sided continuity at a and b is enough (Figure 4.16).

EXAMPLE 2 The function $f(x) = x^2$ (Figure 4.17) is continuous for $0 \le x \le 2$ and differentiable for $0 < x < 2$. Since $f(0) = 0$ and $f(2) = 4$, the Mean Value Theorem says that at some point c in the interval, the derivative $f'(x) = 2x$ must have the value $(4 - 0)/(2 - 0) = 2$. In this case we can identify c by solving the equation $2c = 2$ to get $c = 1$. However, it is not always easy to find c algebraically, even though we know it always exists. ■

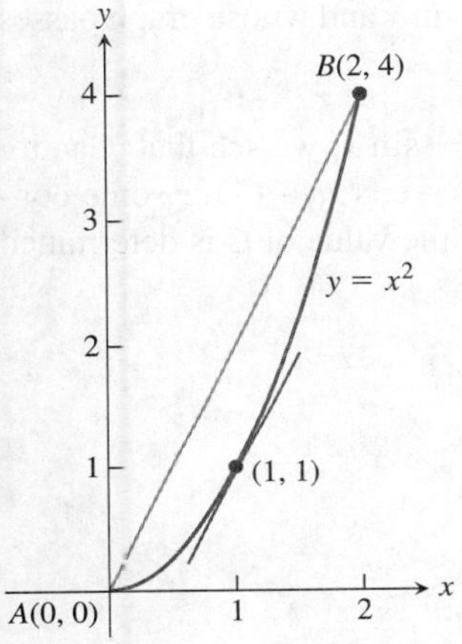

FIGURE 4.17 As we find in Example 1, $c = 1$ is where the tangent line is parallel to the secant line.

A Physical Interpretation

We can think of the number $(f(b) - f(a))/(b - a)$ as the average change in f over $[a, b]$ and can view $f'(c)$ as an instantaneous change. Then the Mean Value Theorem says that the instantaneous change at some interior point is equal to the average change over the entire interval.

EXAMPLE 3 If a car accelerating from zero takes 8 sec to go 352 ft, its average velocity for the 8-sec interval is $352/8 = 44$ ft/sec. The Mean Value Theorem says that at some point during the acceleration, the speedometer must read exactly 30 mph (44 ft/sec) (Figure 4.18). ■

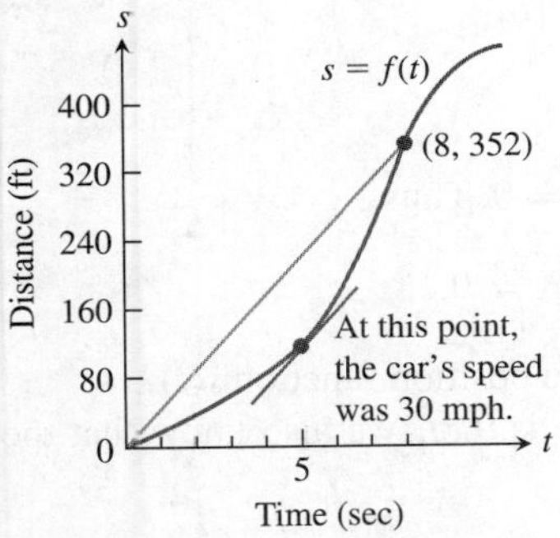

FIGURE 4.18 Distance versus elapsed time for the car in Example 3.

Mathematical Consequences

At the beginning of the section, we asked what kind of function has a zero derivative over an interval. The first corollary of the Mean Value Theorem provides the answer that only constant functions have zero derivatives.

COROLLARY 1 If $f'(x) = 0$ at each point x of an open interval (a, b), then $f(x) = C$ for all $x \in (a, b)$, where C is a constant.

Proof We want to show that f has a constant value on the interval (a, b). We do so by showing that if x_1 and x_2 are any two points in (a, b) with $x_1 < x_2$, then $f(x_1) = f(x_2)$. Now f satisfies the hypotheses of the Mean Value Theorem on $[x_1, x_2]$: It is differentiable at every point of $[x_1, x_2]$ and hence continuous at every point as well. Therefore,

$$\frac{f(x_2) - f(x_1)}{x_2 - x_1} = f'(c)$$

at some point c between x_1 and x_2. Since $f' = 0$ throughout (a, b), this equation implies successively that

$$\frac{f(x_2) - f(x_1)}{x_2 - x_1} = 0, \quad f(x_2) - f(x_1) = 0, \quad \text{and} \quad f(x_1) = f(x_2). \quad ■$$

At the beginning of this section, we also asked about the relationship between two functions that have identical derivatives over an interval. The next corollary tells us that their values on the interval have a constant difference.

COROLLARY 2 If $f'(x) = g'(x)$ at each point x in an open interval (a, b), then there exists a constant C such that $f(x) = g(x) + C$ for all $x \in (a, b)$. That is, $f - g$ is a constant function on (a, b).

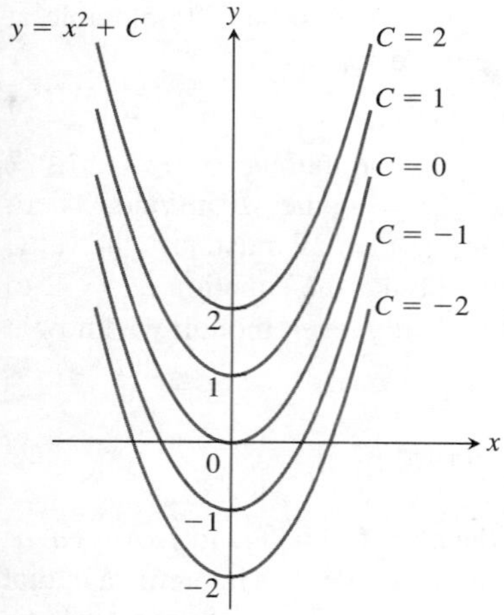

FIGURE 4.19 From a geometric point of view, Corollary 2 of the Mean Value Theorem says that the graphs of functions with identical derivatives on an interval can differ only by a vertical shift. The graphs of the functions with derivative $2x$ are the parabolas $y = x^2 + C$, shown here for several values of C.

Proof At each point $x \in (a, b)$ the derivative of the difference function $h = f - g$ is

$$h'(x) = f'(x) - g'(x) = 0.$$

Thus, $h(x) = C$ on (a, b) by Corollary 1. That is, $f(x) - g(x) = C$ on (a, b), so $f(x) = g(x) + C$. ■

Corollaries 1 and 2 are also true if the open interval (a, b) fails to be finite. That is, they remain true if the interval is (a, ∞), $(-\infty, b)$, or $(-\infty, \infty)$.

Corollary 2 will play an important role when we discuss antiderivatives in Section 4.8. It tells us, for instance, that since the derivative of $f(x) = x^2$ on $(-\infty, \infty)$ is $2x$, any other function with derivative $2x$ on $(-\infty, \infty)$ must have the formula $x^2 + C$ for some value of C (Figure 4.19).

EXAMPLE 4 Find the function $f(x)$ whose derivative is $\sin x$ and whose graph passes through the point $(0, 2)$.

Solution Since the derivative of $g(x) = -\cos x$ is $g'(x) = \sin x$, we see that f and g have the same derivative. Corollary 2 then says that $f(x) = -\cos x + C$ for some constant C. Since the graph of f passes through the point $(0, 2)$, the value of C is determined from the condition that $f(0) = 2$:

$$f(0) = -\cos(0) + C = 2, \quad \text{so} \quad C = 3.$$

The function is $f(x) = -\cos x + 3$. ■

Finding Velocity and Position from Acceleration

We can use Corollary 2 to find the velocity and position functions of an object moving along a vertical line. Assume the object or body is falling freely from rest with acceleration $9.8\ \text{m/sec}^2$. We assume the position $s(t)$ of the body is measured positive downward from the rest position (so the vertical coordinate line points *downward*, in the direction of the motion, with the rest position at 0).

We know that the velocity $v(t)$ is some function whose derivative is 9.8. We also know that the derivative of $g(t) = 9.8t$ is 9.8. By Corollary 2,

$$v(t) = 9.8t + C$$

for some constant C. Since the body falls from rest, $v(0) = 0$. Thus

$$9.8(0) + C = 0, \quad \text{and} \quad C = 0.$$

The velocity function must be $v(t) = 9.8t$. What about the position function $s(t)$?

We know that $s(t)$ is some function whose derivative is $9.8t$. We also know that the derivative of $f(t) = 4.9t^2$ is $9.8t$. By Corollary 2,

$$s(t) = 4.9t^2 + C$$

for some constant C. Since $s(0) = 0$,

$$4.9(0)^2 + C = 0, \quad \text{and} \quad C = 0.$$

The position function is $s(t) = 4.9t^2$ until the body hits the ground.

The ability to find functions from their rates of change is one of the very powerful tools of calculus. As we will see, it lies at the heart of the mathematical developments in Chapter 5.

EXERCISES 4.2

Checking the Mean Value Theorem

Find the value or values of c that satisfy the equation

$$\frac{f(b) - f(a)}{b - a} = f'(c)$$

in the conclusion of the Mean Value Theorem for the functions and intervals in Exercises 1–6.

1. $f(x) = x^2 + 2x - 1,\ [0,1]$ **2.** $f(x) = x^{2/3},\ [0,1]$

3. $f(x) = x + \dfrac{1}{x},\ \left[\dfrac{1}{2}, 2\right]$ **4.** $f(x) = \sqrt{x - 1},\ [1,3]$

5. $f(x) = x^3 - x^2,\ [-1,2]$

6. $g(x) = \begin{cases} x^3, & -2 \le x \le 0 \\ x^2, & 0 < x \le 2 \end{cases}$

Which of the functions in Exercises 7–12 satisfy the hypotheses of the Mean Value Theorem on the given interval, and which do not? Give reasons for your answers.

7. $f(x) = x^{2/3},\ [-1,8]$

8. $f(x) = x^{4/5},\ [0,1]$

9. $f(x) = \sqrt{x(1 - x)},\ [0,1]$

10. $f(x) = \begin{cases} \dfrac{\sin x}{x}, & -\pi \le x < 0 \\ 0, & x = 0 \end{cases}$

11. $f(x) = \begin{cases} x^2 - x, & -2 \le x \le -1 \\ 2x^2 - 3x - 3, & -1 < x \le 0 \end{cases}$

12. $f(x) = \begin{cases} 2x - 3, & 0 \le x \le 2 \\ 6x - x^2 - 7, & 2 < x \le 3 \end{cases}$

13. Prove that $\dfrac{b-a}{b} < \log_e \dfrac{b}{a} < \dfrac{b-a}{a}$ for $0 < a < b$.

Hence, deduce that

(i) $\dfrac{1}{4} < \log_e \dfrac{4}{3} < \dfrac{1}{3}$ **(ii)** $\dfrac{1}{6} < \log_e (1.2) < \dfrac{1}{5}$

(iii) $\dfrac{1}{2} < \log_e 2 < 1$

14. For what values of a, m, and b does the function

$$f(x) = \begin{cases} 3, & x = 0 \\ -x^2 + 3x + a, & 0 < x < 1 \\ mx + b, & 1 \le x \le 2 \end{cases}$$

satisfy the hypotheses of the Mean Value Theorem on the interval $[0, 2]$?

Roots (Zeros)

15. a. Plot the zeros of each polynomial on a line together with the zeros of its first derivative.

i) $y = x^2 - 4$

ii) $y = x^2 + 8x + 15$

iii) $y = x^3 - 3x^2 + 4 = (x + 1)(x - 2)^2$

iv) $y = x^3 - 33x^2 + 216x = x(x - 9)(x - 24)$

b. Use Rolle's Theorem to prove that between every two zeros of $x^n + a_{n-1}x^{n-1} + \cdots + a_1 x + a_0$ there lies a zero of $nx^{n-1} + (n - 1)a_{n-1}x^{n-2} + \cdots + a_1$.

16. Suppose that f'' is continuous on $[a, b]$ and that f has three zeros in the interval. Show that f'' has at least one zero in (a, b). Generalize this result.

17. Show that if $f'' > 0$ throughout an interval $[a, b]$, then f' has at most one zero in $[a, b]$. What if $f'' < 0$ throughout $[a, b]$ instead?

18. Show that a cubic polynomial can have at most three real zeros.

Show that the functions in Exercises 19–26 have exactly one zero in the given interval.

19. $f(x) = x^4 + 3x + 1,\ [-2, -1]$

20. $f(x) = x^3 + \dfrac{4}{x^2} + 7,\ (-\infty, 0)$

21. $g(t) = \sqrt{t} + \sqrt{1 + t} - 4,\ (0, \infty)$

22. $g(t) = \dfrac{1}{1 - t} + \sqrt{1 + t} - 3.1,\ (-1, 1)$

23. $r(\theta) = \theta + \sin^2\left(\dfrac{\theta}{3}\right) - 8,\ (-\infty, \infty)$

24. $r(\theta) = 2\theta - \cos^2\theta + \sqrt{2},\ (-\infty, \infty)$

25. $r(\theta) = \sec^2\theta - \cos(2\theta) - 1,\ (0, \pi/2)$

26. $r(\theta) = 3\tan\theta - \cot\theta - \theta,\ (0, \pi/2)$

Finding Functions from Derivatives

27. Suppose that $f(-1) = 3$ and that $f'(x) = 0$ for all x. Must $f(x) = 3$ for all x? Give reasons for your answer.

28. Suppose that $f(0) = 5$ and that $f'(x) = 2$ for all x. Must $f(x) = 2x + 5$ for all x? Give reasons for your answer.

29. Suppose that $f'(x) = 2x$ for all x. Find $f(2)$ if

a. $f(0) = 0$ **b.** $f(1) = 0$ **c.** $f(-2) = 3$.

30. If $0 < \alpha < \beta < \dfrac{\pi}{2}$, using Cauchy's mean value theorem, prove that $\dfrac{\sin\alpha - \sin\beta}{\cos\beta - \cos\alpha} = \cot\theta$ for some $\theta \in (\alpha, \beta)$.

In Exercises 31–36, find all possible functions with the given derivative.

31. a. $y' = x$ **b.** $y' = x^2$ **c.** $y' = x^3$

32. a. $y' = 2x$ **b.** $y' = 2x - 1$ **c.** $y' = 3x^2 + 2x - 1$

33. a. $y' = -\dfrac{1}{x^2}$ **b.** $y' = 1 - \dfrac{1}{x^2}$ **c.** $y' = 5 + \dfrac{1}{x^2}$

34. a. $y' = \dfrac{1}{2\sqrt{x}}$ **b.** $y' = \dfrac{1}{\sqrt{x}}$ **c.** $y' = 4x - \dfrac{1}{\sqrt{x}}$

35. a. $y' = \sin 2t$ **b.** $y' = \cos\dfrac{t}{2}$ **c.** $y' = \sin 2t + \cos\dfrac{t}{2}$

36. a. $y' = \sec^2\theta$ **b.** $y' = \sqrt{\theta}$ **c.** $y' = \sqrt{\theta} - \sec^2\theta$

In Exercises 37–40, find the function with the given derivative whose graph passes through the point P.

37. $f'(x) = 2x - 1,\ P(0, 0)$

38. $g'(x) = \dfrac{1}{x^2} + 2x,\ P(-1, 1)$

39. $r'(\theta) = 8 - \csc^2\theta,\quad P\left(\frac{\pi}{4}, 0\right)$

40. $r'(t) = \sec t \tan t - 1,\quad P(0, 0)$

Finding Position from Velocity or Acceleration

Exercises 41–44 give the velocity $v = ds/dt$ and initial position of an object moving along a coordinate line. Find the object's position at time t.

41. $v = 9.8t + 5,\quad s(0) = 10$ **42.** $v = 32t - 2,\quad s(0.5) = 4$

43. $v = \sin \pi t,\quad s(0) = 0$ **44.** $v = \frac{2}{\pi}\cos\frac{2t}{\pi},\quad s(\pi^2) = 1$

Exercises 45–48 give the acceleration $a = d^2s/dt^2$, initial velocity, and initial position of an object moving on a coordinate line. Find the object's position at time t.

45. $a = 32,\quad v(0) = 20,\quad s(0) = 5$

46. $a = 9.8,\quad v(0) = -3,\quad s(0) = 0$

47. $a = -4\sin 2t,\quad v(0) = 2,\quad s(0) = -3$

48. $a = \frac{9}{\pi^2}\cos\frac{3t}{\pi},\quad v(0) = 0,\quad s(0) = -1$

Applications

49. Temperature change It took 14 sec for a mercury thermometer to rise from $-19°$C to $100°$C when it was taken from a freezer and placed in boiling water. Show that somewhere along the way, the mercury was rising at the rate of $8.5°$C/sec.

50. A trucker handed in a ticket at a toll booth showing that in 2 hours she had covered 159 mi on a toll road with speed limit 65 mph. The trucker was cited for speeding. Why?

51. Classical accounts tell us that a 170-oar trireme (ancient Greek or Roman warship) once covered 184 sea miles in 24 hours. Explain why at some point during this feat the trireme's speed exceeded 7.5 knots (sea or nautical miles per hour).

52. A marathoner ran the 26.2-mi New York City Marathon in 2.2 hours. Show that at least twice the marathoner was running at exactly 11 mph, assuming the initial and final speeds are zero.

53. Show that at some instant during a 2-hour automobile trip the car's speedometer reading will equal the average speed for the trip.

54. Free fall on the moon On our moon, the acceleration of gravity is 1.6 m/sec^2. If a rock is dropped into a crevasse, how fast will it be going just before it hits bottom 30 sec later?

Theory and Examples

55. The geometric mean of a and b The *geometric mean* of two positive numbers a and b is the number $\sqrt{ab}$. Show that the value of c in the conclusion of the Mean Value Theorem for $f(x) = 1/x$ on an interval of positive numbers $[a, b]$ is $c = \sqrt{ab}$.

56. The arithmetic mean of a and b The *arithmetic mean* of two numbers a and b is the number $(a + b)/2$. Show that the value of c in the conclusion of the Mean Value Theorem for $f(x) = x^2$ on any interval $[a, b]$ is $c = (a + b)/2$.

T **57.** Graph the function

$$f(x) = \sin x \sin(x + 2) - \sin^2(x + 1).$$

What does the graph do? Why does the function behave this way? Give reasons for your answers.

58. Rolle's Theorem

a. Construct a polynomial $f(x)$ that has zeros at $x = -2, -1, 0, 1$, and 2.

b. Graph f and its derivative f' together. How is what you see related to Rolle's Theorem?

c. Do $g(x) = \sin x$ and its derivative g' illustrate the same phenomenon as f and f'?

59. Unique solution Assume that f is continuous on $[a, b]$ and differentiable on (a, b). Also assume that $f(a)$ and $f(b)$ have opposite signs and that $f' \neq 0$ between a and b. Show that $f(x) = 0$ exactly once between a and b.

60. Parallel tangent line Assume that f and g are differentiable on $[a, b]$ and that $f(a) = g(a)$ and $f(b) = g(b)$. Show that there is at least one point between a and b where the tangent lines to the graphs of f and g are parallel or the same line. Illustrate with a sketch.

61. Suppose that $f'(x) \leq 1$ for $1 \leq x \leq 4$. Show that $f(4) - f(1) \leq 3$.

62. Suppose that $0 < f'(x) < 1/2$ for all x-values. Show that $f(-1) < f(1) < 2 + f(-1)$.

63. Show that $|\cos x - 1| \leq |x|$ for all x-values. (*Hint:* Consider $f(t) = \cos t$ on the closed interval with the endpoints 0 and x.)

64. Show that for any numbers a and b, the sine inequality $|\sin b - \sin a| \leq |b - a|$ is true.

65. For any two different real numbers a and b ($a < b$), prove that $a^2 + ab + b^2 = 3c^2$, for some $c \in (a, b)$ using Lagrange's mean value theorem.

66. If $|f(w) - f(x)| \leq |w - x|$ for all values w and x and f is a differentiable function, show that $-1 \leq f'(x) \leq 1$ for all x-values.

67. Assume that f is differentiable on $a \leq x \leq b$ and that $f(b) < f(a)$. Show that f' is negative at some point between a and b.

68. Let f be a function defined on an interval $[a, b]$. What conditions could you place on f to guarantee that

$$\min f' \leq \frac{f(b) - f(a)}{b - a} \leq \max f',$$

where $\min f'$ and $\max f'$ refer to the minimum and maximum values of f' on $[a, b]$? Give reasons for your answers.

T **69.** Use the inequalities in Exercise 68 to estimate $f(0.1)$ if $f'(x) = 1/(1 + x^4\cos x)$ for $0 \leq x \leq 0.1$ and $f(0) = 1$.

T **70.** Use the inequalities in Exercise 68 to estimate $f(0.1)$ if $f'(x) = 1/(1 - x^4)$ for $0 \leq x \leq 0.1$ and $f(0) = 2$.

71. If $f : [0, 4] \to R$ is differentiable, then prove that

$[f(4)]^2 - [f(0)]^2 = 8f'(a)f(b)$ for $a, b \in (0, 4)$.

72. Prove that between any two roots of $e^x \sin x = 1$, there is at least one root of $e^x \cos x + 1 = 0$.

4.3 Monotonic Functions and the First Derivative Test

In sketching the graph of a differentiable function, it is useful to know where it increases (rises from left to right) and where it decreases (falls from left to right) over an interval. This section gives a test to determine where it increases and where it decreases. We also show how to test the critical points of a function to identify whether local extreme values are present.

Increasing Functions and Decreasing Functions

As another corollary to the Mean Value Theorem, we show that functions with positive derivatives are increasing functions and functions with negative derivatives are decreasing functions. A function that is either increasing on an interval or decreasing on an interval is said to be **monotonic** on the interval.

COROLLARY 3 Suppose that f is continuous on $[a, b]$ and differentiable on (a, b).

If $f'(x) > 0$ at each point $x \in (a, b)$, then f is increasing on $[a, b]$.
If $f'(x) < 0$ at each point $x \in (a, b)$, then f is decreasing on $[a, b]$.

Proof Let x_1 and x_2 be any two points in $[a, b]$ with $x_1 < x_2$. The Mean Value Theorem applied to f on $[x_1, x_2]$ says that

$$f(x_2) - f(x_1) = f'(c)(x_2 - x_1)$$

for some c between x_1 and x_2. The sign of the right-hand side of this equation is the same as the sign of $f'(c)$ because $x_2 - x_1$ is positive. Therefore, $f(x_2) > f(x_1)$ if f' is positive on (a, b) and $f(x_2) < f(x_1)$ if f' is negative on (a, b). ■

Corollary 3 tells us that $f(x) = \sqrt{x}$ is increasing on the interval $[0, b]$ for any $b > 0$ because $f'(x) = 1/(2\sqrt{x})$ is positive on $(0, b)$. The derivative does not exist at $x = 0$, but Corollary 3 still applies. The corollary is also valid for open and for infinite intervals, so $f(x) = \sqrt{x}$ is also increasing on $(0, 1)$, on $(0, \infty)$, and on $[0, \infty)$.

To find the intervals where a function f is increasing or decreasing, we first find all of the critical points of f. If $a < b$ are two critical points for f, and if the derivative f' is continuous but never zero on the interval (a, b), then by the Intermediate Value Theorem applied to f', the derivative must be everywhere positive on (a, b), or everywhere negative there. One way we can determine the sign of f' on (a, b) is simply by evaluating the derivative at a single point c in (a, b). If $f'(c) > 0$, then $f'(x) > 0$ for all x in (a, b) so f is increasing on $[a, b]$ by Corollary 3; if $f'(c) < 0$, then f is decreasing on $[a, b]$. It doesn't matter which point c we choose in (a, b) since the sign of $f'(c)$ is the same for all choices. Usually we pick c to be a point where it is easy to evaluate $f'(c)$. The next example illustrates how we use this procedure.

EXAMPLE 1 Find the critical points of $f(x) = x^3 - 12x - 5$ and identify the open intervals on which f is increasing and those on which f is decreasing.

Solution The function f is everywhere continuous and differentiable. The first derivative

$$\begin{aligned} f'(x) &= 3x^2 - 12 = 3(x^2 - 4) \\ &= 3(x + 2)(x - 2) \end{aligned}$$

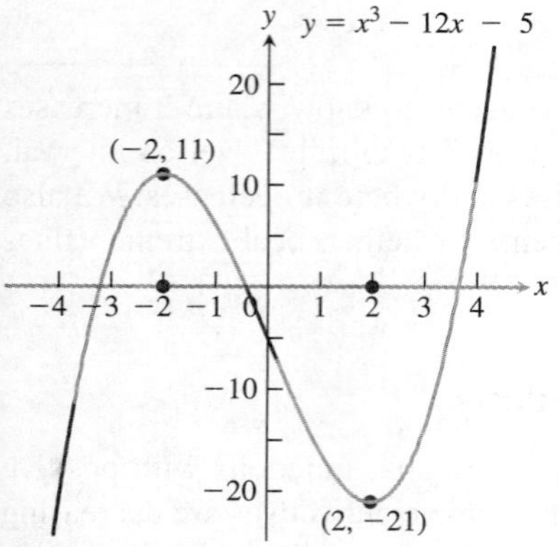

FIGURE 4.20 The function $f(x) = x^3 - 12x - 5$ is monotonic on three separate intervals (Example 1).

is zero at $x = -2$ and $x = 2$. These critical points subdivide the domain of f to create nonoverlapping open intervals $(-\infty, -2), (-2, 2)$, and $(2, \infty)$ on which f' is either positive or negative. We determine the sign of f' by evaluating f' at a convenient point in each subinterval. We evaluate f' at $x = -3$ in the first interval, $x = 0$ in the second interval, and $x = 3$ in the third since f' is relatively easy to compute at these points. The behavior of f is determined by then applying Corollary 3 to each subinterval. The results are summarized in the following table, and the graph of f is given in Figure 4.20.

Interval	$-\infty < x < -2$	$-2 < x < 2$	$2 < x < \infty$
f' evaluated	$f'(-3) = 15$	$f'(0) = -12$	$f'(3) = 15$
Sign of f'	+	−	+
Behavior of f	increasing	decreasing	increasing

∎

We used "strict" less-than inequalities to identify the intervals in the summary table for Example 1, since open intervals were specified. Corollary 3 says that we could use $\le$ inequalities as well. That is, the function f in the example is increasing on $-\infty < x \le -2$, decreasing on $-2 \le x \le 2$, and increasing on $2 \le x < \infty$. We do not talk about whether a function is increasing or decreasing at a single point.

HISTORICAL BIOGRAPHY
Edmund Halley
(1656–1742)
`www.bit.ly/2Ow8sIu`

First Derivative Test for Local Extrema

In Figure 4.21, at the points where f has a minimum value, $f' < 0$ immediately to the left and $f' > 0$ immediately to the right. (If the point is an endpoint, there is only one side to consider.) Thus, the function is decreasing on the left of the minimum value and it is increasing on its right. Similarly, at the points where f has a maximum value, $f' > 0$ immediately to the left and $f' < 0$ immediately to the right. Thus, the function is increasing on the left of the maximum value and decreasing on its right. In summary, at a local extreme point, the sign of $f'(x)$ changes.

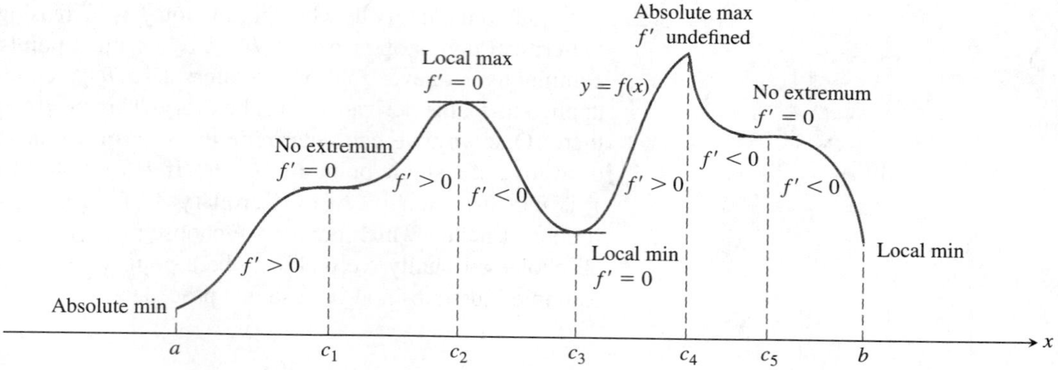

FIGURE 4.21 The critical points of a function locate where it is increasing and where it is decreasing. The first derivative changes sign at a critical point where a local extremum occurs.

These observations lead to a test for the presence and nature of local extreme values of differentiable functions.

First Derivative Test for Local Extrema

Suppose that c is a critical point of a continuous function f, and that f is differentiable at every point in some interval containing c except possibly at c itself. Moving across this interval from left to right,

1. if f' changes from negative to positive at c, then f has a local minimum at c;
2. if f' changes from positive to negative at c, then f has a local maximum at c;
3. if f' does not change sign at c (that is, f' is positive on both sides of c or negative on both sides), then f has no local extremum at c.

The test for local extrema at endpoints is similar, but there is only one side to consider in determining whether f is increasing or decreasing, based on the sign of f'.

Proof of the First Derivative Test Part (1). Since the sign of f' changes from negative to positive at c, there are numbers a and b such that $a < c < b$, $f' < 0$ on (a, c), and $f' > 0$ on (c, b). If $x \in (a, c)$, then $f(c) < f(x)$ because $f' < 0$ implies that f is decreasing on $[a, c]$. If $x \in (c, b)$, then $f(c) < f(x)$ because $f' > 0$ implies that f is increasing on $[c, b]$. Therefore, $f(x) \geq f(c)$ for every $x \in (a, b)$. By definition, f has a local minimum at c.

Parts (2) and (3) are proved similarly. ■

EXAMPLE 2 Find the critical points of

$$f(x) = x^{1/3}(x - 4) = x^{4/3} - 4x^{1/3}.$$

Identify the open intervals on which f is increasing and those on which it is decreasing. Find the function's local and absolute extreme values.

Solution The function f is continuous at all x since it is the product of two continuous functions, $x^{1/3}$ and $(x - 4)$. The first derivative,

$$\begin{aligned} f'(x) &= \frac{d}{dx}\left(x^{4/3} - 4x^{1/3}\right) = \frac{4}{3}x^{1/3} - \frac{4}{3}x^{-2/3} \\ &= \frac{4}{3}x^{-2/3}(x - 1) = \frac{4(x - 1)}{3x^{2/3}}, \end{aligned}$$

is zero at $x = 1$ and undefined at $x = 0$. There are no endpoints in the domain, so the critical points $x = 0$ and $x = 1$ are the only places where f might have an extreme value.

The critical points partition the x-axis into open intervals on which f' is either positive or negative. The sign pattern of f' reveals the behavior of f between and at the critical points, as summarized in the following table.

Interval	$x < 0$	$0 < x < 1$	$x > 1$
Sign of f'	$-$	$-$	$+$
Behavior of f	decreasing	decreasing	increasing

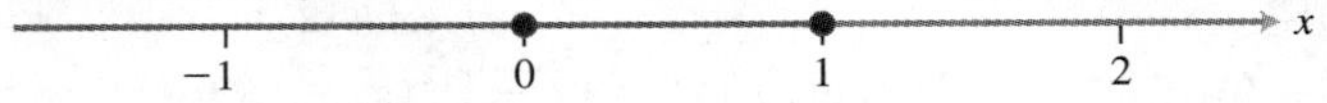

Corollary 3 to the Mean Value Theorem implies that f decreases on $(-\infty, 0)$, decreases on $(0, 1)$, and increases on $(1, \infty)$. The First Derivative Test for Local Extrema tells us that f does not have an extreme value at $x = 0$ (f' does not change sign) and that f has a local minimum at $x = 1$ (f' changes from negative to positive).

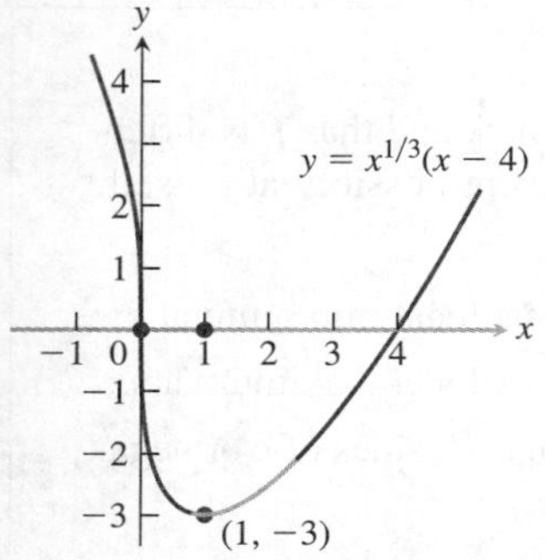

FIGURE 4.22 The function $f(x) = x^{1/3}(x - 4)$ decreases when $x < 1$ and increases when $x > 1$ (Example 2).

The value of the local minimum is $f(1) = 1^{1/3}(1 - 4) = -3$. This is also an absolute minimum since f is decreasing on $(-\infty, 1)$ and increasing on $(1, \infty)$. Figure 4.22 shows this value in relation to the function's graph.

Note that $\lim_{x \to 0} f'(x) = -\infty$, so the graph of f has a vertical tangent line at the origin. ■

EXAMPLE 3 Within the interval $0 \le x \le 2\pi$, find the critical points of

$$f(x) = \sin^2 x - \sin x - 1.$$

Identify the open intervals on which f is increasing and decreasing. Find the function's local and absolute extreme values.

Solution The function f is continuous over $[0, 2\pi]$ and differentiable over $(0, 2\pi)$, so the critical points occur at the zeros of f' in $(0, 2\pi)$. We find

$$f'(x) = 2 \sin x \cos x - \cos x = (2 \sin x - 1)(\cos x).$$

The first derivative is zero if and only if $\sin x = \frac{1}{2}$ or $\cos x = 0$. So the critical points of f in $(0, 2\pi)$ are $x = \pi/6$, $x = 5\pi/6$, $x = \pi/2$, and $x = 3\pi/2$. They partition $[0, 2\pi]$ into open intervals as follows.

Interval	$\left(0, \frac{\pi}{6}\right)$	$\left(\frac{\pi}{6}, \frac{\pi}{2}\right)$	$\left(\frac{\pi}{2}, \frac{5\pi}{6}\right)$	$\left(\frac{5\pi}{6}, \frac{3\pi}{2}\right)$	$\left(\frac{3\pi}{2}, 2\pi\right)$
Sign of f'	−	+	−	+	−
Behavior of f	dec	inc	dec	increasing	decreasing

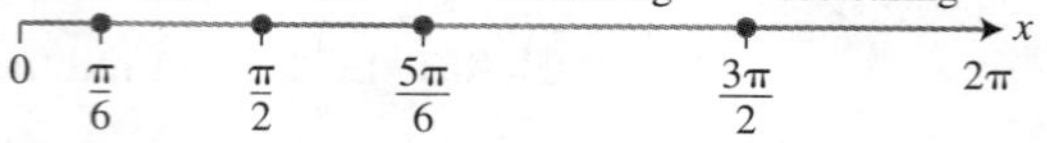

The table displays the open intervals on which f is increasing and decreasing. We can deduce from the table that there is a local minimum value of $f(\pi/6) = \frac{1}{4} - \frac{1}{2} - 1 = -\frac{5}{4}$, a local maximum value of $f(\pi/2) = 1 - 1 - 1 = -1$, another local minimum value of $f(5\pi/6) = -\frac{5}{4}$, and another local maximum value of $f(3\pi/2) = 1 - (-1) - 1 = 1$. The endpoint values are $f(0) = f(2\pi) = -1$. The absolute minimum in $[0, 2\pi]$ is $-\frac{5}{4}$ occurring at $x = \pi/6$ and $x = 5\pi/6$; the absolute maximum is 1 occurring at $x = 3\pi/2$. Figure 4.23 shows the graph. ■

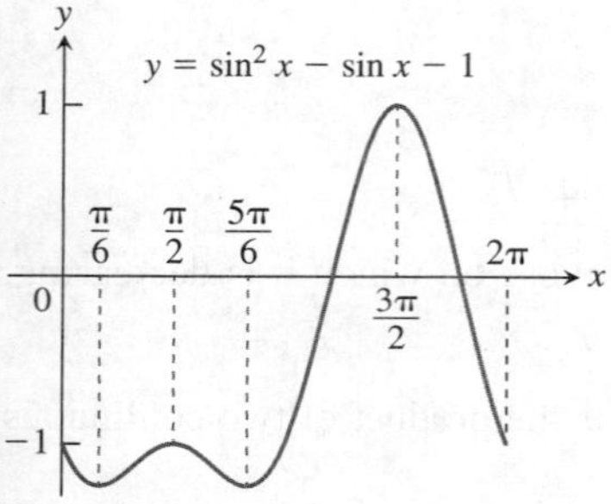

FIGURE 4.23 The graph of the function in Example 3.

EXERCISES 4.3

Analyzing Functions from Derivatives

Answer the following questions about the functions whose derivatives are given in Exercises 1–14:

a. What are the critical points of f?

b. On what open intervals is f increasing or decreasing?

c. At what points, if any, does f assume local maximum or minimum values?

1. $f'(x) = x(x - 1)$

2. $f'(x) = (x - 1)(x + 2)$

3. $f'(x) = (x - 1)^2(x + 2)$

4. $f'(x) = (x - 1)^2(x + 2)^2$

5. $f'(x) = (x - 1)(x + 2)(x - 3)$

6. $f'(x) = (x - 7)(x + 1)(x + 5)$

7. $f'(x) = \dfrac{x^2(x - 1)}{x + 2}, \quad x \ne -2$

8. $f'(x) = \dfrac{(x - 2)(x + 4)}{(x + 1)(x - 3)}, \quad x \ne -1, 3$

9. $f'(x) = 1 - \dfrac{4}{x^2}, \quad x \ne 0$

10. $f'(x) = 3 - \dfrac{6}{\sqrt{x}}, \quad x \ne 0$

11. $f'(x) = x^{-1/3}(x + 2)$

12. $f'(x) = x^{-1/2}(x - 3)$

13. $f'(x) = (\sin x - 1)(2 \cos x + 1), \quad 0 \le x \le 2\pi$

14. $f'(x) = (\sin x + \cos x)(\sin x - \cos x), \quad 0 \le x \le 2\pi$

Identifying Extrema

In Exercises 15–18:

a. Find the open intervals on which the function is increasing and those on which it is decreasing.

b. Identify the function's local and absolute extreme values, if any, saying where they occur.

15.

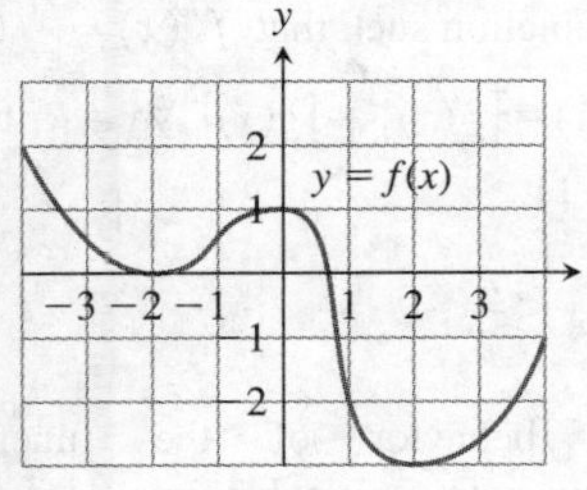

16.

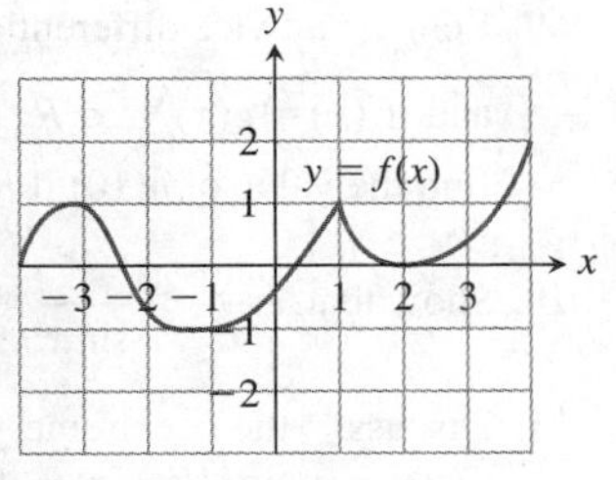

17.

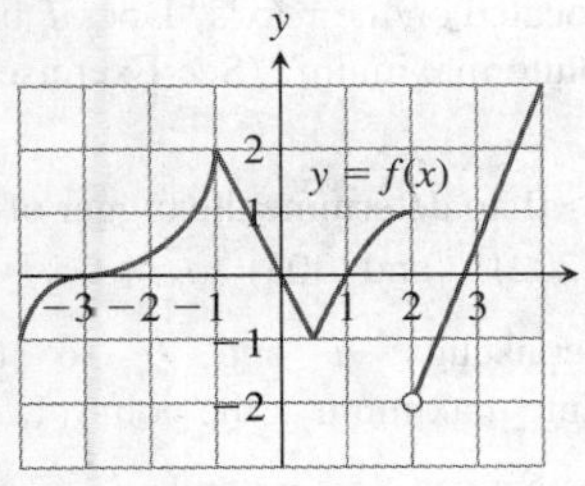

18.

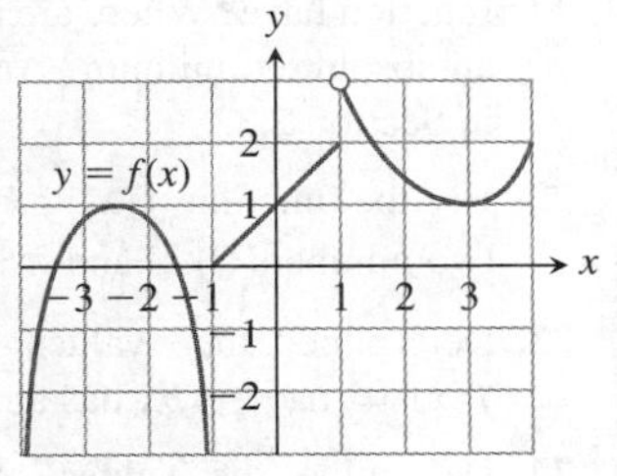

In Exercises 19–40:

a. Find the open intervals on which the function is increasing and those on which it is decreasing.

b. Identify the function's local extreme values, if any, saying where they occur.

19. $g(t) = -t^2 - 3t + 3$

20. $g(t) = -3t^2 + 9t + 5$

21. $h(x) = -x^3 + 2x^2$

22. $h(x) = 2x^3 - 18x$

23. $f(\theta) = 3\theta^2 - 4\theta^3$

24. $f(\theta) = 6\theta - \theta^3$

25. $f(r) = 3r^3 + 16r$

26. $h(r) = (r + 7)^3$

27. $f(x) = x^4 - 8x^2 + 16$

28. $g(x) = x^4 - 4x^3 + 4x^2$

29. $H(t) = \frac{3}{2}t^4 - t^6$

30. $K(t) = 15t^3 - t^5$

31. $f(x) = x - 6\sqrt{x - 1}$

32. $g(x) = 4\sqrt{x} - x^2 + 3$

33. $g(x) = x\sqrt{8 - x^2}$

34. $g(x) = x^2\sqrt{5 - x}$

35. $f(x) = \frac{x^2 - 3}{x - 2}, \quad x \neq 2$

36. $f(x) = \frac{x^3}{3x^2 + 1}$

37. $f(x) = x^{1/3}(x + 8)$

38. $g(x) = x^{2/3}(x + 5)$

39. $h(x) = x^{1/3}(x^2 - 4)$

40. $k(x) = x^{2/3}(x^2 - 4)$

In Exercises 41–52:

a. Identify the function's local extreme values in the given domain, and say where they occur.

T **b.** Graph the function over the given domain. Which of the extreme values, if any, are absolute?

41. $f(x) = 2x - x^2, \quad -\infty < x \leq 2$

42. $f(x) = (x + 1)^2, \quad -\infty < x \leq 0$

43. $g(x) = x^2 - 4x + 4, \quad 1 \leq x < \infty$

44. $g(x) = -x^2 - 6x - 9, \quad -4 \leq x < \infty$

45. $f(t) = 12t - t^3, \quad -3 \leq t < \infty$

46. $f(t) = t^3 - 3t^2, \quad -\infty < t \leq 3$

47. $h(x) = \frac{x^3}{3} - 2x^2 + 4x, \quad 0 \leq x < \infty$

48. $k(x) = x^3 + 3x^2 + 3x + 1, \quad -\infty < x \leq 0$

49. $f(x) = \sqrt{25 - x^2}, \quad -5 \leq x \leq 5$

50. $f(x) = \sqrt{x^2 - 2x - 3}, \quad 3 \leq x < \infty$

51. $g(x) = \frac{x - 2}{x^2 - 1}, \quad 0 \leq x < 1$

52. $g(x) = \frac{x^2}{4 - x^2}, \quad -2 < x \leq 1$

In Exercises 53–60:

a. Find the local extrema of each function on the given interval, and say where they occur.

T **b.** Graph the function and its derivative together. Comment on the behavior of f in relation to the signs and values of f'.

53. $f(x) = \sin 2x, \quad 0 \leq x \leq \pi$

54. $f(x) = \sin x - \cos x, \quad 0 \leq x \leq 2\pi$

55. $f(x) = \sqrt{3}\cos x + \sin x, \quad 0 \leq x \leq 2\pi$

56. $f(x) = -2x + \tan x, \quad \frac{-\pi}{2} < x < \frac{\pi}{2}$

57. $f(x) = \frac{x}{2} - 2\sin\frac{x}{2}, \quad 0 \leq x \leq 2\pi$

58. $f(x) = -2\cos x - \cos^2 x, \quad -\pi \leq x \leq \pi$

59. $f(x) = \csc^2 x - 2\cot x, \quad 0 < x < \pi$

60. $f(x) = \sec^2 x - 2\tan x, \quad \frac{-\pi}{2} < x < \frac{\pi}{2}$

In Exercises 61 and 62, the graph of f' is given. Assume that f is continuous, and determine the x-values corresponding to local minima and local maxima.

61.

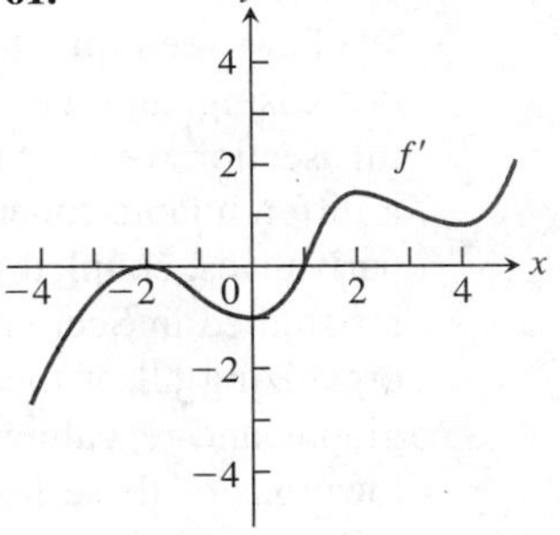

62.

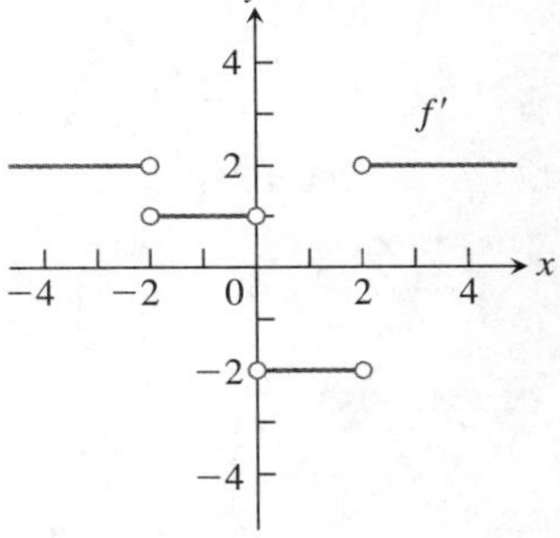

In Exercises 63 and 64, the graph of f' is given. Assume that f has domain $(-2, 2)$.

a. Either use the graph to determine which intervals f is increasing on and which intervals f is decreasing on, or explain why this information cannot be determined from the graph.

b. Either use the graph to determine which intervals f is positive on and which intervals f is negative on, or explain why this information cannot be determined from the graph.

63.

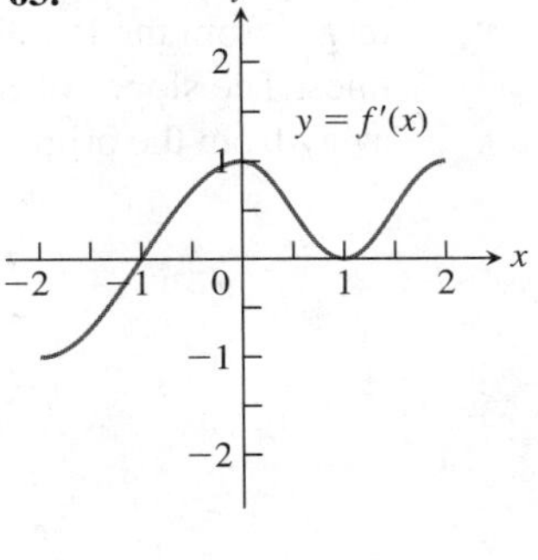

64.

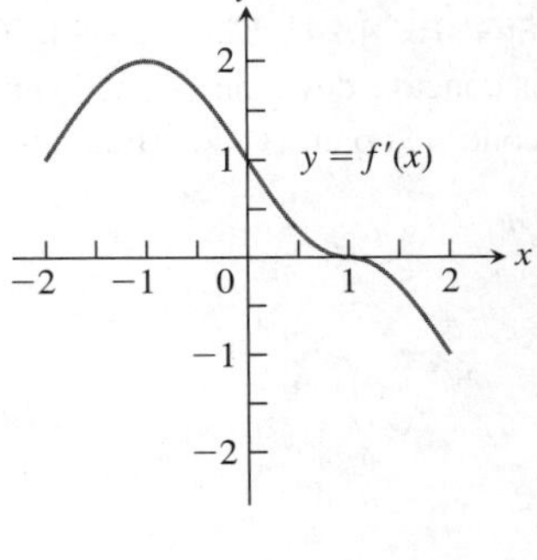

Theory and Examples

Show that the functions in Exercises 65 and 66 have local extreme values at the given values of θ, and say which kind of local extreme the function has.

65. $h(\theta) = 3\cos\frac{\theta}{2}, \quad 0 \le \theta \le 2\pi, \quad \text{at } \theta = 0 \text{ and } \theta = 2\pi$

66. $h(\theta) = 5\sin\frac{\theta}{2}, \quad 0 \le \theta \le \pi, \quad \text{at } \theta = 0 \text{ and } \theta = \pi$

67. Sketch the graph of a differentiable function $y = f(x)$ through the point $(1, 1)$ if $f'(1) = 0$ and

a. $f'(x) > 0$ for $x < 1$ and $f'(x) < 0$ for $x > 1$;

b. $f'(x) < 0$ for $x < 1$ and $f'(x) > 0$ for $x > 1$;

c. $f'(x) > 0$ for $x \ne 1$;

d. $f'(x) < 0$ for $x \ne 1$.

68. Sketch the graph of a differentiable function $y = f(x)$ that has

a. a local minimum at $(1, 1)$ and a local maximum at $(3, 3)$;

b. a local maximum at $(1, 1)$ and a local minimum at $(3, 3)$;

c. local maxima at $(1, 1)$ and $(3, 3)$;

d. local minima at $(1, 1)$ and $(3, 3)$.

69. Let f be a twice differentiable function such that $f''(x) = -f(x)$ and $f'(x) = g(x)\ \forall x \in R$. If $h(x) = [f(x)]^2 + [g(x)]^2\ \forall x \in R$, then find the value of $h(10)$ if $h(5) = 11$.

70. Show that $\dfrac{\tan x}{x} > \dfrac{x}{\sin x}$ if $0 < x < \dfrac{\pi}{2}$.

71. Discuss the extreme-value behavior of the function $f(x) = x\sin(1/x), x \ne 0$. How many critical points does this function have? Where are they located on the x-axis? Does f have an absolute minimum? An absolute maximum? (See Exercise 49 in Section 2.3.)

72. Use the function $f(x) = x^{1/x}, x > 0$, to determine the bigger of the two numbers (i) e^{π} and π^{e} (ii) $(202)^{303}$ and $(303)^{202}$.

73. Determine the values of constants a and b so that $f(x) = ax^2 + bx$ has an absolute maximum at the point $(1, 2)$.

74. Determine the values of constants a, b, c, and d so that $f(x) = ax^3 + bx^2 + cx + d$ has a local maximum at the point $(0, 0)$ and a local minimum at the point $(1, -1)$.

4.4 Concavity and Curve Sketching

We have seen how the first derivative tells us where a function is increasing, where it is decreasing, and whether a local maximum or local minimum occurs at a critical point. In this section we see that the second derivative gives us information about how the graph of a differentiable function bends or turns. With this knowledge about the first and second derivatives, coupled with our previous understanding of symmetry and asymptotic behavior studied in Sections 1.1 and 2.6, we can now draw an accurate graph of a function. By organizing all of these ideas into a coherent procedure, we give a method for sketching graphs and revealing visually the key features of functions. Identifying and knowing the locations of these features is of major importance in mathematics and its applications to science and engineering, especially in the graphical analysis and interpretation of data. When the domain of a function is not a finite closed interval, sketching a graph helps to determine whether absolute maxima or absolute minima exist and, if they do exist, where they are located.

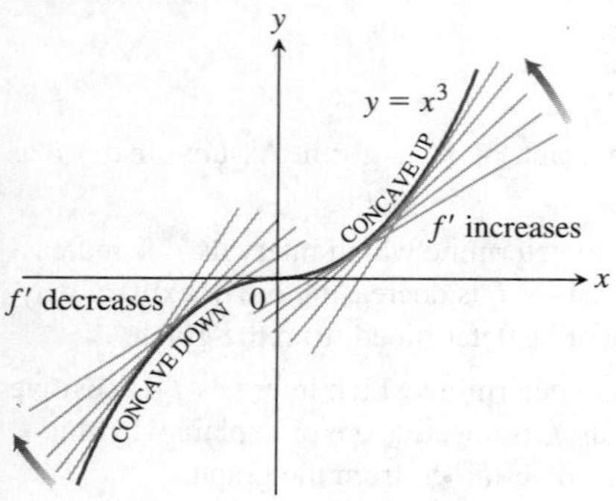

FIGURE 4.24 The graph of $f(x) = x^3$ is concave down on $(-\infty, 0)$ and concave up on $(0, \infty)$ (Example 1a).

Concavity

As you can see in Figure 4.24, the curve $y = x^3$ rises as x increases, but the portions defined on the intervals $(-\infty, 0)$ and $(0, \infty)$ turn in different ways. As we approach the origin from the left along the curve, the curve turns to our right and falls below its tangent lines. The slopes of the tangent lines are decreasing on the interval $(-\infty, 0)$. As we move away from the origin along the curve to the right, the curve turns to our left and rises above its tangent lines. The slopes of the tangent lines are increasing on the interval $(0, \infty)$. This turning or bending behavior defines the *concavity* of the curve.

DEFINITION The graph of a differentiable function $y = f(x)$ is

(a) concave up on an open interval I if f' is increasing on I;

(b) concave down on an open interval I if f' is decreasing on I.

A function whose graph is concave up is also often called **convex**.

If $y = f(x)$ has a second derivative, we can apply Corollary 3 of the Mean Value Theorem to the first derivative function. We conclude that f' increases if $f'' > 0$ on I, and decreases if $f'' < 0$.

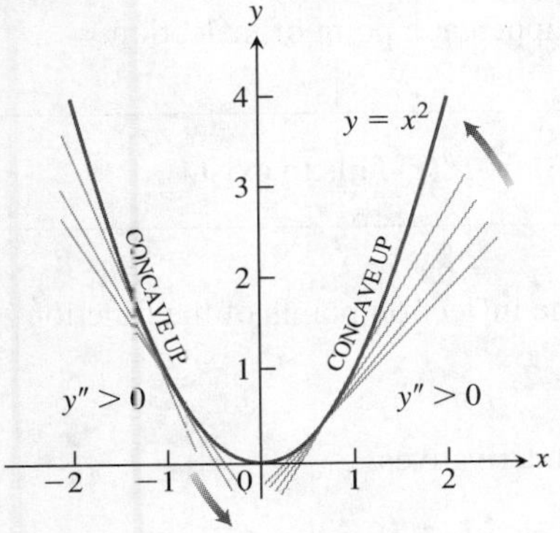

FIGURE 4.25 The graph of $f(x) = x^2$ is concave up on every interval (Example 1b).

The Second Derivative Test for Concavity

Let $y = f(x)$ be twice-differentiable on an interval I.

1. If $f'' > 0$ on I, the graph of f over I is concave up.

2. If $f'' < 0$ on I, the graph of f over I is concave down.

If $y = f(x)$ is twice-differentiable, we will use the notations f'' and y'' interchangeably when denoting the second derivative.

EXAMPLE 1

(a) The curve $y = x^3$ (Figure 4.24) is concave down on $(-\infty, 0)$, where $y'' = 6x < 0$, and concave up on $(0, \infty)$, where $y'' = 6x > 0$.

(b) The curve $y = x^2$ (Figure 4.25) is concave up on $(-\infty, \infty)$ because its second derivative $y'' = 2$ is always positive. ■

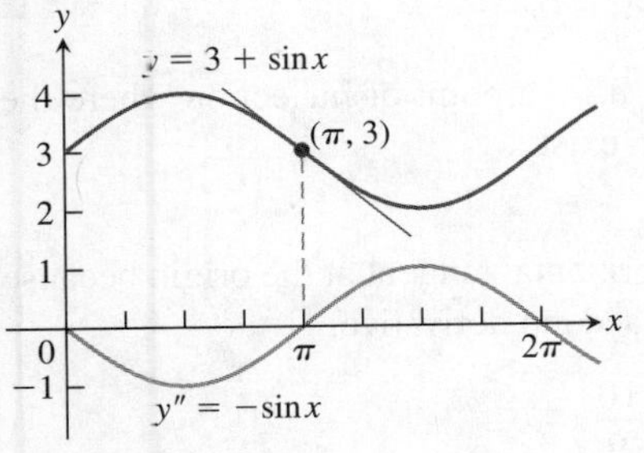

FIGURE 4.26 Using the sign of y'' to determine the concavity of y (Example 2).

EXAMPLE 2 Determine the concavity of $y = 3 + \sin x$ on $[0, 2\pi]$.

Solution The first derivative of $y = 3 + \sin x$ is $y' = \cos x$, and the second derivative is $y'' = -\sin x$. The graph of $y = 3 + \sin x$ is concave down on $(0, \pi)$, where $y'' = -\sin x$ is negative. It is concave up on $(\pi, 2\pi)$, where $y'' = -\sin x$ is positive (Figure 4.26). ■

Points of Inflection

The curve $y = 3 + \sin x$ in Example 2 changes concavity at the point $(\pi, 3)$. Since the first derivative $y' = \cos x$ exists for all x, we see that the curve has a tangent line of slope -1 at the point $(\pi, 3)$. This point is called a *point of inflection* of the curve. Notice from Figure 4.26 that the graph crosses its tangent line at this point and that the second derivative $y'' = -\sin x$ has value 0 when $x = \pi$. In general, we have the following definition.

> **DEFINITION** A point $(c, f(c))$ where the graph of a function has a tangent line and where the concavity changes is a **point of inflection**.

We observed that the second derivative of $f(x) = 3 + \sin x$ is equal to zero at the inflection point $(\pi, 3)$. Generally, if the second derivative exists at a point of inflection $(c, f(c))$, then $f''(c) = 0$. This follows immediately from the Intermediate Value Theorem whenever f'' is continuous over an interval containing $x = c$ because the second derivative changes sign moving across this interval. Even if the continuity assumption is dropped, it is still true that $f''(c) = 0$, provided the second derivative exists (although a more advanced argument is required in this noncontinuous case). Since a tangent line must exist at the point of inflection, either the first derivative $f'(c)$ exists (is finite) or the graph has a vertical tangent line at the point. At a vertical tangent, neither the first nor second derivative exists. In summary, one of two things can happen at a point of inflection.

> At a point of inflection $(c, f(c))$, either $f''(c) = 0$ or $f''(c)$ fails to exist.

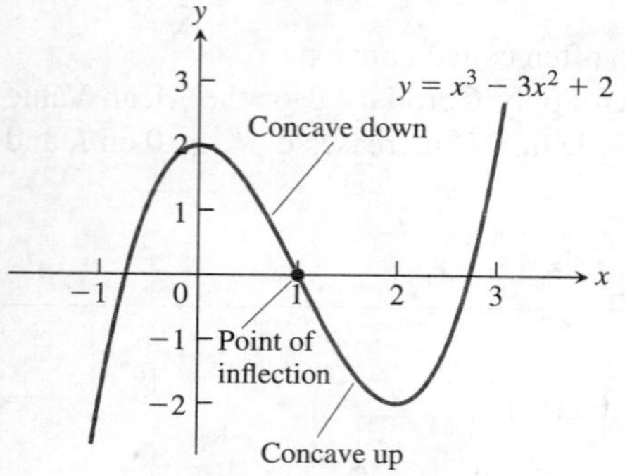

FIGURE 4.27 The concavity of the graph of f changes from concave down to concave up at the inflection point (Example 3).

EXAMPLE 3 Determine the concavity and find the inflection points of the function

$$f(x) = x^3 - 3x^2 + 2.$$

Solution We start by computing the first and second derivatives.

$$f'(x) = 3x^2 - 6x, \qquad f''(x) = 6x - 6.$$

To determine concavity, we look at the sign of the second derivative $f''(x) = 6x - 6$. The sign is negative when $x < 1$, is 0 at $x = 1$, and is positive when $x > 1$. It follows that the graph of f is concave down on $(-\infty, 1)$, is concave up on $(1, \infty)$, and has an inflection point at the point $(1, 0)$ where the concavity changes.

The graph of f is shown in Figure 4.27. Notice that we did not need to know the shape of this graph ahead of time in order to determine its concavity. ■

The next example illustrates that a function can have a point of inflection where the first derivative exists but the second derivative fails to exist.

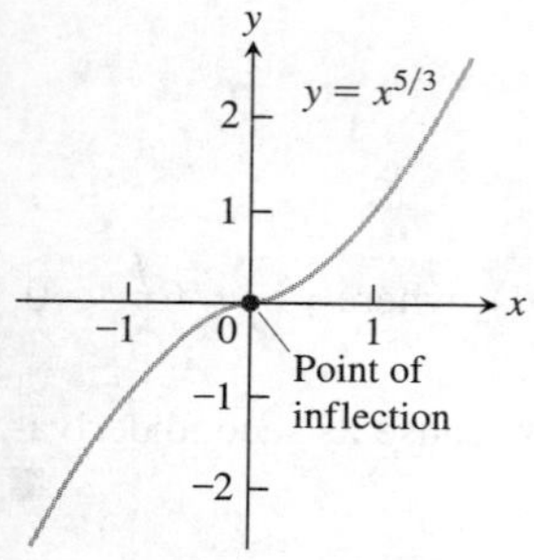

FIGURE 4.28 The graph of $f(x) = x^{5/3}$ has a horizontal tangent at the origin where the concavity changes, although f'' does not exist at $x = 0$ (Example 4).

EXAMPLE 4 The graph of $f(x) = x^{5/3}$ has a horizontal tangent at the origin because $f'(x) = (5/3)x^{2/3} = 0$ when $x = 0$. However, the second derivative,

$$f''(x) = \frac{d}{dx}\left(\frac{5}{3}x^{2/3}\right) = \frac{10}{9}x^{-1/3},$$

fails to exist at $x = 0$. Nevertheless, $f''(x) < 0$ for $x < 0$ and $f''(x) > 0$ for $x > 0$, so the second derivative changes sign at $x = 0$ and there is a point of inflection at the origin. The graph is shown in Figure 4.28. ■

The following example shows that an inflection point need not occur even though both derivatives exist and $f'' = 0$.

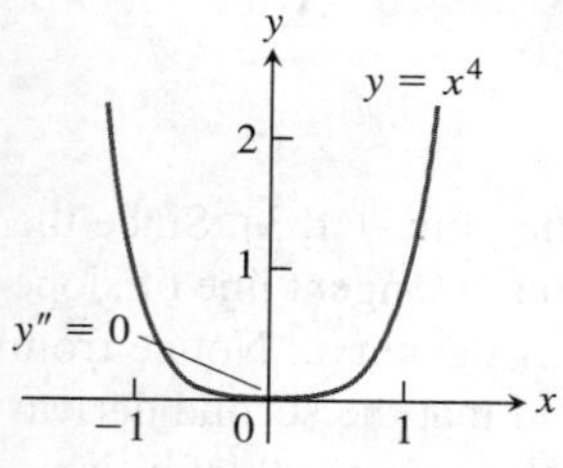

FIGURE 4.29 The graph of $y = x^4$ has no inflection point at the origin, even though $y'' = 0$ there (Example 5).

EXAMPLE 5 The curve $y = x^4$ has no inflection point at $x = 0$ (Figure 4.29). Even though the second derivative $y'' = 12x^2$ is zero there, it does not change sign. The curve is concave up everywhere. ■

In the next example, a point of inflection occurs at a vertical tangent to the curve where neither the first nor the second derivative exists.

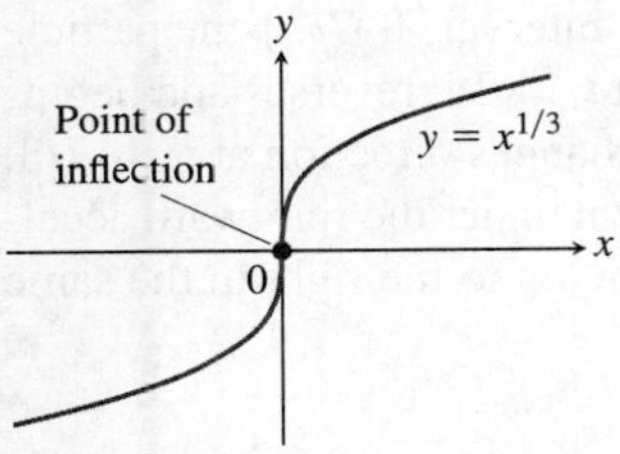

FIGURE 4.30 A point of inflection where y' and y'' fail to exist (Example 6).

EXAMPLE 6 The graph of $y = x^{1/3}$ has a point of inflection at the origin because the second derivative is positive for $x < 0$ and negative for $x > 0$:

$$y'' = \frac{d^2}{dx^2}(x^{1/3}) = \frac{d}{dx}\left(\frac{1}{3}x^{-2/3}\right) = -\frac{2}{9}x^{-5/3}.$$

However, both $y' = x^{-2/3}/3$ and y'' fail to exist at $x = 0$, and there is a vertical tangent there. See Figure 4.30. ■

Caution Example 4 in Section 4.1 (Figure 4.9) shows that the function $f(x) = x^{2/3}$ does not have a second derivative at $x = 0$ and does not have a point of inflection there (there is no change in concavity at $x = 0$). Combined with the behavior of the function in Example 6 above, we see that when the second derivative does not exist at $x = c$, an inflection point may or may not occur there. So we need to be careful about interpreting functional behavior whenever first or second derivatives fail to exist at a point. At such points the graph can have vertical tangent lines, corners, cusps, or various discontinuities. ●

To study the motion of an object moving along a line as a function of time, we often are interested in knowing when the object's acceleration, given by the second derivative, is positive or negative. The points of inflection on the graph of the object's position function reveal where the acceleration changes sign.

EXAMPLE 7 A particle is moving along a horizontal coordinate line (positive to the right) with position function

$$s(t) = 2t^3 - 14t^2 + 22t - 5, \qquad t \geq 0.$$

Find the velocity and acceleration, and describe the motion of the particle.

Solution The velocity is

$$v(t) = s'(t) = 6t^2 - 28t + 22 = 2(t - 1)(3t - 11),$$

and the acceleration is

$$a(t) = v'(t) = s''(t) = 12t - 28 = 4(3t - 7).$$

When the function $s(t)$ is increasing, the particle is moving to the right; when $s(t)$ is decreasing, the particle is moving to the left.

Notice that the first derivative ($v = s'$) is zero at the critical points $t = 1$ and $t = 11/3$.

Interval	$0 < t < 1$	$1 < t < 11/3$	$11/3 < t$
Sign of $v = s'$	+	−	+
Behavior of s	increasing	decreasing	increasing
Particle motion	right	left	right

The particle is moving to the right in the time intervals $[0, 1)$ and $(11/3, \infty)$, and moving to the left in $(1, 11/3)$. It is momentarily stationary (at rest) at $t = 1$ and $t = 11/3$.

The acceleration $a(t) = s''(t) = 4(3t - 7)$ is zero when $t = 7/3$.

Interval	$0 < t < 7/3$	$7/3 < t$
Sign of $a = s''$	−	+
Graph of s	concave down	concave up

Under the influence of the leftward acceleration over the time interval $[0, 7/3)$, the particle starts out moving to the right while slowing down, and then at $t = 1$ it reverses and begins moving to the left while speeding up. The acceleration then changes direction at $t = 7/3$, but the particle continues moving leftward, while slowing down under the rightward acceleration. At $t = 11/3$ the particle reverses direction again: moving to the right in the same direction as the acceleration, so it is speeding up. ■

Second Derivative Test for Local Extrema

Instead of looking for sign changes in f' at critical points, we can sometimes use the following test to determine the presence and nature of local extrema.

> **THEOREM 5—Second Derivative Test for Local Extrema**
> Suppose f'' is continuous on an open interval that contains $x = c$.
> 1. If $f'(c) = 0$ and $f''(c) < 0$, then f has a local maximum at $x = c$.
> 2. If $f'(c) = 0$ and $f''(c) > 0$, then f has a local minimum at $x = c$.
> 3. If $f'(c) = 0$ and $f''(c) = 0$, then the test fails. The function f may have a local maximum, a local minimum, or neither.

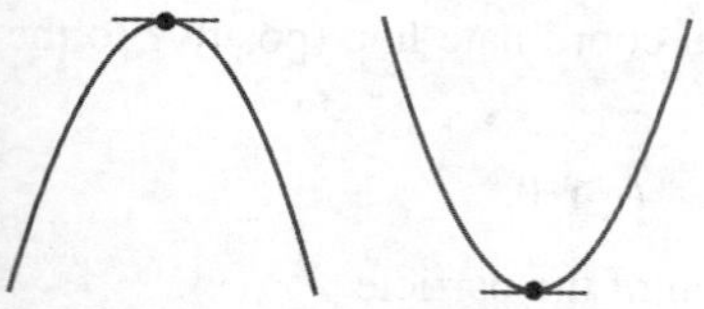

$f' = 0, f'' < 0$ ⇒ local max $f' = 0, f'' > 0$ ⇒ local min

Proof Part (1). If $f''(c) < 0$, then $f''(x) < 0$ on some open interval I containing the point c, since f'' is continuous. Therefore, f' is decreasing on I. Since $f'(c) = 0$, the sign of f' changes from positive to negative at c, so f has a local maximum at c by the First Derivative Test.

The proof of Part (2) is similar.

For Part (3), consider the three functions $y = x^4$, $y = -x^4$, and $y = x^3$. For each function, the first and second derivatives are zero at $x = 0$. Yet the function $y = x^4$ has a local minimum there, $y = -x^4$ has a local maximum, and $y = x^3$ is increasing in any open interval containing $x = 0$ (having neither a maximum nor a minimum there). Thus the test fails. ■

This test requires us to know f'' *only at c itself* and not in an interval about c. This makes the test easy to apply. That's the good news. The bad news is that the test is inconclusive if $f'' = 0$ or if f'' does not exist at $x = c$. When this happens, use the First Derivative Test for local extreme values.

Together f' and f'' tell us the shape of the function's graph—that is, where the critical points are located and what happens at a critical point, where the function is increasing and where it is decreasing, and how the curve is turning or bending as indicated by its concavity. We use this information to sketch a graph of the function that captures its key features.

EXAMPLE 8 Sketch a graph of the function

$$f(x) = x^4 - 4x^3 + 10$$

using the following steps.

(a) Identify where the extrema of f occur.

(b) Find the intervals on which f is increasing and the intervals on which f is decreasing.

(c) Find where the graph of f is concave up and where it is concave down.

(d) Sketch the general shape of the graph for f.

(e) Plot some specific points, such as local maximum and minimum points, points of inflection, and intercepts. Then sketch the curve.

Solution The function f is continuous since $f'(x) = 4x^3 - 12x^2$ exists. The domain of f is $(-\infty, \infty)$, and the domain of f' is also $(-\infty, \infty)$. Thus, the critical points of f occur only at the zeros of f'. Since

$$f'(x) = 4x^3 - 12x^2 = 4x^2(x - 3),$$

the first derivative is zero at $x = 0$ and $x = 3$. We use these critical points to define intervals where f is increasing or decreasing.

Interval	$x < 0$	$0 < x < 3$	$3 < x$
Sign of f'	$-$	$-$	$+$
Behavior of f	decreasing	decreasing	increasing

(a) Using the First Derivative Test for local extrema and the table above, we see that there is no extremum at $x = 0$ and a local minimum at $x = 3$.

(b) Using the table above, we see that f is decreasing on $(-\infty, 0]$ and $[0, 3]$, and increasing on $[3, \infty)$.

(c) $f''(x) = 12x^2 - 24x = 12x(x - 2)$ is zero at $x = 0$ and $x = 2$. We use these points to define intervals where the graph of f is concave up or concave down.

Interval	$x < 0$	$0 < x < 2$	$2 < x$
Sign of f'	$+$	$-$	$+$
Behavior of f	concave up	concave down	concave up

We see that the graph of f is concave up on the intervals $(-\infty, 0)$ and $(2, \infty)$, and concave down on $(0, 2)$.

(d) Summarizing the information in the last two tables, we obtain the following.

$x < 0$	$0 < x < 2$	$2 < x < 3$	$3 < x$
decreasing	decreasing	decreasing	increasing
concave up	concave down	concave up	concave up

The general shape of the curve is shown in the accompanying figure.

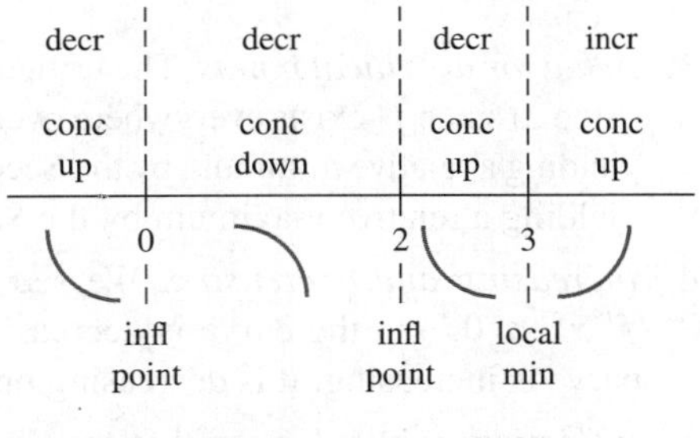

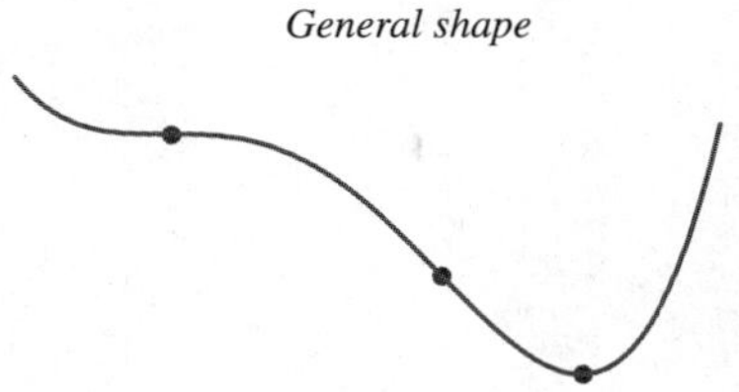

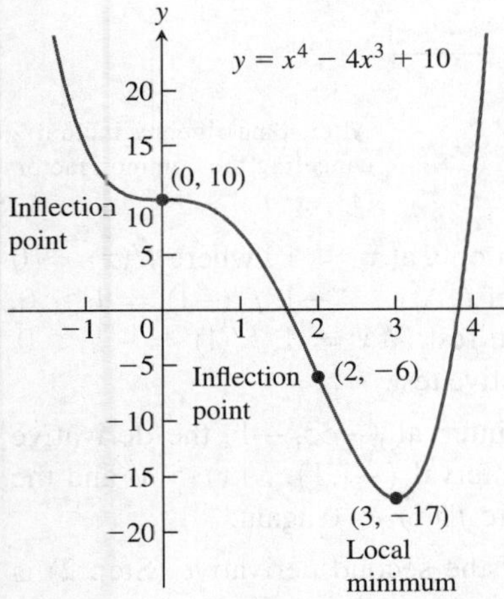

FIGURE 4.31 The graph of $f(x) = x^4 - 4x^3 + 10$ (Example 8).

(e) Plot the curve's intercepts (if possible) and the points where y' and y'' are zero. Indicate any local extreme values and inflection points. Use the general shape as a guide to sketch the curve. (Plot additional points as needed.) Figure 4.31 shows the graph of f. ■

The steps in Example 8 give a procedure for graphing the key features of a function. Asymptotes were defined and discussed in Section 2.6. We can find them for many classes of functions (including rational functions), and the methods in the next section give tools to help find them for even more general functions.

> **Procedure for Graphing $y = f(x)$**
>
> 1. Identify the domain of f and any symmetries the curve may have.
> 2. Find the derivatives y' and y''.
> 3. Find the critical points of f, if any, and identify the function's behavior at each one.
> 4. Find where the curve is increasing and where it is decreasing.
> 5. Find the points of inflection, if any occur, and determine the concavity of the curve.
> 6. Identify any asymptotes that may exist.
> 7. Plot key points, such as the intercepts and the points found in Steps 3–5, and sketch the curve together with any asymptotes that exist.

EXAMPLE 9 Sketch the graph of $f(x) = \dfrac{(x+1)^2}{1+x^2}$.

Solution

1. The domain of f is $(-\infty, \infty)$ and there are no symmetries about either axis or the origin (Section 1.1).
2. *Find f' and f''.*

$$f(x) = \frac{(x+1)^2}{1+x^2}$$

x-intercept at $x = -1$, y-intercept at $y = 1$

$$f'(x) = \frac{(1+x^2)\cdot 2(x+1) - (x+1)^2 \cdot 2x}{(1+x^2)^2}$$

$$= \frac{2(1-x^2)}{(1+x^2)^2}$$

Critical points: $x = -1, x = 1$

$$f''(x) = \frac{(1+x^2)^2 \cdot 2(-2x) - 2(1-x^2)[2(1+x^2)\cdot 2x]}{(1+x^2)^4}$$

$$= \frac{4x(x^2-3)}{(1+x^2)^3}$$

After some algebra, including canceling the common factor $(1 + x^2)$

3. *Behavior at critical points.* The critical points occur only at $x = \pm 1$ where $f'(x) = 0$ (Step 2) since f' exists everywhere over the domain of f. At $x = -1$, $f''(-1) = 1 > 0$, yielding a relative minimum by the Second Derivative Test. At $x = 1$, $f''(1) = -1 < 0$, yielding a relative maximum by the Second Derivative test.
4. *Increasing and decreasing.* We see that on the interval $(-\infty, -1)$ the derivative $f'(x) < 0$, and the curve is decreasing. On the interval $(-1, 1)$, $f'(x) > 0$ and the curve is increasing; it is decreasing on $(1, \infty)$ where $f'(x) < 0$ again.
5. *Inflection points.* Notice that the denominator of the second derivative (Step 2) is always positive. The second derivative f'' is zero when $x = -\sqrt{3}, 0$, and $\sqrt{3}$. The second derivative changes sign at each of these points: negative on $(-\infty, -\sqrt{3})$, positive on $(-\sqrt{3}, 0)$, negative on $(0, \sqrt{3})$, and positive again on $(\sqrt{3}, \infty)$. Thus each point is a point of inflection. The curve is concave down on the interval $(-\infty, -\sqrt{3})$, concave up on $(-\sqrt{3}, 0)$, concave down on $(0, \sqrt{3})$, and concave up again on $(\sqrt{3}, \infty)$.

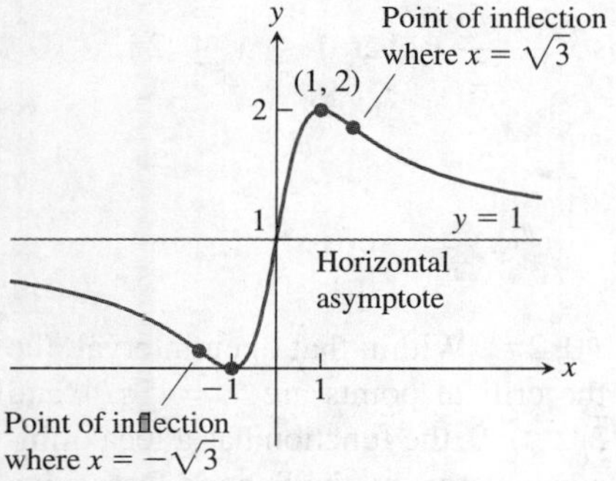

FIGURE 4.32 The graph of $y = \dfrac{(x+1)^2}{1+x^2}$ (Example 9).

6. *Asymptotes.* Expanding the numerator of $f(x)$ and then dividing both numerator and denominator by x^2 gives

$$f(x) = \frac{(x+1)^2}{1+x^2} = \frac{x^2+2x+1}{1+x^2} \qquad \text{Expanding numerator}$$

$$= \frac{1+(2/x)+(1/x^2)}{(1/x^2)+1}. \qquad \text{Dividing by } x^2$$

We see that $f(x) \to 1$ as $x \to \infty$ and that $f(x) \to 1$ as $x \to -\infty$. Thus, the line $y = 1$ is a horizontal asymptote. Since the function is continuous on $(-\infty, \infty)$, there are no vertical asymptotes.

7. The graph of f is sketched in Figure 4.32. Notice how the graph is concave down as it approaches the horizontal asymptote $y = 1$ as $x \to -\infty$, and concave up in its approach to $y = 1$ as $x \to \infty$. ■

EXAMPLE 10 Sketch the graph of $f(x) = \dfrac{x^2+4}{2x}$.

Solution

1. The domain of f is all nonzero real numbers. There are no intercepts because neither x nor $f(x)$ can be zero. Since $f(-x) = -f(x)$, we note that f is an odd function, so the graph of f is symmetric about the origin.

2. We calculate the derivatives of the function, but we first rewrite it in order to simplify our computations:

$$f(x) = \frac{x^2+4}{2x} = \frac{x}{2} + \frac{2}{x} \qquad \text{Function simplified for differentiation}$$

$$f'(x) = \frac{1}{2} - \frac{2}{x^2} = \frac{x^2-4}{2x^2} \qquad \text{Combine fractions to solve easily } f'(x) = 0.$$

$$f''(x) = \frac{4}{x^3} \qquad \text{Exists throughout the entire domain of } f$$

3. The critical points occur at $x = \pm 2$ where $f'(x) = 0$. Since $f''(-2) < 0$ and $f''(2) > 0$, we see from the Second Derivative Test that a relative maximum occurs at $x = -2$ with $f(-2) = -2$, and a relative minimum occurs at $x = 2$ with $f(2) = 2$.

4. On the interval $(-\infty, -2)$ the derivative f' is positive because $x^2 - 4 > 0$ so the graph is increasing; on the interval $(-2, 0)$ the derivative is negative and the graph is decreasing. Similarly, the graph is decreasing on the interval $(0, 2)$ and increasing on $(2, \infty)$.

5. There are no points of inflection because $f''(x) < 0$ whenever $x < 0$, $f''(x) > 0$ whenever $x > 0$, and f'' exists everywhere and is never zero throughout the domain of f. The graph is concave down on the interval $(-\infty, 0)$ and concave up on the interval $(0, \infty)$.

6. From the rewritten formula for $f(x)$, we see that

$$\lim_{x\to 0^+}\left(\frac{x}{2}+\frac{2}{x}\right) = +\infty \quad \text{and} \quad \lim_{x\to 0^-}\left(\frac{x}{2}+\frac{2}{x}\right) = -\infty,$$

so the y-axis is a vertical asymptote. Also, as $x \to \infty$ or as $x \to -\infty$, the graph of $f(x)$ approaches the line $y = x/2$. Thus $y = x/2$ is an oblique asymptote.

7. The graph of f is sketched in Figure 4.33. ■

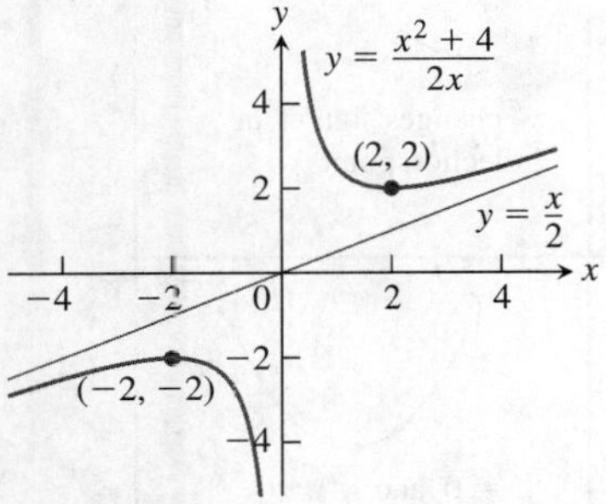

FIGURE 4.33 The graph of $y = \dfrac{x^2+4}{2x}$ (Example 10).

EXAMPLE 11 Sketch the graph of $f(x) = \cos x - \frac{\sqrt{2}}{2}x$ over $0 \leq x \leq 2\pi$.

Solution The derivatives of f are

$$f'(x) = -\sin x - \frac{\sqrt{2}}{2} \quad \text{and} \quad f''(x) = -\cos x.$$

Both derivatives exist everywhere over the interval $(0, 2\pi)$. Within that open interval, the first derivative is zero when $\sin x = -\sqrt{2}/2$, so the critical points are $x = 5\pi/4$ and $x = 7\pi/4$. Since $f''(5\pi/4) = -\cos(5\pi/4) = \sqrt{2}/2 > 0$, the function has a local minimum value of $f(5\pi/4) \approx -3.48$ (evaluated with a calculator) by the Second Derivative Test. Also, $f''(7\pi/4) = -\cos(7\pi/4) = -\sqrt{2}/2 < 0$, so the function has a local maximum value of $f(7\pi/4) \approx -3.18$.

Examining the second derivative, we find that $f'' = 0$ when $x = \pi/2$ or $x = 3\pi/2$. Since f'' changes sign at these two points, we conclude that $(\pi/2, f(\pi/2)) \approx (\pi/2, -1.11)$ and $(3\pi/2, f(3\pi/2)) \approx (3\pi/2, -3.33)$ are points of inflection.

Finally, we evaluate f at the endpoints of the interval to find $f(0) = 1$ and $f(2\pi) \approx -3.44$. Therefore, the values $f(0) = 1$ and $f(5\pi/4) \approx -3.48$ are the absolute maximum and absolute minimum values of f over the closed interval $[0, 2\pi]$. The graph of f is sketched in Figure 4.34. ■

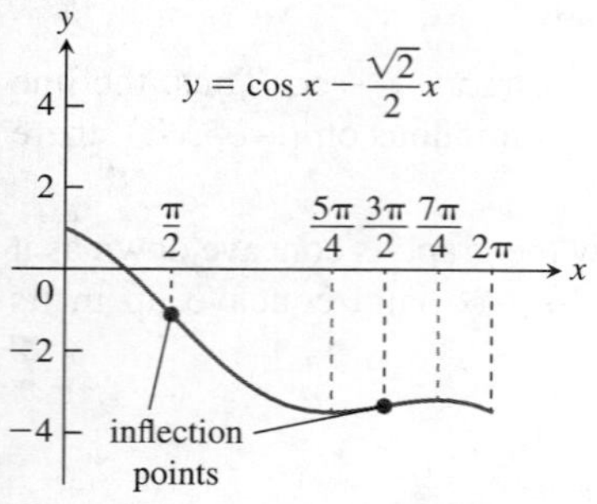

FIGURE 4.34 The graph of the function in Example 11.

Graphical Behavior of Functions from Derivatives

As we saw in Examples 8–11, we can learn much about a twice-differentiable function $y = f(x)$ by examining its first derivative. We can find where the function's graph rises and falls and where any local extrema are located. We can differentiate y' to learn how the graph bends as it passes over the intervals of rise and fall. Together with information about the function's asymptotes and its value at some key points, such as intercepts, this information about the derivatives helps us determine the shape of the function's graph. The following figure summarizes how the first derivative and second derivative affect the shape of a graph.

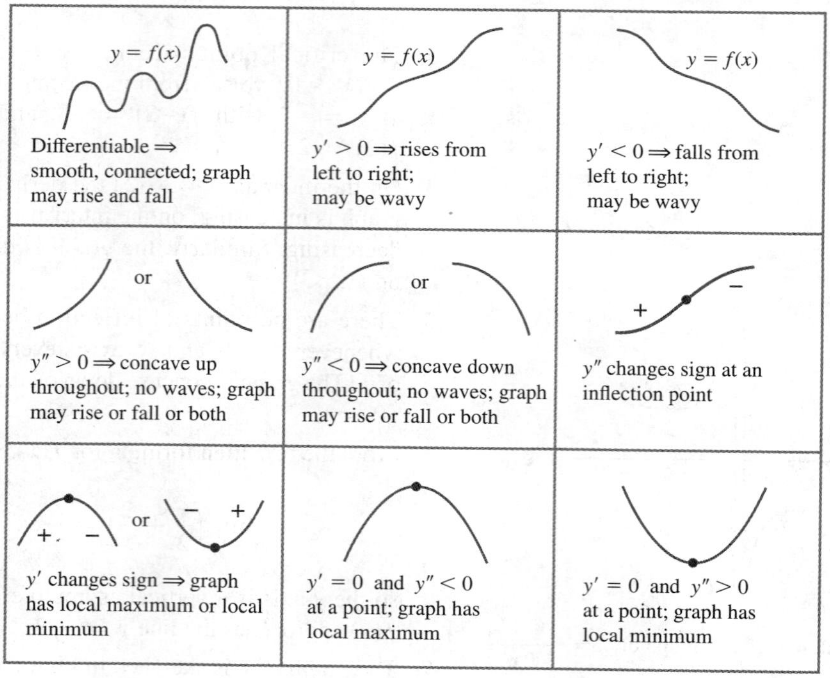

EXERCISES 4.4

Analyzing Functions from Graphs

Identify the inflection points and local maxima and minima of the functions graphed in Exercises 1–8. Identify the open intervals on which the functions are differentiable and the graphs are concave up and concave down.

1. $y = \frac{x^3}{3} - \frac{x^2}{2} - 2x + \frac{1}{3}$

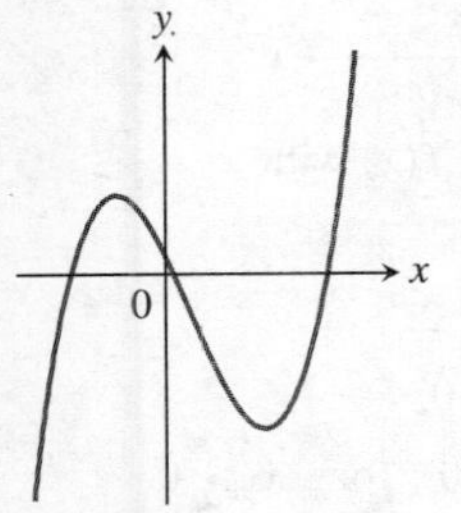

2. $y = \frac{x^4}{4} - 2x^2 + 4$

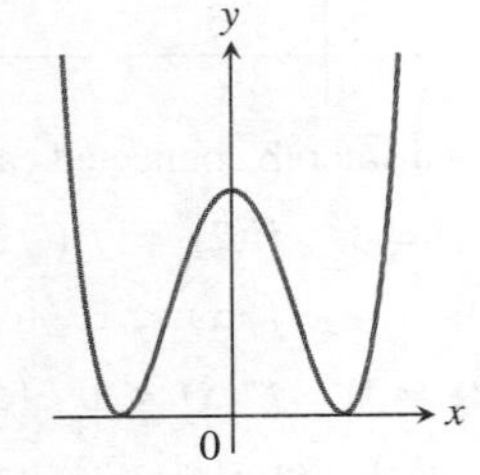

3. $y = \frac{3}{4}(x^2 - 1)^{2/3}$

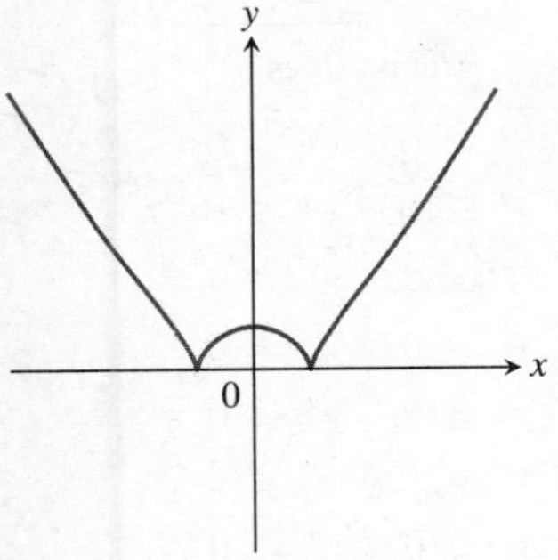

4. $y = \frac{9}{14}x^{1/3}(x^2 - 7)$

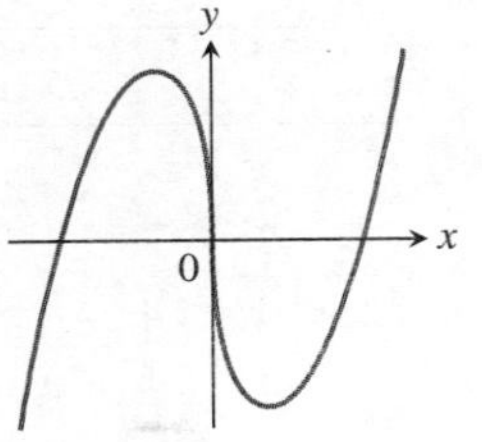

5. $y = x + \sin 2x, \ -\frac{2\pi}{3} \le x \le \frac{2\pi}{3}$

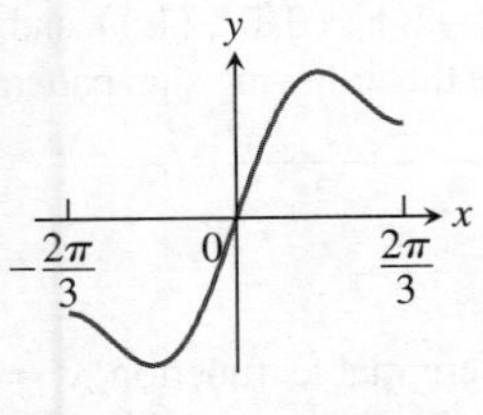

6. $y = \tan x - 4x, \ -\frac{\pi}{2} < x < \frac{\pi}{2}$

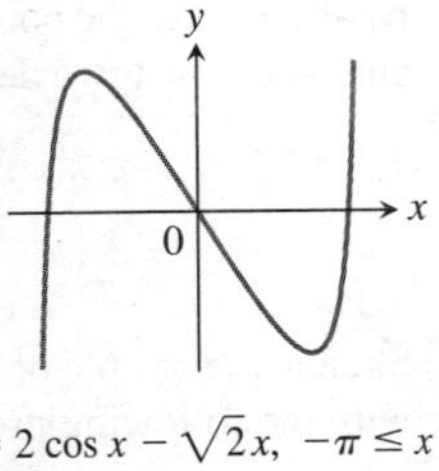

7. $y = \sin|x|, \ -2\pi \le x \le 2\pi$

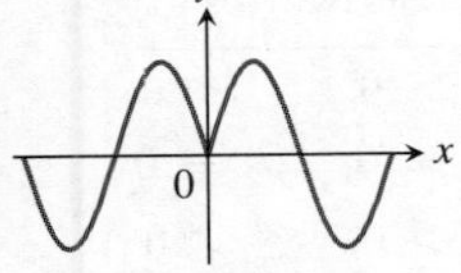

NOT TO SCALE

8. $y = 2\cos x - \sqrt{2}x, \ -\pi \le x \le \frac{3\pi}{2}$

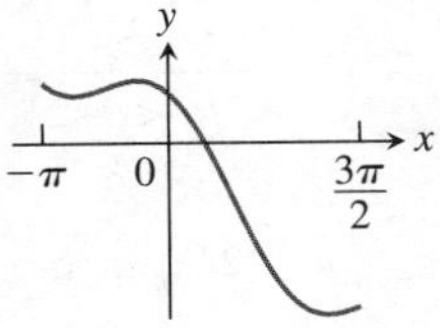

Graphing Functions

In Exercises 9–66, graph the function using appropriate methods from the graphing procedures presented just before Example 9, identifying the coordinates of any local extreme points and inflection points. Then find coordinates of absolute extreme points, if any.

9. $y = x^2 - 4x + 3$

10. $y = 6 - 2x - x^2$

11. $y = x^3 - 3x + 3$

12. $y = x(6 - 2x)^2$

13. $y = -2x^3 + 6x^2 - 3$

14. $y = 1 - 9x - 6x^2 - x^3$

15. $y = (x - 2)^3 + 1$

16. $y = 1 - (x + 1)^3$

17. $y = x^4 - 2x^2 = x^2(x^2 - 2)$

18. $y = -x^4 + 6x^2 - 4 = x^2(6 - x^2) - 4$

19. $y = 4x^3 - x^4 = x^3(4 - x)$

20. $y = x^4 + 2x^3 = x^3(x + 2)$

21. $y = x^5 - 5x^4 = x^4(x - 5)$

22. $y = x\left(\frac{x}{2} - 5\right)^4$

23. $y = \frac{2x^2 + x - 1}{x^2 - 1}$

24. $y = \frac{x^2 - 49}{x^2 + 5x - 14}$

25. $y = \frac{x^4 + 1}{x^2}$

26. $y = \frac{x^2 - 4}{2x}$

27. $y = \frac{x^2 + 1}{x^2 - 1}$

28. $y = \frac{x^2}{x^2 - 1}$

29. $y = -\frac{x^2 - 2}{x^2 - 1}$

30. $y = \frac{x^2 - 4}{x^2 - 2}$

31. $y = \frac{x^2}{x + 1}$

32. $y = -\frac{x^2 - 4}{x + 1}$

33. $y = \frac{x^2 - x + 1}{x - 1}$

34. $y = -\frac{x^2 - x + 1}{x - 1}$

35. $y = \frac{x^3 - 3x^2 + 3x - 1}{x^2 + x - 2}$

36. $y = \frac{x^3 + x - 2}{x - x^2}$

37. $y = \frac{x}{x^2 - 1}$

38. $y = \frac{4x}{x^2 + 4}$ (Newton's serpentine)

39. $y = \frac{8}{x^2 + 4}$ (Agnesi's witch)

40. $y = \frac{x}{\sqrt{x^2 + 1}}$

41. $y = x + \sin x, \quad 0 \le x \le 2\pi$

42. $y = x - \sin x, \quad 0 \le x \le 2\pi$

43. $y = \sqrt{3}x - 2\cos x, \quad 0 \le x \le 2\pi$

44. $y = \frac{4}{3}x - \tan x, \quad \frac{-\pi}{2} < x < \frac{\pi}{2}$

45. $y = \sin x \cos x, \quad 0 \le x \le \pi$

46. $y = \cos x + \sqrt{3}\sin x, \quad 0 \le x \le 2\pi$

47. $y = x^{1/5}$

48. $y = x^{2/5}$

49. $y = 2x - 3x^{2/3}$

50. $y = 5x^{2/5} - 2x$

51. $y = x^{2/3}\left(\frac{5}{2} - x\right)$

52. $y = x^{2/3}(x - 5)$

53. $y = x\sqrt{8 - x^2}$

54. $y = (2 - x^2)^{3/2}$

55. $y = \sqrt{16 - x^2}$

56. $y = x^2 + \frac{2}{x}$

57. $y = \frac{x^2 - 3}{x - 2}$

58. $y = \sqrt[3]{x^3 + 1}$

59. $y = \frac{8x}{x^2 + 4}$

60. $y = \frac{5}{x^4 + 5}$

61. $y = |x^2 - 1|$

62. $y = |x^2 - 2x|$

63. $y = \sqrt{|x|} = \begin{cases} \sqrt{-x}, & x < 0 \\ \sqrt{x}, & x \ge 0 \end{cases}$

64. $y = \sqrt{|x - 4|}$

65. $y = \frac{x}{9 - x^2}$

66. $y = \frac{x^2}{1 - x}$

Sketching the General Shape, Knowing y'

Each of Exercises 67–88 gives the first derivative of a continuous function $y = f(x)$. Find y'' and then use Steps 2–4 of the graphing procedure described in this section to sketch the general shape of the graph of f.

67. $y' = 2 + x - x^2$ **68.** $y' = x^2 - x - 6$

69. $y' = x(x - 3)^2$ **70.** $y' = x^2(2 - x)$

71. $y' = x(x^2 - 12)$ **72.** $y' = (x - 1)^2(2x + 3)$

73. $y' = (8x - 5x^2)(4 - x)^2$ **74.** $y' = (x^2 - 2x)(x - 5)^2$

75. $y' = \sec^2 x, \quad -\frac{\pi}{2} < x < \frac{\pi}{2}$

76. $y' = \tan x, \quad -\frac{\pi}{2} < x < \frac{\pi}{2}$

77. $y' = \cot\frac{\theta}{2}, \quad 0 < \theta < 2\pi$ **78.** $y' = \csc^2\frac{\theta}{2}, \quad 0 < \theta < 2\pi$

79. $y' = \tan^2\theta - 1, \quad -\frac{\pi}{2} < \theta < \frac{\pi}{2}$

80. $y' = 1 - \cot^2\theta, \quad 0 < \theta < \pi$

81. $y' = \cos t, \quad 0 \le t \le 2\pi$ **82.** $y' = \sin t, \quad 0 \le t \le 2\pi$

83. $y' = (x + 1)^{-2/3}$ **84.** $y' = (x - 2)^{-1/3}$

85. $y' = x^{-2/3}(x - 1)$ **86.** $y' = x^{-4/5}(x + 1)$

87. $y' = 2|x| = \begin{cases} -2x, & x \le 0 \\ 2x, & x > 0 \end{cases}$

88. $y' = \begin{cases} -x^2, & x \le 0 \\ x^2, & x > 0 \end{cases}$

Sketching y from Graphs of y' and y''

Each of Exercises 89–92 shows the graphs of the first and second derivatives of a function $y = f(x)$. Copy the picture and add to it a sketch of the approximate graph of f, given that the graph passes through the point P.

89.

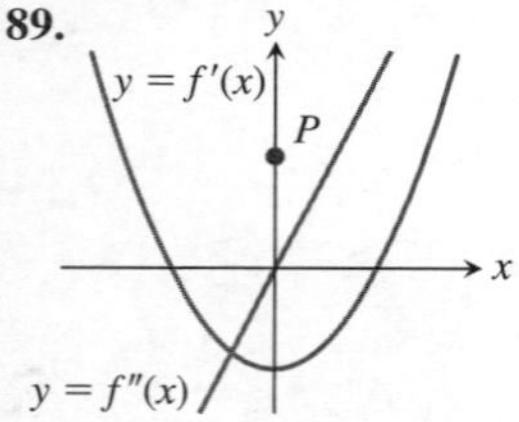

90.

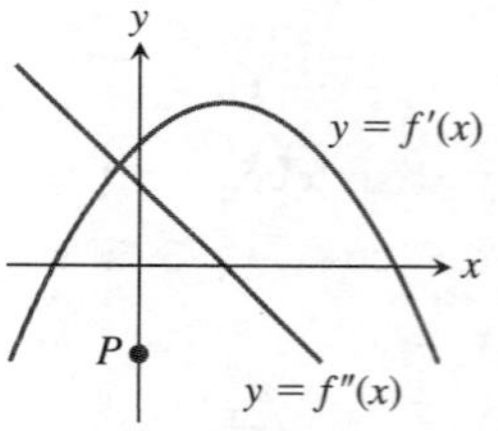

91.

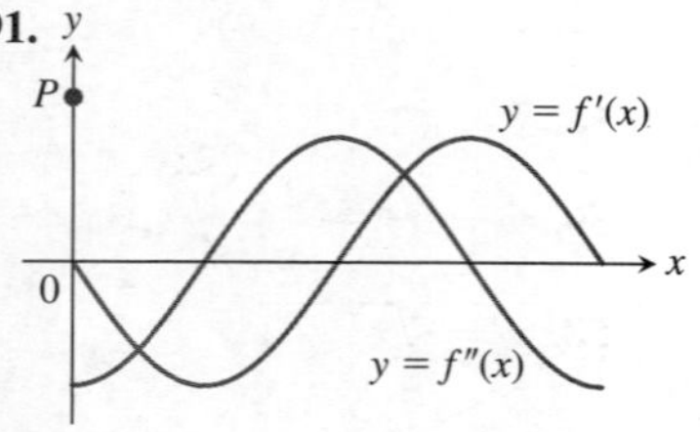

92.

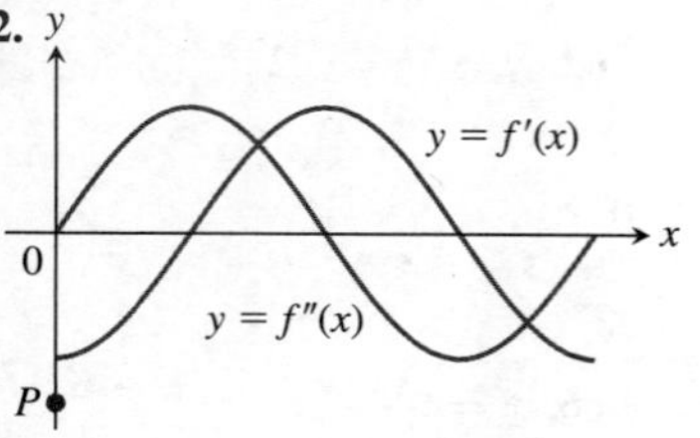

Theory and Examples

93. The accompanying figure shows a portion of the graph of a twice-differentiable function $y = f(x)$. At each of the five labeled points, classify y' and y'' as positive, negative, or zero.

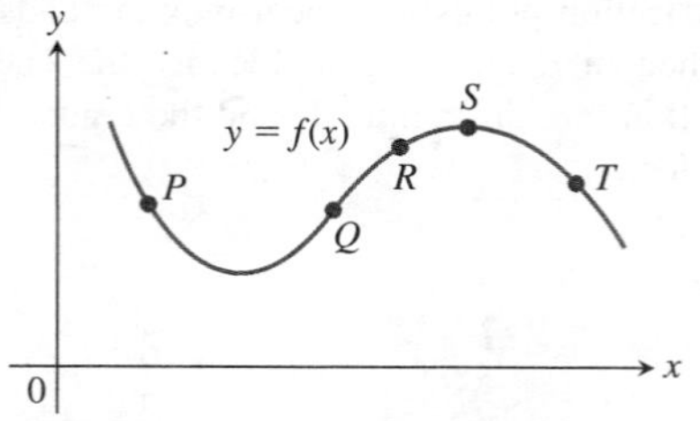

94. Sketch a smooth connected curve $y = f(x)$ with

$$f(-2) = 8, \quad f'(2) = f'(-2) = 0,$$
$$f(0) = 4, \quad f'(x) < 0 \quad \text{for} \quad |x| < 2,$$
$$f(2) = 0, \quad f''(x) < 0 \quad \text{for} \quad x < 0,$$
$$f'(x) > 0 \quad \text{for} \quad |x| > 2, \quad f''(x) > 0 \quad \text{for} \quad x > 0.$$

95. Sketch the graph of a twice-differentiable function $y = f(x)$ with the following properties. Label coordinates where possible.

x	y	Derivatives
$x < 2$		$y' < 0, \quad y'' > 0$
2	1	$y' = 0, \quad y'' > 0$
$2 < x < 4$		$y' > 0, \quad y'' > 0$
4	4	$y' > 0, \quad y'' = 0$
$4 < x < 6$		$y' > 0, \quad y'' < 0$
6	7	$y' = 0, \quad y'' < 0$
$x > 6$		$y' < 0, \quad y'' < 0$

96. Sketch the graph of a twice-differentiable function $y = f(x)$ that passes through the points $(-2, 2)$, $(-1, 1)$, $(0, 0)$, $(1, 1)$, and $(2, 2)$ and whose first two derivatives have the following sign patterns.

y': + (before -2), − (between -2 and 0), + (between 0 and 2), − (after 2)

y'': − (before -1), + (between -1 and 1), − (after 1)

97. Sketch the graph of a twice-differentiable function $y = f(x)$ with the following properties. Label coordinates where possible.

x	y	Derivatives
$x < -2$		$y' > 0, \quad y'' < 0$
-2	-1	$y' = 0, \quad y'' = 0$
$-2 < x < -1$		$y' > 0, \quad y'' > 0$
-1	0	$y' > 0, \quad y'' = 0$
$-1 < x < 0$		$y' > 0, \quad y'' < 0$
0	3	$y' = 0, \quad y'' < 0$
$0 < x < 1$		$y' < 0, \quad y'' < 0$
1	2	$y' < 0, \quad y'' = 0$
$1 < x < 2$		$y' < 0, \quad y'' > 0$
2	0	$y' = 0, \quad y'' > 0$
$x > 2$		$y' > 0, \quad y'' > 0$

98. Sketch the graph of a twice-differentiable function $y = f(x)$ that passes through the points $(-3, -2)$, $(-2, 0)$, $(0, 1)$, $(1, 2)$, and $(2, 3)$ and whose first two derivatives have the following sign patterns.

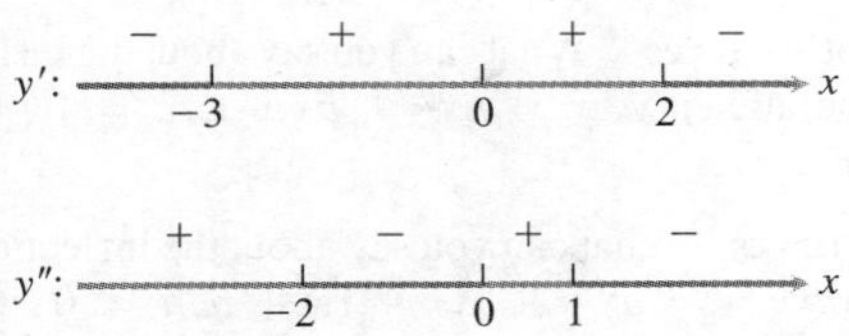

In Exercises 99 and 100, the graph of f' is given. Determine x-values corresponding to inflection points for the graph of f.

99.

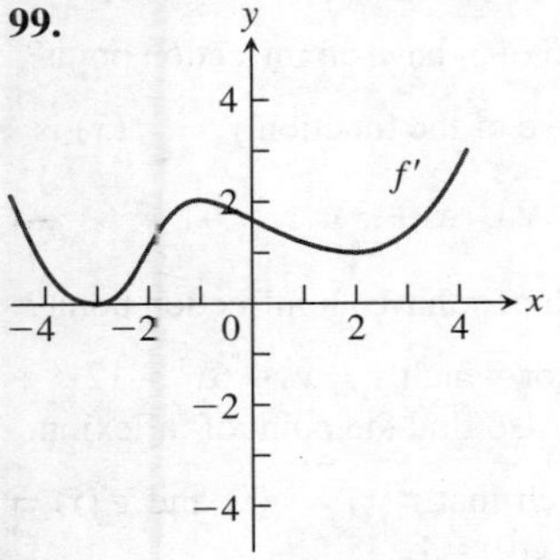

100.

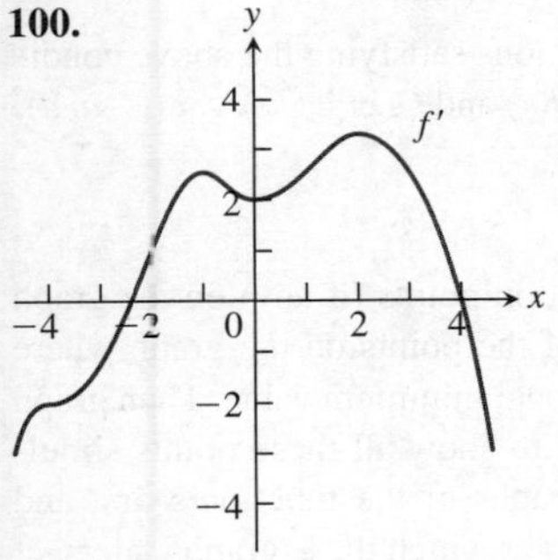

In Exercises 101 and 102, the graph of f' is given. Determine x-values corresponding to local minima, local maxima, and inflection points for the graph of f.

101.

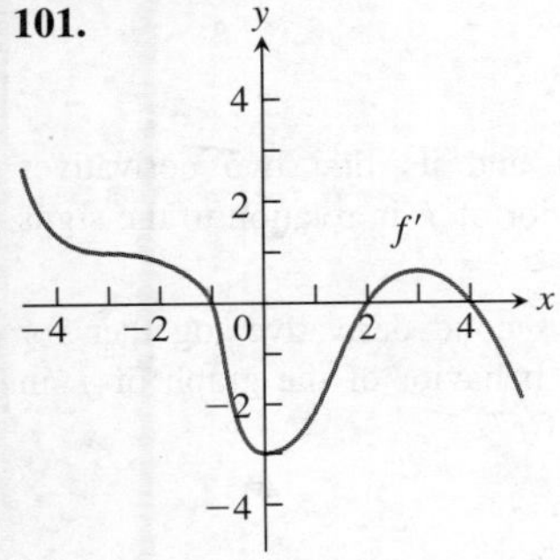

102.

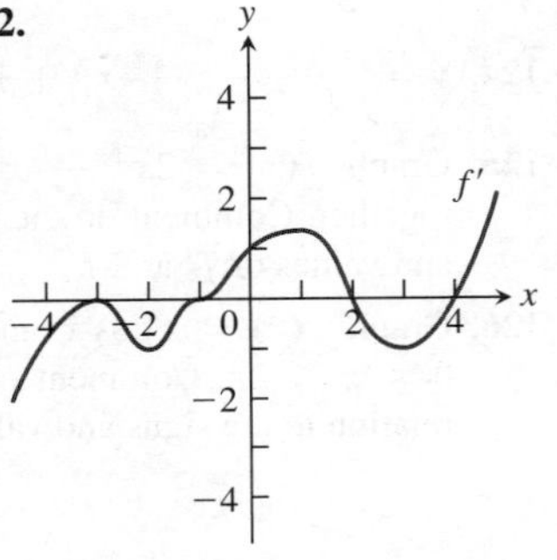

103. A function $f(x)$ has domain $(-2, 2)$. The graph below is a plot of the derivative of f, not a plot of f itself. In other words, this is a graph of $y = f'(x)$. Either use this graph to determine on which intervals the graph of f is concave up and on which intervals the graph of f is concave down, or explain why this information cannot be determined from the graph.

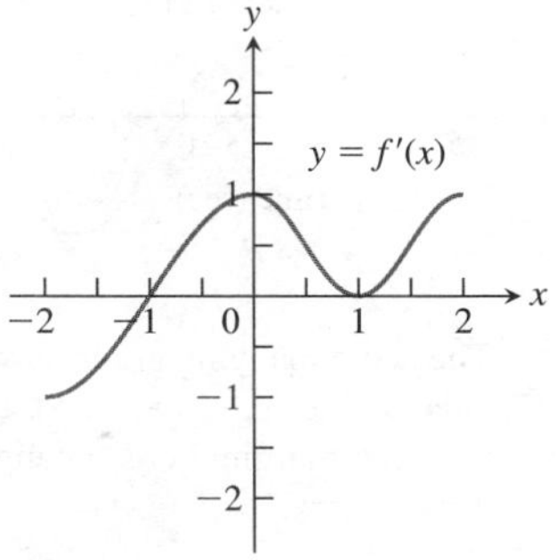

104. A function $f(x)$ has domain $(-2, 2)$. The graph below is a plot of the second derivative of f, not a plot of f itself. In other words, this is a graph of $y = f''(x)$.

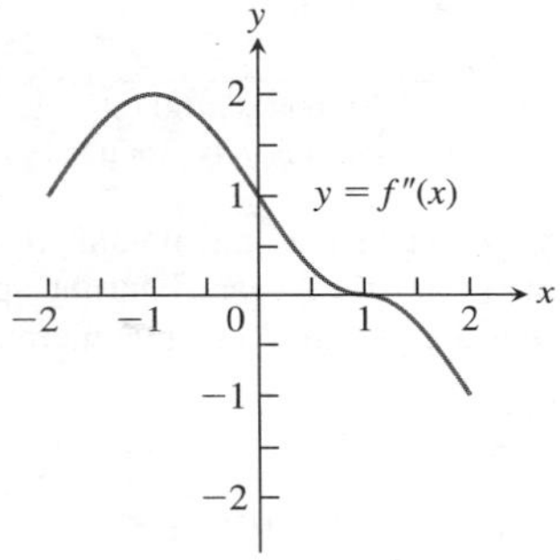

a. Either use the graph above to determine on which intervals the graph of f is concave up and on which intervals the graph of f is concave down and the inflection points of f, or explain why this information cannot be determined from the graph.

b. Either use the graph above to determine on which intervals f is increasing and on which intervals f is decreasing, or explain why this information cannot be determined from the graph.

Motion Along a Line The graphs in Exercises 105 and 106 show the position $s = f(t)$ of an object moving up and down on a coordinate line. **(a)** When is the object moving away from the origin? Toward the origin? At approximately what times is the **(b)** velocity equal to zero? **(c)** Acceleration equal to zero? **(d)** When is the acceleration positive? Negative?

105.

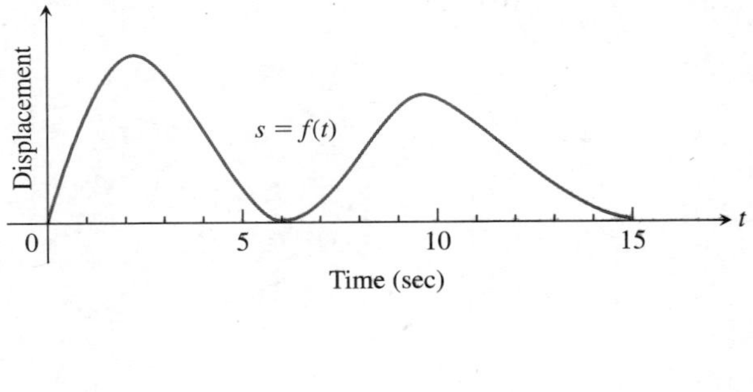

106.

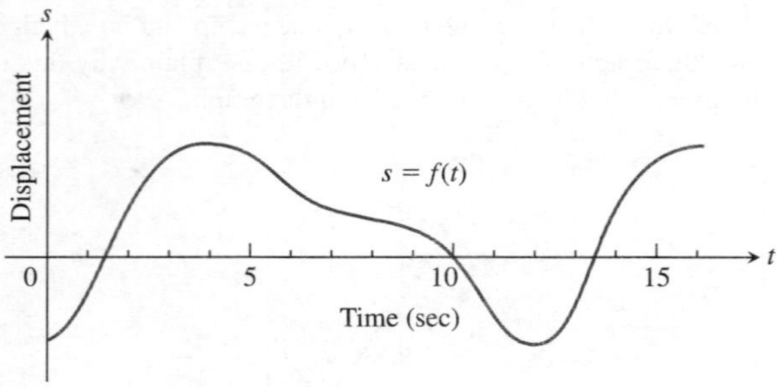

107. Marginal cost The accompanying graph shows the hypothetical cost $c = f(x)$ of manufacturing x items. At approximately what production level does the marginal cost change from decreasing to increasing?

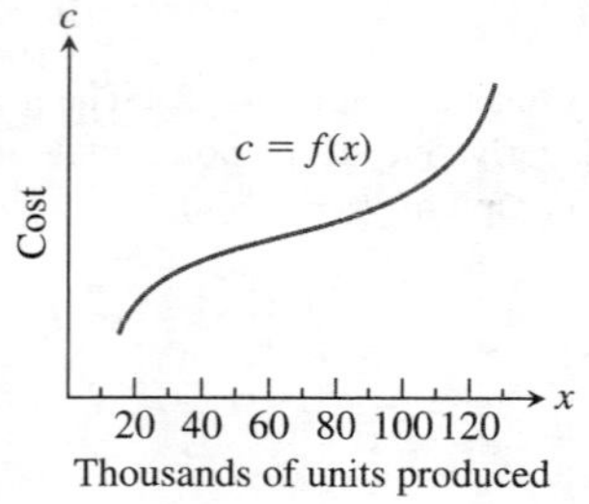

108. The accompanying graph shows the monthly revenue of the Widget Corporation for the past 12 years. During approximately what time intervals was the marginal revenue increasing? Decreasing?

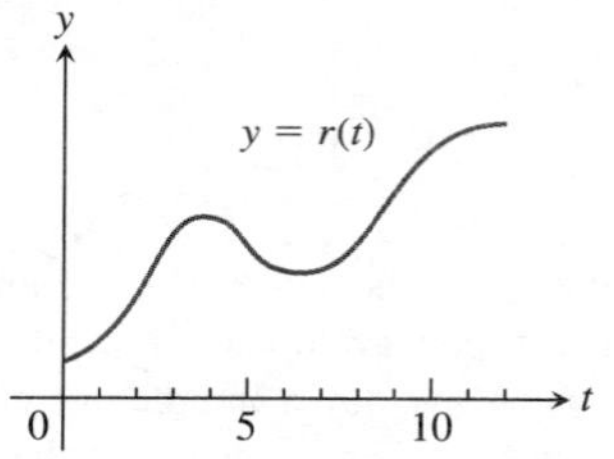

109. Sketch the curve $y^2 = \dfrac{x^2(a^2 - x^2)}{a^2 + x^2}$.

110. Sketch the curve $y^2 = (x-1)(x-2)(x-3)$.

111. For $x > 0$, sketch a curve $y = f(x)$ that has $f(1) = 0$ and $f'(x) = 1/x$. Can anything be said about the concavity of such a curve? Give reasons for your answer.

112. Can anything be said about the graph of a function $y = f(x)$ that has a continuous second derivative that is never zero? Give reasons for your answer.

113. If b, c, and d are constants, for what value of b will the curve $y = x^3 + bx^2 + cx + d$ have a point of inflection at $x = 1$? Give reasons for your answer.

114. Parabolas

a. Find the coordinates of the vertex of the parabola $y = ax^2 + bx + c, a \neq 0$.

b. When is the parabola concave up? Concave down? Give reasons for your answers.

115. Quadratic curves What can you say about the inflection points of a quadratic curve $y = ax^2 + bx + c, a \neq 0$? Give reasons for your answer.

116. Cubic curves What can you say about the inflection points of a cubic curve $y = ax^3 + bx^2 + cx + d, a \neq 0$? Give reasons for your answer.

117. Suppose that the second derivative of the function $y = f(x)$ is

$$y'' = (x + 1)(x - 2).$$

For what x-values does the graph of f have an inflection point?

118. Suppose that the second derivative of the function $y = f(x)$ is

$$y'' = x^2(x - 2)^3(x + 3).$$

For what x-values does the graph of f have an inflection point?

119. Find the ranges of values of x for which $y = x^4 - 6x^3 + 12x^2 + 4x + 10$ is concave up or down. Also find the point of inflexion.

120. Given two functions f and g such that $f'(x) = g(x)$ and $g'(x) = -f(x)\ \forall\ x \in (-a, a)$ and $f(0) = 0$, $g(0) = 1$.

(i) Prove $[f(x)]^2 + [g(x)]^2 = 1\ \forall\ x \in (-a, a)$.

(ii) If F and G are a pair of functions satisfying the above conditions, then prove that $F(x) = f(x)$ and $G(x) = g(x)\ \forall\ x \in (-a, a)$.

COMPUTER EXPLORATIONS

In Exercises 121–124, find the inflection points (if any) on the graph of the function and the coordinates of the points on the graph where the function has a local maximum or local minimum value. Then graph the function in a region large enough to show all these points simultaneously. Add to your picture the graphs of the function's first and second derivatives. How are the values at which these graphs intersect the x-axis related to the graph of the function? In what other ways are the graphs of the derivatives related to the graph of the function?

121. $y = x^5 - 5x^4 - 240$ **122.** $y = x^3 - 12x^2$

123. $y = \dfrac{4}{5}x^5 + 16x^2 - 25$

124. $y = \dfrac{x^4}{4} - \dfrac{x^3}{3} - 4x^2 + 12x + 20$

125. Graph $f(x) = 2x^4 - 4x^2 + 1$ and its first two derivatives together. Comment on the behavior of f in relation to the signs and values of f' and f''.

126. Graph $f(x) = x \cos x$ and its second derivative together for $0 \leq x \leq 2\pi$. Comment on the behavior of the graph of f in relation to the signs and values of f''.

4.5 Applied Optimization

What are the dimensions of a rectangle with fixed perimeter having *maximum area*? What are the dimensions for the *least expensive* cylindrical can of a given volume? How many items should be produced for the *most profitable* production run? Each of these questions asks for the best, or optimal, value of a given function. In this section we use derivatives to solve a variety of optimization problems in mathematics, physics, economics, and business.

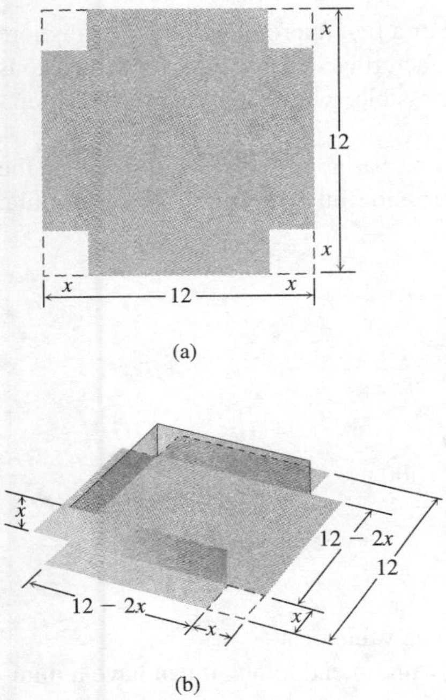

FIGURE 4.35 An open box made by cutting the corners from a square sheet of tin. What size corners maximize the box's volume (Example 1)?

Solving Applied Optimization Problems

1. *Read the problem.* Read the problem until you understand it. What is given? What is the unknown quantity to be optimized (maximized or minimized)?
2. *Introduce variables.* List every relevant relation in the problem as an equation. In most problems it is helpful to draw a picture.
3. *Write an equation for the unknown quantity.* Express the quantity to be optimized as a function of a single variable. This may require considerable manipulation.
4. *Test the critical points and endpoints in the domain of the function found in the previous step.* Use what you know about the shape of the function's graph. Use the first and second derivatives to identify and classify the function's critical points.

EXAMPLE 1 An open-top box is to be made by cutting small congruent squares from the corners of a 12-in.-by-12-in. sheet of tin and bending up the sides. How large should the squares cut from the corners be to make the box hold as much as possible?

Solution We start with a picture (Figure 4.35). In the figure, the corner squares are x in. on a side. The volume of the box is a function of this variable:

$$V(x) = x(12 - 2x)^2 = 144x - 48x^2 + 4x^3. \qquad V = hlw$$

Since the sides of the sheet of tin are only 12 in. long, $x \le 6$ and the domain of V is the interval $0 \le x \le 6$. The graph of V is shown in Figure 4.36.

We examine the first derivative of V with respect to x:

$$\frac{dV}{dx} = 144 - 96x + 12x^2 = 12(12 - 8x + x^2) = 12(2 - x)(6 - x).$$

Of the two zeros, $x = 2$ and $x = 6$, only $x = 2$ lies in the interior of the function's domain and makes the critical-point list. The values of V at this one critical point and two endpoints are

Critical-point value: $V(2) = 128$

Endpoint values: $V(0) = 0, \quad V(6) = 0.$

The maximum volume is 128 in^3. The cutout squares should be 2 in. on a side. ■

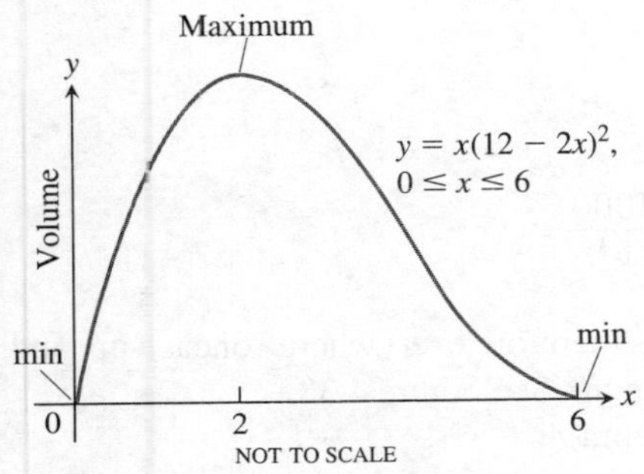

FIGURE 4.36 The volume of the box in Figure 4.35 graphed as a function of x.

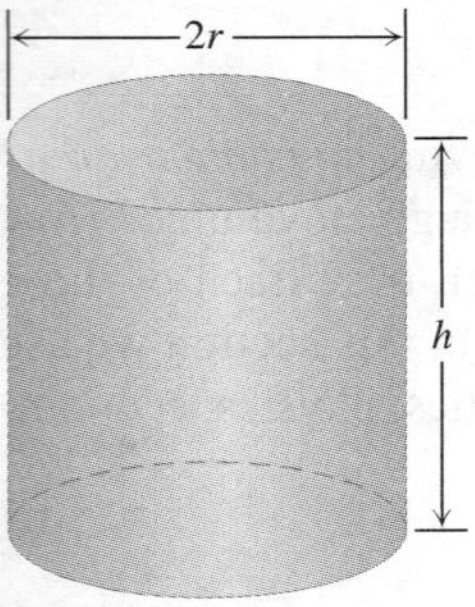

FIGURE 4.37 Example 2 shows that this one-liter can uses the least material when $h = 2r$.

EXAMPLE 2 You have been asked to design a one-liter can shaped like a right circular cylinder (Figure 4.37). What dimensions will use the least material?

Solution *Volume of can:* If r and h are measured in centimeters, then the volume of the can in cubic centimeters is

$$\pi r^2 h = 1000. \qquad \text{1 liter} = 1000 \text{ cm}^3$$

$$\textit{Surface area of can:} \qquad A = \underbrace{2\pi r^2}_{\substack{\text{circular}\\ \text{ends}}} + \underbrace{2\pi r h}_{\substack{\text{cylindrical}\\ \text{wall}}}$$

How can we interpret the phrase "least material"? For a first approximation we can ignore the thickness of the material and the waste in manufacturing. Then we ask for dimensions r and h that make the total surface area as small as possible, while satisfying the constraint $\pi r^2 h = 1000 \text{ cm}^3$.

To express the surface area as a function of one variable, we solve for one of the variables in $\pi r^2 h = 1000$ and substitute that expression into the surface area formula. Solving for h is easier:

$$h = \frac{1000}{\pi r^2}.$$

Thus,

$$\begin{aligned} A &= 2\pi r^2 + 2\pi r h \\ &= 2\pi r^2 + 2\pi r\left(\frac{1000}{\pi r^2}\right) \\ &= 2\pi r^2 + \frac{2000}{r}. \end{aligned}$$

Our goal is to find a value of $r > 0$ that minimizes the value of A.

Since A is differentiable on $r > 0$, an interval with no endpoints, it can have a minimum value only where its first derivative is zero.

$$\begin{aligned} \frac{dA}{dr} &= 4\pi r - \frac{2000}{r^2} \\ 0 &= 4\pi r - \frac{2000}{r^2} && \text{Set } dA/dr = 0. \\ 4\pi r^3 &= 2000 && \text{Multiply by } r^2. \\ r &= \sqrt[3]{\frac{500}{\pi}} \approx 5.42 && \text{Solve for } r. \end{aligned}$$

What happens at $r = \sqrt[3]{500/\pi}$?

The second derivative

$$\frac{d^2A}{dr^2} = 4\pi + \frac{4000}{r^3}$$

is positive throughout the domain of A. The graph is therefore everywhere concave up, and the value of A at $r = \sqrt[3]{500/\pi}$ is an absolute minimum. See Figure 4.38.

The corresponding value of h (after a little algebra) is

$$h = \frac{1000}{\pi r^2} = 2\sqrt[3]{\frac{500}{\pi}} = 2r.$$

The one-liter can that uses the least material has height equal to twice the radius, here with $r \approx 5.42$ cm and $h \approx 10.84$ cm. ■

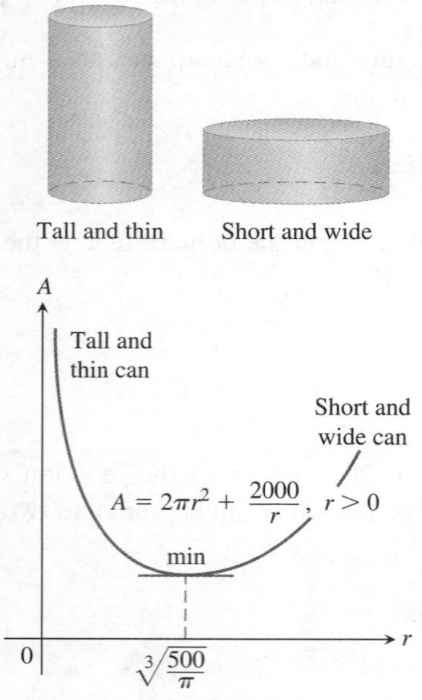

FIGURE 4.38 The graph of $A = 2\pi r^2 + 2000/r$ is concave up.

Examples from Mathematics and Physics

EXAMPLE 3 A rectangle is to be inscribed in a semicircle of radius 2. What is the largest area the rectangle can have, and what are its dimensions?

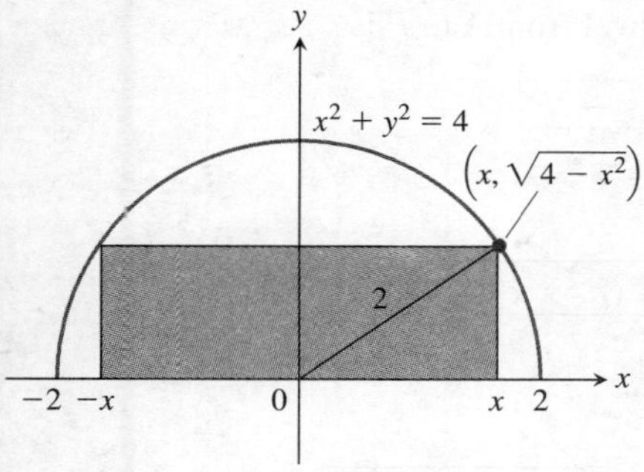

FIGURE 4.39 The rectangle inscribed in the semicircle in Example 3.

Solution Let $\left(x, \sqrt{4 - x^2}\right)$ be the coordinates of the upper right corner of the rectangle obtained by placing the circle and rectangle in the coordinate plane (Figure 4.39). The length, height, and area of the rectangle can then be expressed in terms of the position x of the lower right-hand corner:

$$\text{Length: } 2x, \qquad \text{Height: } \sqrt{4 - x^2}, \qquad \text{Area: } 2x\sqrt{4 - x^2}.$$

Notice that the values of x are to be found in the interval $0 \le x \le 2$, where the selected corner of the rectangle lies.

Our goal is to find the absolute maximum value of the function

$$A(x) = 2x\sqrt{4 - x^2}$$

on the domain $[0, 2]$.

The derivative

$$\frac{dA}{dx} = \frac{-2x^2}{\sqrt{4 - x^2}} + 2\sqrt{4 - x^2}$$

is not defined when $x = 2$ and is equal to zero when

$$\begin{aligned}
\frac{-2x^2}{\sqrt{4 - x^2}} + 2\sqrt{4 - x^2} &= 0 \\
-2x^2 + 2(4 - x^2) &= 0 \\
8 - 4x^2 &= 0 \\
x^2 &= 2 \\
x &= \pm\sqrt{2}.
\end{aligned}$$

Of the two zeros, $x = \sqrt{2}$ and $x = -\sqrt{2}$, only $x = \sqrt{2}$ lies in the interior of A's domain and makes the critical-point list. The values of A at the endpoints and at this one critical point are

$$\begin{aligned}
&\text{Critical-point value: } A\left(\sqrt{2}\right) = 2\sqrt{2}\sqrt{4 - 2} = 4 \\
&\text{Endpoint values: } \quad A(0) = 0, \quad A(2) = 0.
\end{aligned}$$

The area has a maximum value of 4 when the rectangle is $\sqrt{4 - x^2} = \sqrt{2}$ units high and $2x = 2\sqrt{2}$ units long. ■

HISTORICAL BIOGRAPHY
Willebrord Snell van Royen
(1580–1626)
www.bit.ly/2NSNyDX

EXAMPLE 4 The speed of light depends on the medium through which it travels, and is generally slower in denser media.

Fermat's principle in optics states that light travels from one point to another along a path for which the time of travel is a minimum. Describe the path that a ray of light will follow in going from a point A in a medium where the speed of light is c_1 to a point B in a second medium where the speed of light is c_2.

Solution Since light traveling from A to B follows the quickest route, we look for a path that will minimize the travel time. We assume that A and B lie in the xy-plane and that the line separating the two media is the x-axis (Figure 4.40). We place A at coordinates $(0, a)$ and B at coordinates $(d, -b)$ in the xy-plane.

In a uniform medium, where the speed of light remains constant, "shortest time" means "shortest path," and the ray of light will follow a straight line. Thus the path from A

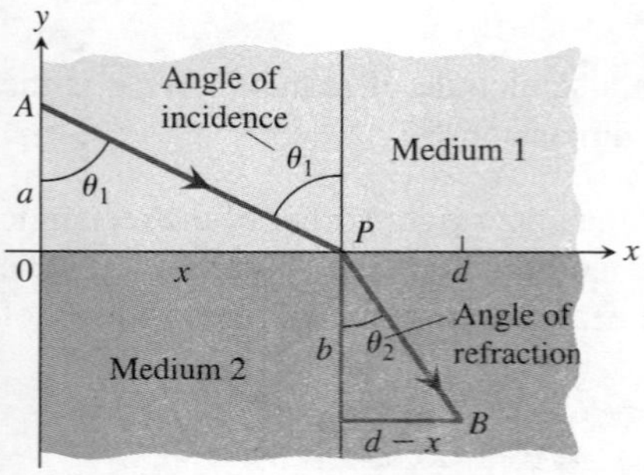

FIGURE 4.40 A light ray refracted (deflected from its path) as it passes from one medium to a denser medium (Example 4).

to B will consist of a line segment from A to a boundary point P, followed by another line segment from P to B. Distance traveled equals rate times time, so

$$\text{Time} = \frac{\text{distance}}{\text{rate}}.$$

From Figure 4.40, the time required for light to travel from A to P is

$$t_1 = \frac{AP}{c_1} = \frac{\sqrt{a^2 + x^2}}{c_1}.$$

From P to B, the time is

$$t_2 = \frac{PB}{c_2} = \frac{\sqrt{b^2 + (d - x)^2}}{c_2}.$$

The time from A to B is the sum of these:

$$t = t_1 + t_2 = \frac{\sqrt{a^2 + x^2}}{c_1} + \frac{\sqrt{b^2 + (d - x)^2}}{c_2}.$$

This equation expresses t as a differentiable function of x whose domain is $[0, d]$. We want to find the absolute minimum value of t on this closed interval. We find the derivative

$$\frac{dt}{dx} = \frac{x}{c_1\sqrt{a^2 + x^2}} - \frac{d - x}{c_2\sqrt{b^2 + (d - x)^2}}$$

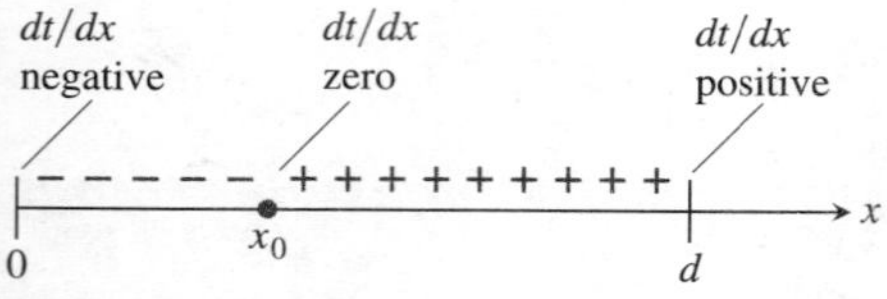

FIGURE 4.41 The sign pattern of dt/dx in Example 4.

and observe that it is continuous. In terms of the angles θ_1 and θ_2 in Figure 4.40,

$$\frac{dt}{dx} = \frac{\sin\theta_1}{c_1} - \frac{\sin\theta_2}{c_2}.$$

The function t has a negative derivative at $x = 0$ and a positive derivative at $x = d$. Since dt/dx is continuous over the interval $[0, d]$, by the Intermediate Value Theorem for continuous functions (Section 2.5), there is a point $x_0 \in [0, d]$ where $dt/dx = 0$ (Figure 4.41). There is only one such point because dt/dx is an increasing function of x (Exercise 68). At this unique point we then have

$$\frac{\sin\theta_1}{c_1} = \frac{\sin\theta_2}{c_2}.$$

This equation is **Snell's Law** or the **Law of Refraction**, and it is an important principle in the theory of optics. It describes the path the ray of light follows. ■

Examples from Economics

Suppose that

$$\begin{aligned} r(x) &= \text{the revenue from selling } x \text{ items} \\ c(x) &= \text{the cost of producing the } x \text{ items} \\ p(x) &= r(x) - c(x) = \text{the profit from producing and selling } x \text{ items.} \end{aligned}$$

Although x is usually an integer in many applications, we can learn about the behavior of these functions by defining them for all nonzero real numbers and by assuming they are differentiable functions. Economists use the terms **marginal revenue**, **marginal cost**, and **marginal profit** to name the derivatives $r'(x)$, $c'(x)$, and $p'(x)$ of the revenue, cost, and profit functions. Let's consider the relationship of the profit p to these derivatives.

If $r(x)$ and $c(x)$ are differentiable for x in some interval of production possibilities, and if $p(x) = r(x) - c(x)$ has a maximum value there, it occurs at a critical point of $p(x)$ or at an endpoint of the interval. If it occurs at a critical point, then $p'(x) = r'(x) - c'(x) = 0$ and we see that $r'(x) = c'(x)$. In economic terms, this last equation means that

At a production level yielding maximum profit, marginal revenue equals marginal cost (Figure 4.42).

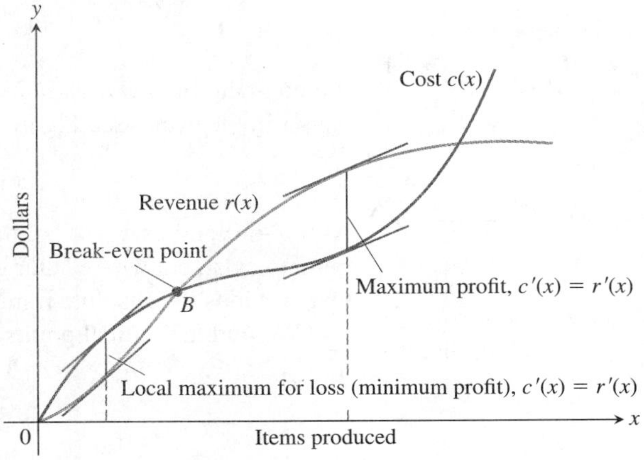

FIGURE 4.42 The graph of a typical cost function starts concave down and later turns concave up. It crosses the revenue curve at the break-even point B. To the left of B, the company operates at a loss. To the right, the company operates at a profit, with the maximum profit occurring where $c'(x) = r'(x)$. Farther to the right, cost exceeds revenue (perhaps because of a combination of rising labor and material costs, and market saturation) and production levels become unprofitable again.

EXAMPLE 5 Suppose that $r(x) = 9x$ and $c(x) = x^3 - 6x^2 + 15x$ are the revenue and the cost functions, given in millions of dollars, where x represents millions of MP3 players produced. Is there a production level that maximizes profit? If so, what is it?

Solution Notice that $r'(x) = 9$ and $c'(x) = 3x^2 - 12x + 15$.

$$3x^2 - 12x + 15 = 9 \qquad \text{Set } c'(x) = r'(x).$$

$$3x^2 - 12x + 6 = 0$$

The two solutions of the quadratic equation are

$$x_1 = \frac{12 - \sqrt{72}}{6} = 2 - \sqrt{2} \approx 0.586 \quad \text{and}$$

$$x_2 = \frac{12 + \sqrt{72}}{6} = 2 + \sqrt{2} \approx 3.414.$$

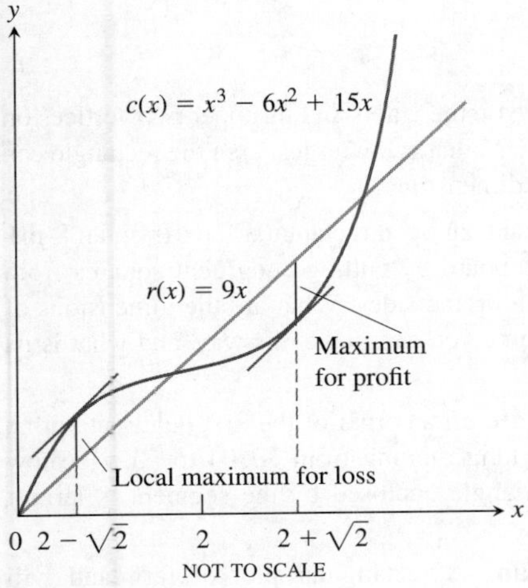

FIGURE 4.43 The cost and revenue curves for Example 5.

The possible production levels for maximum profit are $x \approx 0.586$ million MP3 players or $x \approx 3.414$ million. The second derivative of $p(x) = r(x) - c(x)$ is $p''(x) = -c''(x)$ since $r''(x)$ is everywhere zero. Thus, $p''(x) = 6(2 - x)$, which is negative at $x = 2 + \sqrt{2}$ and positive at $x = 2 - \sqrt{2}$. By the Second Derivative Test, a maximum profit occurs at about $x = 3.414$ (where revenue exceeds costs) and maximum loss occurs at about $x = 0.586$. The graphs of $r(x)$ and $c(x)$ are shown in Figure 4.43. ■

EXAMPLE 6 A cabinetmaker uses cherry wood to produce 5 desks each day. Each delivery of one container of wood is \$5000, whereas the storage of that material is \$10 per day per unit stored, where a unit is the amount of material needed by her to produce 1 desk. How much material should be ordered each time, and how often should the material be delivered, to minimize her average daily cost in the production cycle between deliveries?

Solution If she asks for a delivery every x days, then she must order $5x$ units to have enough material for that delivery cycle. The *average* amount in storage is approximately

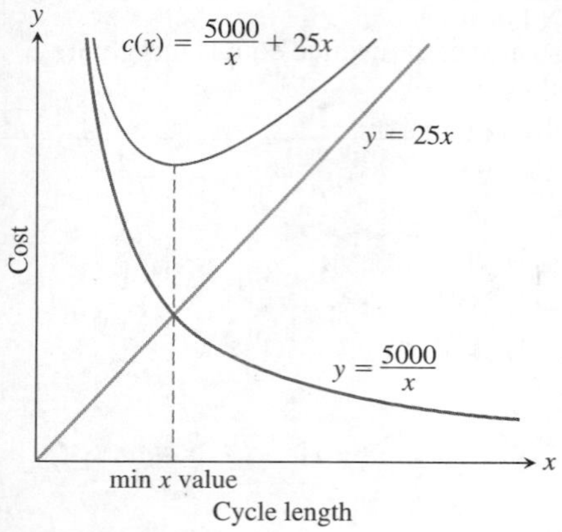

FIGURE 4.44 The average daily cost $c(x)$ is the sum of a hyperbola and a linear function (Example 6).

one-half of the delivery amount, or $5x/2$. Thus, the cost of delivery and storage for each cycle is approximately

$$\text{Cost per cycle} = \text{delivery costs} + \text{storage costs}$$

$$\text{Cost per cycle} = \underbrace{5000}_{\substack{\text{delivery}\\ \text{cost}}} + \underbrace{\left(\frac{5x}{2}\right)}_{\substack{\text{average}\\ \text{amount stored}}} \cdot \underbrace{x}_{\substack{\text{number of}\\ \text{days stored}}} \cdot \underbrace{10}_{\substack{\text{storage cost}\\ \text{per day}}}$$

We compute the *average daily cost* $c(x)$ by dividing the cost per cycle by the number of days x in the cycle (see Figure 4.44).

$$c(x) = \frac{5000}{x} + 25x, \qquad x > 0.$$

As $x \to 0$ and as $x \to \infty$, the average daily cost becomes large. So we expect a minimum to exist, but where? Our goal is to determine the number of days x between deliveries that provides the absolute minimum cost.

We find the critical points by determining where the derivative is equal to zero:

$$c'(x) = -\frac{500}{x^2} + 25 = 0$$

$$x = \pm\sqrt{200} \approx \pm 14.14.$$

Of the two critical points, only $\sqrt{200}$ lies in the domain of $c(x)$. The critical point value of the average daily cost is

$$c\left(\sqrt{200}\right) = \frac{5000}{\sqrt{200}} + 25\sqrt{200} = 500\sqrt{2} \approx \$707.11.$$

We note that $c(x)$ is defined over the open interval $(0, \infty)$ with $c''(x) = 10000/x^3 > 0$. Thus, an absolute minimum exists at $x = \sqrt{200} \approx 14.14$ days.

The cabinetmaker should schedule a delivery of $5(14) = 70$ units of wood every 14 days. ■

EXERCISES 4.5

Mathematical Applications

1. **Minimizing perimeter** What is the smallest perimeter possible for a rectangle whose area is 18 in^2, and what are its dimensions?

2. Show that among all rectangles with an 8-m perimeter, the one with largest area is a square.

3. The figure shows a rectangle inscribed in an isosceles right triangle whose hypotenuse is 2 units long.

   a. Express the y-coordinate of P in terms of x. (*Hint:* Write an equation for the line AB.)

   b. Express the area of the rectangle in terms of x.

   c. What is the largest area the rectangle can have, and what are its dimensions?

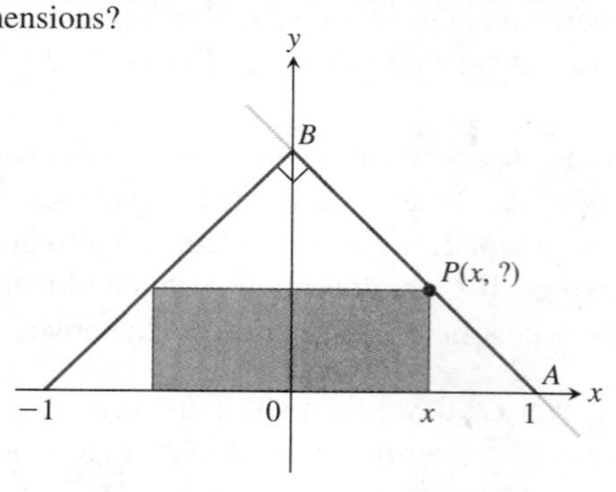

4. A rectangle has its base on the x-axis and its upper two vertices on the parabola $y = 27 - x^2$. What is the largest area the rectangle can have, and what are its dimensions?

5. You are planning to make an open rectangular box from an 8-in.-by-15-in. piece of cardboard by cutting congruent squares from the corners and folding up the sides. What are the dimensions of the box of largest volume you can make this way, and what is its volume?

6. You are planning to close off a corner of the first quadrant with a line segment 20 units long running from $(a, 0)$ to $(0, b)$. Show that the area of the triangle enclosed by the segment is largest when $a = b$.

7. **The best fencing plan** A rectangular plot of farmland will be bounded on one side by a river and on the other three sides by a single-strand electric fence. With 600 m of wire at your disposal, what is the largest area you can enclose, and what are its dimensions?

8. **The shortest fence** A 216 m^2 rectangular pea patch is to be enclosed by a fence and divided into two equal parts by another fence parallel to one of the sides. What dimensions for the outer rectangle will require the smallest total length of fence? How much fence will be needed?

9. **Designing a tank** Your iron works has contracted to design and build a 500 ft^3, square-based, open-top, rectangular steel holding tank for a paper company. The tank is to be made by welding thin stainless steel plates together along their edges. As the production engineer, your job is to find dimensions for the base and height that will make the tank weigh as little as possible.

   a. What dimensions do you tell the shop to use?

   b. Briefly describe how you took weight into account.

10. **Catching rainwater** A 1125 ft^3 open-top rectangular tank with a square base x ft on a side and y ft deep is to be built with its top flush with the ground to catch runoff water. The costs associated with the tank involve not only the material from which the tank is made but also an excavation charge proportional to the product xy.

    a. If the total cost is

    $$c = 5(x^2 + 4xy) + 10xy,$$

    what values of x and y will minimize it?

    b. Give a possible scenario for the cost function in part (a).

11. **Designing a poster** You are designing a rectangular poster to contain 72 in^2 of printing with a 4-in. margin at the top and bottom and a 2-in. margin at each side. What overall dimensions will minimize the amount of paper used?

12. Find the volume of the largest right circular cone that can be inscribed in a sphere of radius 3.

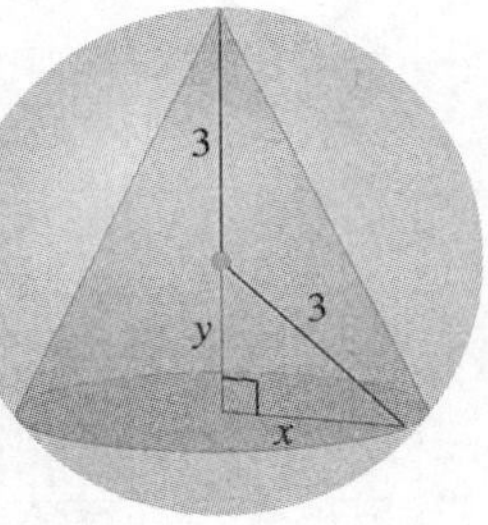

13. Two sides of a triangle have lengths a and b, and the angle between them is θ. What value of θ will maximize the triangle's area? (*Hint:* $A = (1/2)ab \sin \theta$.)

14. **Designing a can** What are the dimensions of the lightest open-top right circular cylindrical can that will hold a volume of 1000 cm^3? Compare the result here with the result in Example 2.

15. **Designing a can** You are designing a 1728 cm^3 right circular cylindrical can whose manufacture will take waste into account. There is no waste in cutting the aluminum for the side, but the top and bottom of radius r will be cut from squares that measure $2r$ units on a side. The total amount of aluminum used up by the can will therefore be rather than the in Example 2. In Example 2, the ratio of h to r for the most economical can was 2 to 1. What is the ratio now?

T 16. **Designing a box with a lid** A piece of cardboard measures 10 in. by 15 in. Two equal squares are removed from the corners of a 10-in. side as shown in the figure. Two equal rectangles are removed from the other corners so that the tabs can be folded to form a rectangular box with lid.

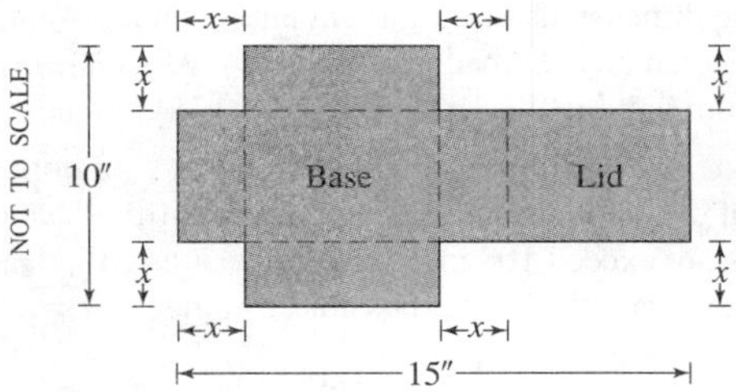

a. Write a formula $V(x)$ for the volume of the box.

b. Find the domain of V for the problem situation and graph V over this domain.

c. Use a graphical method to find the maximum volume and the value of x that gives it.

d. Confirm your result in part (c) analytically.

T 17. **Designing a suitcase** A 24-in.-by-36-in. sheet of cardboard is folded in half to form a 24-in.-by-18-in. rectangle as shown in the accompanying figure. Then four congruent squares of side length x are cut from the corners of the folded rectangle. The sheet is unfolded, and the six tabs are folded up to form a box with sides and a lid.

a. Write a formula $V(x)$ for the volume of the box.

b. Find the domain of V for the problem situation and graph V over this domain.

c. Use a graphical method to find the maximum volume and the value of x that gives it.

d. Confirm your result in part (c) analytically.

e. Find a value of x that yields a volume of 1120 in^3.

f. Write a paragraph describing the issues that arise in part (b).

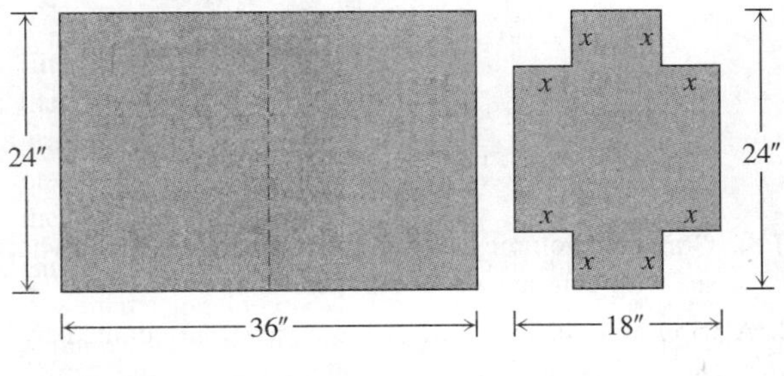

The sheet is then unfolded.

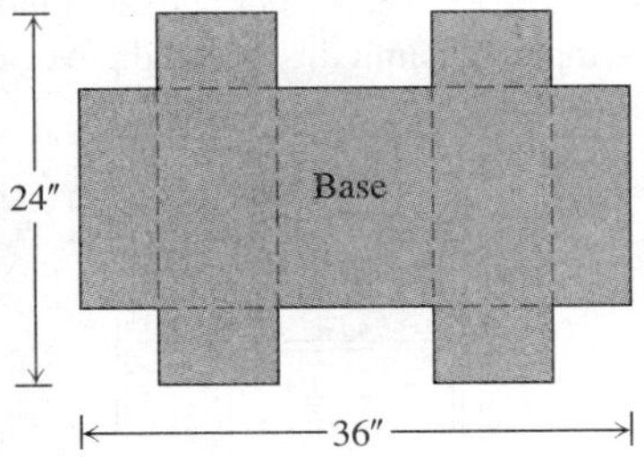

T 18. A rectangle is to be inscribed under the arch of the curve $y = 4\cos(0.5x)$ from $x = -\pi$ to $x = \pi$. What are the dimensions of the rectangle with largest area, and what is the largest area?

19. Find the dimensions of a right circular cylinder of maximum volume that can be inscribed in a sphere of radius 10 cm. What is the maximum volume?

20. a. The U.S. Postal Service will accept a box for domestic shipment only if the sum of its length and girth (distance around) does not exceed 108 in. What dimensions will give a box with a square end the largest possible volume?

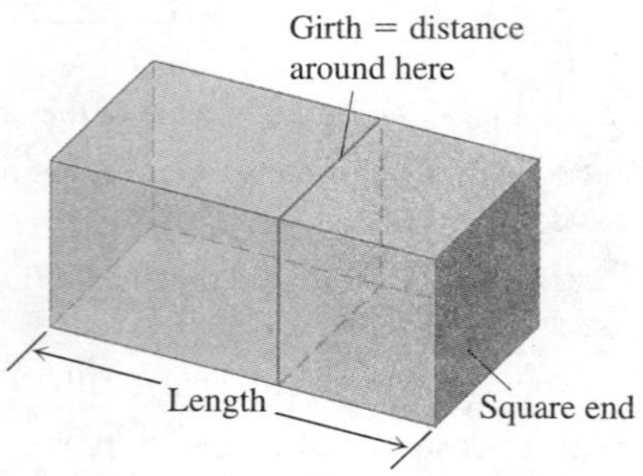

T **b.** Graph the volume of a 108-in. box (length plus girth equals 108 in.) as a function of its length and compare what you see with your answer in part (a).

21. (*Continuation of Exercise 20*)

a. Suppose that instead of having a box with square ends, you have a box with square sides so that its dimensions are h by h by w and the girth is $2h + 2w$. What dimensions will give the box its largest volume now?

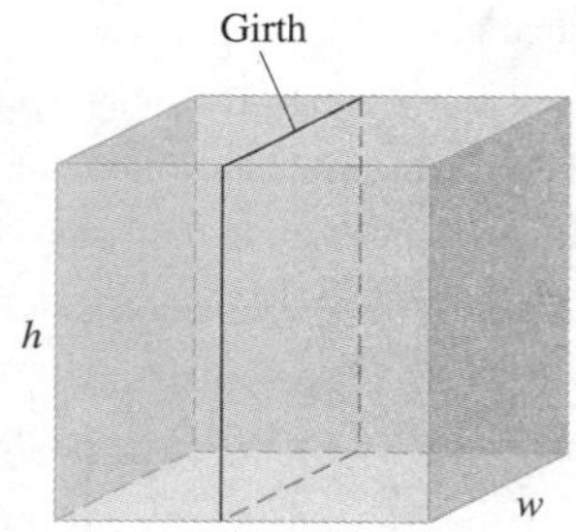

T **b.** Graph the volume as a function of h and compare what you see with your answer in part (a).

22. A window is in the form of a rectangle surmounted by a semicircle. The rectangle is of clear glass, whereas the semicircle is of tinted glass that transmits only half as much light per unit area as clear glass does. The total perimeter is fixed. Find the proportions of the window that will admit the most light. Neglect the thickness of the frame.

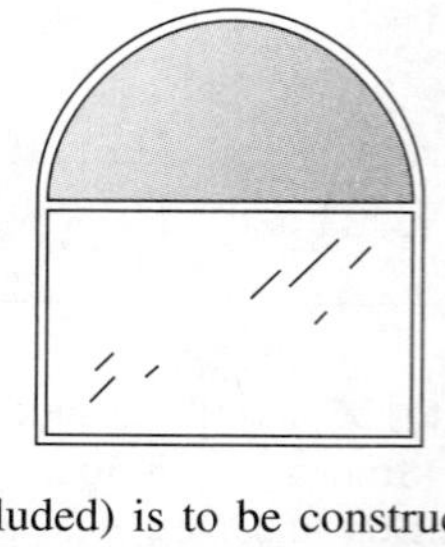

23. A silo (base not included) is to be constructed in the form of a cylinder surmounted by a hemisphere. The cost of construction per square unit of surface area is twice as great for the hemisphere as it is for the cylindrical sidewall. Determine the dimensions to be used if the volume is fixed and the cost of construction is to be kept to a minimum. Neglect the thickness of the silo and waste in construction.

24. The trough in the figure is to be made to the dimensions shown. Only the angle θ can be varied. What value of θ will maximize the trough's volume?

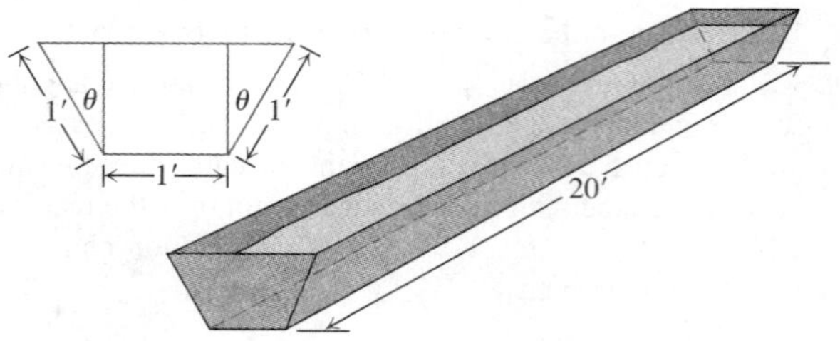

25. Paper folding A rectangular sheet of 8.5-in.-by-11-in. paper is placed on a flat surface. One of the corners is placed on the opposite longer edge, as shown in the figure, and held there as the paper is smoothed flat. The problem is to make the length of the crease as small as possible. Call the length L. Try it with paper.

a. Show that $L^2 = 2x^3/(2x - 8.5)$.

b. What value of x minimizes L^2?

c. What is the minimum value of L?

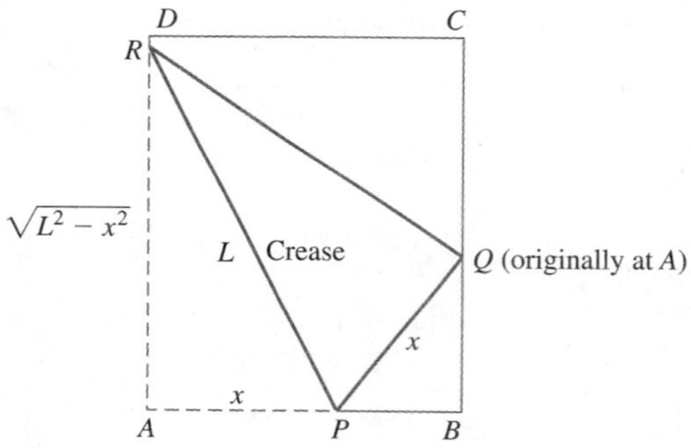

26. Constructing cylinders Compare the answers to the following two construction problems.

a. A rectangular sheet of perimeter 36 cm and dimensions x cm by y cm is to be rolled into a cylinder as shown in part (a) of the figure. What values of x and y give the largest volume?

b. The same sheet is to be revolved about one of the sides of length y to sweep out the cylinder as shown in part (b) of the figure. What values of x and y give the largest volume?

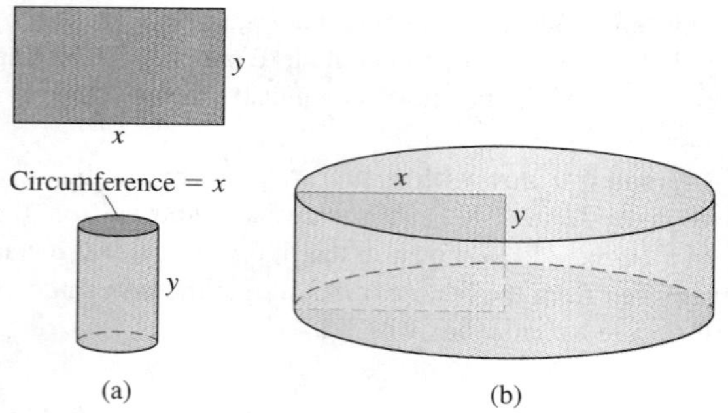

27. Constructing cones A right triangle whose hypotenuse is $\sqrt{3}$ m long is revolved about one of its legs to generate a right circular cone. Find the radius, height, and volume of the cone of greatest volume that can be made this way.

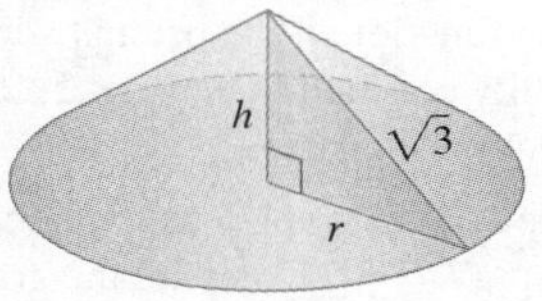

28. Find the point on the line $\frac{x}{a}+\frac{x}{b}=2$ that is closest to the origin.

29. Find a positive number for which the sum of it and its reciprocal is the smallest (least) possible.

30. Find a positive number for which the sum of its reciprocal and thirty two times its square is the smallest possible.

31. A wire b m long is cut into two pieces. One piece is bent into an equilateral triangle and the other is bent into a circle. If the sum of the areas enclosed by each part is a minimum, what is the length of each part?

32. Answer Exercise 31 if one piece is bent into a square and the other into a circle.

33. Suppose a weight D is to be held 10 ft below a horizontal line AB by a wire in the shape of a Y. If the points A and B are 8 ft apart, what is the minimum total length of wire that can be used?

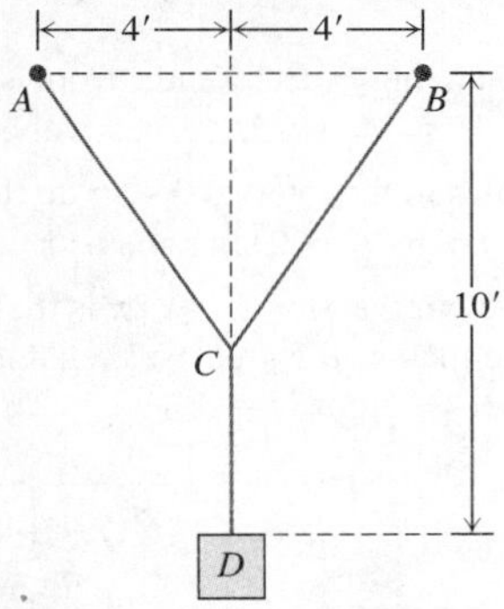

34. Suppose two different gauges of wire must be used to support the weight in Exercise 33: the vertical portion of the wire (the segment CD) costs \$1 per foot, while the remaining wire (the segments AC and CB) must be sturdier and cost \$2 per foot. What is the minimum total cost of the wire that can be used?

35. Determine the dimensions of the rectangle of largest area that can be inscribed in the right triangle shown in the accompanying figure.

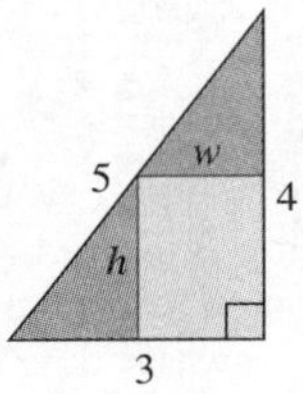

36. Determine the dimensions of the rectangle of largest area that can be inscribed in a semicircle of radius 3. (See the accompanying figure.)

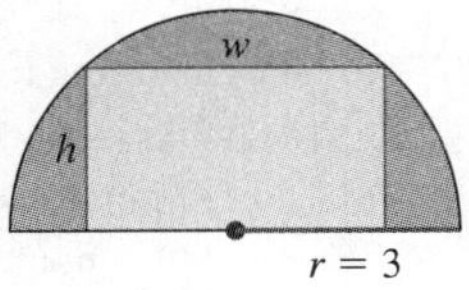

37. What value of a makes $f(x) = x^2 + (a/x)$ have

a. a local minimum at $x = 2$?

b. a point of inflection at $x = 1$?

38. What values of a and b make $f(x) = x^3 + ax^2 + bx$ have

a. a local maximum at $x = -1$ and a local minimum at $x = 3$?

b. a local minimum at $x = 4$ and a point of inflection at $x = 1$?

39. A right circular cone is circumscribed by a sphere of radius 1. Determine the height h and radius r of the cone of maximum volume.

40. Find the point on the graph of $y = 20x^3 + 60x - 3x^5 - 5x^4$ with the largest slope.

41. Among all triangles in the first quadrant formed by the x-axis, the y-axis, and tangent lines to the graph of $y = 3x - x^2$, what is the smallest possible area?

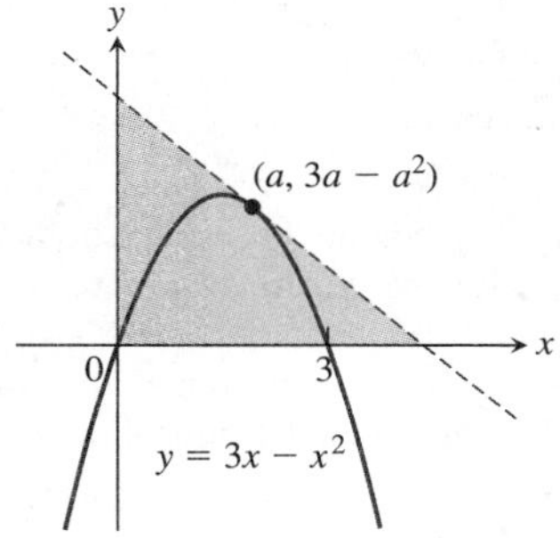

42. A cone is formed from a circular piece of material of radius 1 meter by removing a section of angle θ and then joining the two straight edges. Determine the largest possible volume for the cone.

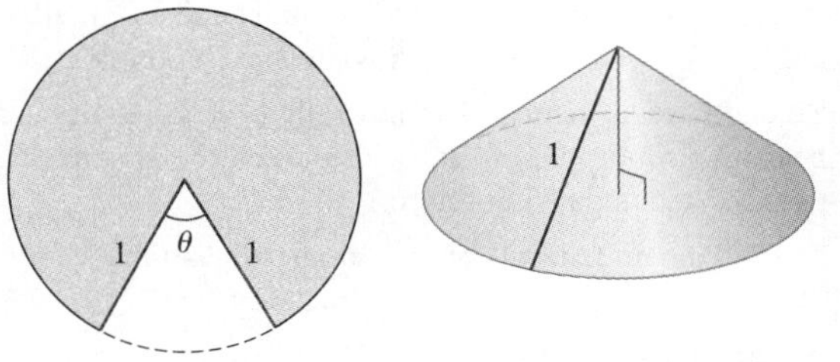

Physical Applications

43. Vertical motion The height above ground of an object moving vertically is given by

$$s = -14t^2 + 56t + 70$$

with s in feet and t in seconds. Find

a. the object's velocity when $t = 0$;

b. its maximum height and when it occurs;

c. its velocity when $s = 0$

44. Quickest route Jane is 2 mi offshore in a boat and wishes to reach a coastal village 6 mi down a straight shoreline from the point nearest the boat. She can row 2 mph and can walk 5 mph. Where should she land her boat to reach the village in the least amount of time?

45. Shortest beam The 8-ft wall shown here stands 27 ft from the building. Find the length of the shortest straight beam that will reach to the side of the building from the ground outside the wall.

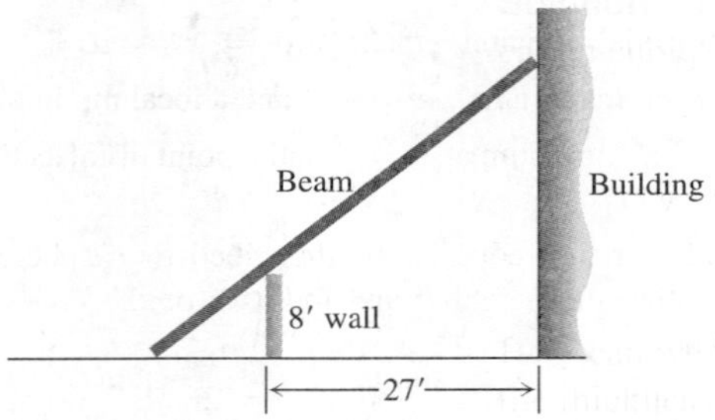

46. Motion on a line The positions of two particles on the s-axis are $s_1 = \sin t$ and $s_2 = \sin(t + \pi/3)$, with s_1 and s_2 in meters and t in seconds.

a. At what time(s) in the interval $0 \le t \le 2\pi$ do the particles meet?

b. What is the farthest apart that the particles ever get?

c. When in the interval $0 \le t \le 2\pi$ is the distance between the particles changing the fastest?

47. The intensity of illumination at any point from a light source is proportional to the square of the reciprocal of the distance between the point and the light source. Two lights, one having an intensity eight times that of the other, are 9 m apart. How far from the stronger light is the total illumination least?

48. Projectile motion The *range* R of a projectile fired from the origin over horizontal ground is the distance from the origin to the point of impact. If the projectile is fired with an initial velocity v_0 at an angle α with the horizontal, then in Chapter 13 we find that

$$R = \frac{v_0^2}{g}\sin 2\alpha,$$

where g is the downward acceleration due to gravity. Find the angle α for which the range R is the largest possible.

T **49. Strength of a beam** The strength S of a rectangular wooden beam is proportional to its width times the square of its depth. (See the accompanying figure.)

a. Find the dimensions of the strongest beam that can be cut from a 12-in.-diameter cylindrical log.

b. Graph S as a function of the beam's width w, assuming the proportionality constant to be $k = 1$. Reconcile what you see with your answer in part (a).

c. On the same screen, graph S as a function of the beam's depth d, again taking $k = 1$. Compare the graphs with one another and with your answer in part (a). What would be the effect of changing to some other value of k? Try it.

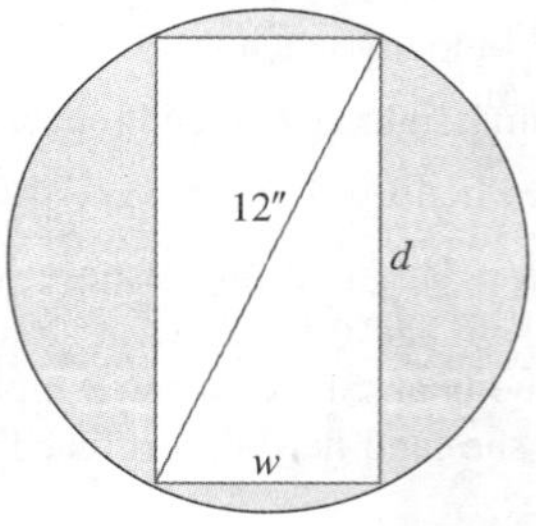

T **50. Stiffness of a beam** The stiffness S of a rectangular beam is proportional to its width times the cube of its depth.

a. Find the dimensions of the stiffest beam that can be cut from a 12-in.-diameter cylindrical log.

b. Graph S as a function of the beam's width w, assuming the proportionality constant to be $k = 1$. Reconcile what you see with your answer in part (a).

c. On the same screen, graph S as a function of the beam's depth d, again taking $k = 1$. Compare the graphs with one another and with your answer in part (a). What would be the effect of changing to some other value of k? Try it.

51. Frictionless cart A small frictionless cart, attached to the wall by a spring, is pulled 10 cm from its rest position and released at time $t = 0$ to roll back and forth for 4 sec. Its position at time t is $s = 10 \cos \pi t$.

a. What is the cart's maximum speed? When is the cart moving that fast? Where is it then? What is the magnitude of the acceleration then?

b. Where is the cart when the magnitude of the acceleration is greatest? What is the cart's speed then?

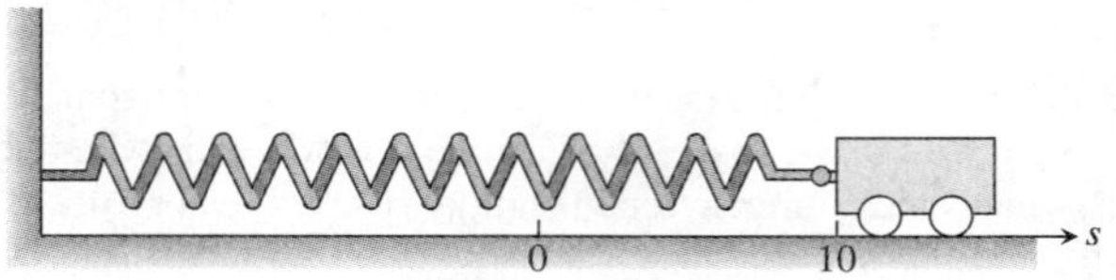

52. Two masses hanging side by side from springs have positions $s_1 = 2 \sin t$ and $s_2 = \sin 2t$, respectively.

a. At what times in the interval $0 < t$ do the masses pass each other? (*Hint:* $\sin 2t = 2 \sin t \cos t$.)

b. When in the interval $0 \le t \le 2\pi$ is the vertical distance between the masses the greatest? What is this distance? (*Hint:* $\cos 2t = 2\cos^2 t - 1$.)

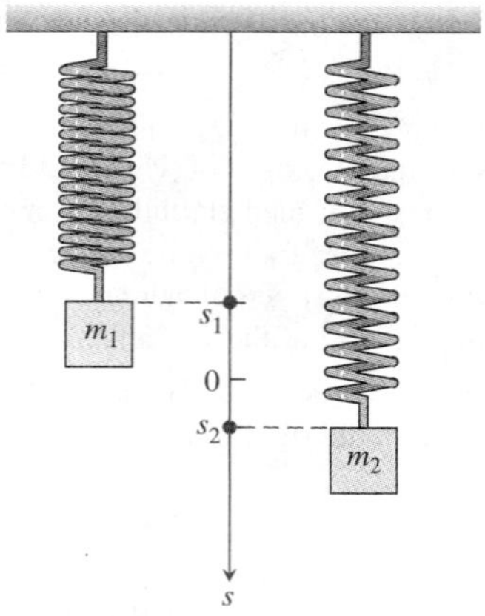

53. Distance between two ships At noon, ship A was 12 nautical miles due north of ship B. Ship A was sailing south at 12 knots (nautical miles per hour; a nautical mile is 2000 yd) and continued to do so all day. Ship B was sailing east at 8 knots and continued to do so all day.

a. Start counting time with $t = 0$ at noon and express the distance s between the ships as a function of t.

b. How rapidly was the distance between the ships changing at noon? One hour later?

c. The visibility that day was 5 nautical miles. Did the ships ever sight each other?

T **d.** Graph s and ds/dt together as functions of t for $-1 \le t \le 3$, using different colors if possible. Compare the graphs and reconcile what you see with your answers in parts (b) and (c).

e. The graph of ds/dt looks as if it might have a horizontal asymptote in the first quadrant. This in turn suggests that ds/dt approaches a limiting value as $t \to \infty$. What is this value? What is its relation to the ships' individual speeds?

54. Fermat's principle in optics Light from a source A is reflected by a plane mirror to a receiver at point B, as shown in the accompanying figure. Show that for the light to obey Fermat's principle, the angle of incidence must equal the angle of reflection, both measured from the line normal to the reflecting surface. (This result can also be derived without calculus. There is a purely geometric argument, which you may prefer.)

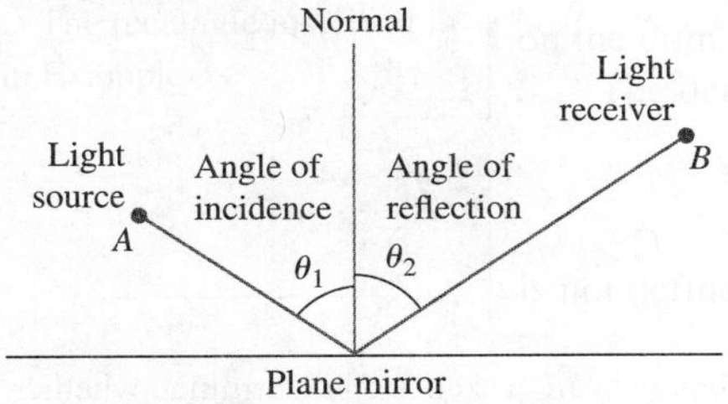

55. Tin pest When metallic tin is kept below 13.2 °C, it slowly becomes brittle and crumbles to a gray powder. Tin objects eventually crumble to this gray powder spontaneously if kept in a cold climate for years. The Europeans who saw tin organ pipes in their churches crumble away years ago called the change *tin pest* because it seemed to be contagious, and indeed it was, for the gray powder is a catalyst for its own formation.

A *catalyst* for a chemical reaction is a substance that controls the rate of reaction without undergoing any permanent change in itself. An *autocatalytic reaction* is one whose product is a catalyst for its own formation. Such a reaction may proceed slowly at first if the amount of catalyst present is small and slowly again at the end, when most of the original substance is used up. But in between, when both the substance and its catalyst product are abundant, the reaction proceeds at a faster pace.

In some cases, it is reasonable to assume that the rate $v = dx/dt$ of the reaction is proportional both to the amount of the original substance present and to the amount of product. That is, v may be considered to be a function of x alone, and

$$v = kx(a - x) = kax - kx^2,$$

where

$x =$ the amount of product,

$a =$ the amount of substance at the beginning, and

$k =$ a positive constant.

At what value of x does the rate v have a maximum? What is the maximum value of v?

56. Airplane landing path An airplane is flying at altitude H when it begins its descent to an airport runway that is at horizontal ground distance L from the airplane, as shown in the accompanying figure. Assume that the landing path of the airplane is the graph of a cubic polynomial function $y = ax^3 + bx^2 + cx + d$, where $y(-L) = H$ and $y(0) = 0$.

a. What is dy/dx at $x = 0$?

b. What is dy/dx at $x = -L$?

c. Use the values for dy/dx at $x = 0$ and $x = -L$ together with $y(0) = 0$ and $y(-L) = H$ to show that

$$y(x) = H\left[2\left(\frac{x}{L}\right)^3 + 3\left(\frac{x}{L}\right)^2\right].$$

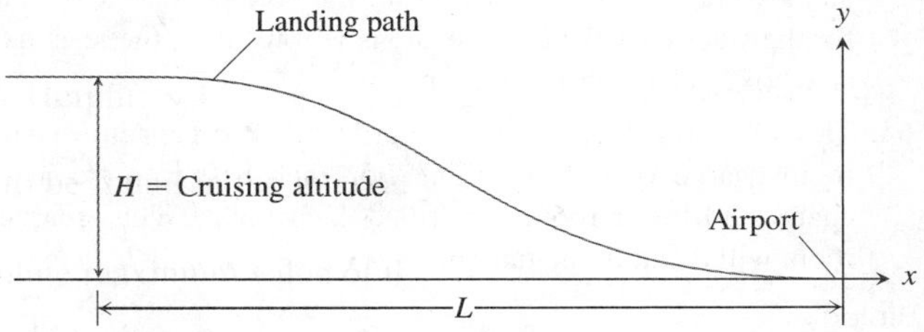

Business and Economics

57. It costs you c dollars each to manufacture and distribute backpacks. If the backpacks sell at x dollars each, the number sold is given by

$$n = \frac{a}{x - c} + b(200 - x)$$

where a and b are positive constants. What selling price will bring a maximum profit?

58. You operate a tour service that offers the following rates:

$300 per person if 60 people (the minimum number to book the tour) go on the tour.

For each additional person, up to a maximum of 80 people total, the rate per person is reduced by $3.

It costs $5000 (a fixed cost) plus $30 per person to conduct the tour. How many people does it take to maximize your profit?

59. Wilson lot size formula One of the formulas for inventory management says that the average weekly cost of ordering, paying for, and holding merchandise is

$$A(q) = \frac{km}{q} + cm + \frac{hq}{2},$$

where q is the quantity you order when things run low (shoes, radios, brooms, or whatever the item might be), k is the cost of placing an order (the same, no matter how often you order), c is the cost of one item (a constant), m is the number of items sold each week (a constant), and h is the weekly holding cost per item (a constant that takes into account things such as space, utilities, insurance, and security).

a. Your job, as the inventory manager for your store, is to find the quantity that will minimize $A(q)$. What is it? (The formula you get for the answer is called the *Wilson lot size formula*.)

b. Shipping costs sometimes depend on order size. When they do, it is more realistic to replace k by $k + bq$, the sum of k and a constant multiple of q. What is the most economical quantity to order now?

60. Production level Prove that the production level (if any) at which average cost is smallest is a level at which the average cost equals marginal cost.

61. Show that if $r(x) = 6x$ and $c(x) = x^3 - 6x^2 + 15x$ are your revenue and cost functions, then the best you can do is break even (have revenue equal cost).

62. Production level Suppose that $c(x) = x^3 - 30x^2 + 10000x$ is the cost of manufacturing x items. Find a production level that will minimize the average cost of making x items.

63. You are to construct an open rectangular box with a square base and a volume of 48. If material for the bottom costs and material for the sides costs what dimensions will result in the least expensive box? What is the minimum cost?

64. The 600-room Mega Motel chain is filled to capacity when the room charge is \$60 per night. For each \$10 increase in room charge, 30 fewer rooms are filled each night. What charge per room will result in the maximum revenue per night?

Biology

65. Sensitivity to medicine *(Continuation of Exercise 60, Section 3.3)* Find the amount of medicine to which the body is most sensitive by finding the value of M that maximizes the derivative dR/dM, where

$$R = M^2\left(\frac{C}{2} - \frac{M}{3}\right)$$

and C is a constant.

66. How we cough

a. When we cough, the trachea (windpipe) contracts to increase the velocity of the air going out. This raises the questions of how much it should contract to maximize the velocity and whether it really contracts that much when we cough.

Under reasonable assumptions about the elasticity of the tracheal wall and about how the air near the wall is slowed by friction, the average flow velocity v can be modeled by the equation

$$v = c(r_0 - r)r^2 \text{ cm/sec}, \qquad \frac{r_0}{2} \le r \le r_0,$$

where r_0 is the rest radius of the trachea in centimeters and c is a positive constant whose value depends in part on the length of the trachea.

Show that v is greatest when $r = (2/3)r_0$, that is, when the trachea is about 33% contracted. The remarkable fact is that X-ray photographs confirm that the trachea contracts about this much during a cough.

T **b.** Take r_0 to be 0.5 and c to be 1, and graph v over the interval $0 \le r \le 0.5$. Compare what you see with the claim that v is at a maximum when $r = (2/3)r_0$.

Theory and Examples

67. An inequality for positive integers Show that if a, b, c, and d are positive integers, then

$$\frac{(a^2+1)(b^2+1)(c^2+1)(d^2+1)}{abcd} \ge 16.$$

68. The derivative dt/dx in Example 4

a. Show that

$$f(x) = \frac{x}{\sqrt{a^2 + x^2}}$$

is an increasing function of x.

b. Show that

$$g(x) = \frac{d - x}{\sqrt{b^2 + (d - x)^2}}$$

is a decreasing function of x.

c. Show that

$$\frac{dt}{dx} = \frac{x}{c_1\sqrt{a^2 + x^2}} - \frac{d - x}{c_2\sqrt{b^2 + (d - x)^2}}$$

is an increasing function of x.

69. Let $f(x)$ and $g(x)$ be the differentiable functions graphed here. Point c is the point where the vertical distance between the curves is the greatest. Is there anything special about the tangent lines to the two curves at c? Give reasons for your answer.

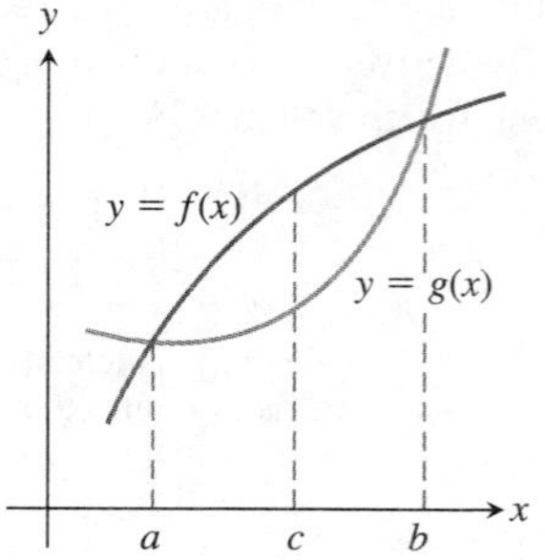

70. You have been asked to determine whether the function $f(x) = 3 + 4\cos x + \cos 2x$ is ever negative.

a. Explain why you need to consider values of x only in the interval $[0, 2\pi]$.

b. Is f ever negative? Explain.

71. a. The function $y = \cot x - \sqrt{2}\csc x$ has an absolute maximum value on the interval $0 < x < \pi$. Find it.

T **b.** Graph the function and compare what you see with your answer in part (a).

72. a. The function $y = \tan x + 3\cot x$ has an absolute minimum value on the interval $0 < x < \pi/2$. Find it.

T **b.** Graph the function and compare what you see with your answer in part (a).

73. a. How close does the curve $y = \sqrt{x}$ come to the point $(3/2, 0)$? (*Hint:* If you minimize the *square* of the distance, you can avoid square roots.)

T **b.** Graph the distance function $D(x)$ and $y = \sqrt{x}$ together and reconcile what you see with your answer in part (a).

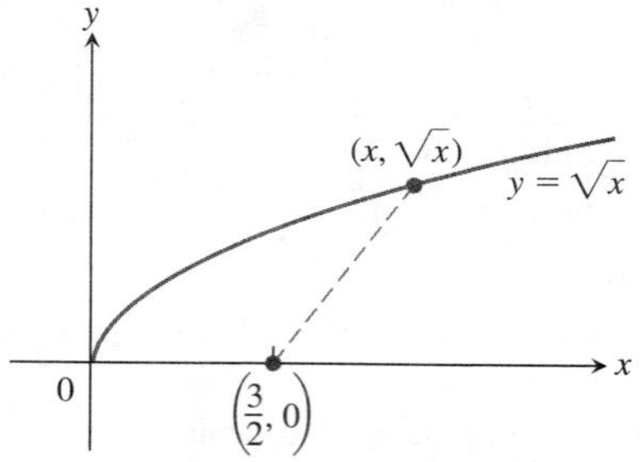

74. a. How close does the semicircle $y = \sqrt{16 - x^2}$ come to the point $(1, \sqrt{3})$?

T **b.** Graph the distance function and $y = \sqrt{16 - x^2}$ together and reconcile what you see with your answer in part (a).

4.6 Newton's Method

For thousands of years, one of the main goals of mathematics has been to find solutions to equations. For linear equations $(ax + b = 0)$, and for quadratic equations $(ax^2 + bx + c = 0)$, we can explicitly solve for a solution. However, for most equations there is no simple formula that gives the solutions.

In this section we study a numerical method called *Newton's method* or the *Newton–Raphson method*, which is a technique to approximate the solutions to an equation $f(x) = 0$. Newton's method estimates the solutions using tangent lines of the graph of $y = f(x)$ near the points where f is zero. A value of x where f is zero is called a *root* of the function f and a *solution* of the equation $f(x) = 0$. Newton's method is both powerful and efficient, and it has numerous applications in engineering and other fields where solutions to complicated equations are needed.

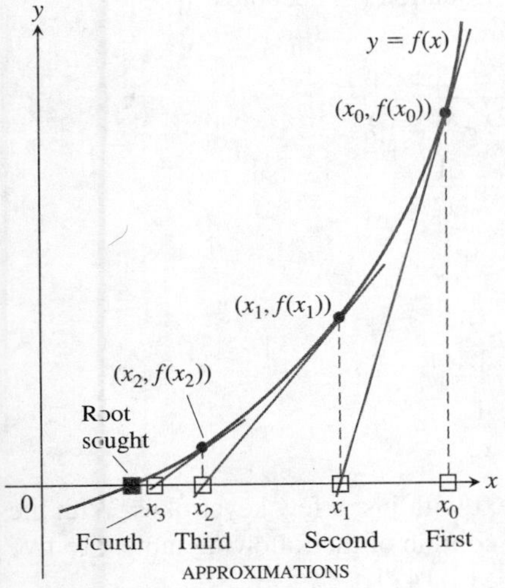

FIGURE 4.45 Newton's method starts with an initial guess x_0 and (under favorable circumstances) improves the guess one step at a time.

Procedure for Newton's Method

The goal of Newton's method for estimating a solution of an equation $f(x) = 0$ is to produce a sequence of approximations that approach the solution. We pick the first number x_0 of the sequence. Then, under favorable circumstances, the method moves step by step toward a point where the graph of f crosses the x-axis (Figure 4.45). At each step the method approximates a zero of f with a zero of one of its linearizations. Here is how it works.

The initial estimate, x_0, may be found by graphing or just plain guessing. The method then uses the tangent to the curve $y = f(x)$ at $(x_0, f(x_0))$ to approximate the curve, calling the point x_1 where the tangent meets the x-axis (Figure 4.45). The number x_1 is usually a better approximation to the solution than is x_0. The point x_2 where the tangent to the curve at $(x_1, f(x_1))$ crosses the x-axis is the next approximation in the sequence. We continue, using each approximation to generate the next, until we are close enough to the root to stop.

We can derive a formula for generating the successive approximations in the following way. Given the approximation x_n, the point-slope equation for the tangent line to the curve at $(x_n, f(x_n))$ is

$$y = f(x_n) + f'(x_n)(x - x_n).$$

We can find where it crosses the x-axis by setting $y = 0$ (Figure 4.46):

$$0 = f(x_n) + f'(x_n)(x - x_n)$$

$$-\frac{f(x_n)}{f'(x_n)} = x - x_n$$

$$x = x_n - \frac{f(x_n)}{f'(x_n)} \qquad \text{If } f'(x_n) \neq 0$$

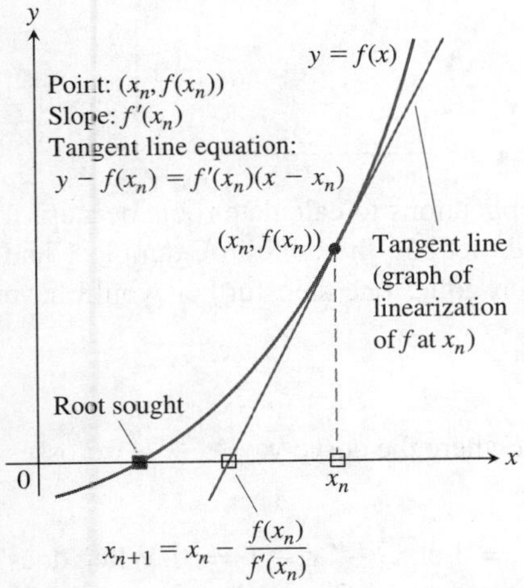

FIGURE 4.46 The geometry of the successive steps of Newton's method. From x_n we go up to the curve and follow the tangent line down to find x_{n+1}.

This value of x is the next approximation x_{n+1}. Here is a summary of Newton's method.

Newton's Method

1. Guess a first approximation to a solution of the equation $f(x) = 0$. A graph of $y = f(x)$ may help.
2. Use the first approximation to get a second, the second to get a third, and so on, using the formula

$$x_{n+1} = x_n - \frac{f(x_n)}{f'(x_n)}, \qquad \text{if } f'(x_n) \neq 0. \tag{1}$$

Applying Newton's Method

Applications of Newton's method generally involve many numerical computations, making them well suited for computers or calculators. Nevertheless, even when the calculations are done by hand (which may be very tedious), they give a powerful way to find solutions of equations.

In our first example, we find decimal approximations to $\sqrt{2}$ by estimating the positive root of the equation $f(x) = x^2 - 2 = 0$.

EXAMPLE 1 Approximate the positive root of the equation

$$f(x) = x^2 - 2 = 0.$$

Solution With $f(x) = x^2 - 2$ and $f'(x) = 2x$, Equation (1) becomes

$$\begin{aligned} x_{n+1} &= x_n - \frac{x_n^{\,2} - 2}{2x_n} \\ &= x_n - \frac{x_n}{2} + \frac{1}{x_n} \\ &= \frac{x_n}{2} + \frac{1}{x_n}. \end{aligned}$$

The equation

$$x_{n+1} = \frac{x_n}{2} + \frac{1}{x_n}$$

enables us to go from each approximation to the next with just a few keystrokes. With the starting value $x_0 = 1$, we get the results in the first column of the following table. (To five decimal places, or, equivalently, to six digits, $\sqrt{2} = 1.41421$.)

	Error	Number of correct digits
$x_0 = 1$	-0.41421	1
$x_1 = 1.5$	0.08579	1
$x_2 = 1.41667$	0.00246	3
$x_3 = 1.41422$	0.00001	5

■

Newton's method is used by many software applications to calculate roots because it converges so fast (more about this later). If the arithmetic in the table in Example 1 had been carried to 13 decimal places instead of 5, then going one step further would have given $\sqrt{2}$ correctly to more than 10 decimal places.

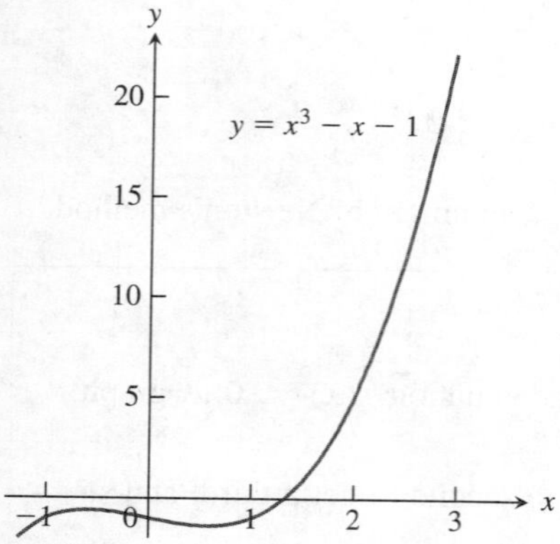

FIGURE 4.47 The graph of $f(x) = x^3 - x - 1$ crosses the x-axis once; this is the root we want to find (Example 2).

EXAMPLE 2 Find the x-coordinate of the point where the curve $y = x^3 - x$ crosses the horizontal line $y = 1$.

Solution The curve crosses the line when $x^3 - x = 1$ or $x^3 - x - 1 = 0$. When does $f(x) = x^3 - x - 1$ equal zero? Since f is continuous on $[1, 2]$ and $f(1) = -1$ while $f(2) = 5$, we know by the Intermediate Value Theorem there is a root in the interval $(1, 2)$ (Figure 4.47).

We apply Newton's method to f with the starting value $x_0 = 1$. The results are displayed in Table 4.1 and Figure 4.48.

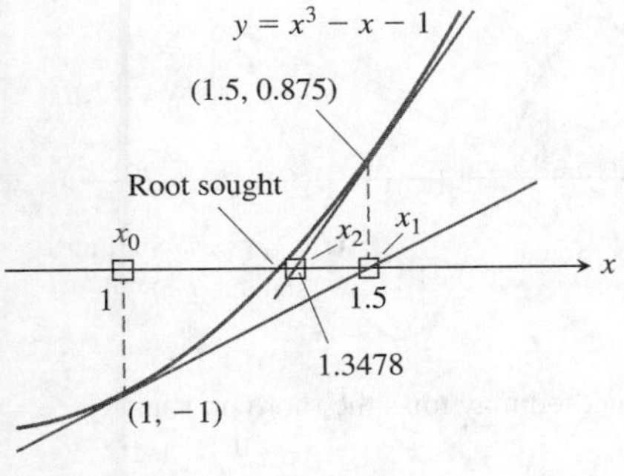

FIGURE 4.48 The first three x-values in Table 4.1 (four decimal places).

At $n = 5$, we come to the result $x_6 = x_5 = 1.3247\,17957$. When $x_{n+1} = x_n$, Equation (1) shows that $f(x_n) = 0$, up to the accuracy of our computation. We have found a solution of $f(x) = 0$ to nine decimals places. ■

TABLE 4.1 The Result of Applying Newton's Method to $f(x) = x^3 - x - 1$ with $x_0 = 1$

n	x_n	$f(x_n)$	$f'(x_n)$	$x_{n+1} = x_n - \dfrac{f(x_n)}{f'(x_n)}$
0	1	−1	2	1.5
1	1.5	0.875	5.75	1.3478 26087
2	1.3478 26087	0.1006 82173	4.4499 05482	1.3252 00399
3	1.3252 00399	0.0020 58362	4.2684 68292	1.3247 18174
4	1.3247 18174	0.0000 00924	4.2646 34722	1.3247 17957
5	1.3247 17957	−1.8672E-13	4.2646 32999	1.3247 17957

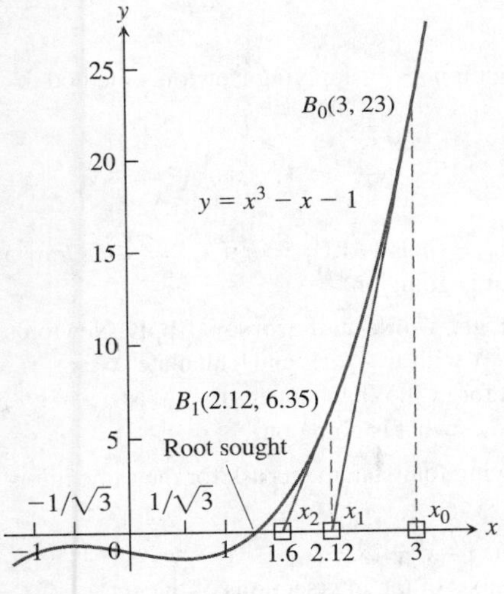

FIGURE 4.49 Any starting value x_0 to the right of $x = 1/\sqrt{3}$ will lead to the root in Example 2.

In Figure 4.49 we have indicated that the process in Example 2 might have started at the point $B_0(3, 23)$ on the curve, with $x_0 = 3$. Point B_0 is quite far from the x-axis, but the tangent at B_0 crosses the x-axis at about (2.12, 0), so x_1 is still an improvement over x_0. If we use Equation (1) repeatedly as before, with $f(x) = x^3 - x - 1$ and $f'(x) = 3x^2 - 1$, we obtain the nine-place solution $x_7 = x_6 = 1.3247\,17957$ in seven steps.

Convergence of the Approximations

In Chapter 10 we define precisely the idea of *convergence* for the approximations x_n in Newton's method. Intuitively, we mean that as the number n of approximations increases without bound, the values x_n get arbitrarily close to the desired root r. (This notion is similar to the idea of the limit of a function $g(t)$ as t approaches infinity, as defined in Section 2.6.)

In practice, Newton's method usually gives convergence with impressive speed, but this is not guaranteed. One way to test convergence is to begin by graphing the function to estimate a good starting value for x_0. You can test that you are getting closer to a zero of the function by checking that $|f(x_n)|$ is approaching zero, and you can check that the approximations are converging by evaluating $|x_n - x_{n+1}|$.

Newton's method does not always converge. For instance, if

$$f(x) = \begin{cases} -\sqrt{r - x}, & x < r \\ \sqrt{x - r}, & x \geq r, \end{cases}$$

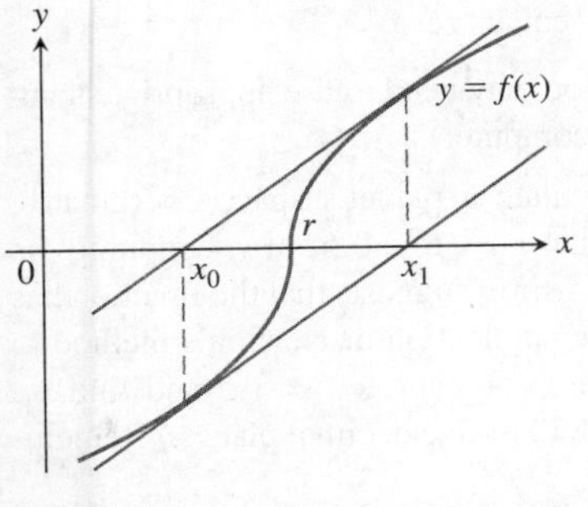

FIGURE 4.50 Newton's method fails to converge. You go from x_0 to x_1 and back to x_0, never getting any closer to r.

the graph will be like the one in Figure 4.50. If we begin with $x_0 = r - h$, we get $x_1 = r + h$, and successive approximations go back and forth between these two values. No amount of iteration brings us closer to the root than our first guess.

If Newton's method does converge, it converges to a root. Be careful, however. There are situations in which the method appears to converge but no root is there. Fortunately, such situations are rare.

When Newton's method converges to a root, it may not be the root you have in mind. Figure 4.51 shows two ways this can happen.

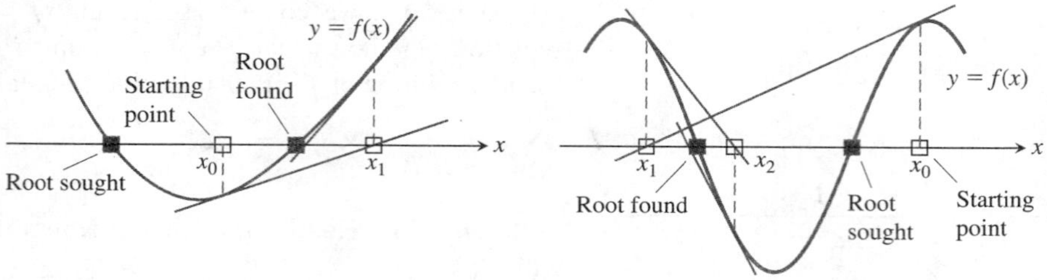

FIGURE 4.51 If you start too far away, Newton's method may miss the root you want.

EXERCISES 4.6

Root Finding

1. Use Newton's method to estimate the solutions of the equation $2x^2 + 2x - 1 = 0$. Start with $x_0 = -1$ for the left hand solution and with $x_0 = 1$ for the solution on the right. Then, in each case, find x_2.
2. Use Newton's method to estimate the one real solution of $x^3 + 3x + 1 = 0$. Start with $x_0 = 0$ and then find x_2.
3. Use Newton's method to estimate the two zeros of the function $f(x) = 2x^4 + 2x - 3 = 0$. Start with $x_0 = -1$ for the left-hand zero and with $x_0 = 1$ for the zero on the right. Then, in each case, find x_2.
4. Use Newton's method to estimate the two zeros of the function $f(x) = 3x - 2x^2 + 1 = 0$. Start with $x_0 = 0$ for the left-hand zero and with $x_0 = 2$ for the zero on the right. Then, in each case, find x_2.
5. Use Newton's method to find the positive fourth root of 2 by solving the equation $x^4 - 2 = 0$. Start with $x_0 = 1$ and find x_2.
6. Use Newton's method to find the negative fourth root of 2 by solving the equation $x^4 - 2 = 0$. Start with $x_0 = -1$ and find x_2.
7. Use Newton's method to find an approximate solution of $5 - 2x = x^3$. Start with $x_0 = 1$ and find x_2.

Dependence on Initial Point

8. Using the function shown in the figure, and, for each initial estimate x_0, determine graphically what happens to the sequence of Newton's method approximations

   a. $x_0 = 0$ **b.** $x_0 = 1$

   c. $x_0 = 2$ **d.** $x_0 = 4$

   e. $x_0 = 5.5$

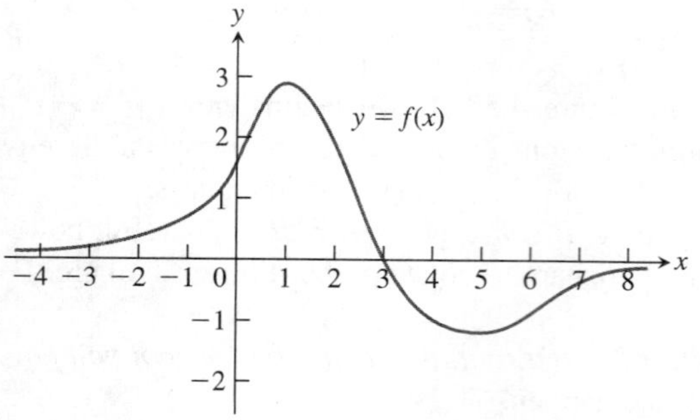

9. **Guessing a root** Suppose that your first guess is lucky, in the sense that x_0 is a root of $f(x) = 0$. Assuming that $f'(x_0)$ is defined and is not 0, what happens to x_1 and later approximations?
10. **Estimating pi** You plan to estimate $\pi/2$ to five decimal places by using Newton's method to solve the equation $\cos x = 0$. Does it matter what your starting value is? Give reasons for your answer.

Theory and Examples

11. **Oscillation** Show that if $h > 0$, applying Newton's method to

$$f(x) = \begin{cases} \sqrt{x}, & x \ge 0 \\ \sqrt{-x}, & x < 0 \end{cases}$$

leads to $x_1 = -h$ if $x_0 = h$ and to $x_1 = h$ if $x_0 = -h$. Draw a picture that shows what is going on.

12. **Approximations that get worse and worse** Apply Newton's method to $f(x) = x^{1/3}$ with $x_0 = 1$ and calculate x_1, x_2, x_3, and x_4. Find a formula for $|x_n|$. What happens to $|x_n|$ as $n \to \infty$? Draw a picture that shows what is going on.
13. Explain why the following four statements ask for the same information:
    i) Find the roots of $f(x) = x^3 - 2x - 5$.
    ii) Find the x-coordinates of the intersections of the curve $y = x^3$ with the line $y = 2x + 5$.
    iii) Find the x-coordinates of the points where the curve $y = x^3 - 2x$ crosses the horizontal line $y = 5$.
    iv) Find the values of x where the derivative of $g(x) = \frac{1}{4}x^3 - x^2 - 5 + 1$ equals zero.

T When solving Exercises 14–30, you may need to use appropriate technology (such as a calculator or a computer).

14. **Locating a planet** To calculate a planet's space coordinates, we have to solve equations like $x = 1 + 0.5 \sin x$. Graphing the function $f(x) = x - 1 - 0.5 \sin x$ suggests that the function has a root near $x = 1.5$. Use one application of Newton's method to improve this estimate. That is, start with $x_0 = 1.5$ and find x_1. (The value of the root is 1.49870 to five decimal places.) Remember to use radians.
15. **Intersecting curves** The curve $y = \tan x$ crosses the line $y = 2x$ between $x = 0$ and $x = \pi/2$. Use Newton's method to find where.

16. **Real solutions of a quartic** Use Newton's method to find the two real solutions of the equation $x^4 - 2x^3 - x^2 - 2x + 2 = 0$.

17. **a.** How many solutions does the equation $\sin 3x = 0.99 - x^2$ have?

 b. Use Newton's method to find them.

18. **Intersection of curves**

 a. Does $\cos 3x$ ever equal x? Give reasons for your answer.

 b. Use Newton's method to find where.

19. Find the four real zeros of the function $f(x) = 2x^4 - 4x^2 + 1$.

20. **Estimating π** Estimate π to as many decimal places as your calculator will display by using Newton's method to solve the equation $2\tan x = 0$ with $x_0 = 3$.

21. **Intersection of curves** At what value(s) of x does $\cos x = 2x$?

22. **Intersection of curves** At what value(s) of x does $\cos x = -x$?

23. The graphs of $y = x^2(x + 1)$ and $y = 1/x$ $(x > 0)$ intersect at one point $x = r$. Use Newton's method to estimate the value of r to four decimal places.

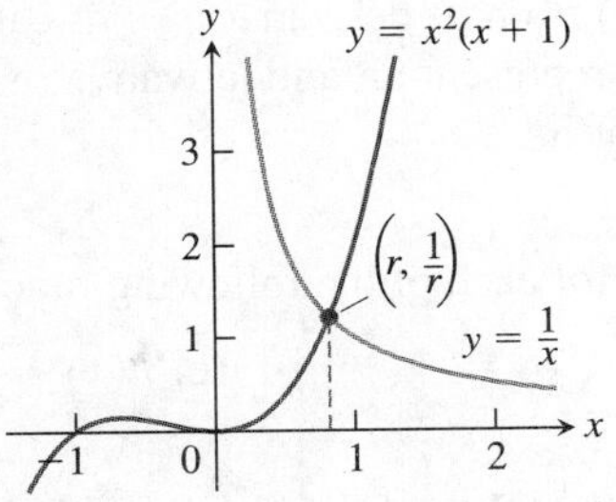

24. The graphs of $y = \sqrt{x}$ and $y = 3 - x^2$ intersect at one point $x = r$. Use Newton's method to estimate the value of r to four decimal places.

25. Use the Intermediate Value Theorem from Section 2.5 to show that $f(x) = x^3 + 2x - 4$ has a root between $x = 1$ and $x = 2$. Then find the root to five decimal places.

26. **Factoring a quartic** Find the approximate values of r_1 through r_4 in the factorization

$$8x^4 - 14x^3 - 9x^2 + 11x - 1 = 8(x - r_1)(x - r_2)(x - r_3)(x - r_4).$$

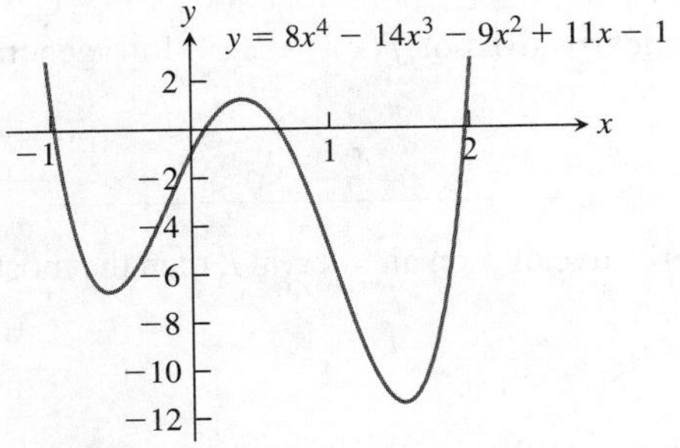

27. **Converging to different zeros** Use Newton's method to find the zeros of $f(x) = 4x^4 - 4x^2$ using the given starting values.

 a. $x_0 = -2$ and $x_0 = -0.8$, lying in $(-\infty, -\sqrt{2}/2)$

 b. $x_0 = -0.5$ and $x_0 = 0.25$, lying in $(-\sqrt{21}/7, \sqrt{21}/7)$

 c. $x_0 = 0.8$ and $x_0 = 2$, lying in $(\sqrt{2}/2, \infty)$

 d. $x_0 = -\sqrt{21}/7$ and $x_0 = \sqrt{21}/7$

28. **The sonobuoy problem** In submarine location problems, it is often necessary to find a submarine's closest point of approach (CPA) to a sonobuoy (sound detector) in the water. Suppose that the submarine travels on the parabolic path $y = x^2$ and that the buoy is located at the point $(2, -1/2)$.

 a. Show that the value of x that minimizes the distance between the submarine and the buoy is a solution of the equation $x = 1/(x^2 + 1)$.

 b. Solve the equation $x = 1/(x^2 + 1)$ with Newton's method.

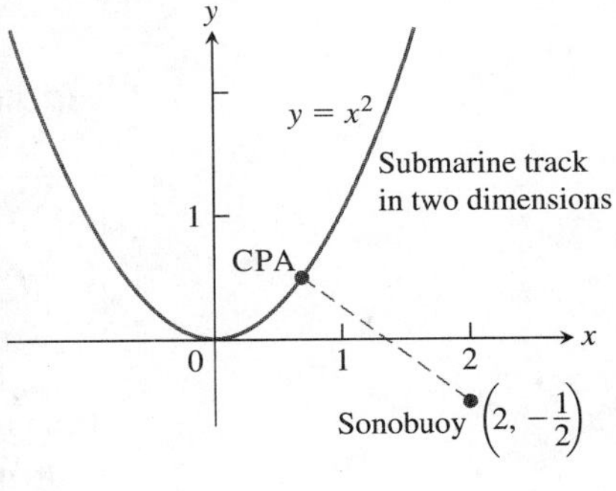

29. **Curves that are nearly flat at the root** Some curves are so flat that, in practice, Newton's method stops too far from the root to give a useful estimate. Try Newton's method on $f(x) = (x - 1)^{40}$ with a starting value of $x_0 = 2$ to see how close your machine comes to the root $x = 1$. See the accompanying graph.

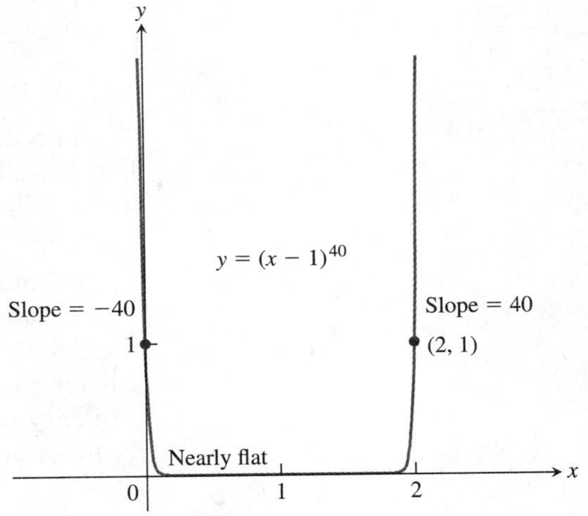

30. The accompanying figure shows a circle of radius r with a chord of length 2 and an arc s of length 3. Use Newton's method to solve for r and θ (radians) to four decimal places. Assume $0 < \theta < \pi$.

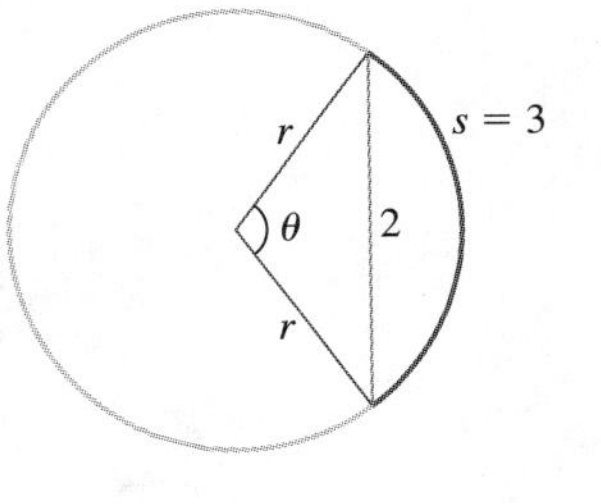

4.7 Antiderivatives

Many problems require that we recover a function from its derivative, or from its rate of change. For instance, the laws of physics tell us the acceleration of an object falling from an initial height, and we can use this to compute its velocity and its height at any time. More generally, starting with a function f, we want to find a function F whose derivative is f. If such a function F exists, it is called an *antiderivative* of f. Antiderivatives are the link connecting the two major elements of calculus: derivatives and definite integrals. Antiderivatives have an important connection to the theory of integrals that is developed in Chapter 5. For this reason the process of taking an antiderivative is also called "integration."

Finding Antiderivatives

DEFINITION A function F is an **antiderivative** of f on an interval I if $F'(x) = f(x)$ for all x in I.

The process of recovering a function $F(x)$ from its derivative $f(x)$ is called *antidifferentiation*. We use capital letters such as F to represent an antiderivative of a function f, G to represent an antiderivative of g, and so forth.

EXAMPLE 1 Find an antiderivative for each of the following functions.

(a) $f(x) = 2x$ **(b)** $g(x) = \cos x$ **(c)** $h(x) = \sec^2 x + \dfrac{1}{2\sqrt{x}}$

Solution We need to think backward here: What function do we know has a derivative equal to the given function?

(a) $F(x) = x^2$ **(b)** $G(x) = \sin x$ **(c)** $H(x) = \tan x + \sqrt{x}$

Each answer can be checked by differentiating. The derivative of $F(x) = x^2$ is $2x$. The derivative of $G(x) = \sin x$ is $\cos x$, and the derivative of $H(x) = \tan x + \sqrt{x}$ is $\sec^2 x + 1/(2\sqrt{x})$. ■

The function $F(x) = x^2$ is not the only function whose derivative is $2x$. The function $x^2 + 1$ has the same derivative. So does $x^2 + C$ for any constant C. Are there others?

Corollary 2 of the Mean Value Theorem in Section 4.2 gives the answer: Any two antiderivatives of a function differ by a constant. So the functions $x^2 + C$, where C is an **arbitrary constant**, form *all* the antiderivatives of $f(x) = 2x$. More generally, we have the following result.

THEOREM 8 If F is an antiderivative of f on an interval I, then the most general antiderivative of f on I is

$$F(x) + C$$

where C is an arbitrary constant.

Thus the most general antiderivative of f on I is a *family* of functions $F(x) + C$ whose graphs are vertical translations of one another. We can select a particular antiderivative from this family by assigning a specific value to C. Here is an example showing how such an assignment might be made.

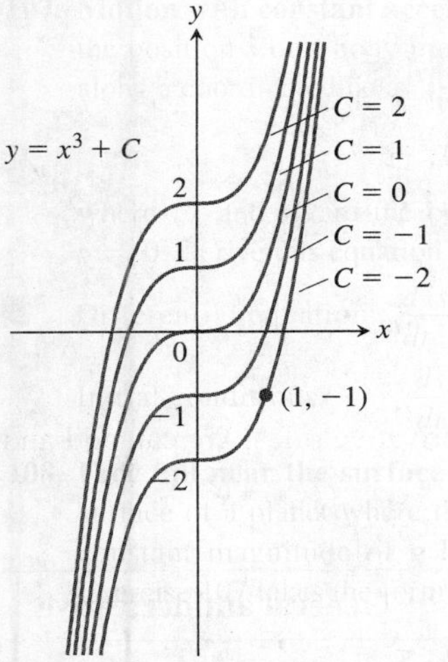

FIGURE 4.52 The curves $y = x^3 + C$ fill the coordinate plane without overlapping. In Example 2, we identify the curve $y = x^3 - 2$ as the one that passes through the given point $(1, -1)$.

EXAMPLE 2 Find an antiderivative of $f(x) = 3x^2$ that satisfies $F(1) = -1$.

Solution Since the derivative of x^3 is $3x^2$, the general antiderivative

$$F(x) = x^3 + C$$

gives all the antiderivatives of $f(x)$. The condition $F(1) = -1$ determines a specific value for C. Substituting $x = 1$ into $f(x) = x^3 + C$ gives

$$F(1) = (1)^3 + C = 1 + C.$$

Since $F(1) = -1$, solving $1 + C = -1$ for C gives $C = -2$. So

$$F(x) = x^3 - 2$$

is the antiderivative satisfying $F(1) = -1$. Notice that this assignment for C selects the particular curve from the family of curves $y = x^3 + C$ that passes through the point $(1, -1)$ in the plane (Figure 4.52). ■

By working backward from assorted differentiation rules, we can derive formulas and rules for antiderivatives. In each case there is an arbitrary constant C in the general expression representing all antiderivatives of a given function. Table 4.2 gives antiderivative formulas for a number of important functions.

The rules in Table 4.2 are easily verified by differentiating the general antiderivative formula to obtain the function to its left. For example, the derivative of $(\tan kx)/k + C$ is $\sec^2 kx$, whatever the value of the constants C or $k \neq 0$, and this shows that Formula 4 gives the general antiderivative of $\sec^2 kx$.

TABLE 4.2 Antiderivative formulas, k a nonzero constant

Function	General antiderivative
1. x^n	$\frac{1}{n+1}x^{n+1} + C, \quad n \neq -1$
2. $\sin kx$	$-\frac{1}{k}\cos kx + C$
3. $\cos kx$	$\frac{1}{k}\sin kx + C$
4. $\sec^2 kx$	$\frac{1}{k}\tan kx + C$
5. $\csc^2 kx$	$-\frac{1}{k}\cot kx + C$
6. $\sec kx \tan kx$	$\frac{1}{k}\sec kx + C$
7. $\csc kx \cot kx$	$-\frac{1}{k}\csc kx + C$

EXAMPLE 3 Find the general antiderivative of each of the following functions.

(a) $f(x) = x^5$ **(b)** $g(x) = \frac{1}{\sqrt{x}}$ **(c)** $h(x) = \sin 2x$

(d) $i(x) = \cos\frac{x}{2}$

Solution In each case, we can use one of the formulas listed in Table 4.2.

(a) $F(x) = \frac{x^6}{6} + C$ Formula 1 with $n = 5$

(b) $g(x) = x^{-1/2}$, so

$$G(x) = \frac{x^{1/2}}{1/2} + C = 2\sqrt{x} + C \qquad \text{Formula 1 with } n = -1/2$$

(c) $$H(x) = \frac{-\cos 2x}{2} + C \qquad \text{Formula 2 with } k = 2$$

(d) $$I(x) = \frac{\sin(x/2)}{1/2} + C = 2\sin\frac{x}{2} + C \qquad \text{Formula 3 with } k = 1/2$$ ■

Other derivative rules also lead to corresponding antiderivative rules. We can add and subtract antiderivatives and multiply them by constants.

TABLE 4.3 Antiderivative linearity rules

	Function	General antiderivative
1. *Constant Multiple Rule:*	$kf(x)$	$kF(x) + C$, k a constant
2. *Sum or Difference Rule:*	$f(x) \pm g(x)$	$F(x) \pm G(x) + C$

The formulas in Table 4.3 are easily proved by differentiating the antiderivatives and verifying that the result agrees with the original function.

EXAMPLE 4 Find the general antiderivative of

$$f(x) = \frac{3}{\sqrt{x}} + \sin 2x.$$

Solution We have that $f(x) = 3g(x) + h(x)$ for the functions g and h in Example 3. Since $G(x) = 2\sqrt{x}$ is an antiderivative of $g(x)$ from Example 3b, it follows from the Constant Multiple Rule for antiderivatives that $3G(x) = 3 \cdot 2\sqrt{x} = 6\sqrt{x}$ is an antiderivative of $3g(x) = 3/\sqrt{x}$. Likewise, from Example 3c we know that $H(x) = (-1/2)\cos 2x$ is an antiderivative of $h(x) = \sin 2x$. From the Sum Rule for antiderivatives, we then get that

$$\begin{aligned} F(x) &= 3G(x) + H(x) + C \\ &= 6\sqrt{x} - \frac{1}{2}\cos 2x + C \end{aligned}$$

is the general antiderivative formula for $f(x)$, where C is an arbitrary constant. ■

Initial Value Problems and Differential Equations

Antiderivatives play several important roles in mathematics and its applications. Methods and techniques for finding them are a major part of calculus, and we take up that study in Chapter 8. Finding an antiderivative for a function $f(x)$ is the same problem as finding a function $y(x)$ that satisfies the equation

$$\frac{dy}{dx} = f(x).$$

This is called a **differential equation**, since it is an equation involving an unknown function y that is being differentiated. To solve it, we need a function $y(x)$ that satisfies the equation. This function is found by taking the antiderivative of $f(x)$. We can fix the arbitrary constant arising in the antidifferentiation process by specifying an initial condition

$$y(x_0) = y_0.$$

This condition means the function $y(x)$ has the value y_0 when $x = x_0$. The combination of a differential equation and an initial condition is called an **initial value problem**. Such problems play important roles in all branches of science.

The most general antiderivative $F(x) + C$ of the function $f(x)$ (such as $x^3 + C$ for the function $3x^2$ in Example 2) gives the **general solution** $y = F(x) + C$ of the differential equation $dy/dx = f(x)$. The general solution gives *all* the solutions of the equation (there are infinitely many, one for each value of C). We **solve** the differential equation by finding its general solution. We then solve the initial value problem by finding the **particular solution** that satisfies the initial condition $y(x_0) = y_0$. In Example 2, the function $y = x^3 - 2$ is the particular solution of the differential equation $dy/dx = 3x^2$ satisfying the initial condition $y(1) = -1$.

Antiderivatives and Motion

We have seen that the derivative of the position function of an object gives its velocity, and the derivative of its velocity function gives its acceleration. If we know an object's acceleration, then by finding an antiderivative we can recover the velocity, and from an antiderivative of the velocity we can recover its position function. This procedure was used as an application of Corollary 2 in Section 4.2. Now that we have a terminology and conceptual framework in terms of antiderivatives, we revisit the problem from the point of view of differential equations.

EXAMPLE 5 A hot-air balloon ascending at the rate of 12 ft/sec is at a height 80 ft above the ground when a package is dropped. How long does it take the package to reach the ground?

Solution Let $v(t)$ denote the velocity of the package at time t, and let $s(t)$ denote its height above the ground. The acceleration of gravity near the surface of the earth is 32 ft/sec^2. Assuming no other forces act on the dropped package, we have

$$\frac{dv}{dt} = -32. \qquad \text{Negative because gravity acts in the direction of decreasing } s$$

This leads to the following initial value problem (Figure 4.53):

Differential equation: $\dfrac{dv}{dt} = -32$

Initial condition: $v(0) = 12.$ Balloon initially rising

This is our mathematical model for the package's motion. We solve the initial value problem to obtain the velocity of the package.

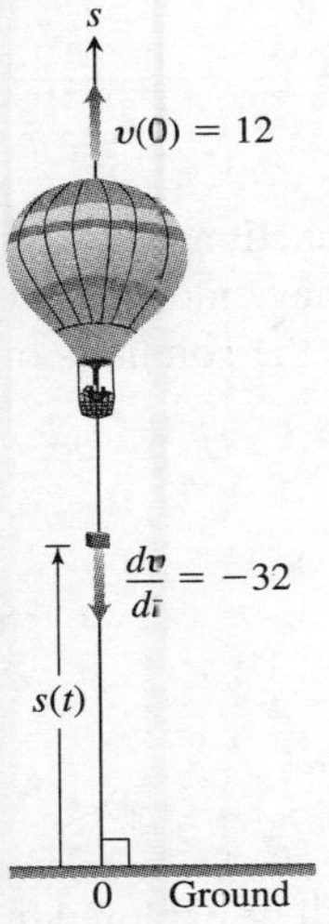

FIGURE 4.53 A package dropped from a rising hot-air balloon (Example 5).

1. *Solve the differential equation*: The general formula for an antiderivative of -32 is

$$v = -32t + C.$$

Having found the general solution of the differential equation, we use the initial condition to find the particular solution that solves our problem.

2. *Evaluate C*:

$$12 = -32(0) + C \qquad \text{Initial condition } v(0) = 12$$
$$C = 12.$$

The solution of the initial value problem is

$$v = -32t + 12.$$

Since velocity is the derivative of height, and the height of the package is 80 ft at time $t = 0$ when it is dropped, we now have a second initial value problem:

Differential equation: $\dfrac{ds}{dt} = -32t + 12$ Set $v = ds/dt$ in the previous equation.

Initial condition: $s(0) = 80.$

We solve this initial value problem to find the height as a function of t.

1. *Solve the differential equation*: Finding the general antiderivative of $-32t + 12$ gives

$$s = -16t^2 + 12t + C.$$

2. *Evaluate C*:

$$\begin{aligned} 80 &= -16(0)^2 + 12(0) + C \qquad \text{Initial condition } s(0) = 80 \\ C &= 80. \end{aligned}$$

The package's height above ground at time t is

$$s = -16t^2 + 12t + 80.$$

Use the solution: To find how long it takes the package to reach the ground, we set s equal to 0 and solve for t:

$$\begin{aligned} -16t^2 + 12t + 80 &= 0 \\ -4t^2 + 3t + 20 &= 0 \\ t &= \frac{-3 \pm \sqrt{329}}{-8} \qquad \text{Quadratic formula} \\ t \approx -1.89, &\qquad t \approx 2.64. \end{aligned}$$

The package hits the ground about 2.64 sec after it is dropped from the balloon. (The negative root has no physical meaning.) ■

Indefinite Integrals

A special symbol is used to denote the collection of all antiderivatives of a function f.

DEFINITION The collection of all antiderivatives of f is called the **indefinite integral** of f with respect to x; it is denoted by

$$\int f(x)\, dx.$$

The symbol $\int$ is an **integral sign**. The function f is the **integrand** of the integral, and x is the **variable of integration**.

After the integral sign in the notation we just defined, the integrand function is always followed by a differential to indicate the variable of integration. We will have more to say about why this is important in Chapter 5. Using this notation, we restate the solutions of Example 1, as follows:

$$\begin{aligned} \int 2x\, dx &= x^2 + C, \\ \int \cos x\, dx &= \sin x + C, \\ \int \left(\sec^2 x + \frac{1}{2\sqrt{x}}\right) dx &= \tan x + \sqrt{x} + C. \end{aligned}$$

This notation is related to the main application of antiderivatives, which will be explored in Chapter 5. Antiderivatives play a key role in computing limits of certain infinite sums, an unexpected and wonderfully useful role that is described in a central result of Chapter 5, the Fundamental Theorem of Calculus.

EXAMPLE 6 Evaluate

$$\int (x^2 - 2x + 5)\,dx.$$

Solution If we recognize that $(x^3/3) - x^2 + 5x$ is an antiderivative of $x^2 - 2x + 5$, we can evaluate the integral as

$$\int (x^2 - 2x + 5)\,dx = \overbrace{\frac{x^3}{3} - x^2 + 5x}^{\text{antiderivative}} + \underbrace{C}_{\text{arbitrary constant}}.$$

If we do not recognize the antiderivative right away, we can generate it term-by-term with the Sum, Difference, and Constant Multiple Rules:

$$\begin{aligned}\int (x^2 - 2x + 5)\,dx &= \int x^2\,dx - \int 2x\,dx + \int 5\,dx \\ &= \int x^2\,dx - 2\int x\,dx + 5\int 1\,dx \\ &= \left(\frac{x^3}{3} + C_1\right) - 2\left(\frac{x^2}{2} + C_2\right) + 5(x + C_3) \\ &= \frac{x^3}{3} + C_1 - x^2 - 2C_2 + 5x + 5C_3.\end{aligned}$$

This formula is more complicated than it needs to be. If we combine $C_1, -2C_2$, and $5C_3$ into a single arbitrary constant $C = C_1 - 2C_2 + 5C_3$, the formula simplifies to

$$\frac{x^3}{3} - x^2 + 5x + C$$

and *still* gives all the possible antiderivatives there are. For this reason, we recommend that you go right to the final form even if you elect to integrate term-by-term. Write

$$\begin{aligned}\int (x^2 - 2x + 5)\,dx &= \int x^2\,dx - \int 2x\,dx + \int 5\,dx \\ &= \frac{x^3}{3} - x^2 + 5x + C.\end{aligned}$$

Find the simplest antiderivative you can for each part, and add the arbitrary constant of integration at the end. ■

We conclude this section with a list of basic antidifferentiation formulas in Table 4.4, using the integral sign to indicate an antiderivative.

TABLE 4.4 Integration formulas

1. $\int x^n\,dx = \frac{x^{n+1}}{n+1} + C \quad (n \neq -1)$
2. $\int \sin x\,dx = -\cos x + C$
3. $\int \cos x\,dx = \sin x + C$
4. $\int \sec^2 x\,dx = \tan x + C$
5. $\int \csc^2 x\,dx = -\cot x + C$
6. $\int \sec x \tan x\,dx = \sec x + C$
7. $\int \csc x \cot x\,dx = -\csc x + C$

EXERCISES 4.7

Finding Antiderivatives

In Exercises 1–16, find an antiderivative for each function. Do as many as you can mentally. Check your answers by differentiation.

1. **a.** $3x$ **b.** x^3 **c.** $x^3 - 2x + 1$
2. **a.** $6x$ **b.** x^7 **c.** $x^7 - 6x + 8$
3. **a.** $-3x^{-2}$ **b.** x^{-2} **c.** $x^{-4} + 5x + 3$
4. **a.** $2x^{-3}$ **b.** $\frac{x^{-3}}{2} + x^2$ **c.** $-x^{-3} + x - 1$
5. **a.** $\frac{1}{2x^2}$ **b.** $\frac{5}{2x^2}$ **c.** $2 - \frac{5}{x\;2\;^{2}}$
6. **a.** $-\frac{2}{x^3}$ **b.** $\frac{1}{2x^3}$ **c.** $x^3 - \frac{1}{x^3}$
7. **a.** $\frac{1}{2}\sqrt{x}$ **b.** $\frac{3}{2\sqrt{x}}$ **c.** $2\sqrt{x} + \frac{1}{2\sqrt{x}}$
8. **a.** $\frac{4}{3}\sqrt[3]{x}$ **b.** $\frac{1}{3\sqrt[3]{x}}$ **c.** $\sqrt[3]{x} + \frac{1}{\sqrt[3]{x}}$
9. **a.** $\frac{2}{5}x^{-1/3}$ **b.** $\frac{2}{3}x^{-2/3}$ **c.** $\frac{2}{3}x^{-4/3}$
10. **a.** $\frac{1}{2}x^{-1/2}$ **b.** $-\frac{1}{2}x^{-3/2}$ **c.** $-\frac{3}{2}x^{-5/2}$
11. **a.** $-\pi \sin \pi x$ **b.** $3 \sin x$ **c.** $\sin \pi x - 3 \sin 3x$
12. **a.** $\pi \cos \pi x$ **b.** $\frac{\pi}{2}\cos\frac{\pi x}{2}$ **c.** $\cos\frac{\pi x}{2} + \pi \cos x$
13. **a.** $\sec^2 x$ **b.** $\frac{2}{3}\sec^2\frac{x}{3}$ **c.** $-\sec^2\frac{3x}{2}$
14. **a.** $\csc^2 x$ **b.** $-\frac{3}{2}\csc^2\frac{3x}{2}$ **c.** $1 - 8\csc^2 2x$
15. **a.** $\csc x \cot x$ **b.** $-\csc 5x \cot 5x$ **c.** $-\pi \csc\frac{\pi x}{2}\cot\frac{\pi x}{2}$
16. **a.** $\sec x \tan x$ **b.** $4 \sec 3x \tan 3x$ **c.** $\sec\frac{\pi x}{2}\tan\frac{\pi x}{2}$

Finding Indefinite Integrals

In Exercises 17–56, find the most general antiderivative or indefinite integral. You may need to try a solution and then adjust your guess. Check your answers by differentiation.

17. $\int (x + 1)\,dx$
18. $\int (2 - 3x)dx$
19. $\int \left(3t^2 + \frac{t}{2}\right)dt$
20. $\int \left(\frac{t^2}{3} + 5t^3\right)dt$
21. $\int (2x^3 - 5x + 7)\,dx$
22. $\int (5 - x^2 - 2x^5)dx$
23. $\int \left(\frac{1}{x^2} - x^2 - \frac{1}{3}\right)dx$
24. $\int \left(\frac{2}{5} - \frac{5}{x^3} + 5x\right)dx$
25. $\int x^{-1/3}\,dx$
26. $\int x^{-3/4}dx$
27. $\int (\sqrt{x} + \sqrt[3]{x})\,dx$
28. $\int \left(\frac{2\sqrt{x}}{3} + \frac{3}{5\sqrt{x}}\right)dx$
29. $\int \left(8y - \frac{2}{y^{1/4}}\right)dy$
30. $\int \left(\frac{1}{7} - \frac{1}{y^{5/4}}\right)dy$
31. $\int 2x(1 - x^{-3})\,dx$
32. $\int x^{-3}(2x + 3)dx$
33. $\int \frac{t\sqrt{t} + \sqrt{t}}{t^2}\,dt$
34. $\int \frac{4 + \sqrt{t}}{t^3}\,dt$
35. $\int (-2\cos t)\,dt$
36. $\int (-5\sin t)\,dt$
37. $\int 7\sin\frac{\theta}{3}\,d\theta$
38. $\int 3\cos 5\theta\,d\theta$
39. $\int (-3\csc^2 x)\,dx$
40. $\int \left(-\frac{\sec^2 x}{3}\right)dx$
41. $\int \frac{\csc\theta\cot\theta}{2}\,d\theta$
42. $\int \frac{2}{5}\sec\theta\tan\theta\,d\theta$
43. $\int (4\sec x\tan x - 2\sec^2 x)\,dx$
44. $\int \frac{1}{2}(\csc^2 x - \csc x\cot x)\,dx$
45. $\int (\sin 2x - \csc^2 x)\,dx$
46. $\int (3\cos 3x - 2\sin 2x)dx$
47. $\int \frac{1 + \cos 4t}{2}\,dt$
48. $\int \left(\frac{1 - \cos 3t}{3}\right)dt$
49. $\int 3x^{\sqrt{3}}\,dx$
50. $\int x^{\sqrt{2}-1}\,dx$
51. $\int (1 + \tan^2\theta)\,d\theta$ (*Hint:* $1 + \tan^2\theta = \sec^2\theta$)
52. $\int (3 + \tan^2\theta)d\theta$
53. $\int \cot^2 x\,dx$ (*Hint:* $1 + \cot^2 x = \csc^2 x$)
54. $\int (1 - \cot^2 x)\,dx$
55. $\int \cos\theta(\tan\theta + \sec\theta)\,d\theta$
56. $\int \frac{\csc\theta}{\csc\theta - \sin\theta}\,d\theta$

Checking Antiderivative Formulas

Verify the formulas in Exercises 57–62 by differentiation.

57. $\int (7x - 2)^3\,dx = \frac{(7x - 2)^4}{28} + C$
58. $\int (2x + 9)^{-4}\,dx = -\frac{(2x + 9)^{-3}}{6} + C$
59. $\int \sec^2(5x - 1)\,dx = \frac{1}{5}\tan(5x - 1) + C$
60. $\int \mathrm{cosec}^2\left(\frac{x - 2}{5}\right)dx = -5\cot\left(\frac{x - 2}{5}\right) + C$
61. $\int \frac{1}{(x + 1)^2}\,dx = -\frac{1}{x + 1} + C$
62. $\int \frac{1}{(x + 1)^2}\,dx = \frac{x}{x + 1} + C$
63. Right, or wrong? Say which for each formula and give a brief reason for each answer.
    - **a.** $\int x\sin x\,dx = \frac{x^2}{2}\sin x + C$
    - **b.** $\int x\sin x\,dx = -x\cos x + C$
    - **c.** $\int x\sin x\,dx = -x\cos x + \sin x + C$
64. Right, or wrong? Say which for each formula and give a brief reason for each answer.
    - **a.** $\int \tan\theta\sec^2\theta\,d\theta = \frac{\sec^3\theta}{3} + C$
    - **b.** $\int \tan\theta\sec^2\theta\,d\theta = \frac{1}{2}\tan^2\theta + C$
    - **c.** $\int \tan\theta\sec^2\theta\,d\theta = \frac{1}{2}\sec^2\theta + C$

65. Right, or wrong? Say which for each formula and give a brief reason for each answer.

a. $\int (2x+1)^2\,dx = \dfrac{(2x+1)^3}{3} + C$

b. $\int 3(2x+1)^2\,dx = \dfrac{(2x+1)^3}{5} + C$

c. $\int 6(2x+1)^2\,dx = (2x+1)^3 + C$

66. Right, or wrong? Say which for each formula and give a brief reason for each answer.

a. $\int \sqrt{2x+1}\,dx = \sqrt{x^2+x+C}$

b. $\int \sqrt{2x+1}\,dx = \sqrt{x^2+x} + C$

c. $\int \sqrt{2x+1}\,dx = \frac{1}{3}\left(\sqrt{2x+1}\right)^3 + C$

67. Right or wrong? Give a brief reason why.

$$\int \frac{-21(x+2)^2}{(x-5)^4}\,dx = \left(\frac{x+2}{x-5}\right)^3$$

68. Right, or wrong? Give a brief reason why.

$$\int \frac{x\cos(x^2) - \sin(x^2)}{x^2}\,dx = \frac{\sin(x^2)}{x} + C$$

Initial Value Problems

69. Which of the following graphs shows the solution of the initial value problem

$$\frac{dy}{dx} = 2x, \quad y = 4 \text{ when } x = 1?$$

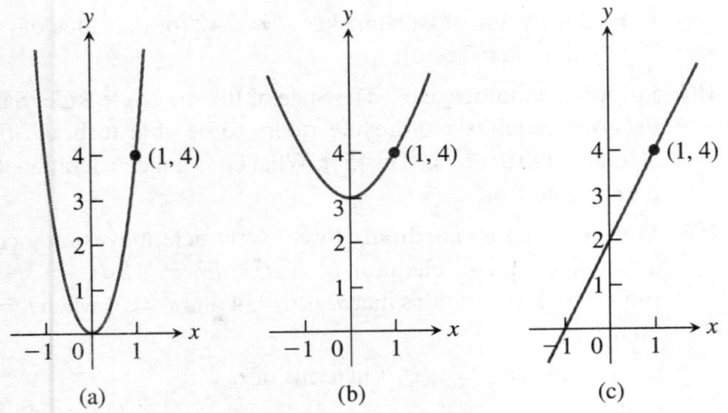

Give reasons for your answer.

70. Which of the following graphs shows the solution of the initial value problem

$$\frac{dy}{dx} = -x, \quad y = 1 \text{ when } x = -1?$$

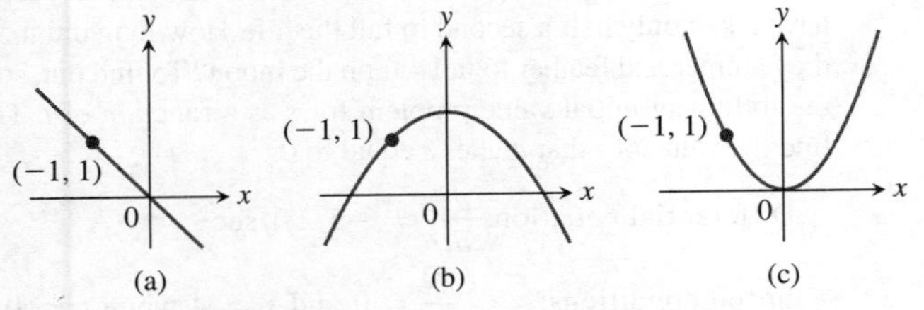

Give reasons for your answer.

Solve the initial value problems in Exercises 71–90.

71. $\dfrac{dy}{dx} = 4x - 12, \quad y(3) = 0$

72. $\dfrac{dy}{dx} = 10 - x, \quad y(0) = -1$

73. $\dfrac{dy}{dx} = \dfrac{1}{x^2} + x, \quad x > 0; \quad y(2) = 1$

74. $\dfrac{dy}{dx} = 12x^2 - 6x + 3, \quad y(-2) = 0$

75. $\dfrac{dy}{dx} = 3x^{-2/3}, \quad y(-1) = -5$

76. $\dfrac{dy}{dx} = \dfrac{1}{2\sqrt{x}}, \quad y(4) = 0$

77. $\dfrac{ds}{dt} = 1 + \sin t, \quad s(0) = 5$

78. $\dfrac{ds}{dt} = \cos t + \sin t, \quad s(\pi) = 1$

79. $\dfrac{dr}{d\theta} = -\pi \sin \pi\theta, \quad r(0) = 0$

80. $\dfrac{dr}{d\theta} = \sin \pi\theta, \quad r\left(\dfrac{1}{2}\right) = 1$

81. $\dfrac{dv}{dt} = \dfrac{1}{2}\sec t \tan t, \quad v(0) = 1$

82. $\dfrac{dv}{dt} = 8t + \csc^2 t, \quad v\left(\dfrac{\pi}{2}\right) = -7$

83. $\dfrac{d^2y}{dx^2} = 5 - 2x, \quad y'(0) = 2, \quad y(0) = 3$

84. $\dfrac{d^2y}{dx^2} = 0; \quad y'(0) = 2, \quad y(0) = 0$

85. $\dfrac{d^2r}{dt^2} = \dfrac{2}{t^3}; \quad \left.\dfrac{dr}{dt}\right|_{t=1} = 1, \quad r(1) = 1$

86. $\dfrac{d^2s}{dt^2} = \dfrac{3t}{8}; \quad \left.\dfrac{ds}{dt}\right|_{t=4} = 3, \quad s(4) = 4$

87. $\dfrac{d^3y}{dx^3} = 12, \quad y''(0) = -5, \quad y'(0) = 1, \quad y(0) = 4$

88. $\dfrac{d^3\theta}{dt^3} = 0; \quad \theta''(0) = -2, \quad \theta'(0) = -\dfrac{1}{2}, \quad \theta(0) = \sqrt{2}$

89. $y^{(4)} = -\sin t + \cos t;$
$y'''(0) = 7, \quad y''(0) = y'(0) = -1, \quad y(0) = 0$

90. $y^{(4)} = -3\cos x + 5\sin 2x, \quad y'''(0) = 1,$
$y''(0) = 2, \quad y'(0) = -1, \quad y(0) = 3$

91. Find the curve $y = f(x)$ in the xy-plane that passes through the point $(9, 4)$ and whose slope at each point is $3\sqrt{x}$.

92. a. Find a curve $y = f(x)$ with the following properties:

i) $\dfrac{d^2y}{dx^2} = 6x$

ii) Its graph passes through the point $(0, 1)$ and has a horizontal tangent line there.

b. How many curves like this are there? How do you know?

In Exercises 93–96, the graph of f' is given. Assume that $f(0) = 1$ and sketch a possible continuous graph of f.

93.

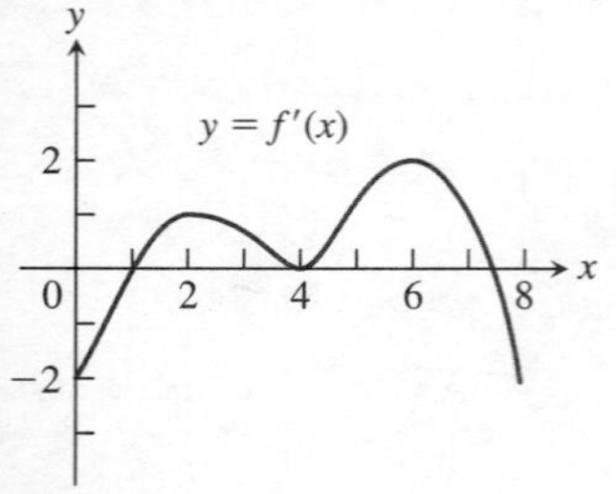

94.

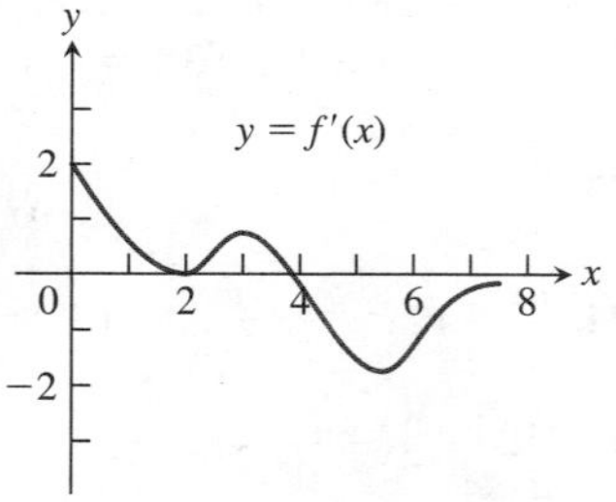

95.

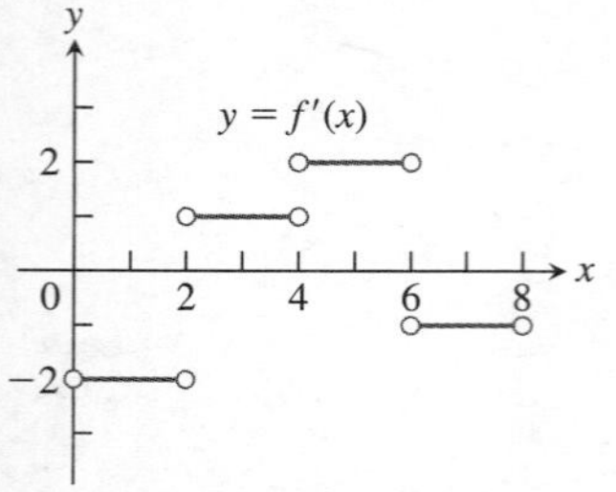

96.

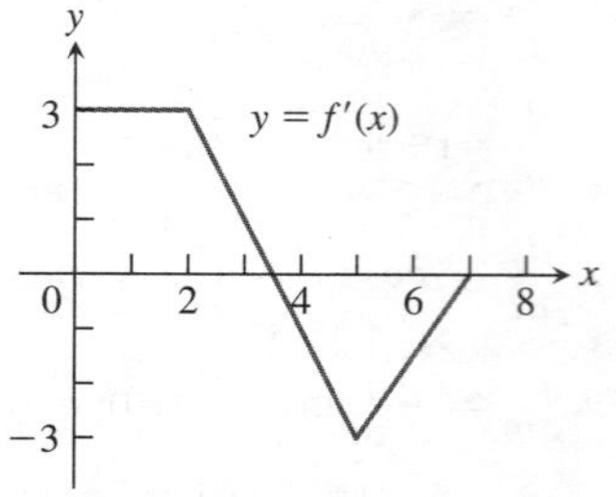

Solution (Integral) Curves

Exercises 97–100 show solution curves of differential equations. In each exercise, find an equation for the curve through the labeled point.

97.

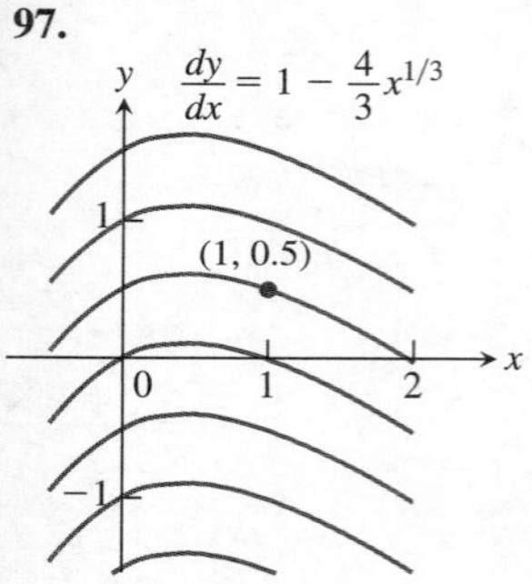

98.

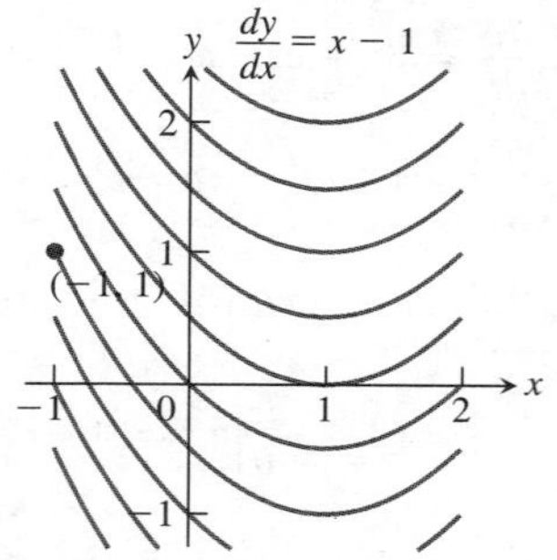

99.

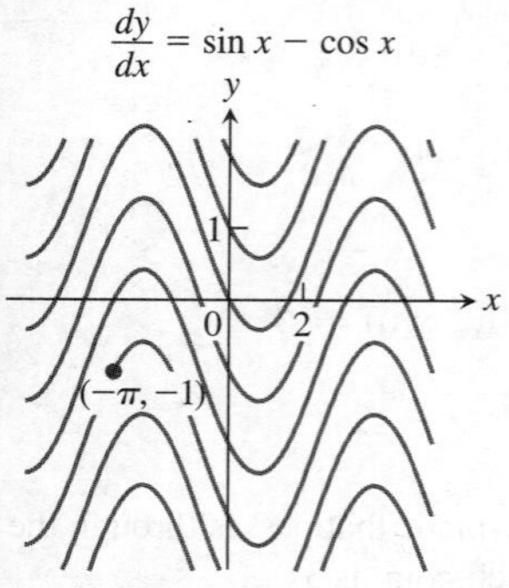

100.

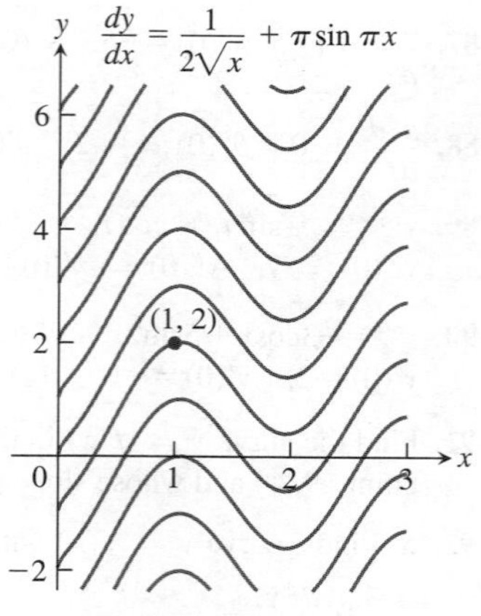

Applications

101. Finding displacement from an antiderivative of velocity

a. Suppose that the velocity of a body moving along the s-axis is

$$\frac{ds}{dt} = v = 9.8t - 3.$$

i) Find the body's displacement over the time interval from $t = 1$ to $t = 3$ given that $s = 5$ when $t = 0$.

ii) Find the body's displacement from $t = 1$ to $t = 3$ given that $s = -2$ when $t = 0$.

iii) Now find the body's displacement from $t = 1$ to $t = 3$ given that $s = s_0$ when $t = 0$.

b. Suppose that the position s of a body moving along a coordinate line is a differentiable function of time t. Is it true that once you know an antiderivative of the velocity function ds/dt, you can find the body's displacement from $t = a$ to $t = b$ even if you do not know the body's exact position at either of those times? Give reasons for your answer.

102. Liftoff from Earth A rocket lifts off the surface of Earth with a constant acceleration of 20 m/sec^2. How fast will the rocket be going 1 min later?

103. Stopping a car in time You are driving along a highway at a steady 60 mph (88 ft/sec) when you see an accident ahead and slam on the brakes. What constant deceleration is required to stop your car in 242 ft? To find out, carry out the following steps.

1. Solve the initial value problem

Differential equation: $\dfrac{d^2s}{dt^2} = -k$ (k constant)

Initial conditions: $\dfrac{ds}{dt} = 88$ and $s = 0$ when $t = 0$.

Measuring time and distance from when the brakes are applied

2. Find the value of t that makes $ds/dt = 0$. (The answer will involve k.)

3. Find the value of k that makes $s = 242$ for the value of t you found in Step 2.

104. Stopping a motorcycle The State of Illinois Cycle Rider Safety Program requires motorcycle riders to be able to brake from 30 mph (44 ft/sec) to 0 in 45 ft. What constant deceleration does it take to do that?

105. Motion along a coordinate line A particle moves on a coordinate line with acceleration $a = d^2s/dt^2 = 15\sqrt{t} - (3/\sqrt{t})$, subject to the conditions that $ds/dt = 4$ and $s = 0$ when $t = 1$. Find

a. the velocity $v = ds/dt$ in terms of t.

b. the position s in terms of t.

T **106. The hammer and the feather** When *Apollo 15* astronaut David Scott dropped a hammer and a feather on the moon to demonstrate that in a vacuum all bodies fall with the same (constant) acceleration, he dropped them from about 4 ft above the ground. The television footage of the event shows the hammer and the feather falling more slowly than on Earth, where, in a vacuum, they would have taken only half a second to fall the 4 ft. How long did it take the hammer and feather to fall 4 ft on the moon? To find out, solve the following initial value problem for s as a function of t. Then find the value of t that makes s equal to 0.

Differential equation: $\dfrac{d^2s}{dt^2} = -5.2$ ft/sec^2

Initial conditions: $\dfrac{ds}{dt} = 0$ and $s = 4$ when $t = 0$

107. Motion with constant acceleration The standard equation for the position s of a body moving with a constant acceleration a along a coordinate line is

$$s = \frac{a}{2}t^2 + v_0 t + s_0, \qquad (1)$$

where v_0 and s_0 are the body's velocity and position at time $t = 0$. Derive this equation by solving the initial value problem

Differential equation: $\dfrac{d^2s}{dt^2} = a$

Initial conditions: $\dfrac{ds}{dt} = v_0$ and $s = s_0$ when $t = 0$.

108. Free fall near the surface of a planet For free fall near the surface of a planet where the acceleration due to gravity has a constant magnitude of g length-units/sec^2, Equation (1) in Exercise 107 takes the form

$$s = -\frac{1}{2}gt^2 + v_0 t + s_0, \qquad (2)$$

where s is the body's height above the surface. The equation has a minus sign because the acceleration acts downward, in the direction of decreasing s. The velocity v_0 is positive if the object is rising at time $t = 0$ and negative if the object is falling.

Instead of using the result of Exercise 107, you can derive Equation (2) directly by solving an appropriate initial value problem. What initial value problem? Solve it to be sure you have the right one, explaining the solution steps as you go along.

109. Suppose that

$$f(x) = \frac{d}{dx}\left(1 - \sqrt{x}\right) \quad \text{and} \quad g(x) = \frac{d}{dx}(x + 2).$$

Find:

a. $\int f(x)\,dx$ **b.** $\int g(x)\,dx$

c. $\int [-f(x)]\,dx$ **d.** $\int [-g(x)]\,dx$

e. $\int [f(x) + g(x)]\,dx$ **f.** $\int [f(x) - g(x)]\,dx$

110. Uniqueness of solutions If differentiable functions $y = F(x)$ and $y = g(x)$ both solve the initial value problem

$$\frac{dy}{dx} = f(x), \qquad y(x_0) = y_0,$$

on an interval I, must $F(x) = G(x)$ for every x in I? Give reasons for your answer.

COMPUTER EXPLORATIONS

Use a CAS to solve the initial value problems in Exercises 111 and 112. Plot the solution curves.

111. $y' = \cos^2 x + \sin x, \quad y(\pi) = 1$

112. $y' = \sqrt{x} + x, \quad y(1) = -1$

CHAPTER 4 Questions to Guide Your Review

1. What can be said about the extreme values of a function that is continuous on a closed interval?
2. What does it mean for a function to have a local extreme value on its domain? An absolute extreme value? How are local and absolute extreme values related, if at all? Give examples.
3. How do you find the absolute extrema of a continuous function on a closed interval? Give examples.
4. What are the hypotheses and conclusion of Rolle's Theorem? Are the hypotheses really necessary? Explain.
5. What are the hypotheses and conclusion of the Mean Value Theorem? What physical interpretations might the theorem have?
6. State the Mean Value Theorem's three corollaries.
7. How can you sometimes identify a function $f(x)$ by knowing f' and knowing the value of f at a point $x = x_0$? Give an example.
8. What is the First Derivative Test for Local Extreme Values? Give examples of how it is applied.
9. How do you test a twice-differentiable function to determine where its graph is concave up or concave down? Give examples.
10. What is an inflection point? Give an example. What physical significance do inflection points sometimes have?
11. What is the Second Derivative Test for Local Extreme Values? Give examples of how it is applied.
12. What do the derivatives of a function tell you about the shape of its graph?
13. List the steps you would take to graph a polynomial function. Illustrate with an example.
14. What is a cusp? Give examples.
15. List the steps you would take to graph a rational function. Illustrate with an example.
16. Outline a general strategy for solving max-min problems. Give examples.
17. Describe Newton's method for solving equations. Give an example. What is the theory behind the method? What are some of the things to watch out for when you use the method?
18. Can a function have more than one antiderivative? If so, how are the antiderivatives related? Explain.
19. What is an indefinite integral? How do you evaluate one? What general formulas do you know for finding indefinite integrals?
20. How can you sometimes solve a differential equation of the form $dy/dx = f(x)$?
21. What is an initial value problem? How do you solve one? Give an example.
22. If you know the acceleration of a body moving along a coordinate line as a function of time, what more do you need to know to find the body's position function? Give an example.

CHAPTER 4 Practice Exercises

Finding Extreme Values

In Exercises 1–10, find the extreme values (absolute and local) of the function over its natural domain, and where they occur.

1. $y = 2x^2 - 8x + 9$

2. $y = x^3 - 2x + 4$

3. $y = x^3 + x^2 - 8x + 5$

4. $y = x^3(x - 5)^2$

5. $y = \sqrt{x^2 - 1}$

6. $y = x - 4\sqrt{x}$

7. $y = \dfrac{1}{\sqrt[3]{1 - x^2}}$

8. $y = \sqrt{3 + 2x - x^2}$

9. $y = \dfrac{x}{x^2 + 1}$

10. $y = \dfrac{x + 1}{x^2 + 2x + 2}$

Extreme Values

11. Does $f(x) = x^3 + 2x + \tan x$ have any local maximum or minimum values? Give reasons for your answer.

12. Does $g(x) = \csc x + 2\cot x$ have any local maximum values? Give reasons for your answer.

13. Does $f(x) = (7 + x)(11 - 3x)^{1/3}$ have an absolute minimum value? An absolute maximum? If so, find them or give reasons why they fail to exist. List all critical points of f.

14. Find values of a and b such that the function

$$f(x) = \frac{ax + b}{x^2 - 1}$$

has a local extreme value of 1 at $x = 3$. Is this extreme value a local maximum or a local minimum? Give reasons for your answer.

15. The greatest integer function $f(x) = \lfloor x \rfloor$, defined for all values of x, assumes a local maximum value of 0 at each point of $[0, 1)$. Could any of these local maximum values also be local minimum values of f? Give reasons for your answer.

16. a. Give an example of a differentiable function f whose first derivative is zero at some point c, even though f has neither a local maximum nor a local minimum at c.

b. How is this consistent with Theorem 2 in Section 4.1? Give reasons for your answer.

17. The function $y = 1/x$ does not take on either a maximum or a minimum on the interval $0 < x < 1$ even though the function is continuous on this interval. Does this contradict the Extreme Value Theorem for continuous functions? Why?

18. What are the maximum and minimum values of the function $y = |x|$ on the interval $-1 \le x < 1$? Notice that the interval is not closed. Is this consistent with the Extreme Value Theorem for continuous functions? Why?

T **19.** A graph that is large enough to show a function's global behavior may fail to reveal important local features. The graph of $f(x) = (x^8/8) - (x^6/2) - x^5 + 5x^3$ is a case in point.

a. Graph f over the interval $-2.5 \le x \le 2.5$. Where does the graph appear to have local extreme values or points of inflection?

b. Now factor $f'(x)$ and show that f has a local maximum at $x = \sqrt[3]{5} \approx 1.70998$ and local minima at $x = \pm\sqrt{3} \approx \pm 1.73205$.

c. Zoom in on the graph to find a viewing window that shows the presence of the extreme values at $x = \sqrt[3]{5}$ and $x = \sqrt{3}$.

The moral here is that without calculus, the existence of two of the three extreme values would probably have gone unnoticed. On any normal graph of the function, the values would lie close enough together to fall within the dimensions of a single pixel on the screen.

(*Source*: *Uses of Technology in the Mathematics Curriculum*, by Benny Evans and Jerry Johnson, Oklahoma State University, published in 1990 under a grant from the National Science Foundation, USE-8950044.)

T **20.** (*Continuation of Exercise 19*)

a. Graph $f(x) = (x^8/8) - (2/5)x^5 - 5x - (5/x^2) + 11$ over the interval $-2 \le x \le 2$. Where does the graph appear to have local extreme values or points of inflection?

b. Show that f has a local maximum value at $x = \sqrt[7]{5} \approx 1.2585$ and a local minimum value at $x = \sqrt[3]{2} \approx 1.2599$.

c. Zoom in to find a viewing window that shows the presence of the extreme values at $x = \sqrt[7]{5}$ and $x = \sqrt[3]{2}$.

The Mean Value Theorem

21. a. Show that $g(t) = \sin^2 t - 3t$ decreases on every interval in its domain.

b. How many solutions does the equation $\sin^2 t - 3t = 5$ have? Give reasons for your answer.

22. a. Show that $y = \tan\theta$ increases on every open interval in its domain.

b. If the conclusion in part (a) is really correct, how do you explain the fact that $\tan\pi = 0$ is less than $\tan(\pi/4) = 1$?

23. a. Show that the equation $x^4 + 2x^2 - 2 = 0$ has exactly one solution on $[0, 1]$.

T **b.** Find the solution to as many decimal places as you can.

24. a. Show that $f(x) = x/(x + 1)$ increases on every open interval in its domain.

b. Show that $f(x) = x^3 + 2x$ has no local maximum or minimum values.

25. Water in a reservoir As a result of a heavy rain, the volume of water in a reservoir increased by 1400 acre-ft in 24 hours. Show that at some instant during that period the reservoir's volume was increasing at a rate in excess of 225,000 gal/min. (An acre-foot is 43,560 ft^3, the volume that would cover 1 acre to the depth of 1 ft. A cubic foot holds 7.48 gal.)

26. The formula $F(x) = 3x + C$ gives a different function for each value of C. All of these functions, however, have the same derivative with respect to x, namely $F'(x) = 3$. Are these the only differentiable functions whose derivative is 3? Could there be any others? Give reasons for your answers.

27. Show that

$$\frac{d}{dx}\left(\frac{x}{x + 1}\right) = \frac{d}{dx}\left(-\frac{1}{x + 1}\right)$$

even though

$$\frac{x}{x+1} \neq -\frac{1}{x+1}.$$

Doesn't this contradict Corollary 2 of the Mean Value Theorem? Give reasons for your answer.

28. Calculate the first derivatives of $f(x) = x^2/(x^2+1)$ and $g(x) = -1/(x^2+1)$. What can you conclude about the graphs of these functions?

Analyzing Graphs

In Exercises 29 and 30, use the graph to answer the questions.

29. Identify any global extreme values of f and the values of x at which they occur.

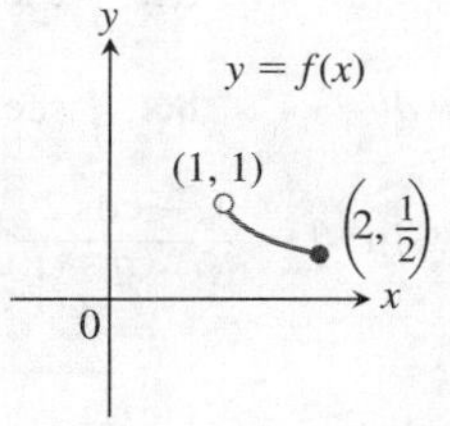

30. Estimate the open intervals on which the function $y = f(x)$ is

a. increasing.

b. decreasing.

c. Use the given graph of f' to indicate where any local extreme values of the function occur, and whether each extreme is a relative maximum or minimum.

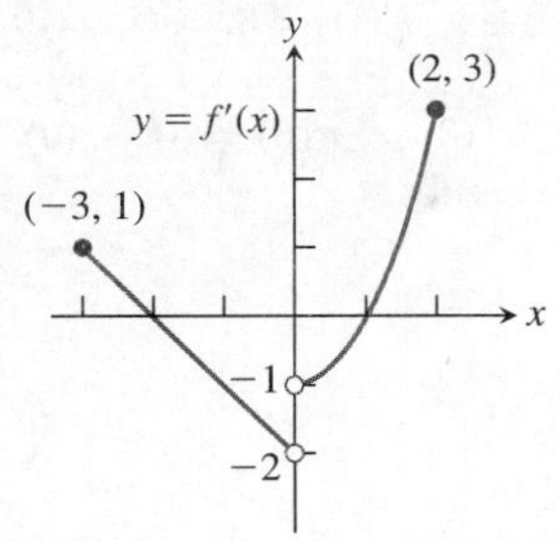

Each of the graphs in Exercises 31 and 32 is the graph of the position function $s = f(t)$ of an object moving on a coordinate line (t represents time). At approximately what times (if any) is each object's **(a)** velocity equal to zero? **(b)** Acceleration equal to zero? During approximately what time intervals does the object move **(c)** forward? **(d)** Backward?

31.

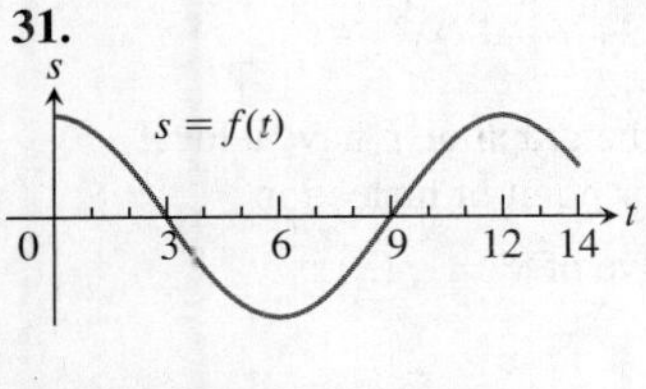

32.

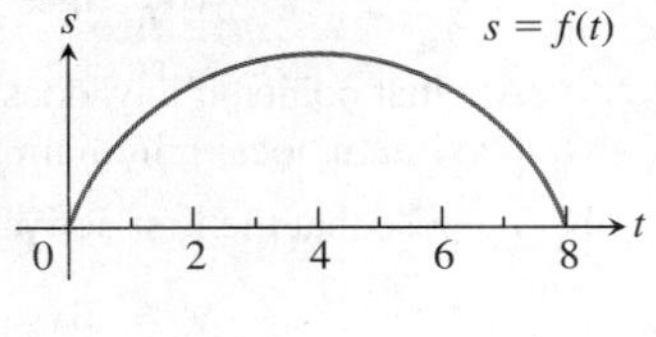

Graphs and Graphing

Graph the curves in Exercises 33–42.

33. $y = x^2 - (x^3/6)$

34. $y = x^3 - 3x^2 + 3$

35. $y = -x^3 + 6x^2 - 9x + 3$

36. $y = (1/8)(x^3 + 3x^2 - 9x - 27)$

37. $y = x^3(8 - x)$

38. $y = x^2(2x^2 - 9)$

39. $y = x - 3x^{2/3}$

40. $y = x^{1/3}(x - 4)$

41. $y = x\sqrt{3 - x}$

42. $y = x\sqrt{4 - x^2}$

Each of Exercises 43–48 gives the first derivative of a function $y = f(x)$. **(a)** At what points, if any, does the graph of f have a local maximum, local minimum, or inflection point? **(b)** Sketch the general shape of the graph.

43. $y' = 16 - x^2$

44. $y' = x^2 - x - 6$

45. $y' = 6x(x + 1)(x - 2)$

46. $y' = x^2(6 - 4x)$

47. $y' = x^4 - 2x^2$

48. $y' = 4x^2 - x^4$

In Exercises 49–52, graph each function. Then use the function's first derivative to explain what you see.

49. $y = x^{2/3} + (x - 1)^{1/3}$

50. $y = x^{2/3} + (x - 1)^{2/3}$

51. $y = x^{1/3} + (x - 1)^{1/3}$

52. $y = x^{2/3} - (x - 1)^{1/3}$

Sketch the graphs of the rational functions in Exercises 53–60.

53. $y = \dfrac{x+1}{x-3}$

54. $y = \dfrac{2x}{x+5}$

55. $y = \dfrac{x^2+1}{x}$

56. $y = \dfrac{x^2 - x + 1}{x}$

57. $y = \dfrac{x^3+2}{2x}$

58. $y = \dfrac{x^4-1}{x^2}$

59. $y = \dfrac{x^2-4}{x^2-3}$

60. $y = \dfrac{x^2}{x^2-4}$

Optimization

61. The sum of two nonnegative numbers is 36. Find the numbers if

a. the difference of their square roots is to be as large as possible.

b. the sum of their square roots is to be as large as possible.

62. The sum of two nonnegative numbers is 20. Find the numbers

a. if the product of one number and the square root of the other is to be as large as possible.

b. if one number plus the square root of the other is to be as large as possible.

63. An isosceles triangle has its vertex at the origin and its base parallel to the x-axis with the vertices above the axis on the curve $y = 27 - x^2$. Find the largest area the triangle can have.

64. A customer has asked you to design an open-top rectangular stainless steel vat. It is to have a square base and a volume of 32 ft^3, to be welded from quarter-inch plate, and to weigh no more than necessary. What dimensions do you recommend?

65. Find the height and radius of the largest right circular cylinder that can be put in a sphere of radius $\sqrt{3}$.

66. The figure here shows two right circular cones, one upside down inside the other. The two bases are parallel, and the vertex of the smaller cone lies at the center of the larger cone's base. What values of r and h will give the smaller cone the largest possible volume?

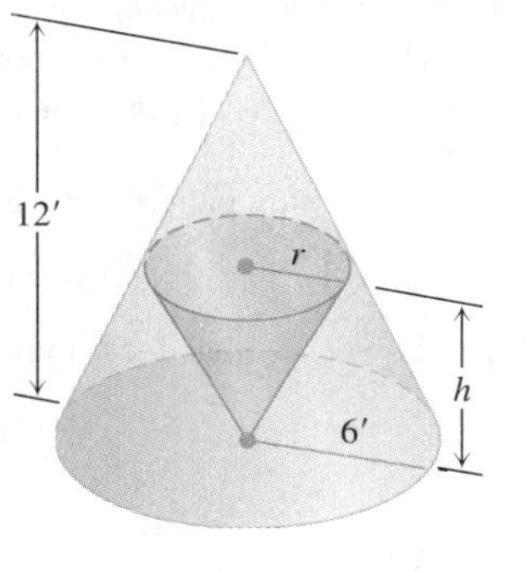

67. Manufacturing tires Your company can manufacture x hundred grade A tires and y hundred grade B tires a day, where $0 \leq x \leq 4$ and

$$y = \frac{40 - 10x}{5 - x}.$$

Your profit on a grade A tire is twice your profit on a grade B tire. What is the most profitable number of each kind to make?

68. Particle motion The positions of two particles on the s-axis are $s_1 = \cos t$ and $s_2 = \cos(t + \pi/4)$.

a. What is the farthest apart the particles ever get?

b. When do the particles collide?

T **69. Open-top box** An open-top rectangular box is constructed from a 10-in.-by-16-in. piece of cardboard by cutting squares of equal side length from the corners and folding up the sides. Find analytically the dimensions of the box of largest volume and the maximum volume. Support your answers graphically.

70. The ladder problem What is the approximate length (in feet) of the longest ladder you can carry horizontally around the corner of the corridor shown here? Round your answer down to the nearest foot.

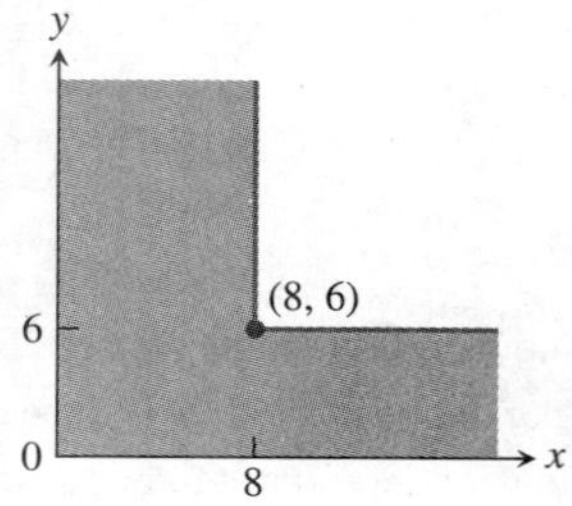

Newton's Method

71. Let $f(x) = 3x - x^3$. Show that the equation $f(x) = -4$ has a solution in the interval $[2, 3]$ and use Newton's method to find it.

72. Let $f(x) = x^4 - x^3$. Show that the equation $f(x) = 75$ has a solution in the interval $[3, 4]$ and use Newton's method to find it.

Finding Indefinite Integrals

Find the indefinite integrals (most general antiderivatives) in Exercises 73–88. You may need to try a solution and then adjust your guess. Check your answers by differentiation.

73. $\int (x^3 + 5x - 7)\, dx$

74. $\int \left(8t^3 - \frac{t^2}{2} + t\right) dt$

75. $\int \left(3\sqrt{t} + \frac{4}{t^2}\right) dt$

76. $\int \left(\frac{1}{2\sqrt{t}} - \frac{3}{t^4}\right) dt$

77. $\int \frac{dr}{(r + 5)^2}$

78. $\int \frac{6\,dr}{\left(r - \sqrt{2}\right)^3}$

79. $\int 3\theta\sqrt{\theta^2 + 1}\, d\theta$

80. $\int \frac{\theta}{\sqrt{7 + \theta^2}}\, d\theta$

81. $\int x^3(1 + x^4)^{-1/4}\, dx$

82. $\int (2 - x)^{3/5}\, dx$

83. $\int \sec^2 \frac{s}{10}\, ds$

84. $\int \csc^2 \pi s\, ds$

85. $\int \csc\sqrt{2}\theta \cot\sqrt{2}\theta\, d\theta$

86. $\int \sec\frac{\theta}{3} \tan\frac{\theta}{3}\, d\theta$

87. $\int \sin^2 \frac{x}{4}\, dx \quad \left(\textit{Hint: } \sin^2\theta = \frac{1 - \cos 2\theta}{2}\right)$

88. $\int \cos^2 \frac{x}{2}\, dx$

Initial Value Problems

Solve the initial value problems in Exercises 89–92.

89. $\frac{dy}{dx} = \frac{x^2 + 1}{x^2}, \quad y(1) = -1$

90. $\frac{dy}{dx} = \left(x + \frac{1}{x}\right)^2, \quad y(1) = 1$

91. $\frac{d^2r}{dt^2} = 15\sqrt{t} + \frac{3}{\sqrt{t}}; \quad r'(1) = 8, \quad r(1) = 0$

92. $\frac{d^3r}{dt^3} = -\cos t; \quad r''(0) = r'(0) = 0, \quad r(0) = -1$

CHAPTER 4 Additional and Advanced Exercises

Functions and Derivatives

1. What can you say about a function whose maximum and minimum values on an interval are equal? Give reasons for your answer.

2. Is it true that a discontinuous function cannot have both an absolute maximum value and an absolute minimum value on a closed interval? Give reasons for your answer.

3. Can you conclude anything about the extreme values of a continuous function on an open interval? On a half-open interval? Give reasons for your answer.

4. Local extrema Use the sign pattern for the derivative

$$\frac{df}{dx} = 6(x - 1)(x - 2)^2(x - 3)^3(x - 4)^4$$

to identify the points where f has local maximum and minimum values.

5. Local extrema

a. Suppose that the first derivative of $y = f(x)$ is

$$y' = 6(x + 1)(x - 2)^2.$$

At what points, if any, does the graph of f have a local maximum, local minimum, or point of inflection?

b. Suppose that the first derivative of $y = f(x)$ is

$$y' = 6x(x + 1)(x - 2).$$

At what points, if any, does the graph of f have a local maximum, local minimum, or point of inflection?

6. If $f'(x) \leq 2$ for all x, what is the most the values of f can increase on $[0, 6]$? Give reasons for your answer.

7. **Bounding a function** Suppose that f is continuous on $[a, b]$ and that c is an interior point of the interval. Show that if $f'(x) \le 0$ on $[a, c)$ and $f'(x) \ge 0$ on $(c, b]$, then $f(x)$ is never less than $f(c)$ on $[a, b]$.
8. **An inequality**
   a. Show that $-1/2 \le x/(1 + x^2) \le 1/2$ for every value of x.
   b. Suppose that f is a function whose derivative is $f'(x) = x/(1 + x^2)$. Use the result in part (a) to show that
   $$|f(b) - f(a)| \le \frac{1}{2}|b - a|$$
   for any a and b.
9. The derivative of $f(x) = x^2$ is zero at $x = 0$, but f is not a constant function. Doesn't this contradict the corollary of the Mean Value Theorem that says that functions with zero derivatives are constant? Give reasons for your answer.
10. **Extrema and inflection points** Let $h = fg$ be the product of two differentiable functions of x.
    a. If f and g are positive, with local maxima at $x = a$, and if f' and g' change sign at a, does h have a local maximum at a?
    b. If the graphs of f and g have inflection points at $x = a$, does the graph of h have an inflection point at a?

In either case, if the answer is yes, give a proof. If the answer is no, give a counterexample.

11. **Finding a function** Use the following information to find the values of a, b, and c in the formula $f(x) = (x + a)/(bx^2 + cx + 2)$.
    a. The values of a, b, and c are either 0 or 1.
    b. The graph of f passes through the point $(-1, 0)$.
    c. The line $y = 1$ is an asymptote of the graph of f.
12. **Horizontal tangent** For what value or values of the constant k will the curve $y = x^3 + kx^2 + 3x - 4$ have exactly one horizontal tangent?

Optimization

13. **Largest inscribed triangle** Points A and B lie at the ends of a diameter of a unit circle and point C lies on the circumference. Is it true that the area of triangle ABC is largest when the triangle is isosceles? How do you know?
14. **Proving the second derivative test** The Second Derivative Test for Local Maxima and Minima (Section 4.4) says:
    a. f has a local maximum value at $x = c$ if $f'(c) = 0$ and $f''(c) < 0$;
    b. f has a local minimum value at $x = c$ if $f'(c) = 0$ and $f''(c) > 0$.
    To prove statement (a), let $\varepsilon = (1/2)|f''(c)|$. Then use the fact that
    $$f''(c) = \lim_{h\to 0} \frac{f'(c + h) - f'(c)}{h} = \lim_{h\to 0} \frac{f'(c + h)}{h}$$
    to conclude that for some $\delta > 0$,
    $$0 < |h| < \delta \quad \Rightarrow \quad \frac{f'(c + h)}{h} < f''(c) + \varepsilon < 0.$$
    Thus, $f'(c + h)$ is positive for $-\delta < h < 0$ and negative for $0 < h < \delta$. Prove statement (b) in a similar way.
15. **Hole in a water tank** You want to bore a hole in the side of the tank shown here at a height that will make the stream of water coming out hit the ground as far from the tank as possible. If you drill the hole near the top, where the pressure is low, the water will exit slowly but spend a relatively long time in the air. If you drill the hole near the bottom, the water will exit at a higher velocity but have only a short time to fall. Where is the best place, if any, for the hole? (*Hint:* How long will it take an exiting droplet of water to fall from height y to the ground?)

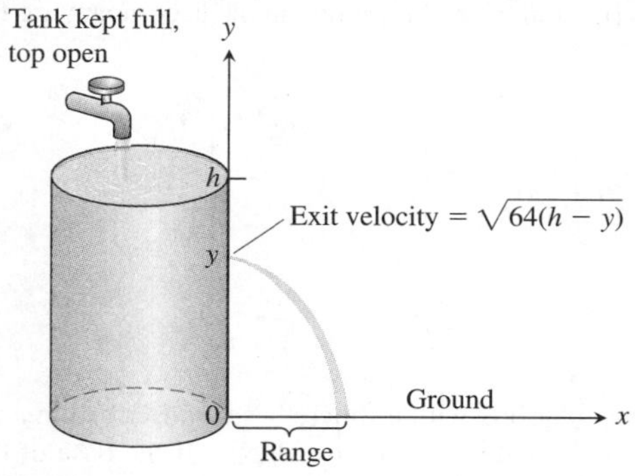

16. **Kicking a field goal** An American football player wants to kick a field goal with the ball being on a right hash mark. Assume that the goal posts are b feet apart and that the hash mark line is a distance $a > 0$ feet from the right goal post. (See the accompanying figure.) Find the distance h from the goal post line that gives the kicker his largest angle β. Assume that the football field is flat.

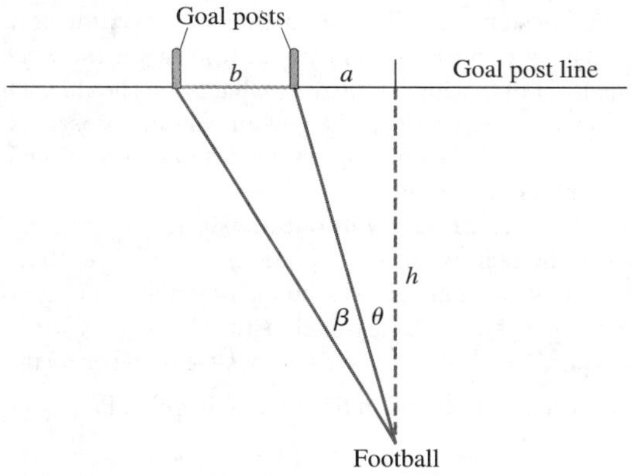

17. **A max-min problem with a variable answer** Sometimes the solution of a max-min problem depends on the proportions of the shapes involved. As a case in point, suppose that a right circular cylinder of radius r and height h is inscribed in a right circular cone of radius R and height H, as shown here. Find the value of r (in terms of R and H) that maximizes the total surface area of the cylinder (including top and bottom). As you will see, the solution depends on whether $H \le 2R$ or $H > 2R$.

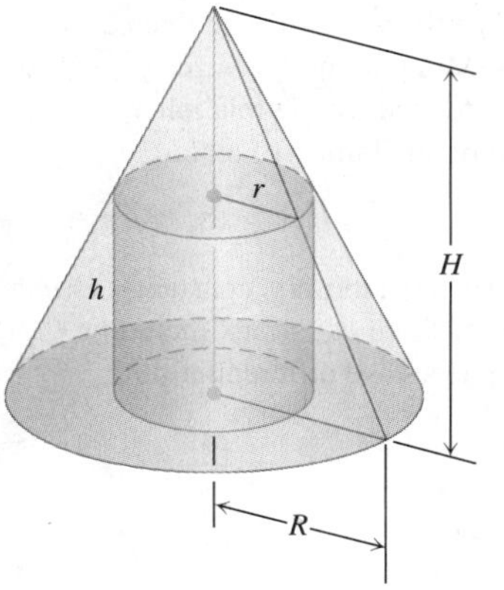

18. **Minimizing a parameter** Find the smallest value of the positive constant m that will make $mx - 1 + (1/x)$ greater than or equal to zero for all positive values of x.
19. Determine the dimensions of the rectangle of largest area that can be inscribed in the right triangle in the accompanying figure.

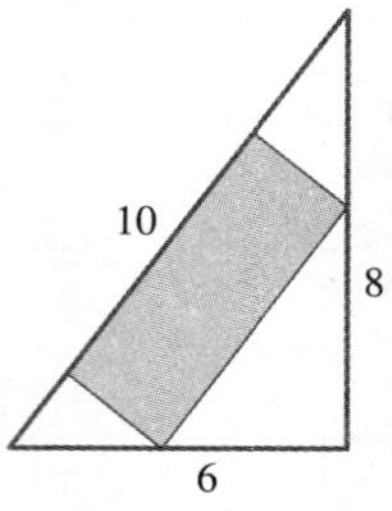

20. A rectangular box with a square base is inscribed in a right circular cone of height 4 and base radius 3. If the base of the box sits on the base of the cone, what is the largest possible volume of the box?

Theory and Examples

21. Suppose that it costs a company $y = a + bx$ dollars to produce x units per week. It can sell x units per week at a price of $P = c - ex$ dollars per unit. Each of a, b, c, and e represents a positive constant. **(a)** What production level maximizes the profit? **(b)** What is the corresponding price? **(c)** What is the weekly profit at this level of production? **(d)** At what price should each item be sold to maximize profits if the government imposes a tax of t dollars per item sold? Comment on the difference between this price and the price before the tax.
22. **Estimating reciprocals without division** You can estimate the value of the reciprocal of a number a without ever dividing by a if you apply Newton's method to the function $f(x) = (1/x) - a$. For example, if $a = 3$, the function involved is $f(x) = (1/x) - 3$.
    a. Graph $y = (1/x) - 3$. Where does the graph cross the x-axis?
    b. Show that the recursion formula in this case is
    $$x_{n+1} = x_n(2 - 3x_n),$$
    so there is no need for division.
23. To find $x = \sqrt[q]{a}$, we apply Newton's method to $f(x) = x^q - a$. Here we assume that a is a positive real number and q is a positive integer. Show that x_1 is a "weighted average" of x_0 and a/x_0^{q-1}, and find the coefficients m_0, m_1 such that
    $$x_1 = m_0x_0 + m_1\left(\frac{a}{x_0^{q-1}}\right), \quad m_0 > 0,\ m_1 > 0,\ m_0 + m_1 = 1.$$
    What conclusion would you reach if x_0 and a/x_0^{q-1} were equal? What would be the value of x_1 in that case?
24. The family of straight lines $y = ax + b$ (a, b arbitrary constants) can be characterized by the relation $y'' = 0$. Find a similar relation satisfied by the family of all circles
    $$(x - h)^2 + (y - h)^2 = r^2,$$
    where h and r are arbitrary constants. (*Hint:* Eliminate h and r from the set of three equations including the given one and two obtained by successive differentiation.)
25. Assume that the brakes of an automobile produce a constant deceleration of k ft/sec^2. **(a)** Determine what k must be to bring an automobile traveling 60 mi/hr (88 ft/sec) to rest in a distance of 100 ft from the point where the brakes are applied. **(b)** With the same k, how far would a car traveling 30 mi/hr go before being brought to a stop?
26. Let $f(x)$ and $g(x)$ be two continuously differentiable functions satisfying the relationships $f'(x) = g(x)$ and $f''(x) = -f(x)$. Let $h(x) = f^2(x) + g^2(x)$. If $h(0) = 5$, find $h(10)$.
27. Can there be a curve satisfying the following conditions? d^2y/dx^2 is everywhere equal to zero and, when $x = 0$, $y = 0$ and $dy/dx = 1$. Give a reason for your answer.
28. Find the equation for the curve in the xy-plane that passes through the point $(1, -1)$ if its slope at x is always $3x^2 + 2$.
29. A particle moves along the x-axis. Its acceleration is $a = -t^2$. At $t = 0$, the particle is at the origin. In the course of its motion, it reaches the point $x = b$, where $b > 0$, but no point beyond b. Determine its velocity at $t = 0$.
30. A particle moves with acceleration $a = \sqrt{t} - (1/\sqrt{t})$. Assuming that the velocity $v = 4/3$ and the position $s = -4/15$ when $t = 0$, find
    a. the velocity v in terms of t.
    b. the position s in terms of t.
31. Given $f(x) = ax^2 + 2bx + c$ with $a > 0$. By considering the minimum, prove that $f(x) \geq 0$ for all real x if and only if $b^2 - ac \leq 0$.
32. **The Cauchy–Schwarz inequality**
    a. In Exercise 31, let
    $$f(x) = (a_1x + b_1)^2 + (a_2x + b_2)^2 + \cdots + (a_nx + b_n)^2,$$
    and deduce The Cauchy–Schwarz inequality:
    $$(a_1b_1 + a_2b_2 + \cdots + a_nb_n)^2 \leq (a_1^2 + a_2^2 + \cdots + a_n^2)(b_1^2 + b_2^2 + \cdots + b_n^2).$$
    b. Show that equality holds in The Cauchy–Schwarz inequality only if there exists a real number x that makes a_ix equal $-b_i$ for every value of i from 1 to n.
33. Consider the unit circle centered at the origin and with a vertical tangent line passing through point A in the accompanying figure. Assume that the lengths of segments AB and AC are equal, and let point D be the intersection of the x-axis with the line passing through points B and C. Find the limit of t as B approaches A.

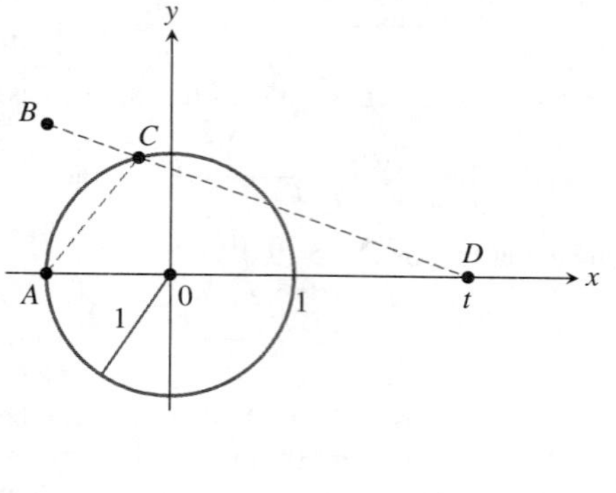

CHAPTER 4 Technology Application Projects

Mathematica/Maple Projects

Projects can be found within MyLab Math.

- ***Motion Along a Straight Line: Position → Velocity → Acceleration***

  You will observe the shape of a graph through dramatic animated visualizations of the derivative relations among the position, velocity, and acceleration. Figures in the text can be animated.

- ***Newton's Method: Estimate π to How Many Places?***

  Plot a function, observe a root, pick a starting point near the root, and use Newton's Iteration Procedure to approximate the root to a desired accuracy. The numbers π, e, and $\sqrt{2}$ are approximated.

5

Integrals

OVERVIEW A great achievement of classical geometry was obtaining formulas for the areas and volumes of triangles, spheres, and cones. In this chapter we develop a method, called *integration*, to calculate the areas and volumes of more general shapes. The *definite integral* is the key tool in calculus for defining and calculating areas and volumes. We also use it to compute quantities such as the lengths of curved paths, probabilities, averages, energy consumption, the mass of an object, and the force against a dam's floodgates.

Like the derivative, the definite integral is defined as a limit. The definite integral is a limit of increasingly fine approximations. The idea is to approximate a quantity (such as the area of a curvy region) by dividing it into many small pieces, each of which we can approximate by something simple (such as a rectangle). Summing the contributions of each of the simple pieces gives us an approximation to the original quantity. As we divide the region into more and more pieces, the approximation given by the sum of the pieces will generally improve, converging to the quantity we are measuring. We take a limit as the number of terms increases to infinity, and when the limit exists, the result is a definite integral. We develop this idea in Section 5.3.

We also show that the process of computing these definite integrals is closely connected to finding antiderivatives. This is one of the most important relationships in calculus; it gives us an efficient way to compute definite integrals, providing a simple and powerful method that eliminates the difficulty of directly computing limits of approximations. This connection is captured in the Fundamental Theorem of Calculus.

5.1 Area and Estimating with Finite Sums

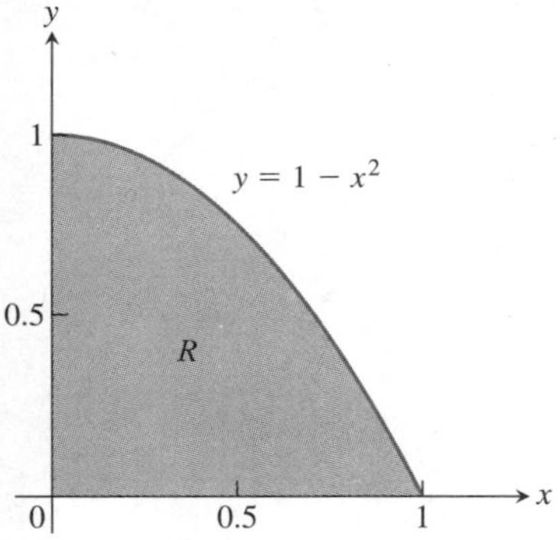

FIGURE 5.1 The area of the shaded region R cannot be found by a simple formula.

The basis for formulating definite integrals is the construction of approximations by finite sums. In this section we consider three examples of this process: finding the area under a graph, the distance traveled by a moving object, and the average value of a function. Although we have yet to define precisely what we mean by the area of a general region in the plane, or the average value of a function over a closed interval, we do have intuitive ideas of what these notions mean. We begin our approach to integration by *approximating* these quantities with simpler finite sums related to these intuitive ideas. We then consider what happens when we take more and more terms in the summation process. In subsequent sections we look at taking the limit of these sums as the number of terms goes to infinity, which leads to a precise definition of the definite integral.

Area

Suppose we want to find the area of the shaded region R that lies above the x-axis, below the graph of $y = 1 - x^2$, and between the vertical lines $x = 0$ and $x = 1$ (see Figure 5.1).

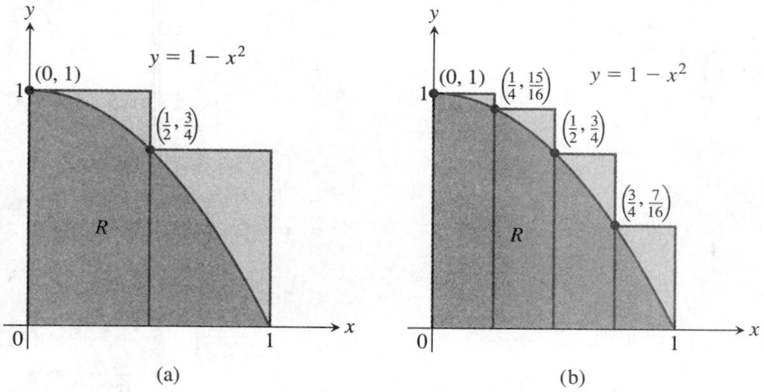

FIGURE 5.2 (a) We get an upper sum approximation of the area of R by using two rectangles containing R. (b) Four rectangles give a better upper sum approximation. Both estimates overshoot the true value for the area by the amount shaded in light gray.

Unfortunately, there is no simple geometric formula for calculating the areas of general shapes having curved boundaries like the region R. How, then, can we find the area of R?

Although we do not yet have a method for determining the exact area of R, we can approximate it in a simple way. Figure 5.2a shows two rectangles that together contain the region R. Each rectangle has width 1/2 and they have heights 1 and 3/4 (left to right). The height of each rectangle is the maximum value of the function f in each subinterval. Because the function f is decreasing, the height is its value at the left endpoint of the subinterval of $[0, 1]$ that forms the base of the rectangle. The total area of the two rectangles approximates the area A of the region R:

$$A \approx 1 \cdot \frac{1}{2} + \frac{3}{4} \cdot \frac{1}{2} = \frac{7}{8} = 0.875.$$

This estimate is larger than the true area A since the two rectangles contain R. We say that 0.875 is an **upper sum** because it is obtained by taking the height of the rectangle corresponding to the maximum (uppermost) value of $f(x)$ over points x lying in the base of each rectangle. In Figure 5.2b, we improve our estimate by using four thinner rectangles, each of width 1/4, which taken together contain the region R. These four rectangles give the approximation

$$A \approx 1 \cdot \frac{1}{4} + \frac{15}{16} \cdot \frac{1}{4} + \frac{3}{4} \cdot \frac{1}{4} + \frac{7}{16} \cdot \frac{1}{4} = \frac{25}{32} = 0.78125,$$

which is still greater than A since the four rectangles contain R.

Suppose instead we use four rectangles contained *inside* the region R to estimate the area, as in Figure 5.3a. Each rectangle has width 1/4, as before, but the rectangles are shorter and lie entirely beneath the graph of f. The function $f(x) = 1 - x^2$ is decreasing on $[0, 1]$, so the height of each of these rectangles is given by the value of f at the right endpoint of the subinterval forming its base. The fourth rectangle has zero height and therefore contributes no area. Summing these rectangles, whose heights are the minimum value of $f(x)$ over points x in the rectangle's base, gives a **lower sum** approximation to the area:

$$A \approx \frac{15}{16} \cdot \frac{1}{4} + \frac{3}{4} \cdot \frac{1}{4} + \frac{7}{16} \cdot \frac{1}{4} + 0 \cdot \frac{1}{4} = \frac{17}{32} = 0.53125.$$

This estimate is smaller than the area A since the rectangles all lie inside of the region R. The true value of A lies somewhere between these lower and upper sums:

$$0.53125 < A < 0.78125.$$

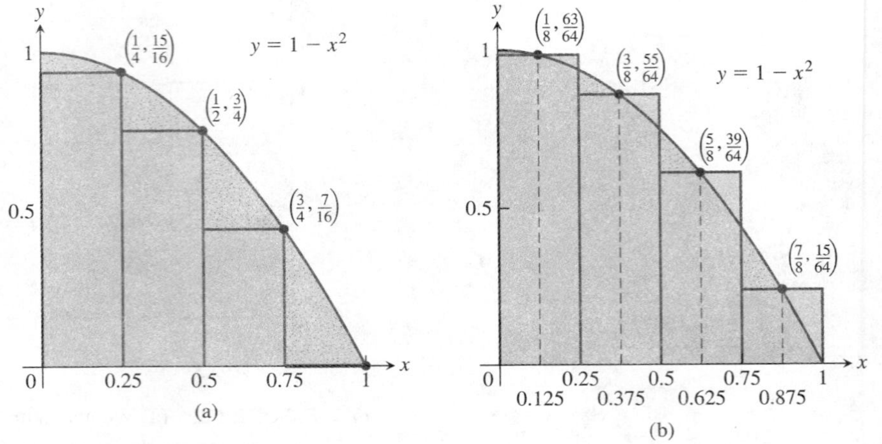

FIGURE 5.3 (a) Rectangles contained in R give a lower sum approximation for the area. This estimate undershoots the true value by the amount shaded in light gray with dots. (b) The midpoint rule uses rectangles whose heights are the values of $y = f(x)$ at the midpoints of their bases. The estimate appears closer to the true value of the area because the light gray overshoot areas roughly balance the dotted light gray undershoot areas.

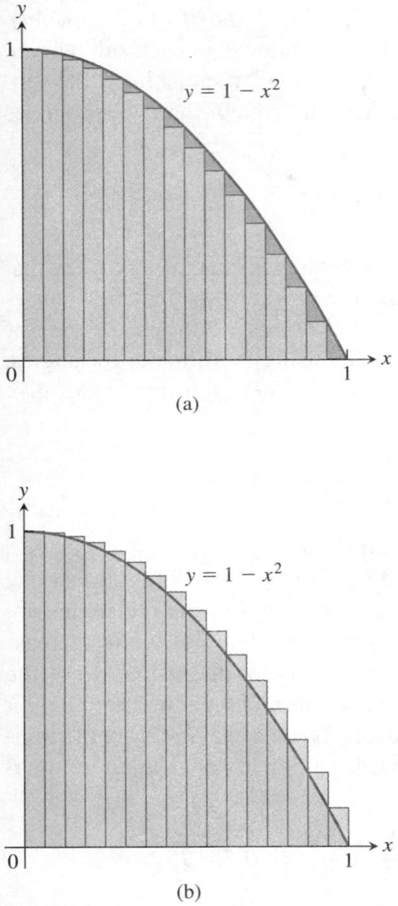

FIGURE 5.4 (a) A lower sum using 16 rectangles of equal width $\Delta x = 1/16$. (b) An upper sum using 16 rectangles.

Considering both lower and upper sum approximations gives us estimates for the area and a bound on the size of the possible error in these estimates since the true value of the area lies somewhere between them. Here the error cannot be greater than the difference $0.78125 - 0.53125 = 0.25$.

Yet another estimate can be obtained by using rectangles whose heights are the values of f at the midpoints of the bases of the rectangles (Figure 5.3b). This method of estimation is called the **midpoint rule** for approximating the area. The midpoint rule gives an estimate that is between a lower sum and an upper sum, but it is not clear whether it overestimates or underestimates the true area. With four rectangles of width 1/4, as before, the midpoint rule estimates the area of R to be

$$A \approx \frac{63}{64}\cdot\frac{1}{4} + \frac{55}{64}\cdot\frac{1}{4} + \frac{39}{64}\cdot\frac{1}{4} + \frac{15}{64}\cdot\frac{1}{4} = \frac{172}{64}\cdot\frac{1}{4} = 0.671875.$$

In each of the sums that we computed, the interval $[a, b]$ over which the function f is defined was subdivided into n subintervals of equal width (or length) $\Delta x = (b - a)/n$, and f was evaluated at a point in each subinterval: c_1 in the first subinterval, c_2 in the second subinterval, and so on. For the upper sum we chose c_k so that $f(c_k)$ was the maximum value of f in the kth subinterval, for the lower sum we chose it so that $f(c_k)$ was the minimum, and for the midpoint rule we chose c_k to be the midpoint of the kth subinterval. In each case the finite sums have the form

$$f(c_1)\,\Delta x + f(c_2)\,\Delta x + f(c_3)\,\Delta x + \cdots + f(c_n)\,\Delta x.$$

As we take more and more rectangles, with each rectangle thinner than before, it appears that these finite sums give better and better approximations to the true area of the region R.

Figure 5.4a shows a lower sum approximation for the area of R using 16 rectangles of equal width. The sum of their areas is 0.634765625, which appears close to the true area but is still somewhat smaller since the rectangles lie inside R.

Figure 5.4b shows an upper sum approximation using 16 rectangles of equal width. The sum of their areas is 0.697265625, which is somewhat larger than the true area because the rectangles taken together contain R. The midpoint rule for 16 rectangles gives a total area approximation of 0.6669921875, but it is not immediately clear whether this estimate is larger or smaller than the true area.

Table 5.1 shows the values of upper and lower sum approximations to the area of R, using up to 1000 rectangles. The values of these approximations appear to be approaching $2/3$. In Section 5.2 we will see how to get an exact value of the area of regions such as R by taking a limit as the base width of each rectangle goes to zero and the number of rectangles goes to infinity. With the techniques developed there, we will be able to show that the area of R is exactly $2/3$.

TABLE 5.1 Finite approximations for the area of R

Number of subintervals	Lower sum	Midpoint sum	Upper sum
2	0.375	0.6875	0.875
4	0.5313	0.6719	0.7813
16	0.6348	0.6670	0.6973
50	0.6566	0.6667	0.6766
100	0.66165	0.666675	0.67165
1000	0.6661665	0.66666675	0.6671665

Distance Traveled

Suppose we know the velocity function $v(t)$ of a car that moves straight down a highway without changing direction, and we want to know how far it traveled between times $t = a$ and $t = b$. The position function $s(t)$ of the car has derivative $v(t)$. If we can find an antiderivative $F(t)$ of $v(t)$, then we can find the car's position function $s(t)$ by setting $s(t) = F(t) + C$. The distance traveled can then be found by calculating the change in position, $s(b) - s(a) = F(b) - F(a)$. However, if the velocity is known only by the readings at various times of a speedometer on the car, then we have no formula for the velocity from which to obtain an antiderivative that gives the position function. So what do we do in this situation?

If we know the velocity at a collection of times $t_1, t_2, \ldots t_n$, we can approximate the distance traveled by using finite sums in a way similar to the area estimates that we discussed before. We first subdivide the interval $[a, b]$ into short time intervals and assume that the velocity on each subinterval is fairly constant. Then we approximate the distance traveled on each time subinterval with the usual distance formula

$$\text{distance} = \text{velocity} \times \text{time}$$

and add the results across $[a, b]$.

Suppose the subdivided interval looks like

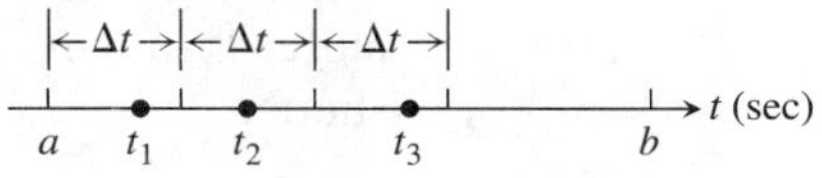

with the subintervals all of equal length Δt. Pick a number t_1 in the first interval. If Δt is so small that the velocity barely changes over a short time interval of duration Δt, then the distance traveled in the first time interval is about $v(t_1)\,\Delta t$. If t_2 is a number in the second interval, the distance traveled in the second time interval is about $v(t_2)\,\Delta t$. The sum of the distances traveled over all the time intervals is approximated by

$$v(t_1)\,\Delta t + v(t_2)\,\Delta t + \cdots + v(t_n)\,\Delta t,$$

where n is the total number of subintervals. This sum is only an approximation to the true distance D, but the approximation increases in accuracy as we take more and more subintervals.

EXAMPLE 1 The velocity function of a projectile fired straight into the air is $f(t) = 160 - 9.8t$ m/sec. Use the summation technique just described to estimate how far the projectile rises during the first 3 sec. How close do the sums come to the exact value of 435.9 m? (We will see how to compute the exact value in Section 5.4.)

Solution We explore the results for different numbers of subintervals and different choices of evaluation points. Notice that $f(t)$ is decreasing, so choosing left endpoints gives an upper sum estimate, and choosing right endpoints gives a lower sum estimate.

(a) *Three subintervals of length* 1, *with f evaluated at left endpoints giving an upper sum*:

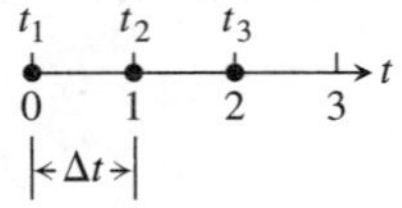

With f evaluated at $t = 0, 1$, and 2, we have

$$\begin{aligned} D &\approx f(t_1)\,\Delta t + f(t_2)\,\Delta t + f(t_3)\,\Delta t \\ &= [160 - 9.8(0)](1) + [160 - 9.8(1)](1) + [160 - 9.8(2)](1) \\ &= 450.6. \end{aligned}$$

(b) *Three subintervals of length* 1, *with f evaluated at right endpoints giving a lower sum*:

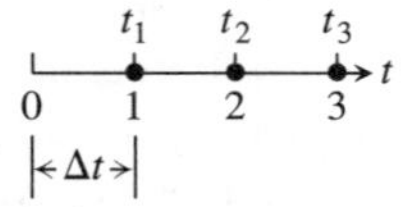

With f evaluated at $t = 1, 2$, and 3, we have

$$\begin{aligned} D &\approx f(t_1)\,\Delta t + f(t_2)\,\Delta t + f(t_3)\,\Delta t \\ &= [160 - 9.8(1)](1) + [160 - 9.8(2)](1) + [160 - 9.8(3)](1) \\ &= 421.2. \end{aligned}$$

(c) *With six subintervals of length* 1/2, *we get*

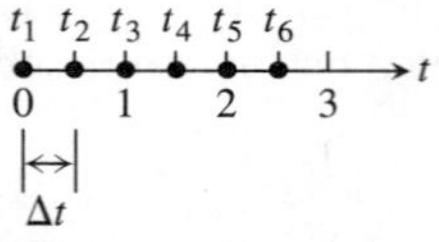

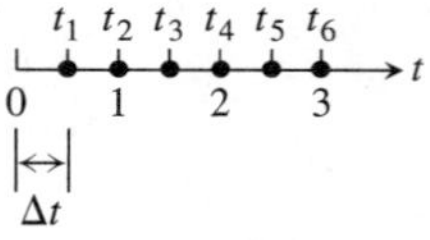

These estimates give an upper sum using left endpoints: $D \approx 443.25$, and a lower sum using right endpoints: $D \approx 428.55$. These six-interval estimates are somewhat closer than the three-interval estimates. The results improve as the subintervals get shorter.

As we can see in Table 5.2, the left-endpoint upper sums approach the true value 435.9 from above, whereas the right-endpoint lower sums approach it from below. The true value lies between these upper and lower sums. The magnitude of the error in the closest entry is 0.23, a small percentage of the true value.

$$\begin{aligned} \text{Error magnitude} &= |\text{true value} - \text{calculated value}| \\ &= |435.9 - 435.67| = 0.23. \\ \text{Error percentage} &= \frac{0.23}{435.9} \approx 0.05\%. \end{aligned}$$

It is reasonable to conclude from the table's last entries that the projectile rose about 436 m during its first 3 sec of flight. ■

TABLE 5.2 Travel-distance estimates

Number of subintervals	Length of each subinterval	Upper sum	Lower sum
3	1	450.6	421.2
6	1/2	443.25	428.55
12	1/4	439.58	432.23
24	1/8	437.74	434.06
48	1/16	436.82	434.98
96	1/32	436.36	435.44
192	1/64	436.13	435.67

Displacement Versus Distance Traveled

If an object with position function $s(t)$ moves along a coordinate line without changing direction, we can calculate the total distance it travels from $t = a$ to $t = b$ by summing the distance traveled over small intervals, as in Example 1. If the object reverses direction one or more times during the trip, then we need to use the object's *speed* $|v(t)|$, which is the absolute value of its velocity function, $v(t)$, to find the total distance traveled. Using the velocity itself, as in Example 1, gives instead an estimate of the object's **displacement**, $s(b) - s(a)$, the difference between its initial and final positions. To see the difference, think about what happens when you walk a mile from your home and then walk back. The total distance traveled is two miles, but your displacement is zero, because you end up back where you started.

To see why using the velocity function in the summation process gives an estimate of the displacement, partition the time interval $[a, b]$ into small enough equal subintervals Δt so that the object's velocity does not change very much from time t_{k-1} to t_k. Then $v(t_k)$ gives a good approximation of the velocity throughout the interval. Accordingly, the change in the object's position coordinate, which is its displacement during the time interval, is about

$$v(t_k)\,\Delta t.$$

The change is positive if $v(t_k)$ is positive and negative if $v(t_k)$ is negative.

In either case, the distance traveled by the object during the subinterval is about

$$|v(t_k)|\,\Delta t.$$

The **total distance traveled** over the time interval is approximately the sum

$$|v(t_1)|\,\Delta t + |v(t_2)|\,\Delta t + \cdots + |v(t_n)|\,\Delta t.$$

We will revisit these ideas in Section 5.4.

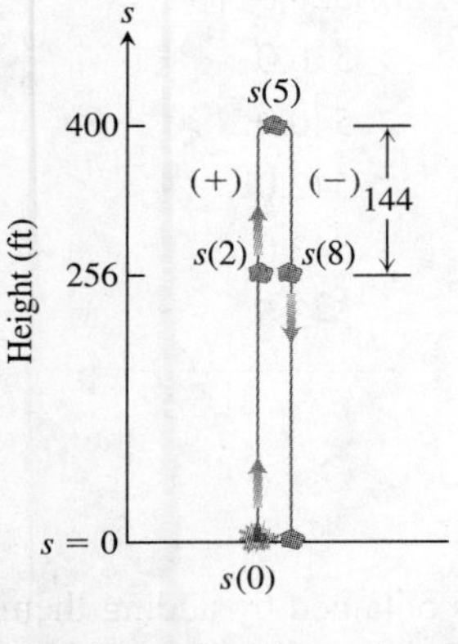

FIGURE 5.5 The rock in Example 2. The height $s = 256$ ft is reached at $t = 2$ sec and $t = 8$ sec. The rock falls 144 ft from its maximum height when $t = 8$.

EXAMPLE 2 In Example 4 in Section 3.4, we analyzed the motion of a heavy rock blown straight up by a dynamite blast. In that example, we found the velocity of the rock at time t was $v(t) = 160 - 32t$ ft/sec. The rock was 256 ft above the ground 2 sec after the explosion, continued upward to reach a maximum height of 400 ft at 5 sec after the explosion, and then fell back down a distance of 144 ft to reach the height of 256 ft again at $t = 8$ sec after the explosion. (See Figure 5.5.) The total distance traveled in these 8 seconds is $400 + 144 = 544$ ft.

If we follow a procedure like the one presented in Example 1, using the velocity function $v(t)$ in the summation process from $t = 0$ to $t = 8$, we obtain an estimate of the rock's height above the ground at time $t = 8$. Starting at time $t = 0$, the rock traveled upward a total of $256 + 144 = 400$ ft, but then it peaked and traveled downward

TABLE 5.3 Velocity function

t	$v(t)$	t	$v(t)$
0	160	4.5	16
0.5	144	5.0	0
1.0	128	5.5	−16
1.5	112	6.0	−32
2.0	96	6.5	−48
2.5	80	7.0	−64
3.0	64	7.5	−80
3.5	48	8.0	−96
4.0	32		

144 ft, ending at a height of 256 ft at time $t = 8$. The velocity $v(t)$ is positive during the upward travel, but negative while the rock falls back down. When we compute the sum $v(t_1)\,\Delta t + v(t_2)\,\Delta t + \cdots + v(t_n)\,\Delta t$, part of the upward positive distance change is canceled by the negative downward movement, giving in the end an approximation of the displacement from the initial position, equal to a positive change of 256 ft.

On the other hand, if we use the speed $|v(t)|$, which is the absolute value of the velocity function, then distances traveled while moving up and distances traveled while moving down are both counted positively. Both the total upward motion of 400 ft and the downward motion of 144 ft are now counted as positive distances traveled, so the sum $|v(t_1)|\,\Delta t + |v(t_2)|\,\Delta t + \cdots + |v(t_n)|\,\Delta t$ gives us an approximation of 544 ft, the total distance that the rock traveled from time $t = 0$ to time $t = 8$.

As an illustration of our discussion, we subdivide the interval $[0, 8]$ into 16 subintervals of length $\Delta t = 1/2$ and take the right endpoint of each subinterval as the value of t_k. Table 5.3 shows the values of the velocity function at these endpoints.

Using $v(t)$ in the summation process, we estimate the displacement at $t = 8$:

$$(144 + 128 + 112 + 96 + 80 + 64 + 48 + 32 + 16$$
$$+ 0 - 16 - 32 - 48 - 64 - 80 - 96) \cdot \frac{1}{2} = 192$$
$$\text{Error magnitude} = 256 - 192 = 64$$

Using $|v(t)|$ in the summation process, we estimate the total distance traveled over the time interval $[0, 8]$:

$$(144 + 128 + 112 + 96 + 80 + 64 + 48 + 32 + 16$$
$$+ 0 + 16 + 32 + 48 + 64 + 80 + 96) \cdot \frac{1}{2} = 528$$
$$\text{Error magnitude} = 544 - 528 = 16$$

If we partition $[0, 8]$ into more and more subintervals, the estimates to the values 256 ft and 544 ft improve, as shown in Table 5.4. ■

TABLE 5.4 Travel estimates for a rock blown straight up during the time interval $[0, 8]$

Number of subintervals	Length of each subinterval	Displacement	Total distance
16	1/2	192.0	528.0
32	1/4	224.0	536.0
64	1/8	240.0	540.0
128	1/16	248.0	542.0
256	1/32	252.0	543.0
512	1/64	254.0	543.5

Average Value of a Nonnegative Continuous Function

The average value of a collection of n numbers $x_1, x_2, \ldots, x_n$ is obtained by adding them together and dividing by n. But what is the average value of a continuous function f on an interval $[a, b]$? Such a function can assume infinitely many values. For example, the temperature at a certain location in a town is a continuous function that goes up and down each day. What does it mean to say that the average temperature in the town over the course of a day is 73 degrees?

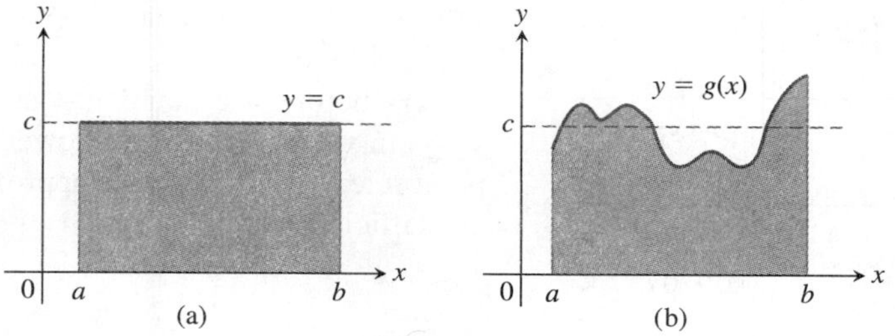

FIGURE 5.6 (a) The average value of $f(x) = c$ on $[a, b]$ is the area of the rectangle divided by $b - a$. (b) The average value of $g(x)$ on $[a, b]$ is the area beneath its graph divided by $b - a$.

When a function is constant, this question is easy to answer. A function with constant value c on an interval $[a, b]$ has average value c. When c is positive, its graph over $[a, b]$ gives a rectangle of height c. The average value of the function can then be interpreted geometrically as the area of this rectangle divided by its width $b - a$ (see Figure 5.6a).

What if we want to find the average value of a nonconstant function, such as the function g in Figure 5.6b? We can think of this graph as a snapshot of the height of some water that is sloshing around in a tank between enclosing walls at $x = a$ and $x = b$. As the water moves, its height over each point changes, but its average height remains the same. To get the average height of the water, we let it settle down until it is level and its height is constant. The resulting height c equals the area under the graph of g divided by $b - a$. We are led to *define* the average value of a nonnegative function on an interval $[a, b]$ to be the area under its graph divided by $b - a$. For this definition to be valid, we need a precise understanding of what is meant by the area under a graph. This will be obtained in Section 5.3, but for now we look at an example.

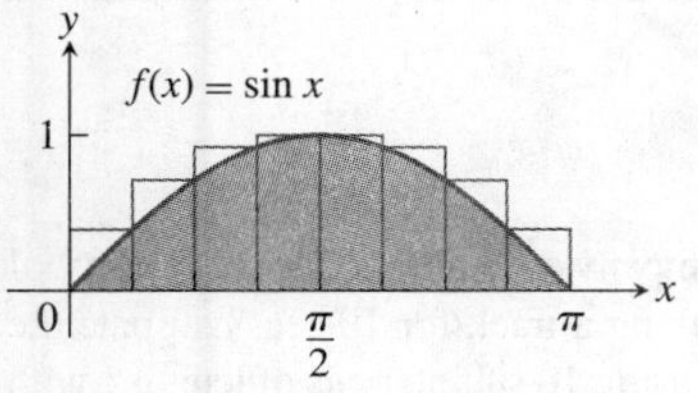

FIGURE 5.7 Approximating the area under $f(x) = \sin x$ between 0 and π to compute the average value of $\sin x$ over $[0, \pi]$, using eight rectangles (Example 3).

EXAMPLE 3 Estimate the average value of the function $f(x) = \sin x$ on the interval $[0, \pi]$.

Solution Looking at the graph of $\sin x$ between 0 and π in Figure 5.7, we can see that its average height is somewhere between 0 and 1. To find the average, we need to calculate the area A under the graph and then divide this area by the length of the interval, $\pi - 0 = \pi$.

We do not have a simple way to determine the area, so we approximate it with finite sums. To get an upper sum approximation, we add the areas of eight rectangles of equal width $\pi/8$ that together contain the region that is beneath the graph of $y = \sin x$ and above the x-axis on $[0, \pi]$. We choose the heights of the rectangles to be the largest value of $\sin x$ on each subinterval. Over a particular subinterval, this largest value may occur at the left endpoint, at the right endpoint, or somewhere between them. We evaluate $\sin x$ at this point to get the height of the rectangle for an upper sum. The sum of the rectangular areas then gives an estimate of the total area (Figure 5.7):

$$A \approx \left(\sin\frac{\pi}{8} + \sin\frac{\pi}{4} + \sin\frac{3\pi}{8} + \sin\frac{\pi}{2} + \sin\frac{\pi}{2} + \sin\frac{5\pi}{8} + \sin\frac{3\pi}{4} + \sin\frac{7\pi}{8}\right) \cdot \frac{\pi}{8}$$

$$\approx (0.38 + 0.71 + 0.92 + 1 + 1 + 0.92 + 0.71 + 0.38) \cdot \frac{\pi}{8} = (6.02) \cdot \frac{\pi}{8} \approx 2.364.$$

To estimate the average value of $\sin x$ on $[0, \pi]$ we divide the estimated area by the length π of the interval and obtain the approximation $2.364/\pi \approx 0.753$.

Since we used an upper sum to approximate the area, this estimate is greater than the actual average value of $\sin x$ over $[0, \pi]$. If we use more and more rectangles, with each rectangle getting thinner and thinner, we get closer and closer to the exact average

TABLE 5.5 Average value of $\sin x$ on $0 \le x \le \pi$

Number of subintervals	Upper sum estimate
8	0.75342
16	0.69707
32	0.65212
50	0.64657
100	0.64161
1000	0.63712

value, as shown in Table 5.5. We will show in Section 5.3 that the true average value is $2/\pi \approx 0.63662$.

As before, we could just as well have used rectangles lying under the graph of $y = \sin x$ and calculated a lower sum approximation, or we could have used the midpoint rule. In each case, the approximations are close to the true area if all the rectangles are sufficiently thin. ■

Summary

The area under the graph of a positive function, the distance traveled by a moving object that doesn't change direction, and the average value of a nonnegative function f over an interval can all be approximated by finite sums constructed in a certain way. First we subdivide the interval into subintervals, treating f as if it were constant over each subinterval. Then we multiply the width of each subinterval by the value of f at some point within it and add these products together. If the interval $[a, b]$ is subdivided into n subintervals of equal widths $\Delta x = (b - a)/n$, and if $f(c_k)$ is the value of f at the chosen point c_k in the kth subinterval, this process gives a finite sum of the form

$$f(c_1)\,\Delta x + f(c_2)\,\Delta x + f(c_3)\,\Delta x + \cdots + f(c_n)\,\Delta x.$$

The choices for the c_k could maximize or minimize the value of f in the kth subinterval, or give some value in between. The true value lies somewhere between the approximations given by upper sums and lower sums. In the examples that we looked at, the finite sum approximations improved as we took more subintervals of smaller width.

EXERCISES 5.1

Area

In Exercises 1–4, apply finite approximations to estimate the area under the graph of the function using

a. a lower sum with two rectangles of equal width.

b. a lower sum with four rectangles of equal width.

c. an upper sum with two rectangles of equal width.

d. an upper sum with four rectangles of equal width.

1. $f(x) = x^2$ between $x = 0$ and $x = 1$.

2. $f(x) = x^3$ between $x = 0$ and $x = 1$.

3. $f(x) = 1/x$ between $x = 1$ and $x = 5$.

4. $f(x) = 4 - x^2$ between $x = -2$ and $x = 2$.

Using rectangles, each of whose height is given by the value of the function at the midpoint of the rectangle's base (*the midpoint rule*), estimate the area under the graphs of the following functions, using first two and then four rectangles.

5. $f(x) = x^2$ between $x = 0$ and $x = 1$.

6. $f(x) = x^3$ between $x = 0$ and $x = 1$.

7. $f(x) = 1/x$ between $x = 1$ and $x = 5$.

8. $f(x) = 4 - x^2$ between $x = -2$ and $x = 2$.

Distance

9. Distance traveled The accompanying table shows the velocity of a model train engine moving along a track for 10 sec. Estimate the distance traveled by the engine using 10 subintervals of length 1 with

a. left-endpoint values.

b. right-endpoint values.

Time (sec)	Velocity (cm/sec)	Time (sec)	Velocity (cm/sec)
0	0	6	28
1	30	7	15
2	56	8	5
3	25	9	15
4	38	10	0
5	33		

10. Distance traveled upstream You are sitting on the bank of a tidal river watching the incoming tide carry a bottle upstream. You record the velocity of the flow every 5 minutes for an hour, with the results shown in the accompanying table. About how far upstream did the bottle travel during that hour? Find an estimate using 12 subintervals of length 5 with

a. left-endpoint values.
b. right-endpoint values.

Time (min)	Velocity (m/sec)	Time (min)	Velocity (m/sec)
0	1	35	1.2
5	1.2	40	1.0
10	1.7	45	1.8
15	2.0	50	1.5
20	1.8	55	1.2
25	1.6	60	0
30	1.4		

11. **Length of a road** You and a companion are about to drive a twisty stretch of dirt road in a car whose speedometer works but whose odometer (mileage counter) is broken. To find out how long this particular stretch of road is, you record the car's velocity at 10-sec intervals, with the results shown in the accompanying table. Estimate the length of the road using
a. left-endpoint values.
b. right-endpoint values.

Time (sec)	Velocity (converted to ft/sec) (30 mi/h = 44 ft/sec)	Time (sec)	Velocity (converted to ft/sec) (30 mi/h = 44 ft/sec)
0	0	70	15
10	44	80	22
20	15	90	35
30	35	100	44
40	30	110	30
50	44	120	35
60	35		

12. **Distance from velocity data** The accompanying table gives data for the velocity of a vintage sports car accelerating from 0 to 142 mi/h in 36 sec (10 thousandths of an hour).

Time (h)	Velocity (mi/h)	Time (h)	Velocity (mi/h)
0.0	0	0.006	116
0.001	40	0.007	125
0.002	62	0.008	132
0.003	82	0.009	137
0.004	96	0.010	142
0.005	108		

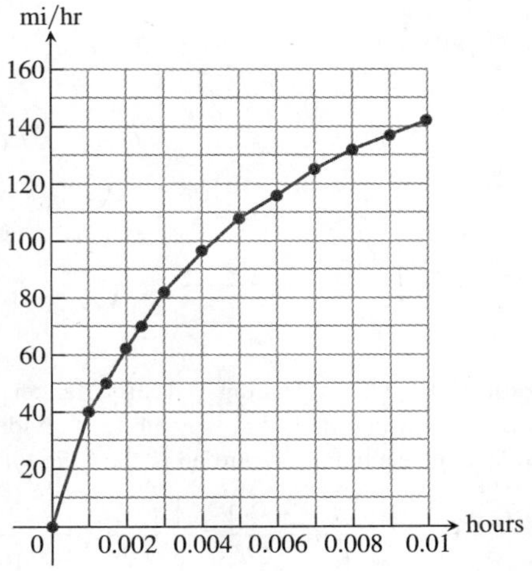

a. Use rectangles to estimate how far the car traveled during the 36 sec it took to reach 142 mi/h.
b. Roughly how many seconds did it take the car to reach the halfway point? About how fast was the car going then?

13. **Free fall with air resistance** An object is dropped straight down from a helicopter. The object falls faster and faster but its acceleration (rate of change of its velocity) decreases over time because of air resistance. The acceleration is measured in ft/sec^2 and recorded every second after the drop for 4 sec, as shown:

t	0	1	2	3	4
a	22.00	15.30	9.21	3.45	2.32

a. Find an upper estimate for the speed when $t = 4$.
b. Find a lower estimate for the speed when $t = 4$.
c. Find an upper estimate for the distance fallen when $t = 2$.

14. **Distance traveled by a projectile** An object is shot straight upward from sea level with an initial velocity of 300 ft/sec.
a. Assuming that gravity is the only force acting on the object, give an upper estimate for its velocity after 3 sec have elapsed. Use $g = 32$ ft/sec^2 for the gravitational acceleration.
b. Find a lower estimate for the height attained after 3 sec.

Average Value of a Function

In Exercises 15–18, use a finite sum to estimate the average value of f on the given interval by partitioning the interval into four subintervals of equal length and evaluating f at the subinterval midpoints.

15. $f(x) = x^3$ on $[0, 2]$
16. $f(x) = 1/x$ on $[1, 5]$
17. $f(t) = (1/2) + \sin^2 \pi t$ on $[0, 2]$

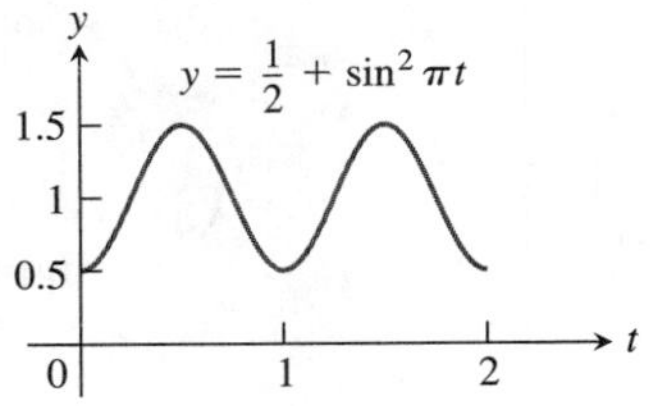

T **18.** $f(t) = 1 - \left(\cos\frac{\pi t}{4}\right)^4$ on $[0, 4]$

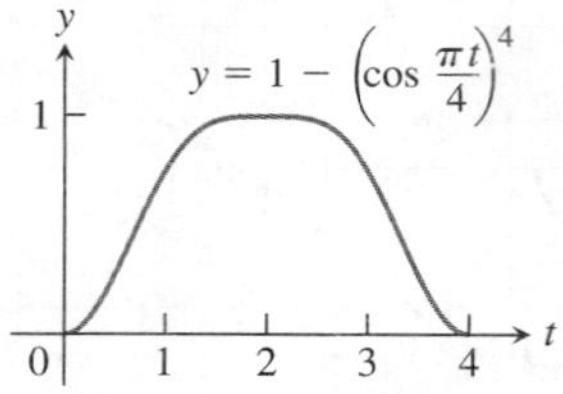

Estimations

19. Water pollution Oil is leaking out of a tanker damaged at sea. The damage to the tanker is worsening as evidenced by the increased leakage each hour, recorded in the following table.

Time (h)	0	1	2	3	4
Leakage (gal/h)	40	60	77	110	155

Time (h)	5	6
Leakage (gal/h)	260	450

a. Give an upper and a lower estimate of the total quantity of oil that has escaped after 3 hours.

b. Repeat part (a) for the quantity of oil that has escaped after 6 hours.

c. The tanker continues to leak 450 gal/h after the first 6 hours. If the tanker originally contained 20,000 gal of oil, approximately how many more hours will elapse in the worst case before all the oil has spilled? In the best case?

20. Air pollution A power plant generates electricity by burning oil. Pollutants produced as a result of the burning process are removed by scrubbers in the smokestacks. Over time, the scrubbers become less efficient and eventually they must be replaced when the amount of pollution released exceeds government standards. Measurements are taken at the end of each month determining the rate at which pollutants are released into the atmosphere, recorded as follows.

Month	Jan	Feb	Mar	Apr	May	Jun
Pollutant release rate (tons/day)	0.10	0.15	0.18	0.24	0.31	0.42

Month	Jul	Aug	Sep	Oct	Nov	Dec
Pollutant release rate (tons/day)	0.50	0.65	0.72	0.80	0.85	0.92

a. Assuming a 30-day month and that new scrubbers allow only 0.06 ton/day to be released, give an upper estimate of the total tonnage of pollutants released by the end of June. What is a lower estimate?

b. In the best case, approximately when will a total of 145 tons of pollutants have been released into the atmosphere?

T **21.** Inscribe a regular n-sided polygon inside a circle of radius 1 and compute the area of the polygon for the following values of n:

a. 4 (square) **b.** 8 (octagon) **c.** 16

d. Compare the areas in parts (a), (b), and (c) with the area inside the circle.

22. (*Continuation of Exercise 21.*)

a. Inscribe a regular n-sided polygon inside a circle of radius 1 and compute the area of one of the n congruent triangles formed by drawing radii to the vertices of the polygon.

b. Compute the limit of the area of the inscribed polygon as $n \to \infty$.

c. Repeat the computations in parts (a) and (b) for a circle of radius r.

COMPUTER EXPLORATIONS

In Exercises 23–26, use a CAS to perform the following steps.

a. Plot the functions over the given interval.

b. Subdivide the interval into $n = 100$, 200, and 1000 subintervals of equal length and evaluate the function at the midpoint of each subinterval.

c. Compute the average value of the function values generated in part (b).

d. Solve the equation $f(x) = (\text{average value})$ for x using the average value calculated in part (c) for the $n = 1000$ partitioning.

23. $f(x) = \sin x$ on $[0, \pi]$ **24.** $f(x) = \sin^2 x$ on $[0, \pi]$

25. $f(x) = x \sin\frac{1}{x}$ on $\left[\frac{\pi}{4}, \pi\right]$ **26.** $f(x) = x \sin^2\frac{1}{x}$ on $\left[\frac{\pi}{4}, \pi\right]$

5.2 Sigma Notation and Limits of Finite Sums

While estimating with finite sums in Section 5.1, we encountered sums that had many terms (up to 1000 terms in Table 5.1). In this section we introduce a notation for sums that have a large number of terms. After describing this notation and its properties, we consider what happens as the number of terms in a sum approaches infinity.

Finite Sums and Sigma Notation

Sigma notation enables us to write a sum with many terms in the compact form

$$\sum_{k=1}^{n} a_k = a_1 + a_2 + a_3 + \cdots + a_{n-1} + a_n.$$

$\sum$ is the capital Greek letter sigma

The Greek letter $\sum$ (capital sigma, corresponding to our letter S), stands for "sum." The **index of summation** k tells us where the sum begins (at the number below the $\sum$ symbol) and where it ends (at the number above $\sum$). Any letter can be used to denote the index, but the letters i, j, k, and n are customary.

The index k ends at $k = n$.

The summation symbol (Greek letter sigma)

$$\sum_{k=1}^{n} a_k$$

a_k is a formula for the kth term.

The index k starts at $k = 1$.

Thus we can write the sum of the squares of the numbers 1 through 11 as

$$1^2 + 2^2 + 3^2 + 4^2 + 5^2 + 6^2 + 7^2 + 8^2 + 9^2 + 10^2 + 11^2 = \sum_{k=1}^{11} k^2,$$

and the sum of $f(i)$ for integers i from 1 to 100 as

$$f(1) + f(2) + f(3) + \cdots + f(100) = \sum_{i=1}^{100} f(i).$$

The starting index does not have to be 1; it can be any integer.

EXAMPLE 1

A sum in sigma notation	The sum written out, one term for each value of k	The value of the sum
$\sum_{k=1}^{5} k$	$1 + 2 + 3 + 4 + 5$	15
$\sum_{k=1}^{3} (-1)^k k$	$(-1)^1(1) + (-1)^2(2) + (-1)^3(3)$	$-1 + 2 - 3 = -2$
$\sum_{k=1}^{2} \frac{k}{k+1}$	$\frac{1}{1+1} + \frac{2}{2+1}$	$\frac{1}{2} + \frac{2}{3} = \frac{7}{6}$
$\sum_{k=4}^{5} \frac{k^2}{k-1}$	$\frac{4^2}{4-1} + \frac{5^2}{5-1}$	$\frac{16}{3} + \frac{25}{4} = \frac{139}{12}$

■

EXAMPLE 2 Express the sum $1 + 3 + 5 + 7 + 9$ in sigma notation.

Solution The formula generating the terms depends on what we choose the lower limit of summation to be, but the terms generated remain the same. It is often simplest to choose the starting index to be $k = 0$ or $k = 1$, but we can start with any integer.

$$\textit{Starting with } k = 0: \quad 1 + 3 + 5 + 7 + 9 = \sum_{k=0}^{4} (2k + 1)$$

$$\textit{Starting with } k = 1: \quad 1 + 3 + 5 + 7 + 9 = \sum_{k=1}^{5} (2k - 1)$$

$$\textit{Starting with } k = 2: \quad 1 + 3 + 5 + 7 + 9 = \sum_{k=2}^{6} (2k - 3)$$

$$\textit{Starting with } k = -3: \quad 1 + 3 + 5 + 7 + 9 = \sum_{k=-3}^{1} (2k + 7)$$

■

When we have a sum such as

$$\sum_{k=1}^{3}(k + k^2),$$

we can rearrange its terms to form two sums:

$$\begin{aligned}\sum_{k=1}^{3}(k + k^2) &= (1 + 1^2) + (2 + 2^2) + (3 + 3^2)\\ &= (1 + 2 + 3) + (1^2 + 2^2 + 3^2) \qquad \text{Regroup terms.}\\ &= \sum_{k=1}^{3} k + \sum_{k=1}^{3} k^2.\end{aligned}$$

This illustrates a general rule for finite sums:

$$\sum_{k=1}^{n}(a_k + b_k) = \sum_{k=1}^{n} a_k + \sum_{k=1}^{n} b_k.$$

This and three other rules are given below. Proofs of these rules can be obtained using mathematical induction (see Appendix A.2).

Algebra Rules for Finite Sums

1. *Sum Rule:* $\sum_{k=1}^{n}(a_k + b_k) = \sum_{k=1}^{n} a_k + \sum_{k=1}^{n} b_k$

2. *Difference Rule:* $\sum_{k=1}^{n}(a_k - b_k) = \sum_{k=1}^{n} a_k - \sum_{k=1}^{n} b_k$

3. *Constant Multiple Rule:* $\sum_{k=1}^{n} ca_k = c \cdot \sum_{k=1}^{n} a_k$ (Any number c)

4. *Constant Value Rule:* $\sum_{k=1}^{n} c = n \cdot c$ (Any number c)

EXAMPLE 3 We demonstrate the use of the algebra rules.

(a) $\sum_{k=1}^{n}(3k - k^2) = 3\sum_{k=1}^{n} k - \sum_{k=1}^{n} k^2$ — Difference Rule and Constant Multiple Rule

(b) $\sum_{k=1}^{n}(-a_k) = \sum_{k=1}^{n}(-1) \cdot a_k = -1 \cdot \sum_{k=1}^{n} a_k = -\sum_{k=1}^{n} a_k$ — Constant Multiple Rule

(c) $\sum_{k=1}^{3}(k + 4) = \sum_{k=1}^{3} k + \sum_{k=1}^{3} 4$ — Sum Rule

$= (1 + 2 + 3) + (3 \cdot 4)$ — Constant Value Rule

$= 6 + 12 = 18$

(d) $\sum_{k=1}^{n}\frac{1}{n} = n \cdot \frac{1}{n} = 1$ — Constant Value Rule ($1/n$ is constant) ■

HISTORICAL BIOGRAPHY
Carl Friedrich Gauss
(1777–1855)
www.bit.ly/2IuVl55

Over the years, people have discovered a variety of formulas for the values of finite sums. The most famous of these are the formula for the sum of the first n positive integers (Gauss is said to have discovered it at age 8) and the formulas for the sums of the squares and cubes of the first n positive integers.

EXAMPLE 4 Show that the sum of the first n positive integers is

$$\sum_{k=1}^{n} k = \frac{n(n+1)}{2}.$$

Solution The formula tells us that the sum of the first 4 positive integers is

$$\frac{(4)(5)}{2} = 10.$$

Addition verifies this prediction:

$$1 + 2 + 3 + 4 = 10.$$

To prove the formula in general, we write out the terms in the sum twice, once forward and once backward.

$$\begin{matrix} 1 & + & 2 & + & 3 & + & \cdots & + & n \\ n & + & (n-1) & + & (n-2) & + & \cdots & + & 1 \end{matrix}$$

If we add the two terms in the first column we get $1 + n = n + 1$. Similarly, if we add the two terms in the second column we get $2 + (n - 1) = n + 1$. The two terms in any column sum to $n + 1$. When we add the n columns together we get n terms, each equal to $n + 1$, for a total of $n(n + 1)$. Since this is twice the desired quantity, the sum of the first n integers is $n(n + 1)/2$. ■

Formulas for the sums of the squares and cubes of the first n positive integers are proved using mathematical induction (see Appendix A.2). We state them here.

Sum of the first n squares: $$\sum_{k=1}^{n} k^2 = \frac{n(n+1)(2n+1)}{6}$$

Sum of the first n cubes: $$\sum_{k=1}^{n} k^3 = \left(\frac{n(n+1)}{2}\right)^2$$

Limits of Finite Sums

The finite sum approximations that we considered in Section 5.1 became more accurate as the number of terms increased and the subinterval widths (lengths) narrowed. The next example shows how to calculate a limiting value as the widths of the subintervals go to zero and the number of subintervals grows to infinity.

EXAMPLE 5 Find the limiting value of lower sum approximations to the area of the region R below the graph of $y = 1 - x^2$ and above the interval $[0, 1]$ on the x-axis using equal-width rectangles whose widths approach zero and whose number approaches infinity. (See Figure 5.4a.)

Solution We compute a lower sum approximation using n rectangles of equal width Δx and then see what happens as $n \to \infty$. We start by subdividing $[0, 1]$ into n equal width subintervals

$$\left[0, \frac{1}{n}\right], \left[\frac{1}{n}, \frac{2}{n}\right], \ldots, \left[\frac{n-1}{n}, \frac{n}{n}\right].$$

Each subinterval has width $\Delta x = 1/n$. The function $1 - x^2$ is decreasing on $[0, 1]$, and its smallest value in a subinterval occurs at the subinterval's right endpoint. So a lower sum is constructed with rectangles whose height over the subinterval $[(k - 1)/n, k/n]$ is $f(k/n) = 1 - (k/n)^2$, giving the sum

$$f\left(\frac{1}{n}\right) \cdot \frac{1}{n} + f\left(\frac{2}{n}\right) \cdot \frac{1}{n} + \cdots + f\left(\frac{k}{n}\right) \cdot \frac{1}{n} + \cdots + f\left(\frac{n}{n}\right) \cdot \frac{1}{n}.$$

We write this in sigma notation and simplify,

$$\begin{aligned}
\sum_{k=1}^{n} f\left(\frac{k}{n}\right)\cdot\frac{1}{n} &= \sum_{k=1}^{n}\left(1-\left(\frac{k}{n}\right)^2\right)\frac{1}{n} \\
&= \sum_{k=1}^{n}\left(\frac{1}{n}-\frac{k^2}{n^3}\right) \\
&= \sum_{k=1}^{n}\frac{1}{n}-\sum_{k=1}^{n}\frac{k^2}{n^3} && \text{Difference Rule} \\
&= n\cdot\frac{1}{n}-\frac{1}{n^3}\sum_{k=1}^{n}k^2 && \text{Constant Value and Constant Multiple Rules} \\
&= 1-\left(\frac{1}{n^3}\right)\frac{n(n+1)(2n+1)}{6} && \text{Sum of the first } n \text{ squares} \\
&= 1-\frac{2n^3+3n^2+n}{6n^3}. && \text{Numerator expanded}
\end{aligned}$$

We have obtained an expression for the lower sum that holds for any n. Taking the limit of this expression as $n \to \infty$, we see that the lower sums converge as the number of subintervals increases and the subinterval widths approach zero:

$$\lim_{n\to\infty}\left(1-\frac{2n^3+3n^2+n}{6n^3}\right)=1-\frac{2}{6}=\frac{2}{3}.$$

The lower sum approximations converge to 2/3. A similar calculation shows that the upper sum approximations also converge to 2/3. Any finite sum approximation $\sum_{k=1}^{n} f(c_k)(1/n)$ also converges to the same value, 2/3. This is because it is possible to show that any finite sum approximation is trapped between the lower and upper sum approximations. For this reason we are led to *define* the area of the region R as this limiting value. In Section 5.3 we study the limits of such finite approximations in a general setting. ■

Riemann Sums

HISTORICAL BIOGRAPHY
Georg Friedrich Bernhard Riemann (1826–1866)
www.bit.ly/2DMOpnO

The theory of limits of finite approximations was made precise by the German mathematician Bernhard Riemann. We now introduce the notion of a *Riemann sum*, which underlies the theory of the definite integral that will be presented in the next section.

We begin with an arbitrary bounded function f defined on a closed interval $[a, b]$. Like the function pictured in Figure 5.8, f may have negative as well as positive values. We subdivide the interval $[a, b]$ into subintervals, not necessarily of equal width (or length), and form sums in the same way as for the finite approximations in Section 5.1. To do so, we choose $n - 1$ points $\{x_1, x_2, x_3, \ldots, x_{n-1}\}$ between a and b that are in increasing order, so that

$$a < x_1 < x_2 < \cdots < x_{n-1} < b.$$

To make the notation consistent, we set $x_0 = a$ and $x_n = b$, so that

$$a = x_0 < x_1 < x_2 < \cdots < x_{n-1} < x_n = b.$$

The set of all of these points,

$$P = \{x_0, x_1, x_2, \ldots, x_{n-1}, x_n\},$$

is called a **partition** of $[a, b]$.

The partition P divides $[a, b]$ into the n closed subintervals

$$[x_0, x_1], [x_1, x_2], \ldots, [x_{n-1}, x_n].$$

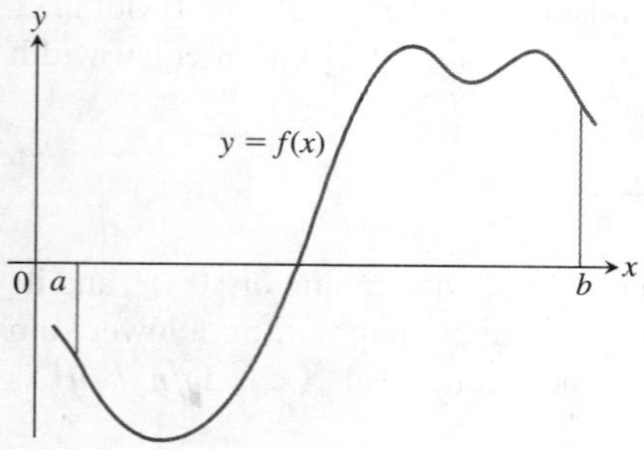

FIGURE 5.8 A typical continuous function $y = f(x)$ over a closed interval $[a, b]$.

HISTORICAL BIOGRAPHY
Richard Dedekind
(1831–1916)
www.bit.ly/2Rdog1r

The first of these subintervals is $[x_0, x_1]$, the second is $[x_1, x_2]$, and the ***k*th subinterval** is $[x_{k-1}, x_k]$ (where k is an integer between 1 and n).

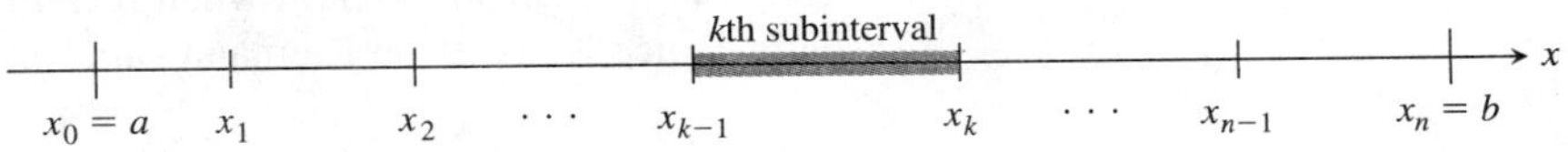

The width of the first subinterval $[x_0, x_1]$ is denoted Δx_1, the width of the second $[x_1, x_2]$ is Δx_2, and the width of the kth subinterval is $\Delta x_k = x_k - x_{k-1}$.

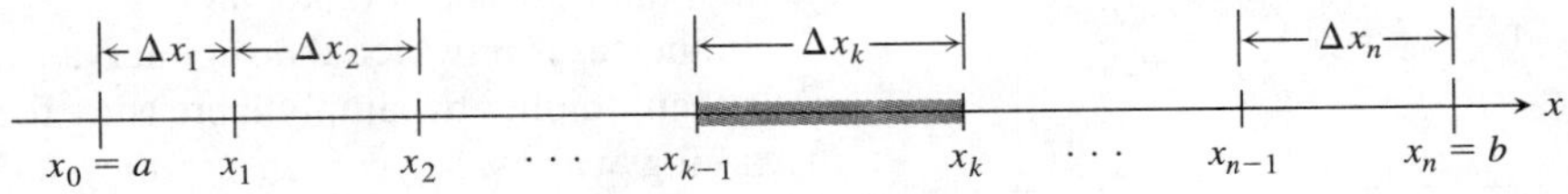

If all n subintervals have equal width, then their common width, which we call Δx, is equal to $(b - a)/n$. Using equal width subintervals is often the simplest choice when doing computations.

In each subinterval we select some point. The point chosen in the kth subinterval $[x_{k-1}, x_k]$ is called c_k. Then on each subinterval, we stand a vertical rectangle that stretches from the x-axis to touch the curve at $(c_k, f(c_k))$. These rectangles can be above or below the x-axis, depending on whether $f(c_k)$ is positive or negative, or on the x-axis if $f(c_k) = 0$ (see Figure 5.9).

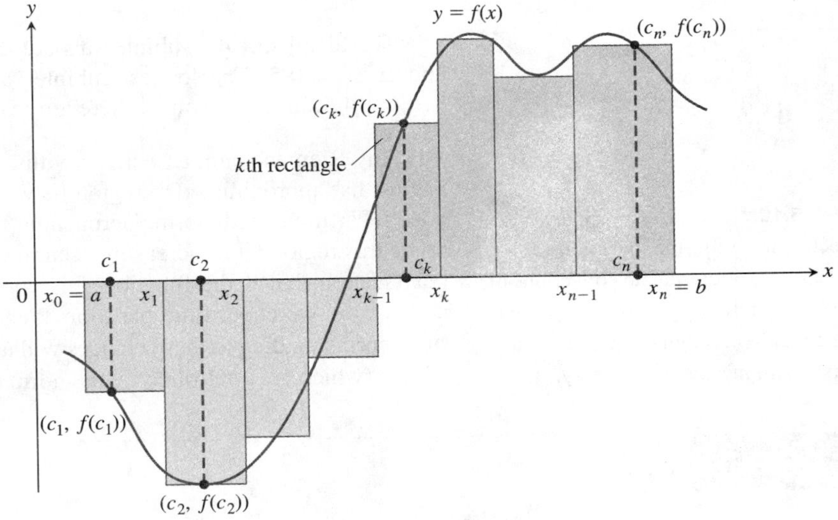

FIGURE 5.9 The rectangles approximate the region between the graph of the function $y = f(x)$ and the x-axis. Figure 5.8 has been repeated and enlarged, the partition of $[a, b]$ and the points c_k have been added, and the corresponding rectangles with heights $f(c_k)$ are shown.

On each subinterval we form the product $f(c_k) \cdot \Delta x_k$. This product is positive, negative, or zero, depending on the sign of $f(c_k)$. When $f(c_k) > 0$, the product $f(c_k) \cdot \Delta x_k$ is the area of a rectangle with height $f(c_k)$ and width Δx_k. When $f(c_k) < 0$, the product $f(c_k) \cdot \Delta x_k$ is a negative number, the negative of the area of a rectangle of width Δx_k that drops from the x-axis to the negative number $f(c_k)$.

Finally, we sum all these products to get

$$S_P = \sum_{k=1}^{n} f(c_k)\,\Delta x_k.$$

The sum S_P is called a **Riemann sum for *f* on the interval $[a, b]$**. There are many such sums, depending on the partition P we choose and on the choices of the points c_k in the subintervals. For instance, we could choose n subintervals all having equal width $\Delta x = (b - a)/n$ to partition $[a, b]$, and then choose the point c_k to be the right-hand endpoint of each subinterval when forming the Riemann sum (as we did in Example 5). This choice leads to the Riemann sum formula

$$S_n = \sum_{k=1}^{n} f\left(a + k\frac{b-a}{n}\right) \cdot \left(\frac{b-a}{n}\right).$$

Similar formulas can be obtained if instead we choose c_k to be the left-hand endpoint, or the midpoint, of each subinterval.

In the cases in which the subintervals all have equal width $\Delta x = (b - a)/n$, we can make them thinner by simply increasing their number n. When a partition has subintervals of varying widths, we can ensure they are all thin by controlling the width of a widest (longest) subinterval. We define the **norm** of a partition P, written $\|P\|$, to be the largest of all the subinterval widths. If $\|P\|$ is a small number, then all of the subintervals in the partition P have a small width.

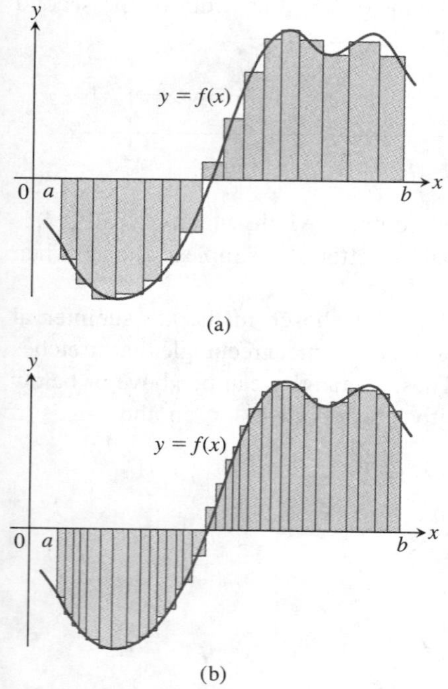

FIGURE 5.10 The curve of Figure 5.9 with rectangles from finer partitions of $[a, b]$. Finer partitions create collections of rectangles with thinner bases that approximate the region between the graph of f and the x-axis with increasing accuracy.

EXAMPLE 6 The set $P = \{0, 0.2, 0.6, 1, 1.5, 2\}$ is a partition of $[0, 2]$. There are five subintervals of P: $[0, 0.2], [0.2, 0.6], [0.6, 1], [1, 1.5]$, and $[1.5, 2]$:

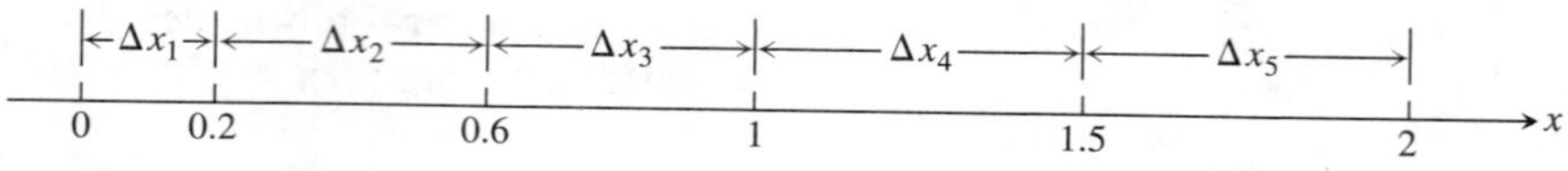

The lengths of the subintervals are $\Delta x_1 = 0.2, \Delta x_2 = 0.4, \Delta x_3 = 0.4, \Delta x_4 = 0.5$, and $\Delta x_5 = 0.5$. The longest subinterval length is 0.5, so the norm of the partition is $\|P\| = 0.5$. In this example, there are two subintervals of this length. ■

Any Riemann sum associated with a partition of a closed interval $[a, b]$ defines rectangles that approximate the region between the graph of a continuous function f and the x-axis. Partitions with norm approaching zero lead to collections of rectangles that approximate this region with increasing accuracy, as suggested by Figure 5.10. We will see in the next section that if the function f is continuous over the closed interval $[a, b]$, then no matter how we choose the partition P and the points c_k in its subintervals, the Riemann sums corresponding to these choices will approach a single limiting value as the subinterval widths (which are controlled by the norm of the partition) approach zero.

EXERCISES 5.2

Sigma Notation

Write the sums in Exercises 1–6 without sigma notation. Then evaluate them.

1. $\displaystyle\sum_{k=1}^{3} \frac{5k}{k+2}$ **2.** $\displaystyle\sum_{k=1}^{3} \frac{k-1}{k}$

3. $\displaystyle\sum_{k=1}^{3} \tan k\pi$ **4.** $\displaystyle\sum_{k=1}^{5} \sin k\pi$

5. $\displaystyle\sum_{k=1}^{3} (-1)^{k+1} \sin\frac{\pi}{k}$ **6.** $\displaystyle\sum_{k=1}^{3} (-1)^k \tan k\pi$

7. Which of the following express $1 + 3 + 9 + 27 + 81 + 243$ in sigma notation?

a. $\displaystyle\sum_{k=1}^{6} 3^{k-1}$ **b.** $\displaystyle\sum_{k=0}^{5} 3^k$ **c.** $\displaystyle\sum_{k=-1}^{4} 3^{k+1}$

8. Which of the following express $1 - 2 + 4 - 8 + 16 - 32$ in sigma notation?

a. $\displaystyle\sum_{k=1}^{6} (-2)^{k-1}$ **b.** $\displaystyle\sum_{k=0}^{5} (-1)^k 2^k$ **c.** $\displaystyle\sum_{k=-2}^{3} (-1)^{k+1} 2^{k+2}$

9. Which formula is not equivalent to the other two?

a. $\displaystyle\sum_{k=2}^{4} \frac{(-1)^{k-1}}{k-1}$ **b.** $\displaystyle\sum_{k=0}^{2} \frac{(-1)^k}{k+1}$ **c.** $\displaystyle\sum_{k=-1}^{1} \frac{(-1)^k}{k+2}$

10. Which formula is not equivalent to the other two?

a. $\displaystyle\sum_{k=1}^{4} (k-1)^2$ **b.** $\displaystyle\sum_{k=-1}^{3} (k+1)^2$ **c.** $\displaystyle\sum_{k=-3}^{-1} k^2$

Express the sums in Exercises 11–16 in sigma notation. The form of your answer will depend on your choice for the starting index.

11. $1 + 2 + 3 + 4 + 5 + 6$ **12.** $4 + 9 + 16 + 25$

13. $\frac{1}{2} + \frac{1}{4} + \frac{1}{8} + \frac{1}{16}$ **14.** $2 + 4 + 6 + 8 + 10$

15. $1 - \frac{1}{2} + \frac{1}{3} - \frac{1}{4} + \frac{1}{5}$

16. $-\frac{1}{5} + \frac{2}{5} - \frac{3}{5} + \frac{4}{5}$

Values of Finite Sums

17. Suppose that $\sum_{k=1}^{n} a_k = -5$ and $\sum_{k=1}^{n} b_k = 6$. Find the values of

a. $\sum_{k=1}^{n} 3a_k$ **b.** $\sum_{k=1}^{n} \frac{b_k}{6}$ **c.** $\sum_{k=1}^{n} (a_k + b_k)$

d. $\sum_{k=1}^{n} (a_k - b_k)$ **e.** $\sum_{k=1}^{n} (b_k - 2a_k)$

18. Suppose that $\sum_{k=1}^{n} a_k = 0$ and $\sum_{k=1}^{n} b_k = 1$. Find the values of

a. $\sum_{k=1}^{n} 5a_k$ **b.** $\sum_{k=1}^{n} 200b_k$

c. $\sum_{k=1}^{n} (a_k + 1)$ **d.** $\sum_{k=1}^{n} (b_k - 1)$

Evaluate the sums in Exercises 19–36.

19. **a.** $\sum_{k=1}^{7} (k+2)$ **b.** $\sum_{k=1}^{7} (k+2)^2$ **c.** $\sum_{k=1}^{7} (k+2)^3$

20. **a.** $\sum_{k=1}^{5} (k+1)$ **b.** $\sum_{k=1}^{5} (k+1)^2$ **c.** $\sum_{k=1}^{5} (k+1)^3$

21. $\sum_{k=1}^{7} (-2k)$

22. $\sum_{k=1}^{7} \frac{\pi k}{14}$

23. $\sum_{k=1}^{6} (3 - k^2)$

24. $\sum_{k=1}^{8} (k^2 - 4)$

25. $\sum_{k=1}^{4} (2k+3)k$

26. $\sum_{k=1}^{7} k(2k + 1)$

27. $\sum_{k=1}^{5} \frac{k^3}{225} + \left(\sum_{k=1}^{5} k\right)^3$

28. $\left(\sum_{k=1}^{7} k\right)^2 - \sum_{k=1}^{7} \frac{k^3}{4}$

29. **a.** $\sum_{k=1}^{5} 7$ **b.** $\sum_{k=1}^{400} 5$ **c.** $\sum_{k=2}^{280} 10$

30. **a.** $\sum_{k=9}^{35} k$ **b.** $\sum_{k=3}^{17} k^2$ **c.** $\sum_{k=18}^{71} k(k - 1)$

31. **a.** $\sum_{k=1}^{n} 3$ **b.** $\sum_{k=2}^{n} 10$ **c.** $\sum_{k=1}^{n} (k+1)$

32. **a.** $\sum_{k=1}^{n} \left(\frac{1}{n} + 2n\right)$ **b.** $\sum_{k=1}^{n} \frac{c}{n}$ **c.** $\sum_{k=1}^{n} \frac{k}{n^2}$

33. $\sum_{k=1}^{50} [(k + 1)^2 - k^2]$

34. $\sum_{k=2}^{20} [\sin(k - 1) - \sin k]$

35. $\sum_{k=7}^{30} (\sqrt{k - 4} - \sqrt{k - 3})$

36. $\sum_{k=1}^{40} \frac{1}{k(k + 1)}$ $\left(\textit{Hint: } \frac{1}{k(k + 1)} = \frac{1}{k} - \frac{1}{k + 1}\right)$

Riemann Sums

In Exercises 37–40, graph each function $f(x)$ over the given interval. Partition the interval into four subintervals of equal length. Then add to your sketch the rectangles associated with the Riemann sum $\sum_{k=1}^{4} f(c_k)\,\Delta x_k$, given that c_k is the **(a)** left-hand endpoint, **(b)** right-hand endpoint, **(c)** midpoint of the kth subinterval. (Make a separate sketch for each set of rectangles.)

37. $f(x) = x^2 - 1, \quad [0, 2]$

38. $f(x) = -x^2, \quad [0, 1]$

39. $f(x) = \sin x, \quad [-\pi, \pi]$

40. $f(x) = \sin x + 1, \quad [-\pi, \pi]$

41. Find the norm of the partition $P = \{0, 1.2, 1.5, 2.3, 2.6, 3\}$.

42. Find the norm of the partition $P = \{-2, -1.6, -0.5, 0, 0.8, 1\}$.

Limits of Riemann Sums

For the functions in Exercises 43–50, find a formula for the Riemann sum obtained by dividing the interval $[a, b]$ into n equal subintervals and using the right-hand endpoint for each c_k. Then take a limit of these sums as $n \to \infty$ to calculate the area under the curve over $[a, b]$.

43. $f(x) = 1 - x^2$ over the interval $[0, 1]$.

44. $f(x) = 2x$ over the interval $[0, 3]$.

45. $f(x) = x^2 + 1$ over the interval $[0, 3]$.

46. $f(x) = 3x^2$ over the interval $[0, 1]$.

47. $f(x) = x + x^2$ over the interval $[0, 1]$.

48. $f(x) = 3x + 2x^2$ over the interval $[0, 1]$.

49. $f(x) = 2x^3$ over the interval $[0, 1]$.

50. $f(x) = x^2 - x^3$ over the interval $[-1, 0]$.

5.3 The Definite Integral

In this section we consider the limit of general Riemann sums as the norm of the partitions of a closed interval $[a, b]$ approaches zero. This limiting process leads us to the definition of the *definite integral* of a function over a closed interval $[a, b]$.

Definition of the Definite Integral

The definition of the definite integral is based on the fact that for some functions, as the norm of the partitions of $[a, b]$ approaches zero, the values of the corresponding Riemann sums approach a limiting value J. In particular, this is true for continuous and piecewise-continuous functions. We again use the symbol ε to represent a small positive number, and use it to specify how close to J the Riemann sum must be. The symbol δ is used for a second small positive number that specifies how small the norm of a partition must be in order for the Riemann sum to differ from J by no more than ε. We now define this limit precisely.

DEFINITION Let $f(x)$ be a function defined on a closed interval $[a, b]$. We say that a number J is the **definite integral of f over $[a, b]$** and that J is the limit of the Riemann sums $\sum_{k=1}^{n} f(c_k)\,\Delta x_k$ if the following condition is satisfied:

Given any number $\varepsilon > 0$, there is a corresponding number $\delta > 0$ such that for every partition $P = \{x_0, x_1, \ldots, x_n\}$ of $[a, b]$ with $\|P\| < \delta$ and any choice of c_k in $[x_{k-1}, x_k]$, we have

$$\left| \sum_{k=1}^{n} f(c_k)\,\Delta x_k - J \right| < \varepsilon.$$

The definition involves a limiting process in which the norm of the partition goes to zero. When this limit exists, the function f is said to be *integrable* over $[a, b]$.

We have many choices for a partition P with norm going to zero, and many choices of points c_k for each partition. The definite integral exists when we always get the same limit J, no matter what choices are made. When the limit exists we write

$$J = \lim_{\|P\| \to 0} \sum_{k=1}^{n} f(c_k)\,\Delta x_k,$$

and we say that the *definite integral exists.*

Leibniz introduced a notation for the definite integral that captures its construction as a limit of Riemann sums. He envisioned the finite sums $\sum_{k=1}^{n} f(c_k)\,\Delta x_k$ becoming an infinite sum of function values $f(x)$ multiplied by "infinitesimal" subinterval widths dx. The sum symbol $\sum$ is replaced in the limit by the integral symbol $\int$, whose origin is in the letter "S" (for sum). The function values $f(c_k)$ are replaced by a continuous selection of function values $f(x)$. The subinterval widths Δx_k become the differential dx. It is as if we were summing all products of the form $f(x) \cdot dx$ as x goes from a to b. While this notation captures the underlying ideas, it is Riemann's definition that gives a precise meaning to the definite integral.

If the definite integral exists, then instead of writing J, we write

$$\int_a^b f(x)\,dx.$$

We read this as "the integral from a to b of f of x dee x" or sometimes as "the integral from a to b of f of x with respect to x." The component parts in the integral symbol also have names:

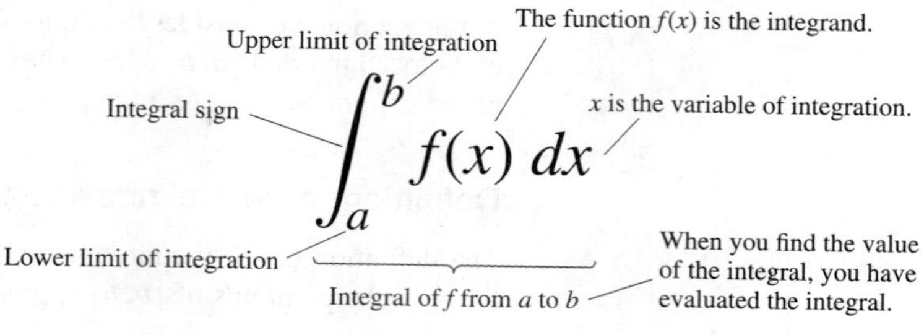

When the definite integral exists, we say that the Riemann sums of f on $[a, b]$ **converge** to the definite integral $J = \int_a^b f(x)\,dx$ and that f is **integrable** over $[a, b]$.

If we choose all the subintervals in a partition to have equal width $\Delta x = (b - a)/n$, the Riemann sums have the form

$$S_n = \sum_{k=1}^{n} f(c_k)\,\Delta x_k = \sum_{k=1}^{n} f(c_k)\left(\frac{b-a}{n}\right), \qquad \Delta x_k = \Delta x = (b-a)/n \text{ for all } k$$

where c_k is any point in the kth subinterval. If the definite integral exists, then these Riemann sums converge to the definite integral of f over $[a, b]$, so

$$J = \int_a^b f(x)\,dx = \lim_{n\to\infty} \sum_{k=1}^{n} f(c_k)\left(\frac{b-a}{n}\right).$$

For equal-width subintervals, the condition $\|P\| \to 0$ is equivalent to $n \to \infty$.

If we pick the point c_k to be the right endpoint of the kth subinterval, so that $c_k = a + k\,\Delta x = a + k(b - a)/n$, then the formula for the definite integral becomes

The Definite Integral as a Limit of Riemann Sums with Equal-Width Subintervals

$$\int_a^b f(x)\,dx = \lim_{n\to\infty} \sum_{k=1}^{n} f\left(a + k\frac{b-a}{n}\right)\left(\frac{b-a}{n}\right) \qquad (1)$$

Equation (1) gives an explicit formula that can be used to compute definite integrals. When the definite integral exists, the Riemann sums coming from other choices of partitions and locations of points c_k will have the same limit as $n \to \infty$, provided that the norms of the partitions approach zero. Another choice of Riemann sums converging to the definite integral is given by the Midpoint Rule, discussed later in this section.

The value of the definite integral of a function over any particular interval depends on the function, not on the letter we choose to represent its independent variable. If we decide to use t or u instead of x, we simply write the integral as

$$\int_a^b f(t)\,dt \qquad \text{or} \qquad \int_a^b f(u)\,du \qquad \text{instead of} \qquad \int_a^b f(x)\,dx.$$

No matter how we write the integral, it is still the same number, the limit of the Riemann sums as the norm of the partition approaches zero. Since it does not matter what letter we use, the variable of integration is called a **dummy variable**. In the three integrals given above, the dummy variables are t, u, and x.

Integrable and Nonintegrable Functions

Not every function defined over a closed interval $[a, b]$ is integrable even if the function is bounded. That is, the Riemann sums for some functions might not converge to the same limiting value, or to any value at all. Understanding which functions defined over $[a, b]$ are integrable and which are not requires advanced mathematical analysis, but fortunately most functions that commonly occur in applications are integrable. In particular, every continuous function over $[a, b]$ is integrable over this interval, and so is every function that has no more than a finite number of jump discontinuities on $[a, b]$. (See Figures 1.9 and 1.10. Such functions are called *piecewise continuous functions*, and they are defined in Additional Exercises 11–18 at the end of this chapter.) The following theorem, which is proved in more advanced courses, establishes these results.

THEOREM 1—Continuous Functions Are Integrable If a function f is continuous over the interval $[a, b]$, or if f has at most finitely many jump discontinuities there, then the definite integral $\int_a^b f(x)\,dx$ exists and f is integrable over $[a, b]$.

The idea behind Theorem 1 for continuous functions is given in Exercises 86 and 87. Briefly, when f is continuous, we can choose each c_k so that $f(c_k)$ gives the maximum value of f on the subinterval $[x_{k-1}, x_k]$, resulting in an upper sum. Likewise, we can choose c_k to give the minimum value of f on $[x_{k-1}, x_k]$ to obtain a lower sum. The upper and lower sums can be shown to converge to the same limiting value as the norm of the partition P tends to zero. Moreover, every Riemann sum is trapped between the values of the upper and lower sums, so every Riemann sum converges to the same limit as well. Therefore, the number J in the definition of the definite integral exists, and the continuous function f is integrable over $[a, b]$.

For integrability to fail, a function needs to be sufficiently discontinuous that the region between its graph and the x-axis cannot be approximated well by increasingly thin rectangles. Our first example is a function that is not integrable over a closed interval.

EXAMPLE 1 The function

$$f(x) = \begin{cases} 1, & \text{if } x \text{ is rational,} \\ 0, & \text{if } x \text{ is irrational,} \end{cases}$$

has no Riemann integral over $[0, 1]$. Underlying this is the fact that between any two numbers there are both a rational number and an irrational number. Thus the function jumps up and down too erratically over $[0, 1]$ to allow the region beneath its graph and above the x-axis to be approximated by rectangles, no matter how thin they are. In fact, we will show that upper sum approximations and lower sum approximations converge to different limiting values.

If we choose a partition P of $[0, 1]$, then the lengths of the intervals in the partition sum to 1; that is, $\sum_{k=1}^{n} \Delta x_k = 1$. In each subinterval $[x_{k-1}, x_k]$ there is a rational point, say c_k. Because c_k is rational, $f(c_k) = 1$. Since 1 is the maximum value that f can take anywhere, the upper sum approximation for this choice of c_k's is

$$U = \sum_{k=1}^{n} f(c_k)\,\Delta x_k = \sum_{k=1}^{n} (1)\,\Delta x_k = 1.$$

As the norm of the partition approaches 0, these upper sum approximations converge to 1 (because each approximation is equal to 1).

On the other hand, we could pick the c_k's differently and get a different result. Each subinterval $[x_{k-1}, x_k]$ also contains an irrational point c_k, and for this choice $f(c_k) = 0$. Since 0 is the minimum value that f can take anywhere, this choice of c_k gives us the minimum value of f on the subinterval. The corresponding lower sum approximation is

$$L = \sum_{k=1}^{n} f(c_k)\,\Delta x_k = \sum_{k=1}^{n} (0)\,\Delta x_k = 0.$$

These lower sum approximations converge to 0 as the norm of the partition converges to 0 (because they each equal 0).

Thus making different choices for the points c_k results in different limits for the corresponding Riemann sums. We conclude that the definite integral of f over the interval $[0, 1]$ does not exist and that f is not integrable over $[0, 1]$. ■

Theorem 1 says nothing about how to *calculate* definite integrals. A method of calculation will be developed in Section 5.4, through a connection of definite integrals to antiderivatives. Meanwhile, finite approximations can be used to calculate an approximation for a definite integral.

The Midpoint Rule

When setting up a Riemann sum

$$\sum_{k=1}^{n} f(c_k)\,\Delta x_k$$

we have many options on how to choose a partition and where to choose a point c_k in the kth interval of the partition. The simplest choice for a partition with n subintervals is to take each subinterval to have equal length $\Delta x = (b - a)/n$. In Equation (1) we obtained a formula by choosing c_k at the right endpoint of the kth interval. Setting c_k to be at the middle of the kth interval is often a better choice for approximating the integral using Riemann sums, as we saw in Figure 5.3. The midpoint is a good choice because on a small interval, the graph of a differentiable function can be approximated by a linear function, and the value of a linear function at an interval's midpoint is its average value over that interval.

With these choices, $[x_{k-1}, x_k] = [a + (k - 1)\Delta x,\ a + k\,\Delta x]$ is the kth interval in the partition, its midpoint is located at $c_k = (x_{k-1} + x_k)/2$, and its width is $\Delta x = (b - a)/n$. The midpoint rule for forming Riemann sums to approximate a definite integral takes the following form.

Midpoint Rule for Approximating a Definite Integral

$$\int_a^b f(x)\,dx \approx \sum_{k=1}^{n} f(c_k)\left(\frac{b-a}{n}\right) = [f(c_1) + f(c_2) + \cdots + f(c_n)]\left(\frac{b-a}{n}\right)$$

with $c_k = \dfrac{x_{k-1} + x_k}{2}$ and $x_k = a + k\left(\dfrac{b-a}{n}\right)$.

EXAMPLE 2 Approximate $\displaystyle\int_0^1 \frac{1}{1+x^2}\,dx$ using the midpoint rule with five intervals of equal length.

Solution The five intervals of the partition are $[0, 0.2]$, $[0.2, 0.4]$, $[0.4, 0.6]$, $[0.6, 0.8]$, and $[0.8, 1.0]$, each of length 1/5. Their midpoints are at 0.1, 0.3, 0.5, 0.7, and 0.9. Applying the midpoint rule gives

$$\begin{aligned}\int_0^1 \frac{1}{1+x^2}\,dx &\approx [f(0.1) + f(0.3) + f(0.5) + f(0.7) + f(0.9)]\left(\frac{1}{5}\right)\\ &= \left[\frac{1}{1.01} + \frac{1}{1.09} + \frac{1}{1.25} + \frac{1}{1.49} + \frac{1}{1.81}\right](0.2)\\ &\approx 0.786.\end{aligned}$$

■

The midpoint rule becomes more powerful when we can combine it with an error bound that tells us how close the approximation it gives with n intervals is to the integral. In Chapter 8 we will examine error bounds in approximations of integrals.

Properties of Definite Integrals

In defining $\int_a^b f(x)\,dx$ as a limit of sums $\sum_{k=1}^{n} f(c_k)\,\Delta x_k$, we moved from left to right across the interval $[a, b]$. What would happen if we instead move right to left, starting with $x_0 = b$ and ending at $x_n = a$? Each Δx_k in the Riemann sum would change its sign, with $x_k - x_{k-1}$ now negative instead of positive. With the same choices of c_k in each subinterval, the sign of any Riemann sum would change, as would the sign of the limit, the integral $\int_b^a f(x)\,dx$. Since we have not previously given a meaning to integrating backward, we are led to define

$$\int_b^a f(x)\,dx = -\int_a^b f(x)\,dx. \qquad a \text{ and } b \text{ interchanged}$$

It is convenient to have a definition for the integral over $[a, b]$ when $a = b$, so that we are computing an integral over an interval of zero width. Since $a = b$ gives $\Delta x = 0$, whenever $f(a)$ exists we define

$$\int_a^a f(x)\,dx = 0. \qquad a \text{ is both the lower and the upper limit of integration.}$$

TABLE 5.6 Rules satisfied by definite integrals

1. *Order of Integration:* $\int_b^a f(x)\,dx = -\int_a^b f(x)\,dx$	A definition
2. *Zero Width Interval:* $\int_a^a f(x)\,dx = 0$	A definition when $f(a)$ exists
3. *Constant Multiple:* $\int_a^b k\,f(x)\,dx = k\int_a^b f(x)\,dx$	Any constant k
4. *Sum and Difference:* $\int_a^b (f(x) \pm g(x))\,dx = \int_a^b f(x)\,dx \pm \int_a^b g(x)\,dx$	
5. *Additivity:* $\int_a^c f(x)\,dx + \int_c^b f(x)\,dx = \int_a^b f(x)\,dx$	$[a, c] \cup [c, b] = [a, b]$
6. *Max-Min Inequality:* If f has maximum value max f and minimum value min f on $[a, b]$, then $(\min f)\cdot(b-a) \le \int_a^b f(x)\,dx \le (\max f)\cdot(b-a).$	
7. *Domination:* If $f(x) \ge g(x)$ on $[a, b]$, then $\int_a^b f(x)\,dx \ge \int_a^b g(x)\,dx.$	
If $f(x) \ge 0$ on $[a, b]$, then $\int_a^b f(x)\,dx \ge 0.$	Special case of Rule 6.

Theorem 2 states some basic properties of integrals, including the two just discussed. These properties, listed in Table 5.6, are very useful for computing integrals. We will refer to them repeatedly to simplify our calculations. Rules 2 through 7 have geometric interpretations, which are shown in Figure 5.11. The graphs in these figures show only positive functions, but the rules apply to general integrable functions, which could take both positive and negative values.

THEOREM 2 When f and g are integrable over the interval $[a, b]$, the definite integral satisfies the rules listed in Table 5.6.

Rules 1 and 2 are definitions, but Rules 3 to 7 of Table 5.6 must be proved. Below we give a proof of Rule 6. Similar proofs can be given to verify the other properties in Table 5.6.

Proof of Rule 6 Rule 6 says that the integral of f over $[a, b]$ is never smaller than the minimum value of f times the length of the interval and never larger than the maximum value of f times the length of the interval. The reason is that for every partition of $[a, b]$ and for every choice of the points c_k,

$$
\begin{aligned}
(\min f)\cdot(b-a) &= (\min f)\cdot\sum_{k=1}^{n}\Delta x_k && \sum_{k=1}^{n}\Delta x_k = b-a\\
&= \sum_{k=1}^{n}(\min f)\cdot\Delta x_k && \text{Constant Multiple Rule}\\
&\le \sum_{k=1}^{n} f(c_k)\,\Delta x_k && \min f \le f(c_k)\\
&\le \sum_{k=1}^{n}(\max f)\cdot\Delta x_k && f(c_k) \le \max f\\
&= (\max f)\cdot\sum_{k=1}^{n}\Delta x_k && \text{Constant Multiple Rule}\\
&= (\max f)\cdot(b-a).
\end{aligned}
$$

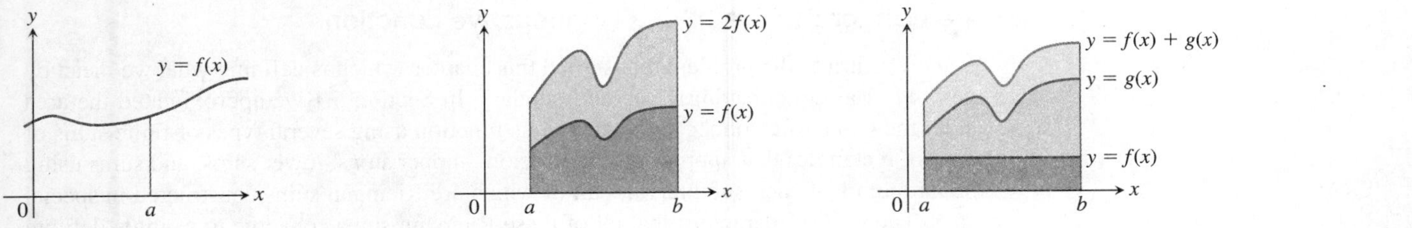

(a) *Zero Width Interval:* $\int_a^a f(x)\,dx = 0$

(b) *Constant Multiple:* $(k = 2)$ $\int_a^b kf(x)\,dx = k\int_a^b f(x)\,dx$

(c) *Sum: (areas add)* $\int_a^b (f(x) + g(x))\,dx = \int_a^b f(x)\,dx + \int_a^b g(x)\,dx$

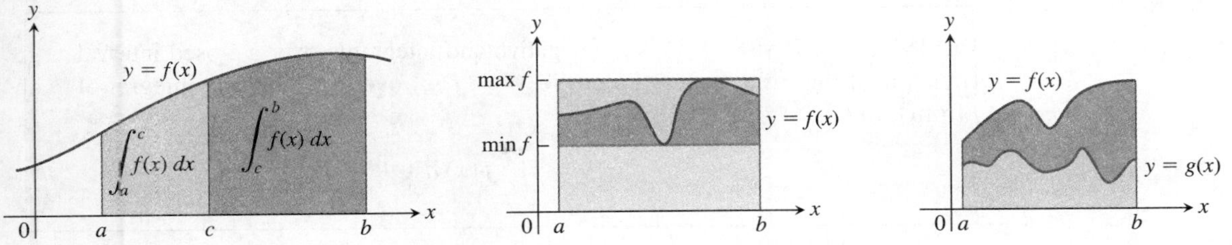

(d) *Additivity for Definite Integrals:* $\int_a^c f(x)\,dx + \int_c^b f(x)\,dx = \int_a^b f(x)\,dx$

(e) *Max-Min Inequality:* $(\min f)\cdot(b - a) \le \int_a^b f(x)\,dx \le (\max f)\cdot(b - a)$

(f) *Domination:* If $f(x) \ge g(x)$ on $[a, b]$, then $\int_a^b f(x)\,dx \ge \int_a^b g(x)\,dx.$

FIGURE 5.11 Geometric interpretations of Rules 2–7 in Table 5.6.

In short, all Riemann sums for f on $[a, b]$ satisfy the inequalities

$$(\min f)\cdot(b - a) \le \sum_{k=1}^{n} f(c_k)\,\Delta x_k \le (\max f)\cdot(b - a).$$

Hence their limit, which is the integral, satisfies the same inequalities. ■

EXAMPLE 3 To illustrate some of the rules, we suppose that

$$\int_{-1}^{1} f(x)\,dx = 5, \qquad \int_{1}^{4} f(x)\,dx = -2, \quad \text{and} \quad \int_{-1}^{1} h(x)\,dx = 7.$$

Then

1. $\int_4^1 f(x)\,dx = -\int_1^4 f(x)\,dx = -(-2) = 2$ Rule 1

2. $\int_{-1}^{1} [2f(x) + 3h(x)]\,dx = 2\int_{-1}^{1} f(x)\,dx + 3\int_{-1}^{1} h(x)\,dx$ Rules 3 and 4
$= 2(5) + 3(7) = 31$

3. $\int_{-1}^{4} f(x)\,dx = \int_{-1}^{1} f(x)\,dx + \int_{1}^{4} f(x)\,dx = 5 + (-2) = 3$ Rule 5 ■

EXAMPLE 4 Show that the value of $\int_0^1 \sqrt{1 + \cos x}\,dx$ is less than or equal to $\sqrt{2}$.

Solution The Max-Min Inequality for definite integrals (Rule 6) says that $(\min f)\cdot(b - a)$ is a *lower bound* for the value of $\int_a^b f(x)\,dx$ and that $(\max f)\cdot(b - a)$ is an *upper bound*. The maximum value of $\sqrt{1 + \cos x}$ on $[0, 1]$ is $\sqrt{1 + 1} = \sqrt{2}$, so

$$\int_0^1 \sqrt{1 + \cos x}\,dx \le \sqrt{2}\cdot(1 - 0) = \sqrt{2}.$$ ■

Area Under the Graph of a Nonnegative Function

We now return to the problem that started this chapter, which is defining what we mean by the *area* of a region having a curved boundary. In Section 5.1 we approximated the area under the graph of a nonnegative continuous function using several types of finite sums of areas of rectangles that approximate the region—upper sums, lower sums, and sums using the midpoints of each subinterval—all of which are Riemann sums constructed in special ways. Theorem 1 guarantees that all of these Riemann sums converge to a single definite integral as the norm of the partitions approaches zero and the number of subintervals goes to infinity. As a result, we can now *define* the area under the graph of a nonnegative integrable function to be the value of that definite integral.

DEFINITION If $y = f(x)$ is nonnegative and integrable over a closed interval $[a, b]$, then the **area under the curve $y = f(x)$ over $[a, b]$** is the integral of f from a to b,

$$A = \int_a^b f(x)\,dx.$$

For the first time, we have a rigorous definition for the area of a region whose boundary is the graph of a continuous function. We now apply this to a simple example, the area under a straight line, and we verify that our new definition agrees with our previous notion of area.

EXAMPLE 5 Compute $\int_0^b x\,dx$ and find the area A under $y = x$ over the interval $[0, b]$, $b > 0$.

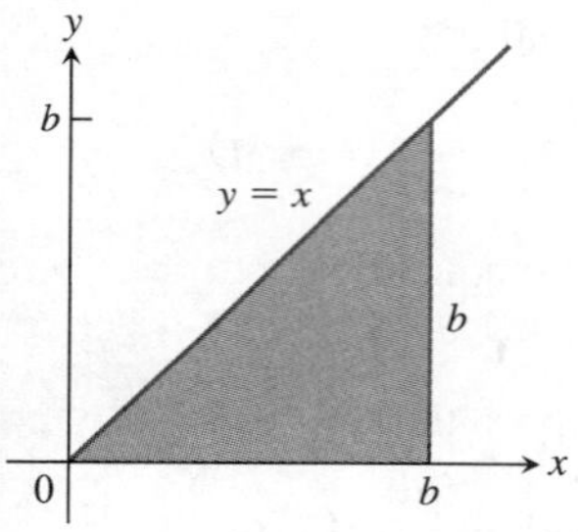

FIGURE 5.12 The region in Example 5 is a triangle.

Solution The region of interest is a triangle (Figure 5.12). We compute the area in two ways.

(a) To compute the definite integral as the limit of Riemann sums, we calculate $\lim_{\|P\|\to 0} \sum_{k=1}^{n} f(c_k)\,\Delta x_k$ for partitions whose norms go to zero. Theorem 1 tells us that it does not matter how we choose the partitions or the points c_k as long as the norms approach zero. All choices give the exact same limit. So we consider the partition P that subdivides the interval $[0, b]$ into n subintervals of equal width $\Delta x = (b - 0)/n = b/n$, and we choose c_k to be the right endpoint in each subinterval as in formula (1). The partition is $P = \left\{0, \frac{b}{n}, \frac{2b}{n}, \frac{3b}{n}, \cdots, \frac{nb}{n}\right\}$ and $c_k = \frac{kb}{n}$. So

$$\begin{aligned}
\sum_{k=1}^{n} f(c_k)\,\Delta x &= \sum_{k=1}^{n} \frac{kb}{n}\cdot\frac{b}{n} && f(c_k) = c_k\\
&= \sum_{k=1}^{n} \frac{kb^2}{n^2}\\
&= \frac{b^2}{n^2}\sum_{k=1}^{n} k && \text{Constant Multiple Rule}\\
&= \frac{b^2}{n^2}\cdot\frac{n(n+1)}{2} && \text{Sum of first } n \text{ integers}\\
&= \frac{b^2}{2}\left(1 + \frac{1}{n}\right).
\end{aligned}$$

As $n \to \infty$ and $\|P\| \to 0$, this last expression on the right has the limit $b^2/2$. Therefore,

$$\int_0^b x\,dx = \frac{b^2}{2}.$$

(b) Since the area equals the definite integral for a nonnegative function, we can quickly derive the definite integral by using the formula for the area of a triangle having base

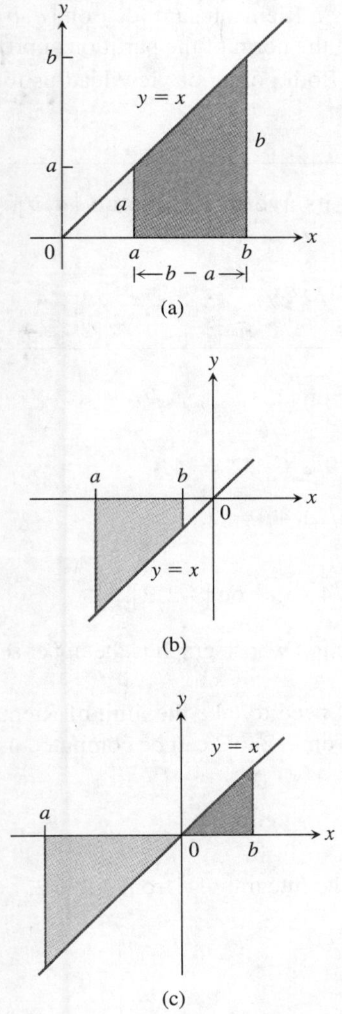

FIGURE 5.13 (a) The area of this trapezoidal region is $A = (b^2 - a^2)/2$. (b) The definite integral in Equation (2) gives the negative of the area of this trapezoidal region. (c) The definite integral in Equation (2) gives the area of the dark gray triangular region added to the negative of the area of the tan triangular region.

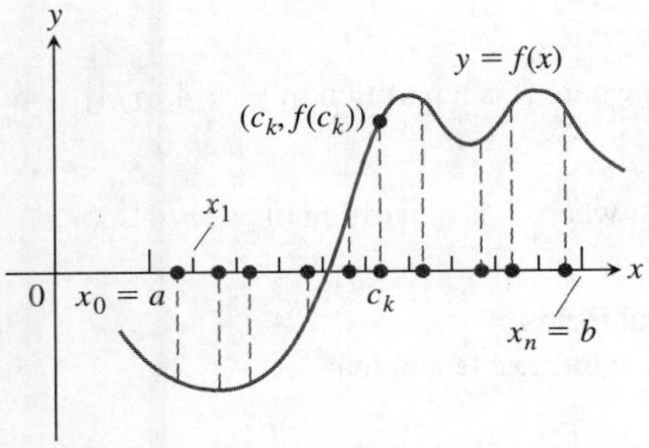

FIGURE 5.14 A sample of values of a function on an interval $[a, b]$.

length b and height $y = b$. The area is $A = (1/2)b \cdot b = b^2/2$. Again we conclude that $\int_0^b x\,dx = b^2/2$. ■

Example 5 can be generalized to integrate $f(x) = x$ over any closed interval $[a, b]$ for which $0 < a < b$.

First write

$$\int_0^b x\,dx = \int_0^a x\,dx + \int_a^b x\,dx \qquad \text{Rule 5}$$

Then, by rearranging this equation and applying Example 5, we obtain

$$\int_a^b x\,dx = \int_0^b x\,dx - \int_0^a x\,dx = \frac{b^2}{2} - \frac{a^2}{2}. \qquad \text{Example 5}$$

In conclusion, we have the following rule for integrating $f(x) = x$ over the interval $[a, b]$:

$$\int_a^b x\,dx = \frac{b^2}{2} - \frac{a^2}{2}, \qquad a < b \qquad (2)$$

This computation gives the area of the trapezoid in Figure 5.13a. Equation (2) remains valid when a and b are negative, but the interpretation of the definite integral changes. When $a < b < 0$, the definite integral value $(b^2 - a^2)/2$ is a negative number, the negative of the area of a trapezoid dropping down to the line $y = x$ below the x-axis (Figure 5.13b). When $a < 0$ and $b > 0$, Equation (2) is still valid and the definite integral gives the difference between two areas, the area under the graph and above $[0, b]$ minus the area below $[a, 0]$ and over the graph (Figure 5.13c).

The following results can also be established by using a Riemann sum calculation similar to the one we used in Example 5 (Exercises 63 and 65).

$$\int_a^b c\,dx = c(b - a), \qquad c \text{ any constant} \qquad (3)$$

$$\int_a^b x^2\,dx = \frac{b^3}{3} - \frac{a^3}{3}, \qquad a < b \qquad (4)$$

Average Value of a Continuous Function Revisited

In Section 5.1 we informally introduced the average value of a nonnegative continuous function f over an interval $[a, b]$, leading us to define this average as the area under the graph of $y = f(x)$ divided by $b - a$. In integral notation we write this as

$$\text{Average} = \frac{1}{b - a}\int_a^b f(x)\,dx.$$

This formula gives us a precise definition of the average value of a continuous (or integrable) function, whether it is positive, negative, or both.

Alternatively, we can justify this formula through the following reasoning. We start with the idea from arithmetic that the average of n numbers is their sum divided by n. A continuous function f on $[a, b]$ may have infinitely many values, but we can still sample them in an orderly way. We divide $[a, b]$ into n subintervals of equal width $\Delta x = (b - a)/n$ and evaluate f at a point c_k in each (Figure 5.14). The average of the n sampled values is

$$\frac{f(c_1) + f(c_2) + \cdots + f(c_n)}{n} = \frac{1}{n}\sum_{k=1}^{n} f(c_k)$$

$$= \frac{\Delta x}{b - a}\sum_{k=1}^{n} f(c_k) \qquad \Delta x = \frac{b-a}{n}, \text{ so } \frac{1}{n} = \frac{\Delta x}{b-a}$$

$$= \frac{1}{b - a}\sum_{k=1}^{n} f(c_k)\,\Delta x. \qquad \text{Constant Multiple Rule}$$

The average of the samples is obtained by dividing a Riemann sum for f on $[a, b]$ by $(b - a)$. As we increase the number of samples and let the norm of the partition approach zero, the average approaches $(1/(b - a))\int_a^b f(x)\,dx$. Both points of view lead us to the following definition.

DEFINITION If f is integrable on $[a, b]$, then its **average value on $[a, b]$**, also called its **mean**, is

$$\text{av}(f) = \frac{1}{b - a}\int_a^b f(x)\,dx.$$

EXAMPLE 6 Find the average value of $f(x) = x$ on $[1, 3]$.

Solution From Equation (2) we see that $\int_1^3 x\,dx = 9/2 - 1/2 = 4$.

So the average value of f on the interval $[1, 3]$ is $(1/2)(4) = 2$. ■

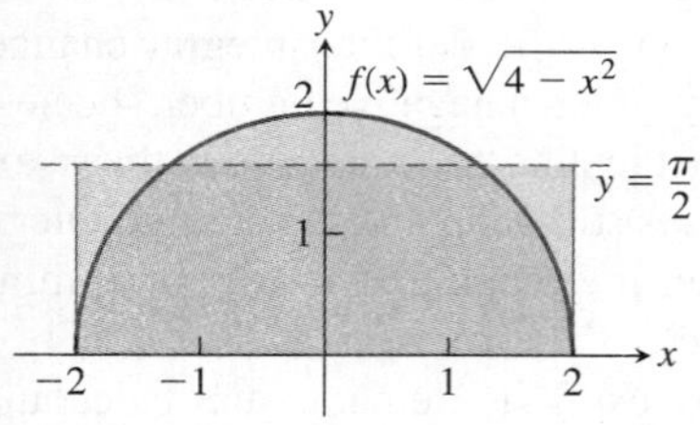

FIGURE 5.15 The average value of $f(x) = \sqrt{4 - x^2}$ on $[-2, 2]$ is $\pi/2$ (Example 7). The area of the rectangle shown here is $4 \cdot (\pi/2) = 2\pi$, which is also the area of the semicircle.

EXAMPLE 7 Find the average value of $f(x) = \sqrt{4 - x^2}$ on $[-2, 2]$.

Solution We recognize $f(x) = \sqrt{4 - x^2}$ as the function whose graph is the upper semicircle of radius 2 centered at the origin (Figure 5.15).

Since we know the area inside a circle, we do not need to take the limit of Riemann sums. The area between the semicircle and the x-axis from -2 to 2 can be computed using the geometry formula

$$\text{Area} = \frac{1}{2} \cdot \pi r^2 = \frac{1}{2} \cdot \pi(2)^2 = 2\pi.$$

Because f is nonnegative, the area is also the value of the integral of f from -2 to 2,

$$\int_{-2}^{2} \sqrt{4 - x^2}\,dx = 2\pi.$$

Therefore, the average value of f is

$$\text{av}(f) = \frac{1}{2 - (-2)}\int_{-2}^{2} \sqrt{4 - x^2}\,dx = \frac{1}{4}(2\pi) = \frac{\pi}{2}.$$

Notice that the average value of f over $[-2, 2]$ is the same as the height of a rectangle over $[-2, 2]$ whose area equals the area of the upper semicircle (see Figure 5.15). ■

EXERCISES 5.3

Interpreting Limits of Sums as Integrals

Express the limits in Exercises 1–8 as definite integrals.

1. $\lim_{\|P\|\to 0} \sum_{k=1}^{n} c_k^2\,\Delta x_k$, where P is a partition of $[0, 3]$

2. $\lim_{\|P\|\to 0} \sum_{k=1}^{n} 2c_k^3\,\Delta x_k$, where P is a partition of $[-1, 0]$

3. $\lim_{\|P\|\to 0} \sum_{k=1}^{n} (2c_k^2 - c_k)\Delta x_k$, where P is a partition of $[-5, 2]$

4. $\lim_{\|P\|\to 0} \sum_{k=1}^{n} \left(\frac{1}{c_k}\right)\Delta x_k$, where P is a partition of $[1, 4]$

5. $\lim_{\|P\|\to 0} \sum_{k=1}^{n} \frac{1}{c_k - 1}\Delta x_k$, where P is a partition of $[3, 5]$

6. $\lim_{\|P\|\to 0} \sum_{k=1}^{n} \sqrt{4 - c_k^2}\,\Delta x_k$, where P is a partition of $[0, 1]$

7. $\lim_{\|P\|\to 0} \sum_{k=1}^{n} (\tan c_k)\Delta x_k$, where P is a partition of $[-\pi/4, \pi/4]$

8. $\lim_{\|P\|\to 0} \sum_{k=1}^{n} (\tan c_k)\Delta x_k$, where P is a partition of $[0, \pi/4]$

Using the Definite Integral Rules

9. Suppose that f and g are integrable and that

$$\int_1^2 f(x)\,dx = -4, \quad \int_1^5 f(x)\,dx = 6, \quad \int_1^5 g(x)\,dx = 8.$$

Use the rules in Table 5.6 to find

a. $\int_3^3 2g(x)dx$ **b.** $\int_5^1 g(x)\,dx$

c. $\int_1^2 3f(x)\,dx$ **d.** $\int_2^5 f(x)\,dx$

e. $\int_1^5 [g(x) - f(x)]dx$ **f.** $\int_1^5 [4f(x) - g(x)]\,dx$

10. Suppose that f and h are integrable and that

$$\int_1^9 f(x)\,dx = -1, \quad \int_7^9 f(x)\,dx = 5, \quad \int_7^9 h(x)\,dx = 4.$$

Use the rules in Table 5.6 to find

a. $\int_1^9 -2f(x)\,dx$ **b.** $\int_7^9 [f(x) + h(x)]\,dx$

c. $\int_7^9 [2f(x) - 3h(x)]\,dx$ **d.** $\int_9^1 f(x)\,dx$

e. $\int_1^7 f(x)\,dx$ **f.** $\int_9^7 [h(x) - f(x)]\,dx$

11. Suppose that $\int_1^2 f(x)\,dx = 5$. Find

a. $\int_1^2 f(u)\,du$ **b.** $\int_2^1 \sqrt{5}\,f(x)dx$

c. $\int_2^1 f(t)\,dt$ **d.** $\int_1^2 [-2f(x)]dx$

12. Suppose that $\int_{-3}^0 g(t)\,dt = \sqrt{2}$. Find

a. $\int_0^{-3} g(t)\,dt$ **b.** $\int_{-3}^0 g(u)\,du$

c. $\int_{-3}^0 [-g(x)]\,dx$ **d.** $\int_{-3}^0 \frac{g(r)}{\sqrt{2}}\,dr$

13. Suppose that f is integrable and that $\int_0^3 f(z)\,dz = 3$ and $\int_0^4 f(z)\,dz = 7$. Find

a. $\int_3^4 [2f(z)]dz$ **b.** $\int_4^3 f(t)\,dt$

14. Suppose that h is integrable and that $\int_{-1}^1 h(r)\,dr = 0$ and $\int_{-1}^3 h(r)\,dr = 6$. Find

a. $\int_1^3 h(r)\,dr$ **b.** $-\int_3^1 h(u)\,du$

Using Known Areas to Find Integrals

In Exercises 15–22, graph the integrands and use known area formulas to evaluate the integrals.

15. $\int_{-2}^4 \left(\frac{x}{2} + 3\right) dx$ **16.** $\int_{1/2}^{3/2} (-2x + 4)\,dx$

17. $\int_{-4}^4 \sqrt{16 - x^2}\,dx$ **18.** $\int_{-4}^0 \sqrt{16 - x^2}\,dx$

19. $\int_{-1}^2 |x + 1|dx$ **20.** $\int_{-1}^1 (3 - |x|)dx$

21. $\int_{-1}^1 (2 - |x|)\,dx$ **22.** $\int_{-1}^1 \left(1 + \sqrt{1 - x^2}\right) dx$

Use known area formulas to evaluate the integrals in Exercises 23–28.

23. $\int_0^b \frac{x}{2}\,dx, \quad b > 0$ **24.** $\int_0^b 4x\,dx, \quad b > 0$

25. $\int_a^b 2s\,ds, \quad 0 < a < b$ **26.** $\int_a^b 3t\,dt, \quad 0 < a < b$

27. $f(x) = \sqrt{16 - x^2}$ on **a.** $[-4, 4]$, **b.** $[0, 4]$

28. $f(x) = 2x + \sqrt{1 - x^2}$ on **a.** $[-1, 0]$, **b.** $[-1, 1]$

Evaluating Definite Integrals

Use the results of Equations (2) and (4) to evaluate the integrals in Exercises 29–40.

29. $\int_1^{\sqrt{2}} x\,dx$ **30.** $\int_{0.4}^{2.4} 2x\,dx$ **31.** $\int_\pi^{2\pi} \theta\,d\theta$

32. $\int_{\sqrt{3}}^{2\sqrt{3}} 2r\,dr$ **33.** $\int_0^{\sqrt{5}} 2x^2\,dx$ **34.** $\int_0^{0.3} s^2\,ds$

35. $\int_0^{1/2} t^2\,dt$ **36.** $\int_0^\pi \theta\,d\theta$ **37.** $\int_a^{2a} x\,dx$

38. $\int_a^{\sqrt{2}} 2x\,dx$ **39.** $\int_0^{\sqrt[3]{b}} x^2\,dx$ **40.** $\int_0^{2b} 2x^2\,dx$

Use the rules in Table 5.6 and Equations (2)–(4) to evaluate the integrals in Exercises 41–50.

41. $\int_3^1 7\,dx$ **42.** $\int_0^2 5x\,dx$

43. $\int_0^3 (3t - 5)\,dt$ **44.** $\int_0^{\sqrt{2}} \left(t - \sqrt{2}\right) dt$

45. $\int_2^1 \left(1 + \frac{z}{2}\right) dz$ **46.** $\int_5^0 (2z - 5)\,dz$

47. $\int_1^3 5u^2\,du$ **48.** $\int_{1/2}^1 15u\,du$

49. $\int_0^2 (3x^2 + x - 5)\,dx$ **50.** $\int_2^0 (5x^2 + 2x - 1)\,dx$

Finding Area by Definite Integrals

In Exercises 51–54, use a definite integral to find the area of the region between the given curve and the x-axis on the interval $[0, b]$.

51. $y = 3x^2$ **52.** $y = \pi x^2$

53. $y = 2x$ **54.** $y = \frac{x}{2} + 1$

Finding Average Value

In Exercises 55–62, graph the function and find its average value over the given interval.

55. $f(x) = x^2 - 1$ on $[0, \sqrt{3}]$ **56.** $f(x) = -\frac{x^2}{2}$ on $[0, 3]$

57. $f(x) = -3x^2 - 1$ on $[0, 1]$ **58.** $f(x) = 3x^2 - 3$ on $[0, 1]$

59. $f(t) = (t - 1)^2$ on $[0, 3]$ **60.** $f(t) = t^2 - t$ on $[-2, 1]$

61. $g(x) = |x| - 1$ on **a.** $[-1, 1]$, **b.** $[1, 3]$, and **c.** $[-1, 3]$

62. $h(x) = -|x|$ on **a.** $[-1, 0]$, **b.** $[0, 1]$, and **c.** $[-1, 1]$

Definite Integrals as Limits of Sums

Use the method of Example 5a or Equation (1) to evaluate the definite integrals in Exercises 63–70.

63. $\int_a^b c\,dx$ **64.** $\int_0^2 (2x + 1)\,dx$

65. $\int_a^b x^2\,dx, \quad a < b$ **66.** $\int_{-1}^0 (x - x^2)\,dx$

67. $\int_{-1}^2 (3x^2 - 2x + 1)\,dx$ **68.** $\int_{-1}^1 x^3\,dx$

69. $\int_a^b x^3\,dx, \quad a < b$ **70.** $\int_0^1 (3x - x^3)\,dx$

Theory and Examples

71. What values of a and b, with $a < b$, maximize the value of

$$\int_a^b (x - x^2)\,dx?$$

(*Hint:* Where is the integrand positive?)

72. What values of a and b, with $a < b$, minimize the value of

$$\int_a^b (x^4 - 2x^2)\,dx?$$

73. Use the Max-Min Inequality to find upper and lower bounds for the value of

$$\int_0^1 \frac{1}{1 + x^2}\,dx.$$

74. (*Continuation of Exercise 73.*) Use the Max-Min Inequality to find upper and lower bounds for

$$\int_0^{0.5} \frac{1}{1 + x^2}\,dx \quad \text{and} \quad \int_{0.5}^1 \frac{1}{1 + x^2}\,dx.$$

Add these to arrive at an improved estimate of

$$\int_0^1 \frac{1}{1 + x^2}\,dx.$$

75. Show that the value of $\int_0^1 \sin(x^2)\,dx$ cannot possibly be 2.

76. Show that the value of $\int_0^1 \sqrt{x + 8}\,dx$ lies between $2\sqrt{2} \approx 2.8$ and 3.

77. Integrals of nonnegative functions Use the Max-Min Inequality to show that if f is integrable, then

$$f(x) \geq 0 \quad \text{on} \quad [a, b] \quad \Rightarrow \quad \int_a^b f(x)\,dx \geq 0.$$

78. Integrals of nonpositive functions Show that if f is integrable, then

$$f(x) \leq 0 \quad \text{on} \quad [a, b] \quad \Rightarrow \quad \int_a^b f(x)\,dx \leq 0.$$

79. Use the inequality $\sin x \leq x$, which holds for $x \geq 0$, to find an upper bound for the value of $\int_0^1 \sin x\,dx$.

80. The inequality $\sec x \geq 1 + (x^2/2)$ holds on $(-\pi/2, \pi/2)$. Use it to find a lower bound for the value of $\int_0^1 \sec x\,dx$.

81. If av(f) really is a typical value of the integrable function $f(x)$ on $[a, b]$, then the constant function av(f) should have the same integral over $[a, b]$ as f. Does it? That is, does

$$\int_a^b \text{av}(f)\,dx = \int_a^b f(x)\,dx?$$

Give reasons for your answer.

82. It would be nice if average values of integrable functions obeyed the following rules on an interval $[a, b]$.

a. $\text{av}(f + g) = \text{av}(f) + \text{av}(g)$

b. $\text{av}(kf) = k\,\text{av}(f)$ (any number k)

c. $\text{av}(f) \leq \text{av}(g)$ if $f(x) \leq g(x)$ on $[a, b]$.

Do these rules ever hold? Give reasons for your answers.

83. Upper and lower sums for increasing functions

a. Suppose the graph of a continuous function $f(x)$ rises steadily as x moves from left to right across an interval $[a, b]$. Let P be a partition of $[a, b]$ into n subintervals of equal length $\Delta x = (b - a)/n$. Show by referring to the accompanying figure that the difference between the upper and lower sums for f on this partition can be represented graphically as the area of a rectangle R whose dimensions are $[f(b) - f(a)]$ by Δx. (*Hint:* The difference $U - L$ is the sum of areas of rectangles whose diagonals $Q_0Q_1, Q_1Q_2, \ldots, Q_{n-1}Q_n$ lie approximately along the curve. There is no overlapping when these rectangles are shifted horizontally onto R.)

b. Suppose that instead of being equal, the lengths Δx_k of the subintervals of the partition of $[a, b]$ vary in size. Show that

$$U - L \leq |f(b) - f(a)|\,\Delta x_{\max},$$

where $\Delta x_{\max}$ is the norm of P, and that hence $\lim_{\|P\| \to 0} (U - L) = 0$.

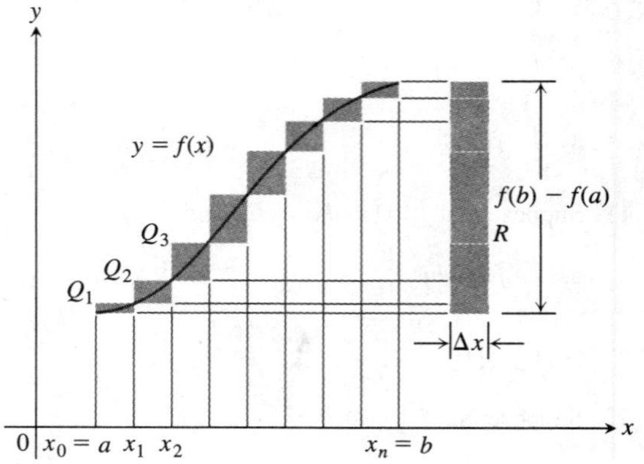

84. Upper and lower sums for decreasing functions (*Continuation of Exercise 83.*)

a. Draw a figure like the one in Exercise 83 for a continuous function $f(x)$ whose values decrease steadily as x moves from left to right across the interval $[a, b]$. Let P be a partition of $[a, b]$ into subintervals of equal length. Find an expression for $U - L$ that is analogous to the one you found for $U - L$ in Exercise 83a.

b. Suppose that instead of being equal, the lengths Δx_k of the subintervals of P vary in size. Show that the inequality

$$U - L \leq |f(b) - f(a)|\,\Delta x_{\max}$$

of Exercise 83b still holds and hence $\lim_{\|P\| \to 0} (U - L) = 0$.

85. Use the formula

$$\sin h + \sin 2h + \sin 3h + \cdots + \sin kh = \frac{\cos(h/2) - \cos((k + (1/2))h)}{2\sin(h/2)}$$

to find the area under the curve $y = \sin x$ from $x = 0$ to $x = \pi/2$ in two steps:

a. Partition the interval $[0, \pi/2]$ into n subintervals of equal length and calculate the corresponding upper sum U; then

b. Find the limit of U as $n \to \infty$ and $\Delta x = (b - a)/n \to 0$.

86. Suppose that f is continuous and nonnegative over $[a, b]$, as in the accompanying figure. By inserting points

$$x_1, x_2, \ldots, x_{k-1}, x_k, \ldots, x_{n-1}$$

as shown, divide $[a, b]$ into n subintervals of lengths $\Delta x_1 = x_1 - a$, $\Delta x_2 = x_2 - x_1, \ldots, \Delta x_n = b - x_{n-1}$, which need not be equal.

a. If $m_k = \min\{f(x) \text{ for } x \text{ in the } k\text{th subinterval}\}$, explain the connection between the **lower sum**

$$L = m_1\,\Delta x_1 + m_2\,\Delta x_2 + \cdots + m_n\,\Delta x_n$$

and the shaded regions in the first part of the figure.

b. If $M_k = \max\{f(x) \text{ for } x \text{ in the } k\text{th subinterval}\}$, explain the connection between the **upper sum**

$$U = M_1\,\Delta x_1 + M_2\,\Delta x_2 + \cdots + M_n\,\Delta x_n$$

and the shaded regions in the second part of the figure.

c. Explain the connection between $U - L$ and the shaded regions along the curve in the third part of the figure.

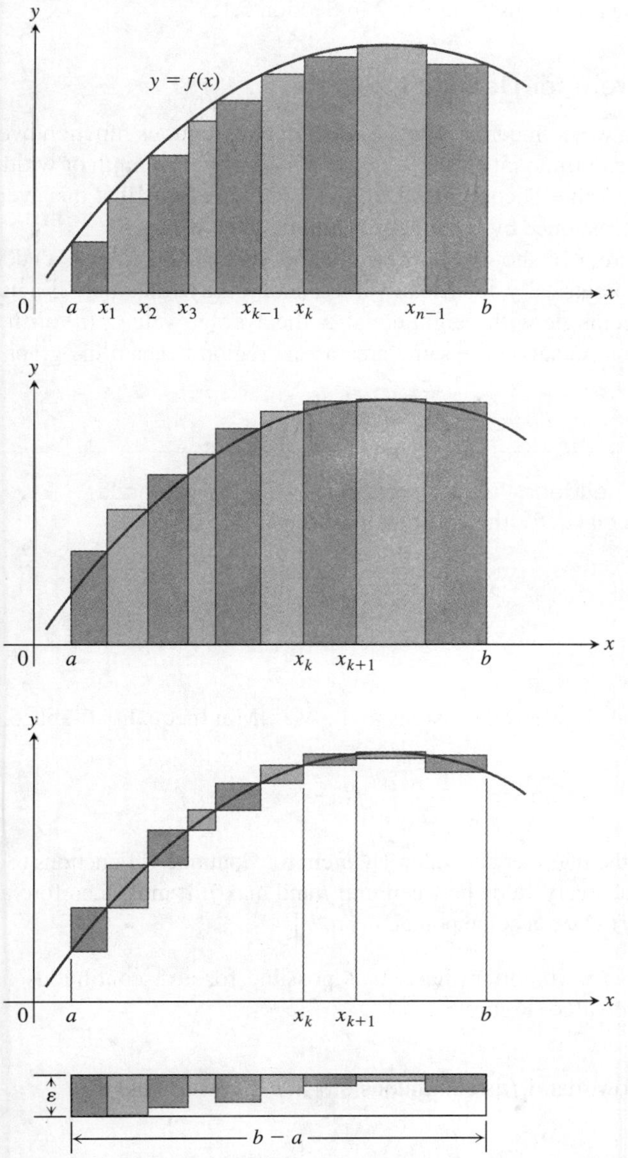

87. We say f is **uniformly continuous** on $[a, b]$ if, given any $\varepsilon > 0$, there is a $\delta > 0$ such that if x_1, x_2 are in $[a, b]$ and $|x_1 - x_2| < \delta$, then $|f(x_1) - f(x_2)| < \varepsilon$. It can be shown that a continuous function on $[a, b]$ is uniformly continuous. Use this and the figure for Exercise 86 to show that if f is continuous and $\varepsilon > 0$ is given, it is possible to make $U - L \le \varepsilon \cdot (b - a)$ by making the largest of the Δx_k's sufficiently small.

88. If you average 30 mi/h on a 150-mi trip and then return over the same 150 mi at the rate of 50 mi/h, what is your average speed for the trip? Give reasons for your answer.

89. Integrals of functions that are equal except at one point Suppose that $f(x)$ is a continuous function over the interval $[a, b]$ and that $g(x)$ is a function on $[a, b]$ such that $g(x) = f(x)$ except at a single point $c \in [a, b]$. Show that $g(x)$ is also integrable over $[a, b]$ and that $\int_a^b g(x)\,dx = \int_a^b f(x)\,dx$. (*Hint:* For a given n, by how much can two different Riemann sums as given in Equation (1) differ?)

90. Some integrable functions that are not continuous

a. The floor function $f(x) = \lfloor x \rfloor$ gives the greatest integer smaller than or equal to x (Example 5 of Section 1.1). This function is not continuous on the interval $[1, 3]$. Show that f is integrable on $[1, 3]$ and that $\int_1^3 \lfloor x \rfloor\,dx = 3$.

b. The function $f(x) = \begin{cases} -1 \text{ if } x < 0, \\ 2 \text{ if } 0 \le x, \end{cases}$ is not continuous on the interval $[-1, 1]$. Show that f is integrable on $[-1, 1]$ and that $\int_{-1}^1 f(x)\,dx = 1$.

COMPUTER EXPLORATIONS

If your CAS can draw rectangles associated with Riemann sums, use it to draw rectangles associated with Riemann sums that converge to the integrals in Exercises 91–96. Use $n = 4$, 10, 20, and 50 subintervals of equal length in each case.

91. $\displaystyle\int_0^1 (1 - x)\,dx = \frac{1}{2}$

92. $\displaystyle\int_0^1 (x^2 + 1)\,dx = \frac{4}{3}$

93. $\displaystyle\int_{-\pi}^{\pi} \cos x\,dx = 0$

94. $\displaystyle\int_0^{\pi/4} \sec^2 x\,dx = 1$

95. $\displaystyle\int_{-1}^{1} |x|\,dx = 1$

96. $\displaystyle\int_1^2 \frac{1}{x}\,dx$ (The integral's value is about 0.693.)

In Exercises 97–100, use a CAS to perform the following steps:

a. Plot the functions over the given interval.

b. Partition the interval into $n = 100$, 200, and 1000 subintervals of equal length, and evaluate the function at the midpoint of each subinterval.

c. Compute the average value of the function values generated in part (b).

d. Solve the equation $f(x) = (\text{average value})$ for x using the average value calculated in part (c) for the $n = 1000$ partitioning.

97. $f(x) = \sin x$ on $[0, \pi]$

98. $f(x) = \sin^2 x$ on $[0, \pi]$

99. $f(x) = x \sin \dfrac{1}{x}$ on $\left[\dfrac{\pi}{4}, \pi\right]$

100. $f(x) = x \sin^2 \dfrac{1}{x}$ on $\left[\dfrac{\pi}{4}, \pi\right]$

5.4 The Fundamental Theorem of Calculus

HISTORICAL BIOGRAPHY
Sir Isaac Newton
(1642–1727)
www.bit.ly/2OjOkcP

In this section we present the Fundamental Theorem of Calculus, which is the central theorem of integral calculus. It connects integration and differentiation, enabling us to compute integrals by using an antiderivative of the integrand function, rather than by taking limits of Riemann sums as we did in Section 5.3. Leibniz and Newton exploited this relationship and started mathematical developments that fueled the scientific revolution for the next 200 years.

Along the way, we will present an integral version of the Mean Value Theorem, which is another important theorem of integral calculus and is used to prove the Fundamental Theorem. We also find that the net change of a function over an interval is the integral of its rate of change, as suggested by Example 2 in Section 5.1.

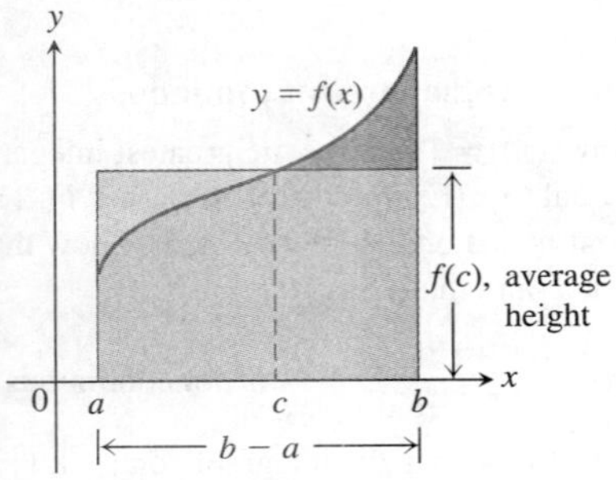

FIGURE 5.16 The value $f(c)$ in the Mean Value Theorem is, in a sense, the average (or *mean*) height of f on $[a, b]$. When $f \geq 0$, the area of the rectangle is the area under the graph of f from a to b,

$$f(c)(b-a) = \int_a^b f(x)\,dx.$$

Mean Value Theorem for Definite Integrals

In the previous section we defined the average value of a continuous function over a closed interval $[a, b]$ to be the definite integral $\int_a^b f(x)\,dx$ divided by the length or width $b - a$ of the interval. The Mean Value Theorem for Definite Integrals asserts that this average value is *always* taken on at least once by the function f in the interval.

The graph in Figure 5.16 shows a *positive* continuous function $y = f(x)$ defined over the interval $[a, b]$. Geometrically, the Mean Value Theorem says that there is a number c in $[a, b]$ such that the rectangle with height equal to the average value $f(c)$ of the function and base width $b - a$ has exactly the same area as the region beneath the graph of f from a to b.

THEOREM 3—The Mean Value Theorem for Definite Integrals
If f is continuous on $[a, b]$, then at some point c in $[a, b]$,

$$f(c) = \frac{1}{b-a}\int_a^b f(x)\,dx.$$

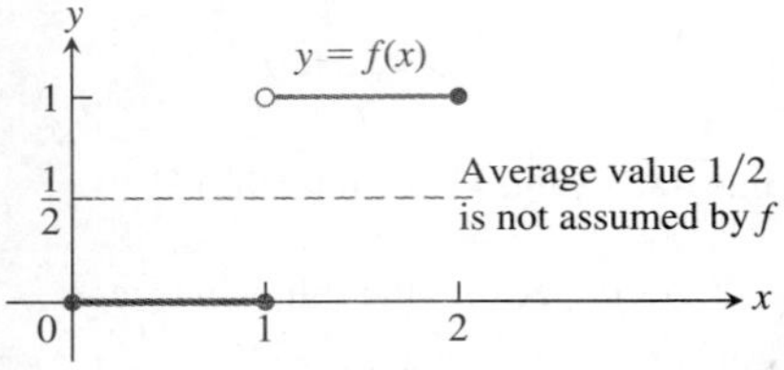

FIGURE 5.17 A discontinuous function need not assume its average value.

Proof If we divide all three expressions in the Max-Min Inequality (Table 5.6, Rule 6) by $(b - a)$, we obtain

$$\min f \leq \frac{1}{b-a}\int_a^b f(x)\,dx \leq \max f.$$

Since f is continuous, the Intermediate Value Theorem for Continuous Functions (Section 2.5) says that f must assume every value between $\min f$ and $\max f$. It must therefore assume the value $(1/(b-a))\int_a^b f(x)\,dx$ at some point c in $[a, b]$. ■

The continuity of f is important here. It is possible for a discontinuous function to never equal its average value (Figure 5.17).

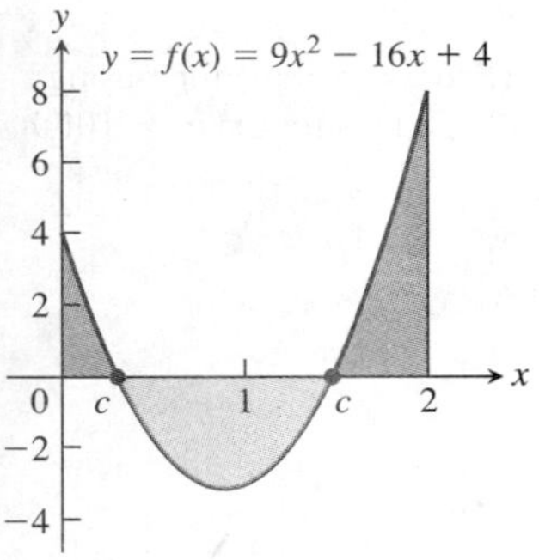

FIGURE 5.18 The function $f(x) = 9x^2 - 16x + 4$ satisfies $\int_0^2 f(x)\,dx = 0$, and there are two values of c in the interval $[0, 2]$ where $f(c) = 0$.

EXAMPLE 1 Show that if f is continuous on $[a, b]$, $a \neq b$, and if

$$\int_a^b f(x)\,dx = 0,$$

then $f(x) = 0$ at least once in $[a, b]$.

Solution The average value of f on $[a, b]$ is

$$\text{av}(f) = \frac{1}{b-a}\int_a^b f(x)\,dx = \frac{1}{b-a}\cdot 0 = 0.$$

By the Mean Value Theorem for Definite Integrals, f assumes this value at some point $c \in [a, b]$. This is illustrated in Figure 5.18 for the function $f(x) = 9x^2 - 16x + 4$ on the interval $[0, 2]$. ■

Fundamental Theorem, Part 1

It can be very difficult to compute definite integrals by taking the limit of Riemann sums. We now develop a powerful new method for evaluating definite integrals, based on using antiderivatives. This method combines the two strands of calculus. One strand involves the idea of taking the limits of finite sums to obtain a definite integral, and the other strand contains derivatives and antiderivatives. They come together in the Fundamental Theorem of Calculus. We begin by considering how to differentiate a certain type of function that is described as an integral.

If $f(t)$ is an integrable function over a finite interval I, then the integral from any fixed number $a \in I$ to another number $x \in I$ defines a new function F whose value at x is

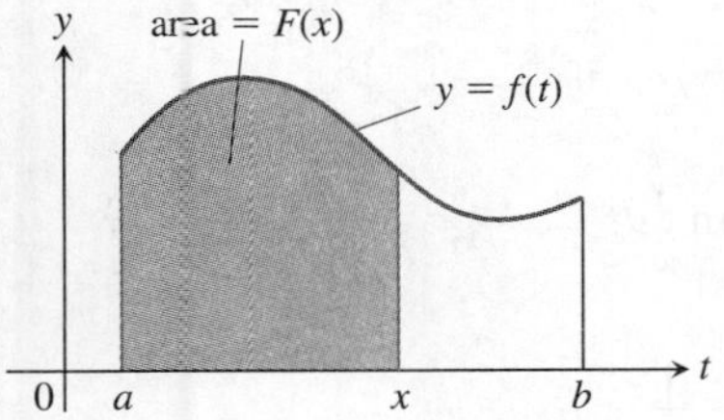

FIGURE 5.19 The function $F(x)$ defined by Equation (1) gives the area under the graph of f from a to x when f is nonnegative and $x > a$.

$$F(x) = \int_a^x f(t)\,dt. \tag{1}$$

For example, if f is nonnegative and x lies to the right of a, then $F(x)$ is the area under the graph from a to x (Figure 5.19). The variable x is the upper limit of integration of an integral, but F is just like any other real-valued function of a real variable. For each value of the input x, there is a single numerical output, in this case the definite integral of f from a to x.

Equation (1) gives a useful way to define new functions (as we will see in Section 7.1), but its key importance is the connection that it makes between integrals and derivatives. If f is a continuous function, then the Fundamental Theorem asserts that F is a differentiable function of x whose derivative is f itself. That is, at each x in the interval $[a, b]$ we have

$$F'(x) = f(x).$$

To gain some insight into why this holds, we look at the geometry behind it.

If $f \geq 0$ on $[a, b]$, then to compute $F'(x)$ from the definition of the derivative, we must take the limit as $h \to 0$ of the difference quotient

$$\frac{F(x+h) - F(x)}{h}.$$

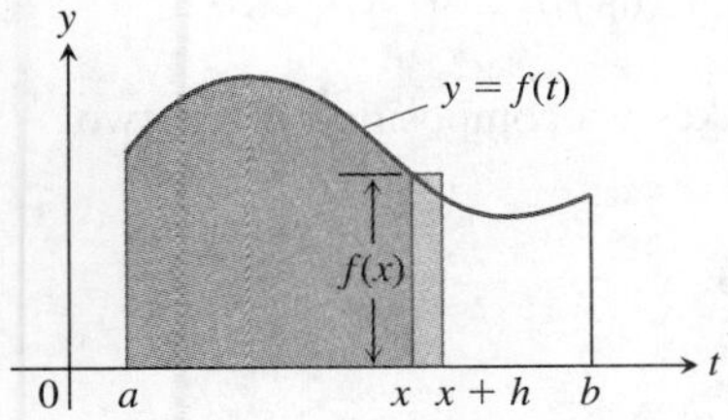

FIGURE 5.20 In Equation (1), $F(x)$ is the area to the left of x. Also, $F(x+h)$ is the area to the left of $x + h$. The difference quotient $[F(x+h) - F(x)]/h$ is then approximately equal to $f(x)$, the height of the rectangle shown here.

If $h > 0$, then $F(x + h)$ is the area under the graph of f from a to $x + h$, while $F(x)$ is the area under the graph of f from a to x. Subtracting the two gives us the area under the graph of f between x and $x + h$ (see Figure 5.20). As shown in Figure 5.20, if h is small, the area under the graph of f from x to $x + h$ is approximated by the area of the rectangle whose height is $f(x)$ and whose base is the interval $[x, x + h]$. That is,

$$F(x+h) - F(x) \approx hf(x).$$

Dividing both sides by h, we see that the value of the difference quotient is very close to the value of $f(x)$:

$$\frac{F(x+h) - F(x)}{h} \approx f(x).$$

This approximation improves as h approaches 0. It is reasonable to expect that $F'(x)$, which is the limit of this difference quotient as $h \to 0$, equals $f(x)$, so that

$$F'(x) = \lim_{h \to 0} \frac{F(x+h) - F(x)}{h} = f(x).$$

This equation is true even if the function f is not positive, and it forms the first part of the Fundamental Theorem of Calculus.

THEOREM 4—The Fundamental Theorem of Calculus, Part 1
If f is continuous on $[a, b]$, then $F(x) = \int_a^x f(t)\,dt$ is continuous on $[a, b]$ and differentiable on (a, b), and its derivative is $f(x)$:

$$F'(x) = \frac{d}{dx}\int_a^x f(t)\,dt = f(x). \tag{2}$$

Before proving Theorem 4, we look at several examples to gain an understanding of what it says. In each of these examples, notice that the independent variable x appears in either the upper or the lower limit of integration (either as part of a formula or by itself). The independent variable on which y depends in these examples is x, while t is merely a dummy variable in the integral.

EXAMPLE 2 Use the Fundamental Theorem to find dy/dx if

(a) $y = \int_a^x (t^3 + 1)\,dt$ **(b)** $y = \int_x^5 3t \sin t\,dt$

(c) $y = \int_1^{x^2} \cos t\,dt$ **(d)** $y = \int_{1+3x^2}^4 \frac{1}{2 + t}\,dt$

Solution We calculate the derivatives with respect to the independent variable x.

(a) $\frac{dy}{dx} = \frac{d}{dx}\int_a^x (t^3 + 1)\,dt = x^3 + 1$ Eq. (2) with $f(t) = t^3 + 1$

(b) $\frac{dy}{dx} = \frac{d}{dx}\int_x^5 3t \sin t\,dt = \frac{d}{dx}\left(-\int_5^x 3t \sin t\,dt\right)$ Table 5.6, Rule 1

$= -\frac{d}{dx}\int_5^x 3t \sin t\,dt$

$= -3x \sin x$ Eq. (2) with $f(t) = 3t \sin t$

(c) The upper limit of integration is not x but x^2. This makes y a composition of the two functions

$$y = \int_1^u \cos t\,dt \qquad \text{and} \qquad u = x^2.$$

We must therefore apply the Chain Rule to find dy/dx:

$$\begin{aligned}
\frac{dy}{dx} &= \frac{dy}{du} \cdot \frac{du}{dx} \\
&= \left(\frac{d}{du}\int_1^u \cos t\,dt\right) \cdot \frac{du}{dx} \\
&= \cos u \cdot \frac{du}{dx} && \text{Eq. (2) with } f(t) = \cos t \\
&= (\cos x^2) \cdot 2x \\
&= 2x \cos x^2
\end{aligned}$$

(d)
$$\begin{aligned}\frac{d}{dx}\int_{1+3x^2}^{4}\frac{1}{2+t}\,dt &= \frac{d}{dx}\left(-\int_{4}^{1+3x^2}\frac{1}{2+t}\,dt\right) && \text{Rule 1}\\ &= -\frac{d}{dx}\int_{4}^{1+3x^2}\frac{1}{2+t}\,dt\\ &= -\frac{1}{2+(1+3x^2)}\cdot\frac{d}{dx}(1+3x^2) && \text{Eq. (2) and the Chain Rule}\\ &= -\frac{2x}{1+x^2}\end{aligned}$$

■

Proof of Theorem 4 We prove the Fundamental Theorem, Part 1, by applying the definition of the derivative directly to the function $F(x)$, when x and $x + h$ are in (a, b). This means writing out the difference quotient

$$\frac{F(x+h)-F(x)}{h} \tag{3}$$

and showing that its limit as $h \to 0$ is the number $f(x)$. Doing so, we find that

$$\begin{aligned}F'(x) &= \lim_{h\to 0}\frac{F(x+h)-F(x)}{h}\\ &= \lim_{h\to 0}\frac{1}{h}\left[\int_a^{x+h} f(t)\,dt - \int_a^x f(t)\,dt\right]\\ &= \lim_{h\to 0}\frac{1}{h}\int_x^{x+h} f(t)\,dt. && \text{Table 5.6, Rule 5}\end{aligned}$$

According to the Mean Value Theorem for Definite Integrals, there is some point c between x and $x + h$ where $f(c)$ equals the average value of f on the interval $[x, x + h]$. That is, there is some number c in $[x, x + h]$ such that

$$\frac{1}{h}\int_x^{x+h} f(t)\,dt = f(c). \tag{4}$$

As $h \to 0$, $x + h$ approaches x, which forces c to approach x also (because c is trapped between x and $x + h$). Since f is continuous at x, $f(c)$ therefore approaches $f(x)$:

$$\lim_{h\to 0} f(c) = f(x). \tag{5}$$

Hence we have shown that, for any x in (a, b),

$$\begin{aligned}F'(x) &= \lim_{h\to 0}\frac{1}{h}\int_x^{x+h} f(t)\,dt\\ &= \lim_{h\to 0} f(c) && \text{Eq. (4)}\\ &= f(x), && \text{Eq. (5)}\end{aligned}$$

and therefore F is differentiable at x. Since differentiability implies continuity, this also shows that F is continuous on the open interval (a, b). To complete the proof, we just have to show that F is also continuous at $x = a$ and $x = b$. To do this, we make a very similar argument, except that at $x = a$ we need only consider the one-sided limit as $h \to 0^+$, and similarly at $x = b$ we need only consider $h \to 0^-$. This shows that F has a one-sided derivative at $x = a$ and at $x = b$, and therefore Theorem 1 in Section 3.2 implies that F is continuous at those two points. ■

Fundamental Theorem, Part 2 (The Evaluation Theorem)

We now come to the second part of the Fundamental Theorem of Calculus. This part describes how to evaluate definite integrals without having to calculate limits of Riemann sums. Instead we find and evaluate an antiderivative at the upper and lower limits of integration.

> **THEOREM 4 (Continued)—The Fundamental Theorem of Calculus, Part 2**
> If f is continuous over $[a, b]$ and F is any antiderivative of f on $[a, b]$, then
> $$\int_a^b f(x)\,dx = F(b) - F(a).$$

Proof Part 1 of the Fundamental Theorem tells us that an antiderivative of f exists, namely

$$G(x) = \int_a^x f(t)\,dt.$$

Thus, if F is *any* antiderivative of f, then $F(x) = G(x) + C$ for some constant C for $a < x < b$ (by Corollary 2 of the Mean Value Theorem for Derivatives, Section 4.2). Since both F and G are continuous on $[a, b]$, we see that the equality $F(x) = G(x) + C$ also holds when $x = a$ and $x = b$ by taking one-sided limits (as $x \to a^+$ and $x \to b^-$).

Evaluating $F(b) - F(a)$, we have

$$\begin{aligned} F(b) - F(a) &= [G(b) + C] - [G(a) + C] \\ &= G(b) - G(a) \\ &= \int_a^b f(t)\,dt - \int_a^a f(t)\,dt \\ &= \int_a^b f(t)\,dt - 0 \\ &= \int_a^b f(t)\,dt. \end{aligned}$$ ■

The Evaluation Theorem is important because it says that to calculate the definite integral of f over an interval $[a, b]$ we need do only two things:

1. Find an antiderivative F of f, and
2. Calculate the number $F(b) - F(a)$, which is equal to $\int_a^b f(x)\,dx$.

This process is much easier than computing Riemann sums and finding their limit. The power of the theorem follows from the realization that the definite integral, which is defined by a complicated process involving all of the values of the function f over $[a, b]$, can be found by knowing the values of *any* antiderivative F at only the two endpoints a and b. The usual notation for the difference $F(b) - F(a)$ is

$$F(x)\Big]_a^b \quad \text{or} \quad \Big[F(x)\Big]_a^b,$$

depending on whether F has one or more terms.

EXAMPLE 3 We calculate several definite integrals using the Evaluation Theorem, rather than by taking limits of Riemann sums.

(a) $$\begin{aligned} \int_0^\pi \cos x\,dx &= \sin x\Big]_0^\pi \\ &= \sin \pi - \sin 0 = 0 - 0 = 0 \end{aligned}$$

$\frac{d}{dx}\sin x = \cos x$

(b) $\displaystyle\int_{-\pi/4}^{0} \sec x \tan x \, dx = \sec x \Big]_{-\pi/4}^{0}$ $\qquad \dfrac{d}{dx}\sec x = \sec x \tan x$

$$= \sec 0 - \sec\left(-\frac{\pi}{4}\right) = 1 - \sqrt{2}$$

(c) $\displaystyle\int_{1}^{4}\left(\frac{3}{2}\sqrt{x} - \frac{4}{x^2}\right) dx = \left[x^{3/2} + \frac{4}{x}\right]_1^4$ $\qquad \dfrac{d}{dx}\left(x^{3/2} + \dfrac{4}{x}\right) = \dfrac{3}{2}x^{1/2} - \dfrac{4}{x^2}$

$$= \left[(4)^{3/2} + \frac{4}{4}\right] - \left[(1)^{3/2} + \frac{4}{1}\right]$$

$$= [8 + 1] - [5] = 4$$

■

Exercise 72 offers another proof of the Evaluation Theorem, bringing together the ideas of Riemann sums, the Mean Value Theorem, and the definition of the definite integral.

The Integral of a Rate

We can interpret Part 2 of the Fundamental Theorem in another way. If F is any antiderivative of f, then $F' = f$. The equation in the theorem can then be rewritten as

$$\int_a^b F'(x)\, dx = F(b) - F(a).$$

Now $F'(x)$ represents the rate of change of the function $F(x)$ with respect to x, so the last equation asserts that the integral of F' is just the *net change* in F as x changes from a to b. Formally, we have the following result.

THEOREM 5—The Net Change Theorem
The net change in a differentiable function $F(x)$ over an interval $a \le x \le b$ is the integral of its rate of change:

$$F(b) - F(a) = \int_a^b F'(x)\, dx. \tag{6}$$

EXAMPLE 4 Here are several interpretations of the Net Change Theorem.

(a) If $c(x)$ is the cost of producing x units of a certain commodity, then $c'(x)$ is the marginal cost (Section 3.4). From Theorem 5,

$$\int_{x_1}^{x_2} c'(x)\, dx = c(x_2) - c(x_1),$$

which is the cost of increasing production from x_1 units to x_2 units.

(b) If an object with position function $s(t)$ moves along a coordinate line, its velocity is $v(t) = s'(t)$. Theorem 5 says that

$$\int_{t_1}^{t_2} v(t)\, dt = s(t_2) - s(t_1),$$

so the integral of velocity is the **displacement** over the time interval $t_1 \le t \le t_2$. On the other hand, the integral of the speed $|v(t)|$ is the **total distance traveled** over the time interval. This is consistent with our discussion in Section 5.1. ■

If we rearrange Equation (6) as

$$F(b) = F(a) + \int_a^b F'(x)\,dx,$$

we see that the Net Change Theorem also says that the final value of a function $F(x)$ over an interval $[a, b]$ equals its initial value $F(a)$ plus its net change over the interval. So if $v(t)$ represents the velocity function of an object moving along a coordinate line, this means that the object's final position $s(t_2)$ over a time interval $t_1 \le t \le t_2$ is its initial position $s(t_1)$ plus its net change in position along the line (see Example 4b).

EXAMPLE 5 Consider again our analysis of a heavy rock blown straight up from the ground by a dynamite blast (Example 2, Section 5.1). The velocity of the rock at any time t during its motion was given as $v(t) = 160 - 32t$ ft/sec.

(a) Find the displacement of the rock during the time period $0 \le t \le 8$.

(b) Find the total distance traveled during this time period.

Solution

(a) From Example 4b, the displacement is the integral

$$\begin{aligned}\int_0^8 v(t)\,dt &= \int_0^8 (160 - 32t)\,dt = \Big[160t - 16t^2\Big]_0^8 \\ &= (160)(8) - (16)(64) = 256.\end{aligned}$$

This means that the rock is 256 ft above the ground 8 sec after the explosion, which agrees with our conclusion in Example 2, Section 5.1.

(b) The velocity function $v(t)$ is positive over the time interval $[0, 5]$ and negative over the interval $[5, 8]$. Therefore, from Example 4b, the total distance traveled is the integral

$$\begin{aligned}\int_0^8 |v(t)|\,dt &= \int_0^5 |v(t)|\,dt + \int_5^8 |v(t)|\,dt \\ &= \int_0^5 (160 - 32t)\,dt - \int_5^8 (160 - 32t)\,dt \qquad |v(t)| = -(160 - 32t) \text{ over } [5,8] \\ &= \Big[160t - 16t^2\Big]_0^5 - \Big[160t - 16t^2\Big]_5^8 \\ &= [(160)(5) - (16)(25)] - [(160)(8) - (16)(64) - ((160)(5) - (16)(25))] \\ &= 400 - (-144) = 544.\end{aligned}$$

This calculation agrees with our conclusion in Example 2, Section 5.1. The total distance of 544 ft traveled by the rock during the time period $0 \le t \le 8$ is (i) the maximum height of 400 ft it reached over the time interval $[0, 5]$ plus (ii) the additional distance of 144 ft the rock fell over the time interval $[5, 8]$. ■

The Relationship Between Integration and Differentiation

The conclusions of the Fundamental Theorem tell us several things. Equation (2) can be rewritten as

$$\frac{d}{dx}\int_a^x f(t)\,dt = f(x),$$

which says that if you first integrate the function f and then differentiate the result, you get the function f back again. Likewise, replacing b by x and x by t in Equation (6) gives

$$\int_a^x F'(t)\,dt = F(x) - F(a),$$

so that if you first differentiate the function F and then integrate the result, you get the function F back (adjusted by an integration constant). In a sense, the processes of integration and differentiation are "inverses" of each other. The Fundamental Theorem also says that every continuous function f has an antiderivative F. It shows the importance of finding antiderivatives in order to evaluate definite integrals easily. Furthermore, it says that the differential equation $dy/dx = f(x)$ has a solution (namely, any of the functions $y = F(x) + C$) when f is a continuous function.

Total Area

Area is always a nonnegative quantity. The Riemann sum approximations contain terms such as $f(c_k)\,\Delta x_k$ that give the area of a rectangle when $f(c_k)$ is positive. When $f(c_k)$ is negative, then the product $f(c_k)\,\Delta x_k$ is the negative of the rectangle's area. When we add up such terms for a negative function, we get the negative of the area between the curve and the x-axis. If we then take the absolute value, we obtain the correct positive area.

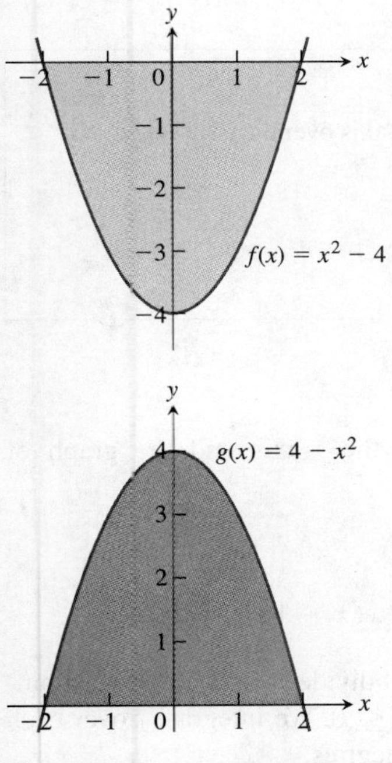

FIGURE 5.21 These graphs enclose the same amount of area with the x-axis, but the definite integrals of the two functions over $[-2, 2]$ differ in sign (Example 6).

EXAMPLE 6 Figure 5.21 shows the graph of $f(x) = x^2 - 4$ and its mirror image $g(x) = 4 - x^2$ reflected across the x-axis. For each function, compute

(a) the definite integral over the interval $[-2, 2]$, and

(b) the area between the graph and the x-axis over $[-2, 2]$.

Solution

(a) $$\int_{-2}^{2} f(x)\,dx = \left[\frac{x^3}{3} - 4x\right]_{-2}^{2} = \left(\frac{8}{3} - 8\right) - \left(-\frac{8}{3} + 8\right) = -\frac{32}{3},$$

and

$$\int_{-2}^{2} g(x)\,dx = \left[4x - \frac{x^3}{3}\right]_{-2}^{2} = \frac{32}{3}.$$

(b) In both cases, the area between the curve and the x-axis over $[-2, 2]$ is $32/3$ square units. Although the definite integral of $f(x)$ is negative, the area is still positive. ■

To compute the area of the region bounded by the graph of a function $y = f(x)$ and the x-axis when the function takes on both positive and negative values, we must be careful to break up the interval $[a, b]$ into subintervals on which the function doesn't change sign. Otherwise, we might get cancelation between positive and negative signed areas, leading to an incorrect total. The correct total area is obtained by adding the absolute value of the definite integral over each subinterval where $f(x)$ does not change sign. The term "area" will be taken to mean this *total area*.

EXAMPLE 7 Figure 5.22 shows the graph of the function $f(x) = \sin x$ between $x = 0$ and $x = 2\pi$. Compute

(a) the definite integral of $f(x)$ over $[0, 2\pi]$,

(b) the area between the graph of $f(x)$ and the x-axis over $[0, 2\pi]$.

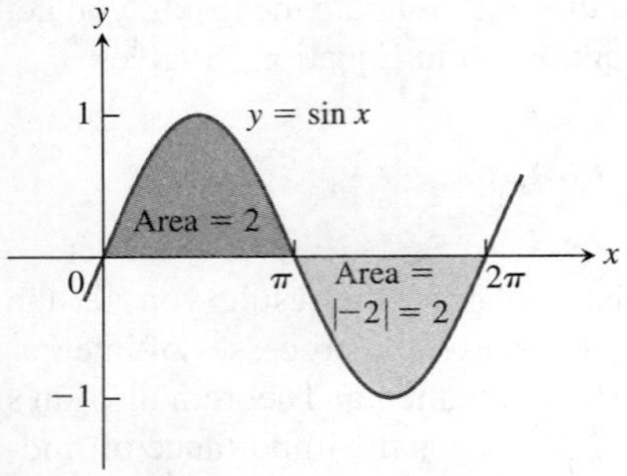

FIGURE 5.22 The total area between $y = \sin x$ and the x-axis for $0 \le x \le 2\pi$ is the sum of the absolute values of two integrals (Example 7).

Solution

(a) The definite integral for $f(x) = \sin x$ is given by

$$\int_0^{2\pi} \sin x \, dx = -\cos x \Big]_0^{2\pi} = -[\cos 2\pi - \cos 0] = -[1 - 1] = 0.$$

The definite integral is zero because the portions of the graph above and below the x-axis make canceling contributions.

(b) The area between the graph of $f(x)$ and the x-axis over $[0, 2\pi]$ is calculated by breaking up the domain of $\sin x$ into two pieces: the interval $[0, \pi]$ over which it is nonnegative and the interval $[\pi, 2\pi]$ over which it is nonpositive.

$$\int_0^{\pi} \sin x \, dx = -\cos x \Big]_0^{\pi} = -[\cos \pi - \cos 0] = -[-1 - 1] = 2$$

$$\int_{\pi}^{2\pi} \sin x \, dx = -\cos x \Big]_{\pi}^{2\pi} = -[\cos 2\pi - \cos \pi] = -[1 - (-1)] = -2$$

The second integral gives a negative value. The area between the graph and the axis is obtained by adding the absolute values,

$$\text{Area} = |2| + |-2| = 4.$$

■

> **Summary:**
> To find the area between the graph of $y = f(x)$ and the x-axis over the interval $[a, b]$:
> 1. Subdivide $[a, b]$ at the zeros of f.
> 2. Integrate f over each subinterval.
> 3. Add the absolute values of the integrals.

EXAMPLE 8 Find the area of the region between the x-axis and the graph of $f(x) = x^3 - x^2 - 2x$, $-1 \le x \le 2$.

Solution First find the zeros of f. Since

$$f(x) = x^3 - x^2 - 2x = x(x^2 - x - 2) = x(x + 1)(x - 2),$$

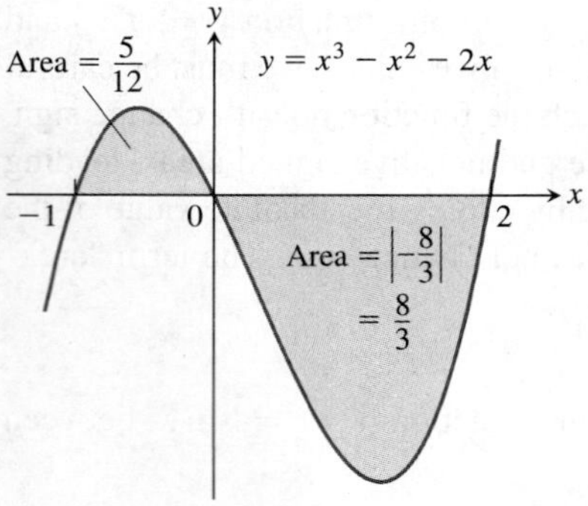

FIGURE 5.23 The region between the curve $y = x^3 - x^2 - 2x$ and the x-axis (Example 8).

the zeros are $x = 0$, -1, and 2 (Figure 5.23). The zeros subdivide $[-1, 2]$ into two subintervals: $[-1, 0]$, on which $f \ge 0$, and $[0, 2]$, on which $f \le 0$. We integrate f over each subinterval and add the absolute values of the calculated integrals.

$$\int_{-1}^{0} (x^3 - x^2 - 2x) \, dx = \left[\frac{x^4}{4} - \frac{x^3}{3} - x^2\right]_{-1}^{0} = 0 - \left[\frac{1}{4} + \frac{1}{3} - 1\right] = \frac{5}{12}$$

$$\int_{0}^{2} (x^3 - x^2 - 2x) \, dx = \left[\frac{x^4}{4} - \frac{x^3}{3} - x^2\right]_{0}^{2} = \left[4 - \frac{8}{3} - 4\right] - 0 = -\frac{8}{3}$$

The total enclosed area is obtained by adding the absolute values of the calculated integrals.

$$\text{Total enclosed area} = \frac{5}{12} + \left|-\frac{8}{3}\right| = \frac{37}{12}$$

■

EXERCISES 5.4

Evaluating Integrals

Evaluate the integrals in Exercises 1–28.

1. $\int_0^2 x(x-3)\,dx$
2. $\int_{-1}^1 (x^2-2x+3)\,dx$
3. $\int_{-2}^2 (x+3)^2\,dx$
4. $\int_{-1}^1 x^{299}\,dx$
5. $\int_1^4 \left(3x^2-\frac{x^3}{4}\right)dx$
6. $\int_{-2}^3 (x^3-2x+3)\,dx$
7. $\int_0^1 (x^2+\sqrt{x})\,dx$
8. $\int_1^{32} x^{-6/5}\,dx$
9. $\int_0^{\pi/3} 2\sec^2 x\,dx$
10. $\int_0^{\pi} (1+\cos x)\,dx$
11. $\int_{\pi/4}^{3\pi/4} \csc\theta\cot\theta\,d\theta$
12. $\int_0^{\pi/3} 4\frac{\sin u}{\cos^2 u}\,du$
13. $\int_{\pi/2}^{0} \frac{1+\cos 2t}{2}\,dt$
14. $\int_{-\pi/3}^{\pi/3} \sin^2 t\,dt$
15. $\int_0^{\pi/4} \tan^2 x\,dx$
16. $\int_0^{\pi/6} (\sec x+\tan x)^2\,dx$
17. $\int_0^{\pi/8} \sin 2x\,dx$
18. $\int_{-\pi/3}^{-\pi/4} \left(4\sec^2 t+\frac{\pi}{t^2}\right)dt$
19. $\int_1^{-1} (r+1)^2\,dr$
20. $\int_{-\sqrt{3}}^{\sqrt{3}} (t+1)(t^2+4)\,dt$
21. $\int_{\sqrt{2}}^{1} \left(\frac{u^7}{2}-\frac{1}{u^5}\right)du$
22. $\int_{-3}^{-1} \frac{y^5-2y}{y^3}\,dy$
23. $\int_1^{\sqrt{2}} \frac{s^2+\sqrt{s}}{s^2}\,ds$
24. $\int_1^8 \frac{(x^{1/3}+1)(2-x^{2/3})}{x^{1/3}}\,dx$
25. $\int_{\pi/6}^{\pi/2} \frac{\sin 2x}{2\sin x}\,dx$
26. $\int_0^{\pi/3} (\cos x+\sec x)^2\,dx$
27. $\int_{-4}^4 |x|\,dx$
28. $\int_0^{\pi} \frac{1}{2}\left(\cos x+|\cos x|\right)dx$

In Exercises 29–32, guess an antiderivative for the integrand function. Validate your guess by differentiation, and then evaluate the given definite integral. (*Hint:* Keep the Chain Rule in mind when trying to guess an antiderivative. You will learn how to find such antiderivatives in the next section.)

29. $\int_0^{\sqrt{\pi/2}} x\cos x^2\,dx$
30. $\int_1^{\pi^2} \frac{\sin\sqrt{x}}{\sqrt{x}}\,dx$
31. $\int_2^5 \frac{x\,dx}{\sqrt{1+x^2}}$
32. $\int_0^{\pi/3} \sin^2 x\cos x\,dx$

Derivatives of Integrals

Find the derivatives in Exercises 33–38.

a. by evaluating the integral and differentiating the result.

b. by differentiating the integral directly.

33. $\frac{d}{dx}\int_0^{\sqrt{x}} \cos t\,dt$
34. $\frac{d}{dx}\int_1^{\sin x} 3t^2\,dt$
35. $\frac{d}{dt}\int_0^{t^4} \sqrt{u}\,du$
36. $\frac{d}{d\theta}\int_0^{\tan\theta} \sec^2 y\,dy$
37. $\frac{d}{dx}\int_1^{1+x^2} t^{-1/2}\,dt$
38. $\frac{d}{dt}\int_0^{\sqrt{t}} (x^4+x^2)\,dx$

Find dy/dx in Exercises 39–46.

39. $y=\int_0^x \sqrt{1+t^2}\,dt$
40. $y=\int_1^{x^2} \frac{t^6}{1+t^4}\,dt$
41. $y=\int_{\sqrt{x}}^0 \sin t^2\,dt$
42. $y=x\int_2^{x^2} \sin t^3\,dt$
43. $y=\int_{-1}^x \frac{t^2}{t^2+4}\,dt-\int_3^x \frac{t^2}{t^2+4}\,dt$
44. $y=\int_0^x (1+t^2)^{-4}\,dt$
45. $y=\int_0^{\sin x} \frac{dt}{\sqrt{1-t^2}},\quad |x|<\frac{\pi}{2}$
46. $y=\int_{\tan x}^0 \frac{dt}{1+t^2}$

Area

In Exercises 47–50, find the total area between the region and the x-axis.

47. $y=-x^2-2x,\quad -3\le x\le 2$
48. $y=3x^2-3,\quad -2\le x\le 2$
49. $y=x^3-3x^2+2x,\quad 0\le x\le 2$
50. $y=x^4-2x^3+x^2+3,\quad 0\le x\le 1$

Find the areas of the shaded regions in Exercises 51–54.

51.

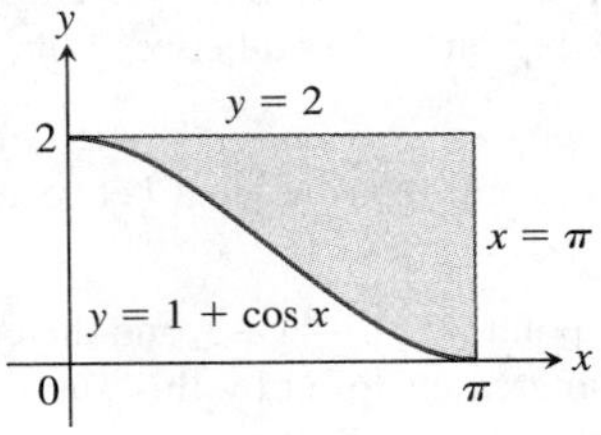

52.

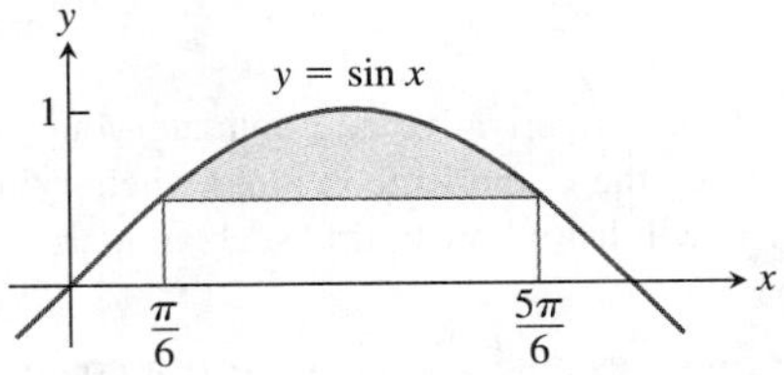

53.

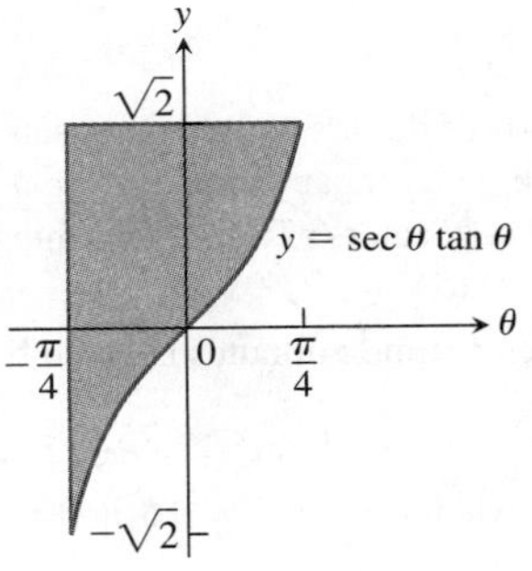

54.

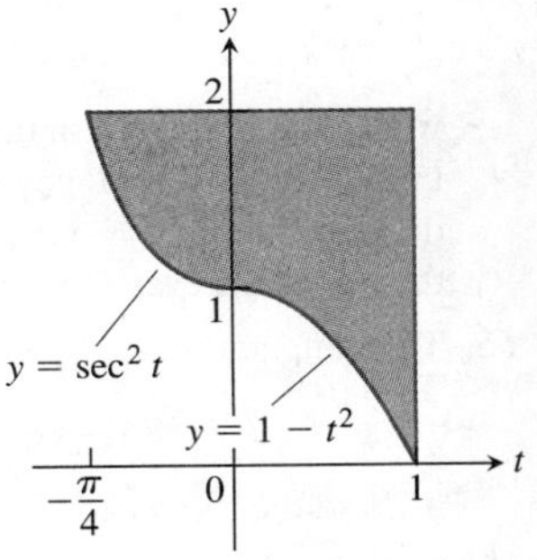

Initial Value Problems

Each of the following functions solves one of the initial value problems in Exercises 55–58. Which function solves which problem? Give brief reasons for your answers.

a. $y = \int_1^x \frac{1}{t}\,dt - 3$ **b.** $y = \int_0^x \sec t\,dt + 4$

c. $y = \int_{-1}^x \sec t\,dt + 4$ **d.** $y = \int_\pi^x \frac{1}{t}\,dt - 3$

55. $\frac{dy}{dx} = \frac{1}{x}, \quad y(\pi) = -3$ **56.** $y' = \sec x, \quad y(-1) = 4$

57. $y' = \sec x, \quad y(0) = 4$ **58.** $y' = \frac{1}{x}, \quad y(1) = -3$

Express the solutions of the initial value problems in Exercises 59 and 60 in terms of integrals.

59. $\frac{dy}{dx} = \sec x, \quad y(2) = 3$ **60.** $\frac{dy}{dx} = \sqrt{1 + x^2}, \quad y(1) = -2$

Theory and Examples

61. Archimedes' area formula for parabolic arches Archimedes (287–212 B.C.), inventor, military engineer, physicist, and the greatest mathematician of classical times in the Western world, discovered that the area under a parabolic arch is two-thirds the base times the height. Sketch the parabolic arch $y = h - (4h/b^2)x^2$, $-b/2 \le x \le b/2$, assuming that h and b are positive. Then use calculus to find the area of the region enclosed between the arch and the x-axis.

62. Show that if k is a positive constant, then the area between the x-axis and one arch of the curve $y = \sin kx$ is $2/k$.

63. The gradient of a curve at any point is $x^2 - 4x + 3$ and the curve passes through (3, 1). Find the area enclosed by this curve, the x-axis and the maximum and minimum ordinates.

In Exercises 64–66, guess an antiderivative and validate your guess by differentiation. (*Hint:* Keep the Chain Rule in mind when trying to guess an antiderivative. You will learn how to find such antiderivatives in the next section.)

64. Revenue from marginal revenue Suppose that a company's marginal revenue from the manufacture and sale of eggbeaters is

$$\frac{dr}{dx} = 2 - 2/(x + 1)^2,$$

where r is measured in thousands of dollars and x in thousands of units. How much money should the company expect from a production run of $x = 3$ thousand eggbeaters? To find out, integrate the marginal revenue from $x = 0$ to $x = 3$.

65. The temperature T(°F) of a room at time t minutes is given by

$$T = 85 - 3\sqrt{25 - t} \quad \text{for} \quad 0 \le t \le 25.$$

a. Find the room's temperature when $t = 0$, $t = 16$, and $t = 25$.

b. Find the room's average temperature for $0 \le t \le 25$.

66. The height H(ft) of a palm tree after growing for t years is given by

$$H = \sqrt{t + 1} + 5t^{1/3} \quad \text{for} \quad 0 \le t \le 8.$$

a. Find the tree's height when $t = 0$, $t = 4$, and $t = 8$.

b. Find the tree's average height for $0 \le t \le 8$.

67. Suppose that $\int_1^x f(t)\,dt = x^2 - 2x + 1$. Find $f(x)$.

68. Find $f(4)$ if $\int_0^x f(t)\,dt = x \cos \pi x$.

69. If $\int_0^x f(t)dt = x + \int_x^0 f(t)dt$,

then find the value of $f(1)$.

70. Suppose $\int_0^x f(t)dt = x^2(1 + x),\ x \ge 0$

find the value of $f(2)$.

71. Suppose that f has a positive derivative for all values of x and that $f(1) = 0$. Which of the following statements must be true of the function

$$g(x) = \int_0^x f(t)\,dt?$$

Give reasons for your answers.

a. g is a differentiable function of x.

b. g is a continuous function of x.

c. The graph of g has a horizontal tangent line at $x = 1$.

d. g has a local maximum at $x = 1$.

e. g has a local minimum at $x = 1$.

f. The graph of g has an inflection point at $x = 1$.

g. The graph of dg/dx crosses the x-axis at $x = 1$.

72. Another proof of the Evaluation Theorem

a. Let $a = x_0 < x_1 < x_2 \cdots < x_n = b$ be any partition of $[a, b]$, and let F be any antiderivative of f. Show that

$$F(b) - F(a) = \sum_{i=1}^{n} [F(x_i) - F(x_{i-1})].$$

b. Apply the Mean Value Theorem to each term to show that $F(x_i) - F(x_{i-1}) = f(c_i)(x_i - x_{i-1})$ for some c_i in the interval (x_{i-1}, x_i). Then show that $F(b) - F(a)$ is a Riemann sum for f on $[a, b]$.

c. From part (b) and the definition of the definite integral, show that

$$F(b) - F(a) = \int_a^b f(x)\,dx.$$

73. Suppose that f is the differentiable function shown in the accompanying graph and that the position at time t (in seconds) of a particle moving along a coordinate axis is

$$s = \int_0^t f(x)\,dx$$

meters. Use the graph to answer the following questions. Give reasons for your answers.

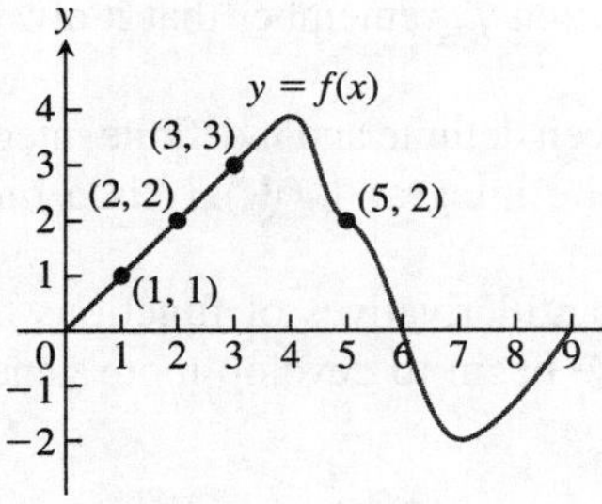

a. What is the particle's velocity at time $t = 5$?

b. Is the acceleration of the particle at time $t = 5$ positive or negative?

c. What is the particle's position at time $t = 3$?

d. At what time during the first 9 sec does s have its largest value?

e. Approximately when is the acceleration zero?

f. When is the particle moving toward the origin? Away from the origin?

g. On which side of the origin does the particle lie at time $t = 9$?

74. Find $\lim_{x\to\infty} \frac{1}{\sqrt{x}} \int_1^x \frac{dt}{\sqrt{t}}$.

COMPUTER EXPLORATIONS

In Exercises 75–78, let $F(x) = \int_a^x f(t)\,dt$ for the specified function f and interval $[a, b]$. Use a CAS to perform the following steps and answer the questions posed.

a. Plot the functions f and F together over $[a, b]$.

b. Solve the equation $F'(x) = 0$. What can you see to be true about the graphs of f and F at points where $F'(x) = 0$? Is your observation borne out by Part 1 of the Fundamental Theorem coupled with information provided by the first derivative? Explain your answer.

c. Over what intervals (approximately) is the function F increasing? Decreasing? What is true about f over those intervals?

d. Calculate the derivative f' and plot it together with F. What can you see to be true about the graph of F at points where $f'(x) = 0$? Is your observation borne out by Part 1 of the Fundamental Theorem? Explain your answer.

75. $f(x) = x^3 - 4x^2 + 3x, \quad [0, 4]$

76. $f(x) = 2x^4 - 17x^3 + 46x^2 - 43x + 12, \quad \left[0, \frac{9}{2}\right]$

77. $f(x) = \sin 2x \cos \frac{x}{3}, \quad [0, 2\pi]$

78. $f(x) = x \cos \pi x, \quad [0, 2\pi]$

In Exercises 79–82, let $F(x) = \int_a^{u(x)} f(t)\,dt$ for the specified a, u, and f. Use a CAS to perform the following steps and answer the questions posed.

a. Find the domain of F.

b. Calculate $F'(x)$ and determine its zeros. For what points in its domain is F increasing? Decreasing?

c. Calculate $F''(x)$ and determine its zero. Identify the local extrema and the points of inflection of F.

d. Using the information from parts (a)–(c), draw a rough hand-sketch of $y = F(x)$ over its domain. Then graph $F(x)$ on your CAS to support your sketch.

79. $a = 1, \quad u(x) = x^2, \quad f(x) = \sqrt{1 - x^2}$

80. $a = 0, \quad u(x) = x^2, \quad f(x) = \sqrt{1 - x^2}$

81. $a = 0, \quad u(x) = 1 - x, \quad f(x) = x^2 - 2x - 3$

82. $a = 0, \quad u(x) = 1 - x^2, \quad f(x) = x^2 - 2x - 3$

In Exercises 83 and 84, assume that f is continuous and $u(x)$ is twice-differentiable.

83. Calculate $\frac{d}{dx} \int_a^{u(x)} f(t)\,dt$ and check your answer using a CAS.

84. Calculate $\frac{d^2}{dx^2} \int_a^{u(x)} f(t)\,dt$ and check your answer using a CAS.

5.5 Indefinite Integrals and the Substitution Method

The Fundamental Theorem of Calculus says that a definite integral of a continuous function can be computed directly if we can find an antiderivative of the function. In Section 4.7 we defined the **indefinite integral** of the function f with respect to x as the set of *all* antiderivatives of f, symbolized by $\int f(x)\,dx$. Since any two antiderivatives of f differ by a constant, the indefinite integral $\int$ notation means that for any antiderivative F of f,

$$\int f(x)\,dx = F(x) + C,$$

where C is any arbitrary constant. The connection between antiderivatives and the definite integral stated in the Fundamental Theorem now explains this notation:

$$\begin{aligned}\int_a^b f(x)\,dx &= F(b) - F(a) = [F(b) + C] - [F(a) + C] \\ &= \Big[F(x) + C\Big]_a^b = \left[\int f(x)\,dx\right]_a^b.\end{aligned}$$

When finding the indefinite integral of a function f, remember that it always includes an arbitrary constant C.

We must keep in mind the difference between definite and indefinite integrals. A definite integral $\int_a^b f(x)\,dx$ is a *number*. An indefinite integral $\int f(x)\,dx$ is a *function* plus an arbitrary constant C.

So far, we have only been able to find antiderivatives of functions that are clearly recognizable as derivatives. In this section we begin to develop more general techniques for finding antiderivatives of functions.

Substitution: Running the Chain Rule Backwards

If u is a differentiable function of x, and n is any number different from -1, the Chain Rule tells us that

$$\frac{d}{dx}\left(\frac{u^{n+1}}{n+1}\right) = u^n \frac{du}{dx}.$$

From another point of view, this same equation says that $u^{n+1}/(n+1)$ is one of the antiderivatives of the function $u^n\,(du/dx)$. Therefore,

$$\int u^n \frac{du}{dx}\,dx = \frac{u^{n+1}}{n+1} + C. \tag{1}$$

The integral in Equation (1) is equal to the simpler integral

$$\int u^n\,du = \frac{u^{n+1}}{n+1} + C,$$

which suggests that we can substitute the simpler expression du for $(du/dx)\,dx$ when computing an integral. Leibniz, one of the founders of calculus, had the insight that indeed this substitution could be done, leading to the *substitution method* for computing integrals. As with differentials, when computing integrals we have

$$du = \frac{du}{dx}\,dx.$$

EXAMPLE 1 Find the integral $\int (x^3 + x)^5 (3x^2 + 1)\,dx$.

Solution We set $u = x^3 + x$. Then

$$du = \frac{du}{dx}\,dx = (3x^2 + 1)\,dx,$$

so that by substitution we have

$$\begin{aligned} \int (x^3 + x)^5 (3x^2 + 1)\,dx &= \int u^5\,du && \text{Let } u = x^3 + x,\ du = (3x^2 + 1)\,dx. \\ &= \frac{u^6}{6} + C && \text{Integrate with respect to } u. \\ &= \frac{(x^3 + x)^6}{6} + C && \text{Substitute } x^3 + x \text{ for } u. \end{aligned}$$

■

EXAMPLE 2 Find $\int \sqrt{2x + 1}\,dx$.

Solution The integral does not fit the formula

$$\int u^n\,du,$$

with $u = 2x + 1$ and $n = 1/2$, because

$$du = \frac{du}{dx}\,dx = 2\,dx,$$

which is not precisely dx. The constant factor 2 is missing from the integral. However, we can introduce this factor after the integral sign if we compensate for it by introducing a factor of 1/2 in front of the integral sign. So we write

$$\begin{aligned}
\int \sqrt{2x+1}\,dx &= \frac{1}{2}\int \underbrace{\sqrt{2x+1}}_{u} \cdot \underbrace{2\,dx}_{du} \\
&= \frac{1}{2}\int u^{1/2}\,du && \text{Let } u = 2x+1,\ du = 2\,dx. \\
&= \frac{1}{2}\frac{u^{3/2}}{3/2} + C && \text{Integrate with respect to } u. \\
&= \frac{1}{3}(2x+1)^{3/2} + C. && \text{Substitute } 2x+1 \text{ for } u.
\end{aligned}$$
■

The substitutions in Examples 1 and 2 are instances of the following general rule.

THEOREM 6—The Substitution Rule
If $u = g(x)$ is a differentiable function whose range is an interval I, and f is continuous on I, then

$$\int f(g(x)) \cdot g'(x)\,dx = \int f(u)\,du.$$

Proof By the Chain Rule, $F(g(x))$ is an antiderivative of $f(g(x)) \cdot g'(x)$ whenever F is an antiderivative of f, because

$$\begin{aligned}
\frac{d}{dx}F(g(x)) &= F'(g(x)) \cdot g'(x) && \text{Chain Rule} \\
&= f(g(x)) \cdot g'(x). && F' = f
\end{aligned}$$

If we make the substitution $u = g(x)$, then

$$\begin{aligned}
\int f(g(x))\,g'(x)\,dx &= \int \frac{d}{dx}F(g(x))\,dx \\
&= F(g(x)) + C && \text{Theorem 8 in Chapter 4} \\
&= F(u) + C && u = g(x) \\
&= \int F'(u)\,du && \text{Theorem 8 in Chapter 4} \\
&= \int f(u)\,du. && F' = f
\end{aligned}$$
■

The use of the variable u in the Substitution Rule is traditional (sometimes it is referred to as u-substitution), but any letter can be used, such as v, t, θ and so forth. The rule provides a method for evaluating an integral of the form $\int f(g(x))\,g'(x)\,dx$, given that the conditions of Theorem 6 are satisfied. The primary challenge is deciding what expression involving x to substitute for in the integrand. The following examples give helpful ideas.

The Substitution Method to evaluate $\int f(g(x))g'(x)\,dx$

1. Substitute $u = g(x)$ and $du = (du/dx)\,dx = g'(x)\,dx$ to obtain $\int f(u)\,du$.
2. Integrate with respect to u.
3. Replace u by $g(x)$.

EXAMPLE 3 Find $\int \sec^2(5x+1)\cdot 5\,dx$.

Solution We substitute $u = 5x+1$ and $du = 5\,dx$. Then

$$\begin{aligned}
\int \sec^2(5x+1)\cdot 5\,dx &= \int \sec^2 u\,du && \text{Let } u = 5x+1,\ du = 5\,dx.\\
&= \tan u + C && \frac{d}{du}\tan u = \sec^2 u\\
&= \tan(5x+1) + C. && \text{Substitute } 5x+1 \text{ for } u.
\end{aligned}$$

■

EXAMPLE 4 Find $\int \cos(7\theta+3)\,d\theta$.

Solution We let $u = 7\theta + 3$ so that $du = 7\,d\theta$. There is a factor of 7 in this formula for du, but there is no corresponding 7 preceding $d\theta$ in the integral. We can compensate for this by multiplying and dividing by 7, using the same procedure as in Example 2. Then

$$\begin{aligned}
\int \cos(7\theta+3)\,d\theta &= \frac{1}{7}\int \cos(7\theta+3)\cdot 7\,d\theta && \text{Place factor } 1/7 \text{ in front of integral.}\\
&= \frac{1}{7}\int \cos u\,du && \text{Substitute } u = 7\theta+3,\ du = 7\,d\theta.\\
&= \frac{1}{7}\sin u + C && \text{Integrate.}\\
&= \frac{1}{7}\sin(7\theta+3) + C. && \text{Replace } u \text{ by } 7\theta+3.
\end{aligned}$$

There is another approach to this problem. With $u = 7\theta + 3$ and $du = 7\,d\theta$ as before, we solve for $d\theta$ to obtain $d\theta = (1/7)\,du$. Then the integral becomes

$$\begin{aligned}
\int \cos(7\theta+3)\,d\theta &= \int \cos u \cdot \frac{1}{7}\,du && \text{Substitute } u = 7\theta+3,\ du = 7\,d\theta, \text{ and } d\theta = (1/7)\,du.\\
&= \frac{1}{7}\sin u + C && \text{Integrate.}\\
&= \frac{1}{7}\sin(7\theta+3) + C. && \text{Replace } u \text{ by } 7\theta+3.
\end{aligned}$$

We can verify this solution by differentiating and checking that we obtain the original function $\cos(7\theta+3)$. ■

EXAMPLE 5 Sometimes we observe that a power of x appears in the integrand that is one less than the power of x appearing in the argument of a function we want to integrate. This observation immediately suggests we try a substitution for the higher power of x. This situation occurs in the following integration.

$$\begin{aligned}
\int x^2 \cos x^3\, dx &= \int \cos x^3 \cdot x^2\, dx \\
&= \int \cos u \cdot \frac{1}{3}\, du && \text{Let } u = x^3,\ du = 3x^2\, dx,\ (1/3)\, du = x^2\, dx. \\
&= \frac{1}{3}\int \cos u\, du \\
&= \frac{1}{3}\sin u + C && \text{Integrate with respect to } u. \\
&= \frac{1}{3}\sin x^3 + C && \text{Replace } u \text{ by } x^3.
\end{aligned}$$

■

HISTORICAL BIOGRAPHY
George David Birkhoff
(1884–1944)
www.bit.ly/2O2MRbz

It may happen that an extra factor of x appears in the integrand when we try a substitution $u = g(x)$. In that case, it may be possible to solve the equation $u = g(x)$ for x in terms of u. Replacing the extra factor of x with that expression may then result in an integral that we can evaluate. Here is an example of this situation.

EXAMPLE 6 Evaluate $\int x\sqrt{2x+1}\, dx$.

Solution Our previous experience with the integral in Example 2 suggests the substitution $u = 2x + 1$ with $du = 2\, dx$. Then

$$\sqrt{2x+1}\, dx = \frac{1}{2}\sqrt{u}\, du.$$

However, in this example the integrand contains an extra factor of x that multiplies the factor $\sqrt{2x+1}$. To adjust for this, we solve the substitution equation $u = 2x + 1$ for x to obtain $x = (u-1)/2$ and find that

$$x\sqrt{2x+1}\, dx = \frac{1}{2}(u-1) \cdot \frac{1}{2}\sqrt{u}\, du.$$

The integration now becomes

$$\begin{aligned}
\int x\sqrt{2x+1}\, dx &= \frac{1}{4}\int (u-1)\sqrt{u}\, du = \frac{1}{4}\int (u-1)\, u^{1/2}\, du && \text{Substitute.} \\
&= \frac{1}{4}\int (u^{3/2} - u^{1/2})\, du && \text{Multiply terms by } u^{1/2}. \\
&= \frac{1}{4}\left(\frac{2}{5}u^{5/2} - \frac{2}{3}u^{3/2}\right) + C && \text{Integrate.} \\
&= \frac{1}{10}(2x+1)^{5/2} - \frac{1}{6}(2x+1)^{3/2} + C. && \text{Replace } u \text{ by } 2x+1.
\end{aligned}$$

■

EXAMPLE 7 Sometimes we can use trigonometric identities to transform an integral we do not know how to evaluate into one that we can evaluate using the Substitution Rule.

(a)
$$\begin{aligned}
\int \sin^2 x\, dx &= \int \frac{1-\cos 2x}{2}\, dx && \sin^2 x = \frac{1-\cos 2x}{2} \\
&= \frac{1}{2}\int (1-\cos 2x)\, dx \\
&= \frac{1}{2}x - \frac{1}{2}\frac{\sin 2x}{2} + C = \frac{x}{2} - \frac{\sin 2x}{4} + C
\end{aligned}$$

(b)
$$\int \cos^2 x\, dx = \int \frac{1+\cos 2x}{2}\, dx = \frac{x}{2} + \frac{\sin 2x}{4} + C \qquad \cos^2 x = \frac{1+\cos 2x}{2}$$

(c) $$\begin{aligned}\int(1-2\sin^2 x)\sin 2x\,dx &= \int(\cos^2 x-\sin^2 x)\sin 2x\,dx \\ &= \int \cos 2x \sin 2x\,dx && \cos 2x=\cos^2 x-\sin^2 x \\ &= \int \frac{1}{2}\sin 4x\,dx = \int \frac{1}{8}\sin u\,du && u=4x,\,du=4\,dx \\ &= -\frac{1}{8}\cos 4x + C.\end{aligned}$$

Trying Different Substitutions

The success of the substitution method depends on finding a substitution that changes an integral we cannot directly evaluate into one that we can. Finding the right substitution gets easier with practice and experience. If your first substitution fails, try another substitution, possibly coupled with other algebraic or trigonometric simplifications to the integrand. Several more complicated types of substitutions will be studied in Chapter 8.

EXAMPLE 8 Evaluate $\int \frac{2z\,dz}{\sqrt[3]{z^2+1}}$.

Solution We use the substitution method of integration as an exploratory tool. First we substitute for the most troublesome part of the integrand and see how things work out. For the integral here, we might try $u = z^2 + 1$, or we might even press our luck and take u to be the entire cube root. In this example both substitutions turn out to be successful, but that is not always the case. If one substitution does not help, a different substitution may work instead.

Method 1: Substitute $u = z^2 + 1$.

$$\begin{aligned}\int \frac{2z\,dz}{\sqrt[3]{z^2+1}} &= \int \frac{du}{u^{1/3}} && \text{Let } u=z^2+1,\ du=2z\,dz. \\ &= \int u^{-1/3}\,du && \text{In the form } \textstyle\int u^n\,du \\ &= \frac{u^{2/3}}{2/3}+C && \text{Integrate.} \\ &= \frac{3}{2}u^{2/3}+C \\ &= \frac{3}{2}(z^2+1)^{2/3}+C && \text{Replace } u \text{ by } z^2+1.\end{aligned}$$

Method 2: Substitute $u = \sqrt[3]{z^2+1}$ instead.

$$\begin{aligned}\int \frac{2z\,dz}{\sqrt[3]{z^2+1}} &= \int \frac{3u^2\,du}{u} && \text{Let } u=\sqrt[3]{z^2+1},\ u^3=z^2+1,\ 3u^2\,du=2z\,dz. \\ &= 3\int u\,du \\ &= 3\cdot\frac{u^2}{2}+C && \text{Integrate.} \\ &= \frac{3}{2}(z^2+1)^{2/3}+C && \text{Replace } u \text{ by } (z^2+1)^{1/3}.\end{aligned}$$

EXERCISES 5.5

Evaluating Indefinite Integrals

In Exercises 1–16, make the given substitutions to evaluate the indefinite integrals.

1. $\int 2(2x+4)^5\,dx, \quad u = 2x+4$

2. $\int 7\sqrt{7x-1}\,dx, \quad u = 7x-1$

3. $\int 2x(x^2+5)^{-4}\,dx, \quad u = x^2+5$

4. $\int \frac{4x^3}{(x^4+1)^2}\,dx, \quad u = x^4+1$

5. $\int (4x^3+3x^4)^5(x^2+x^3)\,dx, \; u = 4x^3+3x^4$

6. $\int \frac{(1+\sqrt{x})^{1/3}}{\sqrt{x}}\,dx, \quad u = 1+\sqrt{x}$

7. $\int \cos 6x\,dx, \; u = 6x$

8. $\int x\sin(2x^2)\,dx, \quad u = 2x^2$

9. $\int \sec 2t\tan 2t\,dt, \quad u = 2t$

10. $\int \left(1-\cos\frac{t}{2}\right)^2 \sin\frac{t}{2}\,dt, \quad u = 1-\cos\frac{t}{2}$

11. $\int \frac{9r^2\,dr}{\sqrt{1-r^3}}, \quad u = 1-r^3$

12. $\int 12(y^4+4y^2+1)^2(y^3+2y)\,dy, \quad u = y^4+4y^2+1$

13. $\int \sqrt{x}\sin^2(x^{3/2}-1)\,dx, \quad u = x^{3/2}-1$

14. $\int \frac{1}{x^2}\cos^2\left(\frac{1}{x}\right)dx, \quad u = -\frac{1}{x}$

15. $\int \csc^2 2\theta\cot 2\theta\,d\theta$

   a. Using $u = \cot 2\theta$ b. Using $u = \csc 2\theta$

16. $\int \frac{dx}{\sqrt{5x+8}}$

   a. Using $u = 5x+8$ b. Using $u = \sqrt{5x+8}$

Evaluate the integrals in Exercises 17–50.

17. $\int \frac{x^2}{\sqrt{x+5}}\,dx$

18. $\int \sqrt{1+\sin 2x}\,dx$

19. $\int \theta\sqrt[4]{1-\theta^2}\,d\theta$

20. $\int 3y\sqrt{7-3y^2}\,dy$

21. $\int \frac{1}{\sqrt{x}(1+\sqrt{x})^2}\,dx$

22. $\int \sqrt{\sin x}\cos^3 x\,dx$

23. $\int \sec^2(3x+2)\,dx$

24. $\int \tan^2 x\sec^2 x\,dx$

25. $\int \sin^5\frac{x}{3}\cos\frac{x}{3}\,dx$

26. $\int \tan^7\frac{x}{2}\sec^2\frac{x}{2}\,dx$

27. $\int r^2\left(\frac{r^3}{18}-1\right)^5 dr$

28. $\int r^4\left(7-\frac{r^5}{10}\right)^3 dr$

29. $\int x^{1/2}\sin(x^{3/2}+1)\,dx$

30. $\int \csc\left(\frac{v-\pi}{2}\right)\cot\left(\frac{v-\pi}{2}\right)dv$

31. $\int \frac{\sin(2t+1)}{\cos^2(2t+1)}\,dt$

32. $\int \frac{\sec z\tan z}{\sqrt{\sec z}}\,dz$

33. $\int \frac{1}{t^2}\cos\left(\frac{1}{t}-1\right)dt$

34. $\int \frac{1}{\sqrt{t}}\cos(\sqrt{t}+3)\,dt$

35. $\int \frac{1}{\theta^2}\sin\frac{1}{\theta}\cos\frac{1}{\theta}\,d\theta$

36. $\int \frac{\cos\sqrt{\theta}}{\sqrt{\theta}\sin^2\sqrt{\theta}}\,d\theta$

37. $\int \frac{x}{\sqrt{1+x}}\,dx$

38. $\int \sqrt{\frac{x-1}{x^5}}\,dx$

39. $\int \frac{1}{x^2}\sqrt{2-\frac{1}{x}}\,dx$

40. $\int \frac{1}{x^3}\sqrt{\frac{x^2-1}{x^2}}\,dx$

41. $\int \sqrt{\frac{x^3-3}{x^{11}}}\,dx$

42. $\int \sqrt{\frac{x^4}{x^3-1}}\,dx$

43. $\int x(x-1)^{10}\,dx$

44. $\int x\sqrt{4-x}\,dx$

45. $\int (x+1)^2(1-x)^5\,dx$

46. $\int (x+5)(x-5)^{1/3}\,dx$

47. $\int x^3\sqrt{x^2+1}\,dx$

48. $\int 3x^5\sqrt{x^3+1}\,dx$

49. $\int \frac{x}{(x^2-4)^3}\,dx$

50. $\int \frac{x}{(2x-1)^{2/3}}\,dx$

If you do not know what substitution to make, try reducing the integral step by step, using a trial substitution to simplify the integral a bit and then another to simplify it some more. You will see what we mean if you try the sequences of substitutions in Exercises 51 and 52.

51. $\int \frac{18\tan^2 x\sec^2 x}{(2+\tan^3 x)^2}\,dx$

   a. $u = \tan x$, followed by $v = u^3$, then by $w = 2+v$

   b. $u = \tan^3 x$, followed by $v = 2+u$

   c. $u = 2+\tan^3 x$

52. $\int \sqrt{1+\sin^2(x-1)}\sin(x-1)\cos(x-1)\,dx$

   a. $u = x-1$, followed by $v = \sin u$, then by $w = 1+v^2$

   b. $u = \sin(x-1)$, followed by $v = 1+u^2$

   c. $u = 1+\sin^2(x-1)$

Evaluate the integrals in Exercises 53 and 54.

53. $\displaystyle\int \frac{(2r-1)\cos\sqrt{3(2r-1)^2+6}}{\sqrt{3(2r-1)^2+6}}\,dr$

54. $\displaystyle\int \frac{\sin\sqrt{\theta}}{\sqrt{\theta}\cos^3\sqrt{\theta}}\,d\theta$

Initial Value Problems

Solve the initial value problems in Exercises 55–60.

55. $\dfrac{ds}{dt} = 12t\,(3t^2-1)^3,\quad s(1) = 3$

56. $\dfrac{dy}{dx} = 4x\,(x^2+8)^{-1/3},\quad y(0) = 0$

57. $\dfrac{ds}{dt} = 8\sin^2\left(t + \dfrac{\pi}{12}\right),\quad s(0) = 8$

58. $\dfrac{dr}{d\theta} = 3\cos^2\left(\dfrac{\pi}{4} - \theta\right),\quad r(0) = \dfrac{\pi}{8}$

59. $\dfrac{d^2s}{dt^2} = -4\sin\left(2t - \dfrac{\pi}{2}\right),\quad s'(0) = 100,\quad s(0) = 0$

60. $\dfrac{d^2y}{dx^2} = 4\sec^2 2x\tan 2x,\quad y'(0) = 4,\quad y(0) = -1$

61. The velocity of a particle moving back and forth on a line is $v = ds/dt = 6\sin 2t$ m/sec for all t. If $s = 0$ when $t = 0$, find the value of s when $t = \pi/2$ sec.

62. The acceleration of a particle moving back and forth on a line is $a = d^2s/dt^2 = \pi^2\cos\pi t$ m/sec^2 for all t. If $s = 0$ and $v = 8$ m/sec when $t = 0$, find s when $t = 1$ sec.

5.6 Definite Integral Substitutions and the Area Between Curves

There are two methods for evaluating a definite integral by substitution. One method is to find an antiderivative using substitution and then to evaluate the definite integral by applying the Evaluation Theorem. The other method extends the process of substitution directly to *definite* integrals by changing the limits of integration. We will use these methods to compute the area between two curves.

The Substitution Formula

The following formula shows how the limits of integration change when we apply a substitution to an integral.

THEOREM 7—Substitution in Definite Integrals
If g' is continuous on the interval $[a, b]$ and f is continuous on the range of $g(x) = u$, then

$$\int_a^b f(g(x))\cdot g'(x)\,dx = \int_{g(a)}^{g(b)} f(u)\,du.$$

Proof Let F denote any antiderivative of f. Then

$$\begin{aligned}\int_a^b f(g(x))\cdot g'(x)\,dx &= F(g(x))\Big]_{x=a}^{x=b} && \frac{d}{dx}F(g(x)) = F'(g(x))g'(x) = f(g(x))g'(x)\\ &= F(g(b)) - F(g(a))\\ &= F(u)\Big]_{u=g(a)}^{u=g(b)}\\ &= \int_{g(a)}^{g(b)} f(u)\,du. && \text{Fundamental Theorem, Part 2}\end{aligned}$$

■

To use Theorem 7, we make the same u-substitution $u = g(x)$ and $du = g'(x)\,dx$ that we would use to evaluate the corresponding indefinite integral. We then integrate the transformed integral with respect to u from the value $g(a)$ (the value of u at $x = a$) to the value $g(b)$ (the value of u at $x = b$).

EXAMPLE 1 Evaluate $\int_{-1}^{1} 3x^2\sqrt{x^3+1}\,dx$.

Solution We will show how to evaluate the integral using Theorem 7, and how to evaluate it using the original limits of integration.

Method 1: Transform the integral and evaluate the transformed integral with the transformed limits given in Theorem 7.

$$\int_{-1}^{1} 3x^2\sqrt{x^3+1}\,dx = \int_0^2 \sqrt{u}\,du$$

Let $u = x^3 + 1$, $du = 3x^2\,dx$.
When $x = -1$, $u = (-1)^3 + 1 = 0$.
When $x = 1$, $u = (1)^3 + 1 = 2$.

$$= \frac{2}{3}u^{3/2}\Big]_0^2$$

Evaluate the new definite integral.

$$= \frac{2}{3}\left[2^{3/2} - 0^{3/2}\right] = \frac{2}{3}\left[2\sqrt{2}\right] = \frac{4\sqrt{2}}{3}$$

Method 2: Transform the integral as an indefinite integral, integrate, change back to x, and use the original x-limits.

$$\int 3x^2\sqrt{x^3+1}\,dx = \int \sqrt{u}\,du$$

Let $u = x^3 + 1$, $du = 3x^2\,dx$.

$$= \frac{2}{3}u^{3/2} + C$$

Integrate with respect to u.

$$= \frac{2}{3}(x^3+1)^{3/2} + C$$

Replace u by $x^3 + 1$.

$$\int_{-1}^{1} 3x^2\sqrt{x^3+1}\,dx = \frac{2}{3}(x^3+1)^{3/2}\Big]_{-1}^{1}$$

Use the integral just found, with limits of integration for x.

$$= \frac{2}{3}\left[\left((1)^3+1\right)^{3/2} - \left((-1)^3+1\right)^{3/2}\right]$$

$$= \frac{2}{3}\left[2^{3/2} - 0^{3/2}\right] = \frac{2}{3}\left[2\sqrt{2}\right] = \frac{4\sqrt{2}}{3}$$ ■

Which method is better—evaluating the transformed definite integral with transformed limits using Theorem 7, or transforming the integral, integrating, and transforming back to use the original limits of integration? In Example 1, the first method seems easier, but that is not always the case. Generally, it is best to know both methods and to use whichever one seems better at the time.

EXAMPLE 2 We use the method of transforming the limits of integration.

(a) $$\int_{\pi/4}^{\pi/2} \cot\theta\csc^2\theta\,d\theta = \int_1^0 u\cdot(-du)$$

Let $u = \cot\theta$, $du = -\csc^2\theta\,d\theta$,
$-du = \csc^2\theta\,d\theta$.
When $\theta = \pi/4$, $u = \cot(\pi/4) = 1$.
When $\theta = \pi/2$, $u = \cot(\pi/2) = 0$.

$$= -\int_1^0 u\,du$$

$$= -\left[\frac{u^2}{2}\right]_1^0$$

$$= -\left[\frac{(0)^2}{2} - \frac{(1)^2}{2}\right] = \frac{1}{2}$$

(b) $$\int_0^{\pi/2} \frac{2\sin x \cos x}{(1+\sin^2 x)^3}\,dx = \int_1^2 \frac{1}{u^3}\,du$$ Let $u = 1+\sin^2 x$, $du = 2\sin x\cos x\,dx$. When $x = 0$, $u = 1$. When $x = \pi/2$, $u = 2$.

$$= -\frac{1}{2u^2}\bigg]_1^2$$

$$= -\frac{1}{8} - \left(-\frac{1}{2}\right) = \frac{3}{8}$$ ■

Definite Integrals of Symmetric Functions

The Substitution Formula in Theorem 7 simplifies the calculation of definite integrals of even and odd functions (Section 1.1) over a symmetric interval $[-a, a]$ (Figure 5.24).

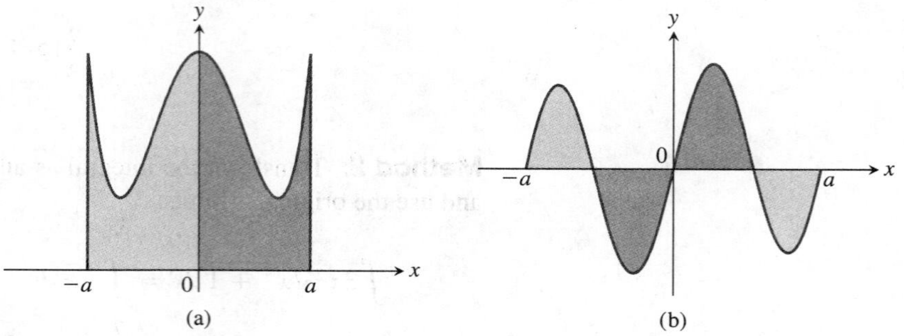

FIGURE 5.24 (a) For f an even function, the integral from $-a$ to a is twice the integral from 0 to a. (b) For f an odd function, the integral from $-a$ to a equals 0.

THEOREM 8 Let f be continuous on the symmetric interval $[-a, a]$.

(a) If f is even, then $\int_{-a}^{a} f(x)\,dx = 2\int_0^a f(x)\,dx$.

(b) If f is odd, then $\int_{-a}^{a} f(x)\,dx = 0$.

Proof of Part (a)

$$\int_{-a}^{a} f(x)\,dx = \int_{-a}^{0} f(x)\,dx + \int_0^a f(x)\,dx$$ Additivity Rule for Definite Integrals

$$= -\int_0^{-a} f(x)\,dx + \int_0^a f(x)\,dx$$ Order of Integration Rule

$$= -\int_0^a f(-u)(-du) + \int_0^a f(x)\,dx$$ Let $u = -x$, $du = -dx$. When $x = 0$, $u = 0$. When $x = -a$, $u = a$.

$$= \int_0^a f(-u)\,du + \int_0^a f(x)\,dx$$

$$= \int_0^a f(u)\,du + \int_0^a f(x)\,dx$$ f is even, so $f(-u) = f(u)$.

$$= 2\int_0^a f(x)\,dx$$

The proof of part (b) is similar, and you are asked to give it in Exercise 90. ■

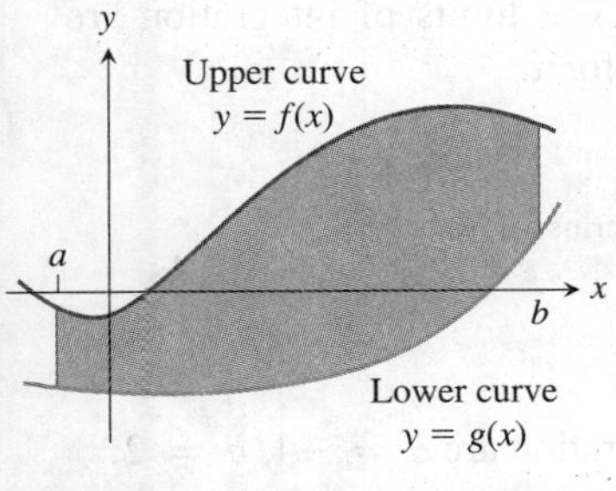

FIGURE 5.25 The region between the curves $y = f(x)$ and $y = g(x)$ and the lines $x = a$ and $x = b$.

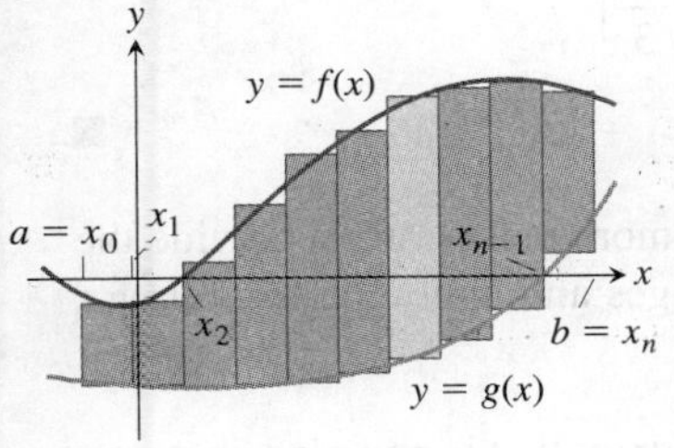

FIGURE 5.26 We approximate the region with rectangles perpendicular to the x-axis.

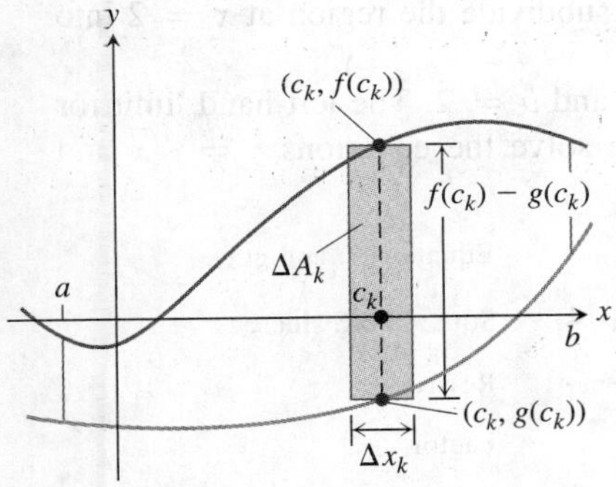

FIGURE 5.27 The area ΔA_k of the kth rectangle is the product of its height, $f(c_k) - g(c_k)$, and its width, Δx_k.

EXAMPLE 3 Evaluate $\int_{-2}^{2} x^4 - 4x^2 + 6\,dx$.

Solution Since $f(x) = x^4 - 4x^2 + 6$ satisfies $f(-x) = f(x)$, it is even on the symmetric interval $[-2, 2]$, so

$$\begin{aligned}\int_{-2}^{2} x^4 - 4x^2 + 6\,dx &= 2\int_{0}^{2} x^4 - 4x^2 + 6\,dx \\ &= 2\left[\frac{x^5}{5} - \frac{4}{3}x^3 + 6x\right]_0^2 \\ &= 2\left(\frac{32}{5} - \frac{32}{3} + 12\right) = \frac{232}{15}.\end{aligned}$$

Areas Between Curves

Suppose we want to find the area of a region that is bounded above by the curve $y = f(x)$, below by the curve $y = g(x)$, and on the left and right by the lines $x = a$ and $x = b$ (Figure 5.25). The region might accidentally have a shape whose area we could find with geometry, but if f and g are arbitrary continuous functions, we usually have to find the area by computing an integral.

To see what the integral should be, we first approximate the region with n vertical rectangles based on a partition $P = \{x_0, x_1, \ldots, x_n\}$ of $[a, b]$ (Figure 5.26). The area of the kth rectangle (Figure 5.27) is

$$\Delta A_k = \text{height} \times \text{width} = [f(c_k) - g(c_k)]\,\Delta x_k.$$

We then approximate the area of the region by adding the areas of the n rectangles:

$$A \approx \sum_{k=1}^{n} \Delta A_k = \sum_{k=1}^{n} [f(c_k) - g(c_k)]\,\Delta x_k. \qquad \text{Riemann sum}$$

As $\|P\| \to 0$, the sums on the right approach the limit $\int_a^b [f(x) - g(x)]\,dx$ because f and g are continuous. The area of the region is defined to be the value of this integral. That is,

$$A = \lim_{\|P\| \to 0} \sum_{k=1}^{n} [f(c_k) - g(c_k)]\,\Delta x_k = \int_a^b [f(x) - g(x)]\,dx.$$

DEFINITION If f and g are continuous with $f(x) \geq g(x)$ throughout $[a, b]$, then the **area of the region between the curves $y = f(x)$ and $y = g(x)$ from a to b** is the integral of $(f - g)$ from a to b:

$$A = \int_a^b [f(x) - g(x)]\,dx.$$

When applying this definition it is usually helpful to graph the curves. The graph reveals which curve is the upper curve f and which is the lower curve g. It also helps you find the limits of integration if they are not given. You may need to find where the curves intersect to determine the limits of integration, and this may involve solving the equation $f(x) = g(x)$ for values of x. Then you can integrate the function $f - g$ for the area between the intersections.

EXAMPLE 4 Find the area of the region enclosed by the parabola $y = 2 - x^2$ and the line $y = -x$.

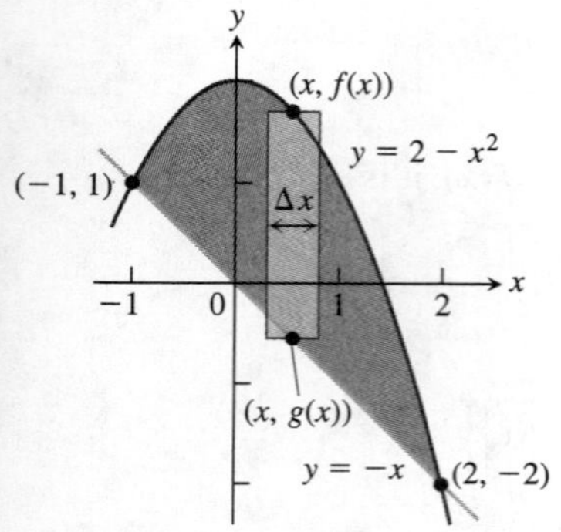

FIGURE 5.28 The region in Example 5 with a typical approximating rectangle from a Riemann sum.

Solution First we sketch the two curves (Figure 5.28). The limits of integration are found by solving $y = 2 - x^2$ and $y = -x$ simultaneously for x.

$$2 - x^2 = -x \qquad \text{Equate } f(x) \text{ and } g(x).$$
$$x^2 - x - 2 = 0 \qquad \text{Rearrange terms.}$$
$$(x + 1)(x - 2) = 0 \qquad \text{Factor.}$$
$$x = -1, \quad x = 2. \qquad \text{Solve.}$$

The region runs from $x = -1$ to $x = 2$. The limits of integration are $a = -1$, $b = 2$.

The area between the curves is

$$A = \int_a^b [f(x) - g(x)]\,dx = \int_{-1}^2 [(2 - x^2) - (-x)]\,dx$$
$$= \int_{-1}^2 (2 + x - x^2)\,dx = \left[2x + \frac{x^2}{2} - \frac{x^3}{3}\right]_{-1}^2$$
$$= \left(4 + \frac{4}{2} - \frac{8}{3}\right) - \left(-2 + \frac{1}{2} + \frac{1}{3}\right) = \frac{9}{2}.$$ ■

If the formula for a bounding curve changes at one or more points, we subdivide the region into subregions that correspond to the formula changes and apply the formula for the area between curves to each subregion.

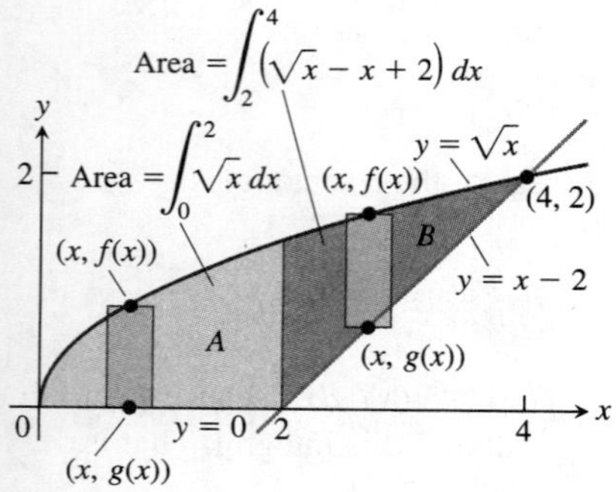

FIGURE 5.29 When the formula for a bounding curve changes, the area integral changes to become the sum of integrals to match, one integral for each of the shaded regions shown here for Example 6.

EXAMPLE 5 Find the area of the region in the first quadrant that is bounded above by $y = \sqrt{x}$ and below by the x-axis and the line $y = x - 2$.

Solution Figure 5.29 shows that the region's upper boundary is the graph of $f(x) = \sqrt{x}$. The lower boundary changes from $g(x) = 0$ for $0 \le x \le 2$ to $g(x) = x - 2$ for $2 \le x \le 4$ (both formulas agree at $x = 2$). We subdivide the region at $x = 2$ into subregions A and B, shown in Figure 5.29.

The limits of integration for region A are $a = 0$ and $b = 2$. The left-hand limit for region B is $a = 2$. To find the right-hand limit, we solve the equations $y = \sqrt{x}$ and $y = x - 2$ simultaneously for x:

$$\sqrt{x} = x - 2 \qquad \text{Equate } f(x) \text{ and } g(x).$$
$$x = (x - 2)^2 = x^2 - 4x + 4 \qquad \text{Square both sides.}$$
$$x^2 - 5x + 4 = 0 \qquad \text{Rewrite.}$$
$$(x - 1)(x - 4) = 0 \qquad \text{Factor.}$$
$$x = 1, \quad x = 4. \qquad \text{Solve.}$$

Only the value $x = 4$ satisfies the equation $\sqrt{x} = x - 2$. The value $x = 1$ is an extraneous root introduced by squaring. The right-hand limit is $b = 4$.

$$\text{For } 0 \le x \le 2: \quad f(x) - g(x) = \sqrt{x} - 0 = \sqrt{x}$$
$$\text{For } 2 \le x \le 4: \quad f(x) - g(x) = \sqrt{x} - (x - 2) = \sqrt{x} - x + 2$$

We add the areas of subregions A and B to find the total area:

$$\text{Total area} = \underbrace{\int_0^2 \sqrt{x}\,dx}_{\text{area of } A} + \underbrace{\int_2^4 (\sqrt{x} - x + 2)\,dx}_{\text{area of } B}$$
$$= \left[\frac{2}{3}x^{3/2}\right]_0^2 + \left[\frac{2}{3}x^{3/2} - \frac{x^2}{2} + 2x\right]_2^4$$
$$= \frac{2}{3}(2)^{3/2} - 0 + \left(\frac{2}{3}(4)^{3/2} - 8 + 8\right) - \left(\frac{2}{3}(2)^{3/2} - 2 + 4\right)$$
$$= \frac{2}{3}(8) - 2 = \frac{10}{3}.$$ ■

Integration with Respect to *y*

If a region's bounding curves are described by functions of y, the approximating rectangles are horizontal instead of vertical, and the basic formula has y in place of x.

To find the areas of regions like these:

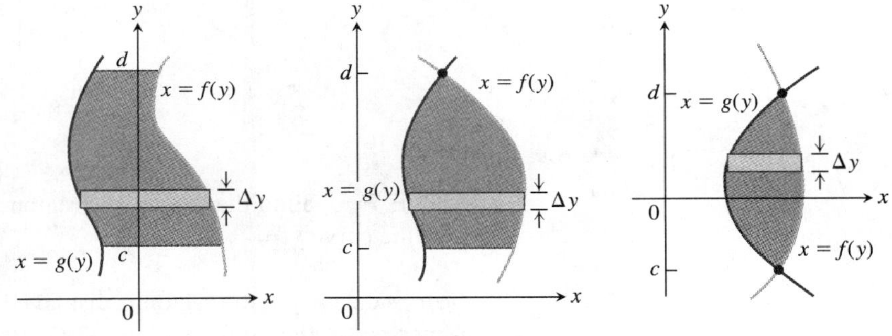

use the formula

$$A = \int_c^d [f(y) - g(y)]\,dy.$$

In this equation f always denotes the right-hand curve and g the left-hand curve, so $f(y) - g(y)$ is nonnegative.

EXAMPLE 6 Find the area of the region in Example 5 by integrating with respect to y.

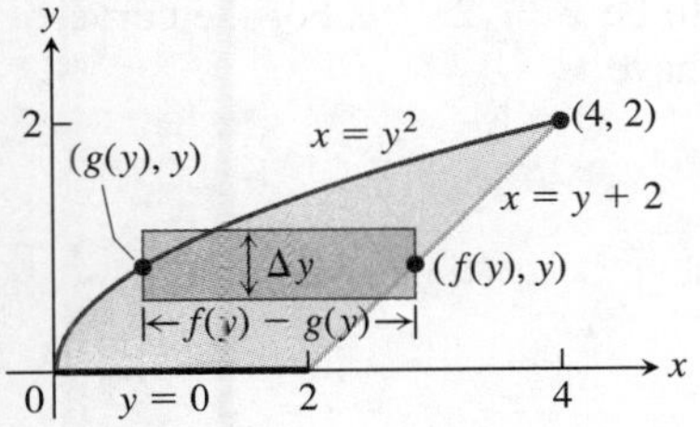

FIGURE 5.30 It takes two integrations to find the area of this region if we integrate with respect to x. It takes only one if we integrate with respect to y (Example 7).

Solution We first sketch the region and a typical *horizontal* rectangle based on a partition of an interval of y-values (Figure 5.30). The region's right-hand boundary is the line $x = y + 2$, so $f(y) = y + 2$. The left-hand boundary is the curve $x = y^2$, so $g(y) = y^2$. The lower limit of integration is $y = 0$. We find the upper limit by solving $x = y + 2$ and $x = y^2$ simultaneously for y:

$$\begin{aligned} y + 2 &= y^2 && \text{Equate } f(y) = y + 2 \text{ and } g(y) = y^2. \\ y^2 - y - 2 &= 0 && \text{Rewrite.} \\ (y + 1)(y - 2) &= 0 && \text{Factor.} \\ y = -1, \quad y &= 2. && \text{Solve.} \end{aligned}$$

The upper limit of integration is $b = 2$. (The value $y = -1$ gives a point of intersection *below* the x-axis.)

The area of the region is

$$\begin{aligned} A &= \int_c^d [f(y) - g(y)]\,dy = \int_0^2 [y + 2 - y^2]\,dy \\ &= \int_0^2 [2 + y - y^2]\,dy \\ &= \left[2y + \frac{y^2}{2} - \frac{y^3}{3}\right]_0^2 \\ &= 4 + \frac{4}{2} - \frac{8}{3} = \frac{10}{3}. \end{aligned}$$

This is the result of Example 5, found with less work. ■

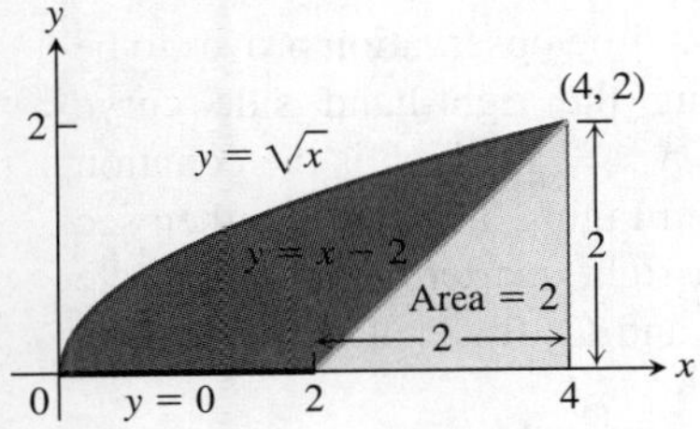

FIGURE 5.31 The area of the black region is the area under the parabola $y = \sqrt{x}$ minus the area of the triangle.

Although it was easier to find the area in Example 5 by integrating with respect to y rather than x (as we did in Example 6), there is an easier way yet. Looking at Figure 5.31, we see that the area we want is the area between the curve $y = \sqrt{x}$ and the x-axis for

$0 \le x \le 4$, *minus* the area of an isosceles triangle of base and height equal to 2. So by combining calculus with some geometry, we find

$$\begin{aligned} \text{Area} &= \int_0^4 \sqrt{x}\, dx - \frac{1}{2}(2)(2) \\ &= \frac{2}{3}x^{3/2}\Big]_0^4 - 2 \\ &= \frac{2}{3}(8) - 0 - 2 = \frac{10}{3}. \end{aligned}$$

EXAMPLE 7 Find the area of the region bounded below by the line $y = 2 - x$ and above by the curve $y = \sqrt{2x - x^2}$.

Solution A sketch of the region is displayed in Figure 5.32, and we see that the line and curve intersect at the points (1, 1) and (2, 0). Using vertical rectangles, the area of the region is given by

$$A = \int_1^2 \left(\sqrt{2x - x^2} + x - 2\right) dx.$$

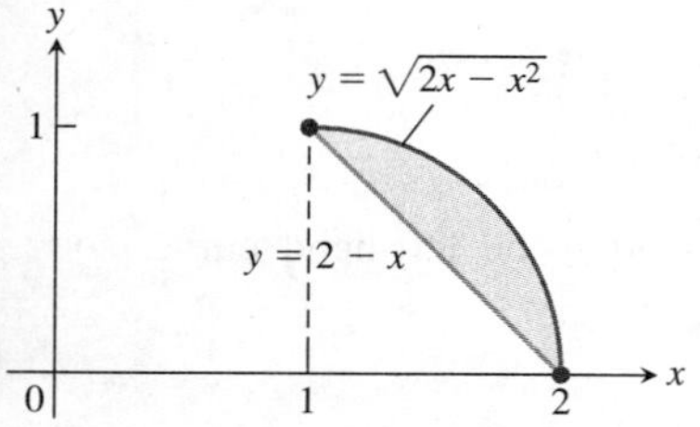

FIGURE 5.32 The region described by the curves in Example 7.

However, we don't know how to find an antiderivative for the term involving the radical, and no simple substitution is apparent.

To use horizontal rectangles, we first need to express each bounding curve as a function of the variable y. The line on the left is easily found to be $x = 2 - y$. For the curve $y = \sqrt{2x - x^2}$ on the right-hand side in Figure 5.32, we have

$$\begin{aligned} y^2 &= 2x - x^2 \\ &= -(x^2 - 2x + 1) + 1 \qquad \text{Complete the square.} \\ &= -(x - 1)^2 + 1. \end{aligned}$$

Solving for x,

$$\begin{aligned} (x - 1)^2 &= 1 - y^2, \\ x &= 1 + \sqrt{1 - y^2}. \qquad x \ge 1, 0 \le y \le 1 \end{aligned}$$

The area of the region is then given by

$$A = \int_0^1 \left[\left(1 + \sqrt{1 - y^2}\right) - (2 - y)\right] dy = \int_0^1 \left(\sqrt{1 - y^2} + y - 1\right) dy.$$

Again, we don't know yet how to integrate the radical term (although we will see how to do that in Section 8.4). We conclude that neither vertical nor horizontal rectangles lead to an integral we currently can evaluate.

Nevertheless, as we found with Example 6, sometimes a little observation proves to be helpful. If we look again at the algebra for expressing the right-hand side curve $y = \sqrt{2x - x^2}$ as a function of y, we see that $(x - 1)^2 + y^2 = 1$, which is the equation of the unit circle with center shifted to the point (1, 0). From Figure 5.32, we can then see that the area of the region we want is the area of the upper right quarter of the unit circle minus the area of the triangle with vertices (1, 1), (1, 0), and (2, 0). That is, the area is given by

$$A = \frac{\pi}{4} - \frac{1}{2} = \frac{\pi - 2}{4} \approx 0.285. \qquad \blacksquare$$

EXERCISES 5.6

Evaluating Definite Integrals

Use the Substitution Formula in Theorem 7 to evaluate the integrals in Exercises 1–24.

1. a. $\int_0^3 \sqrt{y+1}\,dy$ **b.** $\int_{-1}^0 \sqrt{y+1}\,dy$

2. a. $\int_0^1 r\sqrt{1-r^2}\,dr$ **b.** $\int_{-1}^1 r\sqrt{1-r^2}\,dr$

3. a. $\int_0^{\pi/4} \tan x \sec^2 x\,dx$ **b.** $\int_{-\pi/4}^0 \tan x \sec^2 x\,dx$

4. a. $\int_0^{\pi} 3\cos^2 x \sin x\,dx$ **b.** $\int_{2\pi}^{3\pi} 3\cos^2 x \sin x\,dx$

5. a. $\int_0^1 t^3(1+t^4)^3\,dt$ **b.** $\int_{-1}^1 t^3(1+t^4)^3\,dt$

6. a. $\int_0^{\sqrt{7}} t(t^2+1)^{1/3}\,dt$ **b.** $\int_{-\sqrt{7}}^0 t(t^2+1)^{1/3}\,dt$

7. a. $\int_{-1}^1 \frac{5r}{(4+r^2)^2}\,dr$ **b.** $\int_0^1 \frac{5r}{(4+r^2)^2}\,dr$

8. a. $\int_0^1 \frac{10\sqrt{v}}{(1+v^{3/2})^2}\,dv$ **b.** $\int_1^4 \frac{10\sqrt{v}}{(1+v^{3/2})^2}\,dv$

9. a. $\int_0^{\sqrt{3}} \frac{4x}{\sqrt{x^2+1}}\,dx$ **b.** $\int_{-\sqrt{3}}^{\sqrt{3}} \frac{4x}{\sqrt{x^2+1}}\,dx$

10. a. $\int_0^1 \frac{x^3}{\sqrt{x^4+9}}\,dx$ **b.** $\int_{-1}^0 \frac{x^3}{\sqrt{x^4+9}}\,dx$

11. a. $\int_0^1 t\sqrt{4+5t}\,dt$ **b.** $\int_1^9 t\sqrt{4+5t}\,dt$

12. a. $\int_0^{\pi/6} (1-\cos 3t)\sin 3t\,dt$

b. $\int_{\pi/6}^{\pi/3} (1-\cos 3t)\sin 3t\,dt$

13. a. $\int_0^{2\pi} \frac{\cos z}{\sqrt{4+3\sin z}}\,dz$ **b.** $\int_{-\pi}^{\pi} \frac{\cos z}{\sqrt{4+3\sin z}}\,dz$

14. a. $\int_{-\pi/2}^0 \left(2+\tan\frac{t}{2}\right)\sec^2\frac{t}{2}\,dt$

b. $\int_{-\pi/2}^{\pi/2} \left(2+\tan\frac{t}{2}\right)\sec^2\frac{t}{2}\,dt$

15. $\int_0^1 \sqrt{t^5+2t}\,(5t^4+2)\,dt$ **16.** $\int_1^4 \frac{dy}{2\sqrt{y}\,(1+\sqrt{y})^2}$

17. $\int_0^{\pi/6} \cos^{-3} 2\theta \sin 2\theta\,d\theta$ **18.** $\int_{\pi}^{3\pi/2} \cot^5\left(\frac{\theta}{6}\right)\sec^2\left(\frac{\theta}{6}\right)d\theta$

19. $\int_0^{\pi} 5(5-4\cos t)^{1/4}\sin t\,dt$ **20.** $\int_0^{\pi/4} (1-\sin 2t)^{3/2}\cos 2t\,dt$

21. $\int_0^1 (4y-y^2+4y^3+1)^{-2/3}(12y^2-2y+4)\,dy$

22. $\int_0^1 (y^3+6y^2-12y+9)^{-1/2}(y^2+4y-4)\,dy$

23. $\int_0^{\sqrt[3]{\pi^2}} \sqrt{\theta}\cos^2(\theta^{3/2})\,d\theta$ **24.** $\int_{-1}^{-1/2} t^{-2}\sin^2\left(1+\frac{1}{t}\right)dt$

Area

Find the total areas of the shaded regions in Exercises 25–40.

25.

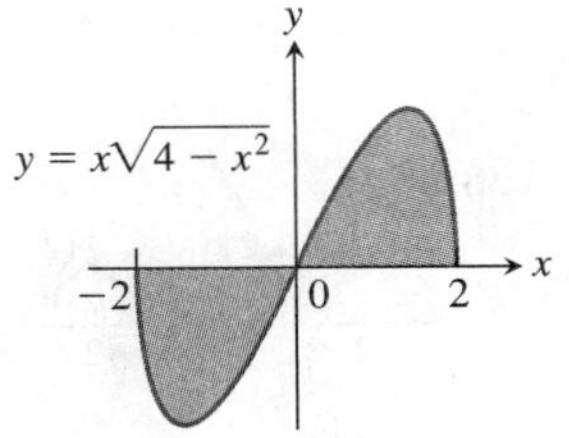

26.

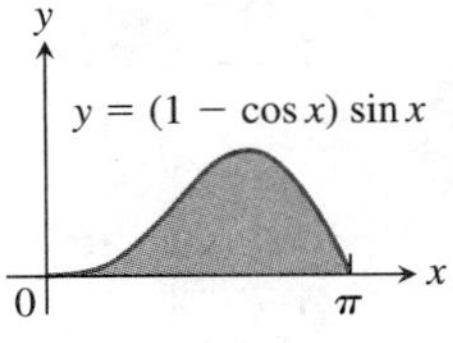

27.

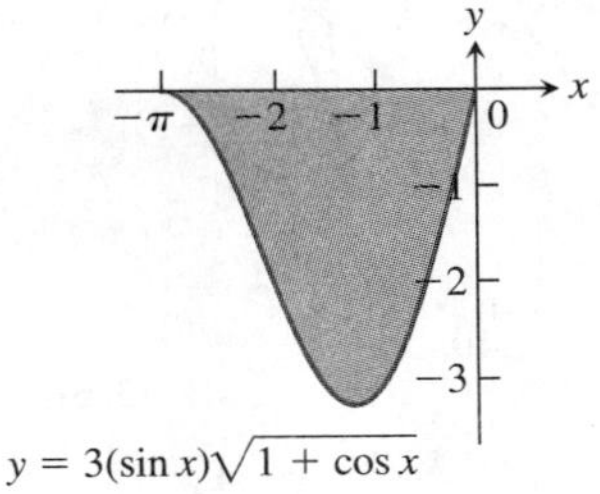

28. $y = \frac{\pi}{2}(\cos x)(\sin(\pi + \pi\sin x))$

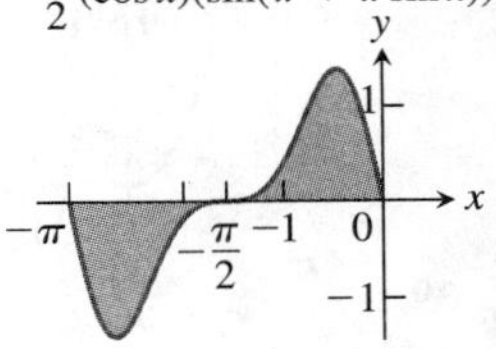

29.

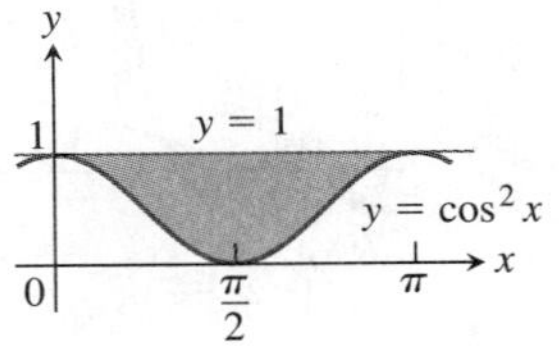

30.

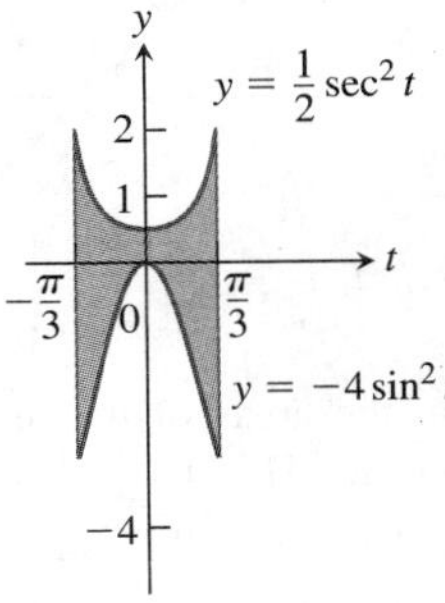

31.

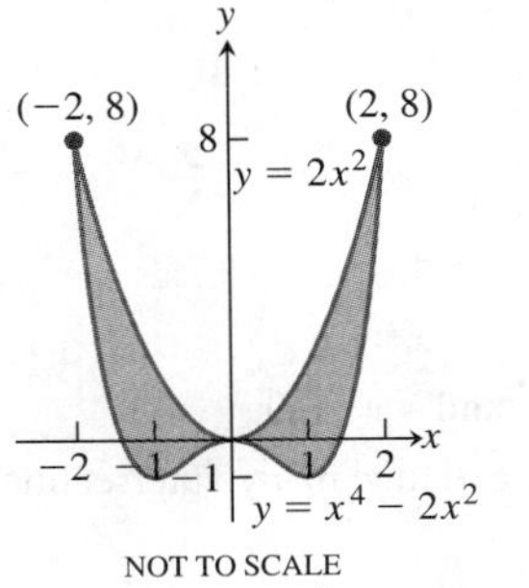

32.

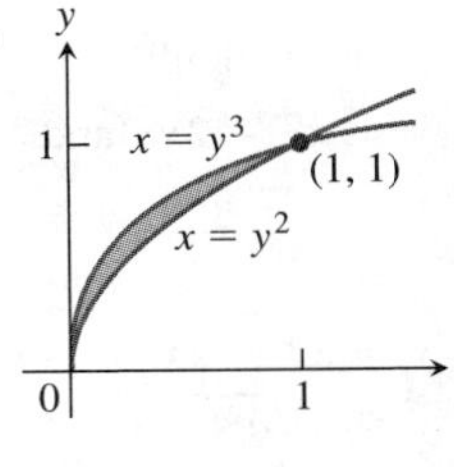

33.

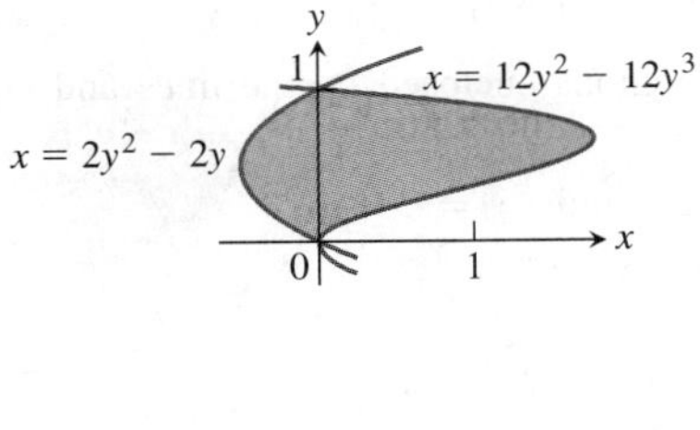

34.

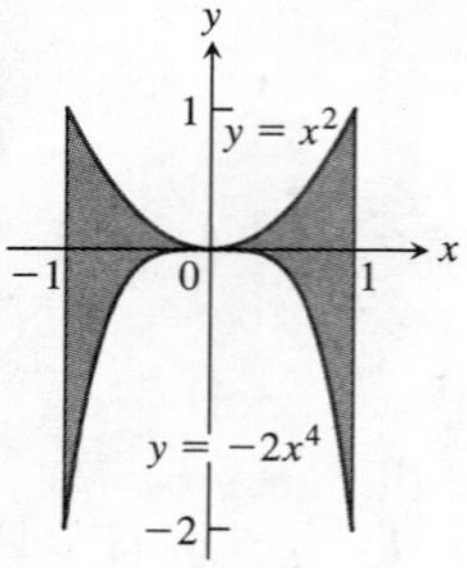

35.

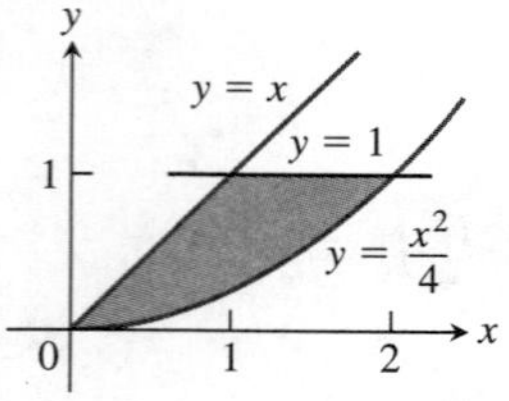

36.

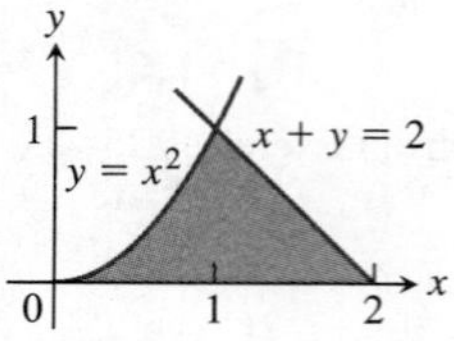

37.

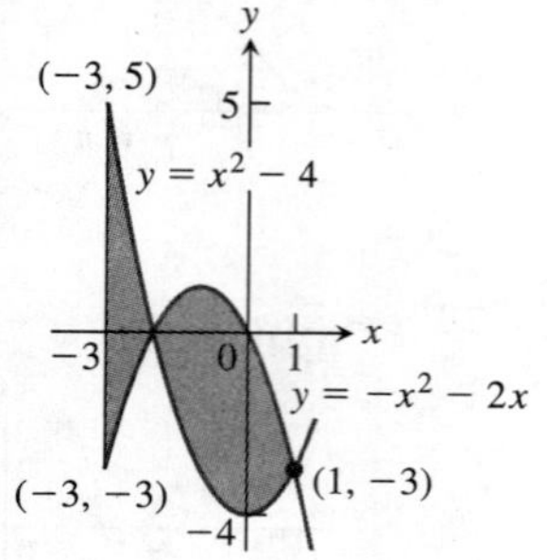

38.

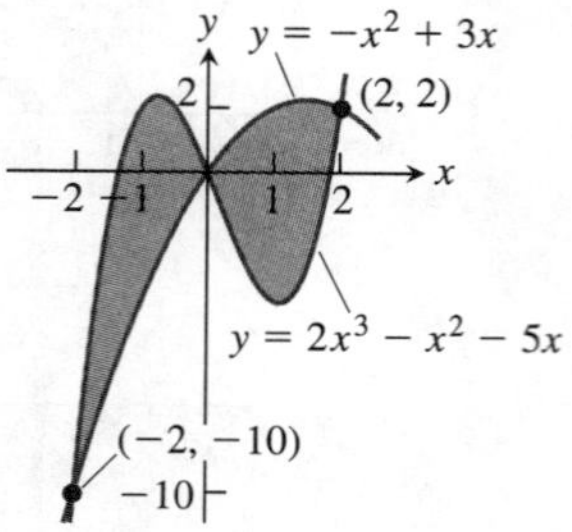

39.

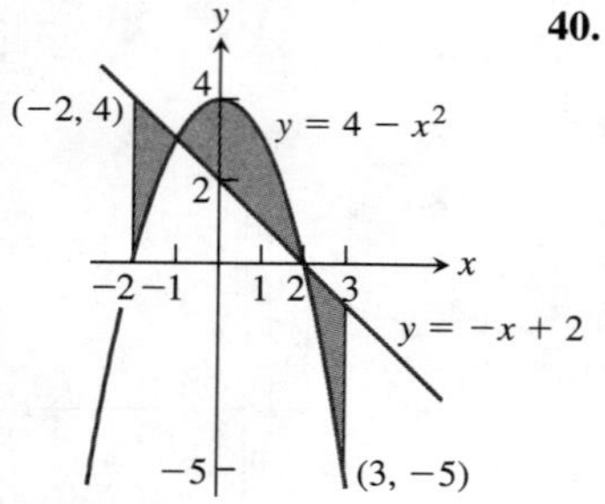

40.

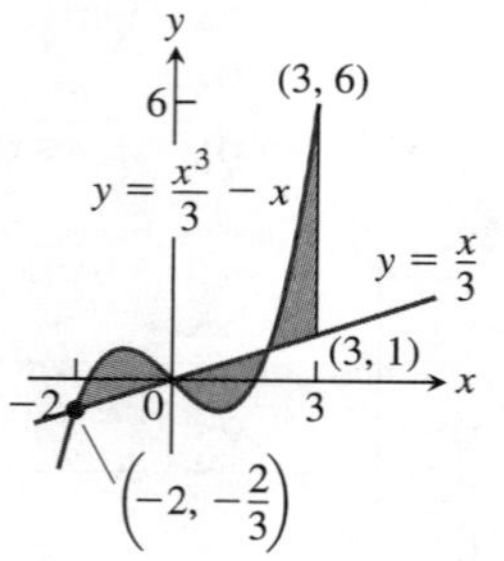

Find the areas of the regions enclosed by the lines and curves in Exercises 41–50.

41. $y = x^2 - 2$ and $y = 2$

42. $y = 2x - x^2$ and $y = -3$

43. $y = x^4$ and $y = 8x$

44. $y = x^2 - 2x$ and $y = x$

45. $y = x^2$ and $y = -x^2 + 4x$

46. $y = 7 - 2x^2$ and $y = x^2 + 4$

47. $y = x^4 - 4x^2 + 4$ and $y = x^2$

48. $y = x\sqrt{a^2 - x^2}$, $a > 0$, and $y = 0$

49. $y = \sqrt{|x|}$ and $5y = x + 6$ (How many intersection points are there?)

50. $y = |x^2 - 4|$ and $y = (x^2/2) + 4$

Find the areas of the regions enclosed by the lines and curves in Exercises 51–58.

51. $x = 2y^2$, $x = 0$, and $y = 3$

52. $x = y^2$ and $x = y + 2$

53. $y^2 - 4x = 4$ and $4x - y = 16$

54. $x - y^2 = 0$ and $x + 2y^2 = 3$

55. $x + y^2 = 0$ and $x + 3y^2 = 2$

56. $x - y^{2/3} = 0$ and $x + y^4 = 2$

57. $x = y^2 - 1$ and $x = |y|\sqrt{1 - y^2}$

58. $x = y^3 - y^2$ and $x = 2y$

Find the areas of the regions enclosed by the curves in Exercises 59–62.

59. $4x^2 + y = 4$ and $x^4 - y = 1$

60. $x^3 - y = 0$ and $3x^2 - y = 4$

61. $x + 4y^2 = 4$ and $x + y^4 = 1$, for $x \ge 0$

62. Find the area between the curves $y = x^4 + x^3 + 16x + 4$ and $y = x^4 + 6x^2 + 8x + 4$.

Find the areas of the regions enclosed by the lines and curves in Exercises 63–70.

63. $y = 2\sin x$ and $y = \sin 2x$, $0 \le x \le \pi$

64. $y = 8\cos x$ and $y = \sec^2 x$, $-\pi/3 \le x \le \pi/3$

65. $y = \cos(\pi x/2)$ and $y = 1 - x^2$

66. $y = \sin(\pi x/2)$ and $y = x$

67. $y = \sec^2 x$, $y = \tan^2 x$, $x = -\pi/4$, and $x = \pi/4$

68. $x = \tan^2 y$ and $x = -\tan^2 y$, $-\pi/4 \le y \le \pi/4$

69. $x = 3\sin y\sqrt{\cos y}$ and $x = 0$, $0 \le y \le \pi/2$

70. $y = \sec^2(\pi x/3)$ and $y = x^{1/3}$, $-1 \le x \le 1$

Area Between Curves

71. Find the area of the propeller-shaped region enclosed by the curve $x - y^3 = 0$ and the line $x - y = 0$.

72. Find the area of the propeller-shaped region enclosed by the curves $x - y^{1/3} = 0$ and $x - y^{1/5} = 0$.

73. For any real t, $x = \dfrac{e^t + e^{-t}}{2}$, $y = \dfrac{e^t - e^{-t}}{2}$ is a point on the hyperbola $x^2 - y^2 = 1$. Find the area bounded by this hyperbola and the lines joining its centre to the points corresponding to t_1 and $-t_1$.

74. Find the area of the "triangular" region in the first quadrant bounded on the left by the y-axis and on the right by the curves $y = \sin x$ and $y = \cos x$.

75. The region bounded below by the parabola $y = x^2$ and above by the line $y = 4$ is to be partitioned into two subsections of equal area by cutting across it with the horizontal line $y = c$.

a. Sketch the region and draw a line $y = c$ across it that looks about right. In terms of c, what are the coordinates of the points where the line and parabola intersect? Add them to your figure.

b. Find c by integrating with respect to y. (This puts c in the limits of integration.)

c. Find c by integrating with respect to x. (This puts c into the integrand as well.)

76. Find the area of the region between the curve $y = 3 - x^2$ and the line $y = -1$ by integrating with respect to **a.** x, **b.** y.

77. Find the area bounded by the curve $x^2 = 4y$ and the straight line $x = 4y - 2$.

78. Find the area of the region in the first quadrant bounded on the left by the y-axis, below by the curve $x = 2\sqrt{y}$, above left by the curve $x = (y-1)^2$, and above right by the line $x = 3 - y$.

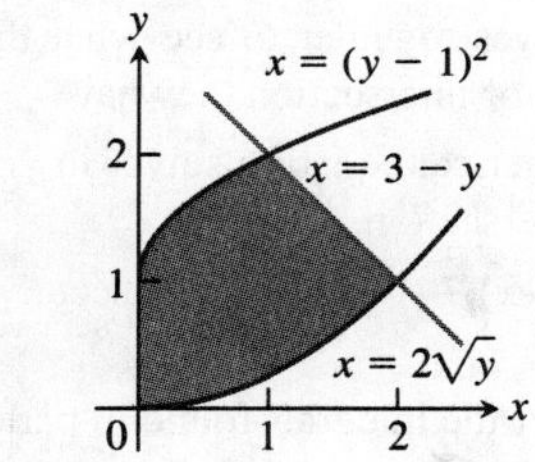

79. The figure here shows triangle AOC inscribed in the region cut from the parabola $y = x^2$ by the line $y = a^2$. Find the limit of the ratio of the area of the triangle to the area of the parabolic region as a approaches zero.

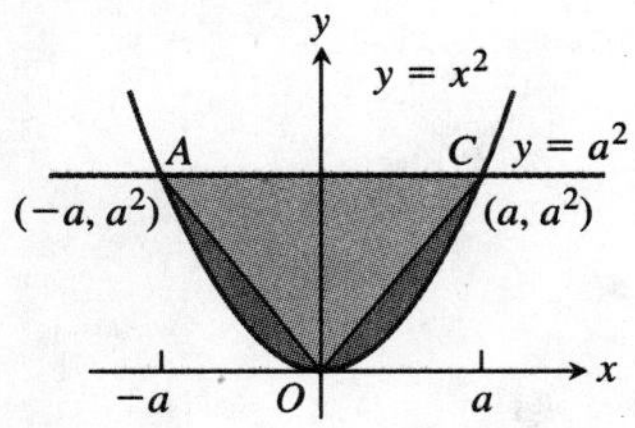

80. Suppose the area of the region between the graph of a positive continuous function f and the x-axis from $x = a$ to $x = b$ is 4 square units. Find the area between the curves $y = f(x)$ and $y = 2f(x)$ from $x = a$ to $x = b$.

81. Which of the following integrals, if either, calculates the area of the shaded region shown here? Give reasons for your answer.

a. $\int_{-1}^{1} (x - (-x))\,dx = \int_{-1}^{1} 2x\,dx$

b. $\int_{-1}^{1} (-x - (x))\,dx = \int_{-1}^{1} -2x\,dx$

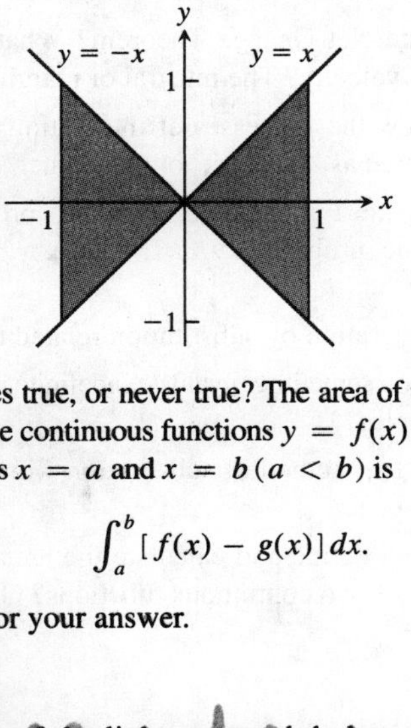

82. True, sometimes true, or never true? The area of the region between the graphs of the continuous functions $y = f(x)$ and $y = g(x)$ and the vertical lines $x = a$ and $x = b$ $(a < b)$ is

$$\int_a^b [f(x) - g(x)]\,dx.$$

Give reasons for your answer.

Comparing Areas

Compute the areas of the light gray and dark gray regions in each of Exercises 83 to 86 and determine which is larger, or show that they have equal area.

83.

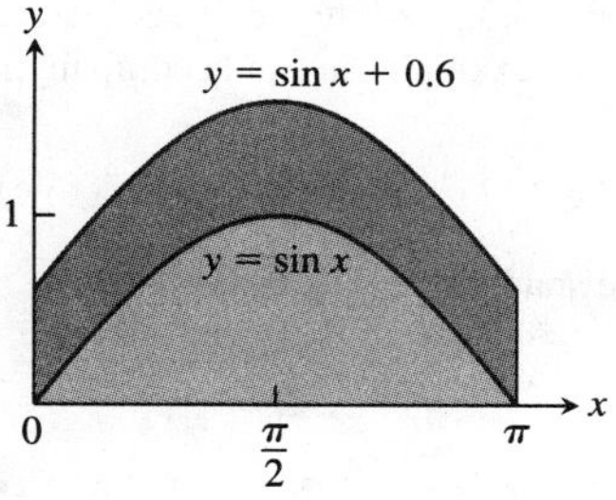

84.

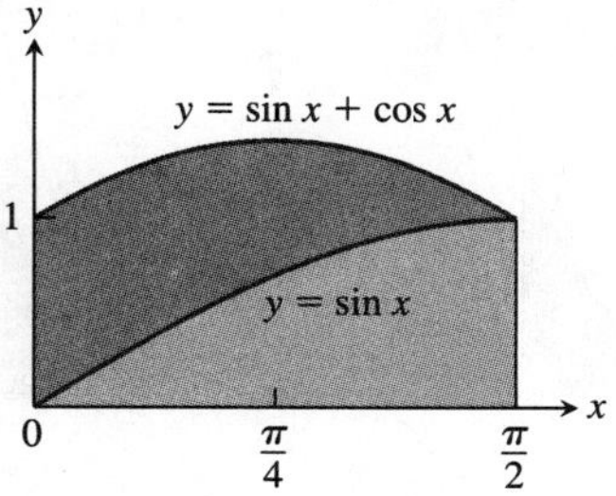

Theory and Examples

85. Suppose that $F(x)$ is an antiderivative of $f(x) = (\sin x)/x$, $x > 0$. Express

$$\int_1^3 \frac{\sin 2x}{x}\,dx$$

in terms of F.

86. Show that if f is continuous, then

$$\int_0^1 f(x)\,dx = \int_0^1 f(1 - x)\,dx.$$

87. Suppose that

$$\int_0^1 f(x)\,dx = 3.$$

Find

$$\int_{-1}^0 f(x)\,dx$$

if **a.** f is odd, **b.** f is even.

88. a. Show that if f is odd on $[-a, a]$, then

$$\int_{-a}^a f(x)\,dx = 0.$$

b. Test the result in part (a) with $f(x) = \sin x$ and $a = \pi/2$.

89. If f is a continuous function, find the value of the integral

$$I = \int_0^a \frac{f(x)\,dx}{f(x) + f(a - x)}$$

by making the substitution $u = a - x$ and adding the resulting integral to I.

90. By using a substitution, prove that for all positive numbers x and y,

$$\int_x^{xy} \frac{1}{t}\,dt = \int_1^y \frac{1}{t}\,dt.$$

The Shift Property for Definite Integrals A basic property of definite integrals is their invariance under translation, as expressed by the equation

$$\int_a^b f(x)\,dx = \int_{a-c}^{b-c} f(x + c)\,dx \qquad (1)$$

The equation holds whenever f is integrable and defined for the necessary values of x. For example, in the accompanying figure, show that

$$\int_{-2}^{-1} (x + 2)^3\, dx = \int_{0}^{1} x^3\, dx$$

because the areas of the shaded regions are congruent.

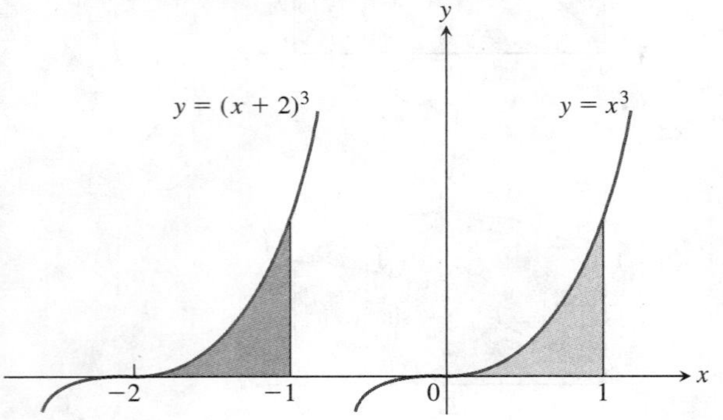

91. Use a substitution to verify Equation (1).

92. For each of the following functions, graph $f(x)$ over $[a, b]$ and $f(x + c)$ over $[a - c, b - c]$ to convince yourself that Equation (1) is reasonable.

a. $f(x) = x^2, \quad a = 0, \quad b = 1, \quad c = 1$

b. $f(x) = \sin x, \quad a = 0, \quad b = \pi, \quad c = \pi/2$

c. $f(x) = \sqrt{x - 4}, \quad a = 4, \quad b = 8, \quad c = 5$

COMPUTER EXPLORATIONS

In Exercises 93–96, you will find the area between curves in the plane when you cannot find their points of intersection using simple algebra. Use a CAS to perform the following steps:

a. Plot the curves together to see what they look like and how many points of intersection they have.

b. Use the numerical equation solver in your CAS to find all the points of intersection.

c. Integrate $|f(x) - g(x)|$ over consecutive pairs of intersection values.

d. Sum together the integrals found in part (c).

93. $f(x) = \dfrac{x^3}{3} - \dfrac{x^2}{2} - 2x + \dfrac{1}{3}, \quad g(x) = x - 1$

94. $f(x) = \dfrac{x^4}{2} - 3x^3 + 10, \quad g(x) = 8 - 12x$

95. $f(x) = x + \sin(2x), \quad g(x) = x^3$

96. $f(x) = x^2 \cos x, \quad g(x) = x^3 - x$

CHAPTER 5 Questions to Guide Your Review

1. How can you sometimes estimate quantities like distance traveled, area, and average value with finite sums? Why might you want to do so?
2. What is sigma notation? What advantage does it offer? Give examples.
3. What is a Riemann sum? Why might you want to consider such a sum?
4. What is the norm of a partition of a closed interval?
5. What is the definite integral of a function f over a closed interval $[a, b]$? When can you be sure it exists?
6. What is the relation between definite integrals and area? Describe some other interpretations of definite integrals.
7. What is the average value of an integrable function over a closed interval? Must the function assume its average value? Explain.
8. Describe the rules for working with definite integrals (Table 5.6). Give examples.
9. What is the Fundamental Theorem of Calculus? Why is it so important? Illustrate each part of the theorem with an example.
10. What is the Net Change Theorem? What does it say about the integral of velocity? The integral of marginal cost?
11. Discuss how the processes of integration and differentiation can be considered as "inverses" of each other.
12. How does the Fundamental Theorem provide a solution to the initial value problem $dy/dx = f(x)$, $y(x_0) = y_0$, when f is continuous?
13. How is integration by substitution related to the Chain Rule?
14. How can you sometimes evaluate indefinite integrals by substitution? Give examples.
15. How does the method of substitution work for definite integrals? Give examples.
16. How do you define and calculate the area of the region between the graphs of two continuous functions? Give an example.

CHAPTER 5 Practice Exercises

Finite Sums and Estimates

1. The accompanying figure shows the graph of the velocity (ft/sec) of a model rocket for the first 8 sec after launch. The rocket accelerated straight up for the first 2 sec and then coasted to reach its maximum height at $t = 8$ sec.

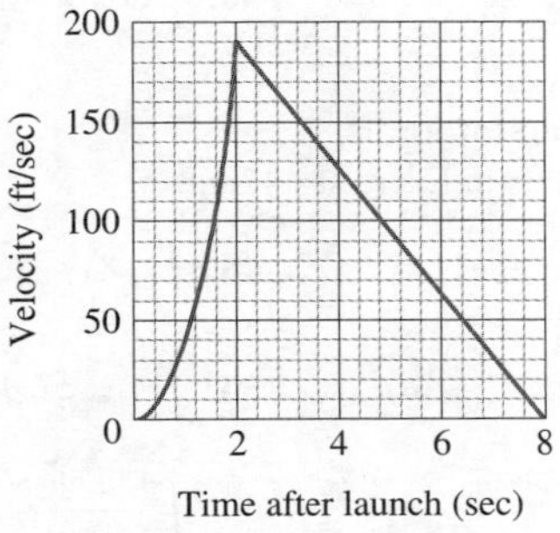

a. Assuming that the rocket was launched from ground level, about how high did it go? (This is the rocket in Section 3.4, Exercise 17, but you do not need to do Exercise 17 to do the exercise here.)

b. Sketch a graph of the rocket's height above ground as a function of time for $0 \le t \le 8$.

2. a. The accompanying figure shows the velocity (m/sec) of a body moving along the s-axis during the time interval from $t = 0$ to $t = 10$ sec. About how far did the body travel during those 10 sec?

b. Sketch a graph of s as a function of t for $0 \le t \le 10$, assuming $s(0) = 0$.

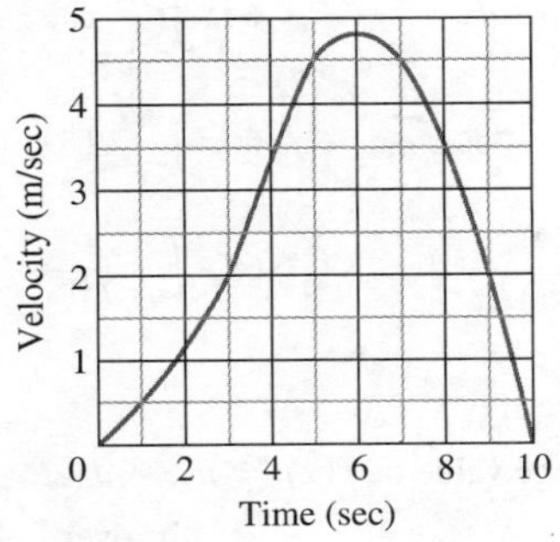

3. Suppose that $\sum_{k=1}^{10} a_k = -2$ and $\sum_{k=1}^{10} b_k = 25$. Find the value of

a. $\sum_{k=1}^{10} \frac{a_k}{4}$ **b.** $\sum_{k=1}^{10} (b_k - 3a_k)$

c. $\sum_{k=1}^{10} (a_k + b_k - 1)$ **d.** $\sum_{k=1}^{10} \left(\frac{5}{2} - b_k\right)$

4. Suppose that $\sum_{k=1}^{20} a_k = 0$ and $\sum_{k=1}^{20} b_k = 7$. Find the value of

a. $\sum_{k=1}^{20} 3a_k$ **b.** $\sum_{k=1}^{20} (a_k + b_k)$

c. $\sum_{k=1}^{20} \left(\frac{1}{2} - \frac{2b_k}{7}\right)$ **d.** $\sum_{k=1}^{20} (a_k - 2)$

Definite Integrals

In Exercises 5–8, express each limit as a definite integral. Then evaluate the integral to find the value of the limit. In each case, P is a partition of the given interval, and the numbers c_k are chosen from the subintervals of P.

5. $\lim_{\|P\|\to 0} \sum_{k=1}^{n} (2c_k - 1)^{-1/2}\,\Delta x_k$, where P is a partition of $[1, 5]$

6. $\lim_{\|P\|\to 0} \sum_{k=1}^{n} c_k(c_k^2 - 1)^{1/3}\,\Delta x_k$, where P is a partition of $[1, 3]$

7. $\lim_{\|P\|\to 0} \sum_{k=1}^{n} \left(\cos\left(\frac{c_k}{2}\right)\right)\Delta x_k$, where P is a partition of $[-\pi, 0]$

8. $\lim_{\|P\|\to 0} \sum_{k=1}^{n} (\sin c_k)(\cos c_k)\,\Delta x_k$, where P is a partition of $[0, \pi/2]$

9. If $\int_{-2}^{2} 3f(x)\,dx = 12$, $\int_{-2}^{5} f(x)\,dx = 6$, and $\int_{-2}^{5} g(x)\,dx = 2$, find the value of each of the following.

a. $\int_{-2}^{2} f(x)\,dx$ **b.** $\int_{2}^{5} f(x)\,dx$

c. $\int_{5}^{-2} g(x)\,dx$ **d.** $\int_{-2}^{5} (-\pi g(x))\,dx$

e. $\int_{-2}^{5} \left(\frac{f(x) + g(x)}{5}\right) dx$

10. If $\int_0^2 f(x)\,dx = \pi$, $\int_0^2 7g(x)\,dx = 7$, and $\int_0^1 g(x)\,dx = 2$, find the value of each of the following.

a. $\int_0^2 g(x)\,dx$ **b.** $\int_1^2 g(x)\,dx$

c. $\int_2^0 f(x)\,dx$ **d.** $\int_0^2 \sqrt{2}\,f(x)\,dx$

e. $\int_0^2 (g(x) - 3f(x))\,dx$

Area

In Exercises 11–14, find the total area of the region between the graph of f and the x-axis.

11. $f(x) = x^2 - 4x + 3, \quad 0 \le x \le 3$

12. $f(x) = 1 - (x^2/4), \quad -2 \le x \le 3$

13. $f(x) = 5 - 5x^{2/3}, \quad -1 \le x \le 8$

14. $f(x) = 1 - \sqrt{x}, \quad 0 \le x \le 4$

Find the areas of the regions enclosed by the curves and lines in Exercises 15–26.

15. $y = x, \quad y = 1/x^2, \quad x = 2$

16. $y = x, \quad y = 1/\sqrt{x}, \quad x = 2$

17. $\sqrt{x} + \sqrt{y} = 1, \quad x = 0, \quad y = 0$

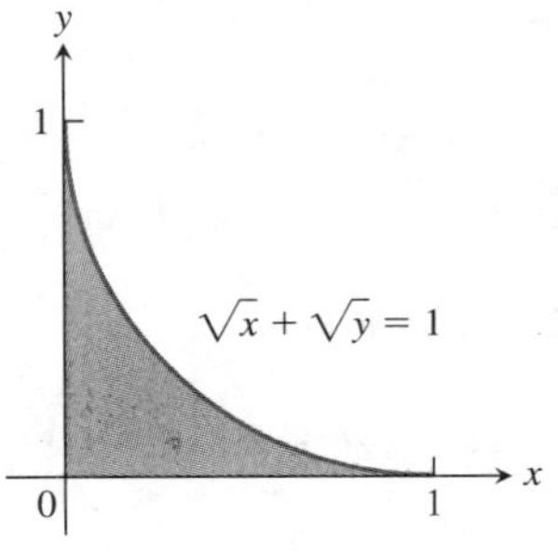

18. $x^3 + \sqrt{y} = 1, \quad x = 0, \quad y = 0, \quad \text{for} \quad 0 \le x \le 1$

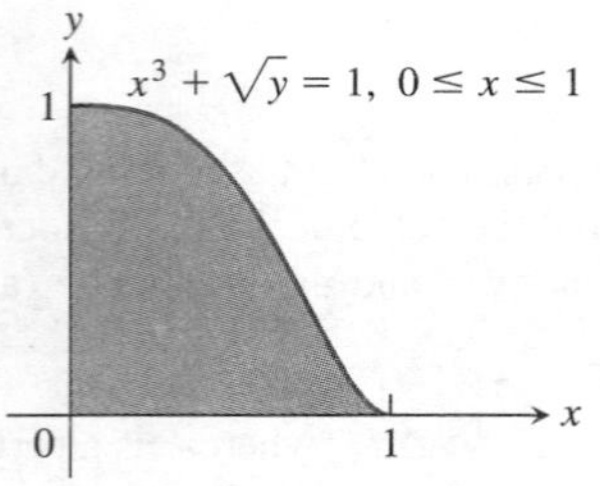

19. $x = 2y^2, \quad x = 0, \quad y = 3$ **20.** $x = 4 - y^2, \quad x = 0$

21. $y^2 = 4x, \quad y = 4x - 2$

22. $y^2 = 4x + 4, \quad y = 4x - 16$

23. $y = \sin x, \quad y = x, \quad 0 \le x \le \pi/4$

24. $y = |\sin x|, \quad y = 1, \quad -\pi/2 \le x \le \pi/2$

25. $y = 2\sin x, \quad y = \sin 2x, \quad 0 \le x \le \pi$

26. $y = 8\cos x, \quad y = \sec^2 x, \quad -\pi/3 \le x \le \pi/3$

27. Find the area of the "triangular" region bounded on the left by $x + y = 2$, on the right by $y = x^2$, and above by $y = 2$.

28. Find the area of the "triangular" region bounded on the left by $y = \sqrt{x}$, on the right by $y = 6 - x$, and below by $y = 1$.

29. Find the extreme values of $f(x) = x^3 - 3x^2$, and find the area of the region enclosed by the graph of f and the x-axis.

30. Find the area of the region cut from the first quadrant by the curve $x^{1/2} + y^{1/2} = a^{1/2}$.

31. Find the area of the curve $x^{2/3} + y^{2/3} = a^{2/3}$.

32. Find the total area of the region between the curves $y = \sin x$ and $y = \cos x$ for $0 \le x \le 3\pi/2$.

Initial Value Problems

33. Show that $y = x^2 + \int_1^x \frac{1}{t}\,dt$ solves the initial value problem

$$\frac{d^2y}{dx^2} = 2 - \frac{1}{x^2}; \quad y'(1) = 3, \quad y(1) = 1.$$

34. Show that $y = \int_0^x \left(1 + 2\sqrt{\sec t}\right) dt$ solves the initial value problem

$$\frac{d^2y}{dx^2} = \sqrt{\sec x}\tan x; \quad y'(0) = 3, \quad y(0) = 0.$$

Express the solutions of the initial value problems in Exercises 35 and 36 in terms of integrals.

35. $\dfrac{dy}{dx} = \dfrac{\sin x}{x}, \quad y(5) = -3$

36. $\dfrac{dy}{dx} = \sqrt{2 - \sin^2 x}, \quad y(-1) = 2$

Evaluating Indefinite Integrals

Evaluate the integrals in Exercises 37–46.

37. $\int 2(\cos x)^{-1/2}\sin x\,dx$ **38.** $\int (\tan x)^{-3/2}\sec^2 x\,dx$

39. $\int (2\theta + 1 + 2\cos(2\theta + 1))\,d\theta$

40. $\int \left(\dfrac{1}{\sqrt{2\theta - \pi}} + 2\sec^2(2\theta - \pi)\right) d\theta$

41. $\int \left(t - \dfrac{2}{t}\right)\left(t + \dfrac{2}{t}\right) dt$

42. $\int \dfrac{(t+1)^2 - 1}{t^4}\,dt$ **43.** $\int \sqrt{t}\sin(2t^{3/2})\,dt$

44. $\int (\sec\theta\tan\theta)\sqrt{1 + \sec\theta}\,d\theta$

45. $\int \dfrac{\sin 2\theta - \cos 2\theta}{(\sin 2\theta + \cos 2\theta)^3}\,d\theta$ **46.** $\int \cos\theta \cdot \sin(\sin\theta)\,d\theta$

Evaluating Definite Integrals

Evaluate the integrals in Exercises 47–68.

47. $\int_{-1}^{1} (3x^2 - 4x + 7)\,dx$ **48.** $\int_0^1 (8s^3 - 12s^2 + 5)\,ds$

49. $\int_1^2 \dfrac{4}{v^2}\,dv$ **50.** $\int_1^{27} x^{-4/3}\,dx$

51. $\int_1^4 \dfrac{dt}{t\sqrt{t}}$ **52.** $\int_1^4 \dfrac{\left(1 + \sqrt{u}\right)^{1/2}}{\sqrt{u}}\,du$

53. $\int_0^1 \dfrac{36\,dx}{(2x+1)^3}$ **54.** $\int_0^1 \dfrac{dr}{\sqrt[3]{(7 - 5r)^2}}$

55. $\int_{1/8}^{1} x^{-1/3}(1 - x^{2/3})^{3/2}\,dx$ **56.** $\int_0^{1/2} x^3(1 + 9x^4)^{-3/2}\,dx$

57. $\int_0^{\pi} \sin^2 5r\,dr$ **58.** $\int_0^{\pi/4} \cos^2\left(4t - \dfrac{\pi}{4}\right) dt$

59. $\int_0^{\pi/3} \sec^2\theta\,d\theta$ **60.** $\int_{\pi/4}^{3\pi/4} \csc^2 x\,dx$

61. $\int_{\pi}^{3\pi} \cot^2\dfrac{x}{6}\,dx$ **62.** $\int_0^{\pi} \tan^2\dfrac{\theta}{3}\,d\theta$

63. $\int_{-\pi/3}^{0} \sec x\tan x\,dx$ **64.** $\int_{\pi/4}^{3\pi/4} \csc z\cot z\,dz$

65. $\int_0^{\pi/2} 5(\sin x)^{3/2}\cos x\,dx$ **66.** $\int_{-\pi/2}^{\pi/2} 15\sin^4 3x\cos 3x\,dx$

67. $\int_0^{\pi/2} \dfrac{3\sin x\cos x}{\sqrt{1 + 3\sin^2 x}}\,dx$ **68.** $\int_0^{\pi/4} \dfrac{\sec^2 x}{(1 + 7\tan x)^{2/3}}\,dx$

Average Values

69. Find the average value of $f(x) = mx + b$

a. over $[-1, 1]$ **b.** over $[-k, k]$

70. Find the average value of

a. $y = \sqrt{3x}$ over $[0, 3]$ **b.** $y = \sqrt{ax}$ over $[0, a]$

71. Let f be a function that is differentiable on $[a, b]$. In Chapter 2 we defined the average rate of change of f over $[a, b]$ to be

$$\frac{f(b) - f(a)}{b - a}$$

and the instantaneous rate of change of f at x to be $f'(x)$. In this chapter we defined the average value of a function. For the new definition of average to be consistent with the old one, we should have

$$\frac{f(b) - f(a)}{b - a} = \text{average value of } f' \text{ on } [a, b].$$

Is this the case? Give reasons for your answer.

72. Is it true that the average value of an integrable function over an interval of length 2 is half the function's integral over the interval? Give reasons for your answer.

T **73.** Compute the average value of the temperature function

$$f(x) = 37\sin\left(\frac{2\pi}{365}(x - 101)\right) + 25$$

for a 365-day year. (See Exercise 86, Section 3.6.) This is one way to estimate the annual mean air temperature in Fairbanks, Alaska. The National Weather Service's official figure, a numerical average of the daily normal mean air temperatures for the year, is 25.7°F, which is slightly higher than the average value of $f(x)$.

T **74. Specific heat of a gas** Specific heat C_v is the amount of heat required to raise the temperature of one mole (gram molecule) of a gas with constant volume by 1°C. The specific heat of oxygen depends on its temperature T and satisfies the formula

$$C_v = 8.27 + 10^{-5}(26T - 1.87T^2).$$

Find the average value of C_v for $20°\text{C} \le T \le 675°\text{C}$ and the temperature at which it is attained.

Differentiating Integrals

In Exercises 75–78, find dy/dx.

75. $y = \int_2^x \sqrt{2 + \cos^3 t}\, dt$ **76.** $y = \int_2^{7x^2} \sqrt{2 + \cos^3 t}\, dt$

77. $y = \int_x^1 \frac{6}{3 + t^4}\, dt$ **78.** $y = \int_{\sec x}^2 \frac{1}{t^2 + 1}\, dt$

Theory and Examples

79. Is it true that every function $y = f(x)$ that is differentiable on $[a, b]$ is itself the derivative of some function on $[a, b]$? Give reasons for your answer.

80. Suppose that $f(x)$ is an antiderivative of $f(x) = \sqrt{1 + x^4}$. Express $\int_0^1 \sqrt{1 + x^4}\, dx$ in terms of F and give a reason for your answer.

81. Find dy/dx if $y = \int_x^1 \sqrt{1 + t^2}\, dt$. Explain the main steps in your calculation.

82. Find dy/dx if $y = \int_{\cos x}^0 (1/(1 - t^2))\, dt$. Explain the main steps in your calculation.

83. A new parking lot To meet the demand for parking, your town has allocated the area shown here. As the town engineer, you have been asked by the town council to find out if the lot can be built for \$10,000. The cost to clear the land will be \$0.10 a square foot, and the lot will cost \$2.00 a square foot to pave. Can the job be done for \$10,000? Use a lower sum estimate to see. (Answers may vary slightly, depending on the estimate used.)

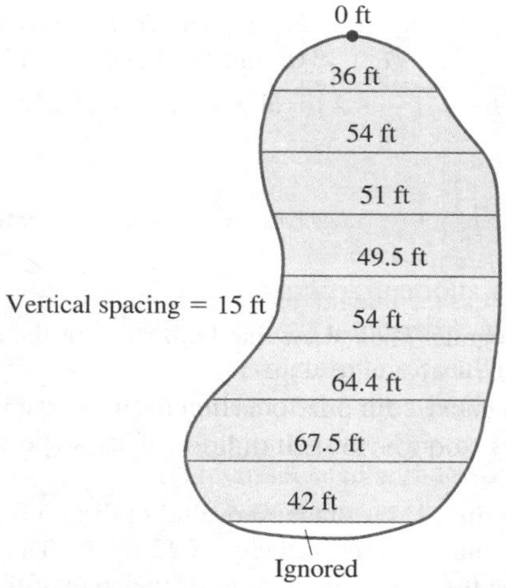

84. Skydivers A and B are in a helicopter hovering at 6400 ft. Skydiver A jumps and descends for 4 sec before opening her parachute. The helicopter then climbs to 7000 ft and hovers there. Forty-five seconds after A leaves the aircraft, B jumps and descends for 13 sec before opening his parachute. Both skydivers descend at 16 ft/sec with parachutes open. Assume that the skydivers fall freely (no effective air resistance) before their parachutes open.

a. At what altitude does A's parachute open?

b. At what altitude does B's parachute open?

c. Which skydiver lands first?

CHAPTER 5 Additional and Advanced Exercises

Theory and Examples

1. a. If $\int_0^1 7f(x)\, dx = 7$, does $\int_0^1 f(x)\, dx = 1$?

b. If $\int_0^1 f(x)\, dx = 4$ and $f(x) \ge 0$, does

$$\int_0^1 \sqrt{f(x)}\, dx = \sqrt{4} = 2?$$

Give reasons for your answers.

2. Suppose $\int_{-2}^2 f(x)\, dx = 4$, $\int_2^5 f(x)\, dx = 3$, $\int_{-2}^5 g(x)\, dx = 2$. Which, if any, of the following statements are true?

a. $\int_5^2 f(x)\, dx = -3$ **b.** $\int_{-2}^5 (f(x) + g(x))\, dx = 9$

c. $f(x) \le g(x)$ on the interval $-2 \le x \le 5$

3. Initial value problem Show that

$$y = \frac{1}{a}\int_0^x f(t)\sin a(x - t)\, dt$$

solves the initial value problem

$$\frac{d^2y}{dx^2} + a^2y = f(x), \qquad \frac{dy}{dx} = 0 \text{ and } y = 0 \text{ when } x = 0.$$

(*Hint:* $\sin(ax - at) = \sin ax \cos at - \cos ax \sin at$.)

4. Proportionality Suppose that x and y are related by the equation

$$x = \int_0^y \frac{1}{\sqrt{1 + 4t^2}}\, dt.$$

Show that d^2y/dx^2 is proportional to y, and find the constant of proportionality.

5. Find $f(4)$ if

a. $\int_0^{x^2} f(t)\,dt = x\cos\pi x$ **b.** $\int_0^{f(x)} t^2\,dt = x\cos\pi x.$

6. Find $f(\pi/2)$ from the following information.

i) f is positive and continuous.

ii) The area under the curve $y = f(x)$ from $x = 0$ to $x = a$ is

$$\frac{a^2}{2} + \frac{a}{2}\sin a + \frac{\pi}{2}\cos a.$$

7. The area of the region in the xy-plane enclosed by the x-axis, the curve $y = f(x)$, $f(x) \geq 0$, and the lines $x = 1$ and $x = b$ is equal to $\sqrt{b^2+1} - \sqrt{2}$ for all $b > 1$. Find $f(x)$.

8. Prove that

$$\int_0^x \left(\int_0^u f(t)\,dt\right) du = \int_0^x f(u)(x-u)\,du.$$

(*Hint:* Express the integral on the right-hand side as the difference of two integrals. Then show that both sides of the equation have the same derivative with respect to x.)

9. Finding a curve Find the equation for the curve in the xy-plane that passes through the point $(1, -1)$ if its slope at x is always $3x^2 + 2$.

10. Shoveling dirt You sling a shovelful of dirt up from the bottom of a hole with an initial velocity of 32 ft/sec. The dirt must rise 17 ft above the release point to clear the edge of the hole. Is that enough speed to get the dirt out, or had you better duck?

Piecewise Continuous Functions

Although we are mainly interested in continuous functions, many functions in applications are piecewise continuous. A function $f(x)$ is **piecewise continuous on a closed interval I** if f has only finitely many discontinuities in I, the limits

$$\lim_{x\to c^-} f(x) \quad \text{and} \quad \lim_{x\to c^+} f(x)$$

exist and are finite at every interior point of I, and the appropriate one-sided limits exist and are finite at the endpoints of I. All piecewise continuous functions are integrable. The points of discontinuity subdivide I into open and half-open subintervals on which f is continuous, and the limit criteria above guarantee that f has a continuous extension to the closure of each subinterval. To integrate a piecewise continuous function, we integrate the individual extensions and add the results. The integral of

$$f(x) = \begin{cases} 1-x, & -1 \leq x < 0 \\ x^2, & 0 \leq x < 2 \\ -1, & 2 \leq x \leq 3 \end{cases}$$

(Figure 5.33) over $[-1, 3]$ is

$$\int_{-1}^{3} f(x)\,dx = \int_{-1}^{0}(1-x)\,dx + \int_0^2 x^2\,dx + \int_2^3(-1)\,dx$$

$$= \left[x - \frac{x^2}{2}\right]_{-1}^{0} + \left[\frac{x^3}{3}\right]_0^2 + \left[-x\right]_2^3$$

$$= \frac{3}{2} + \frac{8}{3} - 1 = \frac{19}{6}.$$

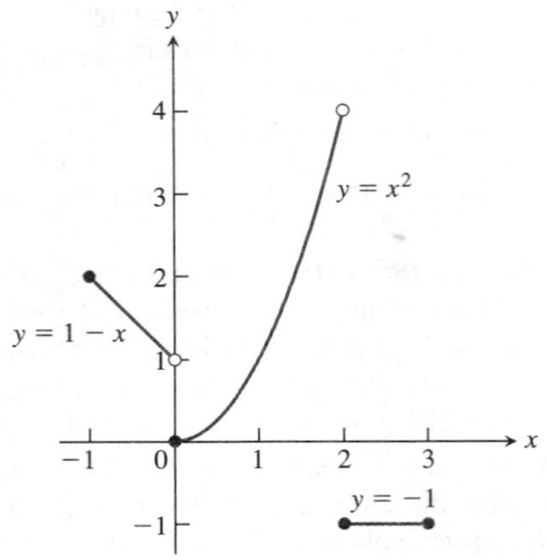

FIGURE 5.33 Piecewise continuous functions like this are integrated piece by piece.

The Fundamental Theorem applies to piecewise continuous functions with the restriction that $(d/dx)\int_a^x f(t)\,dt$ is expected to equal $f(x)$ only at values of x at which f is continuous. There is a similar restriction on Leibniz's Rule (see Exercises 21–23).

Graph the functions in Exercises 11–16 and integrate them over their domains.

11. $f(x) = \begin{cases} x^{2/3}, & -8 \leq x < 0 \\ -4, & 0 \leq x \leq 3 \end{cases}$

12. $f(x) = \begin{cases} \sqrt{-x}, & -4 \leq x < 0 \\ x^2 - 4, & 0 \leq x \leq 3 \end{cases}$

13. $g(t) = \begin{cases} t, & 0 \leq t < 1 \\ \sin\pi t, & 1 \leq t \leq 2 \end{cases}$

14. $h(z) = \begin{cases} \sqrt{1-z}, & 0 \leq z < 1 \\ (7z-6)^{-1/3}, & 1 \leq z \leq 2 \end{cases}$

15. $f(x) = \begin{cases} 1, & -2 \leq x < -1 \\ 1 - x^2, & -1 \leq x < 1 \\ 2, & 1 \leq x \leq 2 \end{cases}$

16. $h(r) = \begin{cases} r, & -1 \leq r < 0 \\ 1 - r^2, & 0 \leq r < 1 \\ 1, & 1 \leq r \leq 2 \end{cases}$

17. Find the area bounded by $y = \sqrt{x}$, $x \in [0,1]$, $y = x^2$, $x \in [1,2]$ and $y = -x^2 + 2x + 4$, $x \in [0,2]$.

18. Find the average value of the function graphed in the accompanying figure.

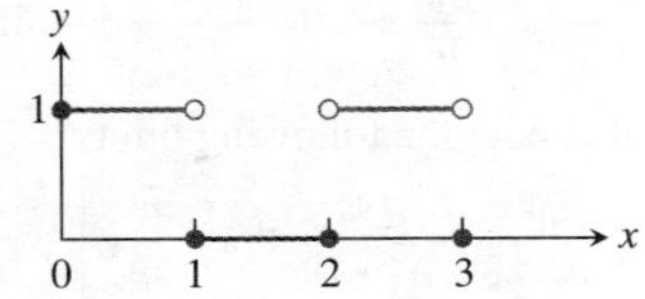

Defining Functions Using the Fundamental Theorem

19. **A function defined by an integral** The graph of a function f consists of a semicircle and two line segments as shown. Let $g(x) = \int_1^x f(t)\,dt$.

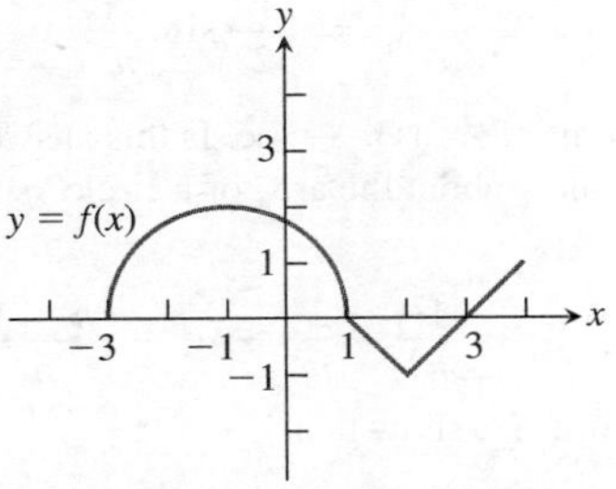

a. Find $g(1)$. **b.** Find $g(3)$. **c.** Find $g(-1)$.

d. Find all values of x on the open interval $(-3, 4)$ at which g has a relative maximum.

e. Write an equation for the line tangent to the graph of g at $x = -1$.

f. Find the x-coordinate of each point of inflection of the graph of g on the open interval $(-3, 4)$.

g. Find the range of g.

20. **A differential equation** Show that both of the following conditions are satisfied by $y = \sin x + \int_x^{\pi} \cos 2t\,dt + 1$:

i) $y'' = -\sin x + 2\sin 2x$

ii) $y = 1$ and $y' = -2$ when $x = \pi$.

Leibniz's Rule In applications, we sometimes encounter functions defined by integrals that have variable upper limits of integration and variable lower limits of integration at the same time. We can find the derivative of such an integral by a formula called **Leibniz's Rule**.

> **Leibniz's Rule**
> If f is continuous on $[a, b]$ and if $u(x)$ and $v(x)$ are differentiable functions of x whose values lie in $[a, b]$, then
> $$\frac{d}{dx}\int_{u(x)}^{v(x)} f(t)\,dt = f(v(x))\frac{dv}{dx} - f(u(x))\frac{du}{dx}.$$

To prove the rule, let F be an antiderivative of f on $[a, b]$. Then

$$\int_{u(x)}^{v(x)} f(t)\,dt = F(v(x)) - F(u(x)).$$

Differentiating both sides of this equation with respect to x gives the equation we want:

$$\begin{aligned}\frac{d}{dx}\int_{u(x)}^{v(x)} f(t)\,dt &= \frac{d}{dx}[F(v(x)) - F(u(x))]\\ &= F'(v(x))\frac{dv}{dx} - F'(u(x))\frac{du}{dx} \qquad \text{Chain Rule}\\ &= f(v(x))\frac{dv}{dx} - f(u(x))\frac{du}{dx}.\end{aligned}$$

Use Leibniz's Rule to find the derivatives of the functions in Exercises 21–23.

21. $f(x) = \int_{x^3}^{x^2} (1+t^2)^{-3}\,dt$

22. $f(x) = \int_{\cos x}^{\sin x} \frac{1}{1-t^2}\,dt$

23. $g(y) = \int_{\sqrt{y}}^{2\sqrt{y}} \sin t^2\,dt$

24. Use Leibniz's Rule to find the value of x that maximizes the value of the integral

$$\int_x^{x+3} t(5-t)\,dt.$$

Approximating Finite Sums with Integrals

In many applications of calculus, integrals are used to approximate finite sums—the reverse of the usual procedure of using finite sums to approximate integrals.

For example, let's estimate the sum of the square roots of the first n positive integers, $\sqrt{1} + \sqrt{2} + \cdots + \sqrt{n}$. The integral

$$\int_0^1 \sqrt{x}\,dx = \frac{2}{3}x^{3/2}\bigg]_0^1 = \frac{2}{3}$$

is the limit of the upper sums

$$\begin{aligned}S_n &= \sqrt{\frac{1}{n}}\cdot\frac{1}{n} + \sqrt{\frac{2}{n}}\cdot\frac{1}{n} + \cdots + \sqrt{\frac{n}{n}}\cdot\frac{1}{n}\\ &= \frac{\sqrt{1}+\sqrt{2}+\cdots+\sqrt{n}}{n^{3/2}}.\end{aligned}$$

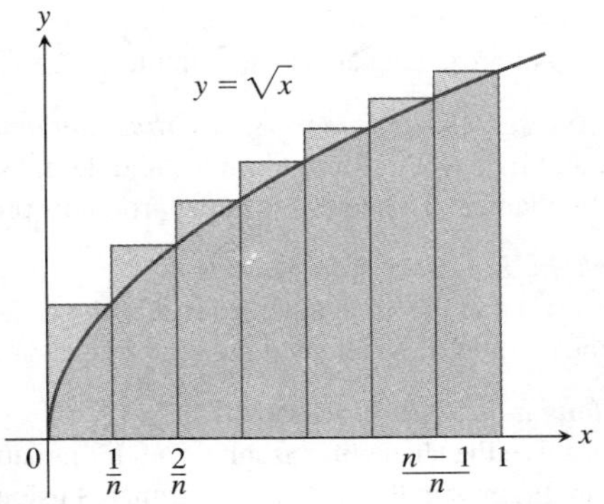

Therefore, when n is large, S_n will be close to $2/3$ and we will have

$$\text{Root sum} = \sqrt{1}+\sqrt{2}+\cdots+\sqrt{n} = S_n \cdot n^{3/2} \approx \frac{2}{3}n^{3/2}.$$

The following table shows how good the approximation can be.

n	Root sum	$(2/3)n^{3/2}$	Relative error
10	22.468	21.082	$1.386/22.468 \approx 6\%$
50	239.04	235.70	1.4%
100	671.46	666.67	0.7%
1000	21,097	21,082	0.07%

25. Evaluate

$$\lim_{n\to\infty} \frac{1^5 + 2^5 + 3^5 + \cdots + n^5}{n^6}$$

by showing that the limit is

$$\int_0^1 x^5\, dx$$

and evaluating the integral.

26. See Exercise 25. Evaluate

$$\lim_{n\to\infty} \frac{1}{n^4}(1^3 + 2^3 + 3^3 + \cdots + n^3).$$

27. Let $f(x)$ be a continuous function. Express

$$\lim_{n\to\infty} \frac{1}{n}\left[f\left(\frac{1}{n}\right) + f\left(\frac{2}{n}\right) + \cdots + f\left(\frac{n}{n}\right)\right]$$

as a definite integral.

28. Use the result of Exercise 27 to evaluate

a. $\lim_{n\to\infty} \frac{1}{n^2}(2 + 4 + 6 + \cdots + 2n)$,

b. $\lim_{n\to\infty} \frac{1}{n^{16}}(1^{15} + 2^{15} + 3^{15} + \cdots + n^{15})$,

c. $\lim_{n\to\infty} \frac{1}{n}\left(\sin\frac{\pi}{n} + \sin\frac{2\pi}{n} + \sin\frac{3\pi}{n} + \cdots + \sin\frac{n\pi}{n}\right)$.

What can be said about the following limits?

d. $\lim_{n\to\infty}\left[\frac{1}{n^3+1} + \frac{4}{n^3+8} + \frac{9}{n^3+27} + \ldots + \frac{n^2}{2n^3}\right]$

e. $\lim_{n\to\infty}\left[\left(1+\frac{1}{n}\right)\left(1+\frac{2}{n}\right)\ldots\left(1+\frac{n}{n}\right)\right]^{\frac{1}{n}}$

29. **a.** Show that the area A_n of an n-sided regular polygon in a circle of radius r is

$$A_n = \frac{nr^2}{2}\sin\frac{2\pi}{n}.$$

b. Find the limit of A_n as $n \to \infty$. Is this answer consistent with what you know about the area of a circle?

30. Let

$$S_n = \frac{1^2}{n^3} + \frac{2^2}{n^3} + \cdots + \frac{(n-1)^2}{n^3}.$$

To calculate $\lim_{n\to\infty} S_n$, show that

$$S_n = \frac{1}{n}\left[\left(\frac{1}{n}\right)^2 + \left(\frac{2}{n}\right)^2 + \cdots + \left(\frac{n-1}{n}\right)^2\right]$$

and interpret S_n as an approximating sum of the integral

$$\int_0^1 x^2\, dx.$$

(*Hint:* Partition $[0,1]$ into n intervals of equal length and write out the approximating sum for inscribed rectangles.)

CHAPTER 5 Technology Application Projects

Mathematica/Maple Projects

Projects can be found within MyLab Math.

- ***Using Riemann Sums to Estimate Areas, Volumes, and Lengths of Curves***
  Visualize and approximate areas and volumes in Part I.
- ***Riemann Sums, Definite Integrals, and the Fundamental Theorem of Calculus***
  Parts I, II, and III develop Riemann sums and definite integrals. Part IV continues the development of the Riemann sum and definite integral using the Fundamental Theorem to solve problems previously investigated.
- ***Rain Catchers, Elevators, and Rockets***
  Part I illustrates that the area under a curve is the same as the area of an appropriate rectangle for examples taken from the chapter. You will compute the amount of water accumulating in basins of different shapes as the basin is filled and drained.
- ***Motion Along a Straight Line, Part II***
  You will observe the shape of a graph through dramatic animated visualizations of the derivative relations among position, velocity, and acceleration. Figures in the text can be animated using this software.
- ***Bending of Beams***
  Study bent shapes of beams, determine their maximum deflections, concavity, and inflection points, and interpret the results in terms of a beam's compression and tension.

6 Applications of Definite Integrals

OVERVIEW In Chapter 5 we saw that a continuous function over a closed interval has a definite integral, which is the limit of Riemann sum approximations for the function. We found a way to evaluate definite integrals using the Fundamental Theorem of Calculus. We saw that the area under a curve and the area between two curves could be defined and computed as definite integrals. In this chapter we will see some of the many additional applications of definite integrals. We will use the definite integral to define and find volumes, lengths of plane curves, and areas of surfaces of revolution. We will see how integrals are used to solve physical problems involving the work done by a force, and how they give the location of an object's center of mass. The integral arises in these and other applications in which we can approximate a desired quantity by Riemann sums. The limit of those Riemann sums, which is the quantity we seek, is given by a definite integral.

6.1 Volumes Using Cross-Sections

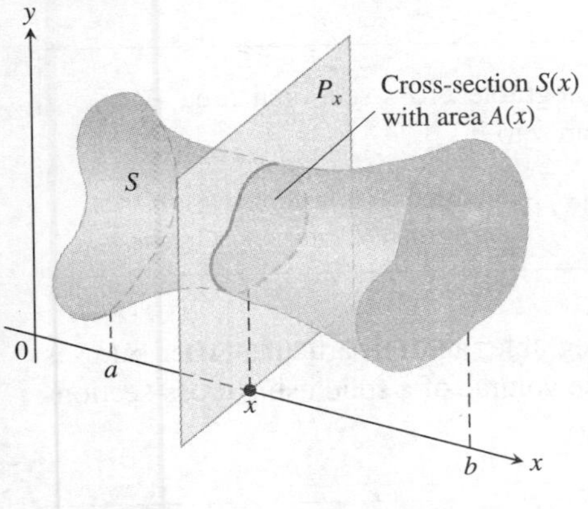

FIGURE 6.1 A cross-section $S(x)$ of the solid S formed by intersecting S with a plane P_x perpendicular to the x-axis through the point x in the interval $[a, b]$.

In this section we define volumes of solids by using the areas of their cross-sections. A **cross-section** of a solid S is the planar region formed by intersecting S with a plane (Figure 6.1). We present three different methods for obtaining the cross-sections appropriate to finding the volume of a particular solid: the method of slicing, the disk method, and the washer method.

Suppose that we want to find the volume of a solid S like the one pictured in Figure 6.1. At each point x in the interval $[a, b]$ we form a cross-section $S(x)$ by intersecting S with a plane perpendicular to the x-axis through the point x, which gives a planar region whose area is $A(x)$. We will show that if A is a continuous function of x, then the volume of the solid S is the definite integral of $A(x)$. This method of computing volumes is known as the **method of slicing**.

Before showing how this method works, we need to extend the definition of a cylinder from the usual cylinders of classical geometry (which have circular, square, or other regular bases) to cylindrical solids that have more general bases. As shown in Figure 6.2, if the

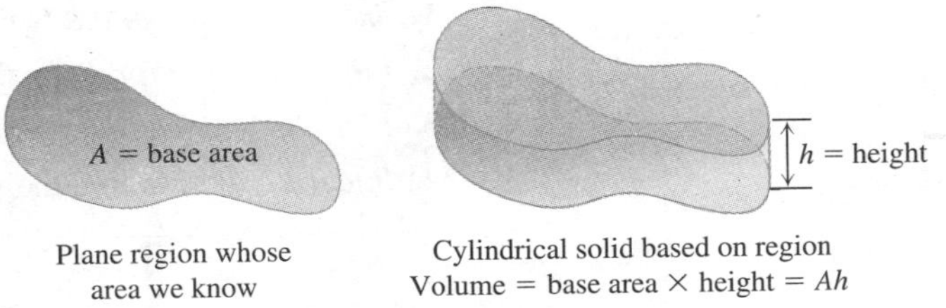

FIGURE 6.2 The volume of a cylindrical solid is equal to its base area times its height.

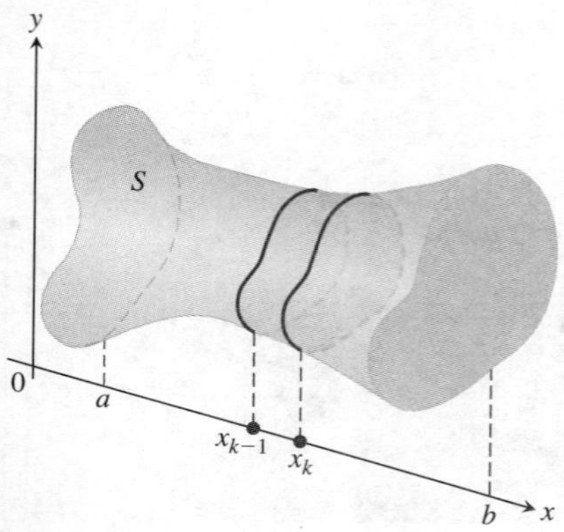

FIGURE 6.3 A typical thin slab in the solid S.

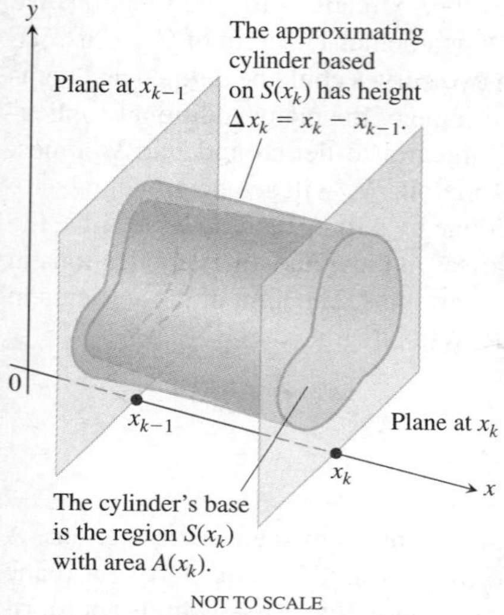

FIGURE 6.4 The solid thin slab in Figure 6.3 is shown enlarged here. It is approximated by the cylindrical solid with base $S(x_k)$ having area $A(x_k)$ and height $\Delta x_k = x_k - x_{k-1}$.

cylindrical solid has a base whose area is A and its height is h, then the volume of the cylindrical solid is

$$\text{Volume} = \text{area} \times \text{height} = A \cdot h.$$

In the method of slicing, the base will be the cross-section of S that has area $A(x)$, and the height will correspond to the width Δx_k of subintervals formed by partitioning the interval $[a, b]$ into finitely many subintervals $[x_{k-1}, x_k]$.

Slicing by Parallel Planes

We partition $[a, b]$ into subintervals of width (length) Δx_k and slice the solid, as we would a loaf of bread, by planes perpendicular to the x-axis at the partition points $a = x_0 < x_1 < \cdots < x_n = b$. These planes slice S into thin "slabs" (like thin slices of a loaf of bread). A typical slab is shown in Figure 6.3. We approximate the slab between the plane at x_{k-1} and the plane at x_k by a cylindrical solid with base area $A(x_k)$ and height $\Delta x_k = x_k - x_{k-1}$ (Figure 6.4). The volume V_k of this cylindrical solid is $A(x_k) \cdot \Delta x_k$, which is approximately the same volume as that of the slab:

$$\text{Volume of the } k\text{th slab} \approx V_k = A(x_k)\,\Delta x_k.$$

The volume V of the entire solid S is therefore approximated by the sum of these cylindrical volumes,

$$V \approx \sum_{k=1}^{n} V_k = \sum_{k=1}^{n} A(x_k)\,\Delta x_k.$$

This is a Riemann sum for the function $A(x)$ on $[a, b]$. The approximation given by this Riemann sum converges to the definite integral of $A(x)$ as $n \to \infty$:

$$\lim_{n\to\infty} \sum_{k=1}^{n} A(x_k)\,\Delta x_k = \int_a^b A(x)\,dx.$$

Therefore, we define this definite integral to be the volume of the solid S.

DEFINITION The **volume** of a solid of integrable cross-sectional area $A(x)$ from $x = a$ to $x = b$ is the integral of A from a to b,

$$V = \int_a^b A(x)\,dx.$$

This definition applies whenever $A(x)$ is integrable, and in particular when $A(x)$ is continuous. To apply this definition to calculate the volume of a solid using cross-sections perpendicular to the x-axis, take the following steps:

Calculating the Volume of a Solid

1. *Sketch the solid and a typical cross-section.*
2. *Find a formula for* $A(x)$, the area of a typical cross-section.
3. *Find the limits of integration.*
4. *Integrate* $A(x)$ to find the volume.

EXAMPLE 1 A pyramid 3 meters high has a square base that is 3 meters on a side. The cross-section of the pyramid perpendicular to the altitude x meters down from the vertex is a square x meters on a side. Find the volume of the pyramid.

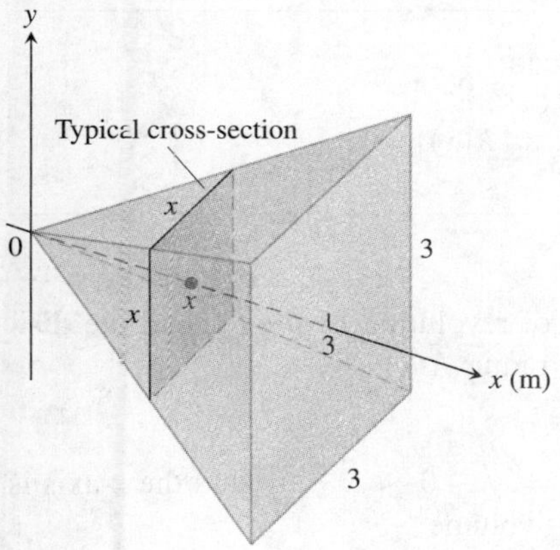

FIGURE 6.5 The cross-sections of the pyramid in Example 1 are squares.

Solution

1. *A sketch.* We draw the pyramid with its altitude along the x-axis and its vertex at the origin and include a typical cross-section (Figure 6.5). Note that by positioning the pyramid in this way, we have vertical cross-sections that are squares, whose areas are easy to calculate.
2. *A formula for $A(x)$.* The cross-section at x is a square x meters on a side, so its area is
$$A(x) = x^2.$$
3. *The limits of integration.* The squares lie on the planes from $x = 0$ to $x = 3$.
4. *Integrate to find the volume:*
$$V = \int_0^3 A(x)\,dx = \int_0^3 x^2\,dx = \frac{x^3}{3}\bigg]_0^3 = 9\text{ m}^3. \qquad ■$$

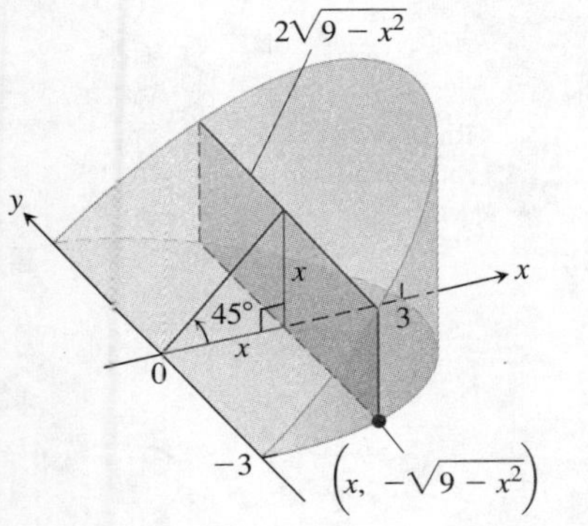

FIGURE 6.6 The wedge of Example 2, sliced perpendicular to the x-axis. The cross-sections are rectangles.

EXAMPLE 2 A curved wedge is cut from a circular cylinder of radius 3 by two planes. One plane is perpendicular to the axis of the cylinder. The second plane crosses the first plane at a 45° angle at the center of the cylinder. Find the volume of the wedge.

Solution We draw the wedge and sketch a typical cross-section perpendicular to the x-axis (Figure 6.6). The base of the wedge in the figure is the semicircle with $x \geq 0$ that is cut from the circle $x^2 + y^2 = 9$ by the 45° plane when it intersects the y-axis. For any x in the interval $[0, 3]$, the y-values in this semicircular base vary from $y = -\sqrt{9 - x^2}$ to $y = \sqrt{9 - x^2}$. When we slice through the wedge by a plane perpendicular to the x-axis, we obtain a cross-section at x, which is a rectangle of height x whose width extends across the semicircular base. The area of this cross-section is

$$\begin{aligned} A(x) &= (\text{height})(\text{width}) = (x)\left(2\sqrt{9 - x^2}\right) \\ &= 2x\sqrt{9 - x^2}. \end{aligned}$$

The rectangles run from $x = 0$ to $x = 3$, so we have

$$\begin{aligned} V &= \int_a^b A(x)\,dx = \int_0^3 2x\sqrt{9 - x^2}\,dx \\ &= -\frac{2}{3}(9 - x^2)^{3/2}\bigg]_0^3 \\ &= 0 + \frac{2}{3}(9)^{3/2} \\ &= 18. \end{aligned}$$

Let $u = 9 - x^2$ and $du = -2x\,dx$, integrate, and substitute back.

■

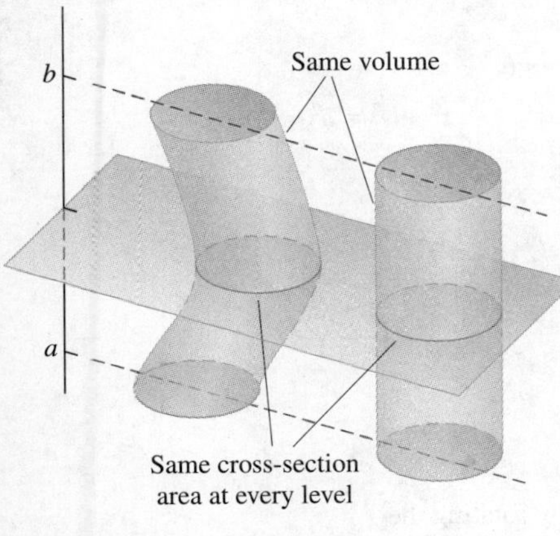

FIGURE 6.7 *Cavalieri's principle:* These solids have the same volume (imagine each solid as a stack of coins).

EXAMPLE 3 Cavalieri's principle says that solids with equal altitudes and identical cross-sectional areas at each height have the same volume (Figure 6.7). This follows immediately from the definition of volume, because the cross-sectional area function $A(x)$ and the interval $[a, b]$ are the same for both solids. ■

Solids of Revolution: The Disk Method

The solid generated by rotating (or revolving) a planar region about an axis in its plane is called a **solid of revolution**. To find the volume of a solid like the one shown in Figure 6.8, we first observe that the cross-sectional area $A(x)$ is the area of a disk of radius $R(x)$, where $R(x)$ is the distance from the axis of revolution to the planar region's boundary. The area is then

$$A(x) = \pi(\text{radius})^2 = \pi[R(x)]^2.$$

Therefore, the definition of volume gives us the following formula.

HISTORICAL BIOGRAPHY
Bonaventura Cavalieri
(1598–1647)
www.bit.ly/2P2HN30

Volume by Disks for Rotation About the x-Axis

$$V = \int_a^b A(x)\,dx = \int_a^b \pi[R(x)]^2\,dx.$$

This method for calculating the volume of a solid of revolution is often called the **disk method** because a cross-section is a circular disk of radius $R(x)$.

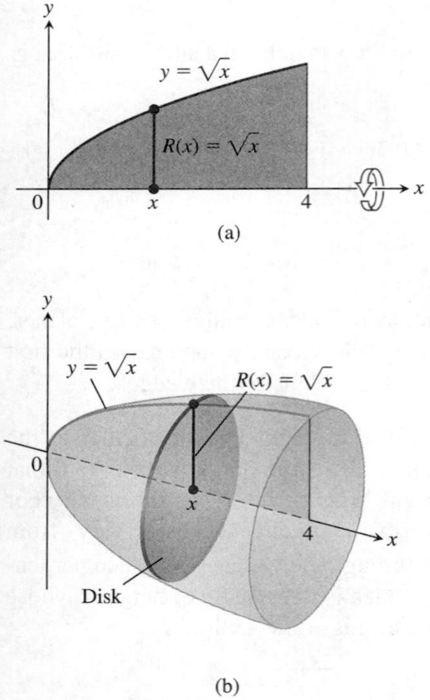

FIGURE 6.8 The region (a) and solid of revolution (b) in Example 4.

EXAMPLE 4 The region between the curve $y = \sqrt{x}$, $0 \le x \le 4$, and the x-axis is revolved about the x-axis to generate a solid. Find its volume.

Solution We draw the region and a typical radius (Figure 6.8a) and the generated solid (Figure 6.8b). The volume is

$$\begin{aligned} V &= \int_a^b \pi[R(x)]^2\,dx \\ &= \int_0^4 \pi\left[\sqrt{x}\right]^2 dx \qquad \text{Radius } R(x) = \sqrt{x} \text{ for rotation around } x\text{-axis.} \\ &= \pi \int_0^4 x\,dx = \pi \frac{x^2}{2}\bigg]_0^4 = \pi \frac{(4)^2}{2} = 8\pi. \end{aligned}$$

■

EXAMPLE 5 The circle

$$x^2 + y^2 = a^2$$

is rotated about the x-axis to generate a sphere. Find its volume.

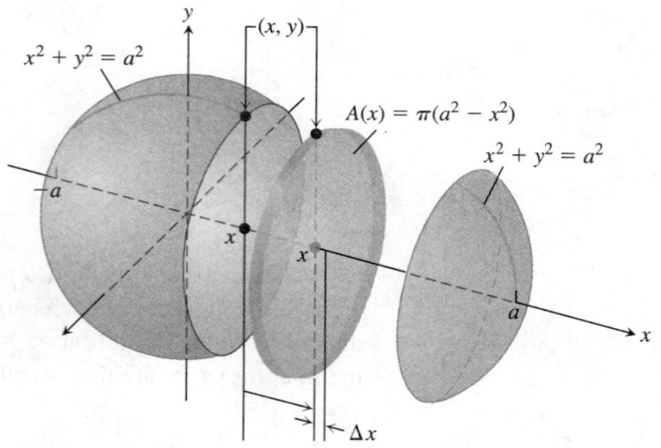

FIGURE 6.9 The sphere generated by rotating the circle $x^2 + y^2 = a^2$ about the x-axis. The radius is $R(x) = y = \sqrt{a^2 - x^2}$ (Example 5).

Solution We imagine the sphere cut into thin slices by planes perpendicular to the x-axis (Figure 6.9). The cross-sectional area at a typical point x between $-a$ and a is

$$A(x) = \pi y^2 = \pi(a^2 - x^2). \qquad R(x) = \sqrt{a^2 - x^2} \text{ for rotation around } x\text{-axis.}$$

Therefore, the volume is

$$V = \int_{-a}^{a} A(x)\,dx = \int_{-a}^{a} \pi(a^2 - x^2)\,dx = \pi\left[a^2x - \frac{x^3}{3}\right]_{-a}^{a} = \frac{4}{3}\pi a^3. \quad \blacksquare$$

The axis of revolution in the next example is not the x-axis, but the rule for calculating the volume is the same: Integrate $\pi(\text{radius})^2$ between appropriate limits.

EXAMPLE 6 Find the volume of the solid generated by revolving the region bounded by $y = \sqrt{x}$ and the lines $y = 1, x = 4$ about the line $y = 1$.

Solution We draw the region and a typical radius (Figure 6.10a), and the generated solid (Figure 6.10b). The volume is

$$\begin{aligned} V &= \int_1^4 \pi[R(x)]^2\,dx \\ &= \int_1^4 \pi[\sqrt{x} - 1]^2\,dx && \text{Radius } R(x) = \sqrt{x} - 1 \text{ for rotation around } y = 1. \\ &= \pi\int_1^4 [x - 2\sqrt{x} + 1]\,dx && \text{Expand integrand.} \\ &= \pi\left[\frac{x^2}{2} - 2\cdot\frac{2}{3}x^{3/2} + x\right]_1^4 = \frac{7\pi}{6}. && \text{Integrate.} \end{aligned}$$

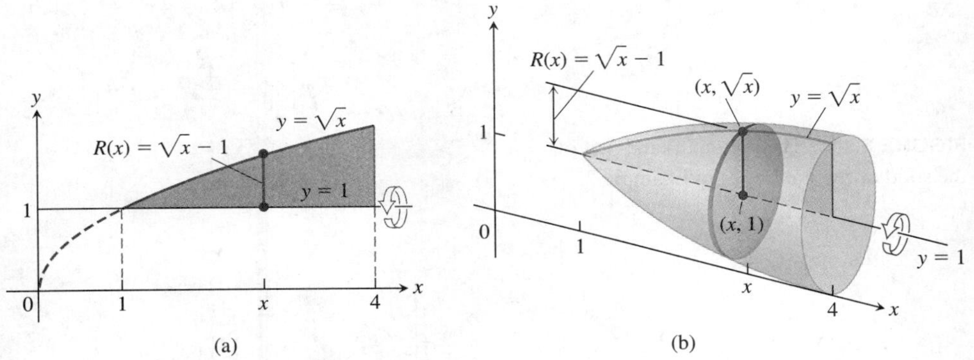

FIGURE 6.10 The region (a) and solid of revolution (b) in Example 6. $\blacksquare$

To find the volume of a solid generated by revolving a region between the y-axis and a curve $x = R(y)$, $c \le y \le d$, about the y-axis, we use the same method with x replaced by y. In this case, the area of the circular cross-section is

$$A(y) = \pi[\text{radius}]^2 = \pi[R(y)]^2,$$

and the definition of volume gives us the following formula.

Volume by Disks for Rotation About the y-Axis

$$V = \int_c^d A(y)\,dy = \int_c^d \pi[R(y)]^2\,dy.$$

EXAMPLE 7 Find the volume of the solid generated by revolving the region between the y-axis and the curve $x = 2/y$, $1 \le y \le 4$, about the y-axis.

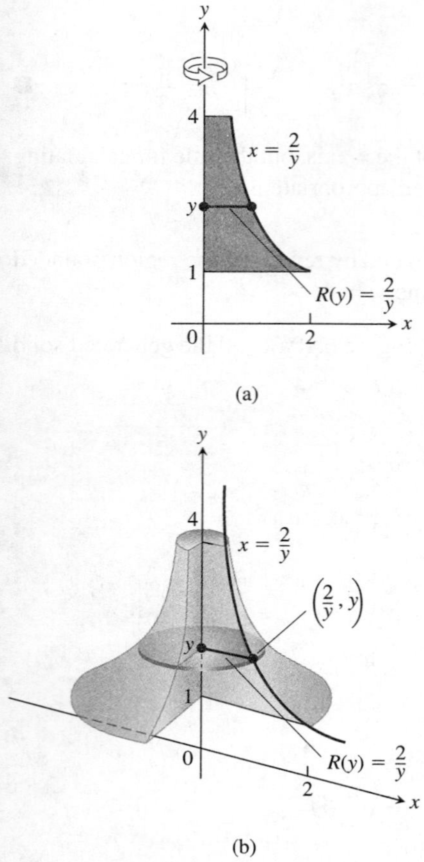

FIGURE 6.11 The region (a) and part of the solid of revolution (b) in Example 7.

Solution We draw the region and a typical radius (Figure 6.11a) and the generated solid (Figure 6.11b). The volume is

$$V = \int_1^4 \pi [R(y)]^2 \, dy$$

$$= \int_1^4 \pi \left(\frac{2}{y}\right)^2 dy \qquad \text{Radius } R(y) = \frac{2}{y} \text{ for rotation around } y\text{-axis}$$

$$= \pi \int_1^4 \frac{4}{y^2} \, dy = 4\pi \left[-\frac{1}{y}\right]_1^4 = 4\pi \left[\frac{3}{4}\right] = 3\pi.$$

■

EXAMPLE 8 Find the volume of the solid generated by revolving the region between the parabola $x = y^2 + 1$ and the line $x = 3$ about the line $x = 3$.

Solution We draw the region and a typical radius (Figure 6.12a) and the generated solid (Figure 6.12b). Note that the cross-sections are perpendicular to the line $x = 3$ and have y-coordinates from $y = -\sqrt{2}$ to $y = \sqrt{2}$. The volume is

$$V = \int_{-\sqrt{2}}^{\sqrt{2}} \pi [R(y)]^2 \, dy \qquad y = \pm\sqrt{2} \text{ when } x = 3$$

$$= \int_{-\sqrt{2}}^{\sqrt{2}} \pi [2 - y^2]^2 \, dy \qquad \text{Radius } R(y) = 3 - (y^2 + 1) \text{ for rotation around axis } x = 3.$$

$$= \pi \int_{-\sqrt{2}}^{\sqrt{2}} [4 - 4y^2 + y^4] \, dy \qquad \text{Expand integrand.}$$

$$= \pi \left[4y - \frac{4}{3}y^3 + \frac{y^5}{5}\right]_{-\sqrt{2}}^{\sqrt{2}} \qquad \text{Integrate.}$$

$$= \frac{64\pi\sqrt{2}}{15}.$$

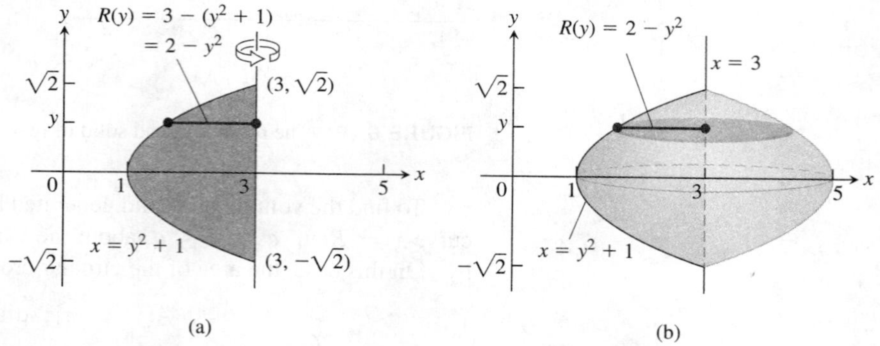

FIGURE 6.12 The region (a) and solid of revolution (b) in Example 8.

■

Solids of Revolution: The Washer Method

If the region we revolve to generate a solid does not border on or cross the axis of revolution, then the solid has a hole in it (Figure 6.13). The cross-sections perpendicular to the axis of revolution are *washers* (the circular surface in Figure 6.13) instead of disks. The dimensions of a typical washer are

Outer radius: $R(x)$

Inner radius: $r(x)$

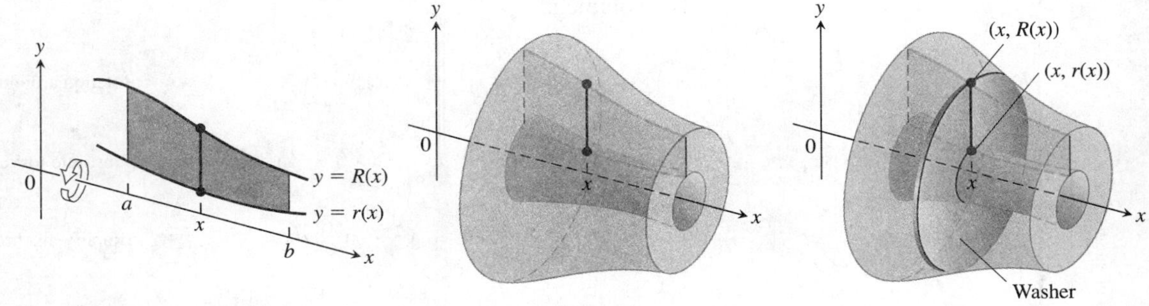

FIGURE 6.13 The cross-sections of the solid of revolution generated here are washers, not disks, so the integral $\int_a^b A(x)\,dx$ leads to a slightly different formula.

The washer's area is the area of a circle of radius $R(x)$ minus the area of a circle of radius $r(x)$:

$$A(x) = \pi[R(x)]^2 - \pi[r(x)]^2 = \pi\left([R(x)]^2 - [r(x)]^2\right).$$

Consequently, the definition of volume in this case gives us the following formula.

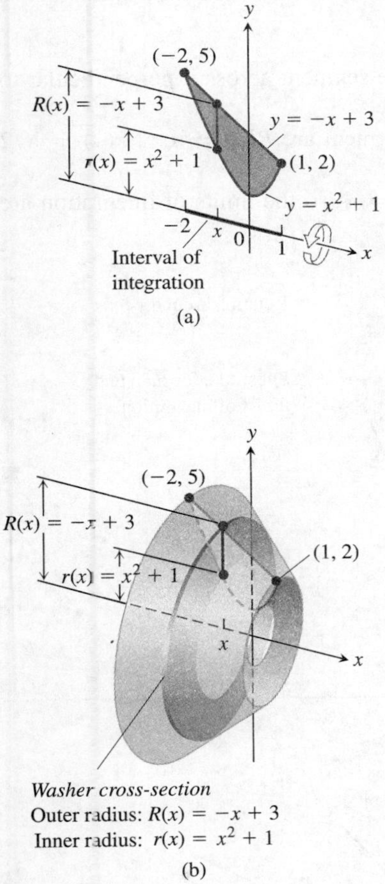

FIGURE 6.14 (a) The region in Example 9 spanned by a line segment perpendicular to the axis of revolution. (b) When the region is revolved about the x-axis, the line segment generates a washer.

Volume by Washers for Rotation About the x-Axis

$$V = \int_a^b A(x)\,dx = \int_a^b \pi\left([R(x)]^2 - [r(x)]^2\right)dx.$$

This method for calculating the volume of a solid of revolution is called the **washer method** because a thin slab of the solid resembles a circular washer with outer radius $R(x)$ and inner radius $r(x)$.

EXAMPLE 9 The region bounded by the curve $y = x^2 + 1$ and the line $y = -x + 3$ is revolved about the x-axis to generate a solid. Find the volume of the solid.

Solution We draw the region and sketch a line segment across it perpendicular to the axis of revolution (Figure 6.14a). We then find the outer and inner radii of the washer that would be swept out by the line segment if it were revolved about the x-axis along with the region. These radii are the distances of the ends of the line segment from the axis of revolution (see Figure 6.14).

$$\text{Outer radius:}\quad R(x) = -x + 3$$

$$\text{Inner radius:}\quad r(x) = x^2 + 1$$

We obtain the limits of integration by finding the x-coordinates of the intersection points of the curve and line in Figure 6.14a.

$$x^2 + 1 = -x + 3$$

$$x^2 + x - 2 = 0$$

$$(x + 2)(x - 1) = 0$$

$$x = -2, \quad x = 1 \qquad \text{Limits of integration}$$

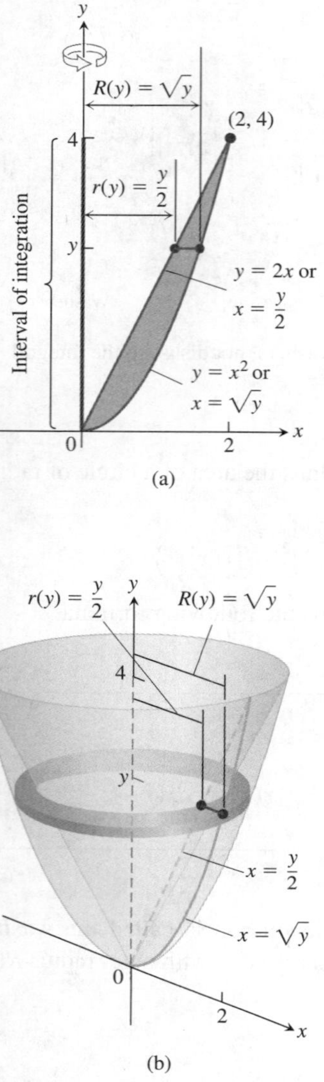

FIGURE 6.15 (a) The region being rotated about the y-axis, the washer radii, and limits of integration in Example 10. (b) The washer swept out by the line segment in part (a).

The volume is

$$V = \int_a^b \pi\left(\left[R(x)\right]^2 - \left[r(x)\right]^2\right) dx \qquad \text{Rotation around } x\text{-axis}$$

$$= \int_{-2}^{1} \pi\left((-x+3)^2 - (x^2+1)^2\right) dx \qquad \text{Substitute for radii and limits of integration.}$$

$$= \pi \int_{-2}^{1} (8 - 6x - x^2 - x^4)\, dx \qquad \text{Simplify algebraically.}$$

$$= \pi\left[8x - 3x^2 - \frac{x^3}{3} - \frac{x^5}{5}\right]_{-2}^{1} = \frac{117\pi}{5}. \qquad \text{Integrate.}$$

■

To find the volume of a solid formed by revolving a region about the y-axis, we use the same procedure as in Example 9, but integrate with respect to y instead of x. In this situation, the line segment sweeping out a typical washer is perpendicular to the y-axis (the axis of revolution), and the outer and inner radii of the washer are functions of y.

EXAMPLE 10 The region bounded by the parabola $y = x^2$ and the line $y = 2x$ in the first quadrant is revolved about the y-axis to generate a solid. Find the volume of the solid.

Solution First we sketch the region and draw a line segment across it perpendicular to the axis of revolution (the y-axis). See Figure 6.15a.

The radii of the washer swept out by the line segment are $R(y) = \sqrt{y}$, $r(y) = y/2$ (Figure 6.15).

The line and parabola intersect at $y = 0$ and $y = 4$, so the limits of integration are $c = 0$ and $d = 4$. We integrate to find the volume:

$$V = \int_c^d \pi\left(\left[R(y)\right]^2 - \left[r(y)\right]^2\right) dy \qquad \text{Rotation around } y\text{-axis}$$

$$= \int_0^4 \pi\left(\left[\sqrt{y}\right]^2 - \left[\frac{y}{2}\right]^2\right) dy \qquad \text{Substitute for radii and limits of integration.}$$

$$= \pi \int_0^4 \left(y - \frac{y^2}{4}\right) dy = \pi\left[\frac{y^2}{2} - \frac{y^3}{12}\right]_0^4 = \frac{8}{3}\pi.$$

■

EXERCISES 6.1

Volumes by Slicing

Find the volumes of the solids in Exercises 1–10.

1. The solid lies between planes perpendicular to the x-axis at $x = 0$ and $x = 4$. The cross-sections perpendicular to the axis on the interval $0 \le x \le 4$ are squares whose diagonals run from the parabola $y = -\sqrt{x}$ to the parabola $y = \sqrt{x}$.

2. The solid lies between planes perpendicular to the x-axis at $x = -1$ and $x = 1$. The cross-sections perpendicular to the x-axis are circular disks whose diameters run from the parabola $y = x^2$ to the parabola $y = 2 - x^2$.

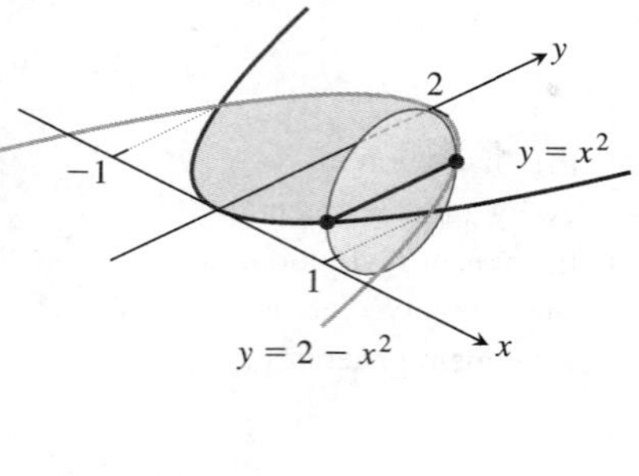

3. The solid lies between planes perpendicular to the x-axis at $x = -1$ and $x = 1$. The cross-sections perpendicular to the x-axis between these planes are squares whose bases run from the semicircle $y = -\sqrt{1 - x^2}$ to the semicircle $y = \sqrt{1 - x^2}$.

4. The solid lies between planes perpendicular to the x-axis at $x = -1$ and $x = 1$. The cross-sections perpendicular to the x-axis between these planes are squares whose bases run from the semicircle $y = -\sqrt{4 - x^2}$ to the semicircle $y = \sqrt{4 - x^2}$.

5. The base of a solid is the region between the curve $y = 2\sqrt{\sin x}$ and the interval $[0, \pi]$ on the x-axis. The cross-sections perpendicular to the x-axis are

a. equilateral triangles with bases running from the x-axis to the curve as shown in the accompanying figure.

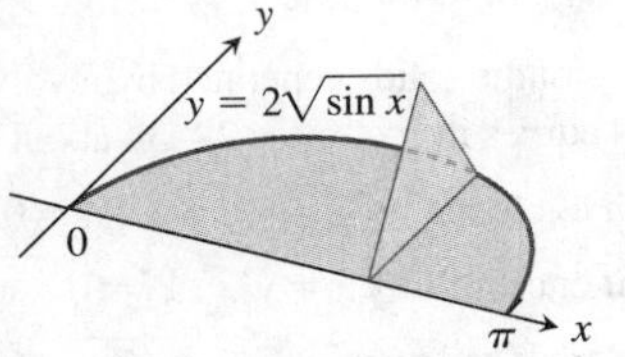

b. squares with bases running from the x-axis to the curve.

6. The solid lies between planes perpendicular to the x-axis at $x = -\pi/3$ and $x = \pi/3$. The cross-sections perpendicular to the x-axis are

a. circular disks with diameters running from the curve $y = \tan x$ to the curve $y = \sec x$.

b. squares whose bases run from the curve $y = \tan x$ to the curve $y = \sec x$.

7. The base of a solid is the region bounded by the graphs of $y = 4x$, $y = 8$, and $x = 0$. The cross-sections perpendicular to the x-axis are

a. rectangles of height 10.

b. rectangles of perimeter 12.

8. The base of a solid is the region bounded by the graphs of $y = \sqrt{x}$ and $y = x/2$. The cross-sections perpendicular to the x-axis are

a. isosceles triangles of height 6.

b. semicircles with diameters running across the base of the solid.

9. The solid lies between planes perpendicular to the y-axis at $y = 0$ and $y = 2$. The cross-sections perpendicular to the y-axis are circular disks with diameters running from the y-axis to the parabola $x = \sqrt{5}y^2$.

10. The base of the solid is the disk $x^2 + y^2 \leq 1$. The cross-sections by planes perpendicular to the y-axis between $y = -1$ and $y = 1$ are isosceles right triangles with one leg in the disk.

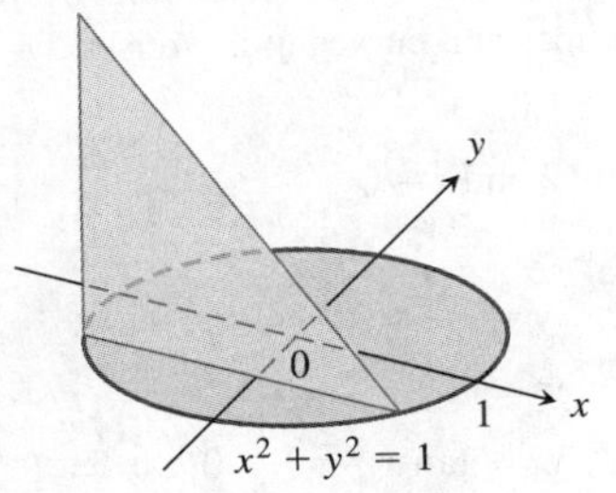

11. Find the volume of the given right tetrahedron. (*Hint:* Consider slices perpendicular to one of the labeled edges.)

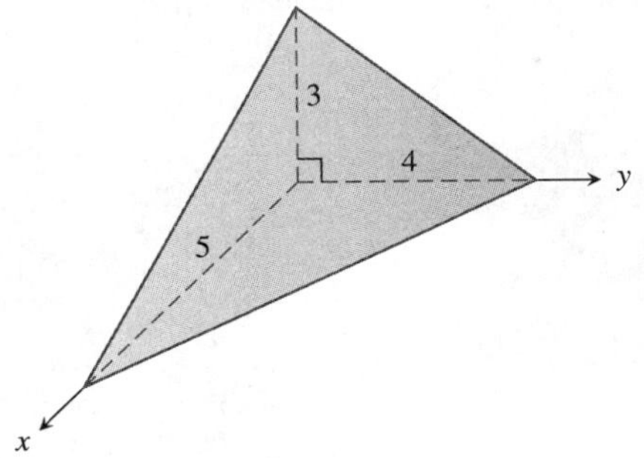

12. Find the volume of the given pyramid, which has a square base of area 9 and height 5.

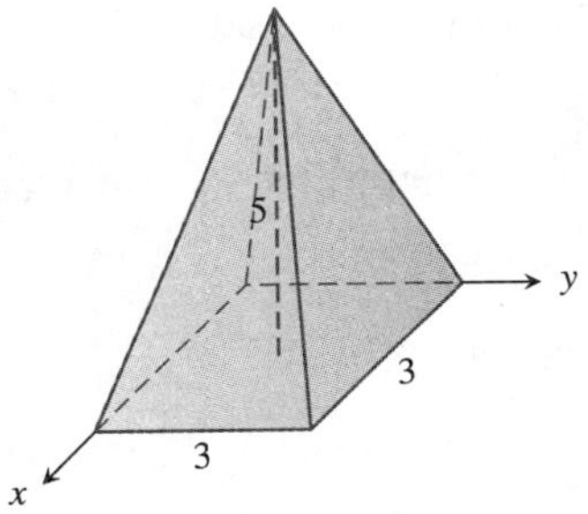

13. A twisted solid A square of side length s lies in a plane perpendicular to a line L. One vertex of the square lies on L. As this square moves a distance h along L, the square turns one revolution about L to generate a corkscrew-like column with square cross-sections.

a. Find the volume of the column.

b. What will the volume be if the square turns twice instead of once? Give reasons for your answer.

14. Cavalieri's principle A solid lies between planes perpendicular to the x-axis at $x = 0$ and $x = 12$. The cross-sections by planes perpendicular to the x-axis are circular disks whose diameters run from the line $y = x/2$ to the line $y = x$ as shown in the accompanying figure. Explain why the solid has the same volume as a right circular cone with base radius 3 and height 12.

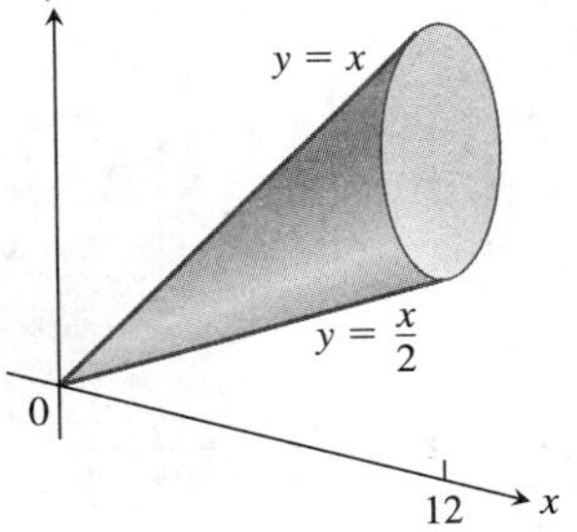

15. **Intersection of two half-cylinders** Two half-cylinders of diameter 2 meet at a right angle in the accompanying figure. Find the volume of the solid region common to both half-cylinders. (*Hint*: Consider slices parallel to the base of the solid.)

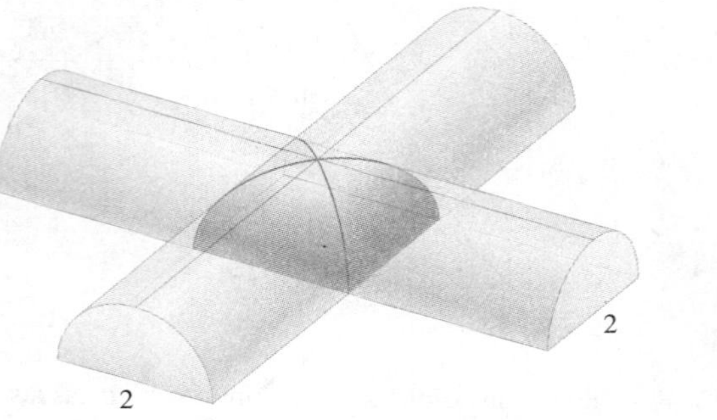

16. **Gasoline in a tank** A gasoline tank is in the shape of a right circular cylinder (lying on its side) of length 10 ft and radius 4 ft. Set up an integral that represents the volume of the gas in the tank if it is filled to a depth of 6 ft. You will learn how to compute this integral in Chapter 8 (or you may use geometry to find its value).

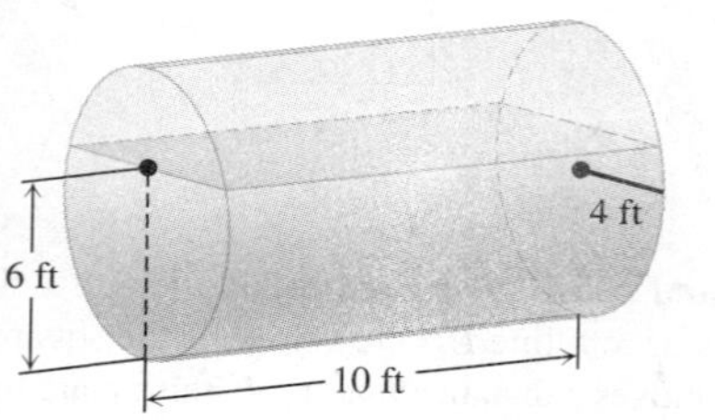

Volumes by the Disk Method

In Exercises 17–20, find the volume of the solid generated by revolving the shaded region about the given axis.

17. About the x-axis

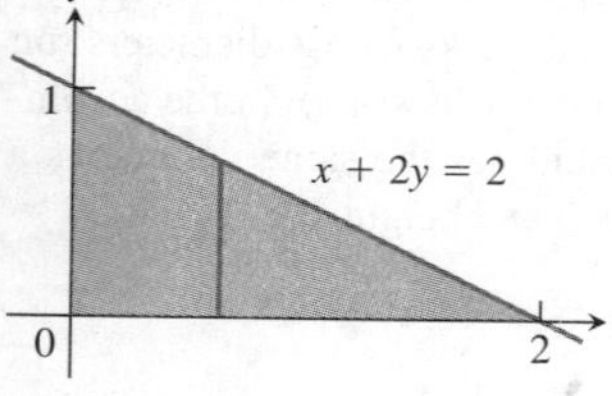

18. About the y-axis

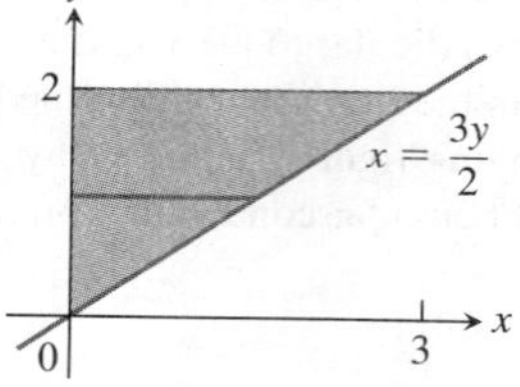

19. About the y-axis

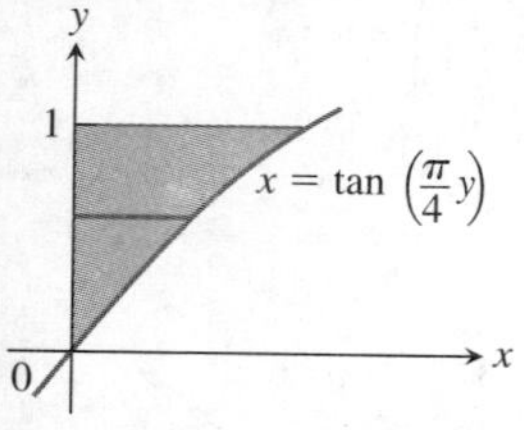

20. About the x-axis

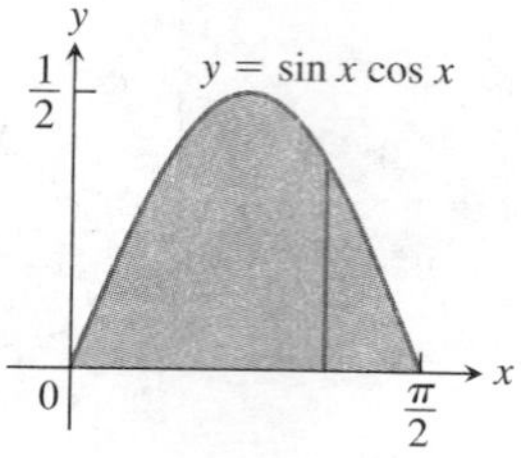

Find the volumes of the solids generated by revolving the regions bounded by the lines and curves in Exercises 21–26 about the x-axis.

21. $y = x^2, \quad y = 0, \quad x = 2$

22. $y = x^4, \quad y = 0, \quad x = 2$

23. $y = \sqrt{9 - x^2}, \quad y = 0$

24. $y = 4x - x^2, \quad y = 0$

25. $y = \sqrt{\cos x}, \quad 0 \le x \le \pi/2, \quad y = 0, \quad x = 0$

26. $y = \sec x, \quad y = 0, \quad x = -\pi/4, \quad x = \pi/4$

In Exercises 27 and 28, find the volume of the solid generated by revolving the region about the given line.

27. The region bounded by the curves $y = x^2$ and $y = \sqrt{x}$ and the lines $x = 0$ and $x = 1$, about the line $y = -1$.

28. The region in the first quadrant bounded above by the line $y = 2$, below by the curve $y = 2\sin x$, $0 \le x \le \pi/2$, and on the left by the y-axis, about the line $y = 2$

Find the volumes of the solids generated by revolving the regions bounded by the lines and curves in Exercises 29–34 about the y-axis.

29. The region enclosed by $x = \sqrt{3}y^2, \quad x = 0, \quad y = -1, \quad y = 1.$

30. The region enclosed by $x = y^2, \quad x = 0, \quad y = 4$

31. The region enclosed by $x = \sqrt{2 \sin 2y}, \quad 0 \le y \le \pi/2, \quad x = 0$

32. The region enclosed by $x = \sqrt{\cos(\pi y/4)}, \quad -2 \le y \le 0, \quad x = 0$

33. $x = 2\sqrt{y + 1}, \quad x = 0, \quad y = 0, \quad y = 3$

34. $x = \sqrt{2y}/(y^2 + 1), \quad x = 0, \quad y = 1$

Volumes by the Washer Method

Find the volumes of the solids generated by revolving the shaded regions in Exercises 35 and 36 about the indicated axes.

35. The x-axis

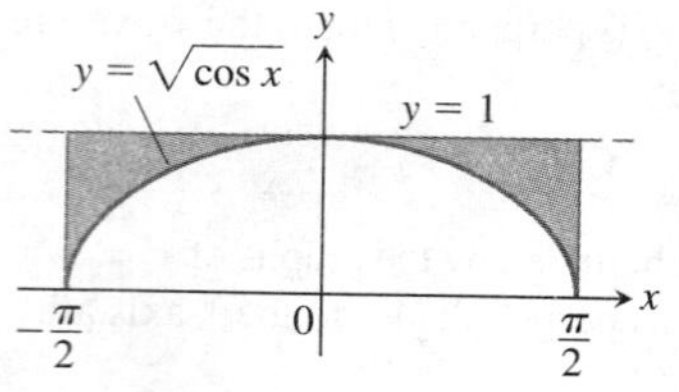

36. The y-axis

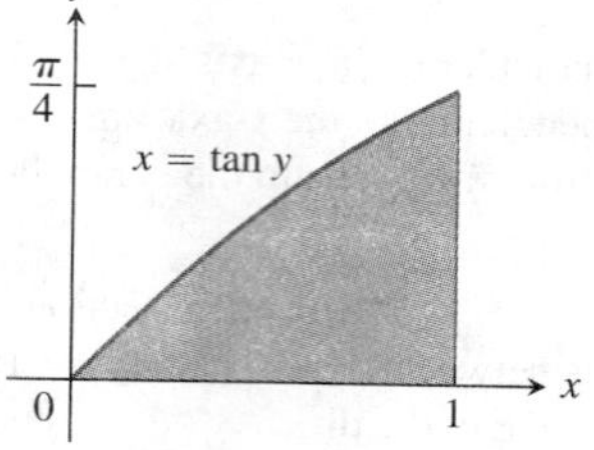

Find the volumes of the solids generated by revolving the regions bounded by the lines and curves in Exercises 37–42 about the x-axis.

37. $y = x, \quad y = 1, \quad x = 0$

38. $y = \sqrt{x}, y = 2$ and $x = 0$

39. $y = x^2 + 1, \quad y = x + 3$

40. $y = x^2 + 2, \quad y = x + 4$

41. $y = \sec x, \quad y = \sqrt{2}, \quad -\pi/4 \le x \le \pi/4$

42. $y = \sec x, \quad y = \tan x, \quad x = 0, \quad x = 1$

In Exercises 43–46, find the volume of the solid generated by revolving each region about the y-axis.

43. The region enclosed by the triangle with vertices (0, 0), (1, 0) and (1, 2).

44. The region enclosed by the triangle with vertices (0, 1), (1, 0), and (1, 1)

45. The region in the first quadrant bounded by the curve $x^2 + 4y^2 = 4$, $x = 0$, $x = 2$ and x axis.

46. The region in the first quadrant bounded on the left by the circle $x^2 - y^2 = 3$, on the right by the line $x = \sqrt{3}$, and above by the line $y = \sqrt{3}$

In Exercises 47 and 48, find the volume of the solid generated by revolving each region about the given axis.

47. The region in the first quadrant bounded by the curve $y = 2\sqrt{x-1}$ and $y = x - 1$, about the line $x = -1$.

48. The region in the first quadrant bounded by the parabola $y = x^2$, the y-axis and the line $y = 1$ is revolved about the line $x = 2$ to generate a solid.

Volumes of Solids of Revolution

49. Find the volume of the solid generated by revolving the region bounded by $y = \sqrt{x}$ and the lines $y = 2$ and $x = 0$ about

a. the x-axis. **b.** the y-axis.

c. the line $y = 2$. **d.** the line $x = 4$.

50. Find the volume of the solid generated by revolving the triangular region bounded by the lines $y = 2x$, $y = 0$, and $x = 1$ about

a. the line $x = 1$. **b.** the line $x = 2$.

51. Find the volume of the solid generated by revolving the region bounded by the parabola $y = x^2$ and the line $y = 1$ about

a. the line $y = 1$. **b.** the line $y = 2$.

c. the line $y = -1$.

52. By integration, find the volume of the solid generated by revolving the triangular region with vertices (0, 0), $(b, 0)$, $(0, h)$ about

a. the x-axis. **b.** the y-axis.

Theory and Applications

53. The volume of a torus The disk $x^2 + y^2 \le a^2$ is revolved about the line $x = b\,(b > a)$ to generate a solid shaped like a doughnut and called a *torus*. Find its volume.

(*Hint:* $\int_{-a}^{a} \sqrt{a^2 - y^2}\,dy = \pi a^2/2$, since it is the area of a semicircle of radius a.)

54. Volume of a bowl A bowl has a shape that can be generated by revolving the graph of $y = x^2/2$ between $y = 0$ and $y = 5$ about the y-axis.

a. Find the volume of the bowl.

b. Related rates If we fill the bowl with water at a constant rate of 3 cubic units per second, how fast will the water level in the bowl be rising when the water is 4 units deep?

55. Volume of a bowl

a. A hemispherical bowl of radius 5 m contains water to a depth 4 m. Find the volume of water in the bowl.

b. Related rates Water runs into a sunken concrete hemispherical bowl of radius 5 m at the rate of $0.2\ \text{m}^3/\text{sec}$. How fast is the water level in the bowl rising when the water is 4 m deep?

56. Explain how you could estimate the volume of a solid of revolution by measuring the shadow cast on a table parallel to its axis of revolution by a light shining directly above it.

57. Volume of a hemisphere Derive the formula $V = (2/3)\pi R^3$ for the volume of a hemisphere of radius R by comparing its cross-sections with the cross-sections of a solid right circular cylinder of radius R and height R from which a solid right circular cone of base radius R and height R has been removed, as suggested by the accompanying figure.

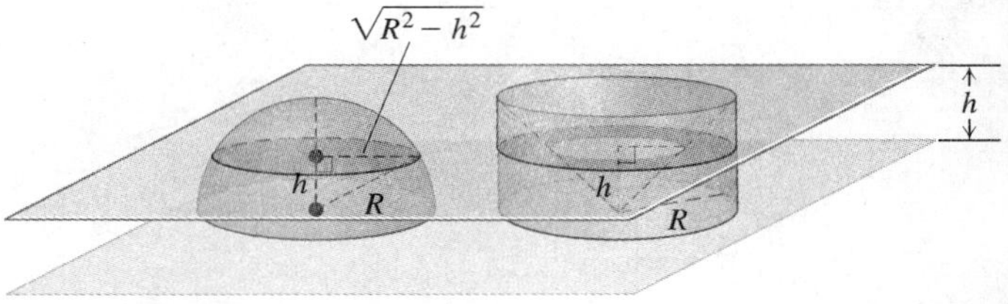

58. Designing a plumb bob Having been asked to design a brass plumb bob that will weigh in the neighborhood of 190 g, you decide to shape it like the solid of revolution shown here. Find the plumb bob's volume. If you specify a brass that weighs $8.5\ \text{g/cm}^3$, how much will the plumb bob weigh (to the nearest gram)?

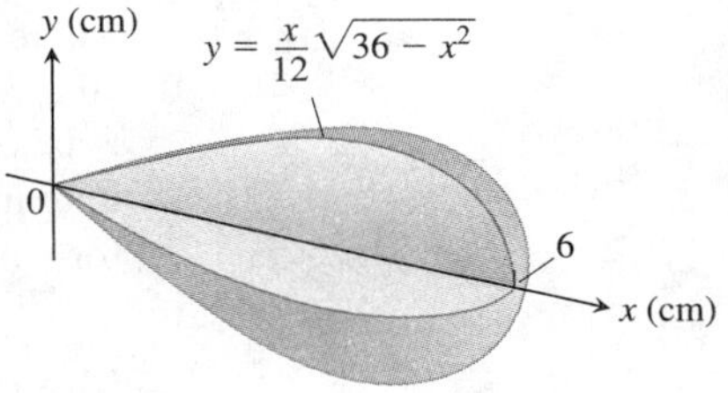

59. Designing a wok You are designing a wok frying pan that will be shaped like a spherical bowl with handles. A bit of experimentation at home persuades you that you can get one that holds about 3 L if you make it 9 cm deep and give the sphere a radius of 16 cm. To be sure, you picture the wok as a solid of revolution, as shown here, and calculate its volume with an integral. To the nearest cubic centimeter, what volume do you really get? ($1\ \text{L} = 1000\ \text{cm}^3$)

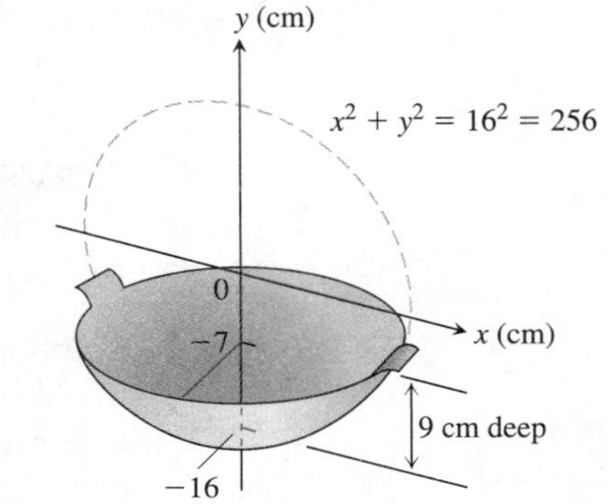

60. Max-min The arch $y = \sin x$, $0 \le x \le \pi$, is revolved about the line $y = c$, $0 \le c \le 1$, to generate the solid in the accompanying figure.

a. Find the value of c that minimizes the volume of the solid. What is the minimum volume?

b. What value of c in $[0, 1]$ maximizes the volume of the solid?

T **c.** Graph the solid's volume as a function of c, first for $0 \le c \le 1$ and then on a larger domain. What happens to the volume of the solid as c moves away from $[0, 1]$? Does this make sense physically? Give reasons for your answers.

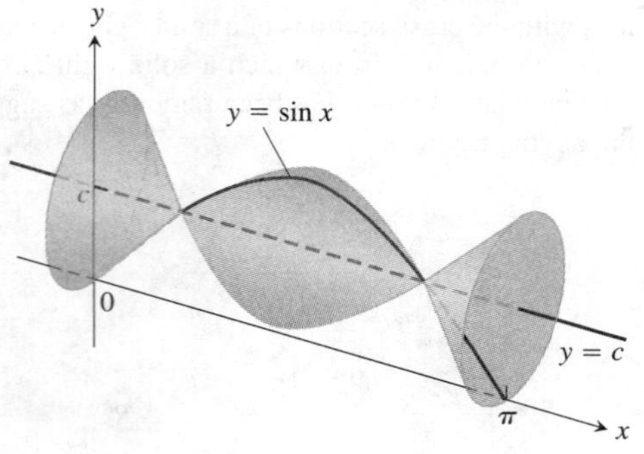

61. Consider the region R bounded by the graphs of $y = f(x) > 0$, $x = a > 0$, $x = b > a$, and $y = 0$ (see accompanying figure). If the volume of the solid formed by revolving R about the x-axis is 4π, and the volume of the solid formed by revolving R about the line $y = -1$ is 8π, find the area of R.

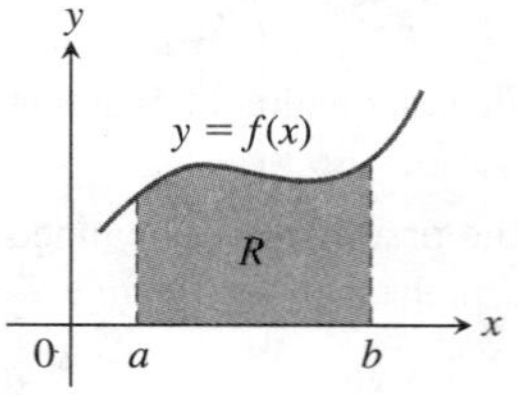

62. Consider the region R given in Exercise 61. If the volume of the solid formed by revolving R around the x-axis is 6π, and the volume of the solid formed by revolving R around the line $y = -2$ is 10π, find the area of R.

6.2 Volumes Using Cylindrical Shells

In Section 6.1 we defined the volume of a solid to be the definite integral $V = \int_a^b A(x)\,dx$, where $A(x)$ is an integrable cross-sectional area of the solid from $x = a$ to $x = b$. The area $A(x)$ was obtained by slicing through the solid with a plane perpendicular to the x-axis. However, this method of slicing is sometimes awkward to apply, as we will illustrate in our first example. To overcome this difficulty, we use the same integral definition for volume, but obtain the area by slicing through the solid in a different way.

Slicing with Cylinders

Suppose we slice through the solid using circular cylinders of increasing radii, like cookie cutters. We slice straight down through the solid so that the axis of each cylinder is parallel to the y-axis. The vertical axis of each cylinder is always the same line, but the radii of the cylinders increase with each slice. In this way the solid is sliced up into thin cylindrical shells of constant thickness that grow outward from their common axis, like circular tree rings. Unrolling a cylindrical shell shows that its volume is approximately that of a rectangular slab with area $A(x)$ and thickness Δx. This slab interpretation allows us to apply the same integral definition for volume as before. The following example provides some insight.

EXAMPLE 1 The region enclosed by the x-axis and the parabola $y = f(x) = 3x - x^2$ is revolved about the vertical line $x = -1$ to generate a solid (see Figure 6.16). Find the volume of the solid.

Solution Using the washer method from Section 6.1 would be awkward here because we would need to express the x-values of the left and right sides of the parabola in Figure 6.16a in terms of y. This is because these x-values, which describe the inner and outer radii of a typical washer, are solutions to the equation $y = 3x - x^2$, and this gives a complicated formula for x. Therefore, instead of rotating a horizontal strip of thickness Δy, we rotate a *vertical strip* of thickness Δx. This rotation produces a *cylindrical shell* of height y_k above a point x_k within the base of the vertical strip and of thickness Δx. An example of a cylindrical shell is shown in Figure 6.17. We can think of the cylindrical shell shown in the figure as approximating a slice of the solid obtained by cutting straight down through it, parallel to the axis of revolution, all the way around.

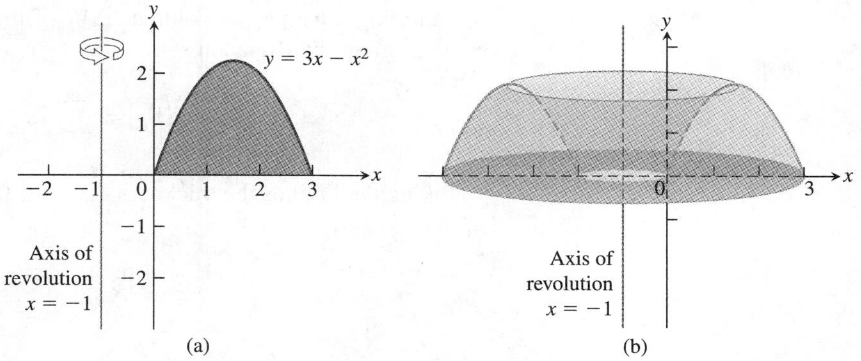

FIGURE 6.16 (a) The graph of the region in Example 1, before revolution. (b) The solid formed when the region in part (a) is revolved about the axis of revolution $x = -1$.

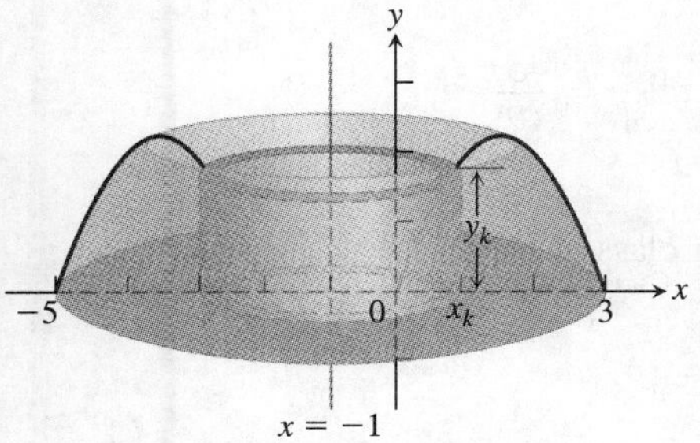

FIGURE 6.17 A cylindrical shell of height y_k obtained by rotating a vertical strip of thickness Δx_k about the line $x = -1$. The outer radius of the cylinder occurs at x_k, where the height of the parabola is $y_k = 3x_k - x_k^2$ (Example 1).

We start by cutting close to the inside hole and then cut another cylindrical slice around the enlarged hole, then another, and so on, obtaining n cylinders. The radii of the cylinders gradually increase, and the heights of the cylinders follow the contour of the parabola: shorter to taller, then back to shorter (Figure 6.16a). The sum of the volumes of the shells is a Riemann sum that approximates the volume of the entire solid.

Each shell sits over a subinterval $[x_{k-1}, x_k]$ in the x-axis. The thickness of the shell is $\Delta x_k = x_k - x_{k-1}$. Because the parabola is rotated around the line $x = -1$, the outer radius of the shell is $1 + x_k$. The height of the shell is the height of the parabola at some point in the interval $[x_{k-1}, x_k]$, or approximately $y_k = f(x_k) = 3x_k - x_k^2$. If we unroll this cylinder and flatten it out, it becomes (approximately) a rectangular slab with thickness Δx_k (see Figure 6.18). The height of the rectangular slab is approximately $y_k = 3x_k - x_k^2$, and its length is the circumference of the shell, which is approximately $2\pi \cdot \text{radius} = 2\pi(1 + x_k)$. Hence the volume of the shell is approximately the volume of the rectangular slab, which is

$$\begin{aligned}\Delta V_k &= \text{circumference} \times \text{height} \times \text{thickness}\\ &= 2\pi(1 + x_k) \cdot (3x_k - x_k^2) \cdot \Delta x_k.\end{aligned}$$

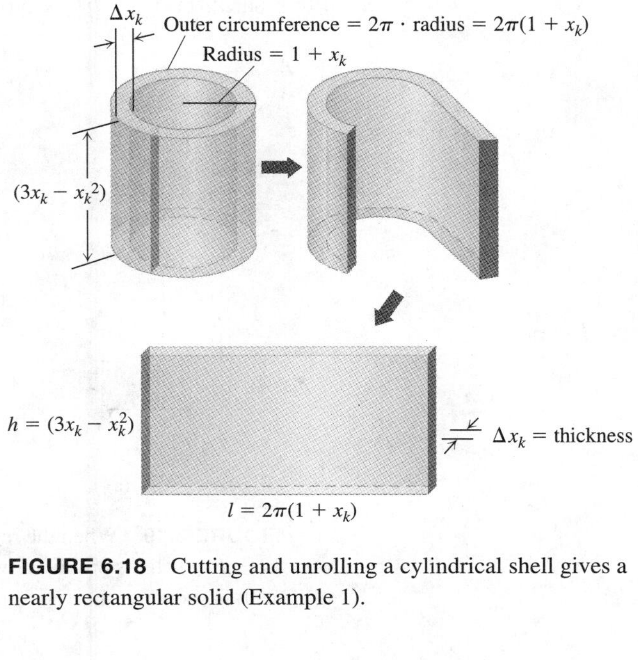

FIGURE 6.18 Cutting and unrolling a cylindrical shell gives a nearly rectangular solid (Example 1).

Summing together the volumes ΔV_k of the individual cylindrical shells over the interval $[0, 3]$ gives the Riemann sum

$$\sum_{k=1}^{n} \Delta V_k = \sum_{k=1}^{n} 2\pi(x_k + 1)(3x_k - x_k^2)\,\Delta x_k.$$

Taking the limit as the thickness $\Delta x_k \to 0$ and $n \to \infty$ gives the volume integral

$$\begin{aligned} V &= \lim_{n\to\infty} \sum_{k=1}^{n} 2\pi(x_k + 1)(3x_k - x_k^2)\,\Delta x_k \\ &= \int_0^3 2\pi(x + 1)(3x - x^2)\,dx \\ &= \int_0^3 2\pi(3x^2 + 3x - x^3 - x^2)\,dx \\ &= 2\pi\int_0^3 (2x^2 + 3x - x^3)\,dx \\ &= 2\pi\left[\frac{2}{3}x^3 + \frac{3}{2}x^2 - \frac{1}{4}x^4\right]_0^3 = \frac{45\pi}{2}. \end{aligned}$$

■

We now generalize this procedure to a broader class of solids.

The Shell Method

Suppose that the region bounded by the graph of a nonnegative continuous function $y = f(x)$ and the x-axis over the finite closed interval $[a, b]$ lies to the right of the vertical line $x = L$ (see Figure 6.19a). We assume $a \geq L$, so the vertical line may touch the region but cannot pass through it. We generate a solid S by rotating this region about the vertical line L.

Let P be a partition of the interval $[a, b]$ by the points $a = x_0 < x_1 < \cdots < x_n = b$. As usual, we choose a point c_k in each subinterval $[x_{k-1}, x_k]$. In Example 1 we chose c_k to be the endpoint x_k, but now it will be more convenient to let c_k be the midpoint of the subinterval $[x_{k-1}, x_k]$. We approximate the region in Figure 6.19a with rectangles based on this partition of $[a, b]$. A typical approximating rectangle has height $f(c_k)$ and width

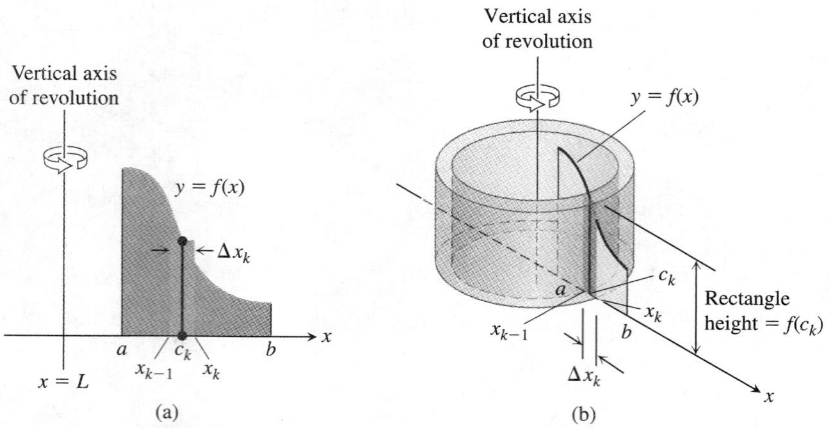

FIGURE 6.19 When the region shown in (a) is revolved about the vertical line $x = L$, a solid is produced which can be sliced into cylindrical shells. A typical shell is shown in (b).

The volume of a cylindrical shell of height h with inner radius r and outer radius R is

$$\pi R^2 h - \pi r^2 h = 2\pi\left(\frac{R+r}{2}\right)(h)(R-r).$$

$\Delta x_k = x_k - x_{k-1}$. If this rectangle is rotated about the vertical line $x = L$, then a shell is swept out, as in Figure 6.19b. A formula from geometry tells us that the volume of the shell swept out by the rectangle is

$$\begin{aligned}\Delta V_k &= 2\pi \times \text{average shell radius} \times \text{shell height} \times \text{thickness}\\ &= 2\pi \cdot (c_k - L) \cdot f(c_k) \cdot \Delta x_k. \qquad R = x_k - L \text{ and } r = x_{k-1} - L\end{aligned}$$

We approximate the volume of the solid S by summing the volumes of the shells swept out by the n rectangles:

$$V \approx \sum_{k=1}^{n} \Delta V_k.$$

The limit of this Riemann sum as each $\Delta x_k \to 0$ and $n \to \infty$ gives the volume of the solid as a definite integral:

$$\begin{aligned}V = \lim_{n\to\infty} \sum_{k=1}^{n} \Delta V_k &= \int_a^b 2\pi(\text{shell radius})(\text{shell height})\, dx\\ &= \int_a^b 2\pi(x - L) f(x)\, dx.\end{aligned}$$

We refer to the variable of integration, here x, as the **thickness variable**. To emphasize the *process* of the shell method, we state the general formula in terms of the shell radius and shell height. This will allow for rotations about a horizontal line $y = L$ as well.

Shell Formula for Revolution About a Vertical Line

The volume of the solid generated by revolving the region between the x-axis and the graph of a continuous function $y = f(x) \geq 0$, $L \leq a \leq x \leq b$, about a vertical line $x = L$ is

$$V = \int_a^b 2\pi \begin{pmatrix}\text{shell}\\ \text{radius}\end{pmatrix}\begin{pmatrix}\text{shell}\\ \text{height}\end{pmatrix} dx.$$

EXAMPLE 2 The region bounded by the curve $y = \sqrt{x}$, the x-axis, and the line $x = 4$ is revolved about the y-axis to generate a solid. Find the volume of the solid.

Solution Sketch the region and draw a line segment across it *parallel* to the axis of revolution (Figure 6.20a). Label the segment's height (shell height) and distance from the axis of revolution (shell radius). (We drew the shell in Figure 6.20b, but you need not do that.)

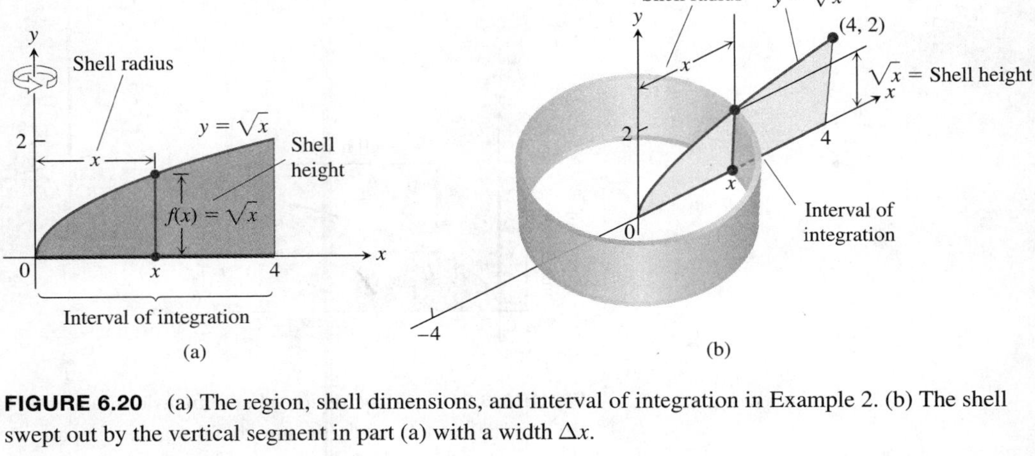

FIGURE 6.20 (a) The region, shell dimensions, and interval of integration in Example 2. (b) The shell swept out by the vertical segment in part (a) with a width Δx.

The shell thickness variable is x, so the limits of integration for the shell formula are $a = 0$ and $b = 4$ (Figure 6.20). The volume is

$$V = \int_a^b 2\pi \begin{pmatrix} \text{shell} \\ \text{radius} \end{pmatrix} \begin{pmatrix} \text{shell} \\ \text{height} \end{pmatrix} dx$$

$$= \int_0^4 2\pi(x)(\sqrt{x})\, dx$$

$$= 2\pi \int_0^4 x^{3/2}\, dx = 2\pi \left[\frac{2}{5}x^{5/2}\right]_0^4 = \frac{128\pi}{5}. \qquad \blacksquare$$

So far, we have used vertical axes of revolution. For horizontal axes, we replace the x's with y's.

EXAMPLE 3 The region bounded by the curve $y = \sqrt{x}$, the x-axis, and the line $x = 4$ is revolved about the x-axis to generate a solid. Find the volume of the solid by the shell method.

Solution This is the solid whose volume was found by the disk method in Example 4 of Section 6.1. Now we find its volume by the shell method. First, sketch the region and draw a line segment across it *parallel* to the axis of revolution (Figure 6.21a). Label the segment's length (shell height) and distance from the axis of revolution (shell radius). (We drew the shell in Figure 6.21b, but you need not do that.)

In this case, the shell thickness variable is y, so the limits of integration for the shell formula method are $a = 0$ and $b = 2$ (along the y-axis in Figure 6.21). The volume of the solid is

$$V = \int_a^b 2\pi \begin{pmatrix} \text{shell} \\ \text{radius} \end{pmatrix} \begin{pmatrix} \text{shell} \\ \text{height} \end{pmatrix} dy$$

$$= \int_0^2 2\pi(y)(4 - y^2)\, dy$$

$$= 2\pi \int_0^2 (4y - y^3)\, dy$$

$$= 2\pi \left[2y^2 - \frac{y^4}{4}\right]_0^2 = 8\pi.$$

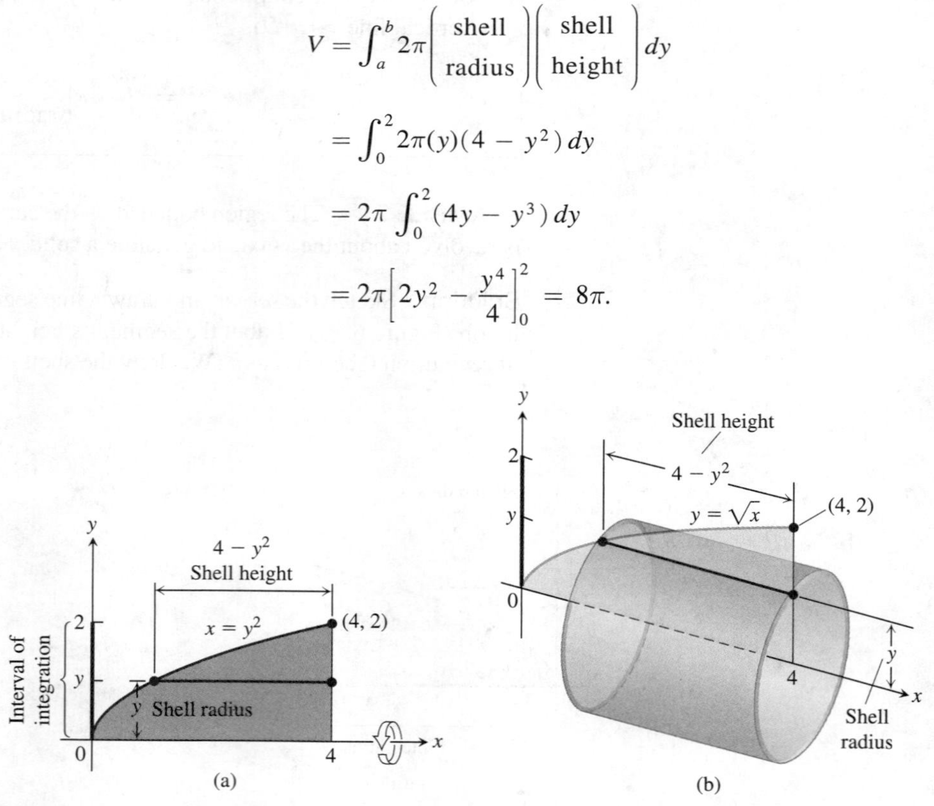

FIGURE 6.21 (a) The region, shell dimensions, and interval of integration in Example 3. (b) The shell swept out by the horizontal segment in part (a) with a width Δy. $\blacksquare$

Summary of the Shell Method

Regardless of the position of the axis of revolution (horizontal or vertical), the steps for implementing the shell method are these.

1. *Draw the region and sketch a line segment* across it *parallel* to the axis of revolution. *Label* the segment's height or length (shell height) and distance from the axis of revolution (shell radius).
2. *Find the limits of integration* for the thickness variable.
3. *Integrate* the product 2π (shell radius) (shell height) with respect to the thickness variable (x or y) to find the volume.

The shell method gives the same answer as the washer method when both are used to calculate the volume of a region. We do not prove that result here, but it is illustrated in Exercises 37 and 38. Both volume formulas are actually special cases of a general volume formula we will look at when studying double and triple integrals in Chapter 15. That general formula also allows for computing volumes of solids other than those swept out by regions of revolution.

EXERCISES 6.2

Revolution About the Axes

In Exercises 1–6, use the shell method to find the volumes of the solids generated by revolving the shaded region about the indicated axis.

1.

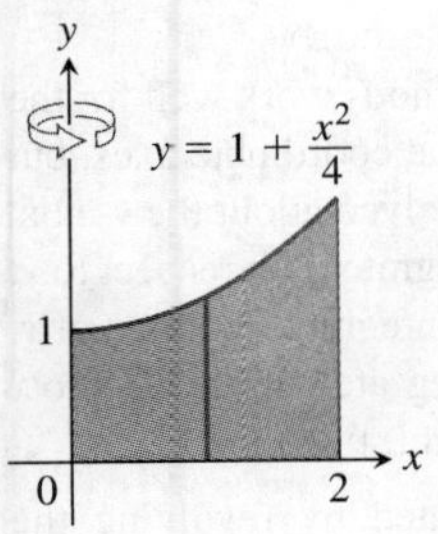

2.

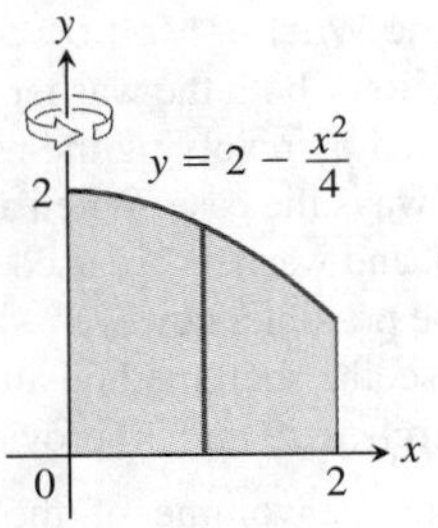

3.

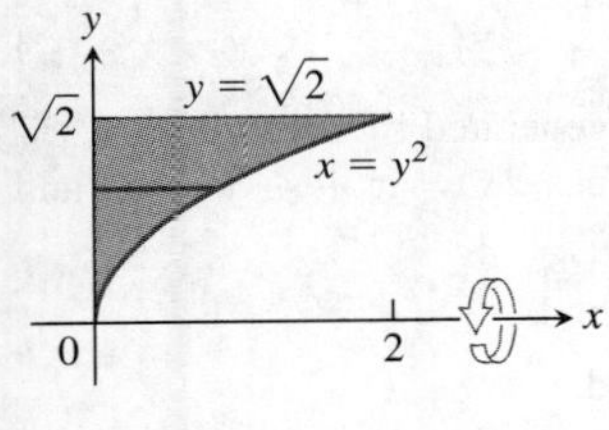

4.

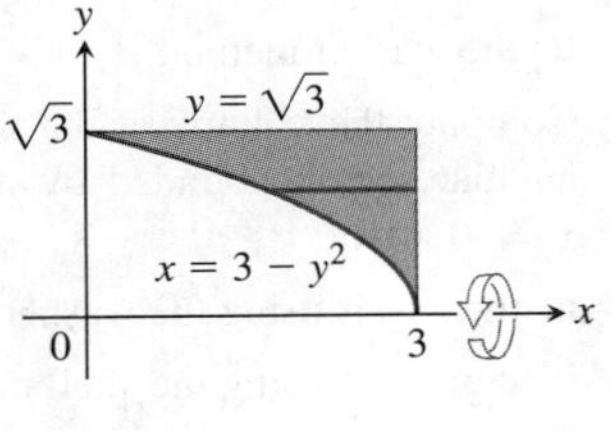

5. The y-axis

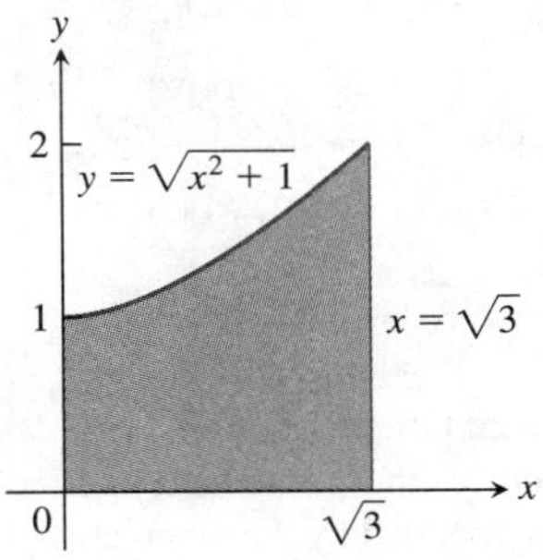

6. The y-axis

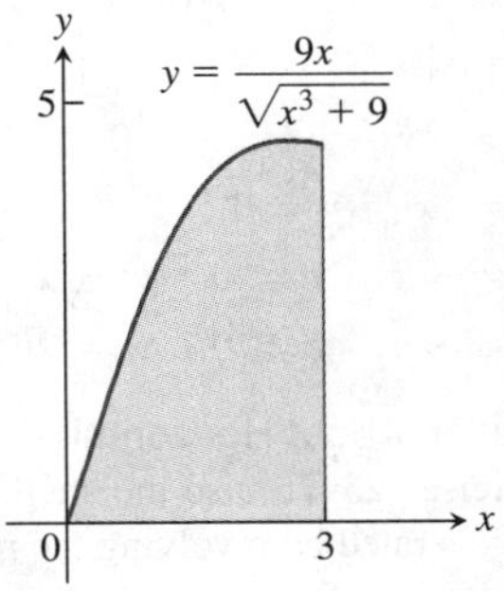

Revolution About the y-Axis

Use the shell method to find the volumes of the solids generated by revolving the regions bounded by the curves and lines in Exercises 7–12 about the y-axis.

7. $y = x, \quad y = -x/2, \quad x = 2$

8. $y = x, \quad y = x^2, \quad x = 1$

9. $y = x^2, \quad y = 2 - x, \quad x = 0, \quad \text{for } x \geq 0$

10. $y = 2x^2 - x^3, \quad y = 0, \quad x = 0, \quad x = 2$

11. $y = 2x - 1, \quad y = \sqrt{x}, \quad x = 0$

12. $y = x^3, \quad y = 0, \quad x = 0, \quad x = 2$

13. Let $f(x) = \begin{cases} (\sin x)/x, & 0 < x \le \pi \\ 1, & x = 0. \end{cases}$

a. Show that $x\,f(x) = \sin x,\ 0 \le x \le \pi$.

b. Find the volume of the solid generated by revolving the shaded region about the y-axis in the accompanying figure.

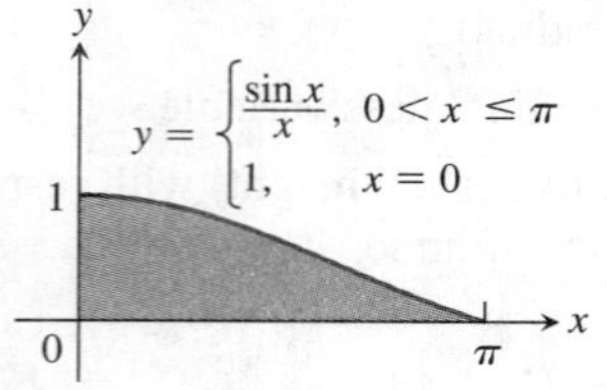

14. Let $g(x) = \begin{cases} (\tan x)^2/x, & 0 < x \le \pi/4 \\ 0, & x = 0. \end{cases}$

a. Show that $x\,g(x) = (\tan x)^2,\ 0 \le x \le \pi/4$.

b. Find the volume of the solid generated by revolving the shaded region about the y-axis in the accompanying figure.

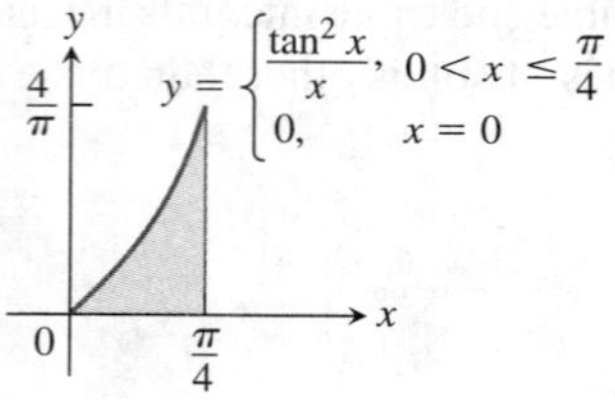

Revolution About the x-Axis

Use the shell method to find the volumes of the solids generated by revolving the regions bounded by the curves and lines in Exercises 15–22 about the x-axis.

15. $x = \sqrt{y},\quad x = -y,\quad y = 2$

16. $x = 1 + y^2,\quad x = 0,\quad y = 1,\quad y = 2$

17. $x = 2y - y^2,\quad x = 0$

18. $x = 2y - y^2,\quad x = y$

19. $y = |x|,\quad y = 1$

20. $y = x,\quad y = 2x,\quad y = 2$

21. $y = \sqrt{x},\quad y = 0,\quad y = x - 2$

22. $y = \sqrt{x},\quad y = 0,\quad y = 2 - x$

Revolution About Horizontal and Vertical Lines

In Exercises 23–26, use the shell method to find the volumes of the solids generated by revolving the regions bounded by the given curves about the given lines.

23. $y = 3x,\quad y = 0,\quad x = 2$

a. The y-axis **b.** The line $x = 4$

c. The line $x = -1$ **d.** The x-axis

e. The line $y = 7$ **f.** The line $y = -2$

24. $y = x^3,\quad y = 8,\quad x = 0$

a. The y-axis **b.** The line $x = 3$

c. The line $x = -2$ **d.** The x-axis

e. The line $y = 8$ **f.** The line $y = -1$

25. $y = x + 2,\quad y = x^2$

a. The line $x = 2$ **b.** The line $x = -1$

c. The x-axis **d.** The line $y = 4$

26. $y = x^4,\quad y = 4 - 3x^2$

a. The line $x = 1$ **b.** The x-axis

In Exercises 27 and 28, use the shell method to find the volumes of the solids generated by revolving the shaded regions about the indicated axes.

27. **a.** The x-axis **b.** The line $y = 1$

c. The line $y = 8/5$ **d.** The line $y = -2/5$

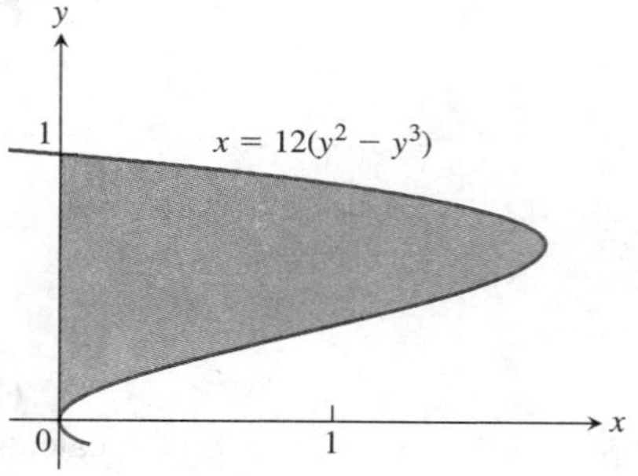

28. **a.** The x-axis **b.** The line $y = 2$

c. The line $y = 5$ **d.** The line $y = -5/8$

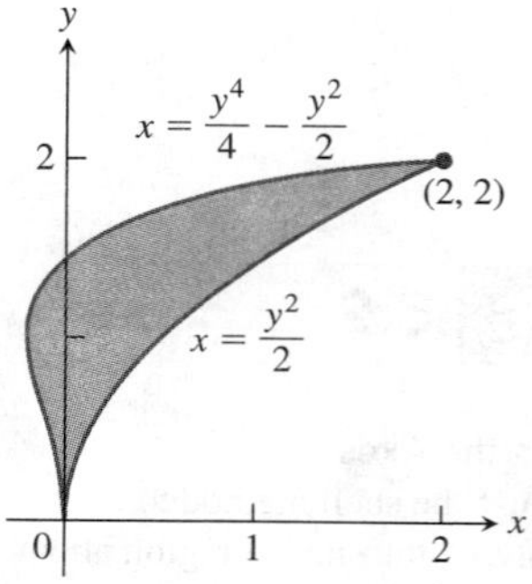

Choosing the Washer Method or the Shell Method

For some regions, both the washer and shell methods work well for the solid generated by revolving the region about the coordinate axes, but this is not always the case. When a region is revolved about the y-axis, for example, and washers are used, we must integrate with respect to y. It may not be possible, however, to express the integrand in terms of y. In such a case, the shell method allows us to integrate with respect to x instead. Exercises 29 and 30 provide some insight.

29. Compute the volume of the solid generated by revolving the region bounded by $y = x$ and $y = x^2$ about each coordinate axis using

a. the shell method.

b. the washer method.

30. Compute the volume of the solid generated by revolving the triangular region bounded by the lines $2y = x + 4$, $y = x$, and $x = 0$ about

a. the x-axis using the washer method.

b. the y-axis using the shell method.

c. the line $x = 4$ using the shell method.

d. the line $y = 8$ using the washer method.

In Exercises 31–36, find the volumes of the solids generated by revolving the regions about the given axes. If you think it would be better to use washers in any given instance, feel free to do so.

31. The triangle with vertices $(1, 1)$, $(1, 2)$, and $(2, 2)$ about

a. the x-axis **b.** the y-axis

c. the line $x = 10/3$ **d.** the line $y = 1$

32. The region bounded by $y = \sqrt{x}$, $y = 2$, $x = 0$ about

a. the x-axis **b.** the y-axis

c. the line $x = 4$ **d.** the line $y = 2$

33. The region in the first quadrant bounded by the curve $x = y - y^3$ and the y-axis about

a. the x-axis **b.** the line $y = 1$

34. The region in the first quadrant bounded by $x = y - y^3$, $x = 1$, and $y = 1$ about

a. the x-axis **b.** the y-axis

c. the line $x = 1$ **d.** the line $y = 1$

35. The region bounded by $y = \sqrt{x}$ and $y = x^2/8$ about

a. the x-axis **b.** the y-axis

36. The region bounded by $y = 2x - x^2$ and $y = x$ about

a. the y-axis **b.** the line $x = 1$

37. The region in the first quadrant that is bounded above by the curve $y = 1/x^{1/4}$, on the left by the line $x = 1/16$, and below by the line $y = 1$ is revolved about the x-axis to generate a solid. Find the volume of the solid by

a. the washer method. **b.** the shell method.

38. The region in the first quadrant that is bounded above by the curve $y = 1/\sqrt{x}$, on the left by the line $x = 1/4$, and below by the line $y = 1$ is revolved about the y-axis to generate a solid. Find the volume of the solid by

a. the washer method. **b.** the shell method.

Theory and Examples

39. The region shown here is to be revolved about the x-axis to generate a solid. Which of the methods (disk, washer, shell) could you use to find the volume of the solid? How many integrals would be required in each case? Explain.

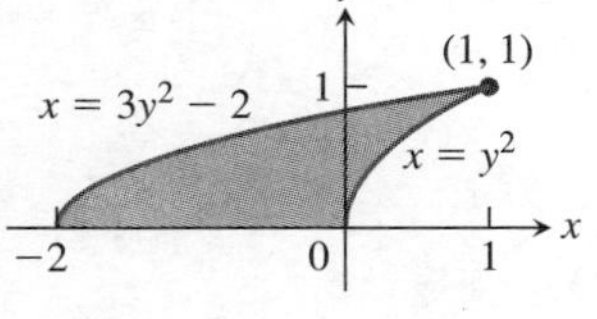

40. The region shown here is to be revolved about the y-axis to generate a solid. Which of the methods (disk, washer, shell) could you use to find the volume of the solid? How many integrals would be required in each case? Give reasons for your answers.

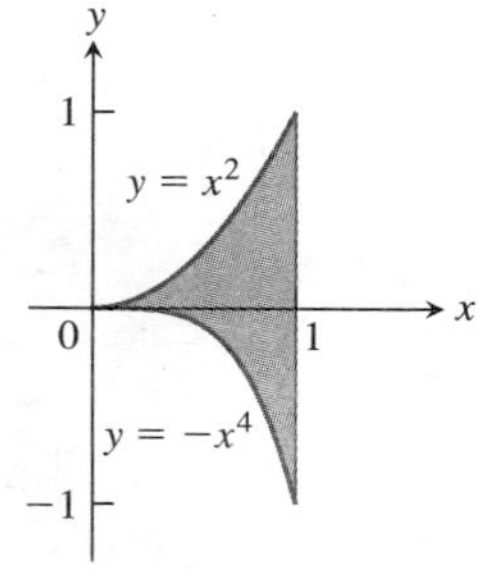

41. A bead is formed from a sphere of radius 5 by drilling through a diameter of the sphere with a drill bit of radius 3.

a. Find the volume of the bead.

b. Find the volume of the removed portion of the sphere.

42. A Bundt cake, well known for having a ringed shape, is formed by revolving around the y-axis the region bounded by the graph of $y = \sin(x^2 - 1)$ and the x-axis over the interval $1 \le x \le \sqrt{1 + \pi}$. Find the volume of the cake.

43. Derive the formula for the volume of a right circular cone of height h and radius r using an appropriate solid of revolution.

44. Derive the equation for the volume of a sphere of radius r using the shell method.

45. Consider the region R bounded by the graphs of $y = f(x) > 0$, $x = a > 0$, and $x = b > a$. If the volume of the solid formed by revolving R about the y-axis is 2π, and the volume formed by revolving R about the line $x = -2$ is 10π, find the area of R.

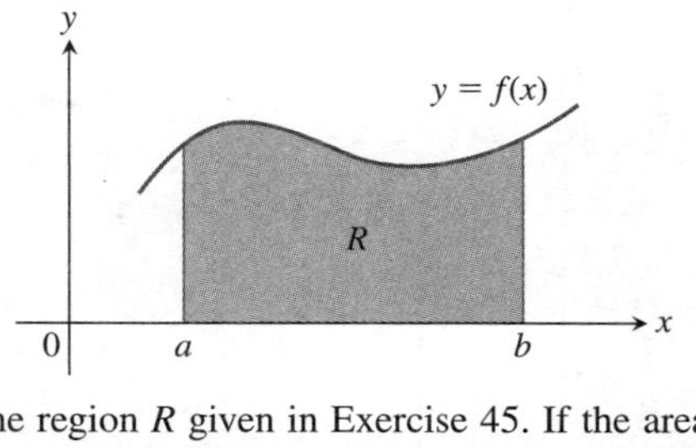

46. Consider the region R given in Exercise 45. If the area of region R is 1, and the volume of the solid formed by revolving R about the line $x = -3$ is 10π, find the volume of the solid formed by revolving R about the y-axis.

6.3 Arc Length

We know what is meant by the length of a straight-line segment, but without calculus, we have no precise definition of the length of a general winding curve. If the curve is the graph of a continuous function defined over an interval, then we can find the length of the curve using a procedure similar to that we used for defining the area between the curve and the x-axis. We divide the curve into many pieces, and we approximate each piece by a straight-line segment. The sum of the lengths of these segments is an approximation to the total curve length that we seek. The total length of the curve is the limiting value of these approximations as the number of segments goes to infinity.

Length of a Curve $y = f(x)$

Suppose the curve whose length we want to find is the graph of the function $y = f(x)$ from $x = a$ to $x = b$. In order to derive an integral formula for the length of the curve, we assume that f has a continuous derivative at every point of $[a, b]$. Such a function is called **smooth**, and its graph is a **smooth curve** because it does not have any breaks, corners, or cusps.

We partition the interval $[a, b]$ into n subintervals with

$$a = x_0 < x_1 < x_2 < \cdots < x_n = b.$$

If $y_k = f(x_k)$, then the corresponding point $P_k(x_k, y_k)$ lies on the curve. Next we connect successive points P_{k-1} and P_k with straight-line segments that, taken together, form a polygonal path whose length approximates the length of the curve (Figure 6.22). If we set $\Delta x_k = x_k - x_{k-1}$ and $\Delta y_k = y_k - y_{k-1}$, then a representative line segment in the path has length

$$L_k = \sqrt{(\Delta x_k)^2 + (\Delta y_k)^2}$$

(see Figure 6.23), so the length of the curve is approximated by the sum

$$\sum_{k=1}^{n} L_k = \sum_{k=1}^{n} \sqrt{(\Delta x_k)^2 + (\Delta y_k)^2}. \tag{1}$$

We expect the approximation to improve as the partition of $[a, b]$ becomes finer. In order to evaluate this limit, we use the Mean Value Theorem, which tells us that there is a point c_k, with $x_{k-1} < c_k < x_k$, such that

$$\Delta y_k = f'(c_k)\,\Delta x_k.$$

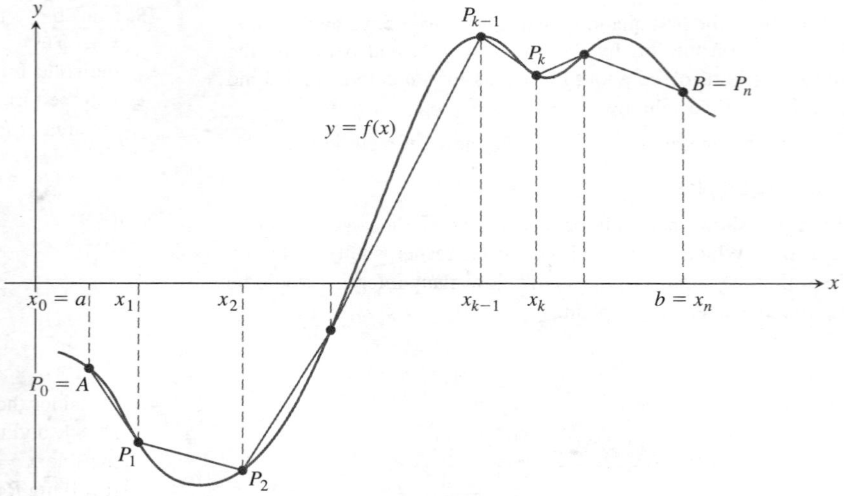

FIGURE 6.22 The length of the polygonal path $P_0P_1P_2 \cdots P_n$ approximates the length of the curve $y = f(x)$ from point A to point B.

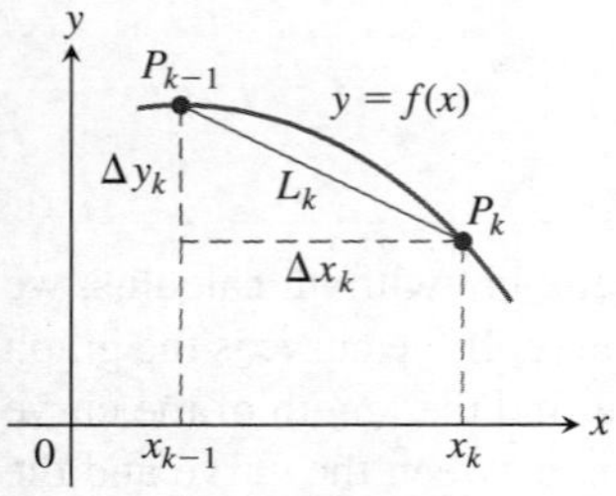

FIGURE 6.23 The arc $P_{k-1}P_k$ of the curve $y = f(x)$ is approximated by the straight-line segment shown here, which has length $L_k = \sqrt{(\Delta x_k)^2 + (\Delta y_k)^2}$.

When this is substituted for Δy_k, the sums in Equation (1) take the form

$$\sum_{k=1}^{n} L_k = \sum_{k=1}^{n} \sqrt{(\Delta x_k)^2 + (f'(c_k)\Delta x_k)^2} = \sum_{k=1}^{n} \sqrt{1 + [f'(c_k)]^2}\,\Delta x_k. \tag{2}$$

This is a Riemann sum whose limit we can evaluate. Because $\sqrt{1 + [f'(x)]^2}$ is continuous on $[a, b]$, the limit of the Riemann sum on the right-hand side of Equation (2) exists and has the value

$$\lim_{n\to\infty} \sum_{k=1}^{n} L_k = \lim_{n\to\infty} \sum_{k=1}^{n} \sqrt{1 + [f'(c_k)]^2}\,\Delta x_k = \int_a^b \sqrt{1 + [f'(x)]^2}\,dx.$$

We define the length of the curve to be this integral.

DEFINITION If f' is continuous on $[a, b]$, then the **length (arc length)** of the curve $y = f(x)$ from the point $A = (a, f(a))$ to the point $B = (b, f(b))$ is the value of the integral

$$L = \int_a^b \sqrt{1 + [f'(x)]^2}\, dx = \int_a^b \sqrt{1 + \left(\frac{dy}{dx}\right)^2}\, dx. \tag{3}$$

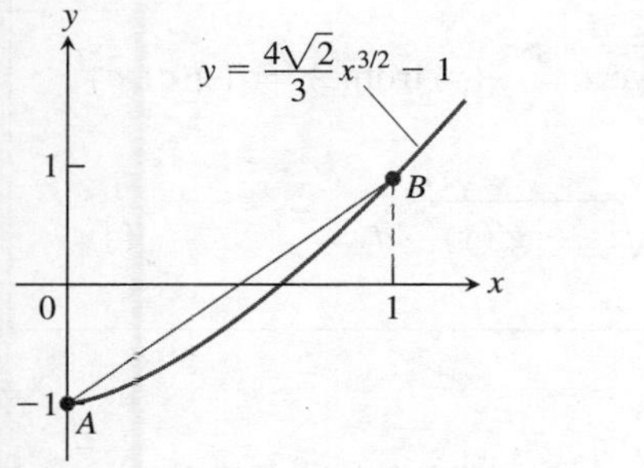

FIGURE 6.24 The length of the curve is slightly larger than the length of the line segment joining points A and B (Example 1).

EXAMPLE 1 Find the length of the curve shown in Figure 6.24, which is the graph of the function

$$y = \frac{4\sqrt{2}}{3}x^{3/2} - 1, \qquad 0 \le x \le 1.$$

Solution We use Equation (3) with $a = 0, b = 1$, and

$$y = \frac{4\sqrt{2}}{3}x^{3/2} - 1 \qquad \text{If } x = 1, \text{ then } y \approx 0.89.$$

$$\frac{dy}{dx} = \frac{4\sqrt{2}}{3} \cdot \frac{3}{2}x^{1/2} = 2\sqrt{2}x^{1/2}$$

$$\left(\frac{dy}{dx}\right)^2 = \left(2\sqrt{2}x^{1/2}\right)^2 = 8x.$$

The length of the curve over $x = 0$ to $x = 1$ is

$$L = \int_0^1 \sqrt{1 + \left(\frac{dy}{dx}\right)^2}\, dx = \int_0^1 \sqrt{1 + 8x}\, dx \qquad \text{Eq. (3) with } a = 0,\ b = 1$$

$$= \frac{2}{3} \cdot \frac{1}{8}(1 + 8x)^{3/2}\Big]_0^1 = \frac{13}{6} \approx 2.17. \qquad \text{Let } u = 1 + 8x, \text{ integrate, and replace } u \text{ by } 1 + 8x.$$

Notice that the length of the curve is slightly larger than the length of the straight-line segment joining the points $A = (0, -1)$ and $B = \left(1, 4\sqrt{2}/3 - 1\right)$ on the curve (see Figure 6.24):

$$2.17 > \sqrt{1^2 + (1.89)^2} \approx 2.14. \qquad \text{Decimal approximations}$$

■

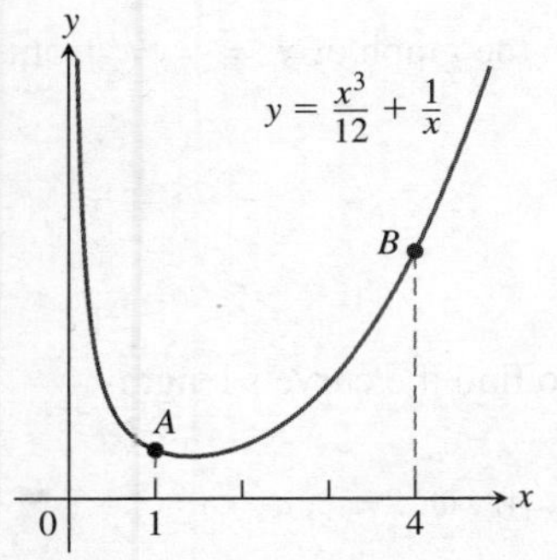

FIGURE 6.25 The curve in Example 2, where $A = (1, 13/12)$ and $B = (4, 67/12)$.

EXAMPLE 2 Find the length of the graph of

$$f(x) = \frac{x^3}{12} + \frac{1}{x}, \qquad 1 \le x \le 4.$$

Solution A graph of the function is shown in Figure 6.25. To use Equation (3), we find

$$f'(x) = \frac{x^2}{4} - \frac{1}{x^2}$$

so

$$1 + [f'(x)]^2 = 1 + \left(\frac{x^2}{4} - \frac{1}{x^2}\right)^2 = 1 + \left(\frac{x^4}{16} - \frac{1}{2} + \frac{1}{x^4}\right)$$

$$= \frac{x^4}{16} + \frac{1}{2} + \frac{1}{x^4} = \left(\frac{x^2}{4} + \frac{1}{x^2}\right)^2$$

and

$$\sqrt{1 + [f'(x)]^2} = \sqrt{\left(\frac{x^2}{4} + \frac{1}{x^2}\right)^2} = \left|\frac{x^2}{4} + \frac{1}{x^2}\right| = \frac{x^2}{4} + \frac{1}{x^2}. \qquad \frac{x^2}{4} + \frac{1}{x^2} > 0$$

The length of the graph over $[1, 4]$ is

$$L = \int_1^4 \sqrt{1 + [f'(x)]^2}\, dx = \int_1^4 \left(\frac{x^2}{4} + \frac{1}{x^2}\right) dx$$

$$= \left[\frac{x^3}{12} - \frac{1}{x}\right]_1^4 = \left(\frac{64}{12} - \frac{1}{4}\right) - \left(\frac{1}{12} - 1\right) = \frac{72}{12} = 6.$$

■

Dealing with Discontinuities in dy/dx

Even if the derivative dy/dx does not exist at some point on a curve, it is possible that dx/dy could exist. This can happen, for example, when a curve has a vertical tangent. In this case, we may be able to find the curve's length by expressing x as a function of y and applying the following analogue of Equation (3).

Formula for the Length of $x = g(y),\ c \le y \le d$
If g' is continuous on $[c, d]$, the length of the curve $x = g(y)$ from $A = (g(c), c)$ to $B = (g(d), d)$ is

$$L = \int_c^d \sqrt{1 + \left(\frac{dx}{dy}\right)^2}\, dy = \int_c^d \sqrt{1 + [g'(y)]^2}\, dy. \tag{4}$$

EXAMPLE 3 Find the length of the curve $y = (x/2)^{2/3}$ from $x = 0$ to $x = 2$.

Solution The derivative

$$\frac{dy}{dx} = \frac{2}{3}\left(\frac{x}{2}\right)^{-1/3}\left(\frac{1}{2}\right) = \frac{1}{3}\left(\frac{2}{x}\right)^{1/3}$$

is not defined at $x = 0$, so we cannot find the curve's length with Equation (3).

We therefore rewrite the equation to express x in terms of y:

$$y = \left(\frac{x}{2}\right)^{2/3}$$

$$y^{3/2} = \frac{x}{2} \qquad \text{Raise both sides to the power } 3/2.$$

$$x = 2y^{3/2}. \qquad \text{Solve for } x.$$

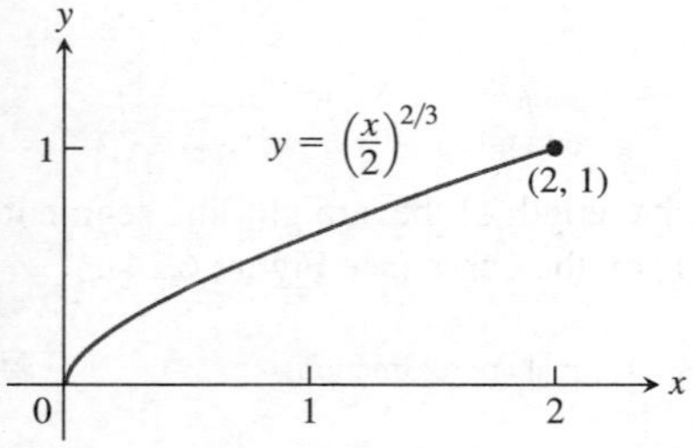

FIGURE 6.26 The graph of $y = (x/2)^{2/3}$ from $x = 0$ to $x = 2$ is also the graph of $x = 2y^{3/2}$ from $y = 0$ to $y = 1$ (Example 3).

From this we see that the curve whose length we want is also the graph of $x = 2y^{3/2}$ from $y = 0$ to $y = 1$ (see Figure 6.26).

The derivative

$$\frac{dx}{dy} = 2\left(\frac{3}{2}\right)y^{1/2} = 3y^{1/2}$$

is continuous on $[0, 1]$. We may therefore use Equation (4) to find the curve's length:

$$L = \int_c^d \sqrt{1 + \left(\frac{dx}{dy}\right)^2}\, dy = \int_0^1 \sqrt{1 + 9y}\, dy \qquad \text{Eq. (4) with } c = 0,\ d = 1$$

$$= \frac{1}{9} \cdot \frac{2}{3}(1 + 9y)^{3/2}\Big]_0^1 \qquad \text{Let } u = 1 + 9y,\ du/9 = dy, \text{ integrate, and substitute back.}$$

$$= \frac{2}{27}\left(10\sqrt{10} - 1\right) \approx 2.27.$$

■

The Differential Formula for Arc Length

If $y = f(x)$ and if f' is continuous on $[a, b]$, then by the Fundamental Theorem of Calculus, we can define a new function

$$s(x) = \int_a^x \sqrt{1 + [f'(t)]^2}\, dt. \tag{5}$$

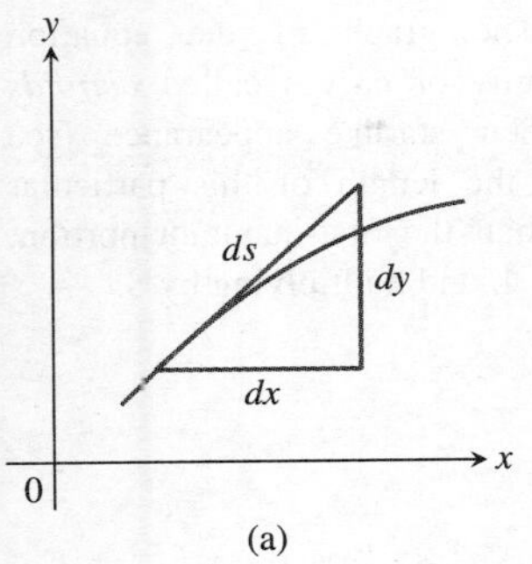

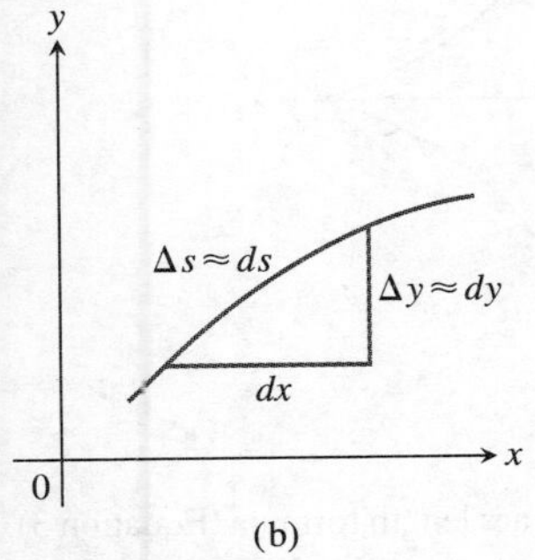

FIGURE 6.27 Diagrams for remembering the equation $ds = \sqrt{dx^2 + dy^2}$.

From Equation (3) and Figure 6.22, we see that this function $s(x)$ is continuous and measures the length along the curve $y = f(x)$ from the initial point $P_0(a, f(a))$ to the point $Q(x, f(x))$ for each $x \in [a, b]$. The function s is called the **arc length function** for $y = f(x)$. From the Fundamental Theorem, the function s is differentiable on (a, b) and

$$\frac{ds}{dx} = \sqrt{1 + [f'(x)]^2} = \sqrt{1 + \left(\frac{dy}{dx}\right)^2}.$$

Then the differential of arc length is

$$ds = \sqrt{1 + \left(\frac{dy}{dx}\right)^2}\, dx. \tag{6}$$

A useful way to remember Equation (6) is to write

$$ds = \sqrt{dx^2 + dy^2}, \tag{7}$$

which can be integrated between appropriate limits to give the total length of a curve. From this point of view, all the arc length formulas are simply different expressions for the equation $L = \int ds$. Figure 6.27a, which corresponds to Equation (7), can be thought of as a simplified approximation of Figure 6.27b. That is, ds is approximately equal to the exact arc length Δs.

EXAMPLE 4 Find the arc length function for the curve in Example 2, taking $A = (1, 13/12)$ as the starting point (see Figure 6.25).

Solution In the solution to Example 2, we found that

$$\sqrt{1 + [f'(x)]^2} = \frac{x^2}{4} + \frac{1}{x^2}.$$

Therefore the arc length function is given by

$$\begin{aligned} s(x) &= \int_1^x \sqrt{1 + [f'(t)]^2}\, dt = \int_1^x \left(\frac{t^2}{4} + \frac{1}{t^2}\right) dt \\ &= \left[\frac{t^3}{12} - \frac{1}{t}\right]_1^x = \frac{x^3}{12} - \frac{1}{x} + \frac{11}{12}. \end{aligned}$$

To compute the arc length along the curve from $A = (1, 13/12)$ to $B = (4, 67/12)$, for instance, we simply calculate

$$s(4) = \frac{4^3}{12} - \frac{1}{4} + \frac{11}{12} = 6.$$

This is the same result we obtained in Example 2. ■

EXERCISES 6.3

Finding Lengths of Curves

Find the lengths of the curves in Exercises 1–12. If you have graphing software, you may want to graph these curves to see what they look like.

1. $y = (1/3)(x^2 + 2)^{3/2}$ from $x = 0$ to $x = 3$
2. $y = x^{3/2}$ from $x = 0$ to $x = 4$
3. $x = (y^3/3) + 1/(4y)$ from $y = 1$ to $y = 3$
4. $x = (y^{3/2}/3) - y^{1/2}$ from $y = 1$ to $y = 9$
5. $x = (y^4/4) + 1/(8y^2)$ from $y = 1$ to $y = 2$
6. $x = (y^3/6) + 1/(2y)$ from $y = 2$ to $y = 3$
7. $y = (3/4)x^{4/3} - (3/8)x^{2/3} + 5, \quad 1 \le x \le 8$
8. $3ay^2 = x(x - a)^2$

9. $y = \frac{x^3}{3} + \frac{1}{4x}, \quad 1 \le x \le 3$

10. $y = \frac{x^5}{5} + \frac{1}{12x^3}, \quad \frac{1}{2} \le x \le 1$

11. $y = \log_e \sec x, \quad 0 \le x \le \frac{\pi}{3}$

12. $y = \int_{-2}^{x} \sqrt{3t^4 - 1}\, dt, \quad -2 \le x \le -1$

T Finding Integrals for Lengths of Curves

In Exercises 13–20, do the following.

a. Set up an integral for the length of the curve.

b. Graph the curve to see what it looks like.

c. Use your grapher's or computer's integral evaluator to find the curve's length numerically.

13. $y = x^2, \quad -1 \le x \le 2$

14. $y = \tan x, \quad -\pi/3 \le x \le 0$

15. $x = \sin y, \quad 0 \le y \le \pi$

16. $27y^2 = 4x^3, \quad 0 \le x \le 6$

17. $y^2 + 2y = 2x + 1$ from $(-1, -1)$ to $(7, 3)$

18. $y = \sin x - x \cos x, \quad 0 \le x \le \pi$

19. $y = \int_0^x \tan t\, dt, \quad 0 \le x \le \pi/6$

20. $x = \int_0^y \sqrt{\sec^2 t - 1}\, dt, \quad -\pi/3 \le y \le \pi/4$

Theory and Examples

21. **a.** Find a curve with a positive derivative through the point $(1, 1)$ whose length integral (Equation 3) is

$$L = \int_1^4 \sqrt{1 + \frac{1}{4x}}\, dx.$$

b. How many such curves are there? Give reasons for your answer.

22. **a.** Find a curve with a positive derivative through the point $(0, 1)$ whose length integral (Equation 4) is

$$L = \int_1^2 \sqrt{1 + \frac{1}{y^4}}\, dy.$$

b. How many such curves are there? Give reasons for your answer.

23. Find the length of the curve

$$y = \int_0^x \sqrt{\cos 2t}\, dt$$

from $x = 0$ to $x = \pi/4$.

24. **The length of an astroid** The graph of the equation $x^{2/3} + y^{2/3} = 1$ is one of a family of curves called *astroids* (not "asteroids") because of their starlike appearance (see the accompanying figure). Find the length of this particular astroid by finding the length of half the first-quadrant portion, $y = (1 - x^{2/3})^{3/2}$, $\sqrt{2}/4 \le x \le 1$, and multiplying by 8.

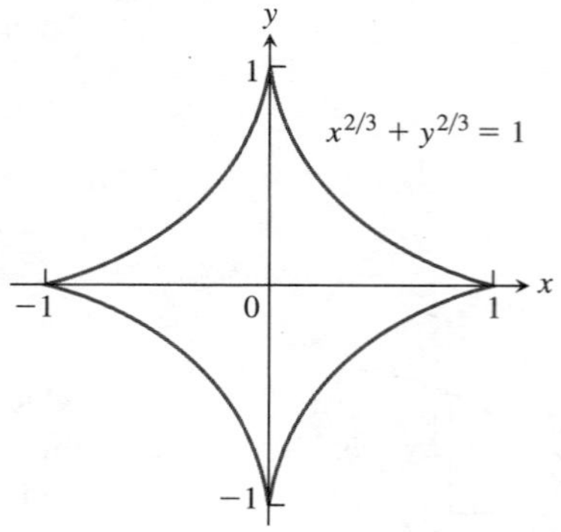

25. **Length of a line segment** Use the arc length formula (Equation 3) to find the length of the line segment $y = 3 - 2x$, $0 \le x \le 2$. Check your answer by finding the length of the segment as the hypotenuse of a right triangle.

26. **Circumference of a circle** Set up an integral to find the circumference of a circle of radius r centered at the origin. You will learn how to evaluate the integral in Section 8.4.

27. If $9x^2 = y(y - 3)^2$, show that

$$ds^2 = \frac{(y + 1)^2}{4y} dy^2.$$

28. If $4x^2 - y^2 = 64$, show that

$$ds^2 = \frac{4}{y^2}(5x^2 - 16)\, dx^2.$$

29. Is there a smooth (continuously differentiable) curve $y = f(x)$ whose length over the interval $0 \le x \le a$ is always $\sqrt{2}a$? Give reasons for your answer.

30. **Using tangent fins to derive the length formula for curves** Assume that f is smooth on $[a, b]$ and partition the interval $[a, b]$ in the usual way. In each subinterval $[x_{k-1}, x_k]$, construct the *tangent fin* at the point $(x_{k-1}, f(x_{k-1}))$, as shown in the accompanying figure.

a. Show that the length of the kth tangent fin over the interval $[x_{k-1}, x_k]$ equals $\sqrt{(\Delta x_k)^2 + (f'(x_{k-1})\,\Delta x_k)^2}$.

b. Show that

$$\lim_{n \to \infty} \sum_{k=1}^{n} (\text{length of } k\text{th tangent fin}) = \int_a^b \sqrt{1 + (f'(x))^2}\, dx,$$

which is the length L of the curve $y = f(x)$ from a to b.

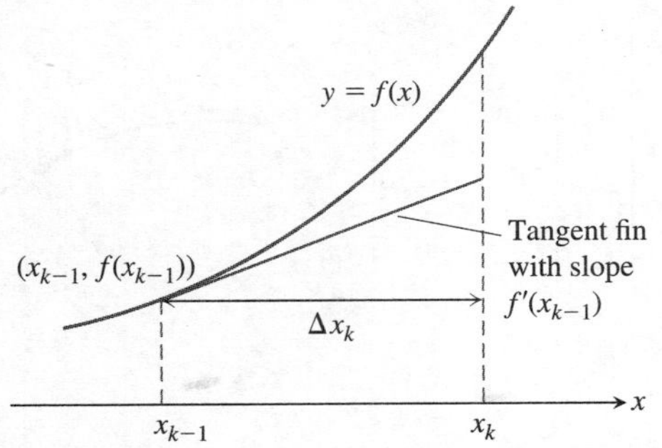

31. Approximate the arc length of one-quarter of the unit circle (which is $\pi/2$) by computing the length of the polygonal approximation with $n = 4$ segments (see accompanying figure).

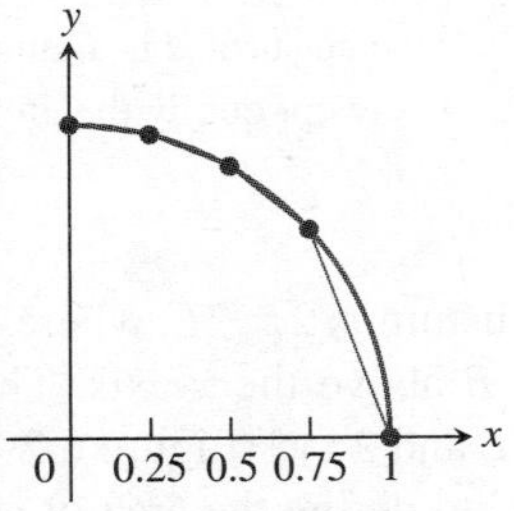

32. Distance between two points Assume that the two points (x_1, y_1) and (x_2, y_2) lie on the graph of the straight line $y = mx + b$. Use the arc length formula (Equation 3) to find the distance between the two points.

33. Find the arc length function for the graph of $f(x) = 2x^{3/2}$ using $(0, 0)$ as the starting point. What is the length of the curve from $(0, 0)$ to $(1, 2)$?

34. Prove that the length of the parabola $y^2 = 4ax$ cut off by the latus rectum is $2a\left[\sqrt{2} + \log(1 + \sqrt{2})\right]$.

COMPUTER EXPLORATIONS

In Exercises 35–40, use a CAS to perform the following steps for the given graph of the function over the closed interval.

a. Plot the curve together with the polygonal path approximations for $n = 2, 4, 8$ partition points over the interval. (See Figure 6.22.)

b. Find the corresponding approximation to the length of the curve by summing the lengths of the line segments.

c. Evaluate the length of the curve using an integral. Compare your approximations for $n = 2, 4, 8$ with the actual length given by the integral. How does the actual length compare with the approximations as n increases? Explain your answer.

35. $f(x) = \sqrt{1 - x^2}, \quad -1 \le x \le 1$

36. $f(x) = x^{1/3} + x^{2/3}, \quad 0 \le x \le 2$

37. $f(x) = \sin(\pi x^2), \quad 0 \le x \le \sqrt{2}$

38. $f(x) = x^2 \cos x, \quad 0 \le x \le \pi$

39. $f(x) = \dfrac{x - 1}{4x^2 + 1}, \quad -\dfrac{1}{2} \le x \le 1$

40. $f(x) = x^3 - x^2, \quad -1 \le x \le 1$

6.4 Areas of Surfaces of Revolution

When you jump rope, the rope sweeps out a surface in the space around you similar to what is called a *surface of revolution*. The surface surrounds a volume of revolution, and many applications require that we know the area of the surface rather than the volume it encloses. In this section we define areas of surfaces of revolution. More general surfaces are treated in Chapter 16.

Defining Surface Area

If you revolve a region in the plane that is bounded by the graph of a function over an interval, it sweeps out a solid of revolution, as we saw earlier in the chapter. However, if you revolve only the bounding curve itself, it does not sweep out any interior volume but rather a surface that surrounds the solid and forms part of its boundary. Just as we were interested in defining and finding the length of a curve in the last section, we are now interested in defining and finding the area of a surface generated by revolving a curve about an axis.

Before considering general curves, we begin by rotating horizontal and slanted line segments about the x-axis. If we rotate the horizontal line segment AB having length Δx about the x-axis (Figure 6.28a), we generate a cylinder with surface area $2\pi y\, \Delta x$. This area is the same as that of a rectangle with side lengths Δx and $2\pi y$ (Figure 6.28b). The length $2\pi y$ is the circumference of the circle of radius y generated by rotating the point (x, y) on the line AB about the x-axis.

Suppose the line segment AB has length L and is slanted rather than horizontal. Now when AB is rotated about the x-axis, it generates a frustum of a cone (Figure 6.29a). From

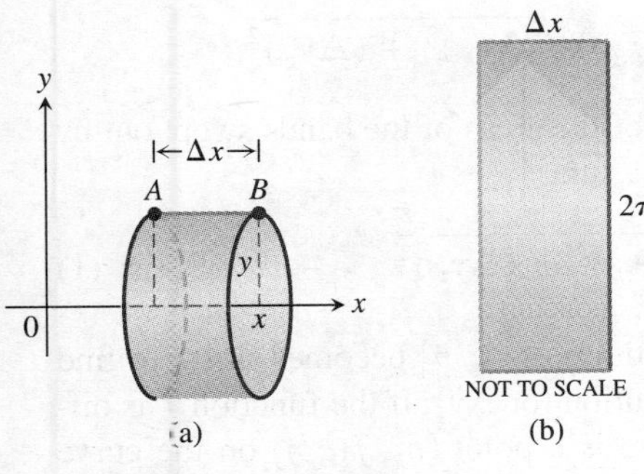

FIGURE 6.28 (a) A cylindrical surface generated by rotating the horizontal line segment AB of length Δx about the x-axis has area $2\pi y\, \Delta x$. (b) The cut and rolled-out cylindrical surface as a rectangle.

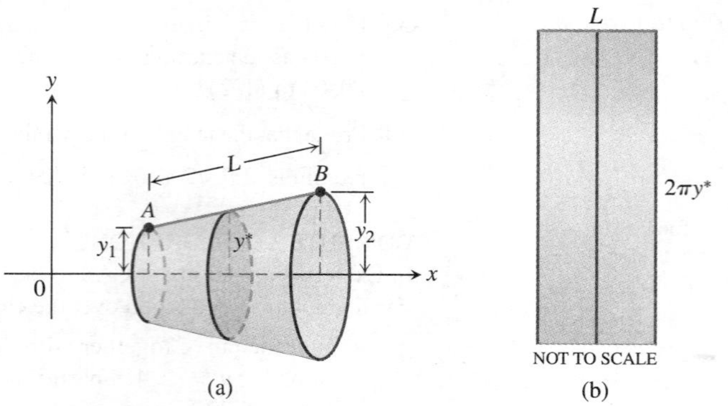

FIGURE 6.29 (a) The frustum of a cone generated by rotating the slanted line segment AB of length L about the x-axis has area $2\pi y^* L$. (b) The area of the rectangle for $y^* = \dfrac{y_1 + y_2}{2}$, the average height of AB above the x-axis.

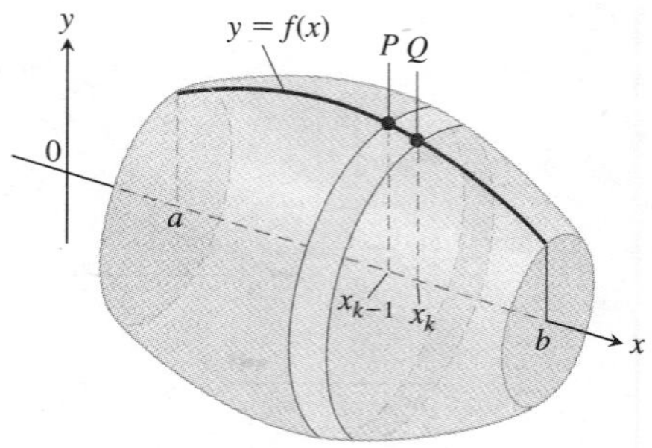

FIGURE 6.30 The surface generated by revolving the graph of a nonnegative function $y = f(x)$, $a \le x \le b$, about the x-axis. The surface is a union of bands like the one swept out by the arc PQ.

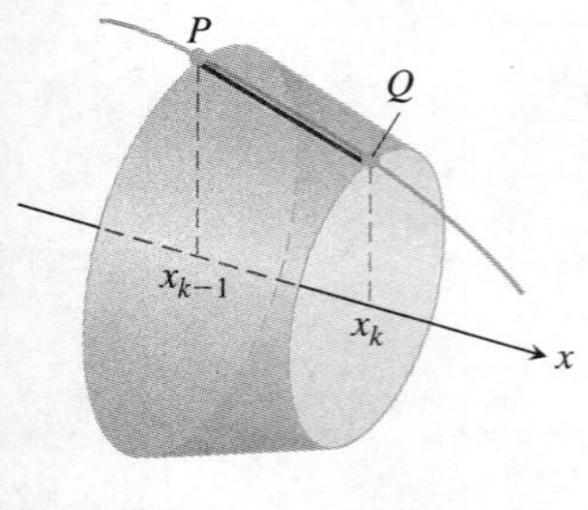

FIGURE 6.31 The line segment joining P and Q sweeps out a frustum of a cone.

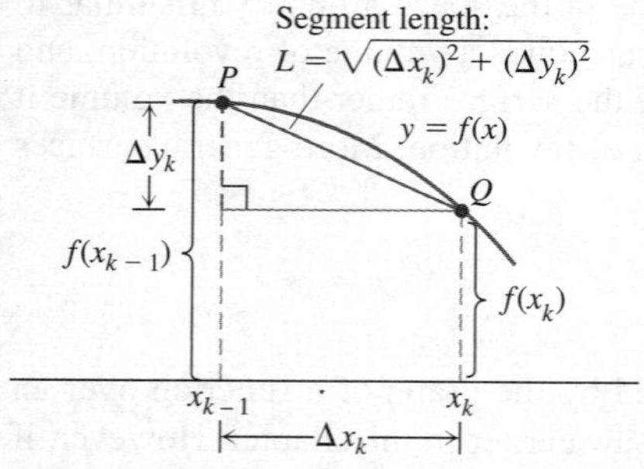

FIGURE 6.32 Dimensions associated with the arc and line segment PQ.

classical geometry, the surface area of this frustum is $2\pi y^* L$, where $y^* = (y_1 + y_2)/2$ is the average height of the slanted segment AB above the x-axis. This surface area is the same as that of a rectangle with side lengths L and $2\pi y^*$ (Figure 6.29b).

Let's build on these geometric principles to define the area of a surface swept out by revolving more general curves about the x-axis. Suppose we want to find the area of the surface swept out by revolving the graph of a nonnegative continuous function $y = f(x)$, $a \le x \le b$, about the x-axis. We partition the closed interval $[a, b]$ in the usual way and use the points in the partition to subdivide the graph into short arcs. Figure 6.30 shows a typical arc PQ and the band it sweeps out as part of the graph of f.

As the arc PQ revolves about the x-axis, the line segment joining P and Q sweeps out a frustum of a cone whose axis lies along the x-axis (Figure 6.31). The surface area of this frustum approximates the surface area of the band swept out by the arc PQ. The surface area of the frustum of the cone shown in Figure 6.31 is $2\pi y^* L$, where y^* is the average height of the line segment joining P and Q, and L is its length (just as before). Since $f \ge 0$, from Figure 6.32 we see that the average height of the line segment is $y^* = (f(x_{k-1}) + f(x_k))/2$, and the slant length is $L = \sqrt{(\Delta x_k)^2 + (\Delta y_k)^2}$. Therefore,

$$\begin{aligned}\text{Frustum surface area} &= 2\pi \cdot \frac{f(x_{k-1}) + f(x_k)}{2} \cdot \sqrt{(\Delta x_k)^2 + (\Delta y_k)^2} \\ &= \pi(f(x_{k-1}) + f(x_k))\sqrt{(\Delta x_k)^2 + (\Delta y_k)^2}.\end{aligned}$$

The area of the original surface, being the sum of the areas of the bands swept out by arcs like arc PQ, is approximated by the frustum area sum

$$\sum_{k=1}^{n} \pi(f(x_{k-1}) + f(x_k))\sqrt{(\Delta x_k)^2 + (\Delta y_k)^2}. \qquad (1)$$

We expect the approximation to improve as the partition of $[a, b]$ becomes finer. To find the limit, we first need to find an appropriate substitution for Δy_k. If the function f is differentiable, then by the Mean Value Theorem, there is a point $(c_k, f(c_k))$ on the curve between P and Q where the tangent is parallel to the segment PQ (Figure 6.33). At this point,

$$\begin{aligned} f'(c_k) &= \frac{\Delta y_k}{\Delta x_k}, \\ \Delta y_k &= f'(c_k)\,\Delta x_k. \end{aligned}$$

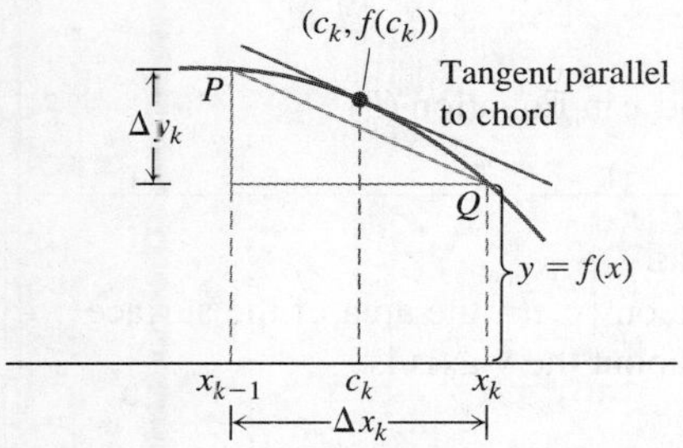

FIGURE 6.33 If f is smooth, the Mean Value Theorem guarantees the existence of a point c_k where the tangent is parallel to segment PQ.

With this substitution for Δy_k, the sums in Equation (1) take the form

$$\sum_{k=1}^{n} \pi(f(x_{k-1}) + f(x_k))\sqrt{(\Delta x_k)^2 + (f'(c_k)\,\Delta x_k)^2}$$

$$= \sum_{k=1}^{n} \pi(f(x_{k-1}) + f(x_k))\sqrt{1 + (f'(c_k))^2}\,\Delta x_k. \quad (2)$$

These sums are not the Riemann sums of any function because the points x_{k-1}, x_k, and c_k are not the same. However, the points x_{k-1}, x_k, and c_k are very close to each other, and so we expect (and it can be proved) that as the norm of the partition of $[a, b]$ goes to zero, the sums in Equation (2) converge to the integral

$$\int_a^b 2\pi f(x)\sqrt{1 + (f'(x))^2}\,dx.$$

We therefore define this integral to be the area of the surface swept out by the graph of f from a to b.

DEFINITION If the function $f(x) \geq 0$ is continuously differentiable on $[a, b]$, the **area of the surface** generated by revolving the graph of $y = f(x)$ about the x-axis is

$$S = \int_a^b 2\pi y\sqrt{1 + \left(\frac{dy}{dx}\right)^2}\,dx = \int_a^b 2\pi f(x)\sqrt{1 + (f'(x))^2}\,dx. \quad (3)$$

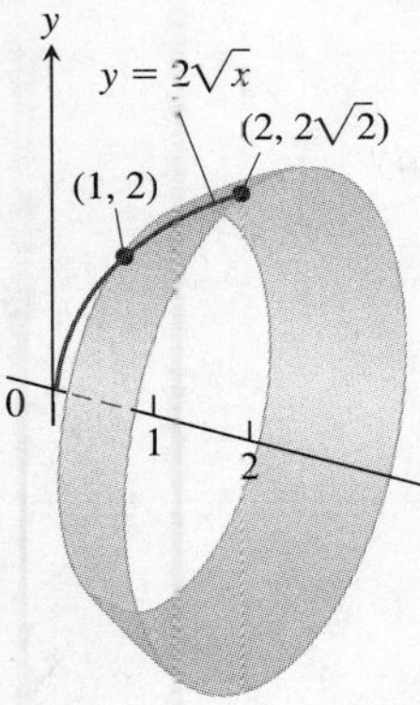

FIGURE 6.34 In Example 1 we calculate the area of this surface.

Note that the square root in Equation (3) is similar to the one that appears in the formula for the arc length of the generating curve in Equation (6) of Section 6.3.

EXAMPLE 1 Find the area of the surface generated by revolving the curve $y = 2\sqrt{x}$, $1 \leq x \leq 2$, about the x-axis (Figure 6.34).

Solution We evaluate the formula

$$S = \int_a^b 2\pi y\sqrt{1 + \left(\frac{dy}{dx}\right)^2}\,dx \qquad \text{Eq. (3)}$$

with

$$a = 1, \qquad b = 2, \qquad y = 2\sqrt{x}, \qquad \frac{dy}{dx} = \frac{1}{\sqrt{x}}.$$

First, we perform some algebraic manipulation on the radical in the integrand to transform it into an expression that is easier to integrate.

$$\sqrt{1 + \left(\frac{dy}{dx}\right)^2} = \sqrt{1 + \left(\frac{1}{\sqrt{x}}\right)^2} = \sqrt{1 + \frac{1}{x}} = \sqrt{\frac{x+1}{x}} = \frac{\sqrt{x+1}}{\sqrt{x}}$$

With these substitutions, we have

$$S = \int_1^2 2\pi \cdot 2\sqrt{x}\,\frac{\sqrt{x+1}}{\sqrt{x}}\,dx = 4\pi\int_1^2 \sqrt{x+1}\,dx$$

$$= 4\pi \cdot \frac{2}{3}(x+1)^{3/2}\Big]_1^2 = \frac{8\pi}{3}\left(3\sqrt{3} - 2\sqrt{2}\right). \quad ■$$

Revolution About the y-Axis

For revolution about the y-axis, we interchange x and y in Equation (3).

Surface Area for Revolution About the y-Axis
If $x = g(y) \geq 0$ is continuously differentiable on $[c, d]$, the area of the surface generated by revolving the graph of $x = g(y)$ about the y-axis is

$$S = \int_c^d 2\pi x\sqrt{1 + \left(\frac{dx}{dy}\right)^2}\, dy = \int_c^d 2\pi g(y)\sqrt{1 + (g'(y))^2}\, dy. \tag{4}$$

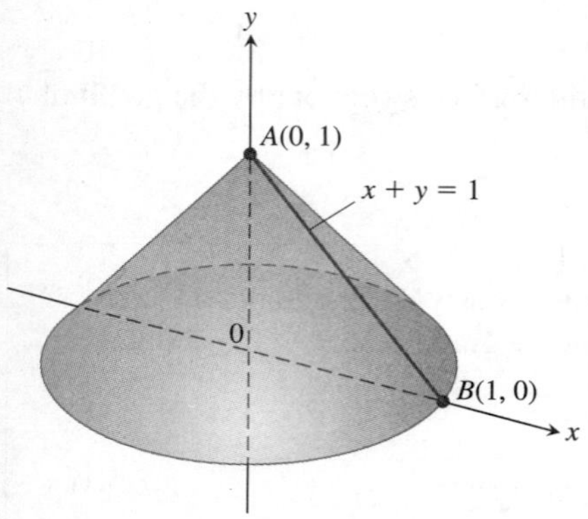

FIGURE 6.35 Revolving line segment AB about the y-axis generates a cone whose lateral surface area we can now calculate in two different ways (Example 2).

EXAMPLE 2 The line segment $x = 1 - y$, $0 \leq y \leq 1$, is revolved about the y-axis to generate the cone in Figure 6.35. Find its lateral surface area (which excludes the base area).

Solution Here we have a calculation we can check with a formula from geometry:

$$\text{Lateral surface area} = \frac{\text{base circumference}}{2} \times \text{slant height} = \pi\sqrt{2}.$$

To see how Equation (4) gives the same result, we take

$$c = 0, \quad d = 1, \quad x = 1 - y, \quad \frac{dx}{dy} = -1,$$

$$\sqrt{1 + \left(\frac{dx}{dy}\right)^2} = \sqrt{1 + (-1)^2} = \sqrt{2}$$

and calculate

$$S = \int_c^d 2\pi x\sqrt{1 + \left(\frac{dx}{dy}\right)^2}\, dy = \int_0^1 2\pi(1 - y)\sqrt{2}\, dy$$

$$= 2\pi\sqrt{2}\left[y - \frac{y^2}{2}\right]_0^1 = 2\pi\sqrt{2}\left(1 - \frac{1}{2}\right) = \pi\sqrt{2}.$$ ■

EXERCISES 6.4

Finding Integrals for Surface Area

In Exercises 1–8:

a. Set up an integral for the area of the surface generated by revolving the given curve about the indicated axis.

T **b.** Graph the curve to see what it looks like. If you can, graph the surface too.

T **c.** Use your utility's integral evaluator to find the surface's area numerically.

1. $y = \tan x, \quad 0 \leq x \leq \pi/4; \quad$ x-axis

2. $y = x^2, \quad 0 \leq x \leq 2; \quad$ x-axis

3. $xy = 1, \quad 1 \leq y \leq 2; \quad$ y-axis

4. $x = \sin y, \quad 0 \leq y \leq \pi; \quad$ y-axis

5. $x^{1/2} + y^{1/2} = 3$ from $(4, 1)$ to $(1, 4)$; x-axis

6. $y = x^2, \quad 0 \leq y \leq 2, \quad$ y-axis

7. $x = \int_0^y \tan t\, dt, \quad 0 \leq y \leq \pi/3; \quad$ y-axis

8. $y = \int_1^x \sqrt{t^2 - 1}\, dt, \quad 1 \leq x \leq \sqrt{5}; \quad$ x-axis

Finding Surface Area

9. Find the lateral (side) surface area of the cone generated by revolving the line segment $y = x/2$, $0 \leq x \leq 4$, about the x-axis. Check your answer with the geometry formula

$$\text{Lateral surface area} = \frac{1}{2} \times \text{base circumference} \times \text{slant height}.$$

10. Find the lateral surface area of the cone generated by revolving the line segment $y = x/2$, $0 \le x \le 4$, about the y-axis. Check your answer with the geometry formula given in Exercise 9.

11. Find the surface area of the cone frustum generated by revolving the line segment $y = (x/2) + (1/2)$, $1 \le x \le 3$, about the x-axis. Check your result with the geometry formula

$$\text{Frustum surface area} = \pi(y_1 + y_2) \times \text{slant height}.$$

12. Find the surface area of the cone frustum generated by revolving the line segment $y = (x/2) + (1/2)$, $1 \le x \le 3$, about the y-axis. Check your result with the geometry formula given in Exercise 11.

Find the areas of the surfaces generated by revolving the curves in Exercises 13–23 about the indicated axes. If you have a grapher, you may want to graph these curves to see what they look like.

13. $y = x^3/9$, $0 \le x \le 2$; x-axis

14. $y = \sqrt{x}$, $3/4 \le x \le 15/4$; x-axis

15. $y = \sqrt{2x - x^2}$, $0.5 \le x \le 1.5$; x-axis

16. $y^2 = 4 + x$, $-4 \le x \le 2$; x-axis

17. $x = y^3/3$, $0 \le y \le 1$; y-axis

18. $3y = x^3$, $-2 \le x \le 2$; x-axis

19. $x = 2\sqrt{4 - y}$, $0 \le y \le 15/4$; y-axis

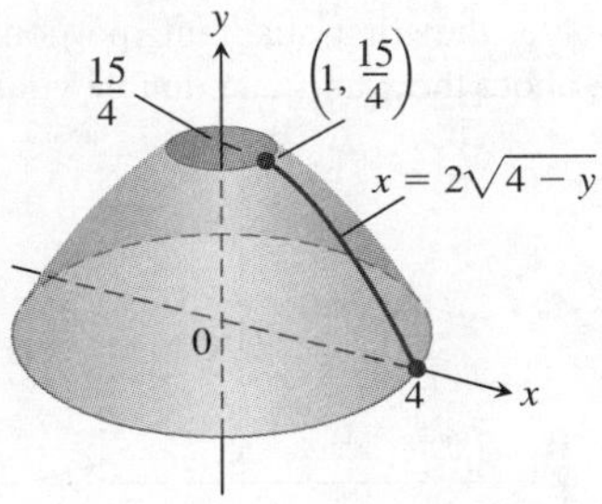

20. $x = \sqrt{2y - 1}$, $5/8 \le y \le 1$; y-axis

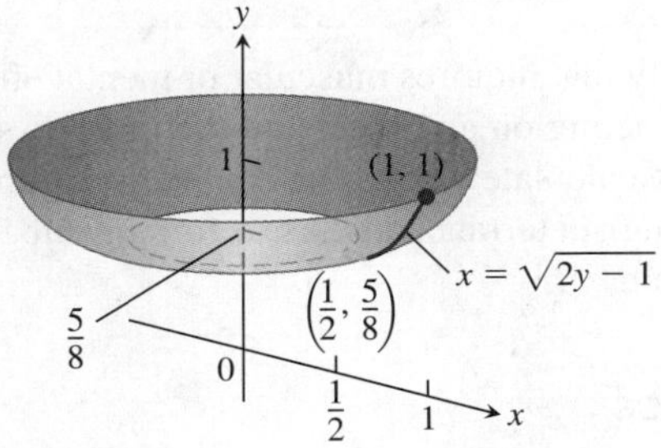

21. $y = (1/2)(x^2 + 1)$, $0 \le x \le 1$; y-axis

22. $y = (1/3)(x^2 + 2)^{3/2}$, $0 \le x \le \sqrt{2}$; y-axis (*Hint:* Express $ds = \sqrt{dx^2 + dy^2}$ in terms of dx, and evaluate the integral $S = \int 2\pi x \, ds$ with appropriate limits.)

23. $x = (y^4/4) + 1/(8y^2)$, $1 \le y \le 2$; x-axis (*Hint:* Express $ds = \sqrt{dx^2 + dy^2}$ in terms of dy, and evaluate the integral $S = \int 2\pi y \, ds$ with appropriate limits.)

24. Write an integral for the area of the surface generated by revolving the curve $y = \cos x$, $-\pi/2 \le x \le \pi/2$, about the x-axis. In Section 8.4 we will see how to evaluate such integrals.

25. **Testing the new definition** Show that the surface area of a sphere of radius a is still $4\pi a^2$ by using Equation (3) to find the area of the surface generated by revolving the curve $y = \sqrt{a^2 - x^2}$, $-a \le x \le a$, about the x-axis.

26. **Testing the new definition** The lateral (side) surface area of a cone of height h and base radius r should be $\pi r\sqrt{r^2 + h^2}$, the semiperimeter of the base times the slant height. Show that this is still the case by finding the area of the surface generated by revolving the line segment $y = (r/h)x$, $0 \le x \le h$, about the x-axis.

T 27. **Enameling woks** Your company decided to put out a deluxe version of a wok you designed. The plan is to coat it inside with white enamel and outside with light gray enamel. Each enamel will be sprayed on 0.5 mm thick before baking. (See accompanying figure.) Your manufacturing department wants to know how much enamel to have on hand for a production run of 5000 woks. What do you tell them? (Neglect waste and unused material and give your answer in liters. Remember that $1 \text{ cm}^3 = 1 \text{ mL}$, so $1 \text{ L} = 1000 \text{ cm}^3$.)

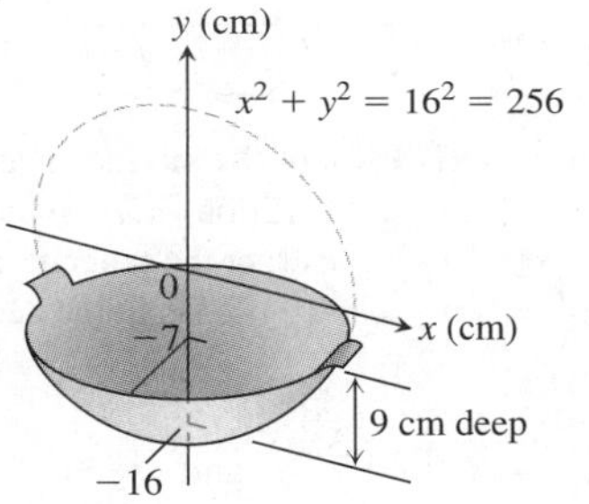

28. Here is a schematic drawing of the 90-ft dome used by the U.S. National Weather Service to house radar in Bozeman, Montana.

a. How much outside surface is there to paint (not counting the bottom)?

T b. Express the answer to the nearest square foot.

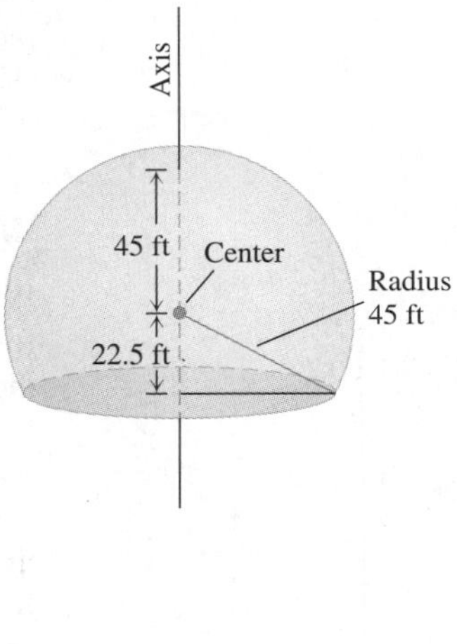

29. The shaded band shown here is cut from a sphere of radius R by parallel planes h units apart. Show that the surface area of the band is $2\pi Rh$.

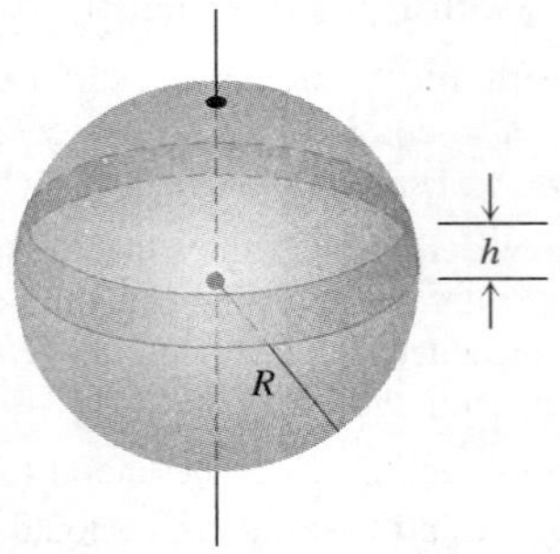

30. Slicing bread Did you know that if you cut a spherical loaf of bread into slices of equal width, each slice will have the same amount of crust? To see why, suppose the semicircle $y = \sqrt{r^2 - x^2}$ shown here is revolved about the x-axis to generate a sphere. Let AB be an arc of the semicircle that lies above an interval of length h on the x-axis. Show that the area swept out by AB does not depend on the location of the interval. (It does depend on the length of the interval.)

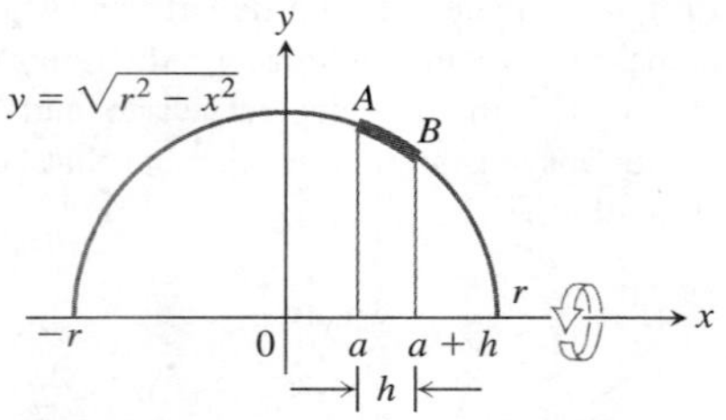

31. An alternative derivation of the surface area formula Assume f is smooth on $[a, b]$ and partition $[a, b]$ in the usual way. In the kth subinterval $[x_{k-1}, x_k]$, construct the tangent line to the curve at the midpoint $m_k = (x_{k-1} + x_k)/2$, as in the accompanying figure.

a. Show that

$$r_1 = f(m_k) - f'(m_k)\frac{\Delta x_k}{2} \quad \text{and} \quad r_2 = f(m_k) + f'(m_k)\frac{\Delta x_k}{2}.$$

b. Show that the length L_k of the tangent line segment in the kth subinterval is $L_k = \sqrt{(\Delta x_k)^2 + (f'(m_k)\,\Delta x_k)^2}$.

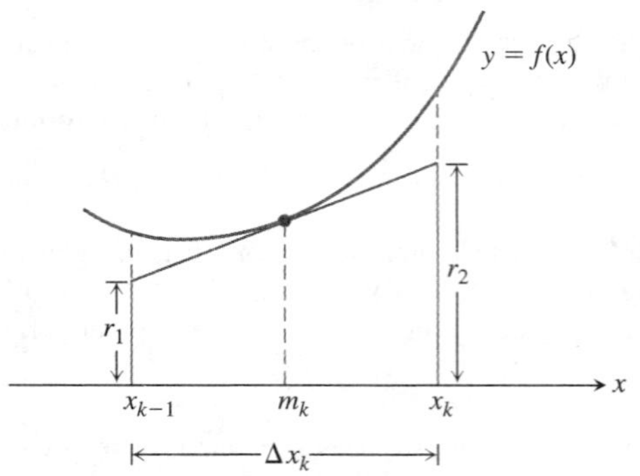

c. Show that the lateral surface area of the frustum of the cone swept out by the tangent line segment as it revolves about the x-axis is $2\pi f(m_k)\sqrt{1 + (f'(m_k))^2}\,\Delta x_k$.

d. Show that the area of the surface generated by revolving $y = f(x)$ about the x-axis over $[a, b]$ is

$$\lim_{n\to\infty} \sum_{k=1}^{n} \left(\begin{array}{c} \text{lateral surface area} \\ \text{of } k\text{th frustum} \end{array} \right) = \int_a^b 2\pi f(x)\sqrt{1 + (f'(x))^2}\,dx.$$

32. The surface of an astroid Find the area of the surface generated by revolving about the x-axis the portion of the astroid $x^{2/3} + y^{2/3} = 1$ shown in the accompanying figure.

(*Hint:* Revolve the first-quadrant portion $y = (1 - x^{2/3})^{3/2}$, $0 \le x \le 1$, about the x-axis and double your result.)

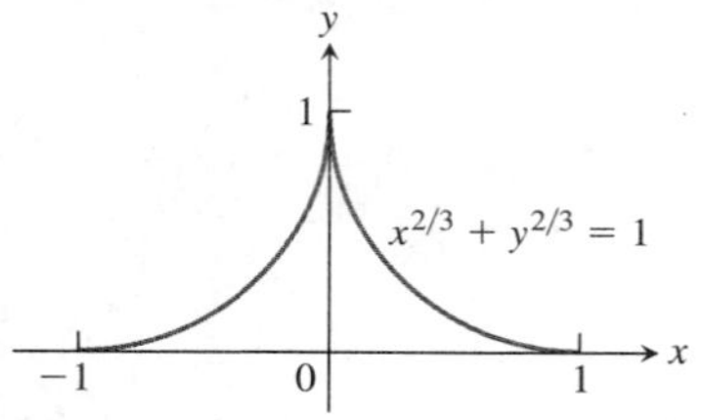

6.5 Work and Fluid Forces

In everyday life, *work* means an activity that requires muscular or mental effort. In science, the term refers specifically to a force acting on an object and the object's subsequent displacement. This section shows how to calculate work. The applications run from compressing railroad car springs and emptying subterranean tanks to forcing subatomic particles to collide and lifting satellites into orbit.

Work Done by a Constant Force

When an object moves a distance d along a straight line as a result of being acted on by a force of constant magnitude F in the direction of motion, we define the **work** W done by the force on the object with the formula

$$W = Fd \qquad \text{(Constant-force formula for work).} \tag{1}$$

From Equation (1) we see that the unit of work in any system is the unit of force multiplied by the unit of distance. In SI units (SI stands for *Système International*, or International System), the unit of force is a newton (N), the unit of distance is a meter (m), and the unit of work is a newton-meter $(\text{N} \cdot \text{m})$. This combination appears so often, it has

Joules
The joule, abbreviated J, is named after the English physicist James Prescott Joule (1818–1889). The defining equation is

$$1 \text{ joule} = (1 \text{ newton})(1 \text{ meter}).$$

In symbols, $1\text{ J} = 1\text{ N} \cdot \text{m}$.

a special name, the **joule** (J). Taking gravitational acceleration at sea level to be 9.8 m/sec^2, to lift one kilogram one meter requires work of 9.8 joules. This is seen by multiplying the force of 9.8 newtons exerted on one kilogram by the one-meter distance moved. In the British system, the unit of work is the foot-pound, a unit sometimes used in applications. It requires one foot-pound of work to lift a one pound weight a distance of one foot.

EXAMPLE 1 Suppose you jack up the side of a 2000-lb car 1.25 ft to change a tire. The jack applies a constant vertical force of about 1000 lb in lifting the side of the car (but because of the mechanical advantage of the jack, the force you apply to the jack itself is only about 30 lb). The total work performed by the jack on the car is $1000 \times 1.25 = 1250$ ft-lb. In SI units, the jack has applied a force of 4448 N through a distance of 0.381 m to do $4448 \times 0.381 \approx 1695$ J of work. ■

Work Done by a Variable Force Along a Line

If the force you apply varies along the way, as it will if you are stretching or compressing a spring, the formula $W = Fd$ has to be replaced by an integral formula that takes the variation in F into account.

Suppose that the force performing the work acts on an object moving along a straight line, which we take to be the x-axis. We assume that the magnitude of the force is a continuous function F of the object's position x. We want to find the work done over the interval from $x = a$ to $x = b$. We partition $[a, b]$ in the usual way and choose an arbitrary point c_k in each subinterval $[x_{k-1}, x_k]$. If the subinterval is short enough, the continuous function F will not vary much from x_{k-1} to x_k. The amount of work done across the interval will be about $F(c_k)$ times the distance Δx_k, the same as it would be if F were constant and we could apply Equation (1). The total work done from a to b is therefore approximated by the Riemann sum

$$\text{Work} \approx \sum_{k=1}^{n} F(c_k)\,\Delta x_k.$$

We expect the approximation to improve as the norm of the partition goes to zero, so we define the work done by the force from a to b to be the integral of F from a to b:

$$\lim_{n\to\infty} \sum_{k=1}^{n} F(c_k)\,\Delta x_k = \int_a^b F(x)\,dx.$$

DEFINITION The **work** done by a variable force $F(x)$ in moving an object along the x-axis from $x = a$ to $x = b$ is

$$W = \int_a^b F(x)\,dx. \tag{2}$$

The units of the integral are joules if F is in newtons and x is in meters, and foot-pounds if F is in pounds and x is in feet. So the work done by a force of $F(x) = 1/x^2$ newtons in moving an object along the x-axis from $x = 1$ m to $x = 10$ m is

$$W = \int_1^{10} \frac{1}{x^2}\,dx = -\frac{1}{x}\bigg]_1^{10} = -\frac{1}{10} + 1 = 0.9 \text{ J}.$$

Hooke's Law for Springs: $F = kx$

One calculation for work arises in finding the work required to stretch or compress a spring. **Hooke's Law** says that the force required to hold a stretched or compressed spring x units from its natural (unstressed) length is proportional to x. In symbols,

$$F = kx. \tag{3}$$

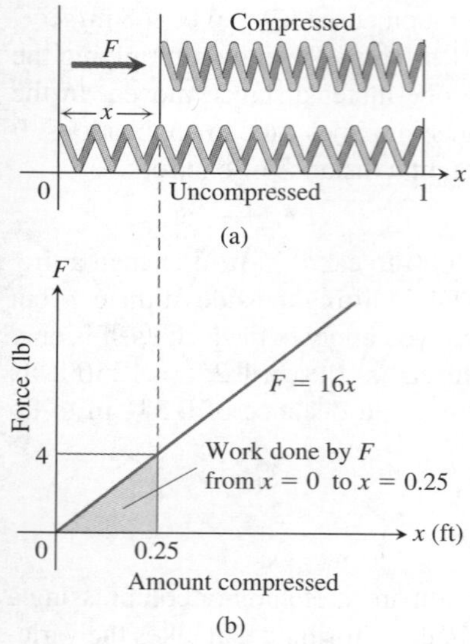

FIGURE 6.36 The force F needed to hold a spring under compression increases linearly as the spring is compressed (Example 2).

The constant k, measured in force units per unit length, is a characteristic of the spring, called the **force constant** (or **spring constant**) of the spring. Hooke's Law, Equation (3), gives good results as long as the force doesn't distort the metal in the spring. We assume that the forces in this section are too small to do that.

EXAMPLE 2 Find the work required to compress a spring from its natural length of 1 ft to a length of 0.75 ft if the force constant is $k = 16$ lb/ft.

Solution We picture the uncompressed spring laid out along the x-axis with its movable end at the origin and its fixed end at $x = 1$ ft (Figure 6.36). This enables us to describe the force required to compress the spring from 0 to x with the formula $F = 16x$. To compress the spring from 0 to 0.25 ft, the force must increase from

$$F(0) = 16 \cdot 0 = 0 \text{ lb} \quad \text{to} \quad F(0.25) = 16 \cdot 0.25 = 4 \text{ lb}.$$

The work done by F over this interval is

$$W = \int_0^{0.25} 16x\, dx = 8x^2\Big]_0^{0.25} = 0.5 \text{ ft-lb.} \qquad \text{Eq. (2) with } a = 0,\ b = 0.25,\ F(x) = 16x$$ ■

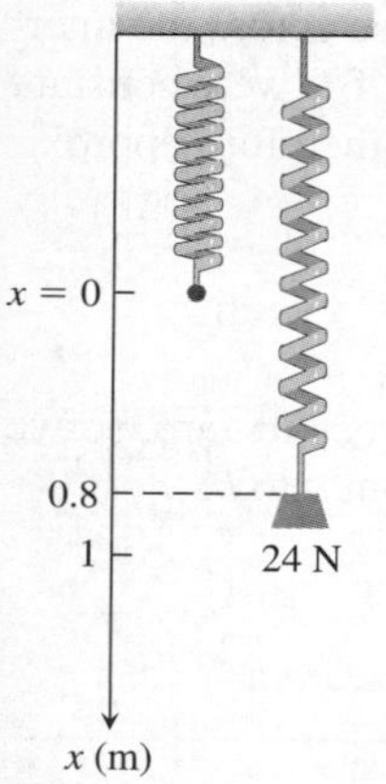

FIGURE 6.37 A 24-N weight stretches this spring 0.8 m beyond its unstressed length (Example 3).

EXAMPLE 3 A spring has a natural length of 1 m. A force of 24 N holds the spring stretched to a total length of 1.8 m.

(a) Find the force constant k.

(b) How much work will it take to stretch the spring from its natural length to a length of 3 m?

(c) How far will a 45-N force stretch the spring?

Solution

(a) *The force constant.* We find the force constant from Equation (3). A force of 24 N maintains the spring at a position where it is stretched 0.8 m from its natural length, so

$$24 = k(0.8) \qquad \text{Eq. (3) with } F = 24,\ x = 0.8$$
$$k = 24/0.8 = 30 \text{ N/m}.$$

(b) *The work to stretch the spring* 2 m. We imagine the unstressed spring hanging along the x-axis with its free end at $x = 0$ (Figure 6.37). The force required to stretch the spring x meters beyond its natural length is the force required to hold the free end of the spring x units from the origin. Hooke's Law with $k = 30$ says that this force is

$$F(x) = 30x.$$

The work done by F on the spring from $x = 0$ m to $x = 2$ m is

$$W = \int_0^2 30x\, dx = 15x^2\Big]_0^2 = 60 \text{ J}.$$

(c) *How far will a 45-N force stretch the spring?* We substitute $F = 45$ in the equation $F = 30x$ to find

$$45 = 30x, \quad \text{or} \quad x = 1.5 \text{ m}.$$

A 45-N force will keep the spring stretched 1.5 m beyond its natural length. ■

Lifting Objects and Pumping Liquids from Containers

The work integral is useful for calculating the work done in lifting objects whose weights vary with their elevation.

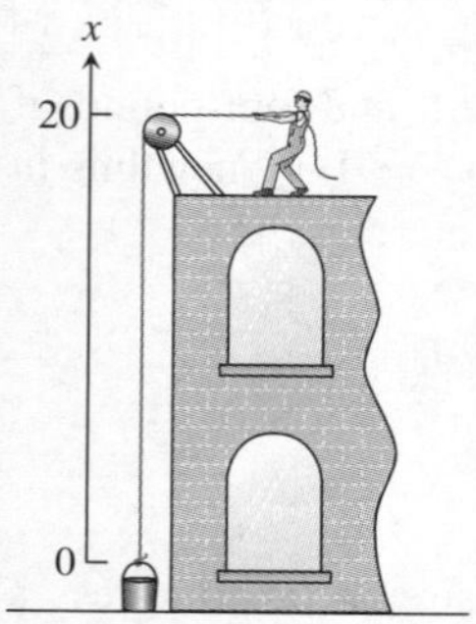

FIGURE 6.38 Lifting the bucket in Example 4.

EXAMPLE 4 A 5-kg bucket is lifted from the ground into the air by pulling in 20 m of rope at a constant speed (Figure 6.38). The rope weighs 0.08 kg/m. How much work is done in lifting the bucket and rope?

Solution The weight of the bucket is obtained by multiplying the mass (5 kg) and the acceleration due to gravity, approximately 9.8 $\mathrm{m/s^2}$. So the bucket's weight is $(5)(9.8) = 49$ N, and the work done in lifting it alone is weight $\times$ distance $= (49)(20) = 980$ J.

The weight of the rope varies with the bucket's elevation, because less of it is freely hanging as the bucket is raised. When the bucket is x meters off the ground, the remaining portion of the rope still being lifted weighs $(0.08)(20 - x)(9.8)$ N. So the work done in lifting the rope is

$$\text{Work on rope} = \int_0^{20} (0.08)(20 - x)(9.8)\,dx = \int_0^{20} (15.68 - 0.784x)\,dx$$

$$= \Big[15.68x - 0.392x^2\Big]_0^{20} = 313.6 - 156.8 = 156.8 \text{ J}.$$

The total work for the bucket and rope combined is

$$980 + 156.8 = 1136.8 \text{ J}.$$

■

How much work does it take to pump all or part of the liquid from a container? Engineers often need to know the answer in order to design or choose the right pump, or to compute the cost to transport water or some other liquid from one place to another. To find out how much work is required to pump the liquid, we imagine lifting the liquid out one thin horizontal slab at a time and applying the equation $W = Fd$ to each slab. We then evaluate the integral that this leads to as the slabs become thinner and more numerous.

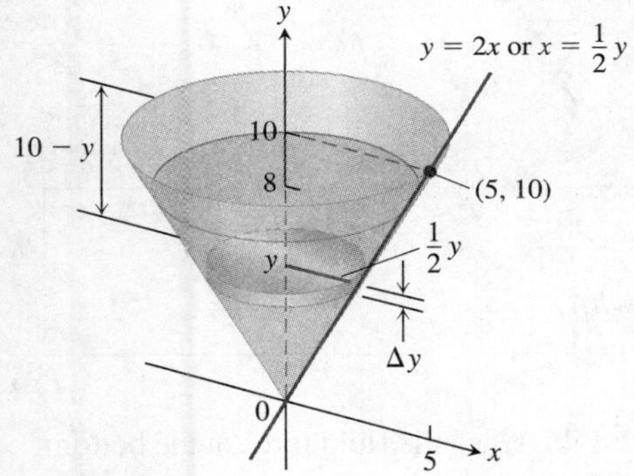

FIGURE 6.39 The olive oil and tank in Example 5.

EXAMPLE 5 The conical tank in Figure 6.39 is filled to within 2 ft of the top with olive oil weighing 57 $\mathrm{lb/ft^3}$. How much work does it take to pump the oil to the rim of the tank?

Solution We imagine the oil divided into thin slabs by planes perpendicular to the y-axis at the points of a partition of the interval $[0, 8]$.

The typical slab between the planes at y and $y + \Delta y$ has a volume of about

$$\Delta V = \pi(\text{radius})^2(\text{thickness}) = \pi\left(\frac{1}{2}y\right)^2 \Delta y = \frac{\pi}{4}y^2\,\Delta y \text{ ft}^3.$$

The force $F(y)$ required to lift this slab is equal to its weight,

$$F(y) = 57\,\Delta V = \frac{57\pi}{4}y^2\,\Delta y \text{ lb.} \qquad \text{Weight} = (\text{weight per unit volume}) \times \text{volume}$$

The distance through which $F(y)$ must act to lift this slab to the level of the rim of the cone is about $(10 - y)$ ft, so the work done in lifting the slab is about

$$\Delta W = \frac{57\pi}{4}(10 - y)y^2\,\Delta y \text{ ft-lb.}$$

Assuming there are n slabs associated with the partition of $[0, 8]$, and that $y = y_k$ denotes height of the plane associated with the kth slab of thickness Δy_k, we can approximate the work done in lifting all of the slabs with the Riemann sum

$$W \approx \sum_{k=1}^{n} \frac{57\pi}{4}(10 - y_k)y_k^2\,\Delta y_k \text{ ft-lb.}$$

The work of pumping the oil to the rim is the limit of these sums as the norm of the partition goes to zero and the number of slabs tends to infinity:

$$W = \lim_{n\to\infty}\sum_{k=1}^{n} \frac{57\pi}{4}(10 - y_k)y_k^2\,\Delta y_k = \int_0^8 \frac{57\pi}{4}(10 - y)y^2\,dy$$

$$= \frac{57\pi}{4}\int_0^8 (10y^2 - y^3)\,dy$$

$$= \frac{57\pi}{4}\left[\frac{10y^3}{3} - \frac{y^4}{4}\right]_0^8 \approx 30{,}561 \text{ ft-lb.}$$

■

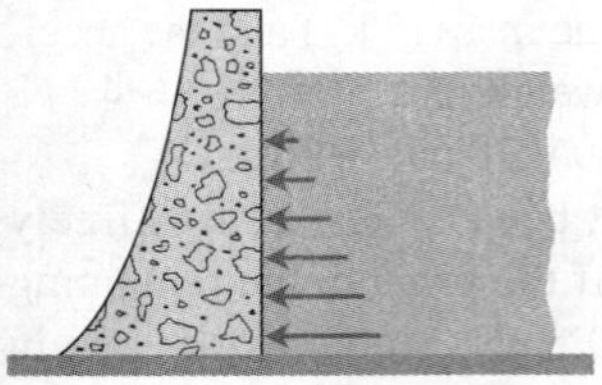

FIGURE 6.40 To withstand the increasing pressure, dams are built thicker as they go down.

Fluid Pressure and Forces

Dams are built thicker at the bottom than at the top (Figure 6.40) because the pressure against them increases with depth. The pressure at any point on a dam depends only on how far below the surface the point is and not on how much the surface of the dam happens to be tilted at that point. The pressure, in pounds per square foot at a point h feet below the surface, is always $62.4h$. The number 62.4 is the weight-density of freshwater in pounds per cubic foot. The pressure h feet below the surface of any fluid is the fluid's *weight-density* times h.

Weight-density

A fluid's weight-density w is its weight per unit volume. Typical values (lb/ft^3) are listed below.

Gasoline	42
Mercury	849
Milk	64.5
Molasses	100
Olive oil	57
Seawater	64
Freshwater	62.4

The Pressure-Depth Equation

In a fluid that is standing still, the pressure p at depth h is the fluid's weight-density w times h:

$$p = wh. \tag{4}$$

In a container of fluid with a flat horizontal base, the total force exerted by the fluid against the base can be calculated by multiplying the area of the base by the pressure at the base. We can do this because total force equals force per unit area (pressure) times area. (See Figure 6.41.) If F, p, and A are the total force, pressure, and area, then

$$\begin{aligned} F &= \text{total force} = \text{force per unit area} \times \text{area} \\ &= \text{pressure} \times \text{area} = pA \qquad\qquad p = wh \text{ from Eq. (4)} \\ &= whA. \end{aligned}$$

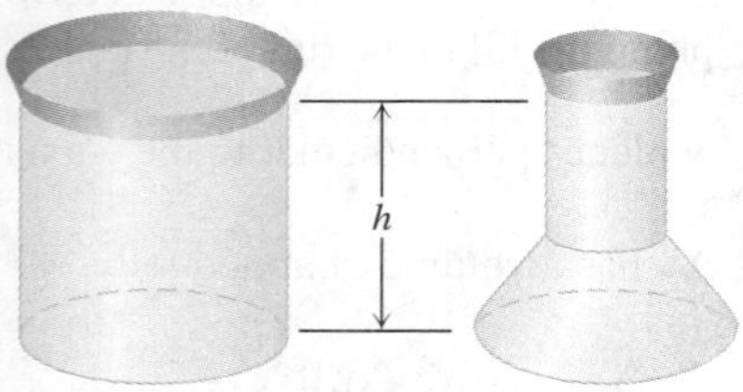

FIGURE 6.41 These containers are filled with water to the same depth and have the same base area. The total force is therefore the same on the bottom of each container. The containers' shapes do not matter here.

Fluid Force on a Constant-Depth Surface

$$F = pA = whA \tag{5}$$

For example, the weight-density of freshwater is 62.4 lb/ft^3, so the fluid force at the bottom of a $10 \text{ ft} \times 20 \text{ ft}$ rectangular swimming pool 3 ft deep is

$$\begin{aligned} F = whA &= (62.4 \text{ lb/ft}^3)(3 \text{ ft})(10 \cdot 20 \text{ ft}^2) \\ &= 37{,}440 \text{ lb}. \end{aligned}$$

For a flat plate submerged *horizontally*, like the bottom of the swimming pool just discussed, the downward force acting on its upper face due to liquid pressure is given by Equation (5). If the plate is submerged *vertically*, however, then the pressure against it will be different at different depths and Equation (5) no longer is usable in that form (because h varies).

Suppose we want to know the force exerted by a fluid against one side of a vertical plate submerged in a fluid of weight-density w. To find it, we model the plate as a region extending from $y = a$ to $y = b$ in the xy-plane (Figure 6.42). We partition $[a, b]$ in the usual way and imagine the region to be cut into thin horizontal strips by planes perpendicular to the y-axis at the partition points. The typical strip from y to $y + \Delta y$ is Δy units wide by $L(y)$ units long. We assume $L(y)$ to be a continuous function of y.

The pressure varies across the strip from top to bottom. If the strip is narrow enough, however, the pressure will remain close to its bottom-edge value of $w \times (\text{strip depth})$. The force exerted by the fluid against one side of the strip will be about

$$\begin{aligned} \Delta F &= (\text{pressure along bottom edge}) \times (\text{area}) \\ &= w \cdot (\text{strip depth}) \cdot L(y)\,\Delta y. \end{aligned}$$

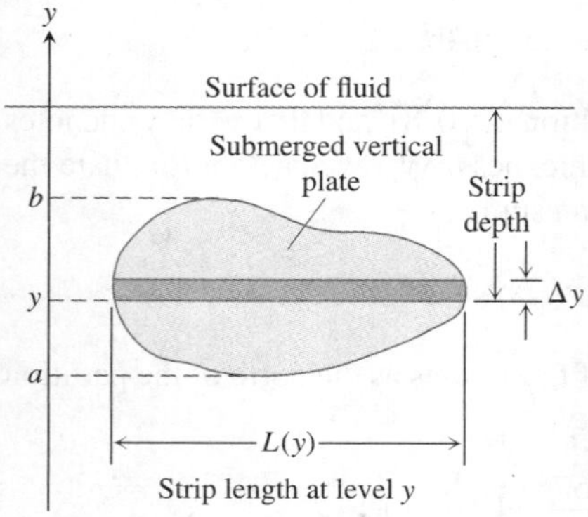

FIGURE 6.42 The force exerted by a fluid against one side of a thin, flat horizontal strip is about $\Delta F = \text{pressure} \times \text{area} = w \times (\text{strip depth}) \times L(y)\,\Delta y$.

Assume there are n strips associated with the partition of $a \le y \le b$ and that y_k is the bottom edge of the kth strip having length $L(y_k)$ and width Δy_k. The force against the entire plate is approximated by summing the forces against each strip, giving the Riemann sum

$$F \approx \sum_{k=1}^{n} w \cdot (\text{strip depth})_k \cdot L(y_k)\,\Delta y_k. \tag{6}$$

The sum in Equation (6) is a Riemann sum for a continuous function on $[a, b]$, and we expect the approximations to improve as the norm of the partition goes to zero. The force against the plate is the limit of these sums:

$$\lim_{n\to\infty} \sum_{k=1}^{n} w \cdot (\text{strip depth})_k \cdot L(y_k)\,\Delta y_k = \int_a^b w \cdot (\text{strip depth}) \cdot L(y)\,dy.$$

The Integral for Fluid Force Against a Vertical Flat Plate
Suppose that a plate submerged vertically in fluid of weight-density w runs from $y = a$ to $y = b$ on the y-axis. Let $L(y)$ be the length of the horizontal strip measured from left to right along the surface of the plate at level y. Then the force exerted by the fluid against one side of the plate is

$$F = \int_a^b w \cdot (\text{strip depth}) \cdot L(y)\,dy. \tag{7}$$

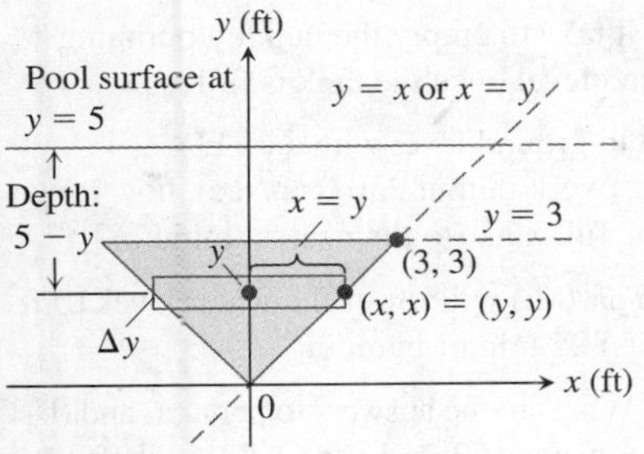

FIGURE 6.43 To find the force on one side of the submerged plate in Example 6, we can use a coordinate system like the one here.

EXAMPLE 6 A flat isosceles right-triangular plate with base 6 ft and height 3 ft is submerged vertically, base up, 2 ft below the surface of a swimming pool. Find the force exerted by the water against one side of the plate.

Solution We establish a coordinate system to work in by placing the origin at the plate's bottom vertex and running the y-axis upward along the plate's axis of symmetry (Figure 6.43). The surface of the pool lies along the line $y = 5$ and the plate's top edge along the line $y = 3$. The plate's right-hand edge lies along the line $y = x$, with the upper-right vertex at $(3, 3)$. The length of a thin strip at level y is

$$L(y) = 2x = 2y.$$

The depth of the strip beneath the surface is $(5 - y)$. The force exerted by the water against one side of the plate is therefore

$$\begin{aligned} F &= \int_a^b w \cdot \begin{pmatrix}\text{strip}\\ \text{depth}\end{pmatrix} \cdot L(y)\,dy \qquad \text{Eq. (7)} \\ &= \int_0^3 62.4(5 - y)2y\,dy \\ &= 124.8 \int_0^3 (5y - y^2)\,dy \\ &= 124.8\left[\frac{5}{2}y^2 - \frac{y^3}{3}\right]_0^3 = 1684.8 \text{ lb.} \end{aligned}$$ ■

EXERCISES 6.5

For some exercises, a calculator may be helpful when expressing answers in decimal form.

Springs

The graphs of force functions (in newtons) are given in Exercises 1 and 2. How much work is done by each force in moving an object 10 m?

1.
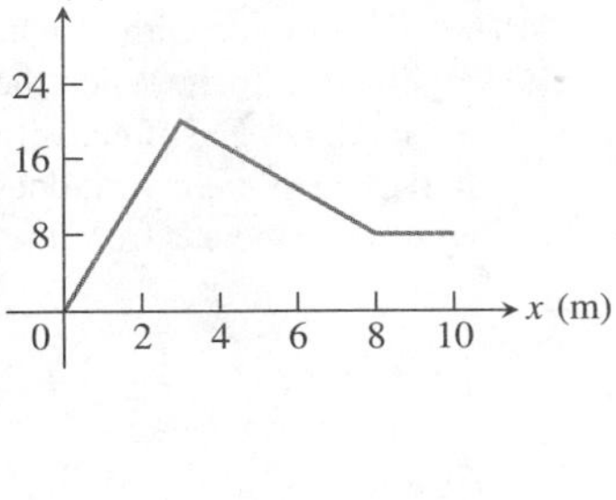

2.

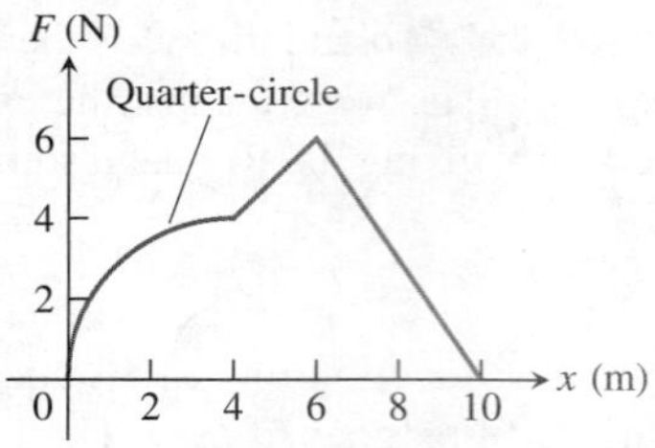

3. **Spring constant** It took 2700 J of work to stretch a spring from its natural length of 3 m to a length of 6 m. Find the spring's force constant.

4. **Stretching a spring** A spring has a natural length of 10 in. An 800-lb force stretches the spring to 14 in.
   a. Find the force constant.
   b. How much work is done in stretching the spring from 10 in. to 12 in.?
   c. How far beyond its natural length will a 1600-lb force stretch the spring?

5. **Stretching a rubber band** A force of 2 N will stretch a rubber band 2 cm (0.02 m). Assuming that Hooke's Law applies, how far will a 4-N force stretch the rubber band? How much work does it take to stretch the rubber band this far?

6. **Stretching a spring** If a force of 90 N stretches a spring 1 m beyond its natural length, how much work does it take to stretch the spring 5 m beyond its natural length?

7. **Subway car springs** It takes a force of 21,714 lb to compress a coil spring assembly on a New York City Transit Authority subway car from its free height of 8 in. to its fully compressed height of 5 in.
   a. What is the assembly's force constant?
   b. How much work does it take to compress the assembly the first half inch? The second half inch? Answer to the nearest in.-lb.

8. **Bathroom scale** A bathroom scale is compressed $1/15$ in. when a 120-lb person stands on it. Assuming that the scale behaves like a spring that obeys Hooke' Law, how much does someone who compresses the scale $1/5$ in. weigh? How much work is done compressing the scale $1/5$ in.?

Work Done by a Variable Force

9. **Lifting a rope** A mountain climber is about to haul up a 50-m length of hanging rope. How much work will it take if the rope weighs 0.624 N/m?

10. **Leaky sandbag** A bag of sand originally weighing 144 lb was lifted at a constant rate. As it rose, sand also leaked out at a constant rate. The sand was half gone by the time the bag had been lifted to 18 ft. How much work was done lifting the sand this far? (Neglect the weight of the bag and lifting equipment.)

11. **Lifting an elevator cable** An electric elevator with a motor at the top has a multistrand cable weighing 5.5 lb/ft. When the car is at the first floor, 200 ft of cable are paid out, and effectively 0 ft are out when the car is at the top floor. How much work does the motor do just lifting the cable when it takes the car from the first floor to the top?

12. **Force of attraction** When a particle of mass m is at $(x, 0)$, it is attracted toward the origin with a force whose magnitude is k/x^2. If the particle starts from rest at $x = b$ and is acted on by no other forces, find the work done on it by the time it reaches $x = a$, $0 < a < b$.

13. **Leaky bucket** Assume the bucket in Example 4 is leaking. It starts with 5 liters of water (5 kg) and leaks at a constant rate. It finishes draining just as it reaches the top. How much work was spent lifting the water alone? (*Hint:* Do not include the rope and bucket, and find the proportion of water left at elevation x m.)

14. (*Continuation of Exercise 13.*) The workers in Example 4 and Exercise 13 changed to a larger bucket that held 10 liters (10 kg) of water, but the new bucket had an even larger leak so that it, too, was empty by the time it reached the top. Assuming that the water leaked out at a steady rate, how much work was done lifting the water alone? (Do not include the rope and bucket.)

Pumping Liquids from Containers

15. **Pumping water** The rectangular tank shown here, with its top at ground level, is used to catch runoff water. Assume that the water weighs $62.4\ \text{lb/ft}^3$.
   a. How much work does it take to empty the tank by pumping the water back to ground level once the tank is full?
   b. If the water is pumped to ground level with a (5/11)-horsepower (hp) motor (work output 250 ft-lb/sec), how long will it take to empty the full tank (to the nearest minute)?
   c. Show that the pump in part (b) will lower the water level 10 ft (halfway) during the first 25 min of pumping.
   d. The weight of water What are the answers to parts (a) and (b) in a location where water weighs $62.26\ \text{lb/ft}^3$? $62.59\ \text{lb/ft}^3$?

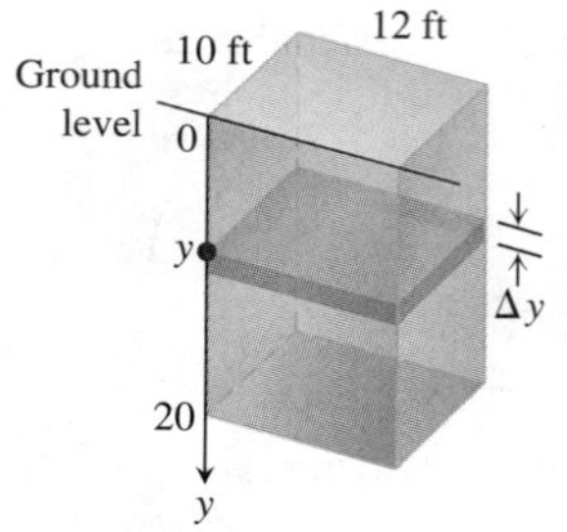

16. **Emptying a cistern** The rectangular cistern (storage tank for rainwater) shown has its top 10 ft below ground level. The cistern, currently full, is to be emptied for inspection by pumping its contents to ground level.
   a. How much work will it take to empty the cistern?
   b. How long will it take a $1/2$-hp pump, rated at 275 ft-lb/sec, to pump the tank dry?
   c. How long will it take the pump in part (b) to empty the tank halfway? (It will be less than half the time required to empty the tank completely.)

d. The weight of water What are the answers to parts (a) through (c) in a location where water weighs 62.26 lb/ft^3? 62.59 lb/ft^3?

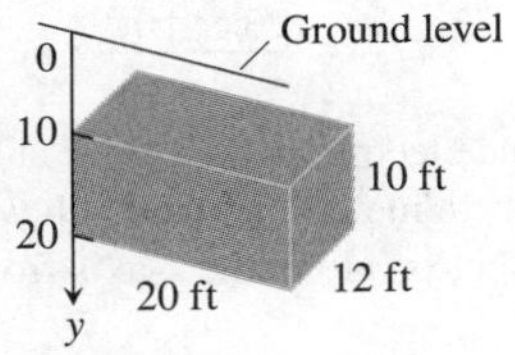

17. Pumping oil How much work would it take to pump oil from the tank in Example 5 to the level of the top of the tank if the tank were completely full?

18. Pumping a half-full tank Suppose that, instead of being full, the tank in Example 5 is only half full. How much work does it take to pump the remaining oil to a level 4 ft above the top of the tank?

19. Emptying a tank A vertical right-circular cylindrical tank measures 30 ft high and 20 ft in diameter. It is full of kerosene weighing 51.5 lb.ft^3. How much work does it take to pump the kerosene to the level of the top of the tank?

20. a. Pumping milk Suppose that the conical container in Example 5 contains milk (weighing 64.5 lb/ft^3) instead of olive oil. How much work will it take to pump the contents to the rim?

b. Pumping oil How much work will it take to pump the oil in Example 5 to a level 3 ft above the cone's rim?

21. The graph of $y = x^2$ on $0 \le x \le 2$ is revolved about the y-axis to form a tank that is then filled with salt water from the Dead Sea (weighing approximately 75 lb/ft^3). How much work does it take to pump all of the water to the top of the tank?

22. A right-circular cylindrical tank of height 10 ft and radius 5 ft is lying horizontally and is full of diesel fuel weighing 53 lb/ft^3. How much work is required to pump all of the fuel to a point 15 ft above the top of the tank?

23. Emptying a water reservoir We model pumping from spherical containers the way we do from other containers, with the axis of integration along the vertical axis of the sphere. Use the figure here to find how much work it takes to empty a full hemispherical water reservoir of radius 5 m by pumping the water to a height of 4 m above the top of the reservoir. Water weighs 9800 N/m^3.

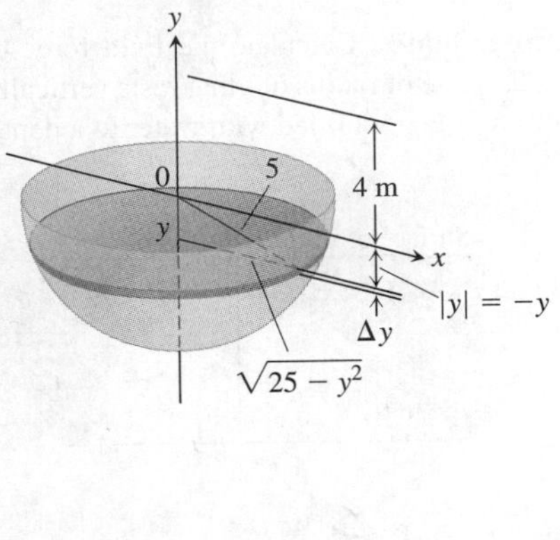

24. You are in charge of the evacuation and repair of the storage tank shown here. The tank is a hemisphere of radius 10 ft and is full of benzene weighing 56 lb/ft^3. A firm you contacted says it can empty the tank for 1/2¢ per foot-pound of work. Find the work required to empty the tank by pumping the benzene to an outlet 2 ft above the top of the tank. If you have $5000 budgeted for the job, can you afford to hire the firm?

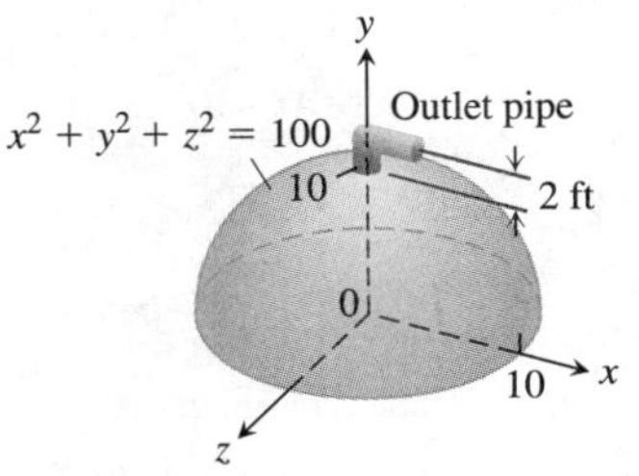

Work and Kinetic Energy

25. Kinetic energy If a variable force of magnitude $F(x)$ moves an object of mass m along the x-axis from x_1 to x_2, the object's velocity v can be written as dx/dt (where t represents time). Use Newton's second law of motion $F = m(dv/dt)$ and the Chain Rule

$$\frac{dv}{dt} = \frac{dv}{dx}\frac{dx}{dt} = v\frac{dv}{dx}$$

to show that the net work done by the force in moving the object from x_1 to x_2 is

$$W = \int_{x_1}^{x_2} F(x)\,dx = \frac{1}{2}mv_2^2 - \frac{1}{2}mv_1^2,$$

where v_1 and v_2 are the object's velocities at x_1 and x_2. In physics, the expression $(1/2)mv^2$ is called the *kinetic energy* of an object of mass m moving with velocity v. Therefore, *the work done by the force equals the change in the object's kinetic energy*, and we can find the work by calculating this change.

In Exercises 26–30, use the result of Exercise 25.

26. Tennis A 2-oz tennis ball was served at 160 ft/sec (about 109 mph). How much work was done on the ball to make it go this fast? (To find the ball's mass from its weight, express the weight in pounds and divide by 32 ft/sec^2, the acceleration of gravity.)

27. Baseball How many foot-pounds of work does it take to throw a baseball 90 mph? A baseball weighs 5 oz, or 0.3125 lb.

28. Golf A 3.2-oz golf ball is driven off the tee at a speed of 260 ft/sec (about 177 mph). How many foot-pounds of work are done on the ball getting it into the air?

29. Tennis At the 2012 Busan Open Challenger Tennis Tournament is Busan, South Korea, the Australian, Samuel Groth, hit a serve measured at 263 kph (163.4 mph). How much work was required by Groth to serve a 0.056699-kg (2-oz) tennis ball at that speed?

30. Softball How much work has to be performed on a 6-oz softball to pitch it 110 ft/sec (75 mph)?

31. Drinking a milkshake The truncated conical container shown here is full of strawberry milkshake that weighs 4/9 oz/in^3. As you can see, the container is 7 in. deep, 2.5 in. across at the base, and 3.5 in. across at the top.

The straw sticks up an inch above the top. About how much work does it take to suck up the milkshake through the straw (neglecting friction)? Answer in inch-ounces.

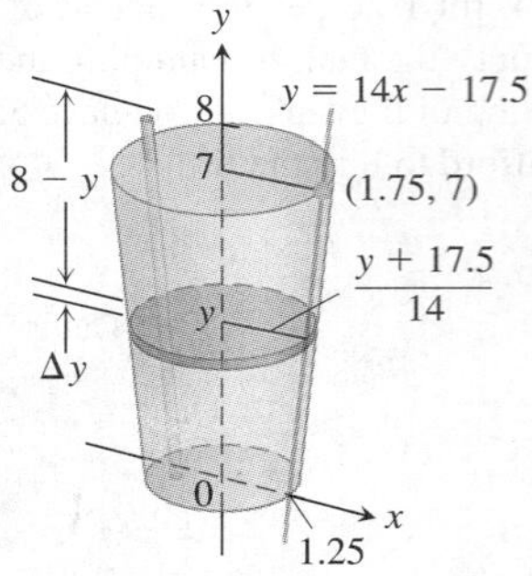

Dimensions in inches

32. Water tower Your town has decided to drill a well to increase its water supply. As the town engineer, you have determined that a water tower will be necessary to provide the pressure needed for distribution, and you have designed the system shown here. The water is to be pumped from a 300-ft well through a vertical 4-in. pipe into the base of a cylindrical tank 20 ft in diameter and 25 ft high. The base of the tank will be 60 ft above ground. The pump is a 3-hp pump, rated at 1650 ft · lb/sec. To the nearest hour, how long will it take to fill the tank the first time? (Include the time it takes to fill the pipe.) Assume that water weighs 62.4 lb/ft^3.

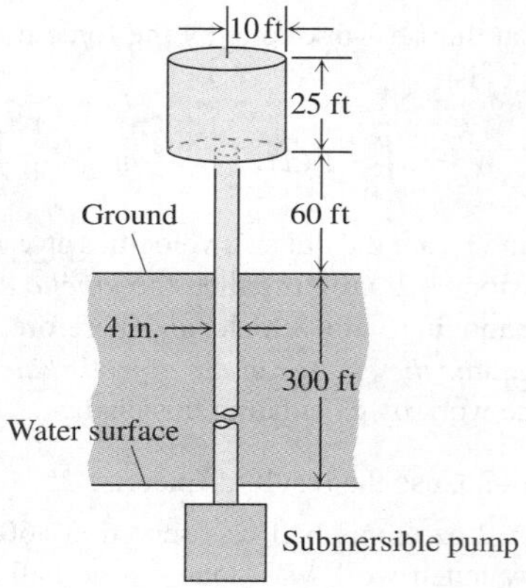

NOT TO SCALE

33. Putting a satellite in orbit The strength of Earth's gravitational field varies with the distance r from Earth's center, and the magnitude of the gravitational force experienced by a satellite of mass m during and after launch is

$$F(r) = \frac{mMG}{r^2}.$$

Here, $M = 5.975 \times 10^{24}$ kg is Earth's mass,

$$G = 6.6720 \times 10^{-11}\ \mathrm{N \cdot m^2\, kg^{-2}}$$

is the universal gravitational constant, and r is measured in meters. The work it takes to lift a 1000-kg satellite from Earth's surface to a circular orbit 35,780 km above Earth's center is therefore given by the integral

$$\text{Work} = \int_{6{,}370{,}000}^{35{,}780{,}000} \frac{1000MG}{r^2}\, dr \text{ joules.}$$

Evaluate the integral. The lower limit of integration is Earth's radius in meters at the launch site. (This calculation does not take into account energy spent lifting the launch vehicle or energy spent bringing the satellite to orbit velocity.)

34. Forcing electrons together Two electrons r meters apart repel each other with a force of

$$F = \frac{23 \times 10^{-29}}{r^2} \text{ newtons.}$$

a. Suppose one electron is held fixed at the point $(1, 0)$ on the x-axis (units in meters). How much work does it take to move a second electron along the x-axis from the point $(-1, 0)$ to the origin?

b. Suppose an electron is held fixed at each of the points $(-1, 0)$ and $(1, 0)$. How much work does it take to move a third electron along the x-axis from $(5, 0)$ to $(3, 0)$?

Finding Fluid Forces

35. Triangular plate Calculate the fluid force on one side of the plate in Example 6 using the coordinate system shown here.

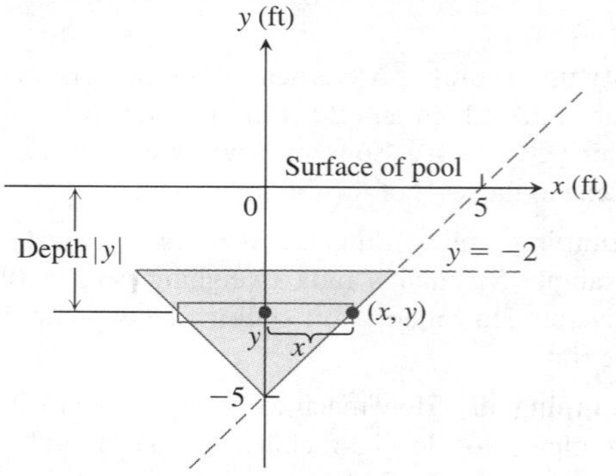

36. Triangular plate Calculate the fluid force on one side of the plate in Example 6 using the coordinate system shown here.

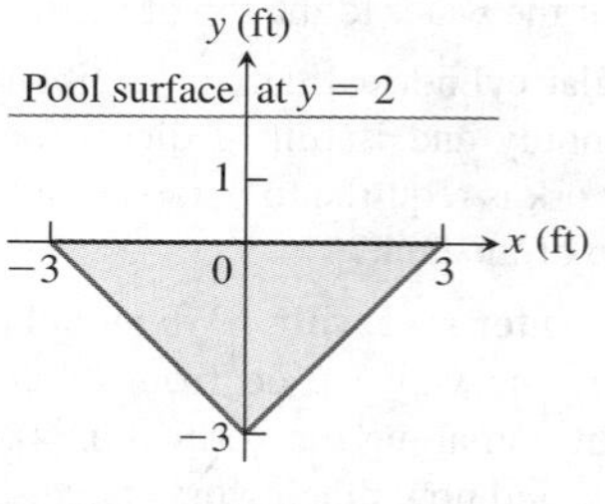

37. Rectangular plate In a pool filled with water to a depth of 10 ft, calculate the fluid force on one side of a 3 ft by 4 ft rectangular plate if the plate rests vertically at the bottom of the pool

a. on its 4-ft edge.

b. on its 3-ft edge.

38. Semicircular plate Calculate the fluid force on one side of a semicircular plate of radius 6 ft that rests vertically on its diameter at the bottom of a pool filled with water to a depth of 8 ft.

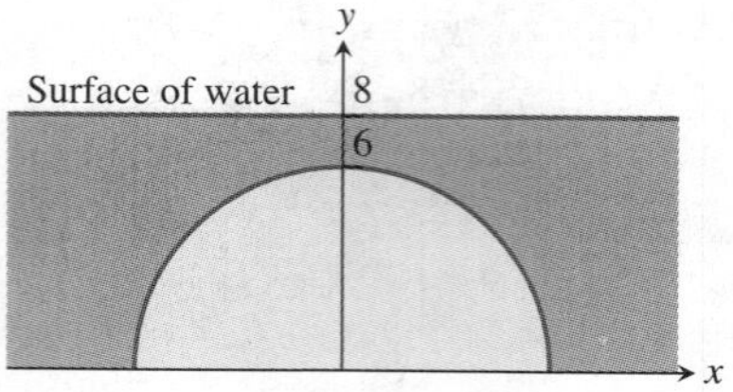

39. Triangular plate The isosceles triangular plate shown here is submerged vertically 1 ft below the surface of a freshwater lake.

a. Find the fluid force against one face of the plate.

b. What would be the fluid force on one side of the plate if the water were seawater instead of freshwater?

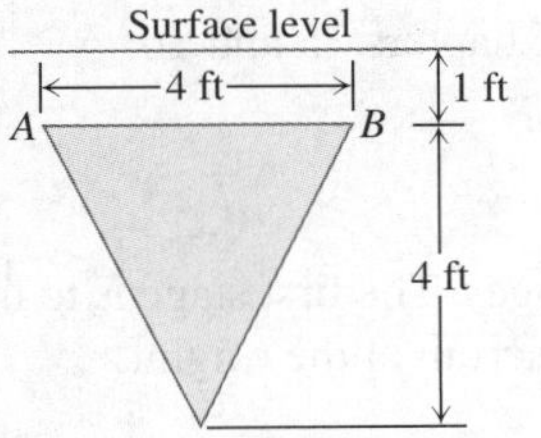

40. Rotated triangular plate The plate in Exercise 39 is revolved 180° about line AB so that part of the plate sticks out of the lake, as shown here. What force does the water exert on one face of the plate now?

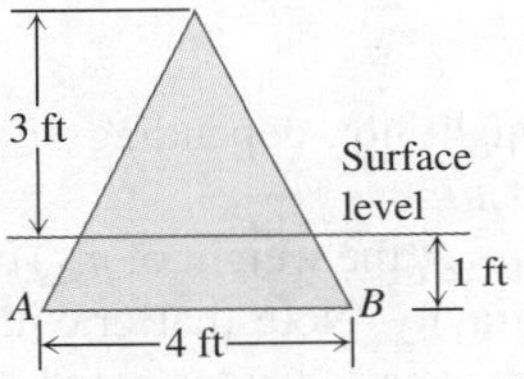

41. New England Aquarium The viewing portion of the rectangular glass window in a typical fish tank at the New England Aquarium in Boston is 63 in. wide and runs from 0.5 in. below the water's surface to 33.5 in. below the surface. Find the fluid force against this portion of the window. The weight-density of seawater is 64 lb/ft^3. (In case you were wondering, the glass is 3/4 in. thick and the tank walls extend 4 in. above the water to keep the fish from jumping out.)

42. Semicircular plate A semicircular plate 4 ft in diameter sticks straight down into freshwater with the diameter along the surface. Find the force exerted by the water on one side of the plate.

43. Tilted plate Calculate the fluid force on one side of a 5 ft by 5 ft square plate if the plate is at the bottom of a pool filled with water to a depth of 8 ft and

a. lying flat on its 5 ft by 5 ft face.

b. resting vertically on a 5-ft edge.

c. resting on a 5-ft edge and tilted at 45° to the bottom of the pool.

44. Tilted plate Calculate the fluid force on one side of a right-triangular plate with edges 3 ft, 4 ft, and 5 ft if the plate sits at the bottom of a pool filled with water to a depth of 6 ft on its 3-ft edge and tilted at 60° to the bottom of the pool.

45. The cubical metal tank shown here has a parabolic gate held in place by bolts and designed to withstand a fluid force of 160 lb without rupturing. The liquid you plan to store has a weight-density of 50 lb/ft^3.

a. What is the fluid force on the gate when the liquid is 2 ft deep?

b. What is the maximum height to which the container can be filled without exceeding the gate's design limitation?

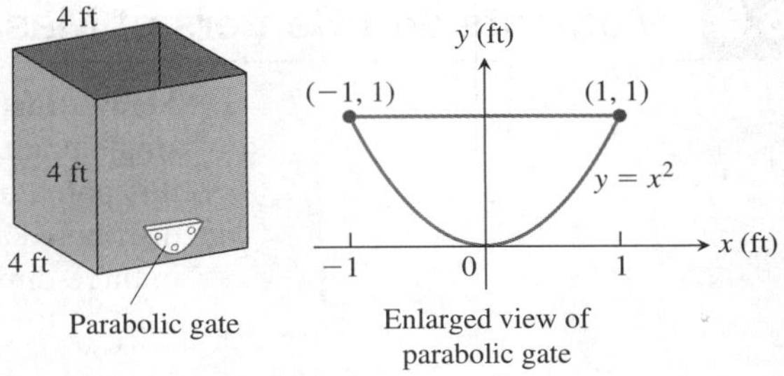

46. The end plates of the trough shown here were designed to withstand a fluid force of 6667 lb. How many cubic feet of water can the tank hold without exceeding this limitation? Round down to the nearest cubic foot. What is the value of h?

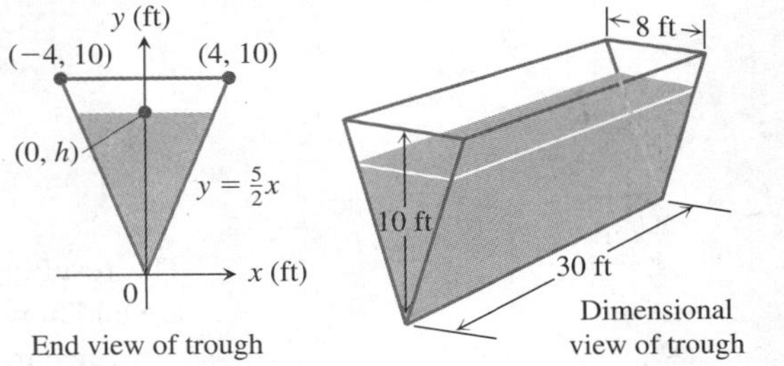

47. A vertical rectangular plate a units long by b units wide is submerged in a fluid of weight-density w with its long edges parallel to the fluid's surface. Find the average value of the pressure along the vertical dimension of the plate. Explain your answer.

48. (*Continuation of Exercise 47.*) Show that the force exerted by the fluid on one side of the plate is the average value of the pressure (found in Exercise 47) times the area of the plate.

49. Water pours into the tank shown here at the rate of 4 ft^3/min. The tank's cross-sections are 4-ft-diameter semicircles. One end of the tank is movable, but moving it to increase the volume compresses a spring. The spring constant is $k = 100$ lb/ft. If the end of the tank moves 5 ft against the spring, the water will drain out of a safety hole in the bottom at the rate of 5 ft^3/min. Will the movable end reach the hole before the tank overflows?

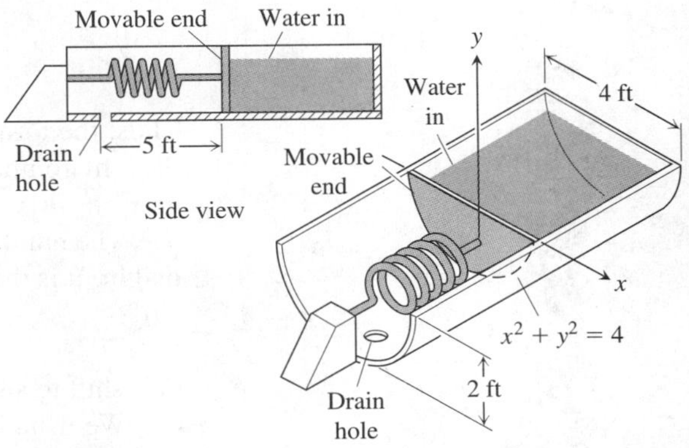

50. Watering trough The vertical ends of a watering trough are squares 3 ft on a side.

a. Find the fluid force against the ends when the trough is full.

b. How many inches do you have to lower the water level in the trough to reduce the fluid force by 25%?

6.6 Moments and Centers of Mass

Many structures and mechanical systems behave as if their masses were concentrated at a single point, called the *center of mass* (Figure 6.44). It is important to know how to locate this point, and doing so is basically a mathematical enterprise. Here we consider masses distributed along a line or region in the plane. Masses distributed across a region or curve in three-dimensional space are treated in Chapters 15 and 16.

Masses Along a Line

We develop our mathematical model in stages. The first stage is to imagine masses m_1, m_2, and m_3 on a rigid x-axis supported by a fulcrum at the origin.

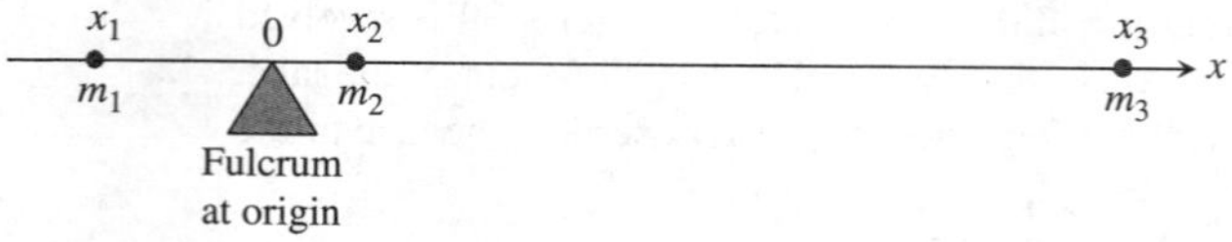

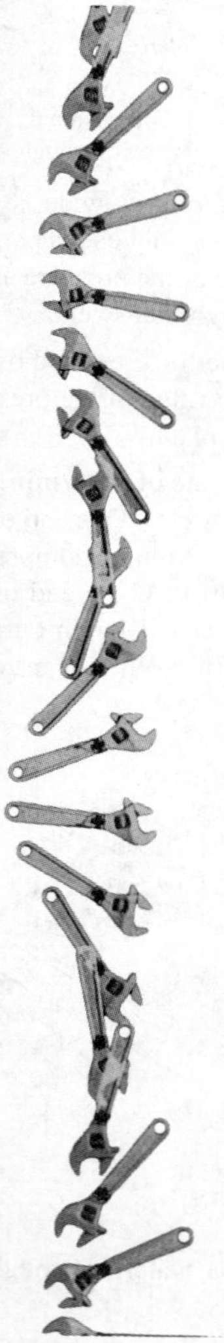

FIGURE 6.44 A wrench gliding on ice turning about its center of mass as the center glides in a vertical line. (*Source:* Berenice Abbott/ScienceSource)

The resulting system might balance, or it might not, depending on how large the masses are and how they are arranged along the x-axis.

Each mass m_k exerts a downward force $m_k g$ (the weight of m_k) equal to the magnitude of the mass times the acceleration due to gravity. Note that gravitational acceleration is downward, hence negative. Each of these forces has a tendency to turn the x-axis about the origin, the way a child turns a seesaw. This turning effect, called a **torque**, is measured by multiplying the force $m_k g$ by the signed distance x_k from the point of application to the origin. By convention, a positive torque induces a counterclockwise turn. Masses to the left of the origin exert positive (counterclockwise) torque. Masses to the right of the origin exert negative (clockwise) torque.

The sum of the torques measures the tendency of a system to rotate about the origin. This sum is called the **system torque**.

$$\text{System torque} = m_1 g x_1 + m_2 g x_2 + m_3 g x_3 \tag{1}$$

The system will balance if and only if its torque is zero.

If we factor out the g in Equation (1), we see that the system torque is

$$\underbrace{g}_{\text{a feature of the environment}} \cdot \underbrace{(m_1 x_1 + m_2 x_2 + m_3 x_3)}_{\text{a feature of the system}}.$$

Thus, the torque is the product of the gravitational acceleration g, which is a feature of the environment in which the system happens to reside, and the number $(m_1 x_1 + m_2 x_2 + m_3 x_3)$, which is a feature of the system itself.

The number $(m_1 x_1 + m_2 x_2 + m_3 x_3)$ is called the **moment of the system about the origin**. It is the sum of the **moments** $m_1 x_1, m_2 x_2, m_3 x_3$ of the individual masses.

$$M_0 = \text{Moment of system about origin} = \sum m_k x_k$$

(We shift to sigma notation here to allow for sums with more terms.)

We usually want to know where to place the fulcrum to make the system balance; that is, we want to know at what point $\bar{x}$ to place the fulcrum to make the torques add to zero.

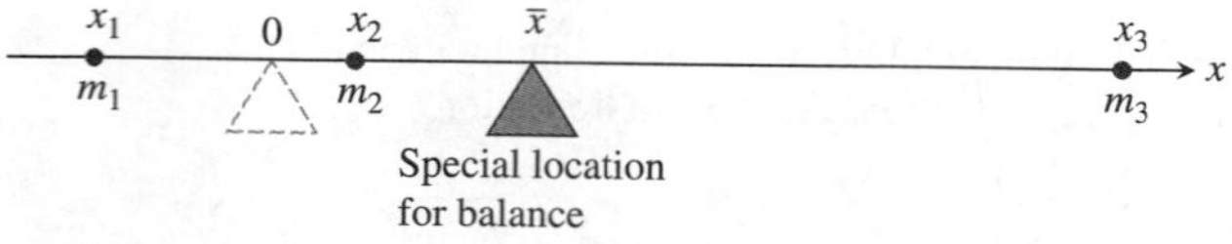

The torque of each mass about the fulcrum in this special location is

$$\text{Torque of } m_k \text{ about } \bar{x} = \begin{pmatrix}\text{signed distance}\\ \text{of } m_k \text{ from } \bar{x}\end{pmatrix}\begin{pmatrix}\text{downward}\\ \text{force}\end{pmatrix} = (x_k - \bar{x})m_k g.$$

When we write the equation that says that the sum of these torques is zero, we get an equation we can solve for $\bar{x}$:

$$\sum (x_k - \bar{x})\, m_k g = 0 \qquad \text{Sum of the torques equals zero.}$$

$$\bar{x} = \frac{\sum m_k x_k}{\sum m_k}. \qquad \text{Solved for } \bar{x}$$

This last equation tells us to find $\bar{x}$ by dividing the system's moment about the origin by the system's total mass:

$$\bar{x} = \frac{\sum m_k x_k}{\sum m_k} = \frac{\text{system moment about origin}}{\text{system mass}}. \tag{2}$$

The point $\bar{x}$ is called the system's **center of mass**.

Thin Wires

Instead of a discrete set of masses arranged in a line, suppose that we have a straight wire or rod located on interval $[a, b]$ on the x-axis. Suppose further that this wire is not homogeneous, but rather the density varies continuously from point to point. If a short segment of a rod containing the point x with length Δx has mass Δm, then the density at x is given by

$$\delta(x) = \lim_{\Delta x \to 0} \Delta m / \Delta x.$$

We often write this formula in one of the alternative forms $\delta = dm/dx$ and $dm = \delta\, dx$.

Partition the interval $[a, b]$ into finitely many subintervals $[x_{k-1}, x_k]$. If we take n subintervals and replace the portion of a wire along a subinterval of length Δx_k containing x_k by a point mass located at x_k with mass $\Delta m_k = \delta(x_k)\,\Delta x_k$, then we obtain a collection of point masses that have approximately the same total mass and same moment as the wire as indicated in Figure 6.45.

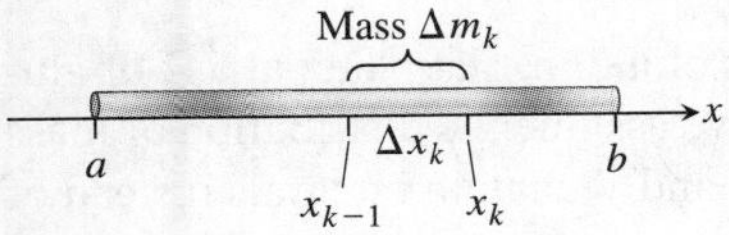

FIGURE 6.45 A rod of varying density can be modeled by a finite number of point masses of mass $\Delta m_k = \delta(x_k)\,\Delta x_k$ located at points x_k along the rod.

The mass M of the wire and the moment M_0 are approximated by the Riemann sums

$$M \approx \sum_{k=1}^{n} \Delta m_k = \sum_{k=1}^{n} \delta(x_k)\,\Delta x_k, \qquad M_0 \approx \sum_{k=1}^{n} x_k \Delta m_k = \sum_{k=1}^{n} x_k \delta(x_k)\,\Delta x_k.$$

By taking a limit of these Riemann sums as the length of the intervals in the partition approaches zero, we get integral formulas for the mass and the moment of the wire about the origin. The mass M, moment about the origin M_0, and center of mass $\bar{x}$ are

$$M = \int_a^b \delta(x)\,dx, \qquad M_0 = \int_a^b x\,\delta(x)\,dx, \qquad \bar{x} = \frac{M_0}{M} = \frac{\int_a^b x\,\delta(x)\,dx}{\int_a^b \delta(x)\,dx}.$$

EXAMPLE 1 Find the mass M and the center of mass $\bar{x}$ of a rod lying on the x-axis over the interval $[1, 2]$ whose density is given by $\delta(x) = 2 + 3x^2$.

Solution The mass of the rod is obtained by integrating the density,

$$M = \int_1^2 (2 + 3x^2)\,dx = \left[2x + x^3\right]_1^2 = (4 + 8) - (2 + 1) = 9,$$

and the center of mass is

$$\bar{x} = \frac{M_0}{M} = \frac{\int_1^2 x(2 + 3x^2)\,dx}{9} = \frac{\left[x^2 + \frac{3x^4}{4}\right]_1^2}{9} = \frac{19}{12}. \qquad \blacksquare$$

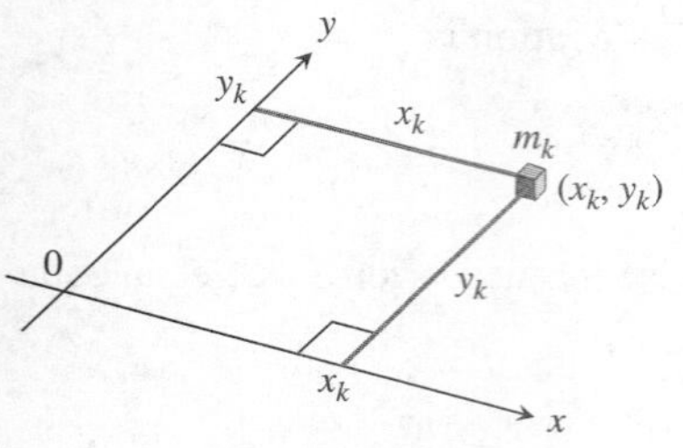

FIGURE 6.46 Each mass m_k has a moment about each axis.

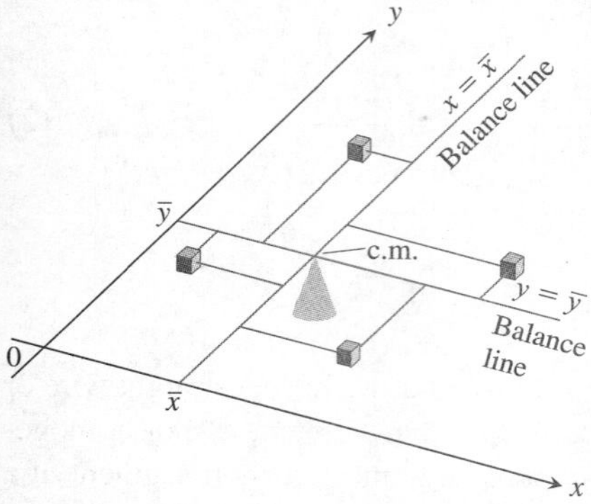

FIGURE 6.47 A two-dimensional array of masses balances on its center of mass.

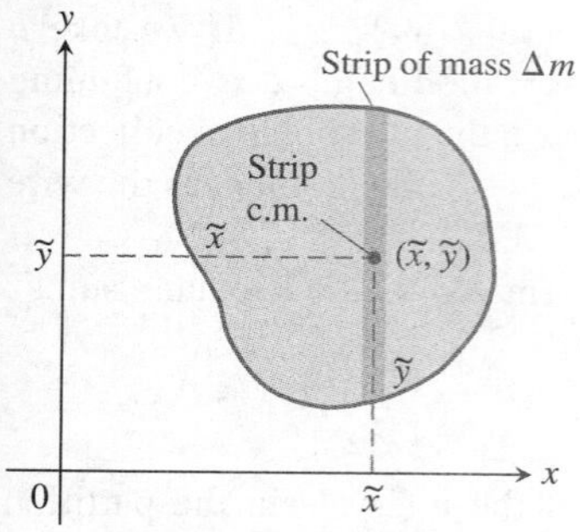

FIGURE 6.48 A plate cut into thin strips parallel to the y-axis. The moment exerted by a typical strip about each axis is the moment its mass Δm would exert if concentrated at the strip's center of mass $(\tilde{x}, \tilde{y})$.

Density of a plate
A material's density is its mass per unit area. For wires, rods, and narrow strips, the density is given in terms of mass per unit length.

Masses Distributed over a Plane Region

Suppose that we have a finite collection of masses located in the plane, with mass m_k at the point (x_k, y_k) (see Figure 6.46). The mass of the system is

$$\text{System mass:} \quad M = \sum m_k.$$

Each mass m_k has a moment about each axis. Its moment about the x-axis is $m_k y_k$, and its moment about the y-axis is $m_k x_k$. The moments of the entire system about the two axes are

$$\text{Moment about } x\text{-axis:} \quad M_x = \sum m_k y_k,$$

$$\text{Moment about } y\text{-axis:} \quad M_y = \sum m_k x_k.$$

The x-coordinate of the system's center of mass is defined to be

$$\bar{x} = \frac{M_y}{M} = \frac{\sum m_k x_k}{\sum m_k}. \tag{3}$$

With this choice of $\bar{x}$, as in the one-dimensional case, the system balances about the line $x = \bar{x}$ (Figure 6.47).

The y-coordinate of the system's center of mass is defined to be

$$\bar{y} = \frac{M_x}{M} = \frac{\sum m_k y_k}{\sum m_k}. \tag{4}$$

With this choice of $\bar{y}$, the system balances about the line $y = \bar{y}$ as well. The torques exerted by the masses about the line $y = \bar{y}$ cancel out. Thus, as far as balance is concerned, the system behaves as if all its mass were at the single point $(\bar{x}, \bar{y})$. We call this point the system's **center of mass (c.m.)**.

Thin, Flat Plates

In many applications, we need to find the center of mass of a thin, flat plate: a disk of aluminum, say, or a triangular sheet of steel. In such cases, we assume the distribution of mass to be continuous, and the formulas we use to calculate $\bar{x}$ and $\bar{y}$ contain integrals instead of finite sums. The integrals arise in the following way.

Imagine that the plate occupying a region in the xy-plane is cut into thin strips parallel to one of the axes (in Figure 6.48, the y-axis). The center of mass of a typical strip is $(\tilde{x}, \tilde{y})$. We treat the strip's mass Δm as if it were concentrated at $(\tilde{x}, \tilde{y})$. The moment of the strip about the y-axis is then $\tilde{x}\,\Delta m$. The moment of the strip about the x-axis is $\tilde{y}\,\Delta m$. Equations (3) and (4) then become

$$\bar{x} = \frac{M_y}{M} = \frac{\sum \tilde{x}\,\Delta m}{\sum \Delta m}, \qquad \bar{y} = \frac{M_x}{M} = \frac{\sum \tilde{y}\,\Delta m}{\sum \Delta m}.$$

These sums are Riemann sums for integrals, and they approach these integrals in the limit as the strips become narrower and narrower. We write these integrals symbolically as

$$\bar{x} = \frac{\int \tilde{x}\,dm}{\int dm} \quad \text{and} \quad \bar{y} = \frac{\int \tilde{y}\,dm}{\int dm}.$$

Moments, Mass, and Center of Mass of a Thin Plate Covering a Region in the *xy*-Plane

$$\text{Moment about the } x\text{-axis:} \quad M_x = \int \tilde{y}\,dm$$

$$\text{Moment about the } y\text{-axis:} \quad M_y = \int \tilde{x}\,dm$$

$$\text{Mass:} \quad M = \int dm \tag{5}$$

$$\text{Center of mass:} \quad \bar{x} = \frac{M_y}{M}, \quad \bar{y} = \frac{M_x}{M}$$

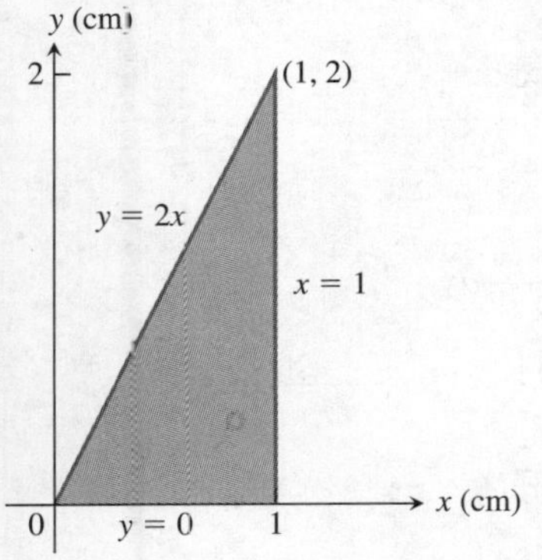

FIGURE 6.49 The plate in Example 2.

Although we do not indicate the limits of integration, the integrals that appear in Equation (5) are definite integrals.

The differential dm in these integrals is the mass of the strip. For this section, we assume the density δ of the plate is a constant or a continuous function of x or of y. Then $dm = \delta\, dA$, which is the mass per unit area δ times the area dA of the strip.

To evaluate the integrals in Equations (5), we picture the plate in the coordinate plane and sketch a strip of mass parallel to one of the coordinate axes. We then express the strip's mass dm and the coordinates $(\tilde{x}, \tilde{y})$ of the strip's center of mass in terms of x or y. Finally, we integrate $\tilde{y}\, dm$, $\tilde{x}\, dm$, and dm between limits of integration determined by the plate's location in the plane.

EXAMPLE 2 The triangular plate shown in Figure 6.49 has a constant density of $\delta = 3\ \text{g/cm}^2$.

(a) Find the plate's moment M_y about the y-axis.

(b) Find the plate's mass M.

(c) Find the x-coordinate of the plate's center of mass (c.m.).

Solution Method 1: Vertical Strips (Figure 6.50)

(a) The moment M_y: The typical vertical strip has the following relevant data.

center of mass (c.m.):	$(\tilde{x}, \tilde{y}) = (x, x)$
length:	$2x$
width:	dx
area:	$dA = 2x\, dx$
mass:	$dm = \delta\, dA = 3 \cdot 2x\, dx = 6x\, dx$
distance of c.m. from y-axis:	$\tilde{x} = x$

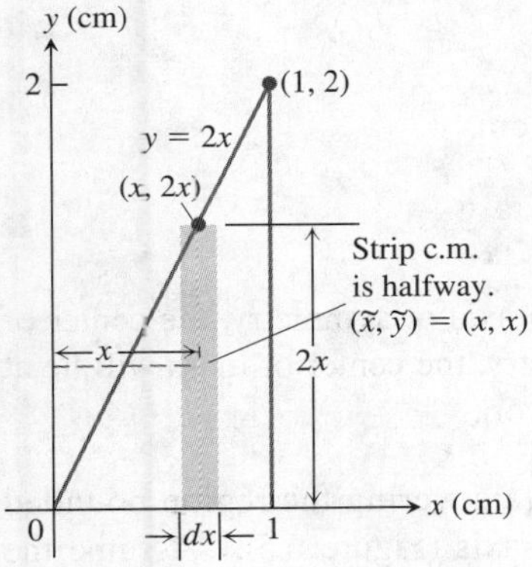

FIGURE 6.50 Modeling the plate in Example 2 with vertical strips.

The moment of the strip about the y-axis is

$$\tilde{x}\, dm = x \cdot 6x\, dx = 6x^2\, dx.$$

The moment of the plate about the y-axis is therefore

$$M_y = \int \tilde{x}\, dm = \int_0^1 6x^2\, dx = 2x^3\Big]_0^1 = 2\ \text{g} \cdot \text{cm}.$$

(b) The plate's mass:

$$M = \int dm = \int_0^1 6x\, dx = 3x^2\Big]_0^1 = 3\ \text{g}.$$

(c) The x-coordinate of the plate's center of mass:

$$\bar{x} = \frac{M_y}{M} = \frac{2\ \text{g} \cdot \text{cm}}{3\ \text{g}} = \frac{2}{3}\ \text{cm}.$$

By a similar computation, we could find M_x and $\bar{y} = M_x/M$.

Method 2: Horizontal Strips (Figure 6.51)

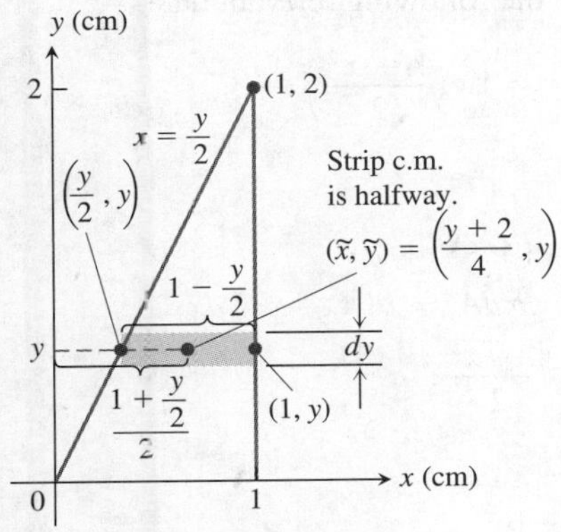

FIGURE 6.51 Modeling the plate in Example 2 with horizontal strips.

(a) The moment M_y: The y-coordinate of the center of mass of a typical horizontal strip is y (see the figure), so

$$\tilde{y} = y.$$

The x-coordinate is the x-coordinate of the point halfway across the triangle. This makes it the average of $y/2$ (the strip's left-hand x-value) and 1 (the strip's right-hand x-value):

$$\tilde{x} = \frac{(y/2) + 1}{2} = \frac{y}{4} + \frac{1}{2} = \frac{y+2}{4}.$$

We also have

$$\begin{aligned}
\text{length:}&\quad 1-\frac{y}{2}=\frac{2-y}{2}\\
\text{width:}&\quad dy\\
\text{area:}&\quad dA=\frac{2-y}{2}\,dy\\
\text{mass:}&\quad dm=\delta\,dA=3\cdot\frac{2-y}{2}\,dy\\
\text{distance of c.m. to } y\text{-axis:}&\quad \tilde{x}=\frac{y+2}{4}.
\end{aligned}$$

The moment of the strip about the y-axis is

$$\tilde{x}\,dm=\frac{y+2}{4}\cdot 3\cdot\frac{2-y}{2}\,dy=\frac{3}{8}(4-y^2)\,dy.$$

The moment of the plate about the y-axis is

$$M_y=\int\tilde{x}\,dm=\int_0^2\frac{3}{8}(4-y^2)\,dy=\frac{3}{8}\left[4y-\frac{y^3}{3}\right]_0^2=\frac{3}{8}\left(\frac{16}{3}\right)=2\text{ g}\cdot\text{cm}.$$

(b) The plate's mass:

$$M=\int dm=\int_0^2\frac{3}{2}(2-y)\,dy=\frac{3}{2}\left[2y-\frac{y^2}{2}\right]_0^2=\frac{3}{2}(4-2)=3\text{ g}.$$

(c) The x-coordinate of the plate's center of mass:

$$\bar{x}=\frac{M_y}{M}=\frac{2\text{ g}\cdot\text{cm}}{3\text{ g}}=\frac{2}{3}\text{ cm}.$$

By a similar computation, we could find M_x and $\bar{y}$. ■

If the distribution of mass in a thin, flat plate has an axis of symmetry, the center of mass will lie on this axis. If there are two axes of symmetry, the center of mass will lie at their intersection. These facts often help to simplify our work.

EXAMPLE 3 Find the center of mass of a thin plate covering the region bounded above by the parabola $y=4-x^2$ and below by the x-axis (Figure 6.52). Assume the density of the plate at the point (x, y) is $\delta=2x^2$, which is twice the square of the distance from the point to the y-axis.

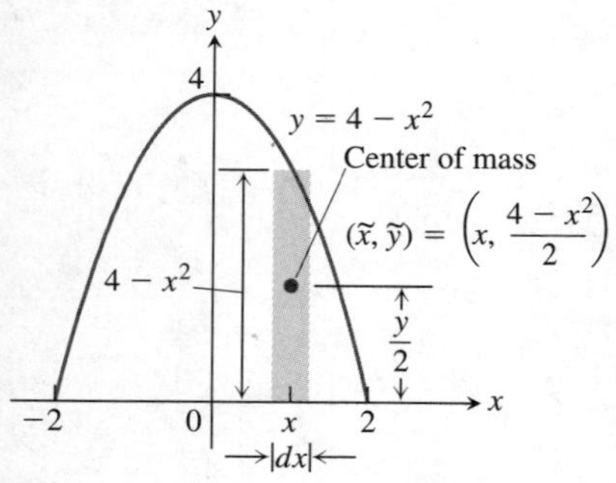

FIGURE 6.52 Modeling the plate in Example 3 with vertical strips.

Solution The mass distribution is symmetric about the y-axis, so $\bar{x}=0$. We model the distribution of mass with vertical strips, since the density is given as a function of the variable x. The typical vertical strip (see Figure 6.52) has the following relevant data.

$$\begin{aligned}
\text{center of mass (c.m.):}&\quad (\tilde{x},\tilde{y})=\left(x,\frac{4-x^2}{2}\right)\\
\text{length:}&\quad 4-x^2\\
\text{width:}&\quad dx\\
\text{area:}&\quad dA=(4-x^2)\,dx\\
\text{mass:}&\quad dm=\delta\,dA=\delta(4-x^2)\,dx\\
\text{distance from c.m. to } x\text{-axis:}&\quad \tilde{y}=\frac{4-x^2}{2}.
\end{aligned}$$

The moment of the strip about the x-axis is

$$\tilde{y}\,dm=\frac{4-x^2}{2}\cdot\delta(4-x^2)\,dx=\frac{\delta}{2}(4-x^2)^2\,dx.$$

The moment of the plate about the x-axis is

$$M_x = \int \tilde{y}\, dm = \int_{-2}^{2} \frac{\delta}{2}(4 - x^2)^2\, dx = \int_{-2}^{2} x^2(4 - x^2)^2\, dx$$

$$= \int_{-2}^{2} (16x^2 - 8x^4 + x^6)\, dx = \frac{2048}{105}.$$

The mass of the plate is

$$M = \int dm = \int_{-2}^{2} \delta(4 - x^2)\, dx = \int_{-2}^{2} 2x^2(4 - x^2)\, dx$$

$$= \int_{-2}^{2} (8x^2 - 2x^4)\, dx = \frac{256}{15}.$$

Therefore,

$$\bar{y} = \frac{M_x}{M} = \frac{2048}{105} \cdot \frac{15}{256} = \frac{8}{7}.$$

The plate's center of mass is

$$(\bar{x}, \bar{y}) = \left(0, \frac{8}{7}\right).$$

■

Plates Bounded by Two Curves

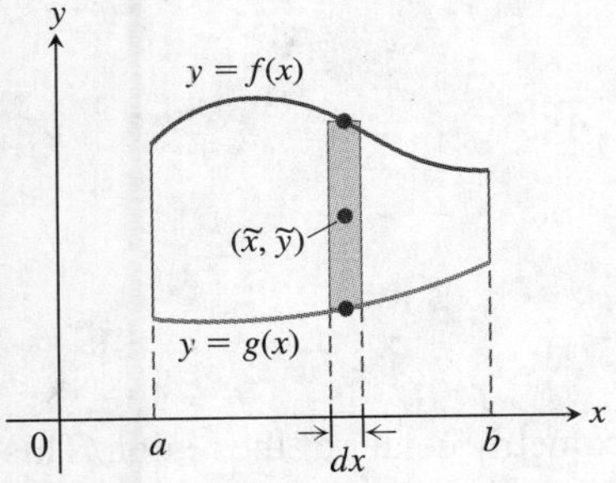

FIGURE 6.53 Modeling the plate bounded by two curves with vertical strips. The strip c.m. is halfway, so $\tilde{y} = \frac{1}{2}[f(x) + g(x)]$.

Suppose a plate covers a region that lies between two curves $y = g(x)$ and $y = f(x)$, where $f(x) \geq g(x)$ and $a \leq x \leq b$. The typical vertical strip (see Figure 6.53) has

center of mass (c.m.):	$(\tilde{x}, \tilde{y}) = \left(x, \frac{1}{2}[f(x) + g(x)]\right)$
length:	$f(x) - g(x)$
width:	dx
area:	$dA = [f(x) - g(x)]\, dx$
mass:	$dm = \delta\, dA = \delta[f(x) - g(x)]\, dx.$

The moment of the plate about the y-axis is

$$M_y = \int \tilde{x}\, dm = \int_a^b x\delta[f(x) - g(x)]\, dx,$$

and the moment about the x-axis is

$$M_x = \int \tilde{y}\, dm = \int_a^b \frac{1}{2}[f(x) + g(x)] \cdot \delta[f(x) - g(x)]\, dx$$

$$= \int_a^b \frac{\delta}{2}[f^2(x) - g^2(x)]\, dx.$$

These moments give us the following formulas.

$$\bar{x} = \frac{1}{M}\int_a^b \delta x[f(x) - g(x)]\, dx \qquad (6)$$

$$\bar{y} = \frac{1}{M}\int_a^b \frac{\delta}{2}[f^2(x) - g^2(x)]\, dx \qquad (7)$$

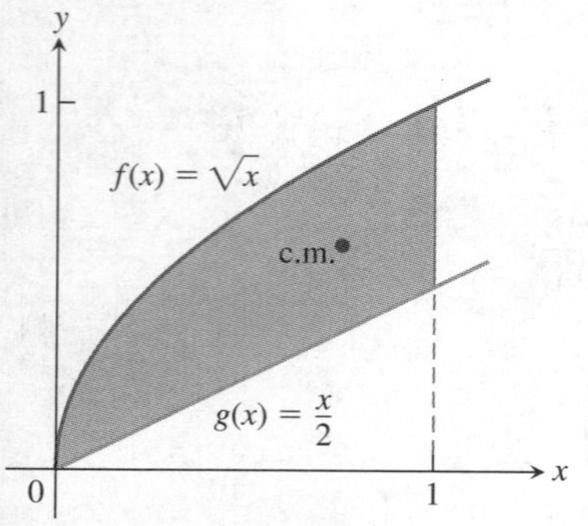

FIGURE 6.54 The region in Example 4.

EXAMPLE 4 Find the center of mass for the thin plate bounded by the curves $g(x) = x/2$ and $f(x) = \sqrt{x}$, $0 \le x \le 1$ (Figure 6.54), using Equations (6) and (7) with the density function $\delta(x) = x^2$.

Solution We first compute the mass of the plate, using $dm = \delta[f(x) - g(x)]\,dx$:

$$M = \int_0^1 x^2\left(\sqrt{x} - \frac{x}{2}\right)dx = \int_0^1 \left(x^{5/2} - \frac{x^3}{2}\right)dx = \left[\frac{2}{7}x^{7/2} - \frac{1}{8}x^4\right]_0^1 = \frac{9}{56}.$$

Then, from Equations (6) and (7) we get

$$\begin{aligned}\bar{x} &= \frac{56}{9}\int_0^1 x^2 \cdot x\left(\sqrt{x} - \frac{x}{2}\right)dx \\ &= \frac{56}{9}\int_0^1 \left(x^{7/2} - \frac{x^4}{2}\right)dx \\ &= \frac{56}{9}\left[\frac{2}{9}x^{9/2} - \frac{1}{10}x^5\right]_0^1 = \frac{308}{405},\end{aligned}$$

and

$$\begin{aligned}\bar{y} &= \frac{56}{9}\int_0^1 \frac{x^2}{2}\left(x - \frac{x^2}{4}\right)dx \\ &= \frac{28}{9}\int_0^1 \left(x^3 - \frac{x^4}{4}\right)dx \\ &= \frac{28}{9}\left[\frac{1}{4}x^4 - \frac{1}{20}x^5\right]_0^1 = \frac{28}{45}.\end{aligned}$$

The center of mass is shown in Figure 6.54. ■

Centroids

The center of mass in Example 4 is not located at the geometric center of the region. This is due to the region's nonuniform density. When the density function is constant, it cancels out of the numerator and denominator of the formulas for $\bar{x}$ and $\bar{y}$. Thus, when the density is constant, the location of the center of mass is a feature of the geometry of the object and not of the material from which it is made. In such cases, engineers may call the center of mass the **centroid** of the shape, as in "Find the centroid of a triangle or a solid cone." To do so, just set δ equal to 1 and proceed to find $\bar{x}$ and $\bar{y}$ as before, by dividing moments by masses.

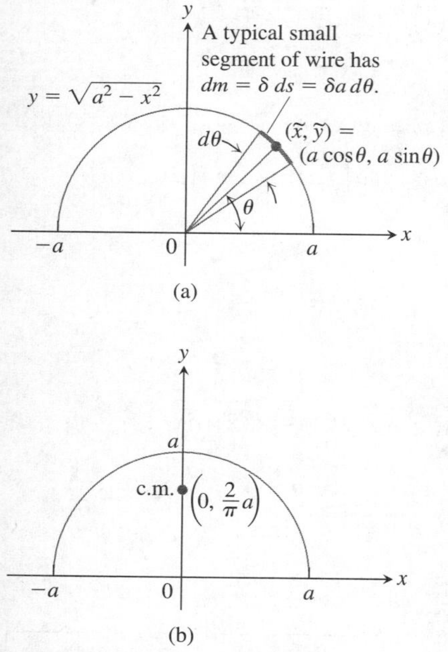

FIGURE 6.55 The semicircular wire in Example 5. (a) The dimensions and variables used in finding the center of mass. (b) The center of mass does not lie on the wire.

EXAMPLE 5 Find the center of mass (centroid) of a thin wire of constant density δ shaped like a semicircle of radius a.

Solution We model the wire with the semicircle $y = \sqrt{a^2 - x^2}$ (Figure 6.55). The distribution of mass is symmetric about the y-axis, so $\bar{x} = 0$. To find $\bar{y}$, we imagine the wire divided into short subarc segments. If $(\tilde{x}, \tilde{y})$ is the center of mass of a subarc and θ is the angle between the x-axis and the radial line joining the origin to $(\tilde{x}, \tilde{y})$, then $\tilde{y} = a\sin\theta$ is a function of the angle θ measured in radians (see Figure 6.55a). The length ds of the subarc containing $(\tilde{x}, \tilde{y})$ subtends an angle of $d\theta$ radians, so $ds = a\,d\theta$. Thus a typical subarc segment has these relevant data for calculating $\bar{y}$:

length:	$ds = a\,d\theta$	
mass:	$dm = \delta\,ds = \delta a\,d\theta$	Mass per unit length times length
distance of c.m. to x-axis:	$\tilde{y} = a\sin\theta$.	

Hence,

$$\bar{y} = \frac{\int \tilde{y}\, dm}{\int dm} = \frac{\int_0^{\pi} a \sin\theta \cdot \delta a\, d\theta}{\int_0^{\pi} \delta a\, d\theta} = \frac{\delta a^2[-\cos\theta]_0^{\pi}}{\delta a \pi} = \frac{2}{\pi} a.$$

The center of mass lies on the axis of symmetry at the point $(0, 2a/\pi)$, about two-thirds of the way up from the origin (Figure 6.55b). Notice how δ cancels in the equation for $\bar{y}$, so we could have set $\delta = 1$ everywhere and obtained the same value for $\bar{y}$. ■

In Example 5 we found the center of mass of a thin wire lying along the graph of a differentiable function in the *xy*-plane. In Chapter 16 we will learn how to find the center of mass of a wire lying along a more general smooth curve in the plane or in space.

Fluid Forces and Centroids

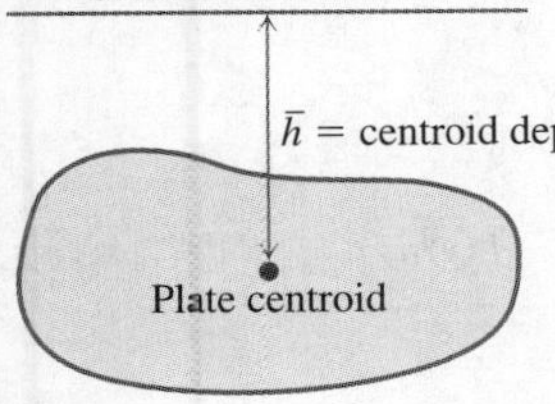

FIGURE 6.56 The force against one side of the plate is $w \cdot \bar{h} \cdot$ plate area.

If we know the location of the centroid of a submerged flat vertical plate (Figure 6.56), we can take a shortcut to find the force against one side of the plate. From Equation (7) in Section 6.5, and the definition of the moment about the *x*-axis, we have

$$\begin{aligned} F &= \int_a^b w \times (\text{strip depth}) \times L(y)\, dy \\ &= w \int_a^b (\text{strip depth}) \times L(y)\, dy \\ &= w \times (\text{moment about surface level line of region occupied by plate}) \\ &= w \times (\text{depth of plate's centroid}) \times (\text{area of plate}). \end{aligned}$$

Fluid Forces and Centroids

The force of a fluid of weight-density w against one side of a submerged flat vertical plate is the product of w, the distance $\bar{h}$ from the plate's centroid to the fluid surface, and the plate's area:

$$F = w\bar{h}A. \qquad (8)$$

EXAMPLE 6 A flat isosceles triangular plate with base 6 ft and height 3 ft is submerged vertically, base up with its vertex at the origin, so that the base is 2 ft below the surface of a swimming pool. (This is Example 6, Section 6.5.) Use Equation (8) to find the force exerted by the water against one side of the plate.

Solution The centroid of the triangle (Figure 6.43) lies on the *y*-axis, one-third of the way from the base to the vertex, so $\bar{h} = 3$ (where $y = 2$), since the pool's surface is $y = 5$. The triangle's area is

$$A = \frac{1}{2}(\text{base})(\text{height}) = \frac{1}{2}(6)(3) = 9.$$

Hence,

$$F = w\bar{h}A = (62.4)(3)(9) = 1684.8 \text{ lb}.$$ ■

The Theorems of Pappus

In the fourth century, an Alexandrian Greek named Pappus discovered two formulas that relate centroids to surfaces and solids of revolution. The formulas provide shortcuts to a number of otherwise lengthy calculations.

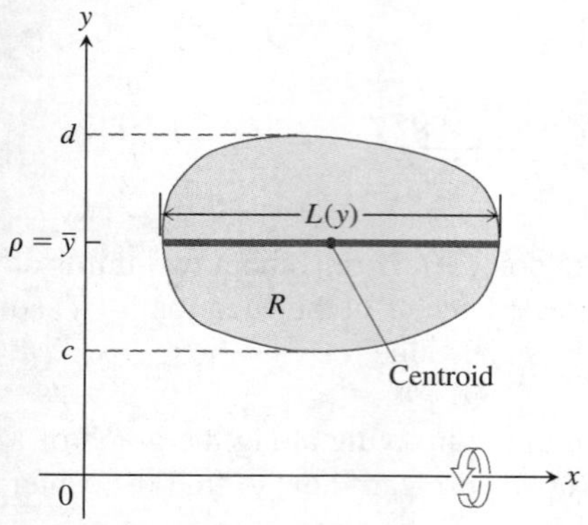

FIGURE 6.57 The region R is to be revolved (once) about the x-axis to generate a solid. A 1700-year-old theorem says that the solid's volume can be calculated by multiplying the region's area by the distance traveled by its centroid during the revolution.

THEOREM 1—Pappus's Theorem for Volumes
If a plane region is revolved once about a line in the plane that does not cut through the region's interior, then the volume of the solid it generates is equal to the region's area times the distance traveled by the region's centroid during the revolution. If ρ is the distance from the axis of revolution to the centroid, then

$$V = 2\pi\rho A. \tag{9}$$

Proof We draw the axis of revolution as the x-axis with the region R in the first quadrant (Figure 6.57). We let $L(y)$ denote the length of the cross-section of R perpendicular to the y-axis at y. We assume $L(y)$ to be continuous.

By the method of cylindrical shells, the volume of the solid generated by revolving the region about the x-axis is

$$V = \int_c^d 2\pi(\text{shell radius})(\text{shell height})\,dy = 2\pi\int_c^d y\,L(y)\,dy. \tag{10}$$

The y-coordinate of R's centroid is

$$\bar{y} = \frac{\int_c^d \tilde{y}\,dA}{A} = \frac{\int_c^d y\,L(y)\,dy}{A}, \qquad \tilde{y} = y,\ dA = L(y)\,dy$$

so that

$$\int_c^d y\,L(y)\,dy = A\bar{y}.$$

Substituting $A\bar{y}$ for the last integral in Equation (10) gives $V = 2\pi\bar{y}A$. With ρ equal to $\bar{y}$, we have $V = 2\pi\rho A$. ■

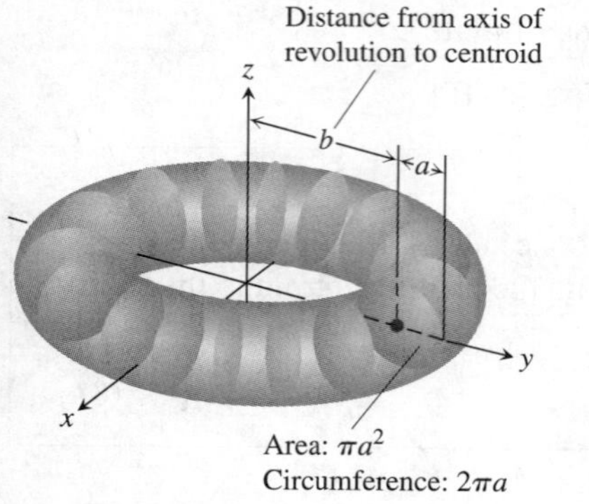

FIGURE 6.58 With Pappus's first theorem, we can find the volume of a torus without having to integrate (Example 7).

EXAMPLE 7 Find the volume of the torus (doughnut) generated by revolving a circular disk of radius a about an axis in its plane at a distance $b \geq a$ from its center (Figure 6.58).

Solution We apply Pappus's Theorem for volumes. The centroid of a disk is located at its center, the area is $A = \pi a^2$, and $\rho = b$ is the distance from the centroid to the axis of revolution (see Figure 6.58). Substituting these values into Equation (9), we find the volume of the torus to be

$$V = 2\pi(b)(\pi a^2) = 2\pi^2 ba^2.$$ ■

The next example shows how we can use Equation (9) in Pappus's Theorem to find one of the coordinates of the centroid of a plane region of known area A when we also know the volume V of the solid generated by revolving the region about the other coordinate axis. That is, if $\bar{y}$ is the coordinate we want to find, we revolve the region around the x-axis so that $\bar{y} = \rho$ is the distance from the centroid to the axis of revolution. The idea is that the rotation generates a solid of revolution whose volume V is an already known quantity. Then we can solve Equation (9) for ρ, which is the value of the centroid's coordinate $\bar{y}$.

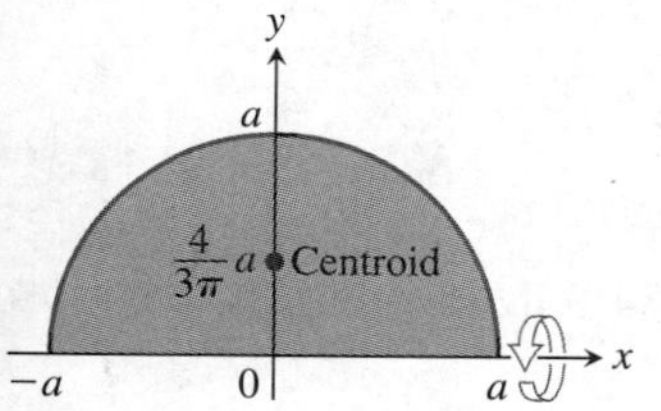

FIGURE 6.59 With Pappus's first theorem, we can locate the centroid of a semicircular region without having to integrate (Example 8).

EXAMPLE 8 Locate the centroid of a semicircular region of radius a.

Solution We consider the region between the semicircle $y = \sqrt{a^2 - x^2}$ (Figure 6.59) and the x-axis and imagine revolving the region about the x-axis to generate a solid sphere. By symmetry, the x-coordinate of the centroid is $\bar{x} = 0$. With $\bar{y} = \rho$ in Equation (9), we have

$$\bar{y} = \frac{V}{2\pi A} = \frac{(4/3)\pi a^3}{2\pi(1/2)\pi a^2} = \frac{4}{3\pi}a.$$ ■

THEOREM 2—Pappus's Theorem for Surface Areas
If an arc of a smooth plane curve is revolved once about a line in the plane that does not cut through the arc's interior, then the area of the surface generated by the arc equals the length L of the arc times the distance traveled by the arc's centroid during the revolution. If ρ is the distance from the axis of revolution to the centroid, then

$$S = 2\pi\rho L. \tag{11}$$

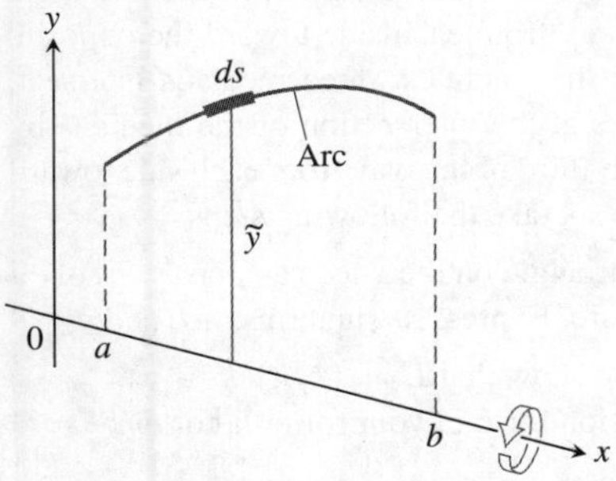

FIGURE 6.60 Figure for proving Pappus's Theorem for surface area. The arc length differential ds is given by Equation (6) in Section 6.3.

The proof we give assumes that we can model the axis of revolution as the x-axis and the arc as the graph of a continuously differentiable function of x.

Proof We draw the axis of revolution as the x-axis with the arc extending from $x = a$ to $x = b$ in the first quadrant (Figure 6.60). The area of the surface generated by the arc is

$$S = \int_{x=a}^{x=b} 2\pi y\, ds = 2\pi \int_{x=a}^{x=b} y\, ds. \tag{12}$$

The y-coordinate of the arc's centroid is

$$\overline{y} = \frac{\displaystyle\int_{x=a}^{x=b} \tilde{y}\, ds}{\displaystyle\int_{x=a}^{x=b} ds} = \frac{\displaystyle\int_{x=a}^{x=b} y\, ds}{L}.$$

$L = \int ds$ is the arc's length and $\tilde{y} = y$.

Hence

$$\int_{x=a}^{x=b} y\, ds = \overline{y}L.$$

Substituting $\overline{y}L$ for the last integral in Equation (12) gives $S = 2\pi\overline{y}L$. With ρ equal to $\overline{y}$, we have $S = 2\pi\rho L$. ■

EXAMPLE 9 Use Pappus's area theorem to find the surface area of the torus in Example 7.

Solution From Figure 6.58, the surface of the torus is generated by revolving a circle of radius a about the z-axis, and $b \geq a$ is the distance from the centroid to the axis of revolution. The arc length of the smooth curve generating this surface of revolution is the circumference of the circle, so $L = 2\pi a$. Substituting these values into Equation (11), we find the surface area of the torus to be

$$S = 2\pi(b)(2\pi a) = 4\pi^2 ba.$$ ■

EXERCISES 6.6

Mass of a wire

In Exercises 1–6, find the mass M and center of mass $\overline{x}$ of the linear wire covering the given interval and having the given density $\delta(x)$.

1. $0 \leq x \leq 4, \quad \delta(x) = \sqrt{x}$

2. $-3 \leq x \leq 3, \quad \delta(x) = 1 + 3x^2$

3. $0 \leq x \leq 3, \quad \delta(x) = x + 1$

4. $1 \leq x \leq 4, \quad \delta(x) = \dfrac{16}{2x^3}$

5. $\delta(x) = \begin{cases} 4, & 0 \leq x \leq 2 \\ 5, & 2 < x \leq 3 \end{cases}$

6. $\delta(x) = \begin{cases} 2 - x, & 0 \leq x < 1 \\ x, & 1 \leq x \leq 2 \end{cases}$

Thin Plates with Constant Density

In Exercises 7–18, find the center of mass of a thin plate of constant density δ covering the given region.

7. The region bounded by the parabola $y = 3x^2$ and the line $y = 12$

8. The region bounded by the parabola $y = 25 - x^2$ and the x-axis

9. The region bounded by the parabola $y = x - x^2$ and the line $y = -x$

10. The region enclosed by the parabolas $y = x^2 - 3$ and $y = -2x^2$

11. The region bounded by the y-axis and the curve $x = y - y^3$, $0 \le y \le 1$

12. The region bounded by the parabola $x = y^2 - y$ and the line $y = x$

13. The region bounded by the x-axis and the curve $y = \cos x$, $-\pi/2 \le x \le \pi/2$

14. The region between the curve $y = \sec^2 x$, $-\pi/4 \le x \le \pi/4$ and the x-axis

15. a. The region cut from the first quadrant by the circle $x^2 + y^2 = 9$

b. The region bounded by the x-axis and the semicircle $y = \sqrt{9 - x^2}$

Compare your answer in part (b) with the answer in part (a).

16. The region bounded by the parabolas $y = x^3$ and $y = \sqrt{x}$

17. The region between the curve $y = \sqrt{x}$ and the x-axis from $x = 1$ to $x = 16$

18. The region bounded above by the curve $y = 1/x^3$, below by the curve $y = -1/x^3$, and on the left and right by the lines $x = 1$ and $x = a > 1$. Also, find $\lim_{a\to\infty} \bar{x}$.

19. Consider a region bounded by the graphs of $y = x^4$ and $y = x^5$. Show that the center of mass lies outside the region.

20. Consider a thin plate of constant density δ lies in the region bounded by the graphs of $y = \sqrt{x}$ and $x = 2y$. Find the plate's

a. moment about the x-axis.

b. moment about the y-axis.

c. moment about the line $x = 5$.

d. moment about the line $x = -1$.

e. moment about the line $y = 2$.

f. moment about the line $y = -3$.

g. mass.

h. center of mass.

Thin Plates with Varying Density

21. Find the center of mass of a thin plate covering the region between the x-axis and the curve $y = 2/x^2$, $1 \le x \le 2$, if the plate's density at the point (x, y) is $\delta(x) = x^2$.

22. Find the center of mass of a thin plate covering the region bounded below by the parabola $y = x^2$ and above by the line $y = x + 2$, if the plate's density at the point (x, y) is $\delta(x) = 9x^2$.

23. The region bounded by the curves $y = \pm 4/\sqrt{x}$ and the lines $x = 1$ and $x = 4$ is revolved about the y-axis to generate a solid.

a. Find the volume of the solid.

b. Find the center of mass of a thin plate covering the region if the plate's density at the point (x, y) is $\delta(x) = 1/x$.

c. Sketch the plate and show the center of mass in your sketch.

24. The region between the curve $y = 2/x$ and the x-axis from $x = 1$ to $x = 4$ is revolved about the x-axis to generate a solid.

a. Find the volume of the solid.

b. Find the center of mass of a thin plate covering the region if the plate's density at the point (x, y) is $\delta(x) = \sqrt{x}$.

c. Sketch the plate and show the center of mass in your sketch.

Centroids of Triangles

25. The centroid of a triangle lies at the intersection of the triangle's medians You may recall that the point inside a triangle that lies one-third of the way from each side toward the opposite vertex is the point where the triangle's three medians intersect. Show that the centroid lies at the intersection of the medians by showing that it too lies one-third of the way from each side toward the opposite vertex. To do so, take the following steps.

i) Stand one side of the triangle on the x-axis as in part (b) of the accompanying figure. Express dm in terms of L and dy.

ii) Use similar triangles to show that $L = (b/h)(h - y)$. Substitute this expression for L in your formula for dm.

iii) Show that $\bar{y} = h/3$.

iv) Extend the argument to the other sides.

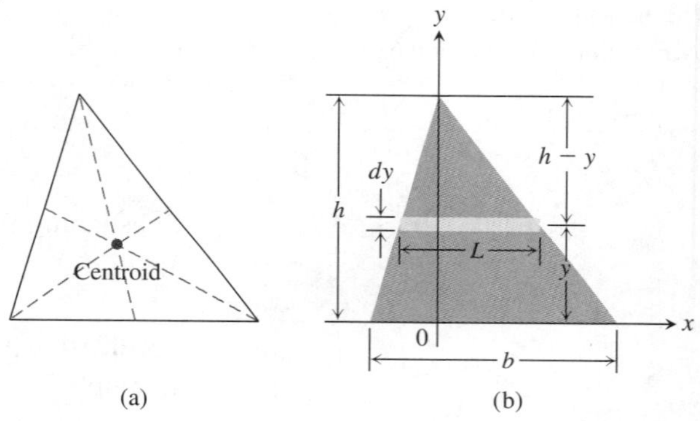

Use the result in Exercise 25 to find the centroids of the triangles whose vertices appear in Exercises 26–30. Assume $a, b > 0$.

26. $(-1, 0), (1, 0), (0, 3)$

27. $(0, 0), (1, 0), (0, 1)$

28. $(0, 0), (a, 0), (0, a)$

29. $(0, 0), (a, 0), (0, b)$

30. $(0, 0), (a, 0), (a/2, b)$

Thin Wires

31. Constant density Find the moment about the x-axis of a wire of constant density that lies along the curve $y = \sqrt{x}$ from $x = 0$ to $x = 2$.

32. Constant density Constant density Find the moment about the y-axis of a wire of constant density that lies along the curve $y = x^2$ from $x = 0$ to $x = \sqrt{2}$

33. Variable density Suppose that the density of the wire in Example 5 is $\delta = k \sin \theta$ (k constant). Find the center of mass.

34. Variable density Suppose that the density of the wire in Example 5 is $\delta = 1 + k|\cos \theta|$ (k constant). Find the center of mass.

Plates Bounded by Two Curves

In Exercises 35–38, find the centroid of the thin plate bounded by the graphs of the given functions. Use Equations (6) and (7) with $\delta = 1$ and M = area of the region covered by the plate.

35. $g(x) = x^2$ and $f(x) = x + 2$

36. $g(x) = x^2$, $f(x) = 4$ and $x = 0$

37. $g(x) = x^2(x - 1)$ and $f(x) = x^2$

38. $g(x) = 0$, $f(x) = 2 + \sin x$, $x = 0$, and $x = 2\pi$

($Hint$: $\int x \sin x\, dx = \sin x - x \cos x + C$.)

Theory and Examples

Verify the statements and formulas in Exercises 39 and 40.

39. The coordinates of the centroid of a differentiable plane curve are

$$\bar{x} = \frac{\int x\, ds}{\text{length}}, \quad \bar{y} = \frac{\int y\, ds}{\text{length}}.$$

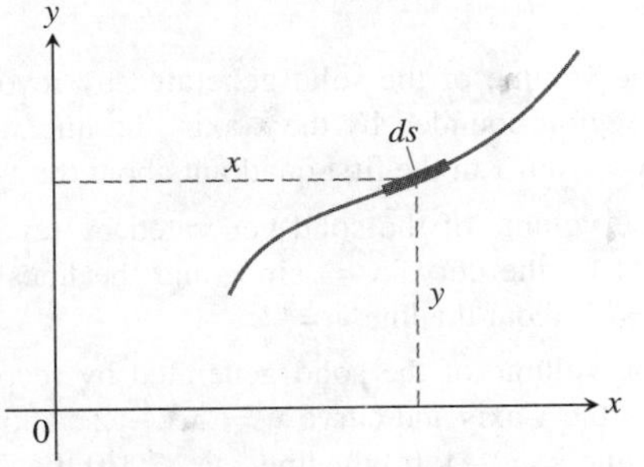

40. Whatever the value of $p > 0$ in the equation $y = x^2/(4p)$, the y-coordinate of the centroid of the parabolic segment shown here is $\bar{y} = (3/5)a$.

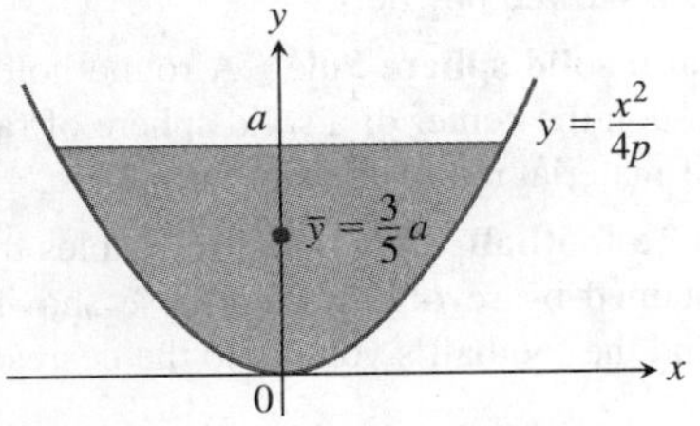

The Theorems of Pappus

41. The square region with vertices (0, 2), (0, 4), (2, 2) and (2, 4) is revolved about the x-axis to generate a solid. Find the volume and the surface area of the solid.

42. Use a theorem of Pappus to find the volume generated by revolving about the line $x = 5$ the triangular region bounded by the coordinate axes and the line $2x + y = 6$ (see Exercise 25).

43. Find the volume of the torus generated by revolving the circle $x^2 + y^2 = 4$ about the line $x = 3$.

44. Use the theorems of Pappus to find the lateral surface area and the volume of a right-circular cone.

45. Use Pappus's Theorem for surface area and the fact that the surface area of a sphere of radius a is $4\pi a^2$ to find the centroid of the semicircle $y = \sqrt{a^2 - x^2}$.

46. As found in Exercise 45, the centroid of the semicircle $y = \sqrt{a^2 - x^2}$ lies at the point $(0, 2a/\pi)$. Find the area of the surface swept out by revolving the semicircle about the line $y = a$.

47. The area of the region R enclosed by the semiellipse $y = (b/a)\sqrt{a^2 - x^2}$ and the x-axis is $(1/2)\pi ab$, and the volume of the ellipsoid generated by revolving R about the x-axis is $(4/3)\pi ab^2$. Find the centroid of R. Notice that the location is independent of a.

48. As found in Example 8, the centroid of the region enclosed by the x-axis and the semicircle $y = \sqrt{a^2 - x^2}$ lies at the point $(0, 4a/3\pi)$. Find the volume of the solid generated by revolving this region about the line $y = -a$.

49. The region of Exercise 48 is revolved about the line $y = x - a$ to generate a solid. Find the volume of the solid.

50. As found in Exercise 45, the centroid of the semicircle $y = \sqrt{a^2 - x^2}$ lies at the point $(0, 2a/\pi)$. Find the area of the surface generated by revolving the semicircle about the line $y = x - a$.

In Exercises 51 and 52, use a theorem of Pappus to find the centroid of the given triangle. Use the fact that the volume of a cone of radius r and height h is $V = \frac{1}{3}\pi r^2 h$.

51.

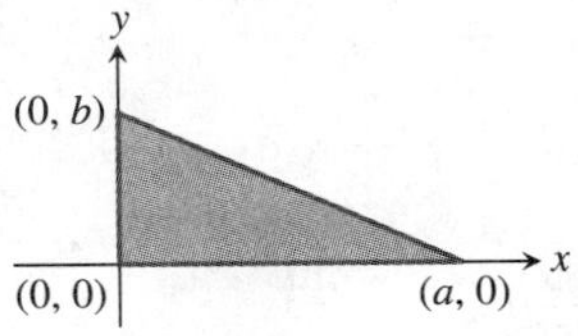

52.

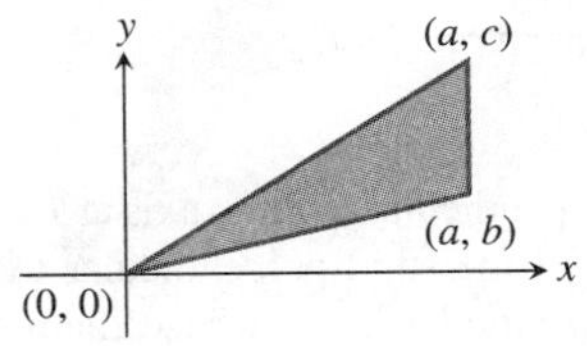

CHAPTER 6 Questions to Guide Your Review

1. How do you define and calculate the volumes of solids by the method of slicing? Give an example.
2. How are the disk and washer methods for calculating volumes derived from the method of slicing? Give examples of volume calculations by these methods.
3. Describe the method of cylindrical shells. Give an example.
4. How do you find the length of the graph of a smooth function over a closed interval? Give an example. What about functions that do not have continuous first derivatives?
5. How do you define and calculate the area of the surface swept out by revolving the graph of a smooth function $y = f(x)$, $a \leq x \leq b$, about the x-axis? Give an example.

6. How do you define and calculate the work done by a variable force directed along a portion of the x-axis? How do you calculate the work it takes to pump a liquid from a tank? Give examples.
7. What is a center of mass? What is a centroid?
8. How do you locate the center of mass of a thin flat plate of material? Give an example.
9. How do you locate the center of mass of a thin plate bounded by two curves $y = f(x)$ and $y = g(x)$ over $a \le x \le b$?

CHAPTER 6 Practice Exercises

Volumes

Find the volumes of the solids in Exercises 1–18.

1. The solid lies between planes perpendicular to the x-axis at $x = 0$ and $x = 1$. The cross-sections perpendicular to the x-axis between these planes are circular disks whose diameters run from the parabola $y = x^2$ to the parabola $y = \sqrt{x}$.
2. The base of the solid is the region in the first quadrant between the line $y = x$ and the parabola $y = 2\sqrt{x}$. The cross-sections of the solid perpendicular to the x-axis are equilateral triangles whose bases stretch from the line to the curve.
3. The solid lies between planes perpendicular to the x-axis at $x = \pi/4$ and $x = 5\pi/4$. The cross-sections between these planes are circular disks whose diameters run from the curve $y = 2\cos x$ to the curve $y = 2\sin x$.
4. The solid lies between planes perpendicular to the x-axis at $x = 0$ and $x = 6$. The cross-sections between these planes are squares whose bases run from the x-axis up to the curve $x^{1/2} + y^{1/2} = \sqrt{6}$.

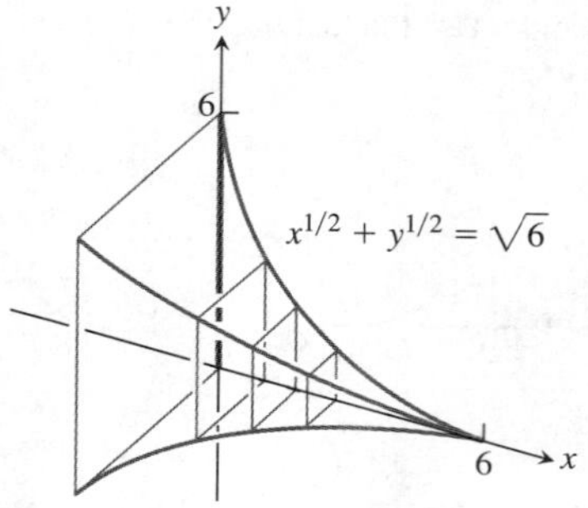

5. The solid lies between planes perpendicular to the x-axis at $x = 0$ and $x = 4$. The cross-sections of the solid perpendicular to the x-axis between these planes are circular disks whose diameters run from the curve $x^2 = 4y$ to the curve $y^2 = 4x$.
6. The base of the solid is the region bounded by the parabola $y^2 = 4x$ and the line $x = 1$ in the xy-plane. Each cross-section perpendicular to the x-axis is an equilateral triangle with one edge in the plane. (The triangles all lie on the same side of the plane.)
7. Find the volume of the solid generated by revolving the region bounded by the x-axis, the curve $y = 3x^4$, and the lines $x = 1$ and $x = -1$ about **(a)** the x-axis; **(b)** the y-axis; **(c)** the line $x = 1$; **(d)** the line $y = 3$.
8. Find the volume of the solid generated by revolving the "triangular" region bounded by the curve $y = 4/x^3$ and the lines $x = 1$ and $y = 1/2$ about **(a)** the x-axis; **(b)** the y-axis; **(c)** the line $x = 2$; **(d)** the line $y = 4$.
9. Find the volume of the solid generated by revolving the region bounded on the left by the parabola $x = y^2 + 1$ and on the right by the line $x = 5$ about **(a)** the x-axis; **(b)** the y-axis; **(c)** the line $x = 5$.
10. Find the volume of the solid generated by revolving the region bounded by the parabola $y^2 = 4x$ and the line $y = x$ about **(a)** the x-axis; **(b)** the y-axis; **(c)** the line $x = 4$; **(d)** the line $y = 4$.
11. Find the volume of the solid generated by revolving the "triangular" region bounded by the x-axis, the line $x = \pi/3$, and the curve $y = \tan x$ in the first quadrant about the x-axis.
12. Find the volume of the solid generated by revolving the region bounded by the curve $y = \sin x$ and the lines $x = 0$, $x = \pi$, and $y = 2$ about the line $y = 2$.
13. Find the volume of the solid generated by revolving the region between the x-axis and curve $y = x^2 - 2x$ about **(a)** the x-axis; **(b)** the line $y = -1$; **(c)** the line $x = 2$; **(d)** the line $y = 2$.
14. Find the volume of the solid generated by revolving about the x-axis the region bounded by $y = 2\tan x$, $y = 0$, $x = -\pi/4$, and $x = \pi/4$. (The region lies in the first and third quadrants and resembles a skewed bowtie.)
15. **Volume of a solid sphere hole** A round hole of radius $\sqrt{3}$ ft is bored through the center of a solid sphere of radius 2 ft. Find the volume of material removed from the sphere.
16. **Volume of a football** A football resembles the surface of revolution obtained by revolving the ellipse shown here around the x-axis. Find the football's volume to the nearest cubic inch.

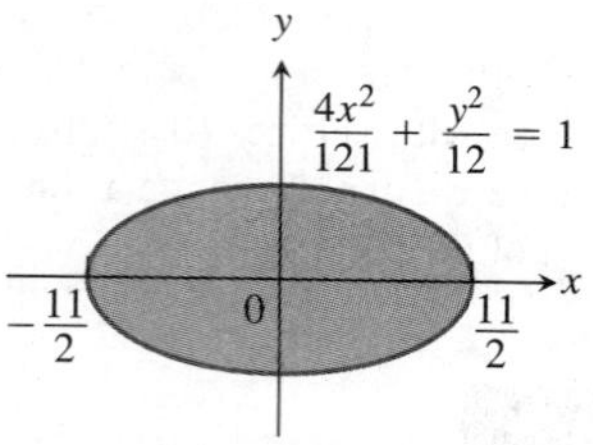

17. Set up and evaluate an integral to find the volume of the given circular frustum of height h and radii a and b.

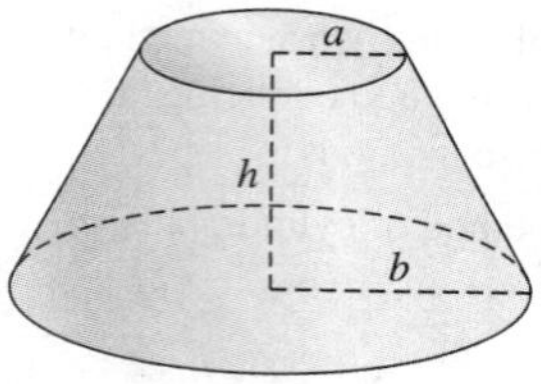

18. The graph of $x^{2/3} + y^{2/3} = 1$ is called an astroid and is given below. Find the volume of the solid formed by revolving the region enclosed by the astroid about the x-axis.

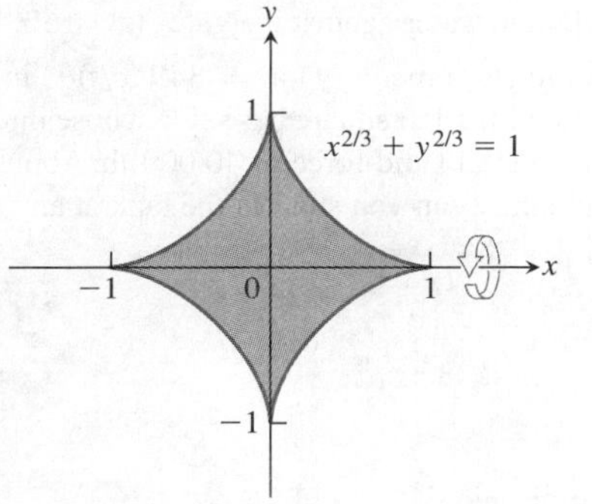

Lengths of Curves

Find the lengths of the curves in Exercises 19–22.

19. $y = x^{1/2} - (1/3)x^{3/2}, \quad 1 \le x \le 4$

20. $x = y^{2/3}, \quad 1 \le y \le 8$

21. $y = (5/12)x^{6/5} - (5/8)x^{4/5}, \quad 1 \le x \le 32$

22. $x = (y^3/12) + (1/y), \quad 1 \le y \le 2$

Areas of Surfaces of Revolution

In Exercises 23–26, find the areas of the surfaces generated by revolving the curves about the given axes.

23. $y = \sqrt{2x + 1}, \quad 0 \le x \le 3; \quad x$-axis

24. $y = x^3/3, \quad 0 \le x \le 1; \quad x$-axis

25. $x = \sqrt{4y - y^2}, \quad 1 \le y \le 2; \quad y$-axis

26. $x = \sqrt{y}, \quad 2 \le y \le 6; \quad y$-axis

Work

27. Lifting equipment A rock climber is about to haul up 100 N (about 22.5 lb) of equipment that has been hanging beneath her on 40 m of rope that weighs 0.8 N/m. How much work will it take? (*Hint:* Solve for the rope and equipment separately, then add.)

28. Leaky tank truck You drove an 800-gal tank truck of water from the base of Mt. Washington to the summit and discovered on arrival that the tank was only half full. You started with a full tank, climbed at a steady rate, and accomplished the 4750-ft elevation change in 50 min. Assuming that the water leaked out at a steady rate, how much work was spent in carrying water to the top? Do not count the work done in getting yourself and the truck there. Water weighs 8 lb/U.S. gal.

29. Earth's attraction The force of attraction on an object below Earth's surface is directly proportional to its distance from Earth's center. Find the work done in moving a weight of w lb located a mi below Earth's surface up to the surface itself. Assume Earth's radius is a constant r mi.

30. Garage door spring A force of 200 N will stretch a garage door spring 0.8 m beyond its unstressed length. How far will a 300-N force stretch the spring? How much work does it take to stretch the spring this far from its unstressed length?

31. Pumping a reservoir A reservoir shaped like a right-circular cone, point down, 20 ft across the top and 8 ft deep, is full of water. How much work does it take to pump the water to a level 6 ft above the top?

32. Pumping a reservoir (*Continuation of Exercise 31.*) The reservoir is filled to a depth of 5 ft, and the water is to be pumped to the same level as the top. How much work does it take?

33. Pumping a conical tank A right-circular conical tank, point down, with top radius 5 ft and height 10 ft, is filled with a liquid whose weight-density is 60 lb/ft^3. How much work does it take to pump the liquid to a point 2 ft above the tank? If the pump is driven by a motor rated at 275 ft-lb/sec (1/2 hp), how long will it take to empty the tank?

34. Pumping a cylindrical tank A storage tank is a right-circular cylinder 20 ft long and 8 ft in diameter with its axis horizontal. If the tank is half full of olive oil weighing 57 lb/ft^3, find the work done in emptying it through a pipe that runs from the bottom of the tank to an outlet that is 6 ft above the top of the tank.

35. Assume that a spring does not follow Hooke's Law. Instead, the force required to stretch the spring x ft from its natural length is $F(x) = 10x^{3/2}$ lb. How much work does it take to

a. stretch the spring 4 ft from its natural length?

b. stretch the spring from an initial 1 ft past its natural length to 5 ft past its natural length?

36. Assume that a spring does not follow Hooke's Law. Instead, the force required to stretch the spring x m from its natural length is $F(x) = k\sqrt{5 + x^2}$ N.

a. If a 3-N force stretches the spring 2 m, find the value of k.

b. How much work is required to stretch the spring 1 m from its natural length?

Centers of Mass and Centroids

37. Find the centroid of a thin, flat plate covering the region enclosed by the parabolas $y = 2x^2$ and $y = 3 - x^2$.

38. Find the centroid of a thin, flat plate covering the region enclosed by the x-axis, the lines $x = 2$ and $x = -2$, and the parabola $y = x^2$.

39. Find the centroid of a thin, flat plate covering the "triangular" region in the first quadrant bounded by the y-axis, the parabola $y = x^2/4$, and the line $y = 4$.

40. Find the centroid of a thin, flat plate covering the region enclosed by the parabola $y^2 = x$ and the line $x = 2y$.

41. Find the center of mass of a thin, flat plate covering the region enclosed by the parabola $y^2 = x$ and the line $x = 2y$ if the density function is $\delta(y) = 1 + y$. (Use horizontal strips.)

42. a. Find the center of mass of a thin plate of constant density covering the region between the curve $y = 3/x^{3/2}$ and the x-axis from $x = 1$ to $x = 9$.

b. Find the plate's center of mass if, instead of being constant, the density is $\delta(x) = x$. (Use vertical strips.)

Fluid Force

43. Trough of water The vertical triangular plate shown here is the end plate of a trough full of water $(w = 62.4)$. What is the fluid force against the plate?

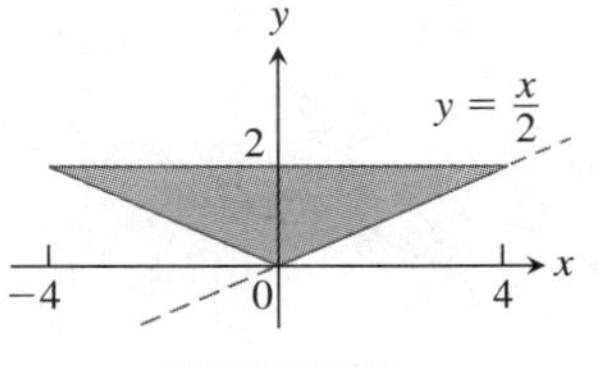

44. Trough of maple syrup The vertical trapezoidal plate shown here is the end plate of a trough full of maple syrup weighing 75 lb/ft^3. What is the force exerted by the syrup against the end plate of the trough when the syrup is 10 in. deep?

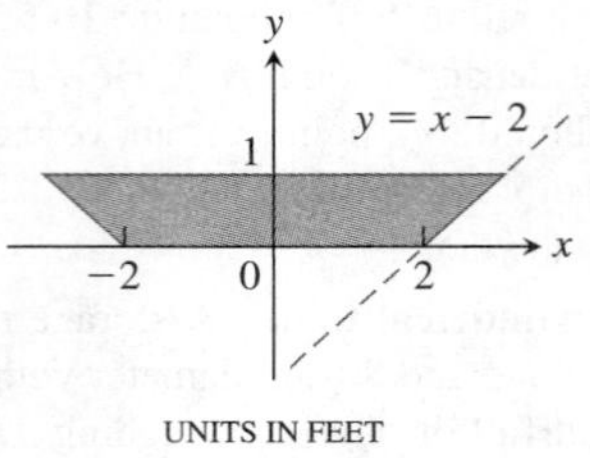

45. Force on a parabolic gate A flat vertical gate in the face of a dam is shaped like the parabolic region between the curve $y = 4x^2$ and the line $y = 4$, with measurements in feet. The top of the gate lies 5 ft below the surface of the water. Find the force exerted by the water against the gate $(w = 62.4)$.

T **46.** You plan to store mercury $(w = 849 \text{ lb/ft}^3)$ in a vertical rectangular tank with a 1 ft square base side whose interior side wall can withstand a total fluid force of 40,000 lb. About how many cubic feet of mercury can you store in the tank at any one time?

CHAPTER 6 Additional and Advanced Exercises

Volume and Length

1. A solid is generated by revolving about the x-axis the region bounded by the graph of the positive continuous function $y = f(x)$, the x-axis, the fixed line $x = a$, and the variable line $x = b$, $b > a$. Its volume, for all b, is $b^2 - ab$. Find $f(x)$.

2. A solid is generated by revolving about the x-axis the region bounded by the graph of the positive continuous function $y = f(x)$, the x-axis, and the lines $x = 0$ and $x = a$. Its volume, for all $a > 0$, is $a^2 + a$. Find $f(x)$.

3. Suppose that the increasing function $f(x)$ is smooth for $x \geq 0$ and that $f(0) = a$. Let $s(x)$ denote the length of the graph of f from $(0, a)$ to $(x, f(x))$, $x > 0$. Find $f(x)$ if $s(x) = Cx$ for some constant C. What are the allowable values for C?

4. a. Show that for $0 < \alpha \leq \pi/2$,

$$\int_0^\alpha \sqrt{1 + \cos^2\theta}\, d\theta > \sqrt{\alpha^2 + \sin^2\alpha}.$$

b. Generalize the result in part (a).

5. Find the volume of the solid formed by revolving the region bounded by the graphs of $y = x$ and $y = x^2$ about the line $y = x$.

6. Consider a right-circular cylinder of diameter 1. Form a wedge by making one slice parallel to the base of the cylinder completely through the cylinder, and another slice at an angle of 45° to the first slice and intersecting the first slice at the opposite edge of the cylinder (see accompanying diagram). Find the volume of the wedge.

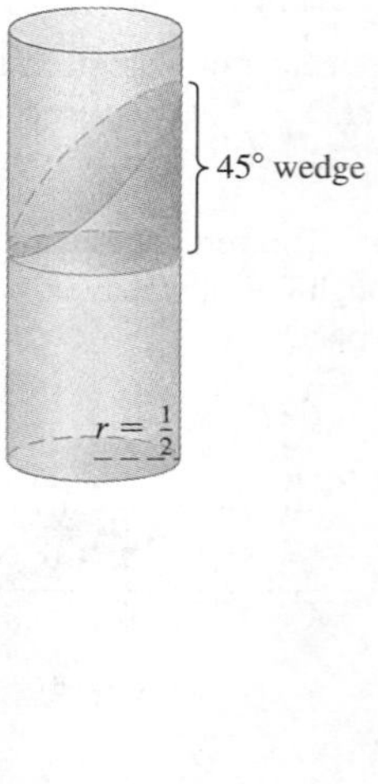

Surface Area

7. At points on the curve $y = 2\sqrt{x}$, line segments of length $h = y$ are drawn perpendicular to the xy-plane. (See accompanying figure.) Find the area of the surface formed by these perpendiculars from $(0, 0)$ to $(3, 2\sqrt{3})$.

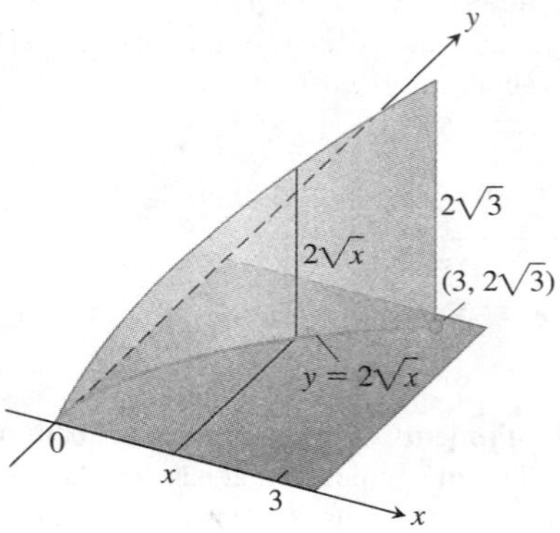

8. At points on a circle of radius a, line segments are drawn perpendicular to the plane of the circle, the perpendicular at each point P being of length ks, where s is the length of the arc of the circle measured counterclockwise from $(a, 0)$ to P, and k is a positive constant, as shown here. Find the area of the surface formed by the perpendiculars along the arc beginning at $(a, 0)$ and extending once around the circle.

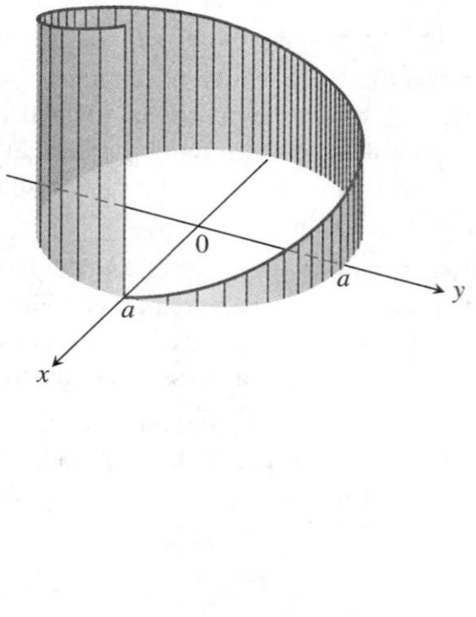

Work

9. A particle of mass m starts from rest at time $t = 0$ and is moved along the x-axis with constant acceleration a from $x = 0$ to $x = h$ against a variable force of magnitude $F(t) = t^2$. Find the work done.
10. **Work and kinetic energy** Suppose a 1.6-oz golf ball is placed on a vertical spring with force constant $k = 2\text{ lb/in}$. The spring is compressed 6 in. and released. About how high does the ball go (measured from the spring's rest position)?

Centers of Mass

11. Find the centroid of the region bounded below by the x-axis and above by the curve $y = 1 - x^n$, n an even positive integer. What is the limiting position of the centroid as $n \to \infty$?
12. If you haul a telephone pole on a two-wheeled carriage behind a truck, you want the wheels to be 3 ft or so behind the pole's center of mass to provide an adequate "tongue" weight. The 40-ft wooden telephone poles used by Verizon have a 27-in. circumference at the top and a 43.5-in. circumference at the base. About how far from the top is the center of mass?
13. Suppose that a thin metal plate of area A and constant density δ occupies a region R in the xy-plane, and let M_y be the plate's moment about the y-axis. Show that the plate's moment about the line $x = b$ is
    a. $M_y - b\delta A$ if the plate lies to the right of the line, and
    b. $b\delta A - M_y$ if the plate lies to the left of the line.
14. Find the center of mass of a thin plate covering the region bounded by the curve $y^2 = 4ax$ and the line $x = a$, $a =$ positive constant, if the density at (x, y) is directly proportional to **(a)** x, **(b)** $|y|$.
15. **a.** Find the centroid of the region in the first quadrant bounded by two concentric circles and the coordinate axes, if the circles have radii a and b, $0 < a < b$, and their centers are at the origin.
    b. Find the limits of the coordinates of the centroid as $a \to b$ and discuss the meaning of the result.
16. A triangular corner is cut from a square 1 ft on a side. The area of the triangle removed is 36 in^2. If the centroid of the remaining region is 7 in. from one side of the original square, how far is it from the remaining sides?

Fluid Force

17. A triangular plate ABC is submerged in water with its plane vertical. The side AB, 4 ft long, is 6 ft below the surface of the water, while the vertex C is 2 ft below the surface. Find the force exerted by the water on one side of the plate.
18. A vertical rectangular plate is submerged in a fluid with its top edge parallel to the fluid's surface. Show that the force exerted by the fluid on one side of the plate equals the average value of the pressure up and down the plate times the area of the plate.

CHAPTER 6 Technology Application Projects

Mathematica/Maple Projects

Projects can be found within MyLab Math.

- ***Using Riemann Sums to Estimate Areas, Volumes, and Lengths of Curves***
  Visualize and approximate areas and volumes in **Part I** and **Part II**: Volumes of Revolution; and **Part III**: Lengths of Curves.
- ***Modeling a Bungee Cord Jump***
  Collect data (or use data previously collected) to build and refine a model for the force exerted by a jumper's bungee cord. Use the work-energy theorem to compute the distance fallen for a given jumper and a given length of bungee cord.

7

Transcendental Functions

OVERVIEW Exponential functions model a wide variety of phenomena of interest in science, engineering, mathematics, and economics. These functions are useful in studying the growth of a biological population, the spread of a disease or of information throughout a human community, drug dosages, radioactive elements and their role in dating fossils, the earth's atmospheric pressure, temperature changes, waves, electrical circuits, the vibrations in bridges, interest rates, and probabilities.

In this chapter, we use the methods of calculus to obtain rigorous and precise definitions and properties of the exponential functions. We first define the natural logarithm function $y = \ln x$ as a certain integral, and then the natural exponential function $y = e^x$ as its inverse function. These two foundational functions form the basis for the full array of logarithmic and exponential functions. We also introduce inverse trigonometric functions, as well as hyperbolic functions and their inverses, and investigate their applications. Along with the trigonometric functions, all of these functions belong to the category of *transcendental functions*.

7.1 Inverse Functions and Their Derivatives

A function that undoes, or inverts, the effect of a function f is called the *inverse* of f. Many common functions, though not all, are paired with an inverse. Important inverse functions often show up in applications. Inverse functions also play a key role in the development and properties of the exponential functions. To have an inverse, a function must possess a special property over its domain.

One-to-One Functions

A function is a rule that assigns a value from its range to each element in its domain. Some functions assign the same range value to more than one element in the domain. The function $f(x) = x^2$ assigns the same value, 1, to both of the numbers -1 and $+1$. Similarly the sines of $\pi/3$ and $2\pi/3$ are both $\sqrt{3}/2$. Other functions assume each value in their range no more than once. The square roots and cubes of different numbers are always different. A function that has distinct values at distinct elements in its domain is called *one-to-one*.

DEFINITION A function $f(x)$ is **one-to-one** on a domain D if $f(x_1) \neq f(x_2)$ whenever $x_1 \neq x_2$ in D.

EXAMPLE 1 Some functions are one-to-one on their entire natural domain. Other functions are not one-to-one on their entire domain, but by restricting the function to a smaller domain we can create a function that is one-to-one. The original and restricted functions are not the same functions because they have different domains. However, the two functions have the same values on the smaller domain.

(a) $f(x) = \sqrt{x}$ is one-to-one on any domain of nonnegative numbers because $\sqrt{x_1} \neq \sqrt{x_2}$ whenever $x_1 \neq x_2$.

(b) $g(x) = \sin x$ is *not* one-to-one on the interval $[0, \pi]$ because $\sin(\pi/6) = \sin(5\pi/6)$. In fact, for each element x_1 in the subinterval $[0, \pi/2)$ there is a corresponding element x_2 in the subinterval $(\pi/2, \pi]$ satisfying $\sin x_1 = \sin x_2$. The sine function *is* one-to-one on $[0, \pi/2]$, however, because it is an increasing function on $[0, \pi/2]$ and hence gives distinct outputs for distinct inputs in that interval. ■

The graph of a one-to-one function $y = f(x)$ can intersect a given horizontal line at most once. If the function intersects the line more than once, then it assumes the same y-value for at least two different x-values and is therefore not one-to-one (Figure 7.1).

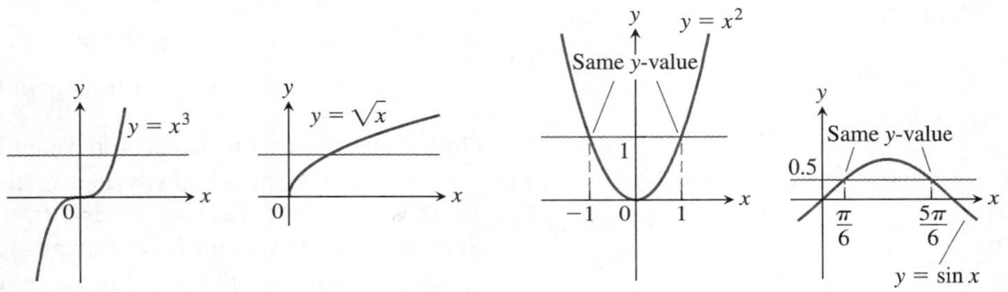

(a) One-to-one: Graph meets each horizontal line at most once.

(b) Not one-to-one: Graph meets one or more horizontal lines more than once.

FIGURE 7.1 (a) $y = x^3$ and $y = \sqrt{x}$ are one-to-one on their domains $(-\infty, \infty)$ and $[0, \infty)$. (b) $y = x^2$ and $y = \sin x$ are not one-to-one on their domains $(-\infty, \infty)$.

> **The Horizontal Line Test for One-to-One Functions**
> A function $y = f(x)$ is one-to-one if and only if its graph intersects each horizontal line at most once.

Inverse Functions

Since each output of a one-to-one function comes from just one input, the effect of the function can be inverted to send each output back to the input from which it came.

> **DEFINITION** Suppose that f is a one-to-one function on a domain D with range R. The **inverse function** f^{-1} is defined by
> $$f^{-1}(b) = a \quad \text{if} \quad f(a) = b.$$
> The domain of f^{-1} is R and the range of f^{-1} is D.

Caution
Do not confuse the inverse function f^{-1} with the reciprocal function $1/f$.

The symbol f^{-1} for the inverse of f is read "f inverse." The "-1" in f^{-1} is *not* an exponent; $f^{-1}(x)$ does not mean $1/f(x)$. Notice that the domains and ranges of f and f^{-1} are interchanged.

EXAMPLE 2 Suppose a one-to-one function $y = f(x)$ is given by a table of values

x	1	2	3	4	5	6	7	8
$f(x)$	3	4.5	7	10.5	15	20.5	27	34.5

A table for the values of $x = f^{-1}(y)$ can then be obtained by simply interchanging the values in each column of the table for f:

y	3	4.5	7	10.5	15	20.5	27	34.5
$f^{-1}(y)$	1	2	3	4	5	6	7	8

■

If we apply f to send an input x to the output $f(x)$ and follow by applying f^{-1} to $f(x)$, we get right back to x, just where we started. Similarly, if we take some number y in the range of f, apply f^{-1} to it, and then apply f to the resulting value $f^{-1}(y)$, we get back the value y from which we began. Composing a function and its inverse has the same effect as doing nothing.

$$(f^{-1} \circ f)(x) = x, \qquad \text{for all } x \text{ in the domain of } f$$
$$(f \circ f^{-1})(y) = y, \qquad \text{for all } y \text{ in the domain of } f^{-1} \text{ (or range of } f)$$

Only a one-to-one function can have an inverse. The reason is that if $f(x_1) = y$ and $f(x_2) = y$ for two distinct inputs x_1 and x_2, then there is no way to assign a value to $f^{-1}(y)$ that satisfies both $f^{-1}(f(x_1)) = x_1$ and $f^{-1}(f(x_2)) = x_2$.

Suppose f is a function whose domain is an interval. If f is increasing, then it satisfies the inequality $f(x_2) > f(x_1)$ when $x_2 > x_1$, so it is one-to-one and has an inverse. Likewise, if f is decreasing, then it also has an inverse. Functions that are neither increasing nor decreasing may still be one-to-one and have an inverse, as with the function $f(x) = 1/x$ for $x \neq 0$ and $f(0) = 0$, defined on $(-\infty, \infty)$ and passing the horizontal line test.

Finding Inverses

The graphs of a function and its inverse are closely related. To read the value of a function from its graph, we start at a point x on the x-axis, go vertically to the graph, and then move horizontally to the y-axis to read the value of y. The inverse function can be read from the graph by reversing this process. Start with a point y on the y-axis, go horizontally to the graph of $y = f(x)$, and then move vertically to the x-axis to read the value of $x = f^{-1}(y)$ (Figure 7.2).

We want to set up the graph of f^{-1} so that its input values lie along the x-axis, as is usually done for functions, rather than on the y-axis. To achieve this we interchange the x- and y-axes by reflecting across the 45° line $y = x$. After this reflection we have a new graph that represents f^{-1}. The value of $f^{-1}(x)$ can now be read from the graph in the usual way, by starting with a point x on the x-axis, going vertically to the graph, and then horizontally to the y-axis to get the value of $f^{-1}(x)$. Figure 7.2 indicates the relationship between the graphs of f and f^{-1}. The graphs are interchanged by reflection through the line $y = x$.

The process of passing from f to f^{-1} can be summarized as a two-step procedure.

1. Solve the equation $y = f(x)$ for x. This gives a formula $x = f^{-1}(y)$ where x is expressed as a function of y.
2. Interchange x and y, obtaining a formula $y = f^{-1}(x)$ where f^{-1} is expressed in the conventional format with x as the independent variable and y as the dependent variable.

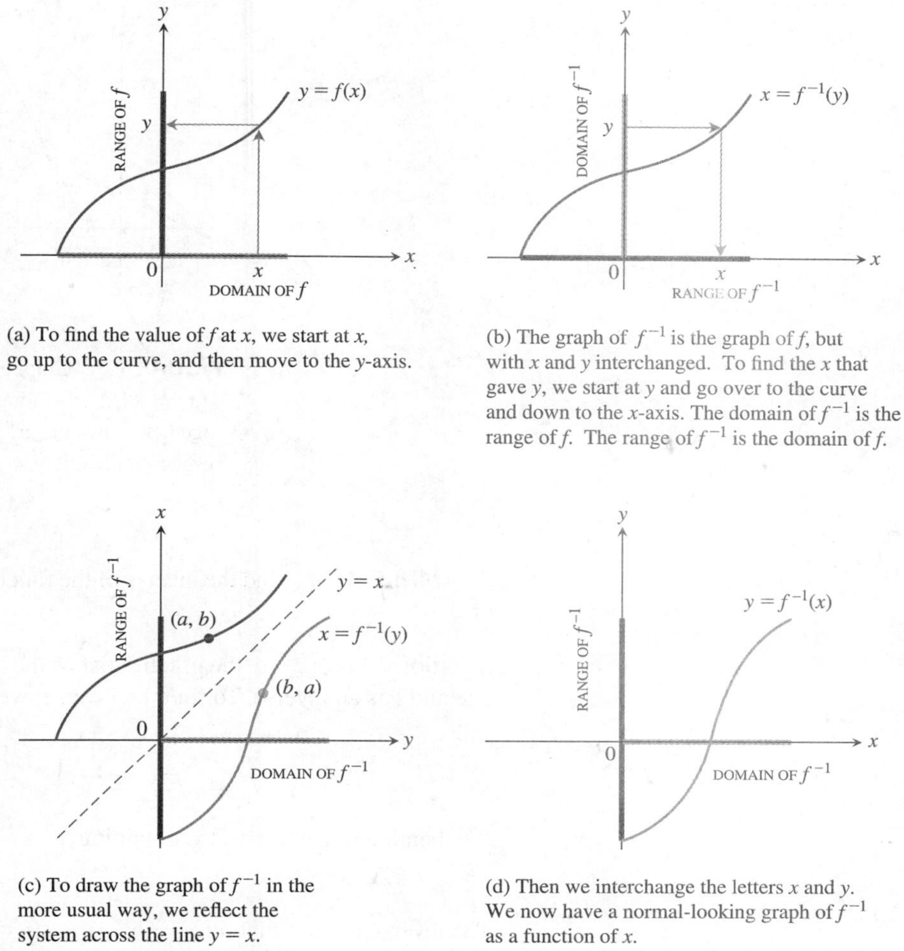

FIGURE 7.2 The graph of $y = f^{-1}(x)$ is obtained by reflecting the graph of $y = f(x)$ about the line $y = x$.

EXAMPLE 3 Find the inverse of $y = \frac{1}{2}x + 1$, expressed as a function of x.

Solution

1. *Solve for x in terms of y:*

$$y = \frac{1}{2}x + 1$$
$$2y = x + 2$$
$$x = 2y - 2.$$

The graph satisfies the horizontal line test, so the function is one-to-one (Figure 7.3).

In this formula, x is a function of y.

2. *Interchange x and y:* $y = 2x - 2.$

Expresses the function in the usual form where y is the dependent variable.

The inverse of the function $f(x) = (1/2)x + 1$ is the function $f^{-1}(x) = 2x - 2$. (See Figure 7.3.) To check, we verify that both compositions give the identity function:

$$f^{-1}(f(x)) = 2\left(\frac{1}{2}x + 1\right) - 2 = x + 2 - 2 = x$$
$$f(f^{-1}(x)) = \frac{1}{2}(2x - 2) + 1 = x - 1 + 1 = x. \quad \blacksquare$$

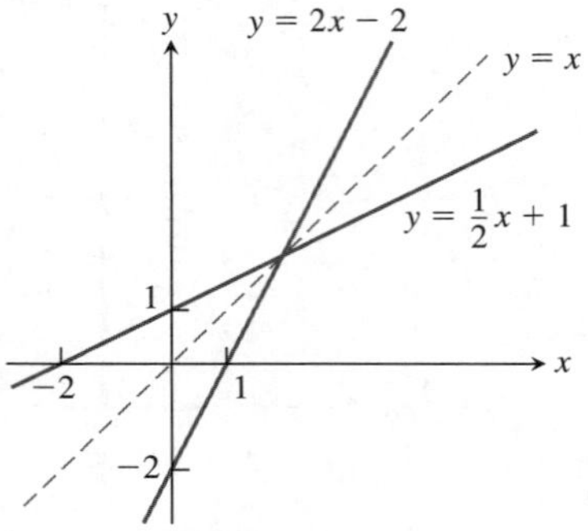

FIGURE 7.3 Graphing the functions $f(x) = (1/2)x + 1$ and $f^{-1}(x) = 2x - 2$ together shows the graphs' symmetry with respect to the line $y = x$ (Example 3).

EXAMPLE 4 Find the inverse of the function $y = x^2, x \geq 0$, expressed as a function of x.

Solution For $x \geq 0$, the graph satisfies the horizontal line test, so the function is one-to-one and has an inverse. To find the inverse, we first solve for x in terms of y:

$$y = x^2$$

$$\sqrt{y} = \sqrt{x^2} = |x| = x \qquad |x| = x \text{ because } x \geq 0$$

We then interchange x and y, obtaining

$$y = \sqrt{x}.$$

The inverse of the function $y = x^2, x \geq 0$, is the function $y = \sqrt{x}$ (Figure 7.4). ■

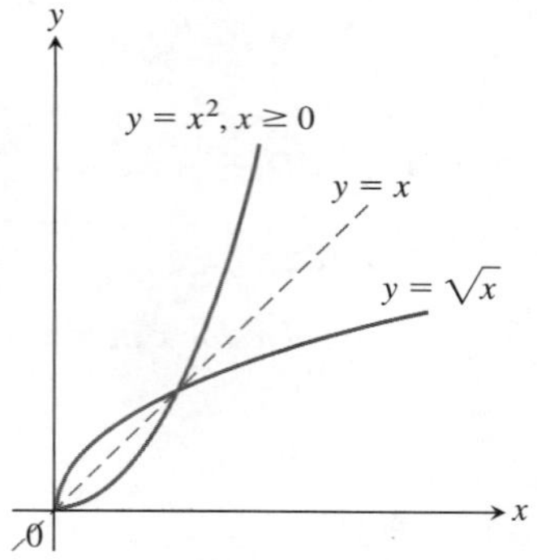

FIGURE 7.4 The functions $y = \sqrt{x}$ and $y = x^2, x \geq 0$, are inverses of one another (Example 4).

Notice that the function $y = x^2, x \geq 0$, with domain *restricted* to the nonnegative real numbers, *is* one-to-one (Figure 7.4) and has an inverse. On the other hand, the function $y = x^2$, with no domain restrictions, *is not* one-to-one (Figure 7.1b) and therefore has no inverse.

Derivatives of Inverses of Differentiable Functions

We calculated the inverse of the function $f(x) = (1/2)x + 1$ to be $f^{-1}(x) = 2x - 2$ in Example 3. If we calculate their derivatives, we see that

$$\frac{d}{dx} f(x) = \frac{d}{dx}\left(\frac{1}{2}x + 1\right) = \frac{1}{2}$$
$$\frac{d}{dx} f^{-1}(x) = \frac{d}{dx}(2x - 2) = 2.$$

The derivatives are reciprocals of one another, so the slope of one line is the reciprocal of the slope of its inverse line. (See Figure 7.3.)

The reciprocal relationship between the slopes of f and f^{-1} holds for other functions as well, but we must be careful to compare slopes at corresponding points. If the slope of $y = f(x)$ at the point $(a, f(a))$ is $f'(a)$ and $f'(a) \neq 0$, then the slope of $y = f^{-1}(x)$ at the point $(f(a), a)$ is the reciprocal $1/f'(a)$ (Figure 7.5). If we set $b = f(a)$, then

$$(f^{-1})'(b) = \frac{1}{f'(a)} = \frac{1}{f'(f^{-1}(b))}.$$

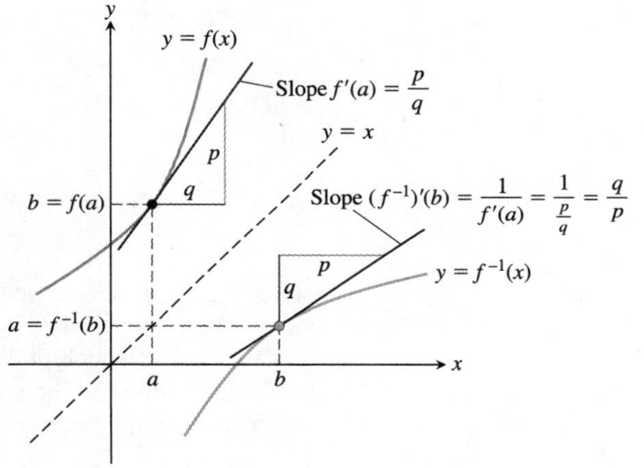

The slopes are reciprocals: $(f^{-1})'(b) = \dfrac{1}{f'(a)}$ or $(f^{-1})'(b) = \dfrac{1}{f'(f^{-1}(b))}$

FIGURE 7.5 The graphs of inverse functions have reciprocal slopes at corresponding points.

If $y = f(x)$ has a horizontal tangent line at $(a, f(a))$, then the inverse function f^{-1} has a vertical tangent line at $(f(a), a)$, so the slope is undefined and f^{-1} is not differentiable at $f(a)$. Theorem 1 gives the conditions under which f^{-1} is differentiable in its domain (which is the same as the range of f).

THEOREM 1—The Derivative Rule for Inverses
If f has an interval I as domain and $f'(x)$ exists and is never zero on I, then f^{-1} is differentiable at every point in its domain (the range of f). The value of $(f^{-1})'$ at a point b in the domain of f^{-1} is the reciprocal of the value of f' at the point $a = f^{-1}(b)$:

$$(f^{-1})'(b) = \frac{1}{f'(f^{-1}(b))} \tag{1}$$

or

$$\left.\frac{df^{-1}}{dx}\right|_{x=b} = \frac{1}{\left.\dfrac{df}{dx}\right|_{x=f^{-1}(b)}}.$$

Theorem 1 makes two assertions. The first of these has to do with the conditions under which f^{-1} is differentiable; the second assertion is a formula for the derivative of f^{-1} when it exists. While we omit the proof of the first assertion, the second one is proved in the following way:

$$
\begin{aligned}
f(f^{-1}(x)) &= x && \text{Inverse function relationship} \\
\frac{d}{dx} f(f^{-1}(x)) &= 1 && \text{Differentiate both sides.} \\
f'(f^{-1}(x)) \cdot \frac{d}{dx} f^{-1}(x) &= 1 && \text{Chain Rule} \\
\frac{d}{dx} f^{-1}(x) &= \frac{1}{f'(f^{-1}(x))}. && \text{Solve for the derivative.}
\end{aligned}
$$

EXAMPLE 5 The function $f(x) = x^2, x > 0$ and its inverse $f^{-1}(x) = \sqrt{x}$ have derivatives $f'(x) = 2x$ and $(f^{-1})'(x) = 1/(2\sqrt{x})$.

Let's verify that Theorem 1 gives the same formula for the derivative of $f^{-1}(x)$:

$$
\begin{aligned}
(f^{-1})'(x) &= \frac{1}{f'(f^{-1}(x))} \\
&= \frac{1}{2(f^{-1}(x))} && f'(x) = 2x \text{ with } x \text{ replaced by } f^{-1}(x) \\
&= \frac{1}{2(\sqrt{x})}. && f^{-1}(x) = \sqrt{x}
\end{aligned}
$$

Theorem 1 gives a derivative that agrees with the known derivative of the square root function.

Let's examine Theorem 1 at a specific point. We pick $x = 2$ (the number a) and $f(2) = 4$ (the value b). Theorem 1 says that the derivative of f at 2, which is $f'(2) = 4$, and the derivative of f^{-1} at $f(2)$, which is $(f^{-1})'(4)$, are reciprocals. It states that

$$
(f^{-1})'(4) = \frac{1}{f'(f^{-1}(4))} = \frac{1}{f'(2)} = \left.\frac{1}{2x}\right|_{x=2} = \frac{1}{4}.
$$

See Figure 7.6. ■

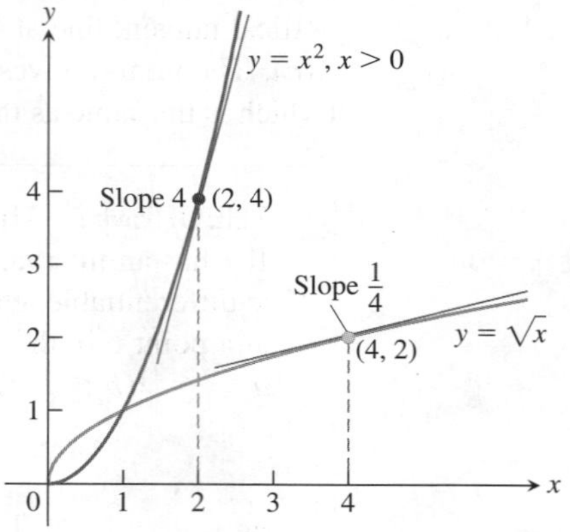

FIGURE 7.6 The derivative of $f^{-1}(x) = \sqrt{x}$ at the point (4, 2) is the reciprocal of the derivative of $f(x) = x^2$ at $(2, 4)$ (Example 5).

We will use the procedure illustrated in Example 5 to calculate formulas for the derivatives of many inverse functions throughout this chapter. Equation (1) sometimes enables us to find specific values of df^{-1}/dx without knowing a formula for f^{-1}.

EXAMPLE 6 Let $f(x) = x^3 - 2, x > 0$. Find the value of df^{-1}/dx at $x = 6 = f(2)$ without finding a formula for $f^{-1}(x)$. See Figure 7.7.

Solution We apply Theorem 1 to obtain the value of the derivative of f^{-1} at $x = 6$:

$$\left.\frac{df}{dx}\right|_{x=2} = \left.3x^2\right|_{x=2} = 12$$

$$\left.\frac{df^{-1}}{dx}\right|_{x=f(2)} = \frac{1}{\left.\frac{df}{dx}\right|_{x=2}} = \frac{1}{12}. \qquad \text{Eq. (1)}$$

■

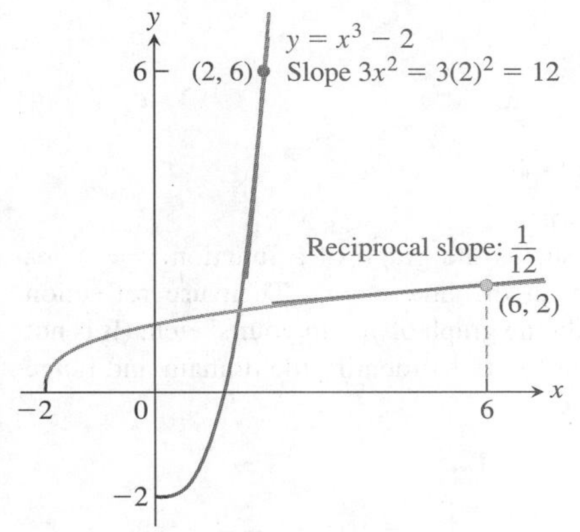

FIGURE 7.7 The derivative of $f(x) = x^3 - 2$ at $x = 2$ tells us the derivative of f^{-1} at $x = 6$ (Example 6).

EXERCISES 7.1

Identifying One-to-One Functions Graphically

Which of the functions graphed in Exercises 1–6 are one-to-one, and which are not?

1.

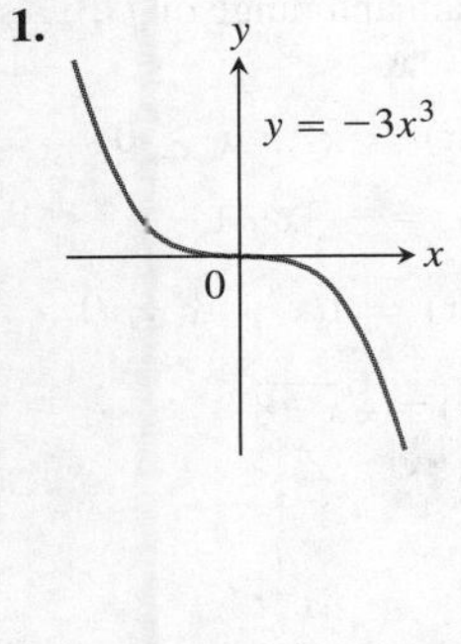

2.

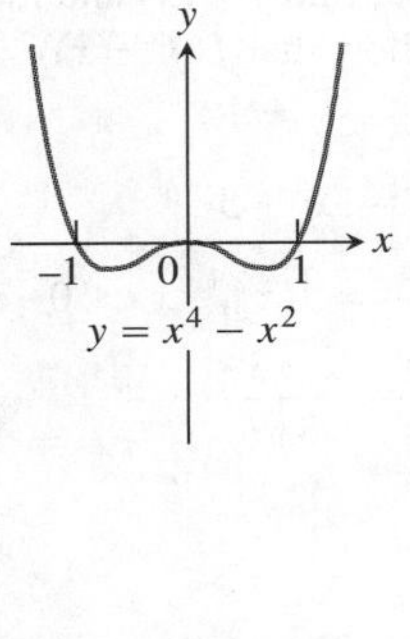

3.

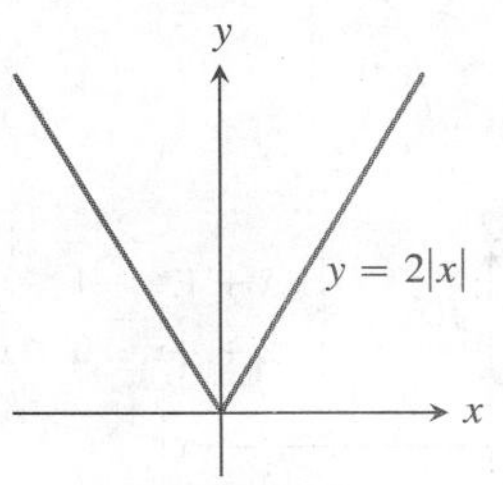

4.

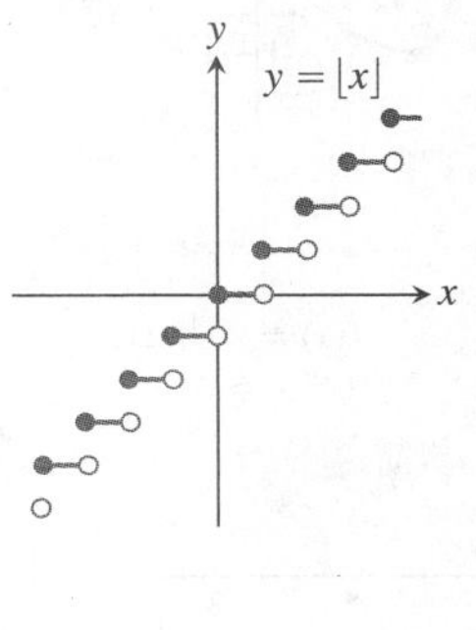

5.

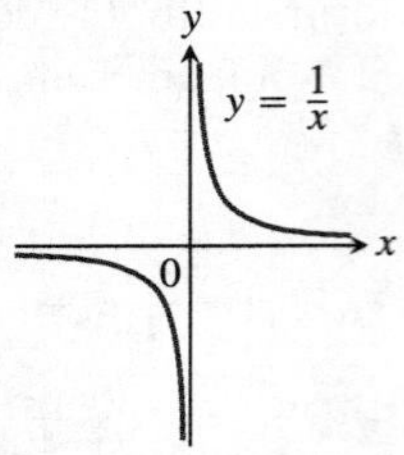

6.

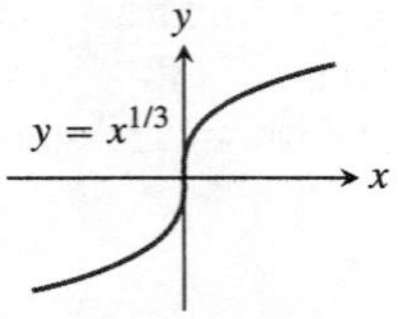

In Exercises 7–10, determine from its graph if the function is one-to-one.

7. $f(x) = \begin{cases} 3 - x, & x < 0 \\ 3, & x \geq 0 \end{cases}$

8. $f(x) = \begin{cases} 2x + 6, & x \leq -3 \\ x + 4, & x > -3 \end{cases}$

9. $f(x) = \begin{cases} 1 - \dfrac{x}{2}, & x \leq 0 \\ \dfrac{x}{x + 2}, & x > 0 \end{cases}$

10. $f(x) = \begin{cases} 2 - x^2, & x \leq 1 \\ x^2, & x > 1 \end{cases}$

Graphing Inverse Functions

Each of Exercises 11–16 shows the graph of a function $y = f(x)$. Copy the graph and draw in the line $y = x$. Then use reflection across the line $y = x$ to add the graph of f^{-1} to your sketch. (It is not necessary to find a formula for f^{-1}.) Identify the domain and range of f^{-1}.

11.

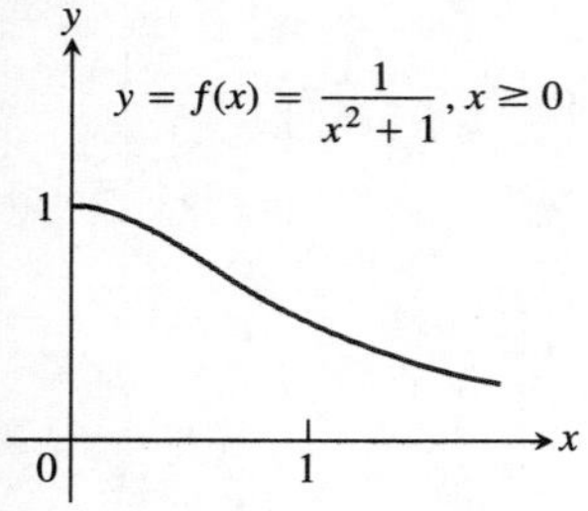

12.

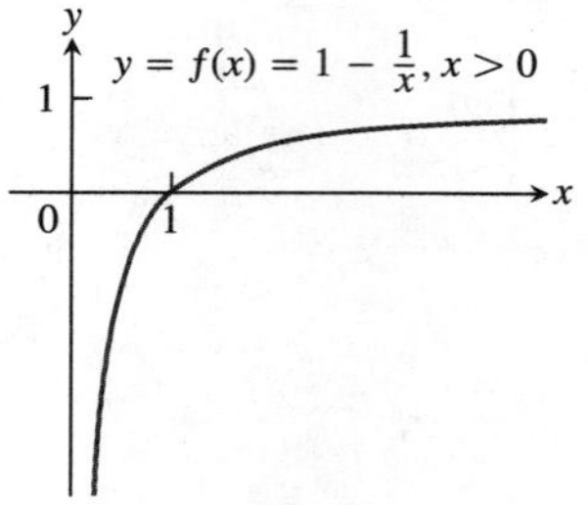

13.

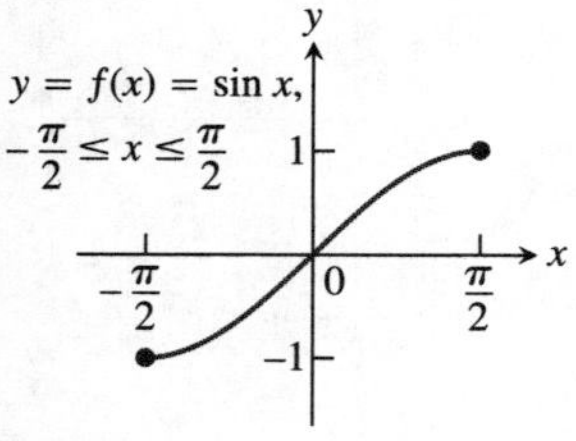

14.

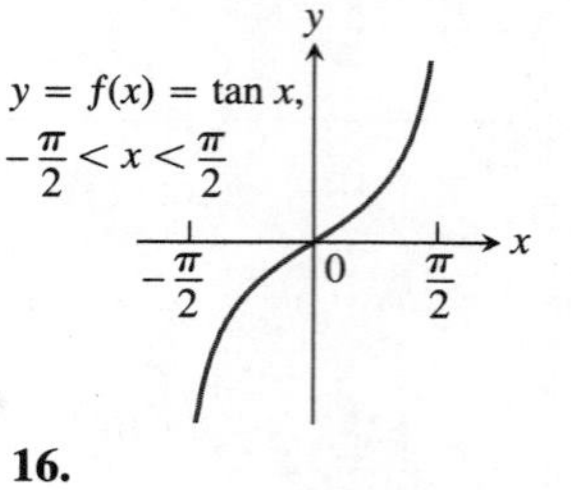

15.

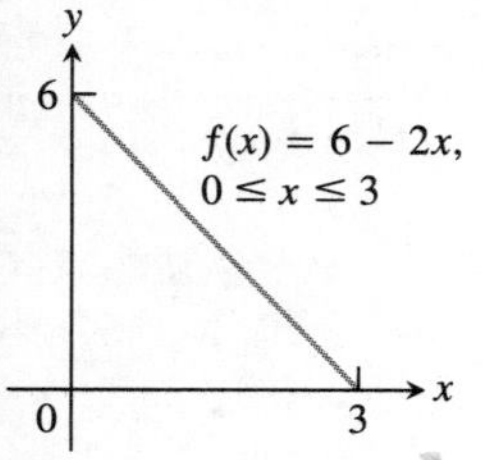

16.

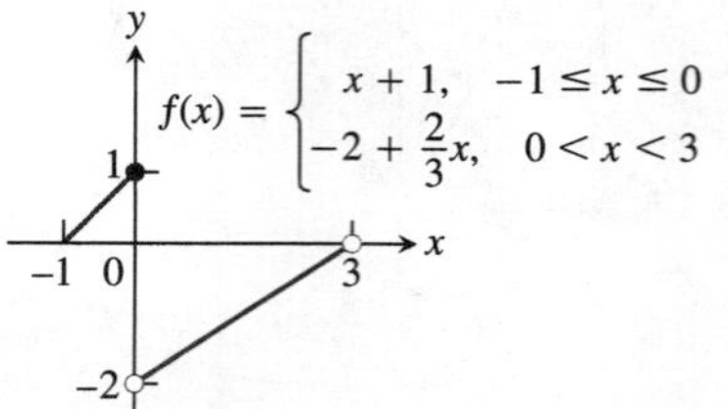

17. **a.** Graph the function $f(x) = \sqrt{1 - x^2}$, $0 \leq x \leq 1$. What symmetry does the graph have?

  b. Show that f is its own inverse. (Remember that $\sqrt{x^2} = x$ if $x \geq 0$.)

18. **a.** Graph the function $f(x) = 1/x$. What symmetry does the graph have?

  b. Show that f is its own inverse.

Formulas for Inverse Functions

Each of Exercises 19–24 gives a formula for a function $y = f(x)$ and shows the graphs of f and f^{-1}. Find a formula for f^{-1} in each case.

19. $f(x) = x^2 + 1, \quad x \geq 0$

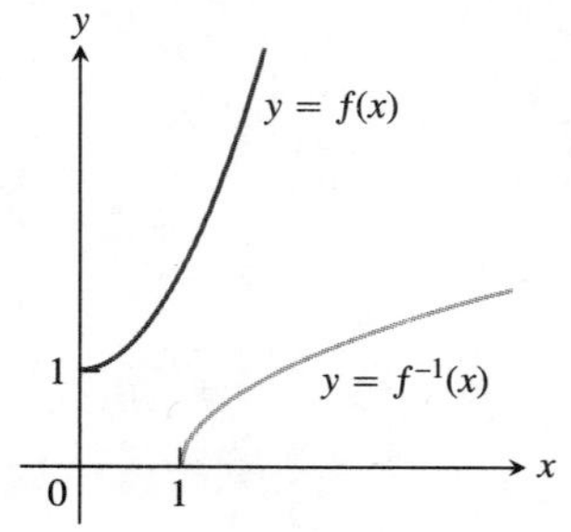

20. $f(x) = x^2, \quad x \leq 0$

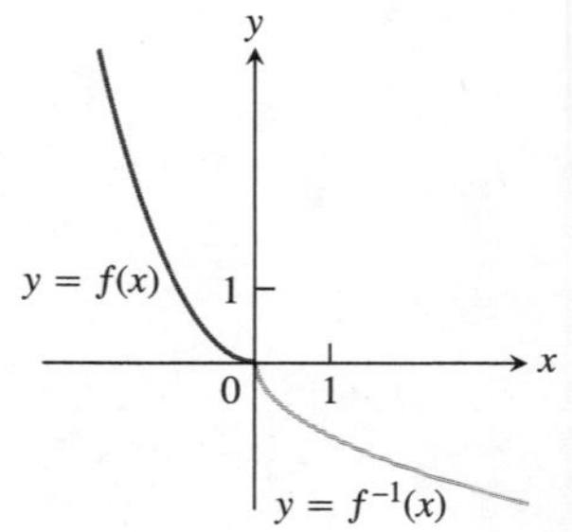

21. $f(x) = x^3 - 1$

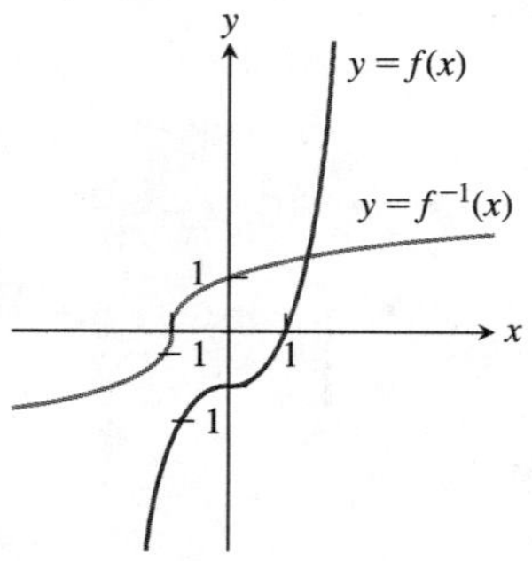

22. $f(x) = x^2 - 2, \quad x \geq 0$

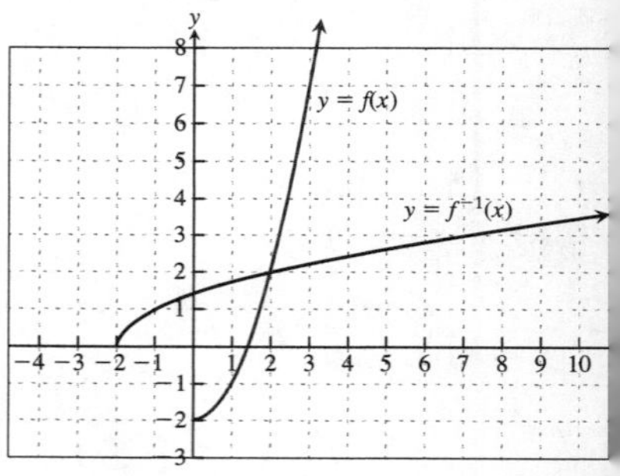

23. $f(x) = (x + 1)^2, \quad x \geq -1$

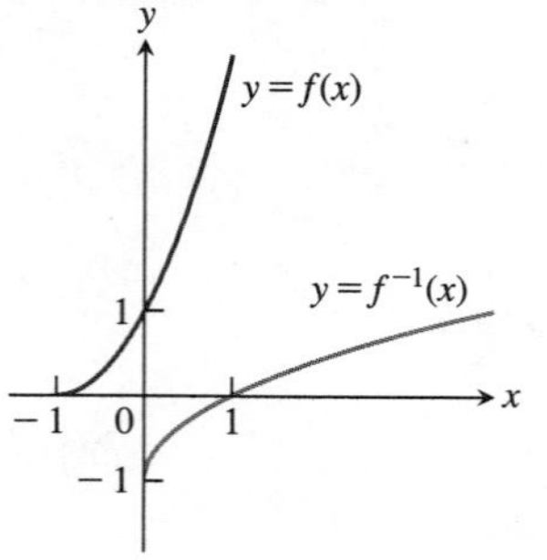

24. $f(x) = x^{2/3}, \quad x \geq 0$

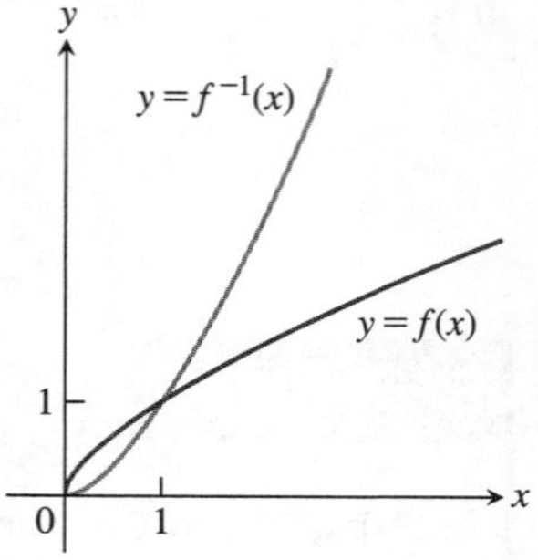

Each of Exercises 25–36 gives a formula for a function $y = f(x)$. In each case, find $f^{-1}(x)$ and identify the domain and range of f^{-1}. As a check, show that $f(f^{-1}(x)) = f^{-1}(f(x)) = x$.

25. $f(x) = 5x$

26. $f(x) = x^4, \quad x \geq 0$

27. $f(x) = 3x^3 + 5$

28. $f(x) = -\dfrac{2}{3}x + 1$

29. $f(x) = 1/x^2, \quad x > 0$

30. $f(x) = 1/x^3, \quad x \neq 0$

31. $f(x) = \dfrac{x - 3}{3x}$

32. $f(x) = \sqrt{x - 8}$

33. $f(x) = x^2 - 2x, \quad x \leq 1$ **34.** $f(x) = \sqrt[5]{x+1} - 3$
(*Hint:* Complete the square.)

35. $f(x) = \dfrac{x+b}{x-2}, \quad b > -2$ and constant

36. $f(x) = x^2 - 2bx, \quad b > 0$ and constant, $x \leq b$

Inverses of Lines

37. a. Find the inverse of the function $f(x) = mx$, where m is a constant different from zero.

b. What can you conclude about the inverse of a function $y = f(x)$ whose graph is a line through the origin with a nonzero slope m?

38. Show that the graph of the inverse of $f(x) = mx + b$, where m and b are constants and $m \neq 0$, is a line with slope $1/m$ and y-intercept $-b/m$.

39. a. Find the inverse of $f(x) = x + 1$. Graph f and its inverse together. Add the line $y = x$ to your sketch, drawing it with dashes or dots for contrast.

b. Find the inverse of $f(x) = x + b$ (b constant). How is the graph of f^{-1} related to the graph of f?

c. What can you conclude about the inverses of functions whose graphs are lines parallel to the line $y = x$?

40. a. Find the inverse of $f(x) = -x + 1$. Graph the line $y = -x + 1$ together with the line $y = x$. At what angle do the lines intersect?

b. Find the inverse of $f(x) = -x + b$ (b constant). What angle does the line $y = -x + b$ make with the line $y = x$?

c. What can you conclude about the inverses of functions whose graphs are lines perpendicular to the line $y = x$?

Derivatives of Inverse Functions

In Exercises 41–44:

a. Find $f^{-1}(x)$.

b. Graph f and f^{-1} together.

c. Evaluate df/dx at $x = a$ and df^{-1}/dx at $x = f(a)$ to show that $(df^{-1}/dx)|_{x=f(a)} = 1/(df/dx)|_{x=a}$.

41. $f(x) = x + 2, \quad a = -1$

42. $f(x) = \dfrac{x+2}{1-x}, \quad a = 1/2$

43. $f(x) = 2 - 2x, \quad a = 1/2$

44. $f(x) = 2x^2, \quad x \geq 0, \quad a = 5$

45. The accompanying figure shows the graph of the function f.

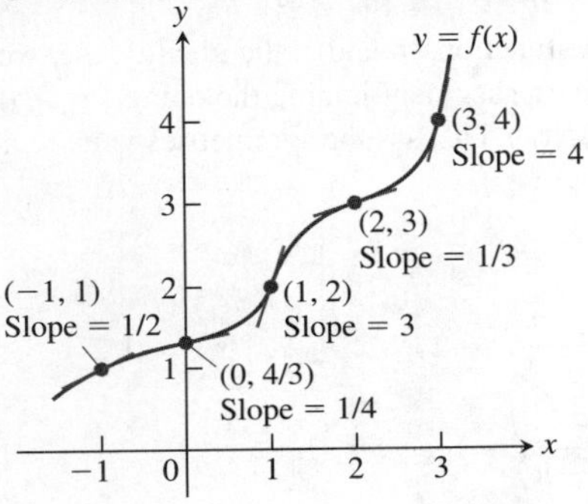

Assuming the inverse function f^{-1} is differentiable, find the slope of $f^{-1}(x)$ at

a. $x = 1$ **b.** $x = 2$ **c.** $x = 3$

46. The accompanying figure shows the graph of the function g.

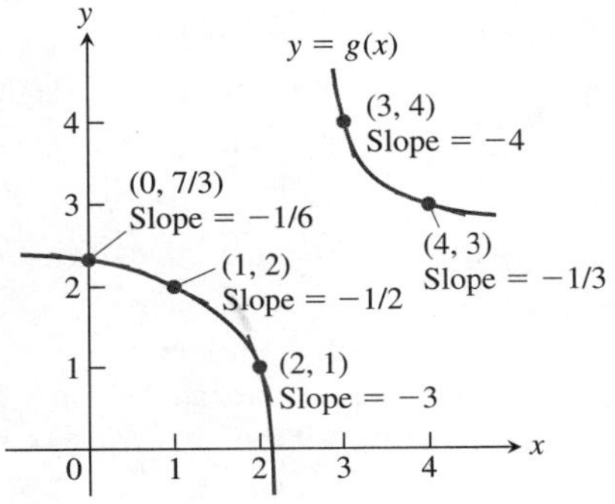

Assuming the inverse function g^{-1} is differentiable, find the slope of $g^{-1}(x)$ at

a. $x = 1$ **b.** $x = 2$ **c.** $x = 3$

47. Suppose that the function f and its derivative with respect to x have the following values at $x = 0, 1, 2, 3,$ and 4.

x	**0**	**1**	**2**	**3**	**4**
$f(x)$	3	6	0	1	2
$f'(x)$	4/3	5	4	1/2	1/7

Assuming the inverse function f^{-1} is differentiable, find the slope of $f^{-1}(x)$ at

a. $x = 1$ **b.** $x = 2$ **c.** $x = 3$

48. Suppose that the function g and its derivative with respect to x have the following values at $x = 0, 1, 2, 3,$ and 4.

x	**0**	**1**	**2**	**3**	**4**
$g(x)$	−4	−1	1	2	3
$g'(x)$	3	2	5/4	2/3	1/5

Assuming the inverse function g^{-1} is differentiable, find the slope of $g^{-1}(x)$ at

a. $x = 1$ **b.** $x = 2$ **c.** $x = 3$

49. a. Show that $f(x) = x^3$ and $g(x) = \sqrt[3]{x}$ are inverses of one another.

b. Graph f and g over an x-interval large enough to show the graphs intersecting at $(1, 1)$ and $(-1, -1)$. Be sure the picture shows the required symmetry about the line $y = x$.

c. Find the slopes of the tangent lines to the graphs of f and g at $(1, 1)$ and $(-1, -1)$ (four tangent lines in all).

d. What lines are tangent to the curves at the origin?

50. a. Show that $h(x) = x^3/4$ and $k(x) = (4x)^{1/3}$ are inverses of one another.

b. Graph h and k over an x-interval large enough to show the graphs intersecting at $(2, 2)$ and $(-2, -2)$. Be sure the picture shows the required symmetry about the line $y = x$.

c. Find the slopes of the tangent lines to the graphs of h and k at $(2, 2)$ and $(-2, -2)$.

d. What lines are tangent to the curves at the origin?

51. Let $f(x) = x^3 - 3x^2$, $x \geq 2$. Find the value of df^{-1}/dx at the point $x = -1 = f(3)$.

52. Let $f(x) = x^2 - 4x$, $x > 2$. Find the value of df^{-1}/dx at the point $x = 0 = f(5)$.

53. Suppose that the differentiable function $y = f(x)$ has an inverse and that the graph of f passes through the point $(2, 4)$ and has a slope of 1/3 there. Find the value of df^{-1}/dx at $x = 4$.

54. Suppose that the differentiable function $y = g(x)$ has an inverse and that the graph of g passes through the origin with slope 2. Find the slope of the graph of g^{-1} at the origin.

Increasing and Decreasing Functions

55. Show that increasing functions and decreasing functions are one-to-one. That is, show that for any x_1 and x_2 in I, $x_2 \neq x_1$ implies $f(x_2) \neq f(x_1)$.

Use the results of Exercise 55 to show that the functions in Exercises 56–60 have inverses over their domains. Find a formula for df^{-1}/dx using Theorem 1.

56. $f(x) = (1/2)x + (3/4)$

57. $f(x) = 27x^3$

58. $f(x) = 1 - 5x^2$

59. $f(x) = (1 - x)^3$

60. $x^{2/3}$

Theory and Applications

61. If $f(x)$ is one-to-one, can anything be said about $g(x) = -f(x)$? Is it also one-to-one? Give reasons for your answer.

62. If $f(x)$ is one-to-one and $f(x)$ is never zero, can anything be said about $h(x) = 1/f(x)$? Is it also one-to-one? Give reasons for your answer.

63. Suppose that the range of g lies in the domain of f so that the composition $f \circ g$ is defined. If f and g are one-to-one, can anything be said about $f \circ g$? Give reasons for your answer.

64. If a composition $f \circ g$ is one-to-one, must g be one-to-one? Give reasons for your answer.

65. If we write $g(x)$ for $f^{-1}(x)$, Equation (1) can be written as

$$g'(f(a)) = \frac{1}{f'(a)}, \quad \text{or} \quad g'(f(a)) \cdot f'(a) = 1.$$

If we then write x for a, we get

$$g'(f(x)) \cdot f'(x) = 1.$$

The latter equation may remind you of the Chain Rule, and indeed there is a connection.

Assume that f and g are differentiable functions that are inverses of one another, so that $(g \circ f)(x) = x$. Differentiate both sides of this equation with respect to x, using the Chain Rule to express $(g \circ f)'(x)$ as a product of derivatives of g and f. What do you find? (This is not a proof of Theorem 1 because we assume here the theorem's conclusion that $g = f^{-1}$ is differentiable.)

66. Equivalence of the washer and shell methods for finding volume Let f be differentiable and increasing on the interval $a \leq x \leq b$, with $a > 0$, and suppose that f has a differentiable inverse, f^{-1}. Revolve about the y-axis the region bounded by the graph of f and the lines $x = a$ and $y = f(b)$ to generate a solid. Then the values of the integrals given by the washer and shell methods for the volume have identical values:

$$\int_{f(a)}^{f(b)} \pi\left((f^{-1}(y))^2 - a^2\right) dy = \int_a^b 2\pi x(f(b) - f(x))\, dx.$$

To prove this equality, define

$$W(t) = \int_{f(a)}^{f(t)} \pi\left((f^{-1}(y))^2 - a^2\right) dy$$

$$S(t) = \int_a^t 2\pi x(f(t) - f(x))\, dx.$$

Then show that the functions W and S agree at a point of $[a, b]$ and have identical derivatives on $[a, b]$. As you saw in Section 4.2, Corollary 2 of the Mean Value Theorem guarantees that $W(t) = S(t)$ for all t in $[a, b]$. In particular, $W(b) = S(b)$. (*Source:* "Disks and Shells Revisited," by Walter Carlip, *American Mathematical Monthly*, Vol. 98, No. 2, Feb. 1991, pp. 154–156.)

COMPUTER EXPLORATIONS

In Exercises 67–72, you will explore some functions and their inverses together with their derivatives and tangent line approximations at specified points. Perform the following steps using your CAS:

a. Plot the function $y = f(x)$ together with its derivative over the given interval. Explain why you know that f is one-to-one over the interval.

b. Solve the equation $y = f(x)$ for x as a function of y, and name the resulting inverse function g.

c. Find the equation for the tangent line to f at the specified point $(x_0, f(x_0))$.

d. Find the equation for the tangent line to g at the point $(f(x_0), x_0)$ located symmetrically across the 45° line $y = x$ (which is the graph of the identity function). Use Theorem 1 to find the slope of this tangent line.

e. Plot the functions f and g, the identity, the two tangent lines, and the line segment joining the points $(x_0, f(x_0))$ and $(f(x_0), x_0)$. Discuss the symmetries you see across the main diagonal $y = x$.

67. $y = \sqrt{3x - 2}, \quad \frac{2}{3} \leq x \leq 4, \quad x_0 = 3$

68. $y = \dfrac{3x + 2}{2x - 11}, \quad -2 \leq x \leq 2, \quad x_0 = 1/2$

69. $y = \dfrac{4x}{x^2 + 1}, \quad -1 \leq x \leq 1, \quad x_0 = 1/2$

70. $y = \dfrac{x^3}{x^2+1}, \quad -1 \le x \le 1, \quad x_0 = 1/2$

71. $y = x^3 - 3x^2 - 1, \quad 2 \le x \le 5, \quad x_0 = \dfrac{27}{10}$

72. $y = 2 - x - x^3, \quad -2 \le x \le 2, \quad x_0 = \dfrac{3}{2}$

In Exercises 73 and 74, repeat the steps above to solve for the functions $y = f(x)$ and $x = f^{-1}(y)$ defined implicitly by the given equations over the interval.

73. $y^{1/3} - 1 = (x+2)^3, \quad -5 \le x \le 5, \quad x_0 = -3/2$

74. $\cos y = x^{1/5}, \quad 0 \le x \le 1, \quad x_0 = 1/2$

7.2 Natural Logarithms

Historically, logarithms played important roles in arithmetic computations, making possible the great seventeenth-century advances in offshore navigation and celestial mechanics. In this section we define the natural logarithm as an integral using the Fundamental Theorem of Calculus. While this indirect approach may at first seem strange, it provides an elegant and rigorous way to obtain the key characteristics of logarithmic and exponential functions.

Definition of the Natural Logarithm Function

The natural logarithm of a positive number x, written as $\ln x$, is the value of an integral.

DEFINITION The **natural logarithm** is the function given by

$$\ln x = \int_1^x \frac{1}{t}\,dt, \qquad x > 0.$$

From the Fundamental Theorem of Calculus, we know that $\ln x$ is a continuous function. Geometrically, if $x > 1$, then $\ln x$ is the area under the curve $y = 1/t$ from $t = 1$ to $t = x$ (Figure 7.8). For $0 < x < 1$, $\ln x$ gives the negative of the area under the curve from x to 1, and the function is not defined for $x \le 0$. From the Zero Width Interval Rule for definite integrals, we also have

$$\ln 1 = \int_1^1 \frac{1}{t}\,dt = 0.$$

Notice that we show the graph of $y = 1/x$ in Figure 7.8 but use $y = 1/t$ in the integral. Using x for everything would have us writing

$$\ln x = \int_1^x \frac{1}{x}\,dx,$$

with x meaning two different things. So we change the variable of integration to t.

By using rectangles to obtain finite approximations of the area under the graph of $y = 1/t$ and over the interval between $t = 1$ and $t = x$, as in Section 5.1, we can approximate the values of the function $\ln x$. Several values are given in Table 7.1. There is an important number between $x = 2$ and $x = 3$ whose natural logarithm equals 1. This number, which we now define, exists because $\ln x$ is a continuous and increasing function and therefore the Intermediate Value Theorem implies that there is exactly one such number in the interval $[2, 3]$.

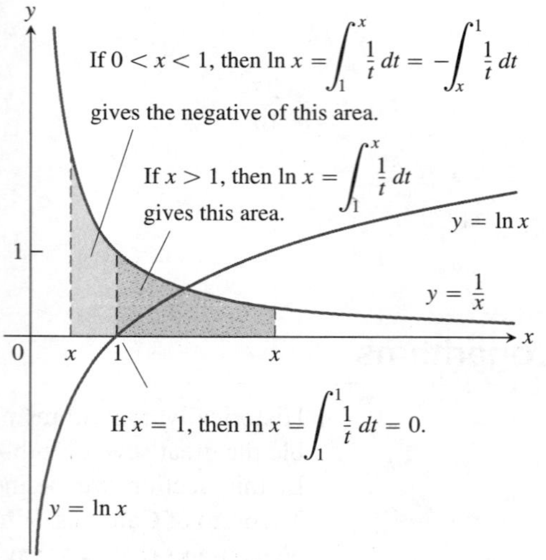

FIGURE 7.8 The graph of $y = \ln x$ and its relation to the function $y = 1/x$, $x > 0$. The graph of the logarithm rises above the x-axis as x moves from 1 to the right, and it falls below the axis as x moves from 1 to the left.

TABLE 7.1 Two-decimal-place approximations of $\ln x$

x	$\ln x$
0	undefined
0.05	-3.00
0.5	-0.69
1	0
2	0.69
3	1.10
4	1.39
10	2.30

DEFINITION The **number e** is the number in the domain of the natural logarithm that satisfies

$$\ln(e) = \int_1^e \frac{1}{t}\,dt = 1.$$

Interpreted geometrically, the number e corresponds to the point on the x-axis for which the area under the graph of $y = 1/t$ and above the interval $[1, e]$ equals the area of the unit square. That is, the area of the region shaded dotted gray in Figure 7.8 is 1 square unit when $x = e$. In the next section, we will see that the number e can be calculated as a limit and has the numerical value $e \approx 2.718281828459045$ to 15 decimal places.

The Derivative of $y = \ln x$

By the first part of the Fundamental Theorem of Calculus (Section 5.4),

$$\frac{d}{dx}\ln x = \frac{d}{dx}\int_1^x \frac{1}{t}\,dt = \frac{1}{x},$$

so we have

$$\frac{d}{dx}\ln x = \frac{1}{x}, \qquad x > 0. \tag{1}$$

Therefore, the function $y = \ln x$ is a solution to the initial value problem $dy/dx = 1/x$, $x > 0$, with $y(1) = 0$. Notice that the derivative is always positive.

If u is a differentiable function of x whose values are positive, so that $\ln u$ is defined, then applying the Chain Rule we obtain

$$\frac{d}{dx}\ln u = \frac{1}{u}\frac{du}{dx}, \qquad u > 0. \tag{2}$$

EXAMPLE 1 We use Equations (1) and (2) to find derivatives.

(a) $\frac{d}{dx}\ln 2x = \frac{1}{2x}\frac{d}{dx}(2x) = \frac{1}{2x}\cdot 2 = \frac{1}{x}, \quad x > 0$

(b) Equation (2) with $u = x^2 + 3$ gives

$$\frac{d}{dx}\ln(x^2+3) = \frac{1}{x^2+3}\cdot\frac{d}{dx}(x^2+3) = \frac{1}{x^2+3}\cdot 2x = \frac{2x}{x^2+3}.$$

(c) Using the Chain Rule and Equation (1), we find

$$\frac{d}{dx}(\ln x)^4 = 4(\ln x)^3\cdot\frac{d}{dx}(\ln x) = 4(\ln x)^3\cdot\frac{1}{x} = \frac{4(\ln x)^3}{x}, \quad x > 0.$$

(d) Equation (2) with $u = |x|$ gives an important derivative:

$$\begin{aligned}
\frac{d}{dx}\ln|x| &= \frac{d}{du}\ln u\cdot\frac{du}{dx} && u = |x|,\ x \neq 0\\
&= \frac{1}{u}\cdot\frac{x}{|x|} && \frac{d}{dx}(|x|) = \frac{x}{|x|}\ \text{(Example 6, Section 3.6)}\\
&= \frac{1}{|x|}\cdot\frac{x}{|x|} && \text{Substitute for } u.\\
&= \frac{x}{x^2}\\
&= \frac{1}{x}.
\end{aligned}$$

So $1/x$ is the derivative of $\ln x$ on the domain $x > 0$, and the derivative of $\ln(-x)$ on the domain $x < 0$. ■

Derivative of $\ln|x|$

$$\frac{d}{dx}\ln|x| = \frac{1}{x}, \ x \neq 0 \tag{3}$$

Notice from Example 1a that the function $y = \ln 2x$ has the same derivative as the function $y = \ln x$. This is true of $y = \ln bx$ for any constant b, provided that $bx > 0$:

$$\frac{d}{dx}\ln bx = \frac{1}{bx}\cdot\frac{d}{dx}(bx) = \frac{1}{bx}(b) = \frac{1}{x}.$$

Properties of Logarithms

HISTORICAL BIOGRAPHY
John Napier
(1550–1617)
bit.ly/2IsZUN5

Logarithms, invented by John Napier, were the single most important improvement in arithmetic calculation before the modern electronic computer. The properties of logarithms reduce multiplication of positive numbers to addition of their logarithms, division of positive numbers to subtraction of their logarithms, and exponentiation of a number to multiplication of its logarithm by the exponent.

THEOREM 2—Algebraic Properties of the Natural Logarithm
For any numbers $b > 0$ and $x > 0$, the natural logarithm satisfies the following rules:

1. *Product Rule:*	$\ln bx = \ln b + \ln x$	
2. *Quotient Rule:*	$\ln \dfrac{b}{x} = \ln b - \ln x$	
3. *Reciprocal Rule:*	$\ln \dfrac{1}{x} = -\ln x$	Rule 2 with $b = 1$
4. *Power Rule:*	$\ln x^r = r \ln x$	For r rational

For now we consider only rational exponents in Rule 4. In Section 7.3 we will see that the rule holds for all real exponents as well.

EXAMPLE 2 We use the properties in Theorem 2 to rewrite three expressions.

(a) $\ln 4 + \ln \sin x = \ln(4 \sin x)$ — Product Rule

(b) $\ln \dfrac{x+1}{2x-3} = \ln(x+1) - \ln(2x-3)$ — Quotient Rule

(c) $\ln \dfrac{1}{8} = -\ln 8$ — Reciprocal Rule

$= -\ln 2^3 = -3 \ln 2$ — Power Rule ■

We now give the proof of Theorem 2. The properties are proved by applying Corollary 2 of the Mean Value Theorem to each rule.

Proof that $\ln bx = \ln b + \ln x$ The argument starts by observing that $\ln bx$ and $\ln x$ have the same derivative:

$$\frac{d}{dx}\ln(bx) = \frac{b}{bx} = \frac{1}{x} = \frac{d}{dx}\ln x.$$

According to Corollary 2 of the Mean Value Theorem, the functions must differ by a constant, which means that

$$\ln bx = \ln x + C$$

for some constant C.

Since this last equation holds for all positive values of x, it must hold for $x = 1$. Hence,

$$\begin{aligned} \ln(b \cdot 1) &= \ln 1 + C \\ \ln b &= 0 + C \qquad \ln 1 = 0 \\ C &= \ln b. \end{aligned}$$

By substituting we conclude that

$$\ln bx = \ln b + \ln x.$$ ■

Proof that $\ln x^r = r \ln x$ (assuming r rational) We use the same-derivative argument again. For all positive values of x,

$$\begin{aligned} \frac{d}{dx} \ln x^r &= \frac{1}{x^r} \frac{d}{dx}(x^r) && \text{Eq. (2) with } u = x^r \\ &= \frac{1}{x^r} r x^{r-1} && \text{General Power Rule for derivatives, } r \text{ rational} \\ &= r \cdot \frac{1}{x} = \frac{d}{dx}(r \ln x). \end{aligned}$$

Since $\ln x^r$ and $r \ln x$ have the same derivative,

$$\ln x^r = r \ln x + C$$

for some constant C. Taking x to be 1 identifies C as zero, and we're done. (Exercise 48 in Section 3.7 indicates a proof of the General Power Rule for derivatives when r is rational.) ■

You are asked to prove Rule 2 in Exercise 90. Rule 3 is a special case of Rule 2, obtained by setting $b = 1$ and noting that $\ln 1 = 0$. This covers all cases of Theorem 2.

We have not yet proved Rule 4 for r irrational; however, the rule does hold for all r, rational or irrational. We will show this in the next section after we define exponential functions and irrational exponents.

The Graph and Range of ln x

The derivative $d(\ln x)/dx = 1/x$ is positive for $x > 0$, so $\ln x$ is an increasing function of x. The second derivative, $-1/x^2$, is negative, so the graph of $\ln x$ is concave down. (See Figure 7.9a.)

We can estimate the value of $\ln 2$ by considering the area under the graph of $y = 1/x$ and above the interval $[1, 2]$. In Figure 7.9b a rectangle of height 1/2 over the interval $[1, 2]$ fits under the graph. Therefore, the area under the graph, which is $\ln 2$, is greater than the area of the rectangle, which is 1/2. So $\ln 2 > 1/2$. Knowing this we have

$$\ln 2^n = n \ln 2 > n\left(\frac{1}{2}\right) = \frac{n}{2}.$$

This result shows that $\ln(2^n) \to \infty$ as $n \to \infty$. Since $\ln x$ is an increasing function, it follows that

$$\lim_{x \to \infty} \ln x = \infty. \qquad \ln x \text{ is increasing and not bounded above.}$$

We also have

$$\lim_{x \to 0^+} \ln x = \lim_{t \to \infty} \ln \frac{1}{t} = \lim_{t \to \infty} (-\ln t) = -\infty. \qquad x = 1/t$$

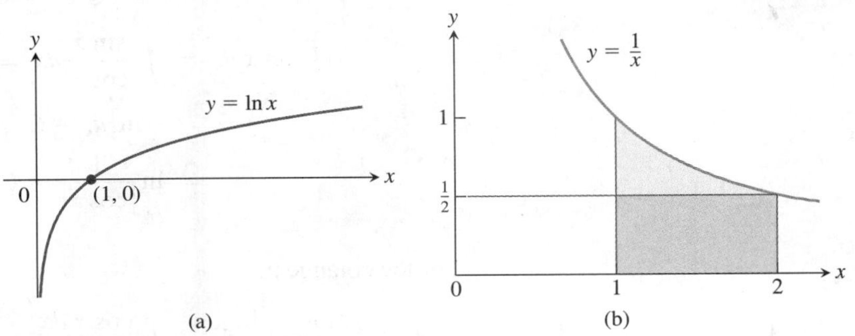

FIGURE 7.9 (a) The graph of the natural logarithm. (b) The rectangle of height $y = 1/2$ fits beneath the graph of $y = 1/x$ for the interval $1 \leq x \leq 2$.

We defined $\ln x$ for $x > 0$, so the domain of $\ln x$ is the set of positive real numbers. The above discussion and the Intermediate Value Theorem show that its range is the entire real line, giving the familiar graph of $y = \ln x$ shown in Figure 7.9a.

The Integral $\int 1/x\, dx$

Equation (3) leads to the following integral formula:

$$\int \frac{1}{x}\, dx = \ln|x| + C. \tag{4}$$

If u is a differentiable function that is never zero, then

$$\int \frac{1}{u}\, du = \ln|u| + C. \tag{5}$$

Equation (5) applies anywhere on the domain of $1/u$, which is the set of points where $u \neq 0$. It says that integrals that have the form $\int \frac{du}{u}$ lead to logarithms. Whenever $u = f(x)$ is a differentiable function that is never zero, we have that $du = f'(x)\, dx$ and

$$\int \frac{f'(x)}{f(x)}\, dx = \ln|f(x)| + C.$$

EXAMPLE 3 We rewrite an integral so that it has the form $\int \frac{du}{u}$.

$$\begin{aligned}\int_{-\pi/2}^{\pi/2} \frac{4\cos\theta}{3 + 2\sin\theta}\, d\theta &= \int_1^5 \frac{2}{u}\, du \qquad \begin{array}{l} u = 3 + 2\sin\theta,\quad du = 2\cos\theta\, d\theta, \\ u(-\pi/2) = 1,\quad u(\pi/2) = 5 \end{array}\\ &= 2\ln|u|\Big]_1^5 \\ &= 2\ln|5| - 2\ln|1| = 2\ln 5\end{aligned}$$

Note that $u = 3 + 2\sin\theta$ is always positive on $[-\pi/2, \pi/2]$, so Equation (5) applies. ■

The Integrals of $\tan x$, $\cot x$, $\sec x$, and $\csc x$

Equation (3) tells us how to integrate these trigonometric functions.

$$\begin{aligned}\int \tan x\, dx &= \int \frac{\sin x}{\cos x}\, dx = \int \frac{-du}{u} \qquad \begin{array}{l} u = \cos x \\ du = -\sin x\, dx \end{array}\\ &= -\ln|u| + C = -\ln|\cos x| + C \\ &= \ln\frac{1}{|\cos x|} + C = \ln|\sec x| + C. \qquad \text{Reciprocal Rule}\end{aligned}$$

For the cotangent,

$$\begin{aligned}\int \cot x\, dx &= \int \frac{\cos x\, dx}{\sin x} = \int \frac{du}{u} \qquad \begin{array}{l} u = \sin x, \\ du = \cos x\, dx \end{array}\\ &= \ln|u| + C = \ln|\sin x| + C = -\ln|\csc x| + C.\end{aligned}$$

To integrate sec x, we multiply and divide by $(\sec x + \tan x)$ as an algebraic form of 1.

$$\int \sec x\,dx = \int \sec x \frac{(\sec x + \tan x)}{(\sec x + \tan x)}\,dx = \int \frac{\sec^2 x + \sec x \tan x}{\sec x + \tan x}\,dx$$

$$= \int \frac{du}{u} = \ln|u| + C = \ln|\sec x + \tan x| + C \qquad \begin{aligned} u &= \sec x + \tan x,\\ du &= (\sec x \tan x + \sec^2 x)\,dx \end{aligned}$$

For csc x, we multiply and divide by $(\csc x + \cot x)$.

$$\int \csc x\,dx = \int \csc x \frac{(\csc x + \cot x)}{(\csc x + \cot x)}\,dx = \int \frac{\csc^2 x + \csc x \cot x}{\csc x + \cot x}\,dx$$

$$= \int \frac{-du}{u} = -\ln|u| + C = -\ln|\csc x + \cot x| + C \qquad \begin{aligned} u &= \csc x + \cot x,\\ du &= (-\csc x \cot x - \csc^2 x)\,dx \end{aligned}$$

In summary, we have the following results.

Integrals of the tangent, cotangent, secant, and cosecant functions

$$\int \tan x\,dx = \ln|\sec x| + C \qquad \int \sec x\,dx = \ln|\sec x + \tan x| + C$$

$$\int \cot x\,dx = \ln|\sin x| + C \qquad \int \csc x\,dx = -\ln|\csc x + \cot x| + C$$

EXAMPLE 4

$$\int_0^{\pi/6} \tan 2x\,dx = \int_0^{\pi/3} \tan u \frac{du}{2} = \frac{1}{2}\int_0^{\pi/3} \tan u\,du \qquad \begin{aligned} &\text{Substitute } u = 2x,\\ &dx = du/2,\\ &u(0) = 0,\\ &u(\pi/6) = \pi/3 \end{aligned}$$

$$= \frac{1}{2}\ln|\sec u|\Big]_0^{\pi/3} = \frac{1}{2}(\ln 2 - \ln 1) = \frac{1}{2}\ln 2$$

■

Logarithmic Differentiation

The derivatives of positive functions given by formulas that involve products, quotients, and powers can often be found more quickly if we take the natural logarithm of both sides before differentiating. This enables us to use the laws of logarithms to simplify the formulas before differentiating. The process, called **logarithmic differentiation**, is illustrated in the next example.

EXAMPLE 5 Find dy/dx if

$$y = \frac{(x^2 + 1)(x + 3)^{1/2}}{x - 1}, \qquad x > 1.$$

Solution We take the natural logarithm of both sides and simplify the result with the algebraic properties of logarithms:

$$\ln y = \ln \frac{(x^2 + 1)(x + 3)^{1/2}}{x - 1}$$

$$= \ln\left((x^2 + 1)(x + 3)^{1/2}\right) - \ln(x - 1) \qquad \text{Quotient Rule}$$

$$= \ln(x^2 + 1) + \ln(x + 3)^{1/2} - \ln(x - 1) \qquad \text{Product Rule}$$

$$= \ln(x^2 + 1) + \frac{1}{2}\ln(x + 3) - \ln(x - 1). \qquad \text{Power Rule}$$

We then take derivatives of both sides with respect to x, using Equation (2):

$$\frac{1}{y}\frac{dy}{dx} = \frac{1}{x^2+1}\cdot 2x + \frac{1}{2}\cdot\frac{1}{x+3} - \frac{1}{x-1}.$$

Next we solve for dy/dx:

$$\frac{dy}{dx} = y\left(\frac{2x}{x^2+1} + \frac{1}{2x+6} - \frac{1}{x-1}\right).$$

Finally, we substitute for y:

$$\frac{dy}{dx} = \frac{(x^2+1)(x+3)^{1/2}}{x-1}\left(\frac{2x}{x^2+1} + \frac{1}{2x+6} - \frac{1}{x-1}\right).$$ ■

The computation in Example 5 would be much longer if we used the product, quotient, and power rules.

EXERCISES 7.2

Using the Algebraic Properties—Theorem 2

1. Express the following logarithms in terms of ln 2 and ln 3.
   a. $\ln 1.5$ **b.** $\ln(4/9)$ **c.** $\ln(1/2)$
   d. $\ln\sqrt[3]{9}$ **e.** $\ln 3\sqrt{2}$ **f.** $\ln\sqrt{13.5}$
2. Express the following logarithms in terms of ln 5 and ln 7.
   a. $\ln(1/125)$ **b.** $\ln 9.8$ **c.** $\ln 7\sqrt{7}$
   d. $\ln 1225$ **e.** $\ln 0.056$
   f. $(\ln 35 + \ln(1/7))/(\ln 25)$

Use the properties of logarithms to write the expressions in Exercises 3 and 4 as a single term.

3. **a.** $\ln\sin\theta - \ln\left(\frac{\sin\theta}{5}\right)$ **b.** $\ln\frac{\sqrt{3x-5}}{7}$
   c. $\frac{1}{2}\ln(4t^4) - \ln 2$
4. **a.** $\ln\sec\theta + \ln\cos\theta$
   b. $\ln(8x+4) - 2\ln 2$
   c. $3\ln\sqrt[3]{t^2-1} - \ln(t+1)$

In Exercises 5 and 6, solve for t.

5. $7 + 3\ln t = 15$
6. $\ln(t-2) = \ln 8 - \ln t$

Finding Derivatives

In Exercises 7–38, find the derivative of y with respect to x, t, or θ, as appropriate.

7. $y = \ln 5x$
8. $y = \ln kx$, k constant
9. $y = \ln(2t^2)$
10. $y = \ln(t^{3/2})$
11. $y = \ln\frac{3}{x}$
12. $y = \ln\left(\frac{5}{x}\right)$
13. $y = \ln(\theta+1)$
14. $y = \ln(2\theta+2)$
15. $y = \ln 2x^3$
16. $y = (\ln x)^3$
17. $y = t(\ln t)^2$
18. $y = 2t\sqrt{\log 2t}$
19. $y = \frac{x^4}{4}\ln x - \frac{x^4}{16}$
20. $y = (x\ln x)^3$
21. $y = \frac{\ln t}{t}$
22. $y = \frac{5+\ln t}{2t}$
23. $y = \frac{\ln x}{1+\ln x}$
24. $y = \frac{x\ln x}{1+\ln x}$
25. $y = 2\ln(\ln 2x)$
26. $y = \ln(\ln(\ln x))$
27. $y = \theta(\sin(\ln\theta) + \cos(\ln\theta))$
28. $y = \ln(\operatorname{cosec}\theta + \cot\theta)$
29. $y = \ln\frac{1}{x\sqrt{x+1}}$
30. $y = \frac{1}{2}\ln\frac{1+x}{1-x}$
31. $y = \frac{5+\ln t}{5-\ln t}$
32. $y = \sqrt{\ln\sqrt{t}}$
33. $y = \ln(\sec(\ln\theta))$
34. $y = \ln\left(\frac{\sqrt{\sin\theta\cos\theta}}{1+2\ln\theta}\right)$
35. $y = \ln\left(\frac{x^2+3}{\sqrt{3-x}}\right)$
36. $y = \ln\sqrt{\frac{(x+1)^5}{(x+2)^{20}}}$
37. $y = \int_{x^2/2}^{x^2}\ln\sqrt{t}\,dt$
38. $y = \int_{\sqrt{x}}^{\sqrt[3]{x}}\ln t\,dt$

Evaluating Integrals

Evaluate the integrals in Exercises 39–56.

39. $\int_{-3}^{-2}\frac{dx}{2x}$
40. $\int_{-1}^{0}\frac{3\,dx}{3x-2}$
41. $\int\frac{2y\,dy}{y^2-25}$
42. $\int\frac{5r\,dr}{3r^2-4}$
43. $\int_0^{\pi}\frac{\sin t}{2-\cos t}\,dt$
44. $\int_0^{\pi/3}\frac{4\sin\theta}{1-4\cos\theta}\,d\theta$
45. $\int_1^2\frac{5\ln x\,dx}{5x}$
46. $\int_2^4\frac{dx}{x\ln x}$

47. $\int_2^4 \frac{dx}{x(\ln x)^2}$

48. $\int_2^{16} \frac{dx}{3x\sqrt{\ln x}}$

49. $\int \frac{3\sec^2 t}{6 + 3\tan t}\,dt$

50. $\int \frac{\sec y \tan y}{2 + \sec y}\,dy$

51. $\int_0^{\pi/2} \tan\frac{x}{3}\,dx$

52. $\int_{\pi/4}^{\pi/2} \cot t\,dt$

53. $\int_{\pi/2}^{\pi} 2\cot\frac{\theta}{3}\,d\theta$

54. $\int_0^{\pi/12} 6\tan 3x\,dx$

55. $\int \frac{dx}{5\sqrt{x}+5x}$

56. $\int \frac{\sec x\,dx}{\sqrt{\ln(\sec x + \tan x)}}$

Logarithmic Differentiation

In Exercises 57–70, use logarithmic differentiation to find the derivative of y with respect to the given independent variable.

57. $y = \sqrt{25 - x^2}$

58. $y = \sqrt{(x^2+1)(x-1)^2}$

59. $y = \sqrt{\frac{t}{t+1}}$

60. $y = \sqrt{\frac{t^2+1}{t^2-1}}$

61. $y = \sqrt{\theta + 3}\sin\theta$

62. $y = (\tan\theta)\sqrt{2\theta + 1}$

63. $y = t(t+1)(t+2)$

64. $y = \frac{(3x+5)^5}{(4x-9)^7}$

65. $y = \frac{\theta + 5}{\theta\cos\theta}$

66. $y = \frac{\theta\sin\theta}{\sqrt{\sec\theta}}$

67. $y = \sqrt{\frac{x-1}{x^4+1}}$

68. $y = \sqrt{\frac{(x+1)^{10}}{(2x+1)^5}}$

69. $y = \sqrt[3]{\frac{x(x-2)}{x^2+1}}$

70. $y = \frac{x^5}{(1-10x)\sqrt{x^2+2}}$

Theory and Applications

71. Locate and identify the absolute extreme values of

 a. $\ln(\cos x)$ on $[-\pi/4, \pi/3]$,

 b. $\cos(\ln x)$ on $[1/2, 2]$.

72. **a.** Prove that $f(x) = x - \ln x$ is increasing for $x > 1$.

 b. Using part (a), show that $\ln x < x$ if $x > 1$.

73. Find the area between the curves $y = \ln x$ and $y = \ln 3x$ from $x = 2$ to $x = 5$.

74. Find the area between the curve $y = \tan x$ and the x-axis from $x = -\pi/4$ to $x = \pi/3$.

In Exercises 75 and 76:

 a. Find the open intervals on which the function is increasing and decreasing.

 b. Identify the function's local and absolute extreme values, if any, saying where they occur.

75. $g(x) = x(\ln x)^2$

76. $g(x) = x^2 - 2x - 4\ln x$

77. The region in the first quadrant bounded by the coordinate axes, the line $y = 4$, and the curve $x = \frac{2}{\sqrt{2y+1}}$ is revolved about the y-axis to generate a solid. Find the volume of the solid.

78. The region between the curve $y = \sqrt{\cot x}$ and the x-axis from $x = \pi/6$ to $x = \pi/2$ is revolved about the x-axis to generate a solid. Find the volume of the solid.

79. The region between the curve $y = 1/x^2$ and the x-axis from $x = 1/2$ to $x = 2$ is revolved about the y-axis to generate a solid. Find the volume of the solid.

80. In Section 6.2, Exercise 6, we revolved about the y-axis the region between the curve $y = 9x/\sqrt{x^3+9}$ and the x-axis from $x = 0$ to $x = 3$ to generate a solid of volume 36π. What volume do you get if you revolve the region about the x-axis instead? (See Section 6.2, Exercise 6, for a graph.)

81. Find the lengths of the following curves.

 a. $y = (x^2/8) - \ln x,\quad 4 \le x \le 8$

 b. $x = (y/4)^2 - 2\ln(y/4),\quad 4 \le y \le 12$

82. Find a curve through the point (1, 0) whose length from $x = 1$ to $x = 2$ is

$$L = \int_1^2 \sqrt{1 + \frac{1}{x^2}}\,dx.$$

T 83. **a.** Find the centroid of the region between the curve $y = 1/x$ and the x-axis from $x = 1$ to $x = 2$. Give the coordinates to two decimal places.

 b. Sketch the region and show the centroid in your sketch.

84. **a.** Find the center of mass of a thin plate of constant density covering the region between the curve $y = 1/\sqrt{x}$ and the x-axis from $x = 1$ to $x = 16$.

 b. Find the center of mass if, instead of being constant, the density function is $\delta(x) = 4/\sqrt{x}$.

85. Use a derivative to show that $f(x) = \ln(x^3 - 1)$ is one-to-one.

86. Use a derivative to show that $g(x) = \sqrt{x^2 + \ln x}$ is one-to-one.

Solve the initial value problems in Exercises 87 and 88.

87. $\frac{dy}{dx} = 2 + \frac{1}{x},\quad y(1) = 3$

88. $\frac{d^2y}{dx^2} = \sec^2 x,\quad y(0) = 0$ and $y'(0) = 1$

T 89. **The linearization of $\ln(1+x)$ at $x = 0$** Instead of approximating $\ln x$ near $x = 1$, we approximate $\ln(1+x)$ near $x = 0$. We get a simpler formula this way.

 a. Derive the linearization $\ln(1+x) \approx x$ at $x = 0$.

 b. Estimate to five decimal places the error involved in replacing $\ln(1+x)$ by x on the interval $[0, 0.1]$.

 c. Graph $\ln(1+x)$ and x together for $0 \le x \le 0.5$. Use different colors, if available. At what points does the approximation of $\ln(1+x)$ seem best? Least good? By reading coordinates from the graphs, find as good an upper bound for the error as your grapher will allow.

90. Use the same-derivative argument, as was done to prove Rules 1 and 4 of Theorem 2, to prove the Quotient Rule property of logarithms.

T 91. **a.** Graph $y = \sin x$ and the curves $y = \ln(a + \sin x)$ for $a = 2, 4, 8, 20$, and 50 together for $0 \le x \le 23$.

 b. Why do the curves flatten as a increases? (*Hint:* Find an a-dependent upper bound for $|y'|$.)

T 92. Does the graph of $y = \sqrt{x} - \ln x$, $x > 0$, have an inflection point? Try to answer this question **(a)** by graphing, **(b)** by using calculus.

7.3 Exponential Functions

Having developed the theory of the function $\ln x$, we introduce its inverse, the exponential function $\exp x = e^x$. We study its properties and compute its derivative and integral. We prove the power rule for derivatives involving general real exponents. Finally, we introduce general exponential functions, a^x, and general logarithmic functions, $\log_a x$.

The Inverse of $\ln x$ and the Number e

The function $\ln x$, being an increasing function of x with domain $(0, \infty)$ and range $(-\infty, \infty)$, has an inverse $\ln^{-1} x$ with domain $(-\infty, \infty)$ and range $(0, \infty)$. The graph of $\ln^{-1} x$ is the graph of $\ln x$ reflected across the line $y = x$. As you can see in Figure 7.10,

$$\lim_{x\to\infty} \ln^{-1} x = \infty \quad \text{and} \quad \lim_{x\to -\infty} \ln^{-1} x = 0.$$

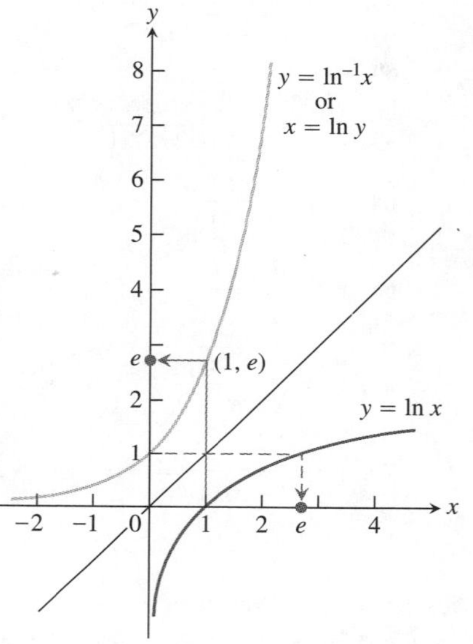

FIGURE 7.10 The graphs of $y = \ln x$ and $y = \ln^{-1} x = \exp x$. The number e is $\ln^{-1} 1 = \exp(1)$.

The inverse function $\ln^{-1} x$ is also denoted by $\exp x$. We have not yet established that $\exp x$ is an exponential function, only that $\exp x$ is the inverse of the function $\ln x$. We will now show that $\ln^{-1} x = \exp x$ is, in fact, the exponential function with base e.

The number e was defined to satisfy the equation $\ln(e) = 1$, so $e = \exp(1)$. We can raise the number e to a rational power r using algebra:

$$e^2 = e \cdot e, \qquad e^{-2} = \frac{1}{e^2}, \qquad e^{1/2} = \sqrt{e}, \qquad e^{2/3} = \sqrt[3]{e^2},$$

and so on. Since e is positive, e^r is positive too, so we can take the logarithm of e^r. When we do, we find that for r rational

$$\ln e^r = r \ln e = r \cdot 1 = r. \qquad \text{Theorem 2, Rule 4}$$

Then applying the function $\ln^{-1}$ to both sides of the equation $\ln e^r = r$, we find that

$$e^r = \exp r \qquad \text{for } r \text{ rational.} \qquad \text{exp is } \ln^{-1}. \tag{1}$$

Thus $\exp r$ coincides with the exponential function e^r for all rational values of r. We have not yet found a way to give an exact meaning to e^x for x irrational, but we can use Equation (1) to do so. The function $\exp x = \ln^{-1} x$ has domain $(-\infty, \infty)$, so it is defined for every x. We have $\exp r = e^r$ for r rational by Equation (1), and we now define e^x to equal $\exp x$ for all x.

The notations $\ln^{-1} x$, $\exp x$, and e^x all refer to the natural exponential function.

DEFINITION For every real number x, we define the **natural exponential function** to be $e^x = \exp x$.

Typical values of e^x

x	e^x (rounded)
−1	0.37
0	1
1	2.72
2	7.39
10	22026
100	2.6881×10^{43}

For the first time we have a precise meaning for a number raised to an irrational power. Usually the exponential function is denoted by e^x rather than $\exp x$. Since $\ln x$ and e^x are inverses of one another, we have the following relations.

Inverse Equations for e^x and $\ln x$

$$e^{\ln x} = x \qquad (\text{all } x > 0)$$
$$\ln(e^x) = x \qquad (\text{all } x)$$

EXAMPLE 1 Solve the equation $e^{2x-6} = 4$ for x.

Solution We take the natural logarithm of both sides of the equation and use the second inverse equation:

$$\begin{aligned} \ln(e^{2x-6}) &= \ln 4 \\ 2x - 6 &= \ln 4 && \text{Inverse relationship} \\ 2x &= 6 + \ln 4 \\ x &= 3 + \frac{1}{2}\ln 4 = 3 + \ln 4^{1/2} \\ x &= 3 + \ln 2. \end{aligned}$$

■

EXAMPLE 2 A line with slope m passes through the origin and is tangent to the graph of $y = \ln x$. What is the value of m?

Solution Suppose the point of tangency occurs at the unknown point $x = a > 0$. Then we know that the point $(a, \ln a)$ lies on the graph and that the tangent line at that point has slope $m = 1/a$ (Figure 7.11). Since the tangent line passes through the origin, its slope is

$$m = \frac{\ln a - 0}{a - 0} = \frac{\ln a}{a}.$$

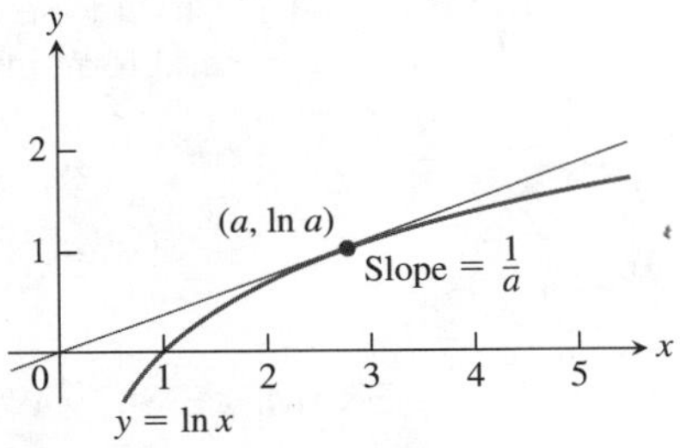

FIGURE 7.11 The tangent line intersects the curve at some point $(a, \ln a)$, where the slope of the curve is $1/a$ (Example 2).

Setting these two formulas for m equal to each other, we have

$$\begin{aligned}
\frac{\ln a}{a} &= \frac{1}{a} \\
\ln a &= 1 \\
e^{\ln a} &= e^1 \\
a &= e \\
m &= \frac{1}{e}.
\end{aligned}$$

■

The Derivative and Integral of e^x

According to Theorem 1, the natural exponential function is differentiable because it is the inverse of a differentiable function whose derivative is never zero. We calculate its derivative using the inverse relationship and the Chain Rule:

$$\begin{aligned}
\ln(e^x) &= x && \text{Inverse relationship} \\
\frac{d}{dx}\ln(e^x) &= 1 && \text{Differentiate both sides.} \\
\frac{1}{e^x}\cdot\frac{d}{dx}(e^x) &= 1 && \text{Eq. (2), Section 7.2, with } u = e^x \\
\frac{d}{dx}e^x &= e^x. && \text{Solve for the derivative.}
\end{aligned}$$

That is, for $y = e^x$, we find that $dy/dx = e^x$ so the natural exponential function e^x is its own derivative.

Derivative of the Natural Exponential Function

$$\frac{d}{dx}e^x = e^x \tag{2}$$

If $f(x) = e^x$, then $f'(0) = e^0 = 1$. This means that the natural exponential function e^x has slope 1 as it crosses the y-axis at $x = 0$.

The Chain Rule extends the derivative result for the natural exponential function to a more general form involving a function $u(x)$: If u is any differentiable function of x, then

$$\frac{d}{dx}e^u = e^u\frac{du}{dx}. \tag{3}$$

EXAMPLE 3 We find derivatives of the exponential using Equations (2) and (3).

(a) $\frac{d}{dx}(5e^x) = 5\frac{d}{dx}e^x = 5e^x$

(b) $\frac{d}{dx}e^{-x} = e^{-x}\frac{d}{dx}(-x) = e^{-x}(-1) = -e^{-x}$ Eq. (3) with $u = -x$

(c) $\frac{d}{dx}e^{\sin x} = e^{\sin x}\frac{d}{dx}(\sin x) = e^{\sin x}\cdot\cos x$ Eq. (3) with $u = \sin x$

(d) $\frac{d}{dx}\left(e^{\sqrt{3x+1}}\right) = e^{\sqrt{3x+1}}\cdot\frac{d}{dx}\left(\sqrt{3x+1}\right)$ Eq. (3) with $u = \sqrt{3x+1}$

$$= e^{\sqrt{3x+1}}\cdot\frac{1}{2}(3x+1)^{-1/2}\cdot 3 = \frac{3}{2\sqrt{3x+1}}e^{\sqrt{3x+1}}$$ ■

Since e^x is its own derivative, it is also its own antiderivative. So the integral equivalent of Equation (2) is the following.

The Antiderivative of the Exponential Function

$$\int e^x\,dx = e^x + C$$

EXAMPLE 4

(a) $\int_0^{\ln 2} e^{3x}\,dx = \int_0^{\ln 8} e^u\cdot\frac{1}{3}\,du$ $u = 3x,\ \frac{1}{3}du = dx,\ u(0) = 0,$ $u(\ln 2) = 3\ln 2 = \ln 2^3 = \ln 8$

$$= \frac{1}{3}\int_0^{\ln 8} e^u\,du$$

$$= \frac{1}{3}e^u\Big]_0^{\ln 8}$$

$$= \frac{1}{3}(8-1) = \frac{7}{3}$$

(b) $\int_0^{\pi/2} e^{\sin x}\cos x\,dx = e^{\sin x}\Big]_0^{\pi/2}$ Antiderivative from Example 3c

$$= e^1 - e^0 = e - 1$$ ■

The derivative of e^x exists and is everywhere positive, confirming that it is a continuous and increasing function as shown in Figure 7.10. Since the second derivative of e^x is also e^x and everywhere positive, the graph is concave up. Moreover, Figure 7.10 shows that the exponential function has the limits

$$\lim_{x\to-\infty} e^x = 0 \quad\text{and}\quad \lim_{x\to\infty} e^x = \infty.$$

From the first of these limits we see that the x-axis is a horizontal asymptote of the graph $y = e^x$.

Laws of Exponents

Even though e^x is defined in a seemingly roundabout way as $\ln^{-1}x$, it obeys the familiar laws of exponents from algebra. The following Theorem 3 shows us that these laws are consequences of the definitions of $\ln x$ and e^x.

THEOREM 3 For all numbers x and y, the natural exponential e^x obeys the following laws:

1. $e^x e^y = e^{x+y}$ **2.** $e^{-x} = \frac{1}{e^x}$

3. $\frac{e^x}{e^y} = e^{x-y}$ **4.** $(e^x)^r = e^{rx}$, if r is rational

Proof of Law 1 Let $a = e^x$ and $b = e^y$. Then

$$\begin{aligned} x = \ln a \quad \text{and} \quad y &= \ln b && \text{Inverse equations} \\ x + y &= \ln a + \ln b \\ &= \ln ab && \text{Product Rule for logarithms} \\ e^{x+y} &= e^{\ln ab} && \text{Exponentiate.} \\ &= ab && e^{\ln u} = u \\ &= e^x e^y. \end{aligned}$$

■

Proof of Law 4 Let $y = (e^x)^r$. Then

$$\begin{aligned} \ln y &= \ln (e^x)^r \\ &= r \ln (e^x) && \text{Power Rule for logarithms, rational } r \\ &= rx. && \ln e^u = u \text{ with } u = x \end{aligned}$$

Exponentiating each side,

$$y = e^{rx}. \qquad e^{\ln y} = y$$

■

Laws 2 and 3 follow from Law 1. Like the Power Rule for logarithms, Law 4 holds for all real numbers r.

The General Exponential Function a^x

Since $a = e^{\ln a}$ for any positive number a, we can express a^x as $(e^{\ln a})^x = e^{x \ln a}$. We therefore use the function e^x to define the other exponential functions, which allow us to raise *any* positive number to an irrational exponent.

> **DEFINITION** For any numbers $a > 0$ and x, the **exponential function with base *a*** is
>
> $$a^x = e^{x \ln a}.$$

When $a = e$, the definition gives $a^x = e^{x \ln a} = e^{x \ln e} = e^{x \cdot 1} = e^x$.

Theorem 3 is also valid for a^x, the exponential function with base a. For example,

$$\begin{aligned} a^x \cdot a^y &= e^{x \ln a} \cdot e^{y \ln a} && \text{Definition of } a^x \\ &= e^{x \ln a + y \ln a} && \text{Law 1} \\ &= e^{(x+y) \ln a} && \text{Factor } \ln a \\ &= a^{x+y}. && \text{Definition of } a^x \end{aligned}$$

In particular, $a^n \cdot a^{-1} = a^{n-1}$ for any real number n.

Proof of the Power Rule (General Version)

The definition of the general exponential function enables us to make sense of raising any positive number to a real power n, rational or irrational. That is, we can define the power function $y = x^n$ for any exponent n.

> **DEFINITION** For any $x > 0$ and for any real number n,
>
> $$x^n = e^{n \ln x}.$$

Because the logarithm and exponential functions are inverses of each other, the definition gives

$$\ln x^n = n \ln x, \qquad \text{for all real numbers } n.$$

That is, the rule for taking the natural logarithm of a power of x holds for *all* real exponents n, not just for rational exponents as previously stated in Theorem 2.

The definition of the power function also enables us to establish the derivative Power Rule for any real power n, as stated in Section 3.3.

General Power Rule for Derivatives

For $x > 0$ and any real number n,

$$\frac{d}{dx}x^n = nx^{n-1}.$$

If $x \leq 0$, then the formula holds whenever the derivative, x^n, and x^{n-1} all exist.

Proof Differentiating x^n with respect to x gives

$$\begin{aligned} \frac{d}{dx}x^n &= \frac{d}{dx}e^{n \ln x} && \text{Definition of } x^n,\ x > 0 \\ &= e^{n \ln x} \cdot \frac{d}{dx}(n \ln x) && \text{Chain Rule for } e^u, \text{ Eq. (3)} \\ &= x^n \cdot \frac{n}{x} && \text{Definition and derivative of } \ln x \\ &= nx^{n-1}. && x^n \cdot x^{-1} = x^{n-1} \end{aligned}$$

In short, whenever $x > 0$,

$$\frac{d}{dx}x^n = nx^{n-1}.$$

For $x < 0$, if $y = x^n$, y', and x^{n-1} all exist, then

$$\ln|y| = \ln|x|^n = n \ln|x|.$$

Using implicit differentiation (which *assumes* the existence of the derivative y') and Equation (3) in Section 7.2, we have

$$\frac{y'}{y} = \frac{n}{x}.$$

Solving for the derivative,

$$y' = n\frac{y}{x} = n\frac{x^n}{x} = nx^{n-1}.$$

It can be shown directly from the definition of the derivative that the derivative equals 0 when $x = 0$ and $n \geq 1$. This completes the proof of the general version of the Power Rule for all values of x. ■

The Number e Expressed as a Limit

We have defined the number e as the number for which $\ln e = 1$, or equivalently, the value $\exp(1)$. We see that e is an important constant for the logarithmic and exponential functions, but what is its numerical value? The next theorem shows one way to calculate e as a limit.

THEOREM 4—The Number e as a Limit
The number e can be calculated as the limit

$$e = \lim_{x\to 0}(1+x)^{1/x}.$$

Proof If $f(x) = \ln x$, then $f'(x) = 1/x$, so $f'(1) = 1$. But, by the definition of derivative,

$$\begin{aligned} f'(1) &= \lim_{h\to 0}\frac{f(1+h)-f(1)}{h} = \lim_{x\to 0}\frac{f(1+x)-f(1)}{x} \\ &= \lim_{x\to 0}\frac{\ln(1+x)-\ln 1}{x} = \lim_{x\to 0}\frac{1}{x}\ln(1+x) && \ln 1 = 0 \\ &= \lim_{x\to 0}\ln(1+x)^{1/x} = \ln\left[\lim_{x\to 0}(1+x)^{1/x}\right]. && \text{ln is continuous; use Theorem 10 in Chapter 2.} \end{aligned}$$

Because $f'(1) = 1$, we have

$$\ln\left[\lim_{x\to 0}(1+x)^{1/x}\right] = 1.$$

Therefore, exponentiating both sides we get (see Figure 7.12)

$$\lim_{x\to 0}(1+x)^{1/x} = e. \quad \blacksquare$$

Approximating the limit in Theorem 4 by taking x very small gives approximations to e. Its value is $e \approx 2.718281828459045$ to 15 decimal places as noted before.

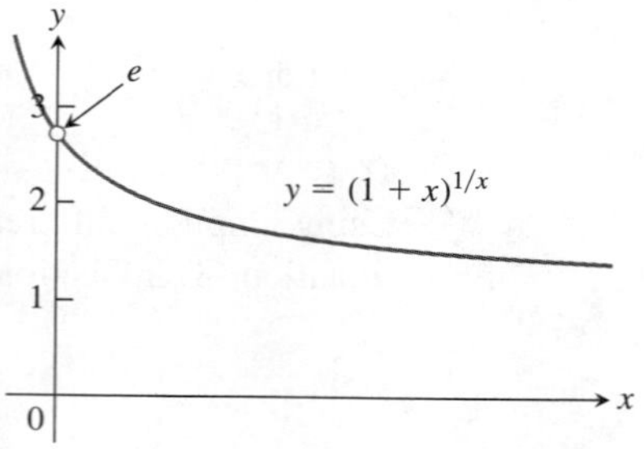

FIGURE 7.12 The number e is the limit of the function graphed here as $x \to 0$.

The Derivative of a^x

To find this derivative, we start with the equation $a^x = e^{x\ln a}$, which defines a^x for any positive a. Then we have

$$\begin{aligned} \frac{d}{dx}a^x &= \frac{d}{dx}e^{x\ln a} \\ &= e^{x\ln a}\cdot\frac{d}{dx}(x\ln a) && \frac{d}{dx}e^u = e^u\frac{du}{dx} \\ &= a^x\ln a. && \ln a \text{ is a constant.} \end{aligned}$$

That is, if $a > 0$, then a^x is differentiable and

$$\frac{d}{dx}a^x = a^x \ln a. \tag{4}$$

We now see why e^x is the exponential function preferred in calculus. If $a = e$, then $\ln a = 1$ and the derivative of a^x simplifies to

$$\frac{d}{dx}e^x = e^x \ln e = e^x. \qquad \ln e = 1$$

If $a > 0$ and u is a differentiable function of x, then, by the Chain Rule, a^u is a differentiable function of x and

$$\frac{d}{dx}a^u = a^u \ln a \frac{du}{dx}. \tag{5}$$

The integral equivalent of Equation (4) gives the antiderivative

$$\int a^x\,dx = \frac{a^x}{\ln a} + C \tag{6}$$

Combining this with the chain rule gives the general antiderivative

$$\int a^u\,du = \frac{a^u}{\ln a} + C. \tag{7}$$

From Equation (4) we see that the derivative of a^x is positive if $\ln a > 0$, or $a > 1$, and negative if $\ln a < 0$, or $0 < a < 1$. Thus, a^x is an increasing function of x if $a > 1$ and a decreasing function of x if $0 < a < 1$. In each case, a^x is one-to-one. The second derivative

$$\frac{d^2}{dx^2}(a^x) = \frac{d}{dx}(a^x \ln a) = (\ln a)^2 a^x$$

is positive for all x, so the graph of a^x is concave up on every interval of the real line. Figure 7.13 displays the graphs of several exponential functions.

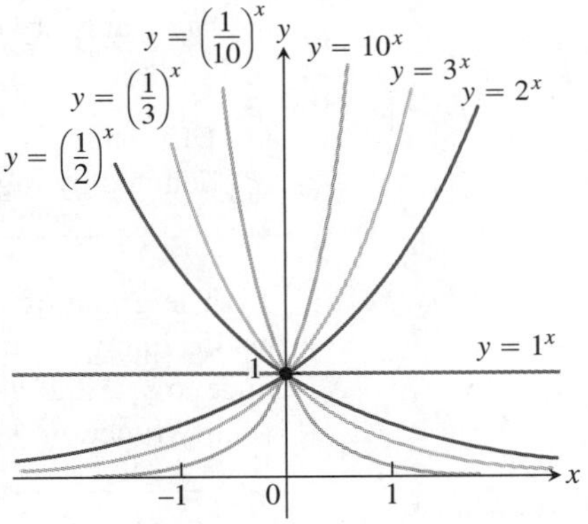

FIGURE 7.13 Exponential functions decrease if $0 < a < 1$ and increase if $a > 1$. As $x \to \infty$, we have $a^x \to 0$ if $0 < a < 1$ and $a^x \to \infty$ if $a > 1$. As $x \to -\infty$, we have $a^x \to \infty$ if $0 < a < 1$ and $a^x \to 0$ if $a > 1$.

EXAMPLE 5 We find derivatives and integrals involving general exponential functions.

(a) $\frac{d}{dx}3^x = 3^x \ln 3$ — Eq. (4) with $a = 3$

(b) $\frac{d}{dx}3^{-x} = 3^{-x}(\ln 3)\frac{d}{dx}(-x) = -3^{-x}\ln 3$ — Eq. (5) with $a = 3$, $u = -x$

(c) $\frac{d}{dx}3^{\sin x} = 3^{\sin x}(\ln 3)\frac{d}{dx}(\sin x) = 3^{\sin x}(\ln 3)\cos x$ — Eq. (5) with $a = 3$, $u = \sin x$

(d) $\int 2^x\,dx = \frac{2^x}{\ln 2} + C$ — Eq. (6) with $a = 2$

(e) $\int 2^{\sin x}\cos x\,dx = \int 2^u\,du = \frac{2^u}{\ln 2} + C$ — $u = \sin x$, $du = \cos x\,dx$, and Eq. (7)

$= \frac{2^{\sin x}}{\ln 2} + C$ — u replaced by $\sin x$ ■

EXAMPLE 6 Differentiate $f(x) = x^x$, $x > 0$.

Solution The Power Rule tells us how to differentiate a function of the form x^a, where a is a fixed real number. However, the exponent in x^x is not a fixed constant, so we cannot use the Power Rule of differentiate x^x. Equation (4) tells us how to differentiate a^x when the base a is constant. However, we cannot use that equation either, because the base x in x^x is not constant. Instead, to find the derivative of this function, we note that $f(x) = x^x = e^{x\ln x}$, so differentiation gives

$$
\begin{aligned}
f'(x) &= \frac{d}{dx}(e^{x\ln x}) && \text{The base } e \text{ is a constant}\\
&= e^{x\ln x}\frac{d}{dx}(x\ln x) && \text{Eq. (3) with } u = x\ln x\\
&= e^{x\ln x}\left(\ln x + x\cdot\frac{1}{x}\right) && \text{Product Rule}\\
&= x^x(\ln x + 1). && x > 0
\end{aligned}
$$

We can also find the derivative of $y = x^x$ using logarithmic differentiation, assuming y' exists. ■

Logarithms with Base a

If a is any positive number other than 1, the function a^x is one-to-one and has a nonzero derivative at every point. It therefore has a differentiable inverse.

> **DEFINITION** For any positive number $a \neq 1$, the **logarithm of x with base a**, denoted by $\log_a x$, is the inverse function of a^x.

The graph of $y = \log_a x$ can be obtained by reflecting the graph of $y = a^x$ across the 45° line $y = x$ (Figure 7.14). When $a = e$, we have $\log_e x =$ inverse of $e^x = \ln x$. Since $\log_a x$ and a^x are inverses of one another, composing them in either order gives the identity function.

> **Inverse Equations for a^x and $\log_a x$**
>
> $$a^{\log_a x} = x \qquad (x > 0)$$
>
> $$\log_a(a^x) = x \qquad (\text{all } x)$$

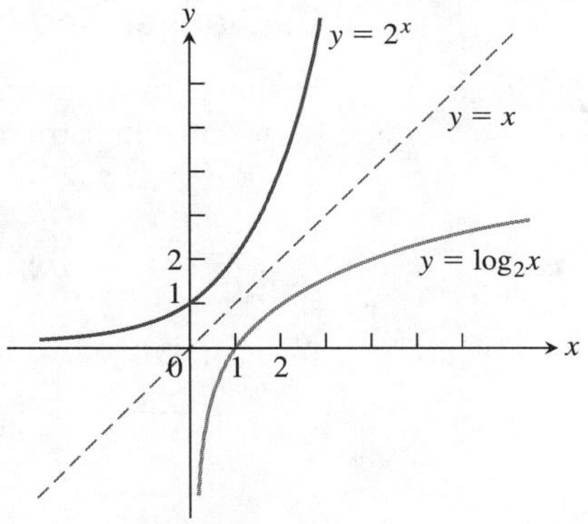

FIGURE 7.14 The graph of 2^x and its inverse, $\log_2 x$.

The function $\log_a x$ is just a numerical multiple of $\ln x$. We see this from the following derivation.

$$
\begin{aligned}
y &= \log_a x && \text{Defining equation for } y \\
a^y &= x && \text{Equivalent equation} \\
\ln a^y &= \ln x && \text{Natural log of both sides} \\
y \ln a &= \ln x && \text{Algebra Rule 4 for natural log} \\
y &= \frac{\ln x}{\ln a} && \text{Solve for } y. \\
\log_a x &= \frac{\ln x}{\ln a} && \text{Substitute for } y.
\end{aligned}
$$

It then follows that the arithmetic rules satisfied by $\log_a x$ are the same as the ones for $\ln x$. These rules, given in Table 7.2, can be proved by dividing the corresponding rules for the natural logarithm function by $\ln a$. For example,

$$
\begin{aligned}
\ln xy &= \ln x + \ln y && \text{Rule 1 for natural logarithms ...} \\
\frac{\ln xy}{\ln a} &= \frac{\ln x}{\ln a} + \frac{\ln y}{\ln a} && \text{... divided by } \ln a \text{ ...} \\
\log_a xy &= \log_a x + \log_a y. && \text{... gives Rule 1 for base } a \text{ logarithms.}
\end{aligned}
$$

TABLE 7.2 Rules for base a logarithms

For any numbers $x > 0$ and $y > 0$,

1. *Product Rule:*
   $\log_a xy = \log_a x + \log_a y$;
2. *Quotient Rule:*
   $\log_a \frac{x}{y} = \log_a x - \log_a y$;
3. *Reciprocal Rule:*
   $\log_a \frac{1}{y} = -\log_a y$;
4. *Power Rule:*
   $\log_a x^y = y \log_a x$.

Derivatives and Integrals Involving $\log_a x$

To find derivatives or integrals involving base a logarithms, we can convert them to natural logarithms. In particular, differentiating $\log_a x$ gives

$$\frac{d}{dx}(\log_a x) = \frac{d}{dx}\left(\frac{\ln x}{\ln a}\right) = \frac{1}{\ln a}\frac{d}{dx}(\ln x) = \frac{1}{\ln a}\cdot\frac{1}{x},$$

so

$$\frac{d}{dx}(\log_a x) = \frac{1}{x \ln a}.$$

If u is a positive differentiable function of x, then

$$\frac{d}{dx}(\log_a u) = \frac{1}{\ln a}\cdot\frac{1}{u}\frac{du}{dx}.$$

EXAMPLE 7 We illustrate the derivative and integral results.

(a) $$\frac{d}{dx}\log_{10}(3x+1) = \frac{1}{\ln 10}\cdot\frac{1}{3x+1}\frac{d}{dx}(3x+1) = \frac{3}{(\ln 10)(3x+1)}$$

(b) $$\int \frac{\log_2 x}{x}\,dx = \frac{1}{\ln 2}\int \frac{\ln x}{x}\,dx \qquad \log_2 x = \frac{\ln x}{\ln 2}$$

$$= \frac{1}{\ln 2}\int u\,du \qquad u = \ln x,\quad du = \frac{1}{x}\,dx$$

$$= \frac{1}{\ln 2}\frac{u^2}{2} + C = \frac{1}{\ln 2}\frac{(\ln x)^2}{2} + C = \frac{(\ln x)^2}{2\ln 2} + C$$ ■

Transcendental Numbers and Transcendental Functions

Numbers that are solutions of polynomial equations with rational coefficients are called **algebraic**: -2 is algebraic because it satisfies the equation $x + 2 = 0$, and $\sqrt{3}$ is algebraic because it satisfies the equation $x^2 - 3 = 0$. Numbers such as e and π that are not algebraic are called **transcendental**.

We call a function $y = f(x)$ algebraic if it satisfies an equation of the form

$$P_n y^n + \cdots + P_1 y + P_0 = 0$$

in which the P's are polynomials in x with rational coefficients. The function $y = 1/\sqrt{x+1}$ is algebraic because it satisfies the equation $(x+1)y^2 - 1 = 0$. Here the polynomials are $P_2 = x + 1$, $P_1 = 0$, and $P_0 = -1$. Functions that are not algebraic are called transcendental.

Summary

We have used calculus to give precise definitions of the logarithmic and exponential functions. This approach is somewhat different from our earlier treatments of the polynomial, rational, and trigonometric functions. There we first defined the function and then we studied its derivatives and integrals. Here we started with an integral from which the functions of interest were obtained. The motivation behind this approach was to address mathematical difficulties that arise when we attempt to define functions such as a^x for any real number x, rational or irrational. Defining $\ln x$ as the integral of the function $1/t$ from $t = 1$ to $t = x$ enabled us to define all of the exponential and logarithmic functions and then to derive their key algebraic and analytic properties.

EXERCISES 7.3

Solving Exponential Equations

In Exercises 1–4, solve for t.

1. **a.** $e^{-0.4t} = 256$ **b.** $e^{kt} = \frac{1}{2}$ **c.** $e^{(\ln 0.2)t} = 0.4$
2. **a.** $e^{-0.001} = 2000$ **b.** $e^{kt} = \frac{1}{10}$ **c.** $e^{(\ln 2)t} = \frac{1}{2}$
3. $e^{\sqrt{t}} = x^2$
4. $e^x e^{x^2} = e^t$
5. $e^{2t} - 3e^t = 0$
6. $e^{-2t} + 6 = 5e^{-t}$

Finding Derivatives

In Exercises 7–26, find the derivative of y with respect to x, t, or θ, as appropriate.

7. $y = e^{-2x}$
8. $y = e^{2x/3}$
9. $y = e^{1-3x}$
10. $y = e^{(4\sqrt{x}+x^2)}$
11. $y = 4xe^{4x} - e^{4x}$
12. $y = (1 + 2x)e^{-2x}$
13. $y = (x^2 - 2x + 2)e^x$
14. $y = (6x^2 - 9x + 2)e^{5x}$

15. $y = e^{\theta}(\sin\theta + \cos\theta)$

16. $y = \ln(3\theta e^{-\theta})$

17. $y = \sin(e^{-\theta^2})$

18. $y = \theta^3 e^{-2\theta}\cos 5\theta$

19. $y = \ln(3te^{-t})$

20. $y = \ln(3e^{-t}\cos t)$

21. $y = \ln\left(\dfrac{e^{\theta}}{1 + e^{\theta}}\right)$

22. $y = \ln\left(\dfrac{\sqrt{\theta}}{1 + \sqrt{\theta}}\right)$

23. $y = e^{-(\sin t + \ln t)}$

24. $y = e^{\sin t}(\ln t^2 + 1)$

25. $y = \int_0^{\ln x} \sin e^t \, dt$

26. $y = \int_{e^{4\sqrt{x}}}^{e^{2x}} \ln t \, dt$

In Exercises 27–32, find dy/dx.

27. $\ln y = e^y \sin x$

28. $\ln xy = e^{x+y}$

29. $e^{2x} = \sin(x + 3y)$

30. $\cot y = e^x + \ln x$

31. $3 + \sin y = y - x^3$

32. $\ln y = xe^y - 2$

Finding Integrals

Evaluate the integrals in Exercises 33–54.

33. $\int (e^{2x} + 4e^{-x})\,dx$

34. $\int (2e^x - 3e^{-2x})\,dx$

35. $\int_{\ln 2}^{\ln 3} e^x \, dx$

36. $\int_{-\ln 2}^{0} e^{-x}\,dx$

37. $\int 5e^{(2x+1)}\,dx$

38. $\int 2e^{(2x-1)}\,dx$

39. $\int_{\ln 4}^{\ln 9} e^{x/2}\,dx$

40. $\int_0^{\ln 16} e^{x/2}\,dx$

41. $\int \dfrac{e^{\sqrt{r}}}{\sqrt{r}}\,dr$

42. $\int \dfrac{e^{-\sqrt{r}}}{\sqrt{r}}\,dr$

43. $\int 2te^{-t^2}\,dt$

44. $\int 2t^3 e^{2t^4}\,dt$

45. $\int \dfrac{e^{1/x}}{x^2}\,dx$

46. $\int \dfrac{e^{-1/x^2}}{x^3}\,dx$

47. $\int_0^{\pi/4} (1 + e^{\tan\theta})\sec^2\theta\,d\theta$

48. $\int_{\pi/4}^{\pi/2} (1 + e^{\cot\theta})\csc^2\theta\,d\theta$

49. $\int e^{\sec \pi t}\sec \pi t \tan \pi t\,dt$

50. $\int e^{\csc(\pi + t)}\csc(\pi + t)\cot(\pi + t)\,dt$

51. $\int_{\ln(\pi/5)}^{\ln(\pi/2)} 2e^v \cos e^v\,dv$

52. $\int_0^{\sqrt{\ln \pi}} 2x\,e^{x^2}\cos(e^{x^2})\,dx$

53. $\int \dfrac{2e^r}{5 - e^r}\,dr$

54. $\int \dfrac{dx}{1 + e^x}$

Initial Value Problems

Solve the initial value problems in Exercises 55–58.

55. $\dfrac{dy}{dt} = e^t \sin(e^t - 2), \quad y(\ln 2) = 0$

56. $\dfrac{dy}{dt} = e^{-t}\sec^2(\pi e^{-t}), \quad y(\ln 4) = 2/\pi$

57. $\dfrac{d^2y}{dx^2} = 3e^{-x}, \quad y(0) = 1$ and $y'(0) = 0$

58. $\dfrac{d^2y}{dt^2} = 1 - e^{2t}, \quad y(1) = -1$ and $y'(1) = 0$

Differentiation

In Exercises 59–86, find the derivative of y with respect to the given independent variable.

59. $y = 4^x$

60. $y = 3^{-x}$

61. $y = 5^{\sqrt{s}}$

62. $y = 2^{(s^2)}$

63. $y = x^{2\pi}$

64. $y = t^{1-e}$

65. $y = (\cos\theta)^{\sqrt{2}}$

66. $y = (\ln\theta)^{\pi}$

67. $y = 7^{\sec\theta}\ln 7$

68. $y = 3^{\tan\theta}\ln 3$

69. $y = 3^{\sin 5t}$

70. $y = 5^{-\cos 2t}$

71. $y = \log_2 3\theta$

72. $y = \log_3(1 + \theta\ln 3)$

73. $y = \log_4 x + \log_4 x^2$

74. $y = \log_{25} e^x - \log_5 \sqrt{x}$

75. $y = x^2 \log_{10} 2x$

76. $y = \log_3 r \cdot \log_9 r$

77. $y = \log_3\left(\left(\dfrac{x+1}{x-1}\right)^{\ln 3}\right)$

78. $y = \log_5 \sqrt{\left(\dfrac{7x}{3x+2}\right)^{\ln 5}}$

79. $y = 2\theta\cos(\log_7 2\theta)$

80. $y = \log_7\left(\dfrac{\sin\theta\cos\theta}{e^{\theta}\,2^{\theta}}\right)$

81. $y = \log_{10} e^x$

82. $y = \dfrac{\theta 5^{\theta}}{2 - \log_5\theta}$

83. $y = 3^{\log_2 x}$

84. $y = 3\log_8(\log_2 t)$

85. $y = \log_3(5t^{\ln 2})$

86. $y = t\log_3(e^{(\sin t)(\ln 3)})$

Integration

Evaluate the integrals in Exercises 87–96.

87. $\int 2^x\,dx$

88. $\int \dfrac{3^x}{3 - 3^x}\,dx$

89. $\int_0^1 2^{-\theta}\,d\theta$

90. $\int_{-3}^{0} 4^{-\theta}\,d\theta$

91. $\int_1^{\sqrt{2}} x2^{(x^2)}\,dx$

92. $\int_1^4 \dfrac{2^{\sqrt{x}}}{\sqrt{x}}\,dx$

93. $\int_0^{\pi/2} 7^{\cos t}\sin t\,dt$

94. $\int_0^{\pi/4} \left(\dfrac{1}{3}\right)^{\tan t}\sec^2 t\,dt$

95. $\int_2^4 x^{2x}(1 + \ln x)\,dx$

96. $\int \dfrac{2x4^{x^2}}{1 + 4^{x^2}}\,dx$

Evaluate the integrals in Exercises 97–110.

97. $\int 3x^{\sqrt{3}}\,dx$

98. $\int x^{\sqrt{4}-1}\,dx$

99. $\int_0^3 (\sqrt{2} + 1)x^{\sqrt{2}}\,dx$

100. $\int_1^e x^{(\ln 2)-1}\,dx$

101. $\int \dfrac{2\log_{10} x}{2x}\,dx$

102. $\int_1^4 \dfrac{\log_2 x}{x}\,dx$

103. $\int_1^4 \dfrac{\ln 2 \log_2 x}{x}\,dx$

104. $\int_1^e \dfrac{2\ln 10 \log_{10} x}{x}\,dx$

105. $\int_0^2 \dfrac{\log_2(x+5)}{x+5}\,dx$

106. $\int_{1/10}^{10} \dfrac{\log_{10}(10x)}{x}\,dx$

107. $\int_0^9 \dfrac{2\log_{10}(x+1)}{x+1}\,dx$

108. $\int_2^3 \dfrac{3\log_3(x-1)}{x-1}\,dx$

109. $\int \dfrac{dx}{x\log_{10} x}$

110. $\int \dfrac{dx}{x(\log_8 x)^2}$

Evaluate the integrals in Exercises 111–114.

111. $\int_1^{\ln x} \frac{1}{t}\,dt,\quad x > 1$ **112.** $\int_1^{e^x} \frac{1}{t}\,dt$

113. $\int_1^{1/x} \frac{1}{t}\,dt,\quad x > 0$ **114.** $\frac{1}{\ln a}\int_1^{x} \frac{1}{t}\,dt,\quad x > 0$

Powers with Variable Bases and Exponents

In Exercises 115–126, use logarithmic differentiation or the method in Example 6 to find the derivative of y with respect to the given independent variable.

115. $y=(2x+1)^{x^2}$ **116.** $y = x^2 + x^{2x}$

117. $y = (\sqrt{t})^t$ **118.** $y = t^{\sqrt{t}}$

119. $y=(\cos x)^x$ **120.** $y = x^{\sin x}$

121. $y=\cos x^x$ **122.** $y = (\ln x)^{\ln x}$

123. $y^x = x^3 y$ **124.** $x^{\sin y} = \ln y$

125. $x = y^{xy}$ **126.** $e^y = y^{\ln x}$

For Exercises 127 and 128 find a function f satisfying each equation.

127. $\int_2^x \sqrt{f(t)}\,dt = x\ln x$ **128.** $f(x) = e^2 + \int_1^x f(t)\,dt$

Theory and Applications

129. Find the absolute maximum and minimum values of $f(x) = e^x - 2x$ on $[0, 1]$.

130. Where does the periodic function $f(x) = 2e^{\sin(x/2)}$ take on its extreme values and what are these values?

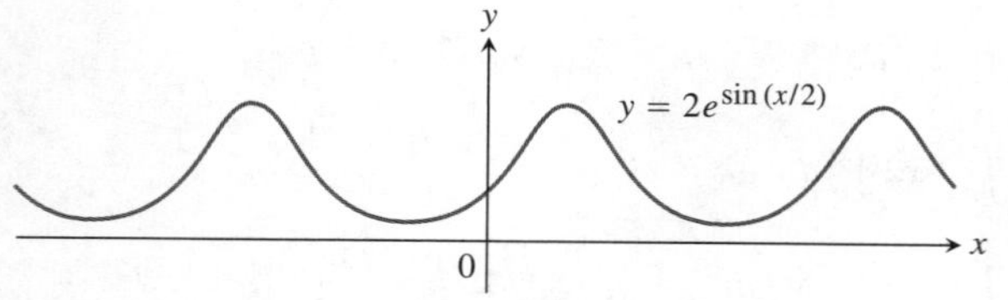

131. Let $f(x) = xe^{-x}$.

a. Find all absolute extreme values for f.

b. Find all inflection points for f.

132. Let $f(x) = \dfrac{e^x}{1+e^{2x}}$.

a. Find all absolute extreme values for f.

b. Find all inflection points for f.

133. Find the absolute maximum value of $f(x) = x^2\ln(1/x)$ and say where it is assumed.

T **134.** Graph $f(x) = (x-3)^2 e^x$ and its first derivative together. Comment on the behavior of f in relation to the signs and values of f'. Identify significant points on the graphs with calculus, as necessary.

135. Find the area of the "triangular" region in the first quadrant that is bounded above by the curve $y=e^{3x}$, below by the curve e^x, and on the right by the line $x=\ln 2$.

136. Find the area of the "triangular" region in the first quadrant that is bounded above by the curve $y=e^{x/3}$, below by the curve $y=e^{-x/3}$, and on the right by the line $x=3\ln 3$.

137. Find a curve through the origin in the xy-plane whose length from $x = 0$ to $x = 1$ is

$$L = \int_0^1 \sqrt{1 + \frac{1}{4}e^x}\,dx.$$

138. Find the area of the surface generated by revolving the curve $x = (e^y + e^{-y})/2,\ 0 \le y \le \ln 2$, about the y-axis.

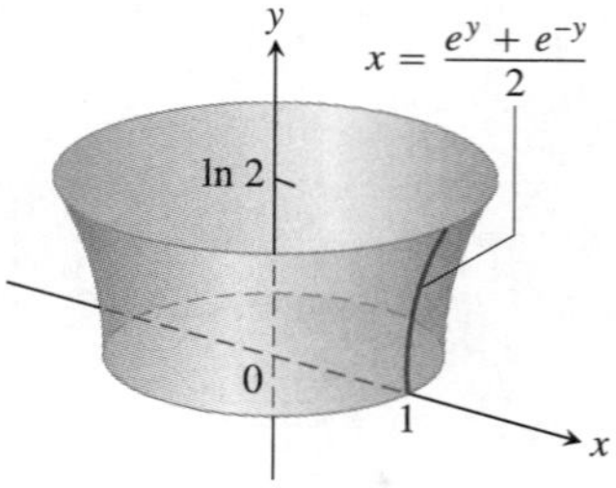

In Exercises 139–142, find the length of each curve.

139. $y = \frac{1}{2}(e^x + e^{-x})$ from $x = 0$ to $x = 1$

140. $y = \ln(e^x - 1) - \ln(e^x + 1)$ from $x = \ln 2$ to $x = \ln 3$

141. $y = \ln(\cos x)$ from $x = 0$ to $x = \pi/4$

142. $y = \ln(\csc x)$ from $x = \pi/6$ to $x = \pi/4$

143. a. Show that $\int \ln x\,dx = x\ln x - x + C$.

b. Find the average value of $\ln x$ over $[1, e]$.

144. Find the average value of $f(x) = 1/x$ on $[1, 2]$.

145. The linearization of e^x at $x = 0$

a. Derive the linear approximation $e^x \approx 1 + x$ at $x = 0$.

T b. Estimate to five decimal places the magnitude of the error involved in replacing e^x by $1 + x$ on the interval $[0, 0.2]$.

T c. Graph e^x and $1 + x$ together for $-2 \le x \le 2$. Use different colors, if available. On what intervals does the approximation appear to overestimate e^x? Underestimate e^x?

146. The geometric, logarithmic, and arithmetic mean inequality

a. Show that the graph of e^x is concave up over every interval of x-values.

b. Show, by reference to the accompanying figure, that if $0 < a < b$, then

$$e^{(\ln a+\ln b)/2}\cdot(\ln b - \ln a) < \int_{\ln a}^{\ln b} e^x\,dx < \frac{e^{\ln a} + e^{\ln b}}{2}\cdot(\ln b - \ln a).$$

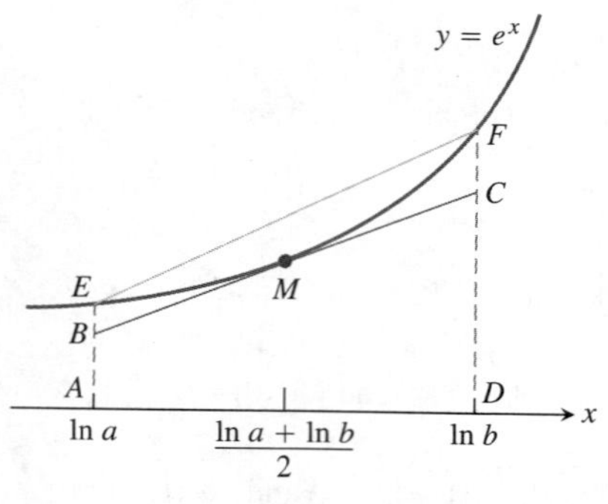

c. Use the inequality in part (b) to conclude that

$$\sqrt{ab} < \frac{b-a}{\ln b - \ln a} < \frac{a+b}{2}.$$

This inequality says that the geometric mean of two positive numbers is less than their logarithmic mean, which in turn is less than their arithmetic mean.

147. Find the area of the region between the curve $y = 2x/(1 + x^2)$ and the interval $-2 \le x \le 2$ of the x-axis.

148. Find the area of the region between the curve $y = 2^{(2-x)}$ and the interval $-2 \le x \le 2$ of the axis.

149. Consider the accompanying graphs of $y = 2x + 3$ and $y = \ln x$. Determine the

a. minimum vertical distance

b. minimum horizontal distance between these graphs.

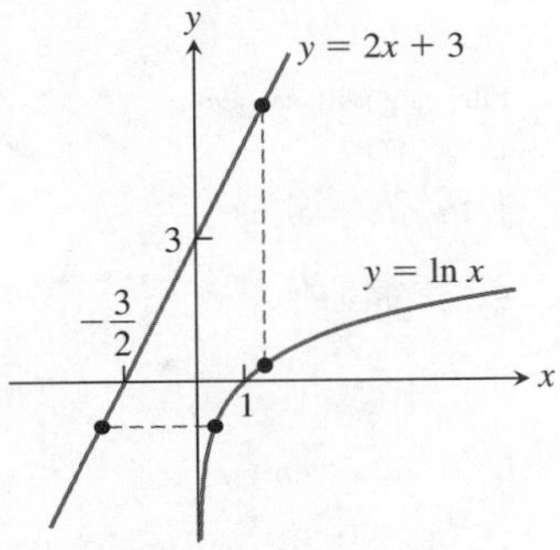

150. Determine the dimensions of the inscribed rectangle of maximum area.

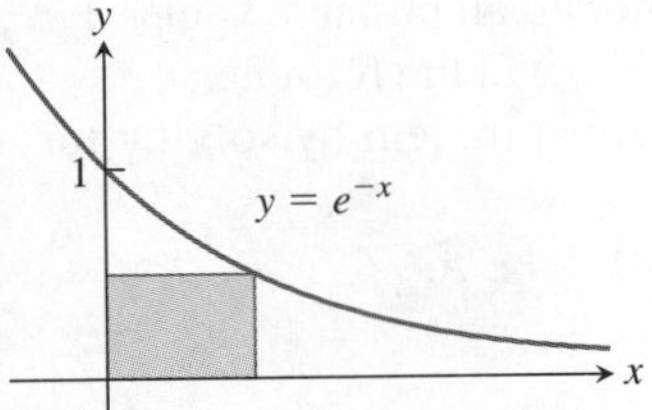

T **151.** The equation $x^2 = 2^x$ has three solutions: $x = 2$, $x = 4$, and one other. Estimate the third solution as accurately as you can by graphing.

T **152.** Could $x^{\ln 2}$ possibly be the same as $2^{\ln x}$ for $x > 0$? Graph the two functions and explain what you see.

153. The linearization of 2^x

a. Find the linearization of $f(x) = 2^x$ at $x = 0$. Then round its coefficients to two decimal places.

T **b.** Graph the linearization and function together for $-3 \le x \le 3$ and $-1 \le x \le 1$.

154. The linearization of $\log_3 x$

a. Find the linearization of $f(x) = \log_3 x$ at $x = 3$. Then round its coefficients to two decimal places.

T **b.** Graph the linearization and function together in the windows $0 \le x \le 8$ and $2 \le x \le 4$.

T **155. Which is bigger, π^e or e^π?** Calculators have taken some of the mystery out of this once-challenging question. (Go ahead and check; you will see that it is a very close call.) You can answer the question without a calculator, though.

a. Find an equation for the line through the origin tangent to the graph of $y = \ln x$.

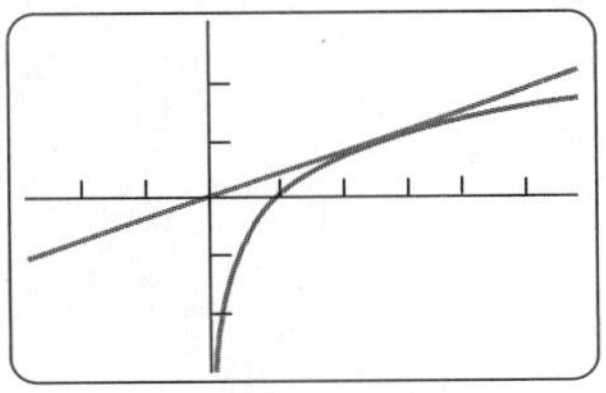

[−3, 6] by [−3, 3]

b. Give an argument based on the graphs of $y = \ln x$ and the tangent line to explain why $\ln x < x/e$ for all positive $x \ne e$.

c. Show that $\ln(x^e) < x$ for all positive $x \ne e$.

d. Conclude that $x^e < e^x$ for all positive $x \ne e$.

e. So which is bigger, π^e or e^π?

T **156. A decimal representation of e** Find e to as many decimal places as your calculator allows by solving the equation $\ln x = 1$ using Newton's method, as in Section 4.6.

7.4 Exponential Change and Separable Differential Equations

Exponential functions increase or decrease very rapidly with changes in the independent variable. They describe growth or decay in many natural and industrial situations. The variety of models based on these functions partly accounts for their importance.

Exponential Change

In modeling many real-world situations, a quantity y increases or decreases at a rate proportional to its size at a given time t. Examples of such quantities include the size of a population, the amount of a decaying radioactive material, and the temperature difference between a hot object and its surrounding medium. Such quantities are said to undergo **exponential change**.

If the amount present at time $t = 0$ is called y_0, then we can find y as a function of t by solving the following initial value problem:

$$\text{Differential equation: } \frac{dy}{dt} = ky \tag{1a}$$

$$\text{Initial condition: } \quad y = y_0 \quad \text{when} \quad t = 0. \tag{1b}$$

If y is positive and increasing, then k is positive, and we use Equation (1a) to say that the rate of growth is proportional to what has already been accumulated. If y is positive and decreasing, then k is negative, and we use Equation (1a) to say that the rate of decay is proportional to the amount still left.

The constant function $y = 0$ is a solution of Equation (1a) if $y_0 = 0$. To find the solutions when y_0 is not zero, we divide Equation (1a) by y:

$$\begin{aligned}
\frac{1}{y} \cdot \frac{dy}{dt} &= k && y \neq 0 \\
\int \frac{1}{y}\frac{dy}{dt}\,dt &= \int k\,dt && \text{Integrate with respect to } t. \\
\ln|y| &= kt + C && \textstyle\int (1/u)\,du = \ln|u| + C \\
|y| &= e^{kt+C} && \text{Exponentiate.} \\
|y| &= e^C \cdot e^{kt} && e^{a+b} = e^a \cdot e^b \\
y &= \pm e^C e^{kt} && \text{If } |y| = r, \text{ then } y = \pm r. \\
y &= Ae^{kt}. && A \text{ is a shorter name for } \pm e^C.
\end{aligned}$$

By allowing A to take on the value 0 in addition to all possible values $\pm e^C$, we can include the solution $y = 0$ (which occurs when $y_0 = 0$) in the formula.

We find the value of A for the initial value problem by solving for A when $y = y_0$ and $t = 0$:

$$y_0 = Ae^{k \cdot 0} = A.$$

> The solution of the initial value problem
>
> $$\frac{dy}{dt} = ky, \qquad y(0) = y_0$$
>
> is
>
> $$y = y_0 e^{kt}. \tag{2}$$

Quantities changing in this way are said to undergo **exponential growth** if $k > 0$ and **exponential decay** if $k < 0$. The number k is called the **rate constant** of the change. (See Figure 7.15.)

The derivation of Equation (2) shows also that the only functions that are their own derivatives ($k = 1$) are constant multiples of the exponential function.

Before presenting several examples of exponential change, let us consider the process we used to derive it.

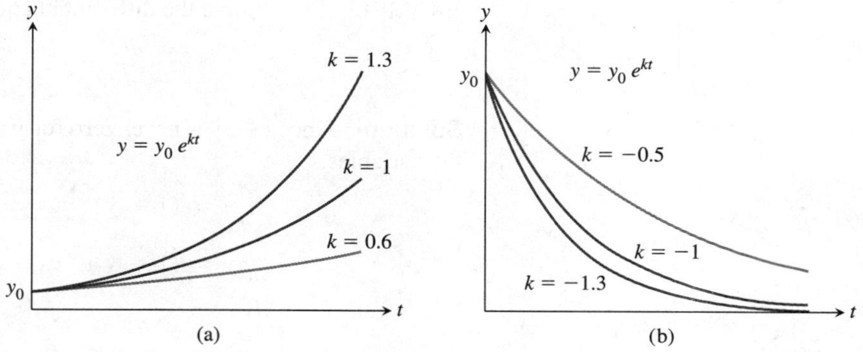

FIGURE 7.15 Graphs of (a) exponential growth and (b) exponential decay. As $|k|$ increases, the growth $(k > 0)$ or decay $(k < 0)$ intensifies.

Separable Differential Equations

Exponential change is modeled by a differential equation of the form $dy/dx = ky$, where k is a nonzero constant. More generally, suppose we have a differential equation of the form

$$\frac{dy}{dx} = f(x, y), \tag{3}$$

where f is a function of *both* the independent and dependent variables. A **solution** of the equation is a differentiable function $y = y(x)$ defined on an interval of x-values (perhaps infinite) such that

$$\frac{d}{dx}y(x) = f(x, y(x))$$

on that interval. That is, when $y(x)$ and its derivative $y'(x)$ are substituted into the differential equation, the resulting equation is true for all x in the solution interval. The **general solution** is a solution $y(x)$ that contains all possible solutions, and it always contains an arbitrary constant.

Equation (3) is **separable** if f can be expressed as a product of a function of x and a function of y. The differential equation then has the form

$$\frac{dy}{dx} = g(x)\,H(y). \qquad \begin{array}{l}g \text{ is a function of } x;\\ H \text{ is a function of } y.\end{array}$$

Then collect all of the y factors so that they are together with dy, and likewise collect all of the x factors together with dx:

$$\frac{1}{H(y)}\,dy = g(x)\,dx.$$

Now we integrate both sides of this equation:

$$\int \frac{1}{H(y)}\,dy = \int g(x)\,dx. \tag{4}$$

After completing the integrations, we obtain the solution y, defined implicitly as a function of x.

The justification that we can integrate both sides in Equation (4) in this way is based on the Substitution Rule (Section 5.5):

$$\begin{aligned}\int \frac{1}{H(y)}\,dy &= \int \frac{1}{H(y(x))}\frac{dy}{dx}\,dx \\ &= \int \frac{1}{H(y(x))}H(y(x))\,g(x)\,dx \qquad \frac{dy}{dx} = H(y(x))\,g(x) \\ &= \int g(x)\,dx.\end{aligned}$$

EXAMPLE 1 Solve the differential equation

$$\frac{dy}{dx} = (1 + y)e^x, \quad y > -1.$$

Solution Since $1 + y$ is never zero for $y > -1$, we can solve the equation by separating the variables.

$$\frac{dy}{dx} = (1 + y)e^x$$

$$dy = (1 + y)e^x\,dx \qquad \text{Treat } dy/dx \text{ as a quotient of differentials and multiply both sides by } dx.$$

$$\frac{dy}{1 + y} = e^x\,dx \qquad \text{Divide by } (1 + y).$$

$$\int \frac{dy}{1 + y} = \int e^x\,dx \qquad \text{Integrate both sides.}$$

$$\ln(1 + y) = e^x + C \qquad C \text{ represents the combined constants of integration.}$$

The last equation gives y as an implicit function of x. ■

EXAMPLE 2 Solve the equation $y(x + 1)\dfrac{dy}{dx} = x(y^2 + 1)$.

Solution We change to differential form, separate the variables, and integrate:

$$y(x + 1)\,dy = x(y^2 + 1)\,dx$$

$$\frac{y\,dy}{y^2 + 1} = \frac{x\,dx}{x + 1} \qquad x \neq -1$$

$$\int \frac{y\,dy}{1 + y^2} = \int \left(1 - \frac{1}{x + 1}\right) dx \qquad \text{Divide } x \text{ by } x + 1.$$

$$\frac{1}{2}\ln(1 + y^2) = x - \ln|x + 1| + C.$$

The last equation gives the solution y as an implicit function of x. ■

The initial value problem

$$\frac{dy}{dt} = ky, \qquad y(0) = y_0$$

involves a separable differential equation, and the solution $y = y_0e^{kt}$ expresses exponential change. We now present several examples of such change.

Unlimited Population Growth

Strictly speaking, the number of individuals in a population (of people, plants, animals, or bacteria, for example) is a discontinuous function of time because it takes on discrete values. However, when the number of individuals becomes large enough, the population can be approximated by a continuous function. Differentiability of the approximating function is another reasonable hypothesis in many settings, allowing for the use of calculus to model and predict population sizes.

If we assume that the proportion of reproducing individuals remains constant and assume a constant fertility, then at any instant t the birth rate is proportional to the number $y(t)$ of individuals present. We likewise assume that the death rate of the population is stable and proportional to $y(t)$. If, further, we neglect departures and arrivals, the growth rate dy/dt is the birth rate minus the death rate, which is the difference of the two proportionalities under our assumptions. In other words, $dy/dt = ky$ so that $y = y_0e^{kt}$, where y_0 is the size of the population at time $t = 0$. As with all kinds of growth, there may be limitations imposed by the surrounding environment, but we will not go into these here. (We treat one model imposing such limitations in Section 9.4.) When k is positive, the proportionality $dy/dt = ky$ models *unlimited population growth.* (See Figure 7.16.)

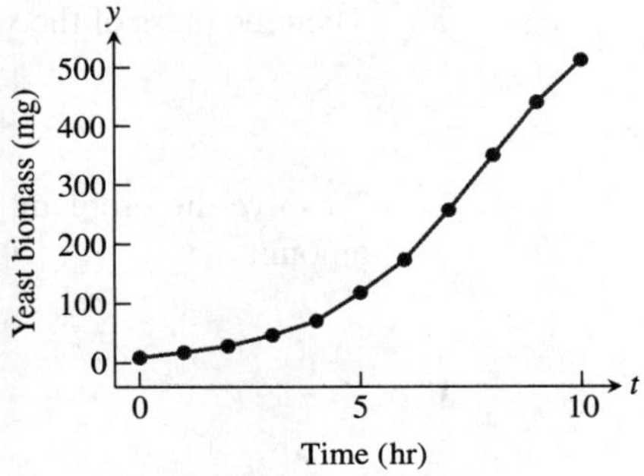

FIGURE 7.16 Graph of the growth of a yeast population over a 10-hour period, based on the data in Table 7.3.

EXAMPLE 3 The biomass of a yeast culture in an experiment is initially 29 grams. After 30 minutes the mass is 37 grams. Assuming that the equation for unlimited population growth gives a good model for the growth of the yeast when the mass is below 100 grams, how long will it take for the mass to double from its initial value?

Solution Let $y(t)$ be the yeast biomass after t minutes. We use the exponential growth model $dy/dt = ky$ for unlimited population growth, with solution $y = y_0e^{kt}$.

We have $y_0 = y(0) = 29$. We are also told that

$$y(30) = 29e^{k(30)} = 37.$$

TABLE 7.3 Population of yeast

Time (hr)	Yeast biomass (mg)
0	9.6
1	18.3
2	29.0
3	47.2
4	71.1
5	119.1
6	174.6
7	257.3
8	350.7
9	441.0
10	513.3

Solving this equation for k, we find

$$\begin{aligned} e^{k(30)} &= \frac{37}{29} \\ 30k &= \ln\left(\frac{37}{29}\right) \\ k &= \frac{1}{30}\ln\left(\frac{37}{29}\right) \approx 0.008118. \end{aligned}$$

Then the mass of the yeast in grams after t minutes is given by the equation

$$y = 29e^{(0.008118)t}.$$

To solve the problem, we find the time t for which $y(t) = 58$, which is twice the initial amount.

$$\begin{aligned} 29e^{(0.008118)t} &= 58 \\ (0.008118)t &= \ln\left(\frac{58}{29}\right) \\ t &= \frac{\ln 2}{0.008118} \approx 85.38 \end{aligned}$$

It takes about 85 minutes for the yeast population to double. ■

In the next example we model the number of people within a given population who are infected by a disease that is being eradicated from the population. Here the constant of proportionality k is negative, and the model describes an exponentially decaying number of infected individuals.

EXAMPLE 4 One model for the way diseases die out when properly treated assumes that the rate dy/dt at which the number of infected people changes is proportional to the number y. The number of people cured is proportional to the number y that are infected with the disease. Suppose that in the course of any given year, the number of cases of a disease is reduced by 20%. If there are 10,000 cases today, how many years will it take to reduce the number to 1000?

Solution We use the equation $y = y_0 e^{kt}$. There are three things to find: the value of y_0, the value of k, and the time t when $y = 1000$.

The value of y_0. We are free to count time beginning anywhere we want. If we count from today, then $y = 10{,}000$ when $t = 0$, so $y_0 = 10{,}000$. Our equation is now

$$y = 10{,}000e^{kt}. \tag{5}$$

The value of k. When $t = 1$ year, the number of cases will be 80% of its present value, or 8000. Hence,

$$\begin{aligned} 8000 &= 10{,}000e^{k(1)} && \text{Eq. (5) with } t = 1 \text{ and } y = 8000 \\ e^k &= 0.8 \\ \ln(e^k) &= \ln 0.8 && \text{Take logs of both sides} \\ k &= \ln 0.8 < 0. && \ln 0.8 \approx -0.223 \end{aligned}$$

At any given time t,

$$y = 10{,}000e^{(\ln 0.8)t}. \tag{6}$$

The value of t that makes $y = 1000$. We set y equal to 1000 in Equation (6) and solve for t:

$$\begin{aligned}
1000 &= 10{,}000e^{(\ln 0.8)t} \\
e^{(\ln 0.8)t} &= 0.1 \\
(\ln 0.8)t &= \ln 0.1 && \text{Logs of both sides} \\
t &= \frac{\ln 0.1}{\ln 0.8} \approx 10.32 \text{ years.}
\end{aligned}$$

It will take a little more than 10 years to reduce the number of cases to 1000. (See Figure 7.17.)

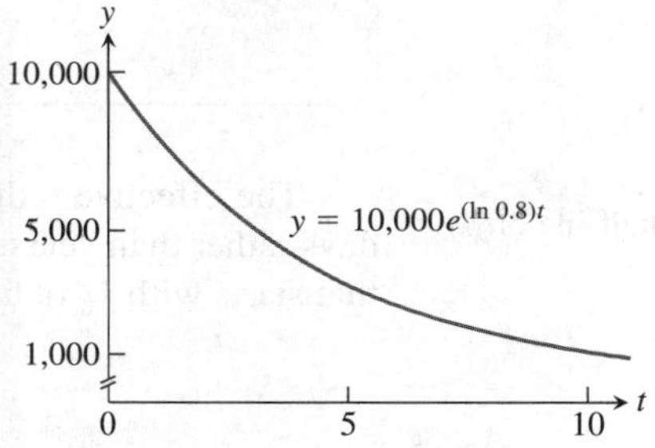

FIGURE 7.17 A graph of the number of people infected by a disease exhibits exponential decay (Example 4). ■

Radioactivity

Some atoms are unstable and can spontaneously emit mass or radiation. This process is called **radioactive decay**, and an element whose atoms go spontaneously through this process is called **radioactive**. Sometimes when an atom emits some of its mass through this process of radioactivity, the remainder of the atom re-forms to make an atom of some new element. For example, radioactive carbon-14 decays into nitrogen; radium, through a number of intermediate radioactive steps, decays into lead.

Experiments have shown that at any given time the rate at which a radioactive element decays (as measured by the number of nuclei that change per unit time) is approximately proportional to the number of radioactive nuclei present. Thus, the decay of a radioactive element is described by the equation $dy/dt = -ky$, $k > 0$. It is conventional to use $-k$, with $k > 0$, to emphasize that y is decreasing. If y_0 is the number of radioactive nuclei present at time zero, the number still present at any later time t will be

$$y = y_0e^{-kt}, \qquad k > 0.$$

For radon-222 gas, t is measured in days and $k = 0.18$. For radium-226, which used to be painted on watch dials to make them glow at night (a dangerous practice), t is measured in years and $k = 4.3 \times 10^{-4}$.

The **half-life** of a radioactive element is the time expected to pass until half of the radioactive nuclei present in a sample decay. The half-life is a constant that does not depend on the number of radioactive nuclei initially present in the sample, but only on the radioactive substance.

To compute the half-life, let y_0 be the number of radioactive nuclei initially present in the sample. Then the number y present at any later time t will be $y = y_0e^{-kt}$. We seek the value of t at which the number of radioactive nuclei present equals half the original number:

$$\begin{aligned} y_0 e^{-kt} &= \frac{1}{2} y_0 \\ e^{-kt} &= \frac{1}{2} \\ -kt &= \ln \frac{1}{2} = -\ln 2 \qquad \text{Reciprocal Rule for logarithms} \\ t &= \frac{\ln 2}{k}. \end{aligned}$$

This value of t is the half-life of the element. It depends only on the value of k; the number y_0 does not have any effect.

$$\text{Half-life} = \frac{\ln 2}{k} \qquad (7)$$

Carbon-14 dating uses the half-life of 5730 years.

The effective radioactive lifetime of polonium-210 is so short that we measure it in days rather than years. The number of radioactive atoms remaining after t days in a sample that starts with y_0 radioactive atoms is

$$y = y_0 e^{-5\times10^{-3} t}.$$

The element's half-life is:

$$\begin{aligned} \text{Half-life} &= \frac{\ln 2}{k} \qquad \text{Eq. (7)} \\ &= \frac{\ln 2}{5 \times 10^{-3}} \qquad \text{The } k \text{ from polonium's decay equation} \\ &\approx 139 \text{ days.} \end{aligned}$$

This means that after 139 days, 1/2 of y_0 radioactive atoms remain; after another 139 days (278 days altogether) half of those remain, or 1/4 of y_0 radioactive atoms remain, and so on (see Figure 7.18).

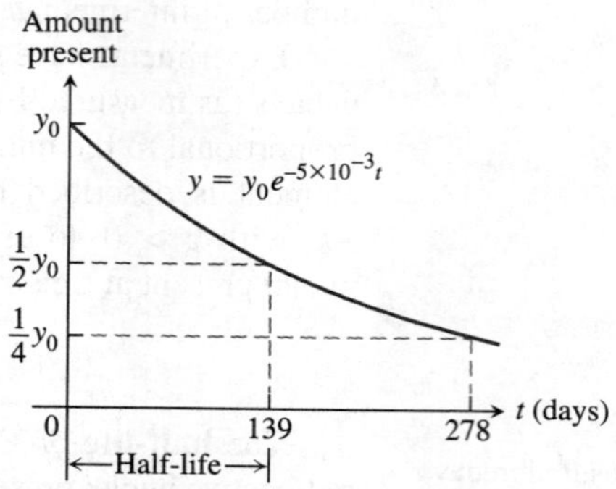

FIGURE 7.18 Amount of polonium-210 present at time t, where y_0 represents the number of radioactive atoms initially present.

EXAMPLE 5 The decay of radioactive elements can sometimes be used to date events from Earth's past. In a living organism, the ratio of radioactive carbon, carbon-14, to ordinary carbon stays fairly constant during the lifetime of the organism, being approximately equal to the ratio in the organism's atmosphere at the time. After the organism's death, however, no new carbon is ingested, and the proportion of carbon-14 in the organism's remains decreases as the carbon-14 decays.

Scientists who do carbon-14 dating often use a figure of 5730 years for its half-life. Find the age of a sample in which 10% of the radioactive nuclei originally present have decayed.

Solution We use the decay equation $y = y_0 e^{-kt}$. There are two things to find: the value of k and the value of t when y is $0.9y_0$ (90% of the radioactive nuclei are still present). That is, find t when $y_0 e^{-kt} = 0.9y_0$, or $e^{-kt} = 0.9$.

The value of k. We use the half-life Equation (7):

$$k = \frac{\ln 2}{\text{half-life}} = \frac{\ln 2}{5730} \qquad (\text{about } 1.2 \times 10^{-4}).$$

The value of t that makes $e^{-kt} = 0.9$.

$$\begin{aligned} e^{-kt} &= 0.9 \\ e^{-(\ln 2/5730)t} &= 0.9 \\ -\frac{\ln 2}{5730}t &= \ln 0.9 && \text{Logs of both sides} \\ t &= -\frac{5730 \ln 0.9}{\ln 2} \approx 871 \text{ years} \end{aligned}$$

The sample is about 871 years old. ■

Heat Transfer: Newton's Law of Cooling

Hot soup left in a tin cup cools to the temperature of the surrounding air. A hot silver bar immersed in a large tub of water cools to the temperature of the surrounding water. In situations like these, the rate at which an object's temperature is changing at any given time is roughly proportional to the difference between its temperature and the temperature of the surrounding medium. This observation is called *Newton's Law of Cooling*, although it applies to warming as well.

If H is the temperature of the object at time t, and H_S is the constant surrounding temperature, then the differential equation is

$$\frac{dH}{dt} = -k(H - H_S). \tag{8}$$

If we substitute y for $(H - H_S)$, then

$$\begin{aligned} \frac{dy}{dt} &= \frac{d}{dt}(H - H_S) = \frac{dH}{dt} - \frac{d}{dt}(H_S) \\ &= \frac{dH}{dt} - 0 && H_S \text{ is a constant.} \\ &= \frac{dH}{dt} \\ &= -k(H - H_S) && \text{Eq. (8)} \\ &= -ky. && H - H_S = y \end{aligned}$$

We know that the solution of the equation $dy/dt = -ky$ is $y = y_0 e^{-kt}$, where $y(0) = y_0$. Substituting $(H - H_S)$ for y, this says that

$$H - H_S = (H_0 - H_S)e^{-kt}, \tag{9}$$

where H_0 is the temperature at $t = 0$. This equation is the solution to Newton's Law of Cooling.

EXAMPLE 6 A hard-boiled egg at 98 °C is put in a sink of 18 °C water. After 5 min, the egg's temperature is 38 °C. Assuming that the water has not warmed appreciably, how much longer will it take the egg to reach 20 °C?

Solution We find how long it would take the egg to cool from 98 °C to 20 °C and subtract the 5 min that have already elapsed. Using Equation (9) with $H_S = 18$ and $H_0 = 98$, the egg's temperature t min after it is put in the sink is

$$H = 18 + (98 - 18)e^{-kt} = 18 + 80e^{-kt}.$$

To find k, we use the information that $H = 38$ when $t = 5$:

$$\begin{aligned} 38 &= 18 + 80e^{-5k} \\ e^{-5k} &= \frac{1}{4} \\ -5k &= \ln\frac{1}{4} = -\ln 4 \\ k &= \frac{1}{5}\ln 4 = 0.2\ln 4 \qquad \text{(about 0.28).} \end{aligned}$$

The egg's temperature at time t is $H = 18 + 80e^{-(0.2\ln 4)t}$. Now find the time t when $H = 20$:

$$\begin{aligned} 20 &= 18 + 80e^{-(0.2\ln 4)t} \\ 80e^{-(0.2\ln 4)t} &= 2 \\ e^{-(0.2\ln 4)t} &= \frac{1}{40} \\ -(0.2\ln 4)t &= \ln\frac{1}{40} = -\ln 40 \\ t &= \frac{\ln 40}{0.2\ln 4} \approx 13 \text{ min.} \end{aligned}$$

The egg's temperature will reach 20 °C about 13 min after it is put in the water to cool. Since it took 5 min to reach 38 °C, it will take about 8 min more to reach 20 °C. ■

EXERCISES 7.4

Verifying Solutions

In Exercises 1–4, show that each function $y = f(x)$ is a solution of the accompanying differential equation.

1. $2y' + 3y = e^{-x}$

a. $y = e^{-x}$ **b.** $y = e^{-x} + e^{-(3/2)x}$

c. $y = e^{-x} + Ce^{-(3/2)x}$

2. $y' = y^2$

a. $-\dfrac{1}{x}$ **b.** $-\dfrac{1}{x+2}$ **c.** $-\dfrac{1}{x+C}$

3. $y = \dfrac{1}{x}\displaystyle\int_1^x \frac{e^t}{t}\,dt,\quad x^2y' + xy = e^x$

4. $y = \dfrac{1}{\sqrt{1+x^4}}\displaystyle\int_1^x \sqrt{1+t^4}\,dt,\quad y' + \frac{2x^3}{1+x^4}y = 1$

Initial Value Problems

In Exercises 5–8, show that each function is a solution of the given initial value problem.

Differential equation	Initial condition	Solution candidate
5. $y' - y = \dfrac{4e^{2x}}{1+4e^{2x}}$	$y(-\ln 2) = \dfrac{\pi}{4}$	$y = 2e^x \tan^{-1}(2e^x)$
6. $y' = e^{-x^2} - 2xy$	$y(2) = 0$	$y = (x-2)e^{-x^2}$
7. $xy' + y = -\sin x,\ x > 0$	$y\left(\dfrac{\pi}{2}\right) = 0$	$y = \dfrac{\cos x}{x}$
8. $x^2y' = xy - y^2,\ x > 1$	$y(e) = e$	$y = \dfrac{x}{\ln x}$

Separable Differential Equations

Solve the differential equation in Exercises 9–22.

9. $2\sqrt{xy}\dfrac{dy}{dx} = 1, \quad x, y > 0$ **10.** $\dfrac{dy}{dx} = x^2 y$

11. $\dfrac{dy}{dx} = e^{x-y}$ **12.** $\dfrac{dy}{dx} = 3x^2 e^{-y}$

13. $\dfrac{dy}{dx} = \sqrt{y}\cos^2\sqrt{y}$ **14.** $3y^2\dfrac{dy}{dx} = 1$

15. $\sqrt{x}\dfrac{dy}{dx} = e^{y+\sqrt{x}}, \quad x > 0$ **16.** $(\sec x)\dfrac{dy}{dx} = e^{y+\sin x}$

17. $\dfrac{dy}{dx} = \sqrt{1-y}$ **18.** $e^x\left(\dfrac{dy}{dx}+1\right) = 1$

19. $y^2\dfrac{dy}{dx} = 3x^2y^3 - 6x^2$ **20.** $\dfrac{dy}{dx} = xy + 3x - 2y - 6$

21. $\dfrac{1}{x}\dfrac{dy}{dx} = ye^{x^2} + 2\sqrt{y}\,e^{x^2}$ **22.** $\dfrac{dy}{dx} = e^{x-y} + e^x + e^{-y} + 1$

Applications and Examples

The answers to most of the following exercises are in terms of logarithms and exponentials. A calculator can be helpful, enabling you to express the answers in decimal form.

23. Human evolution continues The analysis of tooth shrinkage by C. Loring Brace and colleagues at the University of Michigan's Museum of Anthropology indicates that human tooth size is continuing to decrease and that the evolutionary process has not yet come to a halt. In northern Europeans, for example, tooth size reduction now has a rate of 1% per 1000 years.

a. If t represents time in years and y represents tooth size, use the condition that $y = 0.99y_0$ when $t = 1000$ to find the value of k in the equation $y = y_0e^{kt}$. Then use this value of k to answer the following questions.

b. In about how many years will human teeth be 90% of their present size?

c. What will be our descendants' tooth size 20,000 years from now (as a percentage of our present tooth size)?

24. Atmospheric pressure Earth's atmospheric pressure p is often modeled by assuming that the rate dp/dh at which p changes with the altitude h above sea level is proportional to p. Suppose that the pressure at sea level is 1013 millibars (about 14.7 pounds per square inch) and that the pressure at an altitude of 20 km is 90 millibars.

a. Solve the initial value problem

Differential equation: $dp/dh = kp$ (k a constant)
Initial condition: $p = p_0$ when $h = 0$

to express p in terms of h. Determine the values of p_0 and k from the given altitude-pressure data.

b. What is the atmospheric pressure at $h = 50$ km?

c. At what altitude does the pressure equal 900 millibars?

25. Growth rate of rabbits The growth rate of rabbits is given by $\dfrac{dP}{dt} = 0.2P^{3/4}$. If there are 16 rabbits when $t = 0$, how many rabbits will be there after 20 months? (t is measured in months.)

26. The inversion of sugar The processing of raw sugar has a step called "inversion" that changes the sugar's molecular structure. Once the process has begun, the rate of change of the amount of raw sugar is proportional to the amount of raw sugar remaining. If 1000 kg of raw sugar reduces to 800 kg of raw sugar during the first 10 hours, how much raw sugar will remain after another 14 hours?

27. Working underwater The intensity $L(x)$ of light x feet beneath the surface of the ocean satisfies the differential equation

$$\frac{dL}{dx} = -kL.$$

As a diver, you know from experience that diving to 18 ft in the Caribbean Sea cuts the intensity in half. You cannot work without artificial light when the intensity falls below one-tenth of the surface value. About how deep can you expect to work without artificial light?

28. Cell growth The surface area of a biological cell is proportional to its growth rate. If the volume of the cell is V and surface area is given by $S = kV^{2/3}$, then the growth rate is given by

$$\frac{dV}{dt} = kV^{2/3}$$

Solve this differential equation.

29. Cholera bacteria Suppose that the bacteria in a colony can grow unchecked, by the law of exponential change. The colony starts with 1 bacterium and doubles every half-hour. How many bacteria will the colony contain at the end of 24 hours? (Under favorable laboratory conditions, the number of cholera bacteria can double every 30 min. In an infected person, many bacteria are destroyed, but this example helps explain why a person who feels well in the morning may be dangerously ill by evening.)

30. Growth of bacteria A colony of bacteria is grown under ideal conditions in a laboratory so that the population increases exponentially with time. At the end of 5 hours there are 20,000 bacteria. At the end of 7 hours there are 80,000. How many bacteria were present initially?

31. The incidence of a disease (*Continuation of Example 4.*) Suppose that in any given year the number of cases can be reduced by 25% instead of 20%.

a. How long will it take to reduce the number of cases to 1000?

b. How long will it take to eradicate the disease—that is, reduce the number of cases to less than 1?

32. Drug concentration An antibiotic is administered intravenously into the bloodstream at a constant rate r. As the drug flows through the patient's system and acts on the infection that is present, it is removed from the bloodstream at a rate proportional to the amount in the bloodstream at that time. Since the amount of blood in the patient is constant, this means that the concentration $y = y(t)$ of the antibiotic in the bloodstream can be modeled by the differential equation

$$\frac{dy}{dt} = r - ky, \quad k > 0 \text{ and constant.}$$

a. If $y(0) = y_0$, find the concentration $y(t)$ at any time t.

b. Assume that $y_0 < (r/k)$ and find $\lim_{y\to\infty} y(t)$. Sketch the solution curve for the concentration.

33. Endangered species Biologists consider a species of animal or plant to be endangered if it is expected to become extinct within 20 years. If a certain species of wildlife is counted to have 1147 members at the present time, and the population has been steadily declining exponentially at an annual rate averaging 39% over the past 7 years, do you think the species is endangered? Explain your answer.

34. The U.S. population The U.S. Census Bureau keeps a running clock totaling the U.S. population. On April 19, 2021, the total was increasing at the rate of 1 person every 40 sec. The population figure for 1:32 P.M. EST on that day was 330,215,841.

a. Assuming exponential growth at a constant rate, find the rate constant for the population's growth (people per 365-day year).

b. At this rate, what will the U.S. population be at 1:32 P.M. EST on April 19, 2028?

35. Oil depletion Suppose the amount of oil pumped from one of the canyon wells in Whittier, California, decreases at the continuous rate of 10% per year. When will the well's output fall to one-fifth of its present value?

36. Continuous price discounting To encourage buyers to place 100-unit orders, your firm's sales department applies a continuous discount that makes the unit price a function $p(x)$ of the number of units x ordered. The discount decreases the price at the rate of \$0.01 per unit ordered. The price per unit for a 100-unit order is $p(100) = \$20.09$.

a. Find $p(x)$ by solving the following initial value problem.

Differential equation: $\dfrac{dp}{dx} = -\dfrac{1}{100}p$

Initial condition: $p(100) = 20.09$

b. Find the unit price $p(10)$ for a 10-unit order and the unit price $p(90)$ for a 90-unit order.

c. The sales department has asked you to find out if it is discounting so much that the firm's revenue, $r(x) = x \cdot p(x)$, will actually be less for a 100-unit order than, say, for a 90-unit order. Reassure them by showing that r has its maximum value at $x = 100$.

d. Graph the revenue function $r(x) = xp(x)$ for $0 \le x \le 200$.

37. Plutonium-239 The half-life of the plutonium isotope is 24,360 years. If 10 g of plutonium is released into the atmosphere by a nuclear accident, how many years will it take for 80% of the isotope to decay?

38. Polonium-210 The half-life of polonium is 139 days, but your sample will not be useful to you after 95% of the radioactive nuclei present on the day the sample arrives has disintegrated. For about how many days after the sample arrives will you be able to use the polonium?

39. The mean life of a radioactive nucleus Physicists using the radioactivity equation $y = y_0 e^{-kt}$ call the number $1/k$ the *mean life* of a radioactive nucleus. The mean life of a radon nucleus is about $1/0.18 = 5.6$ days. The mean life of a carbon-14 nucleus is more than 8000 years. Show that 95% of the radioactive nuclei originally present in a sample will disintegrate within three mean lifetimes, i.e., by time $t = 3/k$. Thus, the mean life of a nucleus gives a quick way to estimate how long the radioactivity of a sample will last.

40. Californium-252 What costs \$27 million per gram and can be used to treat brain cancer, analyze coal for its sulfur content, and detect explosives in luggage? The answer is californium-252, a radioactive isotope discovered by Glenn Seaborg in 1950. Much less than 1 g of it is produced each year. The half-life of the isotope is 2.645 years—long enough for a useful service life and short enough to have a high radioactivity per unit mass. One microgram of the isotope releases 170 million neutrons per minute.

a. What is the value of k in the decay equation for this isotope?

b. What is the isotope's mean life? (See Exercise 39.)

c. How long will it take 95% of a sample's radioactive nuclei to disintegrate?

41. Cooling soup Suppose that a cup of soup cooled from 95°C to 65°C after 10 min in a room whose temperature was 25°C. Use Newton's law of cooling to answer the following questions.

a. How much longer would it take the soup to cool to 40°C?

b. Instead of being left to stand in the room, the cup of 95°C soup is put in a freezer whose temperature is −10°C. How long will it take the soup to cool from 95°C to 40°C?

42. A beam of unknown temperature An aluminum beam was brought from the outside cold into a machine shop where the temperature was held at 65 °F. After 10 min the beam had warmed to 35 °F, and after another 10 min its temperature was 50 °F. Use Newton's Law of Cooling to estimate the beam's initial temperature.

43. Surrounding medium of unknown temperature A pan of warm water (46 °C) was put in a refrigerator. Ten minutes later, the water's temperature was 39 °C; 10 min after that, it was 33 °C. Use Newton's Law of Cooling to estimate how cold the refrigerator was.

44. Silver cooling in air The temperature of an ingot of silver is 60 °C above room temperature right now. Twenty minutes ago, it was 70 °C above room temperature. How far above room temperature will the silver be

a. 15 min from now?

b. 2 hours from now?

c. When will the silver be 10 °C above room temperature?

45. The age of Crater Lake The charcoal from a tree killed in the volcanic eruption that formed Crater Lake in Oregon contained 44.5% of the carbon-14 found in living matter. About how old is Crater Lake?

46. The sensitivity of carbon-14 dating to measurement To see the effect of a relatively small error in the estimate of the amount of carbon-14 in a sample being dated, consider this hypothetical situation:

a. A bone fragment found in central Illinois in the year 2000 contains 17% of its original carbon-14 content. Estimate the year the animal died.

b. Repeat part (a), assuming 18% instead of 17%.

c. Repeat part (a), assuming 16% instead of 17%.

47. Carbon-14 The oldest known frozen human mummy, discovered in the Schnalstal glacier of the Italian Alps in 1991 and called *Otzi*, was found wearing straw shoes and a leather coat with goat fur, and holding a copper ax and stone dagger. It was estimated that Otzi died 5000 years before he was discovered in the melting glacier. How much of the original carbon-14 remained in Otzi at the time of his discovery?

48. Art forgery A painting attributed to Vermeer (1632–1675), which should contain no more than 96.2% of its original carbon-14, contains 99.5% instead. About how old is the forgery?

49. Lascaux Cave paintings Prehistoric cave paintings of animals were found in the Lascaux Cave in France in 1940. Scientific analysis revealed that only 15% of the original carbon-14 in the paintings remained. What is an estimate of the age of the paintings?

50. Incan mummy The frozen remains of a young Incan woman were discovered by archeologist Johan Reinhard on Mt. Ampato in Peru during an expedition in 1995.

a. How much of the original carbon-14 was present if the estimated age of the "Ice Maiden" was 500 years?

b. If a 1% error can occur in the carbon-14 measurement, what is the oldest possible age for the Ice Maiden?

7.5 Indeterminate Forms and L'Hôpital's Rule

Consider the four limits

$$\lim_{x\to 0}\frac{x^3}{x} = \lim_{x\to 0} x^2 = 0,$$

$$\lim_{x\to 0}\frac{x}{x^3} = \lim_{x\to 0}\frac{1}{x^2} = \infty,$$

$$\lim_{x\to 0}\frac{x}{x} = \lim_{x\to 0} 1 = 1,$$

$$\lim_{x\to 0}\frac{x}{2x} = \lim_{x\to 0}\frac{1}{2} = \frac{1}{2}.$$

In each case both the numerator and the denominator approach zero as $x \to 0$, even though these limits ultimately lead to completely different results: $0, \infty, 1$, and $1/2$. We say the expression

$$\lim_{x\to a}\frac{f(x)}{g(x)} \quad \text{when} \quad \lim_{x\to a} f(x) = 0 \quad \text{and} \quad \lim_{x\to a} g(x) = 0 \tag{1}$$

involves an *indeterminate form* $0/0$. The expression "$0/0$" has the form of a number, but it is not a meaningful quantity. Stating that both the numerator and the denominator approach zero does not provide sufficient information to obtain the limit of the ratio. We have to examine the behavior of the expression in more detail by performing algebraic manipulation or by applying methods that we will introduce in this section.

Other forms exhibit behavior similar to Equation (1). For instance, if both the numerator and the denominator approach $+\infty$ or $-\infty$, then the limit of the ratio leads to an indeterminate form ∞/∞. Additional indeterminate forms we consider in this section are $\infty \cdot 0$, $\infty - \infty$, 1^∞, 0^0, and ∞^0. Their purpose is to summarize the behavior of certain types of limits.

HISTORICAL BIOGRAPHY
Guillaume François Antoine de l'Hôpital
(1661–1704)
www.bit.ly/2QkSnma

Johann Bernoulli
(1667–1748)
www.bit.ly/2OWKvXL

John (Johann) Bernoulli discovered a rule for using derivatives to calculate limits of fractions whose numerators and denominators both approach zero or $\pm\infty$. The rule is known today as **l'Hôpital's Rule**, after Guillaume de l'Hôpital. He was a French nobleman who wrote the first introductory differential calculus text, where the rule first appeared in print. Limits involving transcendental functions often require some use of this rule.

Indeterminate Form 0/0

It is important to understand that the notation "0/0" is *not* intended to imply numerically dividing 0 by 0. Instead, the indeterminate form 0/0 refers to a limit of a ratio of two functions, each of which approaches zero. L'Hôpital's rule can help us evaluate such limits.

THEOREM 5—L'Hôpital's Rule Suppose that $\lim_{x\to a} f(x) = \lim_{x\to a} g(x) = 0$, that f and g are differentiable on an open interval I containing a, and that $g'(x) \neq 0$ on I if $x \neq a$. Then

$$\lim_{x\to a} \frac{f(x)}{g(x)} = \lim_{x\to a} \frac{f'(x)}{g'(x)},$$

assuming that the limit on the right side of this equation exists.

We give a proof of Theorem 5 at the end of this section. Theorem 5 also applies if $x \to \pm\infty$ or when $f'(x)/g'(x) \to \pm\infty$, but we will not prove this.

Caution

To apply l'Hôpital's Rule to f/g, divide the derivative of f by the derivative of g. Do not make the mistake of taking the derivative of f/g. The quotient to use is f'/g', not $(f/g)'$.

EXAMPLE 1 The following limits involve 0/0 indeterminate forms, so we apply l'Hôpital's Rule. In some cases, it must be applied repeatedly.

(a) $\lim_{x\to 0} \dfrac{3x - \sin x}{x}$ The numerator and the denominator are both approaching 0; apply l'Hôpital's Rule.

$= \lim_{x\to 0} \dfrac{3 - \cos x}{1}$ Not 0/0

$= \dfrac{3 - \cos 0}{1} = 2$ Limit is found.

(b) $\lim_{x\to 0} \dfrac{\sqrt{1+x} - 1}{x} = \lim_{x\to 0} \dfrac{\frac{1}{2\sqrt{1+x}}}{1} = \dfrac{1}{2}$

(c) $\lim_{x\to 0} \dfrac{\sqrt{1+x} - 1 - x/2}{x^2}$ $\dfrac{0}{0}$; apply l'Hôpital's Rule.

$= \lim_{x\to 0} \dfrac{(1/2)(1+x)^{-1/2} - 1/2}{2x}$ Still $\dfrac{0}{0}$; apply l'Hôpital's Rule again.

$= \lim_{x\to 0} \dfrac{-(1/4)(1+x)^{-3/2}}{2} = -\dfrac{1}{8}$ Not $\dfrac{0}{0}$; limit is found.

(d) $\lim_{x\to 0} \dfrac{x - \sin x}{x^3}$ $\dfrac{0}{0}$; apply l'Hôpital's Rule.

$= \lim_{x\to 0} \dfrac{1 - \cos x}{3x^2}$ Still $\dfrac{0}{0}$; apply l'Hôpital's Rule again.

$= \lim_{x\to 0} \dfrac{\sin x}{6x}$ Still $\dfrac{0}{0}$; apply l'Hôpital's Rule again.

$= \lim_{x\to 0} \dfrac{\cos x}{6} = \dfrac{1}{6}$ Not $\dfrac{0}{0}$; limit is found.

$$\textbf{(e)}\ \lim_{x\to\infty}\frac{\ln(1+1/x)}{\sin(1/x)} \qquad \frac{0}{0}\text{; apply l'Hôpital's Rule.}$$

$$= \lim_{x\to\infty}\frac{-(1/x^2)\dfrac{1}{1+1/x}}{-(1/x^2)\cos(1/x)} \qquad \text{Still } \frac{0}{0}\text{; simplification is easier than applying l'Hôpital's Rule.}$$

$$= \lim_{x\to\infty}\frac{\dfrac{1}{1+1/x}}{\cos(1/x)} = \frac{\dfrac{1}{1+0}}{\cos 0} = 1 \qquad \text{Not } \frac{0}{0}\text{; limit is found.}$$

■

Here is a summary of the procedure we followed in Example 1.

Using L'Hôpital's Rule To find

$$\lim_{x\to a}\frac{f(x)}{g(x)}$$

by l'Hôpital's Rule, we continue to differentiate f and g, so long as we still get the form $0/0$ as $x \to a$. But as soon as one or the other of these derivatives no longer approaches zero, we stop differentiating. L'Hôpital's Rule does not apply when either the numerator or the denominator has a finite nonzero limit.

EXAMPLE 2 Be careful to apply l'Hôpital's Rule correctly:

$$\lim_{x\to 0}\frac{1-\cos x}{x+x^2} \qquad \frac{0}{0}$$

$$= \lim_{x\to 0}\frac{\sin x}{1+2x} \qquad \text{Not } \frac{0}{0}$$

It is tempting to try to apply l'Hôpital's Rule again, which would result in

$$\lim_{x\to 0}\frac{\cos x}{2} = \frac{1}{2},$$

but this is not the correct limit. l'Hôpital's Rule can be applied only to limits that give indeterminate forms, and $\lim_{x\to 0}(\sin x)/(1+2x)$ does not give an indeterminate form. Instead, this limit is $0/1 = 0$, and the correct answer for the original limit is 0. ■

L'Hôpital's Rule applies to one-sided limits as well.

EXAMPLE 3 In this example the one-sided limits are different.

$$\textbf{(a)}\ \lim_{x\to 0^+}\frac{\sin x}{x^2} \qquad \frac{0}{0}$$

$$= \lim_{x\to 0^+}\frac{\cos x}{2x} = \infty \qquad \text{Positive for } x > 0$$

$$\textbf{(b)}\ \lim_{x\to 0^-}\frac{\sin x}{x^2} \qquad \frac{0}{0}$$

$$= \lim_{x\to 0^-}\frac{\cos x}{2x} = -\infty \qquad \text{Negative for } x < 0$$

■

Indeterminate Forms $\infty/\infty, \infty \cdot 0, \infty - \infty$

Recall that ∞ and $+\infty$ mean the same thing.

Sometimes when we try to evaluate a limit as $x \to a$, we get an indeterminate form like $\infty/\infty, \infty \cdot 0$, or $\infty - \infty$, instead of $0/0$. We first consider the form ∞/∞.

More advanced treatments of calculus prove that l'Hôpital's Rule applies to the indeterminate form ∞/∞, as well as to $0/0$. If $f(x) \to \pm\infty$ and $g(x) \to \pm\infty$ as $x \to a$, then

$$\lim_{x\to a} \frac{f(x)}{g(x)} = \lim_{x\to a} \frac{f'(x)}{g'(x)},$$

provided the limit on the right exists or approaches ∞ or $-\infty$. In the notation $x \to a$, a may be either finite or infinite. Moreover, $x \to a$ may be replaced by the one-sided limits $x \to a^+$ or $x \to a^-$.

EXAMPLE 4 Find the limits of these ∞/∞ forms:

(a) $\displaystyle\lim_{x\to\pi/2} \frac{\sec x}{1 + \tan x}$ **(b)** $\displaystyle\lim_{x\to\infty} \frac{\ln x}{2\sqrt{x}}$ **(c)** $\displaystyle\lim_{x\to\infty} \frac{e^x}{x^2}$.

Solution

(a) The numerator and denominator are discontinuous at $x = \pi/2$, so we investigate the one-sided limits there. To apply l'Hôpital's Rule, we can choose I to be any open interval with $x = \pi/2$ as an endpoint.

$$\lim_{x\to(\pi/2)^-} \frac{\sec x}{1 + \tan x} \qquad \frac{\infty}{\infty} \text{ from left, apply l'Hôpital's Rule}$$

$$= \lim_{x\to(\pi/2)^-} \frac{\sec x \tan x}{\sec^2 x} = \lim_{x\to(\pi/2)^-} \sin x = 1$$

The right-hand limit is 1 also, with $(-\infty)/(-\infty)$ as the indeterminate form. Therefore, the two-sided limit is equal to 1.

(b) $$\lim_{x\to\infty} \frac{\ln x}{2\sqrt{x}} = \lim_{x\to\infty} \frac{1/x}{1/\sqrt{x}} = \lim_{x\to\infty} \frac{1}{\sqrt{x}} = 0 \qquad \frac{1/x}{1/\sqrt{x}} = \frac{\sqrt{x}}{x} = \frac{1}{\sqrt{x}}$$

(c) $$\lim_{x\to\infty} \frac{e^x}{x^2} = \lim_{x\to\infty} \frac{e^x}{2x} = \lim_{x\to\infty} \frac{e^x}{2} = \infty$$ ■

Next we turn our attention to the indeterminate forms $\infty \cdot 0$ and $\infty - \infty$. Sometimes these forms can be handled by using algebra to convert them to a $0/0$ or ∞/∞ form. Here again, we do not mean to suggest that $\infty \cdot 0$ or $\infty - \infty$ is a number. They are only notations for functional behaviors when considering limits. Here are examples of how we might work with these indeterminate forms.

EXAMPLE 5 Find the limits of these $\infty \cdot 0$ forms:

(a) $\displaystyle\lim_{x\to\infty} \left(x \sin\frac{1}{x}\right)$ **(b)** $\displaystyle\lim_{x\to 0^+} \sqrt{x} \ln x$

Solution

(a) $$\lim_{x\to\infty} \left(x \sin\frac{1}{x}\right) = \lim_{x\to\infty} \frac{\sin(1/x)}{1/x} \qquad \infty \cdot 0 \text{ converted to } \frac{0}{0}$$

$$= \lim_{x\to\infty} \frac{(\cos(1/x))(-1/x^2)}{-1/x^2} \qquad \text{L'Hôpital's Rule applied}$$

$$= \lim_{x\to\infty} \left(\cos\frac{1}{x}\right) = 1$$

(See Example 5b in Section 2.6 for an alternative method to solve this problem.)

(b) $$\lim_{x\to 0^+} \sqrt{x}\ln x = \lim_{x\to 0^+} \frac{\ln x}{1/\sqrt{x}} \qquad \infty \cdot 0 \text{ converted to } \infty/\infty$$

$$= \lim_{x\to 0^+} \frac{1/x}{-(1/2)x^{3/2}} \qquad \text{l'Hôpital's Rule applied}$$

$$= \lim_{x\to 0^+} (-2\sqrt{x}) = 0$$

EXAMPLE 6 Find the limit of this $\infty - \infty$ form:

$$\lim_{x\to 0}\left(\frac{1}{\sin x} - \frac{1}{x}\right).$$

Solution If $x \to 0^+$, then $\sin x \to 0^+$ and

$$\frac{1}{\sin x} - \frac{1}{x} \to \infty - \infty.$$

Similarly, if $x \to 0^-$, then $\sin x \to 0^-$ and

$$\frac{1}{\sin x} - \frac{1}{x} \to -\infty - (-\infty) = -\infty + \infty.$$

Neither form reveals what happens in the limit. To find out, we first combine the fractions:

$$\frac{1}{\sin x} - \frac{1}{x} = \frac{x - \sin x}{x \sin x}. \qquad \text{Common denominator is } x \sin x.$$

Then we apply l'Hôpital's Rule to the result:

$$\lim_{x\to 0}\left(\frac{1}{\sin x} - \frac{1}{x}\right) = \lim_{x\to 0} \frac{x - \sin x}{x \sin x} \qquad \frac{0}{0}$$

$$= \lim_{x\to 0} \frac{1 - \cos x}{\sin x + x\cos x} \qquad \text{Still } \frac{0}{0}$$

$$= \lim_{x\to 0} \frac{\sin x}{2\cos x - x \sin x} = \frac{0}{2} = 0. \qquad \blacksquare$$

Indeterminate Powers

Limits that lead to the indeterminate forms 1^∞, 0^0, and ∞^0 can sometimes be handled by first taking the logarithm of the function. We use l'Hôpital's Rule to find the limit of the logarithm expression and then exponentiate the result to find the original function limit. This procedure is justified by the continuity of the exponential function and Theorem 10 in Section 2.5, and it is formulated as follows. (The formula is also valid for one-sided limits.)

> If $\lim_{x\to a} \ln f(x) = L$, then
>
> $$\lim_{x\to a} f(x) = \lim_{x\to a} e^{\ln f(x)} = e^L.$$
>
> Here a may be either finite or infinite.

EXAMPLE 7 Apply l'Hôpital's Rule to show that $\lim_{x\to 0^+} (1 + x)^{1/x} = e$.

Solution The limit leads to the indeterminate form 1^∞. We let $f(x) = (1 + x)^{1/x}$ and find $\lim_{x\to 0^+} \ln f(x)$. Since

$$\ln f(x) = \ln(1 + x)^{1/x} = \frac{1}{x}\ln(1 + x),$$

l'Hôpital's Rule now applies to give

$$\begin{aligned}\lim_{x\to0^+}\ln f(x) &= \lim_{x\to0^+}\frac{\ln(1+x)}{x} && \frac{0}{0}\\ &= \lim_{x\to0^+}\frac{\dfrac{1}{1+x}}{1} && \text{L'Hôpital's Rule applied}\\ &= \frac{1}{1} = 1.\end{aligned}$$

Therefore, $\lim_{x\to0^+}(1+x)^{1/x} = \lim_{x\to0^+} f(x) = \lim_{x\to0^+} e^{\ln f(x)} = e^1 = e.$ ■

EXAMPLE 8 Find $\lim_{x\to\infty} x^{1/x}$.

Solution The limit leads to the indeterminate form ∞^0. We let $f(x) = x^{1/x}$ and find $\lim_{x\to\infty}\ln f(x)$. Since

$$\ln f(x) = \ln x^{1/x} = \frac{\ln x}{x},$$

l'Hôpital's Rule gives

$$\begin{aligned}\lim_{x\to\infty}\ln f(x) &= \lim_{x\to\infty}\frac{\ln x}{x} && \frac{\infty}{\infty}\\ &= \lim_{x\to\infty}\frac{1/x}{1} && \text{L'Hôpital's Rule applied}\\ &= \frac{0}{1} = 0.\end{aligned}$$

Therefore, $\lim_{x\to\infty} x^{1/x} = \lim_{x\to\infty} f(x) = \lim_{x\to\infty} e^{\ln f(x)} = e^0 = 1.$ ■

Proof of L'Hôpital's Rule

Before we prove l'Hôpital's Rule, we consider a special case to provide some geometric insight for its reasonableness. Consider the two functions $f(x)$ and $g(x)$ having *continuous* derivatives and satisfying $f(a) = g(a) = 0$, $g'(a) \neq 0$. The graphs of $f(x)$ and $g(x)$, together with their linearizations $y = f'(a)(x-a)$ and $y = g'(a)(x-a)$, are shown in Figure 7.19. We know that near $x = a$, the linearizations provide good approximations to the functions. In fact,

$$f(x) = f'(a)(x-a) + \varepsilon_1(x-a) \quad\text{and}\quad g(x) = g'(a)(x-a) + \varepsilon_2(x-a),$$

where $\varepsilon_1 \to 0$ and $\varepsilon_2 \to 0$ as $x \to a$. So, as Figure 7.19 suggests,

$$\begin{aligned}\lim_{x\to a}\frac{f(x)}{g(x)} &= \lim_{x\to a}\frac{f'(a)(x-a)+\varepsilon_1(x-a)}{g'(a)(x-a)+\varepsilon_2(x-a)}\\ &= \lim_{x\to a}\frac{f'(a)+\varepsilon_1}{g'(a)+\varepsilon_2} = \frac{f'(a)}{g'(a)} && g'(a)\neq 0\\ &= \lim_{x\to a}\frac{f'(x)}{g'(x)}, && \text{Continuous derivatives}\end{aligned}$$

as asserted by l'Hôpital's Rule. We now proceed to a proof of the rule based on the more general assumptions stated in Theorem 5, which do not require that $g'(a) \neq 0$ or that the two functions have *continuous* derivatives.

The proof of l'Hôpital's Rule is based on Cauchy's Mean Value Theorem, an extension of the Mean Value Theorem that involves two functions instead of one. We prove Cauchy's Theorem first and then show how it leads to l'Hôpital's Rule.

HISTORICAL BIOGRAPHY
Augustin-Louis Cauchy (1789–1857)
www.bit.ly/2OsRgUn

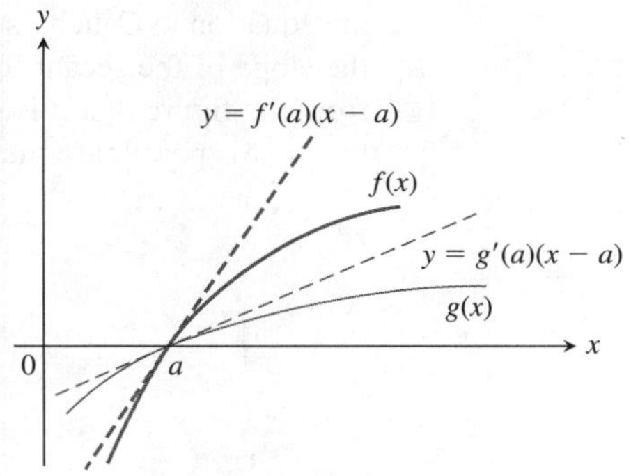

FIGURE 7.19 The two functions in l'Hôpital's Rule, graphed with their linear approximations at $x = a$.

When $g(x) = x$, Theorem 6 is the Mean Value Theorem.

THEOREM 6—Cauchy's Mean Value Theorem
Suppose functions f and g are continuous on $[a, b]$ and differentiable throughout (a, b) and also suppose $g'(x) \neq 0$ throughout (a, b). Then there exists a number c in (a, b) at which

$$\frac{f'(c)}{g'(c)} = \frac{f(b) - f(a)}{g(b) - g(a)}.$$

Proof We apply the Mean Value Theorem of Section 4.2 twice. First we use it to show that $g(a) \neq g(b)$. For if $g(b)$ did equal $g(a)$, then the Mean Value Theorem would give

$$g'(c) = \frac{g(b) - g(a)}{b - a} = 0$$

for some c between a and b, which cannot happen because $g'(x) \neq 0$ in (a, b).

We next apply the Mean Value Theorem to the function

$$F(x) = f(x) - f(a) - \frac{f(b) - f(a)}{g(b) - g(a)}[g(x) - g(a)].$$

This function is continuous and differentiable where f and g are, and $F(b) = F(a) = 0$. Therefore, there is a number c between a and b for which $F'(c) = 0$. When expressed in terms of f and g, this equation becomes

$$F'(c) = f'(c) - \frac{f(b) - f(a)}{g(b) - g(a)}[g'(c)] = 0$$

so that

$$\frac{f'(c)}{g'(c)} = \frac{f(b) - f(a)}{g(b) - g(a)}.$$ ■

Cauchy's Mean Value Theorem has a geometric interpretation for a general winding curve C in the plane joining the two points $A = (g(a), f(a))$ and $B = (g(b), f(b))$. In Chapter 11 you will learn how to describe general curves such as C, along with their tangent lines. There is at least one point P on the curve for which the tangent to the curve at P is parallel to the secant line joining the points A and B. The slope of that tangent line turns out to be the quotient f'/g' evaluated at the number c in the interval (a, b), which is the left-hand side of the equation in Theorem 6. Because the slope of the secant line joining A and B is

$$\frac{f(b) - f(a)}{g(b) - g(a)},$$

the equation in Cauchy's Mean Value Theorem says that the slope of the tangent line equals the slope of the secant line. This geometric interpretation is shown in Figure 7.20. Notice from the figure that it is possible for more than one point on the curve C to have a tangent line that is parallel to the secant line joining A and B.

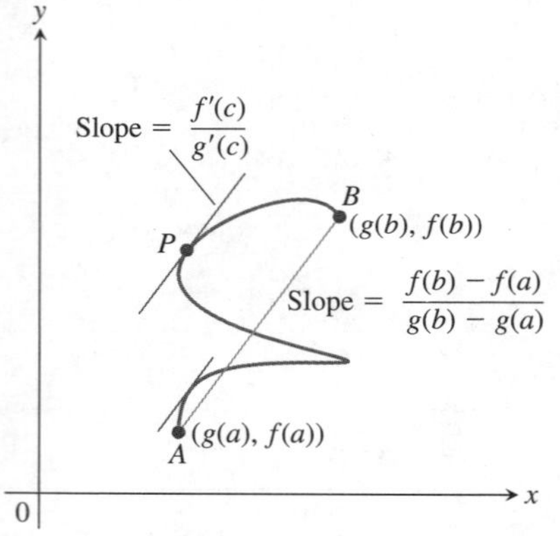

FIGURE 7.20 There is at least one point P on the curve C for which the slope of the tangent line to the curve at P is the same as the slope of the secant line joining the points $A(g(a), f(a))$ and $B(g(b), f(b))$.

Proof of l'Hôpital's Rule Since $\lim_{x\to a} f(x) = \lim_{x\to a} g(x) = 0$ and both f and g are differentiable at a, they must also be continuous at a, hence $f(a) = g(a) = 0$.

We first establish the limit equation for the case $x \to a^+$. The method needs almost no change to apply to $x \to a^-$, and the combination of these two cases establishes the result.

Suppose that x lies in the interval to the right of a. Then $g'(x) \neq 0$, and we can apply Cauchy's Mean Value Theorem to the closed interval from a to x. This step produces a number c between a and x such that

$$\frac{f'(c)}{g'(c)} = \frac{f(x) - f(a)}{g(x) - g(a)}.$$

But $f(a) = g(a) = 0$, so

$$\frac{f'(c)}{g'(c)} = \frac{f(x)}{g(x)}.$$

As x approaches a, c approaches a because it always lies between a and x. Therefore,

$$\lim_{x\to a^+} \frac{f(x)}{g(x)} = \lim_{c\to a^+} \frac{f'(c)}{g'(c)} = \lim_{x\to a^+} \frac{f'(x)}{g'(x)},$$

which establishes l'Hôpital's Rule for the case where x approaches a from above. The case where x approaches a from below is proved by applying Cauchy's Mean Value Theorem to the closed interval $[x, a]$, $x < a$. ■

EXERCISES 7.5

Finding Limits in Two Ways

In Exercises 1–6, use l'Hôpital's Rule to evaluate the limit. Then evaluate the limit using a method studied in Chapter 2.

1. $\lim_{x\to -2} \dfrac{x + 2}{x^2 - 4}$

2. $\lim_{x\to 0} \dfrac{\sin 5x}{x}$

3. $\lim_{x\to\infty} \dfrac{5x^2 - 3x}{7x^2 + 1}$

4. $\lim_{x\to 1} \dfrac{x^3 - 1}{4x^3 - x - 3}$

5. $\lim_{x\to 0} \dfrac{1 - \cos x}{x^2}$

6. $\lim_{x\to\infty} \dfrac{2x^2 + 3x}{x^3 + x + 1}$

Applying l'Hôpital's Rule

Use l'Hôpital's rule to find the limits in Exercises 7–52.

7. $\lim_{x\to 2} \dfrac{x-2}{x^2-4}$

8. $\lim_{x\to -5} \dfrac{x^2-25}{x+5}$

9. $\lim_{t\to -3} \dfrac{t^3-4t+15}{t^2-t-12}$

10. $\lim_{t\to -1} \dfrac{3t^3+3}{4t^3-t+3}$

11. $\lim_{x\to\infty} \dfrac{5x^3-2x}{7x^3+3}$

12. $\lim_{x\to\infty} \dfrac{x-8x^2}{12x^2+5x}$

13. $\lim_{t\to 0} \dfrac{\sin t^2}{t}$

14. $\lim_{x\to 0} \dfrac{x-\tan x}{x^3}$

15. $\lim_{x\to 0} \dfrac{8x^2}{\cos x - 1}$

16. $\lim_{x\to 0} \dfrac{\sin x - x}{x^3}$

17. $\lim_{\theta\to\pi/2} \dfrac{2\theta-\pi}{\cos(2\pi-\theta)}$

18. $\lim_{\theta\to -\pi/3} \dfrac{3\theta+\pi}{\sin(\theta+(\pi/3))}$

19. $\lim_{\theta\to\pi/6} \dfrac{\sin\theta-\frac{1}{2}}{\theta-\frac{\pi}{6}}$

20. $\lim_{\theta\to\pi/4} \dfrac{\tan\theta-1}{\theta-\frac{\pi}{4}}$

21. $\lim_{\theta\to\pi/2} \dfrac{1-\sin\theta}{1+\cos 2\theta}$

22. $\lim_{x\to 1} \dfrac{x-1}{\ln x - \sin\pi x}$

23. $\lim_{x\to 0} \dfrac{x^2}{\ln(\sec x)}$

24. $\lim_{x\to\pi/2} \dfrac{\ln(\csc x)}{(x-(\pi/2))^2}$

25. $\lim_{t\to 0} \dfrac{t(1-\cos t)}{t-\sin t}$

26. $\lim_{t\to 0} \dfrac{t\sin t}{1-\cos t}$

27. $\lim_{x\to(\pi/2)^-} \left(x-\dfrac{\pi}{2}\right)\sec x$

28. $\lim_{x\to(\pi/2)^-} \left(\dfrac{\pi}{2}-x\right)\tan x$

29. $\lim_{\theta\to 0} \dfrac{3^{\sin\theta}-1}{\theta}$

30. $\lim_{\theta\to 0} \dfrac{(1/2)^\theta-1}{\theta}$

31. $\lim_{x\to 0} \dfrac{x2^x}{2^x-1}$

32. $\lim_{x\to 0} \dfrac{3^x-1}{2^x-1}$

33. $\lim_{x\to\infty} \dfrac{\ln(x+1)}{\log_2 x}$

34. $\lim_{x\to\infty} \dfrac{\log_2 x}{\log_3(x+3)}$

35. $\lim_{x\to 0^+} \dfrac{\ln(x^2+2x)}{\ln x}$

36. $\lim_{x\to 0^+} \dfrac{\ln(e^x-1)}{\ln x}$

37. $\lim_{y\to 0} \dfrac{\sqrt{5y+25}-5}{y}$

38. $\lim_{y\to 0} \dfrac{\sqrt{ay+a^2}-a}{y}, \quad a>0$

39. $\lim_{x\to\infty} (\ln 2x - \ln(x+1))$

40. $\lim_{x\to 0^+} (\ln x - \ln\sin x)$

41. $\lim_{x\to 0^+} \dfrac{(\ln x)^2}{\ln(\sin x)}$

42. $\lim_{x\to 0^+} \left(\dfrac{3x+1}{x}-\dfrac{1}{\sin x}\right)$

43. $\lim_{x\to 1^+} \left(\dfrac{1}{x-1}-\dfrac{1}{\ln x}\right)$

44. $\lim_{x\to 0^+} (\csc x - \cot x + \cos x)$

45. $\lim_{\theta\to 0} \dfrac{\cos\theta-1}{e^\theta-\theta-1}$

46. $\lim_{h\to 0} \dfrac{e^h-(1+h)}{h^2}$

47. $\lim_{t\to\infty} \dfrac{e^t+t^2}{e^t-t}$

48. $\lim_{x\to\infty} x^2e^{-x}$

49. $\lim_{x\to 0} \dfrac{x-\sin x}{x\tan x}$

50. $\lim_{x\to 0} \dfrac{(e^x-1)^2}{x\sin x}$

51. $\lim_{\theta\to 0} \dfrac{\theta-\sin\theta\cos\theta}{\tan\theta-\theta}$

52. $\lim_{x\to 0} \dfrac{\sin 3x - 3x + x^2}{\sin x\sin 2x}$

Indeterminate Powers and Products

Find the limits in Exercises 53–68.

53. $\lim_{x\to 1^+} x^{1/(1-x)}$

54. $\lim_{x\to 1^+} x^{1/(x-1)}$

55. $\lim_{x\to 0} \left(\dfrac{\tan x}{x}\right)^{1/x^2}$

56. $\lim_{x\to e^+} (\ln x)^{1/(x-e)}$

57. $\lim_{x\to 0^+} x^{-1/\ln x}$

58. $\lim_{x\to\infty} x^{1/\ln x}$

59. $\lim_{x\to\infty} (1+2x)^{1/(2\ln x)}$

60. $\lim_{x\to 0} (e^x+x)^{1/x}$

61. $\lim_{x\to 0^+} x^x$

62. $\lim_{x\to 0^+} \left(1+\dfrac{1}{x}\right)^x$

63. $\lim_{x\to 0} \left[\dfrac{a^x+b^x+c^x+d^x}{4}\right]^{1/x}$, where a, b, c, d are positive numbers

64. $\lim_{x\to\infty} \left(\dfrac{x^2+1}{x+2}\right)^{1/x}$

65. $\lim_{x\to 0^+} x^2\ln x$

66. $\lim_{x\to 0^+} x(\ln x)^2$

67. $\lim_{x\to 0^+} x\tan\left(\dfrac{\pi}{2}-x\right)$

68. $\lim_{x\to 0^+} \sin x\cdot\ln x$

Theory and Applications

L'Hôpital's Rule does not help with the limits in Exercises 69–76. Try it—you just keep on cycling. Find the limits some other way.

69. $\lim_{x\to\infty} \dfrac{\sqrt{9x+1}}{\sqrt{x+1}}$

70. $\lim_{x\to 0^+} \dfrac{\sqrt{x}}{\sqrt{\sin x}}$

71. $\lim_{x\to(\pi/2)^-} \dfrac{\sec x}{\tan x}$

72. $\lim_{x\to 0^+} \dfrac{\cot x}{\csc x}$

73. $\lim_{x\to\infty} \dfrac{2^x-3^x}{3^x+4^x}$

74. $\lim_{x\to -\infty} \dfrac{2^x+4^x}{5^x-2^x}$

75. $\lim_{x\to\infty} \dfrac{e^{x^2}}{xe^x}$

76. $\lim_{x\to 0^+} \dfrac{x}{e^{-1/x}}$

77. Which one is correct, and which one is wrong? Give reasons for your answers.

a. $\lim_{x\to 3} \dfrac{x-3}{x^2-3} = \lim_{x\to 3} \dfrac{1}{2x} = \dfrac{1}{6}$

b. $\lim_{x\to 3} \dfrac{x-3}{x^2-3} = \dfrac{0}{6} = 0$

78. Which one is correct, and which one is wrong? Give reasons for your answers.

a. $\lim_{x\to 0} \dfrac{x^2-2x}{x^2-\sin x} = \lim_{x\to 0} \dfrac{2x-2}{2x-\cos x}$

$= \lim_{x\to 0} \dfrac{2}{2+\sin x} = \dfrac{2}{2+0} = 1$

b. $\lim_{x\to 0} \dfrac{x^2-2x}{x^2-\sin x} = \lim_{x\to 0} \dfrac{2x-2}{2x-\cos x} = \dfrac{-2}{0-1} = 2$

79. Only one of these calculations is correct. Which one? Why are the others wrong? Give reasons for your answers.

a. $\lim_{x\to 0^+} x\ln x = 0\cdot(-\infty) = 0$

b. $\lim_{x\to 0^+} x\ln x = 0\cdot(-\infty) = -\infty$

c. $\lim_{x\to 0^+} x\ln x = \lim_{x\to 0^+} \dfrac{\ln x}{(1/x)} = \dfrac{-\infty}{\infty} = -1$

d. $\lim_{x\to 0^+} x\ln x = \lim_{x\to 0^+} \dfrac{\ln x}{(1/x)}$

$= \lim_{x\to 0^+} \dfrac{(1/x)}{(-1/x^2)} = \lim_{x\to 0^+} (-x) = 0$

80. Find all values of c that satisfy the conclusion of Cauchy's Mean Value Theorem for the given functions and interval.

a. $f(x) = x, \quad g(x) = x^2, \quad (a, b) = (-2, 0)$

b. $f(x) = x, \quad g(x) = x^2, \quad (a, b)$ arbitrary

c. $f(x) = x^3/3 - 4x, \quad g(x) = x^2, \quad (a, b) = (0, 3)$

81. Continuous extension Find a value of c that makes the function

$$f(x) = \begin{cases} \dfrac{9x - 3\sin 3x}{5x^3}, & x \neq 0 \\ c, & x = 0 \end{cases}$$

continuous at $x = 0$. Explain why your value of c works.

82. For what values of a and b is

$$\lim_{x\to 0}\left(\frac{\tan 2x}{x^3} + \frac{a}{x^2} + \frac{\sin bx}{x}\right) = 0?$$

T **83. $\infty - \infty$ Form**

a. Estimate the value of

$$\lim_{x\to\infty}\left(x - \sqrt{x^2 + x}\right)$$

by graphing $f(x) = x - \sqrt{x^2 + x}$ over a suitably large interval of x-values.

b. Now confirm your estimate by finding the limit with l'Hôpital's Rule. As the first step, multiply $f(x)$ by the fraction $\left(x + \sqrt{x^2 + x}\right)/\left(x + \sqrt{x^2 + x}\right)$ and simplify the new numerator.

84. Find $\lim_{x\to\infty}\left(\sqrt{x^2 + 1} - \sqrt{x}\right)$.

T **85. 0/0 Form** Estimate the value of

$$\lim_{x\to 1}\frac{2x^2 - (3x + 1)\sqrt{x} + 2}{x - 1}$$

by graphing. Then confirm your estimate with l'Hôpital's Rule.

86. This exercise explores the difference between

$$\lim_{x\to\infty}\left(1 + \frac{1}{x^2}\right)^x$$

and

$$\lim_{x\to\infty}\left(1 + \frac{1}{x}\right)^x = e.$$

a. Use l'Hôpital's Rule to show that

$$\lim_{x\to\infty}\left(1 + \frac{1}{x}\right)^x = e.$$

T **b.** Graph

$$f(x) = \left(1 + \frac{1}{x^2}\right)^x \quad \text{and} \quad g(x) = \left(1 + \frac{1}{x}\right)^x$$

together for $x \geq 0$. How does the behavior of f compare with that of g? Estimate the value of $\lim_{x\to\infty} f(x)$.

c. Confirm your estimate of $\lim_{x\to\infty} f(x)$ by calculating it with l'Hôpital's Rule.

87. Show that

$$\lim_{k\to\infty}\left(1 + \frac{r}{k}\right)^k = e^r.$$

88. Given that $x > 0$, find the maximum value, if any, of

a. $x^{1/x}$

b. x^{1/x^2}

c. x^{1/x^n} (n a positive integer)

d. Show that $\lim_{x\to\infty} x^{1/x^n} = 1$ for every positive integer n.

89. Use limits to find horizontal asymptotes for each function.

a. $y = x\tan\left(\dfrac{1}{x}\right)$

b. $y = \dfrac{3x + e^{2x}}{2x + e^{3x}}$

90. Find $f'(0)$ for $f(x) = \begin{cases} e^{-1/x^2}, & x \neq 0 \\ 0, & x = 0. \end{cases}$

T **91. The continuous extension of $(\sin x)^x$ to $[0, \pi]$**

a. Graph $f(x) = (\sin x)^x$ on the interval $0 \leq x \leq \pi$. What value would you assign to f to make it continuous at $x = 0$?

b. Verify your conclusion in part (a) by finding $\lim_{x\to 0^+} f(x)$ with l'Hôpital's Rule.

c. Returning to the graph, estimate the maximum value of f on $[0, \pi]$. About where is max f taken on?

d. Sharpen your estimate in part (c) by graphing f' in the same window to see where its graph crosses the x-axis. To simplify your work, you might want to delete the exponential factor from the expression for f' and graph just the factor that has a zero.

T **92. The function $(\sin x)^{\tan x}$** *(Continuation of Exercise 91)*

a. Graph $f(x) = (\sin x)^{\tan x}$ on the interval $-7 \leq x \leq 7$. How do you account for the gaps in the graph? How wide are the gaps?

b. Now graph f on the interval $0 \leq x \leq \pi$. The function is not defined at $x = \pi/2$, but the graph has no break at this point. What is going on? What value does the graph appear to give for f at $x = \pi/2$? (*Hint:* Use l'Hôpital's Rule to find lim f as $x \to (\pi/2)^-$ and $x \to (\pi/2)^+$.)

c. Continuing with the graphs in part (b), find max f and min f as accurately as you can and estimate the values of x at which they are taken on.

7.6 Inverse Trigonometric Functions

Inverse trigonometric functions arise when we want to calculate angles from side measurements in triangles. They also provide useful antiderivatives and appear frequently in the solutions of differential equations. This section shows how these functions are defined, graphed, and evaluated, how their derivatives are computed, and why they appear as important antiderivatives.

Defining the Inverse Trigonometric Functions

The six basic trigonometric functions are not one-to-one (since their values repeat periodically). However, we can restrict their domains to intervals on which they are one-to-one. The sine function increases from -1 at $x = -\pi/2$ to $+1$ at $x = \pi/2$. By restricting its domain to the interval $[-\pi/2, \pi/2]$ we make it one-to-one, so that it has an inverse which is called $\arcsin x$ (Figure 7.21). Similar domain restrictions can be applied to all six trigonometric functions.

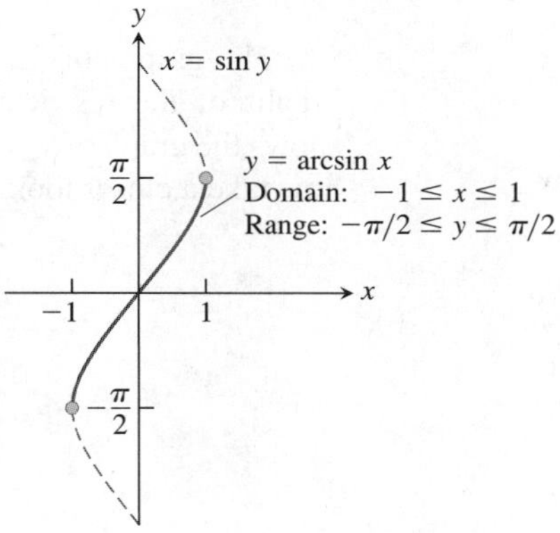

FIGURE 7.21 The graph of $y = \arcsin x$.

Domain restrictions that make the trigonometric functions one-to-one

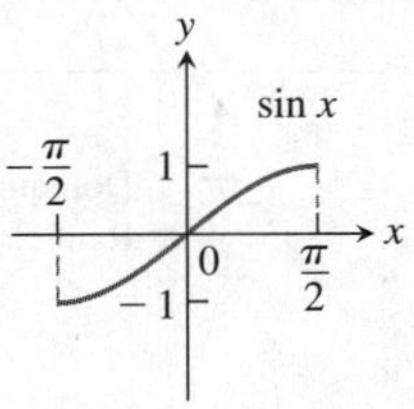

$y = \sin x$
Domain: $[-\pi/2, \pi/2]$
Range: $[-1, 1]$

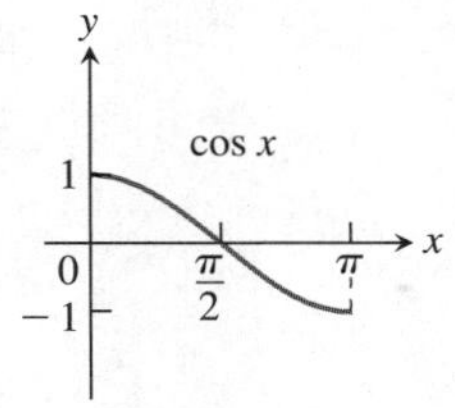

$y = \cos x$
Domain: $[0, \pi]$
Range: $[-1, 1]$

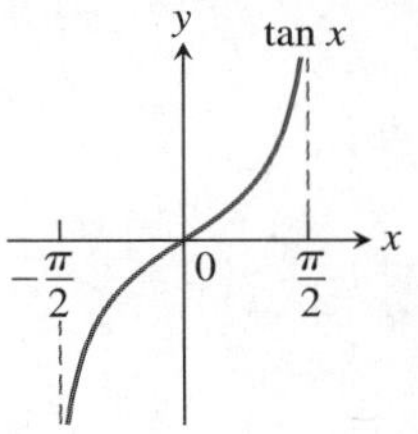

$y = \tan x$
Domain: $(-\pi/2, \pi/2)$
Range: $(-\infty, \infty)$

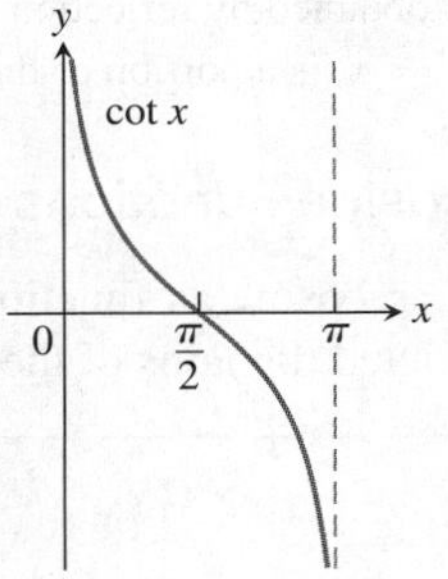

$y = \cot x$
Domain: $(0, \pi)$
Range: $(-\infty, \infty)$

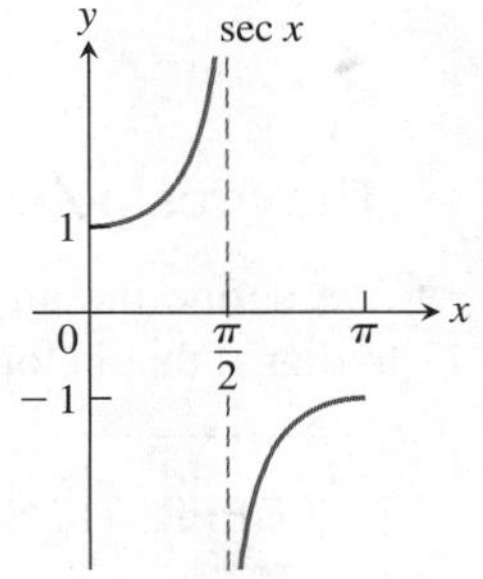

$y = \sec x$
Domain: $[0, \pi/2) \cup (\pi/2, \pi]$
Range: $(-\infty, -1] \cup [1, \infty)$

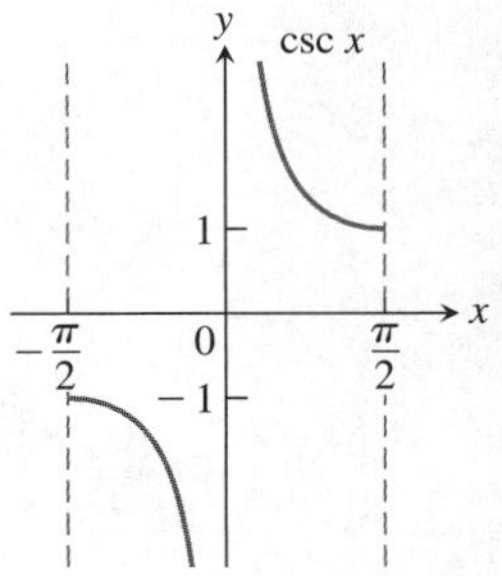

$y = \csc x$
Domain: $[-\pi/2, 0) \cup (0, \pi/2]$
Range: $(-\infty, -1] \cup [1, \infty)$

Since these restricted functions are now one-to-one, they have inverses, which we denote by

$$\begin{array}{llll} y = \sin^{-1} x & \text{or} \quad y = \arcsin x, & y = \cos^{-1} x & \text{or} \quad y = \arccos x \\ y = \tan^{-1} x & \text{or} \quad y = \arctan x, & y = \cot^{-1} x & \text{or} \quad y = \operatorname{arccot} x \\ y = \sec^{-1} x & \text{or} \quad y = \operatorname{arcsec} x, & y = \csc^{-1} x & \text{or} \quad y = \operatorname{arccsc} x \end{array}$$

These equations are read "y equals the arcsine of x" or "y equals arcsin x" and so on.

Caution The -1 in the expressions for the inverse means "inverse." It does *not* mean reciprocal. For example, the *reciprocal* of $\sin x$ is $(\sin x)^{-1} = 1/\sin x = \csc x$. ●

The graphs of the six inverse trigonometric functions are obtained by reflecting the graphs of the restricted trigonometric functions through the line $y = x$. Figure 7.22b shows the graph of $y = \arcsin x$ and Figure 7.23 shows the graphs of all six functions. We now take a closer look at these functions.

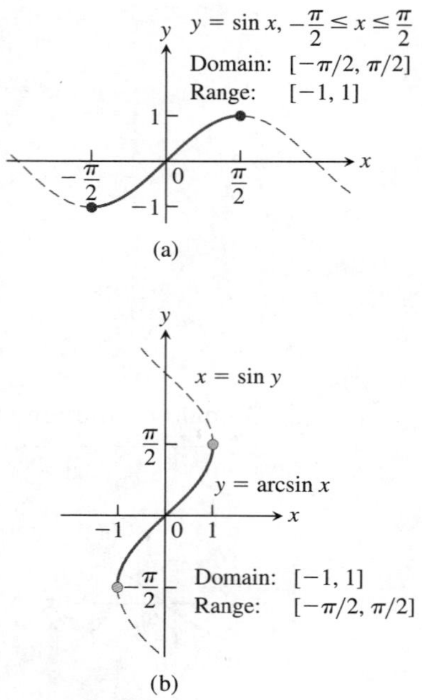

FIGURE 7.22 The graphs of (a) $y = \sin x, -\pi/2 \le x \le \pi/2$, and (b) its inverse, $y = \arcsin x$. The graph of arcsin x, obtained by reflection across the line $y = x$, is a portion of the curve $x = \sin y$.

The Arcsine and Arccosine Functions

We define the arcsine and arccosine as functions whose values are angles (measured in radians) that belong to restricted domains of the sine and cosine functions.

> **DEFINITION**
>
> $y =$ **arcsin** $\boldsymbol{x}$ is the number in $[-\pi/2, \pi/2]$ for which $\sin y = x$.
>
> $y =$ **arccos** $\boldsymbol{x}$ is the number in $[0, \pi]$ for which $\cos y = x$.

The "Arc" in Arcsine and Arccosine
For a unit circle and radian angles, the arc length equation $s = r\theta$ becomes $s = \theta$, so central angles and the arcs they subtend have the same measure. If $x = \sin\theta$, then, in addition to being the angle whose sine is x, θ is also the length of the arc on the unit circle that subtends an angle whose sine is x. So we call θ "the arc whose sine is x."

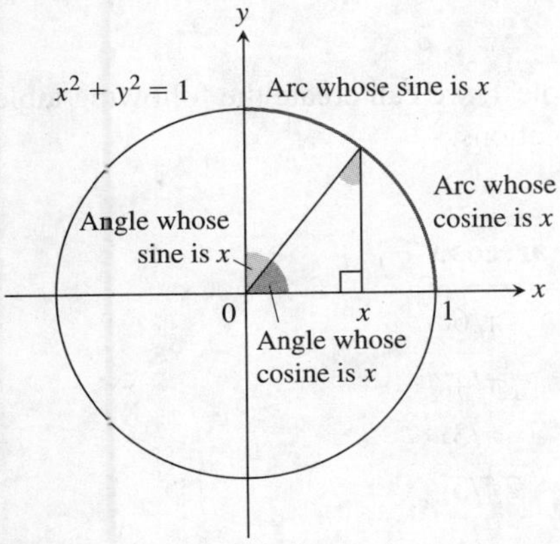

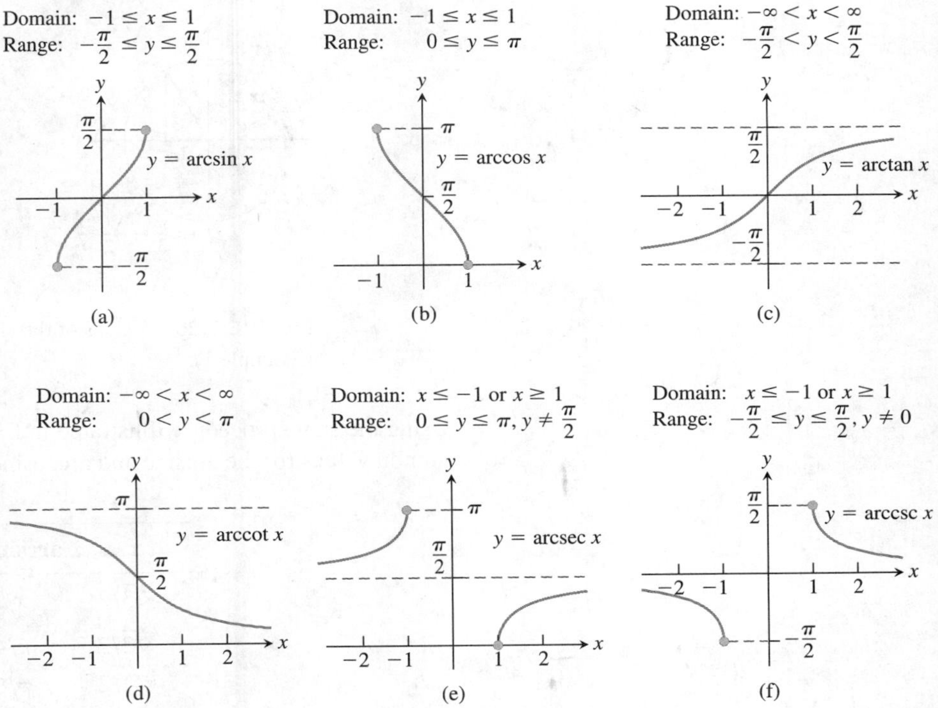

FIGURE 7.23 Graphs of the six basic inverse trigonometric functions.

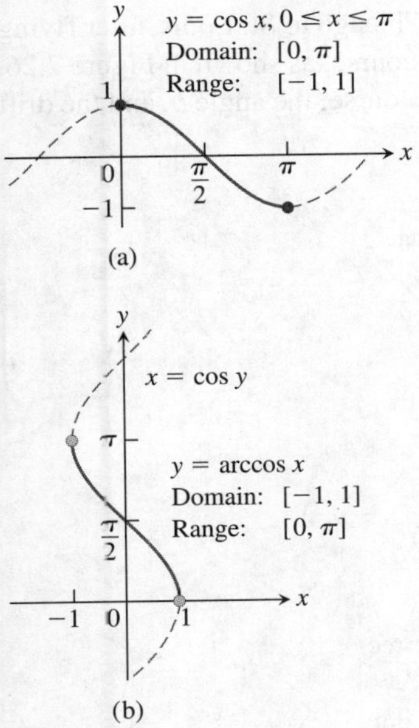

FIGURE 7.24 The graphs of (a) $y = \cos x, 0 \le x \le \pi$, and (b) its inverse, $y = \arccos x$. The graph of $\arccos x$, obtained by reflection across the line $y = x$, is a portion of the curve $x = \cos y$.

The graph of $y = \arcsin x$ (Figure 7.22b) is symmetric about the origin. The arcsine is therefore an odd function:

$$\arcsin(-x) = -\arcsin x. \tag{1}$$

The graph of $y = \arccos x$ (Figure 7.24b) has no such symmetry.

EXAMPLE 1 Evaluate (**a**) $\arcsin\left(\frac{\sqrt{3}}{2}\right)$ and (**b**) $\arccos\left(-\frac{1}{2}\right)$.

Solution

(**a**) We see that

$$\arcsin\left(\frac{\sqrt{3}}{2}\right) = \frac{\pi}{3}$$

because $\sin(\pi/3) = \sqrt{3}/2$ and $\pi/3$ belongs to the range $[-\pi/2, \pi/2]$ of the arcsine function. See Figure 7.25a.

(**b**) We have

$$\arccos\left(-\frac{1}{2}\right) = \frac{2\pi}{3}$$

because $\cos(2\pi/3) = -1/2$ and $2\pi/3$ belongs to the range $[0, \pi]$ of the arccosine function. See Figure 7.25b. ■

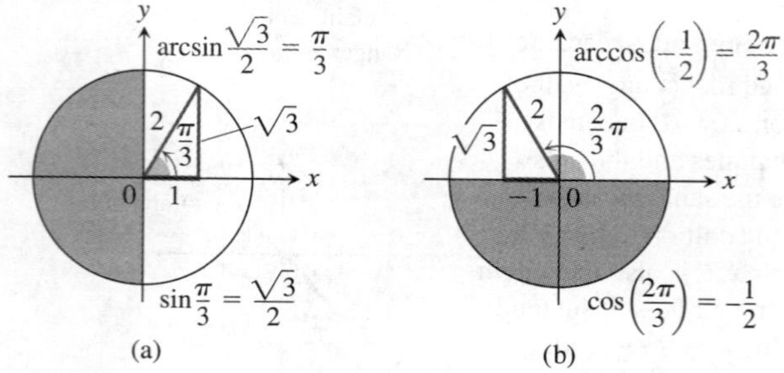

FIGURE 7.25 Values of the arcsine and arccosine functions (Example 1).

Using the same procedure illustrated in Example 1, we can create the following table of common values for the arcsine and arccosine functions.

x	$\arcsin x$	$\arccos x$
$\sqrt{3}/2$	$\pi/3$	$\pi/6$
$\sqrt{2}/2$	$\pi/4$	$\pi/4$
$1/2$	$\pi/6$	$\pi/3$
$-1/2$	$-\pi/6$	$2\pi/3$
$-\sqrt{2}/2$	$-\pi/4$	$3\pi/4$
$-\sqrt{3}/2$	$-\pi/3$	$5\pi/6$

EXAMPLE 2 During a 240-mile airplane flight from Chicago to St. Louis, after flying 180 mi the navigator determines that the plane is 12 mi off course, as shown in Figure 7.26. Find the angle a for a course parallel to the original correct course, the angle b, and the drift correction angle $c = a + b$.

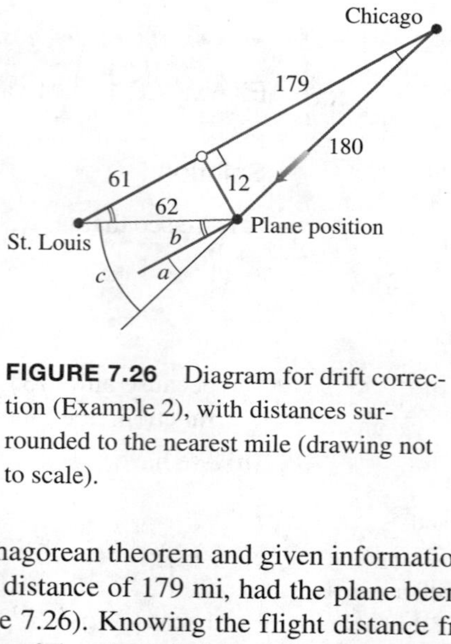

FIGURE 7.26 Diagram for drift correction (Example 2), with distances surrounded to the nearest mile (drawing not to scale).

Solution From the Pythagorean theorem and given information, we compute an approximate hypothetical flight distance of 179 mi, had the plane been flying along the original correct course (see Figure 7.26). Knowing the flight distance from Chicago to St. Louis, we next calculate the remaining leg of the original course to be 61 mi. Applying the Pythagorean theorem again then gives an approximate distance of 62 mi from the position of the plane to St. Louis. Finally, from Figure 7.26, we see that $180 \sin a = 12$ and $62 \sin b = 12$, so

$$a = \arcsin \frac{12}{180} \approx 0.067 \text{ radian} \approx 3.8^\circ$$
$$b = \arcsin \frac{12}{62} \approx 0.195 \text{ radian} \approx 11.2^\circ$$
$$c = a + b \approx 15^\circ.$$

■

Identities Involving Arcsine and Arccosine

As we can see from Figure 7.27, the arccosine of x satisfies the identity

$$\arccos x + \arccos(-x) = \pi, \tag{2}$$

or

$$\arccos(-x) = \pi - \arccos x. \tag{3}$$

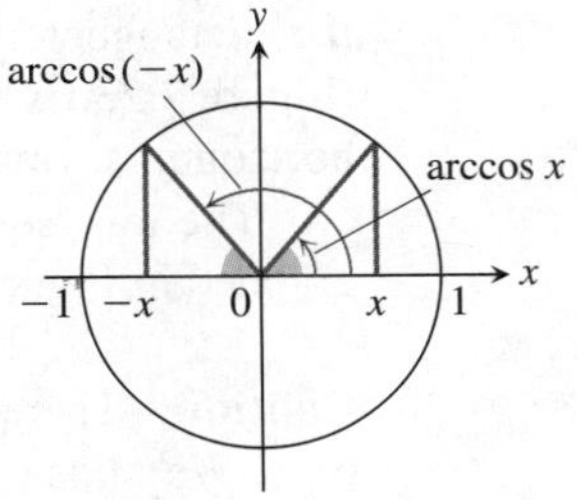

FIGURE 7.27 arccos x and arccos$(-x)$ are supplementary angles (so their sum is π).

Also, we can see from the triangle in Figure 7.28 that for $x > 0$,

$$\arcsin x + \arccos x = \pi/2. \tag{4}$$

Equation (4) holds for the other values of x in $[-1, 1]$ as well, but we cannot conclude this from the triangle in Figure 7.28. It is, however, a consequence of Equations (1) and (3) (Exercise 125).

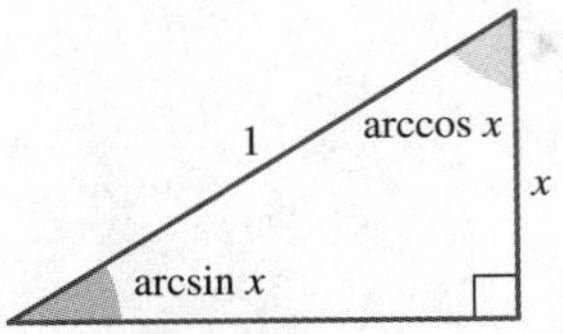

FIGURE 7.28 arcsin x and arccos x are complementary angles (so their sum is $\pi/2$).

Inverses of tan x, cot x, sec x, and csc x

The graphs of these four basic inverse trigonometric functions were shown in Figure 7.23. We obtain these graphs by reflecting the graphs of the restricted trigonometric functions through the line $y = x$. The arctangent of x is a radian angle whose tangent is x. The arccotangent of x is an angle whose cotangent is x, and so forth. The angles belong to the restricted domains of the tangent, cotangent, secant, and cosecant functions.

DEFINITIONS

$y =$ **arctan** $\boldsymbol{x}$ is the number in $(-\pi/2, \pi/2)$ for which $\tan y = x$.

$y =$ **arccot** $\boldsymbol{x}$ is the number in $(0, \pi)$ for which $\cot y = x$.

$y =$ **arcsec** $\boldsymbol{x}$ is the number in $[0, \pi/2) \cup (\pi/2, \pi]$ for which $\sec y = x$.

$y =$ **arccsc** $\boldsymbol{x}$ is the number in $[-\pi/2, 0) \cup (0, \pi/2]$ for which $\csc y = x$.

We use open or half-open intervals to avoid values for which the tangent, cotangent, secant, and cosecant functions are undefined. (See Figure 7.23.)

Just as the inverse sine and the inverse cosine functions are often written as $\sin^{-1} x$ and $\cos^{-1} x$ instead of $\arcsin x$ and $\arccos x$, we often denote the other inverse trigonometric functions by $\tan^{-1} x$, $\cot^{-1} x$, $\sec^{-1} x$ and $\csc^{-1} x$.

The graph of $y = \arctan x$ is symmetric about the origin because it is a branch of the graph $x = \tan y$ that is symmetric about the origin (Figure 7.23c). Algebraically this means that

$$\arctan(-x) = -\arctan x;$$

the arctangent is an odd function. The graph of $y = \operatorname{arccot} x$ has no such symmetry (Figure 7.23d). Notice from Figure 7.23a that the graph of the arctangent function has two horizontal asymptotes: one at $y = \pi/2$ and the other at $y = -\pi/2$.

The inverses of the restricted forms of $\sec x$ and $\csc x$ are chosen to be the functions graphed in Figures 7.23e and 7.23f.

Caution There is no general agreement about how to define $\operatorname{arcsec} x$ for negative values of x. We chose angles in the second quadrant between $\pi/2$ and π. This choice makes $\operatorname{arcsec} x = \arccos(1/x)$. It also makes $\operatorname{arcsec} x$ an increasing function on each interval of its domain. Some tables choose $\operatorname{arcsec} x$ to lie in $[-\pi, -\pi/2)$ for $x < 0$ and some texts choose it to lie in $[\pi, 3\pi/2)$ (Figure 7.29). These choices simplify the formula for the derivative (our formula needs absolute value signs) but fail to satisfy the computational equation $\operatorname{arcsec} x = \arccos(1/x)$. From this, we can derive the identity

$$\operatorname{arcsec} x = \arccos\left(\frac{1}{x}\right) = \frac{\pi}{2} - \arcsin\left(\frac{1}{x}\right) \tag{5}$$

by applying Equation (4). ●

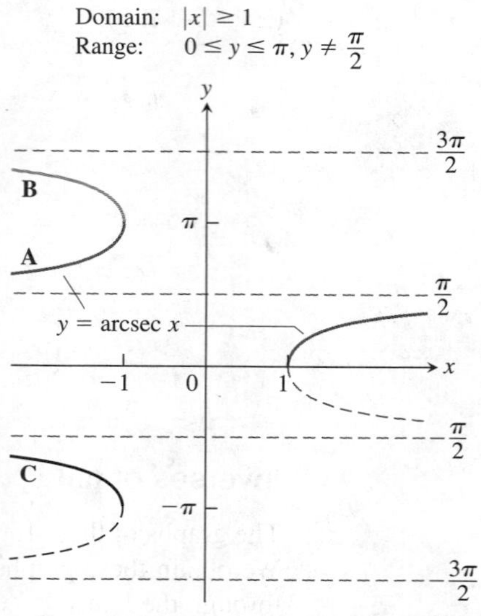

FIGURE 7.29 There are several logical choices for the left-hand branch of $y = \operatorname{arcsec} x$. With choice **A**, $\operatorname{arcsec} x = \arccos(1/x)$, a useful identity employed by many calculators.

EXAMPLE 3 The accompanying figures show two values of arctan x.

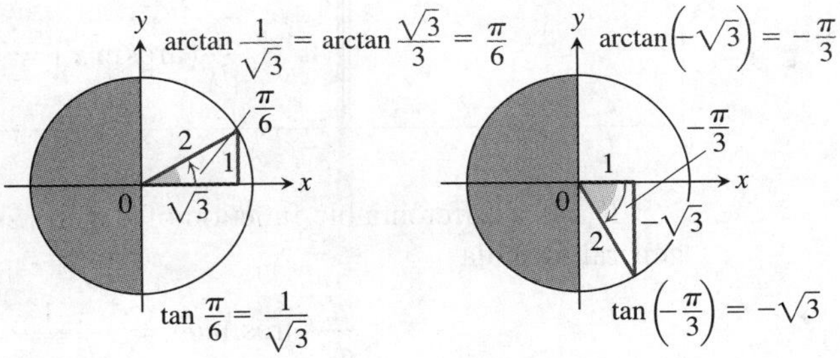

x	**arctan x**
$\sqrt{3}$	$\pi/3$
1	$\pi/4$
$\sqrt{3}/3$	$\pi/6$
$-\sqrt{3}/3$	$-\pi/6$
-1	$-\pi/4$
$-\sqrt{3}$	$-\pi/3$

The angles come from the first and fourth quadrants because the range of arctan x is $(-\pi/2, \pi/2)$. ■

The Derivative of $y = \arcsin x$

We know that the function $x = \sin y$ is differentiable in the interval $-\pi/2 < y < \pi/2$ and that its derivative, the cosine, is positive there. Theorem 1 in Section 7.1 therefore assures us that the inverse function $y = \arcsin x$ is differentiable throughout the interval $-1 < x < 1$. We cannot expect it to be differentiable at $x = 1$ or $x = -1$ because the tangent lines to the graph are vertical at these points (see Figure 7.30).

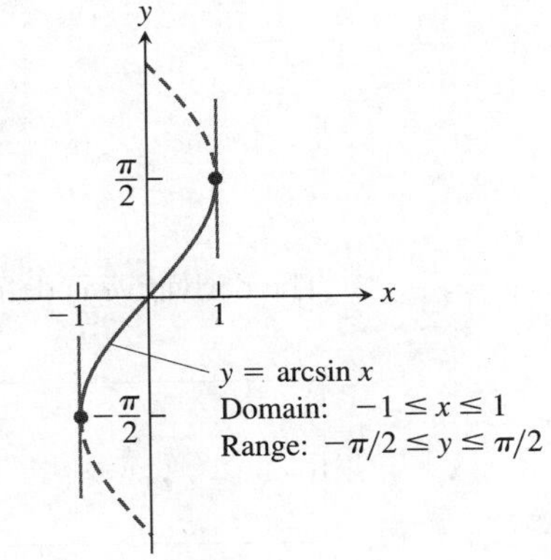

FIGURE 7.30 The graph of $y = \arcsin x$ has vertical tangent lines at $x = -1$ and $x = 1$.

We find the derivative of $y = \arcsin x$ by applying Theorem 1 with $f(x) = \sin x$ and $f^{-1}(x) = \arcsin x$:

$$\begin{aligned}
(f^{-1})'(x) &= \frac{1}{f'(f^{-1}(x))} && \text{Theorem 1} \\
&= \frac{1}{\cos(\arcsin x)} && f'(y) = \cos y \\
&= \frac{1}{\sqrt{1 - \sin^2(\arcsin x)}} && \cos y = \sqrt{1 - \sin^2 y} \\
&= \frac{1}{\sqrt{1 - x^2}}. && \sin(\arcsin x) = x
\end{aligned}$$

For $|x| < 1$,

$$\frac{d}{dx}(\arcsin x) = \frac{1}{\sqrt{1 - x^2}}.$$

If u is a differentiable function of x with $|u| < 1$, we apply the Chain Rule to get the general formula

$$\frac{d}{dx}(\arcsin u) = \frac{1}{\sqrt{1 - u^2}} \frac{du}{dx}, \qquad |u| < 1.$$

EXAMPLE 4 Using the Chain Rule, we calculate the derivative

$$\frac{d}{dx}(\arcsin x^2) = \frac{1}{\sqrt{1 - (x^2)^2}} \cdot \frac{d}{dx}(x^2) = \frac{2x}{\sqrt{1 - x^4}}.$$ ■

The Derivative of $y = \arctan x$

We find the derivative of $y = \arctan x$ by applying Theorem 1 with $f(x) = \tan x$ and $f^{-1}(x) = \arctan x$. Theorem 1 can be applied because the derivative of $\tan x$ is positive for $-\pi/2 < x < \pi/2$:

$$\begin{aligned}
(f^{-1})'(x) &= \frac{1}{f'(f^{-1}(x))} && \text{Theorem 1} \\
&= \frac{1}{\sec^2(\arctan x)} && f'(u) = \sec^2 u \\
&= \frac{1}{1 + \tan^2(\arctan x)} && \sec^2 u = 1 + \tan^2 u \\
&= \frac{1}{1 + x^2}. && \tan(\arctan x) = x
\end{aligned}$$

The derivative is defined for all real numbers:

$$\frac{d}{dx}(\arctan x) = \frac{1}{1 + x^2}.$$

If u is a differentiable function of x, we get the Chain Rule form:

$$\frac{d}{dx}(\arctan u) = \frac{1}{1 + u^2} \frac{du}{dx}.$$

The Chain Rule can also be combined with the arctangent function in other ways, as illustrated by the following example.

EXAMPLE 5

$$\begin{aligned}
\frac{d}{dx}\left(\frac{1}{\arctan x}\right) &= \frac{d}{dx}(\arctan x)^{-1} && \text{Derivative of the reciprocal (not the inverse) of arctangent} \\
&= (-1)(\arctan x)^{-2} \frac{d}{dx}(\arctan x) && \text{Apply the Chain Rule.} \\
&= \frac{-1}{(\arctan x)^2} \cdot \frac{1}{1 + x^2}
\end{aligned}$$ ■

The Derivative of $y = \operatorname{arcsec} x$

Theorem 1 does not apply to the function sec x directly, since its domain is not connected. However, we can apply Theorem 1 to each of the two intervals in its domain to see that the inverse of the one-to-one function sec x is indeed differentiable. The formula for the derivative of arcsec x on its domain $|x| > 1$ can then be found by using implicit differentiation and the Chain Rule as follows:

$$
\begin{aligned}
y &= \operatorname{arcsec} x \\
\sec y &= x && \text{Inverse function relationship} \\
\frac{d}{dx}(\sec y) &= \frac{d}{dx}x && \text{Differentiate both sides.} \\
(\sec y \tan y)\frac{dy}{dx} &= 1 && \text{Chain Rule} \\
\frac{dy}{dx} &= \frac{1}{\sec y \tan y}. && \text{Since } |x| > 1,\ y \text{ lies in } (0, \pi/2) \cup (\pi/2, \pi) \\
& && \text{and } \sec y \tan y \neq 0.
\end{aligned}
$$

To express the result in terms of x, we use the relationships

$$\sec y = x \quad \text{and} \quad \tan y = \pm\sqrt{\sec^2 y - 1} = \pm\sqrt{x^2 - 1}$$

to get

$$\frac{dy}{dx} = \pm\frac{1}{x\sqrt{x^2 - 1}}.$$

Can we do anything about the ± sign? As can be seen in Figure 7.31, the slope of the graph $y = \operatorname{arcsec} x$ is always positive. Thus,

$$\frac{d}{dx}\operatorname{arcsec} x = \begin{cases} +\dfrac{1}{x\sqrt{x^2 - 1}} & \text{if } x > 1 \\ -\dfrac{1}{x\sqrt{x^2 - 1}} & \text{if } x < -1. \end{cases}$$

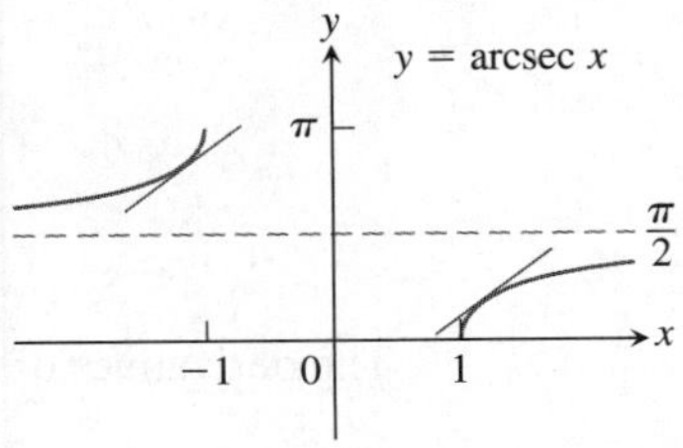

FIGURE 7.31 The slope of the curve $y = \operatorname{arcsec} x$ is positive for both $x < -1$ and $x > 1$.

With the absolute value symbol, we can write a single expression that eliminates the "±" ambiguity:

$$\frac{d}{dx}(\operatorname{arcsec} x) = \frac{1}{|x|\sqrt{x^2 - 1}}, \qquad |x| > 1.$$

If u is a differentiable function of x with $|u| > 1$, we have the formula

$$\frac{d}{dx}(\operatorname{arcsec} u) = \frac{1}{|u|\sqrt{u^2 - 1}}\frac{du}{dx}, \qquad |u| > 1.$$

EXAMPLE 6 Using the Chain Rule and derivative of the arcsecant function, we find

$$\begin{aligned}\frac{d}{dx}\operatorname{arcsec}(5x^4) &= \frac{1}{|5x^4|\sqrt{(5x^4)^2-1}}\frac{d}{dx}(5x^4)\\ &= \frac{1}{5x^4\sqrt{25x^8-1}}(20x^3) \qquad 5x^4 > 1\\ &= \frac{4}{x\sqrt{25x^8-1}}.\end{aligned}$$

■

Derivatives of the Other Three Inverse Trigonometric Functions

We could use the same techniques to find the derivatives of the other three inverse trigonometric functions—arccosine, arccotangent, and arccosecant—but there is an easier way, thanks to the following identities.

Inverse Function–Inverse Cofunction Identities

$$\arccos x = \pi/2 - \arcsin x$$
$$\operatorname{arccot} x = \pi/2 - \arctan x$$
$$\operatorname{arccsc} x = \pi/2 - \operatorname{arcsec} x$$

We saw the first of these identities in Equation (4). The others are derived in a similar way. It follows easily that the derivatives of the inverse cofunctions are the negatives of the derivatives of the corresponding inverse functions. For example, the derivative of arccos x is calculated as follows:

$$\begin{aligned}\frac{d}{dx}(\arccos x) &= \frac{d}{dx}\left(\frac{\pi}{2} - \arcsin x\right) && \text{Identity}\\ &= -\frac{d}{dx}(\arcsin x)\\ &= -\frac{1}{\sqrt{1-x^2}}. && \text{Derivative of arcsine}\end{aligned}$$

The derivatives of the inverse trigonometric functions are summarized in Table 7.4.

TABLE 7.4 **Derivatives of the inverse trigonometric functions**

1. $\frac{d}{dx}(\arcsin x) = \frac{1}{\sqrt{1-x^2}}$ $(\lvert x\rvert < 1)$	4. $\frac{d}{dx}(\arccos x) = -\frac{1}{\sqrt{1-x^2}}$ $(\lvert x\rvert < 1)$
2. $\frac{d}{dx}(\arctan x) = \frac{1}{1+x^2}$	5. $\frac{d}{dx}(\operatorname{arccot} x) = -\frac{1}{1+x^2}$
3. $\frac{d}{dx}(\operatorname{arcsec} x) = \frac{1}{\lvert x\rvert\sqrt{x^2-1}}$ $(\lvert x\rvert > 1)$	6. $\frac{d}{dx}(\operatorname{arccsc} x) = -\frac{1}{\lvert x\rvert\sqrt{x^2-1}}$ $(\lvert x\rvert > 1)$

Integration Formulas

The derivative formulas in Table 7.4 yield three useful integration formulas in Table 7.5. The formulas are readily verified by differentiating the functions on the right-hand sides.

TABLE 7.5 Integrals evaluated with inverse trigonometric functions

The following formulas hold for any constant $a > 0$.

1. $\displaystyle\int \frac{dx}{\sqrt{a^2 - x^2}} = \arcsin\left(\frac{x}{a}\right) + C$ (Valid for $x^2 < a^2$)

2. $\displaystyle\int \frac{dx}{\sqrt{a^2 + x^2}} = \frac{1}{a}\arctan\left(\frac{x}{a}\right) + C$ (Valid for all x)

3. $\displaystyle\int \frac{dx}{x\sqrt{x^2 - a^2}} = \frac{1}{a}\operatorname{arcsec}\left|\frac{x}{a}\right| + C$ (Valid for $|x| > a > 0$)

The derivative formulas in Table 7.4 have $a = 1$, but in most integrations $a \neq 1$, and the formulas in Table 7.5 are more useful.

EXAMPLE 7 These examples illustrate how we use Table 7.5.

(a)
$$\int_{\sqrt{2}/2}^{\sqrt{3}/2} \frac{dx}{\sqrt{1 - x^2}} = \arcsin x\Big]_{\sqrt{2}/2}^{\sqrt{3}/2} \qquad a = 1 \text{ in Table 7.5, Formula 1}$$
$$= \arcsin\left(\frac{\sqrt{3}}{2}\right) - \arcsin\left(\frac{\sqrt{2}}{2}\right) = \frac{\pi}{3} - \frac{\pi}{4} = \frac{\pi}{12}$$

(b)
$$\int \frac{dx}{\sqrt{3 - 4x^2}} = \frac{1}{2}\int \frac{du}{\sqrt{a^2 - u^2}} \qquad a = \sqrt{3},\ u = 2x, \text{ and } du/2 = dx$$
$$= \frac{1}{2}\arcsin\left(\frac{u}{a}\right) + C \qquad \text{Table 7.5, Formula 1}$$
$$= \frac{1}{2}\arcsin\left(\frac{2x}{\sqrt{3}}\right) + C$$

(c)
$$\int \frac{dx}{\sqrt{e^{2x} - 6}} = \int \frac{du/u}{\sqrt{u^2 - a^2}} \qquad u = e^x,\ du = e^x,\ dx = du/e^x = du/u,\ a = \sqrt{6}$$
$$= \int \frac{du}{u\sqrt{u^2 - a^2}}$$
$$= \frac{1}{a}\operatorname{arcsec}\left|\frac{u}{a}\right| + C \qquad \text{Table 7.5, Formula 3}$$
$$= \frac{1}{\sqrt{6}}\operatorname{arcsec}\left(\frac{e^x}{\sqrt{6}}\right) + C$$

■

EXAMPLE 8 Evaluate

(a) $\displaystyle\int \frac{dx}{\sqrt{4x - x^2}}$ **(b)** $\displaystyle\int \frac{dx}{4x^2 + 4x + 2}$

Solution

(a) The expression $\sqrt{4x - x^2}$ does not match any of the formulas in Table 7.5, so we first rewrite $4x - x^2$ by completing the square:

$$4x - x^2 = -(x^2 - 4x) = -(x^2 - 4x + 4) + 4 = 4 - (x - 2)^2.$$

Then we substitute $a = 2$, $u = x - 2$, and $du = dx$ to get

$$\int \frac{dx}{\sqrt{4x - x^2}} = \int \frac{dx}{\sqrt{4 - (x - 2)^2}}$$
$$= \int \frac{du}{\sqrt{a^2 - u^2}} \qquad a = 2,\ u = x - 2, \text{ and } du = dx$$
$$= \arcsin\left(\frac{u}{a}\right) + C \qquad \text{Table 7.5, Formula 1}$$
$$= \arcsin\left(\frac{x - 2}{2}\right) + C$$

(b) We complete the square on the binomial $4x^2 + 4x$:

$$4x^2 + 4x + 2 = 4(x^2 + x) + 2 = 4\left(x^2 + x + \frac{1}{4}\right) + 2 - \frac{4}{4}$$
$$= 4\left(x + \frac{1}{2}\right)^2 + 1 = (2x + 1)^2 + 1.$$

Then,

$$\int \frac{dx}{4x^2 + 4x + 2} = \int \frac{dx}{(2x + 1)^2 + 1} = \frac{1}{2}\int \frac{du}{u^2 + a^2} \qquad a = 1,\ u = 2x + 1, \text{ and } du/2 = dx$$
$$= \frac{1}{2} \cdot \frac{1}{a} \arctan\left(\frac{u}{a}\right) + C \qquad \text{Table 7.5, Formula 2}$$
$$= \frac{1}{2}\arctan(2x + 1) + C \qquad a = 1,\ u = 2x + 1$$

EXERCISES 7.6

Remember that arcsin and $\sin^{-1}$ represent the same function, and similarly for the other trigonometric functions.

Common Values

Use reference triangles in an appropriate quadrant to find the angles in Exercises 1–8.

1. a. $\arctan 1$ **b.** $\arctan(-\sqrt{3})$ **c.** $\tan^{-1}\left(\frac{1}{\sqrt{3}}\right)$

2. a. $\arctan(-1)$ **b.** $\tan^{-1}\sqrt{3}$ **c.** $\arctan\left(\frac{-1}{\sqrt{3}}\right)$

3. a. $\arcsin\left(\frac{-1}{2}\right)$ **b.** $\sin^{-1}\left(\frac{\sqrt{3}}{2}\right)$ **c.** $\sin^{-1}\left(\frac{-\sqrt{3}}{2}\right)$

4. a. $\sin^{-1}\left(\frac{1}{2}\right)$ **b.** $\arcsin\left(\frac{-1}{\sqrt{2}}\right)$ **c.** $\arcsin\left(\frac{\sqrt{3}}{2}\right)$

5. a. $\arccos\left(\frac{1}{2}\right)$ **b.** $\cos^{-1}\left(\frac{-1}{\sqrt{2}}\right)$ **c.** $\cos^{-1}\left(-\frac{\sqrt{3}}{2}\right)$

6. a. $\csc^{-1}\sqrt{2}$ **b.** $\operatorname{arccsc}\left(\frac{-2}{\sqrt{3}}\right)$ **c.** $\operatorname{arccsc} 2$

7. a. $\sec^{-1}(-\sqrt{2})$ **b.** $\operatorname{arcsec}\left(\frac{2}{\sqrt{3}}\right)$ **c.** $\operatorname{arcsec}(-2)$

8. a. $\cot^{-1}(-\sqrt{3})$ **b.** $\operatorname{arccot}(\sqrt{3})$ **c.** $\cot^{-1}\left(\frac{-1}{\sqrt{3}}\right)$

Evaluations

Find the values in Exercises 9–12.

9. $\sin\left(\arccos\left(\frac{\sqrt{2}}{2}\right)\right)$ **10.** $\sec\left(\arccos\frac{1}{2}\right)$

11. $\tan\left(\cos^{-1}\left(-\frac{1}{2}\right)\right)$ **12.** $\cot\left(\arcsin\left(-\frac{\sqrt{3}}{2}\right)\right)$

Limits

Find the limits in Exercises 13–20. (If in doubt, look at the function's graph.)

13. $\lim_{x\to1^-} \cos^{-1} x$ **14.** $\lim_{x\to1^+} \sin^{-1} x$

15. $\lim_{x\to\infty} \arctan x$ **16.** $\lim_{x\to-\infty} \arctan x$

17. $\lim_{x\to\infty} \operatorname{arcsec} x$ **18.** $\lim_{x\to-\infty} \operatorname{arcsec} x$

19. $\lim_{x\to\infty} \operatorname{arccsc} x$ **20.** $\lim_{x\to-\infty} \operatorname{arccsc} x$

Finding Derivatives

In Exercises 21–48, find the derivative of y with respect to the appropriate variable.

21. $y = \cos^{-1}x$ **22.** $y = \arccos(1/x)$

23. $y = \arcsin\sqrt{2}\,t$ **24.** $y = \sin^{-1}(3 - 2t)$

25. $y = \operatorname{arcsec}(2s + 1)$ **26.** $y = \operatorname{arcsec} 5s$

27. $y = \operatorname{arccsc}(x^2 + 1),\quad x > 0$

28. $y = \operatorname{cosec}^{-1}(2x)$

29. $y = \operatorname{arcsec}\frac{1}{t},\quad 0 < t < 1$ **30.** $y = \arcsin\frac{3}{t^2}$

31. $y = \operatorname{arccot}\sqrt{2t}$ **32.** $y = \operatorname{arccot}\sqrt{t - 1}$

33. $y = \ln(\arctan x)$ **34.** $y = \arctan(\ln x)$

35. $y = \sec^{-1}(e^t)$ **36.** $y = \arccos(e^{-t})$

37. $y = s\sqrt{1 - s^2} + \arccos s$ **38.** $y = \sqrt{s^2 - 1} - \operatorname{arcsec} s$

39. $y = \arctan\sqrt{x^2 - 1} + \operatorname{arccsc} x,\quad x > 1$

40. $y = \tan^{-1}\frac{1}{x} + \cot^{-1}x$ **41.** $y = x\arcsin x + \sqrt{1 - x^2}$

42. $y = \ln(x^2 + 4) - x\arctan\left(\frac{x}{2}\right)$

43. $y = \sqrt{\arcsin x}$ **44.** $y = e^{\operatorname{arcsec} x}$

45. $y = \cos(x - \arccos x)$ **46.** $y = \dfrac{x}{1 + \arctan x}$

47. $y = (\operatorname{arccot}(x^3))^3$ **48.** $y = \log_2 \operatorname{arccsc}\sqrt{x}$

For problems 49–52 use implicit differentiation to find $\frac{dy}{dx}$ at the given point P.

49. $3\arctan x + \arcsin y = \frac{\pi}{4};\quad P(1, -1)$

50. $\arcsin(x + y) + \arccos(x - y) = \frac{5\pi}{6};\quad P\left(0, \frac{1}{2}\right)$

51. $y\arccos(xy) = \frac{-3\sqrt{2}}{4}\pi;\quad P\left(\frac{1}{2}, -\sqrt{2}\right)$

52. $16(\arctan 3y)^2 + 9(\arctan 2x)^2 = 2\pi^2;\quad P\left(\frac{\sqrt{3}}{2}, \frac{1}{3}\right)$

Evaluating Integrals

Evaluate the integrals in Exercises 53–76.

53. $\int \frac{dx}{\sqrt{4-x^2}}$

54. $\int \frac{dx}{\sqrt{1-4x^2}}$

55. $\int \frac{dx}{17+x^2}$

56. $\int \frac{dx}{9+3x^2}$

57. $\int \frac{dx}{x\sqrt{25x^2-2}}$

58. $\int \frac{dx}{x\sqrt{5x^2-4}}$

59. $\int_0^1 \frac{4\,ds}{\sqrt{4-s^2}}$

60. $\int_0^{3\sqrt{2}/4} \frac{ds}{\sqrt{9-4s^2}}$

61. $\int_0^2 \frac{dt}{4-t^2}$

62. $\int_{-2}^2 \frac{dt}{4+3t^2}$

63. $\int_{-1}^{-\sqrt{2}/2} \frac{dy}{y\sqrt{4y^2-1}}$

64. $\int_{-2/3}^{-\sqrt{2}/3} \frac{dy}{y\sqrt{9y^2-1}}$

65. $\int \frac{dr}{\sqrt{2-(r-1)^2}}$

66. $\int \frac{6\,dr}{\sqrt{4-(r+1)^2}}$

67. $\int \frac{dx}{5+(x-2)^2}$

68. $\int \frac{dx}{1+(3x+1)^2}$

69. $\int \frac{dx}{(2x-1)\sqrt{(2x-1)^2-4}}$

70. $\int \frac{dx}{(x+5)\sqrt{(x+5)^2-9}}$

71. $\int_{-\pi/2}^{\pi/2} \frac{2\cos\theta\,d\theta}{1+(\sin\theta)^2}$

72. $\int_{\pi/6}^{\pi/4} \frac{\csc^2 x\,dx}{1+(\cot x)^2}$

73. $\int_0^{\ln\sqrt{3}} \frac{e^x\,dx}{1+e^{2x}}$

74. $\int_1^{e^{\pi/4}} \frac{4\,dt}{t(1+\ln^2 t)}$

75. $\int \frac{y\,dy}{\sqrt{4-y^4}}$

76. $\int \frac{\sec^2 y\,dy}{\sqrt{1-\tan^2 y}}$

Evaluate the integrals in Exercises 77–90.

77. $\int \frac{dx}{\sqrt{-x^2+6x-8}}$

78. $\int \frac{dx}{\sqrt{2x-x^2}}$

79. $\int_{-1}^0 \frac{6\,dt}{\sqrt{3-2t-t^2}}$

80. $\int_0^1 \frac{6dt}{\sqrt{3+4t-4t^2}}$

81. $\int \frac{dy}{y^2-2y+5}$

82. $\int \frac{dy}{y^2+6y+10}$

83. $\int_1^2 \frac{5dx}{x^2-4x+5}$

84. $\int_2^4 \frac{2\,dx}{x^2-6x+10}$

85. $\int \frac{x+4}{x^2+4}\,dx$

86. $\int \frac{t-5}{t^2-4t+5}\,dt$

87. $\int \frac{x^2+2x-1}{x^2+9}\,dx$

88. $\int \frac{t^3-2t^2+3t-4}{t^2+1}\,dt$

89. $\int \frac{dx}{(x+1)\sqrt{x^2+2x}}$

90. $\int \frac{dx}{(x-2)\sqrt{x^2-4x+3}}$

Evaluate the integrals in Exercises 91–102.

91. $\int \frac{e^{\arcsin x}\,dx}{\sqrt{1-x^2}}$

92. $\int \frac{e^{\arccos x}\,dx}{\sqrt{1-x^2}}$

93. $\int \frac{(\cos^{-1}x)^2\,dx}{\sqrt{1-x^2}}$

94. $\int \frac{\sqrt{\arctan x}\,dx}{1+x^2}$

95. $\int \frac{dy}{(\arctan y)(1+y^2)}$

96. $\int \frac{dy}{(\cos^{-1}y)\sqrt{1-y^2}}$

97. $\int_{\sqrt{2}}^2 \frac{\sec^2(\operatorname{arcsec} x)\,dx}{x\sqrt{x^2-1}}$

98. $\int_{2/\sqrt{3}}^2 \frac{\cos(\operatorname{arcsec} x)\,dx}{x\sqrt{x^2-1}}$

99. $\int \frac{e^x \sin^{-1} e^x}{\sqrt{1-e^{2x}}}\,dx$

100. $\int \frac{e^x \arcsin e^x}{\sqrt{1-e^{2x}}}\,dx$

101. $\int_0^1 \frac{\arctan x}{1+x^2}\,dx$

102. $\int_{-1/3}^{1/\sqrt{3}} \frac{\cos(\arctan 3x)}{1+9x^2}\,dx$

L'Hôpital's Rule

Find the limits in Exercises 103–110.

103. $\lim_{x\to 0} \frac{\cos^{-1} 5x}{5}$

104. $\lim_{x\to 1^+} \frac{\sqrt{x^2-1}}{\operatorname{arcsec} x}$

105. $\lim_{x\to\infty} x\tan^{-1}\frac{3}{x}$

106. $\lim_{x\to 0} \frac{2\arctan 3x^2}{7x^2}$

107. $\lim_{x\to 0} \frac{\arctan x^2}{x \arcsin x}$

108. $\lim_{x\to\infty} \frac{e^x \arctan e^x}{e^{2x}+x}$

109. $\lim_{x\to 0^+} \frac{(\arctan\sqrt{x})^2}{x\sqrt{x+1}}$

110. $\lim_{x\to 0^-} \frac{\cos^{-1} x^2}{(\cos^{-1} x)^2}$

Integration Formulas

Verify the integration formulas in Exercises 111–114.

111. $\int \frac{\arctan x}{x^2}\,dx = \ln x - \frac{1}{2}\ln(1+x^2) - \frac{\arctan x}{x} + C$

112. $\int x^3 \arccos 5x\,dx = \frac{x^4}{4}\arccos 5x + \frac{5}{4}\int \frac{x^4\,dx}{\sqrt{1-25x^2}}$

113. $\int (\cos^{-1}x)^2\,dx = x(\cos^{-1}x)^2 - 2x - 2\sqrt{1-x^2}\cos^{-1}x$

114. $\int \ln(a^2+x^2)\,dx = x\ln(a^2+x^2) - 2x + 2a\arctan\frac{x}{a} + C$

Initial Value Problems

Solve the initial value problems in Exercises 115–120.

115. $\frac{dy}{dx} = \frac{1}{\sqrt{1-x^2}}, \quad y(0) = 0$

116. $\frac{dy}{dx} = \frac{1}{x^2+1} - 5, \quad y(0) = 1$

117. $\frac{dy}{dx} = \frac{1}{x\sqrt{x^2-1}}, \quad x > 1; \quad y(2) = \pi$

118. $\frac{dy}{dx} = \frac{1}{x^2+1} - \frac{5}{\sqrt{1-x^2}}, \quad y(0) = 2$

Applications and Theory

119. You are sitting in a classroom next to the wall looking at the blackboard at the front of the room. The blackboard is 12 ft long and starts 3 ft from the wall you are sitting next to.

a. Show that your viewing angle is

$$\alpha = \operatorname{arccot}\frac{x}{15} - \operatorname{arccot}\frac{x}{3}$$

if you are x ft from the front wall.

b. Find x so that α is as large as possible.

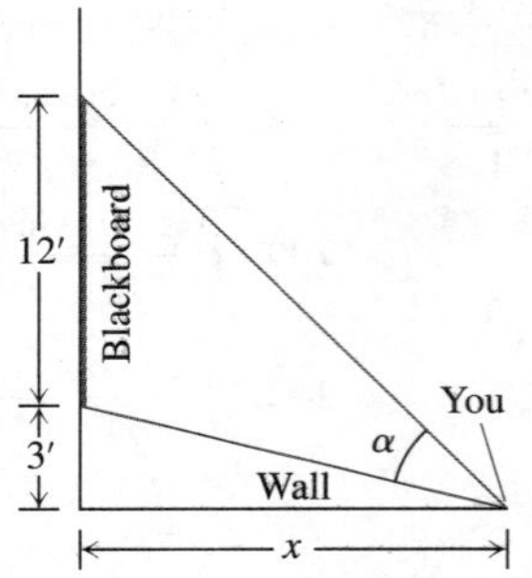

120. The region between the curve $y = \operatorname{arcsec} x$ and the x-axis from $x = 1$ to $x = 2$ (shown here) is revolved about the y-axis to generate a solid. Find the volume of the solid.

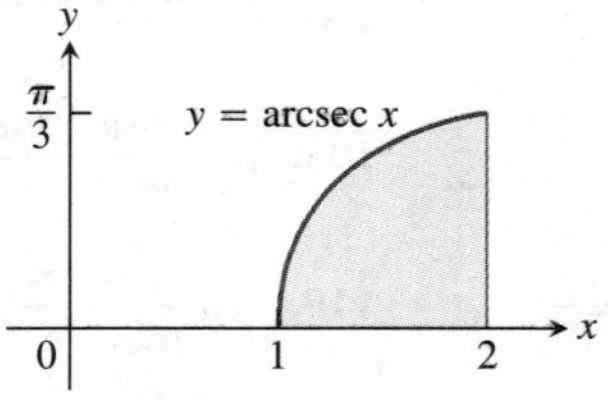

121. The slant height of the cone shown here is 3 m. How large should the indicated angle be to maximize the cone's volume?

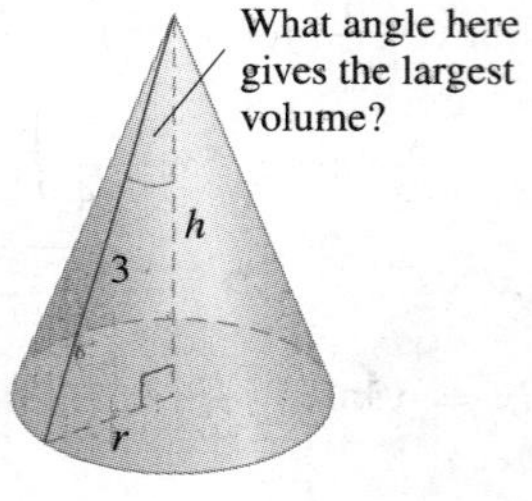

122. Find the angle α.

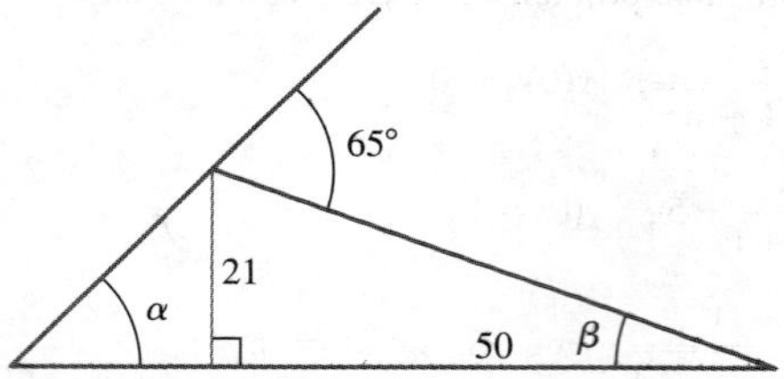

123. Here is an informal proof that $\arctan 1 + \arctan 2 + \arctan 3 = \pi$. Explain what is going on.

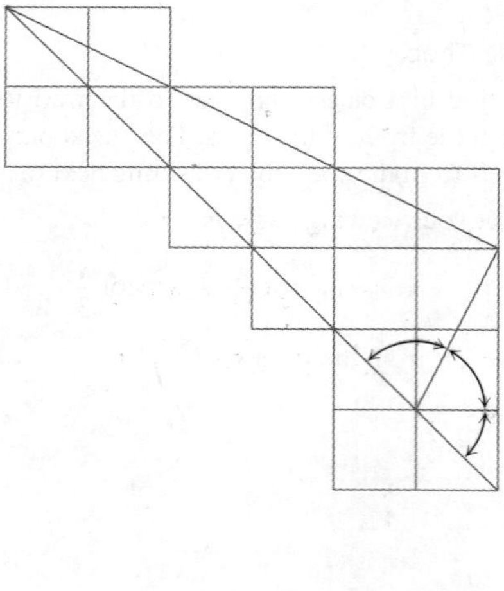

124. Two derivations of the identity $\operatorname{arcsec}(-x) = \pi - \operatorname{arcsec} x$

a. (*Geometric*) Here is a pictorial proof that $\operatorname{arcsec}(-x) = \pi - \operatorname{arcsec} x$. See if you can tell what is going on.

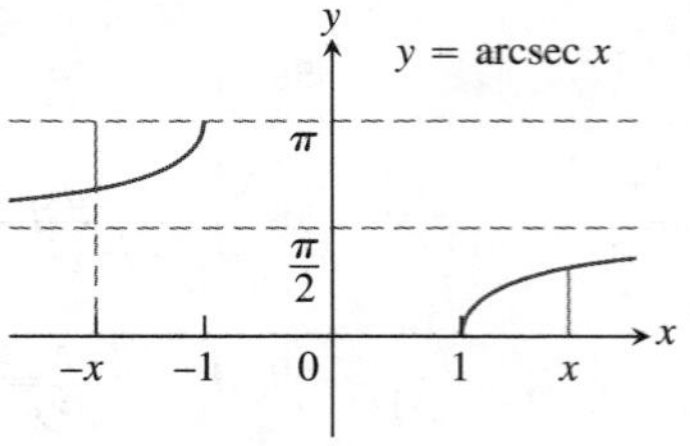

b. (*Algebraic*) Derive the identity $\operatorname{arcsec}(-x) = \pi - \operatorname{arcsec} x$ by combining the following two equations from the text:

$$\arccos(-x) = \pi - \arccos x \qquad \text{Eq. (3)}$$

$$\operatorname{arcsec} x = \arccos(1/x) \qquad \text{Eq. (5)}$$

125. The identity $\arcsin x + \arccos x = \pi/2$ Figure 7.28 establishes the identity for $0 < x < 1$. To establish it for the rest of $[-1, 1]$, verify by direct calculation that it holds for $x = 1, 0$, and -1. Then, for values of x in $(-1, 0)$, let $x = -a$, $a > 0$, and apply Eqs. (1) and (3) to the sum $\arcsin(-a) + \arccos(-a)$.

126. Show that the sum $\arctan x + \arctan(1/x)$ is constant.

127. Use the identity

$$\operatorname{arccsc} u = \frac{\pi}{2} - \operatorname{arcsec} u$$

to derive the formula for the derivative of $\operatorname{arccsc} u$ in Table 7.4 from the formula for the derivative of $\operatorname{arcsec} u$.

128. Derive the formula

$$\frac{dy}{dx} = \frac{1}{1 + x^2}$$

for the derivative of $y = \arctan x$ by differentiating both sides of the equivalent equation $\tan y = x$.

129. Use the Derivative Rule, Theorem 1, to derive

$$\frac{d}{dx}\operatorname{arcsec} x = \frac{1}{|x|\sqrt{x^2 - 1}}, \quad |x| > 1.$$

130. Use the identity

$$\operatorname{arccot} u = \frac{\pi}{2} - \arctan u$$

to derive the formula for the derivative of $\operatorname{arccot} u$ in Table 7.4 from the formula for the derivative of $\arctan u$.

131. What is special about the functions

$$f(x) = \arcsin \frac{x - 1}{x + 1}, \quad x \geq 0, \quad \text{and} \quad g(x) = 2 \arctan \sqrt{x}?$$

Explain.

132. What is special about the functions

$$f(x) = \arcsin \frac{1}{\sqrt{x^2 + 1}} \quad \text{and} \quad g(x) = \arctan \frac{1}{x}?$$

Explain.

133. Find the volume of the solid of revolution shown here.

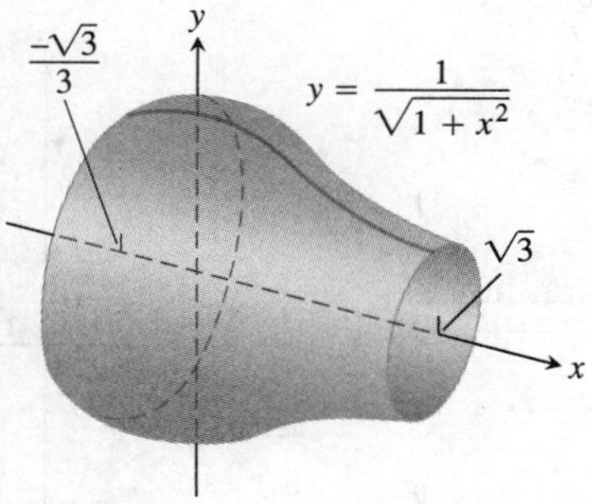

134. Arc length Find the circumference of a circle of radius r using Eq. (3) in Section 6.3.

Find the volumes of the solids in Exercises 135 and 136.

135. The solid lies between planes perpendicular to the x-axis at $x = -1$ and $x = 1$. The cross-sections perpendicular to the x-axis are

a. circles whose diameters stretch from the curve $y = -1/\sqrt{1 + x^2}$ to the curve $y = 1/\sqrt{1 + x^2}$.

b. vertical squares whose base edges run from the curve $y = -1/\sqrt{1 + x^2}$ to the curve $y = 1/\sqrt{1 + x^2}$.

136. The solid lies between planes perpendicular to the x-axis at $x = -\sqrt{2}/2$ and $x = \sqrt{2}/2$. The cross-sections perpendicular to the x-axis are

a. circles whose diameters stretch from the x-axis to the curve $y = 2/\sqrt[4]{1 - x^2}$.

b. squares whose diagonals stretch from the x-axis to the curve $y = 2/\sqrt[4]{1 - x^2}$.

T **137.** Find the values of the following.

a. $\sec^{-1} 2.5$ **b.** $\text{arccsc}(-1.5)$ **c.** $\text{arccot}\, 2$

T **138.** Find the values of the following.

a. $\text{arcsec}(-3)$ **b.** $\text{cosec}^{-1} 2.7$ **c.** $\text{arccot}(-2)$

T In Exercises 139–141, find the domain and range of each composite function. Then graph the compositions on separate screens. Do the graphs make sense in each case? Give reasons for your answers. Comment on any differences you see.

139. a. $y = \arctan(\tan x)$

b. $y = \tan(\arctan x)$

140. a. $y = \arcsin(\sin x)$

b. $y = \sin(\arcsin x)$

141. a. $y = \arccos(\cos x)$

b. $y = \cos(\arccos x)$

T Use graphing software for Exercises 142–146.

142. Graph $y = \sec(\text{arcsec}\, x) = \sec(\arccos(1/x))$. Explain what you see.

143. Newton's serpentine Graph $y = 4x/(x^2 + 1)$, known as Newton's serpentine. Then graph $y = 2\sin(2\arctan x)$ in the same graphing window. What do you see? Explain.

144. Graph the rational function $y = (2 - x^2)/x^2$. Then graph $y = \cos(2\,\text{arcsec}\, x)$ in the same graphing window. What do you see? Explain.

145. Graph $f(x) = \arcsin x$ together with its first two derivatives. Comment on the behavior of f and the shape of its graph in relation to the signs and values of f' and f''.

146. Graph $f(x) = \arctan x$ together with its first two derivatives. Comment on the behavior of f and the shape of its graph in relation to the signs and values of f' and f''.

7.7 Hyperbolic Functions

The hyperbolic functions are formed by taking combinations of the two exponential functions e^x and e^{-x}. The hyperbolic functions simplify many mathematical expressions and occur frequently in mathematical and engineering applications.

Definitions and Identities

The hyperbolic sine and hyperbolic cosine functions are defined by the equations

$$\sinh x = \frac{e^x - e^{-x}}{2} \quad \text{and} \quad \cosh x = \frac{e^x + e^{-x}}{2}.$$

We pronounce sinh x as "cinch x," rhyming with "pinch x," and cosh x as "kosh x," rhyming with "gosh x." From this basic pair, we define the hyperbolic tangent, cotangent, secant, and cosecant functions. The defining equations and graphs of these functions are shown in Table 7.6. We will see that the hyperbolic functions bear many similarities to the trigonometric functions after which they are named.

TABLE 7.6 The six basic hyperbolic functions

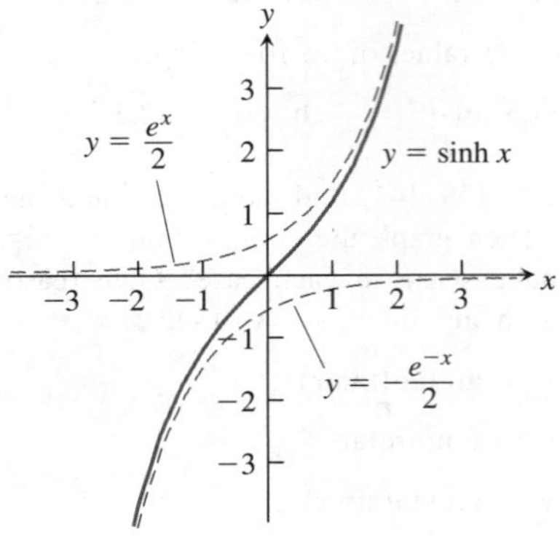

(a)

Hyperbolic sine:

$$\sinh x = \frac{e^x - e^{-x}}{2}$$

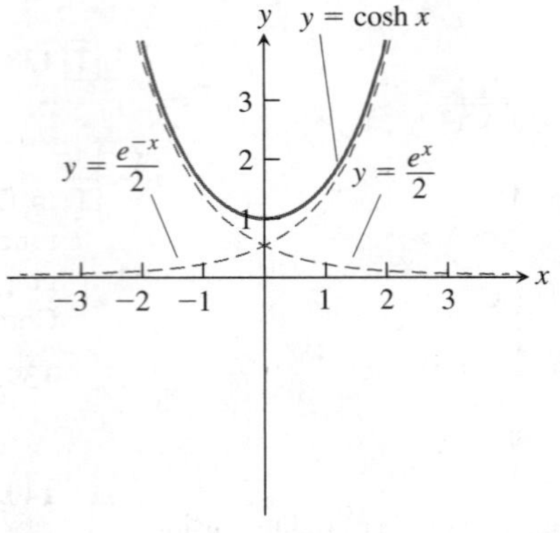

(b)

Hyperbolic cosine:

$$\cosh x = \frac{e^x + e^{-x}}{2}$$

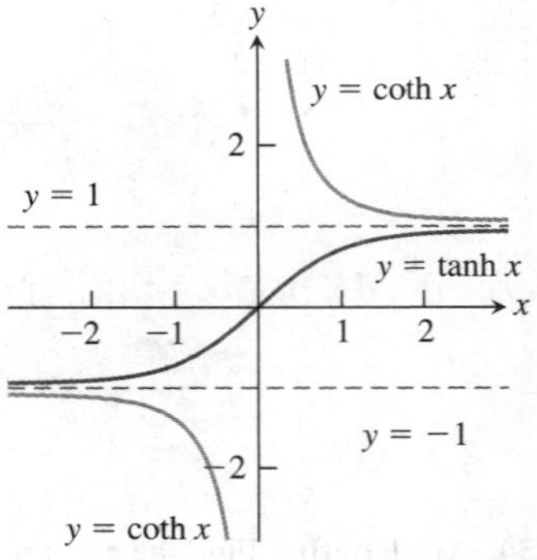

(c)

Hyperbolic tangent:

$$\tanh x = \frac{\sinh x}{\cosh x} = \frac{e^x - e^{-x}}{e^x + e^{-x}}$$

Hyperbolic cotangent:

$$\coth x = \frac{\cosh x}{\sinh x} = \frac{e^x + e^{-x}}{e^x - e^{-x}}$$

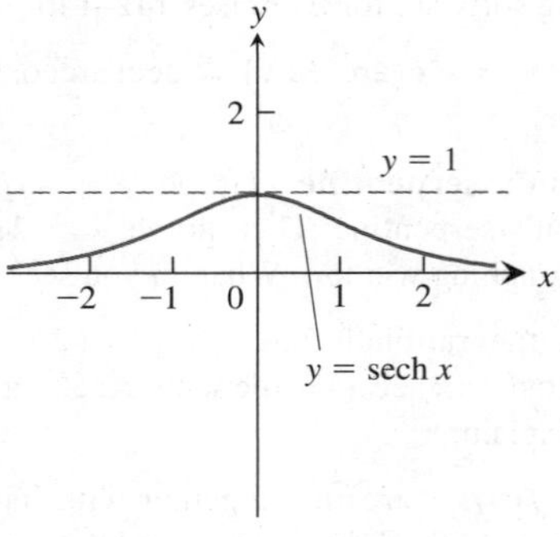

(d)

Hyperbolic secant:

$$\operatorname{sech} x = \frac{1}{\cosh x} = \frac{2}{e^x + e^{-x}}$$

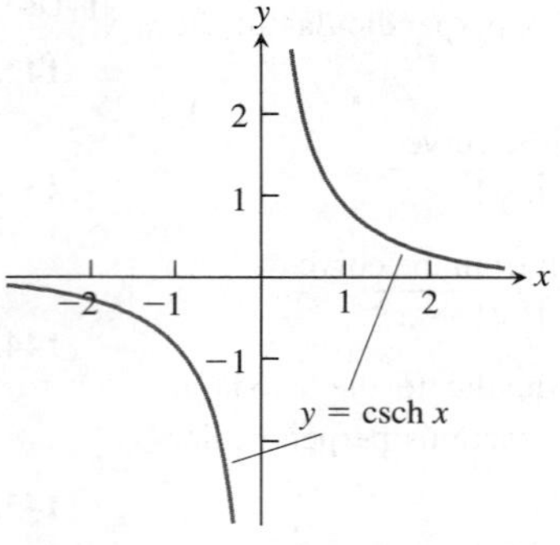

(e)

Hyperbolic cosecant:

$$\operatorname{csch} x = \frac{1}{\sinh x} = \frac{2}{e^x - e^{-x}}$$

Hyperbolic functions satisfy the identities in Table 7.7. Except for differences in sign, these resemble identities we know for the trigonometric functions. The identities are proved directly from the definitions, as we show here for the second one:

$$\begin{aligned}
2 \sinh x \cosh x &= 2\left(\frac{e^x - e^{-x}}{2}\right)\left(\frac{e^x + e^{-x}}{2}\right) \\
&= \frac{e^{2x} - e^{-2x}}{2} && \text{Simplify.} \\
&= \sinh 2x. && \text{Definition of sinh}
\end{aligned}$$

The other identities are obtained similarly, by substituting in the definitions of the hyperbolic functions and using algebra.

For any real number u, we know the point with coordinates $(\cos u, \sin u)$ lies on the unit circle $x^2 + y^2 = 1$. So the trigonometric functions are sometimes called the *circular* functions. Because of the first identity

$$\cosh^2 u - \sinh^2 u = 1,$$

with u substituted for x in Table 7.7, the point having coordinates $(\cosh u, \sinh u)$ lies on the right-hand branch of the hyperbola $x^2 - y^2 = 1$. This is where the *hyperbolic* functions get their names (see Exercise 86).

TABLE 7.7 Identities for hyperbolic functions

$$\cosh^2 x - \sinh^2 x = 1$$

$$\sinh 2x = 2 \sinh x \cosh x$$

$$\cosh 2x = \cosh^2 x + \sinh^2 x$$

$$\cosh^2 x = \frac{\cosh 2x + 1}{2}$$

$$\sinh^2 x = \frac{\cosh 2x - 1}{2}$$

$$\tanh^2 x = 1 - \operatorname{sech}^2 x$$

$$\coth^2 x = 1 + \operatorname{csch}^2 x$$

Hyperbolic functions are useful in finding integrals, which we will see in Chapter 8. They play an important role in science and engineering as well. The hyperbolic cosine describes the shape of a hanging cable or wire that is strung between two points at the same height and hanging freely (see Exercise 83). The shape of the St. Louis Arch is an inverted hyperbolic cosine. The hyperbolic tangent occurs in the formula for the velocity of an ocean wave moving over water having a constant depth, and the inverse hyperbolic tangent describes how relative velocities sum according to Einstein's Law in the Special Theory of Relativity.

Derivatives and Integrals of Hyperbolic Functions

The six hyperbolic functions, being rational combinations of the differentiable functions e^x and e^{-x}, have derivatives at every point at which they are defined (Table 7.8). Again, there are similarities to trigonometric functions.

The derivative formulas are obtained from the derivative of e^x:

$$\begin{aligned} \frac{d}{dx}(\sinh x) &= \frac{d}{dx}\left(\frac{e^x - e^{-x}}{2}\right) && \text{Definition of } \sinh x \\ &= \frac{e^x + e^{-x}}{2} && \text{Derivative of } e^x \\ &= \cosh x. && \text{Definition of } \cosh x \end{aligned}$$

TABLE 7.8 Derivatives of hyperbolic functions

$$\frac{d}{dx}(\sinh x) = \cosh x$$

$$\frac{d}{dx}(\cosh x) = \sinh x$$

$$\frac{d}{dx}(\tanh x) = \operatorname{sech}^2 x$$

$$\frac{d}{dx}(\coth x) = -\operatorname{csch}^2 x$$

$$\frac{d}{dx}(\operatorname{sech} x) = -\operatorname{sech} x \tanh x$$

$$\frac{d}{dx}(\operatorname{csch} x) = -\operatorname{csch} x \coth x$$

This gives the first derivative formula. From the definition, we can calculate the derivative of the hyperbolic cosecant function, as follows:

$$\begin{aligned}
\frac{d}{dx}(\operatorname{csch} x) &= \frac{d}{dx}\left(\frac{1}{\sinh x}\right) && \text{Definition of csch } x \\
&= -\frac{\cosh x}{\sinh^2 x} && \text{Quotient Rule for derivatives} \\
&= -\frac{1}{\sinh x}\frac{\cosh x}{\sinh x} && \text{Rearrange factors.} \\
&= -\operatorname{csch} x \coth x && \text{Definitions of csch } x \text{ and coth } x
\end{aligned}$$

The other formulas in Table 7.8 are obtained similarly.

The derivative formulas lead to the integral formulas in Table 7.9.

TABLE 7.9 Integral formulas for hyperbolic functions

$$\int \sinh x\,dx = \cosh x + C$$

$$\int \cosh x\,dx = \sinh x + C$$

$$\int \operatorname{sech}^2 x\,dx = \tanh x + C$$

$$\int \operatorname{csch}^2 x\,dx = -\coth x + C$$

$$\int \operatorname{sech} x \tanh x\,dx = -\operatorname{sech} x + C$$

$$\int \operatorname{csch} x \coth x\,dx = -\operatorname{csch} x + C$$

EXAMPLE 1 We illustrate the derivative and integral formulas.

(a)
$$\begin{aligned}
\frac{d}{dt}\left(\tanh\sqrt{1+t^2}\right) &= \operatorname{sech}^2\sqrt{1+t^2}\cdot\frac{d}{dt}\left(\sqrt{1+t^2}\right) \\
&= \frac{t}{\sqrt{1+t^2}}\operatorname{sech}^2\sqrt{1+t^2}
\end{aligned}$$

(b)
$$\begin{aligned}
\int \coth 5x\,dx &= \int \frac{\cosh 5x}{\sinh 5x}\,dx = \frac{1}{5}\int\frac{du}{u} && u = \sinh 5x,\ du = 5\cosh 5x\,dx \\
&= \frac{1}{5}\ln|u| + C = \frac{1}{5}\ln|\sinh 5x| + C
\end{aligned}$$

(c)
$$\begin{aligned}
\int_0^1 \sinh^2 x\,dx &= \int_0^1 \frac{\cosh 2x - 1}{2}\,dx && \text{Table 7.7} \\
&= \frac{1}{2}\int_0^1 (\cosh 2x - 1)\,dx = \frac{1}{2}\left[\frac{\sinh 2x}{2} - x\right]_0^1 \\
&= \frac{\sinh 2}{4} - \frac{1}{2} \approx 0.40672 && \text{Evaluate with a calculator.}
\end{aligned}$$

(d)
$$\begin{aligned}
\int_0^{\ln 2} 4e^x \sinh x\,dx &= \int_0^{\ln 2} 4e^x\frac{e^x - e^{-x}}{2}\,dx = \int_0^{\ln 2}(2e^{2x} - 2)\,dx \\
&= \left[e^{2x} - 2x\right]_0^{\ln 2} = (e^{2\ln 2} - 2\ln 2) - (1 - 0) \\
&= 4 - 2\ln 2 - 1 \approx 1.6137
\end{aligned}$$

■

Inverse Hyperbolic Functions

The inverses of the six basic hyperbolic functions are very useful in integration (see Chapter 8). Since $d(\sinh x)/dx = \cosh x > 0$, the hyperbolic sine is an increasing function of x. We denote its inverse by

$$y = \sinh^{-1} x.$$

For every value of x in the interval $-\infty < x < \infty$, the value of $y = \sinh^{-1} x$ is the number whose hyperbolic sine is x. The graphs of $y = \sinh x$ and $y = \sinh^{-1} x$ are shown in Figure 7.32a. Other notations used for the inverse hyperbolic sine function include arcsinh x, arsinh x, and argsinh x.

The function $y = \cosh x$ is not one-to-one because its graph in Table 7.6 does not pass the horizontal line test. The restricted function $y = \cosh x, x \geq 0$, however, is one-to-one and therefore has an inverse, denoted by

$$y = \cosh^{-1} x.$$

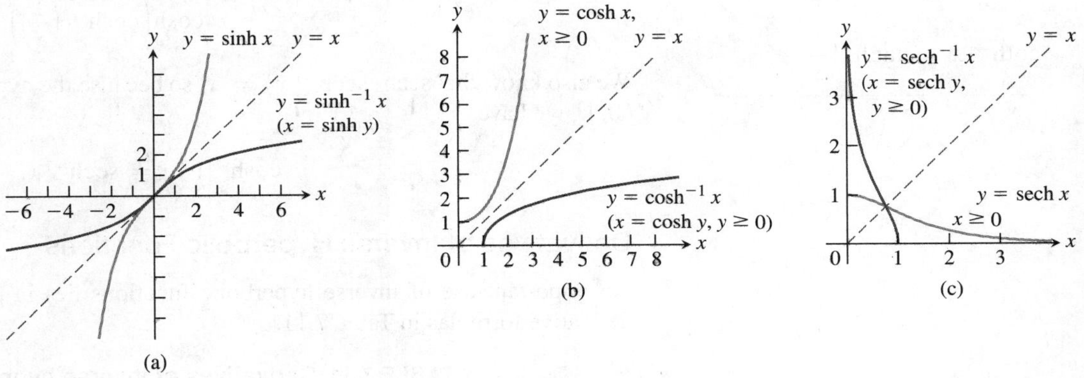

FIGURE 7.32 The graphs of the inverse hyperbolic sine, cosine, and secant of x. Notice the symmetries about the line $y = x$.

For every value of $x \geq 1$, $y = \cosh^{-1} x$ is the number in the interval $0 \leq y < \infty$ whose hyperbolic cosine is x. The graphs of $y = \cosh x, x \geq 0$, and $y = \cosh^{-1} x$ are shown in Figure 7.32b.

Like $y = \cosh x$, the function $y = \operatorname{sech} x = 1/\cosh x$ fails to be one-to-one, but its restriction to nonnegative values of x does have an inverse, denoted by

$$y = \operatorname{sech}^{-1} x.$$

For every value of x in the interval $(0, 1]$, $y = \operatorname{sech}^{-1} x$ is the nonnegative number whose hyperbolic secant is x. The graphs of $y = \operatorname{sech} x, x \geq 0$, and $y = \operatorname{sech}^{-1} x$ are shown in Figure 7.32c.

The hyperbolic tangent, cotangent, and cosecant are one-to-one on their domains and therefore have inverses, denoted by

$$y = \tanh^{-1} x, \qquad y = \coth^{-1} x, \qquad y = \operatorname{csch}^{-1} x.$$

These functions are graphed in Figure 7.33.

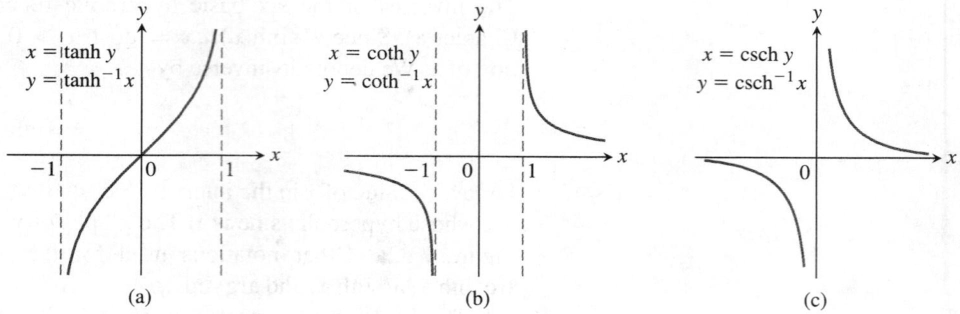

FIGURE 7.33 The graphs of the inverse hyperbolic tangent, cotangent, and cosecant of x.

TABLE 7.10 Identities for inverse hyperbolic functions

$\operatorname{sech}^{-1} x = \cosh^{-1} \frac{1}{x}$
$\operatorname{csch}^{-1} x = \sinh^{-1} \frac{1}{x}$
$\coth^{-1} x = \tanh^{-1} \frac{1}{x}$

Useful Identities

We can use the identities in Table 7.10 to express $\operatorname{sech}^{-1} x$, $\operatorname{csch}^{-1} x$, and $\coth^{-1} x$ in terms of $\cosh^{-1} x$, $\sinh^{-1} x$, and $\tanh^{-1} x$. These identities are direct consequences of the definitions. For example, if $0 < x \le 1$, then

$$\operatorname{sech}\left(\cosh^{-1}\left(\frac{1}{x}\right)\right) = \frac{1}{\cosh\left(\cosh^{-1}\left(\frac{1}{x}\right)\right)} = \frac{1}{\left(\frac{1}{x}\right)} = x.$$

We also know that $\operatorname{sech}(\operatorname{sech}^{-1} x) = x$, so because the hyperbolic secant is one-to-one on $(0, 1]$, we have

$$\cosh^{-1}\left(\frac{1}{x}\right) = \operatorname{sech}^{-1} x.$$

Derivatives of Inverse Hyperbolic Functions

An important use of inverse hyperbolic functions lies in antiderivatives that reverse the derivative formulas in Table 7.11.

TABLE 7.11 Derivatives of inverse hyperbolic functions

$\dfrac{d(\sinh^{-1} x)}{dx} = \dfrac{1}{\sqrt{1 + x^2}}$	
$\dfrac{d(\cosh^{-1} x)}{dx} = \dfrac{1}{\sqrt{x^2 - 1}},$	$x > 1$
$\dfrac{d(\tanh^{-1} x)}{dx} = \dfrac{1}{1 - x^2},$	$\lvert x\rvert < 1$
$\dfrac{d(\coth^{-1} x)}{dx} = \dfrac{1}{1 - x^2},$	$\lvert x\rvert > 1$
$\dfrac{d(\operatorname{sech}^{-1} x)}{dx} = -\dfrac{1}{x\sqrt{1 - x^2}},$	$0 < x < 1$
$\dfrac{d(\operatorname{csch}^{-1} x)}{dx} = -\dfrac{1}{\lvert x\rvert\sqrt{1 + x^2}},$	$x \ne 0$

The restrictions $|x| < 1$ and $|x| > 1$ on the derivative formulas for $\tanh^{-1} x$ and $\coth^{-1} x$ come from the natural restrictions on the values of these functions. (See Figure 7.33a and b.) The distinction between $|x| < 1$ and $|x| > 1$ becomes important when we convert the derivative formulas into integral formulas.

We illustrate how the derivatives of the inverse hyperbolic functions are found in Example 2, where we calculate $d(\cosh^{-1} x)/dx$. The other derivatives are obtained by similar calculations.

EXAMPLE 2 Show that if x is greater than 1, then

$$\frac{d}{dx}(\cosh^{-1} x) = \frac{1}{\sqrt{x^2 - 1}}.$$

Solution We find the derivative of $y = \cosh^{-1} x$ for $x > 1$ by applying Theorem 1 of Section 7.1 with $f(x) = \cosh x$ and $f^{-1}(x) = \cosh^{-1} x$. Theorem 1 can be applied because the derivative of $\cosh x$ is positive when $x > 0$.

$$\begin{aligned}(f^{-1})'(x) &= \frac{1}{f'(f^{-1}(x))} && \text{Theorem 1, Section 7.1}\\ &= \frac{1}{\sinh(\cosh^{-1} x)} && f'(x) = \sinh x\\ &= \frac{1}{\sqrt{\cosh^2(\cosh^{-1} x) - 1}} && \cosh^2 x - \sinh^2 x = 1,\ (x \geq 0),\ \sinh x = \sqrt{\cosh^2 x - 1}\\ &= \frac{1}{\sqrt{x^2 - 1}} && \cosh(\cosh^{-1} x) = x\end{aligned}$$

■

HISTORICAL BIOGRAPHY
Sonya Kovalevsky
(1850–1891)
www.bit.ly/2RbQF8h

With appropriate substitutions, the derivative formulas in Table 7.11 lead to the integration formulas in Table 7.12. Each of the formulas in Table 7.12 can be verified by differentiating the expression on the right-hand side.

TABLE 7.12 Integrals leading to inverse hyperbolic functions

1. $\displaystyle\int \frac{dx}{\sqrt{a^2 + x^2}} = \sinh^{-1}\left(\frac{x}{a}\right) + C, \qquad a > 0$

2. $\displaystyle\int \frac{dx}{\sqrt{x^2 - a^2}} = \cosh^{-1}\left(\frac{x}{a}\right) + C, \qquad x > a > 0$

3. $\displaystyle\int \frac{dx}{a^2 - x^2} = \begin{cases} \dfrac{1}{a}\tanh^{-1}\left(\dfrac{x}{a}\right) + C, & x^2 < a^2 \\ \dfrac{1}{a}\coth^{-1}\left(\dfrac{x}{a}\right) + C, & x^2 > a^2 \end{cases}$

4. $\displaystyle\int \frac{dx}{x\sqrt{a^2 - x^2}} = -\frac{1}{a}\operatorname{sech}^{-1}\left(\frac{x}{a}\right) + C, \qquad 0 < x < a$

5. $\displaystyle\int \frac{dx}{x\sqrt{a^2 + x^2}} = -\frac{1}{a}\operatorname{csch}^{-1}\left|\frac{x}{a}\right| + C, \qquad x \neq 0 \text{ and } a > 0$

EXAMPLE 3 Evaluate

$$\int_0^1 \frac{2\,dx}{\sqrt{3 + 4x^2}}.$$

Solution The indefinite integral is

$$\int \frac{2\,dx}{\sqrt{3+4x^2}} = \int \frac{du}{\sqrt{3+u^2}} \qquad u = 2x, \quad du = 2\,dx$$

$$= \sinh^{-1}\left(\frac{u}{\sqrt{3}}\right) + C \qquad \text{Formula 1 from Table 7.12 with } a^2 = 3$$

$$= \sinh^{-1}\left(\frac{2x}{\sqrt{3}}\right) + C.$$

Therefore,

$$\int_0^1 \frac{2\,dx}{\sqrt{3+4x^2}} = \left[\sinh^{-1}\left(\frac{2x}{\sqrt{3}}\right)\right]_0^1 = \sinh^{-1}\left(\frac{2}{\sqrt{3}}\right) - \sinh^{-1}(0)$$

$$= \sinh^{-1}\left(\frac{2}{\sqrt{3}}\right) - 0 \approx 0.98665.$$

EXERCISES 7.7

Values and Identities

Each of Exercises 1–4 gives a value of $\sinh x$ or $\cosh x$. Use the definitions and the identity $\cosh^2 x - \sinh^2 x = 1$ to find the values of the remaining five hyperbolic functions.

1. $\sinh x = -\dfrac{3}{4}$

2. $\sinh x = \dfrac{4}{3}$

3. $\cosh x = \dfrac{17}{15}, \quad x > 0$

4. $\cosh x = \dfrac{13}{5}, \quad x > 0$

Rewrite the expressions in Exercises 5–10 in terms of exponentials and simplify the results as much as you can.

5. $3\cosh(\ln x)$

6. $\sinh(2\ln x)$

7. $\cosh 7x + \sinh 7x$

8. $\cosh 3x - \sinh 3x$

9. $(\sinh 2x + \cosh 2x)^4$

10. $\ln(\cosh x + \sinh x) + \ln(\cosh x - \sinh x)$

11. Prove the identities

$$\sinh(x+y) = \sinh x \cosh y + \cosh x \sinh y,$$
$$\cosh(x+y) = \cosh x \cosh y + \sinh x \sinh y.$$

Then use them to show that

a. $\sinh 2x = 2\sinh x \cosh x$.

b. $\cosh 2x = \cosh^2 x + \sinh^2 x$.

12. Use the definitions of $\cosh x$ and $\sinh x$ to show that

$$\cosh^2 x - \sinh^2 x = 1.$$

Finding Derivatives

In Exercises 13–24, find the derivative of y with respect to the appropriate variable.

13. $y = 4\sinh x/2$

14. $y = \dfrac{1}{2}\sinh(2x+1)$

15. $y = \sqrt{t}\tanh\sqrt{t}$

16. $y = t^2\tanh\dfrac{1}{t}$

17. $y = \ln(\operatorname{cosech} z)$

18. $y = \ln(\cosh z)$

19. $y = (\tanh\theta)(1 - \ln\tanh\theta)$

20. $y = (\operatorname{csch}\theta)(1 - \ln\operatorname{csch}\theta)$

21. $y = \ln\cosh v - \dfrac{1}{2}\tanh^2 v$

22. $y = \ln\sinh v - \dfrac{1}{2}\coth^2 v$

23. $y = (x^2 - 1)\operatorname{cosech}(\ln x)$

(*Hint:* Before differentiating, express in terms of exponentials and simplify.)

24. $y = (4x^2 - 1)\operatorname{csch}(\ln 2x)$

In Exercises 25–36, find the derivative of y with respect to the appropriate variable.

25. $y = \sinh^{-1}\sqrt{2x}$

26. $y = \cosh^{-1} 2\sqrt{x+1}$

27. $y = (1 - 2\theta)\tanh^{-1} 2\theta$

28. $y = (\theta^2 + 2\theta)\tanh^{-1}(\theta + 1)$

29. $y = (1 - 2t)\coth^{-1}\sqrt{2t}$

30. $y = (1 - t^2)\coth^{-1} t$

31. $y = \sin^{-1} x - x\operatorname{sech}^{-1} x$

32. $y = \ln x + \sqrt{1 - x^2}\operatorname{sech}^{-1} x$

33. $y = \operatorname{csch}^{-1}\left(\dfrac{1}{2}\right)^\theta$

34. $y = \operatorname{csch}^{-1} 2^\theta$

35. $y = \sinh^{-1}(2\tan x)$

36. $y = \cosh^{-1}(\sec 2x), \quad 0 < x < \pi/2$

Integration Formulas

Verify the integration formulas in Exercises 37–40.

37. a. $\int \operatorname{sech} x\, dx = \tan^{-1}(\sinh x) + C$

b. $\int \operatorname{sech} x\, dx = \sin^{-1}(\tanh x) + C$

38. $\int x \operatorname{sech}^{-1} x\, dx = \frac{x^2}{2}\operatorname{sech}^{-1} x - \frac{1}{2}\sqrt{1-x^2} + C$

39. $\int x \coth^{-1} x\, dx = \frac{x^2-1}{2}\coth^{-1} x + \frac{x}{2} + C$

40. $\int \tanh^{-1} x\, dx = x\tanh^{-1} x + \frac{1}{2}\ln(1-x^2) + C$

Evaluating Integrals

Evaluate the integrals in Exercises 41–60.

41. $\int \sinh 3x\, dx$

42. $\int \sinh \frac{x}{5}\, dx$

43. $\int 3\cosh\left(\frac{x}{2} - \ln 4\right) dx$

44. $\int 4\cosh(3x - \ln 2)\, dx$

45. $\int \tanh \frac{x}{7}\, dx$

46. $\int \coth \frac{\theta}{\sqrt{3}}\, d\theta$

47. $\int \operatorname{sech}^2\left(x - \frac{1}{3}\right) dx$

48. $\int \operatorname{csch}^2(5 - x)\, dx$

49. $\int \frac{\operatorname{sech}\sqrt{t}\tanh\sqrt{t}\, dt}{\sqrt{t}}$

50. $\int \frac{\operatorname{csch}(\ln t)\coth(\ln t)\, dt}{t}$

51. $\int_{\ln 3}^{\ln 9} \coth x\, dx$

52. $\int_0^{\ln 2} \tanh 2x\, dx$

53. $\int_{-\ln 4}^{-\ln 2} 2e^{\theta}\cosh\theta\, d\theta$

54. $\int_0^{\ln 2} 4e^{-\theta}\sinh\theta\, d\theta$

55. $\int_{-\pi/4}^{\pi/4} \cosh(\tan\theta)\sec^2\theta\, d\theta$

56. $\int_0^{\pi/2} \sinh(\cos\theta)\sin\theta\, d\theta$

57. $\int_1^3 \frac{2\cosh(\ln t)}{2t}\, dt$

58. $\int_1^4 \frac{8\cosh\sqrt{x}}{\sqrt{x}}\, dx$

59. $\int_{-\ln 2}^{0} \cosh^2\left(\frac{x}{2}\right) dx$

60. $\int_0^{\ln 5} 5\sinh^2\left(\frac{x}{2}\right) dx$

Inverse Hyperbolic Functions and Integrals

Since the hyperbolic functions can be expressed in terms of exponential functions, it is possible to express the inverse hyperbolic functions in terms of logarithms, as shown in the following table.

$\sinh^{-1} x = \ln\left(x + \sqrt{x^2+1}\right),$	$-\infty < x < \infty$
$\cosh^{-1} x = \ln\left(x + \sqrt{x^2-1}\right),$	$x \ge 1$
$\tanh^{-1} x = \frac{1}{2}\ln\frac{1+x}{1-x},$	$\lvert x\rvert < 1$
$\operatorname{sech}^{-1} x = \ln\left(\frac{1+\sqrt{1-x^2}}{x}\right),$	$0 < x \le 1$
$\operatorname{csch}^{-1} x = \ln\left(\frac{1}{x} + \frac{\sqrt{1+x^2}}{\lvert x\rvert}\right),$	$x \ne 0$
$\coth^{-1} x = \frac{1}{2}\ln\frac{x+1}{x-1},$	$\lvert x\rvert > 1$

Use these formulas to express the numbers in Exercises 61–66 in terms of natural logarithms.

61. $\sinh^{-1}(-5/12)$

62. $\cosh^{-1}(7/2)$

63. $\tanh^{-1}(-1/2)$

64. $\coth^{-1}(4/2)$

65. $\operatorname{sech}^{-1}(4/5)$

66. $\operatorname{csch}^{-1}(-1/\sqrt{3})$

Evaluate the integrals in Exercises 67–74 in terms of

a. inverse hyperbolic functions.

b. natural logarithms.

67. $\int_0^{2\sqrt{3}} \frac{dx}{\sqrt{4+x^2}}$

68. $\int_0^{1/3} \frac{6\, dx}{\sqrt{1+9x^2}}$

69. $\int_{5/4}^{2} \frac{dx}{1-x^2}$

70. $\int_0^{1/2} \frac{dx}{1-x^2}$

71. $\int_{1/5}^{3/13} \frac{dx}{x\sqrt{1-16x^2}}$

72. $\int_1^2 \frac{dx}{x\sqrt{4+x^2}}$

73. $\int_0^{\pi} \frac{\cos x\, dx}{\sqrt{1+\sin^2 x}}$

74. $\int_1^e \frac{dx}{x\sqrt{1+(\ln x)^2}}$

Applications and Examples

75. Show that if a function f is defined on an interval symmetric about the origin (so that f is defined at $-x$ whenever it is defined at x), then

$$f(x) = \frac{f(x)+f(-x)}{2} + \frac{f(x)-f(-x)}{2}.$$

Then show that $(f(x)+f(-x))/2$ is even and that $(f(x)-f(-x))/2$ is odd.

76. Derive the formula $\sinh^{-1} x = \ln\left(x + \sqrt{x^2+1}\right)$ for all real x. Explain in your derivation why the plus sign is used with the square root instead of the minus sign.

77. Skydiving If a body of mass m falling from rest under the action of gravity encounters an air resistance proportional to the square of the velocity, then the body's velocity t sec into the fall satisfies the differential equation

$$m\frac{dv}{dt} = mg - kv^2,$$

where k is a constant that depends on the body's aerodynamic properties and the density of the air. (We assume that the fall is short enough so that the variation in the air's density will not affect the outcome significantly.)

a. Show that

$$v = \sqrt{\frac{mg}{k}}\tanh\left(\sqrt{\frac{gk}{m}}\,t\right)$$

satisfies the differential equation and the initial condition that $v = 0$ when $t = 0$.

b. Find the body's *limiting velocity*, $\lim_{t\to\infty} v$.

c. For a 160-lb skydiver ($mg = 160$), with time in seconds and distance in feet, a typical value for k is 0.005. What is the diver's limiting velocity?

78. Acceleration whose magnitudes are proportional to displacement Suppose that the position of a body moving along a coordinate line at time t is

a. $s = a\cos kt + b\sin kt$. **b.** $s = a\cosh kt + b\sinh kt$.

Show in both cases that the acceleration d^2s/dt^2 is proportional to s but that in the first case it is directed toward the origin, whereas in the second case it is directed away from the origin.

79. Volume A region in the first quadrant is bounded above by the curve $y = \coth x$, below by the curve $y = \operatorname{csch} x$, and on the left and right by the y-axis and the line $x = 4$, respectively. Find the volume of the solid generated by revolving the region about the x-axis.

80. Volume The region enclosed by the curve $y = \operatorname{sech} x$, the x-axis, and the lines $x = \pm\ln\sqrt{3}$ is revolved about the x-axis to generate a solid. Find the volume of the solid.

81. Arc length Find the length of the graph of $y = (1/2)\cosh 2x$ from $x = 0$ to $x = \ln\sqrt{5}$.

82. Use the definitions of the hyperbolic functions to find each of the following limits.

a. $\lim_{x\to\infty}\tanh x$ **b.** $\lim_{x\to-\infty}\tanh x$ **c.** $\lim_{x\to\infty}\sinh x$

d. $\lim_{x\to-\infty}\sinh x$ **e.** $\lim_{x\to\infty}\operatorname{sech} x$ **f.** $\lim_{x\to\infty}\coth x$

g. $\lim_{x\to 0^+}\coth x$ **h.** $\lim_{x\to 0^-}\coth x$ **i.** $\lim_{x\to-\infty}\operatorname{csch} x$

83. Hanging cables Imagine a cable, like a telephone line or TV cable, strung from one support to another and hanging freely. The cable's weight per unit length is a constant w, and the horizontal tension at its lowest point is a *vector* of length H. If we choose a coordinate system for the plane of the cable in which the x-axis is horizontal, the force of gravity is straight down, the positive y-axis points straight up, and the lowest point of the cable lies at the point $y = H/w$ on the y-axis (see accompanying figure), then it can be shown that the cable lies along the graph of the hyperbolic cosine

$$y = \frac{H}{w}\cosh\frac{w}{H}x.$$

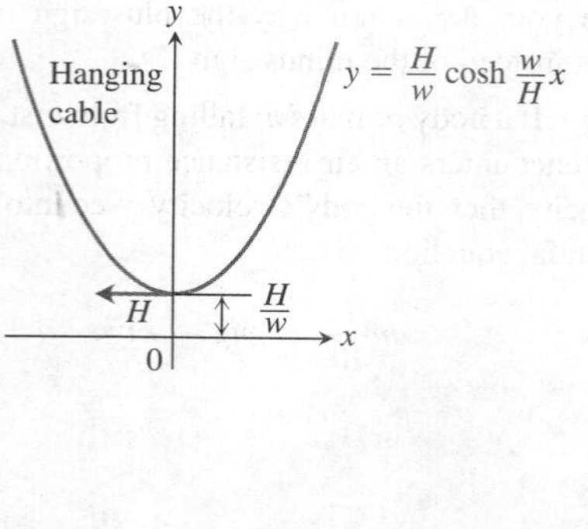

Such a curve is sometimes called a **chain curve** or a **catenary**, the latter deriving from the Latin *catena*, meaning "chain."

a. Let $P(x, y)$ denote an arbitrary point on the cable. The next accompanying figure displays the tension at P as a vector of length (magnitude) T, as well as the tension H at the lowest point A. Show that the cable's slope at P is

$$\tan\phi = \frac{dy}{dx} = \sinh\frac{w}{H}x.$$

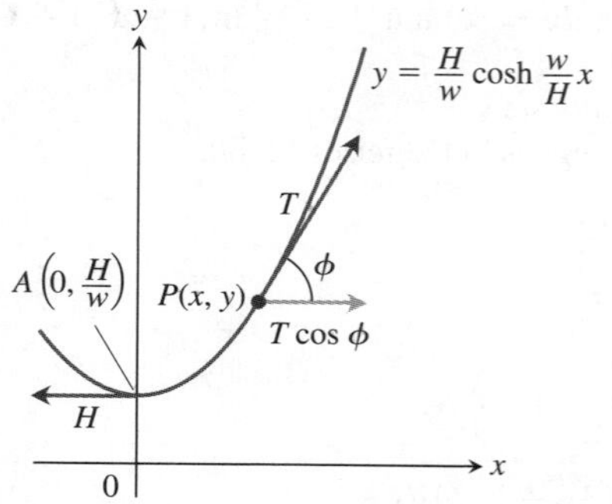

b. Using the result from part (a) and the fact that the horizontal tension at P must equal H (the cable is not moving), show that $T = wy$. Hence, the magnitude of the tension at $P(x, y)$ is exactly equal to the weight of y units of cable.

84. (*Continuation of Exercise 83.*) The length of arc AP in the Exercise 83 figure is $s = (1/a)\sinh ax$, where $a = w/H$. Show that the coordinates of P may be expressed in terms of s as

$$x = \frac{1}{a}\sinh^{-1} as, \qquad y = \sqrt{s^2 + \frac{1}{a^2}}.$$

85. Area Show that the area of the region in the first quadrant enclosed by the curve $y = (1/a)\cosh ax$, the coordinate axes, and the line $x = b$ is the same as the area of a rectangle of height $1/a$ and length s, where s is the length of the curve from $x = 0$ to $x = b$. Draw a figure illustrating this result.

86. The hyperbolic in hyperbolic functions Just as $x = \cos u$ and $y = \sin u$ are identified with points (x, y) on the unit circle, the functions $x = \cosh u$ and $y = \sinh u$ are identified with points (x, y) on the right-hand branch of the unit hyperbola, $x^2 - y^2 = 1$.

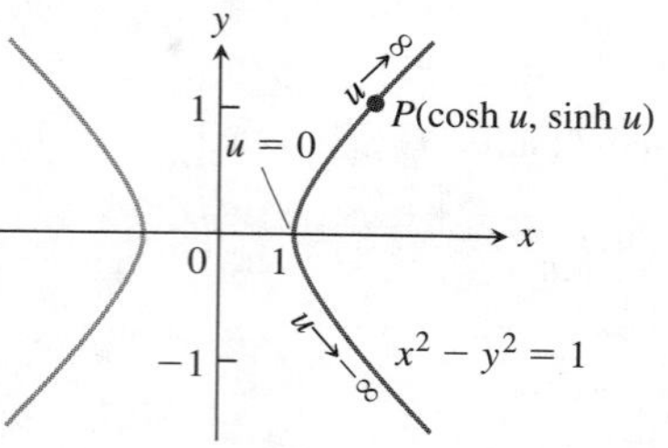

Since $\cosh^2 u - \sinh^2 u = 1$, the point $(\cosh u, \sinh u)$ lies on the right-hand branch of the hyperbola $x^2 - y^2 = 1$ for every value of u.

Another analogy between hyperbolic and circular functions is that the variable u in the coordinates $(\cosh u, \sinh u)$ for the points of the right-hand branch of the hyperbola $x^2 - y^2 = 1$ is twice the area of the sector AOP pictured in the figure following part (c). To see why this is so, carry out the following steps.

a. Show that the area $A(u)$ of sector AOP is

$$A(u) = \frac{1}{2}\cosh u \sinh u - \int_1^{\cosh u} \sqrt{x^2 - 1}\, dx.$$

b. Differentiate both sides of the equation in part (a) with respect to u to show that

$$A'(u) = \frac{1}{2}.$$

c. Solve this last equation for $A(u)$. What is the value of $A(0)$? What is the value of the constant of integration C in your solution? With C determined, what does your solution say about the relationship of u to $A(u)$?

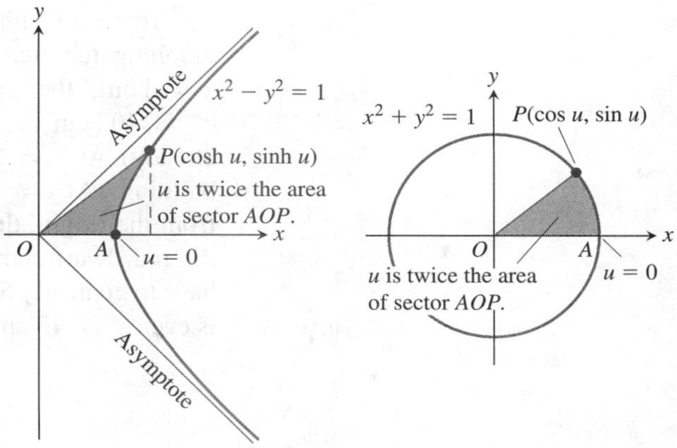

One of the analogies between hyperbolic and circular functions is revealed by these two diagrams (Exercise 86).

7.8 Relative Rates of Growth

It is often important in mathematics, computer science, and engineering to compare the rates at which functions of x grow as x becomes large. Exponential functions are important in these comparisons because of their very fast growth, and logarithmic functions because of their very slow growth. In this section we introduce the *little-oh* and *big-oh* notation used to describe the results of these comparisons. We restrict our attention to functions whose values eventually become and remain positive as $x \to \infty$.

Growth Rates of Functions

You may have noticed that exponential functions like 2^x and e^x seem to grow more rapidly as x gets large than do polynomials and rational functions. These exponentials certainly grow more rapidly than x itself, and you can see 2^x outgrowing x^2 as x increases in Figure 7.34. In fact, as $x \to \infty$, the functions 2^x and e^x grow faster than any power of x, even $x^{1{,}000{,}000}$ (Exercise 19). In contrast, logarithmic functions like $y = \log_2 x$ and $y = \ln x$ grow more slowly as $x \to \infty$ than any positive power of x (Exercise 21).

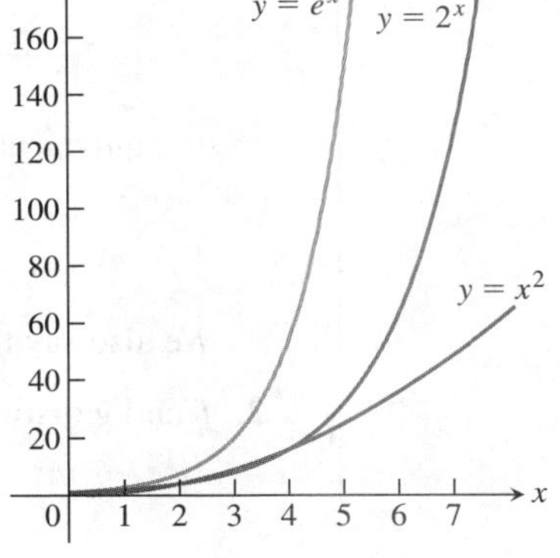

FIGURE 7.34 The graphs of e^x, 2^x, and x^2.

To get a feeling for how rapidly the values of $y = e^x$ grow with increasing x, think of graphing the function on a large blackboard, with the axes scaled in centimeters. At $x = 1$ cm, the graph is $e^1 \approx 3$ cm above the x-axis. At $x = 6$ cm, the graph is $e^6 \approx 403$ cm ≈ 4 m high (it is about to go through the ceiling if it hasn't done so already). At $x = 10$ cm, the graph is $e^{10} \approx 22{,}026$ cm ≈ 220 m high, higher than most buildings. At $x = 24$ cm, the graph is more than halfway to the moon, and at $x = 43$ cm from the origin, the graph is high enough to reach past the sun's closest stellar neighbor, the red dwarf star Proxima Centauri. By contrast, with axes scaled in centimeters, you have to go nearly 5 light-years out on the x-axis to find a point where the graph of $y = \ln x$ is even $y = 43$ cm high. See Figure 7.35.

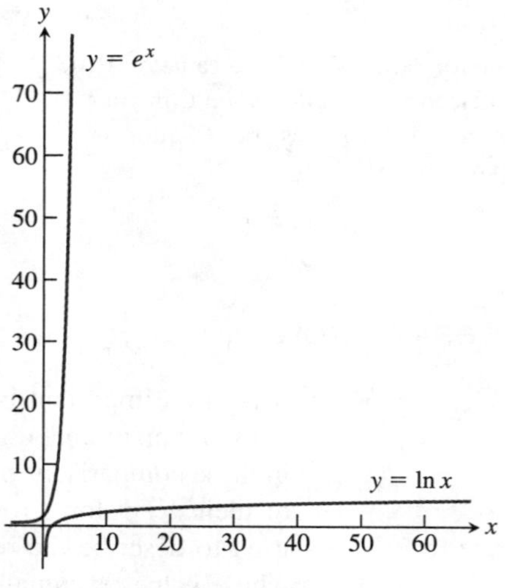

FIGURE 7.35 Scale drawings of the graphs of e^x and $\ln x$.

These important comparisons of exponential, polynomial, and logarithmic functions can be made precise by defining what it means for a function $f(x)$ to grow faster than another function $g(x)$ as $x \to \infty$.

DEFINITION Let $f(x)$ and $g(x)$ be positive for x sufficiently large.

1. f **grows faster than** g as $x \to \infty$ if

$$\lim_{x\to\infty} \frac{f(x)}{g(x)} = \infty$$

or, equivalently, if

$$\lim_{x\to\infty} \frac{g(x)}{f(x)} = 0.$$

We also say that g **grows slower than** f as $x \to \infty$.

2. f and g **grow at the same rate** as $x \to \infty$ if

$$\lim_{x\to\infty} \frac{f(x)}{g(x)} = L$$

where L is finite and positive.

According to these definitions, $y = 2x$ does not grow faster than $y = x$. The two functions grow at the same rate because

$$\lim_{x\to\infty} \frac{2x}{x} = \lim_{x\to\infty} 2 = 2,$$

which is a finite, positive limit. The reason for this departure from more colloquial usage is that we want "f grows faster than g" to mean that for large x-values g is negligible when compared with f.

EXAMPLE 1 We compare the growth rates of several common functions.

(a) e^x grows faster than x^2 as $x \to \infty$ because

$$\underbrace{\lim_{x\to\infty} \frac{e^x}{x^2}}_{\infty/\infty} = \underbrace{\lim_{x\to\infty} \frac{e^x}{2x}}_{\infty/\infty} = \lim_{x\to\infty} \frac{e^x}{2} = \infty. \qquad \text{Using l'Hôpital's Rule twice}$$

(b) 3^x grows faster than 2^x as $x \to \infty$ because

$$\lim_{x\to\infty} \frac{3^x}{2^x} = \lim_{x\to\infty} \left(\frac{3}{2}\right)^x = \infty.$$

(c) x^2 grows faster than $\ln x$ as $x \to \infty$ because

$$\lim_{x\to\infty} \frac{x^2}{\ln x} = \lim_{x\to\infty} \frac{2x}{1/x} = \lim_{x\to\infty} 2x^2 = \infty. \qquad \text{l'Hôpital's Rule}$$

(d) $\ln x$ grows slower than $x^{1/n}$ as $x \to \infty$ for any positive integer n because

$$\lim_{x\to\infty} \frac{\ln x}{x^{1/n}} = \lim_{x\to\infty} \frac{1/x}{(1/n)x^{(1/n)-1}} \qquad \text{l'Hôpital's Rule}$$

$$= \lim_{x\to\infty} \frac{n}{x^{1/n}} = 0. \qquad n \text{ is constant.}$$

(e) As part (b) suggests, exponential functions with different bases never grow at the same rate as $x \to \infty$. If $a > b > 0$, then a^x grows faster than b^x. Since $(a/b) > 1$,

$$\lim_{x\to\infty} \frac{a^x}{b^x} = \lim_{x\to\infty} \left(\frac{a}{b}\right)^x = \infty.$$

(f) In contrast to exponential functions, logarithmic functions with different bases $a > 1$ and $b > 1$ always grow at the same rate as $x \to \infty$:

$$\lim_{x\to\infty} \frac{\log_a x}{\log_b x} = \lim_{x\to\infty} \frac{\ln x/\ln a}{\ln x/\ln b} = \frac{\ln b}{\ln a}.$$

The limiting ratio is always finite and never zero. ■

If f grows at the same rate as g as $x \to \infty$, and g grows at the same rate as h as $x \to \infty$, then f grows at the same rate as h as $x \to \infty$. The reason is that

$$\lim_{x\to\infty} \frac{f}{g} = L_1 \quad \text{and} \quad \lim_{x\to\infty} \frac{g}{h} = L_2$$

together imply

$$\lim_{x\to\infty} \frac{f}{h} = \lim_{x\to\infty} \frac{f}{g} \cdot \frac{g}{h} = L_1 L_2.$$

If L_1 and L_2 are finite and nonzero, then so is $L_1 L_2$.

EXAMPLE 2 Show that $\sqrt{x^2+5}$ and $(2\sqrt{x}-1)^2$ grow at the same rate as $x \to \infty$.

Solution We show that the functions grow at the same rate by showing that they both grow at the same rate as the function $g(x) = x$:

$$\lim_{x\to\infty} \frac{\sqrt{x^2+5}}{x} = \lim_{x\to\infty} \sqrt{1+\frac{5}{x^2}} = 1,$$

$$\lim_{x\to\infty} \frac{(2\sqrt{x}-1)^2}{x} = \lim_{x\to\infty} \left(\frac{2\sqrt{x}-1}{\sqrt{x}}\right)^2 = \lim_{x\to\infty} \left(2-\frac{1}{\sqrt{x}}\right)^2 = 4. \qquad \blacksquare$$

Order and Oh-Notation

The "little-oh" and "big-oh" notation was invented by number theorists over a hundred years ago and is now commonplace in mathematical analysis and computer science. According to this definition, saying $f = o(g)$ as $x \to \infty$ is another way to say that f grows slower than g as $x \to \infty$.

> **DEFINITION** A function f is **of smaller order than** g as $x \to \infty$ if $\lim_{x\to\infty} \frac{f(x)}{g(x)} = 0$. We indicate this by writing $f = o(g)$ ("f is little-oh of g").

EXAMPLE 3 Here we use little-oh notation.

(a) $\ln x = o(x)$ as $x \to \infty$ because $\lim_{x\to\infty} \frac{\ln x}{x} = 0.$

(b) $x^2 = o(x^3+1)$ as $x \to \infty$ because $\lim_{x\to\infty} \frac{x^2}{x^3+1} = 0.$ $\blacksquare$

> **DEFINITION** Let $f(x)$ and $g(x)$ be positive for x sufficiently large. Then f is **of at most the order of** g as $x \to \infty$ if there is a positive integer M for which
>
> $$\frac{f(x)}{g(x)} \le M,$$
>
> for x sufficiently large. We indicate this by writing $f = O(g)$ ("f is big-oh of g").

EXAMPLE 4 Here we use big-oh notation.

(a) $x + \sin x = O(x)$ as $x \to \infty$ because $\frac{x+\sin x}{x} \le 2$ for x sufficiently large.

(b) $e^x + x^2 = O(e^x)$ as $x \to \infty$ because $\frac{e^x+x^2}{e^x} \to 1$ as $x \to \infty$.

(c) $x = O(e^x)$ as $x \to \infty$ because $\frac{x}{e^x} \to 0$ as $x \to \infty$. $\blacksquare$

If you look at the definitions again, you will see that $f = o(g)$ implies $f = O(g)$ for functions that are positive for all sufficiently large x. Also, if f and g grow at the same rate, then $f = O(g)$ and $g = O(f)$ (Exercise 11).

Sequential vs. Binary Search

Computer scientists often measure the efficiency of an algorithm by counting the number of steps a computer must take to execute the algorithm. There can be significant differences in how efficiently algorithms perform, even if they are designed to accomplish the same task. These differences are often described using big-oh notation. Here is an example.

One edition of *Webster's International Dictionary* lists about 26,000 words that begin with the letter *a*. One way to look up a word, or to learn if it is not there, is to read through the list one word at a time until you either find the word or determine that it is not there. This method, called **sequential search**, makes no particular use of the words' alphabetical arrangement in the list. You are sure to get an answer, but it might take 26,000 steps.

Another way to find the word or to learn it is not there is to go straight to the middle of the list (give or take a few words). If you do not find the word, then go to the middle of the half that contains it and forget about the half that does not. (You know which half contains it because you know the list is ordered alphabetically.) This method, called a **binary search**, eliminates roughly 13,000 words in a single step. If you do not find the word on the second try, then jump to the middle of the half that contains it. Continue this way until you have either found the word or divided the list in half so many times there are no words left. How many times do you have to divide the list to find the word or learn that it is not there? At most 15, because

$$(26{,}000/2^{15}) < 1.$$

That certainly beats a possible 26,000 steps.

For a list of length n, a sequential search algorithm takes on the order of n steps to find a word or determine that it is not in the list. A binary search, as the second algorithm is called, takes on the order of $\log_2 n$ steps. The reason is that if $2^{m-1} < n \leq 2^m$, then $m - 1 < \log_2 n \leq m$, and the number of bisections required to narrow the list to one word will be at most $m = \lceil \log_2 n \rceil$, the integer ceiling of the number $\log_2 n$.

Big-oh notation provides a compact way to say all this. The number of steps in a sequential search of an ordered list is $O(n)$; the number of steps in a binary search is $O(\log_2 n)$. In our example, there is a big difference between the two (26,000 versus 15), and the difference can only increase with n because n grows faster than $\log_2 n$ as $n \to \infty$.

EXERCISES 7.8

Comparisons with the Exponential e^x

1. Which of the following functions grow faster than e^x as $x \to \infty$? Which grow at the same rate as e^x? Which grow slower?

a. $x - 3$ **b.** $x^3 + \sin^2 x$

c. $\sqrt{x}$ **d.** 5^x

e. $(3/2)^x$ **f.** $e^{x/2}$

g. $e^x/2$ **h.** $\log_{10} x$

2. Which of the following functions grow faster than e^x as $x \to \infty$? Which grow at the same rate as e^x? Which grow slower?

a. $10x^4 + 30x + 1$ **b.** $x \ln x - x$

c. $\sqrt{1 + x^4}$ **d.** $(5/2)^x$

e. e^{-x} **f.** xe^x

g. $e^{\cos x}$ **h.** e^{x-1}

Comparisons with the Power x^2

3. Which of the following functions grow faster than x^2 as $x \to \infty$? Which grow at the same rate as x^2? Which grow slower?

a. $x^2 + 3x$ **b.** $x^5 - x^2$

c. $\sqrt{x^4 + x^3}$ **d.** $(x + 3)^2$

e. $x \ln x$ **f.** 3^x

g. $x^3 e^{-x}$ **h.** $8x^2$

4. Which of the following functions grow faster than x^2 as $x \to \infty$? Which grow at the same rate as x^2? Which grow slower?

a. $x^2 + \sqrt{x}$ **b.** $10x^2$

c. $x^2 e^{-x}$ **d.** $\log_{10}(x^2)$

e. $x^3 - x^2$ **f.** $(1/10)^x$

g. $(1.1)^x$ **h.** $x^2 + 100x$

Comparisons with the Logarithm $\ln x$

5. Which of the following functions grow faster than $\ln x$ as $x \to \infty$? Which grow at the same rate as $\ln x$? Which grow slower?

a. $\log_3 x$ **b.** $\ln 4x$

c. $\ln \sqrt{x}$ **d.** $\sqrt{x}$

e. $2x$ **f.** $5 \ln x$

g. $1/x$ **h.** e^x

6. Which of the following functions grow faster than $\ln x$ as $x \to \infty$? Which grow at the same rate as $\ln x$? Which grow slower?

a. $\log_2(x^2)$ **b.** $\log_{10} 10x$

c. $1/\sqrt{2x}$ **d.** $1/x^2$

e. $x - 2\ln x$ **f.** e^{-x}

g. $\ln(\ln x)$ **h.** $\ln(2x + 5)$

Ordering Functions by Growth Rates

7. Order the following functions from slowest growing to fastest growing as $x \to \infty$.

a. e^x **b.** x^x

c. $(\ln x)^x$ **d.** $e^{x/2}$

8. Order the following functions from slowest growing to fastest growing as $x \to \infty$.

a. 2^x **b.** x^2

c. $(\ln 2)^x$ **d.** e^x

Big-oh and Little-oh; Order

9. True, or false? As $x \to \infty$,

a. $x = o(x)$ **b.** $x = o(x + 3)$

c. $x = O(x + 5)$ **d.** $x = O(2x)$

e. $e^x = o(e^{2x})$ **f.** $x + \ln x = O(x)$

g. $\ln x = o(\ln 2x)$ **h.** $\sqrt{x^2 + 5} = O(x)$

10. True, or false? As $x \to \infty$,

a. $\dfrac{1}{x + 3} = O\left(\dfrac{1}{x}\right)$ **b.** $\dfrac{1}{x} + \dfrac{1}{x^2} = O\left(\dfrac{1}{x}\right)$

c. $\dfrac{1}{x} - \dfrac{1}{x^2} = o\left(\dfrac{1}{x}\right)$ **d.** $2 + \cos x = O(2)$

e. $e^x + x = O(e^x)$ **f.** $x \ln x = o(x^2)$

g. $\ln(\ln x) = O(\ln x)$ **h.** $\ln(x) = o(\ln(x^2 + 1))$

11. Show that if positive functions $f(x)$ and $g(x)$ grow at the same rate as $x \to \infty$, then $f = O(g)$ and $g = O(f)$.

12. When is a polynomial $f(x)$ of smaller order than a polynomial $g(x)$ as $x \to \infty$? Give reasons for your answer.

13. When is a polynomial $f(x)$ of at most the order of a polynomial $g(x)$ as $x \to \infty$? Give reasons for your answer.

14. What do the conclusions we drew in Section 2.6 about the limits of rational functions tell us about the relative growth of polynomials as $x \to \infty$?

Other Comparisons

T **15.** Investigate

$$\lim_{x\to\infty} \frac{\ln x}{\ln x} \quad \text{and} \quad \lim_{x\to\infty} \frac{\ln(x+100)}{\ln x}$$

Then use l'Hôpital's Rule to explain what you find.

16. (*Continuation of Exercise 15.*) Show that the value of

$$\lim_{x\to\infty} \frac{\ln(x + a)}{\ln x}$$

is the same no matter what value you assign to the constant a. What does this say about the relative rates at which the functions $f(x) = \ln(x + a)$ and $g(x) = \ln x$ grow?

17. Show that $\sqrt{10x + 1}$ and $\sqrt{x + 1}$ grow at the same rate as $x \to \infty$ by showing that they both grow at the same rate as $\sqrt{x}$ as $x \to \infty$.

18. Show that $\sqrt{x^4 + 1}$ and $\sqrt{x^4 - 1}$ grow at the same rate as $x \to \infty$ by showing that they both grow at the same rate as x^2 as $x \to \infty$.

19. Show that e^x grows faster as $x \to \infty$ than x^n for any positive integer n, even $x^{1{,}000{,}000}$. (*Hint:* What is the nth derivative of x^n?)

20. The function e^x outgrows any polynomial Show that e^x grows faster as $x \to \infty$ than any polynomial

$$a_n x^n + a_{n-1} x^{n-1} + \cdots + a_1 x + a_0.$$

21. a. Show that $\ln x$ grows slower as $x \to \infty$ than $x^{1/n}$ for any positive integer n, even $x^{1/1{,}000{,}000}$.

T **b.** Although the values of $x^{1/1{,}000{,}000}$ eventually overtake the values of $\ln x$, you have to go way out on the x-axis before this happens. Find a value of x greater than 1 for which $x^{1/1{,}000{,}000} > \ln x$. You might start by observing that when $x > 1$ the equation $\ln x = x^{1/1{,}000{,}000}$ is equivalent to the equation $\ln(\ln x) = (\ln x)/1{,}000{,}000$.

T **c.** Even $x^{1/10}$ takes a long time to overtake $\ln x$. Experiment with a calculator to find the value of $x > 10$ at which the graphs of $x^{1/10}$ and $\ln x$ cross, or, equivalently, at which $\ln x = 10 \ln(\ln x)$. Bracket the crossing point between powers of 10 and then close in by successive halving.

T **d.** (*Continuation of part (c).*) The value of x at which $\ln x = 10 \ln(\ln x)$ is too far out for some graphers and root finders to identify. Try it on the equipment available to you and see what happens.

22. The function $\ln x$ grows slower than any polynomial Show that $\ln x$ grows slower as $x \to \infty$ than any nonconstant polynomial.

Algorithms and Searches

23. a. Suppose you have three different algorithms for solving the same problem and each algorithm takes a number of steps that is of the order of one of the functions listed here:

$$n \log_2 n, \quad n^{3/2}, \quad n(\log_2 n)^2.$$

Which of the algorithms is the most efficient in the long run? Give reasons for your answer.

T **b.** Graph the functions in part (a) together to get a sense of how rapidly each one grows.

24. Repeat Exercise 23 for the functions

$$n, \quad \sqrt{n} \log_2 n, \quad (\log_2 n)^2.$$

T **25.** Suppose you are looking for an item in an ordered list one million items long. How many steps might it take to find that item with a sequential search? A binary search?

T **26.** You are looking for an item in an ordered list 450,000 items long (the length of *Webster's Third New International Dictionary*). How many steps might it take to find the item with a sequential search? A binary search?

CHAPTER 7 Questions to Guide Your Review

1. What functions have inverses? How do you know if two functions f and g are inverses of one another? Give examples of functions that are (are not) inverses of one another.
2. How are the domains, ranges, and graphs of functions and their inverses related? Give an example.
3. How can you sometimes express the inverse of a function of x as a function of x?
4. Under what circumstances can you be sure that the inverse of a function f is differentiable? How are the derivatives of f and f^{-1} related?
5. What is the natural logarithm function? What are its domain, range, and derivative? What arithmetic properties does it have? Comment on its graph.
6. What is logarithmic differentiation? Give an example.
7. What integrals lead to logarithms? Give examples. What are the integrals of $\tan x$, $\cot x$, $\sec x$, and $\csc x$?
8. How is the exponential function e^x defined? What are its domain, range, and derivative? What laws of exponents does it obey? Comment on its graph.
9. How are the functions a^x and $\log_a x$ defined? Are there any restrictions on a? How is the graph of $\log_a x$ related to the graph of $\ln x$? What truth is there in the statement that there is really only one exponential function and one logarithmic function?
10. How do you solve separable first-order differential equations?
11. What is the law of exponential change? How can it be derived from an initial value problem? What are some of the applications of the law?
12. Describe l'Hôpital's Rule. How do you know when to use the rule and when to stop? Give an example.
13. How can you sometimes handle limits that lead to indeterminate forms ∞/∞, $\infty \cdot 0$, and $\infty - \infty$? Give examples.
14. How can you sometimes handle limits that lead to indeterminate forms 1^∞, 0^0, and ∞^∞? Give examples.
15. How are the inverse trigonometric functions defined? How can you sometimes use right triangles to find values of these functions? Give examples.
16. What are the derivatives of the inverse trigonometric functions? How do the domains of the derivatives compare with the domains of the functions?
17. What integrals lead to inverse trigonometric functions? How do substitution and completing the square broaden the application of these integrals?
18. What are the six basic hyperbolic functions? Comment on their domains, ranges, and graphs. What are some of the identities relating them?
19. What are the derivatives of the six basic hyperbolic functions? What are the corresponding integral formulas? What similarities do you see here with the six basic trigonometric functions?
20. How are the inverse hyperbolic functions defined? Comment on their domains, ranges, and graphs. How can you find values of $\operatorname{sech}^{-1} x$, $\operatorname{csch}^{-1} x$, and $\coth^{-1} x$ using a calculator's keys for $\cosh^{-1} x$, $\sinh^{-1} x$, and $\tanh^{-1} x$?
21. What integrals lead naturally to inverse hyperbolic functions?
22. How do you compare the growth rates of positive functions as $x \to \infty$?
23. What roles do the functions e^x and $\ln x$ play in growth comparisons?
24. Describe big-oh and little-oh notation. Give examples.
25. Which is more efficient—a sequential search or a binary search? Explain.

CHAPTER 7 Practice Exercises

Finding Derivatives

In Exercises 1–24, find the derivative of y with respect to the appropriate variable.

1. $y = 10e^{-x/5}$
2. $y = \sqrt{2}e^{\sqrt{2}x}$
3. $y = \frac{1}{4}xe^{4x} - \frac{1}{16}e^{4x}$
4. $y = x^2e^{-2/x}$
5. $y = \ln(\sin^2\theta)$
6. $y = \ln(\sec^2\theta)$
7. $y = \log_2(x^2/2)$
8. $y = \log_5(3x - 7)$
9. $y = 8^{-t}$
10. $y = 9^{2t}$
11. $y = 5x^{3.6}$
12. $y = \sqrt{2}x^{-\sqrt{2}}$
13. $y = (x + 2)^{x+2}$
14. $y = 2(\ln x)^{x/2}$
15. $y = \sin^{-1}\sqrt{1 - u^2}, \quad 0 < u < 1$
16. $y = \arcsin\left(\frac{1}{\sqrt{v}}\right), \quad v > 1$
17. $y = \ln \arccos x$
18. $y = z \arccos z - \sqrt{1 - z^2}$
19. $y = t \arctan t - \frac{1}{2}\ln t$
20. $y = (1 + t^2)\operatorname{arccot} 2t$
21. $y = z \operatorname{arcsec} z - \sqrt{z^2 - 1}, \quad z > 1$
22. $y = 2\sqrt{x - 1}\operatorname{arcsec}\sqrt{x}$
23. $y = \operatorname{arccsc}(\sec\theta), \quad 0 < \theta < \pi/2$
24. $y = (1 + x^2)e^{\arctan x}$

Logarithmic Differentiation

In Exercises 25–30, use logarithmic differentiation to find the derivative of y with respect to the appropriate variable.

25. $y = \dfrac{2(x^2+1)}{\sqrt{\cos 2x}}$

26. $y = \sqrt[10]{\dfrac{3x+4}{2x-4}}$

27. $y = \left(\dfrac{(t+1)(t-1)}{(t-2)(t+3)}\right)^5, \quad t > 2$

28. $y = \dfrac{2u2^u}{\sqrt{u^2+1}}$

29. $y = (\sin\theta)^{\sqrt{\theta}}$

30. $y = (\ln x)^{1/(\ln x)}$

Evaluating Integrals

Evaluate the integrals in Exercises 31–78.

31. $\int e^x \sin(e^x)\, dx$

32. $\int e^t \cos(3e^t - 2)\, dt$

33. $\int e^x \sec^2(e^x - 7)\, dx$

34. $\int e^y \csc(e^y + 1)\cot(e^y + 1)\, dy$

35. $\int \sec^2 x\, e^{\tan x}\, dx$

36. $\int \csc^2 x\, e^{\cot x}\, dx$

37. $\int_{-1}^{1} \dfrac{dx}{3x-4}$

38. $\int_1^e \dfrac{\sqrt{\ln x}}{x}\, dx$

39. $\int_0^{\pi} \tan\dfrac{x}{3}\, dx$

40. $\int_{1/6}^{1/4} 2\cot \pi x\, dx$

41. $\int_0^4 \dfrac{2t}{t^2-25}\, dt$

42. $\int_{-\pi/2}^{\pi/6} \dfrac{\cos t}{1-\sin t}\, dt$

43. $\int \dfrac{\tan(\ln v)}{v}\, dv$

44. $\int \dfrac{dv}{v \ln v}$

45. $\int \dfrac{(\ln x)^{-3}}{x}\, dx$

46. $\int \dfrac{\ln(x-5)}{x-5}\, dx$

47. $\int \dfrac{1}{r}\csc^2(1 + \ln r)\, dr$

48. $\int \dfrac{\cos(1 - \ln v)}{v}\, dv$

49. $\int x3^{x^2}\, dx$

50. $\int 2^{\tan x}\sec^2 x\, dx$

51. $\int_1^7 \dfrac{3}{x}\, dx$

52. $\int_1^{32} \dfrac{1}{5x}\, dx$

53. $\int_1^4 \left(\dfrac{x}{8} + \dfrac{1}{2x}\right) dx$

54. $\int_1^8 \left(\dfrac{2}{3x} - \dfrac{8}{x^2}\right) dx$

55. $\int_{-2}^{-1} e^{-(x+1)}\, dx$

56. $\int_{-\ln 2}^{0} e^{2w}\, dw$

57. $\int_0^{\ln 5} e^r(3e^r + 1)^{-3/2}\, dr$

58. $\int_0^{\ln 9} e^\theta(e^\theta - 1)^{1/2}\, d\theta$

59. $\int_1^e \dfrac{1}{x}(1 + 7\ln x)^{-1/3}\, dx$

60. $\int_e^{e^2} \dfrac{1}{x\sqrt{\ln x}}\, dx$

61. $\int_1^3 \dfrac{(\ln(v+1))^2}{v+1}\, dv$

62. $\int_2^4 (1 + \ln t)t \ln t\, dt$

63. $\int_1^8 \dfrac{\log_4 \theta}{\theta}\, d\theta$

64. $\int_1^e \dfrac{8 \ln 3 \log_3 \theta}{\theta}\, d\theta$

65. $\int_{-3/4}^{3/4} \dfrac{6\, dx}{\sqrt{9-4x^2}}$

66. $\int_{-1/5}^{1/5} \dfrac{6\, dx}{\sqrt{4-25x^2}}$

67. $\int_{-2}^{2} \dfrac{3\, dt}{4+3t^2}$

68. $\int_{\sqrt{3}}^{3} \dfrac{dt}{3+t^2}$

69. $\int \dfrac{dy}{y\sqrt{4y^2-1}}$

70. $\int \dfrac{24\, dy}{y\sqrt{y^2-16}}$

71. $\int_{\sqrt{2}/3}^{2/3} \dfrac{dy}{|y|\sqrt{9y^2-1}}$

72. $\int_{-2/\sqrt{5}}^{-\sqrt{6}/\sqrt{5}} \dfrac{dy}{|y|\sqrt{5y^2-3}}$

73. $\int \dfrac{dx}{\sqrt{-2x-x^2}}$

74. $\int \dfrac{dx}{\sqrt{-x^2+4x-1}}$

75. $\int_{-2}^{-1} \frac{2\,dv}{v^2 + 4v + 5}$

76. $\int_{-1}^{1} \frac{3\,dv}{4v^2 + 4v + 4}$

77. $\int \frac{dt}{(t + 1)\sqrt{t^2 + 2t - 8}}$

78. $\int \frac{dt}{(3t + 1)\sqrt{9t^2 + 6t}}$

Solving Equations

In Exercises 79–84, solve for y.

79. $3^y = 2^{y+1}$

80. $4^{-y} = 3^{y+2}$

81. $9e^{2y} = x^2$

82. $3^y = 3\ln x$

83. $\ln(y - 1) = x + \ln y$

84. $\ln(10\ln y) = \ln 5x$

Applying L'Hôpital's Rule

Use l'Hôpital's Rule to find the limits in Exercises 85–108.

85. $\lim_{x\to 1} \frac{x^2 + 3x - 4}{x - 1}$

86. $\lim_{x\to 1} \frac{x^a - 1}{x^b - 1}$

87. $\lim_{x\to \pi} \frac{\tan x}{x}$

88. $\lim_{x\to 0} \frac{\tan x}{x + \sin x}$

89. $\lim_{x\to 0} \frac{\sin^2 x}{\tan(x^2)}$

90. $\lim_{x\to 0} \frac{\sin mx}{\sin nx}$

91. $\lim_{x\to \pi/2^-} \sec 7x \cos 3x$

92. $\lim_{x\to 0^+} \sqrt{x} \sec x$

93. $\lim_{x\to 0} (\csc x - \cot x)$

94. $\lim_{x\to 0} \left(\frac{1}{x^4} - \frac{1}{x^2}\right)$

95. $\lim_{x\to \infty} \left(\sqrt{x^2 + x + 1} - \sqrt{x^2 - x}\right)$

96. $\lim_{x\to \infty} \left(\frac{x^3}{x^2 - 1} - \frac{x^3}{x^2 + 1}\right)$

97. $\lim_{x\to 0} \frac{10^x - 1}{x}$

98. $\lim_{\theta\to 0} \frac{3^\theta - 1}{\theta}$

99. $\lim_{x\to 0} \frac{2^{\sin x} - 1}{e^x - 1}$

100. $\lim_{x\to 0} \frac{2^{-\sin x} - 1}{e^x - 1}$

101. $\lim_{x\to 0} \frac{5 - 5\cos x}{e^x - x - 1}$

102. $\lim_{x\to 0} \frac{x \sin x^2}{\tan^3 x}$

103. $\lim_{t\to 0^+} \frac{t - \ln(1 + 2t)}{t^2}$

104. $\lim_{x\to 4} \frac{\sin^2(\pi x)}{e^{x-4} + 3 - x}$

105. $\lim_{t\to 0^+} \left(\frac{e^t}{t} - \frac{1}{t}\right)$

106. $\lim_{y\to 0^+} e^{-1/y} \ln y$

107. $\lim_{x\to \infty} \left(\frac{e^x + 1}{e^x - 1}\right)^{\ln x}$

108. $\lim_{x\to 0^+} \left(1 + \frac{3}{x}\right)^x$

Comparing Growth Rates of Functions

109. Does f grow faster, slower, or at the same rate as g as $x \to \infty$? Give reasons for your answers.

a. $f(x) = \log_2 x$, $\quad g(x) = \log_3 x$

b. $f(x) = x$, $\quad g(x) = x + \frac{1}{x}$

c. $f(x) = x/100$, $\quad g(x) = xe^{-x}$

d. $f(x) = x$, $\quad g(x) = \arctan x$

e. $f(x) = \operatorname{arccsc} x$, $\quad g(x) = 1/x$

f. $f(x) = \sinh x$, $\quad g(x) = e^x$

110. Does f grow faster, slower, or at the same rate as g as $x \to \infty$? Give reasons for your answers.

a. $f(x) = 3^{-x}$, $\quad g(x) = 2^{-x}$

b. $f(x) = \ln 2x$, $\quad g(x) = \ln x^2$

c. $f(x) = 10x^3 + 2x^2$, $\quad g(x) = e^x$

d. $f(x) = \arctan(1/x)$, $\quad g(x) = 1/x$

e. $f(x) = \arcsin(1/x)$, $\quad g(x) = 1/x^2$

f. $f(x) = \operatorname{sech} x$, $\quad g(x) = e^{-x}$

111. True, or false? Give reasons for your answers.

a. $\frac{1}{x^2} + \frac{1}{x^4} = O\left(\frac{1}{x^2}\right)$

b. $\frac{1}{x^2} + \frac{1}{x^4} = O\left(\frac{1}{x^4}\right)$

c. $x = o(x + \ln x)$

d. $\ln(\ln x) = o(\ln x)$

e. $\arctan x = O(1)$

f. $\cosh x = O(e^x)$

112. True, or false? Give reasons for your answers.

a. $\frac{1}{x^4} = O\left(\frac{1}{x^2} + \frac{1}{x^4}\right)$

b. $\frac{1}{x^4} = o\left(\frac{1}{x^2} + \frac{1}{x^4}\right)$

c. $\ln x = o(x + 1)$

d. $\ln 2x = O(\ln x)$

e. $\sec^{-1} x = O(1)$

f. $\sinh x = O(e^x)$

Theory and Applications

113. The function $f(x) = e^x + x$, being differentiable and one-to-one, has a differentiable inverse $f^{-1}(x)$. Find the value of df^{-1}/dx at the point $f(\ln 2)$.

114. Find the inverse of the function $f(x) = 1 + (1/x)$, $x \neq 0$. Then show that $f^{-1}(f(x)) = f(f^{-1}(x)) = x$ and that

$$\left.\frac{df^{-1}}{dx}\right|_{f(x)} = \frac{1}{f'(x)}.$$

In Exercises 115 and 116, find the absolute maximum and minimum values of each function on the given interval.

115. $y = x \ln 2x - x, \quad \left[\frac{1}{2e}, \frac{e}{2}\right]$

116. $y = 10x(2 - \ln x), \quad (0, e^2]$

117. Area Find the area between the curve $y = 2(\ln x)/x$ and the x-axis from $x = 1$ to $x = e$.

a. Show that the area between the curve $y = 1/x$ and the x-axis from $x = 10$ to $x = 20$ is the same as the area between the curve and the x-axis from $x = 1$ to $x = 2$.

b. Show that the area between the curve $y = 1/x$ and the x-axis from ka to kb is the same as the area between the curve and the x-axis from $x = a$ to $x = b (0 < a < b, k > 0)$.

119. A particle is traveling upward and to the right along the curve $y = \ln x$. Its x-coordinate is increasing at the rate $(dx/dt) = \sqrt{x}$ m/sec. At what rate is the y-coordinate changing at the point $(e^2, 2)$?

120. A girl is sliding down a slide shaped like the curve $y = 9e^{-x/3}$. Her y-coordinate is changing at the rate $dy/dt = (-1/4)\sqrt{9 - y}$ ft/sec. At approximately what rate is her x-coordinate changing when she reaches the bottom of the slide at $x = 9$ ft? (Take e^3 to be 20 and round your answer to the nearest ft/sec.)

121. The rectangle shown here has one side on the positive y-axis, one side on the positive x-axis, and its upper right-hand vertex on the curve $y = e^{-x^2}$. What dimensions give the rectangle its largest area, and what is that area?

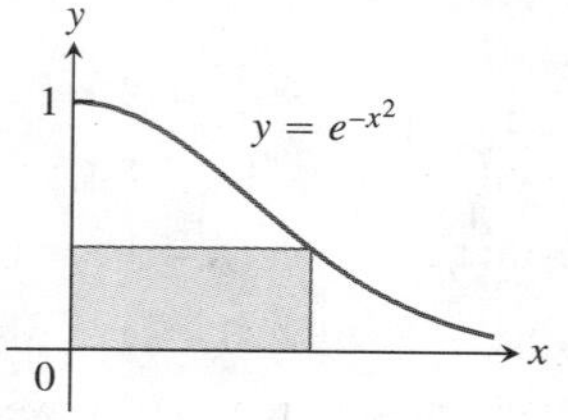

122. The rectangle shown here has one side on the positive y-axis, one side on the positive x-axis, and its upper right-hand vertex on the curve $y = (\ln x)/x^2$. What dimensions give the rectangle its largest area, and what is that area?

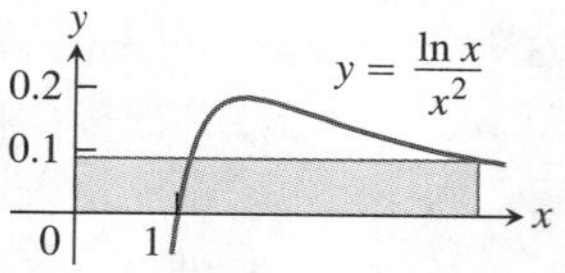

T **123.** Graph the following functions and use what you see to locate and estimate the extreme values, identify the coordinates of the inflection points, and identify the intervals on which the graphs are concave up and concave down. Then confirm your estimates by working with the functions' derivatives.

a. $y = (\ln x)/\sqrt{x}$

b. $y = e^{-x^2}$

c. $y = (1 + x)e^{-x}$

T **124.** Graph $f(x) = x \ln x$. Does the function appear to have an absolute minimum value? Confirm your answer with calculus.

In Exercises 125–128 solve the differential equation.

125. $\dfrac{dy}{dx} = \sqrt{y} \cos^2 \sqrt{y}$

126. $y' = \dfrac{3y(x + 1)^2}{y - 1}$

127. $yy' = \sec y^2 \sec^2 x$

128. $y \cos^2 x \, dy + \sin x \, dx = 0$

In Exercises 129–132 solve the initial value problem.

129. $\dfrac{dy}{dx} = e^{-x-y-2}, \quad y(0) = -2$

130. $\dfrac{dy}{dx} = \dfrac{y \ln y}{1 + x^2}, \quad y(0) = e^2$

131. $x \, dy - \left(y + \sqrt{y}\right) dx = 0, \quad y(1) = 1$

132. $y^{-2} \dfrac{dx}{dy} = \dfrac{e^x}{e^{2x} + 1}, \quad y(0) = 1$

133. What is the age of a sample of charcoal in which 90% of the carbon-14 originally present has decayed?

134. Cooling a pie A deep-dish apple pie, whose internal temperature was 220 °F when removed from the oven, was set out on a breezy 40 °F porch to cool. Fifteen minutes later, the pie's internal temperature was 180 °F. How long did it take the pie to cool from there to 70 °F?

135. Locating a solar station You are under contract to build a solar station at ground level on the east–west line between the two buildings shown here. How far from the taller building should you place the station to maximize the number of hours it will be in the sun on a day when the sun passes directly overhead? Begin by observing that

$$\theta = \pi - \cot^{-1}\frac{x}{60} - \cot^{-1}\frac{50 - x}{30}.$$

Then find the value of x that maximizes θ.

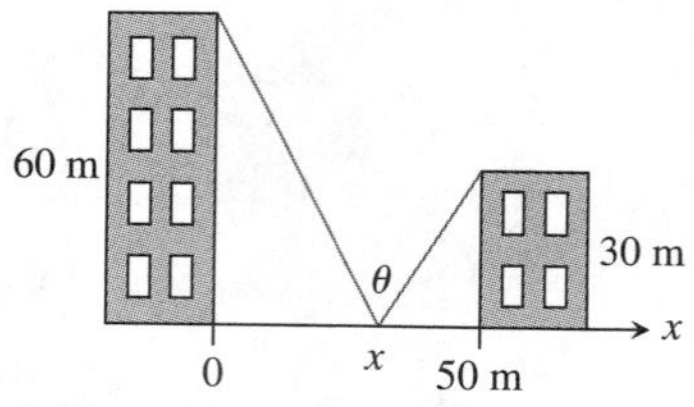

136. A round underwater transmission cable consists of a core of copper wires surrounded by nonconducting insulation. If x denotes the ratio of the radius of the core to the thickness of the insulation, it is known that the speed of the transmission signal is given by the equation $v = x^2 \ln(1/x)$. If the radius of the core is 1 cm, what insulation thickness h will allow the greatest transmission speed?

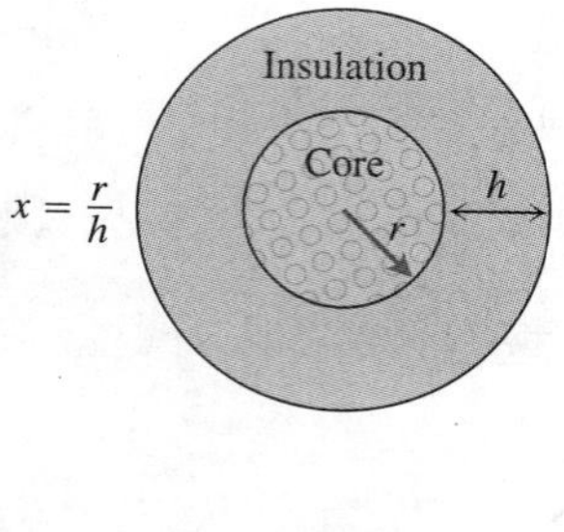

CHAPTER 7 Additional and Advanced Exercises

Limits

Find the limits in Exercises 1–6.

1. $\lim_{b\to 1^-} \int_0^b \frac{dx}{\sqrt{1-x^2}}$

2. $\lim_{x\to\infty} \frac{1}{x}\int_0^x \tan^{-1} t\, dt$

3. $\lim_{x\to 0^+} (\cos\sqrt{x})^{1/x}$

4. $\lim_{x\to\infty} (x + e^x)^{2/x}$

5. $\lim_{n\to\infty} \left(\frac{1}{n+1} + \frac{1}{n+2} + \cdots + \frac{1}{2n}\right)$

6. $\lim_{n\to\infty} \frac{1}{n}\left(e^{1/n} + e^{2/n} + \cdots + e^{(n-1)/n} + e^{n/n}\right)$

7. Let $A(t)$ be the area of the region in the first quadrant enclosed by the coordinate axes, the curve $y = e^{-x}$, and the vertical line $x = t,\ t > 0$. Let $V(t)$ be the volume of the solid generated by revolving the region about the x-axis. Find the following limits.

a. $\lim_{t\to\infty} A(t)$ **b.** $\lim_{t\to\infty} V(t)/A(t)$ **c.** $\lim_{t\to 0^+} V(t)/A(t)$

8. Varying a logarithm's base

a. Find $\lim \log_a 2$ as $a \to 0^+, 1^-, 1^+$, and ∞.

T **b.** Graph $y = \log_a 2$ as a function of a over the interval $0 < a \le 4$.

In Exercises 9 and 10, use implicit differentiation to find $\frac{dy}{dx}$.

9. $y^{e^x} = x^y + 1$

10. $y^{\ln x} = x^{x^y}$

Theory and Examples

11. Find the areas between the curves $y = 2(\log_2 x)/x$ and $y = 2(\log_4 x)/x$ and the x-axis from $x = 1$ to $x = e$. What is the ratio of the larger area to the smaller?

T **12.** Graph $f(x) = \tan^{-1} x + \tan^{-1}(1/x)$ for $-5 \le x \le 5$. Then use calculus to explain what you see. How would you expect f to behave beyond the interval $[-5, 5]$? Give reasons for your answer.

13. For what $x > 0$ does $x^{(x^x)} = (x^x)^x$? Give reasons for your answer.

T **14.** Graph $f(x) = (\sin x)^{\sin x}$ over $[0, 3\pi]$. Explain what you see.

15. Find $f'(2)$ if $f(x) = e^{g(x)}$ and $g(x) = \int_2^x \frac{t}{1+t^4}\, dt$.

16. a. Find df/dx if

$$f(x) = \int_1^{e^x} \frac{2\ln t}{t}\, dt.$$

b. Find $f(0)$.

c. What can you conclude about the graph of f? Give reasons for your answer.

17. Even-odd decompositions

a. Suppose that g is an even function of x and h is an odd function of x. Show that if $g(x) + h(x) = 0$ for all x then $g(x) = 0$ for all x and $h(x) = 0$ for all x.

b. If $f(x) = f_E(x) + f_O(x)$ is the sum of an even function $f_E(x)$ and an odd function $f_O(x)$, then show that

$$f_E(x) = \frac{f(x) + f(-x)}{2} \quad\text{and}\quad f_O(x) = \frac{f(x) - f(-x)}{2}.$$

c. What is the significance of the result in part (b)?

18. Let g be a function that is differentiable throughout an open interval containing the origin. Suppose g has the following properties:

i) $g(x + y) = \frac{g(x) + g(y)}{1 - g(x)g(y)}$ for all real numbers x, y, and $x + y$ in the domain of g.

ii) $\lim_{h\to 0} g(h) = 0$

iii) $\lim_{h\to 0} \frac{g(h)}{h} = 1$

a. Show that $g(0) = 0$.

b. Show that $g'(x) = 1 + [g(x)]^2$.

c. Find $g(x)$ by solving the differential equation in part (b).

19. Center of mass Find the center of mass of a thin plate of constant density covering the region in the first and fourth quadrants enclosed by the curves $y = 1/(1 + x^2)$ and $y = -1/(1 + x^2)$ and by the lines $x = 0$ and $x = 1$.

20. Solid of revolution The region between the curve $y = 1/(2\sqrt{x})$ and the x-axis from $x = 1/4$ to $x = 4$ is revolved about the x-axis to generate a solid.

a. Find the volume of the solid.

b. Find the centroid of the region.

21. The best branching angles for blood vessels and pipes When a smaller pipe branches off from a larger one in a flow system, we may want it to run off at an angle that is best from some energy-saving point of view. We might require, for instance, that energy loss due to friction be minimized along the section AOB shown in the accompanying figure. In this diagram, B is a given point to be reached by the smaller pipe, A is a point in the larger pipe upstream from B, and O is the point where the branching occurs. A law due to Poiseuille states that the loss of energy due to friction in nonturbulent flow is proportional to the length of the path and inversely proportional to the fourth power of the radius. Thus, the loss along AO is $(kd_1)/R^4$ and along OB is $(kd_2)/r^4$, where k is a constant, d_1 is the length of AO, d_2 is the length of OB, R is the radius of the larger pipe, and r is the radius of the smaller pipe. The angle θ is to be chosen to minimize the sum of these two losses:

$$L = k\frac{d_1}{R^4} + k\frac{d_2}{r^4}.$$

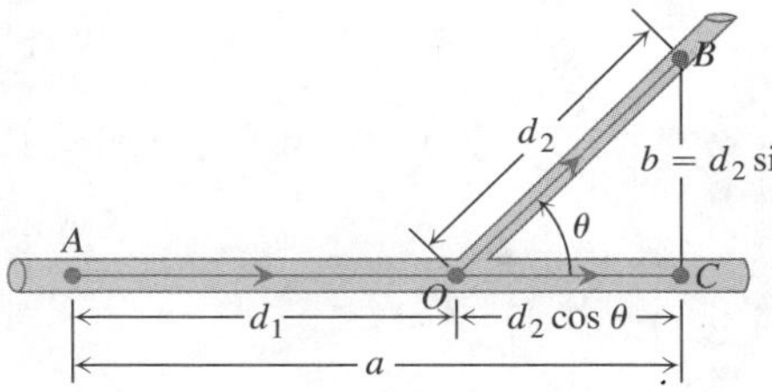

In our model, we assume that $AC = a$ and $BC = b$ are fixed. Thus we have the relations

$$d_1 + d_2\cos\theta = a \qquad d_2\sin\theta = b$$

so that

$$d_2 = b \csc \theta,$$
$$d_1 = a - d_2 \cos \theta = a - b \cot \theta.$$

We can express the total loss L as a function of θ:

$$L = k\left(\frac{a - b \cot \theta}{R^4} + \frac{b \csc \theta}{r^4}\right).$$

a. Show that the critical value of θ for which $dL/d\theta$ equals zero is

$$\theta_c = \cos^{-1} \frac{r^4}{R^4}.$$

b. If the ratio of the pipe radii is $r/R = 5/6$, estimate to the nearest degree the optimal branching angle given in part (a).

The mathematical analysis described here is also used to explain the angles at which arteries branch in an animal's body.

T **22. Urban gardening** A vegetable garden 50 ft wide is to be grown between two buildings, which are 500 ft apart along an east-west line. If the buildings are 200 ft and 350 ft tall, where should the garden be placed in order to receive the maximum number of hours of sunlight exposure? (*Hint:* Determine the value of x in the accompanying figure that maximizes sunlight exposure for the garden.)

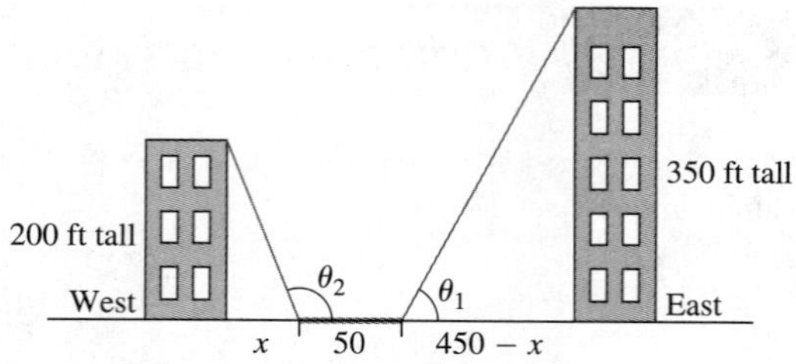

23. Consider point (a, b) on the graph of $y = \ln x$ and triangle ABC formed by the tangent line at (a, b), the y-axis, and the line $y = b$. Show that

$$(\text{area triangle } ABC) = \frac{a}{2}.$$

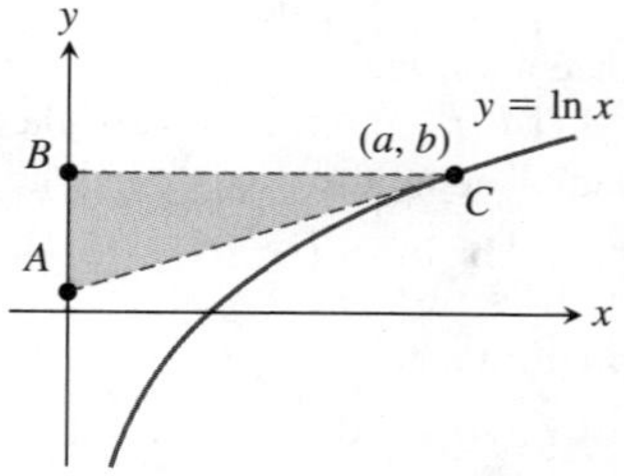

8

Techniques of Integration

OVERVIEW The Fundamental Theorem tells us how to evaluate a definite integral once we have an antiderivative for the integrand function. However, finding antiderivatives (or indefinite integrals) is not as straightforward as finding derivatives. In this chapter we study a number of important techniques that apply to finding integrals for specialized classes of functions such as trigonometric functions, products of certain functions, and rational functions. Since we cannot always find an antiderivative, we develop numerical methods for calculating definite integrals. We also study integrals for which the domain or range is infinite, called *improper integrals*.

8.1 Using Basic Integration Formulas

Table 8.1 summarizes the indefinite integrals of many of the functions we have studied so far, and the substitution method helps us use the table to evaluate more complicated functions involving these basic ones. In this section we combine the Substitution Rules (studied in Chapter 5) with algebraic methods and trigonometric identities to help us use Table 8.1. A more extensive Table of Integrals is given at the back of the chapter, and we discuss its use in Section 8.6.

Sometimes we have to rewrite an integral to match it to a standard form of the type displayed in Table 8.1. We start with an example of this procedure.

EXAMPLE 1 Evaluate the integral

$$\int_3^5 \frac{2x-3}{\sqrt{x^2-3x+1}}\,dx.$$

Solution We rewrite the integral and apply the Substitution Rule for Definite Integrals presented in Section 5.6, to find

$$\begin{aligned}
\int_3^5 \frac{2x-3}{\sqrt{x^2-3x+1}}\,dx &= \int_1^{11} \frac{du}{\sqrt{u}} && u = x^2-3x+1,\ du=(2x-3)\,dx;\ u=1 \text{ when } x=3,\ u=11 \text{ when } x=5\\
&= \int_1^{11} u^{-1/2}\,du\\
&= 2\sqrt{u}\,\Big]_1^{11} = 2\left(\sqrt{11}-1\right) \approx 4.63. && \text{Table 8.1, Formula 2}
\end{aligned}$$

■

TABLE 8.1 Basic integration formulas

1. $\int k\,dx = kx + C \quad$ (any number k)	**12.** $\int \tan x\,dx = \ln\lvert\sec x\rvert + C$
2. $\int x^n\,dx = \dfrac{x^{n+1}}{n+1} + C \quad (n \neq -1)$	**13.** $\int \cot x\,dx = \ln\lvert\sin x\rvert + C$
3. $\int \dfrac{dx}{x} = \ln\lvert x\rvert + C$	**14.** $\int \sec x\,dx = \ln\lvert\sec x + \tan x\rvert + C$
4. $\int e^x\,dx = e^x + C$	**15.** $\int \csc x\,dx = -\ln\lvert\csc x + \cot x\rvert + C$
5. $\int a^x\,dx = \dfrac{a^x}{\ln a} + C \quad (a > 0,\ a \neq 1)$	**16.** $\int \sinh x\,dx = \cosh x + C$
6. $\int \sin x\,dx = -\cos x + C$	**17.** $\int \cosh x\,dx = \sinh x + C$
7. $\int \cos x\,dx = \sin x + C$	**18.** $\int \dfrac{dx}{\sqrt{a^2 - x^2}} = \arcsin\left(\dfrac{x}{a}\right) + C$
8. $\int \sec^2 x\,dx = \tan x + C$	**19.** $\int \dfrac{dx}{a^2 + x^2} = \dfrac{1}{a}\arctan\left(\dfrac{x}{a}\right) + C$
9. $\int \csc^2 x\,dx = -\cot x + C$	**20.** $\int \dfrac{dx}{x\sqrt{x^2 - a^2}} = \dfrac{1}{a}\operatorname{arcsec}\left\lvert\dfrac{x}{a}\right\rvert + C$
10. $\int \sec x \tan x\,dx = \sec x + C$	**21.** $\int \dfrac{dx}{\sqrt{a^2 + x^2}} = \sinh^{-1}\left(\dfrac{x}{a}\right) + C \quad (a > 0)$
11. $\int \csc x \cot x\,dx = -\csc x + C$	**22.** $\int \dfrac{dx}{\sqrt{x^2 - a^2}} = \cosh^{-1}\left(\dfrac{x}{a}\right) + C \quad (x > a > 0)$

EXAMPLE 2 Complete the square to evaluate

$$\int \frac{dx}{\sqrt{8x - x^2}}.$$

Solution We complete the square to simplify the denominator:

$$\begin{aligned} 8x - x^2 &= -(x^2 - 8x) = -(x^2 - 8x + 16 - 16) \\ &= -(x^2 - 8x + 16) + 16 = 16 - (x - 4)^2. \end{aligned}$$

Then

$$\begin{aligned} \int \frac{dx}{\sqrt{8x - x^2}} &= \int \frac{dx}{\sqrt{16 - (x - 4)^2}} \\ &= \int \frac{du}{\sqrt{a^2 - u^2}} && a = 4,\ u = (x - 4),\ du = dx \\ &= \arcsin\left(\frac{u}{a}\right) + C && \text{Table 8.1, Formula 18} \\ &= \arcsin\left(\frac{x - 4}{4}\right) + C. \end{aligned}$$

■

EXAMPLE 3 Evaluate the integral

$$\int (\cos x \sin 2x + \sin x \cos 2x)\, dx.$$

Solution We can replace the integrand with an equivalent trigonometric expression using the Sine Addition Formula to obtain a simple substitution:

$$\begin{aligned}
\int (\cos x \sin 2x + \sin x \cos 2x)\, dx &= \int \sin(x + 2x)\, dx \\
&= \int \sin 3x\, dx \\
&= \int \frac{1}{3} \sin u\, du && u = 3x,\ du = 3\, dx \\
&= -\frac{1}{3} \cos 3x + C. && \text{Table 8.1, Formula 6}
\end{aligned}$$

■

In Section 5.5 we found the indefinite integral of the secant function by multiplying it by a fractional form equal to one, and then integrating the equivalent result. We can use that same procedure in other instances as well, as we illustrate next.

EXAMPLE 4 Find $\displaystyle\int_0^{\pi/4} \frac{dx}{1 - \sin x}$.

Solution We multiply the numerator and denominator of the integrand by $1 + \sin x$. This procedure transforms the integral into one we can evaluate:

$$\begin{aligned}
\int_0^{\pi/4} \frac{dx}{1 - \sin x} &= \int_0^{\pi/4} \frac{1}{1 - \sin x} \cdot \frac{1 + \sin x}{1 + \sin x}\, dx && \text{Multiply and divide by conjugate.} \\
&= \int_0^{\pi/4} \frac{1 + \sin x}{1 - \sin^2 x}\, dx && \text{Simplify.} \\
&= \int_0^{\pi/4} \frac{1 + \sin x}{\cos^2 x}\, dx && 1 - \sin^2 x = \cos^2 x \\
&= \int_0^{\pi/4} (\sec^2 x + \sec x \tan x)\, dx && \text{Use Table 8.1, Formulas 8 and 10} \\
&= \Big[\tan x + \sec x\Big]_0^{\pi/4} = \left(1 + \sqrt{2} - (0 + 1)\right) = \sqrt{2}.
\end{aligned}$$

■

EXAMPLE 5 Evaluate

$$\int \frac{3x^2 - 7x}{3x + 2}\, dx.$$

$$\begin{array}{r}
x - 3 \\
3x + 2 \overline{)\, 3x^2 - 7x} \\
\underline{3x^2 + 2x} \\
-9x \\
\underline{-9x - 6} \\
+6
\end{array}$$

Solution The integrand is an improper fraction since the degree of the numerator is greater than the degree of the denominator. To integrate it, we perform long division to obtain a quotient plus a remainder that is a proper fraction:

$$\frac{3x^2 - 7x}{3x + 2} = x - 3 + \frac{6}{3x + 2}.$$

Therefore,

$$\int \frac{3x^2 - 7x}{3x + 2}\,dx = \int \left(x - 3 + \frac{6}{3x + 2}\right) dx = \frac{x^2}{2} - 3x + 2\ln|3x + 2| + C. \qquad \blacksquare$$

Reducing an improper fraction by long division (Example 5) does not always lead to an expression we can integrate directly. We see what to do about that in Section 8.5.

EXAMPLE 6 Evaluate

$$\int \frac{3x + 2}{\sqrt{1 - x^2}}\,dx.$$

Solution We first separate the integrand to get

$$\int \frac{3x + 2}{\sqrt{1 - x^2}}\,dx = 3\int \frac{x\,dx}{\sqrt{1 - x^2}} + 2\int \frac{dx}{\sqrt{1 - x^2}}.$$

In the first of these new integrals, we substitute

$$u = 1 - x^2, \qquad du = -2x\,dx, \qquad \text{so} \qquad x\,dx = -\frac{1}{2}\,du.$$

Then we obtain

$$\begin{aligned} 3\int \frac{x\,dx}{\sqrt{1 - x^2}} &= 3\int \frac{(-1/2)\,du}{\sqrt{u}} = -\frac{3}{2}\int u^{-1/2}\,du \\ &= -\frac{3}{2} \cdot \frac{u^{1/2}}{1/2} + C_1 = -3\sqrt{1 - x^2} + C_1. \end{aligned}$$

The second of the new integrals is a standard form,

$$2\int \frac{dx}{\sqrt{1 - x^2}} = 2\arcsin x + C_2. \qquad \text{Table 8.1, Formula 18}$$

Combining these results and renaming $C_1 + C_2$ as C gives

$$\int \frac{3x + 2}{\sqrt{1 - x^2}}\,dx = -3\sqrt{1 - x^2} + 2\arcsin x + C. \qquad \blacksquare$$

The question of what to substitute for in an integrand is not always quite so clear. Sometimes we simply proceed by trial-and-error, and if nothing works out, we then try another method altogether. The next several sections of the text present some of these new methods, but substitution works in the following example.

EXAMPLE 7 Evaluate

$$\int \frac{dx}{\left(1 + \sqrt{x}\right)^3}.$$

Solution We might try substituting for the term $\sqrt{x}$, but the derivative factor $1/\sqrt{x}$ is missing from the integrand, so this substitution will not help. The other possibility is to substitute for $\left(1 + \sqrt{x}\right)$, and it turns out this works:

$$\begin{aligned} \int \frac{dx}{\left(1 + \sqrt{x}\right)^3} &= \int \frac{2(u - 1)\,du}{u^3} \qquad u = 1 + \sqrt{x},\ du = \frac{1}{2\sqrt{x}}\,dx; \quad dx = 2\sqrt{x}\,du = 2(u - 1)\,du \\ &= \int \left(\frac{2}{u^2} - \frac{2}{u^3}\right) du \end{aligned}$$

$$= -\frac{2}{u} + \frac{1}{u^2} + C$$

$$= \frac{1 - 2u}{u^2} + C$$

$$= \frac{1 - 2(1 + \sqrt{x})}{(1 + \sqrt{x})^2} + C$$

$$= C - \frac{1 + 2\sqrt{x}}{(1 + \sqrt{x})^2}.$$ ■

When evaluating definite integrals, a property of the integrand may help us in calculating the result.

EXAMPLE 8 Evaluate

$$\int_{-\pi/2}^{\pi/2} x^3 \cos x \, dx.$$

Solution No substitution or algebraic manipulation is clearly helpful here. But we observe that the interval of integration is the symmetric interval $[-\pi/2, \pi/2]$. Moreover, the factor x^3 is an odd function, and $\cos x$ is an even function, so their product is odd. Therefore,

$$\int_{-\pi/2}^{\pi/2} x^3 \cos x \, dx = 0. \qquad \text{Theorem 8, Section 5.6}$$ ■

EXERCISES 8.1

Assorted Integrations

The integrals in Exercises 1–44 are in no particular order. Evaluate each integral using any algebraic method, trigonometric identity, or substitution you think is appropriate.

1. $\int_0^1 \frac{16x}{8x^2 + 2} \, dx$

2. $\int \frac{x^2}{x^2 + 1} \, dx$

3. $\int (\sec x - \tan x)^2 \, dx$

4. $\int_{\pi/4}^{\pi/3} \frac{dx}{\cos^2 x \tan x}$

5. $\int \frac{1 - x}{\sqrt{1 - x^2}} \, dx$

6. $\int \frac{dx}{x - \sqrt{x}}$

7. $\int \frac{e^{-\cot z}}{\sin^2 z} \, dz$

8. $\int \frac{2^{\ln z^3}}{16z} \, dz$

9. $\int \frac{dz}{e^z + e^{-z}}$

10. $\int_1^2 \frac{8 \, dx}{x^2 - 2x + 2}$

11. $\int_{-1}^0 \frac{4 \, dx}{1 + (2x + 1)^2}$

12. $\int_{-1}^3 \frac{4x^2 - 7}{2x + 3} \, dx$

13. $\int \frac{dt}{1 - \sec t}$

14. $\int \csc t \sin 3t \, dt$

15. $\int_0^{\pi/4} \frac{1 + \sin \theta}{\cos^2 \theta} \, d\theta$

16. $\int \frac{d\theta}{\sqrt{2\theta - \theta^2}}$

17. $\int \frac{\ln y}{y + 4y \ln^2 y} \, dy$

18. $\int \frac{2^{\sqrt{y}} \, dy}{2\sqrt{y}}$

19. $\int \frac{d\theta}{\sec \theta + \tan \theta}$

20. $\int \frac{dt}{t\sqrt{3 + t^2}}$

21. $\int \frac{4t^3 - t^2 + 16t}{t^2 + 4} \, dt$

22. $\int \frac{x + 2\sqrt{x - 1}}{2x\sqrt{x - 1}} \, dx$

23. $\int_0^{\pi/2} \sqrt{1 - \cos \theta} \, d\theta$

24. $\int (\sec t + \cot t)^2 \, dt$

25. $\int \frac{dy}{\sqrt{e^{2y} - 1}}$

26. $\int \frac{6 \, dy}{\sqrt{y}(1 + y)}$

27. $\int \frac{2 \, dx}{x\sqrt{1 - 4 \ln^2 x}}$

28. $\int \frac{dx}{(x - 2)\sqrt{x^2 - 4x + 3}}$

29. $\int (\csc x - \sec x)(\sin x + \cos x) \, dx$

30. $\int 3 \sinh\left(\frac{x}{2} + \ln 5\right) dx$

31. $\int_{\sqrt{2}}^3 \frac{2x^3}{x^2 - 1} \, dx$

32. $\int_{-1}^1 \sqrt{1 + x^2} \sin x \, dx$

33. $\int_{-1}^0 \sqrt{\frac{1 + y}{1 - y}} \, dy$

34. $\int e^{z + e^z} \, dz$

35. $\int \frac{7\,dx}{(x-1)\sqrt{x^2-2x-48}}$

36. $\int \frac{dx}{(2x+1)\sqrt{4x+4x^2}}$

37. $\int \frac{4\theta^3-9\theta^2+9\theta}{4\theta-5}\,d\theta$

38. $\int \frac{d\theta}{\cos\theta-1}$

39. $\int \frac{dx}{1-e^{-x}}$

Hint: Use long division.

40. $\int \frac{\sqrt{x}}{1+x^3}\,dx$

Hint: Let $u = x^{3/2}$.

41. $\int \frac{e^{3x}}{e^x+1}\,dx$

42. $\int \frac{4^x-1}{5^x}\,dx$

43. $\int \frac{1}{\sqrt{x}(1+x)}\,dx$

44. $\int \frac{\tan\theta-5}{\cos\theta}\,d\theta$

Theory and Examples

45. **Area** Find the area of the region bounded above by $y = 2\cos x$ and below by $y = \sec x$, $-\pi/4 \le x \le \pi/4$.

46. **Volume** Find the volume of the solid generated by revolving the region in Exercise 45 about the x-axis.

47. **Arc length** Find the length of the curve $y = \ln(\sin x)$, $\pi/3 \le x \le \pi/2$.

48. **Arc length** Find the length of the curve $y = \ln(\sec x)$, $0 \le x \le \pi/4$.

49. **Centroid** Find the centroid of the region bounded by the x-axis, the curve $y = \sec x$, and the lines $x = -\pi/4$, $x = \pi/4$.

50. **Centroid** Find the centroid of the region bounded by the x-axis, the curve $y = \csc x$, and the lines $x = \pi/6$, $x = 5\pi/6$.

51. The functions $y = e^{x^3}$ and $y = x^3 e^{x^3}$ do not have elementary antiderivatives, but $y = (1+3x^3)e^{x^3}$ does. Evaluate

$$\int (1+3x^3)\,e^{x^3}\,dx.$$

52. Use the substitution $u = \tan x$ to evaluate the integral

$$\int \frac{dx}{1+\cos^2 x}.$$

53. Use the substitution $u = x^4+1$ to evaluate the integral

$$\int x^7\sqrt{x^4+1}\,dx.$$

54. **Using different substitutions** Show that the integral

$$\int ((x^2-1)(x+1))^{-2/3}\,dx$$

can be evaluated with any of the following substitutions.

a. $u = 1/(x+1)$

b. $u = ((x-1)/(x+1))^k$ for $k = 1,\ 1/2,\ 1/3,\ -1/3,\ -2/3,$ and -1

c. $u = \arctan x$

d. $u = \tan^{-1}\sqrt{x}$

e. $u = \tan^{-1}((x-1)/2)$

f. $u = \arccos x$

g. $u = \cosh^{-1} x$

What is the value of the integral?

8.2 Integration by Parts

Integration by parts is a technique for simplifying integrals of the form

$$\int u(x)\,v'(x)\,dx.$$

It is useful when u can be differentiated repeatedly and v' can be integrated repeatedly without difficulty. The integrals

$$\int x\cos x\,dx \qquad \text{and} \qquad \int x^2 e^x\,dx$$

are such integrals because $u(x) = x$ or $u(x) = x^2$ can be differentiated repeatedly, and $v'(x) = \cos x$ or $v'(x) = e^x$ can be integrated repeatedly without difficulty. Integration by parts also applies to integrals like

$$\int \ln x\,dx \qquad \text{and} \qquad \int e^x\cos x\,dx.$$

In the first case, the integrand $\ln x$ can be rewritten as $(\ln x)(1)$, and $u(x) = \ln x$ is easy to differentiate while $v'(x) = 1$ easily integrates to x. In the second case, each part of the integrand appears again after repeated differentiation or integration.

Product Rule in Integral Form

If u and v are differentiable functions of x, the Product Rule says that

$$\frac{d}{dx}[u(x)v(x)] = u'(x)v(x) + u(x)v'(x).$$

In terms of indefinite integrals, this equation becomes

$$\int \frac{d}{dx}[u(x)v(x)]\,dx = \int [u'(x)v(x) + u(x)v'(x)]\,dx$$

or

$$\int \frac{d}{dx}[u(x)v(x)]\,dx = \int u'(x)v(x)\,dx + \int u(x)v'(x)\,dx.$$

Rearranging the terms of this last equation, we get

$$\int u(x)v'(x)\,dx = \int \frac{d}{dx}[u(x)v(x)]\,dx - \int v(x)u'(x)\,dx,$$

leading to the following **integration by parts** formula.

Integration by Parts Formula

$$\int u(x)v'(x)\,dx = u(x)v(x) - \int v(x)u'(x)\,dx \tag{1}$$

This formula allows us to exchange the problem of computing the integral $\int u(x)v'(x)\,dx$ for the problem of computing a different integral, $\int v(x)u'(x)\,dx$. In many cases, we can choose the functions u and v so that the second integral is easier to compute than the first. There can be many choices for u and v, and it is not always clear which choice works best, so sometimes we need to try several.

The formula is often given in differential form. With $v'(x)\,dx = dv$ and $u'(x)\,dx = du$, the integration by parts formula becomes

Integration by Parts Formula—Differential Version

$$\int u\,dv = uv - \int v\,du \tag{2}$$

The next examples illustrate the technique.

EXAMPLE 1 Find

$$\int x \cos x\,dx.$$

Solution There is no obvious antiderivative of $x \cos x$, so we use the integration by parts formula

$$\int u(x)\,v'(x)\,dx = u(x)\,v(x) - \int v(x)\,u'(x)\,dx$$

to change this expression to one that is easier to integrate. We first decide how to choose the functions $u(x)$ and $v(x)$. There is more than one way to do this, but here we choose to factor the expression $x \cos x$ into

$$u(x) = x \quad \text{and} \quad v'(x) = \cos x.$$

Next we differentiate $u(x)$ and find an antiderivative of $v'(x)$,

$$u'(x) = 1 \quad \text{and} \quad v(x) = \sin x.$$

When finding an antiderivative for $v'(x)$, we have a choice of how to pick a constant of integration C. We choose the constant $C = 0$, since that makes this antiderivative as simple as possible. We now apply the integration by parts formula:

$$\int \underbrace{x}_{u(x)} \underbrace{\cos x\, dx}_{v'(x)} = \underbrace{x}_{u(x)} \underbrace{\sin x}_{v(x)} - \int \underbrace{\sin x}_{v(x)} \underbrace{(1)}_{u'(x)}\, dx \qquad \text{Integration by parts formula}$$

$$= x \sin x + \cos x + C \qquad \text{Integrate and simplify.}$$

and we have found the integral of the original function. ■

There are at least four apparent choices available for $u(x)$ and $v'(x)$ in Example 1:

1. Let $u(x) = 1$ and $v'(x) = x \cos x$.

2. Let $u(x) = x$ and $v'(x) = \cos x$.

3. Let $u(x) = x \cos x$ and $v'(x) = 1$.

4. Let $u(x) = \cos x$ and $v'(x) = x$.

We used choice 2 in Example 1. The other three choices lead to integrals that we do not know how to evaluate. For instance, Choice 3, with $u'(x) = \cos x - x \sin x$, leads to the integral

$$\int (x \cos x - x^2 \sin x)\, dx.$$

The goal of integration by parts is to go from an integral $\int u(x)v'(x)\, dx$ that we don't see how to evaluate to an integral $\int v(x)u'(x)\, dx$ that we can evaluate. Generally, we choose $v'(x)$ first to be as much of the integrand as we can readily integrate; then we let $u(x)$ be the leftover part. When finding $v(x)$ from $v'(x)$, any antiderivative will work, and we usually pick the simplest one. In particular, no arbitrary constant of integration is needed in $v(x)$ because it would simply cancel out of the right-hand side of Equation (2).

EXAMPLE 2 Find $\int \ln x\, dx$.

Solution We have not yet seen how to find an antiderivative for $\ln x$. If we set $u(x) = \ln x$, then $u'(x)$ is the simpler function $1/x$. It may not appear that a second function $v'(x)$ is multiplying $u(x) = \ln x$, but we can choose $v'(x)$ to be the constant function $v'(x) = 1$. We use the integration by parts formula given in Equation (1), with

$$u(x) = \ln x \quad \text{and} \quad v'(x) = 1.$$

We differentiate $u(x)$ and find an antiderivative of $v'(x)$,

$$u'(x) = \frac{1}{x} \quad \text{and} \quad v(x) = x.$$

Then

$$\int \underbrace{\ln x}_{u(x)} \cdot \underbrace{1\, dx}_{v'(x)} = \underbrace{(\ln x)}_{u(x)}\, \underbrace{x}_{v(x)} - \int \underbrace{x}_{v(x)} \underbrace{\frac{1}{x}}_{u'(x)}\, dx \qquad \text{Integration by parts formula}$$

$$= x \ln x - \int 1\, dx$$

$$= x \ln x - x + C \qquad \text{Simplify and integrate.}$$

■

In the following examples we use the differential form to indicate the process of integration by parts. The computations are the same, with du and dv providing shorter expressions for $u'(x)\, dx$ and $v'(x)\, dx$.

Sometimes we have to use integration by parts more than once, as in the next example.

EXAMPLE 3 Evaluate

$$\int x^2 e^x\, dx.$$

Solution We use the integration by parts formula given in Equation (1), with

$$u(x) = x^2 \quad \text{and} \quad v'(x) = e^x.$$

We differentiate $u(x)$ and find an antiderivative of $v'(x)$,

$$u'(x) = 2x \quad \text{and} \quad v(x) = e^x.$$

We summarize this choice by setting $du = u'(x)\,dx$ and $dv = v'(x)\,dx$, so

$$du = 2x\,dx \quad \text{and} \quad dv = e^x\,dx.$$

We then have

$$\int \underbrace{x^2}_{u}\underbrace{e^x\,dx}_{dv} = \underbrace{x^2}_{u}\underbrace{e^x}_{v} - \int \underbrace{e^x}_{v}\underbrace{2x\,dx}_{du} = x^2e^x - 2\int xe^x\,dx \qquad \text{Integration by parts formula}$$

The new integral is less complicated than the original because the exponent on x is reduced by one. To evaluate the integral on the right, we integrate by parts again with $u = x$, $dv = e^x\,dx$. Then $du = dx$, $v = e^x$, and

$$\int \underbrace{x}_{u}\underbrace{e^x\,dx}_{dv} = \underbrace{x}_{u}\underbrace{e^x}_{v} - \int \underbrace{e^x}_{v}\underbrace{dx}_{du} = xe^x - e^x + C. \qquad \begin{array}{l}\text{Integration by parts Equation (2)}\\ u = x,\ dv = e^x\,dx\\ v = e^x,\ du = dx\end{array}$$

Using this last evaluation, we then obtain

$$\begin{aligned}\int x^2e^x\,dx &= x^2e^x - 2\int xe^x\,dx\\ &= x^2e^x - 2xe^x + 2e^x + C,\end{aligned}$$

where the constant of integration is renamed after substituting for the integral on the right. ■

The technique of Example 3 works for any integral $\int x^ne^x\,dx$ in which n is a positive integer, because differentiating x^n will eventually lead to a constant, and repeatedly integrating e^x is easy.

Integrals like the one in the next example occur in electrical engineering. Their evaluation requires two integrations by parts, followed by solving for the unknown integral.

EXAMPLE 4 Evaluate

$$\int e^x \cos x\,dx.$$

Solution Let $u = e^x$ and $dv = \cos x\,dx$. Then $du = e^x\,dx$, $v = \sin x$, and

$$\int e^x \cos x\,dx = e^x \sin x - \int e^x \sin x\,dx.$$

The second integral is like the first except that it has $\sin x$ in place of $\cos x$. To evaluate it, we use integration by parts with

$$u = e^x, \quad dv = \sin x\,dx, \quad v = -\cos x, \quad du = e^x\,dx.$$

Then

$$\begin{aligned}\int e^x \cos x\,dx &= e^x \sin x - \left(-e^x \cos x - \int(-\cos x)(e^x\,dx)\right)\\ &= e^x \sin x + e^x \cos x - \int e^x \cos x\,dx.\end{aligned}$$

The unknown integral now appears on both sides of the equation, but with opposite signs. Adding the integral to both sides and adding the constant of integration gives

$$2\int e^x \cos x\,dx = e^x \sin x + e^x \cos x + C_1.$$

Dividing by 2 and renaming the constant of integration then gives

$$\int e^x \cos x\,dx = \frac{e^x \sin x + e^x \cos x}{2} + C. \qquad \blacksquare$$

EXAMPLE 5 Obtain a formula that expresses the integral

$$\int \cos^n x\,dx$$

in terms of an integral of a lower power of cos x.

Solution We may think of $\cos^n x$ as $\cos^{n-1} x \cdot \cos x$. Then we let

$$u = \cos^{n-1} x \quad \text{and} \quad dv = \cos x\,dx,$$

so that

$$du = (n-1)(\cos^{n-2} x)(-\sin x\,dx) \quad \text{and} \quad v = \sin x.$$

Integration by parts then gives

$$\begin{aligned}\int \cos^n x\,dx &= \cos^{n-1} x \sin x + (n-1)\int \sin^2 x \cos^{n-2} x\,dx\\ &= \cos^{n-1} x \sin x + (n-1)\int (1-\cos^2 x)\cos^{n-2} x\,dx\\ &= \cos^{n-1} x \sin x + (n-1)\int \cos^{n-2} x\,dx - (n-1)\int \cos^n x\,dx.\end{aligned}$$

If we add

$$(n-1)\int \cos^n x\,dx$$

to both sides of this equation, we obtain

$$n\int \cos^n x\,dx = \cos^{n-1} x \sin x + (n-1)\int \cos^{n-2} x\,dx.$$

We then divide through by n, and the final result is

$$\int \cos^n x\,dx = \frac{\cos^{n-1} x \sin x}{n} + \frac{n-1}{n}\int \cos^{n-2} x\,dx. \qquad \blacksquare$$

The formula found in Example 5 is called a **reduction formula** because it replaces an integral containing some power of a function with an integral of the same form having the power reduced. When n is a positive integer, we may apply the formula repeatedly until the remaining integral is easy to evaluate. For example, the result in Example 5 tells us that

$$\begin{aligned}\int \cos^3 x\,dx &= \frac{\cos^2 x \sin x}{3} + \frac{2}{3}\int \cos x\,dx\\ &= \frac{1}{3}\cos^2 x \sin x + \frac{2}{3}\sin x + C.\end{aligned}$$

Evaluating Definite Integrals by Parts

The integration by parts formula in Equation (1) can be combined with Part 2 of the Fundamental Theorem in order to evaluate definite integrals by parts. Assuming that both u' and v' are continuous over the interval $[a, b]$, Part 2 of the Fundamental Theorem gives

Integration by Parts Formula for Definite Integrals

$$\int_a^b u(x)\,v'(x)\,dx = u(x)\,v(x)\Big]_a^b - \int_a^b v(x)\,u'(x)\,dx \qquad (3)$$

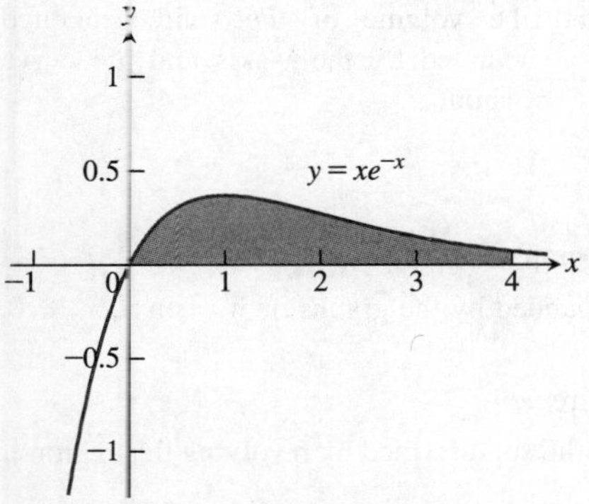

FIGURE 8.1 The region in Example 6.

EXAMPLE 6 Find the area of the region bounded by the curve $y = xe^{-x}$ and the x-axis from $x = 0$ to $x = 4$.

Solution The region is shaded in Figure 8.1. Its area is

$$\int_0^4 xe^{-x}\,dx.$$

Let $u = x$, $dv = e^{-x}\,dx$, $v = -e^{-x}$, and $du = dx$. Then

$$\begin{aligned}
\int_0^4 xe^{-x}\,dx &= -xe^{-x}\Big]_0^4 - \int_0^4 (-e^{-x})\,dx \qquad \text{Integration by parts Formula (3)} \\
&= [-4e^{-4} - (-0e^{-0})] + \int_0^4 e^{-x}\,dx \\
&= -4e^{-4} - e^{-x}\Big]_0^4 \\
&= -4e^{-4} - (e^{-4} - e^{-0}) = 1 - 5e^{-4} \approx 0.91.
\end{aligned}$$

■

EXERCISES 8.2

Integration by Parts

Evaluate the integrals in Exercises 1–24 using integration by parts.

1. $\int x \sin \frac{x}{2}\,dx$
2. $\int \theta \cos \pi\theta\,d\theta$
3. $\int t^2 \cos t\,dt$
4. $\int x^2 \sin x\,dx$
5. $\int_1^2 x \ln x\,dx$
6. $\int_1^e x^3 \ln x\,dx$
7. $\int xe^x\,dx$
8. $\int xe^{3x}\,dx$
9. $\int x^2 e^{-x}\,dx$
10. $\int (x^2 - 2x + 1)e^{2x}\,dx$
11. $\int \tan^{-1} y\,dy$
12. $\int \arcsin y\,dy$
13. $\int x \sec^2 x\,dx$
14. $\int 4x \sec^2 2x\,dx$
15. $\int x^3 e^x\,dx$
16. $\int p^4 e^{-p}\,dp$
17. $\int (x^2 - 5x)e^x\,dx$
18. $\int (r^2 + r + 1)e^r\,dr$
19. $\int x^5 e^x\,dx$
20. $\int t^2 e^{4t}\,dt$
21. $\int e^\theta \sin \theta\,d\theta$
22. $\int e^{-y} \cos y\,dy$
23. $\int e^{2x} \cos 3x\,dx$
24. $\int e^{-2x} \sin 2x\,dx$

Using Substitution

Evaluate the integrals in Exercises 25–30 by using a substitution prior to integration by parts.

25. $\int e^{\sqrt{3s+9}}\,ds$
26. $\int_0^1 x\sqrt{1-x}\,dx$
27. $\int_0^{\pi/3} x \tan^2 x\,dx$
28. $\int \ln(x + x^2)\,dx$
29. $\int \sin(\ln x)\,dx$
30. $\int z(\ln z)^2\,dz$

Evaluating Integrals

Evaluate the integrals in Exercises 31–56. Some integrals do not require integration by parts.

31. $\int x \sec x^2\,dx$
32. $\int \frac{\cos\sqrt{x}}{\sqrt{x}}\,dx$
33. $\int x(\ln x)^2\,dx$
34. $\int \frac{1}{x(\ln x)^2}\,dx$
35. $\int \frac{\ln x}{x^2}\,dx$
36. $\int \frac{(\ln x)^3}{x}\,dx$
37. $\int x^3 e^{x^4}\,dx$
38. $\int x^5 e^{x^3}\,dx$
39. $\int x^3\sqrt{x^2+1}\,dx$
40. $\int x^2 \sin x^3\,dx$
41. $\int \sin 3x \cos 2x\,dx$
42. $\int \sin 2x \cos 4x\,dx$
43. $\int \sqrt{x} \ln x\,dx$
44. $\int \frac{e^{\sqrt{x}}}{\sqrt{x}}\,dx$
45. $\int \cos\sqrt{x}\,dx$
46. $\int \sqrt{x}\,e^{\sqrt{x}}\,dx$
47. $\int_0^{\pi/2} \theta^2 \sin 2\theta\,d\theta$
48. $\int_0^{\pi/2} x^3 \cos 2x\,dx$
49. $\int_{2/\sqrt{3}}^2 t \sec^{-1} t\,dt$
50. $\int_0^{1/\sqrt{2}} 2x \arcsin(x^2)\,dx$
51. $\int x \arctan x\,dx$
52. $\int x^2 \tan^{-1} \frac{x}{2}\,dx$
53. $\int (1 + 2x^2)e^{x^2}\,dx$
54. $\int \frac{xe^x}{(x+1)^2}\,dx$
55. $\int \sqrt{x}(\arcsin\sqrt{x})\,dx$
56. $\int \frac{(\sin^{-1} x)^2}{\sqrt{1-x^2}}\,dx$

Theory and Examples

57. Finding area Find the area of the region enclosed by the curve $y = x \sin x$ and the x-axis (see the accompanying figure) for

a. $0 \le x \le \pi$.

b. $\pi \le x \le 2\pi$.

c. $2\pi \le x \le 3\pi$.

d. What pattern do you see here? What is the area between the curve and the x-axis for $n\pi \le x \le (n + 1)\pi$, n an arbitrary nonnegative integer? Give reasons for your answer.

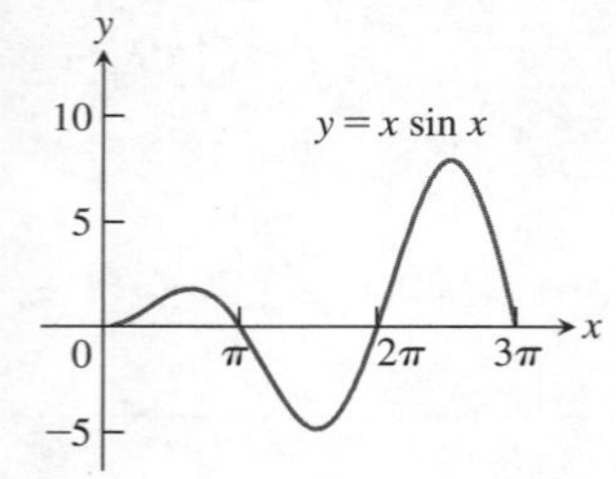

58. Finding area Find the area of the region enclosed by the curve $y = x \cos x$ and the x-axis (see the accompanying figure) for

a. $\pi/2 \le x \le 3\pi/2$.

b. $3\pi/2 \le x \le 5\pi/2$.

c. $5\pi/2 \le x \le 7\pi/2$.

d. What pattern do you see? What is the area between the curve and the x-axis for

$$\left(\frac{2n-1}{2}\right)\pi \le x \le \left(\frac{2n+1}{2}\right)\pi,$$

n an arbitrary positive integer? Give reasons for your answer.

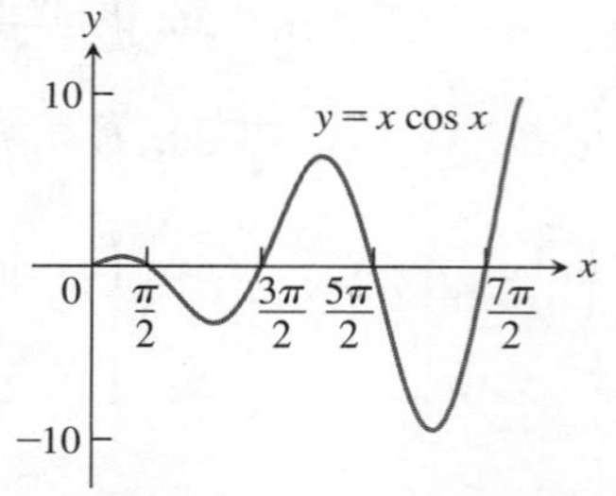

59. Finding volume Find the volume of the solid generated by revolving the region in the first quadrant bounded by the coordinate axes, the curve $y = e^x$, and the line $x = \ln 2$ about the line $x = \ln 2$.

60. Finding volume Find the volume of the solid generated by revolving the region in the first quadrant bounded by the coordinate axes, the curve $y = e^{-x}$, and the line $x = 1$

a. about the y-axis.

b. about the line $x = 1$.

61. Finding volume Find the volume of the solid generated by revolving the region in the first quadrant bounded by the coordinate axes and the curve $y = \cos x$, $0 \le x \le \pi/2$, about

a. the y-axis.

b. the line $x = \pi/2$.

62. Finding volume Find the volume of the solid generated by revolving the region bounded by the x-axis and the curve $y = x \sin x$, $0 \le x \le \pi$, about

a. the y-axis.

b. the line $x = \pi$.

(See Exercise 57 for a graph.)

63. Consider the region bounded by the graphs of $y = \ln x$, $y = 0$, and $x = e$.

a. Find the area of the region.

b. Find the volume of the solid formed by revolving this region about the x-axis.

c. Find the volume of the solid formed by revolving this region about the line $x = -2$.

d. Find the centroid of the region.

64. Consider the region bounded by the graphs of $y = \arctan x$, $y = 0$, and $x = 1$.

a. Find the area of the region.

b. Find the volume of the solid formed by revolving this region about the y-axis.

65. Average value A retarding force, symbolized by the dashpot in the accompanying figure, slows the motion of the weighted spring so that the mass's position at time t is

$$y = 2e^{-t} \cos t, \quad t \ge 0.$$

Find the average value of y over the interval $0 \le t \le 2\pi$.

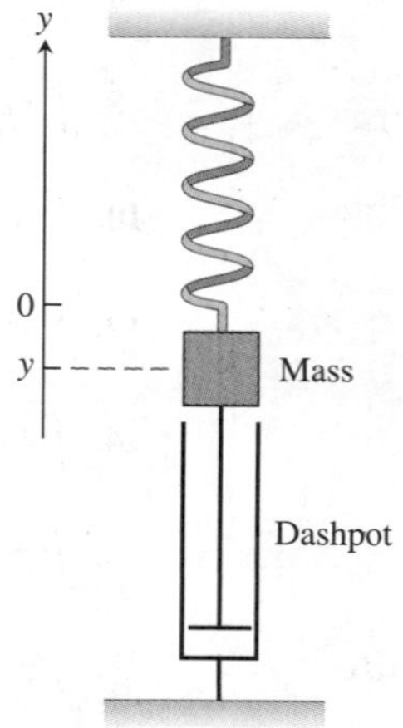

66. Average value In a mass-spring-dashpot system like the one in Exercise 65, the mass's position at time t is

$$y = 4e^{-t}(\sin t - \cos t), \qquad t \ge 0.$$

Find the average value of y over the interval $0 \le t \le 2\pi$.

Reduction Formulas

In Exercises 67–73, use integration by parts to establish the reduction formula.

67. $\int x^n \cos x\, dx = x^n \sin x - n \int x^{n-1} \sin x\, dx$

68. $\int x^n \sin x\, dx = -x^n \cos x + n \int x^{n-1} \cos x\, dx$

69. $\int x^n e^{ax}\, dx = \frac{x^n e^{ax}}{a} - \frac{n}{a}\int x^{n-1} e^{ax}\, dx, \quad a \neq 0$

70. $\int (\ln x)^n\, dx = x(\ln x)^n - n\int (\ln x)^{n-1}\, dx$

71. $\int x^m (\ln x)^n\, dx$

$$= \frac{x^{m+1}}{m+1}(\ln x)^n - \frac{n}{m+1}\int x^m (\ln x)^{n-1}\, dx, \ m \neq -1$$

72. $\int x^n \sqrt{x+1}\, dx$

$$= \frac{2x^n}{2n+3}(x+1)^{3/2} - \frac{2n}{2n+3}\int x^{n-1}\sqrt{x+1}\, dx$$

73. $\int \frac{x^n}{\sqrt{x+1}}\, dx$

$$= \frac{2x^n}{2n+1}\sqrt{x+1} - \frac{2n}{2n+1}\int \frac{x^{n-1}}{\sqrt{x+1}}\, dx$$

74. Use Example 5 to show that

$$\int_0^{\pi/2} \sin^n x\, dx = \int_0^{\pi/2} \cos^n x\, dx = \begin{cases} \left(\frac{\pi}{2}\right)\frac{1\cdot 3\cdot 5\cdots(n-1)}{2\cdot 4\cdot 6\cdots n}, & n \text{ even} \\ \frac{2\cdot 4\cdot 6\cdots(n-1)}{1\cdot 3\cdot 5\cdots n}, & n \text{ odd} \end{cases}$$

75. Show that

$$\int_a^b \left(\int_x^b f(t)\, dt\right) dx = \int_a^b (x-a) f(x)\, dx.$$

76. Use integration by parts to obtain the formula

$$\int \sqrt{1-x^2}\, dx = \frac{1}{2}x\sqrt{1-x^2} + \frac{1}{2}\int \frac{1}{\sqrt{1-x^2}}\, dx.$$

Integrating Inverses of Functions

Integration by parts leads to a rule for integrating inverses that usually gives good results:

$$\begin{aligned} \int f^{-1}(x)\, dx &= \int y f'(y)\, dy && y = f^{-1}(x), \quad x = f(y),\ dx = f'(y)\, dy \\ &= y f(y) - \int f(y)\, dy && \text{Integration by parts with } u = y,\ dv = f'(y)\, dy \\ &= x f^{-1}(x) - \int f(y)\, dy \end{aligned}$$

The idea is to take the most complicated part of the integral, in this case $f^{-1}(x)$, and simplify it first. For the integral of $\ln x$, we get

$$\begin{aligned} \int \ln x\, dx &= \int y e^y\, dy && y = \ln x, \quad x = e^y,\ dx = e^y\, dy \\ &= y e^y - e^y + C \\ &= x \ln x - x + C. \end{aligned}$$

For the integral of $\arccos x$, we get

$$\begin{aligned} \int \arccos x\, dx &= x \arccos x - \int \cos y\, dy && y = \arccos x \\ &= x \arccos x - \sin y + C \\ &= x \arccos x - \sin(\arccos x) + C. \end{aligned}$$

Use the formula

$$\int f^{-1}(x)\, dx = x f^{-1}(x) - \int f(y)\, dy \qquad y = f^{-1}(x) \quad (4)$$

to evaluate the integrals in Exercises 77–80. Express your answers in terms of x.

77. $\int \operatorname{arcsec} x\, dx$

78. $\int \arctan x\, dx$

79. $\int \sec^{-1} x\, dx$

80. $\int \log_2 x\, dx$

Another way to integrate $f^{-1}(x)$ (when f^{-1} is integrable) is to use integration by parts with $u = f^{-1}(x)$ and $dv = dx$ to rewrite the integral of f^{-1} as

$$\int f^{-1}(x)\, dx = x f^{-1}(x) - \int x\left(\frac{d}{dx} f^{-1}(x)\right) dx. \qquad (5)$$

Exercises 81 and 82 compare the results of using Equations (4) and (5).

81. Equations (4) and (5) give different formulas for the integral of $\arccos x$:

a. $\int \arccos x\, dx = x \arccos x - \sin(\arccos x) + C$ Eq. (4)

b. $\int \arccos x\, dx = x \arccos x - \sqrt{1-x^2} + C$ Eq. (5)

Can both integrations be correct? Explain.

82. Equations (4) and (5) lead to different formulas for the integral of $\arctan x$:

a. $\int \arctan x\, dx = x \arctan x - \ln \sec(\arctan x) + C$ Eq. (4)

b. $\int \arctan x\, dx = x \arctan x - \ln \sqrt{1+x^2} + C$ Eq. (5)

Can both integrations be correct? Explain.

Evaluate the integrals in Exercises 83 and 84 with **(a)** Eq. (4) and **(b)** Eq. (5). In each case, check your work by differentiating your answer with respect to x.

83. $\int \sinh^{-1} x\, dx$

84. $\int \tanh^{-1} x\, dx$

8.3 Trigonometric Integrals

Trigonometric integrals involve algebraic combinations of the six basic trigonometric functions. In principle, we can always express such integrals in terms of sines and cosines, but it is often simpler to work with other functions, as in the integral

$$\int \sec^2 x\, dx = \tan x + C.$$

The general idea is to use identities to transform the integrals we must find into integrals that are easier to work with.

Products of Powers of Sines and Cosines

We begin with integrals of the form

$$\int \sin^m x \cos^n x\, dx,$$

where m and n are nonnegative integers (positive or zero). We can divide the appropriate substitution into three cases according to m and n being odd or even.

Case 1 If **m is odd** in $\int \sin^m x \cos^n x\, dx$, we write m as $2k+1$ and use the identity $\sin^2 x = 1 - \cos^2 x$ to obtain

$$\sin^m x = \sin^{2k+1} x = (\sin^2 x)^k \sin x = (1 - \cos^2 x)^k \sin x. \tag{1}$$

Then we substitute $u = \cos x$ and $du = -\sin x\, dx$.

Case 2 If **n is odd** in $\int \sin^m x \cos^n x\, dx$, we write n as $2k+1$ and use the identity $\cos^2 x = 1 - \sin^2 x$ to obtain

$$\cos^n x = \cos^{2k+1} x = (\cos^2 x)^k \cos x = (1 - \sin^2 x)^k \cos x.$$

We then substitute $u = \sin x$ and $du = \cos x\, dx$.

Case 3 If **both m and n are even** in $\int \sin^m x \cos^n x\, dx$, we substitute

$$\sin^2 x = \frac{1 - \cos 2x}{2}, \quad \cos^2 x = \frac{1 + \cos 2x}{2} \tag{2}$$

to reduce the integrand to one in lower powers of $\cos 2x$.

Here are some examples illustrating each case.

EXAMPLE 1 Evaluate $\int \sin^3 x \cos^2 x\, dx$.

Solution This is an example of Case 1.

$$\begin{aligned}
\int \sin^3 x \cos^2 x\, dx &= \int \sin^2 x \cos^2 x \sin x\, dx && m \text{ is odd.}\\
&= \int (1 - \cos^2 x)(\cos^2 x)\sin x\, dx && \sin^2 x = 1 - \cos^2 x\\
&= \int (1 - u^2)(u^2)(-du) && u = \cos x,\ du = -\sin x\, dx\\
&= \int (u^4 - u^2)\, du && \text{Distribute.}\\
&= \frac{u^5}{5} - \frac{u^3}{3} + C = \frac{\cos^5 x}{5} - \frac{\cos^3 x}{3} + C
\end{aligned}$$

■

EXAMPLE 2 Evaluate

$$\int \cos^5 x\, dx.$$

Solution This is an example of Case 2, where $m = 0$ is even and $n = 5$ is odd.

$$\begin{aligned}
\int \cos^5 x\, dx &= \int \cos^4 x \cos x\, dx \\
&= \int (1 - \sin^2 x)^2 \cos x\, dx \\
&= \int (1 - u^2)^2\, du && u = \sin x,\ du = \cos x\, dx \\
&= \int (1 - 2u^2 + u^4)\, du && \text{Square } 1 - u^2. \\
&= u - \frac{2}{3}u^3 + \frac{1}{5}u^5 + C = \sin x - \frac{2}{3}\sin^3 x + \frac{1}{5}\sin^5 x + C
\end{aligned}$$

■

EXAMPLE 3 Evaluate

$$\int \sin^2 x \cos^4 x\, dx.$$

Solution This is an example of Case 3.

$$\begin{aligned}
\int \sin^2 x \cos^4 x\, dx &= \int \left(\frac{1 - \cos 2x}{2}\right)\left(\frac{1 + \cos 2x}{2}\right)^2 dx && m \text{ and } n \text{ both even} \\
&= \frac{1}{8}\int (1 - \cos 2x)(1 + 2\cos 2x + \cos^2 2x)\, dx \\
&= \frac{1}{8}\int (1 + \cos 2x - \cos^2 2x - \cos^3 2x)\, dx \\
&= \frac{1}{8}\left[x + \frac{1}{2}\sin 2x - \int (\cos^2 2x + \cos^3 2x)\, dx\right]
\end{aligned}$$

For the term involving $\cos^2 2x$, we use

$$\begin{aligned}
\int \cos^2 2x\, dx &= \frac{1}{2}\int (1 + \cos 4x)\, dx \\
&= \frac{1}{2}\left(x + \frac{1}{4}\sin 4x\right) + C_1.
\end{aligned}$$

Use the identity $\cos^2 \theta = (1 + \cos 2\theta)/2$, with $\theta = 2x$.

For the $\cos^3 2x$ term, we have

$$\begin{aligned}
\int \cos^3 2x\, dx &= \int (1 - \sin^2 2x)\cos 2x\, dx && u = \sin 2x,\ du = 2\cos 2x\, dx \\
&= \frac{1}{2}\int (1 - u^2)\, du = \frac{1}{2}\left(\sin 2x - \frac{1}{3}\sin^3 2x\right) + C_2.
\end{aligned}$$

Combining everything and simplifying, we get

$$\int \sin^2 x \cos^4 x\, dx = \frac{1}{16}\left(x - \frac{1}{4}\sin 4x + \frac{1}{3}\sin^3 2x\right) + C.$$

■

Eliminating Square Roots

In the next example, we use the identity $\cos^2\theta = (1 + \cos 2\theta)/2$ to eliminate a square root.

EXAMPLE 4 Evaluate

$$\int_0^{\pi/4} \sqrt{1 + \cos 4x}\, dx.$$

Solution To eliminate the square root, we use the identity

$$\cos^2\theta = \frac{1 + \cos 2\theta}{2} \quad \text{or} \quad 1 + \cos 2\theta = 2\cos^2\theta.$$

With $\theta = 2x$, this becomes

$$1 + \cos 4x = 2\cos^2 2x.$$

Therefore,

$$\begin{aligned}
\int_0^{\pi/4} \sqrt{1 + \cos 4x}\, dx &= \int_0^{\pi/4} \sqrt{2\cos^2 2x}\, dx = \int_0^{\pi/4} \sqrt{2}\sqrt{\cos^2 2x}\, dx \\
&= \sqrt{2}\int_0^{\pi/4} |\cos 2x|\, dx = \sqrt{2}\int_0^{\pi/4} \cos 2x\, dx \qquad \begin{matrix}\cos 2x \geq 0 \\ \text{on } [0, \pi/4]\end{matrix} \\
&= \sqrt{2}\left[\frac{\sin 2x}{2}\right]_0^{\pi/4} = \frac{\sqrt{2}}{2}[1 - 0] = \frac{\sqrt{2}}{2}.
\end{aligned}$$

■

Integrals of Powers of tan x and sec x

We know how to integrate the tangent and secant functions and their squares. To integrate higher powers, we use the identities $\tan^2 x = \sec^2 x - 1$ and $\sec^2 x = \tan^2 x + 1$, and integrate by parts when necessary to reduce the higher powers to lower powers.

EXAMPLE 5 Evaluate

$$\int \tan^4 x\, dx.$$

Solution

$$\begin{aligned}
\int \tan^4 x\, dx &= \int \tan^2 x \cdot \tan^2 x\, dx \\
&= \int \tan^2 x \cdot (\sec^2 x - 1)\, dx \qquad \tan^2 x = \sec^2 x - 1 \\
&= \int \tan^2 x \sec^2 x\, dx - \int \tan^2 x\, dx \\
&= \int \tan^2 x \sec^2 x\, dx - \int (\sec^2 x - 1)\, dx \qquad \tan^2 x = \sec^2 x - 1 \\
&= \int \tan^2 x \sec^2 x\, dx - \int \sec^2 x\, dx + \int dx
\end{aligned}$$

In the first integral, we let

$$u = \tan x, \qquad du = \sec^2 x\, dx$$

and have

$$\int u^2\,du = \frac{1}{3}u^3 + C_1.$$

The remaining integrals are standard forms, so

$$\int \tan^4 x\,dx = \frac{1}{3}\tan^3 x - \tan x + x + C.$$ ■

EXAMPLE 6 Evaluate

$$\int \sec^3 x\,dx.$$

Solution We integrate by parts using

$$u = \sec x, \qquad dv = \sec^2 x\,dx, \qquad v = \tan x, \qquad du = \sec x \tan x\,dx.$$

Then

$$\int \sec^3 x\,dx = \sec x \tan x - \int (\tan x)(\sec x \tan x)\,dx \qquad \text{Integrate by parts.}$$

$$= \sec x \tan x - \int (\sec^2 x - 1)\sec x\,dx \qquad \tan^2 x = \sec^2 x - 1$$

$$= \sec x \tan x - \int \sec^3 x\,dx + \int \sec x\,dx$$

Combining the two secant-cubed integrals gives

$$2\int \sec^3 x\,dx = \sec x \tan x + \int \sec x\,dx$$

and therefore

$$\int \sec^3 x\,dx = \frac{1}{2}\sec x \tan x + \frac{1}{2}\ln|\sec x + \tan x| + C.$$ ■

EXAMPLE 7 Evaluate

$$\int \tan^4 x \sec^4 x\,dx.$$

Solution

$$\int (\tan^4 x)(\sec^4 x)\,dx = \int (\tan^4 x)(1 + \tan^2 x)(\sec^2 x)\,dx \qquad \sec^2 x = 1 + \tan^2 x$$

$$= \int (\tan^4 x + \tan^6 x)(\sec^2 x)\,dx \qquad \text{Distribute.}$$

$$= \int (u^4 + u^6)\,du = \frac{u^5}{5} + \frac{u^7}{7} + C \qquad u = \tan x,\ du = \sec^2 x\,dx$$

$$= \frac{\tan^5 x}{5} + \frac{\tan^7 x}{7} + C$$ ■

Products of Sines and Cosines

The integrals

$$\int \sin mx \sin nx\,dx, \qquad \int \sin mx \cos nx\,dx, \qquad \text{and} \qquad \int \cos mx \cos nx\,dx$$

arise in many applications involving periodic functions. We can evaluate these integrals through integration by parts, but two such integrations are required in each case. It is simpler to use the following identities.

$$\sin mx \sin nx = \frac{1}{2}[\cos(m-n)x - \cos(m+n)x] \tag{3}$$

$$\sin mx \cos nx = \frac{1}{2}[\sin(m-n)x + \sin(m+n)x] \tag{4}$$

$$\cos mx \cos nx = \frac{1}{2}[\cos(m-n)x + \cos(m+n)x] \tag{5}$$

These identities come from the angle sum formulas for the sine and cosine functions (Section 1.3). They give functions whose antiderivatives are easily found.

EXAMPLE 8 Evaluate

$$\int \sin 3x \cos 5x\,dx.$$

Solution From Equation (4) with $m = 3$ and $n = 5$, we get

$$\begin{aligned}\int \sin 3x \cos 5x\,dx &= \frac{1}{2}\int [\sin(-2x) + \sin 8x]\,dx \\ &= \frac{1}{2}\int (\sin 8x - \sin 2x)\,dx \\ &= -\frac{\cos 8x}{16} + \frac{\cos 2x}{4} + C.\end{aligned}$$ ■

EXERCISES 8.3

Powers of Sines and Cosines

Evaluate the integrals in Exercises 1–22.

1. $\int \cos 2x\,dx$
2. $\int_0^{\pi} 3 \sin\frac{x}{3}\,dx$
3. $\int \cos^3 x \sin x\,dx$
4. $\int \sin^4 2x \cos 2x\,dx$
5. $\int \sin^3 x\,dx$
6. $\int \cos^3 4x\,dx$
7. $\int \sin^5 x\,dx$
8. $\int_0^{\pi} \sin^5 \frac{x}{2}\,dx$
9. $\int \cos^3 x\,dx$
10. $\int_0^{\pi/6} 3\cos^5 3x\,dx$
11. $\int \sin^3 x \cos^3 x\,dx$
12. $\int \cos^3 2x \sin^5 2x\,dx$
13. $\int \cos^2 x\,dx$
14. $\int_0^{\pi/2} \sin^2 x\,dx$
15. $\int_0^{\pi/2} \sin^7 y\,dy$
16. $\int 7\cos^7 t\,dt$
17. $\int_0^{\pi} 8\sin^4 x\,dx$
18. $\int 8\cos^4 2\pi x\,dx$
19. $\int 16\sin^2 x \cos^2 x\,dx$
20. $\int_0^{\pi} 8\sin^4 y \cos^2 y\,dy$
21. $\int 8\cos^3 2\theta \sin 2\theta\,d\theta$
22. $\int_0^{\pi/2} \sin^2 2\theta \cos^3 2\theta\,d\theta$

Integrating Square Roots

Evaluate the integrals in Exercises 23–32.

23. $\int_0^{2\pi} \sqrt{\frac{1-\cos x}{2}}\, dx$

24. $\int_0^{\pi} \sqrt{1-\cos 2x}\, dx$

25. $\int_0^{\pi} \sqrt{1-\sin^2 t}\, dt$

26. $\int_0^{\pi} \sqrt{1-\cos^2\theta}\, d\theta$

27. $\int_{\pi/3}^{\pi/2} \frac{\sin^2 x}{\sqrt{1-\cos x}}\, dx$

28. $\int_0^{\pi/6} \sqrt{1+\sin x}\, dx$

$\left(\textit{Hint}\text{: Multiply by } \sqrt{\frac{1-\sin x}{1-\sin x}}.\right)$

29. $\int_{5\pi/6}^{\pi} \frac{\cos^4 x}{\sqrt{1-\sin x}}\, dx$

30. $\int_{\pi/2}^{3\pi/4} \sqrt{1-\sin 2x}\, dx$

31. $\int_0^{\pi/2} \theta\sqrt{1-\cos 2\theta}\, d\theta$

32. $\int_{-\pi}^{\pi} (1-\cos^2 t)^{3/2}\, dt$

Powers of Tangents and Secants

Evaluate the integrals in Exercises 33–52.

33. $\int \sec^2 x \tan x\, dx$

34. $\int \sec x \tan^2 x\, dx$

35. $\int \sec^3 x \tan x\, dx$

36. $\int \sec^3 x \tan^3 x\, dx$

37. $\int \sec^2 x \tan^2 x\, dx$

38. $\int \sec^4 x \tan^2 x\, dx$

39. $\int_{-\pi/3}^{0} 2\sec^3 x\, dx$

40. $\int e^x \sec^3 e^x\, dx$

41. $\int \sec^4 \theta\, d\theta$

42. $\int \tan^4 x \sec^3 x\, dx$

43. $\int_{\pi/4}^{\pi/2} \csc^4 \theta\, d\theta$

44. $\int \sec^6 x\, dx$

45. $\int 4\tan^3 x\, dx$

46. $\int_{-\pi/4}^{\pi/4} 6\tan^4 x\, dx$

47. $\int \tan^5 x\, dx$

48. $\int \cot^6 2x\, dx$

49. $\int_{\pi/6}^{\pi/3} \cot^3 x\, dx$

50. $\int 8\cot^4 t\, dt$

51. $\int_{\pi/4}^{\pi/3} \tan^5\theta \sec^4\theta\, d\theta$

52. $\int \cot^3 t \csc^4 t\, dt$

Products of Sines and Cosines

Evaluate the integrals in Exercises 53–58.

53. $\int \sin 3x \cos 2x\, dx$

54. $\int \sin 2x \cos 3x\, dx$

55. $\int_{-\pi}^{\pi} \sin 3x \sin 3x\, dx$

56. $\int_0^{\pi/2} \sin x \cos x\, dx$

57. $\int \cos 3x \cos 4x\, dx$

58. $\int_{-\pi/2}^{\pi/2} \cos x \cos 7x\, dx$

Exercises 59–64 require the use of various trigonometric identities before you evaluate the integrals.

59. $\int \sin^2\theta \cos 3\theta\, d\theta$

60. $\int \cos^2 2\theta \sin\theta\, d\theta$

61. $\int \cos^3\theta \sin 2\theta\, d\theta$

62. $\int \sin^3\theta \cos 2\theta\, d\theta$

63. $\int_{-\pi/2}^{\pi/2} \frac{\cos x}{1+e^x}\, dx$

64. $\int_0^{\pi/2} \frac{\sqrt{\sin x}}{\sqrt{\sin x}+\sqrt{\cos x}}\, dx$

Assorted Integrations

Use any method to evaluate the integrals in Exercises 65–70.

65. $\int_2^3 \frac{\sqrt{x}}{\sqrt{x}+\sqrt{5-x}}\, dx$

66. $\int_0^{\pi/4} \log(1+\tan x)\, dx$

67. $\int \frac{\tan^2 x}{\csc x}\, dx$

68. $\int \frac{\cot x}{\cos^2 x}\, dx$

69. $\int x\sin^2 x\, dx$

70. $\int x\cos^3 x\, dx$

Applications

71. Arc length Find the length of the curve

$$y = \ln(\sin x), \quad \frac{\pi}{6} \le x \le \frac{\pi}{2}$$

72. Center of gravity Find the center of gravity of the region bounded by the x-axis, the curve $y = \sec x$, and the lines $x = -\pi/4$, $x = \pi/4$.

73. Volume Find the volume generated by revolving one arch of the curve $y = \sin x$ about the x-axis.

74. Area Find the area between the x-axis and the curve $y = \sqrt{1+\cos 4x}$, $0 \le x \le \pi$.

75. Centroid Find the centroid of the region bounded by the graphs of $y = x + \cos x$ and $y = 0$ for $0 \le x \le 2\pi$.

76. Volume Find the volume of the solid formed by revolving the region bounded by the graphs of $y = \sin x + \sec x$, $y = 0$, $x = 0$, and $x = \pi/3$ about the x-axis.

77. Volume Find the volume of the solid formed by revolving the region bounded by the graphs of $y = \arctan x$, $x = 0$, and $y = \pi/4$ about the y-axis.

78. Average Value Find the average value of the function $f(x) = \dfrac{1}{1-\sin\theta}$ on $[0, \pi/6]$.

8.4 Trigonometric Substitutions

Trigonometric substitutions occur when we replace the variable of integration by a trigonometric function. The most common substitutions are $x = a\tan\theta$, $x = a\sin\theta$, and $x = a\sec\theta$. These substitutions are effective in transforming integrals involving $\sqrt{a^2 + x^2}$, $\sqrt{a^2 - x^2}$, and $\sqrt{x^2 - a^2}$ into integrals with respect to θ, since they come from the reference right triangles in Figure 8.2.

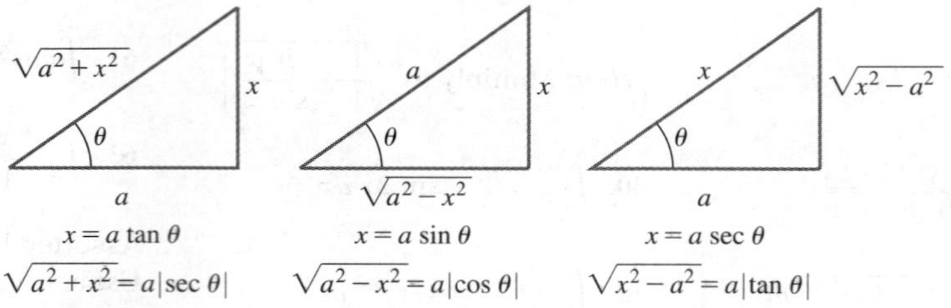

FIGURE 8.2 Reference triangles for the three basic substitutions, identifying the sides labeled x and a for each substitution.

With $x = a\tan\theta$,

$$a^2 + x^2 = a^2 + a^2\tan^2\theta = a^2(1 + \tan^2\theta) = a^2\sec^2\theta.$$

With $x = a\sin\theta$,

$$a^2 - x^2 = a^2 - a^2\sin^2\theta = a^2(1 - \sin^2\theta) = a^2\cos^2\theta.$$

With $x = a\sec\theta$,

$$x^2 - a^2 = a^2\sec^2\theta - a^2 = a^2(\sec^2\theta - 1) = a^2\tan^2\theta.$$

We want any substitution we use in an integration to be reversible so that we can change back to the original variable afterward. For example, if $x = a\tan\theta$, we want to be able to set $\theta = \arctan(x/a)$ after the integration takes place. If $x = a\sin\theta$, we want to be able to set $\theta = \arcsin(x/a)$ when we're done, and similarly for $x = a\sec\theta$.

As we know from Section 7.6, the functions in these substitutions have inverses only for selected values of θ (Figure 8.3). For reversibility,

$$x = a\tan\theta \quad \text{requires} \quad \theta = \arctan\left(\frac{x}{a}\right) \quad \text{with} \quad -\frac{\pi}{2} < \theta < \frac{\pi}{2},$$

$$x = a\sin\theta \quad \text{requires} \quad \theta = \arcsin\left(\frac{x}{a}\right) \quad \text{with} \quad -\frac{\pi}{2} \le \theta \le \frac{\pi}{2},$$

$$x = a\sec\theta \quad \text{requires} \quad \theta = \operatorname{arcsec}\left(\frac{x}{a}\right) \quad \text{with} \quad \begin{cases} 0 \le \theta < \dfrac{\pi}{2} & \text{if } \dfrac{x}{a} \ge 1, \\[2ex] \dfrac{\pi}{2} < \theta \le \pi & \text{if } \dfrac{x}{a} \le -1. \end{cases}$$

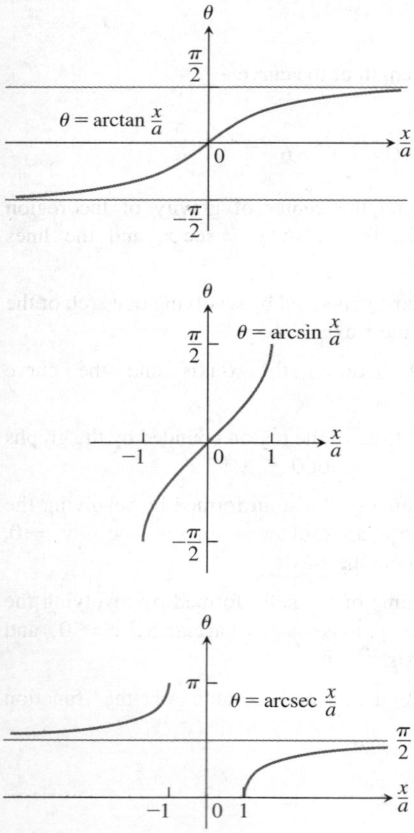

FIGURE 8.3 The arctangent, arcsine, and arcsecant of x/a, graphed as functions of x/a.

To simplify calculations with the substitution $x = a\sec\theta$, we will restrict its use to integrals in which $x/a \ge 1$. This will place θ in $[0, \pi/2)$ and make $\tan\theta \ge 0$. We will then have $\sqrt{x^2 - a^2} = \sqrt{a^2\tan^2\theta} = |a\tan\theta| = a\tan\theta$, free of absolute values, provided $a > 0$.

Procedure for a Trigonometric Substitution

1. Write down the substitution for x, calculate the differential dx, and specify the selected values of θ for the substitution.
2. Substitute the trigonometric expression and the calculated differential into the integrand, and then simplify the results algebraically.
3. Evaluate the trigonometric integral, keeping in mind the restrictions on the angle θ for reversibility.
4. Draw an appropriate reference triangle to reverse the substitution in the integration result and convert it back to the original variable x.

EXAMPLE 1 Evaluate

$$\int \frac{dx}{\sqrt{4 + x^2}}.$$

Solution We set

$$x = 2\tan\theta, \qquad dx = 2\sec^2\theta\, d\theta, \qquad -\frac{\pi}{2} < \theta < \frac{\pi}{2},$$
$$4 + x^2 = 4 + 4\tan^2\theta = 4(1 + \tan^2\theta) = 4\sec^2\theta.$$

Then

$$\int \frac{dx}{\sqrt{4 + x^2}} = \int \frac{2\sec^2\theta\, d\theta}{\sqrt{4\sec^2\theta}} = \int \frac{\sec^2\theta\, d\theta}{|\sec\theta|} \qquad \sqrt{\sec^2\theta} = |\sec\theta|$$

$$= \int \sec\theta\, d\theta \qquad \sec\theta > 0 \text{ for } -\frac{\pi}{2} < \theta < \frac{\pi}{2}$$

$$= \ln|\sec\theta + \tan\theta| + C$$

$$= \ln\left|\frac{\sqrt{4 + x^2}}{2} + \frac{x}{2}\right| + C. \qquad \text{From Fig. 8.4}$$

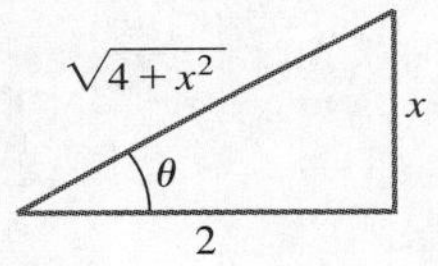

FIGURE 8.4 Reference triangle for $x = 2\tan\theta$ (Example 1):

$$\tan\theta = \frac{x}{2}$$

and

$$\sec\theta = \frac{\sqrt{4 + x^2}}{2}.$$

Notice how we expressed $\ln|\sec\theta + \tan\theta|$ in terms of x: We drew a reference triangle for the original substitution $x = 2\tan\theta$ (Figure 8.4) and read the ratios from the triangle. ■

EXAMPLE 2 Here we find an expression for the inverse hyperbolic sine function in terms of the natural logarithm. Following the same procedure as in Example 1, we find that

$$\int \frac{dx}{\sqrt{a^2 + x^2}} = \int \sec\theta\, d\theta \qquad x = a\tan\theta,\ dx = a\sec^2\theta\, d\theta$$

$$= \ln|\sec\theta + \tan\theta| + C$$

$$= \ln\left|\frac{\sqrt{a^2 + x^2}}{a} + \frac{x}{a}\right| + C \qquad \text{Fig. 8.2}$$

From Table 7.9, $\sinh^{-1}(x/a)$ is also an antiderivative of $1/\sqrt{a^2 + x^2}$, so the two antiderivatives differ by a constant, giving

$$\sinh^{-1}\frac{x}{a} = \ln\left|\frac{\sqrt{a^2 + x^2}}{a} + \frac{x}{a}\right| + C.$$

Setting $x = 0$ in this last equation, we find that $0 = \ln|1| + C$, so $C = 0$. Since $\sqrt{a^2 + x^2} > |x|$, we conclude that $\dfrac{\sqrt{a^2 + x^2}}{a} + \dfrac{x}{a} > 0$, and therefore

$$\sinh^{-1}\frac{x}{a} = \ln\left(\frac{\sqrt{a^2 + x^2}}{a} + \frac{x}{a}\right)$$

(See also Exercise 76 in Section 7.7.) ■

EXAMPLE 3 Evaluate

$$\int \frac{x^2\,dx}{\sqrt{9 - x^2}}.$$

Solution We set

$$x = 3\sin\theta, \qquad dx = 3\cos\theta\,d\theta, \qquad -\frac{\pi}{2} < \theta < \frac{\pi}{2}$$
$$9 - x^2 = 9 - 9\sin^2\theta = 9(1 - \sin^2\theta) = 9\cos^2\theta.$$

Then

$$\int \frac{x^2\,dx}{\sqrt{9 - x^2}} = \int \frac{9\sin^2\theta \cdot 3\cos\theta\,d\theta}{|3\cos\theta|}$$

$$= 9\int \sin^2\theta\,d\theta \qquad \cos\theta > 0 \text{ for } -\frac{\pi}{2} < \theta < \frac{\pi}{2}$$

$$= 9\int \frac{1 - \cos 2\theta}{2}\,d\theta \qquad \sin^2\theta = \frac{1 - \cos 2\theta}{2}$$

$$= \frac{9}{2}\left(\theta - \frac{\sin 2\theta}{2}\right) + C$$

$$= \frac{9}{2}(\theta - \sin\theta\cos\theta) + C \qquad \sin 2\theta = 2\sin\theta\cos\theta$$

$$= \frac{9}{2}\left(\arcsin\frac{x}{3} - \frac{x}{3} \cdot \frac{\sqrt{9 - x^2}}{3}\right) + C \qquad \text{From Fig. 8.5}$$

$$= \frac{9}{2}\arcsin\frac{x}{3} - \frac{x}{2}\sqrt{9 - x^2} + C.$$ ■

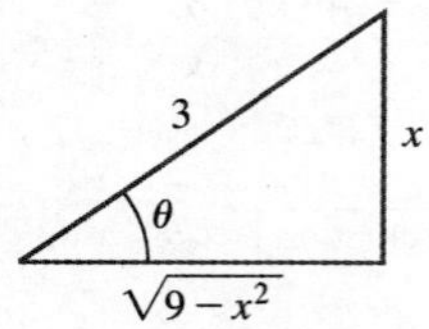

FIGURE 8.5 Reference triangle for $x = 3\sin\theta$ (Example 3):

$$\sin\theta = \frac{x}{3}$$

and

$$\cos\theta = \frac{\sqrt{9 - x^2}}{3}.$$

EXAMPLE 4 Evaluate

$$\int \frac{dx}{\sqrt{25x^2 - 4}}, \qquad x > \frac{2}{5}.$$

Solution We first rewrite the radical as

$$\sqrt{25x^2 - 4} = \sqrt{25\left(x^2 - \frac{4}{25}\right)}$$

$$= 5\sqrt{x^2 - \left(\frac{2}{5}\right)^2} \qquad \sqrt{x^2 - a^2} \text{ with } a = \frac{2}{5}$$

to put the radicand in the form $x^2 - a^2$. We then substitute

$$x = \frac{2}{5}\sec\theta, \quad dx = \frac{2}{5}\sec\theta\tan\theta\,d\theta, \quad 0 < \theta < \frac{\pi}{2}.$$

We then get

$$x^2 - \left(\frac{2}{5}\right)^2 = \frac{4}{25}\sec^2\theta - \frac{4}{25} = \frac{4}{25}(\sec^2\theta - 1) = \frac{4}{25}\tan^2\theta$$

and

$$\sqrt{x^2 - \left(\frac{2}{5}\right)^2} = \frac{2}{5}|\tan\theta| = \frac{2}{5}\tan\theta. \qquad \tan\theta > 0 \text{ for } 0 < \theta < \pi/2$$

With these substitutions, we have

$$\int \frac{dx}{\sqrt{25x^2 - 4}} = \int \frac{dx}{5\sqrt{x^2 - (4/25)}} = \int \frac{(2/5)\sec\theta\tan\theta\,d\theta}{5\cdot(2/5)\tan\theta}$$

$$= \frac{1}{5}\int \sec\theta\,d\theta = \frac{1}{5}\ln|\sec\theta + \tan\theta| + C$$

$$= \frac{1}{5}\ln\left|\frac{5x}{2} + \frac{\sqrt{25x^2 - 4}}{2}\right| + C. \qquad \text{From Fig. 8.6}$$

■

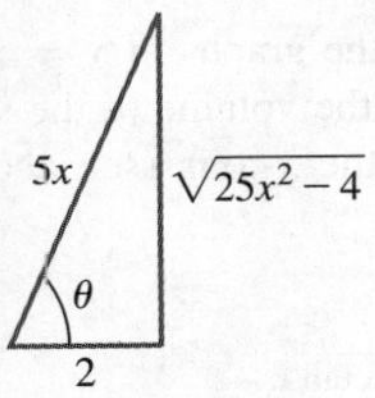

FIGURE 8.6 If $x = (2/5)\sec\theta$, $0 < \theta < \pi/2$, then $\theta = \operatorname{arcsec}(5x/2)$, and we can read the values of the other trigonometric functions of θ from this right triangle (Example 4).

EXERCISES 8.4

Using Trigonometric Substitutions

Evaluate the integrals in Exercises 1–14.

1. $\int \frac{dx}{\sqrt{9 + x^2}}$

2. $\int \frac{3\,dx}{\sqrt{1 + 9x^2}}$

3. $\int_{-2}^{2} \frac{dx}{4 + x^2}$

4. $\int_{0}^{2} \frac{dx}{8 + 2x^2}$

5. $\int_{0}^{3/2} \frac{dx}{\sqrt{9 - x^2}}$

6. $\int_{0}^{1/2\sqrt{2}} \frac{2\,dx}{\sqrt{1 - 4x^2}}$

7. $\int \sqrt{25 - t^2}\,dt$

8. $\int \sqrt{1 - 9t^2}\,dt$

9. $\int \frac{dx}{\sqrt{4x^2 - 49}}, \quad x > \frac{7}{2}$

10. $\int \frac{5\,dx}{\sqrt{25x^2 - 9}}, \quad x > \frac{3}{5}$

11. $\int \frac{\sqrt{y^2 - 49}}{y}\,dy, \quad y > 7$

12. $\int \frac{\sqrt{y^2 - 25}}{y^3}\,dy, \quad y > 5$

13. $\int \frac{dx}{x^2\sqrt{x^2 - 1}}, \quad x > 1$

14. $\int \frac{2\,dx}{x^3\sqrt{x^2 - 1}}, \quad x > 1$

Assorted Integrations

Use any method to evaluate the integrals in Exercises 15–38. Most will require trigonometric substitutions, but some can be evaluated by other methods.

15. $\int \frac{dx}{x\sqrt{x^2 - 1}}$

16. $\int \frac{dx}{1 + x^2}$

17. $\int \frac{x\,dx}{\sqrt{x^2 - 1}}$

18. $\int \frac{dx}{\sqrt{1 - x^2}}$

19. $\int \frac{x}{\sqrt{9 - x^2}}\,dx$

20. $\int \frac{x^2}{4 + x^2}\,dx$

21. $\int \frac{x^3\,dx}{\sqrt{x^2 + 4}}$

22. $\int \frac{dx}{x^2\sqrt{x^2 + 1}}$

23. $\int \frac{8\,dw}{w^2\sqrt{4 - w^2}}$

24. $\int \frac{\sqrt{9 - w^2}}{w^2}\,dw$

25. $\int \sqrt{\frac{x + 1}{1 - x}}\,dx$

26. $\int x\sqrt{x^2 - 4}\,dx$

27. $\int_{0}^{\sqrt{3}/2} \frac{4x^2\,dx}{(1 - x^2)^{3/2}}$

28. $\int_{0}^{1} \frac{dx}{(4 - x^2)^{3/2}}$

29. $\int \frac{dx}{(x^2 - 1)^{3/2}}, \quad x > 1$

30. $\int \frac{x^2\,dx}{(x^2 - 1)^{5/2}}, \quad x > 1$

31. $\int \frac{(1 - x^2)^{3/2}}{x^6}\,dx$

32. $\int \frac{(1 - x^2)^{1/2}}{x^4}\,dx$

33. $\int \frac{1}{x^2(x^4 + 1)^{3/4}}\,dx$

34. $\int \frac{\cos^3 x + \cos^5 x}{\sin^2 x + \sin^4 x}\,dx$

35. $\int \frac{x^3\,dx}{x^2-1}$

36. $\int \frac{x\,dx}{25+4x^2}$

37. $\int \frac{v^2\,dv}{(1-v^2)^{5/2}}$

38. $\int \frac{(1-r^2)^{5/2}}{r^8}\,dr$

In Exercises 39–48, use an appropriate substitution and then a trigonometric substitution to evaluate the integrals.

39. $\int \frac{x^2+1}{x^4+1}\,dx$ $\left[\textit{Hint}: \int \frac{1+\frac{1}{x^2}}{x^2+\frac{1}{x^2}} \text{ put } u=x-\frac{1}{x}\right]$

40. $\int \frac{dx}{x^2+x+1}$

41. $\int_{1/12}^{1/4} \frac{2\,dt}{\sqrt{t}+4t\sqrt{t}}$

42. $\int_1^e \frac{dy}{y\sqrt{1+(\ln y)^2}}$

43. $\int \frac{x\,dx}{\sqrt{1+x^4}}$

44. $\int \frac{\sqrt{1-(\ln x)^2}}{x\ln x}\,dx$

45. $\int \sqrt{\frac{4-x}{x}}\,dx$
(*Hint*: Let $x=u^2$.)

46. $\int \sqrt{\frac{x}{1-x^3}}\,dx$
(*Hint*: Let $u=x^{3/2}$.)

47. $\int \sqrt{x}\sqrt{1-x}\,dx$

48. $\int \frac{\sqrt{x-2}}{\sqrt{x-1}}\,dx$

Complete the Square Before Using Trigonometric Substitutions

For Exercises 49–52, complete the square before using an appropriate trigonometric substitution.

49. $\int \sqrt{8-2x-x^2}\,dx$

50. $\int \frac{1}{\sqrt{x^2-2x+5}}\,dx$

51. $\int \frac{\sqrt{x^2+4x+3}}{x+2}\,dx$

52. $\int \frac{\sqrt{x^2+2x+2}}{x^2+2x+1}\,dx$

Initial Value Problems

Solve the initial value problems in Exercises 53–56 for y as a function of x.

53. $x\frac{dy}{dx} = \sqrt{x^2-4}, \quad x \geq 2, \quad y(2)=0$

54. $\sqrt{x^2-9}\frac{dy}{dx} = 1, \quad x > 3, \quad y(5) = \ln 3$

55. $(x^2+4)\frac{dy}{dx} = 3, \quad y(2)=0$

56. $(x^2+1)^2\frac{dy}{dx} = \sqrt{x^2+1}, \quad y(0)=1$

Applications and Examples

57. **Area** Find the area of the region in the first quadrant that is enclosed by the coordinate axes and the curve $y=\sqrt{9-x^2}/3$.

58. **Area** Find the area enclosed by the ellipse

$$\frac{x^2}{a^2}+\frac{y^2}{b^2}=1.$$

59. Consider the region bounded by the graphs of $y=\sin^{-1}x$, $y=0$, and $x=1/2$.
   a. Find the area of the region.
   b. Find the centroid of the region.

60. Consider the region bounded by the graphs of $y=\sqrt{x}\arctan x$ and $y=0$ for $0\leq x\leq 1$. Find the volume of the solid formed by revolving this region about the x-axis (see accompanying figure).

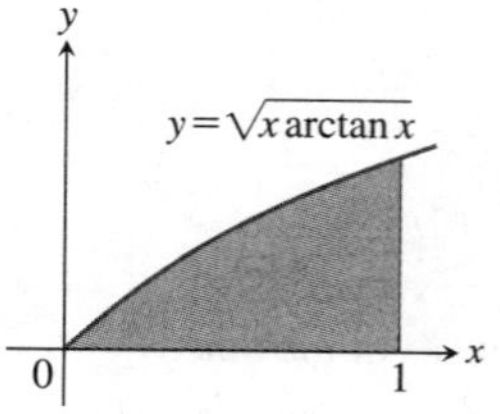

61. Evaluate $\int x^3\sqrt{1-x^2}\,dx$ using
   a. integration by parts.
   b. a u-substitution.
   c. a trigonometric substitution.

62. **Path of a water skier** Suppose that a boat is positioned at the origin with a water skier tethered to the boat at the point $(30, 0)$ on a rope 30 ft long. As the boat travels along the positive y-axis, the skier is pulled behind the boat along an unknown path $y=f(x)$, as shown in the accompanying figure.
   a. Show that $f'(x) = \frac{-\sqrt{900-x^2}}{x}$.
   (*Hint*: Assume that the skier is always pointed directly at the boat and the rope is on a line tangent to the path $y=f(x)$.)
   b. Solve the equation in part (a) for $f(x)$, using $f(30)=0$.

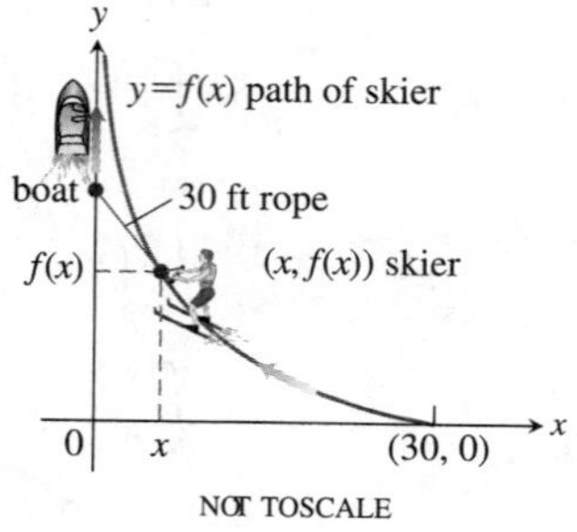

63. Find the average value of $f(x)=\frac{\sqrt{x+1}}{\sqrt{x}}$ on the interval $[1, 3]$.

64. Find the length of the curve $y=1-e^{-x}$, $0\leq x\leq 1$.

8.5 Integration of Rational Functions by Partial Fractions

This section shows how to express a rational function (a quotient of polynomials) as a sum of simpler fractions, called *partial fractions*, which are more easily integrated. For instance, the rational function $(5x - 3)/(x^2 - 2x - 3)$ can be rewritten as

$$\frac{5x-3}{x^2-2x-3} = \frac{2}{x+1} + \frac{3}{x-3}.$$

You can verify this equation algebraically by placing the fractions on the right side over a common denominator $(x + 1)(x - 3)$. The skill acquired in writing rational functions as such a sum is useful in other settings as well (for instance, when using certain transform methods to solve differential equations). To integrate the rational function $(5x - 3)/(x^2 - 2x - 3)$ on the left side of the expression we are considering, we simply sum the integrals of the fractions on the right side:

$$\begin{aligned}\int \frac{5x-3}{(x+1)(x-3)}\,dx &= \int \frac{2}{x+1}\,dx + \int \frac{3}{x-3}\,dx \\ &= 2\ln|x+1| + 3\ln|x-3| + C.\end{aligned}$$

The method for rewriting rational functions as a sum of simpler fractions is called **the method of partial fractions**. In the case of our example, it consists of finding constants A and B such that

$$\frac{5x-3}{x^2-2x-3} = \frac{A}{x+1} + \frac{B}{x-3}. \tag{1}$$

(Pretend for a moment that we do not know that $A = 2$ and $B = 3$ will work.) We call the fractions $A/(x + 1)$ and $B/(x - 3)$ **partial fractions** because their denominators are only part of the original denominator $x^2 - 2x - 3$. We call A and B **undetermined coefficients** until suitable values for them have been found.

To find A and B, we first clear Equation (1) of fractions and regroup in powers of x, obtaining

$$5x - 3 = A(x - 3) + B(x + 1) = (A + B)x - 3A + B.$$

This will be an identity in x if and only if the coefficients of like powers of x on the two sides are equal:

$$A + B = 5, \qquad -3A + B = -3.$$

Solving these equations simultaneously gives $A = 2$ and $B = 3$.

General Description of the Method

Success in writing a rational function $f(x)/g(x)$ as a sum of partial fractions depends on three things:

- *The degree of $f(x)$ must be less than the degree of $g(x)$.* That is, the fraction must be proper. If it isn't, divide $f(x)$ by $g(x)$ and work with the remainder term. Example 3 of this section illustrates such a case.
- We must know the factors of $g(x)$. In theory, any polynomial with real coefficients can be written as a product of real linear factors and real quadratic factors. In practice, the factors may be hard to find.
- The values of the undetermined coefficients form a system of n linear equations in n unknowns. For large n, solving such systems may require linear algebra methods (such as Gaussian Elimination).

Here is how we find the partial fractions of a proper fraction $f(x)/g(x)$ when the factors of g are known. A quadratic polynomial (or factor) is **irreducible** if it cannot be written as the product of two linear factors with real coefficients. That is, the polynomial has no real roots.

Method of Partial Fractions When $f(x)/g(x)$ Is Proper

1. Let $x - r$ be a linear factor of $g(x)$. Suppose that $(x - r)^m$ is the highest power of $x - r$ that divides $g(x)$. Then, to this factor, assign the sum of the m partial fractions:
$$\frac{A_1}{(x-r)} + \frac{A_2}{(x-r)^2} + \cdots + \frac{A_m}{(x-r)^m}.$$
Do this for each distinct linear factor of $g(x)$.
2. Let $x^2 + px + q$ be an irreducible quadratic factor of $g(x)$. In this case, $x^2 + px + q$ has no real roots. Suppose that $(x^2 + px + q)^n$ is the highest power of this factor that divides $g(x)$. Then, to this factor, assign the sum of the n partial fractions:
$$\frac{B_1x + C_1}{(x^2+px+q)} + \frac{B_2x + C_2}{(x^2+px+q)^2} + \cdots + \frac{B_nx + C_n}{(x^2+px+q)^n}.$$
Do this for each distinct quadratic factor of $g(x)$.
3. Set the original fraction $f(x)/g(x)$ equal to the sum of all these partial fractions. Clear the resulting equation of fractions.
4. Find the values of the undetermined coefficients.

There are often multiple ways to find the values of the undetermined coefficients in Step 4. To find the values of the coefficients that satisfy Equation (1), we equated coefficients of like powers of x. In the next example, we instead will assign convenient values of x, leading to simple equations that we can solve for the undetermined coefficients.

EXAMPLE 1 Use partial fractions to evaluate
$$\int \frac{x^2 + 4x + 1}{(x-1)(x+1)(x+3)}\,dx.$$

Solution Note that each of the factors $(x - 1)$, $(x + 1)$, and $(x + 3)$ is raised only to the first power. Therefore, the partial fraction decomposition has the form
$$\frac{x^2 + 4x + 1}{(x-1)(x+1)(x+3)} = \frac{A}{x-1} + \frac{B}{x+1} + \frac{C}{x+3}.$$
To find the values of the undetermined coefficients A, B, and C, we clear fractions and get
$$x^2 + 4x + 1 = A(x+1)(x+3) + B(x-1)(x+3) + C(x-1)(x+1).$$
On the right side, we notice that a factor $(x - 1)$ is present in all terms except for the one containing A. Therefore, letting $x = 1$ allows us to solve for A.
$$x = 1: \qquad 1^2 + 4(1) + 1 = A(2)(4) + B(0) + C(0)$$
$$6 = 8A$$
$$A = \frac{3}{4}$$

In a similar manner, we can let x equal -1 to find B or -3 to find C.

$$x = -1: \qquad (-1)^2 + 4(-1) + 1 = A(0) + B(-2)(2) + C(0)$$
$$-2 = -4B$$
$$B = \frac{1}{2}$$

$$x = -3: \qquad (-3)^2 + 4(-3) + 1 = A(0) + B(0) + C(-4)(-2)$$
$$-2 = 8C$$
$$C = -\frac{1}{4}$$

Hence we have

$$\int \frac{x^2 + 4x + 1}{(x-1)(x+1)(x+3)}\,dx = \int \left[\frac{3}{4}\frac{1}{x-1} + \frac{1}{2}\frac{1}{x+1} - \frac{1}{4}\frac{1}{x+3}\right]dx$$
$$= \frac{3}{4}\ln|x-1| + \frac{1}{2}\ln|x+1| - \frac{1}{4}\ln|x+3| + K,$$

where K is the arbitrary constant of integration (we call it K here to avoid confusion with the undetermined coefficient we labeled as C). ■

You can solve for the undetermined coefficients (A, B, etc.) by equating coefficients of like powers of x or by assigning convenient values to x. You should choose the method that is most convenient for the problem at hand.

EXAMPLE 2 Use partial fractions to evaluate

$$\int \frac{6x + 7}{(x+2)^2}\,dx.$$

Solution First we express the integrand as a sum of partial fractions with undetermined coefficients.

$$\frac{6x+7}{(x+2)^2} = \frac{A}{x+2} + \frac{B}{(x+2)^2} \qquad \text{Two terms because } (x+2) \text{ is squared}$$
$$6x + 7 = A(x+2) + B \qquad \text{Multiply both sides by } (x+2)^2.$$
$$= Ax + (2A + B)$$

Equating coefficients of corresponding powers of x gives

$$A = 6 \quad \text{and} \quad 2A + B = 12 + B = 7, \quad \text{or} \quad A = 6 \quad \text{and} \quad B = -5.$$

Therefore,

$$\int \frac{6x+7}{(x+2)^2}\,dx = \int \left(\frac{6}{x+2} - \frac{5}{(x+2)^2}\right)dx$$
$$= 6\int \frac{dx}{x+2} - 5\int (x+2)^{-2}\,dx$$
$$= 6\ln|x+2| + 5(x+2)^{-1} + C.$$ ■

The next example shows how to handle the case when $f(x)/g(x)$ is an improper fraction. It is a case where the degree of f is larger than the degree of g.

EXAMPLE 3 Use partial fractions to evaluate

$$\int \frac{2x^3 - 4x^2 - x - 3}{x^2 - 2x - 3}\,dx.$$

Solution First we divide the denominator into the numerator to get a polynomial plus a proper fraction.

$$\begin{array}{r} 2x \\ x^2 - 2x - 3\,\overline{)\,2x^3 - 4x^2 - x - 3} \\ \underline{2x^3 - 4x^2 - 6x} \\ 5x - 3 \end{array}$$

Then we write the improper fraction as a polynomial plus a proper fraction.

$$\frac{2x^3 - 4x^2 - x - 3}{x^2 - 2x - 3} = 2x + \frac{5x - 3}{x^2 - 2x - 3}$$

We found the partial fraction decomposition of the fraction on the right in the opening example, so

$$\begin{aligned} \int \frac{2x^3 - 4x^2 - x - 3}{x^2 - 2x - 3}\,dx &= \int 2x\,dx + \int \frac{5x - 3}{x^2 - 2x - 3}\,dx \\ &= \int 2x\,dx + \int \frac{2}{x + 1}\,dx + \int \frac{3}{x - 3}\,dx \\ &= x^2 + 2\ln|x + 1| + 3\ln|x - 3| + C. \end{aligned}$$

■

EXAMPLE 4 Use partial fractions to evaluate

$$\int \frac{-2x + 4}{(x^2 + 1)(x - 1)^2}\,dx.$$

Solution The denominator has an irreducible quadratic factor $x^2 + 1$ as well as a repeated linear factor $(x - 1)^2$, so we write

$$\frac{-2x + 4}{(x^2 + 1)(x - 1)^2} = \frac{Ax + B}{x^2 + 1} + \frac{C}{x - 1} + \frac{D}{(x - 1)^2}. \tag{2}$$

Clearing the equation of fractions gives

$$\begin{aligned} -2x + 4 &= (Ax + B)(x - 1)^2 + C(x - 1)(x^2 + 1) + D(x^2 + 1) \\ &= (A + C)x^3 + (-2A + B - C + D)x^2 \\ &\quad + (A - 2B + C)x + (B - C + D). \end{aligned}$$

Equating coefficients of like terms gives

$$\begin{aligned} &\text{Coefficients of } x^3: & 0 &= A + C \\ &\text{Coefficients of } x^2: & 0 &= -2A + B - C + D \\ &\text{Coefficients of } x^1: & -2 &= A - 2B + C \\ &\text{Coefficients of } x^0: & 4 &= B - C + D \end{aligned}$$

We solve these equations simultaneously to find the values of A, B, C, and D.

$$\begin{aligned} &-4 = -2A, \quad A = 2 && \text{Subtract fourth equation from second.} \\ &C = -A = -2 && \text{From the first equation} \\ &B = (A + C + 2)/2 = 1 && \text{From the third equation and } C = -A \\ &D = 4 - B + C = 1. && \text{From the fourth equation} \end{aligned}$$

We substitute these values into Equation (2), obtaining

$$\frac{-2x+4}{(x^2+1)(x-1)^2} = \frac{2x+1}{x^2+1} - \frac{2}{x-1} + \frac{1}{(x-1)^2}.$$

Finally, using the expansion above, we can integrate:

$$\begin{aligned}\int \frac{-2x+4}{(x^2+1)(x-1)^2}\,dx &= \int\left(\frac{2x+1}{x^2+1} - \frac{2}{x-1} + \frac{1}{(x-1)^2}\right)dx\\ &= \int\left(\frac{2x}{x^2+1} + \frac{1}{x^2+1} - \frac{2}{x-1} + \frac{1}{(x-1)^2}\right)dx\\ &= \ln(x^2+1) + \tan^{-1}x - 2\ln|x-1| - \frac{1}{x-1} + K.\end{aligned}$$

We use the letter K instead of C to represent an arbitrary constant here because we have already used C to represent a variable in the partial fraction representation. ■

EXAMPLE 5 Use partial fractions to evaluate

$$\int \frac{dx}{x(x^2+1)^2}.$$

Solution The form of the partial fraction decomposition is

$$\frac{1}{x(x^2+1)^2} = \frac{A}{x} + \frac{Bx+C}{x^2+1} + \frac{Dx+E}{(x^2+1)^2}.$$

Multiplying by $x(x^2+1)^2$, we have

$$\begin{aligned}1 &= A(x^2+1)^2 + (Bx+C)x(x^2+1) + (Dx+E)x\\ &= A(x^4+2x^2+1) + B(x^4+x^2) + C(x^3+x) + Dx^2 + Ex\\ &= (A+B)x^4 + Cx^3 + (2A+B+D)x^2 + (C+E)x + A.\end{aligned}$$

If we equate coefficients, we get the system

$$A+B=0, \quad C=0, \quad 2A+B+D=0, \quad C+E=0, \quad A=1.$$

Solving this system gives $A=1$, $B=-1$, $C=0$, $D=-1$, and $E=0$. Thus,

$$\begin{aligned}\int \frac{dx}{x(x^2+1)^2} &= \int\left[\frac{1}{x} + \frac{-x}{x^2+1} + \frac{-x}{(x^2+1)^2}\right]dx\\ &= \int\frac{dx}{x} - \int\frac{x\,dx}{x^2+1} - \int\frac{x\,dx}{(x^2+1)^2}\\ &= \int\frac{dx}{x} - \frac{1}{2}\int\frac{du}{u} - \frac{1}{2}\int\frac{du}{u^2} \qquad u = x^2+1,\ du = 2x\,dx\\ &= \ln|x| - \frac{1}{2}\ln|u| + \frac{1}{2u} + K\\ &= \ln|x| - \frac{1}{2}\ln(x^2+1) + \frac{1}{2(x^2+1)} + K\\ &= \ln\frac{|x|}{\sqrt{x^2+1}} + \frac{1}{2(x^2+1)} + K.\end{aligned}$$

■

HISTORICAL BIOGRAPHY
Oliver Heaviside
(1850–1925)
www.bit.ly/2QoRGIV

Determining Coefficients by Differentiating

Another way to determine the constants that appear in partial fractions is to differentiate, as in the next example.

EXAMPLE 6 Find A, B, and C in the equation

$$\frac{x-1}{(x+1)^3} = \frac{A}{x+1} + \frac{B}{(x+1)^2} + \frac{C}{(x+1)^3}.$$

Solution We first clear fractions:

$$x - 1 = A(x+1)^2 + B(x+1) + C.$$

Substituting $x = -1$ shows $C = -2$. We then differentiate both sides with respect to x, obtaining

$$1 = 2A(x+1) + B.$$

Substituting $x = -1$ shows $B = 1$. We differentiate again to get $0 = 2A$, which shows $A = 0$. Hence,

$$\frac{x-1}{(x+1)^3} = \frac{1}{(x+1)^2} - \frac{2}{(x+1)^3}.$$ ■

EXERCISES 8.5

Expanding Quotients into Partial Fractions

Expand the quotients in Exercises 1–8 by partial fractions.

1. $\dfrac{5x-13}{(x-3)(x-2)}$
2. $\dfrac{5x-7}{x^2-3x+2}$
3. $\dfrac{x+4}{(x+1)^2}$
4. $\dfrac{2x+2}{x^2-2x+1}$
5. $\dfrac{z+1}{z^2(z-1)}$
6. $\dfrac{z}{z^3-z^2-6z}$
7. $\dfrac{t^2+8}{t^2-5t+6}$
8. $\dfrac{t^4+9}{t^4+9t^2}$

Nonrepeated Linear Factors

In Exercises 9–16, express the integrand as a sum of partial fractions and evaluate the integrals.

9. $\displaystyle\int \frac{dx}{(x-1)(x-2)}$
10. $\displaystyle\int \frac{dx}{x(x^7+1)}, \; u = x^7$
11. $\displaystyle\int \frac{x+4}{x^2+5x-6}\,dx$
12. $\displaystyle\int \frac{2x+1}{x^2-7x+12}\,dx$
13. $\displaystyle\int_4^8 \frac{y\,dy}{y^2-2y-3}$
14. $\displaystyle\int_0^{\pi/2} \frac{\cos x}{(1+\sin x)(2+\sin x)}\,dx, \; u = \sin x$
15. $\displaystyle\int \frac{dt}{t^3+t^2-2t}$
16. $\displaystyle\int \frac{x+3}{2x^3-8x}\,dx$

Repeated Linear Factors

In Exercises 17–20, express the integrand as a sum of partial fractions and evaluate the integrals.

17. $\displaystyle\int_0^1 \frac{x^3\,dx}{x^2+2x+1}$
18. $\displaystyle\int_{-1}^0 \frac{x^3\,dx}{x^2-2x+1}$
19. $\displaystyle\int \frac{dx}{(x^2-1)^2}$
20. $\displaystyle\int \frac{x^2\,dx}{(x-1)(x^2+2x+1)}$

Irreducible Quadratic Factors

In Exercises 21–32, express the integrand as a sum of partial fractions and evaluate the integrals.

21. $\displaystyle\int_0^1 \frac{dx}{(x+1)(x^2+1)}$
22. $\displaystyle\int_0^1 \frac{dx}{1+x^3}$
23. $\displaystyle\int \frac{y^2+2y+1}{(y^2+1)^2}\,dy$
24. $\displaystyle\int \frac{8x^2+8x+2}{(4x^2+1)^2}\,dx$
25. $\displaystyle\int \frac{2s+2}{(s^2+1)(s-1)^3}\,ds$
26. $\displaystyle\int \frac{s^4+81}{s(s^2+9)^2}\,ds$
27. $\displaystyle\int \frac{x^2-x+2}{x^3-1}\,dx$
28. $\displaystyle\int \frac{1}{x^4+x}\,dx$
29. $\displaystyle\int \frac{x^2}{x^4-1}\,dx$
30. $\displaystyle\int \frac{x^2+x}{x^4-3x^2-4}\,dx$
31. $\displaystyle\int \frac{2\theta^3+5\theta^2+8\theta+4}{(\theta^2+2\theta+2)^2}\,d\theta$
32. $\displaystyle\int \frac{\theta^4-4\theta^3+2\theta^2-3\theta+1}{(\theta^2+1)^3}\,d\theta$

Improper Fractions

In Exercises 33–38, perform long division on the integrand, write the proper fraction as a sum of partial fractions, and then evaluate the integral.

33. $\int \frac{2x^3 - 2x^2 + 1}{x^2 - x}\,dx$ **34.** $\int \frac{x^4}{x^2 - 1}\,dx$

35. $\int \frac{9x^3 - 3x + 1}{x^3 - x^2}\,dx$ **36.** $\int \frac{16x^3}{4x^2 - 4x + 1}\,dx$

37. $\int \frac{y^4 + y^2 - 1}{y^3 + y}\,dy$ **38.** $\int \frac{2y^4}{y^3 - y^2 + y - 1}\,dy$

Evaluating Integrals

Evaluate the integrals in Exercises 39–54.

39. $\int \frac{e^t\,dt}{e^{2t} + 3e^t + 2}$ **40.** $\int \frac{e^{4t} + 2e^{2t} - e^t}{e^{2t} + 1}\,dt$

41. $\int \frac{\cos y\,dy}{\sin^2 y + \sin y - 6}$ **42.** $\int \frac{\sin\theta\,d\theta}{\cos^2\theta + \cos\theta - 2}$

43. $\int \frac{(x-2)^2\tan^{-1}(2x) - 12x^3 - 3x}{(4x^2 + 1)(x - 2)^2}\,dx$

44. $\int \frac{(x+1)^2\tan^{-1}(3x) + 9x^3 + x}{(9x^2 + 1)(x + 1)^2}\,dx$

45. $\int \frac{1}{x^{3/2} - \sqrt{x}}\,dx$ **46.** $\int \frac{1}{(x^{1/3} - 1)\sqrt{x}}\,dx$

(*Hint:* Let $x = u^6$.)

47. $\int \frac{\sqrt{x + 1}}{x}\,dx$ **48.** $\int \frac{1}{x\sqrt{x + 9}}\,dx$

(*Hint:* Let $x + 1 = u^2$.)

49. $\int \frac{1}{x(x^4 + 1)}\,dx$ **50.** $\int \frac{1}{x^6(x^5 + 4)}\,dx$

$\left(\textit{Hint: } \text{Multiply by } \frac{x^3}{x^3}.\right)$

51. $\int \frac{1}{\cos 2\theta \sin\theta}\,d\theta$ **52.** $\int \frac{1}{\cos\theta + \sin 2\theta}\,d\theta$

53. $\int \frac{\sqrt{1 + \sqrt{x}}}{x}\,dx$ **54.** $\int \frac{\sqrt{x}}{\sqrt{2 - \sqrt{x} + \sqrt{x}}}\,dx$

Use any method to evaluate the integrals in Exercises 55–66.

55. $\int \frac{x^3 - 2x^2 - 3x}{x + 2}\,dx$ **56.** $\int \frac{x + 2}{x^3 - 2x^2 - 3x}\,dx$

57. $\int \frac{2^x - 2^{-x}}{2^x + 2^{-x}}\,dx$ **58.** $\int \frac{2^x}{2^{2x} + 2^x - 2}\,dx$

59. $\int \frac{1}{x^4 - 1}\,dx$ **60.** $\int \frac{x^4 - 1}{x^5 - 5x + 1}\,dx$

61. $\int \frac{\ln x + 2}{x(\ln x + 1)(\ln x + 3)}\,dx$

62. $\int \frac{2}{x(\ln x - 2)^3}\,dx$

63. $\int \frac{1}{\sqrt{x^2 - 1}}\,dx$ **64.** $\int \frac{x}{x + \sqrt{x^2 + 2}}\,dx$

65. $\int x^5\sqrt{x^3 + 1}\,dx$ **66.** $\int x^2\sqrt{1 - x^2}\,dx$

Initial Value Problems

Solve the initial value problems in Exercises 67–70 for x as a function of t.

67. $(t^2 - 3t + 2)\frac{dx}{dt} = 1 \quad (t > 2), \quad x(3) = 0$

68. $(3t^4 + 4t^2 + 1)\frac{dx}{dt} = 2\sqrt{3}, \quad x(1) = -\pi\sqrt{3}/4$

69. $(t^2 + 2t)\frac{dx}{dt} = 2x + 2 \quad (t, x > 0), \quad x(1) = 1$

70. $(t + 1)\frac{dx}{dt} = x^2 + 1 \quad (t > -1), \quad x(0) = 0$

Applications and Examples

In Exercises 71 and 72, find the volume of the solid generated by revolving the shaded region about the indicated axis.

71. The x-axis

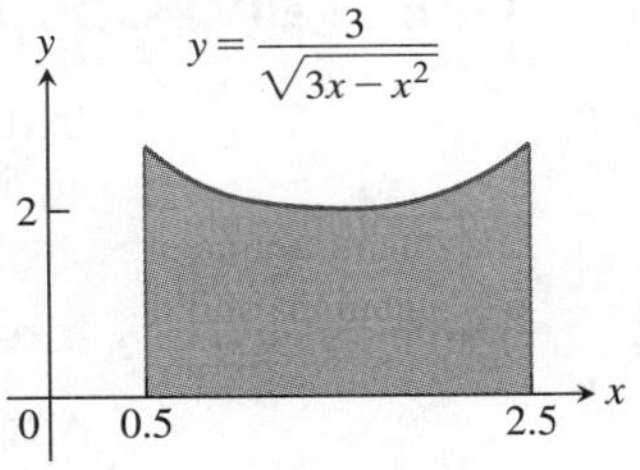

72. The y-axis

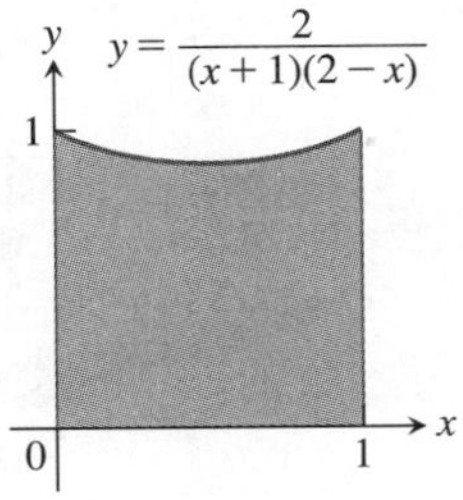

73. Find the length of the curve $y = \ln(1 - x^2)$, $0 \le x \le \frac{1}{2}$.

74. Evaluate $\int \sec\theta\,d\theta$ by

a. multiplying by $\frac{\sec\theta + \tan\theta}{\sec\theta + \tan\theta}$ and then using a u-substitution.

b. writing the integral as $\int \frac{1}{\cos\theta}\,d\theta$. Then multiply by $\frac{\cos\theta}{\cos\theta}$, use a trigonometric identity and a u-substitution, and finally integrate using partial fractions.

T **75.** Find, to two decimal places, the x-coordinate of the centroid of the region in the first quadrant bounded by the x-axis, the curve $y = \arctan x$, and the line $x = \sqrt{3}$.

T **76.** Find the x-coordinate of the centroid of this region to two decimal places.

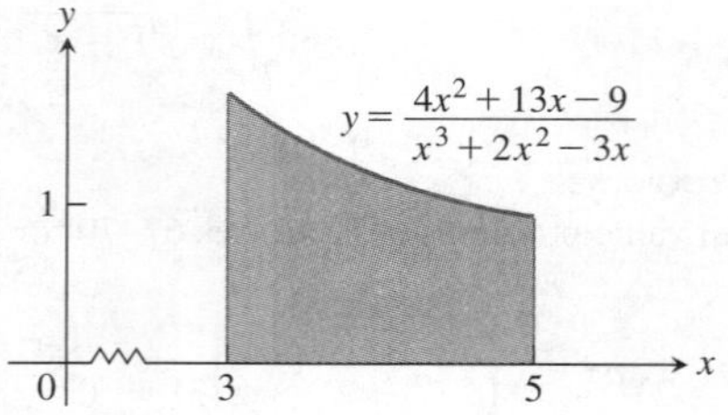

T **77. Social diffusion** Sociologists sometimes use the phrase "social diffusion" to describe the way information spreads through a population. The information might be a rumor, a cultural fad, or news about a technical innovation. In a sufficiently large population, the number of people x who have the information is treated as a differentiable function of time t, and the rate of diffusion, dx/dt, is assumed to be proportional to the number of people who have the information times the number of people who do not. This leads to the equation

$$\frac{dx}{dt} = kx(N - x),$$

where N is the number of people in the population.

Suppose t is in days, $k = 1/250$, and two people start a rumor at time $t = 0$ in a population of $N = 1000$ people.

a. Find x as a function of t.

b. When will half the population have heard the rumor? (This is when the rumor will be spreading the fastest.)

T **78. Second-order chemical reactions** Many chemical reactions are the result of the interaction of two molecules that undergo a change to produce a new product. The rate of the reaction typically depends on the concentrations of the two kinds of molecules. If a is the amount of substance A and b is the amount of substance B at time $t = 0$, and if x is the amount of product at time t, then the rate of formation of x may be given by the differential equation

$$\frac{dx}{dt} = k(a - x)(b - x),$$

or

$$\frac{1}{(a - x)(b - x)}\frac{dx}{dt} = k,$$

where k is a constant for the reaction. Integrate both sides of this equation to obtain a relation between x and t **(a)** if $a = b$, and **(b)** if $a \neq b$. Assume in each case that $x = 0$ when $t = 0$.

8.6 Integral Tables and Computer Algebra Systems

In this section we discuss how to use tables and computer algebra systems (CAS) to evaluate integrals.

Integral Tables

A Brief Table of Integrals is provided at the back of the text, after the index. (More extensive tables appear in compilations such as *CRC Mathematical Tables*, which contain thousands of integrals.) The integration formulas are stated in terms of constants a, b, c, m, n, and so on. These constants can usually assume any real value and need not be integers. Occasional limitations on their values are stated with the formulas. Formula 21 requires $n \neq -1$, for example, and Formula 27 requires $n \neq -2$.

The formulas also assume that the constants do not take on values that require dividing by zero or taking even roots of negative numbers. For example, Formula 24 assumes that $a \neq 0$, and Formulas 29a and 29b cannot be used unless b is positive.

EXAMPLE 1 Find

$$\int x(2x + 5)^{-1}\,dx.$$

Solution We use Formula 24 at the back of the text (not 22, which requires $n \neq -1$):

$$\int x(ax + b)^{-1}\,dx = \frac{x}{a} - \frac{b}{a^2}\ln|ax + b| + C.$$

With $a = 2$ and $b = 5$, we have

$$\int x(2x + 5)^{-1}\,dx = \frac{x}{2} - \frac{5}{4}\ln|2x + 5| + C.$$ ■

EXAMPLE 2 Find

$$\int \frac{dx}{x\sqrt{2x-4}}.$$

Solution We use Formula 29b:

$$\int \frac{dx}{x\sqrt{ax-b}} = \frac{2}{\sqrt{b}}\arctan\sqrt{\frac{ax-b}{b}} + C.$$

With $a = 2$ and $b = 4$, we have

$$\int \frac{dx}{x\sqrt{2x-4}} = \frac{2}{\sqrt{4}}\arctan\sqrt{\frac{2x-4}{4}} + C = \arctan\sqrt{\frac{x-2}{2}} + C. \qquad \blacksquare$$

EXAMPLE 3 Find

$$\int x \arcsin x \, dx.$$

Solution We begin by using Formula 106:

$$\int x^n \arcsin ax \, dx = \frac{x^{n+1}}{n+1}\arcsin ax - \frac{a}{n+1}\int \frac{x^{n+1}\,dx}{\sqrt{1-a^2x^2}}, \qquad n \neq -1.$$

With $n = 1$ and $a = 1$, we have

$$\int x \arcsin x \, dx = \frac{x^2}{2}\arcsin x - \frac{1}{2}\int \frac{x^2\,dx}{\sqrt{1-x^2}}.$$

Next we use Formula 49 to find the integral on the right:

$$\int \frac{x^2}{\sqrt{a^2-x^2}}\,dx = \frac{a^2}{2}\arcsin\left(\frac{x}{a}\right) - \frac{1}{2}x\sqrt{a^2-x^2} + C.$$

With $a = 1$,

$$\int \frac{x^2\,dx}{\sqrt{1-x^2}} = \frac{1}{2}\arcsin x - \frac{1}{2}x\sqrt{1-x^2} + C.$$

The combined result is

$$\begin{aligned}\int x \arcsin x \, dx &= \frac{x^2}{2}\arcsin x - \frac{1}{2}\left(\frac{1}{2}\arcsin x - \frac{1}{2}x\sqrt{1-x^2} + C\right)\\ &= \left(\frac{x^2}{2} - \frac{1}{4}\right)\arcsin x + \frac{1}{4}x\sqrt{1-x^2} + C'.\end{aligned} \qquad \blacksquare$$

Reduction Formulas

The time required for repeated integrations by parts can sometimes be shortened by applying reduction formulas like the following.

$$\int \tan^n x \, dx = \frac{1}{n-1}\tan^{n-1} x - \int \tan^{n-2} x \, dx \tag{1}$$

$$\int (\ln x)^n \, dx = x(\ln x)^n - n\int (\ln x)^{n-1}\, dx \tag{2}$$

$$\int \sin^n x \cos^m x \, dx = -\frac{\sin^{n-1} x \cos^{m+1} x}{m+n} + \frac{n-1}{m+n}\int \sin^{n-2} x \cos^m x \, dx \quad (n \neq -m). \tag{3}$$

By applying such a formula repeatedly, we can eventually express the original integral in terms of a power low enough to be evaluated directly. The next example illustrates this procedure.

EXAMPLE 4 Find

$$\int \tan^5 x\, dx.$$

Solution We apply Equation (1) with $n = 5$ to get

$$\int \tan^5 x\, dx = \frac{1}{4}\tan^4 x - \int \tan^3 x\, dx.$$

We then apply Equation (1) again, with $n = 3$, to evaluate the remaining integral:

$$\int \tan^3 x\, dx = \frac{1}{2}\tan^2 x - \int \tan x\, dx = \frac{1}{2}\tan^2 x + \ln|\cos x| + C_1.$$

The combined result is

$$\int \tan^5 x\, dx = \frac{1}{4}\tan^4 x - \frac{1}{2}\tan^2 x - \ln|\cos x| + C.$$ ■

As their form suggests, reduction formulas are derived using integration by parts. (See Example 5 in Section 8.3.)

Integration with a CAS

A powerful capability of computer algebra systems is their ability to integrate symbolically. This is performed with the **integrate command** specified by the particular system (for example, **int** in Maple, **Integrate** in Mathematica).

EXAMPLE 5 Suppose that you want to evaluate the indefinite integral of the function

$$f(x) = x^2\sqrt{a^2 + x^2}.$$

Using Maple, you first define or name the function:

```
> f:= x^2 * sqrt(a^2 + x^2);
```

Then you use the integrate command on f, identifying the variable of integration:

```
> int(f, x);
```

Maple returns the answer

$$\frac{x(a^2 + x^2)^{3/2}}{4} - \frac{a^2 x\sqrt{a^2 + x^2}}{8} - \frac{a^4 \ln\left(x + \sqrt{a^2 + x^2}\right)}{8}$$

If you want to see whether the answer can be simplified, enter

```
> simplify(%);
```

Maple returns

$$-\frac{a^4 \ln\left(x + \sqrt{a^2 + x^2}\right)}{8} + \frac{x\sqrt{a^2 + x^2}(a^2 + 2x^2)}{8}$$

If you want the definite integral for $0 \le x \le \pi/2$, you can use the format

```
> int(f, x = 0..Pi/2);
```

Maple will return the expression

$$\frac{a^4 \ln(a^2)}{16} + \frac{\pi(\pi^2 + 4a^2)^{3/2}}{64} - \frac{a^2\pi\sqrt{\pi^2 + 4a^2}}{32} + \frac{a^4 \ln(2)}{8}$$
$$- \frac{a^4 \ln\left(\pi + \sqrt{\pi^2 + 4a^2}\right)}{8}$$

You can also find the definite integral for a particular value of the constant a:

$$> a := 1;$$
$$> \text{int}(f, x = 0..1);$$

Maple returns the numerical answer

$$\frac{3}{8}\sqrt{2} + \frac{1}{8}\ln(\sqrt{2} - 1).$$ ■

EXAMPLE 6 Use a CAS to find

$$\int \sin^2 x \cos^3 x \, dx.$$

Solution With Maple, we have the entry

$$> \text{int}((\sin\text{^}2)(x) * (\cos\text{^}3)(x), x);$$

with the immediate return

$$-\frac{1}{5}\sin(x)\cos(x)^4 + \frac{1}{15}\cos(x)^2\sin(x) + \frac{2}{15}\sin(x).$$ ■

Computer algebra systems vary in how they process integrations. We used Maple in Examples 5 and 6. Mathematica would have returned somewhat different results:

1. In Example 5, given

$$In[1] := \text{Integrate}[x\text{^}2 * \text{Sqrt}[a\text{^}2 + x\text{^}2], x]$$

Mathematica returns

$$Out[1] = \frac{1}{8}\sqrt{a^2 + x^2}\left(a^2x + 2x^3 - \frac{a^3 \sinh^{-1}\left(\frac{x}{a}\right)}{\sqrt{1 + \frac{x^2}{a^2}}}\right)$$

without having to simplify an intermediate result. The answer is different from, but equivalent to, Formula 36 in the integral tables.

2. The Mathematica answer to the integral

$$In[2] := \text{Integrate}[\text{Sin}[x]\text{^}2 * \text{Cos}[x]\text{^}3, x]$$

in Example 6 is

$$Out[2] = \frac{\text{Sin}[x]}{8} - \frac{1}{48}\text{Sin}[3x] - \frac{1}{80}\text{Sin}[5x]$$

differing from the Maple answer. Both answers are correct.

Although a CAS is very powerful and can aid us in solving difficult problems, each CAS has its own limitations. There are even situations where a CAS may further complicate a problem (in the sense of producing an answer that is extremely difficult to use or interpret). Note, too, that neither Maple nor Mathematica returns an arbitrary constant $+C$. On the other hand, a little mathematical thinking on your part may reduce the problem to one that is quite easy to handle. We provide an example in Exercise 67.

Nonelementary Integrals

Many functions have antiderivatives that cannot be expressed using the standard functions that we have encountered, such as polynomials, trigonometric functions, and exponential functions. Integrals of functions that do not have elementary antiderivatives are called **nonelementary** integrals. These integrals can sometimes be expressed with infinite series (Chapter 10) or approximated using numerical methods (Section 8.7). Examples of nonelementary integrals include the error function (which measures the probability of random errors)

$$\operatorname{erf}(x) = \frac{2}{\sqrt{\pi}} \int_0^x e^{-t^2}\, dt$$

and integrals such as

$$\int \sin x^2\, dx \qquad \text{and} \qquad \int \sqrt{1 + x^4}\, dx$$

that arise in engineering and physics. These and a number of others, such as

$$\int \frac{e^x}{x}\, dx, \qquad \int e^{(e^x)}\, dx, \qquad \int \frac{1}{\ln x}\, dx, \qquad \int \ln(\ln x)\, dx, \qquad \int \frac{\sin x}{x}\, dx,$$

$$\int \sqrt{1 - k^2 \sin^2 x}\, dx, \qquad 0 < k < 1,$$

look so easy they tempt us to try them just to see how they turn out. It can be proved, however, that there is no way to express any of these integrals as finite combinations of elementary functions. The same applies to integrals that can be changed into these by substitution. The functions in these integrals all have antiderivatives, as a consequence of the Fundamental Theorem of Calculus, Part 1, because they are continuous. However, none of the antiderivatives are elementary. The integrals you are asked to evaluate in this chapter have elementary antiderivatives.

EXERCISES 8.6

Using Integral Tables

Use the table of integrals at the back of the text to evaluate the integrals in Exercises 1–26.

1. $\displaystyle\int \frac{dx}{x\sqrt{3x-5}}$

2. $\displaystyle\int \frac{dx}{x\sqrt{x+4}}$

3. $\displaystyle\int \frac{x\, dx}{\sqrt{x-2}}$

4. $\displaystyle\int \frac{x dx}{(4x+3)^{3/2}}$

5. $\displaystyle\int x\sqrt{2x-3}\, dx$

6. $\displaystyle\int x(7x+5)^{3/2}\, dx$

7. $\displaystyle\int \frac{\sqrt{16-9x}}{x^2}\, dx$

8. $\displaystyle\int \frac{dx}{x^2\sqrt{4x-9}}$

9. $\displaystyle\int x\sqrt{4x-x^2}\, dx$

10. $\displaystyle\int \frac{\sqrt{6x-x^2}}{x}\, dx$

11. $\displaystyle\int \frac{dx}{x\sqrt{7+x^2}}$

12. $\displaystyle\int_2^3 \frac{dx}{(x-1)\sqrt{x^2-2x}} \quad \left[x-1=\frac{1}{u}\right]$

13. $\displaystyle\int \frac{\sqrt{4-x^2}}{x}\, dx$

14. $\displaystyle\int \frac{\sqrt{x^2-9}}{x}\, dx$

15. $\displaystyle\int e^{3t} \cos 2t\, dt$

16. $\displaystyle\int e^{-3t} \sin 4t\, dt$

17. $\displaystyle\int x \arccos x\, dx$

18. $\displaystyle\int_0^{\pi/2} x \tan^2 x\, dx$

19. $\int x^2 \arctan x\, dx$

20. $\int \frac{\tan^{-1} x}{x^3}\, dx$

21. $\int_0^{\pi/2} \sin 3x \cos x\, dx$

22. $\int \sin 2x \cos 3x\, dx$

23. $\int 4 \cos 4t \cos \frac{t}{2}\, dt$

24. $\int \sin \frac{t}{3} \sin \frac{t}{6}\, dt$

25. $\int_0^{\pi/2} \cos 2x \cos 4x\, dx$

26. $\int \sin 7\theta \sin \frac{\theta}{2}\, d\theta$

Substitution and Integral Tables

In Exercises 27–40, use a substitution to change the integral into one you can find in the table. Then evaluate the integral.

27. $\int \frac{x^3 - x + 1}{(x^2 + 1)^2}\, dx$

28. $\int \frac{x^2 + 4x}{(x^2 + 2)^2}\, dx$

29. $\int \arcsin \sqrt{x}\, dx$

30. $\int \frac{\cos^{-1} \sqrt{x}}{\sqrt{x}}\, dx$

31. $\int \frac{\sqrt{x}}{\sqrt{2 - x}}\, dx$

32. $\int \frac{\sqrt{2 - x}}{\sqrt{x}}\, dx$

33. $\int \cot t \sqrt{1 - \sin^2 t}\, dt, \quad 0 < t < \pi/2$

34. $\int \frac{\sin t\, dt}{\cos t \sqrt{4 - \cos^2 t}}$

35. $\int \frac{dy}{y\sqrt{3 + (\ln y)^2}}$

36. $\int \tan^{-1} \sqrt{y}\, dy$

37. $\int \frac{1}{\sqrt{x^2 + 2x + 5}}\, dx$ (*Hint*: Complete the square.)

38. $\int \frac{x^2}{\sqrt{x^2 - 4x + 5}}\, dx$

39. $\int \sqrt{8 - 2x - x^2}\, dx$

40. $\int x^2 \sqrt{2x - x^2}\, dx$

Using Reduction Formulas

Use reduction formulas to evaluate the integrals in Exercises 41–50.

41. $\int \sin^4 3x\, dx$

42. $\int 8 \cos^4 2\pi t\, dt$

43. $\int \sin^2 2\theta \cos^3 2\theta\, d\theta$

44. $\int 2 \sin^2 t \sec^4 t\, dt$

45. $\int 4 \tan^3 2x\, dx$

46. $\int 9 \cot^5 t\, dt$

47. $\int 2 \sec^3 \pi x\, dx$

48. $\int 3 \sec^4 3x\, dx$

49. $\int \csc^5 x\, dx$

50. $\int 9x^2 (\ln x)^3\, dx$

Evaluate the integrals in Exercises 51–56 by making a substitution (possibly trigonometric) and then applying a reduction formula.

51. $\int e^t \sec^3 (e^t - 1)\, dt$

52. $\int \frac{\csc^4 \sqrt{\theta}}{\sqrt{\theta}}\, d\theta$

53. $\int_0^1 2\sqrt{x^2 + 1}\, dx$

54. $\int_0^{\sqrt{3}/2} \frac{dy}{(1 - y^2)^{5/2}}$

55. $\int_1^2 \frac{(r^2 - 1)^{3/2}}{r}\, dr$

56. $\int_0^{1/\sqrt{3}} \frac{dt}{(t^2 + 1)^{5/2}}$

Applications

57. **Surface area** Find the area of the surface generated by revolving the curve $y = \sqrt{x^2 + 2}$, $0 \le x \le \sqrt{2}$, about the x-axis.

58. **Arc length** Find the length of the curve $y = x^2$, $0 \le x \le \sqrt{3}/2$.

59. **Centroid** Find the centroid of the region cut from the first quadrant by the curve $y = 1/\sqrt{x + 1}$ and the line $x = 3$.

60. **Moment about *y*-axis** A thin plate of constant density $\delta = 1$ occupies the region enclosed by the curve $y = 18/(2x + 5)$ and the line $x = 5$ in the first quadrant. Find the moment of the plate about the y-axis.

T 61. Use the integral table and a calculator to find, to two decimal places, the area of the surface generated by revolving the curve $y = x^2$, $-1 \le x \le 1$, about the x-axis.

62. **Volume** The head of your firm's accounting department has asked you to find a formula she can use in a computer program to calculate the year-end inventory of gasoline in the company's tanks. A typical tank is shaped like a right circular cylinder of radius r and length L, mounted horizontally, as shown in the accompanying figure. The data come to the accounting office as depth measurements taken with a vertical measuring stick marked in centimeters.

   a. Show, in the notation of the figure, that the volume of gasoline that fills the tank to a depth d is
   $$V = 2L \int_{-r}^{-r+d} \sqrt{r^2 - y^2}\, dy.$$

   b. Evaluate the integral.

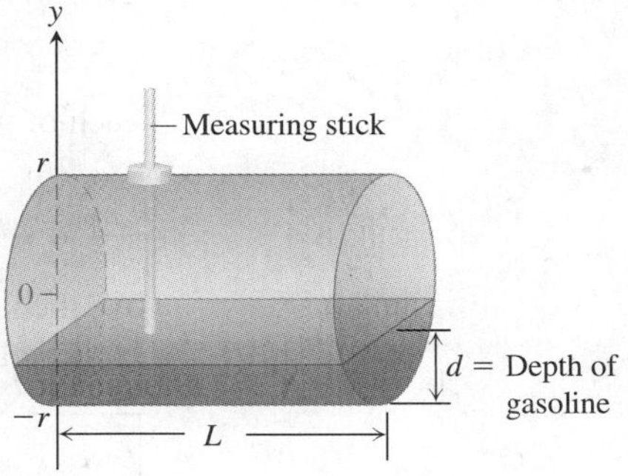

63. What is the largest value
$$\int_a^b \sqrt{2x - x^2}\, dx$$
can have for any a and b? Give reasons for your answer.

64. What is the largest value that

$$\int_a^b x\sqrt{2x - x^2}\, dx$$

can have for any a and b? Give reasons for your answer.

COMPUTER EXPLORATIONS

In Exercises 65 and 66, use a CAS to perform the integrations.

65. Evaluate the integrals

a. $\int x \ln x\, dx$ **b.** $\int x^2 \ln x\, dx$ **c.** $\int x^3 \ln x\, dx.$

d. What pattern do you see? Predict the formula for $\int x^4 \ln x\, dx$ and then see if you are correct by evaluating it with a CAS.

e. What is the formula for $\int x^n \ln x\, dx,\ n \geq 1$? Check your answer using a CAS.

66. Evaluate the integrals

a. $\int \frac{\ln x}{x^2}\, dx$ **b.** $\int \frac{\ln x}{x^3}\, dx$ **c.** $\int \frac{\ln x}{x^4}\, dx.$

d. What pattern do you see? Predict the formula for

$$\int \frac{\ln x}{x^5}\, dx$$

and then see if you are correct by evaluating it with a CAS.

e. What is the formula for

$$\int \frac{\ln x}{x^n}\, dx, \quad n \geq 2?$$

Check your answer using a CAS.

67. a. Use a CAS to evaluate

$$\int_0^{\pi/2} \frac{\sin^n x}{\sin^n x + \cos^n x}\, dx,$$

where n is an arbitrary positive integer. Does your CAS find the result?

b. In succession, find the integral when $n = 1, 2, 3, 5$, and 7. Comment on the complexity of the results.

c. Now substitute $x = (\pi/2) - u$ and add the new and old integrals. What is the value of

$$\int_0^{\pi/2} \frac{\sin^n x}{\sin^n x + \cos^n x}\, dx?$$

This exercise illustrates how a little mathematical ingenuity can sometimes solve a problem not immediately amenable to solution by a CAS.

8.7 Numerical Integration

The antiderivatives of some functions, like $\sin(x^2)$, $1/\ln x$, and $\sqrt{1 + x^4}$, have no elementary formulas. When we cannot find a workable antiderivative for a function f that we have to integrate, we can partition the interval of integration, replace f by a closely fitting polynomial on each subinterval, integrate the polynomials, and add the results to approximate the definite integral of f. This procedure is an example of numerical integration. In this section we start by revisiting the Midpoint Rule, which we studied in Section 5.1. We then study two new methods, the *Trapezoidal Rule* and *Simpson's Rule*. A key goal in our analysis is to control the possible error that is introduced when computing an approximation to an integral.

Approximating Integrals with the Midpoint Rule

In Section 5.1 we introduced the Midpoint Rule to approximate a definite integral over an interval $[a, b]$. The rule is based on subdividing $[a, b]$ into n equal subintervals,

$$[x_0, x_1],\ [x_1, x_2],\ \ldots,\ [x_{n-1}, x_n]$$

each of width

$$\Delta x = \frac{b - a}{n}.$$

We then approximate the integral using n rectangles, where the height of the kth rectangle is the value of f at the midpoint $c_k = (x_{k-1} + x_k)/2$ of the kth subinterval $[x_{k-1}, x_k]$.

Midpoint Rule for Approximating a Definite Integral

$$\int_a^b f(x)\, dx \approx \sum_{k=1}^{n} f(c_k)\left(\frac{b - a}{n}\right) = [f(c_1) + f(c_2) + \cdots + f(c_n)]\left(\frac{b - a}{n}\right)$$

with $c_k = \dfrac{x_{k-1} + x_k}{2}$ and $x_k = a + k\left(\dfrac{b - a}{n}\right)$.

Trapezoidal Approximations

The Trapezoidal Rule for the value of a definite integral is based on approximating the region between a curve and the x-axis with trapezoids instead of rectangles, as in Figure 8.7. It is not necessary for the subdivision points $x_0, x_1, x_2, \ldots, x_n$ in the figure to be evenly spaced, but the resulting formula is simpler if they are. We therefore assume that the length of each subinterval is

$$\Delta x = \frac{b-a}{n}.$$

The length $\Delta x = (b-a)/n$ is called the **step size** or **mesh size**. The area of the trapezoid that lies above the ith subinterval is

$$\Delta x\left(\frac{y_{i-1}+y_i}{2}\right) = \frac{\Delta x}{2}(y_{i-1}+y_i),$$

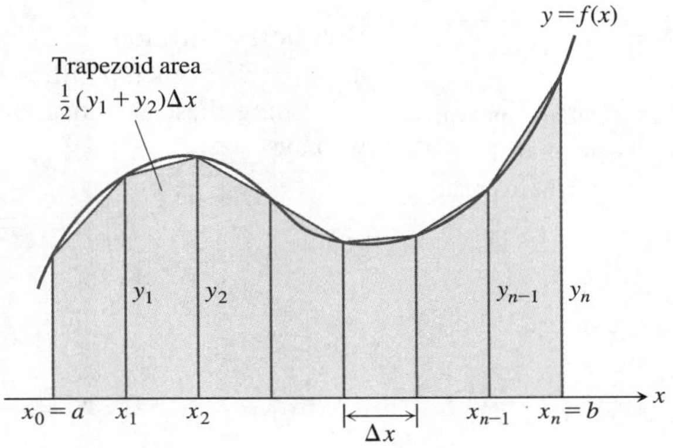

FIGURE 8.7 The Trapezoidal Rule approximates short stretches of the curve $y = f(x)$ with line segments. To approximate the integral of f from a to b, we add the areas of the trapezoids made by vertically joining the ends of the segments to the x-axis.

where $y_{i-1} = f(x_{i-1})$ and $y_i = f(x_i)$. (See Figure 8.7.) The area below the curve $y = f(x)$ and above the x-axis is then approximated by adding the areas of all the trapezoids:

$$\begin{aligned} T &= \frac{1}{2}(y_0+y_1)\Delta x + \frac{1}{2}(y_1+y_2)\Delta x + \cdots \\ &\quad + \frac{1}{2}(y_{n-2}+y_{n-1})\Delta x + \frac{1}{2}(y_{n-1}+y_n)\Delta x \\ &= \Delta x\left(\frac{1}{2}y_0 + y_1 + y_2 + \cdots + y_{n-1} + \frac{1}{2}y_n\right) \\ &= \frac{\Delta x}{2}(y_0 + 2y_1 + 2y_2 + \cdots + 2y_{n-1} + y_n), \end{aligned}$$

where

$$y_0 = f(a), \quad y_1 = f(x_1), \ldots, \quad y_{n-1} = f(x_{n-1}), \quad y_n = f(b).$$

The Trapezoidal Rule says: Use T to estimate the integral of f from a to b.

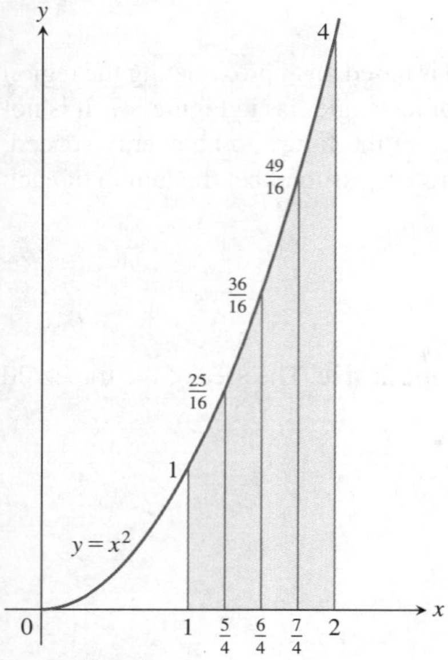

FIGURE 8.8 The trapezoidal approximation of the area under the graph of $y = x^2$ from $x = 1$ to $x = 2$ is a slight overestimate (Example 1).

The Trapezoidal Rule

To approximate $\int_a^b f(x)\,dx$, use

$$T = \frac{\Delta x}{2}(y_0 + 2y_1 + 2y_2 + \cdots + 2y_{n-1} + y_n).$$

The y's are the values of f at the partition points

$$x_0 = a,\ x_1 = a + \Delta x,\ x_2 = a + 2\Delta x,\ \ldots,\ x_{n-1} = a + (n-1)\,\Delta x,\ x_n = b,$$

where $\Delta x = (b - a)/n$.

EXAMPLE 1 Use the Trapezoidal Rule with $n = 4$ to estimate $\int_1^2 x^2\,dx$. Compare the estimate with the exact value.

Solution Partition [1, 2] into four subintervals of equal length (Figure 8.8). Then evaluate $y = x^2$ at each partition point (Table 8.2).

Using these y-values, $n = 4$, and $\Delta x = (2 - 1)/4 = 1/4$ in the Trapezoidal Rule, we have

$$\begin{aligned} T &= \frac{\Delta x}{2}(y_0 + 2y_1 + 2y_2 + 2y_3 + y_4) \\ &= \frac{1}{8}\left(1 + 2\left(\frac{25}{16}\right) + 2\left(\frac{36}{16}\right) + 2\left(\frac{49}{16}\right) + 4\right) \\ &= \frac{75}{32} = 2.34375. \end{aligned}$$

TABLE 8.2

x	$y = x^2$
1	1
$\frac{5}{4}$	$\frac{25}{16}$
$\frac{6}{4}$	$\frac{36}{16}$
$\frac{7}{4}$	$\frac{49}{16}$
2	4

Since the parabola is concave *up*, the approximating segments lie above the curve, giving each trapezoid slightly more area than the corresponding strip under the curve. The exact value of the integral is

$$\int_1^2 x^2\,dx = \frac{x^3}{3}\bigg]_1^2 = \frac{8}{3} - \frac{1}{3} = \frac{7}{3}.$$

The T approximation overestimates the integral by about half a percent of its true value of 7/3. The percentage error is $(2.34375 - 7/3)/(7/3) \approx 0.00446$, or 0.446%. ■

Simpson's Rule: Approximations Using Parabolas

Another rule for approximating the definite integral of a continuous function results from using parabolas instead of the straight-line segments that produced trapezoids. As before, we partition the interval $[a, b]$ into n subintervals of equal length $h = \Delta x = (b - a)/n$, but this time we require that n be an *even* number. On each consecutive pair of intervals we approximate the curve $y = f(x) \geq 0$ by a parabola, as shown in Figure 8.9. A typical parabola passes through three consecutive points (x_{i-1}, y_{i-1}), (x_i, y_i), and (x_{i+1}, y_{i+1}) on the curve.

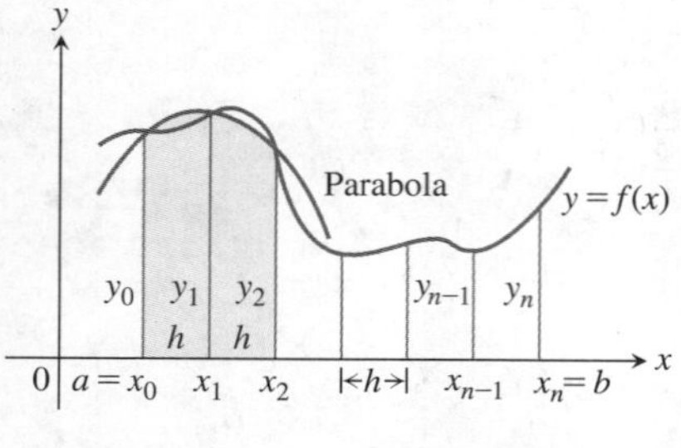

FIGURE 8.9 Simpson's Rule approximates short stretches of the curve with parabolas.

Let's calculate the shaded area beneath a parabola passing through three consecutive points. To simplify our calculations, we first take the case where $x_0 = -h$, $x_1 = 0$, and $x_2 = h$ (Figure 8.10), where $h = \Delta x = (b - a)/n$. The area under the parabola will be the same if we shift the y-axis to the left or right. The parabola has an equation of the form

$$y = Ax^2 + Bx + C,$$

so the area under it from $x = -h$ to $x = h$ is

$$A_p = \int_{-h}^{h} (Ax^2 + Bx + C)\,dx$$

$$= \left[\frac{Ax^3}{3} + \frac{Bx^2}{2} + Cx\right]_{-h}^{h}$$

$$= \frac{2Ah^3}{3} + 2Ch = \frac{h}{3}(2Ah^2 + 6C).$$

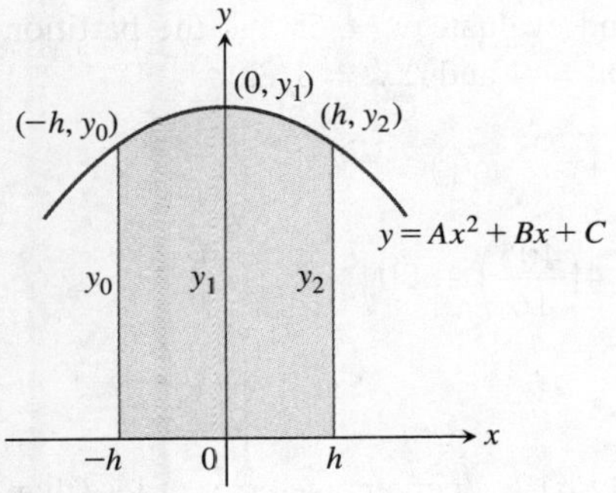

FIGURE 8.10 By integrating from $-h$ to h, we find the shaded area to be

$$\frac{h}{3}(y_0 + 4y_1 + y_2).$$

Since the curve passes through the three points $(-h, y_0), (0, y_1)$, and (h, y_2), we also have

$$y_0 = Ah^2 - Bh + C, \qquad y_1 = C, \qquad y_2 = Ah^2 + Bh + C.$$

After some algebraic manipulation, we find that

$$A_p = \frac{h}{3}(y_0 + 4y_1 + y_2).$$

Now shifting the parabola horizontally to its shaded position in Figure 8.9 does not change the area under it. Thus the area under the parabola through (x_0, y_0), (x_1, y_1), and (x_2, y_2) in Figure 8.9 is still

$$\frac{h}{3}(y_0 + 4y_1 + y_2).$$

Similarly, the area under the parabola through the points (x_2, y_2), (x_3, y_3), and (x_4, y_4) is

$$\frac{h}{3}(y_2 + 4y_3 + y_4).$$

Computing the areas under all the parabolas and adding the results give the approximation

$$\int_a^b f(x)\,dx \approx \frac{h}{3}(y_0 + 4y_1 + y_2) + \frac{h}{3}(y_2 + 4y_3 + y_4) + \cdots$$

$$+ \frac{h}{3}(y_{n-2} + 4y_{n-1} + y_n)$$

$$= \frac{h}{3}(y_0 + 4y_1 + 2y_2 + 4y_3 + 2y_4 + \cdots + 2y_{n-2} + 4y_{n-1} + y_n).$$

The result is known as Simpson's Rule. The function need not be positive, as in our derivation, but the number n of subintervals must be even for us to apply the rule because each parabolic arc uses two subintervals.

HISTORICAL BIOGRAPHY

Thomas Simpson

(1720–1761)

www.bit.ly/2DHSIOI

Simpson's Rule

To approximate $\int_a^b f(x)\,dx$, use

$$S = \frac{\Delta x}{3}(y_0 + 4y_1 + 2y_2 + 4y_3 + \cdots + 2y_{n-2} + 4y_{n-1} + y_n).$$

The y's are the values of f at the partition points

$$x_0 = a,\ x_1 = a + \Delta x,\ x_2 = a + 2\Delta x,\ \ldots,\ x_{n-1} = a + (n-1)\Delta x,\ x_n = b.$$

The number n is even, and $\Delta x = (b - a)/n$.

Note the pattern of the coefficients in the above rule: 1, 4, 2, 4, 2, 4, 2, …, 4, 1.

EXAMPLE 2 Use Simpson's Rule with $n = 4$ to approximate $\int_0^2 5x^4\,dx$.

TABLE 8.3

x	$y = 5x^4$
0	0
$\frac{1}{2}$	$\frac{5}{16}$
1	5
$\frac{3}{2}$	$\frac{405}{16}$
2	80

Solution Partition $[0, 2]$ into four subintervals and evaluate $y = 5x^4$ at the partition points (Table 8.3). Then apply Simpson's Rule with $n = 4$ and $\Delta x = 1/2$:

$$\begin{aligned} S &= \frac{\Delta x}{3}(y_0 + 4y_1 + 2y_2 + 4y_3 + y_4) \\ &= \frac{1}{6}\left(0 + 4\left(\frac{5}{16}\right) + 2(5) + 4\left(\frac{405}{16}\right) + 80\right) \\ &= 32\frac{1}{12}. \end{aligned}$$

This estimate differs from the exact value (32) by only 1/12, a percentage error of less than three-tenths of one percent, and this was with just four subintervals. ■

Error Analysis

Whenever we use an approximation technique, we must consider how accurate the approximation might be. The following theorem gives formulas for estimating the errors when using the Midpoint Rule, the Trapezoidal Rule, and Simpson's Rule. The **error** is the difference between the approximation obtained by using the rule and the actual value of the definite integral $\int_a^b f(x)\,dx$.

THEOREM 1—Error Estimates in the Midpoint, Trapezoidal, and Simpson's Rules

If f'' is continuous and M is any upper bound for the values of $|f''|$ on $[a, b]$, then the error E_M in the Midpoint Rule approximation of the integral of f from a to b for n steps satisfies the inequality

$$|E_M| \le \frac{M(b-a)^3}{24n^2}. \qquad \text{Midpoint Rule}$$

If f'' is continuous and M is any upper bound for the values of $|f''|$ on $[a, b]$, then the error E_T in the Trapezoidal Rule approximation of the integral of f from a to b for n steps satisfies the inequality

$$|E_T| \le \frac{M(b-a)^3}{12n^2}. \qquad \text{Trapezoidal Rule}$$

If $f^{(4)}$ is continuous and M is any upper bound for the values of $|f^{(4)}|$ on $[a, b]$, then the error E_S in the Simpson's Rule approximation of the integral of f from a to b for n steps satisfies the inequality

$$|E_S| \le \frac{M(b-a)^5}{180n^4}. \qquad \text{Simpson's Rule}$$

To give an idea of why Theorem 1 is true in the case of the Trapezoidal Rule, we begin with a result which says that if f'' is continuous on the interval $[a, b]$, then

$$\int_a^b f(x)\,dx = T - \frac{b-a}{12} \cdot f''(c)(\Delta x)^2$$

for some number c between a and b. This result follows from Taylor's Remainder Theorem, which we discuss in Theorem 24 of Section 10.9. It follows that as Δx approaches zero, the error defined by

$$E_T = -\frac{b-a}{12} \cdot f''(c)(\Delta x)^2$$

approaches *zero* at the rate of the *square* of Δx.

The inequality

$$|E_T| \leq \frac{b-a}{12} \max|f''(x)|\,(\Delta x)^2,$$

where "max" refers to the maximum of $|f''(x)|$ over the interval $[a, b]$, gives an upper bound for the magnitude of the error. In practice, we usually cannot find the exact value of $\max|f''(x)|$ and have to estimate an upper bound or "worst case" value for it instead. If M is any upper bound for the values of $|f''(x)|$ on $[a, b]$, so that $|f''(x)| \leq M$ for every x in $[a, b]$, then

$$|E_T| \leq \frac{b-a}{12} M(\Delta x)^2.$$

If we substitute $(b - a)/n$ for Δx, we get

$$|E_T| \leq \frac{M(b-a)^3}{12n^2}.$$

The upper bound for the error in the Midpoint Rule,

$$|E_M| \leq \frac{M(b-a)^3}{24n^2},$$

is based on a similar argument. Taylor's Remainder Theorem implies that the Midpoint Rule approximation and the definite integral differ by at most

$$\frac{|f''(c)|\,(b-a)^3}{24n^2}$$

for some $c \in [a, b]$. Taking M to be at least as large as the maximum of $|f''(c)|$ on $[a, b]$, we obtain the error bound in Theorem 1. Note that this is half as large as the error bound for the Trapezoidal Rule. This does not mean that the Midpoint Rule is always more accurate than the Trapezoidal Rule, but rather that the largest possible error that might occur is only half as large as the largest possible error for the Trapezoidal Rule.

To estimate the error in Simpson's Rule, we start with a result, again following from Taylor's Remainder Theorem, that says that if the fourth derivative $f^{(4)}$ is continuous, then

$$\int_a^b f(x)\,dx = S - \frac{b-a}{180} \cdot f^{(4)}(c)(\Delta x)^4$$

for some point c between a and b. Thus, as Δx approaches zero, the error,

$$E_S = -\frac{b-a}{180} \cdot f^{(4)}(c)(\Delta x)^4,$$

approaches zero as the *fourth power* of Δx. (This helps to explain why Simpson's Rule is likely to give better results than the Trapezoidal Rule.)

The inequality

$$|E_S| \leq \frac{b-a}{180} \max|f^{(4)}(x)|\,(\Delta x)^4,$$

where "max" refers to the maximum of $|f^{(4)}(x)|$ over the interval $[a, b]$, gives an upper bound for the magnitude of the error. As with $\max|f''|$ in the error formula for the Trapezoidal Rule, we usually cannot find the exact value of $\max|f^{(4)}(x)|$ and have to replace it with an upper bound. If M is any upper bound for the values of $|f^{(4)}(x)|$ on $[a, b]$, then

$$|E_S| \leq \frac{b-a}{180} M(\Delta x)^4.$$

Substituting $(b - a)/n$ for Δx in this last expression gives

$$|E_S| \leq \frac{M(b-a)^5}{180n^4}.$$

You might wonder why we don't just take M to be the maximum value of $|f''(x)|$ on $[a, b]$ for the first two rules, or the maximum value of $|f^{(4)}(x)|$ for Simpson's Rule. The reason is that sometimes this maximum is hard to compute, while a less accurate upper bound is easily found. We can certainly set M to the maximum value if we can compute it.

EXAMPLE 3 Find an upper bound for the error in estimating $\int_0^2 5x^4\,dx$ using Simpson's Rule with $n = 4$ (Example 2).

Solution To estimate the error, we first find an upper bound M for the magnitude of the fourth derivative of $f(x) = 5x^4$ on the interval $0 \le x \le 2$. Since the fourth derivative has the constant value $f^{(4)}(x) = 120$, we take $M = 120$. With $b - a = 2$ and $n = 4$, the error estimate for Simpson's Rule gives

$$|E_S| \le \frac{M(b-a)^5}{180n^4} = \frac{120(2)^5}{180 \cdot 4^4} = \frac{1}{12}.$$

This estimate is consistent with the result of Example 2. ■

Theorem 1 can also be used to estimate the number of subintervals required when using the Trapezoidal or Simpson's Rule if we specify a certain tolerance for the error.

EXAMPLE 4 Estimate the minimum number of subintervals needed to approximate the integral in Example 3 using Simpson's Rule with an error of magnitude less than 10^{-4}.

Solution Using the inequality in Theorem 1, if we choose the number of subintervals n to satisfy

$$\frac{M(b-a)^5}{180n^4} < 10^{-4},$$

then the error E_S in Simpson's Rule satisfies $|E_S| < 10^{-4}$, as required.

From the solution in Example 3, we have $M = 120$ and $b - a = 2$, so we want n to satisfy

$$\frac{120(2)^5}{180n^4} < \frac{1}{10^4},$$

or, equivalently,

$$n^4 > \frac{64 \cdot 10^4}{3}.$$

It follows that

$$n > 10\left(\frac{64}{3}\right)^{1/4} \approx 21.5.$$

Since n must be even in Simpson's Rule, we estimate the minimum number of subintervals required for the error tolerance to be $n = 22$. ■

EXAMPLE 5 As we saw in Chapter 7, the value of ln 2 can be calculated from the integral

$$\ln 2 = \int_1^2 \frac{1}{x}\,dx.$$

Table 8.4 shows values of T and S for approximations of $\int_1^2 (1/x)\,dx$ using various values of n. Notice how Simpson's Rule dramatically improves over the Trapezoidal Rule.

TABLE 8.4 Trapezoidal Rule approximations (T_n) and Simpson's Rule approximations (S_n) of $\ln 2 = \int_1^2 (1/x)\,dx$

n	T_n	|Error| less than...	S_n	|Error| less than...
10	0.6937714032	0.0006242227	0.6931502307	0.0000030502
20	0.6933033818	0.0001562013	0.6931473747	0.0000001942
30	0.6932166154	0.0000694349	0.6931472190	0.0000000385
40	0.6931862400	0.0000390595	0.6931471927	0.0000000122
50	0.6931721793	0.0000249988	0.6931471856	0.0000000050
100	0.6931534305	0.0000062500	0.6931471809	0.0000000004

In particular, notice that when we double the value of n (thereby halving the value of $h = \Delta x$), the T error is divided by 2 *squared*, whereas the S error is divided by 2 *to the fourth*.

This has a dramatic effect as $\Delta x = (2 - 1)/n$ gets very small. The Simpson approximation for $n = 50$ rounds accurately to seven places and for $n = 100$ is accurate to nine decimal places (billionths)! ■

If $f(x)$ is a polynomial of degree less than 4, then its fourth derivative is zero, and

$$E_S = -\frac{b-a}{180} f^{(4)}(c)(\Delta x)^4 = -\frac{b-a}{180}(0)(\Delta x)^4 = 0.$$

Thus, there will be no error in the Simpson approximation of any integral of f. In other words, if f is a constant, a linear function, or a quadratic or cubic polynomial, Simpson's Rule will give the value of any integral of f exactly, whatever the number of subdivisions. Similarly, if f is a constant or a linear function, then its second derivative is zero, and

$$E_T = -\frac{b-a}{12} f''(c)(\Delta x)^2 = -\frac{b-a}{12}(0)(\Delta x)^2 = 0.$$

The Trapezoidal Rule will therefore give the exact value of any integral of f. This is no surprise, for the trapezoids fit the graph perfectly.

Although decreasing the step size Δx reduces the error in the Simpson and Trapezoidal approximations in theory, it may fail to do so in practice. When Δx is very small, say $\Delta x = 10^{-8}$, computer or calculator round-off errors in the arithmetic required to evaluate S and T may accumulate to such an extent that the error formulas no longer describe what is going on. Shrinking Δx below a certain size can actually make things worse. You should consult a text on numerical analysis for more sophisticated methods if you are having problems with round-off error using the rules discussed in this section.

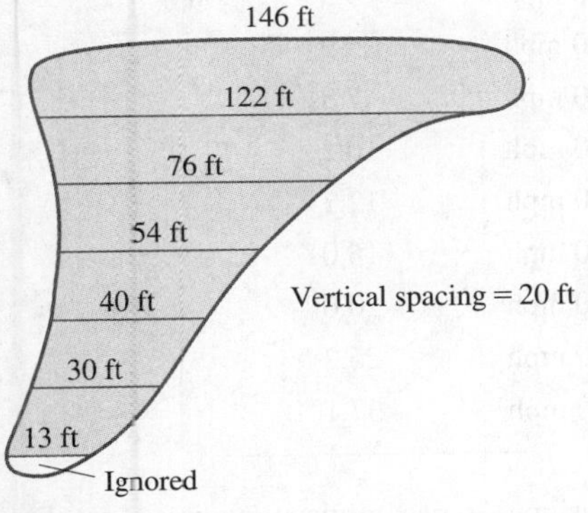

FIGURE 8.11 The dimensions of the swamp in Example 6.

EXAMPLE 6 A town wants to drain and fill a small polluted swamp (Figure 8.11). The swamp averages 5 ft deep. About how many cubic yards of dirt will it take to fill the area after the swamp is drained?

Solution To calculate the volume of the swamp, we estimate the surface area and multiply by 5. To estimate the area, we use Simpson's Rule with $\Delta x = 20$ ft and the y's equal to the distances measured across the swamp, as shown in Figure 8.11.

$$S = \frac{\Delta x}{3}(y_0 + 4y_1 + 2y_2 + 4y_3 + 2y_4 + 4y_5 + y_6)$$

$$= \frac{20}{3}(146 + 488 + 152 + 216 + 80 + 120 + 13) = 8100$$

The volume is about $(8100)(5) = 40{,}500$ ft^3, or 1500 yd^3. ■

EXERCISES 8.7

For some exercises, a calculator may be helpful for expressing answers in decimal form.

Estimating Definite Integrals

The instructions for the integrals in Exercises 1–10 have three parts, one for the Midpoint Rule, one for the Trapezoidal Rule, and one for Simpson's Rule.

I. Using the Midpoint Rule

a. Estimate the integral with $n = 4$ steps and find an upper bound for $|E_M|$.

b. Evaluate the integral directly and find $|E_M|$.

c. Use the formula $(|E_M|/(\text{true value})) \times 100$ to express $|E_M|$ as a percentage of the integral's true value.

II. Using the Trapezoidal Rule

a. Estimate the integral with $n = 4$ steps and find an upper bound for $|E_T|$.

b. Evaluate the integral directly and find $|E_T|$.

c. Use the formula $(|E_T|/(\text{true value})) \times 100$ to express $|E_T|$ as a percentage of the integral's true value.

III. Using Simpson's Rule

a. Estimate the integral with $n = 4$ steps and find an upper bound for $|E_S|$.

b. Evaluate the integral directly and find $|E_S|$.

c. Use the formula $(|E_S|/(\text{true value})) \times 100$ to express $|E_S|$ as a percentage of the integral's true value.

1. $\int_1^2 x\,dx$

2. $\int_1^3 (2x - 1)\,dx$

3. $\int_{-1}^1 (x^2 + 1)\,dx$

4. $\int_{-2}^0 (x^2 - 1)\,dx$

5. $\int_0^2 (t^3 + t)\,dt$

6. $\int_{-1}^1 (t^3 + 1)\,dt$

7. $\int_1^2 \frac{1}{s^2}\,ds$

8. $\int_2^4 \frac{1}{(s-1)^2}\,ds$

9. $\int_0^\pi \sin t\,dt$

10. $\int_0^1 \sin \pi t\,dt$

Estimating the Number of Subintervals

In Exercises 11–22, estimate the minimum number of subintervals needed to approximate the integrals with an error of magnitude less than 10^{-4} by **(a)** the Trapezoidal Rule and **(b)** Simpson's Rule. (The integrals in Exercises 11–18 are the integrals from Exercises 1–8.)

11. $\int_1^2 x\,dx$

12. $\int_1^3 (2x - 1)\,dx$

13. $\int_{-1}^1 (x^2 + 1)\,dx$

14. $\int_{-2}^0 (x^2 - 1)\,dx$

15. $\int_0^2 (t^3 + t)\,dt$

16. $\int_{-1}^1 (t^3 + 1)\,dt$

17. $\int_1^2 \frac{1}{s^2}\,ds$

18. $\int_2^4 \frac{1}{(s-1)^2}\,ds$

19. $\int_0^3 \sqrt{x+1}\,dx$

20. $\int_0^3 \frac{1}{\sqrt{x+1}}\,dx$

21. $\int_0^2 \sin(x + 1)\,dx$

22. $\int_{-1}^1 \cos(x + \pi)\,dx$

Estimates with Numerical Data

23. Volume of water in a swimming pool A rectangular swimming pool is 30 ft wide and 50 ft long. The accompanying table shows the depth $h(x)$ of the water at 5-ft intervals from one end of the pool to the other. Estimate the volume of water in the pool using the Trapezoidal Rule with $n = 10$ applied to the integral

$$V = \int_0^{50} 30 \cdot h(x)\,dx.$$

Position (ft) x	Depth (ft) $h(x)$	Position (ft) x	Depth (ft) $h(x)$
0	6.0	30	11.5
5	8.2	35	11.9
10	9.1	40	12.3
15	9.9	45	12.7
20	10.5	50	13.0
25	11.0		

24. Distance traveled The accompanying table shows time-to-speed data for a sports car accelerating from rest to 130 mph. How far had the car traveled by the time it reached this speed? (Use trapezoids to estimate the area under the velocity curve, but be careful: The time intervals vary in length.)

Speed change	Time (sec)
Zero to 30 mph	2.2
40 mph	3.2
50 mph	4.5
60 mph	5.9
70 mph	7.8
80 mph	10.2
90 mph	12.7
100 mph	16.0
110 mph	20.6
120 mph	26.2
130 mph	37.1

25. Wing design The design of a new airplane requires a gasoline tank of constant cross-sectional area in each wing. A scale drawing of a cross-section is shown here. The tank must hold 5000 lb of gasoline, which has a density of 42 lb/ft^3. Estimate the length of the tank by Simpson's Rule.

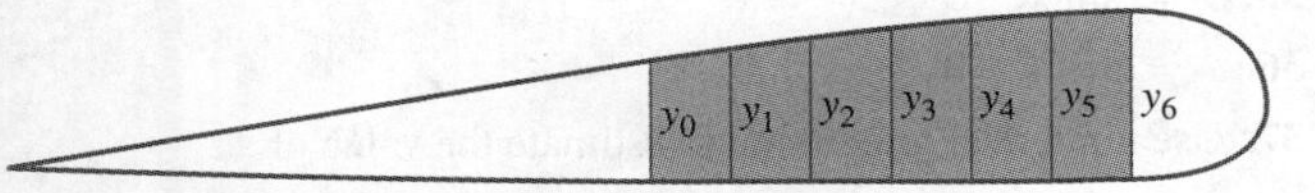

$y_0 = 1.5$ ft, $y_1 = 1.6$ ft, $y_2 = 1.8$ ft, $y_3 = 1.9$ ft,
$y_4 = 2.0$ ft, $y_5 = y_6 = 2.1$ ft Horizontal spacing = 1 ft

26. Oil consumption on Pathfinder Island A diesel generator runs continuously, consuming oil at a gradually increasing rate until it must be temporarily shut down to have the filters replaced. Use the Trapezoidal Rule to estimate the amount of oil consumed by the generator during that week.

Day	**Oil consumption rate (liters/hr)**
Sun	0.019
Mon	0.020
Tue	0.021
Wed	0.023
Thu	0.025
Fri	0.028
Sat	0.031
Sun	0.035

Theory and Examples

27. Usable values of the sine-integral function *The sine-integral function,*

$$\operatorname{Si}(x) = \int_0^x \frac{\sin t}{t}\, dt, \qquad \text{“Sine integral of } x\text{”}$$

is one of the many functions in engineering whose formulas cannot be simplified. There is no elementary formula for the antiderivative of $(\sin t)/t$. The values of Si(x), however, are readily estimated by numerical integration.

Although the notation does not show it explicitly, the function being integrated is

$$f(t) = \begin{cases} \dfrac{\sin t}{t}, & t \neq 0 \\ 1, & t = 0, \end{cases}$$

the continuous extension of $(\sin t)/t$ to the interval $[0, x]$. The function has derivatives of all orders at every point of its domain. Its graph is smooth, and you can expect good results from Simpson's Rule.

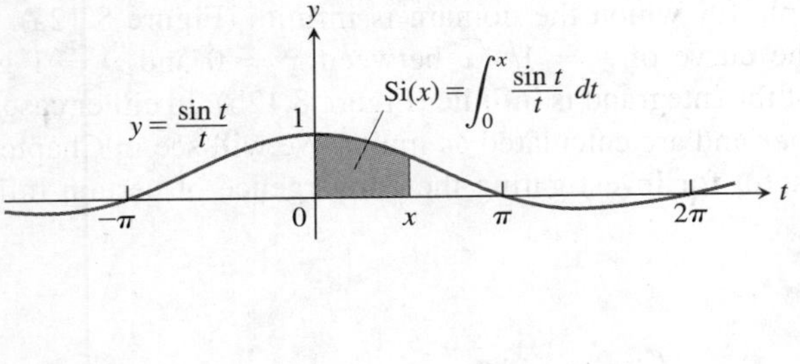

a. Use the fact that $|f^{(4)}| \leq 1$ on $[0, \pi/2]$ to give an upper bound for the error that will occur if

$$\operatorname{Si}\left(\frac{\pi}{2}\right) = \int_0^{\pi/2} \frac{\sin t}{t}\, dt$$

is estimated by Simpson's Rule with $n = 4$.

b. Estimate $\operatorname{Si}(\pi/2)$ by Simpson's Rule with $n = 4$.

c. Express the error bound you found in part (a) as a percentage of the value you found in part (b).

28. The error function *The error function,*

$$\operatorname{erf}(x) = \frac{2}{\sqrt{\pi}} \int_0^x e^{-t^2}\, dt,$$

which is important in probability and in the theories of heat flow and signal transmission, must be evaluated numerically because there is no elementary expression for the antiderivative of e^{-t^2}.

a. Use Simpson's Rule with $n = 10$ to estimate erf (1).

b. In [0, 1],

$$\left|\frac{d^4}{dt^4}\left(e^{-t^2}\right)\right| \leq 12.$$

Give an upper bound for the magnitude of the error of the estimate in part (a).

29. Prove that the sum T in the Trapezoidal Rule for $\int_a^b f(x)\, dx$ is a Riemann sum for f continuous on $[a, b]$. (*Hint:* Use the Intermediate Value Theorem to show the existence of c_k in the subinterval $[x_{k-1}, x_k]$ satisfying $f(c_k) = (f(x_{k-1}) + f(x_k))/2$.)

30. Prove that the sum S in Simpson's Rule for $\int_a^b f(x)\, dx$ is a Riemann sum for f continuous on $[a, b]$. (See Exercise 29.)

31. Elliptic integrals The length of the ellipse

$$\frac{x^2}{a^2} + \frac{y^2}{b^2} = 1$$

turns out to be

$$\text{Length} = 4a \int_0^{\pi/2} \sqrt{1 - e^2 \cos^2 t}\, dt,$$

where $e = \sqrt{a^2 - b^2}/a$ is the ellipse's eccentricity. The integral in this formula, called an *elliptic integral*, is nonelementary except when $e = 0$ or 1.

a. Use the Trapezoidal Rule with $n = 10$ to estimate the length of the ellipse when $a = 1$ and $e = 1/2$.

b. Use the fact that the absolute value of the second derivative of $f(t) = \sqrt{1 - e^2 \cos^2 t}$ is less than 1 to find an upper bound for the error in the estimate you obtained in part (a).

Applications

32. The length of one arch of the curve $y = \sin x$ is given by

$$L = \int_0^{\pi} \sqrt{1 + \cos^2 x}\, dx.$$

Estimate L by Simpson's Rule with $n = 8$.

T When solving Exercises 33-40, you may need to use a calculator or a computer.

33. Your metal fabrication company is bidding for a contract to make sheets of corrugated iron roofing like the one shown here. The cross-sections of the corrugated sheets are to conform to the curve

$$y = \sin \frac{3\pi}{20}x, \quad 0 \le x \le 20 \text{ in.}$$

If the roofing is to be stamped from flat sheets by a process that does not stretch the material, how wide should the original material be? To find out, use numerical integration to approximate the length of the sine curve to two decimal places.

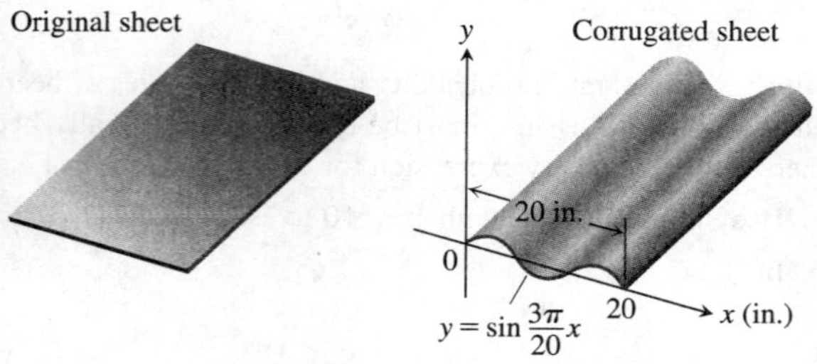

34. Your engineering firm is bidding for the contract to construct the tunnel shown here. The tunnel is 300 ft long and 50 ft wide at the base. The cross-section is shaped like one arch of the curve $y = 25\cos(\pi x/50)$. Upon completion, the tunnel's inside surface (excluding the roadway) will be treated with a waterproof sealer that costs $2.35 per square foot to apply. How much will it cost to apply the sealer? (*Hint:* Use numerical integration to find the length of the cosine curve.)

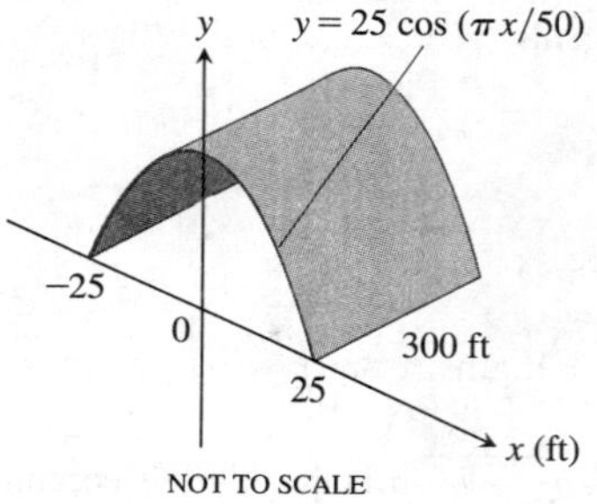

Find, to two decimal places, the areas of the surfaces generated by revolving the curves in Exercises 35 and 36 about the x-axis.

35. $y = \sin x, \quad 0 \le x \le \pi$

36. $y = x^2/4, \quad 0 \le x \le 2$

37. Use numerical integration to estimate the value of

$$\arcsin 0.6 = \int_0^{0.6} \frac{dx}{\sqrt{1 - x^2}}.$$

For reference, arcsin 0.6 = 0.64350 to five decimal places.

38. Use numerical integration to estimate the value of

$$\pi = 4\int_0^1 \frac{1}{1 + x^2}\,dx.$$

39. Drug assimilation An average adult under age 60 years assimilates a 12-hr cold medicine into his or her system at a rate modeled by

$$\frac{dy}{dt} = 6 - \ln(2t^2 - 3t + 3),$$

where y is measured in milligrams and t is the time in hours since the medication was taken. What amount of medicine is absorbed into a person's system over a 12-hr period?

40. Effects of an antihistamine The concentration of an antihistamine in the bloodstream of a healthy adult is modeled by

$$C = 12.5 - 4\ln(t^2 - 3t + 4),$$

where C is measured in grams per liter and t is the time in hours since the medication was taken. What is the average level of concentration in the bloodstream over a 6-hr period?

8.8 Improper Integrals

Up to now, we have required definite integrals to satisfy two properties. First, the domain of integration $[a, b]$ must be finite. Second, the range of the integrand must be finite on this domain. In practice, we may encounter problems that fail to meet one or both of these conditions. The integral for the area under the curve $y = (\ln x)/x^2$ from $x = 1$ to $x = \infty$ is an example for which the domain is infinite (Figure 8.12a). The integral for the area under the curve of $y = 1/\sqrt{x}$ between $x = 0$ and $x = 1$ is an example for which the range of the integrand is infinite (Figure 8.12b). In either case, the integrals are said to be *improper* and are calculated as limits. We will see in Chapter 10 that improper integrals are useful for investigating the convergence of certain infinite series.

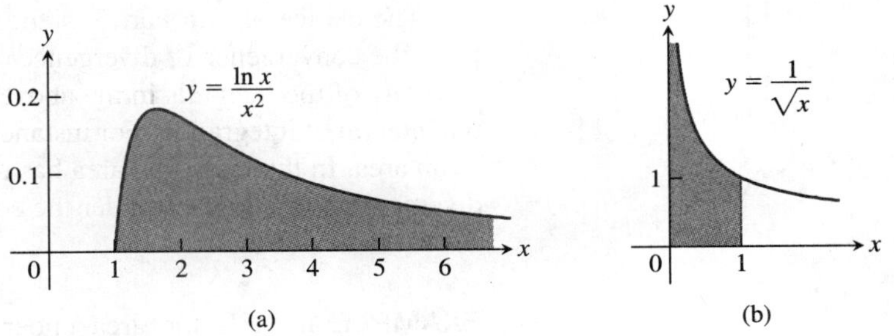

FIGURE 8.12 Are the areas under these infinite curves finite? We will see that the answer is yes for both curves.

Infinite Limits of Integration

Consider the infinite region (unbounded on the right) that lies under the curve $y = e^{-x/2}$ in the first quadrant (Figure 8.13a). You might think this region has infinite area, but we will see that the value is finite. We assign a value to the area in the following way. First find the area $A(b)$ of the portion of the region that is bounded on the right by $x = b$ (Figure 8.13b).

$$A(b) = \int_0^b e^{-x/2}\,dx = -2e^{-x/2}\Big]_0^b = -2e^{-b/2} + 2 = 2 - 2e^{-b/2},$$

which is a little less than 2. Then find the limit of $A(b)$ as $b \to \infty$.

$$\lim_{b\to\infty} A(b) = \lim_{b\to\infty} (2 - 2e^{-b/2}) = 2$$

Therefore, the value we assign to the area under the curve from 0 to ∞ is

$$\int_0^\infty e^{-x/2}\,dx = \lim_{b\to\infty} \int_0^b e^{-x/2}\,dx = 2.$$

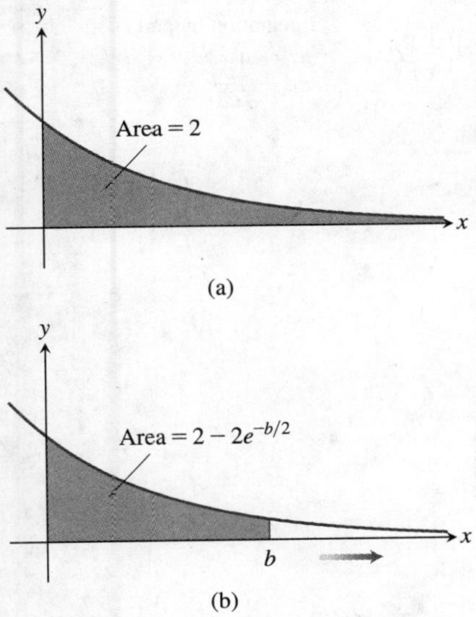

FIGURE 8.13 (a) The area in the first quadrant under the curve $y = e^{-x/2}$. (b) The area is an improper integral of the first type.

DEFINITION Integrals with infinite limits of integration are **improper integrals of Type I**.

1. If $f(x)$ is continuous on $[a, \infty)$, then

$$\int_a^\infty f(x)\,dx = \lim_{b\to\infty} \int_a^b f(x)\,dx.$$

2. If $f(x)$ is continuous on $(-\infty, b]$, then

$$\int_{-\infty}^b f(x)\,dx = \lim_{a\to-\infty} \int_a^b f(x)\,dx.$$

3. If $f(x)$ is continuous on $(-\infty, \infty)$, then

$$\int_{-\infty}^\infty f(x)\,dx = \int_{-\infty}^c f(x)\,dx + \int_c^\infty f(x)\,dx,$$

where c is any real number.

In each case, if the limit exists and is finite, we say that the improper integral **converges** and that the limit is the **value** of the improper integral. If the limit fails to exist, the improper integral **diverges**.

The choice of c in Part 3 of the definition is unimportant. We can evaluate or determine the convergence or divergence of $\int_{-\infty}^{\infty} f(x)\,dx$ with any convenient choice.

Any of the integrals in the above definition can be interpreted as an area if $f \geq 0$ on the interval of integration. For instance, we interpreted the improper integral in Figure 8.13 as an area. In that case, the area has the finite value 2. If $f \geq 0$ and the improper integral diverges, we say the area under the curve is **infinite**.

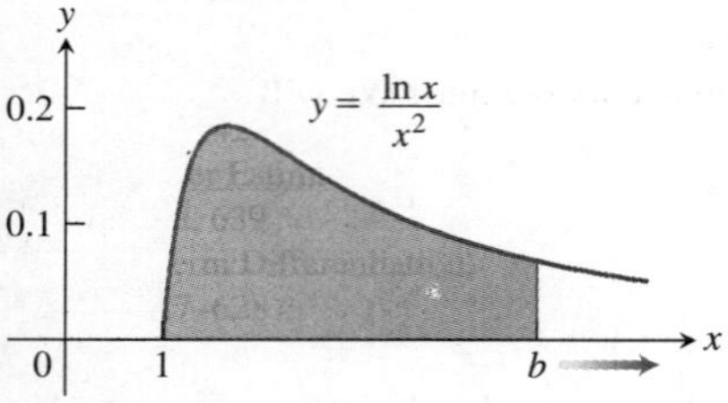

FIGURE 8.14 The area under this curve is an improper integral (Example 1).

EXAMPLE 1 Is the area under the curve $y = (\ln x)/x^2$ from $x = 1$ to $x = \infty$ finite? If so, what is its value?

Solution We find the area under the curve from $x = 1$ to $x = b$ and examine the limit as $b \to \infty$. If the limit is finite, we take it to be the area under the curve (Figure 8.14). The area from 1 to b is

$$\int_1^b \frac{\ln x}{x^2}\,dx = \left[(\ln x)\left(-\frac{1}{x}\right)\right]_1^b - \int_1^b \left(-\frac{1}{x}\right)\left(\frac{1}{x}\right)dx$$

Integration by parts with $u = \ln x$, $dv = dx/x^2$, $du = dx/x$, $v = -1/x$

$$= -\frac{\ln b}{b} - \left[\frac{1}{x}\right]_1^b$$

$$= -\frac{\ln b}{b} - \frac{1}{b} + 1.$$

The limit of the area as $b \to \infty$ is

$$\int_1^\infty \frac{\ln x}{x^2}\,dx = \lim_{b\to\infty} \int_1^b \frac{\ln x}{x^2}\,dx$$

$$= \lim_{b\to\infty}\left[-\frac{\ln b}{b} - \frac{1}{b} + 1\right]$$

$$= -\left[\lim_{b\to\infty}\frac{\ln b}{b}\right] - 0 + 1$$

$$= -\left[\lim_{b\to\infty}\frac{1/b}{1}\right] + 1 = 0 + 1 = 1. \qquad \text{L'Hôpital's Rule}$$

Thus, the improper integral converges and the area has finite value 1. ∎

EXAMPLE 2 Evaluate

$$\int_{-\infty}^{\infty} \frac{dx}{1 + x^2}.$$

Solution According to Part 3 of the definition, we can choose $c = 0$ and write

$$\int_{-\infty}^{\infty} \frac{dx}{1+x^2} = \int_{-\infty}^{0} \frac{dx}{1+x^2} + \int_{0}^{\infty} \frac{dx}{1+x^2}.$$

HISTORICAL BIOGRAPHY
Lejeune Dirichlet
(1805–1859)
www.bit.ly/2Rdogyt

Next we evaluate each improper integral on the right side of the equation above.

$$\int_{-\infty}^{0} \frac{dx}{1+x^2} = \lim_{a\to-\infty} \int_a^0 \frac{dx}{1+x^2}$$

$$= \lim_{a\to-\infty} \tan^{-1} x\Big]_a^0$$

$$= \lim_{a\to-\infty} (\tan^{-1} 0 - \tan^{-1} a) = 0 - \left(-\frac{\pi}{2}\right) = \frac{\pi}{2}$$

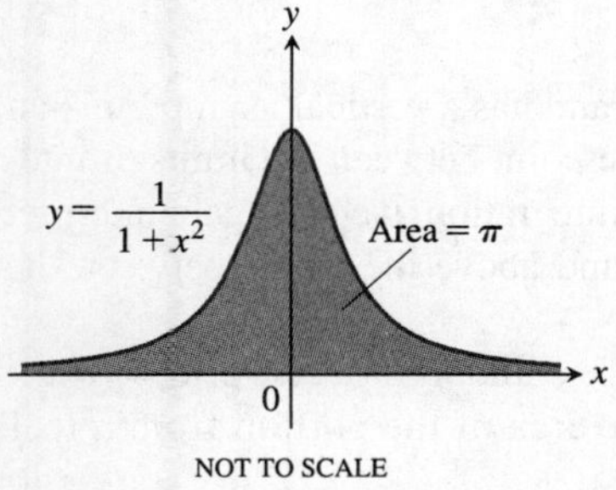

FIGURE 8.15 The area under this curve is finite (Example 2).

$$\begin{aligned}\int_0^{\infty} \frac{dx}{1+x^2} &= \lim_{b\to\infty} \int_0^b \frac{dx}{1+x^2} \\ &= \lim_{b\to\infty} \tan^{-1} x \Big]_0^b \\ &= \lim_{b\to\infty} (\tan^{-1} b - \tan^{-1} 0) = \frac{\pi}{2} - 0 = \frac{\pi}{2}\end{aligned}$$

Thus,

$$\int_{-\infty}^{\infty} \frac{dx}{1+x^2} = \frac{\pi}{2} + \frac{\pi}{2} = \pi.$$

Since $1/(1+x^2) > 0$, the improper integral can be interpreted as the (finite) area beneath the curve and above the x-axis (Figure 8.15). ■

The Integral $\int_1^{\infty} \frac{dx}{x^p}$

The function $y = 1/x$ is the boundary between the convergent and divergent improper integrals with integrands of the form $y = 1/x^p$. As the next example shows, the improper integral converges if $p > 1$ and diverges if $p \le 1$.

EXAMPLE 3 For what values of p does the integral $\int_1^{\infty} dx/x^p$ converge? When the integral does converge, what is its value?

Solution If $p \neq 1$, then

$$\int_1^b \frac{dx}{x^p} = \frac{x^{-p+1}}{-p+1}\Big]_1^b = \left(\frac{1}{1-p}\right)(b^{-p+1} - 1) = \left(\frac{1}{1-p}\right)\left(\frac{1}{b^{p-1}} - 1\right).$$

Thus,

$$\begin{aligned}\int_1^{\infty} \frac{dx}{x^p} &= \lim_{b\to\infty} \int_1^b \frac{dx}{x^p} \\ &= \lim_{b\to\infty} \left[\left(\frac{1}{1-p}\right)\left(\frac{1}{b^{p-1}} - 1\right)\right] = \begin{cases} \frac{1}{p-1}, & p > 1 \\ \infty, & p < 1 \end{cases}\end{aligned}$$

because

$$\lim_{b\to\infty} \frac{1}{b^{p-1}} = \begin{cases} 0, & p > 1 \\ \infty, & p < 1. \end{cases}$$

Therefore, the integral converges to the value $1/(p-1)$ if $p > 1$, and it diverges if $p < 1$.

If $p = 1$, the integral also diverges:

$$\begin{aligned}\int_1^{\infty} \frac{dx}{x^p} &= \int_1^{\infty} \frac{dx}{x} \\ &= \lim_{b\to\infty} \int_1^b \frac{dx}{x} \\ &= \lim_{b\to\infty} \Big[\ln|x|\Big]_1^b \\ &= \lim_{b\to\infty} (\ln b - \ln 1) = \infty.\end{aligned}$$

■

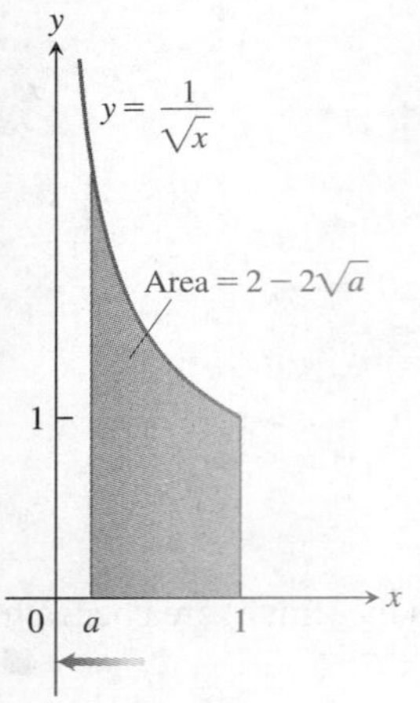

FIGURE 8.16 The area under this curve is an example of an improper integral of the second kind.

Integrands with Vertical Asymptotes

Another type of improper integral arises when the integrand has a vertical asymptote—an infinite discontinuity—at a limit of integration or at some point between the limits of integration. If the integrand f is positive over the interval of integration, we can again interpret the improper integral as the area under the graph of f and above the x-axis between the limits of integration.

Consider the region in the first quadrant that lies under the curve $y = 1/\sqrt{x}$ from $x = 0$ to $x = 1$ (Figure 8.12b). First we find the area of the portion from a to 1 (Figure 8.16):

$$\int_a^1 \frac{dx}{\sqrt{x}} = 2\sqrt{x}\Big]_a^1 = 2 - 2\sqrt{a}.$$

Then we find the limit of this area as $a \to 0^+$:

$$\lim_{a\to 0^+} \int_a^1 \frac{dx}{\sqrt{x}} = \lim_{a\to 0^+} \left(2 - 2\sqrt{a}\right) = 2.$$

Therefore, the area under the curve from 0 to 1 is finite and is defined to be

$$\int_0^1 \frac{dx}{\sqrt{x}} = \lim_{a\to 0^+} \int_a^1 \frac{dx}{\sqrt{x}} = 2.$$

DEFINITION Integrals of functions that become infinite at a point within the interval of integration are **improper integrals of Type II**.

1. If $f(x)$ is continuous on $(a, b]$ and discontinuous at a, then

$$\int_a^b f(x)\,dx = \lim_{c\to a^+} \int_c^b f(x)\,dx.$$

2. If $f(x)$ is continuous on $[a, b)$ and discontinuous at b, then

$$\int_a^b f(x)\,dx = \lim_{c\to b^-} \int_a^c f(x)\,dx.$$

3. If $f(x)$ is discontinuous at c, where $a < c < b$, and continuous on $[a, c) \cup (c, b]$, then

$$\int_a^b f(x)\,dx = \int_a^c f(x)\,dx + \int_c^b f(x)\,dx.$$

In each case, if the limit exists and is finite, we say that the improper integral **converges** and that the limit is the **value** of the improper integral. If the limit does not exist, the integral **diverges**.

In Part 3 of the definition, the integral on the left side of the equation converges if *both* integrals on the right side converge; otherwise, it diverges.

EXAMPLE 4 Investigate the convergence of

$$\int_0^1 \frac{1}{1-x}\,dx.$$

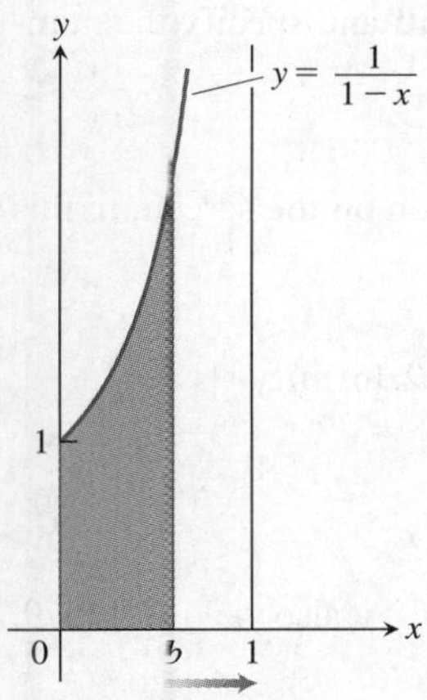

FIGURE 8.17 The area beneath the curve and above the x-axis for $[0, 1)$ is not a real number (Example 4).

Solution The integrand $f(x) = 1/(1 - x)$ is continuous on $[0, 1)$ but is discontinuous at $x = 1$ and becomes infinite as $x \to 1^-$ (Figure 8.17). We evaluate the integral as

$$\lim_{b\to 1^-}\int_0^b \frac{1}{1-x}\,dx = \lim_{b\to 1^-}\Big[-\ln|1-x|\Big]_0^b$$
$$= \lim_{b\to 1^-}[-\ln(1-b) + 0] = \infty.$$

The limit is infinite, so the integral diverges. ■

EXAMPLE 5 Evaluate

$$\int_0^3 \frac{dx}{(x-1)^{2/3}}.$$

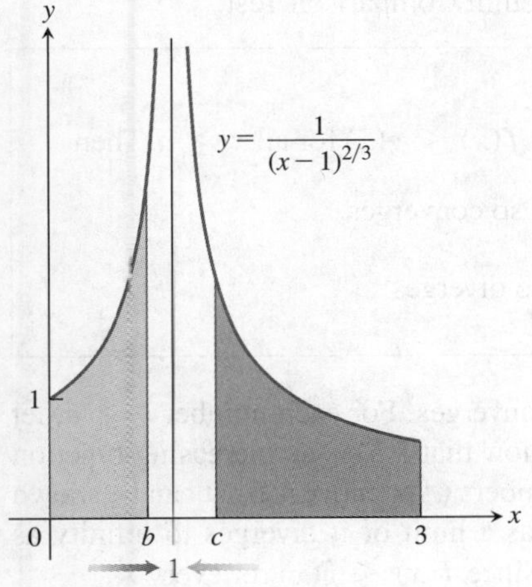

FIGURE 8.18 Example 5 shows that the area under the curve exists (so it is a real number).

Solution The integrand has a vertical asymptote at $x = 1$ and is continuous on $[0, 1)$ and $(1, 3]$ (Figure 8.18). Thus, by Part 3 of the definition above,

$$\int_0^3 \frac{dx}{(x-1)^{2/3}} = \int_0^1 \frac{dx}{(x-1)^{2/3}} + \int_1^3 \frac{dx}{(x-1)^{2/3}}.$$

Next, we evaluate each improper integral on the right-hand side of this equation.

$$\int_0^1 \frac{dx}{(x-1)^{2/3}} = \lim_{b\to 1^-}\int_0^b \frac{dx}{(x-1)^{2/3}}$$
$$= \lim_{b\to 1^-} 3(x-1)^{1/3}\Big]_0^b$$
$$= \lim_{b\to 1^-}\left[3(b-1)^{1/3} + 3\right] = 3$$

$$\int_1^3 \frac{dx}{(x-1)^{2/3}} = \lim_{c\to 1^+}\int_c^3 \frac{dx}{(x-1)^{2/3}}.$$
$$= \lim_{c\to 1^+} 3(x-1)^{1/3}\Big]_c^3$$
$$= \lim_{c\to 1^+}\left[3(3-1)^{1/3} - 3(c-1)^{1/3}\right] = 3\sqrt[3]{2}$$

We conclude that

$$\int_0^3 \frac{dx}{(x-1)^{2/3}} = 3 + 3\sqrt[3]{2}.$$ ■

Improper Integrals with a CAS

Computer algebra systems can evaluate many convergent improper integrals. To evaluate the integral

$$\int_2^\infty \frac{x+3}{(x-1)(x^2+1)}\,dx$$

(which converges) using Maple, enter

```
> f:= (x + 3)/((x - 1)*(x^2 + 1));
```

Then use the integration command

```
> int(f, x = 2..infinity);
```

Maple returns the answer

$$\ln(5) + \arctan(2) - \frac{\pi}{2}$$

To obtain a numerical result, use the evaluation command **evalf** and specify the number of digits as follows:

$$> \text{evalf}(\%, 6);$$

The symbol % instructs the computer to evaluate the last expression on the screen, in this case $(-1/2)\pi + \ln(5) + \arctan(2)$. Maple returns 1.14579.

If you are using Mathematica, entering

$$In[1]{:=}\ \text{Integrate}[(x + 3)/((x - 1)(x\hat{}2 + 1)),\ \{x, 2, \text{Infinity}\}]$$

returns

$$Out[1] = -\frac{\pi}{2} + \text{ArcTan}[2] + \text{Log}[5].$$

To obtain a numerical result with six digits, use the command "N[%, 6]"; it also yields 1.14579.

Tests for Convergence and Divergence

When we cannot evaluate an improper integral directly, we try to determine whether it converges or diverges. If the integral diverges, that's the end of the story. If it converges, we can use numerical methods to approximate its value. The principal tests for convergence or divergence are the Direct Comparison Test and the Limit Comparison Test.

THEOREM 2—Direct Comparison Test
Let f and g be continuous on $[a, \infty)$ with $0 \le f(x) \le g(x)$ for all $x \ge a$. Then

1. if $\displaystyle\int_a^\infty g(x)\,dx$ converges, then $\displaystyle\int_a^\infty f(x)\,dx$ also converges.

2. if $\displaystyle\int_a^\infty f(x)\,dx$ diverges, then $\displaystyle\int_a^\infty g(x)\,dx$ also diverges.

Outline of a Proof Assume that $\int_a^\infty g(x)\,dx$ converges. For each number $b \ge a$, let $I(b) = \int_a^b f(x)\,dx$. Since $f(x)$ is nonnegative, we know that $I(b)$ is an increasing function of b. Using the completeness property of the real numbers (Appendix A.6), it can be shown that since this integral increases with b, it either has a limit or it diverges to infinity as $b \to \infty$. We will not give a proof of this fact here. Since $f(x) \le g(x)$ for every x,

$$I(b) = \int_a^b f(x)\,dx \le \int_a^b g(x)\,dx \le \int_a^\infty g(x)\,dx.$$

Thus, the finite number $M = \int_a^\infty g(x)\,dx$ is an upper bound to the values of $I(b)$. Therefore $I(b)$ cannot diverge to infinity, so it must must converge to a finite value as $b \to \infty$. Hence $\int_a^\infty f(x)\,dx$ converges. This establishes that statement **1** holds. Since statement **2** is the contrapositive form of statement **1**, it must hold as well. ■

HISTORICAL BIOGRAPHY
Karl Weierstrass
(1815–1897)
www.bit.ly/2x0Ijv0

Although the theorem is stated for Type I improper integrals, a similar result is true for integrals of Type II as well.

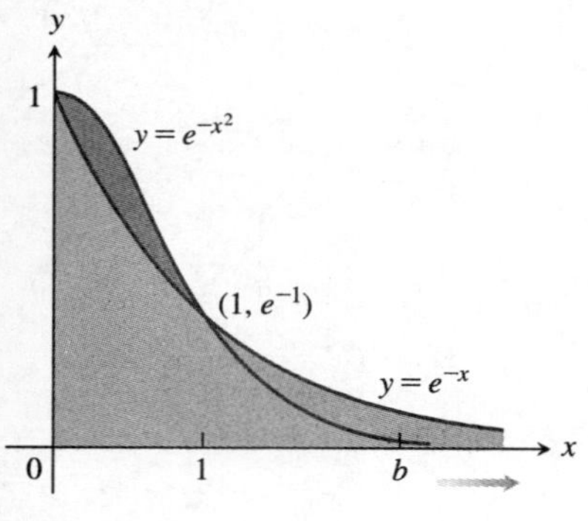

FIGURE 8.19 The graph of e^{-x^2} lies below the graph of e^{-x} for $x > 1$ (Example 6a).

EXAMPLE 6 These examples illustrate how we use Theorem 2.

(a) $\int_1^\infty e^{-x^2}\,dx$ converges because $0 < e^{-x^2} < e^{-x}$ for every $x \ge 1$ (Figure 8.19) and

$$\begin{aligned}
\int_1^\infty e^{-x}\,dx &= \lim_{b\to\infty} \int_1^b e^{-x}\,dx \\
&= \lim_{b\to\infty} \Big[-e^{-x}\,dx\Big]_1^b \\
&= \lim_{b\to\infty} (-e^{-b} + e^{-1}) \\
&= \frac{1}{e}
\end{aligned}$$

converges.

(b) $\displaystyle\int_1^{\infty} \frac{\sin^2 x}{x^2}\,dx$ converges because

$$0 \le \frac{\sin^2 x}{x^2} \le \frac{1}{x^2} \quad \text{on} \quad [1, \infty) \quad \text{and} \quad \int_1^{\infty} \frac{1}{x^2}\,dx \quad \text{converges.} \qquad \text{Example 3}$$

(c) $\displaystyle\int_1^{\infty} \frac{1}{\sqrt{x^2 - 0.1}}\,dx$ diverges because

$$\frac{1}{\sqrt{x^2 - 0.1}} \ge \frac{1}{x} \quad \text{on} \quad [1, \infty) \quad \text{and} \quad \int_1^{\infty} \frac{1}{x}\,dx \quad \text{diverges.} \qquad \text{Example 3}$$

(d) $\displaystyle\int_0^{\pi/2} \frac{\cos x}{\sqrt{x}}\,dx$ converges because

$$0 \le \frac{\cos x}{\sqrt{x}} \le \frac{1}{\sqrt{x}} \quad \text{on} \quad \left[0, \frac{\pi}{2}\right], \qquad 0 \le \cos x \le 1 \text{ on } \left[0, \frac{\pi}{2}\right]$$

and

$$\begin{aligned} \int_0^{\pi/2} \frac{dx}{\sqrt{x}} &= \lim_{a \to 0^+} \int_a^{\pi/2} \frac{dx}{\sqrt{x}} \\ &= \lim_{a \to 0^+} \sqrt{4x}\,\Big]_a^{\pi/2} && 2\sqrt{x} = \sqrt{4x} \\ &= \lim_{a \to 0^+} \left(\sqrt{2\pi} - \sqrt{4a}\right) = \sqrt{2\pi} && \text{converges.} \end{aligned}$$ ■

Although we have shown that the integrals in parts (a), (b), and (d) of Example 6 converge, we do not know the exact values of these integrals. For example, our computations in part (d) imply that

$$0 \le \int_0^{\pi/2} \frac{\cos x}{\sqrt{x}}\,dx \le \int_0^{\pi/2} \frac{dx}{\sqrt{x}} = \sqrt{2\pi},$$

but unless we do further calculations the most we can say is that the integral is some real number between 0 and $\sqrt{2\pi}$.

THEOREM 3—Limit Comparison Test
If the positive functions f and g are continuous on $[a, \infty)$, and if

$$\lim_{x \to \infty} \frac{f(x)}{g(x)} = L, \quad 0 < L < \infty,$$

then

$$\int_a^{\infty} f(x)\,dx \quad \text{and} \quad \int_a^{\infty} g(x)\,dx$$

either *both converge* or *both diverge*.

We omit the proof of Theorem 3, which is similar to that of Theorem 2.

If two functions $f(x)$ and $g(x)$ satisfy the hypotheses of Theorem 3 and their improper integrals both converge, it need not be the case that these two integrals have the same value. This is illustrated in the next example.

EXAMPLE 7 Show that

$$\int_1^{\infty} \frac{dx}{1 + x^2}$$

converges by comparison with $\int_1^{\infty} (1/x^2)\,dx$. Find and compare the values of the two integrals.

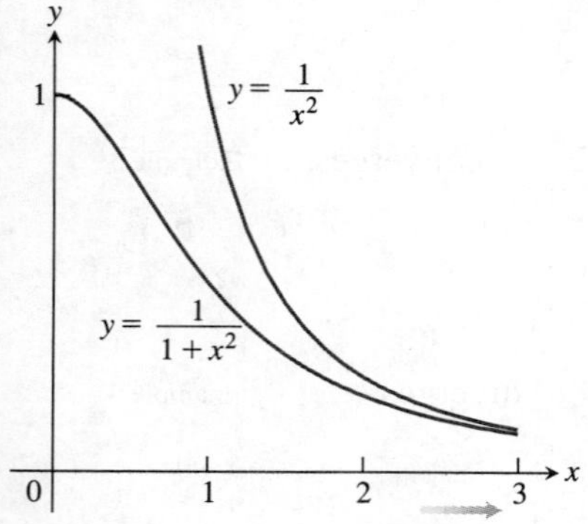

FIGURE 8.20 The functions in Example 7.

Solution The functions $f(x) = 1/x^2$ and $g(x) = 1/(1 + x^2)$ are positive and continuous on $[1, \infty)$. Also,

$$\lim_{x\to\infty} \frac{f(x)}{g(x)} = \lim_{x\to\infty} \frac{1/x^2}{1/(1+x^2)} = \lim_{x\to\infty} \frac{1+x^2}{x^2}$$

$$= \lim_{x\to\infty} \left(\frac{1}{x^2} + 1\right) = 0 + 1 = 1,$$

which is a positive finite limit (Figure 8.20). Therefore, $\int_1^\infty \frac{dx}{1+x^2}$ converges because $\int_1^\infty \frac{dx}{x^2}$ converges.

The integrals converge to different values, however:

$$\int_1^\infty \frac{dx}{x^2} = \frac{1}{2-1} = 1 \qquad \text{Example 3}$$

and

$$\int_1^\infty \frac{dx}{1+x^2} = \lim_{b\to\infty} \int_1^b \frac{dx}{1+x^2} = \lim_{b\to\infty} [\tan^{-1} b - \tan^{-1} 1] = \frac{\pi}{2} - \frac{\pi}{4} = \frac{\pi}{4}. \quad \blacksquare$$

TABLE 8.5

b	$\int_1^b \frac{1-e^{-x}}{x}\,dx$
2	0.5226637569
5	1.3912002736
10	2.0832053156
100	4.3857862516
1000	6.6883713446
10000	8.9909564376
100000	11.2935415306

EXAMPLE 8 Investigate the convergence of $\int_1^\infty \frac{1-e^{-x}}{x}\,dx$.

Solution The integrand suggests a comparison of $f(x) = (1 - e^{-x})/x$ with $g(x) = 1/x$. However, we cannot use the Direct Comparison Test because $f(x) \le g(x)$ and the integral of $g(x)$ *diverges*. On the other hand, using the Limit Comparison Test, we find that

$$\lim_{x\to\infty} \frac{f(x)}{g(x)} = \lim_{x\to\infty} \left(\frac{1-e^{-x}}{x}\right)\left(\frac{x}{1}\right) = \lim_{x\to\infty} (1 - e^{-x}) = 1,$$

which is a positive finite limit. Therefore, $\int_1^\infty \frac{1-e^{-x}}{x}\,dx$ diverges because $\int_1^\infty \frac{dx}{x}$ diverges. Approximations to the improper integral are given in Table 8.5. Note that the values of these approximations do not appear to approach a fixed finite limit as $b \to \infty$. $\blacksquare$

EXERCISES 8.8

Evaluating Improper Integrals

The integrals in Exercises 1–34 converge. Evaluate the integrals without using tables.

1. $\int_0^\infty \frac{dx}{x^2+1}$

2. $\int_1^\infty \frac{dx}{x^{1.005}}$

3. $\int_0^1 \frac{dx}{\sqrt{x}}$

4. $\int_0^4 \frac{dx}{\sqrt{4-x}}$

5. $\int_{-1}^1 \frac{dx}{x^{4/5}}$

6. $\int_{-8}^1 \frac{dx}{x^{1/3}}$

7. $\int_0^1 \frac{dx}{\sqrt{4x-x^2}}$

8. $\int_0^3 \frac{1}{\sqrt{9-x^2}}\,dx$

9. $\int_{-\infty}^{-4} \frac{4dx}{x^2-4}$

10. $\int_{-\infty}^2 \frac{2\,dx}{x^2+4}$

11. $\int_2^\infty \frac{2}{v^2+v}\,dv$

12. $\int_2^\infty \frac{x+3}{(x-1)(x^2+1)}\,dx$

13. $\int_{-\infty}^\infty \frac{2x\,dx}{(x^2+1)^2}$

14. $\int_{-\infty}^\infty \frac{2x\,dx}{(x^2+9)^{3/2}}$

15. $\int_0^1 \frac{\theta+1}{\sqrt{\theta^2+2\theta}}\,d\theta$

16. $\int_0^2 \frac{s+1}{\sqrt{4-s^2}}\,ds$

17. $\int_0^\infty \frac{dx}{(1+x^2)\sqrt{x}}$

18. $\int_1^\infty \frac{1}{x\sqrt{x^2-1}}\,dx$

19. $\int_0^\infty \frac{dv}{(1+v^2)(1+\tan^{-1} v)}$

20. $\int_0^1 \frac{8\sin^{-1} x}{\sqrt{1-x^2}}\,dx$

21. $\int_{-\infty}^0 \theta e^\theta\,d\theta$

22. $\int_0^\infty 2e^{-\theta} \sin\theta\,d\theta$

23. $\int_{-\infty}^0 e^{-|x|}\,dx$

24. $\int_{-\infty}^\infty 2x\,e^{-x^2}\,dx$

25. $\int_0^1 x \ln x\,dx$

26. $\int_0^2 \frac{dx}{(x-1)^{2/3}}$

27. $\int_0^2 \frac{ds}{\sqrt{4 - s^2}}$

28. $\int_0^1 \frac{4r\,dr}{\sqrt{1 - r^4}}$

29. $\int_1^2 \frac{ds}{s\sqrt{s^2 - 1}}$

30. $\int_2^4 \frac{dt}{t\sqrt{t^2 - 4}}$

31. $\int_{-1}^4 \frac{dx}{\sqrt{|x|}}$

32. $\int_0^2 \frac{dx}{\sqrt{|x - 1|}}$

33. $\int_{-1}^{\infty} \frac{d\theta}{\theta^2 + 5\theta + 6}$

34. $\int_0^{\infty} \frac{dx}{(x + 1)(x^2 + 1)}$

Testing for Convergence

In Exercises 35–68, use integration, the Direct Comparison Test, or the Limit Comparison Test to test the integrals for convergence. If more than one method applies, use whatever method you prefer.

35. $\int_{1/2}^2 \frac{dx}{x \ln x}$

36. $\int_{-1}^1 \frac{d\theta}{\theta^2 - 2\theta}$

37. $\int_{1/2}^{\infty} \frac{dx}{x(\ln x)^3}$

38. $\int_0^{\infty} \frac{d\theta}{\theta^2 - 1}$

39. $\int_0^{\pi/2} \tan\theta\, d\theta$

40. $\int_0^{\pi/2} \sec^2\theta\, d\theta$

41. $\int_0^1 \frac{\ln x}{x^2}\, dx$

42. $\int_1^2 \frac{dx}{x \ln x}$

43. $\int_0^{\ln 2} x^{-2} e^{-1/x}\, dx$

44. $\int_0^{\ln 2} x^{-3} e^{-1/x^2}\, dx$

45. $\int_0^1 \frac{x}{\sqrt{1 - x^2}}\, dx$

46. $\int_0^1 \frac{dt}{t - \cos t}$
(*Hint:* $t \geq \cos t$ for $t \geq 0$)

47. $\int_0^2 \frac{dx}{1 - x^2}$

48. $\int_0^1 \frac{dx}{1 + x}$

49. $\int_{-1}^1 \ln|x|\, dx$

50. $\int_{-1}^1 -x \ln|x|\, dx$

51. $\int_1^{\infty} \frac{dx}{x^3 - 1}$

52. $\int_4^{\infty} \frac{dx}{\sqrt{x} - 1}$

53. $\int_2^{\infty} \frac{dv}{\sqrt{v - 1}}$

54. $\int_1^{\infty} \frac{d\theta}{e^{\theta} - 1}$

55. $\int_0^{\infty} \frac{dx}{\sqrt{x^6 + 1}}$

56. $\int_2^{\infty} \frac{dx}{\sqrt{x^2 - 1}}$

57. $\int_1^{\infty} \frac{\sqrt{x^2 + 1}}{x^3}\, dx$

58. $\int_1^{\infty} \frac{\ln x}{x^2}\, dx$

59. $\int_{\pi}^{\infty} \frac{2 + \cos x}{x}\, dx$

60. $\int_2^{\infty} \frac{\cos^2 x}{x^2}\, dx$

61. $\int_4^{\infty} \frac{2\,dt}{t^{3/2} - 1}$

62. $\int_2^{\infty} \frac{1}{\ln x}\, dx$

63. $\int_1^{\infty} \frac{e^x}{x}\, dx$

64. $\int_{e^e}^{\infty} \ln(\ln x)\, dx$

65. $\int_1^{\infty} \frac{1}{\sqrt{e^x - x}}\, dx$

66. $\int_3^{\infty} \frac{1}{x + e^x}\, dx$

67. $\int_{-\infty}^{\infty} \frac{4x}{\sqrt{x^4 + 1}}\, dx$

68. $\int_{-\infty}^{\infty} \frac{dx}{e^x + e^{-x}}$

In Exercises 69–80, determine whether the improper integral converges or diverges. If it converges, evaluate the integral.

69. $\int_0^1 \frac{1}{x\sqrt{x}}\, dx$

70. $\int_2^{\infty} \frac{1}{x\sqrt{x}}\, dx$

71. $\int_0^{32} \frac{1}{\sqrt[5]{x}}\, dx$

72. $\int_1^{\infty} \frac{1}{\sqrt[5]{x}}\, dx$

73. $\int_3^{\infty} \frac{1}{x^4}\, dx$

74. $\int_{-2}^1 \frac{1}{x^4}\, dx$

75. $\int_0^{\infty} x^2 e^{x^3}\, dx$

76. $\int_{-\infty}^0 x^2 e^{x^3}\, dx$

77. $\int_{-3}^0 \frac{1}{x^2 + 3x}\, dx$

78. $\int_1^{\infty} \frac{1}{x^2 + 3x}\, dx$

79. $\int_{-\infty}^4 \frac{x}{(x^2 + 9)^{5/2}}\, dx$

80. $\int_{-\infty}^4 \frac{x}{(x^2 + 9)^{2/5}}\, dx$

Theory and Examples

81. Find the values of p for which each integral converges.

a. $\int_0^1 \frac{\ln(1 + x)}{x^p}\, dx$

b. $\int_0^{\infty} \frac{\ln(1 + x)}{x^p}\, dx$

82. **$\int_{-\infty}^{\infty} f(x)\, dx$ may not equal $\lim_{b \to \infty} \int_{-b}^{b} f(x)\, dx$.** Show that

$$\int_0^{\infty} \frac{2x\,dx}{x^2 + 1}$$

diverges and hence that

$$\int_{-\infty}^{\infty} \frac{2x\,dx}{x^2 + 1}$$

diverges. Then show that

$$\lim_{b \to \infty} \int_{-b}^{b} \frac{2x\,dx}{x^2 + 1} = 0.$$

Exercises 83–86 are about the infinite region in the first quadrant between the curve $y = e^{-x}$ and the x-axis.

83. Find the area of the region.

84. Find the centroid of the region.

85. Find the volume of the solid generated by revolving the region about the y-axis.

86. Find the volume of the solid generated by revolving the region about the x-axis.

87. Find the area of the region that lies between the curves $y = \csc x$ and $y = \cot x$ from $x = 0$ to $x = \pi/2$.

88. The region in Exercise 87 is revolved about the x-axis to generate a solid.

a. Find the volume of the solid.

b. Show that the inner and outer surfaces of the solid have infinite area.

89. Consider the infinite region in the first quadrant bounded by the graphs of $= \dfrac{1}{x^3}$, $y = 0$, and $x = 1$.

a. Find the area of the region.

b. Find the volume of the solid formed by revolving the region (i) about the x-axis; (ii) about the y-axis.

90. Consider the infinite region in the first quadrant bounded by the graphs of $y = \dfrac{1}{\sqrt{x}}$, $y = 0$, $x = 0$, and $x = 1$.

a. Find the area of the region.

b. Find the volume of the solid formed by revolving the region (i) about the x-axis; (ii) about the y-axis.

91. Evaluate the integrals.

a. $\displaystyle\int_0^1 \frac{dt}{\sqrt{t}(1+t)^2}$ **b.** $\displaystyle\int_0^\infty \frac{dt}{\sqrt{t}(1+t)^2}$

92. Evaluate $\displaystyle\int_4^\infty \frac{2dx}{x\sqrt{x^2-16}}$

93. Estimating the value of a convergent improper integral whose domain is infinite

a. Show that

$$\int_3^\infty e^{-3x}\,dx = \frac{1}{3}e^{-9} < 0.000042,$$

and hence that $\int_3^\infty e^{-x^2}\,dx < 0.000042$. Explain why this means that $\int_0^\infty e^{-x^2}\,dx$ can be replaced by $\int_0^3 e^{-x^2}\,dx$ without introducing an error of magnitude greater than 0.000042.

T **b.** Evaluate $\int_0^3 e^{-x^2}\,dx$ numerically.

94. The infinite paint can or Gabriel's horn As Example 3 shows, the integral $\int_1^\infty (dx/x)$ diverges. This means that the integral

$$\int_1^\infty 2\pi\frac{1}{x}\sqrt{1+\frac{1}{x^4}}\,dx,$$

which measures the *surface area* of the solid of revolution traced out by revolving the curve $y = 1/x$, $1 \le x$, about the x-axis, diverges also. By comparing the two integrals, we see that, for every finite value $b > 1$,

$$\int_1^b 2\pi\frac{1}{x}\sqrt{1+\frac{1}{x^4}}\,dx > 2\pi\int_1^b \frac{1}{x}\,dx.$$

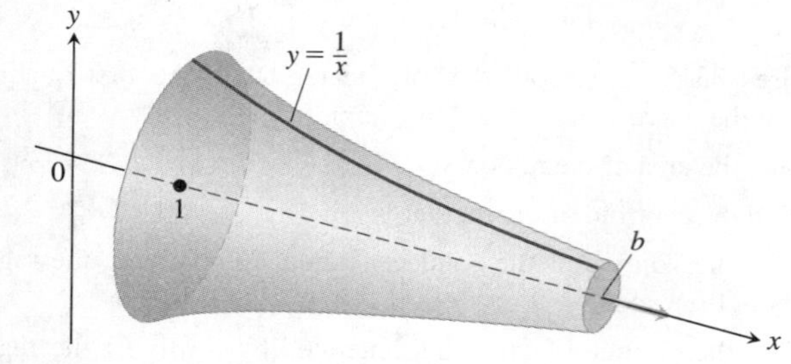

However, the integral

$$\int_1^\infty \pi\left(\frac{1}{x}\right)^2 dx$$

for the *volume* of the solid converges.

a. Calculate it.

b. This solid of revolution is sometimes described as a can that does not hold enough paint to cover its own interior. Think about that for a moment. It is common sense that a finite amount of paint cannot cover an infinite surface. But if we fill the horn with paint (a finite amount), then we *will* have covered an infinite surface. Explain the apparent contradiction.

95. Sine-integral function The integral

$$\text{Si}(x) = \int_0^x \frac{\sin t}{t}\,dt,$$

called the *sine-integral function*, has important applications in optics.

T **a.** Plot the integrand $(\sin t)/t$ for $t > 0$. Is the sine-integral function everywhere increasing or decreasing? Do you think $\text{Si}(x) = 0$ for $x \ge 0$? Check your answers by graphing the function $\text{Si}(x)$ for $0 \le x \le 25$.

b. Explore the convergence of

$$\int_0^\infty \frac{\sin t}{t}\,dt.$$

If it converges, what is its value?

96. Error function The function

$$\text{erf}(x) = \int_0^x \frac{2e^{-t^2}}{\sqrt{\pi}}\,dt,$$

called the *error function*, has important applications in probability and statistics.

T **a.** Plot the error function for $0 \le x \le 25$.

b. Explore the convergence of

$$\int_0^\infty \frac{2e^{-t^2}}{\sqrt{\pi}}\,dt.$$

If it converges, what appears to be its value? You will see how to confirm your estimate in Section 15.4, Exercise 41.

97. Normal probability distribution The function

$$f(x) = \frac{1}{\sigma\sqrt{2\pi}}e^{-\frac{1}{2}\left(\frac{x-\mu}{\sigma}\right)^2}$$

is called the *normal probability density function* with mean μ and standard deviation σ. The number μ tells where the distribution is centered, and σ measures the "scatter" around the mean.

From the theory of probability, it is known that

$$\int_{-\infty}^\infty f(x)\,dx = 1.$$

In what follows, let $\mu = 0$ and $\sigma = 1$.

T **a.** Draw the graph of f. Find the intervals on which f is increasing, the intervals on which f is decreasing, and any local extreme values and where they occur.

b. Evaluate

$$\int_{-n}^n f(x)\,dx$$

for $n = 1, 2$, and 3.

c. Give a convincing argument that

$$\int_{-\infty}^\infty f(x)\,dx = 1.$$

(*Hint:* Show that $0 < f(x) < e^{-x/2}$ for $x > 1$, and for $b > 1$,

$$\int_b^\infty e^{-x/2}\,dx \to 0 \quad \text{as} \quad b \to \infty.)$$

98. Show that if $f(x)$ is integrable on every interval of real numbers, and if a and b are real numbers with $a < b$, then

a. $\int_{-\infty}^{c} f(x)\,dx$ and $\int_{a}^{\infty} f(x)\,dx$ both converge if and only if $\int_{-\infty}^{b} f(x)\,dx$ and $\int_{b}^{\infty} f(x)\,dx$ both converge.

b. $\int_{-\infty}^{c} f(x)\,dx + \int_{a}^{\infty} f(x)\,dx = \int_{-\infty}^{b} f(x)\,dx + \int_{b}^{\infty} f(x)\,dx$ when the integrals involved converge.

COMPUTER EXPLORATIONS

In Exercises 99–102, use a CAS to explore the integrals for various values of p (include noninteger values). For what values of p does the integral converge? What is the value of the integral when it does converge? Plot the integrand for various values of p.

99. $\displaystyle\int_0^e x^p \ln x\,dx$

100. $\displaystyle\int_e^{\infty} x^p \ln x\,dx$

101. $\displaystyle\int_0^{\infty} x^p \ln x\,dx$

102. $\displaystyle\int_{-\infty}^{\infty} x^p \ln|x|\,dx$

Use a CAS to evaluate the integrals.

103. $\displaystyle\int_0^{2/\pi} \sin\frac{1}{x}\,dx$

104. $\displaystyle\int_0^{2/\pi} x\sin\frac{1}{x}\,dx$

8.9 Probability

The outcome of some events, such as a heavy rock falling from a great height, can be modeled so that we can predict with high accuracy what will happen. On the other hand, many events have more than one possible outcome and which one of them will occur is uncertain. If we toss a coin, a head or a tail will result with each outcome being equally likely, but we do not know in advance which one it will be. If we randomly select and then weigh a person from a large population, there are many possible weights the person might have, and it is not certain whether the weight will be between 180 and 190 lb. We are told it is highly likely, but not known for sure, that an earthquake of magnitude 6.0 or greater on the Richter scale will occur near a major population area in California within the next one hundred years. Events having more than one possible outcome are *probabilistic* in nature, and when modeling them we assign a *probability* to the likelihood that a particular outcome may occur. In this section we show how calculus plays a central role in making predictions with probabilistic models.

Random Variables

We begin our discussion with some familiar examples of uncertain events for which the collection of all possible outcomes is finite.

EXAMPLE 1

(a) If we toss a coin once, there are two possible outcomes $\{H, T\}$, where H represents the coin landing head face up and T a tail landing face up. If we toss a coin three times, there are eight possible outcomes, taking into account the order in which a head or tail occurs. The set of outcomes is $\{HHH, HHT, HTH, THH, HTT, THT, TTH, TTT\}$.

(b) If we roll a six-sided die once, the set of possible outcomes is $\{1, 2, 3, 4, 5, 6\}$ representing the six faces of the die.

(c) If we select at random two cards from a 52-card deck, there are 52 possible outcomes for the first card drawn and then 51 possibilities for the second card. Since the order of the cards does not matter, there are $(52 \cdot 51)/2 = 1{,}326$ possible outcomes altogether. ■

It is customary to refer to the set of all possible outcomes as the *sample space* for an event. With an uncertain event we are usually interested in which outcomes, if any, are more likely to occur than others, and to how large an extent. In tossing a coin three times, is it more likely that two heads or that one head will result? To answer such questions, we need a way to quantify the outcomes.

DEFINITION A **random variable** is a function X that assigns a numerical value to each outcome in a sample space.

Random variables that have only finitely many values are called **discrete** random variables. A **continuous random variable** can take on values in an entire interval, and it is associated with a *distribution function*, which we explain later.

EXAMPLE 2

(a) Suppose we toss a coin three times giving the possible outcomes {HHH, HHT, HTH, THH, HTT, THT, TTH, TTT}. Define the random variable X to be the number of heads that appear. So $X(\text{HHT}) = 2, X(\text{THT}) = 1$, and so forth. Since X can only assume the values 0, 1, 2, or 3, it is a discrete random variable.

(b) We spin an arrow anchored by a pin located at the origin. The arrow can wind up pointing in any possible direction and we define the random variable X as the radian angle the arrow makes with the positive x-axis, measured counterclockwise. In this case, X is a continuous random variable that can take on any value in the interval $[0, 2\pi)$.

(c) The weight of a randomly selected person in a given population is a continuous random variable W. The cholesterol level of a randomly chosen person, and the waiting time for service of a person in a queue at a bank, are also continuous random variables.

(d) The scores on the national ACT Examination for college admissions in a particular year are described by a discrete random variable S taking on integer values between 1 and 36. If the number of outcomes is large, or for reasons involving statistical analysis, discrete random variables such as test scores are often modeled as continuous random variables (Example 13).

(e) We roll a pair of dice and define the random variable X to be the sum of the numbers on the top faces. This sum can only assume the integer values from 2 through 12, so X is a discrete random variable.

(f) A tire company produces tires for mid-sized sedans. The tires are guaranteed to last for 30,000 miles, but some will fail sooner and some will last many more miles beyond 30,000. The lifetime in miles of a tire is described by a continuous random variable L. ■

Probability Distributions

A *probability distribution* describes the probabilistic behavior of a random variable. Our chief interest is in probability distributions associated with continuous random variables, but to gain some perspective we first consider a distribution for a discrete random variable.

Suppose we toss a coin three times, with each side H or T equally likely to occur on a given toss. We define the discrete random variable X that assigns the number of heads appearing in each outcome, giving the following.

$$\begin{array}{ccccccccc} & \{\text{HHH}, & \text{HHT}, & \text{HTH}, & \text{HTT}, & \text{THH}, & \text{THT}, & \text{TTH}, & \text{TTT}\} \\ X & \downarrow & \downarrow & \downarrow & \downarrow & \downarrow & \downarrow & \downarrow & \downarrow \\ & 3 & 2 & 2 & 1 & 2 & 1 & 1 & 0 \end{array}$$

Next we count the *frequency* or number of times a specific value of X occurs. Because each of the eight outcomes is equally likely to occur, we can calculate the probability of the random variable X by dividing the frequency of each value by the total number of outcomes. We summarize our results as follows:

Value of X	0	1	2	3
Frequency	1	3	3	1
$P(X)$	1/8	3/8	3/8	1/8

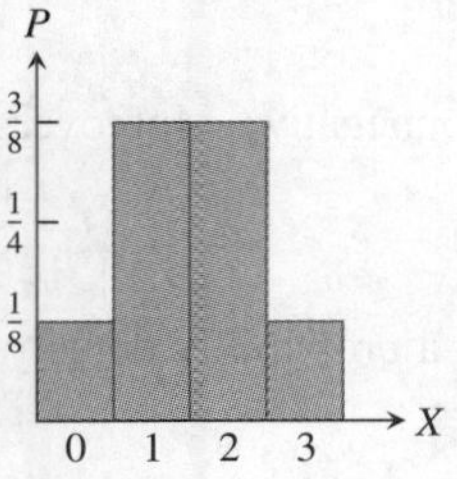

FIGURE 8.21 Probability bar graph for the random variable X when tossing a fair coin three times.

We display this information in a probability bar graph of the discrete random variable X, as shown in Figure 8.21. The values of X are portrayed by intervals of length 1 on the x-axis so the area of each bar in the graph is the probability of the corresponding outcome. For instance, the probability that exactly two heads occurs in the three tosses of the coin is the area of the bar associated with the value $X = 2$, which is 3/8. Similarly, the probability that two or more heads occurs is the sum of areas of the bars associated with the values $X = 2$ and $X = 3$, or 4/8. The probability that either zero or three heads occurs is $\frac{1}{8} + \frac{1}{8} = \frac{1}{4}$, and so forth. Note that the total area of all the bars in the graph is 1, which is the sum of all the probabilities for X.

With a continuous random variable, even when the outcomes are equally likely, we cannot simply count the number of outcomes in the sample space or the frequencies of outcomes that lead to a specific value of X. In fact, the probability that X takes on any particular one of its values is zero. What *is* meaningful to ask is how probable it is that the random variable takes on a value within some specified *interval* of values.

We capture the information we need about the probabilities of X in a function whose graph behaves much like the bar graph in Figure 8.21. That is, we take a nonnegative function f defined over the range of the random variable with the property that the total area beneath the graph of f is 1. The probability that a value of the random variable X lies within some specified interval $[c, d]$ is then the area under the graph of f over that interval. The following definition assumes the range of the continuous random variable X is any real value, but the definition is general enough to account for random variables having a range of finite length.

DEFINITIONS A **probability density function** for a continuous random variable is a function f defined over $(-\infty, \infty)$ and having the following properties:

1. f is continuous, except possibly at a finite number of points.
2. f is nonnegative, so $f \geq 0$.
3. $\int_{-\infty}^{\infty} f(x)\,dx = 1$.

If X is a continuous random variable with probability density function f, the **probability** that X assumes a value in the interval between $X = c$ and $X = d$ is given by the integral

$$P(c \leq X \leq d) = \int_c^d f(x)\,dx.$$

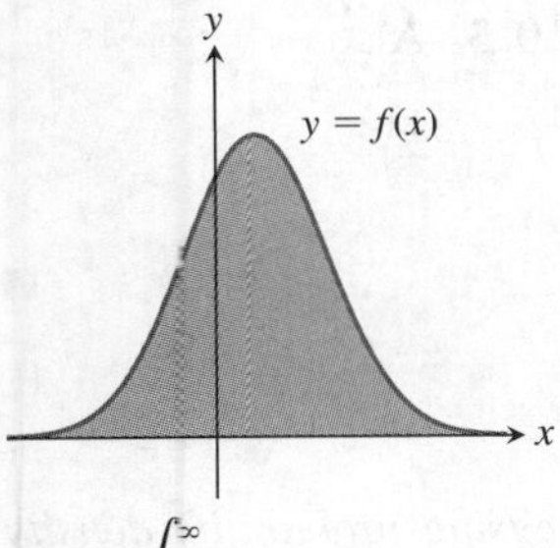

$\int_{-\infty}^{\infty} f(x)\,dx = 1$

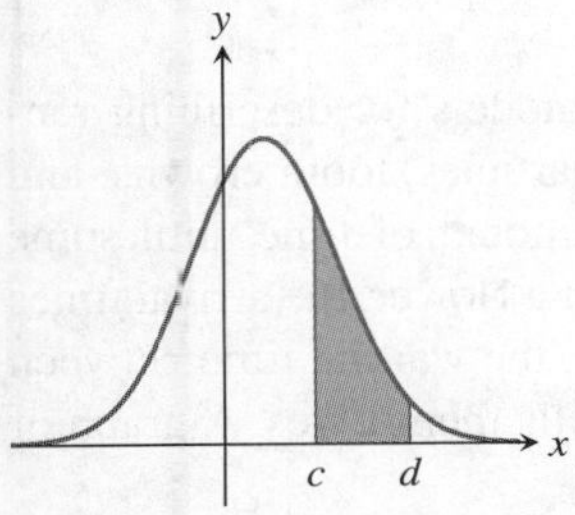

$P(c \leq X \leq d) = \int_c^d f(x)\,dx$

FIGURE 8.22 A probability density function for the continuous random variable X.

The probability that a continuous random variable X assumes a particular real value c is $P(X = c) = \int_c^c f(x)\,dx = 0$, consistent with our previous assertion. Since the area under the graph of f over the interval $[c, d]$ is only a portion of the total area beneath the graph, the probability $P(c \leq X \leq d)$ is always a number between zero and one. Figure 8.22 illustrates a probability density function.

A probability density function for a random variable X resembles the density function for a wire of varying density. To obtain the mass of a segment of the wire, we integrate the density of the wire over an interval. To obtain the probability that a random variable has values in a particular interval, we integrate the probability density function over that interval.

EXAMPLE 3 Let $f(x) = 2e^{-2x}$ if $0 \leq x < \infty$ and $f(x) = 0$ for all negative values of x.

(a) Verify that f is a probability density function.

(b) The time T in hours until a car passes a spot on a remote road is described by the probability density function f. Find the probability $P(T \leq 1)$ that a hitchhiker at that spot will see a car within one hour.

(c) Find the probability $P(T = 1)$ that a car passes by the spot after precisely one hour.

Solution

(a) The function f is continuous except at $x = 0$, and is everywhere nonnegative. Moreover,

$$\int_{-\infty}^{\infty} f(x)\,dx = \int_0^{\infty} 2e^{-2x}\,dx = \lim_{b\to\infty}\int_0^b 2e^{-2x}\,dx = \lim_{b\to\infty}(1 - e^{-2b}) = 1.$$

So all of the conditions are satisfied and we have shown that f is a probability density function.

(b) The probability that a car comes after a time lapse between zero and one hour is given by integrating the probability density function over the interval $[0, 1]$. So

$$P(T \le 1) = \int_0^1 2e^{-2t}\,dt = -e^{-2t}\Big]_0^1 = 1 - e^{-2} \approx 0.865.$$

This result can be interpreted to mean that if 100 people were to hitchhike at that spot, about 87 of them can expect to see a car within one hour.

(c) This probability is the integral $\int_1^1 f(t)\,dt$, which equals zero. We interpret this to mean that a sufficiently accurate measurement of the time until a car comes by the spot would have no possibility of being precisely equal to one hour. It might be very close, perhaps, but it would not be exactly one hour. ■

We can extend the definition to finite intervals. If f is a nonnegative function with at most finitely many discontinuities over the interval $[a, b]$, and its extension F to $(-\infty, \infty)$, obtained by defining F to be 0 outside of $[a, b]$, satisfies the definition for a probability density function, then f is a **probability density function for $[a, b]$**. This means that $\int_a^b f(x)\,dx = 1$. Similar definitions can be made for the intervals (a, b), $(a, b]$, and $[a, b)$.

EXAMPLE 4 Show that $f(x) = \dfrac{4}{27}x^2(3 - x)$ is a probability density function over the interval $[0, 3]$.

Solution The function f is continuous and nonnegative over $[0, 3]$. Also,

$$\int_0^3 \frac{4}{27}x^2(3 - x)\,dx = \frac{4}{27}\left[x^3 - \frac{1}{4}x^4\right]_0^3 = \frac{4}{27}\left(27 - \frac{81}{4}\right) = 1.$$

We conclude that f is a probability density function over $[0, 3]$. ■

Exponentially Decreasing Distributions

The distribution in Example 3 is called an *exponentially decreasing probability density function*. These probability density functions always take on the form

$$f(x) = \begin{cases} 0 & \text{if } x < 0 \\ ce^{-cx} & \text{if } x \ge 0 \end{cases}$$

(see Exercise 23). Exponential density functions can provide models for describing random variables such as the lifetimes of light bulbs, radioactive particles, tooth crowns, and many kinds of electronic components. They also model the amount of time until some specific event occurs, such as the time until a pollinator arrives at a flower, the arrival times of a bus at a stop, the time between individuals joining a queue, the waiting time between phone calls at a help desk, and even the lengths of the phone calls themselves. A graph of an exponential density function is shown in Figure 8.23.

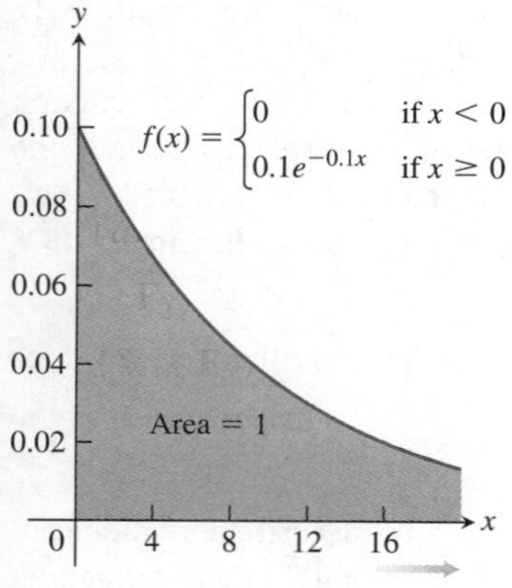

FIGURE 8.23 An exponentially decreasing probability density function.

Random variables with exponential distributions are *memoryless*. If we think of X as describing the lifetime of some object, then the probability that the object survives for at least $s + t$ hours, given that it has survived t hours, is the same as the initial probability that it survives for at least s hours. For instance, the current age t of a radioactive particle does not change the probability that it will survive for at least another time period of length s. Sometimes the exponential distribution is used as a model when the memoryless principle is violated because it provides reasonable approximations that are good enough for their intended use. For instance, this might be the case when predicting the lifetime of an artificial hip replacement or heart valve for a particular patient. Here is an application illustrating the exponential distribution.

EXAMPLE 5 An electronics company models the lifetime T in years of a chip they manufacture with the exponential density function

$$f(t) = \begin{cases} 0 & \text{if } t < 0 \\ 0.1e^{-0.1t} & \text{if } t \geq 0. \end{cases}$$

Using this model,

(a) Find the probability $P(T > 2)$ that a chip will last for more than two years.

(b) Find the probability $P(4 \leq T \leq 5)$ that a chip will fail in the fifth year.

(c) If 1000 chips are shipped to a customer, how many can be expected to fail within three years?

Solution

(a) The probability that a chip lasts at least two years is

$$P(T > 2) = \int_2^\infty 0.1e^{-0.1t}\,dt = \lim_{b\to\infty} \int_2^b 0.1e^{-0.1t}\,dt$$

$$= \lim_{b\to\infty} [e^{-0.2} - e^{-0.1b}] = e^{-0.2} \approx 0.819.$$

That is, about 82% of the chips last more than two years.

(b) The probability is

$$P(4 \leq T \leq 5) = \int_4^5 0.1e^{-0.1t}\,dt = -e^{-0.1t}\Big]_4^5 = e^{-0.4} - e^{-0.5} \approx 0.064$$

which means that about 6% of the chips fail during the fifth year.

(c) We want the probability

$$P(0 \leq T \leq 3) = \int_0^3 0.1e^{-0.1t}\,dt = -e^{-0.1t}\Big]_0^3 = 1 - e^{-0.3} \approx 0.259.$$

We can expect that about 259 of the 1000 chips will fail within three years. ■

Expected Values, Means, and Medians

Suppose the weight in lbs of a steer raised on a cattle ranch is described by a continuous random variable W with probability density function $f(w)$ and that the rancher can sell a steer of weight w for $g(w)$ dollars. How much can the rancher expect to earn for a randomly chosen steer on the ranch?

To answer this question, we consider a small interval $[w_i, w_{i+1}]$ of width Δw_i and note that the probability a steer has weight in this interval is

$$\int_{w_i}^{w_{i+1}} f(w)\,dw \approx f(w_i)\,\Delta w_i.$$

The earning on a steer in this interval is approximately $g(w_i)$. The Riemann sum

$$\sum g(w_i)\,f(w_i)\,\Delta w_i$$

then approximates the amount the rancher would receive for a steer. We assume that steers have a maximum weight, so f is zero outside some finite interval $[0, b]$. Then taking the limit of the Riemann sum as the width of each interval approaches zero gives the integral

$$\int_{-\infty}^{\infty} g(w)\,f(w)\,dw.$$

This integral estimates how much the rancher can expect to earn for a typical steer on the ranch and is the *expected value of the function g*.

The expected values of certain functions of a random variable X have particular importance in probability and statistics. One of the most important of these functions is the expected value of the function $g(x) = x$.

> **DEFINITION** The **expected value** or **mean** of a continuous random variable X with probability density function f is the number
>
> $$\mu = \mathrm{E}(X) = \int_{-\infty}^{\infty} xf(x)\,dx.$$

The expected value $\mathrm{E}(X)$ can be thought of as a weighted average of the random variable X, where each value of X is weighted by $f(X)$. The mean can also be interpreted as the long-run average value of the random variable X, and it is one measure of the centrality of the random variable X.

EXAMPLE 6 Find the mean of the random variable X with exponential probability density function

$$f(x) = \begin{cases} 0 & \text{if } x < 0 \\ ce^{-cx} & \text{if } x \ge 0. \end{cases}$$

Solution From the definition we have

$$\begin{aligned} \mu &= \int_{-\infty}^{\infty} xf(x)\,dx = \int_0^{\infty} xce^{-cx}\,dx \\ &= \lim_{b\to\infty} \int_0^b xce^{-cx}\,dx = \lim_{b\to\infty}\left(-xe^{-cx}\Big]_0^b + \int_0^b e^{-cx}\,dx \right) \\ &= \lim_{b\to\infty}\left(-be^{-cb} - \frac{1}{c}e^{-cb} + \frac{1}{c} \right) = \frac{1}{c}. \end{aligned}$$

l'Hôpital's Rule on first term

Therefore, the mean is $\mu = 1/c$. ■

From the result in Example 6, knowing the mean or expected value μ of a random variable X having an exponential density function allows us to write its entire formula.

> **Exponential Density Function for a Random Variable X with Mean μ**
>
> $$f(x) = \begin{cases} 0 & \text{if } x < 0 \\ \mu^{-1}e^{-x/\mu} & \text{if } x \ge 0 \end{cases}$$

EXAMPLE 7 Suppose the time T before a chip fails in Example 5 is modeled instead by the exponential density function with a mean of eight years. Find the probability that a chip will fail within five years.

Solution The exponential density function with mean $\mu = 8$ is

$$f(t) = \begin{cases} 0 & \text{if } t < 0 \\ \frac{1}{8}e^{-t/8} & \text{if } t \ge 0 \end{cases}$$

Then the probability a chip will fail within five years is the definite integral

$$P(0 \le T \le 5) = \int_0^5 0.125e^{-0.125t}\,dt = -e^{-0.125t}\Big]_0^5 = 1 - e^{-0.625} \approx 0.465$$

so about 47% of the chips can be expected to fail within five years. ■

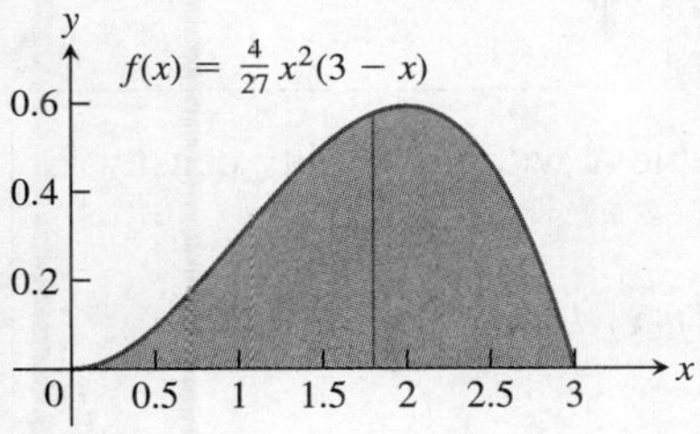

FIGURE 8.24 The expected value of a random variable with this probability density function is $\mu = 1.8$ (Example 8).

EXAMPLE 8 Find the expected value for the random variable X with probability density function given by Example 4.

Solution The expected value is

$$\mu = \mathrm{E}(X) = \int_0^3 \frac{4}{27}x^3(3-x)\,dx = \frac{4}{27}\left[\frac{3}{4}x^4 - \frac{1}{5}x^5\right]_0^3$$

$$= \frac{4}{27}\left(\frac{243}{4} - \frac{243}{5}\right) = 1.8$$

From Figure 8.24, you can see that this expected value is reasonable because the region beneath the probability density function appears to be balanced about the vertical line $x = 1.8$. That is, the horizontal coordinate of the centroid of a plate described by the region is $\bar{x} = 1.8$. ■

There are other ways to measure the centrality of a random variable with a given probability density function.

> **DEFINITION** The **median** of a continuous random variable X with probability density function f is the number m for which
>
> $$\int_{-\infty}^{m} f(x)\,dx = \frac{1}{2} \quad \text{and} \quad \int_{m}^{\infty} f(x)\,dx = \frac{1}{2}.$$

The definition of the median means that there is an equal likelihood that the random variable X will be smaller than m or larger than m.

EXAMPLE 9 Find the median of a random variable X with exponential probability density function

$$f(x) = \begin{cases} 0 & \text{if } x < 0 \\ ce^{-cx} & \text{if } x \ge 0. \end{cases}$$

Solution The median m must satisfy

$$\frac{1}{2} = \int_0^m ce^{-cx}\,dx = -e^{-cx}\Big]_0^m = 1 - e^{-cm}.$$

It follows that

$$e^{-cm} = \frac{1}{2} \quad \text{or} \quad m = \frac{1}{c}\ln 2.$$

Also,

$$\frac{1}{2} = \lim_{b\to\infty}\int_m^b ce^{-cx}\,dx = \lim_{b\to\infty}\Big[-e^{-cx}\Big]_m^b = \lim_{b\to\infty}\left(e^{-cm} - e^{-cb}\right) = e^{-cm}$$

giving the same value for m. Since $1/c$ is the mean μ of X with an exponential distribution, we conclude that the median is $m = \mu \ln 2$. The mean and median differ because the probability density function is skewed and spreads toward the right. ■

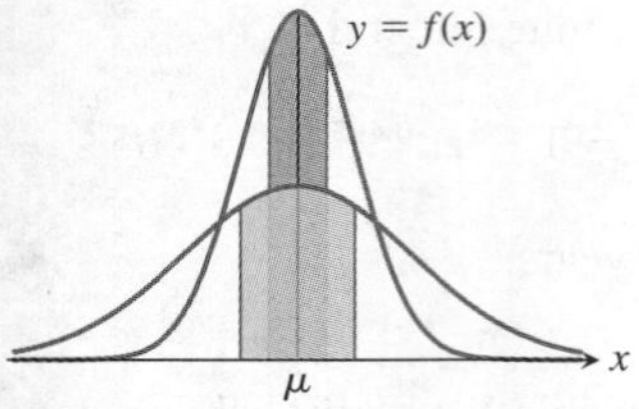

FIGURE 8.25 Probability density functions with the same mean can have different spreads in relation to the mean. The two regions under the curves have equal area.

Variance and Standard Deviation

Random variables with exactly the same mean μ but different distributions can behave very differently (see Figure 8.25). The *variance* of a random variable X measures how spread out the values of X are in relation to the mean, and we measure this dispersion by the expected value of $(X - \mu)^2$. Since the variance measures the expected square of the difference from the mean, we often work instead with its square root.

DEFINITIONS The **variance** of a random variable X with probability density function f is the expected value of $(X - \mu)^2$:

$$\operatorname{Var}(X) = \int_{-\infty}^{\infty} (x - \mu)^2 f(x)\,dx$$

The **standard deviation** of X is

$$\sigma_X = \sqrt{\operatorname{Var}(X)} = \sqrt{\int_{-\infty}^{\infty} (x - \mu)^2 f(x)\,dx}.$$

EXAMPLE 10 Find the standard deviation of the random variable T in Example 5, and find the probability that T lies within one standard deviation of the mean.

Solution The probability density function is the exponential density function with mean $\mu = 10$ by Example 6. To find the standard deviation we first calculate the variance integral:

$$\begin{aligned}
\int_{-\infty}^{\infty} (t - \mu)^2 f(t)\,dt &= \int_0^{\infty} (t - 10)^2 (0.1e^{-0.1t})\,dt \\
&= \lim_{b\to\infty} \int_0^{b} (t - 10)^2 (0.1e^{-0.1t})\,dt \\
&= \lim_{b\to\infty} \left[\left(-(t - 10)^2 - 20(t - 10)\right)e^{-0.1t}\right]_0^b \\
&\quad + \lim_{b\to\infty} \int_0^b 20e^{-0.1t}\,dt \qquad \text{Integrating by parts} \\
&= \left[0 + (-10)^2 + 20(-10)\right] - 20 \lim_{b\to\infty} \left(10e^{-0.1t}\right)\Big]_0^b \\
&= -100 - 200 \lim_{b\to\infty} \left(e^{-0.1b} - 1\right) = 100.
\end{aligned}$$

The standard deviation is the square root of the variance, so $\sigma = 10.0$.

To find the probability that T lies within one standard deviation of the mean, we find the probability $P(\mu - \sigma \le T \le \mu + \sigma)$. For this example, we have

$$P(10 - 10 \le T \le 10 + 10) = \int_0^{20} 0.1e^{-0.1t}\,dt = -e^{-0.1t}\Big]_0^{20} = 1 - e^{-2} \approx 0.865$$

This means that about 87% of the chips will fail within twenty years. ■

Uniform Distributions

The **uniform distribution** is very simple, but it occurs commonly in applications. The probability density function for this distribution on the interval $[a, b]$ is

$$f(x) = \frac{1}{b - a}, \qquad a \le x \le b.$$

If each outcome in the sample space is equally likely to occur, then the random variable X has a uniform distribution. Since f is constant on $[a, b]$, a random variable with a uniform distribution is just as likely to be in one subinterval of a fixed length as in any other of the same length. The probability that X assumes a value in a subinterval of $[a, b]$ is the length of that subinterval divided by $(b - a)$.

EXAMPLE 11 An anchored arrow is spun around the origin, and the random variable X is the radian angle the arrow makes with the positive x-axis, measured within the interval $[0, 2\pi)$. Assuming there is equal probability for the arrow pointing in any direction, find the probability density function and the probability that the arrow ends up pointing between North and East.

Solution We model the probability density function with the uniform distribution $f(x) = 1/2\pi,\ 0 \le x < 2\pi$, and $f(x) = 0$ elsewhere.

The probability that the arrow ends up pointing between North and East is given by

$$P\left(0 \le X \le \frac{\pi}{2}\right) = \int_0^{\pi/2} \frac{1}{2\pi}\, dx = \frac{1}{4}.$$ ■

Normal Distributions

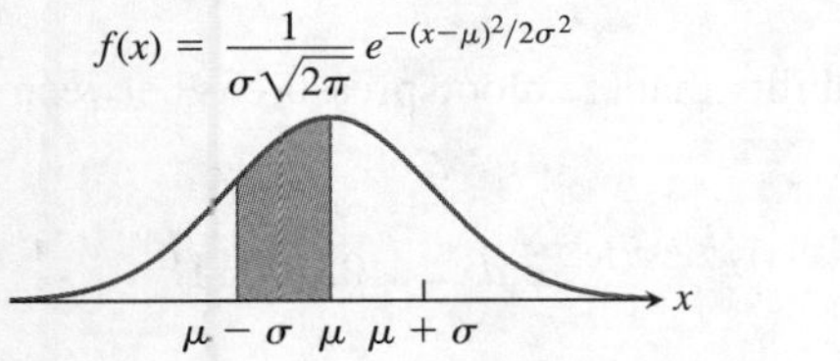

FIGURE 8.26 The normal probability density function with mean μ and standard deviation σ.

Numerous applications use the **normal distribution,** which is defined by the probability density function

$$f(x) = \frac{1}{\sigma\sqrt{2\pi}} e^{-(x-\mu)^2/2\sigma^2}.$$

It can be shown that the mean of a random variable X with this probability density function is μ and its standard deviation is σ. In applications the values of μ and σ are often estimated using large sets of data. The function is graphed in Figure 8.26, and the graph is sometimes called a *bell curve* because of its shape. Since the curve is symmetric about the mean, the median for X is the same as its mean. It is often observed in practice that many random variables have approximately a normal distribution. Some examples illustrating this phenomenon are the height of a man, the annual rainfall in a certain region, an individual's blood pressure, the serum cholesterol level in the blood, the brain weights in a certain population of adults, and the amount of growth in a given period for a population of sunflower seeds.

The normal probability density function does not have an antiderivative expressible in terms of familiar functions. Once μ and σ are fixed, however, an integral involving the normal probability density function can be computed using numerical integration methods. Usually we use the numerical integration capability of a computer or calculator to estimate the values of these integrals. Such computations show that for any normal distribution, we get the following values for the probability that the random variable X lies within $k = 1, 2, 3$, or 4 standard deviations of the mean:

$$P(\mu - \sigma < X < \mu + \sigma) \approx 0.68269$$
$$P(\mu - 2\sigma < X < \mu + 2\sigma) \approx 0.95450$$
$$P(\mu - 3\sigma < X < \mu + 3\sigma) \approx 0.99730$$
$$P(\mu - 4\sigma < X < \mu + 4\sigma) \approx 0.99994$$

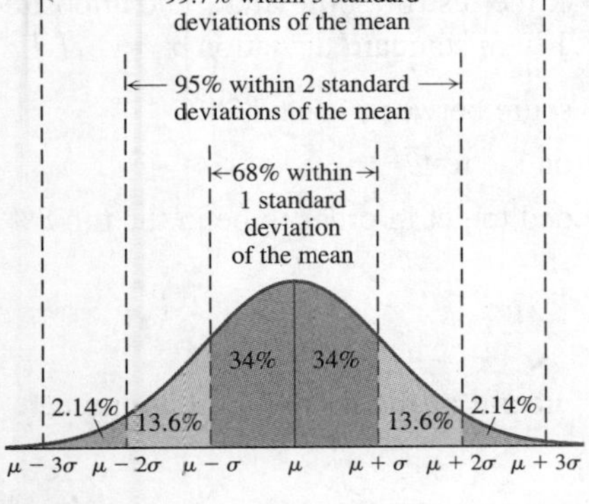

FIGURE 8.27 Probabilities of the normal distribution within its standard deviation bands.

This means, for instance, that the random variable X will take on a value within two standard deviations of the mean about 95% of the time. About 68% of the time, X will lie within one standard deviation of the mean (see Figure 8.27).

EXAMPLE 12 An individual's blood pressure is an important indicator of overall health. A medical study of healthy individuals between 14 and 70 years of age modeled their systolic blood pressure using a normal distribution with mean 119.7 mm Hg and standard deviation 10.9 mm Hg.

(a) Using this model, what percentage of the population has a systolic blood pressure between 140 and 160 mm Hg, the levels set by the American Heart Association for Stage 1 hypertension?

(b) What percentage has a blood pressure between 160 and 180 mm Hg, the levels set by the American Heart Association for Stage 2 hypertension?

(c) What percentage has a blood pressure in the normal range of 90–120, as set by the American Heart Association?

Solution

(a) Since we cannot find an antiderivative, we use a computer to evaluate the probability integral of the normal probability density function with $\mu = 119.7$ and $\sigma = 10.9$:

$$P(140 \le X \le 160) = \int_{140}^{160} \frac{1}{10.9\sqrt{2\pi}} e^{-(x-119.7)^2/2(10.9)^2}\, dx \approx 0.03117.$$

This means that about 3% of the population in the studied age range have Stage 1 hypertension.

(b) Again we use a computer to calculate the probability that the blood pressure is between 160 and 180 mm Hg:

$$P(160 \le X \le 180) = \int_{160}^{180} \frac{1}{10.9\sqrt{2\pi}} e^{-(x-119.7)^2/2(10.9)^2}\, dx \approx 0.00011.$$

We conclude that about 0.011% of the population has Stage 2 hypertension.

(c) The probability that the blood pressure falls in the normal range is

$$P(90 \le X \le 120) = \int_{90}^{120} \frac{1}{10.9\sqrt{2\pi}} e^{-(x-119.7)^2/2(10.9)^2}\, dx \approx 0.50776.$$

That is, about 51% of the population has a normal systolic blood pressure. ■

Many national tests are standardized using the normal distribution. The following example illustrates modeling the discrete random variable for scores on a test using the normal distribution function for a continuous random variable.

EXAMPLE 13 The ACT is a standardized test taken by high school students seeking admission to many colleges and universities. The test measures knowledge, skills, and proficiency in the areas of English, math, and science, with scores ranging over the interval $[1, 36]$. Nearly 1.5 million high school students took the test in 2009, and the composite mean score across the academic areas was $\mu = 21.1$ with standard deviation $\sigma = 5.1$.

(a) What percentage of the population had an ACT score between 18 and 24?

(b) What is the ranking of a student who scored 27 on the test?

(c) What is the minimal integer score a student needed to get in order to be in the top 8% of the scoring population?

Solution

(a) We use a computer to evaluate the probability integral of the normal probability density function with $\mu = 21.1$ and $\sigma = 5.1$:

$$P(18 \le X \le 24) = \int_{18}^{24} \frac{1}{5.1\sqrt{2\pi}} e^{-(x-21.1)^2/2(5.1)^2}\, dx \approx 0.44355.$$

This means that about 44% of the students had an ACT score between 18 and 24.

(b) Again we use a computer to calculate the probability of a student getting a score lower than 27 on the test:

$$P(1 \le X < 27) = \int_1^{27} \frac{1}{5.1\sqrt{2\pi}} e^{-(x-21.1)^2/2(5.1)^2}\, dx \approx 0.87630.$$

We conclude that about 88% of the students scored below a score of 27, so the student ranked in the top 12% of the population.

(c) We look at how many students had a mark higher than 28:

$$P(28 < X \le 36) = \int_{28}^{36} \frac{1}{5.1\sqrt{2\pi}} e^{-(x-21.1)^2/2(5.1)^2}\, dx \approx 0.0863.$$

Since this number gives more than 8% of the students, we look at the next higher integer score:

$$P(29 < X \le 36) = \int_{29}^{36} \frac{1}{5.1\sqrt{2\pi}} e^{-(x-21.1)^2/2(5.1)^2}\, dx \approx 0.0595.$$

Therefore, 29 is the lowest integer score a student could get in order to score in the top 8% of the population (and actually scoring here in the top 6%). ■

The simplest form for a normal distribution of X occurs when its mean is zero and its standard deviation is one. The *standard normal probability density function* f giving mean $\mu = 0$ and standard deviation $\sigma = 1$ is

$$f(x) = \frac{1}{\sqrt{2\pi}} e^{-x^2/2}.$$

Note that the substitution $z = (x - \mu)/\sigma$ gives the equivalent integrals

$$\int_a^b \frac{1}{\sigma\sqrt{2\pi}} e^{-((x-\mu)/\sigma)^2/2}\, dx = \int_\alpha^\beta \frac{1}{\sqrt{2\pi}} e^{-z^2/2}\, dz,$$

where $\alpha = (a - \mu)/\sigma$ and $\beta = (b - \mu)/\sigma$. So we can convert random variable values to the "z-values" to standardize a normal distribution, and then use the integral on the right-hand side of the last equation to calculate probabilities for the original random variable normal distribution with mean μ and standard deviation σ. In a normal distribution, we know that 95.5% of the population lies within two standard deviations of the mean, so a random variable X converted to a z-value has more than a 95% chance of occurring in the interval $[-2, 2]$.

EXERCISES 8.9

Probability Density Functions

In Exercises 1–8, determine which are probability density functions and justify your answer.

1. $f(x) = \frac{1}{18}x$ over $[4, 8]$

2. $f(x) = x^2\left(2x + \frac{3}{2}\right)$ over $[0, 1]$

3. $f(x) = 2^x$ over $\left[0, \frac{\ln(1 + \ln 2)}{\ln 2}\right]$

4. $f(x) = x - 1$ over $[0, 1 + \sqrt{3}]$

5. $f(x) = \begin{cases} \frac{2}{x^3}, & x \ge 1 \\ 0, & x < 1 \end{cases}$

6. $f(x) = \begin{cases} \frac{8}{\pi(4 + x^2)}, & x \ge 0 \\ 0, & x < 0 \end{cases}$

7. $f(x) = 2\sin 2x$ over $\left[0, \frac{\pi}{4}\right]$

8. $f(x) = \frac{1}{x}$ over $(0, e]$

9. Let f be the probability density function for the random variable L in Example 2f. Explain the meaning of each integral.

a. $\int_{25{,}000}^{32{,}000} f(l)\, dl$

b. $\int_{30{,}000}^{\infty} f(l)\, dl$

c. $\int_0^{20{,}000} f(l)\, dl$

d. $\int_{-\infty}^{15{,}000} f(l)\, dl$

10. Let $f(x)$ be the uniform distribution for the random variable X in Example 11. Express the following probabilities as integrals.

a. The probability that the arrow points either between South and West or between North and West.

b. The probability that the arrow makes an angle of at least 2 radians.

Verify that the functions in Exercises 11–16 are probability density functions for a continuous random variable X over the given interval. Determine the specified probability.

11. $f(x)=\frac{1}{2}x^2e^{-x}$ over $[0, \infty)$, $P(1\le X\le 3)$

T **12.** $f(x) = \frac{\ln x}{x^2}$ over $[1, \infty)$, $P(2 < X < 15)$

13. $f(x)=4x\left(1-\frac{3}{4}x\right)$ over $[0, 1]$, $P(0.5<X)$

T **14.** $f(x) = \frac{\sin^2 \pi x}{\pi x^2}$ over $\left[\frac{200}{1059}, \infty\right)$, $P(X < \pi/6)$

15. $f(x)=\begin{cases}\frac{3}{x^4}, & x>1\\ 0, & x\le 1\end{cases}$ over $(-\infty, \infty)$, $P(3\le X<8)$

16. $f(x) = \sin x$ over $[0, \pi/2]$, $P\left(\frac{\pi}{6} < X \le \frac{\pi}{4}\right)$

In Exercises 17–20, find the value of the constant c so that the given function is a probability density function for a random variable over the specified interval.

17. $f(x)=\frac{1}{4}x$ over $[4, c]$

18. $f(x) = \frac{1}{x}$ over $[c, c+1]$

19. $f(x)=6e^{-3x}$ over $[0, c]$

20. $f(x) = cx\sqrt{25-x^2}$ over $[0, 5]$

21. Let $f(x)=cx^2$. Find the value of c so that f is a probability density function over [0, 3]. If f is a probability density function for the random variable X, find the probability $P(1\le X<2)$.

22. Find the value of c so that $f(x) = c\sqrt{x}(1-x)$ is a probability density function for the random variable X over $[0, 1]$, and find the probability $P(0.25 \le X \le 0.5)$.

23. Show that if the exponentially decreasing function

$$f(x) = \begin{cases} 0 & \text{if } x < 0 \\ Ae^{-cx} & \text{if } x \ge 0 \end{cases}$$

is a probability density function, then $A = c$.

24. Suppose f is a probability density function for the random variable X with mean μ. Show that its variance satisfies

$$\text{Var}(X) = \int_{-\infty}^{\infty} x^2 f(x)\, dx - \mu^2.$$

Compute the mean and median for a random variable with the probability density functions in Exercises 25–28.

25. $f(x) = \frac{1}{8}x$ over $[0, 4]$

26. $f(x) = \frac{1}{9}x^2$ over $[0, 3]$

27. $f(x) = \begin{cases}\frac{2}{x^3} & x \ge 1\\ 0 & x < 1\end{cases}$

28. $f(x)=\begin{cases}\frac{1}{5}e^{-x/5}, & x\ge 0\\ 0, & \text{otherwise}\end{cases}$

Exponential Distributions

29. Digestion time The digestion time in hours of a fixed amount of food is exponentially distributed with a mean of 1 hour. What is the probability that the food is digested in less than 30 minutes?

30. Pollinating flowers A biologist models the time in minutes until a bee arrives at a flowering plant with an exponential distribution having a mean of 4 minutes. If 1000 flowers are in a field, how many can be expected to be pollinated within 5 minutes?

31. Lifetime of light bulbs A manufacturer of light bulbs finds that the mean lifetime of a bulb is 1000 hours. Assume the life of a bulb is exponentially distributed.

a. Find the probability that a bulb will last less than its guaranteed lifetime of 800 hours.

b. In a batch of light bulbs, what is the expected time until half the light bulbs in the batch fail?

32. Lifetime of an electronic component The life expectancy in years of a component in a microcomputer is exponentially distributed, and 1/3 of the components fail in the first 3 years. The company that manufactures the component offers a 1 year warranty. What is the probability that a component will fail during the warranty period?

33. Lifetime of an organism A *hydra* is a small fresh-water animal, and studies have shown that its probability of dying does not increase with the passage of time. The lack of influence of age on mortality rates for this species indicates that an exponential distribution is an appropriate model for the mortality of hydra. A biologist studies a population of 500 hydra and observes that 200 of them die within the first 2 years. How many of the hydra would you expect to die within the first six months?

34. Car accidents The number of days that elapse between the beginning of a calendar year and the moment a high-risk driver is involved in an accident is exponentially distributed. Based on historical data, an insurance company expects that 30% of high-risk drivers will be involved in an accident during the first 50 days of the calendar year. In a group of 100 high-risk drivers, how many do you expect to be involved in an accident during the first 80 days of the calendar year?

35. Customer service time The mean waiting time to get served after walking into a bakery is 30 seconds. Assume that an exponential density function describes the waiting times.

a. What is the probability a customer waits 15 seconds or less?

b. What is the probability a customer waits longer than one minute?

c. What is the probability a customer waits exactly 5 minutes?

d. If 200 customers come to the bakery in a day, how many are likely to be served within three minutes?

36. Airport waiting time According to the U.S. Customs and Border Protection Agency, the average airport wait time at Chicago's O'Hare International airport is 16 minutes for a traveler arriving during the hours 7–8 A.M., and 32 minutes for arrival during the hours 4–5 P.M. The wait time is defined as the total processing time from arrival at the airport until the completion of a passenger's security screening. Assume the wait time is exponentially distributed.

a. What is the probability of waiting between 10 and 30 minutes for a traveler arriving during the 7–8 A.M. hour?

b. What is the probability of waiting more than 25 minutes for a traveler arriving during the 7–8 P.M. hour?

c. What is the probability of waiting between 35 and 50 minutes for a traveler arriving during the 4–5 P.M. hour?

d. What is the probability of waiting less than 20 minutes for a traveler arriving during the 4–5 P.M. hour?

37. Printer lifetime The lifetime of a \$300 printer is exponentially distributed with a mean of 2 years. The manufacturer agrees to pay a full refund to a buyer if the printer fails during the first year following its purchase, and a one-half refund if it fails during the second year. If the manufacturer sells 100 printers, how much should it expect to pay in refunds?

38. Failure time The time between failures of a photocopier is exponentially distributed. Half of the copiers at a university require service during the first 2 years of operations. If the university purchased 150 copiers, how many do you expect to require service during the first year of their operation?

T Normal Distributions

39. Cholesterol levels The serum cholesterol levels of children aged 12 to 14 years follows a normal distribution with mean $\mu = 162$ mg/dl and standard deviation $\sigma = 28$ mg/dl. In a population of 1000 of these children, how many would you expect to have serum cholesterol levels between 165 and 193? between 148 and 167?

40. Annual rainfall The annual rainfall in inches for San Francisco, California, is approximately a normal random variable with mean 20.11 in. and standard deviation 4.7 in. What is the probability that next year's rainfall will exceed 17 in.?

41. Manufacturing time The assembly time in minutes for a component at an electronic manufacturing plant is normally distributed with a mean of $\mu = 30$ and standard deviation $\sigma = 4$. What is the probability that a component will be made in less than 40 minutes?

42. Lifetime of a tire Assume the random variable L in Example 2f is normally distributed with mean $\mu = 22{,}000$ miles and $\sigma = 4{,}000$ miles.

a. In a batch of 4000 tires, how many can be expected to last for at least 18,000 miles?

b. What is the minimum number of miles you would expect to find as the lifetime for 90% of the tires?

43. Height The average height of American women aged 18–24 is normally distributed with mean $\mu = 65.5$ inches and $\sigma = 2.5$ inches.

a. What percentage of women are taller than 68 inches?

b. What is the probability a woman is between 5′1″ and 5′4″ tall?

44. Life expectancy At birth, a French citizen has an average life expectancy of 82 years with a standard deviation of 7 years. If 100 newly born French babies are selected at random, how many would you expect to live between 75 and 85 years? Assume life expectancy is normally distributed.

45. Length of pregnancy A team of medical practitioners determines that in a population of 1000 females with ages ranging from 20 to 35 years, the length of pregnancy from conception to birth is approximately normally distributed with a mean of 268 days and a standard deviation of 15 days. How many of these females would you expect to have a pregnancy lasting from 250 to 280 days?

46. Brain weights In a population of 500 adult Swedish men, medical researchers find their brain weights to be approximately normally distributed with mean $\mu = 1400$ gm and standard deviation $\sigma = 100$ gm.

a. What percentage of brain weights are between 1325 and 1450 gm?

b. How many men in the population would you expect to have a brain weight exceeding 1480 gm?

47. Blood pressure Diastolic blood pressure in adults is normally distributed with $\mu = 70$ mm Hg and $\sigma = 10$ mm Hg. In a random sample of 300 adults, how many would be expected to have a diastolic blood pressure below 60 mm Hg?

48. Albumin levels Serum albumin in healthy 20-year-old men is normally distributed with $\mu = 4.4$ and $\sigma = 0.2$. How likely is it for a healthy 20-year-old man to have a level in the range 4.3 to 4.45?

49. Quality control A manufacturer of generator shafts finds that it needs to add additional weight to its shafts in order to achieve proper static and dynamic balance. Based on experimental tests, the average weight it needs to add is $\mu = 35$ gm with $\sigma = 9$ gm. Assuming a normal distribution, from 1000 randomly selected shafts, how many would be expected to need an added weight in excess of 40 gm?

50. Miles driven A taxicab company in New York City analyzed the daily number of miles driven by each of its drivers. It found the average distance was 200 mi with a standard deviation of 30 mi. Assuming a normal distribution, what prediction can we make about the percentage of drivers who will log in either more than 260 mi or less than 170 mi?

51. Germination of sunflower seeds The germination rate of a particular seed is the percentage of seeds in the batch which successfully emerge as plants. Assume that the germination rate for a batch of sunflower seeds is 80%, and that among a large population of n seeds the number of successful germinations is normally distributed with mean $\mu = 0.8n$ and $\sigma = 0.4\sqrt{n}$.

a. In a batch of $n = 2500$ seeds, what is the probability that at least 1960 will successfully germinate?

b. In a batch of $n = 2500$ seeds, what is the probability that at most 1980 will successfully germinate?

c. In a batch of $n = 2500$ seeds, what is the probability that between 1940 and 2020 will successfully germinate?

52. Suppose you toss a fair coin n times and record the number of heads that land. Assume that n is large and approximate the discrete random variable X with a continuous random variable that is normally distributed with $\mu = n/2$ and $\sigma = \sqrt{n}/2$. If $n = 400$, find the given probabilities.

a. $P(190 \le X < 210)$ **b.** $P(X < 170)$

c. $P(X > 220)$ **d.** $P(X = 300)$

Discrete Random Variables

53. A fair coin is tossed four times and the random variable X assigns the number of tails that appear in each outcome.

a. Determine the set of possible outcomes.

b. Find the value of X for each outcome.

c. Create a probability bar graph for X, as in Figure 8.21. What is the probability that at least two heads appear in the four tosses of the coin?

54. You roll a pair of six-sided dice, and the random variable X assigns to each outcome the sum of the number of dots showing on each face, as in Example 2e.

a. Find the set of possible outcomes.

b. Create a probability bar graph for X.

c. What is the probability that $X = 8$?

d. What is the probability that $X \leq 5$? $X > 9$?

55. Three people are asked their opinion in a poll about a particular brand of a common product found in grocery stores. They can answer in one of three ways: "Like the product brand" (L), "Dislike the product brand" (D), or "Undecided" (U). For each outcome, the random variable X assigns the number of L's that appear.

a. Find the set of possible outcomes and the range of X.

b. Create a probability bar graph for X.

c. What is the probability that at least two people like the product brand?

d. What is the probability that no more than one person dislikes the product brand?

56. Spacecraft components A component of a spacecraft has both a main system and a backup system operating throughout a flight. The probability that both systems fail sometime during the flight is 0.0148. Assuming that each system separately has the same failure rate, what is the probability that the main system fails during the flight?

CHAPTER 8 Questions to Guide Your Review

1. What is the formula for integration by parts? Where does it come from? Why might you want to use it?
2. When applying the formula for integration by parts, how do you choose the u and dv? How can you apply integration by parts to an integral of the form $\int f(x)\,dx$?
3. If an integrand is a product of the form $\sin^n x \cos^m x$, where m and n are nonnegative integers, how do you evaluate the integral? Give a specific example of each case.
4. What substitutions are made to evaluate integrals of $\sin mx \sin nx$, $\sin mx \cos nx$, and $\cos mx \cos nx$? Give an example of each case.
5. What substitutions are sometimes used to transform integrals involving $\sqrt{a^2 - x^2}$, $\sqrt{a^2 + x^2}$, and $\sqrt{x^2 - a^2}$ into integrals that can be evaluated directly? Give an example of each case.
6. What restrictions can you place on the variables involved in the three basic trigonometric substitutions to make sure the substitutions are reversible (have inverses)?
7. What is the goal of the method of partial fractions?
8. When the degree of a polynomial $f(x)$ is less than the degree of a polynomial $g(x)$, how do you write $f(x)/g(x)$ as a sum of partial fractions if $g(x)$

   a. is a product of distinct linear factors?

   b. consists of a repeated linear factor?

   c. contains an irreducible quadratic factor?

   What do you do if the degree of f is *not* less than the degree of g?
9. How are integral tables typically used? What do you do if a particular integral you want to evaluate is not listed in the table?
10. What is a reduction formula? How are reduction formulas used? Give an example.
11. How would you compare the relative merits of the Midpoint Rule, the Trapezoidal Rule, and Simpson's Rule?
12. What is an improper integral of Type I? Type II? How are the values of various types of improper integrals defined? Give examples.
13. What tests are available for determining the convergence and divergence of improper integrals that cannot be evaluated directly? Give examples of their use.
14. What is a random variable? What is a continuous random variable? Give some specific examples.
15. What is a probability density function? What is the probability that a continuous random variable has a value in the interval $[c, d]$?
16. What is an exponentially decreasing probability density function? What are some typical events that might be modeled by this distribution? What do we mean when we say such distributions are *memoryless*?
17. What is the expected value of a continuous random variable? What is the expected value of an exponentially distributed random variable?
18. What is the median of a continuous random variable? What is the median of an exponential distribution?
19. What does the variance of a random variable measure? What is the standard deviation of a continuous random variable X?
20. What probability density function describes the normal distribution? What are some examples typically modeled by a normal distribution? How do we usually calculate probabilities for a normal distribution?
21. In a normal distribution, what percentage of the population lies within 1 standard deviation of the mean? Within 2 standard deviations?

CHAPTER 8 Practice Exercises

Integration by Parts

Evaluate the integrals in Exercises 1–8 using integration by parts.

1. $\int \ln(x + 1)\,dx$

2. $\int x^2 \ln x\,dx$

3. $\int \arctan 3x\,dx$

4. $\int \cos^{-1}\left(\frac{x}{2}\right)dx$

5. $\int (x + 1)^2 e^x\,dx$

6. $\int x^2 \sin(1 - x)\,dx$

7. $\int e^x \cos 2x\,dx$

8. $\int 3^x \sin 2x\,dx$

Partial Fractions

Evaluate the integrals in Exercises 9–28. It may be necessary to use a substitution first.

9. $\int \frac{x\,dx}{x^2 - 3x + 2}$

10. $\int \frac{x\,dx}{x^2 + 4x + 3}$

11. $\int \frac{dx}{x(x + 1)^2}$

12. $\int \frac{x + 1}{x^2(x - 1)}\,dx$

13. $\int \frac{\sin\theta\,d\theta}{\cos^2\theta + \cos\theta - 2}$

14. $\int \frac{\cos\theta\,d\theta}{\sin^2\theta + \sin\theta - 6}$

15. $\int \frac{3x^2 + 4x + 4}{x^3 + x}\,dx$

16. $\int \frac{4x\,dx}{x^3 + 4x}$

17. $\int \frac{v + 3}{2v^3 - 8v}\,dv$

18. $\int \frac{(3v - 7)\,dv}{(v - 1)(v - 2)(v - 3)}$

19. $\int \frac{dt}{t^4 + 4t^2 + 3}$

20. $\int \frac{t\,dt}{t^4 - t^2 - 2}$

21. $\int \frac{x^3 + x^2}{x^2 + x - 2}\,dx$

22. $\int \frac{x^3 + 1}{x^3 - x}\,dx$

23. $\int \frac{x^3 + 4x^2}{x^2 + 4x + 3}\,dx$

24. $\int \frac{2x^3 + x^2 - 21x + 24}{x^2 + 2x - 8}\,dx$

25. $\int \frac{dx}{x(3\sqrt{x + 1})}$

26. $\int \frac{dx}{x(1 + \sqrt[3]{x})}$

27. $\int \frac{ds}{e^s - 1}$

28. $\int \frac{ds}{\sqrt{e^s + 1}}$

Trigonometric Substitutions

Evaluate the integrals in Exercises 29–32 **(a)** without using a trigonometric substitution, **(b)** using a trigonometric substitution.

29. $\int \frac{y\,dy}{\sqrt{16 - y^2}}$

30. $\int \frac{x\,dx}{\sqrt{4 + x^2}}$

31. $\int \frac{x\,dx}{4 - x^2}$

32. $\int \frac{t\,dt}{\sqrt{4t^2 - 1}}$

Evaluate the integrals in Exercises 33–36.

33. $\int \frac{x\,dx}{9 - x^2}$

34. $\int \frac{dx}{x(9 - x^2)}$

35. $\int \frac{dx}{9 - x^2}$

36. $\int \frac{dx}{\sqrt{9 - x^2}}$

Trigonometric Integrals

Evaluate the integrals in Exercises 37–44.

37. $\int \sin^3 x \cos^4 x\,dx$

38. $\int \cos^5 x \sin^5 x\,dx$

39. $\int \tan^4 x \sec^2 x\,dx$

40. $\int \tan^3 x \sec^3 x\,dx$

41. $\int \sin 5\theta \cos 6\theta\,d\theta$

42. $\int \sec^2\theta \sin^3\theta\,d\theta$

43. $\int \sqrt{1 + \cos(t/2)}\,dt$

44. $\int e^t \sqrt{\tan^2 e^t + 1}\,dt$

Numerical Integration

45. According to the error-bound formula for Simpson's Rule, how many subintervals should you use to be sure of estimating the value of

$$\ln 3 = \int_1^3 \frac{1}{x}\,dx$$

by Simpson's Rule with an error of no more than 10^{-4} in absolute value? (Remember that for Simpson's Rule, the number of subintervals has to be even.)

46. A brief calculation shows that if $0 \le x \le 1$, then the second derivative of $f(x) = \sqrt{1 + x^4}$ lies between 0 and 8. Based on this, about how many subdivisions would you need to estimate the integral of f from 0 to 1 with an error no greater than 10^{-3} in absolute value using the Trapezoidal Rule?

47. A direct calculation shows that

$$\int_0^{\pi} 2\sin^2 x\,dx = \pi.$$

How close do you come to this value by using the Trapezoidal Rule with $n = 6$? The Midpoint Rule with $n = 6$? Simpson's Rule with $n = 6$? Try them and find out.

48. You are planning to use Simpson's Rule to estimate the value of the integral

$$\int_1^2 f(x)\,dx$$

with an error magnitude less than 10^{-5}. You have determined that $|f^{(4)}(x)| \le 3$ throughout the interval of integration. How many subintervals should you use to ensure the required accuracy? (Remember that for Simpson's Rule, the number has to be even.)

T **49. Mean temperature** Use Simpson's Rule to approximate the average value of the temperature function

$$f(x) = 37\sin\left(\frac{2\pi}{365}(x - 101)\right) + 25$$

for a 365-day year. This is one way to estimate the annual mean air temperature in Fairbanks, Alaska. The National Weather Service's official figure, a numerical average of the daily normal mean air temperatures for the year, is 25.7 °F, which is slightly higher than the average value of $f(x)$.

50. Heat capacity of a gas Heat capacity C_v is the amount of heat required to raise the temperature of a given mass of gas with constant volume by 1 °C, measured in units of cal/deg-mol (calories per degree gram molecular weight). The heat capacity of oxygen depends on its temperature T and satisfies the formula

$$C_v = 8.27 + 10^{-5}(26T - 1.87T^2).$$

Use Simpson's Rule to find the average value of C_v and the temperature at which it is attained for $20\ °C \le T \le 675\ °C$.

51. Fuel efficiency An automobile computer gives a digital readout of fuel consumption in gallons per hour. During a trip, a passenger recorded the fuel consumption every 5 min for a full hour of travel.

Time	Gal/hr	Time	Gal/hr
0	2.5	35	2.5
5	2.4	40	2.4
10	2.3	45	2.3
15	2.4	50	2.4
20	2.4	55	2.4
25	2.5	60	2.3
30	2.6		

a. Use the Trapezoidal Rule to approximate the total fuel consumption during the hour.

b. If the automobile covered 60 mi in the hour, what was its fuel efficiency (in miles per gallon) for that portion of the trip?

52. A new parking lot To meet the demand for parking, your town has allocated the area shown here. As the town engineer, you have been asked by the town council to find out if the lot can be built for \$11,000. The cost to clear the land will be \$0.10 a square foot, and the lot will cost \$2.00 a square foot to pave. Use Simpson's Rule to find out if the job can be done for \$11,000.

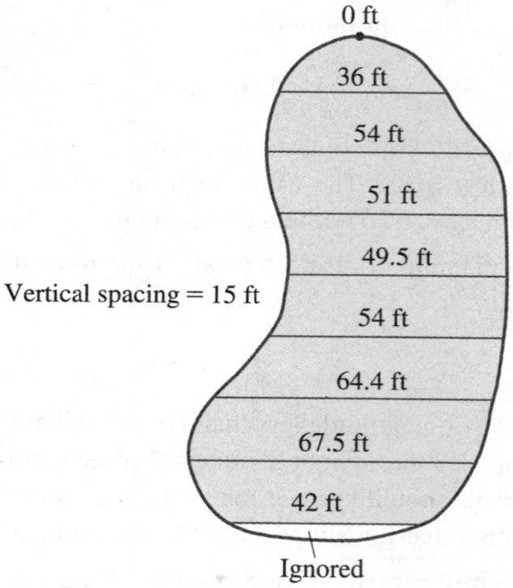

Improper Integrals

Evaluate the improper integrals in Exercises 53–62.

53. $\displaystyle\int_0^3 \frac{dx}{\sqrt{9-x^2}}$

54. $\displaystyle\int_0^1 \ln x\,dx$

55. $\displaystyle\int_0^2 \frac{dy}{(y-1)^{2/3}}$

56. $\displaystyle\int_{-2}^0 \frac{d\theta}{(\theta+1)^{3/5}}$

57. $\displaystyle\int_3^\infty \frac{2\,du}{u^2-2u}$

58. $\displaystyle\int_1^\infty \frac{3v-1}{4v^3-v^2}\,dv$

59. $\displaystyle\int_0^\infty x^2e^{-x}\,dx$

60. $\displaystyle\int_{-\infty}^0 xe^{3x}\,dx$

61. $\displaystyle\int_{-\infty}^\infty \frac{dx}{4x^2+9}$

62. $\displaystyle\int_{-\infty}^\infty \frac{4\,dx}{x^2+16}$

Which of the improper integrals in Exercises 63–68 converge and which diverge?

63. $\displaystyle\int_6^\infty \frac{d\theta}{\sqrt{\theta^2+1}}$

64. $\displaystyle\int_0^\infty e^{-u}\cos u\,du$

65. $\displaystyle\int_1^\infty \frac{\ln z}{z}\,dz$

66. $\displaystyle\int_1^\infty \frac{e^{-t}}{\sqrt{t}}\,dt$

67. $\displaystyle\int_{-\infty}^\infty \frac{2\,dx}{e^x+e^{-x}}$

68. $\displaystyle\int_{-\infty}^\infty \frac{dx}{x^2(1+e^x)}$

Assorted Integrations

Evaluate the integrals in Exercises 69–134. The integrals are listed in random order so you need to decide which integration technique to use.

69. $\displaystyle\int xe^{2x}\,dx$

70. $\displaystyle\int_0^1 x^2e^{x^3}\,dx$

71. $\displaystyle\int (\tan^2 x + \sec^2 x)\,dx$

72. $\displaystyle\int_0^{\pi/4} \cos^2 2x\,dx$

73. $\displaystyle\int x\sec^2 x\,dx$

74. $\displaystyle\int x\sec^2(x^2)\,dx$

75. $\displaystyle\int \sin x\cos^2 x\,dx$

76. $\displaystyle\int \sin 2x\sin(\cos 2x)\,dx$

77. $\displaystyle\int_{-1}^0 \frac{e^x}{e^x+e^{-x}}\,dx$

78. $\displaystyle\int (e^{2x}+e^{-x})^2\,dx$

79. $\displaystyle\int \frac{x+1}{x^4-x^3}\,dx$

80. $\displaystyle\int \frac{e^x+1}{e^x(e^{2x}-4)}\,dx$

81. $\displaystyle\int \frac{e^x+e^{3x}}{e^{2x}}\,dx$

82. $\displaystyle\int (e^x-e^{-x})(e^x+e^{-x})^3\,dx$

83. $\displaystyle\int_0^{\pi/3} \tan^3 x\sec^2 x\,dx$

84. $\displaystyle\int \tan^4 x\sec^4 x\,dx$

85. $\displaystyle\int_0^3 (x+2)\sqrt{x+1}\,dx$

86. $\displaystyle\int (x+1)\sqrt{x^2+2x}\,dx$

87. $\displaystyle\int \cot x\csc^3 x\,dx$

88. $\displaystyle\int \sin x(\tan x - \cot x)^2\,dx$

89. $\displaystyle\int \frac{x\,dx}{1+\sqrt{x}}$

90. $\displaystyle\int \frac{x^3+2}{4-x^2}\,dx$

91. $\displaystyle\int \sqrt{2x-x^2}\,dx$

92. $\displaystyle\int \frac{dx}{\sqrt{-2x-x^2}}$

93. $\displaystyle\int \frac{2-\cos x+\sin x}{\sin^2 x}\,dx$

94. $\displaystyle\int \sin^2\theta\cos^5\theta\,d\theta$

95. $\displaystyle\int \frac{9\,dv}{81-v^4}$

96. $\displaystyle\int_2^\infty \frac{dx}{(x-1)^2}$

97. $\displaystyle\int \theta\cos(2\theta+1)\,d\theta$

98. $\displaystyle\int \frac{x^3\,dx}{x^2-2x+1}$

99. $\displaystyle\int \frac{\sin 2\theta\,d\theta}{(1+\cos 2\theta)^2}$

100. $\displaystyle\int_{\pi/4}^{\pi/2} \sqrt{1+\cos 4x}\,dx$

101. $\displaystyle\int \frac{x\,dx}{\sqrt{2-x}}$

102. $\displaystyle\int \frac{\sqrt{1-v^2}}{v^2}\,dv$

103. $\int \frac{dy}{y^2 - 2y + 2}$

104. $\int \frac{x\,dx}{\sqrt{8 - 2x^2 - x^4}}$

105. $\int \frac{z+1}{z^2(z^2+4)}\,dz$

106. $\int x^2(x-1)^{1/3}\,dx$

107. $\int \frac{t\,dt}{\sqrt{9 - 4t^2}}$

108. $\int \frac{\arctan x}{x^2}\,dx$

109. $\int \frac{e^t\,dt}{e^{2t} + 3e^t + 2}$

110. $\int \tan^3 t\,dt$

111. $\int_1^{\infty} \frac{\ln y}{y^3}\,dy$

112. $\int y^{3/2}(\ln y)^2\,dy$

113. $\int e^{\ln \sqrt{x}}\,dx$

114. $\int e^{\theta}\sqrt{3 + 4e^{\theta}}\,d\theta$

115. $\int \frac{\sin 5t\,dt}{1 + (\cos 5t)^2}$

116. $\int \frac{dv}{\sqrt{e^{2v} - 1}}$

117. $\int \frac{dr}{1 + \sqrt{r}}$

118. $\int \frac{4x^3 - 20x}{x^4 - 10x^2 + 9}\,dx$

119. $\int \frac{x^3}{1 + x^2}\,dx$

120. $\int \frac{x^2}{1 + x^3}\,dx$

121. $\int \frac{1 + x^2}{1 + x^3}\,dx$

122. $\int \frac{1 + x^2}{(1 + x)^3}\,dx$

123. $\int \sqrt{x} \cdot \sqrt{1 + \sqrt{x}}\,dx$

124. $\int \sqrt{1 + \sqrt{1 + x}}\,dx$

125. $\int \frac{1}{\sqrt{x} \cdot \sqrt{1 + x}}\,dx$

126. $\int_0^{1/2} \sqrt{1 + \sqrt{1 - x^2}}\,dx$

127. $\int \frac{\ln x}{x + x\ln x}\,dx$

128. $\int \frac{1}{x \cdot \ln x \cdot \ln(\ln x)}\,dx$

129. Prove that $\int e^x[f(x) + f'(x)]dx = e^x f(x) + c$. Using it evaluate the following integrals.

130. $\int e^x\left(\frac{1-x}{1+x^2}\right)^2 dx$

$$\left[\textit{Hint: } \int e^x\left(\frac{1-x}{1+x^2}\right)^2 dx = \int e^x \frac{1+x^2-2x}{(1+x^2)^2}\,dx = \int\left[e^x\left(\frac{1}{1+x^2} - \frac{2x}{(1+x^2)^2}\right)dx\right]\right]$$

131. $\int e^x \frac{x^3+x+1}{(x^2+1)^{3/2}}\,dx$ $\left[\textit{Hint: } \int e^x\left[\frac{x}{(x^2+1)^{1/2}} + \frac{1}{(x^2+1)^{3/2}}\right]dx\right]$

132. $\int e^{\tan^{-1}x}\left[\frac{1+x+x^2}{1+x^2}\right]dx$ [*Hint:* put $u = \tan^{-1}x$]

133. $\int \frac{\sin^2 x}{1 + \sin^2 x}\,dx$

134. $\int \frac{1 - \cos x}{1 + \cos x}\,dx$

135. Evaluate $\int_0^{\pi/2} \frac{\sin x}{\sin x + \cos x}\,dx$ in two ways:

a. By evaluating $\int \frac{\sin x}{\sin x + \cos x}\,dx$, then using the Evaluation Theorem.

b. By showing that $\int_0^a f(x)\,dx = \int_0^a f(a - x)\,dx$, then using this result.

CHAPTER 8 Additional and Advanced Exercises

Evaluating Integrals

Evaluate the integrals in Exercises 1–6.

1. $\int (\arcsin x)^2\,dx$

2. $\int \frac{dx}{x(x+1)(x+2)\cdots(x+m)}$

3. $\int x \arcsin x\,dx$

4. $\int \sin^{-1}\sqrt{y}\,dy$

5. $\int \frac{dt}{t - \sqrt{1 - t^2}}$

6. $\int \frac{dx}{x^4 + 4}$

Evaluate the limits in Exercise 7 and 8.

7. $\lim_{x\to\infty} \int_{-x}^{x} \sin t\,dt$

8. $\lim_{x\to 0^+} x\int_x^1 \frac{\cos t}{t^2}\,dt$

Evaluate the limits in Exercise 9 and 10 by identifying them with definite integrals and evaluating the integrals.

9. $\lim_{n\to\infty} \sum_{k=1}^{n} \ln \sqrt[n]{1 + \frac{k}{n}}$

10. $\lim_{n\to\infty} \sum_{k=0}^{n-1} \frac{1}{\sqrt{n^2 - k^2}}$

Applications

11. **Finding arc length** Find the length of the curve

$$y = \int_0^x \sqrt{\cos 2t}\,dt, \quad 0 \le x \le \pi/4.$$

12. **Finding arc length** Find the length of the graph of the function $y = \ln(1 - x^2)$, $0 \le x \le 1/2$.

13. **Finding volume** The region in the first quadrant that is enclosed by the x-axis and the curve $y = 3x\sqrt{1 - x}$ is revolved about the y-axis to generate a solid. Find the volume of the solid.

14. **Finding volume** The region in the first quadrant that is enclosed by the x-axis, the curve $y = 5/(x\sqrt{5 - x})$, and the lines $x = 1$ and $x = 4$ is revolved about the x-axis to generate a solid. Find the volume of the solid.

15. **Finding volume** The region in the first quadrant enclosed by the coordinate axes, the curve $y = e^x$, and the line $x = 1$ is revolved about the y-axis to generate a solid. Find the volume of the solid.

16. **Finding volume** The region in the first quadrant that is bounded above by the curve $y = e^x - 1$, below by the x-axis, and on the right by the line $x = \ln 2$ is revolved about the line $x = \ln 2$ to generate a solid. Find the volume of the solid.

17. **Finding volume** Let R be the "triangular" region in the first quadrant that is bounded above by the line $y = 1$, below by the curve $y = \ln x$, and on the left by the line $x = 1$. Find the volume of the solid generated by revolving R about

a. the x-axis. **b.** the line $y = 1$.

18. **Finding volume** (*Continuation of Exercise 17.*) Find the volume of the solid generated by revolving the region R about

a. the y-axis. **b.** the line $x = 1$.

19. Finding volume The region between the x-axis and the curve

$$y = f(x) = \begin{cases} 0, & x = 0 \\ x \ln x, & 0 < x \le 2 \end{cases}$$

is revolved about the x-axis to generate the solid shown here.

a. Show that f is continuous at $x = 0$.

b. Find the volume of the solid.

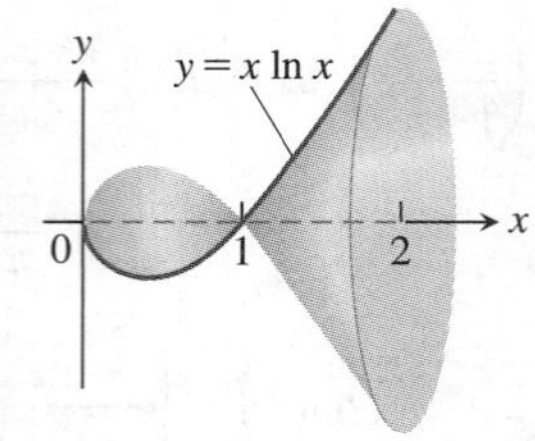

20. Finding volume The infinite region bounded by the coordinate axes and the curve $y = -\ln x$ in the first quadrant is revolved about the x-axis to generate a solid. Find the volume of the solid.

21. Centroid of a region Find the centroid of the region in the first quadrant that is bounded below by the x-axis, above by the curve $y = \ln x$, and on the right by the line $x = e$.

22. Centroid of a region Find the centroid of the region in the plane enclosed by the curves $y = \pm(1 - x^2)^{-1/2}$ and the lines $x = 0$ and $x = 1$.

23. Length of a curve Find the length of the curve $y = \ln x$ from $x = 1$ to $x = e$.

24. Finding surface area Find the area of the surface generated by revolving the curve in Exercise 23 about the y-axis.

25. The surface generated by an astroid The graph of the equation $x^{2/3} + y^{2/3} = 1$ is an *astroid* (see accompanying figure). Find the area of the surface generated by revolving the curve about the x-axis.

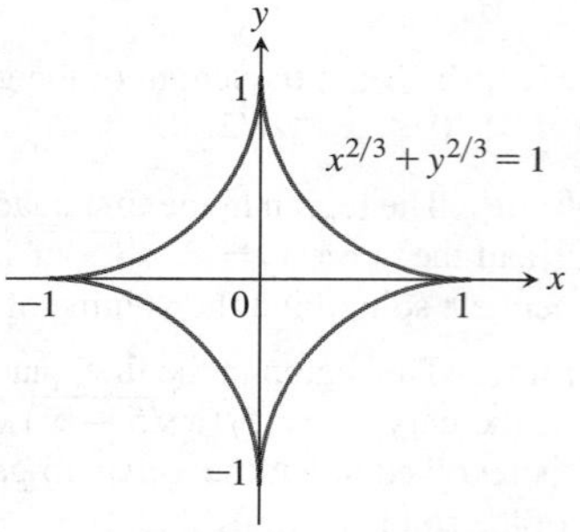

26. Length of a curve Find the length of the curve

$$y = \int_1^x \sqrt{\sqrt{t} - 1}\, dt, \qquad 1 \le x \le 16.$$

27. For what value or values of a does

$$\int_1^{\infty} \left(\frac{ax}{x^2 + 1} - \frac{1}{2x} \right) dx$$

converge? Evaluate the corresponding integral(s).

28. For each $x > 0$, let $G(x) = \int_0^{\infty} e^{-xt}\, dt$. Prove that $xG(x) = 1$ for each $x > 0$.

29. Infinite area and finite volume What values of p have the following property? The area of the region between the curve $y = x^{-p}$, $1 \le x < \infty$, and the x-axis is infinite but the volume of the solid generated by revolving the region about the x-axis is finite.

30. Infinite area and finite volume What values of p have the following property? The area of the region in the first quadrant enclosed by the curve $y = x^{-p}$, the y-axis, the line $x = 1$, and the interval $[0, 1]$ on the x-axis is infinite, but the volume of the solid generated by revolving the region about one of the coordinate axes is finite.

31. Integrating the square of the derivative If f is continuously differentiable on $[0, 1]$, and $f(1) = f(0) = -1/6$, prove that

$$\int_0^1 (f'(x))^2\, dx \ge 2\int_0^1 f(x)\, dx + \frac{1}{4}.$$

Hint: Consider the inequality $0 \le \int_0^1 \left(f'(x) + x - \frac{1}{2} \right)^2 dx$.

Source: Mathematics Magazine, vol. 84, no. 4, Oct. 2011.

32. (*Continuation of Exercise 31.*) If f is continuously differentiable on $[0, a]$ for $a > 0$, and $f(a) = f(0) = b$, prove that

$$\int_0^a (f'(x))^2\, dx \ge 2\int_0^a f(x)\, dx - \left(2ab + \frac{a^3}{12} \right).$$

Hint: Consider the inequality $0 \le \int_0^a \left(f'(x) + x - \frac{a}{2} \right)^2 dx$.

Source: Mathematics Magazine, vol. 84, no. 4, Oct. 2011.

The Substitution $z = \tan(x/2)$

The substitution

$$z = \tan \frac{x}{2} \tag{1}$$

reduces the problem of integrating a rational expression in $\sin x$ and $\cos x$ to a problem of integrating a rational function of z. This in turn can be integrated by partial fractions.

From the accompanying figure

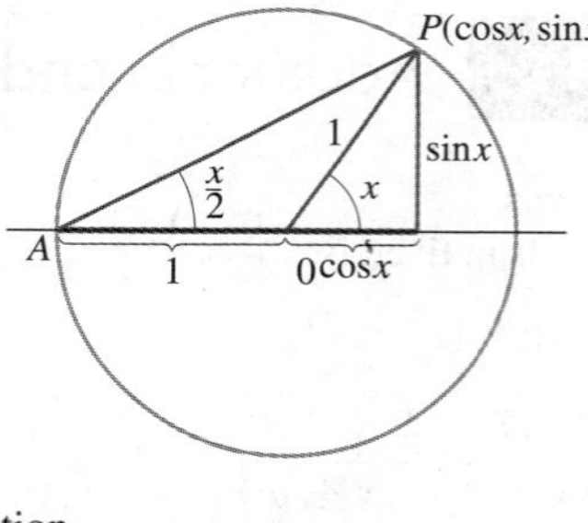

we can read the relation

$$\tan \frac{x}{2} = \frac{\sin x}{1 + \cos x}.$$

To see the effect of the substitution, we calculate

$$\cos x = 2\cos^2\left(\frac{x}{2}\right) - 1 = \frac{2}{\sec^2(x/2)} - 1$$

$$= \frac{2}{1 + \tan^2(x/2)} - 1 = \frac{2}{1 + z^2} - 1$$

$$\cos x = \frac{1 - z^2}{1 + z^2}, \tag{2}$$

and

$$\sin x = 2\sin\frac{x}{2}\cos\frac{x}{2} = 2\frac{\sin(x/2)}{\cos(x/2)} \cdot \cos^2\left(\frac{x}{2}\right)$$

$$= 2\tan\frac{x}{2} \cdot \frac{1}{\sec^2(x/2)} = \frac{2\tan(x/2)}{1 + \tan^2(x/2)}$$

$$\sin x = \frac{2z}{1 + z^2}. \tag{3}$$

Finally, $x = 2\arctan z$, so

$$dx = \frac{2\,dz}{1+z^2}. \qquad (4)$$

Examples

a. $$\int \frac{1}{1+\cos x}\,dx = \int \frac{1+z^2}{2}\frac{2\,dz}{1+z^2} = \int dz = z + C = \tan\left(\frac{x}{2}\right) + C$$

b. $$\int \frac{1}{2+\sin x}\,dx = \int \frac{1+z^2}{2+2z+2z^2}\frac{2\,dz}{1+z^2} = \int \frac{dz}{z^2+z+1} = \int \frac{dz}{(z+(1/2))^2+3/4} = \int \frac{du}{u^2+a^2} = \frac{1}{a}\arctan\left(\frac{u}{a}\right)+C = \frac{2}{\sqrt{3}}\arctan\frac{2z+1}{\sqrt{3}}+C = \frac{2}{\sqrt{3}}\arctan\frac{1+2\tan(x/2)}{\sqrt{3}}+C$$

Use the substitutions in Equations (1)–(4) to evaluate the integrals in Exercises 33–40. Integrals like these arise in calculating the average angular velocity of the output shaft of a universal joint when the input and output shafts are not aligned.

33. $\displaystyle\int \frac{dx}{1-\sin x}$ **34.** $\displaystyle\int \frac{dx}{1+\sin x+\cos x}$

35. $\displaystyle\int_0^{\pi/2} \frac{dx}{1+\sin x}$ **36.** $\displaystyle\int_{\pi/3}^{\pi/2} \frac{dx}{1-\cos x}$

37. $\displaystyle\int_0^{\pi/2} \frac{d\theta}{2+\cos\theta}$ **38.** $\displaystyle\int_{\pi/2}^{2\pi/3} \frac{\cos\theta\,d\theta}{\sin\theta\cos\theta+\sin\theta}$

39. $\displaystyle\int \frac{dt}{\sin t-\cos t}$ **40.** $\displaystyle\int \frac{\cos t\,dt}{1-\cos t}$

Use the substitution $z = \tan(\theta/2)$ to evaluate the integrals in Exercises 41 and 42.

41. $\displaystyle\int \sec\theta\,d\theta$ **42.** $\displaystyle\int \csc\theta\,d\theta$

The Gamma Function and Stirling's Formula

Euler's gamma function $\Gamma(x)$ ("gamma of x"; Γ is a Greek capital g) uses an integral to extend the factorial function from the nonnegative integers to other real values. The formula is

$$\Gamma(x) = \int_0^\infty t^{x-1}e^{-t}\,dt, \quad x > 0.$$

For each positive x, the number $\Gamma(x)$ is the integral of $t^{x-1}e^{-t}$ with respect to t from 0 to ∞. Figure 8.28 shows the graph of Γ near the origin. You will see how to calculate $\Gamma(1/2)$ if you do Additional Exercise 23 in Chapter 15.

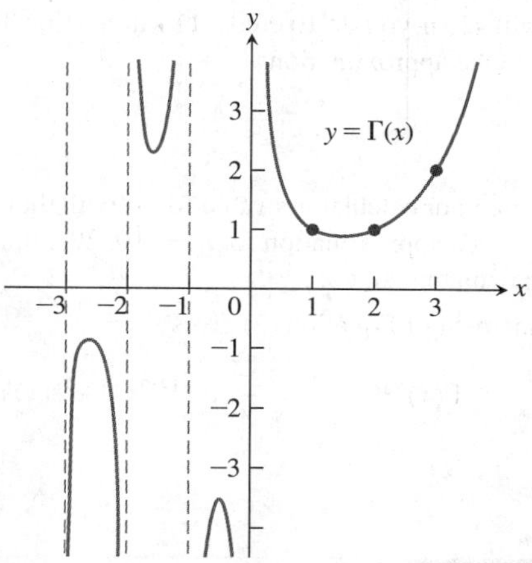

FIGURE 8.28 Euler's gamma function $\Gamma(x)$ is a continuous function of x whose value at each positive integer $n+1$ is $n!$. The defining integral formula for Γ is valid only for $x > 0$, but we can extend Γ to negative noninteger values of x with the formula $\Gamma(x) = (\Gamma(x+1))/x$, which is the subject of Exercise 43.

43. If n is a nonnegative integer, then $\Gamma(n+1) = n!$

a. Show that $\Gamma(1) = 1$.

b. Then apply integration by parts to the integral for $\Gamma(x+1)$ to show that $\Gamma(x+1) = x\Gamma(x)$. This gives

$$\begin{aligned}\Gamma(2) &= 1\Gamma(1) = 1\\ \Gamma(3) &= 2\Gamma(2) = 2\\ \Gamma(4) &= 3\Gamma(3) = 6\\ &\vdots\\ \Gamma(n+1) &= n\,\Gamma(n) = n! \qquad (1)\end{aligned}$$

c. Use mathematical induction to verify Equation (1) for every nonnegative integer n.

44. Stirling's formula Scottish mathematician James Stirling (1692–1770) showed that

$$\lim_{x\to\infty}\left(\frac{e}{x}\right)^x\sqrt{\frac{x}{2\pi}}\,\Gamma(x) = 1,$$

so, for large x,

$$\Gamma(x) = \left(\frac{x}{e}\right)^x\sqrt{\frac{2\pi}{x}}(1+\varepsilon(x)) \qquad \varepsilon(x)\to 0 \text{ as } x\to\infty. \qquad (2)$$

Dropping $\varepsilon(x)$ leads to the approximation

$$\Gamma(x) \approx \left(\frac{x}{e}\right)^x\sqrt{\frac{2\pi}{x}} \qquad \textbf{(Stirling's formula)}. \qquad (3)$$

a. Stirling's approximation for $n!$ Use Equation (3) and the fact that $n! = n\Gamma(n)$ to show that

$$n! \approx \left(\frac{n}{e}\right)^n\sqrt{2n\pi} \qquad \textbf{(Stirling's approximation)}. \qquad (4)$$

As you will see if you do Exercise 114 in Section 10.1, Equation (4) leads to the approximation

$$\sqrt[n]{n!} \approx \frac{n}{e}. \tag{5}$$

T **b.** Compare your calculator's value for $n!$ with the value given by Stirling's approximation for $n = 10, 20, 30, \ldots$, as far as your calculator can go.

T **c.** A refinement of Equation (2) gives

$$\Gamma(x) = \left(\frac{x}{e}\right)^x \sqrt{\frac{2\pi}{x}}\, e^{1/(12x)}(1 + \varepsilon(x))$$

or

$$\Gamma(x) \approx \left(\frac{x}{e}\right)^x \sqrt{\frac{2\pi}{x}}\, e^{1/(12x)},$$

which tells us that

$$n! \approx \left(\frac{n}{e}\right)^n \sqrt{2n\pi}\, e^{1/(12n)}. \tag{6}$$

Compare the values given for 10! by your calculator, Stirling's approximation, and Equation (6).

CHAPTER 8 Technology Application Projects

Mathematica/Maple Projects

Projects can be found within MyLab Math.

- ***Riemann, Trapezoidal, and Simpson Approximations***
  Part I: Visualize the error involved in using Riemann sums to approximate the area under a curve.
  Part II: Build a table of values and compute the relative magnitude of the error as a function of the step size Δx.
  Part III: Investigate the effect of the derivative function on the error.
  Parts IV and V: Trapezoidal Rule approximations.
  Part VI: Simpson's Rule approximations.
- ***Games of Chance: Exploring the Monte Carlo Probabilistic Technique for Numerical Integration***
  Graphically explore the Monte Carlo method for approximating definite integrals.
- ***Computing Probabilities with Improper Integrals***
  More explorations of the Monte Carlo method for approximating definite integrals.

9

First-Order Differential Equations

OVERVIEW Many real-world problems, when formulated mathematically, lead to differential equations. We encountered a number of these equations in previous chapters when studying phenomena such as the motion of an object along a straight line, the decay of a radioactive material, the growth of a population, and the cooling of a heated object placed within a medium of lower temperature.

Section 4.7 introduced differential equations of the form $dy/dx = f(x)$, where f is given and y is an unknown function of x. We learned that when f is continuous over some interval, the general solution $y(x)$ is found directly by integration, $y = \int f(x)\,dx$. In Section 7.4 we investigated differential equations of the form $dy/dx = f(x, y)$, where f is a function of both the independent variable x and the dependent variable y. There we learned how to find the general solution for the special case when the differential equation is separable. In this chapter we further extend our study to include other commonly occurring *first-order* differential equations. These differential equations involve only first derivatives of the unknown function $y(x)$, and they model phenomena varying from simple electrical circuits to the concentration of a chemical in a container. Differential equations involving second derivatives are examined in Chapter 17.

Chapter 17 is available online.
www.bit.ly/3aHAZo5

9.1 Solutions, Slope Fields, and Euler's Method

We begin this section by defining general differential equations involving first derivatives. We then look at slope fields, which give a geometric picture of the solutions to such equations. Many differential equations cannot be solved by obtaining an explicit formula for the solution. However, we can often find numerical approximations to solutions. We present one such method here, called *Euler's method*, which is the basis for many other numerical methods as well.

General First-Order Differential Equations and Solutions

A **first-order differential equation** is an equation

$$\frac{dy}{dx} = f(x, y) \tag{1}$$

in which $f(x, y)$ is a function of two variables defined on a region in the xy-plane. The equation is of *first order* because it involves only the first derivative dy/dx and not higher-order derivatives. In a typical situation y represents an unknown function of x, and $f(x, y)$ is a known function. Some examples of first-order differential equations are

$y' = x + y$, $y' = y/x$, and $y' = 3xy$. In all cases we should think of y as an unknown function of x whose derivative is given by $f(x, y)$. The equations

$$y' = f(x, y) \qquad \text{and} \qquad \frac{d}{dx}y = f(x, y)$$

are equivalent to Equation (1), and all three forms will be used interchangeably in the text.

A **solution** of Equation (1) is a differentiable function $y = y(x)$ defined on an interval I of x-values (possibly an infinite interval) such that

$$\frac{d}{dx}y(x) = f(x, y(x))$$

on that interval. That is, when $y(x)$ and its derivative $y'(x)$ are substituted into Equation (1), the resulting equation is true for all x over the interval I.

EXAMPLE 1 Show that every member of the family of functions

$$y = \frac{C}{x} + 2$$

is a solution of the first-order differential equation

$$\frac{dy}{dx} = \frac{1}{x}(2 - y)$$

on the interval $(0, \infty)$, where C is any constant.

Solution Differentiating $y = C/x + 2$ gives

$$\frac{dy}{dx} = C\frac{d}{dx}\left(\frac{1}{x}\right) + 0 = -\frac{C}{x^2}.$$

We need to show that the differential equation is satisfied when we substitute into it the expressions $(C/x) + 2$ for y, and $-C/x^2$ for dy/dx. That is, we need to verify that for all $x \in (0, \infty)$,

$$-\frac{C}{x^2} = \frac{1}{x}\left[2 - \left(\frac{C}{x} + 2\right)\right].$$

This last equation follows immediately by expanding the expression on the right-hand side:

$$\frac{1}{x}\left[2 - \left(\frac{C}{x} + 2\right)\right] = \frac{1}{x}\left(-\frac{C}{x}\right) = -\frac{C}{x^2}.$$

Therefore, for every value of C, the function $y = C/x + 2$ is a solution of the differential equation. ■

The differential equation in Example 1 has a whole family of solutions, one for each value of C. A natural question to consider is whether this family contains all the solutions to the differential equation, or whether there are others that can somehow arise. The **general solution** to a first-order differential equation is the name given to an expression that contains all possible solutions. The general solution always contains an arbitrary constant, but a solution may contain an arbitrary constant without being the general solution. Establishing when a solution *is* the general solution is left to a more extensive development of the theory of differential equations.

As with antiderivatives, we often need a *particular* rather than the general solution to a first-order differential equation $y' = f(x, y)$. A common way to pick out one of the collection of possible solutions is to specify the value of y at a point $x = x_0$. The **particular solution** satisfying the initial condition $y(x_0) = y_0$ is the solution $y = y(x)$ whose value is y_0 when $x = x_0$. Thus the graph of the particular solution passes through the point (x_0, y_0) in the xy-plane. A **first-order initial value problem** is a differential equation $y' = f(x, y)$ whose solution must satisfy an initial condition $y(x_0) = y_0$.

EXAMPLE 2 Show that the function

$$y = (x + 1) - \frac{1}{3}e^x$$

is a solution to the first-order initial value problem

$$\frac{dy}{dx} = y - x, \qquad y(0) = \frac{2}{3}.$$

Solution The equation

$$\frac{dy}{dx} = y - x$$

is a first-order differential equation with $f(x, y) = y - x$.

On the left side of the equation:

$$\frac{dy}{dx} = \frac{d}{dx}\left(x + 1 - \frac{1}{3}e^x\right) = 1 - \frac{1}{3}e^x.$$

On the right side of the equation:

$$y - x = (x + 1) - \frac{1}{3}e^x - x = 1 - \frac{1}{3}e^x.$$

The function satisfies the initial condition because

$$y(0) = \left[(x + 1) - \frac{1}{3}e^x\right]_{x=0} = 1 - \frac{1}{3} = \frac{2}{3}.$$

The graph of the function is shown in Figure 9.1. ■

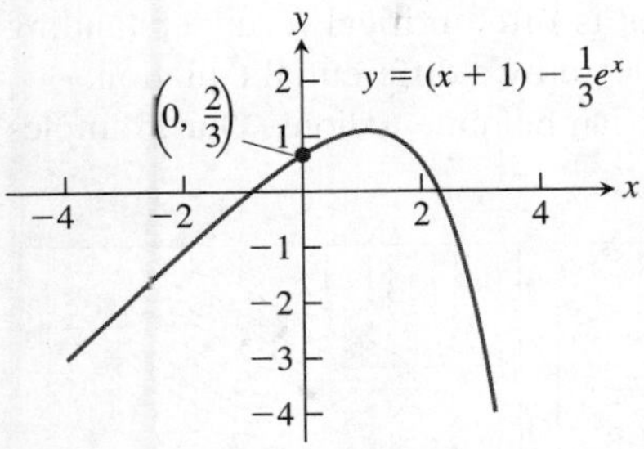

FIGURE 9.1 Graph of the solution to the initial value problem in Example 2.

Slope Fields: Viewing Solution Curves

Each time we specify an initial condition $y(x_0) = y_0$ for the solution of a differential equation $y' = f(x, y)$, the **solution curve** (graph of the solution) is required to pass through the point (x_0, y_0) and to have slope $f(x_0, y_0)$ there. We can picture these slopes graphically by drawing short line segments of slope $f(x, y)$ at selected points (x, y) in the region of the xy-plane that constitutes the domain of f. Each segment has the same slope as the solution curve through (x, y) and so is tangent to the curve there. The resulting picture is called a **slope field** (or **direction field**) and gives a visualization of the general shape of the solution curves. Figure 9.2a shows a slope field, with a particular solution sketched into it in Figure 9.2b. We see how these line segments indicate the direction the solution curve takes at each point it passes through.

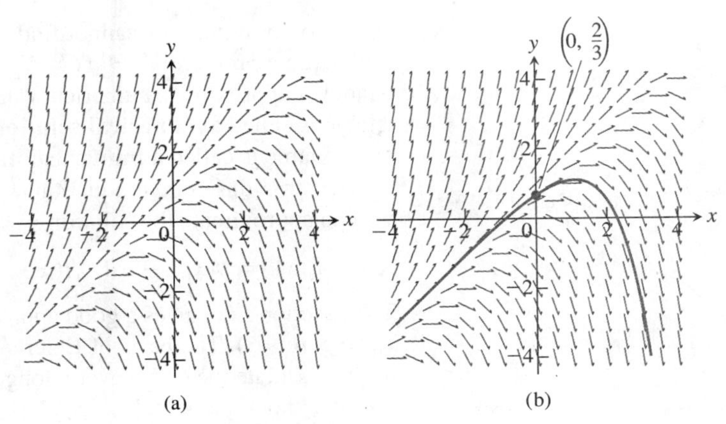

FIGURE 9.2 (a) Slope field for $\frac{dy}{dx} = y - x$. (b) The particular solution curve through the point $\left(0, \frac{2}{3}\right)$ (Example 2).

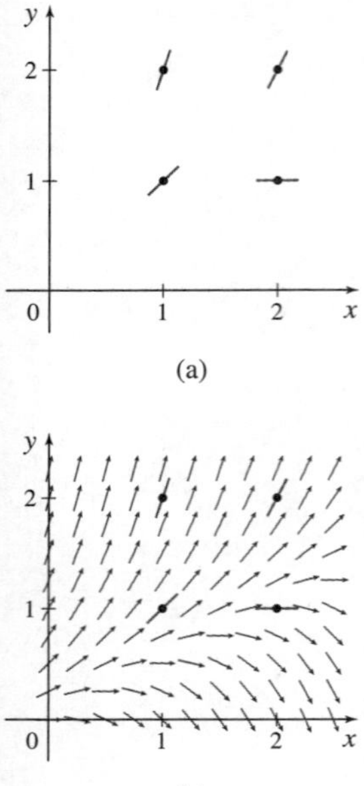

FIGURE 9.3 (a) The slope field for $y' = 2y - x$ is shown at four points. (b) The slope field at several hundred additional points in the plane.

EXAMPLE 3 For the differential equation

$$y' = 2y - x,$$

draw line segments representing the slope field at the points $(1, 2), (1, 1), (2, 2)$, and $(2, 1)$.

Solution At $(1, 2)$ the slope is $2(2) - (1) = 3$. Similarly, the slope at $(2, 2)$ is 2, at $(2, 1)$ is 0, and at $(1, 1)$ is 1. We indicate this on the graph by drawing short line segments of the given slope through each point, as in Figure 9.3. ■

Figure 9.4 shows three slope fields, and we see how the solution curves behave by following the tangent line segments in these fields. Slope fields are useful because they display the overall behavior of the family of solution curves for a given differential equation. For instance, the slope field in Figure 9.4b reveals that every solution $y(x)$ to the differential equation specified in the figure satisfies $\lim_{x\to\pm\infty} y(x) = 0$. We will see that knowing the overall behavior of the solution curves is often critical to understanding and predicting outcomes in a real-world system modeled by a differential equation.

Constructing a slope field with pencil and paper can be quite tedious. Our examples were generated by computer software.

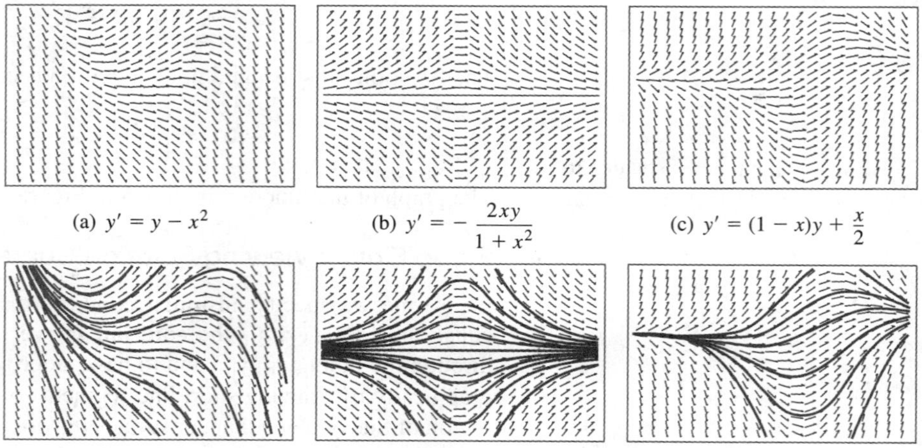

FIGURE 9.4 Slope fields (top row) and selected solution curves (bottom row). In computer renditions, slope segments are sometimes portrayed with arrows, as they are here, but they should be considered as just tangent line segments.

Euler's Method

If we do not require or cannot find an *exact* solution that gives an explicit formula for an initial value problem $y' = f(x, y)$, $y(x_0) = y_0$, we can often use a computer to generate a table of approximate numerical values of y for values of x in an appropriate interval. Such a table is called a **numerical solution** of the problem, and the process by which we generate the table is called a **numerical method**.

Given a differential equation $dy/dx = f(x, y)$ and an initial condition $y(x_0) = y_0$, we can approximate the solution $y = y(x)$ by its linearization

$$L(x) = y(x_0) + y'(x_0)(x - x_0) \quad \text{or} \quad L(x) = y_0 + f(x_0, y_0)(x - x_0).$$

The function $L(x)$ gives a good approximation to the solution $y(x)$ in a short interval about x_0 (Figure 9.5). The basis of Euler's method is to patch together a string of linearizations to approximate the curve over a longer stretch. Here is how the method works.

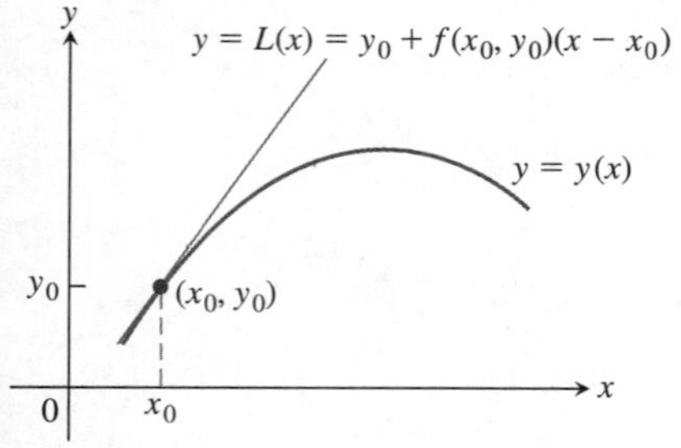

FIGURE 9.5 The linearization $L(x)$ of $y = y(x)$ at $x = x_0$.

We know the point (x_0, y_0) lies on the solution curve. Suppose that we specify a new value for the independent variable to be $x_1 = x_0 + dx$. (Recall that $dx = \Delta x$ in the definition of differentials.) If the increment dx is small, then

$$y_1 = L(x_1) = y_0 + f(x_0, y_0)\,dx$$

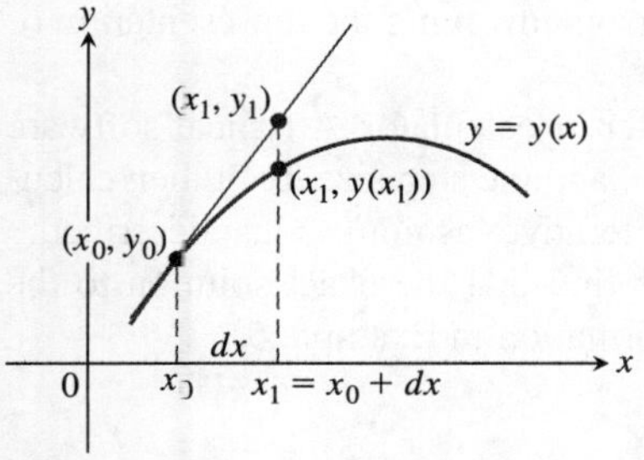

FIGURE 9.6 The first Euler step approximates $y(x_1)$ with $y_1 = L(x_1)$.

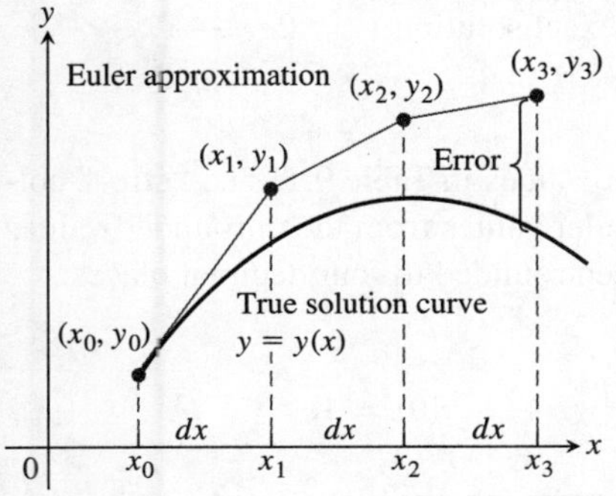

FIGURE 9.7 Three steps in the Euler approximation to the solution of the initial value problem $y' = f(x, y)$, $y(x_0) = y_0$. As we take more steps, the errors involved usually accumulate, but not in the exaggerated way shown here.

is a good approximation to the exact solution value $y = y(x_1)$. So from the point (x_0, y_0), which lies *exactly* on the solution curve, we have obtained the point (x_1, y_1), which lies very close to the point $(x_1, y(x_1))$ on the solution curve (Figure 9.6).

Using the point (x_1, y_1) and the slope $f(x_1, y_1)$ of the solution curve through (x_1, y_1), we take a second step. Setting $x_2 = x_1 + dx$, we use the linearization of the solution curve through (x_1, y_1) to calculate

$$y_2 = y_1 + f(x_1, y_1)\,dx.$$

This gives the next approximation (x_2, y_2) to values along the solution curve $y = y(x)$ (Figure 9.7). Continuing in this fashion, we take a third step from the point (x_2, y_2) with slope $f(x_2, y_2)$ to obtain the third approximation

$$y_3 = y_2 + f(x_2, y_2)\,dx,$$

and so on. We are building an approximation to a solution by following the direction of the slope field of the differential equation.

The steps in Figure 9.7 are drawn large to illustrate the construction process, so the approximation looks crude. In practice, dx would be chosen small enough to make the gray curve hug the black one and give a better approximation.

EXAMPLE 4 Find the first three approximations y_1, y_2, y_3 using Euler's method for the initial value problem

$$y' = 1 + y, \qquad y(0) = 1,$$

starting at $x_0 = 0$ with $dx = 0.1$.

Solution We already have the starting values $x_0 = 0$ and $y_0 = 1$. Next we determine the values of x at which the Euler approximations will take place: $x_1 = x_0 + dx = 0.1$, $x_2 = x_0 + 2\,dx = 0.2$, and $x_3 = x_0 + 3\,dx = 0.3$. Then we find

$$\begin{aligned}
\textit{First:}\quad y_1 &= y_0 + f(x_0, y_0)\,dx \\
&= y_0 + (1 + y_0)\,dx && f(x, y) = 1 + y \\
&= 1 + (1 + 1)(0.1) = 1.2 && y_0 = 1,\ dx = 0.1
\end{aligned}$$

$$\begin{aligned}
\textit{Second:}\quad y_2 &= y_1 + f(x_1, y_1)\,dx \\
&= y_1 + (1 + y_1)\,dx \\
&= 1.2 + (1 + 1.2)(0.1) = 1.42 && y_1 = 1.2,\ dx = 0.1
\end{aligned}$$

$$\begin{aligned}
\textit{Third:}\quad y_3 &= y_2 + f(x_2, y_2)\,dx \\
&= y_2 + (1 + y_2)\,dx \\
&= 1.42 + (1 + 1.42)(0.1) = 1.662 && y_2 = 1.42,\ dx = 0.1
\end{aligned}$$

■

The step-by-step process used in Example 4 can be continued with more points. Using equally spaced values for the independent variable in the table for the numerical solution, and generating n of them, set

$$\begin{aligned}
x_1 &= x_0 + dx \\
x_2 &= x_1 + dx \\
&\vdots \\
x_n &= x_{n-1} + dx.
\end{aligned}$$

Then calculate the approximations to the solution,

$$\begin{aligned}
y_1 &= y_0 + f(x_0, y_0)\,dx \\
y_2 &= y_1 + f(x_1, y_1)\,dx \\
&\vdots \\
y_n &= y_{n-1} + f(x_{n-1}, y_{n-1})\,dx.
\end{aligned}$$

HISTORICAL BIOGRAPHY
Leonhard Euler
(1707–1783)
www.bit.ly/2NcDOzi

The number of steps n can be as large as we like, but errors involving the representation of numbers in software can accumulate if n is too large.

Euler's method is easy to implement on a computer or calculator. A typical software program takes as input x_0 and y_0, the number of steps n, and the step size dx. It then calculates the approximate solution values $y_1, y_2, \ldots, y_n$ in iterative fashion, as just described.

Solving the separable equation in Example 4, we find that the exact solution to the initial value problem is $y = 2e^x - 1$. We use this information in Example 5.

EXAMPLE 5 Use Euler's method to solve

$$y' = 1 + y, \qquad y(0) = 1,$$

on the interval $0 \le x \le 1$, starting at $x_0 = 0$ and taking **(a)** $dx = 0.1$ and **(b)** $dx = 0.05$. Compare the approximations with the values of the exact solution $y = 2e^x - 1$.

Solution

(a) We used a computer to generate the approximate values in Table 9.1. The "error" column is obtained by subtracting the unrounded Euler values from the unrounded values found using the exact solution. All entries are then rounded to four decimal places.

TABLE 9.1 Euler solution of $y' = 1 + y$, $y(0) = 1$, step size $dx = 0.1$

x	y (Euler)	y (exact)	Error
0	1	1	0
0.1	1.2	1.2103	0.0103
0.2	1.42	1.4428	0.0228
0.3	1.662	1.6997	0.0377
0.4	1.9282	1.9836	0.0554
0.5	2.2210	2.2974	0.0764
0.6	2.5431	2.6442	0.1011
0.7	2.8974	3.0275	0.1301
0.8	3.2872	3.4511	0.1639
0.9	3.7159	3.9192	0.2033
1.0	4.1875	4.4366	0.2491

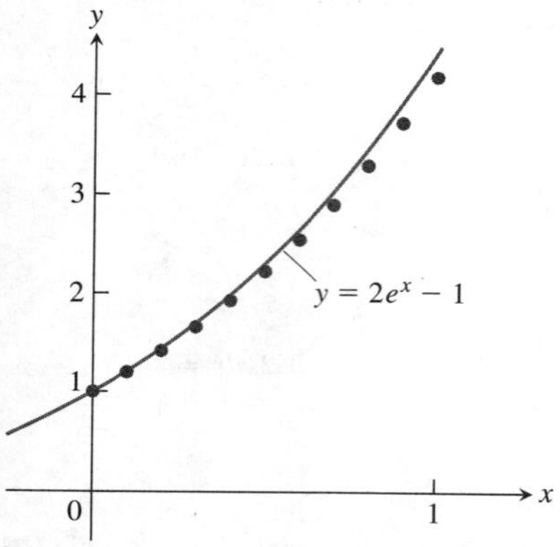

FIGURE 9.8 The graph of $y = 2e^x - 1$ superimposed on a scatterplot of the Euler approximations shown in Table 9.1 (Example 5).

By the time we reach $x = 1$ (after 10 steps), the error is about 5.6% of the exact solution. A plot of the exact solution curve with the scatterplot of Euler solution points from Table 9.1 is shown in Figure 9.8.

(b) One way to try to reduce the error is to decrease the step size. Table 9.2 shows the results and their comparisons with the exact solutions when we decrease the step size to 0.05, doubling the number of steps to 20. As in Table 9.1, all computations are performed before rounding. This time when we reach $x = 1$, the relative error is only about 2.9%. ■

It might be tempting to reduce the step size even further in Example 5 to obtain greater accuracy. Each additional calculation, however, not only requires additional computer time but, more importantly, adds to the buildup of round-off errors due to the approximate representations of numbers inside the computer.

The analysis of error and the investigation of methods to reduce it when making numerical calculations are important. An area of mathematics called *numerical analysis* studies advanced numerical methods that are more accurate than Euler's method.

TABLE 9.2 Euler solution of $y' = 1 + y$, $y(0) = 1$, step size $dx = 0.05$

x	y (Euler)	y (exact)	Error
0	1	1	0
0.05	1.1	1.1025	0.0025
0.10	1.205	1.2103	0.0053
0.15	1.3153	1.3237	0.0084
0.20	1.4310	1.4428	0.0118
0.25	1.5526	1.5681	0.0155
0.30	1.6802	1.6997	0.0195
0.35	1.8142	1.8381	0.0239
0.40	1.9549	1.9836	0.0287
0.45	2.1027	2.1366	0.0340
0.50	2.2578	2.2974	0.0397
0.55	2.4207	2.4665	0.0458
0.60	2.5917	2.6442	0.0525
0.65	2.7713	2.8311	0.0598
0.70	2.9599	3.0275	0.0676
0.75	3.1579	3.2340	0.0761
0.80	3.3657	3.4511	0.0853
0.85	3.5840	3.6793	0.0953
0.90	3.8132	3.9192	0.1060
0.95	4.0539	4.1714	0.1175
1.00	4.3066	4.4366	0.1300

EXERCISES 9.1

Slope Fields

In Exercises 1–4, match the differential equations with their slope fields, graphed here.

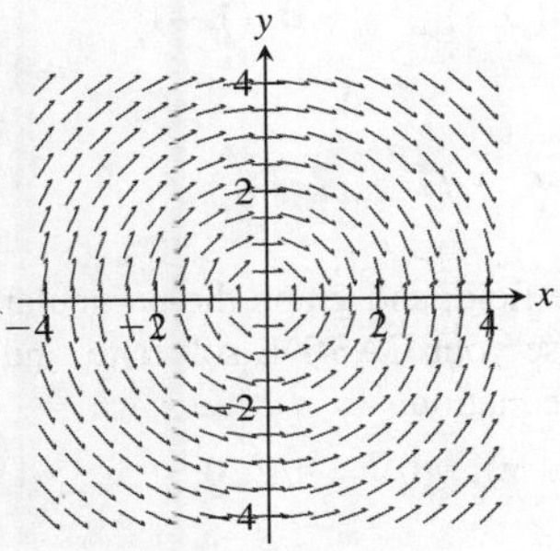

(a)

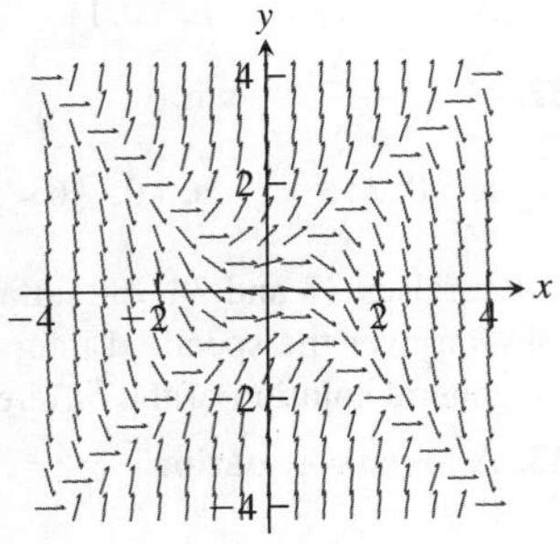

(b)

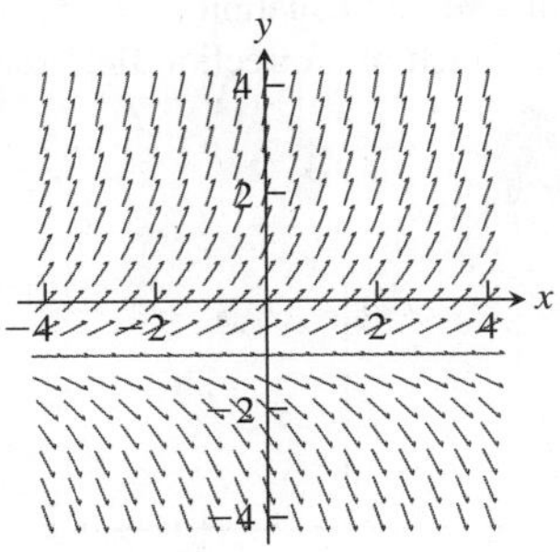

(c)

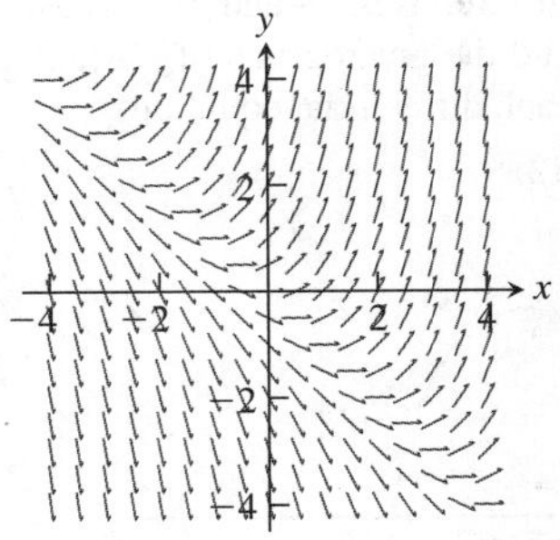

(d)

1. $y' = x + y$

2. $y' = y + 1$

3. $y' = -\dfrac{x}{y}$

4. $y' = y^2 - x^2$

In Exercises 5 and 6, copy the slope fields, and sketch in some of the solution curves.

5. $y' = (y + 1)(y - 1)$

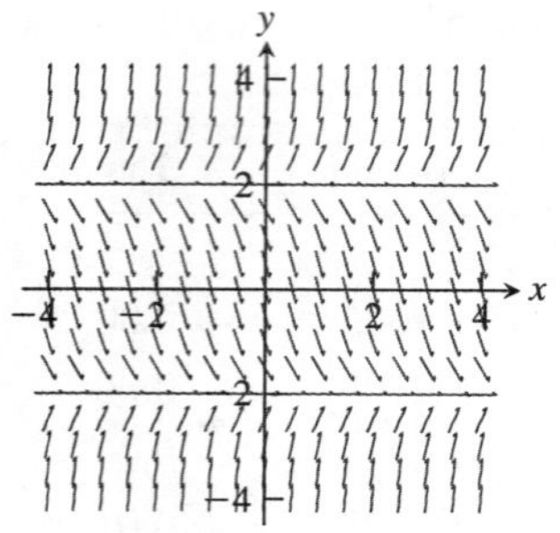

6. $y' = y(y + 2)(y - 2)$

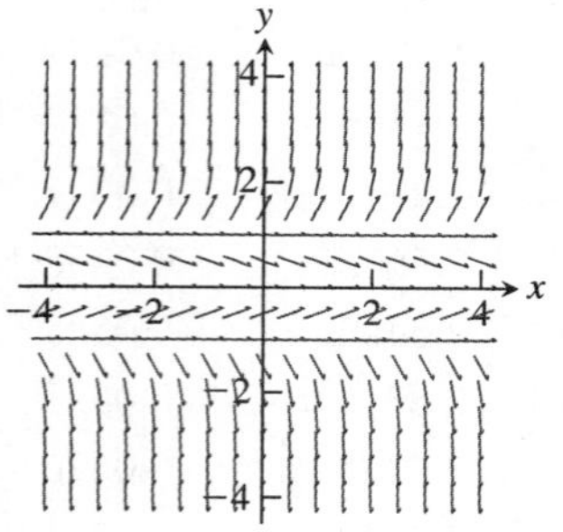

Integral Equations

In Exercises 7–12, write an equivalent first-order differential equation and initial condition for y.

7. $y = -1 + \int_1^x (t - y(t))\,dt$ **8.** $y = \int_1^x \frac{2}{t}\,dt$

9. $y = 2 - \int_0^x (1 + y(t)) \sin t\,dt$

10. $y = 3 + \int_0^x y(t)\,dt$ **11.** $y = x + 4 + \int_{-2}^x te^{y(t)}\,dt$

12. $y = \ln x + \int_x^e \sqrt{t^2 + (y(t))^2}\,dt$

In Exercises 13 and 14, consider the differential equation $y' = f(y)$ and the given graph of f. Make a rough sketch of a direction field for each differential equation.

13.

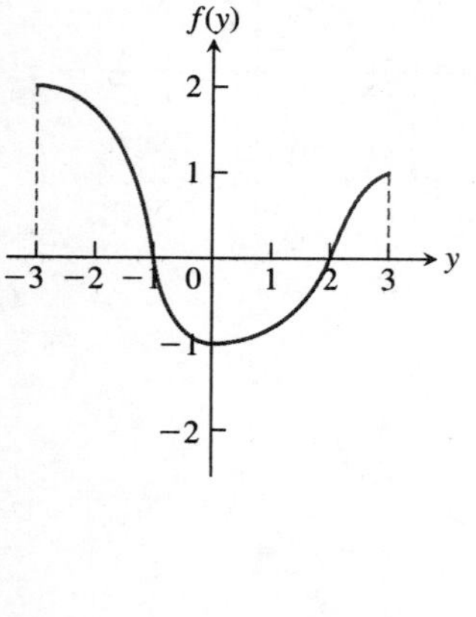

14.

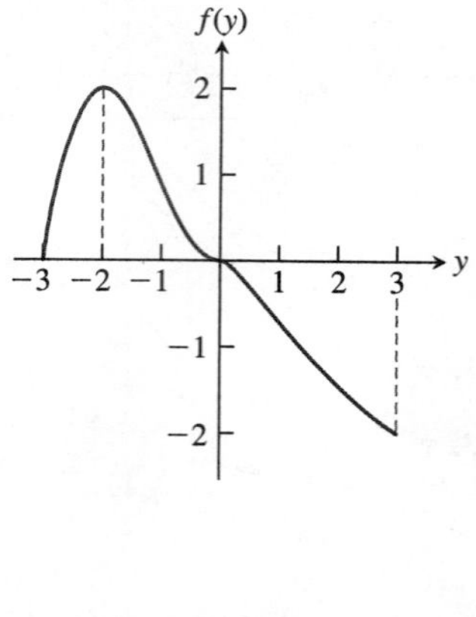

Using Euler's Method

In Exercises 15–20, use Euler's method to calculate the first three approximations to the given initial value problem for the specified increment size. Calculate the exact solution and investigate the accuracy of your approximations. Round your results to four decimal places.

15. $y' = \frac{2y}{x}, \quad y(1) = -1, \quad dx = 0.5$

16. $y' = x(1 - y), \quad y(1) = 0, \quad dx = 0.2$

17. $y' = x + 2y, \quad y(0) = 0, \quad dx = 0.25$

18. $y' = y^2(1 + 2x), \quad y(-1) = 1, \quad dx = 0.5$

T **19.** $y' = 2xe^{x^2}, \quad y(0) = 2, \quad dx = 0.1$

T **20.** $y' = ye^x, \quad y(0) = 2, \quad dx = 0.5$

21. Use the Euler method with $dx = 0.05$ to estimate $y(1)$ if $y' = -50y$ and $y(0) = 1$. What is the exact value of $y(1)$?

22. Use Euler's method with $dx = 0.2$ to estimate $y(2)$ if $y' = y/x$ and $y(1) = 2$. What is the exact value of $y(2)$?

23. Use Euler's method with $dx = 0.5$ to estimate $y(5)$ if $y' = y^2/\sqrt{x}$ and $y(1) = -1$. What is the exact value of $y(5)$?

24. Use Euler's method with $dx = 1/3$ to estimate $y(2)$ if $y' = x \sin y$ and $y(0) = 1$. What is the exact value of $y(2)$?

25. Show that the solution of the initial value problem

$$y' = x + y, \quad y(x_0) = y_0$$

is

$$y = -1 - x + (1 + x_0 + y_0)\,e^{x - x_0}.$$

26. What integral equation is equivalent to the initial value problem $y' = f(x), y(x_0) = y_0$?

COMPUTER EXPLORATIONS

In Exercises 27–32, obtain a slope field and add to it graphs of the solution curves passing through the given points.

27. $y' = y$ with

a. $(0, 1)$ **b.** $(0, 2)$ **c.** $(0, -1)$

28. $y' = 2(y - 4)$ with

a. $(0, 1)$ **b.** $(0, 4)$ **c.** $(0, 5)$

29. $y' = y(x + y)$ with

a. $(0, 1)$ **b.** $(0, -2)$ **c.** $(0, 1/4)$ **d.** $(-1, -1)$

30. $y' = y^2$ with

a. $(0, 1)$ **b.** $(0, 2)$ **c.** $(0, -1)$ **d.** $(0, 0)$

31. $y' = (y - 1)(x + 2)$ with

a. $(0, -1)$ **b.** $(0, 1)$ **c.** $(0, 3)$ **d.** $(1, -1)$

32. $y' = \frac{xy}{x^2 + 4}$ with

a. $(0, 2)$ **b.** $(0, -6)$ **c.** $(-2\sqrt{3}, -4)$

In Exercises 33 and 34, obtain a slope field, and graph the particular solution over the specified interval. Use your CAS DE solver to find the general solution of the differential equation.

33. A logistic equation $y' = y(2 - y), y(0) = 1/2; 0 \le x \le 4, 0 \le y \le 3$

34. $y' = (\sin x)(\sin y), \quad y(0) = 2; \quad -6 \le x \le 6, \quad -6 \le y \le 6$

Exercises 35 and 36 have no explicit solution in terms of elementary functions. Use a CAS to explore graphically each of the differential equations.

35. $y' = \cos(2x - y), \quad y(0) = 2; \quad 0 \le x \le 5, \quad 0 \le y \le 5$

36. A Gompertz equation $y' = y(1/2 - \ln y), \quad y(0) = 1/3;$ $0 \le x \le 4, \quad 0 \le y \le 3$

37. Use a CAS to find the solutions of $y' + y = f(x)$, subject to the initial condition $y(0) = 0$, if $f(x)$ is

a. $2x$ **b.** $\sin 2x$ **c.** $3e^{x/2}$ **d.** $2e^{-x/2}\cos 2x$.

Graph all four solutions over the interval $-2 \le x \le 6$ to compare the results.

38. a. Use a CAS to plot the slope field of the differential equation

$$y' = \frac{3x^2 + 4x + 2}{2(y - 1)}$$

over the region $-3 \le x \le 3$ and $-3 \le y \le 3$.

b. Separate the variables and use a CAS integrator to find the general solution in implicit form.

c. Using a CAS implicit function grapher, plot solution curves for the arbitrary constant values $C = -6, -4, -2, 0, 2, 4, 6$.

d. Find and graph the solution that satisfies the initial condition $y(0) = -1$.

In Exercises 39–42, use Euler's method with the specified step size to estimate the value of the solution at the given point x^*. Find the value of the exact solution at x^*.

39. $y' = 2xe^{x^2}, \quad y(0) = 2, \quad dx = 0.1, \quad x^* = 1$

40. $y' = 2y^2(x - 1), \quad y(2) = -1/2, \quad dx = 0.1, \quad x^* = 3$

41. $y' = \sqrt{x}/y, \quad y > 0, \quad y(0) = 1, \quad dx = 0.1, \quad x^* = 1$

42. $y' = 1 + y^2, \quad y(0) = 0, \quad dx = 0.1, \quad x^* = 1$

Use a CAS to explore graphically each of the differential equations in Exercises 43–46. Perform the following steps to help with your explorations.

a. Plot a slope field for the differential equation in the given xy-window.

b. Find the general solution of the differential equation using your CAS DE solver.

c. Graph the solutions for the values of the arbitrary constant $C = -2, -1, 0, 1, 2$ superimposed on your slope field plot.

d. Find and graph the solution that satisfies the specified initial condition over the interval $[0, b]$.

e. Find the Euler numerical approximation to the solution of the initial value problem with 4 subintervals of the x-interval, and plot the Euler approximation superimposed on the graph produced in part (d).

f. Repeat part (e) for 8, 16, and 32 subintervals. Plot these three Euler approximations superimposed on the graph from part (e).

g. Find the error $(y(\text{exact}) - y(\text{Euler}))$ at the specified point $x = b$ for each of your four Euler approximations. Discuss the improvement in the percentage error.

43. $y' = x + y, \quad y(0) = -7/10; \quad -4 \le x \le 4, \quad -4 \le y \le 4;$ $b = 1$

44. $y' = -x/y, \quad y(0) = 2; \quad -3 \le x \le 3, \quad -3 \le y \le 3; \quad b = 2$

45. $y' = y(2 - y), \quad y(0) = 1/2; \quad 0 \le x \le 4, \quad 0 \le y \le 3;$ $b = 3$

46. $y' = (\sin x)(\sin y), \quad y(0) = 2; \quad -6 \le x \le 6, \quad -6 \le y \le 6;$ $b = 3\pi/2$

9.2 First-Order Linear Equations

A first-order **linear** differential equation is one that can be written in the form

$$\frac{dy}{dx} + P(x)y = Q(x), \tag{1}$$

where P and Q are continuous functions of x. Equation (1) is the linear equation's **standard form**. Since the exponential growth/decay equation $dy/dx = ky$ (Section 7.4) can be put in the standard form

$$\frac{dy}{dx} - ky = 0,$$

we see it is a linear equation with $P(x) = -k$ and $Q(x) = 0$. Equation (1) is *linear* (in y) because y and its derivative dy/dx occur only to the first power, they are not multiplied together, nor do they appear as the argument of a function (such as $\sin y$, e^y, or $\sqrt{dy/dx}$).

EXAMPLE 1 Put the following equation in standard form:

$$x\frac{dy}{dx} = x^2 + 3y, \qquad x > 0.$$

Solution

$$x\frac{dy}{dx} = x^2 + 3y$$

$$\frac{dy}{dx} = x + \frac{3}{x}y \qquad \text{Divide by } x.$$

$$\frac{dy}{dx} - \frac{3}{x}y = x \qquad \text{Standard form with } P(x) = -3/x \text{ and } Q(x) = x$$

Notice that $P(x)$ is $-3/x$, not $+3/x$. The standard form is $y' + P(x)y = Q(x)$, so the minus sign is part of the formula for $P(x)$. ■

Solving Linear Equations

We solve the equation

$$\frac{dy}{dx} + P(x)y = Q(x)$$

by multiplying both sides by a *positive* function $v(x)$ that transforms the left-hand side into the derivative of the product $v(x) \cdot y$. We will show how to find v in a moment, but first we want to show how, once found, it provides the solution we seek.

Here is why multiplying by $v(x)$ works:

$$\frac{dy}{dx} + P(x)y = Q(x) \qquad \text{Original equation is in standard form.}$$

$$v(x)\frac{dy}{dx} + P(x)v(x)y = v(x)Q(x) \qquad \text{Multiply by positive } v(x).$$

$$\frac{d}{dx}(v(x) \cdot y) = v(x)Q(x) \qquad v(x) \text{ is chosen to make } v\frac{dy}{dx} + Pvy = \frac{d}{dx}(v \cdot y).$$

$$v(x) \cdot y = \int v(x)Q(x)\,dx \qquad \text{Integrate with respect to } x.$$

$$y = \frac{1}{v(x)}\int v(x)Q(x)\,dx \qquad \text{Divide by } v(x). \qquad (2)$$

Equation (2) expresses the solution of Equation (1) in terms of the functions $v(x)$ and $Q(x)$. We call $v(x)$ an **integrating factor** for Equation (1) because its presence makes the equation integrable.

Why doesn't the formula for $P(x)$ appear in the solution as well? It does, but indirectly, in the construction of the positive function $v(x)$. We have

$$\frac{d}{dx}(vy) = v\frac{dy}{dx} + Pvy \qquad \text{Condition imposed on } v$$

$$v\frac{dy}{dx} + y\frac{dv}{dx} = v\frac{dy}{dx} + Pvy \qquad \text{Derivative Product Rule}$$

$$y\frac{dv}{dx} = Pvy. \qquad \text{The terms } v\frac{dy}{dx} \text{ cancel.}$$

This last equation will hold if

$$\frac{dv}{dx} = Pv$$

$$\frac{dv}{v} = P\,dx \qquad \text{Variables separated, } v > 0$$

$$\int\frac{dv}{v} = \int P\,dx \qquad \text{Integrate both sides.}$$

$$\ln v = \int P\,dx \qquad \text{Since } v > 0, \text{ we do not need absolute value signs in } \ln v.$$

$$e^{\ln v} = e^{\int P\,dx} \qquad \text{Exponentiate both sides to solve for } v.$$

$$v = e^{\int P\,dx}. \tag{3}$$

Thus a formula for the general solution to Equation (1) is given by Equation (2), where $v(x)$ is given by Equation (3). However, rather than memorizing the formula, just remember how to find the integrating factor once you have the standard form so $P(x)$ is correctly identified. Any antiderivative of P works for Equation (3).

Integrating Factors
To solve the linear equation $y' + P(x)y = Q(x)$, multiply both sides by the integrating factor $v(x) = e^{\int P(x)\,dx}$ and integrate both sides.

When you integrate the product on the left-hand side in this procedure, you always obtain the product $v(x)y$ of the integrating factor and solution function y because of the way v is defined.

EXAMPLE 2 Solve the equation

$$x\frac{dy}{dx} = x^2 + 3y, \qquad x > 0.$$

HISTORICAL BIOGRAPHY
Adrien-Marie Legendre
(1752–1833)
www.bit.ly/2P2HNjw

Solution First we put the equation in standard form (Example 1):

$$\frac{dy}{dx} - \frac{3}{x}y = x,$$

so $P(x) = -3/x$ is identified.

The integrating factor is

$$\begin{aligned} v(x) = e^{\int P(x)\,dx} &= e^{\int(-3/x)\,dx} \\ &= e^{-3\ln|x|} && \text{Constant of integration is 0, so } v \text{ is as simple as possible.} \\ &= e^{-3\ln x} && x > 0 \\ &= e^{\ln x^{-3}} = \frac{1}{x^3}. \end{aligned}$$

Next we multiply both sides of the standard form by $v(x)$ and integrate:

$$\begin{aligned} \frac{1}{x^3}\cdot\left(\frac{dy}{dx} - \frac{3}{x}y\right) &= \frac{1}{x^3}\cdot x \\ \frac{1}{x^3}\frac{dy}{dx} - \frac{3}{x^4}y &= \frac{1}{x^2} \\ \frac{d}{dx}\left(\frac{1}{x^3}y\right) &= \frac{1}{x^2} && \text{Left-hand side is } \frac{d}{dx}(v\cdot y). \\ \frac{1}{x^3}y &= \int \frac{1}{x^2}\,dx && \text{Integrate both sides.} \\ \frac{1}{x^3}y &= -\frac{1}{x} + C. \end{aligned}$$

Solving this last equation for y gives the general solution:

$$y = x^3\left(-\frac{1}{x} + C\right) = -x^2 + Cx^3, \qquad x > 0.$$

■

EXAMPLE 3 Find the particular solution of

$$3xy' - y = \ln x + 1, \qquad x > 0,$$

that satisfying $y(1) = -2$.

Solution With $x > 0$, we write the equation in standard form:

$$y' - \frac{1}{3x}y = \frac{\ln x + 1}{3x}.$$

Then the integrating factor is given by

$$v = e^{\int -1/(3x)\,dx} = e^{(-1/3)\ln x} = x^{-1/3}. \qquad x > 0$$

Thus

$$x^{-1/3}y = \frac{1}{3}\int (\ln x + 1)x^{-4/3}\,dx. \qquad \text{Left-hand side is } vy.$$

Integration by parts of the right-hand side gives

$$x^{-1/3}y = -x^{-1/3}(\ln x + 1) + \int x^{-4/3}\,dx + C.$$

Therefore,

$$x^{-1/3}y = -x^{-1/3}(\ln x + 1) - 3x^{-1/3} + C$$

or, solving for y,

$$y = -(\ln x + 4) + Cx^{1/3}.$$

When $x = 1$ and $y = -2$, this last equation becomes

$$-2 = -(0 + 4) + C,$$

so

$$C = 2.$$

Substitution into the equation for y gives the particular solution

$$y = 2x^{1/3} - \ln x - 4.$$

■

In solving the linear equation in Example 2, we integrated both sides of the equation after multiplying each side by the integrating factor. However, we can shorten the amount of work, as in Example 3, by remembering that the left-hand side *always* integrates into the product $v(x) \cdot y$ of the integrating factor times the solution function. From Equation (2) this means that

$$v(x)y = \int v(x)Q(x)\,dx. \tag{4}$$

We need only integrate the product of the integrating factor $v(x)$ with $Q(x)$ on the right-hand side of Equation (1) and then equate the result with $v(x)y$ to obtain the general solution. Nevertheless, to emphasize the role of $v(x)$ in the solution process, we sometimes follow the complete procedure as illustrated in Example 2.

Observe that if the function $Q(x)$ is identically zero in the standard form given by Equation (1), the linear equation is separable and can be solved by the method of Section 7.4:

$$\frac{dy}{dx} + P(x)y = Q(x)$$

$$\frac{dy}{dx} + P(x)y = 0 \qquad Q(x) = 0$$

$$\frac{dy}{y} = -P(x)\,dx. \qquad \text{Separating the variables}$$

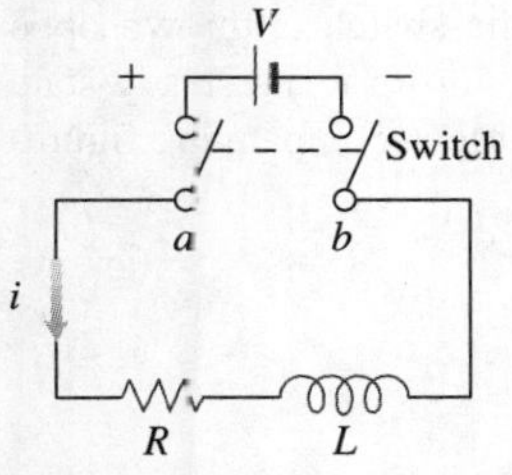

FIGURE 9.9 The *RL* circuit in Example 4.

RL Circuits

The diagram in Figure 9.9 represents an electrical circuit whose total resistance is a constant R ohms and whose self-inductance, shown as a coil, is L henries, also a constant. There is a switch whose terminals at a and b can be closed to connect a constant electrical source of V volts.

Ohm's Law, $V = RI$, has to be augmented for such a circuit. The correct equation accounting for both resistance and inductance is

$$L\frac{di}{dt} + Ri = V, \tag{5}$$

where i is the current in amperes and t is the time in seconds. By solving this equation, we can predict how the current will flow after the switch is closed.

EXAMPLE 4 The switch in the *RL* circuit in Figure 9.9 is closed at time $t = 0$. How will the current flow as a function of time?

Solution Equation (5) is a first-order linear differential equation for i as a function of t. Its standard form is

$$\frac{di}{dt} + \frac{R}{L}i = \frac{V}{L}, \tag{6}$$

and the corresponding solution, given that $i = 0$ when $t = 0$, is

$$i = \frac{V}{R} - \frac{V}{R}e^{-(R/L)t}. \tag{7}$$

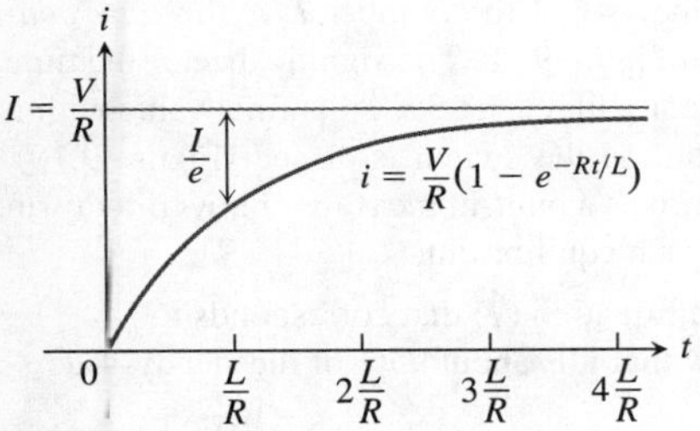

FIGURE 9.10 The growth of the current in the *RL* circuit in Example 4. I is the current's steady-state value. The number $t = L/R$ is the time constant of the circuit. The current gets to within 5% of its steady-state value in 3 time constants (Exercise 27).

(We leave the calculation of the solution to Exercise 28.) Since R and L are positive, $-(R/L)$ is negative and $e^{-(R/L)t} \to 0$ as $t \to \infty$. Thus,

$$\lim_{t\to\infty} i = \lim_{t\to\infty}\left(\frac{V}{R} - \frac{V}{R}e^{-(R/L)t}\right) = \frac{V}{R} - \frac{V}{R}\cdot 0 = \frac{V}{R}.$$

At any given time, the current is less than V/R, but as time passes, the current approaches the **steady-state value** V/R. According to the equation

$$L\frac{di}{dt} + Ri = V,$$

$I = V/R$ is the current that will flow in the circuit if either $L = 0$ (no inductance) or $di/dt = 0$ (steady current, $i =$ constant) (Figure 9.10).

Equation (7) expresses the solution of Equation (6) as the sum of two terms: a steady-state solution V/R and a transient solution $-(V/R)e^{-(R/L)t}$ that tends to zero as $t \to \infty$. ■

EXERCISES 9.2

First-Order Linear Equations

Solve the differential equations in Exercises 1–14.

1. $x\frac{dy}{dx} + y = e^x, \quad x > 0$

2. $e^x\frac{dy}{dx} + 2e^x y = 1$

3. $xy' + 3y = \frac{\sin x}{x^2}, \quad x > 0$

4. $y' + (\tan x)y = \cos^2 x, \quad -\pi/2 < x < \pi/2$

5. $x\frac{dy}{dx} + 2y = 1 - \frac{1}{x}, \quad x > 0$

6. $(1 + x)\, y' + y = \sqrt{x}$

7. $\frac{dy}{dx} - y\tan x = e^x \sec x$

8. $e^{2x}y' + 2e^{2x}y = 2x$

9. $dr + (2r\cot\theta + \sin 2\theta)d\theta = 0$

10. $x\frac{dy}{dx} = \frac{\cos x}{x} - 2y, \quad x > 0$

11. $(t-1)^3 \dfrac{ds}{dt} + 4(t-1)^2 s = t+1, \quad t > 1$

12. $(t+1)\dfrac{ds}{dt} + 2s = 3(t+1) + \dfrac{1}{(t+1)^2}, \quad t > -1$

13. $\sin\theta \dfrac{dr}{d\theta} + (\cos\theta)r = \tan\theta, \quad 0 < \theta < \pi/2$

14. $\tan\theta \dfrac{dr}{d\theta} + r = \sin^2\theta, \quad 0 < \theta < \pi/2$

Solving Initial Value Problems

Solve the initial value problems in Exercises 15–20.

15. $\dfrac{dy}{dt} + 2y = 3, \quad y(0) = 1$

16. $t\dfrac{dy}{dt} + 2y = t^3, \quad t > 0, \quad y(2) = 1$

17. $\theta\dfrac{dy}{d\theta} + y = \sin\theta, \quad \theta > 0, \quad y(\pi/2) = 1$

18. $\theta\dfrac{dy}{d\theta} - 2y = \theta^3\sec\theta\tan\theta, \quad \theta > 0, \quad y(\pi/3) = 2$

19. $(x+1)\dfrac{dy}{dx} - 2(x^2+x)y = \dfrac{e^{x^2}}{x+1}, \quad x > -1, \quad y(0) = 5$

20. $\dfrac{dy}{dx} + xy = x, \quad y(0) = -6$

21. $y - x\dfrac{dy}{dx} - y^2 + \dfrac{dy}{dx} = 0$

$\left[\text{Hint: Rewrite as } \dfrac{dx}{dy} + \dfrac{1}{y(y-1)}x = \dfrac{1}{y(y-1)}, \text{ linear in } x\right]$

22. Solve the following initial value problem for u as a function of t:

$$\frac{du}{dt} + \frac{k}{m}u = 0 \quad (k \text{ and } m \text{ positive constants}), \quad u(0) = u_0$$

a. as a first-order linear equation.

b. as a separable equation.

Theory and Examples

23. Is either of the following equations correct? Give reasons for your answers.

a. $x\displaystyle\int \frac{1}{x}\,dx = x\ln|x| + C$

b. $x\displaystyle\int \frac{1}{x}\,dx = x\ln|x| + Cx$

24. Is either of the following equations correct? Give reasons for your answers.

a. $\dfrac{1}{\cos x}\displaystyle\int \cos x\,dx = \tan x + C$

b. $\dfrac{1}{\cos x}\displaystyle\int \cos x\,dx = \tan x + \frac{C}{\cos x}$

25. **Current in a closed *RL* circuit** How many seconds after the switch in an *RL* circuit is closed will it take the current i to reach half of its steady-state value? Notice that the time depends on R and L, not on how much voltage is applied.

26. **Current in an open RL circuit** If the switch is thrown open after the current in an *RL* circuit has built up to its steady-state value $I = V/R$, the decaying current (see accompanying figure) obeys the equation

$$L\frac{di}{dt} + Ri = 0,$$

which is Equation (5) with $V = 0$.

a. Solve the equation to express i as a function of t.

b. How long after the switch is thrown will it take the current to fall to half its original value?

c. Show that the value of the current when $t = L/R$ is I/e. (The significance of this time is explained in the next exercise.)

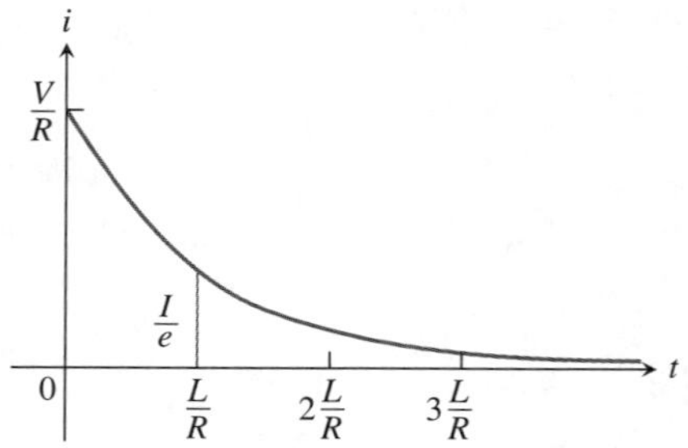

27. **Time constants** Engineers call the number L/R the *time constant* of the *RL* circuit in Figure 9.10. The significance of the time constant is that the current will reach 95% of its final value within 3 time constants of the time the switch is closed (Figure 9.10). Thus, the time constant gives a built-in measure of how rapidly an individual circuit will reach equilibrium.

a. Find the value of i in Equation (7) that corresponds to $t = 3L/R$, and show that it is about 95% of the steady-state value $I = V/R$.

b. Approximately what percentage of the steady-state current will be flowing in the circuit 2 time constants after the switch is closed (i.e., when $t = 2L/R$)?

28. **Derivation of Equation (7) in Example 4**

a. Show that the solution of the equation

$$\frac{di}{dt} + \frac{R}{L}i = \frac{V}{L}$$

is

$$i = \frac{V}{R} + Ce^{-(R/L)t}.$$

b. Then use the initial condition $i(0) = 0$ to determine the value of C. This will complete the derivation of Equation (7).

c. Show that $i = V/R$ is a solution of Equation (6) and that $i = Ce^{-(R/L)t}$ satisfies the equation

$$\frac{di}{dt} + \frac{R}{L}i = 0.$$

HISTORICAL BIOGRAPHY
Jakob Bernoulli
(1654–1705)
www.bit.ly/2NUSHeH

A **Bernoulli differential equation** is of the form

$$\frac{dy}{dx} + P(x)y = Q(x)y^n.$$

Observe that, if $n = 0$ or 1, the Bernoulli equation is linear. For other values of n, the substitution $u = y^{1-n}$ transforms the Bernoulli equation into the linear equation

$$\frac{du}{dx} + (1-n)P(x)u = (1-n)Q(x).$$

For example, in the equation

$$\frac{dy}{dx} - y = e^{-x}y^2,$$

we have $n = 2$, so that $u = y^{1-2} = y^{-1}$ and $du/dx = -y^{-2}\,dy/dx$. Then $dy/dx = -y^2\,du/dx = -u^{-2}\,du/dx$. Substitution into the original equation gives

$$-u^{-2}\frac{du}{dx} - u^{-1} = e^{-x}u^{-2},$$

or, equivalently,

$$\frac{du}{dx} + u = -e^{-x}.$$

This last equation is linear in the (unknown) dependent variable u.

Solve the Bernoulli equations in Exercises 29–32.

29. $y' - y = -y^2$

30. $y' - y = xy^2$

31. $\dfrac{dy}{dx} - y\tan x = \dfrac{\sin x\cos^2 x}{y}$

32. $x^2y' + 2xy = y^3$

9.3 Applications

We now look at four applications of first-order differential equations. The first application analyzes an object moving along a straight line while subject to a force opposing its motion. The second is a model of population growth. The third application considers a curve or curves intersecting each curve in a second family of curves *orthogonally* (that is, at right angles). The final application analyzes chemical concentrations entering and leaving a container. The various models involve separable or linear first-order equations.

Motion with Resistance Proportional to Velocity

In some cases it is reasonable to assume that the resistance encountered by a moving object, such as a car coasting to a stop, is proportional to the object's velocity. The faster the object moves, the more its forward progress is resisted by the air through which it passes. Picture the object as a mass m moving along a coordinate line with position function s and velocity v at time t. From Newton's second law of motion, the resisting force opposing the motion is

$$\text{Force} = \text{mass} \times \text{acceleration} = m\frac{dv}{dt}.$$

If the resisting force is proportional to velocity, we have

$$m\frac{dv}{dt} = -kv \qquad \text{or} \qquad \frac{dv}{dt} = -\frac{k}{m}v \qquad (k > 0).$$

This is a separable differential equation representing exponential change. The solution to the equation with initial condition $v = v_0$ at $t = 0$ is (Section 7.4)

$$v = v_0 e^{-(k/m)t}. \tag{1}$$

What can we learn from Equation (1)? For one thing, we can see that if m is something large, like the mass of a 20,000-ton ore boat in Lake Erie, it will take a long time for the velocity to approach zero (because t must be large in the exponent of the equation in order to make kt/m large enough for v to be small). We can learn even more if we integrate Equation (1) to find the position s as a function of time t.

Suppose that an object is coasting to a stop and the only force acting on it is a resistance proportional to its speed. How far will it coast? To find out, we start with Equation (1) and solve the initial value problem

$$\frac{ds}{dt} = v_0 e^{-(k/m)t}, \qquad s(0) = 0.$$

Integrating with respect to t gives

$$s = -\frac{v_0 m}{k} e^{-(k/m)t} + C.$$

Substituting $s = 0$ when $t = 0$ gives

$$0 = -\frac{v_0 m}{k} + C \qquad \text{and} \qquad C = \frac{v_0 m}{k}.$$

The body's position at time t is therefore

$$s(t) = -\frac{v_0 m}{k} e^{-(k/m)t} + \frac{v_0 m}{k} = \frac{v_0 m}{k}(1 - e^{-(k/m)t}). \tag{2}$$

To find how far the body will coast, we find the limit of $s(t)$ as $t \to \infty$. Since $-(k/m) < 0$, we know that $e^{-(k/m)t} \to 0$ as $t \to \infty$, so that

$$\lim_{t\to\infty} s(t) = \lim_{t\to\infty} \frac{v_0 m}{k}(1 - e^{-(k/m)t}) = \frac{v_0 m}{k}(1 - 0) = \frac{v_0 m}{k}.$$

Thus,

$$\text{Distance coasted} = \frac{v_0 m}{k}. \tag{3}$$

The number $v_0 m/k$ is only an upper bound (albeit a useful one). It is true to life in one respect, at least: If m is large, the body will coast a long way.

In the English system, in which weight is measured in pounds, mass is measured in **slugs**. Thus,

$$\text{Pounds} = \text{slugs} \times 32,$$

assuming that gravitational acceleration is 32 ft/sec^2.

EXAMPLE 1 For a 192-lb ice skater, the k in Equation (1) is about 1/3 slug/sec and $m = 192/32 = 6$ slugs. How long will it take the skater to coast from 11 ft/sec (7.5 mph) to 1 ft/sec? How far will the skater coast before coming to a complete stop?

Solution We answer the first question by solving Equation (1) for t:

$$\begin{aligned} 11e^{-t/18} &= 1 && \text{Eq. (1) with } k = 1/3, \\ e^{-t/18} &= 1/11 && m = 6,\ v_0 = 11,\ v = 1 \\ -t/18 &= \ln(1/11) = -\ln 11 \\ t &= 18 \ln 11 \approx 43 \text{ sec.} \end{aligned}$$

We answer the second question with Equation (3):

$$\text{Distance coasted} = \frac{v_0 m}{k} = \frac{11 \cdot 6}{1/3} = 198 \text{ ft.}$$

■

Inaccuracy of the Exponential Population Growth Model

In Section 7.4 we modeled population growth with the Law of Exponential Change:

$$\frac{dP}{dt} = kP, \qquad P(0) = P_0,$$

where P is the population at time t, $k > 0$ is a constant growth rate, and P_0 is the size of the population at time $t = 0$. In Section 7.4 we found the solution $P = P_0 e^{kt}$ to this model.

To assess the model, notice that the exponential growth differential equation says that

$$\frac{dP/dt}{P} = k \tag{4}$$

is constant. This rate is called the **relative growth rate**. We can use this to predict total future world population based on historical data. Table 9.3 gives the world population at

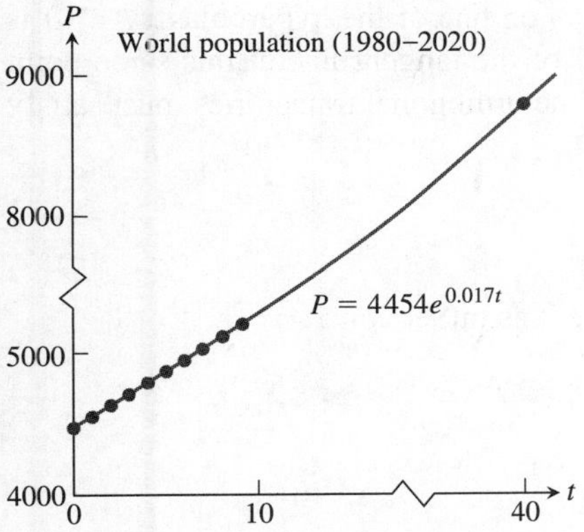

FIGURE 9.11 The value of the solution $P = 4454e^{0.017t}$ is 8792 when $t = 40$, which is nearly 13% more than the actual population in 2020.

TABLE 9.3 World population (midyear)

Year	Population (millions)	$\Delta P/P$
1980	4454	$76/4454 \approx 0.0171$
1981	4530	$80/4530 \approx 0.0177$
1982	4610	$80/4610 \approx 0.0174$
1983	4690	$80/4690 \approx 0.0171$
1984	4770	$81/4770 \approx 0.0170$
1985	4851	$82/4851 \approx 0.0169$
1986	4933	$85/4933 \approx 0.0172$
1987	5018	$87/5018 \approx 0.0173$
1988	5105	$85/5105 \approx 0.0167$
1989	5190	

midyear for the years 1980 to 1989. Taking $dt = 1$ and $dP \approx \Delta P$, we see from the table that the relative growth rate in Equation (4) is approximately equal to 0.017. Thus, based on the tabled data with $t = 0$ representing 1980, $t = 1$ representing 1981, and so forth, the world population could be modeled by the initial value problem

$$\frac{dP}{dt} = 0.017P, \qquad P(0) = 4454.$$

The solution to this initial value problem gives the population function $P = 4454e^{0.017t}$. In year 2008 (so $t = 28$), the solution predicts the world population in midyear to be about 7169 million, or 7.2 billion (Figure 9.11), which is more than the actual population of 6707 million, an error of about 7%. The error grows as the number of years increases. For 2020 ($t = 40$) the model predicts a population of 8792 million. The reported population for 2020 is 7795 million, an overprediction error of about 13%. A more realistic model would consider environmental, economic, and other factors affecting the growth rate, which has been steadily declining. We consider one such model in Section 9.4.

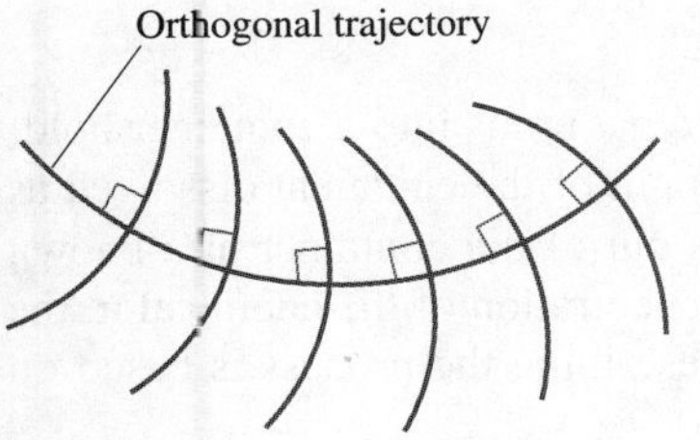

FIGURE 9.12 An orthogonal trajectory intersects the family of curves at right angles, or orthogonally.

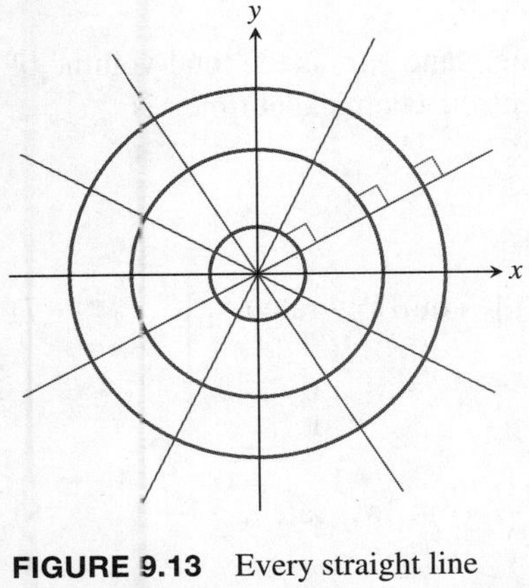

FIGURE 9.13 Every straight line through the origin is orthogonal to the family of circles centered at the origin.

Orthogonal Trajectories

An **orthogonal trajectory** of a family of curves is a curve that intersects each curve of the family at right angles, or *orthogonally* (Figure 9.12). For instance, each straight line through the origin is an orthogonal trajectory of the family of circles $x^2 + y^2 = a^2$, centered at the origin (Figure 9.13). Such mutually orthogonal systems of curves are of particular importance in physical problems related to electrical potential, where the curves in one family correspond to strength of an electric field, and those in the other family correspond to constant electric potential. They also occur in hydrodynamics and heat-flow problems.

EXAMPLE 2 Find the orthogonal trajectories of the family of curves $xy = a$, where $a \neq 0$ is an arbitrary constant.

Solution The curves $xy = a$ form a family of hyperbolas having the coordinate axes as asymptotes. First we find the slopes of each curve in this family, or their dy/dx values. Differentiating $xy = a$ implicitly gives

$$x\frac{dy}{dx} + y = 0 \qquad \text{or} \qquad \frac{dy}{dx} = -\frac{y}{x}.$$

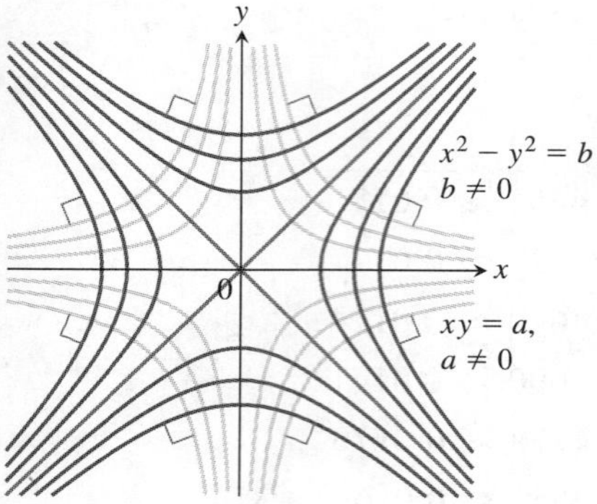

FIGURE 9.14 Each curve is orthogonal to every curve it meets in the other family (Example 2).

Thus the slope of the tangent line at any point (x, y) on one of the hyperbolas $xy = a$ is $y' = -y/x$. On an orthogonal trajectory the slope of the tangent line at this same point must be the negative reciprocal, or x/y. Therefore, the orthogonal trajectories must satisfy the differential equation

$$\frac{dy}{dx} = \frac{x}{y}.$$

This differential equation is separable, and we solve it as in Section 7.4:

$$\begin{aligned} y\,dy &= x\,dx && \text{Separate variables.} \\ \int y\,dy &= \int x\,dx && \text{Integrate both sides.} \\ \frac{1}{2}y^2 &= \frac{1}{2}x^2 + C \\ y^2 - x^2 &= b, \end{aligned} \tag{5}$$

where $b = 2C$ is an arbitrary constant. The orthogonal trajectories are the family of hyperbolas given by Equation (5) and sketched in Figure 9.14. ■

Mixture Problems

Suppose a chemical in a liquid solution (or dispersed in a gas) runs into a container holding the liquid (or the gas) with, possibly, a specified amount of the chemical dissolved as well. The mixture is kept uniform by stirring and flows out of the container at a known rate. In this process, it is often important to know the concentration of the chemical in the container at any given time. The differential equation describing the process is based on the formula

$$\text{Rate of change of amount in container} = \begin{pmatrix}\text{rate at which}\\ \text{chemical}\\ \text{arrives}\end{pmatrix} - \begin{pmatrix}\text{rate at which}\\ \text{chemical}\\ \text{departs.}\end{pmatrix}. \tag{6}$$

If $y(t)$ is the amount of chemical in the container at time t, and $V(t)$ is the total volume of liquid in the container at time t, then the departure rate of the chemical at time t is

$$\begin{aligned} \text{Departure rate} &= \frac{y(t)}{V(t)} \cdot (\text{outflow rate}) \\ &= \begin{pmatrix}\text{concentration in}\\ \text{container at time } t\end{pmatrix} \cdot (\text{outflow rate}). \end{aligned} \tag{7}$$

Accordingly, Equation (6) becomes

$$\frac{dy}{dt} = (\text{chemical's arrival rate}) - \frac{y(t)}{V(t)} \cdot (\text{outflow rate}). \tag{8}$$

If, say, y is measured in pounds, V in gallons, and t in minutes, the units in Equation (8) are

$$\frac{\text{pounds}}{\text{minutes}} = \frac{\text{pounds}}{\text{minutes}} - \frac{\text{pounds}}{\text{gallons}} \cdot \frac{\text{gallons}}{\text{minutes}}.$$

EXAMPLE 3 In an oil refinery, a storage tank contains 2000 gal of gasoline that initially has 100 lb of an additive dissolved in it. In preparation for winter weather, gasoline containing 2 lb of additive per gallon is pumped into the tank at a rate of 40 gal/min.

The well-mixed solution is pumped out at a rate of 45 gal/min. How much of the additive is in the tank 20 min after the pumping process begins (Figure 9.15)?

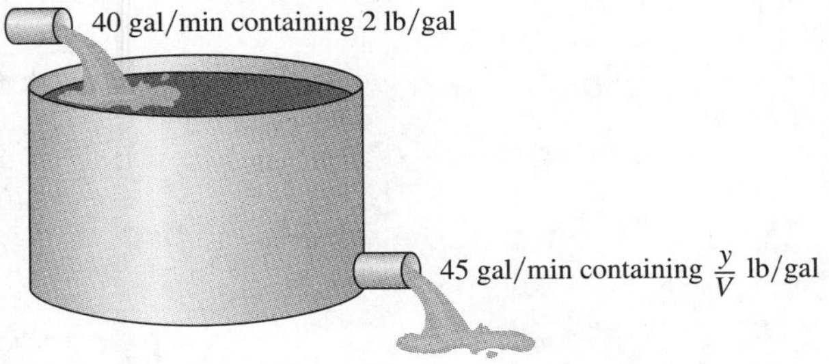

FIGURE 9.15 The storage tank in Example 3 mixes input liquid with stored liquid to produce an output liquid.

Solution Let y be the amount (in pounds) of additive in the tank at time t. We know that $y = 100$ when $t = 0$. The number of gallons of gasoline and additive in solution in the tank at any time t is

$$V(t) = 2000 \text{ gal} + \left(40\frac{\text{gal}}{\text{min}} - 45\frac{\text{gal}}{\text{min}}\right)(t \text{ min}) = (2000 - 5t) \text{ gal}.$$

Therefore,

$$\begin{aligned} \text{Rate out} &= \frac{y(t)}{V(t)} \cdot \text{outflow rate} && \text{Eq. (7)} \\ &= \left(\frac{y}{2000 - 5t}\right)45 && \text{Outflow rate is 45 gal/min and } V = 2000 - 5t. \\ &= \frac{45y}{2000 - 5t}\frac{\text{lb}}{\text{min}}. \end{aligned}$$

Also,

$$\text{Rate in} = \left(2\frac{\text{lb}}{\text{gal}}\right)\left(40\frac{\text{gal}}{\text{min}}\right) = 80\frac{\text{lb}}{\text{min}}.$$

The differential equation modeling the mixture process is

$$\frac{dy}{dt} = 80 - \frac{45y}{2000 - 5t} \qquad \text{Eq. (8)}$$

in pounds per minute.

To solve this differential equation, we first write it in standard linear form:

$$\frac{dy}{dt} + \frac{45}{2000 - 5t}y = 80.$$

Thus, $P(t) = 45/(2000 - 5t)$ and $Q(t) = 80$. The integrating factor is

$$\begin{aligned} v(t) = e^{\int P\,dt} &= e^{\int \frac{45}{200-5t}dt} \\ &= e^{-9\ln(2000-5t)} && 2000 - 5t > 0 \\ &= (2000 - 5t)^{-9}. \end{aligned}$$

Multiplying both sides of the standard equation by $v(t)$ and integrating both sides give

$$(2000 - 5t)^{-9} \cdot \left(\frac{dy}{dt} + \frac{45}{2000 - 5t}y\right) = 80(2000 - 5t)^{-9}$$

$$(2000 - 5t)^{-9}\frac{dy}{dt} + 45(2000 - 5t)^{-10}y = 80(2000 - 5t)^{-9}$$

$$\frac{d}{dt}[(2000 - 5t)^{-9}y] = 80(2000 - 5t)^{-9}$$

$$(2000 - 5t)^{-9}y = \int 80(2000 - 5t)^{-9}\,dt$$

$$(2000 - 5t)^{-9}y = 80 \cdot \frac{(2000 - 5t)^{-8}}{(-8)(-5)} + C.$$

The general solution is

$$y = 2(2000 - 5t) + C(2000 - 5t)^9.$$

Because $y = 100$ when $t = 0$, we can determine the value of C:

$$100 = 2(2000 - 0) + C(2000 - 0)^9$$

$$C = -\frac{3900}{(2000)^9}.$$

The particular solution of the initial value problem is

$$y = 2(2000 - 5t) - \frac{3900}{(2000)^9}(2000 - 5t)^9.$$

The amount of additive in the tank 20 min after the pumping begins is

$$y(20) = 2[2000 - 5(20)] - \frac{3900}{(2000)^9}[2000 - 5(20)]^9 \approx 1342 \text{ lb}.$$

■

EXERCISES 9.3

Motion Along a Line

1. **Coasting bicycle** A 76 kg cyclist on a 10 kg bicycle starts coasting on level ground at 8 m/sec. The k in Equation (1) is about 3.9 kg/sec.
   a. About how far will the cyclist coast before reaching a complete stop?
   b. How long will it take the cyclist's speed to drop to 1 m/sec?
2. **Coasting battleship** An Iowa class battleship has mass around 51,000 metric tons (51,000,000 kg) and a k value in Equation (1) of about 59,000 kg/sec. Assume that the ship loses power when it is moving at a speed of 9 m/sec.
   a. About how far will the ship coast before it is dead in the water?
   b. About how long will it take the ship's speed to drop to 1 m/sec?
3. The data in Table 9.4 were collected with a motion detector and a CBL™ by Valerie Sharritts, then a mathematics teacher at St. Francis DeSales High School in Columbus, Ohio. The table shows the distance s (meters) coasted on inline skates in t sec by her daughter Ashley when she was 10 years old. Find a model for Ashley's position given by the data in Table 9.4 in the form of

TABLE 9.4 Ashley Sharritts skating data

t (sec)	s (m)	t (sec)	s (m)	t (sec)	s (m)
0	0	2.24	3.05	4.48	4.77
0.16	0.31	2.40	3.22	4.64	4.82
0.32	0.57	2.56	3.38	4.80	4.84
0.48	0.80	2.72	3.52	4.96	4.86
0.64	1.05	2.88	3.67	5.12	4.88
0.80	1.28	3.04	3.82	5.28	4.89
0.96	1.50	3.20	3.96	5.44	4.90
1.12	1.72	3.36	4.08	5.60	4.90
1.28	1.93	3.52	4.18	5.76	4.91
1.44	2.09	3.68	4.31	5.92	4.90
1.60	2.30	3.84	4.41	6.08	4.91
1.76	2.53	4.00	4.52	6.24	4.90
1.92	2.73	4.16	4.63	6.40	4.91
2.08	2.89	4.32	4.69	6.56	4.91

Equation (2). Her initial velocity was $v_0 = 2.75$ m/sec, her mass $m = 39.92$ kg (she weighed 88 lb), and her total coasting distance was 4.91 m.

4. **Coasting to a stop** Table 9.5 shows the distance s (meters) coasted on inline skates in terms of time t (seconds) by Kelly Schmitzer. Find a model for her position in the form of Equation (2). Her initial velocity was $v_0 = 0.80$ m/sec, her mass $m = 49.90$ kg (110 lb), and her total coasting distance was 1.32 m.

TABLE 9.5 Kelly Schmitzer skating data

t (sec)	s (m)	t (sec)	s (m)	t (sec)	s (m)
0	0	1.5	0.89	3.1	1.30
0.1	0.07	1.7	0.97	3.3	1.31
0.3	0.22	1.9	1.05	3.5	1.32
0.5	0.36	2.1	1.11	3.7	1.32
0.7	0.49	2.3	1.17	3.9	1.32
0.9	0.60	2.5	1.22	4.1	1.32
1.1	0.71	2.7	1.25	4.3	1.32
1.3	0.81	2.9	1.28	4.5	1.32

Orthogonal Trajectories

In Exercises 5–10, find the orthogonal trajectories of the family of curves. Sketch several members of each family.

5. $y = mx$
6. $y = cx^4$
7. $kx^2 + y^2 = 1$
8. $x^2 + y^2 = 2cx$
9. $y = ce^{-x}$
10. $y = e^{kx}$
11. Show that the curves $2x^2 + 3y^2 = 5$ and $y^2 = x^3$ are orthogonal.
12. Find the family of solutions of the given differential equation and the family of orthogonal trajectories. Sketch both families.
    a. $x\,dx + y\,dy = 0$
    b. $x\,dy - 2y\,dx = 0$

Mixture Problems

13. **Salt mixture** A tank initially contains 200 gal of brine in which 100 lb of salt are dissolved. A brine containing 2 lb gal of salt runs into the tank at the rate of 6 gal min. The mixture is kept uniform by stirring and flows out of the tank at the rate of 5 gal min.
    a. At what rate (pounds per minute) does salt enter the tank at time t?
    b. What is the volume of brine in the tank at time t?
    c. At what rate (pounds per minute) does salt leave the tank at time t?
    d. Write down and solve the initial value problem describing the mixing process.
    e. Find the concentration of salt in the tank 30 min after the process starts.
14. **Mixture problem** A 200-gal tank is half full of distilled water. At time $t = 0$, a solution containing 0.5 lb/gal of concentrate enters the tank at the rate of 5 gal/min, and the well-stirred mixture is withdrawn at the rate of 3 gal/min.
    a. At what time will the tank be full?
    b. At the time the tank is full, how many pounds of concentrate will it contain?
15. **Fertilizer mixture** A tank contains 100 gal of fresh water. A solution containing 1 lb/gal of soluble lawn fertilizer runs into the tank at the rate of 1 gal/min, and the mixture is pumped out of the tank at the rate of 3 gal/min. Find the maximum amount of fertilizer in the tank and the time required to reach the maximum.
16. **Carbon monoxide pollution** An executive conference room of a corporation contains 4500 ft^3 of air initially free of carbon monoxide. Starting at time $t = 0$, cigarette smoke containing 4% carbon monoxide is blown into the room at the rate of 0.3 ft^3/min. A ceiling fan keeps the air in the room well circulated and the air leaves the room at the same rate of 0.3 ft^3/min. Find the time when the concentration of carbon monoxide in the room reaches 0.01%.

9.4 Graphical Solutions of Autonomous Equations

In Chapter 4 we learned that the sign of the first derivative tells where the graph of a function is increasing and where it is decreasing. The sign of the second derivative tells the concavity of the graph. We can build on our knowledge of how derivatives determine the shape of a graph to solve differential equations graphically. We will see that the ability to discern physical behavior from graphs is a powerful tool in understanding real-world systems. The starting ideas for a graphical solution are the notions of *phase line* and *equilibrium value*. We arrive at these notions by investigating, from a point of view quite different from that studied in Chapter 4, what happens when the derivative of a differentiable function is zero.

Equilibrium Values and Phase Lines

When we differentiate implicitly the equation

$$\frac{1}{5}\ln(5y - 15) = x + 1,$$

we obtain

$$\frac{1}{5}\left(\frac{5}{5y - 15}\right)\frac{dy}{dx} = 1.$$

Solving for $y' = dy/dx$, we find $y' = 5y - 15 = 5(y - 3)$. In this case the derivative y' is a function of y only (the dependent variable) and is zero when $y = 3$.

A differential equation for which dy/dx is a function of y only is called an **autonomous** differential equation. Let's investigate what happens when the derivative in an autonomous equation equals zero. We assume any derivatives are continuous.

> **DEFINITION** If $dy/dx = g(y)$ is an autonomous differential equation, then the values of y for which $dy/dx = 0$ are called **equilibrium values** or **rest points**.

Thus, equilibrium values are those at which no change occurs in the dependent variable, so y is at *rest*. The emphasis is on the value of y where $dy/dx = 0$, not the value of x, as we studied in Chapter 4. For example, the equilibrium values for the autonomous differential equation

$$\frac{dy}{dx} = (y + 1)(y - 2)$$

are $y = -1$ and $y = 2$.

To construct a graphical solution to an autonomous differential equation, we first make a **phase line** for the equation, a plot on the y-axis that shows the equation's equilibrium values along with the intervals where dy/dx and d^2y/dx^2 are positive and negative. Then we know where the solutions are increasing and decreasing, and the concavity of the solution curves. These are the essential features we found in Section 4.4, so we can determine the shapes of the solution curves without having to find formulas for them.

EXAMPLE 1 Draw a phase line for the equation

$$\frac{dy}{dx} = (y + 1)(y - 2),$$

and use it to sketch solutions to the equation.

Solution

1. *Draw a number line for y and mark the equilibrium values y* $= -1$ *and y* $= 2$, *where* $dy/dx = 0$.

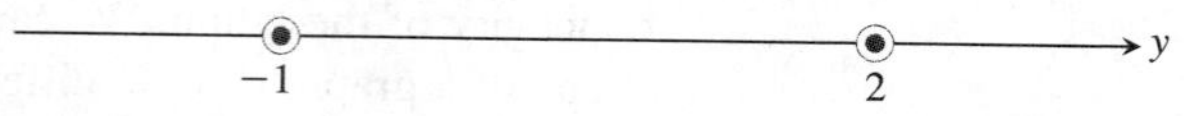

2. *Identify and label the intervals where* $y' > 0$ *and* $y' < 0$. This step resembles what we did in Section 4.3, only now we are marking the y-axis instead of the x-axis.

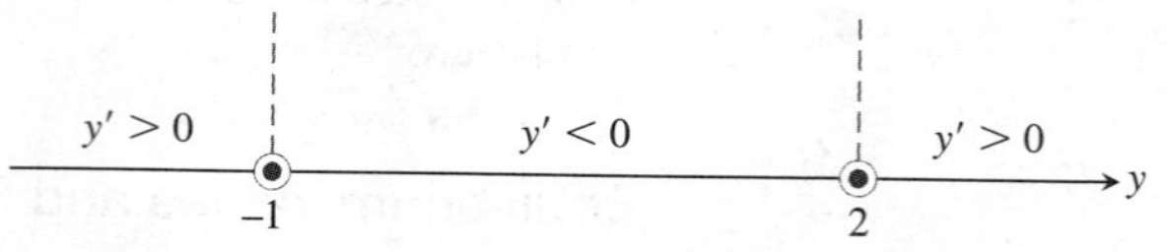

We can encapsulate the information about the sign of y' on the phase line itself. Since $y' > 0$ on the interval to the left of $y = -1$, a solution of the differential equation with a y-value less than -1 will increase from there toward $y = -1$. We display this information by drawing an arrow on the interval pointing to -1.

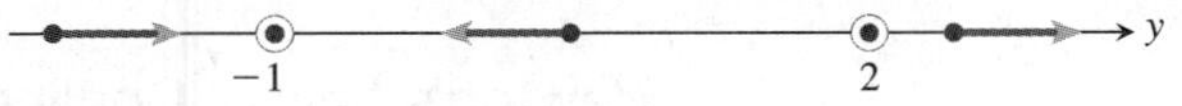

Similarly, $y' < 0$ between $y = -1$ and $y = 2$, so any solution with a value in this interval will decrease toward $y = -1$.

For $y > 2$, we have $y' > 0$, so a solution with a y-value greater than 2 will increase from there without bound.

In short, solution curves below the horizontal line $y = -1$ in the xy-plane rise toward $y = -1$. Solution curves between the lines $y = -1$ and $y = 2$ fall away from $y = 2$ toward $y = -1$. Solution curves above $y = 2$ rise away from $y = 2$ and keep going up.

3. *Calculate y'' and mark the intervals where $y'' > 0$ and $y'' < 0$.* To find y'', we differentiate y' *with respect to* x, using implicit differentiation.

$$y' = (y + 1)(y - 2) = y^2 - y - 2 \qquad \text{Formula for } y'$$

$$\begin{aligned} y'' &= \frac{d}{dx}(y') = \frac{d}{dx}(y^2 - y - 2) && \text{Substitute for } y'. \\ &= 2yy' - y' && \text{Differentiate implicitly with respect to } x. \\ &= (2y - 1)y' \\ &= (2y - 1)(y + 1)(y - 2). \end{aligned}$$

From this formula, we see that y'' changes sign at $y = -1$, $y = 1/2$, and $y = 2$. We add the sign information to the phase line.

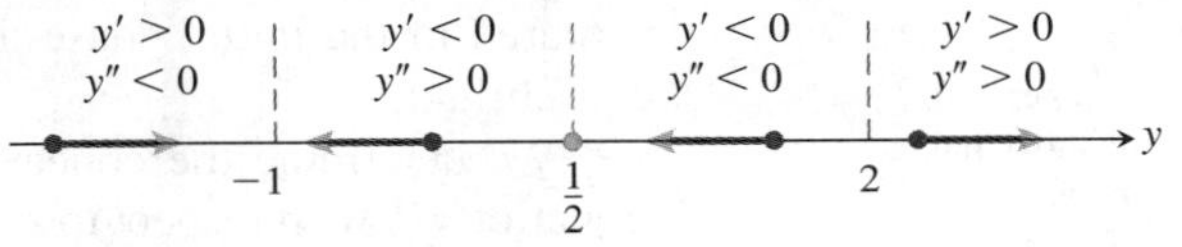

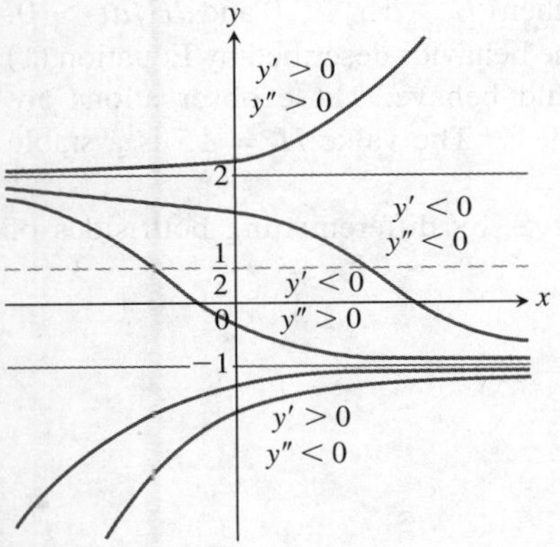

FIGURE 9.16 Graphical solutions from Example 1 include the horizontal lines $y = -1$ and $y = 2$ through the equilibrium values. No two solution curves can ever cross or touch each other.

4. *Sketch an assortment of solution curves in the xy-plane.* The horizontal lines $y = -1$, $y = 1/2$, and $y = 2$ partition the plane into horizontal bands in which we know the signs of y' and y''. In each band, this information tells us whether the solution curves rise or fall and how they bend as x increases (Figure 9.16).

The "equilibrium lines" $y = -1$ and $y = 2$ are also solution curves. (The constant functions $y = -1$ and $y = 2$ satisfy the differential equation.) Solution curves that cross the line $y = 1/2$ have an inflection point there. The concavity changes from concave down (above the line) to concave up (below the line).

As predicted in Step 2, solutions in the middle and lower bands approach the equilibrium value $y = -1$ as x increases. Solutions in the upper band rise steadily away from the value $y = 2$. ■

Stable and Unstable Equilibria

Look at Figure 9.16 once more, in particular at the behavior of the solution curves near the equilibrium values. Once a solution curve has a value near $y = -1$, it tends steadily toward that value; $y = -1$ is a **stable equilibrium**. The behavior near $y = 2$ is just the opposite: All solutions except the equilibrium solution $y = 2$ itself move *away* from it as x increases. We call $y = 2$ an **unstable equilibrium**. If the solution is *at* that value, it stays, but if it is off by any amount, no matter how small, it moves away. (Sometimes an equilibrium value is unstable because a solution moves away from it only on one side of the point.)

Now that we know what to look for, we can already see this behavior on the initial phase line (the second diagram in Step 2 of Example 1). The arrows lead away from $y = 2$ and, once to the left of $y = 2$, toward $y = -1$.

We now present several applied examples for which we can sketch a family of solution curves to the differential equation models using the method in Example 1.

Newton's Law of Cooling

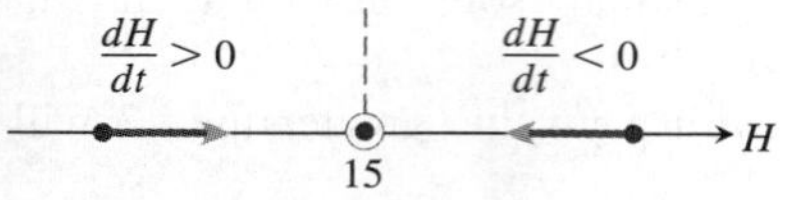

FIGURE 9.17 First step in constructing the phase line for Newton's Law of Cooling. The temperature tends toward the equilibrium (surrounding-medium) value in the long run.

In Section 7.4 we solved analytically the differential equation

$$\frac{dH}{dt} = -k(H - H_S), \qquad k > 0$$

modeling Newton's Law of Cooling. Here H is the temperature of an object at time t, and H_S is the constant temperature of the surrounding medium.

Suppose that the surrounding medium (say, a room in a house) has a constant Celsius temperature of 15 °C. We can then express the difference in temperature as $H(t) - 15$. Assuming H is a differentiable function of time t, by Newton's Law of Cooling, there is a constant of proportionality $k > 0$ such that

$$\frac{dH}{dt} = -k(H - 15) \tag{1}$$

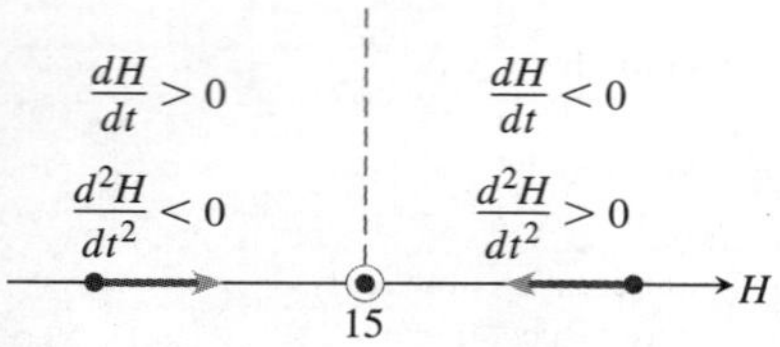

FIGURE 9.18 The complete phase line for Newton's Law of Cooling.

(*minus k* to give a negative derivative when $H > 15$).

Since $dH/dt = 0$ at $H = 15$, the temperature 15 °C is an equilibrium value. If $H > 15$, Equation (1) tells us that $(H - 15) > 0$ and $dH/dt < 0$. If the object is hotter than the room, it will get cooler. Similarly, if $H < 15$, then $(H - 15) < 0$ and $dH/dt > 0$. An object cooler than the room will warm up. Thus, the behavior described by Equation (1) agrees with our intuition of how temperature should behave. These observations are captured in the initial phase line diagram in Figure 9.17. The value $H = 15$ is a stable equilibrium.

We determine the concavity of the solution curves by differentiating both sides of Equation (1) with respect to t:

$$\frac{d}{dt}\left(\frac{dH}{dt}\right) = \frac{d}{dt}(-k(H - 15))$$
$$\frac{d^2H}{dt^2} = -k\frac{dH}{dt}.$$

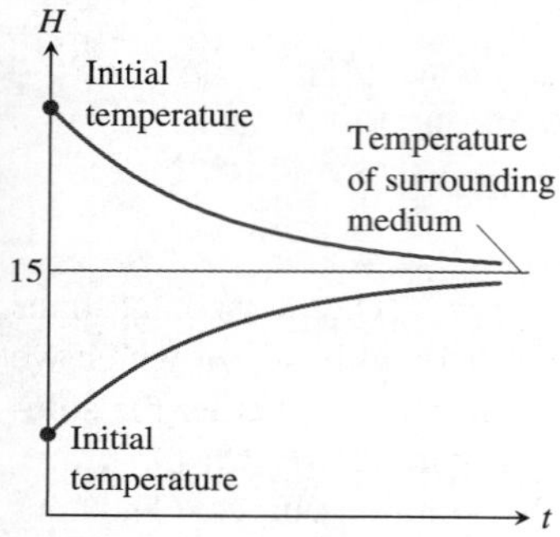

FIGURE 9.19 Temperature versus time. Regardless of initial temperature, the object's temperature $H(t)$ tends toward 15 °C, the temperature of the surrounding medium.

Since $-k$ is negative, we see that d^2H/dt^2 is positive when $dH/dt < 0$ and negative when $dH/dt > 0$. Figure 9.18 adds this information to the phase line.

The completed phase line shows that if the temperature of the object is above the equilibrium value of 15 °C, the graph of $H(t)$ will be decreasing and concave upward. If the temperature is below 15 °C (the temperature of the surrounding medium), the graph of $H(t)$ will be increasing and concave downward. We use this information to sketch typical solution curves (Figure 9.19).

From the upper solution curve in Figure 9.19, we see that as the object cools down, the rate at which it cools slows down because dH/dt approaches zero. This observation is implicit in Newton's Law of Cooling and contained in the differential equation, but the flattening of the graph as time advances gives an immediate visual representation of the phenomenon.

A Falling Body Encountering Resistance

Newton observed that the rate of change of the momentum of a moving object is equal to the net force applied to it. In mathematical terms,

$$F = \frac{d}{dt}(mv), \tag{2}$$

where F is the net force acting on the object, and m and v are the object's mass and velocity. If m varies with time, as it will if the object is a rocket burning fuel, the right-hand side of Equation (2) expands to

$$m\frac{dv}{dt} + v\frac{dm}{dt}$$

using the Derivative Product Rule. In many situations, however, m is constant, $dm/dt = 0$, and Equation (2) takes the simpler form

$$F = m\frac{dv}{dt} \quad \text{or} \quad F = ma, \tag{3}$$

which is known as *Newton's second law of motion* (see Section 9.3).

In free fall, the constant acceleration due to gravity is denoted by g, and the one force propelling the body downward is

$$F_p = mg,$$

the force due to gravity. If, however, we think of a real body falling through the air—say, a penny from a great height or a parachutist from an even greater height—we know that at some point air resistance is a factor in the speed of the fall. A more realistic model of free fall would include air resistance, shown as a force F_r in the schematic diagram in Figure 9.20.

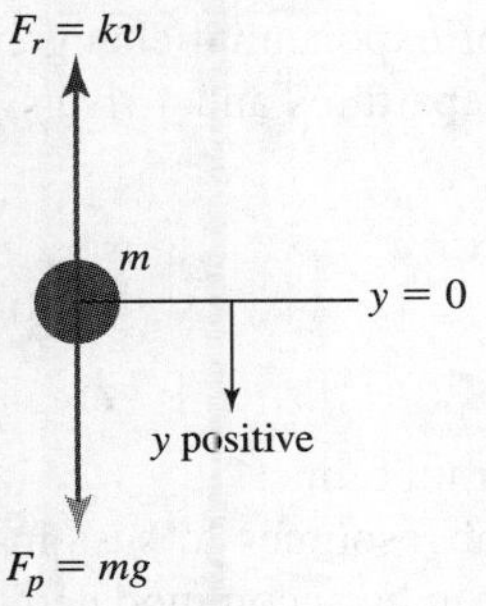

FIGURE 9.20 An object falling under the propulsion due to gravity, with a resistive force assumed to be proportional to the velocity.

For low speeds well below the speed of sound, physical experiments have shown that F_r is approximately proportional to the body's velocity. The net force on the falling body is therefore

$$F = F_p - F_r,$$

giving

$$\begin{aligned} m\frac{dv}{dt} &= mg - kv \\ \frac{dv}{dt} &= g - \frac{k}{m}v. \end{aligned} \tag{4}$$

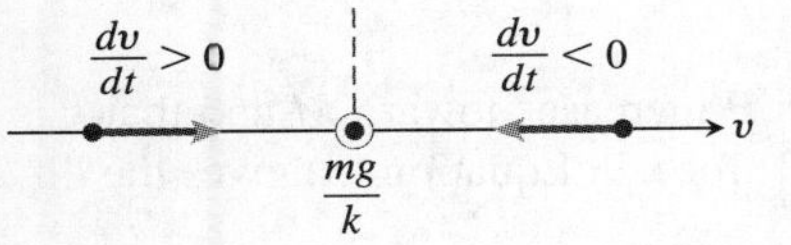

FIGURE 9.21 Initial phase line for the falling body encountering resistance.

We can use a phase line to analyze the velocity functions that solve this differential equation.

The equilibrium point, obtained by setting the right-hand side of Equation (4) equal to zero, is

$$v = \frac{mg}{k}.$$

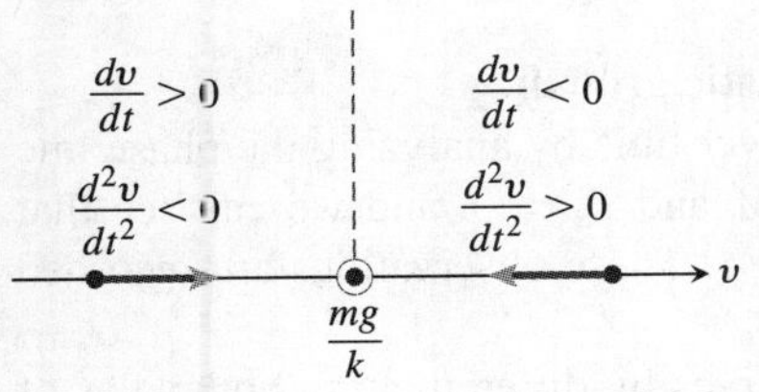

FIGURE 9.22 The completed phase line for the falling body.

If the body is initially moving faster than this, dv/dt is negative and the body slows down. If the body is moving at a velocity below mg/k, then $dv/dt > 0$ and the body speeds up. These observations are captured in the initial phase line diagram in Figure 9.21.

We determine the concavity of the solution curves by differentiating both sides of Equation (4) with respect to t:

$$\frac{d^2v}{dt^2} = \frac{d}{dt}\left(g - \frac{k}{m}v\right) = -\frac{k}{m}\frac{dv}{dt}.$$

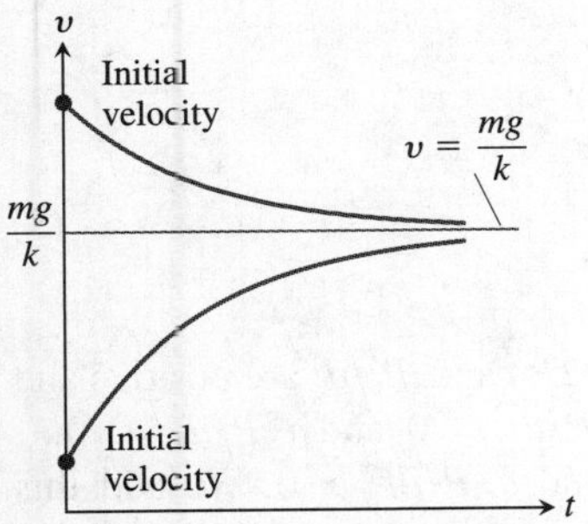

FIGURE 9.23 Typical velocity curves for a falling body encountering resistance. The value $v = mg/k$ is the terminal velocity.

We see that $d^2v/dt^2 < 0$ when $v < mg/k$ and that $d^2v/dt^2 > 0$ when $v > mg/k$. Figure 9.22 adds this information to the phase line. Notice the similarity to the phase line for Newton's Law of Cooling (Figure 9.18). The solution curves are similar as well (Figure 9.23).

Figure 9.23 shows two typical solution curves. Regardless of the initial velocity, we see the body's velocity tending toward the limiting value $v = mg/k$. This value, a stable equilibrium point, is called the body's **terminal velocity**. Skydivers can vary their terminal velocity from 95 mph to 180 mph by changing the amount of body area opposing the fall, which affects the value of k.

The Logistic Model for Population Growth

In Section 9.3 we examined population growth using the model of exponential change. That is, if P represents the number of individuals and we neglect departures and arrivals, then

$$\frac{dP}{dt} = kP, \tag{5}$$

where $k > 0$ is the birth rate minus the death rate per individual per unit time.

Because the natural environment has only a limited number of resources to sustain life, it is reasonable to assume that only a maximum population M can be accommodated. As the population approaches this **limiting population** or **carrying capacity**, resources become less abundant and the growth rate k decreases. A simple relationship exhibiting this behavior is

$$k = r(M - P),$$

where $r > 0$ is a constant. Notice that k decreases as P increases toward M and that k is negative if P is greater than M. Substituting $r(M - P)$ for k in Equation (5) gives the differential equation

$$\frac{dP}{dt} = r(M - P)P = rMP - rP^2. \tag{6}$$

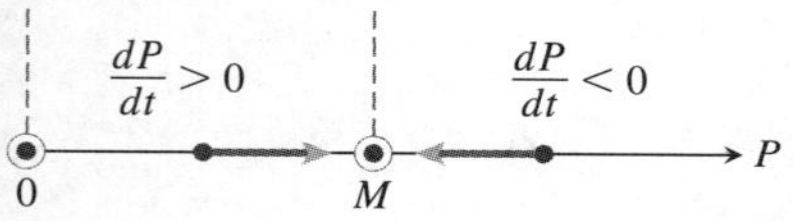

FIGURE 9.24 The initial phase line for logistic growth (Equation 6).

The model given by Equation (6) is referred to as **logistic growth**.

We can forecast the behavior of the population over time by analyzing the phase line for Equation (6). The equilibrium values are $P = M$ and $P = 0$, and we can see that $dP/dt > 0$ if $0 < P < M$ and $dP/dt < 0$ if $P > M$. These observations are recorded on the phase line in Figure 9.24.

We determine the concavity of the population curves by differentiating both sides of Equation (6) with respect to t:

$$\begin{aligned}\frac{d^2P}{dt^2} &= \frac{d}{dt}(rMP - rP^2)\\ &= rM\frac{dP}{dt} - 2rP\frac{dP}{dt}\\ &= r(M - 2P)\frac{dP}{dt}.\end{aligned} \tag{7}$$

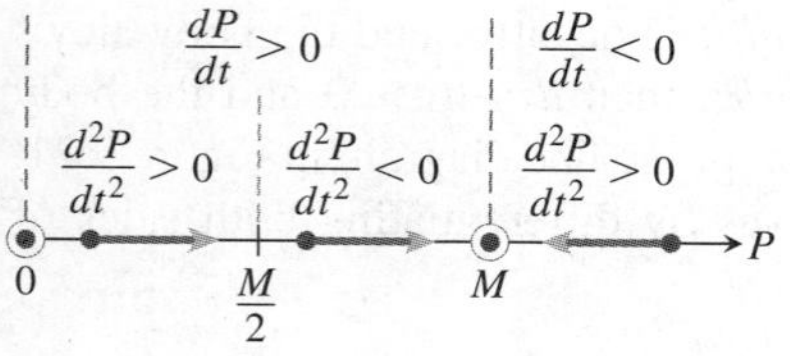

FIGURE 9.25 The completed phase line for logistic growth (Equation 6).

If $P = M/2$, then $d^2P/dt^2 = 0$. If $P < M/2$, then $(M - 2P)$ and dP/dt are positive and $d^2P/dt^2 > 0$. If $M/2 < P < M$, then $(M - 2P) < 0$, $dP/dt > 0$, and $d^2P/dt^2 < 0$. If $P > M$, then $(M - 2P)$ and dP/dt are both negative and $d^2P/dt^2 > 0$. We add this information to the phase line (Figure 9.25).

The lines $P = M/2$ and $P = M$ divide the first quadrant of the tP-plane into horizontal bands in which we know the signs of both dP/dt and d^2P/dt^2. In each band, we know how the solution curves rise and fall, and how they bend as time passes. The equilibrium lines $P = 0$ and $P = M$ are both population curves. Population curves crossing the line

$P = M/2$ have an inflection point there, giving them a **sigmoid** shape (curved in two directions like a letter S). Figure 9.26 displays typical population curves. Notice that each population curve approaches the limiting population M as $t \to \infty$.

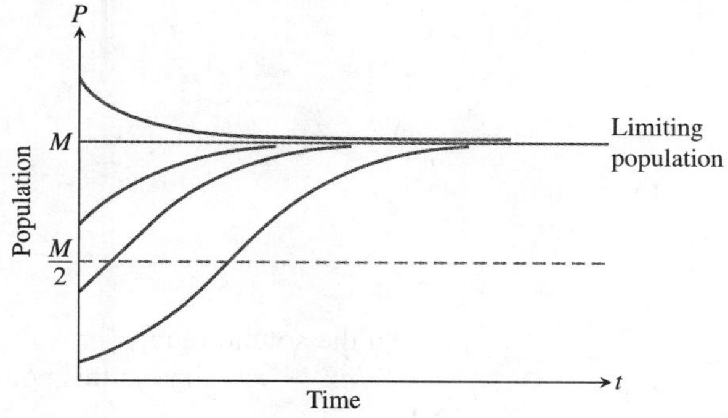

FIGURE 9.26 Population curves for logistic growth.

The Logistic Equation in Neural Networks and Machine Learning

While Figure 9.26 gives a general idea of the behavior of solutions to the Logistic Equation (6), we have not yet found explicit solutions. Exact formulas for solutions of first order differential equations cannot always be found, but they can be derived for the case of the Logistic Equation, where the solutions are called *logistic functions*. In Example 2 we find the solutions lying between $y = 0$ and $y = 1$ for the Logistic Equation in the case where $M = 1$ and r is an arbitrary positive constant.

EXAMPLE 2 Find the solutions to the Logistic Equation $\dfrac{dy}{dx} = ry - ry^2$ that satisfy $0 < y < 1$. Where does a solution cross the horizontal line $y = 1/2$, and what is the slope of its graph at this point?

Solution To solve the differential equation we use the method of separation of variables introduced in Section 7.4.

$$\frac{dy}{dx} = ry - ry^2 = ry(1 - y)$$

$$\frac{1}{y(1-y)}\,dy = r\,dx$$

$$\int \frac{1}{y(1-y)}\,dy = \int r\,dx$$

$$\int \frac{1}{y} + \frac{1}{1-y}\,dy = \int r\,dx \qquad \text{Partial fractions}$$

$$\ln y + C_1 - \ln(1-y) + C_2 = rx + C_3 \qquad |y| = y \text{ and } |1-y| = 1-y, \text{ since } 0 < y < 1.$$

$$\ln y - \ln(1-y) = rx + C \qquad \text{Combine constants, } C = C_3 - C_1 - C_2.$$

$$\ln\frac{y}{1-y} = rx + C$$

$$\frac{y}{1-y} = e^{rx+C} \qquad \text{Exponentiate.}$$

$$y = \frac{e^{rx+C}}{1 + e^{rx+C}} \qquad \text{Use algebra to solve for } y.$$

$$y = \frac{1}{1 + e^{-rx-C}}. \qquad \text{Divide through by } e^{rx+C}.$$

This gives an explicit formula for all the solutions whose graphs lie strictly between $y = 0$ and $y = 1$.

When a solution crosses the line $y = 1/2$ we have

$$\begin{aligned} \frac{1}{1 + e^{-rx-C}} &= \frac{1}{2} \\ 1 + e^{-rx-C} &= 2 \\ e^{-rx-C} &= 1 \\ rx + C &= 0 \\ x &= -C/r. \end{aligned}$$

So the solution graph crosses at the point $(-C/r, 1/2)$. The slope is found by evaluating $dy/dx = ry - ry^2$ at this point,

$$\frac{dy}{dx}(-C/r) = ry - ry^2 = r(1/2) - r(1/2)^2 = \frac{r}{4}. \qquad \blacksquare$$

Figure 9.27 shows the graph of a logistic function with $r = 3$ and $C = -6$.

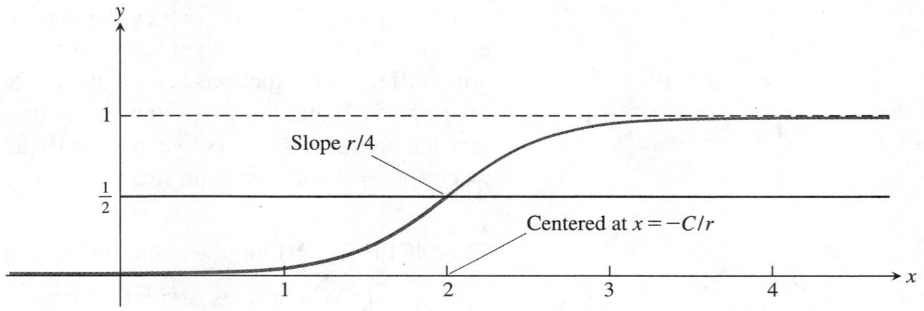

FIGURE 9.27 The constants r and C determine the steepness and horizontal displacement of a solution to the logistic equation. In this example $r = 3$ and $C = -6$.

Logistic functions have applications in many areas beyond the study of population growth. A field of Computer Science called *Machine Learning* develops methods to use a large collection of experimental data, called a training set, to construct a predictor function. The training set might consist of thousands of images of signs, for example, and the predictor function might decide whether a newly obtained image represents a Stop sign.

One highly successful approach to Machine Learning is the method of *Neural Networks*, which creates predictor functions based on a model of interacting neurons. Neural network models are built by taking repeated compositions of linear and logistic functions. Linear functions, such as $L(x) = ax + b$ can give accurate approximations of a function f nearby to a point where f is differentiable, as seen in Chapter 3. The optimal choices for the constants a and b in $L(x)$ are found by minimizing an error function that is calculated using the training set in a process called linear regression. Logistic functions have several features that make them a useful complement to linear functions in constructing predictor functions. They have values lying between 0 and 1 and are well suited to modeling probabilities. They are differentiable and specified by a small number of constants, such as the constants r and C in Example 2. These constants can be adjusted, or tuned, to minimize the error of a prediction. Logistic functions are nonlinear, and taking compositions of linear and logistic functions allows for the approximation of much more complicated functions than linear functions alone. A more complete discussion of the utility of logistic functions involves multivariable functions and their derivatives, which are introduced in Chapter 14.

EXERCISES 9.4

Phase Lines and Solution Curves

In Exercises 1–8,

a. Identify the equilibrium values. Which are stable and which are unstable?

b. Construct a phase line. Identify the signs of y' and y''.

c. Sketch several solution curves.

1. $\frac{dy}{dx} = (y+2)(y-3)$ **2.** $\frac{dy}{dx} = y^2 - 4$

3. $y' = y^3 - 4y^2 + 5y - 2$ **4.** $y' = y(1-y)$

5. $y' = \sqrt{y}, \quad y > 0$ **6.** $y' = y - \sqrt{y}, \quad y > 0$

7. $y' = (y-1)(y-2)(y-3)$ **8.** $y' = y^3 - y^2$

Models of Population Growth

The autonomous differential equations in Exercises 9–12 represent models for population growth. For each exercise, use a phase line analysis to sketch solution curves for $P(t)$, selecting different starting values $P(0)$. Which equilibria are stable, and which are unstable?

9. $\frac{dP}{dt} = 1 - 2P$ **10.** $\frac{dP}{dt} = P(1 - 2P)$

11. $\frac{dP}{dt} = 2P(P-3)$ **12.** $\frac{dP}{dt} = 3P(1-P)\left(P - \frac{1}{2}\right)$

13. Catastrophic change in logistic growth Suppose that a healthy population of some species is growing in a limited environment and that the current population P_0 is fairly close to the carrying capacity M_0. You might imagine a population of fish living in a freshwater lake in a wilderness area. Suddenly a catastrophe such as the Mount St. Helens volcanic eruption contaminates the lake and destroys a significant part of the food and oxygen on which the fish depend. The result is a new environment with a carrying capacity M_1 considerably less than M_0 and, in fact, less than the current population P_0. Starting at some time before the catastrophe, sketch a "before-and-after" curve that shows how the fish population responds to the change in environment.

14. Controlling a population The fish and game department in a certain state is planning to issue hunting permits to control the deer population (one deer per permit). It is known that if the deer population falls below a certain level m, the deer will become extinct. It is also known that if the deer population rises above the carrying capacity M, the population will decrease back to M through disease and malnutrition.

a. Discuss the reasonableness of the following model for the growth rate of the deer population as a function of time:

$$\frac{dP}{dt} = rP(M-P)(P-m),$$

where P is the population of the deer and r is a positive constant of proportionality. Include a phase line.

b. Explain how this model differs from the logistic model $dP/dt = rP(M-P)$. Is it more or less reasonable than the logistic model?

c. Show that if $P > M$ for all t, then $\lim_{t\to\infty} P(t) = M$.

d. What happens if $P < m$ for all t?

e. Discuss the solutions to the differential equation. What are the equilibrium points of the model? Explain the dependence of the steady-state value of P on the initial values of P. About how many permits should be issued?

Applications and Examples

15. Skydiving If a body of mass m falling from rest under the action of gravity encounters an air resistance proportional to the square of velocity, then the body's velocity t seconds into the fall satisfies the equation

$$m\frac{dv}{dt} = mg - kv^2, \qquad k > 0$$

where k is a constant that depends on the body's aerodynamic properties and the density of the air. (We assume that the fall is too short to be affected by changes in the air's density.)

a. Draw a phase line for the equation.

b. Sketch a typical velocity curve.

c. For a 150-lb skydiver and with time in seconds and distance in feet, a typical value of k is 0.005. What is the diver's terminal velocity?

16. Resistance proportional to $\sqrt{v}$ A body of mass m is projected vertically downward with initial velocity v_0. Assume that the resisting force is proportional to the square root of the velocity, and find the terminal velocity from a graphical analysis.

17. Sailing A sailboat is running along a straight course with the wind providing a constant forward force of 75 lb. The only other force acting on the boat is resistance as the boat moves through the water. The resisting force is numerically equal to five times the boat's speed, and the initial velocity is 1 ft/sec. What is the maximum velocity in feet per second of the boat under this wind?

18. The spread of information Sociologists recognize a phenomenon called *social diffusion*, which is the spreading of a piece of information, technological innovation, or cultural fad among a population. The members of the population can be divided into two classes: those who have the information and those who do not. In a fixed population whose size is known, it is reasonable to assume that the rate of diffusion is proportional to the number who have the information times the number yet to receive it. If X denotes the number of individuals who have the information in a population of N people, then a mathematical model for social diffusion is given by

$$\frac{dX}{dt} = kX(N-X),$$

where t represents time in days and k is a positive constant.

a. Discuss the reasonableness of the model.

b. Construct a phase line identifying the signs of X' and X''.

c. Sketch representative solution curves.

d. Predict the value of X for which the information is spreading most rapidly. How many people eventually receive the information?

19. Current in an *RL* circuit The accompanying diagram represents an electrical circuit whose total resistance is a constant R ohms and whose self-inductance, shown as a coil, is L henries, also a constant. There is a switch whose terminals at a and b can be closed to connect a constant electrical source of V volts. From Section 9.2, we have

$$L\frac{di}{dt} + Ri = V,$$

where i is the current in amperes and t is the time in seconds.

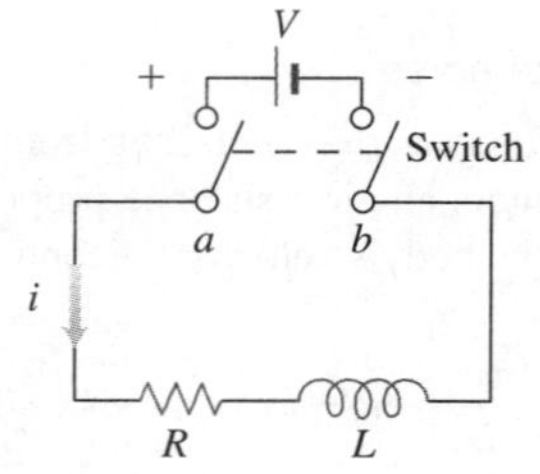

Use a phase line analysis to sketch the solution curve assuming that the switch in the *RL* circuit is closed at time $t = 0$. What happens to the current as $t \to \infty$? This value is called the *steady-state solution.*

20. A pearl in shampoo Suppose that a pearl is sinking in a thick fluid, like shampoo, subject to a frictional force opposing its fall and proportional to its velocity. Suppose that there is also a resistive buoyant force exerted by the shampoo. According to *Archimedes' principle*, the buoyant force equals the weight of the fluid displaced by the pearl. Using m for the mass of the pearl and P for the mass of the shampoo displaced by the pearl as it descends, complete the following steps.

a. Draw a schematic diagram showing the forces acting on the pearl as it sinks, as in Figure 9.20.

b. Using $v(t)$ for the pearl's velocity as a function of time t, write a differential equation modeling the velocity of the pearl as a falling body.

c. Construct a phase line displaying the signs of v' and v''.

d. Sketch typical solution curves.

e. What is the terminal velocity of the pearl?

Logistic Functions

21. Write the formula for a logistic function that has values between $y = 0$ and $y = 1$, crosses the line $y = 1/2$ at $x = 0$, and has slope 5 at this point.

22. Write the formula for a logistic function that has values between $y = 0$ and $y = 1$, crosses the line $y = 1/2$ at $x = 0$, and has slope 1/5 at this point.

9.5 Systems of Equations and Phase Planes

In some situations we are led to consider not one, but several, first-order differential equations. Such a collection is called a **system** of differential equations. In this section we present an approach to understanding systems through a graphical procedure known as a *phase-plane analysis*. We present this analysis in the context of modeling the populations of trout and bass living in a common pond.

Phase Planes

A general system of two first-order differential equations may take the form

$$\frac{dx}{dt} = F(x, y),$$
$$\frac{dy}{dt} = G(x, y).$$

In this system we often think of t as representing time and take $x(t)$ and $y(t)$ to be two functions of t. Such a system of equations is called **autonomous** because dx/dt and dy/dt do not depend on the independent variable time t, but only on the dependent variables x and y. A **solution** of such a system consists of a pair of functions $x(t)$ and $y(t)$ that satisfies both of the differential equations simultaneously for every t over some time interval (finite or infinite).

We cannot look at just one of these equations in isolation to find solutions $x(t)$ or $y(t)$ since each derivative depends on both x and y. To gain insight into the solutions, we look at both dependent variables together by plotting the points $(x(t), y(t))$ in the xy-plane starting at some specified point. Therefore the solution functions define a solution curve through the specified point, called a **trajectory** of the system. The xy-plane itself, in which these trajectories reside, is referred to as the **phase plane**. Thus we consider both solutions together and study the behavior of all the solution trajectories in the phase plane. It can be proved that two trajectories can never cross or touch each other. (Solution trajectories are examples of *parametric curves*, which will be examined in detail in Chapter 10.)

A Competitive-Hunter Model

Imagine two species of fish, say trout and bass, competing for the same limited resources (such as food and oxygen) in a certain pond. We let $x(t)$ represent the number of trout and $y(t)$ the number of bass living in the pond at time t. In reality, $x(t)$ and $y(t)$ are always integer valued, but we will approximate them with real-valued differentiable functions. This allows us to apply the methods of differential equations.

Several factors affect the rates of change of these populations. As time passes, each species breeds, so we assume its population increases proportionally to its size. Taken by itself, this would lead to exponential growth in each of the two populations. However, there is a countervailing effect from the fact that the two species are in competition. A large number of bass tends to cause a decrease in the number of trout, and vice versa. Our model takes the size of this effect to be proportional to the frequency with which the two species interact, which in turn is proportional to xy, the product of the two populations. These considerations lead to the following model for the growth of the trout and bass in the pond:

$$\frac{dx}{dt} = (a - by)x, \tag{1a}$$

$$\frac{dy}{dt} = (m - nx)y. \tag{1b}$$

Here $x(t)$ represents the trout population, $y(t)$ the bass population, and a, b, m, n are positive constants. A solution of this system then consists of a pair of functions $x(t)$ and $y(t)$ that give the population of each fish species at time t. Each equation in (1) contains both of the unknown functions x and y, so we are unable to solve them individually. Instead, we will use a graphical analysis to study the solution trajectories of this **competitive-hunter model**.

We now examine the nature of the phase plane in the trout-bass population model. We will be interested in the 1st quadrant of the xy-plane, where $x \geq 0$ and $y \geq 0$, since populations cannot be negative. First, we determine where the bass and trout populations are both constant. Noting that the $(x(t), y(t))$ values remain unchanged when $dx/dt = 0$ and $dy/dt = 0$, we see that Equations (1a and 1b) then become

$$(a - by)x = 0,$$

$$(m - nx)y = 0.$$

This pair of simultaneous equations has two solutions: $(x, y) = (0, 0)$ and $(x, y) = (m/n, a/b)$. At these (x, y) values, called **equilibrium** or **rest points**, the two populations remain at constant values over all time. The point $(0, 0)$ represents a pond containing no members of either fish species; the point $(m/n, a/b)$ corresponds to a pond with an unchanging number of each fish species.

Next, we note that if $y = a/b$, then Equation (1a) implies $dx/dt = 0$, so the trout population $x(t)$ is constant. Similarly, if $x = m/n$, then Equation (1b) implies $dy/dt = 0$, and the bass population $y(t)$ is constant. This information is recorded in Figure 9.28.

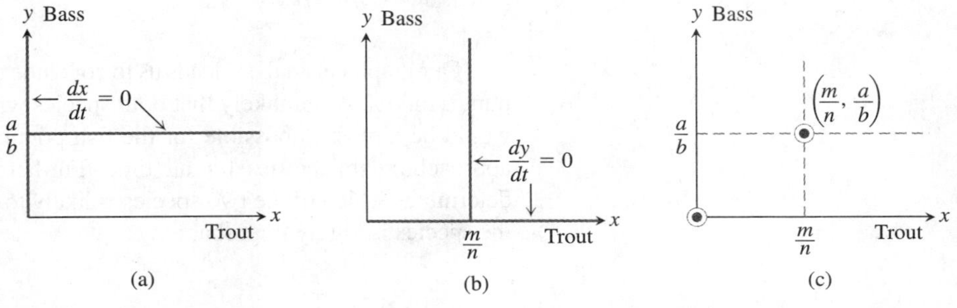

FIGURE 9.28 Rest points in the competitive-hunter model given by Equations (1a) and (1b).

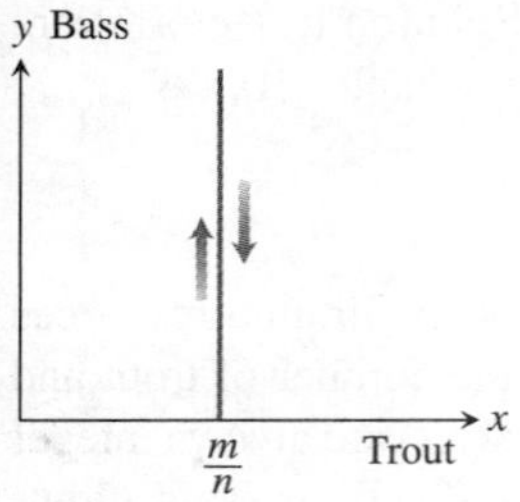

FIGURE 9.29 To the left of the line $x = m/n$ the trajectories move upward, and to the right they move downward.

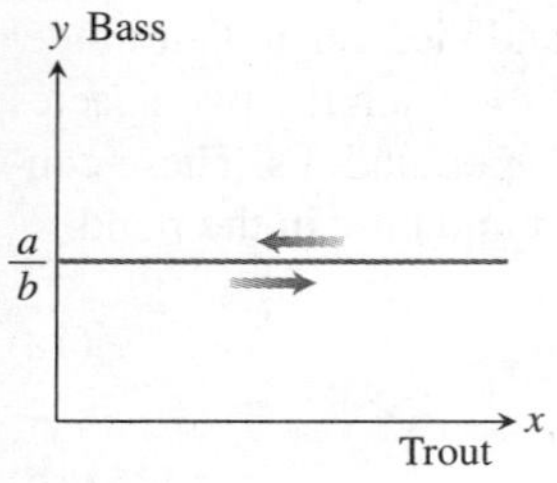

FIGURE 9.30 Above the line $y = a/b$ the trajectories move to the left, and below it they move to the right.

In setting up our competitive-hunter model, we do not generally know precise values of the constants a, b, m, n. Nonetheless, we can analyze the system of Equations (1) to learn the nature of its solution trajectories. We begin by determining the signs of dx/dt and dy/dt throughout the phase plane. Although $x(t)$ represents the number of trout and $y(t)$ the number of bass at time t, we are thinking of the pair of values $(x(t), y(t))$ as a point tracing out a trajectory curve in the phase plane. When dx/dt is positive, $x(t)$ is increasing and the point is moving to the right in the phase plane. If dx/dt is negative, the point is moving to the left. Likewise, the point is moving upward where dy/dt is positive and downward where dy/dt is negative.

We saw that $dy/dt = 0$ along the vertical line $x = m/n$. To the left of this line, dy/dt is positive since $dy/dt = (m - nx)y$ and $x < m/n$. So the trajectories on this side of the line are directed upward. To the right of this line, dy/dt is negative and the trajectories point downward. The directions of the associated trajectories are indicated in Figure 9.29. Similarly, above the horizontal line $y = a/b$, we have $dx/dt < 0$ and the trajectories head leftward; below this line they head rightward, as shown in Figure 9.30. Combining this information gives four distinct regions in the plane A, B, C, D, with their respective trajectory directions shown in Figure 9.31.

Next, we examine what happens near the two equilibrium points. The trajectories near $(0, 0)$ point away from it, upward and to the right. The behavior near the equilibrium point $(m/n, a/b)$ depends on the region in which a trajectory begins. If it starts in region B, for instance, then it will move downward and leftward toward the equilibrium point. Depending on where the trajectory begins, it may move downward into region D, leftward into region A, or perhaps straight into the equilibrium point. If it enters into regions A or D, then it will continue to move away from the rest point. We say that both rest points are **unstable**, meaning (in this setting) there are trajectories near each point that head away from them. These features are indicated in Figure 9.32.

It turns out that in each of the half-planes above and below the line $y = a/b$, there is exactly one trajectory approaching the equilibrium point $(m/n, a/b)$ (see Exercise 7). Above these two trajectories the bass population increases, and below them it decreases. The two trajectories approaching the equilibrium point are suggested in Figure 9.33.

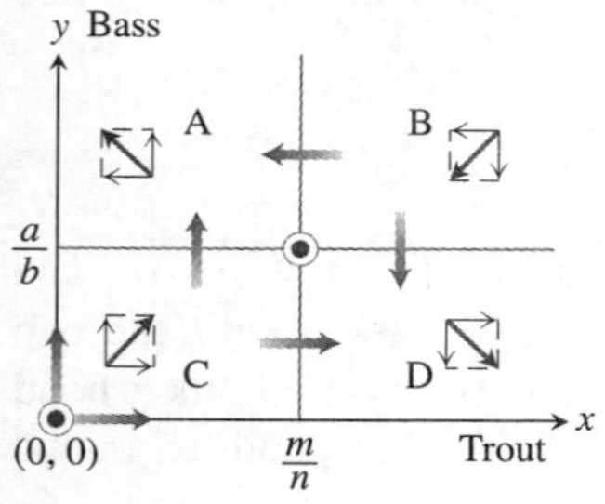

FIGURE 9.31 Composite graphical analysis of the trajectory directions in the four regions determined by $x = m/n$ and $y = a/b$.

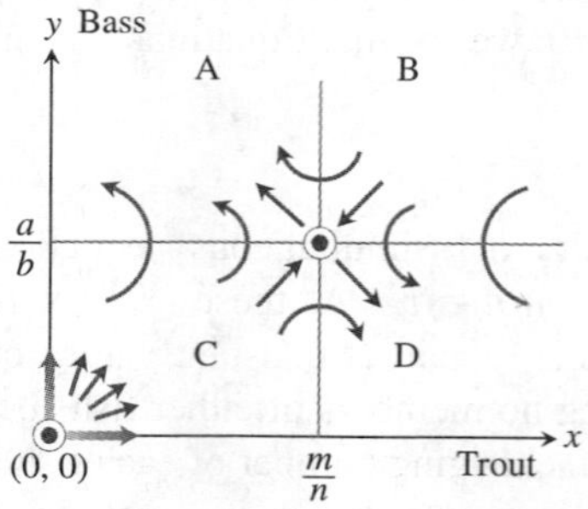

FIGURE 9.32 Motion along the trajectories near the rest points $(0, 0)$ and $(m/n, a/b)$.

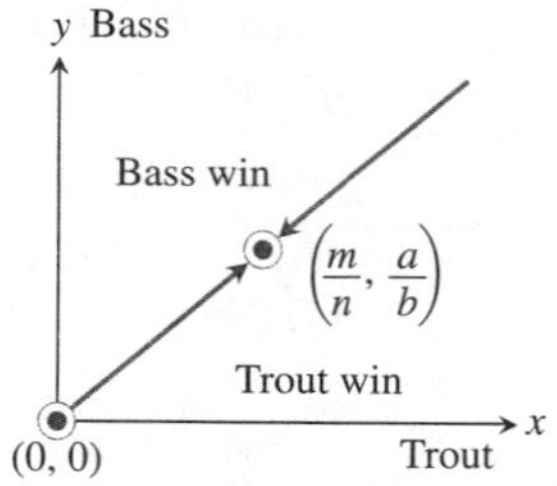

FIGURE 9.33 Qualitative results of analyzing the competitive-hunter model. There are exactly two trajectories approaching the point $(m/n, a/b)$.

Our graphical analysis leads us to conclude that, under the assumptions of the competitive-hunter model, it is unlikely that both species will reach equilibrium levels. This is because it would be almost impossible for the fish populations to move exactly along one of the two approaching trajectories for all time. Furthermore, the initial populations point (x_0, y_0) determines which of the two species is likely to survive over time, and mutual coexistence of the species is highly improbable.

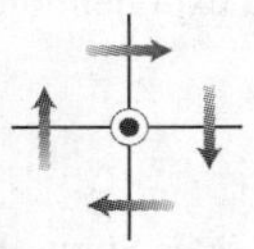

FIGURE 9.34 Trajectory direction near the rest point $(0, 0)$.

Limitations of the Phase-Plane Analysis Method

Unlike the situation for the competitive-hunter model, it is not always possible to determine the behavior of trajectories near a rest point. For example, suppose we know that the trajectories near a rest point, chosen here to be the origin $(0, 0)$, behave as in Figure 9.34. The information provided by Figure 9.34 is not sufficient to distinguish among the three possible trajectories shown in Figure 9.35. Even if we could determine that a trajectory near an equilibrium point resembles that of Figure 9.35c, we would still not know how the other trajectories behave. It could happen that a trajectory closer to the origin behaves like the motions displayed in Figure 9.35a or 9.35b. The spiraling trajectory in Figure 9.35c can never actually reach the rest point in a finite time period.

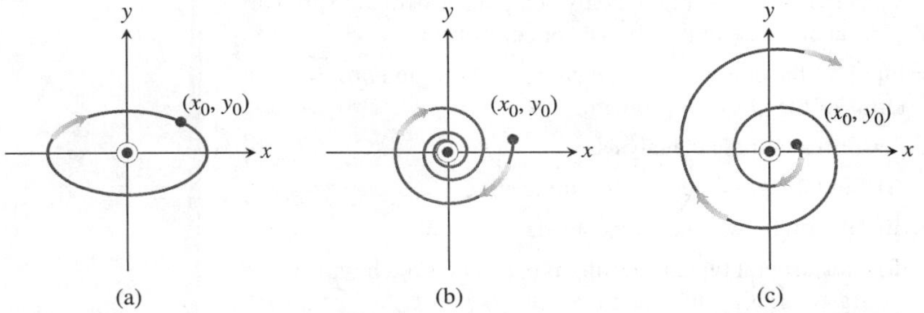

FIGURE 9.35 Three possible trajectory motions: (a) periodic motion, (b) motion toward an asymptotically stable rest point, and (c) motion near an unstable rest point.

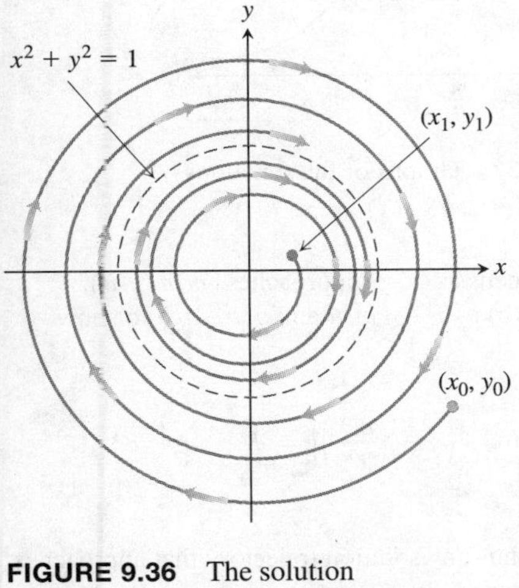

FIGURE 9.36 The solution $x^2 + y^2 = 1$ is a limit cycle.

Another Type of Behavior

The system

$$\frac{dx}{dt} = y + x - x(x^2 + y^2), \tag{2a}$$

$$\frac{dy}{dt} = -x + y - y(x^2 + y^2) \tag{2b}$$

can be shown to have only one equilibrium point at $(0, 0)$. Yet any trajectory starting on the unit circle traverses it clockwise because, when $x^2 + y^2 = 1$, we have $dy/dx = -x/y$ (see Exercise 2). If a trajectory starts inside the unit circle, it spirals outward, asymptotically approaching the circle as $t \to \infty$. If a trajectory starts outside the unit circle, it spirals inward, again asymptotically approaching the circle as $t \to \infty$. The circle $x^2 + y^2 = 1$ is called a **limit cycle** of the system (Figure 9.36). In this system, the values of x and y eventually become periodic.

EXERCISES 9.5

1. List three of the important considerations that are ignored in the competitive-hunter model as presented in the text.
2. For the system (2a) and (2b), show that any trajectory starting on the unit circle $x^2 + y^2 = 1$ will traverse the unit circle in a periodic solution. First introduce polar coordinates and rewrite the system as $dr/dt = r(1 - r^2)$ and $-d\theta/dt = -1$.
3. Develop a model for the growth of trout and bass, assuming that in isolation trout demonstrate exponential decay [so that $a < 0$ in Equations (1a) and (1b)] and that the bass population grows logistically with a population limit M. Analyze graphically the motion in the vicinity of the rest points in your model. Is coexistence possible?

4. How might the competitive-hunter model be validated? Include a discussion of how the various constants a, b, m, and n might be estimated. How could state conservation authorities use the model to ensure the survival of both species?

5. Consider another competitive-hunter model defined by

$$\frac{dx}{dt} = a\left(1 - \frac{x}{k_1}\right)x - bxy,$$

$$\frac{dy}{dt} = m\left(1 - \frac{y}{k_2}\right)y - nxy,$$

where x and y represent trout and bass populations, respectively.

a. What assumptions are implicitly being made about the growth of trout and bass in the absence of competition?

b. Interpret the constants $a, b, m, n, k_1, k_2, a/b$, and m/n in terms of the physical problem.

c. Perform a graphical analysis:

i) Find the possible equilibrium levels.

ii) Determine whether coexistence is possible.

iii) Pick several typical starting points, and sketch typical trajectories in the phase plane.

iv) Interpret the outcomes predicted by your graphical analysis in terms of the constants a, b, m, n, k_1, and k_2.

Note: When you get to part (iii), you should realize that five cases exist. You will need to analyze all five cases.

6. An economic model Consider the following economic model. Let P be the price of a single item on the market. Let Q be the quantity of the item available on the market. Both P and Q are functions of time. If one considers price and quantity as two interacting species, the following model might be proposed:

$$\frac{dP}{dt} = aP\left(\frac{b}{Q} - P\right),$$

$$\frac{dQ}{dt} = cQ(fP - Q),$$

where a, b, c, and f are positive constants. Justify and discuss the adequacy of the model.

a. If $a = 1, b = 20{,}000, c = 1$, and $f = 30$, find the equilibrium points of this system. If possible, classify each equilibrium point with respect to its stability. If a point cannot be readily classified, give some explanation.

b. Perform a graphical stability analysis to determine what will happen to the levels of P and Q as time increases.

c. Give an economic interpretation of the curves that determine the equilibrium points.

7. Two trajectories approach equilibrium Show that the two trajectories leading to $(m/n, a/b)$ shown in Figure 9.33 are unique by carrying out the following steps.

a. From system (1a) and (1b) apply the Chain Rule to derive the following equation:

$$\frac{dy}{dx} = \frac{(m - nx)y}{(a - by)x}.$$

b. Separate the variables, integrate, and exponentiate to obtain

$$y^a e^{-by} = Kx^m e^{-nx},$$

where K is a constant of integration.

c. Let $f(y) = y^a/e^{by}$ and $g(x) = x^m/e^{nx}$. Show that $f(y)$ has a unique maximum of $M_y = (a/eb)^a$ when $y = a/b$ as shown in Figure 9.37. Similarly, show that $g(x)$ has a unique maximum $M_x = (m/en)^m$ when $x = m/n$, also shown in Figure 9.37.

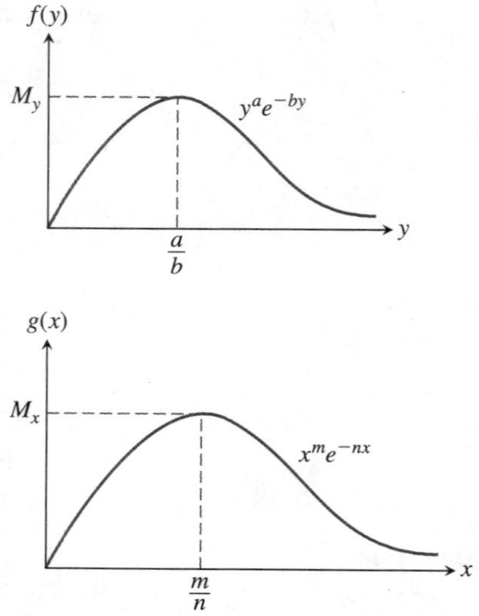

FIGURE 9.37 Graphs of the functions $f(y) = y^a/e^{by}$ and $g(x) = x^m/e^{nx}$.

d. Consider what happens as (x, y) approaches $(m/n, a/b)$. Take limits in part (b) as $x \to m/n$ and $y \to a/b$ to show that either

$$\lim_{\substack{x \to m/n \\ y \to a/b}} \left[\left(\frac{y^a}{e^{by}}\right)\left(\frac{e^{nx}}{x^m}\right)\right] = K$$

or $M_y/M_x = K$. Thus any solution trajectory that approaches $(m/n, a/b)$ must satisfy

$$\frac{y^a}{e^{by}} = \left(\frac{M_y}{M_x}\right)\left(\frac{x^m}{e^{nx}}\right).$$

e. Show that only one trajectory can approach $(m/n, a/b)$ from below the line $y = a/b$. Pick $y_0 < a/b$. From Figure 9.37 you can see that $f(y_0) < M_y$, which implies that

$$\frac{M_y}{M_x}\left(\frac{x^m}{e^{nx}}\right) = y_0^a/e^{by_0} < M_y.$$

This in turn implies that

$$\frac{x^m}{e^{nx}} < M_x.$$

Figure 9.37 tells you that for $g(x)$ there is a unique value $x_0 < m/n$ satisfying this last inequality. That is, for each $y < a/b$ there is a unique value of x satisfying the equation in part (d). Thus there can exist only one trajectory solution approaching $(m/n, a/b)$ from below, as shown in Figure 9.38.

f. Use a similar argument to show that the solution trajectory leading to $(m/n, a/b)$ is unique if $y_0 > a/b$.

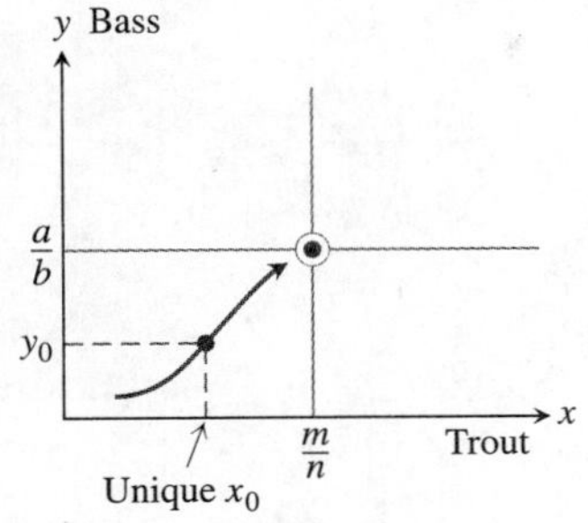

FIGURE 9.38 For any $y < a/b$, only one solution trajectory leads to the rest point $(m/n, a/b)$.

8. Show that the second-order differential equation $y'' = F(x, y, y')$ can be reduced to a system of two first-order differential equations

$$\frac{dy}{dx} = z,$$
$$\frac{dz}{dx} = F(x, y, z).$$

Can something similar be done to the nth-order differential equation $y^{(n)} = F(x, y, y', y'', \ldots, y^{(n-1)})$?

Lotka-Volterra Equations for a Predator-Prey Model

In 1925 Lotka and Volterra introduced the *predator-prey* equations, a system of equations that models the populations of two species, one of which preys on the other. Let $x(t)$ represent the number of rabbits living in a region at time t, and $y(t)$ the number of foxes in the same region. As time passes, the number of rabbits increases at a rate proportional to their population, and decreases at a rate proportional to the number of encounters between rabbits and foxes. The foxes, which compete for food, increase in number at a rate proportional to the number of encounters with rabbits but decrease at a rate proportional to the number of foxes. The number of encounters between rabbits and foxes is assumed to be proportional to the product of the two populations. These assumptions lead to the autonomous system

$$\frac{dx}{dt} = (a - by)x,$$
$$\frac{dy}{dt} = (-c + dx)y,$$

where a, b, c, d are positive constants. The values of these constants vary according to the specific situation being modeled. We can study the nature of the population changes without setting these constants to specific values.

9. What happens to the rabbit population if there are no foxes present?

10. What happens to the fox population if there are no rabbits present?

11. Show that $(0, 0)$ and $(c/d, a/b)$ are equilibrium points. Explain the meaning of each of these points.

12. Show, by differentiating, that the function

$$C(t) = a \ln y(t) - by(t) - dx(t) + c \ln x(t)$$

is constant when $x(t)$ and $y(t)$ are positive and satisfy the predator-prey equations.

While x and y may change over time, $C(t)$ does not. Thus, C is a *conserved quantity* and its existence gives a *conservation law*. A trajectory that begins at a point (x, y) at time $t = 0$ gives a value of C that remains unchanged at future times. Each value of the constant C gives a trajectory for the autonomous system, and these trajectories close up, rather than spiraling inward or outward. The rabbit and fox populations oscillate through repeated cycles along a fixed trajectory. Figure 9.39 shows several trajectories for the predator-prey system.

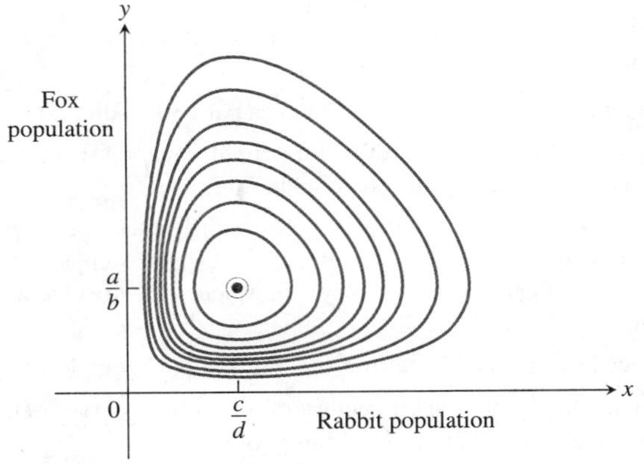

FIGURE 9.39 Some trajectories along which C is conserved.

13. Using a procedure similar to that in the text for the competitive-hunter model, show that each trajectory is traversed in a counterclockwise direction as time t increases.

Along each trajectory, both the rabbit and fox populations fluctuate between their maximum and minimum levels. The maximum and minimum levels for the rabbit population occur where the trajectory intersects the horizontal line $y = a/b$. For the fox population, they occur where the trajectory intersects the vertical line $x = c/d$. When the rabbit population is at its maximum, the fox population is below its maximum value. As the rabbit population declines from this point in time, we move counterclockwise around the trajectory, and the fox population grows until it reaches its maximum value. At this point the rabbit population has declined to $x = c/d$ and is no longer at its peak value. We see that the fox population reaches its maximum value at a later time than the rabbits. The predator population *lags behind* that of the prey in achieving its maximum values. This lag effect is shown in Figure 9.40, which graphs both $x(t)$ and $y(t)$.

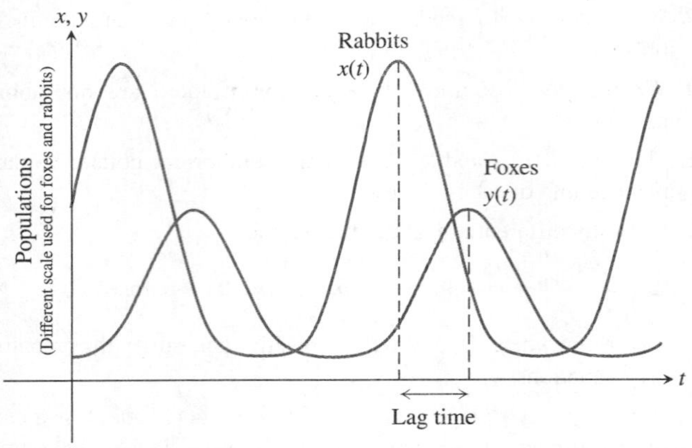

FIGURE 9.40 The fox and rabbit populations oscillate periodically, with the maximum fox population lagging the maximum rabbit population.

14. At some time during a trajectory cycle, a wolf invades the rabbit–fox territory, eats some rabbits, and then leaves. Does this mean that the fox population will from then on have a lower maximum value? Explain your answer.

CHAPTER 9 Questions to Guide Your Review

1. What is a first-order differential equation? When is a function a solution of such an equation?
2. What is a general solution? What is a particular solution?
3. What is the slope field of a differential equation $y' = f(x, y)$? What can we learn from such fields?
4. Describe Euler's method for solving the initial value problem $y' = f(x, y)$, $y(x_0) = y_0$ numerically. Give an example. Comment on the method's accuracy. Why might you want to solve an initial value problem numerically?
5. How do you solve linear first-order differential equations?
6. What is an orthogonal trajectory of a family of curves? Describe how one is found for a given family of curves.
7. What is an autonomous differential equation? What are its equilibrium values? How do they differ from critical points? What is a stable equilibrium value? Unstable?
8. How do you construct the phase line for an autonomous differential equation? How does the phase line help you produce a graph that qualitatively depicts a solution to the differential equation?
9. Why is the exponential model unrealistic for predicting long-term population growth? How does the logistic model correct for the deficiency in the exponential model for population growth? What is the logistic differential equation? What is the form of its solution? Describe the graph of the logistic solution.
10. What is an autonomous system of differential equations? What is a solution to such a system? What is a trajectory of the system?

CHAPTER 9 Practice Exercises

In Exercises 1–22, solve the differential equation.

1. $y' = xe^y\sqrt{x-2}$
2. $y' = xye^{x^2}$
3. $\sec x\,dy + x\cos^2 y\,dx = 0$
4. $2x^2\,dx - 3\sqrt{y}\csc x\,dy = 0$
5. $(1-e^x)\sec^2 y\,dy + 3e^x\tan y\,dx = 0$
6. $y' = xe^{x-y}\csc y$
7. $(x+y)^2\dfrac{dy}{dx} = a^2, \quad [u = x+y]$
8. $y' = (y^2-1)x^{-1}$
9. $2y' - y = xe^{x/2}$
10. $\dfrac{y'}{2} + y = e^{-x}\sin x$
11. $xy' + 2y = 1 - x^{-1}$
12. $xy' - y = 2x\ln x$
13. $(1 + e^x)\,dy + (ye^x + e^{-x})\,dx = 0$
14. $e^{-x}\,dy + (e^{-x}y - 4x)\,dx = 0$
15. $(x + 3y^2)\,dy + y\,dx = 0$ (*Hint:* $d(xy) = y\,dx + x\,dy$)
16. $x\,dy + (3y - x^{-2}\cos x)\,dx = 0, \quad x > 0$
17. $y' = \sin^3 x\cos^2 y$
18. $x\,dy - (x^4 - y)\,dx = 0$
19. $dy + x(2y - e^{x-x^2})\,dx = 0$
20. $y' + 3x^2y = 7x^2$
21. $y' = xy\ln x\ln y$
22. $xy' + 2y\ln x = \ln x$

Initial Value Problems

In Exercises 23–28, solve the initial value problem.

23. $(x+1)\dfrac{dy}{dx} + 2y = x, \quad x > -1, \quad y(0) = 1$
24. $x\dfrac{dy}{dx} + 2y = x^2 + 1, \quad x > 0, \quad y(1) = 1$
25. $\dfrac{dy}{dx} + 3x^2y = x^2, \quad y(0) = -1$
26. $x\,dy + (y - \cos x)\,dx = 0, \quad y\left(\dfrac{\pi}{2}\right) = 0$
27. $xy' + (x-2)y = 3x^3e^{-x}, \quad y(1) = 0$
28. $y\,dx + (3x - xy + 2)\,dy = 0, \quad y(2) = -1, \quad y < 0$

Euler's Method

In Exercises 29 and 30, use Euler's method to solve the initial value problem on the given interval starting at x_0 with $dx = 0.1$.

T 29. $y' = y + \cos x, \quad y(0) = 0; \quad 0 \le x \le 2; \quad x_0 = 0$

T 30. $y' = (2-y)(2x+3), \; y(-3) = 1; \; -3 \le x \le -1; \; x_0 = -3$

In Exercises 31 and 32, use Euler's method with $dx = 0.05$ to estimate $y(c)$, where y is the solution to the given initial value problem.

T **31.** $c = 3; \quad \dfrac{dy}{dx} = \dfrac{x - 2y}{x + 1}, \quad y(0) = 1$

T **32.** $c = 4; \quad \dfrac{dy}{dx} = \dfrac{x^2 - 2y + 1}{x}, \quad y(1) = 1$

In Exercises 33 and 34, use Euler's method to solve the initial value problem graphically, starting at $x_0 = 0$ with

a. $dx = 0.1.$ **b.** $dx = -0.1.$

T **33.** $\dfrac{dy}{dx} = \dfrac{1}{e^{x+y+2}}, \quad y(0) = -2$

T **34.** $\dfrac{dy}{dx} = -\dfrac{x^2 + y}{e^y + x}, \quad y(0) = 0$

Slope Fields

In Exercises 35–38, sketch part of the equation's slope field. Then add to your sketch the solution curve that passes through the point $P(1, -1)$. Use Euler's method with $x_0 = 1$ and $dx = 0.2$ to estimate $y(2)$. Round your answers to four decimal places. Find the exact value of $y(2)$ for comparison.

35. $y' = x$ **36.** $y' = 1/x$

37. $y' = xy$ **38.** $y' = 1/y$

Autonomous Differential Equations and Phase Lines

In Exercises 39 and 40:

a. Identify the equilibrium values. Which are stable and which are unstable?

b. Construct a phase line. Identify the signs of y' and y''.

c. Sketch a representative selection of solution curves.

39. $\dfrac{dy}{dx} = y^2 - 1$ **40.** $\dfrac{dy}{dx} = y - y^2$

Applications

41. Escape velocity The gravitational attraction F exerted by an airless moon on a body of mass m at a distance s from the moon's center is given by the equation $F = -mg\,R^2 s^{-2}$, where g is the acceleration of gravity at the moon's surface and R is the moon's radius (see accompanying figure). The force F is negative because it acts in the direction of decreasing s.

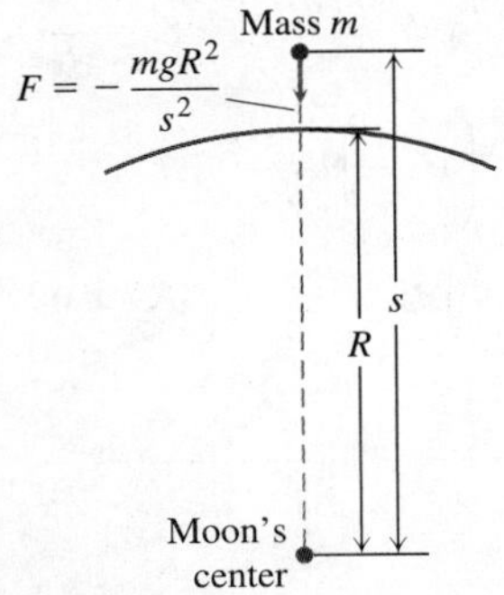

a. If the body is projected vertically upward from the moon's surface with an initial velocity v_0 at time $t = 0$, use Newton's second law, $F = ma$, to show that the body's velocity at position s is given by the equation

$$v^2 = \frac{2gR^2}{s} + v_0^2 - 2gR.$$

Thus, the velocity remains positive as long as $v_0 \geq \sqrt{2gR}$. The velocity $v_0 = \sqrt{2gR}$ is the moon's **escape velocity**. A body projected upward with this velocity or a greater one will escape from the moon's gravitational pull.

b. Show that if $v_0 = \sqrt{2gR}$, then

$$s = R\left(1 + \frac{3v_0}{2R}t\right)^{2/3}.$$

42. Coasting to a stop Table 9.6 shows the distance s (meters) coasted on inline skates in t sec by Johnathon Krueger. Find a model for his position in the form of Equation (2) of Section 9.3. His initial velocity was $v_0 = 0.86$ m/sec, his mass $m = 30.84$ kg (he weighed 68 lb), and his total coasting distance was 0.97 m.

TABLE 9.6 Johnathon Krueger skating data

t(sec)	s(m)	t(sec)	s(m)	t(sec)	s(m)
0	0	0.93	0.61	1.86	0.93
0.13	0.08	1.06	0.68	2.00	0.94
0.27	0.19	1.20	0.74	2.13	0.95
0.40	0.28	1.33	0.79	2.26	0.96
0.53	0.36	1.46	0.83	2.39	0.96
0.67	0.45	1.60	0.87	2.53	0.97
0.80	0.53	1.73	0.90	2.66	0.97

Mixture Problems

In Exercises 43 and 44, let S represent the pounds of salt in a tank at time t minutes. Set up a differential equation representing the given information and the rate at which S changes. Then solve for S and answer the particular questions.

43. A mixture containing $\frac{1}{2}$ lb of salt per gallon flows into a tank at the rate of 6 gal/min, and the well-stirred mixture flows out of the tank at the rate of 4 gal/min. The tank initially holds 100 gal of solution containing 10 lb of salt.

a. How many gallons of solution are in the tank after 1 minute? after 10 minutes? after 1 hour?

b. How many pounds of salt are in the tank after 1 minute? after 10 minutes? after 1 hour?

44. Pure water flows into a tank at the rate of 4 gal/min, and the well-stirred mixture flows out of the tank at the rate of 5 gal/min. The tank initially holds 200 gal of solution containing 50 pounds of salt.

a. How many gallons of solution are in the tank after 1 minute? after 10 minutes? after 200 minutes?

b. How many pounds of salt are in the tank after 1 minute? after 30 minutes?

c. When will the tank have exactly 5 pounds of salt and how many gallons of solution will be in the tank?

CHAPTER 9 Additional and Advanced Exercises

Theory and Applications

1. **Transport through a cell membrane** Under some conditions, the result of the movement of a dissolved substance across a cell's membrane is described by the equation

$$\frac{dy}{dt} = k\frac{A}{V}(c - y).$$

In this equation, y is the concentration of the substance inside the cell, and dy/dt is the rate at which y changes over time. The letters k, A, V, and c stand for constants, k being the *permeability coefficient* (a property of the membrane), A the surface area of the membrane, V the cell's volume, and c the concentration of the substance outside the cell. The equation says that the rate at which the concentration within the cell changes is proportional to the difference between it and the outside concentration.

a. Solve the equation for $y(t)$, using y_0 to denote $y(0)$.

b. Find the steady-state concentration, $\lim_{t\to\infty} y(t)$.

2. **Height of a rocket** If an external force F acts upon a system whose mass varies with time, Newton's law of motion is

$$\frac{d(mv)}{dt} = F + (v + u)\frac{dm}{dt}.$$

In this equation, m is the mass of the system at time t, v is its velocity, and $v + u$ is the velocity of the mass that is entering (or leaving) the system at the rate dm/dt. Suppose that a rocket of initial mass m_0 starts from rest, but is driven upward by firing some of its mass directly backward at the constant rate of $dm/dt = -b$ units per second and at constant speed relative to the rocket $u = -c$. The only external force acting on the rocket is $F = -mg$ due to gravity. Under these assumptions, show that the height of the rocket above the ground at the end of t seconds (t small compared to m_0/b) is

$$y = c\left[t + \frac{m_0 - bt}{b}\ln\frac{m_0 - bt}{m_0}\right] - \frac{1}{2}gt^2.$$

3. **a.** Assume that $P(x)$ and $Q(x)$ are continuous over the interval $[a, b]$. Use the Fundamental Theorem of Calculus, Part 1, to show that any function y satisfying the equation

$$v(x)y = \int v(x)Q(x)\,dx + C$$

for $v(x) = e^{\int P(x)\,dx}$ is a solution to the first-order linear equation

$$\frac{dy}{dx} + P(x)y = Q(x).$$

b. If $C = y_0 v(x_0) - \int_{x_0}^{x} v(t)Q(t)\,dt$, then show that any solution y in part (a) satisfies the initial condition $y(x_0) = y_0$.

4. (*Continuation of Exercise 3.*) Assume the hypotheses of Exercise 3, and assume that $y_1(x)$ and $y_2(x)$ are both solutions to the first-order linear equation satisfying the initial condition $y(x_0) = y_0$.

a. Verify that $y(x) = y_1(x) - y_2(x)$ satisfies the initial value problem

$$y' + P(x)y = 0, \quad y(x_0) = 0.$$

b. For the integrating factor $v(x) = e^{\int P(x)\,dx}$, show that

$$\frac{d}{dx}(v(x)[y_1(x) - y_2(x)]) = 0.$$

Conclude that $v(x)[y_1(x) - y_2(x)] \equiv$ constant.

c. From part (a), we have $y_1(x_0) - y_2(x_0) = 0$. Since $v(x) > 0$ for $a < x < b$, use part (b) to establish that $y_1(x) - y_2(x) \equiv 0$ on the interval (a, b). Thus $y_1(x) = y_2(x)$ for all $a < x < b$.

Homogeneous Equations

A first-order differential equation of the form

$$\frac{dy}{dx} = F\left(\frac{y}{x}\right)$$

is called *homogeneous*. It can be transformed into an equation whose variables are separable by defining the new variable $v = y/x$. Then $y = vx$ and

$$\frac{dy}{dx} = v + x\frac{dv}{dx}.$$

Substituting into the original differential equation and collecting terms with like variables then give the separable equation

$$\frac{dx}{x} + \frac{dv}{v - F(v)} = 0.$$

After solving this separable equation, we obtain the solution of the original equation when we replace v by y/x.

Solve the homogeneous equations in Exercises 5–10. First put the equation in the form of a homogeneous equation.

5. $(x^2 + y^2)\,dx + xy\,dy = 0$

6. $x^2\,dy + (y^2 - xy)\,dx = 0$

7. $(xe^{y/x} + y)\,dx - x\,dy = 0$

8. $\dfrac{y\,dy}{x\,dx} = \sqrt{1 + x^2 + y^2 + x^2y^2}$

9. $(1 + e^{x/y})dy + e^{x/y}\left(1 - \dfrac{x}{y}\right)dy = 0$

10. $\left(x\sin\dfrac{y}{x} - y\cos\dfrac{y}{x}\right)dx + x\cos\dfrac{y}{x}\,dy = 0$

CHAPTER 9 Technology Application Projects

Mathematica/Maple Projects

Projects can be found within MyLab Math.

- ***Drug Dosages: Are They Effective? Are They Safe?***
  Formulate and solve an initial value model for the absorption of a drug in the bloodstream.
- ***First-Order Differential Equations and Slope Fields***
  Plot slope fields and solution curves for various initial conditions to selected first-order differential equations.

10

Infinite Sequences and Series

OVERVIEW In this chapter we introduce the topic of *infinite series*. Such series give us precise ways to express many numbers and functions, both familiar and new, as arithmetic sums with infinitely many terms. For example, we will learn that

$$\frac{\pi}{4} = 1 - \frac{1}{3} + \frac{1}{5} - \frac{1}{7} + \frac{1}{9} - \cdots$$

and

$$\cos x = 1 - \frac{x^2}{2} + \frac{x^4}{24} - \frac{x^6}{720} + \frac{x^8}{40{,}320} - \cdots.$$

We need to develop a method to make sense of such expressions. Everyone knows how to add two numbers together, or even several. But how do you add together infinitely many numbers? Or, when adding together functions, how do you add infinitely many powers of x? In this chapter we answer these questions, which are part of the theory of infinite sequences and series. As with the differential and integral calculus, limits play a major role in the development of infinite series.

One common and important application of series occurs in making computations with complicated functions. A hard-to-compute function is replaced by an expression that looks like an "infinite degree polynomial," an infinite series in powers of x, as we see with the cosine function given above. Using the first few terms of this infinite series can allow for highly accurate approximations of functions by polynomials, enabling us to work with more general functions than those we have encountered before. These new functions are commonly obtained as solutions to differential equations arising in important applications of mathematics to science and engineering.

The terms "sequence" and "series" are sometimes used interchangeably in spoken language. In mathematics, however, each has a distinct meaning. A sequence is a type of infinite list, whereas a series is an infinite sum. To understand the infinite sums described by series, we first must understand infinite sequences.

10.1 Sequences

HISTORICAL ESSAY
Sequences and Series
www.bit.ly/2NSNyUt

Sequences are fundamental to the study of infinite series and to many aspects of mathematics. We saw one example of a sequence when we studied Newton's Method in Section 4.6. Newton's Method produces a sequence of approximations x_n that become closer and closer to the root of a differentiable function. Now we will explore general sequences of numbers and the conditions under which they converge to a finite number.

Representing Sequences

A sequence is a list of numbers

$$a_1, a_2, a_3, \ldots, a_n, \ldots$$

in a given order. Each of a_1, a_2, a_3, and so on represents a number. These are the **terms** of the sequence. For example, the sequence

$$2, 4, 6, 8, 10, 12, \ldots, 2n, \ldots$$

has first term $a_1 = 2$, second term $a_2 = 4$, and nth term $a_n = 2n$. The integer n is called the **index** of a_n and indicates where a_n occurs in the list. Order is important. The sequence $2, 4, 6, 8 \ldots$ is not the same as the sequence $4, 2, 6, 8 \ldots$.

We can think of the sequence

$$a_1, a_2, a_3, \ldots, a_n, \ldots$$

as a function that sends 1 to a_1, 2 to a_2, 3 to a_3, and in general sends the positive integer n to the nth term a_n. More precisely, an **infinite sequence** of numbers is a function whose domain is the set of positive integers. For example, the function associated with the sequence

$$2, 4, 6, 8, 10, 12, \ldots, 2n, \ldots$$

sends 1 to $a_1 = 2$, 2 to $a_2 = 4$, and so on. The general behavior of this sequence is described by the formula $a_n = 2n$.

We can change the index to start at any given number n. For example, the sequence

$$12, 14, 16, 18, 20, 22 \ldots$$

is described by the formula $a_n = 10 + 2n$, if we start with $n = 1$. It can also be described by the simpler formula $b_n = 2n$, where the index n starts at 6 and increases. To allow such simpler formulas, we let the first index of the sequence be any appropriate integer. In the sequence above, $\{a_n\}$ starts with a_1 while $\{b_n\}$ starts with b_6.

Sequences can be described by writing rules that specify their terms, such as

$$a_n = \sqrt{n}, \qquad b_n = (-1)^{n+1}\frac{1}{n}, \qquad c_n = \frac{n-1}{n}, \qquad d_n = (-1)^{n+1},$$

or by listing terms:

$$\{a_n\} = \left\{\sqrt{1}, \sqrt{2}, \sqrt{3}, \ldots, \sqrt{n}, \ldots\right\}$$

$$\{b_n\} = \left\{1, -\frac{1}{2}, \frac{1}{3}, -\frac{1}{4}, \ldots, (-1)^{n+1}\frac{1}{n}, \ldots\right\}$$

$$\{c_n\} = \left\{0, \frac{1}{2}, \frac{2}{3}, \frac{3}{4}, \frac{4}{5}, \ldots, \frac{n-1}{n}, \ldots\right\}$$

$$\{d_n\} = \{1, -1, 1, -1, 1, -1, \ldots, (-1)^{n+1}, \ldots\}.$$

We also sometimes write a sequence using its rule, as with

$$\{a_n\} = \left\{\sqrt{n}\right\}_{n=1}^{\infty}$$

and

$$\{b_n\} = \left\{(-1)^{n+1}\frac{1}{n}\right\}_{n=1}^{\infty}.$$

Figure 10.1 shows two ways to represent sequences graphically. The first marks the first few points from $a_1, a_2, a_3, \ldots, a_n, \ldots$ on the real axis. The second method shows the graph of the function defining the sequence. The function is defined only on integer inputs, and the graph consists of some points in the xy-plane located at $(1, a_1)$, $(2, a_2), \ldots, (n, a_n), \ldots$.

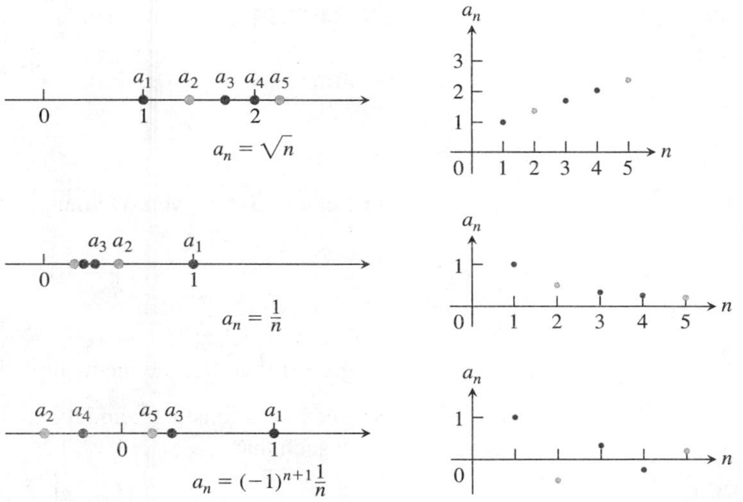

FIGURE 10.1 Sequences can be represented as points on the real line or as points in the plane where the horizontal axis n is the index number of the term and the vertical axis a_n is its value.

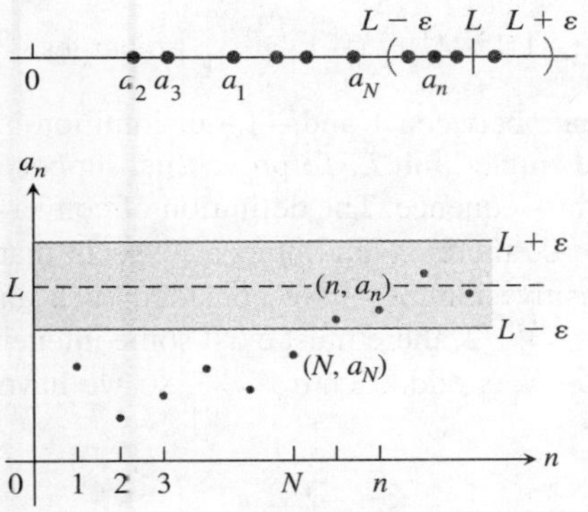

FIGURE 10.2 In the representation of a sequence as points in the plane, $a_n \to L$ if $y = L$ is a horizontal asymptote of the sequence of points $\{(n, a_n)\}$. In this figure, all the a_n's after a_N lie within ε of L.

Convergence and Divergence

Sometimes the numbers in a sequence approach a single value as the index n increases. This happens in the sequence

$$\left\{1, \frac{1}{2}, \frac{1}{3}, \frac{1}{4}, \ldots, \frac{1}{n}, \ldots\right\},$$

whose terms approach 0 as n gets large, and in the sequence

$$\left\{0, \frac{1}{2}, \frac{2}{3}, \frac{3}{4}, \frac{4}{5}, \ldots, 1 - \frac{1}{n}, \ldots\right\},$$

whose terms approach 1. On the other hand, sequences like

$$\{\sqrt{1}, \sqrt{2}, \sqrt{3}, \ldots, \sqrt{n}, \ldots\}$$

have terms that get larger than any number as n increases, and sequences like

$$\{1, -1, 1, -1, 1, -1, \ldots, (-1)^{n+1}, \ldots\}$$

bounce back and forth between 1 and -1, never converging to a single value. The following definition captures the meaning of having a sequence converge to a limiting value. It says that if we specify any number $\varepsilon > 0$, then by going far enough out in the sequence, taking the index n to be larger than some value N, the difference between a_n and the limit of the sequence becomes less than ε.

DEFINITIONS The sequence $\{a_n\}$ **converges** to the number L if for every positive number ε there corresponds an integer N such that

$$|a_n - L| < \varepsilon \qquad \text{whenever} \qquad n > N.$$

If no such number L exists, we say that $\{a_n\}$ **diverges**.

If $\{a_n\}$ converges to L, we write $\lim_{n\to\infty} a_n = L$, or simply $a_n \to L$, and call L the **limit** of the sequence (Figure 10.2).

HISTORICAL BIOGRAPHY
Nicole Oresme
(ca. 1320–1382)
www.bit.ly/2Qng79r

The definition is very similar to the definition of $\lim_{x\to\infty} f(x)$, the limit of a function $f(x)$ as x tends to ∞, discussed in Section 2.6. We will exploit this connection to calculate limits of sequences.

EXAMPLE 1 Show that

(a) $\lim_{n\to\infty} \frac{1}{n} = 0$ **(b)** $\lim_{n\to\infty} k = k$ (where k is a constant)

Solution

(a) Let $\varepsilon > 0$ be given. We must show that there exists an integer N such that

$$\left|\frac{1}{n} - 0\right| < \varepsilon \quad \text{whenever} \quad n > N.$$

The inequality $|1/n - 0| < \varepsilon$ will hold if $1/n < \varepsilon$ or $n > 1/\varepsilon$. If N is any integer greater than $1/\varepsilon$, the inequality will hold for all $n > N$. This proves that $\lim_{n\to\infty} 1/n = 0$.

(b) Let k be a constant, and let $\varepsilon > 0$ be given. We must show that there exists an integer N such that

$$|k - k| < \varepsilon \quad \text{whenever} \quad n > N.$$

Since $k - k = 0$, we can use any positive integer for N and the inequality $|k - k| < \varepsilon$ will hold. This proves that $\lim_{n\to\infty} k = k$. ■

EXAMPLE 2 Show that the sequence $\{1, -1, 1, -1, 1, -1, \ldots, (-1)^{n+1}, \ldots\}$ diverges.

Solution Since the terms in this sequence alternate between 1 and -1, our intuition is that they cannot approach arbitrarily close to some single limit L. To prove this, suppose that there did exist a number L that is the limit of this sequence. The definition of convergence then tells us that for every number $\varepsilon > 0$, there is an integer N such that $|a_n - L| < \varepsilon$ for all $n > N$. No matter what positive number ε we consider, such an integer N must exist. In particular, for the number $\varepsilon = 1/2$, there must exist some integer N such that $|a_n - L| < 1/2$ for all $n > N$. If $n > N$ is odd, then $a_n = 1$, so we have $|1 - L| < 1/2$, or

$$\frac{1}{2} < L < \frac{3}{2}.$$

On the other hand, if $n > N$ is even, then $a_n = -1$, and so we have $|-1 - L| < 1/2$, or

$$-\frac{3}{2} < L < -\frac{1}{2}.$$

However, these two conditions on L cannot both hold, so we have reached a contradiction. Therefore there is no number L to which this sequence converges.

We chose $\varepsilon = 1/2$ in this argument simply for convenience. The same argument would work if we had instead chosen $\varepsilon = 1/4$, or any value less than 1. ■

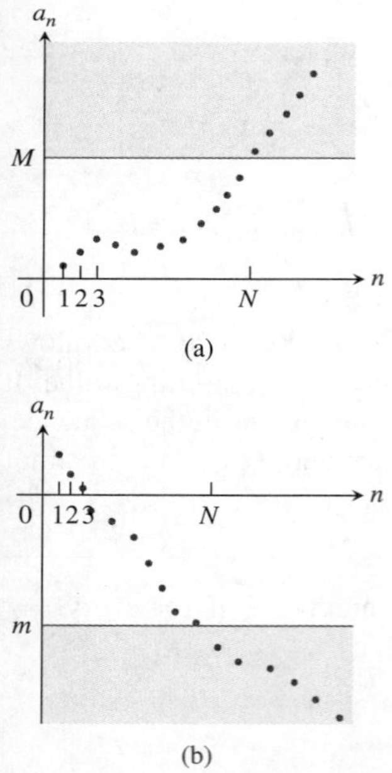

FIGURE 10.3 (a) The sequence diverges to ∞ because no matter what number M is chosen, the terms of the sequence after some index N all lie in the shaded band above M. (b) The sequence diverges to $-\infty$ because all terms after some index N lie below any chosen number m.

The sequence $\{\sqrt{n}\}$ also diverges, but for a different reason. As n increases, its terms become larger than any fixed number. We describe the behavior of this sequence by writing

$$\lim_{n\to\infty} \sqrt{n} = \infty.$$

In writing infinity as the limit of a sequence, we are not saying that the differences between the terms a_n and ∞ become small as n increases, nor are we asserting that there is some number infinity that the sequence approaches. We are merely using a notation that captures the idea that a_n eventually gets and stays larger than any fixed number as n gets large (see Figure 10.3a). The terms of a sequence might also decrease to negative infinity, as in Figure 10.3b.

DEFINITION The sequence $\{a_n\}$ **diverges to infinity** if for every number M there is an integer N such that for all n larger than N, $a_n > M$. If this condition holds, then we write

$$\lim_{n\to\infty} a_n = \infty \quad \text{or} \quad a_n \to \infty.$$

Similarly, if for every number m there is an integer N such that for all $n > N$, we have $a_n < m$, then we say $\{a_n\}$ **diverges to negative infinity** and write

$$\lim_{n\to\infty} a_n = -\infty \quad \text{or} \quad a_n \to -\infty.$$

A sequence may diverge without diverging to infinity or negative infinity, as we saw in Example 2. The sequences $\{1, -2, 3, -4, 5, -6, 7, -8, \ldots\}$ and $\{1, 0, 2, 0, 3, 0, \ldots\}$ are also sequences that diverge but do not diverge to infinity or negative infinity.

The convergence or divergence of a sequence is not affected by the values of any number of its initial terms (whether we omit or change the first 10, the first 1000, or even the first million terms does not matter). From Figure 10.2, we can see that only the part of the sequence that remains after discarding some initial number of terms determines whether the sequence has a limit and the value of that limit when it does exist.

Calculating Limits of Sequences

Since sequences are functions with domain restricted to the positive integers, it is not surprising that the theorems on limits of functions given in Chapter 2 have versions for sequences.

THEOREM 1 Let $\{a_n\}$ and $\{b_n\}$ be sequences of real numbers, and let A and B be real numbers. The following rules hold if $\lim_{n\to\infty} a_n = A$ and $\lim_{n\to\infty} b_n = B$.

1. *Sum Rule:* $\lim_{n\to\infty} (a_n + b_n) = A + B$
2. *Difference Rule:* $\lim_{n\to\infty} (a_n - b_n) = A - B$
3. *Constant Multiple Rule:* $\lim_{n\to\infty} (k \cdot b_n) = k \cdot B$ (any number k)
4. *Product Rule:* $\lim_{n\to\infty} (a_n \cdot b_n) = A \cdot B$
5. *Quotient Rule:* $\lim_{n\to\infty} \frac{a_n}{b_n} = \frac{A}{B}$ if $B \neq 0$

The proof is similar to that of Theorem 1 in Section 2.2 and is omitted.

EXAMPLE 3 By combining Theorem 1 with the limits of Example 1, we have:

(a) $\lim_{n\to\infty}\left(-\frac{1}{n}\right) = -1 \cdot \lim_{n\to\infty}\frac{1}{n} = -1 \cdot 0 = 0$ Constant Multiple Rule and Example 1a

(b) $\lim_{n\to\infty}\left(\frac{n-1}{n}\right) = \lim_{n\to\infty}\left(1 - \frac{1}{n}\right) = \lim_{n\to\infty} 1 - \lim_{n\to\infty}\frac{1}{n} = 1 - 0 = 1$ Difference Rule and Example 1a

(c) $\lim_{n\to\infty}\frac{5}{n^2} = 5 \cdot \lim_{n\to\infty}\frac{1}{n} \cdot \lim_{n\to\infty}\frac{1}{n} = 5 \cdot 0 \cdot 0 = 0$ Product Rule

(d) $\lim_{n\to\infty}\frac{4 - 7n^6}{n^6 + 3} = \lim_{n\to\infty}\frac{(4/n^6) - 7}{1 + (3/n^6)} = \frac{0 - 7}{1 + 0} = -7.$ Divide numerator and denominator by n^6 and use the Sum and Quotient Rules.

■

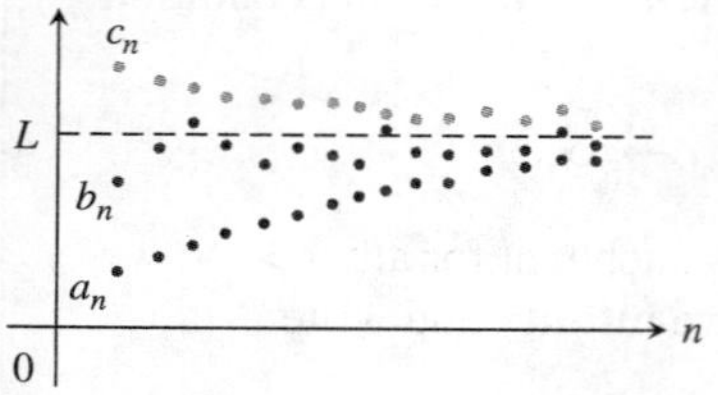

FIGURE 10.4 The terms of sequence $\{b_n\}$ are sandwiched between those of $\{a_n\}$ and $\{c_n\}$, forcing them to the same common limit L.

Be cautious in applying Theorem 1. It does not say, for example, that both of the sequences $\{a_n\}$ and $\{b_n\}$ have limits if their sum $\{a_n + b_n\}$ has a limit. For instance, $\{a_n\} = \{1, 2, 3, \ldots\}$ and $\{b_n\} = \{-1, -2, -3, \ldots\}$ both diverge, but their sum $\{a_n + b_n\} = \{0, 0, 0, \ldots\}$ clearly converges to 0.

One consequence of Theorem 1 is that every nonzero multiple of a divergent sequence $\{a_n\}$ diverges. Suppose, to the contrary, that $\{ca_n\}$ converges for some number $c \neq 0$. Then, by taking $k = 1/c$ in the Constant Multiple Rule in Theorem 1, we see that the sequence

$$\left\{\frac{1}{c} \cdot ca_n\right\} = \{a_n\}$$

converges. Thus, $\{ca_n\}$ cannot converge unless $\{a_n\}$ also converges. If $\{a_n\}$ does not converge, then $\{ca_n\}$ does not converge.

The next theorem is the sequence version of the Sandwich Theorem in Section 2.2. You are asked to prove the theorem in Exercise 119. (See Figure 10.4.)

THEOREM 2—The Sandwich Theorem for Sequences
Let $\{a_n\}$, $\{b_n\}$, and $\{c_n\}$ be sequences of real numbers. If $a_n \leq b_n \leq c_n$ holds for all n beyond some index N, and if $\lim_{n\to\infty} a_n = \lim_{n\to\infty} c_n = L$, then $\lim_{n\to\infty} b_n = L$ also.

An immediate consequence of Theorem 2 is that if $|b_n| \leq c_n$ and $c_n \to 0$, then $b_n \to 0$ because $-c_n \leq b_n \leq c_n$. We use this fact in the next example.

EXAMPLE 4 Since $1/n \to 0$, we know that

(a) $\dfrac{\cos n}{n} \to 0$ because $-\dfrac{1}{n} \leq \dfrac{\cos n}{n} \leq \dfrac{1}{n}$;

(b) $\dfrac{1}{2^n} \to 0$ because $0 \leq \dfrac{1}{2^n} \leq \dfrac{1}{n}$;

(c) $(-1)^n \dfrac{1}{n} \to 0$ because $-\dfrac{1}{n} \leq (-1)^n \dfrac{1}{n} \leq \dfrac{1}{n}$.

(d) If $|a_n| \to 0$, then $a_n \to 0$ because $-|a_n| \leq a_n \leq |a_n|$. ■

The application of Theorems 1 and 2 is broadened by a theorem stating that applying a continuous function to a convergent sequence produces a convergent sequence. We state the theorem, leaving the proof as an exercise (Exercise 120).

THEOREM 3—The Continuous Function Theorem for Sequences
Let $\{a_n\}$ be a sequence of real numbers. If $a_n \to L$ and if f is a function that is continuous at L and defined at all a_n, then $f(a_n) \to f(L)$.

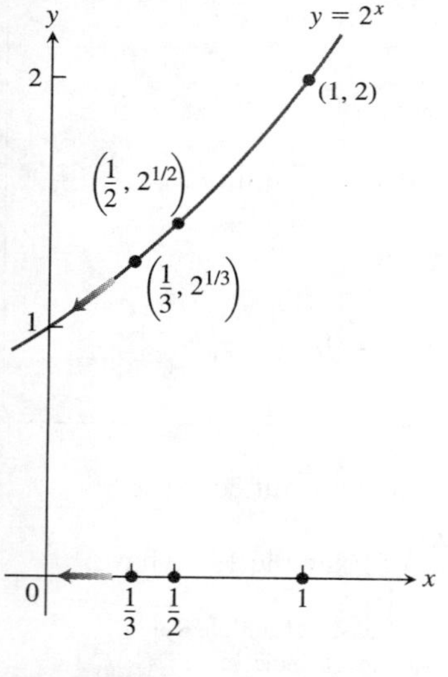

FIGURE 10.5 As $n \to \infty$, $1/n \to 0$ and $2^{1/n} \to 2^0$ (Example 6). The terms of $\{1/n\}$ are shown on the x-axis; the terms of $\{2^{1/n}\}$ are shown as the y-values on the graph of $f(x) = 2^x$.

EXAMPLE 5 Show that $\sqrt{(n+1)/n} \to 1$.

Solution We know that $(n+1)/n \to 1$. Taking $f(x) = \sqrt{x}$ and $L = 1$ in Theorem 3 gives $\sqrt{(n+1)/n} \to \sqrt{1} = 1$. ■

EXAMPLE 6 The sequence $\{1/n\}$ converges to 0. By taking $a_n = 1/n$, $f(x) = 2^x$, and $L = 0$ in Theorem 3, we see that $2^{1/n} = f(1/n) \to f(L) = 2^0 = 1$. The sequence $\{2^{1/n}\}$ converges to 1 (Figure 10.5). ■

Using L'Hôpital's Rule

The next theorem formalizes the connection between $\lim_{n\to\infty} a_n$ and $\lim_{x\to\infty} f(x)$. It enables us to use l'Hôpital's Rule to find the limits of some sequences.

THEOREM 4 Suppose that $f(x)$ is a function defined for all $x \geq n_0$ and that $\{a_n\}$ is a sequence of real numbers such that $a_n = f(n)$ for $n \geq n_0$. Then

$$\lim_{n\to\infty} a_n = L \qquad \text{whenever} \qquad \lim_{x\to\infty} f(x) = L.$$

Proof Suppose that $\lim_{x\to\infty} f(x) = L$. Then for each positive number ε, there is a number M such that

$$|f(x) - L| < \varepsilon \qquad \text{whenever} \qquad x > M.$$

Let N be an integer that is both greater than M and greater than or equal to n_0. Since $a_n = f(n)$, it follows that for all $n > N$, we have

$$|a_n - L| = |f(n) - L| < \varepsilon.$$

■

EXAMPLE 7 Show that

$$\lim_{n\to\infty} \frac{\ln n}{n} = 0.$$

Solution The function $(\ln x)/x$ is defined for all $x \geq 1$ and agrees with the given sequence at positive integers. Therefore, by Theorem 4, $\lim_{n\to\infty} (\ln n)/n$ will equal $\lim_{x\to\infty} (\ln x)/x$ if the latter exists. A single application of l'Hôpital's Rule shows that

$$\lim_{x\to\infty} \frac{\ln x}{x} = \lim_{x\to\infty} \frac{1/x}{1} = \frac{0}{1} = 0.$$

We conclude that $\lim_{n\to\infty} (\ln n)/n = 0$. ■

EXAMPLE 8 Does the sequence whose nth term is

$$a_n = \left(\frac{n+1}{n-1}\right)^n, \quad n \geq 2$$

converge? If so, find $\lim_{n\to\infty} a_n$.

Solution The function $f(x) = \left(\frac{x+1}{x-1}\right)^x$ is defined for all real numbers $x \geq 2$ and agrees with a_n at all integers $n \geq 2$. If we can show that $\lim_{x\to\infty} \left(\frac{x+1}{x-1}\right)^x = L$, then by Theorem 4, $\lim_{n\to\infty} \left(\frac{n+1}{n-1}\right)^n = L$.

The limit $\lim_{x\to\infty} f(x)$ leads to the indeterminate form 1^∞. To evaluate this limit, we apply l'Hôpital's Rule after taking the natural logarithm of $f(x)$:

$$\ln f(x) = \ln\left(\frac{x+1}{x-1}\right)^x = x \ln\left(\frac{x+1}{x-1}\right).$$

Then

$$\begin{aligned}
\lim_{x\to\infty} \ln f(x) &= \lim_{x\to\infty} x \ln\left(\frac{x+1}{x-1}\right) && \infty \cdot 0 \text{ form} \\
&= \lim_{x\to\infty} \frac{\ln\left(\frac{x+1}{x-1}\right)}{1/x} && \frac{0}{0} \text{ form} \\
&= \lim_{x\to\infty} \frac{-2/(x^2-1)}{-1/x^2} && \text{Apply l'Hôpital's Rule.} \\
&= \lim_{x\to\infty} \frac{2x^2}{x^2-1} = 2. && \text{Simplify and evaluate.}
\end{aligned}$$

Therefore, $\lim_{x\to\infty} \left(\frac{x+1}{x-1}\right)^x = \lim_{x\to\infty} f(x) = \lim_{x\to\infty} e^{\ln f(x)} = e^2$. Applying Theorem 4, we conclude that the sequence $\{a_n\}$ also converges to e^2. ■

Commonly Occurring Limits

The next theorem gives some limits that arise frequently.

Factorial Notation

The notation $n!$ ("n factorial") means the product $1 \cdot 2 \cdot 3 \cdots n$ of the integers from 1 to n. Notice that $(n+1)! = (n+1) \cdot n!$. Thus, $4! = 1 \cdot 2 \cdot 3 \cdot 4 = 24$ and $5! = 1 \cdot 2 \cdot 3 \cdot 4 \cdot 5 = 5 \cdot 4! = 120$. We define $0!$ to be 1. Factorials grow even faster than exponentials, as the table suggests. The values in the table are rounded.

n	e^n	$n!$
1	3	1
5	148	120
10	22,026	3,628,800
20	4.9×10^8	2.4×10^{18}

THEOREM 5 The following six sequences converge to the limits listed below.

1. $\lim_{n\to\infty} \frac{\ln n}{n} = 0$
2. $\lim_{n\to\infty} \sqrt[n]{n} = 1$
3. $\lim_{n\to\infty} x^{1/n} = 1 \quad (x > 0)$
4. $\lim_{n\to\infty} x^n = 0 \quad (|x| < 1)$
5. $\lim_{n\to\infty} \left(1 + \frac{x}{n}\right)^n = e^x \quad (\text{any } x)$
6. $\lim_{n\to\infty} \frac{x^n}{n!} = 0 \quad (\text{any } x)$

In Formulas (3) through (6), x remains fixed as $n \to \infty$.

Proof The first limit was computed in Example 7. The next two can be proved by taking logarithms and applying Theorem 4 (Exercises 117 and 118). The remaining proofs are given in Appendix A.5. ■

EXAMPLE 9 These are examples involving the limits in Theorem 5.

(a) $\frac{\ln(n^2)}{n} = \frac{2\ln n}{n} \to 2 \cdot 0 = 0$ Formula 1

(b) $\sqrt[n]{n^2} = n^{2/n} = (n^{1/n})^2 \to (1)^2 = 1$ Formula 2

(c) $\sqrt[n]{3n} = 3^{1/n}(n^{1/n}) \to 1 \cdot 1 = 1$ Formula 3 with $x = 3$ and Formula 2

(d) $\left(-\frac{1}{2}\right)^n \to 0$ Formula 4 with $x = -\frac{1}{2}$

(e) $\left(\frac{n-2}{n}\right)^n = \left(1 + \frac{-2}{n}\right)^n \to e^{-2}$ Formula 5 with $x = -2$

(f) $\frac{100^n}{n!} \to 0$ Formula 6 with $x = 100$ ■

Recursive Definitions

So far, we have calculated each a_n directly from the value of n. But sequences are often defined **recursively** by giving

1. The value(s) of the initial term or terms, and
2. a rule called a **recursion formula** for calculating any later term from terms that precede it.

EXAMPLE 10

(a) The statements $a_1 = 1$ and $a_n = a_{n-1} + 1$ for $n > 1$ define the sequence $1, 2, 3, \ldots, n, \ldots$ of positive integers. With $a_1 = 1$, we have $a_2 = a_1 + 1 = 2$, $a_3 = a_2 + 1 = 3$, and so on.

(b) The statements $a_1 = 1$ and $a_n = n \cdot a_{n-1}$ for $n > 1$ define the sequence $1, 2, 6, 24, \ldots, n!, \ldots$ of factorials. With $a_1 = 1$, we have $a_2 = 2 \cdot a_1 = 2$, $a_3 = 3 \cdot a_2 = 6$, $a_4 = 4 \cdot a_3 = 24$, and so on.

(c) The statements $a_1 = 1, a_2 = 1$, and $a_{n+1} = a_n + a_{n-1}$ for $n > 2$ define the sequence $1, 1, 2, 3, 5, \ldots$ of **Fibonacci numbers**. With $a_1 = 1$ and $a_2 = 1$, we have $a_3 = 1 + 1 = 2$, $a_4 = 2 + 1 = 3$, $a_5 = 3 + 2 = 5$, and so on.

(d) As we can see by applying Newton's method (see Exercise 145), the statements $x_0 = 1$ and $x_{n+1} = x_n - [(\sin x_n - x_n{}^2)/(\cos x_n - 2x_n)]$ for $n > 0$ define a sequence that, when it converges, gives a solution to the equation $\sin x - x^2 = 0$. ■

Bounded Monotonic Sequences

Two concepts that play a key role in determining the convergence of a sequence are those of a *bounded* sequence and a *monotonic* sequence. First we define bounded sequences.

DEFINITION A sequence $\{a_n\}$ is **bounded from above** if there exists a number M such that $a_n \le M$ for all n. The number M is an **upper bound** for $\{a_n\}$. If M is an upper bound for $\{a_n\}$ but no number less than M is an upper bound for $\{a_n\}$, then M is the **least upper bound** for $\{a_n\}$.

A sequence $\{a_n\}$ is **bounded from below** if there exists a number m such that $a_n \ge m$ for all n. The number m is a **lower bound** for $\{a_n\}$. If m is a lower bound for $\{a_n\}$ but no number greater than m is a lower bound for $\{a_n\}$, then m is the **greatest lower bound** for $\{a_n\}$.

If $\{a_n\}$ is bounded from above and below, then $\{a_n\}$ is **bounded**. If $\{a_n\}$ is not bounded, then we say that $\{a_n\}$ is an **unbounded** sequence.

EXAMPLE 11

(a) The sequence $1, 2, 3, \dots, n, \dots$ has no upper bound because it eventually surpasses every number M. However, it is bounded below by every real number less than or equal to 1. The number $m = 1$ is the greatest lower bound of the sequence.

(b) The sequence $\frac{1}{2}, \frac{2}{3}, \frac{3}{4}, \dots, \frac{n}{n+1}, \dots$ is bounded above by every real number greater than or equal to 1. The upper bound $M = 1$ is the least upper bound (Exercise 137). The sequence is also bounded below by every number less than or equal to $\frac{1}{2}$, which is its greatest lower bound. ■

Convergent sequences are bounded.

Suppose that a sequence $\{a_n\}$ converges to a number L. Applying the definition of convergence with the particular value $\varepsilon = 1$, there must exist a number N such that $|a_n - L| < 1$ if $n > N$. That is,

$$L - 1 < a_n < L + 1 \quad \text{for } n > N.$$

If M is a number larger than both $L + 1$ and all of the finitely many numbers $a_1, a_2, \dots, a_N$, then for every index n we have $a_n \le M$, and therefore $\{a_n\}$ is bounded from above. Similarly, if m is a number smaller than both $L - 1$ and all of the numbers $a_1, a_2, \dots, a_N$, then m is a lower bound for the sequence. Therefore, all convergent sequences are bounded.

Although it is true that every convergent sequence is bounded, there are bounded sequences that fail to converge. One example is the bounded sequence $\{(-1)^{n+1}\}$ discussed in Example 2. The problem here is that some bounded sequences bounce around in the band determined by any lower bound m and any upper bound M but do not converge (Figure 10.6). An important type of sequence that does not behave that way is one for which each term is at least as large, or at least as small, as its predecessor.

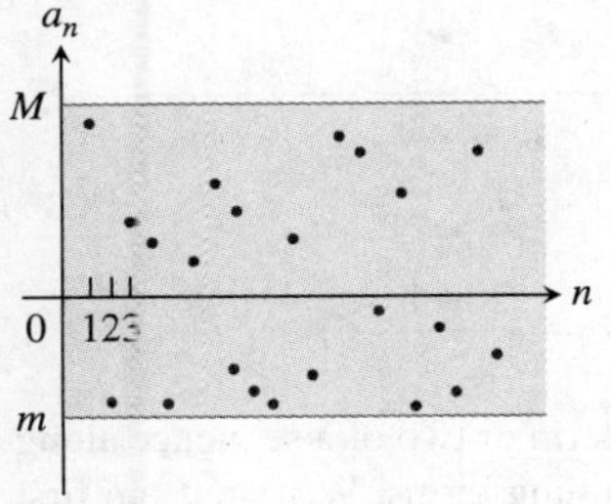

FIGURE 10.6 Some bounded sequences bounce around between their bounds and fail to converge to any limiting value.

DEFINITIONS A sequence $\{a_n\}$ is **nondecreasing** if $a_n \le a_{n+1}$ for all n. That is, $a_1 \le a_2 \le a_3 \le \dots$. The sequence is **nonincreasing** if $a_n \ge a_{n+1}$ for all n. The sequence $\{a_n\}$ is **monotonic** if it is either nondecreasing or nonincreasing.

EXAMPLE 12

(a) The sequence 1, 2, 3, …, n, … is nondecreasing.

(b) The sequence $\frac{1}{2}, \frac{2}{3}, \frac{3}{4}, \ldots, \frac{n}{n+1}, \ldots$ is nondecreasing.

(c) The sequence $1, \frac{1}{2}, \frac{1}{4}, \frac{1}{8}, \ldots, \frac{1}{2^n}, \ldots$ is nonincreasing.

(d) The constant sequence 3, 3, 3, …, 3, … is both nondecreasing and nonincreasing.

(e) The sequence $1, -1, 1, -1, 1, -1, \ldots$ is not monotonic. ■

A sequence that is bounded from above always has a least upper bound. Likewise, a sequence bounded from below always has a greatest lower bound. These results are based on the *completeness property* of the real numbers, discussed in Appendix A.7. We now prove that if L is the least upper bound of a nondecreasing sequence, then the sequence converges to L, and that if L is the greatest lower bound of a nonincreasing sequence, then the sequence converges to L.

THEOREM 6—The Monotonic Sequence Theorem
If a sequence $\{a_n\}$ is both bounded and monotonic, then the sequence converges.

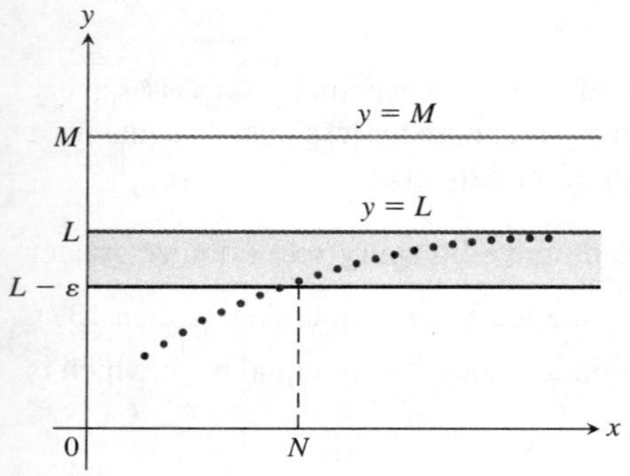

FIGURE 10.7 If the terms of a nondecreasing sequence have an upper bound M, then they have a limit $L \le M$.

Proof Suppose $\{a_n\}$ is nondecreasing, L is its least upper bound, and we plot the points $(1, a_1), (2, a_2), \ldots, (n, a_n), \ldots$ in the xy-plane. If M is an upper bound of the sequence, all these points will lie on or below the line $y = M$ (Figure 10.7). The line $y = L$ is the lowest such line. None of the points (n, a_n) lies above $y = L$, but some do lie above any lower line $y = L - \varepsilon$, if ε is a positive number (because $L - \varepsilon$ is not an upper bound). That is,

a. $a_n \le L$ for *all* values of n, and

b. given any $\varepsilon > 0$, there exists at least one integer N for which $a_N > L - \varepsilon$.

The fact that $\{a_n\}$ is nondecreasing tells us further that

$$a_n \ge a_N > L - \varepsilon \qquad \text{for all } n \ge N.$$

Thus, *all* the numbers a_n beyond the Nth number lie within ε of L. This is precisely the condition for L to be the limit of the sequence $\{a_n\}$.

The proof for nonincreasing sequences bounded from below is similar. ■

It is important to realize that Theorem 6 does not say that convergent sequences are monotonic. The sequence $\{(-1)^{n+1}/n\}$ converges and is bounded, but it is not monotonic since it alternates between positive and negative values as it tends toward zero. What the theorem does say is that a nondecreasing sequence converges when it is bounded from above, but it diverges to infinity otherwise.

EXERCISES 10.1

Finding Terms of a Sequence

Each of Exercises 1–6 gives a formula for the nth term a_n of a sequence $\{a_n\}$. Find the values of a_1, a_2, a_3, and a_4.

1. $a_n = \dfrac{2n-1}{n^2}$

2. $a_n = \dfrac{1}{n!}$

3. $a_n = \dfrac{(-1)^{n+1}}{2n-1}$

4. $a_n = 3 + (-1)^n\left(\dfrac{1}{2}\right)$

5. $a_n = \dfrac{2^n}{2^{n+1}}$

6. $a_n = \dfrac{2^n+1}{2^n}$

Each of Exercises 7–12 gives the first term or two of a sequence along with a recursion formula for the remaining terms. Write out the first ten terms of the sequence.

7. $a_1 = 1, \quad a_{n+1} = a_n + (1/2^n)$

8. $a_1 = 1, \ a_{n+1} = \dfrac{(a_n+1)}{n}$

9. $a_1 = 2, \quad a_{n+1} = (-1)^{n+1} a_n/2$

10. $a_1 = -2, \; a_{n+1} = \dfrac{(n+1)a_n}{n}$

11. $a_1 = a_2 = 1, \quad a_{n+2} = a_{n+1} + a_n$

12. $a_1 = 2, a_2 = -1, a_{n+2} = a_{n+1} \cdot a_n$

Finding a Sequence's Formula

In Exercises 13–30, find a formula for the nth term of the sequence.

13. $1, -1, 1, -1, 1, \dots$	1's with alternating signs
14. $-1, 1, -1, 1, -1, \dots$	1's with alternating signs
15. $-1, 4, -9, 16, -25, \dots$	Squares of the positive integers, with alternating signs
16. $1, -\frac{1}{4}, \frac{1}{9}, -\frac{1}{16}, \frac{1}{25}, \dots$	Reciprocals of squares of the positive integers, with alternating signs
17. $\frac{1}{6}, \frac{3}{8}, \frac{3^2}{10}, \frac{3^3}{12}, \frac{3^4}{14}, \dots$	Powers of 3 divided by multiples of 2
18. $-\frac{3}{2}, -\frac{1}{6}, \frac{1}{12}, \frac{3}{20}, \frac{5}{30}, \dots$	Integers differing by 2 divided by products of consecutive integers
19. $0, 7, 26, 63, 124, \dots$	Cubes of the positive integers diminished by 1
20. $-3, -2, -1, 0, 1, \dots$	Integers, beginning with -3
21. $1, 5, 9, 13, 17, \dots$	Every other odd positive integer
22. $2, 6, 10, 14, 18, \dots$	Every other even positive integer
23. $\frac{5}{1}, \frac{9}{2}, \frac{13}{6}, \frac{17}{24}, \frac{21}{120}, \dots$	Integers differing by 4 divided by factorials
24. $\frac{1}{25}, \frac{8}{125}, \frac{27}{625}, \frac{64}{3125}, \frac{125}{15{,}625}, \dots$	Cubes of positive integers divided by powers of 5
25. $0, 1, 0, 1, 0, \dots$	Alternating 0's and 1's
26. $0, 1, 1, 2, 2, 3, 3, 4, \dots$	Each positive integer repeated

27. $\frac{1}{2}, -\frac{4}{3}, \frac{9}{4}, -\frac{16}{5}, \frac{25}{6}, \dots$

28. $\sqrt{5} - \sqrt{4}, \sqrt{6} - \sqrt{5}, \sqrt{7} - \sqrt{6}, \sqrt{8} - \sqrt{7}, \dots$

29. $\cos\left(\frac{\sqrt{2}}{\sqrt{1+1}}\right), \cos\left(\frac{\sqrt{3}}{\sqrt{1+4}}\right), \cos\left(\frac{\sqrt{4}}{\sqrt{1+9}}\right), \cos\left(\frac{\sqrt{5}}{\sqrt{1+16}}\right), \dots$

30. $\sqrt{\frac{5}{8}}, \sqrt{\frac{7}{11}}, \sqrt{\frac{9}{14}}, \sqrt{\frac{11}{17}}, \dots$

Convergence and Divergence

Which of the sequences $\{a_n\}$ in Exercises 31–100 converge, and which diverge? Find the limit of each convergent sequence.

31. $a_n = 5 + (0.2)^n$

32. $a_n = \dfrac{n + (-1)^n}{n}$

33. $a_n = \dfrac{3n-1}{3n+1}$

34. $a_n = \dfrac{2n-1}{1-2\sqrt{n}}$

35. $a_n = \dfrac{1 - 5n^4}{n^4 + 8n^3}$

36. $a_n = \dfrac{n+3}{n^2 + 5n + 6}$

37. $a_n = \dfrac{n^2 - 1}{n+1}$

38. $a_n = \dfrac{n^3 - 1}{10n + 5n^2}$

39. $a_n = 1 + (-1)^n$

40. $a_n = (-1)^n\left(1 - \dfrac{1}{n}\right)$

41. $a_n = \left(\dfrac{n+1}{2n}\right)\left(1 - \dfrac{1}{n}\right)$

42. $a_n = \left(1 - \dfrac{1}{3^n}\right)\left(4 + \dfrac{1}{3^n}\right)$

43. $a_n = \dfrac{(-1)^{n+1}}{2n - 1}$

44. $a_n = \left(-\dfrac{1}{2}\right)^n$

45. $a_n = \sqrt{\dfrac{3n}{n-1}}$

46. $a_n = \dfrac{1}{(0.9)^n}$

47. $a_n = \cos\left(\pi + \dfrac{1}{n}\right)$

48. $a_n = n\pi \cos(n\pi)$

49. $a_n = \dfrac{\sin n}{n}$

50. $a_n = \dfrac{\cos^2 n}{2^n}$

51. $a_n = \dfrac{n}{2^n}$

52. $a_n = \dfrac{5^n}{2^5}$

53. $a_n = \dfrac{\ln(n+1)}{\sqrt{n}}$

54. $a_n = \dfrac{\ln n}{\ln 2n}$

55. $a_n = 6^{1/n}$

56. $a_n = (0.03)^{1/n}$

57. $a_n = \left(1 + \dfrac{7}{n}\right)^n$

58. $a_n = \left(1 - \dfrac{1}{n}\right)^n$

59. $a_n = \sqrt[n]{10n}$

60. $a_n = \sqrt[n]{n^3}$

61. $a_n = \left(\dfrac{3}{n}\right)^{1/n}$

62. $a_n = (n+4)^{1/(n+4)}$

63. $a_n = n^{1/n} \ln n$

64. $a_n = \ln n - \ln(n+1)$

65. $a_n = \sqrt[n]{2^n n}$

66. $a_n = \sqrt[n]{3^{2n+1}}$

67. $a_n = \dfrac{n!}{n^n}$ (*Hint:* Compare with $1/n$.)

68. $a_n = \dfrac{(-7)^n}{n!}$

69. $a_n = \dfrac{n!}{10^{6n}}$

70. $a_n = \dfrac{8n!}{4^n \cdot 2^n}$

71. $a_n = \left(\dfrac{1}{n}\right)^{1/(\ln n)}$

72. $a_n = \dfrac{(n+1)!}{(n+3)!}$

73. $a_n = \dfrac{(2n+2)!}{(2n-1)!}$

74. $a_n = \dfrac{3e^n + e^{-n}}{e^n + 3e^{-n}}$

75. $a_n = \dfrac{e^{-2n} - 2e^{-3n}}{e^{-2n} - e^{-n}}$

76. $a_n = \left(1 - \dfrac{1}{2}\right) + \left(\dfrac{1}{2} - \dfrac{1}{3}\right) + \left(\dfrac{1}{3} - \dfrac{1}{4}\right) + \cdots + \left(\dfrac{1}{n-2} - \dfrac{1}{n-1}\right) + \left(\dfrac{1}{n-1} - \dfrac{1}{n}\right)$

77. $a_n = (\ln 3 - \ln 2) + (\ln 4 - \ln 3) + (\ln 5 - \ln 4) + \cdots + (\ln(n-1) - \ln(n-2)) + (\ln n - \ln(n-1))$

78. $a_n = \ln\left(1 - \dfrac{1}{n}\right)^n$

79. $a_n = \left(\dfrac{3n+1}{3n-1}\right)^n$

80. $a_n = \left(\dfrac{n}{n+1}\right)^n$

81. $a_n = \left(\dfrac{x^n}{2n+1}\right)^{1/n}, \quad x > 0$

82. $a_n = \left(1 - \dfrac{1}{n^2}\right)^n$

83. $a_n = \dfrac{2^n \cdot 3^n}{5^{-n} \cdot n!}$

84. $a_n = \dfrac{(10/11)^n}{(9/10)^n + (11/12)^n}$

85. $a_n = \tanh n$

86. $a_n = \cosh(\ln n)$

87. $a_n = \dfrac{n^2}{2n-1} \sin\dfrac{1}{n}$

88. $a_n = n\left(1 - \cos\frac{1}{n}\right)$

89. $a_n = \sqrt{n}\sin\frac{1}{\sqrt{n}}$

90. $a_n = \left(2^n + 7^n\right)^{1/n}$

91. $a_n = \tan^{-1} n$

92. $a_n = \frac{1}{\sqrt{n}}\arctan n$

93. $a_n = \left(\frac{1}{5}\right)^n + \frac{1}{\sqrt{3^n}}$

94. $a_n = \sqrt[n]{n^2 + n}$

95. $a_n = \frac{(\ln n)^{100}}{n}$

96. $a_n = \frac{(\ln n)^5}{\sqrt{n}}$

97. $a_n = n - \sqrt{n^2 + n}$

98. $a_n = \frac{1}{\sqrt{n^2+1} + \sqrt{n^2-n}}$

99. $a_n = \frac{1}{n}\int_1^n \frac{1}{x}\,dx$

100. $a_n = \int_1^n \frac{1}{x^p}\,dx, \quad p > 1$

Recursively Defined Sequences

In Exercises 101–108, assume that each sequence converges and find its limit.

101. $a_1 = 2, \quad a_{n+1} = \frac{56}{1 + a_n}$

102. $a_1 = -1, \quad a_{n+1} = \frac{a_n + 6}{a_n + 2}$

103. $a_1 = -4, \quad a_{n+1} = \sqrt{8 + 2a_n}$

104. $a_1 = 0, \quad a_{n+1} = \sqrt{6 + 5a_n}$

105. $a_1 = 5, \quad a_{n+1} = \sqrt{5a_n}$

106. $a_1 = 3, \quad a_{n+1} = 6 - \sqrt{a_n}$

107. $2, 2 + \frac{1}{2}, 2 + \cfrac{1}{2 + \cfrac{1}{2}}, 2 + \cfrac{1}{2 + \cfrac{1}{2 + \cfrac{1}{2}}}, \ldots$

108. $\sqrt{3}, \sqrt{3+\sqrt{3}}, \sqrt{3+\sqrt{3+\sqrt{3}}}, \sqrt{3+\sqrt{3+\sqrt{3+\sqrt{3}}}}, \ldots$

Theory and Examples

109. The first term of a sequence is $x_1 = 1$. Each succeeding term is the sum of all those that come before it:

$$x_{n+1} = x_1 + x_2 + \cdots + x_n.$$

Write out enough early terms of the sequence to deduce a general formula for x_n that holds for $n \geq 2$.

110. A sequence of rational numbers is described as follows:

$$\frac{1}{1}, \frac{3}{2}, \frac{7}{5}, \frac{17}{12}, \ldots, \frac{a}{b}, \frac{a + 2b}{a + b}, \ldots.$$

Here the numerators form one sequence, the denominators form a second sequence, and their ratios form a third sequence. Let x_n and y_n be, respectively, the numerator and the denominator of the nth fraction $r_n = x_n/y_n$.

a. Verify that $x_1^2 - 2y_1^2 = -1$, that $x_2^2 - 2y_2^2 = +1$, and, more generally, that if $a^2 - 2b^2 = -1$ or $+1$, then

$$(a + 2b)^2 - 2(a + b)^2 = +1 \quad \text{or} \quad -1,$$

respectively.

b. The fractions $r_n = x_n/y_n$ approach a limit as n increases. What is that limit? (*Hint:* Use part (a) to show that $r_n^2 - 2 = \pm(1/y_n)^2$ and that y_n is not less than n.)

111. Newton's method The following sequences come from the recursion formula for Newton's method,

$$x_{n+1} = x_n - \frac{f(x_n)}{f'(x_n)}.$$

Do the sequences converge? If so, to what value? In each case, begin by identifying the function f that generates the sequence.

a. $x_0 = 1, \quad x_{n+1} = x_n - \frac{x_n^2 - 2}{2x_n} = \frac{x_n}{2} + \frac{1}{x_n}$

b. $x_0 = 1, \quad x_{n+1} = x_n - \frac{\tan x_n - 1}{\sec^2 x_n}$

c. $x_0 = 1, \quad x_{n+1} = x_n - 1$

112. a. Suppose that $f(x)$ is differentiable for all x in $[0, 1]$ and that $f(0) = 0$. Define sequence $\{a_n\}$ by the rule $a_n = n f(1/n)$. Show that $\lim_{n\to\infty} a_n = f'(0)$. Use the result in part (a) to find the limits of the following sequences $\{a_n\}$.

b. $a_n = n \tan^{-1}\frac{1}{n}$

c. $a_n = n(e^{1/n} - 1)$

d. $a_n = n \ln\left(1 + \frac{2}{n}\right)$

113. Pythagorean triples A triple of positive integers a, b, and c is called a **Pythagorean triple** if $a^2 + b^2 = c^2$. Let a be an odd positive integer and let

$$b = \left\lfloor \frac{a^2}{2} \right\rfloor \quad \text{and} \quad c = \left\lceil \frac{a^2}{2} \right\rceil$$

be, respectively, the integer floor and ceiling for $a^2/2$.

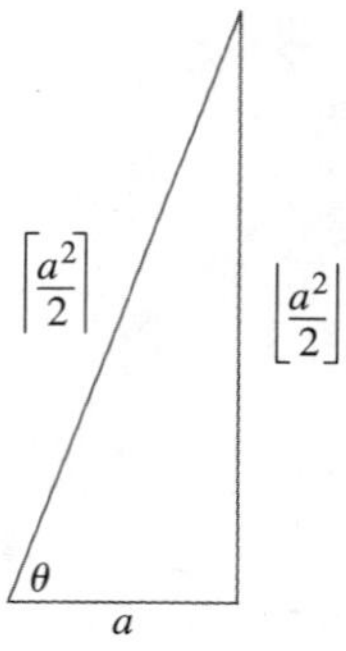

a. Show that $a^2 + b^2 = c^2$. (*Hint:* Let $a = 2n + 1$ and express b and c in terms of n.)

b. By direct calculation, or by appealing to the accompanying figure, find

$$\lim_{a\to\infty} \frac{\left\lfloor \frac{a^2}{2} \right\rfloor}{\left\lceil \frac{a^2}{2} \right\rceil}.$$

114. The *n*th root of *n*!

a. Show that $\lim_{n\to\infty}(2n\pi)^{1/(2n)} = 1$ and hence, using Stirling's approximation (Chapter 8, Additional Exercise 44a), that

$$\sqrt[n]{n!} \approx \frac{n}{e} \quad \text{for large values of } n.$$

T **b.** Test the approximation in part (a) for $n = 40, 50, 60, \ldots$, as far as your calculator will allow.

115. a. Assuming that $\lim_{n\to\infty}(1/n^c) = 0$ if c is any positive constant, show that

$$\lim_{n\to\infty}\frac{\ln n}{n^c} = 0$$

if c is any positive constant.

b. Prove that $\lim_{n\to\infty}(1/n^c) = 0$ if c is any positive constant. (*Hint:* If $\varepsilon = 0.001$ and $c = 0.04$, how large should N be to ensure that $|1/n^c - 0| < \varepsilon$ if $n > N$?)

116. The zipper theorem Prove the "zipper theorem" for sequences: If $\{a_n\}$ and $\{b_n\}$ both converge to L, then the sequence

$$a_1, b_1, a_2, b_2, \ldots, a_n, b_n, \ldots$$

converges to L.

117. Prove that $\lim_{n\to\infty}\sqrt[n]{n} = 1$.

118. Prove that $\lim_{n\to\infty}x^{1/n} = 1, (x > 0)$.

119. Prove Theorem 2.

120. Prove Theorem 3.

In Exercises 121–124, determine whether the sequence is monotonic and whether it is bounded.

121. $a_n = \dfrac{n}{n+1}$

122. $a_n = \dfrac{(2n+3)!}{(n+1)!}$

123. $a_n = \dfrac{2^n 5^n}{n!}$

124. $a_n = 2 - \dfrac{2}{n} - \dfrac{1}{2^n}$

In Exercises 125–134, determine whether the sequence is monotonic, whether it is bounded, and whether it converges.

125. $a_n = 5 - \dfrac{2}{3n}$

126. $a_n = n - \dfrac{1}{n}$

127. $a_n = \dfrac{3^n - 1}{3^n}$

128. $a_n = \dfrac{4^n + 1}{5^n}$

129. $a_n = ((-1)^n + 1)\left(\dfrac{n+1}{n}\right)$

130. The first term of a sequence is $x_1 = \cos(1)$. The next terms are $x_2 = x_1$ or $\cos(2)$, whichever is larger; and $x_3 = x_2$ or $\cos(3)$, whichever is larger (farther to the right). In general,

$$x_{n+1} = \max\{x_n, \cos(n+1)\}.$$

131. $a_n = \dfrac{1+\sqrt{5n}}{\sqrt{n}}$

132. $a_n = \dfrac{n+1}{n}$

133. $a_n = \dfrac{6^n + 7^{n+1}}{7^n}$

134. $a_1 = 1, \quad a_{n+1} = 2a_n - 3$

In Exercises 135–136, use the definition of convergence to prove the given limit.

135. $\lim_{n\to\infty}\dfrac{\sin n}{n} = 0$

136. $\lim_{n\to\infty}\left(1 - \dfrac{1}{n^2}\right) = 1$

137. The sequence $\{n/(n+1)\}$ has a least upper bound of 1 Show that if M is a number less than 1, then the terms of $\{n/(n+1)\}$ eventually exceed M. That is, if $M < 1$, there is an integer N such that $n/(n+1) > M$ whenever $n > N$. Since $n/(n+1) < 1$ for every n, this proves that 1 is a least upper bound for $\{n/(n+1)\}$.

138. Uniqueness of least upper bounds Show that if M_1 and M_2 are least upper bounds for the sequence $\{a_n\}$, then $M_1 = M_2$. That is, a sequence cannot have two different least upper bounds.

139. Is it true that a sequence $\{a_n\}$ of positive numbers must converge if it is bounded from above? Give reasons for your answer.

140. Prove that if $\{a_n\}$ is a convergent sequence, then to every positive number ε there corresponds an integer N such that

$$|a_m - a_n| < \varepsilon \quad \text{whenever} \quad m > N \quad \text{and} \quad n > N.$$

141. Uniqueness of limits Prove that limits of sequences are unique. That is, show that if L_1 and L_2 are numbers such that $a_n \to L_1$ and $a_n \to L_2$, then $L_1 = L_2$.

142. Limits and subsequences If the terms of one sequence appear in another sequence in their given order, we call the first sequence a **subsequence** of the second. Prove that if two subsequences of a sequence $\{a_n\}$ have different limits $L_1 \neq L_2$, then $\{a_n\}$ diverges.

143. For a sequence $\{a_n\}$ the terms of even index are denoted by a_{2k} and the terms of odd index by a_{2k+1}. Prove that if $a_{2k} \to L$ and $a_{2k+1} \to L$, then $a_n \to L$.

144. Prove that a sequence $\{a_n\}$ converges to 0 if and only if the sequence of absolute values $\{|a_n|\}$ converges to 0.

145. Sequences generated by Newton's method Newton's method, applied to a differentiable function $f(x)$, begins with a starting value x_0 and constructs from it a sequence of numbers $\{x_n\}$ that under favorable circumstances converges to a zero of f. The recursion formula for the sequence is

$$x_{n+1} = x_n - \frac{f(x_n)}{f'(x_n)}.$$

a. Show that the recursion formula for $f(x) = x^2 - a, a > 0$, can be written as $x_{n+1} = (x_n + a/x_n)/2$.

T **b.** Starting with $x_0 = 1$ and $a = 3$, calculate successive terms of the sequence until the display begins to repeat. What number is being approximated? Explain.

T **146. A recursive definition of $\pi/2$** If you start with $x_1 = 1$ and if you define the subsequent terms of $\{x_n\}$ by the rule $x_n = x_{n-1} + \cos x_{n-1}$, you generate a sequence that converges rapidly to $\pi/2$. **(a)** Try it. **(b)** Use the accompanying figure to explain why the convergence is so rapid.

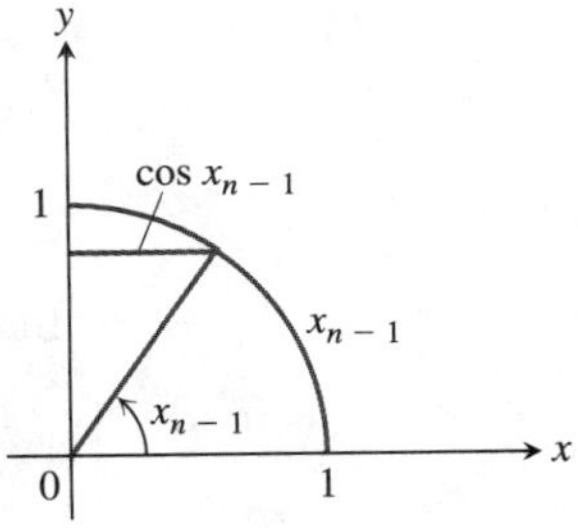

COMPUTER EXPLORATIONS

Use a CAS to perform the following steps for the sequences in Exercises 147–158.

a. Calculate and then plot the first 25 terms of the sequence. Does the sequence appear to be bounded from above or below? Does it appear to converge or diverge? If it does converge, what is the limit L?

b. If the sequence converges, find an integer N such that $|a_n - L| \leq 0.01$ for $n \geq N$. How far in the sequence do you have to get for the terms to lie within 0.0001 of L?

147. $a_n = \sqrt[n]{n}$

148. $a_n = \left(1 + \dfrac{0.5}{n}\right)^n$

149. $a_1 = 1, \quad a_{n+1} = a_n + \dfrac{1}{5^n}$

150. $a_1 = 1, \quad a_{n+1} = a_n + (-2)^n$

151. $a_n = \sin n$

152. $a_n = n \sin \dfrac{1}{n}$

153. $a_n = \dfrac{\sin n}{n}$

154. $a_n = \dfrac{\ln n}{n}$

155. $a_n = (0.9999)^n$

156. $a_n = (123456)^{1/n}$

157. $a_n = \dfrac{8^n}{n!}$

158. $a_n = \dfrac{n^{41}}{19^n}$

10.2 Infinite Series

An *infinite series* is the sum of an infinite sequence of numbers

$$a_1 + a_2 + a_3 + \cdots + a_n + \cdots.$$

The goal of this section is to understand the meaning of such an infinite sum and to develop methods to calculate it. Since there are infinitely many terms to add in an infinite series, we cannot just keep adding to see what comes out. Instead we look at the result of summing just the first n terms of the sequence. The sum of the first n terms

$$s_n = a_1 + a_2 + a_3 + \cdots + a_n$$

is an ordinary finite sum and can be calculated by normal addition. It is called the *nth partial sum*. As n gets larger, we expect the partial sums to get closer and closer to a limiting value in the same sense that the terms of a sequence approach a limit, as discussed in Section 10.1.

For example, to assign meaning to an expression like

$$1 + \frac{1}{2} + \frac{1}{4} + \frac{1}{8} + \frac{1}{16} + \cdots,$$

we add the terms one at a time from the beginning and look for a pattern in how these partial sums grow.

Partial sum		Value	Suggestive expression for partial sum
First:	$s_1 = 1$	1	$2 - 1$
Second:	$s_2 = 1 + \frac{1}{2}$	$\frac{3}{2}$	$2 - \frac{1}{2}$
Third:	$s_3 = 1 + \frac{1}{2} + \frac{1}{4}$	$\frac{7}{4}$	$2 - \frac{1}{4}$
⋮	⋮	⋮	⋮
*n*th:	$s_n = 1 + \frac{1}{2} + \frac{1}{4} + \cdots + \frac{1}{2^{n-1}}$	$\frac{2^n - 1}{2^{n-1}}$	$2 - \frac{1}{2^{n-1}}$

Indeed there is a pattern. The partial sums form a sequence whose nth term is

$$s_n = 2 - \frac{1}{2^{n-1}}.$$

This sequence of partial sums converges to 2 because $\lim_{n\to\infty}(1/2^{n-1}) = 0$. We say

$$\text{“the sum of the infinite series } 1 + \frac{1}{2} + \frac{1}{4} + \cdots + \frac{1}{2^{n-1}} + \cdots \text{ is 2.”}$$

Is the sum of any finite number of terms in this series equal to 2? No. Can we actually add an infinite number of terms one by one? No. But we can still define their sum by defining it to be the limit of the sequence of partial sums as $n \to \infty$, in this case 2 (Figure 10.8). Our knowledge of sequences and limits enables us to break away from the confines of finite sums.

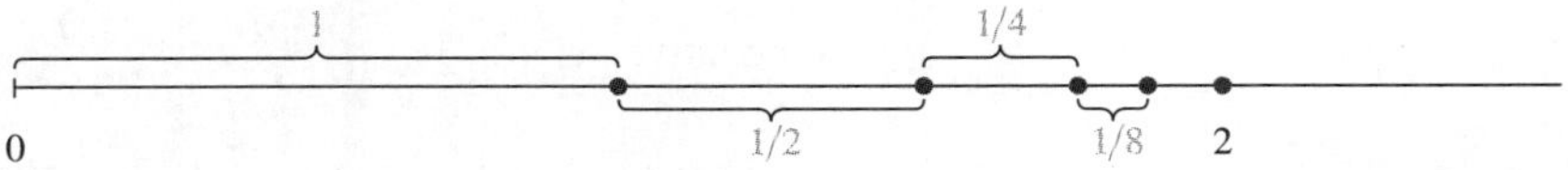

FIGURE 10.8 As the lengths $1, 1/2, 1/4, 1/8, \ldots$ are added one by one, the sum approaches 2.

HISTORICAL BIOGRAPHY
Blaise Pascal
(1623–1662)
www.bit.ly/2ItkU6x

DEFINITIONS Given a sequence of numbers $\{a_n\}$, an expression of the form

$$a_1 + a_2 + a_3 + \cdots + a_n + \cdots$$

is an **infinite series**. The number a_n is the ***n*th term** of the series. The sequence $\{s_n\}$ defined by

$$\begin{aligned} s_1 &= a_1 \\ s_2 &= a_1 + a_2 \\ &\vdots \\ s_n &= a_1 + a_2 + \cdots + a_n = \sum_{k=1}^{n} a_k \\ &\vdots \end{aligned}$$

is the **sequence of partial sums** of the series, the number s_n being the ***n*th partial sum**. If the sequence of partial sums converges to a limit L, we say that the series **converges** and that its **sum** is L. In this case, we also write

$$a_1 + a_2 + \cdots + a_n + \cdots = \sum_{n=1}^{\infty} a_n = L.$$

If the sequence of partial sums of the series does not converge, we say that the series **diverges**.

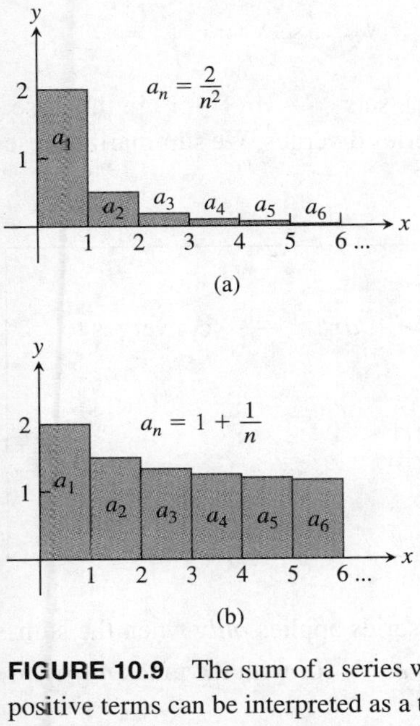

FIGURE 10.9 The sum of a series with positive terms can be interpreted as a total area of an infinite collection of rectangles. The series converges when the total area of the rectangles is finite (a) and diverges when the total area is unbounded (b). Note that the total area can be infinite even if the area of the rectangles is decreasing.

If all the terms a_n in an infinite series are positive, then we can represent each term in the series by the area of a rectangle. The series converges if the total area is finite, and diverges otherwise. Figure 10.9a shows an example where the series converges, and Figure 10.9b shows an example where it diverges. The convergence of the total area is related to the convergence or divergence of improper integrals, as we found in Section 8.8. We make this connection explicit in the next section, where we develop an important test for convergence of series, the Integral Test.

When we begin to study a given series $a_1 + a_2 + \cdots + a_n + \cdots$, we might not know whether it converges or diverges. In either case, it is convenient to use sigma notation to write the series as

$$\sum_{n=1}^{\infty} a_n, \quad \sum_{k=1}^{\infty} a_k, \quad \text{or} \quad \sum a_n$$

A useful shorthand when summation from 1 to ∞ is understood

Geometric Series

Geometric series are series of the form

$$a + ar + ar^2 + \cdots + ar^{n-1} + \cdots = \sum_{n=1}^{\infty} ar^{n-1},$$

in which a and r are fixed real numbers and $a \neq 0$. The series can also be written as $\sum_{n=0}^{\infty} ar^n$. The **ratio** r can be positive, as in

$$1 + \frac{1}{2} + \frac{1}{4} + \cdots + \left(\frac{1}{2}\right)^{n-1} + \cdots, \qquad r = 1/2, a = 1$$

or negative, as in

$$1 - \frac{1}{3} + \frac{1}{9} - \cdots + \left(-\frac{1}{3}\right)^{n-1} + \cdots. \qquad r = -1/3, a = 1$$

If $r = 1$, the nth partial sum of the geometric series is

$$s_n = a + a \cdot 1 + a \cdot 1^2 + \cdots + a \cdot 1^{n-1} = na,$$

and the series diverges because $\lim_{n\to\infty} s_n = \pm\infty$, depending on the sign of a. If $r = -1$, the series diverges because the nth partial sums alternate between a and 0 and never approach a single limit. If $|r| \neq 1$, we can determine the convergence or divergence of the series in the following way:

$$s_n = a + ar + ar^2 + \cdots + ar^{n-1} \qquad \text{Write the } n\text{th partial sum.}$$

$$rs_n = ar + ar^2 + \cdots + ar^{n-1} + ar^n \qquad \text{Multiply } s_n \text{ by } r.$$

$$s_n - rs_n = a - ar^n \qquad \text{Subtract } rs_n \text{ from } s_n\text{. Most of the terms on the right cancel.}$$

$$s_n(1 - r) = a(1 - r^n) \qquad \text{Factor.}$$

$$s_n = \frac{a(1 - r^n)}{1 - r}, \quad (r \neq 1). \qquad \text{We can solve for } s_n \text{ if } r \neq 1.$$

If $|r| < 1$, then $r^n \to 0$ as $n \to \infty$ (as in Section 10.1), so $s_n \to a/(1 - r)$ in this case. On the other hand, if $|r| > 1$, then $|r^n| \to \infty$ and the series diverges. We summarize these results as follows.

Geometric Series

If $|r| < 1$, the geometric series $a + ar + ar^2 + \cdots + ar^{n-1} + \cdots$ converges to $a/(1 - r)$:

$$\sum_{n=1}^{\infty} ar^{n-1} = \frac{a}{1 - r}, \qquad |r| < 1.$$

If $|r| \geq 1$, the series diverges.

The formula $a/(1 - r)$ for the sum of a geometric series applies *only* when the summation index begins with $n = 1$ in the expression $\sum_{n=1}^{\infty} ar^{n-1}$ (or with the index $n = 0$ if we write the series as $\sum_{n=0}^{\infty} ar^n$).

EXAMPLE 1 The geometric series with $a = 1/9$ and $r = 1/3$ is

$$\frac{1}{9} + \frac{1}{27} + \frac{1}{81} + \cdots = \sum_{n=1}^{\infty} \frac{1}{9}\left(\frac{1}{3}\right)^{n-1}.$$

Since $|r| = 1/3 < 1$, the series converges to

$$\frac{1/9}{1 - (1/3)} = \frac{1}{6}. \quad \blacksquare$$

EXAMPLE 2 The series

$$\sum_{n=0}^{\infty} \frac{(-1)^n 5}{4^n} = 5 - \frac{5}{4} + \frac{5}{16} - \frac{5}{64} + \cdots$$

is a geometric series with $a = 5$ and $r = -1/4$. Since $|r| = 1/4 < 1$, the series converges to

$$\frac{a}{1 - r} = \frac{5}{1 + (1/4)} = 4. \quad \blacksquare$$

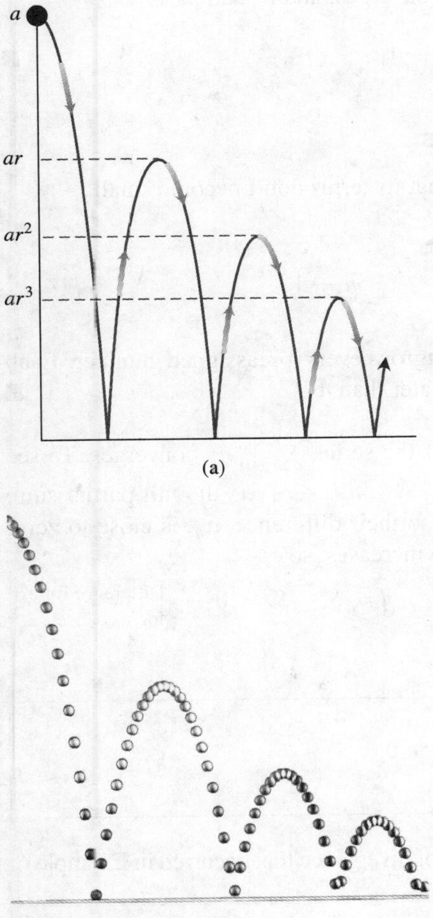

FIGURE 10.10 (a) Example 3 shows how to use a geometric series to calculate the total vertical distance traveled by a bouncing ball if the height of each rebound is reduced by the factor r. (b) A stroboscopic photo of a bouncing ball. (*Source:* Berenice Abbott/Science Source)

EXAMPLE 3 You drop a ball from a meters above a flat surface. Each time the ball hits the surface after falling a distance h, it rebounds a distance rh, where r is positive but less than 1. Find the total distance the ball travels up and down (Figure 10.10).

Solution The total distance is

$$s = a + \underbrace{2ar + 2ar^2 + 2ar^3 + \cdots}_{\text{This sum is } 2ar/(1-r).} = a + \frac{2ar}{1 - r} = a\frac{1 + r}{1 - r}.$$

If $a = 6$ m and $r = 2/3$, for instance, the distance is

$$s = 6 \cdot \frac{1 + (2/3)}{1 - (2/3)} = 6\left(\frac{5/3}{1/3}\right) = 30 \text{ m}. \quad \blacksquare$$

EXAMPLE 4 Express the repeating decimal $5.232323\ldots$ as the ratio of two integers.

Solution From the definition of a decimal number, we get a geometric series.

$$\begin{aligned} 5.232323\ldots &= 5 + \frac{23}{100} + \frac{23}{(100)^2} + \frac{23}{(100)^3} + \cdots \\ &= 5 + \frac{23}{100}\underbrace{\left(1 + \frac{1}{100} + \left(\frac{1}{100}\right)^2 + \cdots\right)}_{1/(1 - 1/100)} \qquad \begin{array}{l} a = 1, \\ r = 1/100 \\ |r| = 1/100 < 1 \end{array} \\ &= 5 + \frac{23}{100}\left(\frac{1}{0.99}\right) = 5 + \frac{23}{99} = \frac{518}{99} \quad \blacksquare \end{aligned}$$

Unfortunately, formulas like the one for the sum of a convergent geometric series are rare, and we usually have to settle for an estimate of a series' sum (more about this later). The next example, however, is another case in which we can find the sum exactly.

EXAMPLE 5 Find the sum of the "telescoping" series $\sum_{n=1}^{\infty} \frac{1}{n(n+1)}$.

Solution We look for a pattern in the sequence of partial sums that might lead to a formula for s_k. The key observation is the partial fraction decomposition

$$\frac{1}{n(n+1)} = \frac{1}{n} - \frac{1}{n+1},$$

so

$$\sum_{n=1}^{k} \frac{1}{n(n+1)} = \sum_{n=1}^{k} \left(\frac{1}{n} - \frac{1}{n+1}\right)$$

and

$$s_k = \left(\frac{1}{1} - \frac{1}{2}\right) + \left(\frac{1}{2} - \frac{1}{3}\right) + \left(\frac{1}{3} - \frac{1}{4}\right) + \cdots + \left(\frac{1}{k} - \frac{1}{k+1}\right).$$

Removing parentheses and canceling adjacent terms of opposite sign collapse the sum to

$$s_k = 1 - \frac{1}{k+1}.$$

We now see that $s_k \to 1$ as $k \to \infty$. The series converges, and its sum is 1:

$$\sum_{n=1}^{\infty} \frac{1}{n(n+1)} = 1.$$

■

The nth-Term Test for a Divergent Series

One reason why a series may fail to converge is that its terms don't become small.

EXAMPLE 6 The series

$$\sum_{n=1}^{\infty} \frac{n+1}{n} = \frac{2}{1} + \frac{3}{2} + \frac{4}{3} + \cdots + \frac{n+1}{n} + \cdots$$

diverges because the partial sums eventually outgrow every preassigned number. Each term is greater than 1, so the sum of n terms is greater than n. ■

We now show that $\lim_{n\to\infty} a_n$ must equal zero if the series $\sum_{n=1}^{\infty} a_n$ converges. To see why, let S represent the series' sum, and $s_n = a_1 + a_2 + \cdots + a_n$ the nth partial sum. When n is large, both s_n and s_{n-1} are close to S, so their difference, a_n, is close to zero. More formally, s_n and s_{n-1} both converge to S as n increases, so

$$\lim_{n\to\infty} a_n = \lim_{n\to\infty} (s_n - s_{n-1}) = \lim_{n\to\infty} s_n - \lim_{n\to\infty} s_{n-1} = S - S = 0.$$ Difference Rule for sequences

This establishes the following theorem.

Caution
Theorem 7 *does not say* that $\sum_{n=1}^{\infty} a_n$ converges if $a_n \to 0$. It is possible for a series to diverge when $a_n \to 0$. (See Example 8.)

THEOREM 7 If $\sum_{n=1}^{\infty} a_n$ converges, then $a_n \to 0$.

Theorem 7 leads to a test for detecting the kind of divergence that occurred in Example 6.

The nth-Term Test for Divergence

$\sum_{n=1}^{\infty} a_n$ diverges if $\lim_{n\to\infty} a_n$ fails to exist or is different from zero.

EXAMPLE 7 The following are all examples of divergent series.

(a) $\sum_{n=1}^{\infty} n^2$ diverges because $n^2 \to \infty$. $\lim_{n\to\infty} a_n$ fails to exist.

(b) $\sum_{n=1}^{\infty} \frac{n+1}{n}$ diverges because $\frac{n+1}{n} \to 1$. $\quad \lim_{n\to\infty} a_n \neq 0$

(c) $\sum_{n=1}^{\infty} (-1)^{n+1}$ diverges because $\lim_{n\to\infty} (-1)^{n+1}$ does not exist.

(d) $\sum_{n=1}^{\infty} \frac{-n}{2n+5}$ diverges because $\lim_{n\to\infty} \frac{-n}{2n+5} = -\frac{1}{2} \neq 0$. ■

EXAMPLE 8 The series

$$1 + \underbrace{\frac{1}{2} + \frac{1}{2}}_{2\text{ terms}} + \underbrace{\frac{1}{4} + \frac{1}{4} + \frac{1}{4} + \frac{1}{4}}_{4\text{ terms}} + \cdots + \underbrace{\frac{1}{2^n} + \frac{1}{2^n} + \cdots + \frac{1}{2^n}}_{2^n\text{ terms}} + \cdots$$

diverges because the terms can be grouped into infinitely many clusters each of which adds to 1, so the partial sums increase without bound. However, the terms of the series form a sequence that converges to 0. Example 1 of Section 10.3 shows that the harmonic series $\sum 1/n$ also behaves in this manner. ■

Combining Series

Whenever we have two convergent series, we can add them term by term, subtract them term by term, or multiply them by constants to make new convergent series.

THEOREM 8 If $\sum a_n = A$ and $\sum b_n = B$ are convergent series, then

1. *Sum Rule:* $\sum (a_n + b_n) = \sum a_n + \sum b_n = A + B$
2. *Difference Rule:* $\sum (a_n - b_n) = \sum a_n - \sum b_n = A - B$
3. *Constant Multiple Rule:* $\sum ka_n = k \sum a_n = kA \quad$ (any number k).

Proof The three rules for series follow from the analogous rules for sequences in Theorem 1, Section 10.1. To prove the Sum Rule for series, let

$$A_n = a_1 + a_2 + \cdots + a_n, \qquad B_n = b_1 + b_2 + \cdots + b_n.$$

Then the partial sums of $\sum (a_n + b_n)$ are

$$\begin{aligned} s_n &= (a_1 + b_1) + (a_2 + b_2) + \cdots + (a_n + b_n) \\ &= (a_1 + \cdots + a_n) + (b_1 + \cdots + b_n) \\ &= A_n + B_n. \end{aligned}$$

Since $A_n \to A$ and $B_n \to B$, we have $s_n \to A + B$ by the Sum Rule for sequences. The proof of the Difference Rule is similar.

To prove the Constant Multiple Rule for series, observe that the partial sums of $\sum ka_n$ form the sequence

$$s_n = ka_1 + ka_2 + \cdots + ka_n = k(a_1 + a_2 + \cdots + a_n) = kA_n,$$

which converges to kA by the Constant Multiple Rule for sequences. ■

As corollaries of Theorem 8, we have the following results. We omit the proofs.

1. Every nonzero constant multiple of a divergent series diverges.
2. If $\sum a_n$ converges and $\sum b_n$ diverges, then $\sum (a_n + b_n)$ and $\sum (a_n - b_n)$ both diverge.

Caution Remember that $\sum(a_n + b_n)$ can converge *even if* both $\sum a_n$ and $\sum b_n$ diverge. For example, both $\sum a_n = 1 + 1 + 1 + \cdots$ and $\sum b_n = (-1) + (-1) + (-1) + \cdots$ diverge, whereas $\sum(a_n + b_n) = 0 + 0 + 0 + \cdots$ converges to 0.

EXAMPLE 9 Find the sums of the following series.

(a)
$$\sum_{n=1}^{\infty} \frac{3^{n-1} - 1}{6^{n-1}} = \sum_{n=1}^{\infty}\left(\frac{1}{2^{n-1}} - \frac{1}{6^{n-1}}\right)$$

$$= \sum_{n=1}^{\infty}\frac{1}{2^{n-1}} - \sum_{n=1}^{\infty}\frac{1}{6^{n-1}} \qquad \text{Difference Rule}$$

$$= \frac{1}{1-(1/2)} - \frac{1}{1-(1/6)} \qquad \text{Geometric series with } a = 1 \text{ and } r = 1/2,\ 1/6$$

Both series converge since $|1/2| < 1$ and $|1/6| < 1$.

$$= 2 - \frac{6}{5} = \frac{4}{5}$$

(b)
$$\sum_{n=0}^{\infty} \frac{4}{2^n} = 4\sum_{n=0}^{\infty}\frac{1}{2^n} \qquad \text{Constant Multiple Rule}$$

$$= 4\left(\frac{1}{1-(1/2)}\right) \qquad \text{Geometric series with } a = 1 \text{ and } r = 1/2$$

Converges since $|r| = 1/2 < 1$.

$$= 8$$

Adding or Deleting Terms

We can add a finite number of terms to a series or delete a finite number of terms without altering the series' convergence or divergence, although in the case of convergence, this will usually change the sum. If $\sum_{n=1}^{\infty} a_n$ converges, then $\sum_{n=k}^{\infty} a_n$ converges for any $k > 1$ and

$$\sum_{n=1}^{\infty} a_n = a_1 + a_2 + \cdots + a_{k-1} + \sum_{n=k}^{\infty} a_n.$$

Conversely, if $\sum_{n=k}^{\infty} a_n$ converges for any $k > 1$, then $\sum_{n=1}^{\infty} a_n$ converges. Thus,

$$\sum_{n=1}^{\infty}\frac{1}{5^n} = \frac{1}{5} + \frac{1}{25} + \frac{1}{125} + \sum_{n=4}^{\infty}\frac{1}{5^n}$$

and

$$\sum_{n=4}^{\infty}\frac{1}{5^n} = \left(\sum_{n=1}^{\infty}\frac{1}{5^n}\right) - \frac{1}{5} - \frac{1}{25} - \frac{1}{125}.$$

The convergence or divergence of a series is not affected by its first few terms. Only the "tail" of the series, the part that remains when we sum beyond some finite number of initial terms, influences whether it converges or diverges.

HISTORICAL BIOGRAPHY
Richard Dedekind
(1831–1916)
www.bit.ly/2Rdog1r

Reindexing

As long as we preserve the order of its terms, we can reindex any series without altering its convergence. To raise the starting value of the index h units, replace the n in the formula for a_n by $n - h$:

$$\sum_{n=1}^{\infty} a_n = \sum_{n=1+h}^{\infty} a_{n-h} = a_1 + a_2 + a_3 + \cdots.$$

To lower the starting value of the index h units, replace the n in the formula for a_n by $n + h$:

$$\sum_{n=1}^{\infty} a_n = \sum_{n=1-h}^{\infty} a_{n+h} = a_1 + a_2 + a_3 + \cdots.$$

We saw this reindexing in starting a geometric series with the index $n = 0$ instead of the index $n = 1$, but we can use any other starting index value as well. We usually give preference to indexings that lead to simple expressions.

EXAMPLE 10 We can write the geometric series

$$\sum_{n=1}^{\infty} \frac{1}{2^{n-1}} = 1 + \frac{1}{2} + \frac{1}{4} + \cdots$$

as

$$\sum_{n=0}^{\infty} \frac{1}{2^n}, \quad \sum_{n=5}^{\infty} \frac{1}{2^{n-5}}, \quad \text{or even} \quad \sum_{n=-4}^{\infty} \frac{1}{2^{n+4}}.$$

The partial sums remain the same no matter what indexing we choose to use. ■

EXERCISES 10.2

Finding *n*th Partial Sums

In Exercises 1–6, find a formula for the nth partial sum of each series and use it to find the series' sum if the series converges.

1. $2 + \frac{2}{3} + \frac{2}{9} + \frac{2}{27} + \cdots + \frac{2}{3^{n-1}} + \cdots$

2. $\frac{11}{100} + \frac{11}{100^2} + \frac{11}{100^3} + \cdots + \frac{11}{100^n} + \cdots$

3. $1 - \frac{1}{2} + \frac{1}{4} - \frac{1}{8} + \cdots + (-1)^{n-1}\frac{1}{2^{n-1}} + \cdots$

4. $1 - 3 + 9 - 27 + \cdots + (-1)^{n-1}3^{n-1} + \cdots$

5. $\frac{1}{2 \cdot 3} + \frac{1}{3 \cdot 4} + \frac{1}{4 \cdot 5} + \cdots + \frac{1}{(n+1)(n+2)} + \cdots$

6. $\frac{7}{1 \cdot 2} + \frac{7}{2 \cdot 3} + \frac{7}{3 \cdot 4} + \cdots + \frac{7}{n(n+1)} + \cdots$

Series with Geometric Terms

In Exercises 7–14, write out the first eight terms of each series to show how the series starts. Then find the sum of the series or show that it diverges.

7. $\sum_{n=0}^{\infty} \frac{(-1)^n}{4^n}$

8. $\sum_{n=2}^{\infty} \frac{1}{2^n}$

9. $\sum_{n=1}^{\infty} \left(1 - \frac{7}{4^n}\right)$

10. $\sum_{n=0}^{\infty} (-1)^n \frac{7}{4^n}$

11. $\sum_{n=0}^{\infty} \left(\frac{5}{2^n} + \frac{1}{3^n}\right)$

12. $\sum_{n=0}^{\infty} \left(\frac{5}{2^n} - \frac{1}{3^n}\right)$

13. $\sum_{n=0}^{\infty} \left(\frac{1}{2^n} + \frac{(-1)^n}{5^n}\right)$

14. $\sum_{n=0}^{\infty} \left(\frac{5^{n+1}}{2^n}\right)$

In Exercises 15–22, determine whether the geometric series converges or diverges. If a series converges, find its sum.

15. $1 + \left(\frac{2}{5}\right) + \left(\frac{2}{5}\right)^2 + \left(\frac{2}{5}\right)^3 + \left(\frac{2}{5}\right)^4 + \cdots$

16. $1 + (-7) + (-7)^2 + (-7)^3 + (-7)^4 + \cdots$

17. $\left(\frac{1}{8}\right) + \left(\frac{1}{8}\right)^2 + \left(\frac{1}{8}\right)^3 + \left(\frac{1}{8}\right)^4 + \left(\frac{1}{8}\right)^5 + \cdots$

18. $\left(\frac{-4}{5}\right)^2 + \left(\frac{-4}{5}\right)^3 + \left(\frac{-4}{5}\right)^4 + \left(\frac{-4}{5}\right)^5 + \left(\frac{-4}{5}\right)^6 + \cdots$

19. $1 - \left(\frac{2}{e}\right) + \left(\frac{2}{e}\right)^2 - \left(\frac{2}{e}\right)^3 + \left(\frac{2}{e}\right)^4 - \cdots$

20. $\left(\frac{1}{3}\right)^{-2} - \left(\frac{1}{3}\right)^{-1} + 1 - \left(\frac{1}{3}\right) + \left(\frac{1}{3}\right)^2 - \cdots$

21. $1 + \left(\frac{7}{6}\right)^2 + \left(\frac{7}{6}\right)^4 + \left(\frac{7}{6}\right)^6 + \left(\frac{7}{6}\right)^8 + \cdots$

22. $\frac{9}{4} - \frac{27}{8} + \frac{81}{16} - \frac{243}{32} + \frac{729}{64} - \cdots$

Repeating Decimals

Express each of the numbers in Exercises 23–30 as the ratio of two integers.

23. $0.\overline{37}$

24. $0.\overline{234} = 0.234\ 234\ 234\ldots$

25. $0.\overline{5}$

26. $0.\overline{d} = 0.dddd\ldots$, where d is a digit

27. $0.0\overline{6} = 0.06666\ldots$

28. $1.\overline{732}$

29. $1.24\overline{123} = 1.24\ 123\ 123\ 123\ldots$

30. $2.\overline{134685}$

Using the *n*th-Term Test

In Exercises 31–38, use the nth-Term Test for divergence to show that the series is divergent, or state that the test is inconclusive.

31. $\sum_{n=1}^{\infty} \frac{2n}{n+4}$

32. $\sum_{n=1}^{\infty} \frac{n(n+1)}{(n+2)(n+3)}$

33. $\sum_{n=0}^{\infty} \frac{1}{n+5}$

34. $\sum_{n=1}^{\infty} \frac{n}{n^2+3}$

35. $\sum_{n=1}^{\infty} \sin\frac{1}{n}$ **36.** $\sum_{n=0}^{\infty} \frac{e^n}{e^n + n}$

37. $\sum_{n=1}^{\infty} \ln\frac{1}{n}$ **38.** $\sum_{n=0}^{\infty} \sin n\pi$

Telescoping Series

In Exercises 39–44, find a formula for the nth partial sum of the series and use it to determine whether the series converges or diverges. If a series converges, find its sum.

39. $\sum_{n=1}^{\infty} \left(\frac{1}{n} - \frac{1}{n+1}\right)$ **40.** $\sum_{n=1}^{\infty} \left(\frac{6}{n^2} - \frac{6}{(n+1)^2}\right)$

41. $\sum_{n=1}^{\infty} (\ln\sqrt{n+1} - \ln\sqrt{n})$ **42.** $\sum_{n=1}^{\infty} (\tan(n) - \tan(n-1))$

43. $\sum_{n=1}^{\infty} \left(\sin^{-1}\left(\frac{1}{n+1}\right) - \sin^{-1}\left(\frac{1}{n+2}\right)\right)$

44. $\sum_{n=1}^{\infty} (\sqrt{n+4} - \sqrt{n+3})$

Find the sum of each series in Exercises 45–52.

45. $\sum_{n=1}^{\infty} \frac{5}{(5n-3)(5n+2)}$ **46.** $\sum_{n=1}^{\infty} \frac{6}{(2n-1)(2n+1)}$

47. $\sum_{n=1}^{\infty} \frac{40n}{(2n-1)^2(2n+1)^2}$ **48.** $\sum_{n=1}^{\infty} \frac{2n+3}{(n+1)^2(n+2)^2}$

49. $\sum_{n=1}^{\infty} \left(\frac{1}{\sqrt{n+1}} - \frac{1}{\sqrt{n}}\right)$ **50.** $\sum_{n=1}^{\infty} \left(\frac{1}{2^{1/n}} - \frac{1}{2^{1/(n+1)}}\right)$

51. $\sum_{n=1}^{\infty} \left(\frac{1}{\ln(n+2)} - \frac{1}{\ln(n+1)}\right)$

52. $\sum_{n=1}^{\infty} \left(\cot^{-1}\left(\frac{1}{n}\right) - \cot^{-1}\left(\frac{1}{n+1}\right)\right)$

Convergence or Divergence

Which series in Exercises 53–76 converge, and which diverge? Give reasons for your answers. If a series converges, find its sum.

53. $\sum_{n=0}^{\infty} \left(\frac{1}{\sqrt{2}}\right)^n$ **54.** $\sum_{n=0}^{\infty} (\sqrt{3})^n$

55. $\sum_{n=1}^{\infty} (-1)^{n+1}\frac{3}{2^n}$ **56.** $\sum_{n=1}^{\infty} (-1)^{n+1} n$

57. $\sum_{n=0}^{\infty} \cos(n\pi)$ **58.** $\sum_{n=0}^{\infty} \frac{\cos n\pi}{5^n}$

59. $\sum_{n=0}^{\infty} e^{-2n}$ **60.** $\sum_{n=1}^{\infty} \ln\frac{1}{3^n}$

61. $\sum_{n=1}^{\infty} \frac{3}{7^n}$ **62.** $\sum_{n=0}^{\infty} \frac{1}{x^n}, \quad |x| > 1$

63. $\sum_{n=0}^{\infty} \frac{2^n - 1}{3^n}$ **64.** $\sum_{n=1}^{\infty} \left(1 - \frac{1}{n}\right)^n$

65. $\sum_{n=0}^{\infty} \frac{n!}{1000^n}$ **66.** $\sum_{n=1}^{\infty} \frac{n^n}{n!}$

67. $\sum_{n=1}^{\infty} \frac{2^n + 3^n}{4^n}$ **68.** $\sum_{n=1}^{\infty} \frac{2^n + 4^n}{3^n + 4^n}$

69. $\sum_{n=1}^{\infty} \ln\left(\frac{n}{n+1}\right)$ **70.** $\sum_{n=1}^{\infty} \ln\left(\frac{n}{2n+1}\right)$

71. $\sum_{n=0}^{\infty} \left(\frac{\pi}{e}\right)^n$ **72.** $\sum_{n=0}^{\infty} \frac{e^{n\pi}}{\pi^{ne}}$

73. $\sum_{n=1}^{\infty} \left(\frac{n}{n+1} - \frac{n+2}{n+3}\right)$

74. $\sum_{n=2}^{\infty} \left(\sin\left(\frac{\pi}{n}\right) - \sin\left(\frac{\pi}{n-1}\right)\right)$

75. $\sum_{n=1}^{\infty} \left(\cos\left(\frac{\pi}{n}\right) + \sin\left(\frac{\pi}{n}\right)\right)$

76. $\sum_{n=0}^{\infty} (\ln(4e^n - 1) - \ln(2e^n + 1))$

Geometric Series with a Variable x

In each of the geometric series in Exercises 77–80, write out the first few terms of the series to find a and r, and find the sum of the series. Then express the inequality $|r| < 1$ in terms of x and find the values of x for which the inequality holds and the series converges.

77. $\sum_{n=0}^{\infty} (-1)^n x^n$ **78.** $\sum_{n=0}^{\infty} (-1)^n x^{3n}$

79. $\sum_{n=0}^{\infty} 3\left(\frac{x-1}{2}\right)^n$ **80.** $\sum_{n=0}^{\infty} \frac{(-1)^n}{2}\left(\frac{1}{3 + \sin x}\right)^n$

In Exercises 81–86, find the values of x for which the given geometric series converges. Also, find the sum of the series (as a function of x) for those values of x.

81. $\sum_{n=0}^{\infty} 2^n x^n$ **82.** $\sum_{n=0}^{\infty} (-1)^n x^{-3n}$

83. $\sum_{n=0}^{\infty} (-1)^n (x+1)^n$ **84.** $\sum_{n=0}^{\infty} \left(-\frac{1}{2}\right)^n (x-3)^n$

85. $\sum_{n=0}^{\infty} \cos^n x$ **86.** $\sum_{n=0}^{\infty} (\ln x)^n$

Theory and Examples

87. The series in Exercise 5 can also be written as

$$\sum_{n=1}^{\infty} \frac{1}{(n+1)(n+2)} \quad \text{and} \quad \sum_{n=-1}^{\infty} \frac{1}{(n+3)(n+4)}.$$

Write this series as a sum beginning with **(a)** $n = -2$, **(b)** $n = 0$, **(c)** $n = 5$.

88. The series in Exercise 6 can also be written as

$$\sum_{n=1}^{\infty} \frac{5}{n(n+1)} \quad \text{and} \quad \sum_{n=0}^{\infty} \frac{5}{(n+1)(n+2)}.$$

Write this series as a sum beginning with **(a)** $n = -1$, **(b)** $n = 3$, **(c)** $n = 20$.

89. Make up an infinite series of nonzero terms whose sum is

a. 1 **b.** −3 **c.** 0.

90. (*Continuation of Exercise 89.*) Can you make an infinite series of nonzero terms that converges to any number you want? Explain.

91. Show by example that $\sum (a_n/b_n)$ may diverge even though $\sum a_n$ and $\sum b_n$ converge and no b_n equals 0.

92. Find convergent geometric series $A = \sum a_n$ and $B = \sum b_n$ that illustrate the fact that $\sum a_n b_n$ may converge without being equal to AB.

93. Show by example that $\sum (a_n/b_n)$ may converge to something other than A/B even when $A = \sum a_n$, $B = \sum b_n \neq 0$, and no b_n equals 0.

94. If $\sum a_n$ converges and $a_n > 0$ for all n, can anything be said about $\sum (1/a_n)$? Give reasons for your answer.

95. What happens if you add a finite number of terms to a divergent series or delete a finite number of terms from a divergent series? Give reasons for your answer.

96. If $\sum a_n$ converges and $\sum b_n$ diverges, can anything be said about their term-by-term sum $\sum (a_n + b_n)$? Give reasons for your answer.

97. Make up a geometric series $\sum ar^{n-1}$ that converges to the number 7 if

a. $a = 3$ **b.** $a = 15/2$.

98. Find the value of b for which

$$1 + e^b + e^{2b} + e^{3b} + \cdots = 9.$$

99. For what values of r does the infinite series

$$1 + 2r + r^2 + 2r^3 + r^4 + 2r^5 + r^6 + \cdots$$

converge? Find the sum of the series when it converges.

100. The accompanying figure shows the first five of a sequence of squares. The outermost square has an area of 4 m^2. Each of the other squares is obtained by joining the midpoints of the sides of the squares before it. Find the sum of the areas of all the squares.

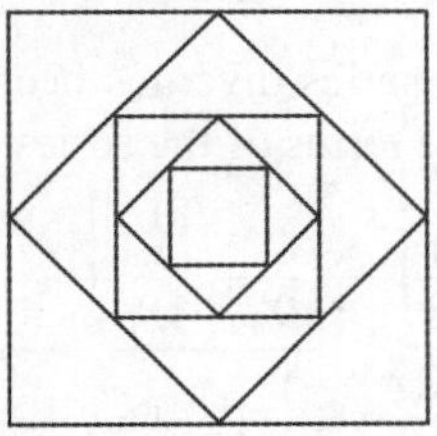

101. Drug dosage A patient takes a 300 mg tablet for the control of high blood pressure every morning at the same time. The concentration of the drug in the patient's system decays exponentially at a constant hourly rate of $k = 0.12$.

a. How many milligrams of the drug are in the patient's system just before the second tablet is taken? Just before the third tablet is taken?

b. After the patient has taken the medication for at least six months, what quantity of drug is in the patient's body just before the next regularly scheduled morning tablet is taken?

102. Show that the error $(L - s_n)$ obtained by replacing a convergent geometric series with one of its partial sums s_n is $ar^n/(1 - r)$.

103. The Cantor set To construct this set, we begin with the closed interval $[0, 1]$. From that interval, we remove the middle open interval $(1/3, 2/3)$, leaving the two closed intervals $[0, 1/3]$ and $[2/3, 1]$. At the second step we remove the open middle third interval from each of those remaining. From $[0, 1/3]$ we remove the open interval $(1/9, 2/9)$, and from $[2/3, 1]$ we remove $(7/9, 8/9)$, leaving behind the four closed intervals $[0, 1/9]$, $[2/9, 1/3]$, $[2/3, 7/9]$, and $[8/9, 1]$. At the next step, we remove the open middle third interval from each closed interval left behind, so $(1/27, 2/27)$ is removed from $[0, 1/9]$, leaving the closed intervals $[0, 1/27]$ and $[2/27, 1/9]$; $(7/27, 8/27)$ is removed from $[2/9, 1/3]$, leaving behind $[2/9, 7/27]$ and $[8/27, 1/3]$, and so forth. We continue this process repeatedly without stopping, at each step removing the open third interval from every closed interval remaining behind from the preceding step. The numbers remaining in the interval $[0, 1]$, after all open middle third intervals have been removed, are the points in the Cantor set (named after Georg Cantor, 1845–1918). The set has some interesting properties.

a. The Cantor set contains infinitely many numbers in $[0, 1]$. List 12 numbers that belong to the Cantor set.

b. Show, by summing an appropriate geometric series, that the total length of all the open middle third intervals that have been removed from $[0, 1]$ is equal to 1.

104. Helge von Koch's snowflake curve Helge von Koch's snowflake is a curve of infinite length that encloses a region of finite area. To see why this is so, suppose the curve is generated by starting with an equilateral triangle whose sides have length 1.

a. Find the length L_n of the nth curve C_n and show that $\lim_{n\to\infty} L_n = \infty$.

b. Find the area A_n of the region enclosed by C_n and show that $\lim_{n\to\infty} A_n = (8/5)A_1$.

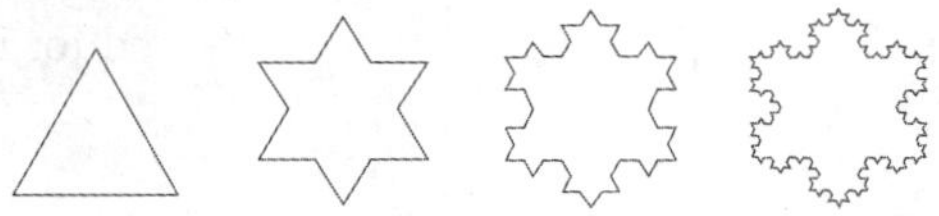

C_1 C_2 C_3 C_4

105. The largest circle in the accompanying figure has radius 1. Consider the sequence of circles of maximum area inscribed in semicircles of diminishing size. What is the sum of the areas of all of the circles?

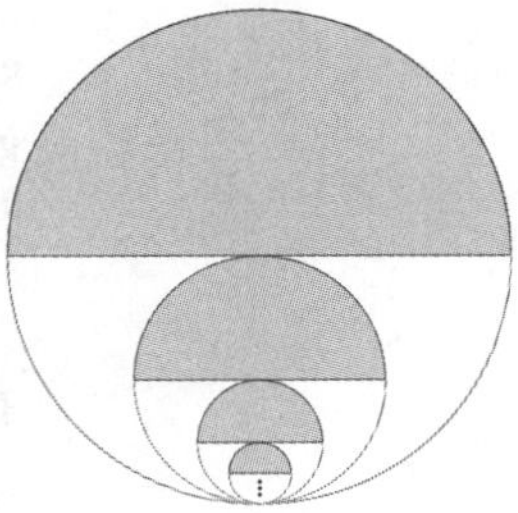

10.3 The Integral Test

The most basic question we can ask about a series is whether it converges. In this section we begin to study this question, starting with series that have nonnegative terms. Such a series converges if its sequence of partial sums is bounded. If we establish that a given series does converge, we generally do not have a formula available for its sum. So to get an estimate for the sum of a convergent series, we investigate the error involved when using a partial sum to approximate the total sum.

Nondecreasing Partial Sums

Suppose that $\sum_{n=1}^{\infty} a_n$ is an infinite series with $a_n \geq 0$ for all n. Then each partial sum is greater than or equal to its predecessor because $s_{n+1} = s_n + a_n$, so

$$s_1 \leq s_2 \leq s_3 \leq \cdots \leq s_n \leq s_{n+1} \leq \cdots.$$

Since the partial sums form a nondecreasing sequence, the Monotonic Sequence Theorem (Theorem 6, Section 10.1) gives the following result.

Corollary of Theorem 6
A series $\sum_{n=1}^{\infty} a_n$ of nonnegative terms converges if and only if its partial sums are bounded from above.

EXAMPLE 1 As an application of the above corollary, consider the **harmonic series**

$$\sum_{n=1}^{\infty} \frac{1}{n} = 1 + \frac{1}{2} + \frac{1}{3} + \cdots + \frac{1}{n} + \cdots.$$

Although the nth term $1/n$ does go to zero, the series diverges because there is no upper bound for its partial sums. To see why, group the terms of the series in the following way:

$$1 + \frac{1}{2} + \underbrace{\left(\frac{1}{3} + \frac{1}{4}\right)}_{>\frac{2}{4}=\frac{1}{2}} + \underbrace{\left(\frac{1}{5} + \frac{1}{6} + \frac{1}{7} + \frac{1}{8}\right)}_{>\frac{4}{8}=\frac{1}{2}} + \underbrace{\left(\frac{1}{9} + \frac{1}{10} + \cdots + \frac{1}{16}\right)}_{>\frac{8}{16}=\frac{1}{2}} + \cdots.$$

The sum of the first two terms is 1.5. The sum of the next two terms is $1/3 + 1/4$, which is greater than $1/4 + 1/4 = 1/2$. The sum of the next four terms is $1/5 + 1/6 + 1/7 + 1/8$, which is greater than $1/8 + 1/8 + 1/8 + 1/8 = 1/2$. The sum of the next eight terms is $1/9 + 1/10 + 1/11 + 1/12 + 1/13 + 1/14 + 1/15 + 1/16$, which is greater than $8/16 = 1/2$. The sum of the next 16 terms is greater than $16/32 = 1/2$, and so on. In general, the sum of 2^m terms ending with $1/2^{m+1}$ is greater than $2^m/2^{m+1} = 1/2$. Therefore, if $n = 2^k$, then the partial sum s_n is greater than $k/2$, so the sequence of partial sums is not bounded from above. The harmonic series diverges. ■

The Integral Test

We introduce the Integral Test with a series that is related to the harmonic series, but whose nth term is $1/n^2$ instead of $1/n$.

EXAMPLE 2 Does the following series converge?

$$\sum_{n=1}^{\infty} \frac{1}{n^2} = 1 + \frac{1}{4} + \frac{1}{9} + \frac{1}{16} + \cdots + \frac{1}{n^2} + \cdots$$

Caution

The series and integral need not have the same value in the convergent case. You will see in Example 6 that

$\sum_{n=1}^{\infty}(1/n^2) \neq \int_1^{\infty}(1/x^2)\,dx = 1.$

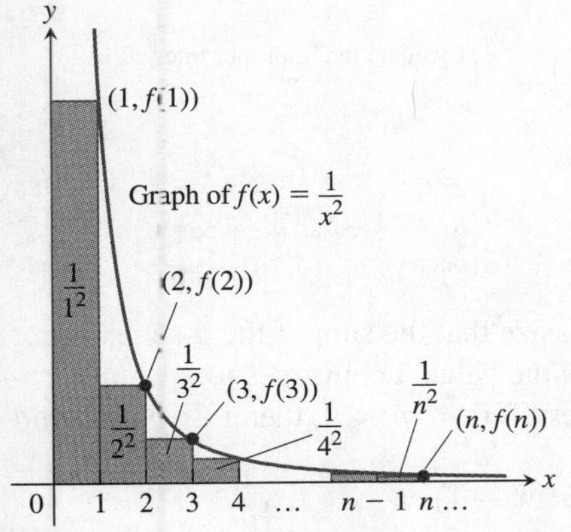

FIGURE 10.11 The sum of the areas of the rectangles under the graph of $f(x) = 1/x^2$ is less than the area under the graph (Example 2).

Solution We determine the convergence of $\sum_{n=1}^{\infty}(1/n^2)$ by comparing it with $\int_1^{\infty}(1/x^2)\,dx$. To carry out the comparison, we think of the terms of the series as values of the function $f(x) = 1/x^2$ and interpret these values as the areas of rectangles under the curve $y = 1/x^2$.

As Figure 10.11 shows,

$$
\begin{aligned}
s_n &= \frac{1}{1^2} + \frac{1}{2^2} + \frac{1}{3^2} + \cdots + \frac{1}{n^2} \\
&= f(1) + f(2) + f(3) + \cdots + f(n) \\
&< f(1) + \int_1^n \frac{1}{x^2}\,dx && \text{Rectangle areas sum to less than area under graph.} \\
&< 1 + \int_1^{\infty} \frac{1}{x^2}\,dx && \int_1^n (1/x^2)\,dx < \int_1^{\infty}(1/x^2)\,dx \\
&= 1 + 1 = 2. && \text{As in Section 8.8, Example 3, } \int_1^{\infty}(1/x^2)\,dx = 1.
\end{aligned}
$$

Thus the partial sums of $\sum_{n=1}^{\infty}(1/n^2)$ are bounded from above (by 2), and the series converges. ■

THEOREM 9—The Integral Test
Let $\{a_n\}$ be a sequence of positive terms. Suppose that $a_n = f(n)$, where f is a continuous, positive, decreasing function of x for all $x \geq N$ (N a positive integer). Then the series $\sum_{n=N}^{\infty} a_n$ and the integral $\int_N^{\infty} f(x)\,dx$ both converge or both diverge.

Proof We establish the test for the case $N = 1$. The proof for general N is similar.

We start with the assumption that f is a decreasing function with $f(n) = a_n$ for every n. This leads us to observe that the rectangles in Figure 10.12a, which have areas $a_1, a_2, \ldots, a_n$, collectively enclose more area than that under the curve $y = f(x)$ from $x = 1$ to $x = n + 1$. That is,

$$\int_1^{n+1} f(x)\,dx \leq a_1 + a_2 + \cdots + a_n.$$

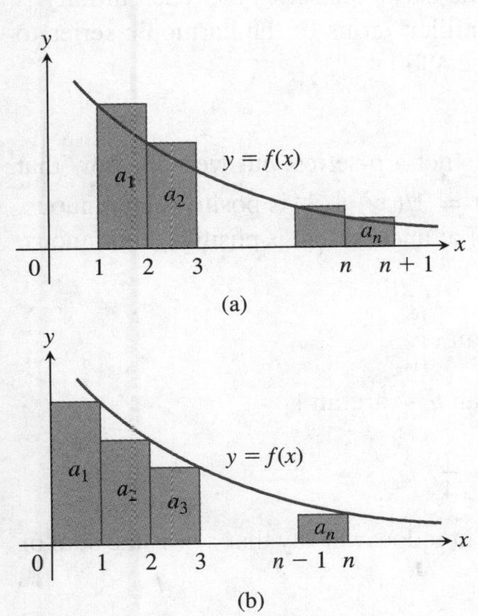

FIGURE 10.12 Subject to the conditions of the Integral Test, the series $\sum_{n=1}^{\infty} a_n$ and the integral $\int_1^{\infty} f(x)\,dx$ both converge or both diverge.

In Figure 10.12b the rectangles have been faced to the left instead of to the right. If we momentarily disregard the first rectangle of area a_1, we see that

$$a_2 + a_3 + \cdots + a_n \leq \int_1^n f(x)\,dx.$$

If we include a_1, we have

$$a_1 + a_2 + \cdots + a_n \leq a_1 + \int_1^n f(x)\,dx.$$

Combining these results gives

$$\int_1^{n+1} f(x)\,dx \leq a_1 + a_2 + \cdots + a_n \leq a_1 + \int_1^n f(x)\,dx.$$

These inequalities hold for each n, and continue to hold as $n \to \infty$.

If $\int_1^{\infty} f(x)\,dx$ is finite, the right-hand inequality shows that $\sum a_n$ is finite. If $\int_1^{\infty} f(x)\,dx$ is infinite, the left-hand inequality shows that $\sum a_n$ is infinite. Hence the series and the integral are either both finite or both infinite. ■

The p-series $\sum_{n=1}^{\infty} \frac{1}{n^p}$ converges if $p > 1$, diverges if $p \leq 1$.

EXAMPLE 3 Show that the ***p*-series**

$$\sum_{n=1}^{\infty} \frac{1}{n^p} = \frac{1}{1^p} + \frac{1}{2^p} + \frac{1}{3^p} + \cdots + \frac{1}{n^p} + \cdots$$

(p a real constant) converges if $p > 1$ and diverges if $p \leq 1$.

Solution If $p > 1$, then $f(x) = 1/x^p$ is a positive, continuous, and decreasing function of x. Since

$$\int_1^{\infty} \frac{1}{x^p}\,dx = \int_1^{\infty} x^{-p}\,dx = \lim_{b\to\infty}\left[\frac{x^{-p+1}}{-p+1}\right]_1^b \qquad \text{Evaluate the improper integral}$$

$$= \frac{1}{1-p}\lim_{b\to\infty}\left(\frac{1}{b^{p-1}} - 1\right)$$

$$= \frac{1}{1-p}(0-1) = \frac{1}{p-1}, \qquad b^{p-1} \to \infty \text{ as } b \to \infty \text{ because } p - 1 > 0.$$

the series converges by the Integral Test. We emphasize that the sum of the p-series is *not* $1/(p-1)$. The series converges, but we don't know the value it converges to.

If $p \leq 0$, the series diverges by the nth-term test. If $0 < p < 1$, then $1 - p > 0$ and

$$\int_1^{\infty} \frac{1}{x^p}\,dx = \frac{1}{1-p}\lim_{b\to\infty}(b^{1-p} - 1) = \infty.$$

Therefore, the series diverges by the Integral Test.

If $p = 1$, we have the (divergent) harmonic series

$$1 + \frac{1}{2} + \frac{1}{3} + \cdots + \frac{1}{n} + \cdots.$$

In summary, we have convergence for $p > 1$ but divergence for all other values of p. ■

The p-series with $p = 1$ is the **harmonic series** (Example 1). The p-Series Test shows that the harmonic series is just *barely* divergent; if we increase p to 1.000000001, for instance, the series converges!

The slowness with which the partial sums of the harmonic series approach infinity is impressive. For instance, it takes more than 178 million terms of the harmonic series to move the partial sums beyond 20. (See also Exercise 49b.)

EXAMPLE 4 The series $\sum_{n=1}^{\infty}(1/(n^2+1))$ is not a p-series, but we will show that it converges by the Integral Test. The function $f(x) = 1/(x^2+1)$ is positive, continuous, and decreasing for $x \geq 1$ since it is the reciprocal of a function that is positive, continuous, and increasing on this interval, and

$$\int_1^{\infty} \frac{1}{x^2+1}\,dx = \lim_{b\to\infty}\Big[\arctan x\Big]_1^b$$

$$= \lim_{b\to\infty}[\arctan b - \arctan 1]$$

$$= \frac{\pi}{2} - \frac{\pi}{4} = \frac{\pi}{4}.$$

The Integral Test tells us that the series converges, but it does *not* say that $\pi/4$ is the sum of the series. ■

EXAMPLE 5 Determine the convergence or divergence of the series.

(a) $\sum_{n=1}^{\infty} ne^{-n^2}$ **(b)** $\sum_{n=1}^{\infty} \frac{1}{2^{\ln n}}$

Solutions

(a) The function $f(x) = xe^{-x^2}$ is continuous and positive for $x \geq 1$. The first derivative

$$f'(x) = e^{-x^2} - 2x^2e^{-x^2} = (1 - 2x^2)e^{-x^2}$$

is negative on that interval (since $x \geq 1$ implies $1 - 2x^2 \leq 1 - 2 < 0$), so f is decreasing and we can apply the Integral Test:

$$\int_1^{\infty} \frac{x}{e^{x^2}}\,dx = \frac{1}{2}\int_1^{\infty} \frac{du}{e^u} \qquad u = x^2, du = 2x\,dx$$

$$= \lim_{b\to\infty}\left[-\frac{1}{2}e^{-u}\right]_1^b$$

$$= \lim_{b\to\infty}\left(-\frac{1}{2e^b} + \frac{1}{2e}\right) = \frac{1}{2e}.$$

Since the integral converges, the series also converges.

(b) The integrand $1/(2^{\ln x})$ is a positive, continuous, and decreasing function on $(1, \infty)$, so we can apply the Integral Test.

$$\int_1^{\infty} \frac{dx}{2^{\ln x}} = \int_0^{\infty} \frac{e^u du}{2^u} \qquad u = \ln x, x = e^u, dx = e^u\,du$$

$$= \int_0^{\infty} \left(\frac{e}{2}\right)^u du$$

$$= \lim_{b\to\infty} \frac{1}{\ln\left(\frac{e}{2}\right)}\left(\left(\frac{e}{2}\right)^b - 1\right) = \infty \qquad (e/2) > 1$$

The improper integral diverges, so the series diverges also. ■

Error Estimation

For some convergent series, such as the geometric series or the telescoping series in Example 5 of Section 10.2, we can actually find the total sum of the series. That is, we can find the limiting value S of the sequence of partial sums. For most convergent series, however, we cannot easily find the total sum. Nevertheless, we can *estimate* the sum by adding the first n terms to get s_n, but we need to know how far off s_n is from the total sum S. An approximation to a function or to a number is more useful when it is accompanied by a bound on the size of the worst possible error that could occur. With such an error bound we can try to make an estimate or approximation that is close enough for the problem at hand. Without a bound on the error size, we are just guessing and hoping that we are close to the actual answer. We now show a way to bound the error size using integrals.

Suppose that a series $\sum a_n$ with positive terms is shown to be convergent by the Integral Test, and we want to estimate the size of the **remainder** R_n measuring the difference between the total sum S of the series and its nth partial sum s_n. That is, we wish to estimate

$$R_n = S - s_n = a_{n+1} + a_{n+2} + a_{n+3} + \cdots.$$

To get a lower bound for the remainder, we compare the sum of the areas of the rectangles with the area under the curve $y = f(x)$ for $x \geq n$ (see Figure 10.13a). We see that

$$R_n = a_{n+1} + a_{n+2} + a_{n+3} + \cdots \geq \int_{n+1}^{\infty} f(x)\,dx.$$

Similarly, from Figure 10.13b, we find an upper bound with

$$R_n = a_{n+1} + a_{n+2} + a_{n+3} + \cdots \leq \int_{n}^{\infty} f(x)\,dx.$$

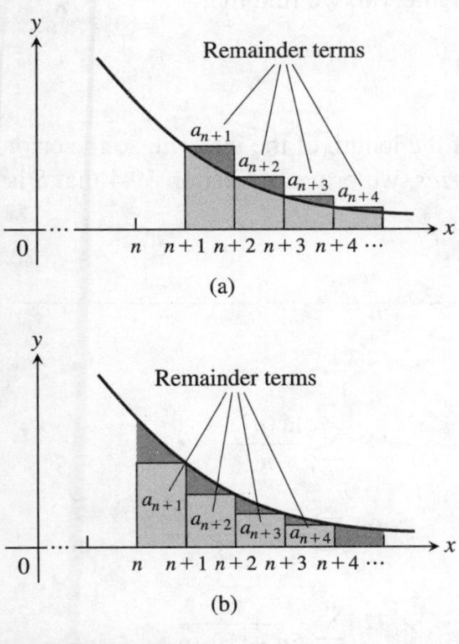

FIGURE 10.13 A geometric interpretation of Remainder Formula (1).

These comparisons prove the following result, giving bounds on the size of the remainder.

Bounds for the Remainder in the Integral Test

Suppose $\{a_k\}$ is a sequence of positive terms with $a_k = f(k)$, where f is a continuous positive decreasing function of x for all $x \geq n$, and suppose that $\sum a_n$ converges to S. Then the remainder $R_n = S - s_n$ satisfies the inequalities

$$\int_{n+1}^{\infty} f(x)\,dx \leq R_n \leq \int_{n}^{\infty} f(x)\,dx. \tag{1}$$

Since $s_n + R_n = S$, if we add the partial sum s_n to each side of the inequalities in (1), we get

$$s_n + \int_{n+1}^{\infty} f(x)\,dx \leq S \leq s_n + \int_{n}^{\infty} f(x)\,dx. \tag{2}$$

The inequalities in (2) are useful for estimating the error in approximating the sum of a series known to converge by the Integral Test. The error can be no larger than the length of the interval containing S, with endpoints given by (2).

EXAMPLE 6 Estimate the sum of the series $\sum (1/n^2)$ using the inequalities in (2) and $n = 10$.

Solution We have that

$$\int_n^{\infty} \frac{1}{x^2}\,dx = \lim_{b\to\infty}\left[-\frac{1}{x}\right]_n^b = \lim_{b\to\infty}\left(-\frac{1}{b} + \frac{1}{n}\right) = \frac{1}{n}.$$

Using this result with the inequalities in (2) gives

$$s_{10} + \frac{1}{11} \leq S \leq s_{10} + \frac{1}{10}.$$

Since $s_{10} = 1 + (1/4) + (1/9) + (1/16) + \cdots + (1/100) \approx 1.54977$, these last inequalities give

$$1.64068 \leq S \leq 1.64977.$$

If we approximate the sum S by the midpoint of this interval, we find that

$$\sum_{n=1}^{\infty} \frac{1}{n^2} \approx 1.6452.$$

The error in this approximation is then less than half the length of the interval, so the error is less than 0.005. Using a trigonometric *Fourier series*, we prove in Section 19.4 that S is equal to $\pi^2/6 \approx 1.64493$. ■

The *p*-series for $p = 2$

$$\sum_{n=1}^{\infty} \frac{1}{n^2} = \frac{\pi^2}{6} \approx 1.64493$$

EXERCISES 10.3

Applying the Integral Test

Use the Integral Test to determine whether the series in Exercises 1–12 converge or diverge. Be sure to check that the conditions of the Integral Test are satisfied.

1. $\sum_{n=1}^{\infty} \frac{1}{n^2}$
2. $\sum_{n=1}^{\infty} \frac{1}{n^{0.2}}$
3. $\sum_{n=1}^{\infty} \frac{1}{n^2 + 4}$
4. $\sum_{n=1}^{\infty} \frac{1}{n + 4}$
5. $\sum_{n=1}^{\infty} e^{-2n}$
6. $\sum_{n=2}^{\infty} \frac{1}{n(\ln n)^2}$
7. $\sum_{n=1}^{\infty} \frac{n}{n^2 + 4}$
8. $\sum_{n=2}^{\infty} \frac{\ln (n^2)}{n}$
9. $\sum_{n=1}^{\infty} \frac{n^2}{e^{n/3}}$
10. $\sum_{n=2}^{\infty} \frac{n - 4}{n^2 - 2n + 1}$
11. $\sum_{n=1}^{\infty} \frac{7}{\sqrt{n + 4}}$
12. $\sum_{n=2}^{\infty} \frac{1}{5n + 10\sqrt{n}}$

Determining Convergence or Divergence

Which of the series in Exercises 13–46 converge, and which diverge? Give reasons for your answers. (When you check an answer, remember that there may be more than one way to determine the series' convergence or divergence.)

13. $\sum_{n=1}^{\infty} \frac{1}{10^n}$ 14. $\sum_{n=1}^{\infty} e^{-n}$ 15. $\sum_{n=1}^{\infty} \frac{n}{n+1}$

16. $\sum_{n=1}^{\infty} \frac{5}{n+1}$ 17. $\sum_{n=1}^{\infty} \frac{3}{\sqrt{n}}$ 18. $\sum_{n=1}^{\infty} \frac{-2}{n\sqrt{n}}$

19. $\sum_{n=1}^{\infty} -\frac{1}{8^n}$ 20. $\sum_{n=1}^{\infty} \frac{\ln n}{n^2}$ 21. $\sum_{n=2}^{\infty} \frac{\ln n}{n}$

22. $\sum_{n=2}^{\infty} \frac{\ln n}{\sqrt{n}}$ 23. $\sum_{n=1}^{\infty} \frac{2^n}{3^n}$ 24. $\sum_{n=1}^{\infty} \frac{5^n}{4^n+3}$

25. $\sum_{n=0}^{\infty} \frac{-2}{n+1}$ 26. $\sum_{n=1}^{\infty} \frac{1}{2n-1}$ 27. $\sum_{n=1}^{\infty} \frac{2^n}{n+1}$

28. $\sum_{n=1}^{\infty} \left(1+\frac{1}{n}\right)^n$ 29. $\sum_{n=2}^{\infty} \frac{\sqrt{n}}{\ln n}$ 30. $\sum_{n=1}^{\infty} \frac{1}{\sqrt{n}(\sqrt{n}+1)}$

31. $\sum_{n=1}^{\infty} \frac{1}{(\ln 2)^n}$ 32. $\sum_{n=2}^{\infty} \frac{1}{\sqrt{n}(\ln n)}$

33. $\sum_{n=3}^{\infty} \frac{(1/n)}{(\ln n)\sqrt{\ln^2 n - 1}}$ 34. $\sum_{n=1}^{\infty} \frac{1}{n(1+\ln^2 n)}$

35. $\sum_{n=1}^{\infty} n \sin\frac{1}{n}$ 36. $\sum_{n=1}^{\infty} n \tan\frac{1}{n}$

37. $\sum_{n=1}^{\infty} \frac{e^n}{1+e^{2n}}$ 38. $\sum_{n=1}^{\infty} \frac{2}{1+e^n}$

39. $\sum_{n=1}^{\infty} \frac{e^n}{10+e^n}$ 40. $\sum_{n=1}^{\infty} \frac{e^n}{(10+e^n)^2}$

41. $\sum_{n=2}^{\infty} \frac{\sqrt{n+2}-\sqrt{n+1}}{\sqrt{n+1}\sqrt{n+2}}$ 42. $\sum_{n=3}^{\infty} \frac{7}{\sqrt{n+1}\ln\sqrt{n+1}}$

43. $\sum_{n=1}^{\infty} \frac{8\tan^{-1} n}{1+n^2}$ 44. $\sum_{n=1}^{\infty} \frac{n}{n^2+1}$

45. $\sum_{n=1}^{\infty} \operatorname{sech} n$ 46. $\sum_{n=1}^{\infty} \operatorname{sech}^2 n$

Theory and Examples

For what values of a, if any, do the series in Exercises 47 and 48 converge?

47. $\sum_{n=1}^{\infty} \left(\frac{a}{n+2} - \frac{1}{n+4}\right)$ 48. $\sum_{n=3}^{\infty} \left(\frac{1}{n-1} - \frac{2a}{n+1}\right)$

49. **a.** Draw illustrations like those in Figures 10.12a and 10.12b to show that the partial sums of the harmonic series satisfy the inequalities

$$\ln(n+1) = \int_1^{n+1} \frac{1}{x}\,dx \le 1 + \frac{1}{2} + \cdots + \frac{1}{n}$$

$$\le 1 + \int_1^{n} \frac{1}{x}\,dx = 1 + \ln n.$$

T **b.** There is absolutely no empirical evidence for the divergence of the harmonic series even though we know it diverges. The partial sums just grow too slowly. To see what we mean, suppose you had started with $s_1 = 1$ the day the universe was formed, 13 billion years ago, and added a new term every *second*. About how large would the partial sum s_n be today, assuming a 365-day year?

50. Are there any values of x for which $\sum_{n=1}^{\infty} (1/nx)$ converges? Give reasons for your answer.

51. Is it true that if $\sum_{n=1}^{\infty} a_n$ is a divergent series of positive numbers, then there is also a divergent series $\sum_{n=1}^{\infty} b_n$ of positive numbers with $b_n < a_n$ for every n? Is there a "smallest" divergent series of positive numbers? Give reasons for your answers.

52. (*Continuation of Exercise 51.*) Is there a "largest" convergent series of positive numbers? Explain.

53. **$\sum_{n=1}^{\infty} (1/\sqrt{n+1})$ diverges**

a. Use the accompanying graph to show that the partial sum $s_{50} = \sum_{n=1}^{50} (1/\sqrt{n+1})$ satisfies

$$\int_1^{51} \frac{1}{\sqrt{x+1}}\,dx < s_{50} < \int_0^{50} \frac{1}{\sqrt{x+1}}\,dx.$$

Conclude that $11.5 < s_{50} < 12.3$.

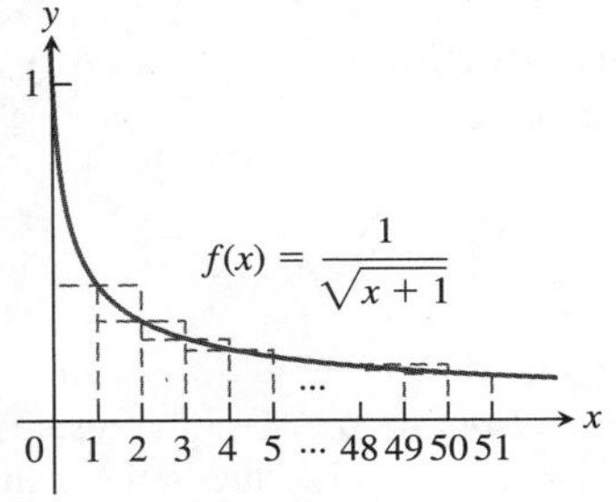

b. What should n be in order that the partial sum

$$s_n = \sum_{i=1}^{n} (1/\sqrt{i+1}) \text{ satisfy } s_n > 1000?$$

54. **$\sum_{n=1}^{\infty} (1/n^4)$ converges**

a. Use the accompanying graph to find an upper bound for the error if $s_{30} = \sum_{n=1}^{30} (1/n^4)$ is used to estimate the value of $\sum_{n=1}^{\infty} (1/n^4)$.

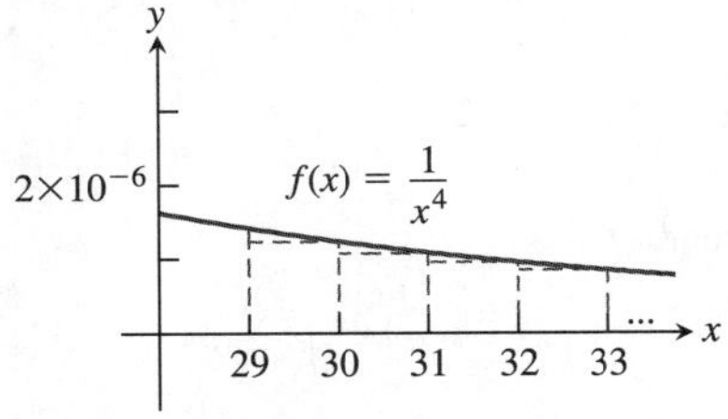

b. Find n so that the partial sum $s_n = \sum_{i=1}^{\infty} (1/i^4)$ estimates the value of $\sum_{n=1}^{\infty} (1/n^4)$ with an error of at most 0.000001. (The exact value of this series is computed in Exercise 20 of Section 19.4.)

55. Estimate the value of $\sum_{n=1}^{\infty} (1/n^3)$ to within 0.01 of its exact value.

56. Estimate the value of $\sum_{n=2}^{\infty} (1/(n^2+4))$ to within 0.1 of its exact value.

57. How many terms of the convergent series $\sum_{n=1}^{\infty} (1/n^{1.1})$ should be used to estimate its value with error at most 0.00001?

58. How many terms of the convergent series $\sum_{n=4}^{\infty} 1/\left(n(\ln n)^3\right)$ should be used to estimate its value with error at most 0.01?

59. The Cauchy condensation test The Cauchy condensation test says: Let $\{a_n\}$ be a nonincreasing sequence ($a_n \geq a_{n+1}$ for all n) of positive terms that converges to 0. Then $\sum a_n$ converges if and only if $\sum 2^n a_{2^n}$ converges. For example, $\sum(1/n)$ diverges because $\sum 2^n \cdot (1/2^n) = \sum 1$ diverges. Show why the test works.

60. Use the Cauchy condensation test from Exercise 59 to show that

a. $\displaystyle\sum_{n=2}^{\infty} \frac{1}{n \ln n}$ diverges;

b. $\displaystyle\sum_{n=1}^{\infty} \frac{1}{n^p}$ converges if $p > 1$ and diverges if $p \leq 1$.

61. Logarithmic *p*-series

a. Show that the improper integral

$$\int_2^{\infty} \frac{dx}{x(\ln x)^p} \quad (p \text{ a positive constant})$$

converges if and only if $p > 1$.

b. What implications does the fact in part (a) have for the convergence of the series

$$\sum_{n=2}^{\infty} \frac{1}{n(\ln n)^p}?$$

Give reasons for your answer.

62. (*Continuation of Exercise 61.*) Use the result in Exercise 61 to determine which of the following series converge and which diverge. Support your answer in each case.

a. $\displaystyle\sum_{n=2}^{\infty} \frac{1}{n(\ln n)}$

b. $\displaystyle\sum_{n=2}^{\infty} \frac{1}{n(\ln n)^{1.01}}$

c. $\displaystyle\sum_{n=2}^{\infty} \frac{1}{n \ln (n^3)}$

d. $\displaystyle\sum_{n=2}^{\infty} \frac{1}{n(\ln n)^3}$

63. Euler's constant Graphs like those in Figure 10.12 suggest that as n increases there is little change in the difference between the sum

$$1 + \frac{1}{2} + \cdots + \frac{1}{n}$$

and the integral

$$\ln n = \int_1^n \frac{1}{x}\, dx.$$

To explore this idea, carry out the following steps.

a. By taking $f(x) = 1/x$ in the proof of Theorem 9, show that

$$\ln(n+1) \leq 1 + \frac{1}{2} + \cdots + \frac{1}{n} \leq 1 + \ln n$$

or

$$0 < \ln(n+1) - \ln n \leq 1 + \frac{1}{2} + \cdots + \frac{1}{n} - \ln n \leq 1.$$

Thus, the sequence

$$a_n = 1 + \frac{1}{2} + \cdots + \frac{1}{n} - \ln n$$

is bounded from below and from above.

b. Show that

$$\frac{1}{n+1} < \int_n^{n+1} \frac{1}{x}\, dx = \ln(n+1) - \ln n,$$

and use this result to show that the sequence $\{a_n\}$ in part (a) is decreasing.

Since a decreasing sequence that is bounded from below converges, the numbers a_n defined in part (a) converge:

$$1 + \frac{1}{2} + \cdots + \frac{1}{n} - \ln n \rightarrow \gamma.$$

The number γ, whose value is $0.5772\ldots$, is called *Euler's constant.*

64. Use the Integral Test to show that the series

$$\sum_{n=0}^{\infty} e^{-n^2}$$

converges.

65. a. For the series $\sum(1/n^3)$, use the inequalities in Equation (2) with $n = 10$ to find an interval containing the sum S.

b. As in Example 5, use the midpoint of the interval found in part (a) to approximate the sum of the series. What is the maximum error for your approximation?

66. Repeat Exercise 65 using the series $\sum(1/n^4)$.

67. Area Consider the sequence $\{1/n\}_{n=1}^{\infty}$. On each subinterval $(1/(n+1), 1/n)$ within the interval $[0, 1]$, erect the rectangle with area a_n having height $1/n$ and width equal to the length of the subinterval. Find the total area $\sum a_n$ of all the rectangles. (*Hint:* Use the result of Example 5 in Section 10.2.)

68. Area Repeat Exercise 67, using trapezoids instead of rectangles. That is, on the subinterval $(1/(n+1), 1/n)$, let a_n denote the area of the trapezoid having heights $y = 1/(n+1)$ at $x = 1/(n+1)$ and $y = 1/n$ at $x = 1/n$.

10.4 Comparison Tests

We have seen how to determine the convergence of geometric series, *p*-series, and a few others. We can test the convergence of many more series by comparing their terms to those of a series whose convergence is already known.

THEOREM 10—Direct Comparison Test

Let $\sum a_n$ and $\sum b_n$ be two series with $0 \le a_n \le b_n$ for all n. Then

1. If $\sum b_n$ converges, then $\sum a_n$ also converges.

2. If $\sum a_n$ diverges, then $\sum b_n$ also diverges.

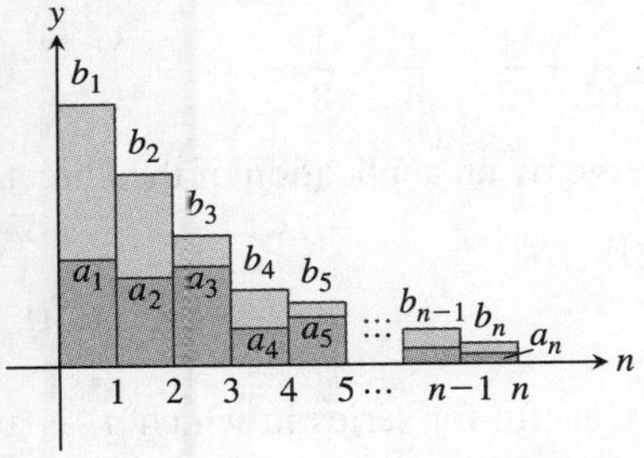

FIGURE 10.14 If the total area $\sum b_n$ of the taller b_n rectangles is finite, then so is the total area $\sum a_n$ of the shorter a_n rectangles.

Proof The series $\sum a_n$ and $\sum b_n$ have nonnegative terms. The Corollary of Theorem 6 stated in Section 10.3 tells us that the series $\sum a_n$ and $\sum b_n$ converge if and only if their partial sums are bounded from above.

In Part (1) we assume that $\sum b_n$ converges to some number M. The partial sums $\sum_{n=1}^{N} a_n$ are all bounded from above by $M = \sum b_n$ because

$$s_N = a_1 + a_2 + \cdots + a_N \le b_1 + b_2 + \cdots + b_N \le \sum_{n=1}^{\infty} b_n = M.$$

Since the partial sums of $\sum a_n$ are bounded from above, the Corollary of Theorem 6 implies that $\sum a_n$ converges. We conclude that if $\sum b_n$ converges, then so does $\sum a_n$. Figure 10.14 illustrates this result, with each term of each series interpreted as the area of a rectangle.

In Part (2), where we assume that $\sum a_n$ diverges, the partial sums of $\sum_{n=1}^{\infty} b_n$ are not bounded from above. If they were, the partial sums for $\sum a_n$ would also be bounded from above, since

$$a_1 + a_2 + \cdots + a_N \le b_1 + b_2 + \cdots + b_N,$$

and this would mean that $\sum a_n$ converges. We conclude that if $\sum a_n$ diverges, then so does $\sum b_n$. ■

EXAMPLE 1 We apply Theorem 10 to several series.

(a) The series

$$\sum_{n=1}^{\infty} \frac{5}{5n-1}$$

diverges because its nth term

$$\frac{5}{5n-1} = \frac{1}{n - \frac{1}{5}} > \frac{1}{n}$$

$\sum_{n=1}^{\infty} \frac{1}{n}$ diverges and $1/n > 0$.

is positive and is greater than the nth term of the (positive) divergent harmonic series.

(b) The series

$$\sum_{n=0}^{\infty} \frac{1}{n!} = 1 + \frac{1}{1!} + \frac{1}{2!} + \frac{1}{3!} + \cdots$$

converges because its terms are all positive and less than or equal to the corresponding terms of

$$1 + \sum_{n=0}^{\infty} \frac{1}{2^n} = 1 + 1 + \frac{1}{2} + \frac{1}{2^2} + \cdots.$$

The geometric series on the left converges (since $|r| = 1/2 < 1$) and we have

$$1 + \sum_{n=0}^{\infty} \frac{1}{2^n} = 1 + \frac{1}{1 - (1/2)} = 3.$$

The fact that 3 is an upper bound for the partial sums of $\sum_{n=0}^{\infty} (1/n!)$ does not mean that the series converges to 3. As we will see in Section 10.9, the series converges to e.

HISTORICAL BIOGRAPHY
Albert of Saxony
(ca. 1316–1390)
www.bit.ly/2DMOp6u

(c) The series

$$5 + \frac{2}{3} + \frac{1}{7} + 1 + \frac{1}{2 + \sqrt{1}} + \frac{1}{4 + \sqrt{2}} + \frac{1}{8 + \sqrt{3}} + \cdots + \frac{1}{2^n + \sqrt{n}} + \cdots$$

converges. To see this, we ignore the first three terms and compare the remaining terms with those of the convergent geometric series $\sum_{n=0}^{\infty}(1/2^n)$. The term $1/(2^n + \sqrt{n})$ of the truncated sequence is positive and is less than the corresponding term $1/2^n$ of the geometric series. We see that term-by-term we have the comparison of positive terms

$$1 + \frac{1}{2 + \sqrt{1}} + \frac{1}{4 + \sqrt{2}} + \frac{1}{8 + \sqrt{3}} + \cdots \le 1 + \frac{1}{2} + \frac{1}{4} + \frac{1}{8} + \cdots.$$

So the truncated series and the original series converge by an application of the Direct Comparison Test. ■

The Limit Comparison Test

We now introduce a comparison test that is particularly useful for series in which a_n is a rational function of n.

THEOREM 11—Limit Comparison Test
Suppose that $a_n > 0$ and $b_n > 0$ for all $n \ge N$ (N an integer).

1. If $\lim_{n\to\infty} \frac{a_n}{b_n} = c$ and $c > 0$, then $\sum a_n$ and $\sum b_n$ both converge or both diverge.

2. If $\lim_{n\to\infty} \frac{a_n}{b_n} = 0$ and $\sum b_n$ converges, then $\sum a_n$ converges.

3. If $\lim_{n\to\infty} \frac{a_n}{b_n} = \infty$ and $\sum b_n$ diverges, then $\sum a_n$ diverges.

Proof We will prove Part 1. Parts 2 and 3 are left as Exercises 57a and b.

We assume that $\lim_{n\to\infty} \frac{a_n}{b_n} = c$ where $c > 0$. Then $\varepsilon = c/2$ is a positive number, so by the definition of convergence there exists an integer N such that

$$\left|\frac{a_n}{b_n} - c\right| < \frac{c}{2} \qquad \text{whenever} \qquad n > N.$$

Limit definition with $\varepsilon = c/2, L = c$, and a_n replaced by a_n/b_n

Thus, for $n > N$,

$$-\frac{c}{2} < \frac{a_n}{b_n} - c < \frac{c}{2},$$

$$\frac{c}{2} < \frac{a_n}{b_n} < \frac{3c}{2},$$

$$\left(\frac{c}{2}\right)b_n < a_n < \left(\frac{3c}{2}\right)b_n.$$

If $\sum b_n$ converges, then $\sum (3c/2)b_n$ converges and $\sum a_n$ converges by the Direct Comparison Test. If $\sum b_n$ diverges, then $\sum (c/2)b_n$ diverges and $\sum a_n$ diverges by the Direct Comparison Test. ■

EXAMPLE 2 Which of the following series converge, and which diverge?

(a) $\frac{3}{4} + \frac{5}{9} + \frac{7}{16} + \frac{9}{25} + \cdots = \sum_{n=1}^{\infty} \frac{2n+1}{(n+1)^2} = \sum_{n=1}^{\infty} \frac{2n+1}{n^2 + 2n + 1}$

(b) $\frac{1}{1} + \frac{1}{3} + \frac{1}{7} + \frac{1}{15} + \cdots = \sum_{n=1}^{\infty} \frac{1}{2^n - 1}$

(c) $\frac{1 + 2\ln 2}{9} + \frac{1 + 3\ln 3}{14} + \frac{1 + 4\ln 4}{21} + \cdots = \sum_{n=2}^{\infty} \frac{1 + n\ln n}{n^2 + 5}$

Solution We apply the Limit Comparison Test to each series.

(a) Let $a_n = (2n + 1)/(n^2 + 2n + 1)$. For large n, we expect a_n to behave like $2n/n^2 = 2/n$ since the leading terms dominate for large n, so we let $b_n = 1/n$. The numbers a_n and b_n are positive for each n, the series

$$\sum_{n=1}^{\infty} b_n = \sum_{n=1}^{\infty} \frac{1}{n} \text{ diverges,}$$

and

$$\lim_{n\to\infty} \frac{a_n}{b_n} = \lim_{n\to\infty} \frac{2n^2 + n}{n^2 + 2n + 1} = 2,$$

so $\sum a_n$ diverges by Part 1 of the Limit Comparison Test. We could just as well have taken $b_n = 2/n$, but $1/n$ is simpler.

(b) Let $a_n = 1/(2^n - 1)$. For large n, we expect a_n to behave like $1/2^n$, so we let $b_n = 1/2^n$. The numbers a_n and b_n are positive for each n, the series

$$\sum_{n=1}^{\infty} b_n = \sum_{n=1}^{\infty} \frac{1}{2^n} \text{ converges,}$$

and

$$\lim_{n\to\infty} \frac{a_n}{b_n} = \lim_{n\to\infty} \frac{2^n}{2^n - 1} = \lim_{n\to\infty} \frac{1}{1 - (1/2^n)} = 1,$$

so $\sum a_n$ converges by Part 1 of the Limit Comparison Test.

(c) Let $a_n = (1 + n\ln n)/(n^2 + 5)$. For large n, we expect a_n to behave like $(n\ln n)/n^2 = (\ln n)/n$, which is greater than $1/n$ for $n \geq 3$, so we let $b_n = 1/n$. The numbers a_n and b_n are positive for each n, the series

$$\sum_{n=2}^{\infty} b_n = \sum_{n=2}^{\infty} \frac{1}{n} \text{ diverges,}$$

and

$$\lim_{n\to\infty} \frac{a_n}{b_n} = \lim_{n\to\infty} \frac{n + n^2 \ln n}{n^2 + 5} = \infty,$$

so $\sum a_n$ diverges by Part 3 of the Limit Comparison Test. ■

EXAMPLE 3 Does $\sum_{n=1}^{\infty} \frac{\ln n}{n^{3/2}}$ converge?

Solution First note that both $\ln n$ and $n^{3/2}$ are positive for $n \geq 3$, so the Limit Comparison Theorem can be applied. Because $\ln n$ grows more slowly than n^c for any positive constant c (Section 10.1, Exercise 115), we can compare the series to a convergent p-series. To get the p-series, we see that

$$\frac{\ln n}{n^{3/2}} < \frac{n^{1/4}}{n^{3/2}} = \frac{1}{n^{5/4}}$$

for n sufficiently large. Then, taking $a_n = (\ln n)/n^{3/2}$ and $b_n = 1/n^{5/4}$, we have

$$\begin{aligned}\lim_{n\to\infty}\frac{a_n}{b_n} &= \lim_{n\to\infty}\frac{\ln n}{n^{1/4}}\\ &= \lim_{n\to\infty}\frac{1/n}{(1/4)n^{-3/4}} \qquad \text{L'Hôpital's Rule}\\ &= \lim_{n\to\infty}\frac{4}{n^{1/4}} = 0.\end{aligned}$$

Since $\sum b_n = \sum (1/n^{5/4})$ is a p-series with $p > 1$, it converges. Therefore, $\sum a_n$ converges by Part 2 of the Limit Comparison Test. ■

EXERCISES 10.4

Direct Comparison Test

In Exercises 1–8, use the Direct Comparison Test to determine whether each series converges or diverges.

1. $\sum_{n=1}^{\infty}\frac{1}{n^2+30}$
2. $\sum_{n=1}^{\infty}\frac{n-1}{n^4+2}$
3. $\sum_{n=2}^{\infty}\frac{1}{\sqrt{n}-1}$
4. $\sum_{n=2}^{\infty}\frac{n+2}{n^2-n}$
5. $\sum_{n=1}^{\infty}\frac{\cos^2 n}{n^{3/2}}$
6. $\sum_{n=1}^{\infty}\frac{1}{n3^n}$
7. $\sum_{n=1}^{\infty}\sqrt{\frac{n+4}{n^4+4}}$
8. $\sum_{n=1}^{\infty}\frac{\sqrt{n}+1}{\sqrt{n^2+3}}$

Limit Comparison Test

In Exercises 9–16, use the Limit Comparison Test to determine whether each series converges or diverges.

9. $\sum_{n=1}^{\infty}\frac{n-2}{n^3-n^2+3}$

(*Hint:* Limit Comparison with $\sum_{n=1}^{\infty}(1/n^2)$)

10. $\sum_{n=1}^{\infty}\sqrt{\frac{n+1}{n^2+2}}$

(*Hint:* Limit Comparison with $\sum_{n=1}^{\infty}(1/\sqrt{n})$)

11. $\sum_{n=2}^{\infty}\frac{n(n+1)}{(n^2+1)(n-1)}$
12. $\sum_{n=1}^{\infty}\frac{2^n}{3+4^n}$
13. $\sum_{n=1}^{\infty}\frac{5^n}{\sqrt{n}\,4^n}$
14. $\sum_{n=1}^{\infty}\left(\frac{2n+3}{5n+4}\right)^n$
15. $\sum_{n=2}^{\infty}\frac{1}{\ln n}$

(*Hint:* Limit Comparison with $\sum_{n=2}^{\infty}(1/n)$)

16. $\sum_{n=1}^{\infty}\ln\left(1+\frac{1}{n^2}\right)$

(*Hint:* Limit Comparison with $\sum_{n=1}^{\infty}(1/n^2)$)

Determining Convergence or Divergence

Which of the series in Exercises 17–56 converge, and which diverge? Use any method, and give reasons for your answers.

17. $\sum_{n=1}^{\infty}\frac{1}{2\sqrt{n}+\sqrt[3]{n}}$
18. $\sum_{n=1}^{\infty}\frac{3}{n+\sqrt{n}}$
19. $\sum_{n=1}^{\infty}\frac{\sin^2 n}{2^n}$
20. $\sum_{n=1}^{\infty}\frac{1+\cos n}{n^2}$
21. $\sum_{n=1}^{\infty}\frac{2n}{3n-1}$
22. $\sum_{n=1}^{\infty}\frac{n+1}{n^2\sqrt{n}}$
23. $\sum_{n=1}^{\infty}\frac{10n+1}{n(n+1)(n+2)}$
24. $\sum_{n=3}^{\infty}\frac{5n^3-3n}{n^2(n-2)(n^2+5)}$
25. $\sum_{n=1}^{\infty}\left(\frac{n}{3n+1}\right)^n$
26. $\sum_{n=1}^{\infty}\frac{1}{\sqrt{n^3+2}}$
27. $\sum_{n=3}^{\infty}\frac{1}{\ln(\ln n)}$
28. $\sum_{n=1}^{\infty}\frac{(\ln n)^2}{n^3}$
29. $\sum_{n=2}^{\infty}\frac{1}{\sqrt{n}\ln n}$
30. $\sum_{n=1}^{\infty}\frac{(\ln n)^2}{n^{3/2}}$
31. $\sum_{n=1}^{\infty}\frac{1}{1+\ln n}$
32. $\sum_{n=2}^{\infty}\frac{\ln(n+1)}{n+1}$
33. $\sum_{n=2}^{\infty}\frac{1}{n\sqrt{n^2-1}}$
34. $\sum_{n=1}^{\infty}\frac{\sqrt{n}}{n^2+1}$
35. $\sum_{n=1}^{\infty}\frac{1-n}{n2^n}$
36. $\sum_{n=1}^{\infty}\frac{n+2^n}{n^2 2^n}$
37. $\sum_{n=1}^{\infty}\left(\sqrt{n^2-1}-n\right)$
38. $\sum_{n=1}^{\infty}\left(\sqrt{n^4+1}-\sqrt{n^4-1}\right)$
39. $\sum_{n=1}^{\infty}\frac{\sqrt{n+1}-1}{(n+2)^3-1}$
40. $\sum_{n=1}^{\infty}\sqrt{\frac{3^n-1}{2^n+1}}$
41. $\sum_{n=1}^{\infty}\frac{2^n-n}{n2^n}$
42. $\sum_{n=1}^{\infty}\frac{\ln n}{\sqrt{n}\,e^n}$
43. $\sum_{n=2}^{\infty}\frac{1}{n!}$

(*Hint:* First show that $(1/n!) \le (1/n(n-1))$ for $n \ge 2$.)

44. $\sum_{n=1}^{\infty}\frac{(n-1)!}{(n+2)!}$
45. $\sum_{n=1}^{\infty}\sin\frac{1}{n}$
46. $\sum_{n=1}^{\infty}\tan\frac{1}{n}$
47. $\sum_{n=1}^{\infty}\frac{\tan^{-1} n}{n^{1.1}}$
48. $\sum_{n=1}^{\infty}\frac{\operatorname{arcsec} n}{n^{1.3}}$
49. $\sum_{n=1}^{\infty}\frac{\coth n}{n^2}$
50. $\sum_{n=1}^{\infty}\frac{\tanh n}{n^2}$
51. $\sum_{n=1}^{\infty}\frac{1}{n\sqrt[n]{n}}$
52. $\sum_{n=1}^{\infty}\frac{\sqrt[n]{n}}{n^2}$
53. $\sum_{n=1}^{\infty}\frac{1}{1+2+3+\cdots+n}$
54. $\sum_{n=1}^{\infty}\frac{1}{1+2^2+3^2+\cdots+n^2}$
55. $\sum_{n=2}^{\infty}\frac{n}{(\ln n)^2}$
56. $\sum_{n=2}^{\infty}\frac{(\ln n)^2}{n}$

Theory and Examples

57. Prove **(a)** Part 2 and **(b)** Part 3 of the Limit Comparison Test.

58. If $\sum_{n=1}^{\infty} a_n$ is a convergent series of nonnegative numbers, can anything be said about $\sum_{n=1}^{\infty} (a_n/n)$? Explain.

59. Show that $\sum_{n=1}^{\infty}\left[\frac{\sqrt{n+1}-\sqrt{n}}{n^p}\right]$ converges for $p>\frac{1}{2}$ and diverges for $0<p\le\frac{1}{2}$.

60. Prove that if $\sum a_n$ is a convergent series of nonnegative terms, then $\sum a_n^2$ converges.

61. Suppose that $a_n>0$ and $\lim_{n\to\infty} a_n = \infty$. Prove that $\sum a_n$ diverges.

62. Suppose that $a_n > 0$ and $\lim_{n\to\infty} n^2 a_n = 0$. Prove that $\sum a_n$ converges.

63. Show that $\sum_{n=2}^{\infty}((\ln n)^q/n^p)$ converges for $-\infty < q < \infty$ and $p > 1$.
(*Hint:* Limit Comparison with $\sum_{n=2}^{\infty} 1/n^r$ for $1 < r < p$.)

64. (*Continuation of Exercise 63.*) Show that $\sum_{n=2}^{\infty}((\ln n)^q/n^p)$ diverges for $-\infty < q < \infty$ and $0 < p < 1$.
(*Hint:* Limit Comparison with an appropriate *p*-series.)

65. Decimal numbers Any real number in the interval $[0,1]$ can be represented by a decimal (not necessarily unique) as

$$0.d_1d_2d_3d_4\cdots = \frac{d_1}{10}+\frac{d_2}{10^2}+\frac{d_3}{10^3}+\frac{d_4}{10^4}+\cdots,$$

where d_i is one of the integers 0, 1, 2, 3, …, 9. Prove that the series on the right-hand side always converges.

66. If $\sum a_n$ is a convergent series of positive terms, prove that $\sum \sin(a_n)$ converges.

In Exercises 67–72, use the results of Exercises 63 and 64 to determine whether each series converges or diverges.

67. $\sum_{n=2}^{\infty}\frac{(\ln n)^3}{n^4}$

68. $\sum_{n=2}^{\infty}\sqrt{\frac{\ln n}{n}}$

69. $\sum_{n=2}^{\infty}\frac{(\ln n)^{1000}}{n^{1.001}}$

70. $\sum_{n=2}^{\infty}\frac{(\ln n)^{1/5}}{n^{0.99}}$

71. $\sum_{n=2}^{\infty}\frac{1}{n^{1.1}(\ln n)^3}$

72. $\sum_{n=2}^{\infty}\frac{1}{\sqrt{n\cdot\ln n}}$

COMPUTER EXPLORATIONS

73. It is not yet known whether the series

$$\sum_{n=1}^{\infty}\frac{1}{n^3\sin^2 n}$$

converges or diverges. Use a CAS to explore the behavior of the series by performing the following steps.

a. Define the sequence of partial sums

$$s_k = \sum_{n=1}^{k}\frac{1}{n^3\sin^2 n}.$$

What happens when you try to find the limit of s_k as $k \to \infty$? Does your CAS find a closed form answer for this limit?

b. Plot the first 100 points (k, s_k) for the sequence of partial sums. Do they appear to converge? What would you estimate the limit to be?

c. Next plot the first 200 points (k, s_k). Discuss the behavior in your own words.

d. Plot the first 400 points (k, s_k). What happens when $k = 355$? Calculate the number 355/113. Explain from your calculation what happened at $k = 355$. For what values of k would you guess this behavior might occur again?

74. a. Use Theorem 8 to show that

$$S = \sum_{n=1}^{\infty}\frac{1}{n(n+1)}+\sum_{n=1}^{\infty}\left(\frac{1}{n^2}-\frac{1}{n(n+1)}\right),$$

where $S = \sum_{n=1}^{\infty}(1/n^2)$, the sum of a convergent *p*-series.

b. From Example 5, Section 10.2, show that

$$S = 1+\sum_{n=1}^{\infty}\frac{1}{n^2(n+1)}.$$

c. Explain why taking the first M terms in the series in part (b) gives a better approximation to S than taking the first M terms in the original series $\sum_{n=1}^{\infty}(1/n^2)$.

d. We know the exact value of S is $\pi^2/6$. Which of these sums,

$$\sum_{n=1}^{1000000}\frac{1}{n^2} \quad\text{or}\quad 1+\sum_{n=1}^{1000}\frac{1}{n^2(n+1)},$$

gives a better approximation to S?

10.5 Absolute Convergence; The Ratio and Root Tests

When some of the terms of a series are positive and others are negative, the series may or may not converge. For example, the geometric series

$$5-\frac{5}{4}+\frac{5}{16}-\frac{5}{64}+\cdots = \sum_{n=0}^{\infty}5\left(\frac{-1}{4}\right)^n \tag{1}$$

converges (since $|r| = \frac{1}{4} < 1$), whereas the different geometric series

$$1-\frac{5}{4}+\frac{25}{16}-\frac{125}{64}+\cdots = \sum_{n=0}^{\infty}\left(\frac{-5}{4}\right)^n \tag{2}$$

diverges (since $|r| = 5/4 > 1$). In series (1), there is some cancelation in the partial sums, which may be assisting the convergence property of the series. However, if we make all of the terms positive in series (1) to form the new series

$$5 + \frac{5}{4} + \frac{5}{16} + \frac{5}{64} + \cdots = \sum_{n=0}^{\infty} \left| 5\left(\frac{-1}{4}\right)^n \right| = \sum_{n=0}^{\infty} 5\left(\frac{1}{4}\right)^n,$$

we see that it still converges. For a general series with both positive and negative terms, we can apply the tests for convergence that we studied before to the series of absolute values of its terms. In doing so, we are led naturally to the following concept.

DEFINITION A series $\sum a_n$ **converges absolutely** (is **absolutely convergent**) if the corresponding series of absolute values, $\sum |a_n|$, converges.

So the geometric series (1) is absolutely convergent. We observed, too, that it is also convergent. This situation is always true: An absolutely convergent series is convergent as well, which we now prove.

Caution
Be careful when using Theorem 12. A convergent series *need not* converge absolutely, as you will see in the next section.

THEOREM 12—The Absolute Convergence Test
If $\sum_{n=1}^{\infty} |a_n|$ converges, then $\sum_{n=1}^{\infty} a_n$ converges.

Proof For each n,

$$-|a_n| \le a_n \le |a_n|, \qquad \text{so} \qquad 0 \le a_n + |a_n| \le 2|a_n|.$$

If $\sum_{n=1}^{\infty} |a_n|$ converges, then $\sum_{n=1}^{\infty} 2|a_n|$ converges and, by the Direct Comparison Test, the nonnegative series $\sum_{n=1}^{\infty} (a_n + |a_n|)$ converges. The equality $a_n = (a_n + |a_n|) - |a_n|$ now lets us express $\sum_{n=1}^{\infty} a_n$ as the difference of two convergent series:

$$\sum_{n=1}^{\infty} a_n = \sum_{n=1}^{\infty} (a_n + |a_n| - |a_n|) = \sum_{n=1}^{\infty} (a_n + |a_n|) - \sum_{n=1}^{\infty} |a_n|.$$

Therefore, $\sum_{n=1}^{\infty} a_n$ converges. ■

EXAMPLE 1 This example gives two series that converge absolutely.

(a) For $\sum_{n=1}^{\infty} (-1)^{n+1} \frac{1}{n^2} = 1 - \frac{1}{4} + \frac{1}{9} - \frac{1}{16} + \cdots$, the corresponding series of absolute values is the convergent series

$$\sum_{n=1}^{\infty} \frac{1}{n^2} = 1 + \frac{1}{4} + \frac{1}{9} + \frac{1}{16} + \cdots.$$

The original series converges because it converges absolutely.

(b) For $\sum_{n=1}^{\infty} \frac{\sin n}{n^2} = \frac{\sin 1}{1} + \frac{\sin 2}{4} + \frac{\sin 3}{9} + \cdots$, which contains both positive and negative terms, the corresponding series of absolute values is

$$\sum_{n=1}^{\infty} \left| \frac{\sin n}{n^2} \right| = \frac{|\sin 1|}{1} + \frac{|\sin 2|}{4} + \cdots,$$

which converges by comparison with $\sum_{n=1}^{\infty} (1/n^2)$ because $|\sin n| \le 1$ for every n. The original series converges absolutely; therefore, it converges. ■

The Ratio Test

The Ratio Test measures the rate of growth (or decline) of a series by examining the ratio a_{n+1}/a_n. For a geometric series $\sum ar^n$, this rate is a constant $((ar^{n+1})/(ar^n) = r)$, and the series converges if and only if its ratio is less than 1 in absolute value. The Ratio Test is a powerful rule extending that result.

ρ is the Greek lowercase letter rho, which is pronounced "row."

THEOREM 13—The Ratio Test
Let $\sum a_n$ be any series and suppose that

$$\lim_{n\to\infty}\left|\frac{a_{n+1}}{a_n}\right| = \rho.$$

Then **(a)** the series *converges absolutely* if $\rho < 1$, **(b)** the series *diverges* if $\rho > 1$ or ρ is infinite, and **(c)** the test is *inconclusive* if $\rho = 1$.

Proof

(a) $\boldsymbol{\rho < 1}$**.** Let r be a number between ρ and 1. Then the number $\varepsilon = r - \rho$ is positive. Since

$$\left|\frac{a_{n+1}}{a_n}\right| \to \rho,$$

$|a_{n+1}/a_n|$ must lie within ε of ρ when n is large enough, say, for all $n \geq N$. In particular,

$$\left|\frac{a_{n+1}}{a_n}\right| < \rho + \varepsilon = r, \qquad \text{when } n \geq N.$$

Hence

$$\begin{aligned}
|a_{N+1}| &< r|a_N|,\\
|a_{N+2}| &< r|a_{N+1}| < r^2|a_N|,\\
|a_{N+3}| &< r|a_{N+2}| < r^3|a_N|,\\
&\ \vdots\\
|a_{N+m}| &< r|a_{N+m-1}| < r^m|a_N|.
\end{aligned}$$

Therefore,

$$\sum_{m=N}^{\infty}|a_m| = \sum_{m=0}^{\infty}|a_{N+m}| \leq \sum_{m=0}^{\infty}|a_N|r^m = |a_N|\sum_{m=0}^{\infty} r^m.$$

The geometric series on the right-hand side converges because $0 < r < 1$, so the series of absolute values $\sum_{m=N}^{\infty}|a_m|$ converges by the Direct Comparison Test. Because adding or deleting finitely many terms in a series does not affect its convergence or divergence property, the series $\sum_{n=1}^{\infty}|a_n|$ also converges. That is, the series $\sum a_n$ is absolutely convergent.

(b) $\boldsymbol{1 < \rho \leq \infty}$**.** From some index M on,

$$\left|\frac{a_{n+1}}{a_n}\right| > 1 \quad \text{and} \quad |a_M| < |a_{M+1}| < |a_{M+2}| < \cdots.$$

The terms of the series do not approach zero as n becomes infinite, so the series diverges by the nth-Term Test.

(c) $\boldsymbol{\rho = 1.}$ The two series

$$\sum_{n=1}^{\infty} \frac{1}{n} \quad \text{and} \quad \sum_{n=1}^{\infty} \frac{1}{n^2}$$

show that some other test for convergence must be used when $\rho = 1$.

$$\text{For } \sum_{n=1}^{\infty} \frac{1}{n}: \quad \left|\frac{a_{n+1}}{a_n}\right| = \frac{1/(n+1)}{1/n} = \frac{n}{n+1} \to 1.$$

$$\text{For } \sum_{n=1}^{\infty} \frac{1}{n^2}: \quad \left|\frac{a_{n+1}}{a_n}\right| = \frac{1/(n+1)^2}{1/n^2} = \left(\frac{n}{n+1}\right)^2 \to 1^2 = 1.$$

In both cases, $\rho = 1$, yet the first series diverges, whereas the second converges. ■

The Ratio Test is often effective when the terms of a series contain factorials of expressions involving n or expressions raised to a power involving n.

EXAMPLE 2 Investigate the convergence of the following series.

(a) $\displaystyle\sum_{n=0}^{\infty} \frac{2^n + 5}{3^n}$ **(b)** $\displaystyle\sum_{n=1}^{\infty} \frac{(2n)!}{n!n!}$ **(c)** $\displaystyle\sum_{n=1}^{\infty} \frac{4^n n! n!}{(2n)!}$

Solution We apply the Ratio Test to each series.

(a) For the series $\sum_{n=0}^{\infty} \dfrac{2^n + 5}{3^n}$,

$$\left|\frac{a_{n+1}}{a_n}\right| = \frac{(2^{n+1} + 5)/3^{n+1}}{(2^n + 5)/3^n} = \frac{1}{3} \cdot \frac{2^{n+1} + 5}{2^n + 5} = \frac{1}{3} \cdot \left(\frac{2 + 5 \cdot 2^{-n}}{1 + 5 \cdot 2^{-n}}\right) \to \frac{1}{3} \cdot \frac{2}{1} = \frac{2}{3}.$$

The series converges absolutely (and thus converges) because $\rho = 2/3$ is less than 1. This does *not* mean that 2/3 is the sum of the series. In fact,

$$\sum_{n=0}^{\infty} \frac{2^n + 5}{3^n} = \sum_{n=0}^{\infty} \left(\frac{2}{3}\right)^n + \sum_{n=0}^{\infty} \frac{5}{3^n} = \frac{1}{1 - (2/3)} + \frac{5}{1 - (1/3)} = \frac{21}{2}.$$

(b) If $a_n = \dfrac{(2n)!}{n!n!}$, then $a_{n+1} = \dfrac{(2n+2)!}{(n+1)!(n+1)!}$ and

$$\left|\frac{a_{n+1}}{a_n}\right| = \frac{n!n!(2n+2)(2n+1)(2n)!}{(n+1)!(n+1)!(2n)!}$$

$$= \frac{(2n+2)(2n+1)}{(n+1)(n+1)} = \frac{4n+2}{n+1} \to 4.$$

The series diverges because $\rho = 4$ is greater than 1.

(c) If $a_n = \dfrac{4^n n! n!}{(2n)!}$, then

$$\left|\frac{a_{n+1}}{a_n}\right| = \frac{4^{n+1}(n+1)!(n+1)!}{(2n+2)(2n+1)(2n)!} \cdot \frac{(2n)!}{4^n n! n!}$$

$$= \frac{4(n+1)(n+1)}{(2n+2)(2n+1)} = \frac{2(n+1)}{2n+1} \to 1.$$

Because the limit is $\rho = 1$, we cannot decide from the Ratio Test whether the series converges. However, when we notice that $a_{n+1}/a_n = (2n+2)/(2n+1)$, we conclude that a_{n+1} is always greater than a_n because $(2n+2)/(2n+1)$ is always greater than 1. Therefore, all terms are greater than or equal to $a_1 = 2$, and the nth term does not approach zero as $n \to \infty$. The series diverges. ■

The Root Test

The convergence tests for $\sum a_n$ that we have studied so far work best when the formula for a_n is relatively simple. However, consider the series with the terms

$$a_n = \begin{cases} n/2^n, & n \text{ odd} \\ 1/2^n, & n \text{ even.} \end{cases}$$

To investigate convergence we write out several terms of the series:

$$\sum_{n=1}^{\infty} a_n = \frac{1}{2^1} + \frac{1}{2^2} + \frac{3}{2^3} + \frac{1}{2^4} + \frac{5}{2^5} + \frac{1}{2^6} + \frac{7}{2^7} + \cdots$$
$$= \frac{1}{2} + \frac{1}{4} + \frac{3}{8} + \frac{1}{16} + \frac{5}{32} + \frac{1}{64} + \frac{7}{128} + \cdots.$$

Clearly, this is not a geometric series. The nth term approaches zero as $n \to \infty$, so the nth-Term Test does not tell us whether the series diverges. The Integral Test does not look promising. The Ratio Test produces

$$\left| \frac{a_{n+1}}{a_n} \right| = \begin{cases} \dfrac{1}{2n}, & n \text{ odd} \\ \dfrac{n+1}{2}, & n \text{ even.} \end{cases}$$

As $n \to \infty$, the ratio is alternately small and large and therefore has no limit. However, we will see that the following test establishes that the series converges.

THEOREM 14—The Root Test
Let $\sum a_n$ be any series and suppose that

$$\lim_{n\to\infty} \sqrt[n]{|a_n|} = \rho.$$

Then **(a)** the series *converges absolutely* if $\rho < 1$, **(b)** the series *diverges* if $\rho > 1$ or ρ is infinite, and **(c)** the test is *inconclusive* if $\rho = 1$.

Proof

(a) $\rho < 1$. Choose an $\varepsilon > 0$ so small that $\rho + \varepsilon < 1$. Since $\sqrt[n]{|a_n|} \to \rho$, the terms $\sqrt[n]{|a_n|}$ eventually get to within ε of ρ. So there exists an index M such that

$$\sqrt[n]{|a_n|} < \rho + \varepsilon \qquad \text{when } n \geq M.$$

Then it is also true that

$$|a_n| < (\rho + \varepsilon)^n \qquad \text{for } n \geq M.$$

Now, $\sum_{n=M}^{\infty} (\rho + \varepsilon)^n$ is a geometric series with ratio $0 < (\rho + \varepsilon) < 1$ and therefore converges. By the Direct Comparison Test, $\sum_{n=M}^{\infty} |a_n|$ converges. Adding finitely many terms to a series does not affect its convergence or divergence, so the series

$$\sum_{n=1}^{\infty} |a_n| = |a_1| + \cdots + |a_{M-1}| + \sum_{n=M}^{\infty} |a_n|$$

also converges. Therefore, $\sum a_n$ converges absolutely.

(b) $1 < \rho \leq \infty$. For all indices beyond some integer M, we have $\sqrt[n]{|a_n|} > 1$, and therefore $|a_n| > 1$ for $n > M$. The terms of the series do not converge to zero. The series diverges by the nth-Term Test.

(c) $\rho = 1$. The series $\sum_{n=1}^{\infty} (1/n)$ and $\sum_{n=1}^{\infty} (1/n^2)$ show that the test is not conclusive when $\rho = 1$. The first series diverges and the second converges, but in both cases $\sqrt[n]{|a_n|} \to 1$. ■

EXAMPLE 3 Consider again the series with terms $a_n = \begin{cases} n/2^n, & n \text{ odd} \\ 1/2^n, & n \text{ even.} \end{cases}$ Does $\sum a_n$ converge?

Solution We apply the Root Test, finding that

$$\sqrt[n]{|a_n|} = \begin{cases} \sqrt[n]{n}/2, & n \text{ odd} \\ 1/2, & n \text{ even.} \end{cases}$$

Therefore,

$$\frac{1}{2} \le \sqrt[n]{|a_n|} \le \frac{\sqrt[n]{n}}{2}.$$

Since $\sqrt[n]{n} \to 1$ (Section 10.1, Theorem 5), we have $\lim_{n\to\infty} \sqrt[n]{|a_n|} = 1/2$ by the Sandwich Theorem. The limit is less than 1, so the series converges absolutely by the Root Test. ■

EXAMPLE 4 Which of the following series converge, and which diverge?

(a) $\sum_{n=1}^{\infty} \frac{n^2}{2^n}$ **(b)** $\sum_{n=1}^{\infty} \frac{2^n}{n^3}$ **(c)** $\sum_{n=1}^{\infty} \left(\frac{1}{1+n}\right)^n$

Solution We apply the Root Test to each series, noting that each series has positive terms.

(a) $\sum_{n=1}^{\infty} \frac{n^2}{2^n}$ converges because $\sqrt[n]{\frac{n^2}{2^n}} = \frac{\sqrt[n]{n^2}}{\sqrt[n]{2^n}} = \frac{\left(\sqrt[n]{n}\right)^2}{2} \to \frac{1^2}{2} < 1.$

(b) $\sum_{n=1}^{\infty} \frac{2^n}{n^3}$ diverges because $\sqrt[n]{\frac{2^n}{n^3}} = \frac{2}{\left(\sqrt[n]{n}\right)^3} \to \frac{2}{1^3} > 1.$

(c) $\sum_{n=1}^{\infty} \left(\frac{1}{1+n}\right)^n$ converges because $\sqrt[n]{\left(\frac{1}{1+n}\right)^n} = \frac{1}{1+n} \to 0 < 1.$ ■

EXERCISES 10.5

Using the Ratio Test

In Exercises 1–8, use the Ratio Test to determine whether each series converges absolutely or diverges.

1. $\sum_{n=1}^{\infty} \frac{2^n}{n!}$

2. $\sum_{n=1}^{\infty} (-1)^n \frac{n+2}{3^n}$

3. $\sum_{n=1}^{\infty} \frac{(n-1)!}{(n+1)^2}$

4. $\sum_{n=1}^{\infty} \frac{2^{n+1}}{n3^{n-1}}$

5. $\sum_{n=1}^{\infty} \frac{n^4}{(-4)^n}$

6. $\sum_{n=2}^{\infty} \frac{3^{n+2}}{\ln n}$

7. $\sum_{n=1}^{\infty} (-1)^n \frac{n^2(n+2)!}{n!\,3^{2n}}$

8. $\sum_{n=1}^{\infty} \frac{n5^n}{(2n+3)\ln(n+1)}$

Using the Root Test

In Exercises 9–16, use the Root Test to determine whether each series converges absolutely or diverges.

9. $\sum_{n=1}^{\infty} \frac{7}{(2n+5)^n}$

10. $\sum_{n=1}^{\infty} \frac{4^n}{(3n)^n}$

11. $\sum_{n=1}^{\infty} \left(\frac{4n+3}{3n-5}\right)^n$

12. $\sum_{n=1}^{\infty} \left(-\ln\left(e^2 + \frac{1}{n}\right)\right)^{n+1}$

13. $\sum_{n=1}^{\infty} \left[\frac{(n+1)^{n+1}}{n^{n+1}} - \frac{n+1}{n}\right]^{-n}$

14. $\sum_{n=1}^{\infty} \sin^n\left(\frac{1}{\sqrt{n}}\right)$

15. $\sum_{n=1}^{\infty} (-1)^n \left(1 - \frac{1}{n}\right)^{n^2}$

(*Hint:* $\lim_{n\to\infty} (1 + x/n)^n = e^x$)

16. $\sum_{n=2}^{\infty} \frac{(-1)^n}{n^{1+n}}$

Determining Convergence or Divergence

In Exercises 17–46, use any method to determine whether the series converges or diverges. Give reasons for your answer.

17. $\sum_{n=1}^{\infty} \frac{n^{\sqrt{2}}}{2^n}$

18. $\sum_{n=1}^{\infty} (-1)^n n^2 e^{-n}$

19. $\sum_{n=1}^{\infty} n!(-e)^{-n}$

20. $\sum_{n=1}^{\infty} \frac{n!}{10^n}$

21. $\sum_{n=1}^{\infty} \frac{n!2^n}{n^n}$

22. $\sum_{n=1}^{\infty} \left(\frac{n-2}{n}\right)^n$

23. $\sum_{n=1}^{\infty} \frac{n^3+1}{2^n+1}$

24. $\sum_{n=1}^{\infty} \frac{(-2)^n}{3^n}$

25. $\sum_{n=1}^{\infty} (-1)^n\left(1-\frac{3}{n}\right)^n$

26. $\sum_{n=1}^{\infty} \left(1-\frac{1}{3n}\right)^n$

27. $\sum_{n=1}^{\infty} \frac{\ln n}{n^3}$

28. $\sum_{n=1}^{\infty} \frac{(-\ln n)^n}{n^n}$

29. $\sum_{n=1}^{\infty} \left(\frac{1}{n}-\frac{1}{n^2}\right)$

30. $\sum_{n=1}^{\infty} \left(\frac{1}{n}-\frac{1}{n^2}\right)^n$

31. $\sum_{n=1}^{\infty} \frac{e^n}{n^e}$

32. $\sum_{n=1}^{\infty} \frac{n \ln n}{(-2)^n}$

33. $\sum_{n=1}^{\infty} \frac{(n+1)(n+2)}{n!}$

34. $\sum_{n=1}^{\infty} e^{-n}(n^3)$

35. $\sum_{n=1}^{\infty} \frac{(n+3)!}{3!n!3^n}$

36. $\sum_{n=1}^{\infty} \frac{n2^n(n+1)!}{3^n n!}$

37. $\sum_{n=1}^{\infty} \frac{n!}{(2n+1)!}$

38. $\sum_{n=1}^{\infty} \frac{n!}{(-n)^n}$

39. $\sum_{n=2}^{\infty} \frac{-n}{(\ln n)^n}$

40. $\sum_{n=2}^{\infty} \frac{n}{(\ln n)^{(n/2)}}$

41. $\sum_{n=1}^{\infty} \frac{n! \ln n}{n(n+2)!}$

42. $\sum_{n=1}^{\infty} \frac{(-3)^n}{n^3 2^n}$

43. $\sum_{n=1}^{\infty} \frac{(n!)^2}{(2n)!}$

44. $\sum_{n=1}^{\infty} \frac{(2n+3)(2^n+3)}{3^n+2}$

45. $\sum_{n=3}^{\infty} \frac{2^n}{n^2}$

46. $\sum_{n=3}^{\infty} \frac{2^{n^2}}{n^{2^n}}$

Recursively Defined Terms Which of the series $\sum_{n=1}^{\infty} a_n$ defined by the formulas in Exercises 47–56 converge, and which diverge? Give reasons for your answers.

47. $a_1 = 2, \quad a_{n+1} = \frac{1+\sin n}{n} a_n$

48. $a_1 = 1, \quad a_{n+1} = \frac{1+\tan^{-1} n}{n} a_n$

49. $a_1 = \frac{1}{3}, \quad a_{n+1} = \frac{3n-1}{2n+5} a_n$

50. $a_1 = 3, \quad a_{n+1} = \frac{n}{n+1} a_n$

51. $a_1 = 2, \quad a_{n+1} = \frac{2}{n} a_n$

52. $a_1 = 5, \quad a_{n+1} = \frac{\sqrt[n]{n}}{2} a_n$

53. $a_1 = 1, \quad a_{n+1} = \frac{1+\ln n}{n} a_n$

54. $a_1 = \frac{1}{2}, \quad a_{n+1} = \frac{n+\ln n}{n+10} a_n$

55. $a_1 = \frac{1}{3}, \quad a_{n+1} = \sqrt[n]{a_n}$

56. $a_1 = \frac{1}{2}, \quad a_{n+1} = (a_n)^{n+1}$

Convergence or Divergence

Which of the series in Exercises 57–64 converge, and which diverge? Give reasons for your answers.

57. $\sum_{n=1}^{\infty} \frac{2^n n! n!}{(2n)!}$

58. $\sum_{n=1}^{\infty} \frac{(-1)^n (3n)!}{n!(n+1)!(n+2)!}$

59. $\sum_{n=1}^{\infty} \frac{(n!)^n}{(n^n)^2}$

60. $\sum_{n=1}^{\infty} (-1)^n \frac{(n!)^n}{n^{(n^2)}}$

61. $\sum_{n=1}^{\infty} \frac{n^n}{2^{(n^2)}}$

62. $\sum_{n=1}^{\infty} \frac{n^n}{(2^n)^2}$

63. $\sum_{n=1}^{\infty} \frac{1 \cdot 3 \cdot \cdots \cdot (2n-1)}{4^n 2^n n!}$

64. $\sum_{n=1}^{\infty} \frac{1 \cdot 3 \cdot \cdots \cdot (2n-1)}{[2 \cdot 4 \cdot \cdots \cdot (2n)](3^n+1)}$

65. Assume that b_n is a sequence of positive numbers converging to 4/5. Determine whether the following series converge or diverge.

a. $\sum_{n=1}^{\infty} (b_n)^{1/n}$

b. $\sum_{n=1}^{\infty} \left(\frac{5}{4}\right)^n (b_n)$

c. $\sum_{n=1}^{\infty} (b_n)^n$

d. $\sum_{n=1}^{\infty} \frac{1000^n}{n!+b_n}$

66. Assume that b_n is a sequence of positive numbers converging to 1/3. Determine whether the following series converge or diverge.

a. $\sum_{n=1}^{\infty} \frac{b_{n+1} b_n}{n 4^n}$

b. $\sum_{n=1}^{\infty} \frac{n^n}{n!\, b_1^2\, b_2^2 \cdots b_n^2}$

Theory and Examples

67. Neither the Ratio Test nor the Root Test helps with p-series. Try them on

$$\sum_{n=1}^{\infty} \frac{1}{n^p}$$

and show that both tests fail to provide information about convergence.

68. Show that neither the Ratio Test nor the Root Test provides information about the convergence of

$$\sum_{n=2}^{\infty} \frac{1}{(\ln n)^p} \qquad (p \text{ constant}).$$

69. Let $a_n = \begin{cases} n/2^n, & \text{if } n \text{ is a prime number} \\ 1/2^n, & \text{otherwise.} \end{cases}$

Does $\sum a_n$ converge? Give reasons for your answer.

70. Show that $\sum_{n=1}^{\infty} 2^{(n^2)}/n!$ diverges. Recall from the Laws of Exponents that $2^{(n^2)} = (2^n)^n$.

71. Determine whether the series $\sum_{n=1}^{\infty} c_n$ converges, where

$$c_n = \begin{cases} -1/n, & \text{if } n \text{ is a perfect square,} \\ 1/n^2, & \text{if } n \text{ is not a perfect square.} \end{cases}$$

10.6 Alternating Series and Conditional Convergence

A series in which the terms are alternately positive and negative is an **alternating series**. Here are three examples:

$$1 - \frac{1}{2} + \frac{1}{3} - \frac{1}{4} + \frac{1}{5} - \cdots + \frac{(-1)^{n+1}}{n} + \cdots \tag{1}$$

$$-2 + 1 - \frac{1}{2} + \frac{1}{4} - \frac{1}{8} + \cdots + \frac{(-1)^n 4}{2^n} + \cdots \tag{2}$$

$$1 - 2 + 3 - 4 + 5 - 6 + \cdots + (-1)^{n+1} n + \cdots \tag{3}$$

We see from these examples that the nth term of an alternating series is of the form

$$a_n = (-1)^{n+1} u_n \qquad \text{or} \qquad a_n = (-1)^n u_n,$$

where $u_n = |a_n|$ is a positive number.

Series (1), called the **alternating harmonic series**, converges, as we will see in a moment. Series (2), which is a geometric series with ratio $r = -1/2$, converges to $-2/[1 + (1/2)] = -4/3$. Series (3) diverges because the nth term does not approach zero.

We prove the convergence of the alternating harmonic series by applying the Alternating Series Test. This test is for *convergence* of an alternating series and cannot be used to conclude that such a series diverges. If we multiply $(u_1 - u_2 + u_3 - u_4 + \cdots)$ by -1, we see that the test is also valid for the alternating series $-u_1 + u_2 - u_3 + u_4 - \cdots$, as with the one in Series (2) given above.

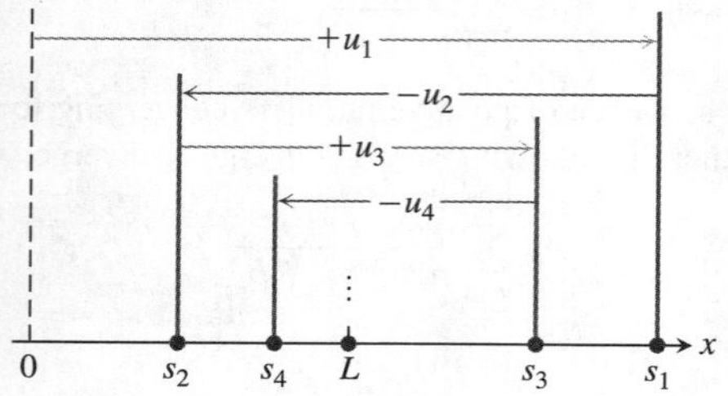

FIGURE 10.15 The partial sums of an alternating series that satisfies the hypotheses of Theorem 15 for $N = 1$ straddle the limit from the beginning.

THEOREM 15—The Alternating Series Test
The series

$$\sum_{n=1}^{\infty} (-1)^{n+1} u_n = u_1 - u_2 + u_3 - u_4 + \cdots$$

converges if the following conditions are satisfied:

1. The u_n's are all positive.
2. The u_n's are eventually nonincreasing: $u_n \geq u_{n+1}$ for all $n \geq N$, for some integer N.
3. $u_n \to 0$.

Proof We look at the case $N = 1$, where we have nonincreasing terms $u_1 \geq u_2 \geq u_3 \geq \cdots$. If n is an even integer, say $n = 2m$, then the sum of the first n terms is

$$s_{2m} = (u_1 - u_2) + (u_3 - u_4) + \cdots + (u_{2m-1} - u_{2m}).$$

Written this way, we see that s_{2m} is the sum of n nonnegative terms, since each term in parentheses is positive or zero. In particular, s_{2m+2} equals s_{2m} plus the nonnegative term $(u_{2m+1} - u_{2m+2})$, so $s_{2m+2} \geq s_{2m}$. This shows that the sequence $\{s_{2m}\}$ is nondecreasing. On the other hand, we can write

$$s_{2m} = u_1 - (u_2 - u_3) - (u_4 - u_5) - \cdots - (u_{2m-2} - u_{2m-1}) - u_{2m}.$$

This shows that $s_{2m} \leq u_1$ for every m. Hence the sequence $\{s_{2m}\}$ is both nondecreasing and bounded from above, so it has a limit, which we call L,

$$\lim_{m\to\infty} s_{2m} = L. \qquad \text{Theorem 6} \tag{4}$$

If n is an odd integer, say $n = 2m + 1$, then the sum of the first n terms is $s_{2m+1} = s_{2m} + u_{2m+1}$. Since $u_n \to 0$,

$$\lim_{m\to\infty} u_{2m+1} = 0.$$

Therefore,

$$\lim_{m\to\infty} s_{2m+1} = \lim_{m\to\infty} s_{2m} + \lim_{m\to\infty} u_{2m+1} = L + 0 = L. \quad (5)$$

Combining the results of Equations (4) and (5) gives $\lim_{n\to\infty} s_n = L$ (Section 10.1, Exercise 143). ■

EXAMPLE 1 The alternating harmonic series

$$\sum_{n=1}^{\infty} (-1)^{n+1}\frac{1}{n} = 1 - \frac{1}{2} + \frac{1}{3} - \frac{1}{4} + \cdots$$

clearly satisfies the three requirements of Theorem 15 with $N = 1$; it therefore converges by the Alternating Series Test. Notice that the test gives no information about what the sum of the series might be. Figure 10.16 shows histograms of the partial sums of the divergent harmonic series and those of the convergent alternating harmonic series. It turns out that the alternating harmonic series converges to ln 2 (Exercise 61 in Section 10.7). ■

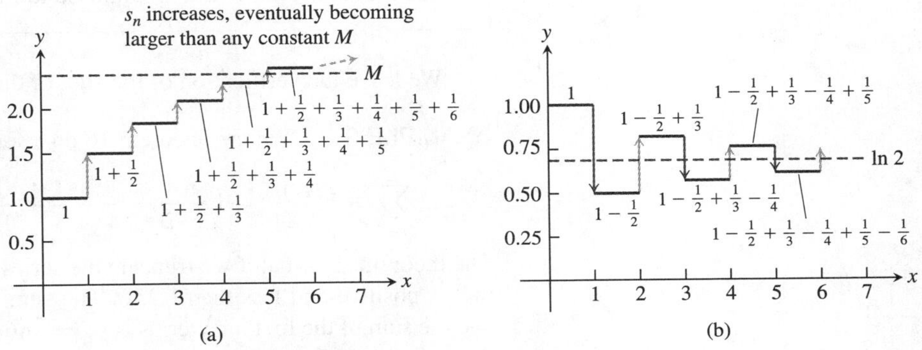

FIGURE 10.16 (a) The harmonic series diverges, with partial sums that eventually exceed any constant. (b) The alternating harmonic series converges to ln 2 ≈ .693.

Rather than directly verifying the definition $u_n \geq u_{n+1}$, a second way to show that the sequence $\{u_n\}$ is nonincreasing is to define a differentiable function $f(x)$ satisfying $f(n) = u_n$. That is, the values of f match the values of the sequence at every positive integer n. If $f'(x) \leq 0$ for all x greater than or equal to some positive integer N, then $f(x)$ is nonincreasing for $x \geq N$. It follows that $f(n) \geq f(n+1)$, or $u_n \geq u_{n+1}$, for $n \geq N$.

EXAMPLE 2 We show that the sequence $u_n = 10n/(n^2 + 16)$ is eventually nonincreasing. Define $f(x) = 10x/(x^2 + 16)$. Then, from the Derivative Quotient Rule,

$$f'(x) = \frac{10(16 - x^2)}{(x^2 + 16)^2} \leq 0 \qquad \text{whenever } x \geq 4.$$

It follows that $u_n \geq u_{n+1}$ for $n \geq 4$. That is, the sequence $\{u_n\}$ is nonincreasing for $n \geq 4$. ■

A graphical interpretation of the partial sums (Figure 10.15) shows how an alternating series converges to its limit L when the three conditions of Theorem 15 are satisfied with $N = 1$. Starting from the origin of the x-axis, we lay off the positive distance $s_1 = u_1$. To find the point corresponding to $s_2 = u_1 - u_2$, we back up a distance equal to u_2. Since $u_2 \leq u_1$, we do not back up any farther than the origin. We continue in this seesaw fashion, backing up or going forward as the signs in the series demand. Each forward or backward step is shorter than (or at most the same size as) the preceding step because $u_{n+1} \leq u_n$. And since the nth term approaches zero as n increases, the size of step we take forward or backward gets smaller and smaller. We oscillate back and forth across the limit L, and the amplitude of oscillation approaches zero. The limit L lies between any two successive sums s_n and s_{n+1} and hence differs from s_n by an amount less than u_{n+1}. If $N > 1$, then the first few forward and backward steps need not get smaller, but from N onward they will decrease. Therefore

$$|L - s_n| < u_{n+1} \qquad \text{for } n \geq N,$$

and we can use this fact to make useful estimates of the sums of convergent alternating series.

THEOREM 16—The Alternating Series Estimation Theorem
If the alternating series $\sum_{n=1}^{\infty} (-1)^{n+1} u_n$ satisfies the three conditions of Theorem 15, then for $n \geq N$,

$$s_n = u_1 - u_2 + \cdots + (-1)^{n+1} u_n$$

approximates the sum L of the series with an error whose absolute value is less than u_{n+1}, the absolute value of the first unused term. Furthermore, the sum L lies between any two successive partial sums s_n and s_{n+1}, and the remainder, $L - s_n$, has the same sign as the first unused term.

We leave the verification of the sign of the remainder for Exercise 87.

EXAMPLE 3 We try Theorem 16 on a series whose sum we know:

$$\sum_{n=0}^{\infty} (-1)^n \frac{1}{2^n} = 1 - \frac{1}{2} + \frac{1}{4} - \frac{1}{8} + \frac{1}{16} - \frac{1}{32} + \frac{1}{64} - \frac{1}{128} + \frac{1}{256} - \cdots.$$

The theorem says that if we truncate the series after the eighth term, we throw away a total that is positive and less than $1/256$. The sum of the first eight terms is $s_8 = 0.6640625$ and the sum of the first nine terms is $s_9 = 0.66796875$. The sum of the geometric series is

$$\frac{1}{1 - (-1/2)} = \frac{1}{3/2} = \frac{2}{3},$$

and we note that $0.6640625 < (2/3) < 0.66796875$. The difference,

$$(2/3) - 0.6640625 = 0.0026041666\ldots,$$

is positive and is less than $(1/256) = 0.00390625$. ■

Conditional Convergence

If we replace all the negative terms in the alternating series in Example 3, changing them to positive terms instead, we obtain the geometric series $\sum 1/2^n$. The original series and the new series of absolute values both converge (although to different sums). For an absolutely convergent series, changing infinitely many of the negative terms in the series to positive values does not change its property of still being a convergent series. Other convergent series may behave differently. The convergent alternating harmonic series has infinitely many negative terms, but if we change its negative terms to positive values, the resulting series is the divergent harmonic series. So the presence of infinitely many negative terms is essential to the convergence of this alternating harmonic series. The following terminology distinguishes these two types of convergent series.

DEFINITION A series that is convergent but not absolutely convergent is called **conditionally convergent**.

The alternating harmonic series is conditionally convergent, or **converges conditionally**. The next example extends that result to the alternating p-series.

EXAMPLE 4 If p is a positive constant, the sequence $\{1/n^p\}$ is a decreasing sequence with limit zero. Therefore, the alternating p-series

$$\sum_{n=1}^{\infty} \frac{(-1)^{n-1}}{n^p} = 1 - \frac{1}{2^p} + \frac{1}{3^p} - \frac{1}{4^p} + \cdots, \qquad p > 0$$

converges.

If $p > 1$, the series converges absolutely as an ordinary p-series. If $0 < p \leq 1$, the series converges conditionally: It converges by the alternating series test, but the corresponding series of absolute values is a divergent p-series. For instance,

$$\text{Absolute convergence } (p = 3/2): \; 1 - \frac{1}{2^{3/2}} + \frac{1}{3^{3/2}} - \frac{1}{4^{3/2}} + \cdots$$

$$\text{Conditional convergence } (p = 1/2): \; 1 - \frac{1}{\sqrt{2}} + \frac{1}{\sqrt{3}} - \frac{1}{\sqrt{4}} + \cdots$$ ■

We need to be careful when using a conditionally convergent series. We have seen with the alternating harmonic series that altering the signs of infinitely many terms of a conditionally convergent series can change its convergence status. Even more, simply changing the order of occurrence of infinitely many of its terms can also have a significant effect, as we now discuss.

Rearranging Series

We can always rearrange the terms of a *finite* collection of numbers without changing their sum. The same result is true for an infinite series that is absolutely convergent (see Exercise 96 for an outline of the proof).

> **THEOREM 17—The Rearrangement Theorem for Absolutely Convergent Series**
> If $\sum_{n=1}^{\infty} a_n$ converges absolutely, and $b_1, b_2, \ldots, b_n, \ldots$ is any arrangement of the sequence $\{a_n\}$, then $\sum_{n=1}^{\infty} b_n$ converges absolutely and
> $$\sum_{n=1}^{\infty} b_n = \sum_{n=1}^{\infty} a_n.$$

On the other hand, if we rearrange the terms of a conditionally convergent series, we can get different results. In fact, for any real number r, a given conditionally convergent series can be rearranged so that its sum is equal to r. (We omit the proof of this.) Here's an example where summing the terms of a conditionally convergent series with different orderings gives different values for the sum.

EXAMPLE 5 We know that the alternating harmonic series $\sum_{n=1}^{\infty} (-1)^{n+1}/n$ converges to some number L. Moreover, by Theorem 16, L lies between the successive partial sums $s_2 = 1/2$ and $s_3 = 5/6$, so $L \neq 0$. If we multiply the series by 2, we obtain

$$\begin{aligned} 2L = 2\sum_{n=1}^{\infty} \frac{(-1)^{n+1}}{n} &= 2\left(1 - \frac{1}{2} + \frac{1}{3} - \frac{1}{4} + \frac{1}{5} - \frac{1}{6} + \frac{1}{7} - \frac{1}{8} + \frac{1}{9} - \frac{1}{10} + \frac{1}{11} - \cdots\right) \\ &= 2 - 1 + \frac{2}{3} - \frac{1}{2} + \frac{2}{5} - \frac{1}{3} + \frac{2}{7} - \frac{1}{4} + \frac{2}{9} - \frac{1}{5} + \frac{2}{11} - \cdots. \end{aligned}$$

Now we change the order of this last sum by grouping each pair of terms with the same odd denominator, but leaving the negative terms with the even denominators as they are placed (so that the denominators are the positive integers in their natural order). This rearrangement gives

$$\begin{aligned}(2-1) - \frac{1}{2} + \left(\frac{2}{3} - \frac{1}{3}\right) - \frac{1}{4} + \left(\frac{2}{5} - \frac{1}{5}\right) - \frac{1}{6} + \left(\frac{2}{7} - \frac{1}{7}\right) - \frac{1}{8} + \cdots \\ = \left(1 - \frac{1}{2} + \frac{1}{3} - \frac{1}{4} + \frac{1}{5} - \frac{1}{6} + \frac{1}{7} - \frac{1}{8} + \frac{1}{9} - \frac{1}{10} + \frac{1}{11} - \cdots\right) \\ = \sum_{n=1}^{\infty} \frac{(-1)^{n+1}}{n} = L.\end{aligned}$$

So when we rearrange the terms of the conditionally convergent series $\sum_{n=1}^{\infty} 2(-1)^{n+1}/n$, the series becomes $\sum_{n=1}^{\infty} (-1)^{n+1}/n$, which is the alternating harmonic series itself. If the two series are the same, it would imply that $2L = L$, which is clearly false since $L \neq 0$. ■

Example 5 shows that we cannot rearrange the terms of a conditionally convergent series and expect the new series to be the same as the original one. When we use a conditionally convergent series, we must add the terms together in the order in which they are given to obtain a correct result. In contrast, Theorem 17 guarantees that the terms of an absolutely convergent series can be summed in any order without affecting the result.

Summary of Tests to Determine Convergence or Divergence

We have developed a variety of tests to determine convergence or divergence for an infinite series of constants. Other tests that we have not presented are sometimes given in more advanced courses. Here is a summary of the tests we have considered.

1. **The *n*th-Term Test for Divergence:** Unless $a_n \to 0$, the series diverges.
2. **Geometric series:** $\sum ar^n$ converges if $|r| < 1$; otherwise, it diverges.
3. ***p*-series:** $\sum 1/n^p$ converges if $p > 1$; otherwise, it diverges.
4. **Series with nonnegative terms:** Try the Integral Test or try comparing to a known series with the Direct Comparison Test or the Limit Comparison Test. Try the Ratio or Root Test.
5. **Series with some negative terms:** Does $\sum |a_n|$ converge by the Ratio or Root Test, or by another of the tests listed above? Remember, absolute convergence implies convergence.
6. **Alternating series:** $\sum a_n$ converges if the series satisfies the conditions of the Alternating Series Test.

EXERCISES 10.6

Convergence of Alternating Series

In Exercises 1–14, determine whether the alternating series converges or diverges. Some of the series do not satisfy the conditions of the Alternating Series Test.

1. $\sum_{n=1}^{\infty} (-1)^{n+1} \frac{1}{\sqrt{n}}$

2. $\sum_{n=1}^{\infty} (-1)^{n+1} \frac{1}{n^{3/2}}$

3. $\sum_{n=1}^{\infty} (-1)^{n+1} \frac{1}{n3^n}$

4. $\sum_{n=2}^{\infty} (-1)^{n} \frac{4}{(\ln n)^2}$

5. $\sum_{n=1}^{\infty} (-1)^{n} \frac{n}{n^2+1}$

6. $\sum_{n=1}^{\infty} (-1)^{n+1} \frac{n^2+5}{n^2+4}$

7. $\sum_{n=1}^{\infty} (-1)^{n+1} \frac{2^n}{n^2}$

8. $\sum_{n=1}^{\infty} (-1)^{n} \frac{10^n}{(n+1)!}$

9. $\sum_{n=1}^{\infty} (-1)^{n+1} \left(\frac{n}{10}\right)^n$

10. $\sum_{n=2}^{\infty} (-1)^{n+1} \frac{1}{\ln n}$

11. $\sum_{n=1}^{\infty}(-1)^{n+1}\frac{\ln n}{n}$

12. $\sum_{n=1}^{\infty}(-1)^{n}\ln\left(1+\frac{1}{n}\right)$

13. $\sum_{n=1}^{\infty}(-1)^{n+1}\frac{\sqrt{n}+1}{n+1}$

14. $\sum_{n=1}^{\infty}(-1)^{n+1}\frac{3\sqrt{n+1}}{\sqrt{n}+1}$

Absolute and Conditional Convergence

Which of the series in Exercises 15–48 converge absolutely, which converge conditionally, and which diverge? Give reasons for your answers.

15. $\sum_{n=1}^{\infty}(-1)^{n+1}(0.1)^{n}$

16. $\sum_{n=1}^{\infty}(-1)^{n+1}\frac{(0.1)^{n}}{n}$

17. $\sum_{n=1}^{\infty}(-1)^{n}\frac{1}{\sqrt{n}}$

18. $\sum_{n=1}^{\infty}\frac{(-1)^{n}}{1+\sqrt{n}}$

19. $\sum_{n=1}^{\infty}\left[\frac{(-1)^{n+1}\sin nx}{n^3}\right]$

20. $\sum_{n=1}^{\infty}(-1)^{n+1}\frac{n!}{2^{n}}$

21. $\sum_{n=1}^{\infty}\left[\frac{(-1)^{n+1}n}{2n-1}\right]$

22. $\sum_{n=1}^{\infty}(-1)^{n}\frac{\sin n}{n^2}$

23. $\sum_{n=1}^{\infty}(-1)^{n+1}\frac{3+n}{5+n}$

24. $\sum_{n=1}^{\infty}\frac{(-2)^{n+1}}{n+5^{n}}$

25. $\sum_{n=1}^{\infty}(-1)^{n+1}\frac{1+n}{n^2}$

26. $\sum_{n=1}^{\infty}(-1)^{n+1}\left(\sqrt[n]{10}\right)$

27. $\sum_{n=1}^{\infty}(-1)^{n}n^2(2/3)^{n}$

28. $\sum_{n=2}^{\infty}(-1)^{n+1}\frac{1}{n\ln n}$

29. $\sum_{n=1}^{\infty}(-1)^{n}\frac{\arctan n}{n^2+1}$

30. $\sum_{n=1}^{\infty}(-1)^{n}\frac{\ln n}{n-\ln n}$

31. $\sum_{n=1}^{\infty}(-1)^{n}\frac{n}{n+1}$

32. $\sum_{n=1}^{\infty}(-5)^{-n}$

33. $\sum_{n=1}^{\infty}\frac{(-100)^{n}}{n!}$

34. $\sum_{n=1}^{\infty}\frac{(-1)^{n-1}}{n^2+2n+1}$

35. $\sum_{n=1}^{\infty}\frac{\cos n\pi}{n\sqrt{n}}$

36. $\sum_{n=1}^{\infty}\frac{\cos n\pi}{n}$

37. $\sum_{n=1}^{\infty}\frac{(-1)^{n}(n+1)^{n}}{(2n)^{n}}$

38. $\sum_{n=1}^{\infty}\frac{(-1)^{n+1}(n!)^2}{(2n)!}$

39. $\sum_{n=1}^{\infty}(-1)^{n}\frac{(2n)!}{2^{n}n!n}$

40. $\sum_{n=1}^{\infty}(-1)^{n}\frac{(n!)^2 3^{n}}{(2n+1)!}$

41. $\sum_{n=1}^{\infty}(-1)^{n}\left(\sqrt{n+1}-\sqrt{n}\right)$

42. $\sum_{n=1}^{\infty}(-1)^{n}\left(\sqrt{n^2+n}-n\right)$

43. $\sum_{n=1}^{\infty}(-1)^{n}\left(\sqrt{n+\sqrt{n}}-\sqrt{n}\right)$

44. $\sum_{n=1}^{\infty}\frac{(-1)^{n}}{\sqrt{n}+\sqrt{n+1}}$

45. $\sum_{n=1}^{\infty}(-1)^{n}\operatorname{sech} n$

46. $\sum_{n=1}^{\infty}(-1)^{n}\operatorname{csch} n$

47. $\frac{1}{4}-\frac{1}{6}+\frac{1}{8}-\frac{1}{10}+\frac{1}{12}-\frac{1}{14}+\cdots$

48. $1+\frac{1}{4}-\frac{1}{9}-\frac{1}{16}+\frac{1}{25}+\frac{1}{36}-\frac{1}{49}-\frac{1}{64}+\cdots$

Error Estimation

In Exercises 49–52, estimate the magnitude of the error involved in using the sum of the first four terms to approximate the sum of the entire series.

49. $\sum_{n=1}^{\infty}(-1)^{n+1}\frac{1}{n}$

50. $\sum_{n=1}^{\infty}(-1)^{n+1}\frac{1}{10^{n}}$

51. $\sum_{n=1}^{\infty}(-1)^{n+1}\frac{(0.01)^{n}}{n}$ As you will see in Section 10.7, the sum is $\ln(1.01)$.

52. $\frac{1}{1+t}=\sum_{n=0}^{\infty}(-1)^{n}t^{n},\quad 0<t<1$

In Exercises 53–56, determine how many terms should be used to estimate the sum of the entire series with an error of less than 0.001.

53. $\sum_{n=1}^{\infty}(-1)^{n}\frac{1}{n^2+3}$

54. $\sum_{n=1}^{\infty}(-1)^{n+1}\frac{n}{n^2+1}$

55. $\sum_{n=1}^{\infty}(-1)^{n+1}\frac{1}{\left(n+3\sqrt{n}\right)^3}$

56. $\sum_{n=1}^{\infty}(-1)^{n}\frac{1}{\ln(\ln(n+2))}$

Determining Convergence or Divergence

In Exercises 57–82, use any method to determine whether the series converges or diverges. Give reasons for your answer.

57. $\sum_{n=1}^{\infty}\frac{3^{n}}{n^{n}}$

58. $\sum_{n=1}^{\infty}\frac{3^{n}}{n^3}$

59. $\sum_{n=1}^{\infty}\left(\frac{1}{n+2}-\frac{1}{n+3}\right)$

60. $\sum_{n=1}^{\infty}\left(\frac{1}{2n+1}-\frac{1}{2n+2}\right)$

61. $\sum_{n=0}^{\infty}(-1)^{n}\frac{(n+2)!}{(2n)!}$

62. $\sum_{n=2}^{\infty}\frac{(3n)!}{(n!)^3}$

63. $\sum_{n=1}^{\infty}n^{-2/\sqrt{5}}$

64. $\sum_{n=2}^{\infty}\frac{3}{10+n^{4/3}}$

65. $\sum_{n=1}^{\infty}\left(1-\frac{2}{n}\right)^{n^2}$

66. $\sum_{n=0}^{\infty}\left(\frac{n+1}{n+2}\right)^{n}$

67. $\sum_{n=1}^{\infty}\frac{n-2}{n^2+3n}\left(-\frac{2}{3}\right)^{n}$

68. $\sum_{n=0}^{\infty}\frac{n+1}{(n+2)!}\left(\frac{3}{2}\right)^{n}$

69. $\frac{1}{2}-\frac{1}{2}+\frac{1}{2}-\frac{1}{2}+\frac{1}{2}-\frac{1}{2}+\cdots$

70. $1-\frac{1}{8}+\frac{1}{64}-\frac{1}{512}+\frac{1}{4096}-\cdots$

71. $\sum_{n=3}^{\infty}\sin\left(\frac{1}{\sqrt{n}}\right)$

72. $\sum_{n=1}^{\infty}\tan(n^{1/n})$

73. $\sum_{n=2}^{\infty}\frac{n}{\ln n}$

74. $\sum_{n=2}^{\infty}\frac{1}{n\sqrt{\ln n}}$

75. $\sum_{n=2}^{\infty}\ln\left(\frac{n+2}{n+1}\right)$

76. $\sum_{n=2}^{\infty}\left(\frac{\ln n}{n}\right)^3$

77. $\sum_{n=2}^{\infty}\frac{1}{1+2+2^2+\cdots+2^{n}}$

78. $\sum_{n=2}^{\infty}\frac{1+3+3^2+\cdots+3^{n-1}}{1+2+3+\cdots+n}$

79. $\sum_{n=0}^{\infty}(-1)^n \frac{e^n}{e^n + e^{n^2}}$

80. $\sum_{n=0}^{\infty}\frac{(2n+3)(2^n+3)}{3^n+2}$

81. $\sum_{n=1}^{\infty}\frac{n^2 3^n}{3\cdot 5\cdot 7\cdots(2n+1)}$

82. $\sum_{n=1}^{\infty}\frac{4\cdot 6\cdot 8\cdots(2n)}{5^{n+1}(n+2)!}$

T Approximate the sums in Exercises 83 and 84 with an error of magnitude less than 5×10^{-6}.

83. $\sum_{n=0}^{\infty}(-1)^n \frac{1}{(2n)!}$ As you will see in Section 10.9, the sum is cos 1, the cosine of 1 radian.

84. $\sum_{n=0}^{\infty}(-1)^n \frac{1}{n!}$ As you will see in Section 10.9 the sum is e^{-1}.

Theory and Examples

85. **a.** The series

$$\frac{1}{3} - \frac{1}{2} + \frac{1}{9} - \frac{1}{4} + \frac{1}{27} - \frac{1}{8} + \cdots + \frac{1}{3^n} - \frac{1}{2^n} + \cdots$$

does not meet one of the conditions of the Alternating Series Test. Which one?

b. Use the Sum Rule for series given in Section 10.2 to find the sum of the series in part (a).

T 86. The limit L of an alternating series that satisfies the conditions of Theorem 15 lies between the values of any two consecutive partial sums. This suggests using the average

$$\frac{s_n + s_{n+1}}{2} = s_n + \frac{1}{2}(-1)^{n+2}a_{n+1}$$

to estimate L. Compute

$$s_{20} + \frac{1}{2}\cdot\frac{1}{21}$$

as an approximation to the sum of the alternating harmonic series. The exact sum is $\ln 2 = 0.69314718\ldots$.

87. **The sign of the remainder of an alternating series that satisfies the conditions of Theorem 15** Prove the assertion in Theorem 16 that whenever an alternating series satisfying the conditions of Theorem 15 is approximated with one of its partial sums, the remainder (the sum of the unused terms) has the same sign as the first unused term. (*Hint:* Group the remainder's terms in consecutive pairs.)

88. Show that the sum of the first $2n$ terms of the series

$$1 - \frac{1}{2} + \frac{1}{2} - \frac{1}{3} + \frac{1}{3} - \frac{1}{4} + \frac{1}{4} - \frac{1}{5} + \frac{1}{5} - \frac{1}{6} + \cdots$$

is the same as the sum of the first n terms of the series

$$\frac{1}{1\cdot 2} + \frac{1}{2\cdot 3} + \frac{1}{3\cdot 4} + \frac{1}{4\cdot 5} + \frac{1}{5\cdot 6} + \cdots.$$

Do these series converge? What is the sum of the first $2n+1$ terms of the first series? If the series converge, what is their sum?

89. Show that if $\sum_{n=1}^{\infty} a_n$ diverges, then $\sum_{n=1}^{\infty} |a_n|$ diverges.

90. Show that if $\sum_{n=1}^{\infty} a_n$ converges absolutely, then

$$\left|\sum_{n=1}^{\infty} a_n\right| \le \sum_{n=1}^{\infty}|a_n|.$$

91. Show that if $\sum_{n=1}^{\infty} a_n$ and $\sum_{n=1}^{\infty} b_n$ both converge absolutely, then so do the following.

a. $\sum_{n=1}^{\infty}(a_n + b_n)$ **b.** $\sum_{n=1}^{\infty}(a_n - b_n)$

c. $\sum_{n=1}^{\infty} ka_n$ (k any number)

92. Show by example that $\sum_{n=1}^{\infty} a_n b_n$ may diverge even if $\sum_{n=1}^{\infty} a_n$ and $\sum_{n=1}^{\infty} b_n$ both converge.

93. Prove that if $\sum a_n$ converges absolutely, then $\sum a_n^2$ converges.

94. Does the series

$$\sum_{n=1}^{\infty}\left(\frac{1}{n} - \frac{1}{n^2}\right)$$

converge or diverge? Justify your answer.

T 95. In the alternating harmonic series, suppose the goal is to arrange the terms to get a new series that converges to $-1/2$. Start the new arrangement with the first negative term, which is $-1/2$. Whenever you have a sum that is less than or equal to $-1/2$, start introducing positive terms, taken in order, until the new total is greater than $-1/2$. Then add negative terms until the total is less than or equal to $-1/2$ again. Continue this process until your partial sums have been above the target at least three times and finish at or below it. If s_n is the sum of the first n terms of your new series, plot the points (n, s_n) to illustrate how the sums are behaving.

96. **Outline of the proof of the Rearrangement Theorem (Theorem 17)**

a. Let ε be a positive real number, let $L = \sum_{n=1}^{\infty} a_n$, and let $s_k = \sum_{n=1}^{k} a_n$. Show that for some index N_1 and for some index $N_2 \ge N_1$,

$$\sum_{n=N_1}^{\infty}|a_n| < \frac{\varepsilon}{2} \quad\text{and}\quad |s_{N_2} - L| < \frac{\varepsilon}{2}.$$

Since all the terms $a_1, a_2, \ldots, a_{N_2}$ appear somewhere in the sequence $\{b_n\}$, there is an index $N_3 \ge N_2$ such that if $n \ge N_3$, then $\left(\sum_{k=1}^{n} b_k\right) - s_{N_2}$ is at most a sum of terms a_m with $m \ge N_1$. Therefore, if $n \ge N_3$, then

$$\left|\sum_{k=1}^{n} b_k - L\right| \le \left|\sum_{k=1}^{n} b_k - s_{N_2}\right| + |s_{N_2} - L|$$
$$\le \sum_{k=N_1}^{\infty}|a_k| + |s_{N_2} - L| < \varepsilon.$$

b. The argument in part (a) shows that if $\sum_{n=1}^{\infty} a_n$ converges absolutely, then $\sum_{n=1}^{\infty} b_n$ converges and $\sum_{n=1}^{\infty} b_n = \sum_{n=1}^{\infty} a_n$. Now show that because $\sum_{n=1}^{\infty} a_n$ converges, $\sum_{n=1}^{\infty} b_n$ converges to $\sum_{n=1}^{\infty} a_n$.

10.7 Power Series

Now that we can test many infinite series of numbers for convergence, we can study sums that look like "infinite polynomials." We call these sums *power series* because they are defined as infinite series of powers of some variable, in our case x. Like polynomials, power series can be added, subtracted, multiplied, differentiated, and integrated to give new power series. With power series we can extend the methods of calculus to a vast array of functions, making the techniques of calculus applicable in an even wider setting.

Power Series and Convergence

We begin with the formal definition, which specifies the notation and terminology used for power series.

DEFINITIONS **A power series about $x = 0$** is a series of the form

$$\sum_{n=0}^{\infty} c_n x^n = c_0 + c_1 x + c_2 x^2 + \cdots + c_n x^n + \cdots. \tag{1}$$

A power series about $x = a$ is a series of the form

$$\sum_{n=0}^{\infty} c_n (x-a)^n = c_0 + c_1(x-a) + c_2(x-a)^2 + \cdots + c_n(x-a)^n + \cdots \tag{2}$$

in which the **center** a and the **coefficients** $c_0, c_1, c_2, \ldots, c_n, \ldots$ are constants.

Equation (1) is the special case obtained by taking $a = 0$ in Equation (2). We will see that a power series defines a function $f(x)$ on a certain interval where it converges. Moreover, this function will be shown to be continuous and differentiable over the interior of that interval.

EXAMPLE 1 Taking all the coefficients to be 1 in Equation (1) gives the geometric power series

$$\sum_{n=0}^{\infty} x^n = 1 + x + x^2 + \cdots + x^n + \cdots.$$

This is the geometric series with first term 1 and ratio x. It converges to $1/(1-x)$ for $|x| < 1$. We express this fact by writing

$$\frac{1}{1-x} = 1 + x + x^2 + \cdots + x^n + \cdots, \qquad -1 < x < 1. \tag{3}$$

■

Power Series for $\frac{1}{1-x}$

$$\frac{1}{1-x} = \sum_{n=0}^{\infty} x^n, \quad |x| < 1$$

Up to now, we have used Equation (3) as a formula for the sum of the series on the right. We now change the focus: We think of the partial sums of the series on the right as polynomials $P_n(x)$ that approximate the function on the left. For values of x near zero, we need take only a few terms of the series to get a good approximation. As we move toward $x = 1$, or -1, we must take more terms. Figure 10.17 shows the graphs of $f(x) = 1/(1-x)$ and the approximating polynomials $y_n = P_n(x)$ for $n = 0, 1, 2,$ and 8. The function $f(x) = 1/(1-x)$ is not continuous on intervals containing $x = 1$, where it has a vertical asymptote. The approximations do not apply when $x \geq 1$.

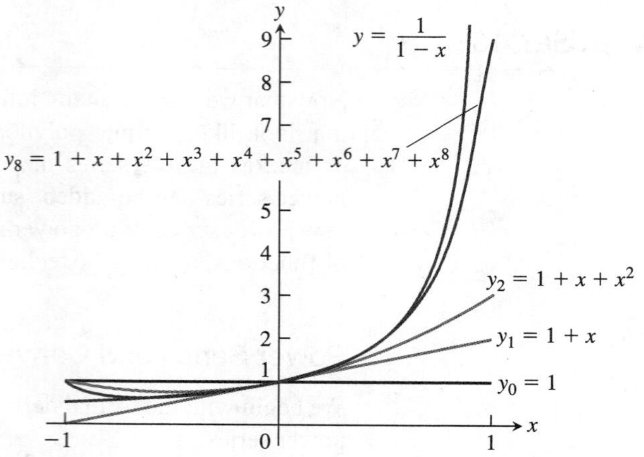

FIGURE 10.17 The graphs of $f(x) = 1/(1 - x)$ in Example 1 and four of its polynomial approximations.

EXAMPLE 2 The power series

$$1 - \frac{1}{2}(x-2) + \frac{1}{4}(x-2)^2 + \cdots + \left(-\frac{1}{2}\right)^n (x-2)^n + \cdots \tag{4}$$

matches Equation (2) with $a = 2, c_0 = 1, c_1 = -1/2, c_2 = 1/4, \ldots, c_n = (-1/2)^n$. This is a geometric series with first term 1 and ratio $r = -\frac{x-2}{2}$. The series converges for $\left|\frac{x-2}{2}\right| < 1$, which simplifies to $0 < x < 4$. The sum is

$$\frac{1}{1-r} = \frac{1}{1 + \frac{x-2}{2}} = \frac{2}{x},$$

so

$$\frac{2}{x} = 1 - \frac{(x-2)}{2} + \frac{(x-2)^2}{4} - \cdots + \left(-\frac{1}{2}\right)^n (x-2)^n + \cdots, \qquad 0 < x < 4.$$

Series (4) generates useful polynomial approximations of $f(x) = 2/x$ for values of x near 2:

$$P_0(x) = 1$$
$$P_1(x) = 1 - \frac{1}{2}(x-2) = 2 - \frac{x}{2}$$
$$P_2(x) = 1 - \frac{1}{2}(x-2) + \frac{1}{4}(x-2)^2 = 3 - \frac{3x}{2} + \frac{x^2}{4},$$

and so on (Figure 10.18). ■

FIGURE 10.18 The graphs of $f(x) = 2/x$ and its first three polynomial approximations (Example 2).

The following example illustrates how we test a power series for convergence by using the Ratio Test to see where it converges and where it diverges.

EXAMPLE 3 For what values of x do the following power series converge?

(a) $\displaystyle\sum_{n=1}^{\infty} (-1)^{n-1} \frac{x^n}{n} = x - \frac{x^2}{2} + \frac{x^3}{3} - \cdots$

(b) $\displaystyle\sum_{n=1}^{\infty} (-1)^{n-1} \frac{x^{2n-1}}{2n-1} = x - \frac{x^3}{3} + \frac{x^5}{5} - \cdots$

(c) $\displaystyle\sum_{n=0}^{\infty} \frac{x^n}{n!} = 1 + x + \frac{x^2}{2!} + \frac{x^3}{3!} + \cdots$

(d) $\displaystyle\sum_{n=0}^{\infty} n!\,x^n = 1 + x + 2!\,x^2 + 3!\,x^3 + \cdots$

Solution Apply the Ratio Test to the series $\sum u_n$, where u_n is the nth term of the power series in question.

(a) $\displaystyle\left|\frac{u_{n+1}}{u_n}\right| = \left|\frac{x^{n+1}}{n+1} \cdot \frac{n}{x}\right| = \frac{n}{n+1}\,|x| \to |x|.$

By the Ratio Test, this series converges absolutely for $|x| < 1$, and it diverges for $|x| > 1$. At $x = 1$, we obtain the alternating harmonic series $1 - 1/2 + 1/3 - 1/4 + \cdots$, which converges (though it does not converge absolutely). At $x = -1$, we get $-1 - 1/2 - 1/3 - 1/4 - \cdots$, the negative of the harmonic series, which diverges. Series (a) converges for $-1 < x \le 1$ and diverges elsewhere. The convergence is absolute for $-1 < x < 1$, but conditional at the point $x = 1$.

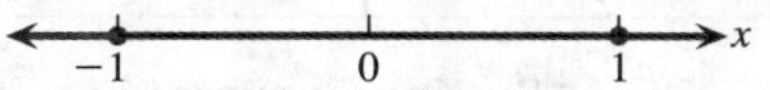

We will see in Example 6 that this series converges to the function $\ln(1 + x)$ on the interval $(-1, 1]$ (see Figure 10.19).

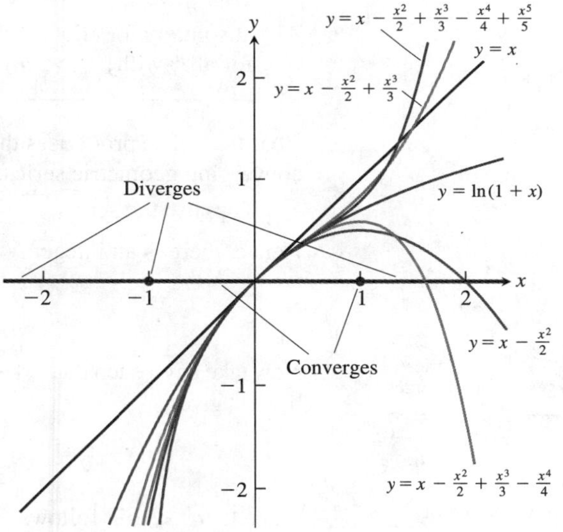

FIGURE 10.19 The power series $x - \dfrac{x^2}{2} + \dfrac{x^3}{3} - \dfrac{x^4}{4} + \cdots$ converges on the interval $(-1, 1]$.

(b) $\displaystyle\left|\frac{u_{n+1}}{u_n}\right| = \left|\frac{x^{2n+1}}{2n+1} \cdot \frac{2n-1}{x^{2n-1}}\right| = \frac{2n-1}{2n+1}x^2 \to x^2.$ $\quad 2(n+1) - 1 = 2n + 1$

By the Ratio Test, the series converges absolutely for $x^2 < 1$ and diverges for $x^2 > 1$. At $x = 1$ the series becomes $1 - 1/3 + 1/5 - 1/7 + \cdots$, which converges by the Alternating Series Theorem. It also converges at $x = -1$ because it is again an alternating series that satisfies the conditions for convergence. The value at $x = -1$ is the negative of the value at $x = 1$. Series (b) converges for $-1 \le x \le 1$ and diverges elsewhere. The convergence is absolute for $-1 < x < 1$, but conditional at the points $x = -1$ and $x = 1$.

(c) $\left|\dfrac{u_{n+1}}{u_n}\right| = \left|\dfrac{x^{n+1}}{(n+1)!} \cdot \dfrac{n!}{x^n}\right| = \dfrac{|x|}{n+1} \rightarrow 0$ for every x. $\quad \dfrac{n!}{(n+1)!} = \dfrac{1 \cdot 2 \cdot 3 \cdots n}{1 \cdot 2 \cdot 3 \cdots n \cdot (n+1)}$

The series converges absolutely for all x.

(d) $\left|\dfrac{u_{n+1}}{u_n}\right| = \left|\dfrac{(n+1)!\,x^{n+1}}{n!\,x^n}\right| = (n+1)|x| \rightarrow \begin{cases} 0 & \text{if } x = 0 \\ \infty & \text{if } x \neq 0. \end{cases}$

The series diverges for all values of x except $x = 0$.

■

The previous example illustrated how a power series might converge. The next result shows that if a power series converges at a nonzero value, then it converges over an entire interval of values. The interval might be finite or infinite and might contain one, both, or none of its endpoints. We will see that each endpoint of a finite interval must be tested independently for convergence or divergence.

THEOREM 18—The Convergence Theorem for Power Series
If the power series

$$\sum_{n=0}^{\infty} a_n x^n = a_0 + a_1 x + a_2 x^2 + \cdots \text{ converges at } x = c \neq 0, \text{ then it converges}$$

absolutely for all x with $|x| < |c|$. If the series diverges at $x = d$, then it diverges for all x with $|x| > |d|$.

Proof The proof uses the Direct Comparison Test, with the given series compared to a converging geometric series.

Suppose the series $\sum_{n=0}^{\infty} a_n c^n$ converges. Then $\lim_{n\to\infty} a_n c^n = 0$ by the nth-Term Test. Hence, there is an integer N such that $|a_n c^n| < 1$ for all $n > N$, so

$$|a_n| < \frac{1}{|c|^n} \qquad \text{for } n > N. \tag{5}$$

Now take any x such that $|x| < |c|$, so that $|x|/|c| < 1$. Multiplying both sides of Equation (5) by $|x|^n$ gives

$$|a_n||x|^n < \frac{|x|^n}{|c|^n} \qquad \text{for } n > N.$$

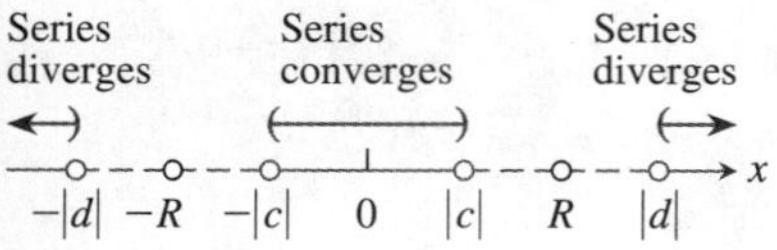

FIGURE 10.20 Convergence of $\sum a_n x^n$ at $x = c$ implies absolute convergence on the interval $-|c| < x < |c|$; divergence at $x = d$ implies divergence for $|x| > |d|$. The corollary to Theorem 18 asserts the existence of a radius of convergence $R \geq 0$. For $|x| < R$ the series converges absolutely, and for $|x| > R$ it diverges.

Since $|x/c| < 1$, it follows that the geometric series $\sum_{n=0}^{\infty} |x/c|^n$ converges. By the Direct Comparison Test (Theorem 10), the series $\sum_{n=0}^{\infty} |a_n||x^n|$ converges, so the original power series $\sum_{n=0}^{\infty} a_n x^n$ converges absolutely for $-|c| < x < |c|$, as claimed by the theorem. (See Figure 10.20.)

Now suppose that the series $\sum_{n=0}^{\infty} a_n x^n$ diverges at $x = d$. If x is a number with $|x| > |d|$ and the series converges at x, then the first half of the theorem shows that the series also converges at d, contrary to our assumption. So the series diverges for all x with $|x| > |d|$. ■

To simplify the notation, Theorem 18 deals with the convergence of series of the form $\sum a_n x^n$. For series of the form $\sum a_n (x - a)^n$, we can replace $x - a$ by t and apply the results to the series $\sum a_n t^n$.

The Radius of Convergence of a Power Series

The theorem we have just proved and the examples we have studied lead to the conclusion that a power series $\sum c_n (x - a)^n$ behaves in one of three possible ways. It might

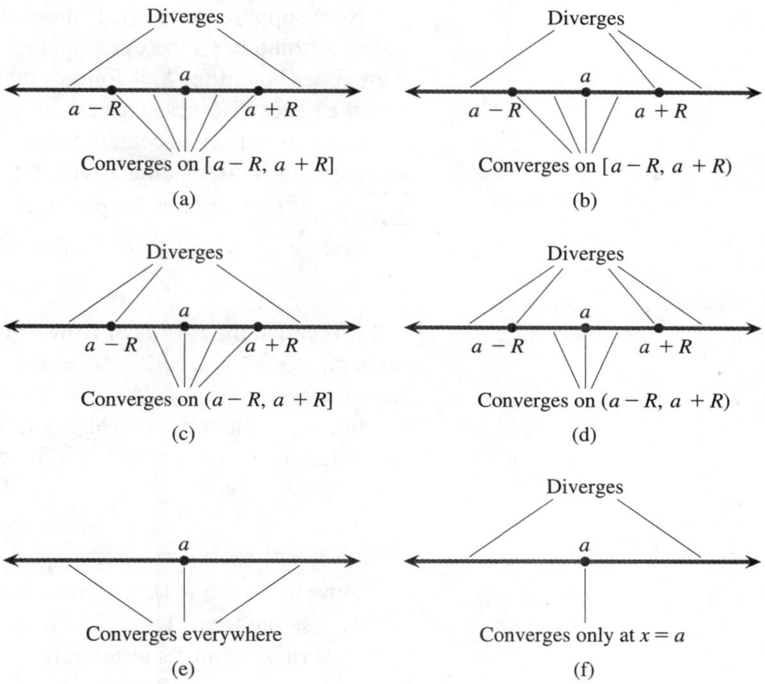

FIGURE 10.21 The six possibilities for an interval of convergence.

converge only at $x = a$, or converge everywhere, or converge on some interval of radius R centered at $x = a$. We prove this as a corollary to Theorem 18. When we also consider the convergence at the endpoints of an interval, we see that there are six different possibilities, shown in Figure 10.21.

Corollary to Theorem 18

The convergence of the series $\sum c_n(x-a)^n$ is described by one of the following three cases:

1. There is a positive number R such that the series diverges for x with $|x - a| > R$ but converges absolutely for x with $|x - a| < R$. The series may or may not converge at either of the endpoints $x = a - R$ and $x = a + R$.
2. The series converges absolutely for every x $(R = \infty)$.
3. The series converges at $x = a$ and diverges elsewhere $(R = 0)$.

Proof We first consider the case where $a = 0$, so that we have a power series $\sum_{n=0}^{\infty} c_n x^n$ centered at 0. If the series converges everywhere we are in Case 2. If it converges only at $x = 0$ then we are in Case 3. Otherwise there is a nonzero number d such that $\sum_{n=0}^{\infty} c_n d^n$ diverges. Let S be the set of values of x for which $\sum_{n=0}^{\infty} c_n x^n$ converges. The set S does not include any x with $|x| > |d|$, since Theorem 18 implies the series diverges at all such values. So the set S is bounded. By the Completeness Property of the Real Numbers (Appendix A.7) S has a least upper bound R. (This is the smallest number with the property that all elements of S are less than or equal to R.) Since we are not in Case 3, the series converges at some number $b \neq 0$ and, by Theorem 18, also on the open interval $(-|b|, |b|)$. Therefore, $R > 0$.

If $|x| < R$ then there is a number c in S with $|x| < c < R$, since otherwise R would not be the least upper bound for S. The series converges at c since $c \in S$, so by Theorem 18 the series converges absolutely at x.

Now suppose $|x| > R$. If the series converges at x, then Theorem 18 implies it converges absolutely on the open interval $(-|x|, |x|)$, so that S contains this interval. Since R is an upper bound for S, it follows that $|x| \leq R$, which is a contradiction. So if $|x| > R$, then the series diverges. This proves the theorem for power series centered at $a = 0$.

For a power series centered at an arbitrary point $x = a$, set $t = x - a$ and repeat the argument above, replacing x with t. Since $t = 0$ when $x = a$, convergence of the series $\sum_{n=0}^{\infty} |c_n t^n|$ on a radius R open interval centered at $t = 0$ corresponds to convergence of the series $\sum_{n=0}^{\infty} |c_n (x - a)^n|$ on a radius R open interval centered at $x = a$. ■

R is called the **radius of convergence** of the power series, and the interval of radius R centered at $x = a$ is called the **interval of convergence**. The interval of convergence may be open, closed, or half-open, depending on the particular series. At points x with $|x - a| < R$, the series converges absolutely. If the series converges for all values of x, we say its radius of convergence is infinite. If it converges only at $x = a$, we say its radius of convergence is zero.

How to Test a Power Series for Convergence

1. Use the Ratio Test or the Root Test to find the largest open interval where the series converges absolutely,

$$|x - a| < R \qquad \text{or} \qquad a - R < x < a + R.$$

2. If R is finite, test for convergence or divergence at each endpoint, as in Examples 3a and b.
3. If R is finite, the series diverges for $|x - a| > R$.

Operations on Power Series

On the intersection of their intervals of convergence, two power series can be added and subtracted term by term just like series of constants (Theorem 8). They can be multiplied just as we multiply polynomials, but we often limit the computation of the product to the first few terms, which are the most important. The following result gives a formula for the coefficients in the product, but we omit the proof. (Power series can also be divided in a way similar to division of polynomials, but we do not give a formula for the general coefficient here.)

THEOREM 19—Series Multiplication for Power Series
If $A(x) = \sum_{n=0}^{\infty} a_n x^n$ and $B(x) = \sum_{n=0}^{\infty} b_n x^n$ converge absolutely for $|x| < R$, and

$$c_n = a_0 b_n + a_1 b_{n-1} + a_2 b_{n-2} + \cdots + a_{n-1} b_1 + a_n b_0 = \sum_{k=0}^{n} a_k b_{n-k},$$

then $\sum_{n=0}^{\infty} c_n x^n$ converges absolutely to $A(x)B(x)$ for $|x| < R$:

$$\left(\sum_{n=0}^{\infty} a_n x^n\right)\left(\sum_{n=0}^{\infty} b_n x^n\right) = \sum_{n=0}^{\infty} c_n x^n.$$

Finding the general coefficient c_n in the product of two power series can be tedious, and the term may be unwieldy. The following computation provides an illustration of a product where we find the first few terms by multiplying the terms of the second series by each term of the first series:

$$\left(\sum_{n=0}^{\infty} x^n\right) \cdot \left(\sum_{n=0}^{\infty} (-1)^n \frac{x^{n+1}}{n+1}\right)$$

$$= (1 + x + x^2 + \cdots)\left(x - \frac{x^2}{2} + \frac{x^3}{3} - \cdots\right) \qquad \text{Multiply second series ...}$$

$$= \underbrace{\left(x - \frac{x^2}{2} + \frac{x^3}{3} - \cdots\right)}_{\text{by } 1} + \underbrace{\left(x^2 - \frac{x^3}{2} + \frac{x^4}{3} - \cdots\right)}_{\text{by } x} + \underbrace{\left(x^3 - \frac{x^4}{2} + \frac{x^5}{3} - \cdots\right)}_{\text{by } x^2} + \cdots$$

$$= x + \frac{x^2}{2} + \frac{5x^3}{6} + \cdots. \qquad \text{and gather the first three powers.}$$

We can also substitute a function $f(x)$ for x in a convergent power series.

THEOREM 20 If $\sum_{n=0}^{\infty} a_n x^n$ converges absolutely for $|x| < R$, and f is a continuous function, then $\sum_{n=0}^{\infty} a_n (f(x))^n$ converges absolutely on the set of points x that satisfy $|f(x)| < R$.

For example, since $1/(1 - x) = \sum_{n=0}^{\infty} x^n$ converges absolutely for $|x| < 1$, it follows from Theorem 20 that $1/(1 - 4x^2) = \sum_{n=0}^{\infty} (4x^2)^n$ converges absolutely when x satisfies $|4x^2| < 1$, or, equivalently, when $|x| < 1/2$.

Theorem 21 says that a power series can be differentiated term by term at each interior point of its interval of convergence. A proof of a restricted case of the theorem is outlined in Exercise 66.

THEOREM 21—Term-by-Term Differentiation
If $\sum c_n (x - a)^n$ has radius of convergence $R > 0$, it defines a function

$$f(x) = \sum_{n=0}^{\infty} c_n (x - a)^n \qquad \text{on the interval} \qquad a - R < x < a + R.$$

This function f has derivatives of all orders inside the interval, and we obtain the derivatives by differentiating the original series term by term:

$$f'(x) = \sum_{n=1}^{\infty} n c_n (x - a)^{n-1},$$

$$f''(x) = \sum_{n=2}^{\infty} n(n - 1) c_n (x - a)^{n-2},$$

and so on. Each of these derived series converges at every point of the interval $a - R < x < a + R$.

EXAMPLE 4 Find series for $f'(x)$ and $f''(x)$ if

$$f(x) = \frac{1}{1 - x} = 1 + x + x^2 + x^3 + x^4 + \cdots + x^n + \cdots$$

$$= \sum_{n=0}^{\infty} x^n, \qquad -1 < x < 1.$$

Solution We differentiate the power series on the right term by term:

$$f'(x) = \frac{1}{(1-x)^2} = 1 + 2x + 3x^2 + 4x^3 + \cdots + nx^{n-1} + \cdots$$
$$= \sum_{n=1}^{\infty} nx^{n-1}, \qquad -1 < x < 1;$$
$$f''(x) = \frac{2}{(1-x)^3} = 2 + 6x + 12x^2 + \cdots + n(n-1)x^{n-2} + \cdots$$
$$= \sum_{n=2}^{\infty} n(n-1)x^{n-2}, \qquad -1 < x < 1. \qquad \blacksquare$$

Caution Term-by-term differentiation might not work for series that are not power series. For example, the trigonometric series

$$\sum_{n=1}^{\infty} \frac{\sin(n!x)}{n^2}$$

converges for all x. But if we differentiate term by term, we get the series

$$\sum_{n=1}^{\infty} \frac{n!\cos(n!x)}{n^2},$$

which diverges for all x. These are not power series since they are not sums of positive integer powers of x. ●

It is also true that a power series can be integrated term by term throughout its interval of convergence. The proof is outlined in Exercise 67.

THEOREM 22—Term-by-Term Integration
Suppose that

$$f(x) = \sum_{n=0}^{\infty} c_n(x-a)^n$$

converges for $a - R < x < a + R$ (where $R > 0$). Then

$$\sum_{n=0}^{\infty} c_n \frac{(x-a)^{n+1}}{n+1}$$

converges for $a - R < x < a + R$ and

$$\int f(x)\,dx = \sum_{n=0}^{\infty} c_n \frac{(x-a)^{n+1}}{n+1} + C$$

for $a - R < x < a + R$.

EXAMPLE 5 Identify the function

$$f(x) = \sum_{n=0}^{\infty} \frac{(-1)^n x^{2n+1}}{2n+1} = x - \frac{x^3}{3} + \frac{x^5}{5} - \cdots, \qquad -1 \le x \le 1.$$

Solution We differentiate the original series term by term and get

$$f'(x) = 1 - x^2 + x^4 - x^6 + \cdots, \qquad -1 < x < 1. \qquad \text{Theorem 21}$$

This is a geometric series with first term 1 and ratio $-x^2$, so

$$f'(x) = \frac{1}{1-(-x^2)} = \frac{1}{1+x^2}.$$

We can now integrate $f'(x) = 1/(1+x^2)$ to get

$$\int f'(x)\,dx = \int \frac{dx}{1+x^2} = \arctan x + C.$$

The series for $f(x)$ is zero when $x = 0$, so $C = 0$. Hence

$$f(x) = x - \frac{x^3}{3} + \frac{x^5}{5} - \frac{x^7}{7} + \cdots = \arctan x, \qquad -1 < x < 1. \tag{6}$$

The Number π as a Series

$$\frac{\pi}{4} = \arctan 1 = \sum_{n=0}^{\infty} \frac{(-1)^n}{2n+1}$$

It can be shown that the series also converges to arctan x at the endpoints $x = \pm 1$, but we omit the proof. ■

Notice that the original series in Example 5 converges at both endpoints of the original interval of convergence, but Theorem 22 can guarantee only the convergence of the integrated series inside the interval.

EXAMPLE 6 The series

$$\frac{1}{1+t} = 1 - t + t^2 - t^3 + \cdots$$

converges on the open interval $-1 < t < 1$. Therefore,

$$\ln(1+x) = \int_0^x \frac{1}{1+t}\,dt = t - \frac{t^2}{2} + \frac{t^3}{3} - \frac{t^4}{4} + \cdots \Big]_0^x \qquad \text{Theorem 22}$$

$$= x - \frac{x^2}{2} + \frac{x^3}{3} - \frac{x^4}{4} + \cdots,$$

or

$$\ln(1+x) = \sum_{n=1}^{\infty} \frac{(-1)^{n-1} x^n}{n}, \qquad -1 < x < 1.$$

Alternating Harmonic Series Sum

$$\ln 2 = \sum_{n=1}^{\infty} \frac{(-1)^{n-1}}{n}$$

It can also be shown that the series converges at $x = 1$ to the number ln 2, but that was not guaranteed by the theorem. A proof of this is outlined in Exercise 63. ■

EXERCISES 10.7

Intervals of Convergence

In Exercises 1–36, **(a)** find the series' radius and interval of convergence. For what values of x does the series converge **(b)** absolutely, **(c)** conditionally?

1. $\sum_{n=0}^{\infty} x^n$

2. $\sum_{n=0}^{\infty} (x+5)^n$

3. $\sum_{n=0}^{\infty} (-1)^n (4x+1)^n$

4. $\sum_{n=1}^{\infty} \frac{(3x-2)^n}{n}$

5. $\sum_{n=0}^{\infty} \frac{(x-2)^n}{10^n}$

6. $\sum_{n=0}^{\infty} (2x)^n$

7. $\sum_{n=0}^{\infty} \frac{nx^n}{n+2}$

8. $\sum_{n=1}^{\infty} \frac{(-1)^n (x+2)^n}{n}$

9. $\sum_{n=1}^{\infty} \frac{x^n}{n\sqrt{n}\,3^n}$

10. $\sum_{n=1}^{\infty} \frac{(x-1)^n}{\sqrt{n}}$

11. $\sum_{n=0}^{\infty} \frac{(-1)^n x^n}{n!}$

12. $\sum_{n=0}^{\infty} \frac{3^n x^n}{n!}$

13. $\sum_{n=1}^{\infty} \frac{4^n x^{2n}}{n}$

14. $\sum_{n=1}^{\infty} \frac{(x-1)^n}{n^3 3^n}$

15. $\sum_{n=0}^{\infty} \frac{x^n}{\sqrt{n^2+3}}$

16. $\sum_{n=0}^{\infty} \frac{(-1)^n x^{n+1}}{\sqrt{n}+3}$

17. $\sum_{n=0}^{\infty} \frac{n(x+3)^n}{5^n}$

18. $\sum_{n=0}^{\infty} \frac{(x-3)^{2n}}{3^n}$

19. $\sum_{n=0}^{\infty} \frac{(-2)^n (2x+1)^n}{n^2}$

20. $\sum_{n=1}^{\infty} \sqrt[n]{n}(2x+5)^n$

21. $\sum_{n=1}^{\infty} \frac{1}{n(1-x)^n}$

22. $\sum_{n=1}^{\infty} \frac{(-1)^n 3^{2n}(x-2)^n}{3n}$

23. $\sum_{n=1}^{\infty} \left(1+\frac{1}{n}\right)^n x^n$

24. $\sum_{n=1}^{\infty} (\ln n) x^n$

25. $\sum_{n=1}^{\infty} n^n x^n$

26. $\sum_{n=0}^{\infty} n!(x-4)^n$

27. $\sum_{n=1}^{\infty} \frac{(-1)^{n+1}(x+2)^n}{n 2^n}$

28. $\sum_{n=0}^{\infty} (-2)^n (n+1)(x-1)^n$

29. $\sum_{n=2}^{\infty} \frac{x^n}{n(\ln n)^2}$ Get the information you need about $\sum 1/\left(n(\ln n)^2\right)$ from Section 10.3, Exercise 61.

30. $\sum_{n=2}^{\infty} \frac{x^n}{n \ln n}$ Get the information you need about $\sum 1/(n \ln n)$ from Section 10.3, Exercise 60.

31. $\sum_{n=1}^{\infty} \frac{(4x-5)^{2n+1}}{n^{3/2}}$

32. $\sum_{n=1}^{\infty} \frac{(3x+1)^{n+1}}{2n+2}$

33. $\sum_{n=1}^{\infty} \frac{1}{2 \cdot 4 \cdot 6 \cdots (2n)} x^n$

34. $\sum_{n=1}^{\infty} \frac{3 \cdot 5 \cdot 7 \cdots (2n+1)}{n^2 \cdot 2^n} x^{n+1}$

35. $\sum_{n=1}^{\infty} \frac{1+2+3+\cdots+n}{1^2+2^2+3^2+\cdots+n^2} x^n$

36. $\sum_{n=1}^{\infty} (\sqrt{n+1}-\sqrt{n})(x-3)^n$

In Exercises 37–42, find the series' radius of convergence.

37. $\sum_{n=1}^{\infty} \frac{n!}{3 \cdot 6 \cdot 9 \cdots 3n} x^n$

38. $\sum_{n=1}^{\infty} \left(\frac{2 \cdot 4 \cdot 6 \cdots (2n)}{2 \cdot 5 \cdot 8 \cdots (3n-1)}\right)^2 x^n$

39. $\sum_{n=1}^{\infty} \frac{(n!)^2}{2^n (2n)!} x^n$

40. $\sum_{n=1}^{\infty} \frac{n! x^n}{n^n}$

41. $\sum_{n=1}^{\infty} \frac{(n+a)!}{n!(n+b)!} x^n$, where a, b are positive integers

42. $\sum_{n=1}^{\infty} \left(\frac{n}{n+1}\right)^{n^2} x^n$

(*Hint:* Apply the Root Test.)

In Exercises 43–50, use Theorem 20 to find the series' interval of convergence and, within this interval, the sum of the series as a function of x.

43. $\sum_{n=0}^{\infty} 3^n x^n$

44. $\sum_{n=0}^{\infty} (e^x - 4)^n$

45. $\sum_{n=0}^{\infty} \frac{(x-1)^{2n}}{4^n}$

46. $\sum_{n=0}^{\infty} \frac{(x+1)^{2n}}{9^n}$

47. $\sum_{n=0}^{\infty} \left(\frac{\sqrt{x}}{2}-1\right)^n$

48. $\sum_{n=0}^{\infty} (\ln x)^n$

49. $\sum_{n=0}^{\infty} \left(\frac{x^2+1}{3}\right)^n$

50. $\sum_{n=0}^{\infty} \left(\frac{x^2-1}{2}\right)^n$

Using the Geometric Series

51. In Example 2 we represented the function $f(x) = 2/x$ as a power series about $x = 2$. Use a geometric series to represent $f(x)$ as a power series about $x = 1$, and find its interval of convergence.

52. Use a geometric series to represent each of the given functions as a power series about $x = 0$, and find their intervals of convergence.

   a. $f(x) = \frac{5}{3-x}$ b. $g(x) = \frac{3}{x-2}$

53. Represent the function $g(x)$ in Exercise 52 as a power series about $x = 5$, and find the interval of convergence.

54. a. Find the interval of convergence of the power series

$$\sum_{n=0}^{\infty} \frac{8}{4^{n+2}} x^n.$$

   b. Represent the power series in part (a) as a power series about $x = 3$ and identify the interval of convergence of the new series. (Later in the chapter you will understand why the new interval of convergence does not necessarily include all of the numbers in the original interval of convergence.)

Theory and Examples

55. For what values of x does the series

$$1 - \frac{1}{2}(x-3) + \frac{1}{4}(x-3)^2 + \cdots + \left(-\frac{1}{2}\right)^n (x-3)^n + \cdots$$

converge? What is its sum? What series do you get if you differentiate the given series term by term? For what values of x does the new series converge? What is its sum?

56. If you integrate the series in Exercise 55 term by term, what new series do you get? For what values of x does the new series converge, and what is another name for its sum?

57. The series

$$\sin x = x - \frac{x^3}{3!} + \frac{x^5}{5!} - \frac{x^7}{7!} + \frac{x^9}{9!} - \frac{x^{11}}{11!} + \cdots$$

converges to $\sin x$ for all x.

   a. Find the first six terms of a series for $\cos x$. For what values of x should the series converge?

   b. By replacing x by $2x$ in the series for $\sin x$, find a series that converges to $\sin 2x$ for all x.

   c. Using the result in part (a) and series multiplication, calculate the first six terms of a series for $2 \sin x \cos x$. Compare your answer with the answer in part (b).

58. The series

$$e^x = 1 + x + \frac{x^2}{2!} + \frac{x^3}{3!} + \frac{x^4}{4!} + \frac{x^5}{5!} + \cdots$$

converges to e^x for all x.

   a. Find a series for $(d/dx)e^x$. Do you get the series for e^x? Explain your answer.

b. Find a series for $\int e^x\,dx$. Do you get the series for e^x? Explain your answer.

c. Replace x by $-x$ in the series for e^x to find a series that converges to e^{-x} for all x. Then multiply the series for e^x and e^{-x} to find the first six terms of a series for $e^{-x}\cdot e^x$.

59. The series

$$\tan x = x + \frac{x^3}{3} + \frac{2}{15}x^5 + \frac{17}{315}x^7 + \frac{62}{2835}x^9 + \cdots$$

converges to $\tan x$ for $-\pi/2 < x < \pi/2$.

a. Find the first five terms of the series for $\ln|\sec x|$. For what values of x should the series converge?

b. Find the first five terms of the series for $\sec^2 x$. For what values of x should this series converge?

c. Check your result in part (b) by squaring the series given for $\sec x$ in Exercise 60.

60. The series

$$\sec x = 1 + \frac{x^2}{2} + \frac{5}{24}x^4 + \frac{61}{720}x^6 + \frac{277}{8064}x^8 + \cdots$$

converges to $\sec x$ for $-\pi/2 < x < \pi/2$.

a. Find the first five terms of a power series for the function $\ln|\sec x + \tan x|$. For what values of x should the series converge?

b. Find the first four terms of a series for $\sec x \tan x$. For what values of x should the series converge?

c. Check your result in part (b) by multiplying the series for $\sec x$ by the series given for $\tan x$ in Exercise 59.

61. Uniqueness of convergent power series

a. Show that if two power series $\sum_{n=0}^{\infty} a_n x^n$ and $\sum_{n=0}^{\infty} b_n x^n$ are convergent and equal for all values of x in an open interval $(-c, c)$, then $a_n = b_n$ for every n. (*Hint:* Let $f(x) = \sum_{n=0}^{\infty} a_n x^n = \sum_{n=0}^{\infty} b_n x^n$. Differentiate term by term to show that a_n and b_n both equal $f^{(n)}(0)/(n!)$.)

b. Show that if $\sum_{n=0}^{\infty} a_n x^n = 0$ for all x in an open interval $(-c, c)$, then $a_n = 0$ for every n.

62. The sum of the series $\sum_{n=0}^{\infty}(n^2/2^n)$ To find the sum of this series, express $1/(1-x)$ as a geometric series, differentiate both sides of the resulting equation with respect to x, multiply both sides of the result by x, differentiate again, multiply by x again, and set x equal to $1/2$. What do you get?

63. The sum of the alternating harmonic series This exercise will show that

$$\sum_{n=1}^{\infty} \frac{(-1)^{n+1}}{n} = \ln 2.$$

Let h_n be the nth partial sum of the harmonic series, and let s_n be the nth partial sum of the alternating harmonic series.

a. Use mathematical induction or algebra to show that

$$s_{2n} = h_{2n} - h_n.$$

b. Use the results in Exercise 63 in Section 10.3 to conclude that

$$\lim_{n\to\infty} (h_n - \ln n) = \gamma$$

and

$$\lim_{n\to\infty} (h_{2n} - \ln 2n) = \gamma,$$

where γ is Euler's constant.

c. Use these facts to show that

$$\sum_{n=1}^{\infty} \frac{(-1)^{n+1}}{n} = \lim_{n\to\infty} s_{2n} = \ln 2.$$

64. Assume that the series $\sum a_n x^n$ converges for $x = 4$ and diverges for $x = 7$. Answer true (T), false (F), or not enough information given (N) for the following statements about the series.

a. Converges absolutely for $x = -4$

b. Diverges for $x = 5$

c. Converges absolutely for $x = -8.5$

d. Converges for $x = -2$

e. Diverges for $x = 8$

f. Diverges for $x = -6$

g. Converges absolutely for $x = 0$

h. Converges absolutely for $x = -7.1$

65. Assume that the series $\sum a_n(x-2)^n$ converges for $x = -1$ and diverges for $x = 6$. Answer true (T), false (F), or not enough information given (N) for the following statements about the series.

a. Converges absolutely for $x = 1$

b. Diverges for $x = -6$

c. Diverges for $x = 2$

d. Converges for $x = 0$

e. Converges absolutely for $x = 5$

f. Diverges for $x = 4.9$

g. Diverges for $x = 5.1$

h. Converges absolutely for $x = 4$

66. Proof of a special case of Theorem 21 Assume that $a = 0$ in Theorem 21, and assume further that $L = \lim_{n\to\infty} \frac{|c_{n+1}|}{|c_n|}$ exists. If $L \neq 0$, then set $R = 1/L$, while if $L = 0$, then set $R = \infty$. The Ratio Test implies that $f(x) = \sum_{n=0}^{\infty} c_n x^n$ converges for $-R < x < R$. Let $g(x) = \sum_{n=1}^{\infty} n c_n x^{n-1}$. This exercise will prove that f is differentiable and that $f'(x) = g(x)$, that is $\lim_{h\to 0} \frac{f(x+h) - f(x)}{h} = g(x)$ for any x with $-R < x < R$. Throughout parts (a)–(h), assume that $-R < x < R$ and that h is small enough that $-R < x + h < R$.

a. Use the Ratio Test to prove that the series defining $g(x)$ converges.

b. Apply the Mean Value Theorem to the function $p_n(x) = x^n$ to show that

$$\frac{(x+h)^n - x^n}{h} = n\,(t_n)^{n-1}$$

for some t_n between x and $x + h$ for $n = 1, 2, 3, \ldots$.

c. Show

$$g(x) - \frac{f(x+h) - f(x)}{h} = \sum_{n=2}^{\infty} n c_n \left(x^{n-1} - (t_n)^{n-1}\right).$$

d. Apply the Mean Value Theorem to the function $p_{n-1}(x) = x^{n-1}$ to show that

$$\frac{x^{n-1} - (t_n)^{n-1}}{x - t_n} = (n-1)(d_{n-1})^{n-2}$$

for some d_{n-1} between x and t_n for $n = 2, 3, 4, \ldots$.

e. Explain why $|x - t_n| < |h|$ and why

$$|d_{n-1}| \le \alpha = \max\{|x|, |x+h|\}.$$

f. Show that

$$\left|g(x) - \frac{f(x+h) - f(x)}{h}\right| \le |h| \sum_{n=2}^{\infty} |n(n-1)c_n \alpha^{n-2}|.$$

Hint: Multiply each term $nc_n(x^{n-1} - (t_n)^{n-1})$ in part (c) by $\dfrac{x - t_n}{x - t_n}$.

g. Show that $\sum_{n=2}^{\infty} n(n-1)c_n \alpha^{n-2}$ converges.

h. Let $h \to 0$ in part (f) to conclude that

$$\lim_{h\to 0} \frac{f(x+h) - f(x)}{h} = g(x).$$

67. Proof of Theorem 22 Assume that $a = 0$ in Theorem 22 and assume that $f(x) = \sum_{n=0}^{\infty} c_n x^n$ converges for $-R < x < R$. Let $g(x) = \sum_{n=0}^{\infty} \frac{c_n}{n+1} x^{n+1}$. This exercise will prove that $g'(x) = f(x)$.

a. Prove that the series defining $g(x)$ converges for $-R < x < R$.

b. Use Theorem 21 to show that $g'(x) = f(x)$, that is,

$$\int f(x)\,dx = g(x) + C.$$

10.8 Taylor and Maclaurin Series

We have seen how geometric series can be used to generate a power series for functions such as $f(x) = 1/(1-x)$ or $g(x) = 3/(x-2)$. Now we expand our capability to represent a function with a power series. This section shows how functions that are infinitely differentiable generate power series called *Taylor series*. In many cases, these series provide useful polynomial approximations of the original functions. Because approximation by polynomials is extremely useful to both mathematicians and scientists, Taylor series are an important application of the theory of infinite series.

Series Representations

We know from Theorem 21 that within its interval of convergence I, the sum of a power series is a continuous function with derivatives of all orders. But what about the other way around? If a function $f(x)$ has derivatives of all orders on an interval, can it be expressed as a power series on at least part of that interval? And if it can, what are its coefficients?

We can answer the last question readily if we assume that $f(x)$ is the sum of a power series about $x = a$,

$$\begin{aligned} f(x) &= \sum_{n=0}^{\infty} a_n (x-a)^n \\ &= a_0 + a_1(x-a) + a_2(x-a)^2 + \cdots + a_n(x-a)^n + \cdots \end{aligned}$$

with a positive radius of convergence. By repeated term-by-term differentiation within the interval of convergence I, we obtain

$$\begin{aligned} f'(x) &= a_1 + 2a_2(x-a) + 3a_3(x-a)^2 + \cdots + na_n(x-a)^{n-1} + \cdots, \\ f''(x) &= 1\cdot 2a_2 + 2\cdot 3a_3(x-a) + 3\cdot 4a_4(x-a)^2 + \cdots, \\ f'''(x) &= 1\cdot 2\cdot 3a_3 + 2\cdot 3\cdot 4a_4(x-a) + 3\cdot 4\cdot 5a_5(x-a)^2 + \cdots, \end{aligned}$$

with the nth derivative being

$$f^{(n)}(x) = n!a_n + \text{a sum of terms with } (x-a) \text{ as a factor.}$$

Since these equations all hold at $x = a$, we have

$$f'(a) = a_1, \qquad f''(a) = 1\cdot 2a_2, \qquad f'''(a) = 1\cdot 2\cdot 3a_3,$$

and, in general,

$$f^{(n)}(a) = n!a_n.$$

These formulas reveal a pattern in the coefficients of any power series $\sum_{n=0}^{\infty} a_n(x-a)^n$ that converges to the values of f on I ("represents f on I"). If there *is* such a series (still an open question), then there is only one such series, and its nth coefficient is

$$a_n = \frac{f^{(n)}(a)}{n!}.$$

If f has a series representation with the center at a, then the series must be

$$f(x) = f(a) + f'(a)(x-a) + \frac{f''(a)}{2!}(x-a)^2 + \cdots + \frac{f^{(n)}(a)}{n!}(x-a)^n + \cdots. \qquad (1)$$

But if we start with an arbitrary function f that is infinitely differentiable on an interval containing $x = a$ and use it to generate the series in Equation (1), does the series converge to $f(x)$ at each x in the interval of convergence? The answer is maybe—for some functions it will, but for other functions it will not (as we will see in Example 4).

HISTORICAL BIOGRAPHIES

Brook Taylor
(1685–1731)
www.bit.ly/2zICD72

Colin Maclaurin
(1698–1746)
www.bit.ly/2NfzJu4

Taylor and Maclaurin Series

The series on the right-hand side of Equation (1) is the most important and useful series we will study in this chapter.

DEFINITIONS Let f be a function with derivatives of all orders throughout some interval containing a as an interior point. Then the **Taylor series generated by f at $x = a$** is

$$\sum_{k=0}^{\infty} \frac{f^{(k)}(a)}{k!}(x-a)^k = f(a) + f'(a)(x-a) + \frac{f''(a)}{2!}(x-a)^2 + \cdots + \frac{f^{(n)}(a)}{n!}(x-a)^n + \cdots.$$

The **Maclaurin series of f** is the Taylor series generated by f at $x = 0$, or

$$\sum_{k=0}^{\infty} \frac{f^{(k)}(0)}{k!}x^k = f(0) + f'(0)x + \frac{f''(0)}{2!}x^2 + \cdots + \frac{f^{(n)}(0)}{n!}x^n + \cdots.$$

The Maclaurin series generated by f is often just called the Taylor series of f.

EXAMPLE 1 Find the Taylor series generated by $f(x) = 1/x$ at $a = 2$. Where, if anywhere, does the series converge to $1/x$?

Solution We need to find $f(2), f'(2), f''(2), \ldots$. Taking derivatives, we get

$$f(x) = x^{-1},\quad f'(x) = -x^{-2},\quad f''(x) = 2!\,x^{-3}, \ldots, f^{(n)}(x) = (-1)^n n!\,x^{-(n+1)},$$

so that

$$f(2) = 2^{-1} = \frac{1}{2},\quad f'(2) = -\frac{1}{2^2},\quad \frac{f''(2)}{2!} = 2^{-3} = \frac{1}{2^3}, \ldots, \quad \frac{f^{(n)}(2)}{n!} = \frac{(-1)^n}{2^{n+1}}.$$

The Taylor series is

$$f(2) + f'(2)(x-2) - \frac{f''(2)}{2!}(x-2)^2 + \cdots + \frac{f^{(n)}(2)}{n!}(x-2)^n + \cdots$$

$$= \frac{1}{2} - \frac{(x-2)}{2^2} + \frac{(x-2)^2}{2^3} - \cdots + (-1)^n\frac{(x-2)^n}{2^{n+1}} + \cdots.$$

This is a geometric series with first term $1/2$ and ratio $r = -(x-2)/2$. It converges absolutely for $|x-2| < 2$ and its sum is

$$\frac{1/2}{1+(x-2)/2} = \frac{1}{2+(x-2)} = \frac{1}{x}.$$

In this example the Taylor series generated by $f(x) = 1/x$ at $a = 2$ converges to $1/x$ for $|x-2| < 2$, or $0 < x < 4$. ■

Taylor Polynomials

The linearization of a differentiable function f at a point a is the polynomial of degree at most 1 given by

$$P_1(x) = f(a) + f'(a)(x-a).$$

In Section 3.9 we used this linearization to approximate $f(x)$ at values of x near a. If f has derivatives of higher order at a, then it has higher-order polynomial approximations as well, one for each available derivative. These polynomials are called the Taylor polynomials of f.

DEFINITION Let f be a function with derivatives of order k for $k = 1, 2, \dots, N$ in some interval containing a as an interior point. Then, for any integer n from 0 through N, the **Taylor polynomial of order n** generated by f at $x = a$ is the polynomial

$$P_n(x) = f(a) + f'(a)(x-a) + \frac{f''(a)}{2!}(x-a)^2 + \cdots + \frac{f^{(k)}(a)}{k!}(x-a)^k + \cdots + \frac{f^{(n)}(a)}{n!}(x-a)^n.$$

We speak of a Taylor polynomial of *order n* rather than *degree n* because $f^{(n)}(a)$ may be zero. The first two Taylor polynomials of $f(x) = \cos x$ at $x = 0$, for example, are $P_0(x) = 1$ and $P_1(x) = 1$. The first-order Taylor polynomial has degree 0, not 1.

Just as the linearization of f at $x = a$ provides the best linear approximation of f in the neighborhood of a, the higher-order Taylor polynomials provide the "best" polynomial approximations of their respective degrees. (See Exercise 51.)

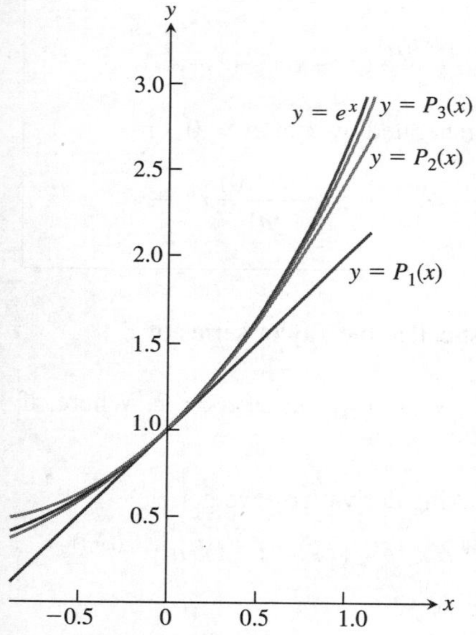

FIGURE 10.22 The graph of $f(x) = e^x$ and its Taylor polynomials

$P_1(x) = 1 + x$

$P_2(x) = 1 + x + (x^2/2!)$

$P_3(x) = 1 + x + (x^2/2!) + (x^3/3!)$.

Notice the very close agreement near the center $x = 0$ (Example 2).

EXAMPLE 2 Find the Taylor series and the Taylor polynomials generated by $f(x) = e^x$ at $x = 0$.

Solution Since $f^{(n)}(x) = e^x$ and $f^{(n)}(0) = 1$ for every $n = 0, 1, 2, \dots$, the Taylor series generated by f at $x = 0$ (see Figure 10.22) is

$$\begin{aligned} f(0) + f'(0)x + \frac{f''(0)}{2!}x^2 + \cdots + \frac{f^{(n)}(0)}{n!}x^n + \cdots &= 1 + x + \frac{x^2}{2} + \cdots + \frac{x^n}{n!} + \cdots \\ &= \sum_{k=0}^{\infty} \frac{x^k}{k!}. \end{aligned}$$

This is also the Maclaurin series for e^x. In the next section we will see that this series converges to e^x at every x.

The Taylor polynomial of order n at $x = 0$ is

$$P_n(x) = 1 + x + \frac{x^2}{2} + \cdots + \frac{x^n}{n!}.$$ ■

EXAMPLE 3 Find the Taylor series and Taylor polynomials generated by $f(x) = \cos x$ at $x = 0$.

Solution The cosine and its derivatives are

$$\begin{aligned} f(x) &= \cos x, & f'(x) &= -\sin x, \\ f''(x) &= -\cos x, & f^{(3)}(x) &= \sin x, \\ &\vdots & &\vdots \\ f^{(2n)}(x) &= (-1)^n \cos x, & f^{(2n+1)}(x) &= (-1)^{n+1} \sin x. \end{aligned}$$

At $x = 0$, the cosines are 1 and the sines are 0, so

$$f^{(2n)}(0) = (-1)^n, \qquad f^{(2n+1)}(0) = 0.$$

The Taylor series generated by f at 0 is

$$\begin{aligned} f(0) + f'(0)x + \frac{f''(0)}{2!}x^2 + \frac{f'''(0)}{3!}x^3 + \cdots + \frac{f^{(n)}(0)}{n!}x^n + \cdots \\ = 1 + 0 \cdot x - \frac{x^2}{2!} + 0 \cdot x^3 + \frac{x^4}{4!} + \cdots + (-1)^n \frac{x^{2n}}{(2n)!} + \cdots \\ = \sum_{k=0}^{\infty} \frac{(-1)^k x^{2k}}{(2k)!}. \end{aligned}$$

This is also the Maclaurin series for $\cos x$. Notice that only even powers of x occur in the Taylor series generated by the cosine function, which is consistent with the fact that it is an even function. In Section 10.9, we will see that the series converges to $\cos x$ at every x.

Because $f^{(2n+1)}(0) = 0$, the Taylor polynomials of orders $2n$ and $2n + 1$ are identical:

$$P_{2n}(x) = P_{2n+1}(x) = 1 - \frac{x^2}{2!} + \frac{x^4}{4!} - \cdots + (-1)^n \frac{x^{2n}}{(2n)!}.$$

Figure 10.23 shows how well these polynomials approximate $f(x) = \cos x$ near $x = 0$. Only the right-hand portions of the graphs are given because the graphs are symmetric about the y-axis. ■

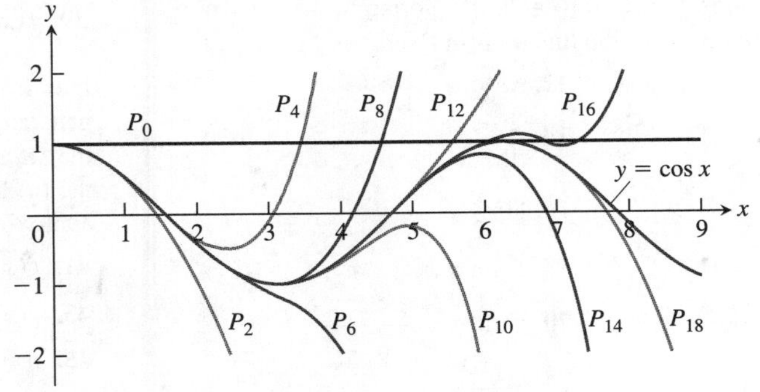

FIGURE 10.23 The polynomials

$$P_{2n}(x) = \sum_{k=0}^{n} \frac{(-1)^k x^{2k}}{(2k)!}$$

converge to $\cos x$ as $n \to \infty$. We can deduce the behavior of $\cos x$ arbitrarily far away solely from knowing the values of the cosine and its derivatives at $x = 0$ (Example 3).

EXAMPLE 4 By using mathematical induction (see Exercise 51), it can be shown that

$$f(x) = \begin{cases} 0, & x = 0 \\ e^{-1/x^2}, & x \neq 0 \end{cases}$$

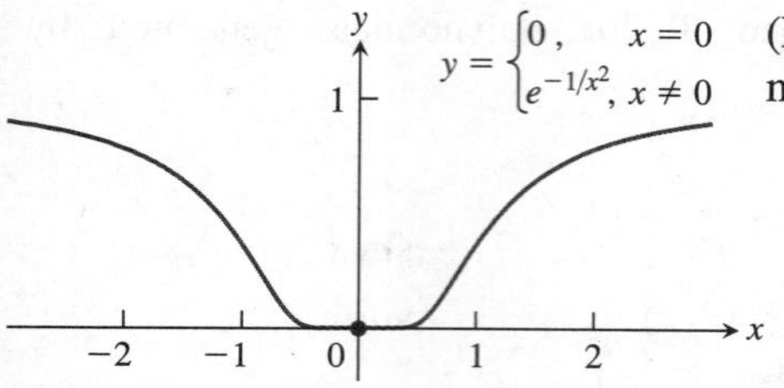

FIGURE 10.24 The graph of the continuous extension of $y = e^{-1/x^2}$ is so flat at the origin that all of its derivatives there are zero (Example 4). Therefore, its Taylor series, which is zero everywhere, is not the function itself.

(Figure 10.24) has derivatives of all orders at $x = 0$ and that $f^{(n)}(0) = 0$ for all n. This means that the Taylor series generated by f at $x = 0$ is

$$\begin{aligned} f(0) + f'(0)x + \frac{f''(0)}{2!}x^2 + \cdots + \frac{f^{(n)}(0)}{n!}x^n + \cdots \\ = 0 + 0 \cdot x + 0 \cdot x^2 + \cdots + 0 \cdot x^n + \cdots \\ = 0 + 0 + \cdots + 0 + \cdots. \end{aligned}$$

The series converges for every x (its sum is 0) but converges to $f(x)$ only at $x = 0$. That is, the Taylor series generated by $f(x)$ in this example is *not* equal to the function $f(x)$ over the entire interval of convergence. ■

Two questions still remain.

1. For what values of x can we normally expect a Taylor series to converge to its generating function?
2. How accurately do a function's Taylor polynomials approximate the function on a given interval?

The answers are provided by a theorem of Taylor in the next section.

EXERCISES 10.8

Finding Taylor Polynomials

In Exercises 1–10, find the Taylor polynomials of orders 0, 1, 2, and 3 generated by f at a.

1. $f(x) = e^{3x}$, $a = 0$
2. $f(x) = \sin x$, $a = 0$
3. $f(x) = \ln\left(\frac{1}{x}\right)$, $a = 1$
4. $f(x) = \ln(1 + x)$, $a = 0$
5. $f(x) = 1/x$, $a = 2$
6. $f(x) = \frac{1}{x+3}$, $a = 0$
7. $f(x) = \sin x$, $a = \pi/4$
8. $f(x) = \tan x$, $a = \pi/4$
9. $f(x) = \sqrt{x}$, $a = 4$
10. $f(x) = \sqrt{x+1}$, $a = 1$

Finding Taylor Series at $x = 0$ (Maclaurin Series)

Find the Maclaurin series for the functions in Exercises 11–24.

11. e^{-x}
12. xe^x
13. $\frac{1}{1-x}$
14. $\frac{2+x}{1-x}$
15. $\cos 3x$
16. $\sin\frac{x}{2}$
17. $7\cos(-x)$
18. $5\sin \pi x$
19. $\cosh x = \frac{e^x + e^{-x}}{2}$
20. $\sinh x = \frac{e^x - e^{-x}}{2}$
21. $x^4 - 3x^3 + 2x + 5$
22. $\frac{x^2}{x+1}$
23. $x \sin x$
24. $(x + 1)\ln(x + 1)$

Finding Taylor and Maclaurin Series

In Exercises 25–34, find the Taylor series generated by f at $x = a$.

25. $f(x) = x^3 + 4x + 5$, $a = 2$
26. $f(x) = 2x^3 + x^2 + 3x - 8$, $a = 1$
27. $f(x) = x^4 + x^2 + 1$, $a = -2$
28. $f(x) = 4x^5 + x^4 - 3x^3 + x^2 - 3$, $a = -1$
29. $f(x) = 1/x^2$, $a = 1$
30. $f(x) = 1/(1-x)^3$, $a = 0$
31. $f(x) = e^x$, $a = 2$
32. $f(x) = 3^x$, $a = 1$
33. $f(x) = \cos(2x + (\pi/2))$, $a = \pi/4$
34. $f(x) = \sqrt{x+1}$, $a = 0$

In Exercises 35–40, find the first three nonzero terms of the Maclaurin series for each function.

35. $f(x) = \sin x - (2/(1-x))$
36. $f(x) = (1 - x + x^2)e^x$
37. $f(x) = (\sin x)\ln(1 + x)$
38. $f(x) = x\cos^2 x$
39. $f(x) = x^4 e^{x^2}$
40. $f(x) = \frac{x^3}{1+2x}$

Quadratic Approximations The Taylor polynomial of order 2 generated by a twice-differentiable function $f(x)$ at $x = a$ is called the *quadratic approximation* of f at $x = a$. In Exercises 41–46, find the **(a)** linearization (Taylor polynomial of order 1) and **(b)** quadratic approximation of f at $x = 0$.

41. $f(x) = \ln(\cos x)$
42. $f(x) = e^{\cos x}$
43. $f(x) = 1/\sqrt{1-x^2}$
44. $f(x) = \cosh x$
45. $f(x) = \cos x$
46. $f(x) = \tan x$
47. If m is a small positive number, then the graphs of $y = \sin x$ and $y = mx$ intersect at a point whose x-coordinate is close to π. Find the quadratic approximation to $f(x) = \sin x - mx$ at $x = \pi$, and use this to show that an approximate solution to the equation $\sin x = mx$ is $x \approx \pi/(1 + m)$.

Theory and Examples

48. Use the Taylor series generated by e^x at $x = a$ to show that
$$e^x = e^a\left[1 + (x - a) + \frac{(x-a)^2}{2!} + \cdots\right].$$
49. (*Continuation of Exercise 48.*) Find the Taylor series generated by e^x at $x = 1$. Compare your answer with the formula in Exercise 48.
50. Let $f(x)$ have derivatives through order n at $x = a$. Show that the Taylor polynomial of order n and its first n derivatives have the same values that f and its first n derivatives have at $x = a$.

51. Approximation properties of Taylor polynomials Suppose that $f(x)$ is differentiable on an interval centered at $x = a$ and that $g(x) = b_0 + b_1(x - a) + \cdots + b_n(x - a)^n$ is a polynomial of degree n with constant coefficients $b_0, \ldots, b_n$. Let $E(x) = f(x) - g(x)$. Show that if we impose on g the conditions

i) $E(a) = 0$ — The approximation error is zero at $x = a$.

ii) $\lim_{x \to a} \frac{E(x)}{(x - a)^n} = 0,$ — The error is negligible when compared to $(x - a)^n$.

then

$$g(x) = f(a) + f'(a)(x - a) + \frac{f''(a)}{2!}(x - a)^2 + \cdots + \frac{f^{(n)}(a)}{n!}(x - a)^n.$$

Thus, the Taylor polynomial $P_n(x)$ is the only polynomial of degree less than or equal to n whose error is both zero at $x = a$ and negligible when compared with $(x - a)^n$.

52. Let f be the function from Example 4. Let $p_1(t) = 2t^3$ and $p_2(t) = 4t^6 - 6t^4$.

a. Show that if $x > 0$, then

$$f'(x) = 2x^{-3}e^{-x^{-2}} = p_1(1/x)f(x),$$

while for $x = 0$,

$$f'(0) = \lim_{h \to 0^+} \frac{f(h) - 0}{h - 0} = \lim_{t \to \infty} \frac{f(1/t)}{1/t} = 0.$$

b. Show that if $x > 0$, then

$$f''(x) = (4x^{-6} - 6x^{-3})e^{-x^{-2}} = p_2(1/x)f(x),$$

and $f''(0) = 0$.

c. For $n \geq 2$, recursively define polynomials p_n by $p_{n+1}(t) = 2t^3 p_n(t) - t^2 p_n'(t)$. Use mathematical induction to prove that p_n has degree $3n$, $f^{(n)}(x) = p_n(1/x)f(x)$ for $x > 0$, and $f^{(n)}(0) = 0$.

10.9 Convergence of Taylor Series

In the last section we asked when a Taylor series for a function can be expected to converge to the function that generates it. The finite-order Taylor polynomials that approximate the Taylor series provide estimates for the generating function. In order for these estimates to be useful, we need a way to control the possible errors we may encounter when approximating a function with its finite-order Taylor polynomials. How do we bound such possible errors? We answer the question in this section with the following theorem.

THEOREM 23—Taylor's Theorem
If f and its first n derivatives $f', f'', \ldots, f^{(n)}$ are continuous on the closed interval between a and b, and $f^{(n)}$ is differentiable on the open interval between a and b, then there exists a number c between a and b such that

$$f(b) = f(a) + f'(a)(b - a) + \frac{f''(a)}{2!}(b - a)^2 + \cdots + \frac{f^{(n)}(a)}{n!}(b - a)^n + \frac{f^{(n+1)}(c)}{(n+1)!}(b - a)^{n+1}.$$

Taylor's Theorem is a generalization of the Mean Value Theorem (Exercise 49), and we omit its proof here.

When we apply Taylor's Theorem, we usually want to hold a fixed and treat b as an independent variable. Taylor's formula is easier to use in circumstances like these if we change b to x. Here is a version of the theorem with this change.

Taylor's Formula
If f has derivatives of all orders in an open interval I containing a, then for each positive integer n and for each x in I,

$$f(x) = f(a) + f'(a)(x - a) + \frac{f''(a)}{2!}(x - a)^2 + \cdots + \frac{f^{(n)}(a)}{n!}(x - a)^n + R_n(x), \tag{1}$$

where

$$R_n(x) = \frac{f^{(n+1)}(c)}{(n+1)!}(x - a)^{n+1} \quad \text{for some } c \text{ between } a \text{ and } x. \tag{2}$$

When we state Taylor's theorem this way, it says that for each $x \in I$,

$$f(x) = P_n(x) + R_n(x).$$

The function $R_n(x)$ is determined by the value of the $(n+1)$st derivative $f^{(n+1)}$ at a point c that depends on both a and x, and that lies somewhere between them. For any value of n we want, the equation gives both a polynomial approximation of f of that order and a formula for the error involved in using that approximation over the interval I.

Equation (1) is called **Taylor's formula**. The function $R_n(x)$ is called the **remainder of order n** or the **error term** for the approximation of f by $P_n(x)$ over I.

> If $R_n(x) \to 0$ as $n \to \infty$ for all $x \in I$, we say that the Taylor series generated by f at $x = a$ **converges** to f on I, and we write
>
> $$f(x) = \sum_{k=0}^{\infty} \frac{f^{(k)}(a)}{k!}(x-a)^k.$$

Often we can estimate R_n without knowing the value of c, as the following example illustrates.

EXAMPLE 1 Show that the Taylor series generated by $f(x) = e^x$ at $x = 0$ converges to $f(x)$ for every real value of x.

Solution The function has derivatives of all orders throughout the interval $I = (-\infty, \infty)$. Equations (1) and (2) with $f(x) = e^x$ and $a = 0$ give

$$e^x = 1 + x + \frac{x^2}{2!} + \cdots + \frac{x^n}{n!} + R_n(x) \qquad \text{Polynomial from Section 10.8, Example 2}$$

and

$$R_n(x) = \frac{e^c}{(n+1)!}x^{n+1} \qquad \text{for some } c \text{ between 0 and } x.$$

Since e^x is an increasing function of x, e^c lies between $e^0 = 1$ and e^x. When x is negative, so is c, and $e^c < 1$. When x is zero, $e^x = 1$ so that $R_n(x) = 0$. When x is positive, so is c, and $e^c < e^x$. Thus, for $R_n(x)$ given as above,

$$|R_n(x)| \le \frac{|x|^{n+1}}{(n+1)!} \quad \text{when } x \le 0, \qquad e^c < 1 \text{ since } c < 0$$

and

$$|R_n(x)| < e^x \frac{x^{n+1}}{(n+1)!} \quad \text{when } x > 0. \qquad e^c < e^x \text{ since } c < x$$

Finally, because

$$\lim_{n\to\infty} \frac{x^{n+1}}{(n+1)!} = 0 \quad \text{for every } x, \qquad \text{Section 10.1, Theorem 5}$$

we see that $\lim_{n\to\infty} R_n(x) = 0$, and therefore the Taylor series converges to e^x for every x. Thus,

$$e^x = \sum_{k=0}^{\infty} \frac{x^k}{k!} = 1 + x + \frac{x^2}{2!} + \cdots + \frac{x^k}{k!} + \cdots. \tag{3}$$

■

The Number e as a Series

$$e = \sum_{n=0}^{\infty} \frac{1}{n!}$$

We can use the result of Example 1 with $x = 1$ to write

$$e = 1 + 1 + \frac{1}{2!} + \cdots + \frac{1}{n!} + R_n(1),$$

where, for some c between 0 and 1,

$$R_n(1) = e^c \frac{1}{(n+1)!} < \frac{3}{(n+1)!}. \qquad e^c < e^1 < 3$$

Estimating the Remainder

It is often possible to estimate $R_n(x)$ as we did in Example 1. This method of estimation is so convenient that we state it as a theorem for future reference.

THEOREM 24—The Remainder Estimation Theorem
If there is a positive constant M such that $|f^{(n+1)}(t)| \leq M$ for all t between x and a, inclusive, then the remainder term $R_n(x)$ in Taylor's Theorem satisfies the inequality

$$|R_n(x)| \leq M \frac{|x-a|^{n+1}}{(n+1)!}.$$

If this inequality holds for every n, and the other conditions of Taylor's Theorem are satisfied by f, then the Taylor series converges to $f(x)$.

The next two examples use Theorem 24 to show that the Taylor series generated by the sine and cosine functions do in fact converge to the functions themselves.

EXAMPLE 2 Show that the Taylor series for $\sin x$ at $x = 0$ converges for all x.

Solution The function and its derivatives are

$$\begin{aligned} f(x) &= \sin x, & f'(x) &= \cos x, \\ f''(x) &= -\sin x, & f'''(x) &= -\cos x, \\ &\vdots & &\vdots \\ f^{(2k)}(x) &= (-1)^k \sin x, & f^{(2k+1)}(x) &= (-1)^k \cos x, \end{aligned}$$

so

$$f^{(2k)}(0) = 0 \quad \text{and} \quad f^{(2k+1)}(0) = (-1)^k.$$

The series has only odd-powered terms, and for $n = 2k + 1$, Taylor's Theorem gives

$$\sin x = x - \frac{x^3}{3!} + \frac{x^5}{5!} - \cdots + \frac{(-1)^k x^{2k+1}}{(2k+1)!} + R_{2k+1}(x).$$

All the derivatives of $\sin x$ have absolute values less than or equal to 1, so we can apply the Remainder Estimation Theorem with $M = 1$ to obtain

$$|R_{2k+1}(x)| \leq 1 \cdot \frac{|x|^{2k+2}}{(2k+2)!}.$$

From Theorem 5, Rule 6, we have $\left(|x|^{2k+2}/(2k+2)!\right) \to 0$ as $k \to \infty$, whatever the value of x, so $R_{2k+1}(x) \to 0$ and the Maclaurin series for $\sin x$ converges to $\sin x$ for every x. Thus,

$$\sin x = \sum_{k=0}^{\infty} \frac{(-1)^k x^{2k+1}}{(2k+1)!} = x - \frac{x^3}{3!} + \frac{x^5}{5!} - \frac{x^7}{7!} + \cdots. \qquad (4)$$

■

$$\sin x = x - \frac{x^3}{3!} + \frac{x^5}{5!} - \frac{x^7}{7!} + \cdots$$

EXAMPLE 3 Show that the Taylor series for $\cos x$ at $x = 0$ converges to $\cos x$ for every value of x.

Solution We add the remainder term to the Taylor polynomial for $\cos x$ (Section 10.8, Example 3) to obtain Taylor's formula for $\cos x$ with $n = 2k$:

$$\cos x = 1 - \frac{x^2}{2!} + \frac{x^4}{4!} - \cdots + (-1)^k \frac{x^{2k}}{(2k)!} + R_{2k}(x).$$

Because the derivatives of the cosine have absolute value less than or equal to 1, the Remainder Estimation Theorem with $M = 1$ gives

$$|R_{2k}(x)| \leq 1 \cdot \frac{|x|^{2k+1}}{(2k+1)!}.$$

For every value of x, $R_{2k}(x) \to 0$ as $k \to \infty$. Therefore, the series converges to $\cos x$ for every value of x. Thus,

$$\cos x = \sum_{k=0}^{\infty} \frac{(-1)^k x^{2k}}{(2k)!} = 1 - \frac{x^2}{2!} + \frac{x^4}{4!} - \frac{x^6}{6!} + \cdots. \tag{5}$$

$$\cos x = 1 - \frac{x^2}{2!} + \frac{x^4}{4!} - \frac{x^6}{6!} + \cdots$$

■

Using Taylor Series

Since every Taylor series is a power series, the operations of adding, subtracting, and multiplying Taylor series are all valid on the intersection of their intervals of convergence.

EXAMPLE 4 Using known series, find the first few terms of the Taylor series for the given function by using power series operations.

(a) $\frac{1}{3}(2x + x\cos x)$ **(b)** $e^x \cos x$

Solution

(a) $$\frac{1}{3}(2x + x\cos x) = \frac{2}{3}x + \frac{1}{3}x\left(1 - \frac{x^2}{2!} + \frac{x^4}{4!} - \cdots + (-1)^k \frac{x^{2k}}{(2k)!} + \cdots\right)$$ Taylor series for $\cos x$

$$= \frac{2}{3}x + \frac{1}{3}x - \frac{x^3}{3!} + \frac{x^5}{3 \cdot 4!} - \cdots = x - \frac{x^3}{6} + \frac{x^5}{72} - \cdots$$

(b) $$e^x \cos x = \left(1 + x + \frac{x^2}{2!} + \frac{x^3}{3!} + \frac{x^4}{4!} + \cdots\right)\left(1 - \frac{x^2}{2!} + \frac{x^4}{4!} - \cdots\right)$$ Multiply the first series by each term of the second series.

$$= \left(1 + x + \frac{x^2}{2!} + \frac{x^3}{3!} + \frac{x^4}{4!} + \cdots\right) - \left(\frac{x^2}{2!} + \frac{x^3}{2!} + \frac{x^4}{2!2!} + \frac{x^5}{2!3!} + \cdots\right)$$
$$+ \left(\frac{x^4}{4!} + \frac{x^5}{4!} + \frac{x^6}{2!4!} + \cdots\right) + \cdots$$

$$= 1 + x - \frac{x^3}{3} - \frac{x^4}{6} + \cdots$$

■

By Theorem 20, if the Taylor series generated by f at $x = 0$,

$$\sum_{k=0}^{\infty} \frac{f^{(k)}(0)}{k!} x^k = f(0) + f'(0)x + \frac{f''(0)}{2!}x^2 + \cdots + \frac{f^{(n)}(0)}{n!}x^n + \cdots, \tag{6}$$

converges absolutely for $|x| < R$, and if u is a continuous function, then the series

$$\sum_{k=0}^{\infty} \frac{f^{(k)}(0)}{k!}(u(x))^k \tag{7}$$

obtained by substituting $u(x)$ for x in the Taylor series of Formula (6) converges absolutely on the set of points x where $|u(x)| < R$.

If $u(x) = cx^m$, where $c \neq 0$ and m is a positive integer, then the series of Formula (7) is a power series and can be shown to be the Taylor series generated by $f(u(x))$ at $x = 0$.

For instance, we can find the Taylor series for $\cos 2x$ by substituting $2x$ for x in the Taylor series for $\cos x$:

$$\begin{aligned}\cos 2x &= \sum_{k=0}^{\infty} \frac{(-1)^k(2x)^{2k}}{(2k)!} = 1 - \frac{(2x)^2}{2!} + \frac{(2x)^4}{4!} - \frac{(2x)^6}{6!} + \cdots \qquad \text{Eq. (5) with } 2x \text{ for } x\\ &= 1 - \frac{2^2x^2}{2!} + \frac{2^4x^4}{4!} - \frac{2^6x^6}{6!} + \cdots\\ &= \sum_{k=0}^{\infty} (-1)^k \frac{2^{2k}x^{2k}}{(2k)!}.\end{aligned}$$

By Theorem 20, this new Taylor series converges for all x.

EXAMPLE 5 For what values of x can we replace $\sin x$ by $x - (x^3/3!)$ and obtain an error whose magnitude is no greater than 3×10^{-4}?

Solution Here we can take advantage of the fact that the Taylor series for $\sin x$ is an alternating series for every nonzero value of x. According to the Alternating Series Estimation Theorem (Section 10.6), the error in truncating

$$\sin x = x - \frac{x^3}{3!} \;\Big\vert\; + \frac{x^5}{5!} - \frac{x^7}{7!} + \cdots$$

after $(x^3/3!)$ is no greater than

$$\left|\frac{x^5}{5!}\right| = \frac{|x|^5}{120}.$$

Therefore, the error will be less than or equal to 3×10^{-4} if

$$\frac{|x|^5}{120} < 3 \times 10^{-4} \qquad \text{or} \qquad |x| < \sqrt[5]{360 \times 10^{-4}} \approx 0.514. \qquad \text{Rounded down, to be safe}$$

The Alternating Series Estimation Theorem tells us something that the Remainder Estimation Theorem does not: The estimate $x - (x^3/3!)$ for $\sin x$ is an underestimate when x is positive, because then $x^5/120$ is positive.

Figure 10.25 shows the graph of $\sin x$, along with the graphs of a number of its approximating Taylor polynomials. The graph of $P_3(x) = x - (x^3/3!)$ is almost indistinguishable from the sine curve when $0 \leq x \leq 1$. ■

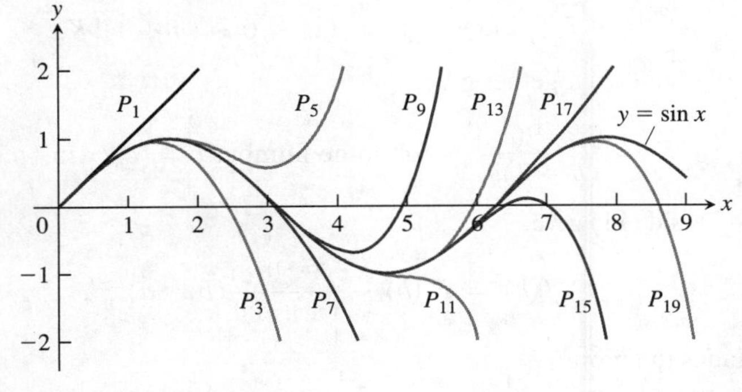

FIGURE 10.25 The polynomials

$$P_{2n+1}(x) = \sum_{k=0}^{n} \frac{(-1)^k x^{2k+1}}{(2k+1)!}$$

converge to $\sin x$ as $n \to \infty$. Notice how closely $P_3(x)$ approximates the sine curve for $x \leq 1$ (Example 5).

A Proof of Taylor's Theorem

We prove Taylor's theorem assuming $a < b$. The proof for $a > b$ is nearly the same.

The Taylor polynomial

$$P_n(x) = f(a) + f'(a)(x - a) + \frac{f''(a)}{2!}(x - a)^2 + \cdots + \frac{f^{(n)}(a)}{n!}(x - a)^n$$

and its first n derivatives match the function f and its first n derivatives at $x = a$. We do not disturb that matching if we add another term of the form $K(x - a)^{n+1}$, where K is any constant, because such a term and its first n derivatives are all equal to zero at $x = a$. The new function

$$\phi_n(x) = P_n(x) + K(x - a)^{n+1}$$

and its first n derivatives still agree with f and its first n derivatives at $x = a$.

We now choose the particular value of K that makes the curve $y = \phi_n(x)$ agree with the original curve $y = f(x)$ at $x = b$. In symbols,

$$f(b) = P_n(b) + K(b - a)^{n+1}, \quad \text{or} \quad K = \frac{f(b) - P_n(b)}{(b - a)^{n+1}}. \tag{8}$$

With K defined by Equation (8), the function

$$F(x) = f(x) - \phi_n(x)$$

measures the difference between the original function f and the approximating function ϕ_n for each x in $[a, b]$.

We now use Rolle's Theorem (Section 4.2). First, because $F(a) = F(b) = 0$ and both F and F' are continuous on $[a, b]$, we know that

$$F'(c_1) = 0 \quad \text{for some } c_1 \text{ in } (a, b).$$

Next, because $F'(a) = F'(c_1) = 0$ and both F' and F'' are continuous on $[a, c_1]$, we know that

$$F''(c_2) = 0 \quad \text{for some } c_2 \text{ in } (a, c_1).$$

Rolle's Theorem, applied successively to $F'', F''', \ldots, F^{(n-1)}$, implies the existence of

$$\begin{aligned} &c_3 \text{ in } (a, c_2) && \text{such that } F'''(c_3) = 0, \\ &c_4 \text{ in } (a, c_3) && \text{such that } F^{(4)}(c_4) = 0, \\ &\vdots \\ &c_n \text{ in } (a, c_{n-1}) && \text{such that } F^{(n)}(c_n) = 0. \end{aligned}$$

Finally, because $F^{(n)}$ is continuous on $[a, c_n]$ and differentiable on (a, c_n), and $F^{(n)}(a) = F^{(n)}(c_n) = 0$, Rolle's Theorem implies that there is a number c_{n+1} in (a, c_n) such that

$$F^{(n+1)}(c_{n+1}) = 0. \tag{9}$$

If we differentiate $F(x) = f(x) - P_n(x) - K(x - a)^{n+1}$ a total of $n + 1$ times, we get

$$F^{(n+1)}(x) = f^{(n+1)}(x) - 0 - (n + 1)!\,K. \tag{10}$$

Equations (9) and (10) together give

$$K = \frac{f^{(n+1)}(c)}{(n + 1)!} \quad \text{for some number } c = c_{n+1} \text{ in } (a, b). \tag{11}$$

Equations (8) and (11) give

$$f(b) = P_n(b) + \frac{f^{(n+1)}(c)}{(n + 1)!}(b - a)^{n+1}.$$

This concludes the proof. ■

EXERCISES 10.9

Finding Taylor Series

Use substitution (as in Formula (7)) to find the Taylor series at $x = 0$ of the functions in Exercises 1–12.

1. e^{-3x} **2.** $e^{-x/2}$ **3.** $5\sin(-x)$

4. $\cos\left(\frac{\pi x}{2}\right)$ **5.** $\cos 5x^2$ **6.** $\cos\left(x^{3/2}/\sqrt{2}\right)$

7. $\ln(1+x^2)$ **8.** $\cot^{-1}(3x^4)$ **9.** $\frac{1}{1+\frac{3}{4}x^3}$

10. $\frac{1}{2-x}$ **11.** $\ln(4+2x)$ **12.** $e^{-x^2+\ln 5}$

Use power series operations to find the Taylor series at $x = 0$ for the functions in Exercises 13–30.

13. xe^{-x} **14.** $x^2\sin x$ **15.** $\cos hx - \frac{x^2}{2} - 1$

16. $\sin x - x + \frac{x^3}{3!}$ **17.** $x\sin\pi x$ **18.** $x^2\cos(x^2)$

19. $\cos^2 x$ (*Hint:* $\cos^2 x = (1+\cos 2x)/2$.)

20. $\sin^2 x$ **21.** $\frac{x^3}{1+2x}$ **22.** $x\ln(1+2x)$

23. $\frac{1}{(1-x)^2}$ **24.** $\frac{2}{(1+x)^3}$ **25.** $x\arctan x^2$

26. $\sin x \cdot \cos x$ **27.** $e^x - \frac{1}{1+x}$ **28.** $\cos x - \sin x$

29. $\frac{x}{5}\ln(1+x^3)$ **30.** $\ln(1+x) - \ln(1-x)$

Find the first four nonzero terms in the Maclaurin series for the functions in Exercises 31–38.

31. $e^x\sin x$ **32.** $\frac{\ln(1-x)}{1+x}$ **33.** $(\arctan x)^2$

34. $\sin^2 x \cdot \cos x$ **35.** $e^{\sin x}$ **36.** $\sin(\tan^{-1} x)$

37. $\sin(e^{-x}-1)$ **38.** $\cos\sqrt{x} + \ln(\cos x)$

Error Estimates

39. Estimate the error if $P_3(x) = x - (x^3/6)$ is used to estimate the value of $\sin x$ at $x = 0.1$.

40. Estimate the error if $P_4(x) = 1 + x + (x^2/2) + (x^3/6) + (x^4/24)$ is used to estimate the value of e^x at $x = 1/2$.

41. For approximately what values of x can you replace $\sin x$ by $x - (x^3/6)$ with an error of magnitude no greater than 4×10^{-4}? Give reasons for your answer.

42. If $\cos x$ is replaced by $1 - (x^2/2)$ and $|x| < 0.5$, what estimate can be made of the error? Does $1 - (x^2/2)$ tend to be too large, or too small? Give reasons for your answer.

43. How close is the approximation $\sin x = x$ when $|x| < 10^{-4}$? For which of these values of x is $x < \sin x$?

44. The estimate $\sqrt{1+x} = 1 + (x/2)$ is used when x is small. Estimate the error when $|x| < 0.01$.

45. The approximation $e^x = 1 + x + (x^2/2)$ is used when x is small. Use the Remainder Estimation Theorem to estimate the error when $|x| < 0.01$.

46. (*Continuation of Exercise 45.*) When $x < 0$, the series for e^x is an alternating series. Use the Alternating Series Estimation Theorem to estimate the error that results from replacing e^x by $1 + x + (x^2/2)$ when $-0.1 < x < 0$. Compare your estimate with the one you obtained in Exercise 45.

Theory and Examples

47. Use the identity $\sin^2 x = (1 - \cos 2x)/2$ to obtain the Maclaurin series for $\sin^2 x$. Then differentiate this series to obtain the Maclaurin series for $2\sin x\cos x$. Check that this is the series for $\sin 2x$.

48. (*Continuation of Exercise 47.*) Use the identity $\cos^2 x = \cos 2x + \sin^2 x$ to obtain a power series for $\cos^2 x$.

49. Taylor's Theorem and the Mean Value Theorem Explain how the Mean Value Theorem (Section 4.2, Theorem 4) is a special case of Taylor's Theorem.

50. Linearizations at inflection points Show that if the graph of a twice-differentiable function $f(x)$ has an inflection point at $x = a$, then the linearization of f at $x = a$ is also the quadratic approximation of f at $x = a$. This explains why tangent lines fit so well at inflection points.

51. The (second) second derivative test Use the equation

$$f(x) = f(a) + f'(a)(x-a) + \frac{f''(c_2)}{2}(x-a)^2$$

to establish the following test.

Let f have continuous first and second derivatives and suppose that $f'(a) = 0$. Then

a. f has a local maximum at a if $f'' \le 0$ throughout an interval whose interior contains a;

b. f has a local minimum at a if $f'' \ge 0$ throughout an interval whose interior contains a.

52. A cubic approximation Use Taylor's formula with $a = 0$ and $n = 3$ to find the standard cubic approximation of $f(x) = 1/(1-x)$ at $x = 0$. Give an upper bound for the magnitude of the error in the approximation when $|x| \le 0.1$.

53. a. Use Taylor's formula with $n = 2$ to find the quadratic approximation of $f(x) = (1+x)^k$ at $x = 0$ (k a constant).

b. If $k = 3$, for approximately what values of x in the interval $[0, 1]$ will the error in the quadratic approximation be less than 1/100?

54. Improving approximations of π

a. Let P be an approximation of π accurate to n decimals. Show that $P + \sin P$ gives an approximation correct to $3n$ decimals. (*Hint:* Let $P = \pi + x$.)

T **b.** Try it with a calculator.

55. The Taylor series generated by $f(x) = \sum_{n=0}^{\infty} a_n x^n$ is $\sum_{n=0}^{\infty} a_n x^n$ A function defined by a power series $\sum_{n=0}^{\infty} a_n x^n$ with a radius of convergence $R > 0$ has a Taylor series that converges to the function at every point of $(-R, R)$. Show this by showing that the Taylor series generated by $f(x) = \sum_{n=0}^{\infty} a_n x^n$ is the series $\sum_{n=0}^{\infty} a_n x^n$ itself.

An immediate consequence of this is that series like

$$x\sin x = x^2 - \frac{x^4}{3!} + \frac{x^6}{5!} - \frac{x^8}{7!} + \cdots$$

and

$$x^2e^x = x^2 + x^3 + \frac{x^4}{2!} + \frac{x^5}{3!} + \cdots,$$

obtained by multiplying Taylor series by powers of x, as well as series obtained by integration and differentiation of convergent power series, are themselves the Taylor series generated by the functions they represent.

56. Taylor series for even functions and odd functions *(Continuation of Section 10.7, Exercise 61.)* Suppose that $f(x) = \sum_{n=0}^{\infty} a_n x^n$ converges for all x in an open interval $(-R, R)$. Show that

a. If f is even, then $a_1 = a_3 = a_5 = \cdots = 0$, i.e., the Taylor series for f at $x = 0$ contains only even powers of x.

b. If f is odd, then $a_0 = a_2 = a_4 = \cdots = 0$, i.e., the Taylor series for f at $x = 0$ contains only odd powers of x.

COMPUTER EXPLORATIONS

Taylor's formula with $n = 1$ and $a = 0$ gives the linearization of a function at $x = 0$. With $n = 2$ and $n = 3$, we obtain the standard quadratic and cubic approximations. In these exercises we explore the errors associated with these approximations. We seek answers to two questions:

a. For what values of x can the function be replaced by each approximation with an error less than 10^{-2}?

b. What is the maximum error we can expect if we replace the function by each approximation over the specified interval?

Using a CAS, perform the following steps to aid in answering questions (a) and (b) for the functions and intervals in Exercises 57–62.

Step 1: Plot the function over the specified interval.

Step 2: Find the Taylor polynomials $P_1(x)$, $P_2(x)$, and $P_3(x)$ at $x = 0$.

Step 3: Calculate the $(n + 1)$st derivative $f^{(n+1)}(c)$ associated with the remainder term for each Taylor polynomial. Plot the derivative as a function of c over the specified interval and estimate its maximum absolute value, M.

Step 4: Calculate the remainder $R_n(x)$ for each polynomial. Using the estimate M from Step 3 in place of $f^{(n+1)}(c)$, plot $R_n(x)$ over the specified interval. Then estimate the values of x that answer question (a).

Step 5: Compare your estimated error with the actual error $E_n(x) = |f(x) - P_n(x)|$ by plotting $E_n(x)$ over the specified interval. This will help you answer question (b).

Step 6: Graph the function and its three Taylor approximations together. Discuss the graphs in relation to the information discovered in Steps 4 and 5.

57. $f(x) = \dfrac{1}{\sqrt{1+x}}, \quad |x| \le \dfrac{3}{4}$

58. $f(x) = (1 + x)^{3/2}, \quad -\dfrac{1}{2} \le x \le 2$

59. $f(x) = \dfrac{x}{x^2 + 1}, \quad |x| \le 2$

60. $f(x) = (\cos x)(\sin 2x), \quad |x| \le 2$

61. $f(x) = e^{-x} \cos 2x, \quad |x| \le 1$

62. $f(x) = e^{x/3} \sin 2x, \quad |x| \le 2$

10.10 Applications of Taylor Series

We can use Taylor series to solve problems that would otherwise be intractable. For example, many functions have antiderivatives that cannot be expressed using familiar functions. In this section we show how to evaluate integrals of such functions by giving them as Taylor series. We also show how to use Taylor series to evaluate limits that lead to indeterminate forms and how Taylor series can be used to extend the exponential function from real to complex numbers. We begin with a discussion of the binomial series, which comes from the Taylor series of the function $f(x) = (1 + x)^m$, and we conclude the section with Table 10.1, which lists some commonly used Taylor series.

The Binomial Series for Powers and Roots

The Taylor series generated by $f(x) = (1 + x)^m$, when m is constant, is

$$1 + mx + \frac{m(m-1)}{2!}x^2 + \frac{m(m-1)(m-2)}{3!}x^3 + \cdots + \frac{m(m-1)(m-2)\cdots(m-k+1)}{k!}x^k + \cdots. \qquad (1)$$

This series, called the **binomial series**, converges absolutely for $|x| < 1$. To derive the series, we first list the function and its derivatives:

$$\begin{aligned} f(x) &= (1 + x)^m \\ f'(x) &= m(1 + x)^{m-1} \\ f''(x) &= m(m - 1)(1 + x)^{m-2} \\ f'''(x) &= m(m - 1)(m - 2)(1 + x)^{m-3} \\ &\vdots \\ f^{(k)}(x) &= m(m - 1)(m - 2)\cdots(m - k + 1)(1 + x)^{m-k}. \end{aligned}$$

We then evaluate these at $x = 0$ and substitute into the Taylor series formula to obtain Series (1).

If m is an integer greater than or equal to zero, the series stops after $(m + 1)$ terms because the coefficients from $k = m + 1$ on are zero.

If m is not a positive integer or zero, the series is infinite and converges for $|x| < 1$. To see why, let u_k be the term involving x^k. Then apply the Ratio Test for absolute convergence to see that

$$\left|\frac{u_{k+1}}{u_k}\right| = \left|\frac{m - k}{k + 1}x\right| \to |x| \qquad \text{as } k \to \infty.$$

Our derivation of the binomial series shows only that it is generated by $(1 + x)^m$ and converges for $|x| < 1$. The derivation does not show that the series converges to $(1 + x)^m$. It does, but we leave the proof to Exercise 58. The following formulation gives a succinct way to express the series.

The Binomial Series

For $-1 < x < 1$,

$$(1 + x)^m = 1 + \sum_{k=1}^{\infty} \binom{m}{k} x^k,$$

where we define

$$\binom{m}{1} = m, \quad \binom{m}{2} = \frac{m(m - 1)}{2!},$$

and

$$\binom{m}{k} = \frac{m(m - 1)(m - 2)\cdots(m - k + 1)}{k!} \qquad \text{for } k \geq 3.$$

EXAMPLE 1 If $m = -1$, then

$$\binom{-1}{1} = -1, \quad \binom{-1}{2} = \frac{-1(-2)}{2!} = 1,$$

and

$$\binom{-1}{k} = \frac{-1(-2)(-3)\cdots(-1 - k + 1)}{k!} = (-1)^k\left(\frac{k!}{k!}\right) = (-1)^k.$$

With these coefficient values and with x replaced by $-x$, the binomial series formula gives the familiar geometric series

$$(1 + x)^{-1} = 1 + \sum_{k=1}^{\infty} (-1)^k x^k = 1 - x + x^2 - x^3 + \cdots + (-1)^k x^k + \cdots. \quad \blacksquare$$

EXAMPLE 2 We know from Section 3.9, Example 1, that $\sqrt{1+x} \approx 1 + (x/2)$ for $|x|$ small. With $m = 1/2$, the binomial series gives quadratic and higher-order approximations as well, along with error estimates that come from the Alternating Series Estimation Theorem:

$$\begin{aligned}(1+x)^{1/2} &= 1 + \frac{x}{2} + \frac{\left(\frac{1}{2}\right)\left(-\frac{1}{2}\right)}{2!}x^2 + \frac{\left(\frac{1}{2}\right)\left(-\frac{1}{2}\right)\left(-\frac{3}{2}\right)}{3!}x^3 \\ &\quad + \frac{\left(\frac{1}{2}\right)\left(-\frac{1}{2}\right)\left(-\frac{3}{2}\right)\left(-\frac{5}{2}\right)}{4!}x^4 + \cdots \\ &= 1 + \frac{x}{2} - \frac{x^2}{8} + \frac{x^3}{16} - \frac{5x^4}{128} + \cdots.\end{aligned}$$

Substitution for x gives still other approximations. For example,

$$\sqrt{1-x^2} \approx 1 - \frac{x^2}{2} - \frac{x^4}{8} \qquad \text{for } |x^2| \text{ small}$$

$$\sqrt{1-\frac{1}{x}} \approx 1 - \frac{1}{2x} - \frac{1}{8x^2} \qquad \text{for } \left|\frac{1}{x}\right| \text{ small, that is, } |x| \text{ large.}$$ ■

Evaluating Nonelementary Integrals

Sometimes we can use a familiar Taylor series to find the sum of a given power series in terms of a known function. For example,

$$x^2 - \frac{x^6}{3!} + \frac{x^{10}}{5!} - \frac{x^{14}}{7!} + \cdots = (x^2) - \frac{(x^2)^3}{3!} + \frac{(x^2)^5}{5!} - \frac{(x^2)^7}{7!} + \cdots = \sin x^2.$$

Additional examples are provided in Exercises 59–62.

Taylor series can be used to express nonelementary integrals in terms of series. Integrals like $\int \sin x^2\, dx$ arise in the study of the diffraction of light.

EXAMPLE 3 Express $\int \sin x^2\, dx$ as a power series.

Solution From the series for $\sin x$, we substitute x^2 for x to obtain

$$\sin x^2 = x^2 - \frac{x^6}{3!} + \frac{x^{10}}{5!} - \frac{x^{14}}{7!} + \frac{x^{18}}{9!} - \cdots.$$

Therefore,

$$\int \sin x^2\, dx = C + \frac{x^3}{3} - \frac{x^7}{7\cdot 3!} + \frac{x^{11}}{11\cdot 5!} - \frac{x^{15}}{15\cdot 7!} + \frac{x^{19}}{19\cdot 9!} - \cdots.$$ ■

EXAMPLE 4 Estimate $\int_0^1 \sin x^2\, dx$ with an error of less than 0.001.

Solution From the indefinite integral in Example 3, we easily find that

$$\int_0^1 \sin x^2\, dx = \frac{1}{3} - \frac{1}{7\cdot 3!} + \frac{1}{11\cdot 5!} - \frac{1}{15\cdot 7!} + \frac{1}{19\cdot 9!} - \cdots.$$

The series on the right-hand side alternates, and we find by numerical evaluations that

$$\frac{1}{11\cdot 5!} \approx 0.00076$$

is the first term to be numerically less than 0.001. The sum of the preceding two terms gives

$$\int_0^1 \sin x^2\,dx \approx \frac{1}{3} - \frac{1}{42} \approx 0.310.$$

With two more terms, we could estimate

$$\int_0^1 \sin x^2\,dx \approx 0.310268$$

with an error of less than 10^{-6}. With only one term beyond that, we have

$$\int_0^1 \sin x^2\,dx \approx \frac{1}{3} - \frac{1}{42} + \frac{1}{1320} - \frac{1}{75600} + \frac{1}{6894720} \approx 0.310268303,$$

with an error of about 1.08×10^{-9}. Guaranteeing this accuracy with the error formula for the Trapezoidal Rule would require using about 8000 subintervals. ■

Arctangents

In Section 10.7, Example 5, we found a series for $\arctan x$ by differentiating to get

$$\frac{d}{dx}\arctan x = \frac{1}{1+x^2} = 1 - x^2 + x^4 - x^6 + \cdots$$

and then integrating to get

$$\arctan x = x - \frac{x^3}{3} + \frac{x^5}{5} - \frac{x^7}{7} + \cdots.$$

However, we did not prove the term-by-term integration theorem on which this conclusion depended. We now derive the series again by integrating both sides of the finite formula

$$\frac{1}{1+t^2} = 1 - t^2 + t^4 - t^6 + \cdots + (-1)^n t^{2n} + \frac{(-1)^{n+1}t^{2n+2}}{1+t^2}, \qquad (2)$$

in which the last term comes from adding the remaining terms as a geometric series with first term $a = (-1)^{n+1}t^{2n+2}$ and ratio $r = -t^2$. Integrating both sides of Equation (2) from $t = 0$ to $t = x$ gives

$$\arctan x = x - \frac{x^3}{3} + \frac{x^5}{5} - \frac{x^7}{7} + \cdots + (-1)^n \frac{x^{2n+1}}{2n+1} + R_n(x),$$

where

$$R_n(x) = \int_0^x \frac{(-1)^{n+1}t^{2n+2}}{1+t^2}\,dt.$$

The denominator of the integrand is greater than or equal to 1; hence

$$|R_n(x)| \le \int_0^{|x|} t^{2n+2}\,dt = \frac{|x|^{2n+3}}{2n+3}.$$

If $|x| \le 1$, the right side of this inequality approaches zero as $n \to \infty$. Therefore, $\lim_{n\to\infty} R_n(x) = 0$ if $|x| \le 1$ and

$$\arctan x = \sum_{n=0}^{\infty} \frac{(-1)^n x^{2n+1}}{2n+1}, \qquad |x| \le 1. \qquad (3)$$

$$\arctan x = x - \frac{x^3}{3} + \frac{x^5}{5} - \frac{x^7}{7} + \cdots, \qquad |x| \le 1.$$

We take this route instead of finding the Taylor series directly because the formulas for the higher-order derivatives of $\arctan x$ are unmanageable. When we put $x = 1$ in Equation (3), we get **Leibniz's formula**:

$$\frac{\pi}{4} = 1 - \frac{1}{3} + \frac{1}{5} - \frac{1}{7} + \frac{1}{9} - \cdots + \frac{(-1)^n}{2n+1} + \cdots.$$

Because this series converges very slowly, it is not used in approximating π to many decimal places. The series for $\arctan x$ converges most rapidly when x is near zero. For that reason, people who use the series for $\arctan x$ to compute π use various trigonometric identities.

For example, if

$$\alpha = \arctan\frac{1}{2} \quad \text{and} \quad \beta = \arctan\frac{1}{3},$$

then

$$\tan(\alpha + \beta) = \frac{\tan\alpha + \tan\beta}{1 - \tan\alpha\tan\beta} = \frac{\frac{1}{2} + \frac{1}{3}}{1 - \frac{1}{6}} = 1 = \tan\frac{\pi}{4},$$

and therefore,

$$\frac{\pi}{4} = \alpha + \beta = \arctan\frac{1}{2} + \arctan\frac{1}{3}.$$

Now Equation (3) may be used with $x = 1/2$ to evaluate $\arctan(1/2)$ and with $x = 1/3$ to give $\arctan(1/3)$. The sum of these results, multiplied by 4, gives π.

Evaluating Indeterminate Forms

We can sometimes evaluate indeterminate forms by expressing the functions involved as Taylor series.

EXAMPLE 5 Evaluate

$$\lim_{x\to 1}\frac{\ln x}{x-1}.$$

Solution We represent $\ln x$ as a Taylor series in powers of $x - 1$. This can be accomplished by calculating the Taylor series generated by $\ln x$ at $x = 1$ directly or by replacing x by $x - 1$ in the series for $\ln(1 + x)$ in Section 10.7, Example 6. Either way, we obtain

$$\ln x = (x-1) - \frac{1}{2}(x-1)^2 + \cdots,$$

from which we find that

$$\lim_{x\to 1}\frac{\ln x}{x-1} = \lim_{x\to 1}\left(1 - \frac{1}{2}(x-1) + \cdots\right) = 1.$$

Of course, this particular limit can be evaluated just as well using l'Hôpital's Rule. ■

EXAMPLE 6 Evaluate

$$\lim_{x\to 0}\frac{\sin x - \tan x}{x^3}.$$

Solution The Taylor series for $\sin x$ and $\tan x$, to terms in x^5, are

$$\sin x = x - \frac{x^3}{3!} + \frac{x^5}{5!} - \cdots, \qquad \tan x = x + \frac{x^3}{3} + \frac{2x^5}{15} + \cdots.$$

Subtracting the series term by term, it follows that

$$\sin x - \tan x = -\frac{x^3}{2} - \frac{x^5}{8} - \cdots = x^3\left(-\frac{1}{2} - \frac{x^2}{8} - \cdots\right).$$

Division of both sides by x^3 and taking limits then give

$$\lim_{x\to 0} \frac{\sin x - \tan x}{x^3} = \lim_{x\to 0}\left(-\frac{1}{2} - \frac{x^2}{8} - \cdots\right) = -\frac{1}{2}. \quad \blacksquare$$

If we apply series to calculate $\lim_{x\to 0}\left((1/\sin x) - (1/x)\right)$, we not only find the limit successfully but also discover an approximation formula for $\csc x$.

EXAMPLE 7 Find $\displaystyle\lim_{x\to 0}\left(\frac{1}{\sin x} - \frac{1}{x}\right)$.

Solution Using algebra and the Taylor series for $\sin x$, we have

$$\frac{1}{\sin x} - \frac{1}{x} = \frac{x - \sin x}{x \sin x} = \frac{x - \left(x - \dfrac{x^3}{3!} + \dfrac{x^5}{5!} - \cdots\right)}{x \cdot \left(x - \dfrac{x^3}{3!} + \dfrac{x^5}{5!} - \cdots\right)}$$

$$= \frac{x^3\left(\dfrac{1}{3!} - \dfrac{x^2}{5!} + \cdots\right)}{x^2\left(1 - \dfrac{x^2}{3!} + \cdots\right)} = x \cdot \frac{\dfrac{1}{3!} - \dfrac{x^2}{5!} + \cdots}{1 - \dfrac{x^2}{3!} + \cdots}.$$

Therefore,

$$\lim_{x\to 0}\left(\frac{1}{\sin x} - \frac{1}{x}\right) = \lim_{x\to 0}\left(x \cdot \frac{\dfrac{1}{3!} - \dfrac{x^2}{5!} + \cdots}{1 - \dfrac{x^2}{3!} + \cdots}\right) = 0.$$

From the quotient on the right, we can see that if $|x|$ is small, then

$$\frac{1}{\sin x} - \frac{1}{x} \approx x \cdot \frac{1}{3!} = \frac{x}{6} \quad \text{or} \quad \csc x \approx \frac{1}{x} + \frac{x}{6}. \quad \blacksquare$$

Euler's Identity

A complex number is a number of the form $a + bi$, where a and b are real numbers and $i = \sqrt{-1}$ (see Chapter 18). If we substitute $x = i\theta$ (with θ real) in the Taylor series for e^x and use the relations

$$i^2 = -1, \qquad i^3 = i^2 i = -i, \qquad i^4 = i^2 i^2 = 1, \qquad i^5 = i^4 i = i,$$

and so on to simplify the result, we obtain

$$\begin{aligned} e^{i\theta} &= 1 + \frac{i\theta}{1!} + \frac{i^2\theta^2}{2!} + \frac{i^3\theta^3}{3!} + \frac{i^4\theta^4}{4!} + \frac{i^5\theta^5}{5!} + \frac{i^6\theta^6}{6!} + \cdots \\ &= \left(1 - \frac{\theta^2}{2!} + \frac{\theta^4}{4!} - \frac{\theta^6}{6!} + \cdots\right) + i\left(\theta - \frac{\theta^3}{3!} + \frac{\theta^5}{5!} - \cdots\right) = \cos\theta + i\sin\theta. \end{aligned}$$

This does not *prove* that $e^{i\theta} = \cos\theta + i\sin\theta$ because we have not yet defined what it means to raise e to an imaginary power. Rather, it tells us how to define $e^{i\theta}$ so that its properties are consistent with the properties of the exponential function for real numbers.

DEFINITION For any real number θ,

$$e^{i\theta} = \cos\theta + i\sin\theta. \tag{4}$$

Equation (4), called **Euler's identity**, enables us to define e^{a+bi} to be $e^a \cdot e^{bi}$ for any complex number $a + bi$. So

$$e^{a+ib} = e^a(\cos b + i\sin b).$$

One consequence of this identity is the equation

$$e^{i\pi} = -1.$$

When written in the form $e^{i\pi} + 1 = 0$, this equation combines five of the most important constants in mathematics.

TABLE 10.1 Frequently Used Taylor Series

$$\frac{1}{1-x} = 1 + x + x^2 + \cdots + x^n + \cdots = \sum_{n=0}^{\infty} x^n, \quad |x| < 1$$

$$\frac{1}{1+x} = 1 - x + x^2 - \cdots + (-x)^n + \cdots = \sum_{n=0}^{\infty} (-1)^n x^n, \quad |x| < 1$$

$$e^x = 1 + x + \frac{x^2}{2!} + \cdots + \frac{x^n}{n!} + \cdots = \sum_{n=0}^{\infty} \frac{x^n}{n!}, \quad |x| < \infty$$

$$\sin x = x - \frac{x^3}{3!} + \frac{x^5}{5!} - \cdots + (-1)^n \frac{x^{2n+1}}{(2n+1)!} + \cdots = \sum_{n=0}^{\infty} \frac{(-1)^n x^{2n+1}}{(2n+1)!}, \quad |x| < \infty$$

$$\cos x = 1 - \frac{x^2}{2!} + \frac{x^4}{4!} - \cdots + (-1)^n \frac{x^{2n}}{(2n)!} + \cdots = \sum_{n=0}^{\infty} \frac{(-1)^n x^{2n}}{(2n)!}, \quad |x| < \infty$$

$$\ln(1+x) = x - \frac{x^2}{2} + \frac{x^3}{3} - \cdots + (-1)^{n-1} \frac{x^n}{n} + \cdots = \sum_{n=1}^{\infty} \frac{(-1)^{n-1} x^n}{n}, \quad -1 < x \le 1$$

$$\arctan x = x - \frac{x^3}{3} + \frac{x^5}{5} - \cdots + (-1)^n \frac{x^{2n+1}}{2n+1} + \cdots = \sum_{n=0}^{\infty} \frac{(-1)^n x^{2n+1}}{2n+1}, \quad |x| \le 1$$

EXERCISES 10.10

Taylor Series

Find the first four nonzero terms of the Taylor series for the functions in Exercises 1–10.

1. $(1+x)^{1/2}$ **2.** $(1+x)^{1/3}$

3. $(1-x)^{-3}$ **4.** $(1-2x)^{1/2}$

5. $\left(1+\frac{x}{2}\right)^{-2}$ **6.** $\left(1-\frac{x}{3}\right)^{4}$

7. $(1+x^3)^{-1/2}$ **8.** $(1+x^2)^{-1/3}$

9. $\left(1+\frac{x^2}{2}\right)^{3/2}$ **10.** $\frac{x}{\sqrt[3]{1+x}}$

Find the binomial series for the functions in Exercises 11–14.

11. $(1+x)^4$ **12.** $(1+x^2)^3$

13. $(1-2x)^3$ **14.** $\left(1-\frac{x}{2}\right)^4$

Approximations and Nonelementary Integrals

T In Exercises 15–18, use series to estimate the integrals' values with an error of magnitude less than 10^{-5}. (The answer section gives the integrals' values rounded to seven decimal places.)

15. $\int_0^{0.6} \sin x^2\, dx$ **16.** $\int_0^{0.4} \frac{e^{-x}-1}{x}\, dx$

17. $\int_0^{0.5} \frac{1}{\sqrt{1+x^4}}\, dx$ **18.** $\int_0^{0.35} \sqrt[3]{1+x^2}\, dx$

T Use series to approximate the values of the integrals in Exercises 19–22 with an error of magnitude less than 10^{-8}.

19. $\int_0^{0.1} \frac{\sin x}{x}\, dx$ **20.** $\int_0^{0.1} e^{-x^2}\, dx$

21. $\int_0^{0.1} \sqrt{1+x^4}\, dx$ **22.** $\int_0^{1} \frac{1-\cos x}{x^2}\, dx$

23. Estimate the error if $\cos t^2$ is approximated by $1-\frac{t^4}{2}+\frac{t^8}{4!}$ in the integral $\int_0^1 \cos t^2\, dt$.

24. Estimate the error if $\cos\sqrt{t}$ is approximated by $1-\frac{t}{2}+\frac{t^2}{4!}-\frac{t^3}{6!}$ in the integral $\int_0^1 \cos\sqrt{t}\, dt$.

In Exercises 25–28, find a polynomial that will approximate $F(x)$ throughout the given interval with an error of magnitude less than 10^{-3}.

25. $F(x)=\int_0^x \sin t^2\, dt,\quad [0,1]$

26. $F(x)=\int_0^x t^2 e^{-t^2}\, dt,\quad [0,1]$

27. $F(x)=\int_0^x \arctan t\, dt,\quad$ **(a)** $[0, 0.5]$ **(b)** $[0,1]$

28. $F(x)=\int_0^x \frac{\ln(1+t)}{t}\, dt,\quad$ **(a)** $[0, 0.5]$ **(b)** $[0,1]$

Indeterminate Forms

Use series to evaluate the limits in Exercises 29–40.

29. $\lim_{x\to 0} \frac{e^x-(1+x)}{x^2}$ **30.** $\lim_{x\to 0} \frac{e^x-e^{-x}}{x}$

31. $\lim_{t\to 0} \frac{1-\cos t-(t^2/2)}{t^4}$ **32.** $\lim_{\theta\to 0} \frac{\sin\theta-\theta+(\theta^3/6)}{\theta^5}$

33. $\lim_{y\to 0} \frac{y-\arctan y}{y^3}$ **34.** $\lim_{y\to 0} \frac{\tan^{-1} y-\sin y}{y^3\cos y}$

35. $\lim_{x\to\infty} x^2\left(e^{-1/x^2}-1\right)$ **36.** $\lim_{x\to\infty} (x+1)\sin\frac{1}{x+1}$

37. $\lim_{x\to 0} \frac{\ln(1+x^2)}{1-\cos x}$ **38.** $\lim_{x\to 2} \frac{x^2-4}{\ln(x-1)}$

39. $\lim_{x\to 0} \frac{\sin 3x^2}{1-\cos 2x}$ **40.** $\lim_{x\to 0} \frac{\ln(1+x^3)}{x\cdot\sin x^2}$

Using Table 10.1

In Exercises 41–52, use Table 10.1 to find the sum of each series.

41. $1+1+\frac{1}{2!}+\frac{1}{3!}+\frac{1}{4!}+\cdots$

42. $\left(\frac{1}{4}\right)^3+\left(\frac{1}{4}\right)^4+\left(\frac{1}{4}\right)^5+\left(\frac{1}{4}\right)^6+\cdots$

43. $1-\frac{3^2}{4^2\cdot 2!}+\frac{3^4}{4^4\cdot 4!}-\frac{3^6}{4^6\cdot 6!}+\cdots$

44. $\frac{1}{2}-\frac{1}{2\cdot 2^2}+\frac{1}{3\cdot 2^3}-\frac{1}{4\cdot 2^4}+\cdots$

45. $\frac{\pi}{3}-\frac{\pi^3}{3^3\cdot 3!}+\frac{\pi^5}{3^5\cdot 5!}-\frac{\pi^7}{3^7\cdot 7!}+\cdots$

46. $\frac{2}{3}-\frac{2^3}{3^3\cdot 3}+\frac{2^5}{3^5\cdot 5}-\frac{2^7}{3^7\cdot 7}+\cdots$

47. $x^3+x^4+x^5+x^6+\cdots$

48. $1-\frac{3^2x^2}{2!}+\frac{3^4x^4}{4!}-\frac{3^6x^6}{6!}+\cdots$

49. $x^3-x^5+x^7-x^9+x^{11}-\cdots$

50. $x^2-2x^3+\frac{2^2x^4}{2!}-\frac{2^3x^5}{3!}+\frac{2^4x^6}{4!}-\cdots$

51. $-1+2x-3x^2+4x^3-5x^4+\cdots$

52. $1+\frac{x}{2}+\frac{x^2}{3}+\frac{x^3}{4}+\frac{x^4}{5}+\cdots$

Theory and Examples

53. Replace x by $-x$ in the Taylor series for $\ln(1+x)$ to obtain a series for $\ln(1-x)$. Then subtract this from the Taylor series for $\ln(1+x)$ to show that for $|x|<1$,

$$\ln\frac{1+x}{1-x}=2\left(x+\frac{x^3}{3}+\frac{x^5}{5}+\cdots\right).$$

54. How many terms of the Taylor series for $\ln(1+x)$ should you add to be sure of calculating $\ln(1.1)$ with an error of magnitude less than 10^{-8}? Give reasons for your answer.

55. Show that $\tan^{-1}\left(\frac{\sqrt{1+x^2}-1}{x}\right)=\frac{1}{2}\left(x-\frac{x^3}{3}+\frac{x^5}{5}-\frac{x^7}{7}+\dots\right)$

56. Show that the Taylor's series for $\tan^{-1}x$ about $x=\frac{\pi}{4}$ is

$$\tan^{-1}x=\tan^{-1}\frac{\pi}{4}+\frac{x-\frac{\pi}{4}}{\left(1+\frac{\pi}{4}\right)^2}-\frac{\pi\left(x-\frac{\pi}{4}\right)^2}{4\left(1+\frac{\pi^2}{16}\right)^2}+\dots$$

T **57. Estimating pi** About how many terms of the Taylor series for $\arctan x$ would you have to use to evaluate each term on the right-hand side of the equation

$$\pi = 48\arctan\frac{1}{18} + 32\arctan\frac{1}{57} - 20\arctan\frac{1}{239}$$

with an error of magnitude less than 10^{-6}? In contrast, the convergence of $\sum_{n=1}^{\infty}(1/n^2)$ to $\pi^2/6$ is so slow that even 50 terms will not yield two-place accuracy.

58. Use the following steps to prove that the binomial series in Equation (1) converges to $(1+x)^m$.

a. Differentiate the series

$$f(x) = 1 + \sum_{k=1}^{\infty}\binom{m}{k}x^k$$

to show that

$$f'(x) = \frac{mf(x)}{1+x}, \quad -1 < x < 1.$$

b. Define $g(x) = (1+x)^{-m}f(x)$ and show that $g'(x) = 0$.

c. From part (b), show that

$$f(x) = (1+x)^m.$$

59. a. Use the binomial series and the fact that

$$\frac{d}{dx}\arcsin x = (1-x^2)^{-1/2}$$

to generate the first four nonzero terms of the Taylor series for $\arcsin x$. What is the radius of convergence?

b. Series for arccos *x* Use your result in part (a) to find the first five nonzero terms of the Taylor series for $\arccos x$.

60. a. Series for $\sinh^{-1}x$ Find the first four nonzero terms of the Taylor series for

$$\sinh^{-1}x = \int_0^x \frac{dt}{\sqrt{1+t^2}}.$$

T **b.** Use the first *three* terms of the series in part (a) to estimate $\sinh^{-1}0.25$. Give an upper bound for the magnitude of the estimation error.

61. Obtain the Taylor series for $1/(1+x)^2$ from the series for $-1/(1+x)$.

62. Use the Taylor series for $1/(1-x^2)$ to obtain a series for $2x/(1-x^2)^2$.

T **63. Estimating pi** The English mathematician Wallis discovered the formula

$$\frac{\pi}{4} = \frac{2\cdot4\cdot4\cdot6\cdot6\cdot8\cdot\cdots}{3\cdot3\cdot5\cdot5\cdot7\cdot7\cdot\cdots}.$$

Find π to two decimal places with this formula.

64. The complete elliptic integral of the first kind is the integral

$$K = \int_0^{\pi/2}\frac{d\theta}{\sqrt{1-k^2\sin^2\theta}},$$

where $0 < k < 1$ is constant.

a. Show that the first four terms of the binomial series for $1/\sqrt{1-x}$ are

$$(1-x)^{-1/2} = 1 + \frac{1}{2}x + \frac{1\cdot3}{2\cdot4}x^2 + \frac{1\cdot3\cdot5}{2\cdot4\cdot6}x^3 + \cdots.$$

b. From part (a) and the reduction integral Formula 67 at the back of the text, show that

$$K = \frac{\pi}{2}\left[1 + \left(\frac{1}{2}\right)^2k^2 + \left(\frac{1\cdot3}{2\cdot4}\right)^2k^4 + \left(\frac{1\cdot3\cdot5}{2\cdot4\cdot6}\right)^2k^6 + \cdots\right].$$

65. Series for arcsin *x* Integrate the binomial series for $(1-x^2)^{-1/2}$ to show that for $|x| < 1$,

$$\arcsin x = x + \sum_{n=1}^{\infty}\frac{1\cdot3\cdot5\cdot\cdots\cdot(2n-1)}{2\cdot4\cdot6\cdot\cdots\cdot(2n)}\frac{x^{2n+1}}{2n+1}.$$

66. Series for arctan *x* for $|x| > 1$ Derive the series

$$\arctan x = \frac{\pi}{2} - \frac{1}{x} + \frac{1}{3x^3} - \frac{1}{5x^5} + \cdots, \quad x > 1$$

$$\arctan x = -\frac{\pi}{2} - \frac{1}{x} + \frac{1}{3x^3} - \frac{1}{5x^5} + \cdots, \quad x < -1,$$

by integrating the series

$$\frac{1}{1+t^2} = \frac{1}{t^2}\cdot\frac{1}{1+(1/t^2)} = \frac{1}{t^2} - \frac{1}{t^4} + \frac{1}{t^6} - \frac{1}{t^8} + \cdots$$

in the first case from x to ∞ and in the second case from $-\infty$ to x.

Euler's Identity

67. Use Equation (4) to write the following powers of e in the form $a + bi$.

a. $e^{-i\pi}$ **b.** $e^{i\pi/4}$ **c.** $e^{-i\pi/2}$

68. Use Equation (4) to show that

$$\cos\theta = \frac{e^{i\theta}+e^{-i\theta}}{2} \quad \text{and} \quad \sin\theta = \frac{e^{i\theta}-e^{-i\theta}}{2i}.$$

69. Establish the equations in Exercise 68 by combining the formal Taylor series for $e^{i\theta}$ and $e^{-i\theta}$.

70. Show that

a. $\cosh i\theta = \cos\theta$, **b.** $\sinh i\theta = i\sin\theta$.

71. By multiplying the Taylor series for e^x and $\sin x$, find the terms through x^5 of the Taylor series for $e^x\sin x$. This series is the imaginary part of the series for

$$e^x\cdot e^{ix} = e^{(1+i)x}.$$

Use this fact to check your answer. For what values of x should the series for $e^x\sin x$ converge?

72. When a and b are real, we define $e^{(a+ib)x}$ with the equation

$$e^{(a+ib)x} = e^{ax}\cdot e^{ibx} = e^{ax}(\cos bx + i\sin bx).$$

Differentiate the right-hand side of this equation to show that

$$\frac{d}{dx}e^{(a+ib)x} = (a + ib)e^{(a+ib)x}.$$

Thus the familiar rule $(d/dx)e^{kx} = ke^{kx}$ holds for k complex as well as real.

73. Use the definition of $e^{i\theta}$ to show that for any real numbers θ, θ_1, and θ_2,

a. $e^{i\theta_1}e^{i\theta_2} = e^{i(\theta_1+\theta_2)}$,

b. $e^{-i\theta} = 1/e^{i\theta}$.

74. Two complex numbers $a + ib$ and $c + id$ are equal if and only if $a = c$ and $b = d$. Use this fact to evaluate

$$\int e^{ax}\cos bx\,dx \quad\text{and}\quad \int e^{ax}\sin bx\,dx$$

from

$$\int e^{(a+ib)x}\,dx = \frac{a - ib}{a^2 + b^2}e^{(a+ib)x} + C,$$

where $C = C_1 + iC_2$ is a complex constant of integration.

CHAPTER 10 Questions to Guide Your Review

1. What is an infinite sequence? What does it mean for such a sequence to converge? To diverge? Give examples.
2. What is a monotonic sequence? Under what circumstances does such a sequence have a limit? Give examples.
3. What theorems are available for calculating limits of sequences? Give examples.
4. What theorem sometimes enables us to use l'Hôpital's Rule to calculate the limit of a sequence? Give an example.
5. What are the six commonly occurring limits in Theorem 5 that arise frequently when you work with sequences and series?
6. What is an infinite series? What does it mean for such a series to converge? To diverge? Give examples.
7. What is a geometric series? When does such a series converge? Diverge? When it does converge, what is its sum? Give examples.
8. Besides geometric series, what other convergent and divergent series do you know?
9. What is the nth-Term Test for Divergence? What is the idea behind the test?
10. What can be said about term-by-term sums and differences of convergent series? About constant multiples of convergent and divergent series?
11. What happens if you add a finite number of terms to a convergent series? A divergent series? What happens if you delete a finite number of terms from a convergent series? A divergent series?
12. How do you reindex a series? Why might you want to do this?
13. Under what circumstances will an infinite series of nonnegative terms converge? Diverge? Why study series of nonnegative terms?
14. What is the Integral Test? What is the reasoning behind it? Give an example of its use.
15. When do p-series converge? Diverge? How do you know? Give examples of convergent and divergent p-series.
16. What are the Direct Comparison Test and the Limit Comparison Test? What is the reasoning behind these tests? Give examples of their use.
17. What are the Ratio and Root Tests? Do they always give you the information you need to determine convergence or divergence? Give examples.
18. What is absolute convergence? Conditional convergence? How are the two related?
19. What is an alternating series? What theorem is available for determining the convergence of such a series?
20. How can you estimate the error involved in approximating the sum of an alternating series with one of the series' partial sums? What is the reasoning behind the estimate?
21. What do you know about rearranging the terms of an absolutely convergent series? Of a conditionally convergent series?
22. What is a power series? How do you test a power series for convergence? What are the possible outcomes?
23. What are the basic facts about
    a. sums, differences, and products of power series?
    b. substitution of a function for x in a power series?
    c. term-by-term differentiation of power series?
    d. term-by-term integration of power series?
    e. Give examples.
24. What is the Taylor series generated by a function $f(x)$ at a point $x = a$? What information do you need about f to construct the series? Give an example.
25. What is a Maclaurin series?
26. Does a Taylor series always converge to its generating function? Explain.
27. What are Taylor polynomials? Of what use are they?
28. What is Taylor's formula? What does it say about the errors involved in using Taylor polynomials to approximate functions? In particular, what does Taylor's formula say about the error in a linearization? A quadratic approximation?
29. What is the binomial series? On what interval does it converge? How is it used?
30. How can you sometimes use power series to estimate the values of nonelementary definite integrals? To find limits?
31. What are the Taylor series for $1/(1 - x)$, $1/(1 + x)$, e^x, $\sin x$, $\cos x$, $\ln(1 + x)$, and $\arctan x$? How do you estimate the errors involved in replacing these series with their partial sums?

CHAPTER 10 Practice Exercises

Determining Convergence of Sequences

Which of the sequences whose nth terms appear in Exercises 1–18 converge, and which diverge? Find the limit of each convergent sequence.

1. $a_n = 1 + \frac{(-1)^n}{n}$

2. $a_n = \frac{1-(-1)^n}{\sqrt{n}}$

3. $a_n = \frac{1-2^n}{2^n}$

4. $a_n = 1 + (0.9)^n$

5. $a_n = \sin\frac{n\pi}{2}$

6. $a_n = \sin n\pi$

7. $a_n = \frac{\ln(n^2)}{n}$

8. $a_n = \frac{\ln(2n+1)}{n}$

9. $a_n = \frac{n+\ln n}{n}$

10. $a_n = \frac{\ln(2n^3+1)}{n}$

11. $a_n = \left(\frac{n-5}{n}\right)^n$

12. $a_n = \left(1+\frac{1}{n}\right)^{-n}$

13. $a_n = \sqrt[n]{\frac{3^n}{n}}$

14. $a_n = \left(\frac{3}{n}\right)^{1/n}$

15. $a_n = n(2^{1/n}-1)$

16. $a_n = \sqrt[n]{2n+1}$

17. $a_n = \frac{(n+1)!}{n!}$

18. $a_n = \frac{(-4)^n}{n!}$

Convergent Series

Find the sums of the series in Exercises 19–24.

19. $\sum_{n=3}^{\infty} \frac{1}{(2n-3)(2n-1)}$

20. $\sum_{n=2}^{\infty} \frac{-2}{n(n+1)}$

21. $\sum_{n=1}^{\infty} \frac{9}{(3n-1)(3n+2)}$

22. $\sum_{n=3}^{\infty} \frac{-8}{(4n-3)(4n+1)}$

23. $\sum_{n=0}^{\infty} e^{-n}$

24. $\sum_{n=1}^{\infty} (-1)^n \frac{3}{4^n}$

Determining Convergence of Series

Which of the series in Exercises 25–44 converge absolutely, which converge conditionally, and which diverge? Give reasons for your answers.

25. $\sum_{n=1}^{\infty} \frac{1}{\sqrt{n}}$

26. $\sum_{n=1}^{\infty} \frac{-5}{n}$

27. $\sum_{n=1}^{\infty} \frac{(-1)^n}{\sqrt{n}}$

28. $\sum_{n=1}^{\infty} \frac{1}{2n^3}$

29. $\sum_{n=1}^{\infty} \frac{(-1)^n}{\ln(n+1)}$

30. $\sum_{n=2}^{\infty} \frac{1}{n(\ln n)^2}$

31. $\sum_{n=1}^{\infty} \frac{\ln n}{n^3}$

32. $\sum_{n=3}^{\infty} \frac{\ln n}{\ln(\ln n)}$

33. $\sum_{n=1}^{\infty} \frac{(-1)^n}{n\sqrt{n^2+1}}$

34. $\sum_{n=1}^{\infty} \frac{(-1)^n 3n^2}{n^3+1}$

35. $\sum_{n=1}^{\infty} \frac{n+1}{n!}$

36. $\sum_{n=1}^{\infty} \frac{(-1)^n(n^2+1)}{2n^2+n-1}$

37. $\sum_{n=1}^{\infty} \frac{(-3)^n}{n!}$

38. $\sum_{n=1}^{\infty} \frac{2^n 3^n}{n^n}$

39. $\sum_{n=1}^{\infty} \frac{1}{\sqrt{n(n+1)(n+2)}}$

40. $\sum_{n=2}^{\infty} \frac{1}{n\sqrt{n^2-1}}$

41. $1 - \left(\frac{1}{\sqrt{3}}\right)^2 + \left(\frac{1}{\sqrt{3}}\right)^4 - \left(\frac{1}{\sqrt{3}}\right)^6 + \left(\frac{1}{\sqrt{3}}\right)^8 - \cdots$

42. $\sum_{n=0}^{\infty} \frac{(-1)^n}{e^{-n}+1}$

43. $\sum_{n=0}^{\infty} \frac{1}{1+r+r^2+\cdots+r^n}$, for $-1 < r < 1$

44. $\sum_{n=1}^{\infty} \frac{(-1)^n}{\sqrt{n+100}-\sqrt{n}}$

45. Prove that $\sum_{n=3}^{\infty} \frac{1}{n(\ln n)(\ln(\ln n))^p}$ converges if and only if $p > 1$.

46. Prove that $\sum_{n=1}^{\infty} \frac{\ln n}{n^p}$ converges if and only if $p > 1$.

Power Series

In Exercises 47–56, **(a)** find the series' radius and interval of convergence. Then identify the values of x for which the series converges **(b)** absolutely and **(c)** conditionally.

47. $\sum_{n=1}^{\infty} \frac{(x+4)^n}{n3^n}$

48. $\sum_{n=1}^{\infty} \frac{(x-1)^{2n-2}}{(2n-1)!}$

49. $\sum_{n=1}^{\infty} \frac{(-1)^{n-1}(3x-1)^n}{n^2}$

50. $\sum_{n=0}^{\infty} \frac{(n+1)(2x+1)^n}{(2n+1)2^n}$

51. $\sum_{n=1}^{\infty} \frac{x^n}{n^n}$

52. $\sum_{n=1}^{\infty} \frac{x^n}{\sqrt{n}}$

53. $\sum_{n=0}^{\infty} \frac{(n+1)x^{2n-1}}{3^n}$

54. $\sum_{n=0}^{\infty} \frac{(-1)^n(x-1)^{2n+1}}{2n+1}$

55. $\sum_{n=1}^{\infty} (\operatorname{csch} n)x^n$

56. $\sum_{n=1}^{\infty} (\coth n)x^n$

Maclaurin Series

Each of the series in Exercises 57–62 is the value of the Taylor series at $x = 0$ of a function $f(x)$ at a particular point. What function and what point? What is the sum of the series?

57. $1 - \frac{1}{4} + \frac{1}{16} - \cdots + (-1)^n \frac{1}{4^n} + \cdots$

58. $\frac{2}{3} - \frac{4}{18} + \frac{8}{81} - \cdots + (-1)^{n-1}\frac{2^n}{n3^n} + \cdots$

59. $\pi - \frac{\pi^3}{3!} + \frac{\pi^5}{5!} - \cdots + (-1)^n \frac{\pi^{2n+1}}{(2n+1)!} + \cdots$

60. $1 - \frac{\pi^2}{9 \cdot 2!} + \frac{\pi^4}{81 \cdot 4!} - \cdots + (-1)^n \frac{\pi^{2n}}{3^{2n}(2n)!} + \cdots$

61. $1 + \ln 2 + \frac{(\ln 2)^2}{2!} + \cdots + \frac{(\ln 2)^n}{n!} + \cdots$

62. $\frac{1}{\sqrt{3}} - \frac{1}{9\sqrt{3}} + \frac{1}{45\sqrt{3}} - \cdots + (-1)^{n-1}\frac{1}{(2n-1)(\sqrt{3})^{2n-1}} + \cdots$

Find Taylor series at $x = 0$ for the functions in Exercises 63–70.

63. $\frac{1}{1-2x}$

64. $\frac{1}{1+x^3}$

65. $\sin \pi x$

66. $\sin \frac{2x}{3}$

67. $\cos(x^{5/3})$

68. $\cos \frac{x^3}{\sqrt{5}}$

69. $e^{(\pi x/2)}$

70. e^{-x^2}

Taylor Series

In Exercises 71–74, find the first four nonzero terms of the Taylor series generated by f at $x = a$.

71. $f(x) = \sqrt{3 + x^2}$ at $x = -1$

72. $f(x) = 1/(1 - x)$ at $x = 2$

73. $f(x) = 1/(x + 1)$ at $x = 3$

74. $f(x) = 1/x$ at $x = a > 0$

Nonelementary Integrals

Use series to approximate the values of the integrals in Exercises 75–78 with an error of magnitude less than 10^{-8}. (The answer section gives the integrals' values rounded to ten decimal places.)

75. $\int_0^{1/2} e^{-x^3}\,dx$

76. $\int_0^1 x \sin(x^3)\,dx$

77. $\int_0^{1/2} \frac{\tan^{-1} x}{x}\,dx$

78. $\int_0^{1/64} \frac{\arctan x}{\sqrt{x}}\,dx$

Using Series to Find Limits

In Exercises 79–84:

a. Use power series to evaluate the limit.

T b. Then use a grapher to support your calculation.

79. $\lim_{x\to 0} \frac{7 \sin x}{e^{2x} - 1}$

80. $\lim_{\theta\to 0} \frac{e^\theta - e^{-\theta} - 2\theta}{\theta - \sin\theta}$

81. $\lim_{t\to 0}\left(\frac{1}{2 - 2\cos t} - \frac{1}{t^2}\right)$

82. $\lim_{h\to 0} \frac{(\sin h)/h - \cos h}{h^2}$

83. $\lim_{z\to 0} \frac{1 - \cos^2 z}{\ln(1 - z) + \sin z}$

84. $\lim_{y\to 0} \frac{y^2}{\cos y - \cosh y}$

Theory and Examples

85. Use a series representation of $\sin 3x$ to find values of r and s for which

$$\lim_{x\to 0}\left(\frac{\sin 3x}{x^3} + \frac{r}{x^2} + s\right) = 0.$$

T 86. Compare the accuracies of the approximations $\sin x \approx x$ and $\sin x \approx 6x/(6 + x^2)$ by comparing the graphs of $f(x) = \sin x - x$ and $g(x) = \sin x - (6x/(6 + x^2))$. Describe what you find.

87. Find the radius of convergence of the series

$$\sum_{n=1}^{\infty} \frac{2 \cdot 5 \cdot 8 \cdot \cdots \cdot (3n-1)}{2 \cdot 4 \cdot 6 \cdot \cdots \cdot (2n)} x^n.$$

88. Find the radius of convergence of the series

$$\sum_{n=1}^{\infty} \frac{3 \cdot 5 \cdot 7 \cdot \cdots \cdot (2n+1)}{4 \cdot 9 \cdot 14 \cdot \cdots \cdot (5n-1)} (x-1)^n.$$

89. Find a closed-form formula for the nth partial sum of the series $\sum_{n=2}^{\infty} \ln(1 - (1/n^2))$ and use it to determine the convergence or divergence of the series.

90. Evaluate $\sum_{k=2}^{\infty} (1/(k^2 - 1))$ by finding the limits as $n \to \infty$ of the series' nth partial sum.

91. a. Find the interval of convergence of the series

$$y = 1 + \frac{1}{6}x^3 + \frac{1}{180}x^6 + \cdots + \frac{1 \cdot 4 \cdot 7 \cdot \cdots \cdot (3n-2)}{(3n)!}x^{3n} + \cdots.$$

b. Show that the function defined by the series satisfies a differential equation of the form

$$\frac{d^2y}{dx^2} = x^a y + b,$$

and find the values of the constants a and b.

92. a. Find the Maclaurin series for the function $x^2/(1 + x)$.

b. Does the series converge at $x = 1$? Explain.

93. If $\sum_{n=1}^{\infty} a_n$ and $\sum_{n=1}^{\infty} b_n$ are convergent series of nonnegative numbers, can anything be said about $\sum_{n=1}^{\infty} a_n b_n$? Give reasons for your answer.

94. If $\sum_{n=1}^{\infty} a_n$ and $\sum_{n=1}^{\infty} b_n$ are divergent series of nonnegative numbers, can anything be said about $\sum_{n=1}^{\infty} a_n b_n$? Give reasons for your answer.

95. Prove that the sequence $\{x_n\}$ and the series $\sum_{k=1}^{\infty} (x_{k+1} - x_k)$ both converge or both diverge.

96. Prove that $\sum_{n=1}^{\infty} (a_n/(1 + a_n))$ converges if $a_n > 0$ for all n and $\sum_{n=1}^{\infty} a_n$ converges.

97. Suppose that $a_1, a_2, a_3, \ldots, a_n$ are positive numbers satisfying the following conditions:

i) $a_1 \ge a_2 \ge a_3 \ge \cdots$;

ii) the series $a_2 + a_4 + a_8 + a_{16} + \cdots$ diverges.

Show that the series

$$\frac{a_1}{1} + \frac{a_2}{2} + \frac{a_3}{3} + \cdots$$

diverges.

98. Use the result in Exercise 97 to show that

$$1 + \sum_{n=2}^{\infty} \frac{1}{n \ln n}$$

diverges.

99. Show that if $a_n > 0$ and $\sum_{n=1}^{\infty} a_n$ converges, then $\sum_{n=1}^{\infty} \frac{\sqrt{a_n}}{n}$ converges.

100. Determine whether $\sum_{n=1}^{\infty} b_n$ converges or diverges.

a. $b_1 = 1, \quad b_{n+1} = (-1)^n \frac{n+1}{3n+2} b_n$

b. $b_1 = 3, \quad b_{n+1} = \frac{n}{\ln n} b_n$

101. Assume that $b_n > 0$ and $\sum_{n=1}^{\infty} b_n$ converges. What, if anything, can be said about the following series?

a. $\sum_{n=1}^{\infty} \tan(b_n)$

b. $\sum_{n=1}^{\infty} \ln(1 + b_n)$

c. $\sum_{n=1}^{\infty} \ln(2 + b_n)$

102. Consider the convergent series $\sum_{n=1}^{\infty} \frac{(-1)^n}{e^n + e^{cn}}$, where c is a constant. What should c be so that the first 10 terms of the series estimate the sum of the entire series with an error of less than 0.00001?

103. Assume that the following sequence has a limit L. Find the value of L.

$$4^{1/3},\ (4(4^{1/3}))^{1/3},\ (4(4(4^{1/3}))^{1/3})^{1/3},\ (4(4(4(4^{1/3}))^{1/3})^{1/3})^{1/3},\ \ldots$$

104. Consider the infinite sequence of shaded right triangles in the accompanying diagram. Compute the total area of the triangles.

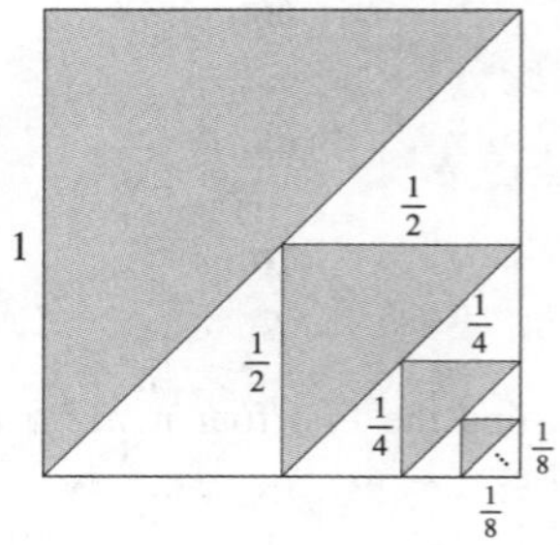

CHAPTER 10 Additional and Advanced Exercises

Determining Convergence of Series

Which of the series $\sum_{n=1}^{\infty} a_n$ defined by the formulas in Exercises 1–4 converge, and which diverge? Give reasons for your answers.

1. $\sum_{n=1}^{\infty} \frac{1}{(3n-2)^{n+(1/2)}}$

2. $\sum_{n=1}^{\infty} \frac{(\tan^{-1} n)^2}{n^2+1}$

3. $\sum_{n=1}^{\infty} (-1)^n \tanh n$

4. $\sum_{n=2}^{\infty} \frac{\log_n (n!)}{n^3}$

Which of the series $\sum_{n=1}^{\infty} a_n$ defined by the formulas in Exercises 5–8 converge, and which diverge? Give reasons for your answers.

5. $a_1 = 1, \quad a_{n+1} = \frac{n(n+1)}{(n+2)(n+3)} a_n$

(*Hint:* Write out several terms, see which factors cancel, and then generalize.)

6. $a_1 = a_2 = 7, \quad a_{n+1} = \frac{n}{(n-1)(n+1)} a_n \quad$ if $n \ge 2$

7. $a_1 = a_2 = 1, \quad a_{n+1} = \frac{1}{1 + a_n} \quad$ if $n \ge 2$

8. $a_n = 1/3^n$ if n is odd, $a_n = n/3^n$ if n is even

Choosing Centers for Taylor Series

Taylor's formula

$$f(x) = f(a) + f'(a)(x-a) + \frac{f''(a)}{2!}(x-a)^2 + \cdots + \frac{f^{(n)}(a)}{n!}(x-a)^n + \frac{f^{(n+1)}(c)}{(n+1)!}(x-a)^{n+1}$$

expresses the value of f at x in terms of the values of f and its derivatives at $x = a$. In numerical computations, we therefore need a to be a point where we know the values of f and its derivatives. We also need a to be close enough to the values of f in which we are interested to make $(x-a)^{n+1}$ so small we can neglect the remainder.

In Exercises 9–14, what Taylor series would you choose to represent the function near the given value of x? (There may be more than one good answer.) Write out the first four nonzero terms of the series you choose.

9. $\cos x$ near $x = 1$

10. $\sin x$ near $x = 6.3$

11. e^x near $x = 0.4$

12. $\ln x$ near $x = 1.3$

13. $\cos x$ near $x = 69$

14. $\arctan x$ near $x = 2$

Theory and Examples

15. Let a and b be constants with $0 < a < b$. Does the sequence $\{(a^n + b^n)^{1/n}\}$ converge? If it does converge, what is the limit?

16. Find the sum of the infinite series

$$1 + \frac{2}{10} + \frac{3}{10^2} + \frac{7}{10^3} + \frac{2}{10^4} + \frac{3}{10^5} + \frac{7}{10^6} + \frac{2}{10^7} + \frac{3}{10^8} + \frac{7}{10^9} + \cdots.$$

17. Evaluate

$$\sum_{n=0}^{\infty} \int_n^{n+1} \frac{1}{1+x^2}\, dx.$$

18. Find all values of x for which

$$\sum_{n=1}^{\infty} \frac{nx^n}{(n+1)(2x+1)^n}$$

converges absolutely.

T **19. a.** Does the value of

$$\lim_{n\to\infty}\left(1-\frac{\cos(a/n)}{n}\right)^n, \quad a \text{ constant},$$

appear to depend on the value of a? If so, how?

b. Does the value of

$$\lim_{n\to\infty}\left(1-\frac{\cos(a/n)}{bn}\right)^n, \quad a \text{ and } b \text{ constant}, b \neq 0,$$

appear to depend on the value of b? If so, how?

c. Use calculus to confirm your findings in parts (a) and (b).

20. Show that if $\sum_{n=1}^{\infty} a_n$ converges, then

$$\sum_{n=1}^{\infty}\left(\frac{1+\sin a_n}{2}\right)^n$$

converges.

21. Find a value for the constant b that will make the radius of convergence of the power series

$$\sum_{n=2}^{\infty}\frac{b^n x^n}{\ln n}$$

equal to 5.

22. Determine a and b so that $\lim_{x\to 0}\frac{a\sin 2x - b\sin x}{x^3} = 1.$

23. Find the value of a for which the limit

$$\lim_{x\to 0}\frac{\sin(ax) - \sin x - x}{x^3}$$

is finite, and evaluate the limit.

24. Find values of a and b for which

$$\lim_{x\to 0}\frac{\cos(ax) - b}{2x^2} = -1.$$

25. Raabe's (or Gauss's) Test The following test, which we state without proof, is an extension of the Ratio Test.

Raabe's Test: If $\sum_{n=1}^{\infty} u_n$ is a series of positive constants and there exist constants C, K, and N such that

$$\frac{u_n}{u_{n+1}} = 1 + \frac{C}{n} + \frac{f(n)}{n^2},$$

where $|f(n)| < K$ for $n \geq N$, then $\sum_{n=1}^{\infty} u_n$ converges if $C > 1$ and diverges if $C \leq 1$.

Show that the results of Raabe's Test agree with what you know about the series $\sum_{n=1}^{\infty}(1/n^2)$ and $\sum_{n=1}^{\infty}(1/n)$.

26. (*Continuation of Exercise 25.*) Suppose that the terms of $\sum_{n=1}^{\infty} u_n$ are defined recursively by the formulas

$$u_1 = 1, \quad u_{n+1} = \frac{(2n-1)^2}{(2n)(2n+1)}u_n.$$

Apply Raabe's Test to determine whether the series converges.

27. Suppose that $\sum_{n=1}^{\infty} a_n$ converges, $a_n \neq 1$, and $a_n > 0$ for all n.

a. Show that $\sum_{n=1}^{\infty} a_n^2$ converges.

b. Does $\sum_{n=1}^{\infty} a_n/(1-a_n)$ converge? Explain.

28. (*Continuation of Exercise 27.*) If $\sum_{n=1}^{\infty} a_n$ converges, and if $0 < a_n < 1$ for all n, show that $\sum_{n=1}^{\infty}\ln(1-a_n)$ converges. (*Hint:* First show that $|\ln(1-a_n)| \leq a_n/(1-a_n)$.)

29. Nicole Oresme's Theorem Prove Nicole Oresme's Theorem:

$$1 + \frac{1}{2}\cdot 2 + \frac{1}{4}\cdot 3 + \cdots + \frac{n}{2^{n-1}} + \cdots = 4.$$

(*Hint:* Differentiate both sides of the equation $1/(1-x) = 1 + \sum_{n=1}^{\infty} x^n$.)

30. a. Find a power series representation of $\frac{e^x - 1}{x}$.

b. By differentiating the series in part (a) term by term, show that $\sum_{n=1}^{\infty}\frac{n}{(n+1)!} = 1.$

31. a. Find a power series representation of e^{-x^2}.

b. By differentiating the series in part (a) twice term-by-term, show that $\sum_{n=1}^{\infty}(-1)^{n+1}\frac{2n+1}{2^n n!} = 1.$

32. a. Show that

$$\sum_{n=1}^{\infty}\frac{n(n+1)}{x^n} = \frac{2x^2}{(x-1)^3}$$

for $|x| > 1$ by differentiating the identity

$$\sum_{n=1}^{\infty} x^{n+1} = \frac{x^2}{1-x}$$

twice, multiplying the result by x, and then replacing x by $1/x$.

b. Use part (a) to find the real solution greater than 1 of the equation

$$x = \sum_{n=1}^{\infty}\frac{n(n+1)}{x^n}.$$

33. Quality control

a. Differentiate the series

$$\frac{1}{1-x} = 1 + x + x^2 + \cdots + x^n + \cdots$$

to obtain a series for $1/(1-x)^2$.

b. In one throw of two dice, the probability of getting a roll of 7 is $p = 1/6$. If you throw the dice repeatedly, the probability that a 7 will appear for the first time at the nth throw is $q^{n-1}p$, where $q = 1 - p = 5/6$. The expected number of throws until a 7 first appears is $\sum_{n=1}^{\infty} nq^{n-1}p$. Find the sum of this series.

c. As an engineer applying statistical control to an industrial operation, you inspect items taken at random from the assembly line. You classify each sampled item as either "good" or "bad." If the probability of an item's being good is p and of an item's being bad is $q = 1 - p$, the probability that the first bad item found is the nth one inspected is $p^{n-1}q$. The average number inspected up to and including the first bad item found is $\sum_{n=1}^{\infty} np^{n-1}q$. Evaluate this sum, assuming $0 < p < 1$.

34. Expected value Suppose that a random variable X may assume the values 1, 2, 3, ..., with probabilities $p_1, p_2, p_3, \ldots$, where p_k is the probability that X equals k $(k = 1, 2, 3, \ldots)$. Suppose also that $p_k \geq 0$ and that $\sum_{k=1}^{\infty} p_k = 1$. The **expected value** of X, denoted by $E(X)$, is the number $\sum_{k=1}^{\infty} kp_k$, provided the series converges. In each of the following cases, show that $\sum_{k=1}^{\infty} p_k = 1$ and find $E(X)$ if it exists. (*Hint:* See Exercise 33.)

a. $p_k = 2^{-k}$ **b.** $p_k = \dfrac{5^{k-1}}{6^k}$

c. $p_k = \dfrac{1}{k(k+1)} = \dfrac{1}{k} - \dfrac{1}{k+1}$

T **35. Safe and effective dosage** The concentration in the blood resulting from a single dose of a drug normally decreases with time as the drug is eliminated from the body. Doses may therefore need to be repeated periodically to keep the concentration from dropping below some particular level. One model for the effect of repeated doses gives the residual concentration just before the $(n+1)$st dose as

$$R_n = C_0e^{-kt_0} + C_0e^{-2kt_0} + \cdots + C_0e^{-nkt_0},$$

where C_0 = the change in concentration achievable by a single dose (mg/mL), k = the *elimination constant* (h^{-1}), and t_0 = time between doses (h). See the accompanying figure.

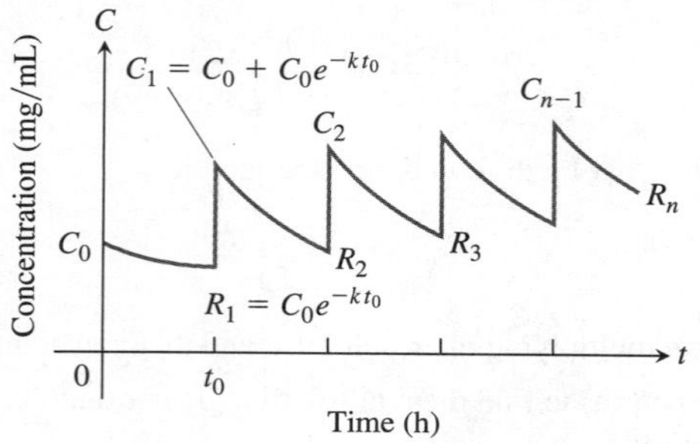

a. Write R_n in closed from as a single fraction, and find $R = \lim_{n\to\infty} R_n$.

b. Calculate R_1 and R_{10} for $C_0 = 1$ mg/mL, $k = 0.1\ \text{h}^{-1}$, and $t_0 = 10$ h. How good an estimate of R is R_{10}?

c. If $k = 0.01\ \text{h}^{-1}$ and $t_0 = 10$ h, find the smallest n such that $R_n > (1/2)R$. Use $C_0 = 1$ mg/mL.

(*Source: Prescribing Safe and Effective Dosage,* B. Horelick and S. Koont, COMAP, Inc., Lexington, MA.)

36. Time between drug doses (*Continuation of Exercise 35.*) If a drug is known to be ineffective below a concentration C_L and harmful above some higher concentration C_H, we need to find values of C_0 and t_0 that will produce a concentration that is safe (not above C_H) but effective (not below C_L). See the accompanying figure. We therefore want to find values for C_0 and t_0 for which

$$R = C_L \quad \text{and} \quad C_0 + R = C_H.$$

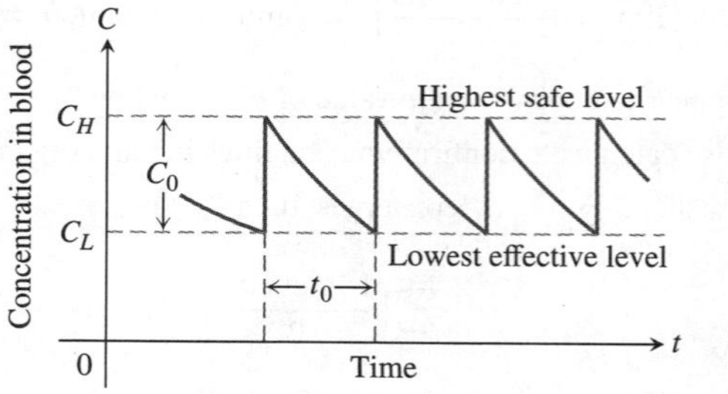

Thus $C_0 = C_H - C_L$. When these values are substituted in the equation for R obtained in part (a) of Exercise 35, the resulting equation simplifies to

$$t_0 = \frac{1}{k}\ln\frac{C_H}{C_L}.$$

To reach an effective level rapidly, one might administer a "loading" dose that would produce a concentration of C_H mg/mL. This could be followed every t_0 hours by a dose that raises the concentration by $C_0 = C_H - C_L$ mg/mL.

a. Verify the preceding equation for t_0.

b. If $k = 0.05\ \text{h}^{-1}$ and the highest safe concentration is e times the lowest effective concentration, find the length of time between doses that will ensure safe and effective concentrations.

c. Given $C_H = 2$ mg/mL, $C_L = 0.5$ mg/mL, and $k = 0.02\ \text{h}^{-1}$, determine a scheme for administering the drug.

d. Suppose that $k = 0.2\ \text{h}^{-1}$ and that the smallest effective concentration is 0.03 mg/mL. A single dose that produces a concentration of 0.1 mg/mL is administered. About how long will the drug remain effective?

CHAPTER 10 Technology Application Projects

Mathematica/Maple Projects

Projects can be found within MyLab Math.

- ***Bouncing Ball***
  The model predicts the height of a bouncing ball, and the time until it stops bouncing.
- ***Taylor Polynomial Approximations of a Function***
  A graphical animation shows the convergence of the Taylor polynomials to functions having derivatives of all orders over an interval in their domains.

11

Parametric Equations and Polar Coordinates

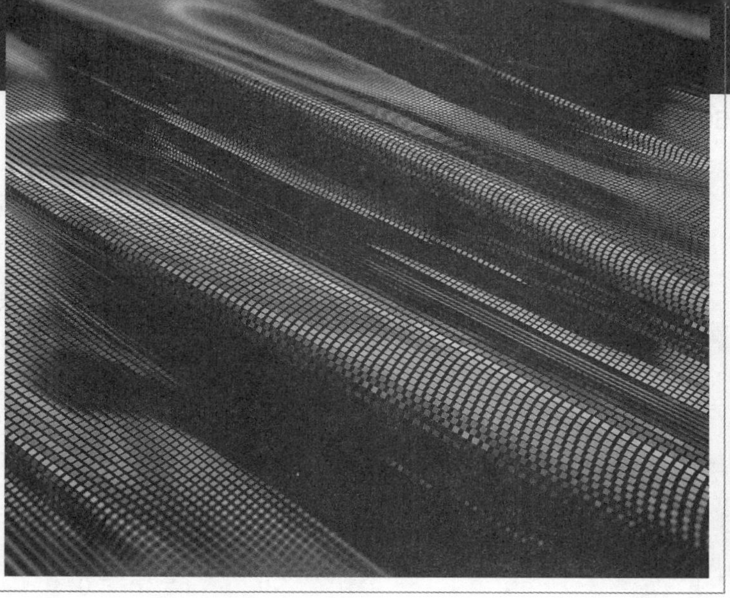

OVERVIEW In this chapter we study new ways to describe curves in the plane. Instead of considering a curve as the graph of a function or equation, we think of it as the path of a moving particle whose position is changing over time. Then each of the x- and y-coordinates of the particle's position becomes a function of a third variable t. We can also change the way in which points in the plane themselves are described by using *polar coordinates* rather than the rectangular or Cartesian system. Both of these new tools are useful for describing motion, like that of planets and satellites, or projectiles moving in the plane or in space.

11.1 Parametrizations of Plane Curves

Parametric Equations

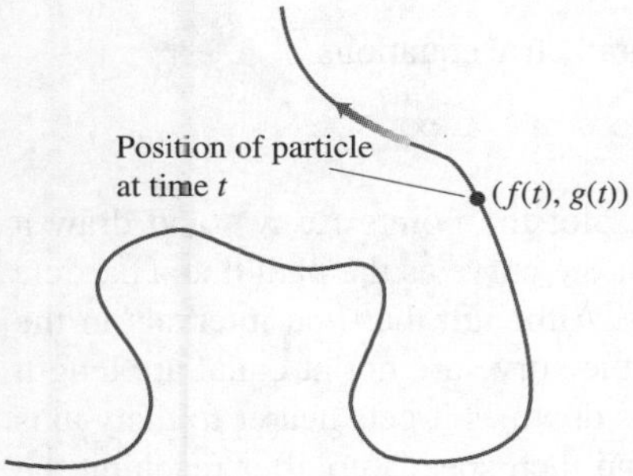

FIGURE 11.1 The curve or path traced by a particle moving in the xy-plane is not always the graph of a function or single equation.

Figure 11.1 shows the path of a moving particle in the xy-plane. Notice that the path fails the vertical line test, so it cannot be described as the graph of a function of the variable x. However, we can describe the path by a pair of equations, $x = f(t)$ and $y = g(t)$, where f and g are continuous functions. In the study of motion, t usually denotes time. Equations like these can describe more general curves than those described by a single function, and they provide not only the graph of the path traced out but also the location of the particle $(x, y) = (f(t), g(t))$ at any time t.

DEFINITION If x and y are given as functions

$$x = f(t), \qquad y = g(t)$$

over an interval I of t-values, then the set of points $(x, y) = (f(t), g(t))$ defined by these equations is a **parametric curve**. The equations are **parametric equations** for the curve.

The variable t is a **parameter** for the curve, and its domain I is the **parameter interval**. If I is a closed interval, $a \le t \le b$, the point $(f(a), g(a))$ is the **initial point** of the curve and $(f(b), g(b))$ is the **terminal point**. When we give parametric equations and a parameter interval for a curve, we say that we have **parametrized** the curve. The equations and interval together constitute a **parametrization** of the curve. A given curve can be represented by different sets of parametric equations. (See Exercises 29 and 30.)

EXAMPLE 1 Sketch the curve defined by the parametric equations

$$x = \sin\frac{\pi t}{2}, \qquad y = t, \qquad 0 \le t \le 6.$$

Solution We make a table of values (Table 11.1), plot the points (x, y), and draw a smooth curve through them (Figure 11.2). If we think of the curve as the path of a moving particle, the particle starts at time $t = 0$ at the initial point $(0, 0)$ and then moves upward in a wavy path until at time $t = 6$ it reaches the terminal point $(0, 6)$. The direction of motion is shown by the arrows in Figure 11.2.

TABLE 11.1 Values of $x = \sin\frac{\pi t}{2}$ and $y = t$ for selected values of t.

t	x	y
0	0	0
1	1	1
2	0	2
3	−1	3
4	0	4
5	1	5
6	0	6

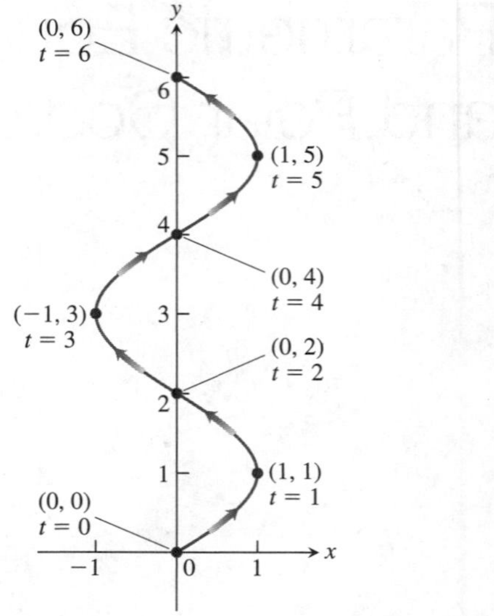

FIGURE 11.2 The curve given by the parametric equations $x = \sin\frac{\pi t}{2}$ and $y = t$ (Example 1). ■

EXAMPLE 2 Sketch the curve defined by the parametric equations

$$x = t^2, \qquad y = t + 1, \qquad -\infty < t < \infty.$$

Solution We make a table of values (Table 11.2), plot the points (x, y), and draw a smooth curve through them (Figure 11.3). We think of the curve as the path that a particle moves along the curve in the direction of the arrows. Although the time intervals in the table are equal, the consecutive points plotted along the curve are not at equal arc length distances. The reason for this is that the particle slows down as it gets nearer to the y-axis along the lower branch of the curve as t increases, and then speeds up after reaching the y-axis at $(0, 1)$ and moving along the upper branch. Since the interval of values for t is all real numbers, there is no initial point or terminal point for the curve.

TABLE 11.2 Values of $x = t^2$ and $y = t + 1$ for selected values of t.

t	x	y
−3	9	−2
−2	4	−1
−1	1	0
0	0	1
1	1	2
2	4	3
3	9	4

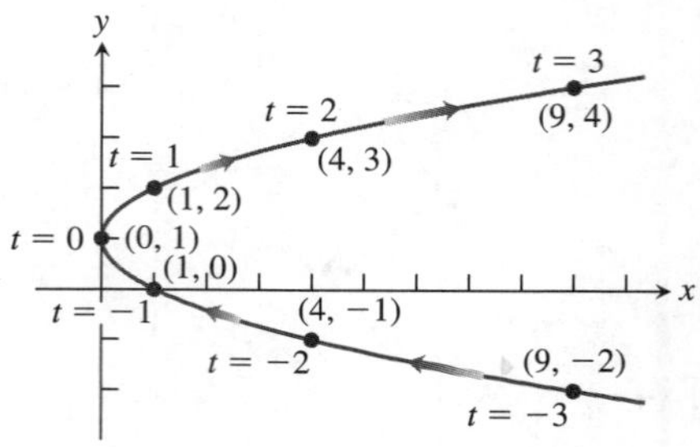

FIGURE 11.3 The curve given by the parametric equations $x = t^2$ and $y = t + 1$ (Example 2).

For this example we can use algebraic manipulation to eliminate the parameter t and obtain an algebraic equation for the curve in terms of x and y alone. We solve $y = t + 1$ for t and substitute the resulting equation $t = y - 1$ into the equation for x, which yields

$$x = t^2 = (y - 1)^2 = y^2 - 2y + 1.$$

The equation $x = y^2 - 2y + 1$ represents a parabola, as displayed in Figure 11.3. It is sometimes quite difficult, or even impossible, to eliminate the parameter from a pair of parametric equations, as we did here. ■

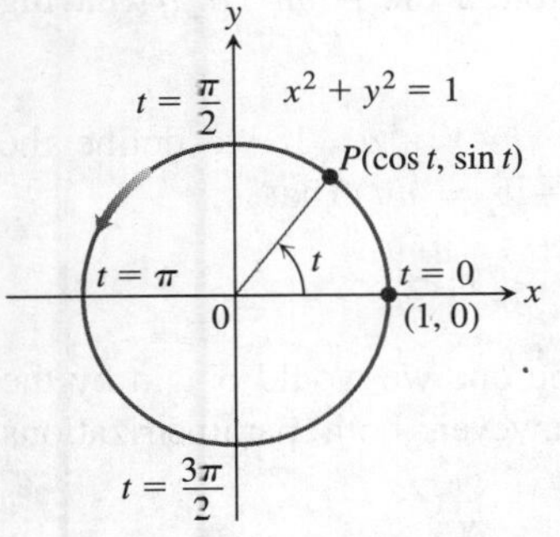

FIGURE 11.4 The equations $x = \cos t$ and $y = \sin t$ describe motion on the circle $x^2 + y^2 = 1$. The arrow shows the direction of increasing t (Example 3).

EXAMPLE 3 Graph the parametric curves

(a) $x = \cos t, \quad y = \sin t, \quad 0 \le t \le 2\pi.$

(b) $x = a\cos t, \quad y = a\sin t, \quad 0 \le t \le 2\pi.$

Solution

(a) Since $x^2 + y^2 = \cos^2 t + \sin^2 t = 1$, the parametric curve lies along the unit circle $x^2 + y^2 = 1$. As t increases from 0 to 2π, the point $(x, y) = (\cos t, \sin t)$ starts at $(1, 0)$ and traces the entire circle once counterclockwise (Figure 11.4).

(b) For $x = a\cos t$, $y = a\sin t$, we have $x^2 + y^2 = a^2\cos^2 t + a^2\sin^2 t = a^2$. The parametrization describes a motion that begins at the point $(a, 0)$ and traverses the circle $x^2 + y^2 = a^2$ once counterclockwise, returning to $(a, 0)$ at $t = 2\pi$. The graph is a circle centered at the origin with radius $r = |a|$ and coordinate points $(a\cos t, a\sin t)$. ■

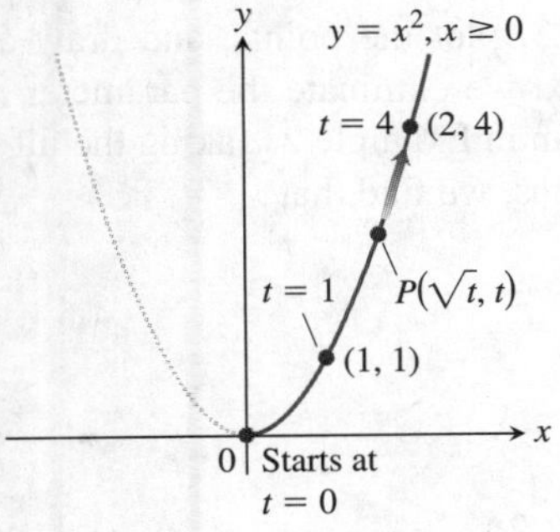

FIGURE 11.5 The equations $x = \sqrt{t}$ and $y = t$ and the interval $t \ge 0$ describe the path of a particle that traces the right-hand half of the parabola $y = x^2$ (Example 4).

EXAMPLE 4 The position $P(x, y)$ of a particle moving in the xy-plane is given by the equations and parameter interval

$$x = \sqrt{t}, \quad y = t, \quad t \ge 0.$$

Identify the path traced by the particle and describe the motion.

Solution We try to identify the path by eliminating t between the equations $x = \sqrt{t}$ and $y = t$, which might produce a recognizable algebraic relation between x and y. We find that

$$y = t = \left(\sqrt{t}\right)^2 = x^2.$$

Thus, the particle's position coordinates satisfy the equation $y = x^2$, so the particle moves along the parabola $y = x^2$.

It would be a mistake, however, to conclude that the particle's path is the entire parabola $y = x^2$; it is only half the parabola. The particle's x-coordinate is never negative. The particle starts at $(0, 0)$ when $t = 0$ and rises into the first quadrant as t increases (Figure 11.5). The parameter interval is $[0, \infty)$, and there is no terminal point. ■

The graph of any function $y = f(x)$ can always be given a **natural parametrization** $x = t$ and $y = f(t)$. The domain of the parameter in this case is the same as the domain of the function f.

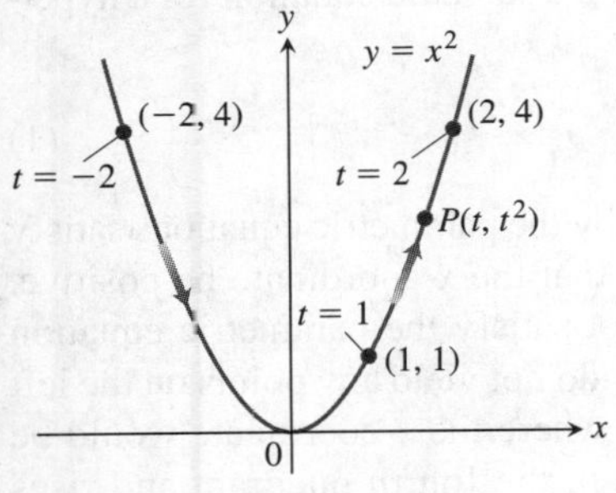

FIGURE 11.6 The path defined by $x = t$, $y = t^2$, $-\infty < t < \infty$ is the entire parabola $y = x^2$ (Example 5).

EXAMPLE 5 A parametrization of the graph of the function $f(x) = x^2$ is given by

$$x = t, \quad y = f(t) = t^2, \quad -\infty < t < \infty.$$

When $t \ge 0$, this parametrization gives the same path in the xy-plane as we had in Example 4. However, since the parameter t here can now also be negative, we obtain the left-hand part of the parabola as well; that is, we have the entire parabolic curve. For this parametrization, there is no starting point and no terminal point (Figure 11.6). ■

Notice that a parametrization also specifies *when* a particle moving along the curve is *located* at a specific point along the curve. In Example 4, the point $(2, 4)$ is reached when $t = 4$; in Example 5, it is reached "earlier" when $t = 2$. You can see the implications of this aspect of parametrizations when considering the possibility of two objects coming into collision: they have to be at the exact same location point $P(x, y)$ for some (possibly different) values of their respective parameters. We will say more about this aspect of parametrizations when we study motion in Chapter 13.

EXAMPLE 6 Find a parametrization for the line through the point (a, b) having slope m.

Solution A Cartesian equation of the line is $y - b = m(x - a)$. If we define the parameter t by $t = x - a$, we find that $x = a + t$ and $y - b = mt$. That is,

$$x = a + t, \qquad y = b + mt, \qquad -\infty < t < \infty$$

parametrizes the line. This parametrization differs from the one we would obtain by the natural parametrization in Example 5 when $t = x$. However, both parametrizations describe the same line. ■

TABLE 11.3 Values of $x = t + (1/t)$ and $y = t - (1/t)$ for selected values of t.

t	$1/t$	x	y
0.1	10.0	10.1	−9.9
0.2	5.0	5.2	−4.8
0.4	2.5	2.9	−2.1
1.0	1.0	2.0	0.0
2.0	0.5	2.5	1.5
5.0	0.2	5.2	4.8
10.0	0.1	10.1	9.9

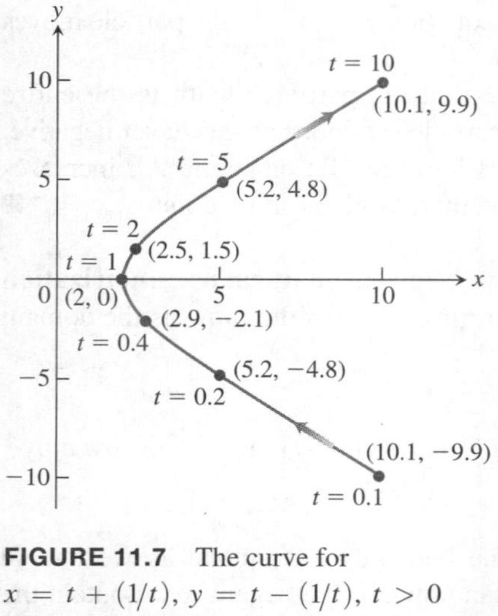

FIGURE 11.7 The curve for $x = t + (1/t)$, $y = t - (1/t)$, $t > 0$ in Example 7. (The part shown is for $0.1 \le t \le 10$.)

EXAMPLE 7 Sketch and identify the path traced by the point $P(x, y)$ if

$$x = t + \frac{1}{t}, \qquad y = t - \frac{1}{t}, \qquad t > 0.$$

Solution We make a brief table of values in Table 11.3, plot the points, and draw a smooth curve through them, as we did in Example 1. Next we eliminate the parameter t from the equations. The procedure is more complicated than in Example 2. Taking the difference between x and y as given by the parametric equations, we find that

$$x - y = \left(t + \frac{1}{t}\right) - \left(t - \frac{1}{t}\right) = \frac{2}{t}.$$

If we add the two parametric equations, we get

$$x + y = \left(t + \frac{1}{t}\right) + \left(t - \frac{1}{t}\right) = 2t.$$

We can then eliminate the parameter t by multiplying these last equations together:

$$(x - y)(x + y) = \left(\frac{2}{t}\right)(2t) = 4.$$

Expanding the expression on the left-hand side, we obtain a standard equation for a hyperbola (Section 11.6):

$$x^2 - y^2 = 4. \tag{1}$$

Thus the coordinates of all the points $P(x, y)$ described by the parametric equations satisfy Equation (1). However, Equation (1) does not require that the x-coordinate be positive. So there are points (x, y) on the hyperbola that do not satisfy the parametric equation $x = t + (1/t)$, $t > 0$. In fact, the parametric equations do not yield any points on the left branch of the hyperbola given by Equation (1), points where the x-coordinate would be negative. For small positive values of t, the path lies in the fourth quadrant and rises into the first quadrant as t increases, crossing the x-axis when $t = 1$ (see Figure 11.7). The parameter domain is $(0, \infty)$ and there is no starting point or terminal point for the path. ■

Examples 4, 5, and 6 illustrate that a given curve, or portion of it, can be represented by different parametrizations. In the case of Example 7, we can also represent the right-hand branch of the hyperbola by the parametrization

$$x = \sqrt{4 + t^2}, \qquad y = t, \qquad -\infty < t < \infty,$$

which is obtained by solving Equation (1) for $x \geq 0$ and letting y be the parameter. Still another parametrization for the right-hand branch of the hyperbola given by Equation (1) is

$$x = 2 \sec t, \qquad y = 2 \tan t, \qquad -\frac{\pi}{2} < t < \frac{\pi}{2}.$$

This parametrization follows from the trigonometric identity $\sec^2 t - \tan^2 t = 1$, because

$$x^2 - y^2 = 4 \sec^2 t - 4 \tan^2 t = 4(\sec^2 t - \tan^2 t) = 4.$$

As t runs between $-\pi/2$ and $\pi/2$, $x = \sec t$ remains positive and $y = \tan t$ runs between $-\infty$ and ∞, so P traverses the hyperbola's right-hand branch. It comes in along the branch's lower half as $t \to 0^-$, reaches $(2, 0)$ at $t = 0$, and moves out into the first quadrant as t increases steadily toward $\pi/2$. This is the same branch of the hyperbola shown in Figure 11.7.

Cycloids

HISTORICAL BIOGRAPHY
Christiaan Huygens
(1629–1695)
www.bit.ly/2P6tPxq

The problem with a pendulum clock whose bob swings in a circular arc is that the frequency of the swing depends on the amplitude of the swing. The wider the swing, the longer it takes the bob to return to center (its lowest position).

This does not happen if the bob can be made to swing in a curve called a *cycloid*. In 1673, Christiaan Huygens designed a pendulum clock whose bob would swing in a cycloid, a curve we define in Example 8. He hung the bob from a fine wire constrained by guards that caused it to draw up as it swung away from center (Figure 11.8). We describe the path parametrically in the next example.

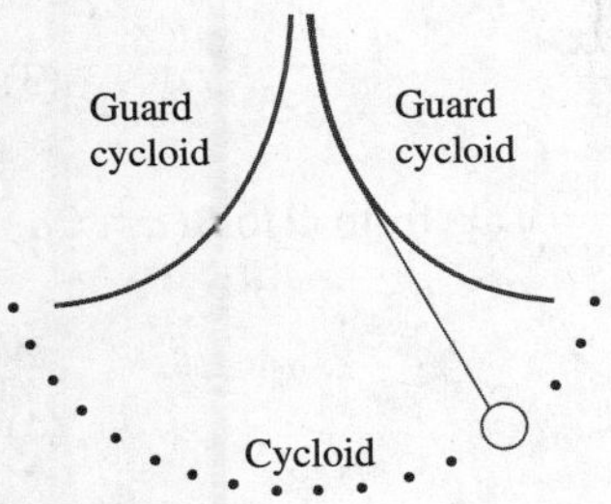

FIGURE 11.8 In Huygens' pendulum clock, the bob swings in a cycloid, so the frequency is independent of the amplitude.

EXAMPLE 8 A wheel of radius a rolls along a horizontal straight line. Find parametric equations for the path traced by a point P on the wheel's circumference. The path is called a **cycloid**.

Solution We take the line to be the x-axis, mark a point P on the wheel, start the wheel with P at the origin, and roll the wheel to the right. As parameter, we use the angle t through which the wheel turns, measured in radians. Figure 11.9 shows the wheel a short while later when its base lies at units from the origin. The wheel's center C lies at (at, a) and the coordinates of P are

$$x = at + a \cos \theta, \qquad y = a + a \sin \theta.$$

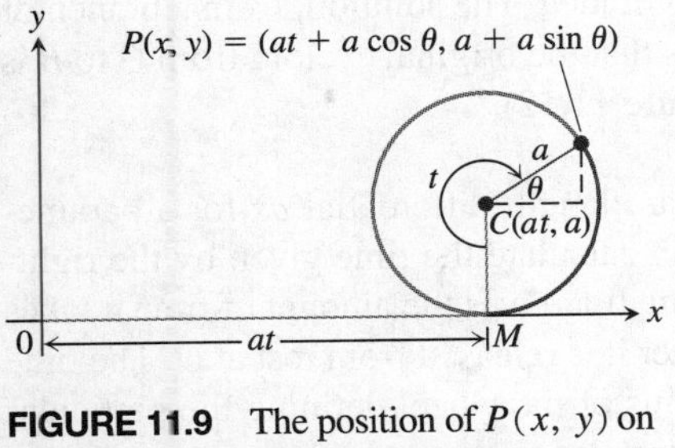

FIGURE 11.9 The position of $P(x, y)$ on the rolling wheel at angle t (Example 8).

To express θ in terms of t, we observe that $t + \theta = 3\pi/2$ in the figure, so that

$$\theta = \frac{3\pi}{2} - t.$$

This makes

$$\cos \theta = \cos\left(\frac{3\pi}{2} - t\right) = -\sin t, \qquad \sin \theta = \sin\left(\frac{3\pi}{2} - t\right) = -\cos t.$$

The equations we seek are

$$x = at - a \sin t, \qquad y = a - a \cos t.$$

These are usually written with the a factored out:

$$x = a(t - \sin t), \qquad y = a(1 - \cos t). \tag{2}$$

Figure 11.10 shows the first arch of the cycloid and part of the next. ■

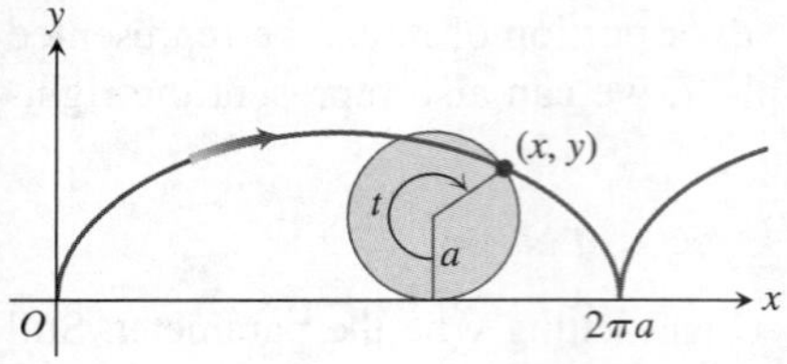

FIGURE 11.10 The cycloid curve $x = a(t - \sin t)$, $y = a(1 - \cos t)$, for $t \geq 0$.

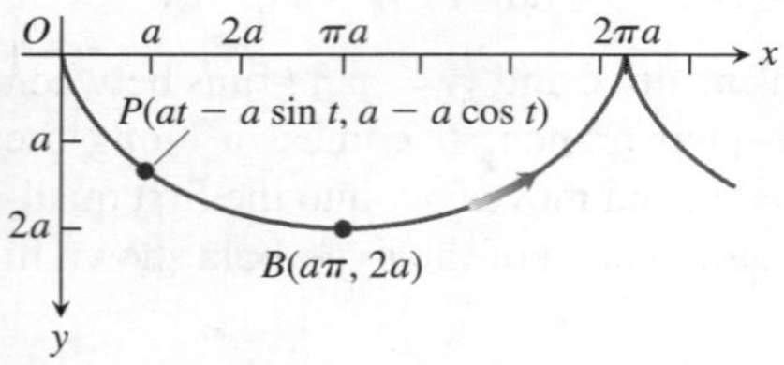

FIGURE 11.11 When Figure 11.10 is turned upside down, the y-axis points downward, indicating the direction of the gravitational force. Equations (2) still describe the curve parametrically.

Brachistochrones and Tautochrones

If we turn Figure 11.10 upside down, Equations (2) still apply and the resulting curve (Figure 11.11) has two interesting physical properties. The first relates to the origin O and the point B at the bottom of the first arch. Among all smooth curves joining these points, the cycloid is the curve along which a frictionless bead, subject only to the force of gravity, will slide from O to B the fastest. This makes the cycloid a **brachistochrone** ("brah-*kiss*-toe-krone"), or shortest-time curve for these points. The second property is that even if you start the bead partway down the curve toward B, it will still take the bead the same amount of time to reach B. This makes the cycloid a **tautochrone** ("*taw*-toe-krone"), or same-time curve for O and B.

Are there any other brachistochrones joining O and B, or is the cycloid the only one? We can formulate this as a mathematical question in the following way. At the start, the kinetic energy of the bead is zero since its velocity (speed) is zero. The work done by gravity in moving the bead from $(0, 0)$ to any other point (x, y) in the plane is mgy, and this must equal the change in kinetic energy. (See Exercise 25 in Section 6.5.) That is,

$$mgy = \frac{1}{2}mv^2 - \frac{1}{2}m(0)^2.$$

Thus, the speed of the bead when it reaches (x, y) has to be $v = \sqrt{2gy}$. That is,

$$\frac{ds}{dT} = \sqrt{2gy}$$

ds is the arc length differential along the bead's path, and T represents time.

or

$$dT = \frac{ds}{\sqrt{2gy}} = \frac{\sqrt{1 + (dy/dx)^2}\, dx}{\sqrt{2gy}}. \tag{3}$$

The time T_f it takes the bead to slide along a particular path $y = f(x)$ from O to $B(a\pi, 2a)$ is

$$T_f = \int_{x=0}^{x=a\pi} \sqrt{\frac{1 + (dy/dx)^2}{2gy}}\, dx. \tag{4}$$

What curves $y = f(x)$, if any, minimize the value of this integral?

At first sight, we might guess that the straight line joining O and B would give the shortest time, but perhaps not. There might be some advantage in having the bead fall vertically at first to build up its speed faster. With a higher speed, the bead could travel a longer path and still reach B first. Indeed, this is the right idea. The solution, from a branch of mathematics known as the *calculus of variations*, is that the original cycloid from O to B is the one and only brachistochrone for O and B (Figure 11.12).

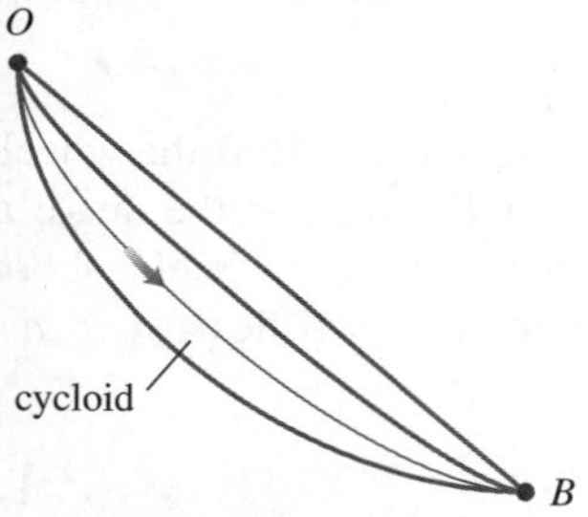

FIGURE 11.12 The cycloid is the unique curve that minimizes the time it takes for a frictionless bead to slide from point O to point B.

In the next section we show how to find the arc length differential ds for a parametrized curve. Once we know how to find ds, we can calculate the time given by the right-hand side of Equation (4) for the cycloid. This calculation gives the amount of time it takes a frictionless bead to slide down the cycloid to B after it is released from rest at O. The time turns out to be equal to $\pi\sqrt{a/g}$, where a is the radius of the wheel defining the particular cycloid. Moreover, if we start the bead at some lower point on the cycloid, corresponding to a parameter value $t_0 > 0$, we can integrate the parametric form of $ds/\sqrt{2gy}$ in Equation (3) over the interval $[t_0, \pi]$ to find the time it takes the bead to reach the point B. That calculation results in the same time $T = \pi\sqrt{a/g}$. It takes the bead the same amount of time to reach B no matter where it starts, which makes the cycloid a tautochrone. Beads starting simultaneously from O, A, and C in Figure 11.13, for instance, will all reach B at exactly the same time. This is the reason why Huygens' pendulum clock in Figure 11.8 is independent of the amplitude of the swing.

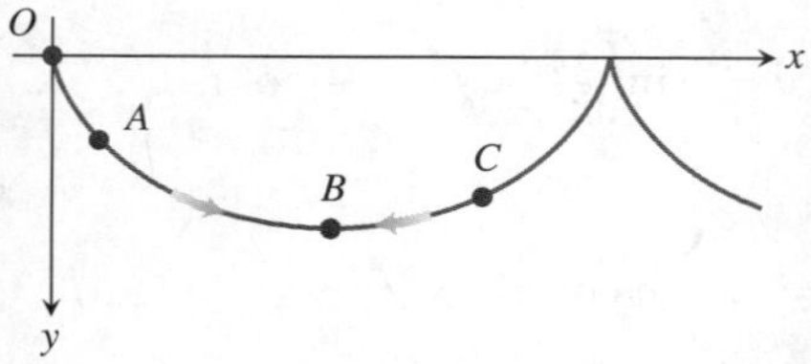

FIGURE 11.13 Beads released simultaneously on the upside-down cycloid at O, A, and C will reach B at the same time.

EXERCISES 11.1

Finding Cartesian from Parametric Equations

Exercises 1–18 give parametric equations and parameter intervals for the motion of a particle in the xy-plane. Identify the particle's path by finding a Cartesian equation for it. Graph the Cartesian equation. (The graphs will vary with the equation used.) Indicate the portion of the graph traced by the particle and the direction of motion.

1. $x = 2t, \quad y = 4t^2, \quad -\infty \le t \le \infty$
2. $x = -\sqrt{t}, \quad y = t, \quad t \ge 0$
3. $x = t - 1, \quad y = 2t - 3, \quad -\infty \le t \le \infty$
4. $x = 2 - 2t, \quad y = 2t, \quad -\infty \le t \le \infty$
5. $x = \cos 4t, \quad y = \sin 4t, \quad 0 \le t \le \pi$
6. $x = \cos(\pi - t), \quad y = \sin(\pi - t), \quad 0 \le t \le \pi$
7. $x = 4\cos t, \quad y = 2\sin t, \quad 0 \le t \le 2\pi$
8. $x = 3\sin t, \quad y = 4\cos t, \quad 0 \le t \le 2\pi$
9. $x = \sin t, \quad y = \cos 2t, \quad -\frac{\pi}{2} \le t \le \frac{\pi}{2}$
10. $x = 1 + \sin t, \quad y = \cos t - 2, \quad 0 \le t \le \pi$
11. $x = t^3, \quad y = t^9 - 2t^6, \quad -\infty \le t \le \infty$
12. $x = \frac{t}{t-1}, \quad y = \frac{t-2}{t+1}, \quad -1 < t < 1$
13. $x = t, \quad y = \sqrt{1 - t^2}, \quad -1 \le t \le 0$
14. $x = \sqrt{t+1}, \quad y = \sqrt{t}, \quad t \ge 0$
15. $x = \sec^2 t - 1, \quad y = \tan t, \quad -\pi/2 < t < \pi/2$
16. $x = -\sec t, \quad y = \tan t, \quad -\pi/2 < t < \pi/2$
17. $x = -\cosh 2t, \quad y = \sinh 2t, \quad -\infty \le t \le \infty$
18. $x = 2\sinh t, \quad y = 2\cosh t, \quad -\infty < t < \infty$

In Exercises 19–24, match the parametric equations with the parametric curves labeled A through F.

19. $x = 1 - \sin t, \quad y = 1 + \cos t$
20. $x = \cos t, \quad y = 2\sin t$
21. $x = \frac{1}{4}t\cos t, \quad y = \frac{1}{4}t\sin t$
22. $x = \sqrt{t}, \quad y = \sqrt{t}\cos t$
23. $x = \ln t, \quad y = 3e^{-t/2}$
24. $x = \cos t, \quad y = \sin 3t$

A.

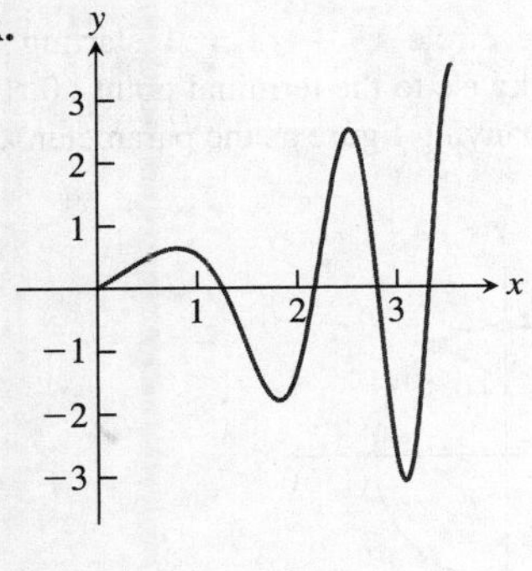

B.

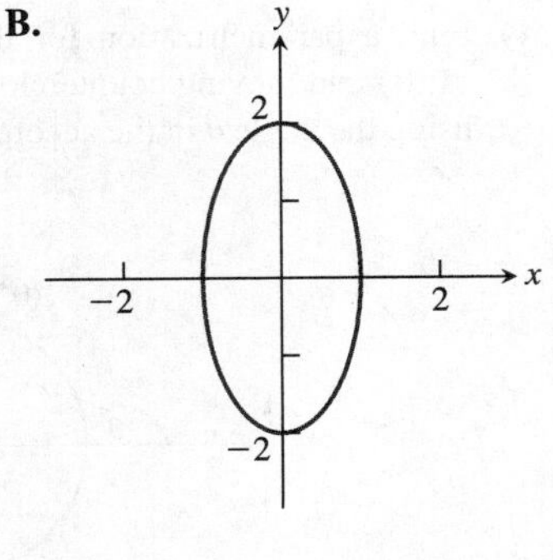

C.

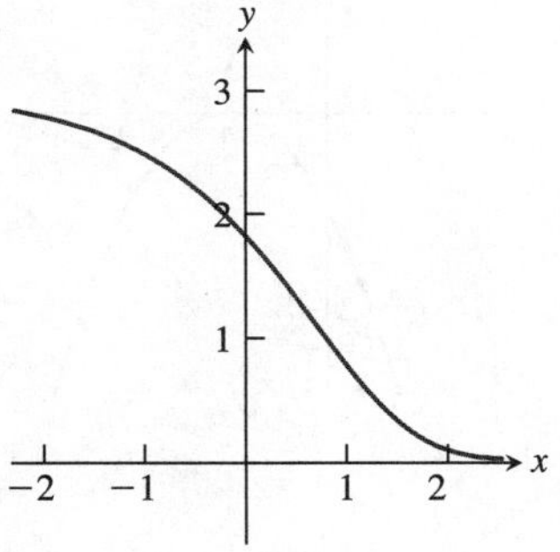

D.

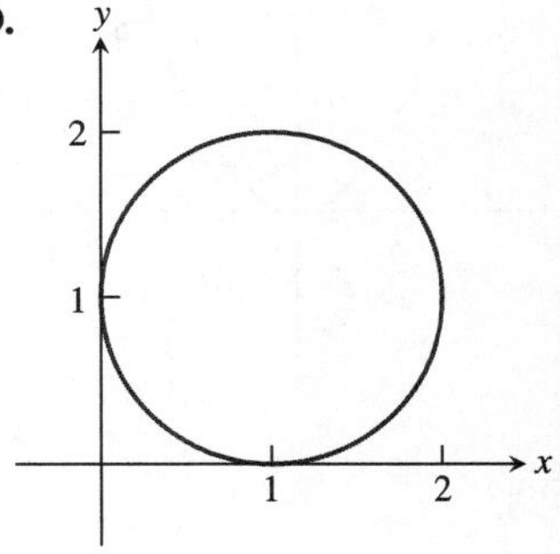

E.

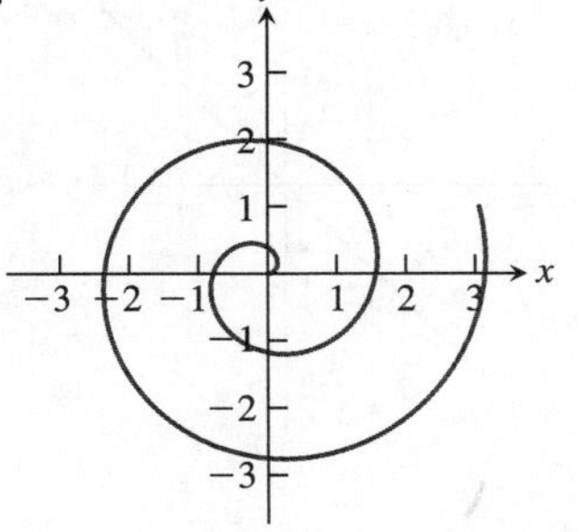

F.

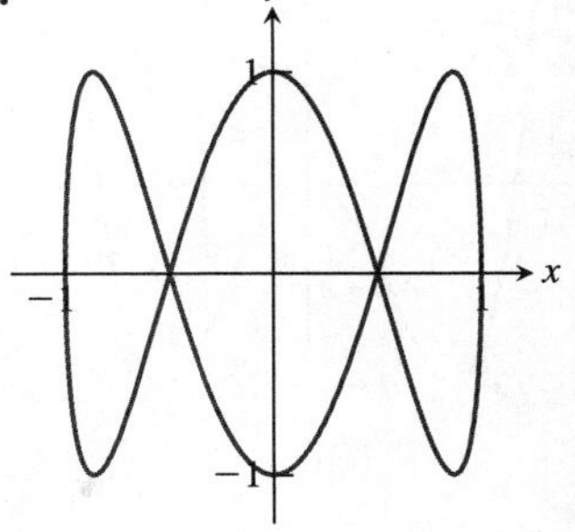

In Exercises 25–28, use the given graphs of $x = f(t)$ and $y = g(t)$ to sketch the corresponding parametric curve in the xy-plane.

25.

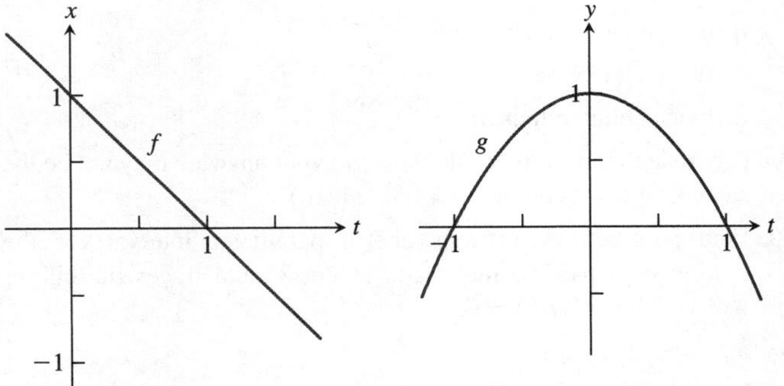

26.

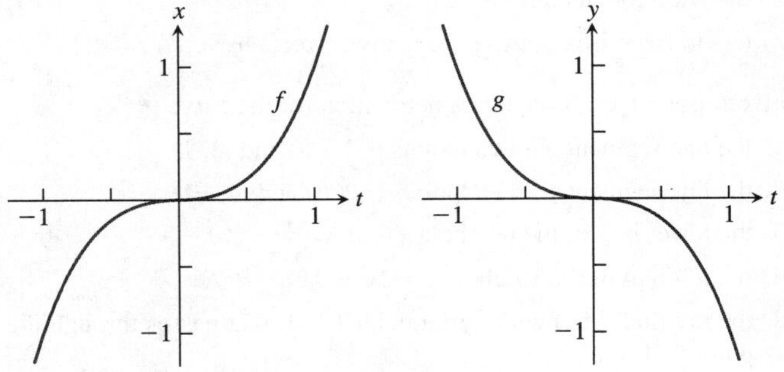

27.

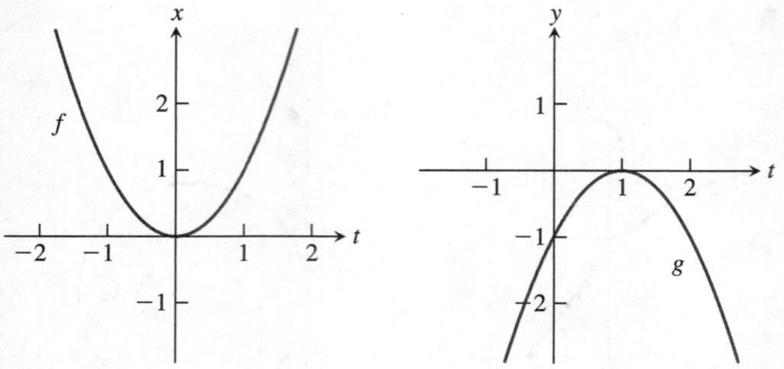

28.

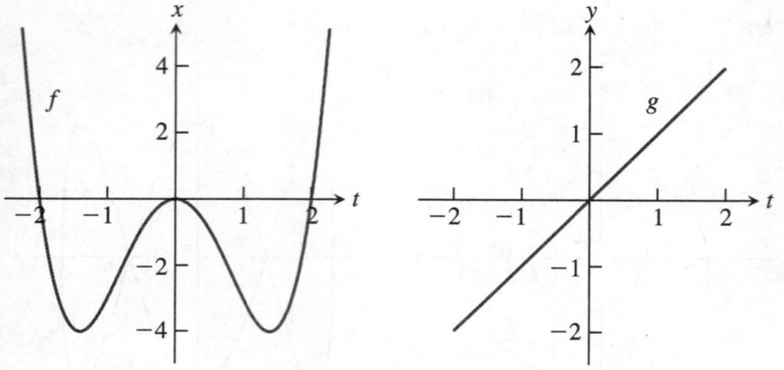

Finding Parametric Equations

29. Find parametric equations and a parameter interval for the motion of a particle that starts at $(a, 0)$ and traces the circle $x^2 + y^2 = a^2$

a. once clockwise.

b. once counterclockwise.

c. twice clockwise.

d. twice counterclockwise.

(There are many ways to do these, so your answers may not be the same as the ones at the back of the text.)

30. Find parametric equations and a parameter interval for the motion of a particle that starts at $(a, 0)$ and traces the ellipse $(x^2/a^2) + (y^2/b^2) = 1$

a. once clockwise.

b. once counterclockwise.

c. twice clockwise.

d. twice counterclockwise.

(As in Exercise 29, there are many correct answers.)

In Exercises 31–36, find a parametrization for the curve.

31. the line segment with endpoints $(-2, -5)$ and $(3, 1)$

32. the line segment with endpoints $(-2, 5)$ and $(4, -1)$

33. the lower half of the parabola $x - 1 = y^2$

34. the left half of the parabola $y = 2x^2 + 4x$

35. the ray (half line) with initial point $(2, 3)$ that passes through the point $(-1, -1)$

36. the ray (half line) with initial point $(-1, 2)$ that passes through the point $(0, 0)$

37. Find parametric equations and a parameter interval for the motion of a particle starting at the point $(2, 0)$ and tracing the top half of the circle $x^2 + y^2 = 4$ four times.

38. Find the parametric equations and a parameter interval for the motion of a particle that moves along the graph of $y = x^2$ in the following way: Beginning at $(0, 0)$ it moves to $(2, 4)$ and then travels back and forth from $(2, 4)$ to $(-2, 4)$ infinitely many times.

39. Find the parametric equations for the semicircle

$$2x^2 + 2y^2 = a^2, \quad y > 0$$

using as parameter the slope $t = dy/dx$ of the tangent to the curve at (x, y)

40. Find parametric equations for the circle

$$x^2 + y^2 = a^2,$$

using as parameter the arc length s measured counterclockwise from the point $(a, 0)$ to the point (x, y).

41. Find a parametrization for the line segment joining points $(0, 2)$ and $(4, 0)$ using the angle θ in the accompanying figure as the parameter.

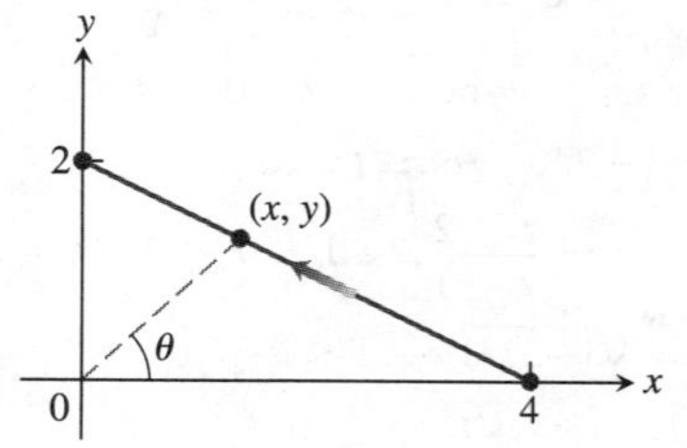

42. Find a parametrization for the curve $y = \sqrt{x}$ with terminal point $(0, 0)$ using the angle θ in the accompanying figure as the parameter.

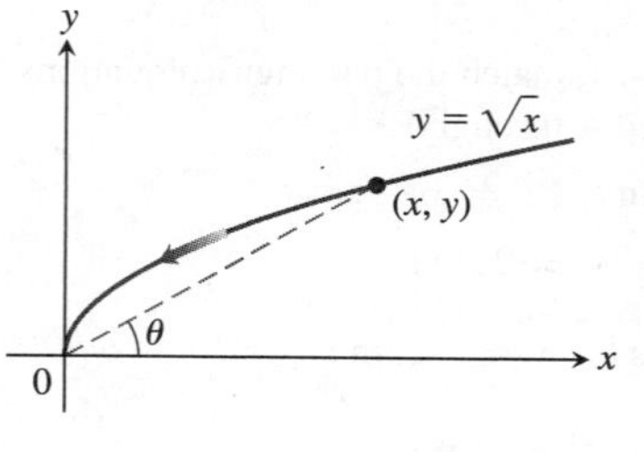

43. Find a parametrization for the circle $(x - 2)^2 + y^2 = 1$ starting at $(1, 0)$ and moving clockwise once around the circle.

44. Find a parametrization for the circle $x^2 + y^2 = 1$ starting at $(1, 0)$ and moving counterclockwise to the terminal point $(0, 1)$, using the angle θ in the accompanying figure as the parameter.

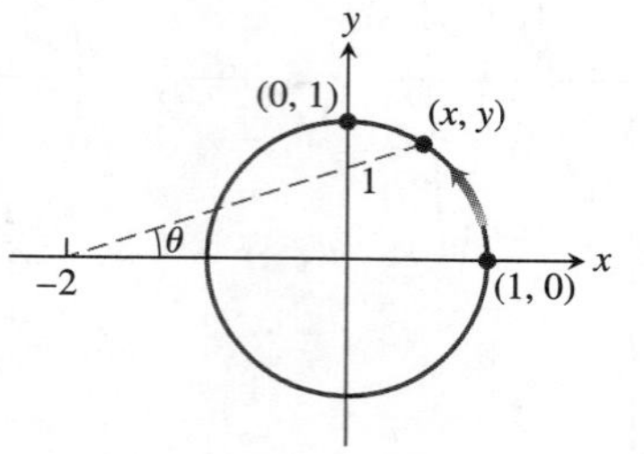

45. The witch of Maria Agnesi The bell-shaped witch of Maria Agnesi can be constructed in the following way. Start with a circle of radius 1, centered at the point $(0, 1)$, as shown in the accompanying figure. Choose a point A on the line $y = 2$ and connect it to the origin with a line segment. Call the point where the segment crosses the circle B. Let P be the point where the vertical line through A crosses the horizontal line through B. The witch is the curve traced by P as A moves along the line $y = 2$. Find parametric equations and a parameter interval for the witch by expressing the coordinates of P in terms of t, the radian measure of the angle that segment OA makes with the positive x-axis. The following equalities (which you may assume) will help.

a. $x = AQ$ **b.** $y = 2 - AB \sin t$

c. $AB \cdot OA = (AQ)^2$

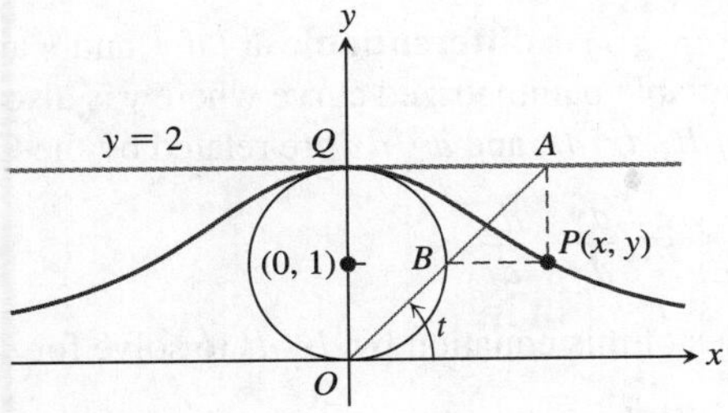

46. Hypocycloid When a circle rolls on the inside of a fixed circle, any point P on the circumference of the rolling circle describes a *hypocycloid*. Let the fixed circle be $x^2 + y^2 = a^2$, let the radius of the rolling circle be b, and let the initial position of the tracing point P be $A(a, 0)$. Find parametric equations for the hypocycloid, using as the parameter the angle θ from the positive x-axis to the line joining the circles' centers. In particular, if $b = a/4$, as in the accompanying figure, show that the hypocycloid is the astroid

$$x = a\cos^3\theta, \quad y = a\sin^3\theta.$$

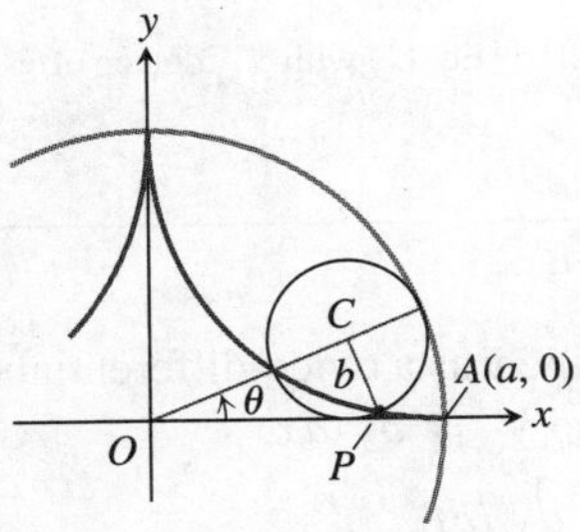

47. As the point N moves along the line $y = a$ in the accompanying figure, P moves in such a way that $OP = MN$. Find parametric equations for the coordinates of P as functions of the angle t that the line ON makes with the positive y-axis.

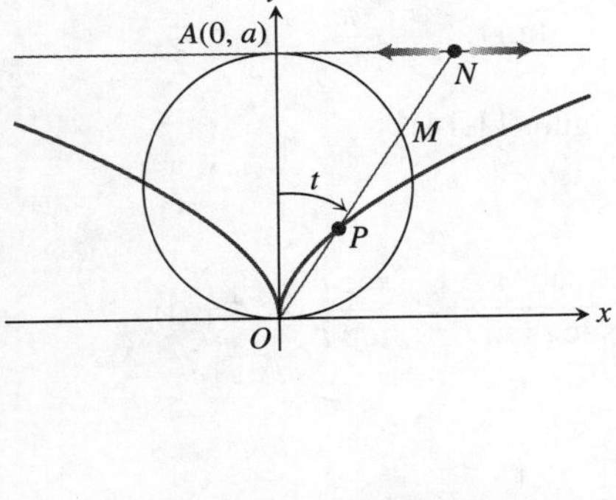

48. Trochoids A wheel of radius a rolls along a horizontal straight line without slipping. Find parametric equations for the curve traced out by a point P on a spoke of the wheel b units from its center. As parameter, use the angle θ through which the wheel turns. The curve is called a *trochoid*, which is a cycloid when $b = a$.

Distance Using Parametric Equations

49. Find the point on the parabola $x = t,\ y = t^2,\ -\infty < t < \infty$, closest to the point $(2, 1/2)$. (*Hint:* Minimize the square of the distance as a function of t.)

50. Find the point on the ellipse $x = 2\cos t,\ y = \sin t,\ 0 \le t \le 2\pi$, closest to the point $(3/4, 0)$. (*Hint:* Minimize the square of the distance as a function of t.)

T GRAPHER EXPLORATIONS

Using a parametric equation grapher, graph the equations over the given intervals in Exercises 51–58.

51. Ellipse $x = 4\cos t, \quad y = 2\sin t,$ over

a. $0 \le t \le 2\pi$

b. $0 \le t \le \pi$

c. $-\pi/2 \le t \le \pi/2$

52. Hyperbola branch $x = \sec t$ (enter as $1/\cos(t)$), $y = \tan t$ (enter as $\sin(t)/\cos(t)$), over

a. $-1.5 \le t \le 1.5$

b. $-0.5 \le t \le 0.5$

c. $-0.1 \le t \le 0.1$

53. Parabola $x = t + 2, \quad y = t^2 + 1,$

54. Cycloid $x = t - \sin t, \quad y = 1 - \cos t,$ over

a. $0 \le t \le 2\pi$

b. $0 \le t \le 4\pi$

c. $\pi \le t \le 3\pi$

55. Deltoid

$$x = 2\cos t + \cos 2t, \quad y = 2\sin t - \sin 2t, \quad 0 \le t \le 2\pi$$

What happens if you replace 2 with -2 in the equations for x and y? Graph the new equations and find out.

56. A nice curve

$$x = 3\cos t + \cos 3t, \quad y = 3\sin t - \sin 3t, \quad 0 \le t \le 2\pi$$

What happens if you replace 3 with -3 in the equations for x and y? Graph the new equations and find out.

57. a. Epicycloid

$$x = 9\cos t - \cos 9t, \quad y = 9\sin t - \sin 9t, \quad 0 \le t \le 2\pi$$

b. Hypocycloid

$$x = 8\cos t + 2\cos 4t, \quad y = 8\sin t - 2\sin 4t, \quad 0 \le t \le 2\pi$$

c. Hypotrochoid

$$x = \cos t + 5\cos 3t, \quad y = 6\cos t - 5\sin 3t, \quad 0 \le t \le 2\pi$$

58. a. $x = 6\cos t + 5\cos 3t, \quad y = 6\sin t - 5\sin 3t,$
$0 \le t \le 2\pi$

b. $x = 6\cos 2t + 5\cos 6t, \quad y = 6\sin 2t - 5\sin 6t,$
$0 \le t \le \pi$

c. $x = 6\cos t + 5\cos 3t, \quad y = 6\sin 2t - 5\sin 3t,$
$0 \le t \le 2\pi$

d. $x = 6\cos 2t + 5\cos 6t, \quad y = 6\sin 4t - 5\sin 6t,$
$0 \le t \le \pi$

11.2 Calculus with Parametric Curves

In this section we use calculus to study parametric curves. Specifically, we find slopes, lengths, and areas associated with parametrized curves.

Tangent Lines and Areas

A parametrized curve $x = f(t)$ and $y = g(t)$ is **differentiable** at t if f and g are each differentiable at t. At a point on a differentiable parametrized curve where y is also a differentiable function of x, the derivatives dy/dt, dx/dt, and dy/dx are related by the Chain Rule:

$$\frac{dy}{dt} = \frac{dy}{dx} \cdot \frac{dx}{dt}.$$

If $dx/dt \neq 0$, we may divide both sides of this equation by dx/dt to solve for dy/dx.

Parametric Formula for dy/dx
If all three derivatives exist and $dx/dt \neq 0$, then

$$\frac{dy}{dx} = \frac{dy/dt}{dx/dt}. \tag{1}$$

If parametric equations define y as a twice-differentiable function of x, we can apply Equation (1) to the function $dy/dx = y'$ to calculate d^2y/dx^2 as a function of t:

$$\frac{d^2y}{dx^2} = \frac{d}{dx}(y') = \frac{dy'/dt}{dx/dt}. \qquad \text{Eq. (1) with } y' \text{ in place of } y$$

Parametric Formula for d^2y/dx^2
If the equations $x = f(t)$, $y = g(t)$ define y as a twice-differentiable function of x, then at any point where $dx/dt \neq 0$ and $y' = dy/dx$,

$$\frac{d^2y}{dx^2} = \frac{dy'/dt}{dx/dt}. \tag{2}$$

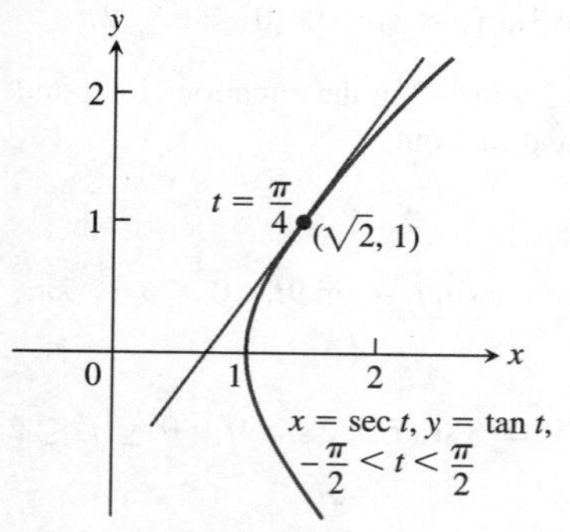

FIGURE 11.14 The curve in Example 1 is the right-hand branch of the hyperbola $x^2 - y^2 = 1$.

EXAMPLE 1 Find the tangent line to the curve

$$x = \sec t, \qquad y = \tan t, \qquad -\frac{\pi}{2} < t < \frac{\pi}{2},$$

at the point $(\sqrt{2}, 1)$, where $t = \pi/4$ (Figure 11.14).

Solution The slope of the curve at t is

$$\frac{dy}{dx} = \frac{dy/dt}{dx/dt} = \frac{\sec^2 t}{\sec t \tan t} = \frac{\sec t}{\tan t}. \qquad \text{Eq. (1)}$$

Setting t equal to $\pi/4$ gives

$$\left.\frac{dy}{dx}\right|_{t=\pi/4} = \frac{\sec(\pi/4)}{\tan(\pi/4)} = \frac{\sqrt{2}}{1} = \sqrt{2}.$$

The tangent line is

$$y - 1 = \sqrt{2}(x - \sqrt{2})$$
$$y = \sqrt{2}x - 2 + 1$$
$$y = \sqrt{2}x - 1.$$

Finding d^2y/dx^2 in Terms of t

1. Express $y' = dy/dx$ in terms of t.
2. Find dy'/dt.
3. Divide dy'/dt by dx/dt.

EXAMPLE 2 Find d^2y/dx^2 as a function of t if $x = t - t^2$ and $y = t - t^3$.

Solution

1. Express $y' = dy/dx$ in terms of t.

$$y' = \frac{dy}{dx} = \frac{dy/dt}{dx/dt} = \frac{1 - 3t^2}{1 - 2t}$$

2. Differentiate y' with respect to t.

$$\frac{dy'}{dt} = \frac{d}{dt}\left(\frac{1 - 3t^2}{1 - 2t}\right) = \frac{2 - 6t + 6t^2}{(1 - 2t)^2} \qquad \text{Derivative Quotient Rule}$$

3. Divide dy'/dt by dx/dt.

$$\frac{d^2y}{dx^2} = \frac{dy'/dt}{dx/dt} = \frac{(2 - 6t + 6t^2)/(1 - 2t)^2}{1 - 2t} = \frac{2 - 6t + 6t^2}{(1 - 2t)^3} \qquad \text{Eq. (2)}$$

EXAMPLE 3 Find the area enclosed by the astroid (Figure 11.15)

$$x = \cos^3 t, \qquad y = \sin^3 t, \qquad 0 \le t \le 2\pi.$$

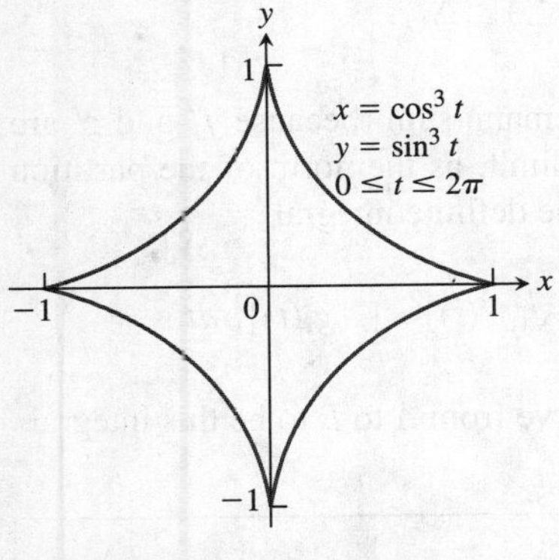

FIGURE 11.15 The astroid in Example 3.

Solution By symmetry, the enclosed area is four times the area beneath the curve in the first quadrant where $0 \le t \le \pi/2$. We can apply the definite integral formula for area studied in Chapter 5, using substitution to express the curve and differential dx in terms of the parameter t. Thus,

$$\begin{aligned}
A &= 4\int_0^1 y\,dx && \text{Four times area under } y \text{ from } x = 0 \text{ to } x = 1 \\
&= 4\int_{\pi/2}^0 (\sin^3 t)(-3\cos^2 t \sin t)\,dt && \cos^3 \pi/2 = 0,\ \cos^3 0 = 1 \\
&= 4\int_0^{\pi/2} (\sin^3 t)(3\cos^2 t \sin t)\,dt && \text{Substitution for } y \text{ and } dx \\
&= 12\int_0^{\pi/2} \left(\frac{1 - \cos 2t}{2}\right)^2 \left(\frac{1 + \cos 2t}{2}\right) dt && \sin^4 t = \left(\frac{1 - \cos 2t}{2}\right)^2 \\
&= \frac{3}{2}\int_0^{\pi/2} (1 - 2\cos 2t + \cos^2 2t)(1 + \cos 2t)\,dt && \text{Expand squared term.} \\
&= \frac{3}{2}\int_0^{\pi/2} (1 - \cos 2t - \cos^2 2t + \cos^3 2t)\,dt && \text{Multiply terms.} \\
&= \frac{3}{2}\left[\int_0^{\pi/2} (1 - \cos 2t)\,dt - \int_0^{\pi/2} \cos^2 2t\,dt + \int_0^{\pi/2} \cos^3 2t\,dt\right] \\
&= \frac{3}{2}\left[\left(t - \frac{1}{2}\sin 2t\right) - \frac{1}{2}\left(t + \frac{1}{4}\sin 4t\right) + \frac{1}{2}\left(\sin 2t - \frac{1}{3}\sin^3 2t\right)\right]_0^{\pi/2} && \text{As in Section 8.3, Example 3} \\
&= \frac{3}{2}\left[\left(\frac{\pi}{2} - 0 - 0 - 0\right) - \frac{1}{2}\left(\frac{\pi}{2} + 0 - 0 - 0\right) + \frac{1}{2}(0 - 0 - 0 + 0)\right] && \text{Evaluate.} \\
&= \frac{3\pi}{8}.
\end{aligned}$$

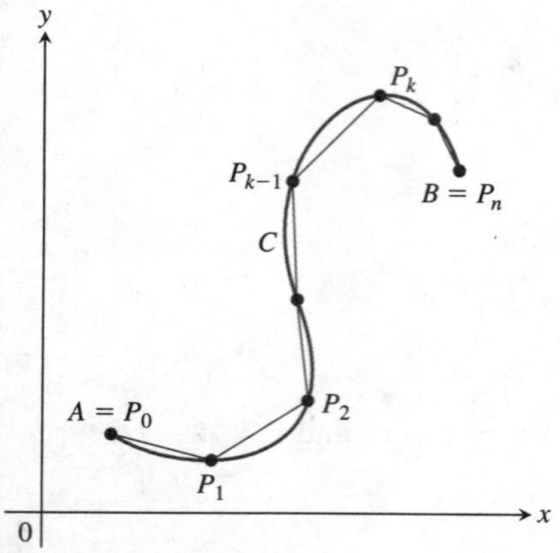

FIGURE 11.16 The length of the smooth curve C from A to B is approximated by the sum of the lengths of the polygonal path (straight-line segments) starting at $A = P_0$, then to P_1, and so on, ending at $B = P_n$.

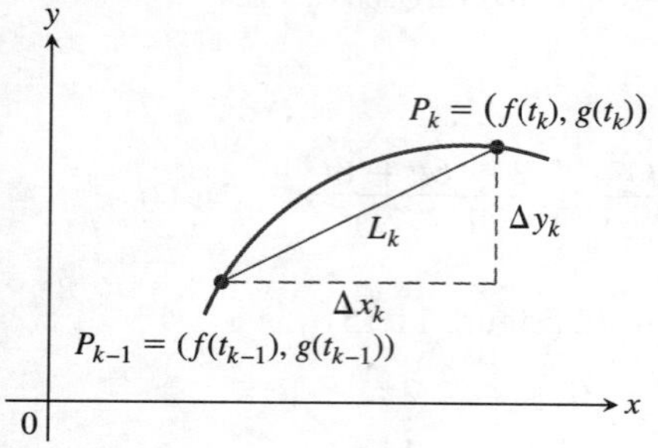

FIGURE 11.17 The arc $P_{k-1}P_k$ is approximated by the straight-line segment shown here, which has length $L_k = \sqrt{(\Delta x_k)^2 + (\Delta y_k)^2}$.

Length of a Parametrically Defined Curve

Let C be a curve given parametrically by the equations

$$x = f(t) \quad \text{and} \quad y = g(t), \qquad a \le t \le b.$$

We assume the functions f and g are **continuously differentiable** (meaning they have continuous first derivatives) on the interval $[a, b]$. We also assume that the derivatives $f'(t)$ and $g'(t)$ are not simultaneously zero, which prevents the curve C from having any corners or cusps. Such a curve is called a **smooth curve**. We subdivide the path (or arc) AB into n pieces at points $A = P_0, P_1, P_2, \ldots, P_n = B$ (Figure 11.16). These points correspond to a partition of the interval $[a, b]$ by $a = t_0 < t_1 < t_2 < \cdots < t_n = b$, where $P_k = (f(t_k), g(t_k))$. Join successive points of this subdivision by straight-line segments (Figure 11.16). A representative line segment has length

$$\begin{aligned} L_k &= \sqrt{(\Delta x_k)^2 + (\Delta y_k)^2} \\ &= \sqrt{[f(t_k) - f(t_{k-1})]^2 + [g(t_k) - g(t_{k-1})]^2} \end{aligned}$$

(see Figure 11.17). If Δt_k is small, the length L_k is approximately the length of arc $P_{k-1}P_k$. By the Mean Value Theorem, there are numbers t_k^* and t_k^{**} in $[t_{k-1}, t_k]$ such that

$$\begin{aligned} \Delta x_k &= f(t_k) - f(t_{k-1}) = f'(t_k^*)\,\Delta t_k, \\ \Delta y_k &= g(t_k) - g(t_{k-1}) = g'(t_k^{**})\,\Delta t_k. \end{aligned}$$

Assuming the path from A to B is traversed exactly once as t increases from $t = a$ to $t = b$, with no doubling back or retracing, an approximation to the (yet to be defined) "length" of the curve AB is the sum of all the lengths L_k:

$$\begin{aligned} \sum_{k=1}^{n} L_k &= \sum_{k=1}^{n} \sqrt{(\Delta x_k)^2 + (\Delta y_k)^2} \\ &= \sum_{k=1}^{n} \sqrt{[f'(t_k^*)]^2 + [g'(t_k^{**})]^2}\,\Delta t_k. \end{aligned}$$

Although this last sum on the right is not exactly a Riemann sum (because f' and g' are evaluated at different points), it can be shown that its limit, as the norm of the partition tends to zero and the number of segments $n \to \infty$, is the definite integral

$$\lim_{\|P\| \to 0} \sum_{k=1}^{n} \sqrt{[f'(t_k^*)]^2 + [g'(t_k^{**})]^2}\,\Delta t_k = \int_a^b \sqrt{[f'(t)]^2 + [g'(t)]^2}\,dt.$$

Therefore, it is reasonable to define the length of the curve from A to B to be this integral.

DEFINITION If a curve C is defined parametrically by $x = f(t)$ and $y = g(t)$, $a \le t \le b$, where f' and g' are continuous and not simultaneously zero on $[a, b]$, and if C is traversed exactly once as t increases from $t = a$ to $t = b$, then **the length of C** is the definite integral

$$L = \int_a^b \sqrt{[f'(t)]^2 + [g'(t)]^2}\,dt.$$

If $x = f(t)$ and $y = g(t)$, then using the Leibniz notation we can write the formula for arc length this way:

$$L = \int_a^b \sqrt{\left(\frac{dx}{dt}\right)^2 + \left(\frac{dy}{dt}\right)^2}\,dt. \tag{3}$$

A smooth curve C does not come to a stop and reverse its direction of motion over the time interval $[a, b]$ since $(f')^2 + (g')^2 > 0$ throughout the interval. At a point where a curve stops and then doubles back on itself, either the curve fails to be differentiable or both derivatives must simultaneously equal zero. We will examine this phenomenon in Chapter 13, where we study tangent vectors to curves.

If there are two different parametrizations for a curve C whose length we want to find, it does not matter which one we use. However, the parametrization we choose must meet the conditions stated in the definition of the length of C (see Exercise 41 for an example).

EXAMPLE 4 Using the definition, find the length of the circle of radius r defined parametrically by

$$x = r\cos t \quad \text{and} \quad y = r\sin t, \qquad 0 \le t \le 2\pi.$$

Solution As t varies from 0 to 2π, the circle is traversed exactly once, so the circumference is

$$L = \int_0^{2\pi} \sqrt{\left(\frac{dx}{dt}\right)^2 + \left(\frac{dy}{dt}\right)^2}\, dt.$$

We find

$$\frac{dx}{dt} = -r\sin t, \qquad \frac{dy}{dt} = r\cos t$$

and

$$\left(\frac{dx}{dt}\right)^2 + \left(\frac{dy}{dt}\right)^2 = r^2(\sin^2 t + \cos^2 t) = r^2.$$

Therefore, the total arc length is

$$L = \int_0^{2\pi} \sqrt{r^2}\, dt = r\Big[t\Big]_0^{2\pi} = 2\pi r. \qquad \blacksquare$$

EXAMPLE 5 Find the length of the astroid (Figure 11.15)

$$x = \cos^3 t, \qquad y = \sin^3 t, \qquad 0 \le t \le 2\pi.$$

Solution Because of the curve's symmetry with respect to the coordinate axes, its length is four times the length of the first-quadrant portion. We have

$$x = \cos^3 t, \qquad\qquad y = \sin^3 t$$

$$\left(\frac{dx}{dt}\right)^2 = [3\cos^2 t(-\sin t)]^2 = 9\cos^4 t \sin^2 t$$

$$\left(\frac{dy}{dt}\right)^2 = [3\sin^2 t(\cos t)]^2 = 9\sin^4 t\cos^2 t$$

$$\begin{aligned}\sqrt{\left(\frac{dx}{dt}\right)^2 + \left(\frac{dy}{dt}\right)^2} &= \sqrt{9\cos^2 t\sin^2 t\underbrace{(\cos^2 t + \sin^2 t)}_{1}}\\ &= \sqrt{9\cos^2 t\sin^2 t}\\ &= 3\,|\cos t\sin t| \qquad \cos t\sin t \ge 0 \text{ for } 0 \le t \le \pi/2\\ &= 3\cos t\sin t.\end{aligned}$$

Therefore,

$$\begin{aligned}\text{Length of first-quadrant portion} &= \int_0^{\pi/2} 3\cos t \sin t\, dt \\ &= \frac{3}{2}\int_0^{\pi/2} \sin 2t\, dt \qquad \cos t \sin t = (1/2)\sin 2t \\ &= -\frac{3}{4}\cos 2t\Big]_0^{\pi/2} = \frac{3}{2}.\end{aligned}$$

The length of the astroid is four times this: $4(3/2) = 6$. ■

EXAMPLE 6 Find the perimeter of the ellipse $\dfrac{x^2}{a^2} + \dfrac{y^2}{b^2} = 1$, where $a > b > 0$.

Solution Parametrically, we represent the ellipse by the equations $x = a \sin t$ and $y = b \cos t$, $0 \le t \le 2\pi$. Then

$$\begin{aligned}\left(\frac{dx}{dt}\right)^2 + \left(\frac{dy}{dt}\right)^2 &= a^2\cos^2 t + b^2 \sin^2 t \\ &= a^2 - (a^2 - b^2)\sin^2 t \\ &= a^2[1 - e^2 \sin^2 t]. \qquad e = \sqrt{1 - \frac{b^2}{a^2}}\end{aligned}$$

(eccentricity, not the number 2.71828 . . .)

From Equation (3), the perimeter is given by

$$P = 4a \int_0^{\pi/2} \sqrt{1 - e^2 \sin^2 t}\, dt.$$

The integral for P is nonelementary and is known as the *complete elliptic integral of the second kind*. We can compute its value to within any degree of accuracy using infinite series in the following way. From the binomial expansion for $\sqrt{1 - x^2}$ in Section 10.10, we have

$$\sqrt{1 - e^2 \sin^2 t} = 1 - \frac{1}{2}e^2 \sin^2 t - \frac{1}{2 \cdot 4}e^4 \sin^4 t - \cdots. \qquad |e \sin t| \le e < 1$$

Then, to each term in this last expression, we apply the integral Formula 157 (at the back of the text) for $\int_0^{\pi/2} \sin^n t\, dt$ when n is even, which yields the perimeter

$$\begin{aligned}P &= 4a \int_0^{\pi/2} \sqrt{1 - e^2 \sin^2 t}\, dt \\ &= 4a\left[\frac{\pi}{2} - \left(\frac{1}{2}e^2\right)\left(\frac{1}{2}\cdot\frac{\pi}{2}\right) - \left(\frac{1}{2\cdot 4}e^4\right)\left(\frac{1\cdot 3}{2\cdot 4}\cdot\frac{\pi}{2}\right) - \left(\frac{1\cdot 3}{2\cdot 4\cdot 6}e^6\right)\left(\frac{1\cdot 3\cdot 5}{2\cdot 4\cdot 6}\cdot\frac{\pi}{2}\right) - \cdots\right] \\ &= 2\pi a\left[1 - \left(\frac{1}{2}\right)^2 e^2 - \left(\frac{1\cdot 3}{2\cdot 4}\right)^2 \frac{e^4}{3} - \left(\frac{1\cdot 3\cdot 5}{2\cdot 4\cdot 6}\right)^2 \frac{e^6}{5} - \cdots\right].\end{aligned}$$

Since $e < 1$, the series on the right-hand side converges by comparison with the geometric series $\sum_{n=1}^{\infty} (e^2)^n$. We do not have an explicit value for P, but we can estimate it as closely as we like by summing finitely many terms from the infinite series. ■

Length of a Curve $y = f(x)$

We will show that the length formula in Section 6.3 is a special case of Equation (3). Given a continuously differentiable function $y = f(x)$, $a \le x \le b$, we can assign $x = t$ as a parameter. The graph of the function f is then the curve C defined parametrically by

$$x = t \quad \text{and} \quad y = f(t), \qquad a \le t \le b,$$

which is a special case of what we have considered in this chapter. We have

$$\frac{dx}{dt} = 1 \quad \text{and} \quad \frac{dy}{dt} = f'(t).$$

From Equation (1),

$$\frac{dy}{dx} = \frac{dy/dt}{dx/dt} = f'(t),$$

giving

$$\begin{aligned} \left(\frac{dx}{dt}\right)^2 + \left(\frac{dy}{dt}\right)^2 &= 1 + [f'(t)]^2 \\ &= 1 + [f'(x)]^2. \qquad t = x \end{aligned}$$

Substitution into Equation (3) gives exactly the arc length formula for the graph of $y = f(x)$ that we found in Section 6.3.

The Arc Length Differential

As in Section 6.3, we define the arc length function for a parametrically defined curve $x = f(t)$ and $y = g(t)$, $a \le t \le b$, by

$$s(t) = \int_a^t \sqrt{[f'(z)]^2 + [g'(z)]^2}\, dz.$$

Then, by the Fundamental Theorem of Calculus,

$$\frac{ds}{dt} = \sqrt{[f'(t)]^2 + [g'(t)]^2} = \sqrt{\left(\frac{dx}{dt}\right)^2 + \left(\frac{dy}{dt}\right)^2}.$$

The differential of arc length is

$$ds = \sqrt{\left(\frac{dx}{dt}\right)^2 + \left(\frac{dy}{dt}\right)^2}\, dt. \tag{4}$$

Equation (4) is often abbreviated as

$$ds = \sqrt{dx^2 + dy^2}.$$

Just as in Section 6.3, we can integrate the differential ds between appropriate limits to find the total length of a curve.

Here's an example where we use the arc length differential to find the centroid of an arc.

EXAMPLE 7 Find the centroid of the first-quadrant arc of the astroid in Example 5.

Solution We take the curve's density to be $\delta = 1$ and calculate the curve's mass and moments about the coordinate axes as we did in Section 6.6.

The distribution of mass is symmetric about the line $y = x$, so $\bar{x} = \bar{y}$. A typical segment of the curve (Figure 11.18) has mass

$$dm = 1 \cdot ds = \sqrt{\left(\frac{dx}{dt}\right)^2 + \left(\frac{dy}{dt}\right)^2}\, dt = 3\cos t \sin t\, dt. \qquad \text{From Example 5}$$

The curve's mass is

$$M = \int_0^{\pi/2} dm = \int_0^{\pi/2} 3\cos t \sin t\, dt = \frac{3}{2}. \qquad \text{Again from Example 5}$$

The curve's moment about the x-axis is

$$M_x = \int \tilde{y}\, dm = \int_0^{\pi/2} \sin^3 t \cdot 3\cos t \sin t\, dt$$

$$= 3\int_0^{\pi/2} \sin^4 t \cos t\, dt = 3 \cdot \frac{\sin^5 t}{5}\bigg]_0^{\pi/2} = \frac{3}{5}.$$

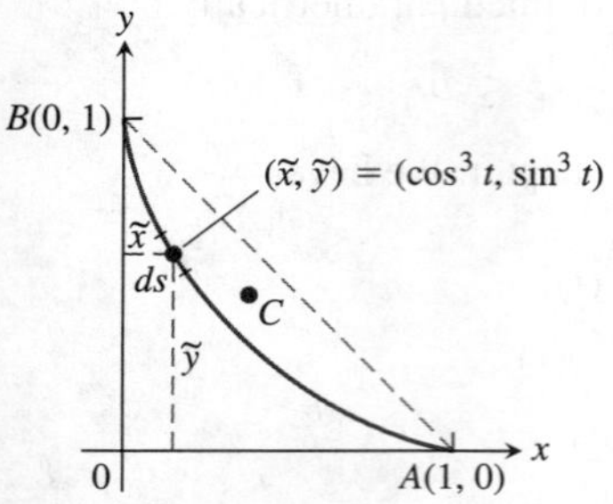

FIGURE 11.18 The centroid C of the astroid arc in Example 7.

It follows that

$$\bar{y} = \frac{M_x}{M} = \frac{3/5}{3/2} = \frac{2}{5}.$$

The centroid is the point $(2/5,\ 2/5)$. ■

EXAMPLE 8 Find the time T_c it takes for a frictionless bead to slide along the cycloid $x = a(t - \sin t)$, $y = a(1 - \cos t)$ from $t = 0$ to $t = \pi$ (see Figure 11.13).

Solution From Equation (3) in Section 11.1, we want to find the time

$$T_c = \int_{t=0}^{t=\pi} \frac{ds}{\sqrt{2gy}}.$$

We need to express ds parametrically in terms of the parameter t. For the cycloid, $dx/dt = a(1 - \cos t)$ and $dy/dt = a\sin t$, so

$$ds = \sqrt{\left(\frac{dx}{dt}\right)^2 + \left(\frac{dy}{dt}\right)^2}\, dt$$

$$= \sqrt{a^2(1 - 2\cos t + \cos^2 t + \sin^2 t)}\, dt$$

$$= \sqrt{a^2(2 - 2\cos t)}\, dt.$$

Substituting for ds and y in the integrand, it follows that

$$T_c = \int_0^{\pi} \sqrt{\frac{a^2(2 - 2\cos t)}{2ga(1 - \cos t)}}\, dt \qquad y = a(1 - \cos t)$$

$$= \int_0^{\pi} \sqrt{\frac{a}{g}}\, dt = \pi\sqrt{\frac{a}{g}}.$$

This is the amount of time it takes the frictionless bead to slide down the cycloid to B after it is released from rest at O (see Figure 11.13). ■

Areas of Surfaces of Revolution

In Section 6.4 we found integral formulas for the area of a surface when a curve is revolved about a coordinate axis. Specifically, we found that the surface area is $S = \int 2\pi y\, ds$ for revolution about the x-axis, and $S = \int 2\pi x\, ds$ for revolution about the y-axis. If the curve is parametrized by the equations $x = f(t)$ and $y = g(t)$, $a \le t \le b$, where f and g are continuously differentiable and $(f')^2 + (g')^2 > 0$ on $[a, b]$, then the arc length differential ds is given by Equation (4). This observation leads to the following formulas for area of surfaces of revolution for smooth parametrized curves.

Area of Surface of Revolution for Parametrized Curves
If a smooth curve $x = f(t)$, $y = g(t)$, $a \leq t \leq b$, is traversed exactly once as t increases from a to b, then the areas of the surfaces generated by revolving the curve about the coordinate axes are as follows.

1. Revolution about the x-axis ($y \geq 0$):

$$S = \int_a^b 2\pi y \sqrt{\left(\frac{dx}{dt}\right)^2 + \left(\frac{dy}{dt}\right)^2}\, dt \tag{5}$$

2. Revolution about the y-axis ($x \geq 0$):

$$S = \int_a^b 2\pi x \sqrt{\left(\frac{dx}{dt}\right)^2 + \left(\frac{dy}{dt}\right)^2}\, dt \tag{6}$$

As with length, we can calculate surface area from any convenient parametrization that meets the stated criteria.

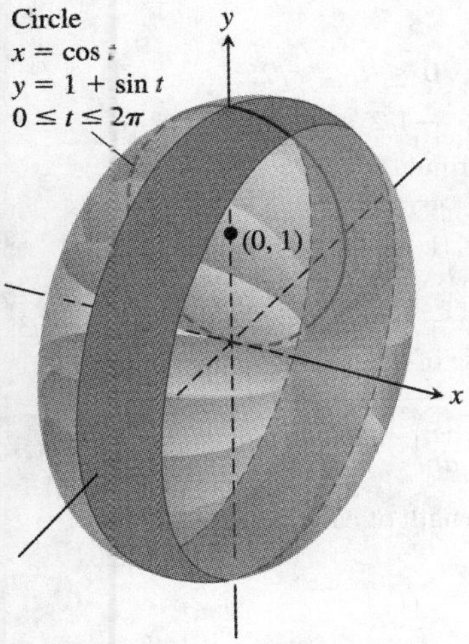

FIGURE 11.19 In Example 9 we calculate the area of the surface of revolution swept out by this parametrized curve.

EXAMPLE 9 The standard parametrization of the circle of radius 1 centered at the point $(0, 1)$ in the xy-plane is

$$x = \cos t, \qquad y = 1 + \sin t, \qquad 0 \leq t \leq 2\pi.$$

Use this parametrization to find the area of the surface swept out by revolving the circle about the x-axis (Figure 11.19).

Solution We evaluate the formula

$$S = \int_a^b 2\pi y \sqrt{\left(\frac{dx}{dt}\right)^2 + \left(\frac{dy}{dt}\right)^2}\, dt$$

Eq. (5) for revolution about the x-axis; $y = 1 + \sin t \geq 0$

$$= \int_0^{2\pi} 2\pi(1 + \sin t)\, \underbrace{\sqrt{(-\sin t)^2 + (\cos t)^2}}_{1}\, dt$$

$$= 2\pi \int_0^{2\pi} (1 + \sin t)\, dt$$

$$= 2\pi \Big[t - \cos t \Big]_0^{2\pi} = 4\pi^2. \qquad \blacksquare$$

EXERCISES 11.2

Tangent Lines to Parametrized Curves

In Exercises 1–14, find an equation for the line tangent to the curve at the point defined by the given value of t. Also, find the value of d^2y/dx^2 at this point.

1. $x = 2\cos t, \quad y = 2\sin t, \quad t = \pi/4$
2. $x = \sin 2\pi t, \quad y = \cos 2\pi t, \quad t = -1/6$
3. $x = 4\sin t, \quad y = 2\cos t, \quad t = \pi/4$
4. $x = \cos t, \quad y = \sqrt{3}\cos t, \quad t = 2\pi/3$
5. $x = t, \quad y = \sqrt{t}, \quad t = 1/4$
6. $x = \sec^2 t - 1, \quad y = \tan t, \quad t = -\pi/4$
7. $x = \sec t, \quad y = \tan t, \quad t = \pi/6$
8. $x = -\sqrt{t+1}, \quad y = \sqrt{3t}, \quad t = 3$
9. $x = \dfrac{t-1}{t+1}, \quad y = \dfrac{t+1}{t-1}, \quad t = 2$
10. $x = 1/t, \quad y = -2 + \ln t, \quad t = 1$
11. $x = t - \sin t, \quad y = 1 - \cos t, \quad t = \pi/3$

12. $x = \cos t, \quad y = 1 + \sin t, \quad t = \pi/2$

13. $x = \dfrac{1}{t+1}, \quad y = \dfrac{t}{t-1}, \quad t = 2$

14. $x = t + e^t, \quad y = 1 - e^t, \quad t = 0$

Implicitly Defined Parametrizations

Assuming that the equations in Exercises 15–20 define x and y implicitly as differentiable functions $x = f(t)$, $y = g(t)$, find the slope of the curve $x = f(t)$, $y = g(t)$ at the given value of t.

15. $x^3 + 2t^2 = 9, \quad 2y^3 - 3t^2 = 4, \quad t = 2$

16. $x = \sqrt{5 - \sqrt{t}}, \quad y(t-1) = \sqrt{t}, \quad t = 4$

17. $x + 2x^{3/2} = t^2 + t, \quad y\sqrt{t+1} + 2t\sqrt{y} = 4, \quad t = 0$

18. $x \sin t + 2x = t, \quad t \sin t - 2t = y, \quad t = \pi$

19. $x = t^3 + t, \quad y + 2t^3 = 2x + t^2, \quad t = 1$

20. $t = \ln(x - t), \quad y = te^t, \quad t = 0$

Area

21. Find the area under one arch of the cycloid

$$x = a(t - \sin t), \quad y = a(1 - \cos t).$$

22. Find the area enclosed by the y-axis and the curve

$$x = t - t^2, \quad y = 1 + e^{-t}.$$

23. Find the area enclosed by the ellipse

$$x = a\cos t, \quad y = b\sin t, \quad 0 \le t \le 2\pi.$$

24. Find the area under $y = x^3$ over $[0, 1]$ using the following parametrizations.

a. $x = t^2, \quad y = t^6$ **b.** $x = t^3, \quad y = t^9$

Lengths of Curves

Find the lengths of the curves in Exercises 25–30.

25. $x = \cos t, \quad y = t + \sin t, \quad 0 \le t \le \pi$

26. $x = t^3, \quad y = 3t^2/2, \quad 0 \le t \le \sqrt{3}$

27. $x = t^2, \quad y = t - t^3/3, \quad 0 \le t \le \sqrt{3}$

28. $x = a(t - \sin t), \; y = a(1 - \cos t), \quad 0 \le t \le 2\pi$

29. $x = 8\cos t + 8t\sin t$, $y = 8\sin t - 8t\cos t$, $0 \le t \le \pi/2$

30. $x = a(\cos\theta + \theta\sin\theta)$, $y = a(\sin\theta - \theta\cos\theta), \; 0 \le \theta \le \pi/2$

Surface Area

Find the areas of the surfaces generated by revolving the curves in Exercises 31–34 about the indicated axes.

31. $x = \cos t, \quad y = 2 + \sin t, \quad 0 \le t \le 2\pi;$ x-axis

32. $x = (2/3)t^{3/2}, \quad y = 2\sqrt{t}, \quad 0 \le t \le \sqrt{3};$ y-axis

33. $x = t + \sqrt{2}, \quad y = (t^2/2) + \sqrt{2}t, \; -\sqrt{2} \le t \le \sqrt{2};$ y-axis

34. $x = \ln(\sec t + \tan t) - \sin t, \; y = \cos t, \; 0 \le t \le \pi/3;$ x-axis

35. A cone frustum The line segment joining the points $(0, 1)$ and $(2, 2)$ is revolved about the x-axis to generate a frustum of a cone. Find the surface area of the frustum using the parametrization $x = 2t, \; y = t + 1, \; 0 \le t \le 1$. Check your result with the geometry formula: Area $= \pi(r_1 + r_2)$(slant height).

36. A cone The line segment joining the origin to the point (h, r) is revolved about the x-axis to generate a cone of height h and base radius r. Find the cone's surface area with the parametric equations $x = ht, \; y = rt, \; 0 \le t \le 1$. Check your result with the geometry formula: Area $= \pi r$(slant height).

Centroids

37. Find the coordinates of the centroid of the curve

$$x = \cos t + t\sin t, \quad y = \sin t - t\cos t, \quad 0 \le t \le \pi/2.$$

38. Find the coordinates of the centroid of the curve

$$x = e^t\cos t, \quad y = e^t\sin t, \quad 0 \le t \le \pi.$$

39. Find the coordinates of the centroid of the curve

$$x = \cos t, \quad y = t + \sin t, \quad 0 \le t \le \pi.$$

T **40.** Most centroid calculations for curves are done with a calculator or computer that has an integral evaluation program. As a case in point, find, to the nearest hundredth, the coordinates of the centroid of the curve

$$x = t^3, \quad y = 3t^2/2, \quad 0 \le t \le \sqrt{3}.$$

Theory and Examples

41. Length is independent of parametrization To illustrate the fact that the numbers we get for length do not depend on the way we parametrize our curves (except for the mild restrictions preventing doubling back mentioned earlier), calculate the length of the semicircle $y = \sqrt{1 - x^2}$ with these two different parametrizations:

a. $x = \cos 2t, \quad y = \sin 2t, \quad 0 \le t \le \pi/2.$

b. $x = \sin \pi t, \quad y = \cos \pi t, \quad -1/2 \le t \le 1/2.$

42. a. Show that the Cartesian formula

$$L = \int_c^d \sqrt{1 + \left(\frac{dx}{dy}\right)^2}\, dy$$

for the length of the curve $x = g(y), \; c \le y \le d$ (Section 6.3, Equation 4), is a special case of the parametric length formula

$$L = \int_a^b \sqrt{\left(\frac{dx}{dt}\right)^2 + \left(\frac{dy}{dt}\right)^2}\, dt.$$

Use this result to find the length of each curve.

b. $x = y^{3/2}, \quad 0 \le y \le 4/3$

c. $x = \dfrac{3}{2}y^{2/3}, \quad 0 \le y \le 1$

43. The curve with parametric equations

$$x = (1 + 2\sin\theta)\cos\theta, \quad y = (1 + 2\sin\theta)\sin\theta$$

is called a *limaçon* and is shown in the accompanying figure. Find the points (x, y) and the slopes of the tangent lines at these points for

a. $\theta = 0$. **b.** $\theta = \pi/2$. **c.** $\theta = 4\pi/3$.

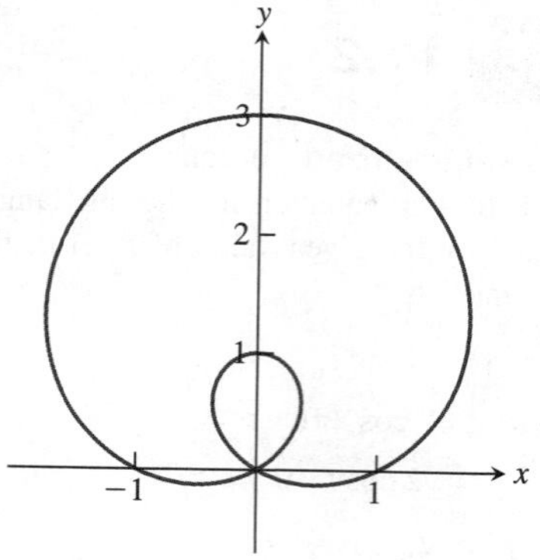

44. The curve with parametric equations

$$x = t, \quad y = 1 - \cos t, \quad 0 \leq t \leq 2\pi$$

is called a *sinusoid* and is shown in the accompanying figure. Find the point (x, y) where the slope of the tangent line is

a. largest. **b.** smallest.

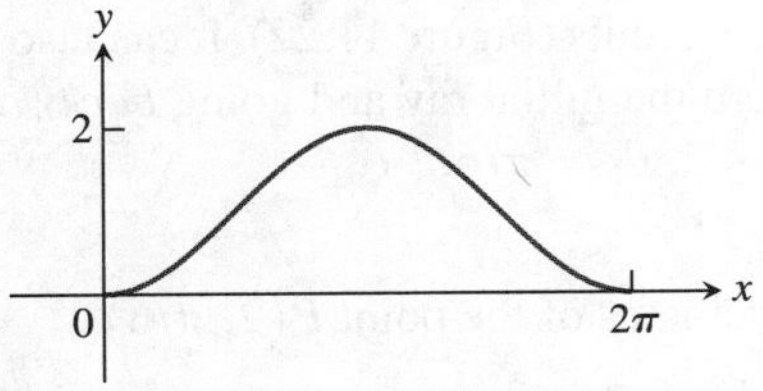

T The curves in Exercises 45 and 46 are called *Bowditch curves* or *Lissajous figures*. In each case, find the point in the interior of the first quadrant where the tangent line to the curve is horizontal, and find the equations of the two tangent lines at the origin.

45.

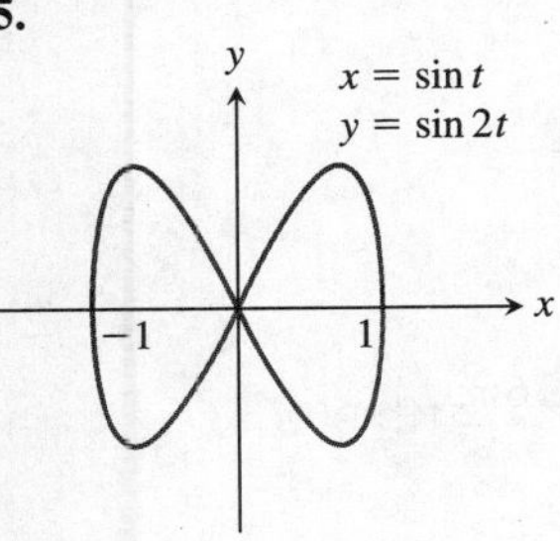

46.

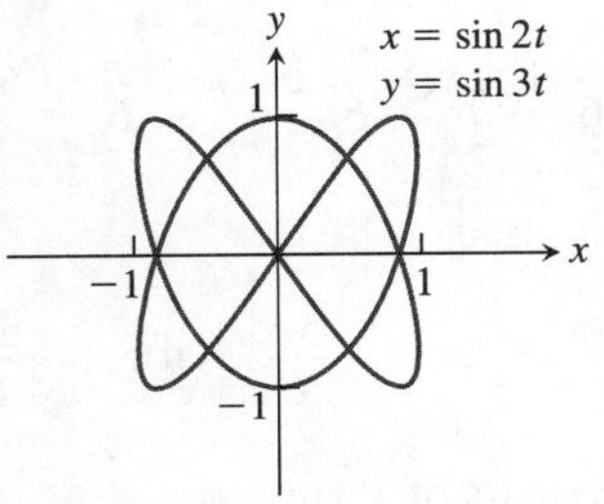

47. Cycloid

a. Find the length of one arch of the cycloid

$$x = a(t - \sin t), \quad y = a(1 - \cos t).$$

b. Find the area of the surface generated by revolving one arch of the cycloid in part (a) about the x-axis for $a = 1$.

48. Volume Find the volume swept out by revolving the region bounded by the x-axis and one arch of the cycloid

$$x = t - \sin t, \quad y = 1 - \cos t$$

about the x-axis.

49. Find the volume swept out by revolving the region bounded by the x-axis and the graph of

$$x = 2t, \quad y = t(2 - t)$$

about the x-axis.

50. Find the volume swept out by revolving the region bounded by the y-axis and the graph of

$$x = t(1 - t), \quad y = 1 + t^2$$

about the y-axis.

COMPUTER EXPLORATIONS

In Exercises 51–54, use a CAS to perform the following steps for the given curve over the closed interval.

a. Plot the curve together with the polygonal path approximations for $n = 2, 4, 8$ partition points over the interval. (See Figure 11.16.)

b. Find the corresponding approximation to the length of the curve by summing the lengths of the line segments.

c. Evaluate the length of the curve using an integral. Compare your approximations for $n = 2, 4, 8$ with the actual length given by the integral. How does the actual length compare with the approximations as n increases? Explain your answer.

51. $x = \frac{1}{3}t^3, \quad y = \frac{1}{2}t^2, \quad 0 \leq t \leq 1$

52. $x = 2t^3 - 16t^2 + 25t + 5, \quad y = t^2 + t - 3, \quad 0 \leq t \leq 6$

53. $x = t - \cos t, \quad y = 1 + \sin t, \quad -\pi \leq t \leq \pi$

54. $x = e^t \cos t, \quad y = e^t \sin t, \quad 0 \leq t \leq \pi$

11.3 Polar Coordinates

In this section we study polar coordinates and their relation to Cartesian coordinates. You will see that polar coordinates are very useful for calculating many multiple integrals studied in Chapter 15. They are also useful in describing the paths of planets and satellites.

Definition of Polar Coordinates

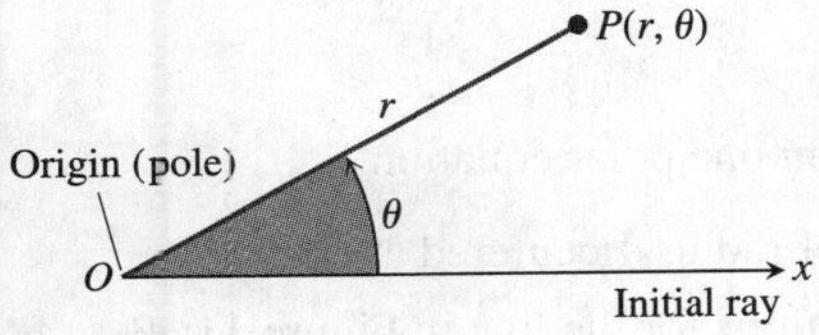

FIGURE 11.20 To define polar coordinates for the plane, we start with an origin, called the pole, and an initial ray.

To define polar coordinates, we first fix an **origin** O (called the **pole**) and an **initial ray** from O (Figure 11.20). Usually the positive x-axis is chosen as the initial ray. Then each point P can be located by assigning to it a **polar coordinate pair** (r, θ) in which r gives the directed distance from O to P, and θ gives the directed angle from the initial ray to ray OP. So we label the point P as

$$P(r, \theta)$$

Directed distance from O to P (for r); Directed angle from initial ray to OP (for θ)

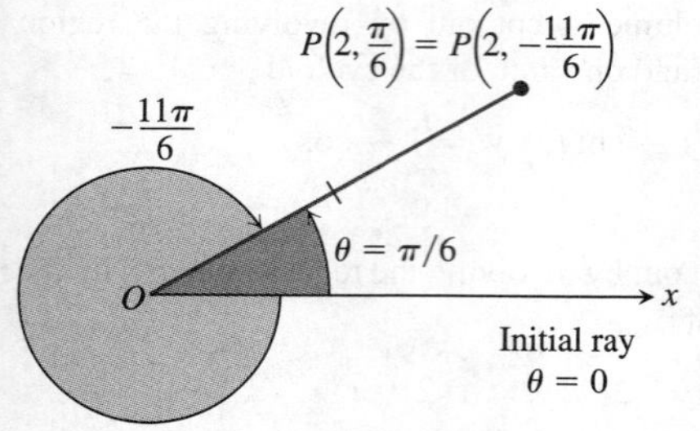

FIGURE 11.21 Polar coordinates are not unique.

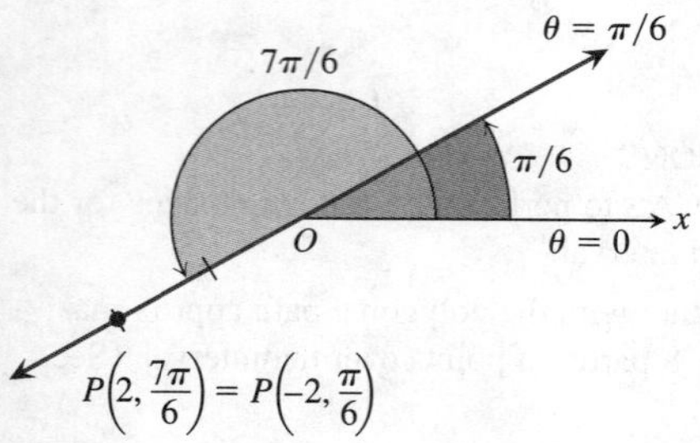

FIGURE 11.22 Polar coordinates can have negative r-values.

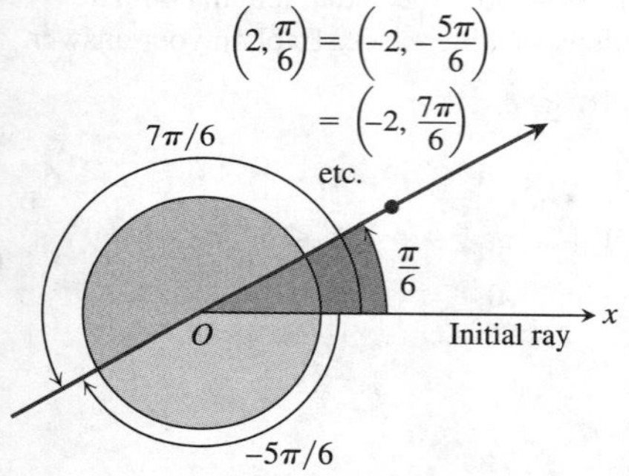

FIGURE 11.23 The point $P(2, \pi/6)$ has infinitely many polar coordinate pairs (Example 1).

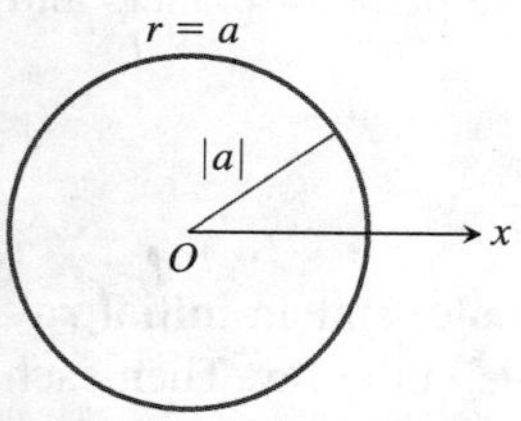

FIGURE 11.24 The polar equation for a circle is $r = a$.

As in trigonometry, θ is positive when measured counterclockwise and negative when measured clockwise. The angle associated with a given point is not unique. A point in the plane has just one pair of Cartesian coordinates, but it has infinitely many pairs of polar coordinates. For instance, the point 2 units from the origin along the ray $\theta = \pi/6$ has polar coordinates $r = 2$, $\theta = \pi/6$. It also has coordinates $r = 2$, $\theta = -11\pi/6$ (Figure 11.21). In some situations we allow r to be negative. That is why we use directed distance in defining $P(r, \theta)$. The point $P(2, 7\pi/6)$ can be reached by turning $7\pi/6$ radians counterclockwise from the initial ray and going forward 2 units (Figure 11.22). It can also be reached by turning $\pi/6$ radians counterclockwise from the initial ray and going *backward* 2 units. So the point also has polar coordinates $r = -2$, $\theta = \pi/6$.

EXAMPLE 1 Find all the polar coordinates of the point $P(2, \pi/6)$.

Solution We sketch the initial ray of the coordinate system, draw the ray from the origin that makes an angle of $\pi/6$ radians with the initial ray, and mark the point $(2, \pi/6)$ (Figure 11.23). We then find the angles for the other coordinate pairs of P in which $r = 2$ and $r = -2$.

For $r = 2$, the complete list of angles is

$$\frac{\pi}{6}, \frac{\pi}{6} \pm 2\pi, \quad \frac{\pi}{6} \pm 4\pi, \quad \frac{\pi}{6} \pm 6\pi, \ldots .$$

For $r = -2$, the angles are

$$-\frac{5\pi}{6}, \quad -\frac{5\pi}{6} \pm 2\pi, \quad -\frac{5\pi}{6} \pm 4\pi, \quad -\frac{5\pi}{6} \pm 6\pi, \ldots .$$

The corresponding coordinate pairs of P are

$$\left(2, \frac{\pi}{6} + 2n\pi\right), \qquad n = 0, \pm 1, \pm 2, \ldots$$

and

$$\left(-2, -\frac{5\pi}{6} + 2n\pi\right), \qquad n = 0, \pm 1, \pm 2, \ldots .$$

When $n = 0$, the formulas give $(2, \pi/6)$ and $(-2, -5\pi/6)$. When $n = 1$, they give $(2, 13\pi/6)$ and $(-2, 7\pi/6)$, and so on. ■

Polar Equations and Graphs

If we hold r fixed at a constant value $r = a \neq 0$, the point $P(r, \theta)$ will lie $|a|$ units from the origin O. As θ varies over any interval of length 2π, P then traces a circle of radius $|a|$ centered at O (Figure 11.24).

If we hold θ fixed at a constant value $\theta = \theta_0$ and let r vary between $-\infty$ and ∞, the point $P(r, \theta)$ traces the line through O that makes an angle of measure θ_0 with the initial ray. (See Figure 11.22 for an example.)

EXAMPLE 2 A circle or line can have more than one polar equation.

(a) $r = 1$ and $r = -1$ are equations for the circle of radius 1 centered at O.

(b) $\theta = \pi/6$, $\theta = 7\pi/6$, and $\theta = -5\pi/6$ are equations for the line in Figure 11.23. ■

Equations of the form $r = a$ and $\theta = \theta_0$, and inequalities such as $r \leq a$ and $0 \leq \theta \leq \pi$, can be combined to define regions, segments, and rays.

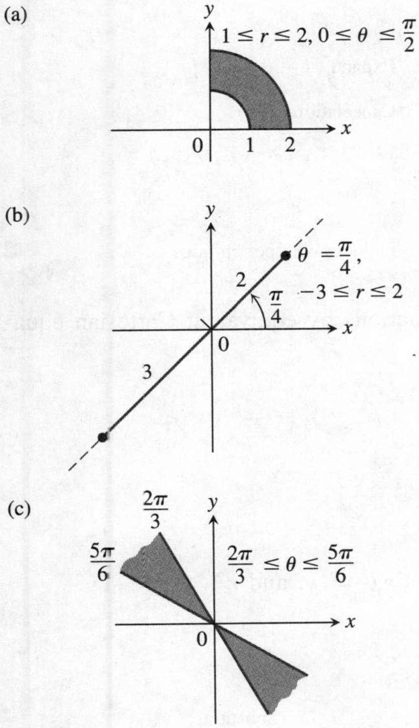

FIGURE 11.25 The graphs of typical inequalities in r and θ (Example 3).

EXAMPLE 3 Graph the sets of points whose polar coordinates satisfy the following conditions.

(a) $1 \le r \le 2$ and $0 \le \theta \le \dfrac{\pi}{2}$

(b) $-3 \le r \le 2$ and $\theta = \dfrac{\pi}{4}$

(c) $\dfrac{2\pi}{3} \le \theta \le \dfrac{5\pi}{6}$ (no restriction on r)

Solution The graphs are shown in Figure 11.25. ■

Relating Polar and Cartesian Coordinates

When we use both polar and Cartesian coordinates in a plane, we place the two origins together and let the initial polar ray be the positive x-axis. The ray $\theta = \pi/2$, $r > 0$, becomes the positive y-axis (Figure 11.26). The two coordinate systems are then related by the following equations.

Equations Relating Polar and Cartesian Coordinates

Polar to Cartesian:

$$x = r\cos\theta, \qquad y = r\sin\theta$$

Cartesian to Polar:

$$r^2 = x^2 + y^2, \qquad \tan\theta = \frac{y}{x}$$

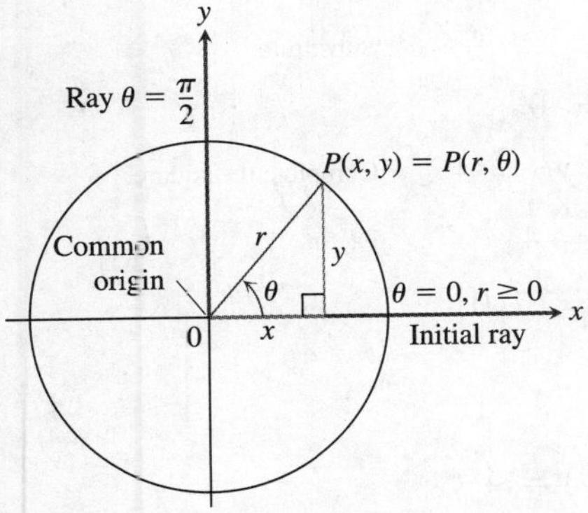

FIGURE 11.26 The usual way to relate polar and Cartesian coordinates.

The first two of these equations uniquely determine the Cartesian coordinates x and y given the polar coordinates r and θ. On the other hand, if x and y are given and $(x, y) \ne (0, 0)$, the third equation gives two possible choices for r (a positive and a negative value). For each of these r values, there is a unique $\theta \in [0, 2\pi)$ satisfying the first two equations, each then giving a polar coordinate representation of the Cartesian point (x, y). The other polar coordinate representations for the point can be determined from these two, as in Example 1.

EXAMPLE 4 Here are some plane curves expressed in terms of both polar coordinate and Cartesian coordinate equations.

Polar equation	Cartesian equivalent
$r\cos\theta = 2$	$x = 2$
$r^2\cos\theta\sin\theta = 4$	$xy = 4$
$r^2\cos^2\theta - r^2\sin^2\theta = 1$	$x^2 - y^2 = 1$
$r = 1 + 2r\cos\theta$	$y^2 - 3x^2 - 4x - 1 = 0$
$r = 1 - \cos\theta$	$x^4 + y^4 + 2x^2y^2 + 2x^3 + 2xy^2 - y^2 = 0$

Some curves are more simply expressed with polar coordinates; others are not. ■

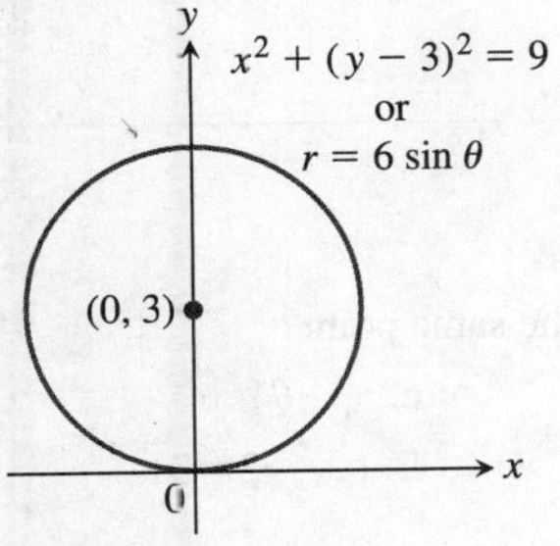

FIGURE 11.27 The circle in Example 5.

EXAMPLE 5 Find a polar equation for the circle $x^2 + (y - 3)^2 = 9$ (Figure 11.27).

Solution We apply the equations relating polar and Cartesian coordinates:

$$
\begin{aligned}
x^2 + (y-3)^2 &= 9 \\
x^2 + y^2 - 6y + 9 &= 9 && \text{Expand } (y-3)^2. \\
x^2 + y^2 - 6y &= 0 && \text{Cancelation} \\
r^2 - 6r\sin\theta &= 0 && x^2 + y^2 = r^2,\ y = r\sin\theta \\
r = 0 \quad \text{or} \quad r - 6\sin\theta &= 0 \\
r &= 6\sin\theta && \text{Includes both possibilities}
\end{aligned}
$$
■

EXAMPLE 6 Replace the following polar equations by equivalent Cartesian equations and identify their graphs.

(a) $r\cos\theta = -4$

(b) $r^2 = 4r\cos\theta$

(c) $r = \dfrac{4}{2\cos\theta - \sin\theta}$

Solution We use the substitutions $r\cos\theta = x$, $r\sin\theta = y$, and $r^2 = x^2 + y^2$.

(a) $r\cos\theta = -4$

The Cartesian equation:
$$
\begin{aligned}
r\cos\theta &= -4 \\
x &= -4 && \text{Substitute.}
\end{aligned}
$$

The graph: Vertical line through $x = -4$ on the x-axis

(b) $r^2 = 4r\cos\theta$

The Cartesian equation:
$$
\begin{aligned}
r^2 &= 4r\cos\theta \\
x^2 + y^2 &= 4x && \text{Substitute.} \\
x^2 - 4x + y^2 &= 0 \\
x^2 - 4x + 4 + y^2 &= 4 && \text{Complete the square.} \\
(x-2)^2 + y^2 &= 4 && \text{Factor.}
\end{aligned}
$$

The graph: Circle, radius 2, center $(h, k) = (2, 0)$

(c) $r = \dfrac{4}{2\cos\theta - \sin\theta}$

The Cartesian equation:
$$
\begin{aligned}
r(2\cos\theta - \sin\theta) &= 4 \\
2r\cos\theta - r\sin\theta &= 4 && \text{Multiply by } r. \\
2x - y &= 4 && \text{Substitute.} \\
y &= 2x - 4 && \text{Solve for } y.
\end{aligned}
$$

The graph: Line, slope $m = 2$, y-intercept $b = -4$ ■

EXERCISES 11.3

Polar Coordinates

1. Which polar coordinate pairs label the same point?

a. $(3, 0)$ **b.** $(-3, 0)$ **c.** $(2, 2\pi/3)$
d. $(2, 7\pi/3)$ **e.** $(-3, \pi)$ **f.** $(2, \pi/3)$
g. $(-3, 2\pi)$ **h.** $(-2, -\pi/3)$

2. Which polar coordinate pairs label the same point?

a. $(-2, \pi/3)$ **b.** $(2, -\pi/3)$ **c.** (r, θ)
d. $(r, \theta + \pi)$ **e.** $(-r, \theta)$ **f.** $(2, -2\pi/3)$
g. $(-r, \theta + \pi)$ **h.** $(-2, 2\pi/3)$

3. Plot the following points, given in polar coordinates. Then find all the polar coordinates of each point.

a. $(2, \pi/2)$ **b.** $(2, 0)$

c. $(-2, \pi/2)$ **d.** $(-2, 0)$

4. Plot the following points, given in polar coordinates. Then find all the polar coordinates of each point.

a. $(3, \pi/4)$ **b.** $(-3, \pi/4)$

c. $(3, -\pi/4)$ **d.** $(-3, -\pi/4)$

Polar to Cartesian Coordinates

5. Find the Cartesian coordinates of the points in Exercise 1.

6. Find the Cartesian coordinates of the following points, given in polar coordinates.

a. $(\sqrt{2}, \pi/4)$ **b.** $(1, 0)$

c. $(0, \pi/5)$ **d.** $(-\sqrt{2}, \pi/4)$

e. $(-3, 5\pi/6)$ **f.** $(5, \tan^{-1}(4/3))$

g. $(-1, 7\pi)$ **h.** $(3\sqrt{2}, 3\pi/2)$

Cartesian to Polar Coordinates

7. Find the polar coordinates, $0 \le \theta < 2\pi$ and $r \ge 0$, of the following points given in Cartesian coordinates.

a. $(2, 2)$ **b.** $(-1, -1)$

c. $(\sqrt{3}, -1)$ **d.** $(-3, 4)$

8. Find the polar coordinates, $-\pi \le \theta < \pi$ and $r \ge 0$, of the following points given in Cartesian coordinates.

a. $(-2, -2)$ **b.** $(0, 3)$

c. $(-\sqrt{3}, 1)$ **d.** $(5, -12)$

9. Find the polar coordinates, $0 \le \theta < 2\pi$ and $r \le 0$, of the following points given in Cartesian coordinates.

a. $(3, 3)$ **b.** $(-1, 0)$

c. $(-1, \sqrt{3})$ **d.** $(4, -3)$

10. Find the polar coordinates, $-\pi \le \theta < \pi$ and $r \le 0$, of the following points given in Cartesian coordinates.

a. $(-2, 0)$ **b.** $(1, 0)$

c. $(0, -3)$ **d.** $\left(\frac{\sqrt{3}}{2}, \frac{1}{2}\right)$

Graphing Sets of Polar Coordinate Points

Graph the sets of points whose polar coordinates satisfy the equations and inequalities in Exercises 11–26.

11. $r = 2$ **12.** $0 \le r \le 2$

13. $r \ge 1$ **14.** $1 \le r \le 2$

15. $0 \le \theta \le \pi/6, \quad r \ge 0$ **16.** $\theta = 2\pi/3, \quad r \le -2$

17. $\theta = \pi/3, \quad -1 \le r \le 3$ **18.** $\theta = 11\pi/4, \quad r \ge -1$

19. $\theta = \pi/2, \quad r \ge 0$ **20.** $\theta = \pi/2, \quad r \le 0$

21. $0 \le \theta \le \pi, \quad r = 1$ **22.** $0 \le \theta \le \pi, \quad r = -1$

23. $\pi/4 \le \theta \le 3\pi/4, \quad 0 \le r \le 1$

24. $-\pi/4 \le \theta \le \pi/4, \quad -1 \le r \le 1$

25. $-\pi/2 \le \theta \le \pi/2, \quad 1 \le r \le 2$

26. $0 \le \theta \le \pi/2, \quad 1 \le |r| \le 2$

Polar to Cartesian Equations

Replace the polar equations in Exercises 27–52 with equivalent Cartesian equations. Then describe or identify the graph.

27. $r \cos \theta = 5$ **28.** $r \sin \theta = -1$

29. $r \sin \theta = 0$ **30.** $r \cos \theta = 0$

31. $r = 3 \csc \theta$ **32.** $r = -3 \sec \theta$

33. $r \cos \theta + r \sin \theta = 1$ **34.** $r \sin \theta = r \cos \theta$

35. $r^2 = 4$ **36.** $r^2 = 4r \sin \theta$

37. $r = \dfrac{5}{\sin \theta - 2 \cos \theta}$ **38.** $r^2 \sin 2\theta = 2$

39. $r = \cot \theta \csc \theta$ **40.** $r = 9 \tan \theta \sec \theta$

41. $r = \sec \theta \, e^{r \sin \theta}$ **42.** $r \cos \theta = \ln r + \ln \sin \theta$

43. $r^2 + 2r^2 \cos \theta \sin \theta = 1$ **44.** $\cos^2 \theta = \sin^2 \theta$

45. $r^2 = -4r \cos \theta$ **46.** $r^2 = -10 r \sin \theta$

47. $r = 8 \sin \theta$ **48.** $r = 3 \cos \theta$

49. $r = 4 \cos \theta + 4 \sin \theta$ **50.** $r = 2 \cos \theta - \sin \theta$

51. $r \sin\left(\theta + \dfrac{\pi}{6}\right) = 2$ **52.** $r \sin\left(\dfrac{2\pi}{3} - \theta\right) = 5$

Cartesian to Polar Equations

Replace the Cartesian equations in Exercises 53–66 with equivalent polar equations.

53. $x = 5$ **54.** $y = 1$ **55.** $x = \sqrt{3}\, y$

56. $x + y = 1$ **57.** $x^2 + y^2 = 4$ **58.** $x^2 - y^2 = 2$

59. $\dfrac{x^2}{4} + \dfrac{y^2}{9} = 1$ **60.** $xy = 2$

61. $y^2 = 9x$ **62.** $x^2 + xy + y^2 = 1$

63. $x^2 + (y - 2)^2 = 4$ **64.** $(x - 6)^2 + y^2 = 36$

65. $(x - 3)^2 + (y + 1)^2 = 4$ **66.** $(x + 1)^2 + (y - 3)^2 = 25$

67. Find all polar coordinates of the origin.

68. Vertical and horizontal lines

a. Show that every vertical line in the xy-plane has a polar equation of the form $r = a \sec \theta$.

b. Find the analogous polar equation for horizontal lines in the xy-plane.

11.4 Graphing Polar Coordinate Equations

It is often helpful to graph an equation expressed in polar coordinates in the Cartesian xy-plane. This section describes some techniques for graphing these equations using symmetries and tangent lines to the graph.

Symmetry

The following list shows how to test for three standard types of symmetries when using polar coordinates. These symmetries are illustrated in Figure 11.28.

Symmetry Tests for Polar Graphs in the Cartesian xy-Plane

1. *Symmetry about the x-axis:* If the point (r, θ) lies on the graph, then the point $(r, -\theta)$ or $(-r, \pi - \theta)$ lies on the graph (Figure 11.28a).
2. *Symmetry about the y-axis:* If the point (r, θ) lies on the graph, then the point $(r, \pi - \theta)$ or $(-r, -\theta)$ lies on the graph (Figure 11.28b).
3. *Symmetry about the origin:* If the point (r, θ) lies on the graph, then the point $(-r, \theta)$ or $(r, \theta + \pi)$ lies on the graph (Figure 11.28c).

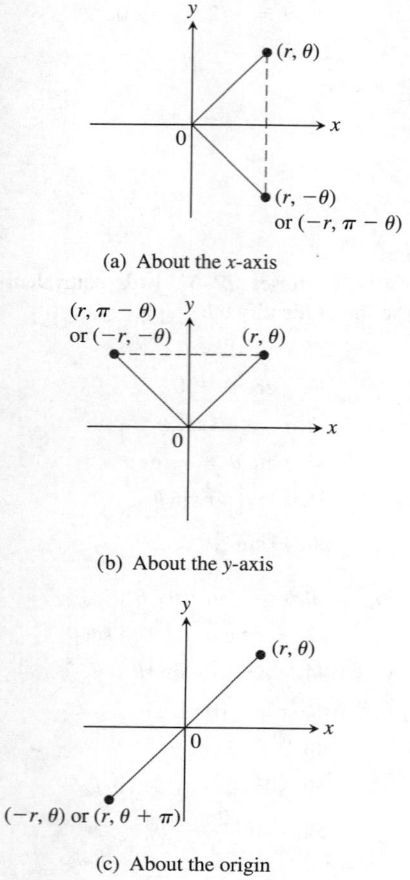

FIGURE 11.28 Three tests for symmetry in polar coordinates.

Slope

The slope of a polar curve $r = f(\theta)$ in the xy-plane is dy/dx, but this is **not** given by the formula $r' = df/d\theta$. To see why, think of the graph of f as the graph of the parametric equations

$$x = r\cos\theta = f(\theta)\cos\theta, \qquad y = r\sin\theta = f(\theta)\sin\theta.$$

If f is a differentiable function of θ, then so are x and y, and when $dx/d\theta \neq 0$, we can calculate dy/dx from the parametric formula

$$\frac{dy}{dx} = \frac{dy/d\theta}{dx/d\theta} \qquad \text{Section 11.2, Eq. (1) with } t = \theta$$

$$= \frac{\frac{d}{d\theta}(f(\theta)\sin\theta)}{\frac{d}{d\theta}(f(\theta)\cos\theta)} \qquad \text{Substitute}$$

$$= \frac{\frac{df}{d\theta}\sin\theta + f(\theta)\cos\theta}{\frac{df}{d\theta}\cos\theta - f(\theta)\sin\theta} \qquad \text{Product Rule for derivatives}$$

Therefore, we see that dy/dx is not the same as $df/d\theta$.

Slope of the Curve $r = f(\theta)$ in the Cartesian xy-Plane

$$\left.\frac{dy}{dx}\right|_{(r,\,\theta)} = \frac{f'(\theta)\sin\theta + f(\theta)\cos\theta}{f'(\theta)\cos\theta - f(\theta)\sin\theta}, \tag{1}$$

provided $dx/d\theta \neq 0$ at (r, θ).

If the curve $r = f(\theta)$ passes through the origin at $\theta = \theta_0$, then $f(\theta_0) = 0$, and the slope equation gives

$$\left.\frac{dy}{dx}\right|_{(0,\,\theta_0)} = \frac{f'(\theta_0)\sin\theta_0}{f'(\theta_0)\cos\theta_0} = \tan\theta_0.$$

That is, the slope at $(0, \theta_0)$ is $\tan\theta_0$. The reason we say "slope at $(0, \theta_0)$" and not just "slope at the origin" is that a polar curve may pass through the origin (or any point) more than once, with different slopes at different θ-values. This is not the case in our first example, however.

θ	$r = 1 - \cos\theta$
0	0
$\frac{\pi}{3}$	$\frac{1}{2}$
$\frac{\pi}{2}$	1
$\frac{2\pi}{3}$	$\frac{3}{2}$
π	2

(a)

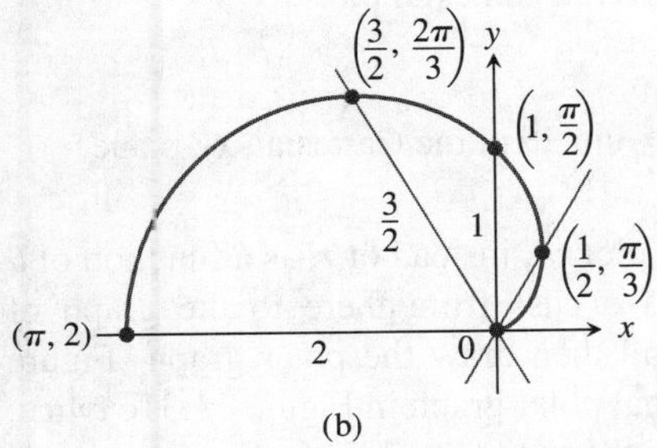

(b)

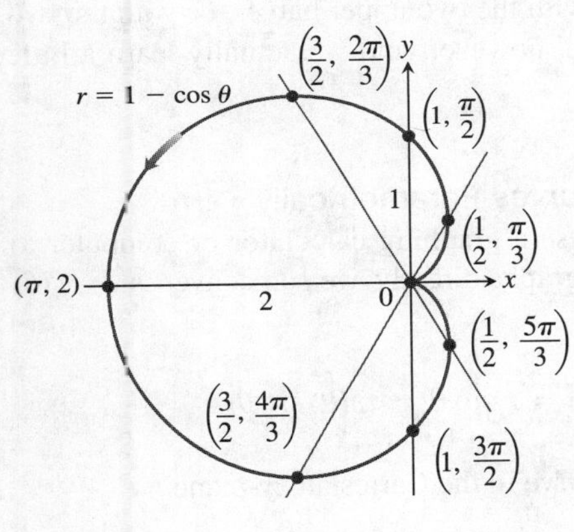

(c)

FIGURE 11.29 The steps in graphing the cardioid $r = 1 - \cos\theta$ (Example 1). The arrow shows the direction of increasing θ.

EXAMPLE 1 Graph the curve $r = 1 - \cos\theta$ in the Cartesian *xy*-plane.

Solution The curve is symmetric about the *x*-axis because

$$(r, \theta) \text{ on the graph} \Rightarrow r = 1 - \cos\theta$$
$$\Rightarrow r = 1 - \cos(-\theta) \qquad \cos\theta = \cos(-\theta)$$
$$\Rightarrow (r, -\theta) \text{ on the graph.}$$

As θ increases from 0 to π, $\cos\theta$ decreases from 1 to -1, and $r = 1 - \cos\theta$ increases from a minimum value of 0 to a maximum value of 2. As θ continues on from π to 2π, $\cos\theta$ increases from -1 back to 1, and r decreases from 2 back to 0. The curve starts to repeat when $\theta = 2\pi$ because the cosine has period 2π.

The curve leaves the origin with slope $\tan(0) = 0$ and returns to the origin with slope $\tan(2\pi) = 0$.

We make a table of values from $\theta = 0$ to $\theta = \pi$, plot the points, draw a smooth curve through them with a horizontal tangent line at the origin, and reflect the curve across the *x*-axis to complete the graph (Figure 11.29). The curve is called a *cardioid* because of its heart shape. ■

EXAMPLE 2 Graph the curve $r^2 = 4\cos\theta$ in the Cartesian *xy*-plane.

Solution The equation $r^2 = 4\cos\theta$ requires $\cos\theta \geq 0$, so we get the entire graph by running θ from $-\pi/2$ to $\pi/2$. The curve is symmetric about the *x*-axis because

$$(r, \theta) \text{ on the graph} \Rightarrow r^2 = 4\cos\theta$$
$$\Rightarrow r^2 = 4\cos(-\theta) \qquad \cos\theta = \cos(-\theta)$$
$$\Rightarrow (r, -\theta) \text{ on the graph.}$$

The curve is also symmetric about the origin because

$$(r, \theta) \text{ on the graph} \Rightarrow r^2 = 4\cos\theta$$
$$\Rightarrow (-r)^2 = 4\cos\theta$$
$$\Rightarrow (-r, \theta) \text{ on the graph.}$$

Together, these two symmetries imply symmetry about the *y*-axis.

The curve passes through the origin when $\theta = -\pi/2$ and $\theta = \pi/2$. It has a vertical tangent line both times because $\tan\theta$ is infinite.

For each value of θ in the interval between $-\pi/2$ and $\pi/2$, the formula $r^2 = 4\cos\theta$ gives two values of r:

$$r = \pm 2\sqrt{\cos\theta}.$$

We make a short table of values, plot the corresponding points, and use information about symmetry and tangent lines to guide us in connecting the points with a smooth curve (Figure 11.30).

θ	$\cos\theta$	$r = \pm 2\sqrt{\cos\theta}$
0	1	± 2
$\pm\frac{\pi}{6}$	$\frac{\sqrt{3}}{2}$	$\approx \pm 1.9$
$\pm\frac{\pi}{4}$	$\frac{1}{\sqrt{2}}$	$\approx \pm 1.7$
$\pm\frac{\pi}{3}$	$\frac{1}{2}$	$\approx \pm 1.4$
$\pm\frac{\pi}{2}$	0	0

(a)

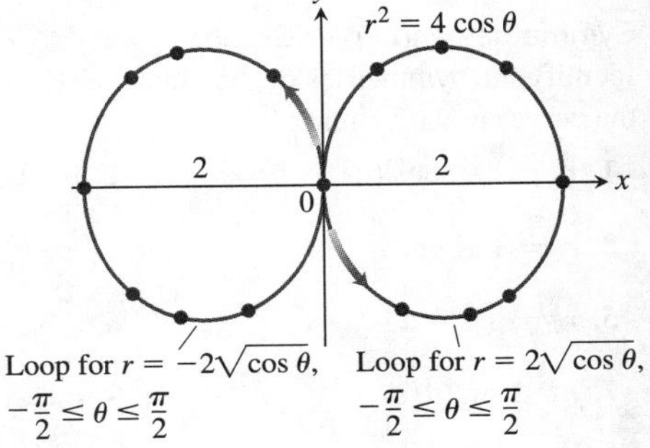

(b)

FIGURE 11.30 The graph of $r^2 = 4\cos\theta$. The arrows show the direction of increasing θ. The values of r in the table are rounded (Example 2). ■

Converting a Graph from the $r\theta$-Plane to the xy-Plane

One way to graph a polar equation $r = f(\theta)$ in the xy-plane is to make a table of (r, θ)-values, plot the corresponding points there, and connect them in order of increasing θ. This can work well if enough points have been plotted to reveal all the loops and dimples in the graph. Another method of graphing follows.

1. First graph the function $r = f(\theta)$ in the *Cartesian* $r\theta$-plane.
2. Then use that Cartesian graph as a "table" and guide to sketch the *polar* coordinate graph in the xy-plane.

This method is sometimes better than simple point plotting because the first Cartesian graph shows at a glance where r is positive, where negative, and where nonexistent, as well as where r is increasing and where it is decreasing. Here is an example.

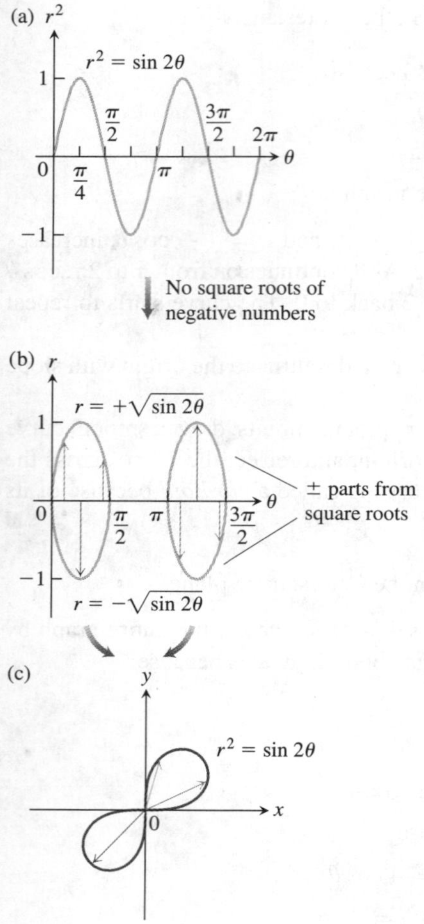

FIGURE 11.31 To plot $r = f(\theta)$ in the Cartesian $r\theta$-plane in (b), we first plot $r^2 = \sin 2\theta$ in the $r^2\theta$-plane in (a) and then ignore the values of θ for which $\sin 2\theta$ is negative. The radii from the sketch in (b) cover the polar graph of the lemniscate in (c) twice (Example 3).

EXAMPLE 3 Graph the *lemniscate* curve $r^2 = \sin 2\theta$ in the Cartesian xy-plane.

Solution For this example it will be easier to first plot r^2, instead of r, as a function of θ in the Cartesian $r^2\theta$-plane (see Figure 11.31a). We pass from there to the graph of $r = \pm\sqrt{\sin 2\theta}$ in the $r\theta$-plane (Figure 11.31b), and then draw the polar graph (Figure 11.31c). The graph in Figure 11.31b "covers" the final polar graph in Figure 11.31c twice. We could have managed with either loop alone, with the two upper halves, or with the two lower halves. The double covering does no harm, however, and we actually learn a little more about the behavior of the function this way. ■

USING TECHNOLOGY Graphing Polar Curves Parametrically
For complicated polar curves, we may need to use a graphing calculator or computer to graph the curve. If the device does not plot polar graphs directly, we can convert $r = f(\theta)$ into parametric form using the equations

$$x = r\cos\theta = f(\theta)\cos\theta, \qquad y = r\sin\theta = f(\theta)\sin\theta.$$

Then we use the device to draw a parametrized curve in the Cartesian xy-plane.

EXERCISES 11.4

Symmetries and Polar Graphs

Identify the symmetries of the curves in Exercises 1–12. Then sketch the curves in the xy-plane.

1. $r = 1 + \cos\theta$	**2.** $r = 1 - 2\cos\theta$
3. $r = 1 - \sin\theta$	**4.** $r = 1 + \sin\theta$
5. $r = 3 + \cos\theta$	**6.** $r = 2 + 3\sin\theta$
7. $r = 2\sin(\theta/2)$	**8.** $r = \cos(\theta/2)$
9. $r^2 = \cos\theta$	**10.** $r^2 = \sin\theta$
11. $r^2 = -\sin\theta$	**12.** $r^2 = -\cos\theta$

Graph the lemniscates in Exercises 13–16. What symmetries do these curves have?

13. $r = 3\cos 2\theta$	**14.** $r^2 = 4\sin 2\theta$
15. $r^2 = -\sin 2\theta$	**16.** $r^2 = -\cos 2\theta$

Slopes of Polar Curves in the xy-Plane

Find the slopes of the curves in Exercises 17–20 at the given points. Sketch the curves along with their tangent lines at these points.

17. Cardioid $r = -1 + \cos\theta$; $\theta = \pm\pi/2$

18. Cardioid $r = -1 + \sin\theta$; $\theta = 0, \pi$

19. Four-leaved rose $r = \sin 2\theta$; $\theta = \pm\pi/4, \pm 3\pi/4$

20. Four-leaved rose $r = \cos 2\theta$; $\theta = 0, \pm\pi/2, \pi$

Concavity of Polar Curves in the xy-Plane

Equation (1) gives the formula for the derivative y' of a polar curve $r = f(\theta)$. The second derivative is $\dfrac{d^2y}{dx^2} = \dfrac{dy'/d\theta}{dx/d\theta}$ (see Equation (2) in Section 11.2). Find the slope and concavity of the curves in Exercises 21–24 at the given points.

21. $r = \sin\theta, \quad \theta = \pi/6, \pi/3$ **22.** $r = e^{\theta}, \quad \theta = 0, \pi$

23. $r = \theta, \quad \theta = 0, \pi/2$ **24.** $r = 1/\theta, \quad \theta = -\pi, 1$

Graphing Limaçons

Graph the limaçons in Exercises 25–28. Limaçon ("*lee*-ma-sahn") is Old French for "snail." You will understand the name when you graph the limaçons in Exercise 25. Equations for limaçons have the form $r = a \pm b\cos\theta$ or $r = a \pm b\sin\theta$. There are four basic shapes.

25. Limaçons with an inner loop

a. $r = 1/3 + \cos\theta$ **b.** $r = \dfrac{1}{2} + \sin\theta$

26. Cardioids

a. $r = 1 - \cos\theta$ **b.** $r = -1 + \sin\theta$

27. Dimpled limaçons

a. $r = \dfrac{3}{2} + \cos\theta$ **b.** $r = 3/2 + \sin\theta$

28. Oval limaçons

a. $r = 2 + \cos\theta$ **b.** $r = -2 - \sin\theta$

Graphing Polar Regions and Curves in the xy-Plane

29. Sketch the region defined by the inequalities $-1 \le r \le 2$ and $-\pi/2 \le \theta \le \pi/2$.

30. Sketch the region defined by the inequalities $0 \le r \le 2\sec\theta$ and $-\pi/4 \le \theta \le \pi/4$.

In Exercises 31 and 32, sketch the region defined by the inequality.

31. $0 \le r \le 2 - 2\cos\theta$ **32.** $0 \le r^2 \le \cos\theta$

T **33.** Which of the following has the same graph as $r = 1 - \cos\theta$?

a. $r = -1 - \cos\theta$ **b.** $r = 1 + \cos\theta$

Confirm your answer with algebra.

T **34.** Which of the following has the same graph as $r = \cos 2\theta$?

a. $r = -\sin(2\theta + \pi/2)$ **b.** $r = -\cos(\theta/2)$

Confirm your answer with algebra.

T **35. A rose within a rose** Graph the equation $r = 1 - 2\cos 3\theta$

T **36. The nephroid of Freeth** Graph the nephroid of Freeth:

$$r = 1 + 2\sin\frac{\theta}{2}.$$

T **37. Roses** Graph the roses $r = \cos m\theta$ for $m = 1/3, 2, 3$, and 7.

T **38. Spirals** Polar coordinates are just the thing for defining spirals. Graph the following spirals.

a. $r = \theta$

b. $r = -\theta$

c. *A logarithmic spiral:* $r = e^{\theta/10}$

d. *A hyperbolic spiral:* $r = 8/\theta$

e. *An equilateral hyperbola:* $r = \pm 10/\sqrt{\theta}$

(Use different colors for the two branches.)

T **39.** Graph the equation $r = \sin\left(\frac{8}{7}\theta\right)$ for $0 \le \theta \le 14\pi$.

T **40.** Graph the equation

$$r = \sin^2(2.3\theta) + \cos^4(2.3\theta)$$

for $0 \le \theta \le 10\pi$.

11.5 Areas and Lengths in Polar Coordinates

This section shows how to calculate areas of plane regions and lengths of curves in polar coordinates.

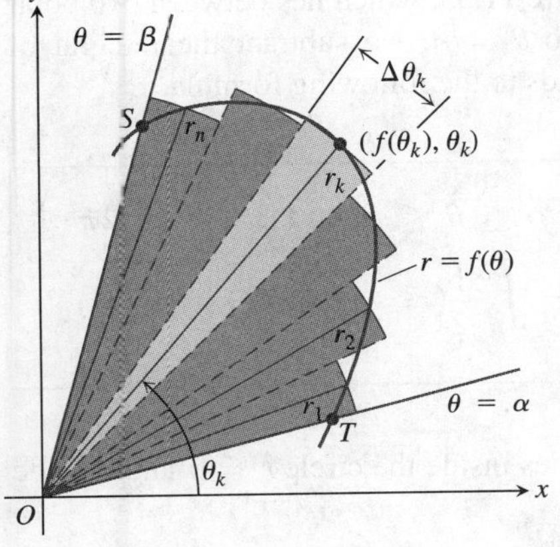

FIGURE 11.32 To derive a formula for the area of region OTS, we approximate the region with fan-shaped circular sectors.

Area in the Plane

The region OTS in Figure 11.32 is bounded by the rays $\theta = \alpha$ and $\theta = \beta$ and the curve $r = f(\theta)$. We approximate the region with n nonoverlapping fan-shaped circular sectors based on a partition P of angle TOS. The typical sector has radius $r_k = f(\theta_k)$ and central angle of radian measure $\Delta\theta_k$. Its area is $\Delta\theta_k/2\pi$ times the area of a circle of radius r_k, or

$$A_k = \frac{1}{2}{r_k}^2\,\Delta\theta_k = \frac{1}{2}\left(f(\theta_k)\right)^2\,\Delta\theta_k.$$

The area of region OTS is approximately

$$\sum_{k=1}^{n} A_k = \sum_{k=1}^{n} \frac{1}{2}\left(f(\theta_k)\right)^2\,\Delta\theta_k.$$

If f is continuous, we expect the approximations to improve as the norm of the partition P goes to zero, where the norm of P is the largest value of $\Delta\theta_k$. We are therefore led to the following formula for the region's area.

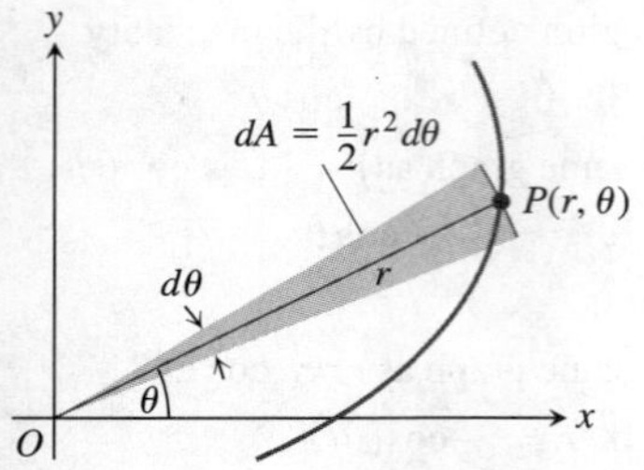

FIGURE 11.33 The area differential dA for the curve $r = f(\theta)$.

$$A = \lim_{\|P\|\to 0} \sum_{k=1}^{n} \frac{1}{2}(f(\theta_k))^2\,\Delta\theta_k = \int_{\alpha}^{\beta} \frac{1}{2}(f(\theta))^2\,d\theta.$$

Area of the Fan-Shaped Region Between the Origin and the Curve $r = f(\theta)$ when $\alpha \le \theta \le \beta$, $r \ge 0$, and $\beta - \alpha \le 2\pi$

$$A = \int_{\alpha}^{\beta} \frac{1}{2} r^2\,d\theta$$

This is the integral of the **area differential** (Figure 11.33)

$$dA = \frac{1}{2} r^2\,d\theta = \frac{1}{2}(f(\theta))^2\,d\theta.$$

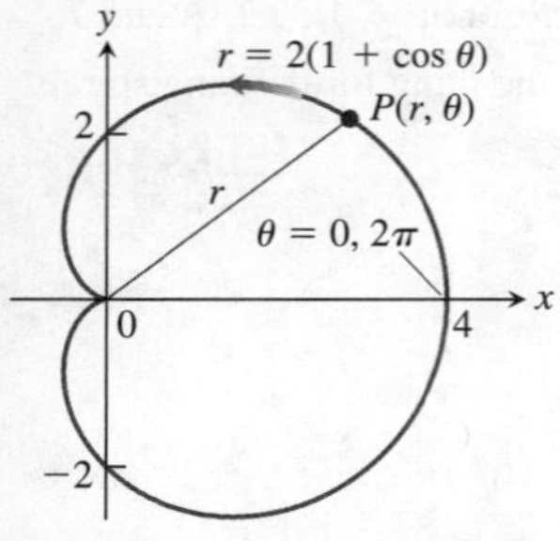

FIGURE 11.34 The cardioid in Example 1.

In the area formula above, we assumed that $r \ge 0$ and that the region does not sweep out an angle of more than 2π. This avoids issues with negatively signed areas or with regions that overlap themselves. More general regions can usually be handled by subdividing them into regions of this type if necessary.

EXAMPLE 1 Find the area of the region in the xy-plane enclosed by the cardioid $r = 2(1 + \cos\theta)$.

Solution We graph the cardioid (Figure 11.34) and determine that the radius OP sweeps out the region exactly once as θ runs from 0 to 2π. The area is therefore

$$\begin{aligned}
\int_{\theta=0}^{\theta=2\pi} \frac{1}{2} r^2\,d\theta &= \int_0^{2\pi} \frac{1}{2}\cdot 4(1+\cos\theta)^2\,d\theta \\
&= \int_0^{2\pi} 2(1 + 2\cos\theta + \cos^2\theta)\,d\theta \\
&= \int_0^{2\pi} \left(2 + 4\cos\theta + 2\cdot\frac{1+\cos 2\theta}{2}\right) d\theta \\
&= \int_0^{2\pi} (3 + 4\cos\theta + \cos 2\theta)\,d\theta \\
&= \left[3\theta + 4\sin\theta + \frac{\sin 2\theta}{2}\right]_0^{2\pi} = 6\pi - 0 = 6\pi.
\end{aligned}$$ ■

To find the area of a region like the one in Figure 11.35, which lies between two polar curves $r_1 = r_1(\theta)$ and $r_2 = r_2(\theta)$ from $\theta = \alpha$ to $\theta = \beta$, we subtract the integral of $(1/2)r_1^2\,d\theta$ from the integral of $(1/2)r_2^2\,d\theta$. This leads to the following formula.

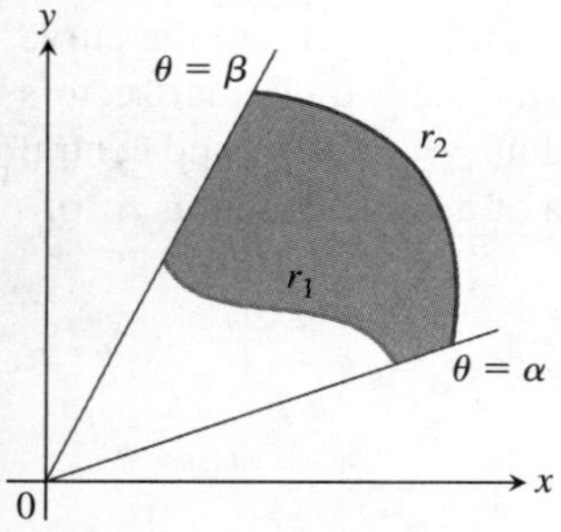

FIGURE 11.35 The area of the shaded region is calculated by subtracting the area of the region between r_1 and the origin from the area of the region between r_2 and the origin.

Area of the Region $0 \le r_1(\theta) \le r \le r_2(\theta)$, $\alpha \le \theta \le \beta$, and $\beta - \alpha \le 2\pi$

$$A = \int_{\alpha}^{\beta} \frac{1}{2} r_2^2\,d\theta - \int_{\alpha}^{\beta} \frac{1}{2} r_1^2\,d\theta = \int_{\alpha}^{\beta} \frac{1}{2}(r_2^2 - r_1^2)\,d\theta \qquad (1)$$

EXAMPLE 2 Find the area of the region that lies inside the circle $r = 1$ and outside the cardioid $r = 1 - \cos\theta$.

Solution We sketch the region to determine its boundaries and find the limits of integration (Figure 11.36). The outer curve is $r_2 = 1$, the inner curve is $r_1 = 1 - \cos\theta$, and θ runs from $-\pi/2$ to $\pi/2$. The area, from Equation (1), is

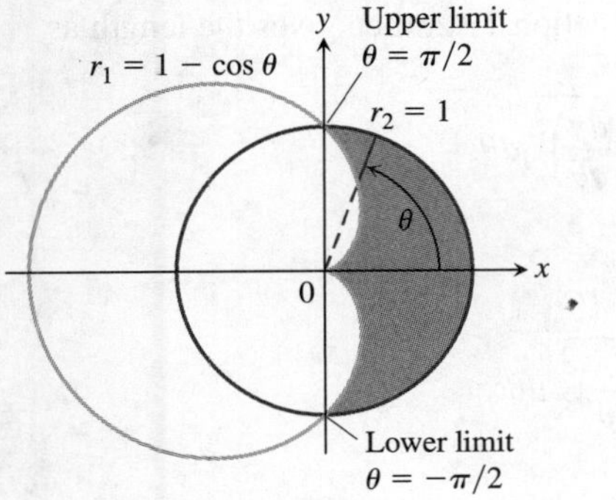

FIGURE 11.36 The region and limits of integration in Example 2.

$$
\begin{aligned}
A &= \int_{-\pi/2}^{\pi/2} \frac{1}{2}(r_2^2 - r_1^2)\,d\theta && \text{Eq. (1)}\\
&= 2\int_{0}^{\pi/2} \frac{1}{2}(r_2^2 - r_1^2)\,d\theta && \text{Symmetry}\\
&= \int_{0}^{\pi/2} (1 - (1 - 2\cos\theta + \cos^2\theta))\,d\theta && r_2 = 1 \text{ and } r_1 = 1 - \cos\theta\\
&= \int_{0}^{\pi/2} (2\cos\theta - \cos^2\theta)\,d\theta = \int_{0}^{\pi/2}\left(2\cos\theta - \frac{1+\cos 2\theta}{2}\right)d\theta\\
&= \left[2\sin\theta - \frac{\theta}{2} - \frac{\sin 2\theta}{4}\right]_0^{\pi/2} = 2 - \frac{\pi}{4}.
\end{aligned}
$$
■

The fact that we can represent a point in different ways in polar coordinates requires that we take extra care in deciding when a point lies on the graph of a polar equation and in determining the points at which polar graphs intersect. (We needed intersection points in Example 2.) In Cartesian coordinates, we can always find the points where two curves cross by solving their equations simultaneously. In polar coordinates, the story is different. Simultaneous solution may reveal some intersection points without revealing others, so it is sometimes difficult to find all points of intersection of two polar curves. One way to identify all the points of intersection is to graph the equations.

EXAMPLE 3 Find all of the points where the curve $r = 2\cos(\theta/3)$ intersects the circle of radius $\sqrt{2}$ centered at the origin.

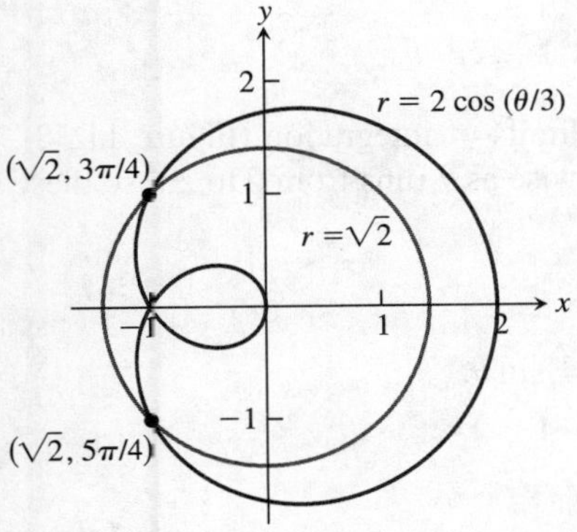

FIGURE 11.37 The curves $r = 2\cos(\theta/3)$ and $r = \sqrt{2}$ intersect at two points (Example 3).

Solution Note that the function $r = 2\cos(\theta/3)$ takes both positive and negative values. Therefore, when we look for the points where this curve intersects the circle, it is important to take into account that the circle is described both by the equation $r = \sqrt{2}$ *and* by the equation $r = -\sqrt{2}$.

Solving $2\cos(\theta/3) = \sqrt{2}$ for θ yields

$$2\cos(\theta/3) = \sqrt{2}, \quad \cos(\theta/3) = \sqrt{2}/2, \quad \theta/3 = \pi/4, \quad \theta = 3\pi/4.$$

This gives us one point, $(\sqrt{2}, 3\pi/4)$, where the two curves intersect. However, as we can see by looking at the graphs in Figure 11.37, there is a second intersection point. To find the second point, we solve $2\cos(\theta/3) = -\sqrt{2}$ for θ:

$$2\cos(\theta/3) = -\sqrt{2}, \quad \cos(\theta/3) = -\sqrt{2}/2, \quad \theta/3 = 3\pi/4, \quad \theta = 9\pi/4.$$

The second intersection point is located at $(-\sqrt{2}, 9\pi/4)$. We can specify this point in polar coordinates using a positive value of r and an angle between 0 and 2π. In polar coordinates, adding multiples of 2π to θ gives a second description of the same point in the plane. Similarly, changing the sign of r, while at the same time adding or subtracting π to or from θ, also gives a description of the same point. So in polar coordinates, $(-\sqrt{2}, 9\pi/4)$ describes the same point in the plane as $(-\sqrt{2}, \pi/4)$ and also as $(\sqrt{2}, 5\pi/4)$. The second intersection point is located at $(\sqrt{2}, 5\pi/4)$. ■

Length of a Polar Curve

We can obtain a polar coordinate formula for the length of a curve $r = f(\theta)$, $\alpha \le \theta \le \beta$, by parametrizing the curve as

$$x = r\cos\theta = f(\theta)\cos\theta, \qquad y = r\sin\theta = f(\theta)\sin\theta, \qquad \alpha \le \theta \le \beta. \qquad (2)$$

The parametric length formula, Equation (3) from Section 11.2, then gives the length as

$$L = \int_{\alpha}^{\beta} \sqrt{\left(\frac{dx}{d\theta}\right)^2 + \left(\frac{dy}{d\theta}\right)^2}\, d\theta.$$

This equation becomes

$$L = \int_{\alpha}^{\beta} \sqrt{r^2 + \left(\frac{dr}{d\theta}\right)^2}\, d\theta$$

when Equations (2) are substituted for x and y (Exercise 29).

Length of a Polar Curve
If $r = f(\theta)$ has a continuous first derivative for $\alpha \le \theta \le \beta$, and if the point $P(r, \theta)$ traces the curve $r = f(\theta)$ exactly once as θ runs from α to β, then the length of the curve is

$$L = \int_{\alpha}^{\beta} \sqrt{r^2 + \left(\frac{dr}{d\theta}\right)^2}\, d\theta. \tag{3}$$

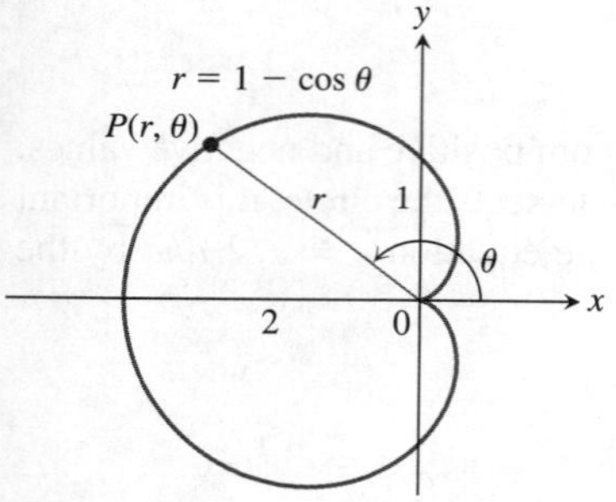

FIGURE 11.38 Calculating the length of a cardioid (Example 4).

EXAMPLE 4 Find the length of the cardioid $r = 1 - \cos\theta$.

Solution We sketch the cardioid to determine the limits of integration (Figure 11.38). The point $P(r, \theta)$ traces the curve once, counterclockwise as θ runs from 0 to 2π, so these are the values we take for α and β.

With

$$r = 1 - \cos\theta, \qquad \frac{dr}{d\theta} = \sin\theta,$$

we have

$$\begin{aligned} r^2 + \left(\frac{dr}{d\theta}\right)^2 &= (1 - \cos\theta)^2 + (\sin\theta)^2 \\ &= 1 - 2\cos\theta + \underbrace{\cos^2\theta + \sin^2\theta}_{1} = 2 - 2\cos\theta \end{aligned}$$

and

$$\begin{aligned} L &= \int_{\alpha}^{\beta} \sqrt{r^2 + \left(\frac{dr}{d\theta}\right)^2}\, d\theta = \int_0^{2\pi} \sqrt{2 - 2\cos\theta}\, d\theta \\ &= \int_0^{2\pi} \sqrt{4\sin^2\frac{\theta}{2}}\, d\theta \qquad 1 - \cos\theta = 2\sin^2(\theta/2) \\ &= \int_0^{2\pi} 2\left|\sin\frac{\theta}{2}\right| d\theta \\ &= \int_0^{2\pi} 2\sin\frac{\theta}{2}\, d\theta \qquad \sin(\theta/2) \ge 0 \quad \text{for} \quad 0 \le \theta \le 2\pi \\ &= \left[-4\cos\frac{\theta}{2}\right]_0^{2\pi} = 4 + 4 = 8. \end{aligned}$$

■

EXERCISES 11.5

Finding Polar Areas

Find the areas of the regions in Exercises 1–8.

1. Bounded by the spiral $r = \theta$ for $0 \le \theta \le \pi$

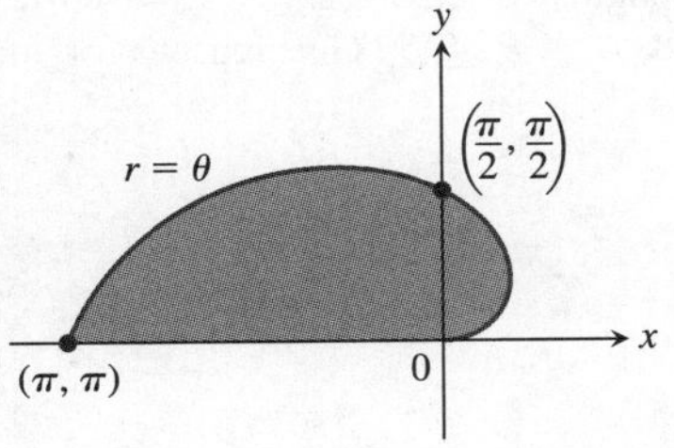

2. Bounded by the circle $r = 2 \sin\theta$ for $\pi/4 \le \theta \le \pi/2$

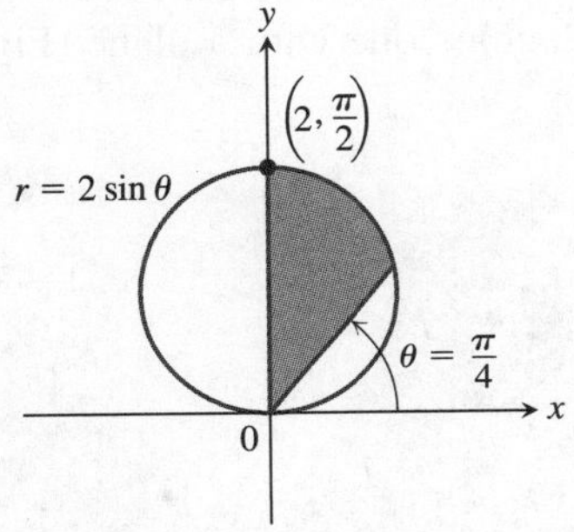

3. Inside the oval limacons $r = 3 + 2\cos\theta$

4. Inside the cardioid $r = a(1 + \cos\theta)$, $a > 0$

5. Inside one leaf of the four-leaved rose $r = \sin 3\theta$

6. Inside one leaf of the three-leaved rose $r = \cos 3\theta$

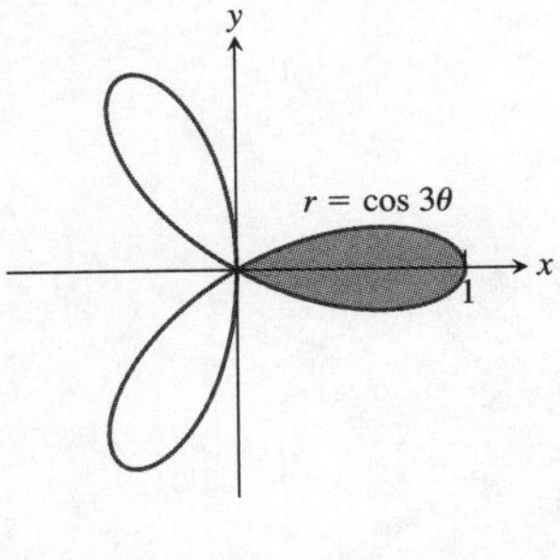

7. Inside one loop of the lemniscate $r^2 = 2 \sin 3\theta$

8. Inside the six-leaved rose $r^2 = 2 \sin 3\theta$

Find the areas of the regions in Exercises 9–18.

9. Shared by the circles $r = 2\cos\theta$ and $r = 2\sin\theta$

10. Shared by the circles $r = 1$ and $r = 2\sin\theta$

11. Shared by the circle $r = 2$ and the cardioid $r = 2(1 - \cos\theta)$

12. Shared by the cardioids $r = 2(1 + \cos\theta)$ and $r = 2(1 - \cos\theta)$

13. Inside the lemniscate $r^2 = 6\cos 2\theta$ and outside the circle $r = \sqrt{3}$

14. Inside the circle $r = 3a\cos\theta$ and outside the cardioid $r = a(1 + \cos\theta)$, $a > 0$

15. Inside the circle $r = -2\cos\theta$ and outside the circle $r = 1$

16. Inside the circle $r = 6$ and above the line $r = 3\csc\theta$

17. Inside the circle $r = 4\cos\theta$ and to the right of the vertical line $r = \sec\theta$

18. Inside the circle $r = 4\sin\theta$ and below the horizontal line $r = 3\csc\theta$

19. a. Find the area of the shaded region in the accompanying figure.

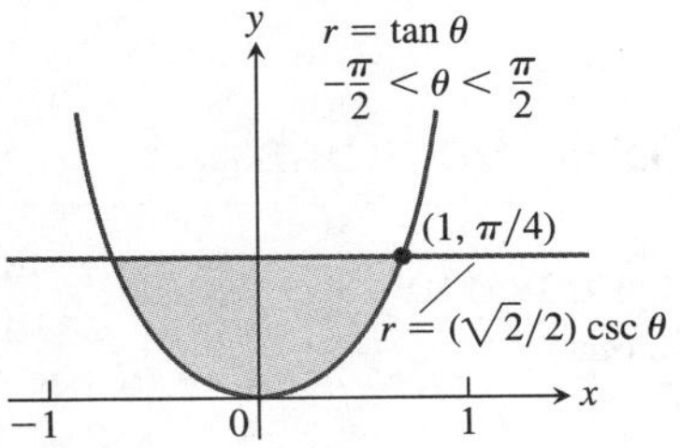

b. It looks as if the graph of $r = \tan\theta$, $-\pi/2 < \theta < \pi/2$, could be asymptotic to the lines $x = 1$ and $x = -1$. Is it? Give reasons for your answer.

20. The area of the region that lies inside the cardioid curve $r = \cos\theta + 1$ and outside the circle $r = \cos\theta$ is not

$$\frac{1}{2}\int_0^{2\pi} \left[(\cos\theta + 1)^2 - \cos^2\theta\right] d\theta = \pi.$$

Why not? What *is* the area? Give reasons for your answers.

Finding Lengths of Polar Curves

Find the lengths of the curves in Exercises 21–28.

21. The spiral $r = \theta^2$, $0 \le \theta \le \sqrt{5}$

22. The spiral $r = e^{\theta}\sqrt{5}$, $0 \le \theta \le \pi$

23. The cardioid $r = 1 + \cos\theta$

24. The curve $r = a\sin^2(\theta/2)$, $0 \le \theta \le \pi$, $a > 0$

25. The parabolic segment $r = 4/(1 + \cos\theta)$, $0 \le \theta \le \pi/2$

26. The parabolic segment $r = 2/(1 - \cos\theta)$, $\pi/2 \le \theta \le \pi$

27. The curve $r = \cos^3(\theta/3)$, $0 \le \theta \le \pi/4$

28. The curve $r = \sqrt{1 + \sin 3\theta}$, $0 \le \theta \le \pi\sqrt{2}$

29. The length of the curve $r = f(\theta)$, $\alpha \le \theta \le \beta$ Assuming that the necessary derivatives are continuous, show how the substitutions

$$x = f(\theta)\cos\theta, \quad y = f(\theta)\sin\theta$$

(Equations 2 in the text) transform

$$L = \int_\alpha^\beta \sqrt{\left(\frac{dx}{d\theta}\right)^2 + \left(\frac{dy}{d\theta}\right)^2}\, d\theta$$

into

$$L = \int_\alpha^\beta \sqrt{r^2 + \left(\frac{dr}{d\theta}\right)^2}\, d\theta.$$

30. Circumferences of circles As usual, when faced with a new formula, it is a good idea to try it on familiar objects to be sure it gives results consistent with past experience. Use the length formula in Equation (3) to calculate the circumferences of the following circles $(a > 0)$.

a. $r = a$ **b.** $r = a\cos\theta$ **c.** $r = a\sin\theta$

Theory and Examples

31. Average value If f is continuous, the average value of the polar coordinate r over the curve $r = f(\theta),\ \alpha \le \theta \le \beta$, with respect to θ is given by the formula

$$r_{\text{av}} = \frac{1}{\beta - \alpha}\int_{\alpha}^{\beta} f(\theta)\, d\theta.$$

Use this formula to find the average value of r with respect to θ over the following curves $(a > 0)$.

a. The cardioid $r = a(1 - \sin\theta)$

b. The circle $r = a$

c. The circle $r = a\cos\theta,\quad -\pi/2 \le \theta \le \pi/2$

32. $\boldsymbol{r = f(\theta)}$ **vs.** $\boldsymbol{r = 2f(\theta)}$ Can anything be said about the relative lengths of the curves $r = f(\theta),\ \alpha \le \theta \le \beta$, and $r = 2f(\theta),\ \alpha \le \theta \le \beta$? Give reasons for your answer.

11.6 Conic Sections

HISTORICAL BIOGRAPHY
Gregory St. Vincent
(1584–1667)
bit.ly/2xLhZ55

In this section we define and review parabolas, ellipses, and hyperbolas geometrically and derive their standard Cartesian equations. These curves are called *conic sections* or *conics* because they are formed by cutting a double cone with a plane (Figure 11.39). This

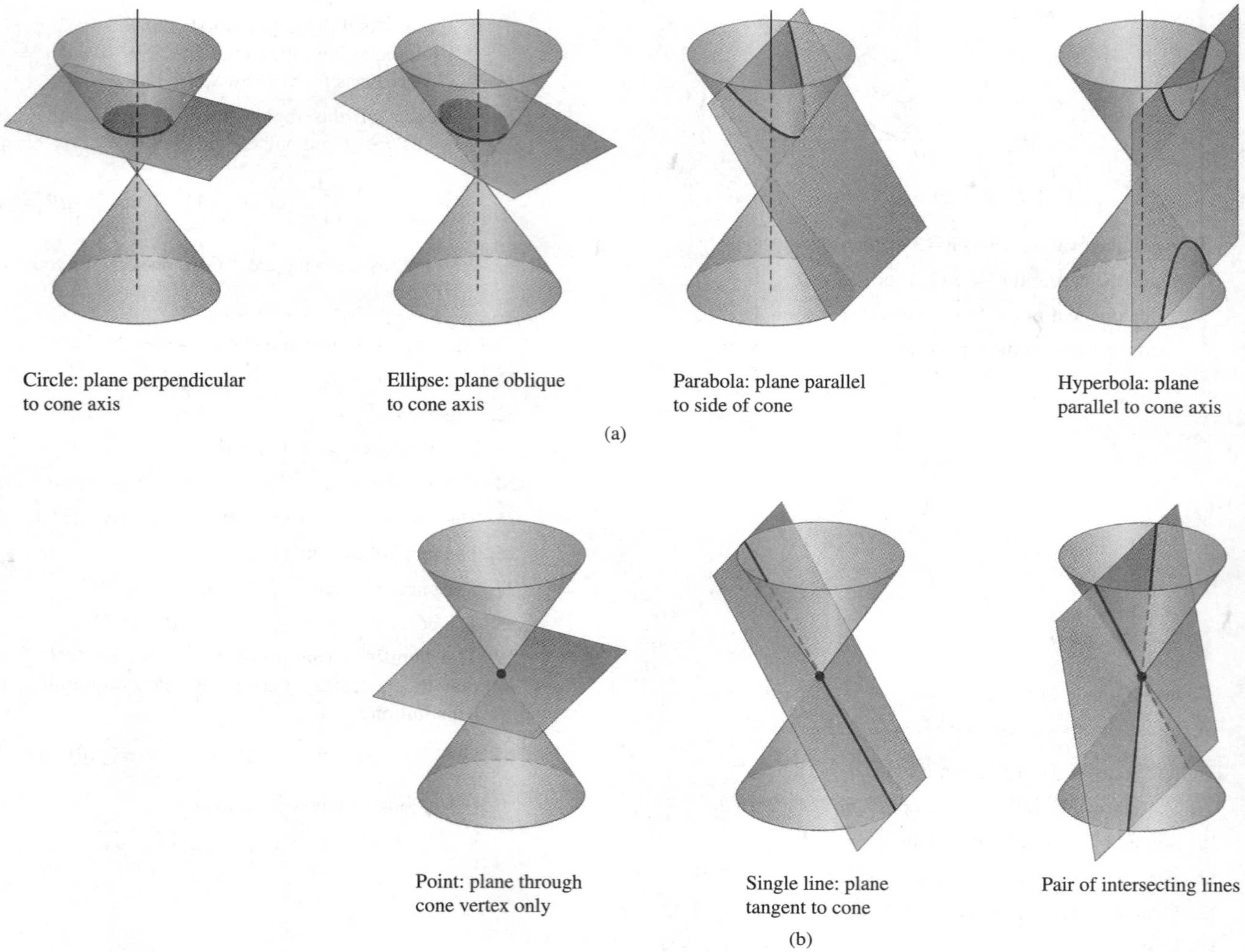

FIGURE 11.39 The standard conic sections (a) are the curves in which a plane cuts a *double* cone. Hyperbolas come in two parts, called *branches*. The point and lines obtained by passing the plane through the cone's vertex (b) are *degenerate* conic sections.

geometric method was the only way that conic sections could be described by Greek mathematicians, since they did not have our tools of Cartesian or polar coordinates. In the next section we express the conics in polar coordinates.

Parabolas

> **DEFINITIONS** A set that consists of all the points in a plane equidistant from a given fixed point and a given fixed line in the plane is a **parabola**. The fixed point is the **focus** of the parabola. The fixed line is the **directrix**.

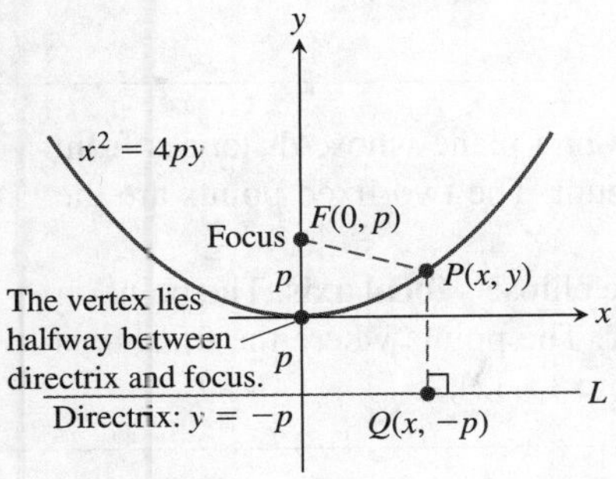

FIGURE 11.40 The standard form of the parabola $x^2 = 4py,\ p > 0$.

If the focus F lies on the directrix L, the parabola is the line through F perpendicular to L. We consider this to be a degenerate case and assume henceforth that F does not lie on L.

A parabola has its simplest equation when its focus and directrix straddle one of the coordinate axes. For example, suppose that the focus lies at the point $F(0, p)$ on the positive y-axis and that the directrix is the line $y = -p$ (Figure 11.40). In the notation of the figure, a point $P(x, y)$ lies on the parabola if and only if $PF = PQ$. From the distance formula,

$$PF = \sqrt{(x-0)^2 + (y-p)^2} = \sqrt{x^2 + (y-p)^2}$$

$$PQ = \sqrt{(x-x)^2 + (y-(-p))^2} = \sqrt{(y+p)^2}.$$

When we equate these expressions, square, and simplify, we get

$$y = \frac{x^2}{4p} \quad \text{or} \quad x^2 = 4py. \qquad \text{Standard form} \tag{1}$$

These equations reveal the parabola's symmetry about the y-axis. We call the y-axis the axis of the parabola (short for "axis of symmetry").

The point where a parabola crosses its axis is the **vertex**. The vertex of the parabola $x^2 = 4py$ lies at the origin (Figure 11.40). The positive number p is the parabola's **focal length**.

If the parabola opens downward, with its focus at $(0, -p)$ and its directrix the line $y = p$, then Equations (1) become

$$y = -\frac{x^2}{4p} \quad \text{and} \quad x^2 = -4py.$$

By interchanging the variables x and y, we obtain similar equations for parabolas opening to the right or to the left (Figure 11.41).

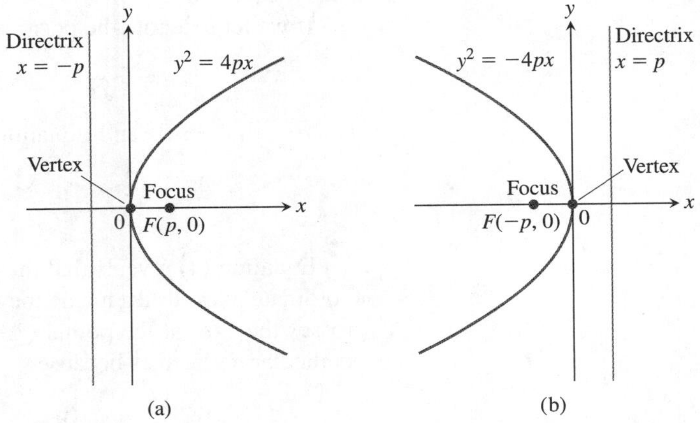

FIGURE 11.41 (a) The parabola $y^2 = 4px$. (b) The parabola $y^2 = -4px$.

EXAMPLE 1 Find the focus and directrix of the parabola $y^2 = 10x$.

Solution We find the value of p in the standard equation $y^2 = 4px$:

$$4p = 10, \quad \text{so} \quad p = \frac{10}{4} = \frac{5}{2}.$$

Then we find the focus and directrix for this value of p:

$$\text{Focus:} \quad (p, 0) = \left(\frac{5}{2}, 0\right)$$

$$\text{Directrix:} \quad x = -p \quad \text{or} \quad x = -\frac{5}{2}.$$

■

Ellipses

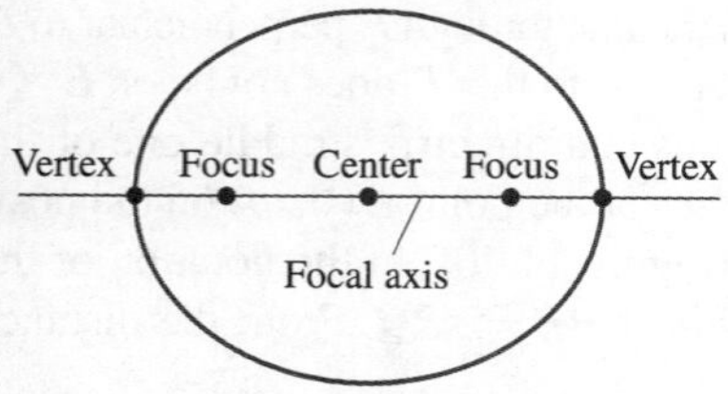

FIGURE 11.42 Points on the focal axis of an ellipse.

DEFINITIONS An **ellipse** is the set of points in a plane whose distances from two fixed points in the plane have a constant sum. The two fixed points are the foci of the ellipse.

The line through the foci of an ellipse is the ellipse's **focal axis**. The point on the axis halfway between the foci is the **center**. The points where the focal axis and ellipse cross are the ellipse's **vertices** (Figure 11.42).

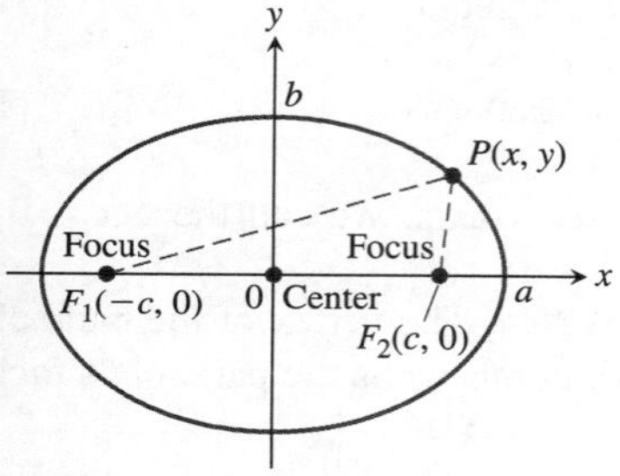

FIGURE 11.43 The ellipse defined by the equation $PF_1 + PF_2 = 2a$ is the graph of the equation $(x^2/a^2) + (y^2/b^2) = 1$, where $b^2 = a^2 - c^2$.

If the foci are $F_1(-c, 0)$ and $F_2(c, 0)$ (Figure 11.43), and $PF_1 + PF_2$ is denoted by $2a$, then the coordinates of a point P on the ellipse satisfy the equation

$$\sqrt{(x + c)^2 + y^2} + \sqrt{(x - c)^2 + y^2} = 2a.$$

To simplify this equation, we move the second radical to the right-hand side, square, isolate the remaining radical, and square again, obtaining

$$\frac{x^2}{a^2} + \frac{y^2}{a^2 - c^2} = 1. \tag{2}$$

Since $PF_1 + PF_2$ is greater than the length F_1F_2 (by the triangle inequality for triangle PF_1F_2), the number $2a$ is greater than $2c$. Accordingly, $a > c$ and the number $a^2 - c^2$ in Equation (2) is positive.

The algebraic steps leading to Equation (2) can be reversed to show that every point P whose coordinates satisfy an equation of this form with $0 < c < a$ also satisfies the equation $PF_1 + PF_2 = 2a$. A point therefore lies on the ellipse if and only if its coordinates satisfy Equation (2).

If we let b denote the positive square root of $a^2 - c^2$,

$$b = \sqrt{a^2 - c^2}, \tag{3}$$

then $a^2 - c^2 = b^2$ and Equation (2) takes the form

$$\frac{x^2}{a^2} + \frac{y^2}{b^2} = 1. \tag{4}$$

Equation (4) reveals that this ellipse is symmetric with respect to the origin and both coordinate axes. It lies inside the rectangle bounded by the lines $x = \pm a$ and $y = \pm b$. It crosses the axes at the points $(\pm a, 0)$ and $(0, \pm b)$. The tangents at these points are perpendicular to the axes because

$$\frac{dy}{dx} = -\frac{b^2x}{a^2y}, \quad \text{Obtained from Eq. (4) by implicit differentiation}$$

which is zero if $x = 0$ and infinite if $y = 0$.

The **major axis** of the ellipse in Equation (4) is the line segment of length $2a$ joining the points $(\pm a, 0)$. The **minor axis** is the line segment of length $2b$ joining the points $(0, \pm b)$. The number a itself is the **semimajor axis**, the number b the **semiminor axis**. The number c, found from Equation (3) as

$$c = \sqrt{a^2 - b^2},$$

is the **center-to-focus distance** of the ellipse. If $a = b$ then the ellipse is a circle.

EXAMPLE 2 The ellipse

$$\frac{x^2}{16} + \frac{y^2}{9} = 1 \tag{5}$$

shown in Figure 11.44 has

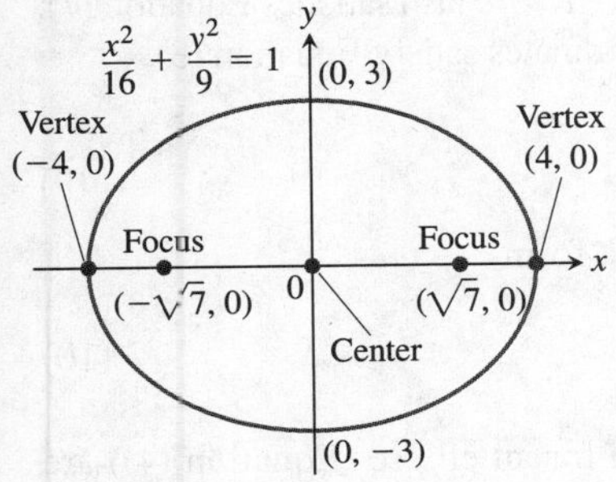

FIGURE 11.44 An ellipse with its major axis horizontal (Example 2).

Semimajor axis: $a = \sqrt{16} = 4$, Semiminor axis: $b = \sqrt{9} = 3$,

Center-to-focus distance: $c = \sqrt{16 - 9} = \sqrt{7}$,

Foci: $(\pm c, 0) = (\pm\sqrt{7}, 0)$,

Vertices: $(\pm a, 0) = (\pm 4, 0)$,

Center: $(0, 0)$. ■

If we interchange x and y in Equation (5), we have the equation

$$\frac{x^2}{9} + \frac{y^2}{16} = 1. \tag{6}$$

The major axis of this ellipse is now vertical instead of horizontal, with the foci and vertices on the y-axis. We can determine which way the major axis runs simply by finding the intercepts of the ellipse with the coordinate axes. The longer of the two axes of the ellipse is the major axis.

Standard-Form Equations for Ellipses Centered at the Origin

Foci on the x-axis: $\frac{x^2}{a^2} + \frac{y^2}{b^2} = 1 \quad (a > b)$

Center-to-focus distance: $c = \sqrt{a^2 - b^2}$

Foci: $(\pm c, 0)$

Vertices: $(\pm a, 0)$

Foci on the y-axis: $\frac{x^2}{b^2} + \frac{y^2}{a^2} = 1 \quad (a > b)$

Center-to-focus distance: $c = \sqrt{a^2 - b^2}$

Foci: $(0, \pm c)$

Vertices: $(0, \pm a)$

In each case, a is the semimajor axis and b is the semiminor axis.

Hyperbolas

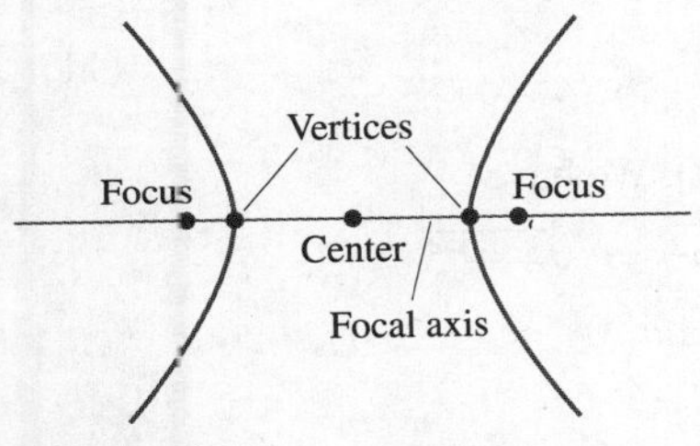

FIGURE 11.45 Points on the focal axis of a hyperbola.

DEFINITIONS A **hyperbola** is the set of points in a plane whose distances from two fixed points in the plane have a constant difference. The two fixed points are the foci of the hyperbola.

The line through the foci of a hyperbola is the **focal axis**. The point on the axis halfway between the foci is the hyperbola's **center**. The points where the focal axis and hyperbola cross are the **vertices** (Figure 11.45).

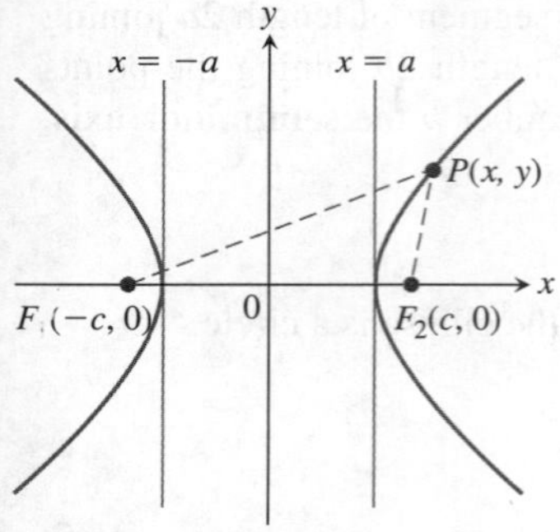

FIGURE 11.46 Hyperbolas have two branches. For points on the right-hand branch of the hyperbola shown here, $PF_1 - PF_2 = 2a$. For points on the left-hand branch, $PF_2 - PF_1 = 2a$. We then let $b = \sqrt{c^2 - a^2}$.

If the foci are $F_1(-c, 0)$ and $F_2(c, 0)$ (Figure 11.46) and the constant difference is $2a$, then a point (x, y) lies on the hyperbola if and only if

$$\sqrt{(x+c)^2 + y^2} - \sqrt{(x-c)^2 + y^2} = \pm 2a. \tag{7}$$

To simplify this equation, we move the second radical to the right-hand side, square, isolate the remaining radical, and square again, obtaining

$$\frac{x^2}{a^2} + \frac{y^2}{a^2 - c^2} = 1. \tag{8}$$

So far, this looks just like the equation for an ellipse. But now $a^2 - c^2$ is negative because $2a$, being the difference of two sides of triangle PF_1F_2, is less than $2c$, the third side.

The algebraic steps leading to Equation (8) can be reversed to show that every point P whose coordinates satisfy an equation of this form with $0 < a < c$ also satisfies Equation (7). A point therefore lies on the hyperbola if and only if its coordinates satisfy Equation (8).

If we let b denote the positive square root of $c^2 - a^2$,

$$b = \sqrt{c^2 - a^2}, \tag{9}$$

then $a^2 - c^2 = -b^2$ and Equation (8) takes the compact form

$$\frac{x^2}{a^2} - \frac{y^2}{b^2} = 1. \tag{10}$$

The differences between Equation (10) and the equation for an ellipse (Equation (4)) are the minus sign and the new relation

$$c^2 = a^2 + b^2. \qquad \text{From Eq. (9)}$$

Like the ellipse, the hyperbola is symmetric with respect to the origin and coordinate axes. It crosses the x-axis at the points $(\pm a, 0)$. The tangents at these points are vertical because

$$\frac{dy}{dx} = \frac{b^2 x}{a^2 y} \qquad \text{Obtained from Eq. (10) by implicit differentiation}$$

and this is infinite when $y = 0$. The hyperbola has no y-intercepts; in fact, no part of the curve lies between the lines $x = -a$ and $x = a$.

The lines

$$y = \pm \frac{b}{a} x$$

are the two **asymptotes** of the hyperbola defined by Equation (10). The fastest way to find the equations of the asymptotes is to replace the 1 in Equation (10) by 0 and solve the new equation for y:

$$\underbrace{\frac{x^2}{a^2} - \frac{y^2}{b^2} = 1}_{\text{hyperbola}} \rightarrow \underbrace{\frac{x^2}{a^2} - \frac{y^2}{b^2} = 0}_{\text{0 for 1}} \rightarrow \underbrace{y = \pm \frac{b}{a} x}_{\text{asymptotes}}.$$

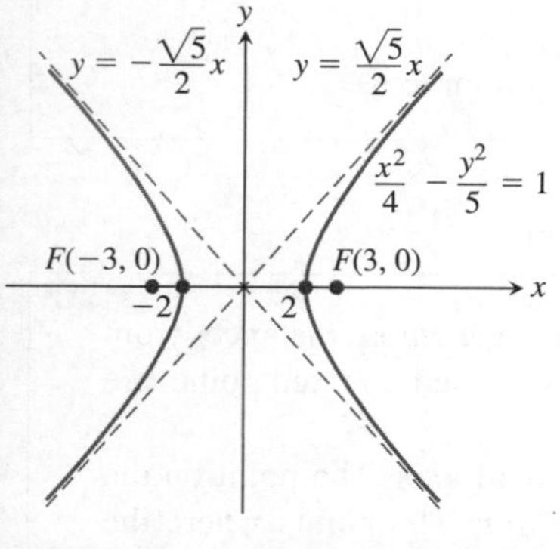

FIGURE 11.47 The hyperbola and its asymptotes in Example 3.

EXAMPLE 3 The equation

$$\frac{x^2}{4} - \frac{y^2}{5} = 1 \tag{11}$$

is Equation (10) with $a^2 = 4$ and $b^2 = 5$ (Figure 11.47). We have

Center-to-focus distance: $c = \sqrt{a^2 + b^2} = \sqrt{4 + 5} = 3,$

Foci: $(\pm c, 0) = (\pm 3, 0)$, Vertices: $(\pm a, 0) = (\pm 2, 0)$,

Center: $(0, 0)$,

Asymptotes: $\frac{x^2}{4} - \frac{y^2}{5} = 0$ or $y = \pm \frac{\sqrt{5}}{2} x.$ ■

If we interchange x and y in Equation (11), the foci and vertices of the resulting hyperbola will lie along the y-axis. We still find the asymptotes in the same way as before, but now their equations will be $y = \pm 2x/\sqrt{5}$.

Standard-Form Equations for Hyperbolas Centered at the Origin

Foci on the x-axis: $\dfrac{x^2}{a^2} - \dfrac{y^2}{b^2} = 1$

Center-to-focus distance: $c = \sqrt{a^2 + b^2}$

Foci: $(\pm c, 0)$

Vertices: $(\pm a, 0)$

Asymptotes: $\dfrac{x^2}{a^2} - \dfrac{y^2}{b^2} = 0$ or $y = \pm \dfrac{b}{a}x$

Foci on the y-axis: $\dfrac{y^2}{a^2} - \dfrac{x^2}{b^2} = 1$

Center-to-focus distance: $c = \sqrt{a^2 + b^2}$

Foci: $(0, \pm c)$

Vertices: $(0, \pm a)$

Asymptotes: $\dfrac{y^2}{a^2} - \dfrac{x^2}{b^2} = 0$ or $y = \pm \dfrac{a}{b}x$

Notice the difference in the asymptote equations (b/a in the first, a/b in the second).

We shift conics using the principles reviewed in Section 1.2, replacing x by $x + h$ and y by $y + k$.

EXAMPLE 4 Show that the equation $x^2 - 4y^2 + 2x + 8y - 7 = 0$ represents a hyperbola. Find its center, asymptotes, and foci.

Solution We reduce the equation to standard form by completing the square in x and y as follows:

$$(x^2 + 2x) - 4(y^2 - 2y) = 7$$

$$(x^2 + 2x + 1) - 4(y^2 - 2y + 1) = 7 + 1 - 4$$

$$\frac{(x+1)^2}{4} - (y-1)^2 = 1.$$

This is the standard form Equation (10) of a hyperbola with x replaced by $x + 1$ and y replaced by $y - 1$. The hyperbola is shifted one unit to the left and one unit upward, and it has center $x + 1 = 0$ and $y - 1 = 0$, or $x = -1$ and $y = 1$. Moreover,

$$a^2 = 4, \qquad b^2 = 1, \qquad c^2 = a^2 + b^2 = 5,$$

so the asymptotes are the two lines

$$\frac{x+1}{2} - (y-1) = 0 \quad \text{and} \quad \frac{x+1}{2} + (y-1) = 0,$$

or

$$y - 1 = \pm\frac{1}{2}(x+1).$$

The shifted foci have coordinates $\left(-1 \pm \sqrt{5}, 1\right)$. ■

EXERCISES 11.6

Identifying Graphs

Match the parabolas in Exercises 1–4 with the following equations:

$$x^2 = 2y, \quad x^2 = -6y, \quad y^2 = 8x, \quad y^2 = -4x.$$

Then find each parabola's focus and directrix.

1.

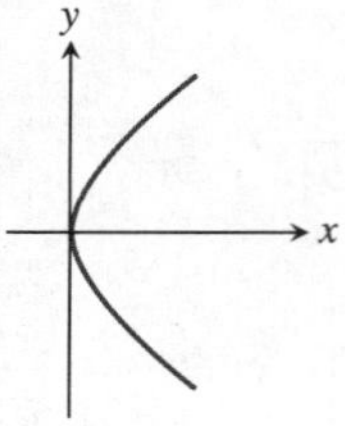

2.

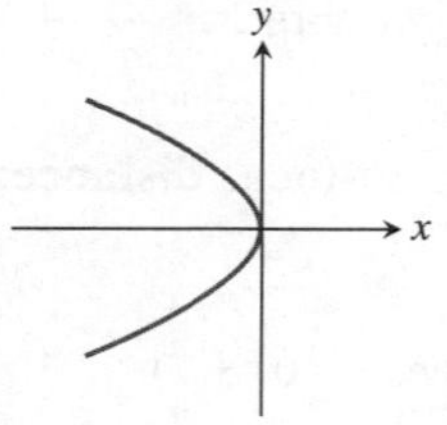

3.

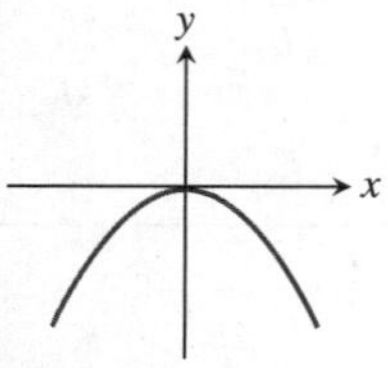

4.

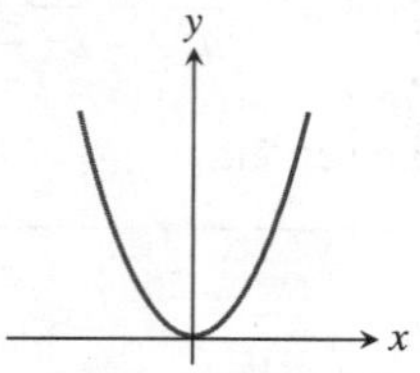

Match each conic section in Exercises 5–8 with one of these equations:

$$\frac{x^2}{4} + \frac{y^2}{9} = 1, \quad \frac{x^2}{2} + y^2 = 1,$$
$$\frac{y^2}{4} - x^2 = 1, \quad \frac{x^2}{4} - \frac{y^2}{9} = 1.$$

Then find the conic section's foci and vertices. If the conic section is a hyperbola, find its asymptotes as well.

5.

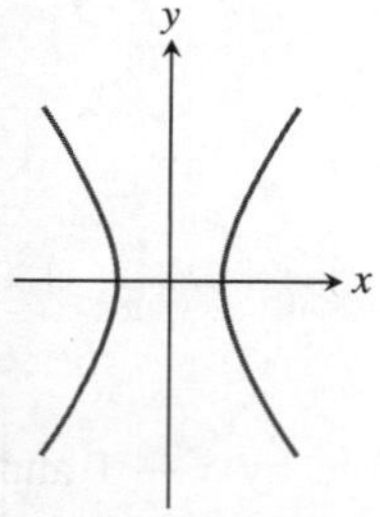

6.

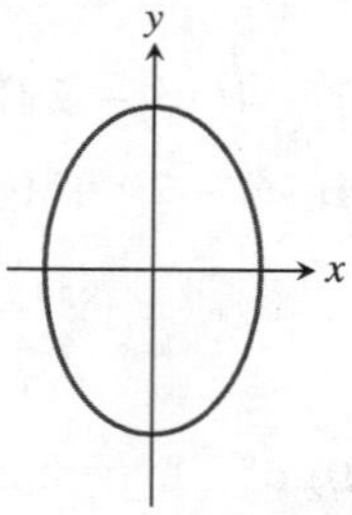

7.

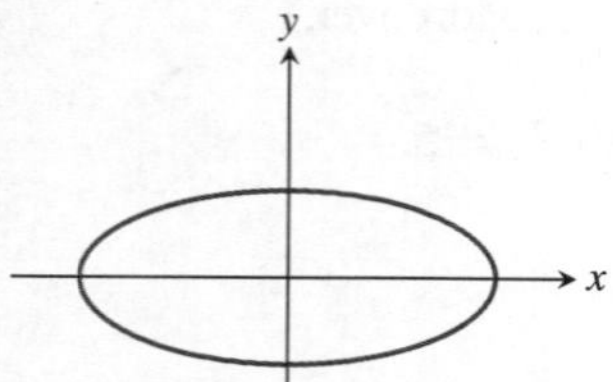

8.

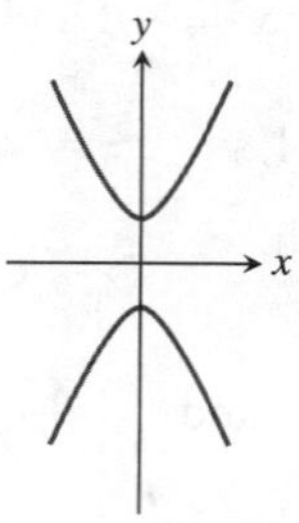

Parabolas

Exercises 9–16 give equations of parabolas. Find each parabola's focus and directrix. Then sketch the parabola. Include the focus and directrix in your sketch.

9. $y^2 = 16x$ **10.** $x^2 = 6y$ **11.** $x^2 = -12y$

12. $y^2 = -4x$ **13.** $y = 4x^2$ **14.** $y = -8x^2$

15. $x = -3y^2$ **16.** $x = y^2$

Ellipses

Exercises 17–24 give equations for ellipses. Put each equation in standard form. Then sketch the ellipse. Include the foci in your sketch.

17. $16x^2 + 25y^2 = 400$ **18.** $5x^2 + 9y^2 = 45$

19. $2x^2 + y^2 = 2$ **20.** $8x^2 + 12y^2 = 96$

21. $7x^2 + 6y^2 = 42$ **22.** $9x^2 + 10y^2 = 90$

23. $6x^2 + 9y^2 = 54$ **24.** $64x^2 + 25y^2 = 1600$

Exercises 25 and 26 give information about the foci and vertices of ellipses centered at the origin of the xy-plane. In each case, find the ellipse's standard-form equation from the given information.

25. Foci: $(\pm\sqrt{2}, 0)$ Vertices: $(\pm 2, 0)$

26. Foci: $(0, \pm 4)$ Vertices: $(0, \pm 5)$

Hyperbolas

Exercises 27–34 give equations for hyperbolas. Put each equation in standard form and find the hyperbola's asymptotes. Then sketch the hyperbola. Include the asymptotes and foci in your sketch.

27. $2x^2 - 2y^2 = 4$ **28.** $9x^2 - 16y^2 = 144$

29. $y^2 - x^2 = 10$ **30.** $y^2 - x^2 = 4$

31. $6x^2 - 2y^2 = 12$ **32.** $y^2 - 3x^2 = 3$

33. $6y^2 - 2x^2 = 12$ **34.** $64x^2 - 36y^2 = 2304$

Exercises 35–38 give information about the foci, vertices, and asymptotes of hyperbolas centered at the origin of the xy-plane. In each case, find the hyperbola's standard-form equation from the information given.

35. Foci: $(0, \pm\sqrt{2})$
Asymptotes: $y = \pm x$

36. Foci: $(\pm 2, 0)$
Asymptotes: $y = \pm \frac{1}{\sqrt{3}}x$

37. Vertices: $(\pm 3, 0)$
Asymptotes: $y = \pm \frac{4}{3}x$

38. Vertices: $(0, \pm 2)$
Asymptotes: $y = \pm \frac{1}{2}x$

Shifting Conic Sections

You may wish to review Section 1.2 before solving Exercises 39–56.

39. The parabola $y^2 = 8x$ is shifted down 2 units and right 1 unit to generate the parabola $(y + 2)^2 = 8(x - 1)$.

a. Find the new parabola's vertex, focus, and directrix.

b. Plot the new vertex, focus, and directrix, and sketch in the parabola.

40. The parabola $x^2 = -4y$ is shifted left 1 unit and up 3 units to generate the parabola $(x + 1)^2 = -4(y - 3)$.

a. Find the new parabola's vertex, focus, and directrix.

b. Plot the new vertex, focus, and directrix, and sketch in the parabola.

41. The ellipse $(x^2/16) + (y^2/9) = 1$ is shifted 4 units to the right and 3 units up to generate the ellipse

$$\frac{(x-4)^2}{16} + \frac{(y-3)^2}{9} = 1.$$

a. Find the foci, vertices, and center of the new ellipse.

b. Plot the new foci, vertices, and center, and sketch in the new ellipse.

42. The ellipse $(x^2/9) + (y^2/25) = 1$ is shifted 3 units to the left and 2 units down to generate the ellipse

$$\frac{(x+3)^2}{9} + \frac{(y+2)^2}{25} = 1.$$

a. Find the foci, vertices, and center of the new ellipse.

b. Plot the new foci, vertices, and center, and sketch in the new ellipse.

43. The hyperbola $(x^2/16) - (y^2/9) = 1$ is shifted 2 units to the right to generate the hyperbola

$$\frac{(x-2)^2}{16} - \frac{y^2}{9} = 1.$$

a. Find the center, foci, vertices, and asymptotes of the new hyperbola.

b. Plot the new center, foci, vertices, and asymptotes, and sketch in the hyperbola.

44. The hyperbola $(y^2/4) - (x^2/5) = 1$ is shifted 2 units down to generate the hyperbola

$$\frac{(y+2)^2}{4} - \frac{x^2}{5} = 1.$$

a. Find the center, foci, vertices, and asymptotes of the new hyperbola.

b. Plot the new center, foci, vertices, and asymptotes, and sketch in the hyperbola.

Exercises 45–48 give equations for parabolas and tell how many units up or down and to the right or left each parabola is to be shifted. Find an equation for the new parabola, and find the new vertex, focus, and directrix.

45. $y^2 = 4x$, left 2, down 3

46. $y^2 = -12x$, right 4, up 3

47. $x^2 = 8y$, right 1, down 7

48. $x^2 = 6y$, left 3, down 2

Exercises 49–52 give equations for ellipses and tell how many units up or down and to the right or left each ellipse is to be shifted. Find an equation for the new ellipse, and find the new foci, vertices, and center.

49. $\frac{x^2}{6} + \frac{y^2}{9} = 1$, left 2, down 1

50. $\frac{x^2}{2} + y^2 = 1$, right 3, up 4

51. $\frac{x^2}{3} + \frac{y^2}{2} = 1$, right 2, up 3

52. $\frac{x^2}{16} + \frac{y^2}{25} = 1$, left 4, down 5

Exercises 53–56 give equations for hyperbolas and tell how many units up or down and to the right or left each hyperbola is to be shifted. Find an equation for the new hyperbola, and find the new center, foci, vertices, and asymptotes.

53. $\frac{x^2}{4} - \frac{y^2}{5} = 1$, right 2, up 2

54. $\frac{x^2}{16} - \frac{y^2}{9} = 1$, left 2, down 1

55. $y^2 - x^2 = 1$, left 1, down 1

56. $\frac{y^2}{3} - x^2 = 1$, right 1, up 3

Find the center, foci, vertices, asymptotes, and radius, as appropriate, of the conic sections in Exercises 57–68.

57. $x^2 + 2x + y^2 = 14$

58. $2x^2 + 2y^2 - 28x + 12y + 114 = 0$

59. $x^2 + 4x + 2y - 3 = 0$

60. $y^2 - 6y - 2x - 14 = 0$

61. $x^2 + 5y^2 + 4x = 1$

62. $9x^2 + 6y^2 + 36y = 0$

63. $x^2 + 3y^2 - 3x - 6y = 9$

64. $4x^2 + y^2 + 8x - 2y = -1$

65. $2x^2 - 2y^2 - x + y = 2$

66. $x^2 - y^2 + 4x - 6y = 6$

67. $x^2 - 2y^2 + 6y = 2$

68. $y^2 - 4x^2 + 16x = 24$

Theory and Examples

69. If lines are drawn parallel to the coordinate axes through a point P on the parabola $y^2 = kx$, $k > 0$, the parabola partitions the rectangular region bounded by these lines and the coordinate axes into two smaller regions, A and B.

a. If the two smaller regions are revolved about the y-axis, show that they generate solids whose volumes have the ratio 4:1.

b. What is the ratio of the volumes generated by revolving the regions about the x-axis?

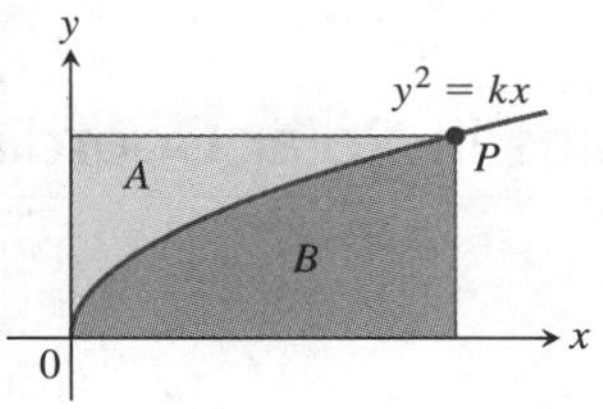

70. **Suspension bridge cables hang in parabolas** The suspension bridge cable shown in the accompanying figure supports a uniform load of w pounds per horizontal foot. It can be shown that if H is the horizontal tension of the cable at the origin, then the curve of the cable satisfies the equation

$$\frac{dy}{dx} = \frac{w}{H}x.$$

Show that the cable hangs in a parabola by solving this differential equation subject to the initial condition that $y = 0$ when $x = 0$.

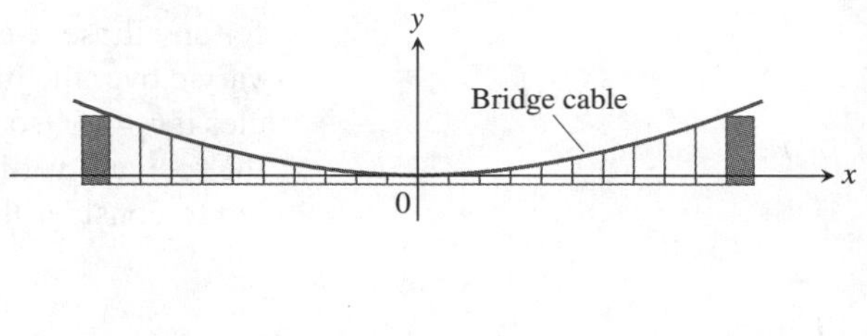

71. **The width of a parabola at the focus** Show that the number $4p$ is the *width* of the parabola $x^2 = 4py (p > 0)$ at the focus by showing that the line $y = p$ cuts the parabola at points that are $4p$ units apart.

72. **The asymptotes of $(x^2/a^2) - (y^2/b^2) = 1$** Show that the vertical distance between the line $y = (b/a)x$ and the upper half of the right-hand branch $y = (b/a)\sqrt{x^2 - a^2}$ of the hyperbola $(x^2/a^2) - (y^2/b^2) = 1$ approaches 0 by showing that

$$\lim_{x\to\infty}\left(\frac{b}{a}x - \frac{b}{a}\sqrt{x^2 - a^2}\right) = \frac{b}{a}\lim_{x\to\infty}\left(x - \sqrt{x^2 - a^2}\right) = 0.$$

Similar results hold for the remaining portions of the hyperbola and the lines $y = \pm(b/a)x$.

73. **Area** Find the dimensions of the rectangle of largest area that can be inscribed in the ellipse $x^2 + 4y^2 = 4$ with its sides parallel to the coordinate axes. What is the area of the rectangle?

74. **Volume** Find the volume of the solid generated by revolving the region enclosed by the ellipse $9x^2 + 4y^2 = 36$ about the **(a)** x-axis, **(b)** y-axis.

75. **Volume** The "triangular region" in the first quadrant bounded by the x-axis, the line $x = 4$, and hyperbola $9x^2 - 9y^2 = 36$ is revolved about x-axis to generate a solid. Find the volume of the solid.

76. **Tangents** Show that the tangents to the curve $y^2 = 4px$ from any point on the line $x = -p$ are perpendicular.

77. **Tangents** Find equations for the tangents to the circle $(x - 2)^2 + (y - 1)^2 = 5$ at the points where the circle crosses the coordinate axes.

78. **Volume** The region bounded on the left by the y-axis, on the right by the hyperbola $2x^2 - 2y^2 = 2$, above and below by the lines $y = \pm 2$ is revolved about the y-axis to generate the solid.

79. **Centroid** Find the centroid of the region that is bounded below by the x-axis and above by the ellipse $(x^2/9) + (y^2/16) = 1$.

80. **Surface area** The curve $y = \sqrt{x^2 + 1}$, $0 \le x \le \sqrt{2}$, which is part of the upper branch of the hyperbola $y^2 - x^2 = 1$, is revolved about the x-axis to generate a surface. Find the area of the surface.

81. **The reflective property of parabolas** The accompanying figure shows a typical point $P(x_0, y_0)$ on the parabola $y^2 = 4px$. The line L is tangent to the parabola at P. The parabola's focus lies at $F(p, 0)$. The ray L' extending from P to the right is parallel to the x-axis. We show that light from F to P will be reflected out along L' by showing that β equals α. Establish this equality by taking the following steps.

a. Show that $\tan\beta = 2p/y_0$.

b. Show that $\tan\phi = y_0/(x_0 - p)$.

c. Use the identity

$$\tan\alpha = \frac{\tan\phi - \tan\beta}{1 + \tan\phi\tan\beta}$$

to show that $\tan\alpha = 2p/y_0$.

Since α and β are both acute, $\tan\beta = \tan\alpha$ implies $\beta = \alpha$.

This reflective property of parabolas is used in applications like car headlights, radio telescopes, and satellite TV dishes.

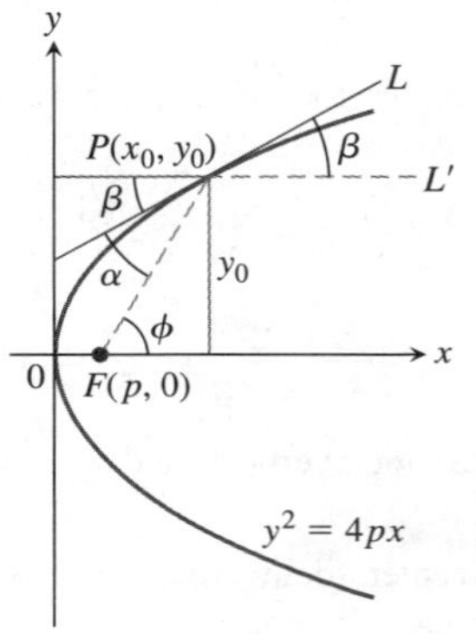

11.7 Conics in Polar Coordinates

Polar coordinates are especially important in astronomy and astronautical engineering because satellites, moons, planets, and comets all move approximately along ellipses, parabolas, and hyperbolas that can be described with a single relatively simple polar coordinate equation. We develop that equation here after first introducing the idea of a conic section's *eccentricity*. The eccentricity reveals the conic section's type (circle, ellipse, parabola, or hyperbola) and the degree to which it is "squashed" or flattened.

Eccentricity

Although the center-to-focus distance c does not appear in the standard Cartesian equation

$$\frac{x^2}{a^2} + \frac{y^2}{b^2} = 1, \qquad (a > b)$$

for an ellipse, we can still determine c from the equation $c = \sqrt{a^2 - b^2}$. If we fix a and vary c over the interval $0 \le c \le a$, the resulting ellipses will vary in shape. They are circles if $c = 0$ (so that $a = b$) and flatten, becoming more oblong, as c increases. If $c = a$, the foci and vertices overlap and the ellipse degenerates into a line segment. Thus we are led to consider the ratio $e = c/a$. We use this ratio for hyperbolas as well, except in this case c equals $\sqrt{a^2 + b^2}$ instead of $\sqrt{a^2 - b^2}$. We refer to this ratio as the *eccentricity* of the ellipse or hyperbola.

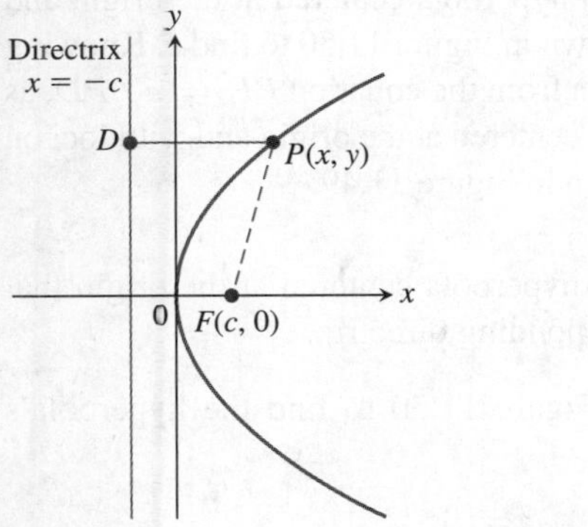

FIGURE 11.48 The distance from the focus F to any point P on a parabola equals the distance from P to the nearest point D on the directrix, so $PF = PD$.

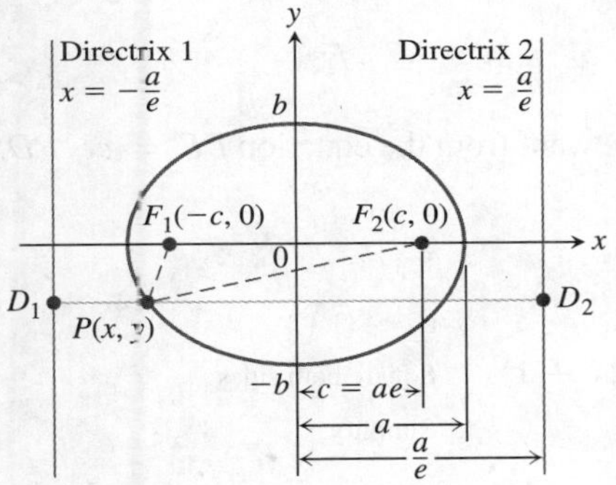

FIGURE 11.49 The foci and directrices of the ellipse $(x^2/a^2) + (y^2/b^2) = 1$. Directrix 1 corresponds to focus F_1 and directrix 2 to focus F_2.

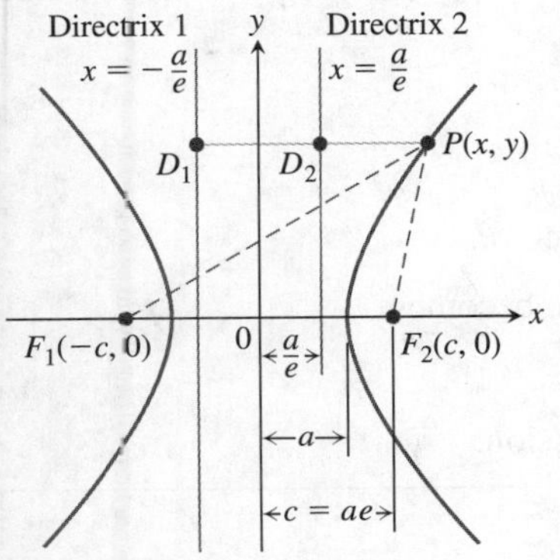

FIGURE 11.50 The foci and directrices of the hyperbola $(x^2/a^2) - (y^2/b^2) = 1$. No matter where P lies on the hyperbola, $PF_1 = e \cdot PD_1$ and $PF_2 = e \cdot PD_2$.

> **DEFINITION**
>
> The **eccentricity** of the ellipse $(x^2/a^2) + (y^2/b^2) = 1 \ (a > b)$ is
>
> $$e = \frac{c}{a} = \frac{\sqrt{a^2 - b^2}}{a}.$$
>
> The **eccentricity** of the hyperbola $(x^2/a^2) - (y^2/b^2) = 1$ is
>
> $$e = \frac{c}{a} = \frac{\sqrt{a^2 + b^2}}{a}.$$
>
> The **eccentricity** of a parabola is $e = 1$.

Whereas a parabola has one focus and one directrix, each **ellipse** has two foci and two **directrices**. These are the lines perpendicular to the major axis at distances $\pm a/e$ from the center. From Figure 11.48 we see that a parabola has the property

$$PF = 1 \cdot PD \tag{1}$$

for any point P on it, where F is the focus and D is the point nearest P on the directrix. For an ellipse, it can be shown that the equations that replace Equation (1) are

$$PF_1 = e \cdot PD_1, \qquad PF_2 = e \cdot PD_2. \tag{2}$$

Here, e is the eccentricity, P is any point on the ellipse, F_1 and F_2 are the foci, and D_1 and D_2 are the points on the directrices nearest P (Figure 11.49).

In both Equations (2) the directrix and focus must correspond; that is, if we use the distance from P to F_1, we must also use the distance from P to the directrix at the same end of the ellipse. The directrix $x = -a/e$ corresponds to $F_1(-c, 0)$, and the directrix $x = a/e$ corresponds to $F_2(c, 0)$.

As with the ellipse, it can be shown that the lines $x = \pm a/e$ act as **directrices** for the **hyperbola** and that

$$PF_1 = e \cdot PD_1 \quad \text{and} \quad PF_2 = e \cdot PD_2. \tag{3}$$

Here P is any point on the hyperbola, F_1 and F_2 are the foci, and D_1 and D_2 are the points nearest P on the directrices (Figure 11.50).

In both the ellipse and the hyperbola, the eccentricity is the ratio of the distance between the foci to the distance between the vertices (because $c/a = 2c/2a$).

$$\text{Eccentricity} = \frac{\text{distance between foci}}{\text{distance between vertices}}$$

In an ellipse, the foci are closer together than the vertices and the ratio is less than 1. In a hyperbola, the foci are farther apart than the vertices and the ratio is greater than 1.

The "focus–directrix" equation $PF = e \cdot PD$ unites the parabola, ellipse, and hyperbola in the following way. Suppose that the distance PF of a point P from a fixed point F (the focus) is a constant multiple of its distance from a fixed line (the directrix). That is, suppose

$$PF = e \cdot PD, \tag{4}$$

where e is the constant of proportionality. Then the path traced by P is

(a) a *parabola* if $e = 1$,

(b) an *ellipse* of eccentricity e if $e < 1$, and

(c) a *hyperbola* of eccentricity e if $e > 1$.

As e increases $(e \to 1^-)$, ellipses become more oblong, and $(e \to \infty)$ hyperbolas flatten toward two lines parallel to the directrix. There are no coordinates in Equation (4), and when we try to translate it into Cartesian coordinate form, it translates in different ways depending on the size of e. However, as we are about to see, in polar coordinates the equation $PF = e \cdot PD$ translates into a single equation regardless of the value of e.

Given the focus and corresponding directrix of a hyperbola centered at the origin and with foci on the x-axis, we can use the dimensions shown in Figure 11.50 to find e. Knowing e, we can derive a Cartesian equation for the hyperbola from the equation $PF = e \cdot PD$, as in the next example. We can find equations for ellipses centered at the origin and with foci on the x-axis in a similar way, using the dimensions shown in Figure 11.49.

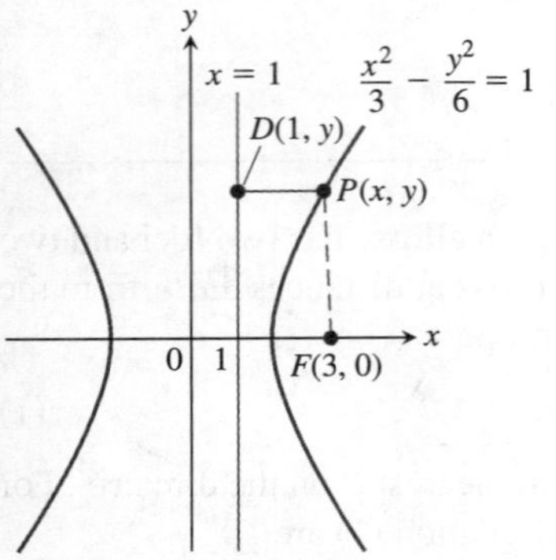

FIGURE 11.51 The hyperbola and directrix in Example 1.

EXAMPLE 1 Find a Cartesian equation for the hyperbola centered at the origin that has a focus at $(3, 0)$ and the line $x = 1$ as the corresponding directrix.

Solution We first use the dimensions shown in Figure 11.50 to find the hyperbola's eccentricity. The focus is (see Figure 11.51)

$$(c, 0) = (3, 0), \qquad \text{so} \qquad c = 3.$$

Again from Figure 11.50, the directrix is the line

$$x = \frac{a}{e} = 1, \qquad \text{so} \qquad a = e.$$

When combined with the equation $e = c/a$ that defines eccentricity, these results give

$$e = \frac{c}{a} = \frac{3}{e}, \qquad \text{so} \qquad e^2 = 3 \quad \text{and} \quad e = \sqrt{3}.$$

Knowing e, we can now derive the equation we want from the equation $PF = e \cdot PD$. In the coordinates of Figure 11.51, we have

$$PF = e \cdot PD \qquad \text{Eq. (4)}$$
$$\sqrt{(x-3)^2 + (y-0)^2} = \sqrt{3}|x - 1| \qquad e = \sqrt{3}$$
$$x^2 - 6x + 9 + y^2 = 3(x^2 - 2x + 1) \qquad \text{Square both sides.}$$
$$2x^2 - y^2 = 6 \qquad \text{Simplify.}$$
$$\frac{x^2}{3} - \frac{y^2}{6} = 1.$$

■

Polar Equations

To find a polar equation for an ellipse, parabola, or hyperbola, we place one focus at the origin and the corresponding directrix to the right of the origin along the vertical line $x = k$ (Figure 11.52). In polar coordinates, this makes

$$PF = r$$

and

$$PD = k - FB = k - r\cos\theta.$$

The conic's focus–directrix equation $PF = e \cdot PD$ then becomes

$$r = e(k - r\cos\theta),$$

which can be solved for r to obtain the following expression.

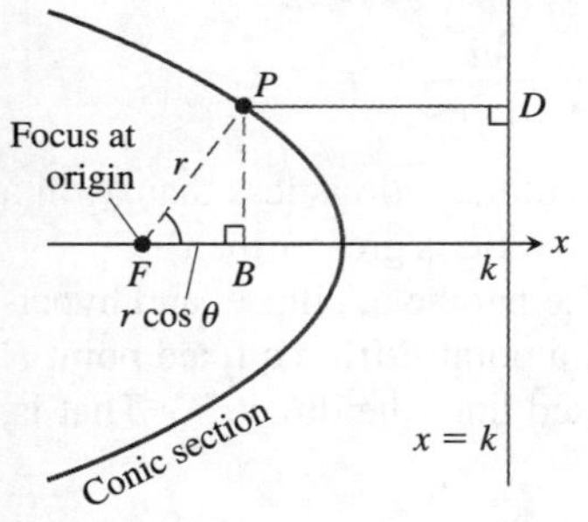

FIGURE 11.52 If a conic section is put in the position with its focus placed at the origin and a directrix perpendicular to the initial ray and right of the origin, we can find its polar equation from the conic's focus–directrix equation.

Polar Equation for a Conic with Eccentricity e

$$r = \frac{ke}{1 + e\cos\theta}, \qquad (5)$$

where $x = k > 0$ is the vertical directrix.

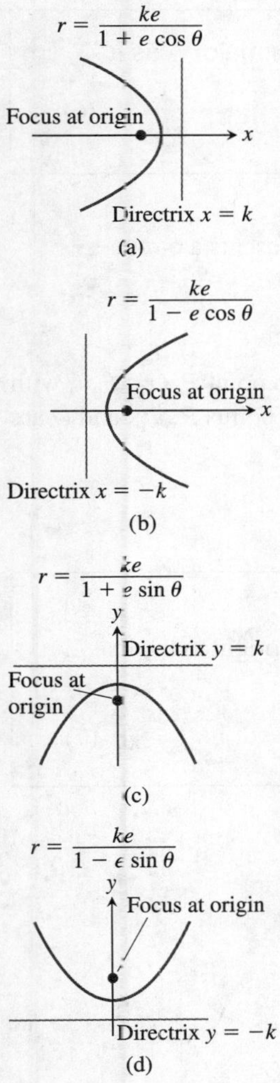

FIGURE 11.53 Equations for conic sections with eccentricity $e > 0$ but different locations of the directrix. The graphs here show a parabola, so $e = 1$.

EXAMPLE 2 Here are polar equations for three conics. The eccentricity values identifying the conic are the same for both polar and Cartesian coordinates.

$$e = \frac{1}{2}: \quad \text{ellipse} \qquad r = \frac{k}{2 + \cos\theta}$$

$$e = 1: \quad \text{parabola} \qquad r = \frac{k}{1 + \cos\theta}$$

$$e = 2: \quad \text{hyperbola} \qquad r = \frac{2k}{1 + 2\cos\theta}$$ ■

You may see variations of Equation (5), depending on the location of the directrix. If the directrix is the line $x = -k$ to the left of the origin (the origin is still a focus), we replace Equation (5) with

$$r = \frac{ke}{1 - e\cos\theta}.$$

The denominator now has a $(-)$ instead of a $(+)$. If the directrix is either of the lines $y = k$ or $y = -k$, the equations have sines in them instead of cosines, as shown in Figure 11.53.

EXAMPLE 3 Find an equation for the hyperbola with eccentricity $3/2$ and directrix $x = 2$.

Solution We use Equation (5) with $k = 2$ and $e = 3/2$:

$$r = \frac{2(3/2)}{1 + (3/2)\cos\theta} \quad \text{or} \quad r = \frac{6}{2 + 3\cos\theta}.$$ ■

EXAMPLE 4 Find the directrix of the parabola $r = \dfrac{25}{10 + 10\cos\theta}$.

Solution We divide the numerator and denominator by 10 to put the equation in standard polar form:

$$r = \frac{5/2}{1 + \cos\theta}.$$

This is the equation

$$r = \frac{ke}{1 + e\cos\theta}$$

with $k = 5/2$ and $e = 1$. The equation of the directrix is $x = 5/2$. ■

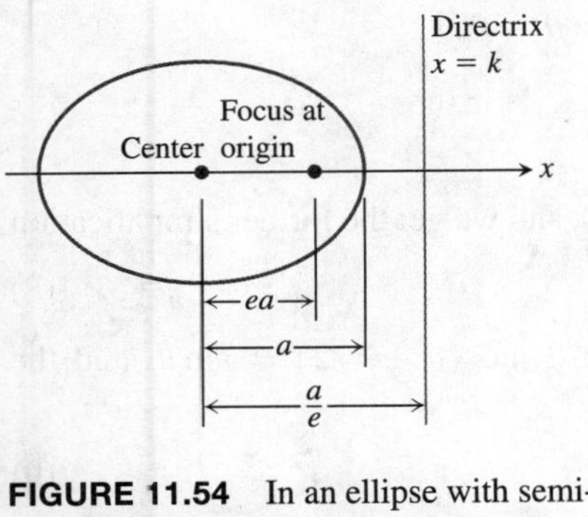

FIGURE 11.54 In an ellipse with semimajor axis a, the focus–directrix distance is $k = (a/e) - ea$, so $ke = a(1 - e^2)$.

From the ellipse diagram in Figure 11.54, we see that k is related to the eccentricity e and the semimajor axis a by the equation

$$k = \frac{a}{e} - ea.$$

From this, we find that $ke = a(1 - e^2)$. Replacing ke in Equation (5) by $a(1 - e^2)$ gives the standard polar equation for an ellipse.

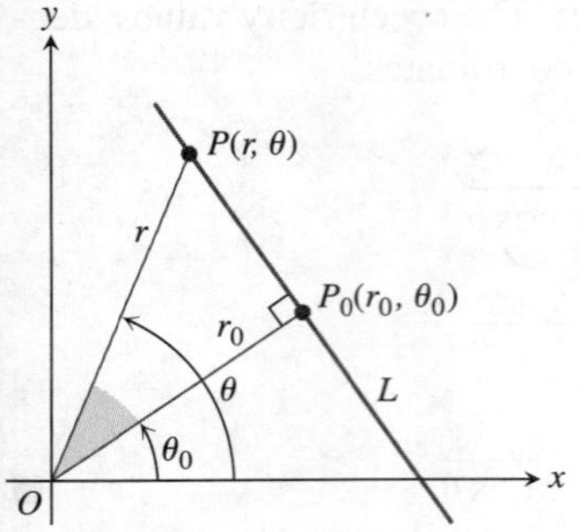

FIGURE 11.55 We can obtain a polar equation for line L by reading the relation $r_0 = r\cos(\theta - \theta_0)$ from the right triangle OP_0P.

Polar Equation for the Ellipse with Eccentricity e and Semimajor Axis a

$$r = \frac{a(1 - e^2)}{1 + e\cos\theta} \tag{6}$$

Notice that when $e = 0$, Equation (6) becomes $r = a$, which represents a circle.

Lines

Suppose the perpendicular from the origin to line L meets L at the point $P_0(r_0, \theta_0)$, with $r_0 \geq 0$ (Figure 11.55). Then, if $P(r, \theta)$ is any other point on L, the points P, P_0, and O are the vertices of a right triangle, from which we can read the relation

$$r_0 = r\cos(\theta - \theta_0).$$

The Standard Polar Equation for Lines

If the point $P_0(r_0, \theta_0)$ is the foot of the perpendicular from the origin to the line L, and $r_0 \geq 0$, then an equation for L is

$$r\cos(\theta - \theta_0) = r_0. \tag{7}$$

For example, if $\theta_0 = \pi/3$ and $r_0 = 2$, we find that

$$r\cos\left(\theta - \frac{\pi}{3}\right) = 2$$

$$r\left(\cos\theta\cos\frac{\pi}{3} + \sin\theta\sin\frac{\pi}{3}\right) = 2$$

$$\frac{1}{2}r\cos\theta + \frac{\sqrt{3}}{2}r\sin\theta = 2, \quad \text{or} \quad x + \sqrt{3}y = 4.$$

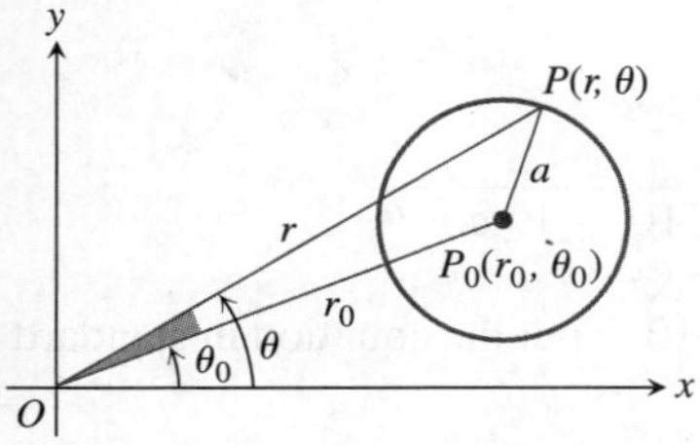

FIGURE 11.56 We can get a polar equation for this circle by applying the Law of Cosines to triangle OP_0P.

Circles

To find a polar equation for the circle of radius a centered at $P_0(r_0, \theta_0)$, we let $P(r, \theta)$ be a point on the circle and apply the Law of Cosines to triangle OP_0P (Figure 11.56). This gives

$$a^2 = r_0{}^2 + r^2 - 2r_0 r\cos(\theta - \theta_0).$$

If the circle passes through the origin, then $r_0 = a$ and this equation simplifies to

$$a^2 = a^2 + r^2 - 2ar\cos(\theta - \theta_0)$$

$$r^2 = 2ar\cos(\theta - \theta_0)$$

$$r = 2a\cos(\theta - \theta_0).$$

If the circle's center lies on the positive x-axis, $\theta_0 = 0$, and we get the further simplification

$$r = 2a\cos\theta. \tag{8}$$

If the center lies on the positive y-axis, $\theta = \pi/2$, $\cos(\theta - \pi/2) = \sin\theta$, and the equation $r = 2a\cos(\theta - \theta_0)$ becomes

$$r = 2a\sin\theta. \tag{9}$$

Equations for circles through the origin centered on the negative x- and y-axes can be obtained by replacing r with $-r$ in the above equations.

EXAMPLE 5 Here are several polar equations given by Equations (8) and (9) for circles through the origin and having centers that lie on the x- or y-axis.

Radius	Center (polar coordinates)	Polar equation
3	$(3, 0)$	$r = 6\cos\theta$
2	$(2, \pi/2)$	$r = 4\sin\theta$
1/2	$(-1/2, 0)$	$r = -\cos\theta$
1	$(-1, \pi/2)$	$r = -2\sin\theta$

■

EXERCISES 11.7

Ellipses and Eccentricity

In Exercises 1–8, find the eccentricity of the ellipse. Then find and graph the ellipse's foci and directrices.

1. $16x^2 + 25y^2 = 400$ **2.** $7x^2 + 16y^2 = 112$

3. $2x^2 + y^2 = 2$ **4.** $2x^2 + y^2 = 4$

5. $3x^2 + 2y^2 = 6$ **6.** $9x^2 + 10y^2 = 90$

7. $6x^2 + 9y^2 = 54$ **8.** $169x^2 + 25y^2 = 4225$

Exercises 9–12 give the foci or vertices and the eccentricities of ellipses centered at the origin of the xy-plane. In each case, find the ellipse's standard-form equation in Cartesian coordinates.

9. Foci: $(0, \pm 2)$
Eccentricity: 0.2

10. Foci: $(\pm 8, 0)$
Eccentricity: 0.2

11. Vertices: $(0, \pm 50)$
Eccentricity: 0.1

12. Vertices: $(\pm 10, 0)$
Eccentricity: 0.24

Exercises 13–16 give foci and corresponding directrices of ellipses centered at the origin of the xy-plane. In each case, use the dimensions in Figure 11.49 to find the eccentricity of the ellipse. Then find the ellipse's standard-form equation in Cartesian coordinates.

13. Focus: $(\sqrt{5}, 0)$
Directrix: $x = \dfrac{9}{\sqrt{5}}$

14. Focus: $(4, 0)$
Directrix: $x = \dfrac{16}{3}$

15. Focus: $(-4, 0)$
Directrix: $x = -16$

16. Focus: $(-\sqrt{2}, 0)$
Directrix: $x = -2\sqrt{2}$

Hyperbolas and Eccentricity

In Exercises 17–24, find the eccentricity of the hyperbola. Then find and graph the hyperbola's foci and directrices.

17. $x^2 - y^2 = 1$ **18.** $9x^2 - 16y^2 = 144$

19. $y^2 - x^2 = 8$ **20.** $y^2 - x^2 = 4$

21. $8x^2 - 2y^2 = 16$ **22.** $y^2 - 3x^2 = 3$

23. $8y^2 - 2x^2 = 16$ **24.** $64x^2 - 36y^2 = 2304$

Exercises 25–28 give the eccentricities and the vertices or foci of hyperbolas centered at the origin of the xy-plane. In each case, find the hyperbola's standard-form equation in Cartesian coordinates.

25. Eccentricity: 4
Vertices: $(0, \pm 2)$

26. Eccentricity: 2
Vertices: $(\pm 2, 0)$

27. Eccentricity: 2
Foci: $(\pm 2, 0)$

28. Eccentricity: 1.25
Foci: $(0, \pm 5)$

Eccentricities and Directrices

Exercises 29–36 give the eccentricities of conic sections with one focus at the origin along with the directrix corresponding to that focus. Find a polar equation for each conic section.

29. $e = 2, \quad x = 1$ **30.** $e = 1, \quad y = 2$

31. $e = 4, \quad y = -2$ **32.** $e = 2, \quad x = 4$

33. $e = 1, \quad x = 1$ **34.** $e = 1/4, \quad x = -2$

35. $e = 1/3, \quad y = -9$ **36.** $e = 1/3, \quad y = 6$

Parabolas and Ellipses

Sketch the parabolas and ellipses in Exercises 37–44. Include the directrix that corresponds to the focus at the origin. Label the vertices with appropriate polar coordinates. Label the centers of the ellipses as well.

37. $r = \dfrac{1}{1 + \cos\theta}$ **38.** $r = \dfrac{6}{2 + \cos\theta}$

39. $r = \dfrac{25}{10 - 5\cos\theta}$ **40.** $r = \dfrac{4}{2 - 2\cos\theta}$

41. $r = \dfrac{400}{16 + 8\sin\theta}$ **42.** $r = \dfrac{12}{3 + 3\sin\theta}$

43. $r = \dfrac{8}{2 - 2\sin\theta}$ **44.** $r = \dfrac{4}{2 - \sin\theta}$

Lines

Sketch the lines in Exercises 45–48 and find Cartesian equations for them.

45. $r\cos\left(\theta-\frac{\pi}{4}\right)=\sqrt{2}$

46. $r\cos\left(\theta+\frac{3\pi}{4}\right)=1$

47. $r\cos\left(\theta-\frac{2\pi}{3}\right)=3$

48. $r\cos\left(\theta+\frac{\pi}{3}\right)=2$

Find a polar equation in the form $r\cos(\theta-\theta_0)=r_0$ for each of the lines in Exercises 49–52.

49. $\sqrt{2}x+\sqrt{2}y=6$

50. $\sqrt{3}x-y=1$

51. $y=-3$

52. $x=-6$

Circles

Sketch the circles in Exercises 53–56. Give polar coordinates for their centers and identify their radii.

53. $r=12\cos\theta$

54. $r=2\sin\theta$

55. $r=-2\cos\theta$

56. $r=-8\sin\theta$

Find polar equations for the circles in Exercises 57–64. Sketch each circle in the coordinate plane and label it with both its Cartesian and polar equations.

57. $(x-3)^2+y^2=9$

58. $(x+2)^2+y^2=4$

59. $x^2+(y-4)^2=16$

60. $x^2+(y+7)^2=49$

61. $x^2+2y+y^2=0$

62. $x^2-16x+y^2=0$

63. $x^2+x+y^2=0$

64. $x^2+y^2-\frac{4}{3}y=0$

Examples of Polar Equations

T Graph the lines and conic sections in Exercises 65–74.

65. $r=2\sec(\theta-\pi/4)$

66. $r=4\sec(\theta+\pi/6)$

67. $r=4\sin\theta$

68. $r=-2\sin\theta$

69. $r=8/(4+\cos\theta)$

70. $r=8/(4+\sin\theta)$

71. $r=1/(1-\cos\theta)$

72. $r=1/(1+\cos\theta)$

73. $r=1/(1+2\sin\theta)$

74. $r=1/(1-2\cos\theta)$

75. Perihelion and aphelion A planet travels about its sun in an ellipse whose semimajor axis has length a. (See accompanying figure.)

a. Show that $r=a(1-e)$ when the planet is closest to the sun and that $r=a(1+e)$ when the planet is farthest from the sun.

b. Use the data in the table in Exercise 76 to find how close each planet in our solar system comes to the sun and how far away each planet gets from the sun.

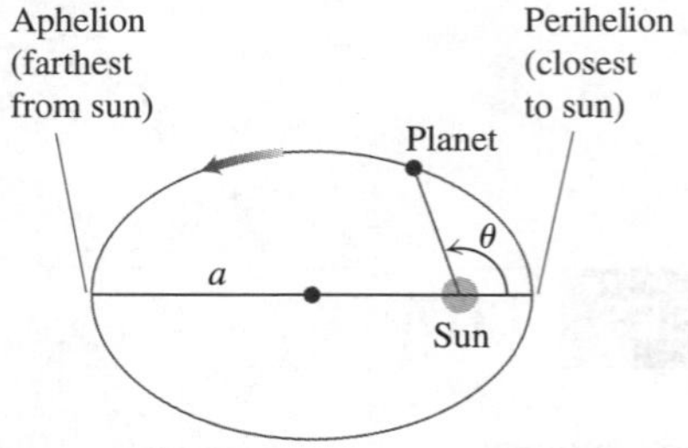

76. Planetary orbits Use the data in the table below and Equation (6) to find polar equations for the orbits of the planets.

Planet	Semimajor axis (astronomical units)	Eccentricity
Mercury	0.3871	0.2056
Venus	0.7233	0.0068
Earth	1.000	0.0167
Mars	1.524	0.0934
Jupiter	5.203	0.0484
Saturn	9.539	0.0543
Uranus	19.18	0.0460
Neptune	30.06	0.0082

CHAPTER 11 Questions to Guide Your Review

1. What is a parametrization of a curve in the xy-plane? Does a function $y=f(x)$ always have a parametrization? Are parametrizations of a curve unique? Give examples.
2. Give some typical parametrizations for lines, circles, parabolas, ellipses, and hyperbolas. How might the parametrized curve differ from the graph of its Cartesian equation?
3. What is a cycloid? What are typical parametric equations for cycloids? What physical properties account for the importance of cycloids?
4. What is the formula for the slope dy/dx of a parametrized curve $x=f(t)$, $y=g(t)$? When does the formula apply? When can you expect to be able to find d^2y/dx^2 as well? Give examples.
5. How can you sometimes find the area bounded by a parametrized curve and one of the coordinate axes?
6. How do you find the length of a smooth parametrized curve $x=f(t)$, $y=g(t)$, $a\le t\le b$? What does smoothness have to do with length? What else do you need to know about the parametrization in order to find the curve's length? Give examples.
7. What is the arc length function for a smooth parametrized curve? What is its arc length differential?
8. Under what conditions can you find the area of the surface generated by revolving a curve $x=f(t)$, $y=g(t)$, $a\le t\le b$, about the x-axis? the y-axis? Give examples.
9. What are polar coordinates? What equations relate polar coordinates to Cartesian coordinates? Why might you want to change from one coordinate system to the other?
10. What consequence does the lack of uniqueness of polar coordinates have for graphing? Give an example.

11. How do you graph equations in polar coordinates? Include in your discussion symmetry, slope, behavior at the origin, and the use of Cartesian graphs. Give examples.
12. How do you find the area of a region $0 \le r_1(\theta) \le r \le r_2(\theta)$, $\alpha \le \theta \le \beta$, in the polar coordinate plane? Give examples.
13. Under what conditions can you find the length of a curve $r = f(\theta)$, $\alpha \le \theta \le \beta$, in the polar coordinate plane? Give an example of a typical calculation.
14. What is a parabola? What are the Cartesian equations for parabolas whose vertices lie at the origin and whose foci lie on the coordinate axes? How can you find the focus and directrix of such a parabola from its equation?
15. What is an ellipse? What are the Cartesian equations for ellipses centered at the origin with foci on one of the coordinate axes? How can you find the foci, vertices, and directrices of such an ellipse from its equation?
16. What is a hyperbola? What are the Cartesian equations for hyperbolas centered at the origin with foci on one of the coordinate axes? How can you find the foci, vertices, and directrices of such an ellipse from its equation?
17. What is the eccentricity of a conic section? How can you classify conic sections by eccentricity? How does eccentricity change the shape of ellipses and hyperbolas?
18. Explain the equation $PF = e \cdot PD$.
19. What are the standard equations for lines and conic sections in polar coordinates? Give examples.

CHAPTER 11 Practice Exercises

Identifying Parametric Equations in the Plane

Exercises 1–6 give parametric equations and parameter intervals for the motion of a particle in the *xy*-plane. Identify the particle's path by finding a Cartesian equation for it. Graph the Cartesian equation and indicate the direction of motion and the portion traced by the particle.

1. $x = t/2, \quad y = t + 1, \quad -\infty < t < \infty$
2. $x = \sqrt{t}, \quad y = 1 - \sqrt{t}, \quad t \ge 0$
3. $x = (1/2)\tan t, \quad y = (1/2)\sec t, \quad -\pi/2 < t < \pi/2$
4. $x = -2\cos t, \quad y = 2\sin t, \quad 0 \le t \le \pi$
5. $x = -\cos t, \quad y = \cos^2 t, \quad 0 \le t \le \pi$
6. $x = 4\cos t, \quad y = 9\sin t, \quad 0 \le t \le 2\pi$

Finding Parametric Equations and Tangent Lines

7. Find parametric equations and a parameter interval for the motion of a particle in the *xy*-plane that traces the ellipse $16x^2 + 9y^2 = 144$ once counterclockwise. (There are many ways to do this.)
8. Find parametric equations and a parameter interval for the motion of a particle that starts at the point $(-2, 0)$ in the *xy*-plane and traces the circle $x^2 + y^2 = 4$ three times clockwise. (There are many ways to do this.)

In Exercises 9 and 10, find an equation for the line in the *xy*-plane that is tangent to the curve at the point corresponding to the given value of *t*. Also, find the value of d^2y/dx^2 at this point.

9. $x = (1/2)\tan t, \quad y = (1/2)\sec t, \quad t = \pi/3$
10. $x = 1 + 1/t^2, \quad y = 1 - 3/t, \quad t = 2$
11. Eliminate the parameter to express the curve in the form $y = f(x)$.
    a. $x = 4t^2, \quad y = t^3 - 1$
    b. $x = \cos t, \quad y = \tan t$
12. Find parametric equations for the given curve.
    a. Line through $(1, -2)$ with slope 3
    b. $(x - 1)^2 + (y + 2)^2 = 9$
    c. $y = 4x^2 - x$
    d. $9x^2 + 4y^2 = 36$

Lengths of Curves

Find the lengths of the curves in Exercises 13–19.

13. $y = x^{1/2} - (1/3)x^{3/2}, \quad 1 \le x \le 4$
14. $x = y^{2/3}, \quad 1 \le y \le 8$
15. $y = (5/12)x^{6/5} - (5/8)x^{4/5}, \quad 1 \le x \le 32$
16. $x = (y^3/12) + (1/y), \quad 1 \le y \le 2$
17. $x = 5\cos t - \cos 5t, \quad y = 5\sin t - \sin 5t, \quad 0 \le t \le \pi/2$
18. $x = t^3 - 6t^2, \quad y = t^3 + 6t^2, \quad 0 \le t \le 1$
19. $x = 3\cos\theta, \quad y = 3\sin\theta, \quad 0 \le \theta \le \dfrac{3\pi}{2}$
20. Find the length of the enclosed loop $x = t^2$, $y = (t^3/3) - t$ shown here. The loop starts at $t = -\sqrt{3}$ and ends at $t = \sqrt{3}$.

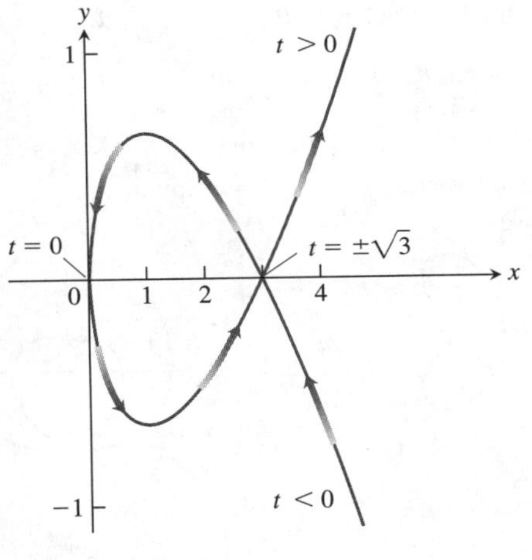

Surface Areas

Find the areas of the surfaces generated by revolving the curves in Exercises 21 and 22 about the indicated axes.

21. $x = t^2/2, \quad y = 2t, \quad 0 \le t \le \sqrt{5}, \quad x\text{-axis}$
22. $x = t^2 + 1/(2t), \quad y = 4\sqrt{t}, \quad 1/\sqrt{2} \le t \le 1, \quad y\text{-axis}$

Polar to Cartesian Equations

Sketch the lines in Exercises 23–28. Also, find a Cartesian equation for each line.

23. $r\cos\left(\theta + \frac{\pi}{3}\right) = 2\sqrt{3}$ **24.** $r\cos\left(\theta - \frac{3\pi}{4}\right) = \frac{\sqrt{2}}{2}$

25. $r = 2\sec\theta$ **26.** $r = -\sqrt{2}\sec\theta$

27. $r = -(3/2)\csc\theta$ **28.** $r = (3\sqrt{3})\csc\theta$

Find Cartesian equations for the circles in Exercises 29–32. Sketch each circle in the coordinate plane and label it with both its Cartesian and polar equations.

29. $r = -4\sin\theta$ **30.** $r = 3\sqrt{3}\sin\theta$

31. $r = 2\sqrt{2}\cos\theta$ **32.** $r = -6\cos\theta$

Cartesian to Polar Equations

Find polar equations for the circles in Exercises 33–36. Sketch each circle in the coordinate plane and label it with both its Cartesian and polar equations.

33. $x^2 + y^2 + 5y = 0$ **34.** $x^2 + y^2 - 2y = 0$

35. $x^2 + y^2 - 3x = 0$ **36.** $x^2 + y^2 + 4x = 0$

Graphs in Polar Coordinates

Sketch the regions defined by the polar coordinate inequalities in Exercises 37 and 38.

37. $0 \le r \le 6\cos\theta$ **38.** $-4\sin\theta \le r \le 0$

Match each graph in Exercises 39–46 with the appropriate equation (a)–(l). There are more equations than graphs, so some equations will not be matched.

a. $r = \cos 2\theta$ **b.** $r\cos\theta = 1$ **c.** $r = \dfrac{6}{1 - 2\cos\theta}$

d. $r = \sin 2\theta$ **e.** $r = \theta$ **f.** $r^2 = \cos 2\theta$

g. $r = 1 + \cos\theta$ **h.** $r = 1 - \sin\theta$ **i.** $r = \dfrac{2}{1 - \cos\theta}$

j. $r^2 = \sin 2\theta$ **k.** $r = -\sin\theta$ **l.** $r = 2\cos\theta + 1$

39. Four-leaved rose

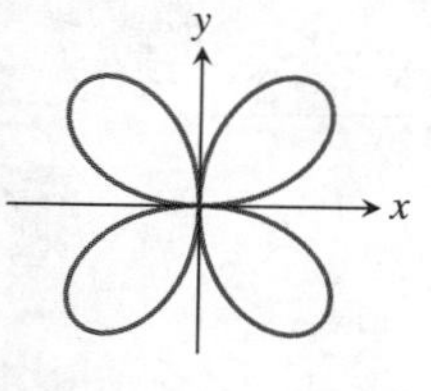

40. Spiral

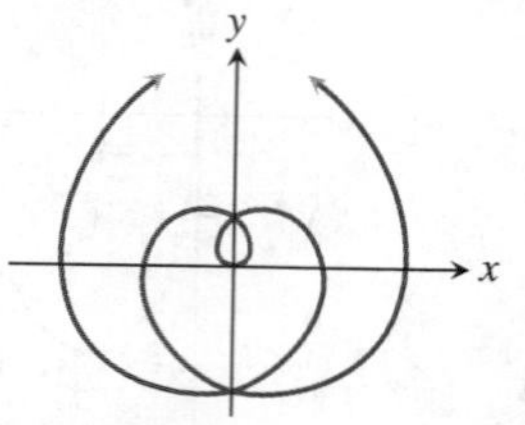

41. Limaçon

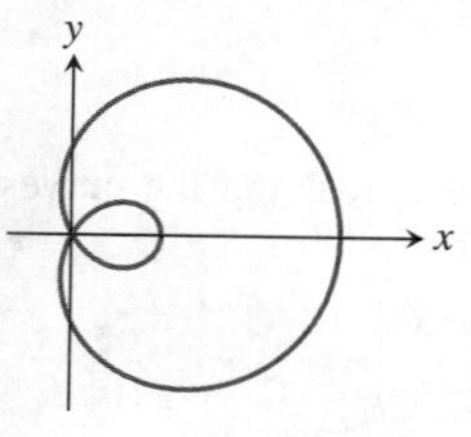

42. Lemniscate

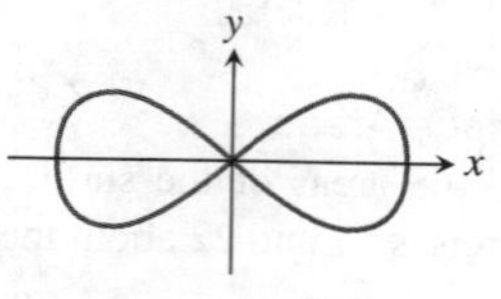

43. Circle

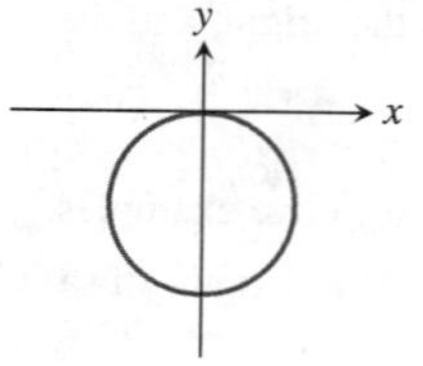

44. Cardioid

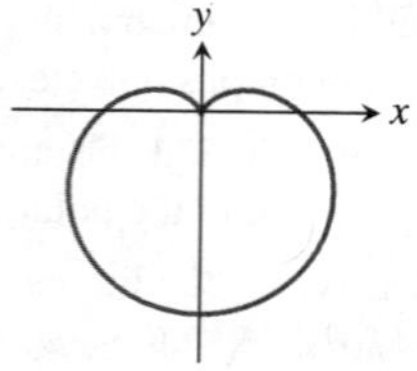

45. Parabola

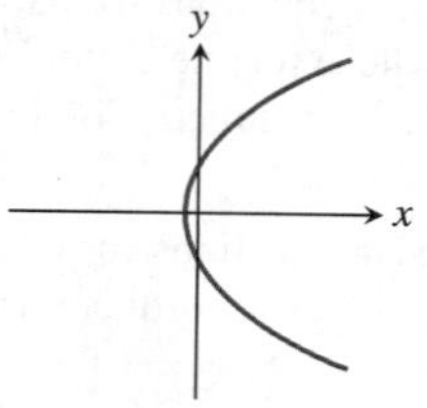

46. Lemniscate

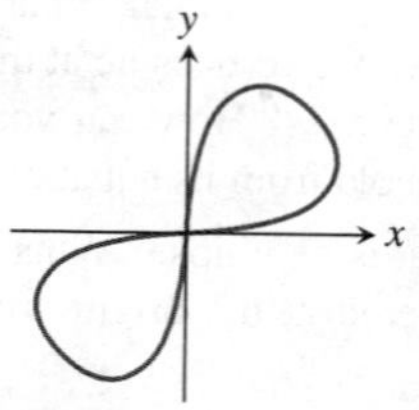

Area in Polar Coordinates

Find the areas of the regions in the polar coordinate plane described in Exercises 47–50.

47. Enclosed by the limaçon $r = 2 - \cos\theta$

48. Enclosed by one leaf of the three-leaved rose $r = \sin 3\theta$

49. Inside the "figure eight" $r = 1 + \cos 2\theta$ and outside the circle $r = 1$

50. Inside the cardioid $r = 2(1 + \sin\theta)$ and outside the circle $r = 2\sin\theta$

Length in Polar Coordinates

Find the lengths of the curves given by the polar coordinate equations in Exercises 51–54.

51. $r = -1 + \cos\theta$

52. $r = 2\sin\theta + 2\cos\theta,\quad 0 \le \theta \le \pi/2$

53. $r = 8\sin^3(\theta/3),\quad 0 \le \theta \le \pi/4$

54. $r = \sqrt{1 + \cos 2\theta},\quad -\pi/2 \le \theta \le \pi/2$

Graphing Conic Sections

Sketch the parabolas in Exercises 55–58. Include the focus and directrix in each sketch.

55. $x^2 = -4y$ **56.** $x^2 = 2y$

57. $y^2 = 3x$ **58.** $y^2 = -(8/3)x$

Find the eccentricities of the ellipses and hyperbolas in Exercises 59–62. Sketch each conic section. Include the foci, vertices, and asymptotes (as appropriate) in your sketch.

59. $16x^2 + 7y^2 = 112$ **60.** $x^2 + 2y^2 = 4$

61. $3x^2 - y^2 = 3$ **62.** $5y^2 - 4x^2 = 20$

Exercises 63–68 give equations for conic sections and tell how many units up or down and to the right or left each curve is to be shifted. Find an equation for the new conic section, and find the new foci, vertices, centers, and asymptotes, as appropriate. If the curve is a parabola, find the new directrix as well.

63. $x^2 = -12y$, right 2, up 3

64. $y^2 = 10x$, left 1/2, down 1

65. $\dfrac{x^2}{9} + \dfrac{y^2}{25} = 1$, left 3, down 5

66. $\frac{x^2}{169} + \frac{y^2}{144} = 1$, right 5, up 12

67. $\frac{y^2}{8} - \frac{x^2}{2} = 1$, right 2, up $2\sqrt{2}$

68. $\frac{x^2}{36} - \frac{y^2}{64} = 1$, left 10, down 3

Identifying Conic Sections

Complete the squares to identify the conic sections in Exercises 69–76. Find their foci, vertices, centers, and asymptotes (as appropriate). If the curve is a parabola, find its directrix as well.

69. $x^2 - 4x - 4y^2 = 0$

70. $4x^2 - y^2 + 4y = 8$

71. $y^2 - 2y + 16x = -49$

72. $x^2 - 2x + 8y = -17$

73. $9x^2 + 16y^2 + 54x - 64y = -1$

74. $25x^2 + 9y^2 - 100x + 54y = 44$

75. $x^2 + y^2 - 2x - 2y = 0$

76. $x^2 + y^2 + 4x + 2y = 1$

Conics in Polar Coordinates

Sketch the conic sections whose polar coordinate equations are given in Exercises 77–80. Give polar coordinates for the vertices and, in the case of ellipses, for the centers as well.

77. $r = \frac{2}{1 + \cos\theta}$

78. $r = \frac{8}{2 + \cos\theta}$

79. $r = \frac{6}{1 - 2\cos\theta}$

80. $r = \frac{12}{3 + \sin\theta}$

Exercises 81–84 give the eccentricities of conic sections with one focus at the origin of the polar coordinate plane, along with the directrix for that focus. Find a polar equation for each conic section.

81. $e = 2, \quad r\cos\theta = 2$

82. $e = 1, \quad r\cos\theta = -4$

83. $e = 1/2, \quad r\sin\theta = 2$

84. $e = 1/3, \quad r\sin\theta = -6$

Theory and Examples

85. Find the volume of the solid generated by revolving the region enclosed by the ellipse $9x^2 + 4y^2 = 36$ about **(a)** the x-axis, **(b)** the y-axis.

86. The "triangular" region in the first quadrant bounded by the x-axis, the line $x = 4$, and the hyperbola $9x^2 - 4y^2 = 36$ is revolved about the x-axis to generate a solid. Find the volume of the solid.

87. Show that the equations $x = r\cos\theta$, $y = r\sin\theta$ transform the polar equation

$$r = \frac{k}{1 + e\cos\theta}$$

into the Cartesian equation

$$(1 - e^2)x^2 + y^2 + 2kex - k^2 = 0.$$

88. **Archimedes spirals** The graph of an equation of the form $r = a\theta$, where a is a nonzero constant, is called an *Archimedes spiral*. Is there anything special about the widths between the successive turns of such a spiral?

CHAPTER 11 Additional and Advanced Exercises

Finding Conic Sections

1. Find an equation for the parabola with focus $(4, 0)$ and directrix $x = 3$. Sketch the parabola together with its vertex, focus, and directrix.

2. Find the vertex, focus, and directrix of the parabola

$$x^2 - 6x - 12y + 9 = 0.$$

3. Find an equation for the curve traced by the point $P(x, y)$ if the distance from P to the vertex of the parabola $x^2 = 4y$ is twice the distance from P to the focus. Identify the curve.

4. A line segment of length $a + b$ runs from the x-axis to the y-axis. The point P on the segment lies a units from one end and b units from the other end. Show that P traces an ellipse as the ends of the segment slide along the axes.

5. The vertices of an ellipse of eccentricity 0.5 lie at the points $(0, \pm 2)$. Where do the foci lie?

6. Find an equation for the ellipse of eccentricity 2/3 that has the line $x = 2$ as a directrix and the point $(4, 0)$ as the corresponding focus.

7. One focus of a hyperbola lies at the point $(0, -7)$ and the corresponding directrix is the line $y = -1$. Find an equation for the hyperbola if its eccentricity is **(a)** 2, **(b)** 5.

8. Find an equation for the hyperbola with foci $(0, -2)$ and $(0, 2)$ that passes through the point $(12, 7)$.

9. Show that the line

$$b^2xx_1 + a^2yy_1 - a^2b^2 = 0$$

is tangent to the ellipse $b^2x^2 + a^2y^2 - a^2b^2 = 0$ at the point (x_1, y_1) on the ellipse.

10. Show that the line

$$b^2xx_1 - a^2yy_1 - a^2b^2 = 0$$

is tangent to the hyperbola $b^2x^2 - a^2y^2 - a^2b^2 = 0$ at the point (x_1, y_1) on the hyperbola.

Equations and Inequalities

What points in the xy-plane satisfy the equations and inequalities in Exercises 11–16? Draw a figure for each exercise.

11. $(x^2 - y^2 - 1)(x^2 + y^2 - 25)(x^2 + 4y^2 - 4) = 0$

12. $(x + y)(x^2 + y^2 - 1) = 0$

13. $(x^2/9) + (y^2/16) \le 1$

14. $(x^2/9) - (y^2/16) \le 1$

15. $(9x^2 + 4y^2 - 36)(4x^2 + 9y^2 - 16) \le 0$

16. $(9x^2 + 4y^2 - 36)(4x^2 + 9y^2 - 16) > 0$

Polar Coordinates

17. a. Find an equation in polar coordinates for the curve

$$x = e^{2t}\cos t, \quad y = e^{2t}\sin t, \quad -\infty < t < \infty.$$

b. Find the length of the curve from $t = 0$ to $t = 2\pi$.

18. Find the length of the curve $r = 2\sin^3(\theta/3)$, $0 \le \theta \le 3\pi$, in the polar coordinate plane.

Exercises 19–22 give the eccentricities of conic sections with one focus at the origin of the polar coordinate plane, along with the directrix for that focus. Find a polar equation for each conic section.

19. $e = 2, \quad r\cos\theta = 2$

20. $e = 1, \quad r\cos\theta = -4$

21. $e = 1/2, \quad r\sin\theta = 2$

22. $e = 1/3, \quad r\sin\theta = -6$

Theory and Examples

23. Epicycloids When a circle rolls externally along the circumference of a second, fixed circle, any point P on the circumference of the rolling circle describes an *epicycloid*, as shown here. Let the fixed circle have its center at the origin O and have radius a.

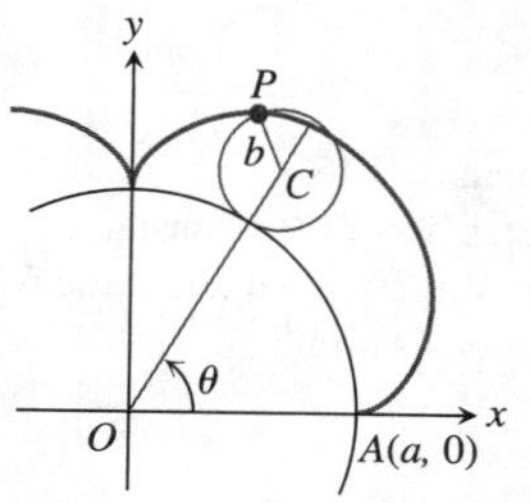

Let the radius of the rolling circle be b and let the initial position of the tracing point P be $A(a, 0)$. Find parametric equations for the epicycloid, using as the parameter the angle θ from the positive x-axis to the line through the circles' centers.

24. Find the centroid of the region enclosed by the x-axis and the cycloid arch

$$x = a(t - \sin t), \quad y = a(1 - \cos t); \quad 0 \le t \le 2\pi.$$

The Angle Between the Radius Vector and the Tangent Line to a Polar Coordinate Curve In Cartesian coordinates, when we want to discuss the direction of a curve at a point, we use the angle ϕ measured counterclockwise from the positive x-axis to the tangent line. In polar coordinates, it is more convenient to calculate the angle ψ from the *radius vector* to the tangent line (see the accompanying figure). The angle ϕ can then be calculated from the relation

$$\phi = \theta + \psi, \tag{1}$$

which comes from applying the Exterior Angle Theorem to the triangle in the accompanying figure.

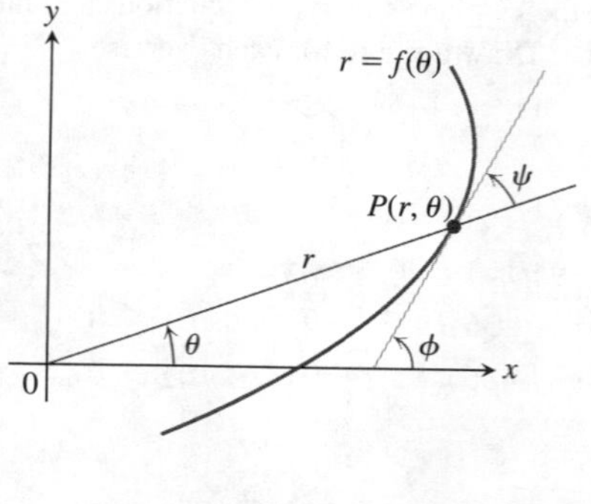

Suppose the equation of the curve is given in the form $r = f(\theta)$, where $f(\theta)$ is a differentiable function of θ. Then

$$x = r\cos\theta \quad \text{and} \quad y = r\sin\theta \tag{2}$$

are differentiable functions of θ with

$$\begin{aligned} \frac{dx}{d\theta} &= -r\sin\theta + \cos\theta\frac{dr}{d\theta}, \\ \frac{dy}{d\theta} &= r\cos\theta + \sin\theta\frac{dr}{d\theta}. \end{aligned} \tag{3}$$

Since $\psi = \phi - \theta$ from (1),

$$\tan\psi = \tan(\phi - \theta) = \frac{\tan\phi - \tan\theta}{1 + \tan\phi\tan\theta}.$$

Furthermore,

$$\tan\phi = \frac{dy}{dx} = \frac{dy/d\theta}{dx/d\theta}$$

because $\tan\phi$ is the slope of the curve at P. Also,

$$\tan\theta = \frac{y}{x}.$$

Hence

$$\tan\psi = \frac{\dfrac{dy/d\theta}{dx/d\theta} - \dfrac{y}{x}}{1 + \dfrac{y}{x}\dfrac{dy/d\theta}{dx/d\theta}} = \frac{x\dfrac{dy}{d\theta} - y\dfrac{dx}{d\theta}}{x\dfrac{dx}{d\theta} + y\dfrac{dy}{d\theta}}. \tag{4}$$

The numerator in the last expression in Equation (4) is found from Equations (2) and (3) to be

$$x\frac{dy}{d\theta} - y\frac{dx}{d\theta} = r^2.$$

Similarly, the denominator is

$$x\frac{dx}{d\theta} + y\frac{dy}{d\theta} = r\frac{dr}{d\theta}.$$

When we substitute these into Equation (4), we obtain

$$\tan\psi = \frac{r}{dr/d\theta}. \tag{5}$$

This is the equation we use for finding ψ as a function of θ.

25. Show, by reference to a figure, that the angle β between the tangents to two curves at a point of intersection may be found from the formula

$$\tan\beta = \frac{\tan\psi_2 - \tan\psi_1}{1 + \tan\psi_2\tan\psi_1}. \tag{6}$$

When will the two curves intersect at right angles?

26. Find the value of $\tan\psi$ for the curve $r = \sin^4(\theta/4)$.

27. Find the angle between the radius vector to the curve $r = 2a \sin 3\theta$ and its tangent when $\theta = \pi/6$.

T **28. a.** Graph the hyperbolic spiral $r\theta = 1$. What appears to happen to ψ as the spiral winds in around the origin?

b. Confirm your finding in part (a) analytically.

29. The circles $r = \sqrt{3} \cos \theta$ and $r = \sin \theta$ intersect at the point $\left(\sqrt{3}/2, \pi/3\right)$. Show that their tangents are perpendicular there.

30. Find the angle at which the cardioid $r = a(1 - \cos \theta)$ crosses the ray $\theta = \pi/2$.

CHAPTER 11 Technology Application Projects

Mathematica/Maple Projects

Projects can be found within MyLab Math.

- ***Radar Tracking of a Moving Object***
  Part I: Convert from polar to Cartesian coordinates.
- ***Parametric and Polar Equations with a Figure Skater***
  Part I: Visualize position, velocity, and acceleration to analyze motion defined by parametric equations.
  Part II: Find and analyze the equations of motion for a figure skater tracing a polar plot.

12

Vectors and the Geometry of Space

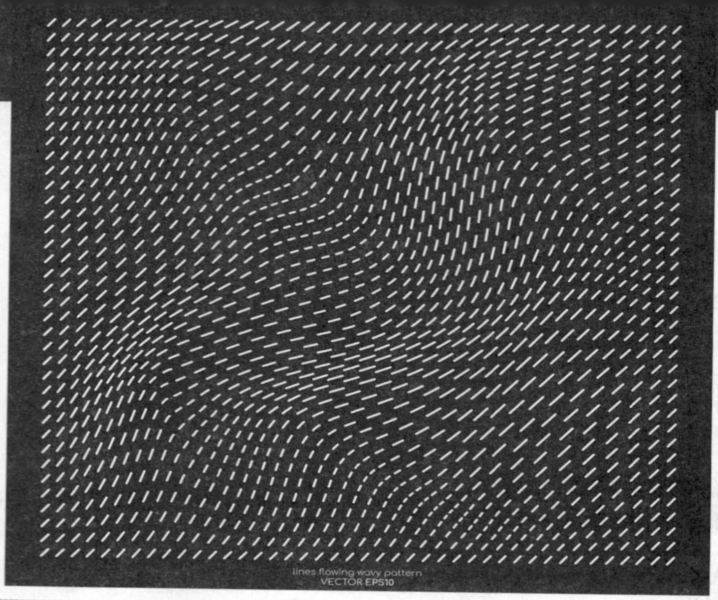

OVERVIEW In this chapter we begin the study of multivariable calculus. To apply calculus in many real-world situations, we introduce three-dimensional coordinate systems and vectors. We establish coordinates in space by adding a third axis that measures distance above and below the xy-plane. Then we define vectors, which provide simple ways to introduce equations for lines, planes, curves, and surfaces in space.

12.1 Three-Dimensional Coordinate Systems

To locate a point in space, we use three mutually perpendicular coordinate axes, arranged as in Figure 12.1. The axes shown there make a *right-handed* coordinate frame. When you hold your right hand so that the fingers curl from the positive x-axis toward the positive y-axis, your thumb points along the positive z-axis. So when you look down on the xy-plane from the positive direction of the z-axis, positive angles in the plane are measured counterclockwise from the positive x-axis and around the positive z-axis. (In a *left-handed* coordinate frame, the z-axis would point downward in Figure 12.1, and angles in the plane would be positive when measured clockwise from the positive x-axis. Right-handed and left-handed coordinate frames are not equivalent.)

The Cartesian coordinates (x, y, z) of a point P in space are the values at which the planes through P perpendicular to the axes cut the axes. Cartesian coordinates for space are also called **rectangular coordinates** because the axes that define them meet at right angles. Points on the x-axis have y- and z-coordinates equal to zero. That is, they have coordinates of the form $(x, 0, 0)$. Similarly, points on the y-axis have coordinates of the form $(0, y, 0)$, and points on the z-axis have coordinates of the form $(0, 0, z)$.

The planes determined by the coordinate axes are the ***xy*-plane**, whose standard equation is $z = 0$; the ***yz*-plane**, whose standard equation is $x = 0$; and the ***xz*-plane**, whose standard equation is $y = 0$. They meet at the **origin** $(0, 0, 0)$ (Figure 12.2), which is also identified by 0 or the letter O.

The three **coordinate planes** $x = 0$, $y = 0$, and $z = 0$ divide space into eight cells called **octants**. The octant in which the point coordinates are all positive is called the **first octant**; there is no convention for numbering the other seven octants.

The points in a plane perpendicular to the x-axis all have the same x-coordinate, this being the number at which that plane cuts the x-axis. The y- and z-coordinates can be any numbers. Similarly, the points in a plane perpendicular to the y-axis have a common y-coordinate, and the points in a plane perpendicular to the z-axis have a common z-coordinate. To write equations for these planes, we name the common coordinate's value. The plane $x = 2$ is the plane perpendicular to the x-axis at $x = 2$. The plane $y = 3$ is the plane perpendicular to the y-axis at $y = 3$. The plane $z = 5$ is the plane perpendicular to the z-axis at $z = 5$. Figure 12.3 shows the planes $x = 2$, $y = 3$, and $z = 5$, together with their intersection point $(2, 3, 5)$.

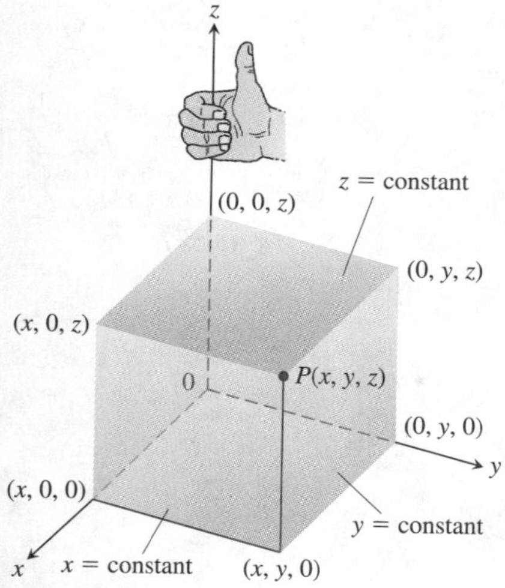

FIGURE 12.1 The Cartesian coordinate system is right-handed.

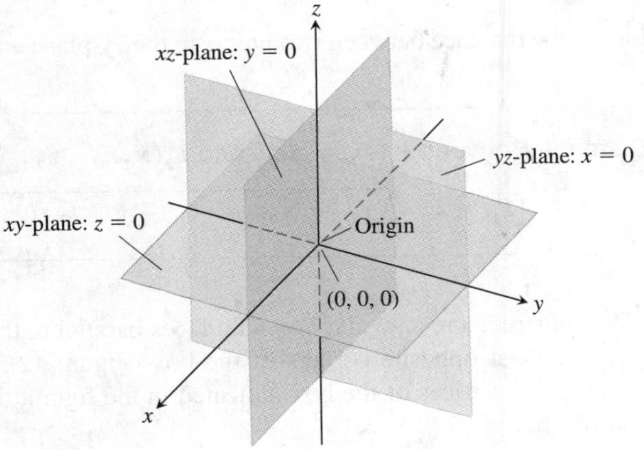

FIGURE 12.2 The planes $x = 0$, $y = 0$, and $z = 0$ divide space into eight octants.

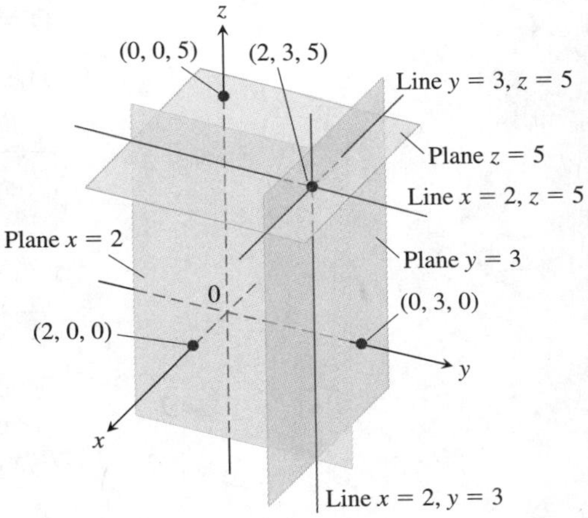

FIGURE 12.3 The planes $x = 2$, $y = 3$, and $z = 5$ determine three lines through the point $(2, 3, 5)$.

The planes $x = 2$ and $y = 3$ in Figure 12.3 intersect in a line parallel to the z-axis. This line is described by the *pair* of equations $x = 2$, $y = 3$. A point (x, y, z) lies on the line if and only if $x = 2$ and $y = 3$. Similarly, the line of intersection of the planes $y = 3$ and $z = 5$ is described by the equation pair $y = 3$, $z = 5$. This line runs parallel to the x-axis. The line of intersection of the planes $x = 2$ and $z = 5$, parallel to the y-axis, is described by the equation pair $x = 2$, $z = 5$.

In the following examples, we match coordinate equations and inequalities with the sets of points they define in space.

EXAMPLE 1 We interpret these equations and inequalities geometrically.

(a) $z \geq 0$	The half-space consisting of the points on and above the xy-plane.
(b) $x = -3$	The plane perpendicular to the x-axis at $x = -3$. This plane lies parallel to the yz-plane and 3 units behind it.
(c) $z = 0, x \leq 0, y \geq 0$	The second quadrant of the xy-plane.
(d) $x \geq 0, y \geq 0, z \geq 0$	The first octant.
(e) $-1 \leq y \leq 1$	The slab between the planes $y = -1$ and $y = 1$ (planes included).
(f) $y = -2, z = 2$	The line in which the planes $y = -2$ and $z = 2$ intersect. Alternatively, the line through the point $(0, -2, 2)$ parallel to the x-axis. ■

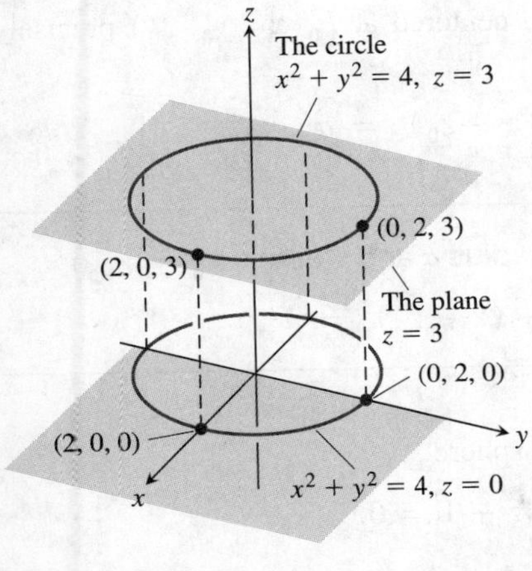

FIGURE 12.4 The circle $x^2 + y^2 = 4$ in the plane $z = 3$ (Example 2).

EXAMPLE 2 What points (x, y, z) satisfy the equations

$$x^2 + y^2 = 4 \quad \text{and} \quad z = 3?$$

Solution The points lie in the horizontal plane $z = 3$ and, in this plane, make up the circle $x^2 + y^2 = 4$. We call this set of points "the circle $x^2 + y^2 = 4$ in the plane $z = 3$" or, more simply, "the circle $x^2 + y^2 = 4$, $z = 3$" (Figure 12.4). ■

Distance and Spheres in Space

The formula for the distance between two points in the xy-plane extends to points in space.

The Distance Between $P_1(x_1, y_1, z_1)$ and $P_2(x_2, y_2, z_2)$

$$|P_1P_2| = \sqrt{(x_2 - x_1)^2 + (y_2 - y_1)^2 + (z_2 - z_1)^2}$$

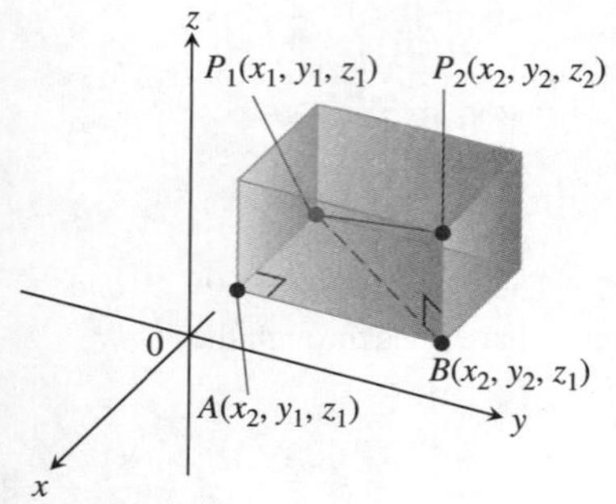

FIGURE 12.5 We find the distance between P_1 and P_2 by applying the Pythagorean theorem to the right triangles P_1AB and P_1BP_2.

Proof We construct a rectangular box with faces parallel to the coordinate planes and the points P_1 and P_2 at opposite corners of the box (Figure 12.5). If $A(x_2, y_1, z_1)$ and $B(x_2, y_2, z_1)$ are the vertices of the box indicated in the figure, then the three box edges P_1A, AB, and BP_2 have lengths

$$|P_1A| = |x_2 - x_1|, \qquad |AB| = |y_2 - y_1|, \qquad |BP_2| = |z_2 - z_1|.$$

Because triangles P_1BP_2 and P_1AB are both right-angled, two applications of the Pythagorean theorem give

$$|P_1P_2|^2 = |P_1B|^2 + |BP_2|^2 \quad \text{and} \quad |P_1B|^2 = |P_1A|^2 + |AB|^2$$

(see Figure 12.5). So

$$\begin{aligned} |P_1P_2|^2 &= |P_1B|^2 + |BP_2|^2 \\ &= |P_1A|^2 + |AB|^2 + |BP_2|^2 \qquad \text{Substitute } |P_1B|^2 = |P_1A|^2 + |AB|^2. \\ &= |x_2 - x_1|^2 + |y_2 - y_1|^2 + |z_2 - z_1|^2 \\ &= (x_2 - x_1)^2 + (y_2 - y_1)^2 + (z_2 - z_1)^2. \end{aligned}$$

Therefore,

$$|P_1P_2| = \sqrt{(x_2 - x_1)^2 + (y_2 - y_1)^2 + (z_2 - z_1)^2}. \qquad \blacksquare$$

EXAMPLE 3 The distance between $P_1(2, 1, 5)$ and $P_2(-2, 3, 0)$ is

$$\begin{aligned} |P_1P_2| &= \sqrt{(-2 - 2)^2 + (3 - 1)^2 + (0 - 5)^2} \\ &= \sqrt{16 + 4 + 25} \\ &= \sqrt{45} \approx 6.708. \end{aligned} \qquad \blacksquare$$

We can use the distance formula to write equations for spheres in space (Figure 12.6). A point $P(x, y, z)$ lies on the sphere of radius a centered at $P_0(x_0, y_0, z_0)$ precisely when $|P_0P| = a$, or

$$(x - x_0)^2 + (y - y_0)^2 + (z - z_0)^2 = a^2.$$

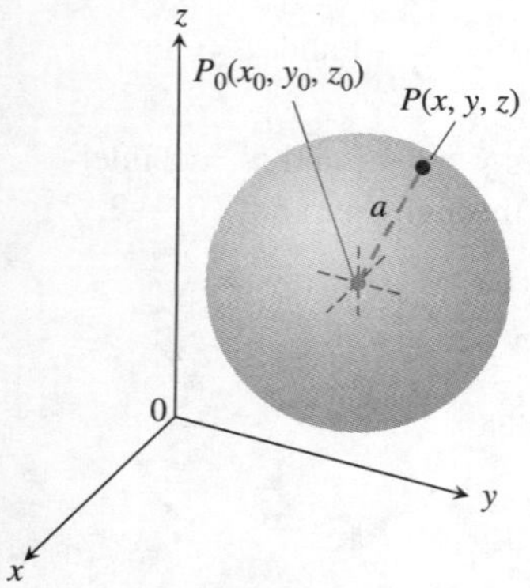

FIGURE 12.6 The sphere of radius a centered at the point (x_0, y_0, z_0).

The Standard Equation for the Sphere of Radius a and Center (x_0, y_0, z_0)

$$(x - x_0)^2 + (y - y_0)^2 + (z - z_0)^2 = a^2$$

EXAMPLE 4 Find the center and radius of the sphere

$$x^2 + y^2 + z^2 + 3x - 4z + 1 = 0.$$

Solution We find the center and radius of a sphere the way we find the center and radius of a circle: Complete the squares on the x-, y-, and z-terms as necessary and write each quadratic as a squared linear expression. Then, from the equation in standard form, read off the center and radius. For this sphere, we have

$$x^2 + y^2 + z^2 + 3x - 4z + 1 = 0$$
$$(x^2 + 3x) + y^2 + (z^2 - 4z) = -1$$
$$\left(x^2 + 3x + \left(\frac{3}{2}\right)^2\right) + y^2 + \left(z^2 - 4z + \left(\frac{-4}{2}\right)^2\right) = -1 + \left(\frac{3}{2}\right)^2 + \left(\frac{-4}{2}\right)^2$$
$$\left(x + \frac{3}{2}\right)^2 + y^2 + (z-2)^2 = \frac{21}{4}.$$

From this standard form, we read that $x_0 = -3/2$, $y_0 = 0$, $z_0 = 2$, and $a = \sqrt{21}/2$. The center is $(-3/2, 0, 2)$. The radius is $\sqrt{21}/2$. ■

EXAMPLE 5 Here are some geometric interpretations of inequalities and equations involving spheres.

(a) $x^2 + y^2 + z^2 < 4$ — The interior of the sphere $x^2 + y^2 + z^2 = 4$.

(b) $x^2 + y^2 + z^2 \le 4$ — The solid ball bounded by the sphere $x^2 + y^2 + z^2 = 4$. Alternatively, the sphere $x^2 + y^2 + z^2 = 4$ together with its interior.

(c) $x^2 + y^2 + z^2 > 4$ — The exterior of the sphere $x^2 + y^2 + z^2 = 4$.

(d) $x^2 + y^2 + z^2 = 4,\ z \le 0$ — The lower hemisphere cut from the sphere $x^2 + y^2 + z^2 = 4$ by the xy-plane (the plane $z = 0$). ■

Just as polar coordinates give another way to locate points in the xy-plane (Section 11.3), alternative coordinate systems, different from the Cartesian coordinate system developed here, exist for three-dimensional space. We examine two of these coordinate systems in Section 15.7.

EXERCISES 12.1

Geometric Interpretations of Equations

In Exercises 1–16, give a geometric description of the set of points in space whose coordinates satisfy the given pairs of equations.

1. $x = 4, \quad y = 5$

2. $x = -1, \quad z = 0$

3. $x = 0, \quad z = 0$

4. $x = 1, \quad y = 0$

5. $x^2 + y^2 = 4, \quad z = 0$

6. $x^2 + y^2 = 9, \quad z = -3$

7. $x^2 + z^2 = 4, \quad y = 0$

8. $x^2 + z^2 = 1, \quad y = 0$

9. $x^2 + y^2 + z^2 = 1, \quad x = 0$

10. $x^2 + y^2 + z^2 = 36, \quad y = -3$

11. $x^2 + y^2 + (z + 3)^2 = 25, \quad z = 0$

12. $(x - 1)^2 + y^2 + z^2 = 9, \quad x = 0$

13. $x^2 + y^2 = 4, \quad z = y$

14. $x^2 + y^2 + z^2 = 4, \quad y = x$

15. $y = x^2, \quad z = 0$

16. $z = y^2, \quad x = 1$

Geometric Interpretations of Inequalities and Equations

In Exercises 17–24, describe the sets of points in space whose coordinates satisfy the given inequalities or combinations of equations and inequalities.

17. a. $x \ge 0, \quad y \ge 0, \quad z = 0$ **b.** $x \ge 0, \quad y \le 0, \quad z = 0$

18. a. $0 \le x \le 1$ **b.** $0 \le x \le 1, \quad 0 \le y \le 1$

c. $0 \le x \le 1, \quad 0 \le y \le 1, \quad 0 \le z \le 1$

19. a. $x^2 + y^2 + z^2 \le 1$ **b.** $x^2 + y^2 + z^2 > 1$

20. a. $x^2 + y^2 \le 1, \quad z = 0$ **b.** $x^2 + y^2 \le 1, \quad z = 3$

c. $x^2 + y^2 \le 1$, no restriction on z

21. a. $1 \le x^2 + y^2 + z^2 \le 4$ **b.** $x^2 + y^2 + z^2 \le 1, \quad z \ge 0$

22. a. $x = y, \quad z = 0$ **b.** $x = y$, no restriction on z

23. a. $y \ge x^2, \quad z \ge 0$ **b.** $x \le y^2, \quad 0 \le z \le 2$

24. a. $z = 1 - y$, no restriction on x

b. $z = y^3, \quad x = 2$

Distance

In Exercises 25–30, find the distance between points P_1 and P_2.

25. $P_1(5, 5, 5)$, $P_2(7, 4, 0)$

26. $P_1(-1, 1, 5)$, $P_2(2, 5, 0)$

27. $P_1(2, 5, 6)$, $P_2(5, -3, 9)$

28. $P_1(3, 4, 5)$, $P_2(2, 3, 4)$

29. $P_1(0, 0, 0)$, $P_2(7, -7, -7)$

30. $P_1(5, 3, -2)$, $P_2(0, 0, 0)$

31. Find the distance from the point $(3, -4, 2)$ to the

a. xy-plane **b.** yz-plane **c.** xz-plane

32. Find the distance from the point $(-2, 1, 4)$ to the

a. plane $x = 3$ **b.** plane $y = -5$ **c.** plane $z = -1$

33. Find the distance from the point $(4, 3, 0)$ to the

a. x-axis **b.** y-axis **c.** z-axis

34. Find the distance from the

a. x-axis to the plane $z = 3$.

b. origin to the plane $2 = z - x$.

c. point $(0, 4, 0)$ to the plane $y = x$.

In Exercises 35–44, describe the given set with a single equation or with a pair of equations.

35. The plane perpendicular to the

a. x-axis at $(3, 0, 0)$ **b.** y-axis at $(0, -1, 0)$

c. z-axis at $(0, 0, -2)$

36. The plane through the point $(3, -1, 2)$ perpendicular to the

a. x-axis **b.** y-axis **c.** z-axis

37. The plane through the point $(3, -1, 1)$ parallel to the

a. xy-plane **b.** yz-plane **c.** xz-plane

38. The circle of radius 2 centered at $(0, 0, 0)$ and lying in the

a. xy-plane **b.** yz-plane **c.** xz-plane

39. The circle of radius 2 centered at $(0, 2, 0)$ and lying in the

a. xy-plane **b.** yz-plane **c.** plane $y = 2$

40. The circle of radius 1 centered at $(-3, 4, 1)$ and lying in a plane parallel to the

a. xy-plane **b.** yz-plane **c.** xz-plane

41. The line through the point $(1, 3, -1)$ parallel to the

a. x-axis **b.** y-axis **c.** z-axis

42. The set of points in space equidistant from the origin and the point $(0, 4, 0)$

43. The circle in which the plane through the point $(1, 1, 5)$ perpendicular to the z-axis meets the sphere of radius 6 centered at the origin

44. The set of points in space that lie 2 units from the point $(0, 0, 1)$ and, at the same time, 2 units from the point $(0, 0, -1)$

Inequalities to Describe Sets of Points

Write inequalities to describe the sets in Exercises 45–50.

45. The slab bounded by the planes $z = 0$ and $z = 1$ (planes included)

46. The solid cube in the first octant bounded by the coordinate planes and the planes $x = 2$, $y = 2$, and $z = 2$

47. The half-space consisting of the points on and below the xy-plane

48. The upper hemisphere of the sphere of radius 3 centered at the origin

49. The (a) interior and (b) exterior of the sphere of radius 3 centered at the point $(3, 3, 3)$

50. The closed region bounded by the spheres of radius 1 and radius 2 centered at the origin. (*Closed* means the spheres are to be included. Had we wanted the spheres left out, we would have asked for the *open* region bounded by the spheres. This is analogous to the way we use *closed* and *open* to describe intervals: *closed* means endpoints included, *open* means endpoints left out. Closed sets include boundaries; open sets leave them out.)

Spheres

Find the center C and the radius a for the spheres in Exercises 51–60.

51. $(x-4)^2 + y^2 + (z+4)^2 = 16$

52. $(x-1)^2 + \left(y + \frac{1}{2}\right)^2 + (z+3)^2 = 25$

53. $(x-\sqrt{8})^2 + (y+\sqrt{8})^2 + (z+\sqrt{8})^2 = 64$

54. $x^2 + \left(y + \frac{1}{3}\right)^2 + \left(z - \frac{1}{3}\right)^2 = \frac{16}{9}$

55. $x^2 + y^2 + z^2 + 14x - 14z = 0$

56. $x^2 + y^2 + z^2 - 6y + 8z = 0$

57. $3x^2 + 3y^2 + 3z^2 + x + y + z = 14$

58. $3x^2 + 3y^2 + 3z^2 + 2y - 2z = 9$

59. $x^2 + y^2 + z^2 - 8x + 10y - 10z = 11$

60. $(x-1)^2 + (y-2)^2 + (z+1)^2 = 103 + 2x + 4y - 2z$

Find equations for the spheres whose centers and radii are given in Exercises 61–64.

	Center	Radius
61.	$(2, 3, 4)$	$\sqrt{15}$
62.	$(0, -1, 5)$	2
63.	$\left(-4, \frac{5}{2}, -\frac{3}{4}\right)$	$\frac{7}{10}$
64.	$(0, -7, 0)$	7

Theory and Examples

65. Find a formula for the distance from the point $P(x, y, z)$ to the

a. x-axis. **b.** y-axis. **c.** z-axis.

66. Find a formula for the distance from the point $P(x, y, z)$ to the

a. xy-plane. **b.** yz-plane. **c.** xz-plane.

67. Perimeter of triangle with vertices $A(-2, 3, 2)$, $B(2, -2, 4)$, and $C(4, 5, 6)$.

68. Show that point $P(5, 3, 4)$ is equidistant from point $A(4, -3, 5)$ and $B(6, -3, 3)$

69. Find an equation for the set of all points equidistant from the planes $y = 3$ and $y = -1$.

70. Find an equation for the set of all points equidistant from the point $(0, 0, 2)$ and the xy-plane.

71. Find the point on the sphere $x^2 + (y - 3)^2 + (z + 5)^2 = 4$ nearest

a. the xy-plane. **b.** the point $(0, 7, -5)$.

72. Find the point equidistant from the points $(0, 0, 0)$, $(0, 4, 0)$, $(3, 0, 0)$, and $(2, 2, -3)$.

73. Find an equation for the set of points equidistant from the point $(0, 0, 2)$ and the x-axis.

74. Find an equation for the set of points equidistant from the y-axis and the plane $z = 6$.

75. Find an equation for the set of points equidistant from the

a. xy-plane and the yz-plane. **b.** x-axis and the y-axis.

76. Find all points that simultaneously lie 3 units from each of the points $(2, 0, 0)$, $(0, 2, 0)$, and $(0, 0, 2)$.

12.2 Vectors

Some of the things we measure are determined simply by their magnitudes. To record mass, length, or time, for example, we need only write down a number and name an appropriate unit of measure. We need more information to describe a force, displacement, or velocity. To describe a force, we need to record the direction in which it acts as well as how large it is. To describe a body's displacement, we have to say in what direction it moved as well as how far. To describe a body's velocity, we have to know its direction of motion, as well as how fast it is going. In this section we show how to represent things that have both magnitude and direction in the plane or in space.

Component Form

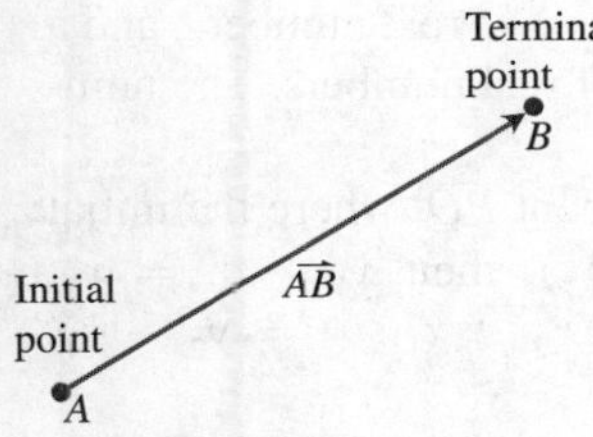

FIGURE 12.7 The directed line segment $\overrightarrow{AB}$ is called a vector.

A quantity such as force, displacement, or velocity is called a **vector** and is represented by a **directed line segment** (Figure 12.7). The arrow points in the direction of the action and its length gives the magnitude of the action in terms of a suitably chosen unit. For example, a force vector points in the direction in which the force acts and its length is a measure of the force's strength; a velocity vector points in the direction of motion and its length is the speed of the moving object. Figure 12.8 displays the velocity vector **v** at a specific location for a particle moving along a path in the plane or in space. (This application of vectors is studied in Chapter 13.)

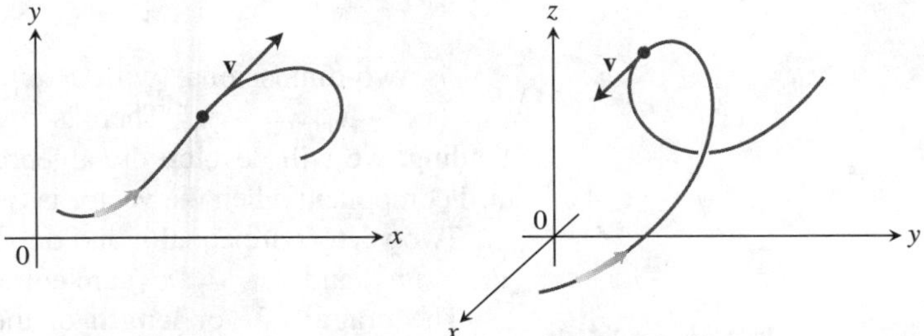

(a) two dimensions (b) three dimensions

FIGURE 12.8 The velocity vector of a particle moving along a path (a) in the plane (b) in space. The arrowhead on the path indicates the direction of motion of the particle.

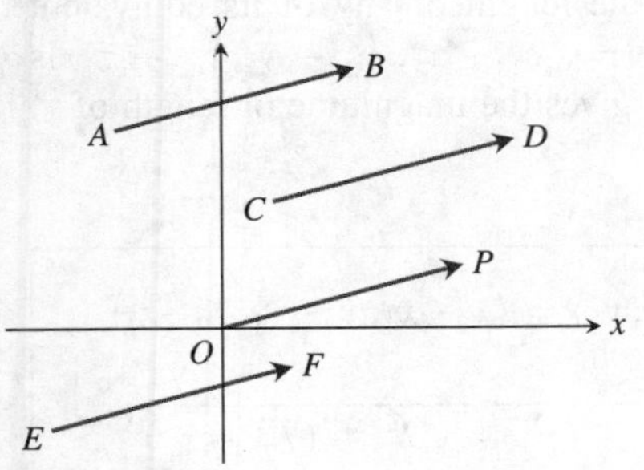

FIGURE 12.9 The four arrows in the plane (directed line segments) shown here have the same length and direction. They therefore represent the same vector, and we write $\overrightarrow{AB} = \overrightarrow{CD} = \overrightarrow{OP} = \overrightarrow{EF}$.

> **DEFINITIONS** The vector represented by the directed line segment $\overrightarrow{AB}$ has **initial point** A and **terminal point** B, and its **length** is denoted by $|\overrightarrow{AB}|$. Two vectors are **equal** if they have the same length and direction.

The arrows we use when we draw vectors are understood to represent the same vector if they have the same length, are parallel, and point in the same direction (Figure 12.9) regardless of the initial point.

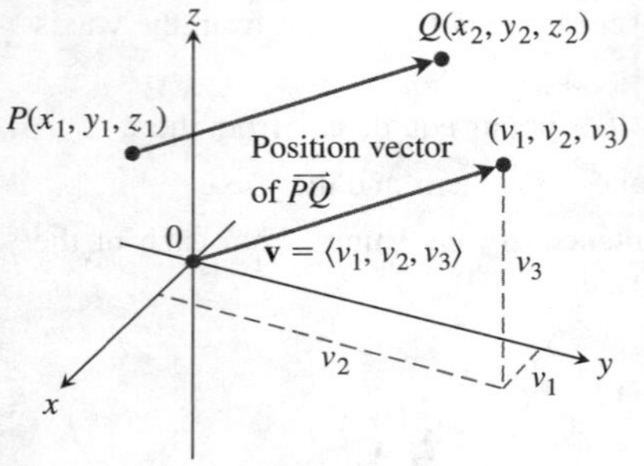

FIGURE 12.10 A vector $\overrightarrow{PQ}$ in standard position has its initial point at the origin. The directed line segments $\overrightarrow{PQ}$ and $\mathbf{v}$ are parallel and have the same length.

In texts, vectors are usually written in lowercase boldface letters—for example **u**, **v**, and **w**. Sometimes we use uppercase boldface letters, such as **F**, to denote a force vector. In handwritten form, it is customary to draw small arrows above the letters—for example $\vec{u}$, $\vec{v}$, $\vec{w}$, and $\vec{F}$.

We need a way to represent vectors algebraically so that we can be more precise about the direction of a vector. Let $\mathbf{v} = \overrightarrow{PQ}$. There is one directed line segment equal to $\overrightarrow{PQ}$ whose initial point is the origin (Figure 12.10). It is the representative of **v** in **standard position** and is the vector we normally use to represent **v**. We can specify **v** by writing the coordinates of its terminal point (v_1, v_2, v_3) when **v** is in standard position. If **v** is a vector in the plane, its terminal point (v_1, v_2) has two coordinates.

DEFINITION If **v** is a **two-dimensional** vector in the plane equal to the vector with initial point at the origin and terminal point (v_1, v_2), then the **component form** of **v** is

$$\mathbf{v} = \langle v_1, v_2 \rangle.$$

If **v** is a **three-dimensional** vector equal to the vector with initial point at the origin and terminal point (v_1, v_2, v_3), then the **component form** of **v** is

$$\mathbf{v} = \langle v_1, v_2, v_3 \rangle.$$

HISTORICAL BIOGRAPHY
Carl Friedrich Gauss
(1777–1855)
www.bit.ly/2xZSPyS

Thus a two-dimensional vector is an ordered pair $\mathbf{v} = \langle v_1, v_2 \rangle$ of real numbers, and a three-dimensional vector is an ordered triple $\mathbf{v} = \langle v_1, v_2, v_3 \rangle$ of real numbers. The numbers v_1, v_2, and v_3 are the **components** of **v**.

If $\mathbf{v} = \langle v_1, v_2, v_3 \rangle$ is represented by the directed line segment $\overrightarrow{PQ}$, where the initial point is $P(x_1, y_1, z_1)$ and the terminal point is $Q(x_2, y_2, z_2)$, then $x_1 + v_1 = x_2$, $y_1 + v_2 = y_2$, and $z_1 + v_3 = z_2$ (see Figure 12.10). Thus $v_1 = x_2 - x_1$, $v_2 = y_2 - y_1$, and $v_3 = z_2 - z_1$ are the components of $\overrightarrow{PQ}$.

In summary, given the points $P(x_1, y_1, z_1)$ and $Q(x_2, y_2, z_2)$, the standard position vector $\mathbf{v} = \langle v_1, v_2, v_3 \rangle$ equal to $\overrightarrow{PQ}$ is

$$\mathbf{v} = \langle x_2 - x_1, y_2 - y_1, z_2 - z_1 \rangle.$$

If **v** is two-dimensional with $P(x_1, y_1)$ and $Q(x_2, y_2)$ as points in the plane, then $\mathbf{v} = \langle x_2 - x_1, y_2 - y_1 \rangle$. There is no third component for planar vectors. With this understanding, we will develop the algebra of three-dimensional vectors and simply drop the third component when the vector is two-dimensional (a planar vector).

Two vectors are equal if and only if their standard position vectors are identical. Thus $\langle u_1, u_2, u_3 \rangle$ and $\langle v_1, v_2, v_3 \rangle$ are **equal** if and only if $u_1 = v_1$, $u_2 = v_2$, and $u_3 = v_3$.

The **magnitude** or **length** of the vector $\overrightarrow{PQ}$ is the length of any of its equivalent directed line segment representations. In particular, if $\mathbf{v} = \langle x_2 - x_1, y_2 - y_1, z_2 - z_1 \rangle$ is the component form of $\overrightarrow{PQ}$, then the distance formula gives the magnitude or length of **v**, denoted by the symbol $|\mathbf{v}|$ or (in some texts) $\|\mathbf{v}\|$.

The **magnitude** or **length** of the vector $\mathbf{v} = \langle v_1, v_2, v_3 \rangle = \langle x_2 - x_1, y_2 - y_1, z_2 - z_1 \rangle$ is the nonnegative number

$$|\mathbf{v}| = \sqrt{v_1^2 + v_2^2 + v_3^2} = \sqrt{(x_2 - x_1)^2 + (y_2 - y_1)^2 + (z_2 - z_1)^2}$$

(see Figure 12.10).

The only vector with length 0 is the **zero vector** $\mathbf{0} = \langle 0, 0 \rangle$ or $\mathbf{0} = \langle 0, 0, 0 \rangle$. This vector is also the only vector with no specific direction.

EXAMPLE 1 Find the **(a)** component form and **(b)** length of the vector with initial point $P(-3, 4, 1)$ and terminal point $Q(-5, 2, 2)$.

Solution

(a) The vector $\mathbf{v} = \overrightarrow{PQ}$ has components

$$v_1 = x_2 - x_1 = -5 - (-3) = -2, \qquad v_2 = y_2 - y_1 = 2 - 4 = -2,$$

and

$$v_3 = z_2 - z_1 = 2 - 1 = 1.$$

The component form of $\overrightarrow{PQ}$ is

$$\mathbf{v} = \langle -2, -2, 1 \rangle.$$

(b) The length, or magnitude, of $\mathbf{v} = \overrightarrow{PQ}$ is

$$|\mathbf{v}| = \sqrt{(-2)^2 + (-2)^2 + (1)^2} = \sqrt{9} = 3.$$ ■

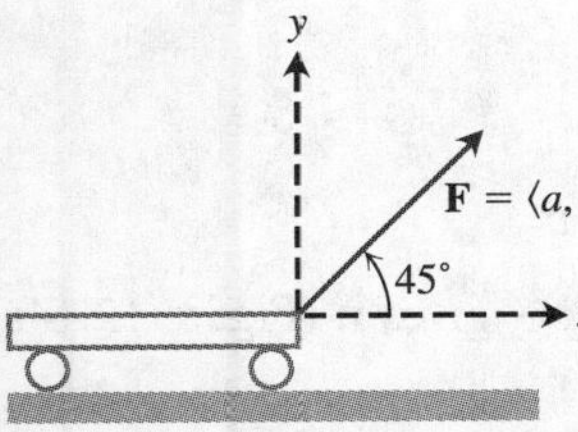

FIGURE 12.11 The force pulling the cart forward is represented by the vector **F** whose horizontal component is the effective force (Example 2).

EXAMPLE 2 A small cart is being pulled along a smooth horizontal floor with a 20-lb force **F** making a 45° angle to the floor (Figure 12.11). What is the *effective* force moving the cart forward?

Solution The effective force is the horizontal component of $\mathbf{F} = \langle a, b \rangle$, given by

$$a = |\mathbf{F}|\cos 45^\circ = (20)\left(\frac{\sqrt{2}}{2}\right) \approx 14.14 \text{ lb}.$$

Notice that **F** is a two-dimensional vector. ■

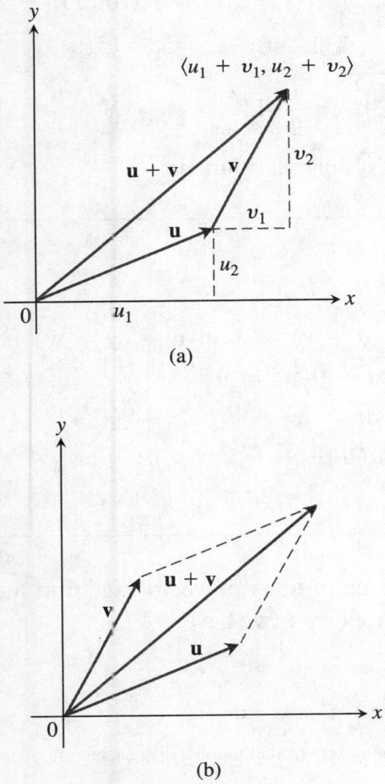

FIGURE 12.12 (a) Geometric interpretation of the vector sum. (b) The parallelogram law of vector addition in which both vectors are in standard position.

Vector Algebra Operations

Two principal operations involving vectors are *vector addition* and *scalar multiplication*. A **scalar** is simply a real number; we call it a scalar when we want to draw attention to the differences between numbers and vectors. Scalars can be positive, negative, or zero and are used to "scale" a vector by multiplication.

> **DEFINITIONS** Let $\mathbf{u} = \langle u_1, u_2, u_3 \rangle$ and $\mathbf{v} = \langle v_1, v_2, v_3 \rangle$ be vectors with k a scalar.
>
> **Addition:** $\mathbf{u} + \mathbf{v} = \langle u_1 + v_1, u_2 + v_2, u_3 + v_3 \rangle$
>
> **Scalar multiplication:** $k\mathbf{u} = \langle ku_1, ku_2, ku_3 \rangle$

We add vectors by adding the corresponding components of the vectors. We multiply a vector by a scalar by multiplying each component by the scalar. The definitions also apply to planar vectors, except in that case there are only two components, $\langle u_1, u_2 \rangle$ and $\langle v_1, v_2 \rangle$.

The definition of vector addition is illustrated geometrically for planar vectors in Figure 12.12a, where the initial point of one vector is placed at the terminal point of the other. Another interpretation is shown in Figure 12.12b. In this **parallelogram law** of addition, the sum, called the **resultant vector**, is the diagonal of the parallelogram. In physics, forces add vectorially, as do velocities, accelerations, and so on. So the force acting on a particle subject to two gravitational forces, for example, is obtained by adding the two force vectors.

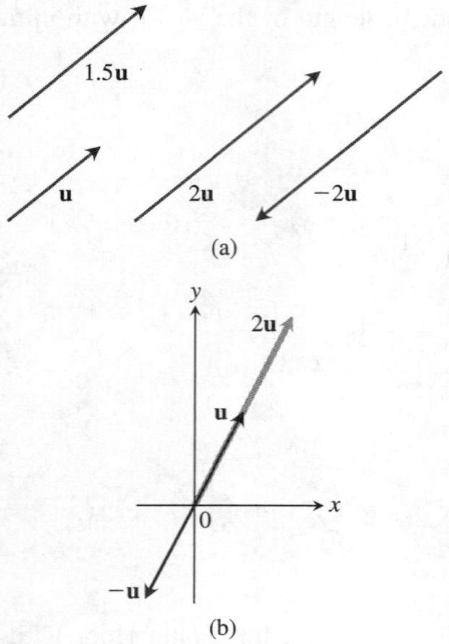

FIGURE 12.13 (a) Scalar multiples of **u**. (b) Scalar multiples of a vector **u** in standard position.

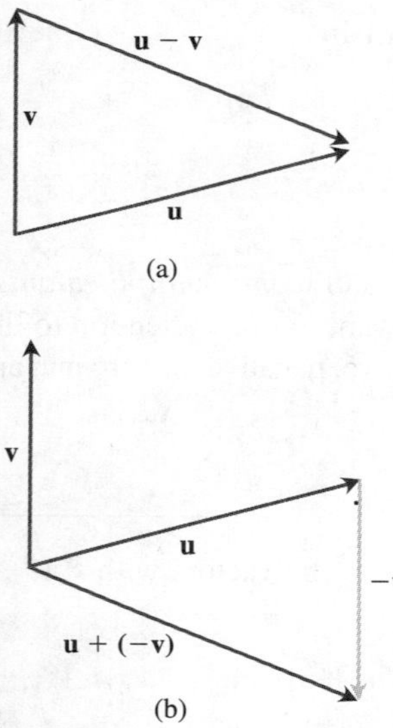

FIGURE 12.14 (a) The vector **u** − **v**, when added to **v**, gives **u**. (b) **u** − **v** = **u** + (−**v**).

Figure 12.13 displays a geometric interpretation of the product $k\mathbf{u}$ of the scalar k and vector **u**. If $k > 0$, then $k\mathbf{u}$ has the same direction as **u**; if $k < 0$, then the direction of $k\mathbf{u}$ is opposite to that of **u**. Comparing the lengths of **u** and $k\mathbf{u}$, we see that

$$
\begin{aligned}
|k\mathbf{u}| &= \sqrt{(ku_1)^2 + (ku_2)^2 + (ku_3)^2} = \sqrt{k^2(u_1{}^2 + u_2{}^2 + u_3{}^2)} \\
&= \sqrt{k^2}\sqrt{u_1{}^2 + u_2{}^2 + u_3{}^2} = |k|\,|\mathbf{u}|.
\end{aligned}
$$

The length of $k\mathbf{u}$ is the absolute value of the scalar k times the length of **u**. The vector $(-1)\mathbf{u} = -\mathbf{u}$ has the same length as **u** but points in the opposite direction. For $k \neq 0$, we often express the scalar multiple $(1/k)\mathbf{u}$ as $\mathbf{u}/k$.

The **difference** $\mathbf{u} - \mathbf{v}$ of two vectors is defined by

$$\mathbf{u} - \mathbf{v} = \mathbf{u} + (-\mathbf{v}).$$

If $\mathbf{u} = \langle u_1, u_2, u_3 \rangle$ and $\mathbf{v} = \langle v_1, v_2, v_3 \rangle$, then

$$\mathbf{u} - \mathbf{v} = \langle u_1 - v_1, u_2 - v_2, u_3 - v_3 \rangle.$$

Note that $(\mathbf{u} - \mathbf{v}) + \mathbf{v} = \mathbf{u}$, so adding the vector $(\mathbf{u} - \mathbf{v})$ to **v** gives **u** (Figure 12.14a). Figure 12.14b shows the difference $\mathbf{u} - \mathbf{v}$ as the sum $\mathbf{u} + (-\mathbf{v})$.

EXAMPLE 3 Let $\mathbf{u} = \langle -1, 3, 1 \rangle$ and $\mathbf{v} = \langle 4, 7, 0 \rangle$. Find the components of

(a) $2\mathbf{u} + 3\mathbf{v}$ **(b)** $\mathbf{u} - \mathbf{v}$ **(c)** $\left|\frac{1}{2}\mathbf{u}\right|$.

Solution

(a) $2\mathbf{u} + 3\mathbf{v} = 2\langle -1, 3, 1 \rangle + 3\langle 4, 7, 0 \rangle = \langle -2, 6, 2 \rangle + \langle 12, 21, 0 \rangle = \langle 10, 27, 2 \rangle$

(b) $\mathbf{u} - \mathbf{v} = \langle -1, 3, 1 \rangle - \langle 4, 7, 0 \rangle = \langle -1 - 4, 3 - 7, 1 - 0 \rangle = \langle -5, -4, 1 \rangle$

(c) $\left|\frac{1}{2}\mathbf{u}\right| = \left|\left\langle -\frac{1}{2}, \frac{3}{2}, \frac{1}{2} \right\rangle\right| = \sqrt{\left(-\frac{1}{2}\right)^2 + \left(\frac{3}{2}\right)^2 + \left(\frac{1}{2}\right)^2} = \frac{1}{2}\sqrt{11}.$ ■

Vector operations have many of the properties of ordinary arithmetic.

> **Properties of Vector Operations**
> Let **u**, **v**, **w** be vectors and a, b be scalars.
>
> 1. $\mathbf{u} + \mathbf{v} = \mathbf{v} + \mathbf{u}$
> 2. $(\mathbf{u} + \mathbf{v}) + \mathbf{w} = \mathbf{u} + (\mathbf{v} + \mathbf{w})$
> 3. $\mathbf{u} + \mathbf{0} = \mathbf{u}$
> 4. $\mathbf{u} + (-\mathbf{u}) = \mathbf{0}$
> 5. $0\mathbf{u} = \mathbf{0}$
> 6. $1\mathbf{u} = \mathbf{u}$
> 7. $a(b\mathbf{u}) = (ab)\mathbf{u}$
> 8. $a(\mathbf{u} + \mathbf{v}) = a\mathbf{u} + a\mathbf{v}$
> 9. $(a + b)\mathbf{u} = a\mathbf{u} + b\mathbf{u}$

These properties are readily verified using the definitions of vector addition and multiplication by a scalar. For instance, to establish Property 1, we have

$$
\begin{aligned}
\mathbf{u} + \mathbf{v} &= \langle u_1, u_2, u_3 \rangle + \langle v_1, v_2, v_3 \rangle \\
&= \langle u_1 + v_1, u_2 + v_2, u_3 + v_3 \rangle && \text{Definition of vector addition} \\
&= \langle v_1 + u_1, v_2 + u_2, v_3 + u_3 \rangle && \text{Commutativity of real numbers (in each component)} \\
&= \langle v_1, v_2, v_3 \rangle + \langle u_1, u_2, u_3 \rangle && \text{Definition of vector addition} \\
&= \mathbf{v} + \mathbf{u}.
\end{aligned}
$$

When three or more space vectors lie in the same plane, we say they are **coplanar** vectors. For example, the vectors **u**, **v**, and $\mathbf{u} + \mathbf{v}$ are always coplanar.

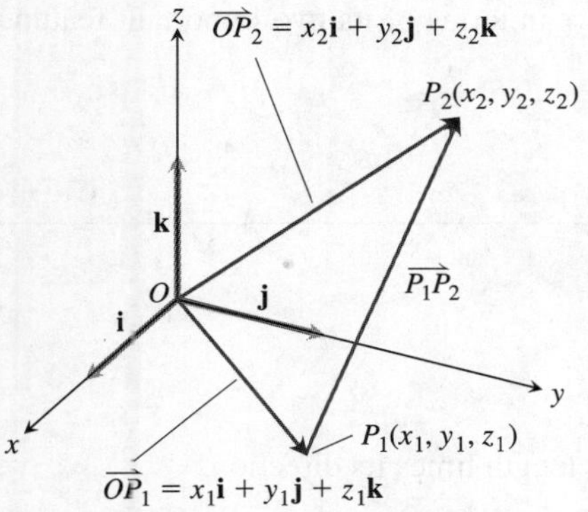

FIGURE 12.15 The vector from P_1 to P_2 is $\overrightarrow{P_1P_2} = (x_2 - x_1)\mathbf{i} + (y_2 - y_1)\mathbf{j} + (z_2 - z_1)\mathbf{k}$.

Unit Vectors

A vector $\mathbf{v}$ of length 1 is called a **unit vector**. The **standard unit vectors** are

$$\mathbf{i} = \langle 1, 0, 0\rangle, \quad \mathbf{j} = \langle 0, 1, 0\rangle, \quad \text{and} \quad \mathbf{k} = \langle 0, 0, 1\rangle.$$

Any vector $\mathbf{v} = \langle v_1, v_2, v_3\rangle$ can be written as a *linear combination* of the standard unit vectors as follows:

$$\begin{aligned}\mathbf{v} = \langle v_1, v_2, v_3\rangle &= \langle v_1, 0, 0\rangle + \langle 0, v_2, 0\rangle + \langle 0, 0, v_3\rangle \\ &= v_1\langle 1, 0, 0\rangle + v_2\langle 0, 1, 0\rangle + v_3\langle 0, 0, 1\rangle \\ &= v_1\mathbf{i} + v_2\mathbf{j} + v_3\mathbf{k}.\end{aligned}$$

We call the scalar (or number) v_1 the **i-component** of the vector $\mathbf{v}$, v_2 the **j-component**, and v_3 the **k-component**. As shown in Figure 12.15, the component form of the vector from $P_1(x_1, y_1, z_1)$ to $P_2(x_2, y_2, z_2)$ is

$$\overrightarrow{P_1P_2} = (x_2 - x_1)\mathbf{i} + (y_2 - y_1)\mathbf{j} + (z_2 - z_1)\mathbf{k}.$$

If $\mathbf{v} \neq \mathbf{0}$, then its length $|\mathbf{v}|$ is not zero and

$$\left|\frac{1}{|\mathbf{v}|}\mathbf{v}\right| = \frac{1}{|\mathbf{v}|}|\mathbf{v}| = 1.$$

That is, if the vector $\mathbf{v}$ is not the zero vector, then $\mathbf{v}/|\mathbf{v}|$ is a **unit vector in the direction of** $\mathbf{v}$, and it is also called the **direction** of $\mathbf{v}$.

EXAMPLE 4 Find a unit vector $\mathbf{u}$ in the direction of the vector from $P_1(1, 0, 1)$ to $P_2(3, 2, 0)$.

Solution We write $\overrightarrow{P_1P_2}$ as a linear combination of the standard unit vectors and then divide it by its length:

$$\overrightarrow{P_1P_2} = (3 - 1)\mathbf{i} + (2 - 0)\mathbf{j} + (0 - 1)\mathbf{k} = 2\mathbf{i} + 2\mathbf{j} - \mathbf{k}$$

$$|\overrightarrow{P_1P_2}| = \sqrt{(2)^2 + (2)^2 + (-1)^2} = \sqrt{4 + 4 + 1} = \sqrt{9} = 3$$

$$\mathbf{u} = \frac{\overrightarrow{P_1P_2}}{|\overrightarrow{P_1P_2}|} = \frac{2\mathbf{i} + 2\mathbf{j} - \mathbf{k}}{3} = \frac{2}{3}\mathbf{i} + \frac{2}{3}\mathbf{j} - \frac{1}{3}\mathbf{k}.$$

This unit vector $\mathbf{u}$ is the direction of $\overrightarrow{P_1P_2}$. ■

EXAMPLE 5 If $\mathbf{v} = 3\mathbf{i} - 4\mathbf{j}$ is a velocity vector, express $\mathbf{v}$ as a product of its magnitude (its speed) times its direction.

Solution Speed is the magnitude (length) of $\mathbf{v}$:

$$|\mathbf{v}| = \sqrt{(3)^2 + (-4)^2} = \sqrt{9 + 16} = 5.$$

The unit vector $\mathbf{v}/|\mathbf{v}|$ is the direction of $\mathbf{v}$:

$$\frac{\mathbf{v}}{|\mathbf{v}|} = \frac{3\mathbf{i} - 4\mathbf{j}}{5} = \frac{3}{5}\mathbf{i} - \frac{4}{5}\mathbf{j}.$$

So

$$\mathbf{v} = 3\mathbf{i} - 4\mathbf{j} = \underbrace{5}_{\text{Length (speed)}}\underbrace{\left(\frac{3}{5}\mathbf{i} - \frac{4}{5}\mathbf{j}\right)}_{\text{Direction of motion}}.$$ ■

HISTORICAL BIOGRAPHY
Hermann Grassmann
(1809–1877)
www.bit.ly/2zIRwGF

In summary, we can express any nonzero vector $\mathbf{v}$ in terms of its two important features, length and direction, by writing $\mathbf{v} = |\mathbf{v}|\dfrac{\mathbf{v}}{|\mathbf{v}|}$.

> If $\mathbf{v} \neq \mathbf{0}$, then
>
> **1.** $\dfrac{\mathbf{v}}{|\mathbf{v}|}$ is a unit vector called the direction of $\mathbf{v}$;
>
> **2.** the equation $\mathbf{v} = |\mathbf{v}|\dfrac{\mathbf{v}}{|\mathbf{v}|}$ expresses $\mathbf{v}$ as its length times its direction.

EXAMPLE 6 A force of 6 newtons is applied in the direction of the vector $\mathbf{v} = 2\mathbf{i} + 2\mathbf{j} - \mathbf{k}$. Express the force $\mathbf{F}$ as a product of its magnitude and direction.

Solution The force vector has magnitude 6 and direction $\dfrac{\mathbf{v}}{|\mathbf{v}|}$, so

$$\mathbf{F} = 6\frac{\mathbf{v}}{|\mathbf{v}|} = 6\frac{2\mathbf{i} + 2\mathbf{j} - \mathbf{k}}{\sqrt{2^2 + 2^2 + (-1)^2}} = 6\frac{2\mathbf{i} + 2\mathbf{j} - \mathbf{k}}{3}$$

$$= 6\left(\frac{2}{3}\mathbf{i} + \frac{2}{3}\mathbf{j} - \frac{1}{3}\mathbf{k}\right).$$ ■

Midpoint of a Line Segment

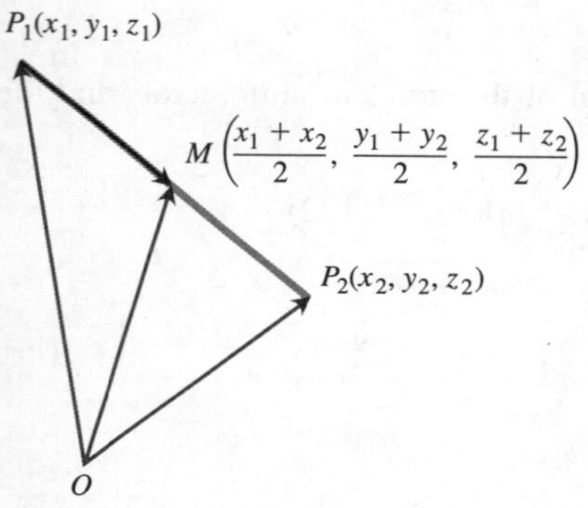

FIGURE 12.16 The coordinates of the midpoint are the averages of the coordinates of P_1 and P_2.

Vectors are often useful in geometry. For example, the coordinates of the midpoint of a line segment are found by averaging.

> The **midpoint** M of the line segment joining points $P_1(x_1, y_1, z_1)$ and $P_2(x_2, y_2, z_2)$ is the point
>
> $$\left(\frac{x_1 + x_2}{2}, \frac{y_1 + y_2}{2}, \frac{z_1 + z_2}{2}\right).$$

To see why, observe (Figure 12.16) that

$$\overrightarrow{OM} = \overrightarrow{OP_1} + \frac{1}{2}(\overrightarrow{P_1P_2}) = \overrightarrow{OP_1} + \frac{1}{2}(\overrightarrow{OP_2} - \overrightarrow{OP_1})$$

$$= \frac{1}{2}(\overrightarrow{OP_1} + \overrightarrow{OP_2})$$

$$= \frac{x_1 + x_2}{2}\mathbf{i} + \frac{y_1 + y_2}{2}\mathbf{j} + \frac{z_1 + z_2}{2}\mathbf{k}.$$

EXAMPLE 7 The midpoint of the segment joining $P_1(3, -2, 0)$ and $P_2(7, 4, 4)$ is

$$\left(\frac{3 + 7}{2}, \frac{-2 + 4}{2}, \frac{0 + 4}{2}\right) = (5, 1, 2).$$ ■

Applications

An important application of vectors occurs in navigation.

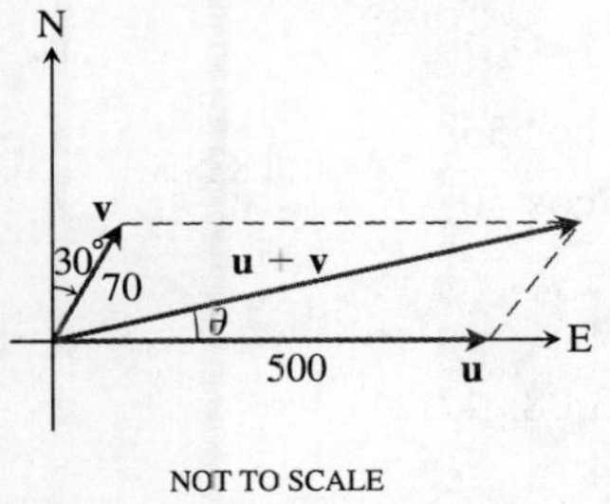

FIGURE 12.17 Vectors representing the velocities of the airplane **u** and tailwind **v** in Example 8.

EXAMPLE 8 A jet airliner, flying due east at 500 mph in still air, encounters a 70-mph tailwind blowing in the direction 60° north of east. The airplane holds its compass heading due east but, because of the wind, acquires a new ground speed and direction. What are they?

Solution If **u** is the velocity of the airplane alone and **v** is the velocity of the tailwind, then $|\mathbf{u}| = 500$ and $|\mathbf{v}| = 70$ (Figure 12.17). The velocity of the airplane with respect to the ground is given by the magnitude and direction of the resultant vector $\mathbf{u} + \mathbf{v}$. If we let the positive x-axis represent east and the positive y-axis represent north, then the component forms of **u** and **v** are

$$\mathbf{u} = \langle 500, 0 \rangle \quad \text{and} \quad \mathbf{v} = \langle 70 \cos 60°, 70 \sin 60° \rangle = \langle 35, 35\sqrt{3} \rangle.$$

Therefore,

$$\mathbf{u} + \mathbf{v} = \langle 535, 35\sqrt{3} \rangle = 535\mathbf{i} + 35\sqrt{3}\mathbf{j}$$

$$|\mathbf{u} + \mathbf{v}| = \sqrt{535^2 + (35\sqrt{3})^2} \approx 538.4$$

and

$$\theta = \arctan \frac{35\sqrt{3}}{535} \approx 6.5°. \qquad \text{Figure 12.17}$$

The new ground speed of the airplane is about 538.4 mph, and its new direction is about 6.5° north of east. ■

Another important application occurs in physics and engineering when several forces are acting on a single object.

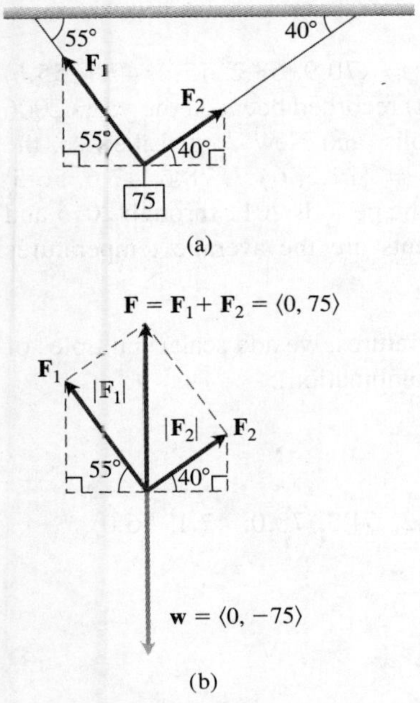

FIGURE 12.18 The suspended weight in Example 9.

EXAMPLE 9 A 75-N weight is suspended by two wires, as shown in Figure 12.18a. Find the forces $\mathbf{F}_1$ and $\mathbf{F}_2$ acting in both wires.

Solution The force vectors $\mathbf{F}_1$ and $\mathbf{F}_2$ have magnitudes $|\mathbf{F}_1|$ and $|\mathbf{F}_2|$ and components that are measured in newtons. The resultant force is the sum $\mathbf{F}_1 + \mathbf{F}_2$, which must be equal in magnitude to the weight vector **w** but acting in the opposite (or upward) direction (see Figure 12.18b). It follows from the figure that

$$\mathbf{F}_1 = \langle -|\mathbf{F}_1| \cos 55°, |\mathbf{F}_1| \sin 55° \rangle \quad \text{and} \quad \mathbf{F}_2 = \langle |\mathbf{F}_2| \cos 40°, |\mathbf{F}_2| \sin 40° \rangle.$$

Since $\mathbf{F}_1 + \mathbf{F}_2 = \langle 0, 75 \rangle$, the resultant vector leads to the system of equations

$$-|\mathbf{F}_1| \cos 55° + |\mathbf{F}_2| \cos 40° = 0$$
$$|\mathbf{F}_1| \sin 55° + |\mathbf{F}_2| \sin 40° = 75.$$

Solving for $|\mathbf{F}_2|$ in the first equation and substituting the result into the second equation, we get

$$|\mathbf{F}_2| = \frac{|\mathbf{F}_1| \cos 55°}{\cos 40°} \quad \text{and} \quad |\mathbf{F}_1| \sin 55° + \frac{|\mathbf{F}_1| \cos 55°}{\cos 40°} \sin 40° = 75.$$

It follows that

$$|\mathbf{F}_1| = \frac{75}{\sin 55° + \cos 55° \tan 40°} \approx 57.67 \text{ N}$$

and

$$|\mathbf{F}_2| = \frac{|\mathbf{F}_1| \cos 55°}{\cos 40°} = \frac{75}{\sin 55° + \cos 55° \tan 40°} \frac{\cos 55°}{\cos 40°} \approx 43.18 \text{ N}.$$

The force vectors are then

$$\mathbf{F}_1 = \langle -|\mathbf{F}_1| \cos 55°, |\mathbf{F}_1| \sin 55° \rangle \approx \langle -33.08, 47.24 \rangle$$

and

$$\mathbf{F}_2 = \langle |\mathbf{F}_2| \cos 40°, |\mathbf{F}_2| \sin 40° \rangle \approx \langle 33.08, 27.76 \rangle.$$ ■

Vectors in *n* Dimensions

So far in this section, we introduced two- and three-dimensional vectors. We extend these notions by considering an n-**dimensional vector** $\mathbf{v} = \langle v_1, v_2, \ldots, v_n \rangle$ (an n-tuple of real numbers). We define

1. the magnitude or length of $\mathbf{v}$: $|\mathbf{v}| = \sqrt{v_1^2 + v_2^2 + \cdots + v_n^2}$;

2. addition of two vectors:

$$\langle u_1, u_2, \ldots, u_n \rangle + \langle v_1, v_2, \ldots, v_n \rangle = \langle u_1 + v_1, u_2 + v_2, \ldots, u_n + v_n \rangle;$$

3. scalar multiplication of a vector by a real number:

$$k \langle v_1, v_2, \ldots, v_n \rangle = \langle kv_1, kv_2, \ldots, kv_n \rangle.$$

The Properties of Vector Operations stated earlier in this section hold for n-dimensional vectors.

When discussing vectors in this text, we will usually focus on the cases where $n = 2$ and $n = 3$ since these correspond to vectors in the plane and vectors in three-dimensional space, which are used in many applications (for instance, such vectors can represent velocity or force). Vectors with more than three components do arise in applications, but in those contexts they do not correspond to line segments in the plane or in three-dimensional space.

EXAMPLE 10 The components of the vector $\mathbf{u} = \langle 70.9, 63.2, 77.2, 47.1, 55.6 \rangle$ contain the average temperatures (in degrees Fahrenheit) recorded between the years 2006 and 2010 in Houston, Los Angeles, Miami, Minneapolis, and New York. Likewise, the vectors $\mathbf{v} = \langle 71.2, 64.3, 78.0, 47.1, 56.0 \rangle$ and $\mathbf{w} = \langle 72.5, 64.7, 78.6, 47.6, 56.5 \rangle$ represent average temperatures in the same cities over the periods 2011 through 2015 and 2016 through 2020. Find the vector whose components are the average temperatures between the year 2006 and the year 2020 in these cities.

Solution To find the vector containing average temperatures, we add scalar multiples of the three vectors (in other words, we evaluate a linear combination):

$$\begin{aligned} \mathbf{a} &= \frac{1}{3}\mathbf{u} + \frac{1}{3}\mathbf{v} + \frac{1}{3}\mathbf{w} \\ &= \frac{1}{3}\langle 70.9, 63.2, 77.2, 47.1, 55.6 \rangle + \frac{1}{3}\langle 71.2, 64.3, 78.0, 47.1, 56.0 \rangle \\ &\quad + \frac{1}{3}\langle 72.5, 64.7, 78.6, 47.6, 56.5 \rangle \\ &\approx \langle 71.5, 64.1, 77.9, 47.3, 56.0 \rangle. \end{aligned}$$ ■

EXAMPLE 11 A rectangular grayscale image m pixels (dots) wide and n pixels tall can be represented on a computer as a vector with $m \cdot n$ components. A common format assigns an integer between 0 and 255 to each pixel, with 255 corresponding to the highest intensity (white), 0 to the lowest (black), and intermediate values to various shades of gray.

Using this format, the 400×400 image containing all white pixels is represented by the 160,000-dimensional vector $\mathbf{w} = \langle 255, \ldots, 255 \rangle$. Figures 12.19a and b can be similarly represented by vectors $\mathbf{u}$ and $\mathbf{v}$, respectively.

In parts c, d, and e of Figure 12.19, we show different linear combinations of the vectors $\mathbf{u}$ and $\mathbf{v}$. Notice how changing the scalar in front of each vector affects the resulting image.

In Figure 12.19f, the image corresponds to the difference $\mathbf{w} - \mathbf{v}$, effectively inverting the grayscale in that image. ■

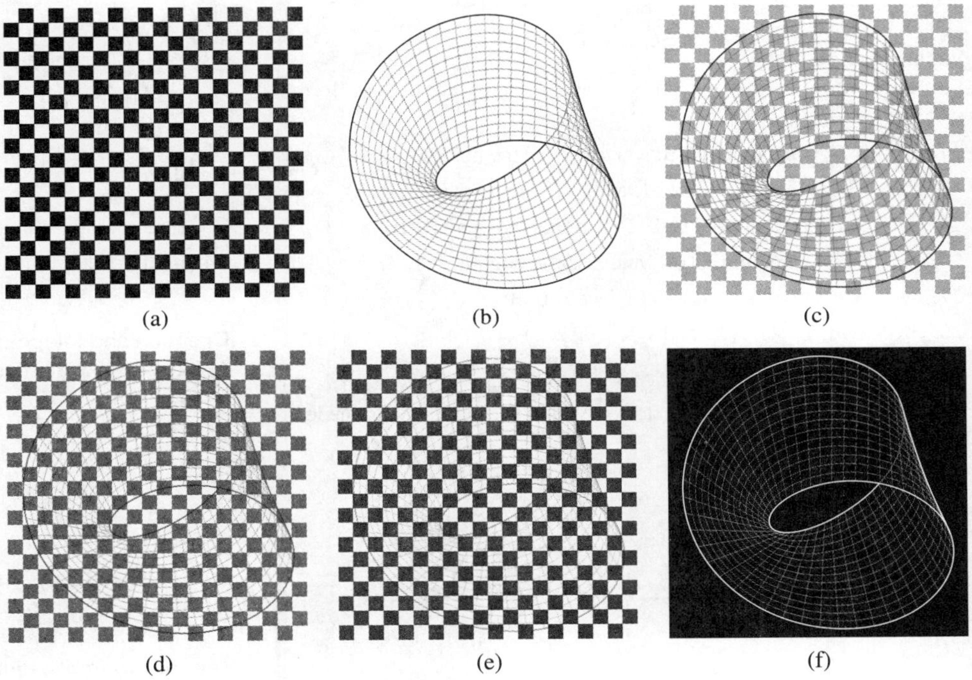

FIGURE 12.19 Each 400×400 pixel image corresponds to a 160,000-dimensional vector: (a) $\mathbf{u}$; (b) $\mathbf{v}$; (c) $\frac{1}{4}\mathbf{u} + \frac{3}{4}\mathbf{v}$; (d) $\frac{1}{2}\mathbf{u} + \frac{1}{2}\mathbf{v}$; (e) $\frac{3}{4}\mathbf{u} + \frac{1}{4}\mathbf{v}$; (f) $\mathbf{w} - \mathbf{v}$.

EXERCISES 12.2

Vectors in the Plane

In Exercises 1–8, let $\mathbf{u} = \langle 3, -2 \rangle$ and $\mathbf{v} = \langle -2, 5 \rangle$. Find the **(a)** component form and **(b)** magnitude (length) of the vector.

1. $3\mathbf{u}$

2. $-7\mathbf{v}$

3. $\mathbf{u} + \mathbf{v}$

4. $\mathbf{u} - \mathbf{v}$

5. $3\mathbf{u} - 2\mathbf{v}$

6. $-2\mathbf{u} + 5\mathbf{v}$

7. $\frac{3}{5}\mathbf{u} + \frac{4}{5}\mathbf{v}$

8. $\frac{-2\mathbf{u}}{13} + \frac{10\mathbf{v}}{13}$

In Exercises 9–16, find the component form of the vector.

9. The vector $\overrightarrow{PQ}$ where $P=(1, 2)$ and $Q=(3, -1)$

10. The vector $\overrightarrow{OP}$, where O is the origin and P is the midpoint of segment RS, where $R = (2, -1)$ and $S = (-4, 3)$

11. The vector from the point $A=(3, 4)$ to the origin

12. The sum of $\overrightarrow{AB}$ and $\overrightarrow{CD}$, where $A = (1, -1)$, $B = (2, 0)$, $C = (-1, 3)$, and $D = (-2, 2)$

13. The unit vector that makes an angle $\theta = \pi/6$ with the positive x-axis

14. The unit vector that makes an angle $\theta = -3\pi/4$ with the positive x-axis

15. The unit vector obtained by rotating the vector $\langle 0, 1\rangle$ by 120° counterclockwise about the origin

16. The unit vector obtained by rotating the vector $\langle 1, 0\rangle$ by 135° counterclockwise about the origin

Vectors in Space

In Exercises 17–22, express each vector in the form $\mathbf{w} = w_1\mathbf{i} + w_2\mathbf{j} + w_3\mathbf{k}$.

17. $\overrightarrow{P_1P_2}$ if P_1 is the point $(2, 3, -1)$ and P_2 is the point $(1, 7, -3)$

18. $\overrightarrow{P_1P_2}$ if P_1 is the point $(1, 2, 0)$ and P_2 is the point $(-3, 0, 5)$

19. $\overrightarrow{AB}$ if A is the point $(-7, -8, 1)$ and B is the point $(-10, 8, 1)$

20. $\overrightarrow{AB}$ if A is the point $(2, 0, 4)$ and B is the point $(-2, 1, 3)$

21. $5\mathbf{u} - \mathbf{v}$ if $\mathbf{u} = \langle 1, 1, -1\rangle$ and $\mathbf{v} = \langle 2, 0, 3\rangle$

22. $-\mathbf{u}+2\mathbf{v}$ if $\mathbf{u}=\langle 1, 0, 1\rangle$ and $\mathbf{v}=\langle -2, 2, 2\rangle$

Geometric Representations

In Exercises 23 and 24, copy vectors **u**, **v**, and **w** head to tail as needed to sketch the indicated vector.

23.

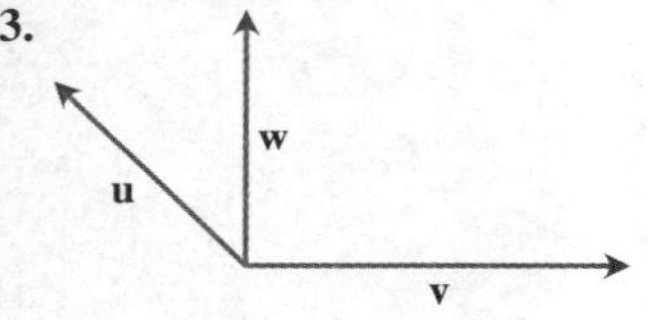

a. $\mathbf{u} + \mathbf{v}$

b. $\mathbf{u} + \mathbf{v} + \mathbf{w}$

c. $\mathbf{u} - \mathbf{v}$

d. $\mathbf{u} - \mathbf{w}$

24.

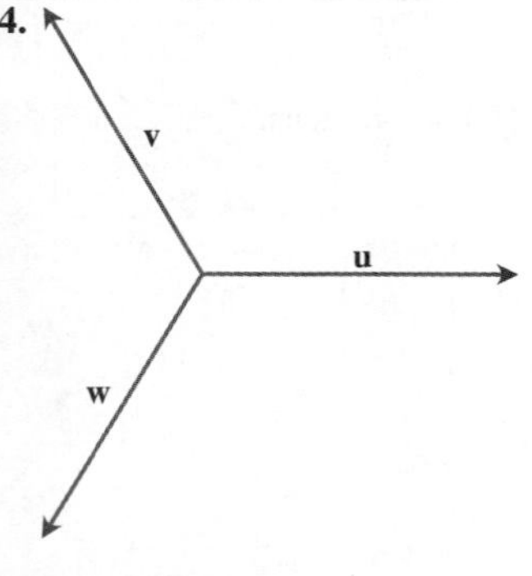

a. $\mathbf{u} - \mathbf{v}$

b. $\mathbf{u} - \mathbf{v} + \mathbf{w}$

c. $2\mathbf{u} - \mathbf{v}$

d. $\mathbf{u} + \mathbf{v} + \mathbf{w}$

Length and Direction

In Exercises 25–30, express each vector as a product of its length and direction.

25. $\mathbf{i}+2\mathbf{j}-2\mathbf{k}$

26. $9\mathbf{i} - 2\mathbf{j} + 6\mathbf{k}$

27. $5\mathbf{k}$

28. $\frac{1}{4}\mathbf{i}+\frac{2}{4}\mathbf{k}$

29. $\frac{1}{\sqrt{6}}\mathbf{i} - \frac{1}{\sqrt{6}}\mathbf{j} - \frac{1}{\sqrt{6}}\mathbf{k}$

30. $\frac{1}{\sqrt{5}}\mathbf{i}+\frac{1}{\sqrt{5}}\mathbf{j}+\frac{1}{\sqrt{5}}\mathbf{k}$

31. Find the vectors whose lengths and directions are given. Try to do the calculations without writing.

Length	Direction
a. 2	$\mathbf{i}$
b. $\sqrt{3}$	$-\mathbf{k}$
c. $\frac{1}{2}$	$\frac{3}{5}\mathbf{j} + \frac{4}{5}\mathbf{k}$
d. 7	$\frac{6}{7}\mathbf{i} - \frac{2}{7}\mathbf{j} + \frac{3}{7}\mathbf{k}$

32. Find the vectors whose lengths and directions are given. Try to do the calculations without writing.

Length	Direction
a. 7	$-\mathbf{j}$
b. $\sqrt{2}$	$-\frac{3}{5}\mathbf{i} - \frac{4}{5}\mathbf{k}$
c. $\frac{13}{12}$	$\frac{3}{13}\mathbf{i} - \frac{4}{13}\mathbf{j} - \frac{12}{13}\mathbf{k}$
d. $a > 0$	$\frac{1}{\sqrt{2}}\mathbf{i} + \frac{1}{\sqrt{3}}\mathbf{j} - \frac{1}{\sqrt{6}}\mathbf{k}$

33. Find a vector of magnitude 5 in the direction of $\mathbf{v}=7\mathbf{i}-4\mathbf{k}$.

34. Find a vector of magnitude 3 in the direction opposite to the direction of $\mathbf{v} = (1/2)\mathbf{i} - (1/2)\mathbf{j} - (1/2)\mathbf{k}$.

Direction and Midpoints

In Exercises 35–38, find **a.** the direction of $\overrightarrow{P_1P_2}$ and **b.** the midpoint of line segment P_1P_2.

35. $P_1(-1, 1, 5)$ $P_2(2, 5, 0)$

36. $P_1(1, 4, 5)$ $P_2(4, -2, 7)$

37. $P_1(3, 4, 5)$ $P_2(2, 3, 4)$

38. $P_1(0, 0, 0)$ $P_2(2, -2, -2)$

39. If $\overrightarrow{AB}=2\mathbf{i}+6\mathbf{j}-4\mathbf{k}$ and B is the point $(2, 3, 1)$, find A.

40. If $\overrightarrow{AB}=-5\mathbf{i}+2\mathbf{j}+7\mathbf{k}$ and A is the point $(-1, -2, 5)$, find B.

Theory and Applications

41. Linear combination Let $\mathbf{u}=3\mathbf{i}+2\mathbf{j}$, $\mathbf{v}=\mathbf{i}+2\mathbf{j}$, $\mathbf{w}=2\mathbf{i}-\mathbf{j}$. Find scalars a and b such that $\mathbf{u}=a\mathbf{v}+b\mathbf{w}$.

42. Linear combination Let $\mathbf{u} = \mathbf{i} - 2\mathbf{j}$, $\mathbf{v} = 2\mathbf{i} + 3\mathbf{j}$, and $\mathbf{w} = \mathbf{i} + \mathbf{j}$. Write $\mathbf{u} = \mathbf{u}_1 + \mathbf{u}_2$, where $\mathbf{u}_1$ is parallel to $\mathbf{v}$ and $\mathbf{u}_2$ is parallel to $\mathbf{w}$. (See Exercise 41.)

43. Linear combination Let $\mathbf{u} = \langle 1, 2, 1\rangle$, $\mathbf{v} = \langle 1, -1, -1\rangle$, $\mathbf{w} = \langle 1, 1, -1\rangle$, and $\mathbf{z} = \langle 2, -3, -4\rangle$. Find scalars a, b, and c such that $\mathbf{z} = a\mathbf{u} + b\mathbf{v} + c\mathbf{w}$.

44. Linear combination Let $\mathbf{u} = \langle 1, 2, 2\rangle$, $\mathbf{v} = \langle 1, -1, -1\rangle$, $\mathbf{w} = \langle 1, 3, -1\rangle$, and $\mathbf{z} = \langle 2, 11, 8\rangle$. Write $\mathbf{z} = \mathbf{u}_1 + \mathbf{u}_2 + \mathbf{u}_3$, where $\mathbf{u}_1$ is parallel to $\mathbf{u}$, $\mathbf{u}_2$ is parallel to $\mathbf{v}$, and $\mathbf{u}_3$ is parallel to $\mathbf{w}$. What are $\mathbf{u}_1$, $\mathbf{u}_2$, $\mathbf{u}_3$?

When solving Exercises 45–50, you may need to use a calculator or a computer.

45. Velocity An airplane is flying in the direction 35° west of north at 900 km/hr. Find the component form of the velocity of the airplane, assuming that the positive x-axis represents due east and the positive y-axis represents due north.

46. (*Continuation of Example 8.*) What speed and direction should the jetliner in Example 8 have in order for the resultant vector to be 500 mph due east?

47. Consider a 100-N weight suspended by two wires as shown in the accompanying figure. Find the magnitudes and components of the force vectors $\mathbf{F}_1$ and $\mathbf{F}_2$.

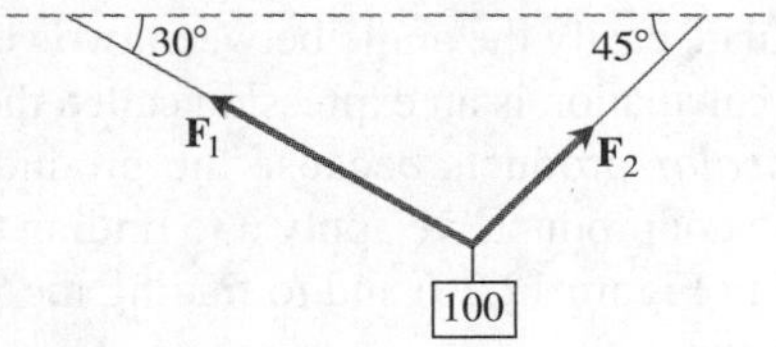

48. Consider a 50-N weight suspended by two wires as shown in the accompanying figure. If the magnitude of vector $\mathbf{F}_1$ is 35 N, find angle α and the magnitude of vector $\mathbf{F}_2$.

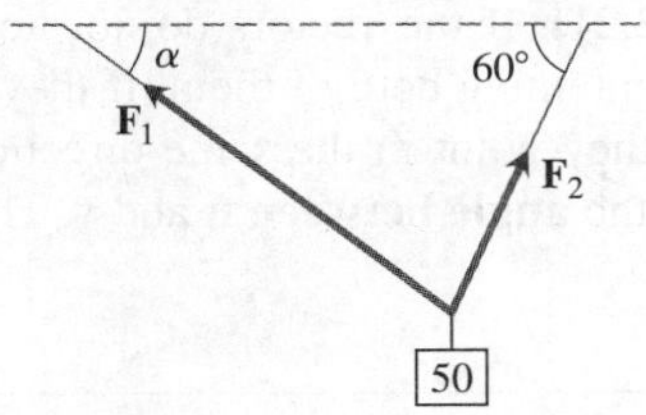

49. Consider a w-N weight suspended by two wires as shown in the accompanying figure. If the magnitude of vector $\mathbf{F}_2$ is 100 N, find w and the magnitude of vector $\mathbf{F}_1$.

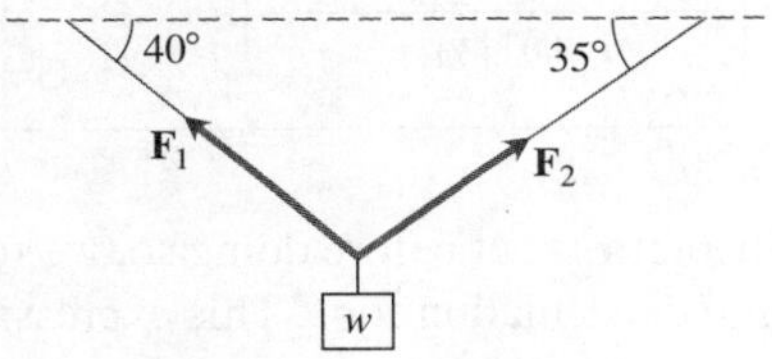

50. Consider a 25-N weight suspended by two wires as shown in the accompanying figure. If the magnitudes of vectors $\mathbf{F}_1$ and $\mathbf{F}_2$ are both 75 N, then angles α and β are equal. Find α.

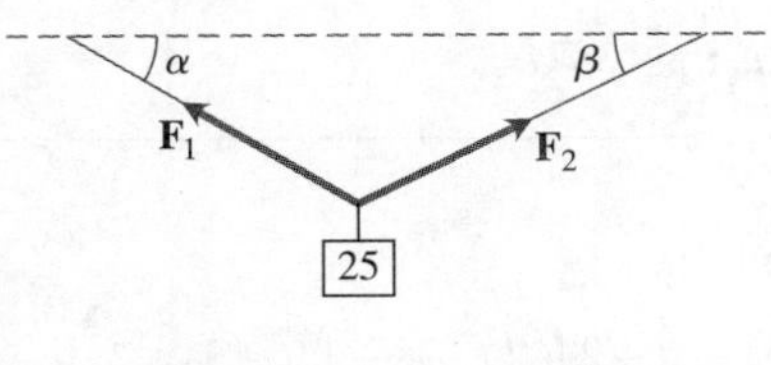

51. **Location** A bird flies from its nest 5 km in the direction 60° north of east, where it stops to rest on a tree. It then flies 10 km in the direction due southeast and lands atop a telephone pole. Place an xy-coordinate system so that the origin is the bird's nest, the x-axis points east, and the y-axis points north.

   a. At what point is the tree located?

   b. At what point is the telephone pole?

52. Use similar triangles to find the coordinates of the point Q that divides the segment from $P_1(x_1, y_1, z_1)$ to $P_2(x_2, y_2, z_2)$ into two lengths whose ratio is $p/q = r$.

53. **Medians of a triangle** Suppose that A, B, and C are the corner points of the thin triangular plate of constant density shown here.

   a. Find the vector from C to the midpoint M of side AB.

   b. Find the vector from C to the point that lies two-thirds of the way from C to M on the median CM.

   c. Find the coordinates of the point in which the medians of ΔABC intersect. According to Exercise 25, Section 6.6, this point is the plate's center of mass. (See the figure.)

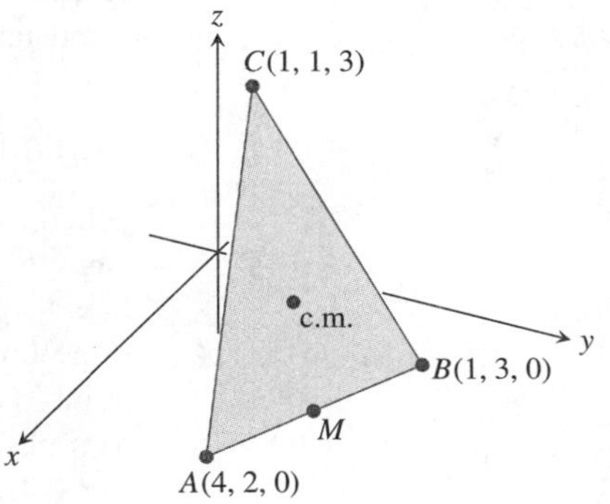

54. Find the vector from the origin to the point of intersection of the medians of the triangle whose vertices are

$$A(1, -1, 2),\quad B(2, 1, 3),\quad \text{and}\quad C(-1, 2, -1).$$

55. Let $ABCD$ be a general, not necessarily planar, quadrilateral in space. Show that the two segments joining the midpoints of opposite sides of $ABCD$ bisect each other. (*Hint:* Show that the segments have the same midpoint.)

56. Vectors are drawn from the center of a regular n-sided polygon in the plane to the vertices of the polygon. Show that the sum of the vectors is zero. (*Hint:* What happens to the sum if you rotate the polygon about its center?)

57. Suppose that A, B, and C are vertices of a triangle and that a, b, and c are, respectively, the midpoints of the opposite sides. Show that $\overrightarrow{Aa} + \overrightarrow{Bb} + \overrightarrow{Cc} = \mathbf{0}$.

58. **Unit vectors in the plane** Show that a unit vector in the plane can be expressed as $\mathbf{u} = (\cos\theta)\mathbf{i} + (\sin\theta)\mathbf{j}$, obtained by rotating $\mathbf{i}$ through an angle θ in the counterclockwise direction. Explain why this form gives *every* unit vector in the plane.

59. Consider a triangle whose vertices are $A(2, -3, 4)$, $B(1, 0, -1)$, and $C(3, 1, 2)$.

   a. Find $\overrightarrow{AB} + \overrightarrow{BC} + \overrightarrow{CA}$. **b.** Find $\overrightarrow{BA} + \overrightarrow{AC} + \overrightarrow{CB}$.

n-Dimensional Vectors

In Exercises 60–65, let $\mathbf{u} = \langle 2, -3, 0, 1\rangle$ and $\mathbf{v} = \langle 0, -4, -1, 3\rangle$. Find the **(a)** component form and **(b)** magnitude (length) of the vector.

60. $2\mathbf{u}$

61. $-4\mathbf{v}$

62. $\mathbf{u} + \mathbf{v}$

63. $\mathbf{v} - \mathbf{u}$

64. $3\mathbf{u} - 2\mathbf{v}$

65. $4\mathbf{u} + 3\mathbf{v}$

12.3 The Dot Product

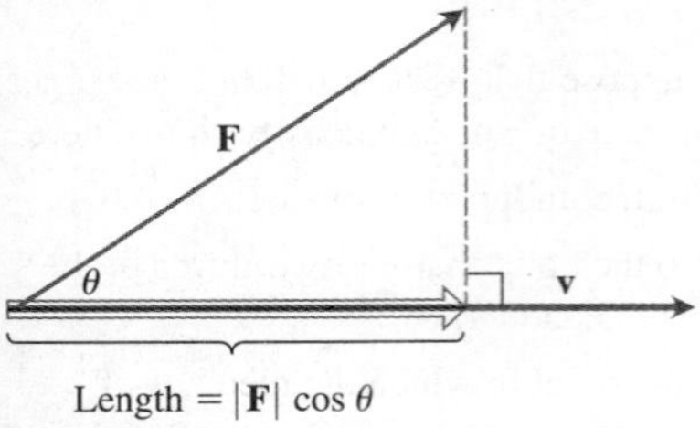

FIGURE 12.20 The magnitude of the force **F** in the direction of vector **v** is the length $|\mathbf{F}|\cos\theta$ of the projection of **F** onto **v**.

If a force **F** is applied to a particle moving along a path, we often need to know the magnitude of the force in the direction of motion. If **v** is parallel to the tangent line to the path at the point where **F** is applied, then we want the magnitude of **F** in the direction of **v**. Figure 12.20 shows that the scalar quantity we seek is the length $|\mathbf{F}|\cos\theta$, where θ is the angle between the two vectors **F** and **v**.

In this section we show how to calculate easily the angle between two vectors directly from their components. A key part of the calculation is an expression called the *dot product*. Dot products are also called *inner* or *scalar* products because the product results in a scalar, not a vector. After investigating the dot product, we apply it to finding the projection of one vector onto another (as displayed in Figure 12.20) and to finding the work done by a constant force acting through a displacement.

Angle Between Vectors

When two nonzero vectors **u** and **v** are placed so their initial points coincide, they form an angle θ of measure $0 \le \theta \le \pi$ (Figure 12.21). If the vectors do not lie along the same line, the angle θ is measured in the plane containing both of them. If they do lie along the same line, the angle between them is 0 if they point in the same direction and π if they point in opposite directions. The angle θ is the **angle between u** and **v**. Theorem 1 gives a formula to determine this angle.

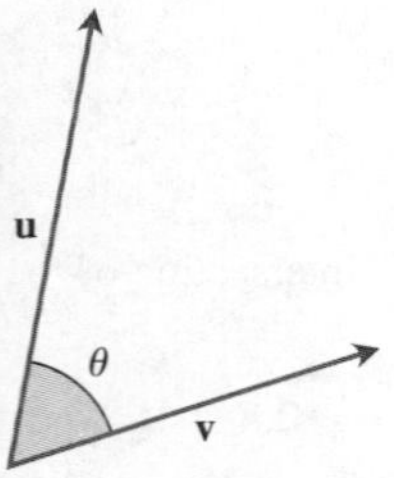

FIGURE 12.21 The angle between **u** and **v** given by Theorem 1 lies in the interval $[0, \pi]$.

THEOREM 1—Angle Between Two Vectors
The angle θ between two nonzero vectors $\mathbf{u} = \langle u_1, u_2, u_3\rangle$ and $\mathbf{v} = \langle v_1, v_2, v_3\rangle$ is given by

$$\theta = \arccos\left(\frac{u_1v_1 + u_2v_2 + u_3v_3}{|\mathbf{u}|\,|\mathbf{v}|}\right).$$

We use the law of cosines to prove Theorem 1, but before doing so, we focus attention on the expression $u_1v_1 + u_2v_2 + u_3v_3$ in the calculation for θ. This expression is the sum of the products of the corresponding components of the vectors **u** and **v**.

DEFINITION The **dot product u · v** (**"u dot v"**) of vectors $\mathbf{u} = \langle u_1, u_2, u_3\rangle$ and $\mathbf{v} = \langle v_1, v_2, v_3\rangle$ is the scalar

$$\mathbf{u}\cdot\mathbf{v} = u_1v_1 + u_2v_2 + u_3v_3.$$

EXAMPLE 1 We illustrate the definition.

(a) $$\begin{aligned}\langle 1, -2, -1\rangle \cdot \langle -6, 2, -3\rangle &= (1)(-6) + (-2)(2) + (-1)(-3)\\ &= -6 - 4 + 3 = -7\end{aligned}$$

(b) $$\left(\frac{1}{2}\mathbf{i} + 3\mathbf{j} + \mathbf{k}\right)\cdot(4\mathbf{i} - \mathbf{j} + 2\mathbf{k}) = \left(\frac{1}{2}\right)(4) + (3)(-1) + (1)(2) = 1$$ ■

The dot product of a pair of two-dimensional vectors is defined in a similar fashion:

$$\langle u_1, u_2\rangle \cdot \langle v_1, v_2\rangle = u_1v_1 + u_2v_2.$$

We will see throughout the remainder of this text that the dot product is a key tool for many important geometric and physical calculations in space (and the plane).

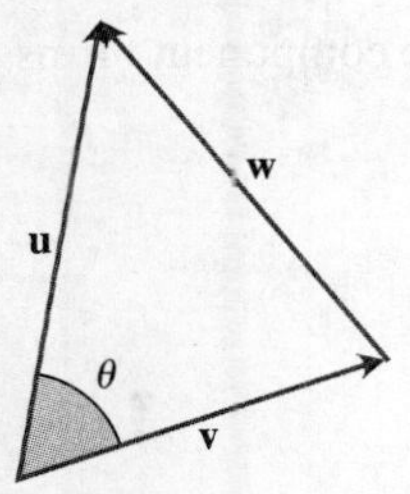

FIGURE 12.22 The parallelogram law of addition of vectors gives $\mathbf{w} = \mathbf{u} - \mathbf{v}$.

Proof of Theorem 1 Applying the law of cosines (Equation (8), Section 1.3) to the triangle in Figure 12.22, we find that

$$|\mathbf{w}|^2 = |\mathbf{u}|^2 + |\mathbf{v}|^2 - 2|\mathbf{u}||\mathbf{v}|\cos\theta \qquad \text{Law of cosines}$$
$$2|\mathbf{u}||\mathbf{v}|\cos\theta = |\mathbf{u}|^2 + |\mathbf{v}|^2 - |\mathbf{w}|^2.$$

Because $\mathbf{w} = \mathbf{u} - \mathbf{v}$, the component form of $\mathbf{w}$ is $\langle u_1 - v_1, u_2 - v_2, u_3 - v_3\rangle$. So

$$|\mathbf{u}|^2 = \left(\sqrt{u_1^2 + u_2^2 + u_3^2}\right)^2 = u_1^2 + u_2^2 + u_3^2$$
$$|\mathbf{v}|^2 = \left(\sqrt{v_1^2 + v_2^2 + v_3^2}\right)^2 = v_1^2 + v_2^2 + v_3^2$$
$$|\mathbf{w}|^2 = \left(\sqrt{(u_1 - v_1)^2 + (u_2 - v_2)^2 + (u_3 - v_3)^2}\right)^2$$
$$= (u_1 - v_1)^2 + (u_2 - v_2)^2 + (u_3 - v_3)^2$$
$$= u_1^2 - 2u_1v_1 + v_1^2 + u_2^2 - 2u_2v_2 + v_2^2 + u_3^2 - 2u_3v_3 + v_3^2$$

and

$$|\mathbf{u}|^2 + |\mathbf{v}|^2 - |\mathbf{w}|^2 = 2(u_1v_1 + u_2v_2 + u_3v_3).$$

Therefore,

$$2|\mathbf{u}||\mathbf{v}|\cos\theta = |\mathbf{u}|^2 + |\mathbf{v}|^2 - |\mathbf{w}|^2 = 2(u_1v_1 + u_2v_2 + u_3v_3)$$
$$|\mathbf{u}||\mathbf{v}|\cos\theta = u_1v_1 + u_2v_2 + u_3v_3$$
$$\cos\theta = \frac{u_1v_1 + u_2v_2 + u_3v_3}{|\mathbf{u}||\mathbf{v}|}.$$

Thus, for $0 \le \theta \le \pi$, we have $\theta = \arccos\left(\frac{u_1v_1 + u_2v_2 + u_3v_3}{|\mathbf{u}||\mathbf{v}|}\right)$. ■

Dot Product and Angles

The angle between two nonzero vectors $\mathbf{u}$ and $\mathbf{v}$ is $\theta = \arccos\left(\frac{\mathbf{u}\cdot\mathbf{v}}{|\mathbf{u}||\mathbf{v}|}\right)$.

The dot product of two vectors $\mathbf{u}$ and $\mathbf{v}$ is given by $\mathbf{u}\cdot\mathbf{v} = |\mathbf{u}||\mathbf{v}|\cos\theta$.

EXAMPLE 2 Find the angle between $\mathbf{u} = \mathbf{i} - 2\mathbf{j} - 2\mathbf{k}$ and $\mathbf{v} = 6\mathbf{i} + 3\mathbf{j} + 2\mathbf{k}$.

Solution We use the formula above:

$$\mathbf{u}\cdot\mathbf{v} = (1)(6) + (-2)(3) + (-2)(2) = 6 - 6 - 4 = -4$$
$$|\mathbf{u}| = \sqrt{(1)^2 + (-2)^2 + (-2)^2} = \sqrt{9} = 3$$
$$|\mathbf{v}| = \sqrt{(6)^2 + (3)^2 + (2)^2} = \sqrt{49} = 7$$
$$\theta = \arccos\left(\frac{\mathbf{u}\cdot\mathbf{v}}{|\mathbf{u}||\mathbf{v}|}\right) = \arccos\left(\frac{-4}{(3)(7)}\right) \approx 1.76 \text{ radians or } 100.98°. \qquad ■$$

The angle formula applies to two-dimensional vectors as well. Note that the angle θ is acute if $\mathbf{u}\cdot\mathbf{v} > 0$ and obtuse if $\mathbf{u}\cdot\mathbf{v} < 0$.

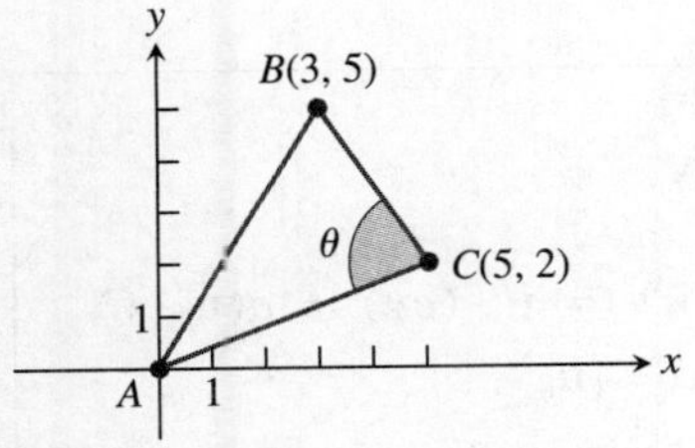

FIGURE 12.23 The triangle in Example 3.

EXAMPLE 3 Find the angle θ in the triangle ABC determined by the vertices $A = (0, 0)$, $B = (3, 5)$, and $C = (5, 2)$ (Figure 12.23).

Solution The angle θ is the angle between the vectors $\overrightarrow{CA}$ and $\overrightarrow{CB}$. The component forms of these two vectors are

$$\overrightarrow{CA} = \langle -5, -2 \rangle \quad \text{and} \quad \overrightarrow{CB} = \langle -2, 3 \rangle.$$

First we calculate the dot product and magnitudes of these two vectors.

$$\begin{aligned} \overrightarrow{CA} \cdot \overrightarrow{CB} &= (-5)(-2) + (-2)(3) = 4 \\ |\overrightarrow{CA}| &= \sqrt{(-5)^2 + (-2)^2} = \sqrt{29} \\ |\overrightarrow{CB}| &= \sqrt{(-2)^2 + (3)^2} = \sqrt{13} \end{aligned}$$

Then, applying the angle formula, we have

$$\begin{aligned} \theta &= \arccos\left(\frac{\overrightarrow{CA} \cdot \overrightarrow{CB}}{|\overrightarrow{CA}|\,|\overrightarrow{CB}|}\right) = \arccos\left(\frac{4}{(\sqrt{29})(\sqrt{13})}\right) \\ &\approx 78.1^\circ \text{ or } 1.36 \text{ radians.} \end{aligned}$$

■

Orthogonal Vectors

Two nonzero vectors $\mathbf{u}$ and $\mathbf{v}$ are perpendicular if the angle between them is $\pi/2$. For such vectors, we have $\mathbf{u} \cdot \mathbf{v} = 0$ because $\cos(\pi/2) = 0$. The converse is also true. If $\mathbf{u}$ and $\mathbf{v}$ are nonzero vectors with $\mathbf{u} \cdot \mathbf{v} = |\mathbf{u}||\mathbf{v}|\cos\theta = 0$, then $\cos\theta = 0$ and $\theta = \arccos 0 = \pi/2$. The following definition also allows for one or both of the vectors to be the zero vector.

DEFINITION Vectors $\mathbf{u}$ and $\mathbf{v}$ are orthogonal if $\mathbf{u} \cdot \mathbf{v} = 0$.

EXAMPLE 4 To determine if two vectors are orthogonal, calculate their dot product.

(a) $\mathbf{u} = \langle 3, -2 \rangle$ and $\mathbf{v} = \langle 4, 6 \rangle$ are orthogonal because $\mathbf{u} \cdot \mathbf{v} = (3)(4) + (-2)(6) = 0$.

(b) $\mathbf{u} = 3\mathbf{i} - 2\mathbf{j} + \mathbf{k}$ and $\mathbf{v} = 2\mathbf{j} + 4\mathbf{k}$ are orthogonal because

$$\mathbf{u} \cdot \mathbf{v} = (3)(0) + (-2)(2) + (1)(4) = 0.$$

(c) $\mathbf{0}$ is orthogonal to every vector $\mathbf{u}$ because

$$\begin{aligned} \mathbf{0} \cdot \mathbf{u} &= \langle 0, 0, 0 \rangle \cdot \langle u_1, u_2, u_3 \rangle \\ &= (0)(u_1) + (0)(u_2) + (0)(u_3) = 0. \end{aligned}$$

■

Dot Product Properties and Vector Projections

The dot product obeys many of the laws that hold for ordinary products of real numbers (scalars).

Properties of the Dot Product

If $\mathbf{u}$, $\mathbf{v}$, and $\mathbf{w}$ are any vectors and c is a scalar, then

1. $\mathbf{u} \cdot \mathbf{v} = \mathbf{v} \cdot \mathbf{u}$

2. $(c\mathbf{u}) \cdot \mathbf{v} = \mathbf{u} \cdot (c\mathbf{v}) = c(\mathbf{u} \cdot \mathbf{v})$

3. $\mathbf{u} \cdot (\mathbf{v} + \mathbf{w}) = \mathbf{u} \cdot \mathbf{v} + \mathbf{u} \cdot \mathbf{w}$

4. $\mathbf{u} \cdot \mathbf{u} = |\mathbf{u}|^2$

5. $\mathbf{0} \cdot \mathbf{u} = 0$.

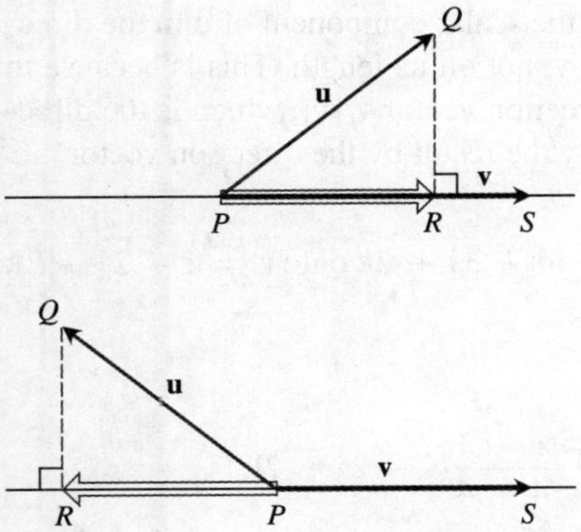

FIGURE 12.24 The vector projection of **u** onto **v**.

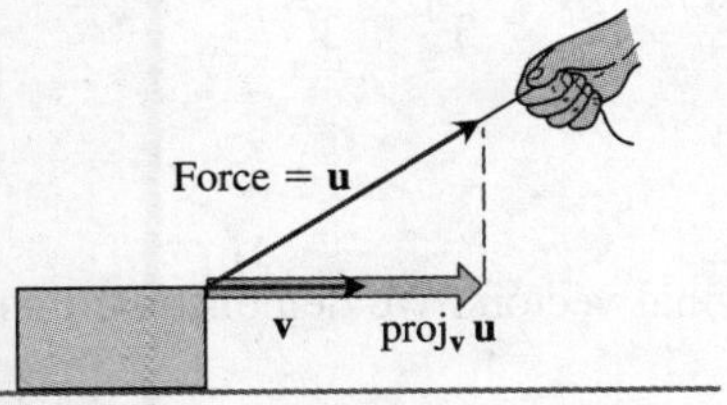

FIGURE 12.25 If we pull on the box with force **u**, the effective force moving the box forward in the direction **v** is the projection of **u** onto **v**.

Proofs of Properties 1 and 3 The properties are easy to prove using the definition. For instance, here are the proofs of Properties 1 and 3.

$$\mathbf{1.}\ \mathbf{u}\cdot\mathbf{v} = u_1v_1 + u_2v_2 + u_3v_3 = v_1u_1 + v_2u_2 + v_3u_3 = \mathbf{v}\cdot\mathbf{u}$$

$$\begin{aligned}\mathbf{3.}\ \mathbf{u}\cdot(\mathbf{v}+\mathbf{w}) &= \langle u_1, u_2, u_3\rangle \cdot \langle v_1 + w_1, v_2 + w_2, v_3 + w_3\rangle \\ &= u_1(v_1 + w_1) + u_2(v_2 + w_2) + u_3(v_3 + w_3) \\ &= u_1v_1 + u_1w_1 + u_2v_2 + u_2w_2 + u_3v_3 + u_3w_3 \\ &= (u_1v_1 + u_2v_2 + u_3v_3) + (u_1w_1 + u_2w_2 + u_3w_3) \\ &= \mathbf{u}\cdot\mathbf{v} + \mathbf{u}\cdot\mathbf{w}\end{aligned}$$

■

We now return to the problem of projecting one vector onto another, posed in the opening to this section. The **vector projection** of $\mathbf{u} = \overrightarrow{PQ}$ onto a nonzero vector $\mathbf{v} = \overrightarrow{PS}$ (Figure 12.24) is the vector $\overrightarrow{PR}$ determined by dropping a perpendicular from Q to the line PS. The notation for this vector is

$$\text{proj}_{\mathbf{v}}\,\mathbf{u} \qquad \text{("the vector projection of } \mathbf{u} \text{ onto } \mathbf{v}\text{")}.$$

If **u** represents a force, then $\text{proj}_{\mathbf{v}}\,\mathbf{u}$ represents the effective force in the direction of **v** (Figure 12.25).

If the angle θ between **u** and **v** is acute, $\text{proj}_{\mathbf{v}}\,\mathbf{u}$ has length $|\mathbf{u}|\cos\theta$ and direction $\mathbf{v}/|\mathbf{v}|$ (Figure 12.26). If θ is obtuse, $\cos\theta < 0$ and $\text{proj}_{\mathbf{v}}\,\mathbf{u}$ has length $-|\mathbf{u}|\cos\theta$ and direction $-\mathbf{v}/|\mathbf{v}|$. In both cases,

$$\begin{aligned}\text{proj}_{\mathbf{v}}\,\mathbf{u} &= (|\mathbf{u}|\cos\theta)\frac{\mathbf{v}}{|\mathbf{v}|} \\ &= \left(\frac{\mathbf{u}\cdot\mathbf{v}}{|\mathbf{v}|}\right)\frac{\mathbf{v}}{|\mathbf{v}|} \qquad |\mathbf{u}|\cos\theta = \frac{|\mathbf{u}||\mathbf{v}|\cos\theta}{|\mathbf{v}|} = \frac{\mathbf{u}\cdot\mathbf{v}}{|\mathbf{v}|} \\ &= \left(\frac{\mathbf{u}\cdot\mathbf{v}}{|\mathbf{v}|^2}\right)\mathbf{v}.\end{aligned}$$

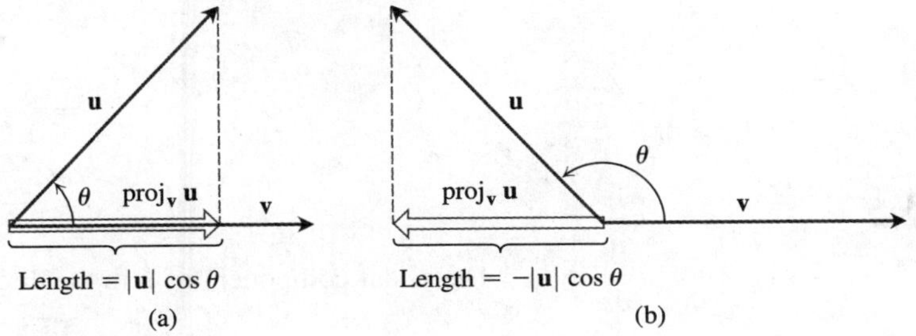

FIGURE 12.26 The length of $\text{proj}_{\mathbf{v}}\,\mathbf{u}$ is (a) $|\mathbf{u}|\cos\theta$ if $\cos\theta \geq 0$ and (b) $-|\mathbf{u}|\cos\theta$ if $\cos\theta < 0$.

The number $|\mathbf{u}|\cos\theta$ is called the **scalar component of u in the direction of v**. To summarize,

> The vector projection of **u** onto **v** is the vector
>
> $$\text{proj}_{\mathbf{v}}\,\mathbf{u} = \left(\frac{\mathbf{u}\cdot\mathbf{v}}{|\mathbf{v}|^2}\right)\mathbf{v} = \left(\frac{\mathbf{u}\cdot\mathbf{v}}{|\mathbf{v}|}\right)\frac{\mathbf{v}}{|\mathbf{v}|}. \tag{1}$$
>
> The scalar component of **u** in the direction of **v** is the scalar
>
> $$|\mathbf{u}|\cos\theta = \frac{\mathbf{u}\cdot\mathbf{v}}{|\mathbf{v}|} = \mathbf{u}\cdot\frac{\mathbf{v}}{|\mathbf{v}|}. \tag{2}$$

Note that both the vector projection of **u** onto **v** and the scalar component of **u** in the direction of **v** depend only on the direction of the vector **v**, not on its length. This is because in both cases we take the dot product of **u** with the direction vector $\mathbf{v}/|\mathbf{v}|$, which is the direction of **v**, and for the projection we go on to multiply the result by the direction vector.

EXAMPLE 5 Find the vector projection of $\mathbf{u} = 6\mathbf{i} + 3\mathbf{j} + 2\mathbf{k}$ onto $\mathbf{v} = \mathbf{i} - 2\mathbf{j} - 2\mathbf{k}$ and the scalar component of **u** in the direction of **v**.

Solution We find $\text{proj}_{\mathbf{v}}\,\mathbf{u}$ from Equation (1):

$$\begin{aligned}\text{proj}_{\mathbf{v}}\,\mathbf{u} &= \frac{\mathbf{u}\cdot\mathbf{v}}{|\mathbf{v}|^2}\mathbf{v} = \frac{\mathbf{u}\cdot\mathbf{v}}{\mathbf{v}\cdot\mathbf{v}}\mathbf{v} = \frac{6-6-4}{1+4+4}(\mathbf{i}-2\mathbf{j}-2\mathbf{k})\\ &= -\frac{4}{9}(\mathbf{i}-2\mathbf{j}-2\mathbf{k}) = -\frac{4}{9}\mathbf{i}+\frac{8}{9}\mathbf{j}+\frac{8}{9}\mathbf{k}.\end{aligned}$$

We find the scalar component of **u** in the direction of **v** from Equation (2):

$$\begin{aligned}|\mathbf{u}|\cos\theta &= \mathbf{u}\cdot\frac{\mathbf{v}}{|\mathbf{v}|} = (6\mathbf{i}+3\mathbf{j}+2\mathbf{k})\cdot\left(\frac{1}{3}\mathbf{i}-\frac{2}{3}\mathbf{j}-\frac{2}{3}\mathbf{k}\right)\\ &= 2-2-\frac{4}{3} = -\frac{4}{3}.\end{aligned}$$

■

Equations (1) and (2) also apply to two-dimensional vectors. We demonstrate this in the next example.

EXAMPLE 6 Find the vector projection of a force $\mathbf{F} = 5\mathbf{i} + 2\mathbf{j}$ onto $\mathbf{v} = \mathbf{i} - 3\mathbf{j}$ and the scalar component of **F** in the direction of **v**.

Solution The vector projection is

$$\begin{aligned}\text{proj}_{\mathbf{v}}\,\mathbf{F} &= \left(\frac{\mathbf{F}\cdot\mathbf{v}}{|\mathbf{v}|^2}\right)\mathbf{v} = \left(\frac{\mathbf{F}\cdot\mathbf{v}}{\mathbf{v}\cdot\mathbf{v}}\right)\mathbf{v}\\ &= \frac{5-6}{1+9}(\mathbf{i}-3\mathbf{j}) = -\frac{1}{10}(\mathbf{i}-3\mathbf{j})\\ &= -\frac{1}{10}\mathbf{i}+\frac{3}{10}\mathbf{j}.\end{aligned}$$

The scalar component of **F** in the direction of **v** is

$$|\mathbf{F}|\cos\theta = \frac{\mathbf{F}\cdot\mathbf{v}}{|\mathbf{v}|} = \frac{5-6}{\sqrt{1+9}} = -\frac{1}{\sqrt{10}}.$$

■

EXAMPLE 7 Verify that the vector $\mathbf{u} - \text{proj}_{\mathbf{v}}\,\mathbf{u}$ is orthogonal to the projection vector $\text{proj}_{\mathbf{v}}\,\mathbf{u}$.

Solution The vector $\text{proj}_{\mathbf{v}}\,\mathbf{u} = \left(\frac{\mathbf{u}\cdot\mathbf{v}}{|\mathbf{v}|^2}\right)\mathbf{v}$ is parallel to **v**. So it suffices to show that the vector $\mathbf{u} - \text{proj}_{\mathbf{v}}\,\mathbf{u}$ is orthogonal to **v**. We verify orthogonality by showing that the dot product of $\mathbf{u} - \text{proj}_{\mathbf{v}}\,\mathbf{u}$ with **v** is zero:

$$\begin{aligned}(\mathbf{u}-\text{proj}_{\mathbf{v}}\,\mathbf{u})\cdot\mathbf{v} &= \mathbf{u}\cdot\mathbf{v} - \left(\frac{\mathbf{u}\cdot\mathbf{v}}{|\mathbf{v}|^2}\mathbf{v}\right)\cdot\mathbf{v} && \text{Definition of } \text{proj}_{\mathbf{v}}\,\mathbf{u}\\ &= \mathbf{u}\cdot\mathbf{v} - \frac{\mathbf{u}\cdot\mathbf{v}}{|\mathbf{v}|^2}(\mathbf{v}\cdot\mathbf{v}) && \text{Dot product property (2)}\\ &= \mathbf{u}\cdot\mathbf{v} - \frac{\mathbf{u}\cdot\mathbf{v}}{|\mathbf{v}|^2}|\mathbf{v}|^2 && \mathbf{v}\cdot\mathbf{v} = |\mathbf{v}|^2\\ &= \mathbf{u}\cdot\mathbf{v} - \mathbf{u}\cdot\mathbf{v} = 0.\end{aligned}$$

■

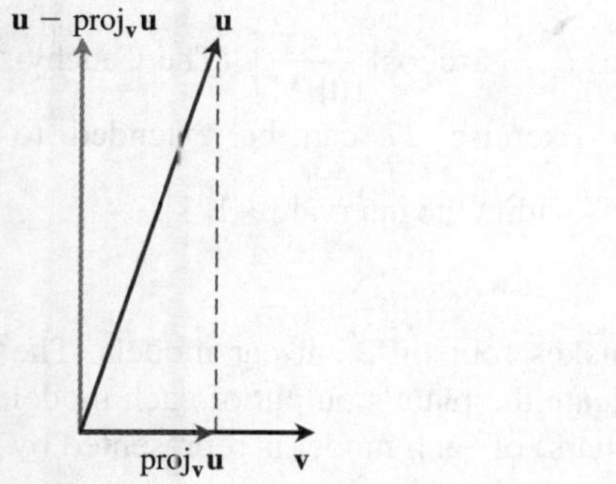

FIGURE 12.27 The vector **u** is the sum of two perpendicular vectors: a vector proj$_v$ **u**, parallel to **v**, and a vector **u** − proj$_v$ **u**, perpendicular to **v**.

Example 7 verifies that the vector **u** − proj$_v$ **u** is orthogonal to the projection vector proj$_v$ **u** (which has the same direction as **v**). So the equation

$$\mathbf{u} = \operatorname{proj}_{\mathbf{v}} \mathbf{u} + (\mathbf{u} - \operatorname{proj}_{\mathbf{v}} \mathbf{u}) = \underbrace{\left(\frac{\mathbf{u}\cdot\mathbf{v}}{|\mathbf{v}|^2}\right)\mathbf{v}}_{\text{Parallel to } \mathbf{v}} + \underbrace{\left(\mathbf{u} - \left(\frac{\mathbf{u}\cdot\mathbf{v}}{|\mathbf{v}|^2}\right)\mathbf{v}\right)}_{\text{Orthogonal to } \mathbf{v}}$$

expresses **u** as a sum of orthogonal vectors (see Figure 12.27).

Work

In Chapter 6, we calculated the work done by a constant force of magnitude F in moving an object through a distance d as $W = Fd$. That formula holds only if the force is directed along the line of motion. If a force **F** moving an object through a displacement $\mathbf{D} = \overrightarrow{PQ}$ has some other direction, the work is performed by the component of **F** in the direction of **D**. If θ is the angle between **F** and **D** (Figure 12.28), then

$$\begin{aligned}\text{Work} &= \begin{pmatrix}\text{scalar component of } \mathbf{F} \\ \text{in the direction of } \mathbf{D}\end{pmatrix}(\text{length of } \mathbf{D}) \\ &= (|\mathbf{F}|\cos\theta)|\mathbf{D}| \\ &= \mathbf{F}\cdot\mathbf{D}.\end{aligned}$$

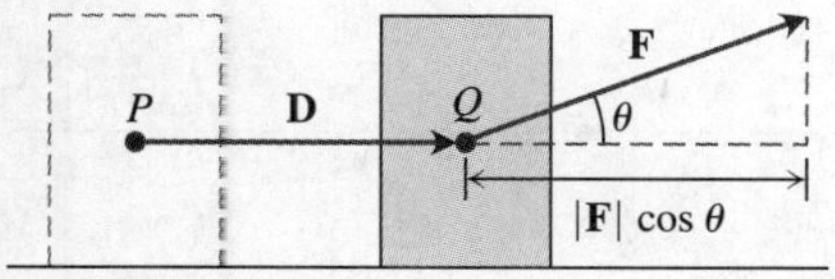

FIGURE 12.28 The work done by a constant force **F** during a displacement **D** is $(|\mathbf{F}|\cos\theta)|\mathbf{D}|$, which is the dot product $\mathbf{F}\cdot\mathbf{D}$.

DEFINITION The **work** done by a constant force **F** acting through a displacement $\mathbf{D} = \overrightarrow{PQ}$ is

$$W = \mathbf{F}\cdot\mathbf{D}.$$

EXAMPLE 8 If $|\mathbf{F}| = 40$ N (newtons), $|\mathbf{D}| = 3$ m, and $\theta = 60°$, the work done by **F** in acting from P to Q is

$$\begin{aligned}\text{Work} &= \mathbf{F}\cdot\mathbf{D} && \text{Definition} \\ &= |\mathbf{F}||\mathbf{D}|\cos\theta \\ &= (40)(3)\cos 60° && \text{Given values} \\ &= (120)(1/2) = 60 \text{ J(joules)}.\end{aligned}$$

■

We encounter more challenging work problems in Chapter 16 when we learn to find the work done by a variable force along a more general *path* in space.

The Dot Product of Two n-Dimensional Vectors

If $\mathbf{u} = \langle u_1, u_2, \dots, u_n\rangle$ and $\mathbf{v} = \langle v_1, v_2, \dots, v_n\rangle$ are n-dimensional vectors, then we define the dot product to be

$$\mathbf{u}\cdot\mathbf{v} = u_1v_1 + u_2v_2 + \cdots + u_nv_n.$$

As for two- and three-dimensional vectors, the dot product is calculated by adding the products of the corresponding components of the two vectors.

This generalized dot product can be shown to satisfy the Properties of the Dot Product that were introduced earlier in this section, and similar terminology is used. If **u** and **v** are n-dimensional vectors, then

1. **u** and **v** are said to be orthogonal if $\mathbf{u}\cdot\mathbf{v} = 0$,

2. the vector projection of **u** onto **v** is $\operatorname{proj}_{\mathbf{v}} \mathbf{u} = \dfrac{\mathbf{u}\cdot\mathbf{v}}{|\mathbf{v}|^2}\mathbf{v}$, and

3. the angle between the vectors **u** and **v** is defined as $\theta = \arccos\left(\frac{\mathbf{u}\cdot\mathbf{v}}{|\mathbf{u}||\mathbf{v}|}\right)$. (The Cauchy-Schwarz inequality, $|\mathbf{u}\cdot\mathbf{v}| \le |\mathbf{u}||\mathbf{v}|$, stated in Exercise 27 can be extended to n-dimensional vectors. This guarantees that $\frac{\mathbf{u}\cdot\mathbf{v}}{|\mathbf{u}||\mathbf{v}|}$ is within the interval $[-1, 1]$.)

EXAMPLE 9 An automobile assembly plant makes four different car models. The components of the vector $\mathbf{u} = \langle 36, 50, 24, 10\rangle$ indicate the plant's output of each model per hour, whereas the revenue per vehicle (in US dollars) of each model is represented by the vector $\mathbf{v} = \langle 24{,}000, 31{,}000, 39{,}000, 52{,}000\rangle$. Calculate the dot product $\mathbf{u}\cdot\mathbf{v}$ and explain the significance of the value that was obtained.

Solution

$$\mathbf{u}\cdot\mathbf{v} = (36)(24{,}000) + (50)(31{,}000) + (24)(39{,}000) + (10)(52{,}000) = 3{,}870{,}000.$$

The value \$3,870,000 represents the total hourly revenue. ■

EXERCISES 12.3

For some exercises, a calculator may be helpful when expressing answers in decimal form.

Dot Product and Projections

In Exercises 1–8, find

a. $\mathbf{v}\cdot\mathbf{u}$, $|\mathbf{v}|$, $|\mathbf{u}|$

b. the cosine of the angle between **v** and **u**

c. the scalar component of **u** in the direction of **v**

d. the vector $\text{proj}_{\mathbf{v}}\,\mathbf{u}$.

1. $\mathbf{v} = \mathbf{i} - 2\mathbf{j} + \sqrt{3}\,\mathbf{k}$, $\mathbf{u} = -\mathbf{i} + 2\mathbf{j} - \sqrt{3}\,\mathbf{k}$
2. $\mathbf{v} = (3/5)\mathbf{i} + (4/5)\mathbf{k}$, $\mathbf{u} = 5\mathbf{i} + 12\mathbf{j}$
3. $\mathbf{v} = 5\mathbf{i} + 12\mathbf{j} - \mathbf{k}$, $\mathbf{u} = 5\mathbf{j} + 3\mathbf{k}$
4. $\mathbf{v} = 2\mathbf{i} + 10\mathbf{j} - 11\mathbf{k}$, $\mathbf{u} = 2\mathbf{i} + 2\mathbf{j} + \mathbf{k}$
5. $\mathbf{v} = 3\mathbf{j} - \mathbf{k}$ $\mathbf{u} = \mathbf{i} - \mathbf{j} - \mathbf{k}$
6. $\mathbf{v} = -\mathbf{i} + \mathbf{j}$, $\mathbf{u} = \sqrt{2}\mathbf{i} + \sqrt{3}\mathbf{j} + 2\mathbf{k}$
7. $\mathbf{v} = 2\mathbf{i} + \mathbf{j}$, $\mathbf{u} = \sqrt{2}\,\mathbf{i} + \sqrt{15}\,\mathbf{j}$
8. $\mathbf{v} = \left\langle \frac{1}{\sqrt{2}}, \frac{1}{\sqrt{3}} \right\rangle$, $\mathbf{u} = \left\langle \frac{1}{\sqrt{2}}, -\frac{1}{\sqrt{3}} \right\rangle$

Angle Between Vectors

Find the angles between the vectors in Exercises 9–12 to the nearest hundredth of a radian.

9. $\mathbf{u} = \mathbf{i} + 2\mathbf{j}$, $\mathbf{v} = 2\mathbf{i} + \mathbf{j} + \mathbf{k}$
10. $\mathbf{u} = 2\mathbf{i} - 2\mathbf{j} + \mathbf{k}$, $\mathbf{v} = 3\mathbf{i} + 4\mathbf{k}$
11. $\mathbf{u} = \sqrt{3}\mathbf{i} - 7\mathbf{j}$, $\mathbf{v} = \sqrt{3}\mathbf{i} + \mathbf{j} - 2\mathbf{k}$
12. $\mathbf{u} = \sqrt{3}\,\mathbf{i} + \mathbf{j} - \sqrt{3}\,\mathbf{k}$, $\mathbf{v} = -3\mathbf{i} + 3\mathbf{j} - 3\mathbf{k}$
13. **Triangle** Find the measures of the angles of the triangle whose vertices are $A = (-2, 0)$, $B = (3, 2)$ and $C = (2, -3)$.
14. **Rectangle** Find the measures of the angles between the diagonals of the rectangle whose vertices are $A = (1, 0)$, $B = (0, 3)$, $C = (3, 4)$, and $D = (4, 1)$.
15. **Direction angles and direction cosines** The *direction angles* α, β, and γ of a vector $\mathbf{v} = a\mathbf{i} + b\mathbf{j} + c\mathbf{k}$ are defined as follows:

α is the angle between **v** and the positive x-axis $(0 \le \alpha \le \pi)$.

β is the angle between **v** and the positive y-axis $(0 \le \beta \le \pi)$.

γ is the angle between **v** and the positive z-axis $(0 \le \gamma \le \pi)$.

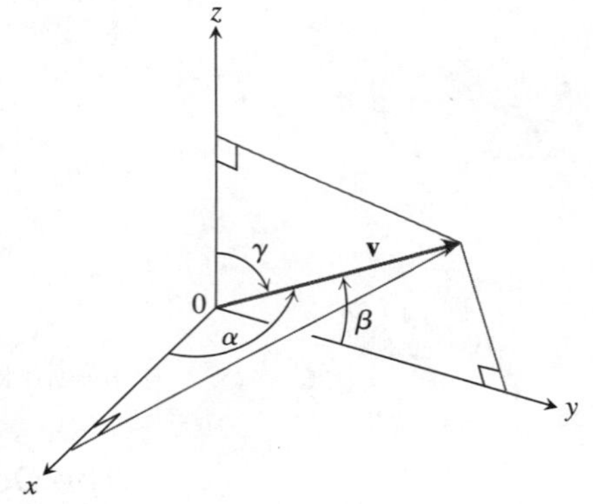

a. Show that

$$\cos\alpha = \frac{a}{|\mathbf{v}|}, \quad \cos\beta = \frac{b}{|\mathbf{v}|}, \quad \cos\gamma = \frac{c}{|\mathbf{v}|},$$

and $\cos^2\alpha + \cos^2\beta + \cos^2\gamma = 1$. These cosines are called the *direction cosines* of **v**.

b. **Unit vectors are built from direction cosines** Show that if $\mathbf{v} = a\mathbf{i} + b\mathbf{j} + c\mathbf{k}$ is a unit vector, then a, b, and c are the direction cosines of **v**.

16. **Water main construction** A water main is to be constructed with a 20% grade in the north direction and a 10% grade in the

east direction. Determine the angle θ required in the water main for the turn from north to east.

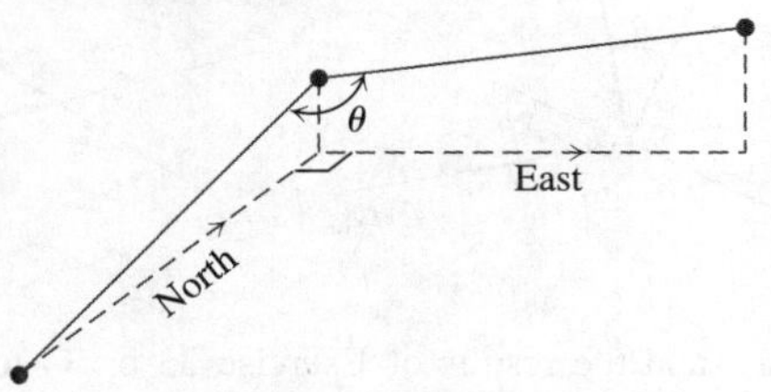

For Exercises 17 and 18, find the acute angle between the given lines by using vectors parallel to the lines.

17. $y = x, \quad y = 2x + 3$

18. $2 - x + 2y = 0, \quad 3x - 4y = -12$

Theory and Examples

19. Sums and differences In the accompanying figure, it looks as if $\mathbf{v}_1 + \mathbf{v}_2$ and $\mathbf{v}_1 - \mathbf{v}_2$ are orthogonal. Is this mere coincidence, or are there circumstances under which we may expect the sum of two vectors to be orthogonal to their difference? Give reasons for your answer.

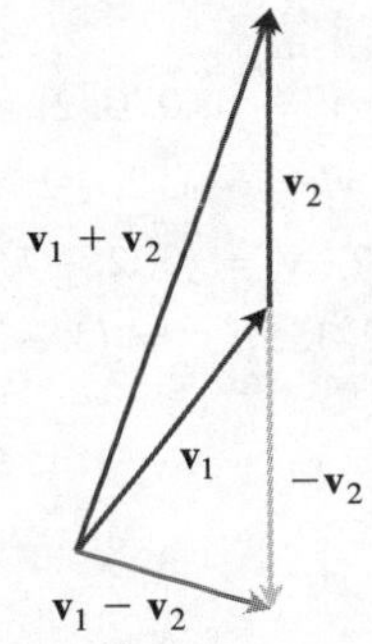

20. Orthogonality on a circle Suppose that AB is the diameter of a circle with center O and that C is a point on one of the two arcs joining A and B. Show that $\overrightarrow{CA}$ and $\overrightarrow{CB}$ are orthogonal.

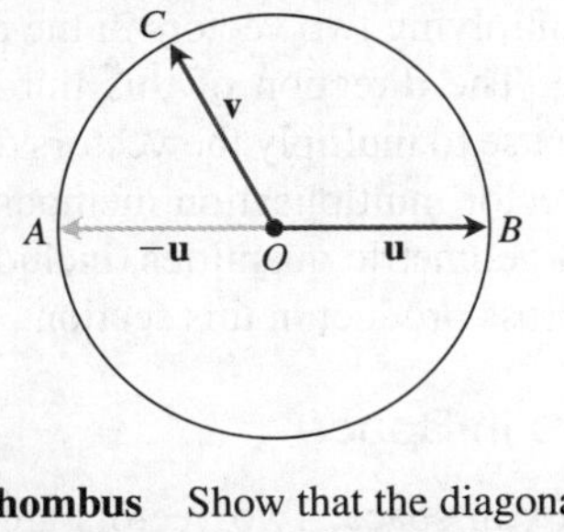

21. Diagonals of a rhombus Show that the diagonals of a rhombus (parallelogram with sides of equal length) are perpendicular.

22. Perpendicular diagonals Show that squares are the only rectangles with perpendicular diagonals.

23. When parallelograms are rectangles Prove that a parallelogram is a rectangle if and only if its diagonals are equal in length. (This fact is often exploited by carpenters.)

24. Diagonal of parallelogram Show that the indicated diagonal of the parallelogram determined by vectors $\mathbf{u}$ and $\mathbf{v}$ bisects the angle between $\mathbf{u}$ and $\mathbf{v}$ if $|\mathbf{u}| = |\mathbf{v}|$.

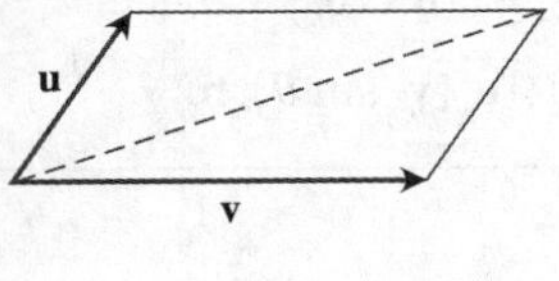

25. Projectile motion A gun with muzzle velocity of 1200 ft/sec is fired at an angle of 8° above the horizontal. Find the horizontal and vertical components of the velocity.

26. Inclined plane Suppose that a box is being towed up an inclined plane as shown in the figure. Find the force $\mathbf{w}$ needed to make the component of the force parallel to the inclined plane equal to 2.5 lb.

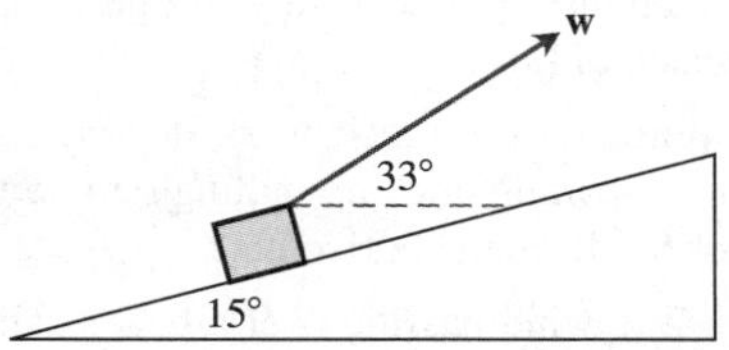

27. a. Cauchy-Schwarz inequality Since $\mathbf{u} \cdot \mathbf{v} = |\mathbf{u}||\mathbf{v}|\cos\theta$, show that the inequality $|\mathbf{u} \cdot \mathbf{v}| \le |\mathbf{u}||\mathbf{v}|$ holds for any vectors $\mathbf{u}$ and $\mathbf{v}$.

b. Under what circumstances, if any, does $|\mathbf{u} \cdot \mathbf{v}|$ equal $|\mathbf{u}||\mathbf{v}|$? Give reasons for your answer.

28. Dot multiplication is positive definite Show that dot multiplication of vectors is *positive definite*; that is, show that $\mathbf{u} \cdot \mathbf{u} \ge 0$ for every vector $\mathbf{u}$ and that $\mathbf{u} \cdot \mathbf{u} = 0$ if and only if $\mathbf{u} = \mathbf{0}$.

29. Orthogonal unit vectors If $\mathbf{u}_1$ and $\mathbf{u}_2$ are orthogonal unit vectors and $\mathbf{v} = a\mathbf{u}_1 + b\mathbf{u}_2$, find $\mathbf{v} \cdot \mathbf{u}_1$.

30. Cancelation in dot products In real-number multiplication, if $uv_1 = uv_2$ and $u \ne 0$, we can cancel the u and conclude that $v_1 = v_2$. Does the same rule hold for the dot product? That is, if $\mathbf{u} \cdot \mathbf{v}_1 = \mathbf{u} \cdot \mathbf{v}_2$ and $\mathbf{u} \ne \mathbf{0}$, can you conclude that $\mathbf{v}_1 = \mathbf{v}_2$? Give reasons for your answer.

31. If $\mathbf{u}$ and $\mathbf{v}$ are orthogonal, show that $\text{proj}_{\mathbf{v}}\,\mathbf{u} = 0$.

32. A force $\mathbf{F} = 3\mathbf{i} + 2\mathbf{j} - \mathbf{k}$ is applied to a spacecraft with velocity vector $\mathbf{v} = 2\mathbf{i} - \mathbf{j}$. Express $\mathbf{F}$ as a sum of a vector parallel to $\mathbf{v}$ and a vector orthogonal to $\mathbf{v}$.

Equations for Lines in the Plane

33. Line perpendicular to a vector Show that $\mathbf{v} = a\mathbf{i} + b\mathbf{j}$ is perpendicular to the line $ax + by = c$. (*Hint:* For a and b nonzero, establish that the slope of the vector $\mathbf{v}$ is the negative reciprocal of the slope of the given line. Also verify the statement when $a = 0$ or $b = 0$.)

34. Line parallel to a vector Show that the vector $\mathbf{v} = a\mathbf{i} + b\mathbf{j}$ is parallel to the line $bx - ay = c$. (*Hint:* For a and b nonzero, establish that the slope of the line segment representing $\mathbf{v}$ is the same as the slope of the given line. Also verify the statement when $a = 0$ or $b = 0$.)

In Exercises 35–38, use the result of Exercise 33 to find an equation for the line through P perpendicular to $\mathbf{v}$. Then sketch the line. Include $\mathbf{v}$ in your sketch *as a vector starting at the origin.*

35. $P(2, 1), \quad \mathbf{v} = \mathbf{i} + 2\mathbf{j}$

36. $P(-1, 2), \quad \mathbf{v} = -2\mathbf{i} - \mathbf{j}$

37. $P(-2, -7), \quad \mathbf{v} = -2\mathbf{i} + \mathbf{j}$

38. $P(11, 10), \quad \mathbf{v} = 2\mathbf{i} - 3\mathbf{j}$

In Exercises 39–42, use the result of Exercise 34 to find an equation for the line through P parallel to $\mathbf{v}$. Then sketch the line. Include $\mathbf{v}$ in your sketch *as a vector starting at the origin.*

39. $P(-2, 1), \quad \mathbf{v} = \mathbf{i} - \mathbf{j}$

40. $P(0, -2), \quad \mathbf{v} = 2\mathbf{i} + 3\mathbf{j}$

41. $P(1, 2), \quad \mathbf{v} = -\mathbf{i} - 2\mathbf{j}$

42. $P(1, 3), \quad \mathbf{v} = 3\mathbf{i} - 2\mathbf{j}$

Work

43. Work along a line Find the work done by a force $\mathbf{F} = 7\mathbf{j}$ (magnitude 7 N) in moving an object along the line from the origin to the point (2, 2) (distance in meters).

44. Locomotive The Union Pacific's *Big Boy* locomotive could pull 6000-ton trains with a tractive effort (pull) of 602,148 N (135,375 lb). At this level of effort, about how much work did *Big Boy* do on the (approximately straight) 605-km journey from San Francisco to Los Angeles?

45. Inclined plane How much work does it take to slide a crate 30 m along a loading dock by pulling on it with a 400 N force at an angle of 60° from the horizontal?

46. Sailboat The wind passing over a boat's sail exerted a 1000-lb magnitude force **F** as shown here. How much work did the wind perform in moving the boat forward 1 mi? Answer in foot-pounds.

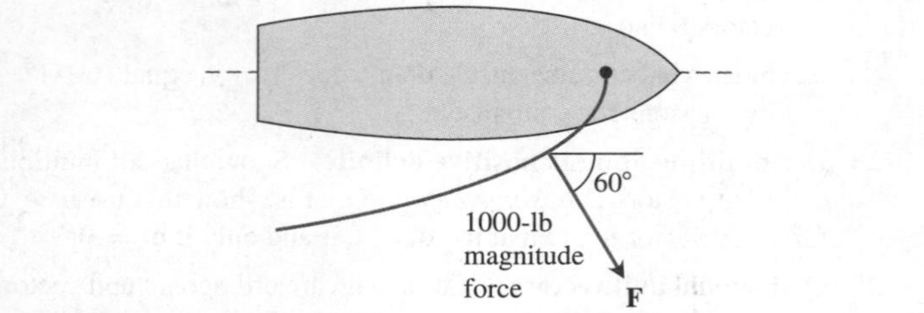

Angles Between Lines in the Plane

The **acute angle between intersecting lines** that do not cross at right angles is the same as the angle determined by vectors normal to the lines or by vectors parallel to the lines.

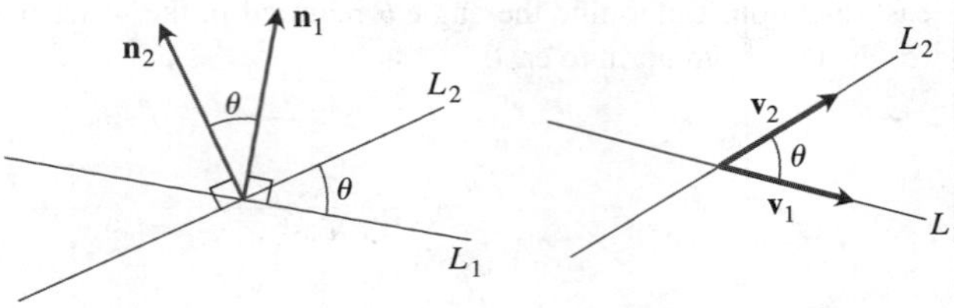

Use this fact and the results of Exercise 33 or 34 to find the acute angles between the lines in Exercises 47–52.

47. $5x + 2y = 7, \quad x - 2y = 1$

48. $y = \sqrt{3}x - 1, \quad y = -\sqrt{3}x + 2$

49. $x - \sqrt{3}y = -3, \quad \sqrt{3}x - y = 3$

50. $2x + \sqrt{3}y = 1, \quad (2 - \sqrt{3})x + (2 + \sqrt{3})y = 7$

51. $3x - 4y = 3, \quad x - y = 7$

52. $10x + 3y = 2, \quad 4x - y = 2$

Dot Products of n-Dimensional Vectors

In Exercises 53–56, **(a)** find $\mathbf{u} \cdot \mathbf{v}$ and **(b)** determine whether the vectors **u** and **v** are orthogonal.

53. $\mathbf{u} = \langle 3, 2, -4, 0\rangle$, $\mathbf{v} = \langle 1, 0, 0, 2\rangle$

54. $\mathbf{u} = \langle -2, 1, 1, 2\rangle$, $\mathbf{v} = \langle -1, 2, -2, -1\rangle$

55. $\mathbf{u} = \langle 6, 3, 0, 1, -2\rangle$, $\mathbf{v} = \langle 0, 2, -7, 0, 3\rangle$

56. $\mathbf{u} = \langle 4, 2, -3, -2, 1, 5\rangle$, $\mathbf{v} = \langle 3, -3, 2, -2, 1, -1\rangle$

12.4 The Cross Product

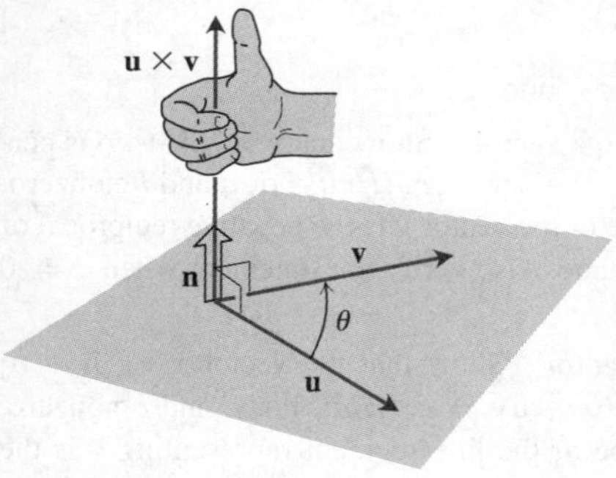

FIGURE 12.29 The construction of $\mathbf{u} \times \mathbf{v}$.

In studying lines in the plane, when we needed to describe how a line was tilting, we used the notions of slope and angle of inclination. In space, we want a way to describe how a *plane* is tilting. We accomplish this by multiplying two vectors in the plane together to get a third vector perpendicular to the plane. The direction of this third vector tells us the "inclination" of the plane. The product we use to multiply the vectors together is the *vector* or *cross product*, the second of the two vector multiplication methods. The cross product gives us a simple way to find a variety of geometric quantities, including volumes, areas, and perpendicular vectors. We study the cross product in this section.

The Cross Product of Two Vectors in Space

We start with two nonzero vectors **u** and **v** in space. Two vectors are *parallel* if one is a nonzero multiple of the other. If **u** and **v** are not parallel, they determine a plane. The vectors in this plane are linear combinations of **u** and **v**, so they can be written as a sum $a\mathbf{u} + b\mathbf{v}$. We select the unit vector **n** perpendicular to the plane by the **right-hand rule**. This means that we choose **n** to be the unit normal vector that points the way your right thumb points when your fingers curl through the angle θ from **u** to **v** (Figure 12.29). Then we define a new vector as follows.

DEFINITION The **cross product** $\mathbf{u} \times \mathbf{v}$ **("u cross v")** is the vector

$$\mathbf{u} \times \mathbf{v} = (|\mathbf{u}||\mathbf{v}| \sin \theta)\, \mathbf{n}.$$

Unlike the dot product, the cross product is a vector. For this reason it is also called the **vector product** of **u** and **v**, and can be applied *only* to vectors in space. The vector $\mathbf{u} \times \mathbf{v}$ is orthogonal to both **u** and **v** because it is a scalar multiple of **n**.

There is a straightforward way to calculate the cross product of two vectors from their components. The method does not require that we know the angle between them (as suggested by the definition), but we postpone that calculation momentarily so we can focus first on the properties of the cross product.

Because the sines of 0 and π are both zero, it makes sense to define the cross product of two parallel nonzero vectors to be **0**. If one or both of **u** and **v** are zero, we also define $\mathbf{u} \times \mathbf{v}$ to be zero. This way, the cross product of two vectors **u** and **v** is zero if and only if **u** and **v** are parallel or one or both of them are zero.

Parallel Vectors

Nonzero vectors **u** and **v** are parallel if and only if $\mathbf{u} \times \mathbf{v} = \mathbf{0}$.

The cross product obeys the following laws.

Properties of the Cross Product

If **u**, **v**, and **w** are any vectors and r, s are scalars, then

1. $(r\mathbf{u}) \times (s\mathbf{v}) = (rs)(\mathbf{u} \times \mathbf{v})$
2. $\mathbf{u} \times (\mathbf{v} + \mathbf{w}) = \mathbf{u} \times \mathbf{v} + \mathbf{u} \times \mathbf{w}$
3. $\mathbf{v} \times \mathbf{u} = -(\mathbf{u} \times \mathbf{v})$
4. $(\mathbf{v} + \mathbf{w}) \times \mathbf{u} = \mathbf{v} \times \mathbf{u} + \mathbf{w} \times \mathbf{u}$
5. $\mathbf{0} \times \mathbf{u} = \mathbf{0}$
6. $\mathbf{u} \times (\mathbf{v} \times \mathbf{w}) = (\mathbf{u} \cdot \mathbf{w})\mathbf{v} - (\mathbf{u} \cdot \mathbf{v})\mathbf{w}$

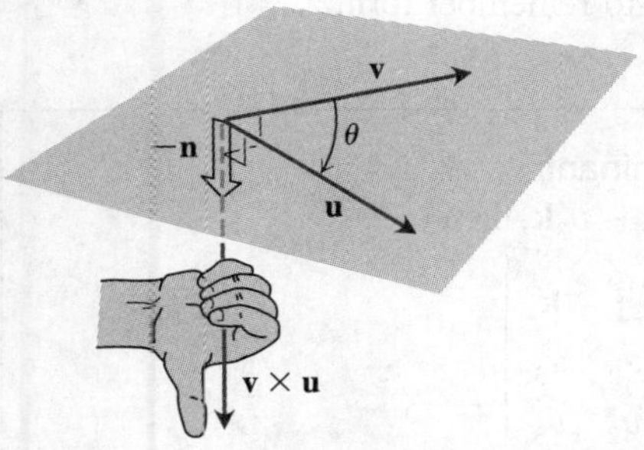

FIGURE 12.30 The construction of $\mathbf{v} \times \mathbf{u}$.

To visualize Property 3, for example, notice that when the fingers of your right hand curl through the angle θ from **v** to **u**, your thumb points the opposite way; the unit vector we choose in forming $\mathbf{v} \times \mathbf{u}$ is the negative of the one we choose in forming $\mathbf{u} \times \mathbf{v}$ (Figure 12.30).

Property 1 can be verified by applying the definition of cross product to both sides of the equation and comparing the results. Property 2 is proved in Appendix A.7. Property 4 follows by multiplying both sides of the equation in Property 2 by –1 and reversing the order of the products using Property 3. Property 5 is a definition. As a rule, cross product multiplication is *not associative* so $(\mathbf{u} \times \mathbf{v}) \times \mathbf{w}$ does not generally equal $\mathbf{u} \times (\mathbf{v} \times \mathbf{w})$. (See Additional Exercise 17.)

When we apply the definition and Property 3 to calculate the pairwise cross products of **i**, **j**, and **k**, we find (Figure 12.31)

$$\mathbf{i} \times \mathbf{j} = -(\mathbf{j} \times \mathbf{i}) = \mathbf{k}$$
$$\mathbf{j} \times \mathbf{k} = -(\mathbf{k} \times \mathbf{j}) = \mathbf{i}$$
$$\mathbf{k} \times \mathbf{i} = -(\mathbf{i} \times \mathbf{k}) = \mathbf{j}$$

and

$$\mathbf{i} \times \mathbf{i} = \mathbf{j} \times \mathbf{j} = \mathbf{k} \times \mathbf{k} = \mathbf{0}.$$

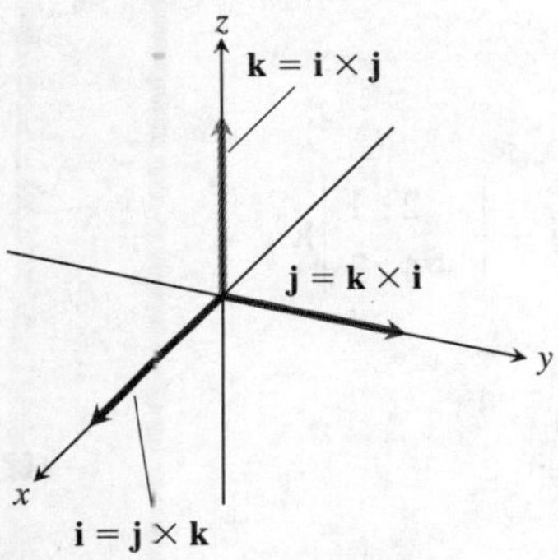

FIGURE 12.31 The pairwise cross products of **i**, **j**, and **k**.

$|\mathbf{u} \times \mathbf{v}|$ Is the Area of a Parallelogram

Because **n** is a unit vector, the magnitude of $\mathbf{u} \times \mathbf{v}$ is

$$|\mathbf{u} \times \mathbf{v}| = |\mathbf{u}||\mathbf{v}||\sin\theta||\mathbf{n}| = |\mathbf{u}||\mathbf{v}|\sin\theta.$$

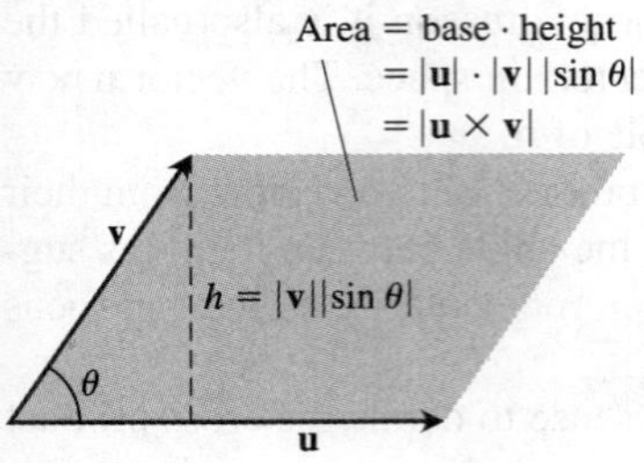

FIGURE 12.32 The parallelogram determined by **u** and **v**.

This is the area of the parallelogram determined by **u** and **v** (Figure 12.32), $|\mathbf{u}|$ being the base of the parallelogram and $|\mathbf{v}||\sin\theta|$ being the height.

Determinant Formula for $\mathbf{u} \times \mathbf{v}$

Our next objective is to calculate $\mathbf{u} \times \mathbf{v}$ from the components of **u** and **v** relative to a Cartesian coordinate system.

Suppose that

$$\mathbf{u} = u_1\mathbf{i} + u_2\mathbf{j} + u_3\mathbf{k} \quad \text{and} \quad \mathbf{v} = v_1\mathbf{i} + v_2\mathbf{j} + v_3\mathbf{k}.$$

Then the distributive laws and the rules for multiplying **i**, **j**, and **k** tell us that

$$\begin{aligned}\mathbf{u} \times \mathbf{v} &= (u_1\mathbf{i} + u_2\mathbf{j} + u_3\mathbf{k}) \times (v_1\mathbf{i} + v_2\mathbf{j} + v_3\mathbf{k}) \\ &= u_1v_1\mathbf{i} \times \mathbf{i} + u_1v_2\mathbf{i} \times \mathbf{j} + u_1v_3\mathbf{i} \times \mathbf{k} \\ &\quad + u_2v_1\mathbf{j} \times \mathbf{i} + u_2v_2\mathbf{j} \times \mathbf{j} + u_2v_3\mathbf{j} \times \mathbf{k} \\ &\quad + u_3v_1\mathbf{k} \times \mathbf{i} + u_3v_2\mathbf{k} \times \mathbf{j} + u_3v_3\mathbf{k} \times \mathbf{k} \\ &= (u_2v_3 - u_3v_2)\mathbf{i} - (u_1v_3 - u_3v_1)\mathbf{j} + (u_1v_2 - u_2v_1)\mathbf{k}.\end{aligned}$$

Determinants

2×2 and 3×3 determinants are evaluated as follows:

$$\begin{vmatrix} a & b \\ c & d \end{vmatrix} = ad - bc$$

$$\begin{vmatrix} a_1 & a_2 & a_3 \\ b_1 & b_2 & b_3 \\ c_1 & c_2 & c_3 \end{vmatrix} = a_1\begin{vmatrix} b_2 & b_3 \\ c_2 & c_3 \end{vmatrix} - a_2\begin{vmatrix} b_1 & b_3 \\ c_1 & c_3 \end{vmatrix} + a_3\begin{vmatrix} b_1 & b_2 \\ c_1 & c_2 \end{vmatrix}$$

The component terms in the last line are hard to remember, but they are the same as the terms in the expansion of the symbolic determinant

$$\begin{vmatrix} \mathbf{i} & \mathbf{j} & \mathbf{k} \\ u_1 & u_2 & u_3 \\ v_1 & v_2 & v_3 \end{vmatrix}.$$

So we restate the calculation in the following easy-to-remember form.

> **Calculating the Cross Product as a Determinant**
> If $\mathbf{u} = u_1\mathbf{i} + u_2\mathbf{j} + u_3\mathbf{k}$ and $\mathbf{v} = v_1\mathbf{i} + v_2\mathbf{j} + v_3\mathbf{k}$, then
>
> $$\mathbf{u} \times \mathbf{v} = \begin{vmatrix} \mathbf{i} & \mathbf{j} & \mathbf{k} \\ u_1 & u_2 & u_3 \\ v_1 & v_2 & v_3 \end{vmatrix}.$$

EXAMPLE 1 Find $\mathbf{u} \times \mathbf{v}$ and $\mathbf{v} \times \mathbf{u}$ if $\mathbf{u} = 2\mathbf{i} + \mathbf{j} + \mathbf{k}$ and $\mathbf{v} = -4\mathbf{i} + 3\mathbf{j} + \mathbf{k}$.

Solution We expand the symbolic determinant.

$$\mathbf{u} \times \mathbf{v} = \begin{vmatrix} \mathbf{i} & \mathbf{j} & \mathbf{k} \\ 2 & 1 & 1 \\ -4 & 3 & 1 \end{vmatrix} = \begin{vmatrix} 1 & 1 \\ 3 & 1 \end{vmatrix}\mathbf{i} - \begin{vmatrix} 2 & 1 \\ -4 & 1 \end{vmatrix}\mathbf{j} + \begin{vmatrix} 2 & 1 \\ -4 & 3 \end{vmatrix}\mathbf{k}$$

$$= -2\mathbf{i} - 6\mathbf{j} + 10\mathbf{k}$$

$$\mathbf{v} \times \mathbf{u} = -(\mathbf{u} \times \mathbf{v}) = 2\mathbf{i} + 6\mathbf{j} - 10\mathbf{k} \qquad \text{Property 3}$$ ■

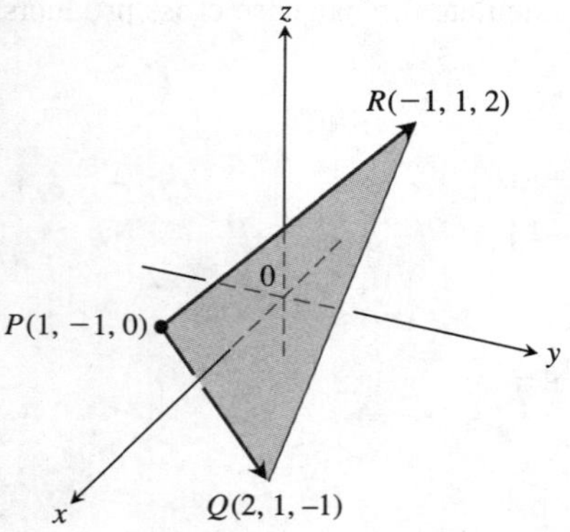

FIGURE 12.33 The vector $\overrightarrow{PQ} \times \overrightarrow{PR}$ is perpendicular to the plane of triangle PQR (Example 2). The area of triangle PQR is half of $|\overrightarrow{PQ} \times \overrightarrow{PR}|$ (Example 3).

EXAMPLE 2 Find a vector perpendicular to the plane of $P(1, -1, 0)$, $Q(2, 1, -1)$, and $R(-1, 1, 2)$ (Figure 12.33).

Solution The vector $\overrightarrow{PQ} \times \overrightarrow{PR}$ is perpendicular to the plane because it is perpendicular to both vectors. In terms of components,

$$\begin{aligned}
\overrightarrow{PQ} &= (2-1)\mathbf{i} + (1+1)\mathbf{j} + (-1-0)\mathbf{k} = \mathbf{i} + 2\mathbf{j} - \mathbf{k} \\
\overrightarrow{PR} &= (-1-1)\mathbf{i} + (1+1)\mathbf{j} + (2-0)\mathbf{k} = -2\mathbf{i} + 2\mathbf{j} + 2\mathbf{k} \\
\overrightarrow{PQ} \times \overrightarrow{PR} &= \begin{vmatrix} \mathbf{i} & \mathbf{j} & \mathbf{k} \\ 1 & 2 & -1 \\ -2 & 2 & 2 \end{vmatrix} = \begin{vmatrix} 2 & -1 \\ 2 & 2 \end{vmatrix}\mathbf{i} - \begin{vmatrix} 1 & -1 \\ -2 & 2 \end{vmatrix}\mathbf{j} + \begin{vmatrix} 1 & 2 \\ -2 & 2 \end{vmatrix}\mathbf{k} \\
&= 6\mathbf{i} + 6\mathbf{k}.
\end{aligned}$$

■

EXAMPLE 3 Find the area of the triangle with vertices $P(1, -1, 0)$, $Q(2, 1, -1)$, and $R(-1, 1, 2)$ (Figure 12.33).

Solution The area of the parallelogram determined by P, Q, and R is

$$\begin{aligned}
|\overrightarrow{PQ} \times \overrightarrow{PR}| &= |6\mathbf{i} + 6\mathbf{k}| && \text{Values from Example 2} \\
&= \sqrt{(6)^2 + (6)^2} = \sqrt{2 \cdot 36} = 6\sqrt{2}.
\end{aligned}$$

The triangle's area is half of this, or $3\sqrt{2}$. ■

EXAMPLE 4 Find a unit vector perpendicular to the plane of $P(1, -1, 0)$, $Q(2, 1, -1)$, and $R(-1, 1, 2)$.

Solution Since $\overrightarrow{PQ} \times \overrightarrow{PR}$ is perpendicular to the plane, its direction $\mathbf{n}$ is a unit vector perpendicular to the plane. Taking values from Examples 2 and 3, we have

$$\mathbf{n} = \frac{\overrightarrow{PQ} \times \overrightarrow{PR}}{|\overrightarrow{PQ} \times \overrightarrow{PR}|} = \frac{6\mathbf{i} + 6\mathbf{k}}{6\sqrt{2}} = \frac{1}{\sqrt{2}}\mathbf{i} + \frac{1}{\sqrt{2}}\mathbf{k}.$$

■

For ease in calculating the cross product using determinants, we usually write vectors in the form $\mathbf{v} = v_1\mathbf{i} + v_2\mathbf{j} + v_3\mathbf{k}$ rather than as ordered triples $\mathbf{v} = \langle v_1, v_2, v_3 \rangle$.

Torque

When we turn a bolt by applying a force $\mathbf{F}$ to a wrench (Figure 12.34), we produce a torque that causes the bolt to rotate. The **torque vector** points in the direction of the axis of the bolt according to the right-hand rule (so the rotation is counterclockwise when viewed from the *tip* of the vector). The magnitude of the torque depends on how far out on the wrench the force is applied and on how much of the force is perpendicular to the wrench at the point of application. The number we use to measure the torque's magnitude is the product of the length of the lever arm $\mathbf{r}$ and the scalar component of $\mathbf{F}$ perpendicular to $\mathbf{r}$. In the notation of Figure 12.34,

$$\text{Magnitude of torque vector} = |\mathbf{r}||\mathbf{F}| \sin\theta,$$

or $|\mathbf{r} \times \mathbf{F}|$. If we let $\mathbf{n}$ be a unit vector along the axis of the bolt in the direction of the torque, then a complete description of the torque vector is $\mathbf{r} \times \mathbf{F}$, or

$$\text{Torque vector} = \mathbf{r} \times \mathbf{F} = (|\mathbf{r}||\mathbf{F}| \sin\theta)\,\mathbf{n}.$$

Recall that we defined $\mathbf{u} \times \mathbf{v}$ to be $\mathbf{0}$ when $\mathbf{u}$ and $\mathbf{v}$ are parallel. This is consistent with the torque interpretation as well. If the force $\mathbf{F}$ in Figure 12.34 is parallel to the wrench, meaning that we are trying to turn the bolt by pushing or pulling along the line of the wrench's handle, the torque produced is zero.

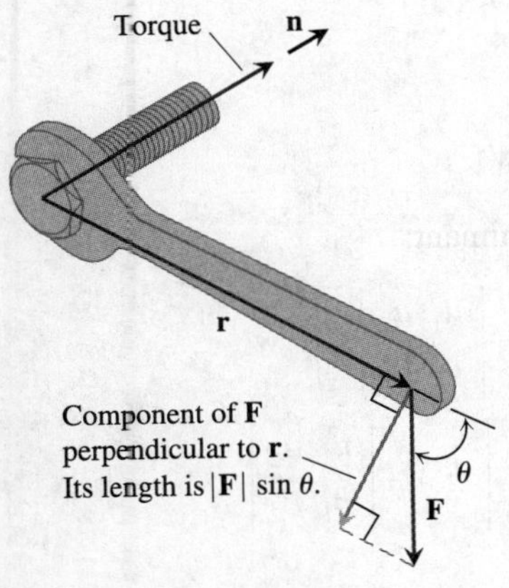

FIGURE 12.34 The torque vector describes the tendency of the force $\mathbf{F}$ to drive the bolt forward.

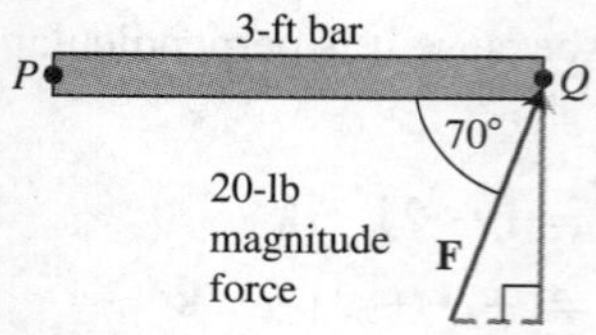

FIGURE 12.35 The magnitude of the torque exerted by **F** at *P* is about 56.4 ft-lb (Example 5). The bar rotates counterclockwise around *P*.

EXAMPLE 5 The magnitude of the torque generated by force **F** at the pivot point *P* in Figure 12.35 is

$$|\overrightarrow{PQ} \times \mathbf{F}| = |\overrightarrow{PQ}||\mathbf{F}| \sin 70^\circ \approx (3)(20)(0.94) \approx 56.4 \text{ ft-lb}.$$

In this example the torque vector is pointing out toward you. ■

Triple Scalar or Box Product

The product $(\mathbf{u} \times \mathbf{v}) \cdot \mathbf{w}$ is called the **triple scalar product** of **u**, **v**, and **w** (in that order). As you can see from the formula

$$|(\mathbf{u} \times \mathbf{v}) \cdot \mathbf{w}| = |\mathbf{u} \times \mathbf{v}||\mathbf{w}||\cos\theta|,$$

the absolute value of this product is the volume of the parallelepiped (parallelogram-sided box) determined by **u**, **v**, and **w** (Figure 12.36). The number $|\mathbf{u} \times \mathbf{v}|$ is the area of the base parallelogram. The number $|\mathbf{w}||\cos\theta|$ is the parallelepiped's height. Because of this geometry, $(\mathbf{u} \times \mathbf{v}) \cdot \mathbf{w}$ is also called the **box product** of **u**, **v**, and **w**.

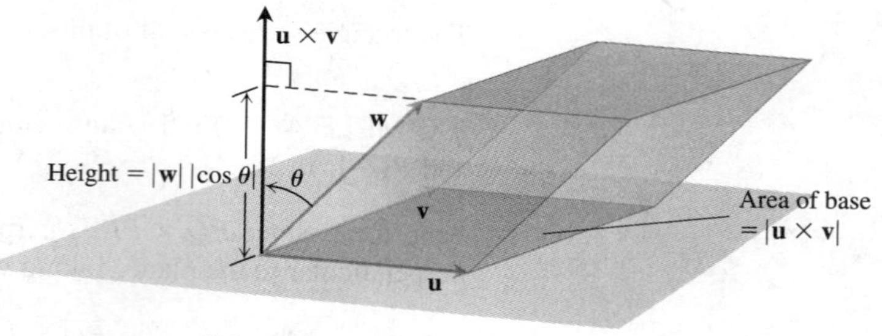

FIGURE 12.36 The number $|(\mathbf{u} \times \mathbf{v}) \cdot \mathbf{w}|$ is the volume of a parallelepiped.

By treating the planes of **v** and **w** and of **w** and **u** as the base planes of the parallelepiped determined by **u**, **v**, and **w**, we see that

$$(\mathbf{u} \times \mathbf{v}) \cdot \mathbf{w} = (\mathbf{v} \times \mathbf{w}) \cdot \mathbf{u} = (\mathbf{w} \times \mathbf{u}) \cdot \mathbf{v}.$$

The dot and cross may be interchanged in a triple scalar product without altering its value.

Since the dot product is commutative, we also have

$$(\mathbf{u} \times \mathbf{v}) \cdot \mathbf{w} = \mathbf{u} \cdot (\mathbf{v} \times \mathbf{w}).$$

The triple scalar product can be evaluated as a determinant:

$$(\mathbf{u} \times \mathbf{v}) \cdot \mathbf{w} = \left(\begin{vmatrix} u_2 & u_3 \\ v_2 & v_3 \end{vmatrix} \mathbf{i} - \begin{vmatrix} u_1 & u_3 \\ v_1 & v_3 \end{vmatrix} \mathbf{j} + \begin{vmatrix} u_1 & u_2 \\ v_1 & v_2 \end{vmatrix} \mathbf{k} \right) \cdot \mathbf{w}$$

$$= w_1 \begin{vmatrix} u_2 & u_3 \\ v_2 & v_3 \end{vmatrix} - w_2 \begin{vmatrix} u_1 & u_3 \\ v_1 & v_3 \end{vmatrix} + w_3 \begin{vmatrix} u_1 & u_2 \\ v_1 & v_2 \end{vmatrix}$$

$$= \begin{vmatrix} u_1 & u_2 & u_3 \\ v_1 & v_2 & v_3 \\ w_1 & w_2 & w_3 \end{vmatrix}.$$

Calculating the Triple Scalar Product as a Determinant

$$(\mathbf{u} \times \mathbf{v}) \cdot \mathbf{w} = \begin{vmatrix} u_1 & u_2 & u_3 \\ v_1 & v_2 & v_3 \\ w_1 & w_2 & w_3 \end{vmatrix}$$

$$\begin{vmatrix} u_1 & u_2 & u_3 \\ v_1 & v_2 & v_3 \\ w_1 & w_2 & w_3 \end{vmatrix} = -\begin{vmatrix} w_1 & w_2 & w_3 \\ v_1 & v_2 & v_3 \\ u_1 & u_2 & u_3 \end{vmatrix}$$

Any two rows of a matrix can be interchanged without changing the *absolute value* of the determinant. So we can take the vectors **u**, **v**, **w** in any order when calculating the absolute value of the triple product.

EXAMPLE 6 Find the volume of the box (parallelepiped) that is determined by $\mathbf{u} = \mathbf{i} + 2\mathbf{j} - \mathbf{k}$, $\mathbf{v} = -2\mathbf{i} + 3\mathbf{k}$, and $\mathbf{w} = 7\mathbf{j} - 4\mathbf{k}$.

Solution Using the rule for calculating a 3×3 determinant, we find

$$(\mathbf{u} \times \mathbf{v}) \cdot \mathbf{w} = \begin{vmatrix} 1 & 2 & -1 \\ -2 & 0 & 3 \\ 0 & 7 & -4 \end{vmatrix} = (1)\begin{vmatrix} 0 & 3 \\ 7 & -4 \end{vmatrix} - (2)\begin{vmatrix} -2 & 3 \\ 0 & -4 \end{vmatrix} + (-1)\begin{vmatrix} -2 & 0 \\ 0 & 7 \end{vmatrix} = -23.$$

The volume is $|(\mathbf{u} \times \mathbf{v}) \cdot \mathbf{w}| = 23$ units cubed. ■

EXERCISES 12.4

Cross Product Calculations

In Exercises 1–8, find the length and direction (when defined) of $\mathbf{u} \times \mathbf{v}$ and $\mathbf{v} \times \mathbf{u}$.

1. $\mathbf{u} = 2\mathbf{i} - 2\mathbf{j} - \mathbf{k}$, $\mathbf{v} = \mathbf{i} - \mathbf{k}$
2. $\mathbf{u} = 5\mathbf{i} + 2\mathbf{j}$, $\mathbf{v} = \mathbf{i} - \mathbf{j}$
3. $\mathbf{u} = 2\mathbf{i} - 2\mathbf{j} + 4\mathbf{k}$, $\mathbf{v} = -\mathbf{i} + \mathbf{j} - 2\mathbf{k}$
4. $\mathbf{u} = 2\mathbf{i} + 2\mathbf{j} - 2\mathbf{k}$, $\mathbf{v} = \mathbf{0}$
5. $\mathbf{u} = -5\mathbf{i}$, $\mathbf{v} = 2\mathbf{j}$
6. $\mathbf{u} = \mathbf{i} \times \mathbf{j}$, $\mathbf{v} = \mathbf{j} \times \mathbf{k}$
7. $\mathbf{u} = -7\mathbf{i} - 3\mathbf{j} - 2\mathbf{k}$, $\mathbf{v} = 3\mathbf{i} + 3\mathbf{j} + \mathbf{k}$
8. $\mathbf{u} = \frac{5}{2}\mathbf{i} - \frac{3}{2}\mathbf{j} + \mathbf{k}$, $\mathbf{v} = -\mathbf{i} - \mathbf{j} - 2\mathbf{k}$

In Exercises 9–14, sketch the coordinate axes and then include the vectors **u**, **v**, and $\mathbf{u} \times \mathbf{v}$ as vectors starting at the origin.

9. $\mathbf{u} = \mathbf{i}$, $\mathbf{v} = \mathbf{j}$
10. $\mathbf{u} = \mathbf{i} - \mathbf{k}$, $\mathbf{v} = \mathbf{j}$
11. $\mathbf{u} = \mathbf{i} - \mathbf{k}$, $\mathbf{v} = \mathbf{j} + \mathbf{k}$
12. $\mathbf{u} = 2\mathbf{i} - \mathbf{j}$, $\mathbf{v} = \mathbf{i} + 2\mathbf{j}$
13. $\mathbf{u} = \mathbf{i} + \mathbf{j}$, $\mathbf{v} = \mathbf{i} - \mathbf{j}$
14. $\mathbf{u} = \mathbf{j} + 2\mathbf{k}$, $\mathbf{v} = \mathbf{i}$

Triangles in Space

In Exercises 15–18,

a. Find the area of the triangle determined by the points P, Q, and R.

b. Find a unit vector perpendicular to plane PQR.

15. $P(-2, -2, 1)$, $Q(-1, 0, 2)$, $R(0, 1, 2)$
16. $P(1, 1, 1)$, $Q(2, 1, 3)$, $R(3, -1, 1)$
17. $P(3, -3, 1)$, $Q(4, -2, 1)$, $R(4, -1, 2)$
18. $P(-2, 2, 0)$, $Q(0, 1, -1)$, $R(-1, 2, -2)$

Triple Scalar Products

In Exercises 19–22, verify that

$$(\mathbf{u} \times \mathbf{v}) \cdot \mathbf{w} = (\mathbf{v} \times \mathbf{w}) \cdot \mathbf{u} = (\mathbf{w} \times \mathbf{u}) \cdot \mathbf{v}$$

and find the volume of the parallelepiped (box) determined by **u**, **v**, and **w**.

	u	v	w
19.	$3\mathbf{i}$	$5\mathbf{j}$	$\mathbf{k}$
20.	$\mathbf{i} - \mathbf{j} + \mathbf{k}$	$2\mathbf{i} + \mathbf{j} - 2\mathbf{k}$	$-\mathbf{i} + 2\mathbf{j} - \mathbf{k}$
21.	$\mathbf{i} + 2\mathbf{j}$	$\mathbf{i} + 2\mathbf{j} - 2\mathbf{k}$	$5\mathbf{i} - 3\mathbf{k}$
22.	$\mathbf{j} + 3\mathbf{k}$	$\mathbf{i} + \mathbf{j}$	$5\mathbf{i} - 2\mathbf{j} + 3\mathbf{k}$

Theory and Examples

23. **Parallel and perpendicular vectors** Let $\mathbf{u} = 3\mathbf{i} + \mathbf{j} - \mathbf{k}$, $\mathbf{v} = -\mathbf{j} + 3\mathbf{k}$, $\mathbf{w} = 12\mathbf{i} + 4\mathbf{j} - 4\mathbf{k}$. Which vectors, if any are **(a)** perpendicular? **(b)** Parallel? Give reasons for your answers.

24. **Parallel and perpendicular vectors** Let $\mathbf{u} = \mathbf{i} + 2\mathbf{j} - \mathbf{k}$, $\mathbf{v} = -\mathbf{i} + \mathbf{j} + \mathbf{k}$, $\mathbf{w} = \mathbf{i} + \mathbf{k}$, $\mathbf{r} = -(\pi/2)\mathbf{i} - \pi\mathbf{j} + (\pi/2)\mathbf{k}$. Which vectors, if any, are **(a)** perpendicular? **(b)** Parallel? Give reasons for your answers.

In Exercises 25 and 26, find the magnitude of the torque exerted by **F** on the bolt at P if $|\overrightarrow{PQ}| = 8$ in. and $|\mathbf{F}| = 30$ lb. Answer in foot-pounds.

25.

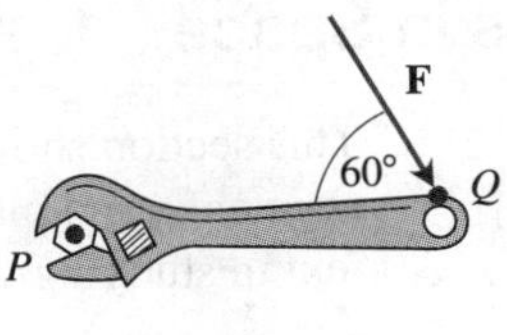

26.

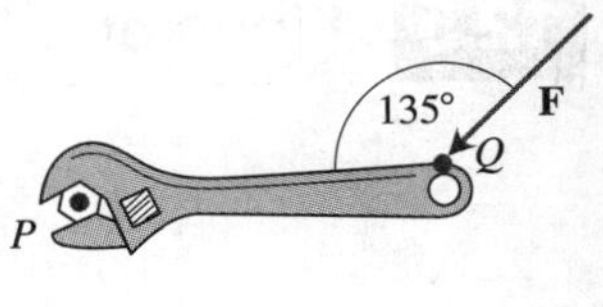

27. Which of the following are *always true*, and which are *not always true*? Give reasons for your answers.

a. $|\mathbf{u}| = \sqrt{\mathbf{u} \cdot \mathbf{u}}$

b. $\mathbf{u} \cdot \mathbf{u} = |\mathbf{u}|$

c. $\mathbf{u} \times \mathbf{0} = \mathbf{0} \times \mathbf{u} = \mathbf{0}$

d. $\mathbf{u} \times (-\mathbf{u}) = \mathbf{0}$

e. $\mathbf{u} \times \mathbf{v} = \mathbf{v} \times \mathbf{u}$

f. $\mathbf{u} \times (\mathbf{v} + \mathbf{w}) = \mathbf{u} \times \mathbf{v} + \mathbf{u} \times \mathbf{w}$

g. $(\mathbf{u} \times \mathbf{v}) \cdot \mathbf{v} = 0$

h. $(\mathbf{u} \times \mathbf{v}) \cdot \mathbf{w} = \mathbf{u} \cdot (\mathbf{v} \times \mathbf{w})$

28. Which of the following are *always true*, and which are *not always true*? Give reasons for your answers.

a. $\mathbf{u} \cdot \mathbf{v} = \mathbf{v} \cdot \mathbf{u}$

b. $\mathbf{u} \times \mathbf{v} = -(\mathbf{v} \times \mathbf{u})$

c. $(-\mathbf{u}) \times \mathbf{v} = -(\mathbf{u} \times \mathbf{v})$

d. $(c\mathbf{u}) \cdot \mathbf{v} = \mathbf{u} \cdot (c\mathbf{v}) = c(\mathbf{u} \cdot \mathbf{v})$ (any number c)

e. $c(\mathbf{u} \times \mathbf{v}) = (c\mathbf{u}) \times \mathbf{v} = \mathbf{u} \times (c\mathbf{v})$ (any number c)

f. $\mathbf{u} \cdot \mathbf{u} = |\mathbf{u}|^2$

g. $(\mathbf{u} \times \mathbf{u}) \cdot \mathbf{u} = 0$

h. $(\mathbf{u} \times \mathbf{v}) \cdot \mathbf{u} = \mathbf{v} \cdot (\mathbf{u} \times \mathbf{v})$

29. Given nonzero vectors **u**, **v**, and **w**, use dot product and cross product notation, as appropriate, to describe the following.

a. The vector projection of **u** onto **v**

b. A vector orthogonal to **u** and **v**

c. A vector orthogonal to $\mathbf{u} \times \mathbf{v}$ and **w**

d. The volume of the parallelepiped determined by **u**, **v**, and **w**

e. A vector orthogonal to $\mathbf{u} \times \mathbf{v}$ and $\mathbf{u} \times \mathbf{w}$

f. A vector of length $|\mathbf{u}|$ in the direction of **v**

30. Compute $(\mathbf{i} \times \mathbf{j}) \times \mathbf{j}$ and $\mathbf{i} \times (\mathbf{j} \times \mathbf{j})$. What can you conclude about the associativity of the cross product?

31. Let **u**, **v**, and **w** be vectors. Which of the following make sense, and which do not? Give reasons for your answers.

a. $(\mathbf{u} \times \mathbf{v}) \cdot \mathbf{w}$

b. $\mathbf{u} \times (\mathbf{v} \cdot \mathbf{w})$

c. $\mathbf{u} \times (\mathbf{v} \times \mathbf{w})$

d. $\mathbf{u} \cdot (\mathbf{v} \cdot \mathbf{w})$

32. Cross products of three vectors Show that except in degenerate cases, $(\mathbf{u} \times \mathbf{v}) \times \mathbf{w}$ lies in the plane of **u** and **v**, whereas $\mathbf{u} \times (\mathbf{v} \times \mathbf{w})$ lies in the plane of **v** and **w**. What *are* the degenerate cases?

33. Cancelation in cross products If $\mathbf{u} \times \mathbf{v} = \mathbf{u} \times \mathbf{w}$ and $\mathbf{u} \neq \mathbf{0}$, then does $\mathbf{v} = \mathbf{w}$? Give reasons for your answer.

34. Double cancelation If $\mathbf{u} \neq \mathbf{0}$ and if $\mathbf{u} \times \mathbf{v} = \mathbf{u} \times \mathbf{w}$ and $\mathbf{u} \cdot \mathbf{v} = \mathbf{u} \cdot \mathbf{w}$, then does $\mathbf{v} = \mathbf{w}$? Give reasons for your answer.

Area of a Parallelogram

Find the areas of the parallelograms whose vertices are given in Exercises 35–40.

35. $A(2, 0)$, $B(0, 2)$, $C(-2, 0)$, $D(0, -2)$

36. $A(0, 0)$, $B(7, 3)$, $C(9, 8)$, $D(2, 5)$

37. $A(-2, 3)$, $B(3, 0)$, $C(5, 2)$, $D(3, 4)$

38. $A(-6, 0)$, $B(1, -4)$, $C(3, 1)$, $D(-4, 5)$

39. $A(2, 0, 0)$, $B(1, 3, 5)$, $C(3, 0, 3)$, $D(5, -2, 0)$

40. $A(1, 0, -1)$, $B(1, 7, 2)$, $C(2, 4, -1)$, $D(0, 3, 2)$

Area of a Triangle

Find the areas of the triangles whose vertices are given in Exercises 41–47.

41. $A(0, 0)$, $B(-2, 3)$, $C(3, 1)$

42. $A(-2, -2)$, $B(5, 5)$, $C(3, 1)$

43. $A(-5, 3)$, $B(1, -2)$, $C(6, -2)$

44. $A(-1, 0)$, $B(5, -20)$, $C(-4, 8)$

45. $A(1, 0, 0)$, $B(0, 2, 0)$, $C(0, 0, -1)$

46. $A(1, 1, 1)$, $B(-2, 2, -2)$, $C(5, 0, 5)$

47. $A(1, -1, 1)$, $B(0, 1, 1)$, $C(1, 0, -1)$

48. Find the volume of a parallelepiped with one of its eight vertices at $A(0, 0, 0)$ and three adjacent vertices at $B(1, 2, 0)$, $C(0, -3, 2)$, and $D(3, -4, 5)$.

49. Triangle area Find a 2×2 determinant formula for the area of the triangle in the xy-plane with vertices at $(0, 0)$, (a_1, a_2), and (b_1, b_2). Explain your work.

50. Triangle area Find a concise 3×3 determinant formula that gives the area of a triangle in the xy-plane having vertices (a_1, a_2), (b_1, b_2), and (c_1, c_2).

Volume of a Tetrahedron

Using the methods of Section 6.1, where volume is computed by integrating cross-sectional area, it can be shown that the volume of a tetrahedron formed by three vectors is equal to $\frac{1}{6}$ the volume of the parallelepiped formed by the three vectors. Find the volumes of the tetrahedra whose vertices are given in Exercises 51–54.

51. $A(0, 0, 0)$, $B(2, 0, 0)$, $C(0, 3, 0)$, $D(0, 0, 4)$

52. $A(0, 0, 0)$, $B(1, 0, 2)$, $C(0, 2, 1)$, $D(3, 4, 0)$

53. $A(1, -1, 0)$, $B(0, 2, -2)$, $C(-3, 0, 3)$, $D(0, 4, 4)$

54. $A(-1, 2, 3)$, $B(2, 0, 1)$, $C(1, -3, 2)$, $D(-2, 1, -1)$

In Exercises 55–57, determine whether the given points are coplanar.

55. $A(1, 1, 1)$, $B(-1, 0, 4)$, $C(0, 2, 1)$, $D(2, -2, 3)$

56. $A(0, 0, 4)$, $B(6, 2, 0)$, $C(2, -1, 1)$, $D(-3, -4, 3)$

57. $A(0, 1, 2)$, $B(-1, 1, 0)$, $C(2, 0, -1)$, $D(1, -1, 1)$

12.5 Lines and Planes in Space

This section shows how to use scalar and vector products to write equations for lines, line segments, and planes in space. We will use these representations throughout the rest of the text in studying the calculus of curves and surfaces in space.

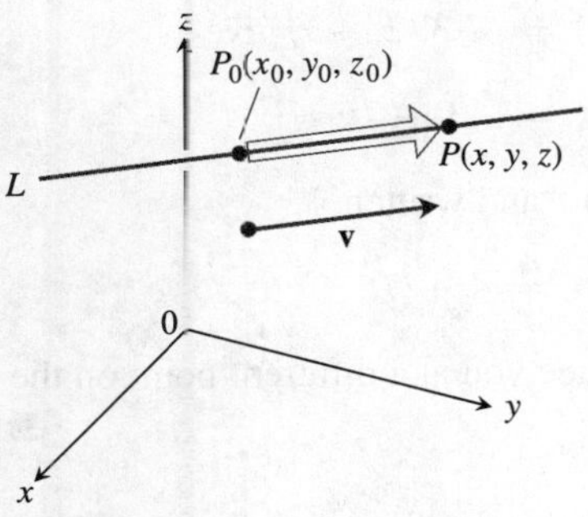

FIGURE 12.37 A point P lies on L through P_0 parallel to $\mathbf{v}$ if and only if $\overrightarrow{P_0P}$ is a scalar multiple of $\mathbf{v}$.

Lines and Line Segments in Space

In the plane, a line is determined by a point and a number giving the slope of the line. In space a line is determined by a point and a *vector* giving the direction of the line.

Suppose that L is a line in space passing through a point $P_0(x_0, y_0, z_0)$ parallel to a vector $\mathbf{v} = v_1\mathbf{i} + v_2\mathbf{j} + v_3\mathbf{k}$. Then L is the set of all points $P(x, y, z)$ for which $\overrightarrow{P_0P}$ is parallel to $\mathbf{v}$ (Figure 12.37). Thus, $\overrightarrow{P_0P} = t\mathbf{v}$ for some scalar parameter t. The value of t depends on the location of the point P along the line, and the domain of t is $(-\infty, \infty)$. The expanded form of the equation $\overrightarrow{P_0P} = t\mathbf{v}$ is

$$(x - x_0)\mathbf{i} + (y - y_0)\mathbf{j} + (z - z_0)\mathbf{k} = t(v_1\mathbf{i} + v_2\mathbf{j} + v_3\mathbf{k}),$$

which can be rewritten as

$$x\mathbf{i} + y\mathbf{j} + z\mathbf{k} = x_0\mathbf{i} + y_0\mathbf{j} + z_0\mathbf{k} + t(v_1\mathbf{i} + v_2\mathbf{j} + v_3\mathbf{k}). \tag{1}$$

If $\mathbf{r}(t)$ is the position vector of a point $P(x, y, z)$ on the line and $\mathbf{r}_0$ is the position vector of the point $P_0(x_0, y_0, z_0)$, then Equation (1) gives the following vector form for the equation of a line in space.

Vector Equation for a Line
A vector equation for the line L through $P_0(x_0, y_0, z_0)$ parallel to a nonzero vector $\mathbf{v}$ is

$$\mathbf{r}(t) = \mathbf{r}_0 + t\mathbf{v}, \qquad -\infty < t < \infty, \tag{2}$$

where $\mathbf{r}$ is the position vector of a point $P(x, y, z)$ on L and $\mathbf{r}_0$ is the position vector of $P_0(x_0, y_0, z_0)$.

Equating the corresponding components of the two sides of Equation (1) gives three scalar equations involving the parameter t:

$$x = x_0 + tv_1, \qquad y = y_0 + tv_2, \qquad z = z_0 + tv_3.$$

These equations give us the standard parametrization of the line for the parameter interval $-\infty < t < \infty$.

Parametric Equations for a Line
The standard parametrization of the line through $P_0(x_0, y_0, z_0)$ parallel to a nonzero vector $\mathbf{v} = v_1\mathbf{i} + v_2\mathbf{j} + v_3\mathbf{k}$ is

$$x = x_0 + tv_1, \quad y = y_0 + tv_2, \quad z = z_0 + tv_3, \quad -\infty < t < \infty. \tag{3}$$

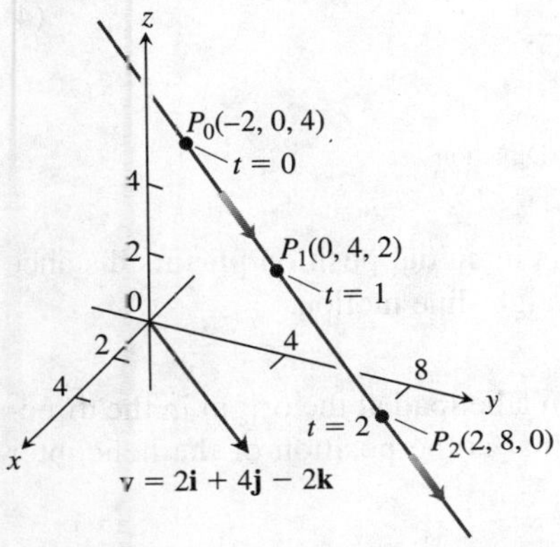

FIGURE 12.38 Selected points and parameter values on the line in Example 1. The arrows show the direction of increasing t.

EXAMPLE 1 Find parametric equations for the line through $(-2, 0, 4)$ parallel to $\mathbf{v} = 2\mathbf{i} + 4\mathbf{j} - 2\mathbf{k}$ (Figure 12.38).

Solution With $P_0(x_0, y_0, z_0)$ equal to $(-2, 0, 4)$ and $v_1\mathbf{i} + v_2\mathbf{j} + v_3\mathbf{k}$ equal to $2\mathbf{i} + 4\mathbf{j} - 2\mathbf{k}$, Equations (3) become

$$x = -2 + 2t, \qquad y = 4t, \qquad z = 4 - 2t. \qquad \blacksquare$$

EXAMPLE 2 Find parametric equations for the line through $P(-3, 2, -3)$ and $Q(1, -1, 4)$.

Solution The vector

$$\overrightarrow{PQ} = (1 - (-3))\mathbf{i} + (-1 - 2)\mathbf{j} + (4 - (-3))\mathbf{k} = 4\mathbf{i} - 3\mathbf{j} + 7\mathbf{k}$$

is parallel to the line, and Equations (3) with $(x_0, y_0, z_0) = (-3, 2, -3)$ give

$$x = -3 + 4t, \qquad y = 2 - 3t, \qquad z = -3 + 7t.$$

We could have chosen $Q(1, -1, 4)$ as the "base point" and written

$$x = 1 + 4t, \qquad y = -1 - 3t, \qquad z = 4 + 7t.$$

These equations serve as well as the first; they simply place you at a different point on the line for a given value of t. ■

Notice that parametrizations are not unique. Not only can the "base point" change, but so can the parameter. The equations $x = -3 + 4t^3$, $y = 2 - 3t^3$, and $z = -3 + 7t^3$ also parametrize the line in Example 2.

To parametrize a line segment joining two points, we first parametrize the line through the points. We then find the t-values for the endpoints and restrict t to lie in the closed interval bounded by these values. The line equations, together with this added restriction, parametrize the segment.

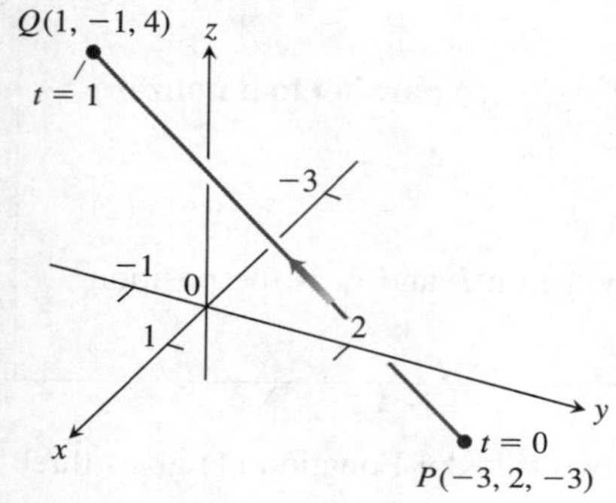

FIGURE 12.39 Example 3 derives a parametrization of line segment PQ. The arrow shows the direction of increasing t.

EXAMPLE 3 Parametrize the line segment joining the points $P(-3, 2, -3)$ and $Q(1, -1, 4)$ (Figure 12.39).

Solution We begin with equations for the line through P and Q, taking them, in this case, from Example 2:

$$x = -3 + 4t, \qquad y = 2 - 3t, \qquad z = -3 + 7t.$$

We observe that the point

$$(x, y, z) = (-3 + 4t, 2 - 3t, -3 + 7t)$$

on the line passes through $P(-3, 2, -3)$ at $t = 0$ and $Q(1, -1, 4)$ at $t = 1$. We add the restriction $0 \le t \le 1$ to parametrize the segment:

$$x = -3 + 4t, \qquad y = 2 - 3t, \qquad z = -3 + 7t, \qquad 0 \le t \le 1.$$ ■

The vector form (Equation (2)) for a line in space is more revealing if we think of a line as the path of a particle starting at position $P_0(x_0, y_0, z_0)$ and moving in the direction of vector $\mathbf{v}$. Rewriting Equation (2), we have

$$\begin{aligned}\mathbf{r}(t) &= \mathbf{r}_0 + t\mathbf{v} \\ &= \underset{\text{Initial position}}{\mathbf{r}_0} + \underset{\text{Time}}{t}\,\underbrace{|\mathbf{v}|}_{\text{Speed}}\,\underbrace{\frac{\mathbf{v}}{|\mathbf{v}|}}_{\text{Direction}}.\end{aligned} \tag{4}$$

In other words, the position of the particle at time t is its initial position plus its distance moved (speed × time) in the direction $\mathbf{v}/|\mathbf{v}|$ of its straight-line motion.

EXAMPLE 4 A helicopter is to fly directly from a helipad at the origin in the direction of the point $(1, 1, 1)$ at a speed of 60 ft/sec. What is the position of the helicopter after 10 sec?

Solution We place the origin at the starting position (helipad) of the helicopter. Then the unit vector

$$\mathbf{u} = \frac{1}{\sqrt{3}}\mathbf{i} + \frac{1}{\sqrt{3}}\mathbf{j} + \frac{1}{\sqrt{3}}\mathbf{k}$$

gives the flight direction of the helicopter. From Equation (4), the position of the helicopter at any time t is

$$\begin{aligned}\mathbf{r}(t) &= \mathbf{r}_0 + t(\text{speed})\,\mathbf{u}\\ &= \mathbf{0} + t(60)\left(\frac{1}{\sqrt{3}}\mathbf{i} + \frac{1}{\sqrt{3}}\mathbf{j} + \frac{1}{\sqrt{3}}\mathbf{k}\right)\\ &= 20\sqrt{3}\,t(\mathbf{i} + \mathbf{j} + \mathbf{k}).\end{aligned}$$

When $t = 10$ sec,

$$\begin{aligned}\mathbf{r}(10) &= 200\sqrt{3}(\mathbf{i} + \mathbf{j} + \mathbf{k})\\ &= \langle 200\sqrt{3}, 200\sqrt{3}, 200\sqrt{3}\rangle.\end{aligned}$$

After 10 sec of flight from the origin toward (1, 1, 1), the helicopter is located at the point $(200\sqrt{3}, 200\sqrt{3}, 200\sqrt{3})$ in space. It has traveled a distance of (60 ft/sec)(10 sec) = 600 ft, which is the length of the vector $\mathbf{r}(10)$. ■

The Distance from a Point to a Line in Space

To find the distance from a point S to a line that passes through a point P parallel to a vector $\mathbf{v}$, we find the absolute value of the scalar component of $\overrightarrow{PS}$ in the direction of a vector normal to the line (Figure 12.40). In the notation of the figure, the absolute value of the scalar component is $|\overrightarrow{PS}|\sin\theta$, which is $\dfrac{|\overrightarrow{PS}|\,|\mathbf{v}|\sin\theta}{|\mathbf{v}|} = \dfrac{|\overrightarrow{PS}\times\mathbf{v}|}{|\mathbf{v}|}$.

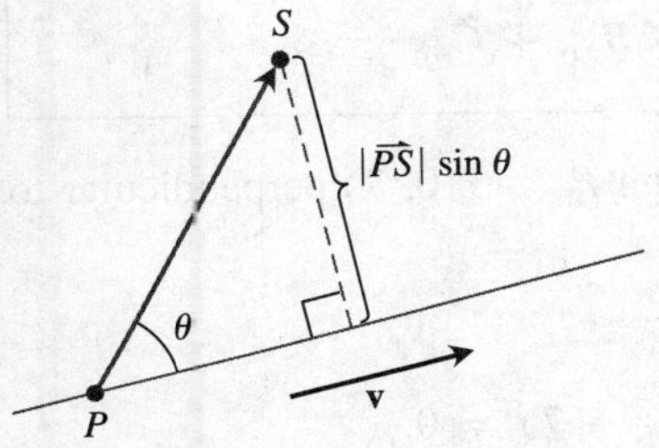

FIGURE 12.40 The distance from S to the line through P parallel to $\mathbf{v}$ is $|\overrightarrow{PS}|\sin\theta$, where θ is the angle between $\overrightarrow{PS}$ and $\mathbf{v}$.

> **Distance from a Point S to a Line Through P Parallel to v**
>
> $$d = \frac{|\overrightarrow{PS}\times\mathbf{v}|}{|\mathbf{v}|} \qquad (5)$$

EXAMPLE 5 Find the distance from the point $S(1, 1, 5)$ to the line

$$L:\quad x = 1 + t,\qquad y = 3 - t,\qquad z = 2t.$$

Solution We see from the equations for L that L passes through $P(1, 3, 0)$ parallel to $\mathbf{v} = \mathbf{i} - \mathbf{j} + 2\mathbf{k}$. With

$$\overrightarrow{PS} = (1-1)\mathbf{i} + (1-3)\mathbf{j} + (5-0)\mathbf{k} = -2\mathbf{j} + 5\mathbf{k}$$

and

$$\overrightarrow{PS}\times\mathbf{v} = \begin{vmatrix}\mathbf{i} & \mathbf{j} & \mathbf{k}\\ 0 & -2 & 5\\ 1 & -1 & 2\end{vmatrix} = \mathbf{i} + 5\mathbf{j} + 2\mathbf{k},$$

Equation (5) gives

$$d = \frac{|\overrightarrow{PS}\times\mathbf{v}|}{|\mathbf{v}|} = \frac{\sqrt{1+25+4}}{\sqrt{1+1+4}} = \frac{\sqrt{30}}{\sqrt{6}} = \sqrt{5}. \quad ■$$

An Equation for a Plane in Space

A plane in space is determined by knowing a point on the plane and its "tilt" or orientation. This "tilt" is defined by specifying a vector that is perpendicular, or normal, to the plane.

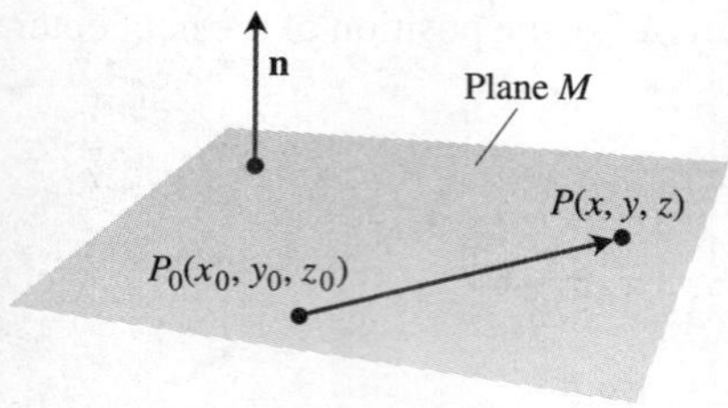

FIGURE 12.41 The standard equation for a plane in space is defined in terms of a vector normal to the plane: A point P lies in the plane through P_0 normal to $\mathbf{n}$ if and only if $\mathbf{n} \cdot \overrightarrow{P_0P} = 0$.

Suppose that plane M passes through a point $P_0(x_0, y_0, z_0)$ and is normal to the nonzero vector $\mathbf{n} = A\mathbf{i} + B\mathbf{j} + C\mathbf{k}$. A vector from P_0 to any point P on the plane is orthogonal to $\mathbf{n}$. Then M is the set of all points $P(x, y, z)$ for which $\overrightarrow{P_0P}$ is orthogonal to $\mathbf{n}$ (Figure 12.41). Thus, the dot product $\mathbf{n} \cdot \overrightarrow{P_0P} = 0$. This equation is equivalent to

$$(A\mathbf{i} + B\mathbf{j} + C\mathbf{k}) \cdot [(x - x_0)\mathbf{i} + (y - y_0)\mathbf{j} + (z - z_0)\mathbf{k}] = 0,$$

so the plane M consists of the points (x, y, z) satisfying

$$A(x - x_0) + B(y - y_0) + C(z - z_0) = 0.$$

Equation for a Plane

The plane through $P_0(x_0, y_0, z_0)$ normal to a nonzero vector $\mathbf{n} = A\mathbf{i} + B\mathbf{j} + C\mathbf{k}$ has

Vector equation:	$\mathbf{n} \cdot \overrightarrow{P_0P} = 0$
Component equation:	$A(x - x_0) + B(y - y_0) + C(z - z_0) = 0$
Component equation simplified:	$Ax + By + Cz = D$, where $D = Ax_0 + By_0 + Cz_0$

EXAMPLE 6 Find an equation for the plane through $P_0(-3, 0, 7)$ perpendicular to $\mathbf{n} = 5\mathbf{i} + 2\mathbf{j} - \mathbf{k}$.

Solution The component equation is

$$5(x - (-3)) + 2(y - 0) + (-1)(z - 7) = 0.$$

Simplifying, we obtain

$$5x + 15 + 2y - z + 7 = 0$$
$$5x + 2y - z = -22.$$

■

Notice in Example 6 how the components of $\mathbf{n} = 5\mathbf{i} + 2\mathbf{j} - \mathbf{k}$ became the coefficients of x, y, and z in the equation $5x + 2y - z = -22$. The vector $\mathbf{n} = A\mathbf{i} + B\mathbf{j} + C\mathbf{k}$ is normal to the plane $Ax + By + Cz = D$.

EXAMPLE 7 Find an equation for the plane through $A(0, 0, 1)$, $B(2, 0, 0)$, and $C(0, 3, 0)$.

Solution We find a vector normal to the plane and use it with one of the points (it does not matter which) to write an equation for the plane.

The cross product

$$\overrightarrow{AB} \times \overrightarrow{AC} = \begin{vmatrix} \mathbf{i} & \mathbf{j} & \mathbf{k} \\ 2 & 0 & -1 \\ 0 & 3 & -1 \end{vmatrix} = 3\mathbf{i} + 2\mathbf{j} + 6\mathbf{k}$$

is normal to the plane. We substitute the components of this vector and the coordinates of $A(0, 0, 1)$ into the component form of the equation to obtain

$$3(x - 0) + 2(y - 0) + 6(z - 1) = 0$$
$$3x + 2y + 6z = 6.$$

■

Lines of Intersection

Just as lines are parallel if and only if they have the same direction, two planes are **parallel** if and only if their normals are parallel, or $\mathbf{n}_1 = k\mathbf{n}_2$ for some scalar k. Two planes that are not parallel intersect in a line.

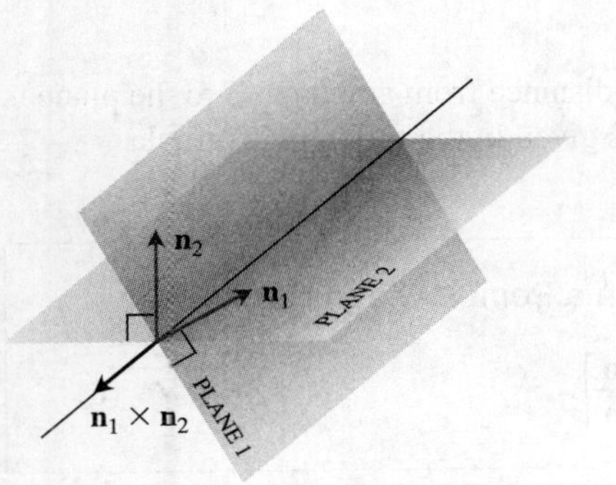

FIGURE 12.42 How the line of intersection of two planes is related to the planes' normal vectors (Example 8).

EXAMPLE 8 Find a vector parallel to the line of intersection of the planes $3x - 6y - 2z = 15$ and $2x + y - 2z = 5$.

Solution The line of intersection of two planes is perpendicular to both planes' normal vectors $\mathbf{n}_1$ and $\mathbf{n}_2$ (Figure 12.42) and therefore parallel to $\mathbf{n}_1 \times \mathbf{n}_2$. Turning this around, $\mathbf{n}_1 \times \mathbf{n}_2$ is a vector parallel to the planes' line of intersection. In our case,

$$\mathbf{n}_1 \times \mathbf{n}_2 = \begin{vmatrix} \mathbf{i} & \mathbf{j} & \mathbf{k} \\ 3 & -6 & -2 \\ 2 & 1 & -2 \end{vmatrix} = 14\mathbf{i} + 2\mathbf{j} + 15\mathbf{k}.$$

Any nonzero scalar multiple of $\mathbf{n}_1 \times \mathbf{n}_2$ will do as well. ■

EXAMPLE 9 Find parametric equations for the line in which the planes $3x - 6y - 2z = 15$ and $2x + y - 2z = 5$ intersect.

Solution We find a vector parallel to the line and a point on the line and use Equations (3).

Example 8 identifies $\mathbf{v} = 14\mathbf{i} + 2\mathbf{j} + 15\mathbf{k}$ as a vector parallel to the line. To find a point on the line, we can take any point common to the two planes. Substituting $z = 0$ in the plane equations and solving for x and y simultaneously identifies one of these points as $(3, -1, 0)$. The line is

$$x = 3 + 14t, \qquad y = -1 + 2t, \qquad z = 15t.$$

The choice $z = 0$ is arbitrary, and we could have chosen $z = 1$ or $z = -1$ just as well. Or we could have let $x = 0$ and solved for y and z. The different choices would simply give different parametrizations of the same line. ■

Sometimes we want to know where a line and a plane intersect. For example, if we are looking at a flat plate and a line segment passes through it, we may be interested in knowing what portion of the line segment is hidden from our view by the plate. This application is used in computer graphics (Exercise 78).

EXAMPLE 10 Find the point where the line

$$x = \frac{8}{3} + 2t, \qquad y = -2t, \qquad z = 1 + t$$

intersects the plane $3x + 2y + 6z = 6$.

Solution The point

$$\left(\frac{8}{3} + 2t, -2t, 1 + t\right)$$

lies in the plane if its coordinates satisfy the equation of the plane—that is, if

$$\begin{aligned} 3\left(\frac{8}{3} + 2t\right) + 2(-2t) + 6(1 + t) &= 6 \\ 8 + 6t - 4t + 6 + 6t &= 6 \\ 8t &= -8 \\ t &= -1. \end{aligned}$$

The point of intersection is

$$(x, y, z)|_{t=-1} = \left(\frac{8}{3} - 2, 2, 1 - 1\right) = \left(\frac{2}{3}, 2, 0\right).$$ ■

The Distance from a Point to a Plane

If P is a point on a plane with a normal $\mathbf{n}$, then the distance from any point S to the plane is the length of the vector projection of $\overrightarrow{PS}$ onto $\mathbf{n}$, as given in the following formula.

Distance from a Point S to a Plane Through a Point P with a Normal $\mathbf{n}$

$$d = \left| \overrightarrow{PS} \cdot \frac{\mathbf{n}}{|\mathbf{n}|} \right| \tag{6}$$

EXAMPLE 11 Find the distance from $S(1, 1, 3)$ to the plane $3x + 2y + 6z = 6$.

Solution We find a point P in the plane and calculate the length of the vector projection of $\overrightarrow{PS}$ onto a vector $\mathbf{n}$ normal to the plane (Figure 12.43). The coefficients in the equation $3x + 2y + 6z = 6$ give

$$\mathbf{n} = 3\mathbf{i} + 2\mathbf{j} + 6\mathbf{k}.$$

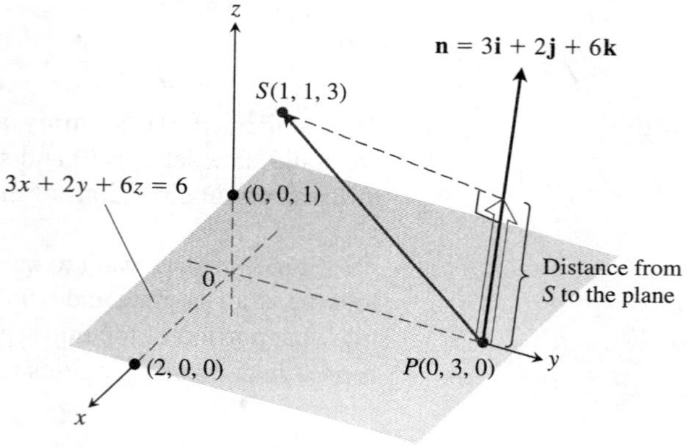

FIGURE 12.43 The distance from S to the plane is the length of the vector projection of $\overrightarrow{PS}$ onto $\mathbf{n}$ (Example 11).

The points on the plane easiest to find from the plane's equation are the intercepts. If we take P to be the y-intercept $(0, 3, 0)$, then

$$\overrightarrow{PS} = (1 - 0)\mathbf{i} + (1 - 3)\mathbf{j} + (3 - 0)\mathbf{k} = \mathbf{i} - 2\mathbf{j} + 3\mathbf{k},$$

$$|\mathbf{n}| = \sqrt{(3)^2 + (2)^2 + (6)^2} = \sqrt{49} = 7.$$

Therefore, the distance from S to the plane is

$$\begin{aligned} d &= \left| \overrightarrow{PS} \cdot \frac{\mathbf{n}}{|\mathbf{n}|} \right| && \text{Length of } \operatorname{proj}_{\mathbf{n}} \overrightarrow{PS} \\ &= \left| (\mathbf{i} - 2\mathbf{j} + 3\mathbf{k}) \cdot \left(\frac{3}{7}\mathbf{i} + \frac{2}{7}\mathbf{j} + \frac{6}{7}\mathbf{k} \right) \right| \\ &= \left| \frac{3}{7} - \frac{4}{7} + \frac{18}{7} \right| = \frac{17}{7}. \end{aligned}$$

■

Angles Between Planes

The angle between two intersecting planes is defined to be the acute angle between their normal vectors (Figure 12.44).

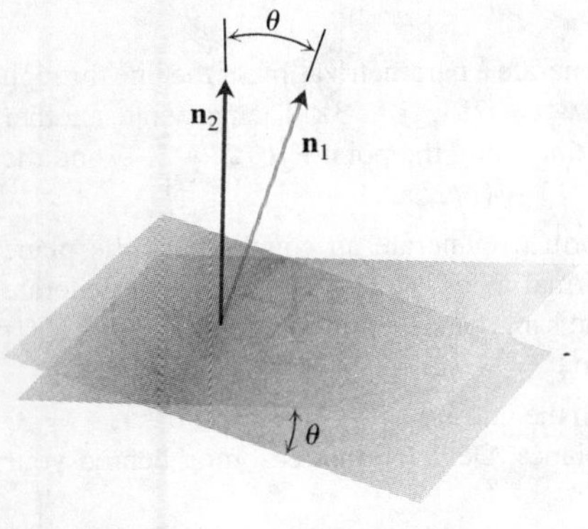

FIGURE 12.44 The angle between two planes is obtained from the angle between their normals.

EXAMPLE 12 Find the angle between the planes $3x - 6y - 2z = 15$ and $2x + y - 2z = 5$.

Solution The vectors

$$\mathbf{n}_1 = 3\mathbf{i} - 6\mathbf{j} - 2\mathbf{k}, \qquad \mathbf{n}_2 = 2\mathbf{i} + \mathbf{j} - 2\mathbf{k}$$

are normals to the planes. The angle between them is

$$\theta = \arccos\left(\frac{\mathbf{n}_1 \cdot \mathbf{n}_2}{|\mathbf{n}_1||\mathbf{n}_2|}\right)$$

$$= \arccos\left(\frac{4}{21}\right) \approx 1.38 \text{ radians.} \qquad \text{About 79 degrees}$$

■

EXERCISES 12.5

Lines and Line Segments

Find parametric equations for the lines in Exercises 1–12.

1. The line through the point $P(3, -4, -1)$ parallel to the vector $\mathbf{i} + \mathbf{j} + \mathbf{k}$
2. The line through $P(1, 2, -1)$ and $Q(-1, 0, 1)$
3. The line through $P(-2, 0, 3)$ and $Q(3, 5, -2)$
4. The line through $P(1, 2, 0)$ and $Q(1, 1, -1)$
5. The line through the origin parallel to the vector $2\mathbf{j} + \mathbf{k}$
6. The line through the point $(3, -2, 1)$ parallel to the line $x = 1 + 2t$, $y = 2 - t$, $z = 3t$
7. The line through $(1, 1, 1)$ parallel to the z-axis
8. The line through $(2, 4, 5)$ perpendicular to the plane $3x + 7y - 5z = 21$
9. The line through (3, 2, –8) and is perpendicular to the plane $3x - y - 2z + 2 = 0$.
10. The line through $(2, 3, 0)$ perpendicular to the vectors $\mathbf{u} = \mathbf{i} + 2\mathbf{j} + 3\mathbf{k}$ and $\mathbf{v} = 3\mathbf{i} + 4\mathbf{j} + 5\mathbf{k}$
11. The line through (2, 1, 0) and perpendicular to the vector $\mathbf{u} = \mathbf{i} + 5\mathbf{j} - \mathbf{k}$ and $\mathbf{v} = 2\mathbf{i} - 5\mathbf{j} + 3\mathbf{k}$.
12. The z-axis

Find parametrizations for the line segments joining the points in Exercises 13–20. Draw coordinate axes and sketch each segment, indicating the direction of increasing t for your parametrization.

13. $(0, 0, 0)$, $(1, 1, 3/2)$
14. $(0, 0, 0)$, $(1, 0, 0)$
15. $(3, 2, 4)$, $(4, 5, 2)$
16. $(4, 5, 2)$, $(5, 8, 0)$
17. $(0, 1, 1)$, $(0, -1, 1)$
18. $(0, 2, 0)$, $(3, 0, 0)$
19. $(2, 0, 2)$, $(0, 2, 0)$
20. $(1, 0, -1)$, $(0, 3, 0)$

Planes

Find equations for the planes in Exercises 21–26.

21. The plane through $P_0(0, 2, -1)$ normal to $\mathbf{n} = 3\mathbf{i} - 2\mathbf{j} - \mathbf{k}$
22. The plane through $(1, -1, 3)$ parallel to the plane
$$3x + y + z = 7$$
23. The plane through $(1, 1, -1)$, $(2, 0, 2)$, and $(0, -2, 1)$
24. The plane through the point (2, –3, 1) and perpendicular to the line joining the points (3, 4, –1) and (2, –1, 5).
25. The plane through $P_0(2, 4, 5)$ perpendicular to the line
$$x = 5 + t, \quad y = 1 + 3t, \quad z = 4t$$
26. The plane through $A(1, -2, 1)$ perpendicular to the vector from the origin to A.
27. Find the point of intersection of the lines $x = 2t + 1$, $y = 3t + 2$, $z = 4t + 3$, and $x = s + 2$, $y = 2s + 4$, $z = -4s - 1$, and then find the plane determined by these lines.
28. Find the point of intersection of the lines $x = 2t - 3$, $y = 3t - 5$, $z = -3t + 7$ and $x = 4s - 1$, $y = 5s - 1$, $z = -s - 1$ and then find the plane determined by these lines.

In Exercises 29 and 30, find the plane containing the intersecting lines.

29. $L1$: $x = -1 + t$, $y = 2 + t$, $z = 1 - t$; $-\infty < t < \infty$
$L2$: $x = 1 - 4s$, $y = 1 + 2s$, $z = 2 - 2s$; $-\infty < s < \infty$
30. $L1: x = 2t + 1$, $y = 3t + 2$, $z = 4t + 3$; $-\infty < t < \infty$
$L2: x = 3s + 2$, $y = 4s + 3$, $z = 5s + 4$; $-\infty < s < \infty$
31. Find a plane through $P_0(2, 1, -1)$ and perpendicular to the line of intersection of the planes $2x + y - z = 3$, $x + 2y + z = 2$.
32. Find a plane through the points $P_1(1, 2, 3)$, and $P_2(3, 2, 1)$ and perpendicular to the plane $4x - y + 2z = 7$.

Distances

In Exercises 33–38, find the distance from the point to the line.

33. $(0, 0, 12)$; $x = 4t$, $y = -2t$, $z = 2t$
34. $(0, 0, 0)$; $x = 5 + 3t$, $y = 5 + 4t$, $z = -3 - 5t$
35. $(4, 6, 2)$; $x = 3t + 2$, $y = 2t + 2$, $z = t + 2$
36. $(3, -1, 11)$; $x = 2t$, $y = 3t + 2$, $z = 4t + 3$
37. $(3, -1, 4)$; $x = 4 - t$, $y = 3 + 2t$, $z = -5 + 3t$
38. $(-1, 4, 3)$; $x = 10 + 4t$, $y = -3$, $z = 4t$

In Exercises 39–44, find the distance from the point to the plane.

39. $(2, -3, 4)$, $x + 2y + 2z = 13$
40. $(0, 0, 0)$, $3x + 2y + 6z = 6$
41. $(1, 3, 4)$, $2x - y + z = 0$

42. $(2, -3, 3), \quad x - 2y - z + 1 = 0$

43. $(0, -1, 0), \quad 2x + y + 2z = 4$

44. $(1, 0, -1), \quad -4x + y + z = 4$

45. Find the distance from the plane $x + 2y + 6z = 1$ to the plane $x + 2y + 6z = 10$.

46. Find the distance from the line $x = 2 + t$, $y = 1 + t$, $z = -(1/2) - (1/2)t$ to the plane $x + 2y + 6z = 10$.

Angles

In Exercises 47 and 48, find the angles between the planes.

47. $x + y = 1, \quad 2x + y - 2z = 2$

48. $x + 2y + z - 1 = 0, \quad 2x + y - z + 1 = 0$

In Exercises 49 and 50, find the acute angles between the intersecting lines.

49. $x = t,\ y = 2t,\ z = -t$ and $x = 1 - t,\ y = 5 + t,\ z = 2t$

50. $x = 2 + t,\ y = 4t + 2,\ z = 1 + t$ and $x = 3t - 2,\ y = -2,\ z = 2 - 2t$

In Exercises 51 and 52, find the acute angles between the lines and planes.

51. $x = 3t + 1, \quad y = 5t + 3, \quad z = 2t + 4; \quad 2x - y + z + 3 = 0$

52. $x = 2,\ y = 3 + 2t,\ z = 1 - 2t; \quad x - y + z = 0$

T Use a calculator to find the acute angles between the planes in Exercises 53–56 to the nearest hundredth of a radian.

53. $x + 2y + 3z + 6 = 0, \quad x - 2y + z + 1 = 0$

54. $x + y + z = 1, \quad z = 0$ (the xy-plane)

55. $2x + 2y - z = 3, \quad x + 2y + z = 2$

56. $4y + 3z = -12, \quad 3x + 2y + 6z = 6$

Intersecting Lines and Planes

In Exercises 57–60, find the point in which the line meets the plane.

57. $x = 1 - t, \quad y = 3t, \quad z = 1 + t; \quad 2x - y + 3z = 6$

58. $x = 2, \quad y = 3 + 2t, \quad z = -2 - 2t; \quad 6x + 3y - 4z = -12$

59. $x = 1 + 2t, \quad y = 1 + 5t, \quad z = 3t; \quad x + y + z = 2$

60. $x = -1 + 3t, \quad y = -2, \quad z = 5t; \quad 2x - 3z = 7$

Find parametrizations for the lines in which the planes in Exercises 61–64 intersect.

61. $x + y + z = 1, \quad x + y = 2$

62. $x - 3y + 2z + 4 = 0, \quad 2x + y + 4z + 1 = 0$

63. $x - 2y + 4z = 2, \quad x + y - 2z = 5$

64. $x + y + z - 3 = 0, \quad 2x + 3y + 4z - 5 = 0$

Given two lines in space, either they are parallel, they intersect, or they are skew (lie in parallel planes). In Exercises 65 and 66, determine whether the lines, taken two at a time, are parallel, intersect, or are skew. If they intersect, find the point of intersection. Otherwise, find the distance between the two lines.

65. L1: $x = 3 + 2t,\ y = -1 + 4t,\ z = 2 - t; \quad -\infty < t < \infty$

L2: $x = 1 + 4s,\ y = 1 + 2s,\ z = -3 + 4s; \quad -\infty < s < \infty$

L3: $x = 3 + 2r,\ y = 2 + r,\ z = -2 + 2r; \quad -\infty < r < \infty$

66. L1: $x = 1 + 2t, \quad y = -1 - t, \quad z = 3t; \quad -\infty < t < \infty$

L2: $x = 2 - s, \quad y = 3s, \quad z = 1 + s; \quad -\infty < s < \infty$

L3: $x = 5 + 2r, \quad y = 1 - r, \quad z = 8 + 3r; \quad -\infty < r < \infty$

Theory and Examples

67. Use Equations (3) to generate a parametrization of the line through $P(2, -4, 7)$ parallel to $\mathbf{v}_1 = 2\mathbf{i} - \mathbf{j} + 3\mathbf{k}$. Then generate another parametrization of the line using the point $P_2(-2, -2, 1)$ and the vector $\mathbf{v}_2 = -\mathbf{i} + (1/2)\mathbf{j} - (3/2)\mathbf{k}$.

68. Use the component form to generate an equation for the plane through $P_1(4, 1, 5)$ normal to $\mathbf{n}_1 = \mathbf{i} - 2\mathbf{j} + \mathbf{k}$. Then generate another equation for the same plane using the point $P_2(3, -2, 0)$ and the normal vector $\mathbf{n}_2 = -\sqrt{2}\mathbf{i} + 2\sqrt{2}\mathbf{j} - \sqrt{2}\mathbf{k}$.

69. Find the points in which the line $x = 1 + 2t,\ y = -1 - t,\ z = 3t$ meets the coordinate planes. Describe the reasoning behind your answer.

70. Find equations for the line in the plane $z = 3$ that makes an angle of $\pi/6$ rad with $\mathbf{i}$ and an angle of $\pi/3$ rad with $\mathbf{j}$. Describe the reasoning behind your answer.

71. Is the line $x = 1 - 2t,\ y = 2 + 5t,\ z = -3t$ parallel to the plane $2x + y - z = 8$? Give reasons for your answer.

72. How can you tell when two planes $A_1x + B_1y + C_1z = D_1$ and $A_2x + B_2y + C_2z = D_2$ are parallel? Perpendicular? Give reasons for your answer.

73. Find two different planes whose intersection is the line $x = 1 + t,\ y = 2 - t,\ z = 3 + 2t$. Write equations for each plane in the form $Ax + By + Cz = D$.

74. Find a plane through the origin that is perpendicular to the plane $M: 2x + 3y + z = 12$ in a right angle. How do you know that your plane is perpendicular to M?

75. The graph of $(x/a) + (y/b) + (z/c) = 1$ is a plane for any nonzero numbers a, b, and c. Which planes have an equation of this form?

76. Suppose L_1 and L_2 are disjoint (nonintersecting) nonparallel lines. Is it possible for a nonzero vector to be perpendicular to both L_1 and L_2? Give reasons for your answer.

77. Perspective in computer graphics In computer graphics and perspective drawing, we need to represent objects seen by the eye in space as images on a two-dimensional plane. Suppose that the eye is at $E(x_0, 0, 0)$ as shown here and that we want to represent a point $P_1(x_1, y_1, z_1)$ as a point on the yz-plane. We do this by projecting P_1 onto the plane with a ray from E. The point P_1 will be portrayed as the point $P(0, y, z)$. The problem for us as graphics designers is to find y and z given E and P_1.

a. Write a vector equation that holds between $\overrightarrow{EP}$ and $\overrightarrow{EP_1}$. Use the equation to express y and z in terms of x_0, x_1, y_1, and z_1.

b. Test the formulas obtained for y and z in part (a) by investigating their behavior at $x_1 = 0$ and $x_1 = x_0$ and by seeing what happens as $x_0 \to \infty$. What do you find?

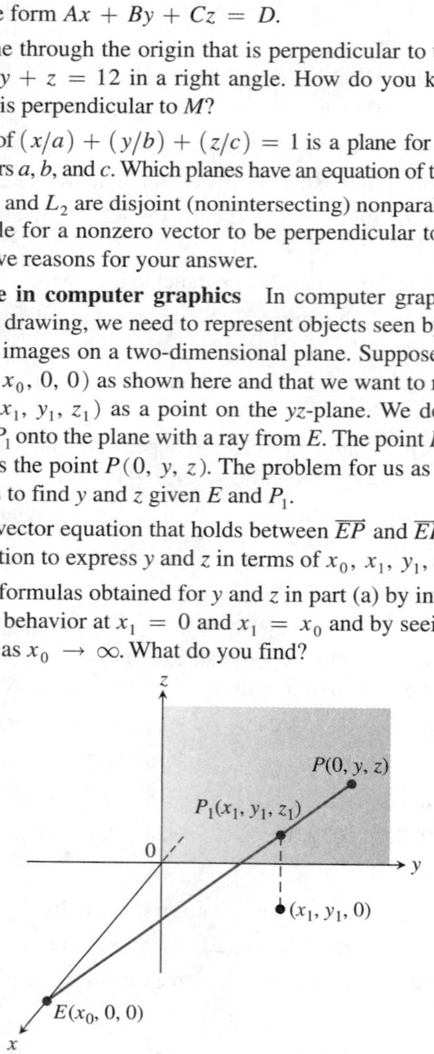

78. Hidden lines in computer graphics Here is another typical problem in computer graphics. Your eye is at (4, 0, 0). You are looking at a triangular plate whose vertices are at (1, 0, 1), (1, 1, 0), and (−2, 2, 2). The line segment from (1, 0, 0) to (0, 2, 2) passes through the plate. What portion of the line segment is hidden from your view by the plate? (This is an exercise in finding intersections of lines and planes.)

12.6 Cylinders and Quadric Surfaces

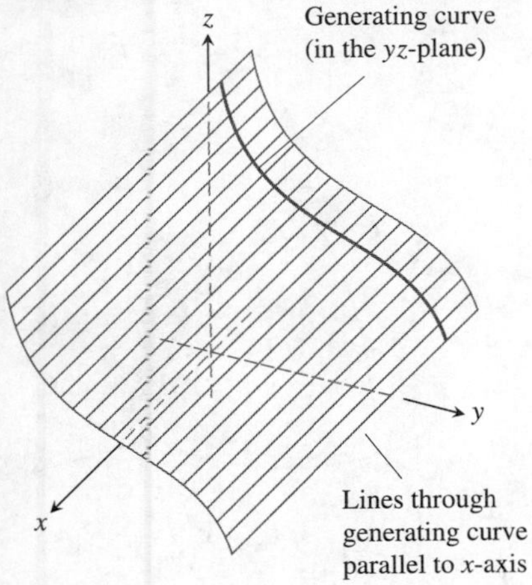

FIGURE 12.45 A cylinder and generating curve.

Up to now, we have studied two special types of surfaces: spheres and planes. In this section, we extend our inventory to include a variety of cylinders and quadric surfaces. Quadric surfaces are surfaces defined by second-degree equations in x, y, and z. Spheres are quadric surfaces, but there are others of equal interest that will be needed in Chapters 14–16.

Cylinders

Suppose we are given a plane in space that contains a curve, and in addition we are given a line that is not parallel to this plane. A **cylinder** is a surface that is generated by moving a line that is parallel to the given line along the curve, while keeping it parallel to the given line. The curve is called a **generating curve** for the cylinder (Figure 12.45 illustrates this when the given plane is the yz-plane and the given line is the x-axis). In solid geometry, where *cylinder* means *circular cylinder*, the generating curves are circles, but now we allow generating curves of any kind. The cylinder in our first example is generated by a parabola.

EXAMPLE 1 Find an equation for the cylinder made by the lines parallel to the z-axis that pass through the parabola $y = x^2$, $z = 0$ (Figure 12.46).

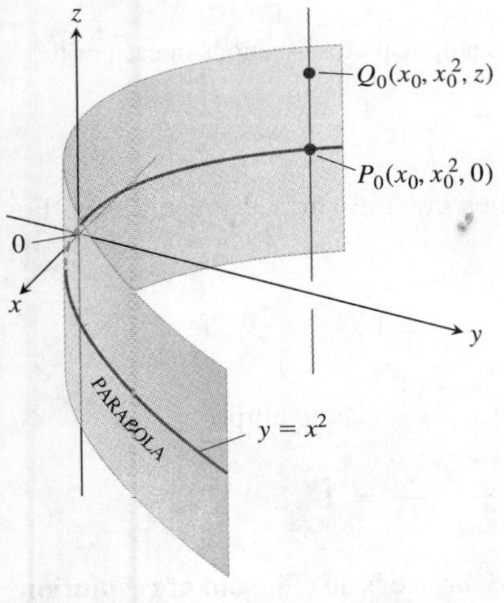

FIGURE 12.46 Every point of the cylinder in Example 1 has coordinates of the form (x_0, x_0^2, z).

Solution The point $P_0(x_0, x_0^2, 0)$ lies on the parabola $y = x^2$ in the xy-plane. Then, for any value of z, the point $Q(x_0, x_0^2, z)$ lies on the cylinder because it lies on the line $x = x_0$, $y = x_0^2$ through P_0 parallel to the z-axis. Conversely, any point $Q(x_0, x_0^2, z)$ whose y-coordinate is the square of its x-coordinate lies on the cylinder because it lies on the line $x = x_0$, $y = x_0^2$ through P_0 parallel to the z-axis (Figure 12.46).

Regardless of the value of z, therefore, the points on the surface are the points whose coordinates satisfy the equation $y = x^2$. This makes $y = x^2$ an equation for the cylinder. ■

As Example 1 suggests, any curve $f(x, y) = c$ in the xy-plane generates a cylinder parallel to the z-axis whose equation is also $f(x, y) = c$. For instance, the equation $x^2 + y^2 = 1$ corresponds to the circular cylinder made by the lines parallel to the z-axis that pass through the circle $x^2 + y^2 = 1$ in the xy-plane.

In a similar way, any curve $g(x, z) = c$ in the xz-plane generates a cylinder parallel to the y-axis whose space equation is also $g(x, z) = c$. Any curve $h(y, z) = c$ generates a cylinder parallel to the x-axis whose space equation is also $h(y, z) = c$. The axis of a cylinder need not be parallel to a coordinate axis, however.

Quadric Surfaces

A **quadric surface** is the graph in space of a second-degree equation in x, y, and z. We first focus on quadric surfaces given by the equation

$$Ax^2 + By^2 + Cz^2 + Dz = E,$$

where A, B, C, D, and E are constants. The basic quadric surfaces are **ellipsoids**, **paraboloids**, **elliptical cones**, and **hyperboloids**. Spheres are special cases of ellipsoids. We present a few examples illustrating how to sketch a quadric surface, and then we give a summary table of graphs of the basic types.

EXAMPLE 2 The **ellipsoid**

$$\frac{x^2}{a^2} + \frac{y^2}{b^2} + \frac{z^2}{c^2} = 1$$

(Figure 12.47) cuts the coordinate axes at $(\pm a, 0, 0)$, $(0, \pm b, 0)$, and $(0, 0, \pm c)$. It lies within the rectangular box defined by the inequalities $|x| \leq a$, $|y| \leq b$, and $|z| \leq c$. The surface is symmetric with respect to each of the coordinate planes because each variable in the defining equation is squared.

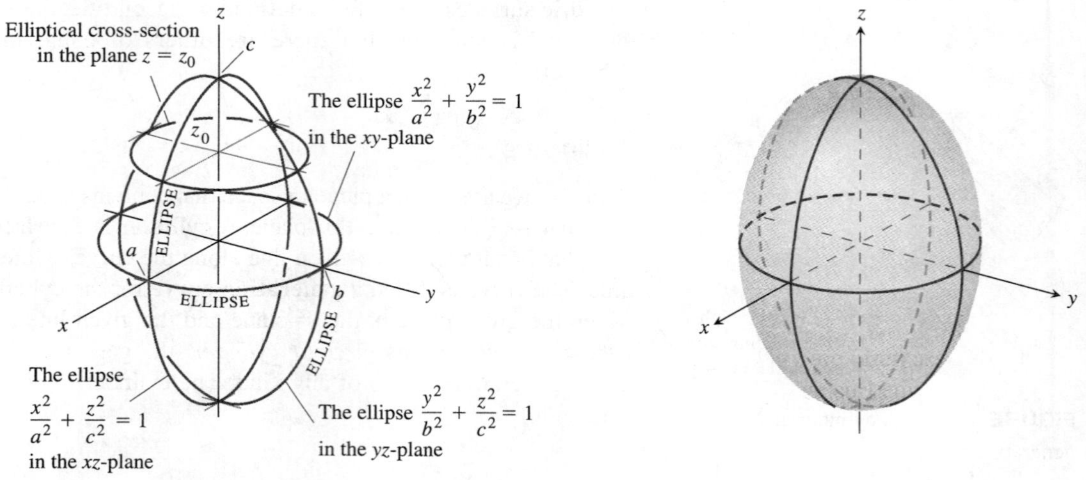

FIGURE 12.47 The ellipsoid $\frac{x^2}{a^2} + \frac{y^2}{b^2} + \frac{z^2}{c^2} = 1$ in Example 2 has elliptical cross-sections in each of the three coordinate planes.

The curves in which the three coordinate planes cut the surface are ellipses. For example,

$$\frac{x^2}{a^2} + \frac{y^2}{b^2} = 1 \quad \text{when} \quad z = 0.$$

The curve cut from the surface by the plane $z = z_0$, $|z_0| < c$, is the ellipse

$$\frac{x^2}{a^2\left(1 - (z_0/c)^2\right)} + \frac{y^2}{b^2\left(1 - (z_0/c)^2\right)} = 1.$$

If any two of the semiaxes a, b, and c are equal, the surface is an **ellipsoid of revolution**. If all three are equal, the surface is a sphere. ■

EXAMPLE 3 The **hyperbolic paraboloid**

$$\frac{y^2}{b^2} - \frac{x^2}{a^2} = \frac{z}{c}, \qquad c > 0$$

has symmetry with respect to the planes $x = 0$ and $y = 0$ (Figure 12.48). The cross-sections in these planes are

$$x = 0: \quad \text{the parabola } z = \frac{c}{b^2}y^2. \tag{1}$$

$$y = 0: \quad \text{the parabola } z = -\frac{c}{a^2}x^2. \tag{2}$$

In the plane $x = 0$, the parabola opens upward from the origin. The parabola in the plane $y = 0$ opens downward.

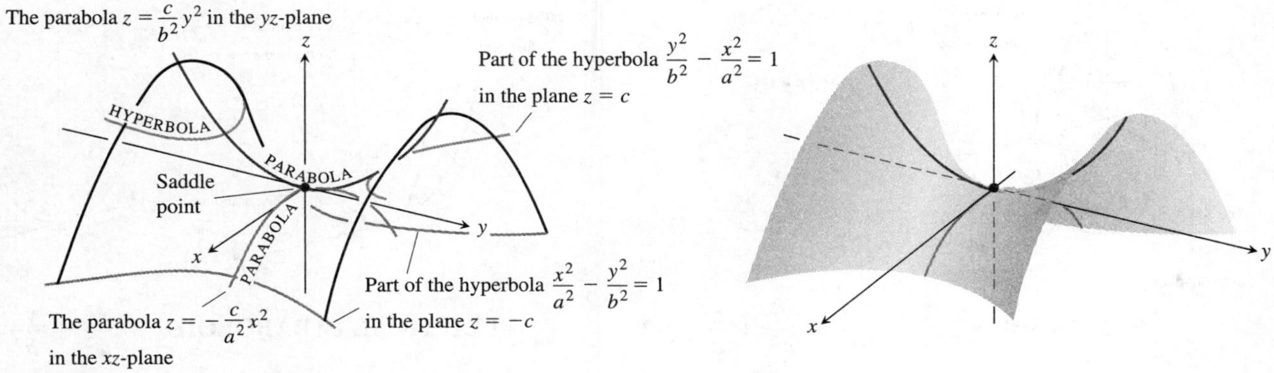

FIGURE 12.48 The hyperbolic paraboloid $(y^2/b^2) - (x^2/a^2) = z/c$, $c > 0$. The cross-sections in planes perpendicular to the z-axis above and below the xy-plane are hyperbolas. The cross-sections in planes perpendicular to the other axes are parabolas.

If we cut the surface by a plane $z = z_0 > 0$, the cross-section is a hyperbola,

$$\frac{y^2}{b^2} - \frac{x^2}{a^2} = \frac{z_0}{c},$$

with its focal axis parallel to the y-axis and its vertices on the parabola in Equation (1). If z_0 is negative, the focal axis is parallel to the x-axis and the vertices lie on the parabola in Equation (2).

Near the origin, the surface is shaped like a saddle or mountain pass. To a person traveling along the surface in the yz-plane the origin looks like a minimum. To a person traveling the xz-plane the origin looks like a maximum. Such a point is called a **saddle point** of a surface. We will say more about saddle points in Section 14.7. ■

Table 12.1 shows graphs of the six basic types of quadric surfaces. Each surface shown is symmetric with respect to the z-axis, but other coordinate axes can serve as well (with appropriate changes to the equation).

General Quadric Surfaces

The quadric surfaces we have considered have symmetries relative to the x-, y-, or z-axes. The general equation of second degree in three variables x, y, z is

$$Ax^2 + By^2 + Cz^2 + Dxy + Exz + Fyz + Gz + Hy + Iz + J = 0,$$

where $A, B, C, D, E, F, G, H, I$, and J are constants. This equation leads to surfaces similar to those in Table 12.1, but in general these surfaces might be translated and rotated relative to the x-, y-, and z-axes. Terms of the type Gx, Hy, or Iz in the above formula lead to translations, which can be seen by a process of completing the squares.

EXAMPLE 4 Identify the surface given by the equation

$$x^2 + y^2 + 4z^2 - 2x + 4y + 1 = 0.$$

Solution We complete the squares to simplify the expression:

$$\begin{aligned} x^2 + y^2 + 4z^2 - 2x + 4y + 1 &= (x-1)^2 - 1 + (y+2)^2 - 4 + 4z^2 + 1 \\ &= (x-1)^2 + (y+2)^2 + 4z^2 - 4. \end{aligned}$$

TABLE 12.1 Graphs of Quadric Surfaces

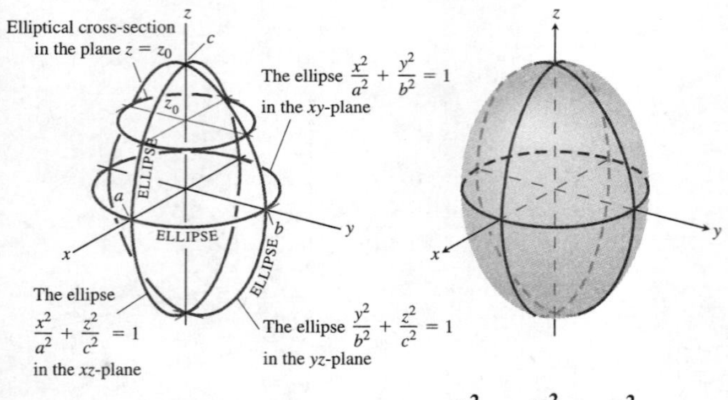

ELLIPSOID $$\frac{x^2}{a^2} + \frac{y^2}{b^2} + \frac{z^2}{c^2} = 1$$

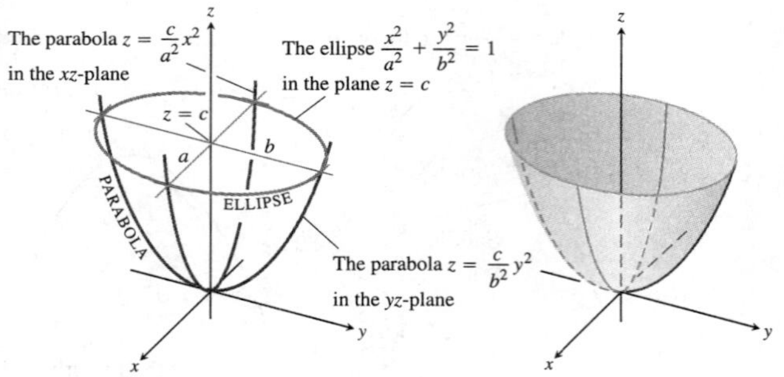

ELLIPTICAL PARABOLOID $$\frac{x^2}{a^2} + \frac{y^2}{b^2} = \frac{z}{c}$$

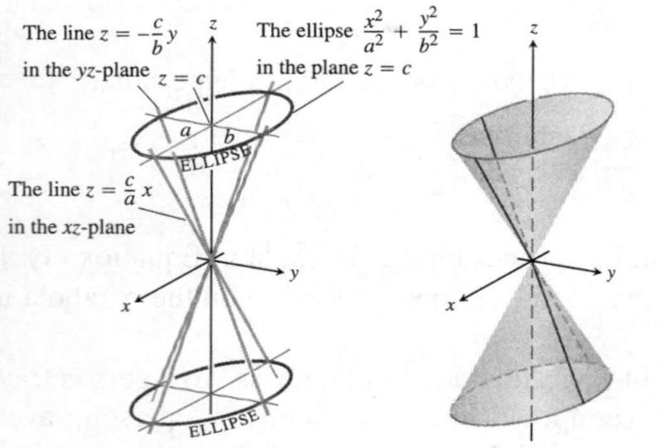

ELLIPTICAL CONE $$\frac{x^2}{a^2} + \frac{y^2}{b^2} = \frac{z^2}{c^2}$$

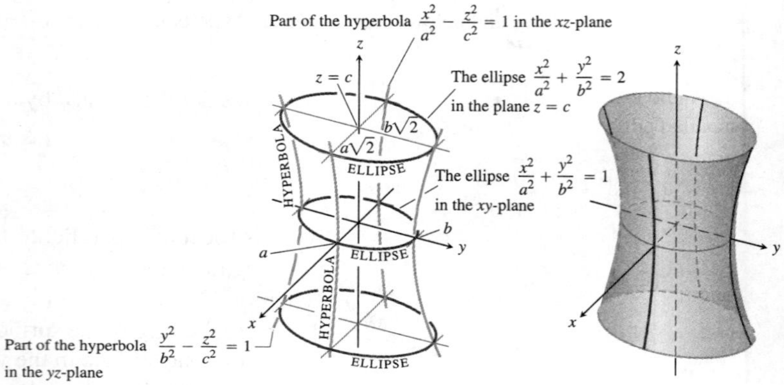

HYPERBOLOID OF ONE SHEET $$\frac{x^2}{a^2} + \frac{y^2}{b^2} - \frac{z^2}{c^2} = 1$$

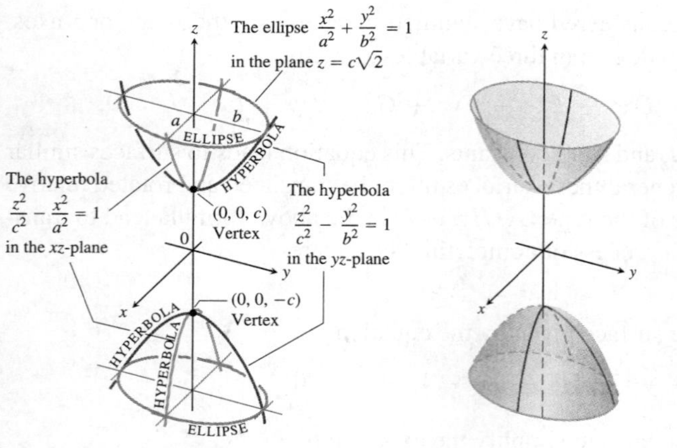

HYPERBOLOID OF TWO SHEETS $$\frac{z^2}{c^2} - \frac{x^2}{a^2} - \frac{y^2}{b^2} = 1$$

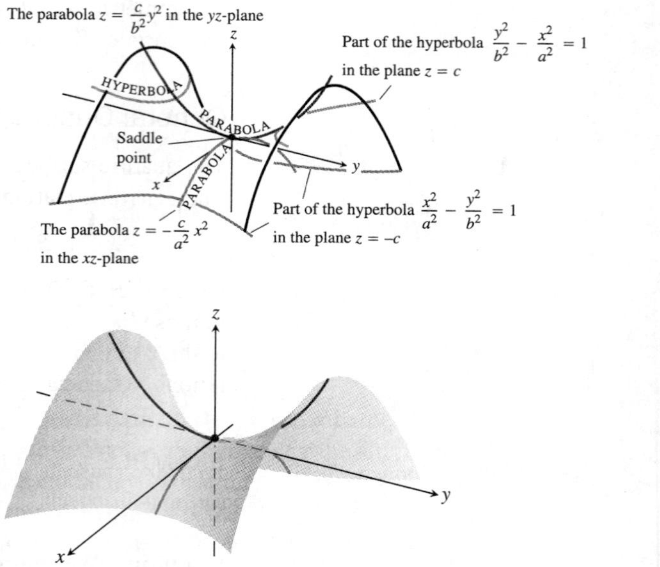

HYPERBOLIC PARABOLOID $$\frac{y^2}{b^2} - \frac{x^2}{a^2} = \frac{z}{c},\ c > 0$$

We can rewrite the original equation as

$$\frac{(x-1)^2}{4} + \frac{(y+2)^2}{4} + \frac{z^2}{1} = 1.$$

This is the equation of an ellipsoid whose three semiaxes have lengths 2, 2, and 1 and which is centered at the point $(1, -2, 0)$, as shown in Figure 12.49. ■

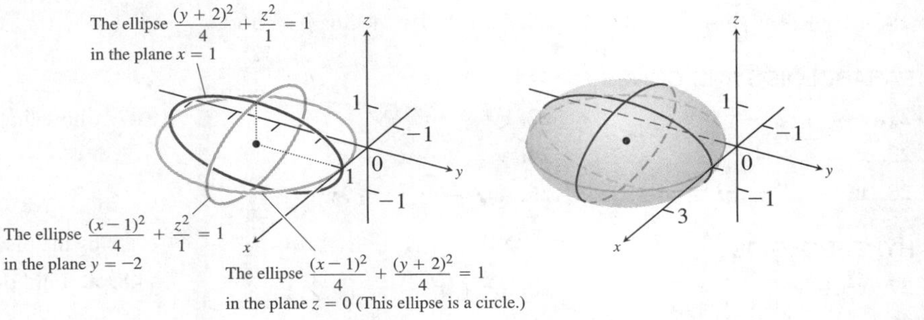

FIGURE 12.49 An ellipsoid centered at the point $(1, -2, 0)$.

EXERCISES 12.6

Matching Equations with Surfaces

In Exercises 1–12, match the equation with the surface it defines. Also, identify each surface by type (paraboloid, ellipsoid, etc.). The surfaces are labeled (a)–(l).

1. $x^2 + y^2 + 4z^2 = 10$ **2.** $z^2 + 4y^2 - 4x^2 = 4$

3. $4y^2 + z^2 = 9$ **4.** $2y^2 + z^2 = 2x^2$

5. $3x = y^2 - 3z^2$ **6.** $x = -2y^2 - 2z^2$

7. $x^2 + 2z^2 = 8$ **8.** $z^2 + x^2 - y^2 = 1$

9. $x = z^2 - y^2$ **10.** $z = -4x^2 - y^2$

11. $x^2 + 4z^2 = y^2$ **12.** $8x^2 + 2y^2 + 4z^2 = 32$

a.

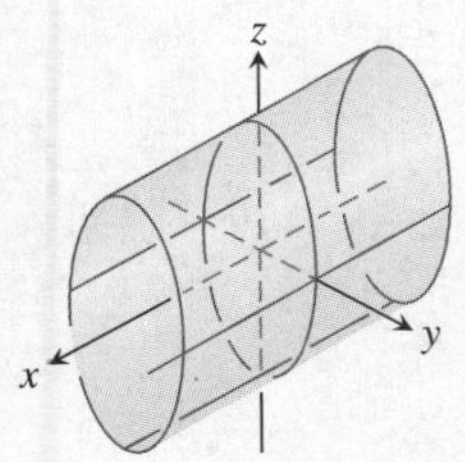

b.

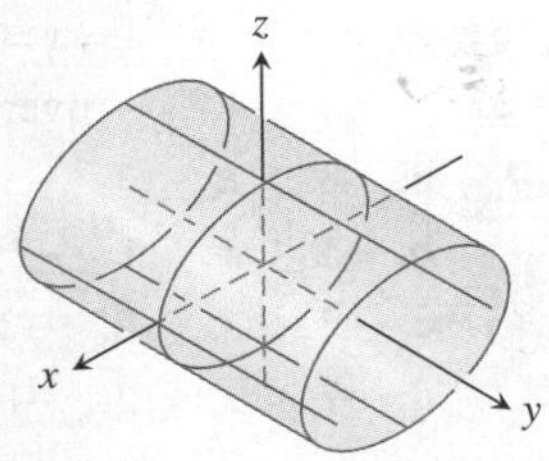

c.

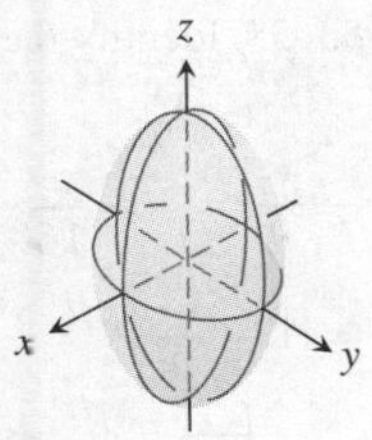

d.

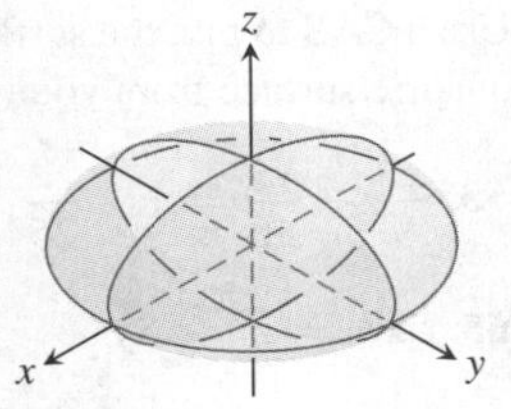

e.

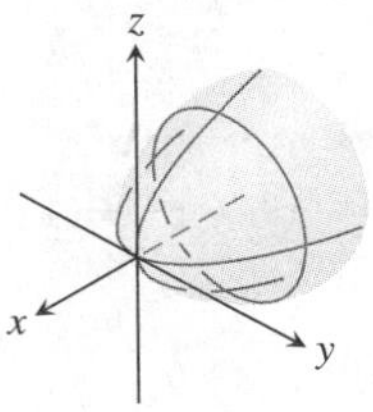

f.

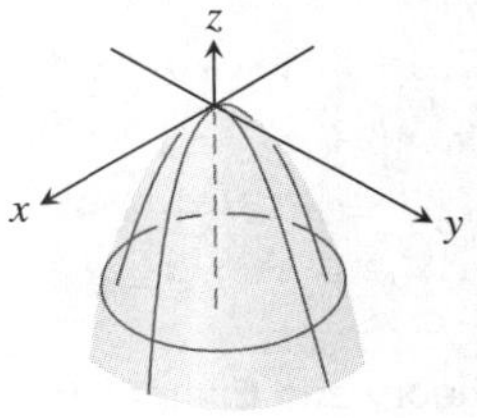

g.

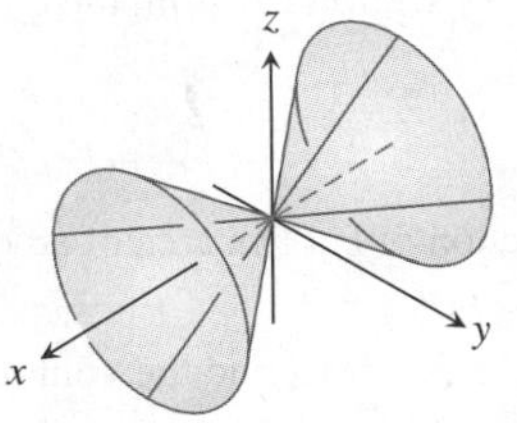

h.

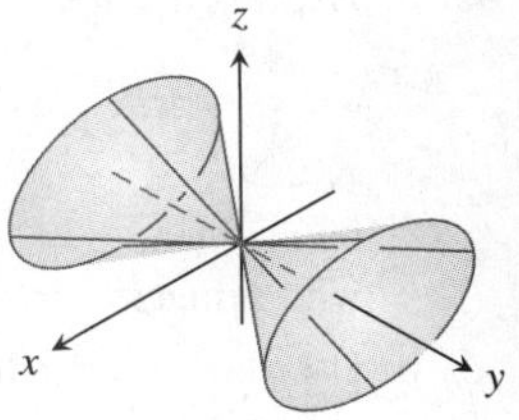

i.

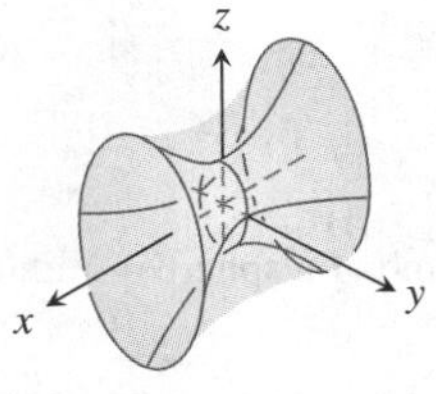

j.

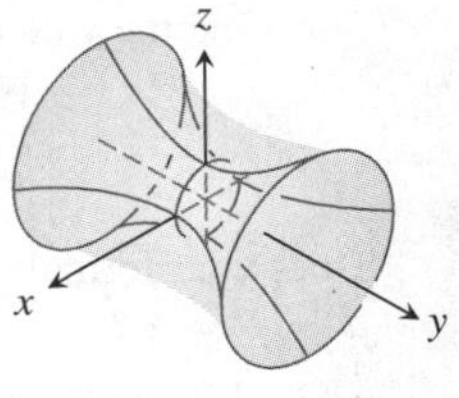

k.

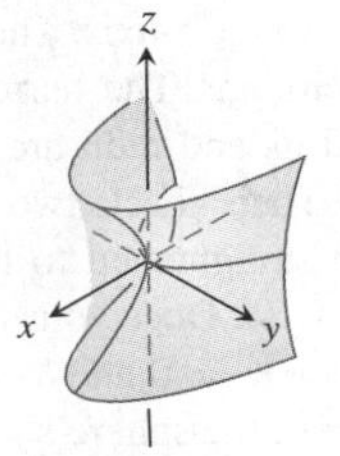

l.

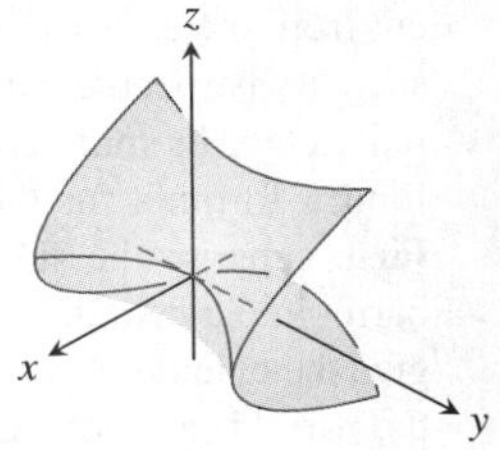

Drawing

Sketch the surfaces in Exercises 13–44.

CYLINDERS

13. $x^2 + y^2 = 9$

14. $z = y^2 - 1$

15. $x^2 + 4z^2 = 16$

16. $4x^2 + y^2 = 36$

ELLIPSOIDS

17. $4x^2 + y^2 + z^2 = 4$

18. $4x^2 + 4y^2 + z^2 = 16$

19. $4x^2 + 9y^2 + 4z^2 = 36$

20. $9x^2 + 4y^2 + 36z^2 = 36$

PARABOLOIDS AND CONES

21. $z = x^2 + 4y^2$

22. $z = 9 - x^2 - y^2$

23. $x = 4 - 4y^2 - z^2$

24. $y = 1 - x^2 - z^2$

25. $x^2 + y^2 = z^2$

26. $4x^2 + 9z^2 = 9y^2$

HYPERBOLOIDS

27. $x^2 + y^2 - z^2 = 1$

28. $y^2 + z^2 - x^2 = 1$

29. $z^2 - x^2 - y^2 = 1$

30. $(y^2/4) - (x^2/4) - z^2 = 1$

HYPERBOLIC PARABOLOIDS

31. $y^2 - x^2 = z$

32. $x^2 - y^2 = z$

ASSORTED

33. $z = 1 + y^2 - x^2$

34. $4x^2 + 4y^2 = z^2$

35. $y = -(x^2 + z^2)$

36. $16x^2 + 4y^2 = 1$

37. $x^2 + y^2 - z^2 = 4$

38. $x^2 + z^2 = y$

39. $x^2 + z^2 = 1$

40. $25y^2 + 16z^2 = 9x^2$

41. $z = -(x^2 + y^2)$

42. $y^2 - x^2 - z^2 = 1$

43. $9y^2 + z^2 - 9x^2 = 9$

44. $x^2 + y^2 = z$

Theory and Examples

45. a. Express the area A of the cross-section cut from the ellipsoid

$$x^2 + \frac{y^2}{4} + \frac{z^2}{9} = 1$$

by the plane $z = c$ as a function of c. (The area of an ellipse with semiaxes a and b is πab.)

b. Use slices perpendicular to the z-axis to find the volume of the ellipsoid in part (a).

c. Now find the volume of the ellipsoid

$$\frac{x^2}{a^2} + \frac{y^2}{b^2} + \frac{z^2}{c^2} = 1.$$

Does your formula give the volume of a sphere of radius a if $a = b = c$?

46. The barrel shown here is shaped like an ellipsoid with equal pieces cut from the ends by planes perpendicular to the z-axis. The cross-sections perpendicular to the z-axis are circular. The barrel is $2h$ units high, its midsection radius is R, and its end radii are both r. Find a formula for the barrel's volume. Then check two things. First, suppose the sides of the barrel are straightened to turn the barrel into a cylinder of radius R and height $2h$. Does your formula give the cylinder's volume? Second, suppose $r = 0$ and $h = R$ so the barrel is a sphere. Does your formula give the sphere's volume?

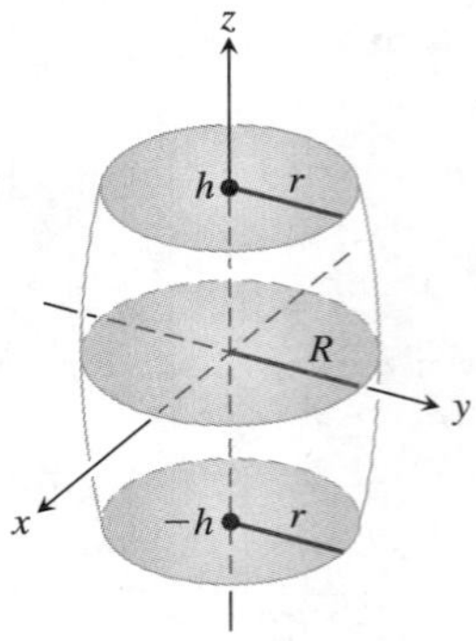

47. Show that the volume of the segment cut from the paraboloid

$$\frac{x^2}{a^2} + \frac{y^2}{b^2} = \frac{z}{c}$$

by the plane $z = h$ equals half the segment's base times its altitude.

48. a. Find the volume of the solid bounded by the hyperboloid

$$\frac{x^2}{a^2} + \frac{y^2}{b^2} - \frac{z^2}{c^2} = 1$$

and the planes $z = 0$ and $z = h$, $h > 0$.

b. Express your answer in part (a) in terms of h and the areas A_0 and A_h of the regions cut by the hyperboloid from the planes $z = 0$ and $z = h$.

c. Show that the volume in part (a) is also given by the formula

$$V = \frac{h}{6}(A_0 + 4A_m + A_h),$$

where A_m is the area of the region cut by the hyperboloid from the plane $z = h/2$.

Viewing Surfaces

T Plot the surfaces in Exercises 49–52 over the indicated domains. If you can, rotate the surface into different viewing positions.

49. $z = -y^2$, $-4 \le x \le 4$, $-.5 \le y \le 4$

50. $z = 1 + y^2$, $-4 \le x \le 4$, $-4 \le y \le 4$

51. $z = x^2 - y^2$, $-4 \le x \le 4$, $-.5 \le y \le 4$

52. $z = x^2 + 2y^2$ over

a. $-3 \le x \le 3$, $-3 \le y \le 3$

b. $-1 \le x \le 1$, $-2 \le y \le 3$

c. $-2 \le x \le 2$, $-2 \le y \le 2$

d. $-2 \le x \le 2$, $-1 \le y \le 1$

COMPUTER EXPLORATIONS

Use a CAS to plot the surfaces in Exercises 53–58. Identify the type of quadric surface from your graph.

53. $\frac{x^2}{9} + \frac{y^2}{36} = 1 - \frac{z^2}{25}$

54. $\frac{x^2}{9} - \frac{z^2}{9} = 1 - \frac{y^2}{16}$

55. $5x^2 = z^2 - 3y^2$

56. $\frac{y^2}{16} = 1 - \frac{x^2}{9} + z$

57. $\frac{x^2}{9} - 1 = \frac{y^2}{16} + \frac{z^2}{2}$

58. $y - \sqrt{4 - z^2} = 0$

CHAPTER 12 Questions to Guide Your Review

1. When do directed line segments in the plane represent the same vector?
2. How are vectors added and subtracted geometrically? How are they added and subtracted algebraically?
3. How do you find a vector's magnitude and direction?
4. If a vector is multiplied by a positive scalar, how is the result related to the original vector? What if the scalar is zero? Negative?
5. Define the *dot product (scalar product)* of two vectors. Which algebraic laws are satisfied by dot products? Give examples. When is the dot product of two vectors equal to zero?
6. What geometric interpretation does the dot product have? Give examples.
7. What is the vector projection of a vector **u** onto a vector **v**? Give an example of a useful application of a vector projection.
8. Define the *cross product (vector product)* of two vectors. Which algebraic laws are satisfied by cross products, and which are not? Give examples. When is the cross product of two vectors equal to zero?
9. What geometric or physical interpretations do cross products have? Give examples.
10. What is the determinant formula for calculating the cross product of two vectors relative to the Cartesian **i**, **j**, **k**-coordinate system? Use it in an example.
11. How do you find equations for lines, line segments, and planes in space? Give examples. Can you express a line in space by a single equation? A plane?
12. How do you find the distance from a point to a line in space? From a point to a plane? Give examples.
13. What are box products? What significance do they have? How are they evaluated? Give an example.
14. How do you find equations for spheres in space? Give examples.
15. How do you find the intersection of two lines in space? A line and a plane? Two planes? Give examples.
16. What is a cylinder? Give examples of equations that define cylinders in Cartesian coordinates.
17. What are quadric surfaces? Give examples of different kinds of ellipsoids, paraboloids, cones, and hyperboloids (equations and sketches).

CHAPTER 12 Practice Exercises

Vector Calculations in Two Dimensions

In Exercises 1–4, let $\mathbf{u} = \langle -3, 4\rangle$ and $\mathbf{v} = \langle 2, -5\rangle$. Find **(a)** the component form of the vector and **(b)** its magnitude.

1. $3\mathbf{u} - 4\mathbf{v}$
2. $\mathbf{u} + \mathbf{v}$
3. $-2\mathbf{u}$
4. $5\mathbf{v}$

In Exercises 5–8, find the component form of the vector.

5. The vector obtained by rotating $\langle 0, 1\rangle$ through an angle of $2\pi/3$ radians
6. The unit vector that makes an angle of $\pi/6$ radian with the positive x-axis
7. The vector 2 units long in the direction $4\mathbf{i} - \mathbf{j}$
8. The vector 5 units long in the direction opposite to the direction of $(3/5)\mathbf{i} + (4/5)\mathbf{j}$

Express the vectors in Exercises 9–12 in terms of their lengths and directions.

9. $\sqrt{2}\mathbf{i} + \sqrt{2}\mathbf{j}$
10. $-\mathbf{i} - \mathbf{j}$
11. Velocity vector $\mathbf{v} = (-2\sin t)\mathbf{i} + (2\cos t)\mathbf{j}$ when $t = \pi/2$.
12. Velocity vector $\mathbf{v} = (e^t\cos t - e^t\sin t)\mathbf{i} + (e^t\sin t + e^t\cos t)\mathbf{j}$ when $t = \ln 2$.

Vector Calculations in Three Dimensions

Express the vectors in Exercises 13 and 14 in terms of their lengths and directions.

13. $2\mathbf{i} - 3\mathbf{j} + 6\mathbf{k}$
14. $\mathbf{i} + 2\mathbf{j} - \mathbf{k}$
15. Find a vector 2 units long in the direction of $\mathbf{v} = 4\mathbf{i} - \mathbf{j} + 4\mathbf{k}$.
16. Find a vector 5 units long in the direction opposite to the direction of $\mathbf{v} = (3/5)\mathbf{i} + (4/5)\mathbf{k}$.

In Exercises 17 and 18, find $|\mathbf{v}|$, $|\mathbf{u}|$, $\mathbf{v}\cdot\mathbf{u}$, $\mathbf{u}\cdot\mathbf{v}$, $\mathbf{v}\times\mathbf{u}$, $\mathbf{u}\times\mathbf{v}$, $|\mathbf{v}\times\mathbf{u}|$, the angle between **v** and **u**, the scalar component of **u** in the direction of **v**, and the vector projection of **u** onto **v**.

17. $\mathbf{v} = \mathbf{i} + \mathbf{j}$
    $\mathbf{u} = 2\mathbf{i} + \mathbf{j} - 2\mathbf{k}$
18. $\mathbf{v} = \mathbf{i} + \mathbf{j} + 2\mathbf{k}$
    $\mathbf{u} = -\mathbf{i} - \mathbf{k}$

In Exercises 19 and 20, find $\text{proj}_{\mathbf{v}}\,\mathbf{u}$.

19. $\mathbf{v} = 2\mathbf{i} + \mathbf{j} - \mathbf{k}$
    $\mathbf{u} = \mathbf{i} + \mathbf{j} - 5\mathbf{k}$
20. $\mathbf{u} = \mathbf{i} - 2\mathbf{j}$
    $\mathbf{v} = \mathbf{i} + \mathbf{j} + \mathbf{k}$

In Exercises 21 and 22, draw coordinate axes and then sketch **u**, **v**, and $\mathbf{u}\times\mathbf{v}$ as vectors at the origin.

21. $\mathbf{u} = \mathbf{i}$, $\mathbf{v} = \mathbf{i} + \mathbf{j}$
22. $\mathbf{u} = \mathbf{i} - \mathbf{j}$, $\mathbf{v} = \mathbf{i} + \mathbf{j}$
23. If $|\mathbf{v}| = 2$, $|\mathbf{w}| = 3$, and the angle between **v** and **w** is $\pi/3$, find $|\mathbf{v} - 2\mathbf{w}|$.
24. For what value or values of a will the vectors $\mathbf{u} = 2\mathbf{i} + 4\mathbf{j} - 5\mathbf{k}$ and $\mathbf{v} = -4\mathbf{i} - 8\mathbf{j} + a\mathbf{k}$ be parallel?

In Exercises 25 and 26, find (a) the area of the parallelogram determined by vectors **u** and **v** and (b) the volume of the parallelepiped determined by the vectors **u**, **v**, and **w**.

25. $\mathbf{u} = \mathbf{i} + \mathbf{j} - \mathbf{k}, \quad \mathbf{v} = 2\mathbf{i} + \mathbf{j} + \mathbf{k}, \quad \mathbf{w} = -\mathbf{i} - 2\mathbf{j} + 3\mathbf{k}$

26. $\mathbf{u} = \mathbf{i} + \mathbf{j}, \quad \mathbf{v} = \mathbf{j}, \quad \mathbf{w} = \mathbf{i} + \mathbf{j} + \mathbf{k}$

Lines, Planes, and Distances

27. Suppose that **n** is normal to a plane and that **v** is parallel to the plane. Describe how you would find a vector **n** that is both perpendicular to **v** and parallel to the plane.

28. Find a vector in the plane parallel to the line $ax + by = c$.

In Exercises 29 and 30, find the distance from the point to the line.

29. $(2, 2, 0)$; $x = -t, \quad y = t, \quad z = -1 + t$

30. $(0, 4, 1)$; $x = 2 + t, \quad y = 2 + t, \quad z = t$

31. Parametrize the line that passes through the point $(1, 2, 3)$ parallel to the vector $\mathbf{v} = -3\mathbf{i} + 7\mathbf{k}$.

32. Parametrize the line segment joining the points $P(1, 2, 0)$ and $Q(1, 3, -1)$.

In Exercises 33 and 34, find the distance from the point to the plane.

33. $(6, 0, -6), \quad x - y = 4$

34. $(3, 0, 10), \quad 2x + 3y + z = 2$

35. Find an equation for the plane that passes through the point $(3, -2, 1)$ normal to the vector $\mathbf{n} = 2\mathbf{i} + \mathbf{j} + \mathbf{k}$.

36. Find an equation for the plane that passes through the point $(-1, 6, 0)$ perpendicular to the line $x = -1 + t$, $y = 6 - 2t$, $z = 3t$.

In Exercises 37 and 38, find an equation for the plane through points P, Q, and R.

37. $P(1, -1, 2), \quad Q(2, 1, 3), \quad R(-1, 2, -1)$

38. $P(1, 0, 0), \quad Q(0, 1, 0), \quad R(0, 0, 1)$

39. Find the points in which the line $x = 1 + 2t$, $y = -1 - t$, $z = 3t$ meets the three coordinate planes.

40. Find the point in which the line through the origin perpendicular to the plane $2x - y - z = 4$ meets the plane $3x - 5y + 2z = 6$.

41. Find the acute angle between the planes $x = 7$ and $x + y + \sqrt{2}z = -3$.

42. Find the acute angle between the planes $x + y = 1$ and $y + z = 1$.

43. Find parametric equations for the line in which the planes $x + 2y + z = 1$ and $x - y + 2z = -8$ intersect.

44. Show that the line in which the planes

$$x + 2y - 2z = 5 \quad \text{and} \quad 5x - 2y - z = 0$$

intersect is parallel to the line

$$x = -3 + 2t, \quad y = 3t, \quad z = 1 + 4t.$$

45. The planes $3x + 6z = 1$ and $2x + 2y - z = 3$ intersect in a line.

a. Show that the planes are orthogonal.

b. Find equations for the line of intersection.

46. Find an equation for the plane that passes through the point $(1, 2, 3)$ parallel to $\mathbf{u} = 2\mathbf{i} + 3\mathbf{j} + \mathbf{k}$ and $\mathbf{v} = \mathbf{i} - \mathbf{j} + 2\mathbf{k}$.

47. Is $\mathbf{v} = 2\mathbf{i} - 4\mathbf{j} + \mathbf{k}$ related in any special way to the plane $2x + y = 5$? Give reasons for your answer.

48. The equation $\mathbf{n} \cdot \overrightarrow{P_0P} = 0$ represents the plane through P_0 normal to **n**. What set does the inequality $\mathbf{n} \cdot \overrightarrow{P_0P} > 0$ represent?

49. Find the distance from the point $P(1, 4, 0)$ to the plane through $A(0, 0, 0)$, $B(2, 0, -1)$, and $C(2, -1, 0)$.

50. Find the distance from the point $(2, 2, 3)$ to the plane $2x + 3y + 5z = 0$.

51. Find a vector parallel to the plane $2x - y - z = 4$ and orthogonal to $\mathbf{i} + \mathbf{j} + \mathbf{k}$.

52. Find a unit vector orthogonal to **A** in the plane of **B** and **C** if $\mathbf{A} = 2\mathbf{i} - \mathbf{j} + \mathbf{k}$, $\mathbf{B} = \mathbf{i} + 2\mathbf{j} + \mathbf{k}$, and $\mathbf{C} = \mathbf{i} + \mathbf{j} - 2\mathbf{k}$.

53. Find a vector of magnitude 2 parallel to the line of intersection of the planes $x + 2y + z - 1 = 0$ and $x - y + 2z + 7 = 0$.

54. Find the point in which the line through the origin perpendicular to the plane $2x - y - z = 4$ meets the plane $3x - 5y + 2z = 6$.

55. Find the point in which the line through $P(3, 2, 1)$ normal to the plane $2x - y + 2z = -2$ meets the plane.

56. What angle does the line of intersection of the planes $2x + y - z = 0$ and $x + y + 2z = 0$ make with the positive x-axis?

57. The line

$$L: \quad x = 3 + 2t, \quad y = 2t, \quad z = t$$

intersects the plane $x + 3y - z = -4$ in a point P. Find the coordinates of P and find equations for the line in the plane through P perpendicular to L.

58. Show that for every real number k, the plane

$$x - 2y + z + 3 + k(2x - y - z + 1) = 0$$

contains the line of intersection of the planes

$$x - 2y + z + 3 = 0 \quad \text{and} \quad 2x - y - z + 1 = 0.$$

59. Find an equation for the plane through $A(-2, 0, -3)$ and $B(1, -2, 1)$ that lies parallel to the line through $C(-2, -13/5, 26/5)$ and $D(16/5, -13/5, 0)$.

60. Is the line $x = 1 + 2t$, $y = -2 + 3t$, $z = -5t$ related in any way to the plane $-4x - 6y + 10z = 9$? Give reasons for your answer.

61. Which of the following are equations for the plane through the points $P(1, 1, -1)$, $Q(3, 0, 2)$, and $R(-2, 1, 0)$?

a. $(2\mathbf{i} - 3\mathbf{j} + 3\mathbf{k}) \cdot ((x + 2)\mathbf{i} + (y - 1)\mathbf{j} + z\mathbf{k}) = 0$

b. $x = 3 - t, \quad y = -11t, \quad z = 2 - 3t$

c. $(x + 2) + 11(y - 1) = 3z$

d. $(2\mathbf{i} - 3\mathbf{j} + 3\mathbf{k}) \times ((x + 2)\mathbf{i} + (y - 1)\mathbf{j} + z\mathbf{k}) = \mathbf{0}$

e. $(2\mathbf{i} - \mathbf{j} + 3\mathbf{k}) \times (-3\mathbf{i} + \mathbf{k}) \cdot ((x + 2)\mathbf{i} + (y - 1)\mathbf{j} + z\mathbf{k}) = 0$

62. The parallelogram shown here has vertices at $A(2, -1, 4)$, $B(1, 0, -1)$, $C(1, 2, 3)$, and D. Find

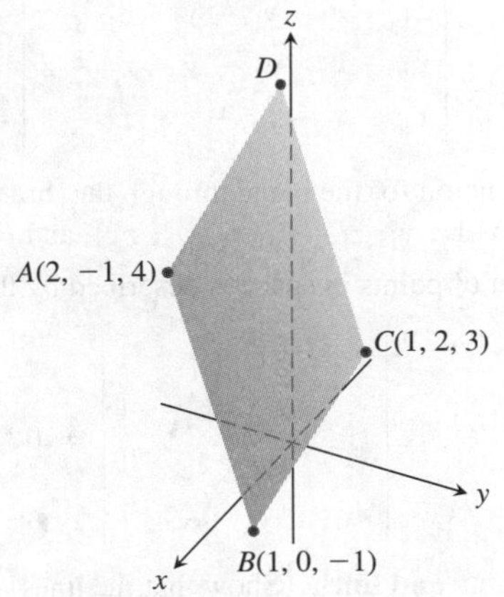

a. the coordinates of D.

b. the cosine of the interior angle at B.

c. the vector projection of $\overrightarrow{BA}$ onto $\overrightarrow{BC}$.

d. the area of the parallelogram.

e. an equation for the plane of the parallelogram.

f. the areas of the orthogonal projections of the parallelogram on the three coordinate planes.

63. Distance between skew lines Find the distance between the line L_1 through the points $A(1, 0, -1)$ and $B(-1, 1, 0)$ and the line L_2 through the points $C(3, 1, -1)$ and $D(4, 5, -2)$. The distance is to be measured along the line perpendicular to the two lines. First find a vector **n** perpendicular to both lines. Then project $\overrightarrow{AC}$ onto **n**.

64. (*Continuation of Exercise 63.*) Find the distance between the line through $A(4, 0, 2)$ and $B(2, 4, 1)$ and the line through $C(1, 3, 2)$ and $D(2, 2, 4)$.

Quadric Surfaces

Identify and sketch the surfaces in Exercises 65–76.

65. $x^2 + y^2 + z^2 = 4$

66. $x^2 + (y - 1)^2 + z^2 = 1$

67. $4x^2 + 4y^2 + z^2 = 4$

68. $36x^2 + 9y^2 + 4z^2 = 36$

69. $z = -(x^2 + y^2)$

70. $y = -(x^2 + z^2)$

71. $x^2 + y^2 = z^2$

72. $x^2 + z^2 = y^2$

73. $x^2 + y^2 - z^2 = 4$

74. $4y^2 + z^2 - 4x^2 = 4$

75. $y^2 - x^2 - z^2 = 1$

76. $z^2 - x^2 - y^2 = 1$

CHAPTER 12 Additional and Advanced Exercises

1. Submarine hunting Two surface ships on maneuvers are trying to determine a submarine's course and speed to prepare for an aircraft intercept. As shown here, ship A is located at $(4, 0, 0)$, whereas ship B is located at $(0, 5, 0)$. All coordinates are given in thousands of feet. Ship A locates the submarine in the direction of the vector $2\mathbf{i} + 3\mathbf{j} - (1/3)\mathbf{k}$, and ship B locates it in the direction of the vector $18\mathbf{i} - 6\mathbf{j} - \mathbf{k}$. Four minutes ago, the submarine was located at $(2, -1, -1/3)$. The aircraft is due in 20 min. Assuming that the submarine moves in a straight line at a constant speed, to what position should the surface ships direct the aircraft?

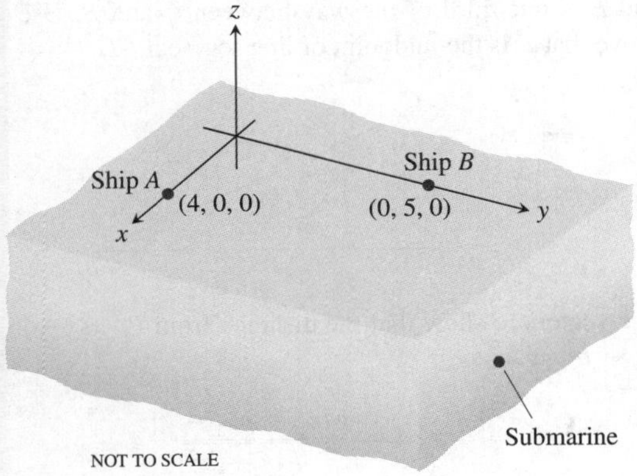

2. A helicopter rescue Two helicopters, H_1 and H_2, are traveling together. At time $t = 0$, they separate and follow different straight-line paths given by

$$H_1: \quad x = 6 + 40t, \quad y = -3 + 10t, \quad z = -3 + 2t$$

$$H_2: \quad x = 6 + 110t, \quad y = -3 + 4t, \quad z = -3 + t.$$

Time t is measured in hours, and all coordinates are measured in miles. Due to system malfunctions, H_2 stops its flight at $(446, 13, 1)$ and, in a negligible amount of time, lands at $(446, 13, 0)$. Two hours later, H_1 is advised of this fact and heads toward H_2 at 150 mph. How long will it take H_1 to reach H_2?

3. Torque The operator's manual for the Toro® 21-in. lawnmower says, "tighten the spark plug to 15 ft-lb (20.4 N · m)." If you are installing the plug with a 10.5-in. socket wrench that places the center of your hand 9 in. from the axis of the spark plug, about how hard should you pull? Answer in pounds.

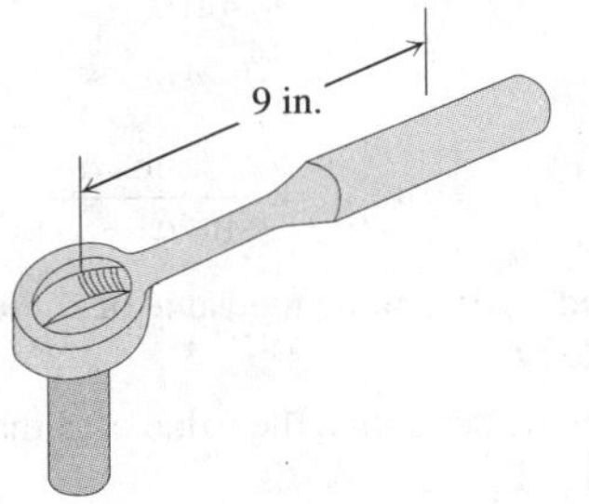

4. Rotating body The line through the origin and the point $A(1, 1, 1)$ is the axis of rotation of a rigid body rotating with a constant angular speed of $3/2$ rad/sec. The rotation appears to be clockwise when we look toward the origin from A. Find the velocity $\mathbf{v}$ of the point of the body that is at the position $B(1, 3, 2)$.

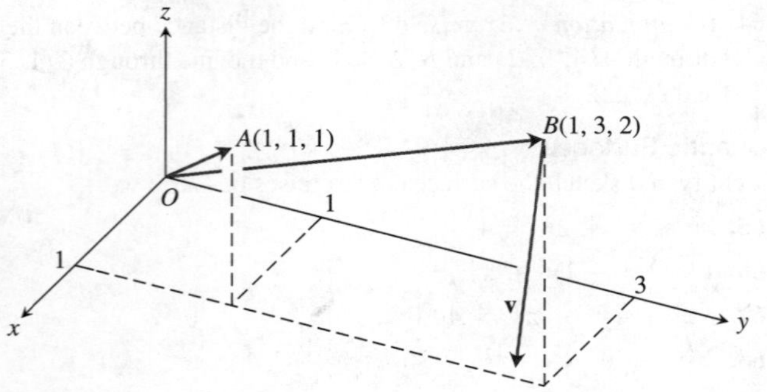

5. Consider the weight suspended by two wires in each diagram. Find the magnitudes and components of vectors $\mathbf{F}_1$ and $\mathbf{F}_2$, and angles α and β.

a.

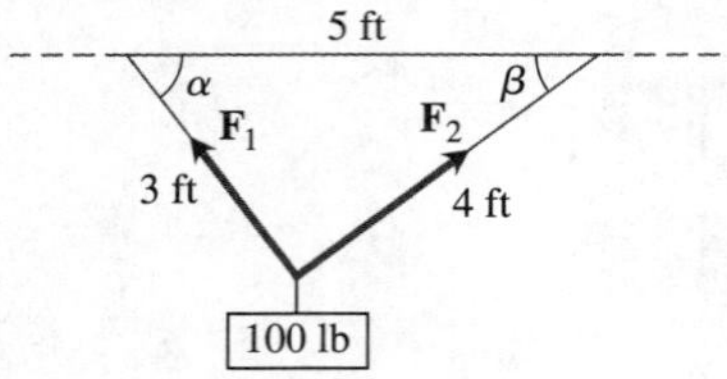

b.

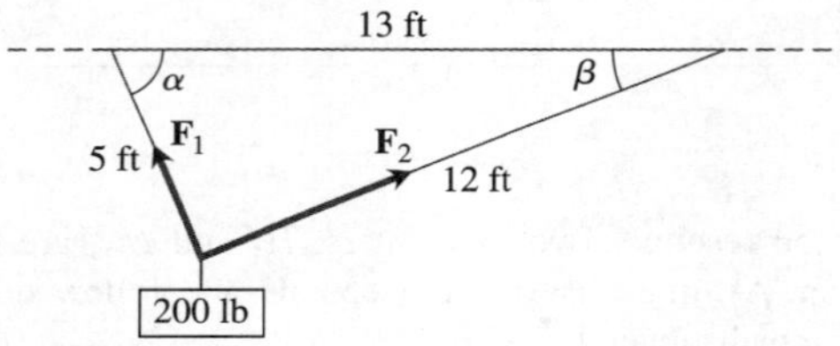

(*Hint:* This triangle is a right triangle.)

6. Consider a weight of w N suspended by two wires in the diagram, where $\mathbf{T}_1$ and $\mathbf{T}_2$ are force vectors directed along the wires.

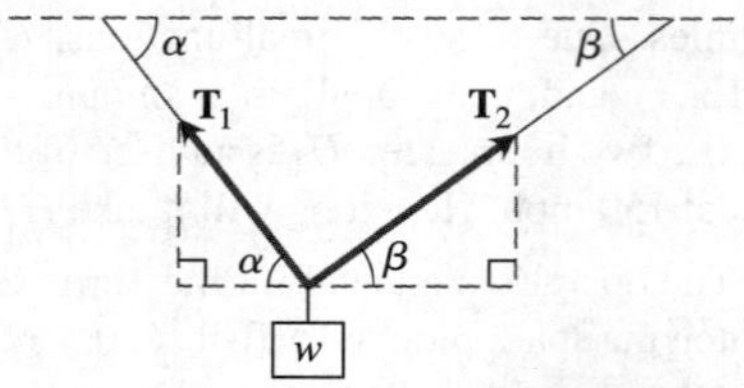

a. Find the vectors $\mathbf{T}_1$ and $\mathbf{T}_2$ and show that their magnitudes are

$$|\mathbf{T}_1| = \frac{w \cos \beta}{\sin(\alpha + \beta)}$$

and

$$|\mathbf{T}_2| = \frac{w \cos \alpha}{\sin(\alpha + \beta)}.$$

b. For a fixed β, determine the value of α that minimizes the magnitude $|\mathbf{T}_1|$.

c. For a fixed α, determine the value of β that minimizes the magnitude $|\mathbf{T}_2|$.

7. Determinants and planes

a. Show that

$$\begin{vmatrix} x_1 - x & y_1 - y & z_1 - z \\ x_2 - x & y_2 - y & z_2 - z \\ x_3 - x & y_3 - y & z_3 - z \end{vmatrix} = 0$$

is an equation for the plane through the three noncollinear points $P_1(x_1, y_1, z_1)$, $P_2(x_2, y_2, z_2)$, and $P_3(x_3, y_3, z_3)$.

b. What set of points in space is described by the equation

$$\begin{vmatrix} x & y & z & 1 \\ x_1 & y_1 & z_1 & 1 \\ x_2 & y_2 & z_2 & 1 \\ x_3 & y_3 & z_3 & 1 \end{vmatrix} = 0?$$

8. Determinants and lines Show that the lines

$$x = a_1 s + b_1, \quad y = a_2 s + b_2, \quad z = a_3 s + b_3, \quad -\infty < s < \infty$$

and

$$x = c_1 t + d_1, \quad y = c_2 t + d_2, \quad z = c_3 t + d_3, \quad -\infty < t < \infty$$

intersect or are parallel if and only if

$$\begin{vmatrix} a_1 & c_1 & b_1 - d_1 \\ a_2 & c_2 & b_2 - d_2 \\ a_3 & c_3 & b_3 - d_3 \end{vmatrix} = 0.$$

9. Consider a regular tetrahedron of side length 2.

a. Use vectors to find the angle θ formed by the base of the tetrahedron and any one of its other edges.

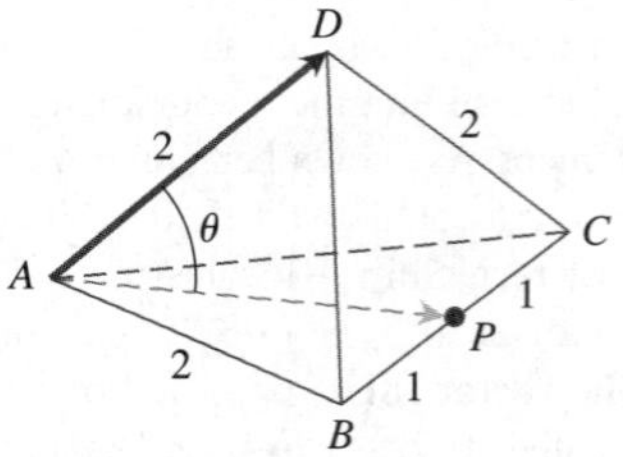

b. Use vectors to find the angle θ formed by any two adjacent faces of the tetrahedron. This angle is commonly referred to as a dihedral angle.

10. In the figure here, D is the midpoint of side AB of triangle ABC, and E is one-third of the way between C and B. Use vectors to prove that F is the midpoint of line segment CD.

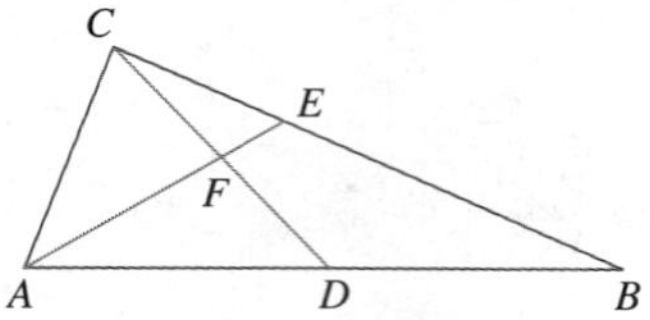

11. Use vectors to show that the distance from $P_1(x_1, y_1)$ to the line $ax + by = c$ is

$$d = \frac{|ax_1 + by_1 - c|}{\sqrt{a^2 + b^2}}.$$

12. a. Use vectors to show that the distance from $P_1(x_1, y_1, z_1)$ to the plane $Ax + By + Cz = D$ is

$$d = \frac{|Ax_1 + By_1 + Cz_1 - D|}{\sqrt{A^2 + B^2 + C^2}}.$$

b. Find an equation for the sphere that is tangent to the planes $x + y + z = 3$ and $x + y + z = 9$ if the planes $2x - y = 0$ and $3x - z = 0$ pass through the center of the sphere.

13. a. **Distance between parallel planes** Show that the distance between the parallel planes $Ax + By + Cz = D_1$ and $Ax + By + Cz = D_2$ is

$$d = \frac{|D_1 - D_2|}{|A\mathbf{i} + B\mathbf{j} + C\mathbf{k}|}.$$

b. Find the distance between the planes $2x + 3y - z = 6$ and $2x + 3y - z = 12$.

c. Find an equation for the plane parallel to the plane $2x - y + 2z = -4$ if the point $(3, 2, -1)$ is equidistant from the two planes.

d. Write equations for the planes that lie parallel to, and 5 units away from, the plane $x - 2y + z = 3$.

14. Prove that four points A, B, C, and D are coplanar (lie in a common plane) if and only if $\overrightarrow{AD} \cdot (\overrightarrow{AB} \times \overrightarrow{BC}) = 0$.

15. **The projection of a vector on a plane** Let P be a plane in space and let $\mathbf{v}$ be a vector. The vector projection of $\mathbf{v}$ onto the plane P, $\text{proj}_P\,\mathbf{v}$, can be defined informally as follows. Suppose the sun is shining so that its rays are normal to the plane P. Then $\text{proj}_P\,\mathbf{v}$ is the "shadow" of $\mathbf{v}$ onto P. If P is the plane $x + 2y + 6z = 6$ and $\mathbf{v} = \mathbf{i} + \mathbf{j} + \mathbf{k}$, find $\text{proj}_P\,\mathbf{v}$.

16. The accompanying figure shows nonzero vectors $\mathbf{v}$, $\mathbf{w}$, and $\mathbf{z}$, with $\mathbf{z}$ orthogonal to the line L, and $\mathbf{v}$ and $\mathbf{w}$ making equal angles β with L. Assuming $|\mathbf{v}| = |\mathbf{w}|$, find $\mathbf{w}$ in terms of $\mathbf{v}$ and $\mathbf{z}$.

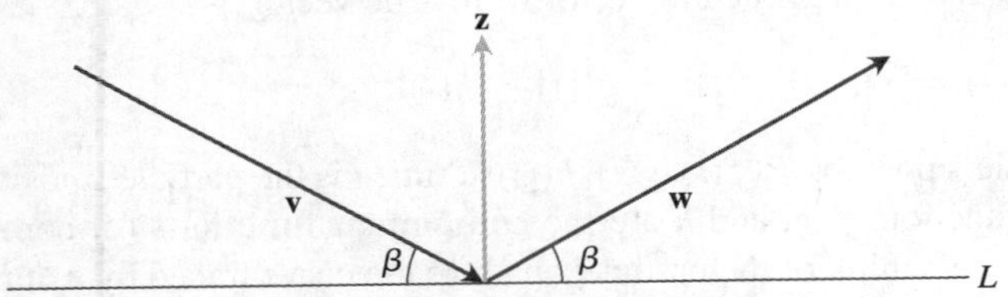

17. **Triple vector products** The *triple vector products* $(\mathbf{u} \times \mathbf{v}) \times \mathbf{w}$ and $\mathbf{u} \times (\mathbf{v} \times \mathbf{w})$ are usually not equal, although the formulas for evaluating them from components are similar:

$$(\mathbf{u} \times \mathbf{v}) \times \mathbf{w} = (\mathbf{u} \cdot \mathbf{w})\mathbf{v} - (\mathbf{v} \cdot \mathbf{w})\mathbf{u}.$$

$$\mathbf{u} \times (\mathbf{v} \times \mathbf{w}) = (\mathbf{u} \cdot \mathbf{w})\mathbf{v} - (\mathbf{u} \cdot \mathbf{v})\mathbf{w}.$$

Verify each formula for the following vectors by evaluating its two sides and comparing the results.

	u	v	w
a.	$2\mathbf{i}$	$2\mathbf{j}$	$2\mathbf{k}$
b.	$\mathbf{i} - \mathbf{j} + \mathbf{k}$	$2\mathbf{i} + \mathbf{j} - 2\mathbf{k}$	$-\mathbf{i} + 2\mathbf{j} - \mathbf{k}$
c.	$2\mathbf{i} + \mathbf{j}$	$2\mathbf{i} - \mathbf{j} + \mathbf{k}$	$\mathbf{i} + 2\mathbf{k}$
d.	$\mathbf{i} + \mathbf{j} - 2\mathbf{k}$	$-\mathbf{i} - \mathbf{k}$	$2\mathbf{i} + 4\mathbf{j} - 2\mathbf{k}$

18. **Cross and dot products** Show that if $\mathbf{u}$, $\mathbf{v}$, $\mathbf{w}$, and $\mathbf{r}$ are any vectors, then

a. $\mathbf{u} \times (\mathbf{v} \times \mathbf{w}) + \mathbf{v} \times (\mathbf{w} \times \mathbf{u}) + \mathbf{w} \times (\mathbf{u} \times \mathbf{v}) = \mathbf{0}$

b. $\mathbf{u} \times \mathbf{v} = (\mathbf{u} \cdot \mathbf{v} \times \mathbf{i})\mathbf{i} + (\mathbf{u} \cdot \mathbf{v} \times \mathbf{j})\mathbf{j} + (\mathbf{u} \cdot \mathbf{v} \times \mathbf{k})\mathbf{k}$

c. $(\mathbf{u} \times \mathbf{v}) \cdot (\mathbf{w} \times \mathbf{r}) = \begin{vmatrix} \mathbf{u} \cdot \mathbf{w} & \mathbf{v} \cdot \mathbf{w} \\ \mathbf{u} \cdot \mathbf{r} & \mathbf{v} \cdot \mathbf{r} \end{vmatrix}$.

19. **Cross and dot products** Prove or disprove the formula

$$\mathbf{u} \times (\mathbf{u} \times (\mathbf{u} \times \mathbf{v})) \cdot \mathbf{w} = -|\mathbf{u}|^2 \mathbf{u} \cdot \mathbf{v} \times \mathbf{w}.$$

20. By forming the cross product of two appropriate vectors, derive the trigonometric identity

$$\sin(A - B) = \sin A \cos B - \cos A \sin B.$$

21. Use vectors to prove that

$$(a^2 + b^2)(c^2 + d^2) \geq (ac + bd)^2$$

for any four numbers a, b, c, and d. (*Hint:* Let $\mathbf{u} = a\mathbf{i} + b\mathbf{j}$ and $\mathbf{v} = c\mathbf{i} + d\mathbf{j}$.)

22. **Dot multiplication is positive definite** Show that dot multiplication of vectors is *positive definite;* that is, show that $\mathbf{u} \cdot \mathbf{u} \geq 0$ for every vector $\mathbf{u}$ and that $\mathbf{u} \cdot \mathbf{u} = 0$ if and only if $\mathbf{u} = \mathbf{0}$.

23. Show that $|\mathbf{u} + \mathbf{v}| \leq |\mathbf{u}| + |\mathbf{v}|$ for any vectors $\mathbf{u}$ and $\mathbf{v}$.

24. Show that $\mathbf{w} = |\mathbf{v}|\mathbf{u} + |\mathbf{u}|\mathbf{v}$ bisects the angle between $\mathbf{u}$ and $\mathbf{v}$.

25. Show that $|\mathbf{v}|\mathbf{u} + |\mathbf{u}|\mathbf{v}$ and $|\mathbf{v}|\mathbf{u} - |\mathbf{u}|\mathbf{v}$ are orthogonal.

CHAPTER 12 Technology Application Projects

Mathematica/Maple Projects

Projects can be found within MyLab Math.

- ***Using Vectors to Represent Lines and Find Distances***
  Parts I and II: Learn the advantages of interpreting lines as vectors.
  Part III: Use vectors to find the distance from a point to a line.
- ***Putting a Scene in Three Dimensions onto a Two-Dimensional Canvas***
  Use the concept of planes in space to obtain a two-dimensional image.
- ***Getting Started in Plotting in 3D***
  Part I: Use the vector definition of lines and planes to generate graphs and equations, and to compare different forms for the equations of a single line.
  Part II: Plot functions that are defined implicitly.

13

Vector-Valued Functions and Motion in Space

OVERVIEW In this chapter we introduce the calculus of vector-valued functions. The domains of these functions are sets of real numbers, as before, but their ranges consist of vectors instead of scalars. When a vector-valued function changes, the change can occur in both magnitude and direction, so the derivative is itself a vector. The integral of a vector-valued function is also a vector. We use the calculus of these functions to describe the paths and motions of objects moving in a plane or in space, so their velocities and accelerations are given by vectors.

13.1 Curves in Space and Their Tangents

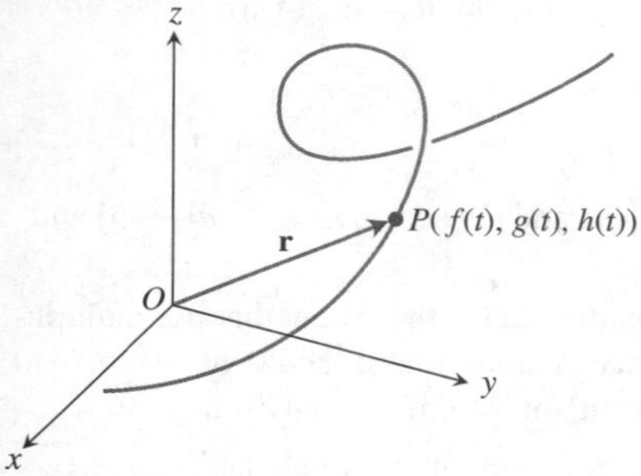

FIGURE 13.1 The position vector $\mathbf{r} = \overrightarrow{OP}$ of a particle moving through space is a function of time.

When a particle moves through space during a time interval I, we think of the particle's coordinates as functions defined on I:

$$x = f(t), \qquad y = g(t), \qquad z = h(t), \qquad t \in I. \tag{1}$$

The points $(x, y, z) = (f(t), g(t), h(t))$, $t \in I$, make up the **curve** in space that we call the particle's **path**. The equations and interval in Equation (1) parametrize the curve.

A curve in space can also be represented in vector form. The vector

$$\mathbf{r}(t) = \overrightarrow{OP} = f(t)\mathbf{i} + g(t)\mathbf{j} + h(t)\mathbf{k} \tag{2}$$

from the origin to the particle's position $P(f(t), g(t), h(t))$ at time t is the particle's position vector (Figure 13.1). The functions f, g, and h are the **component functions** (or components) of the position vector. We think of the particle's path as the curve traced by $\mathbf{r}$ during the time interval I. Figure 13.2 displays several space curves generated by a computer graphing program.

Equation (2) defines $\mathbf{r}$ as a vector function of the real variable t on the interval I. More generally, a **vector-valued function** or **vector function** on a domain set D is a rule that assigns a vector in space to each element in D. For now, the domains will be intervals of real numbers, and the graph of the function represents a curve in space. Vector functions on a domain in the plane or in space give rise to "vector fields," which are important to the study of fluid flows, gravitational fields, and electromagnetic phenomena. We investigate vector fields and their applications in Chapter 16.

Real-valued functions are often called **scalar functions** to distinguish them from vector functions. The components of $\mathbf{r}$ in Equation (2) are scalar functions of t. The domain of a vector-valued function is the common domain of its components.

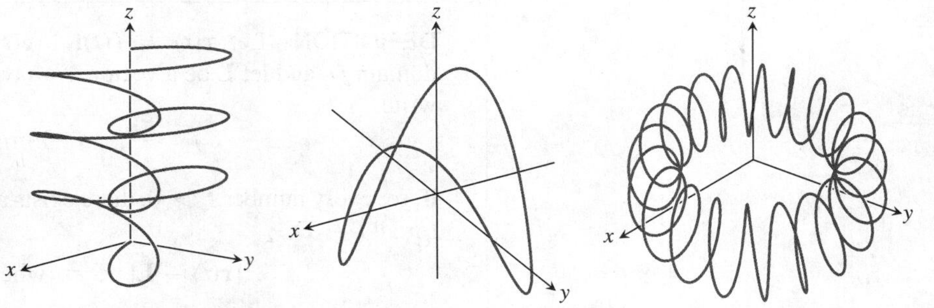

$\mathbf{r}(t) = (\sin 3t)(\cos t)\mathbf{i} + (\sin 3t)(\sin t)\mathbf{j} + t\mathbf{k}$

(a)

$\mathbf{r}(t) = (\cos t)\mathbf{i} + (\sin t)\mathbf{j} + (\sin 2t)\mathbf{k}$

(b)

$\mathbf{r}(t) = (4 + \sin 20t)(\cos t)\mathbf{i} + (4 + \sin 20t)(\sin t)\mathbf{j} + (\cos 20t)\mathbf{k}$

(c)

FIGURE 13.2 Space curves are defined by the position vectors $\mathbf{r}(t)$.

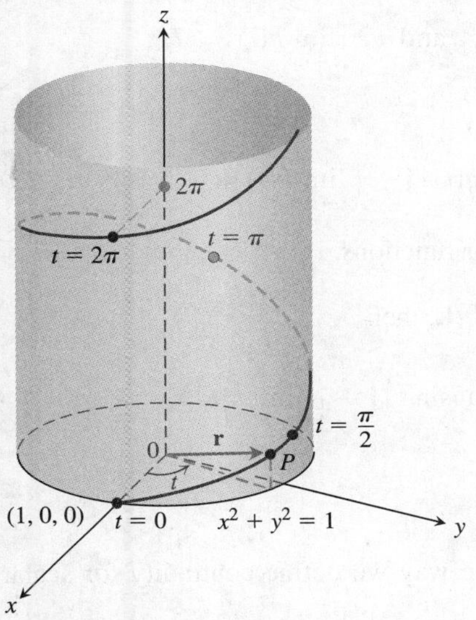

FIGURE 13.3 The helix $\mathbf{r}(t) = (\cos t)\mathbf{i} + (\sin t)\mathbf{j} + t\mathbf{k}$ (Example 1).

EXAMPLE 1 Graph the vector function

$$\mathbf{r}(t) = (\cos t)\mathbf{i} + (\sin t)\mathbf{j} + t\mathbf{k}.$$

Solution This vector function $\mathbf{r}(t)$ is defined for all real values of t. The curve traced by $\mathbf{r}$ winds around the circular cylinder $x^2 + y^2 = 1$ (Figure 13.3). The curve lies on the cylinder because the **i**- and **j**-components of **r**, being the x- and y-coordinates of the tip of **r**, satisfy the cylinder's equation:

$$x^2 + y^2 = (\cos t)^2 + (\sin t)^2 = 1.$$

The curve rises as the **k**-component $z = t$ increases. Each time t increases by 2π, the curve completes one turn around the cylinder. The curve is called a **helix** (from an ancient Greek word for "spiral"). The equations

$$x = \cos t, \qquad y = \sin t, \qquad z = t$$

parametrize the helix. The domain is the largest set of points t for which all three equations are defined, or $-\infty < t < \infty$ for this example. Figure 13.4 shows more helices. ■

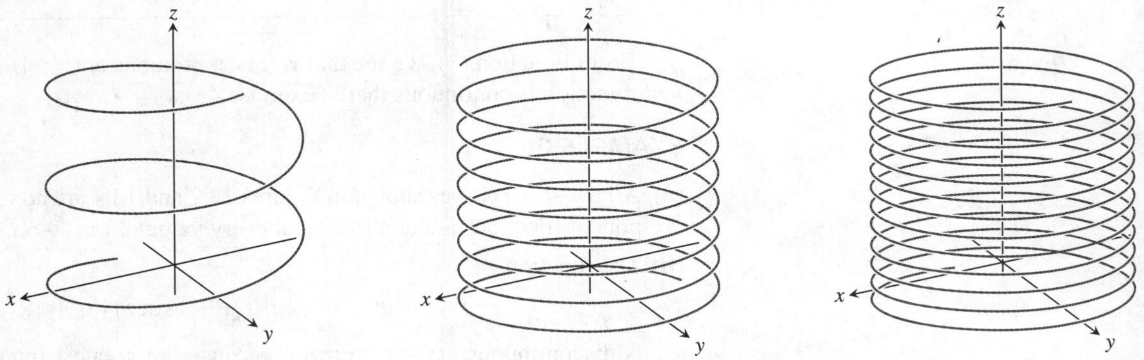

$\mathbf{r}(t) = (\cos t)\mathbf{i} + (\sin t)\mathbf{j} + t\mathbf{k}$

$\mathbf{r}(t) = (\cos t)\mathbf{i} + (\sin t)\mathbf{j} + 0.3t\mathbf{k}$

$\mathbf{r}(t) = (\cos 5t)\mathbf{i} + (\sin 5t)\mathbf{j} + t\mathbf{k}$

FIGURE 13.4 Helices spiral upward around a cylinder, like coiled springs.

Limits and Continuity

The way we define limits of vector-valued functions is similar to the way we define limits of real-valued functions.

DEFINITION Let $\mathbf{r}(t) = f(t)\mathbf{i} + g(t)\mathbf{j} + h(t)\mathbf{k}$ be a vector function with domain D, and let $\mathbf{L}$ be a vector. We say that $\mathbf{r}$ has **limit L** as t approaches t_0 and write

$$\lim_{t\to t_0} \mathbf{r}(t) = \mathbf{L}$$

if, for every number $\varepsilon > 0$, there exists a corresponding number $\delta > 0$ such that, for all $t \in D$,

$$|\mathbf{r}(t) - \mathbf{L}| < \varepsilon \quad \text{whenever} \quad 0 < |t - t_0| < \delta.$$

If $\mathbf{L} = L_1\mathbf{i} + L_2\mathbf{j} + L_3\mathbf{k}$, then it can be shown that $\lim_{t\to t_0} \mathbf{r}(t) = \mathbf{L}$ precisely when

$$\lim_{t\to t_0} f(t) = L_1, \qquad \lim_{t\to t_0} g(t) = L_2, \qquad \text{and} \qquad \lim_{t\to t_0} h(t) = L_3.$$

We omit the proof. The equation

To calculate the limit of a vector function, we find the limit of each component scalar function.

$$\lim_{t\to t_0} \mathbf{r}(t) = \left(\lim_{t\to t_0} f(t)\right)\mathbf{i} + \left(\lim_{t\to t_0} g(t)\right)\mathbf{j} + \left(\lim_{t\to t_0} h(t)\right)\mathbf{k} \tag{3}$$

provides a practical way to calculate limits of vector functions.

EXAMPLE 2 If $\mathbf{r}(t) = (\cos t)\mathbf{i} + (\sin t)\mathbf{j} + t\mathbf{k}$, then

$$\begin{aligned} \lim_{t\to\pi/4} \mathbf{r}(t) &= \left(\lim_{t\to\pi/4} \cos t\right)\mathbf{i} + \left(\lim_{t\to\pi/4} \sin t\right)\mathbf{j} + \left(\lim_{t\to\pi/4} t\right)\mathbf{k} \\ &= \frac{\sqrt{2}}{2}\mathbf{i} + \frac{\sqrt{2}}{2}\mathbf{j} + \frac{\pi}{4}\mathbf{k}. \end{aligned}$$

■

We define continuity for vector functions the same way we define continuity for scalar functions defined over an interval.

DEFINITION A vector function $\mathbf{r}(t)$ is **continuous at a point** $t = t_0$ in its domain if $\lim_{t\to t_0} \mathbf{r}(t) = \mathbf{r}(t_0)$. The function is **continuous** if it is continuous at every point in its domain.

From Equation (3), we see that $\mathbf{r}(t)$ is continuous at $t = t_0$ if and only if each component function is continuous there (Exercise 45).

EXAMPLE 3

(a) All the space curves shown in Figures 13.2 and 13.4 are continuous because their component functions are continuous at every value of t in $(-\infty, \infty)$.

(b) The function

$$\mathbf{g}(t) = (\cos t)\mathbf{i} + (\sin t)\mathbf{j} + \lfloor t \rfloor\mathbf{k}$$

is discontinuous at every integer, because the greatest integer function $\lfloor t \rfloor$ is discontinuous at every integer. ■

Derivatives and Motion

Suppose that $\mathbf{r}(t) = f(t)\mathbf{i} + g(t)\mathbf{j} + h(t)\mathbf{k}$ is the position vector of a particle moving along a curve in space and that f, g, and h are differentiable functions of t. Then the difference between the particle's positions at time t and time $t + \Delta t$ is the vector

$$\Delta\mathbf{r} = \mathbf{r}(t + \Delta t) - \mathbf{r}(t)$$

(Figure 13.5a). In terms of components,

$$\begin{aligned}\Delta\mathbf{r} &= \mathbf{r}(t+\Delta t) - \mathbf{r}(t)\\ &= [f(t+\Delta t)\mathbf{i} + g(t+\Delta t)\mathbf{j} + h(t+\Delta t)\mathbf{k}] - [f(t)\mathbf{i} + g(t)\mathbf{j} + h(t)\mathbf{k}]\\ &= [f(t+\Delta t) - f(t)]\mathbf{i} + [g(t+\Delta t) - g(t)]\mathbf{j} + [h(t+\Delta t) - h(t)]\mathbf{k}.\end{aligned}$$

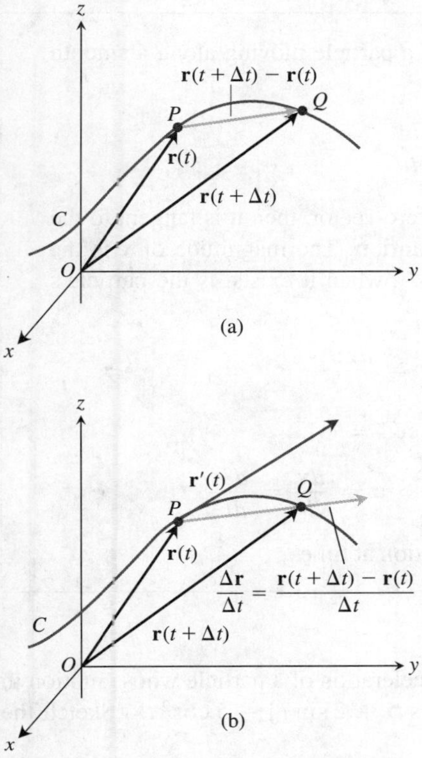

FIGURE 13.5 As $\Delta t \to 0$, the point Q approaches the point P along the curve C. In the limit, the vector $\overrightarrow{PQ}/\Delta t$ becomes the tangent vector $\mathbf{r}'(t)$.

As Δt approaches zero, three things seem to happen simultaneously. First, Q approaches P along the curve. Second, the secant line PQ seems to approach a limiting position tangent to the curve at P. Third, the quotient $\Delta\mathbf{r}/\Delta t$ (Figure 13.5b) approaches the limit

$$\begin{aligned}\lim_{\Delta t\to 0}\frac{\Delta\mathbf{r}}{\Delta t} &= \left[\lim_{\Delta t\to 0}\frac{f(t+\Delta t)-f(t)}{\Delta t}\right]\mathbf{i} + \left[\lim_{\Delta t\to 0}\frac{g(t+\Delta t)-g(t)}{\Delta t}\right]\mathbf{j}\\ &\quad + \left[\lim_{\Delta t\to 0}\frac{h(t+\Delta t)-h(t)}{\Delta t}\right]\mathbf{k}\\ &= \left[\frac{df}{dt}\right]\mathbf{i} + \left[\frac{dg}{dt}\right]\mathbf{j} + \left[\frac{dh}{dt}\right]\mathbf{k}.\end{aligned}$$

These observations lead us to the following definition.

DEFINITION The vector function $\mathbf{r}(t) = f(t)\mathbf{i} + g(t)\mathbf{j} + h(t)\mathbf{k}$ has a **derivative (is differentiable) at** t if f, g, and h have derivatives at t. The derivative is the vector function

$$\mathbf{r}'(t) = \frac{d\mathbf{r}}{dt} = \lim_{\Delta t\to 0}\frac{\mathbf{r}(t+\Delta t)-\mathbf{r}(t)}{\Delta t} = \frac{df}{dt}\mathbf{i} + \frac{dg}{dt}\mathbf{j} + \frac{dh}{dt}\mathbf{k}.$$

A vector function $\mathbf{r}$ is **differentiable** if it is differentiable at every point of its domain.

The geometric significance of the definition of derivative is shown in Figure 13.5. The points P and Q have position vectors $\mathbf{r}(t)$ and $\mathbf{r}(t+\Delta t)$, and the vector $\overrightarrow{PQ}$ is represented by $\mathbf{r}(t+\Delta t) - \mathbf{r}(t)$. For $\Delta t > 0$, the scalar multiple $(1/\Delta t)(\mathbf{r}(t+\Delta t) - \mathbf{r}(t))$ points in the same direction as the vector $\overrightarrow{PQ}$. As $\Delta t \to 0$, this vector approaches the vector $\mathbf{r}'(t)$, which is a vector tangent to the curve at P, as long as it is different from the zero vector $\mathbf{0}$ (Figure 13.5b).

The curve traced by $\mathbf{r}$ is **smooth** if $d\mathbf{r}/dt$ is continuous and never $\mathbf{0}$, that is, if f, g, and h have continuous first derivatives that are not simultaneously 0. We require $d\mathbf{r}/dt \neq 0$ for a smooth curve to make sure the curve has a continuously turning tangent at each point. On a smooth curve, there are no sharp corners or cusps.

A curve that is made up of a finite number of smooth curves pieced together in a continuous fashion is called **piecewise smooth** (Figure 13.6).

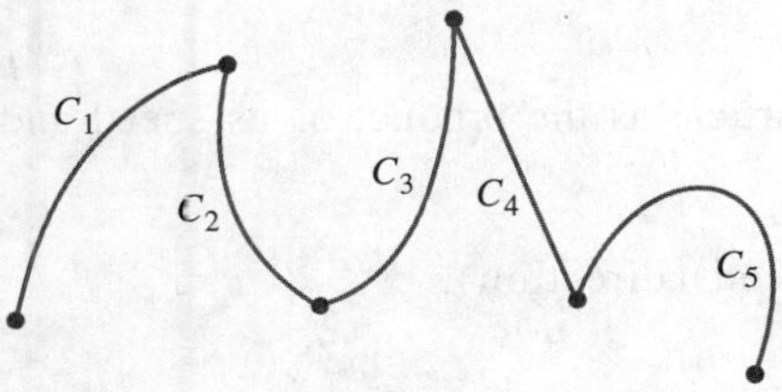

FIGURE 13.6 A piecewise smooth curve made up of five smooth curves connected end to end in a continuous fashion. The curve here is not smooth at the points joining the five smooth curves.

Look once again at Figure 13.5. We drew the figure for Δt positive, so $\Delta\mathbf{r}$ points forward, in the direction of the motion. The vector $\Delta\mathbf{r}/\Delta t$, having the same direction as $\Delta\mathbf{r}$, points forward too. Had Δt been negative, $\Delta\mathbf{r}$ would have pointed backward, against the direction of motion. The quotient $\Delta\mathbf{r}/\Delta t$, however, being a negative scalar multiple of $\Delta\mathbf{r}$, would once again have pointed forward. No matter how $\Delta\mathbf{r}$ points, $\Delta\mathbf{r}/\Delta t$ points forward, and we expect the vector $d\mathbf{r}/dt = \lim_{\Delta t\to 0}\Delta\mathbf{r}/\Delta t$, when different from $\mathbf{0}$, to do the same. This means that the derivative $d\mathbf{r}/dt$, which is the rate of change of position with respect to time, always points in the direction of motion. For a smooth curve, $d\mathbf{r}/dt$ is never zero; the particle does not stop or reverse direction.

DEFINITIONS If $\mathbf{r}$ is the position vector of a particle moving along a smooth curve in space, then

$$\mathbf{v}(t) = \frac{d\mathbf{r}}{dt}$$

is the particle's **velocity vector**. If $\mathbf{v}$ is a nonzero vector, then it is tangent to the curve, and its direction is the **direction of motion**. The magnitude of $\mathbf{v}$ is the particle's **speed**, and the derivative $\mathbf{a} = d\mathbf{v}/dt$, when it exists, is the particle's **acceleration vector**. In summary,

1. Velocity is the derivative of position: $\mathbf{v} = \dfrac{d\mathbf{r}}{dt}$.
2. Speed is the magnitude of velocity: Speed $= |\mathbf{v}|$.
3. Acceleration is the derivative of velocity: $\mathbf{a} = \dfrac{d\mathbf{v}}{dt} = \dfrac{d^2\mathbf{r}}{dt^2}$.
4. The unit vector $\mathbf{v}/|\mathbf{v}|$ is the direction of motion at time t.

EXAMPLE 4 Find the velocity, speed, and acceleration of a particle whose motion in space is given by the position vector $\mathbf{r}(t) = 2\cos t\mathbf{i} + 2\sin t\mathbf{j} + 5\cos^2 t\mathbf{k}$. Sketch the velocity vector $\mathbf{v}(7\pi/4)$.

Solution The velocity and acceleration vectors at time t are

$$\begin{aligned}\mathbf{v}(t) = \mathbf{r}'(t) &= -2\sin t\mathbf{i} + 2\cos t\mathbf{j} - 10\cos t\sin t\mathbf{k}\\ &= -2\sin t\mathbf{i} + 2\cos t\mathbf{j} - 5\sin 2t\mathbf{k},\\ \mathbf{a}(t) = \mathbf{r}''(t) &= -2\cos t\mathbf{i} - 2\sin t\mathbf{j} - 10\cos 2t\mathbf{k},\end{aligned}$$

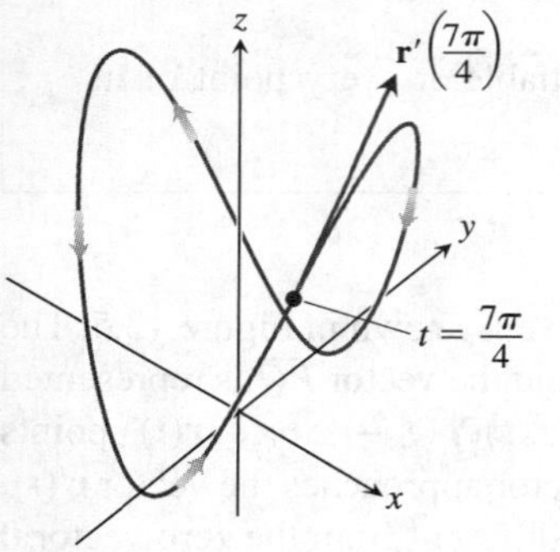

FIGURE 13.7 The curve and the velocity vector when $t = 7\pi/4$ for the motion given in Example 4.

and the speed is

$$|\mathbf{v}(t)| = \sqrt{(-2\sin t)^2 + (2\cos t)^2 + (-5\sin 2t)^2} = \sqrt{4 + 25\sin^2 2t}.$$

When $t = 7\pi/4$, we have

$$\mathbf{v}\left(\frac{7\pi}{4}\right) = \sqrt{2}\mathbf{i} + \sqrt{2}\mathbf{j} + 5\mathbf{k}, \quad \mathbf{a}\left(\frac{7\pi}{4}\right) = -\sqrt{2}\mathbf{i} + \sqrt{2}\mathbf{j}, \quad \left|\mathbf{v}\left(\frac{7\pi}{4}\right)\right| = \sqrt{29}.$$

A sketch of the curve of motion, and the velocity vector when $t = 7\pi/4$, can be seen in Figure 13.7. ■

We can express the velocity of a moving particle as the product of its speed and direction:

$$\text{Velocity} = |\mathbf{v}|\left(\frac{\mathbf{v}}{|\mathbf{v}|}\right) = (\text{speed})(\text{direction}).$$

Differentiation Rules

Because the derivatives of vector functions may be computed component by component, the rules for differentiating vector functions have the same form as the rules for differentiating scalar functions.

Differentiation Rules for Vector Functions

Let **u** and **v** be differentiable vector functions of t, **C** a constant vector, c any scalar, and f any differentiable scalar function.

1. *Constant Function Rule:* $\frac{d}{dt}\mathbf{C} = \mathbf{0}$
2. *Scalar Multiple Rules:* $\frac{d}{dt}[c\mathbf{u}(t)] = c\mathbf{u}'(t)$

   $\frac{d}{dt}[f(t)\,\mathbf{u}(t)] = f'(t)\,\mathbf{u}(t) + f(t)\,\mathbf{u}'(t)$
3. *Sum Rule:* $\frac{d}{dt}[\mathbf{u}(t) + \mathbf{v}(t)] = \mathbf{u}'(t) + \mathbf{v}'(t)$
4. *Difference Rule:* $\frac{d}{dt}[\mathbf{u}(t) - \mathbf{v}(t)] = \mathbf{u}'(t) - \mathbf{v}'(t)$
5. *Dot Product Rule:* $\frac{d}{dt}[\mathbf{u}(t)\cdot\mathbf{v}(t)] = \mathbf{u}'(t)\cdot\mathbf{v}(t) + \mathbf{u}(t)\cdot\mathbf{v}'(t)$
6. *Cross Product Rule:* $\frac{d}{dt}[\mathbf{u}(t)\times\mathbf{v}(t)] = \mathbf{u}'(t)\times\mathbf{v}(t) + \mathbf{u}(t)\times\mathbf{v}'(t)$
7. *Chain Rule:* $\frac{d}{dt}[\mathbf{u}(f(t))] = f'(t)\mathbf{u}'(f(t))$

When you use the Cross Product Rule, remember to preserve the order of the factors. If **u** comes first on the left side of the equation, it must also come first on the right, or the signs will be wrong.

We will prove the product rules and the Chain Rule but will leave the rules for constants, scalar multiples, sums, and differences as exercises.

Proof of the Dot Product Rule Suppose that

$$\mathbf{u} = u_1(t)\mathbf{i} + u_2(t)\mathbf{j} + u_3(t)\mathbf{k}$$

and

$$\mathbf{v} = v_1(t)\mathbf{i} + v_2(t)\mathbf{j} + v_3(t)\mathbf{k}.$$

Then

$$\begin{aligned}\frac{d}{dt}(\mathbf{u}\cdot\mathbf{v}) &= \frac{d}{dt}(u_1v_1 + u_2v_2 + u_3v_3)\\ &= \underbrace{u_1'v_1 + u_2'v_2 + u_3'v_3}_{\mathbf{u}'\cdot\mathbf{v}} + \underbrace{u_1v_1' + u_2v_2' + u_3v_3'}_{\mathbf{u}\cdot\mathbf{v}'}.\end{aligned}$$

■

Proof of the Cross Product Rule We model the proof after the proof of the Product Rule for scalar functions. According to the definition of derivative,

$$\frac{d}{dt}(\mathbf{u}\times\mathbf{v}) = \lim_{h\to 0}\frac{\mathbf{u}(t+h)\times\mathbf{v}(t+h) - \mathbf{u}(t)\times\mathbf{v}(t)}{h}.$$

To change this fraction into an equivalent one that contains the difference quotients for the derivatives of **u** and **v**, we subtract and add $\mathbf{u}(t)\times\mathbf{v}(t+h)$ in the numerator. Then

$$\begin{aligned}&\frac{d}{dt}(\mathbf{u}\times\mathbf{v})\\ &= \lim_{h\to 0}\frac{\mathbf{u}(t+h)\times\mathbf{v}(t+h) - \mathbf{u}(t)\times\mathbf{v}(t+h) + \mathbf{u}(t)\times\mathbf{v}(t+h) - \mathbf{u}(t)\times\mathbf{v}(t)}{h}\\ &= \lim_{h\to 0}\left[\frac{\mathbf{u}(t+h) - \mathbf{u}(t)}{h}\times\mathbf{v}(t+h) + \mathbf{u}(t)\times\frac{\mathbf{v}(t+h) - \mathbf{v}(t)}{h}\right]\\ &= \lim_{h\to 0}\frac{\mathbf{u}(t+h) - \mathbf{u}(t)}{h}\times\lim_{h\to 0}\mathbf{v}(t+h) + \lim_{h\to 0}\mathbf{u}(t)\times\lim_{h\to 0}\frac{\mathbf{v}(t+h) - \mathbf{v}(t)}{h}.\end{aligned}$$

The last of these equalities holds because the limit of the cross product of two vector functions is the cross product of their limits if the latter exist (Exercise 46). As h approaches zero, $\mathbf{v}(t+h)$ approaches $\mathbf{v}(t)$ because $\mathbf{v}$, being differentiable at t, is continuous at t (Exercise 47). The two fractions approach the values of $d\mathbf{u}/dt$ and $d\mathbf{v}/dt$ at t. In short,

$$\frac{d}{dt}(\mathbf{u}\times\mathbf{v}) = \left(\frac{d\mathbf{u}}{dt}\times\mathbf{v}\right) + \left(\mathbf{u}\times\frac{d\mathbf{v}}{dt}\right). \quad \blacksquare$$

Proof of the Chain Rule Suppose that $\mathbf{u}(s) = a(s)\mathbf{i} + b(s)\mathbf{j} + c(s)\mathbf{k}$ is a differentiable vector function of s and that $s = f(t)$ is a differentiable scalar function of t. Then a, b, and c are differentiable functions of t, and the Chain Rule for differentiable real-valued functions gives

$$\begin{aligned}
\frac{d}{dt}[\mathbf{u}(s)] &= \frac{da}{dt}\mathbf{i} + \frac{db}{dt}\mathbf{j} + \frac{dc}{dt}\mathbf{k} \\
&= \frac{da}{ds}\frac{ds}{dt}\mathbf{i} + \frac{db}{ds}\frac{ds}{dt}\mathbf{j} + \frac{dc}{ds}\frac{ds}{dt}\mathbf{k} \\
&= \frac{ds}{dt}\left(\frac{da}{ds}\mathbf{i} + \frac{db}{ds}\mathbf{j} + \frac{dc}{ds}\mathbf{k}\right) \\
&= \frac{ds}{dt}\frac{d\mathbf{u}}{ds} \\
&= f'(t)\,\mathbf{u}'(f(t)). && s = f(t) \quad \blacksquare
\end{aligned}$$

As an algebraic convenience, we sometimes write the product of a scalar c and a vector $\mathbf{v}$ as $\mathbf{v}c$ instead of $c\mathbf{v}$. This permits us, for instance, to write the Chain Rule in a familiar form:

$$\frac{d\mathbf{u}}{dt} = \frac{d\mathbf{u}}{ds}\frac{ds}{dt},$$

where $s = f(t)$.

Vector Functions of Constant Length

When we track a particle moving on a sphere centered at the origin (Figure 13.8), the position vector has a constant length equal to the radius of the sphere. The velocity vector $d\mathbf{r}/dt$, tangent to the path of motion, is tangent to the sphere and hence perpendicular to $\mathbf{r}$. This is always the case for a differentiable vector function of constant length: The vector and its first derivative are orthogonal. By direct calculation,

$$\begin{aligned}
\mathbf{r}(t)\cdot\mathbf{r}(t) &= |\mathbf{r}(t)|^2 = c^2 && |\mathbf{r}(t)| = c \text{ is constant.} \\
\frac{d}{dt}[\mathbf{r}(t)\cdot\mathbf{r}(t)] &= 0 && \text{Differentiate both sides.} \\
\mathbf{r}'(t)\cdot\mathbf{r}(t) + \mathbf{r}(t)\cdot\mathbf{r}'(t) &= 0 && \text{Rule 5 with } \mathbf{r}(t) = \mathbf{u}(t) = \mathbf{v}(t) \\
2\mathbf{r}'(t)\cdot\mathbf{r}(t) &= 0.
\end{aligned}$$

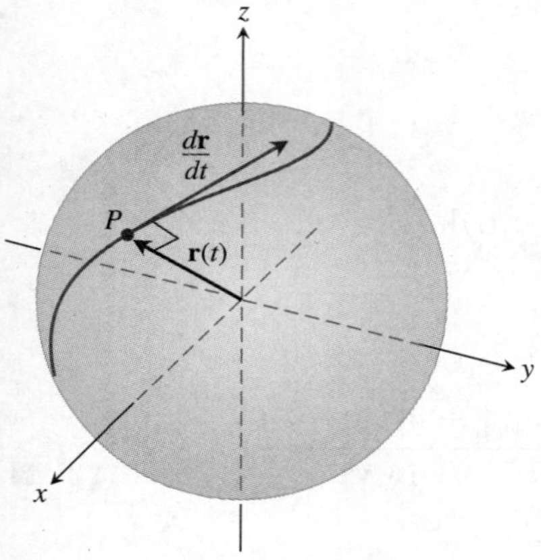

FIGURE 13.8 If a particle moves on a sphere in such a way that its position $\mathbf{r}$ is a differentiable function of time, then $\mathbf{r}\cdot(d\mathbf{r}/dt) = 0$.

Thus the vectors $\mathbf{r}'(t)$ and $\mathbf{r}(t)$ are orthogonal because their dot product is 0. In summary, the following holds.

> If $\mathbf{r}$ is a differentiable vector function of t and the length of $\mathbf{r}(t)$ is constant, then
> $$\mathbf{r}\cdot\frac{d\mathbf{r}}{dt} = 0. \qquad (4)$$

We will use this observation repeatedly in Section 13.4. The converse is also true (see Exercise 41).

EXERCISES 13.1

In Exercises 1–4, find the given limits.

1. $\lim_{t\to\pi}\left[\left(\sin\frac{t}{2}\right)\mathbf{i}+\left(\cos\frac{2}{3}t\right)\mathbf{j}+\left(\tan\frac{5}{4}t\right)\mathbf{k}\right]$

2. $\lim_{t\to-1}\left[t^3\mathbf{i}+\left(\sin\frac{\pi}{2}t\right)\mathbf{j}+(\ln(t+2))\mathbf{k}\right]$

3. $\lim_{t\to1}\left[\left(\frac{t^2-1}{\ln t}\right)\mathbf{i}-\left(\frac{\sqrt{t}-1}{1-t}\right)\mathbf{j}+(\arctan t)\mathbf{k}\right]$

4. $\lim_{t\to0}\left[\left(\frac{\sin t}{t}\right)\mathbf{i}+\left(\frac{\tan^2 t}{\sin 2t}\right)\mathbf{j}-\left(\frac{t^3-8}{t+2}\right)\mathbf{k}\right]$

Motion in the Plane

In Exercises 5–8, $\mathbf{r}(t)$ is the position of a particle in the xy-plane at time t. Find an equation in x and y whose graph is the path of the particle. Then find the particle's velocity and acceleration vectors at the given value of t.

5. $\mathbf{r}(t)=(t^2+1)\mathbf{i}+(2t-1)\mathbf{j},\quad t=2$

6. $\mathbf{r}(t)=\frac{2t}{t-1}\mathbf{i}+\frac{2}{t^2}\mathbf{j},\quad t=\frac{1}{2}$

7. $\mathbf{r}(t)=e^t\mathbf{i}+\frac{2}{9}e^{2t}\mathbf{j},\quad t=\ln 3$

8. $\mathbf{r}(t)=(\sin 4t)\mathbf{i}+(3\cos 4t)\mathbf{j},\quad t=0$

Exercises 9–12 give the position vectors of particles moving along various curves in the xy-plane. In each case, find the particle's velocity and acceleration vectors at the stated times and sketch them as vectors on the curve.

9. **Motion on the circle $x^2+y^2=1$**

$$\mathbf{r}(t)=(\sin t)\mathbf{i}+(\cos t)\mathbf{j};\quad t=\pi/4 \text{ and } \pi/2$$

10. **Motion on the circle $x^2+y^2=16$**

$$\mathbf{r}(t)=\left(4\cos\frac{t}{2}\right)\mathbf{i}+\left(4\sin\frac{t}{2}\right)\mathbf{j};\quad t=\pi \text{ and } 3\pi/2$$

11. **Motion on the cycloid $x=t-\sin t,\ y=1-\cos t$**

$$\mathbf{r}(t)=(t-\sin t)\mathbf{i}+(1-\cos t)\mathbf{j};\quad t=\pi \text{ and } 3\pi/2$$

12. **Motion on the parabola $y=x^2+1$**

$$\mathbf{r}(t)=t\mathbf{i}+(t^2+1)\mathbf{j};\quad t=-1, 0, \text{ and } 1$$

Motion in Space

In Exercises 13–18, $\mathbf{r}(t)$ is the position of a particle in space at time t. Find the particle's velocity and acceleration vectors. Then find the particle's speed and direction of motion at the given value of t. Write the particle's velocity at that time as the product of its speed and direction.

13. $\mathbf{r}(t)=(t^2+1)\mathbf{i}+(2t-1)\mathbf{j}-2t\mathbf{k},\quad t=2$

14. $\mathbf{r}(t)=(1-t)\mathbf{i}+\frac{2t^2}{5}\mathbf{j}+5t^3\mathbf{k},\quad t=3$

15. $\mathbf{r}(t)=(2\cos t)\mathbf{i}+(3\sin t)\mathbf{j}+4t\mathbf{k},\quad t=\pi/2$

16. $\mathbf{r}(t)=(\tan 2t)\mathbf{i}+(2\sec 2t)\mathbf{j}+\mathbf{k},\quad t=\frac{\pi}{2}$

17. $\mathbf{r}(t)=(2\ln(t+1))\mathbf{i}+t^2\mathbf{j}+\frac{t^2}{2}\mathbf{k},\quad t=1$

18. $\mathbf{r}(t)=2e^{-t}\mathbf{i}+(3\sin 2t)\mathbf{j}+(3\cos 2t)\mathbf{k},\quad t=0$

In Exercises 19–22, $\mathbf{r}(t)$ is the position of a particle in space at time t. Find the angle between the velocity and acceleration vectors at time $t=0$.

19. $\mathbf{r}(t)=(3t+1)\mathbf{i}+\sqrt{3}t\mathbf{j}+t^2\mathbf{k}$

20. $\mathbf{r}(t)=\frac{\sqrt{3}}{3}t\mathbf{i}+\left(\frac{\sqrt{3}}{3}t-9t^2\right)\mathbf{j}$

21. $\mathbf{r}(t)=\ln(2t^2+1)\mathbf{i}+(\tan^{-1}2t)\mathbf{j}+\left(\sqrt{2t^2+1}\right)\mathbf{k}$

22. $\mathbf{r}(t)=\frac{4}{9}(1+t)^{3/2}\mathbf{i}+\frac{4}{9}(1-t)^{3/2}\mathbf{j}+\frac{1}{3}t\mathbf{k}$

Tangents to Curves

As mentioned in the text, the **tangent line** to a smooth curve $\mathbf{r}(t)=f(t)\mathbf{i}+g(t)\mathbf{j}+h(t)\mathbf{k}$ at $t=t_0$ is the line that passes through the point $(f(t_0), g(t_0), h(t_0))$ parallel to $\mathbf{v}(t_0)$, the curve's velocity vector at t_0. In Exercises 23–26, find parametric equations for the line that is tangent to the given curve at the given parameter value $t=t_0$.

23. $\mathbf{r}(t)=(\cos t)\mathbf{i}+(2t^2-\sin t)\mathbf{j}+e^{-t}\mathbf{k},\quad t_0=0$

24. $\mathbf{r}(t)=2t^2\mathbf{i}+(3t+1)\mathbf{j}+3t^3\mathbf{k},\quad t_0=3$

25. $\mathbf{r}(t)=\ln t\,\mathbf{i}+\frac{t-1}{t+2}\mathbf{j}+t\ln t\,\mathbf{k},\quad t_0=1$

26. $\mathbf{r}(t)=(\cos 2t)\mathbf{i}+(\sin 2t)\mathbf{j}+(\cos 2t)\mathbf{k},\quad t_0=\pi$

In Exercises 27–30, find the value(s) of t so that the tangent line to the given curve contains the given point.

27. $\mathbf{r}(t)=t^2\mathbf{i}+(1+t)\mathbf{j}+(2t-3)\mathbf{k};\quad(-8,2,-1)$

28. $\mathbf{r}(t)=t\mathbf{i}+3\mathbf{j}+\left(\frac{2}{3}t^{3/2}\right)\mathbf{k};\quad(0,3,-8/3)$

29. $\mathbf{r}(t)=2t\mathbf{i}+t^2\mathbf{j}-t^2\mathbf{k};\quad(0,-4,4)$

30. $\mathbf{r}(t)=-t\mathbf{i}+t^2\mathbf{j}+(\ln t)\mathbf{k};\quad(2,-5,-3)$

In Exercises 31–36, $\mathbf{r}(t)$ is the position of a particle in space at time t. Match each position function with one of the graphs A–F.

31. $\mathbf{r}(t)=(t\cos t)\mathbf{i}+(t\sin t)\mathbf{j}+t\mathbf{k}$

32. $\mathbf{r}(t)=(\cos t)\mathbf{i}+(\sin t)\mathbf{j}+(\sin 2t)\mathbf{k}$

33. $\mathbf{r}(t)=t^2\mathbf{i}+(t^2+1)\mathbf{j}+t^4\mathbf{k}$

34. $\mathbf{r}(t)=t\mathbf{i}+(\ln t)\mathbf{j}+(\sin t)\mathbf{k}$

35. $\mathbf{r}(t) = t\mathbf{i} + (\cos t)\mathbf{j} + (\sin t)\mathbf{k}$

36. $\mathbf{r}(t) = (t\sin t)\mathbf{i} + (t\cos t)\mathbf{j} + \left(\dfrac{t}{t^2+1}\right)\mathbf{k}$

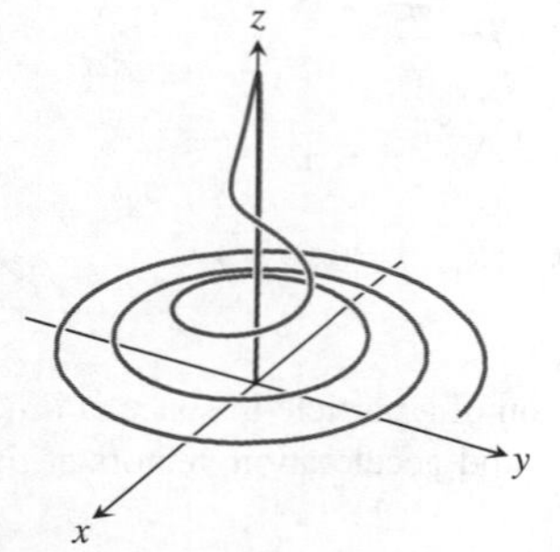

A.

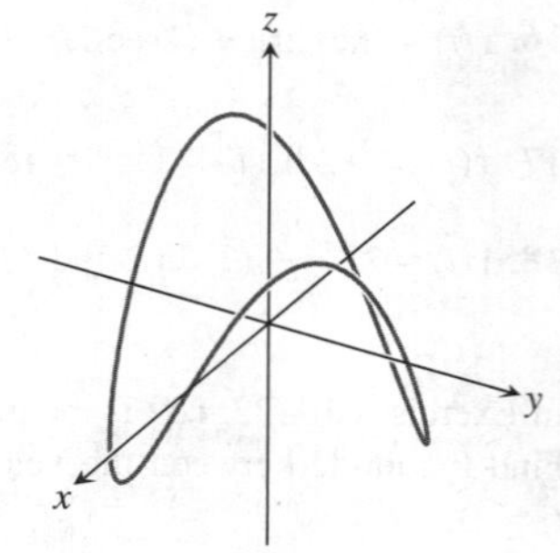

B.

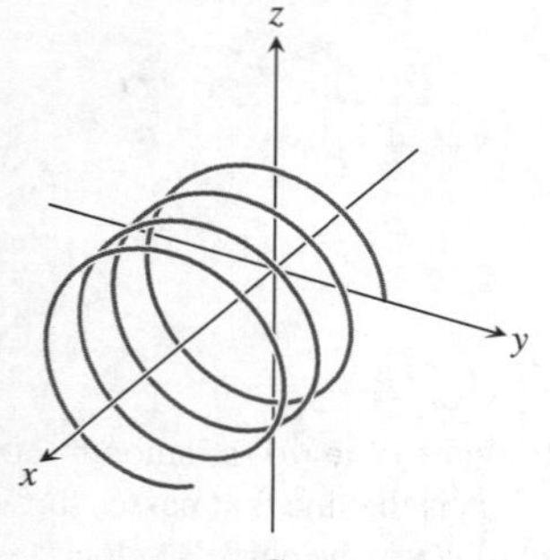

C.

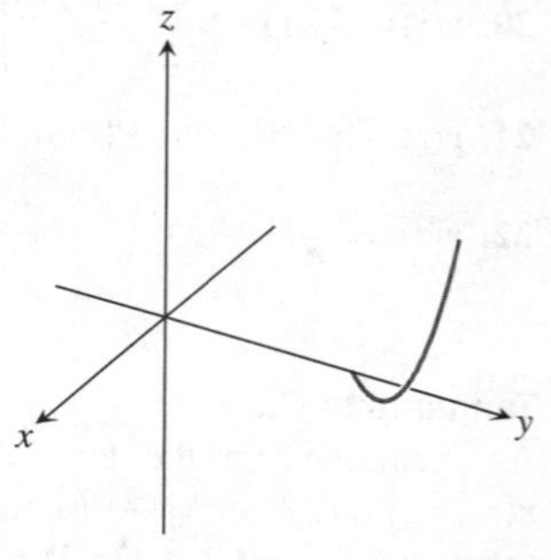

D.

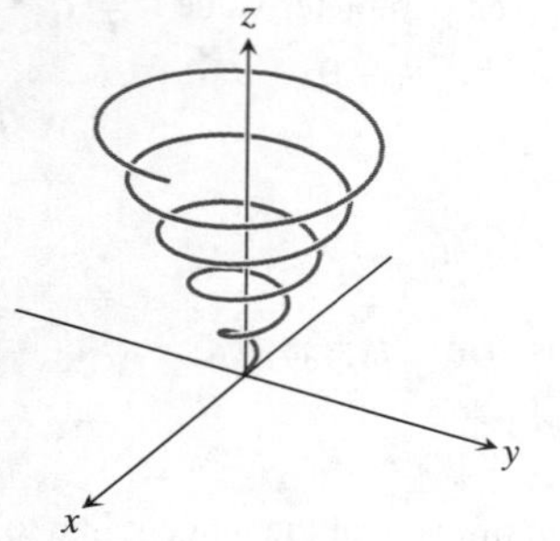

E.

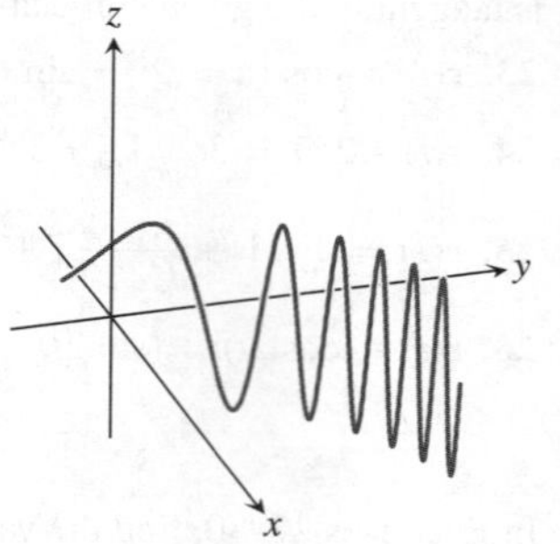

F.

Theory and Examples

37. Motion along a circle Each of the following equations in parts (a)–(e) describes the motion of a particle having the same path, namely the unit circle $x^2 + y^2 = 1$. Although the path of each particle in parts (a)–(e) is the same, the behavior, or "dynamics," of each particle is different. For each particle, answer the following questions.

i) Does the particle have constant speed? If so, what is its constant speed?

ii) Is the particle's acceleration vector always orthogonal to its velocity vector?

iii) Does the particle move clockwise or counterclockwise around the circle?

iv) Is the particle initially located at the point $(1, 0)$?

a. $\mathbf{r}(t) = (\sin t)\mathbf{i} + (\cos t)\mathbf{j}, \quad t \geq 0$

b. $\mathbf{r}(t) = \cos(2t)\mathbf{i} + \sin(2t)\mathbf{j}, \quad t \geq 0$

c. $\mathbf{r}(t) = \cos(t - \pi/2)\mathbf{i} + \sin(t - \pi/2)\mathbf{j}, \quad t \geq 0$

d. $\mathbf{r}(t) = (\sin t)\mathbf{i} - (\cos t)\mathbf{j}, \quad t \geq 0$

e. $\mathbf{r}(t) = \cos(t^2)\mathbf{i} + \sin(t^2)\mathbf{j}, \quad t \geq 0$

38. Motion along a circle Show that the vector-valued function

$$\mathbf{r}(t) = (2\mathbf{i} + 2\mathbf{j} + \mathbf{k}) + \cos t\left(\frac{1}{\sqrt{2}}\mathbf{i} - \frac{1}{\sqrt{2}}\mathbf{j}\right) + \sin t\left(\frac{1}{\sqrt{3}}\mathbf{i} + \frac{1}{\sqrt{3}}\mathbf{j} + \frac{1}{\sqrt{3}}\mathbf{k}\right)$$

describes the motion of a particle moving in the circle of radius 1 centered at the point $(2, 2, 1)$ and lying in the plane $x + y - 2z = 2$.

39. Motion along a parabola A particle moves along the top of the parabola $y^2 = 3x$ from left to right at a constant speed of 6 units per second. Find the velocity of the particle as it moves through the point (3, 3).

40. Motion along a cycloid A particle moves in the xy-plane in such a way that its position at time t is

$$\mathbf{r}(t) = (t - \sin t)\mathbf{i} + (1 - \cos t)\mathbf{j}.$$

T **a.** Graph $\mathbf{r}(t)$. The resulting curve is a cycloid.

b. Find the maximum and minimum values of $|\mathbf{v}|$ and $|\mathbf{a}|$. (*Hint:* Find the extreme values of $|\mathbf{v}|^2$ and $|\mathbf{a}|^2$ first, and take square roots later.)

41. Let $\mathbf{r}$ be a differentiable vector function of t. Show that if $\mathbf{r} \cdot (d\mathbf{r}/dt) = 0$ for all t, then $|\mathbf{r}|$ is constant.

42. Derivatives of triple scalar products

a. Show that if $\mathbf{u}$, $\mathbf{v}$, and $\mathbf{w}$ are differentiable vector functions of t, then

$$\frac{d}{dt}(\mathbf{u} \cdot \mathbf{v} \times \mathbf{w}) = \frac{d\mathbf{u}}{dt} \cdot \mathbf{v} \times \mathbf{w} + \mathbf{u} \cdot \frac{d\mathbf{v}}{dt} \times \mathbf{w} + \mathbf{u} \cdot \mathbf{v} \times \frac{d\mathbf{w}}{dt}.$$

b. Show that

$$\frac{d}{dt}\left(\mathbf{r} \cdot \frac{d\mathbf{r}}{dt} \times \frac{d^2\mathbf{r}}{dt^2}\right) = \mathbf{r} \cdot \left(\frac{d\mathbf{r}}{dt} \times \frac{d^3\mathbf{r}}{dt^3}\right).$$

(*Hint:* Differentiate on the left and look for vectors whose products are zero.)

43. Prove the two Scalar Multiple Rules for vector functions.

44. Prove the Sum and Difference Rules for vector functions.

45. Component test for continuity at a point Show that the vector function $\mathbf{r}$ defined by $\mathbf{r}(t) = f(t)\mathbf{i} + g(t)\mathbf{j} + h(t)\mathbf{k}$ is continuous at $t = t_0$ if and only if f, g, and h are continuous at t_0.

46. Limits of cross products of vector functions Suppose that $\mathbf{r}_1(t) = f_1(t)\mathbf{i} + f_2(t)\mathbf{j} + f_3(t)\mathbf{k}$, $\mathbf{r}_2(t) = g_1(t)\mathbf{i} + g_2(t)\mathbf{j} + g_3(t)\mathbf{k}$, $\lim_{t\to t_0} \mathbf{r}_1(t) = \mathbf{A}$, and $\lim_{t\to t_0} \mathbf{r}_2(t) = \mathbf{B}$. Use the determinant formula for cross products and the Limit Product Rule for scalar functions to show that

$$\lim_{t\to t_0} (\mathbf{r}_1(t) \times \mathbf{r}_2(t)) = \mathbf{A} \times \mathbf{B}.$$

47. Differentiable vector functions are continuous Show that if $\mathbf{r}(t) = f(t)\mathbf{i} + g(t)\mathbf{j} + h(t)\mathbf{k}$ is differentiable at $t = t_0$, then it is continuous at t_0 as well.

48. Constant Function Rule Prove that if $\mathbf{u}$ is the vector function with the constant value $\mathbf{C}$, then $d\mathbf{u}/dt = \mathbf{0}$.

COMPUTER EXPLORATIONS

Use a CAS to perform the following steps in Exercises 49–52.

a. Plot the space curve traced out by the position vector $\mathbf{r}$.

b. Find the components of the velocity vector $d\mathbf{r}/dt$.

c. Evaluate $d\mathbf{r}/dt$ at the given point t_0 and determine the equation of the tangent line to the curve at $\mathbf{r}(t_0)$.

d. Plot the tangent line together with the curve over the given interval.

49. $\mathbf{r}(t) = (\sin t - t\cos t)\mathbf{i} + (\cos t + t\sin t)\mathbf{j} + t^2\mathbf{k}$, $0 \le t \le 6\pi$, $t_0 = 3\pi/2$

50. $\mathbf{r}(t) = \sqrt{2}t\mathbf{i} + e^t\mathbf{j} + e^{-t}\mathbf{k}$, $-2 \le t \le 3$, $t_0 = 1$

51. $\mathbf{r}(t) = (\sin 2t)\mathbf{i} + (\ln(1 + t))\mathbf{j} + t\mathbf{k}$, $0 \le t \le 4\pi$, $t_0 = \pi/4$

52. $\mathbf{r}(t) = (\ln(t^2 + 2))\mathbf{i} + (\arctan 3t)\mathbf{j} + \sqrt{t^2 + 1}\,\mathbf{k}$, $-3 \le t \le 5$, $t_0 = 3$

In Exercises 53 and 54, you will explore graphically the behavior of the helix

$$\mathbf{r}(t) = (\cos at)\mathbf{i} + (\sin at)\mathbf{j} + bt\mathbf{k}$$

as you change the values of the constants a and b. Use a CAS to perform the steps in each exercise.

53. Set $b = 1$. Plot the helix $\mathbf{r}(t)$ together with the tangent line to the curve at $t = 3\pi/2$ for $a = 1, 2, 4$, and 6 over the interval $0 \le t \le 4\pi$. Describe in your own words what happens to the graph of the helix and the position of the tangent line as a increases through these positive values.

54. Set $a = 1$. Plot the helix $\mathbf{r}(t)$ together with the tangent line to the curve at $t = 3\pi/2$ for $b = 1/4, 1/2, 2$, and 4 over the interval $0 \le t \le 4\pi$. Describe in your own words what happens to the graph of the helix and the position of the tangent line as b increases through these positive values.

13.2 Integrals of Vector Functions; Projectile Motion

In this section we investigate integrals of vector functions and their application to motion along a path in space or in the plane.

Integrals of Vector Functions

A differentiable vector function $\mathbf{R}(t)$ is an **antiderivative** of a vector function $\mathbf{r}(t)$ on an interval I if $d\mathbf{R}/dt = \mathbf{r}$ at each point of I. If $\mathbf{R}$ is an antiderivative of $\mathbf{r}$ on I, it can be shown, working one component at a time, that every antiderivative of $\mathbf{r}$ on I has the form $\mathbf{R} + \mathbf{C}$ for some constant vector $\mathbf{C}$ (Exercise 45). The set of all antiderivatives of $\mathbf{r}$ on I is the **indefinite integral** of $\mathbf{r}$ on I.

> DEFINITION The **indefinite integral** of $\mathbf{r}$ with respect to t is the set of all antiderivatives of $\mathbf{r}$, denoted by
>
> $$\int \mathbf{r}(t)\,dt.$$

The usual arithmetic rules for indefinite integrals apply.

EXAMPLE 1 To integrate a vector function, we integrate each of its components.

$$\int ((\cos t)\mathbf{i} + \mathbf{j} - 2t\mathbf{k})\,dt = \left(\int \cos t\,dt\right)\mathbf{i} + \left(\int dt\right)\mathbf{j} - \left(\int 2t\,dt\right)\mathbf{k} \quad (1)$$

$$= (\sin t + C_1)\mathbf{i} + (t + C_2)\mathbf{j} - (t^2 + C_3)\mathbf{k} \quad (2)$$

$$= (\sin t)\mathbf{i} + t\mathbf{j} - t^2\mathbf{k} + \mathbf{C} \qquad C = C_1\mathbf{i} + C_2\mathbf{j} - C_3\mathbf{k}$$

As in the integration of scalar functions, we recommend that you skip the steps in Equations (1) and (2) and go directly to the final form. Find an antiderivative for each component and add a *constant vector* at the end. ■

Definite integrals of vector functions are best defined in terms of components. The definition is consistent with how we compute limits and derivatives of vector functions.

DEFINITION If the components of $\mathbf{r}(t) = f(t)\mathbf{i} + g(t)\mathbf{j} + h(t)\mathbf{k}$ are integrable over $[a, b]$, then so is $\mathbf{r}$, and the **definite integral** of $\mathbf{r}$ from a to b is

$$\int_a^b \mathbf{r}(t)\,dt = \left(\int_a^b f(t)\,dt\right)\mathbf{i} + \left(\int_a^b g(t)\,dt\right)\mathbf{j} + \left(\int_a^b h(t)\,dt\right)\mathbf{k}.$$

EXAMPLE 2 As in Example 1, we integrate each component.

$$\begin{aligned}
\int_0^\pi ((\cos t)\mathbf{i} + \mathbf{j} - 2t\mathbf{k})\,dt &= \left(\int_0^\pi \cos t\,dt\right)\mathbf{i} + \left(\int_0^\pi dt\right)\mathbf{j} - \left(\int_0^\pi 2t\,dt\right)\mathbf{k} \\
&= \Big[\sin t\Big]_0^\pi \mathbf{i} + \Big[t\Big]_0^\pi \mathbf{j} - \Big[t^2\Big]_0^\pi \mathbf{k} \\
&= [0 - 0]\mathbf{i} + [\pi - 0]\mathbf{j} - [\pi^2 - 0^2]\mathbf{k} \\
&= \pi\mathbf{j} - \pi^2\mathbf{k}
\end{aligned}$$

■

The Fundamental Theorem of Calculus for continuous vector functions says that

$$\int_a^b \mathbf{r}(t)\,dt = \mathbf{R}(t)\Big]_a^b = \mathbf{R}(b) - \mathbf{R}(a),$$

where $\mathbf{R}$ is any antiderivative of $\mathbf{r}$, so that $\mathbf{R}'(t) = \mathbf{r}(t)$ (Exercise 46). Notice that an antiderivative of a vector function is also a vector function, whereas a definite integral of a vector function is a single constant vector.

EXAMPLE 3 Suppose we do not know the path of a hang glider, but only its acceleration vector $\mathbf{a}(t) = -(3\cos t)\mathbf{i} - (3\sin t)\mathbf{j} + 2\mathbf{k}$. We also know that initially (at time $t = 0$) the glider departed from the point $(4, 0, 0)$ with velocity $\mathbf{v}(0) = 3\mathbf{j}$. Find the glider's position as a function of t.

Solution Our goal is to find $\mathbf{r}(t)$ knowing

The differential equation: $\mathbf{a} = \dfrac{d^2\mathbf{r}}{dt^2} = -(3\cos t)\mathbf{i} - (3\sin t)\mathbf{j} + 2\mathbf{k}$

The initial conditions: $\mathbf{v}(0) = 3\mathbf{j}$ and $\mathbf{r}(0) = 4\mathbf{i} + 0\mathbf{j} + 0\mathbf{k}$.

Integrating both sides of the differential equation with respect to t gives

$$\mathbf{v}(t) = -(3\sin t)\mathbf{i} + (3\cos t)\mathbf{j} + 2t\mathbf{k} + \mathbf{C}_1.$$

We use $\mathbf{v}(0) = 3\mathbf{j}$ to find $\mathbf{C}_1$:

$$\begin{aligned}
3\mathbf{j} &= -(3\sin 0)\mathbf{i} + (3\cos 0)\mathbf{j} + (0)\mathbf{k} + \mathbf{C}_1 \\
3\mathbf{j} &= 3\mathbf{j} + \mathbf{C}_1 \\
\mathbf{C}_1 &= 0.
\end{aligned}$$

The glider's velocity as a function of time is

$$\frac{d\mathbf{r}}{dt} = \mathbf{v}(t) = -(3\sin t)\mathbf{i} + (3\cos t)\mathbf{j} + 2t\mathbf{k}.$$

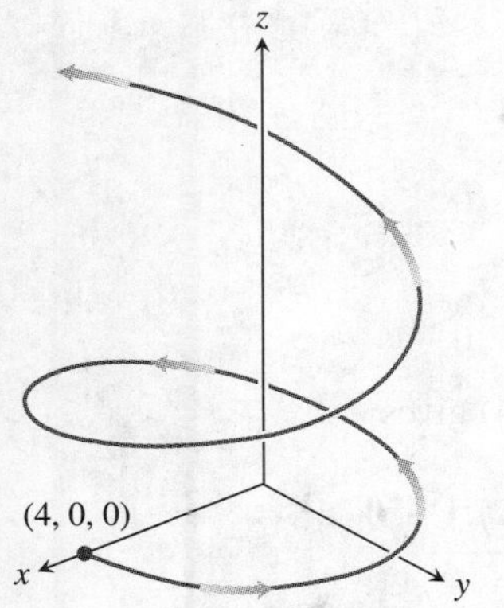

FIGURE 13.9 The path of the hang glider in Example 3. Although the path spirals around the z-axis, it is not a helix.

Integrating both sides of this last differential equation gives

$$\mathbf{r}(t) = (3\cos t)\mathbf{i} + (3\sin t)\mathbf{j} + t^2\mathbf{k} + \mathbf{C}_2.$$

We then use the initial condition $\mathbf{r}(0) = 4\mathbf{i}$ to find $\mathbf{C}_2$:

$$\begin{aligned} 4\mathbf{i} &= (3\cos 0)\mathbf{i} + (3\sin 0)\mathbf{j} + (0^2)\mathbf{k} + \mathbf{C}_2 \\ 4\mathbf{i} &= 3\mathbf{i} + (0)\mathbf{j} + (0)\mathbf{k} + \mathbf{C}_2 \\ \mathbf{C}_2 &= \mathbf{i}. \end{aligned}$$

The glider's position as a function of t is

$$\mathbf{r}(t) = (1 + 3\cos t)\mathbf{i} + (3\sin t)\mathbf{j} + t^2\mathbf{k}.$$

This is the path of the glider shown in Figure 13.9. Although the path resembles that of a helix due to its spiraling nature around the z-axis, it is not a helix because of the way it is rising. (We say more about this in Section 13.5.) ■

The Vector and Parametric Equations for Ideal Projectile Motion

A classic example of integrating vector functions is the derivation of the equations for the motion of a projectile. In physics, projectile motion describes how an object fired at some angle from an initial position, and acted upon by only the force of gravity, moves in a vertical coordinate plane. In the classic example, we ignore the effects of any frictional drag on the object, which may vary with its speed and altitude, and also the fact that the force of gravity changes slightly with the projectile's changing height. In addition, we ignore the long-distance effects of Earth turning beneath the projectile, such as in a rocket launch or the firing of a projectile from a cannon. Ignoring these effects gives us a reasonable approximation of the motion in most cases.

To derive equations for projectile motion, we assume that the projectile behaves like a particle moving in a vertical coordinate plane and that the only force acting on the projectile during its flight is the constant force of gravity, which always points straight down. The magnitude of the gravitational acceleration g is approximately 9.8 m/sec^2 at sea level, or 32 ft/sec^2. We assume that the projectile is launched from the origin at time $t = 0$ into the first quadrant with an initial velocity $\mathbf{v}_0$ (Figure 13.10). If $\mathbf{v}_0$ makes an angle α with the horizontal, then

$$\mathbf{v}_0 = (|\mathbf{v}_0|\cos\alpha)\mathbf{i} + (|\mathbf{v}_0|\sin\alpha)\mathbf{j}.$$

If we use the simpler notation v_0 for the initial speed $|\mathbf{v}_0|$, then

$$\mathbf{v}_0 = (v_0\cos\alpha)\mathbf{i} + (v_0\sin\alpha)\mathbf{j}. \tag{3}$$

The projectile's initial position is

$$\mathbf{r}_0 = 0\mathbf{i} + 0\mathbf{j} = \mathbf{0}. \tag{4}$$

Newton's second law of motion says that the force acting on the projectile is equal to the projectile's mass m times its acceleration, or $m(d^2\mathbf{r}/dt^2)$, if $\mathbf{r}$ is the projectile's position vector and t is time. If the force is solely the gravitational force $-mg\mathbf{j}$, then

$$m\frac{d^2\mathbf{r}}{dt^2} = -mg\mathbf{j} \quad \text{and} \quad \frac{d^2\mathbf{r}}{dt^2} = -g\mathbf{j},$$

where g is the acceleration due to gravity. We find $\mathbf{r}$ as a function of t by solving the following initial value problem.

$$\text{Differential equation:} \quad \frac{d^2\mathbf{r}}{dt^2} = -g\mathbf{j}$$

$$\text{Initial conditions:} \quad \mathbf{r} = \mathbf{r}_0 \quad \text{and} \quad \frac{d\mathbf{r}}{dt} = \mathbf{v}_0 \quad \text{when } t = 0$$

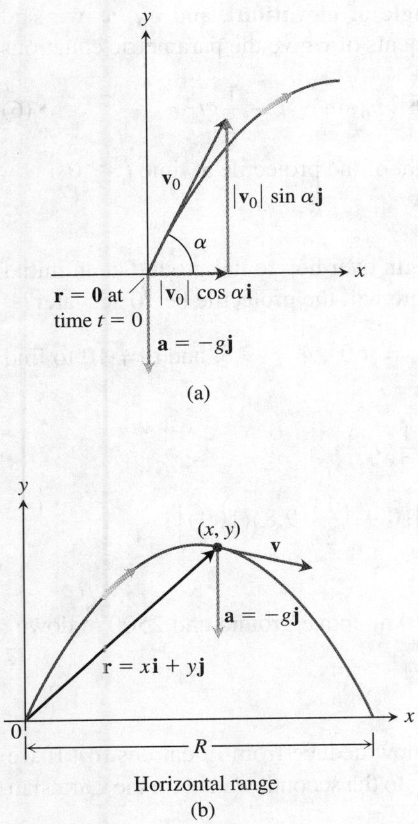

FIGURE 13.10 (a) Position, velocity, acceleration, and launch angle at $t = 0$. (b) Position, velocity, and acceleration at a later time t.

The first integration gives

$$\frac{d\mathbf{r}}{dt} = -(gt)\mathbf{j} + \mathbf{v}_0.$$

A second integration gives

$$\mathbf{r} = -\frac{1}{2}gt^2\mathbf{j} + \mathbf{v}_0 t + \mathbf{r}_0.$$

Substituting the values of $\mathbf{v}_0$ and $\mathbf{r}_0$ from Equations (3) and (4) gives

$$\mathbf{r} = -\frac{1}{2}gt^2\mathbf{j} + \underbrace{(v_0 \cos\alpha)t\mathbf{i} + (v_0 \sin\alpha)t\mathbf{j}}_{\mathbf{v}_0 t} + \mathbf{0}.$$

Collecting terms, we obtain the following.

Ideal Projectile Motion Equation

$$\mathbf{r} = (v_0 \cos\alpha)t\mathbf{i} + \left((v_0 \sin\alpha)t - \frac{1}{2}gt^2\right)\mathbf{j}. \tag{5}$$

Equation (5) is the *vector equation* of the path for ideal projectile motion. The angle α is the projectile's **launch angle (firing angle, angle of elevation)**, and v_0, as we said before, is the projectile's **initial speed**. The components of $\mathbf{r}$ give the parametric equations

$$x = (v_0 \cos\alpha)t \quad \text{and} \quad y = (v_0 \sin\alpha)t - \frac{1}{2}gt^2, \tag{6}$$

where x is the distance downrange and y is the height of the projectile at time $t \geq 0$.

EXAMPLE 4 A projectile is fired from the origin over horizontal ground at an initial speed of 500 m/sec and a launch angle of 60°. Where will the projectile be 10 sec later?

Solution We use Equation (5) with $v_0 = 500$, $\alpha = 60°$, $g = 9.8$, and $t = 10$ to find the projectile's components 10 sec after firing.

$$\begin{aligned}\mathbf{r} &= (v_0 \cos\alpha)t\mathbf{i} + \left((v_0 \sin\alpha)t - \frac{1}{2}gt^2\right)\mathbf{j} \\ &= (500)\left(\frac{1}{2}\right)(10)\mathbf{i} + \left[(500)\left(\frac{\sqrt{3}}{2}\right)10 - \left(\frac{1}{2}\right)(9.8)(100)\right]\mathbf{j} \\ &\approx 2500\mathbf{i} + 3840\mathbf{j}\end{aligned}$$

Ten seconds after firing, the projectile is about 3840 m above ground and 2500 m downrange from the origin. ■

Ideal projectiles move along parabolas, as we now deduce from Equations (6). If we substitute $t = x/(v_0 \cos\alpha)$ from the first equation into the second, we obtain the Cartesian coordinate equation

$$y = -\left(\frac{g}{2v_0{}^2 \cos^2\alpha}\right)x^2 + (\tan\alpha)x.$$

This equation has the form $y = ax^2 + bx$, so its graph is a parabola.

A projectile reaches its highest point when its vertical velocity component is zero. When fired over horizontal ground, the projectile lands when its vertical component equals zero in Equation (5), and the **range** R is the distance from the origin to the point of impact. We summarize the results here, which you are asked to verify in Exercise 31.

Height, Flight Time, and Range for Ideal Projectile Motion

For ideal projectile motion when an object is launched from the origin over a horizontal surface with initial speed v_0 and launch angle α:

Maximum height: $$y_{\max} = \frac{(v_0 \sin \alpha)^2}{2g}$$

Flight time: $$t = \frac{2v_0 \sin \alpha}{g}$$

Range: $$R = \frac{v_0{}^2}{g} \sin 2\alpha$$

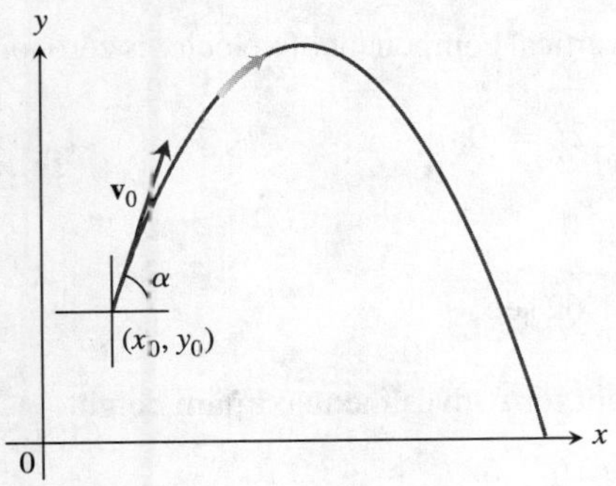

FIGURE 13.11 The path of a projectile fired from (x_0, y_0) with an initial velocity $\mathbf{v}_0$ at an angle of α degrees with the horizontal.

If we fire our ideal projectile from the point (x_0, y_0) instead of the origin (Figure 13.11), the position vector for the path of motion is

$$\mathbf{r} = (x_0 + (v_0 \cos \alpha)t)\mathbf{i} + \left(y_0 + (v_0 \sin \alpha)t - \frac{1}{2}gt^2\right)\mathbf{j}, \tag{7}$$

as you are asked to show in Exercise 33.

Projectile Motion with Wind Gusts

The next example shows how to account for another force acting on a projectile due to a gust of wind. We assume that the path of the baseball in Example 5 lies in a vertical plane.

EXAMPLE 5 A baseball is hit when it is 3 ft above the ground. It leaves the bat with initial speed of 152 ft/sec, making an angle of 20° with the horizontal. At the instant the ball is hit, an instantaneous gust of wind blows in the horizontal direction directly opposite the direction the ball is taking toward the outfield, adding a component of $-8.8\mathbf{i}$ (ft/sec) to the ball's initial velocity (8.8 ft/sec = 6 mph).

(a) Find a vector equation (position vector) for the path of the baseball.

(b) How high does the baseball go, and when does it reach maximum height?

(c) Assuming that the ball is not caught, find its range and flight time.

Solution

(a) Using Equation (3) and accounting for the gust of wind, the initial velocity of the baseball is

$$\begin{aligned}\mathbf{v}_0 &= (v_0 \cos \alpha)\mathbf{i} + (v_0 \sin \alpha)\mathbf{j} - 8.8\mathbf{i} \\ &= (152 \cos 20°)\mathbf{i} + (152 \sin 20°)\mathbf{j} - (8.8)\mathbf{i} \\ &= (152 \cos 20° - 8.8)\mathbf{i} + (152 \sin 20°)\mathbf{j}.\end{aligned}$$

The initial position is $\mathbf{r}_0 = 0\mathbf{i} + 3\mathbf{j}$. Integration of $d^2\mathbf{r}/dt^2 = -g\mathbf{j}$ gives

$$\frac{d\mathbf{r}}{dt} = -(gt)\mathbf{j} + \mathbf{v}_0.$$

A second integration gives

$$\mathbf{r} = -\frac{1}{2}gt^2\mathbf{j} + \mathbf{v}_0t + \mathbf{r}_0.$$

Substituting the values of $\mathbf{v}_0$ and $\mathbf{r}_0$ into the last equation gives the position vector of the baseball.

$$\begin{aligned}\mathbf{r} &= -\frac{1}{2}gt^2\mathbf{j} + \mathbf{v}_0t + \mathbf{r}_0 \\ &= -16t^2\mathbf{j} + (152\cos 20° - 8.8)t\mathbf{i} + (152\sin 20°)t\mathbf{j} + 3\mathbf{j} \\ &= (152\cos 20° - 8.8)t\mathbf{i} + (3 + (152\sin 20°)t - 16t^2)\mathbf{j}.\end{aligned}$$

(b) The baseball reaches its highest point when the vertical component of velocity is zero, or

$$\frac{dy}{dt} = 152\sin 20° - 32t = 0.$$

Solving for t we find

$$t = \frac{152\sin 20°}{32} \approx 1.62 \text{ sec}.$$

Substituting this time into the vertical component for $\mathbf{r}$ gives the maximum height

$$y_{\max} = 3 + (152\sin 20°)(1.62) - 16(1.62)^2 \approx 45.2 \text{ ft}.$$

That is, the maximum height of the baseball is about 45.2 ft, reached about 1.6 sec after leaving the bat.

(c) To find when the baseball lands, we set the vertical component for $\mathbf{r}$ equal to 0 and solve for t:

$$\begin{aligned}3 + (152\sin 20°)t - 16t^2 &= 0 \\ 3 + (51.99)t - 16t^2 &= 0.\end{aligned}$$

The solution values are about $t = 3.3$ sec and $t = -0.06$ sec. Substituting the positive time into the horizontal component for $\mathbf{r}$, we find the range

$$R = (152\cos 20° - 8.8)(3.3) \approx 442 \text{ ft}.$$

Thus, the horizontal range is about 442 ft, and the flight time is about 3.3 sec. ■

In Exercises 41 and 42, we consider projectile motion when there is air resistance slowing down the flight.

EXERCISES 13.2

Integrating Vector-Valued Functions

Evaluate the integrals in Exercises 1–10.

1. $\int_0^1 [2t^3\mathbf{i} + 5\mathbf{j} + (t-1)\mathbf{k}]\,dt$

2. $\int_1^2 \left[(6-6t)\mathbf{i} + 3\sqrt{t}\mathbf{j} + \left(\frac{4}{t^2}\right)\mathbf{k}\right]dt$

3. $\int_0^{\pi} [(\cos t)\mathbf{i} + (1+\sin t)\mathbf{j} + (\sec^2 t)\mathbf{k}]\,dt$

4. $\int_0^{\pi/3} [(\sec t\tan t)\mathbf{i} + (\tan t)\mathbf{j} + (2\sin t\cos t)\mathbf{k}]\,dt$

5. $\int_1^2 \left[\frac{2}{t}\mathbf{i} + \frac{1}{4-t}\mathbf{j} + \frac{1}{4t}\mathbf{k}\right]dt$

6. $\int_0^1 \left[\frac{2}{\sqrt{1-t^2}}\mathbf{i} + \frac{\sqrt{3}}{1+t^2}\mathbf{k}\right]dt$

7. $\int_0^2 [2te^{t^2}\mathbf{i} + 2e^{-t}\mathbf{j} + 2\mathbf{k}]\,dt$

8. $\int_1^{\ln 3} [2te^t\mathbf{i} + 2e^t\mathbf{j} + \ln t\,\mathbf{k}]\,dt$

9. $\int_0^{\pi/2} [\cos t\,\mathbf{i} - \sin 2t\,\mathbf{j} + \sin^2 t\,\mathbf{k}]\,dt$

10. $\int_{-\pi/4}^{\pi/4} \left[\tan t\,\mathbf{i} + \sec^2 t\,\mathbf{j} - t\cos t\,\mathbf{k}\right]dt$

Initial Value Problems

Solve the initial value problems in Exercises 11–20 for $\mathbf{r}$ as a vector function of t.

11. Differential equation: $\dfrac{d\mathbf{r}}{dt} = 2t\mathbf{i} + 2t\mathbf{j} + 2t\mathbf{k}$

Initial condition: $\mathbf{r}(0) = 2\mathbf{i} + \mathbf{j} + 5\mathbf{k}$

12. Differential equation: $\dfrac{d\mathbf{r}}{dt} = (180t)\mathbf{i} + (180t - 16t^2)\mathbf{j}$

Initial condition: $\mathbf{r}(0) = 100\mathbf{j}$

13. Differential equation: $\dfrac{d\mathbf{r}}{dt} = \dfrac{3}{2}(t+1)^{1/2}\mathbf{i} + e^{-t}\mathbf{j} + \dfrac{1}{t+1}\mathbf{k}$

Initial condition: $\mathbf{r}(0) = \mathbf{k}$

14. Differential equation: $\dfrac{d\mathbf{r}}{dt} = (2t^3 + t)\mathbf{i} + t^2\mathbf{j} + t\mathbf{k}$

Initial condition: $\mathbf{r}(0) = \mathbf{j} + \mathbf{k}$

15. Differential equation:

$$\frac{d\mathbf{r}}{dt} = (\tan t)\mathbf{i} + \left(\cos\left(\frac{1}{2}t\right)\right)\mathbf{j} - (\sec 2t)\mathbf{k},\ -\frac{\pi}{4} < t < \frac{\pi}{4}$$

Initial condition: $\mathbf{r}(0) = 3\mathbf{i} - 2\mathbf{j} + \mathbf{k}$

16. Differential equation:

$$\frac{d\mathbf{r}}{dt} = \left(\frac{t}{t^2+2}\right)\mathbf{i} - \left(\frac{t^2+1}{t-2}\right)\mathbf{j} + \left(\frac{t^2+4}{t^2+3}\right)\mathbf{k},\ t < 2$$

Initial condition: $\mathbf{r}(0) = \mathbf{i} - \mathbf{j} + \mathbf{k}$

17. Differential equation: $\dfrac{d^2\mathbf{r}}{dt^2} = -32\mathbf{k}$

Initial conditions: $\mathbf{r}(0) = 100\mathbf{k}$ and

$\left.\dfrac{d\mathbf{r}}{dt}\right|_{t=0} = 8\mathbf{i} + 8\mathbf{j}$

18. Differential equation: $\dfrac{d^2\mathbf{r}}{dt^2} = -(\mathbf{i} + \mathbf{j} + \mathbf{k})$

Initial conditions: $\mathbf{r}(0) = 10\mathbf{i} + 10\mathbf{j} + 10\mathbf{k}$ and

$\left.\dfrac{d\mathbf{r}}{dt}\right|_{t=0} = \mathbf{0}$

19. Differential equation: $\dfrac{d^2\mathbf{r}}{dt^2} = e^t\mathbf{i} - e^{-t}\mathbf{j} + 4e^{2t}\mathbf{k}$

Initial conditions: $\mathbf{r}(0) = 3\mathbf{i} + \mathbf{j} + 2\mathbf{k}$ and

$\left.\dfrac{d\mathbf{r}}{dt}\right|_{t=0} = -\mathbf{i} + 4\mathbf{j}$

20. Differential equation:

$$\frac{d^2\mathbf{r}}{dt^2} = (\sin t)\mathbf{i} - (\cos t)\mathbf{j} + (4\sin t\cos t)\mathbf{k}$$

Initial conditions: $\mathbf{r}(0) = \mathbf{i} - \mathbf{k}$ and

$\left.\dfrac{d\mathbf{r}}{dt}\right|_{t=0} = \mathbf{i}$

Motion Along a Straight Line

21. At time $t = 0$, a particle is located at the point $(1, 2, 3)$. It travels in a straight line to the point $(4, 1, 4)$, has speed 2 at $(1, 2, 3)$, and has constant acceleration $3\mathbf{i} - \mathbf{j} + \mathbf{k}$. Find an equation for the position vector $\mathbf{r}(t)$ of the particle at time t.

22. A particle traveling in a straight line is located at the point $(1, -1, 2)$ and has speed 2 at time $t = 0$. The particle moves toward the point $(3, 0, 3)$ with constant acceleration $2\mathbf{i} + \mathbf{j} + \mathbf{k}$. Find its position vector $\mathbf{r}(t)$ at time t.

Projectile Motion

Projectile flights in Exercises 23–40 are to be treated as ideal unless stated otherwise. All launch angles are assumed to be measured from the horizontal. All projectiles are assumed to be launched from the origin over a horizontal surface unless stated otherwise. For some exercises, a calculator may be helpful.

23. Travel time A projectile is fired at a speed of 960 m/sec at an angle of 60°. How long will it take to get 30 km downrange?

24. Range and height versus speed

a. Show that doubling a projectile's initial speed at a given launch angle multiplies its range by 4.

b. By about what percentage should you increase the initial speed to double the height and range?

25. Flight time and height A projectile is fired with an initial speed of 600 m/sec at an angle of elevation of 45°.

a. When and how far away will the projectile strike?

b. How high overhead will projectile be when it is 6 km downrange?

c. What is the greatest height reached by the projectile?

26. Throwing a baseball A baseball is thrown from the stands 32 ft above the field at an angle of 30° up from the horizontal. When and how far away will the ball strike the ground if its initial speed is 32 ft sec?

27. Firing golf balls A spring gun at ground level fires a golf ball at an angle of 45°. The ball lands 15 m away.

a. What was the ball's initial speed?

b. For the same initial speed, find the two firing angles that make the range 5 m.

28. Beaming electrons An electron in a TV tube is beamed horizontally at a speed of 6×10^6 m/sec toward the face of the tube 50 cm away. About how far will the electron drop before it hits?

29. Equal-range firing angles What two angles of elevation will enable a projectile to reach a target 25 km downrange on the same level as the gun if the projectile's initial speed is 500 m/sec?

30. Finding muzzle speed Find the muzzle speed of a gun whose maximum range is 32 km.

31. Verify the results given in the text (following Example 4) for the maximum height, flight time, and range for ideal projectile motion.

32. Colliding marbles The accompanying figure shows an experiment with two marbles. Marble A was launched toward marble B with launch angle α and initial speed v_0. At the same instant, marble B was released to fall from rest at $R \tan\alpha$ units directly above a spot R units downrange from A. The marbles were found to collide regardless of the value of v_0. Was this mere coincidence, or must this happen? Give reasons for your answer.

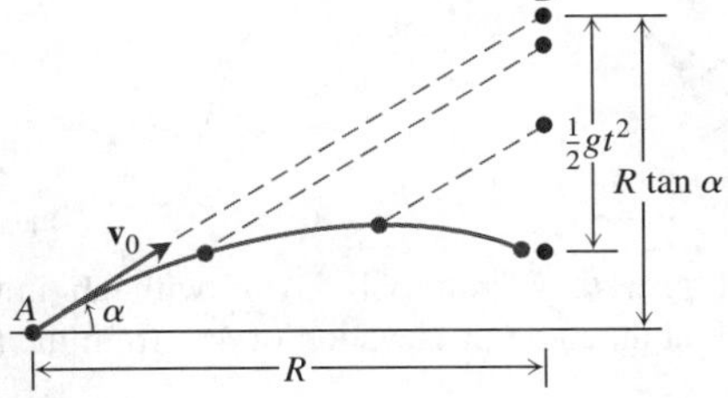

33. Firing from (x_0, y_0) Derive the equations

$$x = x_0 + (v_0 \cos \alpha)t,$$
$$y = y_0 + (v_0 \sin \alpha)t - \frac{1}{2}gt^2$$

(see Equation (7) in the text) by solving the following initial value problem for a vector **r** in the plane.

Differential equation: $\dfrac{d^2\mathbf{r}}{dt^2} = -g\mathbf{j}$

Initial conditions: $\mathbf{r}(0) = x_0\mathbf{i} + y_0\mathbf{j}$

$$\frac{d\mathbf{r}}{dt}(0) = (v_0 \cos \alpha)\mathbf{i} + (v_0 \sin \alpha)\mathbf{j}$$

34. Where trajectories crest For a projectile fired from the ground at launch angle α with initial speed v_0, consider α as a variable and v_0 as a fixed constant. For each α, $0 < \alpha < \pi/2$, we obtain a parabolic trajectory as shown in the accompanying figure. Show that the points in the plane that give the maximum heights of these parabolic trajectories all lie on the ellipse

$$x^2 + 4\left(y - \frac{v_0^2}{4g}\right)^2 = \frac{v_0^4}{4g^2},$$

where $x \geq 0$.

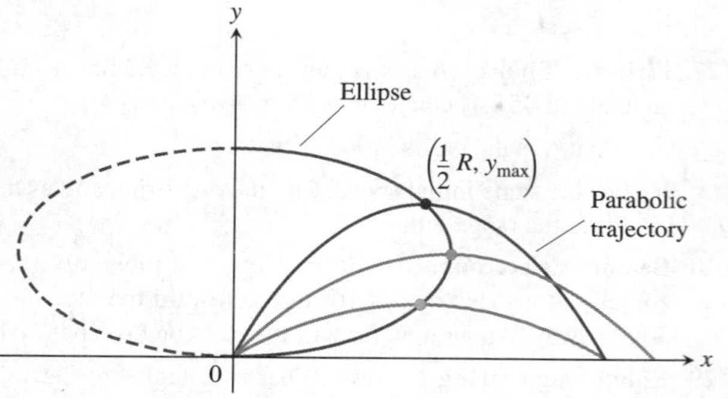

35. Launching downhill An ideal projectile is launched straight down an inclined plane as shown in the accompanying figure.

a. Show that the greatest downhill range is achieved when the initial velocity vector bisects angle *AOR*.

b. If the projectile were fired uphill instead of down, what launch angle would maximize its range? Give reasons for your answer.

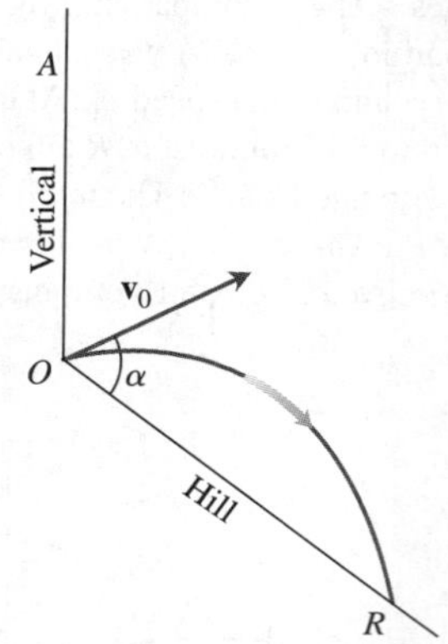

36. Elevated green A golf ball is hit with an initial speed of 116 ft/sec at an angle of elevation of 45° from the tee to a green that is elevated 45 ft above the tee as shown in the diagram. Assuming that the pin, 369 ft downrange, does not get in the way, where will the ball land in relation to the pin?

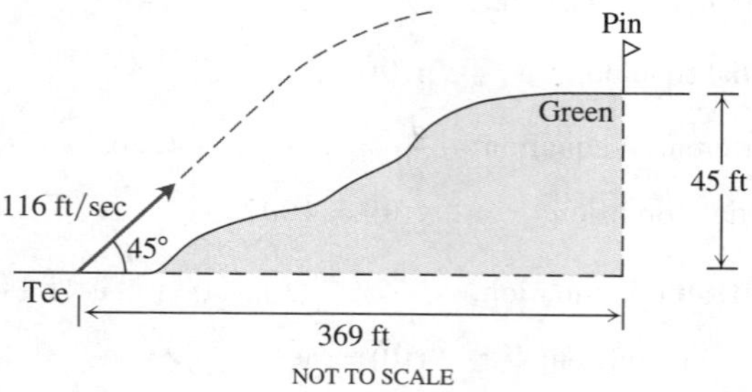

37. Volleyball A volleyball is hit when it is 4 ft above the ground and 12 ft from a 6-ft-high net. It leaves the point of impact with an initial velocity of 35 ft/sec at an angle of 27° and slips by the opposing team untouched.

a. Find a vector equation for the path of the volleyball.

b. How high does the volleyball go, and when does it reach maximum height?

c. Find its range and flight time.

d. When is the volleyball 7 ft above the ground? How far (ground distance) is the volleyball from where it will land?

e. Suppose that the net is raised to 8 ft. Does this change things? Explain.

38. Shot put In Moscow in 1987, Natalya Lisovskaya set a women's world record by putting an 8 lb 13 oz shot 73 ft 10 in. Assuming that she launched the shot at a 40° angle to the horizontal from 6.5 ft above the ground, what was the shot's initial speed?

39. A child throws a ball with an initial speed of 60 ft/sec at an angle of elevation of 60° toward a tall building that is 25 ft from the child. If the child's hand is 5 ft from the ground, show that the ball hits the building, and find the height above the ground of the point where the ball hits the building.

40. Hitting a baseball under a wind gust A baseball is hit when it is 2.5 ft above the ground. It leaves the bat with an initial velocity of 145 ft/sec at a launch angle of 23°. At the instant the ball is hit, an instantaneous gust of wind blows against the ball, adding a component of $-14\mathbf{i}$ (ft/sec) to the ball's initial velocity. A 15-ft-high fence lies 300 ft from home plate in the direction of the flight.

a. Find a vector equation for the path of the baseball.

b. How high does the baseball go, and when does it reach maximum height?

c. Find the range and flight time of the baseball, assuming that the ball is not caught.

d. When is the baseball 20 ft high? How far (ground distance) is the baseball from home plate at that height?

e. Has the batter hit a home run? Explain.

Projectile Motion with Linear Drag

The main force affecting the motion of a projectile, other than gravity, is air resistance. This slowing down force is **drag force**, and it acts in a direction *opposite* to the velocity of the projectile (see accompanying figure). For projectiles moving through the air at relatively low speeds, however, the drag force is (very nearly) proportional to the speed (to the first power) and so is called **linear**.

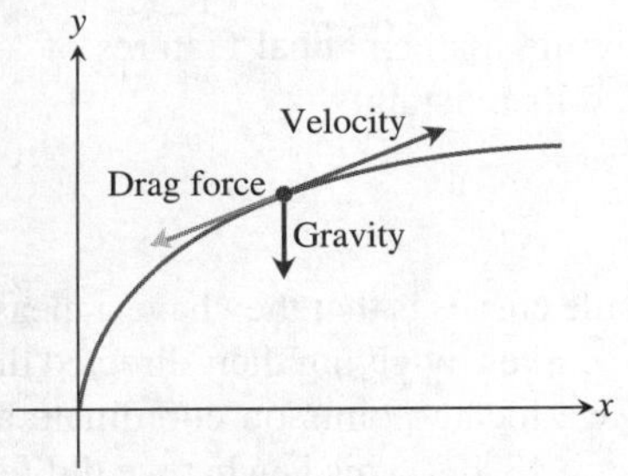

41. Linear drag Derive the equations

$$x = \frac{v_0}{k}(1 - e^{-kt})\cos\alpha$$

$$y = \frac{v_0}{k}(1 - e^{-kt})(\sin\alpha) + \frac{g}{k^2}(1 - kt - e^{-kt})$$

by solving the following initial value problem for a vector $\mathbf{r}$ in the plane.

Differential equation: $\frac{d^2\mathbf{r}}{dt^2} = -g\mathbf{j} - k\mathbf{v} = -g\mathbf{j} - k\frac{d\mathbf{r}}{dt}$

Initial conditions: $\mathbf{r}(0) = \mathbf{0}$

$$\left.\frac{d\mathbf{r}}{dt}\right|_{t=0} = \mathbf{v}_0 = (v_0\cos\alpha)\mathbf{i} + (v_0\sin\alpha)\mathbf{j}$$

The **drag coefficient** k is a positive constant representing resistance due to air density, v_0 and α are the projectile's initial speed and launch angle, and g is the acceleration of gravity.

42. Hitting a baseball with linear drag Consider the baseball problem in Example 5 when there is linear drag (see Exercise 41). Assume a drag coefficient $k = 0.12$, but no gust of wind.

a. From Exercise 41, find a vector form for the path of the baseball.

b. How high does the baseball go, and when does it reach maximum height?

c. Find the range and flight time of the baseball.

d. When is the baseball 30 ft high? How far (ground distance) is the baseball from home plate at that height?

e. A 10-ft-high outfield fence is 340 ft from home plate in the direction of the flight of the baseball. The outfielder can jump and catch any ball up to 11 ft off the ground to stop it from going over the fence. Has the batter hit a home run?

Theory and Examples

43. Establish the following properties of integrable vector functions.

a. The *Constant Scalar Multiple Rule:*

$$\int_a^b k\mathbf{r}(t)\,dt = k\int_a^b \mathbf{r}(t)\,dt \quad (\text{any scalar } k)$$

The *Rule for Negatives,*

$$\int_a^b (-\mathbf{r}(t))\,dt = -\int_a^b \mathbf{r}(t)\,dt,$$

is obtained by taking $k = -1$.

b. The *Sum and Difference Rules:*

$$\int_a^b (\mathbf{r}_1(t) \pm \mathbf{r}_2(t))\,dt = \int_a^b \mathbf{r}_1(t)\,dt \pm \int_a^b \mathbf{r}_2(t)\,dt$$

c. The *Constant Vector Multiple Rules:*

$$\int_a^b \mathbf{C}\cdot\mathbf{r}(t)\,dt = \mathbf{C}\cdot\int_a^b \mathbf{r}(t)\,dt \quad (\text{any constant vector } \mathbf{C})$$

and

$$\int_a^b \mathbf{C}\times\mathbf{r}(t)\,dt = \mathbf{C}\times\int_a^b \mathbf{r}(t)\,dt \quad (\text{any constant vector } \mathbf{C})$$

44. Products of scalar and vector functions Suppose that the scalar function $u(t)$ and the vector function $\mathbf{r}(t)$ are both defined for $a \le t \le b$.

a. Show that $u\mathbf{r}$ is continuous on $[a, b]$ if u and $\mathbf{r}$ are continuous on $[a, b]$.

b. If u and $\mathbf{r}$ are both differentiable on $[a, b]$, show that $u\mathbf{r}$ is differentiable on $[a, b]$ and that

$$\frac{d}{dt}(u\mathbf{r}) = u\frac{d\mathbf{r}}{dt} + \frac{du}{dt}\mathbf{r}.$$

45. Antiderivatives of vector functions

a. Use Corollary 2 of the Mean Value Theorem for scalar functions to show that if two vector functions $\mathbf{R}_1(t)$ and $\mathbf{R}_2(t)$ have identical derivatives on an interval I, then the functions differ by a constant vector value throughout I.

b. Use the result in part (a) to show that if $\mathbf{R}(t)$ is any antiderivative of $\mathbf{r}(t)$ on I, then any other antiderivative of $\mathbf{r}$ on I equals $\mathbf{R}(t) + \mathbf{C}$ for some constant vector $\mathbf{C}$.

46. The Fundamental Theorem of Calculus The Fundamental Theorem of Calculus for scalar functions of a real variable holds for vector functions of a real variable as well. Prove this by using the theorem for scalar functions to show first that if a vector function $\mathbf{r}(t)$ is continuous for $a \le t \le b$, then

$$\frac{d}{dt}\int_a^t \mathbf{r}(\tau)\,d\tau = \mathbf{r}(t)$$

at every point t of (a, b). Then use the conclusion in part (b) of Exercise 45 to show that if $\mathbf{R}$ is any antiderivative of $\mathbf{r}$ on $[a, b]$, then

$$\int_a^b \mathbf{r}(t)\,dt = \mathbf{R}(b) - \mathbf{R}(a).$$

47. Hitting a baseball with linear drag under a wind gust Consider again the baseball problem in Example 5. This time assume a drag coefficient of 0.08 *and* an instantaneous gust of wind that adds a component of $-17.6\mathbf{i}$ (ft/sec) to the initial velocity at the instant the baseball is hit.

a. Find a vector equation for the path of the baseball.

b. How high does the baseball go, and when does it reach maximum height?

c. Find the range and flight time of the baseball.

d. When is the baseball 35 ft high? How far (ground distance) is the baseball from home plate at that height?

e. A 20-ft-high outfield fence is 380 ft from home plate in the direction of the flight of the baseball. Has the batter hit a home run? If "yes," what change in the horizontal component of the ball's initial velocity would have kept the ball in the park? If "no," what change would have allowed it to be a home run?

48. Height versus time Show that a projectile attains three-quarters of its maximum height in half the time it takes to reach the maximum height.

13.3 Arc Length in Space

In this and the next two sections, we study the mathematical features of a curve's shape that describe the sharpness of its turning and its twisting.

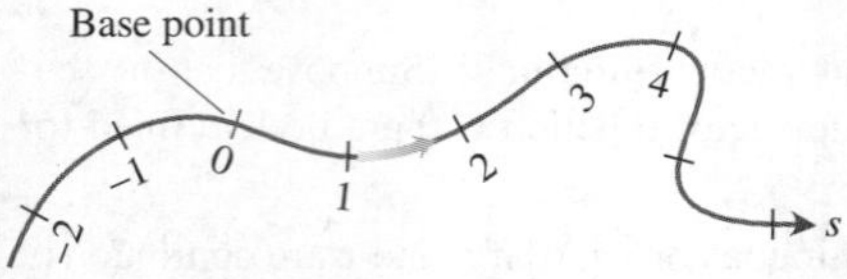

FIGURE 13.12 Smooth curves can be scaled like number lines, the coordinate of each point being its directed distance along the curve from a preselected base point.

Arc Length Along a Space Curve

One of the features of smooth space and plane curves is that they have a measurable length. This enables us to locate points along these curves by giving their directed distance s along the curve from some base point, the way we locate points on coordinate axes by giving their directed distance from the origin (Figure 13.12). This is what we did for plane curves in Section 11.2.

To measure distance along a smooth curve in space, we add a z-term to the formula we use for curves in the plane.

DEFINITION The **length** of a smooth curve $\mathbf{r}(t) = x(t)\mathbf{i} + y(t)\mathbf{j} + z(t)\mathbf{k}$, $a \le t \le b$, that is traced exactly once as t increases from $t = a$ to $t = b$ is

$$L = \int_a^b \sqrt{\left(\frac{dx}{dt}\right)^2 + \left(\frac{dy}{dt}\right)^2 + \left(\frac{dz}{dt}\right)^2}\, dt. \tag{1}$$

Just as for plane curves, we can calculate the length of a curve in space from any convenient parametrization that meets the stated conditions. We omit the proof.

The square root in Equation (1) is $|\mathbf{v}|$, the length of a velocity vector $d\mathbf{r}/dt$. This enables us to write the formula for length a shorter way.

Arc Length Formula

$$L = \int_a^b |\mathbf{v}|\, dt \tag{2}$$

EXAMPLE 1 A glider is soaring upward along the helix

$$\mathbf{r}(t) = (\cos t)\mathbf{i} + (\sin t)\mathbf{j} + t\mathbf{k}.$$

How long is the glider's path from $t = 0$ to $t = 2\pi$?

Solution The path segment during this time corresponds to one full turn of the helix (Figure 13.13). The length of this portion of the curve is

$$\begin{aligned} L = \int_a^b |\mathbf{v}|\, dt &= \int_0^{2\pi} \sqrt{(-\sin t)^2 + (\cos t)^2 + (1)^2}\, dt \\ &= \int_0^{2\pi} \sqrt{2}\, dt = 2\pi\sqrt{2} \text{ units of length.} \end{aligned}$$

This is $\sqrt{2}$ times the circumference of the circle in the xy-plane over which the helix stands. ■

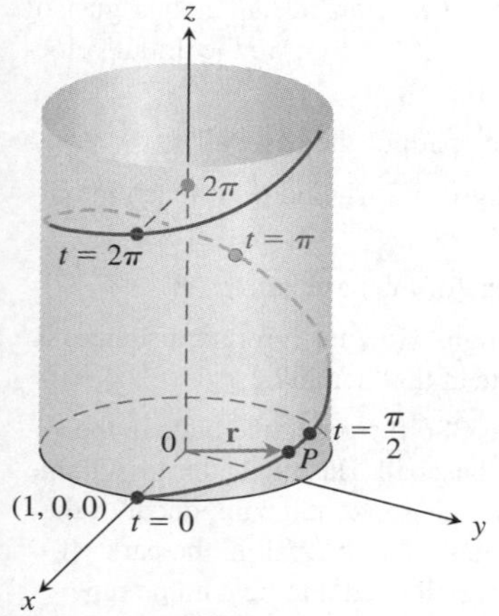

FIGURE 13.13 The helix in Example 1, $\mathbf{r}(t) = (\cos t)\mathbf{i} + (\sin t)\mathbf{j} + t\mathbf{k}$.

If we choose a base point $P(t_0)$ on a smooth curve C parametrized by t, each value of t determines a point $P(t) = (x(t), y(t), z(t))$ on C and a "directed distance"

$$s(t) = \int_{t_0}^{t} |\mathbf{v}(\tau)|\, d\tau,$$

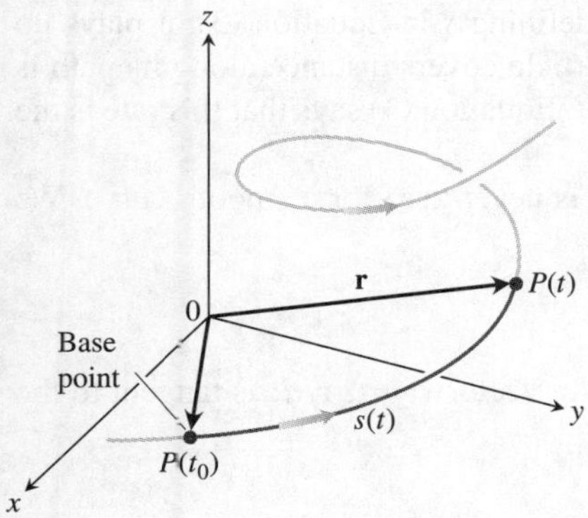

FIGURE 13.14 The directed distance along the curve from $P(t_0)$ to any point $P(t)$ is $s(t) = \int_{t_0}^{t} |\mathbf{v}(\tau)|\, d\tau$.

measured along C from the base point (Figure 13.14). This is the arc length function we defined in Section 11.2 for plane curves that have no z-component. If $t > t_0$, $s(t)$ is the distance along the curve from $P(t_0)$ to $P(t)$. If $t < t_0$, $s(t)$ is the negative of the distance. Each value of s determines a point on C, and this parametrizes C with respect to s. We call s an **arc length parameter** for the curve. The parameter's value increases in the direction of increasing t. We will see that the arc length parameter is particularly effective for investigating the turning and twisting nature of a space curve.

Arc Length Parameter with Base Point $P(t_0)$

$$s(t) = \int_{t_0}^{t} \sqrt{[x'(\tau)]^2 + [y'(\tau)]^2 + [z'(\tau)]^2}\, d\tau = \int_{t_0}^{t} |\mathbf{v}(\tau)|\, d\tau \qquad (3)$$

We use the Greek letter τ ("tau") as the variable of integration in Equation (3) because the letter t is already in use as the upper limit.

τ is the Greek letter tau (rhymes with "now")

If a curve $\mathbf{r}(t)$ is already given in terms of some parameter t, and $s(t)$ is the arc length function given by Equation (3), then we may be able to solve for t as a function of s: $t = t(s)$. Then the curve can be reparametrized in terms of s by substituting for t: $\mathbf{r} = \mathbf{r}(t(s))$. The new parametrization identifies a point on the curve with its directed distance along the curve from the base point.

EXAMPLE 2 This is an example for which we can actually find the arc length parametrization of a curve. If $t_0 = 0$, then the arc length parameter along the helix

$$\mathbf{r}(t) = (\cos t)\mathbf{i} + (\sin t)\mathbf{j} + t\mathbf{k}$$

from t_0 to t is

$$\begin{aligned} s(t) &= \int_{t_0}^{t} |\mathbf{v}(\tau)|\, d\tau && \text{Eq. (3)} \\ &= \int_{0}^{t} \sqrt{2}\, d\tau && \text{Value from Example 1} \\ &= \sqrt{2}\, t. \end{aligned}$$

Solving this equation for t gives $t = s/\sqrt{2}$. Substituting into the position vector $\mathbf{r}$ gives the following arc length parametrization for the helix:

$$\mathbf{r}(t(s)) = \left(\cos\frac{s}{\sqrt{2}}\right)\mathbf{i} + \left(\sin\frac{s}{\sqrt{2}}\right)\mathbf{j} + \frac{s}{\sqrt{2}}\mathbf{k}. \qquad \blacksquare$$

Unlike the case that appears in Example 2, the arc length parametrization is generally difficult to find analytically for a curve already given in terms of some other parameter t. Fortunately, however, we rarely need an exact formula for $s(t)$ or its inverse $t(s)$.

Speed on a Smooth Curve

HISTORICAL BIOGRAPHY
Josiah Willard Gibbs
(1839–1903)
www.bit.ly/2xZS8pg

Since the derivatives beneath the radical in Equation (3) are continuous (the curve is smooth), the Fundamental Theorem of Calculus tells us that s is a differentiable function of t with derivative

$$\frac{ds}{dt} = |\mathbf{v}(t)|. \qquad (4)$$

Although the base point $P(t_0)$ plays a role in defining s in Equation (3), it plays no role in Equation (4). The rate at which a moving particle covers distance along its path is independent of how far away it is from the base point. Equation (4) says that this rate is the magnitude of **v**.

Notice that $ds/dt > 0$ since, by definition, $|\mathbf{v}|$ is never zero for a smooth curve. We see once again that s is an increasing function of t.

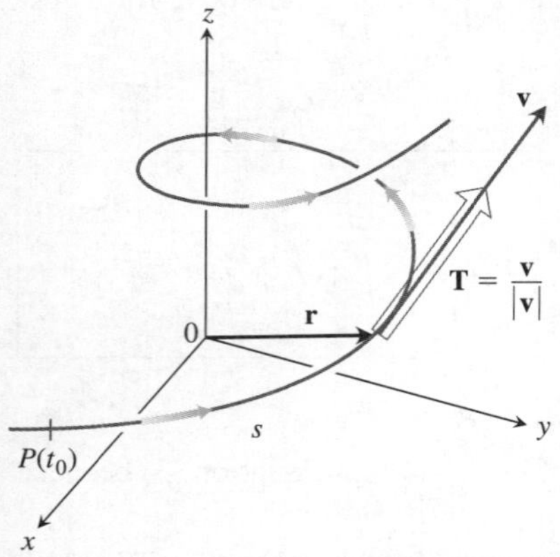

FIGURE 13.15 We find the unit tangent vector **T** by dividing **v** by its length $|\mathbf{v}|$.

Unit Tangent Vector

On a smooth curve, we already know that the velocity vector $\mathbf{v} = d\mathbf{r}/dt$ is tangent to the curve $\mathbf{r}(t)$ and that the vector

$$\mathbf{T} = \frac{\mathbf{v}}{|\mathbf{v}|}$$

is therefore a unit vector tangent to the curve, called the **unit tangent vector** (Figure 13.15). The unit tangent vector **T** for a smooth curve is a differentiable function of t whenever **v** is a differentiable function of t. As we will see in Section 13.5, **T** is one of three unit vectors in a traveling reference frame that is used to describe the motion of objects traveling in three dimensions.

EXAMPLE 3 Find the unit tangent vector of the curve

$$\mathbf{r}(t) = (1 + 3\cos t)\mathbf{i} + (3\sin t)\mathbf{j} + t^2\mathbf{k}$$

representing the path of the glider in Example 3, Section 13.2.

Solution In that example, we found

$$\mathbf{v} = \frac{d\mathbf{r}}{dt} = -(3\sin t)\mathbf{i} + (3\cos t)\mathbf{j} + 2t\mathbf{k}$$

and

$$|\mathbf{v}| = \sqrt{9 + 4t^2}.$$

Thus,

$$\mathbf{T} = \frac{\mathbf{v}}{|\mathbf{v}|} = -\frac{3\sin t}{\sqrt{9+4t^2}}\mathbf{i} + \frac{3\cos t}{\sqrt{9+4t^2}}\mathbf{j} + \frac{2t}{\sqrt{9+4t^2}}\mathbf{k}. \quad \blacksquare$$

For the counterclockwise motion

$$\mathbf{r}(t) = (\cos t)\mathbf{i} + (\sin t)\mathbf{j}$$

around the unit circle, we see that

$$\mathbf{v} = (-\sin t)\mathbf{i} + (\cos t)\mathbf{j}$$

is already a unit vector, so $\mathbf{T} = \mathbf{v}$ and **T** is orthogonal to **r** (Figure 13.16).

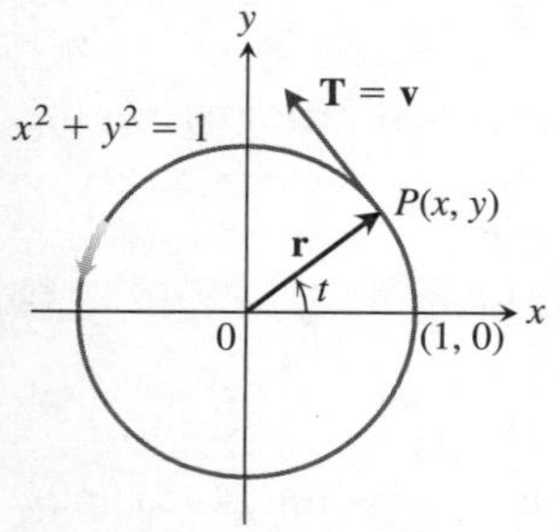

FIGURE 13.16 Counterclockwise motion around the unit circle.

The velocity vector is the change in the position vector **r** with respect to time t, but how does the position vector change with respect to arc length? More precisely, what is the derivative $d\mathbf{r}/ds$? Since $ds/dt > 0$ for the curves we are considering, s is one-to-one and has an inverse that gives t as a differentiable function of s (Section 7.1). The derivative of the inverse is

$$\frac{dt}{ds} = \frac{1}{ds/dt} = \frac{1}{|\mathbf{v}|}.$$

This makes **r** a differentiable function of s whose derivative can be calculated with the Chain Rule to be

$$\frac{d\mathbf{r}}{ds} = \frac{d\mathbf{r}}{dt}\frac{dt}{ds} = \mathbf{v}\frac{1}{|\mathbf{v}|} = \frac{\mathbf{v}}{|\mathbf{v}|} = \mathbf{T}. \tag{5}$$

This equation says that $d\mathbf{r}/ds$ is the unit tangent vector in the direction of the velocity vector **v** (Figure 13.15).

EXERCISES 13.3

Finding Tangent Vectors and Lengths

In Exercises 1–8, find the curve's unit tangent vector. Also, find the length of the indicated portion of the curve.

1. $\mathbf{r}(t) = (1 + 2\cos t)\mathbf{i} + (2\sin t)\mathbf{j} + \sqrt{3}\,t\mathbf{k}, \quad 0 \le t \le \pi$

2. $\mathbf{r}(t) = (6\sin 2t)\mathbf{i} + (6\cos 2t)\mathbf{j} + 5t\mathbf{k}, \quad 0 \le t \le \pi$

3. $\mathbf{r}(t) = 2t\mathbf{i} + (2/5)t^{3/2}\mathbf{k}, \quad 0 \le t \le 5$

4. $\mathbf{r}(t) = (4 + t)\mathbf{i} - (t + 3)\mathbf{j} + t^2\mathbf{k}, \quad 0 \le t \le 4$

5. $\mathbf{r}(t) = (\cos^3 t)\mathbf{j} + (\sin^3 t)\mathbf{k}, \quad 0 \le t \le \pi/2$

6. $\mathbf{r}(t) = 5t^3\mathbf{i} + 3t^3\mathbf{j} + t^3\mathbf{k}, \quad 1 \le t \le 3$

7. $\mathbf{r}(t) = (t\cos t)\mathbf{i} + (t\sin t)\mathbf{j} + \left(2\sqrt{2}/3\right)t^{3/2}\mathbf{k}, \quad 0 \le t \le \pi$

8. $\mathbf{r}(t) = (t\sin t + \cos t)\mathbf{i} + (t\cos t - \sin t)\mathbf{j}, \quad \sqrt{2} \le t \le 2$

9. Find the point on the curve

$$\mathbf{r}(t) = 12t\mathbf{i} + 5\sin t\mathbf{j} + 5\cos t\mathbf{k}$$

at a distance 13π units along the curve from the point $(0, 0, 5)$ in the direction of increasing arc length.

10. Find the point on the curve

$$\mathbf{r}(t) = (12\sin t)\mathbf{i} - (12\cos t)\mathbf{j} + 5t\mathbf{k}$$

at a distance 13π units along the curve from the point $(0, -12, 0)$ in the direction corresponding to decreasing t values.

Arc Length Parameter

In Exercises 11–14, find the arc length parameter along the curve from the point where $t = 0$ by evaluating the integral

$$s(t) = \int_0^t |\mathbf{v}(\tau)|\,d\tau$$

from Equation (3). Then use the formula for $s(t)$ to find the length of the indicated portion of the curve.

11. $\mathbf{r}(t) = (\cos t)\mathbf{i} + (\sin t)\mathbf{j} + t\mathbf{k}, \quad 0 \le t \le \pi$

12. $\mathbf{r}(t) = (\cos t + t\sin t)\mathbf{i} + (\sin t - t\cos t)\mathbf{j}, \quad \pi/2 \le t \le \pi$

13. $\mathbf{r}(t) = (e^t\cos t)\mathbf{i} + (e^t\sin t)\mathbf{j} + e^t\,\mathbf{k}, \quad -\ln 4 \le t \le 0$

14. $\mathbf{r}(t) = (1 - 3t)\mathbf{i} + (1 - 2t)\mathbf{j} + 6t\mathbf{k}, \quad 0 \le t \le 1$

Theory and Examples

15. Arc length Find the length of the curve

$$\mathbf{r}(t) = \left(\sqrt{2}t\right)\mathbf{i} + \left(\sqrt{2}t\right)\mathbf{j} + (1 - t^2)\mathbf{k}$$

from $(0, 0, 1)$ to $\left(\sqrt{2}, \sqrt{2}, 0\right)$.

16. Length of helix The length $2\pi\sqrt{2}$ of the turn of the helix in Example 1 is also the length of the diagonal of a square 2π units on a side. Show how to obtain this square by cutting away and flattening a portion of the cylinder around which the helix winds.

17. Ellipse

a. Show that the curve $\mathbf{r}(t) = (\cos t)\mathbf{i} + (\sin t)\mathbf{j} + (1 - \cos t)\mathbf{k}$, $0 \le t \le 2\pi$, is an ellipse by showing that it is the intersection of a right circular cylinder and a plane. Find equations for the cylinder and plane.

b. Sketch the ellipse on the cylinder. Add to your sketch the unit tangent vectors at $t = 0, \pi/2, \pi$, and $3\pi/2$.

c. Show that the acceleration vector always lies parallel to the plane (orthogonal to a vector normal to the plane). Thus, if you draw the acceleration as a vector attached to the ellipse, it will lie in the plane of the ellipse. Add the acceleration vectors for $t = 0, \pi/2, \pi$, and $3\pi/2$ to your sketch.

d. Write an integral for the length of the ellipse. Do not try to evaluate the integral; it is nonelementary.

T **e. Numerical integrator** Estimate the length of the ellipse to two decimal places.

18. Length is independent of parametrization To illustrate that the length of a smooth space curve does not depend on the parametrization you use to compute it, calculate the length of one turn of the helix in Example 1 with the following parametrizations.

a. $\mathbf{r}(t) = (\cos 4t)\mathbf{i} + (\sin 4t)\mathbf{j} + 4t\mathbf{k}, \quad 0 \le t \le \pi/2$

b. $\mathbf{r}(t) = [\cos(t/2)]\mathbf{i} + [\sin(t/2)]\mathbf{j} + (t/2)\mathbf{k}, \quad 0 \le t \le 4\pi$

c. $\mathbf{r}(t) = (\cos t)\mathbf{i} - (\sin t)\mathbf{j} - t\mathbf{k}, \quad -2\pi \le t \le 0$

19. The involute of a circle If a string wound around a fixed circle is unwound while held taut in the plane of the circle, its end P traces an *involute* of the circle. In the accompanying figure, the circle in question is the circle $x^2 + y^2 = 1$ and the tracing point starts at $(1, 0)$. The unwound portion of the string is tangent to the circle at Q, and t is the radian measure of the angle from the positive x-axis to segment OQ. Derive the parametric equations

$$x = \cos t + t\sin t, \quad y = \sin t - t\cos t, \quad t > 0$$

of the point $P(x, y)$ for the involute.

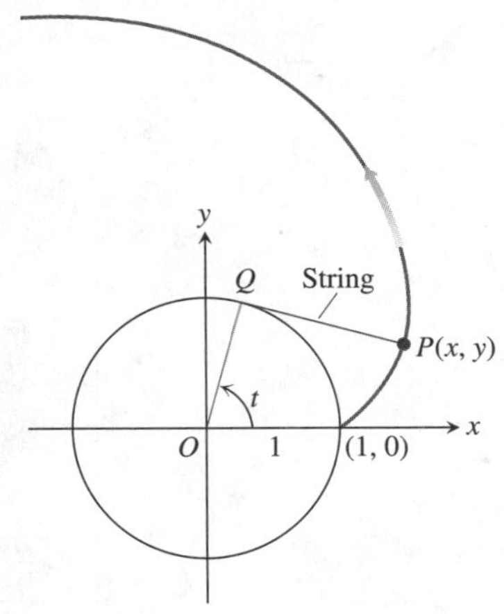

20. (*Continuation of Exercise 19.*) Find the unit tangent vector to the involute of the circle at the point $P(x, y)$.

21. Distance along a line Show that if $\mathbf{u}$ is a unit vector, then the arc length parameter along the line $\mathbf{r}(t) = P_0 + t\mathbf{u}$ from the point $P_0(x_0, y_0, z_0)$ where $t = 0$, is t itself.

22. Use Simpson's Rule with $n = 10$ to approximate the length of arc of $\mathbf{r}(t) = t\mathbf{i} + t^2\mathbf{j} + t^3\mathbf{k}$ from the origin to the point $(2, 4, 8)$.

13.4 Curvature and Normal Vectors of a Curve

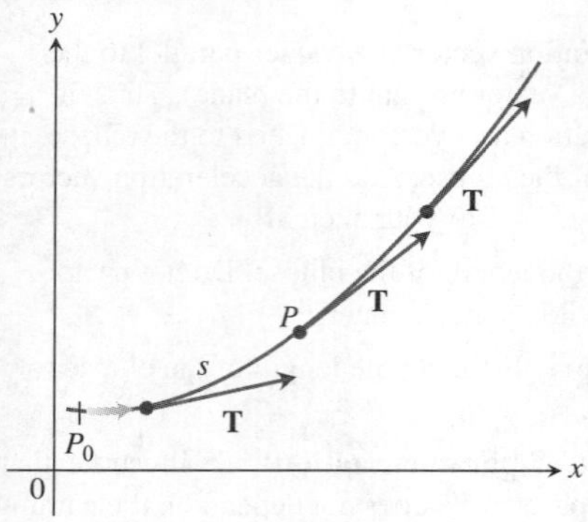

FIGURE 13.17 As P moves along the curve in the direction of increasing arc length, the unit tangent vector turns. The value of $|d\mathbf{T}/ds|$ at P is called the *curvature* of the curve at P.

In this section we study how a curve turns or bends. To gain perspective, we look first at curves in the coordinate plane. Then we consider curves in space.

Curvature of a Plane Curve

As a particle moves along a smooth curve in the plane, $\mathbf{T} = d\mathbf{r}/ds$ turns as the curve bends. Since $\mathbf{T}$ is a unit vector, its length remains constant and only its direction changes as the particle moves along the curve. The rate at which $\mathbf{T}$ turns per unit of length along the curve is called the *curvature* (Figure 13.17). The traditional symbol for the curvature function is the Greek letter κ ("kappa").

DEFINITION If $\mathbf{T}$ is the unit tangent vector of a smooth curve in the plane, then the **curvature** function of the curve is

$$\kappa = \left|\frac{d\mathbf{T}}{ds}\right|.$$

If $|d\mathbf{T}/ds|$ is large, $\mathbf{T}$ turns sharply as the particle passes through P, and the curvature at P is large. If $|d\mathbf{T}/ds|$ is close to zero, $\mathbf{T}$ turns more slowly, and the curvature at P is smaller.

If a smooth curve $\mathbf{r}(t)$ is already given in terms of some parameter t other than the arc length parameter s, we can calculate the curvature as

κ is the Greek letter kappa.

$$\begin{aligned}\kappa &= \left|\frac{d\mathbf{T}}{ds}\right| = \left|\frac{d\mathbf{T}}{dt}\frac{dt}{ds}\right| && \text{Chain Rule}\\ &= \frac{1}{|ds/dt|}\left|\frac{d\mathbf{T}}{dt}\right| \\ &= \frac{1}{|\mathbf{v}|}\left|\frac{d\mathbf{T}}{dt}\right|. && \frac{ds}{dt} = |\mathbf{v}|\end{aligned}$$

Formula for Calculating Curvature

If $\mathbf{r}(t)$ is a smooth curve in the plane, then the curvature is the scalar function

$$\kappa = \frac{1}{|\mathbf{v}|}\left|\frac{d\mathbf{T}}{dt}\right|, \qquad (1)$$

where $\mathbf{T} = \mathbf{v}/|\mathbf{v}|$ is the unit tangent vector.

Testing the definition, we see in Examples 1 and 2 below that the curvature is constant for straight lines and circles.

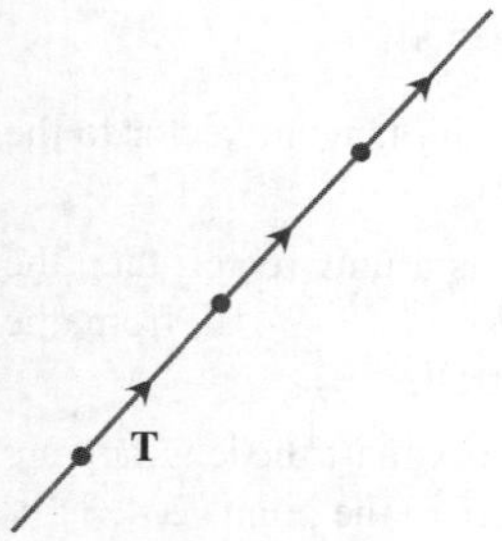

FIGURE 13.18 Along a straight line, $\mathbf{T}$ always points in the same direction. The curvature, $|d\mathbf{T}/ds|$, is zero (Example 1).

EXAMPLE 1 A straight line is parametrized by $\mathbf{r}(t) = \mathbf{C} + t\mathbf{v}$ for constant vectors $\mathbf{C}$ and $\mathbf{v}$. Thus, $\mathbf{r}'(t) = \mathbf{v}$, and the unit tangent vector $\mathbf{T} = \mathbf{v}/|\mathbf{v}|$ is a constant vector that always points in the same direction and has derivative $\mathbf{0}$ (Figure 13.18). It follows that, for any value of the parameter t, the curvature of the straight line is zero:

$$\kappa = \frac{1}{|\mathbf{v}|}\left|\frac{d\mathbf{T}}{dt}\right| = \frac{1}{|\mathbf{v}|}|\mathbf{0}| = 0. \quad \blacksquare$$

EXAMPLE 2 Here we find the curvature of a circle. We begin with the parametrization

$$\mathbf{r}(t) = (a\cos t)\mathbf{i} + (a\sin t)\mathbf{j}$$

of a circle of radius a. Then

$$\mathbf{v} = \frac{d\mathbf{r}}{dt} = -(a\sin t)\mathbf{i} + (a\cos t)\mathbf{j}$$

$$|\mathbf{v}| = \sqrt{(-a\sin t)^2 + (a\cos t)^2} = \sqrt{a^2} = |a| = a. \qquad \text{Since } a > 0,\ |a| = a.$$

From this we find

$$\mathbf{T} = \frac{\mathbf{v}}{|\mathbf{v}|} = -(\sin t)\mathbf{i} + (\cos t)\mathbf{j}$$

$$\frac{d\mathbf{T}}{dt} = -(\cos t)\mathbf{i} - (\sin t)\mathbf{j}$$

$$\left|\frac{d\mathbf{T}}{dt}\right| = \sqrt{\cos^2 t + \sin^2 t} = 1.$$

Hence, for any value of the parameter t, the curvature of the circle is

$$\kappa = \frac{1}{|\mathbf{v}|}\left|\frac{d\mathbf{T}}{dt}\right| = \frac{1}{a}(1) = \frac{1}{a} = \frac{1}{\text{radius}}.$$ ■

Among the vectors orthogonal to the unit tangent vector $\mathbf{T}$, there is one of particular significance because it points in the direction in which the curve is turning. Since $\mathbf{T}$ has constant length (because its length is always 1), the derivative $d\mathbf{T}/ds$ is orthogonal to $\mathbf{T}$ (Equation 4, Section 13.1). Therefore, if we divide $d\mathbf{T}/ds$ by its length κ, we obtain a *unit* vector $\mathbf{N}$ orthogonal to $\mathbf{T}$ (Figure 13.19).

DEFINITION At a point where $\kappa \neq 0$, the **principal unit normal** vector for a smooth curve in the plane is

$$\mathbf{N} = \frac{1}{\kappa}\frac{d\mathbf{T}}{ds}.$$

The vector $d\mathbf{T}/ds$ points in the direction in which $\mathbf{T}$ turns as the curve bends. Therefore, if we face in the direction of increasing arc length, the vector $d\mathbf{T}/ds$ points toward the right if $\mathbf{T}$ turns clockwise and toward the left if $\mathbf{T}$ turns counterclockwise. In other words, the principal normal vector $\mathbf{N}$ will point toward the concave side of the curve (Figure 13.19).

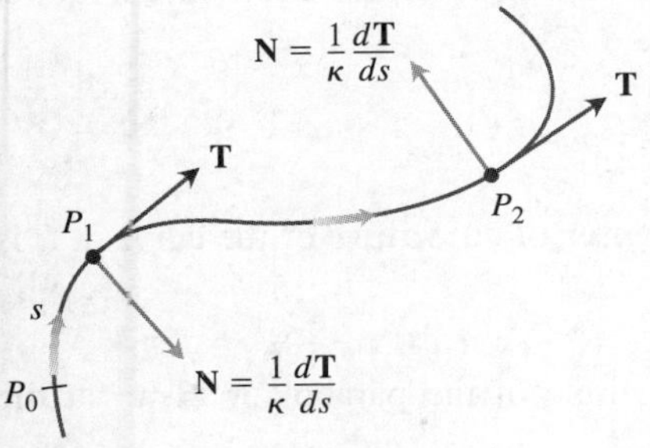

FIGURE 13.19 The vector $d\mathbf{T}/ds$, normal to the curve, always points in the direction in which $\mathbf{T}$ is turning. The unit normal vector $\mathbf{N}$ is the direction of $d\mathbf{T}/ds$.

If a smooth curve $\mathbf{r}(t)$ is already given in terms of some parameter t other than the arc length parameter s, we can use the Chain Rule to calculate $\mathbf{N}$ directly:

$$\mathbf{N} = \frac{d\mathbf{T}/ds}{|d\mathbf{T}/ds|}$$

$$= \frac{(d\mathbf{T}/dt)(dt/ds)}{|d\mathbf{T}/dt||dt/ds|}$$

$$= \frac{d\mathbf{T}/dt}{|d\mathbf{T}/dt|}. \qquad \frac{dt}{ds} = \frac{1}{ds/dt} > 0 \text{ cancels.}$$

This formula enables us to find $\mathbf{N}$ without having to find κ and s first.

Formula for Calculating N

If $\mathbf{r}(t)$ is a smooth curve in the plane, then the principal unit normal is

$$\mathbf{N} = \frac{d\mathbf{T}/dt}{|d\mathbf{T}/dt|}, \qquad (2)$$

where $\mathbf{T} = \mathbf{v}/|\mathbf{v}|$ is the unit tangent vector.

EXAMPLE 3 Find $\mathbf{T}$ and $\mathbf{N}$ for the circular motion

$$\mathbf{r}(t) = (\cos 2t)\mathbf{i} + (\sin 2t)\mathbf{j}.$$

Solution We first find $\mathbf{T}$:

$$\mathbf{v} = -(2\sin 2t)\mathbf{i} + (2\cos 2t)\mathbf{j}$$

$$|\mathbf{v}| = \sqrt{4\sin^2 2t + 4\cos^2 2t} = 2$$

$$\mathbf{T} = \frac{\mathbf{v}}{|\mathbf{v}|} = -(\sin 2t)\mathbf{i} + (\cos 2t)\mathbf{j}.$$

From this we find

$$\frac{d\mathbf{T}}{dt} = -(2\cos 2t)\mathbf{i} - (2\sin 2t)\mathbf{j}$$

$$\left|\frac{d\mathbf{T}}{dt}\right| = \sqrt{4\cos^2 2t + 4\sin^2 2t} = 2$$

and

$$\mathbf{N} = \frac{d\mathbf{T}/dt}{|d\mathbf{T}/dt|} = -(\cos 2t)\mathbf{i} - (\sin 2t)\mathbf{j}. \qquad \text{Eq. (2)}$$

Notice that $\mathbf{T} \cdot \mathbf{N} = 0$, verifying that $\mathbf{N}$ is orthogonal to $\mathbf{T}$. Notice too, that for the circular motion here, $\mathbf{N}$ points from $\mathbf{r}(t)$ toward the circle's center at the origin. ■

Circle of Curvature for Plane Curves

The **circle of curvature** or **osculating circle** at a point P on a plane curve where $\kappa \neq 0$ is the circle in the plane of the curve that

1. is tangent to the curve at P (has the same tangent line the curve has)
2. has the same curvature the curve has at P
3. has center that lies toward the concave or inner side of the curve (as in Figure 13.20).

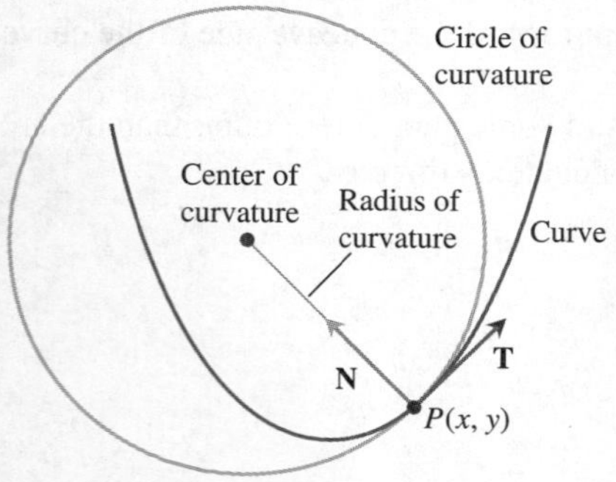

FIGURE 13.20 The center of the osculating circle at $P(x, y)$ lies toward the inner side of the curve.

The **radius of curvature** of the curve at P is the radius of the circle of curvature, which, according to Example 2, is

$$\text{Radius of curvature} = \rho = \frac{1}{\kappa}.$$

To find ρ, we find κ and take the reciprocal. The **center of curvature** of the curve at P is the center of the circle of curvature.

EXAMPLE 4 Find and graph the osculating circle of the parabola $y = x^2$ at the origin.

Solution We parametrize the parabola using the parameter $t = x$ (Section 11.1, Example 5):

$$\mathbf{r}(t) = t\mathbf{i} + t^2\mathbf{j}.$$

First we find the curvature of the parabola at the origin, using Equation (1):

$$\mathbf{v} = \frac{d\mathbf{r}}{dt} = \mathbf{i} + 2t\mathbf{j}$$
$$|\mathbf{v}| = \sqrt{1 + 4t^2}$$

so that

$$\mathbf{T} = \frac{\mathbf{v}}{|\mathbf{v}|} = (1 + 4t^2)^{-1/2}\mathbf{i} + 2t(1 + 4t^2)^{-1/2}\mathbf{j}.$$

From this we find

$$\frac{d\mathbf{T}}{dt} = -4t(1 + 4t^2)^{-3/2}\mathbf{i} + \left[2(1 + 4t^2)^{-1/2} - 8t^2(1 + 4t^2)^{-3/2}\right]\mathbf{j}.$$

At the origin, $t = 0$, so the curvature is

$$\begin{aligned}\kappa(0) &= \frac{1}{|\mathbf{v}(0)|}\left|\frac{d\mathbf{T}}{dt}(0)\right| && \text{Eq. (1)}\\ &= \frac{1}{\sqrt{1}}|0\mathbf{i} + 2\mathbf{j}|\\ &= (1)\sqrt{0^2 + 2^2} = 2.\end{aligned}$$

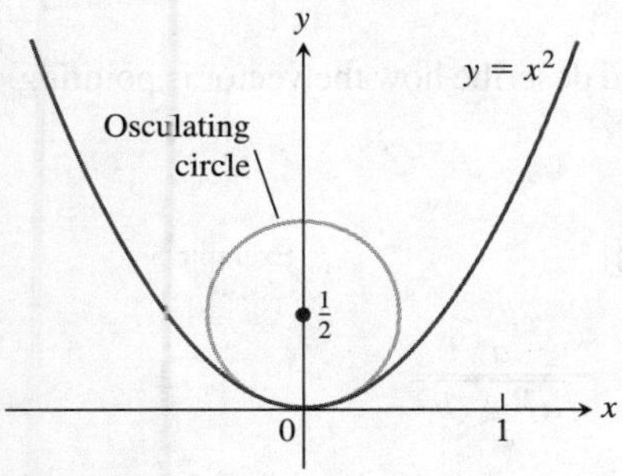

FIGURE 13.21 The osculating circle for the parabola $y = x^2$ at the origin (Example 4).

Therefore, the radius of curvature is $1/\kappa = 1/2$. At the origin we have $t = 0$ and $\mathbf{T} = \mathbf{i}$, so $\mathbf{N} = \mathbf{j}$. Thus the center of the circle is $(0, 1/2)$. The equation of the osculating circle is

$$(x - 0)^2 + \left(y - \frac{1}{2}\right)^2 = \left(\frac{1}{2}\right)^2.$$

You can see from Figure 13.21 that the osculating circle is a better approximation to the parabola at the origin than is the tangent line approximation $y = 0$. ■

Curvature and Normal Vectors for Space Curves

If a smooth curve in space is specified by the position vector $\mathbf{r}(t)$ as a function of some parameter t, and if s is the arc length parameter of the curve, then the unit tangent vector $\mathbf{T}$ is $d\mathbf{r}/ds = \mathbf{v}/|\mathbf{v}|$. The **curvature** in space is then defined to be

$$\kappa = \left|\frac{d\mathbf{T}}{ds}\right| = \frac{1}{|\mathbf{v}|}\left|\frac{d\mathbf{T}}{dt}\right| \qquad (3)$$

just as for plane curves. The vector $d\mathbf{T}/ds$ is orthogonal to $\mathbf{T}$, and we define the **principal unit normal** to be

$$\mathbf{N} = \frac{1}{\kappa}\frac{d\mathbf{T}}{ds} = \frac{d\mathbf{T}/dt}{|d\mathbf{T}/dt|}. \qquad (4)$$

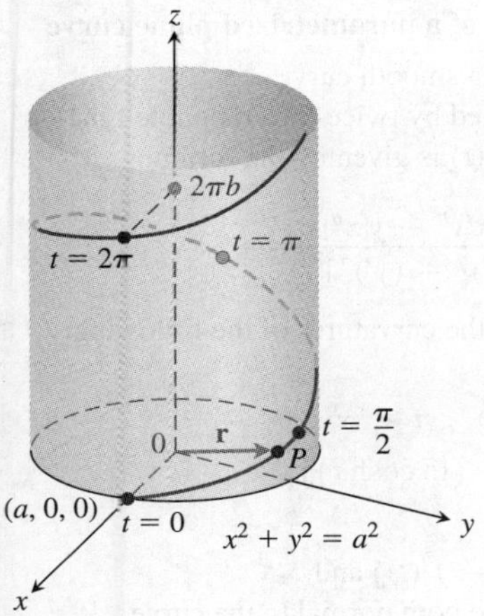

FIGURE 13.22 The helix

$$\mathbf{r}(t) = (a\cos t)\mathbf{i} + (a\sin t)\mathbf{j} + bt\mathbf{k},$$

drawn with a and b positive and $t \geq 0$ (Example 5).

EXAMPLE 5 Find the curvature for the helix (Figure 13.22)

$$\mathbf{r}(t) = (a\cos t)\mathbf{i} + (a\sin t)\mathbf{j} + bt\mathbf{k}, \qquad a, b \geq 0, \qquad a^2 + b^2 \neq 0.$$

Solution We calculate $\mathbf{T}$ from the velocity vector $\mathbf{v}$:

$$\begin{aligned}\mathbf{v} &= -(a\sin t)\mathbf{i} + (a\cos t)\mathbf{j} + b\mathbf{k}\\ |\mathbf{v}| &= \sqrt{a^2\sin^2 t + a^2\cos^2 t + b^2} = \sqrt{a^2 + b^2}\\ \mathbf{T} &= \frac{\mathbf{v}}{|\mathbf{v}|} = \frac{1}{\sqrt{a^2 + b^2}}[-(a\sin t)\mathbf{i} + (a\cos t)\mathbf{j} + b\mathbf{k}].\end{aligned}$$

Then we use Equation (3):

$$\begin{aligned}\kappa &= \frac{1}{|\mathbf{v}|}\left|\frac{d\mathbf{T}}{dt}\right| \\ &= \frac{1}{\sqrt{a^2+b^2}}\left|\frac{1}{\sqrt{a^2+b^2}}[-(a\cos t)\mathbf{i}-(a\sin t)\mathbf{j}]\right| \\ &= \frac{a}{a^2+b^2}\left|-(\cos t)\mathbf{i}-(\sin t)\mathbf{j}\right| \\ &= \frac{a}{a^2+b^2}\sqrt{(\cos t)^2+(\sin t)^2} = \frac{a}{a^2+b^2}.\end{aligned}$$

From this equation, we see that increasing b for a fixed a decreases the curvature. Decreasing a for a fixed b eventually decreases the curvature as well.

If $b = 0$, the helix reduces to a circle of radius a, and its curvature reduces to $1/a$, as it should. If $a = 0$, the helix becomes the z-axis, and its curvature reduces to 0, again as it should. ■

EXAMPLE 6 Find $\mathbf{N}$ for the helix in Example 5 and describe how the vector is pointing.

Solution We have

$$\frac{d\mathbf{T}}{dt} = -\frac{1}{\sqrt{a^2+b^2}}[(a\cos t)\mathbf{i}+(a\sin t)\mathbf{j}] \qquad \text{Example 5}$$

$$\left|\frac{d\mathbf{T}}{dt}\right| = \frac{1}{\sqrt{a^2+b^2}}\sqrt{a^2\cos^2 t+a^2\sin^2 t} = \frac{a}{\sqrt{a^2+b^2}}$$

$$\begin{aligned}\mathbf{N} &= \frac{d\mathbf{T}/dt}{|d\mathbf{T}/dt|} \qquad \text{Eq. (4)} \\ &= -\frac{\sqrt{a^2+b^2}}{a}\cdot\frac{1}{\sqrt{a^2+b^2}}[(a\cos t)\mathbf{i}+(a\sin t)\mathbf{j}] \\ &= -(\cos t)\mathbf{i}-(\sin t)\mathbf{j}.\end{aligned}$$

Thus, $\mathbf{N}$ is parallel to the xy-plane and always points toward the z-axis. ■

EXERCISES 13.4

Plane Curves

Find $\mathbf{T}$, $\mathbf{N}$, and κ for the plane curves in Exercises 1–4.

1. $\mathbf{r}(t) = t\mathbf{i} + \ln(\sin t)\mathbf{j} \quad -\pi < t < \pi$
2. $\mathbf{r}(t) = \ln(\cos t)\mathbf{i} + t\mathbf{j} \quad -\pi < t < \pi$
3. $r(t) = (3t+5)\mathbf{i} + (2-t^2)\mathbf{j}$
4. $\mathbf{r}(t) = (\cos t + t\sin t)\mathbf{i} + (\sin t - t\cos t)\mathbf{j}, \quad t > 0$
5. **A formula for the curvature of the graph of a function in the xy-plane**
   a. The graph $y = f(x)$ in the xy-plane automatically has the parametrization $x = x$, $y = f(x)$, and the vector formula $\mathbf{r}(x) = x\mathbf{i} + f(x)\mathbf{j}$. Use this formula to show that if f is a twice-differentiable function of x, then
   $$\kappa(x) = \frac{|f''(x)|}{\left[1+(f'(x))^2\right]^{3/2}}.$$
   b. Use the formula for κ in part (a) to find the curvature of $y = \ln(\cos x)$, $-\pi/2 < x < \pi/2$. Compare your answer with the answer in Exercise 1.
   c. Show that the curvature is zero at a point of inflection.
6. **A formula for the curvature of a parametrized plane curve**
   a. Show that the curvature of a smooth curve $\mathbf{r}(t) = f(t)\mathbf{i} + g(t)\mathbf{j}$ defined by twice-differentiable functions $x = f(t)$ and $y = g(t)$ is given by the formula
   $$\kappa = \frac{|x'y'' - y'x''|}{[(x')^2+(y')^2]^{3/2}}.$$
   Apply this formula to find the curvatures of the following curves.
   b. $\mathbf{r}(t) = t\mathbf{i} + (\ln\sin t)\mathbf{j}, \quad 0 < t < \pi$
   c. $\mathbf{r}(t) = [\arctan(\sinh t)]\mathbf{i} + (\ln\cosh t)\mathbf{j}$
7. **Normals to plane curves**
   a. Show that $\mathbf{n}(t) = -g'(t)\mathbf{i} + f'(t)\mathbf{j}$ and $-\mathbf{n}(t) = g'(t)\mathbf{i} - f'(t)\mathbf{j}$ are both normal to the curve $\mathbf{r}(t) = f(t)\mathbf{i} + g(t)\mathbf{j}$ at the point $(f(t), g(t))$.

   To obtain $\mathbf{N}$ for a particular plane curve, we can choose the one of $\mathbf{n}$ or $-\mathbf{n}$ from part (a) that points toward the concave side of the curve, and make it into a unit vector. (See Figure 13.19.) Apply this method to find $\mathbf{N}$ for the following curves.

b. $\mathbf{r}(t) = t\mathbf{i} + e^{2t}\mathbf{j}$

c. $\mathbf{r}(t) = \sqrt{4 - t^2}\,\mathbf{i} + t\mathbf{j}, \quad -2 \le t \le 2$

8. (*Continuation of Exercise 7*)

a. Use the method of Exercise 7 to find **N** for the curve $\mathbf{r}(t) = t\mathbf{i} + (1/3)t^3\mathbf{j}$ when $t < 0$; when $t > 0$.

b. Calculate **N** for $t \ne 0$ directly from **T** using Equation (4) for the curve in part (a). Does **N** exist at $t = 0$? Graph the curve and explain what is happening to **N** as t passes from negative to positive values.

Space Curves

Find **T**, **N**, and κ for the space curves in Exercises 9–16.

9. $\mathbf{r}(t) = (3\sin t)\mathbf{i} + (3\cos t)\mathbf{j} + 4t\mathbf{k}$

10. $\mathbf{r}(t) = (\sin t - t\cos t)\mathbf{i} + (\cos t + t\sin t)\mathbf{j} - 4\mathbf{k}$

11. $\mathbf{r}(t) = (2e^t \sin t)\mathbf{i} + (2\,e^t \cos t)\mathbf{j} + 3\mathbf{k}$

12. $\mathbf{r}(t) = (2\sin 4t)\mathbf{i} + (2\cos 4t)\mathbf{j} + 3t\mathbf{k}$

13. $\mathbf{r}(t) = \left(\dfrac{2t^3}{3}\right)\mathbf{i} + \left(\dfrac{3t^2}{2}\right)\mathbf{j}, \quad t > 0$

14. $\mathbf{r}(t) = (\cos^3 t)\mathbf{j} + (\sin^3 t)\mathbf{k}, \quad 0 < t < \pi/2$

15. $\mathbf{r}(t) = t\mathbf{i} + (a\cosh(t/a))\mathbf{k}, \quad a > 0$

16. $\mathbf{r}(t) = (\cosh t)\mathbf{i} - (\sinh t)\mathbf{j} + t\mathbf{k}$

More on Curvature

17. Show that the parabola $y = ax^2$, $a \ne 0$, has its largest curvature at its vertex and has no minimum curvature. (*Note:* Since the curvature of a curve remains the same if the curve is translated or rotated, this result is true for any parabola.)

18. Show that the ellipse $x = a\cos t$, $y = b\sin t$, $a > b > 0$, has its largest curvature on its major axis and its smallest curvature on its minor axis. (The same is true for any ellipse.)

19. Maximizing the curvature of a helix In Example 5, we found the curvature of the helix $\mathbf{r}(t) = (a\cos t)\mathbf{i} + (a\sin t)\mathbf{j} + bt\mathbf{k}$ $(a, b \ge 0)$ to be $\kappa = a/(a^2 + b^2)$. What is the largest value κ can have for a given value of b? Give reasons for your answer.

20. Total curvature We find the **total curvature** of the portion of a smooth curve that runs from $s = s_0$ to $s = s_1 > s_0$ by integrating κ from s_0 to s_1. If the curve has some other parameter, say t, then the total curvature is

$$K = \int_{s_0}^{s_1} \kappa\, ds = \int_{t_0}^{t_1} \kappa \frac{ds}{dt}\, dt = \int_{t_0}^{t_1} \kappa |\mathbf{v}|\, dt,$$

where t_0 and t_1 correspond to s_0 and s_1. Find the total curvatures of

a. The portion of the helix $\mathbf{r}(t) = (3\cos t)\mathbf{i} + (3\sin t)\mathbf{j} + t\mathbf{k}$, $0 \le t \le 4\pi$.

b. The parabola $y = x^2$, $-\infty < x < \infty$.

21. Find an equation for the circle of curvature of the curve $\mathbf{r}(t) = t\mathbf{i} + (\sin t)\mathbf{j}$ at the point $(\pi/2, 1)$. (The curve parametrizes the graph of $y = \sin x$ in the xy-plane.)

22. Find an equation for the circle of curvature of the curve $\mathbf{r}(t) = (2\ln t)\mathbf{i} - [t + (1/t)]\mathbf{j}, e \quad \le t \le e$, at the point $(0, -2)$, where $t = 1$.

T The formula

$$\kappa(x) = \frac{|f''(x)|}{\left[1 + (f'(x))^2\right]^{3/2}},$$

derived in Exercise 5, expresses the curvature $\kappa(x)$ of a twice-differentiable plane curve $y = f(x)$ as a function of x. Find the curvature function of each of the curves in Exercises 23–26. Then graph $f(x)$ together with $\kappa(x)$ over the given interval. You will find some surprises.

23. $y = 2x^2, \quad -2 \le x \le 2$

24. $y = x^4/4, \quad -2 \le x \le 2$

25. $y = 2\sin x, \quad 0 \le x \le 2\pi$

26. $y = e^x, \quad -1 \le x \le 2$

In Exercises 27 and 28, determine the maximum curvature for the graph of each function.

27. $f(x) = \ln x$

28. $f(x) = \dfrac{x}{x+1}$ for $x > -1$

29. Osculating circle Show that the center of the osculating circle for the parabola $y = x^2$ at the point (a, a^2) is located at $\left(-4a^3, 3a^2 + \dfrac{1}{2}\right)$.

30. Osculating circle Find a parametrization of the osculating circle for the parabola $y = x^2$ when $x = 1$.

COMPUTER EXPLORATIONS

In Exercises 31–38 you will use a CAS to explore the osculating circle at a point P on a plane curve where $\kappa \ne 0$. Use a CAS to perform the following steps:

a. Plot the plane curve given in parametric or function form over the specified interval to see what it looks like.

b. Calculate the curvature κ of the curve at the given value t_0 using the appropriate formula from Exercise 5 or 6. Use the parametrization $x = t$ and $y = f(t)$ if the curve is given as a function $y = f(x)$.

c. Find the unit normal vector **N** at t_0. Notice that the signs of the components of **N** depend on whether the unit tangent vector **T** is turning clockwise or counterclockwise at $t = t_0$. (See Exercise 7.)

d. If $\mathbf{C} = a\mathbf{i} + b\mathbf{j}$ is the vector from the origin to the center (a, b) of the osculating circle, find the center **C** from the vector equation

$$\mathbf{C} = \mathbf{r}(t_0) + \frac{1}{\kappa(t_0)}\mathbf{N}(t_0).$$

The point $P(x_0, y_0)$ on the curve is given by the position vector $\mathbf{r}(t_0)$.

e. Plot implicitly the equation $(x - a)^2 + (y - b)^2 = 1/\kappa^2$ of the osculating circle. Then plot the curve and osculating circle together. You may need to experiment with the size of the viewing window, but be sure the axes are equally scaled.

31. $\mathbf{r}(t) = (3\cos t)\mathbf{i} + (5\sin t)\mathbf{j}, \quad 0 \le t \le 2\pi, \quad t_0 = \pi/4$

32. $\mathbf{r}(t) = (\cos^3 t)\mathbf{i} + (\sin^3 t)\mathbf{j}, \quad 0 \le t \le 2\pi, \quad t_0 = \pi/4$

33. $\mathbf{r}(t) = t^2\mathbf{i} + (t^3 - 3t)\mathbf{j}, \quad -4 \le t \le 4, \quad t_0 = 3/5$

34. $\mathbf{r}(t) = (t^3 - 2t^2 - t)\mathbf{i} + \dfrac{3t}{\sqrt{1 + t^2}}\mathbf{j}, \quad -2 \le t \le 5, \quad t_0 = 1$

35. $\mathbf{r}(t) = (2t - \sin t)\mathbf{i} + (2 - 2\cos t)\mathbf{j}, \quad 0 \le t \le 3\pi$, $t_0 = 3\pi/2$

36. $\mathbf{r}(t) = (e^{-t}\cos t)\mathbf{i} + (e^{-t}\sin t)\mathbf{j}, \quad 0 \le t \le 6\pi, \quad t_0 = \pi/4$

37. $y = x^2 - x, \quad -2 \le x \le 5, \quad x_0 = 1$

38. $y = x(1 - x)^{2/5}, \quad -1 \le x \le 2, \quad x_0 = 1/2$

13.5 Tangential and Normal Components of Acceleration

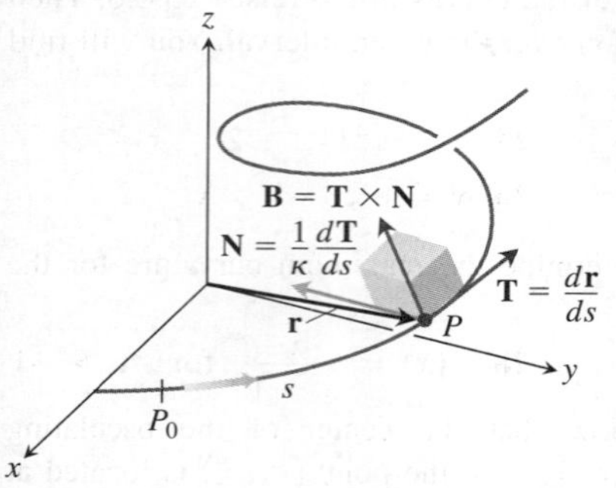

FIGURE 13.23 The **TNB** frame of mutually orthogonal unit vectors traveling along a curve in space.

If you are flying in an airplane that is traveling along a curve in space, the Cartesian **i**, **j**, and **k** coordinate system for representing the vectors describing your motion may not be very relevant to you. Vectors that are likely to be more important are those representing your forward direction (the unit tangent vector **T**) and the direction in which your path is turning (the unit normal vector **N**), along with a third unit vector perpendicular to the other two. Expressing the acceleration vector along the curve as a linear combination of these three mutually orthogonal unit vectors traveling with the motion (Figure 13.23) can reveal much about the nature of your path and your motion along it.

The TNB Frame

The **binormal vector** of a curve in space is $\mathbf{B} = \mathbf{T} \times \mathbf{N}$, which is a unit vector that is orthogonal to both **T** and **N** (Figure 13.24). Together **T**, **N**, and **B** define a moving right-handed vector frame that plays a significant role in analyzing the paths of particles moving through space. It is called the **Frenet** ("fre-*nay*") **frame** (after Jean-Frédéric Frenet, 1816–1900), or the **TNB frame**.

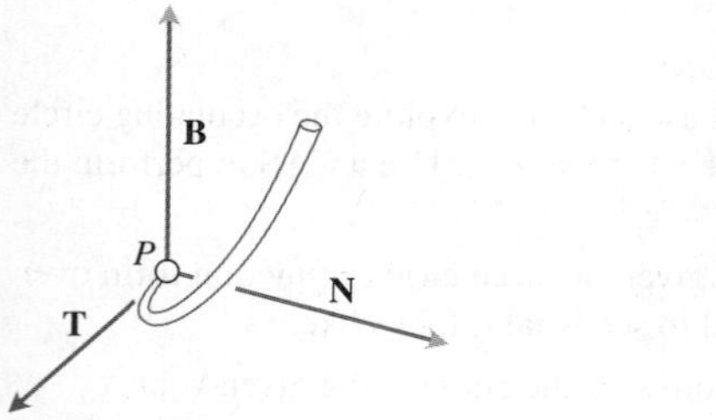

FIGURE 13.24 The vectors **T**, **N**, and **B** (in that order) make a right-handed frame of mutually orthogonal unit vectors in space.

Tangential and Normal Components of Acceleration

When an object is accelerated by gravity, brakes, or rocket motors, we often need to know how much of the acceleration acts in the direction of motion, which is the direction of the tangent vector **T**. We can calculate this using the Chain Rule to rewrite **v** as

$$\mathbf{v} = \frac{d\mathbf{r}}{dt} = \frac{d\mathbf{r}}{ds}\frac{ds}{dt} = \mathbf{T}\frac{ds}{dt}.$$

Then we differentiate both ends of this string of equalities to get

$$\begin{aligned}\mathbf{a} &= \frac{d\mathbf{v}}{dt} = \frac{d}{dt}\left(\mathbf{T}\frac{ds}{dt}\right) = \frac{d^2s}{dt^2}\mathbf{T} + \frac{ds}{dt}\frac{d\mathbf{T}}{dt}\\ &= \frac{d^2s}{dt^2}\mathbf{T} + \frac{ds}{dt}\left(\frac{d\mathbf{T}}{ds}\frac{ds}{dt}\right) = \frac{d^2s}{dt^2}\mathbf{T} + \frac{ds}{dt}\left(\kappa\mathbf{N}\frac{ds}{dt}\right) \qquad \frac{d\mathbf{T}}{ds} = \kappa\mathbf{N}\\ &= \frac{d^2s}{dt^2}\mathbf{T} + \kappa\left(\frac{ds}{dt}\right)^2\mathbf{N}.\end{aligned}$$

DEFINITION If the acceleration vector is written as

$$\mathbf{a} = a_{\mathrm{T}}\mathbf{T} + a_{\mathrm{N}}\mathbf{N}, \tag{1}$$

then

$$a_{\mathrm{T}} = \frac{d^2s}{dt^2} = \frac{d}{dt}|\mathbf{v}| \quad \text{and} \quad a_{\mathrm{N}} = \kappa\left(\frac{ds}{dt}\right)^2 = \kappa|\mathbf{v}|^2 \tag{2}$$

are the **tangential** and the **normal** scalar components of acceleration.

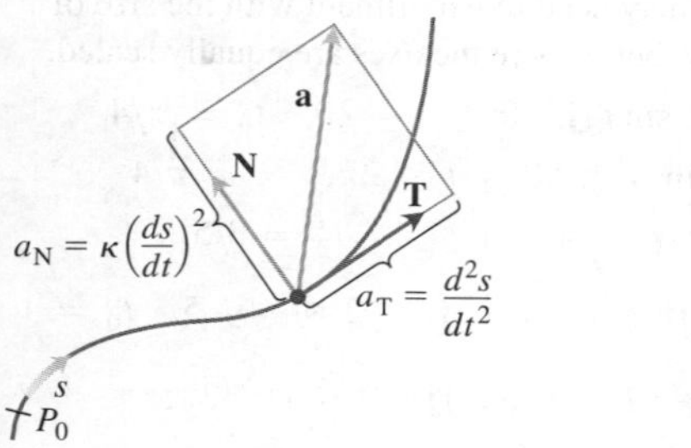

FIGURE 13.25 The tangential and normal components of acceleration. The acceleration **a** always lies in the plane of **T** and **N** and is orthogonal to **B**.

Notice that the binormal vector **B** does not appear in Equation (1). No matter how the path of the moving object we are watching may appear to twist and turn in space, the acceleration **a** *always lies in the plane of* **T** and **N** and therefore is orthogonal to **B**. The equation also tells us exactly how much of the acceleration takes place tangent to the motion (d^2s/dt^2) and how much takes place normal to the motion $\left[\kappa(ds/dt)^2\right]$ (Figure 13.25).

What information can we discover from Equations (2)? By definition, acceleration **a** is the rate of change of velocity **v**, and in general, both the length and direction of **v** change as an object moves along its path. The tangential component of acceleration a_{T} measures

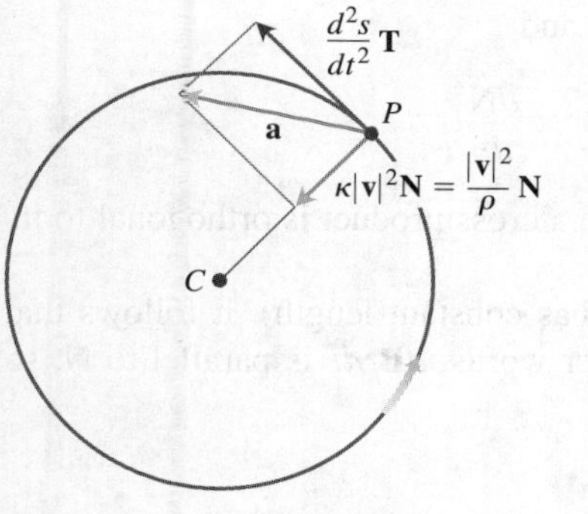

FIGURE 13.26 The tangential and normal components of the acceleration of an object that is speeding up as it moves counterclockwise around a circle of radius ρ.

the rate of change of the *length* of **v** (that is, the change in the speed). The normal component of acceleration a_N is proportional to the rate of change of the *direction* of **v**.

Notice that the normal scalar component of the acceleration is the curvature times the *square* of the speed. This explains why you have to hold on when your car makes a sharp (large κ), high-speed (large $|\mathbf{v}|$) turn. If you double the speed of your car, you will experience four times the normal component of acceleration for the same curvature.

If an object moves in a circle at a constant speed, d^2s/dt^2 is zero and all the acceleration points along **N** toward the circle's center. If the object is speeding up or slowing down, **a** has a positive or negative tangential component (Figure 13.26).

To calculate a_N, we usually use the formula $a_N = \sqrt{|\mathbf{a}|^2 - a_T^2}$, which comes from solving the equation $|\mathbf{a}|^2 = \mathbf{a}\cdot\mathbf{a} = a_T^2 + a_N^2$ for a_N (which, unlike the tangential component, cannot be negative). With this formula, we can find a_N without having to calculate κ first.

Formula for Calculating the Normal Component of Acceleration

$$a_N = \sqrt{|\mathbf{a}|^2 - a_T^2} \qquad (3)$$

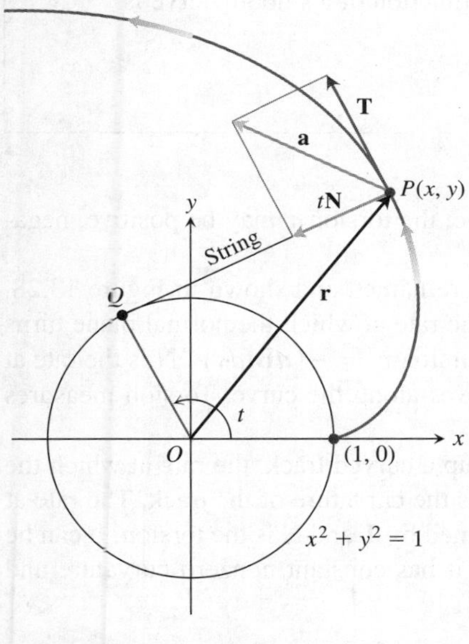

FIGURE 13.27 The tangential and normal components of the acceleration of the motion $\mathbf{r}(t) = (\cos t + t\sin t)\mathbf{i} + (\sin t - t\cos t)\mathbf{j}$, for $t > 0$. If a string wound around a fixed circle is unwound while held taut in the plane of the circle, its end P traces an involute of the circle (Example 1).

EXAMPLE 1 Without finding **T** and **N**, write the acceleration of the motion

$$\mathbf{r}(t) = (\cos t + t\sin t)\mathbf{i} + (\sin t - t\cos t)\mathbf{j}, \qquad t > 0$$

in the form $\mathbf{a} = a_T\mathbf{T} + a_N\mathbf{N}$. (The path of the motion is the involute of the circle in Figure 13.27. See also Section 13.3, Exercise 19.)

Solution We use the first of Equations (2) to find a_T:

$$\mathbf{v} = \frac{d\mathbf{r}}{dt} = (-\sin t + \sin t + t\cos t)\mathbf{i} + (\cos t - \cos t + t\sin t)\mathbf{j}$$
$$= (t\cos t)\mathbf{i} + (t\sin t)\mathbf{j}$$
$$|\mathbf{v}| = \sqrt{t^2\cos^2 t + t^2\sin^2 t} = \sqrt{t^2} = |t| = t \qquad t > 0$$
$$a_T = \frac{d}{dt}|\mathbf{v}| = \frac{d}{dt}(t) = 1. \qquad \text{Eq. (2)}$$

Knowing a_T, we use Equation (3) to find a_N:

$$\mathbf{a} = (\cos t - t\sin t)\mathbf{i} + (\sin t + t\cos t)\mathbf{j}$$
$$|\mathbf{a}|^2 = t^2 + 1 \qquad \text{After some algebra}$$
$$a_N = \sqrt{|\mathbf{a}|^2 - a_T^2}$$
$$= \sqrt{(t^2 + 1) - (1)} = \sqrt{t^2} = t.$$

We then use Equation (1) to write

$$\mathbf{a} = a_T\mathbf{T} + a_N\mathbf{N} = (1)\mathbf{T} + (t)\mathbf{N} = \mathbf{T} + t\mathbf{N}.$$ ■

Torsion

How does $d\mathbf{B}/ds$ behave in relation to **T**, **N**, and **B**? From the rule for differentiating a cross product in Section 13.1, we have

$$\frac{d\mathbf{B}}{ds} = \frac{d(\mathbf{T}\times\mathbf{N})}{ds} = \frac{d\mathbf{T}}{ds}\times\mathbf{N} + \mathbf{T}\times\frac{d\mathbf{N}}{ds}.$$

Since $\mathbf{N}$ is the direction of $d\mathbf{T}/ds$, $(d\mathbf{T}/ds) \times \mathbf{N} = \mathbf{0}$ and

$$\frac{d\mathbf{B}}{ds} = \mathbf{0} + \mathbf{T} \times \frac{d\mathbf{N}}{ds} = \mathbf{T} \times \frac{d\mathbf{N}}{ds}.$$

From this we see that $d\mathbf{B}/ds$ is orthogonal to $\mathbf{T}$, since a cross product is orthogonal to its factors.

Since $d\mathbf{B}/ds$ is also orthogonal to $\mathbf{B}$ (the latter has constant length), it follows that $d\mathbf{B}/ds$ is orthogonal to the plane of $\mathbf{B}$ and $\mathbf{T}$. In other words, $d\mathbf{B}/ds$ is parallel to $\mathbf{N}$, so $d\mathbf{B}/ds$ is a scalar multiple of $\mathbf{N}$. In symbols,

$$\frac{d\mathbf{B}}{ds} = -\tau\mathbf{N}.$$

The negative sign in this equation is traditional. The scalar τ is called the *torsion* along the curve. Notice that

$$\frac{d\mathbf{B}}{ds} \cdot \mathbf{N} = -\tau\mathbf{N} \cdot \mathbf{N} = -\tau(1) = -\tau.$$

We use this equation for our next definition.

DEFINITION Let $\mathbf{B} = \mathbf{T} \times \mathbf{N}$. The **torsion** function of a smooth curve is

$$\tau = -\frac{d\mathbf{B}}{ds} \cdot \mathbf{N}. \tag{4}$$

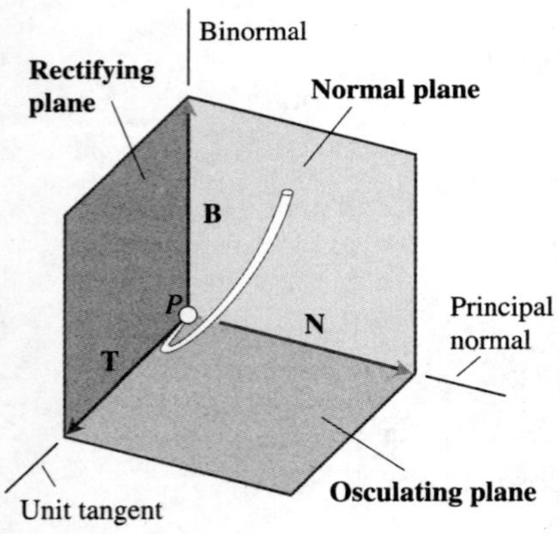

FIGURE 13.28 The names of the three planes determined by **T**, **N**, and **B**.

Unlike the curvature κ, which is never negative, the torsion τ may be positive, negative, or zero.

The three planes determined by $\mathbf{T}$, $\mathbf{N}$, and $\mathbf{B}$ are named and shown in Figure 13.28. The curvature $\kappa = |d\mathbf{T}/ds|$ can be thought of as the rate at which the normal plane turns as the point P moves along its path. Similarly, the torsion $\tau = -(d\mathbf{B}/ds) \cdot \mathbf{N}$ is the rate at which the osculating plane turns about $\mathbf{T}$ as P moves along the curve. Torsion measures how the curve twists.

Look at Figure 13.29. If P is a train climbing up a curved track, the rate at which the headlight turns from side to side per unit distance is the curvature of the track. The rate at which the engine tends to twist out of the plane formed by $\mathbf{T}$ and $\mathbf{N}$ is the torsion. It can be shown that a space curve is a helix if and only if it has constant nonzero curvature and constant nonzero torsion.

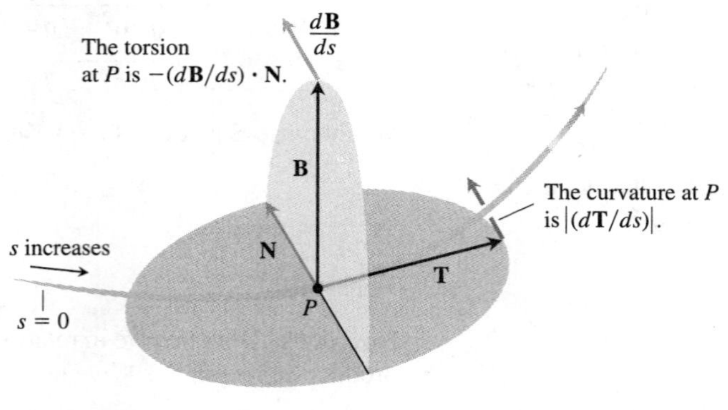

FIGURE 13.29 Every moving body travels with a **TNB** frame that characterizes the geometry of its path of motion.

Formulas for Computing Curvature and Torsion

We now give easy-to-use formulas for computing the curvature and torsion of a smooth curve. From Equations (1) and (2), we have

$$\begin{aligned}\mathbf{v}\times\mathbf{a} &= \left(\frac{ds}{dt}\mathbf{T}\right)\times\left[\frac{d^2s}{dt^2}\mathbf{T}+\kappa\left(\frac{ds}{dt}\right)^2\mathbf{N}\right] && \mathbf{v}=d\mathbf{r}/dt=(ds/dt)\mathbf{T}\\ &= \left(\frac{ds}{dt}\frac{d^2s}{dt^2}\right)(\mathbf{T}\times\mathbf{T})+\kappa\left(\frac{ds}{dt}\right)^3(\mathbf{T}\times\mathbf{N})\\ &= \kappa\left(\frac{ds}{dt}\right)^3\mathbf{B}. && \mathbf{T}\times\mathbf{T}=0 \text{ and } \mathbf{T}\times\mathbf{N}=\mathbf{B}\end{aligned}$$

It follows that

$$|\mathbf{v}\times\mathbf{a}| = \kappa\left|\frac{ds}{dt}\right|^3|\mathbf{B}| = \kappa|\mathbf{v}|^3. \qquad \frac{ds}{dt}=|\mathbf{v}| \text{ and } |\mathbf{B}|=1$$

Solving for κ gives the following formula.

Vector Formula for Curvature

$$\kappa = \frac{|\mathbf{v}\times\mathbf{a}|}{|\mathbf{v}|^3} \tag{5}$$

Equation (5) calculates the curvature, a geometric property of the curve, from the velocity and acceleration of any vector representation of the curve in which $|\mathbf{v}|$ is different from zero. From any formula for motion along a curve, no matter how variable the motion may be (as long as $\mathbf{v}$ is never zero), we can calculate a geometric property of the curve that seems to have nothing to do with the way the curve is parametrically defined.

The most widely used formula for torsion, derived in more advanced texts, is given in a determinant form.

Formula for Torsion

$$\tau = \frac{\begin{vmatrix}\dot{x} & \dot{y} & \dot{z}\\ \ddot{x} & \ddot{y} & \ddot{z}\\ \dddot{x} & \dddot{y} & \dddot{z}\end{vmatrix}}{|\mathbf{v}\times\mathbf{a}|^2} \qquad (\text{if } \mathbf{v}\times\mathbf{a}\neq\mathbf{0}) \tag{6}$$

This formula calculates the torsion directly from the derivatives of the component functions $x = f(t)$, $y = g(t)$, $z = h(t)$ that make up $\mathbf{r}$. The determinant's first row comes from $\mathbf{v}$, the second row comes from $\mathbf{a}$, and the third row comes from $\dot{\mathbf{a}} = d\mathbf{a}/dt$. This formula for torsion is traditionally written using Newton's dot notation for derivatives.

Newton's Dot Notation for Derivatives

The dots in Equation (6) denote differentiation with respect to t, one derivative for each dot. Thus, $\dot{x}$ ("x dot") means dx/dt, $\ddot{x}$ ("x double dot") means d^2x/dt^2, and $\dddot{x}$ ("x triple dot") means d^3x/dt^3. Similarly, $\dot{y} = dy/dt$, and so on.

EXAMPLE 2 Use Equations (5) and (6) to find the curvature κ and torsion τ for the helix

$$\mathbf{r}(t) = (a\cos t)\mathbf{i} + (a\sin t)\mathbf{j} + bt\mathbf{k}, \qquad a, b \geq 0, \qquad a^2+b^2\neq 0.$$

Solution We calculate the curvature with Equation (5):

$$\mathbf{v} = -(a\sin t)\mathbf{i} + (a\cos t)\mathbf{j} + b\mathbf{k}$$
$$\mathbf{a} = -(a\cos t)\mathbf{i} - (a\sin t)\mathbf{j}$$
$$\mathbf{v}\times\mathbf{a} = \begin{vmatrix} \mathbf{i} & \mathbf{j} & \mathbf{k} \\ -a\sin t & a\cos t & b \\ -a\cos t & -a\sin t & 0 \end{vmatrix}$$
$$= (ab\sin t)\mathbf{i} - (ab\cos t)\mathbf{j} + a^2\mathbf{k}$$
$$\kappa = \frac{|\mathbf{v}\times\mathbf{a}|}{|\mathbf{v}|^3} = \frac{\sqrt{a^2b^2+a^4}}{(a^2+b^2)^{3/2}} = \frac{a\sqrt{a^2+b^2}}{(a^2+b^2)^{3/2}} = \frac{a}{a^2+b^2}. \tag{7}$$

Notice that Equation (7) agrees with the result in Example 5 in Section 13.4, where we calculated the curvature directly from its definition.

To evaluate Equation (6) for the torsion, we find the entries in the determinant by differentiating **r** with respect to t. We already have **v** and **a**, and

$$\dot{\mathbf{a}} = \frac{d\mathbf{a}}{dt} = (a\sin t)\mathbf{i} - (a\cos t)\mathbf{j}.$$

Hence,

$$\tau = \frac{\begin{vmatrix} \dot{x} & \dot{y} & \dot{z} \\ \ddot{x} & \ddot{y} & \ddot{z} \\ \dddot{x} & \dddot{y} & \dddot{z} \end{vmatrix}}{|\mathbf{v}\times\mathbf{a}|^2} = \frac{\begin{vmatrix} -a\sin t & a\cos t & b \\ -a\cos t & -a\sin t & 0 \\ a\sin t & -a\cos t & 0 \end{vmatrix}}{\left(a\sqrt{a^2+b^2}\right)^2} \qquad \text{Value of } |\mathbf{v}\times\mathbf{a}| \text{ from Eq. (7)}$$
$$= \frac{b(a^2\cos^2 t + a^2\sin^2 t)}{a^2(a^2+b^2)}$$
$$= \frac{b}{a^2+b^2}.$$

From this last equation we see that the torsion of a helix about a circular cylinder is constant. In fact, constant curvature and constant torsion characterize the helix among all curves in space. ■

Computation Formulas for Curves in Space

Unit tangent vector: $\mathbf{T} = \dfrac{\mathbf{v}}{|\mathbf{v}|}$

Principal unit normal vector: $\mathbf{N} = \dfrac{d\mathbf{T}/dt}{|d\mathbf{T}/dt|}$

Binormal vector: $\mathbf{B} = \mathbf{T}\times\mathbf{N}$

Curvature: $\kappa = \left|\dfrac{d\mathbf{T}}{ds}\right| = \dfrac{|\mathbf{v}\times\mathbf{a}|}{|\mathbf{v}|^3}$

Torsion: $\tau = -\dfrac{d\mathbf{B}}{ds}\cdot\mathbf{N} = \dfrac{\begin{vmatrix} \dot{x} & \dot{y} & \dot{z} \\ \ddot{x} & \ddot{y} & \ddot{z} \\ \dddot{x} & \dddot{y} & \dddot{z} \end{vmatrix}}{|\mathbf{v}\times\mathbf{a}|^2}$

Tangential and normal scalar components of acceleration:

$$\mathbf{a} = a_{\mathrm{T}}\mathbf{T} + a_{\mathrm{N}}\mathbf{N}$$
$$a_{\mathrm{T}} = \frac{d}{dt}|\mathbf{v}|$$
$$a_{\mathrm{N}} = \kappa|\mathbf{v}|^2 = \sqrt{|\mathbf{a}|^2 - a_{\mathrm{T}}^2}$$

EXERCISES 13.5

Finding Tangential and Normal Components

In Exercises 1–4, write $\mathbf{a}$ in the form $\mathbf{a} = a_T\mathbf{T} + a_N\mathbf{N}$ without finding $\mathbf{T}$ and $\mathbf{N}$.

1. $\mathbf{r}(t) = (2t + 3)\mathbf{i} + (t^2 - 1)\mathbf{j}$
2. $\mathbf{r}(t) = (1 + 5t)\mathbf{i} + (t + 1)\mathbf{j} - 2t\mathbf{k}$
3. $\mathbf{r}(t) = (a\cos t)\mathbf{i} + (a\sin t)\mathbf{j} + bt\mathbf{k}$
4. $\mathbf{r}(t) = (2t\sin t)\mathbf{i} + (2t\cos t)\mathbf{j} + 2t^2\mathbf{k}, \quad t = 0$

In Exercises 5–10, write $\mathbf{a}$ in the form $\mathbf{a} = a_T\mathbf{T} + a_N\mathbf{N}$ at the given value of t without finding $\mathbf{T}$ and $\mathbf{N}$.

5. $\mathbf{r}(t) = t^3\mathbf{i} + (t + \frac{1}{2}t^2)\mathbf{j} + (t - \frac{1}{2}t^2)\mathbf{k}, \quad t = 0$
6. $\mathbf{r}(t) = (2e^t\sin t)\mathbf{i} + (2e^t\cos t)\mathbf{j} + 2e^t\mathbf{k}, \quad t = 0$
7. $\mathbf{r}(t) = (t + 1)\mathbf{i} + 2t\mathbf{j} + t^2\mathbf{k}, \quad t = 1$
8. $\mathbf{r}(t) = (t\cos t)\mathbf{i} + (t\sin t)\mathbf{j} + t^2\mathbf{k}, \quad t = 0$
9. $\mathbf{r}(t) = t^2\mathbf{i} + (t + (1/3)t^3)\mathbf{j} + (t - (1/3)t^3)\mathbf{k}, \quad t = 0$
10. $\mathbf{r}(t) = (e^t\cos t)\mathbf{i} + (e^t\sin t)\mathbf{j} + \sqrt{2}e^t\mathbf{k}, \quad t = 0$

Finding the TNB Frame

In Exercises 11 and 12, find $\mathbf{r}$, $\mathbf{T}$, $\mathbf{N}$, and $\mathbf{B}$ at the given value of t. Then find equations for the osculating, normal, and rectifying planes at that value of t.

11. $\mathbf{r}(t) = (\sin t)\mathbf{i} - t\mathbf{j} + (\cos t)\mathbf{k}$
12. $\mathbf{r}(t) = (\cos t)\mathbf{i} + (\sin t)\mathbf{j} + t\mathbf{k}, \quad t = 0$

In Exercises 9–16 of Section 13.4, you found $\mathbf{T}$, $\mathbf{N}$, and κ. Now, in the following Exercises 13–20, find $\mathbf{B}$ and τ for these space curves.

13. $\mathbf{r}(t) = (\sin t - t\cos t)\mathbf{i} + (\cos t + t\sin t)\mathbf{j} - 4\mathbf{k}$
14. $\mathbf{r}(t) = (2e^t\sin t)\mathbf{i} + (2e^t\cos t)\mathbf{j} + 3\mathbf{k}$
15. $\mathbf{r}(t) = (2\sin 4t)\mathbf{i} + (2\cos 4t)\mathbf{j} + 3t\mathbf{k}$
16. $\mathbf{r}(t) = \left(\frac{2t^3}{3}\right)\mathbf{i} + \left(\frac{3t^2}{2}\right)\mathbf{j}, \quad t > 0$
17. $\mathbf{r}(t) = (t^3/3)\mathbf{i} + (t^2/2)\mathbf{j} + \mathbf{k}, \quad t > 0$
18. $\mathbf{r}(t) = (\cos^3 t)\mathbf{j} + (\sin^3 t)\mathbf{k}, \quad 0 < t < \pi/2$
19. $\mathbf{r}(t) = t\mathbf{i} + (a\cosh(t/a))\mathbf{k}, \quad a > 0$
20. $\mathbf{r}(t) = (\cosh t)\mathbf{i} - (\sinh t)\mathbf{j} + t\mathbf{k}$

Physical Applications

21. The speedometer on your car reads a steady 35 mph. Could you be accelerating? Explain.
22. Can anything be said about the acceleration of a particle that is moving at a constant speed? Give reasons for your answer.
23. Can anything be said about the speed of a particle whose acceleration is always orthogonal to its velocity? Give reasons for your answer.
24. An object of mass m travels along the parabola $y = x^2$ with a constant speed of 10 units/sec. What is the force on the object due to its acceleration at $(0, 0)$? at $(2^{1/2}, 2)$? Write your answers in terms of $\mathbf{i}$ and $\mathbf{j}$. (Remember Newton's law, $\mathbf{F} = m\mathbf{a}$.)

Theory and Examples

25. Show that κ and τ are both zero for the line

$$\mathbf{r}(t) = (x_0 + At)\mathbf{i} + (y_0 + Bt)\mathbf{j} + (z_0 + Ct)\mathbf{k}.$$

26. Show that a moving particle will move in a straight line if the normal component of its acceleration is zero.
27. **A sometime shortcut to curvature** If you already know $|a_N|$ and $|\mathbf{v}|$, then the formula $a_N = \kappa|\mathbf{v}|^2$ gives a convenient way to find the curvature. Use it to find the curvature and radius of curvature of the curve

$$\mathbf{r}(t) = (\cos t + t\sin t)\mathbf{i} + (\sin t - t\cos t)\mathbf{j}, \quad t > 0.$$

(Take a_N and $|\mathbf{v}|$ from Example 1.)

28. What can be said about the torsion of a smooth plane curve $\mathbf{r}(t) = f(t)\mathbf{i} + g(t)\mathbf{j}$? Give reasons for your answer.
29. **Differentiable curves with zero torsion lie in planes** That a sufficiently differentiable curve with zero torsion lies in a plane is a special case of the fact that a particle whose velocity remains perpendicular to a fixed vector $\mathbf{C}$ moves in a plane perpendicular to $\mathbf{C}$. This, in turn, can be viewed as the following result.

Suppose $\mathbf{r}(t) = f(t)\mathbf{i} + g(t)\mathbf{j} + h(t)\mathbf{k}$ is twice differentiable for all t in an interval $[a, b]$, that $\mathbf{r} = 0$ when $t = a$, and that $\mathbf{v} \cdot \mathbf{k} = 0$ for all t in $[a, b]$. Show that $h(t) = 0$ for all t in $[a, b]$. (*Hint:* Start with $\mathbf{a} = d^2\mathbf{r}/dt^2$ and apply the initial conditions in reverse order.)

30. **A formula that calculates τ from B and v** If we start with the definition $\tau = -(d\mathbf{B}/ds) \cdot \mathbf{N}$ and apply the Chain Rule to rewrite $d\mathbf{B}/ds$ as

$$\frac{d\mathbf{B}}{ds} = \frac{d\mathbf{B}}{dt}\frac{dt}{ds} = \frac{d\mathbf{B}}{dt}\frac{1}{|\mathbf{v}|},$$

we arrive at the formula

$$\tau = -\frac{1}{|\mathbf{v}|}\left(\frac{d\mathbf{B}}{dt} \cdot \mathbf{N}\right).$$

Use the formula to find the torsion of the helix in Example 2.

COMPUTER EXPLORATIONS

Rounding the answers to four decimal places, use a CAS to find $\mathbf{v}$, $\mathbf{a}$, speed, $\mathbf{T}$, $\mathbf{N}$, $\mathbf{B}$, κ, τ, and the tangential and normal components of acceleration for the curves in Exercises 31–34 at the given values of t.

31. $\mathbf{r}(t) = (t\cos t)\mathbf{i} + (t\sin t)\mathbf{j} + t\mathbf{k}, \quad t = \sqrt{3}$
32. $\mathbf{r}(t) = (e^t\cos t)\mathbf{i} + (e^t\sin t)\mathbf{j} + e^t\mathbf{k}, \quad t = \ln 2$
33. $\mathbf{r}(t) = (t - \sin t)\mathbf{i} + (1 - \cos t)\mathbf{j} + \sqrt{-t}\mathbf{k}, \quad t = -3\pi$
34. $\mathbf{r}(t) = (3t - t^2)\mathbf{i} + (3t^2)\mathbf{j} + (3t + t^3)\mathbf{k}, \quad t = 1$

13.6 Velocity and Acceleration in Polar Coordinates

In this section we derive equations for velocity and acceleration in polar coordinates. These equations are useful for calculating the paths of planets and satellites in space, and we use them to examine Kepler's three laws of planetary motion.

Motion in Polar and Cylindrical Coordinates

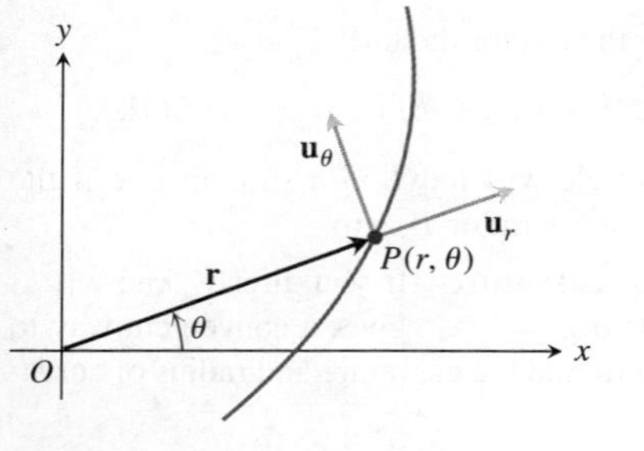

FIGURE 13.30 The length of $\mathbf{r}$ is the positive polar coordinate r of the point P. Thus $\mathbf{u}_r$, which is $\mathbf{r}/|\mathbf{r}|$, is also $\mathbf{r}/r$. Equations (1) express $\mathbf{u}_r$ and $\mathbf{u}_\theta$ in terms of $\mathbf{i}$ and $\mathbf{j}$.

When a particle at $P(r, \theta)$ moves along a curve in the polar coordinate plane, we express its position, velocity, and acceleration in terms of the moving unit vectors

$$\mathbf{u}_r = (\cos\theta)\mathbf{i} + (\sin\theta)\mathbf{j}, \qquad \mathbf{u}_\theta = -(\sin\theta)\mathbf{i} + (\cos\theta)\mathbf{j}, \tag{1}$$

shown in Figure 13.30. The vector $\mathbf{u}_r$ points along the position vector $\overrightarrow{OP}$, so $\mathbf{r} = r\mathbf{u}_r$. The vector $\mathbf{u}_\theta$, orthogonal to $\mathbf{u}_r$, points in the direction of increasing θ.

We find from Equations (1) that

$$\frac{d\mathbf{u}_r}{d\theta} = -(\sin\theta)\mathbf{i} + (\cos\theta)\mathbf{j} = \mathbf{u}_\theta$$

$$\frac{d\mathbf{u}_\theta}{d\theta} = -(\cos\theta)\mathbf{i} - (\sin\theta)\mathbf{j} = -\mathbf{u}_r.$$

We next differentiate $\mathbf{u}_r$ and $\mathbf{u}_\theta$ with respect to t to find how they change with time. The Chain Rule gives

$$\dot{\mathbf{u}}_r = \frac{d\mathbf{u}_r}{d\theta}\dot{\theta} = \dot{\theta}\mathbf{u}_\theta, \qquad \dot{\mathbf{u}}_\theta = \frac{d\mathbf{u}_\theta}{d\theta}\dot{\theta} = -\dot{\theta}\mathbf{u}_r. \tag{2}$$

Hence, we can express the velocity vector in terms of $\mathbf{u}_r$ and $\mathbf{u}_\theta$ as

$$\mathbf{v} = \dot{\mathbf{r}} = \frac{d}{dt}(r\mathbf{u}_r) = \dot{r}\mathbf{u}_r + r\dot{\mathbf{u}}_r = \dot{r}\mathbf{u}_r + r\dot{\theta}\mathbf{u}_\theta.$$

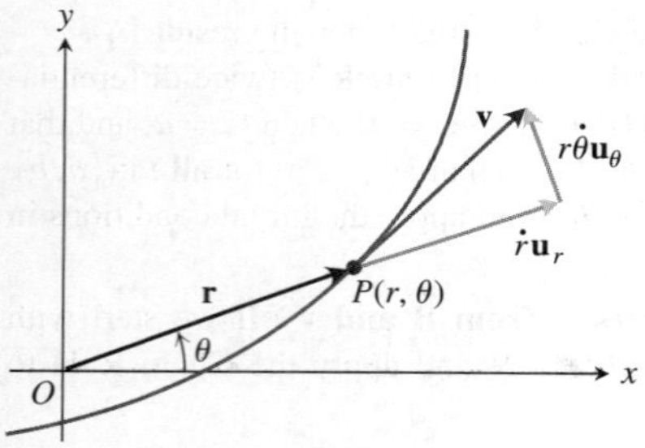

FIGURE 13.31 In polar coordinates, the velocity vector is $\mathbf{v} = \dot{r}\mathbf{u}_r + r\dot{\theta}\mathbf{u}_\theta$.

See Figure 13.31. As in the previous section, we use Newton's dot notation for time derivatives to keep the formulas as simple as we can: $\dot{\mathbf{u}}_r$ means $d\mathbf{u}_r/dt$, $\dot{\theta}$ means $d\theta/dt$, and so on.

The acceleration is

$$\mathbf{a} = \dot{\mathbf{v}} = (\ddot{r}\mathbf{u}_r + \dot{r}\dot{\mathbf{u}}_r) + (\dot{r}\dot{\theta}\mathbf{u}_\theta + r\ddot{\theta}\mathbf{u}_\theta + r\dot{\theta}\dot{\mathbf{u}}_\theta).$$

When Equations (2) are used to evaluate $\dot{\mathbf{u}}_r$ and $\dot{\mathbf{u}}_\theta$ and the components are separated, the equation for acceleration in terms of $\mathbf{u}_r$ and $\mathbf{u}_\theta$ becomes

$$\mathbf{a} = (\ddot{r} - r\dot{\theta}^2)\mathbf{u}_r + (r\ddot{\theta} + 2\dot{r}\dot{\theta})\mathbf{u}_\theta.$$

To extend these equations of motion to space, we add $z\mathbf{k}$ to the right-hand side of the equation $\mathbf{r} = r\mathbf{u}_r$. Then, in these *cylindrical coordinates*, we have

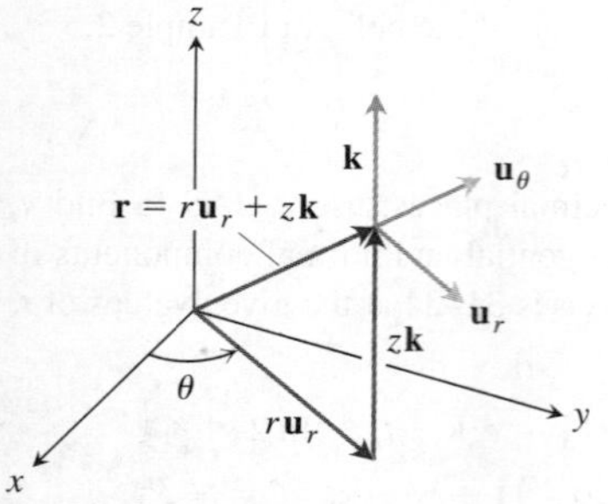

FIGURE 13.32 Position vector and basic unit vectors in cylindrical coordinates. Notice that $|\mathbf{r}| \neq r$ if $z \neq 0$ since $|\mathbf{r}| = \sqrt{r^2 + z^2}$.

Position:	$\mathbf{r} = r\mathbf{u}_r + z\mathbf{k}$	
Velocity:	$\mathbf{v} = \dot{r}\mathbf{u}_r + r\dot{\theta}\mathbf{u}_\theta + \dot{z}\mathbf{k}$	(3)
Acceleration:	$\mathbf{a} = (\ddot{r} - r\dot{\theta}^2)\mathbf{u}_r + (r\ddot{\theta} + 2\dot{r}\dot{\theta})\mathbf{u}_\theta + \ddot{z}\mathbf{k}$	

The vectors $\mathbf{u}_r$, $\mathbf{u}_\theta$, and $\mathbf{k}$ make a right-handed frame (Figure 13.32) in which

$$\mathbf{u}_r \times \mathbf{u}_\theta = \mathbf{k}, \qquad \mathbf{u}_\theta \times \mathbf{k} = \mathbf{u}_r, \qquad \mathbf{k} \times \mathbf{u}_r = \mathbf{u}_\theta.$$

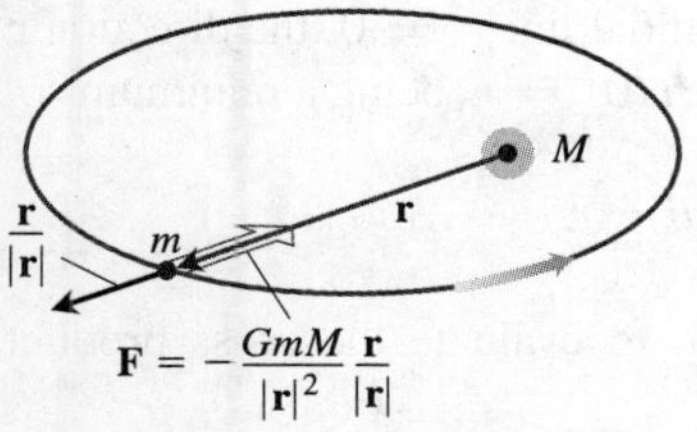

FIGURE 13.33 The force of gravity is directed along the line joining the centers of mass.

Planets Move in Planes

Newton's law of gravitation says that if $\mathbf{r}$ is the radius vector from the center of a sun of mass M to the center of a planet of mass m, then the force $\mathbf{F}$ of the gravitational attraction between the planet and sun is

$$\mathbf{F} = -\frac{GmM}{|\mathbf{r}|^2}\frac{\mathbf{r}}{|\mathbf{r}|}$$

(Figure 13.33). The number G is the **universal gravitational constant**. If we measure mass in kilograms, force in newtons, and distance in meters, G is about $6.6738 \times 10^{-11}\ \mathrm{Nm}^2\ \mathrm{kg}^{-2}$.

Combining the gravitation law with Newton's second law, $\mathbf{F} = m\ddot{\mathbf{r}}$, for the force acting on the planet gives

$$m\ddot{\mathbf{r}} = -\frac{GmM}{|\mathbf{r}|^2}\frac{\mathbf{r}}{|\mathbf{r}|},$$

$$\ddot{\mathbf{r}} = -\frac{GM}{|\mathbf{r}|^2}\frac{\mathbf{r}}{|\mathbf{r}|}.$$

The planet is therefore accelerated toward the sun's center of mass at all times.

Since $\ddot{\mathbf{r}}$ is a scalar multiple of $\mathbf{r}$, we have

$$\mathbf{r} \times \ddot{\mathbf{r}} = \mathbf{0}.$$

From this last equation,

$$\frac{d}{dt}(\mathbf{r} \times \dot{\mathbf{r}}) = \underbrace{\dot{\mathbf{r}} \times \dot{\mathbf{r}}}_{0} + \mathbf{r} \times \ddot{\mathbf{r}} = \mathbf{r} \times \ddot{\mathbf{r}} = \mathbf{0}.$$

It follows that

$$\mathbf{r} \times \dot{\mathbf{r}} = \mathbf{C} \tag{4}$$

for some constant vector $\mathbf{C}$.

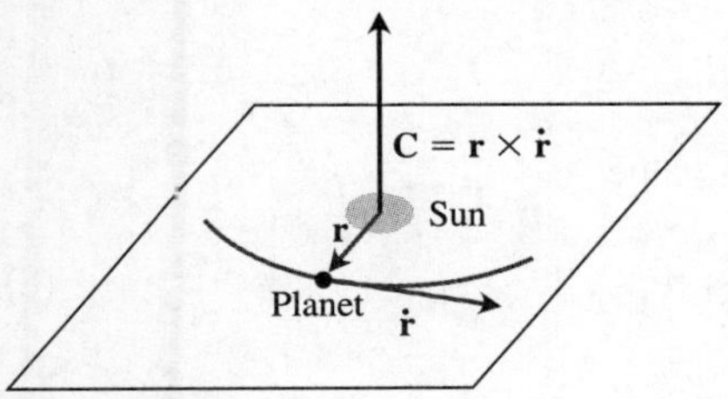

FIGURE 13.34 A planet that obeys Newton's laws of gravitation and motion travels in the plane through its sun's center of mass perpendicular to $\mathbf{C} = \mathbf{r} \times \dot{\mathbf{r}}$.

Equation (4) tells us that $\mathbf{r}$ and $\dot{\mathbf{r}}$ always lie in a plane perpendicular to $\mathbf{C}$. Hence, the planet moves in a fixed plane through the center of mass of its sun (Figure 13.34). We next see how Kepler's laws describe the motion in a precise way.

Kepler's First Law (Ellipse Law)

Kepler's first law says that a planet's path is an ellipse with its sun at one focus. The eccentricity of the ellipse is

$$e = \frac{r_0 v_0^{\,2}}{GM} - 1, \tag{5}$$

and the polar equation (see Section 11.7 Equation (5)) is

$$r = \frac{(1 + e)r_0}{1 + e\cos\theta}. \tag{6}$$

Here v_0 is the speed when the planet is positioned at its minimum distance r_0 from the sun. We omit the lengthy proof. The sun's mass M is 1.99×10^{30} kg.

Kepler's Second Law (Equal Area Law)

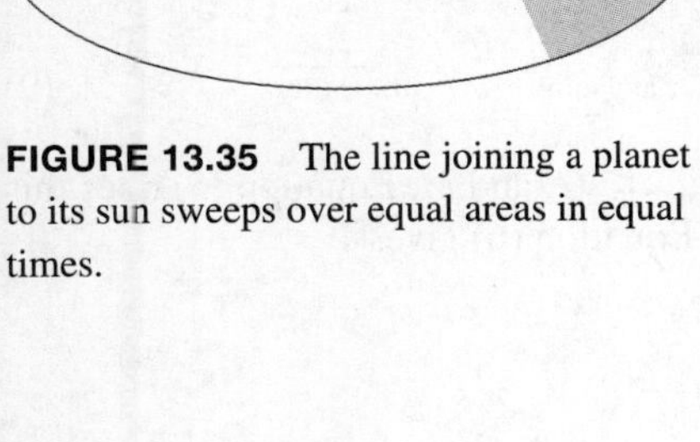

FIGURE 13.35 The line joining a planet to its sun sweeps over equal areas in equal times.

Kepler's second law says that the radius vector from the sun to a planet (the vector $\mathbf{r}$ in our model) sweeps out equal areas in equal times, as displayed in Figure 13.35. In that figure, we assume the plane of the planet is the xy-plane, so the unit vector in the direction of $\mathbf{C}$ is $\mathbf{k}$.

We introduce polar coordinates in the plane, choosing as initial line $\theta = 0$, the direction $\mathbf{r}$ when $|\mathbf{r}| = r$ is a minimum value. Then at $t = 0$, we have $r(0) = r_0$ being a minimum so,

$$\dot{r}\big|_{t=0} = \frac{dr}{dt}\bigg|_{t=0} = 0 \quad \text{and} \quad v_0 = |\mathbf{v}|_{t=0} = \left[r\dot{\theta}\right]_{t=0}. \qquad \text{Eq. (3), } \dot{z} = 0$$

To derive Kepler's second law, we use Equation (3) to evaluate the cross product $\mathbf{C} = \mathbf{r} \times \dot{\mathbf{r}}$ from Equation (4):

$$\begin{aligned} \mathbf{C} &= \mathbf{r} \times \dot{\mathbf{r}} = \mathbf{r} \times \mathbf{v} \\ &= r\mathbf{u}_r \times (\dot{r}\mathbf{u}_r + r\dot{\theta}\mathbf{u}_\theta) && \text{Eq. (3), } \dot{z} = 0 \\ &= r\dot{r}\underbrace{(\mathbf{u}_r \times \mathbf{u}_r)}_{0} + r(r\dot{\theta})\underbrace{(\mathbf{u}_r \times \mathbf{u}_\theta)}_{\mathbf{k}} \\ &= r(r\dot{\theta})\mathbf{k}. \end{aligned} \tag{7}$$

Setting t equal to zero shows that

$$\mathbf{C} = \left[r(r\dot{\theta})\right]_{t=0}\mathbf{k} = r_0 v_0 \mathbf{k}.$$

Substituting this value for $\mathbf{C}$ in Equation (7) gives

$$r_0 v_0 \mathbf{k} = r^2\dot{\theta}\mathbf{k}, \quad \text{or} \quad r^2\dot{\theta} = r_0 v_0.$$

This is where the area comes in. The area differential in polar coordinates is

$$dA = \frac{1}{2}r^2\, d\theta$$

(Section 11.5). Accordingly, dA/dt has the constant value

$$\frac{dA}{dt} = \frac{1}{2}r^2\dot{\theta} = \frac{1}{2}r_0 v_0. \tag{8}$$

So dA/dt is constant, giving Kepler's second law.

HISTORICAL BIOGRAPHY
Johannes Kepler
(1571–1630)
www.bit.ly/2zIRwXb

Kepler's Third Law (Time–Distance Law)

The time T it takes a planet to go around its sun once is the planet's **orbital period**. *Kepler's third law* says that T and the orbit's semimajor axis a are related by the equation

$$\frac{T^2}{a^3} = \frac{4\pi^2}{GM}.$$

Since the right-hand side of this equation is constant within a given solar system, the ratio of T^2 to a^3 *is the same for every planet in the system.*

Here is a partial derivation of Kepler's third law. The area enclosed by the planet's elliptical orbit is calculated as follows:

$$\begin{aligned} \text{Area} &= \int_0^T dA \\ &= \int_0^T \frac{1}{2} r_0 v_0\, dt && \text{Eq. (8)} \\ &= \frac{1}{2} T r_0 v_0. \end{aligned}$$

If b is the semiminor axis, the area of the ellipse is πab, so

$$T = \frac{2\pi ab}{r_0 v_0} = \frac{2\pi a^2}{r_0 v_0}\sqrt{1 - e^2}. \qquad \text{For any ellipse, } b = a\sqrt{1 - e^2}. \tag{9}$$

It remains only to express a and e in terms of r_0, v_0, G, and M. Equation (5) does this for e. For a, we observe that setting θ equal to π in Equation (6) gives

$$r_{\max} = r_0 \frac{1 + e}{1 - e}.$$

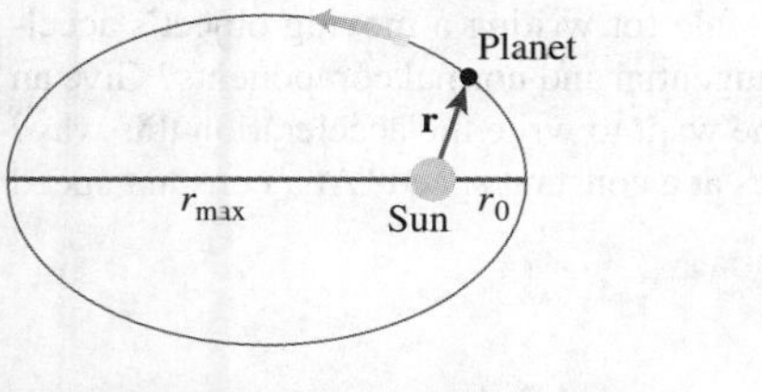

FIGURE 13.36 The length of the major axis of the ellipse is $2a = r_0 + r_{max}$.

Hence, from Figure 13.36,

$$2a = r_0 + r_{max} = \frac{2r_0}{1 - e} = \frac{2r_0 GM}{2GM - r_0 v_0^{\,2}}. \tag{10}$$

Squaring both sides of Equation (9) and substituting the results of Equations (5) and (10) produce Kepler's third law (Exercise 11).

EXERCISES 13.6

In Exercises 1–7, find the velocity and acceleration vectors in terms of $\mathbf{u}_r$ and $\mathbf{u}_\theta$.

1. $r = 2\theta$ and $\dfrac{d\theta}{dt} = 4$

2. $r = \dfrac{1}{6}$ and $\dfrac{d\theta}{dt} = t^2$

3. $r = a(1 + \sin\theta)$ and $\dfrac{d\theta}{dt} = 5$

4. $r = a\sin 2\theta$ and $\dfrac{d\theta}{dt} = 2t$

5. $r = 2e^{\theta}$ and $\dfrac{d\theta}{dt} = 3$

6. $r = a(1 + \sin t)$ and $\theta = 1 - e^{-t}$

7. $r = \cos 6t$ and $\theta = 3t$

8. Type of orbit For what values of v_0 in Equation (5) is the orbit in Equation (6) a circle? An ellipse? A parabola? A hyperbola?

9. Circular orbits Show that a planet in a circular orbit moves with a constant speed. (*Hint:* This is a consequence of one of Kepler's laws.)

10. Suppose that $\mathbf{r}$ is the position vector of a particle moving along a plane curve and dA/dt is the rate at which the vector sweeps out area. Without introducing coordinates, and assuming the necessary derivatives exist, give a geometric argument based on increments and limits for the validity of the equation

$$\frac{dA}{dt} = \frac{1}{2}|\mathbf{r} \times \dot{\mathbf{r}}|.$$

11. Kepler's third law Complete the derivation of Kepler's third law (the part following Equation (10)).

12. Do the data in the accompanying table support Kepler's third law? Give reasons for your answer.

Planet	Semimajor axis a (10^{10} m)	Period T (years)
Mercury	5.79	0.241
Venus	10.81	0.615
Mars	22.78	1.881
Saturn	142.70	29.457

13. Earth's major axis Estimate the length of the major axis of Earth's orbit if its orbital period is 365.256 days.

14. Estimate the length of the major axis of the orbit of Uranus if its orbital period is 84 years.

15. The eccentricity of Earth's orbit is $e = 0.0167$, so the orbit is nearly circular, with radius approximately 150×10^6 km. Find, in units of km^2/sec, the rate dA/dt satisfying Kepler's second law.

16. Jupiter's orbital period Estimate the orbital period of Jupiter, assuming that $a = 77.8 \times 10^{10}$ m.

17. Mass of Jupiter Europa is one of the moons of Jupiter. It has semi major axis of 0.067×10^{10} m and an orbital period of 3.551 days. Use these data to estimate the mass of Jupiter.

18. Distance from Earth to the moon The period of the moon's rotation around Earth is 2.36055×10^6 sec. Estimate the distance to the moon.

CHAPTER 13 Questions to Guide Your Review

1. State the rules for differentiating and integrating vector functions. Give examples.

2. How do you define and calculate the velocity, speed, direction of motion, and acceleration of a body moving along a sufficiently differentiable space curve? Give an example.

3. What is special about the derivatives of vector functions of constant length? Give an example.

4. What are the vector and parametric equations for ideal projectile motion? How do you find a projectile's maximum height, flight time, and range? Give examples.

5. How do you define and calculate the length of a segment of a smooth space curve? Give an example. What mathematical assumptions are involved in the definition?

6. How do you measure distance along a smooth curve in space from a preselected base point? Give an example.

7. What is a differentiable curve's unit tangent vector? Give an example.

8. Define curvature, circle of curvature (osculating circle), center of curvature, and radius of curvature for twice-differentiable curves in the plane. Give examples. What curves have zero curvature? Constant curvature?

9. What is a plane curve's principal normal vector? When is it defined? Which way does it point? Give an example.

10. How do you define **N** and κ for curves in space? How are these quantities related? Give examples.

11. What is a curve's binormal vector? Give an example. How is this vector related to the curve's torsion? Give an example.

12. What formulas are available for writing a moving object's acceleration as a sum of its tangential and normal components? Give an example. Why might one want to write the acceleration this way? What if the object moves at a constant speed? At a constant speed around a circle?

13. State Kepler's laws.

CHAPTER 13 Practice Exercises

Motion in the Plane

In Exercises 1 and 2, graph the curves and sketch their velocity and acceleration vectors at the given values of t. Then write **a** in the form $\mathbf{a} = a_T\mathbf{T} + a_N\mathbf{N}$ without finding **T** and **N**, and find the value of κ at the given values of t.

1. $\mathbf{r}(t) = (4\cos t)\mathbf{i} + (\sqrt{2}\sin t)\mathbf{j}, \quad t = 0$ and $\pi/4$

2. $\mathbf{r}(t) = (\sqrt{3}\sec t)\mathbf{i} + (\sqrt{3}\tan t)\mathbf{j}, \quad t = 0$

3. The position of a particle in the plane at time t is

$$\mathbf{r} = \frac{1}{\sqrt{1+t^2}}\mathbf{i} + \frac{t}{\sqrt{1+t^2}}\mathbf{j}.$$

Find the particle's highest speed.

4. Suppose $\mathbf{r}(t) = (e^t\cos t)\mathbf{i} + (e^t\sin t)\mathbf{j}$. Show that the angle between **r** and **a** never changes. What *is* the angle?

5. Finding curvature At point P, the velocity and acceleration of a particle moving in the plane are $\mathbf{v} = 3\mathbf{i} + 4\mathbf{j}$ and $\mathbf{a} = 5\mathbf{i} + 15\mathbf{j}$. Find the curvature of the particle's path at P.

6. Find the point on the curve $y = e^x$ where the curvature is greatest.

7. A particle moves around the unit circle in the xy-plane. Its position at time t is $\mathbf{r} = x\mathbf{i} + y\mathbf{j}$, where x and y are differentiable functions of t. Find dy/dt if $\mathbf{v}\cdot\mathbf{i} = y$. Is the motion clockwise or counterclockwise?

8. You send a message through a pneumatic tube that follows the curve $9y = x^3$ (distance in meters). At the point $(3, 3)$, $\mathbf{v}\cdot\mathbf{i} = 4$ and $\mathbf{a}\cdot\mathbf{i} = -2$. Find the values of $\mathbf{v}\cdot\mathbf{j}$ and $\mathbf{a}\cdot\mathbf{j}$ at $(3, 3)$.

9. Characterizing circular motion A particle moves in the plane so that its velocity and position vectors are always orthogonal. Show that the particle moves in a circle centered at the origin.

10. Speed along a cycloid A circular wheel with radius 1 ft and center C rolls to the right along the x-axis at a half-turn per second. (See the accompanying figure.) At time t seconds, the position vector of the point P on the wheel's circumference is

$$\mathbf{r} = (\pi t - \sin\pi t)\mathbf{i} + (1 - \cos\pi t)\mathbf{j}.$$

a. Sketch the curve traced by P during the interval $0 \le t \le 3$.

b. Find **v** and **a** at $t = 0$, 1, 2, and 3 and add these vectors to your sketch.

c. At any given time, what is the forward speed of the topmost point of the wheel? Of C?

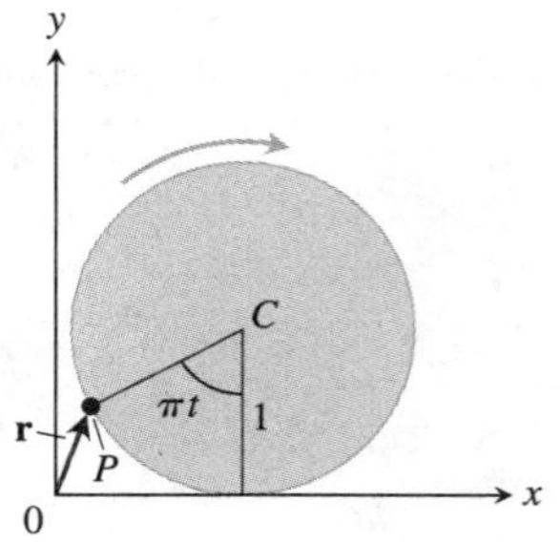

Projectile Motion

11. Shot put A shot leaves the thrower's hand 6.5 ft above the ground at a 45° angle at 44 ft/sec. Where is it 3 sec later?

12. Javelin A javelin leaves the thrower's hand 7 ft above the ground at a 45° angle at 80 ft/sec. How high does it go?

13. A golf ball is hit with an initial speed v_0 at an angle α to the horizontal from a point that lies at the foot of a straight-sided hill that is inclined at an angle ϕ to the horizontal, where

$$0 < \phi < \alpha < \frac{\pi}{2}.$$

Show that the ball lands at a distance

$$\frac{2v_0^2\cos\alpha}{g\cos^2\phi}\sin(\alpha - \phi),$$

measured up the face of the hill. Hence, show that the greatest range that can be achieved for a given v_0 occurs when $\alpha = (\phi/2) + (\pi/4)$, that is, when the initial velocity vector bisects the angle between the vertical and the hill.

T **14. Javelin** In Potsdam in 1988, Petra Felke of (then) East Germany set a women's world record by throwing a javelin 262 ft 5 in.

a. Assuming that Felke launched the javelin at a 40° angle to the horizontal 6.5 ft above the ground, what was the javelin's initial speed?

b. How high did the javelin go?

Motion in Space

Find the lengths of the curves in Exercises 15 and 16.

15. $\mathbf{r}(t) = (2\cos t)\mathbf{i} + (2\sin t)\mathbf{j} + t^2\mathbf{k}, \quad 0 \le t \le \pi/4$

16. $\mathbf{r}(t) = (3\cos t)\mathbf{i} + (3\sin t)\mathbf{j} + 2t^{3/2}\mathbf{k}, \quad 0 \le t \le 3$

In Exercises 17–20, find $\mathbf{T}$, $\mathbf{N}$, $\mathbf{B}$, κ, and τ at the given value of t.

17. $\mathbf{r}(t) = \frac{4}{9}(1+t)^{3/2}\mathbf{i} + \frac{4}{9}(1-t)^{3/2}\mathbf{j} + \frac{1}{3}t\mathbf{k}, \quad t = 0$

18. $\mathbf{r}(t) = (e^t \sin 2t)\mathbf{i} + (e^t \cos 2t)\mathbf{j} + 2e^t\mathbf{k}, \quad t = 0$

19. $\mathbf{r}(t) = t\mathbf{i} + \frac{1}{2}e^{2t}\mathbf{j}, \quad t = \ln 2$

20. $\mathbf{r}(t) = (3\cosh 2t)\mathbf{i} + (3\sinh 2t)\mathbf{j} + 6t\mathbf{k}, \quad t = \ln 2$

In Exercises 21 and 22, write $\mathbf{a}$ in the form $\mathbf{a} = a_{\mathrm{T}}\mathbf{T} + a_{\mathrm{N}}\mathbf{N}$ at $t = 0$ without finding $\mathbf{T}$ and $\mathbf{N}$.

21. $\mathbf{r}(t) = (2 + 3t + 3t^2)\mathbf{i} + (4t + 4t^2)\mathbf{j} - (6\cos t)\mathbf{k}$

22. $\mathbf{r}(t) = (2 + t)\mathbf{i} + (t + 2t^2)\mathbf{j} + (1 + t^2)\mathbf{k}$

23. Find $\mathbf{T}$, $\mathbf{N}$, $\mathbf{B}$, κ, and τ as functions of t if

$$\mathbf{r}(t) = (\sin t)\mathbf{i} + (\sqrt{2}\cos t)\mathbf{j} + (\sin t)\mathbf{k}.$$

24. At what times in the interval $0 \le t \le \pi$ are the velocity and acceleration vectors of the motion $\mathbf{r}(t) = \mathbf{i} + (5\cos t)\mathbf{j} + (3\sin t)\mathbf{k}$ orthogonal?

25. The position of a particle moving in space at time $t \ge 0$ is

$$\mathbf{r}(t) = 2\mathbf{i} + \left(4\sin\frac{t}{2}\right)\mathbf{j} + \left(3 - \frac{t}{\pi}\right)\mathbf{k}.$$

Find the first time $\mathbf{r}$ is orthogonal to the vector $\mathbf{i} - \mathbf{j}$.

26. Find equations for the osculating, normal, and rectifying planes of the curve $\mathbf{r}(t) = t\mathbf{i} + t^2\mathbf{j} + t^3\mathbf{k}$ at the point $(1, 1, 1)$.

27. Find parametric equations for the line that is tangent to the curve $\mathbf{r}(t) = e^t\mathbf{i} + (\sin t)\mathbf{j} + \ln(1 - t)\mathbf{k}$ at $t = 0$.

28. Find parametric equations for the line that is tangent to the helix $\mathbf{r}(t) = (\sqrt{2}\cos t)\mathbf{i} + (\sqrt{2}\sin t)\mathbf{j} + t\mathbf{k}$ at the point where $t = \pi/4$.

Theory and Examples

29. Synchronous curves By eliminating α from the ideal projectile equations

$$x = (v_0\cos\alpha)t, \quad y = (v_0\sin\alpha)t - \frac{1}{2}gt^2,$$

show that $x^2 + (y + gt^2/2)^2 = v_0{}^2t^2$. This shows that projectiles launched simultaneously from the origin at the same initial speed will, at any given instant, all lie on the circle of radius v_0t centered at $(0, -gt^2/2)$, regardless of their launch angle. These circles are the *synchronous curves* of the launching.

30. Radius of curvature Show that the radius of curvature of a twice-differentiable plane curve $\mathbf{r}(t) = f(t)\mathbf{i} + g(t)\mathbf{j}$ is given by the formula

$$\rho = \frac{\dot{x}^2 + \dot{y}^2}{\sqrt{\ddot{x}^2 + \ddot{y}^2 - \ddot{s}^2}}, \quad \text{where} \quad \ddot{s} = \frac{d}{dt}\sqrt{\dot{x}^2 + \dot{y}^2}.$$

31. An alternative definition of curvature in the plane An alternative definition gives the curvature of a sufficiently differentiable plane curve to be $|d\phi/ds|$, where ϕ is the angle between $\mathbf{T}$ and $\mathbf{i}$ (Figure 13.37a). Figure 13.37b shows the distance s measured counterclockwise around the circle $x^2 + y^2 = a^2$ from the point $(a, 0)$ to a point P, along with the angle ϕ at P. Calculate the circle's curvature using the alternative definition. (*Hint:* $\phi = \theta + \pi/2$.)

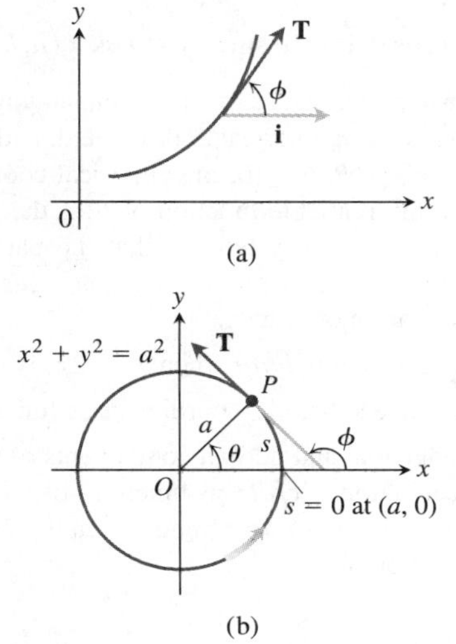

FIGURE 13.37 Figures for Exercise 31.

32. The view from *Skylab 4* What percentage of Earth's surface area could the astronauts see when *Skylab 4* was at its apogee height, 437 km above the surface? To find out, model the visible surface as the surface generated by revolving the circular arc GT, shown here, about the y-axis. Then carry out these steps:

1. Use similar triangles in the figure to show that $y_0/6380 = 6380/(6380 + 437)$. Solve for y_0.
2. To four significant digits, calculate the visible area as

$$VA = \int_{y_0}^{6380} 2\pi x\sqrt{1 + \left(\frac{dx}{dy}\right)^2}\,dy.$$

3. Express the result as a percentage of Earth's surface area.

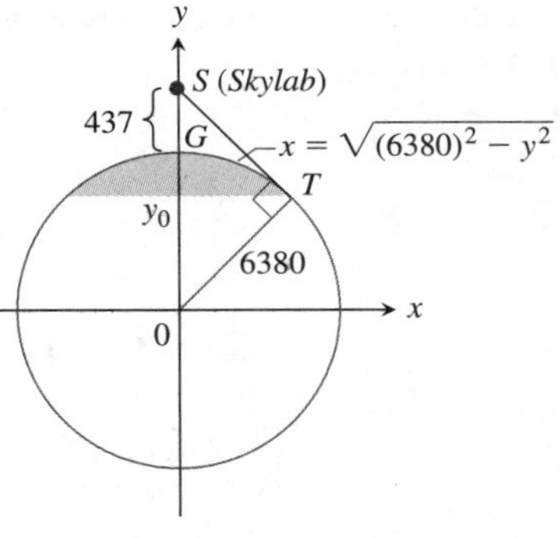

CHAPTER 13 Additional and Advanced Exercises

Applications

1. A frictionless particle P, starting from rest at time $t = 0$ at the point $(a, 0, 0)$, slides down the helix

$$\mathbf{r}(\theta) = (a\cos\theta)\mathbf{i} + (a\sin\theta)\mathbf{j} + b\theta\mathbf{k} \quad (a, b > 0)$$

under the influence of gravity, as in the accompanying figure. The θ in this equation is the cylindrical coordinate θ, and the helix is the curve $r = a$, $z = b\theta$, $\theta \geq 0$, in cylindrical coordinates. We assume θ to be a differentiable function of t for the motion. The law of conservation of energy tells us that the particle's speed after it has fallen straight down a distance z is $\sqrt{2gz}$, where g is the constant acceleration of gravity.

a. Find the angular velocity $d\theta/dt$ when $\theta = 2\pi$.

b. Express the particle's θ- and z-coordinates as functions of t.

c. Express the tangential and normal components of the velocity $d\mathbf{r}/dt$ and acceleration $d^2\mathbf{r}/dt^2$ as functions of t. Does the acceleration have any nonzero component in the direction of the binormal vector $\mathbf{B}$?

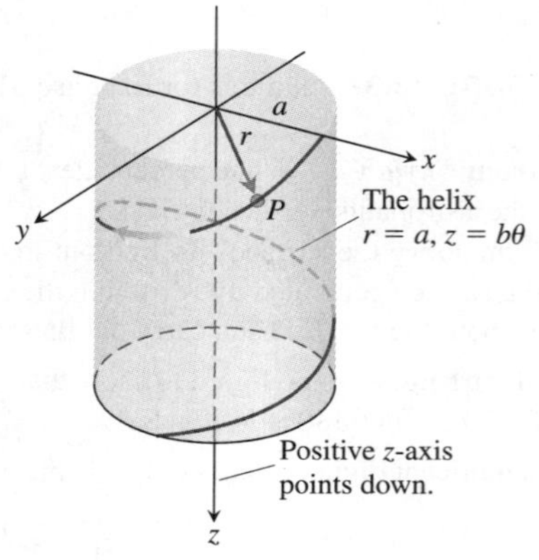

2. Suppose the curve in Exercise 1 is replaced by the conical helix $r = a\theta$, $z = b\theta$ shown in the accompanying figure.

a. Express the angular velocity $d\theta/dt$ as a function of θ.

b. Express the distance the particle travels along the helix as a function of θ.

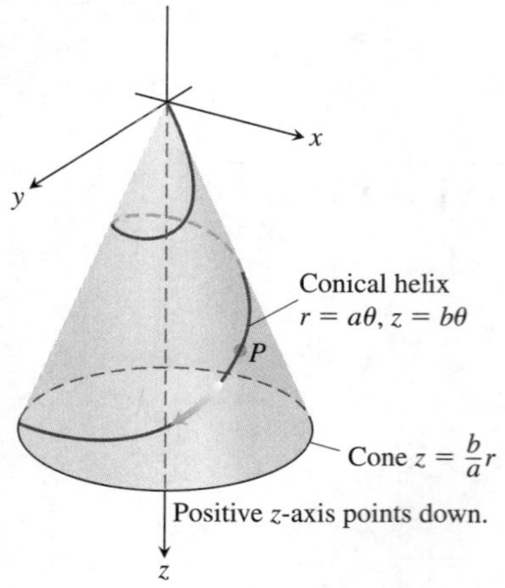

Motion in Polar and Cylindrical Coordinates

3. Deduce from the orbit equation

$$r = \frac{(1 + e)r_0}{1 + e\cos\theta}$$

that a planet is closest to its sun when $\theta = 0$, and show that $r = r_0$ at that time.

T 4. **A Kepler equation** The problem of locating a planet in its orbit at a given time and date eventually leads to solving "Kepler" equations of the form

$$f(x) = x - 1 - \frac{1}{2}\sin x = 0.$$

a. Show that this particular equation has a solution between $x = 0$ and $x = 2$.

b. With your computer or calculator in radian mode, use Newton's method to find the solution to as many places as you can.

5. In Section 13.6, we found the velocity of a particle moving in the plane to be

$$\mathbf{v} = \dot{x}\mathbf{i} + \dot{y}\mathbf{j} = \dot{r}\mathbf{u}_r + r\dot{\theta}\mathbf{u}_\theta.$$

a. Express $\dot{x}$ and $\dot{y}$ in terms of $\dot{r}$ and $r\dot{\theta}$ by evaluating the dot products $\mathbf{v} \cdot \mathbf{i}$ and $\mathbf{v} \cdot \mathbf{j}$.

b. Express $\dot{r}$ and $r\dot{\theta}$ in terms of $\dot{x}$ and $\dot{y}$ by evaluating the dot products $\mathbf{v} \cdot \mathbf{u}_r$ and $\mathbf{v} \cdot \mathbf{u}_\theta$.

6. Express the curvature of a twice-differentiable curve $r = f(\theta)$ in the polar coordinate plane in terms of f and its derivatives.

7. A slender rod through the origin of the polar coordinate plane rotates (in the plane) about the origin at the rate of 3 rad/min. A beetle starting from the point $(2, 0)$ crawls along the rod toward the origin at the rate of 1 in./min.

a. Find the beetle's acceleration and velocity in polar form when it is halfway to (1 in. from) the origin.

T **b.** To the nearest tenth of an inch, what will be the length of the path the beetle has traveled by the time it reaches the origin?

8. **Arc length in cylindrical coordinates**

a. Show that when you express $ds^2 = dx^2 + dy^2 + dz^2$ in terms of cylindrical coordinates, you get $ds^2 = dr^2 + r^2 d\theta^2 + dz^2$.

b. Interpret this result geometrically in terms of the edges and a diagonal of a box. Sketch the box.

c. Use the result in part (a) to find the length of the curve $r = e^\theta$, $z = e^\theta$, $0 \leq \theta \leq \ln 8$.

9. **Unit vectors for position and motion in cylindrical coordinates** When the position of a particle moving in space is given in cylindrical coordinates, the unit vectors we use to describe its position and motion are

$$\mathbf{u}_r = (\cos\theta)\mathbf{i} + (\sin\theta)\mathbf{j}, \qquad \mathbf{u}_\theta = -(\sin\theta)\mathbf{i} + (\cos\theta)\mathbf{j},$$

and **k** (see accompanying figure). The particle's position vector is then $\mathbf{r} = r\mathbf{u}_r + z\mathbf{k}$, where r is the positive polar distance coordinate of the particle's position.

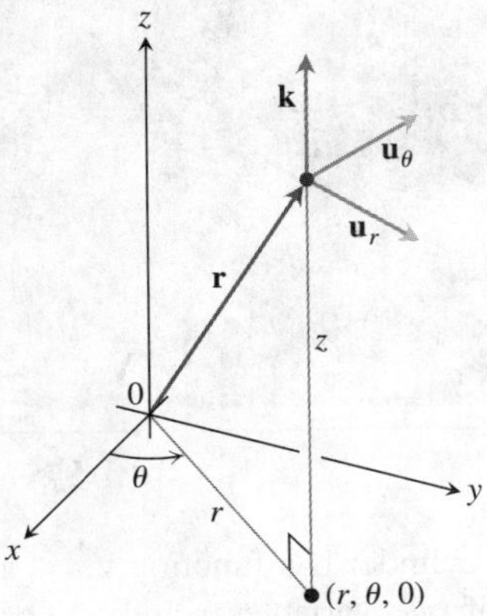

a. Show that $\mathbf{u}_r$, $\mathbf{u}_\theta$, and **k**, in this order, form a right-handed frame of unit vectors.

b. Show that

$$\frac{d\mathbf{u}_r}{d\theta} = \mathbf{u}_\theta \quad \text{and} \quad \frac{d\mathbf{u}_\theta}{d\theta} = -\mathbf{u}_r.$$

c. Assuming that the necessary derivatives with respect to t exist, express $\mathbf{v} = \dot{\mathbf{r}}$ and $\mathbf{a} = \ddot{\mathbf{r}}$ in terms of $\mathbf{u}_r$, $\mathbf{u}_\theta$, **k**, $\dot{r}$, and $\dot{\theta}$.

d. Conservation of angular momentum Let $\mathbf{r}(t)$ denote the position in space of a moving object at time t. Suppose the force acting on the object at time t is

$$\mathbf{F}(t) = -\frac{c}{|\mathbf{r}(t)|^3}\mathbf{r}(t),$$

where c is a constant. In physics the **angular momentum** of an object at time t is defined to be $\mathbf{L}(t) = \mathbf{r}(t) \times m\mathbf{v}(t)$, where m is the mass of the object and $\mathbf{v}(t)$ is the velocity. Prove that angular momentum is a conserved quantity; that is, prove that $\mathbf{L}(t)$ is a constant vector, independent of time. Remember Newton's law $\mathbf{F} = m\mathbf{a}$. (This is a calculus problem, not a physics problem.)

CHAPTER 13 Technology Application Projects

Mathematica/Maple Projects

Projects can be found within MyLab Math.

- ***Radar Tracking of a Moving Object***
  Visualize position, velocity, and acceleration vectors to analyze motion.
- ***Parametric and Polar Equations with a Figure Skater***
  Visualize position, velocity, and acceleration vectors to analyze motion.
- ***Moving in Three Dimensions***
  Compute distance traveled, speed, curvature, and torsion for motion along a space curve. Visualize and compute the tangential, normal, and binormal vectors associated with motion along a space curve.

14

Partial Derivatives

OVERVIEW The volume of a right circular cylinder is a function $V = \pi r^2 h$ of its radius and its height, so it is a function $V(r, h)$ of two variables r and h. The speed of sound through seawater is primarily a function of salinity S and temperature T. The monthly payment on a home mortgage is a function of the principal borrowed P, the interest rate i, and the term t of the loan. These are examples of functions that depend on more than one independent variable. In this chapter we extend the ideas of single-variable differential calculus to functions of several variables.

14.1 Functions of Several Variables

Real-valued functions of several independent real variables are defined analogously to functions of a single variable. Points in the domain are now ordered pairs (or triples, quadruples, n-tuples) of real numbers, and values in the range are real numbers.

DEFINITIONS Suppose D is a set of n-tuples of real numbers $(x_1, x_2, \dots, x_n)$. A **real-valued function** f on D is a rule that assigns a real number

$$w = f(x_1, x_2, \dots, x_n)$$

to each element in D. The set D is the function's **domain**. The set of w-values taken on by f is the function's **range**. The symbol w is the **dependent variable** of f, and f is said to be a function of the n **independent variables** x_1 to x_n. We also call the x_j's the function's **input variables** and call w the function's **output variable**.

If f is a function of two independent variables, we usually call the independent variables x and y and the dependent variable z, and we picture the domain of f as a region in the xy-plane (Figure 14.1). If f is a function of three independent variables, we call the independent variables x, y, and z and the dependent variable w, and we picture the domain as a region in space.

In applications, we tend to use letters that remind us of what the variables stand for. To say that the volume of a right circular cylinder is a function of its radius and height, we might write $V = f(r, h)$. To be more specific, we might replace the notation $f(r, h)$ by the formula that calculates the value of V from the values of r and h, and write $V = \pi r^2 h$. In either case, r and h would be the independent variables and V the dependent variable of the function.

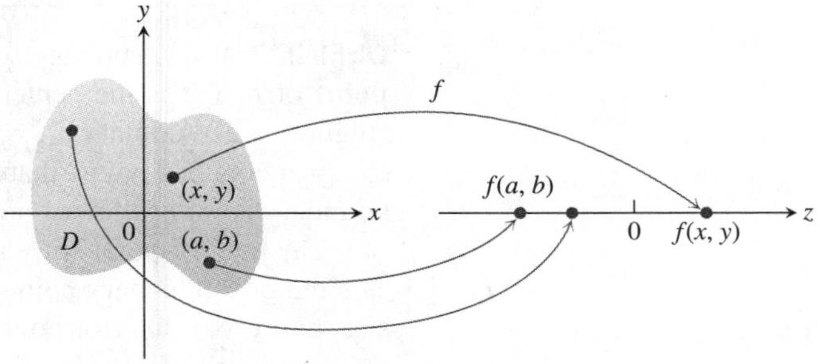

FIGURE 14.1 An arrow diagram for the function $z = f(x, y)$.

As usual, we evaluate functions defined by formulas by substituting the values of the independent variables in the formula and calculating the corresponding value of the dependent variable. For example, the value of $f(x, y, z) = \sqrt{x^2 + y^2 + z^2}$ at the point $(3, 0, 4)$ is

$$f(3, 0, 4) = \sqrt{(3)^2 + (0)^2 + (4)^2} = \sqrt{25} = 5.$$

Domains and Ranges

In defining a function of more than one variable, we follow the usual practice of excluding inputs that lead to complex numbers or division by zero. If $f(x, y) = \sqrt{y - x^2}$, then y cannot be less than x^2. If $f(x, y) = 1/(xy)$, then xy cannot be zero. The domain of a function is assumed to be the largest set for which the defining rule generates real numbers, unless the domain is otherwise specified explicitly. The range consists of the set of output values for the dependent variable.

EXAMPLE 1

(a) These are functions of two variables. Note the restrictions that apply to their domains in order to obtain a real value for the dependent variable z.

Function	Domain	Range
$z = \sqrt{y - x^2}$	$y \geq x^2$	$[0, \infty)$
$z = \dfrac{1}{xy}$	$xy \neq 0$	$(-\infty, 0) \cup (0, \infty)$
$z = \sin xy$	Entire plane	$[-1, 1]$

(b) These are functions of three variables with restrictions on some of their domains.

Function	Domain	Range
$w = \sqrt{x^2 + y^2 + z^2}$	Entire space	$[0, \infty)$
$w = \dfrac{1}{x^2 + y^2 + z^2}$	$(x, y, z) \neq (0, 0, 0)$	$(0, \infty)$
$w = xy \ln z$	Half-space $z > 0$	$(-\infty, \infty)$

■

Functions of Two Variables

On the real line, closed intervals $[a, b]$ include their boundary points while open intervals (a, b) do not. Intervals such as $[a, b)$, which includes only one of its two boundary points, are neither open nor closed. Regions in the plane can also be open, closed, or neither.

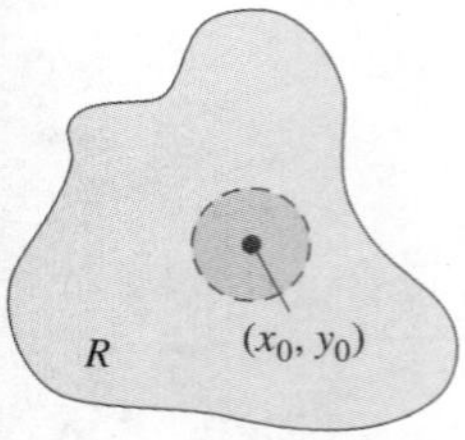

(a) Interior point

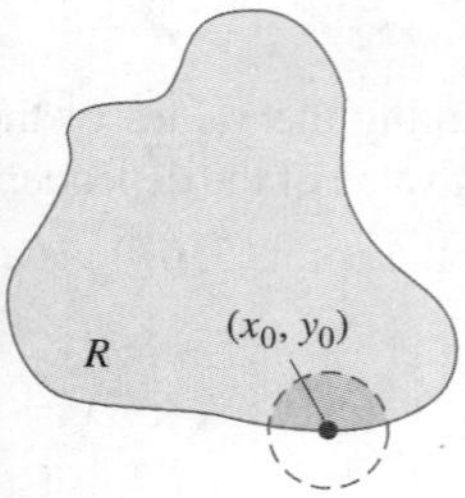

(b) Boundary point

FIGURE 14.2 Interior points and boundary points of a plane region R. An interior point is necessarily a point of R. A boundary point of R need not belong to R.

DEFINITIONS A point (x_0, y_0) in a region (set) R in the xy-plane is an **interior point** of R if it is the center of a disk of positive radius that lies entirely in R (Figure 14.2). A point (x_0, y_0) is a **boundary point** of R if every disk centered at (x_0, y_0) contains points that lie outside of R as well as points that lie in R. (The boundary point itself need not belong to R.)

The interior points of a region, as a set, make up the **interior** of the region. The region's boundary points make up its **boundary**. A region is **open** if it consists entirely of interior points. A region is **closed** if it contains all its boundary points (Figure 14.3).

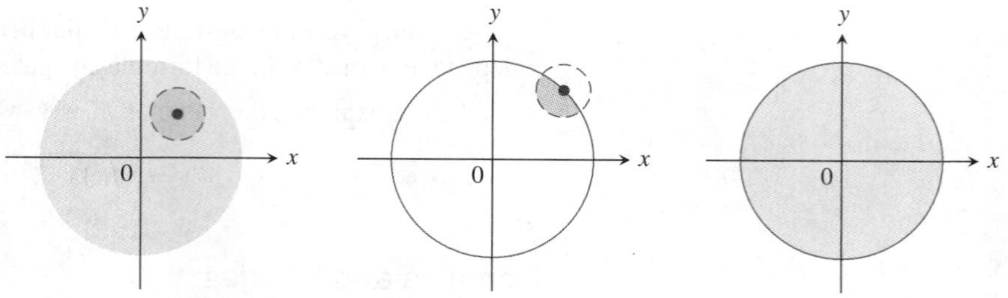

$\{(x, y) \mid x^2 + y^2 < 1\}$
Open unit disk.
Every point an interior point.

$\{(x, y) \mid x^2 + y^2 = 1\}$
Boundary of unit disk. (The unit circle.)

$\{(x, y) \mid x^2 + y^2 \le 1\}$
Closed unit disk.
Contains all boundary points.

FIGURE 14.3 Interior points and boundary points of the unit disk in the plane.

As with a half-open interval of real numbers $[a, b)$, some regions in the plane are neither open nor closed. If you start with the open disk in Figure 14.3 and add to it some, but not all, of its boundary points, the resulting set is neither open nor closed. The boundary points that *are* there keep the set from being open. The absence of the remaining boundary points keeps the set from being closed. Two interesting examples are the empty set and the entire plane. The empty set has no interior points and no boundary points. This implies that the empty set is open (because it does not contain points that are not interior points), and at the same time it is closed (because there are no boundary points that it fails to contain). The entire xy-plane is also both open and closed: open because every point in the plane is an interior point, and closed because it has no boundary points. The empty set and the entire plane are the only subsets of the plane that are both open and closed. Other sets may be open, or closed, or neither.

DEFINITIONS A region in the plane is **bounded** if it lies inside a disk of finite radius. A region is **unbounded** if it is not bounded.

Examples of *bounded* sets in the plane include line segments, triangles, interiors of triangles, rectangles, circles, and disks. Examples of *unbounded* sets in the plane include lines, coordinate axes, the graphs of functions defined on infinite intervals, quadrants, half-planes, and the plane itself.

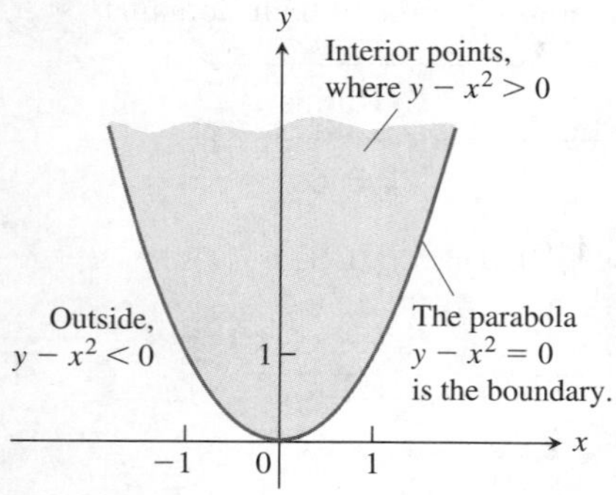

FIGURE 14.4 The domain of $f(x, y)$ in Example 2 consists of the shaded region and its bounding parabola.

EXAMPLE 2 Describe the domain of the function $f(x, y) = \sqrt{y - x^2}$.

Solution Since f is defined only where $y - x^2 \ge 0$, the domain is the closed, unbounded region shown in Figure 14.4. The parabola $y = x^2$ is the boundary of the domain. The points above the parabola make up the domain's interior. ■

Graphs, Level Curves, and Contours of Functions of Two Variables

There are two standard ways to picture the values of a function $f(x, y)$. One is to draw and label curves in the domain on which f has a constant value. The other is to sketch the surface $z = f(x, y)$ in space.

> **DEFINITIONS** The set of points in the plane where a function $f(x, y)$ has a constant value $f(x, y) = c$ is called a **level curve** of f. The set of all points $(x, y, f(x, y))$ in space, for (x, y) in the domain of f, is called the **graph** of f.

The graph of f is often called the **surface** $z = \boldsymbol{f(x, y)}$.

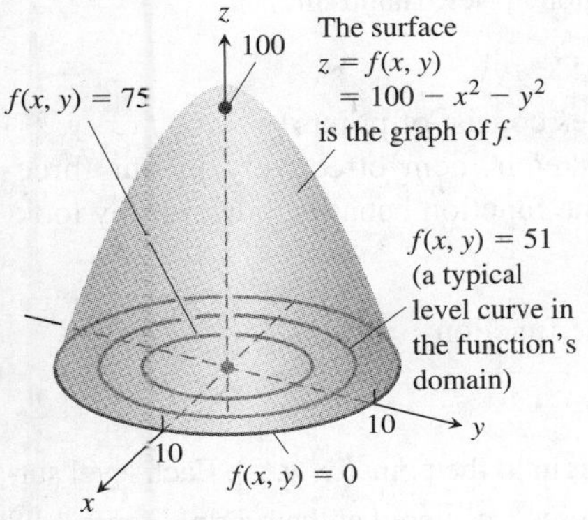

FIGURE 14.5 The graph and selected level curves of the function $f(x, y)$ in Example 3. The level curves lie in the xy-plane, which is the domain of the function $f(x, y)$.

EXAMPLE 3 Graph $f(x, y) = 100 - x^2 - y^2$ and plot the level curves $f(x, y) = 0$, $f(x, y) = 51$, and $f(x, y) = 75$ in the domain of f in the plane.

Solution The domain of f is the entire xy-plane, and the range of f is the set of real numbers less than or equal to 100. The graph is the paraboloid $z = 100 - x^2 - y^2$, the positive portion of which is shown in Figure 14.5.

The level curve $f(x, y) = 0$ is the set of points in the xy-plane at which

$$f(x, y) = 100 - x^2 - y^2 = 0, \qquad \text{or} \qquad x^2 + y^2 = 100,$$

which is the circle of radius 10 centered at the origin. Similarly, the level curves $f(x, y) = 51$ and $f(x, y) = 75$ (Figure 14.5) are the circles

$$f(x, y) = 100 - x^2 - y^2 = 51, \qquad \text{or} \qquad x^2 + y^2 = 49$$

$$f(x, y) = 100 - x^2 - y^2 = 75, \qquad \text{or} \qquad x^2 + y^2 = 25.$$

The level curve $f(x, y) = 100$ consists of the origin alone. (It is still a level curve.)

If $x^2 + y^2 > 100$, then the values of $f(x, y)$ are negative. For example, the circle $x^2 + y^2 = 144$, which is the circle centered at the origin with radius 12, gives the constant value $f(x, y) = -44$ and is a level curve of f. ■

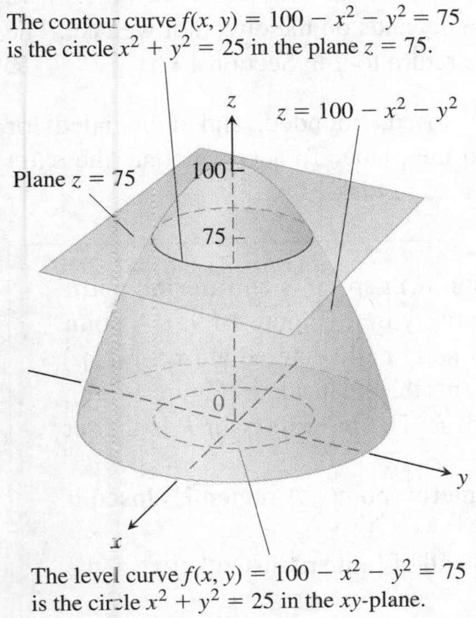

FIGURE 14.6 A plane $z = c$ parallel to the xy-plane intersecting a surface $z = f(x, y)$ produces a contour curve.

The curve in space in which the plane $z = c$ cuts a surface $z = f(x, y)$ is made up of the points that represent the function value $f(x, y) = c$. It is called the **contour curve** $f(x, y) = c$ to distinguish it from the level curve $f(x, y) = c$ in the domain of f. Figure 14.6 shows the contour curve $f(x, y) = 75$ on the surface $z = 100 - x^2 - y^2$ defined by the function $f(x, y) = 100 - x^2 - y^2$. The contour curve lies directly above the circle $x^2 + y^2 = 25$, which is the level curve $f(x, y) = 75$ in the function's domain.

The distinction between level curves and contour curves is often overlooked, and it is common to call both types of curves by the same name, relying on context to make it clear which type of curve is meant. On most maps, for example, the curves that represent constant elevation (height above sea level) are called contours, not level curves (Figure 14.7).

Functions of Three Variables

In the plane, the points where a function of two independent variables has a constant value $f(x, y) = c$ make a curve in the function's domain. In space, the points where a function of three independent variables has a constant value $f(x, y, z) = c$ make a surface in the function's domain.

> **DEFINITION** The set of points (x, y, z) in space where a function of three independent variables has a constant value $f(x, y, z) = c$ is called a **level surface** of f.

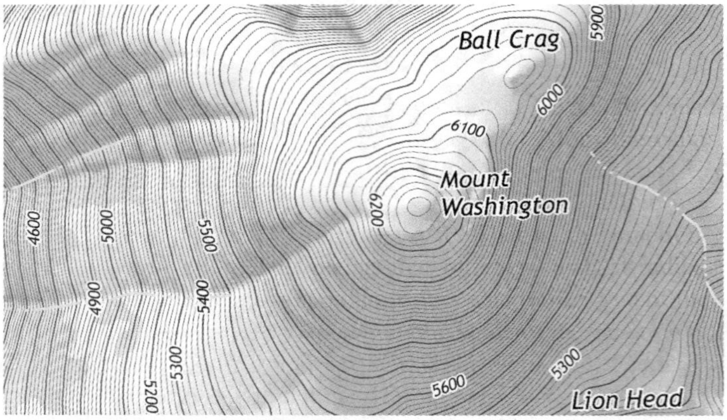

FIGURE 14.7 Contours on Mt. Washington in New Hampshire. (*Source:* United States Geological Survey)

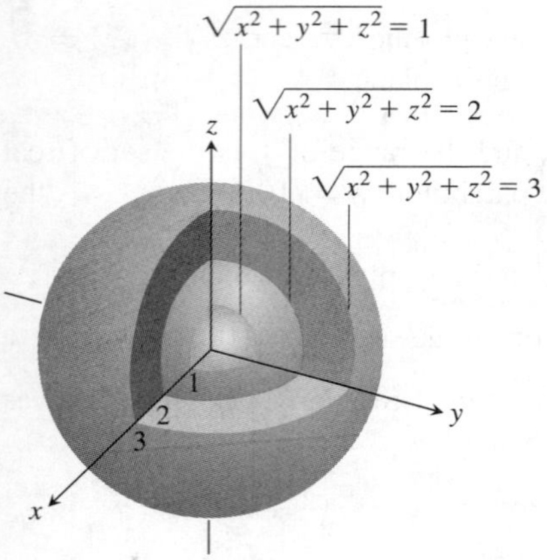

FIGURE 14.8 The level surfaces of $f(x, y, z) = \sqrt{x^2 + y^2 + z^2}$ are concentric spheres (Example 4).

Since the graphs of functions of three variables consist of points $(x, y, z, f(x, y, z))$ lying in a four-dimensional space, we cannot sketch them effectively in our three-dimensional frame of reference. We can see how the function behaves, however, by looking at its three-dimensional level surfaces.

EXAMPLE 4 Describe the level surfaces of the function

$$f(x, y, z) = \sqrt{x^2 + y^2 + z^2}.$$

Solution The value of f is the distance from the origin to the point (x, y, z). Each level surface $\sqrt{x^2 + y^2 + z^2} = c, c > 0$, is a sphere of radius c centered at the origin. Figure 14.8 shows a cutaway view of three of these spheres. The level surface $\sqrt{x^2 + y^2 + z^2} = 0$ consists of the origin alone.

We are not graphing the function here; we are looking at level surfaces in the function's domain. The level surfaces show how the function's values change as we move through its domain. If we remain on a sphere of radius c centered at the origin, the function maintains a constant value, namely c. If we move from a point on one sphere to a point on another, the function's value changes. It increases if we move away from the origin and decreases if we move toward the origin. The way the values change depends on the direction we take. The dependence of change on direction is important. We return to it in Section 14.5. ■

The definitions of interior, boundary, open, closed, bounded, and unbounded for regions in space are similar to those for regions in the plane. To accommodate the extra dimension, we use solid balls of positive radius instead of disks.

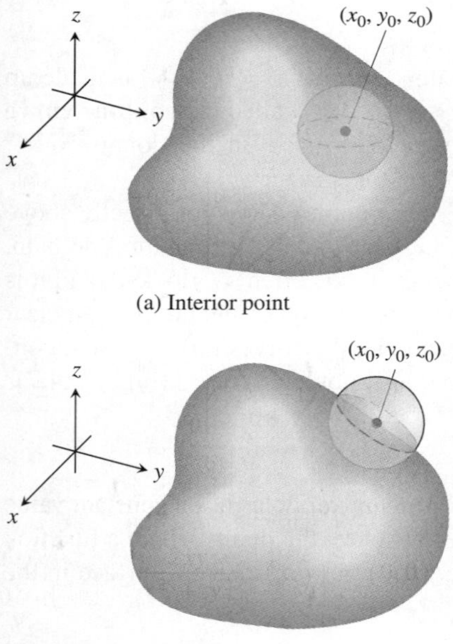

(a) Interior point

(b) Boundary point

FIGURE 14.9 Interior points and boundary points of a region in space. As with regions in the plane, a boundary point need not belong to the space region R.

DEFINITIONS A point (x_0, y_0, z_0) in a region R in space is an **interior point** of R if it is the center of a solid ball that lies entirely in R (Figure 14.9a). A point (x_0, y_0, z_0) is a **boundary point** of R if every solid ball centered at (x_0, y_0, z_0) contains points that lie outside of R as well as points that lie inside R (Figure 14.9b). The **interior** of R is the set of interior points of R. The **boundary** of R is the set of boundary points of R.

A region is **open** if it consists entirely of interior points. A region is **closed** if it contains its entire boundary.

A region is **bounded** if it lies inside a solid ball of finite radius; otherwise, the region is **unbounded**.

Examples of *open* sets in space include the interior of a sphere, the open half-space $z > 0$, the first octant (where x, y, and z are all positive), and space itself. Examples of *closed* sets in space include lines, planes, and the closed half-space $z \geq 0$. A solid sphere with part of its boundary removed or a solid cube with a missing face, edge, or corner point is *neither open nor closed*.

Functions of more than three independent variables are also important. For example, a model that measures temperature in the atmosphere may depend not only on the location of the point $P(x, y, z)$ in space, but also on the time t when it is measured, so we would write $T = f(x, y, z, t)$.

Computer Graphing

Three-dimensional graphing software makes it possible to graph functions of two variables. We can often get information more quickly from a graph than from a formula, since the surfaces reveal increasing and decreasing behavior, and high points or low points.

EXAMPLE 5 The temperature w beneath Earth's surface is a function of the depth x beneath the surface and the time t of the year. If we measure x in feet and t as the number of days elapsed from the expected date of the yearly highest surface temperature, we can model the variation in temperature with the function

$$w = \cos(1.7 \times 10^{-2}t - 0.2x)e^{-0.2x}.$$

(The temperature at 0 ft is scaled to vary from +1 to −1, so that the variation at x feet can be interpreted as a fraction of the variation at the surface.)

Figure 14.10 shows a graph of the function. At a depth of 15 ft, the variation (change in vertical amplitude in the figure) is about 5% of the surface variation. At 25 ft, there is almost no variation during the year.

The graph also shows that the temperature 15 ft below the surface is about half a year out of phase with the surface temperature. When the temperature is lowest on the surface (late January, say), it is at its highest 15 ft below. Fifteen feet below the ground, the seasons are reversed. ■

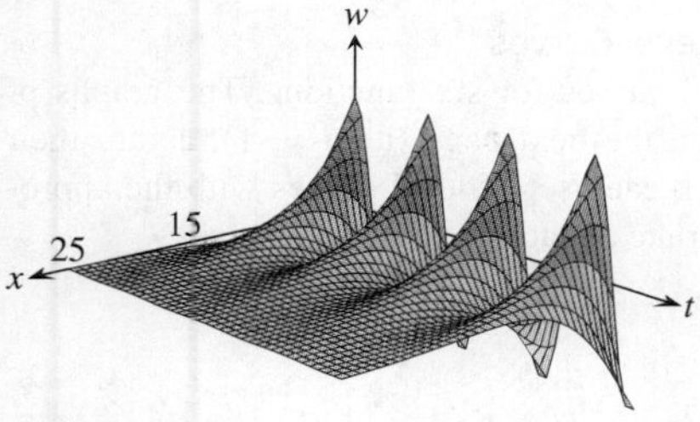

FIGURE 14.10 This graph shows the seasonal variation of the temperature below ground as a fraction of surface temperature (Example 5).

Figure 14.11 shows computer-generated graphs of a number of functions of two variables together with their level curves.

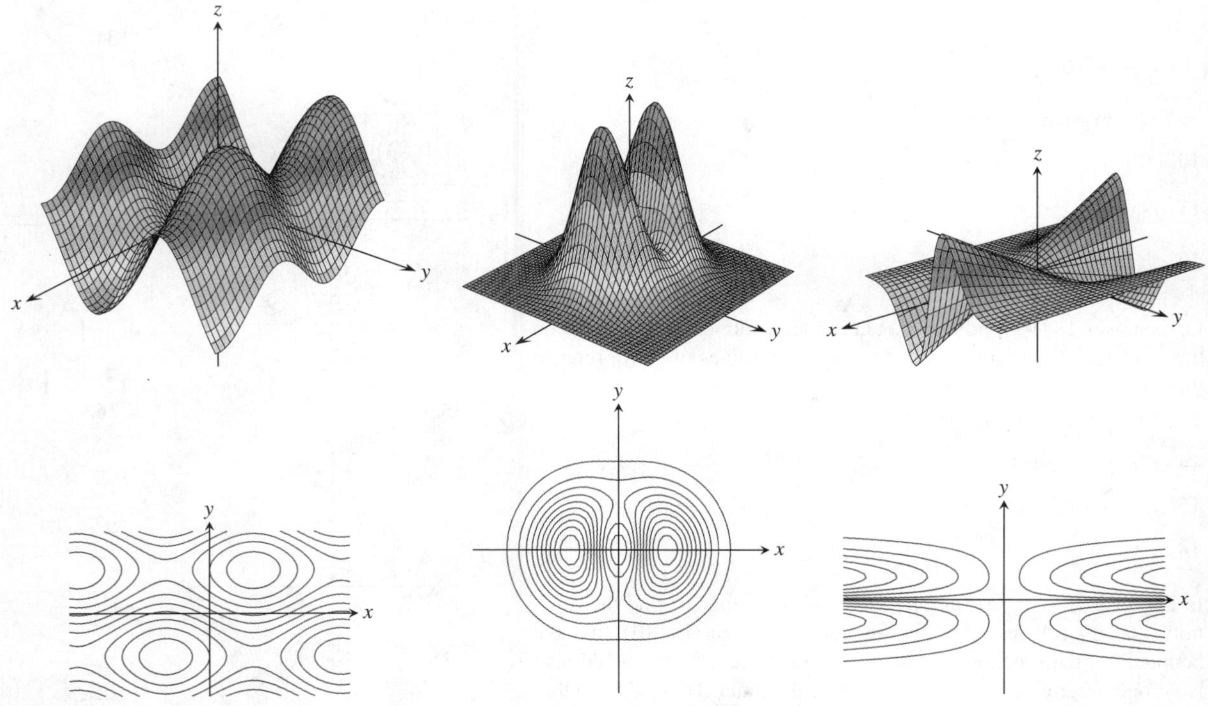

(a) $z = \sin x + 2\sin y$ (b) $z = (4x^2 + y^2)e^{-x^2-y^2}$ (c) $z = xye^{-y^2}$

FIGURE 14.11 Computer-generated graphs and level curves of typical functions of two variables.

EXERCISES 14.1

Domain, Range, and Level Curves

In Exercises 1–4, find the specific function values.

1. $f(x, y) = x^2 + xy^3$

a. $f(0, 0)$ **b.** $f(-1, 1)$

c. $f(2, 3)$ **d.** $f(-3, -2)$

2. $f(x, y) = \cos(xy)$

a. $f\left(2, \frac{\pi}{12}\right)$ **b.** $f\left(-3, \frac{\pi}{9}\right)$

c. $f\left(\pi, \frac{1}{4}\right)$ **d.** $f\left(-\frac{\pi}{2}, -3\right)$

3. $f(x, y, z) = \dfrac{x - y}{y^2 + z^2}$

a. $f(3, -1, 2)$ **b.** $f\left(1, \frac{1}{2}, -\frac{1}{4}\right)$

c. $f\left(0, -\frac{1}{3}, 0\right)$ **d.** $f(2, 2, 100)$

4. $f(x, y, z) = \sqrt{36 + x^2 - y^2 - z^2}$

a. $f(0, 0, 0)$ **b.** $f(2, -5, 3)$

c. $f(1, 2, -3)$ **d.** $f\left(\frac{3}{\sqrt{2}}, \frac{4}{\sqrt{2}}, \frac{5}{\sqrt{2}}\right)$

In Exercises 5–12, find and sketch the domain for each function.

5. $f(x, y) = \sqrt{y - x - 2}$

6. $f(x, y) = \ln(9 - x^2 - 9y^2)$

7. $f(x, y) = \dfrac{(x - 1)(y + 2)}{(y - x)(y - x^3)}$

8. $f(x, y) = \dfrac{\sin(xy)}{x^2 + y^2 - 25}$

9. $f(x, y) = \sin^{-1}(y - x^2)$

10. $f(x, y) = \ln(xy + x - y - 1)$

11. $f(x, y) = \sqrt{x^2 - y^2}$

12. $f(x, y) = \dfrac{1}{\ln(4 - x^2 - y^2)}$

In Exercises 13–16, find and sketch the level curves $f(x, y) = c$ on the same set of coordinate axes for the given values of c. We refer to these level curves as a contour map.

13. $f(x, y) = x + y - 1, \quad c = -3, -2, -1, 0, 1, 2, 3$

14. $f(x, y) = x^2 + 4y^2, \quad c = 0, 1, 4, 9, 16, 25$

15. $f(x, y) = xy, \quad c = -3, -2, -1, 0, 1, 2, 3$

16. $f(x, y) = \sqrt{25 - x^2 - y^2}, \quad c = 0, 1, 2, 3, 4$

In Exercises 17–30, **(a)** find the function's domain, **(b)** find the function's range, **(c)** describe the function's level curves, **(d)** find the boundary of the function's domain, **(e)** determine whether the domain is an open region, a closed region, or neither, and **(f)** decide whether the domain is bounded or unbounded.

17. $f(x, y) = y - x$

18. $f(x, y) = \sqrt{y - x}$

19. $f(x, y) = 9x^2 + 4y^2$

20. $f(x, y) = x^2 - y^2$

21. $f(x, y) = xy$

22. $f(x, y) = x^2/y$

23. $f(x, y) = \dfrac{1}{\sqrt{16 - x^2 - y^2}}$

24. $f(x, y) = \sqrt{16 - x^2 - y^2}$

25. $f(x, y) = \ln(x^2 + y^2)$

26. $f(x, y) = e^{-(x^2 + y^2)}$

27. $f(x, y) = \cos^{-1}(y - x)$

28. $f(x, y) = \tan^{-1}\left(\frac{y}{x}\right)$

29. $f(x, y) = \ln(x^2 + y^2 - 1)$

30. $f(x, y) = \ln(4 - x^2 - y^2)$

Matching Surfaces with Level Curves

Exercises 31–36 show level curves for six functions. The graphs of these functions are given on the next page (items a–f), as are their equations (items g–l). Match each set of level curves with the appropriate graph and the appropriate equation.

31.

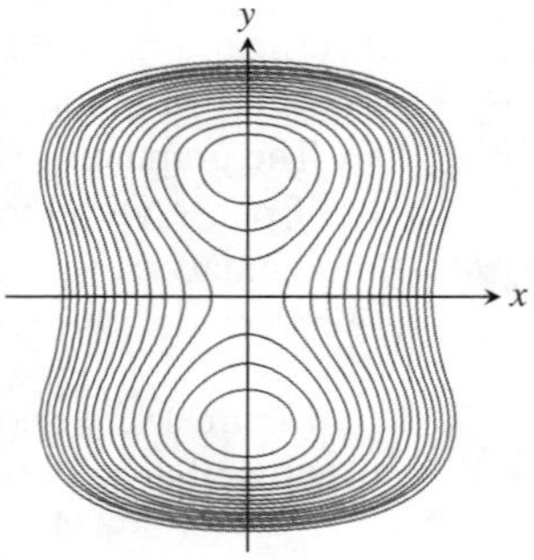

32.

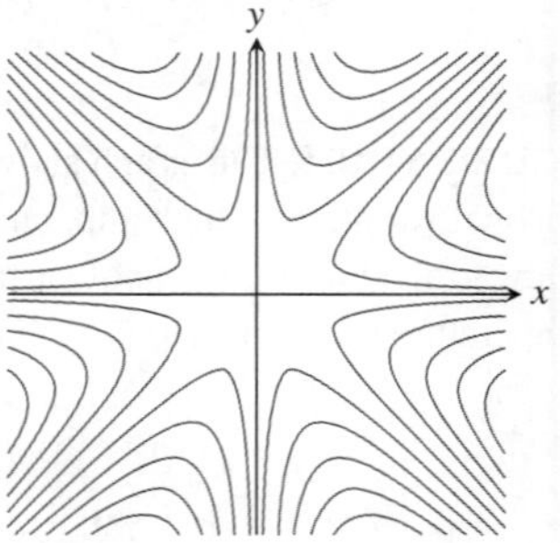

33.

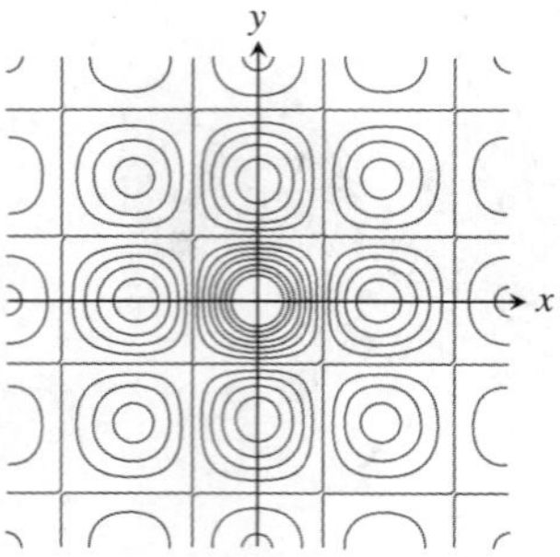

34.

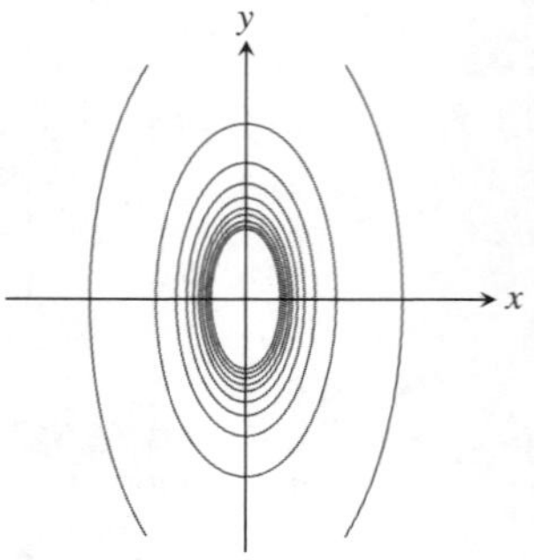

35.

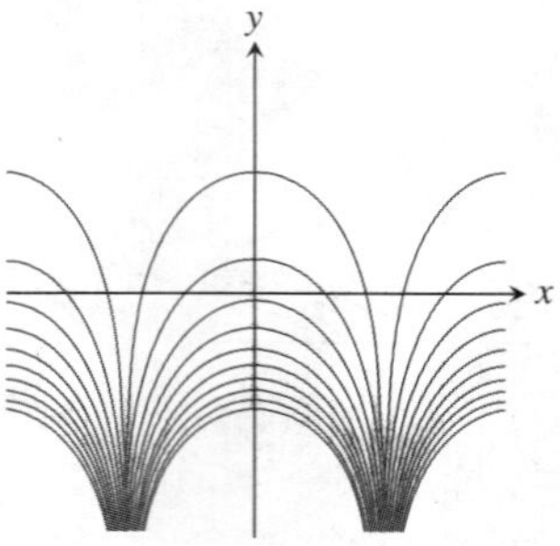

36.

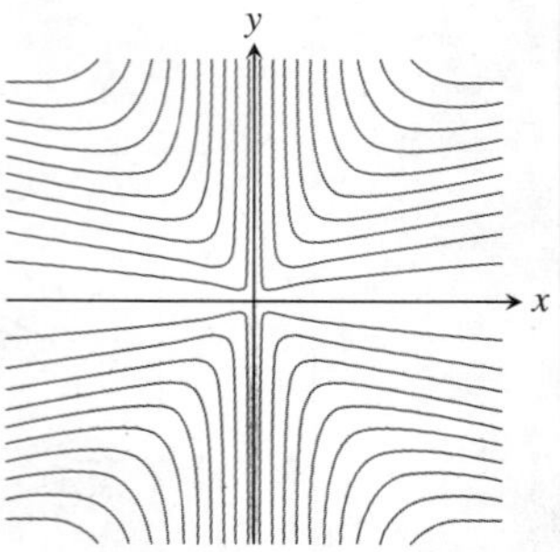

a.

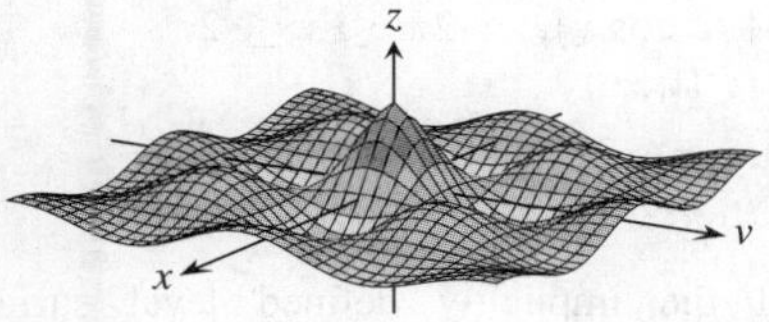

b.

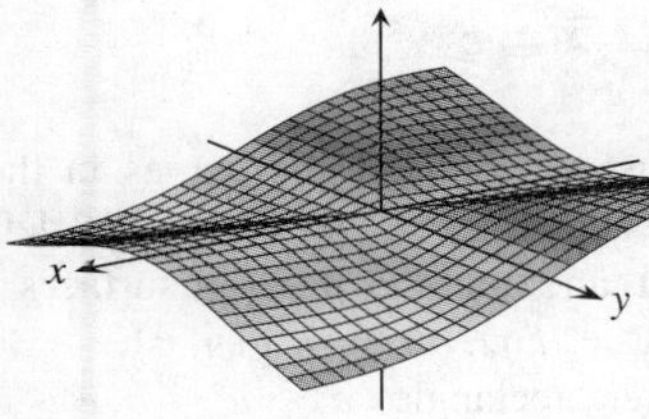

c.

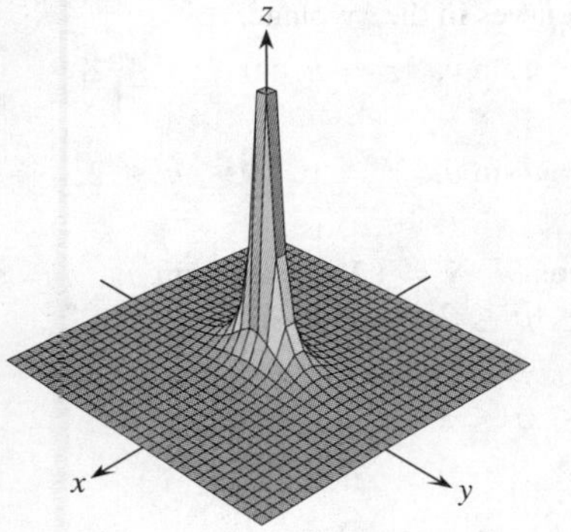

d.

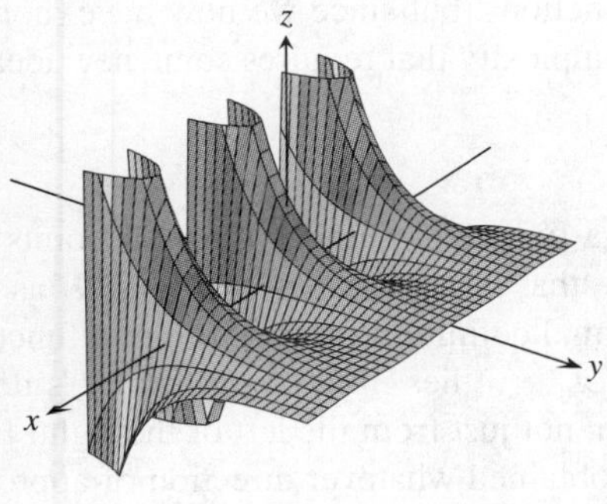

e.

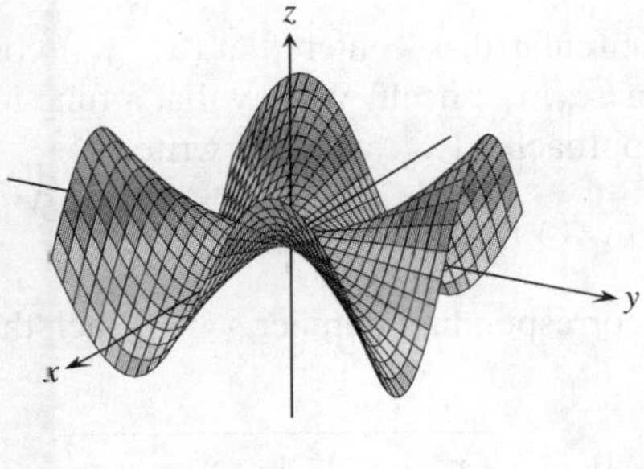

f.

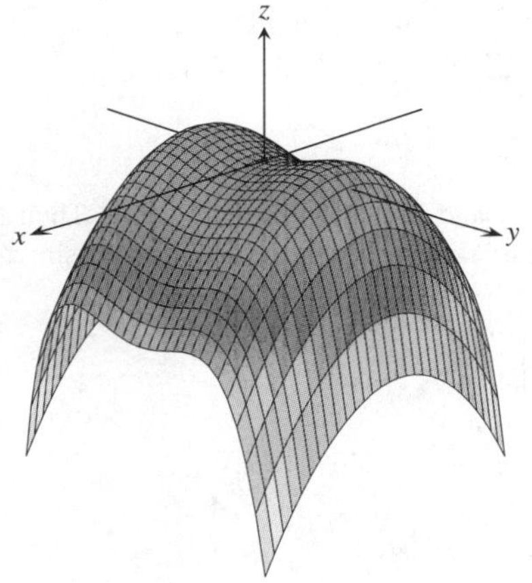

g. $z = -\dfrac{xy^2}{x^2 + y^2}$

h. $z = y^2 - y^4 - x^2$

i. $z = (\cos x)(\cos y)e^{-\sqrt{x^2+y^2}/4}$

j. $z = e^{-y}\cos x$

k. $z = \dfrac{1}{4x^2 + y^2}$

l. $z = \dfrac{xy(x^2 - y^2)}{x^2 + y^2}$

Functions of Two Variables

Display the values of the functions in Exercises 37–48 in two ways: **(a)** by sketching the surface $z = f(x, y)$ and **(b)** by drawing an assortment of level curves in the function's domain. Label each level curve with its function value.

37. $f(x, y) = y^2$

38. $f(x, y) = \sqrt{x}$

39. $f(x, y) = x^2 - y^2$

40. $f(x, y) = \sqrt{x^2 + y^2}$

41. $f(x, y) = x^2 - y$

42. $f(x, y) = 4 - x^2 - y^2$

43. $f(x, y) = x^2 + 4y^2$

44. $f(x, y) = 6 - 2x - 3y$

45. $f(x, y) = 1 - |y|$

46. $f(x, y) = 1 - |x| - |y|$

47. $f(x, y) = \sqrt{x^2 + y^2 + 4}$

48. $f(x, y) = \sqrt{x^2 + y^2 - 4}$

Finding Level Curves

In Exercises 49–52, find an equation for, and sketch the graph of, the level curve of the function $f(x, y)$ that passes through the given point.

49. $f(x, y) = 25 - x^2 - y^2, \quad \left(\sqrt{2}, 3\sqrt{2}\right)$

50. $f(x, y) = \sqrt{x^2 - 1}, \quad (1, 0)$

51. $f(x, y) = \sqrt{x + y^2 - 3}, \quad (3, -1)$

52. $f(x, y) = \dfrac{2x - 2y}{x - y - 1}, \quad (2, -1)$

Sketching Level Surfaces

In Exercises 53–60, sketch a typical level surface for the function.

53. $f(x, y, z) = x^2 + y^2 + z^2$

54. $f(x, y, z) = \ln(x^2 + y^2 + z^2)$

55. $f(x, y, z) = x + z$

56. $f(x, y, z) = z$

57. $f(x, y, z) = x^2 + y^2$

58. $f(x, y, z) = y^2 + z^2$

59. $f(x, y, z) = z - x^2 - y^2$

60. $f(x, y, z) = (x^2/25) + (y^2/16) + (z^2/9)$

Finding Level Surfaces

In Exercises 61–64, find an equation for the level surface of the function through the given point.

61. $f(x, y, z) = \sqrt{x - y} - \ln z, \quad (3, -1, 1)$

62. $f(x, y, z) = \ln(x^2 - 2y + z^2)$, $(-2, 1, 2)$

63. $g(x, y, z) = \sqrt{x^2 + y^2 + z^2}$, $(1, -1, \sqrt{2})$

64. $g(x, y, z) = \dfrac{2x - y + z}{x + y + z}$, $(-1, 0, -2)$

In Exercises 65–68, find and sketch the domain of f. Then find an equation for the level curve or surface of the function passing through the given point.

65. $f(x, y) = \displaystyle\sum_{n=0}^{\infty}\left(\frac{x}{y}\right)^n$, $(1, 2)$

66. $g(x, y, z) = \displaystyle\sum_{n=0}^{\infty}\frac{(x + y)^n}{n!z^n}$, $(\ln 4, \ln 9, 2)$

67. $f(x, y) = \displaystyle\int_x^y \frac{d\theta}{\sqrt{1 - \theta^2}}$, $(0, 1)$

68. $g(x, y, z) = \displaystyle\int_x^y \frac{dt}{1 + t^2} + \int_0^z \frac{d\theta}{\sqrt{4 - \theta^2}}$, $(0, 1, \sqrt{3})$

COMPUTER EXPLORATIONS

Use a CAS to perform the following steps for each of the functions in Exercises 69–72.

a. Plot the surface over the given rectangle.

b. Plot several level curves in the rectangle.

c. Plot the level curve of f through the given point.

69. $f(x, y) = x \sin\frac{y}{2} + y \sin 2x$, $0 \le x \le 5\pi$, $0 \le y \le 5\pi$, $P(3\pi, 3\pi)$

70. $f(x, y) = (\sin x)(\cos y)e^{\sqrt{x^2+y^2}/8}$, $0 \le x \le 5\pi$, $0 \le y \le 5\pi$, $P(4\pi, 4\pi)$

71. $f(x, y) = \sin(x + 2\cos y)$, $-2\pi \le x \le 2\pi$, $-2\pi \le y \le 2\pi$, $P(\pi, \pi)$

72. $f(x, y) = e^{(x^{0.1} - y)} \sin(x^2 + y^2)$, $0 \le x \le 2\pi$, $-2\pi \le y \le \pi$, $P(\pi, -\pi)$

Use a CAS to plot the implicitly defined level surfaces in Exercises 73–76.

73. $4\ln(x^2 + y^2 + z^2) = 1$

74. $x^2 + z^2 = 1$

75. $x + y^2 - 3z^2 = 1$

76. $\sin\left(\frac{x}{2}\right) - (\cos y)\sqrt{x^2 + z^2} = 2$

Parametrized Surfaces Just as you describe curves in the plane parametrically with a pair of equations $x = f(t)$, $y = g(t)$ defined on some parameter interval I, you can sometimes describe surfaces in space with a triple of equations $x = f(u, v)$, $y = g(u, v)$, $z = h(u, v)$ defined on some parameter rectangle $a \le u \le b$, $c \le v \le d$. Many computer algebra systems permit you to plot such surfaces in *parametric mode*. (Parametrized surfaces are discussed in detail in Section 16.5.) Use a CAS to plot the surfaces in Exercises 77–80. Also plot several level curves in the xy-plane.

77. $x = u\cos v$, $y = u \sin v$, $z = u$, $0 \le u \le 2$, $0 \le v \le 2\pi$

78. $x = u\cos v$, $y = u\sin v$, $z = v$, $0 \le u \le 2$, $0 \le v \le 2\pi$

79. $x = (2 + \cos u)\cos v$, $y = (2 + \cos u)\sin v$, $z = \sin u$, $0 \le u \le 2\pi$, $0 \le v \le 2\pi$

80. $x = 2\cos u \cos v$, $y = 2\cos u \sin v$, $z = 2\sin u$, $0 \le u \le 2\pi$, $0 \le v \le \pi$

14.2 Limits and Continuity in Higher Dimensions

In this section we develop limits and continuity for multivariable functions. The theory is similar to that developed for single-variable functions, but since we now have more than one independent variable, there is additional complexity that requires some new ideas.

Limits for Functions of Two Variables

If the values of $f(x, y)$ lie arbitrarily close to a fixed real number L for all points (x, y) sufficiently close to a point (x_0, y_0), we say that f approaches the limit L as (x, y) approaches (x_0, y_0). This is similar to the informal definition for the limit of a function of a single variable. Notice, however, that when (x_0, y_0) lies in the interior of f's domain, (x, y) can approach (x_0, y_0) from any direction, not just from the left or the right. For the limit to exist, the same limiting value must be obtained whatever direction of approach is taken. We illustrate this issue in several examples following the definition.

DEFINITION Suppose that every open circular disk centered at (x_0, y_0) contains a point in the domain of f other than (x_0, y_0) itself. We say that a function $f(x, y)$ approaches the **limit** L as (x, y) approaches (x_0, y_0), and write

$$\lim_{(x,y)\to(x_0,y_0)} f(x, y) = L,$$

if, for every number $\varepsilon > 0$, there exists a corresponding number $\delta > 0$ such that for all (x, y) in the domain of f,

$$|f(x, y) - L| < \varepsilon \quad \text{whenever} \quad 0 < \sqrt{(x - x_0)^2 + (y - y_0)^2} < \delta.$$

The definition of limit says that the distance between $f(x, y)$ and L becomes arbitrarily small whenever the distance from (x, y) to (x_0, y_0) is made sufficiently small (but not 0). The definition applies to interior points (x_0, y_0) as well as boundary points of the domain of f, although a boundary point need not lie within the domain. The points (x, y) that approach (x_0, y_0) are always taken to be in the domain of f. See Figure 14.12.

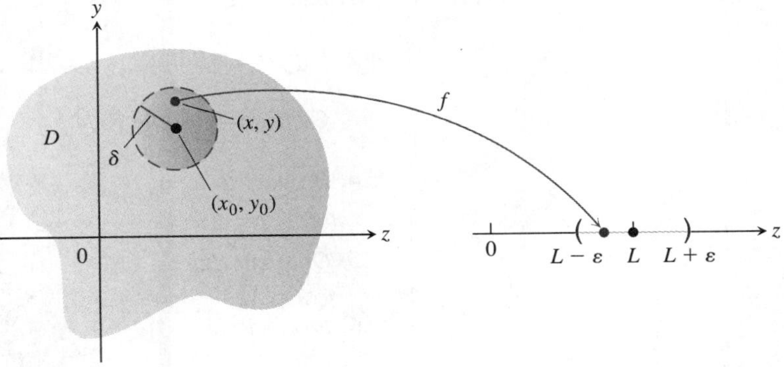

FIGURE 14.12 In the limit definition, δ is the radius of a disk centered at (x_0, y_0). For all points (x, y) within this disk, the function values $f(x, y)$ lie inside the corresponding interval $(L - \varepsilon, L + \varepsilon)$.

As for functions of a single variable, it can be shown that

$$\lim_{(x,y)\to(x_0,y_0)} x = x_0 \tag{1}$$

$$\lim_{(x,y)\to(x_0,y_0)} y = y_0 \tag{2}$$

$$\lim_{(x,y)\to(x_0,y_0)} k = k \quad \text{(any number } k\text{)}. \tag{3}$$

For example, in the first limit statement above, $f(x, y) = x$ and $L = x_0$. Using the definition of limit, suppose that $\varepsilon > 0$ is chosen. If we let δ equal this ε, we see that if

$$0 < \sqrt{(x - x_0)^2 + (y - y_0)^2} < \delta = \varepsilon,$$

then

$$\begin{aligned} \sqrt{(x - x_0)^2} &< \varepsilon & \quad & (x - x_0)^2 \le (x - x_0)^2 + (y - y_0)^2 \\ |x - x_0| &< \varepsilon & & \sqrt{a^2} = |a| \\ |f(x, y) - x_0| &< \varepsilon. & & x = f(x, y) \end{aligned}$$

That is,

$$|f(x, y) - x_0| < \varepsilon \quad \text{whenever} \quad 0 < \sqrt{(x - x_0)^2 + (y - y_0)^2} < \delta.$$

So a δ has been found satisfying the requirement of the definition, and therefore we have proved that

$$\lim_{(x,y)\to(x_0,y_0)} f(x, y) = \lim_{(x,y)\to(x_0,y_0)} x = x_0.$$

Equation (1) is a special case of the more general formula

$$\lim_{(x,y)\to(x_0,y_0)} g(x) = \lim_{x\to x_0} g(x), \tag{4}$$

according to which, if $f(x, y)$ can be expressed as a function g of a single variable x, then $\lim_{(x,y)\to(x_0,y_0)} f(x, y)$ depends only on what happens to g as x approaches x_0. Similarly, the following formula generalizes Equation (2):

$$\lim_{(x,y)\to(x_0,y_0)} h(y) = \lim_{y\to y_0} h(y) \tag{5}$$

As with single-variable functions, the limit of the sum of two functions is the sum of their limits (when they both exist), with similar results for the limits of the differences, constant multiples, products, quotients, powers, and roots. These facts are summarized in Theorem 1.

THEOREM 1—Properties of Limits of Functions of Two Variables
The following rules hold if L, M, and k are real numbers and

$$\lim_{(x,y)\to(x_0,y_0)} f(x,y) = L \quad \text{and} \quad \lim_{(x,y)\to(x_0,y_0)} g(x,y) = M.$$

1. *Sum Rule:* $\lim_{(x,y)\to(x_0,y_0)} [f(x,y) + g(x,y)] = L + M$
2. *Difference Rule:* $\lim_{(x,y)\to(x_0,y_0)} [f(x,y) - g(x,y)] = L - M$
3. *Constant Multiple Rule:* $\lim_{(x,y)\to(x_0,y_0)} kf(x,y) = kL \quad \text{(any number } k)$
4. *Product Rule:* $\lim_{(x,y)\to(x_0,y_0)} [f(x,y) \cdot g(x,y)] = L \cdot M$
5. *Quotient Rule:* $\lim_{(x,y)\to(x_0,y_0)} \dfrac{f(x,y)}{g(x,y)} = \dfrac{L}{M}, \quad M \neq 0$
6. *Power Rule:* $\lim_{(x,y)\to(x_0,y_0)} [f(x,y)]^n = L^n$, n a positive integer
7. *Root Rule:* $\lim_{(x,y)\to(x_0,y_0)} \sqrt[n]{f(x,y)} = \sqrt[n]{L} = L^{1/n}$,
   n a positive integer, and if n is even, we assume that $L > 0$.
8. *Composition Rule:* If $h(z)$ is continuous at $z = L$, then
   $\lim_{(x,y)\to(x_0,y_0)} h(f(x,y)) = h(L)$.

Although we will not prove Theorem 1 here, we give an informal discussion of why it is true. If (x, y) is sufficiently close to (x_0, y_0), then $f(x, y)$ is close to L and $g(x, y)$ is close to M (from the informal interpretation of limits). It is then reasonable that $f(x, y) + g(x, y)$ is close to $L + M$; $f(x, y) - g(x, y)$ is close to $L - M$; $kf(x, y)$ is close to kL; $f(x, y)g(x, y)$ is close to LM; and $f(x, y)/g(x, y)$ is close to L/M if $M \neq 0$. Similarly, powers and roots of f are close to those of L, and a continuous function h composed with f has a value close to its value $h(L)$ when applied to L.

When we apply Theorem 1 and Equations (1)–(3) to polynomials and rational functions, we obtain the useful result that the limits of these functions as $(x, y) \to (x_0, y_0)$ can be calculated by evaluating the functions at (x_0, y_0). The only requirement is that the rational functions be defined at (x_0, y_0).

EXAMPLE 1 In this example, we combine Equations (1)–(5) with the results in Theorem 1 to calculate the limits.

(a) $$\lim_{(x,y)\to(0,1)} \frac{x - xy + 3}{x^2y + 5xy - y^3} = \frac{0 - (0)(1) + 3}{(0)^2(1) + 5(0)(1) - (1)^3} = -3$$

(b) $$\begin{aligned} \lim_{(x,y)\to(3,-4)} \sqrt{x^2 + y^2} &= \sqrt{\lim_{(x,y)\to(3,-4)} (x^2 + y^2)} && \text{Rule 7} \\ &= \sqrt{3^2 + (-4)^2} && \text{Rules 1 and 6 and Eq. (1) and (2)} \\ &= \sqrt{25} = 5 \end{aligned}$$

(c) $$\begin{aligned} \lim_{(x,y)\to(\pi/2,0)} \left(\frac{x}{\sin x} - \frac{\sin y}{y} \right) &= \lim_{(x,y)\to(\pi/2,0)} \frac{x}{\sin x} - \lim_{(x,y)\to(\pi/2,0)} \frac{\sin y}{y} && \text{Rule 2} \\ &= \lim_{x\to\pi/2} \frac{x}{\sin x} - \lim_{y\to 0} \frac{\sin y}{y} && \text{Eq. (4) and (5)} \\ &= \frac{\pi}{2} - 1 && \text{Theorem 6, Section 2.4} \end{aligned}$$ ■

EXAMPLE 2 Find $\lim_{(x,y)\to(0,0)} \dfrac{x^2 - xy}{\sqrt{x} - \sqrt{y}}$.

Solution Since the denominator $\sqrt{x} - \sqrt{y}$ approaches 0 as $(x, y) \to (0, 0)$, we cannot use the Quotient Rule from Theorem 1. If we multiply numerator and denominator by $\sqrt{x} + \sqrt{y}$, however, we produce an equivalent fraction whose limit we *can* find:

$$\begin{aligned}
\lim_{(x,y)\to(0,0)} \frac{x^2 - xy}{\sqrt{x} - \sqrt{y}} &= \lim_{(x,y)\to(0,0)} \frac{(x^2 - xy)(\sqrt{x} + \sqrt{y})}{(\sqrt{x} - \sqrt{y})(\sqrt{x} + \sqrt{y})} && \text{Multiply by a form equal to 1.}\\
&= \lim_{(x,y)\to(0,0)} \frac{x(x - y)(\sqrt{x} + \sqrt{y})}{x - y} && \text{Algebra}\\
&= \lim_{(x,y)\to(0,0)} x(\sqrt{x} + \sqrt{y}) && \text{Cancel the nonzero factor } (x - y).\\
&= \left(\lim_{(x,y)\to(0,0)} x\right)\left[\left(\lim_{(x,y)\to(0,0)} \sqrt{x}\right) + \left(\lim_{(x,y)\to(0,0)} \sqrt{y}\right)\right] && \text{Rules 4 and 1}\\
&= \left(\lim_{(x,y)\to(0,0)} x\right)\left[\sqrt{\lim_{(x,y)\to(0,0)} x} + \sqrt{\lim_{(x,y)\to(0,0)} y}\right] && \text{Rule 7}\\
&= (0)\left[\sqrt{0} + \sqrt{0}\right] = 0 && \text{Eq. (1) and (2)}
\end{aligned}$$

We can cancel the factor $(x - y)$ because the path $y = x$ (where we would have $x - y = 0$) is *not* in the domain of the function

$$f(x, y) = \frac{x^2 - xy}{\sqrt{x} - \sqrt{y}}.$$

■

EXAMPLE 3 Find $\lim_{(x,y)\to(0,0)} \dfrac{4xy^2}{x^2 + y^2}$ if it exists.

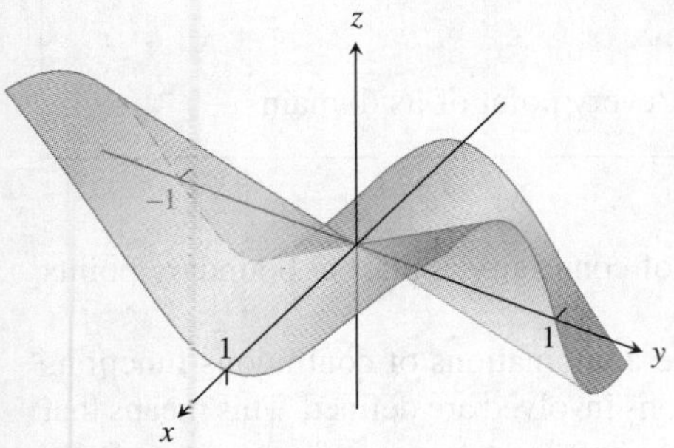

FIGURE 14.13 The surface graph suggests that the limit of the function in Example 3 must be 0, if it exists.

Solution We first observe that along the line $x = 0$, the function always has value 0 when $y \neq 0$. Likewise, along the line $y = 0$, the function has value 0 provided $x \neq 0$. So if the limit does exist as (x, y) approaches $(0, 0)$, the value of the limit must be 0 (see Figure 14.13). To see whether this is true, we apply the definition of limit.

Let $\varepsilon > 0$ be given, but arbitrary. We want to find a $\delta > 0$ such that

$$\left|\frac{4xy^2}{x^2 + y^2} - 0\right| < \varepsilon \quad \text{whenever} \quad 0 < \sqrt{x^2 + y^2} < \delta$$

or

$$\frac{4\,|x|\,y^2}{x^2 + y^2} < \varepsilon \quad \text{whenever} \quad 0 < \sqrt{x^2 + y^2} < \delta.$$

Since $y^2 \le x^2 + y^2$, we have that

$$\frac{4\,|x|\,y^2}{x^2 + y^2} \le 4\,|x| = 4\sqrt{x^2} \le 4\sqrt{x^2 + y^2}. \qquad \frac{y^2}{x^2 + y^2} \le 1$$

So if we choose $\delta = \varepsilon/4$ and let $0 < \sqrt{x^2 + y^2} < \delta$, we get

$$\left|\frac{4xy^2}{x^2 + y^2} - 0\right| \le 4\sqrt{x^2 + y^2} < 4\delta = 4\left(\frac{\varepsilon}{4}\right) = \varepsilon.$$

It follows from the definition that

$$\lim_{(x,y)\to(0,0)} \frac{4xy^2}{x^2 + y^2} = 0.$$

■

EXAMPLE 4 If $f(x, y) = \dfrac{y}{x}$, does $\lim\limits_{(x,y)\to(0,0)} f(x, y)$ exist?

Solution The domain of f does not include the y-axis, so we do not consider any points (x, y) where $x = 0$ in the approach toward the origin $(0, 0)$. Along the x-axis, the value of the function is $f(x, 0) = 0$ for all $x \neq 0$. So if the limit does exist as $(x, y) \to (0, 0)$, the value of the limit must be $L = 0$. On the other hand, along the line $y = x$, the value of the function is $f(x, x) = x/x = 1$ for all $x \neq 0$. That is, the function f approaches the value 1 along the line $y = x$. This means that for every disk of radius δ centered at $(0, 0)$, the disk will contain points $(x, 0)$ on the x-axis where the value of the function is 0, and also points (x, x) along the line $y = x$ where the value of the function is 1. So no matter how small we choose δ as the radius of the disk in Figure 14.12, there will be points within the disk for which the function values differ by 1. Therefore, the limit cannot exist because we can take ε to be any number less than 1 in the limit definition and deny that $L = 0$ or 1, or any other real number. The limit does not exist because we have different limiting values along different paths approaching the point $(0, 0)$. ■

Continuity

As with functions of a single variable, continuity is defined in terms of limits.

DEFINITION Suppose that every open circular disk centered at (x_0, y_0) contains a point in the domain of f other than (x_0, y_0) itself. Then a function $f(x, y)$ is **continuous at the point** (x_0, y_0) if

1. f is defined at (x_0, y_0),
2. $\lim\limits_{(x,y)\to(x_0,y_0)} f(x, y)$ exists, and
3. $\lim\limits_{(x,y)\to(x_0,y_0)} f(x, y) = f(x_0, y_0)$.

A function is **continuous** if it is continuous at every point of its domain.

As with the definition of limit, the definition of continuity applies at boundary points as well as interior points of the domain of f.

A consequence of Theorem 1 is that algebraic combinations of continuous functions are continuous at every point at which all the functions involved are defined. This means that sums, differences, constant multiples, products, quotients, and powers of continuous functions are continuous where defined. In particular, polynomials and rational functions of two variables are continuous at every point at which they are defined.

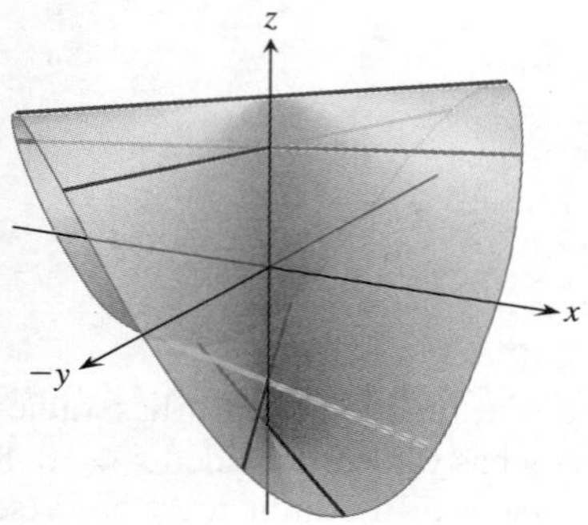

(a)

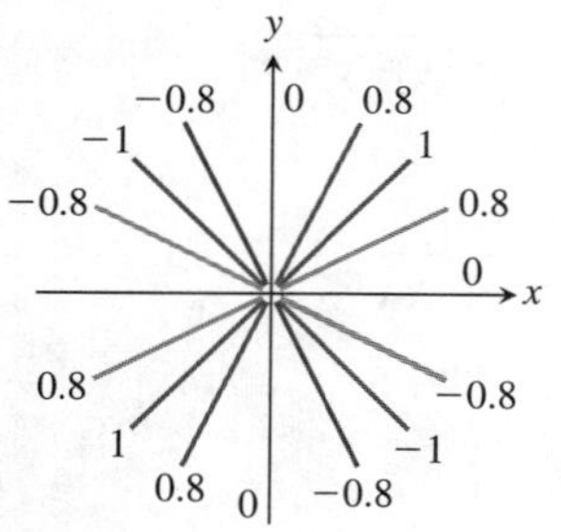

(b)

FIGURE 14.14 (a) The graph of

$$f(x, y) = \begin{cases} \dfrac{2xy}{x^2 + y^2}, & (x, y) \neq (0, 0) \\ 0, & (x, y) = (0, 0). \end{cases}$$

The function is continuous at every point except the origin. (b) The value of f along each line $y = mx$, $x \neq 0$, is constant but varies with m (Example 5).

EXAMPLE 5 Show that

$$f(x, y) = \begin{cases} \dfrac{2xy}{x^2 + y^2}, & (x, y) \neq (0, 0) \\ 0, & (x, y) = (0, 0). \end{cases}$$

is continuous at every point except the origin (Figure 14.14).

Solution The function f is continuous at every point (x, y) except $(0, 0)$ because its values at points other than $(0, 0)$ are given by a rational function of x and y, and therefore at those points the limiting value is simply obtained by substituting the values of x and y into that rational expression.

At $(0, 0)$, the value of f is defined, but f has no limit as $(x, y) \to (0, 0)$. The reason is that different paths of approach to the origin can lead to different results, as we now see.

For every value of m, the function f has a constant value on the "punctured" line $y = mx, x \neq 0$, because

$$f(x, y)\bigg|_{y=mx} = \frac{2xy}{x^2 + y^2}\bigg|_{y=mx} = \frac{2x(mx)}{x^2 + (mx)^2} = \frac{2mx^2}{x^2 + m^2x^2} = \frac{2m}{1 + m^2}.$$

Therefore, f has this number as its limit as (x, y) approaches $(0, 0)$ along the line:

$$\lim_{\substack{(x,y)\to(0,0)\\ \text{along } y=mx}} f(x, y) = \lim_{(x,y)\to(0,0)} \left[f(x, y)\bigg|_{y=mx} \right] = \frac{2m}{1 + m^2}.$$

This limit changes with each value of the slope m. There is therefore no single number we may call the limit of f as (x, y) approaches the origin. The limit fails to exist, and the function is not continuous at the origin. ■

Examples 4 and 5 illustrate an important point about limits of functions of two or more variables. For a limit to exist at a point, the limit must be the same along every approach path. This result is analogous to the single-variable case where both the left- and right-sided limits had to have the same value. For functions of two or more variables, if we ever find paths with different limits, we know the function has no limit at the point they approach.

Two-Path Test for Nonexistence of a Limit

If a function $f(x, y)$ has different limits along two different paths in the domain of f as (x, y) approaches (x_0, y_0), then $\lim_{(x,y)\to(x_0,y_0)} f(x, y)$ does not exist.

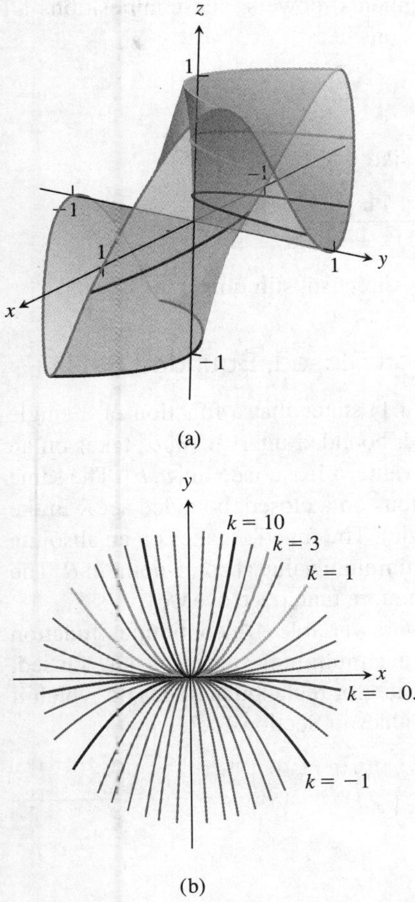

FIGURE 14.15 (a) The graph of $f(x, y) = 2x^2y/(x^4 + y^2)$. (b) Along each path $y = kx^2, x \neq 0$, the value of f is constant, but varies with k (Example 6).

EXAMPLE 6 Show that the function

$$f(x, y) = \frac{2x^2y}{x^4 + y^2}$$

(Figure 14.15) has no limit as (x, y) approaches $(0, 0)$.

Solution As (x, y) approaches $(0, 0)$, both the numerator and the denominator approach 0, which gives the indeterminate form 0/0. We examine the values of f along parabolic curves that end at $(0, 0)$. Along the curve $y = kx^2, x \neq 0$, the function has the constant value

$$f(x, y)\bigg|_{y=kx^2} = \frac{2x^2y}{x^4 + y^2}\bigg|_{y=kx^2} = \frac{2x^2(kx^2)}{x^4 + (kx^2)^2} = \frac{2kx^4}{x^4 + k^2x^4} = \frac{2k}{1 + k^2}.$$

Therefore,

$$\lim_{\substack{(x,y)\to(0,0)\\ \text{along } y=kx^2}} f(x, y) = \lim_{(x,y)\to(0,0)} \left[f(x, y)\bigg|_{y=kx^2} \right] = \frac{2k}{1 + k^2}.$$

This limit varies with the path of approach. If (x, y) approaches $(0, 0)$ along the parabola $y = x^2$, for instance, $k = 1$ and the limit is 1. If (x, y) approaches $(0, 0)$ along the x-axis, $k = 0$ and the limit is 0. By the two-path test, f has no limit as (x, y) approaches $(0, 0)$. ■

It can be shown that the function in Example 6 has limit 0 along every straight line path $y = mx$ (Exercise 57). This implies the following observation:

Having the same limit along all straight lines approaching (x_0, y_0) does not imply that a limit exists at (x_0, y_0).

Whenever it is correctly defined, the composition of continuous functions is also continuous. The only requirement is that each function be continuous where it is applied. The proof, omitted here, is similar to that for functions of a single variable (Theorem 9 in Section 2.5).

Continuity of Compositions
If f is continuous at (x_0, y_0) and g is a single-variable function continuous at $f(x_0, y_0)$, then the composition $h = g \circ f$ defined by $h(x, y) = g(f(x, y))$ is also continuous at (x_0, y_0).

For example, the composite functions

$$e^{x-y}, \qquad \cos\frac{xy}{x^2+1}, \qquad \ln(1 + x^2y^2)$$

are continuous at every point (x, y).

Functions of More Than Two Variables

The definitions of limit and continuity for functions of two variables and the conclusions about limits and continuity for sums, products, quotients, powers, and compositions all extend to functions of three or more variables. Functions like

$$\ln(x + y + z) \qquad \text{and} \qquad \frac{y \sin z}{x - 1}$$

are continuous throughout their domains, and limits like

$$\lim_{P \to (1,0,-1)} \frac{e^{x+z}}{z^2 + \cos\sqrt{xy}} = \frac{e^{1-1}}{(-1)^2 + \cos 0} = \frac{1}{2},$$

where P denotes the point (x, y, z), may be found by direct substitution.

Extreme Values of Continuous Functions on Closed, Bounded Sets

The Extreme Value Theorem (Theorem 1, Section 4.1) states that a function of a single variable that is continuous at every point of a closed, bounded interval $[a, b]$ takes on an absolute maximum value and an absolute minimum value at least once in $[a, b]$. The same holds true of a function $z = f(x, y)$ that is continuous on a closed, bounded set R in the plane (like a line segment, a disk, or a filled-in triangle). The function takes on an absolute maximum value at some point in R and an absolute minimum value at some point in R. The function may take on a maximum or minimum value more than once over R.

Similar results hold for functions of three or more variables. A continuous function $w = f(x, y, z)$ must take on absolute maximum and minimum values on any closed, bounded set (such as a solid ball or cube, spherical shell, or rectangular solid) on which it is defined. We will learn how to find these extreme values in Section 14.7.

EXERCISES 14.2

Limits with Two Variables

Find the limits in Exercises 1–12.

1. $\displaystyle\lim_{(x,y)\to(0,0)} \frac{3x^2 - y^2 + 5}{x^2 + y^2 + 2}$

2. $\displaystyle\lim_{(x,y)\to(4,2)} \frac{\sqrt{x}}{y}$

3. $\displaystyle\lim_{(x,y)\to(3,4)} \sqrt{x^2 + y^2 - 1}$

4. $\displaystyle\lim_{(x,y)\to(2,-3)} \left(\frac{1}{x} + \frac{1}{y}\right)^2$

5. $\displaystyle\lim_{(x,y)\to\left(\frac{\pi}{4},\frac{\pi}{2}\right)} \cot x \csc y$

6. $\displaystyle\lim_{(x,y)\to(0,0)} \cos\frac{x^2 + y^3}{x + y + 1}$

7. $\lim\limits_{(x,y)\to(0,\ln 3)} e^{x+y}$

8. $\lim\limits_{(x,y)\to(1,1)} \ln|1+x^2y^2|$

9. $\lim\limits_{(x,y)\to(0,0)} \dfrac{e^y \sin x}{x}$

10. $\lim\limits_{(x,y)\to\left(\frac{1}{64},\pi^3\right)} \sin\sqrt[3]{xy}$

11. $\lim\limits_{(x,y)\to(1,\pi/6)} \dfrac{x\sin y}{x^2+1}$

12. $\lim\limits_{(x,y)\to(\pi,0)} \dfrac{1-\sin y}{y+\cos x}$

Limits of Quotients

Find the limits in Exercises 13–24 by rewriting the fractions first.

13. $\lim\limits_{\substack{(x,y)\to(1,1)\\ x\neq y}} \dfrac{x^2-2xy+y^2}{x-y}$

14. $\lim\limits_{\substack{(x,y)\to(1,1)\\ x\neq y}} \dfrac{x^2-y^2}{x+y}$

15. $\lim\limits_{\substack{(x,y)\to(1,1)\\ x\neq 1}} \dfrac{xy-y-2x+2}{x-1}$

16. $\lim\limits_{\substack{(x,y)\to(2,-4)\\ x\neq -4,\, x\neq x^2}} \dfrac{y+4}{x^2y-xy+4x^2-4x}$

17. $\lim\limits_{\substack{(x,y)\to(0,0)\\ x\neq y}} \dfrac{\sqrt{x}+\sqrt{y}}{x+2\sqrt{x}-y+2\sqrt{y}}$

18. $\lim\limits_{\substack{(x,y)\to(2,2)\\ x+y\neq 4}} \dfrac{x+y-4}{\sqrt{x+y}-2}$

19. $\lim\limits_{\substack{(x,y)\to(3,0)\\ 3x-y\neq 9}} \dfrac{\sqrt{3x-y}-3}{3x-y-9}$

20. $\lim\limits_{\substack{(x,y)\to(4,3)\\ x\neq y+1}} \dfrac{\sqrt{x}-\sqrt{y+1}}{x-y-1}$

21. $\lim\limits_{(x,y)\to(0,0)} \dfrac{\sin(x-y)}{x-y}$

22. $\lim\limits_{(x,y)\to(0,0)} \dfrac{1-\cos(xy)}{xy}$

23. $\lim\limits_{(x,y)\to(1,-1)} \dfrac{x^3+y^3}{x+y}$

24. $\lim\limits_{(x,y)\to(3,3)} \dfrac{x^2-y^2}{x^4-y^4}$

Limits with Three Variables

Find the limits in Exercises 25–30.

25. $\lim\limits_{P\to(1,3,4)} \left(\dfrac{1}{x}+\dfrac{1}{y}+\dfrac{1}{z}\right)$

26. $\lim\limits_{(x,y,z)\to(-1,1,-1)} \dfrac{3xy+2yz}{x^3+z^3}$

27. $\lim\limits_{P\to(\pi,\pi,0)} (\sin^2 x+\cos^2 y+\sec^2 z)$

28. $\lim\limits_{P\to(-1/4,\pi/2,2)} \tan^{-1} xyz$

29. $\lim\limits_{P\to(\pi,0,3)} ze^{-2y}\cos 2x$

30. $\lim\limits_{(x,y,z)\to(6,-4,2)} \ln\sqrt{x^2-y^2-z^2}$

Continuity for Two Variables

At what points (x, y) in the plane are the functions in Exercises 31–34 continuous?

31. **a.** $f(x,y)=\sin(x+y)$ **b.** $f(x,y)=\ln(x^2+y^2)$

32. **a.** $f(x,y)=\dfrac{x^2+y^2}{x^2-y^2}$ **b.** $f(x,y)=\dfrac{x^2y}{x^2+y^2}$

33. **a.** $g(x,y)=\sin\dfrac{1}{xy}$ **b.** $g(x,y)=\dfrac{x+y}{2+\cos x}$

34. **a.** $f(x,y)=\dfrac{x^2+y^2}{x^2-5x+6}$ **b.** $f(x,y)=\dfrac{1}{x+y}$

Continuity for Three Variables

At what points (x, y, z) in space are the functions in Exercises 35–40 continuous?

35. **a.** $f(x,y,z)=x^2+y^2-2z^2$
b. $f(x,y,z)=\sqrt{x^2+y^2-1}$

36. **a.** $f(x,y,z)=\ln xyz$ **b.** $f(x,y,z)=e^{x+y}\cos z$

37. **a.** $h(x,y,z)=xy\cos\dfrac{1}{z}$ **b.** $h(x,y,z)=\dfrac{1}{y^2+z^2-4}$

38. **a.** $h(x,y,z)=\dfrac{1}{|y|+|z|}$ **b.** $h(x,y,z)=\dfrac{1}{|xy|+|z|}$

39. **a.** $h(x,y,z)=\ln(z-x^2-y^2-1)$
b. $h(x,y,z)=\dfrac{1}{z-\sqrt{x^2+y^2}}$

40. **a.** $h(x,y,z)=\sqrt{9-x^2-y^2-z^2}$
b. $h(x,y,z)=\dfrac{1}{5-\sqrt{x^2+y^2+z^2-16}}$

No Limit Exists at the Origin

By considering different paths of approach, show that the functions in Exercises 41–48 have no limit as $(x, y)\to(0,0)$.

41. $f(x,y)=-\dfrac{x}{\sqrt{x^2+y^2}}$

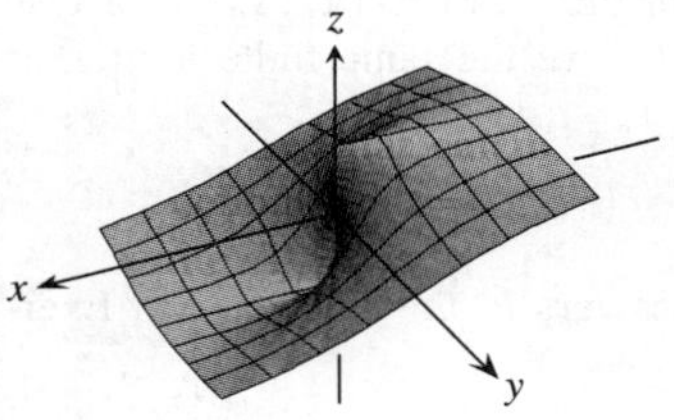

42. $f(x,y)=\dfrac{x^4}{x^4+y^2}$

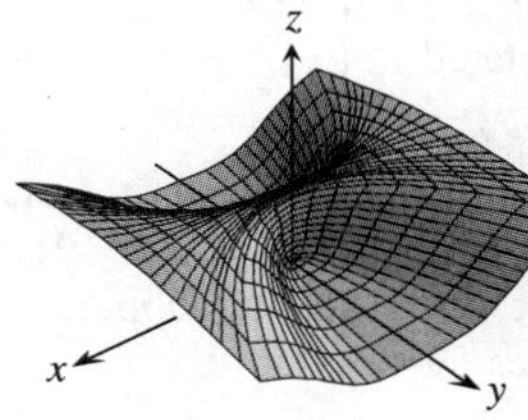

43. $f(x,y)=\dfrac{x^2-y^2}{x^2+y^2}$

44. $f(x,y)=\dfrac{xy}{|xy|}$

45. $g(x,y)=\dfrac{x-y}{x+y}$

46. $g(x,y)=\dfrac{x^2-y}{x-y}$

47. $h(x,y)=\dfrac{x^2+y}{y}$

48. $h(x,y)=\dfrac{x^2y}{x^4+y^2}$

Theory and Examples

In Exercises 49–54, show that the limits do not exist.

49. $\lim\limits_{(x,y)\to(1,1)} \dfrac{x^2y-1}{x-1}$

50. $\lim\limits_{(x,y)\to(1,-1)} \dfrac{xy+1}{x^2-y^2}$

51. $\lim\limits_{(x,y)\to(0,1)} \dfrac{x\ln y}{x^2+(\ln y)^2}$

52. $\lim\limits_{(x,y)\to(1,0)} \dfrac{xe^y-1}{xe^y-1+y}$

53. $\lim\limits_{(x,y)\to(0,0)} \dfrac{y+\sin x}{x+\sin y}$

54. $\lim\limits_{(x,y)\to(1,1)} \dfrac{\tan y-y\tan x}{y-x}$

55. Let $f(x, y) = \begin{cases} 1, & y \geq x^4 \\ 1, & y \leq 0 \\ 0, & \text{otherwise.} \end{cases}$

Find each of the following limits, or explain that the limit does not exist.

a. $\lim_{(x, y)\to(0,1)} f(x, y)$

b. $\lim_{(x, y)\to(2,3)} f(x, y)$

c. $\lim_{(x, y)\to(0,0)} f(x, y)$

56. Let $f(x, y) = \begin{cases} x^2, & x \geq 0 \\ x, & x < 0 \end{cases}.$

Find the following limits.

a. $\lim_{(x, y)\to(2,-1)} f(x,y)$

b. $\lim_{(x, y)\to(-3,2)} f(x, y)$

c. $\lim_{(x, y)\to(0,0)} f(x, y)$

57. Show that the function in Example 6 has limit 0 along every straight line approaching $(0, 0)$.

58. If $f(x_0, y_0) = 5$, what can you say about

$$\lim_{(x, y)\to(x_0, y_0)} f(x, y)$$

if f is continuous at (x_0, y_0)? If f is not continuous at (x_0, y_0)? Give reasons for your answers.

The Sandwich Theorem for functions of two variables states that if $g(x, y) \leq f(x, y) \leq h(x, y)$ for all $(x, y) \neq (x_0, y_0)$ in a disk centered at (x_0, y_0) and if g and h have the same finite limit L as $(x, y) \to (x_0, y_0)$, then

$$\lim_{(x, y)\to(x_0, y_0)} f(x, y) = L.$$

Use this result to support your answers to the questions in Exercises 59–62.

59. Does knowing that

$$1 - \frac{x^2y^2}{3} < \frac{\tan^{-1} xy}{xy} < 1$$

tell you anything about

$$\lim_{(x, y)\to(0,0)} \frac{\tan^{-1} xy}{xy}?$$

Give reasons for your answer.

60. Does knowing that

$$2|xy| - \frac{x^2y^2}{6} < 4 - 4\cos\sqrt{|xy|} < 2|xy|$$

tell you anything about

$$\lim_{(x, y)\to(0,0)} \frac{4 - 4\cos\sqrt{|xy|}}{|xy|}?$$

Give reasons for your answer.

61. Does knowing that $|\sin(1/x)| \leq 1$ tell you anything about

$$\lim_{(x, y)\to(0,0)} y \sin\frac{1}{x}?$$

Give reasons for your answer.

62. Does knowing that $|\cos(1/y)| \leq 1$ tell you anything about

$$\lim_{(x, y)\to(0,0)} x \cos\frac{1}{y}?$$

Give reasons for your answer.

63. (*Continuation of Example 5.*)

a. Reread Example 5. Then substitute $m = \tan\theta$ into the formula

$$f(x, y)\Big|_{y=mx} = \frac{2m}{1 + m^2}$$

and simplify the result to show how the value of f varies with the line's angle of inclination.

b. Use the formula you obtained in part (a) to show that the limit of f as $(x, y) \to (0, 0)$ along the line $y = mx$ varies from -1 to 1, depending on the angle of approach.

64. Continuous extension Define $f(0, 0)$ in a way that extends

$$f(x, y) = xy\frac{x^2 - y^2}{x^2 + y^2}$$

to be continuous at the origin.

Changing Variables to Polar Coordinates

If you cannot make any headway with $\lim_{(x, y)\to(0,0)} f(x, y)$ in rectangular coordinates, try changing to polar coordinates. Substitute $x = r\cos\theta$, $y = r\sin\theta$, and investigate the limit of the resulting expression as $r \to 0$. In other words, try to decide whether there exists a number L satisfying the following criterion:

Given $\varepsilon > 0$, there exists a $\delta > 0$ such that for all r and θ,

$$|r| < \delta \;\Rightarrow\; |f(r, \theta) - L| < \varepsilon. \tag{1}$$

If such an L exists, then

$$\lim_{(x, y)\to(0,0)} f(x, y) = \lim_{r\to 0} f(r\cos\theta, r\sin\theta) = L.$$

For instance,

$$\lim_{(x, y)\to(0,0)} \frac{x^3}{x^2 + y^2} = \lim_{r\to 0} \frac{r^3\cos^3\theta}{r^2} = \lim_{r\to 0} r\cos^3\theta = 0.$$

To verify the last of these equalities, we need to show that Equation (1) is satisfied with $f(r, \theta) = r\cos^3\theta$ and $L = 0$. That is, we need to show that given any $\varepsilon > 0$, there exists a $\delta > 0$ such that for all r and θ,

$$|r| < \delta \;\Rightarrow\; |r\cos^3\theta - 0| < \varepsilon.$$

Since

$$|r\cos^3\theta| = |r|\,|\cos^3\theta| \leq |r| \cdot 1 = |r|,$$

the implication holds for all r and θ if we take $\delta = \varepsilon$.

In contrast,

$$\frac{x^2}{x^2 + y^2} = \frac{r^2\cos^2\theta}{r^2} = \cos^2\theta$$

takes on all values from 0 to 1 regardless of how small $|r|$ is, so that $\lim_{(x,y)\to(0,0)} x^2/(x^2+y^2)$ does not exist.

In each of these instances, the existence or nonexistence of the limit as $r \to 0$ is fairly clear. Shifting to polar coordinates does not always help, however, and may even tempt us to false conclusions. For example, the limit may exist along every straight line (or ray) $\theta =$ constant and yet fail to exist in the broader sense. Example 5 illustrates this point. In polar coordinates, $f(x, y) = (2x^2y)/(x^4 + y^2)$ becomes

$$f(r\cos\theta, r\sin\theta) = \frac{r\cos\theta\sin 2\theta}{r^2\cos^4\theta + \sin^2\theta}$$

for $r \neq 0$. If we hold θ constant and let $r \to 0$, the limit is 0. On the path $y = x^2$, however, we have $r\sin\theta = r^2\cos^2\theta$ and

$$f(r\cos\theta, r\sin\theta) = \frac{r\cos\theta\sin 2\theta}{r^2\cos^4\theta + (r\cos^2\theta)^2}$$

$$= \frac{2r\cos^2\theta\sin\theta}{2r^2\cos^4\theta} = \frac{r\sin\theta}{r^2\cos^2\theta} = 1.$$

In Exercises 65–70, find the limit of f as $(x, y) \to (0,0)$ or show that the limit does not exist.

65. $f(x, y) = \dfrac{y^3 - x^2y}{x^2 + y^2}$

66. $f(x, y) = \cos\left(\dfrac{x^3 - y^3}{x^2 + y^2}\right)$

67. $f(x, y) = \dfrac{y^2}{x^2 + y^2}$

68. $f(x, y) = \dfrac{2y}{x^2 + y + y^2}$

69. $f(x, y) = \tan^{-1}\left(\dfrac{|x| + |y|}{x^2 + y^2}\right)$

70. $f(x, y) = \dfrac{x^2 - y^2}{x^2 + y^2}$

In Exercises 71 and 72, define $f(0,0)$ in a way that extends f to be continuous at the origin.

71. $f(x, y) = \ln\left(\dfrac{5x^2 - 2x^2y^2 + 5y^2}{x^2 + y^2}\right)$

72. $f(x, y) = \dfrac{3x^2y}{x^2 + y^2}$

Using the Limit Definition

Each of Exercises 73–78 gives a function $f(x, y)$ and a positive number ε. In each exercise, show that there exists a $\delta > 0$ such that for all (x, y),

$$\sqrt{x^2 + y^2} < \delta \Rightarrow |f(x, y) - f(0,0)| < \varepsilon.$$

73. $f(x, y) = x^2 + y^2, \quad \varepsilon = 0.01$

74. $f(x, y) = \dfrac{x}{(y^2 + 1)}$

75. $f(x, y) = (x + y)/(x^2 + 1), \quad \varepsilon = 0.01$

76. $f(x, y) = (x + y)/(2 + \cos x), \quad \varepsilon = 0.02$

77. $f(x, y) = \dfrac{xy^2}{x^2 + y^2}$ and $f(0,0) = 0, \quad \varepsilon = 0.04$

78. $f(x, y) = \dfrac{x^3 + y^4}{x^2 + y^2}$ and $f(0,0) = 0, \quad \varepsilon = 0.02$

Each of Exercises 79–82 gives a function $f(x, y, z)$ and a positive number ε. In each exercise, show that there exists a $\delta > 0$ such that for all (x, y, z),

$$\sqrt{x^2 + y^2 + z^2} < \delta \Rightarrow |f(x, y, z) - f(0,0,0)| < \varepsilon.$$

79. $f(x, y, z) = x^2 + y^2 + z^2, \quad \varepsilon = 0.015$

80. $f(x, y, z) = xyz, \quad \varepsilon = 0.008$

81. $f(x, y, z) = \dfrac{x + y + z}{x^2 + y^2 + z^2 + 1}, \quad \varepsilon = 0.015$

82. $f(x, y, z) = \tan^2 x + \tan^2 y + \tan^2 z, \quad \varepsilon = 0.03$

83. Show that $f(x, y, z) = x + y - z$ is continuous at every point (x_0, y_0, z_0).

84. Show that $f(x, y, z) = x^2 + y^2 + z^2$ is continuous at the origin.

14.3 Partial Derivatives

The calculus of several variables is similar to single-variable calculus applied to several variables, one at a time. When we hold all but one of the independent variables of a function constant and differentiate with respect to that one variable, we get a "partial" derivative. This section shows how partial derivatives are defined and interpreted geometrically, and how to calculate them by applying the familiar rules for differentiating functions of a single variable. The idea of *differentiability* for functions of several variables requires more than the existence of the partial derivatives, because a point can be approached from many different directions. However, we will see that differentiable functions of several variables behave similarly to differentiable single-variable functions. In particular, they are continuous and can be well approximated by linear functions.

Partial Derivatives of a Function of Two Variables

If (x_0, y_0) is a point in the domain of a function $f(x, y)$, the vertical plane $y = y_0$ will cut the surface $z = f(x, y)$ in the curve $z = f(x, y_0)$ (Figure 14.16). This curve is the graph

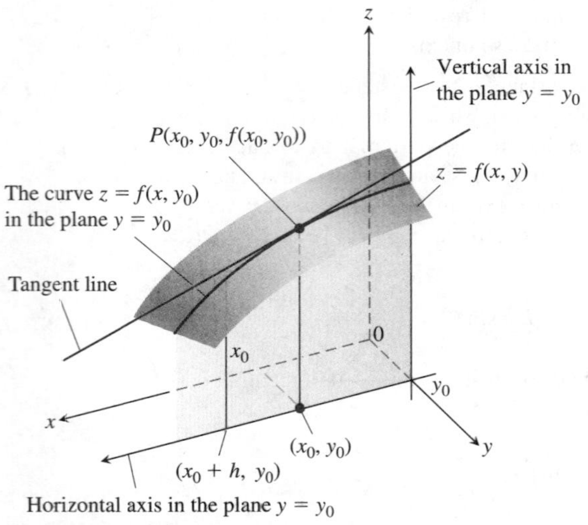

FIGURE 14.16 The intersection of the plane $y = y_0$ with the surface $z = f(x, y)$, viewed from above the first quadrant of the xy-plane.

of the function $z = f(x, y_0)$ in the plane $y = y_0$. The horizontal coordinate in this plane is x; the vertical coordinate is z. The y-value is held constant at y_0, so y is not a variable.

We define the partial derivative of f with respect to x at the point (x_0, y_0) as the ordinary derivative of $f(x, y_0)$ with respect to x at the point $x = x_0$. To distinguish partial derivatives from ordinary derivatives, we use the symbol ∂ rather than the d previously used. In the definition, h represents a real number, positive or negative.

DEFINITION The **partial derivative of $f(x, y)$ with respect to x at the point (x_0, y_0)** is

$$\left.\frac{\partial f}{\partial x}\right|_{(x_0, y_0)} = \lim_{h \to 0} \frac{f(x_0 + h, y_0) - f(x_0, y_0)}{h},$$

provided the limit exists.

The partial derivative of $f(x, y)$ with respect to x at the point (x_0, y_0) is the same as the ordinary derivative of $f(x, y_0)$ at the point x_0:

$$\left.\frac{\partial f}{\partial x}\right|_{(x_0, y_0)} = \left.\frac{d}{dx} f(x, y_0)\right|_{x = x_0}.$$

A variety of notations are used to denote the partial derivative at a point (x_0, y_0), including

$$\frac{\partial f}{\partial x}(x_0, y_0), \qquad f_x(x_0, y_0), \qquad \text{and} \qquad \left.\frac{\partial z}{\partial x}\right|_{(x_0, y_0)}.$$

When we do not specify a specific point (x_0, y_0) at which the partial derivative is being evaluated, then the partial derivative becomes a *function* whose domain is the points where the partial derivative exists. Notations for this function include

$$\frac{\partial f}{\partial x}, \qquad f_x, \qquad \text{and} \qquad \frac{\partial z}{\partial x}.$$

The slope of the curve $z = f(x, y_0)$ at the point $P(x_0, y_0, f(x_0, y_0))$ in the plane $y = y_0$ is the value of the partial derivative of f with respect to x at (x_0, y_0). (In Figure 14.16 this slope is negative.) The tangent line to the curve at P is the line in the plane $y = y_0$ that passes through P with this slope. The partial derivative $\partial f/\partial x$ at (x_0, y_0) gives the rate of change of f with respect to x when y is held fixed at the value y_0.

The definition of the partial derivative of $f(x, y)$ with respect to y at a point (x_0, y_0) is similar to the definition of the partial derivative of f with respect to x. We hold x fixed at the value x_0 and take the ordinary derivative of $f(x_0, y)$ with respect to y at y_0.

DEFINITION The **partial derivative of $f(x, y)$ with respect to y at the point (x_0, y_0)** is

$$\left.\frac{\partial f}{\partial y}\right|_{(x_0, y_0)} = \left.\frac{d}{dy} f(x_0, y)\right|_{y=y_0} = \lim_{h \to 0} \frac{f(x_0, y_0 + h) - f(x_0, y_0)}{h},$$

provided the limit exists.

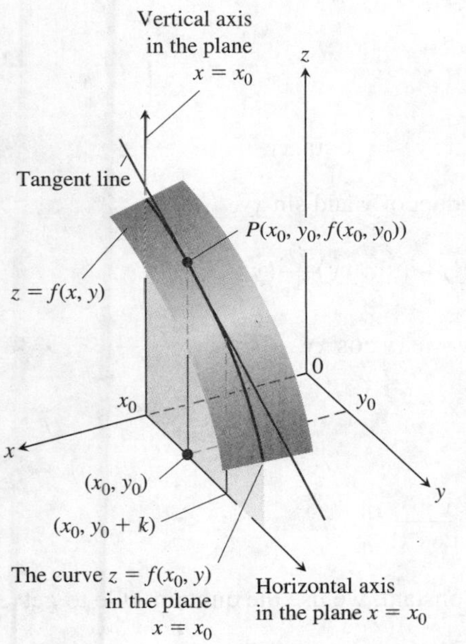

FIGURE 14.17 The intersection of the plane $x = x_0$ with the surface $z = f(x, y)$, viewed from above the first quadrant of the xy-plane.

The slope of the curve $z = f(x_0, y)$ at the point $P(x_0, y_0, f(x_0, y_0))$ in the vertical plane $x = x_0$ (Figure 14.17) is the partial derivative of f with respect to y at (x_0, y_0). The tangent line to the curve at P is the line in the plane $x = x_0$ that passes through P with this slope. The partial derivative gives the rate of change of f with respect to y at (x_0, y_0) when x is held fixed at the value x_0.

The partial derivative with respect to y is denoted the same way as the partial derivative with respect to x:

$$\frac{\partial f}{\partial y}(x_0, y_0), \qquad f_y(x_0, y_0), \qquad \frac{\partial f}{\partial y}, \qquad f_y.$$

Notice that we now have two tangent lines associated with the surface $z = f(x, y)$ at the point $P(x_0, y_0, f(x_0, y_0))$ (Figure 14.18). Is the plane they determine tangent to the surface at P? We will see that it is for the *differentiable* functions defined at the end of this section, and we will learn how to find the tangent plane in Section 14.6. First we have to better understand partial derivatives.

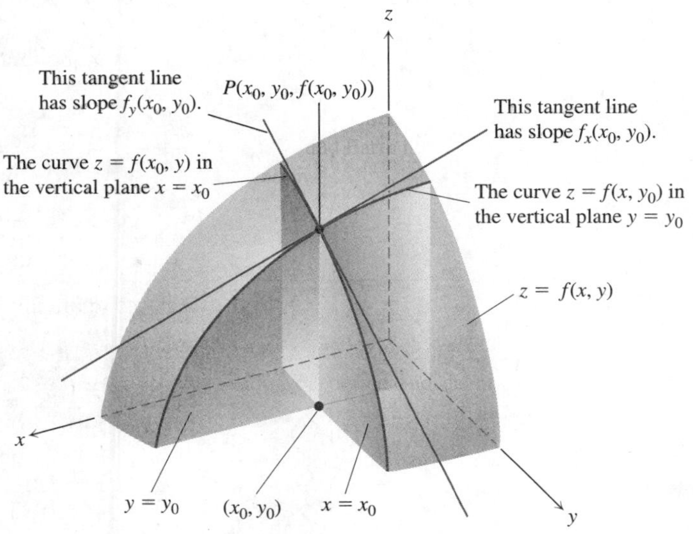

FIGURE 14.18 Figures 14.16 and 14.17 combined. The tangent lines at the point $(x_0, y_0, f(x_0, y_0))$ determine a plane that, in this picture at least, appears to be tangent to the surface.

Calculations

The definitions of $\partial f/\partial x$ and $\partial f/\partial y$ give us two different ways of differentiating f at a point: with respect to x in the usual way while treating y as a constant, and with respect to y in the usual way while treating x as a constant. As the following examples show, the values of these partial derivatives are usually different at a given point (x_0, y_0).

EXAMPLE 1 Find the values of $\partial f/\partial x$ and $\partial f/\partial y$ at the point $(4, -5)$ if

$$f(x, y) = x^2 + 3xy + y - 1.$$

Solution To find $\partial f/\partial x$, we treat y as a constant and differentiate with respect to x:

$$\frac{\partial f}{\partial x} = \frac{\partial}{\partial x}(x^2 + 3xy + y - 1) = 2x + 3 \cdot 1 \cdot y + 0 - 0 = 2x + 3y.$$

The value of $\partial f/\partial x$ at $(4, -5)$ is $2(4) + 3(-5) = -7$.

To find $\partial f/\partial y$, we treat x as a constant and differentiate with respect to y:

$$\frac{\partial f}{\partial y} = \frac{\partial}{\partial y}(x^2 + 3xy + y - 1) = 0 + 3 \cdot x \cdot 1 + 1 - 0 = 3x + 1.$$

The value of $\partial f/\partial y$ at $(4, -5)$ is $3(4) + 1 = 13$. ■

EXAMPLE 2 Find $\partial f/\partial y$ as a function if $f(x, y) = y \sin xy$.

Solution We treat x as a constant and f as a product of y and $\sin xy$:

$$\begin{aligned}\frac{\partial f}{\partial y} &= \frac{\partial}{\partial y}(y \sin xy) = y\frac{\partial}{\partial y}\sin xy + (\sin xy)\frac{\partial}{\partial y}(y)\\ &= (y \cos xy)\frac{\partial}{\partial y}(xy) + \sin xy = xy \cos xy + \sin xy.\end{aligned}$$ ■

EXAMPLE 3 Find f_x and f_y as functions if

$$f(x, y) = \frac{2y}{y + \cos x}.$$

Solution We treat f as a quotient. With y held constant, we use the quotient rule to get

$$\begin{aligned}f_x &= \frac{\partial}{\partial x}\left(\frac{2y}{y + \cos x}\right) = \frac{(y + \cos x)\frac{\partial}{\partial x}(2y) - 2y\frac{\partial}{\partial x}(y + \cos x)}{(y + \cos x)^2}\\ &= \frac{(y + \cos x)(0) - 2y(-\sin x)}{(y + \cos x)^2} = \frac{2y \sin x}{(y + \cos x)^2}.\end{aligned}$$

With x held constant and again applying the quotient rule, we get

$$\begin{aligned}f_y &= \frac{\partial}{\partial y}\left(\frac{2y}{y + \cos x}\right) = \frac{(y + \cos x)\frac{\partial}{\partial y}(2y) - 2y\frac{\partial}{\partial y}(y + \cos x)}{(y + \cos x)^2}\\ &= \frac{(y + \cos x)(2) - 2y(1)}{(y + \cos x)^2} = \frac{2 \cos x}{(y + \cos x)^2}.\end{aligned}$$ ■

Implicit differentiation works for partial derivatives the way it works for ordinary derivatives, as the next example illustrates.

EXAMPLE 4 Find $\partial z/\partial x$ assuming that the equation

$$yz - \ln z = x + y$$

defines z as a function of the two independent variables x and y and the partial derivative exists.

Solution We differentiate both sides of the equation with respect to x, holding y constant and treating z as a differentiable function of x:

$$\frac{\partial}{\partial x}(yz) - \frac{\partial}{\partial x}\ln z = \frac{\partial x}{\partial x} + \frac{\partial y}{\partial x}$$

$$y\frac{\partial z}{\partial x} - \frac{1}{z}\frac{\partial z}{\partial x} = 1 + 0 \qquad \text{With } y \text{ constant, } \frac{\partial}{\partial x}(yz) = y\frac{\partial z}{\partial x}.$$

$$\left(y - \frac{1}{z}\right)\frac{\partial z}{\partial x} = 1$$

$$\frac{\partial z}{\partial x} = \frac{z}{yz - 1}.$$

EXAMPLE 5 The plane $x = 1$ intersects the paraboloid $z = x^2 + y^2$ in a parabola. Find the slope of the tangent line to the parabola at $(1, 2, 5)$ (Figure 14.19).

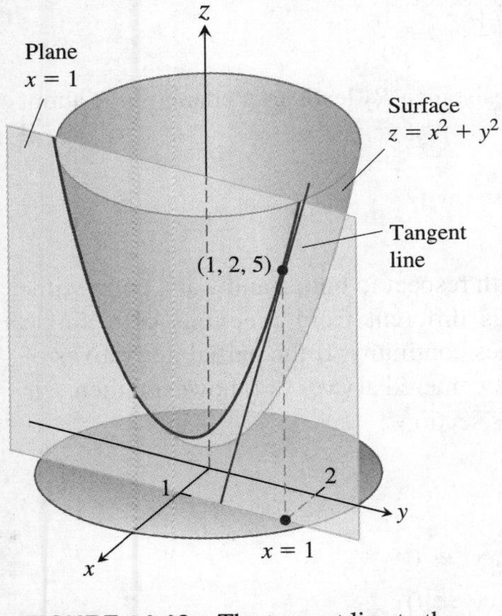

FIGURE 14.19 The tangent line to the curve of intersection of the plane $x = 1$ and the surface $z = x^2 + y^2$ at the point $(1, 2, 5)$ (Example 5).

Solution The parabola lies in a plane parallel to the yz-plane, and the slope is the value of the partial derivative $\partial z/\partial y$ at $(1, 2)$:

$$\left.\frac{\partial z}{\partial y}\right|_{(1,2)} = \left.\frac{\partial}{\partial y}(x^2 + y^2)\right|_{(1,2)} = \left.2y\right|_{(1,2)} = 2(2) = 4.$$

As a check, we can treat the parabola as the graph of the single-variable function $z = (1)^2 + y^2 = 1 + y^2$ in the plane $x = 1$ and ask for the slope at $y = 2$. The slope, calculated now as an ordinary derivative, is

$$\left.\frac{dz}{dy}\right|_{y=2} = \left.\frac{d}{dy}(1 + y^2)\right|_{y=2} = \left.2y\right|_{y=2} = 4.$$

Functions of More Than Two Variables

The definitions of the partial derivatives of functions of more than two independent variables are similar to the definitions for functions of two variables. They are ordinary derivatives with respect to one variable, taken while the other independent variables are held constant.

EXAMPLE 6 If x, y, and z are independent variables and

$$f(x, y, z) = x\sin(y + 3z),$$

then

$$\frac{\partial f}{\partial z} = \frac{\partial}{\partial z}[x\sin(y + 3z)] = x\frac{\partial}{\partial z}\sin(y + 3z) \qquad x \text{ held constant}$$

$$= x\cos(y + 3z)\frac{\partial}{\partial z}(y + 3z) \qquad \text{Chain rule}$$

$$= 3x\cos(y + 3z). \qquad y \text{ held constant}$$

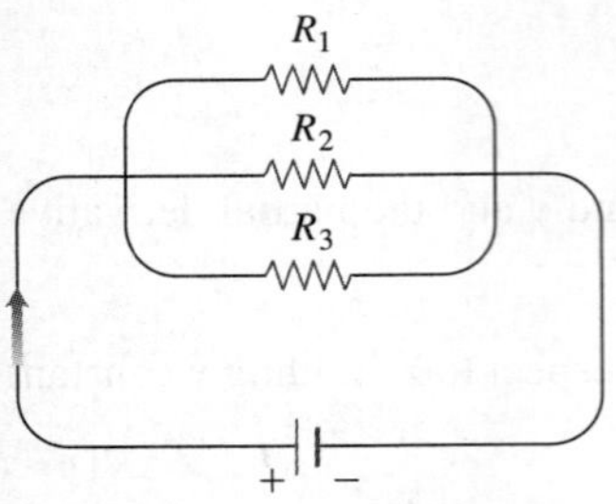

FIGURE 14.20 Resistors arranged this way are said to be connected in parallel (Example 7). Each resistor lets a portion of the current through. Their equivalent resistance R is calculated with the formula

$$\frac{1}{R} = \frac{1}{R_1} + \frac{1}{R_2} + \frac{1}{R_3}.$$

EXAMPLE 7 If resistors of R_1, R_2, and R_3 ohms are connected in parallel to make an R-ohm resistor, the value of R can be found from the equation

$$\frac{1}{R} = \frac{1}{R_1} + \frac{1}{R_2} + \frac{1}{R_3}$$

(Figure 14.20). Find the value of $\partial R/\partial R_2$ when $R_1 = 30$, $R_2 = 45$, and $R_3 = 90$ ohms.

Solution To find $\partial R/\partial R_2$, we treat R_1 and R_3 as constants and, using implicit differentiation, differentiate both sides of the equation with respect to R_2:

$$\frac{\partial}{\partial R_2}\left(\frac{1}{R}\right) = \frac{\partial}{\partial R_2}\left(\frac{1}{R_1} + \frac{1}{R_2} + \frac{1}{R_3}\right)$$

$$-\frac{1}{R^2}\frac{\partial R}{\partial R_2} = 0 - \frac{1}{R_2{}^2} + 0$$

$$\frac{\partial R}{\partial R_2} = \frac{R^2}{R_2{}^2} = \left(\frac{R}{R_2}\right)^2.$$

When $R_1 = 30$, $R_2 = 45$, and $R_3 = 90$,

$$\frac{1}{R} = \frac{1}{30} + \frac{1}{45} + \frac{1}{90} = \frac{3 + 2 + 1}{90} = \frac{6}{90} = \frac{1}{15},$$

so $R = 15$ and

$$\frac{\partial R}{\partial R_2} = \left(\frac{15}{45}\right)^2 = \left(\frac{1}{3}\right)^2 = \frac{1}{9}.$$

Thus at the given values, a small change in the resistance R_2 leads to a change in R about one-ninth as large. ■

Partial Derivatives and Continuity

A function $f(x, y)$ can have partial derivatives with respect to both x and y at a point without the function being continuous there. This is different from functions of a single variable, where the existence of a derivative implies continuity. If the partial derivatives of $f(x, y)$ exist and are continuous throughout a disk centered at (x_0, y_0), however, then f is continuous at (x_0, y_0), as we see at the end of this section.

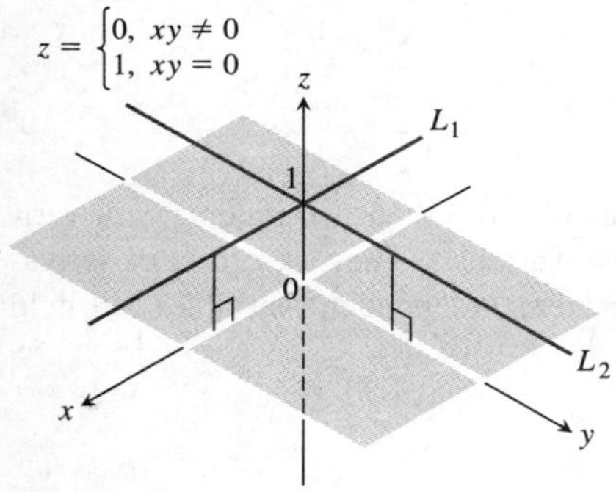

FIGURE 14.21 The graph of

$$f(x, y) = \begin{cases} 0, & xy \neq 0 \\ 1, & xy = 0 \end{cases}$$

consists of the lines L_1 and L_2 (lying 1 unit above the xy-plane) and the four open quadrants of the xy-plane. The function has partial derivatives at the origin but is not continuous there (Example 8).

EXAMPLE 8 Let

$$f(x, y) = \begin{cases} 0, & xy \neq 0 \\ 1, & xy = 0 \end{cases}$$

(Figure 14.21).

(a) Find the limit of f as (x, y) approaches $(0, 0)$ along the line $y = x$.
(b) Find the limit of f as (x, y) approaches $(0, 0)$ along the line $y = 0$.
(c) Prove that f is not continuous at the origin.
(d) Show that both partial derivatives $\partial f/\partial x$ and $\partial f/\partial y$ exist at the origin.

Solution

(a) Since $f(x, y)$ is zero at every point on the line $y = x$ (except at the origin), we have

$$\lim_{(x, y)\to(0, 0)} f(x, y)\Big|_{y=x} = \lim_{(x, y)\to(0, 0)} 0 = 0.$$

(b) Since $f(x, y)$ takes the constant value 1 at every point on the line $y = 0$, we have

$$\lim_{(x, y)\to(0, 0)} f(x, y)\Big|_{y=0} = \lim_{(x, y)\to(0, 0)} 1 = 1.$$

(c) By the two-path test, f has no limit as (x, y) approaches $(0, 0)$. Consequently, f is not continuous at $(0, 0)$.

(d) To find $\partial f/\partial x$ at $(0, 0)$, we hold y fixed at $y = 0$. Then $f(x, y) = 1$ for all x, and the graph of f is the line L_1 in Figure 14.21. The slope of this line at any x is $\partial f/\partial x = 0$. In particular, $\partial f/\partial x = 0$ at $(0, 0)$. Similarly, $\partial f/\partial y$ is the slope of line L_2 at any y, so $\partial f/\partial y = 0$ at $(0, 0)$. ■

What Example 8 suggests is that we need a stronger requirement for differentiability in higher dimensions than the mere existence of the partial derivatives. We define differentiability for functions of two variables (which is somewhat more complicated than for single-variable functions) at the end of this section and then revisit the connection to continuity.

Second-Order Partial Derivatives

When we differentiate a function $f(x, y)$ twice, we produce its second-order derivatives. These derivatives are usually denoted by

$$\frac{\partial^2 f}{\partial x^2} \text{ or } f_{xx}, \qquad \frac{\partial^2 f}{\partial y^2} \text{ or } f_{yy},$$

$$\frac{\partial^2 f}{\partial x\,\partial y} \text{ or } f_{yx}, \qquad \text{and} \qquad \frac{\partial^2 f}{\partial y\,\partial x} \text{ or } f_{xy}.$$

The defining equations are

$$\frac{\partial^2 f}{\partial x^2} = \frac{\partial}{\partial x}\left(\frac{\partial f}{\partial x}\right), \qquad \frac{\partial^2 f}{\partial x\,\partial y} = \frac{\partial}{\partial x}\left(\frac{\partial f}{\partial y}\right),$$

and so on. Notice the order in which the mixed partial derivatives are taken:

$$\frac{\partial^2 f}{\partial x\,\partial y} \qquad \text{Differentiate first with respect to } y\text{, then with respect to } x.$$

$$f_{yx} = (f_y)_x \qquad \text{Means the same thing}$$

HISTORICAL BIOGRAPHY
Pierre-Simon Laplace
(1749–1827)
www.bit.ly/2NdfIV5

EXAMPLE 9 If $f(x, y) = x \cos y + ye^x$, find the second-order derivatives

$$\frac{\partial^2 f}{\partial x^2}, \quad \frac{\partial^2 f}{\partial y\,\partial x}, \quad \frac{\partial^2 f}{\partial y^2}, \quad \text{and} \quad \frac{\partial^2 f}{\partial x\,\partial y}.$$

Solution The first step is to calculate both first partial derivatives.

$$\frac{\partial f}{\partial x} = \frac{\partial}{\partial x}(x \cos y + ye^x) = \cos y + ye^x \qquad \frac{\partial f}{\partial y} = \frac{\partial}{\partial y}(x \cos y + ye^x) = -x \sin y + e^x$$

Now we find both partial derivatives of each first partial:

$$\frac{\partial^2 f}{\partial y\,\partial x} = \frac{\partial}{\partial y}\left(\frac{\partial f}{\partial x}\right) = -\sin y + e^x \qquad \frac{\partial^2 f}{\partial x\,\partial y} = \frac{\partial}{\partial x}\left(\frac{\partial f}{\partial y}\right) = -\sin y + e^x$$

$$\frac{\partial^2 f}{\partial x^2} = \frac{\partial}{\partial x}\left(\frac{\partial f}{\partial x}\right) = ye^x. \qquad \frac{\partial^2 f}{\partial y^2} = \frac{\partial}{\partial y}\left(\frac{\partial f}{\partial y}\right) = -x \cos y.$$ ■

The Mixed Derivative Theorem

You may have noticed that the "mixed" second-order partial derivatives

$$\frac{\partial^2 f}{\partial y\,\partial x} \quad \text{and} \quad \frac{\partial^2 f}{\partial x\,\partial y}$$

in Example 9 are equal. This is not a coincidence. They must be equal whenever f, f_x, f_y, f_{xy}, and f_{yx} are continuous, as stated in the following theorem. However, the mixed derivatives can be different when the continuity conditions are not satisfied (see Exercise 82).

THEOREM 2—The Mixed Derivative Theorem
If $f(x, y)$ and its partial derivatives f_x, f_y, f_{xy}, and f_{yx} are defined throughout an open region containing a point (a, b) and are all continuous at (a, b), then

$$f_{xy}(a, b) = f_{yx}(a, b).$$

HISTORICAL BIOGRAPHY
Alexis Clairaut
(1713–1765)
www.bit.ly/2Rfl7yc

Theorem 2 is also known as Clairaut's Theorem, after the French mathematician Alexis Clairaut, who discovered it. A proof is given in Appendix A.8. Theorem 2 says that to calculate a mixed second-order derivative, we may differentiate in either order, provided the continuity conditions are satisfied. This ability to proceed in different order sometimes simplifies our calculations.

EXAMPLE 10 Find $\dfrac{\partial^2 w}{\partial x\, \partial y}$ if

$$w = xy + \frac{e^y}{y^2 + 1}.$$

Solution The symbol $\partial^2 w/\partial x\, \partial y$ tells us to differentiate first with respect to y and then with respect to x. However, if we interchange the order of differentiation and differentiate first with respect to x, we get the answer more quickly. In two steps,

$$\frac{\partial w}{\partial x} = y \qquad \text{and} \qquad \frac{\partial^2 w}{\partial y\, \partial x} = 1.$$

If we differentiate first with respect to y, we obtain $\partial^2 w/\partial x\, \partial y = 1$ as well, but with more work. We can differentiate in either order because the conditions of Theorem 2 hold for w at all points (x_0, y_0). ■

Partial Derivatives of Still Higher Order

Although we will deal mostly with first- and second-order partial derivatives, because these appear the most frequently in applications, there is no theoretical limit to how many times we can differentiate a function as long as the derivatives involved exist. Thus, we get third- and fourth-order derivatives denoted by symbols like

$$\frac{\partial^3 f}{\partial x\, \partial y^2} = f_{yyx},$$

$$\frac{\partial^4 f}{\partial x^2\, \partial y^2} = f_{yyxx},$$

and so on. As with second-order derivatives, the order of differentiation is immaterial as long as all the derivatives through the order in question are continuous.

EXAMPLE 11 Find f_{yxyz} if $f(x, y, z) = 1 - 2xy^2z + x^2y$.

Solution We first differentiate with respect to the variable y, then x, then y again, and finally with respect to z:

$$\begin{aligned} f_y &= -4xyz + x^2 \\ f_{yx} &= -4yz + 2x \\ f_{yxy} &= -4z \\ f_{yxyz} &= -4. \end{aligned}$$

■

Differentiability

The concept of differentiability for functions of several variables is more complicated than for single-variable functions, because a point in the domain can be approached from many directions and along any path, not just from the left or from the right. The existence of both partial derivatives at a point (x_0, y_0) is not by itself even enough to show continuity at (x_0, y_0), as we saw in Example 8. The differentiability of f is instead based on the idea that a linear function gives a good model of a differentiable function near a point.

In Section 3.9, we saw that a differentiable function f can be approximated near a point x_0 by its linearization,

$$L(x) = f(x_0) + f'(x_0)(x - x_0).$$

This formula allows us to find a linear function L, a function whose graph is a straight line, such that L closely approximates f near x_0. This can be done whenever f is differentiable, even when f itself is described by a very complicated formula. Approximations are much more useful and meaningful when they are accompanied by information on their accuracy. In Section 3.9, Equation (1), we saw that a differentiable function f satisfies

$$f(x) - f(x_0) = f'(x_0)(x - x_0) + \varepsilon(x - x_0),$$

where $\varepsilon \to 0$ as $x \to x_0$. Framed in terms of approximating f by L, this becomes

$$f(x) - L(x) = \varepsilon(x - x_0), \tag{1}$$

where again $\varepsilon \to 0$ as $x \to x_0$.

Rather than being a consequence of the definition, the differentiability for a function of two variables $f(x, y)$ is defined to mean that f can be approximated by a linear function. The approximating linear function $L(x, y)$ for $f(x, y)$ near the point (x_0, y_0) takes the form

$$L(x, y) = f(x_0, y_0) + f_x(x_0, y_0)(x - x_0) + f_y(x_0, y_0)(y - y_0),$$

and the graph of L is a plane, called the *tangent plane*, that approximates the graph of f near (x_0, y_0). Notice that $L(x_0, y_0) = f(x_0, y_0)$, so the functions L and f coincide at (x_0, y_0). Moreover the partial derivatives of L and f are also equal at (x_0, y_0). We will study tangent planes in detail in Section 14.6.

We now specify how closely f is approximated by L at (x_0, y_0). Extending the formula for single variable functions in Equation (1), we require that the difference between f and L satisfies

$$f(x, y) - L(x, y) = \varepsilon_1(x - x_0) + \varepsilon_2(y - y_0), \tag{2}$$

where both $\varepsilon_1 \to 0$ and $\varepsilon_2 \to 0$ as $(x, y) \to (x_0, y_0)$.

If we insert the formula for $L(x, y)$ into Equation (2) we see that

$$f(x, y) - f(x_0, y_0) = f_x(x_0, y_0)(x - x_0) + f_y(x_0, y_0)(y - y_0) + \varepsilon_1(x - x_0) + \varepsilon_2(y - y_0).$$

Setting $\Delta x = x - x_0$, $\Delta y = y - y_0$, and $\Delta z = f(x, y) - f(x_0, y_0)$, we get

$$\Delta z = f_x(x_0, y_0)\Delta x + f_y(x_0, y_0)\Delta y + \varepsilon_1 \Delta x + \varepsilon_2 \Delta y.$$

Based on these ideas, we now state the formal definition of differentiability, which captures the idea that f is well approximated by L.

> **DEFINITION** A function $z = f(x, y)$ is **differentiable at** (x_0, y_0) if both $f_x(x_0, y_0)$ and $f_y(x_0, y_0)$ exist and if $\Delta z = f(x, y) - f(x_0, y_0)$ satisfies
>
> $$\Delta z = f_x(x_0, y_0)\Delta x + f_y(x_0, y_0)\Delta y + \varepsilon_1 \Delta x + \varepsilon_2 \Delta y,$$
>
> where $\Delta x = x - x_0$, $\Delta y = y - y_0$, and both $\varepsilon_1 \to 0$ and $\varepsilon_2 \to 0$ as $(x, y) \to (x_0, y_0)$. We call the function f **differentiable** if it is differentiable at every point in its domain, and we then say that its graph is a **smooth surface**.

Differentiability is defined in a similar way for functions of more than two variables.

The following theorem (proved in Appendix A.8) and its accompanying corollary tell us that functions with *continuous* first partial derivatives at (x_0, y_0) are differentiable there, and they are closely approximated locally by a linear function. We study this approximation in Section 14.6.

THEOREM 3—The Increment Theorem for Functions of Two Variables
Suppose that the first partial derivatives of $f(x, y)$ are defined throughout an open region R containing the point (x_0, y_0) and that f_x and f_y are continuous at (x_0, y_0). Then the change

$$\Delta z = f(x_0 + \Delta x, y_0 + \Delta y) - f(x_0, y_0)$$

in the value of f that results from moving from (x_0, y_0) to another point $(x_0 + \Delta x, y_0 + \Delta y)$ in R satisfies an equation of the form

$$\Delta z = f_x(x_0, y_0)\Delta x + f_y(x_0, y_0)\Delta y + \varepsilon_1 \Delta x + \varepsilon_2 \Delta y,$$

in which each of $\varepsilon_1, \varepsilon_2 \to 0$ as both $\Delta x, \Delta y \to 0$.

In many cases the partial derivatives are defined and continuous at every point in the domain of f. We then have the following Corollary.

Corollary of Theorem 3
If the partial derivatives f_x and f_y of a function $f(x, y)$ are continuous throughout an open region R, then f is differentiable at every point of R.

If $z = f(x, y)$ is differentiable, then the definition of differentiability ensures that $\Delta z = f(x_0 + \Delta x, y_0 + \Delta y) - f(x_0, y_0)$ approaches 0 as Δx and Δy approach 0. This tells us that a function of two variables is continuous at every point where it is differentiable.

THEOREM 4—Differentiable Implies Continuous
If a function $f(x, y)$ is differentiable at (x_0, y_0), then f is continuous at (x_0, y_0).

As we can see from Corollary 3 and Theorem 4, a function $f(x, y)$ must be continuous at a point (x_0, y_0) if f_x and f_y are continuous throughout an open region containing (x_0, y_0). Remember, however, that it is still possible for a function of two variables to be discontinuous at a point where its first partial derivatives exist, as we saw in Example 8. Existence alone of the partial derivatives at that point is not enough, but continuity of the partial derivatives guarantees differentiability.

EXERCISES 14.3

Calculating First-Order Partial Derivatives

In Exercises 1–22, find $\partial f/\partial x$ and $\partial f/\partial y$.

1. $f(x, y) = 2x^2 - 3y - 4$

2. $f(x, y) = x^2 - xy + y^2$

3. $f(x, y) = (x^2 - 1)(y + 2)$

4. $f(x, y) = \dfrac{xy}{x + y}$

5. $f(x, y) = y^x$

6. $f(x, y) = (2x - 3y)^3$

7. $f(x, y) = \sqrt{x^2 + y^2}$

8. $f(x, y) = (x^3 + (y/2))^{2/3}$

9. $f(x, y) = 1/(x + y)$

10. $f(x, y) = x/(x^2 + y^2)$

11. $f(x, y) = (x + y)/(xy - 1)$

12. $f(x, y) = \tan^{-1}(y/x)$

13. $f(x, y) = e^{(x+y+1)}$

14. $f(x, y) = e^{-x}\sin(x + y)$

15. $f(x, y) = \ln(x^2 + xy + y^2)$

16. $f(x, y) = \ln(x^2 + y^2) + \tan^{-1} y/x$

17. $f(x, y) = \sin^2(x - 3y)$

18. $f(x, y) = \ln\left[\dfrac{\sqrt{x^2 + y^2} - x}{\sqrt{x^2 + y^2} + x}\right]$

19. $f(x, y) = x^y$

20. $f(x, y) = \log_y x$

21. $f(x, y) = \int_x^y g(t)\,dt$ (g continuous for all t)

22. $f(x, y) = \sum_{n=0}^{\infty} (xy)^n$ $(|xy| < 1)$

In Exercises 23–34, find f_x, f_y, and f_z.

23. $f(x, y, z) = 1 + xy^2 - 2z^2$

24. $f(x, y, z) = (x - y)(y - z)(z - x)$

25. $f(x, y, z) = x - \sqrt{y^2 + z^2}$

26. $f(x, y, z) = (x^2 + y^2 + z^2)^{-1/2}$

27. $f(x, y, z) = \arcsin(xyz)$

28. $f(x, y, z) = \dfrac{x}{y} + \dfrac{y}{z} + \dfrac{z}{x}$

29. $f(x, y, z) = \ln(x + 2y + 3z)$

30. $f(x, y, z) = yz \ln(xy)$

31. $f(x, y, z) = e^{-(x^2+y^2+z^2)}$

32. $f(x, y, z) = e^{-xyz}$

33. $f(x, y, z) = e^{x/y} + e^{z/y}$

34. $f(x, y, z) = \sinh(xy - z^2)$

In Exercises 35–40, find the partial derivative of the function with respect to each variable.

35. $f(t, \alpha) = \cos(2\pi t - \alpha)$

36. $g(u, v) = v^2 e^{(2u/v)}$

37. $h(\rho, \phi, \theta) = \rho \sin\phi \cos\theta$

38. $g(r, \theta, z) = r(1 - \cos\theta) - z$

39. **Work done by the heart** (Section 3.9, Exercise 51)

$$W(P, V, \delta, v, g) = PV + \frac{V\delta v^2}{2g}$$

40. **Wilson lot size formula** (Section 4.5, Exercise 59)

$$A(c, h, k, m, q) = \frac{km}{q} + cm + \frac{hq}{2}$$

Calculating Second-Order Partial Derivatives

Find all the second-order partial derivatives of the functions in Exercises 41–54.

41. $f(x, y) = x + y + xy$

42. $f(x, y) = \sin xy$

43. $g(x, y) = x^2 y + \cos y + y \sin x$

44. $h(x, y) = xe^y + y + 1$

45. $r(x, y) = \ln(x^2 + y^2)$

46. $s(x, y) = \arctan(y/x)$

47. $w = x^2 \tan(xy)$

48. $w(x, y) = e^x \ln y + \sin y \ln x$

49. $w = x \sin(x^2 y)$

50. $w = \dfrac{x - y}{x^2 + y}$

51. $f(x, y) = x^2 y^3 - x^4 + y^5$

52. $g(x, y) = \cos x^2 - \sin 3y$

53. $z = x \sin(2x - y^2)$

54. $z = xe^{x/y^2}$

Mixed Partial Derivatives

In Exercises 55–60, verify that $w_{xy} = w_{yx}$.

55. $w = \ln(2x + 3y)$

56. $w = e^x + x \ln y + y \ln x$

57. $w(x, y) = (x^2 + y^2)^{3/2}$

58. $w = x \sin y + y \sin x + xy$

59. $w = \dfrac{x^2}{y^3}$

60. $w = \dfrac{3x - y}{x + y}$

61. Which order of differentiation enables one to calculate f_{xy} faster: x first or y first? Try to answer without writing anything down.

a. $f(x, y) = x \sin y + e^y$

b. $f(x, y) = 1/x$

c. $f(x, y) = y + (x/y)$

d. $f(x, y) = y + x^2 y + 4y^3 - \ln(y^2 + 1)$

e. $f(x, y) = x^2 + 5xy + \sin x + 7e^x$

f. $f(x, y) = x \ln xy$

62. The fifth-order partial derivative $\partial^5 f/\partial x^2 \partial y^3$ is zero for each of the following functions. To show this as quickly as possible, which variable would you differentiate with respect to first: x or y? Try to answer without writing anything down.

a. $f(x, y) = y^2 x^4 e^x + 2$

b. $f(x, y) = y^2 + y(\sin x - x^4)$

c. $f(x, y) = x^2 + 5xy + \sin x + 7e^x$

d. $f(x, y) = xe^{y^2/2}$

Using the Partial Derivative Definition

In Exercises 63–66, use the limit definition of partial derivative to compute the partial derivatives of the functions at the specified points.

63. $f(x, y) = 1 - x + y - 3x^2 y,\quad \dfrac{\partial f}{\partial x}$ and $\dfrac{\partial f}{\partial y}$ at $(1, 2)$

64. $f(x, y) = 4 + 2x - 3y - xy^2,\quad \dfrac{\partial f}{\partial x}$ and $\dfrac{\partial f}{\partial y}$ at $(-2, 1)$

65. $f(x, y) = \sqrt{2x + 3y - 1},\quad \dfrac{\partial f}{\partial x}$ and $\dfrac{\partial f}{\partial y}$ at $(-2, 3)$

66. $f(x, y) = \begin{cases} \dfrac{\sin(x^3 + y^4)}{x^2 + y^2}, & (x, y) \neq (0, 0) \\ 0, & (x, y) = (0, 0), \end{cases}$

$\dfrac{\partial f}{\partial x}$ and $\dfrac{\partial f}{\partial y}$ at $(0, 0)$

67. **Three variables** Let $w = f(x, y, z)$ be a function of three independent variables and write the formal definition of the partial derivative $\partial f/\partial z$ at (x_0, y_0, z_0). Use this definition to find $\partial f/\partial z$ at $(1, 2, 3)$ for $f(x, y, z) = x^2 yz^2$.

68. **Three variables** Let $w = f(x, y, z)$ be a function of three independent variables and write the formal definition of the partial derivative $\partial f/\partial y$ at (x_0, y_0, z_0). Use this definition to find $\partial f/\partial y$ at $(-1, 0, 3)$ for $f(x, y, z) = -2xy^2 + yz^2$.

Differentiating Implicitly

69. Find the value of $\partial z/\partial x$ at the point $(1, 1, 1)$ if the equation

$$xy + z^3 x - 2yz = 0$$

defines z as a function of the two independent variables x and y and the partial derivative exists.

70. Find the value of $\partial x/\partial z$ at the point $(1, -1, -3)$ if the equation

$$xz + y \ln x - x^2 + 4 = 0$$

defines x as a function of the two independent variables y and z and the partial derivative exists.

Exercises 71 and 72 are about the triangle shown here.

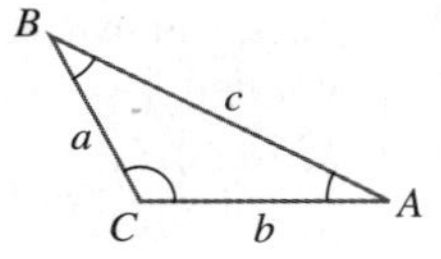

71. Express A implicitly as a function of a, b, and c and calculate $\partial A/\partial a$ and $\partial A/\partial b$.

72. Express a implicitly as a function of A, b, and B and calculate $\partial a/\partial A$ and $\partial a/\partial B$.

73. Two dependent variables Express v_x in terms of u and y if the equations $x = v \ln u$ and $y = u \ln v$ define u and v as functions of the independent variables x and y, and if v_x exists. (*Hint:* Differentiate both equations with respect to x and solve for v_x by eliminating u_x.)

74. Two dependent variables Find $\partial x/\partial u$ and $\partial y/\partial u$ if the equations $u = x^2 - y^2$ and $v = x^2 - y$ define x and y as functions of the independent variables u and v, and the partial derivatives exist. (See the hint in Exercise 73.) Then let $s = x^2 + y^2$ and find $\partial s/\partial u$.

Theory and Examples

75. Let $f(x, y) = 2x + 3y - 4$. Find the slope of the line tangent to this surface at the point $(2, -1)$ and lying in **a.** the plane $x = 2$ **b.** the plane $y = -1$.

76. Let $f(x, y) = x^2 + y^3$. Find the slope of the line tangent to this surface at the point $(-1, 1)$ and lying in **a.** the plane $x = -1$ **b.** the plane $y = 1$.

In Exercises 77–80, find a function $z = f(x, y)$ whose partial derivatives are as given, or explain why this is impossible.

77. $\dfrac{\partial f}{\partial x} = 3x^2y^2 - 2x, \quad \dfrac{\partial f}{\partial y} = 2x^3y + 6y$

78. $\dfrac{\partial f}{\partial x} = 2xe^{xy^2} + x^2y^2e^{xy^2} + 3, \quad \dfrac{\partial f}{\partial y} = 2x^3ye^{xy^2} - e^y$

79. $\dfrac{\partial f}{\partial x} = \dfrac{2y}{(x+y)^2}, \quad \dfrac{\partial f}{\partial y} = \dfrac{2x}{(x+y)^2}$

80. $\dfrac{\partial f}{\partial x} = xy\cos(xy) + \sin(xy), \quad \dfrac{\partial f}{\partial y} = x\cos(xy)$

81. Let $f(x, y) = \begin{cases} y^3, & y \ge 0 \\ -y^2, & y < 0. \end{cases}$

Find f_x, f_y, f_{xy}, and f_{yx}, and state the domain for each partial derivative.

82. Let $f(x, y) = \begin{cases} xy\dfrac{x^2 - y^2}{x^2 + y^2}, & \text{if } (x, y) \ne 0, \\ 0, & \text{if } (x, y) = 0. \end{cases}$

a. Show that $\dfrac{\partial f}{\partial y}(x, 0) = x$ for all x, and $\dfrac{\partial f}{\partial x}(0, y) = -y$ for all y.

b. Show that $\dfrac{\partial^2 f}{\partial y\,\partial x}(0,0) \ne \dfrac{\partial^2 f}{\partial x\,\partial y}(0,0)$.

The **three-dimensional Laplace equation**

$$\frac{\partial^2 f}{\partial x^2} + \frac{\partial^2 f}{\partial y^2} + \frac{\partial^2 f}{\partial z^2} = 0$$

is satisfied by steady-state temperature distributions $T = f(x, y, z)$ in space, by gravitational potentials, and by electrostatic potentials. The **two-dimensional Laplace equation**

$$\frac{\partial^2 f}{\partial x^2} + \frac{\partial^2 f}{\partial y^2} = 0,$$

obtained by dropping the $\partial^2 f/\partial z^2$ term from the previous equation, describes potentials and steady-state temperature distributions in a plane. The plane may be treated as a thin slice of the solid perpendicular to the z-axis.

Show that each function in Exercises 83–90 satisfies a Laplace equation.

83. $f(x, y, z) = x^2 + y^2 - 2z^2$

84. $f(x, y, z) = 2z^3 - 3(x^2 + y^2)z$

85. $f(x, y) = e^{-2y}\cos 2x$

86. $f(x, y) = 3x^2y + 2x^2 - y^3 - 2y^2$

87. $f(x, y) = e^x(x\cos y - y\sin y)$

88. $f(x, y) = \arctan\dfrac{x}{y}$

89. $f(x, y, z) = (x^2 + y^2 + z^2)^{-1/2}$

90. $f(x, y, z) = e^{3x+4y}\cos 5z$

The wave equation If we stand on an ocean shore and take a snapshot of the waves, the picture shows a regular pattern of peaks and valleys in an instant of time. We see periodic vertical motion in space, with respect to distance. If we stand in the water, we can feel the rise and fall of the water as the waves go by. We see periodic vertical motion in time. In physics, this beautiful symmetry is expressed by the **one-dimensional wave equation**

$$\frac{\partial^2 w}{\partial t^2} = c^2\frac{\partial^2 w}{\partial x^2},$$

where w is the wave height, x is the distance variable, t is the time variable, and c is the velocity with which the waves are propagated.

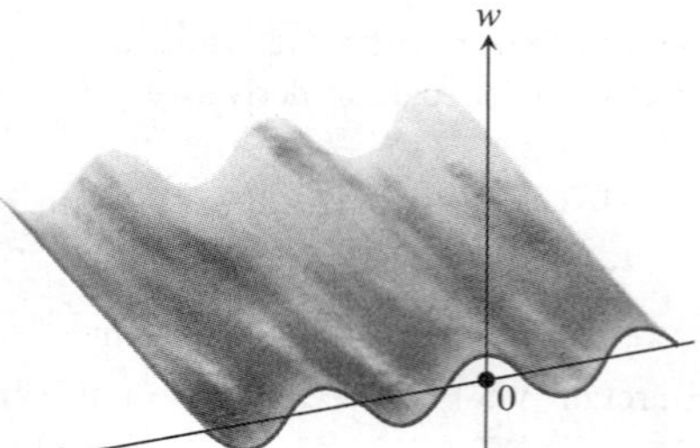

In our example, x is the distance across the ocean's surface, but in other applications, x might be the distance along a vibrating string, distance through air (sound waves), or distance through space (light waves). The number c varies with the medium and type of wave.

Show that the functions in Exercises 91–97 are all solutions of the wave equation.

91. $w = \sin(x + ct)$

92. $w = \cos(2x + 2ct)$

93. $w = \sin(x + ct) + \cos(2x + 2ct)$

94. $w = \ln(2x + 2ct)$

95. $w = \tan(2x - 2ct)$

96. $w = 5\cos(3x + 3ct) + e^{x+ct}$

97. $w = f(u)$, where f is a differentiable function of u, and $u = a(x + ct)$, where a is a constant

98. Does a function $f(x, y)$ with continuous first partial derivatives throughout an open region R have to be continuous on R? Give reasons for your answer.

99. If a function $f(x, y)$ has continuous second partial derivatives throughout an open region R, must the first-order partial derivatives of f be continuous on R? Give reasons for your answer.

100. The heat equation An important partial differential equation that describes the distribution of heat in a region at time t can be represented by the *one-dimensional heat equation*

$$\frac{\partial f}{\partial t} = \frac{\partial^2 f}{\partial x^2}.$$

Show that $u(x, t) = \sin(\alpha x) \cdot e^{-\beta t}$ satisfies the heat equation for constants α and β. What is the relationship between α and β for this function to be a solution?

101. Let $f(x, y) = \begin{cases} \dfrac{xy^2}{x^2 + y^4}, & (x, y) \neq (0, 0) \\ 0, & (x, y) = (0, 0). \end{cases}$

Show that $f_x(0, 0)$ and $f_y(0, 0)$ exist, but f is not differentiable at $(0, 0)$. (*Hint:* Use Theorem 4 and show that f is not continuous at $(0, 0)$.)

102. Let $f(x, y) = \begin{cases} 0, & x^2 < y < 2x^2 \\ 1, & \text{otherwise.} \end{cases}$

Show that $f_x(0, 0)$ and $f_y(0, 0)$ exist, but f is not differentiable at $(0, 0)$.

103. The Korteweg–de Vries equation
This nonlinear differential equation, which describes wave motion on shallow water surfaces, is given by

$$u_t + u_{xxx} + 12uu_x = 0.$$

Show that $u(x, t) = \operatorname{sech}^2(x - t)$ satisfies the Korteweg–de Vries equation.

104. Show that $T = \dfrac{1}{\sqrt{x^2 + y^2}}$ satisfies the equation $T_{xx} + T_{yy} = T^3$.

14.4 The Chain Rule

The Chain Rule for functions of a single variable studied in Section 3.6 says that if $w = f(x)$ is a differentiable function of x, and $x = g(t)$ is a differentiable function of t, then w is a differentiable function of t, and dw/dt can be calculated by the formula

$$\frac{dw}{dt} = \frac{dw}{dx}\frac{dx}{dt}.$$

To find dw/dt, we read down the route from w to t, multiplying derivatives along the way.

Chain Rule

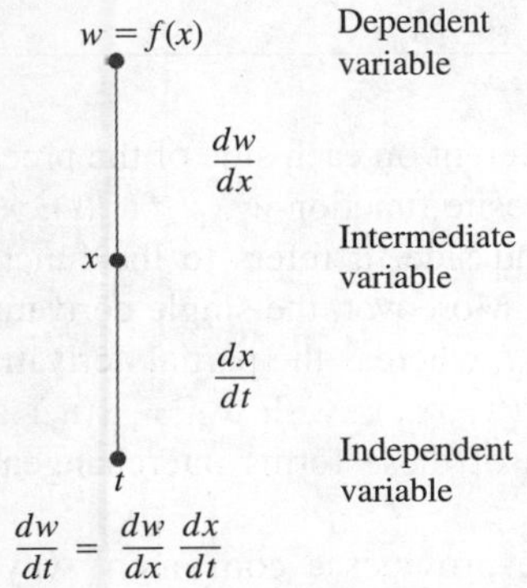

For this composite function $w(t) = f(g(t))$, we can think of t as the independent variable and $x = g(t)$ as the "intermediate variable" because t determines the value of x that in turn gives the value of w from the function f. We display the Chain Rule in a "dependency diagram" in the margin. Such diagrams capture which variables depend on which.

For functions of several variables the Chain Rule has more than one form, which depends on how many independent and intermediate variables are involved. However, once the variables are taken into account, the Chain Rule works in the same way we just discussed.

Functions of Two Variables

The Chain Rule formula for a differentiable function $w = f(x, y)$ when $x = x(t)$ and $y = y(t)$ are both differentiable functions of t is given in the following theorem.

THEOREM 5—Chain Rule for Functions of One Independent Variable and Two Intermediate Variables
If $w = f(x, y)$ is differentiable and if $x = x(t)$, $y = y(t)$ are differentiable functions of t, then the composition $w = f(x(t), y(t))$ is a differentiable function of t and

$$\frac{dw}{dt} = f_x(x(t), y(t))x'(t) + f_y(x(t), y(t))y'(t),$$

or

$$\frac{dw}{dt} = \frac{\partial f}{\partial x}\frac{dx}{dt} + \frac{\partial f}{\partial y}\frac{dy}{dt}.$$

Each of $\frac{\partial f}{\partial x}, \frac{\partial w}{\partial x}, f_x$ indicates the partial derivative of f with respect to x.

Proof The proof consists of showing that if x and y are differentiable at $t = t_0$, then w is differentiable at t_0 and

$$\frac{dw}{dt}(t_0) = \frac{\partial w}{\partial x}(P_0)\frac{dx}{dt}(t_0) + \frac{\partial w}{\partial y}(P_0)\frac{dy}{dt}(t_0),$$

where $P_0 = (x(t_0), y(t_0))$.

Let Δx, Δy, and Δw be the increments that result from changing t from t_0 to $t_0 + \Delta t$. Since f is differentiable (see the definition in Section 14.3),

$$\Delta w = \frac{\partial w}{\partial x}(P_0)\,\Delta x + \frac{\partial w}{\partial y}(P_0)\,\Delta y + \varepsilon_1 \Delta x + \varepsilon_2 \Delta y,$$

where $\varepsilon_1, \varepsilon_2 \to 0$ as $\Delta x, \Delta y \to 0$. To find dw/dt, we divide this equation through by Δt and let Δt approach zero (therefore, Δx and Δy approach zero as well since the fact that $x(t)$ and $y(t)$ are differentiable implies that they are continuous). The division gives

$$\frac{\Delta w}{\Delta t} = \frac{\partial w}{\partial x}(P_0)\frac{\Delta x}{\Delta t} + \frac{\partial w}{\partial y}(P_0)\frac{\Delta y}{\Delta t} + \varepsilon_1 \frac{\Delta x}{\Delta t} + \varepsilon_2 \frac{\Delta y}{\Delta t}.$$

Letting Δt approach zero gives

$$\begin{aligned}\frac{dw}{dt}(t_0) &= \lim_{\Delta t \to 0} \frac{\Delta w}{\Delta t} \\ &= \frac{\partial w}{\partial x}(P_0)\frac{dx}{dt}(t_0) + \frac{\partial w}{\partial y}(P_0)\frac{dy}{dt}(t_0) + 0 \cdot \frac{dx}{dt}(t_0) + 0 \cdot \frac{dy}{dt}(t_0).\end{aligned}$$ ■

To remember the Chain Rule, picture the diagram below. To find dw/dt, start at w and read down each route to t, multiplying derivatives along the way. Then add the products.

Chain Rule

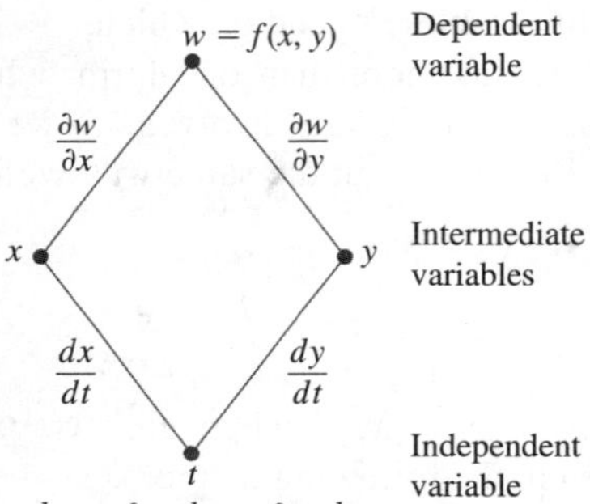

$$\frac{dw}{dt} = \frac{\partial w}{\partial x}\frac{dx}{dt} + \frac{\partial w}{\partial y}\frac{dy}{dt}$$

Often we write $\partial w/\partial x$ for the partial derivative $\partial f/\partial x$, so we can rewrite the Chain Rule in Theorem 5 in the form

$$\frac{dw}{dt} = \frac{\partial w}{\partial x}\frac{dx}{dt} + \frac{\partial w}{\partial y}\frac{dy}{dt}.$$

However, the meaning of the dependent variable w is different on each side of the preceding equation. On the left-hand side, it refers to the composite function $w = f(x(t), y(t))$ as a function of the single variable t. On the right-hand side, it refers to the function $w = f(x, y)$ as a function of the two variables x and y. Moreover, the single derivatives dw/dt, dx/dt, and dy/dt are being evaluated at a point t_0, whereas the partial derivatives $\partial w/\partial x$ and $\partial w/\partial y$ are being evaluated at the point (x_0, y_0), with $x_0 = x(t_0)$ and $y_0 = y(t_0)$. With that understanding, we will use both of these forms interchangeably throughout the text whenever no confusion will arise.

The **dependency diagram** on the preceding page provides a convenient way to remember the Chain Rule. The "true" independent variable in the composite function is t, whereas x and y are *intermediate variables* (controlled by t) and w is the dependent variable.

A more precise notation for the Chain Rule shows where the various derivatives in Theorem 5 are evaluated:

$$\frac{dw}{dt}(t_0) = \frac{\partial f}{\partial x}(x_0, y_0)\frac{dx}{dt}(t_0) + \frac{\partial f}{\partial y}(x_0, y_0)\frac{dy}{dt}(t_0),$$

or, using another notation,

$$\left.\frac{dw}{dt}\right|_{t_0} = \left.\frac{\partial f}{\partial x}\right|_{(x_0, y_0)} \left.\frac{dx}{dt}\right|_{t_0} + \left.\frac{\partial f}{\partial y}\right|_{(x_0, y_0)} \left.\frac{dy}{dt}\right|_{t_0}.$$

EXAMPLE 1 Use the Chain Rule to find the derivative of

$$w = xy$$

with respect to t along the path $x = \cos t$, $y = \sin t$. What is the derivative's value at $t = \pi/2$?

Solution We apply the Chain Rule to find dw/dt as follows:

$$\begin{aligned}\frac{dw}{dt} &= \frac{\partial w}{\partial x}\frac{dx}{dt} + \frac{\partial w}{\partial y}\frac{dy}{dt} \\ &= \frac{\partial(xy)}{\partial x}\frac{d}{dt}(\cos t) + \frac{\partial(xy)}{\partial y}\frac{d}{dt}(\sin t) \\ &= (y)(-\sin t) + (x)(\cos t) \\ &= (\sin t)(-\sin t) + (\cos t)(\cos t) \\ &= -\sin^2 t + \cos^2 t \\ &= \cos 2t.\end{aligned}$$

In this example, we can check the result with a more direct calculation. As a function of t,

$$w = xy = \cos t \sin t = \frac{1}{2}\sin 2t,$$

so

$$\frac{dw}{dt} = \frac{d}{dt}\left(\frac{1}{2}\sin 2t\right) = \frac{1}{2}(2\cos 2t) = \cos 2t.$$

In either case, at the given value of t,

$$\left.\frac{dw}{dt}\right|_{t=\pi/2} = \cos\left(2\frac{\pi}{2}\right) = \cos \pi = -1.$$ ■

Functions of Three Variables

You can probably predict the Chain Rule for functions of three intermediate variables, as it involves adding the expected third term to the two-variable formula.

Here we have three routes from w to t instead of two, but finding dw/dt is still the same. Read down each route, multiplying derivatives along the way; then add.

Chain Rule

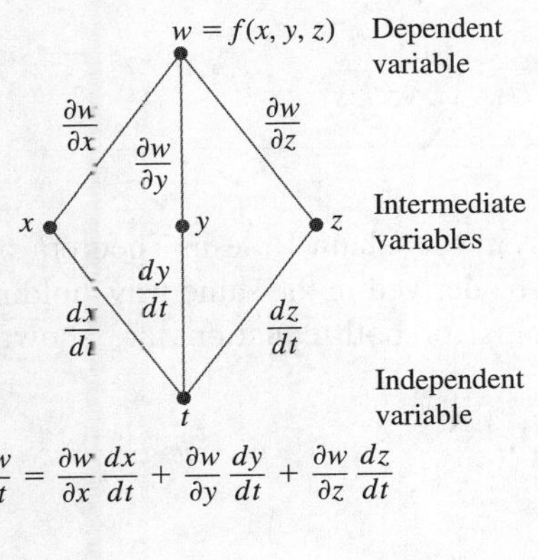

$$\frac{dw}{dt} = \frac{\partial w}{\partial x}\frac{dx}{dt} + \frac{\partial w}{\partial y}\frac{dy}{dt} + \frac{\partial w}{\partial z}\frac{dz}{dt}$$

THEOREM 6—Chain Rule for Functions of One Independent Variable and Three Intermediate Variables
If $w = f(x, y, z)$ is differentiable and x, y, and z are differentiable functions of t, then w is a differentiable function of t, and

$$\frac{dw}{dt} = \frac{\partial w}{\partial x}\frac{dx}{dt} + \frac{\partial w}{\partial y}\frac{dy}{dt} + \frac{\partial w}{\partial z}\frac{dz}{dt}.$$

The proof is identical to the proof of Theorem 5, except that there are now three intermediate variables instead of two. The dependency diagram we use for remembering the new equation is similar as well, with three routes from w to t.

EXAMPLE 2 Find dw/dt if

$$w = xy + z, \qquad x = \cos t, \qquad y = \sin t, \qquad z = t.$$

In this example the values of $w(t)$ are changing along the path of a helix (Section 13.1) as t changes. What is the derivative's value at $t = 0$?

Solution Using the Chain Rule for three intermediate variables, we have

$$\begin{aligned}\frac{dw}{dt} &= \frac{\partial w}{\partial x}\frac{dx}{dt} + \frac{\partial w}{\partial y}\frac{dy}{dt} + \frac{\partial w}{\partial z}\frac{dz}{dt} \\ &= (y)(-\sin t) + (x)(\cos t) + (1)(1) \\ &= (\sin t)(-\sin t) + (\cos t)(\cos t) + 1 \\ &= -\sin^2 t + \cos^2 t + 1 = 1 + \cos 2t,\end{aligned}$$

Substitute for intermediate variables.

so

$$\left.\frac{dw}{dt}\right|_{t=0} = 1 + \cos(0) = 2.$$

For a physical interpretation of change along a curve, think of an object whose position is changing with time t. If $w = T(x, y, z)$ is the temperature at each point (x, y, z) along a curve C with parametric equations $x = x(t)$, $y = y(t)$, and $z = z(t)$, then the composite function $w = T(x(t), y(t), z(t))$ represents the temperature relative to t along the curve. The derivative dw/dt is then the instantaneous rate of change of temperature due to the motion along the curve, as calculated in Theorem 6.

Functions Defined on Surfaces

If we are interested in the temperature $w = f(x, y, z)$ at points (x, y, z) on Earth's surface, we might prefer to think of x, y, and z as functions of the variables r and s that give the points' longitudes and latitudes. If $x = g(r, s)$, $y = h(r, s)$, and $z = k(r, s)$, we could then express the temperature as a function of r and s with the composite function

$$w = f(g(r, s), h(r, s), k(r, s)).$$

Under the conditions stated below, w has partial derivatives with respect to both r and s that can be calculated in the following way.

THEOREM 7—Chain Rule for Two Independent Variables and Three Intermediate Variables
Suppose that $w = f(x, y, z)$, $x = g(r, s)$, $y = h(r, s)$, and $z = k(r, s)$. If all four functions are differentiable, then w has partial derivatives with respect to r and s, given by the formulas

$$\frac{\partial w}{\partial r} = \frac{\partial w}{\partial x}\frac{\partial x}{\partial r} + \frac{\partial w}{\partial y}\frac{\partial y}{\partial r} + \frac{\partial w}{\partial z}\frac{\partial z}{\partial r}$$

$$\frac{\partial w}{\partial s} = \frac{\partial w}{\partial x}\frac{\partial x}{\partial s} + \frac{\partial w}{\partial y}\frac{\partial y}{\partial s} + \frac{\partial w}{\partial z}\frac{\partial z}{\partial s}.$$

The first of these equations can be derived from the Chain Rule in Theorem 6 by holding s fixed and treating r as t. The second can be derived in the same way, holding r fixed and treating s as t. The dependency diagrams for both equations are shown in Figure 14.22.

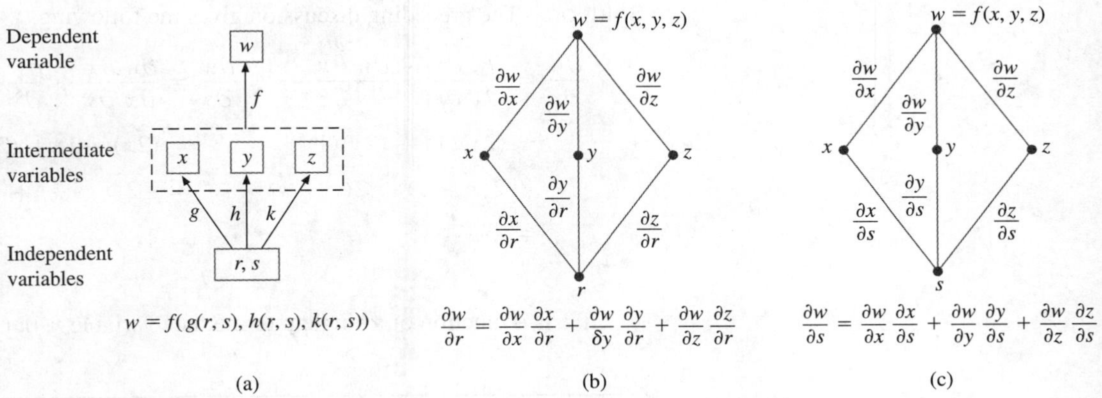

FIGURE 14.22 Composite function and dependency diagrams for Theorem 7.

EXAMPLE 3 Express $\partial w/\partial r$ and $\partial w/\partial s$ in terms of r and s if

$$w = x + 2y + z^2, \qquad x = \frac{r}{s}, \qquad y = r^2 + \ln s, \qquad z = 2r.$$

Solution Using the formulas in Theorem 7, we find

$$\begin{aligned}
\frac{\partial w}{\partial r} &= \frac{\partial w}{\partial x}\frac{\partial x}{\partial r} + \frac{\partial w}{\partial y}\frac{\partial y}{\partial r} + \frac{\partial w}{\partial z}\frac{\partial z}{\partial r} \\
&= (1)\left(\frac{1}{s}\right) + (2)(2r) + (2z)(2) \\
&= \frac{1}{s} + 4r + (4r)(2) = \frac{1}{s} + 12r \qquad \text{Substitute for intermediate variable } z.
\end{aligned}$$

$$\begin{aligned}
\frac{\partial w}{\partial s} &= \frac{\partial w}{\partial x}\frac{\partial x}{\partial s} + \frac{\partial w}{\partial y}\frac{\partial y}{\partial s} + \frac{\partial w}{\partial z}\frac{\partial z}{\partial s} \\
&= (1)\left(-\frac{r}{s^2}\right) + (2)\left(\frac{1}{s}\right) + (2z)(0) = \frac{2}{s} - \frac{r}{s^2}.
\end{aligned}$$ ■

If f is a function of two intermediate variables instead of three, each equation in Theorem 7 becomes correspondingly one term shorter.

> If $w = f(x, y)$, $x = g(r, s)$, and $y = h(r, s)$, then
>
> $$\frac{\partial w}{\partial r} = \frac{\partial w}{\partial x}\frac{\partial x}{\partial r} + \frac{\partial w}{\partial y}\frac{\partial y}{\partial r} \quad \text{and} \quad \frac{\partial w}{\partial s} = \frac{\partial w}{\partial x}\frac{\partial x}{\partial s} + \frac{\partial w}{\partial y}\frac{\partial y}{\partial s}.$$

Figure 14.23 shows the dependency diagram for the first of these equations. The diagram for the second equation is similar; just replace r with s.

Chain Rule

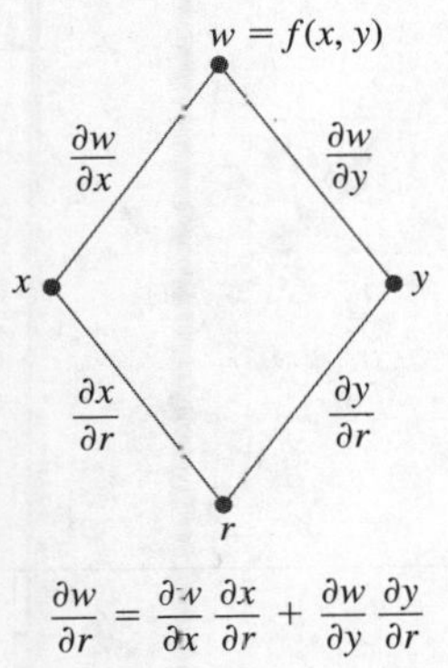

FIGURE 14.23 Dependency diagram for the equation

$$\frac{\partial w}{\partial r} = \frac{\partial w}{\partial x}\frac{\partial x}{\partial r} + \frac{\partial w}{\partial y}\frac{\partial y}{\partial r}.$$

EXAMPLE 4 Express $\partial w/\partial r$ and $\partial w/\partial s$ in terms of r and s if

$$w = x^2 + y^2, \qquad x = r - s, \qquad y = r + s.$$

Solution The preceding discussion gives the following.

$$\begin{aligned}\frac{\partial w}{\partial r} &= \frac{\partial w}{\partial x}\frac{\partial x}{\partial r} + \frac{\partial w}{\partial y}\frac{\partial y}{\partial r} & \frac{\partial w}{\partial s} &= \frac{\partial w}{\partial x}\frac{\partial x}{\partial s} + \frac{\partial w}{\partial y}\frac{\partial y}{\partial s}\\ &= (2x)(1) + (2y)(1) & &= (2x)(-1) + (2y)(1)\\ &= 2(r - s) + 2(r + s) & &= -2(r - s) + 2(r + s)\\ &= 4r & &= 4s\end{aligned}$$

Substitute for the intermediate variables.

■

If f is a function of a single intermediate variable x, our equations are even simpler.

> If $w = f(x)$ and $x = g(r, s)$, then
>
> $$\frac{\partial w}{\partial r} = \frac{dw}{dx}\frac{\partial x}{\partial r} \quad \text{and} \quad \frac{\partial w}{\partial s} = \frac{dw}{dx}\frac{\partial x}{\partial s}.$$

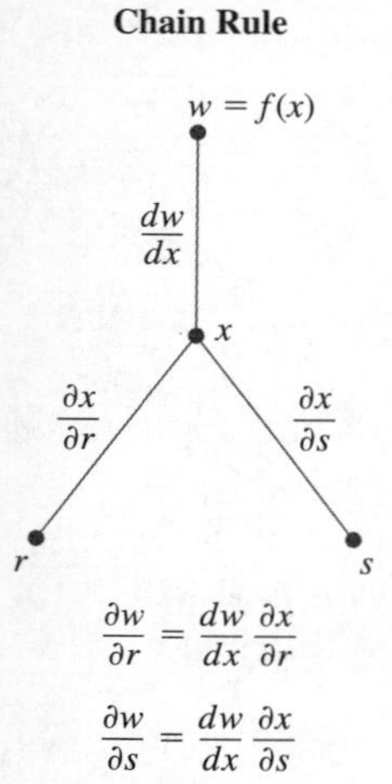

FIGURE 14.24 Dependency diagram for differentiating f as a composite function of r and s with one intermediate variable.

In this case, we use the ordinary (single-variable) derivative, dw/dx. The dependency diagram is shown in Figure 14.24.

Implicit Differentiation Revisited

The two-variable Chain Rule in Theorem 5 leads to a formula that takes some of the algebra out of implicit differentiation. Suppose that

1. The function $F(x, y)$ is differentiable and
2. The equation $F(x, h(x)) = 0$ defines y implicitly as a differentiable function of x, say $y = h(x)$.

Since $w = F(x, h(x)) = 0$, the derivative dw/dx must be zero. Computing the derivative from the Chain Rule (dependency diagram in Figure 14.25), we find

$$\begin{aligned}0 = \frac{dw}{dx} &= F_x\frac{dx}{dx} + F_y\frac{dy}{dx} \qquad \text{Theorem 5 with } t = x \text{ and } f = F\\ &= F_x \cdot 1 + F_y \cdot \frac{dy}{dx}.\end{aligned}$$

If $F_y = \partial w/\partial y \neq 0$, we can solve this equation for dy/dx to get

$$\frac{dy}{dx} = -\frac{F_x}{F_y}.$$

We state this result formally.

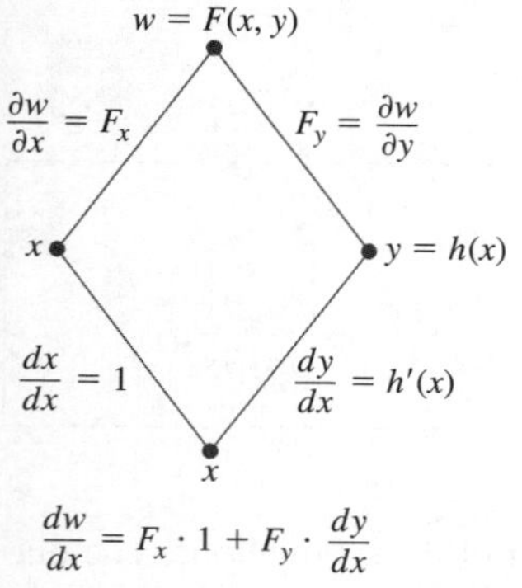

FIGURE 14.25 Dependency diagram for differentiating $w = F(x, y)$ with respect to x. Setting $dw/dx = 0$ leads to a simple computational formula for implicit differentiation (Theorem 8).

> **THEOREM 8—A Formula for Implicit Differentiation**
> Suppose that $F(x, y)$ is differentiable and that the equation $F(x, y) = 0$ defines y as a differentiable function of x. Then, at any point where $F_y \neq 0$,
>
> $$\frac{dy}{dx} = -\frac{F_x}{F_y}. \qquad (1)$$

EXAMPLE 5 Use Theorem 8 to find dy/dx if $y^2 - x^2 - \sin xy = 0$.

Solution Take $F(x, y) = y^2 - x^2 - \sin xy$. Then

$$\frac{dy}{dx} = -\frac{F_x}{F_y} = -\frac{-2x - y\cos xy}{2y - x\cos xy} = \frac{2x + y\cos xy}{2y - x\cos xy}.$$

This calculation is significantly shorter than a single-variable calculation using implicit differentiation. ■

The result in Theorem 8 is easily extended to three variables. Suppose that the equation $F(x, y, z) = 0$ defines the variable z implicitly as a function $z = f(x, y)$. Then, for all (x, y) in the domain of f, we have $F(x, y, f(x, y)) = 0$. Assuming that F and f are differentiable functions, we can use the Chain Rule to differentiate the equation $F(x, y, z) = 0$ with respect to the independent variable x:

$$\begin{aligned} 0 &= \frac{\partial F}{\partial x}\frac{\partial x}{\partial x} + \frac{\partial F}{\partial y}\frac{\partial y}{\partial x} + \frac{\partial F}{\partial z}\frac{\partial z}{\partial x} \\ &= F_x \cdot 1 + F_y \cdot 0 + F_z \cdot \frac{\partial z}{\partial x}, \end{aligned}$$

y is constant when we differentiate with respect to x.

so

$$F_x + F_z\frac{\partial z}{\partial x} = 0.$$

A similar calculation for differentiating with respect to the independent variable y gives

$$F_y + F_z\frac{\partial z}{\partial y} = 0.$$

Whenever $F_z \neq 0$, we can solve these last two equations for the partial derivatives of $z = f(x, y)$ to obtain

$$\frac{\partial z}{\partial x} = -\frac{F_x}{F_z} \quad \text{and} \quad \frac{\partial z}{\partial y} = -\frac{F_y}{F_z}. \tag{2}$$

An important result from advanced calculus, called the **Implicit Function Theorem**, states the conditions for which our results in Equations (2) are valid. If the partial derivatives F_x, F_y, and F_z are continuous throughout an open region R in space containing the point (x_0, y_0, z_0), and if for some constant c, $F(x_0, y_0, z_0) = c$ and $F_z(x_0, y_0, z_0) \neq 0$, then the equation $F(x, y, z) = c$ defines z implicitly as a differentiable function of x and y near (x_0, y_0, z_0), and the partial derivatives of z are given by Equations (2).

EXAMPLE 6 Find $\dfrac{\partial z}{\partial x}$ and $\dfrac{\partial z}{\partial y}$ at $(0, 0, 0)$ if $x^3 + z^2 + ye^{xz} + z\cos y = 0$.

Solution Let $F(x, y, z) = x^3 + z^2 + ye^{xz} + z\cos y$. Then

$$F_x = 3x^2 + zye^{xz}, \qquad F_y = e^{xz} - z\sin y, \qquad \text{and} \qquad F_z = 2z + xye^{xz} + \cos y.$$

Since $F(0,0,0) = 0, F_z(0,0,0) = 1 \neq 0$, and all first partial derivatives are continuous, the Implicit Function Theorem says that $F(x, y, z) = 0$ defines z as a differentiable function of x and y near the point $(0, 0, 0)$. From Equations (2),

$$\frac{\partial z}{\partial x} = -\frac{F_x}{F_z} = -\frac{3x^2 + zye^{xz}}{2z + xye^{xz} + \cos y} \quad \text{and} \quad \frac{\partial z}{\partial y} = -\frac{F_y}{F_z} = -\frac{e^{xz} - z\sin y}{2z + xye^{xz} + \cos y}.$$

At $(0, 0, 0)$ we find

$$\frac{\partial z}{\partial x} = -\frac{0}{1} = 0 \quad \text{and} \quad \frac{\partial z}{\partial y} = -\frac{1}{1} = -1.$$

■

Functions of Many Variables

We have seen several different forms of the Chain Rule in this section, but each one is just a special case of one general formula. When solving particular problems, it may help to draw the appropriate dependency diagram by placing the dependent variable on top, the intermediate variables in the middle, and the selected independent variable at the bottom. To find the derivative of the dependent variable with respect to the selected independent variable, start at the dependent variable and read down each route of the dependency diagram to the independent variable, calculating and multiplying the derivatives along each route. Then add the products found for the different routes.

In general, suppose that $w = f(x_1, x_2, \ldots, x_n)$ is a differentiable function of the intermediate variables $x_1, x_2, \ldots, x_n$ (a finite set) and that $x_1, x_2, \ldots, x_n$ are differentiable functions of the independent variables $t_1, t_2, \ldots, t_m$ (another finite set). Then w is a differentiable function of the variables $t_1, t_2, \ldots, t_m$, and the partial derivatives of w with respect to these variables are given by equations of the form

$$\frac{\partial w}{\partial t_i} = \frac{\partial w}{\partial x_1}\frac{\partial x_1}{\partial t_i} + \frac{\partial w}{\partial x_2}\frac{\partial x_2}{\partial t_i} + \cdots + \frac{\partial w}{\partial x_n}\frac{\partial x_n}{\partial t_i} \quad \text{for } i = 1, 2, \ldots, m.$$

One way to remember this equation is to think of the right-hand side as the dot product of two n-dimensional vectors:

$$\frac{\partial w}{\partial t_i} = \underbrace{\left\langle \frac{\partial w}{\partial x_1}, \frac{\partial w}{\partial x_2}, \ldots, \frac{\partial w}{\partial x_n} \right\rangle}_{\substack{\text{Derivatives of } w \text{ with}\\ \text{respect to the}\\ \text{intermediate variables}}} \cdot \underbrace{\left\langle \frac{\partial x_1}{\partial t_i}, \frac{\partial x_2}{\partial t_i}, \ldots, \frac{\partial x_n}{\partial t_i} \right\rangle}_{\substack{\text{Derivatives of the}\\ \text{intermediate variables with}\\ \text{respect to the selected}\\ \text{independent variable}}}.$$

The first vector describes how w changes in various directions, while the second vector indicates the velocity vector of $\mathbf{x}(t_i) = \langle x_1(t_i), x_2(t_i), \ldots, x_n(t_i) \rangle$. These concepts will be studied further in the next section.

EXERCISES 14.4

Chain Rule: One Independent Variable

In Exercises 1–6, **(a)** express dw/dt as a function of t, both by using the Chain Rule and by expressing w in terms of t and differentiating directly with respect to t. Then **(b)** evaluate dw/dt at the given value of t.

1. $w = x^2 + y^2, \quad x = \cos t, \quad y = \sin t; \quad t = \pi$

2. $w = x \ \ + y \ , \quad x = at \ , \quad y = 2at$

3. $w = \frac{x}{z} + \frac{y}{z}, \quad x = \cos^2 t, \quad y = \sin^2 t, \quad z = 1/t; \quad t = 3$

4. $w = \ln(x^2 + y^2 + z^2), \quad x = \cos t, \quad y = \sin t, \quad z = 4\sqrt{t}; \quad t = 3$

5. $w = 2ye^x - \ln z, \quad x = \ln(t^2 + 1), \quad y = \tan^{-1} t, \quad z = e^t; \quad t = 1$

6. $w = x^2 + y^2 + z^2, \quad x = e^t, \quad y = e^t \cos t, \quad z = e^t \sin t, \quad t = 0$

Chain Rule: Two and Three Independent Variables

In Exercises 7 and 8, **(a)** express $\partial z/\partial u$ and $\partial z/\partial v$ as functions of u and v both by using the Chain Rule and by expressing z directly in terms of u and v before differentiating. Then **(b)** evaluate $\partial z/\partial u$ and $\partial z/\partial v$ at the given point (u, v).

7. $z = 4e^x \ln y, \quad x = \ln(u \cos v), \quad y = u \sin v; \quad (u, v) = (2, \pi/4)$

8. $z = \tan^{-1}(x/y), \quad x = u\cos v, \quad y = u\sin v;$
$(u, v) = (1.3, \pi/6)$

In Exercises 9 and 10, **(a)** express $\partial w/\partial u$ and $\partial w/\partial v$ as functions of u and v both by using the Chain Rule and by expressing w directly in terms of u and v before differentiating. Then **(b)** evaluate $\partial w/\partial u$ and $\partial w/\partial v$ at the given point (u, v).

9. $w = xy + yz + xz, \quad x = u + v, \quad y = u - v, \quad z = uv;$
$(u, v) = (1/2, 1)$

10. $w = \ln(x^2 + y^2 + z^2), \quad x = ue^v \sin u, \quad y = ue^v \cos u,$
$z = ue^v; \quad (u, v) = (-2, 0)$

In Exercises 11 and 12, **(a)** express $\partial u/\partial x, \partial u/\partial y$, and $\partial u/\partial z$ as functions of x, y, and z both by using the Chain Rule and by expressing u directly in terms of x, y, and z before differentiating. Then **(b)** evaluate $\partial u/\partial x, \partial u/\partial y$, and $\partial u/\partial z$ at the given point (x, y, z).

11. $u = \dfrac{p - q}{q - r}, \quad p = x + y + z, \quad q = x - y + z,$
$r = x + y - z; \quad (x, y, z) = (\sqrt{3}, 2, 1)$

12. $u = e^{qr} \sin^{-1} p, \quad p = \sin x, \quad q = z^2 \ln y, \quad r = 1/z;$
$(x, y, z) = (\pi/4, 1/2, -1/2)$

Using a Dependency Diagram

In Exercises 13–24, draw a dependency diagram and write a Chain Rule formula for each derivative.

13. $\dfrac{dz}{dt}$ for $z = f(x, y), \quad x = g(t), \quad y = h(t)$

14. $\dfrac{dz}{dt}$ for $z = f(u, v, w), \quad u = g(t), \quad v = h(t), \quad w = k(t)$

15. $\dfrac{\partial w}{\partial u}$ and $\dfrac{\partial w}{\partial v}$ for $w = h(x, y, z), \quad x = f(u, v), \quad y = g(u, v),$
$z = k(u, v)$

16. $\dfrac{\partial w}{\partial x}$ and $\dfrac{\partial w}{\partial y}$ for $w = f(r, s, t), \quad r = g(x, y), \quad s = h(x, y),$
$t = k(x, y)$

17. $\dfrac{\partial w}{\partial u}$ and $\dfrac{\partial w}{\partial v}$ for $w = g(x, y), \quad x = h(u, v), \quad y = k(u, v)$

18. $\dfrac{\partial w}{\partial x}$ and $\dfrac{\partial w}{\partial y}$ for $w = g(u, v), \quad u = h(x, y), \quad v = k(x, y)$

19. $\dfrac{\partial z}{\partial t}$ and $\dfrac{\partial z}{\partial s}$ for $z = f(x, y), \quad x = g(t, s), \quad y = h(t, s)$

20. $\dfrac{\partial y}{\partial r}$ for $y = f(u), \quad u = g(r, s)$

21. $\dfrac{\partial w}{\partial s}$ and $\dfrac{\partial w}{\partial t}$ for $w = g(u), \quad u = h(s, t)$

22. $\dfrac{\partial w}{\partial p}$ for $w = f(x, y, z, v), \quad x = g(p, q), \quad y = h(p, q),$
$z = j(p, q), \quad v = k(p, q)$

23. $\dfrac{\partial w}{\partial r}$ and $\dfrac{\partial w}{\partial s}$ for $w = f(x, y), \quad x = g(r), \quad y = h(s)$

24. $\dfrac{\partial w}{\partial s}$ for $w = g(x, y), \quad x = h(r, s, t), \quad y = k(r, s, t)$

Implicit Differentiation

Assuming that the equations in Exercises 25–30 define y as a differentiable function of x, use Theorem 8 to find the value of dy/dx at the given point.

25. $x^3 - 2y^2 + xy = 0, \quad (1, 1)$

26. $xy + y^2 - 3x - 3 = 0, \quad (-1, 1)$

27. $x^2 + xy + y^2 - 7 = 0, \quad (1, 2)$

28. $xe^y + \sin xy + y - \ln 2 = 0, \quad (0, \ln 2)$

29. $(x^3 - y^4)^6 + \ln(x^2 + y) = 1, \quad (-1, 0)$

30. $xe^{x^2y} - ye^x = x + y - 2, \quad (1, 1)$

Find the values of $\partial z/\partial x$ and $\partial z/\partial y$ at the points in Exercises 31–34.

31. $z^3 - xy + yz + y^3 - 2 = 0, \quad (1, 1, 1)$

32. $\dfrac{1}{x} + \dfrac{1}{y} + \dfrac{1}{z} - 1 = 0, \quad (2, 3, 6)$

33. $\sin(x + y) + \sin(y + z) + \sin(x + z) = 0, \quad (\pi, \pi, \pi)$

34. $xe^y + ye^z + 2\ln x - 2 - 3\ln 2 = 0, \quad (1, \ln 2, \ln 3)$

Finding Partial Derivatives at Specified Points

35. Find $\partial w/\partial r$ when $r = 1, s = -1$ if $w = (x + y + z)^2$, $x = r - s, y = \cos(r + s), z = \sin(r + s)$.

36. Find $\partial w/\partial v$ when $u = -1, v = 2$ if $w = xy + \ln z$, $x = v^2/u, y = u + v, z = \cos u$.

37. Find $\partial w/\partial v$ when $u = 0, v = 0$ if $w = x^2 + (y/x)$, $x = u - 2v + 1, y = 2u + v - 2$.

38. Find $\partial z/\partial u$ when $u = 0, v = 1$ if $z = \sin xy + x\sin y$, $x = u^2 + v^2, y = uv$.

39. Find $\partial z/\partial u$ and $\partial z/\partial v$ when $u = \ln 2, v = 1$ if $z = 5\tan^{-1} x$ and $x = e^u + \ln v$.

40. Find $\partial z/\partial u$ and $\partial z/\partial v$ when $u = 1, v = -2$ if $z = \ln q$ and $q = \sqrt{v + 3}\tan^{-1} u$.

Theory and Examples

41. Assume that $w = f(s^3 + t^2)$ and $f'(x) = e^x$. Find $\dfrac{\partial w}{\partial t}$ and $\dfrac{\partial w}{\partial s}$.

42. Assume that $w = f\left(ts^2, \dfrac{s}{t}\right)$, $\dfrac{\partial f}{\partial x}(x, y) = xy$, and $\dfrac{\partial f}{\partial y}(x, y) = \dfrac{x^2}{2}$. Find $\dfrac{\partial w}{\partial t}$ and $\dfrac{\partial w}{\partial s}$.

43. Assume that $z = f(x, y), x = g(t), y = h(t), f_x(2, -1) = 3$, and $f_y(2, -1) = -2$. If $g(0) = 2, h(0) = -1, g'(0) = 5$, and $h'(0) = -4$, find $\left.\dfrac{dz}{dt}\right|_{t=0}$.

44. Assume that $z = f(x, y)^2, x = g(t), y = h(t), f_x(1, 0) = -1$, $f_y(1, 0) = 1$, and $f(1, 0) = 2$. If $g(3) = 1, h(3) = 0$, $g'(3) = -3$, and $h'(3) = 4$, find $\left.\dfrac{dz}{dt}\right|_{t=3}$.

45. Assume that $z = f(w), w = g(x, y), x = 2r^3 - s^2$, and $y = re^s$. If $g_x(2, 1) = -3, g_y(2, 1) = 2, f'(7) = -1$, and $g(2, 1) = 7$, find $\left.\dfrac{\partial z}{\partial r}\right|_{r=1, s=0}$ and $\left.\dfrac{\partial z}{\partial s}\right|_{r=1, s=0}$.

46. Assume that $z = \ln(f(w)), w = g(x, y), x = \sqrt{r - s}$, and $y = r^2 s$. If $g_x(2, -9) = -1, g_y(2, -9) = 3, f'(-2) = 2$, $f(-2) = 5$, and $g(2, -9) = -2$, find $\left.\dfrac{\partial z}{\partial r}\right|_{r=3, s=-1}$ and $\left.\dfrac{\partial z}{\partial s}\right|_{r=3, s=-1}$.

47. **Changing voltage in a circuit** The voltage V in a circuit that satisfies the law $V = IR$ is slowly dropping as the battery wears out. At the same time, the resistance R is increasing as the resistor heats up. Use the equation

$$\frac{dV}{dt} = \frac{\partial V}{\partial I}\frac{dI}{dt} + \frac{\partial V}{\partial R}\frac{dR}{dt}$$

to find how the current is changing at the instant when $R = 600$ ohms, $I = 0.04$ amp, $dR/dt = 0.5$ ohm/sec, and $dV/dt = -0.01$ volt/sec.

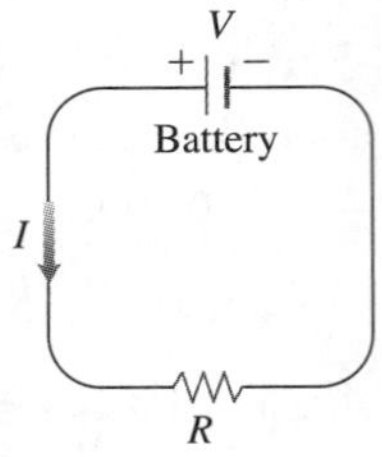

48. Changing dimensions in a box The lengths a, b, and c of the edges of a rectangular box are changing with time. At the instant in question, $a = 1$ m, $b = 2$ m, $c = 3$ m, $da/dt = db/dt = 1$ m/sec, and $dc/dt = -3$ m/sec. At what rates are the box's volume V and surface area S changing at that instant? Are the box's interior diagonals increasing in length or decreasing?

49. If $f(u, v, w)$ is differentiable and $u = x - y$, $v = y - z$, and $w = z - x$, show that

$$\frac{\partial f}{\partial x} + \frac{\partial f}{\partial y} + \frac{\partial f}{\partial z} = 0.$$

50. Polar coordinates Suppose that we substitute polar coordinates $x = r\cos\theta$ and $y = r\sin\theta$ in a differentiable function $w = f(x, y)$.

a. Show that

$$\frac{\partial w}{\partial r} = f_x \cos\theta + f_y \sin\theta$$

and

$$\frac{1}{r}\frac{\partial w}{\partial \theta} = -f_x \sin\theta + f_y \cos\theta.$$

b. Solve the equations in part (a) to express f_x and f_y in terms of $\partial w/\partial r$ and $\partial w/\partial \theta$.

c. Show that

$$(f_x)^2 + (f_y)^2 = \left(\frac{\partial w}{\partial r}\right)^2 + \frac{1}{r^2}\left(\frac{\partial w}{\partial \theta}\right)^2.$$

51. Laplace equations Show that if $w = f(u, v)$ satisfies the Laplace equation $f_{uu} + f_{vv} = 0$ and if $u = (x^2 - y^2)/2$ and $v = xy$, then w satisfies the Laplace equation $w_{xx} + w_{yy} = 0$.

52. Laplace equations Let $w = f(u) + g(v)$, where $u = x + iy$, $v = x - iy$, and $i = \sqrt{-1}$. Show that w satisfies the Laplace equation $w_{xx} + w_{yy} = 0$ if all the necessary functions are differentiable.

53. Extreme values on a helix Suppose that the partial derivatives of a function $f(x, y, z)$ at points on the helix $x = \cos t$, $y = \sin t$, $z = t$ are

$$f_x = \cos t, \quad f_y = \sin t, \quad f_z = t^2 + t - 2.$$

At what points on the curve, if any, can f take on extreme values?

54. A space curve Let $w = x^2 e^{2y} \cos 3z$. Find the value of dw/dt at the point $(1, \ln 2, 0)$ on the curve $x = \cos t$, $y = \ln(t + 2)$, $z = t$.

55. Temperature on a circle Let $T = f(x, y)$ be the temperature at the point (x, y) on the circle $x = \cos t$, $y = \sin t$, $0 \le t \le 2\pi$, and suppose that

$$\frac{\partial T}{\partial x} = 8x - 4y, \qquad \frac{\partial T}{\partial y} = 8y - 4x.$$

a. Find where the maximum and minimum temperatures on the circle occur by examining the derivatives dT/dt and d^2T/dt^2.

b. Suppose that $T = 4x^2 - 4xy + 4y^2$. Find the maximum and minimum values of T on the circle.

56. Temperature on an ellipse Let $T = g(x, y)$ be the temperature at the point (x, y) on the ellipse

$$x = 2\sqrt{2}\cos t, \qquad y = \sqrt{2}\sin t, \qquad 0 \le t \le 2\pi,$$

and suppose that

$$\frac{\partial T}{\partial x} = y, \qquad \frac{\partial T}{\partial y} = x.$$

a. Locate the maximum and minimum temperatures on the ellipse by examining dT/dt and d^2T/dt^2.

b. Suppose that $T = xy - 2$. Find the maximum and minimum values of T on the ellipse.

57. The temperature $T = T(x, y)$ in °C at point (x, y) satisfies $T_x(1, 2) = 3$ and $T_y(1, 2) = -1$. If $x = e^{2t-2}$ cm and $y = 2 + \ln t$ cm, find the rate at which the temperature T changes when $t = 1$ sec.

58. A bug crawls on the surface $z = x^2 - y^2$ directly above a path in the xy-plane given by $x = f(t)$ and $y = g(t)$. If $f(2) = 4$, $f'(2) = -1$, $g(2) = -2$, and $g'(2) = -3$, then at what rate is the bug's elevation z changing when $t = 2$?

Differentiating Integrals Under mild continuity restrictions, it is true that if

$$F(x) = \int_a^b g(t, x)\, dt,$$

then $F'(x) = \int_a^b g_x(t, x)\, dt$. Using this fact and the Chain Rule, we can find the derivative of

$$F(x) = \int_a^{f(x)} g(t, x)\, dt$$

by letting

$$G(u, x) = \int_a^u g(t, x)\, dt,$$

where $u = f(x)$. Find the derivatives of the functions in Exercises 59 and 60.

59. $F(x) = \displaystyle\int_0^{x^2} \sqrt{t^4 + x^3}\, dt$

60. $F(x) = \displaystyle\int_{x^2}^{1} \sqrt{t^3 + x^2}\, dt$

61. Water is flowing into a tank in the form of a right-circular cylinder at the rate of $(4/5)\pi$ ft³/min. The tank is stretching in such a way that even though it remains cylindrical, its radius is increasing at the rate of 0.002 ft/min. How fast is the surface of the water rising when the radius is 2 ft and the volume of water in the tank is 20π ft³?

62. Suppose f is a differentiable function of x, y, and z and $u = f(x, y, z)$. Then if $x = r\sin\phi\cos\theta$, $y = r\sin\phi\sin\theta$, and $z = r\cos\phi$, express $\partial u/\partial r$, $\partial u/\partial \phi$, and $\partial u/\partial \theta$ in terms of $\partial u/\partial x$, $\partial u/\partial y$, and $\partial u/\partial z$.

63. At a given instant, the length of one leg of a right triangle is 10 ft, and it is increasing at the rate of 1 ft/min, and the length of the other leg of the right triangle is 12 ft, and it is decreasing at the rate of 2 ft/min. Find the rate of change of the measure of the acute angle opposite the leg of length 12 ft at the given instant.

14.5 Directional Derivatives and Gradient Vectors

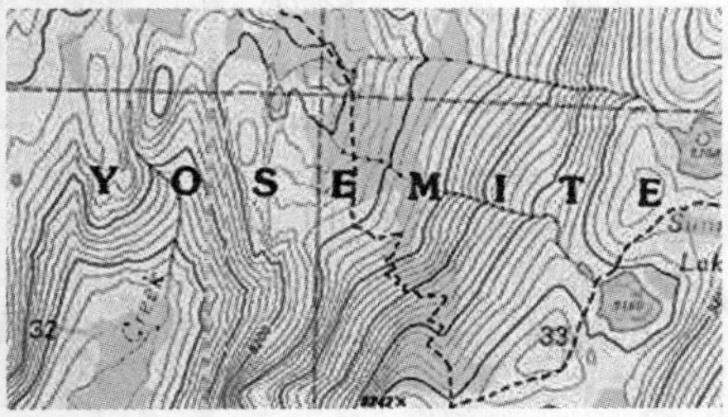

FIGURE 14.26 Contours within Yosemite National Park in California show streams, which follow paths of steepest descent, running perpendicular to the contours. (*Source:* Yosemite National Park Map from U.S. Geological Survey, http://www.usgs.gov)

If you look at the map (Figure 14.26) showing contours within Yosemite National Park in California, you will notice that the streams flow perpendicular to the contours. The streams are following paths of steepest descent so the waters reach lower elevations as quickly as possible. Therefore, the fastest instantaneous rate of change in a stream's elevation above sea level has a particular direction. In this section, you will see why this direction, called the "downhill" direction, is perpendicular to the contours.

Directional Derivatives in the Plane

We know from Section 14.4 that if $f(x, y)$ is differentiable, then the rate at which f changes with respect to t along a differentiable curve $x = g(t)$, $y = h(t)$ is

$$\frac{df}{dt} = \frac{\partial f}{\partial x}\frac{dx}{dt} + \frac{\partial f}{\partial y}\frac{dy}{dt}.$$

At any point $P_0(x_0, y_0) = P_0(g(t_0), h(t_0))$, this equation gives the rate of change of f with respect to increasing t and therefore depends, among other things, on the direction of motion along the curve. If the curve is a straight line and t is the arc length parameter along the line measured from P_0 in the direction of a given unit vector $\mathbf{u}$, then df/dt is the rate of change of f with respect to distance in its domain in the direction of $\mathbf{u}$. By varying $\mathbf{u}$, we find the rates at which f changes with respect to distance as we move through P_0 in different directions. We now define this idea more precisely.

Suppose that the function $f(x, y)$ is defined throughout a region R in the xy-plane, that $P_0(x_0, y_0)$ is a point in R, and that $\mathbf{u} = u_1\mathbf{i} + u_2\mathbf{j}$ is a unit vector. Then the equations

$$x = x_0 + su_1, \quad y = y_0 + su_2$$

parametrize the line through P_0 parallel to $\mathbf{u}$. If the parameter s measures arc length from P_0 in the direction of $\mathbf{u}$, we find the rate of change of f at P_0 in the direction of $\mathbf{u}$ by calculating df/ds at P_0 (Figure 14.27).

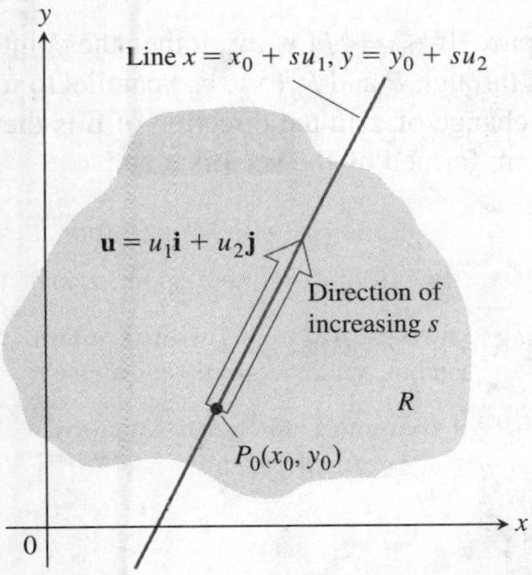

FIGURE 14.27 The rate of change of f in the direction of $\mathbf{u}$ at a point P_0 is the rate at which f changes along this line at P_0.

DEFINITION The **derivative of f at $P_0(x_0, y_0)$ in the direction of the unit vector $\mathbf{u} = u_1\mathbf{i} + u_2\mathbf{j}$** is the number

$$\left(\frac{df}{ds}\right)_{\mathbf{u}, P_0} = \lim_{s \to 0} \frac{f(x_0 + su_1, y_0 + su_2) - f(x_0, y_0)}{s}, \qquad (1)$$

provided the limit exists.

The **directional derivative** defined by Equation (1) is also denoted by

$$D_{\mathbf{u}}f(P_0) \qquad \text{or} \qquad D_{\mathbf{u}}f|_{P_0}.$$

"The derivative of f in the direction of $\mathbf{u}$, evaluated at P_0"

The partial derivatives $f_x(x_0, y_0)$ and $f_y(x_0, y_0)$ are the directional derivatives of f at P_0 in the $\mathbf{i}$ and $\mathbf{j}$ directions. This observation can be seen by comparing Equation (1) to the definitions of the two partial derivatives given in Section 14.3.

EXAMPLE 1 Using the definition, find the derivative of

$$f(x, y) = x^2 + xy$$

at $P_0(1, 2)$ in the direction of the unit vector $\mathbf{u} = (1/\sqrt{2})\mathbf{i} + (1/\sqrt{2})\mathbf{j}$.

Solution Applying the definition in Equation (1), we obtain

$$\left(\frac{df}{ds}\right)_{\mathbf{u},P_0} = \lim_{s\to 0}\frac{f(x_0+su_1, y_0+su_2)-f(x_0,y_0)}{s} \qquad \text{Eq. (1)}$$

$$= \lim_{s\to 0}\frac{f\left(1+s\cdot\frac{1}{\sqrt{2}}, 2+s\cdot\frac{1}{\sqrt{2}}\right)-f(1,2)}{s} \qquad \text{Substitute.}$$

$$= \lim_{s\to 0}\frac{\left(1+\frac{s}{\sqrt{2}}\right)^2+\left(1+\frac{s}{\sqrt{2}}\right)\left(2+\frac{s}{\sqrt{2}}\right)-(1^2+1\cdot 2)}{s}$$

$$= \lim_{s\to 0}\frac{\left(1+\frac{2s}{\sqrt{2}}+\frac{s^2}{2}\right)+\left(2+\frac{3s}{\sqrt{2}}+\frac{s^2}{2}\right)-3}{s}$$

$$= \lim_{s\to 0}\frac{\frac{5s}{\sqrt{2}}+s^2}{s} = \lim_{s\to 0}\left(\frac{5}{\sqrt{2}}+s\right) = \frac{5}{\sqrt{2}}.$$

The rate of change of $f(x, y) = x^2 + xy$ at $P_0(1, 2)$ in the direction $\mathbf{u}$ is $5/\sqrt{2}$. ■

Interpretation of the Directional Derivative

The equation $z = f(x, y)$ represents a surface S in space. If $z_0 = f(x_0, y_0)$, then the point $P(x_0, y_0, z_0)$ lies on S. The vertical plane that passes through P and $P_0(x_0, y_0)$ parallel to $\mathbf{u}$ intersects S in a curve C (Figure 14.28). The rate of change of f in the direction of $\mathbf{u}$ is the slope of the tangent to C at P in the right-handed system formed by the vectors $\mathbf{u}$ and $\mathbf{k}$.

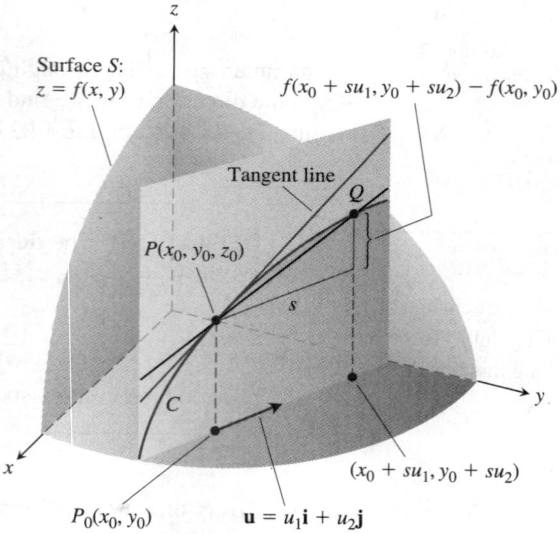

FIGURE 14.28 The slope of the trace curve C at P_0 is $\lim_{Q\to P}$ slope (PQ); this is the directional derivative

$$\left(\frac{df}{ds}\right)_{\mathbf{u},P_0} = D_{\mathbf{u}}f\Big|_{P_0}.$$

When $\mathbf{u} = \mathbf{i}$, the directional derivative at P_0 is $\partial f/\partial x$ evaluated at (x_0, y_0). When $\mathbf{u} = \mathbf{j}$, the directional derivative at P_0 is $\partial f/\partial y$ evaluated at (x_0, y_0). The directional derivative generalizes the two partial derivatives. We can now ask for the rate of change of f in any direction $\mathbf{u}$, not just in the directions $\mathbf{i}$ and $\mathbf{j}$.

For a physical interpretation of the directional derivative, suppose that $T = f(x, y)$ is the temperature at each point (x, y) over a region in the plane. Then $f(x_0, y_0)$ is the temperature at the point $P_0(x_0, y_0)$, and $D_{\mathbf{u}}f|_{P_0}$ is the instantaneous rate of change of the temperature at P_0 stepping off in the direction $\mathbf{u}$.

Calculation and Gradients

We now develop an efficient formula to calculate the directional derivative for a differentiable function f. We begin with the line

$$x = x_0 + su_1, \quad y = y_0 + su_2, \tag{2}$$

through $P_0(x_0, y_0)$, parametrized with the arc length parameter s increasing in the direction of the unit vector $\mathbf{u} = u_1\mathbf{i} + u_2\mathbf{j}$. Then, by the Chain Rule we find

$$\left(\frac{df}{ds}\right)_{\mathbf{u},P_0} = \left.\frac{\partial f}{\partial x}\right|_{P_0}\frac{dx}{ds} + \left.\frac{\partial f}{\partial y}\right|_{P_0}\frac{dy}{ds} \qquad \text{Chain Rule for differentiable } f$$

$$= \left.\frac{\partial f}{\partial x}\right|_{P_0}u_1 + \left.\frac{\partial f}{\partial y}\right|_{P_0}u_2 \qquad \text{From Eqs. (2), } dx/ds = u_1 \text{ and } dy/ds = u_2$$

$$= \underbrace{\left[\left.\frac{\partial f}{\partial x}\right|_{P_0}\mathbf{i} + \left.\frac{\partial f}{\partial y}\right|_{P_0}\mathbf{j}\right]}_{\text{Gradient of } f \text{ at } P_0} \cdot \underbrace{\left[u_1\mathbf{i} + u_2\mathbf{j}\right]}_{\text{Direction } \mathbf{u}}. \tag{3}$$

Equation (3) says that the derivative of a differentiable function f in the direction of $\mathbf{u}$ at P_0 is the dot product of $\mathbf{u}$ with a special vector, which we now define.

DEFINITION The **gradient vector** (or **gradient**) of $f(x, y)$ is the vector

$$\nabla f = \frac{\partial f}{\partial x}\mathbf{i} + \frac{\partial f}{\partial y}\mathbf{j}.$$

The value of the gradient vector obtained by evaluating the partial derivatives at a point $P_0(x_0, y_0)$ is written

$$\nabla f|_{P_0} \quad \text{or} \quad \nabla f(x_0, y_0).$$

The notation ∇f is read "grad f" as well as "gradient of f" and "del f." The symbol ∇ by itself is read "del." Another notation for the gradient is grad f. Using the gradient notation, we restate Equation (3) as a theorem.

THEOREM 9—The Directional Derivative Is a Dot Product
If $f(x, y)$ is differentiable in an open region containing $P_0(x_0, y_0)$, then

$$\left(\frac{df}{ds}\right)_{\mathbf{u},P_0} = \nabla f|_{P_0} \cdot \mathbf{u}, \tag{4}$$

the dot product of the gradient ∇f at P_0 with the vector $\mathbf{u}$. In brief, $D_{\mathbf{u}}f = \nabla f \cdot \mathbf{u}$.

EXAMPLE 2 Find the derivative of $f(x, y) = xe^y + \cos(xy)$ at the point $(2, 0)$ in the direction of $\mathbf{v} = 3\mathbf{i} - 4\mathbf{j}$.

Solution Recall that the direction of a vector $\mathbf{v}$ is the unit vector obtained by dividing $\mathbf{v}$ by its length:

$$\mathbf{u} = \frac{\mathbf{v}}{|\mathbf{v}|} = \frac{\mathbf{v}}{5} = \frac{3}{5}\mathbf{i} - \frac{4}{5}\mathbf{j}.$$

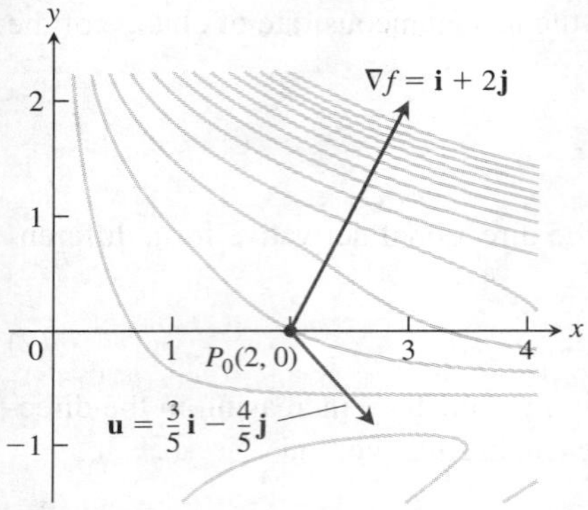

FIGURE 14.29 Picture ∇f as a vector in the domain of f. The figure shows a number of level curves of f. The rate at which f changes at $(2, 0)$ in the direction $\mathbf{u}$ is $\nabla f \cdot \mathbf{u} = -1$, which is the component of ∇f in the direction of unit vector $\mathbf{u}$ (Example 2).

The partial derivatives of f are everywhere continuous and at $(2, 0)$ are given by

$$f_x(2,0) = (e^y - y\sin(xy))\Big|_{(2,0)} = e^0 - 0 = 1$$

$$f_y(2,0) = (xe^y - x\sin(xy))\Big|_{(2,0)} = 2e^0 - 2\cdot 0 = 2.$$

The gradient of f at $(2, 0)$ is

$$\nabla f|_{(2,0)} = f_x(2,0)\mathbf{i} + f_y(2,0)\mathbf{j} = \mathbf{i} + 2\mathbf{j}$$

(Figure 14.29). The derivative of f at $(2, 0)$ in the direction of $\mathbf{v}$ is therefore

$$D_{\mathbf{u}}f|_{(2,0)} = \nabla f|_{(2,0)} \cdot \mathbf{u} \qquad \text{Eq. (4) with the } D_{\mathbf{u}}f|_{P_0} \text{ notation}$$

$$= (\mathbf{i} + 2\mathbf{j}) \cdot \left(\frac{3}{5}\mathbf{i} - \frac{4}{5}\mathbf{j}\right) = \frac{3}{5} - \frac{8}{5} = -1.$$

■

Evaluating the dot product in the brief version of Equation (4) gives

$$D_{\mathbf{u}}f = \nabla f \cdot \mathbf{u} = |\nabla f||\mathbf{u}|\cos\theta = |\nabla f|\cos\theta,$$

where θ is the angle between the vectors $\mathbf{u}$ and ∇f, and reveals the following properties.

Properties of the Directional Derivative $D_{\mathbf{u}}f = \nabla f \cdot \mathbf{u} = |\nabla f|\cos\theta$

1. The function f increases most rapidly when $\cos\theta = 1$, which means that $\theta = 0$ and $\mathbf{u}$ is the direction of ∇f. That is, at each point P in its domain, f increases most rapidly in the direction of the gradient vector ∇f at P. The derivative in this direction is
$$D_{\mathbf{u}}f = |\nabla f|\cos(0) = |\nabla f|.$$
2. Similarly, f decreases most rapidly in the direction of $-\nabla f$. The derivative in this direction is $D_{\mathbf{u}}f = |\nabla f|\cos(\pi) = -|\nabla f|$.
3. Any direction $\mathbf{u}$ orthogonal to a gradient $\nabla f \neq 0$ is a direction of zero change in f because θ then equals $\pi/2$ and
$$D_{\mathbf{u}}f = |\nabla f|\cos(\pi/2) = |\nabla f| \cdot 0 = 0.$$

As we discuss later, these properties hold in three dimensions as well as two.

EXAMPLE 3 Find the directions in which $f(x, y) = (x^2/2) + (y^2/2)$

(a) increases most rapidly at the point $(1, 1)$, and

(b) decreases most rapidly at $(1, 1)$.

(c) What are the directions of zero change in f at $(1, 1)$?

Solution

(a) The function increases most rapidly in the direction of ∇f at $(1, 1)$. The gradient there is

$$\nabla f|_{(1,1)} = (x\mathbf{i} + y\mathbf{j})\Big|_{(1,1)} = \mathbf{i} + \mathbf{j}.$$

Its direction is

$$\mathbf{u} = \frac{\mathbf{i} + \mathbf{j}}{|\mathbf{i} + \mathbf{j}|} = \frac{\mathbf{i} + \mathbf{j}}{\sqrt{(1)^2 + (1)^2}} = \frac{1}{\sqrt{2}}\mathbf{i} + \frac{1}{\sqrt{2}}\mathbf{j}.$$

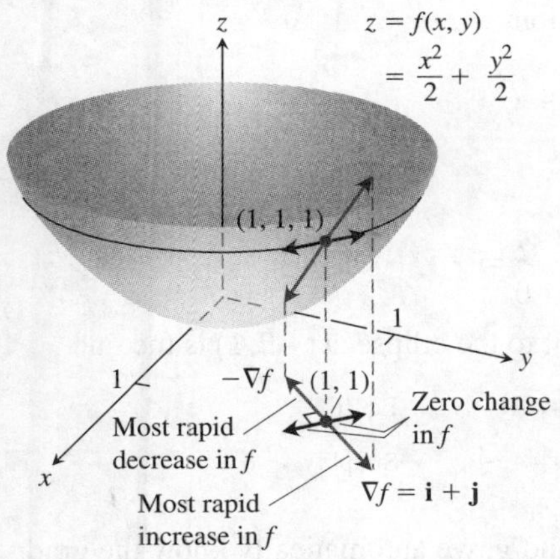

FIGURE 14.30 The direction in which $f(x, y)$ increases most rapidly at $(1, 1)$ is the direction of $\nabla f|_{(1,1)} = \mathbf{i} + \mathbf{j}$. It corresponds to the direction of steepest ascent on the surface at $(1, 1, 1)$ (Example 3).

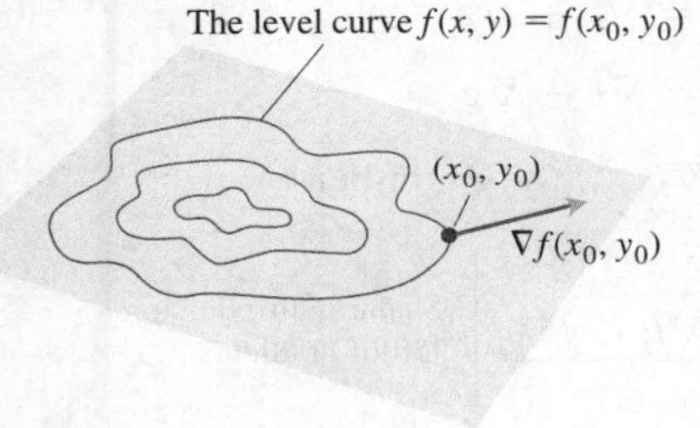

FIGURE 14.31 When it is nonzero, the gradient of a differentiable function of two variables at a point is always normal to the function's level curve through that point.

(b) The function decreases most rapidly in the direction of $-\nabla f$ at $(1, 1)$, which is

$$-\mathbf{u} = -\frac{1}{\sqrt{2}}\mathbf{i} - \frac{1}{\sqrt{2}}\mathbf{j}.$$

(c) The directions of zero change at $(1, 1)$ are the directions orthogonal to ∇f:

$$\mathbf{n} = -\frac{1}{\sqrt{2}}\mathbf{i} + \frac{1}{\sqrt{2}}\mathbf{j} \quad \text{and} \quad -\mathbf{n} = \frac{1}{\sqrt{2}}\mathbf{i} - \frac{1}{\sqrt{2}}\mathbf{j}.$$

See Figure 14.30. ■

Gradients and Tangents to Level Curves

If a differentiable function $f(x, y)$ has a constant value c along a smooth curve $\mathbf{r} = g(t)\mathbf{i} + h(t)\mathbf{j}$ (making the curve part of a level curve of f), then $f(g(t), h(t)) = c$. Differentiating both sides of this equation with respect to t leads to the equations

$$\frac{d}{dt}f(g(t), h(t)) = \frac{d}{dt}(c)$$

$$\frac{\partial f}{\partial x}\frac{dg}{dt} + \frac{\partial f}{\partial y}\frac{dh}{dt} = 0 \qquad \text{Chain Rule} \qquad (5)$$

$$\underbrace{\left(\frac{\partial f}{\partial x}\mathbf{i} + \frac{\partial f}{\partial y}\mathbf{j}\right)}_{\nabla f} \cdot \underbrace{\left(\frac{dg}{dt}\mathbf{i} + \frac{dh}{dt}\mathbf{j}\right)}_{\frac{d\mathbf{r}}{dt}} = 0.$$

Assuming the gradient of f is a nonzero vector, Equation (5) says that ∇f is normal to the tangent vector $d\mathbf{r}/dt$, so it is normal to the curve. This is seen in Figure 14.31.

> At every point (x_0, y_0) in the domain of a differentiable function $f(x, y)$ where the gradient of f is a nonzero vector, this vector is normal to the level curve through (x_0, y_0) (Figure 14.31).

Equation (5) validates our observation that streams flow perpendicular to the contours in topographical maps (see Figure 14.26). Since the downflowing stream will reach its destination in the fastest way, it must flow in the direction of the negative gradient vectors from Property 2 for the directional derivative. Equation (5) tells us these directions are perpendicular to the level curves.

This observation also enables us to find equations for tangent lines to level curves. They are the lines normal to the gradients. The line through a point $P_0(x_0, y_0)$ normal to a nonzero vector $\mathbf{N} = A\mathbf{i} + B\mathbf{j}$ has the equation

$$A(x - x_0) + B(y - y_0) = 0$$

(Exercise 39). If $\mathbf{N}$ is the gradient $\nabla f|_{(x_0, y_0)} = f_x(x_0, y_0)\mathbf{i} + f_y(x_0, y_0)\mathbf{j}$, and this gradient is not the zero vector, then this equation gives the following formula.

> **Equation for the Tangent Line to a Level Curve**
>
> $$f_x(x_0, y_0)(x - x_0) + f_y(x_0, y_0)(y - y_0) = 0 \qquad (6)$$

EXAMPLE 4 Find an equation for the tangent to the ellipse

$$\frac{x^2}{4} + y^2 = 2$$

(Figure 14.32) at the point $(-2, 1)$.

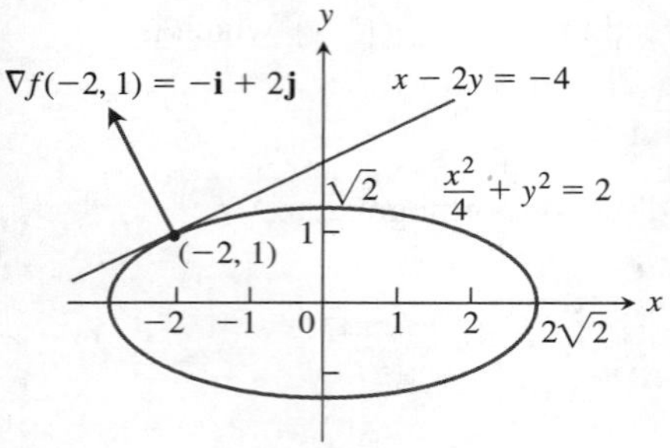

FIGURE 14.32 We can find the tangent to the ellipse $(x^2/4) + y^2 = 2$ by treating the ellipse as a level curve of the function $f(x, y) = (x^2/4) + y^2$ (Example 4).

Solution The ellipse is a level curve of the function

$$f(x, y) = \frac{x^2}{4} + y^2.$$

The gradient of f at $(-2, 1)$ is

$$\nabla f|_{(-2,1)} = \left(\frac{x}{2}\mathbf{i} + 2y\mathbf{j}\right)\bigg|_{(-2,1)} = -\mathbf{i} + 2\mathbf{j}.$$

Because this gradient vector is nonzero, the tangent to the ellipse at $(-2, 1)$ is the line

$$(-1)(x + 2) + (2)(y - 1) = 0 \qquad \text{Eq. (6)}$$
$$x - 2y = -4. \qquad \text{Simplify.}$$

■

If we know the gradients of two functions f and g, we automatically know the gradients of their sum, difference, constant multiples, product, and quotient. You are asked to establish the following rules in Exercise 40. Notice that these rules have the same form as the corresponding rules for derivatives of single-variable functions.

Algebra Rules for Gradients

1. *Sum Rule:* $\nabla(f + g) = \nabla f + \nabla g$
2. *Difference Rule:* $\nabla(f - g) = \nabla f - \nabla g$
3. *Constant Multiple Rule:* $\nabla(kf) = k\nabla f$ (any number k)
4. *Product Rule:* $\nabla(fg) = f\nabla g + g\nabla f$
5. *Quotient Rule:* $\nabla\left(\frac{f}{g}\right) = \frac{g\nabla f - f\nabla g}{g^2}$

(Rules 4 and 5: Scalar multipliers on left of gradients)

EXAMPLE 5 We illustrate two of the rules with

$$f(x, y) = x - y \qquad g(x, y) = 3y$$
$$\nabla f = \mathbf{i} - \mathbf{j} \qquad \nabla g = 3\mathbf{j}.$$

We have

1. $\nabla(f - g) = \nabla(x - 4y) = \mathbf{i} - 4\mathbf{j} = \nabla f - \nabla g$ Rule 2
2. $\nabla(fg) = \nabla(3xy - 3y^2) = 3y\mathbf{i} + (3x - 6y)\mathbf{j}$

and

$$f\nabla g + g\nabla f = (x - y)3\mathbf{j} + 3y(\mathbf{i} - \mathbf{j}) \qquad \text{Substitute.}$$
$$= 3y\mathbf{i} + (3x - 6y)\mathbf{j}. \qquad \text{Simplify.}$$

We have therefore verified that for this example, $\nabla(fg) = f\nabla g + g\nabla f$. ■

Functions of Three Variables

For a differentiable function $f(x, y, z)$ and a unit vector $\mathbf{u} = u_1\mathbf{i} + u_2\mathbf{j} + u_3\mathbf{k}$ in space, we have

$$\nabla f = \frac{\partial f}{\partial x}\mathbf{i} + \frac{\partial f}{\partial y}\mathbf{j} + \frac{\partial f}{\partial z}\mathbf{k}$$

and

$$D_{\mathbf{u}}f = \nabla f \cdot \mathbf{u} = \frac{\partial f}{\partial x}u_1 + \frac{\partial f}{\partial y}u_2 + \frac{\partial f}{\partial z}u_3.$$

The directional derivative can once again be written in the form

$$D_{\mathbf{u}}f = \nabla f \cdot \mathbf{u} = |\nabla f||\mathbf{u}| \cos\theta = |\nabla f| \cos\theta,$$

so the properties listed earlier for functions of two variables extend to three variables. At any given point, f increases most rapidly in the direction of ∇f and decreases most rapidly in the direction of $-\nabla f$. In any direction orthogonal to ∇f, the derivative is zero.

EXAMPLE 6

(a) Find the derivative of $f(x, y, z) = x^3 - xy^2 - z$ at $P_0(1, 1, 0)$ in the direction of $\mathbf{v} = 2\mathbf{i} - 3\mathbf{j} + 6\mathbf{k}$.

(b) In what directions does f change most rapidly at P_0, and what are the rates of change in these directions?

Solution

(a) The direction of $\mathbf{v}$ is obtained by dividing $\mathbf{v}$ by its length:

$$|\mathbf{v}| = \sqrt{(2)^2 + (-3)^2 + (6)^2} = \sqrt{49} = 7$$

$$\mathbf{u} = \frac{\mathbf{v}}{|\mathbf{v}|} = \frac{2}{7}\mathbf{i} - \frac{3}{7}\mathbf{j} + \frac{6}{7}\mathbf{k}.$$

The partial derivatives of f at P_0 are

$$f_x = (3x^2 - y^2)\Big|_{(1,1,0)} = 2, \qquad f_y = -2xy\Big|_{(1,1,0)} = -2, \qquad f_z = -1\Big|_{(1,1,0)} = -1.$$

The gradient of f at P_0 is

$$\nabla f|_{(1,1,0)} = 2\mathbf{i} - 2\mathbf{j} - \mathbf{k}.$$

The derivative of f at P_0 in the direction of $\mathbf{v}$ is therefore

$$D_{\mathbf{u}}f|_{(1,1,0)} = \nabla f|_{(1,1,0)} \cdot \mathbf{u} = (2\mathbf{i} - 2\mathbf{j} - \mathbf{k}) \cdot \left(\frac{2}{7}\mathbf{i} - \frac{3}{7}\mathbf{j} + \frac{6}{7}\mathbf{k}\right)$$

$$= \frac{4}{7} + \frac{6}{7} - \frac{6}{7} = \frac{4}{7}.$$

(b) The function increases most rapidly in the direction of $\nabla f = 2\mathbf{i} - 2\mathbf{j} - \mathbf{k}$ and decreases most rapidly in the direction of $-\nabla f$. The rates of change in the directions are, respectively,

$$|\nabla f| = \sqrt{(2)^2 + (-2)^2 + (-1)^2} = \sqrt{9} = 3 \quad \text{and} \quad -|\nabla f| = -3. \qquad \blacksquare$$

Functions of More Than Three Variables

The gradient of a differentiable function of n variables $f(x_1, x_2, \ldots, x_n)$ is

$$\nabla f = \left\langle \frac{\partial f}{\partial x_1}, \frac{\partial f}{\partial x_2}, \ldots, \frac{\partial f}{\partial x_n} \right\rangle.$$

If $\mathbf{u} = \langle u_1, u_2, \ldots, u_n \rangle$ is an n-dimensional vector such that $u_1^2 + u_2^2 + \cdots + u_n^2 = 1$ (so $\mathbf{u}$ is a unit vector since $|\mathbf{u}| = \sqrt{u_1^2 + u_2^2 + \cdots + u_n^2} = 1$), then the directional derivative of f in the direction of $\mathbf{u}$ is

$$D_u f = \nabla f \cdot \mathbf{u} = \frac{\partial f}{\partial x_1}u_1 + \frac{\partial f}{\partial x_2}u_2 + \cdots + \frac{\partial f}{\partial x_n}u_n.$$

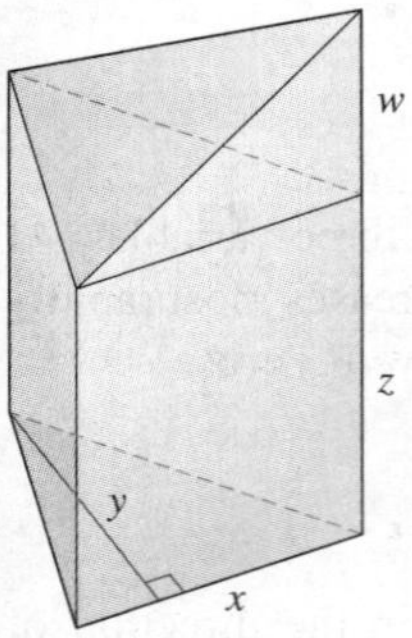

FIGURE 14.33 A tetrahedron on top of a triangular prism (Example 7).

EXAMPLE 7 The volume of the solid shown in Figure 14.33 consisting of a tetrahedron on top of a triangular prism is given by $f(x, y, z, w) = \frac{xy}{2}\left(z + \frac{w}{3}\right)$.

(a) Calculate the derivative of $f(x, y, z, w)$ at the point $P_0(6, 5, 8, 4)$ in the direction of $\mathbf{v} = \langle 1, -1, -1, 1 \rangle$.

(b) What is the geometric significance of the value obtained in part (a)?

Solution

(a) The direction of $\mathbf{v}$ is the unit vector

$$\begin{aligned}\mathbf{u} &= \frac{1}{|\mathbf{v}|}\mathbf{v}\\ &= \frac{1}{\sqrt{1^2 + (-1)^2 + (-1)^2 + 1^2}}\langle 1, -1, -1, 1 \rangle\\ &= \frac{1}{2}\langle 1, -1, -1, 1 \rangle\\ &= \left\langle \frac{1}{2}, -\frac{1}{2}, -\frac{1}{2}, \frac{1}{2} \right\rangle.\end{aligned}$$

The four partial derivatives of f at the point P_0 are

$$\begin{aligned}f_x &= \frac{y}{2}\left(z + \frac{w}{3}\right)\bigg|_{(6,5,8,4)} = \frac{70}{3},\\ f_y &= \frac{x}{2}\left(z + \frac{w}{3}\right)\bigg|_{(6,5,8,4)} = 28,\\ f_z &= \frac{xy}{2}\bigg|_{(6,5,8,4)} = 15,\\ f_w &= \frac{xy}{6}\bigg|_{(6,5,8,4)} = 5.\end{aligned}$$

The gradient of f at P_0 is

$$\nabla f|_{(6,5,8,4)} = \left\langle \frac{70}{3}, 28, 15, 5 \right\rangle.$$

The derivative of f at $(6, 5, 8, 4)$ in the direction of $\mathbf{v}$ is

$$\begin{aligned}D_{\mathbf{u}}f|_{P_0} &= \nabla f|_{P_0} \cdot \mathbf{u}\\ &= \left\langle \frac{70}{3}, 28, 15, 5 \right\rangle \cdot \left\langle \frac{1}{2}, -\frac{1}{2}, -\frac{1}{2}, \frac{1}{2} \right\rangle\\ &= \left(\frac{70}{3}\right)\left(\frac{1}{2}\right) + (28)\left(-\frac{1}{2}\right) + (15)\left(-\frac{1}{2}\right) + (5)\left(\frac{1}{2}\right)\\ &= -\frac{22}{3}.\end{aligned}$$

(b) Geometrically, this means that if the dimensions are $x = 6, y = 5, z = 8$, and $w = 4$, and the dimensions are changed by moving at unit speed so that x and w increase at the same rate while both y and z decrease at that rate, then the volume of the solid decreases at the rate of 7 1/3. ■

The Chain Rule for Paths

If $\mathbf{r}(t) = x(t)\mathbf{i} + y(t)\mathbf{j} + z(t)\mathbf{k}$ is a smooth path C, and $w = f(\mathbf{r}(t))$ is a scalar function evaluated along C, then according to the Chain Rule, Theorem 6 in Section 14.4,

$$\frac{dw}{dt} = \frac{\partial w}{\partial x}\frac{dx}{dt} + \frac{\partial w}{\partial y}\frac{dy}{dt} + \frac{\partial w}{\partial z}\frac{dz}{dt}.$$

The partial derivatives on the right-hand side of the above equation are evaluated along the curve $\mathbf{r}(t)$, and the derivatives of the intermediate variables are evaluated at t. If we express this equation using vector notation, we have

The Derivative Along a Path

$$\frac{d}{dt}f(\mathbf{r}(t)) = \nabla f(\mathbf{r}(t)) \cdot \mathbf{r}'(t). \tag{7}$$

What Equation (7) says is that the derivative of the composite function $f(\mathbf{r}(t))$ is the "derivative" (gradient) of the outside function f, evaluated at $\mathbf{r}(t)$, "times" (dot product) the derivative of the inside function $\mathbf{r}$. This is analogous to the "Outside-Inside" Rule for derivatives of composite functions studied in Section 3.6. That is, the multivariable Chain Rule for paths has exactly *the same form* as the rule for single-variable differential calculus when appropriate interpretations are given to the meanings of the terms and operations involved.

EXERCISES 14.5

Calculating Gradients

In Exercises 1–6, find the gradient of the function at the given point. Then sketch the gradient, together with the level curve that passes through the point.

1. $f(x, y) = y - x, \quad (2, 1)$

2. $f(x, y) = \ln(x^2 + y^2), \quad (1, 1)$

3. $g(x, y) = xy^2, \quad (2, -1)$

4. $g(x, y) = \frac{x^2}{2} - \frac{y^2}{2}, \quad (\sqrt{2}, 1)$

5. $f(x, y) = \sqrt{2x + 3y}, \quad (-1, 2)$

6. $f(x, y) = \tan^{-1}\frac{\sqrt{x}}{y}, \quad (4, -2)$

In Exercises 7–10, find ∇f at the given point.

7. $f(x, y, z) = x^2 + y^2 - 2z^2 + z\ln x, \quad (1, 1, 1)$

8. $f(x, y, z) = \log_e(x^2 + y^2 + z^2), \quad (1, 2, 1)$

9. $f(x, y, z) = (x^2 + y^2 + z^2)^{-1/2} + \ln(xyz), \quad (-1, 2, -2)$

10. $f(x, y, z) = e^{x+y}\cos z + (y + 1)\arcsin x, \quad (0, 0, \pi/6)$

Finding Directional Derivatives

In Exercises 11–18, find the derivative of the function at P_0 in the direction of $\mathbf{v}$.

11. $f(x, y) = 2xy - 3y^2, \quad P_0(5, 5), \quad \mathbf{v} = 4\mathbf{i} + 3\mathbf{j}$

12. $f(x, y) = 2x^2 + y^2, \quad P_0(-1, 1), \quad \mathbf{v} = 3\mathbf{i} - 4\mathbf{j}$

13. $g(x, y) = \frac{x - y}{xy + 2}, \quad P_0(1, -1), \quad \mathbf{v} = 12\mathbf{i} + 5\mathbf{j}$

14. $h(x, y) = \arctan(y/x) + \sqrt{3}\arcsin(xy/2), \quad P_0(1, 1),$
$\mathbf{v} = 3\mathbf{i} - 2\mathbf{j}$

15. $f(x, y, z) = xy + yz + zx, \quad P_0(1, -1, 2), \quad \mathbf{v} = 3\mathbf{i} + 6\mathbf{j} - 2\mathbf{k}$

16. $f(x, y, z) = x^2yz + 4xz^2, \quad P_0(1, -2, -1), \quad \mathbf{u} = 2\mathbf{i} + \mathbf{j} - 2\mathbf{k}$

17. $g(x, y, z) = 3e^x\cos yz, \quad P_0(0, 0, 0), \quad \mathbf{v} = 2\mathbf{i} + \mathbf{j} - 2\mathbf{k}$

18. $h(x, y, z) = \cos xy + e^{yz} + \ln zx, \quad P_0(1, 0, 1/2),$
$\mathbf{v} = \mathbf{i} + 2\mathbf{j} + 2\mathbf{k}$

In Exercises 19–24, find the directions in which the functions increase most rapidly, and the directions in which they decrease most rapidly, at P_0. Then find the derivatives of the functions in these directions.

19. $f(x, y) = x^2 + xy + y^2, \quad P_0(-1, 1)$

20. $f(x, y) = x^2y + e^{xy}\sin y, \quad P_0(1, 0)$

21. $f(x, y, z) = (x/y) - yz, \quad P_0(4, 1, 1)$

22. $g(x, y, z) = x^3yz, \quad P_0(1, 4, 1)$

23. $f(x, y, z) = \ln xy + \ln yz + \ln xz, \quad P_0(1, 1, 1)$

24. $h(x, y, z) = x^2 + y^2 - z, \quad P_0(4, 4, 2)$

Tangent Lines to Level Curves

In Exercises 25–28, sketch the curve $f(x, y) = c$, together with ∇f and the tangent line at the given point. Then write an equation for the tangent line.

25. $x^2 + y^2 = 4, \quad (\sqrt{2}, \sqrt{2})$

26. $x^2 - y = 1, \quad (\sqrt{2}, 1)$

27. $xy = -4, \quad (2, -2)$

28. $x^2 - xy + y^2 = 7, \quad (-1, 2)$

Theory and Examples

29. Let $f(x, y) = x^2 - xy + y^2 - y$. Find the directions $\mathbf{u}$ and the values of $D_{\mathbf{u}}f(1, -1)$ for which

a. $D_{\mathbf{u}}f(1, -1)$ is largest

b. $D_{\mathbf{u}}f(1, -1)$ is smallest

c. $D_{\mathbf{u}}f(1, -1) = 0$

d. $D_{\mathbf{u}}f(1, -1) = 4$

e. $D_{\mathbf{u}}f(1, -1) = -3$

30. Let $f(x, y) = \frac{(x - y)}{(x + y)}$. Find the directions $\mathbf{u}$ and the values of $D_{\mathbf{u}}f\left(-\frac{1}{2}, \frac{3}{2}\right)$ for which

a. $D_{\mathbf{u}}f\left(-\frac{1}{2}, \frac{3}{2}\right)$ is largest

b. $D_{\mathbf{u}}f\left(-\frac{1}{2}, \frac{3}{2}\right)$ is smallest

c. $D_{\mathbf{u}}f\left(-\frac{1}{2}, \frac{3}{2}\right) = 0$

d. $D_{\mathbf{u}}f\left(-\frac{1}{2}, \frac{3}{2}\right) = -2$

e. $D_{\mathbf{u}}f\left(-\frac{1}{2}, \frac{3}{2}\right) = 1$

31. Zero directional derivative In what direction is the derivative of $f(x, y) = xy + y^2$ at $P(3, 2)$ equal to zero?

32. Zero directional derivative In what directions is the derivative of $f(x, y) = (x^2 - y^2)/(x^2 + y^2)$ at $P(1, 1)$ equal to zero?

33. Is there a direction $\mathbf{u}$ in which the rate of change of $f(x, y) = x^2 - 3xy + 4y^2$ at $P(1, 2)$ equals 14? Give reasons for your answer.

34. Changing temperature along a circle Is there a direction $\mathbf{u}$ in which the rate of change of the temperature function $T(x, y, z) = 2xy - yz$ (temperature in degrees Celsius, distance in feet) at $P(1, -1, 1)$ is $-3\,°\mathrm{C/ft}$? Give reasons for your answer.

35. The derivative of $f(x, y)$ at $P_0(1, 2)$ in the direction of $\mathbf{i} + \mathbf{j}$ is $2\sqrt{2}$ and in the direction of $-2\mathbf{j}$ is -3. What is the derivative of f in the direction of $-\mathbf{i} - 2\mathbf{j}$? Give reasons for your answer.

36. The derivative of $f(x, y, z)$ at a point P is greatest in the direction of $\mathbf{v} = \mathbf{i} + \mathbf{j} - \mathbf{k}$. In this direction, the value of the derivative is $2\sqrt{3}$.

a. What is ∇f at P? Give reasons for your answer.

b. What is the derivative of f at P in the direction of $\mathbf{i} + \mathbf{j}$?

37. Directional derivatives and scalar components How is the derivative of a differentiable function $f(x, y, z)$ at a point P_0 in the direction of a unit vector $\mathbf{u}$ related to the scalar component of $\nabla f|_{P_0}$ in the direction of $\mathbf{u}$? Give reasons for your answer.

38. Directional derivatives and partial derivatives Assuming that the necessary derivatives of $f(x, y, z)$ are defined, how are $D_{\mathbf{i}}f$, $D_{\mathbf{j}}f$, and $D_{\mathbf{k}}f$ related to f_x, f_y, and f_z? Give reasons for your answer.

39. Lines in the xy-plane Show that $A(x - x_0) + B(y - y_0) = 0$ is an equation for the line in the xy-plane through the point (x_0, y_0) normal to the vector $\mathbf{N} = A\mathbf{i} + B\mathbf{j}$.

40. The algebra rules for gradients Given a constant k and the gradients

$$\nabla f = \frac{\partial f}{\partial x}\mathbf{i} + \frac{\partial f}{\partial y}\mathbf{j} + \frac{\partial f}{\partial z}\mathbf{k},$$

$$\nabla g = \frac{\partial g}{\partial x}\mathbf{i} + \frac{\partial g}{\partial y}\mathbf{j} + \frac{\partial g}{\partial z}\mathbf{k},$$

establish the algebra rules for gradients.

In Exercises 41–44, find a parametric equation for the line that is perpendicular to the graph of the given equation at the given point.

41. $x^2 + y^2 = 25,\quad (-3, 4)$

42. $x^2 + xy + y^2 = 3,\quad (2, -1)$

43. $x^2 + y^2 + z^2 = 14,\quad (3, -2, 1)$

44. $z = x^3 - xy^2,\quad (-1, 1, 0)$

Gradients and Directional Derivatives for Functions of More Than Three Variables

In Exercises 45–48, find ∇f at the given point.

45. $f(x, y, z, w) = \dfrac{x\sqrt{y}}{w} - x^2z^3, (2, 4, -1, 3)$

46. $f(x, y, z, w) = x^3 \sin y + w^2 \cos z, (-2, \pi, 0, 3)$

47. $f(x, y, z, s, t) = e^y \ln s + x^2 t \tan z, \left(3, 0, \frac{\pi}{4}, e, 5\right)$

48. $f(x, y, z, s, t) = \dfrac{(x^2 + y^2)\arctan t}{z^2 s}, (-2, 1, -1, 2, 1)$

In Exercises 49–52, find the derivative of the function at P_0 in the direction of $\mathbf{v}$.

49. $f(x, y, z, w) = \dfrac{w \ln x}{y^2 z^3},\quad P_0(e^2, -2, 1, -3),\quad \mathbf{v} = \langle -1, 2, -2, 4\rangle$

50. $f(x, y, z, w) = (x - y)^2 + e^{z-w}, P_0(4, 2, 3, 1)$, $\mathbf{v} = \langle 1, 0, -2, 2\rangle$

51. $f(x, y, z, s, t) = s \arcsin(x + y) - t^2 \arctan(x - z)$, $P_0\left(0, \frac{1}{2}, -1, 1, -1\right), \mathbf{v} = \langle -1, 1, 0, 3, 5\rangle$

52. $f(x, y, z, s, t) = \sin tx + \cos sy - \dfrac{st}{z}, P_0\left(\frac{\pi}{4}, \frac{\pi}{6}, 2, 5, 1\right)$, $\mathbf{v} = \langle -3, 2, -2, 2, 2\rangle$

14.6 Tangent Planes and Differentials

In single-variable differential calculus, we saw how the derivative defined the tangent line to the graph of a differentiable function at a point on the graph. The tangent line then provided for a linearization of the function at the point. In this section, we will see analogously how the gradient defines the *tangent plane* to the level surface of a function $w = f(x, y, z)$ at a point on the surface. The tangent plane then provides for a linearization of f at the point and defines the total differential of the function.

Tangent Planes and Normal Lines

If $\mathbf{r}(t) = x(t)\mathbf{i} + y(t)\mathbf{j} + z(t)\mathbf{k}$ is a smooth curve on the level surface $f(x, y, z) = c$ of a differentiable function f, we found in Equation (7) of the last section that

$$\frac{d}{dt} f(\mathbf{r}(t)) = \nabla f(\mathbf{r}(t)) \cdot \mathbf{r}'(t).$$

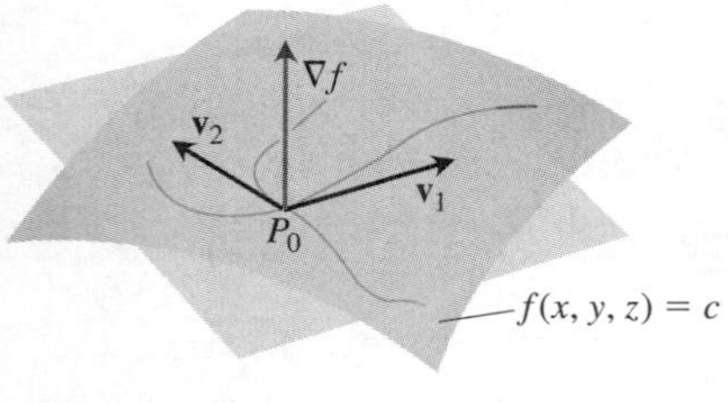

FIGURE 14.34 The gradient ∇f is orthogonal to the velocity vector of every smooth curve in the surface through P_0. The velocity vectors at P_0 therefore lie in a common plane, which we call the tangent plane at P_0.

Since f is constant along the curve $\mathbf{r}$, the derivative on the left-hand side of the equation is 0, so the gradient ∇f is orthogonal to the curve's velocity vector $\mathbf{r}'$.

Now let us restrict our attention to the curves that pass through a point P_0 (Figure 14.34). All the velocity vectors at P_0 are orthogonal to ∇f at P_0, so the curves' tangent lines all lie in the plane through P_0 normal to ∇f. (assuming it is a nonzero vector). We now define this plane.

DEFINITIONS The **tangent plane** to the level surface $f(x, y, z) = c$ of a differentiable function f at a point P_0 where the gradient is not zero is the plane through P_0 normal to $\nabla f|_{P_0}$.

The **normal line** of the surface at P_0 is the line through P_0 parallel to $\nabla f|_{P_0}$.

The results of Section 12.5 imply that the tangent plane and normal line satisfy the following equations, as long as the gradient at the point P_0 is not the zero vector.

Tangent Plane to $f(x, y, z) = c$ at $P_0(x_0, y_0, z_0)$

$$f_x(P_0)(x - x_0) + f_y(P_0)(y - y_0) + f_z(P_0)(z - z_0) = 0 \qquad (1)$$

Normal Line to $f(x, y, z) = c$ at $P_0(x_0, y_0, z_0)$

$$x = x_0 + f_x(P_0)t, \qquad y = y_0 + f_y(P_0)t, \qquad z = z_0 + f_z(P_0)t \qquad (2)$$

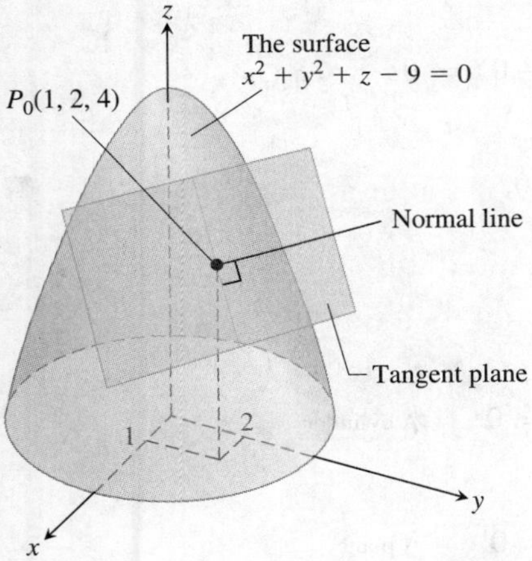

FIGURE 14.35 The tangent plane and normal line to this level surface at P_0 (Example 1).

EXAMPLE 1 Find the tangent plane and normal line of the level surface

$$f(x, y, z) = x^2 + y^2 + z - 9 = 0 \qquad \text{A circular paraboloid}$$

at the point $P_0(1, 2, 4)$.

Solution The surface is shown in Figure 14.35.

The tangent plane is the plane through P_0 perpendicular to the gradient of f at P_0. The gradient is

$$\nabla f|_{P_0} = (2x\mathbf{i} + 2y\mathbf{j} + \mathbf{k})\Big|_{(1,2,4)} = 2\mathbf{i} + 4\mathbf{j} + \mathbf{k}.$$

The tangent plane is therefore the plane

$$2(x - 1) + 4(y - 2) + (z - 4) = 0, \qquad \text{or} \qquad 2x + 4y + z = 14.$$

The line normal to the surface at P_0 is

$$x = 1 + 2t, \qquad y = 2 + 4t, \qquad z = 4 + t.$$

■

To find an equation for the plane tangent to a smooth surface $z = f(x, y)$ at a point $P_0(x_0, y_0, z_0)$ where $z_0 = f(x_0, y_0)$, we first observe that the equation $z = f(x, y)$ is equivalent to $f(x, y) - z = 0$. The surface $z = f(x, y)$ is therefore the zero level surface of the function $F(x, y, z) = f(x, y) - z$. The partial derivatives of F are

$$F_x = \frac{\partial}{\partial x}(f(x, y) - z) = f_x - 0 = f_x$$

$$F_y = \frac{\partial}{\partial y}(f(x, y) - z) = f_y - 0 = f_y$$

$$F_z = \frac{\partial}{\partial z}(f(x, y) - z) = 0 - 1 = -1.$$

The formula

$$F_x(P_0)(x - x_0) + F_y(P_0)(y - y_0) + F_z(P_0)(z - z_0) = 0$$

for the plane tangent to the level surface at P_0 therefore reduces to

$$f_x(x_0, y_0)(x - x_0) + f_y(x_0, y_0)(y - y_0) - (z - z_0) = 0.$$

Plane Tangent to a Surface $z = f(x, y)$ at $(x_0, y_0, f(x_0, y_0))$
The plane tangent to the surface $z = f(x, y)$ of a differentiable function f at the point $P_0(x_0, y_0, z_0) = (x_0, y_0, f(x_0, y_0))$ is

$$f_x(x_0, y_0)(x - x_0) + f_y(x_0, y_0)(y - y_0) - (z - z_0) = 0. \qquad (3)$$

EXAMPLE 2 Find the plane tangent to the surface $z = x\cos y - ye^x$ at $(0, 0, 0)$.

Solution We calculate the partial derivatives of $f(x, y) = x\cos y - ye^x$ and use Equation (3):

$$f_x(0,0) = (\cos y - ye^x)\Big|_{(0,0)} = 1 - 0 \cdot 1 = 1$$

$$f_y(0,0) = (-x\sin y - e^x)\Big|_{(0,0)} = 0 - 1 = -1.$$

The tangent plane is therefore

$$1 \cdot (x - 0) - 1 \cdot (y - 0) - (z - 0) = 0, \qquad \text{Eq. (3)}$$

or

$$x - y - z = 0.$$ ■

EXAMPLE 3 The surfaces

$$f(x, y, z) = x^2 + y^2 - 2 = 0 \qquad \text{A cylinder}$$

and

$$g(x, y, z) = x + z - 4 = 0 \qquad \text{A plane}$$

meet in an ellipse E (Figure 14.36). Find parametric equations for the line tangent to E at the point $P_0(1, 1, 3)$.

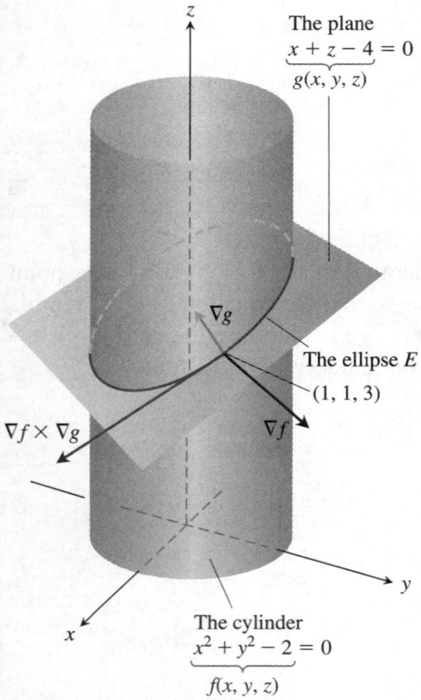

FIGURE 14.36 This cylinder and plane intersect in an ellipse E (Example 3).

Solution The tangent line is orthogonal to both ∇f and ∇g at P_0, and therefore parallel to $\mathbf{v} = \nabla f \times \nabla g$. The components of $\mathbf{v}$ and the coordinates of P_0 give us equations for the line. We have

$$\nabla f\big|_{(1,1,3)} = (2x\mathbf{i} + 2y\mathbf{j})\Big|_{(1,1,3)} = 2\mathbf{i} + 2\mathbf{j}$$

$$\nabla g\big|_{(1,1,3)} = (\mathbf{i} + \mathbf{k})\Big|_{(1,1,3)} = \mathbf{i} + \mathbf{k}$$

$$\mathbf{v} = (2\mathbf{i} + 2\mathbf{j}) \times (\mathbf{i} + \mathbf{k}) = \begin{vmatrix} \mathbf{i} & \mathbf{j} & \mathbf{k} \\ 2 & 2 & 0 \\ 1 & 0 & 1 \end{vmatrix} = 2\mathbf{i} - 2\mathbf{j} - 2\mathbf{k}.$$

The tangent line to the ellipse of intersection is

$$x = 1 + 2t, \qquad y = 1 - 2t, \qquad z = 3 - 2t.$$ ■

Estimating Change in a Specific Direction

The directional derivative plays a role similar to that of an ordinary derivative when we want to estimate how much the value of a function f changes if we move a small distance ds from a point P_0 to another point nearby. If f were a function of a single variable, we would have

$$df = f'(P_0)\,ds. \qquad \text{Ordinary derivative} \times \text{increment}$$

For a function of two or more variables, we use the formula

$$df = \left(\nabla f|_{P_0} \cdot \mathbf{u}\right) ds, \qquad \text{Directional derivative} \times \text{increment}$$

where $\mathbf{u}$ is the direction of the motion away from P_0.

Estimating the Change in f in a Direction u
To estimate the change in the value of a differentiable function f when we move a small distance ds from a point P_0 in a particular direction $\mathbf{u}$, use this formula:

$$df = \underbrace{\left(\nabla f|_{P_0} \cdot \mathbf{u}\right)}_{\substack{\text{Directional}\\ \text{derivative}}} \underbrace{ds}_{\substack{\text{Distance}\\ \text{increment}}}$$

EXAMPLE 4 Estimate how much the value of

$$f(x, y, z) = y \sin x + 2yz$$

will change if the point $P(x, y, z)$ moves 0.1 unit from $P_0(0, 1, 0)$ straight toward $P_1(2, 2, -2)$.

Solution We first find the derivative of f at P_0 in the direction of the vector $\overrightarrow{P_0P_1} = 2\mathbf{i} + \mathbf{j} - 2\mathbf{k}$. The direction of this vector is

$$\mathbf{u} = \frac{\overrightarrow{P_0P_1}}{|\overrightarrow{P_0P_1}|} = \frac{\overrightarrow{P_0P_1}}{3} = \frac{2}{3}\mathbf{i} + \frac{1}{3}\mathbf{j} - \frac{2}{3}\mathbf{k}.$$

The gradient of f at P_0 is

$$\nabla f|_{(0,1,0)} = \left((y \cos x)\mathbf{i} + (\sin x + 2z)\mathbf{j} + 2y\mathbf{k}\right)\Big|_{(0,1,0)} = \mathbf{i} + 2\mathbf{k}.$$

Therefore,

$$\nabla f|_{P_0} \cdot \mathbf{u} = (\mathbf{i} + 2\mathbf{k}) \cdot \left(\frac{2}{3}\mathbf{i} + \frac{1}{3}\mathbf{j} - \frac{2}{3}\mathbf{k}\right) = \frac{2}{3} - \frac{4}{3} = -\frac{2}{3}.$$

The change df in f that results from moving $ds = 0.1$ unit away from P_0 in the direction of $\mathbf{u}$ is approximately

$$df = \left(\nabla f|_{P_0} \cdot \mathbf{u}\right)(ds) = \left(-\frac{2}{3}\right)(0.1) \approx -0.067 \text{ unit.}$$

See Figure 14.37. ■

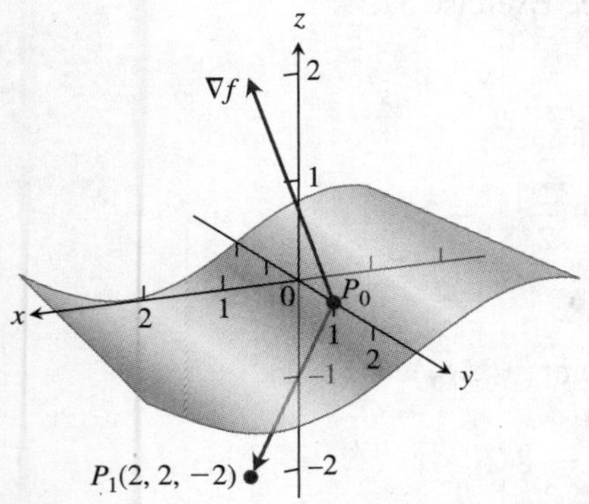

FIGURE 14.37 As $P(x, y, z)$ moves off the level surface at P_0 by 0.1 unit directly toward P_1, the function f changes value by approximately -0.067 unit (Example 4).

How to Linearize a Function of Two Variables

Functions of two variables can be quite complicated, and we sometimes need to approximate them with simpler ones that give the accuracy required for specific applications without being so difficult to work with. We do this in a way that is similar to the way we find linear replacements for functions of a single variable (Section 3.9).

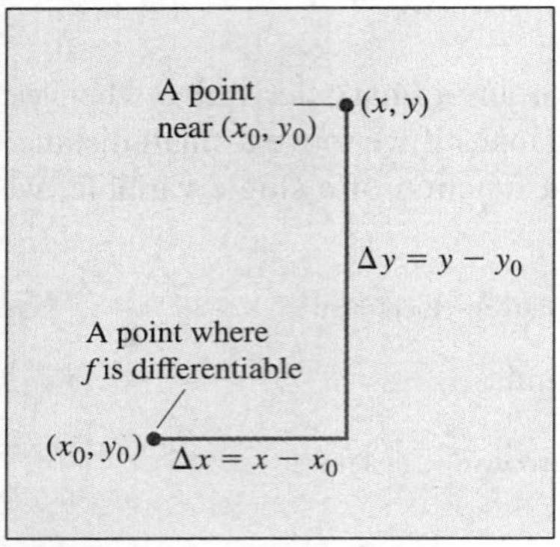

FIGURE 14.38 If f is differentiable at (x_0, y_0), then the value of f at point (x, y) nearby is approximately $f(x_0, y_0) + f_x(x_0, y_0)\Delta x + f_y(x_0, y_0)\Delta y$.

Suppose the function we wish to approximate is $z = f(x, y)$ near a point (x_0, y_0) at which we know the values of f, f_x, and f_y and at which f is differentiable. If we move from (x_0, y_0) to a nearby point (x, y) by increments $\Delta x = x - x_0$ and $\Delta y = y - y_0$ (see Figure 14.38), then the definition of differentiability from Section 14.3 shows that the change

$$f(x, y) - f(x_0, y_0) = f_x(x_0, y_0)\Delta x + f_y(x_0, y_0)\Delta y + \varepsilon_1 \Delta x + \varepsilon_2 \Delta y,$$

where $\varepsilon_1, \varepsilon_2 \to 0$ as $\Delta x, \Delta y \to 0$. If the increments Δx and Δy are small, the products $\varepsilon_1 \Delta x$ and $\varepsilon_2 \Delta y$ will eventually be smaller still, and we have the approximation

$$f(x, y) \approx \underbrace{f(x_0, y_0) + f_x(x_0, y_0)(x - x_0) + f_y(x_0, y_0)(y - y_0)}_{L(x, y)}.$$

In other words, as long as Δx and Δy are small, f will have approximately the same value as the linear function L.

DEFINITIONS The **linearization** of a function $f(x, y)$ at a point (x_0, y_0) where f is differentiable is the function

$$L(x, y) = f(x_0, y_0) + f_x(x_0, y_0)(x - x_0) + f_y(x_0, y_0)(y - y_0).$$

The approximation

$$f(x, y) \approx L(x, y)$$

is the **standard linear approximation** of f at (x_0, y_0).

From Equation (3), we find that the plane $z = L(x, y)$ is tangent to the surface $z = f(x, y)$ at the point (x_0, y_0). Thus, the linearization of a function of two variables is a tangent-*plane* approximation in the same way that the linearization of a function of a single variable is a tangent-*line* approximation. (See Exercise 57.)

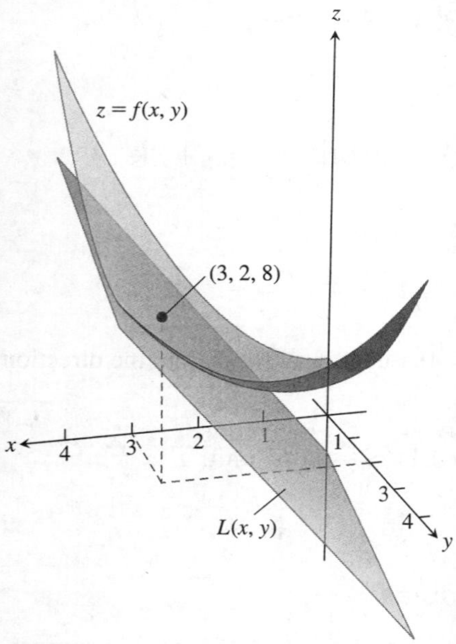

FIGURE 14.39 The tangent plane $L(x, y)$ represents the linearization of $f(x, y)$ in Example 5.

EXAMPLE 5 Find the linearization of

$$f(x, y) = x^2 - xy + \frac{1}{2}y^2 + 3$$

at the point $(3, 2)$.

Solution We first evaluate f, f_x, and f_y at the point $(x_0, y_0) = (3, 2)$:

$$f(3, 2) = \left(x^2 - xy + \frac{1}{2}y^2 + 3\right)\bigg|_{(3,2)} = 8$$

$$f_x(3, 2) = \frac{\partial}{\partial x}\left(x^2 - xy + \frac{1}{2}y^2 + 3\right)\bigg|_{(3,2)} = (2x - y)\bigg|_{(3,2)} = 4$$

$$f_y(3, 2) = \frac{\partial}{\partial y}\left(x^2 - xy + \frac{1}{2}y^2 + 3\right)\bigg|_{(3,2)} = (-x + y)\bigg|_{(3,2)} = -1,$$

which yields

$$\begin{aligned} L(x, y) &= f(x_0, y_0) + f_x(x_0, y_0)(x - x_0) + f_y(x_0, y_0)(y - y_0) \\ &= 8 + (4)(x - 3) + (-1)(y - 2) = 4x - y - 2. \end{aligned}$$

The linearization of f at $(3, 2)$ is $L(x, y) = 4x - y - 2$ (see Figure 14.39). ■

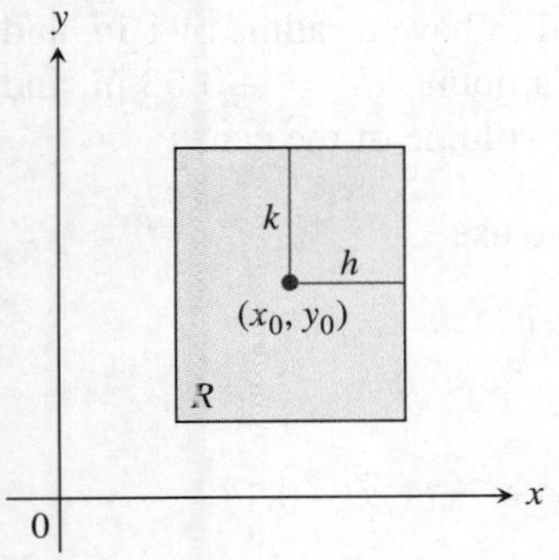

FIGURE 14.40 The rectangular region R: $|x - x_C| \le h, |y - y_0| \le k$ in the xy-plane.

When we approximate a differentiable function $f(x, y)$ by its linearization $L(x, y)$ at (x_0, y_0), an important question is how accurate the approximation might be.

If we can find a common upper bound M for $|f_{xx}|, |f_{yy}|$, and $|f_{xy}|$ on a rectangle R centered at (x_0, y_0) (Figure 14.40), then we can bound the error E throughout R by using a simple formula. The **error** is defined by $E(x, y) = f(x, y) - L(x, y)$.

The Error in the Standard Linear Approximation

If f has continuous first and second partial derivatives throughout an open set containing a rectangle R centered at (x_0, y_0), and if M is any upper bound for the values of $|f_{xx}|, |f_{yy}|$, and $|f_{xy}|$ on R, then the error $E(x, y)$ incurred in replacing $f(x, y)$ on R by its linearization

$$L(x, y) = f(x_0, y_0) + f_x(x_0, y_0)(x - x_0) + f_y(x_0, y_0)(y - y_0)$$

satisfies the inequality

$$|E(x, y)| \le \frac{1}{2}M(|x - x_0| + |y - y_0|)^2.$$

To make $|E(x, y)|$ small for a given M, we just make $|x - x_0|$ and $|y - y_0|$ small.

Differentials

Recall from Section 3.9 that for a function of a single variable, $y = f(x)$, we defined the change in f as x changes from a to $a + \Delta x$ by

$$\Delta f = f(a + \Delta x) - f(a)$$

and the differential of f as

$$df = f'(a)\Delta x.$$

We now consider the differential of a function of two variables.

Suppose a differentiable function $f(x, y)$ and its partial derivatives exist at a point (x_0, y_0). If we move to a nearby point $(x_0 + \Delta x, y_0 + \Delta y)$, the change in f is

$$\Delta f = f(x_0 + \Delta x, y_0 + \Delta y) - f(x_0, y_0).$$

A straightforward calculation based on the definition of $L(x, y)$, using the notation $x - x_0 = \Delta x$ and $y - y_0 = \Delta y$, shows that the corresponding change in L is

$$\Delta L = L(x_0 + \Delta x, y_0 + \Delta y) - L(x_0, y_0) = f_x(x_0, y_0)\Delta x + f_y(x_0, y_0)\Delta y.$$

The **differentials** dx and dy are independent variables, so they can be assigned any values. Often we take $dx = \Delta x = x - x_0$, and $dy = \Delta y = y - y_0$. We then have the following definition of the differential or *total* differential of f.

DEFINITION If we move from (x_0, y_0) to a point $(x_0 + dx, y_0 + dy)$ nearby, the resulting change

$$df = f_x(x_0, y_0)\,dx + f_y(x_0, y_0)\,dy$$

in the linearization of f is called the **total differential of f**.

EXAMPLE 6 Suppose that a cylindrical can is designed to have a radius of 1 in. and a height of 5 in., but that the radius and height are off by the amounts $dr = +0.03$ in. and $dh = -0.1$ in. Estimate the resulting absolute change in the volume of the can.

Solution To estimate the absolute change in $V = \pi r^2 h$, we use

$$\Delta V \approx dV = V_r(r_0, h_0)\,dr + V_h(r_0, h_0)\,dh.$$

With $V_r = 2\pi rh$ and $V_h = \pi r^2$, we get

$$\begin{aligned} dV &= 2\pi r_0 h_0\,dr + \pi r_0^2\,dh = 2\pi(1)(5)(0.03) + \pi(1)^2(-0.1) \\ &= 0.3\pi - 0.1\pi = 0.2\pi \approx 0.63\ \text{in}^3. \end{aligned}$$

■

EXAMPLE 7 A company manufactures stainless steel right circular cylindrical molasses storage tanks that are 25 ft high with a radius of 5 ft. How sensitive are the tanks' volumes to small variations in height and radius?

Solution With $V = \pi r^2 h$, the total differential gives the approximation for the change in volume as

$$\begin{aligned} dV &= V_r(5, 25)\,dr + V_h(5, 25)\,dh \\ &= (2\pi rh)\Big|_{(5,25)} dr + (\pi r^2)\Big|_{(5,25)} dh \\ &= 250\pi\,dr + 25\pi\,dh. \end{aligned}$$

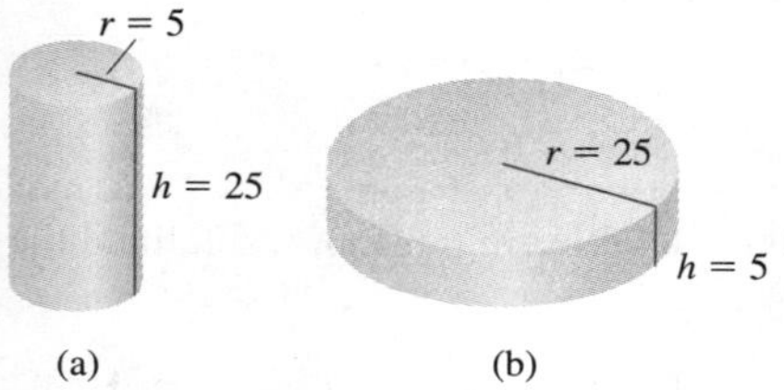

FIGURE 14.41 The volume of cylinder (a) is more sensitive to a small change in r than it is to an equally small change in h. The volume of cylinder (b) is more sensitive to small changes in h than it is to small changes in r (Example 7).

Thus, a 1-unit change in r will change V by about 250π units. A 1-unit change in h will change V by about 25π units. The tank's volume is 10 times more sensitive to a small change in r than it is to a small change of equal size in h. As a quality control engineer concerned with being sure the tanks have the correct volume, you would want to pay special attention to their radii.

In contrast, if the values of r and h are reversed to make $r = 25$ and $h = 5$, then the total differential in V becomes

$$dV = (2\pi rh)\Big|_{(25,5)} dr + (\pi r^2)\Big|_{(25,5)} dh = 250\pi\,dr + 625\pi\,dh.$$

Now the volume is more sensitive to changes in h than to changes in r (Figure 14.41).

The general rule is that functions are most sensitive to small changes in the variables that generate the largest partial derivatives. ■

Functions of More Than Two Variables

Analogous results hold for differentiable functions of more than two variables.

1. The **linearization** of $f(x, y, z)$ at a point $P_0(x_0, y_0, z_0)$ is

$$L(x, y, z) = f(P_0) + f_x(P_0)(x - x_0) + f_y(P_0)(y - y_0) + f_z(P_0)(z - z_0).$$

2. Suppose that R is a closed rectangular solid centered at P_0 and lying in an open region on which the second partial derivatives of f are continuous. Suppose also that $|f_{xx}|, |f_{yy}|, |f_{zz}|, |f_{xy}|, |f_{xz}|$, and $|f_{yz}|$ are all less than or equal to M throughout R. Then the **error** $E(x, y, z) = f(x, y, z) - L(x, y, z)$ in the approximation of f by L is bounded throughout R by the inequality

$$|E| \le \frac{1}{2}M(|x - x_0| + |y - y_0| + |z - z_0|)^2.$$

3. If the second partial derivatives of f are continuous and if x, y, and z change from x_0, y_0, and z_0 by small amounts dx, dy, and dz, the **total differential**

$$df = f_x(P_0)\,dx + f_y(P_0)\,dy + f_z(P_0)\,dz$$

gives a good approximation of the resulting change in f.

EXAMPLE 8 Find the linearization $L(x, y, z)$ of

$$f(x, y, z) = x^2 - xy + 3\sin z$$

at the point $(x_0, y_0, z_0) = (2, 1, 0)$. Find an upper bound for the error incurred in replacing f by L on the rectangular region

$$R:\ |x - 2| \le 0.01, \qquad |y - 1| \le 0.02, \qquad |z| \le 0.01.$$

Solution Routine calculations give

$$f(2,1,0) = 2, \qquad f_x(2,1,0) = 3, \qquad f_y(2,1,0) = -2, \qquad f_z(2,1,0) = 3.$$

Thus,

$$L(x, y, z) = 2 + 3(x - 2) + (-2)(y - 1) + 3(z - 0) = 3x - 2y + 3z - 2.$$

Since

$$f_{xx} = 2, \qquad f_{yy} = 0, \qquad f_{zz} = -3\sin z, \qquad f_{xy} = -1, \qquad f_{xz} = 0, \qquad f_{yz} = 0,$$

and $|-3\sin z| \le 3\sin 0.01 \approx 0.03$, we may take $M = 2$ as a bound on the second partials. Hence, the error incurred by replacing f by L on R satisfies

$$|E| \le \frac{1}{2}(2)(0.01 + 0.02 + 0.01)^2 = 0.0016.$$ ■

EXERCISES 14.6

Tangent Planes and Normal Lines to Surfaces

In Exercises 1–10, find equations for the

(a) tangent plane and

(b) normal line at the point P_0 on the given surface.

1. $x^2 + y^2 + z^2 = 3, \quad P_0(1, 1, 1)$
2. $x^2 - 4y^2 + 3z^2 + 4 = 0, \quad P_0(3, 2, 1)$
3. $x^2 y - 2xz^2 = 8, \quad P_0(1, 0, 2)$
4. $x^2 + 2xy - y^2 + z^2 = 7, \quad P_0(1, -1, 3)$
5. $\cos \pi x - x^2 y + e^{xz} + yz = 4, \quad P_0(0, 1, 2)$
6. $x^2 - xy - y^2 - z = 0, \quad P_0(1, 1, -1)$
7. $x + y + z = 1, \quad P_0(0, 1, 0)$
8. $2xz^2 - 3xy - 4x = 7, \quad P_0(1, -1, -2)$
9. $x \ln y + y \ln z = x, \quad P_0(1, 1, e)$
10. $ye^x + ze^{y^2} = z, \quad P_0(0, 0, 1)$

In Exercises 11–14, find an equation for the plane that is tangent to the given surface at the given point.

11. $z = \ln(x^2 + y^2), \quad (1, 0, 0)$
12. $z = e^{-(x^2+y^2)}, \quad (0, 0, 1)$
13. $z = \sqrt{y - x}, \quad (1, 2, 1)$
14. $z = 4x^2 + y^2, \quad (1, 1, 5)$

Tangent Lines to Intersecting Surfaces

In Exercises 15–20, find parametric equations for the line tangent to the curve of intersection of the surfaces at the given point.

15. Surfaces: $x^2 - y^2 - z^2 = 11, \quad xy + yz - zx = 18$
    Point: $(6, 4, 3)$
16. Surfaces: $xyz = 1, \quad x^2 + 2y^2 + 3z^2 = 6$
    Point: $(1, 1, 1)$
17. Surfaces: $x^2 + 2y + 2z = 4, \quad y = 1$
    Point: $(1, 1, 1/2)$
18. Surfaces: $5x^2 - 2yz - 9x = 0$ and $4x^2 y + z^3 - 4 = 0$
    Point: $(1, -1, 2)$
19. Surfaces: $x^3 + 3x^2y^2 + y^3 + 4xy - z^2 = 0,$
    $x^2 + y^2 + z^2 = 11$
    Point: $(1, 1, 3)$
20. Surfaces: $x^2 + y^2 = 4, \quad x^2 + y^2 - z = 0$
    Point: $(\sqrt{2}, \sqrt{2}, 4)$

Estimating Change

21. By about how much will

$$f(x, y, z) = \ln\sqrt{x^2 + y^2 + z^2}$$

change if the point $P(x, y, z)$ moves from $P_0(3, 4, 12)$ a distance of $ds = 0.1$ unit in the direction of $3\mathbf{i} + 6\mathbf{j} - 2\mathbf{k}$?

22. By about how much will

$$f(x, y, z) = e^x \cos yz$$

change as the point $P(x, y, z)$ moves from the origin a distance of $ds = 0.1$ unit in the direction of $2\mathbf{i} + 2\mathbf{j} - 2\mathbf{k}$?

23. By about how much will

$$g(x, y, z) = x + x\cos z - y\sin z + y$$

change if the point $P(x, y, z)$ moves from $P_0(2, -1, 0)$ a distance of $ds = 0.2$ unit toward the point $P_1(0, 1, 2)$?

24. By about how much will

$$h(x, y, z) = \cos(\pi xy) + xz^2$$

change if the point $P(x, y, z)$ moves from $P_0(-1, -1, -1)$ a distance of $ds = 0.1$ unit toward the origin?

25. Temperature change along a circle Suppose that the Celsius temperature at the point (x, y) in the xy-plane is $T(x, y) = x\sin 2y$ and that distance in the xy-plane is measured in meters. A particle is moving *clockwise* around the circle of radius 1 m centered at the origin at the constant rate of 2 m/sec.

a. How fast is the temperature experienced by the particle changing in degrees Celsius per meter at the point $P(1/2, \sqrt{3}/2)$?

b. How fast is the temperature experienced by the particle changing in degrees Celsius per second at P?

26. Changing temperature along a space curve The Celsius temperature in a region in space is given by $T(x, y, z) = 2x^2 - xyz$. A particle is moving in this region and its position at time t is given by $x = 2t^2$, $y = 3t$, $z = -t^2$, where time is measured in seconds and distance in meters.

a. How fast is the temperature experienced by the particle changing in degrees Celsius per meter when the particle is at the point $P(8, 6, -4)$?

b. How fast is the temperature experienced by the particle changing in degrees Celsius per second at P?

Finding Linearizations

In Exercises 27–32, find the linearization $L(x, y)$ of the function at each point.

27. $f(x, y) = x^2 + y^2 + 1$ at **a.** $(0, 0)$, **b.** $(1, 1)$

28. $f(x, y) = (x + y + 2)^2$ at **a.** $(0, 0)$, **b.** $(1, 2)$

29. $f(x, y) = 3x - 4y + 5$ at **a.** $(0, 0)$, **b.** $(1, 1)$

30. $f(x, y) = x^3y^4$ at **a.** $(1, 1)$, **b.** $(0, 0)$

31. $f(x, y) = e^x \cos y$ at **a.** $(0, 0)$, **b.** $(0, \pi/2)$

32. $f(x, y) = e^{2y-x}$ at **a.** $(0, 0)$, **b.** $(1, 2)$

33. Wind chill factor Wind chill, a measure of the apparent temperature felt on exposed skin, is a function of air temperature and wind speed. The precise formula, updated by the National Weather Service in 2001 and based on modern heat transfer theory, a human face model, and skin tissue resistance, is

$$W = W(v, T) = 35.74 + 0.6215\,T - 35.75\,v^{0.16} + 0.4275\,T \cdot v^{0.16},$$

where T is air temperature in °F and v is wind speed in mph. A partial wind chill chart follows.

***T*(°F)**

v (mph)	30	25	20	15	10	5	0	−5	−10
5	25	19	13	7	1	−5	−11	−16	−22
10	21	15	9	3	−4	−10	−16	−22	−28
15	19	13	6	0	−7	−13	−19	−26	−32
20	17	11	4	−2	−9	−15	−22	−29	−35
25	16	9	3	−4	−11	−17	−24	−31	−37
30	15	8	1	−5	−12	−19	−26	−33	−39
35	14	7	0	−7	−14	−21	−27	−34	−41

a. Use the table to find $W(20, 25)$, $W(30, -10)$, and $W(15, 15)$.

b. Use the formula to find $W(10, -40)$, $W(50, -40)$, and $W(60, 30)$.

c. Find the linearization $L(v, T)$ of the function $W(v, T)$ at the point $(25, 5)$.

d. Use $L(v, T)$ in part (c) to estimate the following wind chill values.

i) $W(24, 6)$ **ii)** $W(27, 2)$

iii) $W(5, -10)$ (Explain why this value is much different from the value found in the table.)

34. Find the linearization $L(v, T)$ of the function $W(v, T)$ in Exercise 33 at the point $(50, -20)$. Use it to estimate the following wind chill values.

a. $W(49, -22)$

b. $W(53, -19)$

c. $W(60, -30)$

Bounding the Error in Linear Approximations

In Exercises 35–40, find the linearization $L(x, y)$ of the function $f(x, y)$ at P_0. Then find an upper bound for the magnitude $|E|$ of the error in the approximation $f(x, y) \approx L(x, y)$ over the rectangle R.

35. $f(x, y) = x^2 - 3xy + 5$ at $P_0(2, 1)$,
R: $|x - 2| \le 0.1$, $|y - 1| \le 0.1$

36. $f(x, y) = (1/2)x^2 + xy + (1/4)y^2 + 3x - 3y + 4$ at $P_0(2, 2)$,
R: $|x - 2| \le 0.1$, $|y - 2| \le 0.1$

37. $f(x, y) = 1 + y + x\cos y$ at $P_0(0, 0)$,
R: $|x| \le 0.2$, $|y| \le 0.2$
(Use $|\cos y| \le 1$ and $|\sin y| \le 1$ in estimating E.)

38. $f(x, y) = xy^2 + y\cos(x - 1)$ at $P_0(1, 2)$,
R: $|x - 1| \le 0.1$, $|y - 2| \le 0.1$

39. $f(x, y) = e^x \cos y$ at $P_0(0, 0)$,
R: $|x| \leq 0.1$, $|y| \leq 0.1$
(Use $e^x \leq 1.11$ and $|\cos y| \leq 1$ in estimating E.)

40. $f(x, y) = \ln x + \ln y$ at $P_0(1, 1)$,
R: $|x - 1| \leq 0.2$, $|y - 1| \leq 0.2$

Linearizations for Three Variables

Find the linearizations $L(x, y, z)$ of the functions in Exercises 41–46 at the given points.

41. $f(x, y, z) = xy + yz + xz$ at
a. $(1, 1, 1)$ **b.** $(1, 0, 0)$ **c.** $(0, 0, 0)$

42. $f(x, y, z) = x^2 + y^2 + z^2$ at
a. $(1, 1, 1)$ **b.** $(0, 1, 0)$ **c.** $(1, 0, 0)$

43. $f(x, y, z) = \sqrt{x^2 + y^2 + z^2}$ at
a. $(1, 0, 0)$ **b.** $(1, 1, 0)$ **c.** $(1, 2, 2)$

44. $f(x, y, z) = (\sin xy)/z$ at
a. $(\pi/2, 1, 1)$ **b.** $(2, 0, 1)$

45. $f(x, y, z) = e^x + \cos(y + z)$ at
a. $(0, 0, 0)$ **b.** $\left(0, \frac{\pi}{2}, 0\right)$ **c.** $\left(0, \frac{\pi}{4}, \frac{\pi}{4}\right)$

46. $f(x, y, z) = \tan^{-1}(xyz)$ at
a. $(1, 0, 0)$ **b.** $(1, 1, 0)$ **c.** $(1, 1, 1)$

In Exercises 47–50, find the linearization $L(x, y, z)$ of the function $f(x, y, z)$ at P_0. Then find an upper bound for the magnitude of the error E in the approximation $f(x, y, z) \approx L(x, y, z)$ over the region R.

47. $f(x, y, z) = xz - 3yz + 2$ at $P_0(1, 1, 2)$,
R: $|x - 1| \leq 0.01$, $|y - 1| \leq 0.01$, $|z - 2| \leq 0.02$

48. $f(x, y, z) = x^2 + xy + yz + (1/4)z^2$ at $P_0(1, 1, 2)$,
R: $|x - 1| \leq 0.01$, $|y - 1| \leq 0.01$, $|z - 2| \leq 0.08$

49. $f(x, y, z) = xy + 2yz - 3xz$ at $P_0(1, 1, 0)$,
R: $|x - 1| \leq 0.01$, $|y - 1| \leq 0.01$, $|z| \leq 0.01$

50. $f(x, y, z) = \sqrt{2} \cos x \sin(y + z)$ at $P_0(0, 0, \pi/4)$,
R: $|x| \leq 0.01$, $|y| \leq 0.01$, $|z - \pi/4| \leq 0.01$

Estimating Error; Sensitivity to Change

51. Estimating maximum error Suppose that T is to be found from the formula $T = x(e^y + e^{-y})$, where x and y are found to be 2 and ln 2 with maximum possible errors of $|dx| = 0.1$ and $|dy| = 0.02$. Estimate the maximum possible error in the computed value of T.

52. Variation in electrical resistance The resistance R produced by wiring resistors of R_1 and R_2 ohms in parallel (see accompanying figure) can be calculated from the formula

$$\frac{1}{R} = \frac{1}{R_1} + \frac{1}{R_2}.$$

a. Show that

$$dR = \left(\frac{R}{R_1}\right)^2 dR_1 + \left(\frac{R}{R_2}\right)^2 dR_2.$$

b. You have designed a two-resistor circuit, like the one shown, to have resistances of $R_1 = 100$ ohms and $R_2 = 400$ ohms, but there is always some variation in manufacturing, and the resistors received by your firm will probably not have these exact values. Will the value of R be more sensitive to variation in R_1 or to variation in R_2? Give reasons for your answer.

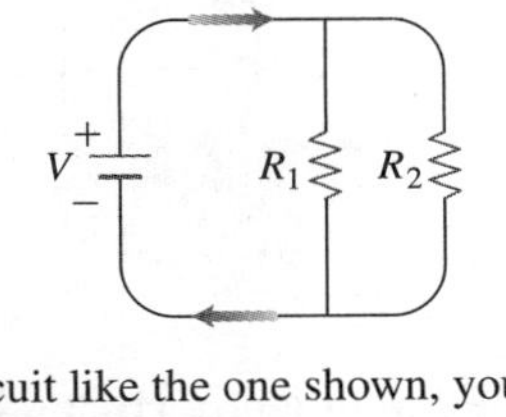

c. In another circuit like the one shown, you plan to change R_1 from 20 to 20.1 ohms and R_2 from 25 to 24.9 ohms. By about what percentage will this change R?

53. You plan to calculate the area of a long, thin rectangle from measurements of its length and width. Which dimension should you measure more carefully? Give reasons for your answer.

54. a. Around the point $(1, 0)$, is $f(x, y) = x^2(y + 1)$ more sensitive to changes in x or to changes in y? Give reasons for your answer.

b. What ratio of dx to dy will make df equal zero at $(1, 0)$?

55. Value of a 2 × 2 determinant If $|a|$ is much greater than $|b|$, $|c|$, and $|d|$, to which of a, b, c, and d is the value of the determinant

$$f(a, b, c, d) = \begin{vmatrix} a & b \\ c & d \end{vmatrix}$$

most sensitive? Give reasons for your answer.

56. The Wilson lot size formula The Wilson lot size formula in economics says that the most economical quantity Q of goods (radios, shoes, brooms, whatever) for a store to order is given by the formula $Q = \sqrt{2KM/h}$, where K is the cost of placing the order, M is the number of items sold per week, and h is the weekly holding cost for each item (cost of space, utilities, security, and so on). To which of the variables K, M, and h is Q most sensitive near the point $(K_0, M_0, h_0) = (2, 20, 0.05)$? Give reasons for your answer.

Theory and Examples

57. The linearization of $f(x, y)$ is a tangent-plane approximation Show that the tangent plane at the point $P_0(x_0, y_0, f(x_0, y_0))$ on the surface $z = f(x, y)$ defined by a differentiable function f is the plane

$$f_x(x_0, y_0)(x - x_0) + f_y(x_0, y_0)(y - y_0) - (z - f(x_0, y_0)) = 0,$$

or

$$z = f(x_0, y_0) + f_x(x_0, y_0)(x - x_0) + f_y(x_0, y_0)(y - y_0).$$

Thus, the tangent plane at P_0 is the graph of the linearization of f at P_0 (see accompanying figure).

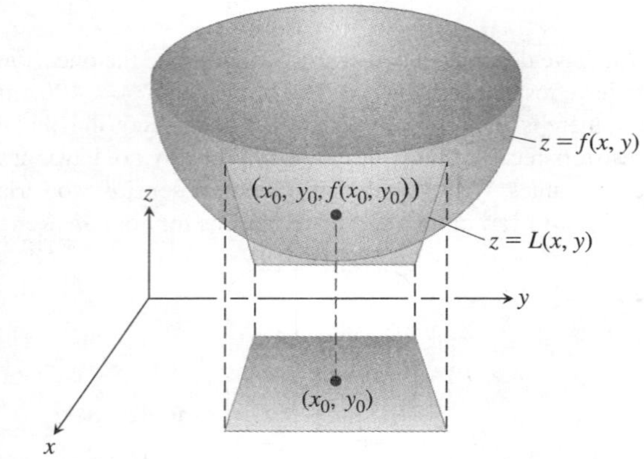

58. Change along the involute of a circle Find the derivative of $f(x, y) = x^2 + y^2$ in the direction of the unit tangent vector of the curve

$$\mathbf{r}(t) = (\cos t + t \sin t)\mathbf{i} + (\sin t - t \cos t)\mathbf{j}, \qquad t > 0.$$

59. Tangent curves A smooth curve is *tangent* to the surface at a point of intersection if its velocity vector is orthogonal to ∇f there.

Show that the curve

$$\mathbf{r}(t) = \sqrt{t}\mathbf{i} + \sqrt{t}\mathbf{j} + (2t - 1)\mathbf{k}$$

is tangent to the surface $x^2 + y^2 - z = 1$ when $t = 1$.

60. Normal curves A smooth curve is *normal* to a surface $f(x, y, z) = c$ at a point of intersection if the curve's velocity vector is a nonzero scalar multiple of ∇f at the point.

Show that the curve

$$\mathbf{r}(t) = \sqrt{t}\mathbf{i} + \sqrt{t}\,\mathbf{j} - \frac{1}{4}(t + 3)\mathbf{k}$$

is normal to the surface $x^2 + y^2 - z = 3$ when $t = 1$.

61. Consider a closed rectangular box with a square base, as shown in the figure. Assume x is measured with an error of at most 0.5% and y is measured with an error of at most 0.75%, so we have $|dx|/x < 0.005$ and $|dy|/y < 0.0075$.

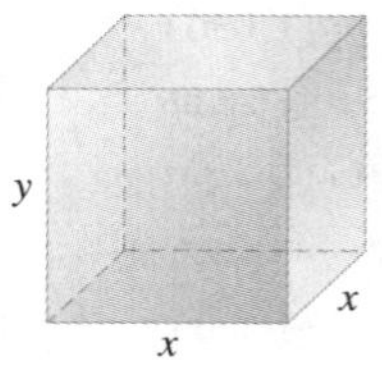

a. Use a differential to estimate the relative error $|dV|/V$ in computing the box's volume V.

b. Use a differential to estimate the relative error $|dS|/S$ in computing the box's surface area S.

Hint for **b**: $\dfrac{4x^2 + 4xy}{2x^2 + 4xy} \le \dfrac{4x^2 + 8xy}{2x^2 + 4xy} = 2$ and $\dfrac{4xy}{2x^2 + 4xy} \le \dfrac{2x^2 + 4xy}{2x^2 + 4xy} = 1.$

14.7 Extreme Values and Saddle Points

HISTORICAL BIOGRAPHY
Siméon-Denis Poisson
(1781–1840)
www.bit.ly/2y3g8rw

Continuous functions of two variables assume extreme values on closed, bounded domains (see Figures 14.42 and 14.43). We see in this section that we can narrow the search for these extreme values by examining the functions' first partial derivatives. A function of two variables can assume extreme values only at boundary points of the domain or at interior domain points where both first partial derivatives are zero or where one or both of the first partial derivatives fail to exist. However, the vanishing of derivatives at an interior point (a, b) does not always signal the presence of an extreme value. The surface that is the graph of the function might be shaped like a saddle right above (a, b) and cross its tangent plane there.

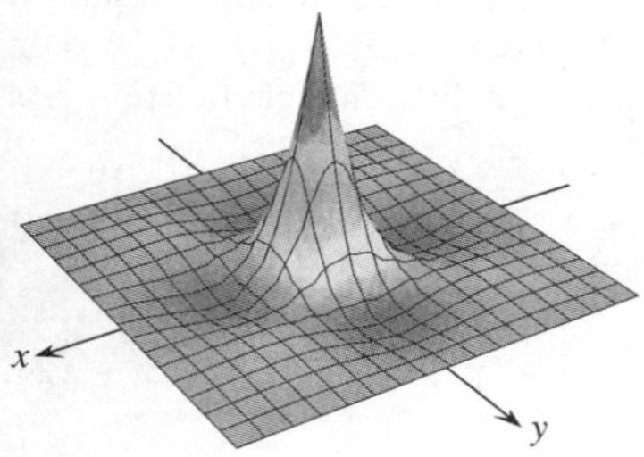

FIGURE 14.42 The function

$$z = (\cos x)(\cos y)e^{-\sqrt{x^2+y^2}}$$

has a maximum value of 1 and a minimum value of about -0.067 on the square region $|x| \le 3\pi/2, |y| \le 3\pi/2$.

Local Extreme Values for Functions of Two Variables

To find the local extreme values of a function of a single variable, we look for points where the graph has a horizontal tangent line. At such points, we then look for local maxima, local minima, and points of inflection. For a function $f(x, y)$ of two variables, we look for points where the surface $z = f(x, y)$ has a horizontal tangent *plane*. At such points, we then look for local maxima, local minima, and saddle points. We begin by defining maxima and minima.

DEFINITIONS Let $f(x, y)$ be defined on a region R containing the point (a, b). Then

1. $f(a, b)$ is a **local maximum** value of f if $f(a, b) \ge f(x, y)$ for all domain points (x, y) in an open disk centered at (a, b). $f(a, b)$ is an **absolute maximum** value of f on R if $f(a, b) \ge f(x, y)$ for all domain points (x, y) in R.
2. $f(a, b)$ is a **local minimum** value of f if $f(a, b) \le f(x, y)$ for all domain points (x, y) in an open disk centered at (a, b). $f(a, b)$ is an **absolute minimum** value of f on R if $f(a, b) \le f(x, y)$ for all domain points (x, y) in R.

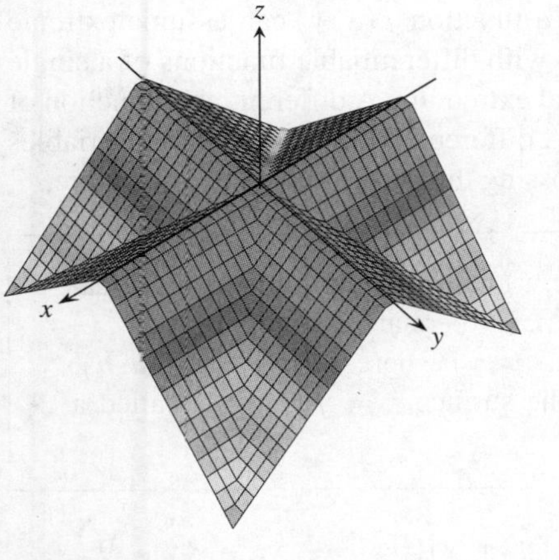

FIGURE 14.43 The "roof surface"

$$z = \frac{1}{2}(||x| - |y|| - |x| - |y|)$$

has a maximum value of 0 and a minimum value of $-a$ on the square region $|x| \le a$, $|y| \le a$.

Local maxima correspond to mountain peaks on the surface $z = f(x, y)$, and local minima correspond to valley bottoms (Figure 14.44). At such points the tangent planes, when they exist, are horizontal. Local extrema are also called **relative extrema**.

As with functions of a single variable, the key to identifying the local extrema is the First Derivative Theorem, which we next state and prove.

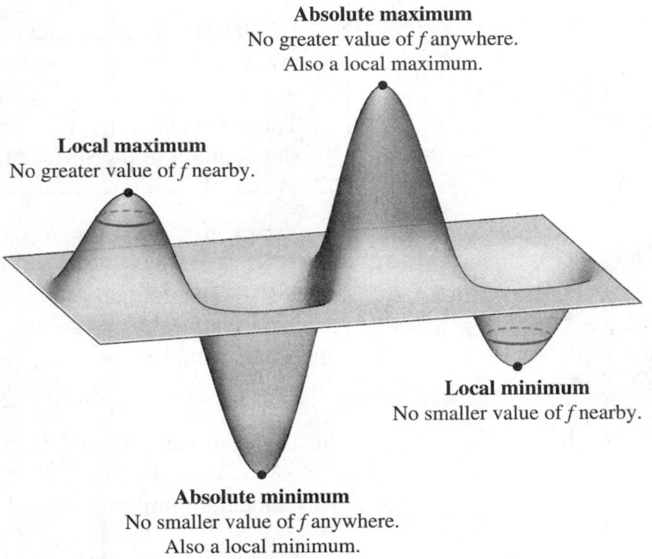

FIGURE 14.44 A local maximum occurs at a mountain peak, and a local minimum occurs at a valley low point.

THEOREM 10—First Derivative Theorem for Local Extreme Values
If $f(x, y)$ has a local maximum or minimum value at an interior point (a, b) of its domain and if the first partial derivatives exist there, then $f_x(a, b) = 0$ and $f_y(a, b) = 0$.

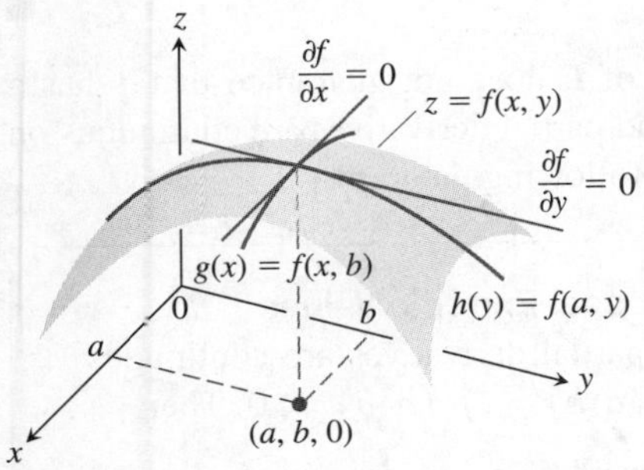

FIGURE 14.45 If a local maximum of f occurs at $x = a$, $y = b$, then the first partial derivatives $f_x(a, b)$ and $f_y(a, b)$ are both zero.

Proof If f has a local extremum at (a, b), then the function $g(x) = f(x, b)$ has a local extremum at $x = a$ (Figure 14.45). Therefore, $g'(a) = 0$ (Chapter 4, Theorem 2). Now $g'(a) = f_x(a, b)$, so $f_x(a, b) = 0$. A similar argument with the function $h(y) = f(a, y)$ shows that $f_y(a, b) = 0$. ■

If we substitute the values $f_x(a, b) = 0$ and $f_y(a, b) = 0$ into the equation

$$f_x(a, b)(x - a) + f_y(a, b)(y - b) - (z - f(a, b)) = 0$$

for the tangent plane to the surface $z = f(x, y)$ at (a, b), the equation reduces to

$$0 \cdot (x - a) + 0 \cdot (y - b) - z + f(a, b) = 0,$$

or

$$z = f(a, b).$$

Thus, Theorem 10 says that the surface does indeed have a horizontal tangent plane at a local extremum, provided there is a tangent plane there.

DEFINITION An interior point of the domain of a function $f(x, y)$ where both f_x and f_y are zero or where one or both of f_x and f_y do not exist is a **critical point** of f.

Theorem 10 says that the only points where a function $f(x, y)$ can assume extreme values are critical points and boundary points. As with differentiable functions of a single variable, not every critical point gives rise to a local extremum. A differentiable function of a single variable might have a point of inflection. A differentiable function of two variables might have a *saddle point*, with the graph of f crossing the tangent plane defined there.

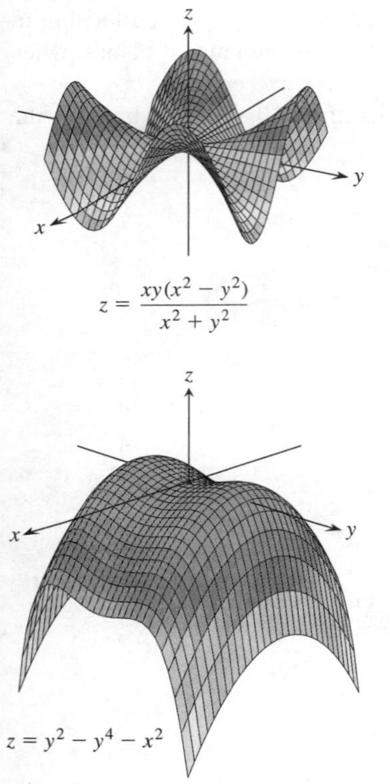

FIGURE 14.46 Saddle points at the origin.

> **DEFINITION** A differentiable function $f(x, y)$ has a **saddle point** at a critical point (a, b) if in every open disk centered at (a, b) there are domain points (x, y) where $f(x, y) > f(a, b)$ and domain points (x, y) where $f(x, y) < f(a, b)$. The corresponding point $(a, b, f(a, b))$ on the surface $z = f(x, y)$ is called a saddle point of the surface (Figure 14.46).

EXAMPLE 1 Find the local extreme values of $f(x, y) = x^2 + y^2 - 4y + 9$.

Solution The domain of f is the entire plane (so there are no boundary points) and the partial derivatives $f_x = 2x$ and $f_y = 2y - 4$ exist everywhere. Therefore, local extreme values can occur only where

$$f_x = 2x = 0 \quad \text{and} \quad f_y = 2y - 4 = 0.$$

The only possibility is the point $(0, 2)$, where the value of f is 5. Since $f(x, y) = x^2 + (y - 2)^2 + 5$ is never less than 5, we see that the critical point $(0, 2)$ gives a local minimum (Figure 14.47). ■

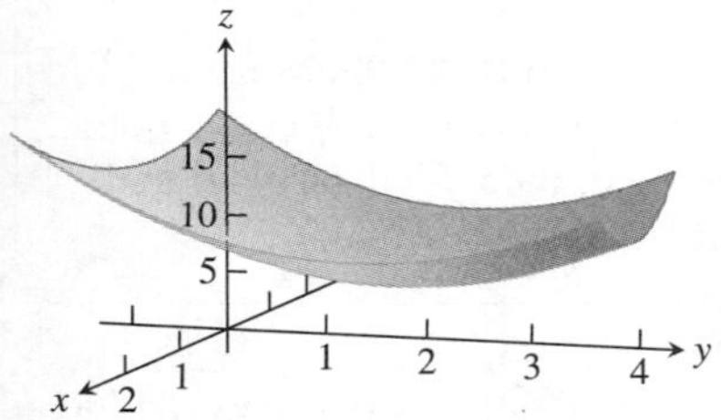

FIGURE 14.47 The graph of the function $f(x, y) = x^2 + y^2 - 4y + 9$ is a paraboloid which has a local minimum value of 5 at the point $(0, 2)$ (Example 1).

EXAMPLE 2 Find the local extreme values (if any) of $f(x, y) = y^2 - x^2$.

Solution The domain of f is the entire plane (so there are no boundary points) and the partial derivatives $f_x = -2x$ and $f_y = 2y$ exist everywhere. Therefore, local extrema can occur only at the origin $(0, 0)$, where $f_x = 0$ and $f_y = 0$. The value of f at the origin is 0. However, away from the origin along the positive x-axis, f has the value $f(x, 0) = -x^2 < 0$; along the positive y-axis, f has the value $f(0, y) = y^2 > 0$. Therefore, every open disk in the xy-plane centered at $(0, 0)$ contains points where the function is positive and points where it is negative. The function has a saddle point at the origin and no local extreme values (Figure 14.48a). Figure 14.48b displays the level curves (they are hyperbolas) of f and shows the function decreasing and increasing in an alternating fashion among the groupings of hyperbolas. ■

That $f_x = f_y = 0$ at an interior point (a, b) of R does not guarantee that f has a local extreme value there. If f and its first and second partial derivatives are continuous on R, however, we may be able to learn more from the following theorem.

> **THEOREM 11—Second Derivative Test for Local Extreme Values**
> Suppose that $f(x, y)$ and its first and second partial derivatives are continuous throughout a disk centered at (a, b) and that $f_x(a, b) = f_y(a, b) = 0$. Then
>
> **i)** f has a **local maximum** at (a, b) if $f_{xx} < 0$ and $f_{xx}f_{yy} - f_{xy}^2 > 0$ at (a, b).
> **ii)** f has a **local minimum** at (a, b) if $f_{xx} > 0$ and $f_{xx}f_{yy} - f_{xy}^2 > 0$ at (a, b).
> **iii)** f has a **saddle point** at (a, b) if $f_{xx}f_{yy} - f_{xy}^2 < 0$ at (a, b).
> **iv)** **the test is inconclusive** at (a, b) if $f_{xx}f_{yy} - f_{xy}^2 = 0$ at (a, b). In this case, we must find some other way to determine the behavior of f at (a, b).

The expression $f_{xx}f_{yy} - f_{xy}^2$ is called the **discriminant** or **Hessian** of f. It is sometimes easier to remember it in determinant form,

$$f_{xx}f_{yy} - f_{xy}^2 = \begin{vmatrix} f_{xx} & f_{xy} \\ f_{xy} & f_{yy} \end{vmatrix}.$$

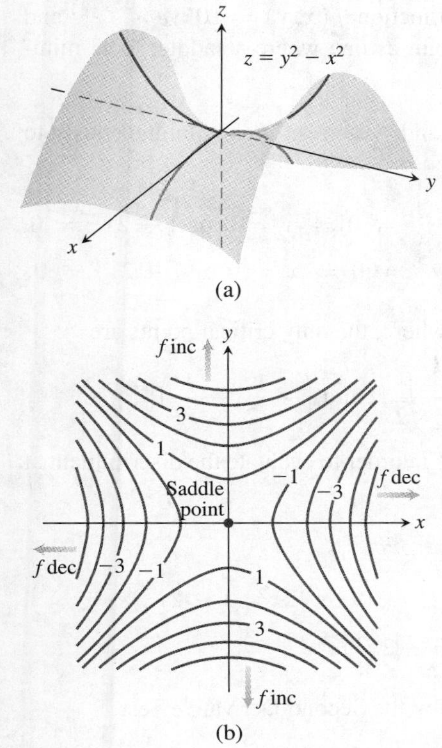

FIGURE 14.48 (a) The origin is a saddle point of the function $f(x, y) = y^2 - x^2$. There are no local extreme values (Example 2). (b) Level curves for the function f in Example 2.

The discriminant is the determinant of the **Hessian matrix** of f,

$$Hf(x, y) = \begin{bmatrix} f_{xx} & f_{xy} \\ f_{yx} & f_{yy} \end{bmatrix}.$$

Note that by Theorem 2, we have $f_{xy}(a, b) = f_{yx}(a, b)$ at any point (a, b) satisfying the assumptions of Theorem 11.

Theorem 11 says that if the discriminant is positive at the point (a, b), then the surface curves the same way in all directions: downward if $f_{xx} < 0$, giving rise to a local maximum, and upward if $f_{xx} > 0$, giving a local minimum. On the other hand, if the discriminant is negative at (a, b), then the surface curves up in some directions and down in others, so we have a saddle point.

EXAMPLE 3 Find the local extreme values of the function

$$f(x, y) = xy - x^2 - y^2 - 2x - 2y + 4.$$

Solution The function is defined and differentiable for all x and y, and its domain has no boundary points. The function therefore has extreme values only at the points where f_x and f_y are simultaneously zero. This leads to

$$f_x = y - 2x - 2 = 0, \qquad f_y = x - 2y - 2 = 0,$$

or

$$x = y = -2.$$

Therefore, the point $(-2, -2)$ is the only point where f may take on an extreme value. To see whether it does so, we calculate

$$f_{xx} = -2, \qquad f_{yy} = -2, \qquad f_{xy} = 1.$$

The discriminant of f at $(a, b) = (-2, -2)$ is

$$f_{xx}f_{yy} - f_{xy}{}^2 = (-2)(-2) - (1)^2 = 4 - 1 = 3.$$

The combination

$$f_{xx} < 0 \quad \text{and} \quad f_{xx}f_{yy} - f_{xy}{}^2 > 0$$

tells us that f has a local maximum at $(-2, -2)$. The value of f at this point is $f(-2, -2) = 8$. ■

EXAMPLE 4 Find the local extreme values of $f(x, y) = 3y^2 - 2y^3 - 3x^2 + 6xy$.

Solution Since f is differentiable everywhere, it can assume extreme values only where

$$f_x = 6y - 6x = 0 \quad \text{and} \quad f_y = 6y - 6y^2 + 6x = 0.$$

From the first of these equations we find $x = y$, and substitution for y into the second equation then gives

$$6x - 6x^2 + 6x = 0 \quad \text{or} \quad 6x(2 - x) = 0.$$

The two critical points are therefore $(0, 0)$ and $(2, 2)$.

To classify the critical points, we calculate the second derivatives:

$$f_{xx} = -6, \qquad f_{yy} = 6 - 12y, \qquad f_{xy} = 6.$$

The discriminant is given by

$$f_{xx}f_{yy} - f_{xy}{}^2 = (-36 + 72y) - 36 = 72(y - 1).$$

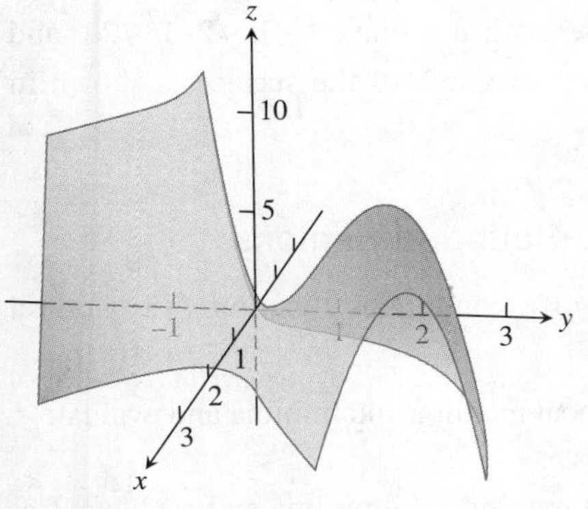

FIGURE 14.49 The surface $z = 3y^2 - 2y^3 - 3x^2 + 6xy$ has a saddle point at the origin and a local maximum at the point $(2, 2)$ (Example 4).

At the critical point $(0, 0)$ we see that the value of the discriminant is the negative number -72, so the function has a saddle point at the origin. At the critical point $(2, 2)$ we see that the discriminant has the positive value 72. Combining this result with the negative value of the second partial $f_{xx} = -6$, Theorem 11 says that the critical point $(2, 2)$ gives a local maximum value of $f(2, 2) = 12 - 16 - 12 + 24 = 8$. A graph of the surface is shown in Figure 14.49. ■

EXAMPLE 5 Find the critical points of the function $f(x, y) = 10xye^{-(x^2+y^2)}$ and use the Second Derivative Test to classify each point as one where a saddle, local minimum, or local maximum occurs.

Solution First we find the partial derivatives f_x and f_y and set them simultaneously to zero in seeking the critical points:

$$f_x = 10ye^{-(x^2+y^2)} - 20x^2ye^{-(x^2+y^2)} = 10y(1 - 2x^2)e^{-(x^2+y^2)} = 0 \Rightarrow y = 0 \text{ or } 1 - 2x^2 = 0,$$
$$f_y = 10xe^{-(x^2+y^2)} - 20xy^2e^{-(x^2+y^2)} = 10x(1 - 2y^2)e^{-(x^2+y^2)} = 0 \Rightarrow x = 0 \text{ or } 1 - 2y^2 = 0.$$

Since both partial derivatives are continuous everywhere, the only critical points are

$$(0, 0), \left(\frac{1}{\sqrt{2}}, \frac{1}{\sqrt{2}}\right), \left(-\frac{1}{\sqrt{2}}, \frac{1}{\sqrt{2}}\right), \left(\frac{1}{\sqrt{2}}, -\frac{1}{\sqrt{2}}\right), \text{ and } \left(-\frac{1}{\sqrt{2}}, -\frac{1}{\sqrt{2}}\right).$$

Next we calculate the second partial derivatives in order to evaluate the discriminant at each critical point:

$$f_{xx} = -20xy(1 - 2x^2)e^{-(x^2+y^2)} - 40xye^{-(x^2+y^2)} = -20xy(3 - 2x^2)e^{-(x^2+y^2)},$$
$$f_{xy} = f_{yx} = 10(1 - 2x^2)e^{-(x^2+y^2)} - 20y^2(1 - 2x^2)e^{-(x^2+y^2)} = 10(1 - 2x^2)(1 - 2y^2)e^{-(x^2+y^2)},$$
$$f_{yy} = -20xy(1 - 2y^2)e^{-(x^2+y^2)} - 40xye^{-(x^2+y^2)} = -20xy(3 - 2y^2)e^{-(x^2+y^2)}.$$

The following table summarizes the values needed by the Second Derivative Test.

Critical Point	f_{xx}	f_{xy}	f_{yy}	Discriminant D
$(0, 0)$	0	10	0	-100
$\left(\frac{1}{\sqrt{2}}, \frac{1}{\sqrt{2}}\right)$	$-\frac{20}{e}$	0	$-\frac{20}{e}$	$\frac{400}{e^2}$
$\left(-\frac{1}{\sqrt{2}}, \frac{1}{\sqrt{2}}\right)$	$\frac{20}{e}$	0	$\frac{20}{e}$	$\frac{400}{e^2}$
$\left(\frac{1}{\sqrt{2}}, -\frac{1}{\sqrt{2}}\right)$	$\frac{20}{e}$	0	$\frac{20}{e}$	$\frac{400}{e^2}$
$\left(-\frac{1}{\sqrt{2}}, -\frac{1}{\sqrt{2}}\right)$	$-\frac{20}{e}$	0	$-\frac{20}{e}$	$\frac{400}{e^2}$

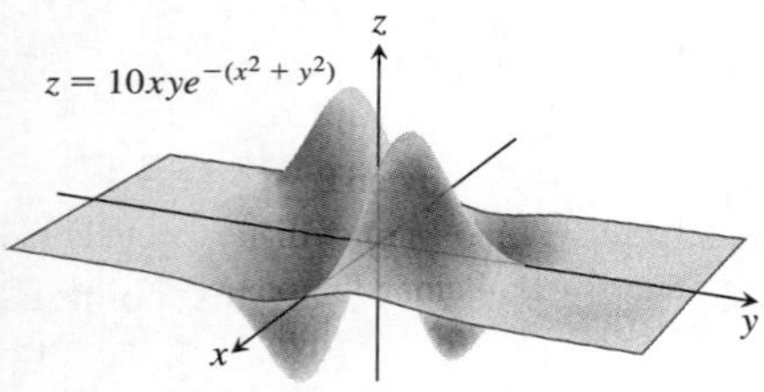

FIGURE 14.50 A graph of the function in Example 5.

From the table we find that $D < 0$ at the critical point $(0, 0)$, giving a saddle; $D > 0$ and $f_{xx} < 0$ at the critical points $(1/\sqrt{2}, 1/\sqrt{2})$ and $(-1/\sqrt{2}, -1/\sqrt{2})$, giving local maximum values there; and $D > 0$ and $f_{xx} > 0$ at the critical points $(-1/\sqrt{2}, 1/\sqrt{2})$ and $(1/\sqrt{2}, -1/\sqrt{2})$, each giving local minimum values. A graph of the surface is shown in Figure 14.50. ■

Absolute Maxima and Minima on Closed Bounded Regions

We organize the search for the absolute extrema of a continuous function $f(x, y)$ on a closed and bounded region R into three steps.

1. *List the interior points of R* where f may have local maxima and minima and evaluate f at these points. These are the critical points of f.
2. *List the boundary points of R* where f has local maxima and minima and evaluate f at these points. We show how to do this in the next example.
3. *Look through the lists* for the maximum and minimum values of f. These will be the absolute maximum and minimum values of f on R.

EXAMPLE 6 Find the absolute maximum and minimum values of

$$f(x, y) = 2 + 2x + 4y - x^2 - y^2$$

on the triangular region in the first quadrant bounded by the lines $x = 0$, $y = 0$, and $y = 9 - x$.

Solution Since f is differentiable, the only places where f can assume these values are points inside the triangle where $f_x = f_y = 0$ and points on the boundary (Figure 14.51a).

(a) Interior points. For these we have

$$f_x = 2 - 2x = 0, \qquad f_y = 4 - 2y = 0,$$

yielding the single point $(x, y) = (1, 2)$. The value of f there is

$$f(1, 2) = 7.$$

(b) Boundary points. We take the triangle one side at a time:

i) On the segment OA we always have $y = 0$. Therefore, we can regard $f(x, y)$ as being solely a function of x on this segment. That is, on this segment we want to consider the function

$$g(x) = f(x, 0) = 2 + 2x - x^2$$

for $0 \le x \le 9$. Its extreme values (as we know from Chapter 4) may occur at the endpoints

$$x = 0 \quad \text{where} \quad g(0) = f(0, 0) = 2$$
$$x = 9 \quad \text{where} \quad g(9) = f(9, 0) = 2 + 18 - 81 = -61$$

or at the interior points where $g'(x) = 2 - 2x = 0$. The only interior point where $g'(x) = 0$ is $x = 1$, where

$$g(1) = f(1, 0) = 3.$$

ii) On the segment OB we always have $x = 0$. Therefore, on this segment we can regard $f(x, y)$ as being solely a function of y, and so we consider the function

$$h(y) = f(0, y) = 2 + 4y - y^2$$

on the closed interval $[0, 9]$. Its extreme values can occur at the endpoints or at interior points where $h'(y) = 0$. Since $h'(y) = 4 - 2y$, the only interior point where $h'(y) = 0$ occurs at $(0, 2)$, with $h(2) = 6$. So the candidates for this segment are

$$h(0) = f(0, 0) = 2, \quad h(9) = f(0, 9) = -43, \quad \text{and} \quad h(2) = f(0, 2) = 6.$$

iii) We have already accounted for the values of f at the endpoints of AB, so we need only look at the interior points of the line segment AB. On this segment we have $y = 9 - x$, so we consider the function

$$k(x) = f(x, 9 - x) = 2 + 2x + 4(9 - x) - x^2 - (9 - x)^2 = -43 + 16x - 2x^2.$$

Setting $k'(x) = 16 - 4x = 0$ gives

$$x = 4.$$

At this value of x,

$$y = 9 - 4 = 5 \quad \text{and} \quad k(4) = f(4, 5) = -11.$$

Summary We list all the function value candidates: $7, 2, -61, 3, -43, 6, -11$. The maximum is 7, which f assumes at $(1, 2)$. The minimum is -61, which f assumes at $(9, 0)$. See Figure 14.51b. ■

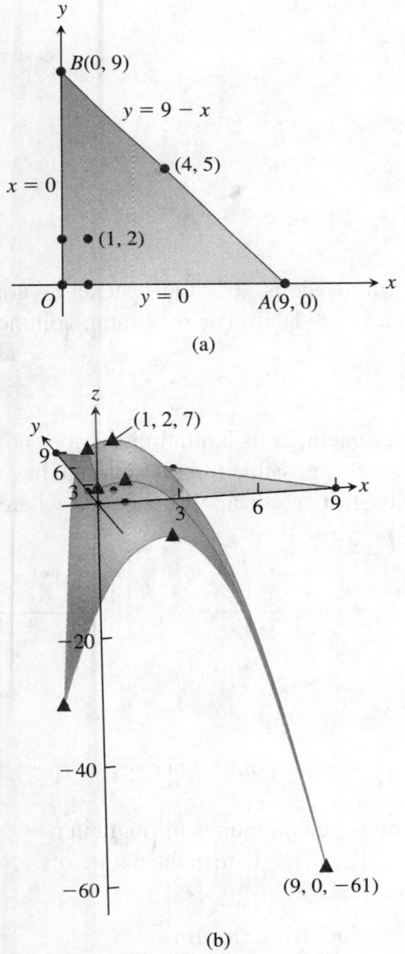

FIGURE 14.51 (a) This triangular region is the domain of the function in Example 6. (b) The graph of the function in Example 6. The triangular points are the candidates for maxima or minima.

Solving extreme value problems with algebraic constraints on the variables usually requires the method of Lagrange multipliers, which is introduced in the next section. But sometimes we can solve such problems directly, as in the next example.

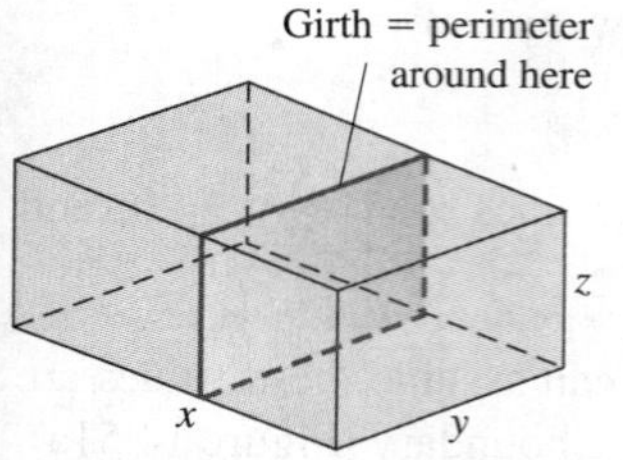

FIGURE 14.52 The box in Example 7.

EXAMPLE 7 A delivery company accepts only rectangular boxes the sum of whose length and girth (perimeter of a cross-section) does not exceed 108 in. Find the dimensions of an acceptable box of largest volume.

Solution Let x, y, and z represent the length, width, and height of the rectangular box, respectively. Then the girth is $2y + 2z$. We want to maximize the volume $V = xyz$ of the box (Figure 14.52) satisfying $x + 2y + 2z = 108$ (the largest box accepted by the delivery company). Thus, we can write the volume of the box as a function of two variables:

$$\begin{aligned} V(y, z) &= (108 - 2y - 2z)yz \qquad \text{$V = xyz$ and $x = 108 - 2y - 2z$} \\ &= 108yz - 2y^2z - 2yz^2. \end{aligned}$$

Setting the first partial derivatives equal to zero,

$$\begin{aligned} V_y(y, z) &= 108z - 4yz - 2z^2 = (108 - 4y - 2z)z = 0 \\ V_z(y, z) &= 108y - 2y^2 - 4yz = (108 - 2y - 4z)y = 0, \end{aligned}$$

gives the critical points $(0, 0)$, $(0, 54)$, $(54, 0)$, and $(18, 18)$. The volume is zero at $(0, 0)$, $(0, 54)$, and $(54, 0)$, which are not maximum values. At the point $(18, 18)$, we apply the Second Derivative Test (Theorem 11):

$$V_{yy} = -4z, \qquad V_{zz} = -4y, \qquad V_{yz} = 108 - 4y - 4z.$$

Then

$$V_{yy}V_{zz} - V_{yz}^2 = 16yz - 16(27 - y - z)^2.$$

Thus,

$$V_{yy}(18, 18) = -4(18) < 0$$

and

$$\left.\left(V_{yy}V_{zz} - V_{yz}^2\right)\right|_{(18,18)} = 16(18)(18) - 16(-9)^2 > 0,$$

so $(18, 18)$ gives a maximum volume. The dimensions of the package are $x = 108 - 2(18) - 2(18) = 36$ in., $y = 18$ in., and $z = 18$ in. The maximum volume is $V = (36)(18)(18) = 11{,}664\ \text{in}^3$, or $6.75\ \text{ft}^3$. ■

Despite the power of Theorem 11, we urge you to remember its limitations. It does not apply to boundary points of a function's domain, where it is possible for a function to have extreme values along with nonzero derivatives. Also, it does not apply to points where either f_x or f_y fails to exist.

Summary of Max-Min Tests

The extreme values of $f(x, y)$ can occur only at

i) boundary points of the domain of f

ii) critical points (interior points where $f_x = f_y = 0$ or points where f_x or f_y fails to exist)

If the first- and second-order partial derivatives of f are continuous throughout a disk centered at a point (a, b) and if $f_x(a, b) = f_y(a, b) = 0$, then the nature of $f(a, b)$ can be tested with the **Second Derivative Test**:

i) $f_{xx} < 0$ and $f_{xx}f_{yy} - f_{xy}^2 > 0$ at (a, b) $\Rightarrow$ **local maximum**

ii) $f_{xx} > 0$ and $f_{xx}f_{yy} - f_{xy}^2 > 0$ at (a, b) $\Rightarrow$ **local minimum**

iii) $f_{xx}f_{yy} - f_{xy}^2 < 0$ at (a, b) $\Rightarrow$ **saddle point**

iv) $f_{xx}f_{yy} - f_{xy}^2 = 0$ at (a, b) $\Rightarrow$ **test is inconclusive**

Finding maximum and minimum values for functions of more than two variables is an important problem with many important applications, from machine learning to making economic predictions. The problem becomes much harder as the number of variables increases. The process of finding extrema for functions with high-dimensional domains is discussed in Appendices B.2 and B.3.

EXERCISES 14.7

Finding Local Extrema

Find all the local maxima, local minima, and saddle points of the functions in Exercises 1–30.

1. $f(x, y) = x^2 + xy + y^2 + 3x - 3y + 4$
2. $f(x, y) = x^3 + y^3 - 12x - 3y + 20$
3. $f(x, y) = x^2 + xy + 3x + 2y + 5$
4. $f(x, y) = x^4 + y^4 - 2x^2 - 2y^2 + 4xy$
5. $f(x, y) = 2xy - x^2 - 2y^2 + 3x + 4$
6. $f(x, y) = x^2 - 4xy + y^2 + 6y + 2$
7. $f(x, y) = 2x^2 + 3xy + 4y^2 - 5x + 2y$
8. $f(x, y) = x^2 - 2xy + 2y^2 - 2x + 2y + 1$
9. $f(x, y) = x^2 - y^2 - 2x + 4y + 6$
10. $f(x, y) = x^4 + y^2 + x^2y$
11. $f(x, y) = \sqrt{56x^2 - 8y^2 - 16x - 31} + 1 - 8x$
12. $f(x, y) = x^3 + y^3 - 3xy$
13. $f(x, y) = x^3 - y^3 - 2xy + 6$
14. $f(x, y) = x^3 + 3xy + y^3$
15. $f(x, y) = 6x^2 - 2x^3 + 3y^2 + 6xy$
16. $f(x, y) = x^3 + y^3 + 3x^2 - 3y^2 - 8$
17. $f(x, y) = x^3 + 3xy^2 - 15x + y^3 - 15y$
18. $f(x, y) = 2x^3 + 2y^3 - 9x^2 + 3y^2 - 12y$
19. $f(x, y) = 4xy - x^4 - y^4$
20. $f(x, y) = x^4 + y^4 + 4xy$
21. $f(x, y) = \sin x \sin y \sin(x + y)$, $0 < x, y < \pi$
22. $f(x, y) = \dfrac{1}{x} + xy + \dfrac{1}{y}$
23. $f(x, y) = y \sin x$
24. $f(x, y) = e^{2x} \cos y$
25. $f(x, y) = e^{x^2+y^2-4x}$
26. $f(x, y) = e^y - ye^x$
27. $f(x, y) = e^{-y}(x^2 + y^2)$
28. $f(x\ y) = e^x(x^2 - y^2)$
29. $f(x, y) = 2 \ln x + \ln y - 4x - y$
30. $f(x, y) = \ln(x + y) + x^2 - y$

Finding Absolute Extrema

In Exercises 31–38, find the absolute maxima and minima of the functions on the given domains.

31. $f(x, y) = 2x^2 - 4x + y^2 - 4y + 1$ on the closed triangular plate bounded by the lines $x = 0$, $y = 2$, $y = 2x$ in the first quadrant
32. $D(x, y) = x^2 - xy + y^2 + 1$ on the closed triangular plate in the first quadrant bounded by the lines $x = 0$, $y = 4$, $y = x$
33. $f(x, y) = x^2 + y^2$ on the closed triangular plate bounded by the lines $x = 0$, $y = 0$, $y + 2x = 2$ in the first quadrant
34. $T(x, y) = x^2 + xy + y^2 - 6x$ on the rectangular plate $0 \le x \le 5, -3 \le y \le 3$
35. $T(x, y) = x^2 + xy + y^2 - 6x + 2$ on the rectangular plate $0 \le x \le 5, -3 \le y \le 0$
36. $f(x, y) = 48xy - 32x^3 - 24y^2$ on the rectangular plate $0 \le x \le 1, 0 \le y \le 1$
37. $f(x, y) = (4x - x^2)\cos y$ on the rectangular plate $1 \le x \le 3, -\pi/4 \le y \le \pi/4$
38. $f(x, y) = 4x - 8xy + 2y + 1$ on the triangular plate bounded by the lines $x = 0$, $y = 0$, $x + y = 1$ in the first quadrant
39. Find two numbers a and b with $a \le b$ such that
$$\int_a^b (6 - x - x^2)\, dx$$
has its largest value.
40. Find two numbers a and b with $a \le b$ such that
$$\int_a^b (24 - 2x - x^2)^{1/3}\, dx$$
has its largest value.
41. **Temperatures** A flat circular plate has the shape of the region $x^2 + y^2 \le 1$. The plate, including the boundary where $x^2 + y^2 = 1$, is heated so that the temperature at the point (x, y) is
$$T(x, y) = x^2 + 2y^2 - x.$$
Find the temperatures at the hottest and coldest points on the plate.
42. Find the critical point of
$$f(x, y) = xy + 2x - \ln x^2y$$
in the open first quadrant $(x > 0, y > 0)$ and show that f takes on a minimum there.

Theory and Examples

43. Find the maxima, minima, and saddle points of $f(x, y)$, if any, given that
    a. $f_x = 2x - 4y$ and $f_y = 2y - 4x$
    b. $f_x = 2x - 2$ and $f_y = 2y - 4$
    c. $f_x = 9x^2 - 9$ and $f_y = 2y + 4$

    Describe your reasoning in each case.
44. The discriminant $f_{xx}f_{yy} - f_{xy}^2$ is zero at the origin for each of the following functions, so the Second Derivative Test fails there. Determine whether the function has a maximum, a minimum, or neither at the origin by imagining what the surface $z = f(x, y)$ looks like. Describe your reasoning in each case.
    a. $f(x, y) = x^2y^2$
    b. $f(x, y) = 1 - x^2y^2$
    c. $f(x, y) = xy^2$
    d. $f(x, y) = x^3y^2$
    e. $f(x, y) = x^3y^3$
    f. $f(x, y) = x^4y^4$
45. Show that $(0, 0)$ is a critical point of $f(x, y) = x^2 + kxy + y^2$ no matter what value the constant k has. (*Hint:* Consider two cases: $k = 0$ and $k \ne 0$.)

46. For what values of the constant k does the Second Derivative Test guarantee that $f(x, y) = x^2 + kxy + y^2$ will have a saddle point at $(0, 0)$? A local minimum at $(0, 0)$? For what values of k is the Second Derivative Test inconclusive? Give reasons for your answers.

47. If $f_x(a, b) = f_y(a, b) = 0$, must f have a local maximum or minimum value at (a, b)? Give reasons for your answer.

48. Can you conclude anything about $f(a, b)$ if f and its first and second partial derivatives are continuous throughout a disk centered at the critical point (a, b) and $f_{xx}(a, b)$ and $f_{yy}(a, b)$ differ in sign? Give reasons for your answer.

49. Among all the points on the graph of $z = 10 - x^2 - y^2$ that lie above the plane $x + 2y + 3z = 0$, find the point farthest from the plane.

50. Find the minimum value of $S : xy + 2xz + 2yz$ where $xyz = 32$, $x, y, z > 0$.

51. Find the point on the plane $3x + 2y + z = 6$ that is nearest the origin.

52. Find the minimum distance from the point $(2, -1, 1)$ to the plane $x + y - z = 2$.

53. Find three numbers whose sum is 9 and whose sum of squares is a minimum.

54. Find three positive numbers whose sum is 3 and whose product is a maximum.

55. Find the maximum value of $s = xy + yz + xz$ where $x + y + z = 6$.

56. Find the minimum distance from the cone $z = \sqrt{x^2 + y^2}$ to the point $(-6, 4, 0)$.

57. Find the dimensions of the rectangular box of maximum volume that can be inscribed inside the sphere $x^2 + y^2 + z^2 = 4$.

58. Among all closed rectangular boxes of volume 27 cm^3, what is the smallest surface area?

59. You are to construct an open rectangular box from 12 ft^2 of material. What dimensions will result in a box of maximum volume?

60. Consider the function $f(x, y) = x^2 + y^2 + 2xy - x - y + 1$ over the square $0 \le x \le 1$ and $0 \le y \le 1$.

a. Show that f has an absolute minimum along the line segment $2x + 2y = 1$ in this square. What *is* the absolute minimum value?

b. Find the absolute maximum value of f over the square.

61. Find the point on the graph of $y^2 - xz^2 = 4$ nearest the origin.

62. A rectangular box is inscribed in the region in the first octant bounded above by the plane with x-intercept 6, y-intercept 6, and z-intercept 6.

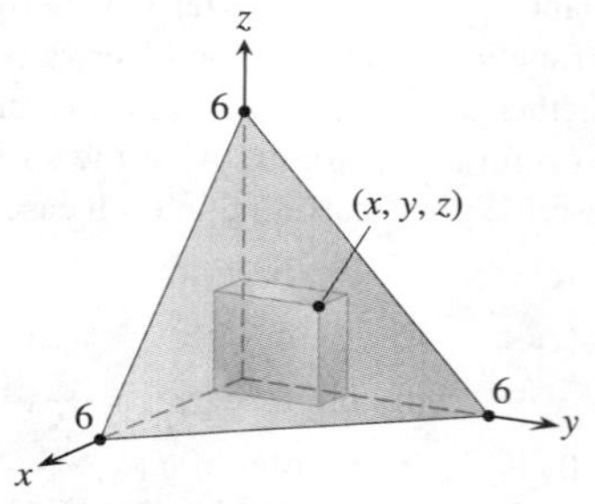

a. Find an equation for the plane.

b. Find the dimensions of the box of maximum volume.

Extreme Values on Parametrized Curves To find the extreme values of a function $f(x, y)$ on a curve $x = x(t)$, $y = y(t)$, we treat f as a function of the single variable t and use the Chain Rule to find where df/dt is zero. As in any other single-variable case, the extreme values of f are then found among the values at

a. The critical points (points where df/dt is zero or fails to exist), and

b. The endpoints of the parameter domain.

In Exercises 63–66, find the absolute maximum and minimum values of the following functions on the given curves.

63. Functions:

a. $f(x, y) = x + y$ **b.** $g(x, y) = xy$

c. $h(x, y) = 2x^2 + y^2$

Curves:

i) The semicircle $x^2 + y^2 = 4$, $y \ge 0$

ii) The quarter circle $x^2 + y^2 = 4$, $x \ge 0$, $y \ge 0$

Use the parametric equations $x = 2\cos t$, $y = 2\sin t$.

64. Functions:

a. $f(x, y) = 2x + 3y$ **b.** $g(x, y) = xy$

c. $h(x, y) = x^2 + 3y^2$

Curves:

i) The semiellipse $(x^2/9) + (y^2/4) = 1$, $y \ge 0$

ii) The quarter ellipse $(x^2/9) + (y^2/4) = 1$, $x \ge 0$, $y \ge 0$

Use the parametric equations $x = 3\cos t$, $y = 2\sin t$.

65. Function: $f(x, y) = xy$

Curves:

i) The line $x = 2t$, $y = t + 1$

ii) The line segment $x = 2t$, $y = t + 1$, $-1 \le t \le 0$

iii) The line segment $x = 2t$, $y = t + 1$, $0 \le t \le 1$

66. Functions:

a. $f(x, y) = x^2 + y^2$ **b.** $g(x, y) = 1/(x^2 + y^2)$

Curves:

i) The line $x = t$, $y = 2 - 2t$

ii) The line segment $x = t$, $y = 2 - 2t$, $0 \le t \le 1$

67. Least squares and regression lines When we try to fit a line $y = mx + b$ to a set of numerical data points $(x_1, y_1), (x_2, y_2), \dots, (x_n, y_n)$, we usually choose the line that minimizes the sum of the squares of the vertical distances from the points to the line. In theory, this means finding the values of m and b that minimize the value of the function

$$w = (mx_1 + b - y_1)^2 + \cdots + (mx_n + b - y_n)^2. \quad (1)$$

(See the accompanying figure.) Show that the values of m and b that do this are

$$m = \frac{\left(\sum x_k\right)\left(\sum y_k\right) - n\sum x_k y_k}{\left(\sum x_k\right)^2 - n\sum x_k^{\,2}}, \quad (2)$$

$$b = \frac{1}{n}\left(\sum y_k - m\sum x_k\right), \quad (3)$$

with all sums running from $k = 1$ to $k = n$. Many scientific calculators have these formulas built in, enabling you to find m and b with only a few keystrokes after you have entered the data.

The line $y = mx + b$ determined by these values of m and b is called the **least squares line, regression line,** or **trend line** for the data under study. Finding a least squares line lets you

1. summarize data with a simple expression,
2. predict values of y for other, experimentally untried values of x,
3. handle data analytically.

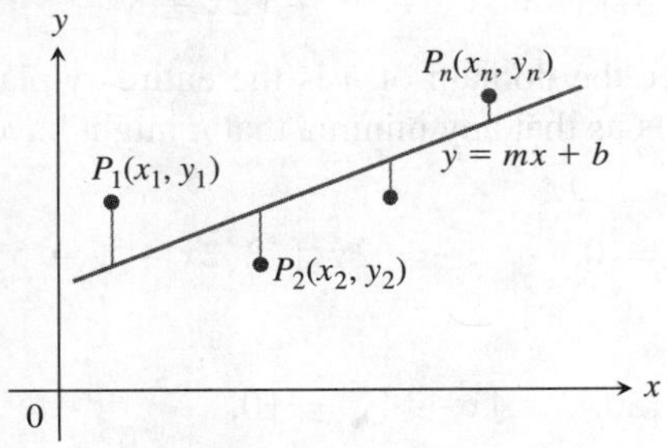

In Exercises 68–70, use Equations (2) and (3) to find the least squares line for each set of data points. Then use the linear equation you obtain to predict the value of y that would correspond to $x = 4$.

68. $(-2, 0), (0, 2), (2, 3)$
69. $(-1, 2), (0, 1), (3, -4)$
70. $(0, 0), (1, 2), (2, 3)$

COMPUTER EXPLORATIONS

In Exercises 71–76, you will explore functions to identify their local extrema. Use a CAS to perform the following steps:

a. Plot the function over the given rectangle.

b. Plot some level curves in the rectangle.

c. Calculate the function's first partial derivatives and use the CAS equation solver to find the critical points. How are the critical points related to the level curves plotted in part (b)? Which critical points, if any, appear to give a saddle point? Give reasons for your answer.

d. Calculate the function's second partial derivatives and find the discriminant $f_{xx}f_{yy} - f_{xy}{}^2$.

e. Using the max-min tests, classify the critical points found in part (c). Are your findings consistent with your discussion in part (c)?

71. $f(x, y) = x^2 + y^3 - 3xy, \quad -5 \le x \le 5, \quad -5 \le y \le 5$

72. $f(x, y) = x^3 - 3xy^2 + y^2, \quad -2 \le x \le 2, \quad -2 \le y \le 2$

73. $f(x, y) = x^4 + y^2 - 8x^2 - 6y + 16, \quad -3 \le x \le 3, \quad -6 \le y \le 6$

74. $f(x, y) = 2x^4 + y^4 - 2x^2 - 2y^2 + 3, \quad -3/2 \le x \le 3/2, \quad -3/2 \le y \le 3/2$

75. $f(x, y) = 5x^6 + 18x^5 - 30x^4 + 30xy^2 - 120x^3, \quad -4 \le x \le 3, \quad -2 \le y \le 2$

76. $f(x, y) = \begin{cases} x^5 \ln(x^2 + y^2), & (x, y) \ne (0, 0) \\ 0, & (x, y) = (0, 0), \end{cases}$
$-2 \le x \le 2, \quad -2 \le y \le 2$

14.8 Lagrange Multipliers

HISTORICAL BIOGRAPHY
Joseph Louis Lagrange
(1736–1813)
www.bit.ly/2NdfJs7

Sometimes we need to find the extreme values of a function whose domain is constrained to lie within some particular subset of the plane—for example, a disk, a closed triangular region, or along a curve. We saw an instance of this situation in Example 6 of the previous section. Here we explore a powerful method for finding extreme values of constrained functions: the method of *Lagrange multipliers*.

Constrained Maxima and Minima

To gain some insight, we first consider a problem where a constrained minimum can be found by eliminating a variable.

EXAMPLE 1 Find the point $p(x, y, z)$ on the plane $2x + y - z - 5 = 0$ that is closest to the origin.

Solution The problem asks us to find the minimum value of the function

$$|\overrightarrow{OP}| = \sqrt{(x - 0)^2 + (y - 0)^2 + (z - 0)^2} = \sqrt{x^2 + y^2 + z^2}$$

subject to the constraint that

$$2x + y - z - 5 = 0.$$

Since $|\overrightarrow{OP}|$ has a minimum value wherever the function

$$f(x, y, z) = x^2 + y^2 + z^2$$

has a minimum value, we may solve the problem by finding the minimum value of $f(x, y, z)$ subject to the constraint $2x + y - z - 5 = 0$ (thus avoiding square roots). If we regard x and y as the independent variables in this equation and write z as

$$z = 2x + y - 5,$$

our problem reduces to finding the points (x, y) at which the function

$$h(x, y) = f(x, y, 2x + y - 5) = x^2 + y^2 + (2x + y - 5)^2$$

has its minimum value or values. Since the domain of h is the entire xy-plane, the First Derivative Theorem of Section 14.7 tells us that any minima that h might have must occur at points where

$$h_x = 2x + 2(2x + y - 5)(2) = 0, \qquad h_y = 2y + 2(2x + y - 5) = 0.$$

This leads to

$$10x + 4y = 20, \qquad 4x + 4y = 10,$$

which has the solution

$$x = \frac{5}{3}, \qquad y = \frac{5}{6}.$$

We may apply a geometric argument together with the Second Derivative Test to show that these values minimize h. The z-coordinate of the corresponding point on the plane $z = 2x + y - 5$ is

$$z = 2\left(\frac{5}{3}\right) + \frac{5}{6} - 5 = -\frac{5}{6}.$$

Therefore, the point we seek is

$$\text{Closest point:} \qquad P\left(\frac{5}{3}, \frac{5}{6}, -\frac{5}{6}\right).$$

The distance from P to the origin is $5/\sqrt{6} \approx 2.04$. ■

Attempts to solve a constrained maximum or minimum problem by substitution, as we might call the method of Example 1, do not always go smoothly.

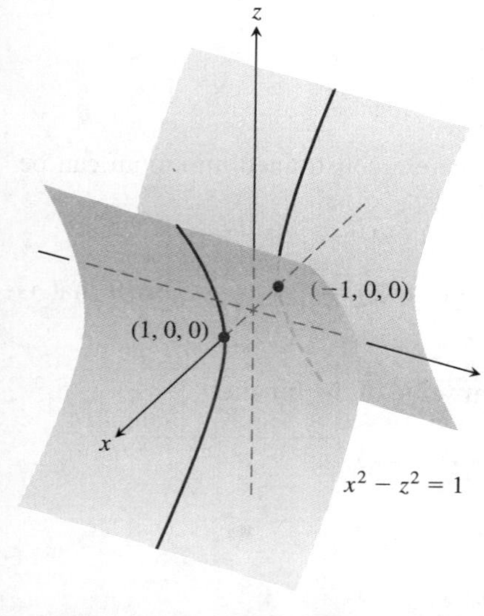

FIGURE 14.53 The hyperbolic cylinder $x^2 - z^2 - 1 = 0$ in Example 2.

EXAMPLE 2 Find the points on the hyperbolic cylinder $x^2 - z^2 - 1 = 0$ that are closest to the origin.

Solution 1 The cylinder is shown in Figure 14.53. We seek the points on the cylinder closest to the origin. These are the points whose coordinates minimize the value of the function

$$f(x, y, z) = x^2 + y^2 + z^2 \qquad \text{Square of the distance}$$

subject to the constraint that $x^2 - z^2 - 1 = 0$. If we regard x and y as independent variables in the constraint equation, then

$$z^2 = x^2 - 1,$$

and the values of $f(x, y, z) = x^2 + y^2 + z^2$ on the cylinder are given by the function

$$h(x, y) = x^2 + y^2 + (x^2 - 1) = 2x^2 + y^2 - 1.$$

To find the points on the cylinder whose coordinates minimize f, we look for the points in the xy-plane whose coordinates minimize h. The only extreme value of h occurs where

$$h_x = 4x = 0 \quad \text{and} \quad h_y = 2y = 0,$$

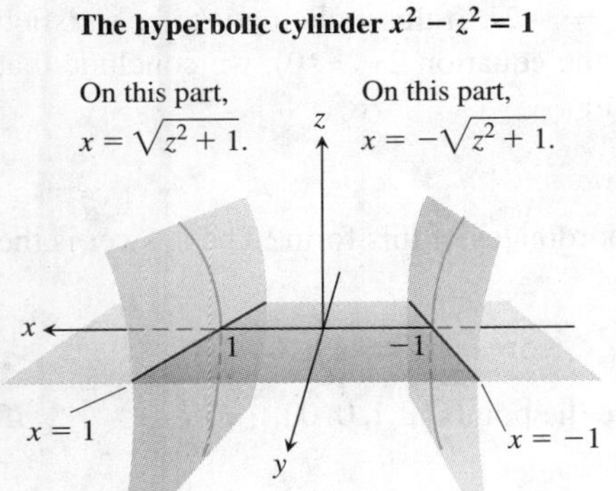

FIGURE 14.54 The region in the xy-plane from which the first two coordinates of the points (x, y, z) on the hyperbolic cylinder $x^2 - z^2 = 1$ are selected excludes the band $-1 < x < 1$ in the xy-plane (Example 2).

that is, at the point $(0, 0)$. But there are no points on the cylinder where both x and y are zero. What went wrong?

What happened is that the First Derivative Theorem found (as it should have) the point *in the domain of h* where h has a minimum value. We, on the other hand, want the points *on the cylinder* where h has a minimum value. Although the domain of h is the entire xy-plane, the domain from which we can select the first two coordinates of the points (x, y, z) on the cylinder is restricted to the projection, or "shadow" of the cylinder on the xy-plane; it does not include the band between the lines $x = -1$ and $x = 1$ (Figure 14.54).

We can avoid this problem if we treat y and z as independent variables (instead of x and y) and express x in terms of y and z as

$$x^2 = z^2 + 1.$$

With this substitution, $f(x, y, z) = x^2 + y^2 + z^2$ becomes

$$k(y, z) = (z^2 + 1) + y^2 + z^2 = 1 + y^2 + 2z^2$$

and we look for the points where k takes on its smallest value. The domain of k in the yz-plane now matches the domain from which we select the y- and z-coordinates of the points (x, y, z) on the cylinder. Hence, the points that minimize k in the plane will have corresponding points on the cylinder. The smallest values of k occur where

$$k_y = 2y = 0 \quad \text{and} \quad k_z = 4z = 0,$$

or where $y = z = 0$. This leads to

$$x^2 = z^2 + 1 = 1, \qquad x = \pm 1.$$

The corresponding points on the cylinder are $(\pm 1, 0, 0)$. We can see from the inequality

$$k(y, z) = 1 + y^2 + 2z^2 \geq 1$$

that the points $(\pm 1, 0, 0)$ give a minimum value for k. We can also see that the minimum distance from the origin to a point on the cylinder is 1 unit.

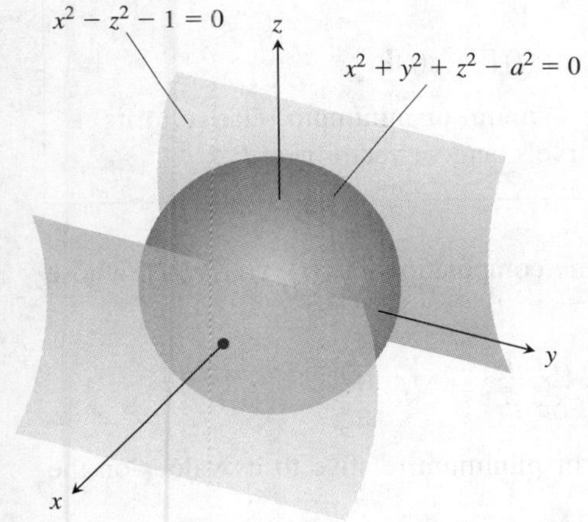

FIGURE 14.55 A sphere expanding like a soap bubble centered at the origin until it just touches the hyperbolic cylinder $x^2 - z^2 - 1 = 0$ (Example 2).

Solution 2 Another way to find the points on the cylinder closest to the origin is to imagine a small sphere centered at the origin expanding like a soap bubble until it just touches the cylinder (Figure 14.55). At each point of contact, the cylinder and sphere have the same tangent plane and normal line. Therefore, if the sphere and cylinder are represented as the level surfaces obtained by setting

$$f(x, y, z) = x^2 + y^2 + z^2 - a^2 \quad \text{and} \quad g(x, y, z) = x^2 - z^2 - 1$$

equal to 0, then the gradients ∇f and ∇g will be parallel where the surfaces touch. At any point of contact, we should therefore be able to find a scalar λ ("lambda") such that

$$\nabla f = \lambda \nabla g,$$

or

$$2x\mathbf{i} + 2y\mathbf{j} + 2z\mathbf{k} = \lambda(2x\mathbf{i} - 2z\mathbf{k}).$$

λ is the Greek letter lambda.

Thus, the coordinates x, y, and z of any point of tangency will have to satisfy the three scalar equations

$$2x = 2\lambda x, \qquad 2y = 0, \qquad 2z = -2\lambda z.$$

For what values of λ will a point (x, y, z) whose coordinates satisfy these scalar equations also lie on the surface $x^2 - z^2 - 1 = 0$? To answer this question, we use our knowledge that no point on the surface has a zero x-coordinate to conclude that $x \neq 0$. Hence, $2x = 2\lambda x$ only if

$$2 = 2\lambda, \qquad \text{or} \qquad \lambda = 1.$$

For $\lambda = 1$, the equation $2z = -2\lambda z$ becomes $2z = -2z$. If this equation is to be satisfied as well, z must be zero. Since $y = 0$ also (from the equation $2y = 0$), we conclude that the points we seek all have coordinates of the form

$$(x, 0, 0).$$

What points on the surface $x^2 - z^2 = 1$ have coordinates of this form? The answer is the points $(x, 0, 0)$ for which

$$x^2 - (0)^2 = 1, \qquad x^2 = 1, \qquad \text{or} \qquad x = \pm 1.$$

The points on the cylinder closest to the origin are the points $(\pm 1, 0, 0)$. ■

The Method of Lagrange Multipliers

In Solution 2 of Example 2, we used the **method of Lagrange multipliers**. The method says that the local extreme values of a function $f(x, y, z)$ whose variables are subject to a constraint $g(x, y, z) = 0$ are to be found on the surface $g = 0$ among the points where

$$\nabla f = \lambda \nabla g$$

for some scalar λ (called a **Lagrange multiplier**).

To explore the method further and see why it works, we first make the following observation, which we state as a theorem.

THEOREM 12—The Orthogonal Gradient Theorem
Suppose that $f(x, y, z)$ is differentiable in a region whose interior contains a smooth curve

$$C: \quad \mathbf{r}(t) = x(t)\mathbf{i} + y(t)\mathbf{j} + z(t)\mathbf{k}.$$

If P_0 is a point on C where f has a local maximum or minimum relative to its values on C, then ∇f is orthogonal to the curve's tangent vector $\mathbf{r}'$ at P_0.

Proof The values of f on C are given by the composition $f(x(t), y(t), z(t))$, whose derivative with respect to t is

$$\frac{df}{dt} = \frac{\partial f}{\partial x}\frac{dx}{dt} + \frac{\partial f}{\partial y}\frac{dy}{dt} + \frac{\partial f}{\partial z}\frac{dz}{dt} = \nabla f \cdot \mathbf{r}'.$$

At any point P_0 where f has a local maximum or minimum relative to its values on the curve, $df/dt = 0$, so

$$\nabla f \cdot \mathbf{r}' = 0.$$ ■

By dropping the z-terms in Theorem 12, we obtain a similar result for functions of two variables.

COROLLARY At the points on a smooth curve $\mathbf{r}(t) = x(t)\mathbf{i} + y(t)\mathbf{j}$ where a differentiable function $f(x, y)$ takes on its local maxima or minima relative to its values on the curve, we have $\nabla f \cdot \mathbf{r}' = 0$.

Theorem 12 is the key to the method of Lagrange multipliers. Suppose that $f(x, y, z)$ and $g(x, y, z)$ are differentiable and that P_0 is a point on the surface $g(x, y, z) = 0$ where f has a local maximum or minimum value relative to its other values on the surface. We assume also that $\nabla g \neq \mathbf{0}$ at points on the surface $g(x, y, z) = 0$. Then f takes on a local maximum or minimum at P_0 relative to its values on every differentiable curve through P_0 on the surface $g(x, y, z) = 0$. Therefore, ∇f is orthogonal to the tangent vector of every such differentiable curve through P_0. Moreover, so is ∇g (because ∇g is perpendicular to the level surface $g = 0$, as we saw in Section 14.5). Therefore, at P_0, ∇f is some scalar multiple λ of ∇g.

The Method of Lagrange Multipliers

Suppose that $f(x, y, z)$ and $g(x, y, z)$ are differentiable and $\nabla g \neq \mathbf{0}$ when $g(x, y, z) = 0$. To find the local maximum and minimum values of f subject to the constraint $g(x, y, z) = 0$ (if these exist), find the values of x, y, z, and λ that simultaneously satisfy the equations

$$\nabla f = \lambda \nabla g \quad \text{and} \quad g(x, y, z) = 0. \tag{1}$$

If they exist, absolute extrema can be found by comparing these values of f at each critical point satisfying Equation (1). For functions of two independent variables, the condition is similar, but without the variable z.

Some care must be used in applying this method. An extreme value may not actually exist (Exercise 45).

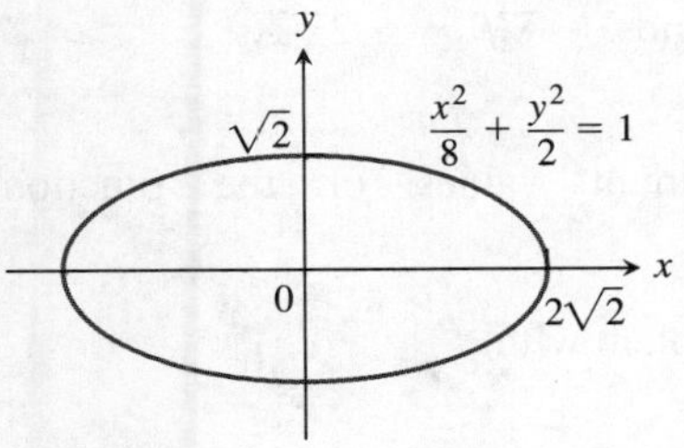

FIGURE 14.56 Example 3 shows how to find the largest and smallest values of the product xy on this ellipse.

EXAMPLE 3 Find the largest and smallest values that the function

$$f(x, y) = xy$$

takes on the ellipse (Figure 14.56)

$$\frac{x^2}{8} + \frac{y^2}{2} = 1.$$

Solution We want to find the extreme values of $f(x, y) = xy$ subject to the constraint

$$g(x, y) = \frac{x^2}{8} + \frac{y^2}{2} - 1 = 0.$$

To do so, we first find the values of x, y, and λ for which

$$\nabla f = \lambda \nabla g \quad \text{and} \quad g(x, y) = 0.$$

The gradient equation in Equations (1) gives

$$y\mathbf{i} + x\mathbf{j} = \frac{\lambda}{4} x\mathbf{i} + \lambda y\mathbf{j},$$

from which we find

$$y = \frac{\lambda}{4} x, \qquad x = \lambda y,$$

and

$$y = \frac{\lambda}{4}(\lambda y) = \frac{\lambda^2}{4} y,$$

Caution: Don't cancel y without considering the case where $y = 0$.

so that

$$y = 0 \text{ or } \lambda = \pm 2.$$

We now consider these two cases.

Case 1: If $y = 0$, then $x = y = 0$. But $(0, 0)$ is not on the ellipse. Hence, $y \neq 0$.

Case 2: If $y \neq 0$, then $\lambda = \pm 2$ and $x = \pm 2y$. Substituting this in the equation $g(x, y) = 0$ gives

$$\frac{(\pm 2y)^2}{8} + \frac{y^2}{2} = 1, \qquad 4y^2 + 4y^2 = 8 \quad \text{and} \quad y = \pm 1.$$

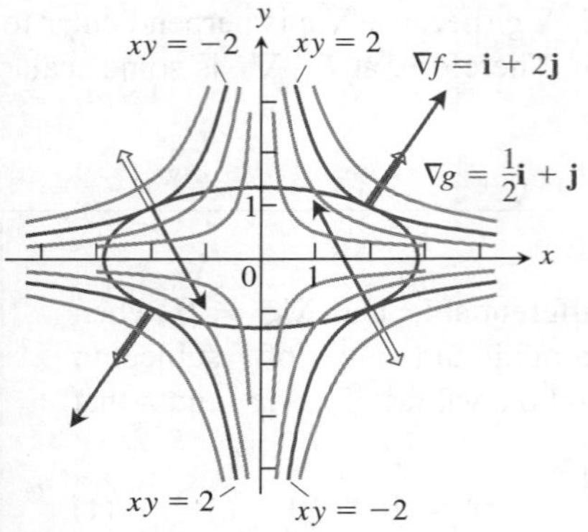

FIGURE 14.57 When subjected to the constraint $g(x, y) = x^2/8 + y^2/2 - 1 = 0$, the function $f(x, y) = xy$ takes on extreme values at the four points $(\pm 2, \pm 1)$. These are the points on the ellipse where ∇f is a scalar multiple of ∇g (Example 3).

The function $f(x, y) = xy$ therefore has critical points on the ellipse at the four points $(\pm 2, 1), (\pm 2, -1)$. The extreme values are found by examining the values of f at these four points. The absolute maximum is $f(2, 1) = f(-2, -1) = 2$, and the absolute minimum is $f(-2, 1) = f(2, -1) = -2$.

The Geometry of the Solution The level curves of the function $f(x, y) = xy$ are the hyperbolas $xy = c$ (Figure 14.57). The farther the hyperbolas lie from the origin, the larger the absolute value of f. We want to find the extreme values of $f(x, y)$, given that the point (x, y) also lies on the ellipse $x^2 + 4y^2 = 8$. Which hyperbolas intersecting the ellipse lie farthest from the origin? The hyperbolas that just graze the ellipse, the ones that are tangent to it, are farthest. At these points, any vector normal to the hyperbola is normal to the ellipse, so $\nabla f = y\mathbf{i} + x\mathbf{j}$ is a multiple $(\lambda = \pm 2)$ of $\nabla g = (x/4)\mathbf{i} + y\mathbf{j}$. At the point $(2, 1)$, for example,

$$\nabla f = \mathbf{i} + 2\mathbf{j}, \qquad \nabla g = \frac{1}{2}\mathbf{i} + \mathbf{j}, \qquad \text{and} \qquad \nabla f = 2\nabla g.$$

At the point $(-2, 1)$,

$$\nabla f = \mathbf{i} - 2\mathbf{j}, \qquad \nabla g = -\frac{1}{2}\mathbf{i} + \mathbf{j}, \qquad \text{and} \qquad \nabla f = -2\nabla g.$$

■

EXAMPLE 4 Find the maximum and minimum values of the function $f(x, y) = 3x + 4y$ on the circle $x^2 + y^2 = 1$.

Solution We model this as a Lagrange multiplier problem with

$$f(x, y) = 3x + 4y, \qquad g(x, y) = x^2 + y^2 - 1$$

and look for the values of x, y, and λ that satisfy the equations

$$\nabla f = \lambda \nabla g: \quad 3\mathbf{i} + 4\mathbf{j} = 2x\lambda\mathbf{i} + 2y\lambda\mathbf{j}$$
$$g(x, y) = 0: \quad x^2 + y^2 - 1 = 0.$$

The gradient equation implies that $\lambda \neq 0$ and gives

$$x = \frac{3}{2\lambda}, \qquad y = \frac{2}{\lambda}.$$

These equations tell us, among other things, that x and y have the same sign. With these values for x and y, the equation $g(x, y) = 0$ gives

$$\left(\frac{3}{2\lambda}\right)^2 + \left(\frac{2}{\lambda}\right)^2 - 1 = 0,$$

so

$$\frac{9}{4\lambda^2} + \frac{4}{\lambda^2} = 1, \qquad 9 + 16 = 4\lambda^2, \qquad 4\lambda^2 = 25, \qquad \text{and} \qquad \lambda = \pm\frac{5}{2}.$$

Thus,

$$x = \frac{3}{2\lambda} = \pm\frac{3}{5}, \qquad y = \frac{2}{\lambda} = \pm\frac{4}{5},$$

and $f(x, y) = 3x + 4y$ has critical points at $(x, y) = \pm(3/5, 4/5)$.

By calculating the value of $3x + 4y$ at the points $\pm(3/5, 4/5)$, we see that its maximum and minimum values on the circle $x^2 + y^2 = 1$ are

$$3\left(\frac{3}{5}\right) + 4\left(\frac{4}{5}\right) = \frac{25}{5} = 5 \qquad \text{and} \qquad 3\left(-\frac{3}{5}\right) + 4\left(-\frac{4}{5}\right) = -\frac{25}{5} = -5.$$

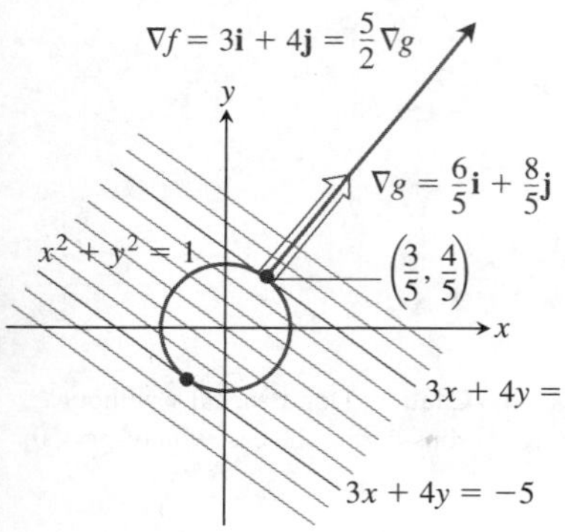

FIGURE 14.58 The function $f(x, y) = 3x + 4y$ takes on its largest value on the unit circle $g(x, y) = x^2 + y^2 - 1 = 0$ at the point $(3/5, 4/5)$ and its smallest value at the point $(-3/5, -4/5)$ (Example 4). At each of these points, ∇f is a scalar multiple of ∇g. The figure shows the gradients at the first point but not at the second.

The Geometry of the Solution The level curves of $f(x, y) = 3x + 4y$ are the lines $3x + 4y = c$ (Figure 14.58). The farther the lines lie from the origin, the larger the absolute value of f. We want to find the extreme values of $f(x, y)$ given that the point (x, y)

also lies on the circle $x^2 + y^2 = 1$. Which lines intersecting the circle lie farthest from the origin? The lines tangent to the circle are farthest. At the points of tangency, any vector normal to the line is normal to the circle, so the gradient $\nabla f = 3\mathbf{i} + 4\mathbf{j}$ is a multiple $(\lambda = \pm 5/2)$ of the gradient $\nabla g = 2x\mathbf{i} + 2y\mathbf{j}$. At the point $(3/5, 4/5)$, for example,

$$\nabla f = 3\mathbf{i} + 4\mathbf{j}, \qquad \nabla g = \frac{6}{5}\mathbf{i} + \frac{8}{5}\mathbf{j}, \qquad \text{and} \qquad \nabla f = \frac{5}{2}\nabla g. \qquad \blacksquare$$

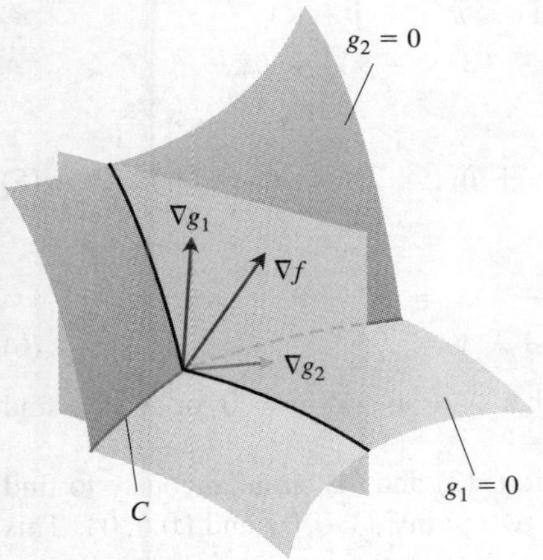

FIGURE 14.59 The vectors ∇g_1 and ∇g_2 lie in a plane perpendicular to the curve C, because ∇g_1 is normal to the surface $g_1 = 0$ and ∇g_2 is normal to the surface $g_2 = 0$.

μ is the Greek letter mu, pronounced "mew".

Lagrange Multipliers with Two Constraints

Many problems require us to find the extreme values of a differentiable function $f(x, y, z)$ whose variables are subject to two constraints. If the constraints are

$$g_1(x, y, z) = 0 \qquad \text{and} \qquad g_2(x, y, z) = 0$$

and g_1 and g_2 are differentiable, with ∇g_1 not parallel to ∇g_2, we find the constrained local maxima and minima of f by introducing two Lagrange multipliers λ and μ (mu, pronounced "mew"). That is, we locate the points $P(x, y, z)$ where f takes on its constrained extreme values by finding the values of x, y, z, λ, and μ that simultaneously satisfy the three equations

$$\nabla f = \lambda \nabla g_1 + \mu \nabla g_2, \qquad g_1(x, y, z) = 0, \qquad g_2(x, y, z) = 0 \tag{2}$$

Equations (2) have a nice geometric interpretation. The surfaces $g_1 = 0$ and $g_2 = 0$ (usually) intersect in a smooth curve, say C (Figure 14.59). Along this curve we seek the points where f has local maximum and minimum values relative to its other values on the curve. These are the points where ∇f is normal to C, as we saw in Theorem 12. But ∇g_1 and ∇g_2 are also normal to C at these points because C lies in the surfaces $g_1 = 0$ and $g_2 = 0$. Therefore, ∇f lies in the plane determined by ∇g_1 and ∇g_2, which means that $\nabla f = \lambda \nabla g_1 + \mu \nabla g_2$ for some λ and μ. Since the points we seek also lie in both surfaces, their coordinates must satisfy the equations $g_1(x, y, z) = 0$ and $g_2(x, y, z) = 0$, which are the remaining requirements in Equations (2).

EXAMPLE 5 The plane $x + y + z = 1$ cuts the cylinder $x^2 + y^2 = 1$ in an ellipse (Figure 14.60). Find the points on the ellipse that lie closest to and farthest from the origin.

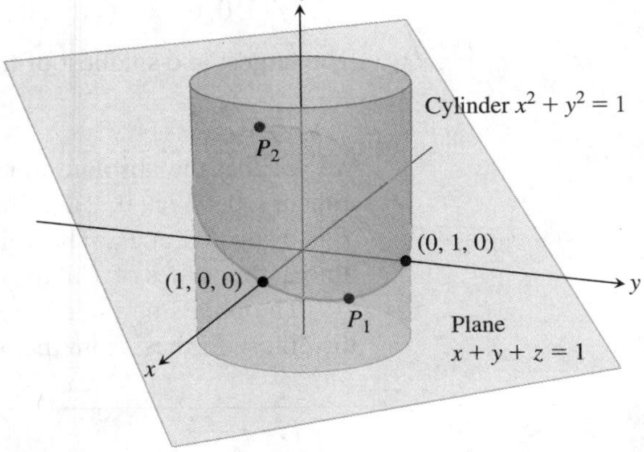

FIGURE 14.60 On the ellipse where the plane and cylinder meet, we find the points closest to and farthest from the origin (Example 5).

Solution We find the extreme values of

$$f(x, y, z) = x^2 + y^2 + z^2$$

(the square of the distance from (x, y, z) to the origin) subject to the constraints

$$g_1(x, y, z) = x^2 + y^2 - 1 = 0 \tag{3}$$

$$g_2(x, y, z) = x + y + z - 1 = 0. \tag{4}$$

The gradient equation in Equations (2) then gives

$$\begin{aligned} \nabla f &= \lambda \nabla g_1 + \mu \nabla g_2 \\ 2x\mathbf{i} + 2y\mathbf{j} + 2z\mathbf{k} &= \lambda(2x\mathbf{i} + 2y\mathbf{j}) + \mu(\mathbf{i} + \mathbf{j} + \mathbf{k}) \\ 2x\mathbf{i} + 2y\mathbf{j} + 2z\mathbf{k} &= (2\lambda x + \mu)\mathbf{i} + (2\lambda y + \mu)\mathbf{j} + \mu\mathbf{k}, \end{aligned}$$

or

$$2x = 2\lambda x + \mu, \qquad 2y = 2\lambda y + \mu, \qquad 2z = \mu. \tag{5}$$

The scalar equations in Equations (5) yield

$$\begin{aligned} 2x &= 2\lambda x + 2z \Rightarrow (1 - \lambda)x = z, \\ 2y &= 2\lambda y + 2z \Rightarrow (1 - \lambda)y = z. \end{aligned} \tag{6}$$

Equations (6) are satisfied simultaneously if either $\lambda = 1$ and $z = 0$ or $\lambda \neq 1$ and $x = y = z/(1 - \lambda)$.

In the first case, where $z = 0$, solving Equations (3) and (4) simultaneously to find the corresponding points on the ellipse gives the two points $(1, 0, 0)$ and $(0, 1, 0)$. This makes sense when you look at Figure 14.60.

In the second case, where $x = y$, Equations (3) and (4) give

$$\begin{aligned} x^2 + x^2 - 1 &= 0 \qquad & x + x + z - 1 &= 0 \\ 2x^2 &= 1 & z &= 1 - 2x \\ x &= \pm\frac{\sqrt{2}}{2} & z &= 1 \mp \sqrt{2}. \end{aligned}$$

The corresponding points on the ellipse are

$$P_1 = \left(\frac{\sqrt{2}}{2}, \frac{\sqrt{2}}{2}, 1 - \sqrt{2}\right) \quad \text{and} \quad P_2 = \left(-\frac{\sqrt{2}}{2}, -\frac{\sqrt{2}}{2}, 1 + \sqrt{2}\right).$$

To find the points at maximum and minimum distance from the origin, we evaluate f at the four critical points $(1, 0, 0)$, $(0, 1, 0)$, P_1, and P_2. We see that

$$f(1, 0, 0) = f(0, 1, 0) = 1, \quad f(P_1) = 4 - 2\sqrt{2}, \quad \text{and} \quad f(P_2) = 4 + 2\sqrt{2}.$$

The largest and smallest of these give the absolute extrema. Since

$$1 < 4 - 2\sqrt{2} < 4 + 2\sqrt{2},$$

we see that the absolute minimum value of f is 1 and is attained when f is evaluated at either $(1, 0, 0)$ or $(0, 1, 0)$. The absolute maximum value of f is $4 + 2\sqrt{2}$ and occurs when f is evaluated at P_2. The value $f(P_1) = 4 - 2\sqrt{2}$ is neither the largest nor the smallest among the values of f at the critical points, so f does not have an absolute extremum at P_1.

The points on the ellipse closest to the origin are $(1, 0, 0)$ and $(0, 1, 0)$. The point on the ellipse farthest from the origin is P_2. (See Figure 14.60.) ■

EXERCISES 14.8

Two Independent Variables with One Constraint

1. **Extrema on an ellipse** Find the points on the ellipse $x^2 + 2y^2 = 1$ where $f(x, y) = xy$ has its extreme values.
2. **Extrema on a circle** Find the extreme values of $f(x, y) = xy$ subject to the constraint $g(x, y) = x^2 + y^2 - 10 = 0$.
3. **Maximum on a line** Find the maximum value of $f(x, y) = 49 - x^2 - y^2$ on the line $x + 3y = 10$.
4. **Extrema on a line** Find the local extreme values of $f(x, y) = x^2 y$ on the line $x + y = 3$.

5. Constrained minimum Find the points on the curve $xy^2 = 54$ nearest the origin.

6. Constrained minimum Find the points on the curve $y^2 = 4x$ which is nearest to the point (2, –8).

7. Use the method of Lagrange multipliers to find

a. Minimum on a hyperbola The minimum value of $x + y$, subject to the constraints $xy = 16$, $x > 0$, $y > 0$.

b. Maximum on a line The maximum value of xy, subject to the constraint $x + y = 16$.

Comment on the geometry of each solution.

8. Extrema on a curve Prove that among all the right angle triangles of given hypotenuse, the isosceles triangle has the maximum area.

9. Minimum surface area with fixed volume Find the dimensions of the closed right circular cylindrical can of smallest surface area whose volume is 16π cm^3.

10. Cylinder in a sphere Find the radius and height of the open right circular cylinder of largest surface area that can be inscribed in a sphere of radius a. What *is* the largest surface area?

11. Rectangle of greatest area in an ellipse Use the method of Lagrange multipliers to find the dimensions of the rectangle of greatest area that can be inscribed in the ellipse $x^2/16 + y^2/9 = 1$ with sides parallel to the coordinate axes.

12. Rectangle of longest perimeter in an ellipse Given the slant height of a right circular cone, show that the volume of the cone is maximum when semi-vertical angle is $\tan^{-1}\sqrt{2}$.

13. Extrema on a circle Find the maximum and minimum values of $x^2 + y^2$ subject to the constraint $x^2 - 2x + y^2 - 4y = 0$.

14. Extrema on a circle Find the maximum and minimum values of $3x - y + 6$ subject to the constraint $x^2 + y^2 = 4$.

15. Ant on a metal plate The temperature at a point (x, y) on a metal plate is $T(x, y) = 4x^2 - 4xy + y^2$. An ant on the plate walks around the circle of radius 5 centered at the origin. What are the highest and lowest temperatures encountered by the ant?

16. Cheapest storage tank Your firm has been asked to design a storage tank for liquid petroleum gas. The customer's specifications call for a cylindrical tank with hemispherical ends, and the tank is to hold 8000 m^3 of gas. The customer also wants to use the smallest amount of material possible in building the tank. What radius and height do you recommend for the cylindrical portion of the tank?

Three Independent Variables with One Constraint

17. Minimum distance to a point Find the point on the plane $x + 2y + 3z = 13$ closest to the point $(1, 1, 1)$.

18. Maximum distance to a point Find the point on the sphere $x^2 + y^2 + z^2 = 4$ farthest from the point $(1, -1, 1)$.

19. Minimum distance to the origin Find the minimum distance from the surface $x^2 - y^2 - z^2 = 1$ to the origin.

20. Minimum distance to the origin Find the point on the surface $z = xy + 1$ nearest the origin.

21. Minimum distance to the origin Find the points on the surface $z^2 = xy + 4$ closest to the origin.

22. Minimum distance to the origin Find the minimum value of x^2yz^3 subject to $2x + y + 3z = 3$.

23. Extrema on a sphere Find the maximum and minimum values of

$$f(x, y, z) = x - 2y + 5z$$

on the sphere $x^2 + y^2 + z^2 = 30$.

24. Extrema on a sphere Find the points on the sphere $x^2 + y^2 + z^2 = 24$ that are farthest and nearest from the point (1, 2, –1).

25. Minimizing a sum of squares Find three real numbers whose sum is 9 and the sum of whose squares is as small as possible.

26. Maximizing a product Find the largest product the positive numbers x, y, and z can have if $x + y + z^2 = 16$.

27. Rectangular box of largest volume in a sphere Find the dimensions of the closed rectangular box with maximum volume that can be inscribed in the unit sphere.

28. Box with vertex on a plane Find the volume of the largest closed rectangular box in the first octant having three faces in the coordinate planes and a vertex on the plane $x/a + y/b + z/c = 1$, where $a > 0$, $b > 0$, and $c > 0$.

29. Hottest point on a space probe Find the volume of the largest rectangular parallelepiped that can be inscribed in the ellipsoid

$$\frac{x^2}{a^2} + \frac{y^2}{b^2} + \frac{z^2}{c^2} = 1.$$

30. Extreme temperatures on a sphere Suppose that the Celsius temperature at the point (x, y, z) on the sphere $x^2 + y^2 + z^2 = 1$ is $T = 400xyz^2$. Locate the highest and lowest temperatures on the sphere.

31. Cobb–Douglas production function During the 1920s, Charles Cobb and Paul Douglas modeled total production output P (of a firm, industry, or entire economy) as a function of labor hours involved x and capital invested y (which includes the monetary worth of all buildings and equipment). The Cobb–Douglas production function is given by

$$P(x, y) = kx^{\alpha}y^{1-\alpha},$$

where k and α are constants representative of a particular firm or economy.

a. Show that a doubling of both labor and capital results in a doubling of production P.

b. Suppose a particular firm has the production function for $k = 120$ and $\alpha = 3/4$. Assume that each unit of labor costs \$250 and each unit of capital costs \$400, and that the total expenses for all costs cannot exceed \$100,000. Find the maximum production level for the firm.

32. (*Continuation of Exercise 31.*) If the cost of a unit of labor is c_1 and the cost of a unit of capital is c_2, and if the firm can spend only B dollars as its total budget, then production P is constrained by $c_1x + c_2y = B$. Show that the maximum production level subject to the constraint occurs at the point

$$x = \frac{\alpha B}{c_1} \quad \text{and} \quad y = \frac{(1 - \alpha)B}{c_2}.$$

33. Maximizing a utility function: an example from economics In economics, the usefulness or *utility* of amounts x and y of two capital goods G_1 and G_2 is sometimes measured by a function $U(x, y)$. For example, G_1 and G_2 might be two chemicals a pharmaceutical company needs to have on hand, and $U(x, y)$ might be the gain from manufacturing a product whose synthesis requires different amounts of the chemicals depending on the process used. If G_1 costs a dollars per kilogram, G_2 costs b dollars per kilogram, and the total amount allocated for the purchase of G_1 and G_2 together is c dollars, then the company's managers want to maximize $U(x, y)$ given that $ax + by = c$. Thus, they need to solve a typical Lagrange multiplier problem.

Suppose that

$$U(x, y) = xy + 2x$$

and that the equation $ax + by = c$ simplifies to

$$2x + y = 30.$$

Find the maximum value of U and the corresponding values of x and y subject to this latter constraint.

34. Blood types Human blood types are classified by three gene forms A, B, and O. Blood types AA, BB, and OO are *homozygous*, and blood types AB, AO, and BO are *heterozygous*. If p, q, and r represent the proportions of the three gene forms to the population, respectively, then the *Hardy–Weinberg Law* asserts that the proportion Q of heterozygous persons in any specific population is modeled by

$$Q(p, q, r) = 2(pq + pr + qr),$$

subject to $p + q + r = 1$. Find the maximum value of Q.

35. Length of a beam In Section 4.5, Exercise 45, we posed a problem of finding the length L of the shortest beam that can reach over a wall of height h to a tall building located k units from the wall. Use Lagrange multipliers to show that

$$L = (h^{2/3} + k^{2/3})^{3/2}.$$

36. Locating a radio telescope You are in charge of erecting a radio telescope on a newly discovered planet. To minimize interference, you want to place it where the magnetic field of the planet is weakest. The planet is spherical, with a radius of 6 units. Based on a coordinate system whose origin is at the center of the planet, the strength of the magnetic field is given by $M(x, y, z) = 6x - y^2 + xz + 60$. Where should you locate the radio telescope?

Extreme Values Subject to Two Constraints

37. Maximize the function $f(x, y, z) = x^2 + 2y - z^2$ subject to the constraints $2x - y = 0$ and $y + z = 0$.

38. Minimize the function $f(x, y, z) = x^2 + y^2 + z^2$ subject to the constraints $x + 2y + 3z = 6$ and $x + 3y + 9z = 9$.

39. Minimum distance to the origin Find the point closest to the origin on the line of intersection of the planes $y + 2z = 12$ and $x + y = 6$.

40. Find the extreme values of $f(x, y, z) = 2x^2 + yz$ on the intersection of the cylinder $x^2 + z^2 = 9$ and the plane $y - z = 4$.

41. Extrema on a curve of intersection Find the extreme values of $f(x, y, z) = x^2yz + 1$ on the intersection of the plane $z = 1$ with the sphere $x^2 + y^2 + z^2 = 10$.

42. a. Maximum on line of intersection Find the maximum value of $w = xyz$ on the line of intersection of the two planes $x + y + z = 40$ and $x + y - z = 0$.

b. Give a geometric argument to support your claim that you have found a maximum, and not a minimum, value of w.

43. Extrema on a circle of intersection Find the extreme values of the function $f(x, y, z) = xy + z^2$ on the circle in which the plane $y - x = 0$ intersects the sphere $x^2 + y^2 + z^2 = 4$.

44. Minimum distance to the origin Find the point closest to the origin on the curve of intersection of the plane $2y + 4z = 5$ and the cone $z^2 = 4x^2 + 4y^2$.

Theory and Examples

45. The condition $\nabla f = \lambda \nabla g$ is not sufficient Even though $\nabla f = \lambda \nabla g$ is a necessary condition for the occurrence of an extreme value of $f(x, y)$ subject to the conditions $g(x, y) = 0$ and $\nabla g \neq \mathbf{0}$, it does not in itself guarantee that one exists. As a case in point, try using the method of Lagrange multipliers to find a maximum value of $f(x, y) = x + y$ subject to the constraint that $xy = 16$. The method will identify the two points $(4, 4)$ and $(-4, -4)$ as candidates for the location of extreme values. Yet the sum $x + y$ has no maximum value on the hyperbola $xy = 16$. The farther you go from the origin on this hyperbola in the first quadrant, the larger the sum $f(x, y) = x + y$ becomes.

46. A least squares plane The plane $z = Ax + By + C$ is to be "fitted" to the following points (x_k, y_k, z_k):

$$(0, 0, 0), \quad (0, 1, 1), \quad (1, 1, 1), \quad (1, 0, -1).$$

Find the values of A, B, and C that minimize

$$\sum_{k=1}^{4} (Ax_k + By_k + C - z_k)^2,$$

the sum of the squares of the deviations.

47. a. Maximum on a sphere Show that the maximum value of $a^2b^2c^2$ on a sphere of radius r centered at the origin of a Cartesian abc-coordinate system is $(r^2/3)^3$.

b. Geometric and arithmetic means Using part (a), show that for nonnegative numbers a, b, and c,

$$(abc)^{1/3} \leq \frac{a + b + c}{3};$$

that is, the *geometric mean* of three nonnegative numbers is less than or equal to their *arithmetic mean*.

48. Sum of products Let $a_1, a_2, \ldots, a_n$ be n positive numbers. Find the maximum of $\sum_{i=1}^{n} a_i x_i$ subject to the constraint $\sum_{i=1}^{n} x_i^2 = 1$.

COMPUTER EXPLORATIONS

In Exercises 49–54, use a CAS to perform the following steps implementing the method of Lagrange multipliers for finding constrained extrema:

a. Form the function $h = f - \lambda_1 g_1 - \lambda_2 g_2$, where f is the function to optimize subject to the constraints $g_1 = 0$ and $g_2 = 0$.

b. Determine all the first partial derivatives of h, including the partials with respect to λ_1 and λ_2, and set them equal to 0.

c. Solve the system of equations found in part (b) for all the unknowns, including λ_1 and λ_2.

d. Evaluate f at each of the solution points found in part (c), and select the extreme value subject to the constraints asked for in the exercise.

49. Minimize $f(x, y, z) = xy + yz$ subject to the constraints $x^2 + y^2 - 2 = 0$ and $x^2 + z^2 - 2 = 0$.

50. Minimize $f(x, y, z) = xyz$ subject to the constraints $x^2 + y^2 - 1 = 0$ and $x - z = 0$.

51. Maximize $f(x, y, z) = x^2 + y^2 + z^2$ subject to the constraints $2y + 4z - 5 = 0$ and $4x^2 + 4y^2 - z^2 = 0$.

52. Minimize $f(x, y, z) = x^2 + y^2 + z^2$ subject to the constraints $x^2 - xy + y^2 - z^2 - 1 = 0$ and $x^2 + y^2 - 1 = 0$.

53. Minimize $f(x, y, z, w) = x^2 + y^2 + z^2 + w^2$ subject to the constraints $2x - y + z - w - 1 = 0$ and $x + y - z + w - 1 = 0$.

54. Determine the distance from the line $y = x + 1$ to the parabola $y^2 = x$. (*Hint:* Let (x, y) be a point on the line and (w, z) a point on the parabola. You want to minimize $(x - w)^2 + (y - z)^2$.)

14.9 Taylor's Formula for Two Variables

In this section we use Taylor's formula to derive the Second Derivative Test for local extreme values (Section 14.7) and the error formula for linearizations of functions of two independent variables (Section 14.6). The use of Taylor's formula in these derivations leads to an extension of the formula that provides polynomial approximations of all orders for functions of two independent variables.

Derivation of the Second Derivative Test

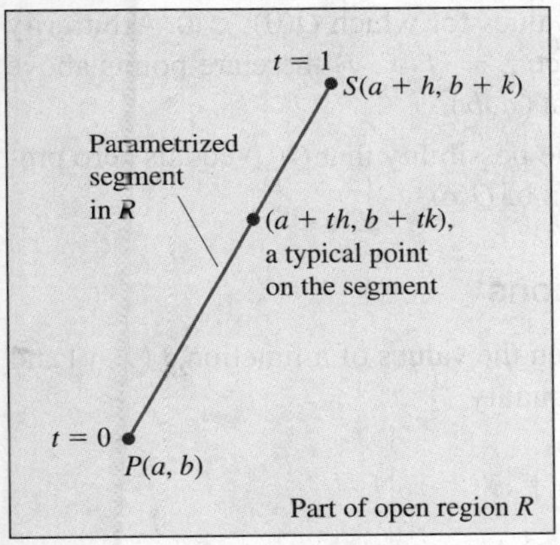

FIGURE 14.61 We begin the derivation of the Second Derivative Test at $P(a, b)$ by parametrizing a typical line segment from P to a point S nearby.

Let $f(x, y)$ have continuous first and second partial derivatives in an open region R containing a point $P(a, b)$ where $f_x = f_y = 0$ (Figure 14.61). Let h and k be increments small enough to put the point $S(a + h, b + k)$ and the line segment joining it to P inside R. We parametrize the segment PS as

$$x = a + th, \quad y = b + tk, \quad 0 \le t \le 1.$$

If $F(t) = f(a + th, b + tk)$, the Chain Rule gives

$$F'(t) = f_x \frac{dx}{dt} + f_y \frac{dy}{dt} = hf_x + kf_y.$$

Since f_x and f_y are differentiable (because they have continuous partial derivatives), F' is a differentiable function of t and

$$F'' = \frac{\partial F'}{\partial x}\frac{dx}{dt} + \frac{\partial F'}{\partial y}\frac{dy}{dt} = \frac{\partial}{\partial x}(hf_x + kf_y)\cdot h + \frac{\partial}{\partial y}(hf_x + kf_y)\cdot k$$
$$= h^2 f_{xx} + 2hk f_{xy} + k^2 f_{yy}. \qquad f_{xy} = f_{yx}$$

Since F and F' are continuous on $[0, 1]$ and F' is differentiable on $(0, 1)$, we can apply Taylor's formula with $n = 2$ and $a = 0$ to obtain

$$\begin{aligned} F(1) &= F(0) + F'(0)(1 - 0) + F''(c)\frac{(1-0)^2}{2} \\ &= F(0) + F'(0) + \frac{1}{2}F''(c) \end{aligned} \tag{1}$$

for some c between 0 and 1. Writing Equation (1) in terms of f gives

$$\begin{aligned} f(a + h, b + k) &= f(a, b) + hf_x(a, b) + kf_y(a, b) \\ &\quad + \frac{1}{2}\left(h^2 f_{xx} + 2hk f_{xy} + k^2 f_{yy}\right)\Big|_{(a+ch,\, b+ck)}. \end{aligned} \tag{2}$$

Since $f_x(a, b) = f_y(a, b) = 0$, this reduces to

$$f(a + h, b + k) - f(a, b) = \frac{1}{2}\left(h^2 f_{xx} + 2hk f_{xy} + k^2 f_{yy}\right)\Big|_{(a+ch,\, b+ck)}. \tag{3}$$

To determine whether f has an extremum at (a, b), we examine the sign of the difference $f(a + h, b + k) - f(a, b)$. By Equation (3), this is the same as the sign of

$$Q(c) = \left(h^2 f_{xx} + 2hk f_{xy} + k^2 f_{yy}\right)\Big|_{(a+ch,\, b+ck)}.$$

Now, if $Q(0) \neq 0$, the sign of $Q(c)$ will be the same as the sign of $Q(0)$ for sufficiently small values of h and k. We can predict the sign of

$$Q(0) = h^2 f_{xx}(a, b) + 2hk f_{xy}(a, b) + k^2 f_{yy}(a, b) \tag{4}$$

from the signs of f_{xx} and $f_{xx} f_{yy} - f_{xy}{}^2$ at (a, b). Multiply both sides of Equation (4) by f_{xx} and rearrange the right-hand side to get

$$f_{xx} Q(0) = \left(h f_{xx} + k f_{xy}\right)^2 + \left(f_{xx} f_{yy} - f_{xy}{}^2\right) k^2. \tag{5}$$

From Equation (5) we see that

1. If $f_{xx} < 0$ and $f_{xx} f_{yy} - f_{xy}{}^2 > 0$ at (a, b), then $Q(0) < 0$ for all sufficiently small nonzero values of h and k, and f has a *local maximum* value at (a, b).
2. If $f_{xx} > 0$ and $f_{xx} f_{yy} - f_{xy}{}^2 > 0$ at (a, b), then $Q(0) > 0$ for all sufficiently small nonzero values of h and k, and f has a *local minimum* value at (a, b).
3. If $f_{xx} f_{yy} - f_{xy}{}^2 < 0$ at (a, b), there are combinations of arbitrarily small nonzero values of h and k for which $Q(0) > 0$, and other values for which $Q(0) < 0$. Arbitrarily close to the point $P_0(a, b, f(a, b))$ on the surface $z = f(x, y)$ there are points above P_0 and points below P_0, so f has a *saddle point* at (a, b).
4. If $f_{xx} f_{yy} - f_{xy}{}^2 = 0$, another test is needed. The possibility that $Q(0)$ equals zero prevents us from drawing conclusions about the sign of $Q(c)$.

The Error Formula for Linear Approximations

We want to show that the difference $E(x, y)$ between the values of a function $f(x, y)$ and its linearization $L(x, y)$ at (x_0, y_0) satisfies the inequality

$$|E(x, y)| \leq \frac{1}{2} M \left(|x - x_0| + |y - y_0|\right)^2.$$

The function f is assumed to have continuous second partial derivatives throughout an open set containing a closed rectangular region R centered at (x_0, y_0). The number M is an upper bound for $|f_{xx}|$, $|f_{yy}|$, and $|f_{xy}|$ on R.

The inequality we want comes from Equation (2). We substitute x_0 and y_0 for a and b, and $x - x_0$ and $y - y_0$ for h and k, respectively, and rearrange the result as

$$f(x, y) = \underbrace{f(x_0, y_0) + f_x(x_0, y_0)(x - x_0) + f_y(x_0, y_0)(y - y_0)}_{\text{linearization } L(x, y)}$$

$$+ \underbrace{\frac{1}{2}\left((x - x_0)^2 f_{xx} + 2(x - x_0)(y - y_0) f_{xy} + (y - y_0)^2 f_{yy}\right)\Big|_{(x_0 + c(x - x_0),\, y_0 + c(y - y_0))}}_{\text{error } E(x, y)}.$$

This equation reveals that

$$|E| \leq \frac{1}{2}\left(|x - x_0|^2 |f_{xx}| + 2|x - x_0||y - y_0||f_{xy}| + |y - y_0|^2 |f_{yy}|\right).$$

Hence, if M is an upper bound for the values of $|f_{xx}|$, $|f_{xy}|$, and $|f_{yy}|$ on R, then

$$\begin{aligned} |E| &\leq \frac{1}{2}\left(|x - x_0|^2 M + 2|x - x_0||y - y_0| M + |y - y_0|^2 M\right) \\ &= \frac{1}{2} M \left(|x - x_0| + |y - y_0|\right)^2. \end{aligned}$$

Taylor's Formula for Functions of Two Variables

The formulas derived earlier for F' and F'' can be obtained by applying to $f(x, y)$ the differentiation operators

$$\left(h\frac{\partial}{\partial x} + k\frac{\partial}{\partial y}\right) \quad \text{and} \quad \left(h\frac{\partial}{\partial x} + k\frac{\partial}{\partial y}\right)^2 = h^2\frac{\partial^2}{\partial x^2} + 2hk\frac{\partial^2}{\partial x\,\partial y} + k^2\frac{\partial^2}{\partial y^2}.$$

These are the first two instances of a more general formula,

$$F^{(n)}(t) = \frac{d^n}{dt^n}F(t) = \left(h\frac{\partial}{\partial x} + k\frac{\partial}{\partial y}\right)^n f(x, y), \tag{6}$$

which says that applying d^n/dt^n to $F(t)$ gives the same result as applying the operator

$$\left(h\frac{\partial}{\partial x} + k\frac{\partial}{\partial y}\right)^n$$

to $f(x, y)$ after expanding it by the Binomial Theorem.

If the partial derivatives of f through order $n + 1$ are continuous throughout a rectangular region centered at (a, b), we may extend the Taylor formula for $F(t)$ to

$$F(t) = F(0) + F'(0)t + \frac{F''(0)}{2!}t^2 + \cdots + \frac{F^{(n)}(0)}{n!}t^{(n)} + \text{remainder},$$

and take $t = 1$ to obtain

$$F(1) = F(0) + F'(0) + \frac{F''(0)}{2!} + \cdots + \frac{F^{(n)}(0)}{n!} + \text{remainder}.$$

When we replace the first n derivatives on the right of this last series by their equivalent expressions from Equation (6) evaluated at $t = 0$ and add the appropriate remainder term, we arrive at the following formula.

Taylor's Formula for $f(x, y)$ at the Point (a, b)

Suppose $f(x, y)$ and its partial derivatives through order $n + 1$ are continuous throughout an open rectangular region R centered at a point (a, b). Then, throughout R,

$$\begin{aligned} f(a + h, b + k) = {} & f(a, b) + \left(hf_x + kf_y\right)\Big|_{(a,b)} + \frac{1}{2!}\left(h^2 f_{xx} + 2hkf_{xy} + k^2 f_{yy}\right)\Big|_{(a,b)} \\ & + \frac{1}{3!}\left(h^3 f_{xxx} + 3h^2kf_{xxy} + 3hk^2 f_{xyy} + k^3 f_{yyy}\right)\Big|_{(a,b)} + \cdots + \frac{1}{n!}\left(h\frac{\partial}{\partial x} + k\frac{\partial}{\partial y}\right)^n f\Big|_{(a,b)} \\ & + \frac{1}{(n+1)!}\left(h\frac{\partial}{\partial x} + k\frac{\partial}{\partial y}\right)^{n+1} f\Big|_{(a+ch,\, b+ck)}. \end{aligned} \tag{7}$$

The first n derivative terms are evaluated at (a, b). The last term is evaluated at some point $(a + ch, b + ck)$ on the line segment joining (a, b) and $(a + h, b + k)$.

If $(a, b) = (0, 0)$ and we treat h and k as independent variables (denoting them now by x and y), then Equation (7) assumes the following form.

Taylor's Formula for $f(x, y)$ at the Origin

$$f(x, y) = f(0,0) + xf_x + yf_y + \frac{1}{2!}\left(x^2 f_{xx} + 2xyf_{xy} + y^2 f_{yy}\right)$$
$$+ \frac{1}{3!}\left(x^3 f_{xxx} + 3x^2 y f_{xxy} + 3xy^2 f_{xyy} + y^3 f_{yyy}\right) + \cdots + \frac{1}{n!}\left(x^n \frac{\partial^n f}{\partial x^n} + nx^{n-1}y\frac{\partial^n f}{\partial x^{n-1}\,\partial y} + \cdots + y^n \frac{\partial^n f}{\partial y^n}\right)$$
$$+ \frac{1}{(n+1)!}\left(x^{n+1}\frac{\partial^{n+1} f}{\partial x^{n+1}} + (n+1)x^n y \frac{\partial^{n+1} f}{\partial x^n\,\partial y} + \cdots + y^{n+1}\frac{\partial^{n+1} f}{\partial y^{n+1}}\right)\Bigg|_{(cx,cy)} \quad (8)$$

The first n derivative terms are evaluated at $(0, 0)$. The last term is evaluated at a point on the line segment joining the origin and (x, y).

Taylor's formula provides polynomial approximations of two-variable functions. The first n derivative terms give the polynomial; the last term gives the approximation error. The first three terms of Taylor's formula give the function's linearization. To improve on the linearization, we add higher-power terms.

EXAMPLE 1 Find a quadratic approximation to $f(x, y) = \sin x \sin y$ near the origin. How accurate is the approximation if $|x| \le 0.1$ and $|y| \le 0.1$?

Solution We take $n = 2$ in Equation (8):

$$f(x, y) = f(0,0) + (xf_x + yf_y) + \frac{1}{2}(x^2 f_{xx} + 2xyf_{xy} + y^2 f_{yy})$$
$$+ \frac{1}{6}(x^3 f_{xxx} + 3x^2 y f_{xxy} + 3xy^2 f_{xyy} + y^3 f_{yyy})\Big|_{(cx,cy)}.$$

Calculating the values of the partial derivatives,

$$f(0,0) = \sin x \sin y\Big|_{(0,0)} = 0, \quad f_{xx}(0,0) = -\sin x \sin y\Big|_{(0,0)} = 0,$$
$$f_x(0,0) = \cos x \sin y\Big|_{(0,0)} = 0, \quad f_{xy}(0,0) = \cos x \cos y\Big|_{(0,0)} = 1,$$
$$f_y(0,0) = \sin x \cos y\Big|_{(0,0)} = 0, \quad f_{yy}(0,0) = -\sin x \sin y\Big|_{(0,0)} = 0,$$

we have the result

$$\sin x \sin y \approx 0 + 0 + 0 + \frac{1}{2}(x^2(0) + 2xy(1) + y^2(0)), \quad \text{or} \quad \sin x \sin y \approx xy.$$

The error in the approximation is

$$E(x, y) = \frac{1}{6}(x^3 f_{xxx} + 3x^2 y f_{xxy} + 3xy^2 f_{xyy} + y^3 f_{yyy})\Big|_{(cx,cy)}.$$

The third derivatives never exceed 1 in absolute value because they are products of sines and cosines. Also, $|x| \le 0.1$ and $|y| \le 0.1$. Hence

$$|E(x, y)| \le \frac{1}{6}((0.1)^3 + 3(0.1)^3 + 3(0.1)^3 + (0.1)^3) = \frac{8}{6}(0.1)^3 \le 0.00134$$

(rounded up). The error will not exceed 0.00134 if $|x| \le 0.1$ and $|y| \le 0.1$. ■

EXERCISES 14.9

Finding Quadratic and Cubic Approximations

In Exercises 1–10, use Taylor's formula for $f(x, y)$ at the origin to find quadratic and cubic approximations of f near the origin.

1. $f(x, y) = xe^y$

2. $f(x, y) = e^x \cos y$

3. $f(x, y) = x \cos y$

4. $f(x, y) = \sin x \cos y$

5. $f(x, y) = e^{x^2 - y}$

6. $f(x, y) = \ln(2x + y + 1)$

7. $f(x, y) = \sin(x^2 + y^2)$

8. $f(x, y) = \sin 2x + \cos y$

9. $f(x, y) = \dfrac{1}{1 - x - y}$

10. $f(x, y) = \dfrac{1}{1 - x - y + xy}$

11. Use Taylor's formula to find a quadratic approximation of $f(x, y) = \cos x \cos y$ at the origin. Estimate the error in the approximation if $|x| \le 0.1$ and $|y| \le 0.1$.

12. $f(x, y) = e^y \sin x$

14.10 Partial Derivatives with Constrained Variables

In finding partial derivatives of functions like $w = f(x, y)$, we have assumed x and y to be independent. In many applications, however, this is not the case. For example, the internal energy U of a gas may be expressed as a function $U = f(P, V, T)$ of pressure P, volume V, and temperature T. If the individual molecules of the gas do not interact, however, P, V, and T obey (and are constrained by) the ideal gas law

$$PV = nRT \qquad (n \text{ and } R \text{ constant}),$$

and fail to be independent. In this section we learn how to find partial derivatives in situations like this, which occur in economics, engineering, and physics.

Decide Which Variables Are Dependent and Which Are Independent

If the variables in a function $w = f(x, y, z)$ are constrained by a relation like the one imposed on x, y, and z by the equation $z = x^2 + y^2$, the geometric meanings and the numerical values of the partial derivatives of f will depend on which variables are chosen to be dependent and which are chosen to be independent. To see how this choice can affect the outcome, we consider the calculation of $\partial w/\partial x$ when $w = x^2 + y^2 + z^2$ and $z = x^2 + y^2$.

EXAMPLE 1 Find $\partial w/\partial x$ if $w = x^2 + y^2 + z^2$ and $z = x^2 + y^2$.

Solution We are given two equations in the four unknowns x, y, z, and w. Like many such systems, this one can be solved for two of the unknowns (the dependent variables) in terms of the others (the independent variables). In being asked for $\partial w/\partial x$, we are told that w is to be a dependent variable and x an independent variable. The possible choices for the other variables come down to

	Dependent	*Independent*
Choice 1:	w, z	x, y
Choice 2:	w, y	x, z

In either case, we can express w explicitly in terms of the selected independent variables. We do this by using the second equation $z = x^2 + y^2$ to eliminate the remaining dependent variable in the first equation.

In the first case, the remaining dependent variable is z. We eliminate it from the first equation by replacing it by $x^2 + y^2$. The resulting expression for w is

$$\begin{aligned} w &= x^2 + y^2 + z^2 = x^2 + y^2 + (x^2 + y^2)^2 \\ &= x^2 + y^2 + x^4 + 2x^2y^2 + y^4 \end{aligned}$$

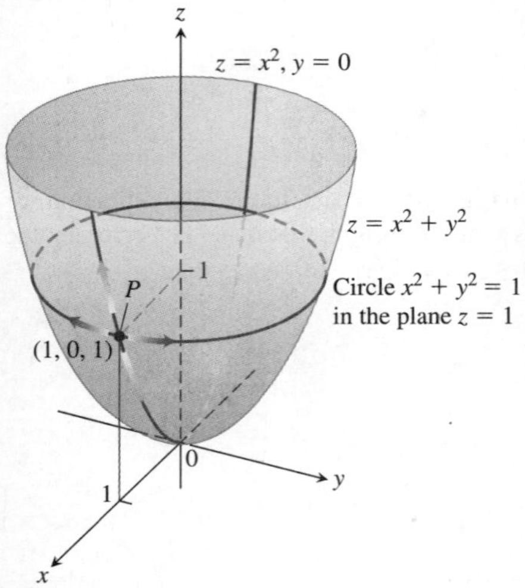

FIGURE 14.62 If P is constrained to lie on the paraboloid $z = x^2 + y^2$, the value of the partial derivative of $w = x^2 + y^2 + z^2$ with respect to x at P depends on the direction of motion (Example 1). (1) As x changes, with $y = 0$, P moves up or down the surface on the parabola $z = x^2$ in the xz-plane with $\partial w/\partial x = 2x + 4x^3$. (2) As x changes, with $z = 1$, P moves on the circle $x^2 + y^2 = 1, z = 1$, and $\partial w/\partial x = 0$.

and therefore

$$\frac{\partial w}{\partial x} = 2x + 4x^3 + 4xy^2. \tag{1}$$

This is the formula for $\partial w/\partial x$ when x and y are the independent variables.

In the second case, where the independent variables are x and z and the remaining dependent variable is y, we eliminate the dependent variable y in the expression for w by replacing y^2 in the second equation by $z - x^2$. This gives

$$w = x^2 + y^2 + z^2 = x^2 + (z - x^2) + z^2 = z + z^2$$

and therefore

$$\frac{\partial w}{\partial x} = 0. \tag{2}$$

This is the formula for $\partial w/\partial x$ when x and z are the independent variables.

The formulas for $\partial w/\partial x$ in Equations (1) and (2) are genuinely different. We cannot change either formula into the other by using the relation $z = x^2 + y^2$. There is not just one $\partial w/\partial x$, there are two, and we see that the original instruction to find $\partial w/\partial x$ was incomplete. *Which* $\partial w/\partial x$? we ask.

The geometric interpretations of Equations (1) and (2) help to explain why the equations differ. The function $w = x^2 + y^2 + z^2$ measures the square of the distance from the point (x, y, z) to the origin. The condition $z = x^2 + y^2$ says that the point (x, y, z) lies on the paraboloid of revolution shown in Figure 14.62. What does it mean to calculate $\partial w/\partial x$ at a point $P(x, y, z)$ that can move only on this surface? What is the value of $\partial w/\partial x$ when the coordinates of P are, say, $(1, 0, 1)$?

If we take x and y to be independent, then we find $\partial w/\partial x$ by holding y fixed (at $y = 0$ in this case) and letting x vary. Hence, P moves along the parabola $z = x^2$ in the xz-plane. As P moves on this parabola, w, which is the square of the distance from P to the origin, changes. We calculate $\partial w/\partial x$ in this case (our first solution above) to be

$$\frac{\partial w}{\partial x} = 2x + 4x^3 + 4xy^2.$$

At the point $P(1, 0, 1)$, the value of this derivative is

$$\frac{\partial w}{\partial x} = 2 + 4 + 0 = 6.$$

If we take x and z to be independent, then we find $\partial w/\partial x$ by holding z fixed while x varies. Since the z-coordinate of P is 1, varying x moves P along a circle in the plane $z = 1$. As P moves along this circle, its distance from the origin remains constant, and w, being the square of this distance, does not change. That is,

$$\frac{\partial w}{\partial x} = 0,$$

as we found in our second solution. ■

How to Find $\partial w/\partial x$ When the Variables in $w = f(x, y, z)$ Are Constrained by Another Equation

As we saw in Example 1, a typical routine for finding $\partial w/\partial x$ when the variables in the function $w = f(x, y, z)$ are related by another equation has three steps. These steps apply to finding $\partial w/\partial y$ and $\partial w/\partial z$ as well.

1. *Decide* which variables are to be dependent and which are to be independent. (In practice, the decision is based on the physical or theoretical context of our work. In the exercises at the end of this section, we say which variables are which.)
2. *Eliminate* the other dependent variable(s) in the expression for w.
3. *Differentiate* as usual.

If we cannot carry out Step 2 after deciding which variables are dependent, we differentiate the equations as they are and try to solve for $\partial w/\partial x$ afterward. The next example shows how this is done.

EXAMPLE 2 Find $\partial w/\partial x$ at the point $(x, y, z) = (2, -1, 1)$ if

$$w = x^2 + y^2 + z^2, \qquad z^3 - xy + yz + y^3 = 1,$$

and x and y are the independent variables.

Solution It is not convenient to eliminate z in the expression for w. We therefore differentiate both equations implicitly with respect to x, treating x and y as independent variables and w and z as dependent variables. This gives

$$\frac{\partial w}{\partial x} = 2x + 2z\frac{\partial z}{\partial x} \tag{3}$$

and

$$3z^2\frac{\partial z}{\partial x} - y + y\frac{\partial z}{\partial x} + 0 = 0. \tag{4}$$

These equations may now be combined to express $\partial w/\partial x$ in terms of x, y, and z. We solve Equation (4) for $\partial z/\partial x$ to get

$$\frac{\partial z}{\partial x} = \frac{y}{y + 3z^2}$$

and substitute into Equation (3) to get

$$\frac{\partial w}{\partial x} = 2x + \frac{2yz}{y + 3z^2}.$$

The value of this derivative at $(x, y, z) = (2, -1, 1)$ is

$$\left.\frac{\partial w}{\partial x}\right|_{(2,-1,1)} = 2(2) + \frac{2(-1)(1)}{-1 + 3(1)^2} = 4 + \frac{-2}{2} = 3.$$ ■

Notation

To show what variables are assumed to be independent in calculating a derivative, we can use the following notation:

$\left(\frac{\partial w}{\partial x}\right)_y$ $\quad \partial w/\partial x$ with x and y independent

$\left(\frac{\partial f}{\partial y}\right)_{x,t}$ $\quad \partial f/\partial y$ with y, x, and t independent.

HISTORICAL BIOGRAPHY
Sonya Kovalevsky
(1850–1891)
bit.ly/2RgmUCZ

EXAMPLE 3 Find $(\partial w/\partial x)_{y,z}$ if $w = x^2 + y - z + \sin t$ and $x + y = t$.

Solution With x, y, z independent, we have

$$t = x + y, \qquad w = x^2 + y - z + \sin(x + y)$$

$$\begin{aligned}\left(\frac{\partial w}{\partial x}\right)_{y,z} &= 2x + 0 - 0 + \cos(x + y)\frac{\partial}{\partial x}(x + y)\\ &= 2x + \cos(x + y).\end{aligned}$$

■

Arrow Diagrams

In solving problems like the one in Example 3, it often helps to start with an arrow diagram that shows how the variables and functions are related. If

$$w = x^2 + y - z + \sin t \qquad \text{and} \qquad x + y = t$$

and we are asked to find $\partial w/\partial x$ when x, y, and z are independent, the appropriate diagram is one like this:

$$\begin{pmatrix} x \\ y \\ z \end{pmatrix} \rightarrow \begin{pmatrix} x \\ y \\ z \\ t \end{pmatrix} \rightarrow w \tag{5}$$

Independent variables | Intermediate variables | Dependent variable

To avoid confusion between the independent and intermediate variables with the same symbolic names in the diagram, it is helpful to rename the intermediate variables (so they are seen as *functions* of the independent variables). Thus, let $u = x$, $v = y$, and $s = z$ denote the renamed intermediate variables. With this notation, the arrow diagram becomes

$$\begin{pmatrix} x \\ y \\ z \end{pmatrix} \rightarrow \begin{pmatrix} u \\ v \\ s \\ t \end{pmatrix} \rightarrow w \tag{6}$$

Independent variables | Intermediate variables and relations $u = x$, $v = y$, $s = z$, $t = x + y$ | Dependent variable

The diagram shows the independent variables on the left, the intermediate variables and their relation to the independent variables in the middle, and the dependent variable on the right. The function w now becomes

$$w = u^2 + v - s + \sin t,$$

where

$$u = x, \qquad v = y, \qquad s = z, \qquad \text{and} \qquad t = x + y.$$

To find $\partial w/\partial x$, we apply the four-variable form of the Chain Rule to w, guided by the arrow diagram in Equation (6):

$$\begin{aligned}\frac{\partial w}{\partial x} &= \frac{\partial w}{\partial u}\frac{\partial u}{\partial x} + \frac{\partial w}{\partial v}\frac{\partial v}{\partial x} + \frac{\partial w}{\partial s}\frac{\partial s}{\partial x} + \frac{\partial w}{\partial t}\frac{\partial t}{\partial x}\\ &= (2u)(1) + (1)(0) + (-1)(0) + (\cos t)(1)\\ &= 2u + \cos t\\ &= 2x + \cos(x + y).\end{aligned}$$

Substitute the original independent variables $u = x$ and $t = x + y$

EXERCISES 14.10

Finding Partial Derivatives with Constrained Variables

In Exercises 1–3, begin by drawing a diagram that shows the relations among the variables.

1. If $w = 2x^2 + 3y^2 + z^2$ and $z = x^2 + 2y^2$, find

a. $\left(\frac{\partial w}{\partial y}\right)_z$ **b.** $\left(\frac{\partial w}{\partial z}\right)_x$ **c.** $\left(\frac{\partial w}{\partial z}\right)_y$.

2. If $w = x^2 + y - z + \sin t$ and $x + y = t$, find

a. $\left(\frac{\partial w}{\partial y}\right)_{x,z}$ **b.** $\left(\frac{\partial w}{\partial y}\right)_{z,t}$ **c.** $\left(\frac{\partial w}{\partial z}\right)_{x,y}$

d. $\left(\frac{\partial w}{\partial z}\right)_{y,t}$ **e.** $\left(\frac{\partial w}{\partial t}\right)_{x,z}$ **f.** $\left(\frac{\partial w}{\partial t}\right)_{y,z}$.

3. Let $U = f(P, V, T)$ be the internal energy of a gas that obeys the ideal gas law $PV = nRT$ (n and R constant). Find

a. $\left(\frac{\partial U}{\partial P}\right)_V$ **b.** $\left(\frac{\partial U}{\partial T}\right)_V$.

4. Find

a. $\left(\frac{\partial w}{\partial x}\right)_y$ **b.** $\left(\frac{\partial w}{\partial z}\right)_y$

at the point $(x, y, z) = \left(\frac{\pi}{2}, 1, \pi\right)$ if

$w = x^2 + y^2 - z^2$ and $y \cos z - z \cos x = 0$.

5. Find

a. $\left(\frac{\partial w}{\partial y}\right)_x$ **b.** $\left(\frac{\partial w}{\partial y}\right)_z$

at the point $(w, x, y, z) = (4, 2, 1, -1)$ if

$$w = x^2y^2 + yz - z^3 \quad \text{and} \quad x^2 + y^2 + z^2 = 6.$$

6. Find $\left(\frac{\partial u}{\partial x}\right)_y$ at the point $(u, v) = \left(\sqrt{2}, 1\right)$, if $y = u^2 + v^2$ and $x = uv$.

7. Suppose that $x^2 + y^2 = r^2$ and $y = r \sin \theta$, as in polar coordinates. Find

$$\left(\frac{\partial y}{\partial r}\right)_\theta \quad \text{and} \quad \left(\frac{\partial r}{\partial y}\right)_x.$$

8. Suppose that

$$w = x^2 - y^2 + 4z + t \quad \text{and} \quad x + 2z + t = 25.$$

Show that the equations

$$\frac{\partial w}{\partial x} = 2x - 1 \quad \text{and} \quad \frac{\partial w}{\partial x} = 2x - 2$$

each give $\partial w/\partial x$, depending on which variables are chosen to be dependent and which variables are chosen to be independent. Identify the independent variables in each case.

Theory and Examples

9. Establish the fact, widely used in hydrodynamics, that if $f(x, y, z) = 0$, then

$$\left(\frac{\partial x}{\partial y}\right)_z \left(\frac{\partial y}{\partial z}\right)_x \left(\frac{\partial z}{\partial x}\right)_y = -1.$$

(*Hint:* Express all the derivatives in terms of the formal partial derivatives $\partial f/\partial x$, $\partial f/\partial y$, and $\partial f/\partial z$.)

10. If $z = x + f(u)$, where $u = xy$, show that

$$x\frac{\partial z}{\partial x} - y\frac{\partial z}{\partial y} = x.$$

11. Suppose that the equation $g(x, y, z) = 0$ determines z as a differentiable function of the independent variables x and y and that $g_z \neq 0$. Show that

$$\left(\frac{\partial z}{\partial y}\right)_x = -\frac{\partial g/\partial y}{\partial g/\partial z}.$$

12. Suppose that $f(x, y, z, w) = 0$ and $g(x, y, z, w) = 0$ determine z and w as differentiable functions of the independent variables x and y, and suppose that

$$\frac{\partial f}{\partial z}\frac{\partial g}{\partial w} - \frac{\partial f}{\partial w}\frac{\partial g}{\partial z} \neq 0.$$

Show that

$$\left(\frac{\partial z}{\partial x}\right)_y = -\frac{\dfrac{\partial f}{\partial x}\dfrac{\partial g}{\partial w} - \dfrac{\partial f}{\partial w}\dfrac{\partial g}{\partial x}}{\dfrac{\partial f}{\partial z}\dfrac{\partial g}{\partial w} - \dfrac{\partial f}{\partial w}\dfrac{\partial g}{\partial z}}$$

and

$$\left(\frac{\partial w}{\partial y}\right)_x = -\frac{\dfrac{\partial f}{\partial z}\dfrac{\partial g}{\partial y} - \dfrac{\partial f}{\partial y}\dfrac{\partial g}{\partial z}}{\dfrac{\partial f}{\partial z}\dfrac{\partial g}{\partial w} - \dfrac{\partial f}{\partial w}\dfrac{\partial g}{\partial z}}.$$

CHAPTER 14 Questions to Guide Your Review

1. What is a real-valued function of two independent variables? Three independent variables? Give examples.
2. What does it mean for sets in the plane or in space to be open? Closed? Give examples. Give examples of sets that are neither open nor closed.
3. How can you display the values of a function $f(x, y)$ of two independent variables graphically? How do you do the same for a function $f(x, y, z)$ of three independent variables?
4. What does it mean for a function $f(x, y)$ to have limit L as $(x, y) \to (x_0, y_0)$? What are the basic properties of limits of functions of two independent variables?
5. When is a function of two (three) independent variables continuous at a point in its domain? Give examples of functions that are continuous at some points but not others.
6. What can be said about algebraic combinations and compositions of continuous functions?

7. Explain the two-path test for nonexistence of limits.
8. How are the partial derivatives $\partial f/\partial x$ and $\partial f/\partial y$ of a function $f(x, y)$ defined? How are they interpreted and calculated?
9. How does the relation between first partial derivatives and continuity of functions of two independent variables differ from the relation between first derivatives and continuity for real-valued functions of a single independent variable? Give an example.
10. What is the Mixed Derivative Theorem for mixed second-order partial derivatives? How can it help in calculating partial derivatives of second and higher orders? Give examples.
11. What does it mean for a function $f(x, y)$ to be differentiable? What does the Increment Theorem say about differentiability?
12. How can you sometimes decide from examining f_x and f_y that a function $f(x, y)$ is differentiable? What is the relation between the differentiability of f and the continuity of f at a point?
13. What is the general Chain Rule? What form does it take for functions of two independent variables? Three independent variables? Functions defined on surfaces? How do you diagram these different forms? Give examples. What pattern enables one to remember all the different forms?
14. What is the derivative of a function $f(x, y)$ at a point P_0 in the direction of a unit vector **u**? What rate does it describe? What geometric interpretation does it have? Give examples.
15. What is the gradient vector of a differentiable function $f(x, y)$? How is it related to the function's directional derivatives? State the analogous results for functions of three independent variables.
16. How do you find the tangent line at a point on a level curve of a differentiable function $f(x, y)$? How do you find the tangent plane and normal line at a point on a level surface of a differentiable function $f(x, y, z)$? Give examples.
17. How can you use directional derivatives to estimate change?
18. How do you linearize a function $f(x, y)$ of two independent variables at a point (x_0, y_0)? Why might you want to do this? How do you linearize a function of three independent variables?
19. What can you say about the accuracy of linear approximations of functions of two (three) independent variables?
20. If (x, y) moves from (x_0, y_0) to a point $(x_0 + dx, y_0 + dy)$ nearby, how can you estimate the resulting change in the value of a differentiable function $f(x, y)$? Give an example.
21. How do you define local maxima, local minima, and saddle points for a differentiable function $f(x, y)$? Give examples.
22. What derivative tests are available for determining the local extreme values of a function $f(x, y)$? How do they enable you to narrow your search for these values? Give examples.
23. How do you find the extrema of a continuous function $f(x, y)$ on a closed bounded region of the *xy*-plane? Give an example.
24. Describe the method of Lagrange multipliers and give examples.
25. How does Taylor's formula for a function $f(x, y)$ generate polynomial approximations and error estimates?
26. If $w = f(x, y, z)$, where the variables x, y, and z are constrained by an equation $g(x, y, z) = 0$, what is the meaning of the notation $(\partial w/\partial x)_y$? How can an arrow diagram help you calculate this partial derivative with constrained variables? Give examples.

CHAPTER 14 Practice Exercises

Domain, Range, and Level Curves

In Exercises 1–4, find the domain and range of the given function and identify its level curves. Sketch a typical level curve.

1. $f(x, y) = 9x^2 + y^2$
2. $f(x, y) = e^{x+y}$
3. $g(x, y) = 1/xy$
4. $g(x, y) = \sqrt{x^2 - y}$

In Exercises 5–8, find the domain and range of the given function and identify its level surfaces. Sketch a typical level surface.

5. $f(x, y, z) = x^2 + y^2 - z$
6. $g(x, y, z) = x^2 + 4y^2 + 9z^2$
7. $h(x, y, z) = \dfrac{1}{x^2 + y^2 + z^2}$
8. $k(x, y, z) = \dfrac{1}{x^2 + y^2 + z^2 + 1}$

Evaluating Limits

Find the limits in Exercises 9–14.

9. $\displaystyle\lim_{(x, y)\to(\pi, \ln 2)} e^y \cos x$
10. $\displaystyle\lim_{(x, y)\to(0, 0)} \frac{2 + y}{x + \cos y}$
11. $\displaystyle\lim_{(x, y)\to(1, 1)} \frac{x - y}{x^2 - y^2}$
12. $\displaystyle\lim_{(x, y)\to(1, 1)} \frac{x^3y^3 - 1}{xy - 1}$
13. $\displaystyle\lim_{P\to(1, -1, e)} \ln|x + y + z|$
14. $\displaystyle\lim_{P\to(1, -1, -1)} \arctan(x + y + z)$

By considering different paths of approach, show that the limits in Exercises 15 and 16 do not exist.

15. $\displaystyle\lim_{\substack{(x, y)\to(0, 0)\\ y\neq x^2}} \frac{y}{x^2 - y}$
16. $\displaystyle\lim_{\substack{(x, y)\to(0, 0)\\ xy\neq 0}} \frac{x^2 + y^2}{xy}$

17. **Continuous extension** Let $f(x, y) = (x^2 - y^2)/(x^2 + y^2)$ for $(x, y) \neq (0, 0)$. Is it possible to define $f(0, 0)$ in a way that makes f continuous at the origin? Why?
18. **Continuous extension** Let

$$f(x, y) = \begin{cases} \dfrac{\sin(x - y)}{|x| + |y|}, & |x| + |y| \neq 0 \\ 0, & (x, y) = (0, 0). \end{cases}$$

Is f continuous at the origin? Why?

Partial Derivatives

In Exercises 19–24, find the partial derivative of the function with respect to each variable.

19. $g(r, \theta) = r \cos\theta + r \sin\theta$
20. $f(x, y) = \dfrac{1}{2}\ln(x^2 + y^2) + \arctan\dfrac{y}{x}$

21. $f(R_1, R_2, R_3) = \dfrac{1}{R_1} + \dfrac{1}{R_2} + \dfrac{1}{R_3}$

22. $h(x, y, z) = \sin(2\pi x + y - 3z)$

23. $P(n, R, T, V) = \dfrac{nRT}{V}$ (the ideal gas law)

24. $f(r, l, T, w) = \dfrac{1}{2rl}\sqrt{\dfrac{T}{\pi w}}$

Second-Order Partials

Find the second-order partial derivatives of the functions in Exercises 25–28.

25. $g(x, y) = y + \dfrac{x}{y}$

26. $g(x, y) = e^x + y \sin x$

27. $f(x, y) = x + xy - 5x^3 + \ln(x^2 + 1)$

28. $f(x, y) = y^2 - 3xy + \cos y + 7e^y$

Chain Rule Calculations

29. Find dw/dt at $t = 0$ if $w = \sin(xy + \pi)$, $x = e^t$, and $y = \ln(t + 1)$.

30. Find dw/dt at $t = 1$ if $w = xe^y + y \sin z - \cos z$, $x = 2\sqrt{t}$, $y = t - 1 + \ln t$, and $z = \pi t$.

31. Find $\partial w/\partial r$ and $\partial w/\partial s$ when $r = \pi$ and $s = 0$ if $w = \sin(2x - y)$, $x = r + \sin s$, $y = rs$.

32. Find $\partial w/\partial u$ and $\partial w/\partial v$ when $u = v = 0$ if $w = \ln\sqrt{1 + x^2} - \tan^{-1} x$ and $x = 2e^u \cos v$.

33. Find the value of the derivative of $f(x, y, z) = xy + yz + xz$ with respect to t on the curve $x = \cos t$, $y = \sin t$, $z = \cos 2t$ at $t = 1$.

34. Show that if $w = f(s)$ is any differentiable function of s and if $s = y + 5x$, then

$$\frac{\partial w}{\partial x} - 5\frac{\partial w}{\partial y} = 0.$$

Implicit Differentiation

Assuming that the equations in Exercises 35 and 36 define y as a differentiable function of x, find the value of dy/dx at point P.

35. $1 - x - y^2 - \sin xy = 0$, $P(0, 1)$

36. $2xy + e^{x+y} - 2 = 0$, $P(0, \ln 2)$

Directional Derivatives

In Exercises 37–40, find the directions in which f increases and decreases most rapidly at P_0 and find the derivative of f in each direction. Also, find the derivative of f at P_0 in the direction of the vector $\mathbf{v}$.

37. $f(x, y) = \cos x \cos y$, $P_0(\pi/4, \pi/4)$, $\mathbf{v} = 3\mathbf{i} + 4\mathbf{j}$

38. $f(x, y) = x^2 e^{-2y}$, $P_0(1, 0)$, $\mathbf{v} = \mathbf{i} + \mathbf{j}$

39. $f(x, y, z) = \ln(2x + 3y + 6z)$, $P_0(-1, -1, 1)$, $\mathbf{v} = 2\mathbf{i} + 3\mathbf{j} + 6\mathbf{k}$

40. $f(x, y, z) = x^2 + 3xy - z^2 + 2y + z + 4$, $P_0(0, 0, 0)$, $\mathbf{v} = \mathbf{i} + \mathbf{j} + \mathbf{k}$

41. **Derivative in velocity direction** Find the derivative of $f(x, y, z) = xyz$ in the direction of the velocity vector of the helix

$$\mathbf{r}(t) = (\cos 3t)\mathbf{i} + (\sin 3t)\mathbf{j} + 3t\mathbf{k}$$

at $t = \pi/3$.

42. **Maximum directional derivative** What is the largest value that the directional derivative of $f(x, y, z) = xyz$ can have at the point $(1, 1, 1)$?

43. **Directional derivatives with given values** At the point $(1, 2)$, the function $f(x, y)$ has a derivative of 2 in the direction toward $(2, 2)$ and a derivative of -2 in the direction toward $(1, 1)$.

  a. Find $f_x(1, 2)$ and $f_y(1, 2)$.

  b. Find the derivative of f at $(1, 2)$ in the direction toward the point $(4, 6)$.

44. Which of the following statements are true if $f(x, y)$ is differentiable at (x_0, y_0)? Give reasons for your answers.

  a. If $\mathbf{u}$ is a unit vector, the derivative of f at (x_0, y_0) in the direction of $\mathbf{u}$ is $(f_x(x_0, y_0)\mathbf{i} + f_y(x_0, y_0)\mathbf{j}) \cdot \mathbf{u}$.

  b. The derivative of f at (x_0, y_0) in the direction of $\mathbf{u}$ is a vector.

  c. The directional derivative of f at (x_0, y_0) has its greatest value in the direction of ∇f.

  d. At (x_0, y_0), vector ∇f is normal to the curve $f(x, y) = f(x_0, y_0)$.

Gradients, Tangent Planes, and Normal Lines

In Exercises 45 and 46, sketch the surface $f(x, y, z) = c$ together with ∇f at the given points.

45. $x^2 + y + z^2 = 0$; $(0, -1, \pm 1)$, $(0, 0, 0)$

46. $y^2 + z^2 = 4$; $(2, \pm 2, 0)$, $(2, 0, \pm 2)$

In Exercises 47 and 48, find an equation for the plane tangent to the level surface $f(x, y, z) = c$ at the point P_0. Also, find parametric equations for the line that is normal to the surface at P_0.

47. $x^2 - y - 5z = 0$, $P_0(2, -1, 1)$

48. $x^2 + y^2 + z = 4$, $P_0(1, 1, 2)$

In Exercises 49 and 50, find an equation for the plane tangent to the surface $z = f(x, y)$ at the given point.

49. $z = \ln(x^2 + y^2)$, $(0, 1, 0)$

50. $z = 1/(x^2 + y^2)$, $(1, 1, 1/2)$

In Exercises 51 and 52, find equations for the lines that are tangent and normal to the level curve $f(x, y) = c$ at the point P_0. Then sketch the lines and level curve together with ∇f at P_0.

51. $y - \sin x = 1$, $P_0(\pi, 1)$

52. $\dfrac{y^2}{2} - \dfrac{x^2}{2} = \dfrac{3}{2}$, $P_0(1, 2)$

Tangent Lines to Curves

In Exercises 53 and 54, find parametric equations for the line that is tangent to the curve of intersection of the surfaces at the given point.

53. Surfaces: $x^2 + 2y + 2z = 4$, $y = 1$
  Point: $(1, 1, 1/2)$

54. Surfaces: $x + y^2 + z = 2$, $y = 1$
  Point: $(1/2, 1, 1/2)$

Linearizations

In Exercises 55 and 56, find the linearization $L(x, y)$ of the function $f(x, y)$ at the point P_0. Then find an upper bound for the magnitude of the error E in the approximation $f(x, y) \approx L(x, y)$ over the rectangle R.

55. $f(x, y) = \sin x \cos y$, $P_0(\pi/4, \pi/4)$
  R: $\left|x - \dfrac{\pi}{4}\right| \le 0.1$, $\left|y - \dfrac{\pi}{4}\right| \le 0.1$

56. $f(x, y) = xy - 3y^2 + 2$, $P_0(1, 1)$
  R: $|x - 1| \le 0.1$, $|y - 1| \le 0.2$

Find the linearizations of the functions in Exercises 57 and 58 at the given points.

57. $f(x, y, z) = xy + 2yz - 3xz$ at $(1, 0, 0)$ and $(1, 1, 0)$

58. $f(x, y, z) = \sqrt{2} \cos x \sin(y + z)$ at $(0, 0, \pi/4)$ and $(\pi/4, \pi/4, 0)$

Estimates and Sensitivity to Change

59. Measuring the volume of a pipeline You plan to calculate the volume inside a stretch of pipeline that is about 36 in. in diameter and 1 mile long. With which measurement should you be more careful, the length or the diameter? Why?

60. Sensitivity to change Is $f(x, y) = x^2 - xy + y^2 - 3$ more sensitive to changes in x or to changes in y when it is near the point $(1, 2)$? How do you know?

61. Change in an electrical circuit Suppose that the current I (amperes) in an electrical circuit is related to the voltage V (volts) and the resistance R (ohms) by the equation $I = V/R$. If the voltage drops from 24 to 23 volts and the resistance drops from 100 to 80 ohms, will I increase or decrease? By about how much? Is the change in I more sensitive to change in the voltage or to change in the resistance? How do you know?

62. Maximum error in estimating the area of an ellipse If $a = 10$ cm and $b = 16$ cm to the nearest millimeter, what should you expect the maximum percentage error to be in the calculated area $A = \pi ab$ of the ellipse $x^2/a^2 + y^2/b^2 = 1$?

63. Error in estimating a product Let $y = uv$ and $z = u + v$, where u and v are positive independent variables.

a. If u is measured with an error of 2% and v with an error of 3%, about what is the percentage error in the calculated value of y?

b. Show that the percentage error in the calculated value of z is less than the percentage error in the value of y.

64. Cardiac index To make different people comparable in studies of cardiac output, researchers divide the measured cardiac output by the body surface area to find the *cardiac index* C:

$$C = \frac{\text{cardiac output}}{\text{body surface area}}.$$

The body surface area B of a person with weight w and height h is approximated by the formula

$$B = 71.84 w^{0.425} h^{0.725},$$

which gives B in square centimeters when w is measured in kilograms and h in centimeters. You are about to calculate the cardiac index of a person 180 cm tall, weighing 70 kg, with cardiac output of 7 L/min. Which will have a greater effect on the calculation, a 1-kg error in measuring the weight or a 1-cm error in measuring the height?

Local Extrema

Test the functions in Exercises 65–70 for local maxima and minima and saddle points. Find each function's value at these points.

65. $f(x, y) = x^2 - xy + y^2 + 2x + 2y - 4$

66. $f(x, y) = 5x^2 + 4xy - 2y^2 + 4x - 4y$

67. $f(x, y) = 2x^3 + 3xy + 2y^3$

68. $f(x, y) = x^3 + y^3 - 3xy + 15$

69. $f(x, y) = x^3 + y^3 + 3x^2 - 3y^2$

70. $f(x, y) = x^4 - 8x^2 + 3y^2 - 6y$

Absolute Extrema

In Exercises 71–78, find the absolute maximum and minimum values of f on the region R.

71. $f(x, y) = x^2 + xy + y^2 - 3x + 3y$

R: The triangular region cut from the first quadrant by the line $x + y = 4$

72. $f(x, y) = x^2 - y^2 - 2x + 4y + 1$

R: The rectangular region in the first quadrant bounded by the coordinate axes and the lines $x = 4$ and $y = 2$

73. $f(x, y) = y^2 - xy - 3y + 2x$

R: The square region enclosed by the lines $x = \pm 2$ and $y = \pm 2$

74. $f(x, y) = 2x + 2y - x^2 - y^2$

R: The square region bounded by the coordinate axes and the lines $x = 2$, $y = 2$ in the first quadrant

75. $f(x, y) = x^2 - y^2 - 2x + 4y$

R: The triangular region bounded below by the x-axis, above by the line $y = x + 2$, and on the right by the line $x = 2$

76. $f(x, y) = 4xy - x^4 - y^4 + 16$

R: The triangular region bounded below by the line $y = -2$, above by the line $y = x$, and on the right by the line $x = 2$

77. $f(x, y) = x^3 + y^3 + 3x^2 - 3y^2$

R: The square region enclosed by the lines $x = \pm 1$ and $y = \pm 1$

78. $f(x, y) = x^3 + 3xy + y^3 + 1$

R: The square region enclosed by the lines $x = \pm 1$ and $y = \pm 1$

Lagrange Multipliers

79. Extrema on a circle Find the extreme values of $f(x, y) = x^3 + y^2$ on the circle $x^2 + y^2 = 1$.

80. Extrema on a circle Find the extreme values of $f(x, y) = xy$ on the circle $x^2 + y^2 = 1$.

81. Extrema in a disk Find the extreme values of $f(x, y) = x^2 + 3y^2 + 2y$ on the unit disk $x^2 + y^2 \le 1$.

82. Extrema in a disk Find the extreme values of $f(x, y) = x^2 + y^2 - 3x - xy$ on the disk $x^2 + y^2 \le 9$.

83. Extrema on a sphere Find the extreme values of $f(x, y, z) = x - y + z$ on the unit sphere $x^2 + y^2 + z^2 = 1$.

84. Minimum distance to origin Find the points on the surface $x^2 - zy = 4$ closest to the origin.

85. Minimizing cost of a box A closed rectangular box is to have volume V cm^3. The cost of the material used in the box is a cents/cm^2 for top and bottom, b cents/cm^2 for front and back, and c cents/cm^2 for the remaining sides. What dimensions minimize the total cost of materials?

86. Least volume Find the plane $x/a + y/b + z/c = 1$ that passes through the point $(2, 1, 2)$ and cuts off the least volume from the first octant.

87. Extrema on curve of intersecting surfaces Find the extreme values of $f(x, y, z) = x(y + z)$ on the curve of intersection of the right circular cylinder $x^2 + y^2 = 1$ and the hyperbolic cylinder $xz = 1$.

88. Minimum distance to origin on curve of intersecting plane and cone Find the point closest to the origin on the curve of intersection of the plane $x + y + z = 1$ and the cone $z^2 = 2x^2 + 2y^2$.

Theory and Examples

89. Let $w = f(r, \theta)$, $r = \sqrt{x^2 + y^2}$, and $\theta = \tan^{-1}(y/x)$. Find $\partial w/\partial x$ and $\partial w/\partial y$, and express your answers in terms of r and θ.

90. Let $z = f(u, v)$, $u = ax + by$, and $v = ax - by$. Express z_x and z_y in terms of f_u, f_v, and the constants a and b.

91. If a and b are constants, $w = u^3 + \tanh u + \cos u$, and $u = ax + by$, show that

$$a\frac{\partial w}{\partial y} = b\frac{\partial w}{\partial x}.$$

92. Using the Chain Rule If $w = \ln(x^2 + y^2 + 2z)$, $x = r + s$, $y = r - s$, and $z = 2rs$, find w_r and w_s by the Chain Rule. Then check your answer another way.

93. Angle between vectors The equations $e^u \cos v - x = 0$ and $e^u \sin v - y = 0$ define u and v as differentiable functions of x and y. Show that the angle between the vectors

$$\frac{\partial u}{\partial x}\mathbf{i} + \frac{\partial u}{\partial y}\mathbf{j} \quad \text{and} \quad \frac{\partial v}{\partial x}\mathbf{i} + \frac{\partial v}{\partial y}\mathbf{j}$$

is constant.

94. Polar coordinates and second derivatives Introducing polar coordinates $x = r\cos\theta$ and $y = r\sin\theta$ changes $f(x, y)$ to $g(r, \theta)$. Find the value of $\partial^2 g/\partial\theta^2$ at the point $(r, \theta) = (2, \pi/2)$, given that

$$\frac{\partial f}{\partial x} = \frac{\partial f}{\partial y} = \frac{\partial^2 f}{\partial x^2} = \frac{\partial^2 f}{\partial y^2} = 1$$

at that point.

95. Normal line parallel to a plane Find the points on the surface

$$(y + z)^2 + (z - x)^2 = 16$$

where the normal line is parallel to the yz-plane.

96. Tangent plane parallel to xy-plane Find the points on the surface

$$xy + yz + zx - x - z^2 = 0$$

where the tangent plane is parallel to the xy-plane.

97. When gradient is parallel to position vector Suppose that $\nabla f(x, y, z)$ is always parallel to the position vector $x\mathbf{i} + y\mathbf{j} + z\mathbf{k}$. Show that $f(0, 0, a) = f(0, 0, -a)$ for any a.

98. One-sided directional derivative in all directions, but no gradient The *one-sided directional derivative of f at $P(x_0, y_0, z_0)$ in the direction* $\mathbf{u} = u_1\mathbf{i} + u_2\mathbf{j} + u_3\mathbf{k}$ is the number

$$\lim_{s\to 0^+} \frac{f(x_0 + su_1, y_0 + su_2, z_0 + su_3) - f(x_0, y_0, z_0)}{s}.$$

Show that the one-sided directional derivative of

$$f(x, y, z) = \sqrt{x^2 + y^2 + z^2}$$

at the origin equals 1 in any direction but that f has no gradient vector at the origin.

99. Normal line through origin Show that the line normal to the surface $xy + z = 2$ at the point $(1, 1, 1)$ passes through the origin.

100. Tangent plane and normal line

a. Sketch the surface $x^2 - y^2 + z^2 = 4$.

b. Find a vector normal to the surface at $(2, -3, 3)$. Add the vector to your sketch.

c. Find equations for the tangent plane and the normal line at $(2, -3, 3)$.

Partial Derivatives with Constrained Variables

In Exercises 101 and 102, begin by drawing a diagram that shows the relations among the variables.

101. If $w = x^2 e^{yz}$ and $z = x^2 - y^2$ find

a. $\left(\frac{\partial w}{\partial y}\right)_z$ **b.** $\left(\frac{\partial w}{\partial z}\right)_x$ **c.** $\left(\frac{\partial w}{\partial z}\right)_y$.

102. Let $U = f(P, V, T)$ be the internal energy of a gas that obeys the ideal gas law $PV = nRT$ (n and R constant). Find

a. $\left(\frac{\partial U}{\partial T}\right)_P$ **b.** $\left(\frac{\partial U}{\partial V}\right)_T$.

CHAPTER 14 Additional and Advanced Exercises

Partial Derivatives

1. Function with saddle at the origin If you did Exercise 64 in Section 14.2, you know that the function

$$f(x, y) = \begin{cases} xy\dfrac{x^2 - y^2}{x^2 + y^2}, & (x, y) \neq (0, 0) \\ 0, & (x, y) = (0, 0) \end{cases}$$

(see the accompanying figure) is continuous at $(0, 0)$. Find $f_{xy}(0, 0)$ and $f_{yx}(0, 0)$.

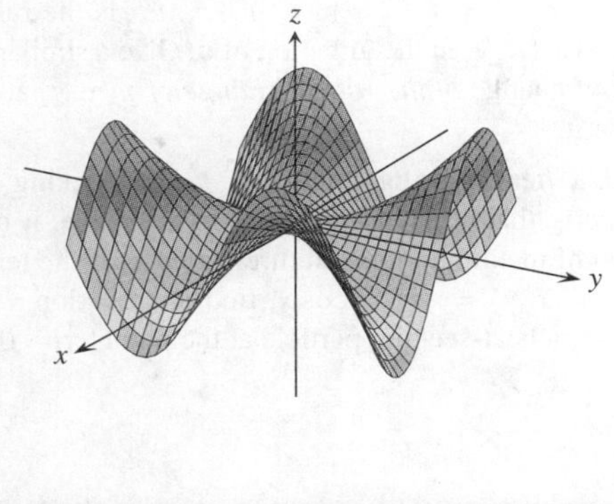

2. Finding a function from second partials Find a function $w = f(x, y)$ whose first partial derivatives are $\partial w/\partial x = 1 + e^x \cos y$ and $\partial w/\partial y = 2y - e^x \sin y$ and whose value at the point $(\ln 2, 0)$ is $\ln 2$.

3. A proof of Leibniz's Rule Leibniz's Rule says that if f is continuous on $[a, b]$ and if $u(x)$ and $v(x)$ are differentiable functions of x whose values lie in $[a, b]$, then

$$\frac{d}{dx}\int_{u(x)}^{v(x)} f(t)\,dt = f(v(x))\frac{dv}{dx} - f(u(x))\frac{du}{dx}.$$

Prove the rule by setting

$$g(u, v) = \int_u^v f(t)\,dt, \qquad u = u(x), \qquad v = v(x)$$

and calculating dg/dx with the Chain Rule.

4. Finding a function with constrained second partials Suppose that f is a twice-differentiable function of r, that $r = \sqrt{x^2 + y^2 + z^2}$, and that

$$f_{xx} + f_{yy} + f_{zz} = 0.$$

Show that for some constants a and b,

$$f(r) = \frac{a}{r} + b.$$

5. **Homogeneous functions** A function $f(x, y)$ is *homogeneous of degree n* (n a nonnegative integer) if $f(tx, ty) = t^n f(x, y)$ for all t, x, and y. For such a function (sufficiently differentiable), prove that

   a. $x\frac{\partial f}{\partial x} + y\frac{\partial f}{\partial y} = nf(x, y)$

   b. $x^2\left(\frac{\partial^2 f}{\partial x^2}\right) + 2xy\left(\frac{\partial^2 f}{\partial x\, \partial y}\right) + y^2\left(\frac{\partial^2 f}{\partial y^2}\right) = n(n-1)f.$

6. **Surface in polar coordinates** Let

$$f(r, \theta) = \begin{cases} \dfrac{\sin 6r}{6r}, & r \neq 0 \\ 1, & r = 0, \end{cases}$$

   where r and θ are polar coordinates. Find

   a. $\lim_{r \to 0} f(r, \theta)$

   b. $f_r(0, 0)$

   c. $f_\theta(r, \theta), \quad r \neq 0.$

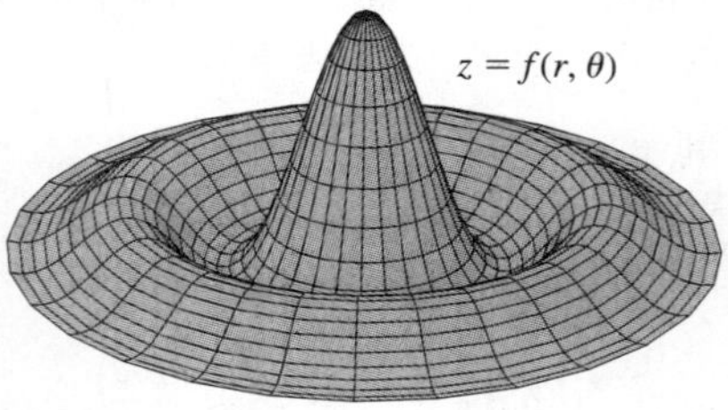

Gradients and Tangents

7. **Properties of position vectors** Let $\mathbf{r} = x\mathbf{i} + y\mathbf{j} + z\mathbf{k}$ and let $r = |\mathbf{r}|$.

   a. Show that $\nabla r = \mathbf{r}/r$.

   b. Show that $\nabla(r^n) = nr^{n-2}\mathbf{r}$.

   c. Find a function whose gradient equals $\mathbf{r}$.

   d. Show that $\mathbf{r} \cdot d\mathbf{r} = r\, dr$.

   e. Show that $\nabla(\mathbf{A} \cdot \mathbf{r}) = \mathbf{A}$ for any constant vector $\mathbf{A}$.

8. **Gradient orthogonal to tangent** Suppose that a differentiable function $f(x, y)$ has the constant value c along the differentiable curve $x = g(t)$, $y = h(t)$; that is,

$$f(g(t), h(t)) = c$$

   for all values of t. Differentiate both sides of this equation with respect to t to show that ∇f is orthogonal to the curve's tangent vector at every point on the curve.

9. **Curve tangent to a surface** Show that the curve

$$\mathbf{r}(t) = (\ln t)\mathbf{i} + (t \ln t)\mathbf{j} + t\mathbf{k}$$

   is tangent to the surface

$$xz^2 - yz + \cos xy = 1$$

   at $(0, 0, 1)$.

10. **Curve tangent to a surface** Show that the curve

$$\mathbf{r}(t) = \left(\frac{t^3}{4} - 2\right)\mathbf{i} + \left(\frac{4}{t} - 3\right)\mathbf{j} + \cos(t - 2)\mathbf{k}$$

   is tangent to the surface

$$x^3 + y^3 + z^3 - xyz = 0$$

   at $(0, -1, 1)$.

Extreme Values

11. **Extrema on a surface** Show that the only possible maxima and minima of z on the surface $z = x^3 + y^3 - 9xy + 27$ occur at $(0, 0)$ and $(3, 3)$. Show that neither a maximum nor a minimum occurs at $(0, 0)$. Determine whether z has a maximum or a minimum at $(3, 3)$.

12. **Maximum in closed first quadrant** Find the maximum value of $f(x, y) = 6xye^{-(2x+3y)}$ in the closed first quadrant (includes the nonnegative axes).

13. **Minimum volume cut from first octant** Find the minimum volume for a region bounded by the planes $x = 0$, $y = 0$, $z = 0$ and a plane tangent to the ellipsoid

$$\frac{x^2}{a^2} + \frac{y^2}{b^2} + \frac{z^2}{c^2} = 1$$

   at a point in the first octant.

14. **Minimum distance from a line to a parabola in *xy*-plane** By minimizing the function $f(x, y, u, v) = (x - u)^2 + (y - v)^2$ subject to the constraints $y = x + 1$ and $u = v^2$, find the minimum distance in the xy-plane from the line $y = x + 1$ to the parabola $y^2 = x$.

Theory and Examples

15. **Boundedness of first partials implies continuity** Prove the following theorem: If $f(x, y)$ is defined in an open region R of the xy-plane and if f_x and f_y are bounded on R, then $f(x, y)$ is continuous on R. (The assumption of boundedness is essential.)

16. Suppose that $\mathbf{r}(t) = g(t)\mathbf{i} + h(t)\mathbf{j} + k(t)\mathbf{k}$ is a smooth curve in the domain of a differentiable function $f(x, y, z)$. Describe the relation among df/dt, ∇f, and $\mathbf{v} = d\mathbf{r}/dt$. What can be said about ∇f and $\mathbf{v}$ at interior points of the curve where f has extreme values relative to its other values on the curve? Give reasons for your answer.

17. **Finding functions from partial derivatives** Suppose that f and g are functions of x and y such that

$$\frac{\partial f}{\partial y} = \frac{\partial g}{\partial x} \quad \text{and} \quad \frac{\partial f}{\partial x} = \frac{\partial g}{\partial y},$$

   and suppose that

$$\frac{\partial f}{\partial x} = 0, \quad f(1, 2) = g(1, 2) = 5, \quad \text{and} \quad f(0, 0) = 4.$$

   Find $f(x, y)$ and $g(x, y)$.

18. **Rate of change of the rate of change** We know that if $f(x, y)$ is a function of two variables and if $\mathbf{u} = a\mathbf{i} + b\mathbf{j}$ is a unit vector, then $D_\mathbf{u} f(x, y) = f_x(x, y)a + f_y(x, y)b$ is the rate of change of $f(x, y)$ at (x, y) in the direction of $\mathbf{u}$. Give a similar formula for the rate of change *of the rate of change of* $f(x, y)$ at (x, y) in the direction $\mathbf{u}$.

19. **Path of a heat-seeking particle** A heat-seeking particle has the property that at any point (x, y) in the plane, it moves in the direction of maximum temperature increase. If the temperature at (x, y) is $T(x, y) = -e^{-2y} \cos x$, find an equation $y = f(x)$ for the path of a heat-seeking particle at the point $(\pi/4, 0)$.

20. Velocity after a ricochet A particle traveling in a straight line with constant velocity $\mathbf{i} + \mathbf{j} - 5\mathbf{k}$ passes through the point $(0, 0, 30)$ and hits the surface $z = 2x^2 + 3y^2$. The particle ricochets off the surface, the angle of reflection being equal to the angle of incidence. Assuming no loss of speed, what is the velocity of the particle after the ricochet? Simplify your answer.

21. Directional derivatives tangent to a surface Let S be the surface that is the graph of $f(x, y) = 10 - x^2 - y^2$. Suppose that the temperature in space at each point (x, y, z) is $T(x, y, z) = x^2y + y^2z + 4x + 14y + z$.

a. Among all the possible directions tangential to the surface S at the point $(0, 0, 10)$, which direction will make the rate of change of temperature at $(0, 0, 10)$ a maximum?

b. Which direction tangential to S at the point $(1, 1, 8)$ will make the rate of change of temperature a maximum?

22. Drilling another borehole On a flat surface of land, geologists drilled a borehole straight down and hit a mineral deposit at 1000 ft. They drilled a second borehole 100 ft to the north of the first and hit the mineral deposit at 950 ft. A third borehole 100 ft east of the first borehole struck the mineral deposit at 1025 ft. The geologists have reasons to believe that the mineral deposit is in the shape of a dome, and for the sake of economy, they would like to find where the deposit is closest to the surface. Assuming the surface to be the xy-plane, in what direction from the first borehole would you suggest the geologists drill their fourth borehole?

The one-dimensional heat equation If $w(x, t)$ represents the temperature at position x at time t in a uniform wire with perfectly insulated sides, then the partial derivatives w_{xx} and w_t satisfy a differential equation of the form

$$w_{xx} = \frac{1}{c^2} w_t.$$

This equation is called the *one-dimensional heat equation*. The value of the positive constant c^2 is determined by the material from which the wire is made.

23. Find all solutions of the one-dimensional heat equation of the form $w = e^{rt} \sin \pi x$, where r is a constant.

24. Find all solutions of the one-dimensional heat equation that have the form $w = e^{rt} \sin kx$ and satisfy the conditions that $w(0, t) = 0$ and $w(L, t) = 0$. What happens to these solutions as $t \to \infty$?

CHAPTER 14 Technology Application Projects

Mathematica/Maple Projects

Projects can be found within MyLab Math.

- ***Plotting Surfaces***
  Efficiently generate plots of surfaces, contours, and level curves.
- ***Exploring the Mathematics Behind Skateboarding: Analysis of the Directional Derivative***
  The path of a skateboarder is introduced, first on a level plane, then on a ramp, and finally on a paraboloid. Compute, plot, and analyze the directional derivative in terms of the skateboarder.
- ***Looking for Patterns and Applying the Method of Least Squares to Real Data***
  Fit a line to a set of numerical data points by choosing the line that minimizes the sum of the squares of the vertical distances from the points to the line.
- ***Lagrange Goes Skateboarding: How High Does He Go?***
  Revisit and analyze the skateboarders' adventures for maximum and minimum heights from both a graphical and analytic perspective using Lagrange multipliers.

15 Multiple Integrals

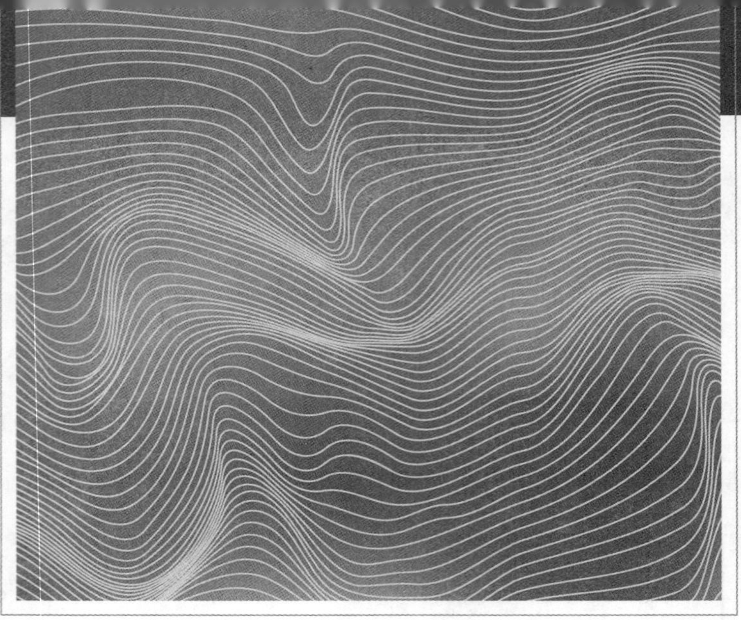

OVERVIEW In this chapter we define the *double integral* of a function of two variables $f(x, y)$ over a region in the plane as the limit of approximating Riemann sums. Just as a single integral can represent signed area, so can a double integral represent signed volume. Double integrals can be evaluated using the Fundamental Theorem of Calculus studied in Section 5.4, but now the evaluations are done twice by integrating with respect to each of the variables x and y in turn. Double integrals can be used to find areas of more general regions in the plane than those encountered in Chapter 5. Moreover, just as the Substitution Rule could simplify finding single integrals, we can sometimes use polar coordinates to simplify computing a double integral. We study more general substitutions for evaluating double integrals as well.

We also define the *triple integral* of a function of three variables $f(x, y, z)$ over a region in space. Triple integrals can be used to find volumes of still more general regions in space, and their evaluation is like that of double integrals with yet a third evaluation. *Cylindrical* or *spherical coordinates* can sometimes be used to simplify the calculation of a triple integral, and we investigate those techniques. Double and triple integrals have a number of applications, such as calculating the average value of a multivariable function, and finding moments and centers of mass.

15.1 Double and Iterated Integrals over Rectangles

In Chapter 5 we defined the definite integral of a function $f(x)$ over an interval $[a, b]$ as a limit of Riemann sums. In this section we extend this idea to define the *double integral* of a function of two variables $f(x, y)$ over a bounded rectangle R in the plane. The Riemann sums for the integral of a single-variable function $f(x)$ are obtained by partitioning a finite interval into thin subintervals, multiplying the width of each subinterval by the value of f at a point c_k inside that subinterval, and then adding together all the products. A similar method of partitioning, multiplying, and summing is used to construct double integrals as limits of approximating Riemann sums.

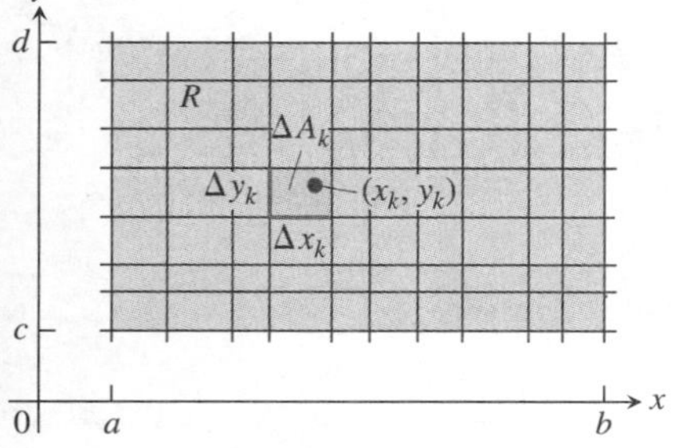

FIGURE 15.1 Rectangular grid partitioning the region R into small rectangles of area $\Delta A_k = \Delta x_k \, \Delta y_k$.

Double Integrals

We begin our investigation of double integrals by considering the simplest type of planar region, a rectangle. We consider a function $f(x, y)$ defined on a rectangular region R,

$$R: \quad a \leq x \leq b, \quad c \leq y \leq d.$$

We subdivide R into small rectangles using a network of lines parallel to the x- and y-axes (Figure 15.1). The lines divide R into n rectangular pieces, where the number of such pieces n gets large as the width and height of each piece gets small. These rectangles form a **partition** of R. A small rectangular piece of width Δx and height Δy has area

$\Delta A = \Delta x\,\Delta y$. If we number the small pieces partitioning R in some order, then their areas are given by numbers $\Delta A_1, \Delta A_2, \ldots, \Delta A_n$, where ΔA_k is the area of the kth small rectangle.

To form a Riemann sum over R, we choose a point (x_k, y_k) in the kth small rectangle, multiply the value of f at that point by the area ΔA_k, and add together the products:

$$S_n = \sum_{k=1}^{n} f(x_k, y_k)\,\Delta A_k.$$

Depending on how we pick (x_k, y_k) in the kth small rectangle, we may get different values for S_n.

We are interested in what happens to these Riemann sums as the widths and heights of all the small rectangles in the partition of R approach zero. The **norm** of a partition P, written $\|P\|$, is the largest width or height of any rectangle in the partition. If $\|P\| = 0.1$, then all the rectangles in the partition of R have width at most 0.1 and height at most 0.1. Sometimes the Riemann sums converge as the norm of P goes to zero, which is written $\|P\| \to 0$. The resulting limit is then written as

$$\lim_{\|P\|\to 0} \sum_{k=1}^{n} f(x_k, y_k)\,\Delta A_k.$$

As $\|P\| \to 0$ and the rectangles get narrow and short, their number n increases, so we can also write this limit as

$$\lim_{n\to\infty} \sum_{k=1}^{n} f(x_k, y_k)\,\Delta A_k,$$

with the understanding that $\|P\| \to 0$, and hence $\Delta A_k \to 0$, as $n \to \infty$.

Many choices are involved in a limit of this kind. The collection of small rectangles is determined by the grid of vertical and horizontal lines that determine a rectangular partition of R. In each of the resulting small rectangles there is a choice of an arbitrary point (x_k, y_k) at which f is evaluated. These choices together determine a single Riemann sum. To form a limit, we repeat the whole process again and again, choosing partitions whose rectangle widths and heights both go to zero and whose number goes to infinity.

When a limit of the sums S_n exists, giving the same limiting value no matter what choices are made, then the function f is said to be **integrable** and the limit is called the **double integral** of f over R, which is written as

$$\iint_R f(x, y)\,dA \quad \text{or} \quad \iint_R f(x, y)\,dx\,dy.$$

It can be shown that if $f(x, y)$ is a continuous function throughout R, then f is integrable, as in the single-variable case discussed in Chapter 5. Many discontinuous functions are also integrable, including functions that are discontinuous only on a finite number of points or smooth curves. We leave the proof of these facts to a more advanced text.

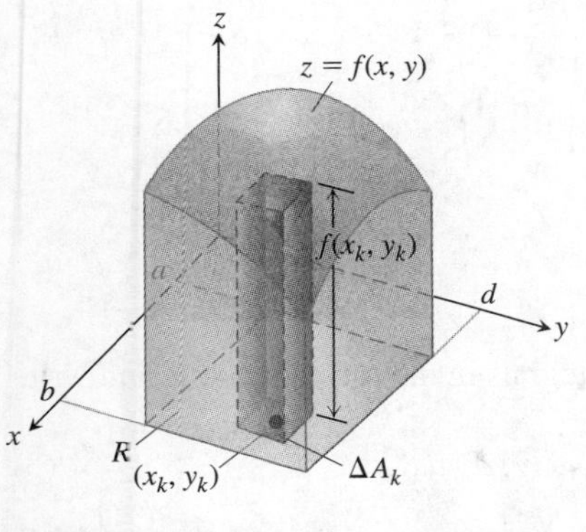

FIGURE 15.2 Approximating solids with rectangular boxes leads us to define the volumes of more general solids as double integrals. The volume of the solid shown here is the double integral of $f(x, y)$ over the base region R.

Double Integrals as Volumes

When $f(x, y)$ is a positive function over a rectangular region R in the xy-plane, we may interpret the double integral of f over R as the volume of the three-dimensional solid region over the xy-plane bounded below by R and above by the surface $z = f(x, y)$ (Figure 15.2). Each term $f(x_k, y_k)\,\Delta A_k$ in the sum $S_n = \sum f(x_k, y_k)\,\Delta A_k$ is the volume of a vertical rectangular box that approximates the volume of the portion of the solid that stands directly above the base ΔA_k. The sum S_n thus approximates what we want to call the total volume of the solid. We *define* this volume to be

$$\text{Volume} = \lim_{n\to\infty} S_n = \iint_R f(x, y)\,dA,$$

where $\Delta A_k \to 0$ as $n \to \infty$.

As you might expect, this more general method of calculating volume agrees with the methods in Chapter 6, but we do not prove this here. Figure 15.3 shows Riemann sum approximations to the volume becoming more accurate as the number n of boxes increases.

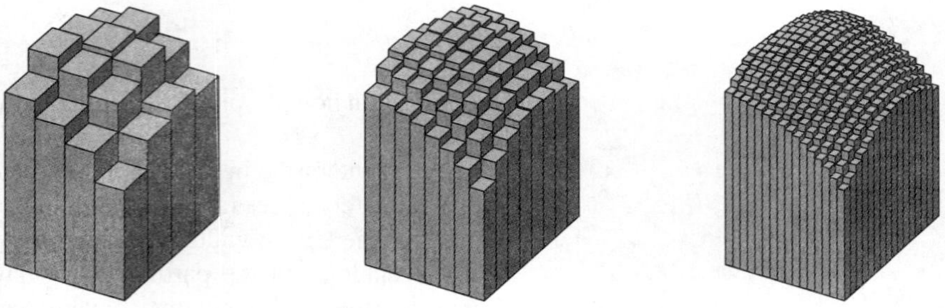

(a) $n = 16$ (b) $n = 64$ (c) $n = 256$

FIGURE 15.3 As n increases, the Riemann sum approximations approach the total volume of the solid shown in Figure 15.2.

Fubini's Theorem for Calculating Double Integrals

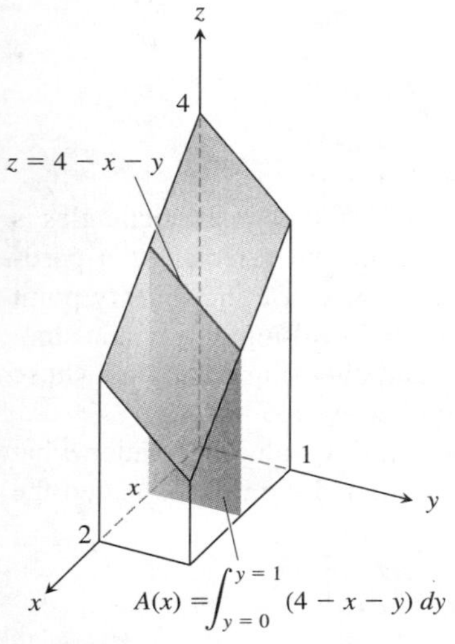

FIGURE 15.4 To obtain the cross-sectional area $A(x)$, we hold x fixed and integrate with respect to y.

Suppose that we wish to calculate the volume under the plane $z = 4 - x - y$ over the rectangular region $R: 0 \le x \le 2,\ 0 \le y \le 1$ in the xy-plane. If we apply the method of slicing from Section 6.1, with slices perpendicular to the x-axis (Figure 15.4), then the volume is

$$\int_{x=0}^{x=2} A(x)\,dx, \tag{1}$$

where $A(x)$ is the cross-sectional area at x. For each value of x, we may calculate $A(x)$ as the integral

$$A(x) = \int_{y=0}^{y=1} (4 - x - y)\,dy, \tag{2}$$

which is the area under the curve $z = 4 - x - y$ in the plane of the cross-section at x. In calculating $A(x)$, x is held fixed and the integration takes place with respect to y. Combining Equations (1) and (2), we see that the volume of the entire solid is

$$\begin{aligned}
\text{Volume} &= \int_{x=0}^{x=2} A(x)\,dx = \int_{x=0}^{x=2}\left(\int_{y=0}^{y=1}(4 - x - y)\,dy\right)dx \\
&= \int_{x=0}^{x=2}\left[4y - xy - \frac{y^2}{2}\right]_{y=0}^{y=1} dx = \int_{x=0}^{x=2}\left(\frac{7}{2} - x\right)dx \\
&= \left[\frac{7}{2}x - \frac{x^2}{2}\right]_0^2 = 5.
\end{aligned}$$

We often omit parentheses separating the two integrals in the formula above and write

$$\text{Volume} = \int_0^2\int_0^1 (4 - x - y)\,dy\,dx. \tag{3}$$

The expression on the right, called an **iterated** or **repeated integral**, says that the volume is obtained by integrating $4 - x - y$ with respect to y from $y = 0$ to $y = 1$ while holding x fixed, and then integrating the resulting expression in x from $x = 0$ to $x = 2$. The limits of integration 0 and 1 are associated with y, so they are placed on the integral closest to dy. The other limits of integration, 0 and 2, are associated with the variable x, so they are placed on the outside integral symbol that is paired with dx.

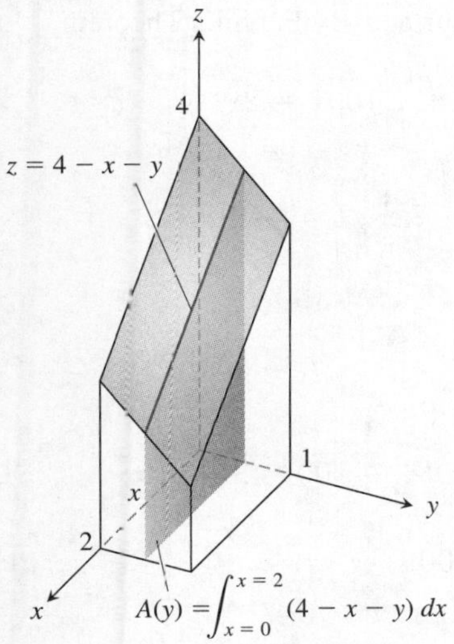

FIGURE 15.5 To obtain the cross-sectional area $A(y)$, we hold y fixed and integrate with respect to x.

What would have happened if we had calculated the volume by slicing with planes perpendicular to the y-axis (Figure 15.5)? As a function of y, the typical cross-sectional area is

$$A(y) = \int_{x=0}^{x=2} (4 - x - y)\, dx = \left[4x - \frac{x^2}{2} - xy\right]_{x=0}^{x=2} = 6 - 2y. \qquad (4)$$

The volume of the entire solid is therefore

$$\text{Volume} = \int_{y=0}^{y=1} A(y)\, dy = \int_{y=0}^{y=1} (6 - 2y)\, dy = \left[6y - y^2\right]_0^1 = 5,$$

in agreement with our earlier calculation.

Again, we may give a formula for the volume as an iterated integral by writing

$$\text{Volume} = \int_0^1 \int_0^2 (4 - x - y)\, dx\, dy.$$

The expression on the right says we can find the volume by integrating $4 - x - y$ with respect to x from $x = 0$ to $x = 2$ as in Equation (4) and integrating the result with respect to y from $y = 0$ to $y = 1$. In this iterated integral, the order of integration is first x and then y, the reverse of the order in Equation (3).

What do these two volume calculations with iterated integrals have to do with the double integral

$$\iint_R (4 - x - y)\, dA$$

over the rectangle $R: 0 \le x \le 2,\ 0 \le y \le 1$? The answer is that both iterated integrals give the value of the double integral. This is what we would reasonably expect, since the double integral measures the volume of the same region as the two iterated integrals. A theorem published in 1907 by Guido Fubini says that the double integral of any continuous function over a rectangle can be calculated as an iterated integral in either order of integration. (Fubini proved his theorem in greater generality, but this is what it says in our setting.)

HISTORICAL BIOGRAPHY
Guido Fubini
(1879–1943)
www.bit.ly/2xOIk22

THEOREM 1—Fubini's Theorem (First Form)
If $f(x, y)$ is continuous throughout the rectangular region $R: a \le x \le b$, $c \le y \le d$, then

$$\iint_R f(x, y)\, dA = \int_c^d \int_a^b f(x, y)\, dx\, dy = \int_a^b \int_c^d f(x, y)\, dy\, dx.$$

Fubini's Theorem says that double integrals over rectangles can be calculated as iterated integrals. Thus, we can evaluate a double integral by integrating with respect to one variable at a time using the Fundamental Theorem of Calculus.

Fubini's Theorem also says that we may calculate the double integral by integrating in *either* order, a genuine convenience. When we calculate a volume by slicing, we may use either planes perpendicular to the x-axis or planes perpendicular to the y-axis.

EXAMPLE 1 Calculate $\iint_R f(x, y)\, dA$ for

$$f(x, y) = 100 - 6x^2y \quad \text{and} \quad R:\ 0 \le x \le 2,\ -1 \le y \le 1.$$

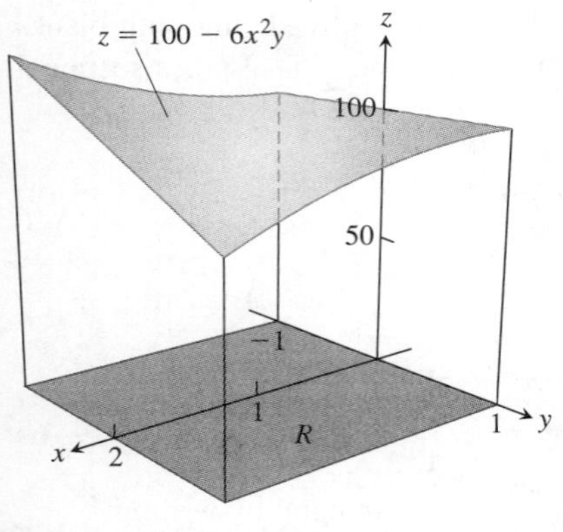

FIGURE 15.6 The double integral $\iint_R f(x, y)\,dA$ gives the volume under this surface over the rectangular region R (Example 1).

Solution Figure 15.6 displays the volume beneath the surface. By Fubini's Theorem,

$$\iint_R f(x, y)\,dA = \int_{-1}^{1}\int_{0}^{2}(100 - 6x^2y)\,dx\,dy = \int_{-1}^{1}\Big[100x - 2x^3y\Big]_{x=0}^{x=2}dy$$

$$= \int_{-1}^{1}(200 - 16y)\,dy = \Big[200y - 8y^2\Big]_{-1}^{1} = 400.$$

Reversing the order of integration gives the same answer:

$$\int_{0}^{2}\int_{-1}^{1}(100 - 6x^2y)\,dy\,dx = \int_{0}^{2}\Big[100y - 3x^2y^2\Big]_{y=-1}^{y=1}dx$$

$$= \int_{0}^{2}[(100 - 3x^2) - (-100 - 3x^2)]\,dx$$

$$= \int_{0}^{2}200\,dx = 400.$$

■

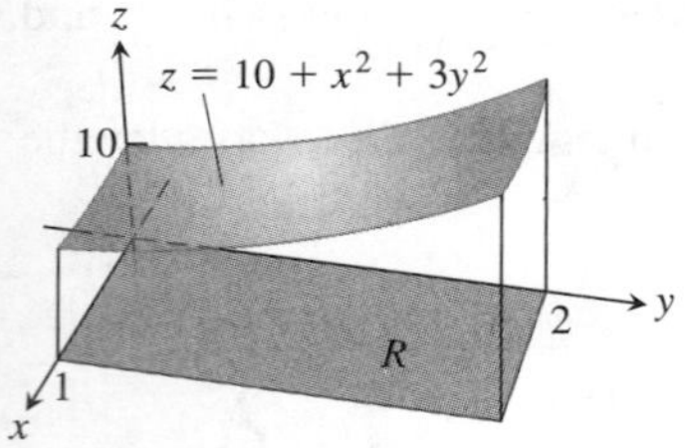

FIGURE 15.7 The double integral $\iint_R f(x, y)\,dA$ gives the volume under this surface over the rectangular region R (Example 2).

EXAMPLE 2 Find the volume of the region bounded above by the elliptical paraboloid $z = 10 + x^2 + 3y^2$ and below by the rectangle $R\!: 0 \le x \le 1,\ 0 \le y \le 2$.

Solution The surface and volume are shown in Figure 15.7. The volume is given by the double integral

$$V = \iint_R (10 + x^2 + 3y^2)\,dA = \int_{0}^{1}\int_{0}^{2}(10 + x^2 + 3y^2)\,dy\,dx$$

$$= \int_{0}^{1}\Big[10y + x^2y + y^3\Big]_{y=0}^{y=2}dx$$

$$= \int_{0}^{1}(28 + 2x^2)\,dx = \Big[28x + \frac{2}{3}x^3\Big]_{0}^{1} = \frac{86}{3}.$$

■

EXERCISES 15.1

Evaluating Iterated Integrals

In Exercises 1–14, evaluate the iterated integral.

1. $\int_{1}^{2}\int_{0}^{4} 2xy\,dy\,dx$
2. $\int_{0}^{1}\int_{0}^{2} x(x + y)\,dy\,dx$
3. $\int_{-1}^{0}\int_{-1}^{1}(x + y + 1)\,dx\,dy$
4. $\int_{0}^{1}\int_{0}^{1}\left(1 - \frac{x^2 + y^2}{2}\right)dx\,dy$
5. $\int_{0}^{3}\int_{0}^{2}(4 - y^2)\,dy\,dx$
6. $\int_{0}^{3}\int_{-2}^{0}(x^2y - 2xy)\,dy\,dx$
7. $\int_{0}^{1}\int_{0}^{1}\frac{y}{1 + xy}\,dx\,dy$
8. $\int_{1}^{4}\int_{0}^{4}\left(\frac{x}{2} + \sqrt{y}\right)dx\,dy$
9. $\int_{0}^{\ln 2}\int_{1}^{\ln 5} e^{2x+y}\,dy\,dx$
10. $\int_{0}^{-1}\int_{0}^{-1}\frac{dx\,dy}{\sqrt{1-x^2}\sqrt{1-y^2}}$
11. $\int_{-1}^{2}\int_{0}^{\pi/2} y\sin x\,dx\,dy$
12. $\int_{\pi}^{2\pi}\int_{0}^{\pi}(\sin x + \cos y)\,dx\,dy$
13. $\int_{1}^{4}\int_{1}^{e}\frac{\ln x}{xy}\,dx\,dy$
14. $\int_{-1}^{2}\int_{1}^{2} x\ln y\,dy\,dx$
15. Find all values of the constant c so that $\int_{0}^{1}\int_{0}^{c}(2x + y)\,dx\,dy = 3$.
16. Find all values of the constant c so that
$$\int_{-1}^{c}\int_{0}^{2}(xy + 1)\,dy\,dx = 4 + 4c.$$

Evaluating Double Integrals over Rectangles

In Exercises 17–24, evaluate the double integral over the given region R.

17. $\iint_R (x^2 + y^2)\,dx\,dy,\quad R\!: 1 \le x \le 2,\quad 0 \le y \le 1$
18. $\iint_R \frac{1}{(x + y)^2}\,dx\,dy,\quad R\!: 1 \le x \le 2,\quad 3 \le y \le 4$

19. $\iint_R xy\cos y\, dA, \quad R: -1 \le x \le 1,\ 0 \le y \le \pi$

20. $\iint_R y\sin(x+y)\, dA, \quad R: -\pi \le x \le 0,\ 0 \le y \le \pi$

21. $\iint_R e^{x-y}\, dA, \quad R: 0 \le x \le \ln 2,\ 0 \le y \le \ln 2$

22. $\iint_R xye^{xy^2}\, dA, \quad R: 0 \le x \le 2,\ 0 \le y \le 1$

23. $\iint_R \frac{xy^3}{x^2+1}\, dA, \quad R: 0 \le x \le 1,\ 0 \le y \le 2$

24. $\iint_R \frac{y}{x^2y^2+1}\, dA, \quad R: 0 \le x \le 1,\ 0 \le y \le 1$

In Exercises 25 and 26, integrate f over the given region.

25. **Square** $f(x, y) = 1/(xy)$ over the square $1 \le x \le 2$, $1 \le y \le 2$

26. **Rectangle** $f(x, y) = y\cos xy$ over the rectangle $0 \le x \le \pi$, $0 \le y \le 1$

In Exercises 27 and 28, sketch the solid whose volume is given by the specified integral.

27. $\int_0^1\int_0^2 (9 - x^2 - y^2)\, dy\, dx$ 28. $\int_0^3\int_1^4 (7 - x - y)\, dx\, dy$

29. Find the volume of the region bounded above by the paraboloid $z = x^2 + y^2$ and below by the square $R: -1 \le x \le 1$, $-1 \le y \le 1$.

30. Find the volume of the region bounded above by the elliptical paraboloid $z = 16 - x^2 - y^2$ and below by the square $R: 0 \le x \le 2$, $0 \le y \le 2$.

31. Find the volume of the region bounded above by the plane $z = 2 - x - y$ and below by the square $R: 0 \le x \le 1$, $0 \le y \le 1$.

32. Find the volume of the region bounded above by the plane $z = y/2$ and below by the rectangle $R: 0 \le x \le 4, 0 \le y \le 2$.

33. Find the volume of the region bounded above by the surface $z = 2\sin x\cos y$ and below by the rectangle $R: 0 \le x \le \pi/2$, $0 \le y \le \pi/4$.

34. Find the volume of the region bounded above by the surface $z = 4 - y^2$ and below by the rectangle $R: 0 \le x \le 1$, $0 \le y \le 2$.

35. Find a value of the constant k so that $\int_1^2\int_0^3 kx^2y\, dx\, dy = 1$.

36. Evaluate $\int_{-1}^1\int_0^{\pi/2} x\sin\sqrt{y}\, dy\, dx$.

37. Use Fubini's Theorem to evaluate

$$\int_0^2\int_0^1 \frac{x}{1+xy}\, dx\, dy.$$

38. Use Fubini's Theorem to evaluate

$$\int_0^1\int_0^3 xe^{xy}\, dx\, dy.$$

T 39. Use a software application to compute the integrals

a. $\int_0^1\int_0^2 \frac{y-x}{(x+y)^3}\, dx\, dy$ **b.** $\int_0^2\int_0^1 \frac{y-x}{(x+y)^3}\, dy\, dx$

Explain why your results do not contradict Fubini's Theorem.

40. If $f(x, y)$ is continuous over $R: a \le x \le b,\ c \le y \le d$ and

$$F(x, y) = \int_a^x\int_c^y f(u, v)\, dv\, du$$

on the interior of R, find the second partial derivatives F_{xy} and F_{yx}.

15.2 Double Integrals over General Regions

In this section we define and evaluate double integrals over bounded regions in the plane that are more general than rectangles. These double integrals are also evaluated as iterated integrals, with the main practical problem being that of determining the limits of integration. Since the region of integration may have boundaries other than line segments parallel to the coordinate axes, the limits of integration often involve variables, not just constants.

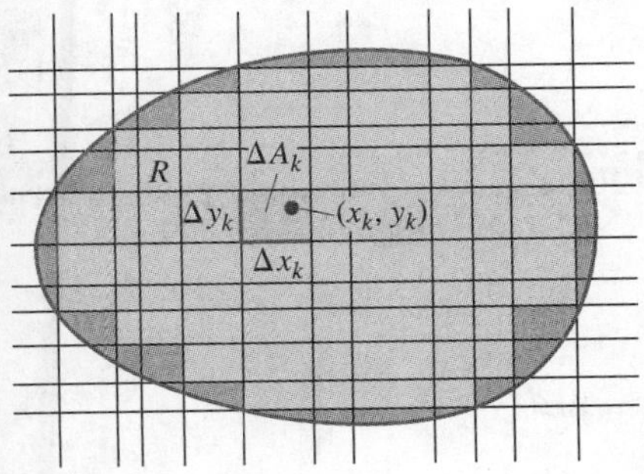

FIGURE 15.8 A rectangular grid partitioning a bounded, nonrectangular region into rectangular cells.

Double Integrals over Bounded, Nonrectangular Regions

To define the double integral of a function $f(x, y)$ over a bounded, nonrectangular region R, such as the one in Figure 15.8, we again begin by covering R with a grid of small rectangular cells whose union contains all points of R. This time, however, we cannot exactly fill R with a finite number of rectangles lying inside R since its boundary is curved, and some of the small rectangles in the grid lie partly outside R. A partition of R is formed by taking the rectangles that lie completely inside it, not using any that are either partly or completely outside. For commonly arising regions, more and more of R is included as the norm of a partition (the largest width or height of any rectangle used) approaches zero.

Once we have a partition of R, we number the rectangles in some order from 1 to n and let ΔA_k be the area of the kth rectangle. We then choose a point (x_k, y_k) in the kth rectangle and form the Riemann sum

$$S_n = \sum_{k=1}^{n} f(x_k, y_k)\,\Delta A_k.$$

As the norm of the partition forming S_n goes to zero, $\|P\| \to 0$, the width and height of each enclosed rectangle go to zero, their area ΔA_k goes to zero, and their number goes to infinity. If $f(x, y)$ is a continuous function, then these Riemann sums converge to a limiting value that is not dependent on any of the choices we made. This limit is called the **double integral** of $f(x, y)$ over R:

$$\lim_{\|P\|\to 0} \sum_{k=1}^{n} f(x_k, y_k)\,\Delta A_k = \iint_R f(x, y)\,dA.$$

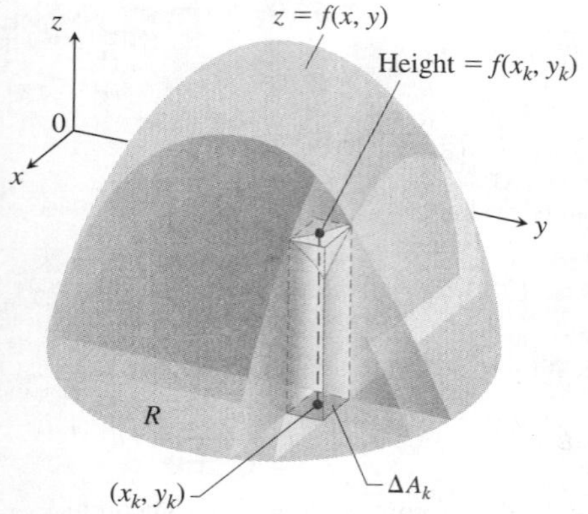

Volume $= \lim \sum f(x_k, y_k)\,\Delta A_k = \iint_R f(x, y)\,dA$

FIGURE 15.9 We define the volume of a solid with a curved base as a limit of the sums of volumes of approximating rectangular boxes.

The nature of the boundary of R introduces issues not found in integrals over an interval. When R has a curved boundary, the n rectangles of a partition lie inside R but do not cover all of R. In order for a partition to approximate R well, the parts of R covered by small rectangles lying partly outside R must become negligible as the norm of the partition approaches zero. This property of being nearly filled in by a partition of small norm is satisfied by all the regions that we will encounter. There is no problem with boundaries made from polygons, circles, and ellipses or from continuous graphs over an interval, joined end to end. A curve with a "fractal" type of shape would be problematic, but such curves arise rarely in most applications. A careful discussion of which types of regions R can be used for computing double integrals is left to a more advanced text.

Volumes

If $f(x, y)$ is positive and continuous over R, we define the volume of the solid region between R and the surface $z = f(x, y)$ to be $\iint_R f(x, y)\,dA$, as before (Figure 15.9).

If R is a region like the one shown in the xy-plane in Figure 15.10, bounded "above" and "below" by the curves $y = g_2(x)$ and $y = g_1(x)$ and on the sides by the lines $x = a$, $x = b$, we may again calculate the volume by the method of slicing. We first calculate the cross-sectional area

$$A(x) = \int_{y=g_1(x)}^{y=g_2(x)} f(x, y)\,dy$$

and then integrate $A(x)$ from $x = a$ to $x = b$ to get the volume as an iterated integral:

$$V = \int_a^b A(x)\,dx = \int_a^b \int_{g_1(x)}^{g_2(x)} f(x, y)\,dy\,dx. \tag{1}$$

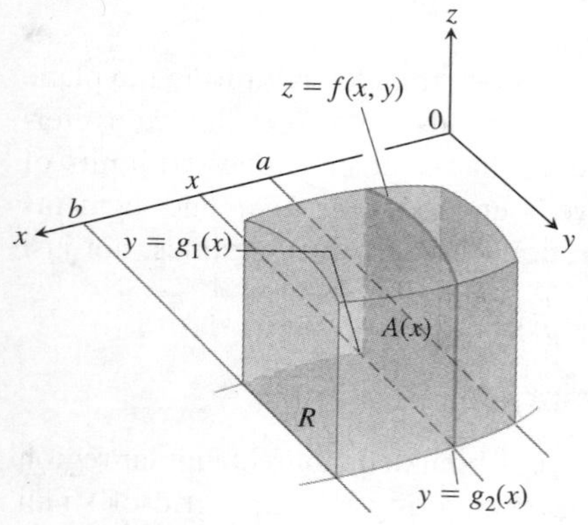

FIGURE 15.10 The area of the vertical slice shown here is $A(x)$. To calculate the volume of the solid, we integrate this area from $x = a$ to $x = b$:

$$\int_a^b A(x)\,dx = \int_a^b \int_{g_1(x)}^{g_2(x)} f(x, y)\,dy\,dx.$$

Similarly, if R is a region like the one shown in Figure 15.11, bounded by the curves $x = h_2(y)$ and $x = h_1(y)$ and the lines $y = c$ and $y = d$, then the volume calculated by slicing is given by the iterated integral

$$\text{Volume} = \int_c^d \int_{h_1(y)}^{h_2(y)} f(x, y)\,dx\,dy. \tag{2}$$

That the iterated integrals in Equations (1) and (2) both give the volume that we defined to be the double integral of f over R is a consequence of the following stronger form of Fubini's Theorem.

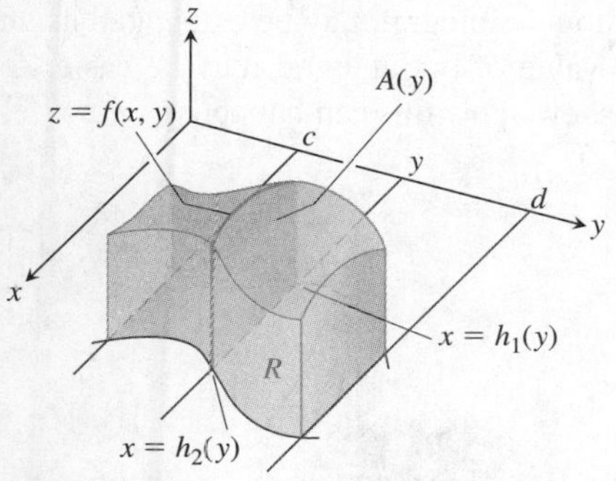

FIGURE 15.11 The volume of the solid shown here is

$$\int_c^d A(y)\,dy = \int_c^d \int_{h_1(y)}^{h_2(y)} f(x, y)\,dx\,dy.$$

For a given solid, Theorem 2 says we can calculate the volume as in Figure 15.10 or in the way shown here. Both calculations have the same result.

THEOREM 2—Fubini's Theorem (Stronger Form)
Let $f(x, y)$ be continuous on a region R.

1. If R is defined by $a \le x \le b,\ g_1(x) \le y \le g_2(x)$, with g_1 and g_2 continuous on $[a, b]$, then

$$\iint_R f(x, y)\,dA = \int_a^b \int_{g_1(x)}^{g_2(x)} f(x, y)\,dy\,dx.$$

2. If R is defined by $c \le y \le d,\ h_1(y) \le x \le h_2(y)$, with h_1 and h_2 continuous on $[c, d]$, then

$$\iint_R f(x, y)\,dA = \int_c^d \int_{h_1(y)}^{h_2(y)} f(x, y)\,dx\,dy.$$

Some iterated double integrals we will encounter later in this text will use variables of integration other than x and y. For instance, we may write

$$\int_{u=p}^{u=q} \int_{v=G_1(u)}^{v=G_2(u)} F(u, v)\,dv\,du = \int_p^q \int_{G_1(u)}^{G_2(u)} F(u, v)\,dv\,du.$$

Regardless of which specific variables of integration are used, the limits of an iterated double integral always satisfy these properties:

- The limits of the *outside* integral are *constants* (they do not depend on either variable of integration), and
- the limits of the *inside* integral are *functions* that may depend on the variable of the *outside* integral.

EXAMPLE 1 Find the volume of the right prism whose base is the triangle in the xy-plane bounded by the x-axis and the lines $y = x$ and $x = 1$ and whose top lies in the plane

$$z = f(x, y) = 3 - x - y.$$

Solution See Figure 15.12a. For any x between 0 and 1, y may vary from $y = 0$ to $y = x$ (Figure 15.12b). Hence,

$$V = \int_0^1 \int_0^x (3 - x - y)\,dy\,dx = \int_0^1 \left[3y - xy - \frac{y^2}{2}\right]_{y=0}^{y=x} dx$$

$$= \int_0^1 \left(3x - \frac{3x^2}{2}\right) dx = \left[\frac{3x^2}{2} - \frac{x^3}{2}\right]_{x=0}^{x=1} = 1.$$

When the order of integration is reversed (Figure 15.12c), the integral for the volume is

$$V = \int_0^1 \int_y^1 (3 - x - y)\,dx\,dy = \int_0^1 \left[3x - \frac{x^2}{2} - xy\right]_{x=y}^{x=1} dy$$

$$= \int_0^1 \left(3 - \frac{1}{2} - y - 3y + \frac{y^2}{2} + y^2\right) dy$$

$$= \int_0^1 \left(\frac{5}{2} - 4y + \frac{3}{2}y^2\right) dy = \left[\frac{5}{2}y - 2y^2 + \frac{y^3}{2}\right]_{y=0}^{y=1} = 1.$$

The two integrals are equal, as they should be. ■

Although Fubini's Theorem assures us that a double integral may be calculated as an iterated integral in either order of integration, the value of one integral may be easier to find than the value of the other. The next example shows how this can happen.

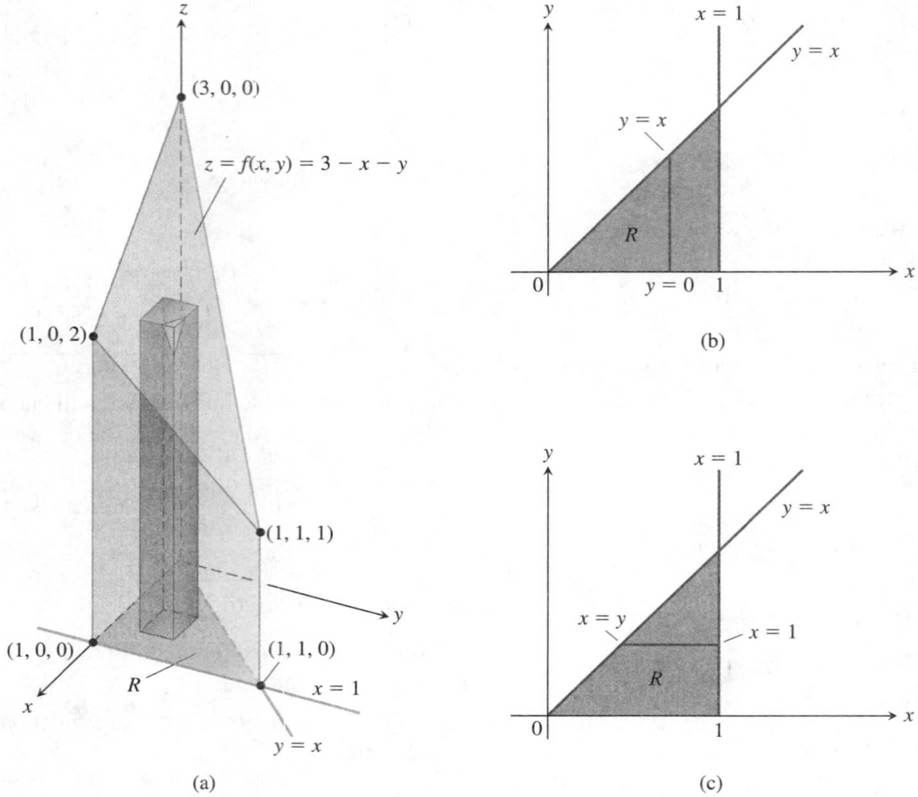

FIGURE 15.12 (a) Prism with a triangular base in the *xy*-plane. The volume of this prism is defined as a double integral over R. To evaluate it as an iterated integral, we may integrate first with respect to y and then with respect to x, or the other way around (Example 1). (b) Integration limits of

$$\int_{x=0}^{x=1}\int_{y=0}^{y=x} f(x, y)\,dy\,dx.$$

If we integrate first with respect to y, we integrate along a vertical line through R and then integrate from left to right to include all the vertical lines in R. (c) Integration limits of

$$\int_{y=0}^{y=1}\int_{x=y}^{x=1} f(x, y)\,dx\,dy.$$

If we integrate first with respect to x, we integrate along a horizontal line through R and then integrate from bottom to top to include all the horizontal lines in R.

EXAMPLE 2 Calculate

$$\iint_R \frac{\sin x}{x}\,dA,$$

where R is the triangle in the *xy*-plane bounded by the x-axis, the line $y = x$, and the line $x = 1$.

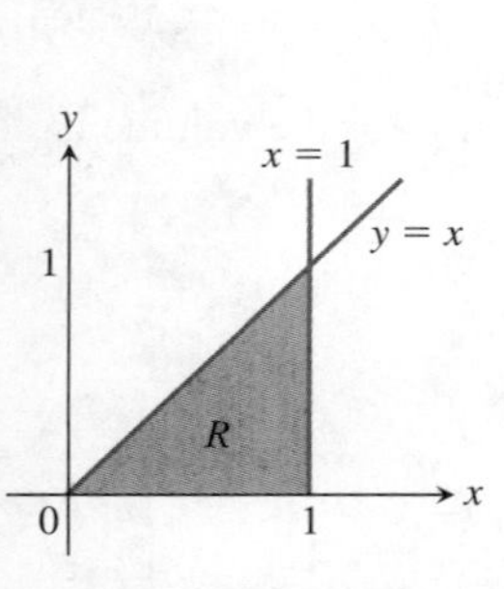

FIGURE 15.13 The region of integration in Example 2.

Solution The region of integration is shown in Figure 15.13. If we integrate first with respect to y and next with respect to x, then because x is held fixed in the first integration, we find

$$\int_0^1\left(\int_0^x \frac{\sin x}{x}\,dy\right)dx = \int_0^1\left[y\frac{\sin x}{x}\right]_{y=0}^{y=x} dx = \int_0^1 \sin x\,dx = -\cos(1) + 1 \approx 0.46.$$

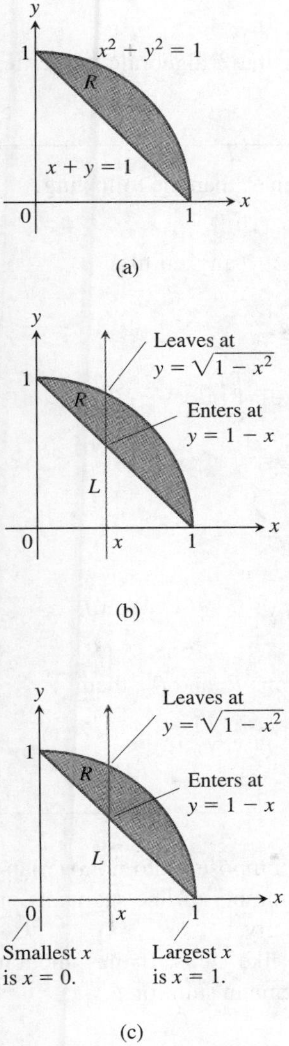

FIGURE 15.14 Finding the limits of integration when integrating first with respect to y and then with respect to x.

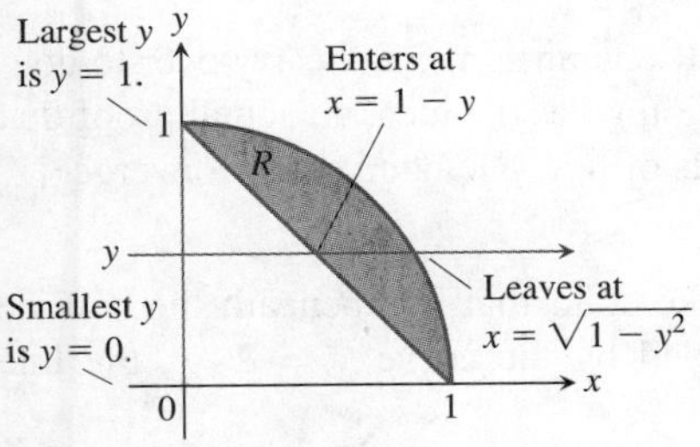

FIGURE 15.15 Finding the limits of integration when integrating first with respect to x and then with respect to y.

If we reverse the order of integration and attempt to calculate

$$\int_0^1 \int_y^1 \frac{\sin x}{x}\, dx\, dy,$$

we run into a problem because $\int ((\sin x)/x)\, dx$ cannot be expressed in terms of elementary functions (there is no simple antiderivative).

There is no general rule for predicting which order of integration will be the good one in circumstances like these. If the order you first choose doesn't work, try the other. Sometimes neither order will work, and then we may need to use numerical approximations. ■

Finding Limits of Integration

We now give a procedure for finding limits of integration that applies for many regions in the plane. Regions that are more complicated, and for which this procedure fails, can often be split up into pieces on which the procedure works.

Using Vertical Cross-Sections When faced with evaluating $\iint_R f(x, y)\, dA$, integrating first with respect to y and then with respect to x, do the following three steps:

1. *Sketch.* Sketch the region of integration and label the bounding curves (Figure 15.14a).
2. *Find the y-limits of integration.* Imagine a vertical line L cutting through R in the direction of increasing y. Mark the y-values where L enters and leaves. These are the y-limits of integration and are usually functions of x (instead of constants) (Figure 15.14b).
3. *Find the x-limits of integration.* Choose x-limits that include all the vertical lines through R. These must be constants. The integral whose region of integration is shown in Figure 15.14c is

$$\iint_R f(x, y)\, dA = \int_{x=0}^{x=1} \int_{y=1-x}^{y=\sqrt{1-x^2}} f(x, y)\, dy\, dx.$$

Using Horizontal Cross-Sections To evaluate the same double integral as an iterated integral with the order of integration reversed, use horizontal lines instead of vertical lines in Steps 2 and 3 (see Figure 15.15). The integral is

$$\iint_R f(x, y)\, dA = \int_0^1 \int_{1-y}^{\sqrt{1-y^2}} f(x, y)\, dx\, dy.$$

EXAMPLE 3 Sketch the region of integration for the integral

$$\int_0^2 \int_{x^2}^{2x} (4x + 2)\, dy\, dx$$

and write an equivalent integral with the order of integration reversed.

Solution The region of integration is given by the inequalities $x^2 \le y \le 2x$ and $0 \le x \le 2$. It is therefore the region bounded by the curves $y = x^2$ and $y = 2x$ between $x = 0$ and $x = 2$ (Figure 15.16a).

To find limits for integrating in the reverse order, we imagine a horizontal line passing from left to right through the region. It enters at $x = y/2$ and leaves at $x = \sqrt{y}$. To include all such lines, we let y run from $y = 0$ to $y = 4$ (Figure 15.16b). The integral is

$$\int_0^4 \int_{y/2}^{\sqrt{y}} (4x + 2)\, dx\, dy.$$

The common value of these integrals is 8. ■

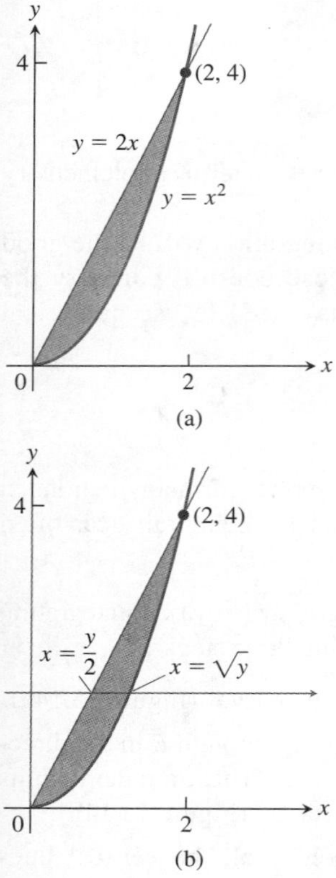

FIGURE 15.16 Region of integration for Example 3.

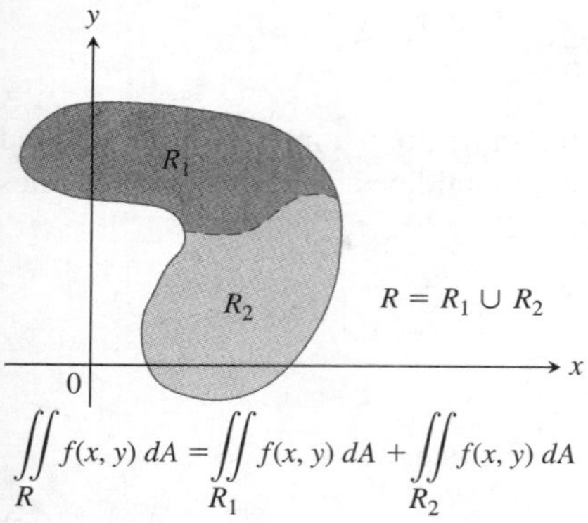

FIGURE 15.17 The Additivity Property for rectangular regions holds for regions bounded by smooth curves.

Properties of Double Integrals

Like single integrals, double integrals of continuous functions have algebraic properties that are useful in computations and applications.

If $f(x, y)$ and $g(x, y)$ are continuous on the bounded region R, then the following properties hold.

1. *Constant Multiple:* $\displaystyle\iint_R cf(x, y)\, dA = c\iint_R f(x, y)\, dA$ (any number c)

2. *Sum and Difference:*

$$\iint_R (f(x, y) \pm g(x, y))\, dA = \iint_R f(x, y)\, dA \pm \iint_R g(x, y)\, dA$$

3. *Domination:*

(a) $\displaystyle\iint_R f(x, y)\, dA \geq 0 \quad \text{if} \quad f(x, y) \geq 0 \text{ on } R$

(b) $\displaystyle\iint_R f(x, y)\, dA \geq \iint_R g(x, y)\, dA \quad \text{if} \quad f(x, y) \geq g(x, y) \text{ on } R$

4. *Additivity:* If R is the union of two nonoverlapping regions R_1 and R_2, then

$$\iint_R f(x, y)\, dA = \iint_{R_1} f(x, y)\, dA + \iint_{R_2} f(x, y)\, dA$$

Property 4 assumes that the region of integration R is decomposed into nonoverlapping regions R_1 and R_2 with boundaries consisting of a finite number of line segments or smooth curves. Figure 15.17 illustrates an example of this property.

The idea behind these properties is that integrals behave like sums. If the function $f(x, y)$ is replaced by its constant multiple $cf(x, y)$, then a Riemann sum for f,

$$S_n = \sum_{k=1}^{n} f(x_k, y_k)\,\Delta A_k,$$

is replaced by a Riemann sum for cf:

$$\sum_{k=1}^{n} cf(x_k, y_k)\,\Delta A_k = c\sum_{k=1}^{n} f(x_k, y_k)\,\Delta A_k = cS_n.$$

Taking limits as $n \to \infty$ shows that $c \lim_{n\to\infty} S_n = c\iint_R f\, dA$ and $\lim_{n\to\infty} cS_n = \iint_R cf\, dA$ are equal. It follows that the Constant Multiple Property carries over from sums to double integrals.

The other properties are also easy to verify for Riemann sums, and carry over to double integrals for the same reason. While this discussion gives the idea, an actual proof that these properties hold requires a more careful analysis of how Riemann sums converge.

EXAMPLE 4 Find the volume of the wedgelike solid that lies beneath the surface $z = 16 - x^2 - y^2$ and above the region R bounded by the curve $y = 2\sqrt{x}$, the line $y = 4x - 2$, and the x-axis.

Solution Figure 15.18a shows the surface and the "wedgelike" solid whose volume we want to calculate. Figure 15.18b shows the region of integration in the xy-plane. If we integrate in the order $dy\,dx$ (first with respect to y and then with respect to x), two integrations

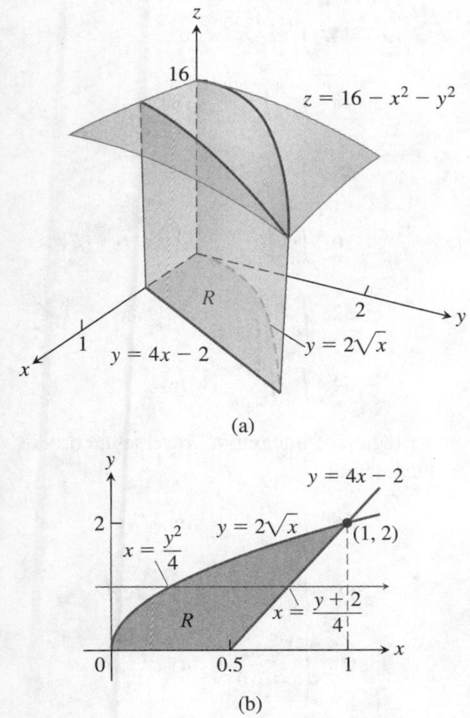

FIGURE 15.18 (a) The solid "wedge-like" region whose volume is found in Example 4. (b) The region of integration R showing the order $dx\,dy$.

will be required because y varies from $y = 0$ to $y = 2\sqrt{x}$ for $0 \le x \le 0.5$, and then varies from $y = 4x - 2$ to $y = 2\sqrt{x}$ for $0.5 \le x \le 1$. So we choose to integrate in the order $dx\,dy$, which requires only one double integral whose limits of integration are indicated in Figure 15.18b. The volume is then calculated as the iterated integral:

$$\iint_R (16 - x^2 - y^2)\,dA$$

$$= \int_0^2 \int_{y^2/4}^{(y+2)/4} (16 - x^2 - y^2)\,dx\,dy$$

$$= \int_0^2 \left[16x - \frac{x^3}{3} - xy^2\right]_{x=y^2/4}^{x=(y+2)/4} dx$$

$$= \int_0^2 \left[4(y+2) - \frac{(y+2)^3}{3 \cdot 64} - \frac{(y+2)y^2}{4} - 4y^2 + \frac{y^6}{3 \cdot 64} + \frac{y^4}{4}\right] dy$$

$$= \left[\frac{191y}{24} + \frac{63y^2}{32} - \frac{145y^3}{96} - \frac{49y^4}{768} + \frac{y^5}{20} + \frac{y^7}{1344}\right]_0^2 = \frac{20803}{1680} \approx 12.4. \quad ■$$

Our development of the double integral has focused on its representation of the volume of the solid region between R and the surface $z = f(x, y)$ of a positive continuous function. Just as we saw with signed area in the case of single integrals, when $f(x_k, y_k)$ is negative, the product $f(x_k, y_k)\,\Delta A_k$ is the negative of the volume of the rectangular box shown in Figure 15.9 that was used to form the approximating Riemann sum. So for an arbitrary continuous function f defined over R, the limit of any Riemann sum represents the *signed* volume (not the total volume) of the solid region between R and the surface. The double integral has other interpretations as well, and in the next section we will see how it is used to calculate the area of a general region in the plane.

EXERCISES 15.2

Sketching Regions of Integration

In Exercises 1–8, sketch the regions of integration associated with the given double integrals.

1. $\int_0^3 \int_0^{2x} f(x, y)\,dy\,dx$

2. $\int_{-1}^2 \int_{x-1}^{x^2} f(x, y)\,dy\,dx$

3. $\int_{-2}^2 \int_{y^2}^4 f(x, y)\,dx\,dy$

4. $\int_0^1 \int_y^{2y} f(x, y)\,dx\,dy$

5. $\int_0^1 \int_{e^x}^e f(x, y)\,dy\,dx$

6. $\int_1^{e^2} \int_0^{\ln x} f(x, y)\,dy\,dx$

7. $\int_0^1 \int_0^{\arcsin y} f(x, y)\,dx\,dy$

8. $\int_0^8 \int_{y/4}^{y^{1/3}} f(x, y)\,dx\,dy$

Finding Limits of Integration

In Exercises 9–18, write an iterated integral for $\iint_R dA$ over the described region R using (a) vertical cross-sections, (b) horizontal cross-sections.

9.

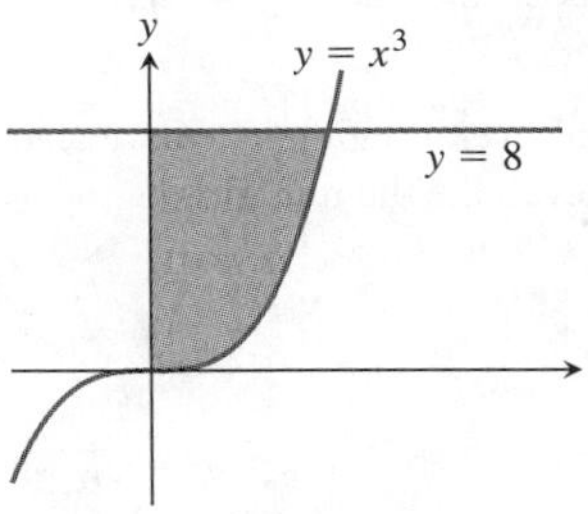

10.

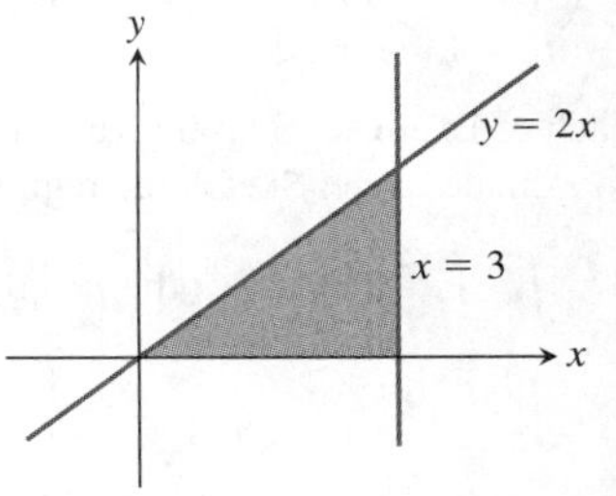

11.

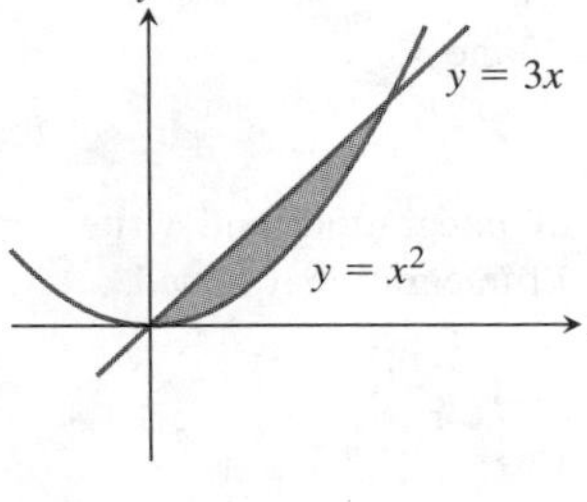

12.

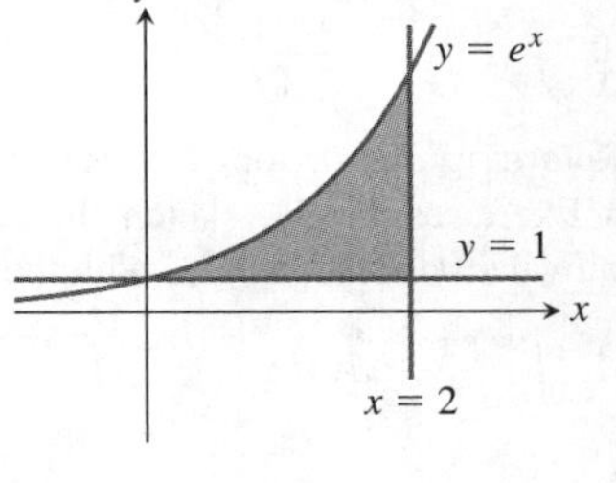

13. Bounded by $y = \sqrt{x}$, $y = 0$, and $x = 9$

14. Bounded by $y = \tan x$, $x = 0$, and $y = 1$

15. Bounded by $y = e^{-x}$, $y = 1$, and $x = \ln 3$

16. Bounded by $y = 0$, $x = 0$, $y = 1$, and $y = \ln x$

17. Bounded by $y = 3 - 2x$, $y = x$, and $x = 0$

18. Bounded by $y = x^2$ and $y = x + 2$

Evaluating Iterated Integrals

In Exercises 19–26, evaluate the integral.

19. $\int_1^2 \int_0^{2x} xy^3\,dy\,dx$ **20.** $\int_1^3 \int_y^{2y} y\,dx\,dy$

21. $\int_0^1 \int_y^1 (\sqrt{x} + xy)\,dx\,dy$ **22.** $\int_0^a \int_0^{\sqrt{a^2-x^2}} xy$

23. $\int_0^1 \int_{\sqrt{x}}^{2-x} x\,dy\,dx$ **24.** $\int_0^1 \int_0^{\arctan y} \frac{1}{1+y^2}\,dx\,dy$

25. $\int_1^4 \int_y^{y^2} \sqrt{\frac{y}{x}}\,dx\,dy$ **26.** $\int_3^5 \int_1^{e^x} \frac{1}{xy}\,dy\,dx$

Finding Regions of Integration and Double Integrals

In Exercises 27–32, sketch the region of integration and evaluate the integral.

27. $\int_0^\pi \int_0^x x \sin y\,dy\,dx$ **28.** $\int_0^\pi \int_0^{\sin x} y\,dy\,dx$

29. $\int_1^{\ln 8} \int_1^{\ln y} e^{x+y}\,dx\,dy$ **30.** $\int_1^2 \int_y^{y^2} dx\,dy$

31. $\int_0^1 \int_0^{y^2} 3y^3 e^{xy}\,dx\,dy$ **32.** $\int_1^4 \int_0^{\sqrt{x}} \frac{3}{2} e^{y/\sqrt{x}}\,dy\,dx$

In Exercises 33–36, integrate f over the given region.

33. Quadrilateral $\iint_R xy\,dx\,dy$, where R is the region bounded by $x = 2a$ and $x^2 = 4ay$.

34. Triangle $f(x, y) = x^2 + y^2$ over the triangular region with vertices $(0, 0)$, $(1, 0)$, and $(0, 1)$

35. Triangle $f(x, y) = x^2 + y^2$ over the triangular region enclosed by the lines $x = 0$, $y = 0$, $x + y = 1$.

36. Curved region $f(s, t) = e^s \ln t$ over the region in the first quadrant of the st-plane that lies above the curve $s = \ln t$ from $t = 1$ to $t = 2$

Each of Exercises 37–40 gives an integral over a region in a Cartesian coordinate plane. Sketch the region and evaluate the integral.

37. $\int_{-2}^0 \int_v^{-v} 2\,dp\,dv$ (the pv-plane)

38. $\int_0^1 \int_0^{\sqrt{1+y^2}} \frac{1}{+x^2+y^2}\,dx\,dy$

39. $\int_{-\pi/3}^{\pi/3} \int_0^{\sec t} 3 \cos t\,du\,dt$ (the tu-plane)

40. $\int_0^{3/2} \int_1^{4-2u} \frac{4-2u}{v^2}\,dv\,du$ (the uv-plane)

Reversing the Order of Integration

In Exercises 41–54, sketch the region of integration, and write an equivalent double integral with the order of integration reversed.

41. $\int_0^1 \int_2^{4-2x} dy\,dx$ **42.** $\int_0^2 \int_{y-2}^0 dx\,dy$

43. $\int_0^1 \int_y^{\sqrt{y}} dx\,dy$ **44.** $\int_0^1 \int_{1-x}^{1-x^2} dy\,dx$

45. $\int_0^1 \int_1^{e^x} dy\,dx$ **46.** $\int_0^{4a} \int_{x^2/4a}^{2\sqrt{ax}} dy\,dx$

47. $\int_0^{3/2} \int_0^{9-4x^2} 16x\,dy\,dx$ **48.** $\int_0^2 \int_0^{4-y^2} y\,dx\,dy$

49. $\int_0^a \int_{a-\sqrt{a^2-y^2}}^{a+\sqrt{a^2-y^2}} xy\,dx\,dy$ **50.** $\int_0^2 \int_{-\sqrt{4-x^2}}^{\sqrt{4-x^2}} 6x\,dy\,dx$

51. $\int_0^3 \int_1^{\sqrt{4-y}} xy\,dy\,dx$ **52.** $\int_0^{\pi/6} \int_{\sin x}^{1/2} xy^2\,dy\,dx$

53. $\int_0^3 \int_1^{e^y} (x + y)\,dx\,dy$ **54.** $\int_0^1 \int_0^{2-x} xy\,dy\,dx$

In Exercises 55–64, sketch the region of integration, reverse the order of integration, and evaluate the integral.

55. $\int_0^\pi \int_x^\pi \frac{\sin y}{y}\,dy\,dx$ **56.** $\int_0^2 \int_x^2 2y^2 \sin xy\,dy\,dx$

57. $\int_0^1 \int_y^1 x^2 e^{xy}\,dx\,dy$ **58.** $\int_0^2 \int_0^{4-x^2} \frac{xe^{2y}}{4-y}\,dy\,dx$

59. $\int_0^1 \int_y^{2-y} xy\,dx\,dy$ **60.** $\int_0^3 \int_{\sqrt{x/3}}^1 e^{y^3}\,dy\,dx$

61. $\int_0^{1/16} \int_{y^{1/4}}^{1/2} \cos(16\pi x^5)\,dx\,dy$ **62.** $\int_0^8 \int_{\sqrt[3]{x}}^2 \frac{dy\,dx}{y^4+1}$

63. Square region $\iint_R \sqrt{xy - y^2}\,dx\,dy$, where R is the triangle with vertices (0, 0), (10, 1), (1, 1)

64. Triangular region $\iint_R xy\,dA$ where R is the region bounded by the lines $y = x$, $y = 2x$, and $x + y = 2$

Volume Beneath a Surface $z = f(x, y)$

65. Find the volume of the region bounded above by the paraboloid $z = x^2 + y^2$ and below by the triangle enclosed by the lines $y = x$, $x = 0$, and $x + y = 2$ in the xy-plane.

66. Find the volume of the solid that is bounded above by the cylinder $z = x^2$ and below by the region enclosed by the parabola $y = 2 - x^2$ and the line $y = x$ in the xy-plane.

67. Find the volume of the solid whose base is the region in the xy-plane that is bounded by the parabola $y = 4 - x^2$ and the line $y = 3x$, while the top of the solid is bounded by the plane $z = x + 4$.

68. Find the volume of the solid in the first octant bounded by the coordinate planes, the cylinder $x^2 + y^2 = 4$, and the plane $z + y = 3$.

69. Find the volume of the solid in the first octant bounded by the coordinate planes, the plane $x = 3$, and the parabolic cylinder $z = 4 - y^2$.

70. Find the volume of the solid cut from the first octant by the surface $z = 4 - x^2 - y$.

71. Find the volume of the wedge cut from the first octant by the cylinder $z = 12 - 3y^2$ and the plane $x + y = 2$.

72. Find the volume of the solid cut from the square column $|x| + |y| \leq 1$ by the planes $z = 0$ and $3x + z = 3$.

73. Find the volume of the solid that is bounded on the front and back by the planes $x = 2$ and $x = 1$, on the sides by the cylinders $y = \pm 1/x$, and above and below by the planes $z = x + 1$ and $z = 0$.

74. Find the volume of the solid bounded on the front and back by the planes $x = \pm\pi/3$, on the sides by the cylinders $y = \pm \sec x$, above by the cylinder $z = 1 + y^2$, and below by the xy-plane.

In Exercises 75 and 76, sketch the region of integration and the solid whose volume is given by the double integral.

75. $\int_0^3 \int_0^{2-2x/3} \left(1 - \frac{1}{3}x - \frac{1}{2}y\right) dy\, dx$

76. $\int_0^4 \int_{-\sqrt{16-y^2}}^{\sqrt{16-y^2}} \sqrt{25 - x^2 - y^2}\, dx\, dy$

Integrals over Unbounded Regions

Improper double integrals can often be computed similarly to improper integrals of one variable. The first iteration of the following improper integrals is conducted just as if they were proper integrals. One then evaluates an improper integral of a single variable by taking appropriate limits, as in Section 8.8. Evaluate the improper integrals in Exercises 77–80 as iterated integrals.

77. $\int_1^\infty \int_{e^{-x}}^1 \frac{1}{x^3 y}\, dy\, dx$

78. $\int_{-1}^1 \int_{-1/\sqrt{1-x^2}}^{1/\sqrt{1-x^2}} (2y + 1)\, dy\, dx$

79. $\int_{-\infty}^\infty \int_{-\infty}^\infty \frac{1}{(x^2 + 1)(y^2 + 1)}\, dx\, dy$

80. $\int_0^\infty \int_0^\infty xe^{-(x+2y)}\, dx\, dy$

Approximating Integrals with Finite Sums

In Exercises 81 and 82, approximate the double integral of $f(x, y)$ over the region R partitioned by the given vertical lines $x = a$ and horizontal lines $y = c$. In each subrectangle, use (x_k, y_k) as indicated for your approximation.

$$\iint_R f(x, y)\, dA \approx \sum_{k=1}^n f(x_k, y_k)\, \Delta A_k$$

81. $f(x, y) = x + y$ over the region R bounded above by the semicircle $y = \sqrt{1 - x^2}$ and below by the x-axis, using the partition $x = -1, -1/2, 0, 1/4, 1/2, 1$ and $y = 0, 1/2, 1$ with (x_k, y_k) the lower left corner in the kth subrectangle (provided the subrectangle lies within R)

82. $f(x, y) = x + 2y$ over the region R inside the circle $(x - 2)^2 + (y - 3)^2 = 1$ using the partition $x = 1, 3/2, 2, 5/2, 3$ and $y = 2, 5/2, 3, 7/2, 4$ with (x_k, y_k) the center (centroid) in the kth subrectangle (provided the subrectangle lies within R)

Theory and Examples

83. Circular sector Integrate $f(x, y) = \sqrt{4 - x^2}$ over the smaller sector cut from the disk $x^2 + y^2 \le 4$ by the rays $\theta = \pi/6$ and $\theta = \pi/2$.

84. Unbounded region Integrate $f(x, y) = 1/\left[(x^2 - x)(y - 1)^{2/3}\right]$ over the infinite rectangle $2 \le x < \infty$, $0 \le y \le 2$.

85. Noncircular cylinder A solid right (noncircular) cylinder has its base R in the xy-plane and is bounded above by the paraboloid $z = x^2 + y^2$. The cylinder's volume is

$$V = \int_0^1 \int_0^y (x^2 + y^2)\, dx\, dy + \int_1^2 \int_0^{2-y} (x^2 + y^2)\, dx\, dy.$$

Sketch the base region R, and express the cylinder's volume as a single iterated integral with the order of integration reversed. Then evaluate the integral to find the volume.

86. Converting to a double integral Evaluate the integral

$$\int_0^2 (\arctan \pi x - \arctan x)\, dx.$$

(*Hint:* Write the integrand as an integral.)

87. Maximizing a double integral What region R in the xy-plane maximizes the value of

$$\iint_R (4 - x^2 - 2y^2)\, dA?$$

Give reasons for your answer.

88. Minimizing a double integral What region R in the xy-plane minimizes the value of

$$\iint_R (x^2 + y^2 - 9)\, dA?$$

Give reasons for your answer.

89. Evaluate $\int_0^\infty \int_x^\infty \frac{e^{-y}}{y}\, dy\, dx$ by changing the order of integration.

Area as double integral.

90. How would you evaluate the double integral of a continuous function $f(x, y)$ over the region R in the xy-plane enclosed by the triangle with vertices $(0, 1)$, $(2, 0)$, and $(1, 2)$? Give reasons for your answer.

91. Unbounded region Prove that

$$\int_{-\infty}^\infty \int_{-\infty}^\infty e^{-x^2-y^2}\, dx\, dy = \lim_{b\to\infty} \int_{-b}^b \int_{-b}^b e^{-x^2-y^2}\, dx\, dy$$
$$= 4\left(\int_0^\infty e^{-x^2}\, dx\right)^2.$$

92. Improper double integral Evaluate the improper integral

$$\int_0^1 \int_0^3 \frac{x^2}{(y - 1)^{2/3}}\, dy\, dx.$$

COMPUTER EXPLORATIONS

Use a CAS double-integral evaluator to estimate the values of the integrals in Exercises 93–96.

93. $\int_1^3 \int_1^x \frac{1}{xy}\, dy\, dx$

94. $\int_0^1 \int_0^1 e^{-(x^2+y^2)}\, dy\, dx$

95. $\int_0^1 \int_0^1 \arctan xy\, dy\, dx$

96. $\int_{-1}^1 \int_0^{\sqrt{1-x^2}} 3\sqrt{1 - x^2 - y^2}\, dy\, dx$

Use a CAS double-integral evaluator to find the integrals in Exercises 97–102. Then reverse the order of integration and evaluate, again with a CAS.

97. $\int_0^1 \int_{2y}^4 e^{x^2}\, dx\, dy$

98. $\int_0^3 \int_{x^2}^9 x \cos(y^2)\, dy\, dx$

99. $\int_0^2 \int_{y^3}^{4\sqrt{2y}} (x^2y - xy^2)\, dx\, dy$

100. $\int_0^2 \int_0^{4-y^2} e^{xy}\, dx\, dy$

101. $\int_1^2 \int_0^{x^2} \frac{1}{x+y}\, dy\, dx$

102. $\int_1^2 \int_{y^3}^8 \frac{1}{\sqrt{x^2+y^2}}\, dx\, dy$

15.3 Area by Double Integration

In this section we show how to use double integrals to calculate the areas of bounded regions in the plane, and to find the average value of a function of two variables.

Areas of Bounded Regions in the Plane

If we take $f(x, y) = 1$ in the definition of the double integral over a region R in the preceding section, the Riemann sums reduce to

$$S_n = \sum_{k=1}^{n} f(x_k, y_k)\,\Delta A_k = \sum_{k=1}^{n} \Delta A_k. \tag{1}$$

This is simply the sum of the areas of the small rectangles in the partition of R, and it approximates what we would like to call the area of R. As the norm of a partition of R approaches zero, the height and width of all rectangles in the partition approach zero, and the coverage of R becomes increasingly complete (Figure 15.8). We define the area of R to be the limit

$$\lim_{\|P\|\to 0} \sum_{k=1}^{n} \Delta A_k = \iint_R dA. \tag{2}$$

DEFINITION The **area** of a closed, bounded plane region R is

$$A = \iint_R dA.$$

As with the other definitions in this chapter, the definition here applies to a greater variety of regions than does the earlier single-variable definition of area, but it agrees with the earlier definition on regions to which they both apply. To evaluate the integral in the definition of area, we integrate the constant function $f(x, y) = 1$ over R.

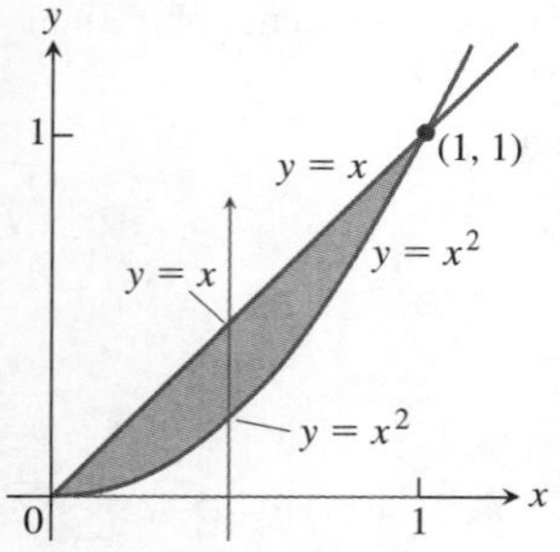

FIGURE 15.19 The region in Example 1.

EXAMPLE 1 Find the area of the region R bounded by $y = x$ and $y = x^2$ in the first quadrant.

Solution We sketch the region (Figure 15.19), noting where the two curves intersect at the origin and $(1, 1)$, and calculate the area as

$$A = \int_0^1 \int_{x^2}^{x} dy\, dx = \int_0^1 \left[y \right]_{y=x^2}^{y=x} dx = \int_0^1 (x - x^2)\, dx = \left[\frac{x^2}{2} - \frac{x^3}{3} \right]_0^1 = \frac{1}{6}.$$

Notice that the single-variable integral $\int_0^1 (x - x^2)\, dx$, obtained from evaluating the inside iterated integral, is the integral for the area between these two curves using the method of Section 5.6. ■

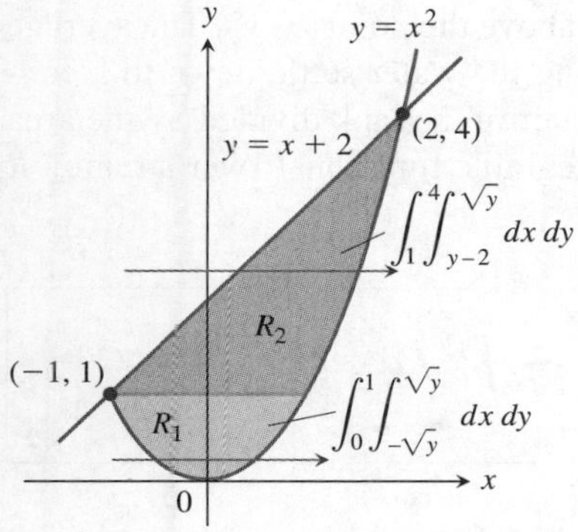

(a)

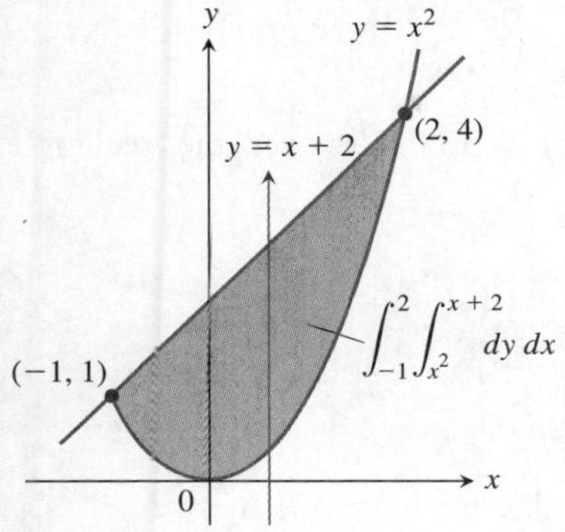

(b)

FIGURE 15.20 Calculating this area takes (a) two double integrals if the first integration is with respect to x, but (b) only one if the first integration is with respect to y (Example 2).

EXAMPLE 2 Find the area of the region R enclosed by the parabola $y = x^2$ and the line $y = x + 2$.

Solution If we divide R into the regions R_1 and R_2 shown in Figure 15.20a, we may calculate the area as

$$A = \iint_{R_1} dA + \iint_{R_2} dA = \int_0^1 \int_{-\sqrt{y}}^{\sqrt{y}} dx\, dy + \int_1^4 \int_{y-2}^{\sqrt{y}} dx\, dy.$$

On the other hand, reversing the order of integration (Figure 15.20b) gives

$$A = \int_{-1}^{2} \int_{x^2}^{x+2} dy\, dx.$$

This second result, which requires only one integral, is simpler to evaluate, giving

$$A = \int_{-1}^{2} \Big[y \Big]_{y=x^2}^{y=x+2} dx = \int_{-1}^{2} (x + 2 - x^2)\, dx = \left[\frac{x^2}{2} + 2x - \frac{x^3}{3} \right]_{-1}^{2} = \frac{9}{2}.$$ ■

EXAMPLE 3 Find the area of the playing field described by R: $-2 \le x \le 2,\ -1 - \sqrt{4 - x^2} \le y \le 1 + \sqrt{4 - x^2}$, using

(a) Fubini's Theorem **(b)** simple geometry.

Solution The region R is shown in Figure 15.21a.

(a) From the symmetries observed in the figure, we see that the area of R is 4 times its area in the first quadrant. As shown in Figure 15.21b, a vertical line at x enters this part of the region at $y = 0$ and exits at $y = 1 + \sqrt{4 - x^2}$. Therefore, using Fubini's Theorem, we have

$$\begin{aligned} A &= \iint_R dA = 4 \int_0^2 \int_0^{1+\sqrt{4-x^2}} dy\, dx \\ &= 4 \int_0^2 \left(1 + \sqrt{4 - x^2}\right) dx \\ &= 4 \left[x + \frac{x}{2}\sqrt{4 - x^2} + \frac{4}{2}\sin^{-1}\frac{x}{2} \right]_0^2 && \text{Integral Table Formula 45} \\ &= 4\left(2 + 0 + 2 \cdot \frac{\pi}{2} - 0\right) = 8 + 4\pi. \end{aligned}$$

(b) The region R consists of a rectangle mounted on two sides by half disks of radius 2. The area can be computed by summing the area of the 4×2 rectangle and the area of a circle of radius 2, so

$$A = 8 + \pi 2^2 = 8 + 4\pi.$$ ■

Average Value

The average value of an integrable function of one variable on a closed interval is the integral of the function over the interval divided by the length of the interval. For an integrable function of two variables defined on a bounded region in the plane, the average value is the integral over the region divided by the area of the region. This can be visualized by thinking of the region as being the base of a tank with vertical walls around the boundary of the region, and imagining that the tank is filled with water that is sloshing around. The value

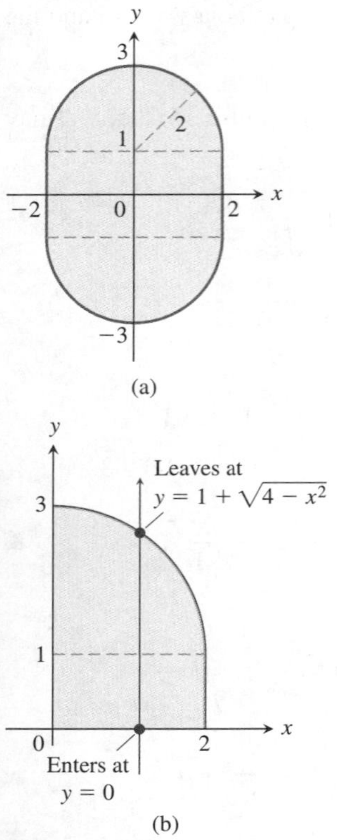

FIGURE 15.21 (a) The playing field described by the region R in Example 3. (b) First quadrant of the playing field.

$f(x, y)$ is then the height of the water that is directly above the point (x, y). The average height of the water in the tank can be found by letting the water settle down to a constant height. This height is equal to the volume of water in the tank divided by the area of R. We therefore define the average value of an integrable function f over a region R as follows:

$$\textbf{Average value} \text{ of } f \text{ over } R = \frac{1}{\text{area of } R} \iint_R f\, dA. \tag{3}$$

If f is the temperature of a thin plate covering R, then the double integral of f over R divided by the area of R is the plate's average temperature. If $f(x, y)$ is the distance from the point (x, y) to a fixed point P, then the average value of f over R is the average distance of points in R from P.

EXAMPLE 4 Find the average value of $f(x, y) = x \cos xy$ over the rectangle $R: 0 \le x \le \pi,\ 0 \le y \le 1$.

Solution The value of the integral of f over R is

$$\int_0^{\pi}\int_0^1 x\cos xy\, dy\, dx = \int_0^{\pi}\left[\sin xy\right]_{y=0}^{y=1} dx \qquad \int x\cos xy\, dy = \sin xy + C$$

$$= \int_0^{\pi}(\sin x - 0)\, dx = -\cos x\Big]_0^{\pi} = 1 + 1 = 2.$$

The area of R is π. The average value of f over R is $2/\pi$. ■

EXERCISES 15.3

Area by Double Integrals

In Exercises 1–12, sketch the region bounded by the given lines and curves. Then express the region's area as an iterated double integral and evaluate the integral.

1. The coordinate axes and the line $x + y = 2$
2. The lines $x = 0,\ y = 2x$, and $y = 4$
3. The parabola $x = -y^2$ and the line $y = x + 2$
4. The parabola $y = 4x - x^2$ and the line $y = 2$
5. The parabola $y^2 = 4 - x,\ y^2 = x$
6. The curves $y = \ln x$ and $y = 2\ln x$ and the line $x = e$, in the first quadrant
7. The parabolas $x = y^2$ and $x = 2y - y^2$
8. The parabolas $x = y^2 - 1$ and $x = 2y^2 - 2$
9. The lines $y = x,\ y = x/3$, and $y = 2$
10. The parabola $y = 2x^2$ and $y^2 = 4x$
11. The ellipse $\dfrac{x^2}{a^2} + \dfrac{y^2}{b^2} = 1$
12. The lines $y = x - 2$ and $y = -x$ and the curve $y = \sqrt{x}$

Identifying the Region of Integration

The integrals and sums of integrals in Exercises 13–18 give the areas of regions in the xy-plane. Sketch each region, label each bounding curve with its equation, and give the coordinates of the points where the curves intersect. Then find the area of the region.

13. $\displaystyle\int_0^6\int_{y^2/3}^{2y} dx\, dy$

14. $\displaystyle\int_0^3\int_{-x}^{x(2-x)} dy\, dx$

15. $\displaystyle\int_0^{\pi/4}\int_{\sin x}^{\cos x} dy\, dx$

16. $\displaystyle\int_0^1\int_0^{\sqrt{y}} dx\, dy + \int_1^2\int_0^{2-y} dx\, dy$

17. $\displaystyle\int_{-1}^0\int_{-2x}^{1-x} dy\, dx + \int_0^2\int_{-x/2}^{1-x} dy\, dx$

18. $\displaystyle\int_0^2\int_{x^2-4}^0 dy\, dx + \int_0^4\int_0^{\sqrt{x}} dy\, dx$

Finding Average Values

19. Find the average value of $f(x, y) = \sin(x + y)$ over
    a. the rectangle $0 \le x \le \pi,\quad 0 \le y \le \pi$.
    b. the rectangle $0 \le x \le \pi,\quad 0 \le y \le \pi/2$.
20. Which do you think will be larger, the average value of $f(x, y) = xy$ over the square $0 \le x \le 1,\ 0 \le y \le 1$, or the average value of f over the quarter circle $x^2 + y^2 \le 1$ in the first quadrant? Calculate them to find out.
21. Find the average height of the paraboloid $z = x^2 + y^2$ over the square $0 \le x \le 2,\ 0 \le y \le 2$.

22. Find the average value of $f(x, y) = 1/(xy)$ over the square $\ln 2 \leq x \leq 2 \ln 2$, $\ln 2 \leq y \leq 2 \ln 2$.

Theory and Examples

23. **Geometric area** Find the area of the region

$$R: 0 \leq x \leq 2, \; 2 - x \leq y \leq \sqrt{4 - x^2},$$

using **(a)** Fubini's Theorem, **(b)** simple geometry.

24. **Geometric area** Find the area of the circular washer with outer radius 2 and inner radius 1, using **(a)** Fubini's Theorem, **(b)** simple geometry.

25. **Bacterium population** If $f(x, y) = (10{,}000e^y)/(1 + |x|/2)$ represents the "population density" of a certain bacterium on the xy-plane, where x and y are measured in centimeters, find the total population of bacteria within the rectangle $-5 \leq x \leq 5$ and $-2 \leq y \leq 0$.

26. **Regional population** If $f(x, y) = 100(y + 1)$ represents the population density of a planar region on Earth, where x and y are measured in miles, find the number of people in the region bounded by the curves $x = y^2$ and $x = 2y - y^2$.

27. **Average temperature in Texas** According to the *Texas Almanac*, Texas has 254 counties and a National Weather Service station in each county. Assume that at time t_0, each of the 254 weather stations recorded the local temperature. Find a formula that would give a reasonable approximation of the average temperature in Texas at time t_0. Your answer should involve information that you would expect to be readily available in the *Texas Almanac*.

28. If $y = f(x)$ is a nonnegative continuous function over the closed interval $a \leq x \leq b$, show that the double integral definition of area for the closed plane region bounded by the graph of f, the vertical lines $x = a$ and $x = b$, and the x-axis agrees with the definition for area beneath the curve in Section 5.3.

29. Suppose $f(x, y)$ is continuous over a region R in the plane and that the area $A(R)$ of the region is defined. If there are constants m and M such that $m \leq f(x, y) \leq M$ for all $(x, y) \in R$, prove that

$$mA(R) \leq \iint_R f(x, y)\, dA \leq MA(R).$$

30. Suppose $f(x, y)$ is continuous and nonnegative over a region R in the plane with a defined area $A(R)$. If $\iint_R f(x, y)\, dA = 0$, prove that $f(x, y) = 0$ at every point $(x, y) \in R$.

15.4 Double Integrals in Polar Form

Double integrals are sometimes easier to evaluate if we change to polar coordinates. This section shows how to accomplish the change and how to evaluate double integrals over regions whose boundaries are given by polar equations.

Integrals in Polar Coordinates

When we defined the double integral of a function over a region R in the xy-plane, we began by cutting R into rectangles whose sides were parallel to the coordinate axes. These were the natural shapes to use because their sides have either constant x-values or constant y-values. In polar coordinates, the natural shape is a "polar rectangle" whose sides have constant r- and θ-values. To avoid ambiguities when describing the region of integration with polar coordinates, we use polar coordinate points (r, θ) where $r \geq 0$.

Suppose that a function $f(r, \theta)$ is defined over a region R that is bounded by the rays $\theta = \alpha$ and $\theta = \beta$ and by the continuous curves $r = g_1(\theta)$ and $r = g_2(\theta)$. Suppose also that $0 \leq g_1(\theta) \leq g_2(\theta) \leq a$ for every value of θ between α and β. Then R lies in a fan-shaped region Q defined by the inequalities $0 \leq r \leq a$ and $\alpha \leq \theta \leq \beta$, where $0 \leq \beta - \alpha \leq 2\pi$. See Figure 15.22.

We cover Q by a grid of circular arcs and rays. The arcs are cut from circles centered at the origin, with radii $\Delta r, 2\Delta r, \ldots, m\Delta r$, where $\Delta r = a/m$. The rays are given by

$$\theta = \alpha, \quad \theta = \alpha + \Delta\theta, \quad \theta = \alpha + 2\Delta\theta, \quad \ldots, \quad \theta = \alpha + m'\Delta\theta = \beta,$$

where $\Delta\theta = (\beta - \alpha)/m'$. The arcs and rays partition Q into small patches called "polar rectangles."

We number the polar rectangles that lie inside R (the order does not matter), calling their areas $\Delta A_1, \Delta A_2, \ldots, \Delta A_n$. We let (r_k, θ_k) be any point in the polar rectangle whose area is ΔA_k. We then form the sum

$$S_n = \sum_{k=1}^{n} f(r_k, \theta_k)\, \Delta A_k.$$

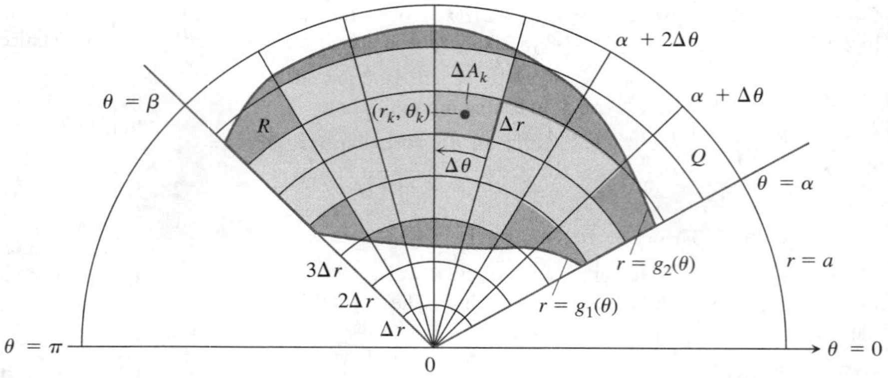

FIGURE 15.22 The region R: $g_1(\theta) \le r \le g_2(\theta)$, $\alpha \le \theta \le \beta$, is contained in the fan-shaped region Q: $0 \le r \le a$, $\alpha \le \theta \le \beta$, where $0 \le \beta - \alpha \le 2\pi$. The partition of Q by circular arcs and rays induces a partition of R.

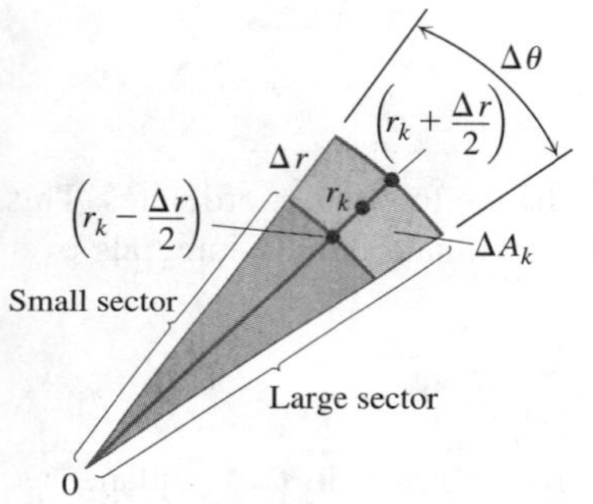

FIGURE 15.23 The observation that

$$\Delta A_k = \begin{pmatrix}\text{area of}\\ \text{large sector}\end{pmatrix} - \begin{pmatrix}\text{area of}\\ \text{small sector}\end{pmatrix}$$

leads to the formula $\Delta A_k = r_k\,\Delta r\,\Delta\theta$.

If f is continuous throughout R, this sum will approach a limit as we refine the grid to make Δr and $\Delta\theta$ go to zero. The limit is the double integral of f over R. In symbols,

$$\lim_{n\to\infty} S_n = \iint_R f(r,\theta)\,dA.$$

To evaluate this limit, we first have to write the sum S_n in a way that expresses ΔA_k in terms of Δr and $\Delta\theta$. For convenience we choose r_k to be the average of the radii of the inner and outer arcs bounding the kth polar rectangle ΔA_k. The radius of the inner arc bounding ΔA_k is then $r_k - (\Delta r/2)$ (Figure 15.23). The radius of the outer arc is $r_k + (\Delta r/2)$.

The area of a wedge-shaped sector of a circle having radius r and central angle $\Delta\theta$ is

$$A = \frac{1}{2}\Delta\theta \cdot r^2,$$

as can be seen by multiplying πr^2, the area of the circle, by $\Delta\theta/2\pi$, the fraction of the circle's area contained in the wedge. So the areas of the circular sectors subtended by these arcs at the origin are

$$\text{Area of small sector:} \quad \frac{1}{2}\left(r_k - \frac{\Delta r}{2}\right)^2 \Delta\theta$$

$$\text{Area of large sector:} \quad \frac{1}{2}\left(r_k + \frac{\Delta r}{2}\right)^2 \Delta\theta.$$

Therefore,

$$\begin{aligned}\Delta A_k &= \text{area of large sector} - \text{area of small sector}\\ &= \frac{\Delta\theta}{2}\left[\left(r_k + \frac{\Delta r}{2}\right)^2 - \left(r_k - \frac{\Delta r}{2}\right)^2\right] = \frac{\Delta\theta}{2}(2r_k\,\Delta r) = r_k\,\Delta r\,\Delta\theta.\end{aligned}$$

Combining this result with the sum defining S_n gives

$$S_n = \sum_{k=1}^{n} f(r_k, \theta_k) r_k\,\Delta r\,\Delta\theta.$$

As $n \to \infty$ and the values of Δr and $\Delta\theta$ approach zero, these sums converge to the double integral

$$\lim_{n\to\infty} S_n = \iint_R f(r,\theta)\,r\,dr\,d\theta.$$

A version of Fubini's Theorem says that the limit approached by these sums can be evaluated by repeated single integrations with respect to r and θ as

$$\iint_R f(r, \theta)\, dA = \int_{\theta=\alpha}^{\theta=\beta} \int_{r=g_1(\theta)}^{r=g_2(\theta)} f(r, \theta)\, r\, dr\, d\theta.$$

Finding Limits of Integration

The procedure for finding limits of integration in rectangular coordinates also works for polar coordinates. We illustrate this using the region R shown in Figure 15.24. To evaluate $\iint_R f(r, \theta)\, dA$ in polar coordinates, integrating first with respect to r and then with respect to θ, take the following steps.

1. *Sketch.* Sketch the region and label the bounding curves (Figure 15.24a).
2. *Find the r-limits of integration.* Imagine a ray L from the origin cutting through R in the direction of increasing r. Mark the r-values where L enters and leaves R. These are the r-limits of integration. They usually depend on the angle θ that L makes with the positive x-axis (Figure 15.24b).
3. *Find the θ-limits of integration.* Find the smallest and largest θ-values that bound R. These are the θ-limits of integration (Figure 15.24c). The polar iterated integral is

$$\iint_R f(r, \theta)\, dA = \int_{\theta=\pi/4}^{\theta=\pi/2} \int_{r=\sqrt{2}\csc\theta}^{r=2} f(r, \theta)\, r\, dr\, d\theta.$$

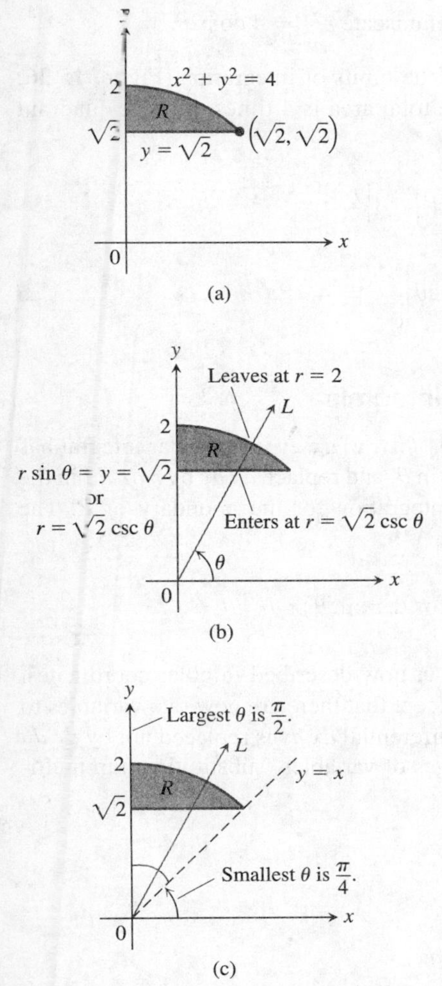

FIGURE 15.24 Finding the limits of integration in polar coordinates.

EXAMPLE 1 Find the limits of integration for integrating $f(r, \theta)$ over the region R that lies inside the cardioid $r = 1 + \cos\theta$ and outside the circle $r = 1$.

Solution

1. We first sketch the region and label the bounding curves (Figure 15.25).
2. Next we find the *r-limits of integration*. A typical ray from the origin enters R where $r = 1$ and leaves where $r = 1 + \cos\theta$.
3. Finally, we find the *θ-limits of integration*. The rays from the origin that intersect R run from $\theta = -\pi/2$ to $\theta = \pi/2$. The integral is

$$\int_{-\pi/2}^{\pi/2} \int_{1}^{1+\cos\theta} f(r, \theta) r\, dr\, d\theta. \qquad \blacksquare$$

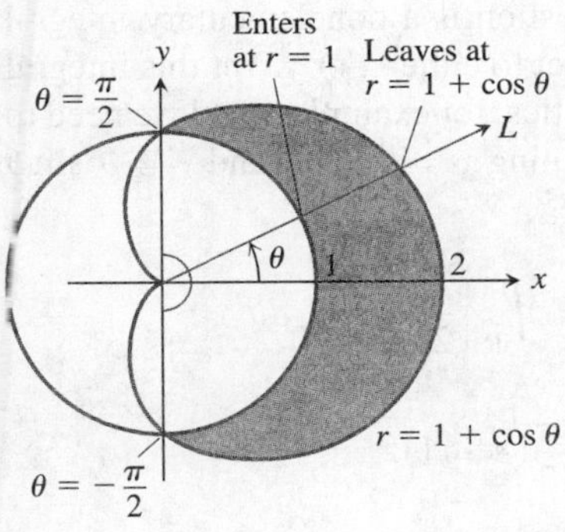

FIGURE 15.25 Finding the limits of integration in polar coordinates for the region in Example 1.

If $f(r, \theta)$ is the constant function whose value is 1, then the integral of f over R is the area of R.

> **Area in Polar Coordinates**
> The area of a closed and bounded region R in the polar coordinate plane is
>
> $$A = \iint_R r\, dr\, d\theta.$$

> **Area Differential in Polar Coordinates**
>
> $$dA = r\, dr\, d\theta$$

This formula for area is consistent with all earlier formulas.

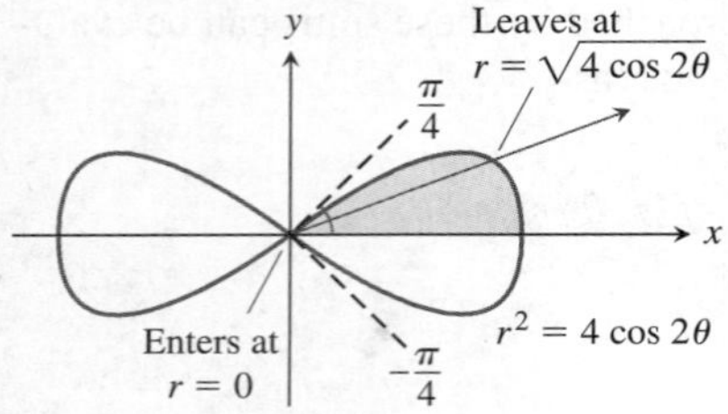

FIGURE 15.26 To integrate over the shaded region, we run r from 0 to $\sqrt{4\cos 2\theta}$ and θ from 0 to $\pi/4$ (Example 2).

EXAMPLE 2 Find the area enclosed by the lemniscate $r^2 = 4\cos 2\theta$.

Solution We graph the lemniscate to determine the limits of integration (Figure 15.26) and see from the symmetry of the region that the total area is 4 times the first-quadrant portion.

$$A = 4\int_0^{\pi/4}\int_0^{\sqrt{4\cos 2\theta}} r\,dr\,d\theta = 4\int_0^{\pi/4}\left[\frac{r^2}{2}\right]_{r=0}^{r=\sqrt{4\cos 2\theta}} d\theta$$

$$= 4\int_0^{\pi/4} 2\cos 2\theta\,d\theta = 4\sin 2\theta\Big]_0^{\pi/4} = 4.$$ ■

Changing Cartesian Integrals into Polar Integrals

The procedure for changing a Cartesian integral $\iint_R f(x, y)\,dx\,dy$ into a polar integral has two steps. First substitute $x = r\cos\theta$ and $y = r\sin\theta$, and replace $dx\,dy$ by $r\,dr\,d\theta$ in the Cartesian integral. Then supply polar limits of integration for the boundary of R. The Cartesian integral then becomes

$$\iint_R f(x, y)\,dx\,dy = \iint_G f(r\cos\theta, r\sin\theta)\,r\,dr\,d\theta,$$

where G denotes the same region of integration, but now described in polar coordinates. This is like the substitution method in Chapter 5 except that there are now two variables to substitute for instead of one. Notice that the area differential $dx\,dy$ is replaced not by $dr\,d\theta$ but by $r\,dr\,d\theta$. A more general discussion of changes of variables (substitutions) in multiple integrals is given in Section 15.8.

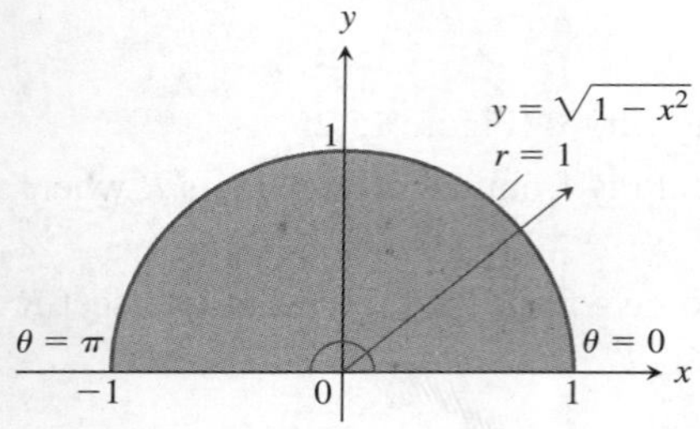

FIGURE 15.27 The semicircular region in Example 3 is the region

$$0 \le r \le 1, \quad 0 \le \theta \le \pi.$$

EXAMPLE 3 Evaluate

$$\iint_R e^{x^2+y^2}\,dy\,dx,$$

where R is the semicircular region bounded by the x-axis and the curve $y = \sqrt{1 - x^2}$ (Figure 15.27).

Solution In Cartesian coordinates, the integral in question is a nonelementary integral and there is no direct way to integrate $e^{x^2+y^2}$ with respect to either x or y. Yet this integral and others like it are important in mathematics—in statistics, for example—and we need to evaluate it. Polar coordinates make this possible. Substituting $x = r\cos\theta$ and $y = r\sin\theta$ and replacing $dy\,dx$ by $r\,dr\,d\theta$ give

$$\iint_R e^{x^2+y^2}\,dy\,dx = \int_0^{\pi}\int_0^1 e^{r^2} r\,dr\,d\theta = \int_0^{\pi}\left[\frac{1}{2}e^{r^2}\right]_{r=0}^{r=1} d\theta$$

$$= \int_0^{\pi}\frac{1}{2}(e - 1)\,d\theta = \frac{\pi}{2}(e - 1).$$

The r in the $r\,dr\,d\theta$ is what allowed us to integrate e^{r^2}. Without it, we would have been unable to find an antiderivative for the first (innermost) iterated integral. ■

EXAMPLE 4 Evaluate the integral

$$\int_0^1\int_0^{\sqrt{1-x^2}} (x^2 + y^2)\,dy\,dx.$$

Solution Integration with respect to y gives

$$\int_0^1 \left(x^2 \sqrt{1-x^2} + \frac{(1-x^2)^{3/2}}{3} \right) dx,$$

which is difficult to evaluate without tables. Things go better if we change the original integral to polar coordinates. The region of integration in Cartesian coordinates is given by the inequalities $0 \le y \le \sqrt{1-x^2}$ and $0 \le x \le 1$, which correspond to the interior of the unit quarter circle $x^2 + y^2 = 1$ in the first quadrant. (See Figure 15.27, first quadrant.) Substituting the polar coordinates $x = r\cos\theta$, $y = r\sin\theta$, $0 \le \theta \le \pi/2$, and $0 \le r \le 1$, and replacing $dy\,dx$ by $r\,dr\,d\theta$ in the double integral, we get

$$\begin{aligned}\int_0^1 \int_0^{\sqrt{1-x^2}} (x^2 + y^2)\,dy\,dx &= \int_0^{\pi/2} \int_0^1 (r^2)\,r\,dr\,d\theta \\ &= \int_0^{\pi/2} \left[\frac{r^4}{4} \right]_{r=0}^{r=1} d\theta = \int_0^{\pi/2} \frac{1}{4}\,d\theta = \frac{\pi}{8}.\end{aligned}$$

The polar coordinate transformation is effective here because $x^2 + y^2$ simplifies to r^2 and the limits of integration become constants. ■

EXAMPLE 5 Find the volume of the solid region bounded above by the paraboloid $z = 9 - x^2 - y^2$ and below by the unit circle in the xy-plane.

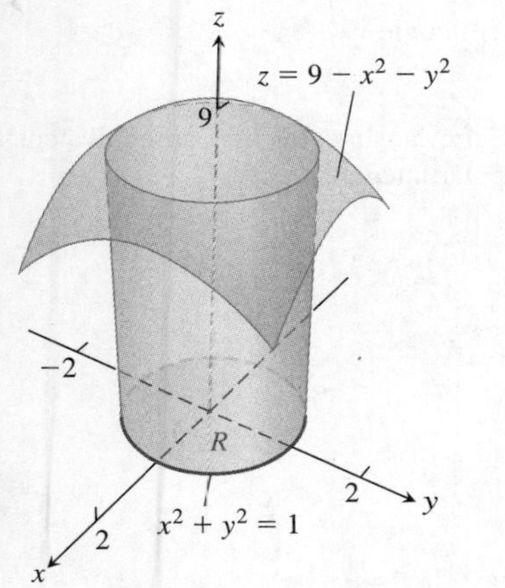

FIGURE 15.28 The solid region in Example 5.

Solution The region of integration R is bounded by the unit circle $x^2 + y^2 = 1$, which is described in polar coordinates by $r = 1$, $0 \le \theta \le 2\pi$. The solid region is shown in Figure 15.28. The volume is given by the double integral

$$\begin{aligned}\iint_R (9 - x^2 - y^2)\,dA &= \int_0^{2\pi} \int_0^1 (9 - r^2)\,r\,dr\,d\theta \qquad r^2 = x^2 + y^2,\ dA = r\,dr\,d\theta. \\ &= \int_0^{2\pi} \int_0^1 (9r - r^3)\,dr\,d\theta \\ &= \int_0^{2\pi} \left[\frac{9}{2}r^2 - \frac{1}{4}r^4 \right]_{r=0}^{r=1} d\theta \\ &= \frac{17}{4} \int_0^{2\pi} d\theta = \frac{17\pi}{2}.\end{aligned}$$

■

EXAMPLE 6 Using polar integration, find the area of the region R enclosed by the circle $x^2 + y^2 = 4$, above the line $y = 1$, and below the line $y = \sqrt{3}x$.

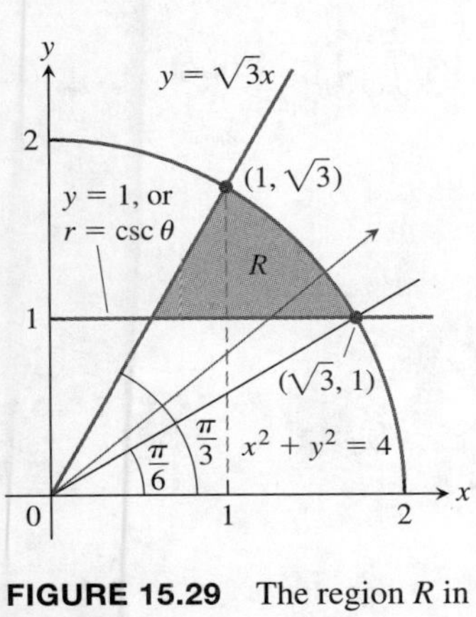

FIGURE 15.29 The region R in Example 6.

Solution A sketch of the region R is shown in Figure 15.29. First we note that the line $y = \sqrt{3}x$ has slope $\sqrt{3} = \tan\theta$, so $\theta = \pi/3$. Next we observe that the line $y = 1$ intersects the circle $x^2 + y^2 = 4$ when $x^2 + 1 = 4$, or $x = \sqrt{3}$. Moreover, the radial line from the origin through the point $(\sqrt{3}, 1)$ has slope $1/\sqrt{3} = \tan\theta$, giving its angle of inclination as $\theta = \pi/6$. This information is shown in Figure 15.29.

Now, for the region R, as θ varies from $\pi/6$ to $\pi/3$, the polar coordinate r varies from the horizontal line $y = 1$ to the circle $x^2 + y^2 = 4$. Substituting $r\sin\theta$ for y in the equation for the horizontal line, we have $r\sin\theta = 1$, or $r = \csc\theta$, which is the polar equation of the line. The polar equation for the circle is $r = 2$. So in polar coordinates, for $\pi/6 \le \theta \le \pi/3$, r varies from $r = \csc\theta$ to $r = 2$. It follows that the iterated integral for the area is

$$\iint_R dA = \int_{\pi/6}^{\pi/3}\int_{\csc\theta}^{2} r\,dr\,d\theta$$

$$= \int_{\pi/6}^{\pi/3}\left[\frac{1}{2}r^2\right]_{r=\csc\theta}^{r=2} d\theta$$

$$= \int_{\pi/6}^{\pi/3}\frac{1}{2}[4 - \csc^2\theta]\,d\theta$$

$$= \frac{1}{2}\Big[4\theta + \cot\theta\Big]_{\pi/6}^{\pi/3}$$

$$= \frac{1}{2}\left(\frac{4\pi}{3} + \frac{1}{\sqrt{3}}\right) - \frac{1}{2}\left(\frac{4\pi}{6} + \sqrt{3}\right) = \frac{\pi - \sqrt{3}}{3}. \quad \blacksquare$$

EXERCISES 15.4

Regions in Polar Coordinates

In Exercises 1–8, describe the given region in polar coordinates.

1.

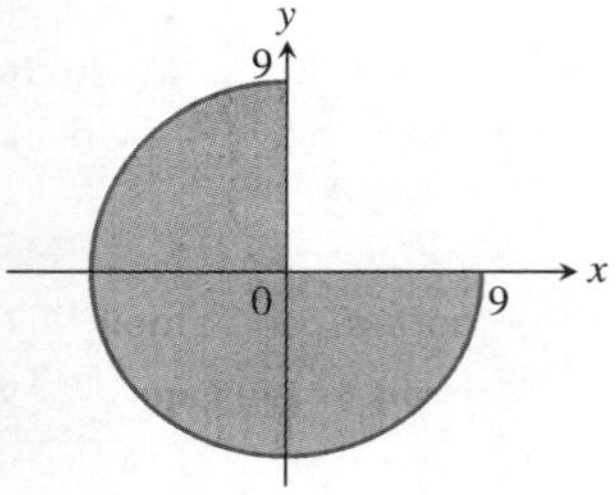

2.

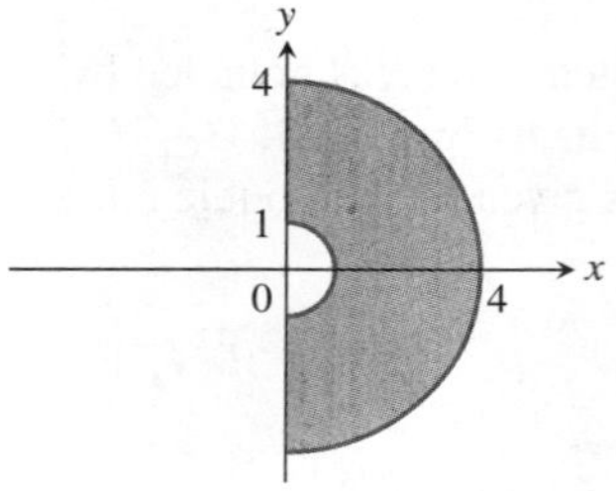

3.

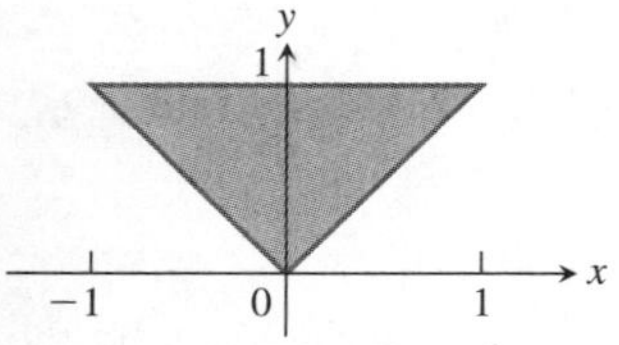

4.

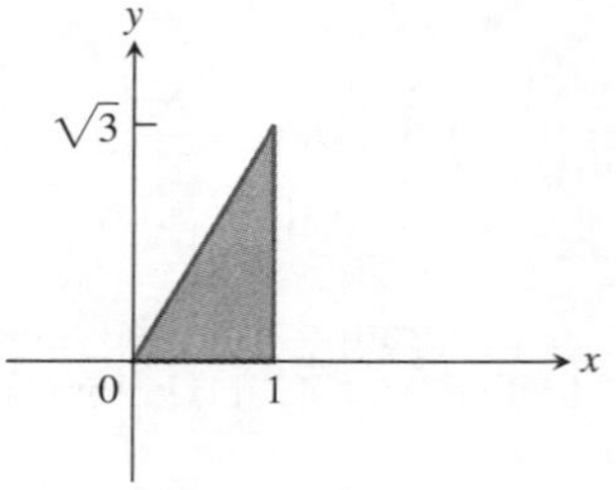

5.

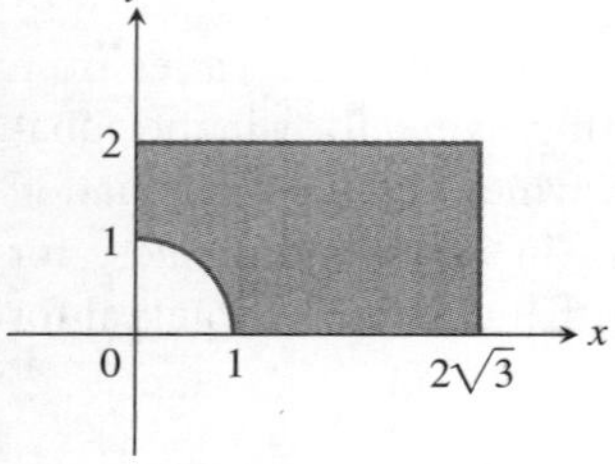

6.

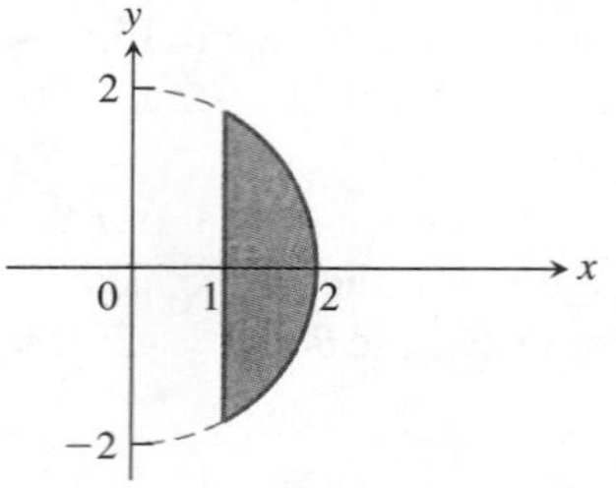

7. The region enclosed by the circle $x^2 + y^2 = 2x$

8. The region enclosed by the semicircle $x^2 + y^2 = 2y,\ y \geq 1$

Evaluating Polar Integrals

In Exercises 9–22, change the Cartesian integral into an equivalent polar integral. Then evaluate the polar integral.

9. $\displaystyle\int_{-1}^{1}\int_{0}^{\sqrt{1-x^2}} dy\,dx$

10. $\displaystyle\int_{0}^{1}\int_{0}^{\sqrt{1-y^2}} (x^2 + y^2)\,dx\,dy$

11. $\displaystyle\int_{0}^{2}\int_{0}^{\sqrt{4-y^2}} (x^2 + y^2)\,dx\,dy$

12. $\displaystyle\int_{-a}^{a}\int_{-\sqrt{a^2-x^2}}^{\sqrt{a^2-x^2}} dy\,dx$

13. $\displaystyle\int_{0}^{2}\int_{0}^{\sqrt{2x-x^2}} (x^2 + y^2)\,dy\,dx$

14. $\displaystyle\int_{0}^{a}\int_{y}^{a} \frac{x^2}{\sqrt{x^2 + y^2}}\,dx\,dy$

15. $\displaystyle\int_{1}^{\sqrt{3}}\int_{1}^{x} dy\,dx$

16. $\displaystyle\int_{\sqrt{2}}^{2}\int_{\sqrt{4-y^2}}^{y} dx\,dy$

17. $\displaystyle\int_{-1}^{0}\int_{-\sqrt{1-x^2}}^{0} \frac{2}{1 + \sqrt{x^2 + y^2}}\,dy\,dx$

18. $\displaystyle\int_{0}^{a}\int_{0}^{\sqrt{a^2-y^2}} (x^2 + y^2)\,dx\,dy$

19. $\displaystyle\int_{0}^{2}\int_{0}^{\sqrt{4-x^2}} y(x^2 + y^2)\,dx\,dy$

20. $\displaystyle\int_{0}^{\infty}\int_{0}^{\infty} e^{-(x^2+y^2)}\,dx\,dy$

21. $\displaystyle\int_{0}^{1}\int_{x}^{\sqrt{2-x^2}} (x + 2y)\,dy\,dx$

22. $\int_1^2 \int_0^{\sqrt{2x-x^2}} \frac{1}{(x^2+y^2)^2}\, dy\, dx$

In Exercises 23–26, sketch the region of integration, and convert each polar integral or sum of integrals into a Cartesian integral or sum of integrals. Do not evaluate the integrals.

23. $\int_0^{\pi/2} \int_0^1 r^3 \sin\theta \cos\theta\, dr\, d\theta$

24. $\int_{\pi/6}^{\pi/2} \int_1^{\csc\theta} r^2 \cos\theta\, dr\, d\theta$

25. $\int_0^{\pi/4} \int_0^{2\sec\theta} r^5 \sin^2\theta\, dr\, d\theta$

26. $\int_0^{\arctan\frac{4}{3}} \int_0^{3\sec\theta} r^7\, dr\, d\theta + \int_{\arctan\frac{4}{3}}^{\pi/2} \int_0^{4\csc\theta} r^7\, dr\, d\theta$

Area in Polar Coordinates

27. Find the area of the region cut from the first quadrant by the curve $r = 2(2 - \sin 2\theta)^{1/2}$.

28. **Cardioid overlapping a circle** Find the area of the region that lies inside the cardioid $r = 1 + \cos\theta$ and outside the circle $r = 1$.

29. **One leaf of a rose** Find the area enclosed by one leaf of the rose $r = 12\cos 3\theta$.

30. **Snail shell** Find the area of the region enclosed by the positive x-axis and spiral $r = 4\theta/3$, $0 \le \theta \le 2\pi$. The region looks like a snail shell.

31. **Cardioid in the first quadrant** Find the area of the region cut from the first quadrant by the cardioid $r = 1 + \sin\theta$.

32. **Overlapping cardioids** Find the area of the region common to the interiors of the cardioids $r = 1 + \cos\theta$ and $r = 1 - \cos\theta$.

Average Values

In polar coordinates, the **average value** of a function over a region R (Section 15.3) is given by

$$\frac{1}{\text{Area}(R)} \iint_R f(r,\theta)\, r\, dr\, d\theta.$$

33. **Average height of a hemisphere** Find the average height of the hemispherical surface $z = \sqrt{a^2 - x^2 - y^2}$ above the disk $x^2 + y^2 \le a^2$ in the xy-plane.

34. **Average height of a cone** Find the average height of the (single) cone $z = \sqrt{x^2 + y^2}$ above the disk $x^2 + y^2 \le a^2$ in the xy-plane.

35. **Average distance from interior of disk to center** Find the average distance from a point $P(x, y)$ in the disk $x^2 + y^2 \le a^2$ to the origin.

36. **Average distance squared from a point in a disk to a point in its boundary** Find the average value of the *square* of the distance from the point $P(x, y)$ in the disk $x^2 + y^2 \le 1$ to the boundary point $A(1, 0)$.

Theory and Examples

37. **Converting to a polar integral** Integrate $f(x, y) = [\ln(x^2 + y^2)]/\sqrt{x^2 + y^2}$ over the region $1 \le x^2 + y^2 \le e$.

38. **Converting to a polar integral** Integrate $f(x, y) = [\ln(x^2 + y^2)]/(x^2 + y^2)$ over the region $1 \le x^2 + y^2 \le e^2$.

39. **Volume of noncircular right cylinder** The region that lies inside the cardioid $r = 1 + \cos\theta$ and outside the circle $r = 1$ is the base of a solid right cylinder. The top of the cylinder lies in the plane $z = x$. Find the cylinder's volume.

40. **Volume of noncircular right cylinder** The region enclosed by the lemniscate $r^2 = 2\cos 2\theta$ is the base of a solid right cylinder whose top is bounded by the sphere $z = \sqrt{2 - r^2}$. Find the cylinder's volume.

41. **Converting to polar integrals**

a. The usual way to evaluate the improper integral $I = \int_0^\infty e^{-x^2}\, dx$ is first to calculate its square:

$$I^2 = \left(\int_0^\infty e^{-x^2}\, dx\right)\left(\int_0^\infty e^{-y^2}\, dy\right) = \int_0^\infty \int_0^\infty e^{-(x^2+y^2)}\, dx\, dy.$$

Evaluate the last integral using polar coordinates and solve the resulting equation for I.

b. Evaluate

$$\lim_{x\to\infty} \operatorname{erf}(x) = \lim_{x\to\infty} \int_0^x \frac{2e^{-t^2}}{\sqrt{\pi}}\, dt.$$

42. **Converting to a polar integral** Evaluate the integral

$$\int_0^\infty \int_0^\infty \frac{1}{(1 + x^2 + y^2)^2}\, dx\, dy.$$

43. **Existence** Integrate the function $f(x, y) = 1/(1 - x^2 - y^2)$ over the disk $x^2 + y^2 \le 3/4$. Does the integral of $f(x, y)$ over the disk $x^2 + y^2 \le 1$ exist? Give reasons for your answer.

44. **Area formula in polar coordinates** Use the double integral in polar coordinates to derive the formula

$$A = \int_\alpha^\beta \frac{1}{2} r^2\, d\theta$$

for the area of the fan-shaped region between the origin and the polar curve $r = f(\theta)$, $\alpha \le \theta \le \beta$.

45. **Average distance to a given point inside a disk** Let P_0 be a point inside a circle of radius a and let h denote the distance from P_0 to the center of the circle. Let d denote the distance from an arbitrary point P to P_0. Find the average value of d^2 over the region enclosed by the circle. (*Hint:* Simplify your work by placing the center of the circle at the origin and P_0 on the x-axis.)

46. **Area** Suppose that the area of a region in the polar coordinate plane is

$$A = \int_{\pi/4}^{3\pi/4} \int_{\csc\theta}^{2\sin\theta} r\, dr\, d\theta.$$

Sketch the region and find its area.

47. Evaluate the integral $\iint_R \sqrt{x^2 + y^2}\, dA$, where R is the region inside the upper semicircle of radius 2 centered at the origin, but outside the circle $x^2 + (y - 1)^2 = 1$.

48. Evaluate the integral $\iint_R (x^2 + y^2)^{-2}\, dA$, where R is the region inside the circle $x^2 + y^2 = 2$ for $x \le -1$.

COMPUTER EXPLORATIONS

In Exercises 49–52, use a CAS to change the Cartesian integrals into an equivalent polar integral and evaluate the polar integral. Perform the following steps in each exercise.

a. Plot the Cartesian region of integration in the *xy*-plane.

b. Change each boundary curve of the Cartesian region in part (a) to its polar representation by solving its Cartesian equation for r and θ.

c. Using the results in part (b), plot the polar region of integration in the $r\theta$-plane.

d. Change the integrand from Cartesian to polar coordinates. Determine the limits of integration from your plot in part (c) and evaluate the polar integral using the CAS integration utility.

49. $\int_0^1 \int_x^1 \frac{y}{x^2 + y^2}\, dy\, dx$

50. $\int_0^1 \int_0^{x/2} \frac{x}{x^2 + y^2}\, dy\, dx$

51. $\int_0^1 \int_{-y/3}^{y/3} \frac{y}{\sqrt{x^2 + y^2}}\, dx\, dy$

52. $\int_0^1 \int_y^{2-y} \sqrt{x + y}\, dx\, dy$

15.5 Triple Integrals in Rectangular Coordinates

Just as double integrals allow us to deal with more general situations than could be handled by single integrals, triple integrals enable us to solve still more general problems. We use triple integrals to calculate the volumes of three-dimensional shapes and the average value of a function over a three-dimensional region. Triple integrals also arise in the study of vector fields and fluid flow in three dimensions, as we will see in Chapter 16.

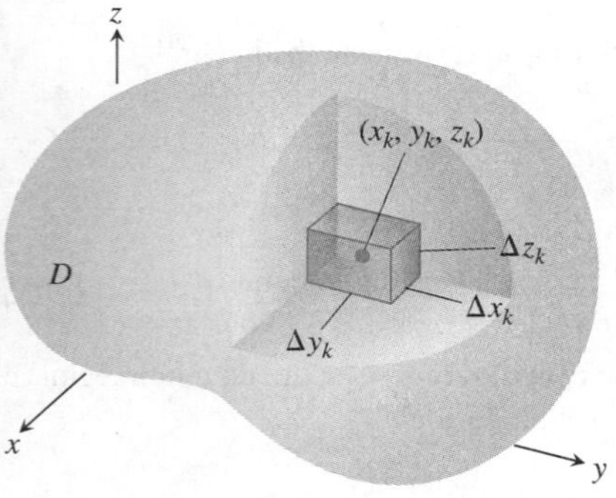

FIGURE 15.30 Partitioning a solid with rectangular cells of volume ΔV_k.

Triple Integrals

If $F(x, y, z)$ is a function defined on a closed bounded solid region D in space, such as the region occupied by a solid ball or a lump of clay, then the integral of F over D may be defined in the following way. We partition a rectangular boxlike region containing D into rectangular cells by planes parallel to the coordinate axes (Figure 15.30). We number the cells that lie completely inside D from 1 to n in some order, the kth cell having dimensions Δx_k by Δy_k by Δz_k and volume $\Delta V_k = \Delta x_k \Delta y_k \Delta z_k$. We choose a point (x_k, y_k, z_k) in each cell and form the sum

$$S_n = \sum_{k=1}^{n} F(x_k, y_k, z_k)\,\Delta V_k. \tag{1}$$

We are interested in what happens as D is partitioned by smaller and smaller cells, so that Δx_k, Δy_k, Δz_k, and the norm of the partition $\|P\|$, the largest value among Δx_k, Δy_k, Δz_k, all approach zero. When a single limiting value is attained, no matter how the partitions and points (x_k, y_k, z_k) are chosen, we say that F is **integrable** over D. As before, it can be shown that when F is continuous and the bounding surface of D is formed from finitely many smooth surfaces joined together along finitely many smooth curves, then F is integrable. In this case, as $\|P\| \to 0$ and the number of cells n goes to ∞, the sums S_n approach a limit. We call this limit the **triple integral of F over D** and write

$$\lim_{n\to\infty} S_n = \iiint_D F(x, y, z)\, dV \quad \text{or} \quad \lim_{\|P\|\to 0} S_n = \iiint_D F(x, y, z)\, dx\, dy\, dz.$$

The regions D over which continuous functions are integrable are those having "reasonably smooth" boundaries.

Volume of a Solid Region in Space

If F is the constant function whose value is 1, then the sums in Equation (1) reduce to

$$S_n = \sum_{k=1}^{n} F(x_k, y_k, z_k)\,\Delta V_k = \sum_{k=1}^{n} 1 \cdot \Delta V_k = \sum_{k=1}^{n} \Delta V_k.$$

As Δx_k, Δy_k, and Δz_k approach zero, the cells ΔV_k become smaller and more numerous and fill up more and more of D. We therefore define the volume of D to be the triple integral

$$\lim_{n\to\infty} \sum_{k=1}^{n} \Delta V_k = \iiint_D dV.$$

DEFINITION The **volume** of a closed and bounded solid region D in space is

$$V = \iiint_D dV.$$

This definition is in agreement with our previous definitions of volume, although we omit the verification of this fact. As we will see in a moment, this integral enables us to calculate the volumes of solids enclosed by curved surfaces. These are more general solids than the ones encountered before (Chapter 6 and Section 15.2).

Iterated Integrals

We evaluate a triple integral by applying a three-dimensional version of Fubini's Theorem (Section 15.2) to evaluate it by three repeated single integrations. As with double integrals, there is a geometric procedure for finding the limits of integration for these iterated integrals.

To evaluate

$$\iiint_D F(x, y, z)\, dV$$

over a solid region D, integrate first with respect to z, then with respect to y, and finally with respect to x. (You might choose a different order of integration, but the procedure is similar, as we illustrate in Example 2.)

1. *Sketch.* Sketch the solid region D along with its "shadow" R (vertical projection) in the xy-plane. Label the upper and lower bounding surfaces of D and the upper and lower bounding curves of R.

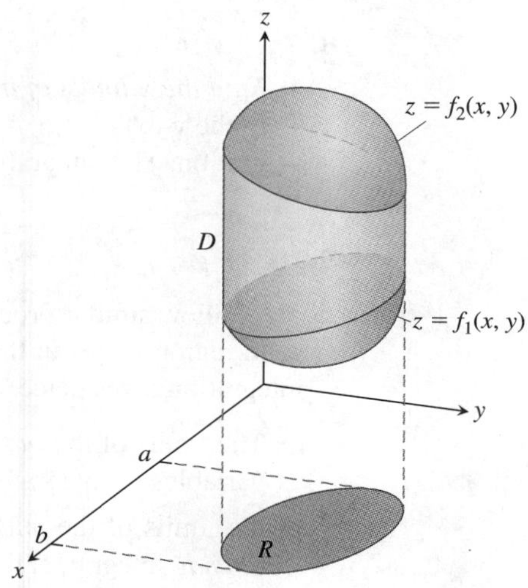

2. *Find the z-limits of integration.* Draw a line M passing through a typical point (x, y) in R parallel to the z-axis. As z increases, M enters D at $z = f_1(x, y)$ and leaves at $z = f_2(x, y)$. These are the z-limits of integration.

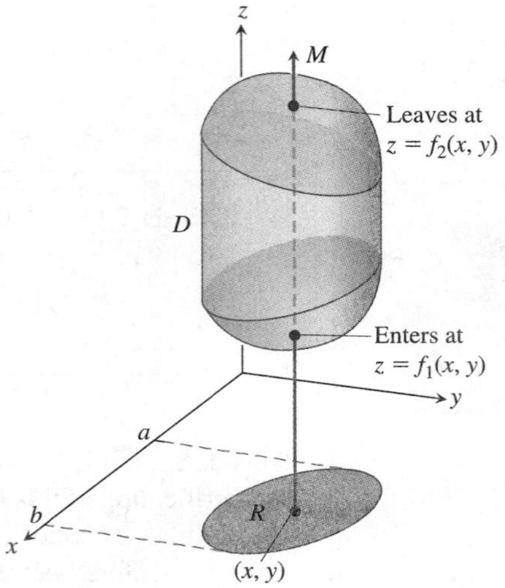

3. *Find the y-limits of integration.* Draw a line L through (x, y) parallel to the y-axis. As y increases, L enters R at $y = g_1(x)$ and leaves at $y = g_2(x)$. These are the y-limits of integration.

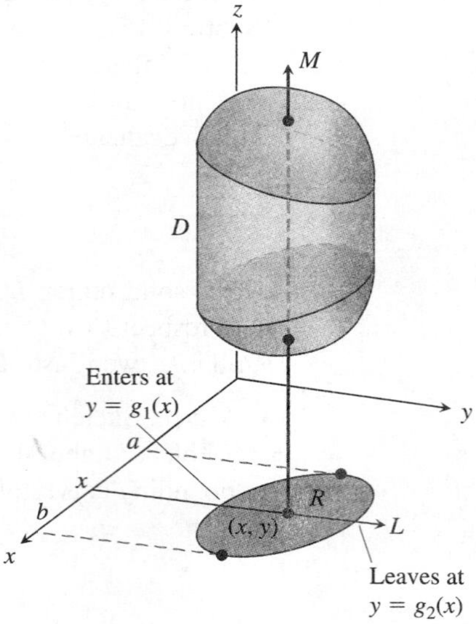

4. *Find the x-limits of integration.* Choose x-limits that include all lines through R parallel to the y-axis ($x = a$ and $x = b$ in the preceding figure). These are the x-limits of integration. The integral is

$$\int_{x=a}^{x=b} \int_{y=g_1(x)}^{y=g_2(x)} \int_{z=f_1(x, y)}^{z=f_2(x, y)} F(x, y, z)\, dz\, dy\, dx.$$

Follow similar procedures if you change the order of integration. The "shadow" of the solid region D lies in the plane of the last two variables with respect to which the iterated integration takes place. The limits of an iterated triple integral satisfy these properties:

- The limits of the *outside* integral are *constants* (they do not depend on any of the three variables of integration),
- the limits of the *middle* integral are *functions* that may depend on the variable of the *outside* integral, and
- the limits of the *inside* integral are *functions* that may depend on two variables: the *middle* integration variable and the *outside* integration variable.

The preceding procedure applies whenever a solid region D is bounded above and below by a surface, and when the "shadow" region R is bounded by a lower and upper curve. It does not apply to regions with more complicated shapes (such as regions containing holes); although, sometimes such regions can be subdivided into simpler regions for which the procedure does apply.

We illustrate this method of finding the limits of integration in our first example.

EXAMPLE 1 Let S be the sphere of radius 5 centered at the origin, and let D be the solid region under the sphere that lies above the plane $z = 3$. Set up the limits of integration for evaluating the triple integral of a function $F(x, y, z)$ over the region D.

Solution The solid region under the sphere that lies above the plane $z = 3$ is enclosed by the surfaces $x^2 + y^2 + z^2 = 25$ and $z = 3$.

To find the limits of integration, we first sketch the solid region, as shown in Figure 15.31. The "shadow region" R in the *xy*-plane is a circle of some radius centered at the origin. By considering a side view of the region D, we can determine that the radius of this circle is 4; see Figure 15.32a.

If we fix a point (x, y) in R and draw a vertical line M above (x, y), then we see that this line enters the region D at the height $z = 3$ and leaves the region at the height $z = \sqrt{25 - x^2 - y^2}$; see Figure 15.31. This gives us the z-limits of integration.

To find the y-limits of integration, we consider a line L that lies in the region R, passes through the point (x, y), and is parallel to the y-axis. For clarity we have separately pictured the region R and the line L in Figure 15.32b. The line L enters R when $y = -\sqrt{16 - x^2}$ and exits when $y = \sqrt{16 - x^2}$. This gives us the y-limits of integration.

Finally, as L sweeps across R from left to right, the value of x varies from $x = -4$ to $x = 4$. This gives us the x-limits of integration. Therefore, the triple integral of F over the region D is given by

$$\iiint\limits_D F(x, y, z)\, dz\, dy\, dx = \int_{-4}^{4} \int_{-\sqrt{16-x^2}}^{\sqrt{16-x^2}} \int_{3}^{\sqrt{25-x^2-y^2}} F(x, y, z)\, dz\, dy\, dx. \qquad \blacksquare$$

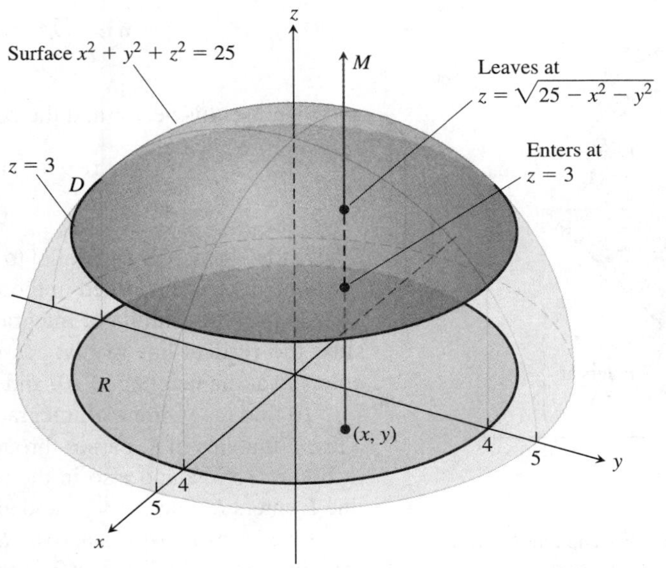

FIGURE 15.31 Finding the limits of integration for evaluating the triple integral of a function defined over the portion of the sphere of radius 5 that lies above the plane $z = 3$ (Example 1).

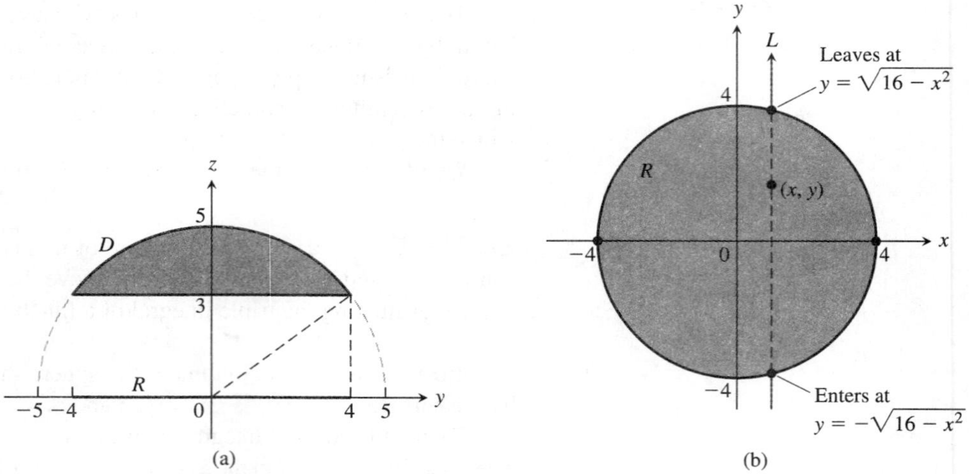

FIGURE 15.32 (a) Side view of the solid region from Example 1, looking down the x-axis. The dashed right triangle has a hypotenuse of length 5 and sides of lengths 3 and 4. In this side view, the shadow region R lies between -4 and 4 on the y-axis. (b) The "shadow region" R shown face-on in the xy-plane.

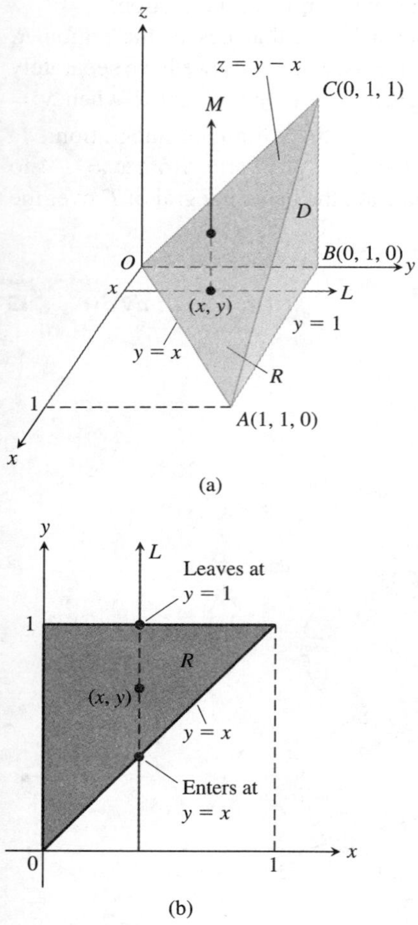

FIGURE 15.33 (a) The tetrahedron in Example 2, showing how the limits of integration are found for the order $dz\,dy\,dx$. (b) The "shadow region" R shown face-on in the xy-plane.

The region D in Example 1 has a great deal of symmetry, which makes visualization easier. Even without symmetry, the steps in finding the limits of integration are the same, as shown in the next example.

EXAMPLE 2 Set up the limits of integration for evaluating the triple integral of a function $F(x, y, z)$ over the tetrahedron D whose vertices are $O(0,0,0)$, $A(1,1,0)$, $B(0,1,0)$, and $C(0,1,1)$. Use the order of integration $dz\,dy\,dx$.

Solution The solid region D and its "shadow" R in the xy-plane are shown in Figure 15.33a. The "top" face is contained in the plane through the points O, A, and C. Following the procedure introduced in Example 7 of Section 12.5, we first form a normal vector to that plane:

$$\mathbf{n} = \overrightarrow{OA} \times \overrightarrow{OC} = \begin{vmatrix} \mathbf{i} & \mathbf{j} & \mathbf{k} \\ 1 & 1 & 0 \\ 0 & 1 & 1 \end{vmatrix} = \mathbf{i} - \mathbf{j} + \mathbf{k},$$

and then use this vector and the coordinates of O to set up an equation for the plane:

$$1(x - 0) - 1(y - 0) + 1(z - 0) = 0$$
$$x - y + z = 0.$$

The "side" face of D is parallel to the xz-plane, the "back" face lies in the yz-plane, and the "bottom" face is contained in the xy-plane.

To find the z-limits of integration, fix a point (x, y) in the shadow region R, and consider the vertical line M that passes through (x, y) and is parallel to the z-axis. This line enters D at the height $z = 0$, and it exits at height $z = y - x$.

To find the y-limits of integration we again fix a point (x, y) in R, but now we consider a line L that lies in R, passes through (x, y), and is parallel to the y-axis. This line is shown in Figure 15.33a and also in the face-on view of R that is pictured in Figure 15.33b. The line L enters R when $y = x$ and exits when $y = 1$.

Finally, as L sweeps across R, the value of x varies from $x = 0$ to $x = 1$. Therefore, the triple integral of F over the region D is given by

$$\iiint_D F(x, y, z)\,dz\,dy\,dx = \int_0^1 \int_x^1 \int_0^{y-x} F(x, y, z)\,dz\,dy\,dx. \qquad \blacksquare$$

In the next example we project the region D onto the xz-plane instead of the xy-plane, to show how to use a different order of integration.

EXAMPLE 3 Find the volume of the tetrahedron D from Example 2 by integrating $F(x, y, z) = 1$ over the region using the order $dz\, dy\, dx$. Then do the same calculation using the order $dy\, dz\, dx$.

Solution Using the limits of integration that we found in Example 2, we calculate the volume of the tetrahedron as follows:

$$V = \int_0^1 \int_x^1 \int_0^{y-x} dz\, dy\, dx \qquad \text{Integrand is 1 when computing volume.}$$

$$= \int_0^1 \int_x^1 (y - x)\, dy\, dx \qquad \text{Integrate over } z \text{ and evaluate.}$$

$$= \int_0^1 \left[\frac{1}{2}y^2 - xy\right]_{y=x}^{y=1} dx \qquad \text{Integrate over } y.$$

$$= \int_0^1 \left(\frac{1}{2} - x + \frac{1}{2}x^2\right) dx \qquad \text{Evaluate.}$$

$$= \left[\frac{1}{2}x - \frac{1}{2}x^2 + \frac{1}{6}x^3\right]_0^1 \qquad \text{Integrate over } x.$$

$$= \frac{1}{6}. \qquad \text{Evaluate.}$$

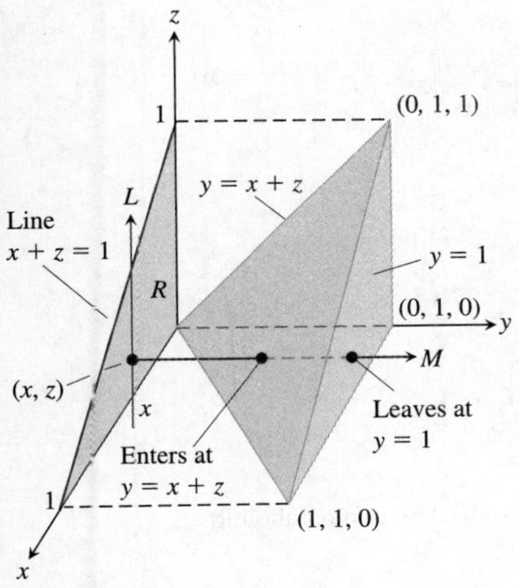

FIGURE 15.34 Finding the limits of integration for evaluating the triple integral of a function defined over the tetrahedron D (Example 3).

Now we will compute the volume using the order of integration $dy\, dz\, dx$. The procedure for finding the limits of integration is similar, except that we find the limits for y first, then for z, and then for x. The region D is the same tetrahedron as before, but now the "shadow region" R lies in the xz-plane, as shown in Figure 15.34.

To find the y-limits of integration, we fix a point (x, z) in the shadow R and consider the line M that passes through (x, z) and is parallel to the y-axis. As shown in Figure 15.34, this line enters D when $y = x + z$, and it leaves when $y = 1$.

Next we find the z-limits of integration. The line L that passes through a point (x, z) in R and is parallel to the z-axis enters R when $z = 0$ and exits when $z = 1 - x$ (see Figure 15.34).

Finally, as L sweeps across R, the value of x varies from $x = 0$ to $x = 1$. Therefore, the volume of the tetrahedron is

$$V = \int_0^1 \int_0^{1-x} \int_{x+z}^1 dy\, dz\, dx$$

$$= \int_0^1 \int_0^{1-x} (1 - x - z)\, dz\, dx$$

$$= \int_0^1 \left[(1 - x)z - \frac{1}{2}z^2\right]_{z=0}^{z=1-x} dx$$

$$= \int_0^1 \left[(1 - x)^2 - \frac{1}{2}(1 - x)^2\right] dx$$

$$= \frac{1}{2}\int_0^1 (1 - x)^2\, dx$$

$$= -\frac{1}{6}(1 - x)^3\Big]_0^1 = \frac{1}{6}.$$

■

Next we set up and evaluate a triple integral over a more complicated region.

EXAMPLE 4 Find the volume of the solid region D enclosed by the surfaces $z = x^2 + 3y^2$ and $z = 8 - x^2 - y^2$.

Solution The volume is

$$V = \iiint_D dz\,dy\,dx,$$

the integral of $F(x, y, z) = 1$ over D. To find the limits of integration for evaluating the integral, we first sketch the region. The surfaces (Figure 15.35) intersect on the elliptical cylinder $x^2 + 3y^2 = 8 - x^2 - y^2$ or $x^2 + 2y^2 = 4$, $z > 0$. The boundary of the region R, the projection of D onto the xy-plane, is an ellipse with the same equation: $x^2 + 2y^2 = 4$. The "upper" boundary of R is the curve $y = \sqrt{(4 - x^2)/2}$. The lower boundary is the curve $y = -\sqrt{(4 - x^2)/2}$.

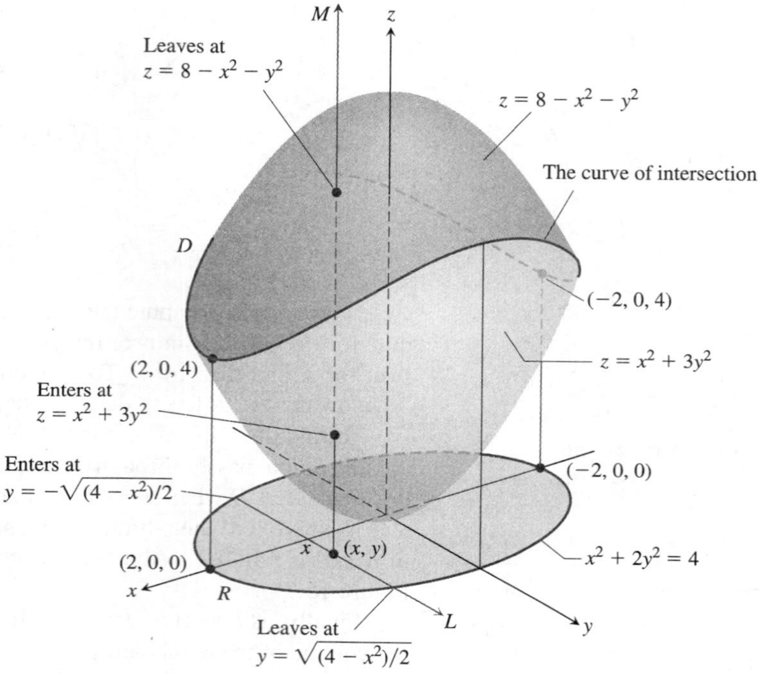

FIGURE 15.35 The volume of the region enclosed by two paraboloids, calculated in Example 4.

Now we find the z-limits of integration. The line M passing through a typical point (x, y) in R parallel to the z-axis enters D at $z = x^2 + 3y^2$ and leaves at $z = 8 - x^2 - y^2$.

Next we find the y-limits of integration. The line L through (x, y) that lies parallel to the y-axis enters the region R when $y = -\sqrt{(4 - x^2)/2}$ and leaves when $y = \sqrt{(4 - x^2)/2}$.

Finally, we find the x-limits of integration. As L sweeps across R, the value of x varies from $x = -2$ at $(-2, 0, 0)$ to $x = 2$ at $(2, 0, 0)$. The volume of D is

$$V = \iiint_D dz\,dy\,dx \qquad \text{Integrand is 1 when computing volume.}$$

$$= \int_{-2}^{2}\int_{-\sqrt{(4-x^2)/2}}^{\sqrt{(4-x^2)/2}}\int_{x^2+3y^2}^{8-x^2-y^2} dz\,dy\,dx \qquad \text{Form an iterated integral.}$$

$$= \int_{-2}^{2}\int_{-\sqrt{(4-x^2)/2}}^{\sqrt{(4-x^2)/2}} (8 - 2x^2 - 4y^2)\,dy\,dx \qquad \text{Integrate over } z \text{ and evaluate.}$$

$$= \int_{-2}^{2}\left[(8 - 2x^2)y - \frac{4}{3}y^3\right]_{y=-\sqrt{(4-x^2)/2}}^{y=\sqrt{(4-x^2)/2}} dx \qquad \text{Integrate over } y.$$

$$= \int_{-2}^{2}\left(2(8 - 2x^2)\sqrt{\frac{4 - x^2}{2}} - \frac{8}{3}\left(\frac{4 - x^2}{2}\right)^{3/2}\right) dx \qquad \text{Evaluate.}$$

$$= \int_{-2}^{2}\left[8\left(\frac{4 - x^2}{2}\right)^{3/2} - \frac{8}{3}\left(\frac{4 - x^2}{2}\right)^{3/2}\right] dx$$

$$= \frac{4\sqrt{2}}{3}\int_{-2}^{2}(4 - x^2)^{3/2}\,dx$$

$$= 8\pi\sqrt{2}. \qquad \text{After integration with the substitution } x = 2\sin\theta$$

■

Average Value of a Function in Space

The average value of a function F over a solid region D in space is defined by the formula

$$\textbf{Average value} \text{ of } F \text{ over } D = \frac{1}{\text{volume of } D}\iiint_D F\,dV. \tag{2}$$

For example, if $F(x, y, z) = \sqrt{x^2 + y^2 + z^2}$, then the average value of F over D is the average distance of points in D from the origin. If $F(x, y, z)$ is the temperature at (x, y, z) on a solid that occupies a region D in space, then the average value of F over D is the average temperature of the solid.

EXAMPLE 5 Find the average value of $F(x, y, z) = xyz$ throughout the cubical region D bounded by the coordinate planes and the planes $x = 2$, $y = 2$, and $z = 2$ in the first octant.

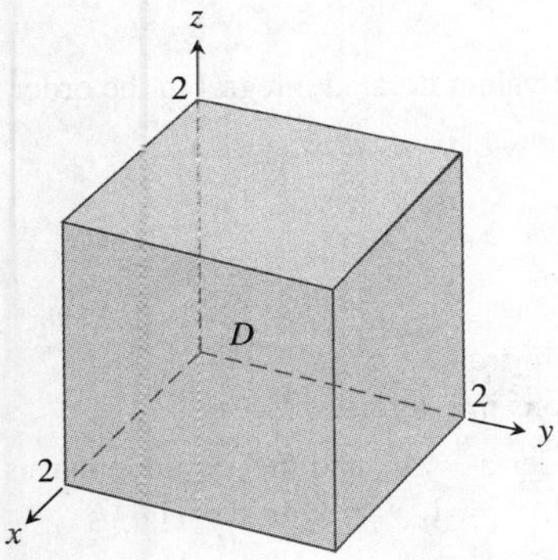

FIGURE 15.36 The region of integration in Example 5.

Solution We sketch the cube with enough detail to show the limits of integration (Figure 15.36). We then use Equation (2) to calculate the average value of F over the cube.

The volume of the region D is $(2)(2)(2) = 8$. The value of the integral of F over the cube is

$$\int_0^2\int_0^2\int_0^2 xyz\,dx\,dy\,dz = \int_0^2\int_0^2\left[\frac{x^2}{2}yz\right]_{x=0}^{x=2} dy\,dz = \int_0^2\int_0^2 2yz\,dy\,dz$$

$$= \int_0^2\left[y^2z\right]_{y=0}^{y=2} dz = \int_0^2 4z\,dz = \left[2z^2\right]_0^2 = 8.$$

With these values, Equation (2) gives

$$\begin{matrix}\text{Average value of} \\ xyz \text{ over the cube}\end{matrix} = \frac{1}{\text{volume}}\iiint_{\text{cube}} xyz\,dV = \left(\frac{1}{8}\right)(8) = 1.$$

In evaluating the integral, we chose the order $dx\,dy\,dz$, but any of the other five possible orders would have done as well. ■

Properties of Triple Integrals

Triple integrals have the same algebraic properties as double and single integrals. Simply replace the double integrals in the four properties given in Section 15.2, page 904, with triple integrals.

EXERCISES 15.5

Triple Integrals in Different Iteration Orders

1. Evaluate the integral in Example 3, taking $F(x, y, z) = 1$ to find the volume of the tetrahedron in the order $dz\,dx\,dy$.

2. **Volume of rectangular solid** Write six different iterated triple integrals for the volume of the rectangular solid in the first octant bounded by the coordinate planes and the planes $x = 1$, $y = 2$, and $z = 3$. Evaluate one of the integrals.

3. **Volume of tetrahedron** Write six different iterated triple integrals for the volume of the tetrahedron cut from the first octant by the plane $6x + 3y + 2z = 6$. Evaluate one of the integrals.

4. **Volume of solid** Write six different iterated triple integrals for the volume of the solid region in the first octant enclosed by the cylinder $x^2 + z^2 = 4$ and the plane $y = 3$. Evaluate one of the integrals.

5. **Volume enclosed by paraboloids** Let D be the solid region bounded by the paraboloids $z = 8 - x^2 - y^2$ and $z = x^2 + y^2$. Write six different triple iterated integrals for the volume of D. Evaluate one of the integrals.

6. **Volume inside paraboloid beneath a plane** Let D be the solid region bounded by the paraboloid $z = x^2 + y^2$ and the plane $z = 2y$. Write triple iterated integrals in the order $dz\,dx\,dy$ and $dz\,dy\,dx$ that give the volume of D. Do not evaluate either integral.

Evaluating Triple Iterated Integrals

Evaluate the integrals in Exercises 7–20.

7. $\int_0^1\int_0^1\int_0^1 (x^2 + y^2 + z^2)\,dz\,dy\,dx$

8. $\int_0^{\sqrt{2}}\int_0^{3y}\int_{x^2+3y^2}^{8-x^2-y^2} dz\,dx\,dy$

9. $\int_0^{\ln 2}\int_0^{x}\int_0^{x+y} e^{x+y+z}\,dz\,dy\,dx$

10. $\int_0^1\int_0^{3-3x}\int_0^{3-3x-y} dz\,dy\,dx$

11. $\int_0^{\pi/6}\int_0^1\int_{-2}^3 y\sin z\,dx\,dy\,dz$

12. $\int_{-1}^1\int_0^1\int_0^2 (x + y + z)\,dy\,dx\,dz$

13. $\int_0^3\int_0^{\sqrt{9-x^2}}\int_0^{\sqrt{9-x^2}} dz\,dy\,dx$

14. $\int_0^2\int_{-\sqrt{4-y^2}}^{\sqrt{4-y^2}}\int_0^{2x+y} dz\,dx\,dy$

15. $\int_0^1\int_0^{2-x}\int_0^{2-x-y} dz\,dy\,dx$

16. $\int_0^1\int_0^{1-x^2}\int_3^{4-x^2-y} x\,dz\,dy\,dx$

17. $\int_0^1\int_0^{1-x}\int_0^{1-x-y} xy\,dz\,dy\,dx$

18. $\int_0^1\int_0^{\sqrt{1-x^2}}\int_0^{\sqrt{1-x^2-y^2}} \dfrac{1}{\sqrt{1-x^2-y^2-z^2}}\,dz\,dy\,dx$

19. $\int_0^{\pi/4}\int_0^{\ln \sec v}\int_{-\infty}^{2t} e^x\,dx\,dt\,dv$ (tvx-space)

20. $\int_0^7\int_0^2\int_0^{\sqrt{4-q^2}} \dfrac{q}{r+1}\,dp\,dq\,dr$ (pqr-space)

Finding Equivalent Iterated Integrals

21. Here is the region of integration of the integral

$$\int_{-1}^1\int_{x^2}^1\int_0^{1-y} dz\,dy\,dx.$$

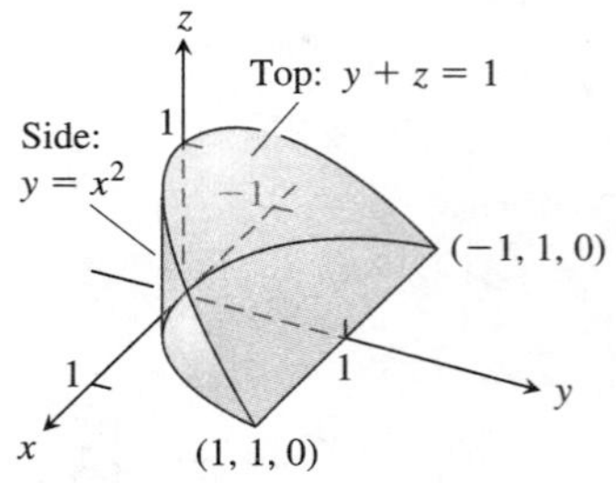

Rewrite the integral as an equivalent iterated integral in the order

a. $dy\,dz\,dx$
b. $dy\,dx\,dz$
c. $dx\,dy\,dz$
d. $dx\,dz\,dy$
e. $dz\,dx\,dy$.

22. Here is the region of integration of the integral

$$\int_0^1\int_{-1}^0\int_0^{y^2} dz\,dy\,dx$$

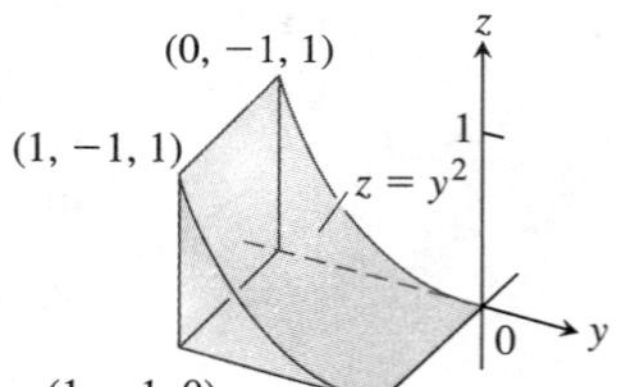

Rewrite the integral as an equivalent iterated integral in the order

a. $dy\,dz\,dx$
b. $dy\,dx\,dz$
c. $dx\,dy\,dz$
d. $dx\,dz\,dy$
e. $dz\,dx\,dy$.

Finding Volumes Using Triple Integrals

Find the volumes of the solid regions in Exercises 23–36.

23. The region between the cylinder $z = y^2$ and the xy-plane that is bounded by the planes $x = 0$, $x = 1$, $y = -1$, $y = 1$

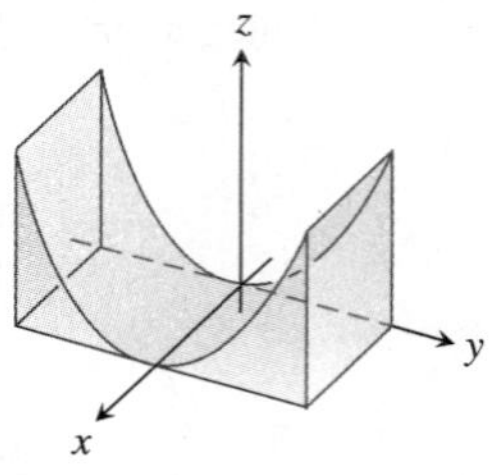

24. The region in the first octant bounded by the coordinate planes and the planes $x + z = 1$, $y + 2z = 2$

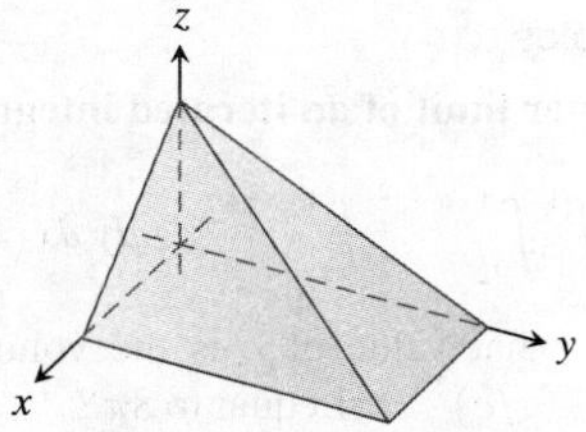

25. The region in the first octant bounded by the coordinate planes, the plane $y + z = 2$, and the cylinder $x = 4 - y^2$

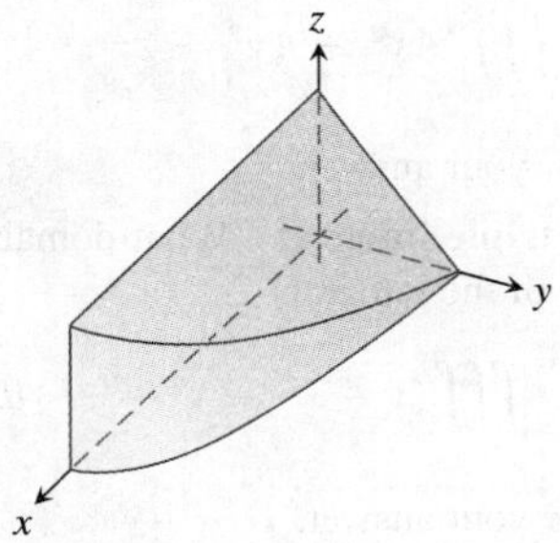

26. The wedge cut from the cylinder $x^2 + y^2 = 1$ with $z \geq 0$ by the planes $z = -y$ and $z = 0$

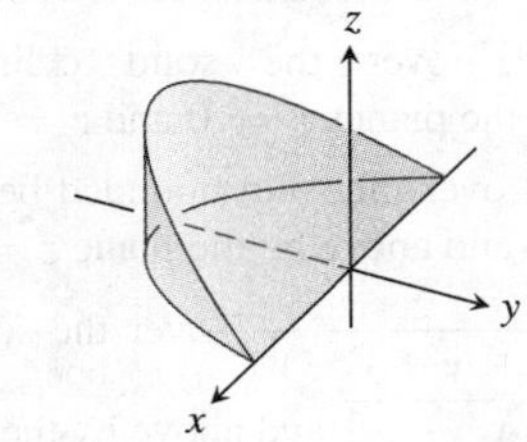

27. The tetrahedron in the first octant bounded by the coordinate planes and the plane passing through $(1, 0, 0)$, $(0, 2, 0)$, and $(0, 0, 3)$

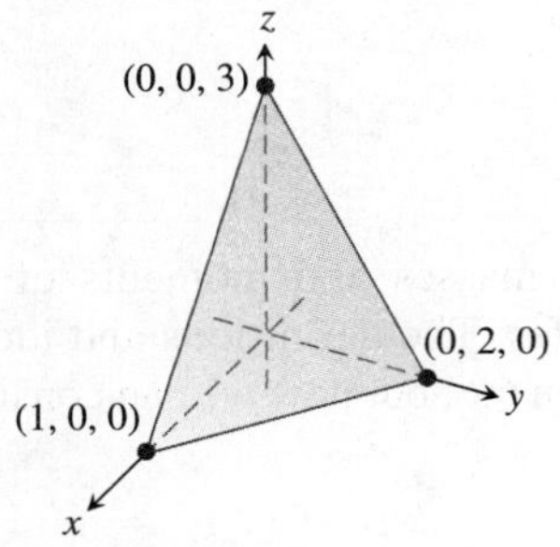

28. The region in the first octant bounded by the coordinate planes, the plane $y = 1 - x$, and the surface $z = \cos(\pi x/2)$, $0 \leq x \leq 1$

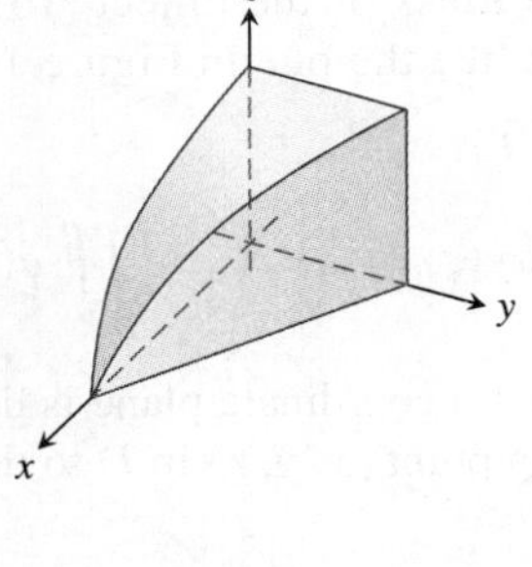

29. The region common to the interiors of the cylinders $x^2 + y^2 = 1$ and $x^2 + z^2 = 1$, one-eighth of which is shown in the accompanying figure

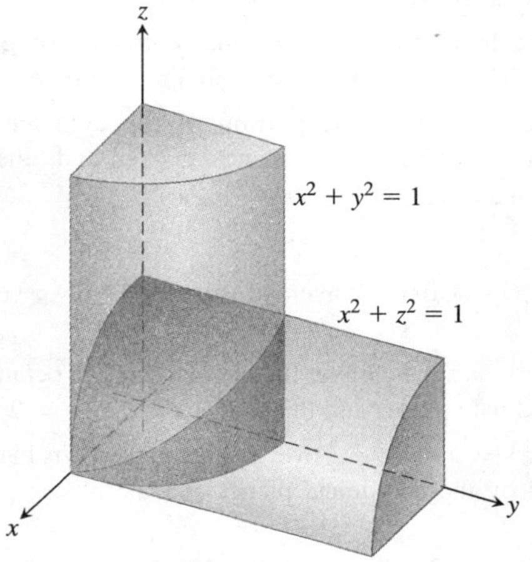

30. The region in the first octant bounded by the coordinate planes and the surface $z = 4 - x^2 - y$

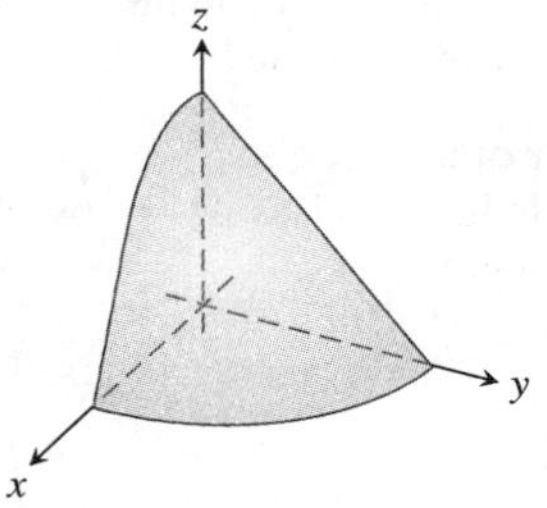

31. The region in the first octant bounded by the coordinate planes, the plane $x + y = 4$, and the cylinder $y^2 + 4z^2 = 16$

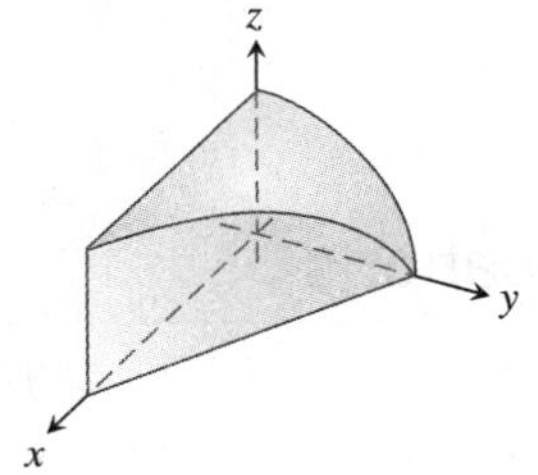

32. The region cut from the cylinder $x^2 + y^2 = 4$ by the plane $z = 0$ and the plane $x + z = 3$

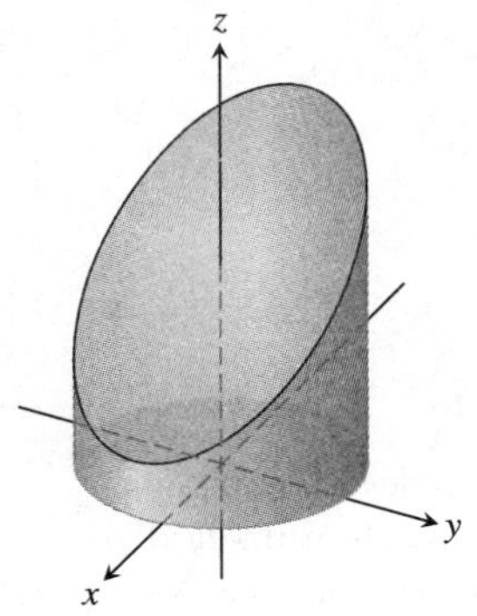

33. The region between the planes $x + y + 2z = 2$ and $2x + 2y + z = 4$ in the first octant

34. The finite region bounded by the planes $z = x$, $x + z = 8$, $z = y$, $y = 8$, and $z = 0$

35. A circular hole of radius b is made centrally through a sphere of radius a. Find the volume of the remaining sphere.

36. Find the volume of the portion of the cylinder $x^2 + y^2 = 1$ intercepted between the plane $z = 0$ and the paraboloid $x^2 + y^2 = 4 - z$.

Average Values

In Exercises 37–40, find the average value of $F(x, y, z)$ over the given region.

37. $F(x, y, z) = x^2 + 9$ over the cube in the first octant bounded by the coordinate planes and the planes $x = 2$, $y = 2$, and $z = 2$

38. $F(x, y, z) = x + y - z$ over the rectangular box in the first octant bounded by the coordinate planes and the planes $x = 1$, $y = 1$, and $z = 2$

39. $F(x, y, z) = x^2 + y^2 + z^2$ over the cube in the first octant bounded by the coordinate planes and the planes $x = 1$, $y = 1$, and $z = 1$

40. $F(x, y, z) = xyz$ over the cube in the first octant bounded by the coordinate planes and the planes $x = 2$, $y = 2$, and $z = 2$

Changing the Order of Integration

Evaluate the integrals in Exercises 41–44 by changing the order of integration in an appropriate way.

41. $\displaystyle\int_0^4\int_0^1\int_{2y}^2 \frac{4\cos(x^2)}{2\sqrt{z}}\,dx\,dy\,dz$

42. $\displaystyle\int_0^1\int_0^1\int_{x^2}^1 12xze^{zy^2}\,dy\,dx\,dz$

43. $\displaystyle\int_0^1\int_{\sqrt[3]{z}}^1\int_0^{\ln 3} \frac{\pi e^{2x}\sin \pi y^2}{y^2}\,dx\,dy\,dz$

44. $\displaystyle\int_0^2\int_0^{4-x^2}\int_0^x \frac{\sin 2z}{4-z}\,dy\,dz\,dx$

Theory and Examples

45. Finding an upper limit of an iterated integral Solve for a:

$$\int_0^1\int_0^{4-a-x^2}\int_a^{4-x^2-y} dz\,dy\,dx = \frac{4}{15}.$$

46. Ellipsoid For what value of c is the volume of the ellipsoid $x^2 + (y/2)^2 + (z/c)^2 = 1$ equal to 8π?

47. Minimizing a triple integral What domain D in space minimizes the value of the integral

$$\iiint_D (4x^2 + 4y^2 + z^2 - 4)\,dV?$$

Give reasons for your answer.

48. Maximizing a triple integral What domain D in space maximizes the value of the integral

$$\iiint_D (1 - x^2 - y^2 - z^2)\,dV?$$

Give reasons for your answer.

COMPUTER EXPLORATIONS

In Exercises 49–52, use a CAS integration utility to evaluate the triple integral of the given function over the specified solid region.

49. $F(x, y, z) = x^2y^2z$ over the solid cylinder bounded by $x^2 + y^2 = 1$ and the planes $z = 0$ and $z = 1$

50. $F(x, y, z) = |xyz|$ over the solid bounded below by the paraboloid $z = x^2 + y^2$ and above by the plane $z = 1$

51. $F(x, y, z) = \dfrac{z}{(x^2 + y^2 + z^2)^{3/2}}$ over the solid bounded below by the cone $z = \sqrt{x^2 + y^2}$ and above by the plane $z = 1$

52. $F(x, y, z) = x^4 + y^2 + z^2$ over the solid sphere $x^2 + y^2 + z^2 \le 1$

15.6 Applications

This section shows how to calculate the masses and moments of two- and three-dimensional objects in Cartesian coordinates. The definitions and ideas are similar to the single-variable case we studied in Section 6.6, but now we can consider more general situations.

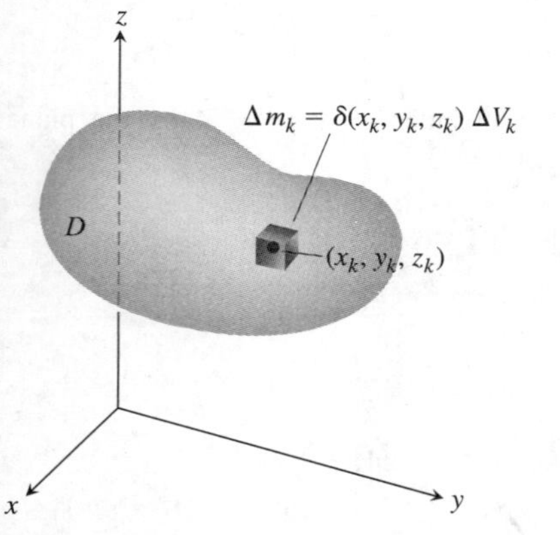

FIGURE 15.37 To define an object's mass, we first imagine it to be partitioned into a finite number of mass elements Δm_k.

Masses and First Moments

If $\delta(x, y, z)$ is the density (mass per unit volume) of an object occupying a solid region D in space, the integral of δ over D gives the **mass** of the object. To see why, imagine partitioning the object into n mass elements like the one in Figure 15.37. The object's mass is the limit

$$M = \lim_{n\to\infty}\sum_{k=1}^{n}\Delta m_k = \lim_{n\to\infty}\sum_{k=1}^{n}\delta(x_k, y_k, z_k)\,\Delta V_k = \iiint_D \delta(x, y, z)\,dV.$$

The *first moment* of a solid region D about a coordinate plane is defined as the triple integral over D of the (signed) distance from a point (x, y, z) in D to the plane multiplied by the density of the solid at that point. For instance, the first moment about the yz-plane is the integral

$$M_{yz} = \iiint_D x\,\delta(x, y, z)\,dV.$$

The *center of mass* is found from the first moments. For instance, the x-coordinate of the center of mass is $\bar{x} = M_{yz}/M$.

For a two-dimensional object, such as a thin, flat plate, we calculate first moments about the coordinate axes by simply dropping the z-coordinate. So the first moment about the y-axis is the double integral over the region R forming the plate of the (signed) distance from the axis multiplied by the density, or

$$M_y = \iint_R x\,\delta(x, y)\,dA.$$

Table 15.1 summarizes the formulas.

TABLE 15.1 Mass and first moment formulas

THREE-DIMENSIONAL SOLID

Mass: $M = \iiint_D \delta\,dV \qquad \delta = \delta(x, y, z)$ is the density at (x, y, z).

First moments about the coordinate planes:

$$M_{yz} = \iiint_D x\,\delta\,dV, \qquad M_{xz} = \iiint_D y\,\delta\,dV, \qquad M_{xy} = \iiint_D z\,\delta\,dV$$

Center of mass: $\bar{x} = \dfrac{M_{yz}}{M}, \quad \bar{y} = \dfrac{M_{xz}}{M}, \quad \bar{z} = \dfrac{M_{xy}}{M}$

TWO-DIMENSIONAL PLATE

Mass: $M = \iint_R \delta\,dA \qquad \delta = \delta(x, y)$ is the density at (x, y).

First moments: $M_y = \iint_R x\,\delta\,dA, \quad M_x = \iint_R y\,\delta\,dA$

Center of mass: $\bar{x} = \dfrac{M_y}{M}, \quad \bar{y} = \dfrac{M_x}{M}$

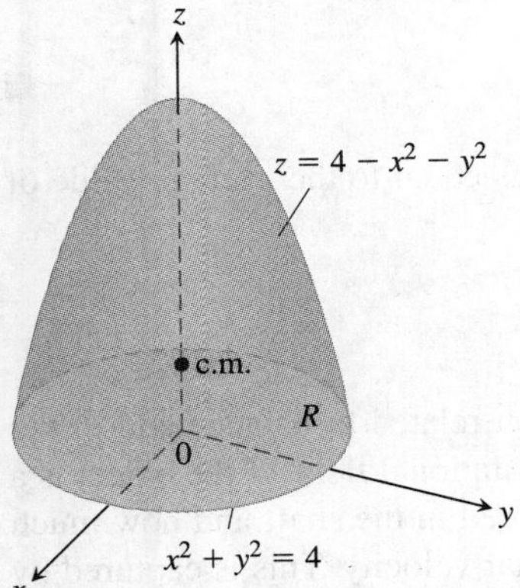

FIGURE 15.38 Finding the center of mass of a solid (Example 1).

EXAMPLE 1 Find the center of mass of a solid of constant density δ bounded below by the disk R: $x^2 + y^2 \leq 4$ in the plane $z = 0$ and above by the paraboloid $z = 4 - x^2 - y^2$ (Figure 15.38).

Solution By symmetry $\bar{x} = \bar{y} = 0$. To find $\bar{z}$, we first calculate

$$M_{xy} = \iint_R \int_{z=0}^{z=4-x^2-y^2} z\,\delta\,dz\,dy\,dx = \iint_R \left[\frac{z^2}{2}\right]_{z=0}^{z=4-x^2-y^2} \delta\,dy\,dx$$

$$= \frac{\delta}{2}\iint_R (4 - x^2 - y^2)^2\,dy\,dx$$

$$= \frac{\delta}{2}\int_0^{2\pi}\int_0^2 (4 - r^2)^2\, r\, dr\, d\theta \qquad \text{Polar coordinates simplify the integration.}$$

$$= \frac{\delta}{2}\int_0^{2\pi}\left[-\frac{1}{6}(4 - r^2)^3\right]_{r=0}^{r=2} d\theta = \frac{16\delta}{3}\int_0^{2\pi} d\theta = \frac{32\pi\delta}{3}.$$

A similar calculation gives the mass:

$$M = \iint_R \int_0^{4-x^2-y^2} \delta\, dz\, dy\, dx = 8\pi\delta.$$

Therefore, $\bar{z} = (M_{xy}/M) = 4/3$ and the center of mass is $(\bar{x}, \bar{y}, \bar{z}) = (0, 0, 4/3)$. ■

When the density of a solid object or plate is constant (as in Example 1), the center of mass is called the **centroid** of the object. To find a centroid, we set δ equal to 1 and proceed to find $\bar{x}$, $\bar{y}$, and $\bar{z}$ as before, by dividing first moments by masses. These calculations are also valid for two-dimensional objects.

EXAMPLE 2 Find the centroid of the region in the first quadrant that is bounded above by the line $y = x$ and below by the parabola $y = x^2$.

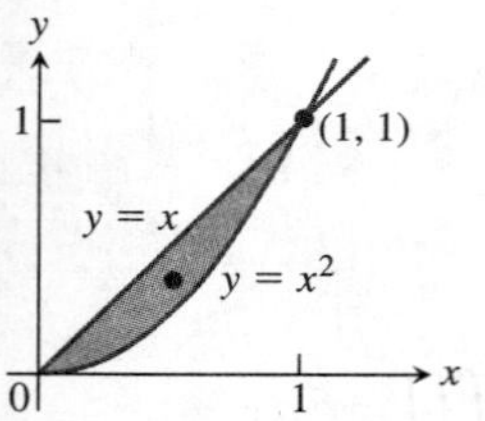

FIGURE 15.39 The centroid of this region is found in Example 2.

Solution We sketch the region and include enough detail to determine the limits of integration (Figure 15.39). We then set δ equal to 1 and evaluate the appropriate formulas from Table 15.1:

$$M = \int_0^1\int_{x^2}^{x} 1\, dy\, dx = \int_0^1 \Big[y\Big]_{y=x^2}^{y=x} dx = \int_0^1 (x - x^2)\, dx = \left[\frac{x^2}{2} - \frac{x^3}{3}\right]_0^1 = \frac{1}{6}$$

$$M_x = \int_0^1\int_{x^2}^{x} y\, dy\, dx = \int_0^1 \left[\frac{y^2}{2}\right]_{y=x^2}^{y=x} dx$$

$$= \int_0^1 \left(\frac{x^2}{2} - \frac{x^4}{2}\right) dx = \left[\frac{x^3}{6} - \frac{x^5}{10}\right]_0^1 = \frac{1}{15}$$

$$M_y = \int_0^1\int_{x^2}^{x} x\, dy\, dx = \int_0^1 \Big[xy\Big]_{y=x^2}^{y=x} dx = \int_0^1 (x^2 - x^3)\, dx = \left[\frac{x^3}{3} - \frac{x^4}{4}\right]_0^1 = \frac{1}{12}.$$

From these values of M, M_x, and M_y, we find

$$\bar{x} = \frac{M_y}{M} = \frac{1/12}{1/6} = \frac{1}{2} \quad \text{and} \quad \bar{y} = \frac{M_x}{M} = \frac{1/15}{1/6} = \frac{2}{5}.$$

The centroid is the point $(1/2, 2/5)$. ■

Note that each coordinate of the centroid of a region is equal to the average value of the corresponding variable over the region.

Moments of Inertia

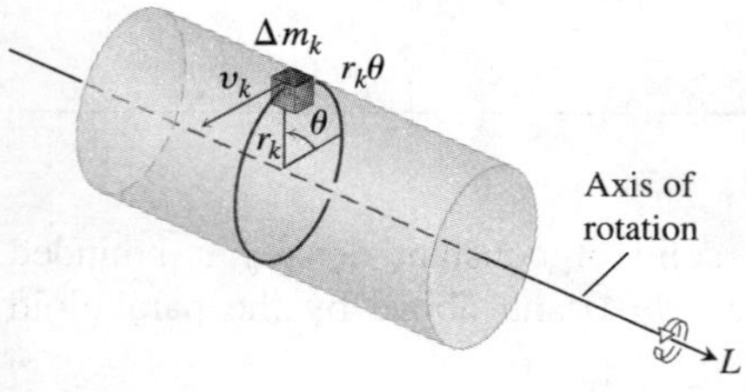

FIGURE 15.40 To find an integral for the amount of energy stored in a rotating shaft, we first imagine the shaft to be partitioned into small blocks. Each block has its own kinetic energy. We add the contributions of the individual blocks to find the kinetic energy of the shaft.

An object's first moments (Table 15.1) give us information related to balance and to the torque the object experiences about different axes in a gravitational field. If the object is a rotating shaft, we are interested in how much energy is stored in the shaft and how much energy is generated by a shaft rotating at a particular angular velocity. This is captured by the second moment or moment of inertia.

Think of partitioning the shaft into small blocks of mass Δm_k and let r_k denote the distance from the kth block's center of mass to the axis of rotation (Figure 15.40). If the shaft rotates at a constant angular velocity of $\omega = d\theta/dt$ radians per second, the block's center of mass will trace its orbit at a linear speed of

$$v_k = \frac{d}{dt}(r_k\theta) = r_k\frac{d\theta}{dt} = r_k\omega.$$

The block's kinetic energy will be approximately

$$\frac{1}{2}\Delta m_k v_k^2 = \frac{1}{2}\Delta m_k (r_k\omega)^2 = \frac{1}{2}\omega^2 r_k^{\,2}\,\Delta m_k.$$

The kinetic energy of the shaft will be approximately

$$\sum \frac{1}{2}\omega^2 r_k^{\,2}\,\Delta m_k.$$

The integral approached by these sums as the shaft is partitioned into smaller and smaller blocks gives the shaft's kinetic energy:

$$\text{KE}_{\text{shaft}} = \int \frac{1}{2}\omega^2 r^2\,dm = \frac{1}{2}\omega^2 \int r^2\,dm. \tag{1}$$

The factor

$$I = \int r^2\,dm$$

is the *moment of inertia* of the shaft about its axis of rotation, and we see from Equation (1) that the shaft's kinetic energy is

$$\text{KE}_{\text{shaft}} = \frac{1}{2}I\omega^2.$$

The moment of inertia of a shaft resembles in some ways the inertial mass of a locomotive. To start a locomotive with mass m moving at a linear velocity v, we need to provide a kinetic energy of $\text{KE} = (1/2)mv^2$. To stop the locomotive we have to remove this amount of energy. To start a shaft with moment of inertia I rotating at an angular velocity ω, we need to provide a kinetic energy of $\text{KE} = (1/2)I\omega^2$. To stop the shaft we have to take this amount of energy back out. The shaft's moment of inertia is analogous to the locomotive's mass. What makes the locomotive hard to start or stop is its mass. What makes the shaft hard to start or stop is its moment of inertia. The moment of inertia depends not only on the mass of the shaft but also on its distribution. Mass that is farther away from the axis of rotation contributes more to the moment of inertia.

We now derive a formula for the moment of inertia for a solid in space. If $r(x, y, z)$ is the distance from the point (x, y, z) in D to a line L, then the moment of inertia of the mass $\Delta m_k = \delta(x_k, y_k, z_k)\,\Delta V_k$ about the line L (as in Figure 15.40) is approximately $\Delta I_k = r^2(x_k, y_k, z_k)\,\Delta m_k$. **The moment of inertia about** L of the entire object is

$$I_L = \lim_{n\to\infty}\sum_{k=1}^{n}\Delta I_k = \lim_{n\to\infty}\sum_{k=1}^{n} r^2(x_k, y_k, z_k)\,\delta(x_k, y_k, z_k)\,\Delta V_k = \iiint_D r^2\delta\,dV.$$

If L is the x-axis, then $r^2 = y^2 + z^2$ (Figure 15.41) and

$$I_x = \iiint_D (y^2 + z^2)\,\delta(x, y, z)\,dV.$$

Similarly, if L is the y-axis or the z-axis, we have

$$I_y = \iiint_D (x^2 + z^2)\,\delta(x, y, z)\,dV \qquad \text{and} \qquad I_z = \iiint_D (x^2 + y^2)\,\delta(x, y, z)\,dV.$$

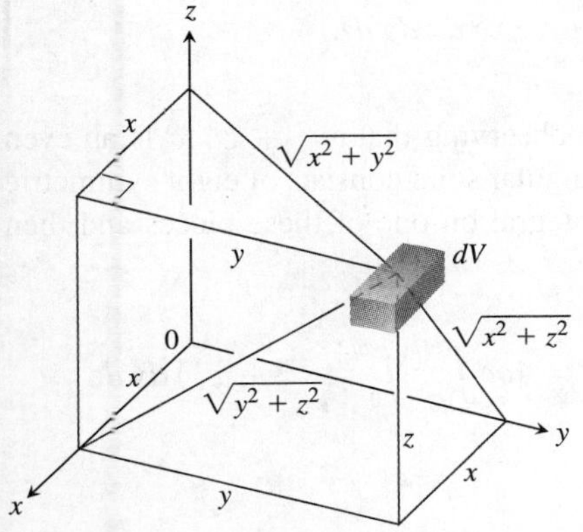

FIGURE 15.41 Distances from dV to the axes.

Table 15.2 summarizes the formulas for these moments of inertia (second moments because they invoke the *squares* of the distances). It shows the definition of the *polar moment* about the origin as well.

TABLE 15.2 Moments of inertia (second moments) formulas

THREE-DIMENSIONAL SOLID

About the x-axis:	$I_x = \iiint_D (y^2 + z^2)\,\delta\,dV$	$\delta = \delta(x, y, z)$
About the y-axis:	$I_y = \iiint_D (x^2 + z^2)\,\delta\,dV$	
About the z-axis:	$I_z = \iiint_D (x^2 + y^2)\,\delta\,dV$	
About a line L:	$I_L = \iiint_D r^2(x, y, z)\,\delta\,dV$	$r(x, y, z) =$ distance from the point (x, y, z) to line L

TWO-DIMENSIONAL PLATE

About the x-axis:	$I_x = \iint_R y^2\,\delta\,dA$	$\delta = \delta(x, y)$
About the y-axis:	$I_y = \iint_R x^2\,\delta\,dA$	
About a line L:	$I_L = \iint_R r^2(x, y)\,\delta\,dA$	$r(x, y) =$ distance from (x, y) to L
About the origin (polar moment):	$I_0 = \iint_R (x^2 + y^2)\,\delta\,dA = I_x + I_y$	

EXAMPLE 3 Find I_x, I_y, I_z for the rectangular solid of constant density δ shown in Figure 15.42.

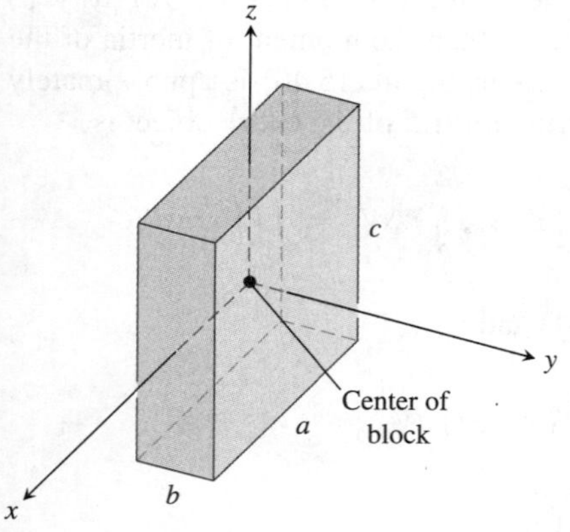

FIGURE 15.42 Finding I_x, I_y, and I_z for the block shown here. The origin lies at the center of the block (Example 3).

Solution The formula for I_x gives

$$I_x = \int_{-c/2}^{c/2}\int_{-b/2}^{b/2}\int_{-a/2}^{a/2} (y^2 + z^2)\,\delta\,dx\,dy\,dz.$$

We can avoid some of the work of integration by observing that $(y^2 + z^2)\delta$ is an even function of x, y, and z since δ is constant. The rectangular solid consists of eight symmetric pieces, one in each octant. We can evaluate the integral on one of these pieces and then multiply by 8 to get the total value:

$$I_x = 8\int_0^{c/2}\int_0^{b/2}\int_0^{a/2} (y^2 + z^2)\,\delta\,dx\,dy\,dz = 4a\delta\int_0^{c/2}\int_0^{b/2} (y^2 + z^2)\,dy\,dz$$

$$= 4a\delta\int_0^{c/2}\left[\frac{y^3}{3} + z^2 y\right]_{y=0}^{y=b/2} dz$$

$$= 4a\delta\int_0^{c/2}\left(\frac{b^3}{24} + \frac{z^2 b}{2}\right) dz$$

$$= 4a\delta\left(\frac{b^3c}{48} + \frac{c^3b}{48}\right) = \frac{abc\delta}{12}(b^2 + c^2) = \frac{M}{12}(b^2 + c^2). \qquad M = abc\delta$$

Similarly,

$$I_y = \frac{M}{12}(a^2 + c^2) \quad \text{and} \quad I_z = \frac{M}{12}(a^2 + b^2).$$ ■

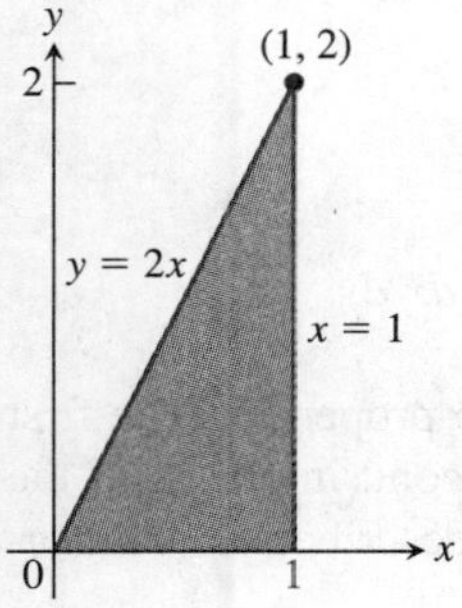

FIGURE 15.43 The triangular region covered by the plate in Example 4.

EXAMPLE 4 A thin plate covers the triangular region bounded by the x-axis and the lines $x = 1$ and $y = 2x$ in the first quadrant. The plate's density at the point (x, y) is $\delta(x, y) = 6x + 6y + 6$. Find the plate's moments of inertia about the coordinate axes and the origin.

Solution We sketch the plate and put in enough detail to determine the limits of integration for the integrals we have to evaluate (Figure 15.43). The moment of inertia about the x-axis is

$$I_x = \int_0^1 \int_0^{2x} y^2\,\delta(x, y)\,dy\,dx = \int_0^1 \int_0^{2x} (6xy^2 + 6y^3 + 6y^2)\,dy\,dx$$

$$= \int_0^1 \left[2xy^3 + \frac{3}{2}y^4 + 2y^3\right]_{y=0}^{y=2x} dx = \int_0^1 (40x^4 + 16x^3)\,dx$$

$$= \left[8x^5 + 4x^4\right]_0^1 = 12.$$

Similarly, the moment of inertia about the y-axis is

$$I_y = \int_0^1 \int_0^{2x} x^2\,\delta(x, y)\,dy\,dx = \frac{39}{5}.$$

Notice that we integrate y^2 times density in calculating I_x, and x^2 times density to find I_y.

Since we know I_x and I_y, we do not need to evaluate an integral to find I_0; we can use the equation $I_0 = I_x + I_y$ from Table 15.2 instead:

$$I_0 = 12 + \frac{39}{5} = \frac{60 + 39}{5} = \frac{99}{5}.$$ ■

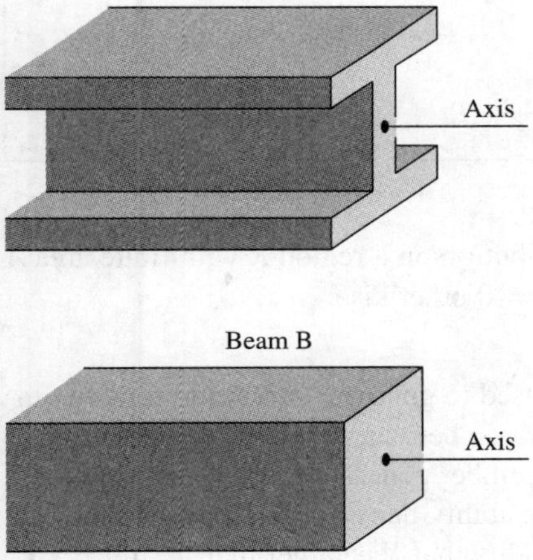

FIGURE 15.44 The greater the polar moment of inertia of the cross-section of a beam about the beam's longitudinal axis, the stiffer the beam. Beams A and B have the same cross-sectional area, but A is stiffer.

The moment of inertia also plays a role in determining how much a horizontal metal beam will bend under a load. The stiffness of the beam is a constant times I, the moment of inertia of a typical cross-section of the beam about the beam's longitudinal axis. The greater the value of I, the stiffer the beam and the less it will bend under a given load. That is why we use I-beams instead of beams whose cross-sections are square. The flanges at the top and bottom of the beam hold most of the beam's mass away from the longitudinal axis to increase the value of I (Figure 15.44).

Probability

In Chapter 8 we saw that the probability that a continuous random variable X takes values between a and b is found by integrating a probability density function f,

$$P(a \le X \le b) = \int_a^b f(x)\,dx.$$

A similar process applies to probabilities involving two continuous random variables. The probability that a pair of random variables (X, Y) takes values lying within a particular region is determined by a *joint probability density function* f. Integrating the joint probability density function over a region R in the plane gives the probability that the pair of random variables take values in that region:

$$P((X, Y) \in R) = \iint_R f(x, y)\, dx\, dy.$$

If the region is a rectangle, then this expression has the simple form

$$P(a \le X \le b \text{ and } c \le Y \le d) = \int_c^d \int_a^b f(x, y)\, dx\, dy.$$

A joint probability density function f is defined by three basic properties. The first property ensures that there are no negative probabilities, and the second implies that the total probability of all possible outcomes is one. The final property describes the connection of f to probabilities.

DEFINITION A **joint probability density function** f is a function that satisfies three conditions:

1. $f(x, y) \ge 0$
2. $\int_{-\infty}^{\infty} \int_{-\infty}^{\infty} f(x, y)\, dx\, dy = 1$
3. $P((X, Y) \in R) = \iint_R f(x, y)\, dx\, dy.$

A pair of random variables has a *uniform* distribution on a region R with finite area A if $f(x, y) = 1/A$ for any $(x, y) \in R$, and $f(x, y) = 0$ otherwise.

EXAMPLE 5 A random number generator is used to generate two random real numbers X and Y in succession. The first number X is chosen between 0 and 10, and the second number Y is chosen between 0 and 5. The random number generation is done by a process that gives a uniform distribution. Find the joint probability density function f for the pair of numbers (X, Y) and use it to compute the probability that X is larger than Y.

Solution The joint probability density function f is constant on the rectangle $0 \le x \le 10$, $0 \le y \le 5$, because (X, Y) is uniformly distributed. The area of the rectangle is 50, so f takes the value $1/50$ inside this rectangle:

$$f(x, y) = \begin{cases} 1/50, & \text{if } 0 \le x \le 10 \text{ and } 0 \le y \le 5, \\ 0, & \text{otherwise.} \end{cases}$$

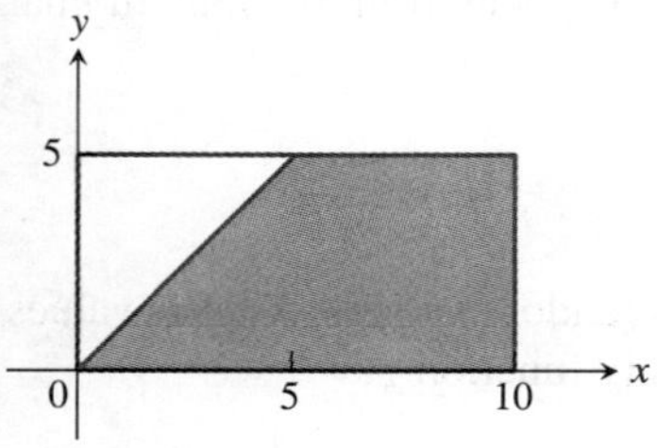

FIGURE 15.45 The pair of random variables X and Y take values anywhere in this rectangle with equal probability. In the shaded region we have $X > Y$.

To compute the probability that $X > Y$, we integrate the joint probability density function f over the region in the rectangle where $X > Y$. This region is bounded on the left by the line $x = y$ and on the right by the line $x = 10$. An integral over this region has limits of integration given by $y \le x \le 10$, $0 \le y \le 5$ (see Figure 15.45). The probability is given by

$$P(X > Y) = \int_0^5 \int_y^{10} \frac{1}{50}\, dx\, dy = \frac{3}{4}.$$

There is a 75% probability that the first number is larger than the second. ■

EXAMPLE 6 Using the joint probability density function

$$f(x, y) = \begin{cases} e^{-(x+y)}, & \text{if } 0 < x \text{ and } 0 < y \\ 0, & \text{otherwise} \end{cases}$$

find the probability that $1 < X < 2$ and $2 < Y < 3$.

Solution

$$P(1 < X < 2,\ 2 < Y < 3) = \int_2^3 \int_1^2 e^{-(x+y)}\,dx\,dy = e^{-5} + e^{-3} - 2e^{-4} \approx 0.01989.$$

There is slightly less than a 2% probability that X and Y fall within these bounds. ■

Means and Expected Values

The mean, or expected value, of a random variable was seen in Section 8.9 to be

$$\mu = \int_{-\infty}^{\infty} x f(x)\,dx.$$

When X and Y have joint probability density function f, the expected value of X and the expected value of Y are

$$\mu_X = \int_{-\infty}^{\infty}\int_{-\infty}^{\infty} x f(x, y)\,dx\,dy \quad \text{and} \quad \mu_Y = \int_{-\infty}^{\infty}\int_{-\infty}^{\infty} y f(x, y)\,dx\,dy.$$

These indicate the average value expected for each of X and Y. The expected values μ_X and μ_Y are sometimes called the first moments of the distribution, because their defining formulas have the same form as those seen in Table 15.1 for the moments of a two-dimensional plate. The joint probability density function plays the role in computing μ_X that the mass density function plays in computing the x-coordinate $\bar{x}$ of the center of mass, and the same applies to μ_Y and $\bar{y}$. One can roughly think of the joint probability density function as measuring the probability concentration per unit area on the plane, just as density measures the mass per unit area for a plate.

EXAMPLE 7 Find the expected values μ_X and μ_Y for the joint probability density function in Example 5.

Solution For the joint probability density function in Example 5, we compute

$$\mu_X = \int_0^5 \int_0^{10} x(1/50)\,dx\,dy = 5$$

and

$$\mu_Y = \int_0^5 \int_0^{10} y(1/50)\,dx\,dy = 2.5.$$

The expected value of X is 5 and that of Y is 2.5. ■

EXERCISES 15.6

Plates of Constant Density

1. **Finding a center of mass** Find the center of mass of thin plate of density $\delta = 5$ bounded by the lines $x = 0$, $y = x$, and the parabola $y = 2 + x^2$ in the first quadrant.

2. **Finding moments of inertia** Find the moments of inertia about the coordinate axes of a thin rectangular plate of constant density δ gm/cm^2 bounded by the lines $x = 5$ and $y = 5$ in the first quadrant.

3. **Finding a centroid** Find the centroid of the region in the first quadrant bounded by $y^2 = 6x$, $y = 0$ and $x = 6$.

4. **Finding a centroid** Find the centroid of the triangular region cut from the first quadrant by the line $x + y = 3$.
5. **Finding a centroid** Find the centroid of the region cut from the first quadrant by the circle $x^2 + y^2 = a^2$.
6. **Finding a centroid** Find the moments of inertia of the triangular plate bounded by $3x + 4y = 24$, $x = 0$ and $y = 0$ and having density 1.
7. **Finding moments of inertia** Find the moment of inertia about the x-axis of a thin plate of density $\delta = 1$ gm/cm^2 bounded by the circle $x^2 + y^2 = 9$. Then use your result to find I_y and I_0 for the plate.
8. **Finding a moment of inertia** Find the moment of inertia with respect to y-axis of a thin sheet of constant density $\delta = 1$ gm/cm^2 bounded by the curve $y = (\sin^2 x)/x^2$ and interval $-\pi \leq x \leq \pi$ of the x-axis.
9. **The centroid of an infinite region** Find the centroid of the infinite region in the second quadrant enclosed by the coordinate axes and the curve $y = e^x$. (Use improper integrals in the mass-moment formulas.)
10. **The first moment of an infinite plate** Find the first moment about the y-axis of a thin plate of density $\delta(x, y) = 1$ covering the infinite region under the curve $y = e^{-x^2}$ in the first quadrant.

Plates with Varying Density

11. **Finding a moment of inertia** Find the moment of inertia about the x-axis of a thin plate bounded by the parabola $x = y - y^2$ and the line $x + y = 0$ if $\delta(x, y) = x + y$.
12. **Finding mass** Find the mass of a thin plate occupying the smaller region cut from the ellipse $x^2 + 4y^2 = 12$ by the parabola $x = 4y^2$ if $\delta(x, y) = 5x$ kg/m^2.
13. **Finding a center of mass** Find the center of mass of a thin triangular plate bounded by the y-axis and the lines $y = x$ and $y = 2 + x$ if $\delta(x, y) = 3x + 2y + 1$.
14. **Finding a center of mass and moment of inertia** Find the center of mass and moment of inertia about the x-axis of a thin plate bounded by the curves $x = y^2$ and $x = 2y - y^2$ if the density at the point (x, y) is $\delta(x, y) = y + 1$.
15. **Center of mass, moment of inertia** Find the center of mass and the moment of inertia about the y-axis of a thin rectangular plate cut from the first quadrant by the lines $x = 6$ and $y = 1$ if $\delta(x, y) = x + y + 1$.
16. **Center of mass, moment of inertia** Find the center of mass and the moment of inertia about the y-axis of a thin plate bounded by the line $y = 1$ and the parabola $y = x^2$ if the density is $\delta(x, y) = y + 1$.
17. **Center of mass, moment of inertia** Find the center of mass and moment of inertia about the y-axis of a thin plate bounded by the x-axis, the lines $x = \pm 2$, and the parabola $y = 2x^2$ if $\delta(x, y) = 2y + 1$.
18. **Center of mass, moments of inertia** Find the center of mass and the moments of inertia about the x-axis of a thin rectangular plate bounded by the lines $x = 0$, $x = 20$, $y = -1$, and $y = 1$ if $\delta(x, y) = 1 + (x/20)$.
19. **Center of mass, moments of inertia** Find the center of mass, the moment of inertia about the coordinate axes, and the polar moment of inertia of a thin triangular plate bounded by the lines $y = x$, $y = -x$, and $y = 1$ if $\delta(x, y) = y + 1$ kg/m^2.
20. **Center of mass, moments of inertia** Repeat Exercise 19 for $\delta(x, y) = 3x^2 + 1$ kg/m^2.

Solids with Constant Density

21. **Moments of inertia** Find the moments of inertia of the rectangular box of constant density $\delta(x, y, z) = 1$ shown here with respect to its edges by calculating I_x, I_y, and I_z.

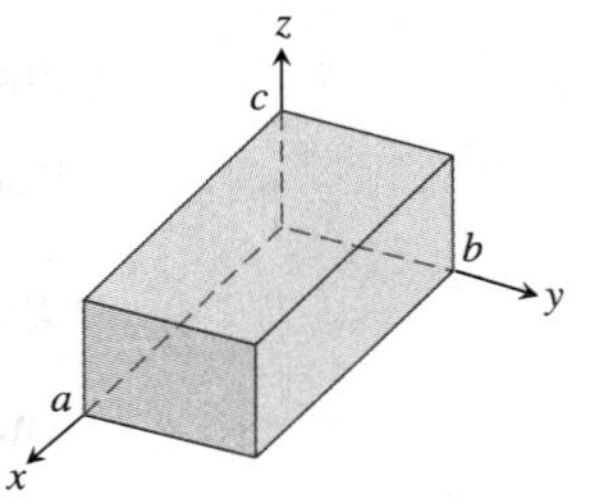

22. **Moments of inertia** The coordinate axes in the figure run through the centroid of a solid wedge parallel to the labeled edges. Find I_x, I_y, and I_z if $a = b = 6$, $c = 4$, and the density is $\delta(x, y, z) = 1$.

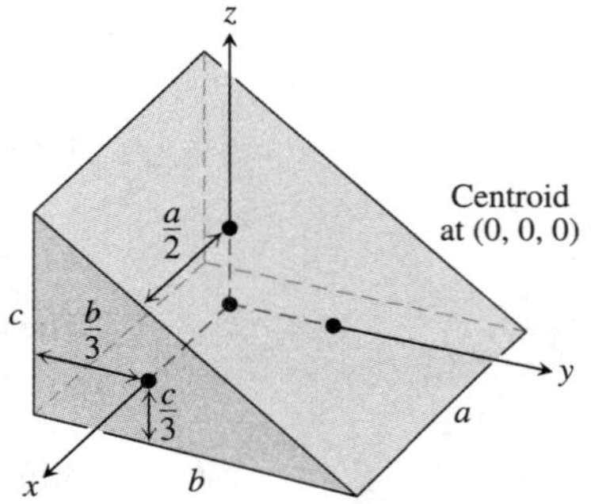

23. **Center of mass and moments of inertia** A solid "trough" of constant density $\delta(x, y, z) = 1$ is bounded below by the surface $z = 4y^2$, above by the plane $z = 4$, and on the ends by the planes $x = 1$ and $x = -1$. Find the center of mass and the moments of inertia with respect to the three axes.
24. **Center of mass** A solid of constant density is bounded below by the plane $z = 0$, on the sides by the elliptical cylinder $x^2 + 4y^2 = 4$, and above by the plane $z = 2 - x$ (see the accompanying figure).
    a. Find $\bar{x}$ and $\bar{y}$.
    b. Evaluate the integral

$$M_{xy} = \int_{-2}^{2} \int_{-(1/2)\sqrt{4-x^2}}^{(1/2)\sqrt{4-x^2}} \int_{0}^{2-x} z \, dz \, dy \, dx,$$

using integral tables to carry out the final integration with respect to x. Then divide M_{xy} by M to verify that $\bar{z} = 5/4$.

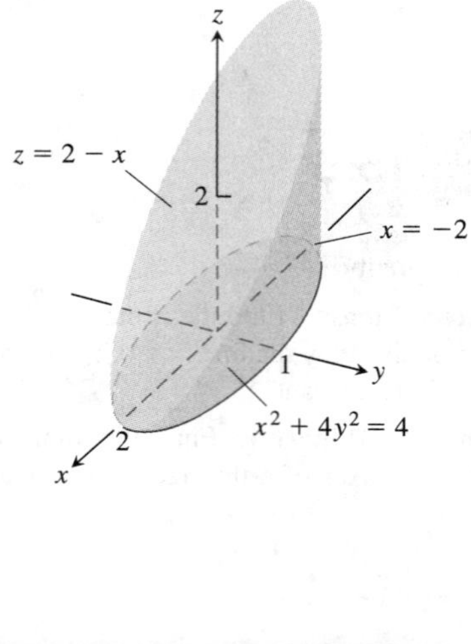

25. a. Center of mass Find the center of mass of a solid of constant density bounded below by the paraboloid $z = x^2 + y^2$ and above by the plane $z = 4$.

b. Find the plane $z = c$ that divides the solid into two parts of equal volume. This plane does not pass through the center of mass.

26. Moments A solid cube of constant density $\delta(x, y, z) = 1$, 2 units on a side, is bounded by the planes $x = \pm 1$, $z = \pm 1$, $y = 3$, and $y = 5$. Find the center of mass and the moments of inertia about the coordinate axes.

27. Moment of inertia about a line A wedge like the one in Exercise 22 has $a = 4$, $b = 6$, $c = 3$, and a constant density $\delta(x, y, z) = 1$. Make a quick sketch to check for yourself that the square of the distance from a typical point (x, y, z) of the wedge to the line L: $z = 0$, $y = 6$ is $r^2 = (y - 6)^2 + z^2$. Then calculate the moment of inertia of the wedge about L.

28. Moment of inertia about a line A wedge like the one in Exercise 22 has $a = 4$, $b = 6$, $c = 3$, and a constant density $\delta(x, y, z) = 1$. Make a quick sketch to check for yourself that the square of the distance from a typical point (x, y, z) of the wedge to the line L: $x = 4$, $y = 0$ is $r^2 = (x - 4)^2 + y^2$. Then calculate the moment of inertia of the wedge about L.

Solids with Varying Density

In Exercises 29 and 30, find

a. the mass of the solid. **b.** the center of mass.

29. A solid region in the first octant is bounded by the coordinate planes and the plane $x + y + z = 3$. The density of the solid is $\delta(x, y, z) = 3x$ gm/cm^3.

30. A solid in the first octant is bounded by the planes $y = 0$ and $z = 0$ and by the surfaces $z = 4 - x^2$ and $x = y^2$ (see the accompanying figure). Its density function is $\delta(x, y, z) = kxy$, k a constant.

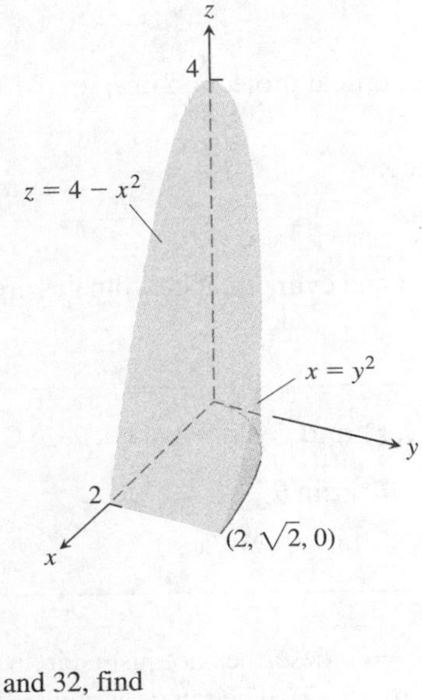

In Exercises 31 and 32, find

a. the mass of the solid.

b. the center of mass.

c. the moments of inertia about the coordinate axes.

31. A solid cube in the first octant is bounded by the coordinate planes and by the planes $x = 2$, $y = 2$, and $z = 2$. The density of the cube is $\delta(x, y, z) = x + y + z + 2$.

32. A wedge like the one in Exercise 22 has dimensions $a = 2$, $b = 6$, and $c = 3$. The density is $\delta(x, y, z) = x + 1$. Notice that if the density is constant, the center of mass will be $(0, 0, 0)$.

33. Mass Find the mass of the solid bounded by the planes $x + z = 1$, $x - z = -1$, $y = 0$, and the surface $y = \sqrt{z}$. The density of the solid is $\delta(x, y, z) = 2y + 5$ kg/m^3.

34. Mass Find the mass of the solid region bounded by the parabolic surfaces $z = 16 - 2x^2 - 2y^2$ and $z = 2x^2 + 2y^2$ if the density of the solid is $\delta(x, y, z) = \sqrt{x^2 + y^2}$.

Theory and Examples

The Parallel Axis Theorem Let $L_{\text{c.m.}}$ be a line through the center of mass of a body of mass m and let L be a parallel line h units away from $L_{\text{c.m.}}$. The *Parallel Axis Theorem* says that the moments of inertia $I_{\text{c.m.}}$ and I_L of the body about $L_{\text{c.m.}}$ and L satisfy the equation

$$I_L = I_{\text{c.m.}} + mh^2. \tag{2}$$

As in the two-dimensional case, the theorem gives a quick way to calculate one moment when the other moment and the mass are known.

35. Proof of the Parallel Axis Theorem

a. Show that the first moment of a body in space about any plane through the body's center of mass is zero. (*Hint:* Place the body's center of mass at the origin and let the plane be the yz-plane. What does the formula $\bar{x} = M_{yz}/M$ then tell you?)

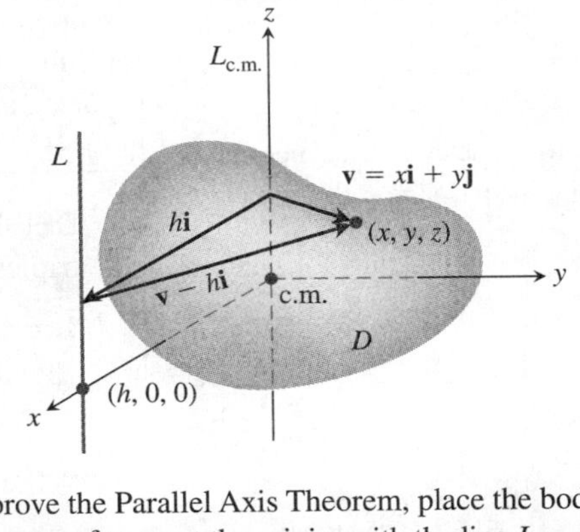

b. To prove the Parallel Axis Theorem, place the body with its center of mass at the origin, with the line $L_{\text{c.m.}}$ along the z-axis and the line L perpendicular to the xy-plane at the point $(h, 0, 0)$. Let D be the region of space occupied by the body. Then, in the notation of the figure,

$$I_L = \iiint_D |\mathbf{v} - h\mathbf{i}|^2 \, dm.$$

Expand the integrand in this integral and complete the proof.

36. The moment of inertia about a diameter of a solid sphere of constant density and radius a is $(2/5)ma^2$, where m is the mass of the sphere. Find the moment of inertia about a line tangent to the sphere.

37. The moment of inertia of the solid in Exercise 21 about the z-axis is $I_z = abc(a^2 + b^2)/3$.

a. Use Equation (2) to find the moment of inertia of the solid about the line parallel to the z-axis through the solid's center of mass.

b. Use Equation (2) and the result in part (a) to find the moment of inertia of the solid about the line $x = 0,\ y = 2b$.

38. If $a = b = 6$ and $c = 4$, the moment of inertia of the solid wedge in Exercise 22 about the x-axis is $I_x = 208$. Find the moment of inertia of the wedge about the line $y = 4,\ z = -4/3$ (the edge of the wedge's narrow end).

Joint Probability Density Functions

For Exercises 39–42, verify that f gives a joint probability density function. Then find the expected values μ_X and μ_Y.

39. $f(x, y) = \begin{cases} x + y, & \text{if } 0 \le x \le 1 \text{ and } 0 \le y \le 1, \\ 0, & \text{otherwise.} \end{cases}$

40. $f(x, y) = \begin{cases} 4xy, & \text{if } 0 \le x \le 1 \text{ and } 0 \le y \le 1, \\ 0, & \text{otherwise.} \end{cases}$

41. $f(x, y) = \begin{cases} 6x^2y, & \text{if } 0 \le x \le 1 \text{ and } 0 \le y \le 1, \\ 0, & \text{otherwise.} \end{cases}$

42. $f(x, y) = \begin{cases} \frac{3}{2}(x^2 + y^2), & \text{if } 0 \le x \le 1 \text{ and } 0 \le y \le 1, \\ 0, & \text{otherwise.} \end{cases}$

43. Suppose that f is a uniform joint probability density function on $0 \le x < 2,\ 0 \le y < 3$. What is the formula for f? What is the probability that $X < Y$?

44. The following formula defines a joint probability density function. What is the value of C? What are the expected values μ_X and μ_Y?

$$f(x, y) = \begin{cases} Cxy, & \text{if } 0 \le x \le 2 \text{ and } 0 \le y \le 3, \\ 0, & \text{otherwise.} \end{cases}$$

15.7 Triple Integrals in Cylindrical and Spherical Coordinates

When a calculation in physics, engineering, or geometry involves a cylinder, cone, or sphere, we can often simplify our work by using cylindrical or spherical coordinates, which are introduced in this section. The procedure for transforming to these coordinates and evaluating the resulting triple integrals is similar to the transformation to polar coordinates in the plane discussed in Section 15.4.

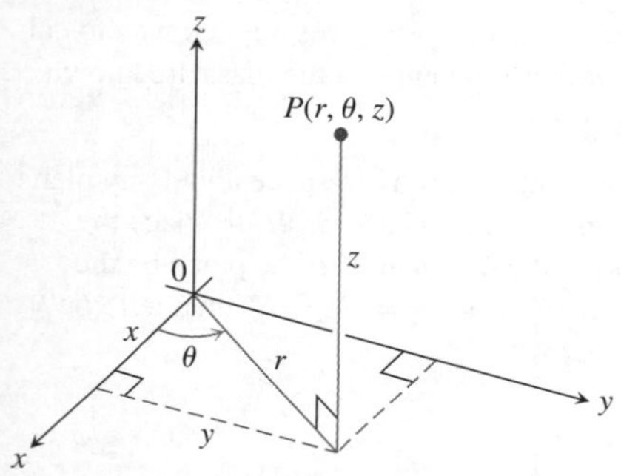

FIGURE 15.46 The cylindrical coordinates of a point in space are r, θ, and z.

Integration in Cylindrical Coordinates

We obtain cylindrical coordinates for space by combining polar coordinates in the xy-plane with the usual z-axis. This assigns to every point in space coordinate triples of the form (r, θ, z), as shown in Figure 15.46. Here we require $r \ge 0$.

> **DEFINITION** **Cylindrical coordinates** represent a point P in space by ordered triples (r, θ, z) in which
>
> 1. r and θ are polar coordinates for the vertical projection of P on the xy-plane, with $r \ge 0$, and
> 2. z is the rectangular vertical coordinate.

The values of x, y, r, and θ in rectangular and cylindrical coordinates are related by the usual equations.

> **Equations Relating Rectangular (x, y, z) and Cylindrical (r, θ, z) Coordinates**
>
> $$x = r\cos\theta, \quad y = r\sin\theta, \quad z = z,$$
> $$r^2 = x^2 + y^2, \quad \tan\theta = y/x$$

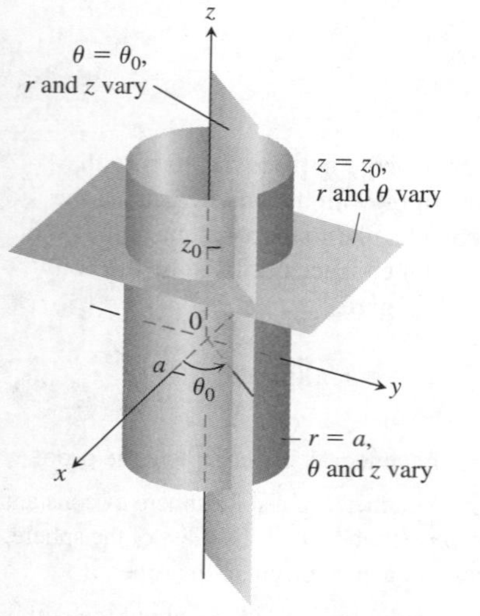

FIGURE 15.47 Constant-coordinate equations in cylindrical coordinates yield cylinders and planes.

In cylindrical coordinates, the equation $r = a$ describes not just a circle in the xy-plane but an entire cylinder about the z-axis (Figure 15.47). The z-axis is given by $r = 0$. The equation $\theta = \theta_0$ describes the half-plane that contains the z-axis and makes an angle θ_0 with the positive x-axis. And, just as in rectangular coordinates, the equation $z = z_0$ describes a plane perpendicular to the z-axis.

Cylindrical coordinates are good for describing cylinders whose axes run along the z-axis and planes that either contain the z-axis or lie perpendicular to the z-axis. Surfaces like these have equations of constant coordinate value:

$$r = 4 \qquad \text{Cylinder, radius 4, axis the } z\text{-axis}$$

$$\theta = \frac{\pi}{3} \qquad \text{Half-plane containing the } z\text{-axis}$$

$$z = 2. \qquad \text{Plane perpendicular to the } z\text{-axis}$$

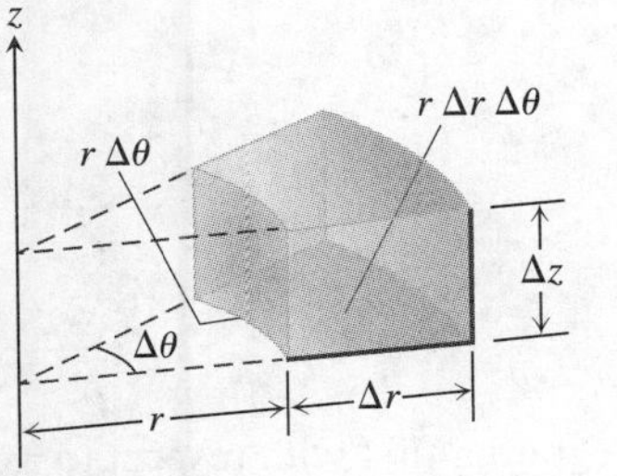

FIGURE 15.48 In cylindrical coordinates the volume of the wedge is approximated by the product $\Delta V = r\,\Delta z\,\Delta r\,\Delta\theta$.

When computing triple integrals over a solid region D in cylindrical coordinates, we partition the region into n small cylindrical wedges, rather than into rectangular boxes. In the kth cylindrical wedge, r, θ, and z change by Δr_k, $\Delta\theta_k$, and Δz_k, and the largest of these numbers among all the cylindrical wedges is called the **norm** of the partition. We express the triple integral as a limit of Riemann sums using these wedges. The volume of such a cylindrical wedge ΔV_k is obtained by taking the area ΔA_k of its base in the $r\theta$-plane and multiplying by the height Δz_k (Figure 15.48).

For a point (r_k, θ_k, z_k) in the center of the kth wedge, we calculated in polar coordinates that $\Delta A_k = r_k\,\Delta r_k\,\Delta\theta_k$. So $\Delta V_k = \Delta z_k\,r_k\,\Delta r_k\,\Delta\theta_k = r_k\,\Delta z_k\,\Delta r_k\,\Delta\theta_k$, and a Riemann sum for f over D has the form

$$S_n = \sum_{k=1}^{n} f(r_k, \theta_k, z_k)\,r_k\,\Delta z_k\,\Delta r_k\,\Delta\theta_k.$$

The triple integral of a function f over D is obtained by taking a limit of such Riemann sums with partitions whose norms approach zero:

Volume Differential in Cylindrical Coordinates

$$dV = r\,dz\,dr\,d\theta$$

$$\lim_{n\to\infty} S_n = \iiint_D f\,dV = \iiint_D f\,r\,dz\,dr\,d\theta.$$

Triple integrals in cylindrical coordinates are then evaluated as iterated integrals, as in the following example. Although the definition of cylindrical coordinates makes sense without any restrictions on θ, in most situations when integrating, we will need to restrict θ to an interval of length 2π. So we impose the requirement that $\alpha \le \theta \le \beta$, where $0 \le \beta - \alpha \le 2\pi$.

EXAMPLE 1 Find the limits of integration in cylindrical coordinates for integrating a function $f(r, \theta, z)$ over the solid region D bounded below by the plane $z = 0$, laterally by the circular cylinder $x^2 + (y-1)^2 = 1$, and above by the paraboloid $z = x^2 + y^2$.

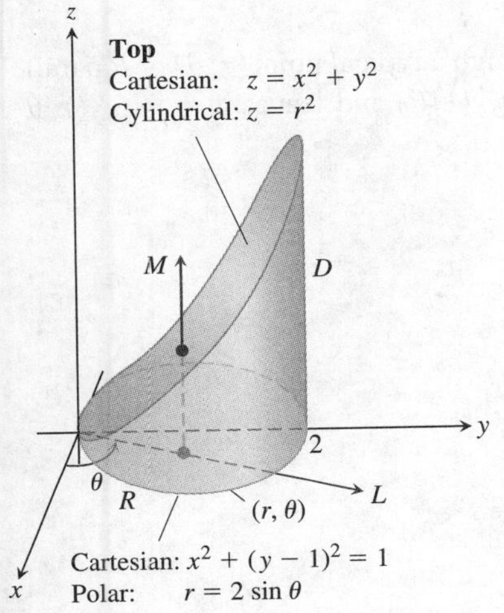

FIGURE 15.49 Finding the limits of integration for evaluating an integral in cylindrical coordinates (Example 1).

Solution The base of D is also the region's projection R on the xy-plane. The boundary of R is the circle $x^2 + (y-1)^2 = 1$. Its polar coordinate equation is

$$\begin{aligned} x^2 + (y-1)^2 &= 1 \\ x^2 + y^2 - 2y + 1 &= 1 \\ r^2 - 2r\sin\theta &= 0 \\ r &= 2\sin\theta. \end{aligned}$$

The region is sketched in Figure 15.49.

We find the limits of integration, starting with the z-limits. A line M through a typical point (r, θ) in R parallel to the z-axis enters D at $z = 0$ and leaves at $z = x^2 + y^2 = r^2$.

Next we find the r-limits of integration. A ray L through (r, θ) from the origin enters R at $r = 0$ and leaves at $r = 2\sin\theta$.

Finally, we find the θ-limits of integration. As L sweeps across R, the angle θ it makes with the positive x-axis runs from $\theta = 0$ to $\theta = \pi$. The integral is

$$\iiint_D f(r, \theta, z)\, dV = \int_0^{\pi} \int_0^{2\sin\theta} \int_0^{r^2} f(r, \theta, z)\, r\, dz\, dr\, d\theta. \qquad \blacksquare$$

Example 1 illustrates a good procedure for finding limits of integration in cylindrical coordinates. The procedure is summarized as follows.

How to Integrate in Cylindrical Coordinates

To evaluate

$$\iiint_D f(r, \theta, z)\, dV$$

over a solid region D in space in cylindrical coordinates, integrating first with respect to z, then with respect to r, and finally with respect to θ, take the following steps.

1. *Sketch.* Sketch the solid region D along with its projection R on the xy-plane. Label the surfaces and curves that bound D and R.

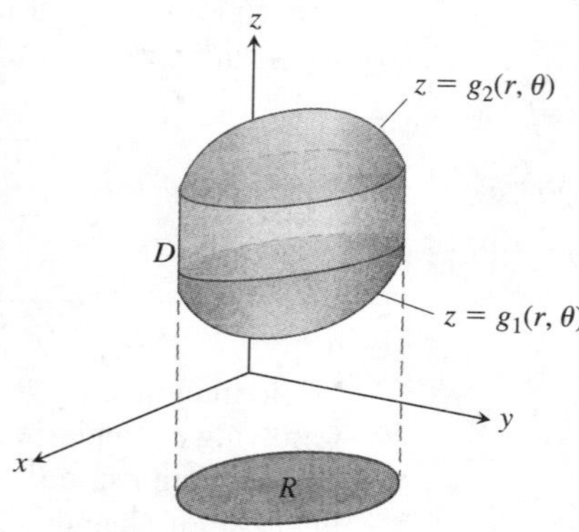

2. *Find the z-limits of integration.* Draw a line M through a typical point (r, θ) of R parallel to the z-axis. As z increases, M enters D at $z = g_1(r, \theta)$ and leaves at $z = g_2(r, \theta)$. These are the z-limits of integration.

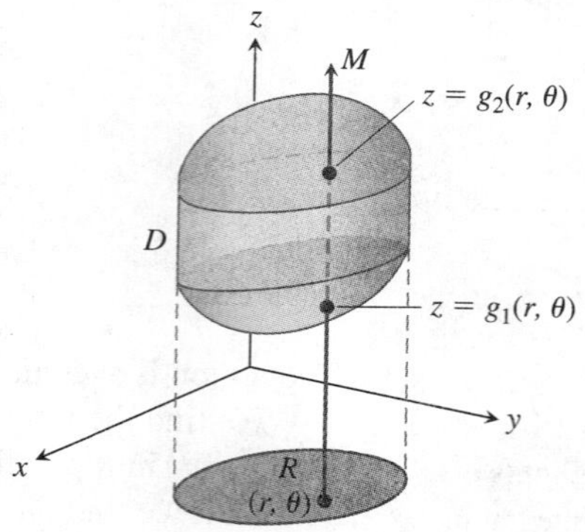

3. *Find the r-limits of integration.* Draw a ray L through (r, θ) from the origin. The ray enters R at $r = h_1(\theta)$ and leaves at $r = h_2(\theta)$. These are the r-limits of integration.

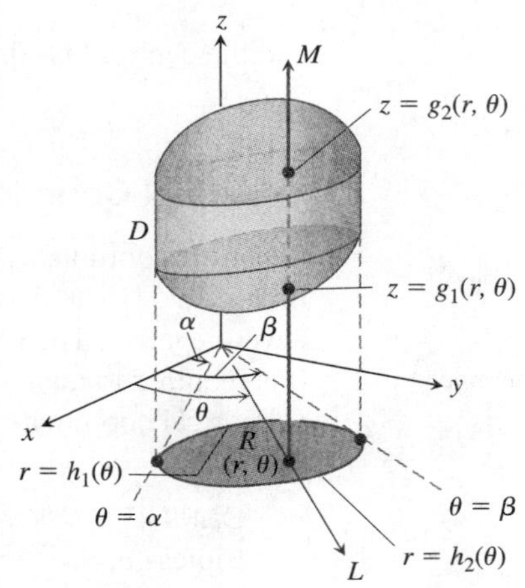

4. *Find the θ-limits of integration.* As L sweeps across R, the angle θ it makes with the positive x-axis runs from $\theta = \alpha$ to $\theta = \beta$. These are the θ-limits of integration. The integral is

$$\iiint_D f(r, \theta, z)\, dV = \int_{\theta=\alpha}^{\theta=\beta} \int_{r=h_1(\theta)}^{r=h_2(\theta)} \int_{z=g_1(r,\theta)}^{z=g_2(r,\theta)} f(r, \theta, z)\, r\, dz\, dr\, d\theta.$$

EXAMPLE 2 Find the centroid ($\delta = 1$) of the solid enclosed by the cylinder $x^2 + y^2 = 4$, bounded above by the paraboloid $z = x^2 + y^2$, and bounded below by the xy-plane.

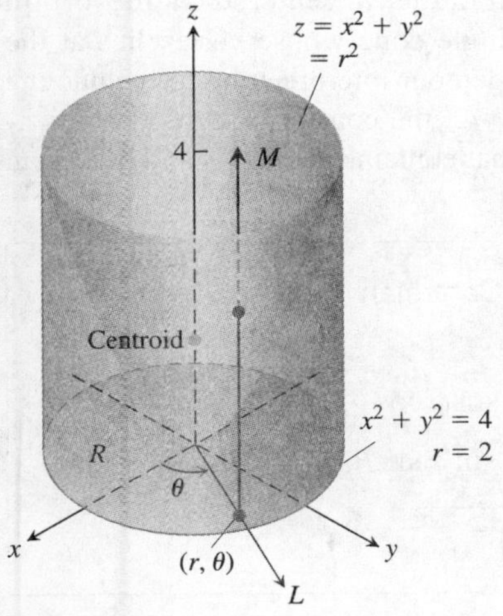

FIGURE 15.50 Example 2 shows how to find the centroid of this solid.

Solution We sketch the solid, bounded above by the paraboloid $z = r^2$ and below by the plane $z = 0$ (Figure 15.50). Its base R is the disk $0 \le r \le 2$ in the xy-plane.

The solid's centroid $(\bar{x}, \bar{y}, \bar{z})$ lies on its axis of symmetry, here the z-axis. This makes $\bar{x} = \bar{y} = 0$. To find $\bar{z}$, we divide the first moment M_{xy} by the mass M.

To find the limits of integration for the mass and moment integrals, we continue with the four basic steps. We completed our initial sketch. The remaining steps give the limits of integration.

The z-limits. A line M through a typical point (r, θ) in the base parallel to the z-axis enters the solid at $z = 0$ and leaves at $z = r^2$.

The r-limits. A ray L through (r, θ) from the origin enters R at $r = 0$ and leaves at $r = 2$.

The θ-limits. As L sweeps over the base like a clock hand, the angle θ it makes with the positive x-axis runs from $\theta = 0$ to $\theta = 2\pi$. The value of M_{xy} is

$$M_{xy} = \int_0^{2\pi} \int_0^2 \int_0^{r^2} z\, r\, dz\, dr\, d\theta = \int_0^{2\pi} \int_0^2 \left[\frac{z^2}{2}\right]_{z=0}^{z=r^2} r\, dr\, d\theta$$

$$= \int_0^{2\pi} \int_0^2 \frac{r^5}{2}\, dr\, d\theta = \int_0^{2\pi} \left[\frac{r^6}{12}\right]_{r=0}^{r=2} d\theta = \int_0^{2\pi} \frac{16}{3}\, d\theta = \frac{32\pi}{3}.$$

The value of M is

$$M = \int_0^{2\pi} \int_0^2 \int_0^{r^2} r\, dz\, dr\, d\theta = \int_0^{2\pi} \int_0^2 \left[z\right]_{z=0}^{z=r^2} r\, dr\, d\theta$$

$$= \int_0^{2\pi} \int_0^2 r^3\, dr\, d\theta = \int_0^{2\pi} \left[\frac{r^4}{4}\right]_{r=0}^{r=2} d\theta = \int_0^{2\pi} 4\, d\theta = 8\pi.$$

Therefore,

$$\bar{z} = \frac{M_{xy}}{M} = \frac{32\pi}{3}\frac{1}{8\pi} = \frac{4}{3},$$

and the centroid is $(0, 0, 4/3)$. Notice that the centroid lies on the z-axis, outside the solid. ■

Spherical Coordinates and Integration

Spherical coordinates locate points in space with two angles and one distance, as shown in Figure 15.51. The first coordinate, $\rho = |\overrightarrow{OP}|$, is the point's distance from the origin and is never negative. The second coordinate, ϕ, is the angle $\overrightarrow{OP}$ makes with the positive z-axis. It is required to lie in the interval $[0, \pi]$. The third coordinate is the angle θ as measured in cylindrical coordinates.

ϕ is the Greek letter phi, pronounced "fee."

DEFINITION Spherical coordinates represent a point P in space by ordered triples (ρ, ϕ, θ) in which

1. ρ is the distance from P to the origin $(\rho \geq 0)$.
2. ϕ is the angle $\overrightarrow{OP}$ makes with the positive z-axis $(0 \leq \phi \leq \pi)$.
3. θ is the angle from cylindrical coordinates.

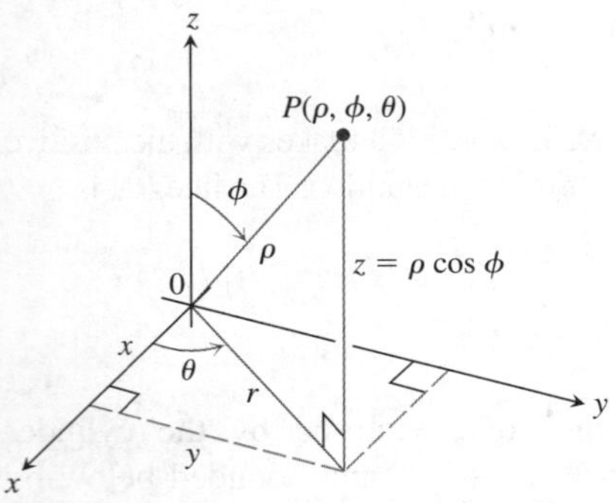

FIGURE 15.51 The spherical coordinates ρ, ϕ, and θ and their relation to x, y, z, and r.

On maps of Earth, θ is related to the longitude of a point on the planet and ϕ to its latitude, while ρ is related to elevation above Earth's surface.

The equation $\rho = a$ describes the sphere of radius a centered at the origin (Figure 15.52). The equation $\phi = \phi_0$ describes a single cone whose vertex lies at the origin and whose axis lies along the z-axis. (We broaden our interpretation to include the xy-plane as the cone $\phi = \pi/2$.) If ϕ_0 is greater than $\pi/2$, the cone $\phi = \phi_0$ opens downward. The equation $\theta = \theta_0$ describes the half-plane that contains the z-axis and makes an angle θ_0 with the positive x-axis.

Equations Relating Spherical Coordinates to Cartesian and Cylindrical Coordinates

$$\begin{aligned} r &= \rho \sin\phi, & x &= r\cos\theta = \rho\sin\phi\cos\theta, \\ z &= \rho\cos\phi, & y &= r\sin\theta = \rho\sin\phi\sin\theta, \end{aligned} \tag{1}$$
$$\rho = \sqrt{x^2 + y^2 + z^2} = \sqrt{r^2 + z^2}.$$

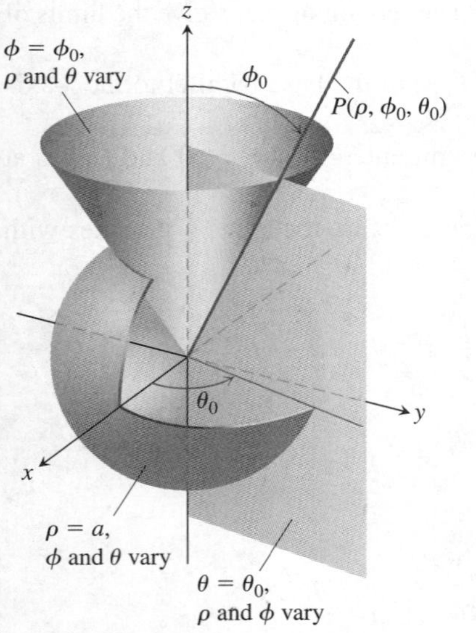

FIGURE 15.52 Constant-coordinate equations in spherical coordinates yield spheres, single cones, and half-planes.

EXAMPLE 3 Find a spherical coordinate equation for the sphere $x^2 + y^2 + (z - 1)^2 = 1$.

Solution We use Equations (1) to substitute for x, y, and z:

$$\begin{aligned} x^2 + y^2 + (z-1)^2 &= 1 \\ \rho^2\sin^2\phi\cos^2\theta + \rho^2\sin^2\phi\sin^2\theta + (\rho\cos\phi - 1)^2 &= 1 \qquad \text{Eqs. (1)} \\ \rho^2\sin^2\phi\underbrace{(\cos^2\theta + \sin^2\theta)}_{1} + \rho^2\cos^2\phi - 2\rho\cos\phi + 1 &= 1 \\ \rho^2\underbrace{(\sin^2\phi + \cos^2\phi)}_{1} &= 2\rho\cos\phi \\ \rho^2 &= 2\rho\cos\phi \\ \rho &= 2\cos\phi. \qquad \text{Includes } \rho = 0 \end{aligned}$$

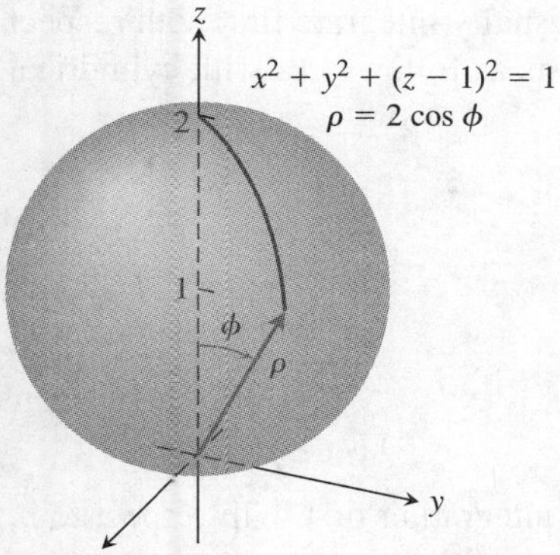

FIGURE 15.53 The sphere in Example 3.

The angle ϕ varies from 0 at the north pole of the sphere to $\pi/2$ at the south pole; the angle θ does not appear in the expression for ρ, reflecting the symmetry about the z-axis (see Figure 15.53). ■

EXAMPLE 4 Find a spherical coordinate equation for the cone $z = \sqrt{x^2 + y^2}$.

Solution 1 *Use geometry.* The cone is symmetric with respect to the z-axis and cuts the first quadrant of the yz-plane along the line $z = y$. The angle between the cone and the positive z-axis is therefore $\pi/4$ radians. The cone consists of the points whose spherical coordinates have ϕ equal to $\pi/4$, so its equation is $\phi = \pi/4$. (See Figure 15.54.)

Solution 2 *Use algebra.* If we use Equations (1) to substitute for x, y, and z, we obtain the same result:

$$\begin{aligned}
z &= \sqrt{x^2 + y^2} \\
\rho \cos \phi &= \sqrt{\rho^2 \sin^2 \phi} && \text{Example 3} \\
\rho \cos \phi &= \rho \sin \phi && \rho \geq 0, \sin \phi \geq 0 \\
\cos \phi &= \sin \phi && \text{Includes } \rho = 0 \text{ (the origin)} \\
\phi &= \frac{\pi}{4}. && 0 \leq \phi \leq \pi
\end{aligned}$$

■

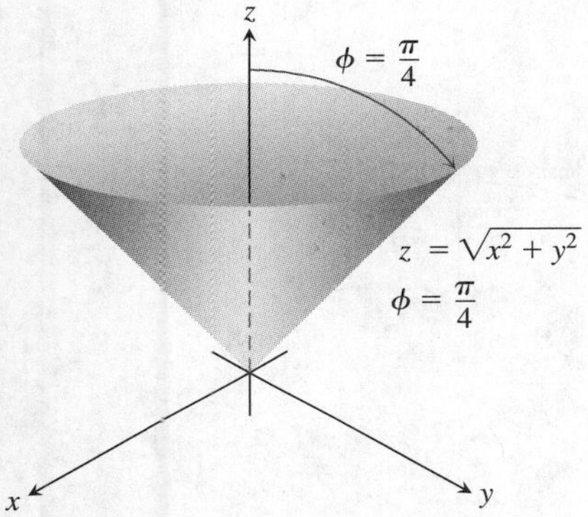

FIGURE 15.54 The cone in Example 4.

Spherical coordinates are useful for describing spheres centered at the origin, half-planes hinged along the z-axis, and cones whose vertices lie at the origin and whose axes lie along the z-axis. Surfaces like these have equations of constant coordinate value:

$\rho = 4$	Sphere, radius 4, center at origin
$\phi = \frac{2\pi}{3}$	Cone opening down from the origin, making an angle of $2\pi/3$ radians with the positive z-axis
$\theta = \frac{\pi}{3}$.	Half-plane, hinged along the z-axis, making an angle of $\pi/3$ radians with the positive x-axis

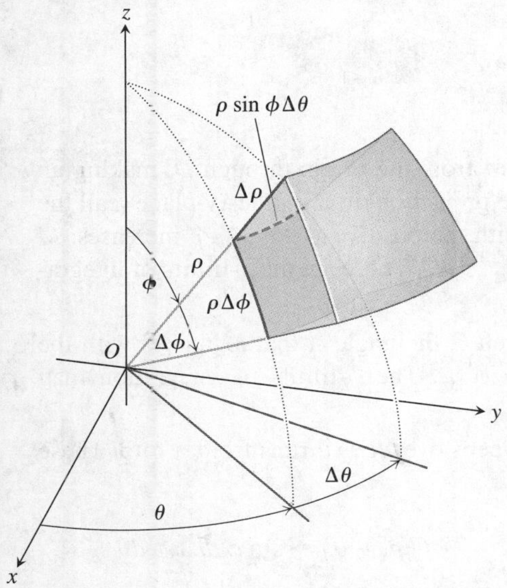

FIGURE 15.55 In spherical coordinates we use the volume of a spherical wedge, which closely approximates that of a rectangular box.

When computing triple integrals over a solid region D in spherical coordinates, we partition the region into n spherical wedges. The size of the kth spherical wedge, which contains a point $(\rho_k, \phi_k, \theta_k)$, is given by the changes $\Delta\rho_k$, $\Delta\phi_k$, and $\Delta\theta_k$ in ρ, ϕ, and θ. Such a spherical wedge has one edge a circular arc of length $\rho_k \Delta\phi_k$, another edge a circular arc of length $\rho_k \sin \phi_k \Delta\theta_k$, and thickness $\Delta\rho_k$. The spherical wedge closely approximates a rectangular box of these dimensions when $\Delta\rho_k$, $\Delta\phi_k$, and $\Delta\theta_k$ are all small (Figure 15.55). It can be shown that the volume of this spherical wedge ΔV_k is $\Delta V_k = \rho_k^2 \sin \phi_k \Delta\rho_k \Delta\phi_k \Delta\theta_k$ for $(\rho_k, \phi_k, \theta_k)$, a point chosen inside the wedge.

The corresponding Riemann sum for a function $f(\rho, \phi, \theta)$ is

$$S_n = \sum_{k=1}^{n} f(\rho_k, \phi_k, \theta_k)\, \rho_k^2 \sin \phi_k \,\Delta\rho_k \,\Delta\phi_k \,\Delta\theta_k.$$

As the norm of a partition approaches zero, and the spherical wedges get smaller, the limit of the Riemann sums is the triple integral:

$$\lim_{n \to \infty} S_n = \iiint_D f(\rho, \phi, \theta)\, dV = \iiint_D f(\rho, \phi, \theta)\, \rho^2 \sin \phi \, d\rho \, d\phi \, d\theta.$$

Volume Differential in Spherical Coordinates

$$dV = \rho^2 \sin \phi \, d\rho \, d\phi \, d\theta$$

To evaluate integrals in spherical coordinates, we usually integrate first with respect to ρ. The procedure for finding the limits of integration is as follows. As with cylindrical coordinates, we restrict θ in the form $\alpha \leq \theta \leq \beta$ and $0 \leq \beta - \alpha \leq 2\pi$.

How to Integrate in Spherical Coordinates

To evaluate

$$\iiint_D f(\rho, \phi, \theta)\, dV$$

over a solid region D in space in spherical coordinates, integrating first with respect to ρ, then with respect to ϕ, and finally with respect to θ, take the following steps.

1. *Sketch.* Sketch the solid region D along with its projection R on the xy-plane. Label the surfaces that bound D.

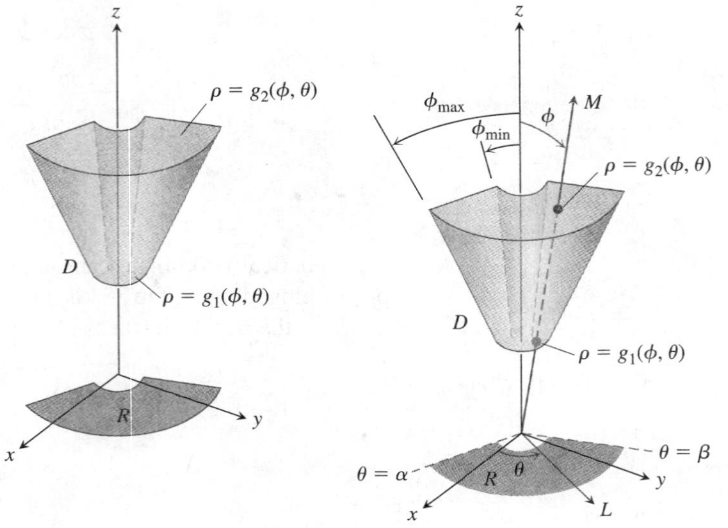

2. *Find the ρ-limits of integration.* Draw a ray M from the origin through D, making an angle ϕ with the positive z-axis. Also draw the projection of M on the xy-plane (call the projection L). The ray L makes an angle θ with the positive x-axis. As ρ increases, M enters D at $\rho = g_1(\phi, \theta)$ and leaves at $\rho = g_2(\phi, \theta)$. These are the ρ-limits of integration shown in the above figure.
3. *Find the ϕ-limits of integration.* For any given θ, the angle ϕ that M makes with the positive z-axis runs from $\phi = \phi_{\min}$ to $\phi = \phi_{\max}$. The ϕ-limits of integration may depend on θ, but they are often constant.
4. *Find the θ-limits of integration.* The ray L sweeps over R as θ runs from α to β. These are the θ-limits of integration. The integral is

$$\iiint_D f(\rho, \phi, \theta)\, dV = \int_{\theta=\alpha}^{\theta=\beta} \int_{\phi=\phi_{\min}}^{\phi=\phi_{\max}} \int_{\rho=g_1(\phi,\theta)}^{\rho=g_2(\phi,\theta)} f(\rho, \phi, \theta)\, \rho^2 \sin\phi\, d\rho\, d\phi\, d\theta.$$

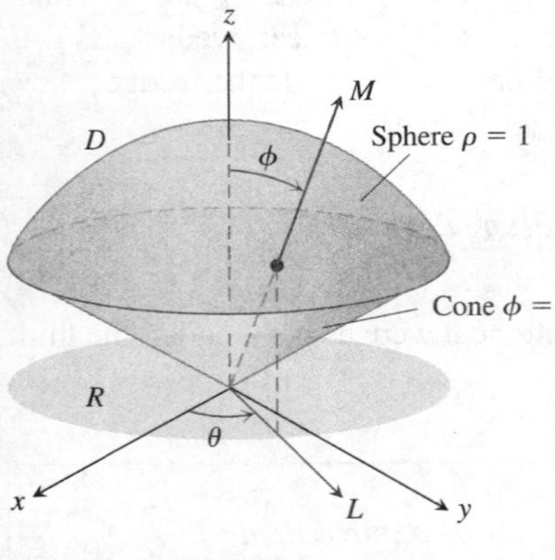

FIGURE 15.56 The ice cream cone in Example 5.

EXAMPLE 5 Find the volume of the "ice cream cone" D bounded above by the sphere $\rho = 1$ and bounded below by the cone $\phi = \pi/3$.

Solution The volume is $V = \iiint_D \rho^2 \sin\phi\, d\rho\, d\phi\, d\theta$, the integral of $f(\rho, \phi, \theta) = 1$ over D.

To find the limits of integration for evaluating the integral, we begin by sketching D and its projection R on the xy-plane (Figure 15.56).

The ρ-limits of integration. We draw a ray M from the origin through D, making an angle ϕ with the positive z-axis. We also draw L, the projection of M on the xy-plane, along with the angle θ that L makes with the positive x-axis. Ray M enters D at $\rho = 0$ and leaves at $\rho = 1$.

The ϕ-limits of integration. The cone $\phi = \pi/3$ makes an angle of $\pi/3$ with the positive z-axis. For any given θ, the angle ϕ can run from $\phi = 0$ to $\phi = \pi/3$.

The θ-limits of integration. The ray L sweeps over R as θ runs from 0 to 2π. The volume is

$$
\begin{aligned}
V &= \iiint_D \rho^2 \sin\phi \, d\rho \, d\phi \, d\theta = \int_0^{2\pi} \int_0^{\pi/3} \int_0^1 \rho^2 \sin\phi \, d\rho \, d\phi \, d\theta \\
&= \int_0^{2\pi} \int_0^{\pi/3} \left[\frac{\rho^3}{3}\right]_{\rho=0}^{\rho=1} \sin\phi \, d\phi \, d\theta = \int_0^{2\pi} \int_0^{\pi/3} \frac{1}{3} \sin\phi \, d\phi \, d\theta \\
&= \int_0^{2\pi} \left[-\frac{1}{3}\cos\phi\right]_{\phi=0}^{\phi=\pi/3} d\theta = \int_0^{2\pi} \left(-\frac{1}{6} + \frac{1}{3}\right) d\theta = \frac{1}{6}(2\pi) = \frac{\pi}{3}.
\end{aligned}
$$

■

EXAMPLE 6 A solid of constant density $\delta = 1$ occupies the solid region D in Example 5. Find the solid's moment of inertia about the z-axis.

Solution In rectangular coordinates, the moment is

$$I_z = \iiint_D (x^2 + y^2)\, dV.$$

In spherical coordinates, $x^2 + y^2 = (\rho \sin\phi \cos\theta)^2 + (\rho \sin\phi \sin\theta)^2 = \rho^2 \sin^2\phi$. Hence,

$$I_z = \iiint_D (\rho^2 \sin^2\phi)\rho^2 \sin\phi \, d\rho \, d\phi \, d\theta = \iiint_D \rho^4 \sin^3\phi \, d\rho \, d\phi \, d\theta.$$

For the region D in Example 5, this becomes

$$
\begin{aligned}
I_z &= \int_0^{2\pi} \int_0^{\pi/3} \int_0^1 \rho^4 \sin^3\phi \, d\rho \, d\phi \, d\theta = \int_0^{2\pi} \int_0^{\pi/3} \left[\frac{\rho^5}{5}\right]_{\rho=0}^{\rho=1} \sin^3\phi \, d\phi \, d\theta \\
&= \frac{1}{5} \int_0^{2\pi} \int_0^{\pi/3} (1 - \cos^2\phi) \sin\phi \, d\phi \, d\theta = \frac{1}{5} \int_0^{2\pi} \left[-\cos\phi + \frac{\cos^3\phi}{3}\right]_{\phi=0}^{\phi=\pi/3} d\theta \\
&= \frac{1}{5} \int_0^{2\pi} \left(-\frac{1}{2} + \frac{1}{24} + 1 - \frac{1}{3}\right) d\theta = \frac{1}{5} \int_0^{2\pi} \frac{5}{24}\, d\theta = \frac{1}{24}(2\pi) = \frac{\pi}{12}.
\end{aligned}
$$

■

Coordinate Conversion Formulas

Cylindrical to Rectangular	Spherical to Rectangular	Spherical to Cylindrical
$x = r\cos\theta$	$x = \rho \sin\phi \cos\theta$	$r = \rho \sin\phi$
$y = r\sin\theta$	$y = \rho \sin\phi \sin\theta$	$z = \rho \cos\phi$
$z = z$	$z = \rho \cos\phi$	$\theta = \theta$

Corresponding formulas for dV in triple integrals:

$$
\begin{aligned}
dV &= dx\, dy\, dz \\
&= r\, dz\, dr\, d\theta \\
&= \rho^2 \sin\phi \, d\rho \, d\phi \, d\theta
\end{aligned}
$$

In the next section we offer a more general procedure for determining dV in cylindrical and spherical coordinates. The results, of course, will be the same.

EXERCISES 15.7

In Exercises 1–12, sketch the region described by the following cylindrical coordinates in three-dimensional space.

1. $r = 2$

2. $\theta = \frac{\pi}{4}$

3. $z = -1$

4. $z = r$

5. $r = \theta$

6. $z = r \sin \theta$

7. $r^2 + z^2 = 4$

8. $1 \leq r \leq 2, \quad 0 \leq \theta \leq \frac{\pi}{3}$

9. $r \leq z \leq \sqrt{9 - r^2}$

10. $0 \leq r \leq 2 \sin \theta, \quad 1 \leq z \leq 3$

11. $0 \leq r \leq 4 \cos \theta, \quad 0 \leq \theta \leq \frac{\pi}{2}, \quad 0 \leq z \leq 5$

12. $0 \leq r \leq 3, \quad \frac{-\pi}{2} \leq \theta \leq \frac{\pi}{2}, \quad 0 \leq z \leq r \cos \theta$

In Exercises 13–22, sketch the region described by the following spherical coordinates in three-dimensional space.

13. $\rho = 3$

14. $\phi = \frac{\pi}{6}$

15. $\theta = \frac{2}{3}\pi$

16. $\rho = \csc \phi$

17. $\rho \cos \phi = 4$

18. $1 \leq \rho \leq 2 \sec \phi, \quad 0 \leq \phi \leq \frac{\pi}{4}$

19. $0 \leq \rho \leq 3 \csc \phi$

20. $0 \leq \rho \leq 1, \quad \frac{\pi}{2} \leq \phi \leq \pi, \quad 0 \leq \theta \leq \pi$

21. $0 \leq \rho \cos \theta \sin \phi \leq 2, \quad 0 \leq \rho \sin \theta \sin \phi \leq 3,$
$0 \leq \rho \cos \phi \leq 4$

22. $4 \sec \phi \leq \rho \leq 5, 0 \leq \phi \leq \frac{\pi}{2}$

Evaluating Integrals in Cylindrical Coordinates

Evaluate the cylindrical coordinate integrals in Exercises 23–28.

23. $\int_0^{\pi} \int_0^1 \int_0^{\sqrt{3-r^2}} dz\, r\, dr\, d\theta$

24. $\int_0^{2\pi} \int_0^3 \int_{r^2/3}^{\sqrt{18-r^2}} r\, dz\, dr\, d\theta$

25. $\int_0^{\pi} \int_0^{\theta} \int_0^{2+16r^2} dz\, r\, dr\, d\theta$

26. $\int_0^{\pi} \int_0^{\theta/\pi} \int_{-\sqrt{4-r^2}}^{3\sqrt{4-r^2}} z\, r\, dz\, dr\, d\theta$

27. $\int_0^{\pi} \int_0^1 \int_0^{1/\sqrt{1-r^2}} 2\, dz\, r\, dr\, d\theta$

28. $\int_0^{\pi} \int_0^1 \int_{-1/4}^{1/4} (r \sin\theta + z)\, dz\, r\, dr\, d\theta$

Changing the Order of Integration in Cylindrical Coordinates

The integrals we have seen so far suggest that there are preferred orders of integration for cylindrical coordinates, but other orders usually work well and are occasionally easier to evaluate. Evaluate the integrals in Exercises 29–32.

29. $\int_0^{2\pi} \int_0^3 \int_0^{z/3} r^3\, dr\, dz\, d\theta$

30. $\int_{-1}^1 \int_0^{\pi} \int_0^{1+\sin\theta} 2r\, dr\, d\theta dz$

31. $\int_0^1 \int_0^{\sqrt{z}} \int_0^{2\pi} (r^2 \cos^2 \theta + z^2) r\, d\theta\, dr\, dz$

32. $\int_0^1 \int_r^{\sqrt{2-r^2}} \int_0^{\pi} (r \sin \theta + 1) r\, d\theta\, dz\, dr$

33. Let D be the solid region bounded below by the plane $z = 0$, above by the sphere $x^2 + y^2 + z^2 = 4$, and on the sides by the cylinder $x^2 + y^2 = 1$. Set up the triple integrals in cylindrical coordinates that give the volume of D using the following orders of integration.

a. $dz\, dr\, d\theta$ **b.** $dr\, dz\, d\theta$ **c.** $d\theta\, dz\, dr$

34. Let D be the solid region bounded below by the cone $z = \sqrt{x^2 + y^2}$ and above by the paraboloid $z = 2 - x^2 - y^2$. Set up the triple integrals in cylindrical coordinates that give the volume of D using the following orders of integration.

a. $dz\, dr\, d\theta$ **b.** $dr\, dz\, d\theta$ **c.** $d\theta\, dz\, dr$

Finding Iterated Integrals in Cylindrical Coordinates

35. Give the limits of integration for evaluating the integral

$$\iiint_D f(r, \theta, z)\, r\, dz\, dr\, d\theta$$

as an iterated integral over the solid region D that is bounded below by the plane $z = 0$, on the side by the cylinder $r = \cos \theta$, and on top by the paraboloid $z = 3r^2$.

36. Convert the integral

$$\int_{-1}^1 \int_0^{\sqrt{1-x^2}} \int_0^y x^2\, dz\, dy\, d.$$

to an equivalent integral in cylindrical coordinates and evaluate the result.

In Exercises 37–42, set up the iterated integral for evaluating $\iiint_D f(r, \theta, z)\, r\, dz\, dr\, d\theta$ over the given solid region D.

37. D is the right circular cylinder whose base is the circle $r = 2 \sin \theta$ in the xy-plane and whose top lies in the plane $z = 4 - y$.

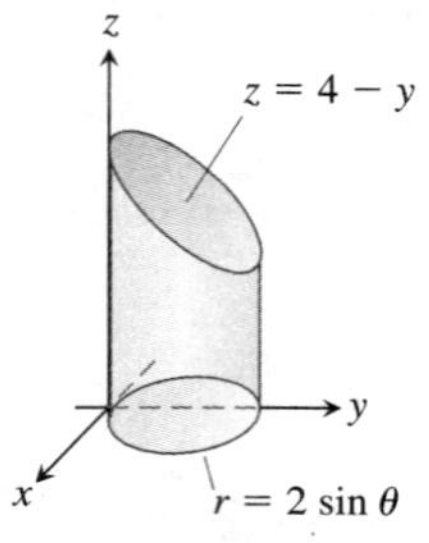

38. D is the right circular cylinder whose base is the circle $r = 3\cos\theta$ and whose top lies in the plane $z = 5 - x$.

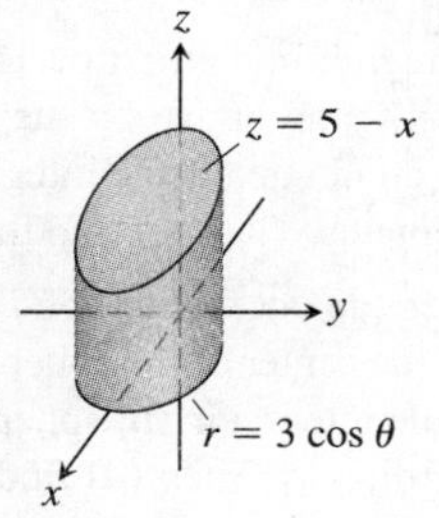

39. D is the solid right cylinder whose base is the region in the xy-plane that lies inside the cardioid $r = 1 + \cos\theta$ and outside the circle $r = 1$ and whose top lies in the plane $z = 4$.

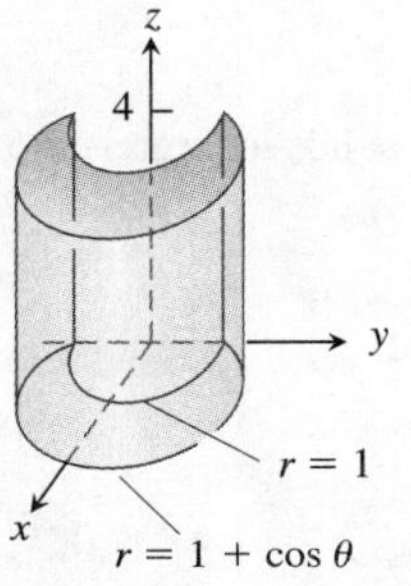

40. D is the solid right cylinder whose base is the region between the circles $r = \cos\theta$ and $r = 2\cos\theta$ and whose top lies in the plane $z = 3 - y$.

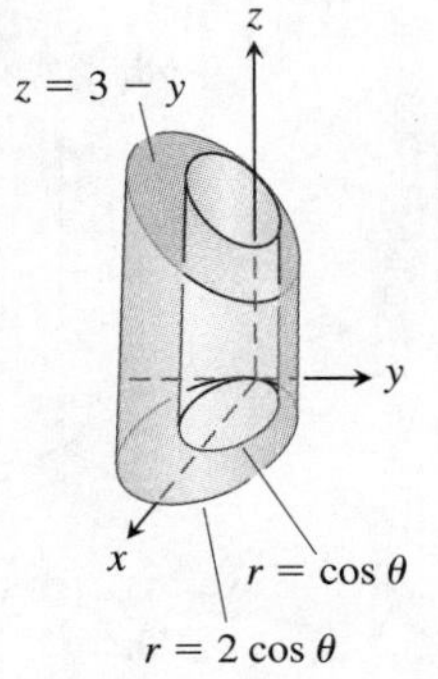

41. D is the right prism whose base is the triangle in the xy-plane bounded by the x-axis and the lines $y = x$ and $x = 1$ and whose top lies in the plane $z = 2 - y$.

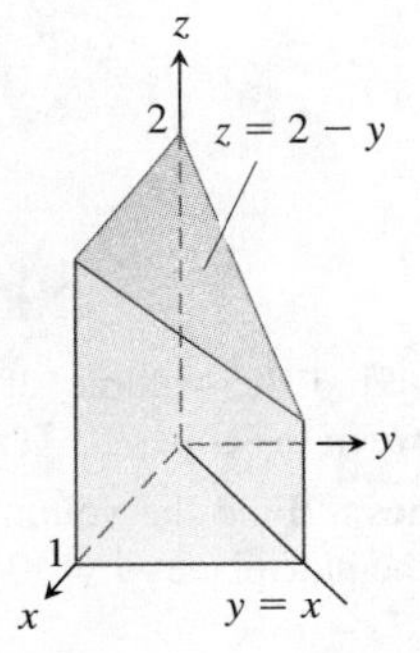

42. D is the right prism whose base is the triangle in the xy-plane bounded by the y-axis and the lines $y = x$ and $y = 1$ and whose top lies in the plane $z = 2 - x$.

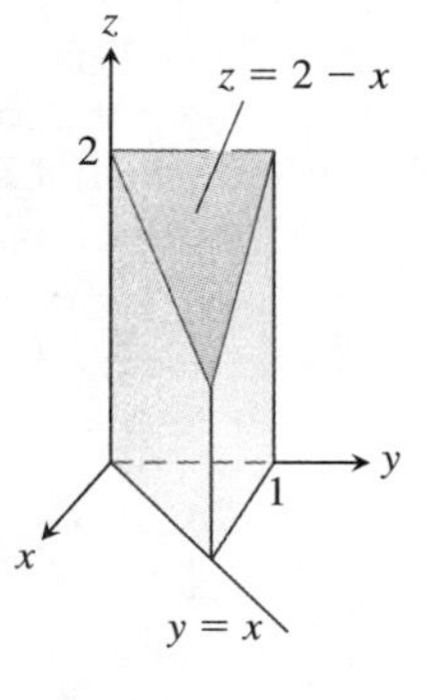

Evaluating Integrals in Spherical Coordinates

Evaluate the spherical coordinate integrals in Exercises 43–48.

43. $\displaystyle\int_0^{2\pi}\int_0^{2\pi}\int_0^{3\sin\phi} \rho^2 \sin\phi \, d\rho d\phi d\theta$

44. $\displaystyle\int_0^{2\pi}\int_0^{\pi/4}\int_0^{2} (\rho\cos\phi)\,\rho^2 \sin\phi \, d\rho\, d\phi\, d\theta$

45. $\displaystyle\int_0^{\pi}\int_0^{\pi}\int_0^{\frac{1-\sin\phi}{2}} \rho^2 \sin\phi \, d\rho\, d\phi d\theta$

46. $\displaystyle\int_0^{3\pi/2}\int_0^{\pi}\int_0^{1} 5\rho^3 \sin^3\phi \, d\rho\, d\phi\, d\theta$

47. $\displaystyle\int_0^{\pi}\int_0^{\pi/2}\int_{\cos\phi}^{2} \rho^3 \sin\phi \, d\rho\, d\phi d\theta$

48. $\displaystyle\int_0^{2\pi}\int_0^{\pi/4}\int_0^{\sec\phi} (\rho\cos\phi)\,\rho^2 \sin\phi \, d\rho\, d\phi\, d\theta$

Changing the Order of Integration in Spherical Coordinates

The previous integrals suggest there are preferred orders of integration for spherical coordinates, but other orders give the same value and are occasionally easier to evaluate. Evaluate the integrals in Exercises 49–52.

49. $\displaystyle\int_0^{2}\int_{-\pi}^{0}\int_{\pi/4}^{\pi/2} \rho^3 \sin 2\phi \, d\phi\, d\theta\, d\rho$

50. $\displaystyle\int_{\pi/6}^{\pi/2}\int_{\mathrm{cosec}\,\phi}^{3\,\mathrm{cosec}\,\phi}\int_0^{\pi} \rho^2 \sin\phi \, d\theta d\rho d\phi$

51. $\displaystyle\int_0^{1}\int_0^{\pi}\int_0^{\pi/4} 12\rho \sin^3\phi \, d\phi\, d\theta\, d\rho$

52. $\displaystyle\int_{\pi/6}^{\pi/3}\int_{-\pi}^{\pi/2}\int_{\cos\phi}^{2} 10\rho^4 \sin^3\phi \, d\rho\, d\theta d\phi$

53. Let D be the region in Exercise 33. Set up the triple integrals in spherical coordinates that give the volume of D using the following orders of integration.

a. $d\rho\, d\phi\, d\theta$

b. $d\phi\, d\rho\, d\theta$

54. Let D be the solid region bounded below by the cone $z = \sqrt{x^2 + y^2}$ and above by the plane $z = 1$. Set up the triple integrals in spherical coordinates that give the volume of D using the following orders of integration.

a. $d\rho\, d\phi\, d\theta$ **b.** $d\phi\, d\rho\, d\theta$

Finding Iterated Integrals in Spherical Coordinates

In Exercises 55–60, **(a)** find the spherical coordinate limits for the integral that calculates the volume of the given solid and then **(b)** evaluate the integral.

55. The solid between the sphere $\rho = \cos\phi$ and the hemisphere $\rho = 6$, $z \geq 0$

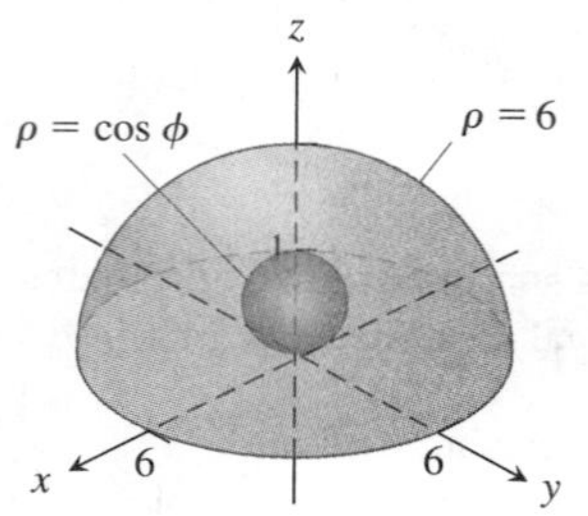

56. The solid bounded below by the hemisphere $\rho = 1$, $z \geq 0$, and above by the surface $\rho = 1 + \cos\phi$

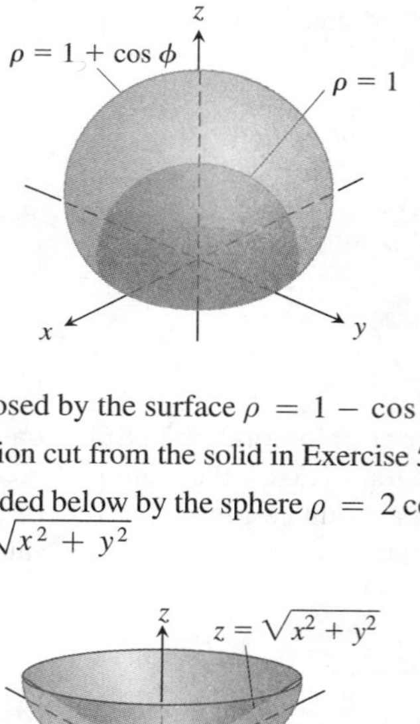

57. The solid enclosed by the surface $\rho = 1 - \cos\phi$

58. The upper portion cut from the solid in Exercise 57 by the xy-plane

59. The solid bounded below by the sphere $\rho = 2\cos\phi$ and above by the cone $z = \sqrt{x^2 + y^2}$

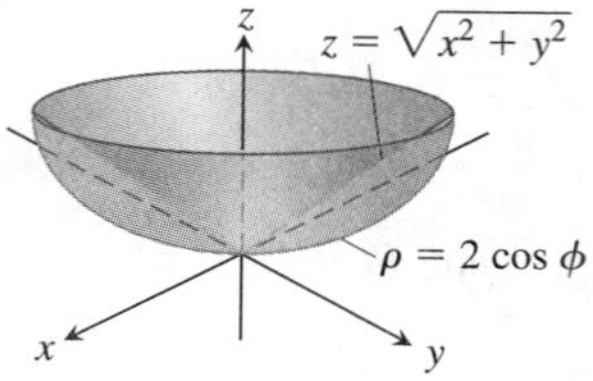

60. The solid bounded by the cone $z = \sqrt{3(x^2 + y^2)}$ and the hemisphere $z = \sqrt{4 - x^2 - y^2}$

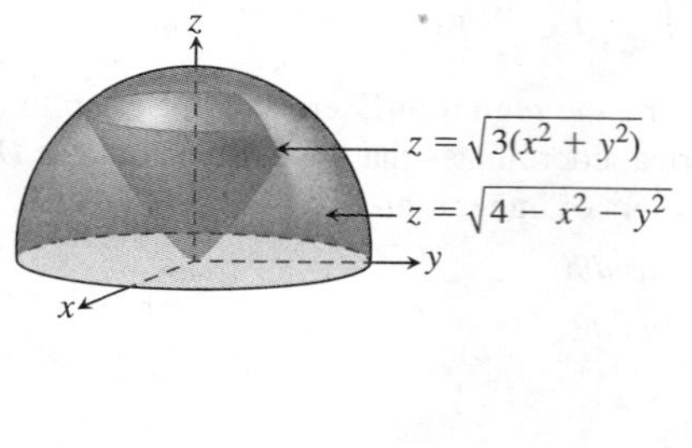

Finding Triple Integrals

61. Set up triple integrals for the volume of the sphere $\rho = 2$ in **(a)** spherical, **(b)** cylindrical, and **(c)** rectangular coordinates.

62. Let D be the solid region in the first octant that is bounded below by the cone $\phi = \pi/4$ and above by the sphere $\rho = 3$. Express the volume of D as an iterated triple integral in **(a)** cylindrical and **(b)** spherical coordinates. Then **(c)** find the volume.

63. Let D be the smaller cap cut from a solid ball of radius 2 units by a plane 1 unit from the center of the sphere. Express the volume of D as an iterated triple integral in **(a)** spherical, **(b)** cylindrical, and **(c)** rectangular coordinates. Then **(d)** find the volume by evaluating one of the three triple integrals.

64. Let D be the solid hemisphere $x^2 + y^2 + z^2 \leq 1$, $z \geq 0$. If the density is $\delta(x, y, z) = 1$, express the moment of intertia I_z as an iterated integral in **(a)** cylindrical and **(b)** spherical coordinates. Then **(c)** find I_z.

Volumes

Find the volumes of the solids in Exercises 65–70.

65.

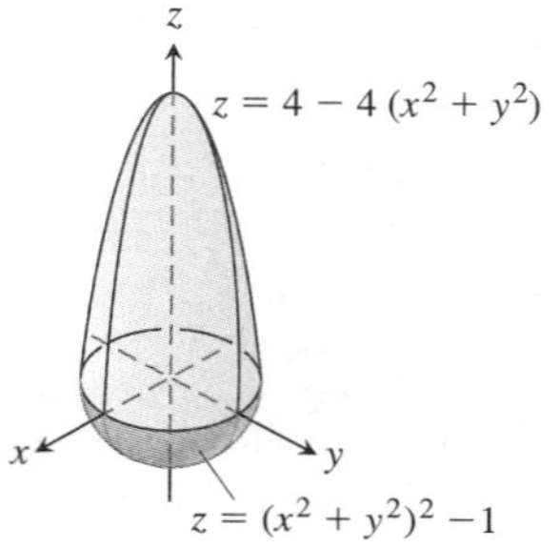

66.

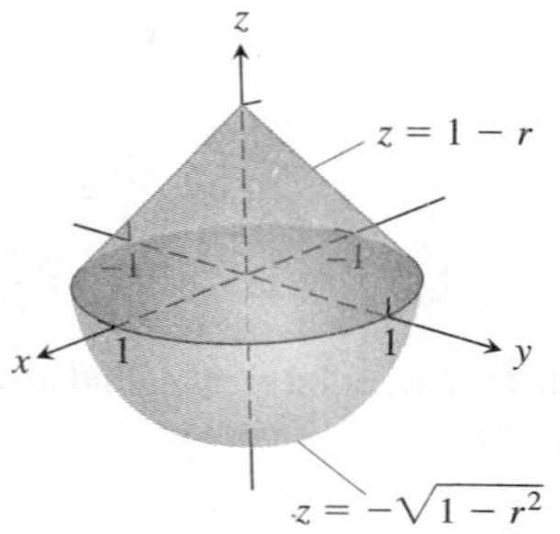

67.

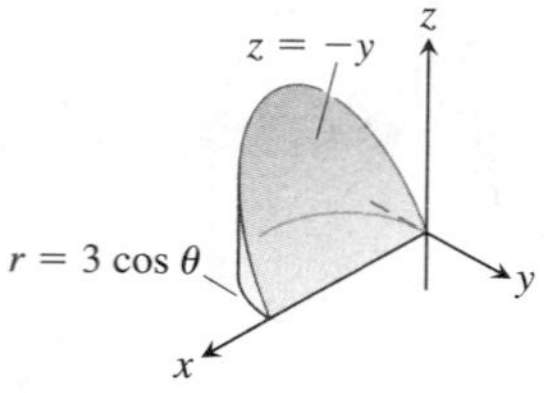

68.

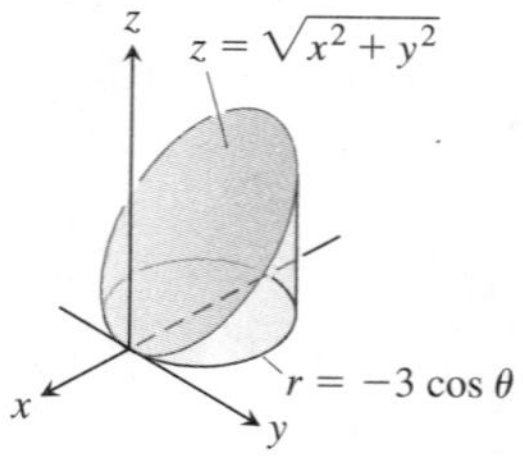

69.

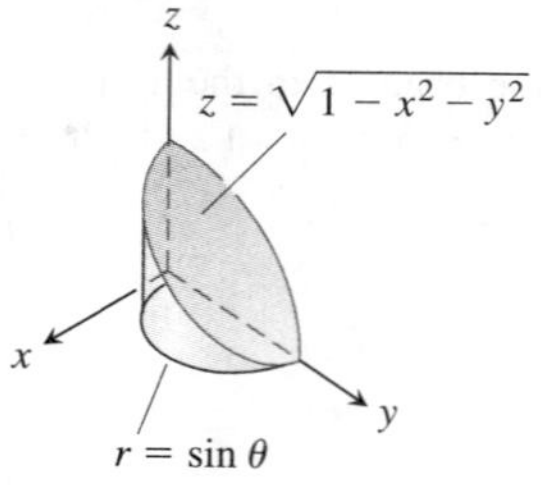

70.

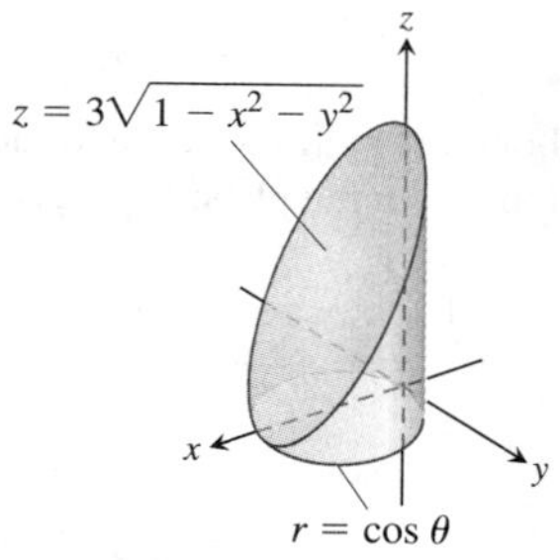

71. Ball and cones Find the volume of the portion of the ball $\rho \leq a$ that lies between the cones $\phi = \pi/3$ and $\phi = 2\pi/3$.

72. Ball and half-planes Find the volume of the region cut from the ball $\rho \leq a$ by the half-planes $\theta = 0$ and $\theta = \pi/6$ in the first octant.

73. **Ball and plane** Find the volume of the smaller region cut from the ball $\rho \leq \sqrt{2}$ by the plane $z = 1$.

74. **Cone and planes** Find the volume of the solid enclosed by the cone $z = \sqrt{x^2 + y^2}$ between the planes $z = 1$ and $z = 2$.

75. **Cone and paraboloid** Find the volume of solid bounded by the paraboloid $z = 2 - x^2 - y^2$ and conic surface $z = \sqrt{x^2 + y^2}$

76. **Cylinder and paraboloids** Find the volume of the solid region bounded below by the paraboloid $z = x^2 + y^2$, laterally by the cylinder $x^2 + y^2 = 1$, and above by the paraboloid $z = x^2 + y^2 + 1$.

77. **Cylinder and cones** Find the volume of the solid cut from the thick-walled cylinder $1 \leq x^2 + y^2 \leq 2$ by the cones $z = \pm\sqrt{x^2 + y^2}$.

78. **Sphere and cylinder** Find the volume of the solid region that lies inside the sphere $x^2 + y^2 + z^2 = 2$ and outside the cylinder $x^2 + y^2 = 1$.

79. **Cylinders** Find the volume of the solid, the region common to the interiors of the cylinder $x^2 + y^2 = 1$ and $x^2 + z^2 = 1$.

80. **Cylinder and planes** Find the volume of the region cut from the cylinder $x^2 + y^2 = 4$ by the plane $x + z = 3$.

81. **Region trapped by paraboloids** Find the volume of the solid formed by two paraboloids $z = x^2 + y^2$ and $z = 1 - x^2 - y^2$.

82. **Paraboloid and cylinder** Find the volume of the region bounded in the back by the plane $x = 0$, on the front and sides by the parabolic cylinder $x = 1 - y^2$, on the top by the paraboloid $z = x^2 + y^2$, on the bottom by the xy-plane.

83. **Cylinder and sphere** Find the volume of the region cut from the solid cylinder $x^2 + y^2 \leq 1$ by the sphere $x^2 + y^2 + z^2 = 4$.

84. **Sphere and paraboloid** Find the volume of the solid bounded by the sphere $x^2 + y^2 + z^2 = 6$ and paraboloid $x^2 + y^2 = z$.

Average Values

85. Find the average value of the function $f(r, \theta, z) = r$ over the solid region bounded by the cylinder $r = 1$ between the planes $z = -1$ and $z = 1$.

86. Find the average value of the function $f(r, \theta, z) = r$ over the solid ball bounded by the sphere $r^2 + z^2 = 1$. (This is the sphere $x^2 + y^2 + z^2 = 1$.)

87. Find the average value of the function $f(\rho, \phi, \theta) = \rho$ over the solid ball $\rho \leq 1$.

88. Find the average value of the function $f(\rho, \phi, \theta) = \rho \cos \phi$ over the upper half of the solid ball $\rho \leq 1$, $0 \leq \phi \leq \pi/2$.

Masses, Moments, and Centroids

89. **Center of mass** Find the center of mass of solid right circular cone of height h and radius of base b, if the density is equal to the distance from the base.

90. **Centroid** Find the centroid of the solid under $z^2 = xy$ and above the triangle bounded by $y = x$, $y = 0$ and $x = 4$.

91. **Centroid** Find the centroid of the solid in Exercise 60.

92. **Centroid** Find the centroid of the solid bounded above by the sphere $\rho = a$ and below by the cone $\phi = \pi/4$.

93. **Centroid** Find the centroid of the solid region that is bounded above by the surface $z = \sqrt{r}$, on the sides by the cylinder $r = 4$, and below by the xy-plane.

94. **Centroid** Find the centroid of the region cut from the solid ball $r^2 + z^2 \leq 1$ by the half-planes $\theta = -\pi/3$, $r \geq 0$, and $\theta = \pi/3$, $r \geq 0$.

95. **Moment of inertia of solid cone** Find the moment of inertia of a solid right circular cone of base radius 1 and height 1 about an axis through the vertex parallel to the base if the density is $\delta = 1$.

96. **Moment of inertia of ball** Find the moment of inertia of a ball of radius a about a diameter if the density is $\delta = 1$.

97. **Moment of inertia of solid cone** Find the moment of inertia of a solid right circular cone of base radius a and height h about its axis if the density is $\delta = 1$. (*Hint:* Place the cone with its vertex at the origin and its axis along the z-axis.)

98. **Variable density** A solid is bounded on the top by the paraboloid $z = r^2$, on the bottom by the plane $z = 0$, and on the sides by the cylinder $r = 1$. Find the center of mass and the moment of inertia about the z-axis if the density is

    a. $\delta(r, \theta, z) = z$

    b. $\delta(r, \theta, z) = r$.

99. **Variable density** A solid is bounded below by the cone $z = \sqrt{x^2 + y^2}$ and above by the plane $z = 1$. Find the center of mass and the moment of inertia about the z-axis if the density is

    a. $\delta(r, \theta, z) = z$

    b. $\delta(r, \theta, z) = z^2$.

100. **Variable density** A solid ball is bounded by the sphere $\rho = a$. Find the moment of inertia about the z-axis if the density is

    a. $\delta(\rho, \phi, \theta) = \rho^2$

    b. $\delta(\rho, \phi, \theta) = r = \rho \sin \phi$.

101. **Centroid of solid semi-ellipsoid** Show that the centroid of the solid semi-ellipsoid of revolution $(r^2/a^2) + (z^2/h^2) \leq 1$, $z \geq 0$, lies on the z-axis three-eighths of the way from the base to the top. The special case $h = a$ gives a solid hemisphere. Thus, the centroid of a solid hemisphere lies on the axis of symmetry three-eighths of the way from the base to the top.

102. **Centroid of solid cone** Show that the centroid of a solid right circular cone is one-fourth of the way from the base to the vertex. (In general, the centroid of a solid cone or pyramid is one-fourth of the way from the centroid of the base to the vertex.)

103. **Density of center of a planet** A planet is in the shape of a sphere of radius R and total mass M with spherically symmetric density distribution that increases linearly as one approaches its center. What is the density at the center of this planet if the density at its edge (surface) is taken to be zero?

104. **Mass of planet's atmosphere** A spherical planet of radius R has an atmosphere whose density is $\mu = \mu_0 e^{-ch}$, where h is the altitude above the surface of the planet, μ_0 is the density at sea level, and c is a positive constant. Find the mass of the planet's atmosphere.

Theory and Examples

105. Vertical planes in cylindrical coordinates

a. Show that planes perpendicular to the x-axis have equations of the form $r = a \sec \theta$ in cylindrical coordinates.

b. Show that planes perpendicular to the y-axis have equations of the form $r = b \csc \theta$.

106. (*Continuation of Exercise 105.*) Find an equation of the form $r = f(\theta)$ in cylindrical coordinates for the plane $ax + by = c$, $c \neq 0$.

107. Symmetry What symmetry will you find in a surface that has an equation of the form $r = f(z)$ in cylindrical coordinates? Give reasons for your answer.

108. Symmetry What symmetry will you find in a surface that has an equation of the form $\rho = f(\phi)$ in spherical coordinates? Give reasons for your answer.

15.8 Substitutions in Multiple Integrals

This section introduces the ideas involved in coordinate transformations to evaluate multiple integrals by substitution. The method replaces complicated integrals by ones that are easier to evaluate. Substitutions accomplish this by simplifying the integrand, the limits of integration, or both. A thorough discussion of multivariable transformations and substitutions is best left to a more advanced course, but our introduction here shows how the substitutions just studied reflect the general idea derived for single integral calculus.

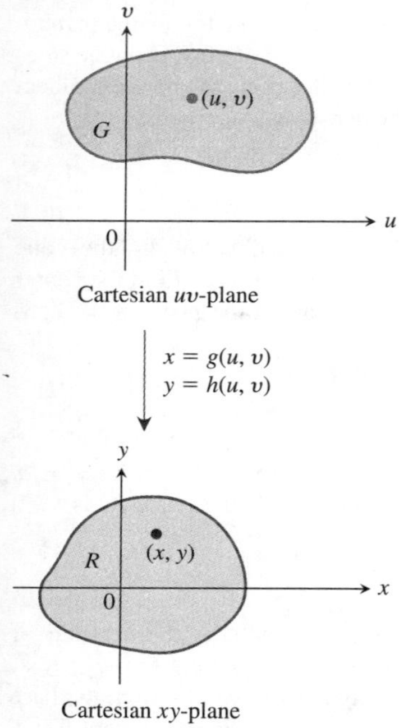

FIGURE 15.57 The equations $x = g(u, v)$ and $y = h(u, v)$ allow us to change an integral over a region R in the xy-plane into an integral over a region G in the uv-plane.

Substitutions in Double Integrals

The polar coordinate substitution of Section 15.4 is a special case of a more general substitution method for double integrals, a method that pictures changes in variables as transformations of regions.

Suppose that a region G in the uv-plane is transformed into the region R in the xy-plane by equations of the form

$$x = g(u, v), \qquad y = h(u, v),$$

as suggested in Figure 15.57. We assume the transformation is one-to-one on the interior of G. We call R the **image** of G under the transformation, and G the **preimage** of R. Any function $f(x, y)$ defined on R can be thought of as a function $f(g(u, v), h(u, v))$ defined on G as well. How is the integral of $f(x, y)$ over R related to the integral of $f(g(u, v), h(u, v))$ over G?

To gain some insight into the question, we look again at the single variable case. To be consistent with how we are using them now, we interchange the variables x and u used in the substitution method for single integrals in Chapter 5, so the equation is

$$\int_{g(a)}^{g(b)} f(x)\,dx = \int_a^b f(g(u))\,g'(u)\,du. \qquad x = g(u),\ dx = g'(u)du$$

To propose an analogue for substitution in a double integral $\iint_R f(x, y)\,dx\,dy$, we need a derivative factor like $g'(u)$ as a multiplier that transforms the area element $du\,dv$ in the region G to its corresponding area element $dx\,dy$ in the region R. We denote this factor by J. In continuing with our analogy, it is reasonable to assume that J is a function of both variables u and v, just as g' is a function of the single variable u. Moreover, J should register instantaneous change, so partial derivatives are going to be involved in its expression. Since four partial derivatives are associated with the transforming equations $x = g(u, v)$ and $y = h(u, v)$, it is also reasonable to assume that the factor $J(u, v)$ we seek includes them all. These features are captured in the following definition, which is constructed from the partial derivatives and is named after the German mathematician Carl Jacobi.

HISTORICAL BIOGRAPHY
Carl Gustav Jacob Jacobi
(1804–1851)
www.bit.ly/2xZS8Wi

DEFINITION The **Jacobian determinant** or **Jacobian** of the coordinate transformation $x = g(u, v)$, $y = h(u, v)$ is

$$J(u, v) = \begin{vmatrix} \dfrac{\partial x}{\partial u} & \dfrac{\partial x}{\partial v} \\ \dfrac{\partial y}{\partial u} & \dfrac{\partial y}{\partial v} \end{vmatrix} = \frac{\partial x}{\partial u}\frac{\partial y}{\partial v} - \frac{\partial y}{\partial u}\frac{\partial x}{\partial v}. \qquad (1)$$

The Jacobian can also be denoted by

$$J(u, v) = \frac{\partial(x, y)}{\partial(u, v)}$$

Differential Area Change Substituting $x = g(u, v)$, $y = h(u, v)$

$$dx\,dy = \left|\frac{\partial(x, y)}{\partial(u, v)}\right| du\,dv$$

to help us remember how the determinant in Equation (1) is constructed from the partial derivatives of x and y. The array of partial derivatives in Equation (1) behaves just like the derivative g' in the single variable situation. The Jacobian measures how much the transformation is expanding or contracting the area around the point (u, v). Effectively, the factor $|J|$ converts the area of the differential rectangle $du\,dv$ in G to match its corresponding differential area $dx\,dy$ in R. We note that, in general, the value of the scaling factor $|J|$ depends on the point (u, v) in G; that is, the scaling changes as the point (u, v) varies through the region G. Our examples to follow will show how it scales the differential area $du\,dv$ for specific transformations.

Now we can answer our original question concerning the relationship of the integral of $f(x, y)$ over the region R to the integral of $f(g(u, v), h(u, v))$ over G.

THEOREM 3—Substitution for Double Integrals
Suppose that $f(x, y)$ is continuous over the region R. Let G be the preimage of R under the transformation $x = g(u, v)$, $y = h(u, v)$, which is assumed to be one-to-one on the interior of G. If the functions g and h have continuous first partial derivatives within the interior of G, then

$$\iint_R f(x, y)\,dx\,dy = \iint_G f(g(u, v), h(u, v))\left|\frac{\partial(x, y)}{\partial(u, v)}\right| du\,dv. \qquad (2)$$

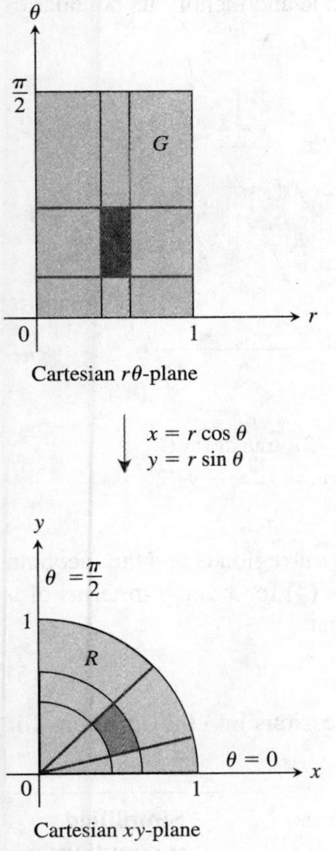

FIGURE 15.58 The equations $x = r\cos\theta$, $y = r\sin\theta$ transform G into R. The Jacobian factor r, calculated in Example 1, scales the differential rectangle $dr\,d\theta$ in G to match the differential area element $dx\,dy$ in R.

The derivation of Equation (2) is intricate and properly belongs to a course in advanced calculus, so we do not include it here. We now present examples illustrating the substitution method defined by the equation.

EXAMPLE 1 Find the Jacobian for the polar coordinate transformation $x = r\cos\theta$, $y = r\sin\theta$, and use Equation (2) to write the Cartesian integral $\iint_R f(x, y)\,dx\,dy$ as a polar integral.

Solution Figure 15.58 shows how the equations $x = r\cos\theta$, $y = r\sin\theta$ transform the rectangle $G: 0 \le r \le 1$, $0 \le \theta \le \pi/2$, into the quarter of a circular disk R bounded by $x^2 + y^2 = 1$ in the first quadrant of the xy-plane.

For polar coordinates, we have r and θ in place of u and v. With $x = r\cos\theta$ and $y = r\sin\theta$, the Jacobian is

$$J(r, \theta) = \begin{vmatrix} \dfrac{\partial x}{\partial r} & \dfrac{\partial x}{\partial \theta} \\ \dfrac{\partial y}{\partial r} & \dfrac{\partial y}{\partial \theta} \end{vmatrix} = \begin{vmatrix} \cos\theta & -r\sin\theta \\ \sin\theta & r\cos\theta \end{vmatrix} = r(\cos^2\theta + \sin^2\theta) = r.$$

Since we assume $r \geq 0$ when integrating in polar coordinates, $|J(r, \theta)| = |r| = r$ so that Equation (2) gives

$$\iint_R f(x, y)\, dx\, dy = \iint_G f(r\cos\theta, r\sin\theta)\, r\, dr\, d\theta. \tag{3}$$

This is the same formula we derived independently using a geometric argument for polar area in Section 15.4. ■

Here is an example of a substitution in which the image of a rectangle under the coordinate transformation is a trapezoid. Transformations like this one are called **linear transformations**, and their Jacobians are constant throughout G.

EXAMPLE 2 Evaluate

$$\int_0^4 \int_{x=y/2}^{x=(y/2)+1} \frac{2x - y}{2}\, dx\, dy$$

by applying the transformation

$$u = \frac{2x - y}{2}, \qquad v = \frac{y}{2} \tag{4}$$

and integrating over an appropriate region in the uv-plane.

Solution We sketch the region R of integration in the xy-plane and identify its boundaries (Figure 15.59).

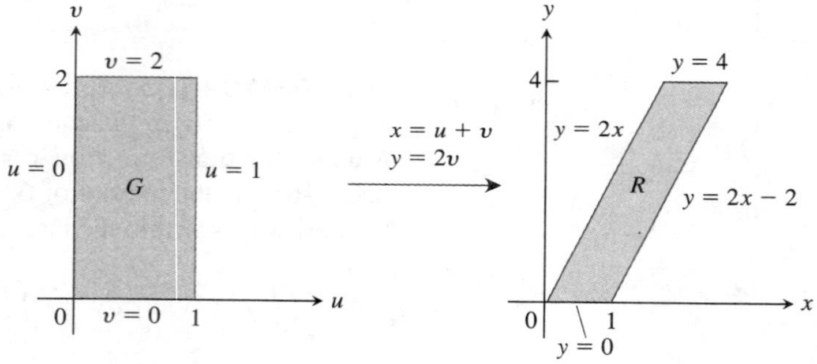

FIGURE 15.59 The equations $x = u + v$ and $y = 2v$ transform G into R. Reversing the transformation by the equations $u = (2x - y)/2$ and $v = y/2$ transforms R into G (Example 2).

To apply Equation (2), we need to find the corresponding uv-region G and the Jacobian of the transformation. To find them, we first solve Equations (4) for x and y in terms of u and v. From those equations it is easy to find algebraically that

$$x = u + v, \qquad y = 2v. \tag{5}$$

We then find the boundaries of G by substituting these expressions into the equations for the boundaries of R (Figure 15.59)

xy*-equations for the boundary of *R	**Corresponding *uv*-equations for the boundary of *G***	**Simplified *uv*-equations**
$x = y/2$	$u + v = 2v/2 = v$	$u = 0$
$x = (y/2) + 1$	$u + v = (2v/2) + 1 = v + 1$	$u = 1$
$y = 0$	$2v = 0$	$v = 0$
$y = 4$	$2v = 4$	$v = 2$

From Equations (5) the Jacobian of the transformation is

$$J(u,v) = \begin{vmatrix} \dfrac{\partial x}{\partial u} & \dfrac{\partial x}{\partial v} \\ \dfrac{\partial y}{\partial u} & \dfrac{\partial y}{\partial v} \end{vmatrix} = \begin{vmatrix} \dfrac{\partial}{\partial u}(u+v) & \dfrac{\partial}{\partial v}(u+v) \\ \dfrac{\partial}{\partial u}(2v) & \dfrac{\partial}{\partial v}(2v) \end{vmatrix} = \begin{vmatrix} 1 & 1 \\ 0 & 2 \end{vmatrix} = 2.$$

We now have everything we need to apply Equation (2):

$$\int_0^4 \int_{x=y/2}^{x=(y/2)+1} \frac{2x-y}{2}\,dx\,dy = \int_{v=0}^{v=2}\int_{u=0}^{u=1} u\,|J(u,v)|\,du\,dv$$

$$= \int_0^2 \int_0^1 (u)(2)\,du\,dv = \int_0^2 \Big[u^2\Big]_{u=0}^{u=1} dv = \int_0^2 dv = 2. \quad \blacksquare$$

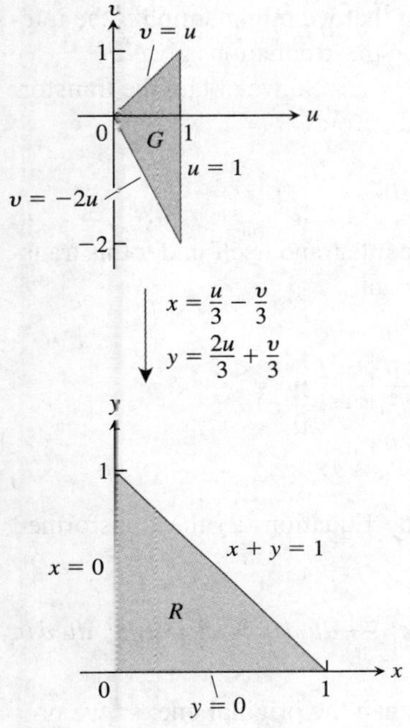

FIGURE 15.60 The equations $x = (u/3) - (v/3)$ and $y = (2u/3) + (v/3)$ transform G into R. Reversing the transformation by the equations $u = x + y$ and $v = y - 2x$ transforms R into G (Example 3).

EXAMPLE 3 Evaluate

$$\int_0^1 \int_0^{1-x} \sqrt{x+y}\,(y-2x)^2\,dy\,dx.$$

Solution We sketch the region R of integration in the xy-plane and identify its boundaries (Figure 15.60). The integrand suggests the transformation $u = x + y$ and $v = y - 2x$. Routine algebra produces x and y as functions of u and v:

$$x = \frac{u}{3} - \frac{v}{3}, \qquad y = \frac{2u}{3} + \frac{v}{3}. \tag{6}$$

From Equations (6), we can find the boundaries of the uv-region G (Figure 15.60).

xy*-equations for the boundary of *R	**Corresponding *uv*-equations for the boundary of *G***	**Simplified *uv*-equations**
$x + y = 1$	$\left(\frac{u}{3} - \frac{v}{3}\right) + \left(\frac{2u}{3} + \frac{v}{3}\right) = 1$	$u = 1$
$x = 0$	$\frac{u}{3} - \frac{v}{3} = 0$	$v = u$
$y = 0$	$\frac{2u}{3} + \frac{v}{3} = 0$	$v = -2u$

The Jacobian of the transformation in Equations (6) is

$$J(u,v) = \begin{vmatrix} \dfrac{\partial x}{\partial u} & \dfrac{\partial x}{\partial v} \\ \dfrac{\partial y}{\partial u} & \dfrac{\partial y}{\partial v} \end{vmatrix} = \begin{vmatrix} \dfrac{1}{3} & -\dfrac{1}{3} \\ \dfrac{2}{3} & \dfrac{1}{3} \end{vmatrix} = \frac{1}{3}.$$

Applying Equation (2), we evaluate the integral:

$$\int_0^1 \int_0^{1-x} \sqrt{x+y}\,(y-2x)^2\,dy\,dx = \int_{u=0}^{u=1}\int_{v=-2u}^{v=u} u^{1/2}v^2\,|J(u,v)|\,dv\,du$$

$$= \int_0^1 \int_{-2u}^{u} u^{1/2}v^2\left(\frac{1}{3}\right)dv\,du = \frac{1}{3}\int_0^1 u^{1/2}\left[\frac{1}{3}v^3\right]_{v=-2u}^{v=u} du$$

$$= \frac{1}{9}\int_0^1 u^{1/2}(u^3 + 8u^3)\,du = \int_0^1 u^{7/2}\,du = \frac{2}{9}u^{9/2}\Big]_0^1 = \frac{2}{9}. \quad \blacksquare$$

In the next example we illustrate a nonlinear transformation of coordinates resulting from simplifying the form of the integrand. Like the polar coordinates' transformation, nonlinear transformations can map a straight-line boundary of a region into a curved boundary (or vice versa with the inverse transformation). In general, nonlinear transformations are more complex to analyze than linear ones, and a complete treatment is left to a more advanced course.

EXAMPLE 4 Evaluate the integral

$$\int_1^2 \int_{1/y}^{y} \sqrt{\frac{y}{x}}\, e^{\sqrt{xy}}\, dx\, dy.$$

Solution The square root terms in the integrand suggest that we might simplify the integration by substituting $u = \sqrt{xy}$ and $v = \sqrt{y/x}$. Squaring these equations gives $u^2 = xy$ and $v^2 = y/x$, which imply that $u^2v^2 = y^2$ and $u^2/v^2 = x^2$. So we obtain the transformation (in the same ordering of the variables as discussed before)

$$x = \frac{u}{v} \quad \text{and} \quad y = uv,$$

with $u > 0$ and $v > 0$. Let's first see what happens to the integrand itself under this transformation. The Jacobian of the transformation is not constant:

$$J(u, v) = \begin{vmatrix} \dfrac{\partial x}{\partial u} & \dfrac{\partial x}{\partial v} \\[2mm] \dfrac{\partial y}{\partial u} & \dfrac{\partial y}{\partial v} \end{vmatrix} = \begin{vmatrix} \dfrac{1}{v} & \dfrac{-u}{v^2} \\[2mm] v & u \end{vmatrix} = \frac{2u}{v}.$$

If G is the region of integration in the uv-plane, then by Equation (2) the transformed double integral under the substitution is

$$\iint\limits_R \sqrt{\frac{y}{x}}\, e^{\sqrt{xy}}\, dx\, dy = \iint\limits_G ve^u |J(u, v)|\, du\, dv = \iint\limits_G ve^u \frac{2u}{v}\, du\, dv = \iint\limits_G 2ue^u\, du\, dv.$$

The transformed integrand function is easier to integrate than the original one, so we proceed to determine the limits of integration for the transformed integral.

The region of integration R of the original integral in the xy-plane is shown in Figure 15.61. From the substitution equations $u = \sqrt{xy}$ and $v = \sqrt{y/x}$, we see that the image of the left-hand boundary $xy = 1$ for R is the vertical line segment $u = 1$, $2 \geq v \geq 1$, in G (see Figure 15.62). Likewise, the right-hand boundary $y = x$ of R maps to the horizontal line segment $v = 1$, $1 \leq u \leq 2$, in G. Finally, the horizontal top boundary $y = 2$ of R maps to $uv = 2$, $1 \leq v \leq 2$, in G. As we move counterclockwise around the boundary of the region R, we also move counterclockwise around the boundary of G, as shown in Figure 15.62. Knowing the region of integration G in the uv-plane, we can now write equivalent iterated integrals:

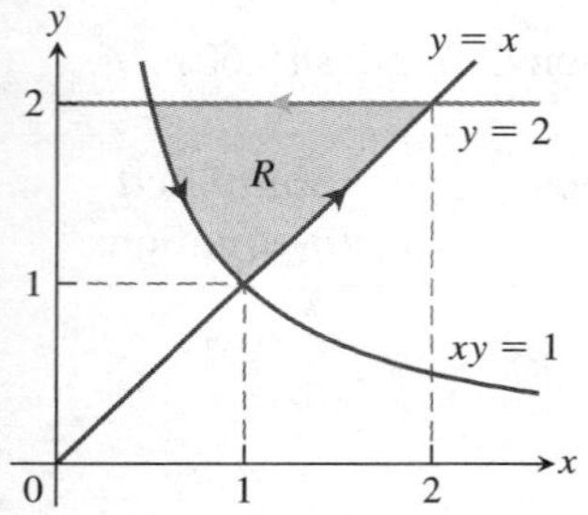

FIGURE 15.61 The region of integration R in Example 4.

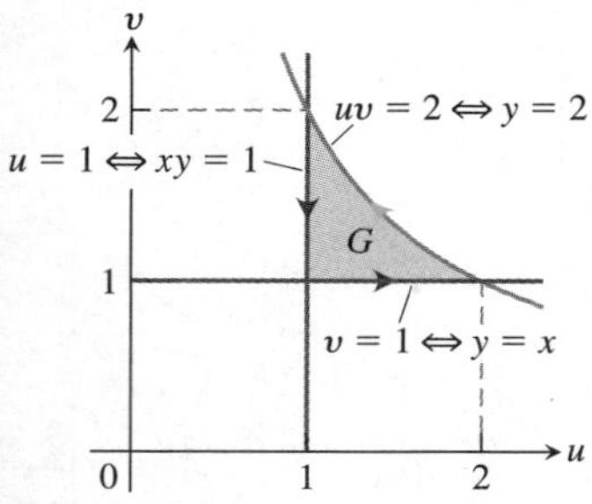

FIGURE 15.62 The boundaries of the region G correspond to those of region R in Figure 15.61. Notice that as we move counterclockwise around the region R, we move counterclockwise around the region G as well. The inverse transformation equations $u = \sqrt{xy}$, $v = \sqrt{y/x}$ produce the region G from the region R.

$$\int_1^2 \int_{1/y}^{y} \sqrt{\frac{y}{x}}\, e^{\sqrt{xy}}\, dx\, dy = \int_1^2 \int_1^{2/u} 2ue^u\, dv\, du. \qquad \text{Note the order of integration.}$$

We now evaluate the transformed integral on the right-hand side:

$$\begin{aligned} \int_1^2 \int_1^{2/u} 2ue^u\, dv\, du &= 2\int_1^2 \Big[vue^u \Big]_{v=1}^{v=2/u} du \\ &= 2\int_1^2 (2e^u - ue^u)\, du \\ &= 2\int_1^2 (2 - u)e^u\, du \\ &= 2\Big[(2 - u)e^u + e^u \Big]_{u=1}^{u=2} \qquad \text{Integrate by parts.} \\ &= 2(e^2 - (e + e)) = 2e(e - 2). \end{aligned}$$

■

Substitutions in Triple Integrals

The cylindrical and spherical coordinate substitutions in Section 15.7 are special cases of a substitution method that pictures changes of variables in triple integrals as transformations of solid regions. The method is like the method for double integrals given by Equation (2) except that now we work in three dimensions instead of two.

Suppose that a solid region G in uvw-space is transformed one-to-one into the solid region D in xyz-space by differentiable equations of the form

$$x = g(u, v, w), \qquad y = h(u, v, w), \qquad z = k(u, v, w),$$

as suggested in Figure 15.63. Then any function $F(x, y, z)$ defined on D can be thought of as a function

$$F(g(u, v, w), h(u, v, w), k(u, v, w)) = H(u, v, w)$$

defined on G. If g, h, and k have continuous first partial derivatives, then the integral of $F(x, y, z)$ over D is related to the integral of $H(u, v, w)$ over G by the equation

$$\iiint\limits_D F(x, y, z)\, dx\, dy\, dz = \iiint\limits_G H(u, v, w)|J(u, v, w)|\, du\, dv\, dw. \tag{7}$$

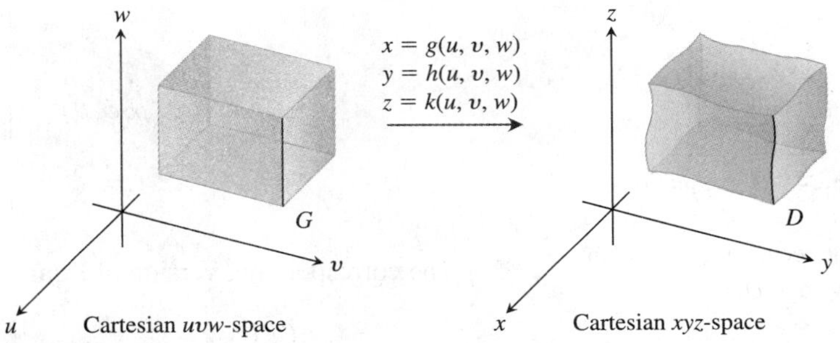

FIGURE 15.63 The equations $x = g(u, v, w)$, $y = h(u, v, w)$, and $z = k(u, v, w)$ allow us to change an integral over a region D in Cartesian xyz-space into an integral over a region G in Cartesian uvw-space using Equation (7).

The factor $J(u, v, w)$, whose absolute value appears in this equation, is the **Jacobian determinant**

$$J(u, v, w) = \begin{vmatrix} \frac{\partial x}{\partial u} & \frac{\partial x}{\partial v} & \frac{\partial x}{\partial w} \\ \frac{\partial y}{\partial u} & \frac{\partial y}{\partial v} & \frac{\partial y}{\partial w} \\ \frac{\partial z}{\partial u} & \frac{\partial z}{\partial v} & \frac{\partial z}{\partial w} \end{vmatrix} = \frac{\partial(x, y, z)}{\partial(u, v, w)}.$$

This determinant measures how much the volume near a point in G is being expanded or contracted by the transformation from (u, v, w) to (x, y, z) coordinates. As in the two-dimensional case, the derivation of the change-of-variable formula in Equation (7) is omitted.

For cylindrical coordinates, r, θ, and z take the place of u, v, and w. The transformation from Cartesian $r\theta z$-space to Cartesian xyz-space is given by the equations

$$x = r\cos\theta, \qquad y = r\sin\theta, \qquad z = z$$

Determinants

2×2 and 3×3 determinants are evaluated as follows:

$$\begin{vmatrix} a & b \\ c & d \end{vmatrix} = ad - bc$$

$$\begin{vmatrix} a_1 & a_2 & a_3 \\ b_1 & b_2 & b_3 \\ c_1 & c_2 & c_3 \end{vmatrix} = a_1 \begin{vmatrix} b_2 & b_3 \\ c_2 & c_3 \end{vmatrix} - a_2 \begin{vmatrix} b_1 & b_3 \\ c_1 & c_3 \end{vmatrix} + a_3 \begin{vmatrix} b_1 & b_2 \\ c_1 & c_2 \end{vmatrix}$$

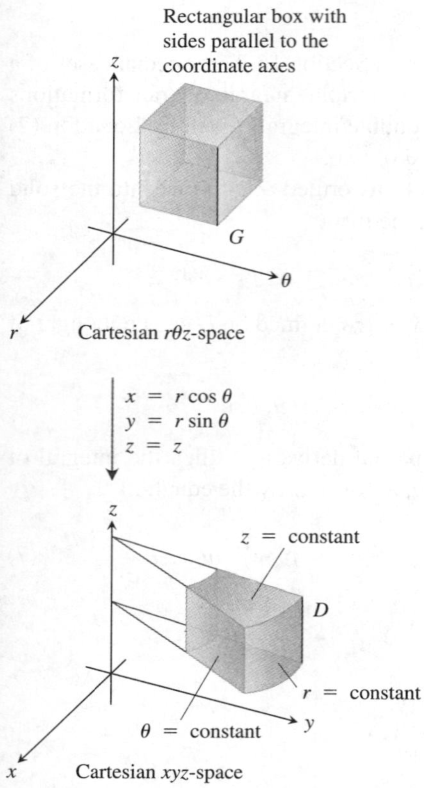

FIGURE 15.64 The equations $x = r\cos\theta$, $y = r\sin\theta$, and $z = z$ transform the rectangular box G into a cylindrical wedge D.

(Figure 15.64). The Jacobian of the transformation is

$$J(r,\theta,z) = \begin{vmatrix} \dfrac{\partial x}{\partial r} & \dfrac{\partial x}{\partial \theta} & \dfrac{\partial x}{\partial z} \\ \dfrac{\partial y}{\partial r} & \dfrac{\partial y}{\partial \theta} & \dfrac{\partial y}{\partial z} \\ \dfrac{\partial z}{\partial r} & \dfrac{\partial z}{\partial \theta} & \dfrac{\partial z}{\partial z} \end{vmatrix} = \begin{vmatrix} \cos\theta & -r\sin\theta & 0 \\ \sin\theta & r\cos\theta & 0 \\ 0 & 0 & 1 \end{vmatrix} = r\cos^2\theta + r\sin^2\theta = r.$$

The corresponding version of Equation (7) is

$$\iiint_D F(x,y,z)\,dx\,dy\,dz = \iiint_G H(r,\theta,z)\,|r|\,dr\,d\theta\,dz.$$

We can drop the absolute value signs because $r \geq 0$.

For spherical coordinates, ρ, ϕ, and θ take the place of u, v, and w. The transformation from Cartesian $\rho\phi\theta$-space to Cartesian xyz-space is given by

$$x = \rho\sin\phi\cos\theta, \qquad y = \rho\sin\phi\sin\theta, \qquad z = \rho\cos\phi$$

(Figure 15.65). The Jacobian of the transformation (see Exercise 23) is

$$J(\rho,\phi,\theta) = \begin{vmatrix} \dfrac{\partial x}{\partial \rho} & \dfrac{\partial x}{\partial \phi} & \dfrac{\partial x}{\partial \theta} \\ \dfrac{\partial y}{\partial \rho} & \dfrac{\partial y}{\partial \phi} & \dfrac{\partial y}{\partial \theta} \\ \dfrac{\partial z}{\partial \rho} & \dfrac{\partial z}{\partial \phi} & \dfrac{\partial z}{\partial \theta} \end{vmatrix} = \rho^2\sin\phi.$$

The corresponding version of Equation (7) is

$$\iiint_D F(x,y,z)\,dx\,dy\,dz = \iiint_G H(\rho,\phi,\theta)\,|\rho^2\sin\phi|\,d\rho\,d\phi\,d\theta.$$

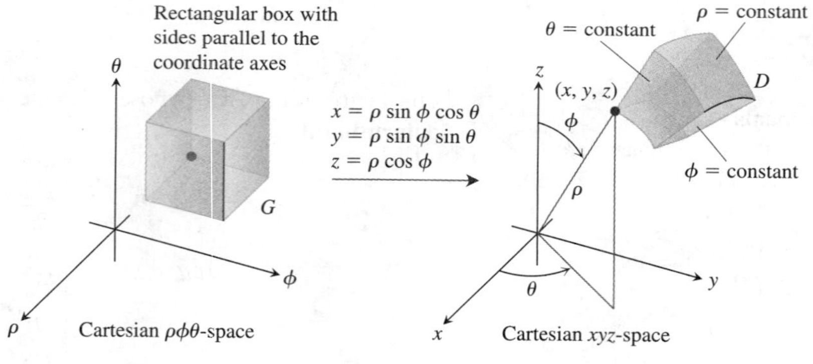

FIGURE 15.65 The equations $x = \rho\sin\phi\cos\theta$, $y = \rho\sin\phi\sin\theta$, and $z = \rho\cos\phi$ transform the rectangular box G into the spherical wedge D.

We can drop the absolute value signs because $\sin\phi$ is never negative for $0 \leq \phi \leq \pi$. Note that this is the same result we obtained in Section 15.7.

Here is an example of another substitution. Although we could evaluate the integral in this example directly, we have chosen it to illustrate the substitution method in a simple (and fairly intuitive) setting.

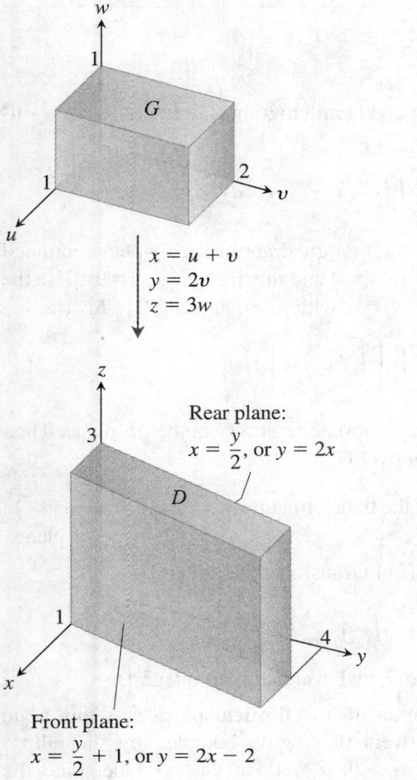

FIGURE 15.66 The equations $x = u + v$, $y = 2v$, and $z = 3w$ transform G into D. Reversing the transformation by the equations $u = (2x - y)/2$, $v = y/2$, and $w = z/3$ transforms D into G (Example 5).

EXAMPLE 5 Evaluate

$$\int_0^3 \int_0^4 \int_{x=y/2}^{x=(y/2)+1} \left(\frac{2x - y}{2} + \frac{z}{3} \right) dx\, dy\, dz$$

by applying the transformation

$$u = (2x - y)/2, \qquad v = y/2, \qquad w = z/3 \tag{8}$$

and integrating over an appropriate region in uvw-space.

Solution We sketch the solid region D of integration in xyz-space and identify its boundaries (Figure 15.66). In this case, the bounding surfaces are planes.

To apply Equation (7), we need to find the corresponding uvw-region G and the Jacobian of the transformation. To find them, we first solve Equations (8) for x, y, and z in terms of u, v, and w. Routine algebra gives

$$x = u + v, \qquad y = 2v, \qquad z = 3w. \tag{9}$$

We then find the boundaries of G by substituting these expressions into the equations for the boundaries of D.

xyz*-equations for the boundary of *D	**Corresponding *uvw*-equations for the boundary of *G***	**Simplified *uvw*-equations**
$x = y/2$	$u + v = 2v/2 = v$	$u = 0$
$x = (y/2) + 1$	$u + v = (2v/2) + 1 = v + 1$	$u = 1$
$y = 0$	$2v = 0$	$v = 0$
$y = 4$	$2v = 4$	$v = 2$
$z = 0$	$3w = 0$	$w = 0$
$z = 3$	$3w = 3$	$w = 1$

The Jacobian of the transformation, again from Equations (9), is

$$J(u, v, w) = \begin{vmatrix} \dfrac{\partial x}{\partial u} & \dfrac{\partial x}{\partial v} & \dfrac{\partial x}{\partial w} \\ \dfrac{\partial y}{\partial u} & \dfrac{\partial y}{\partial v} & \dfrac{\partial y}{\partial w} \\ \dfrac{\partial z}{\partial u} & \dfrac{\partial z}{\partial v} & \dfrac{\partial z}{\partial w} \end{vmatrix} = \begin{vmatrix} 1 & 1 & 0 \\ 0 & 2 & 0 \\ 0 & 0 & 3 \end{vmatrix} = 6.$$

We now have everything we need to apply Equation (7):

$$\begin{aligned}
&\int_0^3 \int_0^4 \int_{x=y/2}^{x=(y/2)+1} \left(\frac{2x - y}{2} + \frac{z}{3} \right) dx\, dy\, dz \\
&= \int_0^1 \int_0^2 \int_0^1 (u + w)\,|J(u, v, w)|\, du\, dv\, dw \\
&= \int_0^1 \int_0^2 \int_0^1 (u + w)(6)\, du\, dv\, dw = 6 \int_0^1 \int_0^2 \left[\frac{u^2}{2} + uw \right]_{u=0}^{u=1} dv\, dw \\
&= 6 \int_0^1 \int_0^2 \left(\frac{1}{2} + w \right) dv\, dw = 6 \int_0^1 \left[\frac{v}{2} + vw \right]_{v=0}^{v=2} dw = 6 \int_0^1 (1 + 2w)\, dw \\
&= 6 \left[w + w^2 \right]_0^1 = 6(2) = 12.
\end{aligned}$$

■

EXERCISES 15.8

Jacobians and Transformed Regions in the Plane

1. a. Solve the system

$$u = x - y, \qquad v = 2x + y$$

for x and y in terms of u and v. Then find the value of the Jacobian $\partial(x, y)/\partial(u, v)$.

b. Find the image under the transformation $u = x - y$, $v = 2x + y$ of the triangular region with vertices $(0, 0)$, $(1, 1)$, and $(1, -2)$ in the xy-plane. Sketch the transformed region in the uv-plane.

2. a. Solve the system

$$u = x + 2y, \qquad v = x - y$$

for x and y in terms of u and v. Then find the value of the Jacobian $\partial(x, y)/\partial(u, v)$.

b. Find the image under the transformation $u = x + 2y$, $v = x - y$ of the triangular region in the xy-plane bounded by the lines $y = 0$, $y = x$, and $x + 2y = 2$. Sketch the transformed region in the uv-plane.

3. a. Solve the system

$$u = 3x + 2y, \qquad v = x + 4y$$

for x and y in terms of u and v. Then find the value of the Jacobian $\partial(x, y)/\partial(u, v)$.

b. Find the image under the transformation $u = 3x + 2y$, $v = x + 4y$ of the triangular region in the xy-plane bounded by the x-axis, the y-axis, and the line $x + y = 1$. Sketch the transformed region in the uv-plane.

4. a. Solve the system

$$u = 2x - 3y, \qquad v = -x + y$$

for x and y in terms of u and v. Then find the value of the Jacobian $\partial(x, y)/\partial(u, v)$.

b. Find the image under the transformation $u = 2x - 3y$, $v = -x + y$ of the parallelogram R in the xy-plane with boundaries $x = -3$, $x = 0$, $y = x$, and $y = x + 1$. Sketch the transformed region in the uv-plane.

Substitutions in Double Integrals

5. Evaluate the integral

$$\int_0^4 \int_{x=y/2}^{x=(y/2)+1} \frac{2x - y}{2}\, dx\, dy$$

from Example 1 directly by integration with respect to x and y to confirm that its value is 2.

6. Use the transformation in Exercise 1 to evaluate the integral

$$\iint_R (2x^2 - xy - y^2)\, dx\, dy$$

for the region R in the first quadrant bounded by the lines $y = -2x + 4$, $y = -2x + 7$, $y = x - 2$, and $y = x + 1$.

7. Use the transformation in Exercise 3 to evaluate the integral

$$\iint_R (3x^2 + 14xy + 8y^2)\, dx\, dy$$

for the region R bounded by the lines $y = -(3/2)x + 1$, $y = -(3/2)x + 3$, $y = -(1/4)x$, and $y = -(1/4)x + 1$.

8. Use the transformation and parallelogram R in Exercise 4 to evaluate the integral

$$\iint_R 2(x - y)\, dx\, dy.$$

9. Let R be the region in the first quadrant of the xy-plane bounded by the hyperbolas $xy = 1$, $xy = 4$ and the lines $y = x$, $y = 9x$. Use the transformation $x = u/v$, $y = uv$ with $u > 0$ and $v > 0$ to rewrite

$$\iint_R \left(\sqrt{\frac{y}{x}} + \sqrt{xy}\right) dx\, dy$$

as an integral over an appropriate region G in the uv-plane. Then evaluate the uv-integral over G.

10. a. Find the Jacobian of the transformation $x = u$, $y = uv$ and sketch the region G: $1 \le u \le 2$, $1 \le uv \le 2$, in the uv-plane.

b. Then use Equation (2) to transform the integral

$$\int_1^2 \int_1^2 \frac{y}{x}\, dy\, dx$$

into an integral over G, and evaluate both integrals.

11. Polar moment of inertia of an elliptical plate A thin plate of constant density covers the region bounded by the ellipse $x^2/a^2 + y^2/b^2 = 1$, $a > 0$, $b > 0$, in the xy-plane. Find the first moment of the plate about the origin. (*Hint:* Use the transformation $x = ar\cos\theta$, $y = br\sin\theta$.)

12. The area of an ellipse The area πab of the ellipse $x^2/a^2 + y^2/b^2 = 1$ can be found by integrating the function $f(x, y) = 1$ over the region bounded by the ellipse in the xy-plane. Evaluating the integral directly requires a trigonometric substitution. An easier way to evaluate the integral is to use the transformation $x = au$, $y = bv$ and evaluate the transformed integral over the disk G: $u^2 + v^2 \le 1$ in the uv-plane. Find the area this way.

13. Use the transformation in Exercise 2 to evaluate the integral

$$\int_0^{2/3} \int_y^{2-2y} (x + 2y)\, e^{(y-x)}\, dx\, dy$$

by first writing it as an integral over a region G in the uv-plane.

14. Use the transformation $x = u + (1/2)v$, $y = v$ to evaluate the integral

$$\int_0^2 \int_{y/2}^{(y+4)/2} y^3 (2x - y)\, e^{(2x-y)^2}\, dx\, dy$$

by first writing it as an integral over a region G in the uv-plane.

15. Use the transformation $u = \dfrac{2x - y}{2}$, $v = \dfrac{y}{2}$ and $w = \dfrac{z}{3}$ to evaluate the integral

$$\int_0^3 \int_0^4 \int_{x=y/2}^{x=(y/2)+1} \left(\frac{2x - y}{2} + \frac{z}{3}\right) dx\, dy\, dz$$

16. Use the transformation $x = u^2 - v^2$, $y = 2uv$ to evaluate the integral

$$\int_0^1 \int_0^{2\sqrt{1-x}} \sqrt{x^2 + y^2}\, dy\, dx.$$

(*Hint:* Show that the image of the triangular region G with vertices $(0, 0)$, $(1, 0)$, $(1, 1)$ in the uv-plane is the region of integration R in the xy-plane defined by the limits of integration.)

Substitutions in Triple Integrals

17. Evaluate the integral in Example 5 by integrating with respect to x, y, and z.

18. Volume of a solid ellipsoid Find the volume of the solid ellipsoid

$$\frac{x^2}{a^2} + \frac{y^2}{b^2} + \frac{z^2}{c^2} \leq 1.$$

(*Hint:* Let $x = au$, $y = bv$, and $z = cw$. Then find the volume of an appropriate region in uvw-space.)

19. Evaluate

$$\iiint_D |xyz|\, dx\, dy\, dz$$

over the solid ellipsoid D,

$$\frac{x^2}{a^2} + \frac{y^2}{b^2} + \frac{z^2}{c^2} \leq 1.$$

(*Hint:* Let $x = au$, $y = bv$, and $z = cw$. Then integrate over an appropriate region in uvw-space.)

20. Let D be the region in xyz-space defined by the inequalities

$$0 \leq x \leq 1, \quad 0 \leq xy \leq 2, \quad 1 \leq z \leq 2.$$

Evaluate

$$\iiint_D (x^2y + 3xyz)\, dx\, dy\, dz$$

by applying the transformation

$$u = 3x, \quad v = xy, \quad w = z$$

and integrating over an appropriate region G in uvw-space.

Theory and Examples

21. Find the Jacobian $\partial(x, y)/\partial(u, v)$ of the transformation

a. $x = v \cos u, \quad y = v \sin u$

b. $x = u \sin v, \quad y = u \cos v.$

22. Find the Jacobian $\partial(x, y, z)/\partial(u, v, w)$ of the transformation

a. $x = u \cos v, \quad y = u \sin v, \quad z = w$

b. $x = 3u + 1, \quad y = 2v - 3, \quad z = 1/3\ (w)$

23. Evaluate the appropriate determinant to show that the Jacobian of the transformation from Cartesian $\rho\phi\theta$-space to Cartesian xyz-space is $\rho^2 \sin \phi$.

24. Substitutions in single integrals How can substitutions in single definite integrals be viewed as transformations of regions? What is the Jacobian in such a case? Illustrate with an example.

25. Centroid of a solid semi-ellipsoid Assuming the result that the centroid of a solid hemisphere lies on the axis of symmetry three-eighths of the way from the base toward the top, show, by transforming the appropriate integrals, that the center of mass of a solid semi-ellipsoid $(x^2/a^2) + (y^2/b^2) + (z^2/c^2) \leq 1$, $z \geq 0$, lies on the z-axis three-eighths of the way from the base toward the top. (You can do this without evaluating any of the integrals.)

26. Cylindrical shells In Section 6.2, we learned how to find the volume of a solid of revolution using the shell method. Specifically, if the region between the curve $y = f(x)$ and the x-axis from a to b $(0 < a < b)$ is revolved about the y-axis, the volume of the resulting solid is $\int_a^b 2\pi x f(x)\, dx$. Prove that finding volumes by using triple integrals gives the same result. (*Hint:* Use cylindrical coordinates with the roles of y and z changed.)

27. Inverse transform The equations $x = g(u, v)$, $y = h(u, v)$ in Figure 15.57 transform the region G in the uv-plane into the region R in the xy-plane. Since the substitution transformation is one-to-one with continuous first partial derivatives, it has an inverse transformation, and there are equations $u = \alpha(x, y)$, $v = \beta(x, y)$ with continuous first partial derivatives transforming R back into G. Moreover, the Jacobian determinants of the transformations are related reciprocally by

$$\frac{\partial(x, y)}{\partial(u, v)} = \left(\frac{\partial(u, v)}{\partial(x, y)}\right)^{-1}. \tag{10}$$

Equation (10) is proved in advanced calculus. Use it to find the area of the region R in the first quadrant of the xy-plane bounded by the lines $y = 2x$, $2y = x$, and the curves $xy = 2$, $2xy = 1$ for $u = xy$ and $v = y/x$.

28. (*Continuation of Exercise 27.*) For the region R described in Exercise 27, evaluate the integral $\iint_R y^2\, dA$.

CHAPTER 15 Questions to Guide Your Review

1. Define the double integral of a function of two variables over a bounded region in the coordinate plane.
2. How are double integrals evaluated as iterated integrals? Does the order of integration matter? How are the limits of integration determined? Give examples.
3. How are double integrals used to calculate areas and average values. Give examples.
4. How can you change a double integral in rectangular coordinates into a double integral in polar coordinates? Why might it be worthwhile to do so? Give an example.
5. Define the triple integral of a function $f(x, y, z)$ over a bounded solid region in space.
6. How are triple integrals in rectangular coordinates evaluated? How are the limits of integration determined? Give an example.
7. How are double and triple integrals in rectangular coordinates used to calculate volumes, average values, masses, moments, and centers of mass? Give examples.
8. How are triple integrals defined in cylindrical and spherical coordinates? Why might one prefer working in one of these coordinate systems to working in rectangular coordinates?
9. How are triple integrals in cylindrical and spherical coordinates evaluated? How are the limits of integration found? Give examples.
10. How are substitutions in double integrals pictured as transformations of regions in the plane? Give a sample calculation.
11. How are substitutions in triple integrals pictured as transformations of solid regions? Give a sample calculation.

CHAPTER 15 Practice Exercises

Evaluating Double Iterated Integrals

In Exercises 1–4, sketch the region of integration and evaluate the double integral.

1. $\int_1^{10} \int_0^{1/y} ye^{xy}\, dx\, dy$

2. $\int_0^1 \int_0^{x^3} e^{y/x}\, dy\, dx$

3. $\int_0^{3/2} \int_{-\sqrt{9-4t^2}}^{\sqrt{9-4t^2}} t\, ds\, dt$

4. $\int_0^1 \int_{\sqrt{y}}^{2-\sqrt{y}} xy\, dx\, dy$

In Exercises 5–8, sketch the region of integration and write an equivalent integral with the order of integration reversed. Then evaluate both integrals.

5. $\int_0^4 \int_{-\sqrt{4-y}}^{(y-4)/2} dx\, dy$

6. $\int_0^1 \int_{x^2}^{x} \sqrt{x}\, dy\, dx$

7. $\int_0^{3/2} \int_{-\sqrt{9-4y^2}}^{\sqrt{9-4y^2}} y\, dx\, dy$

8. $\int_0^2 \int_0^{4-x^2} 2x\, dy\, dx$

Evaluate the integrals in Exercises 9–12.

9. $\int_0^1 \int_{2y}^{2} 4\cos(x^2)\, dx\, dy$

10. $\int_0^2 \int_{y/2}^{1} e^{x^2}\, dx\, dy$

11. $\int_0^8 \int_{\sqrt[3]{x}}^{2} \frac{dy\, dx}{y^4 + 1}$

12. $\int_0^1 \int_{\sqrt[3]{y}}^{1} \frac{2\pi \sin \pi x^2}{x^2}\, dx\, dy$

Areas and Volumes Using Double Integrals

13. **Area between line and parabola** Find the area of the region enclosed by the line $y = 2x + 4$ and the parabola $y = 4 - x^2$ in the xy-plane.

14. **Area bounded by lines and parabola** Find the area of the "triangular" region in the xy-plane that is bounded on the right by the parabola $y = x^2$, on the left by the line $x + y = 2$, and above by the line $y = 4$.

15. **Volume of the region under a paraboloid** Find the volume under the paraboloid $z = x^2 + y^2$ above the triangle enclosed by the lines $y = x$, $x = 0$, and $x + y = 2$ in the xy-plane.

16. **Volume of the region under a parabolic cylinder** Find the volume under the parabolic cylinder $z = x^2$ above the region enclosed by the parabola $y = 6 - x^2$ and the line $y = x$ in the xy-plane.

Average Values

Find the average value of $f(x, y) = xy$ over the regions in Exercises 17 and 18.

17. The square bounded by the lines $x = 1$, $y = 1$ in the first quadrant

18. The quarter circle $x^2 + y^2 \le 1$ in the first quadrant

Polar Coordinates

Evaluate the integrals in Exercises 19 and 20 by changing to polar coordinates.

19. $\int_{-1}^1 \int_{-\sqrt{1-x^2}}^{\sqrt{1-x^2}} \frac{2\, dy\, dx}{(1 + x^2 + y^2)^2}$

20. $\int_{-1}^1 \int_{-\sqrt{1-y^2}}^{\sqrt{1-y^2}} \ln(x^2 + y^2 + 1)\, dx\, dy$

21. **Integrating over a lemniscate** Integrate the function $f(x, y) = 1/(1 + x^2 + y^2)^2$ over the region enclosed by one loop of the lemniscate $(x^2 + y^2)^2 - (x^2 - y^2) = 0$.

22. Integrate $f(x, y) = 1/(1 + x^2 + y^2)^2$ over

    a. **Triangular region** The triangle with vertices $(0, 0)$, $(1, 0)$, and $(1, \sqrt{3})$.

    b. **First quadrant** The first quadrant of the xy-plane.

Evaluating Triple Iterated Integrals

Evaluate the integrals in Exercises 23–26.

23. $\int_0^\pi \int_0^\pi \int_0^\pi \cos(x + y + z)\, dx\, dy\, dz$

24. $\int_{\ln 6}^{\ln 7} \int_0^{\ln 2} \int_{\ln 4}^{\ln 5} e^{(x+y+z)}\, dz\, dy\, dx$

25. $\int_0^1 \int_0^{x^2} \int_0^{x+y} (2x - y - z)\, dz\, dy\, dx$

26. $\int_1^e \int_1^x \int_0^z \frac{2y}{z^3}\, dy\, dz\, dx$

Volumes and Average Values Using Triple Integrals

27. **Volume** Find the volume of the wedge-shaped solid region enclosed on the side by the cylinder $x = -\cos y$, $-\pi/2 \le y \le \pi/2$, on the top by the plane $z = -2x$, and below by the xy-plane.

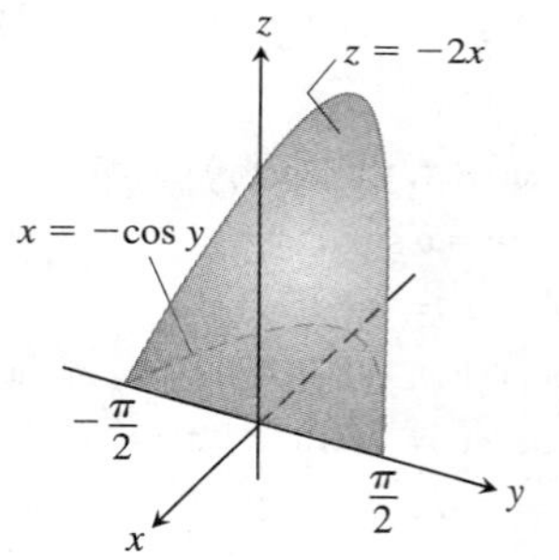

28. **Volume** Find the volume of the solid that is bounded above by the cylinder $z = 4 - x^2$, on the sides by the cylinder $x^2 + y^2 = 4$, and below by the xy-plane.

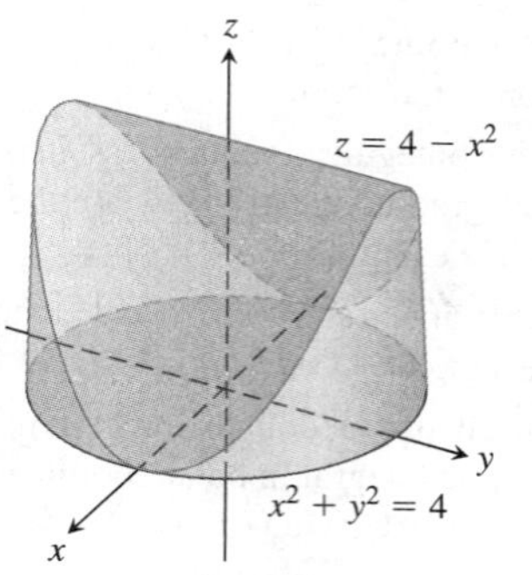

29. **Average value** Find the average value of $f(x, y, z) = 30xz\sqrt{x^2 + y}$ over the rectangular solid in the first octant bounded by the coordinate planes and the planes $x = 1$, $y = 3$, $z = 1$.

30. Average value Find the average value of ρ over the ball $\rho \le a$ (spherical coordinates).

Cylindrical and Spherical Coordinates

31. Cylindrical to rectangular coordinates Convert

$$\int_0^{2\pi}\int_0^{\sqrt{2}}\int_r^{\sqrt{4-r^2}} 3r\,dz\,dr\,d\theta, \qquad r \ge 0$$

to **(a)** rectangular coordinates with the order of integration $dz\,dx\,dy$ and **(b)** spherical coordinates. Then **(c)** evaluate one of the integrals.

32. Rectangular to cylindrical coordinates **(a)** Convert to cylindrical coordinates. Then **(b)** evaluate the new integral.

$$\int_0^1\int_{-\sqrt{1-x^2}}^{\sqrt{1-x^2}}\int_{-(x^2+y^2)}^{(x^2+y^2)} 21xy^2\,dz\,dy\,dx$$

33. Rectangular to spherical coordinates **(a)** Convert to spherical coordinates. Then **(b)** evaluate the new integral.

$$\int_{-1}^1\int_{-\sqrt{1-x^2}}^{\sqrt{1-x^2}}\int_{\sqrt{x^2+y^2}}^1 dz\,dy\,dx$$

34. Rectangular, cylindrical, and spherical coordinates Write an iterated triple integral for the integral of $f(x, y, z) = 6 + 4y$ over the region in the first octant bounded by the cone $z = \sqrt{x^2 + y^2}$, the cylinder $x^2 + y^2 = 1$, and the coordinate planes in **(a)** rectangular coordinates, **(b)** cylindrical coordinates, and **(c)** spherical coordinates. Then **(d)** find the integral of f by evaluating one of the triple integrals.

35. Cylindrical to rectangular coordinates Set up an integral in rectangular coordinates equivalent to the integral

$$\int_0^{\pi/2}\int_1^{\sqrt{3}}\int_1^{\sqrt{4-r^2}} r^3(\sin\theta\cos\theta)z^2\,dz\,dr\,d\theta.$$

Arrange the order of integration to be z first, then y, then x.

36. Rectangular to cylindrical coordinates The volume of a solid is

$$\int_0^2\int_0^{\sqrt{2x-x^2}}\int_{-\sqrt{4-x^2-y^2}}^{\sqrt{4-x^2-y^2}} dz\,dy\,dx.$$

a. Describe the solid by giving equations for the surfaces that form its boundary.

b. Convert the integral to cylindrical coordinates, but do not evaluate the integral.

37. Spherical versus cylindrical coordinates Triple integrals involving spherical shapes do not always require spherical coordinates for convenient evaluation. Some calculations may be accomplished more easily with cylindrical coordinates. As a case in point, find the volume of the solid region bounded above by the sphere $x^2 + y^2 + z^2 = 8$ and below by the plane $z = 2$ by using **(a)** cylindrical coordinates and **(b)** spherical coordinates.

Masses and Moments

38. Finding I_z in spherical coordinates Find the moment of inertia about the z-axis of a solid of constant density $\delta = 1$ that is bounded above by the sphere $\rho = 2$ and below by the cone $\phi = \pi/3$ (spherical coordinates).

39. Moment of inertia of a "thick" sphere Find the moment of inertia of a solid of constant density δ bounded by two concentric spheres of radii a and $b(a < b)$ about a diameter.

40. Moment of inertia of an apple Find the moment of inertia about the z-axis of a solid of density $\delta = 1$ enclosed by the spherical coordinate surface $\rho = 1 - \cos\phi$. The solid is the curve rotated about the z-axis in the accompanying figure.

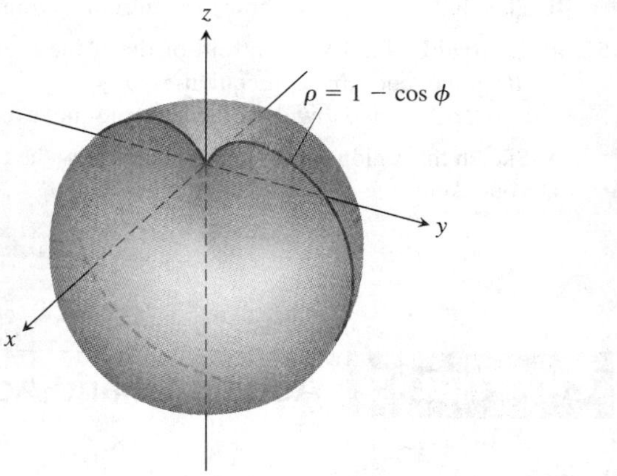

41. Centroid Find the centroid of the "triangular" region bounded by the lines $x = 2$, $y = 2$ and the hyperbola $xy = 2$ in the xy-plane.

42. Centroid Find the centroid of the region between the parabola $x + y^2 - 2y = 0$ and the line $x + 2y = 0$ in the xy-plane.

43. Polar moment Find the polar moment of inertia about the origin of a thin triangular plate of constant density $\delta = 3$ bounded by the y-axis and the lines $y = 2x$ and $y = 4$ in the xy-plane.

44. Polar moment Find the polar moment of inertia about the center of a thin rectangular sheet of constant density $\delta = 1$ bounded by the lines

a. $x = \pm 2$, $y = \pm 1$ in the xy-plane

b. $x = \pm a$, $y = \pm b$ in the xy-plane.

(*Hint:* Find I_x. Then use the formula for I_x to find I_y, and add the two to find I_0.)

45. Inertial moment Find the moment of inertia about the x-axis of a thin plate of constant density δ covering the triangle with vertices $(0, 0)$, $(3, 0)$, and $(3, 2)$ in the xy-plane.

46. Plate with variable density Find the center of mass and the moments of inertia about the coordinate axes of a thin plate bounded by the line $y = x$ and the parabola $y = x^2$ in the xy-plane if the density is $\delta(x, y) = x + 1$.

47. Plate with variable density Find the mass and first moments about the coordinate axes of a thin square plate bounded by the lines $x = \pm 1$, $y = \pm 1$ in the xy-plane if the density is $\delta(x, y) = x^2 + y^2 + 1/3$.

48. Triangles with same inertial moment Find the moment of inertia about the x-axis of a thin triangular plate of constant density δ whose base lies along the interval $[0, b]$ on the x-axis and whose vertex lies on the line $y = h$ above the x-axis. As you will see, it does not matter where on the line this vertex lies. All such triangles have the same moment of inertia about the x-axis.

49. Centroid Find the centroid of the region in the polar coordinate plane defined by the inequalities $0 \le r \le 3$, $-\pi/3 \le \theta \le \pi/3$.

50. Centroid Find the centroid of the region in the first quadrant bounded by the rays $\theta = 0$ and $\theta = \pi/2$ and the circles $r = 1$ and $r = 3$.

51. a. Centroid Find the centroid of the region in the polar coordinate plane that lies inside the cardioid $r = 1 + \cos\theta$ and outside the circle $r = 1$.

b. Sketch the region and show the centroid in your sketch.

52. a. Centroid Find the centroid of the plane region defined by the polar coordinate inequalities $0 \le r \le a, -\alpha \le \theta \le \alpha$ $(0 < \alpha \le \pi)$. How does the centroid move as $\alpha \to \pi^-$?

b. Sketch the region for $\alpha = 5\pi/6$ and show the centroid in your sketch.

Substitutions

53. Show that if $u = x - y$ and $v = y$, then for any continuous f,

$$\int_0^\infty \int_0^x e^{-sx} f(x - y, y)\, dy\, dx = \int_0^\infty \int_0^\infty e^{-s(u+v)} f(u, v)\, du\, dv.$$

54. What relationship must hold between the constants a, b, and c to make

$$\int_{-\infty}^\infty \int_{-\infty}^\infty e^{-(ax^2+2bxy+cy^2)}\, dx\, dy = 1?$$

(*Hint:* Let $s = \alpha x + \beta y$ and $t = \gamma x + \delta y$, where $(\alpha\delta - \beta\gamma)^2 = ac - b^2$. Then $ax^2 + 2bxy + cy^2 = s^2 + t^2$.)

CHAPTER 15 Additional and Advanced Exercises

Volumes

1. Sand pile: double and triple integrals The base of a sand pile covers the region in the xy-plane that is bounded by the parabola $x^2 + y = 6$ and the line $y = x$. The height of the sand above the point (x, y) is x^2. Express the volume of sand as **(a)** a double integral and **(b)** a triple integral. Then **(c)** find the volume.

2. Water in a hemispherical bowl A hemispherical bowl of radius 5 cm is filled with water to within 3 cm of the top. Find the volume of water in the bowl.

3. Solid cylindrical region between two planes Find the volume of the portion of the solid cylinder $x^2 + y^2 \le 1$ that lies between the planes $z = 0$ and $x + y + z = 2$.

4. Sphere and paraboloid Find the volume of the solid region bounded above by the sphere $x^2 + y^2 + z^2 = 2$ and below by the paraboloid $z = x^2 + y^2$.

5. Two paraboloids Find the volume of the solid region bounded above by the paraboloid $z = 3 - x^2 - y^2$ and below by the paraboloid $z = 2x^2 + 2y^2$.

6. Spherical coordinates Find the volume of the solid region enclosed by the spherical coordinate surface $\rho = 2\sin\phi$ (see accompanying figure).

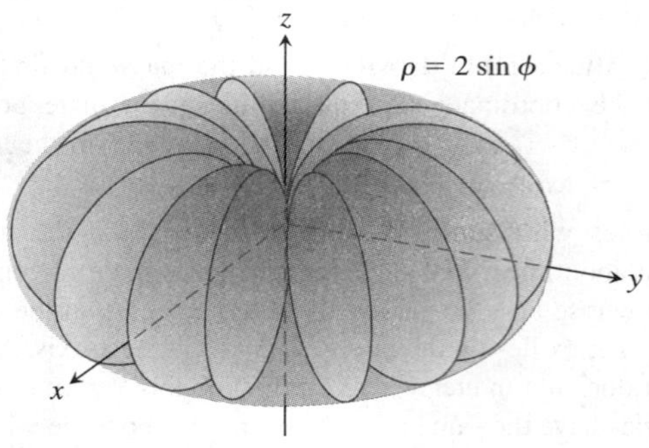

7. Hole in solid ball A circular cylindrical hole is bored through a ball, the axis of the hole being a diameter of the sphere. The volume of the remaining solid is

$$V = 2\int_0^{2\pi}\int_0^{\sqrt{3}}\int_1^{\sqrt{4-z^2}} r\, dr\, dz\, d\theta.$$

a. Find the radius of the hole and the radius of the sphere.

b. Evaluate the integral.

8. Ball and cylinder Find the volume of material cut from the ball $r^2 + z^2 \le 9$ by the cylinder $r = 3\sin\theta$.

9. Two paraboloids Find the volume of the solid region enclosed by the surfaces $z = x^2 + y^2$ and $z = (x^2 + y^2 + 1)/2$.

10. Cylinder and surface $z = xy$ Find the volume of the solid region in the first octant that lies between the cylinders $r = 1$ and $r = 2$ and is bounded below by the xy-plane and above by the surface $z = xy$.

Changing the Order of Integration

11. Evaluate the integral

$$\int_0^\infty \frac{e^{-ax} - e^{-bx}}{x}\, dx.$$

(*Hint:* Use the relation

$$\frac{e^{-ax} - e^{-bx}}{x} = \int_a^b e^{-xy}\, dy$$

to form a double integral, and evaluate the integral by changing the order of integration.)

12. a. Polar coordinates Show, by changing to polar coordinates, that

$$\int_0^{a\sin\beta}\int_{y\cot\beta}^{\sqrt{a^2-y^2}} \ln(x^2 + y^2)\, dx\, dy = a^2\beta\left(\ln a - \frac{1}{2}\right),$$

where $a > 0$ and $0 < \beta < \pi/2$.

b. Rewrite the Cartesian integral with the order of integration reversed.

13. Reducing a double to a single integral By changing the order of integration, show that the following double integral can be reduced to a single integral:

$$\int_0^x \int_0^u e^{m(x-t)} f(t)\, dt\, du = \int_0^x (x-t)e^{m(x-t)} f(t)\, dt.$$

Similarly, it can be shown that

$$\int_0^x \int_0^v \int_0^u e^{m(x-t)} f(t)\, dt\, du\, dv = \int_0^x \frac{(x-t)^2}{2} e^{m(x-t)} f(t)\, dt.$$

14. Transforming a double integral to obtain constant limits Sometimes a multiple integral with variable limits can be changed into one with constant limits. By changing the order of integration, show that

$$\int_0^1 f(x)\left(\int_0^x g(x-y) f(y)\, dy\right) dx$$
$$= \int_0^1 f(y)\left(\int_y^1 g(x-y) f(x)\, dx\right) dy$$
$$= \frac{1}{2}\int_0^1 \int_0^1 g(|x-y|) f(x) f(y)\, dx\, dy.$$

Masses and Moments

15. Minimizing polar inertia A thin plate of constant density is to occupy the triangular region in the first quadrant of the xy-plane having vertices $(0, 0)$, $(a, 0)$, and $(a, 1/a)$. What value of a will minimize the plate's polar moment of inertia about the origin?

16. Polar inertia of triangular plate Find the polar moment of inertia about the origin of a thin triangular plate of constant density $\delta = 3$ bounded by the y-axis and the lines $y = 2x$ and $y = 4$ in the xy-plane.

17. Mass and polar inertia of a counterweight The counterweight of a flywheel of constant density 1 has the form of the smaller segment cut from a circle of radius a by a chord at a distance b from the center $(b < a)$. Find the mass of the counterweight and its polar moment of inertia about the center of the wheel.

18. Centroid of a boomerang Find the centroid of the boomerang-shaped region between the parabolas $y^2 = -4(x-1)$ and $y^2 = -2(x-2)$ in the xy-plane.

Theory and Examples

19. Evaluate

$$\int_0^a \int_0^b e^{\max(b^2x^2,\, a^2y^2)}\, dy\, dx,$$

where a and b are positive numbers and

$$\max(b^2x^2, a^2y^2) = \begin{cases} b^2x^2 & \text{if } b^2x^2 \ge a^2y^2 \\ a^2y^2 & \text{if } b^2x^2 < a^2y^2. \end{cases}$$

20. Show that

$$\iint \frac{\partial^2 F(x,y)}{\partial x\, \partial y}\, dx\, dy$$

over the rectangle $x_0 \le x \le x_1$, $y_0 \le y \le y_1$ is

$$F(x_1, y_1) - F(x_0, y_1) - F(x_1, y_0) + F(x_0, y_0).$$

21. Suppose that $f(x, y)$ can be written as a product $f(x, y) = F(x)G(y)$ of a function of x and a function of y. Then the integral of f over the rectangle $R: a \le x \le b,\ c \le y \le d$ can be evaluated as a product as well, by the formula

$$\iint_R f(x,y)\, dA = \left(\int_a^b F(x)\, dx\right)\left(\int_c^d G(y)\, dy\right). \qquad (1)$$

The argument is that

$$\iint_R f(x,y)\, dA = \int_c^d \left(\int_a^b F(x)\, G(y)\, dx\right) dy \qquad \text{(i)}$$
$$= \int_c^d \left(G(y) \int_a^b F(x)\, dx\right) dy \qquad \text{(ii)}$$
$$= \int_c^d \left(\int_a^b F(x)\, dx\right) G(y)\, dy \qquad \text{(iii)}$$
$$= \left(\int_a^b F(x)\, dx\right) \int_c^d G(y)\, dy. \qquad \text{(iv)}$$

a. Give reasons for Steps (i) through (iv).

When it applies, Equation (1) can be a time-saver. Use it to evaluate the following integrals.

b. $\displaystyle\int_0^{\ln 2} \int_0^{\pi/2} e^x \cos y\, dy\, dx$ **c.** $\displaystyle\int_1^2 \int_{-1}^1 \frac{x}{y^2}\, dx\, dy$

22. Let $D_{\mathbf{u}}f$ denote the derivative of $f(x, y) = (x^2 + y^2)/2$ in the direction of the unit vector $\mathbf{u} = u_1\mathbf{i} + u_2\mathbf{j}$.

a. Finding average value Find the average value of $D_{\mathbf{u}}f$ over the triangular region cut from the first quadrant by the line $x + y = 1$.

b. Average value and centroid Show in general that the average value of $D_{\mathbf{u}}f$ over a region in the xy-plane is the value of $D_{\mathbf{u}}f$ at the centroid of the region.

23. The value of $\Gamma(1/2)$ The gamma function,

$$\Gamma(x) = \int_0^\infty t^{x-1} e^{-t}\, dt,$$

extends the factorial function from the nonnegative integers to other real values. Of particular interest in the theory of differential equations is the number

$$\Gamma\left(\frac{1}{2}\right) = \int_0^\infty t^{(1/2)-1} e^{-t}\, dt = \int_0^\infty \frac{e^{-t}}{\sqrt{t}}\, dt. \qquad (2)$$

a. If you have not yet done Exercise 41 in Section 15.4, do it now to show that

$$I = \int_0^\infty e^{-y^2}\, dy = \frac{\sqrt{\pi}}{2}.$$

b. Substitute $y = \sqrt{t}$ in Equation (2) to show that $\Gamma(1/2) = 2I = \sqrt{\pi}$.

24. Total electrical charge over circular plate The electrical charge distribution on a circular plate of radius R meters is $\sigma(r, \theta) = kr(1 - \sin\theta)$ coulomb/m^2 (k a constant). Integrate σ over the plate to find the total charge Q.

25. A parabolic rain gauge A bowl is in the shape of the graph of $z = x^2 + y^2$ from $z = 0$ to $z = 10$ in. You plan to calibrate the bowl to make it into a rain gauge. What height in the bowl would correspond to 1 in. of rain? 3 in. of rain?

26. Water in a satellite dish A parabolic satellite dish is 2 m wide and 1/2 m deep. Its axis of symmetry is tilted 30 degrees from the vertical.

a. Set up, but do not evaluate, a triple integral in rectangular coordinates that gives the amount of water the satellite dish will hold. (*Hint:* Put your coordinate system so that the satellite dish is in "standard position" and the plane of the water level is slanted.) (*Caution:* The limits of integration are not "nice.")

b. What would be the smallest tilt of the satellite dish so that it holds no water?

27. An infinite half-cylinder Let D be the interior of the infinite right circular half-cylinder of radius 1 with its single-end face suspended 1 unit above the origin and its axis the ray from $(0, 0, 1)$ to ∞. Use cylindrical coordinates to evaluate

$$\iiint_D z(r^2 + z^2)^{-5/2}\, dV.$$

28. Hypervolume We have learned that $\int_a^b 1\, dx$ is the length of the interval $[a, b]$ on the number line (one-dimensional space), $\iint_R 1\, dA$ is the area of region R in the xy-plane (two-dimensional space), and $\iiint_D 1\, dV$ is the volume of the region D in three-dimensional space (xyz-space). We could continue: If Q is a region in 4-space ($xyzw$-space), then $\iiiint_Q 1\, dV$ is the "hyper-volume" of Q. Use your generalizing abilities and a Cartesian coordinate system of 4-space to find the hypervolume inside the unit four-dimensional sphere $x^2 + y^2 + z^2 + w^2 = 1$.

CHAPTER 15 Technology Application Projects

Mathematica/Maple Projects

Projects can be found within MyLab Math.

- ***Take Your Chances: Try the Monte Carlo Technique for Numerical Integration in Three Dimensions***
  Use the Monte Carlo technique to integrate numerically in three dimensions.
- ***Means and Moments and Exploring New Plotting Techniques, Part II***
  Use the method of moments in a form that makes use of geometric symmetry as well as multiple integration.

16

Integrals and Vector Fields

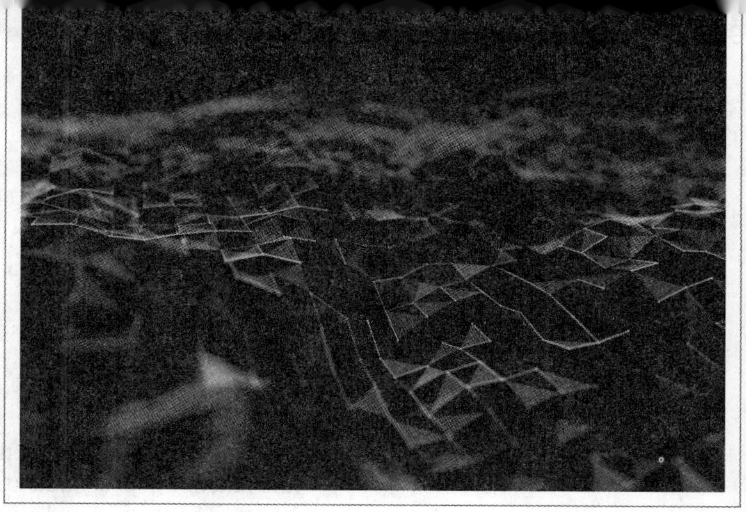

OVERVIEW In this chapter we extend the theory of integration to functions whose domains are curves and surfaces in space. The resulting line and surface integrals give powerful mathematical tools for science and engineering. Line integrals are used to find the work done by a force in moving an object along a path and to find the mass of a curved wire with variable density. Surface integrals are used to find the rate of flow of a fluid across a surface and to describe the interactions of electric and magnetic forces. We present the fundamental theorems of vector integral calculus and discuss their mathematical consequences and physical applications. The theorems of vector calculus are then shown to be generalized versions of the Fundamental Theorem of Calculus.

16.1 Line Integrals of Scalar Functions

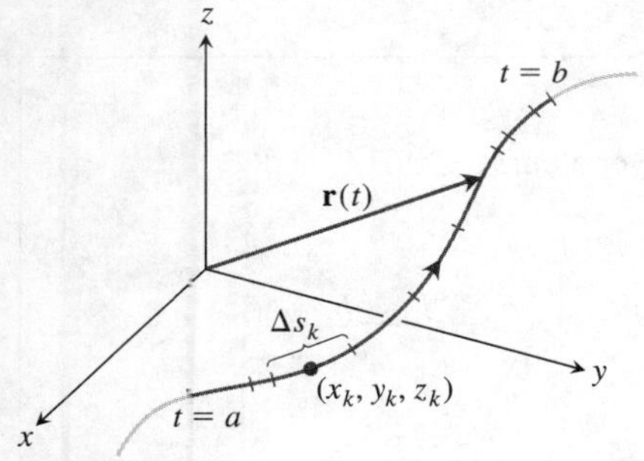

FIGURE 16.1 The curve $\mathbf{r}(t)$ partitioned into small arcs from $t = a$ to $t = b$. The length of a typical subarc is Δs_k.

To calculate the total mass of a wire lying along a curve in space, or to find the work done by a variable force acting along such a curve, we need a more general notion of integral than was defined in Chapter 5. We need to integrate over a curve C rather than over an interval $[a, b]$. These more general integrals are called *line integrals* (although *path integrals* might be more descriptive). We make our definitions for space curves, with curves in the xy-plane being the special case with z-coordinate identically zero.

Suppose that $f(x, y, z)$ is a real-valued function we wish to integrate over the curve C lying within the domain of f and parametrized by $\mathbf{r}(t) = g(t)\mathbf{i} + h(t)\mathbf{j} + k(t)\mathbf{k}$, $a \le t \le b$. The values of f along the curve are given by the composite function $f(g(t), h(t), k(t))$. We are going to integrate this composition with respect to arc length from $t = a$ to $t = b$. To begin, we first partition the curve C into a finite number n of subarcs (Figure 16.1). The typical subarc has length Δs_k. In each subarc we choose a point (x_k, y_k, z_k) and form the sum

$$S_n = \sum_{k=1}^{n} \underbrace{f(x_k, y_k, z_k)}_{\text{value of } f \text{ at a point on the subarc}} \underbrace{\Delta s_k}_{\text{length of a small subarc of the curve}},$$

which is similar to a Riemann sum. Depending on how we partition the curve C and pick (x_k, y_k, z_k) in the kth subarc, we may get different values for S_n. If f is continuous and the functions g, h, and k have continuous first derivatives, then these sums approach a limit as n increases and the lengths Δs_k approach zero. This leads to the following definition, which is similar to that for a single integral. In the definition, we assume that the norm of the partition approaches zero as $n \to \infty$, so that the length of the longest subarc approaches zero.

DEFINITION If f is defined on a curve C given parametrically by $\mathbf{r}(t) = g(t)\mathbf{i} + h(t)\mathbf{j} + k(t)\mathbf{k}$, $a \le t \le b$, then the **line integral of f over C** is

$$\int_C f(x, y, z)\, ds = \lim_{n\to\infty} \sum_{k=1}^{n} f(x_k, y_k, z_k)\,\Delta s_k, \tag{1}$$

provided this limit exists.

If the curve C is smooth for $a \le t \le b$ (so $\mathbf{v} = d\mathbf{r}/dt$ is continuous and never $\mathbf{0}$) and the function f is continuous on C, then the limit in Equation (1) can be shown to exist. We can then apply the Fundamental Theorem of Calculus to differentiate the arc length equation,

$$s(t) = \int_a^t |\mathbf{v}(\tau)|\, d\tau, \qquad \text{Eq. (3) of Section 13.3 with } t_0 = a$$

to express ds in Equation (1) as $ds = |\mathbf{v}(t)|\, dt$ and evaluate the integral of f over C as

$$\frac{ds}{dt} = |\mathbf{v}| = \sqrt{\left(\frac{dx}{dt}\right)^2 + \left(\frac{dy}{dt}\right)^2 + \left(\frac{dz}{dt}\right)^2}$$

$$\int_C f(x, y, z)\, ds = \int_a^b f(g(t), h(t), k(t))|\mathbf{v}(t)|\, dt. \tag{2}$$

The integral on the right side of Equation (2) is just an ordinary definite integral, as defined in Chapter 5, where we are integrating with respect to the parameter t. The formula evaluates the line integral on the left side correctly no matter what smooth parametrization is used. Note that the parameter t defines a direction along the path. The starting point on C is the position $\mathbf{r}(a)$, and movement along the path is in the direction of increasing t (see Figure 16.1).

How to Evaluate a Line Integral

To integrate a continuous function $f(x, y, z)$ over a curve C:

1. Find a smooth parametrization of C,
$$\mathbf{r}(t) = g(t)\mathbf{i} + h(t)\mathbf{j} + k(t)\mathbf{k}, \qquad a \le t \le b.$$
2. Evaluate the integral as
$$\int_C f(x, y, z)\, ds = \int_a^b f(g(t), h(t), k(t))|\mathbf{v}(t)|\, dt.$$

$$f(\mathbf{r}(t)) = f(g(t), h(t), k(t))$$

If f has the constant value 1, then the integral of f over C gives the length of C from $t = a$ to $t = b$. We also write $f(\mathbf{r}(t))$ for the evaluation $f(g(t), h(t), k(t))$ along the curve $\mathbf{r}$.

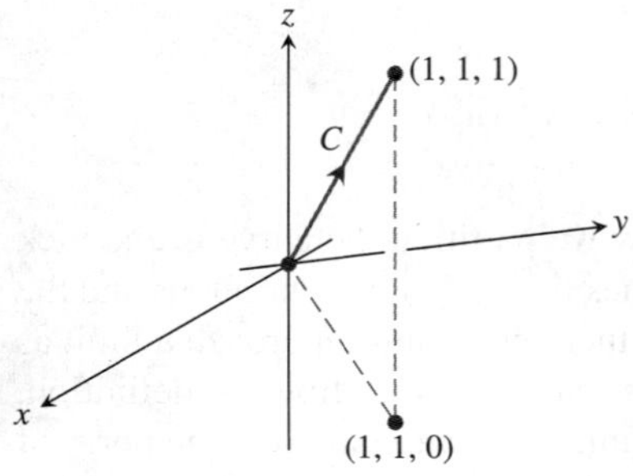

FIGURE 16.2 The integration path in Example 1.

EXAMPLE 1 Integrate $f(x, y, z) = x - 3y^2 + z$ over the line segment C joining the origin to the point $(1, 1, 1)$ (Figure 16.2).

Solution Since any choice of parametrization will give the same answer, we choose the simplest parametrization we can think of:

$$\mathbf{r}(t) = t\mathbf{i} + t\mathbf{j} + t\mathbf{k}, \qquad 0 \le t \le 1.$$

The components have continuous first derivatives, and $|\mathbf{v}(t)| = |\mathbf{i} + \mathbf{j} + \mathbf{k}| = \sqrt{1^2 + 1^2 + 1^2} = \sqrt{3}$ is never 0, so the parametrization is smooth. The integral of f over C is

$$\int_C f(x, y, z)\, ds = \int_0^1 f(t, t, t)\sqrt{3}\, dt \qquad \text{Eq. (2), } ds = |\mathbf{v}(t)|\, dt = \sqrt{3}\, dt$$

$$= \int_0^1 (t - 3t^2 + t)\sqrt{3}\, dt$$

$$= \sqrt{3}\int_0^1 (2t - 3t^2)\, dt = \sqrt{3}\Big[t^2 - t^3\Big]_0^1 = 0. \qquad \blacksquare$$

Additivity

Line integrals have the useful property that if a piecewise smooth curve C is made by joining a finite number of smooth curves $C_1, C_2, \ldots, C_n$ end to end (Section 13.1), then the integral of a function over C is the sum of the integrals over the curves that make it up:

$$\int_C f\, ds = \int_{C_1} f\, ds + \int_{C_2} f\, ds + \cdots + \int_{C_n} f\, ds. \tag{3}$$

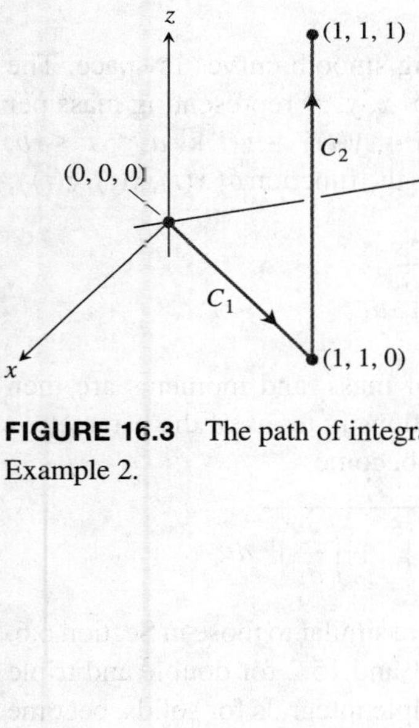

FIGURE 16.3 The path of integration in Example 2.

EXAMPLE 2 Figure 16.3 shows another path from the origin to $(1, 1, 1)$, formed from two line segments C_1 and C_2. Integrate $f(x, y, z) = x - 3y^2 + z$ over $C_1 \cup C_2$.

Solution We choose the simplest parametrizations for C_1 and C_2 we can find, calculating the lengths of the velocity vectors as we go along:

$$C_1: \ \mathbf{r}(t) = t\mathbf{i} + t\mathbf{j}, \quad 0 \le t \le 1; \quad |\mathbf{v}| = \sqrt{1^2 + 1^2} = \sqrt{2}$$

$$C_2: \ \mathbf{r}(t) = \mathbf{i} + \mathbf{j} + t\mathbf{k}, \quad 0 \le t \le 1; \quad |\mathbf{v}| = \sqrt{0^2 + 0^2 + 1^2} = 1.$$

With these parametrizations we find that

$$\int_{C_1 \cup C_2} f(x, y, z)\, ds = \int_{C_1} f(x, y, z)\, ds + \int_{C_2} f(x, y, z)\, ds \qquad \text{Eq. (3)}$$

$$= \int_0^1 f(t, t, 0)\sqrt{2}\, dt + \int_0^1 f(1, 1, t)(1)\, dt \qquad \text{Eq. (2)}$$

$$= \int_0^1 (t - 3t^2 + 0)\sqrt{2}\, dt + \int_0^1 (1 - 3 + t)(1)\, dt$$

$$= \sqrt{2}\left[\frac{t^2}{2} - t^3\right]_0^1 + \left[\frac{t^2}{2} - 2t\right]_0^1 = -\frac{\sqrt{2}}{2} - \frac{3}{2}. \qquad \blacksquare$$

Notice three things about the integrations in Examples 1 and 2. First, as soon as the components of the appropriate curve were substituted into the formula for f, the integration became a standard integration with respect to t. Second, the integral of f over $C_1 \cup C_2$ was obtained by integrating f over each section of the path and adding the results. Third, the integrals of f over C and $C_1 \cup C_2$ had different values. We investigate this third observation in Section 16.3.

> The value of a line integral along a path joining two points can change if you change the path between them.

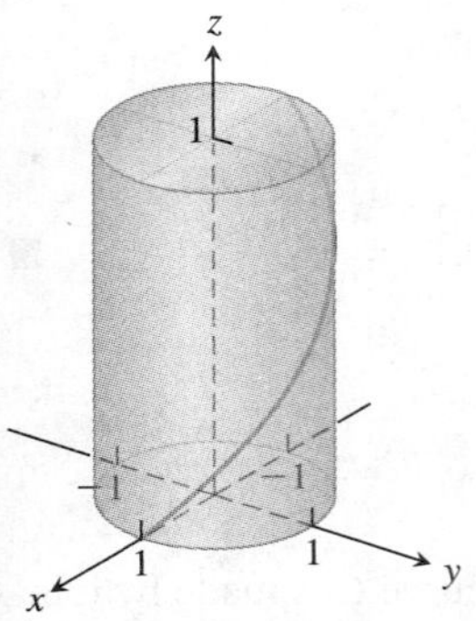

FIGURE 16.4 A line integral is taken over a curve such as this helix from Example 3.

EXAMPLE 3 Find the line integral of $f(x, y, z) = 2xy + \sqrt{z}$ over the helix $\mathbf{r}(t) = \cos t\mathbf{i} + \sin t\mathbf{j} + t\mathbf{k},\ 0 \le t \le \pi$.

Solution For the helix (Figure 16.4) we find $\mathbf{v}(t) = \mathbf{r}'(t) = -\sin t\mathbf{i} + \cos t\mathbf{j} + \mathbf{k}$ and $|\mathbf{v}(t)| = \sqrt{(-\sin t)^2 + (\cos t)^2 + 1} = \sqrt{2}$. Evaluating the function f at the point $\mathbf{r}(t)$, we obtain

$$f(\mathbf{r}(t)) = f(\cos t, \sin t, t) = 2\cos t \sin t + \sqrt{t} = \sin 2t + \sqrt{t}.$$

The line integral is given by

$$\begin{aligned}\int_C f(x, y, z)\, ds &= \int_0^{\pi} (\sin 2t + \sqrt{t})\sqrt{2}\, dt \\ &= \sqrt{2}\left[-\frac{1}{2}\cos 2t + \frac{2}{3}t^{3/2}\right]_0^{\pi} \\ &= \frac{2\sqrt{2}}{3}\pi^{3/2} \approx 5.25.\end{aligned}$$

■

Mass and Moment Calculations

We treat coil springs and wires as masses distributed along smooth curves in space. The distribution is described by a continuous density function $\delta(x, y, z)$ representing mass per unit length. When a curve C is parametrized by $\mathbf{r}(t) = x(t)\mathbf{i} + y(t)\mathbf{j} + z(t)\mathbf{k},\ a \le t \le b$, then x, y, and z are functions of the parameter t, the density is the function $\delta(x(t), y(t), z(t))$, and the arc length differential is given by

$$ds = \sqrt{\left(\frac{dx}{dt}\right)^2 + \left(\frac{dy}{dt}\right)^2 + \left(\frac{dz}{dt}\right)^2}\, dt.$$

(See Section 13.3.) The spring's or wire's mass, center of mass, and moments are then calculated using the formulas in Table 16.1, with the integrations in terms of the parameter t over the interval $[a, b]$. For example, the formula for mass becomes

$$M = \int_a^b \delta(x(t), y(t), z(t))\sqrt{\left(\frac{dx}{dt}\right)^2 + \left(\frac{dy}{dt}\right)^2 + \left(\frac{dz}{dt}\right)^2}\, dt.$$

These formulas also apply to thin rods, and their derivations are similar to those in Section 6.6. Notice how similar the formulas are to those in Tables 15.1 and 15.2 for double and triple integrals. The double integrals for planar regions, and the triple integrals for solids, become line integrals for coil springs, wires, and thin rods.

Notice that the element of mass dm is equal to $\delta\, ds$ in the table, rather than to $\delta\, dV$ as in Table 15.1, and that the integrals are taken over the curve C.

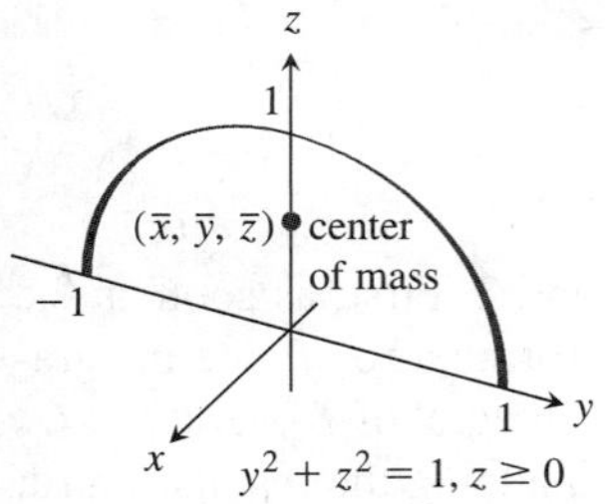

FIGURE 16.5 Example 4 shows how to find the center of mass of a circular arch of variable density.

EXAMPLE 4 A slender metal arch, denser at the bottom than at the top, lies along the semicircle $y^2 + z^2 = 1,\ z \ge 0$, in the yz-plane (Figure 16.5). Find the center of the arch's mass if the density at the point (x, y, z) on the arch is $\delta(x, y, z) = 2 - z$.

Solution We know that $\bar{x} = 0$ and $\bar{y} = 0$ because the arch lies in the yz-plane with its mass distributed symmetrically about the z-axis. To find $\bar{z}$, we parametrize the circle as

$$\mathbf{r}(t) = (\cos t)\mathbf{j} + (\sin t)\mathbf{k}, \qquad 0 \le t \le \pi.$$

TABLE 16.1 Mass and moment formulas for coil springs, wires, and thin rods lying along a smooth curve C in space

Mass: $M = \int_C \delta\, ds$ $\quad \delta = \delta(x, y, z)$ is the density at (x, y, z).

First moments about the coordinate planes:

$$M_{yz} = \int_C x\,\delta\, ds, \qquad M_{xz} = \int_C y\,\delta\, ds, \qquad M_{xy} = \int_C z\,\delta\, ds$$

Coordinates of the center of mass:

$$\bar{x} = M_{yz}/M, \qquad \bar{y} = M_{xz}/M, \qquad \bar{z} = M_{xy}/M$$

Moments of inertia about axes and other lines:

$$I_x = \int_C (y^2 + z^2)\delta\, ds, \quad I_y = \int_C (x^2 + z^2)\delta\, ds, \quad I_z = \int_C (x^2 + y^2)\delta\, ds,$$

$$I_L = \int_C r^2\delta\, ds$$ $r = r(x, y, z)$ is the distance from the point (x, y, z) to line L.

For this parametrization,

$$|\mathbf{v}(t)| = \sqrt{\left(\frac{dx}{dt}\right)^2 + \left(\frac{dy}{dt}\right)^2 + \left(\frac{dz}{dt}\right)^2} = \sqrt{(0)^2 + (-\sin t)^2 + (\cos t)^2} = 1,$$

so $ds = |\mathbf{v}|\, dt = dt$.

The formulas in Table 16.1 then give

$$M = \int_C \delta\, ds = \int_C (2 - z)\, ds = \int_0^{\pi} (2 - \sin t)\, dt = 2\pi - 2$$

$$M_{xy} = \int_C z\delta\, ds = \int_C z(2 - z)\, ds = \int_0^{\pi} (\sin t)(2 - \sin t)\, dt$$

$$= \int_0^{\pi} (2 \sin t - \sin^2 t)\, dt = \frac{8 - \pi}{2} \qquad \text{Routine integration}$$

$$\bar{z} = \frac{M_{xy}}{M} = \frac{8 - \pi}{2} \cdot \frac{1}{2\pi - 2} = \frac{8 - \pi}{4\pi - 4} \approx 0.57.$$

With $\bar{z}$ to the nearest hundredth, the center of mass is $(0, 0, 0.57)$. ■

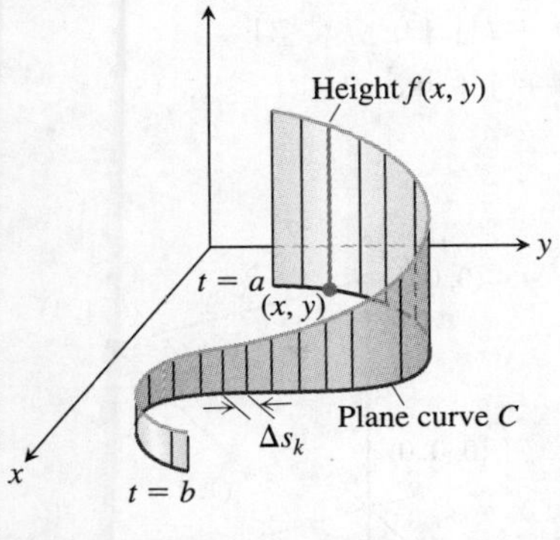

FIGURE 16.6 The line integral $\int_C f\, ds$ gives the area of the portion of the cylindrical surface or "wall" beneath $z = f(x, y) \geq 0$.

Line Integrals in the Plane

Line integrals for curves in the plane have a natural geometric interpretation. If C is a smooth curve in the xy-plane parametrized by $\mathbf{r}(t) = x(t)\mathbf{i} + y(t)\mathbf{j}$, $a \leq t \leq b$, we generate a cylindrical surface by moving a straight line along C perpendicular to the plane, holding the line parallel to the z-axis, as in Figure 16.6. If $z = f(x, y)$ is a nonnegative continuous function over a region in the plane containing the curve C, then the graph of f is a surface that lies above the plane. The cylinder cuts through this surface, forming a curve on it that lies above the curve C and follows its winding nature. The part of the cylindrical surface that lies beneath the surface curve and above the xy-plane forms a "curved wall" or "fence" standing on the curve C and orthogonal to the plane. At any point (x, y) along the curve, the height of the wall is $f(x, y)$. From the definition

$$\int_C f\,ds = \lim_{n\to\infty} \sum_{k=1}^{n} f(x_k, y_k)\,\Delta s_k,$$

where $\Delta s_k \to 0$ as $n \to \infty$, we see that the line integral $\int_C f\,ds$ is the area of the wall shown in the figure.

EXERCISES 16.1

Graphs of Vector Equations

Match the vector equations in Exercises 1–8 with the graphs (a)–(h) given here.

a.

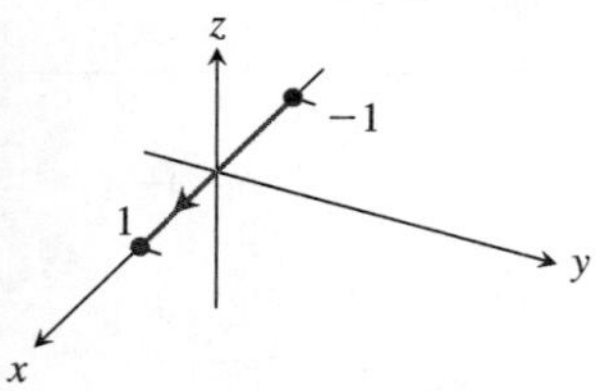

b.

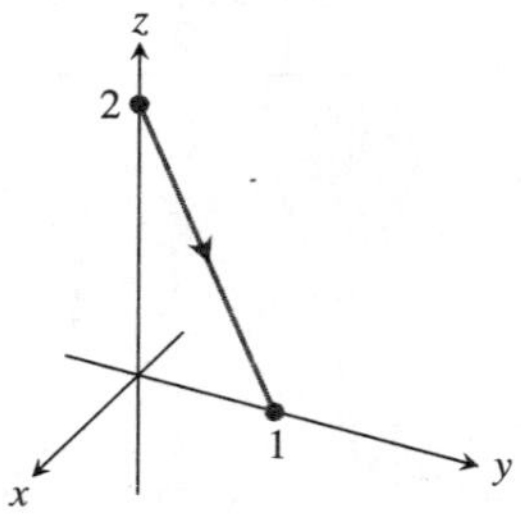

c.

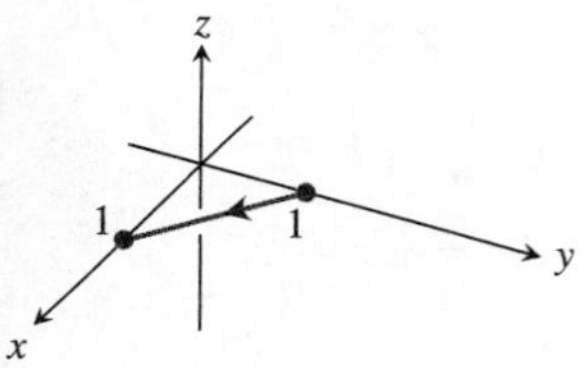

d.

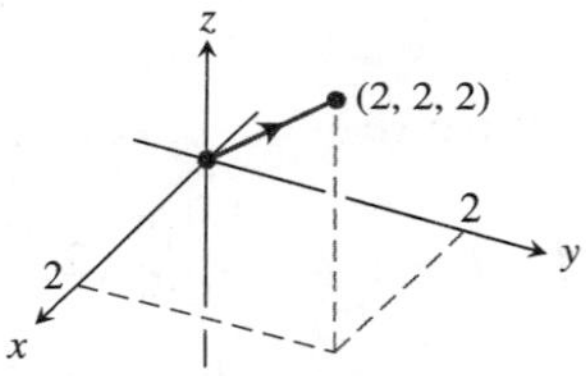

e.

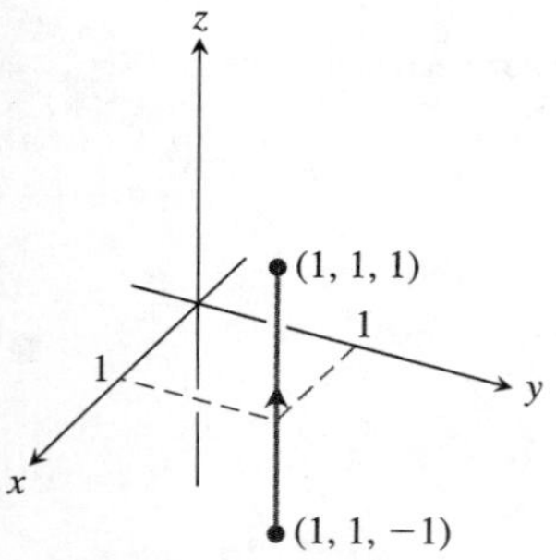

f.

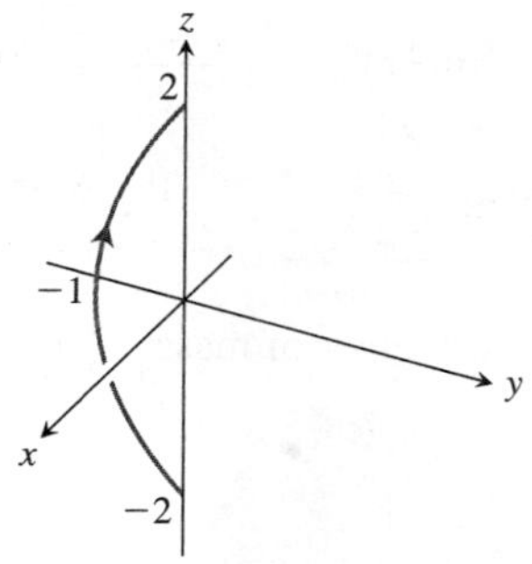

g.

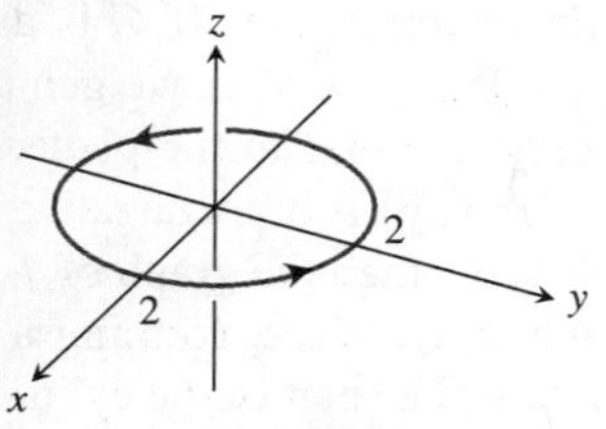

h.

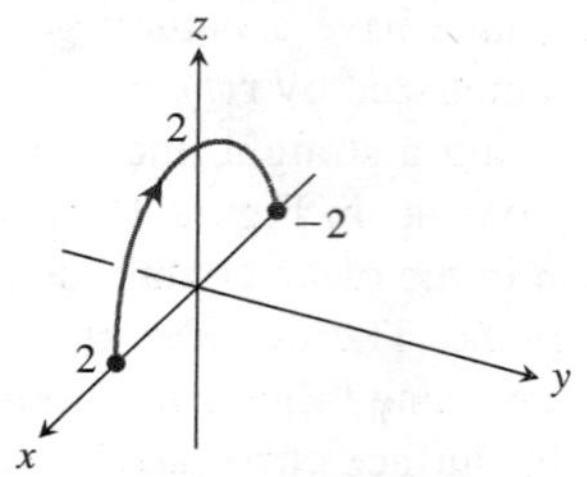

1. $\mathbf{r}(t) = t\mathbf{i} + (1 - t)\mathbf{j},\quad 0 \le t \le 1$
2. $\mathbf{r}(t) = \mathbf{i} + \mathbf{j} + t\mathbf{k},\quad -1 \le t \le 1$
3. $\mathbf{r}(t) = (2\cos t)\mathbf{i} + (2\sin t)\mathbf{j},\quad 0 \le t \le 2\pi$
4. $\mathbf{r}(t) = t\mathbf{i},\quad -1 \le t \le 1$
5. $\mathbf{r}(t) = t\mathbf{i} + t\mathbf{j} + t\mathbf{k},\quad 0 \le t \le 2$
6. $\mathbf{r}(t) = t\mathbf{j} + (2 - 2t)\mathbf{k},\quad 0 \le t \le 1$
7. $\mathbf{r}(t) = (t^2 - 1)\mathbf{j} + 2t\mathbf{k},\quad -1 \le t \le 1$
8. $\mathbf{r}(t) = (2\cos t)\mathbf{i} + (2\sin t)\mathbf{k},\quad 0 \le t \le \pi$

Evaluating Line Integrals over Space Curves

9. Evaluate $\int_C (x + y)\,ds$, where C is the straight-line segment $x = t,\ y = (1 - t),\ z = 0$, from $(0, 1, 0)$ to $(1, 0, 0)$.
10. Evaluate $\int_C (x - y + z - 2)\,ds$, where C is the straight-line segment $x = t,\ y = (1 - t),\ z = 1$, from $(0, 1, 1)$ to $(1, 0, 1)$.
11. Evaluate $\int_C (x^2 - y^2 + z)\,ds$ along the curve $\mathbf{r}(t) = t\mathbf{i} + 2t\mathbf{j} + (1 - 2t)\mathbf{k}, 0 \le t \le 1$.
12. Evaluate $\int_C \sqrt{x^2 + y^2}\,ds$ along the curve $\mathbf{r}(t) = (4\cos t)\mathbf{i} + (4\sin t)\mathbf{j} + 3t\mathbf{k},\ -2\pi \le t \le 2\pi$.
13. Find the line integral of $f(x, y, z) = x + y + z$ over the straight-line segment from $(1, 2, 3)$ to $(0, -1, 1)$.
14. Find the line integral of $f(x, y, z) = 2x + 3y + z$ over the straight line segment from $(0, 1, -1)$ to $(1, 1, 2)$.
15. Integrate $f(x, y, z) = x + \sqrt{y} - z^2$ over the path C_1 followed by C_2 from $(0, 0, 0)$ to $(1, 1, 1)$ (see accompanying figure) given by

$$C_1:\ \mathbf{r}(t) = t\mathbf{i} + t^2\mathbf{j},\quad 0 \le t \le 1$$
$$C_2:\ \mathbf{r}(t) = \mathbf{i} + \mathbf{j} + t\mathbf{k},\quad 0 \le t \le 1.$$

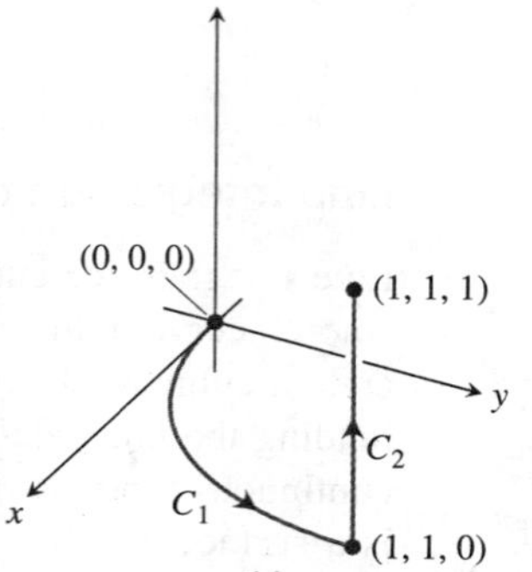

(a)

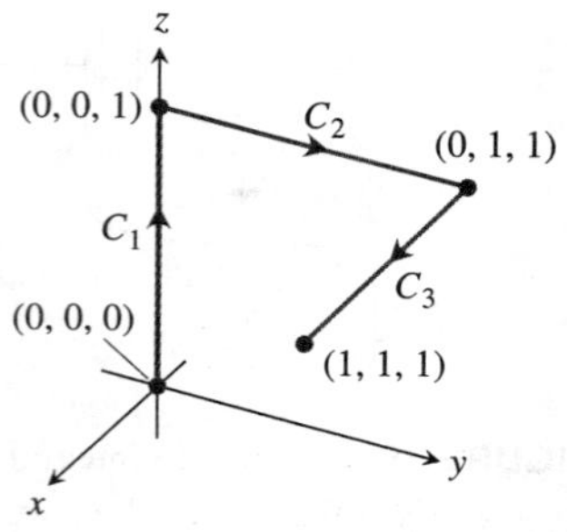

(b)

The paths of integration for Exercises 15 and 16.

16. Integrate $f(x, y, z) = x + \sqrt{y} - z^2$ over the path C_1 followed by C_2 followed by C_3 from $(0, 0, 0)$ to $(1, 1, 1)$ (see accompanying figure) given by

$$C_1: \ \mathbf{r}(t) = t\mathbf{k}, \quad 0 \le t \le 1$$
$$C_2: \ \mathbf{r}(t) = t\mathbf{j} + \mathbf{k}, \quad 0 \le t \le 1$$
$$C_3: \ \mathbf{r}(t) = t\mathbf{i} + \mathbf{j} + \mathbf{k}, \quad 0 \le t \le 1.$$

17. Integrate $f(x, y, z) = x + y - z^2$ over the path from (0, 0, 0) to (2, 4, 8) given by C_1: $2t\mathbf{i}$, C_2: $2\mathbf{i} + 4\mathbf{j}$, C_3: $2\mathbf{i} + 4\mathbf{j} + 8\mathbf{k}$, $0 \le t \le 1$.

18. Integrate $f(x, y, z) = -\sqrt{x^2 + z^2}$ over the circle

$$\mathbf{r}(t) = (a \cos t)\mathbf{j} + (a \sin t)\mathbf{k}, \qquad 0 \le t \le 2\pi.$$

Line Integrals over Plane Curves

19. Evaluate $\int_C x \, ds$, where C is
    a. the straight-line segment $x = t, \ y = t/2$, from $(0, 0)$ to $(4, 2)$.
    b. the parabolic curve $x = t, \ y = t^2$, from $(0, 0)$ to $(2, 4)$.

20. Evaluate $\int_C \sqrt{x + 2y} \, ds$, where C is
    a. the straight-line segment $x = t, \ y = 4t$, from $(0, 0)$ to $(1, 4)$.
    b. $C_1 \cup C_2$; C_1 is the line segment from $(0, 0)$ to $(1, 0)$ and C_2 is the line segment from $(1, 0)$ to $(1, 2)$.

21. Find the line integral of $f(x, y) = ye^{x^2}$ along the curve $\mathbf{r}(t) = 4t\mathbf{i} - 3t\mathbf{j}$, $-1 \le t \le 2$.

22. Find the line integral of $f(x, y) = x - y + 3$ along the curve $\mathbf{r}(t) = (\cos t)\mathbf{i} + (\sin t)\mathbf{j}$, $0 \le t \le 2\pi$.

23. Evaluate $\displaystyle\int_C \frac{x^2}{y^{4/3}} \, ds$, where C is the curve $x = t^2, \ y = t^3$, for $1 \le t \le 2$.

24. Find the line integral of $f(x, y) = \sqrt{y}/x$ along the curve $\mathbf{r}(t) = t^3\mathbf{i} + t^4\mathbf{j}$, $1/2 \le t \le 1$.

25. Evaluate $\int_C (x + \sqrt{y}) \, ds$, where C is given in the accompanying figure.

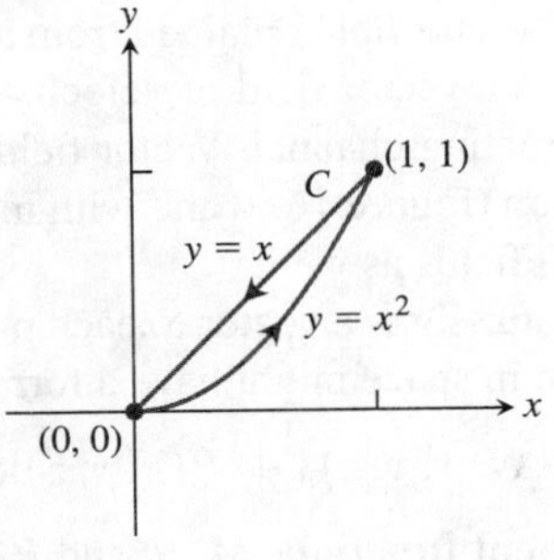

26. Evaluate $\displaystyle\int_C \frac{1}{x^2 + y^2 + 1} \, ds$, where C is given in the accompanying figure.

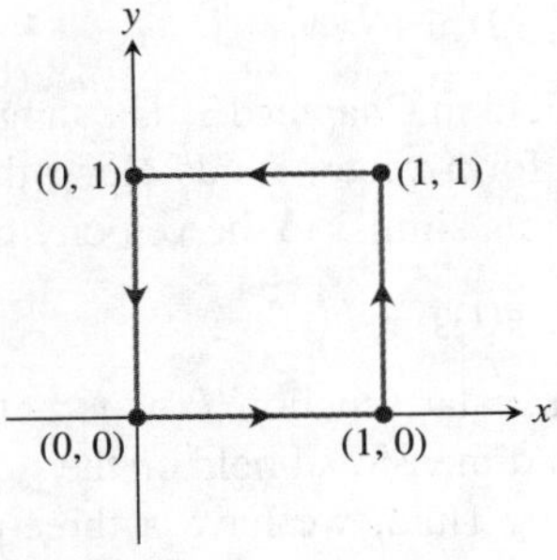

In Exercises 27–30, integrate f over the given curve.

27. $f(x, y) = x^3/y, \quad C: \ y = x^2/2, \quad 0 \le x \le 2$

28. $f(x, y) = (x + y^2)/\sqrt{1 + x^2}, \quad C: \ y = x^2/2$ from $(1, 1/2)$ to $(0, 0)$

29. $f(x, y) = x + y, \quad C: \ x^2 + y^2 = 4$ in the first quadrant from $(2, 0)$ to $(0, 2)$

30. $f(x, y) = x^2 - y, \quad C: \ x^2 + y^2 = 4$ in the first quadrant from $(0, 2)$ to $(\sqrt{2}, \sqrt{2})$

31. Find the area of one side of the "winding wall" standing perpendicularly on the curve $y = x^2$, $0 \le x \le 2$, and beneath the curve on the surface $f(x, y) = x + \sqrt{y}$.

32. Find the area of one side of the "wall" standing perpendicularly on the curve $2x + 3y = 6$, $0 \le x \le 6$, and beneath the curve on the surface $f(x, y) = 4 + 3x + 2y$.

Masses and Moments

33. **Mass of a wire** Find the mass of a wire that lies along the curve $\mathbf{r}(t) = (t^2 - 1)\mathbf{j} + 2t\mathbf{k}$, $0 \le t \le 1$, if the density is $\delta = (3/2)t$.

34. **Center of mass of a curved wire** A wire of density $\delta(x, y, z) = 15\sqrt{y + 2}$ lies along the curve $\mathbf{r}(t) = (t^2 - 1)\mathbf{j} + 2t\mathbf{k}$, $-1 \le t \le 1$. Find its center of mass. Then sketch the curve and center of mass together.

35. **Mass of wire with variable density** Find the mass of a thin wire lying along the curve $\mathbf{r}(t) = \sqrt{2}t\mathbf{i} + \sqrt{2}t\mathbf{j} + (4 - t^2)\mathbf{k}$, $0 \le t \le 1$, if the density is **(a)** $\delta = 3t$ and **(b)** $\delta = 1$.

36. **Center of mass of wire with variable density** Find the center of mass of a thin wire lying along the curve $\mathbf{r}(t) = t\mathbf{i} + 2t\mathbf{j} + (2/3)t^{3/2}\mathbf{k}$, $0 \le t \le 2$, if the density is $\delta = 3\sqrt{5 + t}$.

37. **Moment of inertia of wire hoop** A circular wire hoop of constant density δ lies along the circle $x^2 + y^2 = a^2$ in the xy-plane. Find the hoop's moment of inertia about the z-axis.

38. **Inertia of a slender rod** A slender rod of constant density lies along the line segment $\mathbf{r}(t) = t\mathbf{j} + (2 - 2t)\mathbf{k}$, $0 \le t \le 1$, in the yz-plane. Find the moments of inertia of the rod about the three coordinate axes.

39. **Two springs of constant density** A spring of constant density δ lies along the helix

$$\mathbf{r}(t) = (\cos t)\mathbf{i} + (\sin t)\mathbf{j} + t\mathbf{k}, \qquad 0 \le t \le 2\pi.$$

    a. Find I_z.
    b. Suppose that you have another spring of constant density δ that is twice as long as the spring in part (a) and lies along the helix for $0 \le t \le 4\pi$. Do you expect I_z for the longer spring to be the same as that for the shorter one, or should it be different? Check your prediction by calculating I_z for the longer spring.

40. **Wire of constant density** A wire of constant density $\delta = 1$ lies along the curve

$$\mathbf{r}(t) = (t \cos t)\mathbf{i} + (t \sin t)\mathbf{j} + (2\sqrt{2}/3)t^{3/2}\mathbf{k}, \qquad 0 \le t \le 1.$$

Find $\bar{z}$ and I_z.

41. **The arch in Example 4** Find I_x for the arch in Example 4.

42. Center of mass and moments of inertia for wire with variable density Find the center of mass and the moments of inertia about the coordinate axes of a thin wire lying along the curve

$$\mathbf{r}(t) = t\mathbf{i} + \frac{2\sqrt{2}}{3}t^{3/2}\mathbf{j} + \frac{t^2}{2}\mathbf{k}, \qquad 0 \le t \le 2,$$

if the density is $\delta = 1/(t + 1)$.

COMPUTER EXPLORATIONS

In Exercises 43–46, use a CAS to perform the following steps to evaluate the line integrals.

a. Find $ds = |\mathbf{v}(t)|\,dt$ for the path $\mathbf{r}(t) = g(t)\mathbf{i} + h(t)\mathbf{j} + k(t)\mathbf{k}$.

b. Express the integrand $f(g(t), h(t), k(t))\,|\mathbf{v}(t)|$ as a function of the parameter t.

c. Evaluate $\int_C f\,ds$ using Equation (2) in the text.

43. $f(x, y, z) = \sqrt{1 + 30x^2 + 10y}$; $\mathbf{r}(t) = t\mathbf{i} + t^2\mathbf{j} + 3t^2\mathbf{k}$, $0 \le t \le 2$

44. $f(x, y, z) = \sqrt{1 + x^3 + 5y^3}$; $\mathbf{r}(t) = t\mathbf{i} + \frac{1}{3}t^2\mathbf{j} + \sqrt{t}\mathbf{k}$, $0 \le t \le 2$

45. $f(x, y, z) = x\sqrt{y} - 3z^2$; $\mathbf{r}(t) = (\cos 2t)\mathbf{i} + (\sin 2t)\mathbf{j} + 5t\mathbf{k}$, $0 \le t \le 2\pi$

46. $f(x, y, z) = \left(1 + \frac{9}{4}z^{1/3}\right)^{1/4}$; $\mathbf{r}(t) = (\cos 2t)\mathbf{i} + (\sin 2t)\mathbf{j} + t^{5/2}\mathbf{k}$, $0 \le t \le 2\pi$

16.2 Vector Fields and Line Integrals: Work, Circulation, and Flux

Gravitational and electric forces have both a direction and a magnitude. They are represented by a vector at each point in their domain, producing a *vector field.* In this section we show how to compute the work done in moving an object through such a field by using a line integral involving the vector field. We also discuss velocity fields, such as the vector field representing the velocity of a flowing fluid in its domain. A line integral can be used to find the rate at which the fluid flows along or across a curve within the domain.

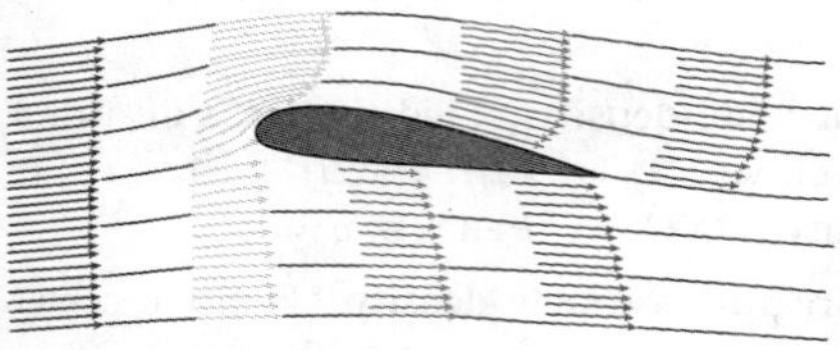

FIGURE 16.7 Velocity vectors of a flow around an airfoil.

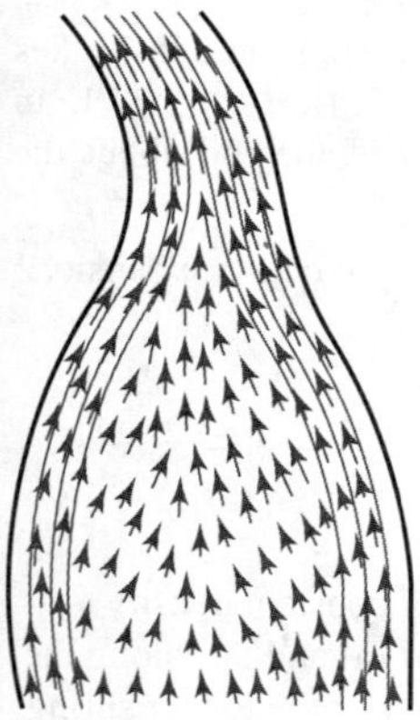

FIGURE 16.8 Streamlines in a contracting channel. The water speeds up as the channel narrows, and the velocity vectors increase in length.

Vector Fields

Suppose a region in the plane or in space is occupied by a moving fluid, such as air or water. The fluid is made up of a large number of particles, and at any instant of time, a particle has a velocity $\mathbf{v}$. At different points of the region at a given (same) time, these velocities can vary. We can think of a velocity vector being attached to each point of the fluid, representing the velocity of a particle at that point. Such a fluid flow is an example of a *vector field.* Figure 16.7 shows a velocity vector field obtained from air flowing around an airfoil in a wind tunnel. Figure 16.8 shows a vector field of velocity vectors along the streamlines of water moving through a contracting channel. Vector fields are also associated with forces such as gravitational attraction (Figure 16.9) and with magnetic fields and electric fields. There are purely mathematical fields as well.

Generally, a **vector field** is a function that assigns a vector to each point in its domain. A vector field on a three-dimensional domain in space might have a formula like

$$\mathbf{F}(x, y, z) = M(x, y, z)\mathbf{i} + N(x, y, z)\mathbf{j} + P(x, y, z)\mathbf{k}.$$

The vector field is **continuous** if the **component functions** M, N, and P are continuous; it is **differentiable** if each of the component functions is differentiable. The formula for a field of two-dimensional vectors could look like

$$\mathbf{F}(x, y) = M(x, y)\mathbf{i} + N(x, y)\mathbf{j}.$$

We encountered another type of vector field in Chapter 13. The tangent vectors $\mathbf{T}$ and normal vectors $\mathbf{N}$ for a curve in space both form vector fields along the curve. Along a curve $\mathbf{r}(t)$ they might have a component formula similar to the velocity field expression

$$\mathbf{v}(t) = f(t)\mathbf{i} + g(t)\mathbf{j} + h(t)\mathbf{k}.$$

If we attach the gradient vector ∇f of a scalar function $f(x, y, z)$ to each point of a level surface of the function, we obtain a three-dimensional field on the surface. If we attach the velocity vector to each point of a flowing fluid, we have a three-dimensional field defined on a region in space. These and other fields are illustrated in Figures 16.7–16.16. To sketch the fields, we picked a representative selection of domain points and drew the vectors attached to them. The arrows are drawn with their tails, not their heads, attached to the points where the vector functions are evaluated.

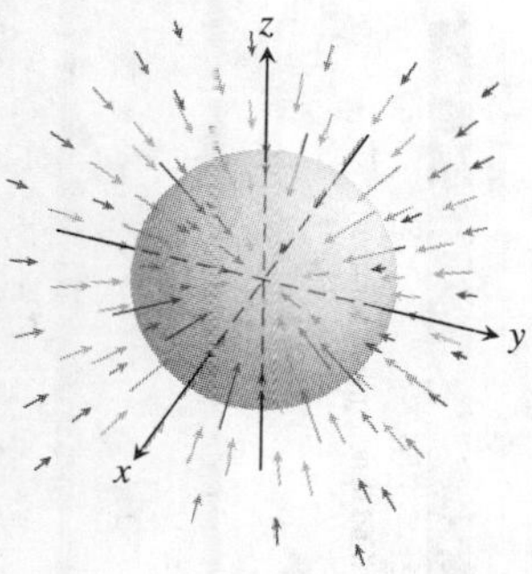

FIGURE 16.9 Vectors in a gravitational field point toward the center of mass that gives the source of the field.

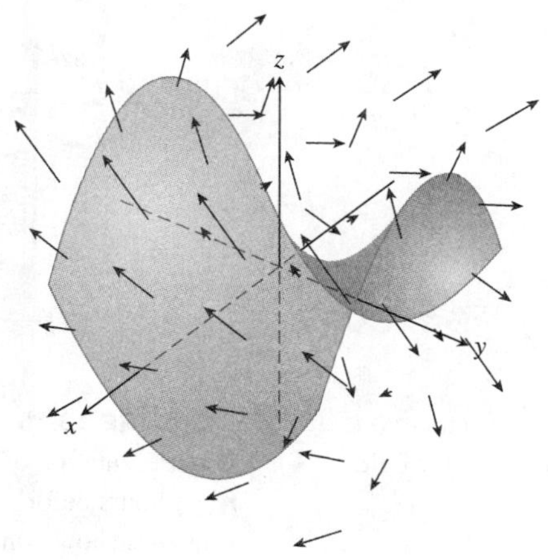

FIGURE 16.10 A surface might represent a filter (or a net or a parachute) in a vector field representing water or wind flow velocity vectors. The arrows show the direction of fluid flow, and their lengths indicate speed.

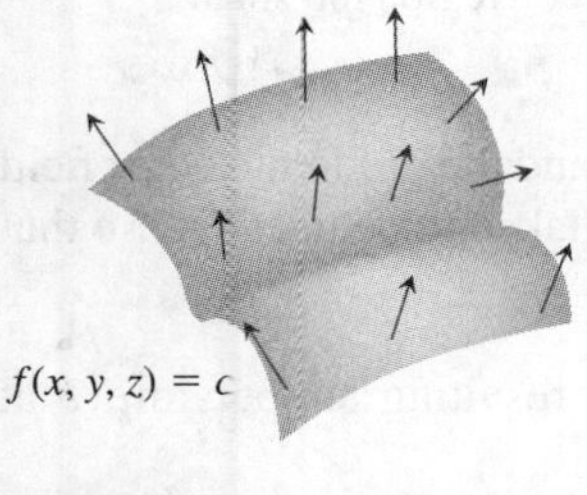

FIGURE 16.11 The field of gradient vectors ∇f on a level surface $f(x, y, z) = c$. The function f is constant on the surface, and each vector points in the direction where f is increasing fastest.

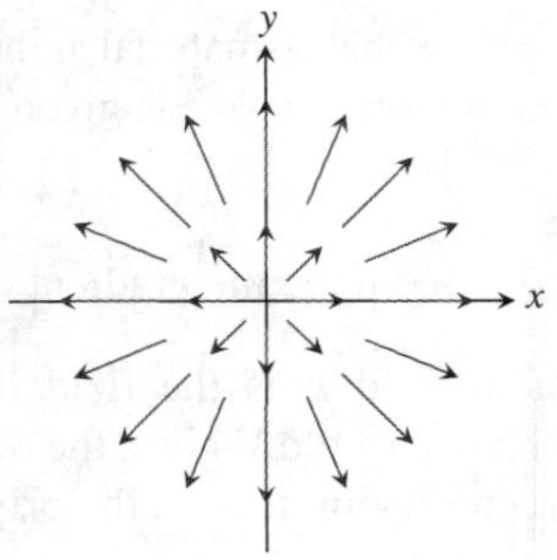

FIGURE 16.12 The radial field $\mathbf{F} = x\mathbf{i} + y\mathbf{j}$ formed by the position vectors of points in the plane. Notice the convention that an arrow is drawn with its tail, not its head, at the point where $\mathbf{F}$ is evaluated.

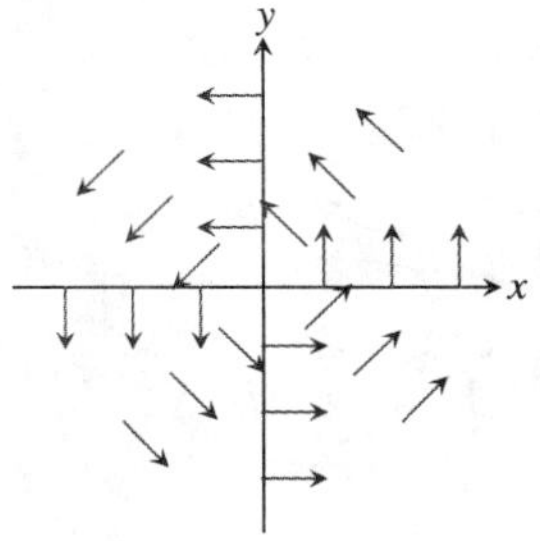

FIGURE 16.13 A "spin" field of rotating unit vectors

$$\mathbf{F} = (-y\mathbf{i} + x\mathbf{j})/(x^2 + y^2)^{1/2}$$

in the plane. The field is not defined at the origin.

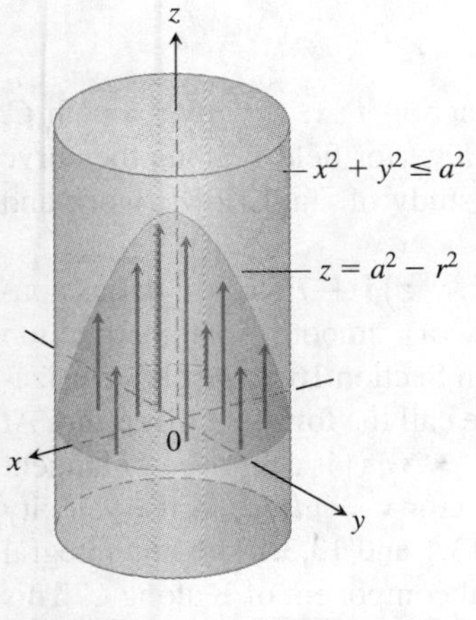

FIGURE 16.14 The flow of fluid in a long cylindrical pipe. The vectors $\mathbf{v} = (a^2 - r^2)\mathbf{k}$ inside the cylinder that have their bases in the xy-plane have their tips on the paraboloid $z = a^2 - r^2$.

Gradient Fields

The gradient vector of a differentiable scalar-valued function at a point gives the direction of greatest increase of the function. An important type of vector field is formed by all the gradient vectors of the function (see Section 14.5). We define the **gradient field** of a differentiable function $f(x, y, z)$ to be the field of gradient vectors

$$\nabla f = \frac{\partial f}{\partial x}\mathbf{i} + \frac{\partial f}{\partial y}\mathbf{j} + \frac{\partial f}{\partial z}\mathbf{k}.$$

At each point (x, y, z), the gradient field gives a vector pointing in the direction of greatest increase of f, with magnitude being the value of the directional derivative in that direction. The gradient field might represent a force field, or a velocity field that gives the motion of a fluid, or the flow of heat through a medium, depending on the application being considered.

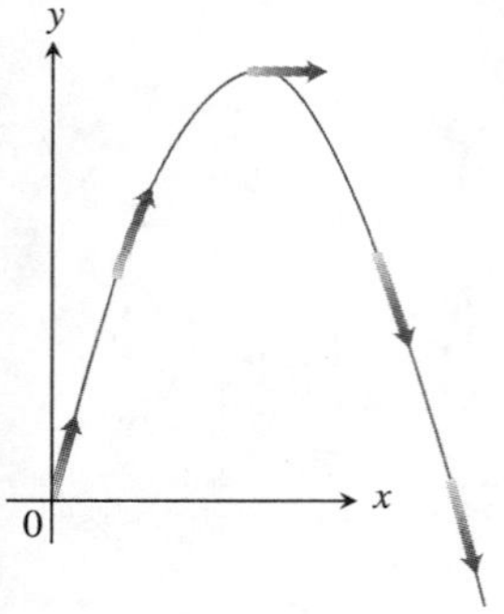

FIGURE 16.15 The velocity vectors $\mathbf{v}(t)$ of a projectile's motion make a vector field along the trajectory.

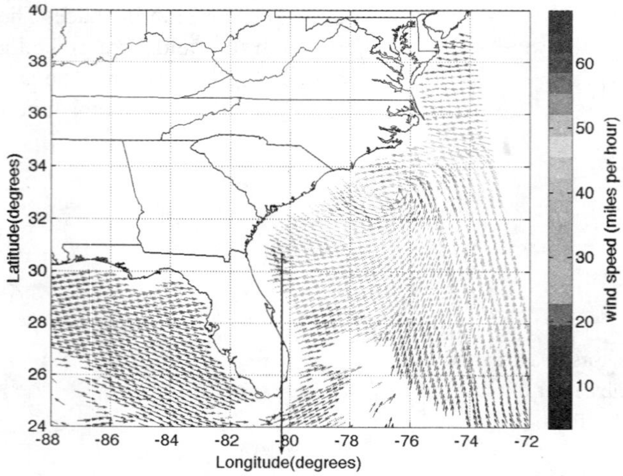

FIGURE 16.16 Data from NASA's QuikSCAT satellite were used to create this representation of wind speed and wind direction in Hurricane Irene approximately six hours before it made landfall in North Carolina on August 27, 2011. The arrows show wind direction, and speed is indicated by color (rather than length). The maximum wind speeds (over 130 km/hour) occurred over a region too small to see in this illustration. (*Source:* JPL-Caltech/ISRO/NASA)

In many physical applications, f represents a potential energy, and the gradient vector field indicates the corresponding force. In such situations, f is often taken to be negative, so that the force gives the direction of decreasing potential energy.

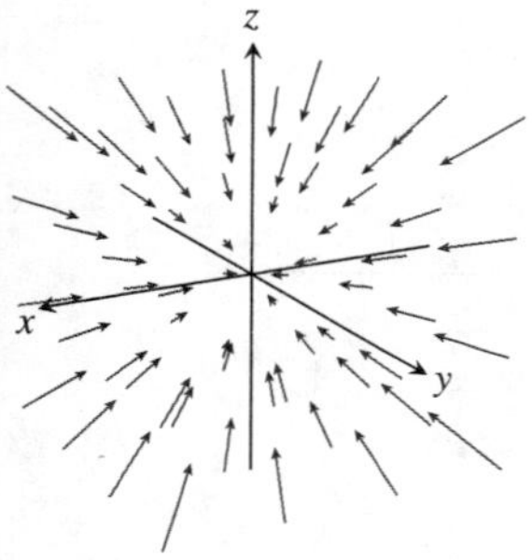

FIGURE 16.17 The vectors in a temperature gradient field point in the direction of greatest increase in temperature. In this case they are pointing toward the origin.

EXAMPLE 1 Suppose that a material is heated, that the resulting temperature T at each point (x, y, z) in a region of space is given by

$$T = 100 - x^2 - y^2 - z^2,$$

and that $\mathbf{F}(x, y, z)$ is defined to be the gradient of T. Find the vector field $\mathbf{F}$.

Solution The gradient field $\mathbf{F}$ is the field $\mathbf{F} = \nabla T = -2x\mathbf{i} - 2y\mathbf{j} - 2z\mathbf{k}$. At each point in the region, the vector field $\mathbf{F}$ gives the direction for which the increase in temperature is greatest. The vectors point toward the origin, where the temperature is greatest. See Figure 16.17. ■

Line Integrals of Vector Fields

In Section 16.1 we defined the line integral of a scalar function $f(x, y, z)$ over a path C. We turn our attention now to the idea of a line integral of a vector field $\mathbf{F}$ along the curve C. Such line integrals have important applications in the study of fluid flows, work and energy, and electrical or gravitational fields.

Assume that the vector field $\mathbf{F} = M(x, y, z)\mathbf{i} + N(x, y, z)\mathbf{j} + P(x, y, z)\mathbf{k}$ has continuous components, and that the curve C has a smooth parametrization $\mathbf{r}(t) = g(t)\mathbf{i} + h(t)\mathbf{j} + k(t)\mathbf{k}$, $a \le t \le b$. As discussed in Section 16.1, the parametrization $\mathbf{r}(t)$ defines a direction (or orientation) along C that we call the **forward direction**. At each point along the path C, the tangent vector $\mathbf{T} = d\mathbf{r}/ds = \mathbf{v}/|\mathbf{v}|$ is a unit vector tangent to the path and pointing in this forward direction. (The vector $\mathbf{v} = d\mathbf{r}/dt$ is the velocity vector tangent to C at the point, as discussed in Sections 13.1 and 13.3.) The line integral of the vector field is the line integral of the scalar tangential component of $\mathbf{F}$ along C. This tangential component is given by the dot product

$$\mathbf{F} \cdot \mathbf{T} = \mathbf{F} \cdot \frac{d\mathbf{r}}{ds},$$

so we are led to the following definition.

> **DEFINITION** Let $\mathbf{F}$ be a vector field with continuous components defined along a smooth curve C parametrized by $\mathbf{r}(t)$, $a \le t \le b$. Then the **line integral of F along *C*** is
>
> $$\int_C \mathbf{F} \cdot \mathbf{T}\, ds = \int_C \left(\mathbf{F} \cdot \frac{d\mathbf{r}}{ds}\right) ds = \int_C \mathbf{F} \cdot d\mathbf{r}. \tag{1}$$

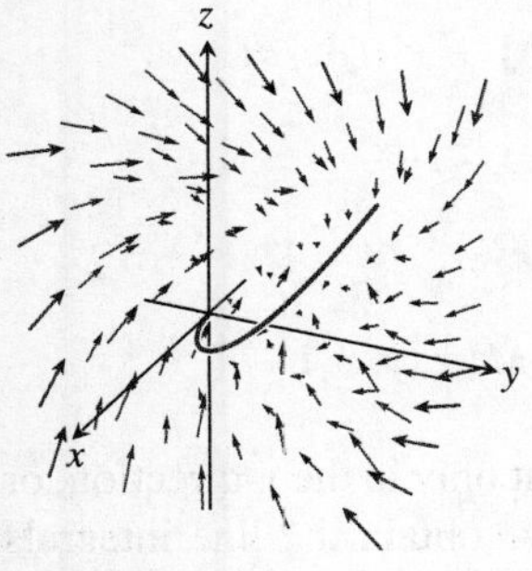

FIGURE 16.18 A curve winds through a vector field as in Example 2. The line integral is determined by the vectors that lie along the curve.

We evaluate line integrals of vector fields in a way similar to the way we evaluate line integrals of scalar functions (Section 16.1). The vector field may also be defined on points not meeting the curve, but only the vectors along the curve play a role in the line integral. See Figure 16.18.

> **Evaluating the Line Integral of $\mathbf{F} = M\mathbf{i} + N\mathbf{j} + P\mathbf{k}$ Along $C: \mathbf{r}(t) = g(t)\mathbf{i} + h(t)\mathbf{j} + k(t)\mathbf{k}$**
>
> 1. Express the vector field $\mathbf{F}$ along the parametrized curve C as $\mathbf{F}(\mathbf{r}(t))$ by substituting the components $x = g(t)$, $y = h(t)$, $z = k(t)$ of $\mathbf{r}$ into the scalar components $M(x, y, z)$, $N(x, y, z)$, $P(x, y, z)$ of $\mathbf{F}$.
> 2. Find the derivative (velocity) vector $d\mathbf{r}/dt$.
> 3. Evaluate the line integral with respect to the parameter t, $a \le t \le b$, to obtain
>
> $$\int_C \mathbf{F} \cdot d\mathbf{r} = \int_a^b \mathbf{F}(\mathbf{r}(t)) \cdot \frac{d\mathbf{r}}{dt}\, dt. \tag{2}$$

EXAMPLE 2 Evaluate $\int_C \mathbf{F} \cdot d\mathbf{r}$, where $\mathbf{F}(x, y, z) = z\mathbf{i} + xy\mathbf{j} - y^2\mathbf{k}$ along the curve C given by $\mathbf{r}(t) = t^2\mathbf{i} + t\mathbf{j} + \sqrt{t}\mathbf{k}$, $0 \le t \le 1$.

Solution We have

$$\mathbf{F}(\mathbf{r}(t)) = \sqrt{t}\mathbf{i} + t^3\mathbf{j} - t^2\mathbf{k} \qquad z = \sqrt{t},\ xy = t^3,\ -y^2 = -t^2$$

and

$$\frac{d\mathbf{r}}{dt} = 2t\mathbf{i} + \mathbf{j} + \frac{1}{2\sqrt{t}}\mathbf{k}.$$

Thus,

$$\begin{aligned}\int_C \mathbf{F} \cdot d\mathbf{r} &= \int_0^1 \mathbf{F}(\mathbf{r}(t)) \cdot \frac{d\mathbf{r}}{dt}\, dt \qquad \text{Eq. (2)}\\ &= \int_0^1 \left(2t^{3/2} + t^3 - \frac{1}{2}t^{3/2}\right) dt\\ &= \left[\left(\frac{3}{2}\right)\left(\frac{2}{5}t^{5/2}\right) + \frac{1}{4}t^4\right]_0^1 = \frac{17}{20}.\end{aligned}$$

■

Line Integrals with Respect to *dx*, *dy*, or *dz*

When analyzing forces or flows, it is often useful to consider each component direction separately. For example, when analyzing the effect of a gravitational force, we might want to consider motion and forces in the vertical direction, while ignoring horizontal motions. Or we might be interested only in the force exerted horizontally by water pushing against the face of a dam or in wind affecting the course of a plane. In such situations we want to evaluate a line integral of a scalar function with respect to only one of the coordinates, such as $\int_C M\,dx$. This type of integral is not the same as the arc length line integral $\int_C M\,ds$ we defined in Section 16.1, since it picks out displacement in the direction of only one coordinate. To define the integral $\int_C M\,dx$ for the scalar function $M(x, y, z)$, we specify a vector field $\mathbf{F} = M(x, y, z)\mathbf{i}$ having a component only in the x-direction, and none in the y- or the z-direction. Then, over the curve C parametrized by $\mathbf{r}(t) = g(t)\mathbf{i} + h(t)\mathbf{j} + k(t)\mathbf{k}$ for $a \le t \le b$, we have $x = g(t)$, $dx = g'(t)\,dt$, and

$$\begin{aligned}\mathbf{F}\cdot d\mathbf{r} = \mathbf{F}\cdot\frac{d\mathbf{r}}{dt}\,dt &= M(x, y, z)\mathbf{i}\cdot\left(g'(t)\mathbf{i} + h'(t)\mathbf{j} + k'(t)\mathbf{k}\right)dt\\ &= M(x, y, z)g'(t)\,dt = M(x, y, z)\,dx.\end{aligned}$$

As in the definition of the line integral of $\mathbf{F}$ along C, we define

$$\int_C M(x, y, z)\,dx = \int_C \mathbf{F}\cdot d\mathbf{r}, \quad \text{where} \quad \mathbf{F} = M(x, y, z)\mathbf{i}.$$

In the same way, by defining $\mathbf{F} = N(x, y, z)\mathbf{j}$ with a component only in the y-direction, or $\mathbf{F} = P(x, y, z)\mathbf{k}$ with a component only in the z-direction, we obtain the line integrals $\int_C N\,dy$ and $\int_C P\,dz$. Expressing everything in terms of the parameter t along the curve C, we have the following formulas for these three integrals:

$$\int_C M(x, y, z)\,dx = \int_a^b M(g(t), h(t), k(t))\,g'(t)\,dt \tag{3}$$

$$\int_C N(x, y, z)\,dy = \int_a^b N(g(t), h(t), k(t))\,h'(t)\,dt \tag{4}$$

$$\int_C P(x, y, z)\,dz = \int_a^b P(g(t), h(t), k(t))\,k'(t)\,dt \tag{5}$$

Line Integral Notation
The commonly occurring expression

$$\int_C M\,dx + N\,dy + P\,dz$$

is a short way of expressing the sum of three line integrals, one for each coordinate direction:

$$\int_C M(x, y, z)\,dx + \int_C N(x, y, z)\,dy + \int_C P(x, y, z)\,dz.$$

To evaluate these integrals, we parametrize C as $g(t)\mathbf{i} + h(t)\mathbf{j} + k(t)\mathbf{k}$ and use Equations (3), (4), and (5).

It often happens that these line integrals occur in combination, and we abbreviate the notation by writing

$$\int_C M(x, y, z)\,dx + \int_C N(x, y, z)\,dy + \int_C P(x, y, z)\,dz = \int_C M\,dx + N\,dy + P\,dz.$$

EXAMPLE 3 Evaluate the line integral $\int_C -y\,dx + z\,dy + 2x\,dz$, where C is the helix $\mathbf{r}(t) = (\cos t)\mathbf{i} + (\sin t)\mathbf{j} + t\mathbf{k}$, $0 \le t \le 2\pi$.

Solution We express everything in terms of the parameter t, so $x = \cos t$, $y = \sin t$, $z = t$, and $dx = -\sin t\,dt$, $dy = \cos t\,dt$, $dz = dt$. Then

$$\begin{aligned}\int_C -y\,dx + z\,dy + 2x\,dz &= \int_0^{2\pi} [(-\sin t)(-\sin t) + t\cos t + 2\cos t]\,dt\\ &= \int_0^{2\pi} [2\cos t + t\cos t + \sin^2 t]\,dt\\ &= \left[2\sin t + (t\sin t + \cos t) + \left(\frac{t}{2} - \frac{\sin 2t}{4}\right)\right]_0^{2\pi}\\ &= [0 + (0 + 1) + (\pi - 0)] - [0 + (0 + 1) + (0 - 0)]\\ &= \pi.\end{aligned}$$

■

Work Done by a Force over a Curve in Space

Suppose that the vector field $\mathbf{F} = M(x, y, z)\mathbf{i} + N(x, y, z)\mathbf{j} + P(x, y, z)\mathbf{k}$ represents a force throughout a region in space (it might be the force of gravity or an electromagnetic force) and that

$$\mathbf{r}(t) = g(t)\mathbf{i} + h(t)\mathbf{j} + k(t)\mathbf{k}, \qquad a \le t \le b,$$

represents a smooth curve C in the region. The formula for the work done by the force in moving an object along the curve is motivated by the same kind of reasoning we used in Chapter 6 to derive the ordinary single integral for the work done by a continuous force of magnitude $F(x)$ directed along an interval of the x-axis. For the curve C in space, we define the work done by a continuous force field $\mathbf{F}$ to move an object along C from a point A to another point B as follows.

We divide C into n subarcs $P_{k-1}P_k$ with lengths Δs_k, starting at A and ending at B. We choose any point (x_k, y_k, z_k) in the subarc $P_{k-1}P_k$ and let $\mathbf{T}(x_k, y_k, z_k)$ be the unit tangent vector at the chosen point. The work W_k done to move the object along the subarc $P_{k-1}P_k$ is approximated by the tangential component of the force $\mathbf{F}(x_k, y_k, z_k)$ times the arc length Δs_k, the distance the object moves along the subarc (see Figure 16.19). The total work done in moving the object from point A to point B is then obtained by summing the work done along each of the subarcs, so

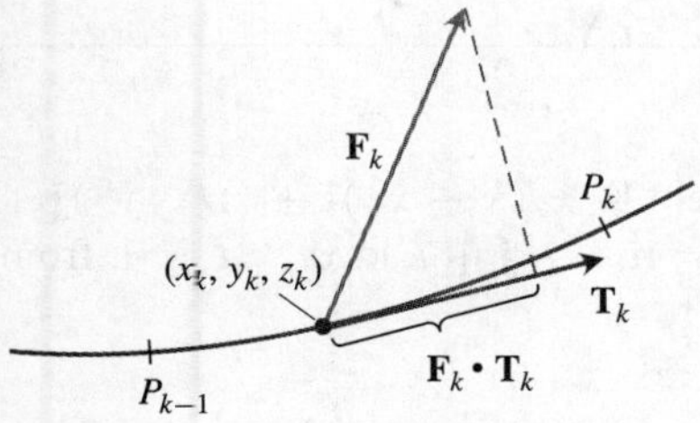

FIGURE 16.19 The work done along the subarc shown here is approximately $\mathbf{F}_k \cdot \mathbf{T}_k \, \Delta s_k$, where $\mathbf{F}_k = \mathbf{F}(x_k, y_k, z_k)$ and $\mathbf{T}_k = \mathbf{T}(x_k, y_k, z_k)$.

$$W = \sum_{k=1}^{n} W_k \approx \sum_{k=1}^{n} \mathbf{F}(x_k, y_k, z_k) \cdot \mathbf{T}(x_k, y_k, z_k)\, \Delta s_k.$$

For any subdivision of C into n subarcs, and for any choice of the points (x_k, y_k, z_k) within each subarc, as $n \to \infty$ and $\Delta s_k \to 0$, these sums approach the line integral

$$\int_C \mathbf{F} \cdot \mathbf{T}\, ds.$$

This is the line integral of $\mathbf{F}$ along C, which now defines the total work done.

DEFINITION Let C be a smooth curve parametrized by $\mathbf{r}(t)$, $a \le t \le b$, and let $\mathbf{F}$ be a continuous force field over a region containing C. Then the **work** done in moving an object from the point $A = \mathbf{r}(a)$ to the point $B = \mathbf{r}(b)$ along C is

$$W = \int_C \mathbf{F} \cdot \mathbf{T}\, ds = \int_a^b \mathbf{F}(\mathbf{r}(t)) \cdot \frac{d\mathbf{r}}{dt}\, dt. \qquad (6)$$

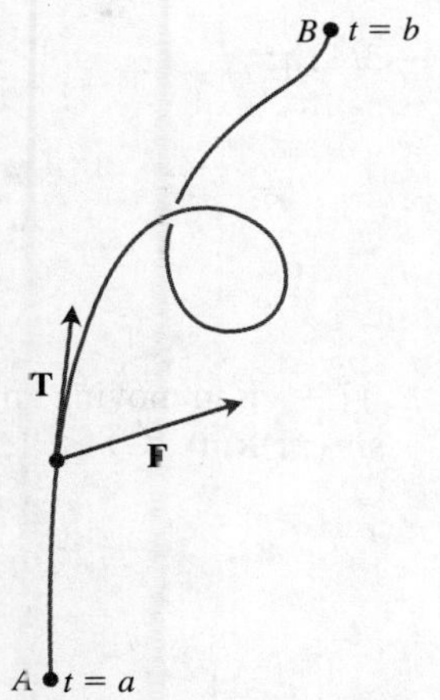

FIGURE 16.20 The work done by a force $\mathbf{F}$ is the line integral of the scalar component $\mathbf{F} \cdot \mathbf{T}$ over the smooth curve from A to B.

The sign of the number we calculate with this integral depends on the direction in which the curve is traversed. If we reverse the direction of motion, then we reverse the direction of $\mathbf{T}$ in Figure 16.20 and change the sign of $\mathbf{F} \cdot \mathbf{T}$ and its integral.

Using the notations we have presented, we can express the work integral in a variety of ways, depending upon what seems most suitable or convenient for a particular discussion. Table 16.2 shows five ways we can write the work integral in Equation (6). In the table, the field components M, N, and P are functions of the intermediate variables x, y, and z, which in turn are functions of the independent variable t along the curve C in the vector field. So along the curve, $x = g(t)$, $y = h(t)$, and $z = k(t)$ with $dx = g'(t)\,dt$, $dy = h'(t)\,dt$, and $dz = k'(t)\,dt$.

TABLE 16.2 Different ways to write the work integral for $\mathbf{F} = M\mathbf{i} + N\mathbf{j} + P\mathbf{k}$ over the curve C: $\mathbf{r}(t) = g(t)\mathbf{i} + h(t)\mathbf{j} + k(t)\mathbf{k}$, $a \le t \le b$

$W = \int_C \mathbf{F} \cdot \mathbf{T}\, ds$	The definition
$= \int_C \mathbf{F} \cdot d\mathbf{r}$	Vector differential form
$= \int_a^b \mathbf{F} \cdot \frac{d\mathbf{r}}{dt}\, dt$	Parametric vector evaluation
$= \int_a^b (Mg'(t) + Nh'(t) + Pk'(t))\, dt$	Parametric scalar evaluation
$= \int_C M\,dx + N\,dy + P\,dz$	Scalar differential form

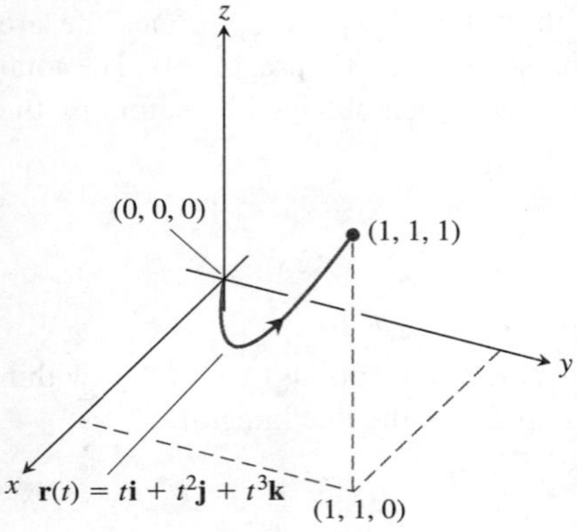

FIGURE 16.21 The curve in Example 4.

EXAMPLE 4 Find the work done by the force field $\mathbf{F} = (y - x^2)\mathbf{i} + (z - y^2)\mathbf{j} + (x - z^2)\mathbf{k}$ in moving an object along the curve $\mathbf{r}(t) = t\mathbf{i} + t^2\mathbf{j} + t^3\mathbf{k}$, $0 \le t \le 1$, from $(0, 0, 0)$ to $(1, 1, 1)$ (Figure 16.21).

Solution First we evaluate $\mathbf{F}$ on the curve $\mathbf{r}(t)$:

$$\begin{aligned}\mathbf{F} &= (y - x^2)\mathbf{i} + (z - y^2)\mathbf{j} + (x - z^2)\mathbf{k} \\ &= \underbrace{(t^2 - t^2)}_{0}\mathbf{i} + (t^3 - t^4)\mathbf{j} + (t - t^6)\mathbf{k}. \qquad \text{Substitute } x = t,\ y = t^2,\ z = t^3.\end{aligned}$$

Then we find $d\mathbf{r}/dt$:

$$\frac{d\mathbf{r}}{dt} = \frac{d}{dt}(t\mathbf{i} + t^2\mathbf{j} + t^3\mathbf{k}) = \mathbf{i} + 2t\mathbf{j} + 3t^2\mathbf{k}.$$

Finally, we find $\mathbf{F} \cdot d\mathbf{r}/dt$ and integrate from $t = 0$ to $t = 1$:

$$\begin{aligned}\mathbf{F} \cdot \frac{d\mathbf{r}}{dt} &= [(t^3 - t^4)\mathbf{j} + (t - t^6)\mathbf{k}] \cdot (\mathbf{i} + 2t\mathbf{j} + 3t^2\mathbf{k}) \\ &= (t^3 - t^4)(2t) + (t - t^6)(3t^2) = 2t^4 - 2t^5 + 3t^3 - 3t^8. \qquad \text{Evaluate dot product.}\end{aligned}$$

Thus

$$\begin{aligned}\text{Work} &= \int_a^b \mathbf{F} \cdot \frac{d\mathbf{r}}{dt}\, dt = \int_0^1 (2t^4 - 2t^5 + 3t^3 - 3t^8)\, dt \\ &= \left[\frac{2}{5}t^5 - \frac{2}{6}t^6 + \frac{3}{4}t^4 - \frac{3}{9}t^9\right]_0^1 = \frac{29}{60}.\end{aligned}$$

■

EXAMPLE 5 Find the work done by the force field $\mathbf{F} = x\mathbf{i} + y\mathbf{j} + z\mathbf{k}$ in moving an object along the curve C parametrized by $\mathbf{r}(t) = \cos(\pi t)\mathbf{i} + t^2\mathbf{j} + \sin(\pi t)\mathbf{k}$, $0 \le t \le 1$.

Solution We begin by writing $\mathbf{F}$ along C as a function of t:

$$\mathbf{F}(\mathbf{r}(t)) = \cos(\pi t)\mathbf{i} + t^2\mathbf{j} + \sin(\pi t)\mathbf{k}.$$

Next we compute $d\mathbf{r}/dt$:

$$\frac{d\mathbf{r}}{dt} = -\pi\sin(\pi t)\mathbf{i} + 2t\mathbf{j} + \pi\cos(\pi t)\mathbf{k}.$$

We then calculate the dot product:

$$\mathbf{F}(\mathbf{r}(t)) \cdot \frac{d\mathbf{r}}{dt} = -\pi \sin(\pi t)\cos(\pi t) + 2t^3 + \pi \sin(\pi t)\cos(\pi t) = 2t^3.$$

The work done is the line integral

$$\int_a^b \mathbf{F}(\mathbf{r}(t)) \cdot \frac{d\mathbf{r}}{dt}\, dt = \int_0^1 2t^3\, dt = \frac{t^4}{2}\bigg]_0^1 = \frac{1}{2}.$$ ■

Flow Integrals and Circulation for Velocity Fields

Suppose that **F** represents the velocity field of a fluid flowing through a region in space (a tidal basin or the turbine chamber of a hydroelectric generator, for example). Under these circumstances, the integral of $\mathbf{F} \cdot \mathbf{T}$ along a curve in the region gives the fluid's flow along, or *circulation* around, the curve. For instance, the vector field in Figure 16.12 gives zero circulation around the unit circle in the plane. By contrast, the vector field in Figure 16.13 gives nonzero circulation around the unit circle.

DEFINITION If $\mathbf{r}(t)$ parametrizes a smooth curve C in the domain of a continuous velocity field **F**, the **flow** along the curve from $A = \mathbf{r}(a)$ to $B = \mathbf{r}(b)$ is

$$\text{Flow} = \int_C \mathbf{F} \cdot \mathbf{T}\, ds. \tag{7}$$

The integral is called a **flow integral**. If the curve starts and ends at the same point, so that $A = B$, the flow is called the **circulation** around the curve.

The direction we travel along C matters. If we reverse the direction, then **T** is replaced by $-\mathbf{T}$ and the sign of the integral changes. We evaluate flow integrals the same way we evaluate work integrals.

EXAMPLE 6 A fluid's velocity field is $\mathbf{F} = x\mathbf{i} + z\mathbf{j} + y\mathbf{k}$. Find the flow along the helix $\mathbf{r}(t) = (\cos t)\mathbf{i} + (\sin t)\mathbf{j} + t\mathbf{k}$, $0 \le t \le \pi/2$.

Solution We evaluate **F** on the curve $\mathbf{r}(t)$:

$$\mathbf{F} = x\mathbf{i} + z\mathbf{j} + y\mathbf{k} = (\cos t)\mathbf{i} + t\mathbf{j} + (\sin t)\mathbf{k} \qquad \text{Substitute } x = \cos t,\ z = t,\ y = \sin t.$$

and then find $d\mathbf{r}/dt$:

$$\frac{d\mathbf{r}}{dt} = (-\sin t)\mathbf{i} + (\cos t)\mathbf{j} + \mathbf{k}.$$

The dot product of **F** with $d\mathbf{r}/dt$ is

$$\begin{aligned} \mathbf{F} \cdot \frac{d\mathbf{r}}{dt} &= (\cos t)(-\sin t) + (t)(\cos t) + (\sin t)(1) \\ &= -\sin t \cos t + t \cos t + \sin t. \end{aligned}$$

Finally, we integrate $\mathbf{F} \cdot (d\mathbf{r}/dt)$ from $t = 0$ to $t = \dfrac{\pi}{2}$:

$$\begin{aligned} \text{Flow} &= \int_{t=a}^{t=b} \mathbf{F} \cdot \frac{d\mathbf{r}}{dt}\, dt = \int_0^{\pi/2} (-\sin t \cos t + t \cos t + \sin t)\, dt \\ &= \left[\frac{\cos^2 t}{2} + t \sin t\right]_0^{\pi/2} = \left(0 + \frac{\pi}{2}\right) - \left(\frac{1}{2} + 0\right) = \frac{\pi}{2} - \frac{1}{2}. \end{aligned}$$ ■

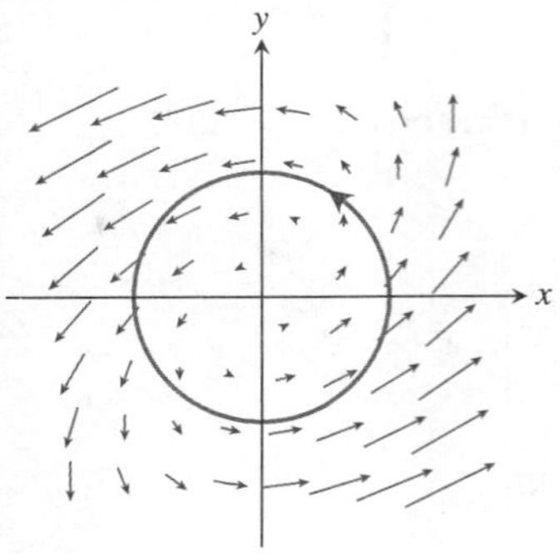

FIGURE 16.22 The vector field **F** and curve $\mathbf{r}(t)$ in Example 7.

EXAMPLE 7 Find the circulation of the field $\mathbf{F} = (x - y)\mathbf{i} + x\mathbf{j}$ around the circle $\mathbf{r}(t) = (\cos t)\mathbf{i} + (\sin t)\mathbf{j}$, $0 \le t \le 2\pi$ (Figure 16.22).

Solution On the circle, $\mathbf{F} = (x - y)\mathbf{i} + x\mathbf{j} = (\cos t - \sin t)\mathbf{i} + (\cos t)\mathbf{j}$, and

$$\frac{d\mathbf{r}}{dt} = (-\sin t)\mathbf{i} + (\cos t)\mathbf{j}.$$

Then

$$\mathbf{F} \cdot \frac{d\mathbf{r}}{dt} = -\sin t \cos t + \underbrace{\sin^2 t + \cos^2 t}_{1}$$

gives

$$\begin{aligned}\text{Circulation} &= \int_0^{2\pi} \mathbf{F} \cdot \frac{d\mathbf{r}}{dt}\, dt = \int_0^{2\pi} (1 - \sin t \cos t)\, dt \\ &= \left[t - \frac{\sin^2 t}{2}\right]_0^{2\pi} = 2\pi.\end{aligned}$$

As Figure 16.22 suggests, a fluid with this velocity field is circulating *counterclockwise* around the circle. The circle is also traversed counterclockwise as t increases from 0 to 2π, so the circulation is positive. ■

Flux Across a Simple Closed Plane Curve

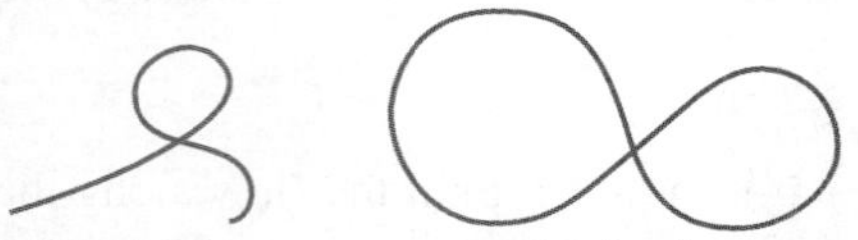

FIGURE 16.23 Distinguishing between curves that are simple and curves that are closed. Closed curves are also called loops.

A curve in the xy-plane is **simple** if it does not cross itself (Figure 16.23). When a curve starts and ends at the same point, it is a **closed curve** or **loop**. To find the rate at which a fluid is entering or leaving a region enclosed by a smooth simple closed curve C in the xy-plane, we calculate the line integral over C of $\mathbf{F} \cdot \mathbf{n}$, the scalar component of the fluid's velocity field in the direction of the curve's outward-pointing normal vector. We use only the normal component of **F**, while ignoring the tangential component, because the normal component leads to the flow across C. The value of this integral is the *flux* of **F** across C. *Flux* is Latin for flow, but many flux calculations involve no motion at all. When **F** is an electric or magnetic field, for instance, the integral of $\mathbf{F} \cdot \mathbf{n}$ is still called the flux of the field across C.

DEFINITION If C is a smooth simple closed curve in the domain of a continuous vector field $\mathbf{F} = M(x, y)\mathbf{i} + N(x, y)\mathbf{j}$ in the plane, and if **n** is the outward-pointing unit normal vector on C, the **flux** of **F** across C is

$$\text{Flux of } \mathbf{F} \text{ across } C = \int_C \mathbf{F} \cdot \mathbf{n}\, ds. \tag{8}$$

Notice the difference between flux and circulation. The flux of **F** across C is the line integral with respect to arc length of $\mathbf{F} \cdot \mathbf{n}$, the scalar component of **F** in the direction of the outward normal. The circulation of **F** around C is the line integral with respect to arc length of $\mathbf{F} \cdot \mathbf{T}$, the scalar component of **F** in the direction of the unit tangent vector. Flux is the integral of the normal component of **F**; circulation is the integral of the tangential component of **F**. In Section 16.6 we will define flux across a surface.

To evaluate the integral for flux in Equation (8), we begin with a smooth parametrization

$$x = g(t), \qquad y = h(t), \qquad a \le t \le b,$$

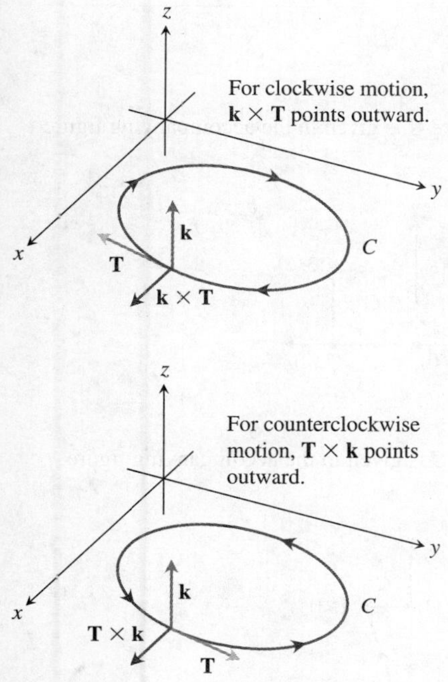

FIGURE 16.24 To find an outward unit normal vector for a smooth simple curve C in the xy-plane that is traversed counterclockwise as t increases, we take $\mathbf{n} = \mathbf{T} \times \mathbf{k}$. For clockwise motion, we take $\mathbf{n} = \mathbf{k} \times \mathbf{T}$.

that traces the curve C exactly once as t increases from a to b. We can find the outward unit normal vector $\mathbf{n}$ by crossing the curve's unit tangent vector $\mathbf{T}$ with the vector $\mathbf{k}$. But which order do we choose, $\mathbf{T} \times \mathbf{k}$ or $\mathbf{k} \times \mathbf{T}$? Which one points outward? It depends on which way C is traversed as t increases. If the motion is clockwise, $\mathbf{k} \times \mathbf{T}$ points outward; if the motion is counterclockwise, $\mathbf{T} \times \mathbf{k}$ points outward (Figure 16.24). The usual choice is $\mathbf{n} = \mathbf{T} \times \mathbf{k}$, the choice that assumes counterclockwise motion. Thus, even though the value of the integral in Equation (8) does not depend on which way C is traversed, the formulas we are about to derive for computing $\mathbf{n}$ and evaluating the integral assume counterclockwise motion.

In terms of components,

$$\mathbf{n} = \mathbf{T} \times \mathbf{k} = \left(\frac{dx}{ds}\mathbf{i} + \frac{dy}{ds}\mathbf{j}\right) \times \mathbf{k} = \frac{dy}{ds}\mathbf{i} - \frac{dx}{ds}\mathbf{j}. \qquad \begin{vmatrix} \mathbf{i} & \mathbf{j} & \mathbf{k} \\ \frac{dx}{ds} & \frac{dy}{ds} & 0 \\ 0 & 0 & 1 \end{vmatrix}$$

If $\mathbf{F} = M(x, y)\mathbf{i} + N(x, y)\mathbf{j}$, then

$$\mathbf{F} \cdot \mathbf{n} = M(x, y)\frac{dy}{ds} - N(x, y)\frac{dx}{ds}.$$

Hence,

$$\int_C \mathbf{F} \cdot \mathbf{n}\, ds = \int_C \left(M\frac{dy}{ds} - N\frac{dx}{ds}\right) ds = \oint_C M\, dy - N\, dx.$$

We put a directed circle ⟲ on the last integral as a reminder that the integration around the closed curve C is to be in the counterclockwise direction. To evaluate this integral, we express M, dy, N, and dx in terms of the parameter t and integrate from $t = a$ to $t = b$. We do not need to know $\mathbf{n}$ or ds explicitly to find the flux.

> **Calculating Flux Across a Smooth Closed Plane Curve**
>
> $$(\text{Flux of } \mathbf{F} = M\mathbf{i} + N\mathbf{j} \text{ across } C) = \oint_C M\, dy - N\, dx \qquad (9)$$
>
> The integral can be evaluated from any smooth parametrization $x = g(t)$, $y = h(t)$, $a \le t \le b$, that traces C counterclockwise exactly once.

EXAMPLE 8 Find the flux of $\mathbf{F} = (x - y)\mathbf{i} + x\mathbf{j}$ across the circle $x^2 + y^2 = 1$ in the xy-plane. (The vector field and curve were shown in Figure 16.22.)

Solution The parametrization $\mathbf{r}(t) = (\cos t)\mathbf{i} + (\sin t)\mathbf{j}$, $0 \le t \le 2\pi$, traces the circle counterclockwise exactly once. We can therefore use this parametrization in Equation (9). With

$$M = x - y = \cos t - \sin t, \qquad dy = d(\sin t) = \cos t\, dt,$$
$$N = x = \cos t, \qquad dx = d(\cos t) = -\sin t\, dt,$$

we find

$$\text{Flux} = \oint_C M\, dy - N\, dx = \int_0^{2\pi} (\cos^2 t - \sin t \cos t + \cos t \sin t)\, dt \qquad \text{Eq. (9)}$$

$$= \int_0^{2\pi} \cos^2 t\, dt = \int_0^{2\pi} \frac{1 + \cos 2t}{2}\, dt = \left[\frac{t}{2} + \frac{\sin 2t}{4}\right]_0^{2\pi} = \pi.$$

The flux of $\mathbf{F}$ across the circle is π. Since the answer is positive, the net flow across the curve is outward. A net inward flow would have given a negative flux. ■

EXERCISES 16.2

Vector Fields

Find the gradient fields of the functions in Exercises 1–4.

1. $f(x, y, z) = (x^2 + y^2 + z^2)^{-1/2}$

2. $f(x, y, z) = x^2y + y^2x + z^2$

3. $g(x, y, z) = e^z - \ln(x^2 + y^2)$

4. $g(x, y, z) = xy + yz + xz$

5. Give a formula $\mathbf{F} = M(x, y)\mathbf{i} + N(x, y)\mathbf{j}$ for the vector field in the plane that has the property that $\mathbf{F}$ points toward the origin with magnitude inversely proportional to the square of the distance from (x, y) to the origin. (The field is not defined at $(0, 0)$.)

6. Give a formula $\mathbf{F} = M(x, y)\mathbf{i} + N(x, y)\mathbf{j}$ for the vector field in the plane that has the properties that $\mathbf{F} = \mathbf{0}$ at $(0, 0)$ and that at any other point (a, b), $\mathbf{F}$ is tangent to the circle $x^2 + y^2 = a^2 + b^2$ and points in the clockwise direction with magnitude $|\mathbf{F}| = \sqrt{a^2 + b^2}$.

Line Integrals of Vector Fields

In Exercises 7–12, find the line integrals of $\mathbf{F}$ from $(0, 0, 0)$ to $(1, 1, 1)$ over each of the following paths in the accompanying figure.

a. The straight-line path C_1: $\mathbf{r}(t) = t\mathbf{i} + t\mathbf{j} + t\mathbf{k}$, $0 \le t \le 1$

b. The curved path C_2: $\mathbf{r}(t) = t\mathbf{i} + t^2\mathbf{j} + t^4\mathbf{k}$, $0 \le t \le 1$

c. The path $C_3 \cup C_4$ consisting of the line segment from $(0, 0, 0)$ to $(1, 1, 0)$ followed by the segment from $(1, 1, 0)$ to $(1, 1, 1)$

7. $\mathbf{F} = 3y\mathbf{i} + 2x\mathbf{j} + 4z\mathbf{k}$ **8.** $\mathbf{F} = [1/(x^2 + 1)]\mathbf{j}$

9. $\mathbf{F} = \sqrt{z}\mathbf{i} - 2x\mathbf{j} + \sqrt{y}\mathbf{k}$ **10.** $\mathbf{F} = xy\mathbf{i} + yz\mathbf{j} + xz\mathbf{k}$

11. $\mathbf{F} = (3x^2 - 3x)\mathbf{i} + 3z\mathbf{j} + \mathbf{k}$

12. $\mathbf{F} = (y + z)\mathbf{i} + (z + x)\mathbf{j} + (x + y)\mathbf{k}$

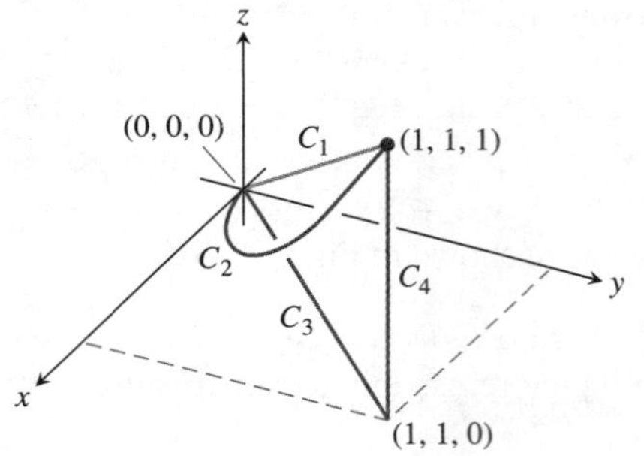

Line Integrals with Respect to x, y, and z

In Exercises 13–16, find the line integrals along the given path C.

13. $\int_C (x - y)\,dx$, where C: $x = t$, $y = 2t + 1$, for $0 \le t \le 3$

14. $\int_C \frac{x}{y}\,dy$, where C: $x = t$, $y = t^2$, for $1 \le t \le 2$

15. $\int_C (x^2 + y^2)\,dy$, where C is given in the accompanying figure

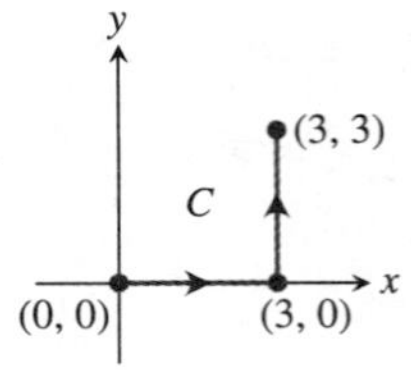

16. $\int_C \sqrt{x + y}\,dx$, where C is given in the accompanying figure

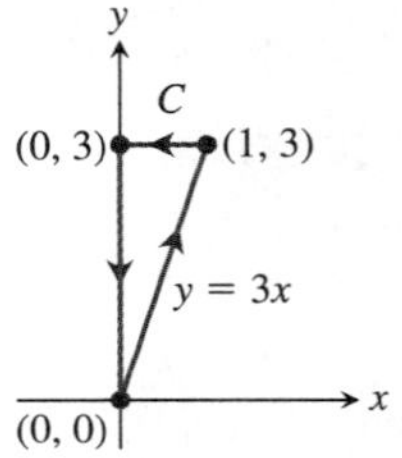

17. Along the curve $\mathbf{r}(t) = t\mathbf{i} - \mathbf{j} + t^2\mathbf{k}$, $0 \le t \le 1$, evaluate each of the following integrals.

a. $\int_C (x + y - z)\,dx$ **b.** $\int_C (x + y - z)\,dy$

c. $\int_C (x + y - z)\,dz$

18. Along the curve $\mathbf{r}(t) = (\cos t)\mathbf{i} + (\sin t)\mathbf{j} - (\cos t)\mathbf{k}$, $0 \le t \le \pi$, evaluate each of the following integrals.

a. $\int_C xz\,dx$ **b.** $\int_C xz\,dy$

c. $\int_C xyz\,dz$

Work

In Exercises 19–22, find the work done by $\mathbf{F}$ over the curve in the direction of increasing t.

19. $\mathbf{F} = xy\mathbf{i} + y\mathbf{j} - yz\mathbf{k}$

$\mathbf{r}(t) = t\mathbf{i} + t^2\mathbf{j} + t\mathbf{k}$, $0 \le t \le 1$

20. $\mathbf{F} = 2y\mathbf{i} + 3x\mathbf{j} + (x + y)\mathbf{k}$

$\mathbf{r}(t) = (\cos t)\mathbf{i} + (\sin t)\mathbf{j} + (t/6)\mathbf{k}$, $0 \le t \le 2\pi$

21. $\mathbf{F} = 3xy\mathbf{i} - 52\mathbf{j} + 10x\mathbf{k}$

$\mathbf{r}(t) = (t^2 + 1)\mathbf{i} + 2t^2\mathbf{j} + t^3\mathbf{k}$, $1 \le t \le 2$

22. $\mathbf{F} = yz\mathbf{i} + zx\mathbf{j} - xy\mathbf{k}$

$\mathbf{r}(t) = t\mathbf{i} + t^2\mathbf{j} + t^3\mathbf{k}$, $0 \le t \le 2$

Line Integrals in the Plane

23. $\int y^2 dx + 2xy\,dy$ along the circle $x^2 + y^2 = a^2$ in the counter clockwise-sense.

24. Evaluate $\int_C (x - y)\,dx + (x + y)\,dy$ counterclockwise around the triangle with vertices $(0,0)$, $(1,0)$, and $(0,1)$.

25. Evaluate $\int_C \mathbf{F} \cdot \mathbf{T}\,ds$ for the vector field $\mathbf{F} = x^2\mathbf{i} - y\mathbf{j}$ along the curve $x = y^2$ from $(4,2)$ to $(1,-1)$.

26. Evaluate $\int_C \mathbf{F} \cdot d\mathbf{r}$ for the vector field $\mathbf{F} = y\mathbf{i} - x\mathbf{j}$ counterclockwise along the unit circle $x^2 + y^2 = 1$ from $(1,0)$ to $(0,1)$.

Work, Circulation, and Flux in the Plane

27. **Work** Find the work done by the force $\mathbf{F} = xy\mathbf{i} + (y - x)\mathbf{j}$ over the straight line from $(1,1)$ to $(2,3)$.

28. **Work** Find the work done by the gradient of $f(x, y) = (x + y)^2$ counterclockwise around the circle $x^2 + y^2 = 4$ from $(2,0)$ to itself.

29. **Circulation and flux** Find the circulation and flux of the fields

$$\mathbf{F}_1 = x\mathbf{i} + y\mathbf{j} \quad \text{and} \quad \mathbf{F}_2 = -y\mathbf{i} + x\mathbf{j}$$

around and across each of the following curves.

a. The circle $\mathbf{r}(t) = (\cos t)\mathbf{i} + (\sin t)\mathbf{j}, \quad 0 \le t \le 2\pi$

b. The ellipse $\mathbf{r}(t) = (\cos t)\mathbf{i} + (4\sin t)\mathbf{j}, \quad 0 \le t \le 2\pi$

30. **Flux across a circle** Find the flux of the fields

$$\mathbf{F}_1 = 2x\mathbf{i} - 3y\mathbf{j} \quad \text{and} \quad \mathbf{F}_2 = 2x\mathbf{i} + (x - y)\mathbf{j}$$

across the circle

$$\mathbf{r}(t) = (a\cos t)\mathbf{i} + (a\sin t)\mathbf{j}, \quad 0 \le t \le 2\pi.$$

In Exercises 31–34, find the circulation and flux of the field $\mathbf{F}$ around and across the closed semicircular path that consists of the semicircular arch $\mathbf{r}_1(t) = (a\cos t)\mathbf{i} + (a\sin t)\mathbf{j}$, $0 \le t \le \pi$, followed by the line segment $\mathbf{r}_2(t) = t\mathbf{i}$, $-a \le t \le a$.

31. $\mathbf{F} = x\mathbf{i} + y\mathbf{j}$

32. $\mathbf{F} = x^2\mathbf{i} + y^2\mathbf{j}$

33. $\mathbf{F} = -y\mathbf{i} + x\mathbf{j}$

34. $\mathbf{F} = -y^2\mathbf{i} + x^2\mathbf{j}$

35. **Flow integrals** Find the flow of the velocity field $\mathbf{F} = (x + y)\mathbf{i} - (x^2 + y^2)\mathbf{j}$ along each of the following paths from $(1,0)$ to $(-1,0)$ in the xy-plane.

a. The upper half of the circle $x^2 + y^2 = 1$

b. The line segment from $(1,0)$ to $(-1,0)$

c. The line segment from $(1,0)$ to $(0,-1)$ followed by the line segment from $(0,-1)$ to $(-1,0)$

36. **Flux across a triangle** Find the flux of the field $\mathbf{F}$ in Exercise 35 outward across the triangle with vertices $(1,0)$, $(0,1)$, $(-1,0)$.

37. The flow of a gas with a density of $\delta = 0.001\ \text{kg/m}^2$ over the closed curve $\mathbf{r}(t) = (-\sin t)\mathbf{i} + (\cos t)\mathbf{j}$, $0 \le t \le 2\pi$, is given by the vector field $\mathbf{F} = \delta\mathbf{v}$, where $\mathbf{v} = x\mathbf{i} + y^2\mathbf{j}$ is a velocity field measured in meters per second. Find the flux of $\mathbf{F}$ across the curve $\mathbf{r}(t)$.

38. The flow of a gas with a density of $\delta = 0.3\ \text{kg/m}^2$ over the closed curve $\mathbf{r}(t) = (\cos t)\mathbf{i} + (\sin t)\mathbf{j}$, $0 \le t \le 2\pi$, is given by the vector field $\mathbf{F} = \delta\mathbf{v}$, where $\mathbf{v} = x^2\mathbf{i} - y\mathbf{j}$ is a velocity field measured in meters per second. Find the flux of $\mathbf{F}$ across the curve $\mathbf{r}(t)$.

39. Find the flow of the velocity field $\mathbf{F} = y^2\mathbf{i} + 2xy\mathbf{j}$ along each of the following paths from $(0,0)$ to $(2,4)$.

a.

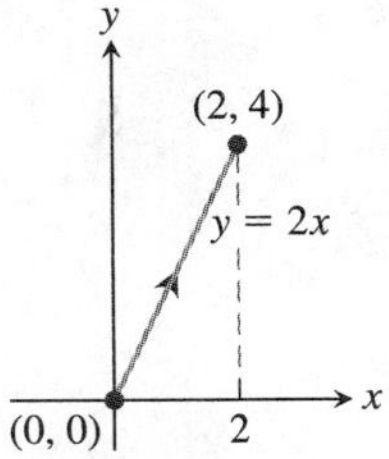

b.

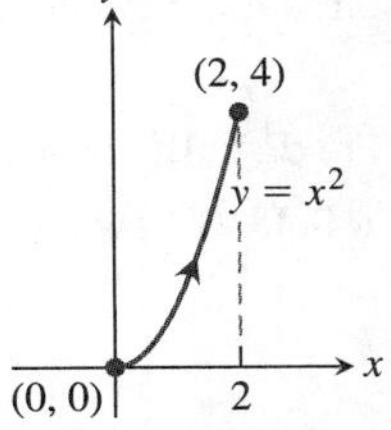

c. Use any path from $(0,0)$ to $(2,4)$ different from parts (a) and (b).

40. Find the circulation of the field $\mathbf{F} = y\mathbf{i} + (x + 2y)\mathbf{j}$ around each of the following closed paths.

a.

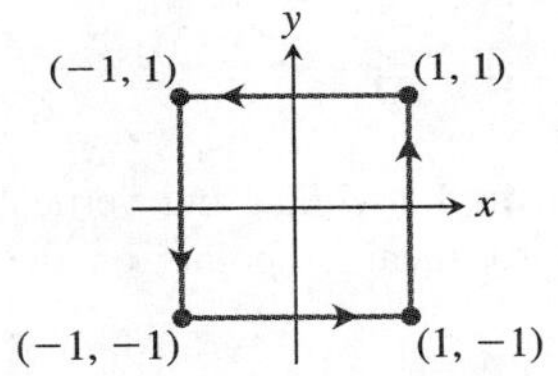

b.

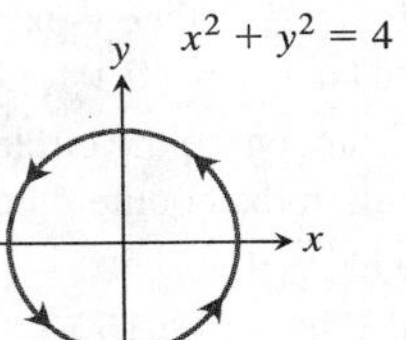

c. Use any closed path different from parts (a) and (b).

41. Find the work done by the force $\mathbf{F} = y^2\mathbf{i} + x^3\mathbf{j}$, where force is measured in newtons, in moving an object over the curve $\mathbf{r}(t) = 2t\mathbf{i} + t^2\mathbf{j}$, $0 \le t \le 2$, where distance is measured in meters.

42. Find the work done by the force $\mathbf{F} = e^y\mathbf{i} + (\ln x)\mathbf{j} + 3z\mathbf{k}$, where force is measured in newtons, in moving an object over the curve $\mathbf{r}(t) = e^t\mathbf{i} + (\ln t)\mathbf{j} + t^2\mathbf{k}$, $1 \le t \le e$, where distance is measured in meters.

43. Find the flow of the velocity field $\mathbf{F} = \dfrac{x}{y+1}\mathbf{i} + \dfrac{y}{x+1}\mathbf{j}$, where velocity is measured in meters per second, over the curve $\mathbf{r}(t) = t^2\mathbf{i} + t\mathbf{j}$, $0 \le t \le 1$.

44. Find the flow of the velocity field $\mathbf{F} = (y + z)\mathbf{i} + x\mathbf{j} - y\mathbf{k}$, where velocity is measured in meters per second, over the curve $\mathbf{r}(t) = e^t\mathbf{i} - e^{2t}\mathbf{j} + e^{-t}\mathbf{k}$, $0 \le t \le \ln 2$.

45. Salt water with a density of $\delta = 0.25\ \text{g/cm}^2$ flows over the curve $\mathbf{r}(t) = \sqrt{t}\mathbf{i} + t\mathbf{j}$, $0 \le t \le 4$, according to the vector field $\mathbf{F} = \delta\mathbf{v}$, where $\mathbf{v} = xy\mathbf{i} + (y - x)\mathbf{j}$ is a velocity field measured in centimeters per second. Find the flow of $\mathbf{F}$ over the curve $\mathbf{r}(t)$.

46. Propyl alcohol with a density of $\delta = 0.2\ \text{g/cm}^2$ flows over the closed curve $\mathbf{r}(t) = (\sin t)\mathbf{i} - (\cos t)\mathbf{j}$, $0 \le t \le 2\pi$, according to the vector field $\mathbf{F} = \delta\mathbf{v}$, where $\mathbf{v} = (x - y)\mathbf{i} + x^2\mathbf{j}$ is a velocity field measured in centimeters per second. Find the circulation of $\mathbf{F}$ around the curve $\mathbf{r}(t)$.

Vector Fields in the Plane

47. Spin field Draw the spin field

$$\mathbf{F} = -\frac{y}{\sqrt{x^2 + y^2}}\mathbf{i} + \frac{x}{\sqrt{x^2 + y^2}}\mathbf{j}$$

(see Figure 16.13) along with its horizontal and vertical components at a representative assortment of points on the circle $x^2 + y^2 = 4$.

48. Radial field Draw the radial field

$$\mathbf{F} = x\mathbf{i} + y\mathbf{j}$$

(see Figure 16.12) along with its horizontal and vertical components at a representative assortment of points on the circle $x^2 + y^2 = 1$.

49. A field of tangent vectors

a. Find a field $\mathbf{G} = P(x, y)\mathbf{i} + Q(x, y)\mathbf{j}$ in the xy-plane with the property that at any point $(a, b) \ne (0, 0)$, $\mathbf{G}$ is a vector of magnitude $\sqrt{a^2 + b^2}$ tangent to the circle $x^2 + y^2 = a^2 + b^2$ and pointing in the counterclockwise direction. (The field is undefined at $(0, 0)$.)

b. How is $\mathbf{G}$ related to the spin field $\mathbf{F}$ in Figure 16.13?

50. A field of tangent vectors

a. Find a field $\mathbf{G} = P(x, y)\mathbf{i} + Q(x, y)\mathbf{j}$ in the xy-plane with the property that at any point $(a, b) \ne (0, 0)$, $\mathbf{G}$ is a unit vector tangent to the circle $x^2 + y^2 = a^2 + b^2$ and pointing in the clockwise direction.

b. How is $\mathbf{G}$ related to the spin field $\mathbf{F}$ in Figure 16.13?

51. Unit vectors pointing toward the origin Find a field $\mathbf{F} = M(x, y)\mathbf{i} + N(x, y)\mathbf{j}$ in the xy-plane with the property that at each point $(x, y) \ne (0, 0)$, $\mathbf{F}$ is a unit vector pointing toward the origin. (The field is undefined at $(0, 0)$.)

52. Two "central" fields Find a field $\mathbf{F} = M(x, y)\mathbf{i} + N(x, y)\mathbf{j}$ in the xy-plane with the property that at each point $(x, y) \ne (0, 0)$, $\mathbf{F}$ points toward the origin and $|\mathbf{F}|$ is **(a)** the distance from (x, y) to the origin, **(b)** inversely proportional to the distance from (x, y) to the origin. (The field is undefined at $(0, 0)$.)

53. Work and area Suppose that $f(t)$ is differentiable and positive for $a \le t \le b$. Let C be the path $\mathbf{r}(t) = t\mathbf{i} + f(t)\mathbf{j}$, $a \le t \le b$, and $\mathbf{F} = y\mathbf{i}$. Is there any relation between the value of the work integral

$$\int_C \mathbf{F} \cdot d\mathbf{r}$$

and the area of the region bounded by the t-axis, the graph of f, and the lines $t = a$ and $t = b$? Give reasons for your answer.

54. Work done by a radial force with constant magnitude A particle moves along the smooth curve $y = f(x)$ from $(a, f(a))$ to $(b, f(b))$. The force moving the particle has constant magnitude k and always points away from the origin. Show that the work done by the force is

$$\int_C \mathbf{F} \cdot \mathbf{T}\, ds = k\left[\left(b^2 + (f(b))^2\right)^{1/2} - \left(a^2 + (f(a))^2\right)^{1/2}\right].$$

Flow Integrals in Space

In Exercises 55–58, $\mathbf{F}$ is the velocity field of a fluid flowing through a region in space. Find the flow along the given curve in the direction of increasing t.

55. $\mathbf{F} = -4xy\mathbf{i} + 8y\mathbf{j} + 2\mathbf{k}$
$\mathbf{r}(t) = t\mathbf{i} + t^2\mathbf{j} + \mathbf{k}, \quad 0 \le t \le 2$

56. $\mathbf{F} = x^2\mathbf{i} + yz\mathbf{j} + y^2\mathbf{k}$
$\mathbf{r}(t) = 3t\mathbf{j} + 4t\mathbf{k}, \quad 0 \le t \le 1$

57. $\mathbf{F} = (x - z)\mathbf{i} + x\mathbf{k}$
$\mathbf{r}(t) = (\cos t)\mathbf{i} + (\sin t)\mathbf{k}, \quad 0 \le t \le \pi$

58. $\mathbf{F} = -y\mathbf{i} + x\mathbf{j} + 2\mathbf{k}$
$\mathbf{r}(t) = (-2\cos t)\mathbf{i} + (2\sin t)\mathbf{j} + 2t\mathbf{k}, \quad 0 \le t \le 2\pi$

59. Circulation Find the circulation of $\mathbf{F} = 2x\mathbf{i} + 2z\mathbf{j} + 2y\mathbf{k}$ around the closed path consisting of the following three curves traversed in the direction of increasing t.

C_1: $\mathbf{r}(t) = (\cos t)\mathbf{i} + (\sin t)\mathbf{j} + t\mathbf{k}, \quad 0 \le t \le \pi/2$
C_2: $\mathbf{r}(t) = \mathbf{j} + (\pi/2)(1 - t)\mathbf{k}, \quad 0 \le t \le 1$
C_3: $\mathbf{r}(t) = t\mathbf{i} + (1 - t)\mathbf{j}, \quad 0 \le t \le 1$

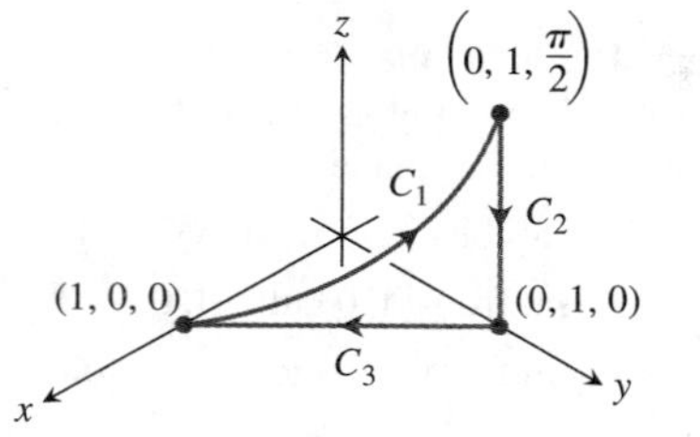

60. Zero circulation Let C be the ellipse in which the plane $2x + 3y - z = 0$ meets the cylinder $x^2 + y^2 = 12$. Show, without evaluating either line integral directly, that the circulation of the field $\mathbf{F} = x\mathbf{i} + y\mathbf{j} + z\mathbf{k}$ around C in either direction is zero.

61. Flow along a curve The field $\mathbf{F} = xy\mathbf{i} + y\mathbf{j} - yz\mathbf{k}$ is the velocity field of a flow in space. Find the flow from $(0, 0, 0)$ to $(1, 1, 1)$ along the curve of intersection of the cylinder $y = x^2$ and the plane $z = x$. (*Hint:* Use $t = x$ as the parameter.)

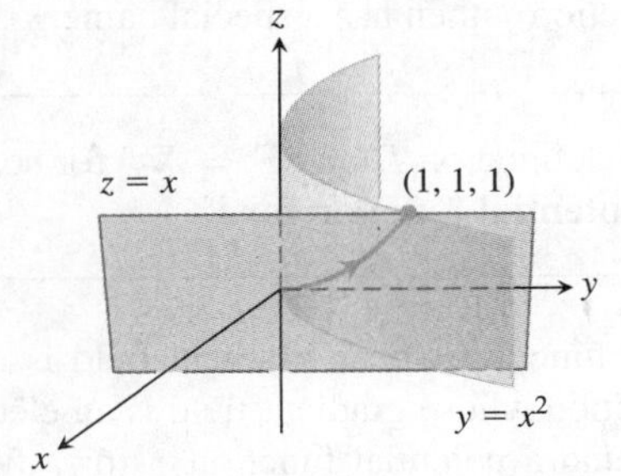

62. Flow of a gradient field Find the flow of the field $\mathbf{F} = \nabla(xy^2z^3)$:

a. Once around the curve C in Exercise 58, clockwise as viewed from above;

b. Along the line segment from $(1, 1, 1)$ to $(2, 1, -1)$.

COMPUTER EXPLORATIONS

In Exercises 63–68, use a CAS to perform the following steps for finding the work done by force $\mathbf{F}$ over the given path:

a. Find $d\mathbf{r}$ for the path $\mathbf{r}(t) = g(t)\mathbf{i} + h(t)\mathbf{j} + k(t)\mathbf{k}$.

b. Evaluate the force $\mathbf{F}$ along the path.

c. Evaluate $\int_C \mathbf{F} \cdot d\mathbf{r}$.

63. $\mathbf{F} = xy^6\mathbf{i} + 3x(xy^5 + 2)\mathbf{j}$; $\mathbf{r}(t) = (2\cos t)\mathbf{i} + (\sin t)\mathbf{j}$, $0 \le t \le 2\pi$

64. $\mathbf{F} = \dfrac{3}{1 + x^2}\mathbf{i} + \dfrac{2}{1 + y^2}\mathbf{j}$; $\mathbf{r}(t) = (\cos t)\mathbf{i} + (\sin t)\mathbf{j}$, $0 \le t \le \pi$

65. $\mathbf{F} = (y + yz\cos xyz)\mathbf{i} + (x^2 + xz\cos xyz)\mathbf{j} + (z + xy\cos xyz)\mathbf{k}$; $\mathbf{r}(t) = (2\cos t)\mathbf{i} + (3\sin t)\mathbf{j} + \mathbf{k}$, $0 \le t \le 2\pi$

66. $\mathbf{F} = 2xy\mathbf{i} - y^2\mathbf{j} + ze^x\mathbf{k}$; $\mathbf{r}(t) = -t\mathbf{i} + \sqrt{t}\mathbf{j} + 3t\mathbf{k}$, $1 \le t \le 4$

67. $\mathbf{F} = (2y + \sin x)\mathbf{i} + (z^2 + (1/3)\cos y)\mathbf{j} + x^4\mathbf{k}$; $\mathbf{r}(t) = (\sin t)\mathbf{i} + (\cos t)\mathbf{j} + (\sin 2t)\mathbf{k}$, $-\pi/2 \le t \le \pi/2$

68. $\mathbf{F} = (x^2y)\mathbf{i} + \dfrac{1}{3}x^3\mathbf{j} + xy\mathbf{k}$; $\mathbf{r}(t) = (\cos t)\mathbf{i} + (\sin t)\mathbf{j} + (2\sin^2 t - 1)\mathbf{k}$, $0 \le t \le 2\pi$

16.3 Path Independence, Conservative Fields, and Potential Functions

A **gravitational field G** is a vector field that represents the effect of gravity at a point in space due to the presence of a massive object. The gravitational force on a body of mass m placed in the field is given by $\mathbf{F} = m\mathbf{G}$. Similarly, an **electric field E** is a vector field in space that represents the effect of electric forces on a charged particle placed within it. The force on a body of charge q placed in the field is given by $\mathbf{F} = q\mathbf{E}$. In gravitational and electric fields, the amount of work it takes to move a mass or charge from one point to another depends on the initial and final positions of the object—not on which path is taken between these positions. In this section we study vector fields with this independence-of-path property and the calculation of work integrals associated with them.

Path Independence

If A and B are two points in an open region D in space, the line integral of $\mathbf{F}$ along C from A to B for a field $\mathbf{F}$ defined on D usually depends on the path C taken, as we saw in Section 16.1. For some special fields, however, the integral's value is the same for all paths from A to B.

> **DEFINITIONS** Let $\mathbf{F}$ be a vector field defined on an open region D in space, and suppose that for any two points A and B in D, the line integral $\int_C \mathbf{F} \cdot d\mathbf{r}$ along a path C from A to B in D is the same over all paths from A to B. Then the integral $\int_C \mathbf{F} \cdot d\mathbf{r}$ is **path independent in D** and the field $\mathbf{F}$ is **conservative on D**.

The word *conservative* comes from physics, where it refers to fields in which the principle of conservation of energy holds. When a line integral is independent of the path C from point A to point B, we sometimes represent the integral by the symbol $\int_A^B$ rather than the usual line integral symbol $\int_C$. This substitution helps us remember the path-independence

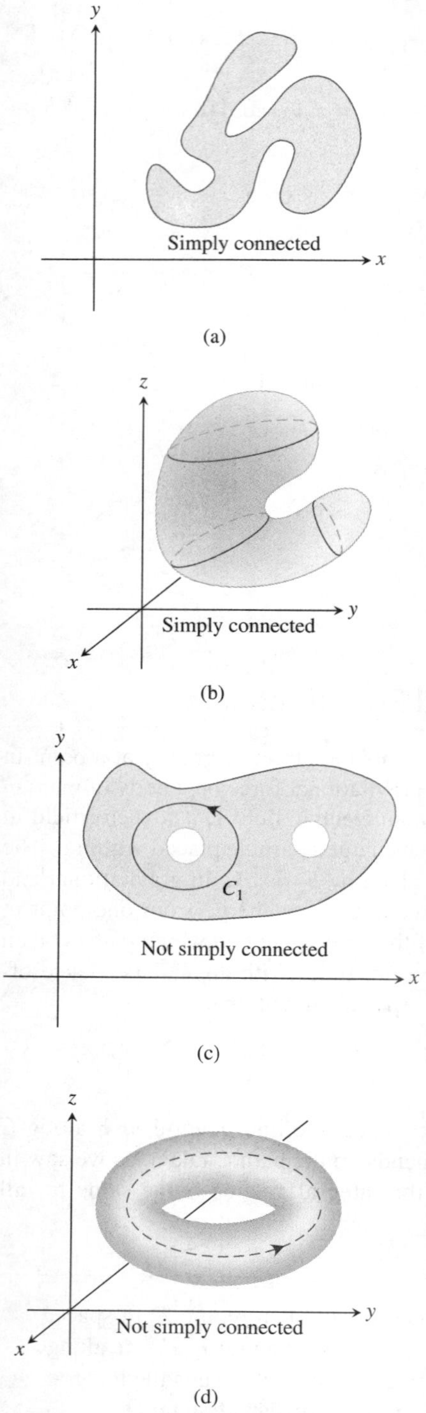

FIGURE 16.25 Four connected regions. In (a) and (b), the regions are simply connected. In (c) and (d), the regions are not simply connected because the curves C_1 and C_2 cannot be contracted to a point inside the regions containing them.

property by indicating that the integral depends only on the initial and final points, not on the path connecting them.

Under reasonable differentiability conditions that we will specify, we will show that a field $\mathbf{F}$ is conservative if and only if it is the gradient field of a scalar function f—that is, if and only if $\mathbf{F} = \nabla f$ for some f. The function f then has a special name.

> **DEFINITION** If $\mathbf{F}$ is a vector field defined on D and $\mathbf{F} = \nabla f$ for some scalar function f on D, then f is called a **potential function for F**.

A gravitational potential is a scalar function whose gradient field is a gravitational field, an electric potential is a scalar function whose gradient field is an electric field, and so on. As we will see, once we have found a potential function f for a field $\mathbf{F}$, we can evaluate all the line integrals in the domain of $\mathbf{F}$ over any path between A and B by

$$\int_A^B \mathbf{F} \cdot d\mathbf{r} = \int_A^B \nabla f \cdot d\mathbf{r} = f(B) - f(A). \tag{1}$$

If you think of ∇f for functions of several variables as analogous to the derivative f' for functions of a single variable, then you see that Equation (1) is the vector calculus rendition of the Fundamental Theorem of Calculus formula

$$\int_a^b f'(x)\,dx = f(b) - f(a).$$

Conservative fields have other important properties. For example, saying that $\mathbf{F}$ is conservative on D is equivalent to saying that the integral of $\mathbf{F}$ around every closed path in D is zero. Certain conditions on the curves, fields, and domains must be satisfied for Equation (1) to be valid. We discuss these conditions next.

Assumptions on Curves, Vector Fields, and Domains

In order for the computations and results we derive below to be valid, we must assume certain properties for the curves, surfaces, domains, and vector fields we consider. We give these assumptions in the statements of theorems, and they also apply to the examples and exercises unless otherwise stated.

The curves we consider are **piecewise smooth**. Such curves are made up of finitely many smooth pieces connected end to end, as discussed in Section 13.1. For such curves we can compute lengths and, except at finitely many points where the smooth pieces connect, tangent vectors. We consider vector fields $\mathbf{F}$ whose components have continuous first partial derivatives.

The domains D we consider are **connected**. For an open region, this means that any two points in D can be joined by a smooth curve that lies in the region. Some results require D to be **simply connected**, which means that every loop in D can be contracted to a point in D without ever leaving D. The plane with a disk removed is a two-dimensional region that is *not* simply connected; a loop in the plane that goes around the disk cannot be contracted to a point without going into the "hole" left by the removed disk (see Figure 16.25c). Similarly, if we remove a line from space, the remaining region D is *not* simply connected. A curve encircling the line cannot be shrunk to a point while remaining inside D.

Connectivity and simple connectivity are not the same, and neither property implies the other. Think of connected regions as being in "one piece" and of simply connected regions as not having any "loop-catching holes." All of space itself is both connected and simply connected. Figure 16.25 illustrates some of these properties.

Caution Some of the results in this chapter can fail to hold if applied to situations where the conditions we've imposed are not met. In particular, the component test for conservative fields, given later in this section, is not valid on domains that are not simply connected (see Example 5). The condition will be stated when needed. ●

Line Integrals in Conservative Fields

A gradient field $\mathbf{F}$ is obtained by differentiating a scalar function f. A theorem analogous to the Fundamental Theorem of Calculus gives a way to evaluate the line integrals of gradient fields.

Like the Fundamental Theorem of Calculus, Theorem 1 gives a direct way to evaluate line integrals without having to take limits of Riemann sums and without needing to compute a line integral by the procedure used in Section 16.2. Before proving Theorem 1, we give an example.

THEOREM 1—Fundamental Theorem of Line Integrals
Let C be a smooth curve joining the point A to the point B in the plane or in space and parametrized by $\mathbf{r}(t)$. Let f be a differentiable function with a continuous gradient vector $\mathbf{F} = \nabla f$ on a domain D containing C. Then

$$\int_C \mathbf{F} \cdot d\mathbf{r} = f(B) - f(A).$$

EXAMPLE 1 Suppose the force field $\mathbf{F} = \nabla f$ is the gradient of the function

$$f(x, y, z) = -\frac{1}{x^2 + y^2 + z^2}.$$

Find the work done by $\mathbf{F}$ in moving an object along a smooth curve C joining $(1, 0, 0)$ to $(0, 0, 2)$ that does not pass through the origin.

Solution An application of Theorem 1 shows that the work done by $\mathbf{F}$ along any smooth curve C joining the two points and not passing through the origin is

$$\int_C \mathbf{F} \cdot d\mathbf{r} = f(0, 0, 2) - f(1, 0, 0) = -\frac{1}{4} - (-1) = \frac{3}{4}.$$ ■

The gravitational force due to a planet, and the electric force associated with a charged particle, can both be modeled by the field $\mathbf{F}$ given in Example 1 up to a constant that depends on the units of measurement. When used to model gravity, the function f in Example 1 represents gravitational potential energy. The sign of f is negative, and f approaches $-\infty$ near the origin. This choice ensures that the gravitational force $\mathbf{F}$, the gradient of f, points toward the origin, so that objects fall down rather than up.

Proof of Theorem 1 Suppose that A and B are two points in the region D and that C: $\mathbf{r}(t) = g(t)\mathbf{i} + h(t)\mathbf{j} + k(t)\mathbf{k}$, $a \le t \le b$, is a smooth curve in D joining A to B. In Section 14.5 we found that the derivative of a scalar function f along a path C is the dot product $\nabla f(\mathbf{r}(t)) \cdot \mathbf{r}'(t)$, so we have

$$\begin{aligned}
\int_C \mathbf{F} \cdot d\mathbf{r} &= \int_C \nabla f \cdot d\mathbf{r} && \mathbf{F} = \nabla f \\
&= \int_{t=a}^{t=b} \nabla f(\mathbf{r}(t)) \cdot \mathbf{r}'(t)\,dt && \text{Eq. (2) of Section 16.2 for computing } d\mathbf{r} \\
&= \int_a^b \frac{d}{dt} f(\mathbf{r}(t))\,dt && \text{Eq. (7) of Section 14.5 giving derivative along a path} \\
&= f(\mathbf{r}(b)) - f(\mathbf{r}(a)) && \text{Fundamental Theorem of Calculus} \\
&= f(B) - f(A). && \mathbf{r}(a) = A,\ \mathbf{r}(b) = B
\end{aligned}$$ ■

We see from Theorem 1 that the line integral of a gradient field $\mathbf{F} = \nabla f$ is straightforward to compute once we know the function f. Many important vector fields arising in applications are indeed gradient fields. The next result, which follows from Theorem 1, shows that any conservative field is of this type.

THEOREM 2—Conservative Fields Are Gradient Fields
Let $\mathbf{F} = M\mathbf{i} + N\mathbf{j} + P\mathbf{k}$ be a vector field whose components are continuous throughout an open connected region D in space. Then $\mathbf{F}$ is conservative if and only if $\mathbf{F}$ is a gradient field ∇f for a differentiable function f.

Theorem 2 says that $\mathbf{F} = \nabla f$ if and only if, for any two points A and B in the region D, the value of the line integral $\int_C \mathbf{F} \cdot d\mathbf{r}$ is independent of the path C joining A to B in D.

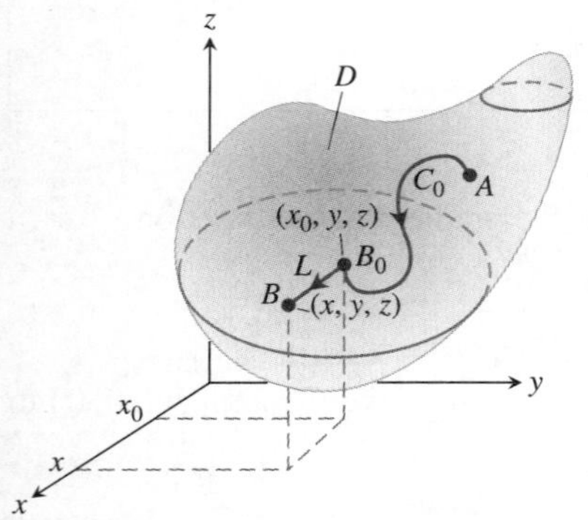

FIGURE 16.26 The function $f(x, y, z)$ in the proof of Theorem 2 is computed by a line integral $\int_{C_0} \mathbf{F} \cdot d\mathbf{r} = f(B_0)$ from A to B_0, plus a line integral $\int_L \mathbf{F} \cdot d\mathbf{r}$ along a line segment L parallel to the x-axis and joining B_0 to B located at (x, y, z). The value of f at A is $f(A) = 0$.

Proof of Theorem 2 If $\mathbf{F}$ is a gradient field, then $\mathbf{F} = \nabla f$ for a differentiable function f, and Theorem 1 shows that $\int_C \mathbf{F} \cdot d\mathbf{r} = f(B) - f(A)$. The value of the line integral does not depend on C, but only on its endpoints A and B. So the line integral is path independent and $\mathbf{F}$ satisfies the definition of a conservative field.

On the other hand, suppose that $\mathbf{F}$ is a conservative vector field. We want to find a function f on D satisfying $\nabla f = \mathbf{F}$. First, pick a point A in D and set $f(A) = 0$. For any other point B in D define $f(B)$ to equal $\int_C \mathbf{F} \cdot d\mathbf{r}$, where C is *any* smooth path in D from A to B. The value of $f(B)$ does not depend on the choice of C, since $\mathbf{F}$ is conservative. To show that $\nabla f = \mathbf{F}$, we need to demonstrate that $\partial f/\partial x = M$, $\partial f/\partial y = N$, and $\partial f/\partial z = P$.

Suppose that B has coordinates (x, y, z). By the definition of f, the value of the function f at a nearby point B_0 located at (x_0, y, z) is $\int_{C_0} \mathbf{F} \cdot d\mathbf{r}$, where C_0 is any path from A to B_0. We take a path $C = C_0 \cup L$ from A to B formed by first traveling along C_0 to arrive at B_0 and then traveling along the line segment L from B_0 to B (Figure 16.26). When B_0 is close to B, the segment L lies in D and, since the value $f(B)$ is independent of the path from A to B,

$$f(x, y, z) = \int_{C_0} \mathbf{F} \cdot d\mathbf{r} + \int_L \mathbf{F} \cdot d\mathbf{r}.$$

Differentiating, we have

$$\frac{\partial}{\partial x} f(x, y, z) = \frac{\partial}{\partial x}\left(\int_{C_0} \mathbf{F} \cdot d\mathbf{r} + \int_L \mathbf{F} \cdot d\mathbf{r}\right).$$

Only the last term on the right depends on x, so

$$\frac{\partial}{\partial x} f(x, y, z) = \frac{\partial}{\partial x} \int_L \mathbf{F} \cdot d\mathbf{r}.$$

Now we parametrize L as $\mathbf{r}(t) = t\mathbf{i} + y\mathbf{j} + z\mathbf{k}$, $x_0 \le t \le x$. Then $d\mathbf{r}/dt = \mathbf{i}$, and since $\mathbf{F} = M\mathbf{i} + N\mathbf{j} + P\mathbf{k}$, it follows that $\mathbf{F} \cdot d\mathbf{r}/dt = M$ and $\int_L \mathbf{F} \cdot d\mathbf{r} = \int_{x_0}^{x} M(t, y, z)\, dt$. Differentiating then gives

$$\frac{\partial}{\partial x} f(x, y, z) = \frac{\partial}{\partial x} \int_{x_0}^{x} M(t, y, z)\, dt = M(x, y, z)$$

by the Fundamental Theorem of Calculus. The partial derivatives $\partial f/\partial y = N$ and $\partial f/\partial z = P$ follow similarly, showing that $\mathbf{F} = \nabla f$. ■

EXAMPLE 2 Find the work done by the conservative field

$$\mathbf{F} = yz\mathbf{i} + xz\mathbf{j} + xy\mathbf{k} = \nabla f, \quad \text{where} \quad f(x, y, z) = xyz,$$

in moving an object along any smooth curve C joining the point $A(-1, 3, 9)$ to $B(1, 6, -4)$.

Solution With $f(x, y, z) = xyz$, we have

$$\begin{aligned}\int_C \mathbf{F} \cdot d\mathbf{r} &= \int_A^B \nabla f \cdot d\mathbf{r} && \mathbf{F} = \nabla f \text{ and path independence}\\ &= f(B) - f(A) && \text{Theorem 1}\\ &= xyz|_{(1,6,-4)} - xyz|_{(-1,3,9)}\\ &= (1)(6)(-4) - (-1)(3)(9)\\ &= -24 + 27 = 3.\end{aligned}$$

■

A very useful property of line integrals in conservative fields comes into play when the path of integration is a closed curve, or loop. We often use the notation $\oint_C$ for integration around a closed path (discussed with more detail in the next section).

THEOREM 3—Loop Property of Conservative Fields
The following statements are equivalent.

1. $\oint_C \mathbf{F} \cdot d\mathbf{r} = 0$ around every loop (that is, closed curve C) in D.
2. The field $\mathbf{F}$ is conservative on D.

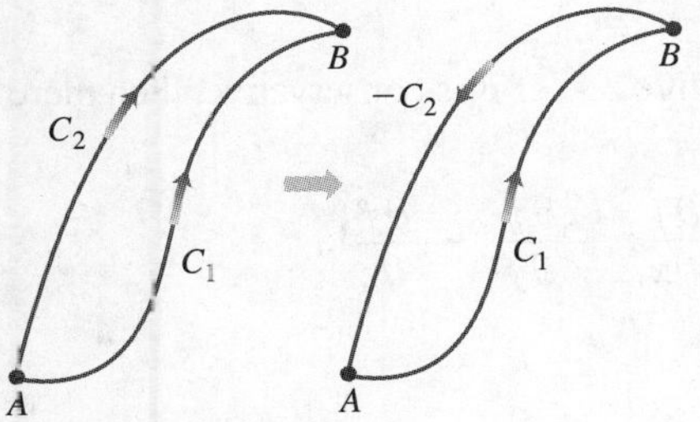

FIGURE 16.27 If we have two paths from A to B, one of them can be reversed to make a loop.

Proof that Part 1 ⇒ Part 2 We want to show that for any two points A and B in D, the integral of $\mathbf{F} \cdot d\mathbf{r}$ has the same value over any two paths C_1 and C_2 from A to B. We reverse the direction on C_2 to make a path $-C_2$ from B to A (Figure 16.27). Together, C_1 and $-C_2$ make a closed loop C, and by assumption,

$$\int_{C_1} \mathbf{F} \cdot d\mathbf{r} - \int_{C_2} \mathbf{F} \cdot d\mathbf{r} = \int_{C_1} \mathbf{F} \cdot d\mathbf{r} + \int_{-C_2} \mathbf{F} \cdot d\mathbf{r} = \int_C \mathbf{F} \cdot d\mathbf{r} = 0.$$

Thus, the integrals over C_1 and C_2 give the same value. Note that the definition of $\mathbf{F} \cdot d\mathbf{r}$ shows that changing the direction along a curve reverses the sign of the line integral.

Proof that Part 2 ⇒ Part 1 We want to show that the integral of $\mathbf{F} \cdot d\mathbf{r}$ is zero over any closed loop C. We pick two points A and B on C and use them to break C into two pieces: C_1 from A to B followed by C_2 from B back to A (Figure 16.28). Then

$$\oint_C \mathbf{F} \cdot d\mathbf{r} = \int_{C_1} \mathbf{F} \cdot d\mathbf{r} + \int_{C_2} \mathbf{F} \cdot d\mathbf{r} = \int_A^B \mathbf{F} \cdot d\mathbf{r} - \int_A^B \mathbf{F} \cdot d\mathbf{r} = 0.$$

■

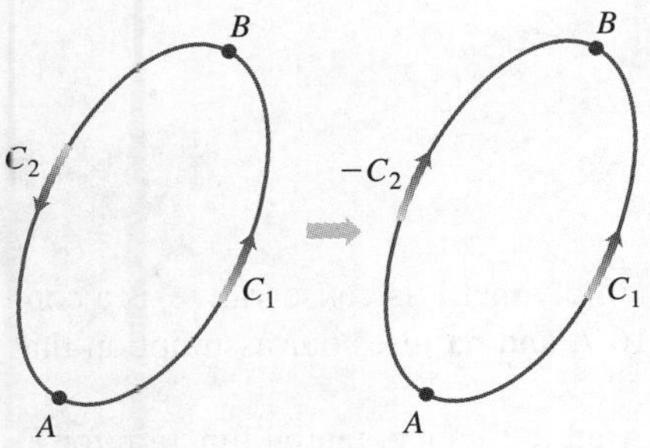

FIGURE 16.28 If A and B lie on a loop, we can reverse part of the loop to make two paths from A to B.

The following diagram summarizes the results of Theorems 2 and 3.

$$\mathbf{F} = \nabla f \text{ on } D \quad \overset{\text{Theorem 2}}{\Leftrightarrow} \quad \mathbf{F} \text{ conservative on } D \quad \overset{\text{Theorem 3}}{\Leftrightarrow} \quad \oint_C \mathbf{F} \cdot d\mathbf{r} = 0 \text{ over any loop in } D$$

Two questions arise:

1. How do we know whether a given vector field **F** is conservative?
2. If **F** is in fact conservative, how do we find a potential function f (so that $\mathbf{F} = \nabla f$)?

Finding Potentials for Conservative Fields

The test for a vector field being conservative involves the equivalence of certain first partial derivatives of the field components.

> **Component Test for Conservative Fields**
> Let $\mathbf{F} = M(x, y, z)\mathbf{i} + N(x, y, z)\mathbf{j} + P(x, y, z)\mathbf{k}$ be a field on an open simply connected domain whose component functions have continuous first partial derivatives. Then **F** is conservative if and only if
>
> $$\frac{\partial P}{\partial y} = \frac{\partial N}{\partial z}, \quad \frac{\partial M}{\partial z} = \frac{\partial P}{\partial x}, \quad \text{and} \quad \frac{\partial N}{\partial x} = \frac{\partial M}{\partial y}. \tag{2}$$

We can view the component test as saying that on a simply connected region, the vector

$$\left(\frac{\partial P}{\partial y} - \frac{\partial N}{\partial z}\right)\mathbf{i} + \left(\frac{\partial M}{\partial z} - \frac{\partial P}{\partial x}\right)\mathbf{j} + \left(\frac{\partial N}{\partial x} - \frac{\partial M}{\partial y}\right)\mathbf{k} \tag{3}$$

is zero if and only if **F** is conservative. This interesting vector curl **F** is called the **curl** of **F**. We study it in Sections 16.4 and 16.7.

Proof that Equations (2) hold if F is conservative If **F** is conservative, then there is a potential function f such that

$$\mathbf{F} = M\mathbf{i} + N\mathbf{j} + P\mathbf{k} = \nabla f = \frac{\partial f}{\partial x}\mathbf{i} + \frac{\partial f}{\partial y}\mathbf{j} + \frac{\partial f}{\partial z}\mathbf{k}.$$

Hence,

$$\begin{aligned}
\frac{\partial P}{\partial y} &= \frac{\partial}{\partial y}\left(\frac{\partial f}{\partial z}\right) = \frac{\partial^2 f}{\partial y\, \partial z} \\
&= \frac{\partial^2 f}{\partial z\, \partial y} && \text{Mixed Derivative Theorem, Section 14.3} \\
&= \frac{\partial}{\partial z}\left(\frac{\partial f}{\partial y}\right) = \frac{\partial N}{\partial z}.
\end{aligned}$$

The others in Equations (2) are proved similarly. ■

The second half of the proof, that Equations (2) imply that **F** is conservative, is a consequence of Stokes' Theorem, taken up in Section 16.7, and requires our assumption that the domain of **F** be simply connected.

Once we know that **F** is conservative, we often want to find a potential function for **F**. This requires solving the equation $\nabla f = \mathbf{F}$ or

$$\frac{\partial f}{\partial x}\mathbf{i} + \frac{\partial f}{\partial y}\mathbf{j} + \frac{\partial f}{\partial z}\mathbf{k} = M\mathbf{i} + N\mathbf{j} + P\mathbf{k}$$

for f. We accomplish this by integrating the three equations

$$\frac{\partial f}{\partial x} = M, \qquad \frac{\partial f}{\partial y} = N, \qquad \frac{\partial f}{\partial z} = P,$$

as illustrated in the next example.

EXAMPLE 3 Show that $\mathbf{F} = (e^x \cos y + yz)\mathbf{i} + (xz - e^x \sin y)\mathbf{j} + (xy + z)\mathbf{k}$ is conservative over its natural domain, and find a potential function for it.

Solution The natural domain of $\mathbf{F}$ is all of space, which is open and simply connected. We apply the test in Equations (2) to

$$M = e^x \cos y + yz, \qquad N = xz - e^x \sin y, \qquad P = xy + z$$

and calculate

$$\frac{\partial P}{\partial y} = x = \frac{\partial N}{\partial z}, \qquad \frac{\partial M}{\partial z} = y = \frac{\partial P}{\partial x}, \qquad \frac{\partial N}{\partial x} = -e^x \sin y + z = \frac{\partial M}{\partial y}.$$

The partial derivatives are continuous, so these equalities tell us that $\mathbf{F}$ is conservative, so there is a function f with $\nabla f = \mathbf{F}$ (Theorem 2).

We find f by integrating the equations

$$\frac{\partial f}{\partial x} = e^x \cos y + yz, \qquad \frac{\partial f}{\partial y} = xz - e^x \sin y, \qquad \frac{\partial f}{\partial z} = xy + z. \tag{4}$$

We integrate the first equation with respect to x, holding y and z fixed, to get

$$f(x, y, z) = e^x \cos y + xyz + g(y, z).$$

We write the constant of integration as a function of y and z because its value may depend on y and z, though not on x. We then calculate $\partial f/\partial y$ from this equation and match it with the expression for $\partial f/\partial y$ in Equations (4). This gives

$$-e^x \sin y + xz + \frac{\partial g}{\partial y} = xz - e^x \sin y,$$

so $\partial g/\partial y = 0$. Therefore, g is a function of z alone, and

$$f(x, y, z) = e^x \cos y + xyz + h(z).$$

We now calculate $\partial f/\partial z$ from this equation and match it to the formula for $\partial f/\partial z$ in Equations (4). This gives

$$xy + \frac{dh}{dz} = xy + z, \qquad \text{or} \qquad \frac{dh}{dz} = z,$$

so

$$h(z) = \frac{z^2}{2} + C.$$

Hence,

$$f(x, y, z) = e^x \cos y + xyz + \frac{z^2}{2} + C.$$

We found infinitely many potential functions of $\mathbf{F}$, one for each value of C. ■

EXAMPLE 4 Show that $\mathbf{F} = (2x - 3)\mathbf{i} - z\mathbf{j} + (\cos z)\mathbf{k}$ is not conservative.

Solution We apply the Component Test in Equations (2) and find immediately that

$$\frac{\partial P}{\partial y} = \frac{\partial}{\partial y}(\cos z) = 0, \qquad \frac{\partial N}{\partial z} = \frac{\partial}{\partial z}(-z) = -1.$$

The two are unequal, so **F** is not conservative. No further testing is required. ■

EXAMPLE 5 Show that the vector field

$$\mathbf{F} = \frac{-y}{x^2 + y^2}\mathbf{i} + \frac{x}{x^2 + y^2}\mathbf{j} + 0\mathbf{k}$$

satisfies the equations in the Component Test but is not conservative over its natural domain. Explain why this is possible.

Solution We have $M = -y/(x^2 + y^2)$, $N = x/(x^2 + y^2)$, and $P = 0$. If we apply the Component Test, we find

$$\frac{\partial P}{\partial y} = 0 = \frac{\partial N}{\partial z}, \qquad \frac{\partial P}{\partial x} = 0 = \frac{\partial M}{\partial z}, \qquad \text{and} \qquad \frac{\partial M}{\partial y} = \frac{y^2 - x^2}{(x^2 + y^2)^2} = \frac{\partial N}{\partial x}.$$

So it may appear that the field **F** passes the Component Test. However, the test assumes that the domain of **F** is simply connected, which is not the case here. Since $x^2 + y^2$ cannot equal zero, the natural domain is the complement of the z-axis and contains loops that cannot be contracted to a point. One such loop is the unit circle C in the xy-plane. The circle is parametrized by $\mathbf{r}(t) = (\cos t)\mathbf{i} + (\sin t)\mathbf{j}$, $0 \le t \le 2\pi$. This loop wraps around the z-axis and cannot be contracted to a point while staying within the complement of the z-axis.

To show that **F** is not conservative, we compute the line integral $\oint_C \mathbf{F} \cdot d\mathbf{r}$ around the loop C. First we write the field in terms of the parameter t:

$$\mathbf{F} = \frac{-y}{x^2 + y^2}\mathbf{i} + \frac{x}{x^2 + y^2}\mathbf{j} = \frac{-\sin t}{\sin^2 t + \cos^2 t}\mathbf{i} + \frac{\cos t}{\sin^2 t + \cos^2 t}\mathbf{j} = (-\sin t)\mathbf{i} + (\cos t)\mathbf{j}.$$

Next we find $d\mathbf{r}/dt = (-\sin t)\mathbf{i} + (\cos t)\mathbf{j}$ and then calculate the line integral as

$$\oint_C \mathbf{F} \cdot d\mathbf{r} = \oint_C \mathbf{F} \cdot \frac{d\mathbf{r}}{dt}\, dt = \int_0^{2\pi} (\sin^2 t + \cos^2 t)\, dt = 2\pi.$$

Since the line integral of **F** around the loop C is not zero, the field **F** is not conservative, by Theorem 3. The field **F** is displayed in Figure 16.31d in the next section. ■

Example 5 shows that the Component Test does not apply when the domain of the field is not simply connected. However, if we change the domain in the example so that it is restricted to the ball of radius 1 centered at the point $(2, 2, 2)$, or to any similar ball-shaped region that does not contain a piece of the z-axis, then this new domain D *is* simply connected. Now the partial derivative Equations (2), as well as all the assumptions of the Component Test, are satisfied. In this new situation, the field **F** in Example 5 is conservative on D. Just as we must be careful with a function when determining whether it satisfies a property throughout its domain (such as continuity, which is required for the Intermediate Value Property), so must we also be careful with a vector field in determining the properties it may or may not have over its assigned domain.

Exact Differential Forms

It is often convenient to express work and circulation integrals in the differential form

$$\int_C M\,dx + N\,dy + P\,dz$$

discussed in Section 16.2. Such line integrals are relatively easy to evaluate if $M\,dx + N\,dy + P\,dz$ is the total differential of a function f and if C is any path joining the point A to the point B, for then

$$\begin{aligned}\int_C M\,dx + N\,dy + P\,dz &= \int_C \frac{\partial f}{\partial x}dx + \frac{\partial f}{\partial y}dy + \frac{\partial f}{\partial z}dz \\ &= \int_A^B \nabla f \cdot d\mathbf{r} && \nabla f \text{ is conservative.} \\ &= f(B) - f(A). && \text{Theorem 1}\end{aligned}$$

Thus,

$$\int_A^B df = f(B) - f(A),$$

just as with differentiable functions of a single variable.

DEFINITIONS Any expression $M(x, y, z)\,dx + N(x, y, z)\,dy + P(x, y, z)\,dz$ is a **differential form**. A differential form is **exact** on a domain D in space if

$$M\,dx + N\,dy + P\,dz = \frac{\partial f}{\partial x}dx + \frac{\partial f}{\partial y}dy + \frac{\partial f}{\partial z}dz = df$$

for some scalar function f throughout D.

Notice that if $M\,dx + N\,dy + P\,dz = df$ on D, then $\mathbf{F} = M\mathbf{i} + N\mathbf{j} + P\mathbf{k}$ is the gradient field of f on D. Conversely, if $\mathbf{F} = \nabla f$, then the form $M\,dx + N\,dy + P\,dz$ is exact. The test for the form being exact is therefore the same as the test for $\mathbf{F}$ being conservative.

Component Test for Exactness of $M\,dx + N\,dy + P\,dz$
The differential form $M\,dx + N\,dy + P\,dz$ is exact on an open simply connected domain if and only if

$$\frac{\partial P}{\partial y} = \frac{\partial N}{\partial z}, \quad \frac{\partial M}{\partial z} = \frac{\partial P}{\partial x}, \quad \text{and} \quad \frac{\partial N}{\partial x} = \frac{\partial M}{\partial y}.$$

This is equivalent to saying that the field $\mathbf{F} = M\mathbf{i} + N\mathbf{j} + P\mathbf{k}$ is conservative.

EXAMPLE 6 Show that $y\,dx + x\,dy + 4\,dz$ is exact, and evaluate the integral

$$\int_{(1,1,1)}^{(2,3,-1)} y\,dx + x\,dy + 4\,dz$$

over any path from $(1, 1, 1)$ to $(2, 3, -1)$.

Solution Note that the domain of **F** fills all of three-dimensional space, so it is simply connected. We let $M = y$, $N = x$, $P = 4$ and apply the Component Test for Exactness:

$$\frac{\partial P}{\partial y} = 0 = \frac{\partial N}{\partial z}, \qquad \frac{\partial M}{\partial z} = 0 = \frac{\partial P}{\partial x}, \qquad \frac{\partial N}{\partial x} = 1 = \frac{\partial M}{\partial y}.$$

These equalities tell us that $y\,dx + x\,dy + 4\,dz$ is exact, so

$$y\,dx + x\,dy + 4\,dz = df$$

for some function f, and the integral's value is $f(2, 3, -1) - f(1, 1, 1)$.

We find f up to a constant by integrating the equations

$$\frac{\partial f}{\partial x} = y, \qquad \frac{\partial f}{\partial y} = x, \qquad \frac{\partial f}{\partial z} = 4. \tag{5}$$

From the first equation we get

$$f(x, y, z) = xy + g(y, z).$$

The second equation tells us that

$$\frac{\partial f}{\partial y} = x + \frac{\partial g}{\partial y} = x, \qquad \text{or} \qquad \frac{\partial g}{\partial y} = 0.$$

Hence, g is a function of z alone, and

$$f(x, y, z) = xy + h(z).$$

The third of Equations (5) tells us that

$$\frac{\partial f}{\partial z} = 0 + \frac{dh}{dz} = 4, \qquad \text{or} \qquad h(z) = 4z + C.$$

Therefore,

$$f(x, y, z) = xy + 4z + C.$$

The value of the line integral is independent of the path taken from $(1, 1, 1)$ to $(2, 3, -1)$ and equals

$$f(2, 3, -1) - f(1, 1, 1) = 2 + C - (5 + C) = -3.$$ ■

EXERCISES 16.3

Testing for Conservative Fields

Which fields in Exercises 1–6 are conservative, and which are not?

1. $\mathbf{F} = yz\mathbf{i} + xz\mathbf{j} + xy\mathbf{k}$
2. $\mathbf{F} = (y \sin z)\mathbf{i} + (x \sin z)\mathbf{j} + (xy \cos z)\mathbf{k}$
3. $\mathbf{F} = (4xy - 3x^2z^2)\mathbf{i} + 2x^2\mathbf{j} - 2x^3z\mathbf{k}$
4. $\mathbf{F} = (3x^2 + 6xy)\mathbf{i} + (3x^2 - y^3)\mathbf{j}$
5. $\mathbf{F} = (z + y)\mathbf{i} + z\mathbf{j} + (y + x)\mathbf{k}$
6. $\mathbf{F} = (e^x \cos y)\mathbf{i} - (e^x \sin y)\mathbf{j} + z\mathbf{k}$

Finding Potential Functions

In Exercises 7–12, find a potential function f for the field $\mathbf{F}$.

7. $\mathbf{F} = 2x\mathbf{i} + 3y\mathbf{j} + 4z\mathbf{k}$
8. $\mathbf{F} = (y + z)\mathbf{i} + (x + z)\mathbf{j} + (x + y)\mathbf{k}$
9. $\mathbf{F} = e^{y+2z}(\mathbf{i} + x\mathbf{j} + 2x\mathbf{k})$
10. $\mathbf{F} = (y \sin z)\mathbf{i} + (x \sin z)\mathbf{j} + (xy \cos z)\mathbf{k}$
11. $\mathbf{F} = (2xy + z^3)\mathbf{i} + x^2\mathbf{j} + 3xz^2\mathbf{k}$
12. $\mathbf{F} = (y + y^2 + z^2)\mathbf{i} + (x + z + 2xy)\mathbf{j} + (y + 2zx)\mathbf{k}$

Exact Differential Forms

In Exercises 13–17, show that the differential forms in the integrals are exact. Then evaluate the integrals.

13. $\displaystyle\int_{(0,0,0)}^{(2,3,-6)} 2x\,dx + 2y\,dy + 2z\,dz$
14. $\displaystyle\int_{(1,1,2)}^{(3,5,0)} yz\,dx + xz\,dy + xy\,dz$

15. $\int_{(1,-2,1)}^{(3,1,4)} (2xy + z^3)dx + x^2 dy + 3xz^2 dz$

16. $\int_{(0,0,0)}^{(3,3,1)} 2x\,dx - y^2\,dy - \frac{4}{1+z^2}dz$

17. $\int_{(1,0,0)}^{(0,1,1)} \sin y \cos x\,dx + \cos y \sin x\,dy + dz$

Finding Potential Functions to Evaluate Line Integrals

Although they are not defined on all of space R^3, the fields associated with Exercises 18–22 are conservative. Find a potential function for each field, and evaluate the integrals as in Example 6.

18. $\int_{(0,2,1)}^{(1,\pi/2,2)} 2\cos y\,dx + \left(\frac{1}{y} - 2x\sin y\right)dy + \frac{1}{z}dz$

19. $\int_{(1,1,1)}^{(1,2,3)} 3x^2\,dx + \frac{z^2}{y}dy + 2z\ln y\,dz$

20. $\int_{(0,1,-1)}^{(2,3,0)} (ze^x - 2xy)dx - (x^2 - 1)dy + (e^x + z)dz$

21. $\int_{(1,1,1)}^{(2,2,2)} \frac{1}{y}dx + \left(\frac{1}{z} - \frac{x}{y^2}\right)dy - \frac{y}{z^2}dz$

22. $\int_{(-1,-1,-1)}^{(2,2,2)} \frac{2x\,dx + 2y\,dy + 2z\,dz}{x^2 + y^2 + z^2}$

Applications and Examples

23. Revisiting Example 6 Evaluate the integral

$$\int_{(1,1,1)}^{(2,3,-1)} y\,dx + x\,dy + 4\,dz$$

from Example 6 by finding parametric equations for the line segment from $(1,1,1)$ to $(2,3,-1)$ and evaluating the line integral of $\mathbf{F} = y\mathbf{i} + x\mathbf{j} + 4\mathbf{k}$ along the segment. Since $\mathbf{F}$ is conservative, the integral is independent of the path.

24. Evaluate

$$\int_C x^2\,dx + yz\,dy + (y^2/2)\,dz$$

along the line segment C joining $(0,0,0)$ to $(0,3,4)$.

Independence of path Show that the values of the integrals in Exercises 25 and 26 do not depend on the path taken from A to B.

25. $\int_A^B z^2\,dx + 2y\,dy + 2xz\,dz$

26. $\int_A^B \frac{x\,dx + y\,dy + z\,dz}{\sqrt{x^2+y^2+z^2}}$

In Exercises 27 and 28, find a potential function for $\mathbf{F}$.

27. $\mathbf{F} = \frac{2x}{y}\mathbf{i} + \left(\frac{1-x^2}{y^2}\right)\mathbf{j}, \quad \{(x,y): y > 0\}$

28. $\mathbf{F} = (e^x \ln y)\mathbf{i} + \left(\frac{e^x}{y} + \sin z\right)\mathbf{j} + (y\cos z)\mathbf{k}$

29. Work along different paths Find the work done by $\mathbf{F} = (x^2 + y)\mathbf{i} + (y^2 + x)\mathbf{j} + ze^z\mathbf{k}$ over the following paths from $(1,0,0)$ to $(1,0,1)$.

a. The line segment $x = 1,\ y = 0,\ 0 \le z \le 1$

b. The helix $\mathbf{r}(t) = (\cos t)\mathbf{i} + (\sin t)\mathbf{j} + (t/2\pi)\mathbf{k},\ 0 \le t \le 2\pi$

c. The x-axis from $(1,0,0)$ to $(0,0,0)$ followed by the parabola $z = x^2,\ y = 0$ from $(0,0,0)$ to $(1,0,1)$

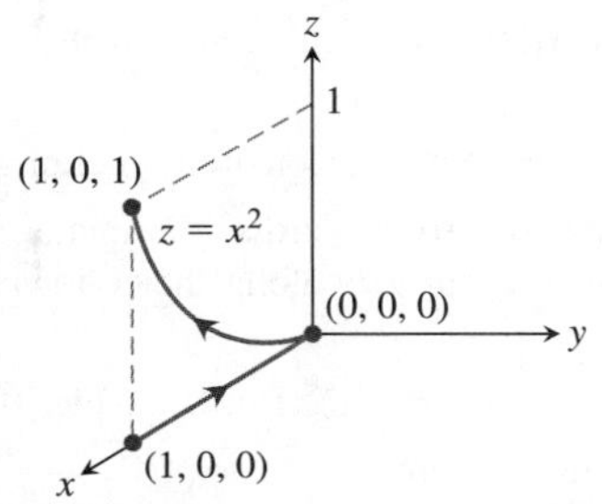

30. Work along different paths Find the work done by $\mathbf{F} = e^{yz}\mathbf{i} + (xze^{yz} + z\cos y)\mathbf{j} + (xye^{yz} + \sin y)\mathbf{k}$ over the following paths from $(1,0,1)$ to $(1,\pi/2,0)$.

a. The line segment $x = 1,\ y = \pi t/2,\ z = 1 - t,\ 0 \le t \le 1$

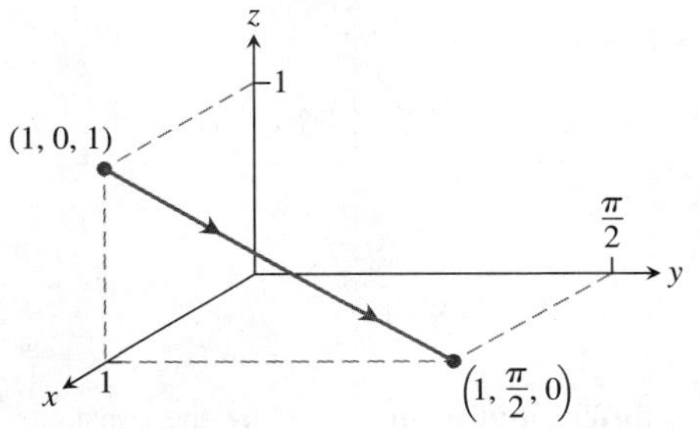

b. The line segment from $(1,0,1)$ to the origin followed by the line segment from the origin to $(1,\pi/2,0)$

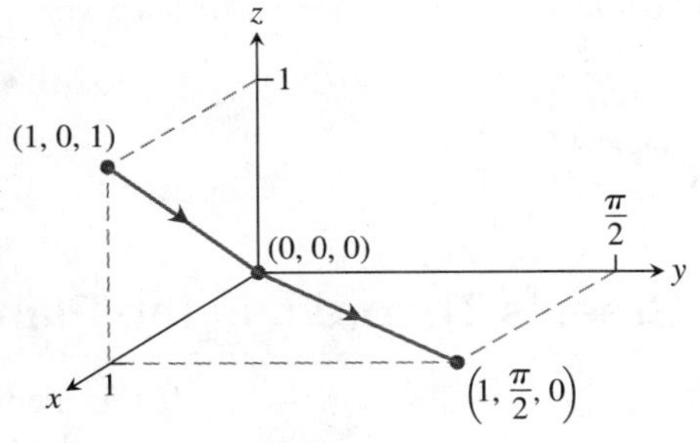

c. The line segment from $(1,0,1)$ to $(1,0,0)$, followed by the x-axis from $(1,0,0)$ to the origin, followed by the parabola $y = \pi x^2/2,\ z = 0$ from there to $(1,\pi/2,0)$

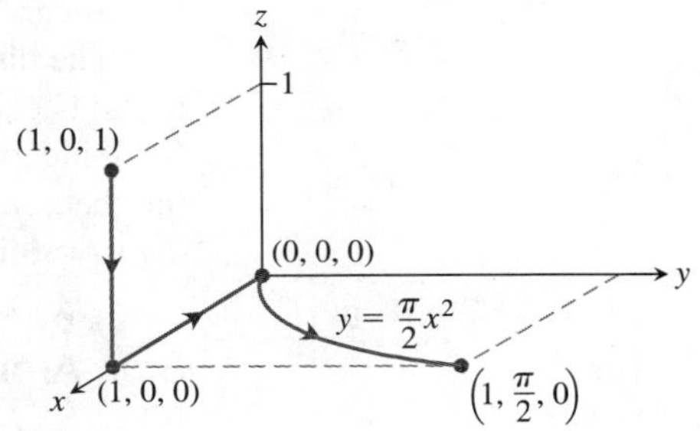

31. Evaluating a work integral two ways Let $\mathbf{F} = \nabla(x^3y^2)$ and let C be the path in the xy-plane from $(-1, 1)$ to $(1, 1)$ that consists of the line segment from $(-1, 1)$ to $(0, 0)$ followed by the line segment from $(0, 0)$ to $(1, 1)$. Evaluate $\int_C \mathbf{F} \cdot d\mathbf{r}$ in two ways.

a. Find parametrizations for the segments that make up C and evaluate the integral.

b. Use $f(x, y) = x^3y^2$ as a potential function for $\mathbf{F}$.

32. Integral along different paths Evaluate the line integral $\int_C 2x \cos y\, dx - x^2 \sin y\, dy$ along the following paths C in the xy-plane.

a. The parabola $y = (x - 1)^2$ from $(1, 0)$ to $(0, 1)$

b. The line segment from $(-1, \pi)$ to $(1, 0)$

c. The x-axis from $(-1, 0)$ to $(1, 0)$

d. The astroid $\mathbf{r}(t) = (\cos^3 t)\mathbf{i} + (\sin^3 t)\mathbf{j},\ 0 \le t \le 2\pi$, counterclockwise from $(1, 0)$ back to $(1, 0)$

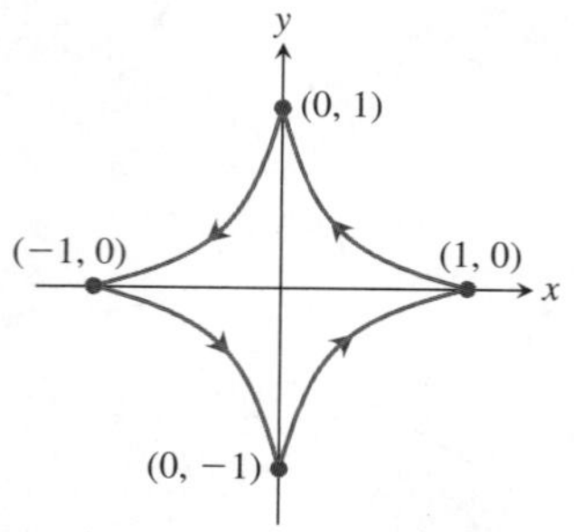

33. a. Exact differential form How are the constants a, b, and c related if the following differential form is exact?

$$(ay^2 + 2czx)\,dx + y(bx + cz)\,dy + (ay^2 + cx^2)\,dz$$

b. Gradient field For what values of b and c will

$$\mathbf{F} = (y^2 + 2czx)\mathbf{i} + y(bx + cz)\mathbf{j} + (y^2 + cx^2)\mathbf{k}$$

be a gradient field?

34. Gradient of a line integral Suppose that $\mathbf{F} = \nabla f$ is a conservative vector field and

$$g(x, y, z) = \int_{(0,0,0)}^{(x,y,z)} \mathbf{F} \cdot d\mathbf{r}.$$

Show that $\nabla g = \mathbf{F}$.

35. Path of least work You have been asked to find the path along which a force field $\mathbf{F}$ will perform the least work in moving a particle between two locations. A quick calculation on your part shows $\mathbf{F}$ to be conservative. How should you respond? Give reasons for your answer.

36. A revealing experiment By experiment, you find that a force field $\mathbf{F}$ performs only half as much work in moving an object along path C_1 from A to B as it does in moving the object along path C_2 from A to B. What can you conclude about $\mathbf{F}$? Give reasons for your answer.

37. Work by a constant force Show that the work done by a constant force field $\mathbf{F} = a\mathbf{i} + b\mathbf{j} + c\mathbf{k}$ in moving a particle along any path from A to B is $W = \mathbf{F} \cdot \overrightarrow{AB}$.

38. Gravitational field

a. Find a potential function for the gravitational field

$$\mathbf{F} = -GmM\frac{x\mathbf{i} + y\mathbf{j} + z\mathbf{k}}{(x^2 + y^2 + z^2)^{3/2}}$$

(G, m, and M are constants).

b. Let P_1 and P_2 be points at distances s_1 and s_2 from the origin. Show that the work done by the gravitational field in part (a) in moving a particle from P_1 to P_2 is

$$GmM\left(\frac{1}{s_2} - \frac{1}{s_1}\right).$$

16.4 Green's Theorem in the Plane

If $\mathbf{F}$ is a conservative field, then we know $\mathbf{F} = \nabla f$ for a differentiable function f, and we can calculate the line integral of $\mathbf{F}$ over any path C joining point A to point B as $\int_C \mathbf{F} \cdot d\mathbf{r} = f(B) - f(A)$. In this section we derive a method for computing a work or flux integral over a *closed* curve C in the plane. This method, which can be used even when the field $\mathbf{F}$ is not conservative, comes from Green's Theorem. It enables us to convert the line integral into a double integral over the region enclosed by C.

The discussion is given in terms of velocity fields of fluid flows (a fluid is a liquid or a gas) because they are easy to visualize. However, Green's Theorem applies to any vector field, independent of any particular interpretation of the field, provided the assumptions of the theorem are satisfied. We introduce two new ideas for Green's Theorem: *circulation density* around an axis perpendicular to the plane and *divergence* (or *flux density*).

Spin Around an Axis: The k-Component of Curl

Suppose that $\mathbf{F}(x, y) = M(x, y)\mathbf{i} + N(x, y)\mathbf{j}$ is the velocity field of a fluid flowing in the plane and that the first partial derivatives of M and N are continuous at each point of a region R. Let (x, y) be a point in R, and let A be a small rectangle with one corner at (x, y)

that, along with its interior, lies entirely in R. The sides of the rectangle, parallel to the coordinate axes, have lengths of Δx and Δy. Assume that the components M and N do not change sign throughout a small region containing the rectangle A. The first idea we use to convey Green's Theorem quantifies the rate at which a floating paddle wheel, with axis perpendicular to the plane, spins at a point in a fluid flowing in a plane region. This idea gives some sense of how the fluid is circulating around axes located at different points and perpendicular to the plane. Physicists sometimes refer to this as the *circulation density* of a vector field **F** at a point. To obtain it, we consider the velocity field

$$\mathbf{F}(x, y) = M(x, y)\mathbf{i} + N(x, y)\mathbf{j}$$

and the rectangle A in Figure 16.29 (where we assume both components of **F** are positive).

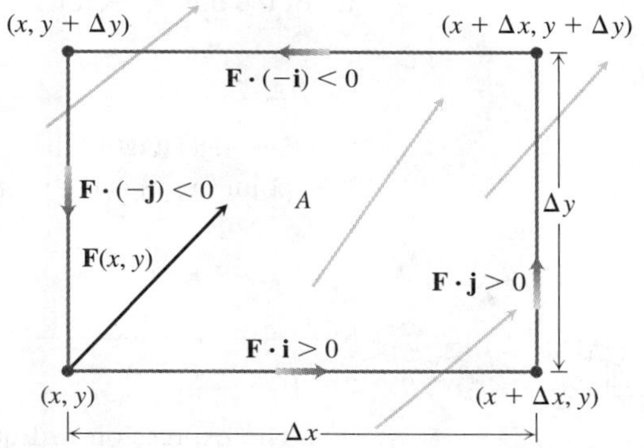

FIGURE 16.29 The rate at which a fluid flows along the bottom edge of a rectangular region A in the direction **i** is approximately $\mathbf{F}(x, y) \cdot \mathbf{i}\,\Delta x$, which is positive for the vector field **F** shown here. To approximate the rate of circulation at the point (x, y), we calculate the (approximate) flow rates along each edge in the directions of the shaded gray arrows, sum these rates, and then divide the sum by the area of A. Taking the limit as $\Delta x \to 0$ and $\Delta y \to 0$ gives the rate of the circulation per unit area.

The circulation rate of **F** around the boundary of A is the sum of flow rates along the sides in the tangential direction. For the bottom edge, the flow rate is approximately

$$\mathbf{F}(x, y) \cdot \mathbf{i}\,\Delta x = M(x, y)\,\Delta x.$$

This is the scalar component of the velocity $\mathbf{F}(x, y)$ in the tangent direction **i** times the length of the segment. The flow rates may be positive or negative, depending on the components of **F**. We approximate the net circulation rate around the rectangular boundary of A by summing the flow rates along the four edges as defined by the following dot products.

$$\begin{aligned}
&\text{Top:} && \mathbf{F}(x, y + \Delta y) \cdot (-\mathbf{i})\,\Delta x = -M(x, y + \Delta y)\,\Delta x \\
&\text{Bottom:} && \mathbf{F}(x, y) \cdot \mathbf{i}\,\Delta x = M(x, y)\,\Delta x \\
&\text{Right:} && \mathbf{F}(x + \Delta x, y) \cdot \mathbf{j}\,\Delta y = N(x + \Delta x, y)\,\Delta y \\
&\text{Left:} && \mathbf{F}(x, y) \cdot (-\mathbf{j})\,\Delta y = -N(x, y)\,\Delta y
\end{aligned}$$

We sum opposite pairs to get

$$\text{Top and bottom:} \quad -(M(x, y + \Delta y) - M(x, y))\,\Delta x \approx -\left(\frac{\partial M}{\partial y}\Delta y\right)\Delta x$$

$$\text{Right and left:} \quad (N(x + \Delta x, y) - N(x, y))\,\Delta y \approx \left(\frac{\partial N}{\partial x}\Delta x\right)\Delta y.$$

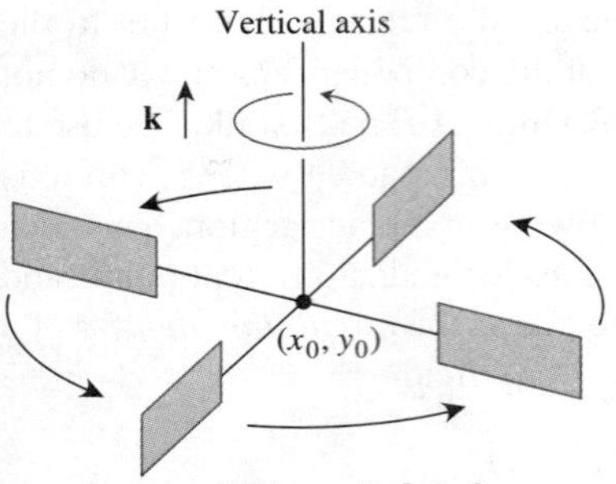

curl $\mathbf{F}(x_0, y_0) \cdot \mathbf{k} > 0$
Counterclockwise circulation

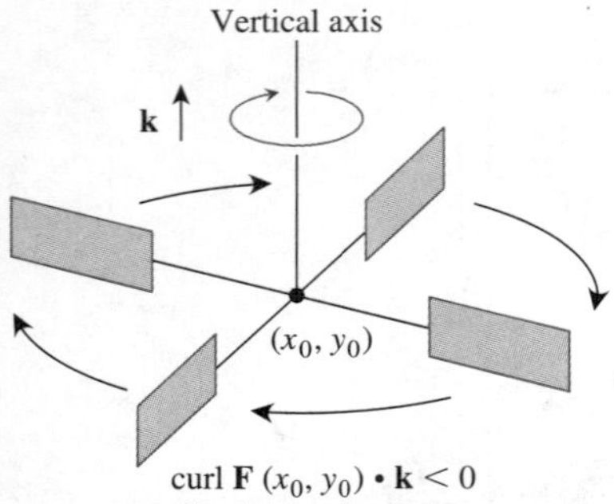

curl $\mathbf{F}(x_0, y_0) \cdot \mathbf{k} < 0$
Clockwise circulation

FIGURE 16.30 In the flow of an incompressible fluid over a plane region, the **k**-component of the curl measures the rate of the fluid's rotation at a point. The **k**-component of the curl is positive at points where the rotation is counterclockwise and negative where the rotation is clockwise.

Adding these last two equations gives the net circulation rate relative to the counterclockwise orientation,

$$\text{Circulation rate around rectangle} \approx \left(\frac{\partial N}{\partial x} - \frac{\partial M}{\partial y}\right)\Delta x\,\Delta y.$$

We now divide by $\Delta x\,\Delta y$ to estimate the circulation rate per unit area, or *circulation density*, for the rectangle:

$$\frac{\text{Circulation around rectangle}}{\text{rectangle area}} \approx \frac{\partial N}{\partial x} - \frac{\partial M}{\partial y}.$$

We let Δx and Δy approach zero to define the *circulation density* of **F** at the point (x, y).

If we see a counterclockwise rotation looking downward onto the *xy*-plane from the tip of the unit **k** vector, then the circulation density is positive (Figure 16.30).

DEFINITION The **circulation density** of a vector field $\mathbf{F} = M\mathbf{i} + N\mathbf{j}$ at the point (x, y) is the scalar expression

$$\frac{\partial N}{\partial x} - \frac{\partial M}{\partial y}. \qquad (1)$$

The expression in Equation (1) is the the **k**-component of the curl of **F**, which was introduced in Equation (3) of Section 16.3:

$$\frac{\partial N}{\partial x} - \frac{\partial M}{\partial y} = (\text{curl } \mathbf{F}) \cdot \mathbf{k}.$$

If water is moving about a region in the *xy*-plane in a thin layer, then the **k**-component of the curl at a point (x_0, y_0) gives a way to measure how fast and in what direction a small paddle wheel spins if it is put into the water at (x_0, y_0) with its axis perpendicular to the plane, parallel to **k** (Figure 16.30). Looking downward onto the *xy*-plane, it spins counterclockwise when $(\text{curl } \mathbf{F}) \cdot \mathbf{k}$ is positive and clockwise when the **k**-component is negative.

EXAMPLE 1 The following vector fields represent the velocity of a gas flowing in the *xy*-plane. Find the circulation density of each vector field and interpret its physical meaning. Figure 16.31 displays the vector fields.

(a) *Uniform expansion or compression:* $\mathbf{F}(x, y) = cx\mathbf{i} + cy\mathbf{j}$ $\quad c$ a constant

(b) *Uniform rotation:* $\mathbf{F}(x, y) = -cy\mathbf{i} + cx\mathbf{j}$

(c) *Shearing flow:* $\mathbf{F}(x, y) = y\mathbf{i}$

(d) *Whirlpool effect:* $\mathbf{F}(x, y) = \dfrac{-y}{x^2 + y^2}\mathbf{i} + \dfrac{x}{x^2 + y^2}\mathbf{j}$

Solution

(a) *Uniform expansion:* $(\text{curl } \mathbf{F}) \cdot \mathbf{k} = \dfrac{\partial}{\partial x}(cy) - \dfrac{\partial}{\partial y}(cx) = 0$. The gas is not circulating at very small scales.

(b) *Rotation:* $(\text{curl } \mathbf{F}) \cdot \mathbf{k} = \dfrac{\partial}{\partial x}(cx) - \dfrac{\partial}{\partial y}(-cy) = 2c$. The constant circulation density indicates rotation around every point. If $c > 0$, the rotation is counterclockwise; if $c < 0$, the rotation is clockwise.

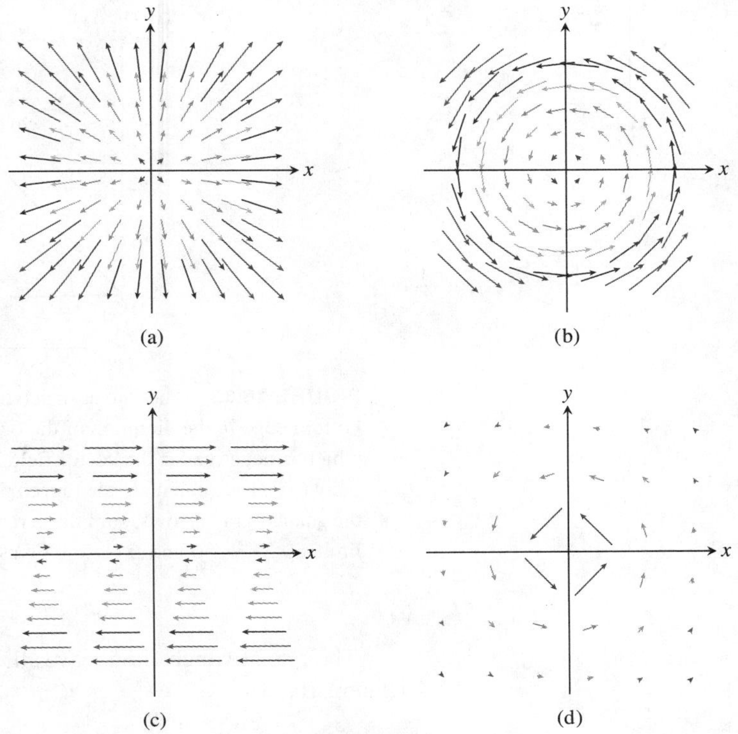

FIGURE 16.31 Velocity fields of a gas flowing in the plane (Example 1).

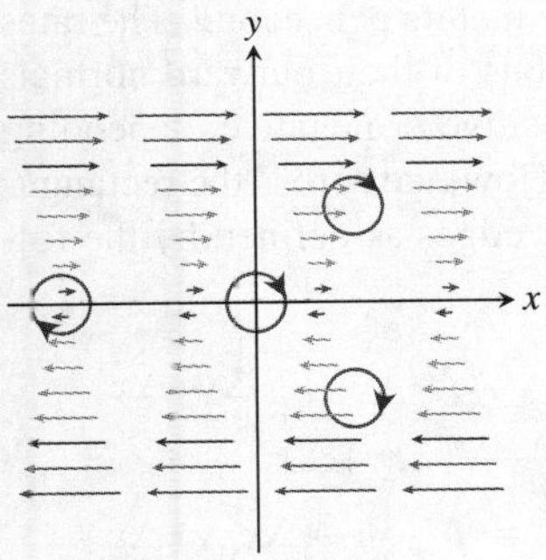

FIGURE 16.32 A shearing flow pushes the fluid clockwise around each point (Example 1c).

(c) *Shear:* $(\text{curl } \mathbf{F}) \cdot \mathbf{k} = -\frac{\partial}{\partial y}(y) = -1$. The circulation density is constant and negative, so a paddle wheel floating in water undergoing such a shearing flow spins clockwise. The rate of rotation is the same at each point. The average rotational effect of the fluid flow is to push fluid clockwise around each of the small circles shown in Figure 16.32.

(d) *Whirlpool:*

$$(\text{curl } \mathbf{F}) \cdot \mathbf{k} = \frac{\partial}{\partial x}\left(\frac{x}{x^2 + y^2}\right) - \frac{\partial}{\partial y}\left(\frac{-y}{x^2 + y^2}\right) = \frac{y^2 - x^2}{(x^2 + y^2)^2} - \frac{y^2 - x^2}{(x^2 + y^2)^2} = 0.$$

The circulation density is 0 at every point away from the origin (where the vector field is undefined and the whirlpool effect blows up), and the gas is not circulating at any point for which the vector field is defined. ■

One form of Green's Theorem tells us how circulation density can be used to calculate the line integral for flow in the xy-plane. (The flow integral was defined in Section 16.2.) A second form of the theorem tells us how we can calculate the flux integral, which gives the flow across the boundary, from *flux density*. We define this idea next and then present both versions of the theorem.

Divergence

Consider again the velocity field $\mathbf{F}(x, y) = M(x, y)\mathbf{i} + N(x, y)\mathbf{j}$ in a domain containing the rectangle A, as shown in Figure 16.33. As before, we assume the field components do not change sign throughout a small region containing the rectangle A. Our interest now is to determine the rate at which the fluid leaves A by flowing across its boundary.

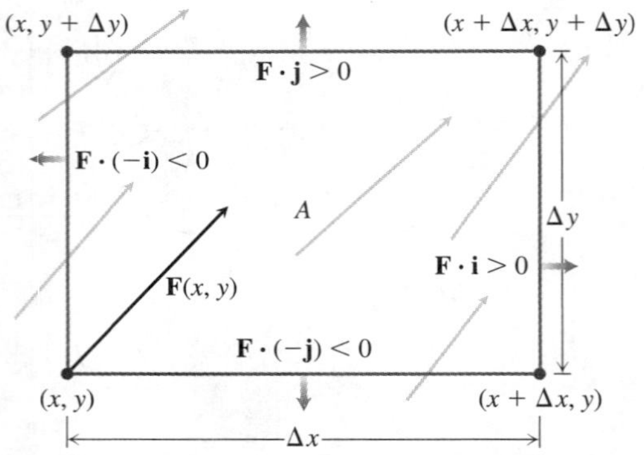

FIGURE 16.33 The rate at which the fluid leaves the rectangular region A across the bottom edge in the direction of the outward normal $-\mathbf{j}$ is approximately $\mathbf{F}(x, y) \cdot (-\mathbf{j})\,\Delta x$, which is negative for the vector field $\mathbf{F}$ shown here. To approximate the flow rate at the point (x, y), we calculate the (approximate) flow rates across each edge in the directions of the shaded gray arrows, sum these rates, and then divide the sum by the area of A. Taking the limit as $\Delta x \to 0$ and $\Delta y \to 0$ gives the flow rate per unit area.

The rate at which fluid leaves the rectangle across the bottom edge is approximately (Figure 16.33)

$$\mathbf{F}(x, y) \cdot (-\mathbf{j})\,\Delta x = -N(x, y)\,\Delta x.$$

This is the scalar component of the velocity at (x, y) in the direction of the outward normal times the length of the segment. If the velocity is in meters per second, for example, the flow rate will be in meters per second times meters, or square meters per second. The rates at which the fluid crosses the other three sides in the directions of their outward normals can be estimated in a similar way. The flow rates may be positive or negative, depending on the signs of the components of $\mathbf{F}$. We approximate the net flow rate across the rectangular boundary of A by summing the flow rates across the four edges as defined by the following dot products.

Fluid Flow Rates:

Top:	$\mathbf{F}(x, y + \Delta y) \cdot \mathbf{j}\,\Delta x$	$= N(x, y + \Delta y)\,\Delta x$
Bottom:	$\mathbf{F}(x, y) \cdot (-\mathbf{j})\,\Delta x$	$= -N(x, y)\,\Delta x$
Right:	$\mathbf{F}(x + \Delta x, y) \cdot \mathbf{i}\,\Delta y$	$= M(x + \Delta x, y)\,\Delta y$
Left:	$\mathbf{F}(x, y) \cdot (-\mathbf{i})\,\Delta y$	$= -M(x, y)\,\Delta y$

Summing opposite pairs gives

$$\text{Top and bottom:}\quad (N(x, y + \Delta y) - N(x, y))\,\Delta x \approx \left(\frac{\partial N}{\partial y}\Delta y\right)\Delta x,$$

$$\text{Right and left:}\quad (M(x + \Delta x, y) - M(x, y))\,\Delta y \approx \left(\frac{\partial M}{\partial x}\Delta x\right)\Delta y.$$

Adding these last two equations gives the net effect of the flow rates, or the

$$\text{Flux across rectangle boundary} \approx \left(\frac{\partial M}{\partial x} + \frac{\partial N}{\partial y}\right)\Delta x\,\Delta y.$$

We now divide by $\Delta x\,\Delta y$ to estimate the total flux per unit area, or *flux density*, for the rectangle:

$$\frac{\text{Flux across rectangle boundary}}{\text{rectangle area}} \approx \left(\frac{\partial M}{\partial x} + \frac{\partial N}{\partial y}\right).$$

div **F** is the symbol for divergence.

Finally, we let Δx and Δy approach zero to define the flux density of **F** at the point (x, y). The mathematical term for the flux density is the *divergence* of **F**. The symbol for it is div **F**, which is pronounced "divergence of **F**" or "div **F**."

DEFINITION The **divergence (flux density)** of a vector field $\mathbf{F} = M\mathbf{i} + N\mathbf{j}$ at the point (x, y) is

$$\text{div}\,\mathbf{F} = \frac{\partial M}{\partial x} + \frac{\partial N}{\partial y}. \tag{2}$$

Source: div **F** $(x_0, y_0) > 0$

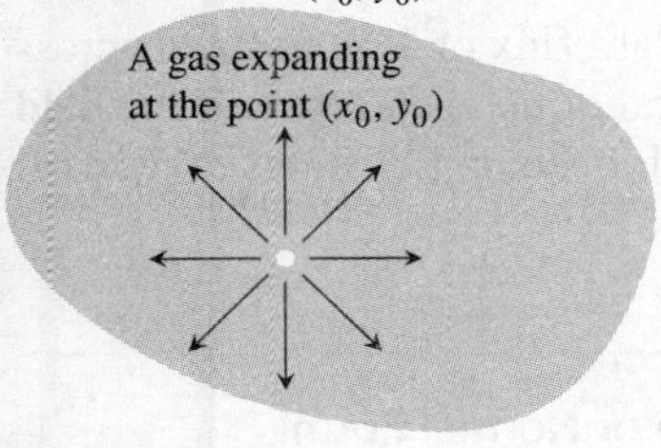

Sink: div **F** $(x_0, y_0) < 0$

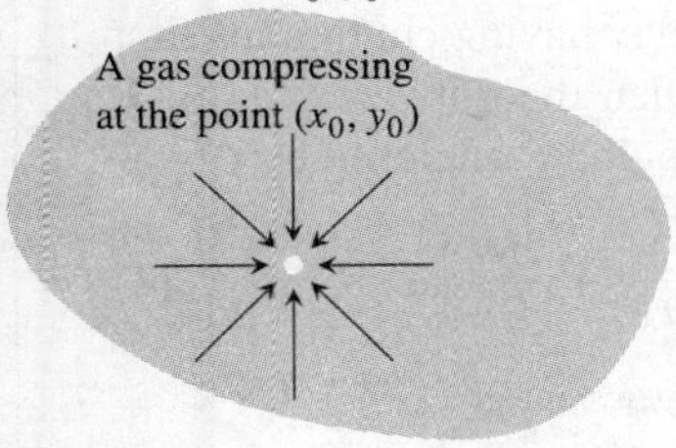

FIGURE 16.34 If a gas is expanding at a point (x_0, y_0), the lines of flow have positive divergence; if the gas is compressing, the divergence is negative.

A gas is compressible, unlike a liquid, and the divergence of its velocity field measures to what extent it is expanding or compressing at each point. Intuitively, if a gas is expanding at the point (x_0, y_0), the lines of flow would diverge there (hence the name), and since the gas would be flowing out of a small rectangle about (x_0, y_0), the divergence of **F** at (x_0, y_0) would be positive. If the gas were compressing instead of expanding, the divergence would be negative (Figure 16.34).

EXAMPLE 2 Find the divergence, and interpret what it means, for each vector field in Example 1 representing the velocity of a gas flowing in the *xy*-plane.

Solution

(a) $\text{div}\,\mathbf{F} = \frac{\partial}{\partial x}(cx) + \frac{\partial}{\partial y}(cy) = 2c$: If $c > 0$, the gas is undergoing uniform expansion; if $c < 0$, it is undergoing uniform compression.

(b) $\text{div}\,\mathbf{F} = \frac{\partial}{\partial x}(-cy) + \frac{\partial}{\partial y}(cx) = 0$: The gas is neither expanding nor compressing.

(c) $\text{div}\,\mathbf{F} = \frac{\partial}{\partial x}(y) = 0$: The gas is neither expanding nor compressing.

(d) $\text{div}\,\mathbf{F} = \frac{\partial}{\partial x}\left(\frac{-y}{x^2 + y^2}\right) + \frac{\partial}{\partial y}\left(\frac{x}{x^2 + y^2}\right) = \frac{2xy}{(x^2 + y^2)^2} - \frac{2xy}{(x^2 + y^2)^2} = 0$: Again, the divergence is zero at all points in the domain of the velocity field. ■

Cases (b), (c), and (d) of Figure 16.31 are plausible models for the two-dimensional flow of a liquid. In fluid dynamics, when the velocity field of a flowing fluid always has divergence equal to zero, as in those cases, the flow is said to be **incompressible**.

Two Forms for Green's Theorem

A simple closed curve C can be traversed in two possible directions. (Recall that a curve is simple if it does not cross itself.) The curve is traversed counterclockwise, and said to be *positively oriented*, if the region it encloses is always to the left when moving along the curve. If the curve is traversed clockwise, then the enclosed region is on the right when moving along the curve, and the curve is said to be *negatively oriented*. The line integral of a vector field **F** along C reverses sign if we change the orientation. We use the notation

$$\oint_C \mathbf{F}(x, y) \cdot d\mathbf{r}$$

for the line integral when the simple closed curve C is traversed counterclockwise, with its positive orientation.

In one form, Green's Theorem says that the counterclockwise circulation of a vector field around a simple closed curve is the double integral of the **k**-component of the curl of the field over the region enclosed by the curve. Recall the defining Equation (5) for circulation in Section 16.2.

Circulation around $C = \oint_C \mathbf{F} \cdot \mathbf{T}\, ds$

$(\text{curl } \mathbf{F}) \cdot \mathbf{k} = \dfrac{\partial N}{\partial x} - \dfrac{\partial M}{\partial y}$

THEOREM 4—Green's Theorem (Circulation-Curl or Tangential Form) Let C be a piecewise smooth, simple closed curve enclosing a region R in the plane. Let $\mathbf{F} = M\mathbf{i} + N\mathbf{j}$ be a vector field with M and N having continuous first partial derivatives in an open region containing R. Then the counterclockwise circulation of $\mathbf{F}$ around C equals the double integral of $(\text{curl } \mathbf{F}) \cdot \mathbf{k}$ over R.

$$\underbrace{\oint_C \mathbf{F} \cdot \mathbf{T}\, ds = \oint_C M\,dx + N\,dy}_{\text{Counterclockwise circulation}} = \underbrace{\iint_R \left(\frac{\partial N}{\partial x} - \frac{\partial M}{\partial y}\right) dx\,dy}_{\text{Curl integral}} \tag{3}$$

A second form of Green's Theorem says that the outward flux of a vector field across a simple closed curve in the plane equals the double integral of the divergence of the field over the region enclosed by the curve. Recall the formulas for flux in Equations (8) and (9) in Section 16.2.

Flux of $\mathbf{F}$ across $C = \oint_C \mathbf{F} \cdot \mathbf{n}\, ds$

$\text{div } \mathbf{F} = \dfrac{\partial M}{\partial x} + \dfrac{\partial N}{\partial y}$

THEOREM 5—Green's Theorem (Flux-Divergence or Normal Form) Let C be a piecewise smooth, simple closed curve enclosing a region R in the plane. Let $\mathbf{F} = M\mathbf{i} + N\mathbf{j}$ be a vector field with M and N having continuous first partial derivatives in an open region containing R. Then the outward flux of $\mathbf{F}$ across C equals the double integral of div $\mathbf{F}$ over the region R enclosed by C.

$$\underbrace{\oint_C \mathbf{F} \cdot \mathbf{n}\, ds = \oint_C M\,dy - N\,dx}_{\text{Outward flux}} = \underbrace{\iint_R \left(\frac{\partial M}{\partial x} + \frac{\partial N}{\partial y}\right) dx\,dy}_{\text{Divergence integral}} \tag{4}$$

The two forms of Green's Theorem are equivalent. Applying Equation (3) to the field $\mathbf{G}_1 = -N\mathbf{i} + M\mathbf{j}$ gives Equation (4), and applying Equation (4) to $\mathbf{G}_2 = N\mathbf{i} - M\mathbf{j}$ gives Equation (3).

Both forms of Green's Theorem can be viewed as two-dimensional generalizations of the Fundamental Theorem of Calculus from Section 5.4. The counterclockwise circulation of $\mathbf{F}$ around C, defined by the line integral on the left-hand side of Equation (3), is the integral of its rate of change (circulation density) over the region R enclosed by C, which is the double integral on the right-hand side of Equation (3). Likewise, the outward flux of $\mathbf{F}$ across C, defined by the line integral on the left-hand side of Equation (4), is the integral of its rate of change (flux density) over the region R enclosed by C, which is the double integral on the right-hand side of Equation (4).

EXAMPLE 3 Verify both forms of Green's Theorem for the vector field

$$\mathbf{F}(x, y) = (x - y)\mathbf{i} + x\mathbf{j}$$

and the region R bounded by the unit circle

$$C: \quad \mathbf{r}(t) = (\cos t)\mathbf{i} + (\sin t)\mathbf{j}, \quad 0 \le t \le 2\pi.$$

Solution First we evaluate the counterclockwise circulation of $\mathbf{F} = M\mathbf{i} + N\mathbf{j}$ around C. On the curve C we have $x = \cos t$ and $y = \sin t$. Evaluating $\mathbf{F}(\mathbf{r}(t))$ and computing the derivatives of the components of $\mathbf{r}$, we have

$$\begin{aligned} M &= x - y = \cos t - \sin t, \qquad & dx &= d(\cos t) = -\sin t\,dt, \\ N &= x = \cos t, & dy &= d(\sin t) = \cos t\,dt. \end{aligned}$$

Therefore,

$$\oint_C \mathbf{F}\cdot\mathbf{T}\,ds = \oint_C M\,dx + N\,dy$$

$$= \int_{t=0}^{t=2\pi} (\cos t - \sin t)(-\sin t)\,dt + (\cos t)(\cos t)\,dt$$

$$= \int_0^{2\pi} (-\sin t\cos t + 1)\,dt = 2\pi.$$

This gives the left side of Equation (3). Next we find the curl integral, the right side of Equation (3). Since $M = x - y$ and $N = x$, we have

$$\frac{\partial M}{\partial x} = 1, \qquad \frac{\partial M}{\partial y} = -1, \qquad \frac{\partial N}{\partial x} = 1, \qquad \frac{\partial N}{\partial y} = 0.$$

Therefore,

$$\iint_R \left(\frac{\partial N}{\partial x} - \frac{\partial M}{dy}\right) dx\,dy = \iint_R (1-(-1))\,dx\,dy$$

$$= 2\iint_R dx\,dy = 2(\text{area inside the unit circle}) = 2\pi.$$

Thus, the right and left sides of Equation (3) both equal 2π, as asserted by the circulation-curl version of Green's Theorem.

Figure 16.35 displays the vector field and circulation around C.

Now we compute the two sides of Equation (4) in the flux-divergence form of Green's Theorem, starting with the outward flux:

$$\oint_C M\,dy - N\,dx = \int_{t=0}^{t=2\pi} (\cos t - \sin t)(\cos t\,dt) - (\cos t)(-\sin t\,dt)$$

$$= \int_0^{2\pi} \cos^2 t\,dt = \pi.$$

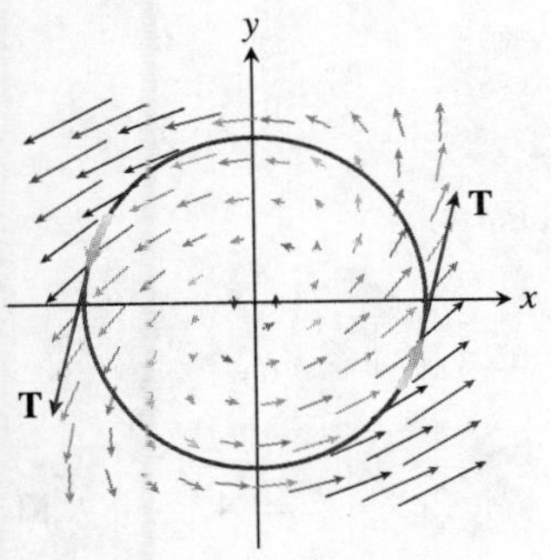

FIGURE 16.35 The vector field in Example 3 has a counterclockwise circulation of 2π around the unit circle.

Next we compute the divergence integral:

$$\iint_R \left(\frac{\partial M}{\partial x} + \frac{\partial N}{\partial y}\right) dx\,dy = \iint_R (1+0)\,dx\,dy = \iint_R dx\,dy = \pi.$$

Hence the right and left sides of Equation (4) both equal π, as asserted by the flux-divergence version of Green's Theorem. ■

Using Green's Theorem to Evaluate Line Integrals

If we construct a closed curve C by piecing together a number of different curves end to end, the process of evaluating a line integral over C can be lengthy because there are so many different integrals to evaluate. If C bounds a region R to which Green's Theorem applies, however, we can use Green's Theorem to change the line integral around C into one double integral over R.

EXAMPLE 4 Evaluate the line integral

$$\oint_C xy\,dy - y^2\,dx,$$

where C is the boundary of the square $0 \le x \le 1,\ 0 \le y \le 1$.

Solution We can use either form of Green's Theorem to change the line integral into a double integral over the square, where C is the square's boundary and R is its interior.

1. *With the Tangential Form* Equation (3): Taking $M = -y^2$ and $N = xy$ gives the result:

$$\oint_C -y^2\,dx + xy\,dy = \iint_R \left(\frac{\partial N}{\partial x} - \frac{\partial M}{\partial y}\right) dx\,dy = \iint_R (y - (-2y))\,dx\,dy$$

$$= \int_0^1 \int_0^1 3y\,dx\,dy = \int_0^1 \left[3xy\right]_{x=0}^{x=1} dy = \int_0^1 3y\,dy = \frac{3}{2}y^2\Big]_0^1 = \frac{3}{2}.$$

2. *With the Normal Form* Equation (4): Taking $M = xy$, $N = y^2$, gives the same result:

$$\oint_C xy\,dy - y^2\,dx = \iint_R \left(\frac{\partial M}{\partial x} + \frac{\partial N}{\partial y}\right) dx\,dy = \iint_R (y + 2y)\,dx\,dy = \frac{3}{2}.$$ ■

EXAMPLE 5 Calculate the outward flux of the vector field $\mathbf{F}(x, y) = 2e^{xy}\mathbf{i} + y^3\mathbf{j}$ across the boundary of the square $-1 \le x \le 1$, $-1 \le y \le 1$.

Solution Calculating the flux with a line integral would take four integrations, one for each side of the square. With Green's Theorem, we can change the line integral to one double integral. With $M = 2e^{xy}$, $N = y^3$, C the square's boundary, and R the square's interior, we have

$$\text{Flux} = \oint_C \mathbf{F}\cdot\mathbf{n}\,ds = \oint_C M\,dy - N\,dx$$

$$= \iint_R \left(\frac{\partial M}{\partial x} + \frac{\partial N}{\partial y}\right) dx\,dy \qquad \text{Green's Theorem, Eq. (4)}$$

$$= \int_{-1}^1 \int_{-1}^1 (2ye^{xy} + 3y^2)\,dx\,dy = \int_{-1}^1 \left[2e^{xy} + 3xy^2\right]_{x=-1}^{x=1} dy$$

$$= \int_{-1}^1 (2e^y + 6y^2 - 2e^{-y})\,dy = \left[2e^y + 2y^3 + 2e^{-y}\right]_{-1}^1 = 4.$$ ■

Proof of Green's Theorem for Special Regions

Let C be a smooth simple closed curve in the xy-plane with the property that lines parallel to the axes cut it at no more than two points. Let R be the region enclosed by C and suppose that M, N, and their first partial derivatives are continuous at every point of some open region containing C and R. We want to prove the circulation-curl form of Green's Theorem,

$$\oint_C M\,dx + N\,dy = \iint_R \left(\frac{\partial N}{\partial x} - \frac{\partial M}{\partial y}\right) dx\,dy. \tag{5}$$

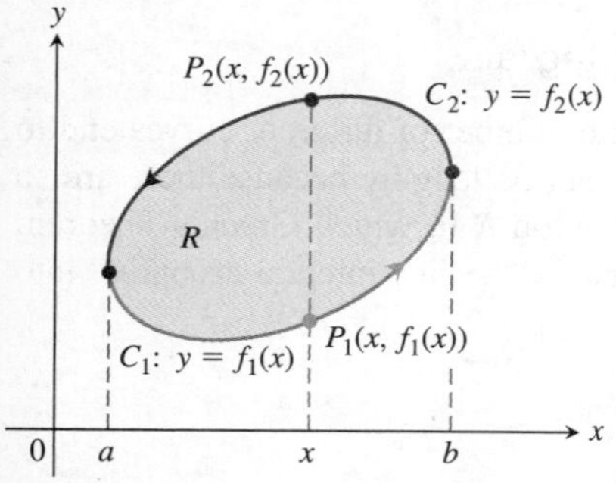

FIGURE 16.36 The boundary curve C is made up of C_1, the graph of $y = f_1(x)$, and C_2, the graph of $y = f_2(x)$.

Figure 16.36 shows C made up of two directed parts:

$$C_1:\ y = f_1(x),\ a \le x \le b, \qquad C_2:\ y = f_2(x),\ b \ge x \ge a.$$

For any x between a and b, we can integrate $\partial M/\partial y$ with respect to y from $y = f_1(x)$ to $y = f_2(x)$ and obtain

$$\int_{f_1(x)}^{f_2(x)} \frac{\partial M}{\partial y}\,dy = M(x, y)\Big]_{y=f_1(x)}^{y=f_2(x)} = M(x, f_2(x)) - M(x, f_1(x)).$$

We can then integrate this with respect to x from a to b:

$$\begin{aligned}\int_a^b \int_{f_1(x)}^{f_2(x)} \frac{\partial M}{\partial y}\,dy\,dx &= \int_a^b [M(x, f_2(x)) - M(x, f_1(x))]\,dx \\ &= -\int_b^a M(x, f_2(x))\,dx - \int_a^b M(x, f_1(x))\,dx \\ &= -\int_{C_2} M\,dx - \int_{C_1} M\,dx \\ &= -\oint_C M\,dx.\end{aligned}$$

Therefore, reversing the order of the equations, we have

$$\oint_C M\,dx = \iint_R \left(-\frac{\partial M}{\partial y}\right) dx\,dy. \tag{6}$$

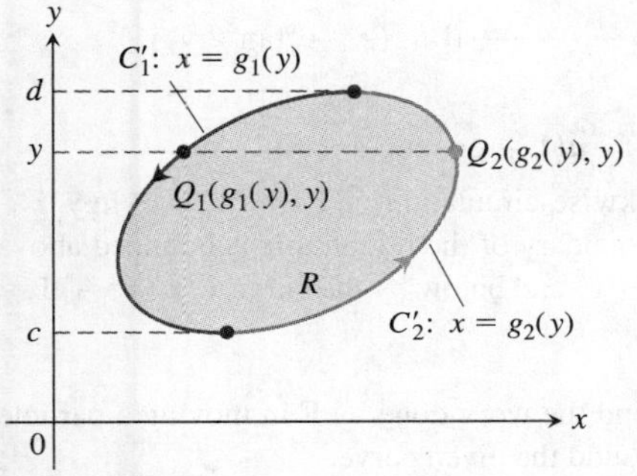

FIGURE 16.37 The boundary curve C is made up of C_1', the graph of $x = g_1(y)$, and C_2', the graph of $x = g_2(y)$.

Equation (6) is half the result we need for Equation (5). We derive the other half by integrating $\partial N/\partial x$ first with respect to x and then with respect to y, as suggested by Figure 16.37. This shows the curve C of Figure 16.36 decomposed into the two directed parts C_1': $x = g_1(y)$, $d \ge y \ge c$ and C_2': $x = g_2(y)$, $c \le y \le d$. The result of this double integration is

$$\oint_C N\,dy = \iint_R \frac{\partial N}{\partial x}\,dx\,dy. \tag{7}$$

Summing Equations (6) and (7) gives Equation (5). This concludes the proof. ■

Green's Theorem also holds for more general regions, such as those shown in Figure 16.38. Notice that the region in Figure 16.38c is not simply connected. The curves C_1 and C_h on its boundary are oriented so that the region R is always on the left-hand side as the curves are traversed in the directions shown, and cancelation occurs over common boundary arcs traversed in opposite directions. With this convention, Green's Theorem is valid for regions that are not simply connected. The proof proceeds by summing the contributions to the integral of a collection of special regions, which overlap along their boundaries. Cancelation occurs along arcs that are traversed twice, once in each direction, as in Figure 16.38c. We do not give the full proof here.

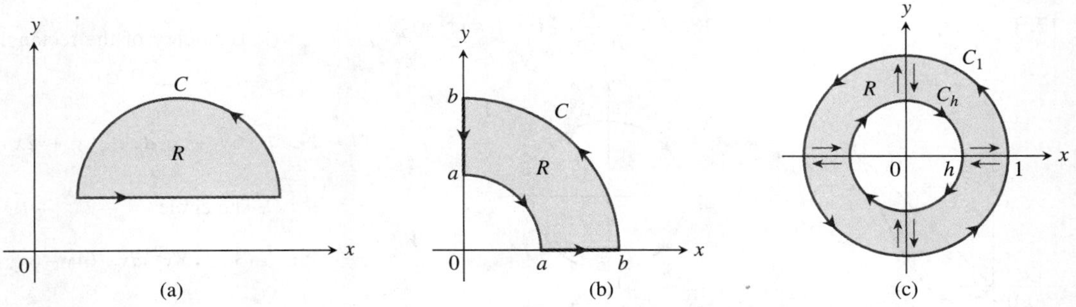

FIGURE 16.38 Other regions to which Green's Theorem applies. In (c) the axes convert the region into four simply connected regions, and we sum the line integrals along the oriented boundaries.

EXERCISES 16.4

Computing the k-Component of Curl(F)

In Exercises 1–6, find the **k**-component of curl(**F**) for the following vector fields on the plane.

1. $\mathbf{F} = (x + y)\mathbf{i} + (2xy)\mathbf{j}$
2. $\mathbf{F} = (x^2 - y)\mathbf{i} + (y^2)\mathbf{j}$
3. $\mathbf{F} = (xe^y)\mathbf{i} + (ye^x)\mathbf{j}$
4. $\mathbf{F} = (x^2 y)\mathbf{i} + (xy^2)\mathbf{j}$
5. $\mathbf{F} = (y \sin x)\mathbf{i} + (x \sin y)\mathbf{j}$
6. $\mathbf{F} = (x/y)\mathbf{i} - (y/x)\mathbf{j}$

Verifying Green's Theorem

In Exercises 7–10, verify the conclusion of Green's Theorem by evaluating both sides of Equations (3) and (4) for the field $\mathbf{F} = M\mathbf{i} + N\mathbf{j}$. Take the domains of integration in each case to be the disk R: $x^2 + y^2 \le a^2$ and its bounding circle C: $\mathbf{r} = (a \cos t)\mathbf{i} + (a \sin t)\mathbf{j}$, $0 \le t \le 2\pi$.

7. $\mathbf{F} = -y\mathbf{i} + x\mathbf{j}$
8. $\mathbf{F} = (x - y)\mathbf{i} + 2xy\,\mathbf{j}$

   C: Boundary of the region formed by the curves $y = x^2$ and $x = y^2$
9. $\mathbf{F} = 2x\mathbf{i} - 3y\mathbf{j}$
10. $\mathbf{F} = -x^2 y\mathbf{i} + xy^2\mathbf{j}$

Circulation and Flux

In Exercises 11–20, use Green's Theorem to find the counterclockwise circulation and outward flux for the field **F** and the curve C.

11. $\mathbf{F} = (x - y)\mathbf{i} + (y - x)\mathbf{j}$

    C: The square bounded by $x = 0$, $x = 1$, $y = 0$, and $y = 1$
12. $\mathbf{F} = (x^2 + 4y)\mathbf{i} + (x + y^2)\mathbf{j}$

    C: The square bounded by $x = 0$, $x = 1$, $y = 0$, and $y = 1$
13. $\mathbf{F} = (y^2 - x^2)\mathbf{i} + (x^2 + y^2)\mathbf{j}$

    C: The triangle bounded by $y = 0$, $x = 3$, and $y = x$
14. $\mathbf{F} = (x + y)\mathbf{i} - (x^2 + y^2)\mathbf{j}$

    C: The triangle bounded by $y = 0$, $x = 1$, and $y = x$
15. $\mathbf{F} = (xy + y^2)\mathbf{i} + (x - y)\mathbf{j}$

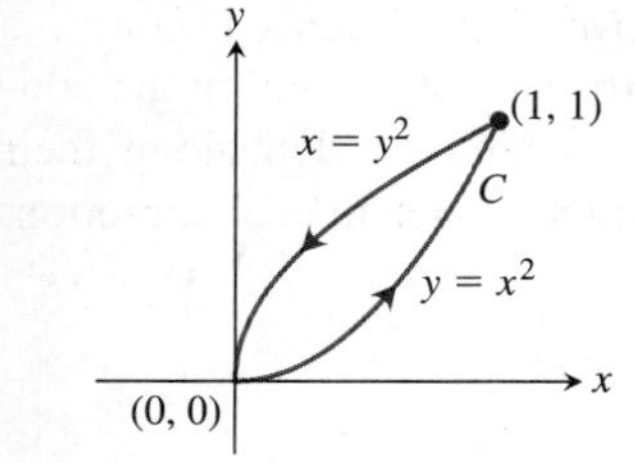

16. $\mathbf{F} = (x + 3y)\mathbf{i} + (2x - y)\mathbf{j}$

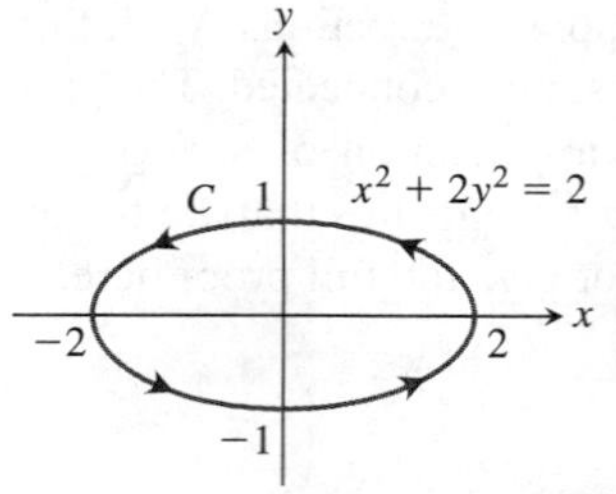

17. $\mathbf{F} = x^3 y^2\mathbf{i} + \frac{1}{2}x^4 y\mathbf{j}$

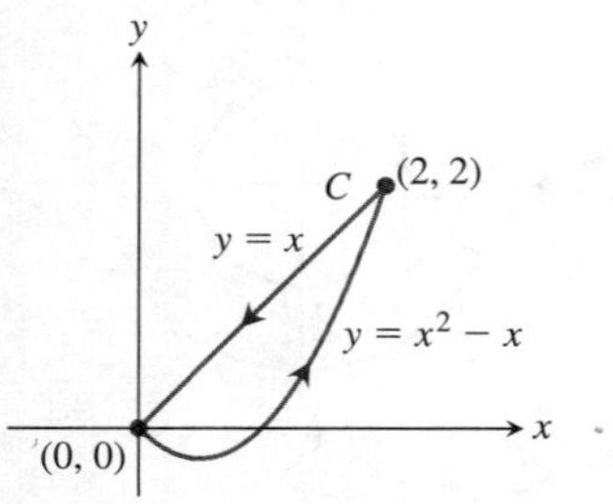

18. $\mathbf{F} = \dfrac{x}{1 + y^2}\mathbf{i} + (\tan^{-1} y)\mathbf{j}$

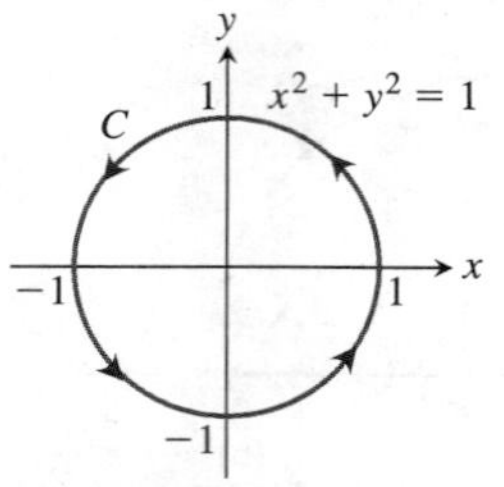

19. $\mathbf{F} = (x + e^x \sin y)\mathbf{i} + (x + e^x \cos y)\mathbf{j}$

    C: The right-hand loop of the lemniscate $r^2 = \cos 2\theta$
20. $\mathbf{F} = \left(\tan^{-1}\dfrac{y}{x}\right)\mathbf{i} + \ln(x^2 + y^2)\mathbf{j}$

    C: The boundary of the region defined by the polar coordinate inequalities $1 \le r \le 2$, $0 \le \theta \le \pi$
21. Find the counterclockwise circulation and outward flux of the field $\mathbf{F} = xy\mathbf{i} + y^2\mathbf{j}$ around and over the boundary of the region enclosed by the curve $y = x^2$ and the line $y = x$.
22. Find the counterclockwise circulation and the outward flux of the field $\mathbf{F} = (-\sin y)\mathbf{i} + (x \cos y)\mathbf{j}$ around and over the boundary of the square $0 \le x \le \pi/2$, $0 \le y \le \pi/2$.
23. Find the outward flux of the field

    $$\mathbf{F} = \left(3xy - \frac{x}{1 + y^2}\right)\mathbf{i} + (e^x + \tan^{-1} y)\mathbf{j}$$

    across the cardioid $r = a(1 + \cos\theta)$, $a > 0$.
24. Find the counterclockwise circulation of $\mathbf{F} = (y + e^x \ln y)\mathbf{i} + (e^x/y)\mathbf{j}$ around the boundary of the region that is bounded above by the curve $y = 3 - x^2$ and below by the curve $y = x^4 + 1$.

Work

In Exercises 25 and 26, find the work done by **F** in moving a particle once counterclockwise around the given curve.

25. $\mathbf{F} = 2xy^3\mathbf{i} + 4x^2y^2\mathbf{j}$

    C: The boundary of the "triangular" region in the first quadrant enclosed by the x-axis, the line $x = 1$, and the curve $y = x^3$
26. $\mathbf{F} = (4x - 2y)\mathbf{i} + (2x - 4y)\mathbf{j}$

    C: The circle $(x - 2)^2 + (y - 2)^2 = 4$

Using Green's Theorem

Apply Green's Theorem to evaluate the integrals in Exercises 27–30.

27. $\oint_C (y^2\,dx + x^2\,dy)$

    C: The boundary of the triangle enclosed by the lines $x = 0$, $x + y = 1$, and $y = 0$
28. $\oint_C e^{-x}[\sin y\,dx + \cos y\,dy]$

    C: Boundary of the rectangle formed by the lines $x = 0$, $x = \pi$, $y = 0$, $y = \dfrac{\pi}{2}$
29. $\oint_C (6y + x)\,dx + (y + 2x)\,dy$

    C: The circle $(x - 2)^2 + (y - 3)^2 = 4$
30. $\oint_C (3x^2 - 8y^2)dx + (4y - 6xy)dy$

    C: The boundary of the triangle formed by the lines $x = 0$, $y = 0$, $x + y = 1$

Calculating Area with Green's Theorem If a simple closed curve C in the plane and the region R it encloses satisfy the hypotheses of Green's Theorem, the area of R is given by

Green's Theorem Area Formula

$$\text{Area of } R = \frac{1}{2}\oint_C x\,dy - y\,dx$$

The reason is that, by Equation (4) run backward,

$$\text{Area of } R = \iint_R dy\,dx = \iint_R \left(\frac{1}{2} + \frac{1}{2}\right) dy\,dx$$
$$= \oint_C \frac{1}{2}x\,dy - \frac{1}{2}y\,dx.$$

Use the Green's Theorem area formula given above to find the areas of the regions enclosed by the curves in Exercises 31–34.

31. The circle $\mathbf{r}(t) = (a\cos t)\mathbf{i} + (a\sin t)\mathbf{j},\quad 0 \le t \le 2\pi$

32. The ellipse $\mathbf{r}(t) = (a\cos t)\mathbf{i} + (b\sin t)\mathbf{j},\quad 0 \le t \le 2\pi$

33. The astroid $\mathbf{r}(t) = (\cos^3 t)\mathbf{i} + (\sin^3 t)\mathbf{j},\quad 0 \le t \le 2\pi$

34. One arch of the cycloid $x = t - \sin t,\quad y = 1 - \cos t$

35. Find the area of the region bounded by $y^2 = 4x$ and $x^2 = 4y$ by Green's theorem.

36. Integral dependent only on area Show that the value of

$$\oint_C xy^2\,dx + (x^2y + 2x)\,dy$$

around any square depends only on the area of the square and not on its location in the plane.

37. Evaluate the integral

$$\oint_C 4x^3y\,dx + x^4\,dy$$

for any closed path C.

38. Evaluate the integral

$$\oint_C -y^3\,dy + x^3\,dx$$

for any closed path C.

39. Area as a line integral Show that if R is a region in the plane bounded by a piecewise smooth, simple closed curve C, then

$$\text{Area of } R = \oint_C x\,dy = -\oint_C y\,dx.$$

40. Definite integral as a line integral Suppose that a nonnegative function $y = f(x)$ has a continuous first derivative on $[a, b]$. Let C be the boundary of the region in the xy-plane that is bounded below by the x-axis, above by the graph of f, and on the sides by the lines $x = a$ and $x = b$. Show that

$$\int_a^b f(x)\,dx = -\oint_C y\,dx.$$

41. Area and the centroid Let $\bar{x}$ be the x-coordinate of the centroid of a region R that is bounded by a piecewise smooth, simple closed curve C in the xy-plane. If A is the area of R, show that

$$\frac{1}{2}\oint_C x^2\,dy = -\oint_C xy\,dx = \frac{1}{3}\oint_C x^2\,dy - xy\,dx = A\bar{x}.$$

42. Moment of inertia Let I_y be the moment of inertia about the y-axis of the region in Exercise 41. Show that

$$\frac{1}{3}\oint_C x^3\,dy = -\oint_C x^2y\,dx = \frac{1}{4}\oint_C x^3\,dy - x^2y\,dx = I_y.$$

43. Green's Theorem and Laplace's equation Assuming that all the necessary derivatives exist and are continuous, show that if $f(x, y)$ satisfies the Laplace equation

$$\frac{\partial^2 f}{\partial x^2} + \frac{\partial^2 f}{\partial y^2} = 0,$$

then

$$\oint_C \frac{\partial f}{\partial y}dx - \frac{\partial f}{\partial x}dy = 0$$

for all closed curves C to which Green's Theorem applies. (The converse is also true: If the line integral is always zero, then f satisfies the Laplace equation.)

44. Maximizing work Among all smooth, simple closed curves in the plane, oriented counterclockwise, find the one along which the work done by

$$\mathbf{F} = \left(\frac{1}{4}x^2y + \frac{1}{3}y^3\right)\mathbf{i} + x\mathbf{j}$$

is greatest. (*Hint:* Where is $(\text{curl } \mathbf{F}) \cdot \mathbf{k}$ positive?)

45. Regions with many holes Green's Theorem holds for a region R with any finite number of holes as long as the bounding curves are smooth, simple, and closed and we integrate over each component of the boundary in the direction that keeps R on our immediate left as we proceed along the curve (see accompanying figure).

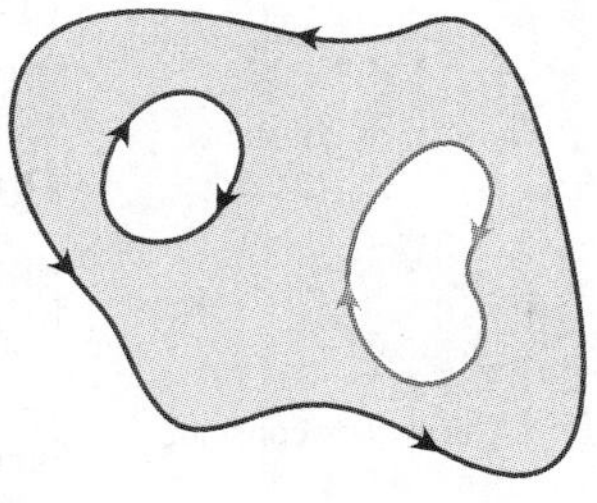

a. Let $f(x, y) = \ln(x^2 + y^2)$ and let C be the circle $x^2 + y^2 = a^2$. Evaluate the flux integral

$$\oint_C \nabla f \cdot \mathbf{n}\,ds.$$

b. Let K be an arbitrary smooth, simple closed curve in the plane that does not pass through $(0, 0)$. Use Green's Theorem to show that

$$\oint_K \nabla f \cdot \mathbf{n}\, ds$$

has two possible values, depending on whether $(0, 0)$ lies inside K or outside K.

46. Bendixson's criterion The *streamlines* of a planar fluid flow are the smooth curves traced by the fluid's individual particles. The vectors $\mathbf{F} = M(x, y)\mathbf{i} + N(x, y)\mathbf{j}$ of the flow's velocity field are the tangent vectors of the streamlines. Show that if the flow takes place over a simply connected region R (no holes or missing points) and that if $M_x + N_y \neq 0$ throughout R, then none of the streamlines in R is closed. In other words, no particle of fluid ever has a closed trajectory in R. The criterion $M_x + N_y \neq 0$ is called **Bendixson's criterion** for the nonexistence of closed trajectories.

47. Establish Equation (7) to finish the proof of the special case of Green's Theorem.

48. Curl component of conservative fields Can anything be said about the curl component of a conservative two-dimensional vector field? Give reasons for your answer.

COMPUTER EXPLORATIONS

In Exercises 49–52, use a CAS and Green's Theorem to find the counterclockwise circulation of the field $\mathbf{F}$ around the simple closed curve C. Perform the following CAS steps.

a. Plot C in the xy-plane.

b. Determine the integrand $(\partial N/\partial x) - (\partial M/\partial y)$ for the tangential form of Green's Theorem.

c. Determine the (double integral) limits of integration from your plot in part (a), and evaluate the curl integral for the circulation.

49. $\mathbf{F} = (2x - y)\mathbf{i} + (x + 3y)\mathbf{j}$, C: The ellipse $x^2 + 4y^2 = 4$

50. $\mathbf{F} = (2x^3 - y^3)\mathbf{i} + (x^3 + y^3)\mathbf{j}$, C: The ellipse $\frac{x^2}{4} + \frac{y^2}{9} = 1$

51. $\mathbf{F} = x^{-1}e^y\mathbf{i} + (e^y \ln x + 2x)\mathbf{j}$,
C: The boundary of the region defined by $y = 1 + x^4$ (below) and $y = 2$ (above)

52. $\mathbf{F} = xe^y\mathbf{i} + (4x^2 \ln y)\mathbf{j}$,
C: The triangle with vertices $(0, 0)$, $(2, 0)$, and $(0, 4)$

16.5 Surfaces and Area

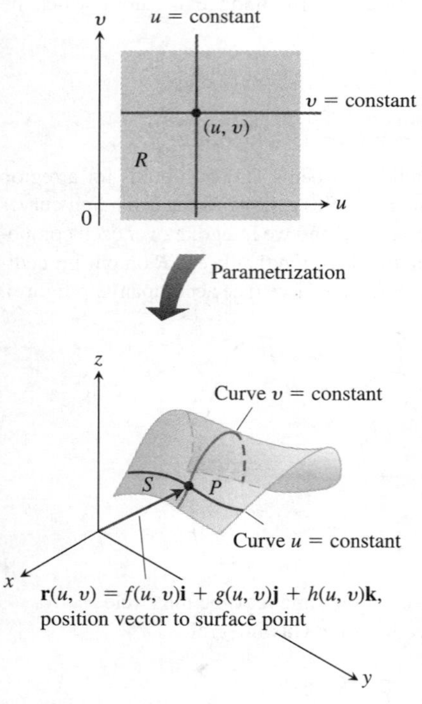

FIGURE 16.39 A parametrized surface S expressed as a vector function of two variables defined on a region R.

We have described curves in the plane in three different ways.

Explicit form:	$y = f(x)$
Implicit form:	$F(x, y) = 0$
Parametric vector form:	$\mathbf{r}(t) = f(t)\mathbf{i} + g(t)\mathbf{j}, \quad a \le t \le b.$

We have analogous descriptions of surfaces in space.

Explicit form:	$z = f(x, y)$
Implicit form:	$F(x, y, z) = 0.$

There is also a parametric form for surfaces that gives the position of a point on the surface as a vector function of two variables. We discuss this new form in this section and apply the form to obtain the area of a surface as a double integral. Double integral formulas for areas of surfaces given in implicit and explicit forms are then obtained as special cases of the more general parametric formula.

Parametrizations of Surfaces

Suppose

$$\mathbf{r}(u, v) = f(u, v)\mathbf{i} + g(u, v)\mathbf{j} + h(u, v)\mathbf{k} \tag{1}$$

is a continuous vector function that is defined on a region R in the uv-plane and is one-to-one on the interior of R (Figure 16.39). We call the range of $\mathbf{r}$ the **surface** S defined or traced by $\mathbf{r}$. Equation (1) together with the domain R constitutes a **parametrization** of the surface. The variables u and v are the **parameters**, and R is the **parameter domain**. To simplify our discussion, we take R to be a rectangle defined by inequalities of the form $a \le u \le b$, $c \le v \le d$. The requirement that $\mathbf{r}$ be one-to-one on the interior of R ensures that S does not cross itself. Notice that Equation (1) is the vector equivalent of *three* parametric equations:

$$x = f(u, v), \qquad y = g(u, v), \qquad z = h(u, v).$$

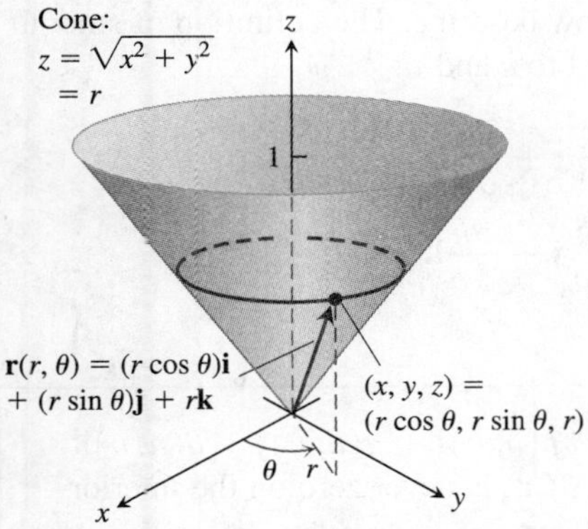

FIGURE 16.40 The cone in Example 1 can be parametrized using cylindrical coordinates.

EXAMPLE 1 Find a parametrization of the cone

$$z = \sqrt{x^2 + y^2}, \qquad 0 \le z \le 1.$$

Solution Here, cylindrical coordinates provide a parametrization. A typical point (x, y, z) on the cone (Figure 16.40) has $x = r \cos \theta$, $y = r \sin \theta$, and $z = \sqrt{x^2 + y^2} = r$, with $0 \le r \le 1$ and $0 \le \theta \le 2\pi$. Taking $u = r$ and $v = \theta$ in Equation (1) gives the parametrization

$$\mathbf{r}(r, \theta) = (r \cos \theta)\mathbf{i} + (r \sin \theta)\mathbf{j} + r\mathbf{k}, \qquad 0 \le r \le 1, \; 0 \le \theta \le 2\pi.$$

The parametrization is one-to-one on the interior of the domain, though not on the boundary where $r = 0$ (mapped to the tip of the cone) or where $\theta = 0$ or $\theta = 2\pi$ (where the cone glues together along a seam above the x-axis). ■

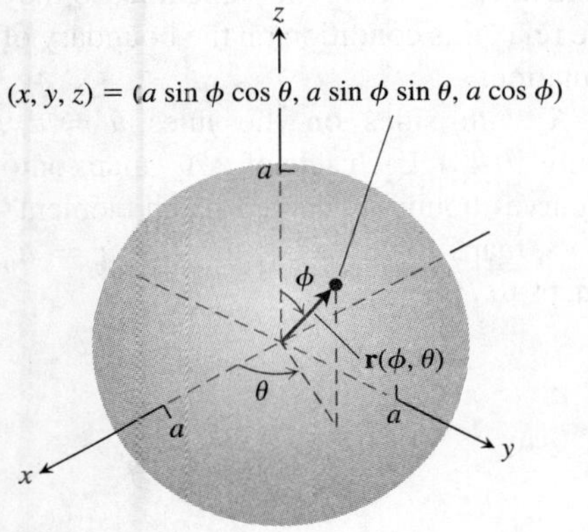

FIGURE 16.41 The sphere in Example 2 can be parametrized using spherical coordinates.

EXAMPLE 2 Find a parametrization of the sphere $x^2 + y^2 + z^2 = a^2$.

Solution Spherical coordinates provide what we need. A typical point (x, y, z) on the sphere (Figure 16.41) has $x = a \sin \phi \cos \theta$, $y = a \sin \phi \sin \theta$, and $z = a \cos \phi$, $0 \le \phi \le \pi$, $0 \le \theta \le 2\pi$. Taking $u = \phi$ and $v = \theta$ in Equation (1) gives the parametrization

$$\mathbf{r}(\phi, \theta) = (a \sin \phi \cos \theta)\mathbf{i} + (a \sin \phi \sin \theta)\mathbf{j} + (a \cos \phi)\mathbf{k},$$
$$0 \le \phi \le \pi, \qquad 0 \le \theta \le 2\pi.$$

Again, the parametrization is one-to-one on the interior of the domain, though not on its boundary. ■

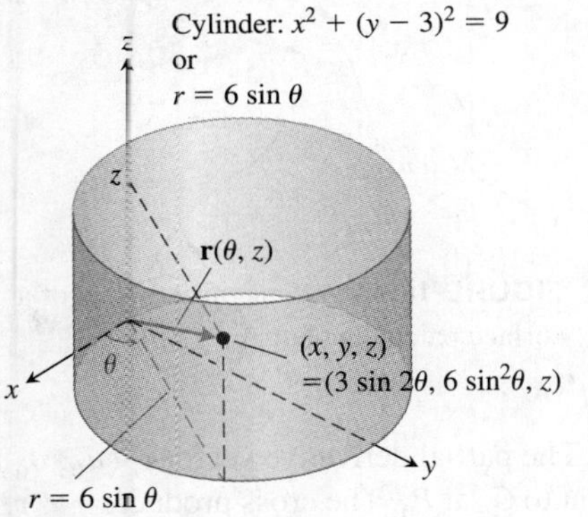

FIGURE 16.42 The cylinder in Example 3 can be parametrized using cylindrical coordinates.

EXAMPLE 3 Find a parametrization of the cylinder

$$x^2 + (y - 3)^2 = 9, \qquad 0 \le z \le 5.$$

Solution In cylindrical coordinates, a point (x, y, z) has $x = r \cos \theta$, $y = r \sin \theta$, and $z = z$. For points on the cylinder $x^2 + (y - 3)^2 = 9$ (Figure 16.42), the equation is the same as the polar equation for the cylinder's base in the xy-plane:

$$x^2 + (y^2 - 6y + 9) = 9$$
$$r^2 - 6r \sin \theta = 0 \qquad x^2 + y^2 = r^2,\; y = r \sin \theta$$

or

$$r = 6 \sin \theta, \qquad 0 \le \theta \le \pi.$$

A typical point on the cylinder therefore has

$$x = r \cos \theta = 6 \sin \theta \cos \theta = 3 \sin 2\theta$$
$$y = r \sin \theta = 6 \sin^2 \theta$$
$$z = z.$$

Taking $u = \theta$ and $v = z$ in Equation (1) gives the parametrization

$$\mathbf{r}(\theta, z) = (3 \sin 2\theta)\mathbf{i} + (6 \sin^2 \theta)\mathbf{j} + z\mathbf{k}, \quad 0 \le \theta \le \pi, \quad 0 \le z \le 5,$$

which is one-to-one on the interior of the domain. ■

Surface Area

Our goal is to find a double integral that gives the area of a curved surface S based on the parametrization

$$\mathbf{r}(u, v) = f(u, v)\mathbf{i} + g(u, v)\mathbf{j} + h(u, v)\mathbf{k}, \quad a \le u \le b, \quad c \le v \le d.$$

We need S to be smooth for the construction we now describe. The definition of smoothness involves the partial derivatives of $\mathbf{r}$ with respect to u and v:

$$\mathbf{r}_u = \frac{\partial \mathbf{r}}{\partial u} = \frac{\partial f}{\partial u}\mathbf{i} + \frac{\partial g}{\partial u}\mathbf{j} + \frac{\partial h}{\partial u}\mathbf{k}$$

$$\mathbf{r}_v = \frac{\partial \mathbf{r}}{\partial v} = \frac{\partial f}{\partial v}\mathbf{i} + \frac{\partial g}{\partial v}\mathbf{j} + \frac{\partial h}{\partial v}\mathbf{k}.$$

DEFINITION A parametrized surface $\mathbf{r}(u, v) = f(u, v)\mathbf{i} + g(u, v)\mathbf{j} + h(u, v)\mathbf{k}$ is **smooth** if $\mathbf{r}_u$ and $\mathbf{r}_v$ are continuous and if $\mathbf{r}_u \times \mathbf{r}_v$ is never zero on the interior of the parameter domain.

The condition that $\mathbf{r}_u \times \mathbf{r}_v$ is never the zero vector in the definition of smoothness means that the two vectors $\mathbf{r}_u$ and $\mathbf{r}_v$ are nonzero and never lie along the same line, so they always determine a plane tangent to the surface. We relax this condition on the boundary of the domain, but this does not affect the area computations.

Now consider a small rectangle ΔA_{uv} in R with sides on the lines $u = u_0$, $u = u_0 + \Delta u$, $v = v_0$, and $v = v_0 + \Delta v$ (Figure 16.43). Each side of ΔA_{uv} maps onto a curve on the surface S, and together these four curves bound a "curved patch element" $\Delta\sigma_{uv}$. In the notation of the figure, the side $v = v_0$ maps to curve C_1, the side $u = u_0$ maps onto C_2, and their common vertex (u_0, v_0) maps to P_0.

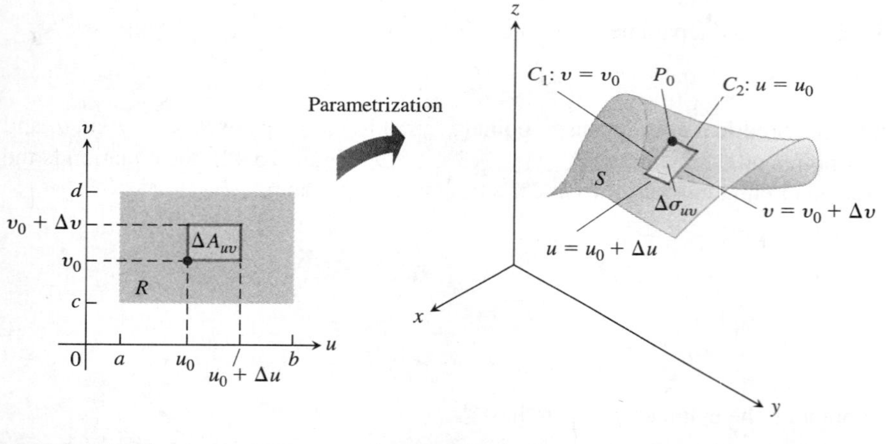

FIGURE 16.43 A rectangular area element ΔA_{uv} in the uv-plane maps onto a curved patch element $\Delta\sigma_{uv}$ on S.

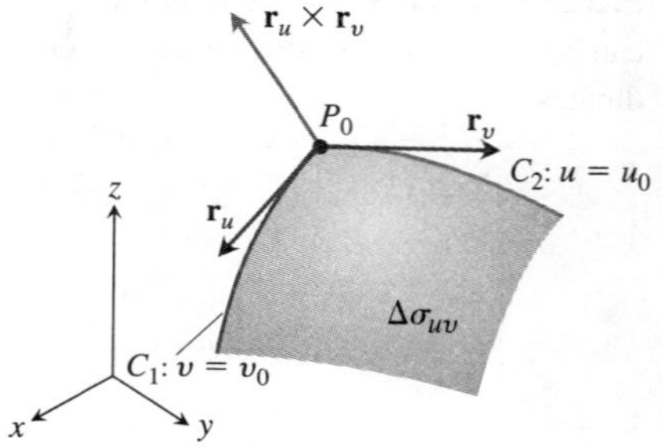

FIGURE 16.44 A magnified view of a surface patch element $\Delta\sigma_{uv}$.

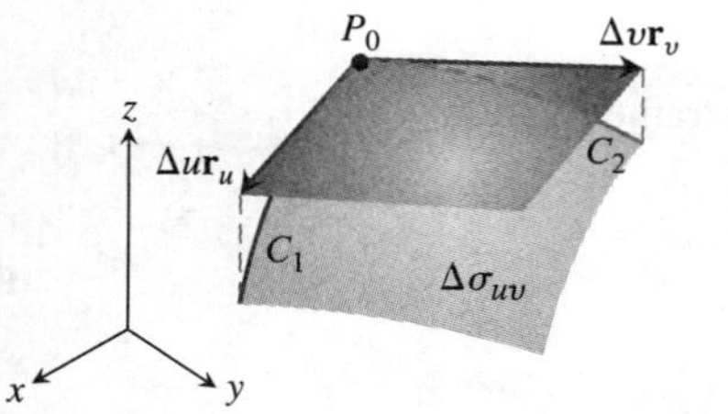

FIGURE 16.45 The area of the parallelogram determined by the vectors $\Delta u\mathbf{r}_u$ and $\Delta v\mathbf{r}_v$ approximates the area of the surface patch element $\Delta\sigma_{uv}$.

Figure 16.44 shows an enlarged view of $\Delta\sigma_{uv}$. The partial derivative vector $\mathbf{r}_u(u_0, v_0)$ is tangent to C_1 at P_0. Likewise, $\mathbf{r}_v(u_0, v_0)$ is tangent to C_2 at P_0. The cross product $\mathbf{r}_u \times \mathbf{r}_v$ is normal to the surface at P_0. (Here is where we begin to use the assumption that S is smooth. We want to be sure that $\mathbf{r}_u \times \mathbf{r}_v \neq \mathbf{0}$.)

We next approximate the surface patch element $\Delta\sigma_{uv}$ by the parallelogram on the tangent plane whose sides are determined by the vectors $\Delta u\mathbf{r}_u$ and $\Delta v\mathbf{r}_v$ (Figure 16.45). The area of this parallelogram is

$$|\Delta u\mathbf{r}_u \times \Delta v\mathbf{r}_v| = |\mathbf{r}_u \times \mathbf{r}_v|\,\Delta u\,\Delta v. \tag{2}$$

A partition of the region R in the uv-plane by rectangular regions ΔA_{uv} induces a partition of the surface S into surface patch elements $\Delta\sigma_{uv}$. We approximate the area of each surface patch element $\Delta\sigma_{uv}$ by the parallelogram area in Equation (2) and sum these areas together to obtain an approximation of the surface area of S:

$$\sum_n |\mathbf{r}_u \times \mathbf{r}_v|\, \Delta u\, \Delta v. \tag{3}$$

As Δu and Δv approach zero independently, the number of area elements n approaches ∞ and the continuity of $\mathbf{r}_u$ and $\mathbf{r}_v$ guarantees that the sum in Equation (3) approaches the double integral $\int_c^d \int_a^b |\mathbf{r}_u \times \mathbf{r}_v|\, du\, dv$. This double integral over the region R is used to define the area of the surface S.

DEFINITION The **area** of the smooth surface

$$\mathbf{r}(u, v) = f(u, v)\mathbf{i} + g(u, v)\mathbf{j} + h(u, v)\mathbf{k}, \qquad a \le u \le b, \;\; c \le v \le d$$

is

$$A = \iint_R |\mathbf{r}_u \times \mathbf{r}_v|\, dA = \int_c^d \int_a^b |\mathbf{r}_u \times \mathbf{r}_v|\, du\, dv. \tag{4}$$

We can abbreviate the integral in Equation (4) by writing $d\sigma$ for $|\mathbf{r}_u \times \mathbf{r}_v|\, du\, dv$. The surface area differential $d\sigma$ is analogous to the arc length differential ds in Section 13.3.

Surface Area Differential for a Parametrized Surface

$$d\sigma = |\mathbf{r}_u \times \mathbf{r}_v|\, du\, dv \qquad\qquad \iint_S d\sigma \tag{5}$$

Surface area differential, also called surface area element — Differential formula for surface area

EXAMPLE 4 Find the surface area of the cone in Example 1 (Figure 16.40).

Solution In Example 1, we found the parametrization

$$\mathbf{r}(r, \theta) = (r\cos\theta)\mathbf{i} + (r\sin\theta)\mathbf{j} + r\mathbf{k}, \qquad 0 \le r \le 1, \;\; 0 \le \theta \le 2\pi.$$

To apply Equation (4), we first find $\mathbf{r}_r \times \mathbf{r}_\theta$:

$$\mathbf{r}_r \times \mathbf{r}_\theta = \begin{vmatrix} \mathbf{i} & \mathbf{j} & \mathbf{k} \\ \cos\theta & \sin\theta & 1 \\ -r\sin\theta & r\cos\theta & 0 \end{vmatrix}$$

$$= -(r\cos\theta)\mathbf{i} - (r\sin\theta)\mathbf{j} + \underbrace{(r\cos^2\theta + r\sin^2\theta)}_{r}\mathbf{k}.$$

Thus, $|\mathbf{r}_r \times \mathbf{r}_\theta| = \sqrt{r^2\cos^2\theta + r^2\sin^2\theta + r^2} = \sqrt{2r^2} = \sqrt{2}r$. The area of the cone is

$$A = \int_0^{2\pi}\int_0^1 |\mathbf{r}_r \times \mathbf{r}_\theta|\, dr\, d\theta \qquad \text{Eq. (4) with } u = r,\ v = \theta$$

$$= \int_0^{2\pi}\int_0^1 \sqrt{2}r\, dr\, d\theta = \int_0^{2\pi} \frac{\sqrt{2}}{2}\, d\theta = \frac{\sqrt{2}}{2}(2\pi) = \pi\sqrt{2} \text{ square units.}$$ ■

EXAMPLE 5 Find the surface area of a sphere of radius a.

Solution We use the parametrization from Example 2:

$$\mathbf{r}(\phi,\theta) = (a\sin\phi\cos\theta)\mathbf{i} + (a\sin\phi\sin\theta)\mathbf{j} + (a\cos\phi)\mathbf{k},$$
$$0 \le \phi \le \pi, \quad 0 \le \theta \le 2\pi.$$

For $\mathbf{r}_\phi \times \mathbf{r}_\theta$, we get

$$\mathbf{r}_\phi \times \mathbf{r}_\theta = \begin{vmatrix} \mathbf{i} & \mathbf{j} & \mathbf{k} \\ a\cos\phi\cos\theta & a\cos\phi\sin\theta & -a\sin\phi \\ -a\sin\phi\sin\theta & a\sin\phi\cos\theta & 0 \end{vmatrix}$$
$$= (a^2\sin^2\phi\cos\theta)\mathbf{i} + (a^2\sin^2\phi\sin\theta)\mathbf{j} + (a^2\sin\phi\cos\phi)\mathbf{k}.$$

Thus,

$$|\mathbf{r}_\phi \times \mathbf{r}_\theta| = \sqrt{a^4\sin^4\phi\cos^2\theta + a^4\sin^4\phi\sin^2\theta + a^4\sin^2\phi\cos^2\phi}$$
$$= \sqrt{a^4\sin^4\phi + a^4\sin^2\phi\cos^2\phi} = \sqrt{a^4\sin^2\phi(\sin^2\phi + \cos^2\phi)}$$
$$= a^2\sqrt{\sin^2\phi} = a^2\sin\phi$$

because $\sin\phi \ge 0$ for $0 \le \phi \le \pi$. Therefore, the area of the sphere is

$$A = \int_0^{2\pi}\int_0^{\pi} a^2\sin\phi\, d\phi\, d\theta$$
$$= \int_0^{2\pi}\Big[-a^2\cos\phi\Big]_{\phi=0}^{\phi=\pi} d\theta = \int_0^{2\pi} 2a^2\, d\theta = 4\pi a^2 \quad \text{square units.}$$

This gives the well-known formula for the surface area of a sphere. ■

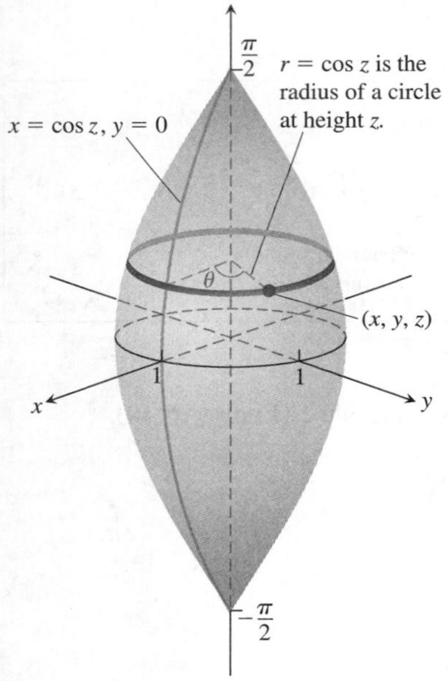

FIGURE 16.46 The "football" surface in Example 6 obtained by rotating the curve $x = \cos z$ about the z-axis.

EXAMPLE 6 Let S be the "football" surface formed by rotating the curve $x = \cos z$, $y = 0$, $-\pi/2 \le z \le \pi/2$ around the z-axis (see Figure 16.46). Find a parametrization for S and compute its surface area.

Solution Example 2 suggests finding a parametrization of S based on its rotation around the z-axis. If we rotate a point $(x, 0, z)$ on the curve $x = \cos z$, $y = 0$ about the z-axis, we obtain a circle at height z above the xy-plane that is centered on the z-axis and has radius $r = \cos z$ (see Figure 16.46). The point sweeps out the circle through an angle of rotation θ, $0 \le \theta \le 2\pi$. We let (x, y, z) be an arbitrary point on this circle, and define the parameters $u = z$ and $v = \theta$. Then we have $x = r\cos\theta = \cos u\cos v$, $y = r\sin\theta = \cos u\sin v$, and $z = u$, giving a parametrization for S as

$$\mathbf{r}(u,v) = \cos u\cos v\,\mathbf{i} + \cos u\sin v\,\mathbf{j} + u\mathbf{k}, \quad -\frac{\pi}{2} \le u \le \frac{\pi}{2}, \quad 0 \le v \le 2\pi.$$

Next we use Equation (5) to find the surface area of S. Differentiation of the parametrization gives

$$\mathbf{r}_u = -\sin u\cos v\,\mathbf{i} - \sin u\sin v\,\mathbf{j} + \mathbf{k}$$

and

$$\mathbf{r}_v = -\cos u\sin v\,\mathbf{i} + \cos u\cos v\,\mathbf{j}.$$

Computing the cross product, we have

$$\mathbf{r}_u \times \mathbf{r}_v = \begin{vmatrix} \mathbf{i} & \mathbf{j} & \mathbf{k} \\ -\sin u\cos v & -\sin u\sin v & 1 \\ -\cos u\sin v & \cos u\cos v & 0 \end{vmatrix}$$
$$= -\cos u\cos v\,\mathbf{i} - \cos u\sin v\,\mathbf{j} - (\sin u\cos u\cos^2 v + \cos u\sin u\sin^2 v)\mathbf{k}.$$

Taking the magnitude of the cross product gives

$$\begin{aligned}|\mathbf{r}_u \times \mathbf{r}_v| &= \sqrt{\cos^2 u\,(\cos^2 v + \sin^2 v) + \sin^2 u \cos^2 u} \\ &= \sqrt{\cos^2 u\,(1 + \sin^2 u)} \\ &= \cos u\,\sqrt{1 + \sin^2 u}. \qquad \cos u \geq 0 \text{ for } -\frac{\pi}{2} \leq u \leq \frac{\pi}{2}\end{aligned}$$

From Equation (4) the surface area is given by the integral

$$A = \int_0^{2\pi} \int_{-\pi/2}^{\pi/2} \cos u\,\sqrt{1 + \sin^2 u}\, du\, dv.$$

To evaluate the integral, we substitute $w = \sin u$ and $dw = \cos u\, du$, $-1 \leq w \leq 1$. Since the surface S is symmetric across the xy-plane, we need only integrate with respect to w from 0 to 1 and multiply the result by 2. In summary, we have

$$\begin{aligned}A &= 2\int_0^{2\pi} \int_0^1 \sqrt{1 + w^2}\, dw\, dv \\ &= 2\int_0^{2\pi} \left[\frac{w}{2}\sqrt{1 + w^2} + \frac{1}{2}\ln\left(w + \sqrt{1 + w^2}\right)\right]_{w=0}^{w=1} dv \qquad \text{Integral Table Formula 35} \\ &= \int_0^{2\pi} 2\left[\frac{1}{2}\sqrt{2} + \frac{1}{2}\ln\left(1 + \sqrt{2}\right)\right] dv \\ &= 2\pi\left[\sqrt{2} + \ln\left(1 + \sqrt{2}\right)\right].\end{aligned}$$

■

Implicit Surfaces

Surfaces are often presented as level sets of a function, described by an equation such as

$$F(x, y, z) = c,$$

for some constant c. Such a level surface does not come with an explicit parametrization and is called an *implicitly defined surface.* Implicit surfaces arise, for example, as equipotential surfaces in electric or gravitational fields. Figure 16.47 shows a piece of such a surface. It may be difficult to find explicit formulas for the functions f, g, and h that describe the surface in the form $\mathbf{r}(u, v) = f(u, v)\mathbf{i} + g(u, v)\mathbf{j} + h(u, v)\mathbf{k}$. We now show how to compute the surface area differential $d\sigma$ for implicit surfaces.

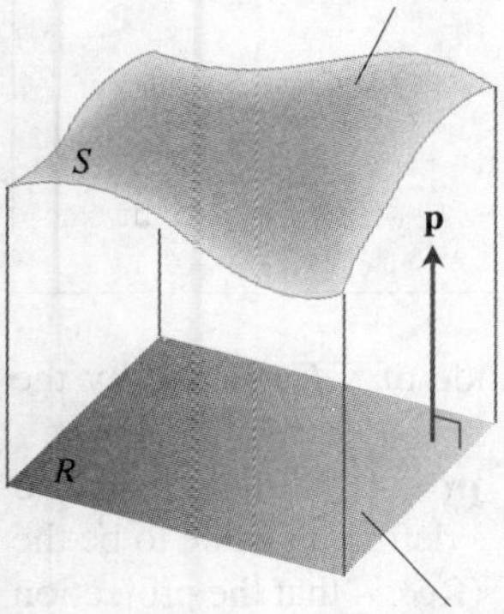

FIGURE 16.47 As we soon see, the area of a surface S in space can be calculated by evaluating a related double integral over the vertical projection or "shadow" of S on a coordinate plane. The unit vector $\mathbf{p}$ is normal to the plane.

Figure 16.47 shows a piece of an implicit surface S that lies above its "shadow" region R in the plane beneath it. The surface is defined by the equation $F(x, y, z) = c$, and we choose $\mathbf{p}$ to be a unit vector normal to the plane region R. We assume that the surface is **smooth** (F is differentiable and ∇F is nonzero and continuous on S) and that $\nabla F \cdot \mathbf{p} \neq 0$, so the surface never folds back over itself.

Assume that the normal vector $\mathbf{p}$ is the unit vector $\mathbf{k}$, so the region R in Figure 16.47 lies in the xy-plane. By assumption, we then have $\nabla F \cdot \mathbf{p} = \nabla F \cdot \mathbf{k} = F_z \neq 0$ on S. The Implicit Function Theorem (see Section 14.4) implies that S is then the graph of a differentiable function $z = h(x, y)$, although the function $h(x, y)$ is not explicitly known. Define the parameters u and v by $u = x$ and $v = y$. Then $z = h(u, v)$ and

$$\mathbf{r}(u, v) = u\mathbf{i} + v\mathbf{j} + h(u, v)\mathbf{k} \tag{6}$$

gives a parametrization of the surface S. We use Equation (4) to find the area of S.

Calculating the partial derivatives of $\mathbf{r}$, we find

$$\mathbf{r}_u = \mathbf{i} + \frac{\partial h}{\partial u}\mathbf{k} \quad \text{and} \quad \mathbf{r}_v = \mathbf{j} + \frac{\partial h}{\partial v}\mathbf{k}.$$

Applying the Chain Rule for implicit differentiation (see Equation (2) in Section 14.4) to $F(x, y, z) = c$, where $x = u$, $y = v$, and $z = h(u, v)$, we obtain the partial derivatives

$$\frac{\partial h}{\partial u} = -\frac{F_x}{F_z} \quad \text{and} \quad \frac{\partial h}{\partial v} = -\frac{F_y}{F_z}. \qquad F_z \neq 0$$

Substitution of these derivatives into the derivatives of $\mathbf{r}$ gives

$$\mathbf{r}_u = \mathbf{i} - \frac{F_x}{F_z}\mathbf{k} \quad \text{and} \quad \mathbf{r}_v = \mathbf{j} - \frac{F_y}{F_z}\mathbf{k}.$$

From a routine calculation of the cross product, we find

$$\begin{aligned} \mathbf{r}_u \times \mathbf{r}_v &= \frac{F_x}{F_z}\mathbf{i} + \frac{F_y}{F_z}\mathbf{j} + \mathbf{k} \qquad \begin{vmatrix} \mathbf{i} & \mathbf{j} & \mathbf{k} \\ 1 & 0 & -F_x/F_z \\ 0 & 1 & -F_y/F_z \end{vmatrix} \quad \text{cross product of } \mathbf{r}_u, \mathbf{r}_v \\ &= \frac{1}{F_z}(F_x\mathbf{i} + F_y\mathbf{j} + F_z\mathbf{k}) \\ &= \frac{\nabla F}{F_z} = \frac{\nabla F}{\nabla F \cdot \mathbf{k}} \\ &= \frac{\nabla F}{\nabla F \cdot \mathbf{p}}. \qquad \mathbf{p} = \mathbf{k} \end{aligned}$$

Therefore, the surface area differential is given by

$$d\sigma = |\mathbf{r}_u \times \mathbf{r}_v|\, du\, dv = \frac{|\nabla F|}{|\nabla F \cdot \mathbf{p}|}\, dx\, dy. \qquad u = x \text{ and } v = y$$

We obtain similar calculations if instead the vector $\mathbf{p} = \mathbf{j}$ is normal to the xz-plane when $F_y \neq 0$ on S, or if $\mathbf{p} = \mathbf{i}$ is normal to the yz-plane when $F_x \neq 0$ on S. Combining these results with Equation (4) then gives the following general formula.

Formula for the Surface Area of an Implicit Surface

The area of the surface $F(x, y, z) = c$ over a closed and bounded plane region R is

$$\text{Surface area} = \iint_R \frac{|\nabla F|}{|\nabla F \cdot \mathbf{p}|}\, dA, \qquad (7)$$

where $\mathbf{p} = \mathbf{i}$, $\mathbf{j}$, or $\mathbf{k}$ is normal to R and $\nabla F \cdot \mathbf{p} \neq 0$.

Thus, the area is the double integral over R of the magnitude of ∇F divided by the magnitude of the scalar component of ∇F normal to R.

We reached Equation (7) under the assumption that $\nabla F \cdot \mathbf{p} \neq 0$ throughout R and that ∇F is continuous. Whenever the integral exists, however, we define its value to be the area of the portion of the surface $F(x, y, z) = c$ that lies over R. (Recall that the projection is assumed to be one-to-one.)

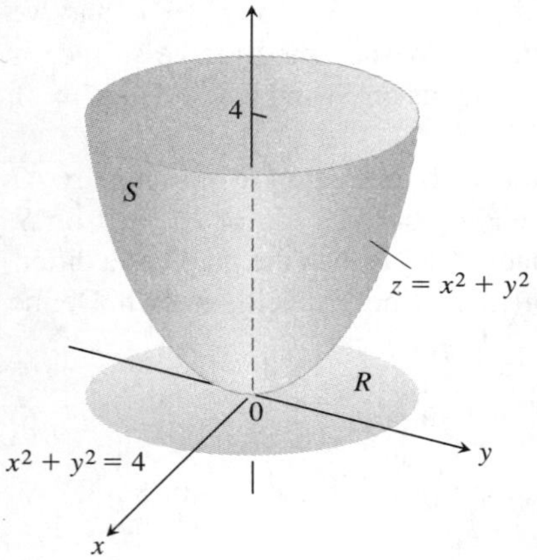

FIGURE 16.48 The area of this parabolic surface is calculated in Example 7.

EXAMPLE 7 Find the area of the surface cut from the bottom of the paraboloid $x^2 + y^2 - z = 0$ by the plane $z = 4$.

Solution We sketch the surface S and the region R below it in the xy-plane (Figure 16.48). The surface S is part of the level surface $F(x, y, z) = x^2 + y^2 - z = 0$, and R is the disk $x^2 + y^2 \leq 4$ in the xy-plane. To get a unit vector normal to the plane of R, we can take $\mathbf{p} = \mathbf{k}$.

At any point (x, y, z) on the surface, we have

$$F(x, y, z) = x^2 + y^2 - z$$
$$\nabla F = 2x\mathbf{i} + 2y\mathbf{j} - \mathbf{k}$$
$$|\nabla F| = \sqrt{(2x)^2 + (2y)^2 + (-1)^2} = \sqrt{4x^2 + 4y^2 + 1}$$
$$|\nabla F \cdot \mathbf{p}| = |\nabla F \cdot \mathbf{k}| = |-1| = 1.$$

In the region R, $dA = dx\,dy$. Therefore,

$$\begin{aligned}
\text{Surface area} &= \iint_R \frac{|\nabla F|}{|\nabla F \cdot \mathbf{p}|}\,dA && \text{Eq. (7)} \\
&= \iint_{x^2+y^2 \le 4} \sqrt{4x^2 + 4y^2 + 1}\,dx\,dy \\
&= \int_0^{2\pi}\int_0^2 \sqrt{4r^2 + 1}\,r\,dr\,d\theta && \text{Polar coordinates} \\
&= \int_0^{2\pi}\left[\frac{1}{12}(4r^2 + 1)^{3/2}\right]_{r=0}^{r=2} d\theta \\
&= \int_0^{2\pi} \frac{1}{12}(17^{3/2} - 1)\,d\theta = \frac{\pi}{6}(17\sqrt{17} - 1). \qquad \blacksquare
\end{aligned}$$

Example 7 illustrates how to find the surface area for a function $z = f(x, y)$ over a region R in the xy-plane. Actually, the surface area differential can be obtained in two ways, and we show this in the next example.

EXAMPLE 8 Derive the surface area differential $d\sigma$ of the surface $z = f(x, y)$ over a region R in the xy-plane **(a)** parametrically using Equation (5), and **(b)** implicitly, as in Equation (7).

Solution

(a) We parametrize the surface by taking $x = u$, $y = v$, and $z = f(x, y)$ over R. This gives the parametrization

$$\mathbf{r}(u, v) = u\mathbf{i} + v\mathbf{j} + f(u, v)\mathbf{k}.$$

Computing the partial derivatives gives $\mathbf{r}_u = \mathbf{i} + f_u\mathbf{k}$, $\mathbf{r}_v = \mathbf{j} + f_v\mathbf{k}$ and

$$\mathbf{r}_u \times \mathbf{r}_v = -f_u\mathbf{i} - f_v\mathbf{j} + \mathbf{k}. \qquad \begin{vmatrix} \mathbf{i} & \mathbf{j} & \mathbf{k} \\ 1 & 0 & f_u \\ 0 & 1 & f_v \end{vmatrix}$$

Then $|\mathbf{r}_u \times \mathbf{r}_v|\,du\,dv = \sqrt{f_u^2 + f_v^2 + 1}\,du\,dv$. Substituting for u and v. then gives the surface area differential

$$d\sigma = \sqrt{f_x^2 + f_y^2 + 1}\,dx\,dy.$$

(b) We define the implicit function $F(x, y, z) = f(x, y) - z$. Since (x, y) belongs to the region R, the unit normal to the plane of R is $\mathbf{p} = \mathbf{k}$. Then $\nabla F = f_x\mathbf{i} + f_y\mathbf{j} - \mathbf{k}$ so that $|\nabla F \cdot \mathbf{p}| = |-1| = 1, |\nabla F| = \sqrt{f_x^2 + f_y^2 + 1}$, and $|\nabla F|/|\nabla F \cdot \mathbf{p}| = |\nabla F|$. The surface area differential is again given by

$$d\sigma = \sqrt{f_x^2 + f_y^2 + 1}\, dx\, dy.$$

The surface area differential derived in Example 8 gives the following formula for calculating the surface area of the graph of a function defined explicitly as $z = f(x, y)$.

Formula for the Surface Area of a Graph $z = f(x, y)$
For a graph $z = f(x, y)$ over a region R in the xy-plane, the surface area formula is

$$A = \iint_R \sqrt{f_x^2 + f_y^2 + 1}\, dx\, dy. \qquad (8)$$

EXERCISES 16.5

Finding Parametrizations

In Exercises 1–16, find a parametrization of the surface. (There are many correct ways to do these, so your answers may not be the same as those in the back of the text.)

1. The paraboloid $z = x^2 + y^2, z \le 4$
2. The paraboloid $z = 9 - x^2 - y^2, z \ge 0$
3. **Cone frustum** The first-octant portion of the cone $z = \sqrt{x^2 + y^2}/2$ between the planes $z = 0$ and $z = 3$
4. **Cone frustum** The portion of the cone $z = 2\sqrt{x^2 + y^2}$ between the planes $z = 2$ and $z = 4$
5. **Spherical cap** The cap cut from the sphere $x^2 + y^2 + z^2 = 9$ by the cone $z = \sqrt{x^2 + y^2}$
6. **Spherical cap** The portion of the sphere $x^2 + y^2 + z^2 = 4$ in the first octant between the xy-plane and the cone $z = \sqrt{x^2 + y^2}$
7. **Spherical band** The portion of the sphere $x^2 + y^2 + z^2 = 3$ between the planes $z = \sqrt{3}/2$ and $z = -\sqrt{3}/2$
8. **Spherical cap** The upper portion cut from the sphere $x^2 + y^2 + z^2 = 8$ by the plane $z = -2$
9. **Parabolic cylinder between planes** The surface cut from the parabolic cylinder $z = 4 - y^2$ by the planes $x = 0, x = 2$, and $z = 0$
10. **Parabolic cylinder between planes** The surface cut from the parabolic cylinder $y = x^2$ by the planes $z = 0, z = 3$, and $y = 2$
11. **Circular cylinder band** The portion of the cylinder $y^2 + z^2 = 9$ between the planes $x = 0$ and $x = 3$
12. **Circular cylinder band** The portion of the cylinder $x^2 + z^2 = 4$ above the xy-plane between the planes $y = -2$ and $y = 2$
13. **Tilted plane inside cylinder** The portion of the plane $x + y + z = 1$
    a. Inside the cylinder $x^2 + y^2 = 9$
    b. Inside the cylinder $y^2 + z^2 = 9$
14. **Tilted plane inside cylinder** The portion of the plane $x - y + 2z = 2$
    a. Inside the cylinder $x^2 + z^2 = 3$
    b. Inside the cylinder $y^2 + z^2 = 2$
15. **Circular cylinder band** The portion of the cylinder $(x - 2)^2 + z^2 = 4$ between the planes $y = 0$ and $y = 3$
16. **Circular cylinder band** The portion of the cylinder $y^2 + (z - 5)^2 = 25$ between the planes $x = 0$ and $x = 10$

Surface Area of Parametrized Surfaces

In Exercises 17–26, use a parametrization to express the area of the surface as a double integral. Then evaluate the integral. (There are many correct ways to set up the integrals, so your integrals may not be the same as those in the back of the text. They should have the same values, however.)

17. **Tilted plane inside cylinder** The portion of the plane $y + 2z = 2$ inside the cylinder $x^2 + y^2 = 1$
18. **Plane inside cylinder** The portion of the plane $z = -x$ inside the cylinder $x^2 + y^2 = 4$
19. **Cone frustum** The portion of the cone $z = 2\sqrt{x^2 + y^2}$ between the planes $z = 2$ and $z = 6$
20. **Cone frustum** The portion of the cone $z = \sqrt{x^2 + y^2}/3$ between the planes $z = 1$ and $z = 4/3$
21. **Circular cylinder band** The portion of the cylinder $x^2 + y^2 = 1$ between the planes $z = 1$ and $z = 4$
22. **Circular cylinder band** The portion of the cylinder $x^2 + z^2 = 10$ between the planes $y = -1$ and $y = 1$
23. **Parabolic cap** The cap cut from the paraboloid $z = 2 - x^2 - y^2$ by the cone $z = \sqrt{x^2 + y^2}$
24. **Parabolic band** The portion of the paraboloid $z = x^2 + y^2$ between the planes $z = 1$ and $z = 4$
25. **Sawed-off sphere** The lower portion cut from the sphere $x^2 + y^2 + z^2 = 2$ by the cone $z = \sqrt{x^2 + y^2}$
26. **Spherical band** The portion of the sphere $x^2 + y^2 + z^2 = 4$ between the planes $z = -1$ and $z = \sqrt{3}$

Planes Tangent to Parametrized Surfaces

The tangent plane at a point $P_0(f(u_0, v_0), g(u_0, v_0), h(u_0, v_0))$ on a parametrized surface $\mathbf{r}(u, v) = f(u, v)\mathbf{i} + g(u, v)\mathbf{j} + h(u, v)\mathbf{k}$ is the plane through P_0 normal to the vector $\mathbf{r}_u(u_0, v_0) \times \mathbf{r}_v(u_0, v_0)$, the cross product of the tangent vectors $\mathbf{r}_u(u_0, v_0)$ and $\mathbf{r}_v(u_0, v_0)$ at P_0. In Exercises 27–30, find an equation for the plane tangent to the surface at P_0. Then find a Cartesian equation for the surface, and sketch the surface and tangent plane together.

27. Cone The cone $\mathbf{r}(r, \theta) = (r\cos\theta)\mathbf{i} + (r\sin\theta)\mathbf{j} + r\mathbf{k}$, $r \geq 0$, $0 \leq \theta \leq 2\pi$ at the point $P_0(\sqrt{2}, \sqrt{2}, 2)$ corresponding to $(r, \theta) = (2, \pi/4)$

28. Hemisphere The hemisphere surface $\mathbf{r}(\phi, \theta) = (4\sin\phi\cos\theta)\mathbf{i} + (4\sin\phi\sin\theta)\mathbf{j} + (4\cos\phi)\mathbf{k}$, $0 \leq \phi \leq \pi/2$, $0 \leq \theta \leq 2\pi$, at the point $P_0(\sqrt{2}, \sqrt{2}, 2\sqrt{3})$ corresponding to $(\phi, \theta) = (\pi/6, \pi/4)$

29. Circular cylinder The circular cylinder $\mathbf{r}(\theta, z) = (3\sin 2\theta)\mathbf{i} + (6\sin^2\theta)\mathbf{j} + z\mathbf{k}$, $0 \leq \theta \leq \pi$, at the point $P_0(3\sqrt{3}/2, 9/2, 0)$ corresponding to $(\theta, z) = (\pi/3, 0)$ (See Example 3.)

30. Parabolic cylinder The parabolic cylinder surface $\mathbf{r}(x, y) = x\mathbf{i} + y\mathbf{j} - x^2\mathbf{k}$, $-\infty < x < \infty$, $-\infty < y < \infty$, at the point $P_0(1, 2, -1)$ corresponding to $(x, y) = (1, 2)$

More Parametrizations of Surfaces

31. a. A *torus of revolution* (doughnut) is obtained by rotating a circle C in the xz-plane about the z-axis in space. (See the accompanying figure.) If C has radius $r > 0$ and center $(R, 0, 0)$, show that a parametrization of the torus is

$$\mathbf{r}(u, v) = ((R + r\cos u)\cos v)\mathbf{i} + ((R + r\cos u)\sin v)\mathbf{j} + (r\sin u)\mathbf{k},$$

where $0 \leq u \leq 2\pi$ and $0 \leq v \leq 2\pi$ are the angles in the figure.

b. Show that the surface area of the torus is $A = 4\pi^2 Rr$.

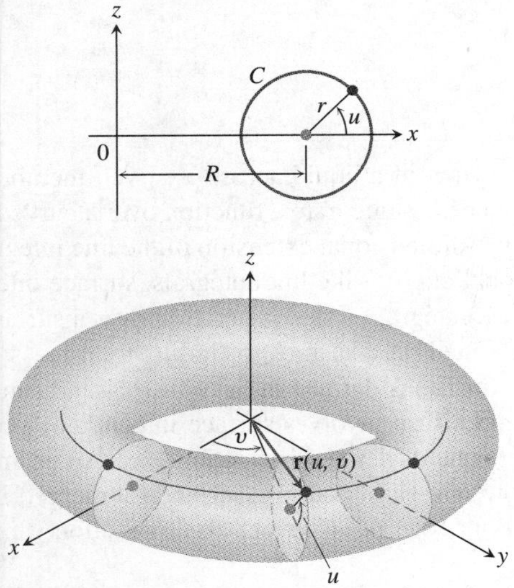

32. Parametrization of a surface of revolution Suppose that the parametrized curve C: $(f(u), g(u))$ is revolved about the x-axis, where $g(u) > 0$ for $a \leq u \leq b$.

a. Show that

$$\mathbf{r}(u, v) = f(u)\mathbf{i} + (g(u)\cos v)\mathbf{j} + (g(u)\sin v)\mathbf{k}$$

is a parametrization of the resulting surface of revolution, where $0 \leq v \leq 2\pi$ is the angle from the xy-plane to the point $\mathbf{r}(u, v)$ on the surface. (See the accompanying figure.) Notice that $f(u)$ measures distance *along* the axis of revolution and $g(u)$ measures distance *from* the axis of revolution.

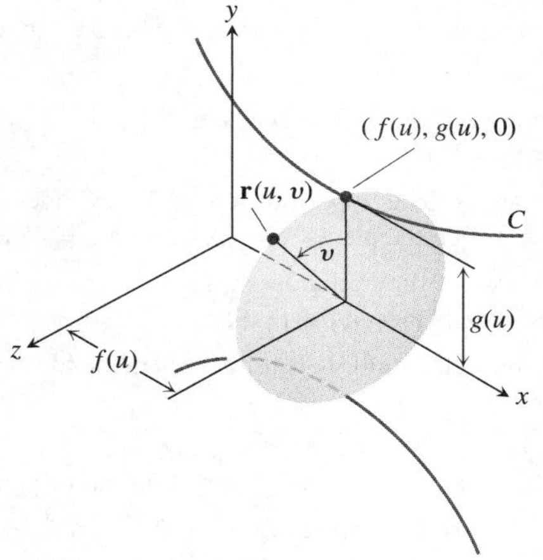

b. Find a parametrization for the surface obtained by revolving the curve $x = y^2$, $y \geq 0$, about the x-axis.

33. a. Parametrization of an ellipsoid The parametrization $x = a\cos\theta$, $y = b\sin\theta$, $0 \leq \theta \leq 2\pi$ gives the ellipse $(x^2/a^2) + (y^2/b^2) = 1$. Using the angles θ and ϕ in spherical coordinates, show that

$$\mathbf{r}(\theta, \phi) = (a\cos\theta\sin\phi)\mathbf{i} + (b\sin\theta\sin\phi)\mathbf{j} + (c\cos\phi)\mathbf{k}$$

is a parametrization of the ellipsoid

$$(x^2/a^2) + (y^2/b^2) + (z^2/c^2) = 1.$$

b. Write an integral for the surface area of the ellipsoid, but do not evaluate the integral.

34. Hyperboloid of one sheet

a. Find a parametrization for the hyperboloid of one sheet $x^2 + y^2 - z^2 = 1$ in terms of the angle θ associated with the circle $x^2 + y^2 = r^2$ and the hyperbolic parameter u associated with the hyperbolic function $r^2 - z^2 = 1$. (*Hint:* $\cosh^2 u - \sinh^2 u = 1$.)

b. Generalize the result in part (a) to the hyperboloid $(x^2/a^2) + (y^2/b^2) - (z^2/c^2) = 1$.

35. (*Continuation of Exercise 34.*) Find a Cartesian equation for the plane tangent to the hyperboloid $x^2 + y^2 - z^2 = 25$ at the point $(x_0, y_0, 0)$, where $x_0{}^2 + y_0{}^2 = 25$.

36. Hyperboloid of two sheets Find a parametrization of the hyperboloid of two sheets $(z^2/c^2) - (x^2/a^2) - (y^2/b^2) = 1$.

Surface Area for Implicit and Explicit Forms

37. Find the surface area of the sphere $x^2 + y^2 + z^2 = a^2$

38. Find the area of the band cut from the paraboloid $x^2 + y^2 - z = 0$ by the planes $z = 2$ and $z = 6$.

39. Find the area of the region cut from the plane $x + 2y + 2z = 5$ by the cylinder whose walls are $x = y^2$ and $x = 2 - y^2$.

40. Find the area of the portion of the surface $x^2 - 2z = 0$ that lies above the triangle bounded by the lines $x = \sqrt{3}$, $y = 0$, and $y = x$ in the xy-plane.

41. Find the area of the surface $x^2 - 2y - 2z = 0$ that lies above the triangle bounded by the lines $x = 2$, $y = 0$, and $y = 3x$ in the xy-plane.

42. Find the area of the cap cut from the sphere $x^2 + y^2 + z^2 = 2$ by the cone $z = \sqrt{x^2 + y^2}$.

43. Find the area of the ellipse cut from the plane $z = cx$ (c a constant) by the cylinder $x^2 + y^2 = 1$.

44. Find the area of the upper portion of the cylinder $x^2 + z^2 = 1$ that lies between the planes $x = \pm 1/2$ and $y = \pm 1/2$.

45. Find the area of the triangular surface cut from $x + y + z = a$ by the coordinate planes in the first octant.

46. Find the area of the surface cut from the paraboloid $x^2 + y + z^2 = 2$ by the plane $y = 0$.

47. Find the area of the surface $x^2 - 2\ln x + \sqrt{15}y - z = 0$ above the square R: $1 \le x \le 2$, $0 \le y \le 1$, in the xy-plane.

48. Find the area of the surface $2x^{3/2} + 2y^{3/2} - 3z = 0$ above the square R: $0 \le x \le 1$, $0 \le y \le 1$, in the xy-plane.

Find the area of the surfaces in Exercises 49–54.

49. The surface cut from the bottom of the paraboloid $z = x^2 + y^2$ by the plane $z = 3$

50. The surface cut from the "nose" of the paraboloid $x = 1 - y^2 - z^2$ by the yz-plane

51. The portion of the cone $z = \sqrt{x^2 + y^2}$ that lies over the region between the circle $x^2 + y^2 = 1$ and the ellipse $9x^2 + 4y^2 = 36$ in the xy-plane. (*Hint:* A formula from geometry states that the area inside the ellipse $x^2/a^2 + y^2/b^2 = 1$ is πab.)

52. The triangle cut from the plane $2x + 6y + 3z = 6$ by the bounding planes of the first octant. Calculate the area three ways, using different explicit forms.

53. The surface in the first octant cut from the cylinder $y = (2/3)z^{3/2}$ by the planes $x = 1$ and $y = 16/3$

54. The portion of the plane $y + z = 4$ that lies above the region cut from the first quadrant of the xz-plane by the parabola $x = 4 - z^2$

55. Use the parametrization

$$\mathbf{r}(x, z) = x\mathbf{i} + f(x, z)\mathbf{j} + z\mathbf{k}$$

and Equation (5) to derive a formula for $d\sigma$ associated with the explicit form $y = f(x, z)$.

56. Let S be the surface obtained by rotating the smooth curve $y = f(x)$, $a \le x \le b$, about the x-axis, where $f(x) \ge 0$.

a. Show that the vector function

$$\mathbf{r}(x, \theta) = x\mathbf{i} + f(x)\cos\theta\mathbf{j} + f(x)\sin\theta\mathbf{k}$$

is a parametrization of S, where θ is the angle of rotation around the x-axis (see the accompanying figure).

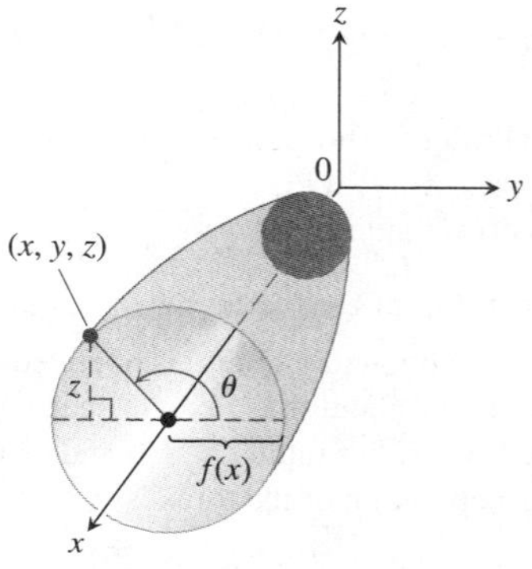

b. Use Equation (4) to show that the surface area of this surface of revolution is given by

$$A = \int_a^b 2\pi f(x)\sqrt{1 + [f'(x)]^2}\,dx.$$

16.6 Surface Integrals

To compute the mass of a surface, the flow of a liquid across a curved membrane, or the total electrical charge on a surface, we need to integrate a function over a curved surface in space. Such a *surface integral* is the two-dimensional extension of the line integral concept used to integrate over a one-dimensional curve. Like line integrals, surface integrals arise in two forms. The first occurs when we integrate a scalar function over a surface, such as integrating a mass density function defined on a surface to find its total mass. This form corresponds to line integrals of scalar functions defined in Section 16.1 and can be used to find the mass of a thin wire. The second form involves surface integrals of vector fields, analogous to the line integrals for vector fields defined in Section 16.2. An example occurs when we want to measure the net flow of a fluid across a surface submerged in the fluid (just as we previously defined the flux of **F** across a curve). In this section we investigate these ideas and their applications.

Surface Integrals

Suppose that the function $G(x, y, z)$ gives the *mass density* (mass per unit area) at each point on a surface S. Then we can calculate the total mass of S as an integral in the following way.

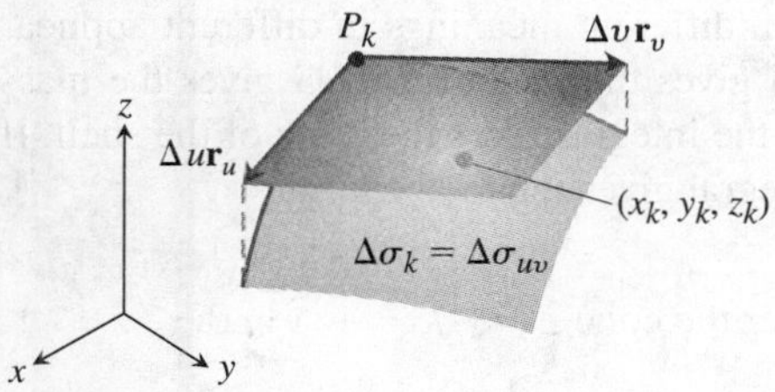

FIGURE 16.49 The area of the patch $\Delta\sigma_k$ is approximated by the area of the tangent parallelogram determined by the vectors $\Delta u\, \mathbf{r}_u$ and $\Delta v\, \mathbf{r}_v$. The point (x_k, y_k, z_k) lies on the surface patch, beneath the parallelogram shown here.

Assume, as in Section 16.5, that the surface S is defined parametrically on a region R in the uv-plane,

$$\mathbf{r}(u, v) = f(u, v)\mathbf{i} + g(u, v)\mathbf{j} + h(u, v)\mathbf{k}, \qquad (u, v) \in R.$$

In Figure 16.49, we see how a subdivision of R (considered as a rectangle for simplicity) divides the surface S into corresponding curved surface elements, or patches, of area

$$\Delta\sigma_{uv} \approx |\mathbf{r}_u \times \mathbf{r}_v|\, du\, dv.$$

As we did for the subdivisions when defining double integrals in Section 15.2, we number the surface element patches in some order with their areas given by $\Delta\sigma_1, \Delta\sigma_2, \ldots, \Delta\sigma_n$. To form a Riemann sum over S, we choose a point (x_k, y_k, z_k) in the kth patch, multiply the value of the function G at that point by the area $\Delta\sigma_k$, and add together the products:

$$\sum_{k=1}^{n} G(x_k, y_k, z_k)\, \Delta\sigma_k.$$

Depending on how we pick (x_k, y_k, z_k) in the kth patch, we may get different values for this Riemann sum. Then we take the limit as the number of surface patches increases, their areas shrink to zero, and both $\Delta u \to 0$ and $\Delta v \to 0$. This limit, whenever it exists independent of all choices made, defines the **surface integral of G over the surface S** as

$$\iint_S G(x, y, z)\, d\sigma = \lim_{n\to\infty} \sum_{k=1}^{n} G(x_k, y_k, z_k)\, \Delta\sigma_k. \tag{1}$$

Notice the analogy with the definition of the double integral (Section 15.2) and with the line integral (Section 16.1). If S is a piecewise smooth surface, and G is continuous over S, then the surface integral defined by Equation (1) can be shown to exist.

The formula for evaluating the surface integral depends on the manner in which S is described—parametrically, implicitly, or explicitly—as discussed in Section 16.5.

Formulas for a Surface Integral of a Scalar Function

1. For a smooth surface S defined **parametrically** as

$$\mathbf{r}(u, v) = f(u, v)\mathbf{i} + g(u, v)\mathbf{j} + h(u, v)\mathbf{k}, \quad (u, v) \in R,$$

and a continuous function $G(x, y, z)$ defined on S, the surface integral of G over S is given by the double integral over R,

$$\iint_S G(x, y, z)\, d\sigma = \iint_R G(f(u, v), g(u, v), h(u, v))\, |\mathbf{r}_u \times \mathbf{r}_v|\, du\, dv. \tag{2}$$

2. For a surface S given **implicitly** by $F(x, y, z) = c$, where F is a continuously differentiable function, with S lying above its closed and bounded shadow region R in the coordinate plane beneath it, the surface integral of the continuous function G over S is given by the double integral over R,

$$\iint_S G(x, y, z)\, d\sigma = \iint_R G(x, y, z) \frac{|\nabla F|}{|\nabla F \cdot \mathbf{p}|}\, dA, \tag{3}$$

where $\mathbf{p}$ is a unit vector normal to R and $\nabla F \cdot \mathbf{p} \neq 0$.

3. For a surface S given **explicitly** as the graph of $z = f(x, y)$, where f is a continuously differentiable function over a region R in the xy-plane, the surface integral of the continuous function G over S is given by the double integral over R,

$$\iint_S G(x, y, z)\, d\sigma = \iint_R G(x, y, f(x, y)) \sqrt{f_x^{\,2} + f_y^{\,2} + 1}\, dx\, dy. \tag{4}$$

The surface integral in Equation (1) takes on different meanings in different applications. If G has the constant value 1, the integral gives the area of S. If G gives the mass density of a thin shell of material modeled by S, the integral gives the mass of the shell. If G gives the charge density of a thin shell, the integral gives the total charge.

EXAMPLE 1 Integrate $G(x, y, z) = x^2$ over the cone $z = \sqrt{x^2 + y^2}$, $0 \le z \le 1$.

Solution Using Equation (2) and the calculations from Example 4 in Section 16.5, we have $|\mathbf{r}_r \times \mathbf{r}_\theta| = \sqrt{2}r$ and

$$\iint_S x^2\, d\sigma = \int_0^{2\pi}\int_0^1 (r^2\cos^2\theta)\left(\sqrt{2}r\right) dr\, d\theta \qquad x = r\cos\theta$$

$$= \sqrt{2}\int_0^{2\pi}\int_0^1 r^3\cos^2\theta\, dr\, d\theta$$

$$= \frac{\sqrt{2}}{4}\int_0^{2\pi}\cos^2\theta\, d\theta = \frac{\sqrt{2}}{4}\left[\frac{\theta}{2} + \frac{1}{4}\sin 2\theta\right]_0^{2\pi} = \frac{\pi\sqrt{2}}{4}. \qquad \blacksquare$$

Surface integrals behave like other double integrals, the integral of the sum of two functions being the sum of their integrals and so on. The domain Additivity Property takes the form

$$\iint_S G\, d\sigma = \iint_{S_1} G\, d\sigma + \iint_{S_2} G\, d\sigma + \cdots + \iint_{S_n} G\, d\sigma.$$

When S is partitioned by smooth curves into a finite number of smooth patches with nonoverlapping interiors (i.e., if S is piecewise smooth), then the integral over S is the sum of the integrals over the patches. Thus, the integral of a function over the surface of a cube is the sum of the integrals over the faces of the cube. We integrate over a "turtle shell" of welded plates by integrating over one plate at a time and adding the results.

EXAMPLE 2 Integrate $G(x, y, z) = xyz$ over the surface of the cube cut from the first octant by the planes $x = 1$, $y = 1$, and $z = 1$ (Figure 16.50).

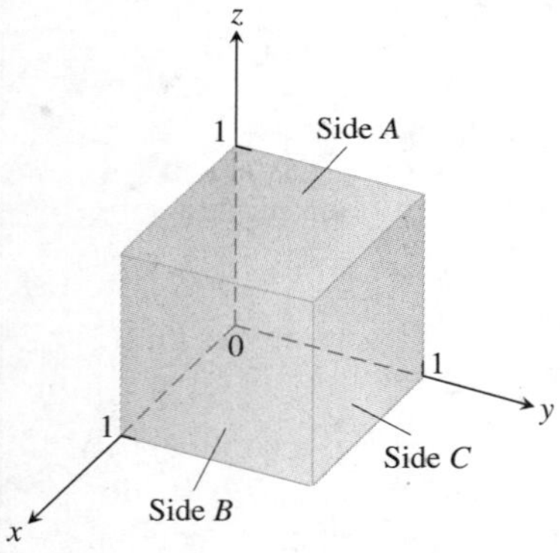

FIGURE 16.50 The cube in Example 2.

Solution We integrate xyz over each of the six sides and add the results. Since $xyz = 0$ on the sides that lie in the coordinate planes, the integral over the surface of the cube reduces to

$$\iint_{\substack{\text{Cube}\\\text{surface}}} xyz\, d\sigma = \iint_{\text{Side }A} xyz\, d\sigma + \iint_{\text{Side }B} xyz\, d\sigma + \iint_{\text{Side }C} xyz\, d\sigma.$$

Side A is the surface $f(x, y, z) = z = 1$ over the square region R_{xy}: $0 \le x \le 1$, $0 \le y \le 1$, in the xy-plane. For this surface and region,

$$\mathbf{p} = \mathbf{k}, \quad \nabla f = \mathbf{k}, \quad |\nabla f| = 1, \quad |\nabla f \cdot \mathbf{p}| = |\mathbf{k}\cdot\mathbf{k}| = 1$$

$$d\sigma = \frac{|\nabla f|}{|\nabla f \cdot \mathbf{p}|}dA = \frac{1}{1}dx\, dy = dx\, dy \qquad \text{Eq. (3)}$$

$$xyz = xy(1) = xy$$

and

$$\iint_{\text{Side }A} xyz\, d\sigma = \iint_{R_{xy}} xy\, dx\, dy = \int_0^1\int_0^1 xy\, dx\, dy = \int_0^1 \frac{y}{2}\, dy = \frac{1}{4}.$$

Symmetry tells us that the integrals of xyz over sides B and C are also 1/4. Hence,

$$\iint_{\substack{\text{Cube}\\\text{surface}}} xyz\, d\sigma = \frac{1}{4} + \frac{1}{4} + \frac{1}{4} = \frac{3}{4}. \qquad \blacksquare$$

EXAMPLE 3 Integrate $G(x, y, z) = \sqrt{1 - x^2 - y^2}$ over the "football" surface S formed by rotating the curve $x = \cos z$, $y = 0$, $-\pi/2 \le z \le \pi/2$, around the z-axis.

Solution The surface is displayed in Figure 16.46, and in Example 6 of Section 16.5 we found the parametrization

$$x = \cos u \cos v, \quad y = \cos u \sin v, \quad z = u, \quad -\frac{\pi}{2} \le u \le \frac{\pi}{2} \quad \text{and} \quad 0 \le v \le 2\pi,$$

where v represents the angle of rotation from the xz-plane about the z-axis. Substituting this parametrization into the expression for G gives

$$\sqrt{1 - x^2 - y^2} = \sqrt{1 - (\cos^2 u)(\cos^2 v + \sin^2 v)} = \sqrt{1 - \cos^2 u} = |\sin u|.$$

The surface area differential for the parametrization was found to be (Example 6, Section 16.5)

$$d\sigma = \cos u \sqrt{1 + \sin^2 u}\, du\, dv.$$

These calculations give the surface integral

$$\iint_S \sqrt{1 - x^2 - y^2}\, d\sigma = \int_0^{2\pi} \int_{-\pi/2}^{\pi/2} |\sin u| \cos u \sqrt{1 + \sin^2 u}\, du\, dv \qquad \begin{array}{l} |\sin u| = \sin(-u) \\ \text{for } -\pi/2 < u < 0 \end{array}$$

$$= 2 \int_0^{2\pi} \int_0^{\pi/2} \sin u \cos u \sqrt{1 + \sin^2 u}\, du\, dv$$

$$= \int_0^{2\pi} \int_1^2 \sqrt{w}\, dw\, dv \qquad \begin{array}{l} w = 1 + \sin^2 u, \\ dw = 2 \sin u \cos u\, du \\ \text{When } u = 0,\ w = 1. \\ \text{When } u = \pi/2,\ w = 2. \end{array}$$

$$= 2\pi \cdot \frac{2}{3} w^{3/2} \Big]_1^2 = \frac{4\pi}{3}(2\sqrt{2} - 1).$$

■

EXAMPLE 4 Evaluate $\iint_S \sqrt{x(1 + 2z)}\, d\sigma$ on the portion of the cylinder $z = y^2/2$ over the triangular region $R: x \ge 0,\ y \ge 0,\ x + y \le 1$, in the xy-plane (Figure 16.51).

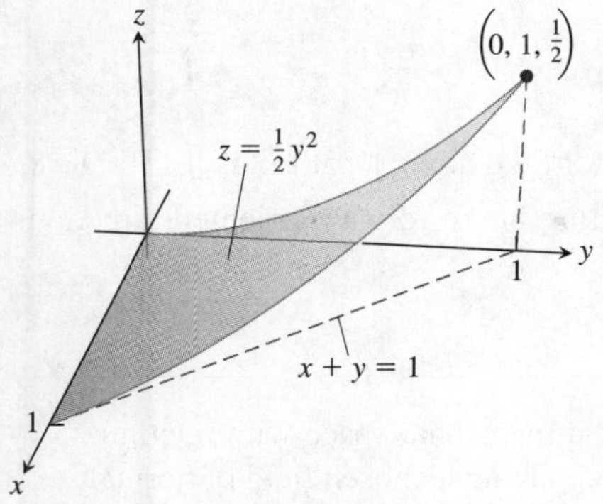

FIGURE 16.51 The surface S in Example 4.

Solution The function G on the surface S is given by

$$G(x, y, z) = \sqrt{x(1 + 2z)} = \sqrt{x}\sqrt{1 + y^2}.$$

With $z = f(x, y) = y^2/2$, we use Equation (4) to evaluate the surface integral:

$$d\sigma = \sqrt{f_x{}^2 + f_y{}^2 + 1}\, dx\, dy = \sqrt{0 + y^2 + 1}\, dx\, dy$$

and

$$\iint_S G(x, y, z)\, d\sigma = \iint_R \left(\sqrt{x}\sqrt{1 + y^2}\right)\sqrt{1 + y^2}\, dx\, dy$$

$$= \int_0^1 \int_0^{1-x} \sqrt{x}\,(1 + y^2)\, dy\, dx$$

$$= \int_0^1 \sqrt{x}\left[(1 - x) + \frac{1}{3}(1 - x)^3\right] dx \qquad \text{Integrate and evaluate.}$$

$$= \int_0^1 \left(\frac{4}{3}x^{1/2} - 2x^{3/2} + x^{5/2} - \frac{1}{3}x^{7/2}\right) dx \qquad \text{Routine algebra}$$

$$= \left[\frac{8}{9}x^{3/2} - \frac{4}{5}x^{5/2} + \frac{2}{7}x^{7/2} - \frac{2}{27}x^{9/2}\right]_0^1$$

$$= \frac{8}{9} - \frac{4}{5} + \frac{2}{7} - \frac{2}{27} = \frac{284}{945} \approx 0.30.$$

■

Orientation of a Surface

A curve C with a parametrization $\mathbf{r}(t)$ has a natural orientation, or direction, that comes from the direction of increasing t. The unit tangent vector $\mathbf{T}$ along C points in this forward direction at each point on the curve. There are two possible orientations for a curve, corresponding to whether we follow the direction of the tangent vector $\mathbf{T}$ at each point, or the direction of $-\mathbf{T}$.

To specify an orientation on a surface in space S, we do something similar, but this time we specify a normal vector at each point on the surface. A parametrization of a surface $\mathbf{r}(u, v)$ gives a vector $\mathbf{r}_u \times \mathbf{r}_v$ that is normal to the surface, and so gives an orientation wherever the parametrization applies. A second choice of orientation is found by taking $-(\mathbf{r}_u \times \mathbf{r}_v)$, giving a vector that points to the opposite side of the surface at each point. In essence, an orientation is a way of consistently choosing one of the two sides of a surface. Not all surfaces have orientations, but a surface that does have one also has a second, opposite orientation.

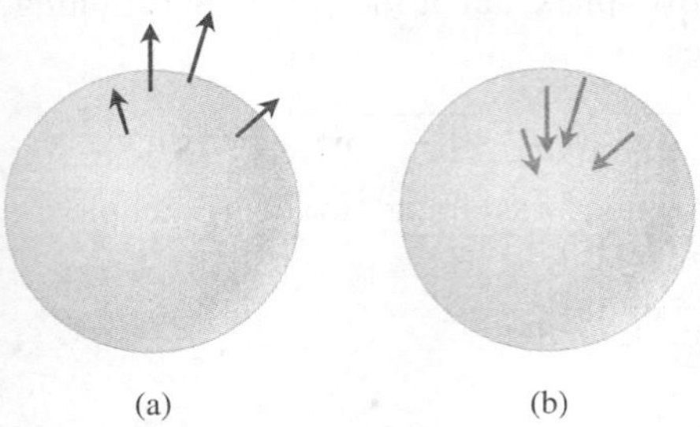

FIGURE 16.52 (a) An outward-pointing vector field and (b) an inward-pointing vector field give the two possible orientations of a sphere.

Each point on the sphere in Figure 16.52 has one normal vector pointing inward, toward the center of the sphere, and another opposite normal vector pointing outward. We specify one of two possible orientations for the sphere by choosing either the inward vector at each point, or alternatively the outward vector at each point.

When we can choose a continuous field of unit normal vectors $\mathbf{n}$ on a smooth surface S, we say that S is *orientable* (or *two-sided*). Spheres and other smooth surfaces that are the boundaries of solid regions in space are orientable since we can choose an outward-pointing unit vector $\mathbf{n}$ at each point to specify an orientation.

A surface together with its normal field $\mathbf{n}$, or, equivalently, a surface with a consistent choice of sides, is called an oriented surface. The vector $\mathbf{n}$ at any point gives the positive direction or positively oriented side at that point (Figure 16.52). Not all surfaces can be oriented. The Möbius band in Figure 16.53 is an example of a surface that is not orientable. No matter how you try to construct a continuous unit normal vector field (shown as the shafts of thumbtacks in the figure), starting at one point and moving the vector continuously around the surface in the manner shown will return it to the starting point, but pointing in the opposite direction. No choice of a vectors can give a continuous normal vector field on the Möbius band, so the Möbius band is not orientable.

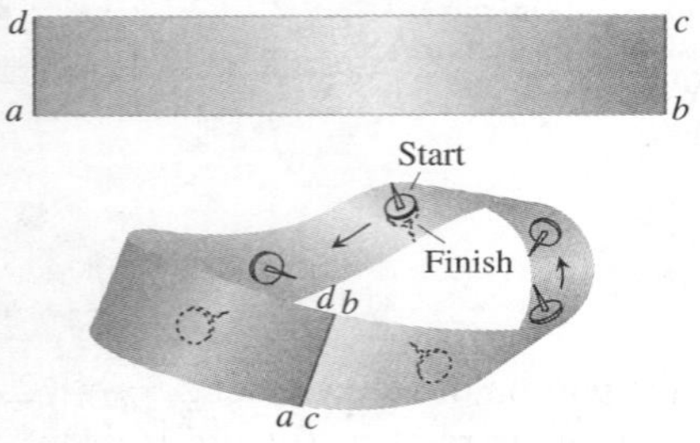

FIGURE 16.53 To make a Möbius band, take a rectangular strip of paper *abcd*, give the end *bc* a single twist, and paste the ends of the strip together to match *a* with *c* and *b* with *d*. The Möbius band is a nonorientable, or one-sided, surface.

Surface Integrals of Vector Fields

In Section 16.2 we defined the line integral of a vector field along a path C as $\int_C \mathbf{F} \cdot \mathbf{T}\, ds$, where $\mathbf{T}$ is the unit tangent vector to the path pointing in the forward-oriented direction. We have a similar definition for surface integrals.

DEFINITION Let $\mathbf{F}$ be a vector field in three-dimensional space with continuous components defined over a smooth surface S having a chosen field of normal unit vectors $\mathbf{n}$ orienting S. Then the **surface integral of F over** S is

$$\iint_S \mathbf{F} \cdot \mathbf{n}\, d\sigma. \tag{5}$$

This integral is also called the **flux** of the vector field $\mathbf{F}$ across S.

If $\mathbf{F}$ is the velocity field of a three-dimensional fluid flow, then the flux of $\mathbf{F}$ across S is the net rate at which fluid is crossing S per unit time in the chosen positive direction $\mathbf{n}$ defined by the orientation of S. Fluid flows are discussed in more detail in Section 16.7.

Computing a Surface Integral for a Parametrized Surface

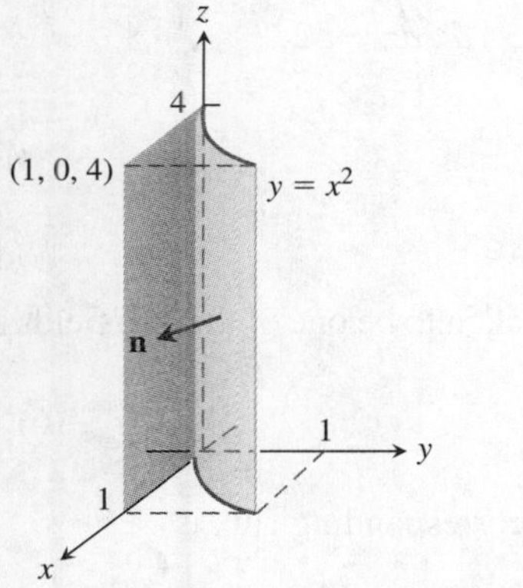

FIGURE 16.54 Finding the flux through the surface of a parabolic cylinder (Example 5).

EXAMPLE 5 Find the flux of $\mathbf{F} = yz\mathbf{i} + x\mathbf{j} - z^2\mathbf{k}$ through the parabolic cylinder $y = x^2$, $0 \le x \le 1$, $0 \le z \le 4$, in the direction $\mathbf{n}$ indicated in Figure 16.54.

Solution On the surface we have $x = x$, $y = x^2$, and $z = z$, so we have the parametrization $\mathbf{r}(x, z) = x\mathbf{i} + x^2\mathbf{j} + z\mathbf{k}$, $0 \le x \le 1$, $0 \le z \le 4$. The cross product of tangent vectors,

$$\mathbf{r}_x \times \mathbf{r}_z = \begin{vmatrix} \mathbf{i} & \mathbf{j} & \mathbf{k} \\ 1 & 2x & 0 \\ 0 & 0 & 1 \end{vmatrix} = 2x\mathbf{i} - \mathbf{j}, \qquad \begin{aligned} \mathbf{r}_x &= \mathbf{i} + 2x\mathbf{j} \\ \mathbf{r}_z &= \mathbf{k} \end{aligned}$$

can be used to find unit normal vectors to the surface,

$$\mathbf{n} = \frac{\mathbf{r}_x \times \mathbf{r}_z}{|\mathbf{r}_x \times \mathbf{r}_z|} = \frac{2x\mathbf{i} - \mathbf{j}}{\sqrt{4x^2 + 1}}.$$

We can equally well choose the unit normal vectors $-\mathbf{n}$ that point in the opposite direction. The first choice is shown in Figure 16.54.

On the surface we have $y = x^2$, so the vector field there is

$$\mathbf{F} = yz\mathbf{i} + x\mathbf{j} - z^2\mathbf{k} = x^2z\mathbf{i} + x\mathbf{j} - z^2\mathbf{k}.$$

Thus,

$$\mathbf{F} \cdot \mathbf{n} = \frac{1}{\sqrt{4x^2 + 1}}\left((x^2z)(2x) + (x)(-1) + (-z^2)(0)\right) = \frac{2x^3z - x}{\sqrt{4x^2 + 1}}.$$

The flux of $\mathbf{F}$ outward through the surface is

$$\begin{aligned} \iint_S \mathbf{F} \cdot \mathbf{n}\, d\sigma &= \int_0^4 \int_0^1 \frac{2x^3z - x}{\sqrt{4x^2 + 1}} |\mathbf{r}_x \times \mathbf{r}_z|\, dx\, dz \qquad d\sigma = |\mathbf{r}_x \times \mathbf{r}_z|\, dx\, dz \\ &= \int_0^4 \int_0^1 \frac{2x^3z - x}{\sqrt{4x^2 + 1}} \sqrt{4x^2 + 1}\, dx\, dz \\ &= \int_0^4 \int_0^1 (2x^3z - x)\, dx\, dz = \int_0^4 \left[\frac{1}{2}x^4z - \frac{1}{2}x^2\right]_{x=0}^{x=1} dz \\ &= \int_0^4 \frac{1}{2}(z - 1)\, dz = \frac{1}{4}(z - 1)^2\Big]_0^4 \\ &= \frac{1}{4}(9) - \frac{1}{4}(1) = 2. \end{aligned}$$

■

There is a simple formula for the flux of $\mathbf{F}$ across a parametrized surface $\mathbf{r}(u, v)$. Since

$$d\sigma = |\mathbf{r}_u \times \mathbf{r}_v|\, du\, dv$$

and

$$\mathbf{n} = \frac{\mathbf{r}_u \times \mathbf{r}_v}{|\mathbf{r}_u \times \mathbf{r}_v|},$$

it follows that

$$\iint_S \mathbf{F} \cdot \mathbf{n}\, d\sigma = \iint_R \mathbf{F} \cdot \frac{\mathbf{r}_u \times \mathbf{r}_v}{|\mathbf{r}_u \times \mathbf{r}_v|} |\mathbf{r}_u \times \mathbf{r}_v|\, du\, dv = \iint_R \mathbf{F} \cdot (\mathbf{r}_u \times \mathbf{r}_v)\, du\, dv.$$

Flux Across a Parametrized Surface

$$\text{Flux} = \pm\iint_R \mathbf{F} \cdot (\mathbf{r}_u \times \mathbf{r}_v)\, du\, dv$$

The other choice of unit normal vector, $-\mathbf{n}$, would add a negative sign to this formula. The choice of $\mathbf{n}$ or $-\mathbf{n}$ depends on the direction in which we choose to measure the flux across the surface.

This integral for flux simplifies the computation in Example 5 by eliminating the need to compute the canceled factor $|\mathbf{r}_u \times \mathbf{r}_v|$. Since

$$\begin{aligned} \mathbf{F} \cdot (\mathbf{r}_x \times \mathbf{r}_z) &= (yz\mathbf{i} + x\mathbf{j} - z^2\mathbf{k} = x^2z\mathbf{i} + x\mathbf{j} - z^2\mathbf{k}) \cdot (2x\mathbf{i} - \mathbf{j}) \\ &= (x^2z)(2x) + (x)(-1) = 2x^3z - x, \end{aligned}$$

we obtain directly

$$\text{Flux} = \iint_S \mathbf{F} \cdot \mathbf{n}\, d\sigma = \int_0^4 \int_0^1 (2x^3 z - x)\, dx\, dz = 2$$

in Example 5.

Computing a Surface Integral for a Level Surface

If S is part of a level surface $g(x, y, z) = c$, then $\mathbf{n}$ may be taken to be one of the two fields

$$\mathbf{n} = \pm \frac{\nabla g}{|\nabla g|}, \tag{6}$$

depending on which one gives the preferred direction. The corresponding flux is

$$\begin{aligned} \text{Flux} &= \iint_S \mathbf{F} \cdot \mathbf{n}\, d\sigma \\ &= \iint_R \left(\mathbf{F} \cdot \frac{\pm \nabla g}{|\nabla g|} \right) \frac{|\nabla g|}{|\nabla g \cdot \mathbf{p}|}\, dA \qquad \text{Eqs. (6) and (3)} \\ &= \iint_R \mathbf{F} \cdot \frac{\pm \nabla g}{|\nabla g \cdot \mathbf{p}|}\, dA. \end{aligned} \tag{7}$$

EXAMPLE 6 Find the flux of $\mathbf{F} = yz\mathbf{j} + z^2\mathbf{k}$ through the surface S cut from the cylinder $y^2 + z^2 = 1, z \geq 0$, by the planes $x = 0$ and $x = 1$, in the direction away from the x-axis.

Solution The normal field on S (Figure 16.55) in the specified direction may be calculated from the gradient of $g(x, y, z) = y^2 + z^2$ to be

$$\mathbf{n} = +\frac{\nabla g}{|\nabla g|} = \frac{2y\mathbf{j} + 2z\mathbf{k}}{\sqrt{4y^2 + 4z^2}} = \frac{2y\mathbf{j} + 2z\mathbf{k}}{2\sqrt{1}} = y\mathbf{j} + z\mathbf{k}.$$

With $\mathbf{p} = \mathbf{k}$, we also have

$$d\sigma = \frac{|\nabla g|}{|\nabla g \cdot \mathbf{k}|}\, dA = \frac{2}{|2z|}\, dA = \frac{1}{z}\, dA. \qquad \text{Eq. (3)}$$

We can drop the absolute value bars because $z \geq 0$ on S.

The value of $\mathbf{F} \cdot \mathbf{n}$ on the surface is

$$\begin{aligned} \mathbf{F} \cdot \mathbf{n} &= (yz\mathbf{j} + z^2\mathbf{k}) \cdot (y\mathbf{j} + z\mathbf{k}) \\ &= y^2 z + z^3 = z(y^2 + z^2) \\ &= z. \qquad y^2 + z^2 = 1 \text{ on } S. \end{aligned}$$

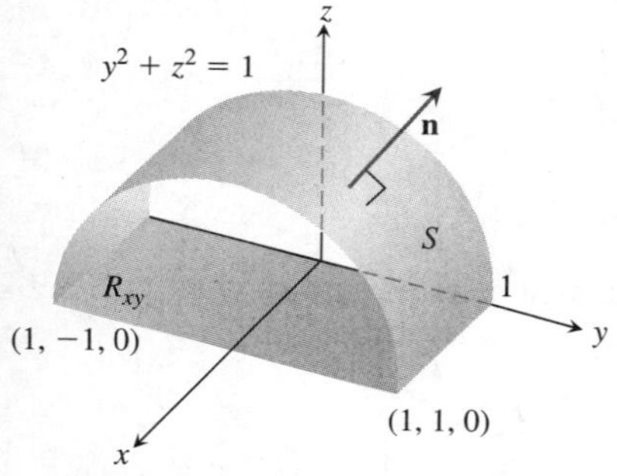

FIGURE 16.55 Calculating the flux of a vector field through the surface S. The area of the shadow region R_{xy} is 2 (Example 6).

The surface projects onto the shadow region R_{xy}, which is the rectangle in the xy-plane shown in Figure 16.55. Therefore, the flux of $\mathbf{F}$ through S in the direction away from the x-axis is

$$\iint_S \mathbf{F} \cdot \mathbf{n}\, d\sigma = \iint_{R_{xy}} (z)\left(\frac{1}{z}\, dA\right) = \iint_{R_{xy}} dA = \text{area}(R_{xy}) = 2. \qquad \blacksquare$$

Moments and Masses of Thin Shells

Thin shells of material like bowls, metal drums, and domes are modeled with surfaces. Their moments and masses are calculated with the formulas in Table 16.3. The derivations are similar to those in Section 6.6. The formulas resemble those for line integrals in Table 16.1, Section 16.1.

TABLE 16.3 Mass and moment formulas for very thin shells

Mass: $M = \iint_S \delta\, d\sigma$ $\qquad \delta = \delta(x, y, z) =$ density at (x, y, z) is mass per unit area.

First moments about the coordinate planes:

$$M_{yz} = \iint_S x\,\delta\, d\sigma, \quad M_{xz} = \iint_S y\,\delta\, d\sigma, \quad M_{xy} = \iint_S z\,\delta\, d\sigma$$

Coordinates of center of mass:

$$\bar{x} = M_{yz}/M, \quad \bar{y} = M_{xz}/M, \quad \bar{z} = M_{xy}/M$$

Moments of inertia about coordinate axes:

$$I_x = \iint_S (y^2 + z^2)\,\delta\, d\sigma, \quad I_y = \iint_S (x^2 + z^2)\,\delta\, d\sigma, \quad I_z = \iint_S (x^2 + y^2)\,\delta\, d\sigma,$$

$$I_L = \iint_S r^2 \delta\, d\sigma \qquad r(x, y, z) = \text{distance from point } (x, y, z) \text{ to line } L$$

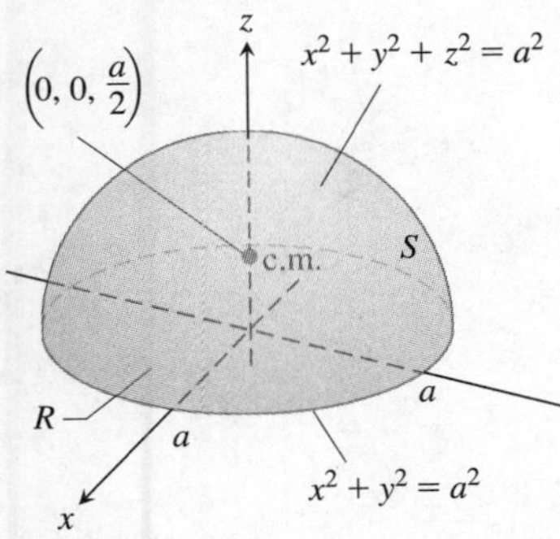

FIGURE 16.56 The center of mass of a thin hemispherical shell of constant density lies on the axis of symmetry halfway from the base to the top (Example 7).

EXAMPLE 7 Find the center of mass of a thin hemispherical shell of radius a and constant density δ.

Solution We model the shell with the hemisphere

$$f(x, y, z) = x^2 + y^2 + z^2 = a^2, \qquad z \geq 0$$

(Figure 16.56). The symmetry of the surface about the z-axis tells us that $\bar{x} = \bar{y} = 0$. It remains only to find $\bar{z}$ from the formula $\bar{z} = M_{xy}/M$.

The mass of the shell is

$$M = \iint_S \delta\, d\sigma = \delta \iint_S d\sigma = (\delta)(\text{area of } S) = 2\pi a^2 \delta. \qquad \delta = \text{constant}$$

To evaluate the integral for M_{xy}, we take $\mathbf{p} = \mathbf{k}$ and calculate

$$|\nabla f| = |2x\mathbf{i} + 2y\mathbf{j} + 2z\mathbf{k}| = 2\sqrt{x^2 + y^2 + z^2} = 2a$$

$$|\nabla f \cdot \mathbf{p}| = |\nabla f \cdot \mathbf{k}| = |2z| = 2z$$

$$d\sigma = \frac{|\nabla f|}{|\nabla f \cdot \mathbf{p}|}\, dA = \frac{a}{z}\, dA. \qquad \text{Eq. (3)}$$

Then

$$M_{xy} = \iint_S z\delta\, d\sigma = \delta \iint_R z\frac{a}{z}\, dA = \delta a \iint_R dA = \delta a(\pi a^2) = \delta \pi a^3$$

$$\bar{z} = \frac{M_{xy}}{M} = \frac{\pi a^3 \delta}{2\pi a^2 \delta} = \frac{a}{2}.$$

The shell's center of mass is the point $(0, 0, a/2)$. ■

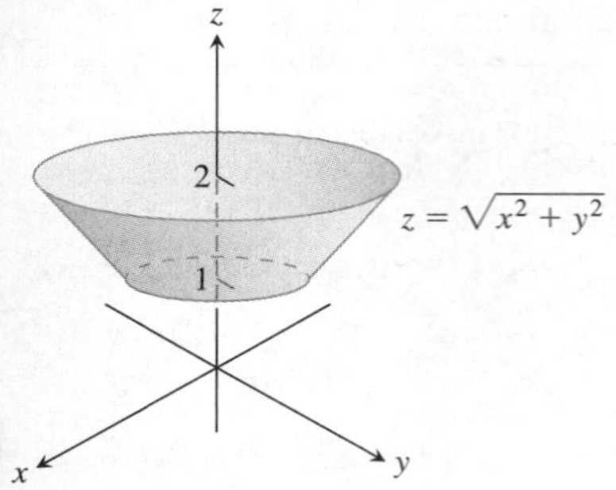

FIGURE 16.57 The cone frustum formed when the cone $z = \sqrt{x^2 + y^2}$ is cut by the planes $z = 1$ and $z = 2$ (Example 8).

EXAMPLE 8 Find the center of mass of a thin shell of density $\delta = 1/z^2$ cut from the cone $z = \sqrt{x^2 + y^2}$ by the planes $z = 1$ and $z = 2$ (Figure 16.57).

Solution Since the surface and the density function δ are symmetric about the z-axis, we have $\bar{x} = \bar{y} = 0$. We now proceed to find $\bar{z} = M_{xy}/M$. Working as in Example 4 of Section 16.5, we have

$$\mathbf{r}(r, \theta) = (r\cos\theta)\mathbf{i} + (r\sin\theta)\mathbf{j} + r\mathbf{k}, \qquad 1 \le r \le 2, \quad 0 \le \theta \le 2\pi,$$

and

$$|\mathbf{r}_r \times \mathbf{r}_\theta| = \sqrt{2}r.$$

Therefore,

$$\begin{aligned}
M &= \iint_S \delta\, d\sigma = \int_0^{2\pi}\int_1^2 \frac{1}{r^2}\sqrt{2}r\, dr\, d\theta \\
&= \sqrt{2}\int_0^{2\pi}\Big[\ln r\Big]_1^2 d\theta = \sqrt{2}\int_0^{2\pi} \ln 2\, d\theta \\
&= 2\pi\sqrt{2}\ln 2, \\
M_{xy} &= \iint_S \delta z\, d\sigma = \int_0^{2\pi}\int_1^2 \frac{1}{r^2} r\sqrt{2}r\, dr\, d\theta \\
&= \sqrt{2}\int_0^{2\pi}\int_1^2 dr\, d\theta \\
&= \sqrt{2}\int_0^{2\pi} d\theta = 2\pi\sqrt{2}, \\
\bar{z} &= \frac{M_{xy}}{M} = \frac{2\pi\sqrt{2}}{2\pi\sqrt{2}\ln 2} = \frac{1}{\ln 2}.
\end{aligned}$$

The shell's center of mass is the point $(0, 0, 1/\ln 2)$. ■

EXERCISES 16.6

Surface Integrals of Scalar Functions

In Exercises 1–8, integrate the given function over the given surface.

1. **Parabolic cylinder** $G(x, y, z) = x$, over the parabolic cylinder $y = x^2$, $0 \le x \le 2$, $0 \le z \le 3$
2. **Circular cylinder** $G(x, y, z) = z$, over the cylindrical surface $y^2 + z^2 = 4$, $z \ge 0$, $1 \le x \le 4$
3. **Sphere** $G(x, y, z) = x^2$, over the unit sphere $x^2 + y^2 + z^2 = 1$
4. **Hemisphere** $G(x, y, z) = z^2$, over the hemisphere $x^2 + y^2 + z^2 = a^2$, $z \ge 0$
5. **Portion of plane** $F(x, y, z) = z$, over the portion of the plane $x + y + z = 4$ that lies above the square $0 \le x \le 1$, $0 \le y \le 1$, in the xy-plane
6. **Cone** $F(x, y, z) = z - x$, over the cone $z = \sqrt{x^2 + y^2}$, $0 \le z \le 1$
7. **Parabolic dome** $H(x, y, z) = x^2\sqrt{5 - 4z}$, over the parabolic dome $z = 1 - x^2 - y^2$, $z \ge 0$
8. **Spherical cap** $H(x, y, z) = yz$, over the part of the sphere $x^2 + y^2 + z^2 = 4$ that lies above the cone $z = \sqrt{x^2 + y^2}$
9. Integrate $G(x, y, z) = x + y + z$ over the surface of the cube cut from the first octant by the planes $x = a$, $y = a$, $z = a$.
10. Integrate $G(x, y, z) = y + z$ over the surface of the wedge in the first octant bounded by the coordinate planes and the planes $x = 2$ and $y + z = 1$.
11. Integrate $G(x, y, z) = xyz$ over the surface of the rectangular solid cut from the first octant by the planes $x = a$, $y = b$, and $z = c$.
12. Integrate $G(x, y, z) = xyz$ over the surface of the rectangular solid bounded by the planes $x = \pm a$, $y = \pm b$, and $z = \pm c$.
13. Integrate $G(x, y, z) = x + y + z$ over the portion of the plane $2x + 2y + z = 2$ that lies in the first octant.
14. Integrate $G(x, y, z) = x\sqrt{y^2 + 4}$ over the surface cut from the parabolic cylinder $y^2 + 4z = 16$ by the planes $x = 0$, $x = 1$, and $z = 0$.

15. Integrate $G(x, y, z) = z - x$ over the portion of the graph of $z = x + y^2$ above the triangle in the xy-plane having vertices $(0, 0, 0)$, $(1, 1, 0)$, and $(0, 1, 0)$. (See accompanying figure.)

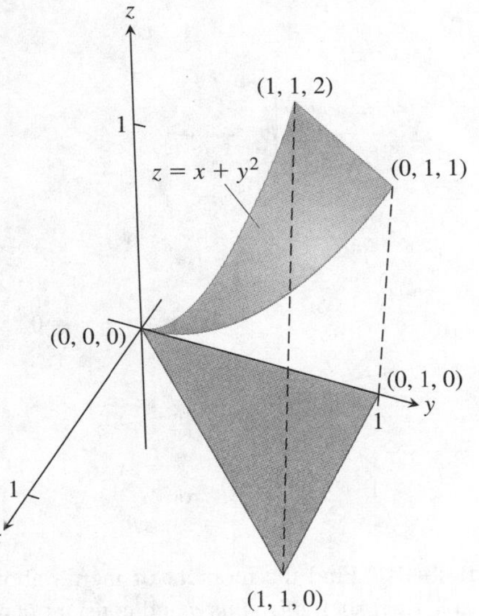

16. Integrate $G(x, y, z) = x$ over the surface given by

$$z = x^2 + y \quad \text{for} \quad 0 \le x \le 1, \quad -1 \le y \le 1.$$

17. Integrate $G(x, y, z) = xyz$ over the triangular surface with vertices $(1, 0, 0)$, $(0, 2, 0)$, and $(0, 1, 1)$.

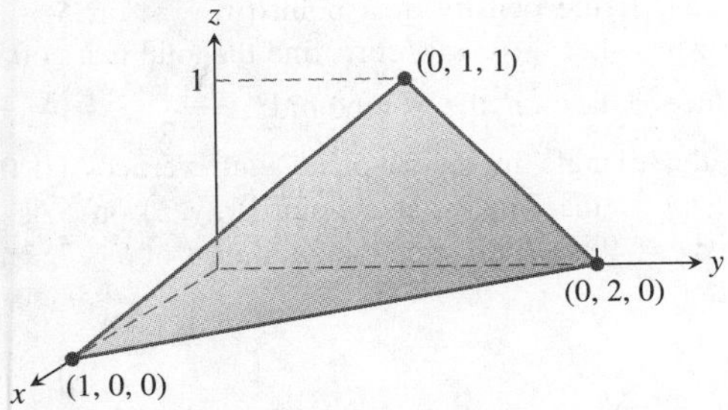

18. Integrate $G(x, y, z) = x - y - z$ over the portion of the plane $x + y = 1$ in the first octant between $z = 0$ and $z = 1$ (see the figure below).

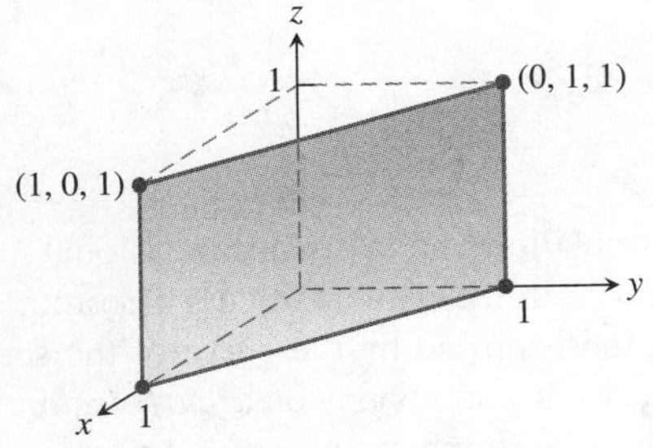

Finding Flux or Surface Integrals of Vector Fields

In Exercises 19–28, use a parametrization to find the flux $\iint_S \mathbf{F} \cdot \mathbf{n}\, d\sigma$ across the surface in the specified direction.

19. **Parabolic cylinder** $\mathbf{F} = z^2\mathbf{i} + x\mathbf{j} - 3z\mathbf{k}$ through the surface cut from the parabolic cylinder $z = 4 - y^2$ by the planes $x = 0$, $x = 1$, and $z = 0$ in the direction away from the x-axis

20. **Parabolic cylinder** $\mathbf{F} = 4y\mathbf{i} + 18z\mathbf{j} - x\mathbf{k}$ outward normal to the surfaces cut from the plane $3x + 2y + 6z = 6$ by the coordinate planes in the first octant.

21. **Sphere** $\mathbf{F} = z\mathbf{k}$ across the portion of the sphere $x^2 + y^2 + z^2 = a^2$ in the first octant in the direction away from the origin

22. **Sphere** $\mathbf{F} = x\mathbf{i} + y\mathbf{j} + z\mathbf{k}$ across the sphere $x^2 + y^2 + z^2 = a^2$ in the direction away from the origin

23. **Plane** $\mathbf{F} = 2xy\mathbf{i} + 2yz\mathbf{j} + 2xz\mathbf{k}$ upward across the portion of the plane $x + y + z = 2a$ that lies above the square $0 \le x \le a$, $0 \le y \le a$, in the xy-plane

24. **Cylinder** $\mathbf{F} = x\mathbf{i} + y\mathbf{j} + z\mathbf{k}$ through the portion of the cylinder $x^2 + y^2 = 1$ cut by the planes $z = 0$ and $z = a$ in the direction away from the z-axis

25. **Cone** $\mathbf{F} = xy\mathbf{i} - z\mathbf{k}$ through the cone $z = \sqrt{x^2 + y^2}$, $0 \le z \le 1$, in the direction away from the z-axis

26. **Cone** $\mathbf{F} = y^2\mathbf{i} + xz\mathbf{j} - \mathbf{k}$ through the cone $z = 2\sqrt{x^2 + y^2}$, $0 \le z \le 2$, in the direction away from the z-axis

27. **Cone frustum** $\mathbf{F} = -x\mathbf{i} - y\mathbf{j} + z^2\mathbf{k}$ through the portion of the cone $z = \sqrt{x^2 + y^2}$ between the planes $z = 1$ and $z = 2$ in the direction away from the z-axis

28. **Paraboloid** $\mathbf{F} = 4x\mathbf{i} + 4y\mathbf{j} + 2\mathbf{k}$ through the surface cut from the bottom of the paraboloid $z = x^2 + y^2$ by the plane $z = 1$ in the direction away from the z-axis

In Exercises 29 and 30, find the surface integral of the field $\mathbf{F}$ over the portion of the given surface in the specified direction.

29. $\mathbf{F}(x, y, z) = -\mathbf{i} + 2\mathbf{j} + 3\mathbf{k}$

S: rectangular surface $z = 0$, $0 \le x \le 2$, $0 \le y \le 3$, direction $\mathbf{k}$

30. $\mathbf{F}(x, y, z) = yx^2\mathbf{i} - 2\mathbf{j} + xz\mathbf{k}$

S: rectangular surface $y = 0$, $-1 \le x \le 2$, $2 \le z \le 7$, direction $-\mathbf{j}$

In Exercises 31–36, use Equation (7) to find the surface integral of the field $\mathbf{F}$ over the portion of the sphere $x^2 + y^2 + z^2 = a^2$ in the first octant in the direction away from the origin.

31. $\mathbf{F}(x, y, z) = z\mathbf{k}$
32. $\mathbf{F}(x, y, z) = -y\mathbf{i} + x\mathbf{j}$
33. $\mathbf{F}(x, y, z) = y\mathbf{i} - x\mathbf{j} + \mathbf{k}$
34. $\mathbf{F}(x, y, z) = zx\mathbf{i} + zy\mathbf{j} + z^2\mathbf{k}$
35. $\mathbf{F}(x, y, z) = x\mathbf{i} + y\mathbf{j} + z\mathbf{k}$
36. $\mathbf{F}(x, y, z) = yz\mathbf{i} + zx\mathbf{j} - xy\mathbf{k}$

37. Find the flux of the field $\mathbf{F}(x, y, z) = z^2\mathbf{i} + x\mathbf{j} - 3z\mathbf{k}$ through the surface cut from the parabolic cylinder $z = 4 - y^2$ by the planes $x = 0$, $x = 1$, and $z = 0$ in the direction away from the x-axis.

38. Find the flux of the field $\mathbf{F}(x, y, z) = 4x\mathbf{i} + 4y\mathbf{j} + 2\mathbf{k}$ through the surface cut from the bottom of the paraboloid $z = x^2 + y^2$ by the plane $z = 1$ in the direction away from the z-axis.

39. Let S be the portion of the cylinder $y = e^x$ in the first octant that projects parallel to the x-axis onto the rectangle $R_{yz}: 1 \le y \le 2$, $0 \le z \le 1$, in the yz-plane (see the accompanying figure). Let $\mathbf{n}$ be the unit vector normal to S that points away from the yz-plane. Find the flux of the field $\mathbf{F}(x, y, z) = -2\mathbf{i} + 2y\mathbf{j} + z\mathbf{k}$ across S in the direction of $\mathbf{n}$.

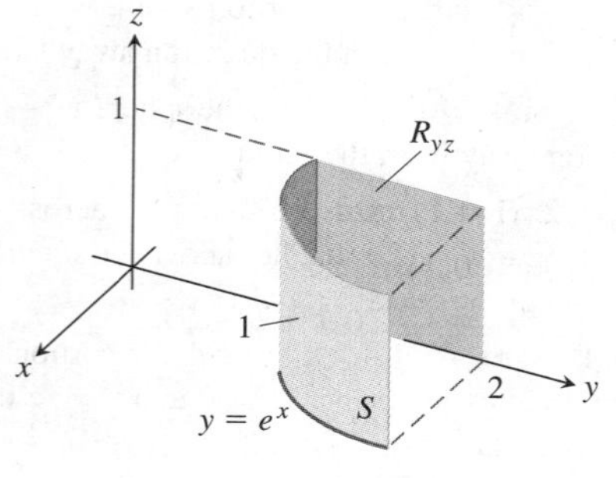

40. Let S be the surface of the parabolic cylinder $y^2 = 4x$ in the first octant bounded by the planes $x = 4$, $z = 3$, $\mathbf{n}$ is the outward normal. Find the flux of the vector $\mathbf{F} = y\mathbf{i} - x\mathbf{j} + z\mathbf{k}$ through S in the direction of $\mathbf{n}$.

41. Find the outward flux of the field $\mathbf{F} = 2xy\mathbf{i} + 2yz\mathbf{j} + 2xz\mathbf{k}$ across the surface of the cube cut from the first octant by the planes $x = a$, $y = a$, and $z = a$.

42. Find the outward flux of the field $\mathbf{F} = xz\mathbf{i} + yz\mathbf{j} + \mathbf{k}$ across the surface of the upper cap cut from the ball $x^2 + y^2 + z^2 \le 25$ by the plane $z = 3$.

Moments and Masses

43. Centroid Find the centroid of the portion of the sphere $x^2 + y^2 + z^2 = a^2$ that lies in the first octant.

44. Centroid Find the centroid of the surface cut from the cylinder $y^2 + z^2 = 9$, $z \ge 0$, by the planes $x = 0$ and $x = 3$ (resembles the surface in Example 6).

45. Thin shell of constant density Find the center of mass and the moment of inertia about the z-axis of a thin shell of constant density δ cut from the cone $x^2 + y^2 - z^2 = 0$ by the planes $z = 1$ and $z = 2$.

46. Conical surface of constant density Find the moment of inertia about the z-axis of a thin shell of constant density δ cut from the cone $4x^2 + 4y^2 - z^2 = 0$, $z \ge 0$, by the circular cylinder $x^2 + y^2 = 2x$ (see the accompanying figure).

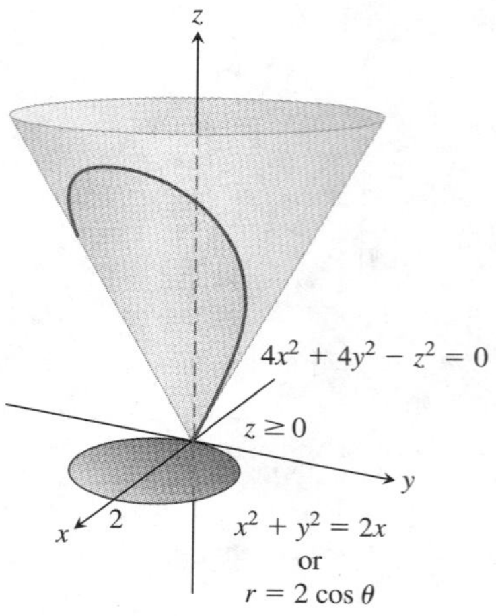

47. Spherical shells Find the moment of inertia about a diameter of a thin spherical shell of radius a and constant density δ. (Work with a hemispherical shell and double the result.)

48. Conical Surface Find the centroid of the lateral surface of a solid cone of base radius a and height h (cone surface minus the base).

49. A surface S lies on the plane $2x + 3y + 6z = 12$ directly above the rectangle in the xy-plane with vertices $(0, 0)$, $(1, 0)$, $(0, 2)$, and $(1, 2)$. If the density at a point (x, y, z) on S is given by $\delta(x, y, z) = 4xy + 6z$ mg/cm^2, find the total mass of S.

50. A surface S lies on the paraboloid $z = \frac{1}{2}x^2 + \frac{1}{2}y^2$ directly above the triangle in the xy-plane with vertices $(0, 0)$, $(2, 0)$, and $(2, 4)$. If the density at a point (x, y, z) on S is given by $\delta(x, y, z) = 9xy$ g/cm^2, find the total mass of S.

16.7 Stokes' Theorem

To calculate the counterclockwise circulation of a two-dimensional vector field $\mathbf{F} = M\mathbf{i} + N\mathbf{j}$ around a simple closed curve in the plane, Green's Theorem says we can compute the double integral over the region enclosed by the curve of the scalar quantity $(\partial N/\partial x - \partial M/\partial y)$. This expression is the $\mathbf{k}$-component of a *curl vector* field, and it measures the rate of rotation of $\mathbf{F}$ at each point in the region around an axis parallel to $\mathbf{k}$. For a vector field in three-dimensional space, the rotation at each point is around an axis that is parallel to the curl vector at that point. When a closed curve C in space is the boundary of an oriented surface, we will see that the circulation of $\mathbf{F}$ around C is equal to the surface integral of the curl vector field. This result extends Green's Theorem from regions in the plane to general surfaces in space having a smooth boundary curve.

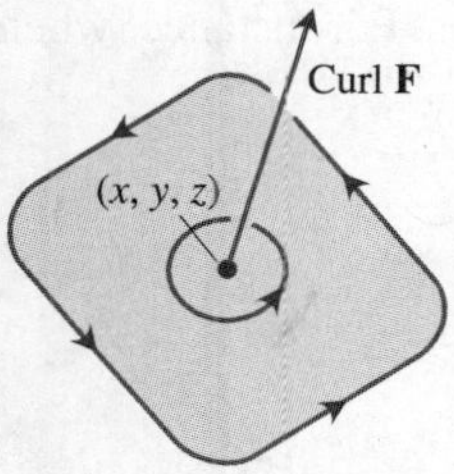

FIGURE 16.58 The circulation vector at a point (x, y, z) in a plane in a three-dimensional fluid flow. Notice its right-hand relation to the rotating particles in the fluid.

The Curl Vector Field

Suppose that $\mathbf{F}$ is the velocity field of a fluid flowing in space. Particles near the point (x, y, z) in the fluid tend to rotate around an axis through (x, y, z) that is parallel to a certain vector we are about to identify. This vector points in the direction for which the rotation is counterclockwise when viewed looking down onto the plane of the circulation from the tip of the arrow representing the vector. This is the direction your right-hand thumb points when your fingers curl around the axis of rotation in the way consistent with the rotating motion of the particles in the fluid (see Figure 16.58). The length of the vector measures the rate of rotation. The vector, introduced in Equation (3) of Section 16.3, is called the **curl vector** for the vector field $\mathbf{F} = M\mathbf{i} + N\mathbf{j} + P\mathbf{k}$, and it is given by

$$\text{curl } \mathbf{F} = \left(\frac{\partial P}{\partial y} - \frac{\partial N}{\partial z}\right)\mathbf{i} + \left(\frac{\partial M}{\partial z} - \frac{\partial P}{\partial x}\right)\mathbf{j} + \left(\frac{\partial N}{\partial x} - \frac{\partial M}{\partial y}\right)\mathbf{k}. \tag{1}$$

This information is a consequence of Stokes' Theorem, the generalization to space of the circulation-curl form of Green's Theorem.

Notice that $(\text{curl } \mathbf{F}) \cdot \mathbf{k} = (\partial N/\partial x - \partial M/\partial y)$, which is consistent with our discussion in Section 16.4 when $\mathbf{F} = M(x, y)\mathbf{i} + N(x, y)\mathbf{j}$. The formula for curl $\mathbf{F}$ in Equation (1) is often expressed using the symbol

$$\nabla = \mathbf{i}\frac{\partial}{\partial x} + \mathbf{j}\frac{\partial}{\partial y} + \mathbf{k}\frac{\partial}{\partial z}. \tag{2}$$

∇ is the symbol "del."

The symbol ∇ is pronounced "del," and we can use this symbol to express the curl of $\mathbf{F}$ with the formula

$$\begin{aligned}\nabla \times \mathbf{F} &= \begin{vmatrix} \mathbf{i} & \mathbf{j} & \mathbf{k} \\ \dfrac{\partial}{\partial x} & \dfrac{\partial}{\partial y} & \dfrac{\partial}{\partial z} \\ M & N & P \end{vmatrix} \\ &= \left(\frac{\partial P}{\partial y} - \frac{\partial N}{\partial z}\right)\mathbf{i} + \left(\frac{\partial M}{\partial z} - \frac{\partial P}{\partial x}\right)\mathbf{j} + \left(\frac{\partial N}{\partial x} - \frac{\partial M}{\partial y}\right)\mathbf{k}.\end{aligned}$$

We often use this cross product notation to write the curl symbolically as "del cross $\mathbf{F}$."

$$\text{curl } \mathbf{F} = \nabla \times \mathbf{F} \tag{3}$$

EXAMPLE 1 Find the curl of $\mathbf{F} = (x^2 - z)\mathbf{i} + xe^z\mathbf{j} + xy\mathbf{k}$.

Solution We use Equation (3) and the determinant form for the cross product, which gives,

$$\begin{aligned}\text{curl } \mathbf{F} &= \nabla \times \mathbf{F} \\ &= \begin{vmatrix} \mathbf{i} & \mathbf{j} & \mathbf{k} \\ \dfrac{\partial}{\partial x} & \dfrac{\partial}{\partial y} & \dfrac{\partial}{\partial z} \\ x^2 - z & xe^z & xy \end{vmatrix} \\ &= \left(\frac{\partial}{\partial y}(xy) - \frac{\partial}{\partial z}(xe^z)\right)\mathbf{i} - \left(\frac{\partial}{\partial x}(xy) - \frac{\partial}{\partial z}(x^2 - z)\right)\mathbf{j} \\ &\quad + \left(\frac{\partial}{\partial x}(xe^z) - \frac{\partial}{\partial y}(x^2 - z)\right)\mathbf{k} \\ &= (x - xe^z)\mathbf{i} - (y + 1)\mathbf{j} + (e^z - 0)\mathbf{k} \\ &= x(1 - e^z)\mathbf{i} - (y + 1)\mathbf{j} + e^z\mathbf{k}.\end{aligned}$$

Curl $\mathbf{F}$ is a vector, not a scalar.

■

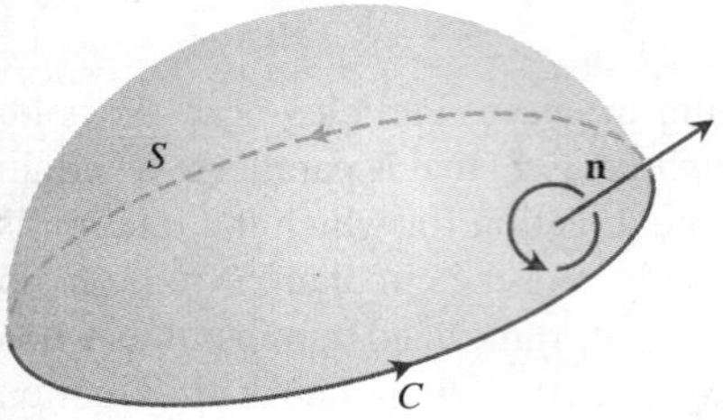

FIGURE 16.59 The orientation of the bounding curve C gives it a right-hand relation to the normal field $\mathbf{n}$. If the thumb of a right hand points along $\mathbf{n}$, the fingers curl in the direction of C.

As we will see, the operator ∇ has a number of other applications. For instance, when applied to a scalar function $f(x, y, z)$, it gives the gradient of f:

$$\nabla f = \frac{\partial f}{\partial x}\mathbf{i} + \frac{\partial f}{\partial y}\mathbf{j} + \frac{\partial f}{\partial z}\mathbf{k}.$$

In this setting it is read sometimes as "del f" and sometimes as "grad f."

Stokes' Theorem

Stokes' Theorem generalizes Green's Theorem to three dimensions. The circulation-curl form of Green's Theorem relates the counterclockwise circulation of a vector field around a simple closed curve C in the xy-plane to a double integral over the plane region R enclosed by C. Stokes' Theorem relates the circulation of a vector field around the boundary C of an oriented surface S in space (Figure 16.59) to a surface integral over the surface S. We require that the surface be **piecewise smooth**, which means that it is a finite union of smooth surfaces joining along smooth curves.

THEOREM 6—Stokes' Theorem
Let S be a piecewise smooth oriented surface having a piecewise smooth boundary curve C. Let $\mathbf{F} = M\mathbf{i} + N\mathbf{j} + P\mathbf{k}$ be a vector field whose components have continuous first partial derivatives on an open region containing S. Then the circulation of $\mathbf{F}$ around C in the direction counterclockwise with respect to the surface's unit normal vector $\mathbf{n}$ equals the integral of the curl vector field $\nabla \times \mathbf{F}$ over S:

$$\underbrace{\oint_C \mathbf{F} \cdot d\mathbf{r}}_{\text{Counterclockwise circulation}} = \underbrace{\iint_S (\nabla \times \mathbf{F}) \cdot \mathbf{n}\, d\sigma}_{\text{Curl integral}} \qquad (4)$$

Notice from Equation (4) that if two different oriented surfaces S_1 and S_2 have the same boundary C, their curl integrals are equal:

$$\iint_{S_1} (\nabla \times \mathbf{F}) \cdot \mathbf{n}_1\, d\sigma = \iint_{S_2} (\nabla \times \mathbf{F}) \cdot \mathbf{n}_2\, d\sigma.$$

Both curl integrals equal the counterclockwise circulation integral on the left side of Equation (4) as long as the unit normal vectors $\mathbf{n}_1$ and $\mathbf{n}_2$ correctly orient the surfaces. So the curl integral is independent of the surface and depends only on circulation along the boundary curve. This independence of surface resembles the path independence for the flow integral of a conservative velocity field along a curve, where the value of the flow integral depends only on the endpoints (that is, the boundary points) of the path. In that sense, the curl field $\nabla \times \mathbf{F}$ is analogous to the gradient field ∇f of a scalar function f.

If C is a curve in the xy-plane, oriented counterclockwise, and R is the region in the xy-plane bounded by C, then $d\sigma = dx\,dy$ and

$$(\nabla \times \mathbf{F}) \cdot \mathbf{n} = (\nabla \times \mathbf{F}) \cdot \mathbf{k} = \left(\frac{\partial N}{\partial x} - \frac{\partial M}{\partial y}\right).$$

Under these conditions, Stokes' equation becomes

$$\oint_C \mathbf{F} \cdot d\mathbf{r} = \iint_R \left(\frac{\partial N}{\partial x} - \frac{\partial M}{\partial y}\right) dx\,dy,$$

which is the circulation-curl form of the equation in Green's Theorem. Conversely, by reversing these steps we can rewrite the circulation-curl form of Green's Theorem for two-dimensional fields in del notation as

$$\oint_C \mathbf{F} \cdot d\mathbf{r} = \iint_R (\nabla \times \mathbf{F}) \cdot \mathbf{k}\, dA. \tag{5}$$

See Figure 16.60.

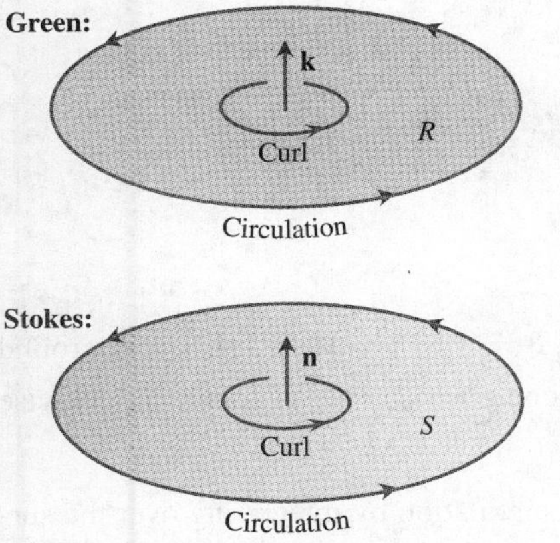

FIGURE 16.60 When applied to curves and surfaces in the plane, Stokes' Theorem gives the circulation-curl version of Green's Theorem. But Stokes' Theorem also applies more generally, to curves and surfaces not lying in the plane.

EXAMPLE 2 Evaluate both sides of Equation (4) for the hemisphere S: $x^2 + y^2 + z^2 = 9$, $z \geq 0$; its bounding circle C: $x^2 + y^2 = 9$, $z = 0$, traversed counterclockwise (when viewed from above); and the field $\mathbf{F} = y\mathbf{i} - x\mathbf{j}$.

Solution The hemisphere looks much like the surface in Figure 16.59 with the bounding circle C in the xy-plane (see Figure 16.61). We calculate the counterclockwise circulation around C (as viewed from above) using the parametrization $\mathbf{r}(\theta) = (3\cos\theta)\mathbf{i} + (3\sin\theta)\mathbf{j}$, $0 \leq \theta \leq 2\pi$:

$$\begin{aligned}
d\mathbf{r} &= (-3\sin\theta\, d\theta)\mathbf{i} + (3\cos\theta\, d\theta)\mathbf{j} \\
\mathbf{F} &= y\mathbf{i} - x\mathbf{j} = (3\sin\theta)\mathbf{i} - (3\cos\theta)\mathbf{j} \\
\mathbf{F} \cdot d\mathbf{r} &= -9\sin^2\theta\, d\theta - 9\cos^2\theta\, d\theta = -9\, d\theta \\
\oint_C \mathbf{F} \cdot d\mathbf{r} &= \int_0^{2\pi} -9\, d\theta = -18\pi.
\end{aligned}$$

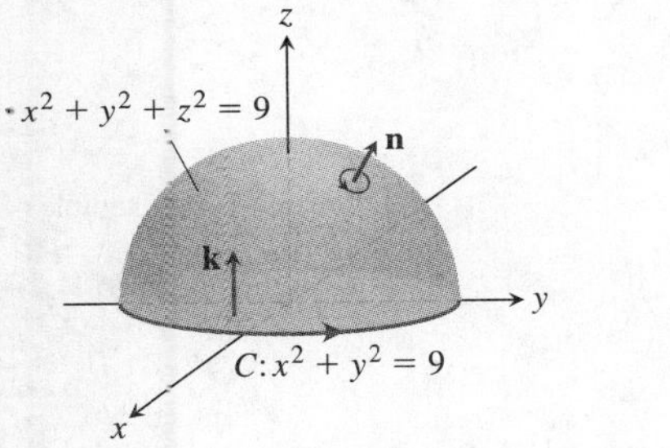

FIGURE 16.61 A hemisphere and a disk, each with boundary C (Examples 2 and 3).

When evaluating the right side of Equation (4), we choose the orientation of the unit normal vector so that it points away from the origin, giving it a right-hand relation to the prescribed orientation of the curve C (see Figure 16.61). We have

$$\begin{aligned}
\nabla \times \mathbf{F} &= \left(\frac{\partial P}{\partial y} - \frac{\partial N}{\partial z}\right)\mathbf{i} + \left(\frac{\partial M}{\partial z} - \frac{\partial P}{\partial x}\right)\mathbf{j} + \left(\frac{\partial N}{\partial x} - \frac{\partial M}{\partial y}\right)\mathbf{k} \\
&= (0 - 0)\mathbf{i} + (0 - 0)\mathbf{j} + (-1 - 1)\mathbf{k} = -2\mathbf{k}
\end{aligned}$$

$$\mathbf{n} = \frac{x\mathbf{i} + y\mathbf{j} + z\mathbf{k}}{\sqrt{x^2 + y^2 + z^2}} = \frac{x\mathbf{i} + y\mathbf{j} + z\mathbf{k}}{3} \qquad \text{Unit normal vector}$$

$$d\sigma = \frac{3}{z}\, dA \qquad \text{Section 16.6, Example 7, with } a = 3$$

$$(\nabla \times \mathbf{F}) \cdot \mathbf{n}\, d\sigma = (-2\mathbf{k}) \times \left(\frac{x\mathbf{i} + y\mathbf{j} + z\mathbf{k}}{3}\right) d\sigma = -\frac{2z}{3}\frac{3}{z}\, dA = -2\, dA$$

and

$$\iint_S (\nabla \times \mathbf{F}) \cdot \mathbf{n}\, d\sigma = \iint_{x^2+y^2 \leq 9} -2\, dA = -18\pi.$$

The circulation around the circle equals the integral of the curl over the hemisphere, as it should from Stokes' Theorem. ■

The surface integral in Stokes' Theorem can be computed using any surface having boundary curve C, provided the surface is properly oriented and lies within the domain of the field $\mathbf{F}$. The next example illustrates this fact for the circulation around the curve C in Example 2.

EXAMPLE 3 Calculate the circulation around the bounding circle C in Example 2, using the disk of radius 3 centered at the origin in the xy-plane as the surface S (instead of the hemisphere). See Figure 16.61.

Solution As in Example 2, $\nabla \times \mathbf{F} = -2\mathbf{k}$. When the surface is the described disk in the xy-plane, we have the normal vector $\mathbf{n} = \mathbf{k}$, chosen to give a counterclockwise direction for C as required by Stokes' Theorem, so that

$$(\nabla \times \mathbf{F}) \cdot \mathbf{n}\, d\sigma = -2\mathbf{k} \cdot \mathbf{k}\, dA = -2\, dA$$

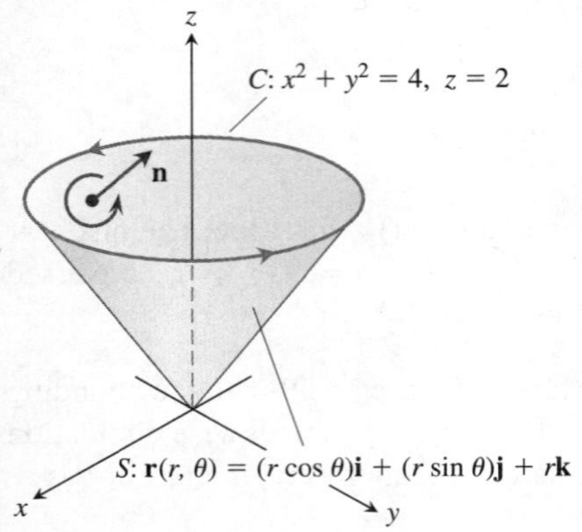

FIGURE 16.62 The curve C and cone S in Example 4.

and

$$\iint_S (\nabla \times \mathbf{F}) \cdot \mathbf{n}\, d\sigma = \iint_{x^2+y^2 \le 9} -2\, dA = -18\pi,$$

a simpler calculation than before. ■

EXAMPLE 4 Find the circulation of the field $\mathbf{F} = (x^2 - y)\mathbf{i} + 4z\mathbf{j} + x^2\mathbf{k}$ around the curve C in which the plane $z = 2$ meets the cone $z = \sqrt{x^2 + y^2}$, counterclockwise as viewed from above (Figure 16.62).

Solution Stokes' Theorem enables us to find the circulation by integrating over the surface of the cone. Traversing C in the counterclockwise direction viewed from above corresponds to taking the *inner* normal $\mathbf{n}$ to the cone, the normal with a positive $\mathbf{k}$-component.

We parametrize the cone as

$$\mathbf{r}(r, \theta) = (r\cos\theta)\mathbf{i} + (r\sin\theta)\mathbf{j} + r\mathbf{k}, \qquad 0 \le r \le 2, \quad 0 \le \theta \le 2\pi.$$

We then have

$$\mathbf{n} = \frac{\mathbf{r}_r \times \mathbf{r}_\theta}{|\mathbf{r}_r \times \mathbf{r}_\theta|} = \frac{-(r\cos\theta)\mathbf{i} - (r\sin\theta)\mathbf{j} + r\mathbf{k}}{r\sqrt{2}} \qquad \text{Section 16.5, Example 4}$$

$$= \frac{1}{\sqrt{2}}(-(\cos\theta)\mathbf{i} - (\sin\theta)\mathbf{j} + \mathbf{k})$$

$$d\sigma = r\sqrt{2}\, dr\, d\theta \qquad \text{Section 16.5, Example 4}$$

$$\nabla \times \mathbf{F} = -4\mathbf{i} - 2x\mathbf{j} + \mathbf{k} \qquad \text{Computation of curl}$$

$$= -4\mathbf{i} - 2r\cos\theta\mathbf{j} + \mathbf{k}. \qquad x = r\cos\theta$$

Accordingly,

$$(\nabla \times \mathbf{F}) \cdot \mathbf{n} = \frac{1}{\sqrt{2}}(4\cos\theta + 2r\cos\theta\sin\theta + 1)$$

$$= \frac{1}{\sqrt{2}}(4\cos\theta + r\sin 2\theta + 1),$$

and the circulation is

$$\oint_C \mathbf{F} \cdot d\mathbf{r} = \iint_S (\nabla \times \mathbf{F}) \cdot \mathbf{n}\, d\sigma \qquad \text{Stokes' Theorem, Eq. (4)}$$

$$= \int_0^{2\pi}\int_0^2 \frac{1}{\sqrt{2}}(4\cos\theta + r\sin 2\theta + 1)\left(r\sqrt{2}\, dr\, d\theta\right) = 4\pi.$$ ■

EXAMPLE 5 The cone used in Example 4 is not the easiest surface to use for calculating the circulation around the bounding circle C lying in the plane $z = 2$. If instead we use the flat disk of radius 2 centered on the z-axis and lying in the plane $z = 2$, then the normal vector to the surface S is $\mathbf{n} = \mathbf{k}$ (chosen to give a counterclockwise direction for the curve C). Just as in the computation for Example 4, we still have $\nabla \times \mathbf{F} = -4\mathbf{i} - 2x\mathbf{j} + \mathbf{k}$. However, now we get $(\nabla \times \mathbf{F}) \cdot \mathbf{n} = 1$, so that

$$\iint_S (\nabla \times \mathbf{F}) \cdot \mathbf{n}\, d\sigma = \iint_{x^2+y^2 \le 4} 1\, dA = 4\pi. \qquad \text{The shadow is the disk of radius 2 in the } xy\text{-plane.}$$

This result agrees with the circulation value found in Example 4. ■

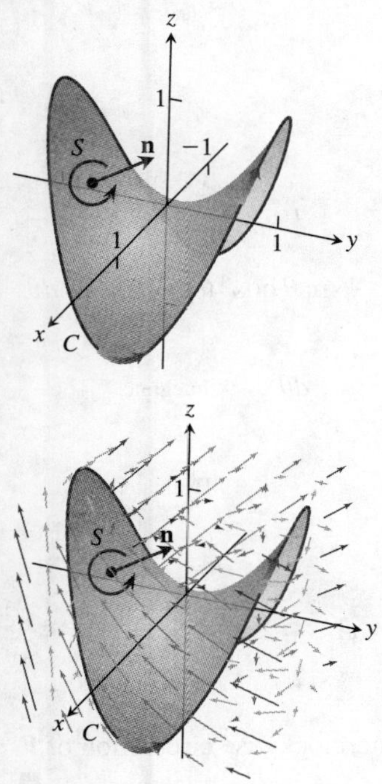

FIGURE 16.63 The surface and vector field for Example 6.

EXAMPLE 6 Find a parametrization for the surface S formed by the part of the hyperbolic paraboloid $z = y^2 - x^2$ lying inside the cylinder of radius one around the z-axis and for the boundary curve C of S. (See Figure 16.63.) Then verify Stokes' Theorem for S using the normal having positive $\mathbf{k}$-component and the vector field $\mathbf{F} = y\mathbf{i} - x\mathbf{j} + x^2\mathbf{k}$.

Solution As the unit circle is traversed in the xy-plane, the z-coordinate of the surface with the curve C as boundary is given by $y^2 - x^2$. We choose the orientation for the curve C to be counterclockwise when viewed from above (see Figure 16.63). A parametrization of C is given by

$$\mathbf{r}(t) = (\cos t)\mathbf{i} + (\sin t)\mathbf{j} + (\sin^2 t - \cos^2 t)\mathbf{k}, \quad 0 \le t \le 2\pi$$

with

$$\frac{d\mathbf{r}}{dt} = (-\sin t)\mathbf{i} + (\cos t)\mathbf{j} + (4 \sin t \cos t)\mathbf{k}, \quad 0 \le t \le 2\pi.$$

Along the curve $\mathbf{r}(t)$ the formula for the vector field $\mathbf{F}$ is

$$\mathbf{F} = (\sin t)\mathbf{i} - (\cos t)\mathbf{j} + (\cos^2 t)\mathbf{k}.$$

The counterclockwise circulation along C is the value of the line integral

$$\begin{aligned}\int_0^{2\pi} \mathbf{F} \cdot \frac{d\mathbf{r}}{dt}\,dt &= \int_0^{2\pi} (-\sin^2 t - \cos^2 t + 4 \sin t \cos^3 t)\,dt \\ &= \int_0^{2\pi} (4 \sin t \cos^3 t - 1)\,dt \\ &= \Big[-\cos^4 t - t\Big]_0^{2\pi} = -2\pi.\end{aligned}$$

We now compute the same quantity by integrating $(\nabla \times \mathbf{F}) \cdot \mathbf{n}$ over the surface S. We use polar coordinates and parametrize S by noting that above the point (r, θ) in the plane, the z–coordinate of S is $y^2 - x^2 = r^2 \sin^2 \theta - r^2 \cos^2 \theta$. A parametrization of S is

$$\mathbf{r}(r, \theta) = (r \cos \theta)\mathbf{i} + (r \sin \theta)\mathbf{j} + r^2(\sin^2 \theta - \cos^2 \theta)\mathbf{k}, \quad 0 \le r \le 1, \quad 0 \le \theta \le 2\pi.$$

We next compute $(\nabla \times \mathbf{F}) \cdot \mathbf{n}\,d\sigma$. We have

$$\nabla \times \mathbf{F} = \begin{vmatrix} \mathbf{i} & \mathbf{j} & \mathbf{k} \\ \dfrac{\partial}{\partial x} & \dfrac{\partial}{\partial y} & \dfrac{\partial}{\partial z} \\ y & -x & x^2 \end{vmatrix} = -2x\mathbf{j} - 2\mathbf{k} = -(2r \cos \theta)\mathbf{j} - 2\mathbf{k}$$

and

$$\begin{aligned}\mathbf{r}_r &= (\cos \theta)\mathbf{i} + (\sin \theta)\mathbf{j} + 2r(\sin^2 \theta - \cos^2 \theta)\mathbf{k} \\ \mathbf{r}_\theta &= (-r \sin \theta)\mathbf{i} + (r \cos \theta)\mathbf{j} + 4r^2(\sin \theta \cos \theta)\mathbf{k}\end{aligned}$$

$$\begin{aligned}\mathbf{r}_r \times \mathbf{r}_\theta &= \begin{vmatrix} \mathbf{i} & \mathbf{j} & \mathbf{k} \\ \cos \theta & \sin \theta & 2r(\sin^2 \theta - \cos^2 \theta) \\ -r \sin \theta & r \cos \theta & 4r^2(\sin \theta \cos \theta) \end{vmatrix} \\ &= 2r^2(2 \sin^2 \theta \cos \theta - \sin^2 \theta \cos \theta + \cos^3 \theta)\mathbf{i} \\ &\quad -2r^2(2 \sin \theta \cos^2 \theta + \sin^3 \theta + \sin \theta \cos^2 \theta)\mathbf{j} + r\mathbf{k}.\end{aligned}$$

Note that the $\mathbf{k}$-component is always nonnegative. Therefore, we take

$$\mathbf{n} = +\frac{(\mathbf{r}_r \times \mathbf{r}_\theta)}{|\mathbf{r}_r \times \mathbf{r}_\theta|}.$$

We now obtain

$$\iint_S (\nabla \times \mathbf{F}) \cdot \mathbf{n}\, d\sigma = \int_0^{2\pi}\int_0^1 (\nabla \times \mathbf{F}) \cdot \frac{\mathbf{r}_r \times \mathbf{r}_\theta}{|\mathbf{r}_r \times \mathbf{r}_\theta|}|\mathbf{r}_r \times \mathbf{r}_\theta|\, dr\, d\theta$$

$$= \int_0^{2\pi}\int_0^1 (\nabla \times \mathbf{F}) \cdot (\mathbf{r}_r \times \mathbf{r}_\theta)\, dr\, d\theta$$

$$= \int_0^{2\pi}\int_0^1 [4r^3(2\sin\theta\cos^3\theta + \sin^3\theta\cos\theta + \sin\theta\cos^3\theta) - 2r]\, dr\, d\theta$$

$$= \int_0^{2\pi} \left[r^4(3\sin\theta\cos^3\theta + \sin^3\theta\cos\theta) - r^2\right]_{r=0}^{r=1} d\theta \qquad \text{Integrate.}$$

$$= \int_0^{2\pi} (3\sin\theta\cos^3\theta + \sin^3\theta\cos\theta - 1)\, d\theta \qquad \text{Evaluate.}$$

$$= \left[-\frac{3}{4}\cos^4\theta + \frac{1}{4}\sin^4\theta - \theta\right]_0^{2\pi}$$

$$= \left(-\frac{3}{4} + 0 - 2\pi + \frac{3}{4} - 0 + 0\right) = -2\pi.$$

So the surface integral of $(\nabla \times \mathbf{F}) \cdot \mathbf{n}$ over S equals the counterclockwise circulation of $\mathbf{F}$ along C, as asserted by Stokes' Theorem. ■

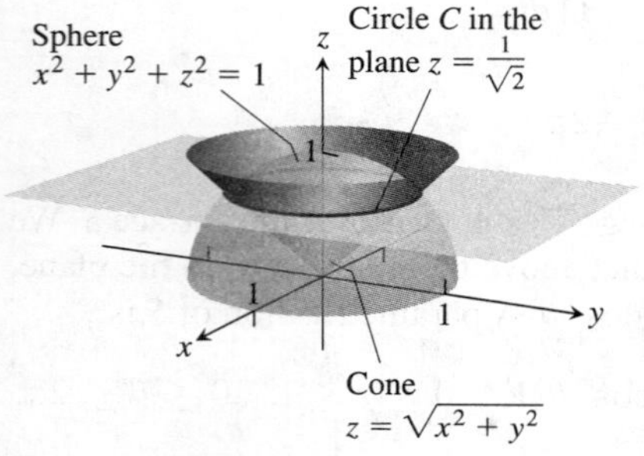

FIGURE 16.64 Circulation curve C in Example 7.

EXAMPLE 7 Calculate the circulation of the vector field

$$\mathbf{F} = (x^2 + z)\mathbf{i} + (y^2 + 2x)\mathbf{j} + (z^2 - y)\mathbf{k}$$

along the curve of intersection of the sphere $x^2 + y^2 + z^2 = 1$ with the cone $z = \sqrt{x^2 + y^2}$ traversed in the counterclockwise direction around the z-axis when viewed from above.

Solution The sphere and cone intersect when $1 = (x^2 + y^2) + z^2 = z^2 + z^2 = 2z^2$, or $z = 1/\sqrt{2}$ (see Figure 16.64). We apply Stokes' Theorem to the curve of intersection $x^2 + y^2 = 1/2$ considered as the boundary of the enclosed disk in the plane $z = 1/\sqrt{2}$. The normal vector to the surface that gives a counterclockwise orientation to the boundary curve is then $\mathbf{n} = \mathbf{k}$. We calculate the curl vector as

$$\nabla \times \mathbf{F} = \begin{vmatrix} \mathbf{i} & \mathbf{j} & \mathbf{k} \\ \dfrac{\partial}{\partial x} & \dfrac{\partial}{\partial y} & \dfrac{\partial}{\partial z} \\ x^2 + z & y^2 + 2x & z^2 - y \end{vmatrix} = -\mathbf{i} + \mathbf{j} + 2\mathbf{k},$$

so that $(\nabla \times \mathbf{F}) \cdot \mathbf{k} = 2$. The circulation around the disk is

$$\oint_C \mathbf{F} \cdot d\mathbf{r} = \iint_S (\nabla \times \mathbf{F}) \cdot \mathbf{k}\, d\sigma$$

$$= \iint_S 2\, d\sigma = 2 \cdot \text{area of disk} = 2 \cdot \pi\left(\frac{1}{\sqrt{2}}\right)^2 = \pi. \qquad ■$$

Paddle Wheel Interpretation of $\nabla \times \mathbf{F}$

Suppose that $\mathbf{F}$ is the velocity field of a fluid moving in a region R in space containing the closed curve C. Then

$$\oint_C \mathbf{F} \cdot d\mathbf{r}$$

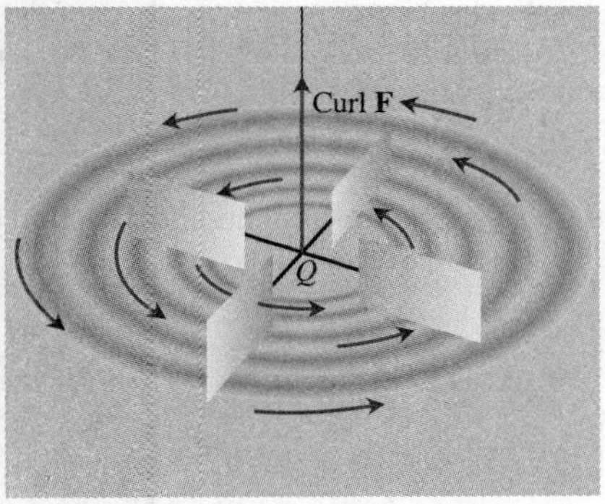

FIGURE 16.65 A small paddle wheel in a fluid spins fastest at point Q when its axle points in the direction of curl $\mathbf{F}$.

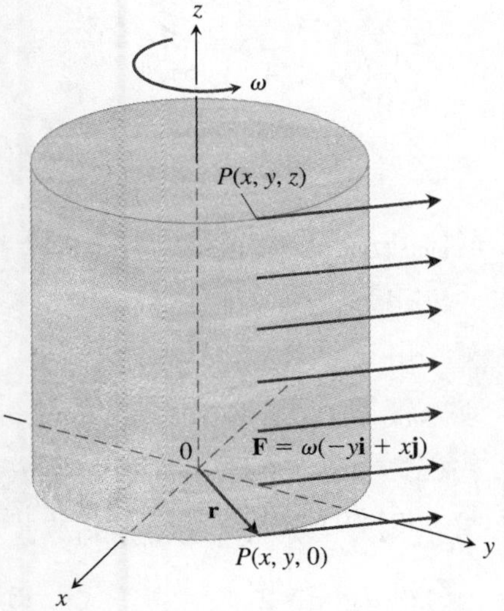

FIGURE 16.66 A steady rotational flow parallel to the xy-plane, with constant angular velocity ω in the positive (counterclockwise) direction (Example 8).

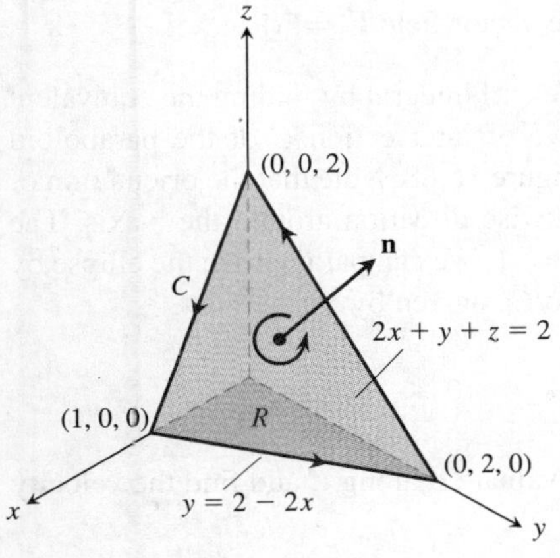

FIGURE 16.67 The planar surface in Example 9.

is the circulation of the fluid around C. By Stokes' Theorem, the circulation is equal to the flux of $\nabla \times \mathbf{F}$ through any suitably oriented surface S with boundary C:

$$\oint_C \mathbf{F} \cdot d\mathbf{r} = \iint_S (\nabla \times \mathbf{F}) \cdot \mathbf{n}\, d\sigma.$$

Suppose we fix a point Q in the region R and a direction $\mathbf{u}$ at Q. Take C to be a circle of radius ρ, with center at Q, whose plane is normal to $\mathbf{u}$. If $\nabla \times \mathbf{F}$ is continuous at Q, the average value of the $\mathbf{u}$-component of $\nabla \times \mathbf{F}$ over the circular disk S bounded by C approaches the $\mathbf{u}$-component of $\nabla \times \mathbf{F}$ at Q as the radius $\rho \to 0$:

$$\left((\nabla \times \mathbf{F}) \cdot \mathbf{u}\right)\Big|_Q = \lim_{\rho \to 0} \frac{1}{\pi\rho^2} \iint_S (\nabla \times \mathbf{F}) \cdot \mathbf{u}\, d\sigma.$$

If we apply Stokes' Theorem and replace the surface integral by a line integral over C, we get

$$\left((\nabla \times \mathbf{F}) \cdot \mathbf{u}\right)\Big|_Q = \lim_{\rho \to 0} \frac{1}{\pi\rho^2} \oint_C \mathbf{F} \cdot d\mathbf{r}. \tag{6}$$

The left-hand side of Equation (6) has its maximum value when $\mathbf{u}$ is the direction of $\nabla \times \mathbf{F}$. When ρ is small, the limit on the right-hand side of Equation (6) is approximately

$$\frac{1}{\pi\rho^2} \oint_C \mathbf{F} \cdot d\mathbf{r},$$

which is the circulation around C divided by the area of the disk (circulation density). Suppose that a small paddle wheel of radius ρ is introduced into the fluid at Q, with its axle directed along $\mathbf{u}$ (Figure 16.65). The circulation of the fluid around C affects the rate of spin of the paddle wheel. The wheel spins fastest when the circulation integral is maximized; therefore, it spins fastest when the axle of the paddle wheel points in the direction of $\nabla \times \mathbf{F}$.

EXAMPLE 8 A fluid of constant density rotates around the z-axis with velocity $\mathbf{F} = \omega(-y\mathbf{i} + x\mathbf{j})$, where ω is a positive constant called the *angular velocity* of the rotation (Figure 16.66). Find $\nabla \times \mathbf{F}$ and relate it to the circulation density.

Solution With $\mathbf{F} = -\omega y\mathbf{i} + \omega x\mathbf{j}$, we find the curl

$$\begin{aligned}\nabla \times \mathbf{F} &= \left(\frac{\partial P}{\partial y} - \frac{\partial N}{\partial z}\right)\mathbf{i} + \left(\frac{\partial M}{\partial z} - \frac{\partial P}{\partial x}\right)\mathbf{j} + \left(\frac{\partial N}{\partial x} - \frac{\partial M}{\partial y}\right)\mathbf{k} \\ &= (0 - 0)\mathbf{i} + (0 - 0)\mathbf{j} + (\omega - (-\omega))\mathbf{k} = 2\omega\mathbf{k},\end{aligned}$$

and therefore $(\nabla \times \mathbf{F}) \cdot \mathbf{k} = 2\omega$. By Stokes' Theorem, the circulation of $\mathbf{F}$ around a circle C of radius ρ (traversed counterclockwise when viewed from above) bounding a disk S in a plane normal to $\nabla \times \mathbf{F}$, say the xy-plane, is

$$\oint_C \mathbf{F} \cdot d\mathbf{r} = \iint_S (\nabla \times \mathbf{F}) \cdot \mathbf{n}\, d\sigma = \iint_S 2\omega\mathbf{k} \cdot \mathbf{k}\, dx\, dy = (2\omega)(\pi\rho^2).$$

Solving this last equation for 2ω, we see that

$$(\nabla \times \mathbf{F}) \cdot \mathbf{k} = 2\omega = \frac{1}{\pi\rho^2} \oint_C \mathbf{F} \cdot d\mathbf{r},$$

which is consistent with Equation (6) when $\mathbf{u} = \mathbf{k}$. ■

EXAMPLE 9 Use Stokes' Theorem to evaluate $\int_C \mathbf{F} \cdot d\mathbf{r}$, if $\mathbf{F} = xz\mathbf{i} + xy\mathbf{j} + 3xz\mathbf{k}$ and C is the boundary of the portion of the plane $2x + y + z = 2$ in the first octant, traversed counterclockwise as viewed from above (Figure 16.67).

Solution The plane is the level surface $f(x, y, z) = 2$ of the function $f(x, y, z) = 2x + y + z$. The unit normal vector

$$\mathbf{n} = \frac{\nabla f}{|\nabla f|} = \frac{(2\mathbf{i} + \mathbf{j} + \mathbf{k})}{|2\mathbf{i} + \mathbf{j} + \mathbf{k}|} = \frac{1}{\sqrt{6}}(2\mathbf{i} + \mathbf{j} + \mathbf{k})$$

is consistent with the counterclockwise motion around C. To apply Stokes' Theorem, we find

$$\text{curl } \mathbf{F} = \nabla \times \mathbf{F} = \begin{vmatrix} \mathbf{i} & \mathbf{j} & \mathbf{k} \\ \dfrac{\partial}{\partial x} & \dfrac{\partial}{\partial y} & \dfrac{\partial}{\partial z} \\ xz & xy & 3xz \end{vmatrix} = (x - 3z)\mathbf{j} + y\mathbf{k}.$$

On the plane, z equals $2 - 2x - y$, so

$$\nabla \times \mathbf{F} = (x - 3(2 - 2x - y))\mathbf{j} + y\mathbf{k} = (7x + 3y - 6)\mathbf{j} + y\mathbf{k}$$

and

$$(\nabla \times \mathbf{F}) \cdot \mathbf{n} = \frac{1}{\sqrt{6}}(7x + 3y - 6 + y) = \frac{1}{\sqrt{6}}(7x + 4y - 6).$$

The surface area differential is

$$d\sigma = \frac{|\nabla f|}{|\nabla f \cdot \mathbf{k}|}\, dA = \frac{\sqrt{6}}{1}\, dx\, dy. \qquad \text{Formula (7) in Section 16.5}$$

The circulation is

$$\begin{aligned} \oint_C \mathbf{F} \cdot d\mathbf{r} &= \iint_S (\nabla \times \mathbf{F}) \cdot \mathbf{n}\, d\sigma \qquad \text{Stokes' Theorem, Eq. (4)} \\ &= \int_0^1 \int_0^{2-2x} \frac{1}{\sqrt{6}}(7x + 4y - 6)\sqrt{6}\, dy\, dx \\ &= \int_0^1 \int_0^{2-2x} (7x + 4y - 6)\, dy\, dx = -1. \end{aligned}$$

■

EXAMPLE 10 Let the surface S be the elliptic paraboloid $z = x^2 + 4y^2$ lying beneath the plane $z = 1$ (Figure 16.68). We define the orientation of S by taking the *inner* normal vector $\mathbf{n}$ to the surface, which is the normal having a positive $\mathbf{k}$-component. Find the flux of $\nabla \times \mathbf{F}$ across S in the direction $\mathbf{n}$ for the vector field $\mathbf{F} = y\mathbf{i} - xz\mathbf{j} + xz^2\mathbf{k}$.

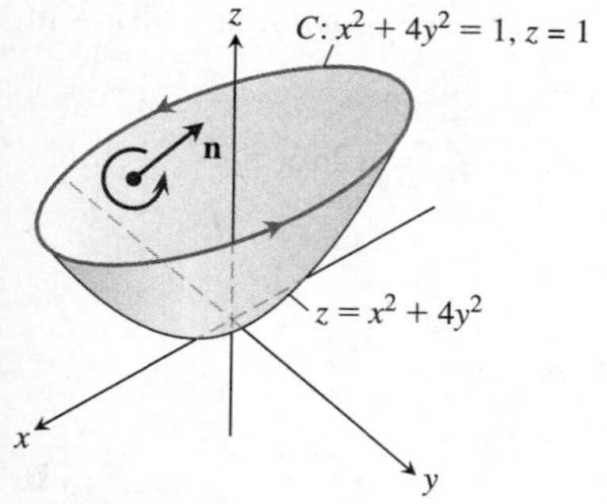

FIGURE 16.68 The portion of the elliptic paraboloid in Example 10, showing its curve of intersection C with the plane $z = 1$ and its inner normal orientation by $\mathbf{n}$.

Solution We use Stokes' Theorem to calculate the curl integral by finding the equivalent counterclockwise circulation of $\mathbf{F}$ around the curve of intersection C of the paraboloid $z = x^2 + 4y^2$ and the plane $z = 1$, as shown in Figure 16.68. Note that the orientation of S is consistent with traversing C in a counterclockwise direction around the z-axis. The curve C is the ellipse $x^2 + 4y^2 = 1$ in the plane $z = 1$. We can parametrize the ellipse by $x = \cos t,\ y = \frac{1}{2}\sin t,\ z = 1$ for $0 \le t \le 2\pi$, so C is given by

$$\mathbf{r}(t) = (\cos t)\mathbf{i} + \frac{1}{2}(\sin t)\mathbf{j} + \mathbf{k}, \qquad 0 \le t \le 2\pi.$$

To compute the circulation integral $\oint_C \mathbf{F} \cdot d\mathbf{r}$, we evaluate $\mathbf{F}$ along C and find the velocity vector $d\mathbf{r}/dt$:

$$\mathbf{F}(\mathbf{r}(t)) = \frac{1}{2}(\sin t)\mathbf{i} - (\cos t)\mathbf{j} + (\cos t)\mathbf{k}$$

and

$$\frac{d\mathbf{r}}{dt} = -(\sin t)\mathbf{i} + \frac{1}{2}(\cos t)\mathbf{j}.$$

Then

$$\begin{aligned}\oint_C \mathbf{F} \cdot d\mathbf{r} &= \int_0^{2\pi} \mathbf{F}(\mathbf{r}(t)) \cdot \frac{d\mathbf{r}}{dt}\, dt \\ &= \int_0^{2\pi} \left(-\frac{1}{2}\sin^2 t - \frac{1}{2}\cos^2 t\right) dt \\ &= -\frac{1}{2}\int_0^{2\pi} dt = -\pi.\end{aligned}$$

Therefore, by Stokes' Theorem the flux of the curl across S in the direction $\mathbf{n}$ for the field $\mathbf{F}$ is

$$\iint_S (\nabla \times \mathbf{F}) \cdot \mathbf{n}\, d\sigma = -\pi. \qquad \blacksquare$$

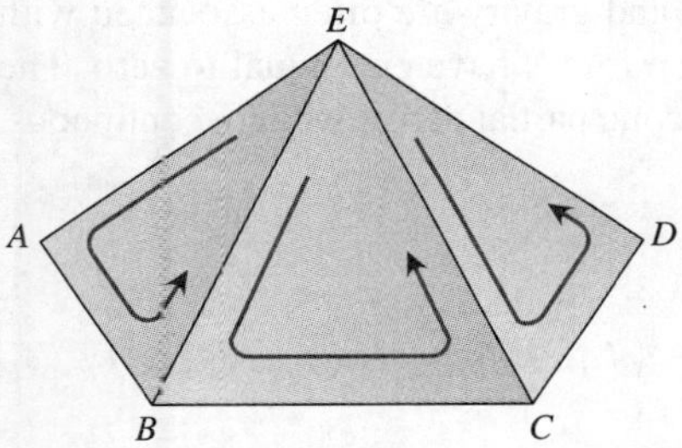

(a)

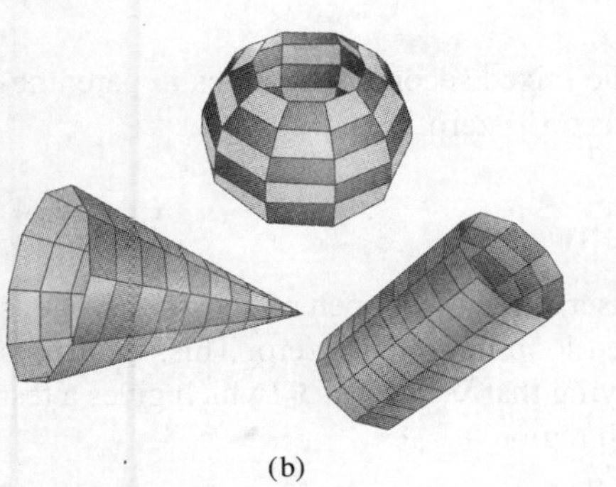
(b)

FIGURE 16.69 (a) Part of a polyhedral surface. (b) Other polyhedral surfaces.

Proof Outline of Stokes' Theorem for Polyhedral Surfaces

Let S be a polyhedral surface consisting of a finite number of plane regions or faces. (See Figure 16.69 for examples.) We apply Green's Theorem to each separate face of S. There are two types of faces:

1. Those that are surrounded on all sides by other faces.
2. Those that have one or more edges that are not adjacent to other faces.

The boundary of S consists of those edges of the type 2 faces that are not adjacent to other faces. In Figure 16.69a, the triangles EAB, BCE, and CDE represent a part of S, with $ABCD$ part of the boundary of the surface, boundary(S). Although Green's Theorem was stated for curves in the xy-plane, a generalized form applies to curves that lie in a plane in space. In the generalized form, the theorem asserts that the line integral of $\mathbf{F}$ around the curve enclosing the plane region R normal to $\mathbf{n}$ equals the double integral of (curl $\mathbf{F}$) $\cdot$ $\mathbf{n}$ over R. Applying this generalized form to the three triangles of Figure 16.69a in turn, and adding the results, gives

$$\left(\oint_{EAB} + \oint_{BCE} + \oint_{CDE}\right)\mathbf{F} \cdot d\mathbf{r} = \left(\iint_{EAB} + \iint_{BCE} + \iint_{CDE}\right)(\nabla \times \mathbf{F}) \cdot \mathbf{n}\, d\sigma. \tag{7}$$

The three line integrals on the left-hand side of Equation (7) combine into a single line integral taken around the periphery $ABCDE$ because the integrals along interior segments cancel in pairs. For example, the integral along segment BE in triangle ABE is opposite in sign to the integral along the same segment in triangle EBC. The same holds for segment CE. Hence, Equation (7) reduces to

$$\oint_{ABCDE} \mathbf{F} \cdot d\mathbf{r} = \iint_{ABCDE} (\nabla \times \mathbf{F}) \cdot \mathbf{n}\, d\sigma.$$

When we apply Green's Theorem to all the faces and add the results, we get

$$\oint_{\text{boundary}(S)} \mathbf{F} \cdot d\mathbf{r} = \iint_S (\nabla \times \mathbf{F}) \cdot \mathbf{n}\, d\sigma.$$

This is Stokes' Theorem for the polyhedral surface S in Figure 16.69a. More general polyhedral surfaces are shown in Figure 16.69b, and the proof can be extended to them. General smooth surfaces can be obtained as limits of polyhedral surfaces.

Stokes' Theorem for Surfaces with Holes

Stokes' Theorem holds for an oriented surface S that has one or more holes (Figure 16.70). The surface integral over S of the normal component of $\nabla \times \mathbf{F}$ equals the sum of the line integrals around all the boundary curves of the tangential component of $\mathbf{F}$, where the curves are to be traced in the direction induced by the orientation of S. For such surfaces the theorem is unchanged, but C is considered as a union of simple closed curves.

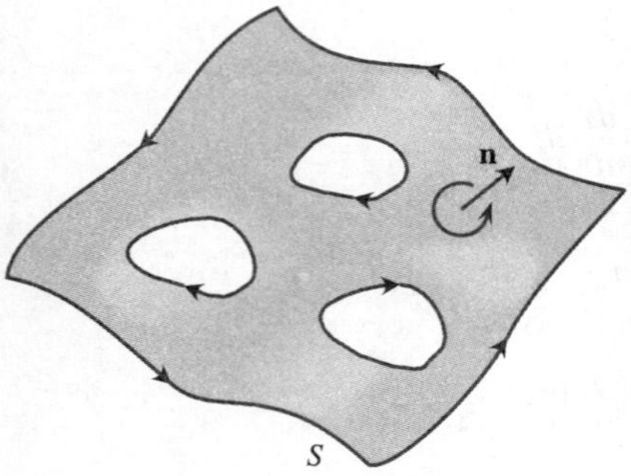

FIGURE 16.70 Stokes' Theorem also holds for oriented surfaces with holes. Consistent with the orientation of S, the outer curve is traversed counterclockwise around $\mathbf{n}$, and the inner curves surrounding the holes are traversed clockwise.

An Important Identity

The following identity arises frequently in mathematics and the physical sciences.

$$\text{curl grad } f = \mathbf{0} \quad \text{or} \quad \nabla \times \nabla f = \mathbf{0} \tag{8}$$

Forces arising in the study of electromagnetism and gravity are often associated with a potential function f. The identity (8) says that these forces have curl equal to zero. The identity (8) holds for any function $f(x, y, z)$ whose second partial derivatives are continuous. The proof goes like this:

$$\nabla \times \nabla f = \begin{vmatrix} \mathbf{i} & \mathbf{j} & \mathbf{k} \\ \dfrac{\partial}{\partial x} & \dfrac{\partial}{\partial y} & \dfrac{\partial}{\partial z} \\ \dfrac{\partial f}{\partial x} & \dfrac{\partial f}{\partial y} & \dfrac{\partial f}{\partial z} \end{vmatrix} = (f_{zy} - f_{yz})\mathbf{i} - (f_{zx} - f_{xz})\mathbf{j} + (f_{yx} - f_{xy})\mathbf{k}.$$

If the second partial derivatives are continuous, the mixed second derivatives in parentheses are equal (Theorem 2, Section 14.3) and the vector is zero.

Conservative Fields and Stokes' Theorem

In Section 16.3, we found that a field $\mathbf{F}$ being conservative in an open region D in space is equivalent to the integral of $\mathbf{F}$ around every closed loop in D being zero. This, in turn, is equivalent in *simply connected* open regions to saying that $\nabla \times \mathbf{F} = \mathbf{0}$ (which gives a test for determining whether $\mathbf{F}$ is conservative for such regions).

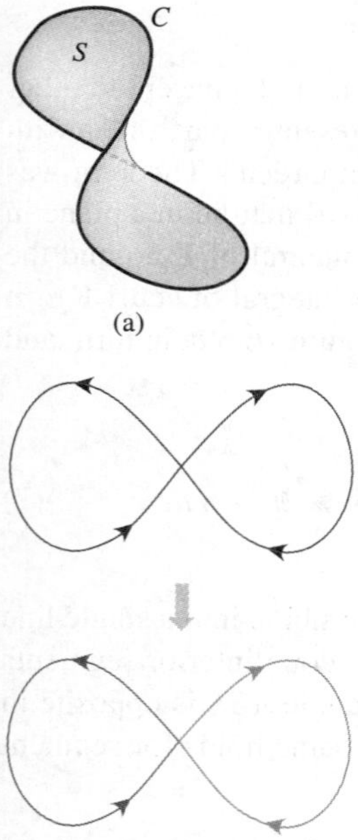

FIGURE 16.71 (a) In a simply connected open region in space, a simple closed curve C is the boundary of a smooth surface S. (b) Smooth curves that cross themselves can be divided into loops to which Stokes' Theorem applies.

THEOREM 7—Curl $\mathbf{F} = \mathbf{0}$ Related to the Closed-Loop Property
If $\nabla \times \mathbf{F} = \mathbf{0}$ at every point of a simply connected open region D in space, then on any piecewise-smooth closed path C in D,

$$\oint_C \mathbf{F} \cdot d\mathbf{r} = 0.$$

Sketch of a Proof Theorem 7 can be proved in two steps. The first step is for simple closed curves (loops that do not cross themselves), like the one in Figure 16.71a. A theorem from topology, a branch of advanced mathematics, states that every smooth simple closed curve C in a simply connected open region D is the boundary of a smooth two-sided surface S that also lies in D. Hence, by Stokes' Theorem,

$$\oint_C \mathbf{F}\cdot d\mathbf{r} = \iint_S (\nabla \times \mathbf{F})\cdot \mathbf{n}\, d\sigma = 0.$$

The second step is for curves that cross themselves, like the one in Figure 16.71b. The idea is to break these into simple loops spanned by orientable surfaces, apply Stokes' Theorem one loop at a time, and add the results.

The following diagram summarizes the results for conservative fields defined on connected, simply connected open regions. For such regions, the four statements are equivalent to each other.

$\mathbf{F}$ conservative on D	$\Longleftrightarrow$ (Theorem 2, Section 16.3)	$\mathbf{F} = \nabla f$ on D
$\Updownarrow$ Theorem 3, Section 16.3		$\Downarrow$ Vector identity (Eq. 8) (continuous second partial derivatives)
$\oint_C \mathbf{F}\cdot d\mathbf{r} = 0$ over any closed path in D	$\Longleftarrow$ (Theorem 7 Domain's simple connectivity and Stokes' Theorem)	$\nabla \times \mathbf{F} = \mathbf{0}$ throughout D

EXERCISES 16.7

In Exercises 1–6, find the curl of each vector field $\mathbf{F}$.

1. $\mathbf{F} = (x + y - z)\mathbf{i} + (2x - y + 3z)\mathbf{j} + (3x + 2y + z)\mathbf{k}$
2. $\mathbf{F} = (x^2 - y)\mathbf{i} + (y^2 - z)\mathbf{j} + (z^2 - x)\mathbf{k}$
3. $\mathbf{F} = (xy + z)\mathbf{i} + (yz + x)\mathbf{j} + (xz + y)\mathbf{k}$
4. $\mathbf{F} = ye^z\mathbf{i} + ze^x\mathbf{j} - xe^y\mathbf{k}$
5. $\mathbf{F} = x^2yz\mathbf{i} + xy^2z\mathbf{j} + xyz^2\mathbf{k}$
6. $\mathbf{F} = \dfrac{x}{yz}\mathbf{i} - \dfrac{y}{xz}\mathbf{j} + \dfrac{z}{xy}\mathbf{k}$

Using Stokes' Theorem to Find Line Integrals

In Exercises 7–12, use the surface integral in Stokes' Theorem to calculate the circulation of the field $\mathbf{F}$ around the curve C in the indicated direction.

7. $\mathbf{F} = x^2\mathbf{i} + 2x\mathbf{j} + z^2\mathbf{k}$

   C: The ellipse $4x^2 + y^2 = 4$ in the xy-plane, counterclockwise when viewed from above
8. $\mathbf{F} = 2y\mathbf{i} + 3x\mathbf{j} - z^2\mathbf{k}$

   C: The circle $x^2 + y^2 = 9$ in the xy-plane, counterclockwise when viewed from above
9. $\mathbf{F} = y\mathbf{i} + xz\mathbf{j} + x^2\mathbf{k}$

   C: The boundary of the triangle cut from the plane $x + y + z = 1$ by the first octant, counterclockwise when viewed from above
10. $\mathbf{F} = (y^2 + z^2)\mathbf{i} + (x^2 + z^2)\mathbf{j} + (x^2 + y^2)\mathbf{k}$

    C: The boundary of the triangle cut from the plane $x + y + z = 1$ by the first octant, counterclockwise when viewed from above
11. $\mathbf{F} = (y^2 + z^2)\mathbf{i} + (x^2 + y^2)\mathbf{j} + (x^2 + y^2)\mathbf{k}$

    C: The square bounded by the lines $x = \pm 1$ and $y = \pm 1$ in the xy-plane, counterclockwise when viewed from above
12. $\mathbf{F} = y^2\mathbf{i} + x^2\mathbf{j} - (x + z)\mathbf{k}$

    C: The boundary of the triangle with vertices at (0, 0, 0), (1, 0, 0), (1, 1, 0).

Integral of the Curl Vector Field

13. Let $\mathbf{n}$ be the unit normal in the direction away from the origin of the elliptic shell

    $$S:\ 4x^2 + 9y^2 + 36z^2 = 36, \qquad z \ge 0,$$

    and let

    $$\mathbf{F} = y\mathbf{i} + x^2\mathbf{j} + (x^2 + y^4)^{3/2} \sin e^{\sqrt{xyz}}\,\mathbf{k}.$$

    Find the value of

    $$\iint_S (\nabla \times \mathbf{F})\cdot \mathbf{n}\, d\sigma.$$

    (*Hint:* One parametrization of the ellipse at the base of the shell is $x = 3\cos t$, $y = 2\sin t$, $0 \le t \le 2\pi$.)

14. $\mathbf{F} = 2(x-y)\mathbf{i} - yz^2\mathbf{j} - y^2z\mathbf{k}$. Let $\mathbf{n}$ be the outer unit normal to the upper half of the surface $x^2+y^2+z^2=1$ and C is the circle $x^2+y^2=1$. Evaluate $\oint_C \mathbf{F}\cdot d\mathbf{r}$ by Stokes theorem.

15. Let S be the cylinder $x^2 + y^2 = a^2$, $0 \le z \le h$, together with its top, $x^2 + y^2 \le a^2$, $z = h$. Let $\mathbf{F} = -y\mathbf{i} + x\mathbf{j} + x^2\mathbf{k}$. Use Stokes' Theorem to find the flux of $\nabla \times \mathbf{F}$ through S in the direction away from the origin.

16. Evaluate

$$\iint_S (\nabla \times (y\mathbf{i})) \cdot \mathbf{n}\, d\sigma,$$

where S is the hemisphere $x^2 + y^2 + z^2 = 1$, $z \ge 0$, in the direction away from the origin.

17. Suppose $\mathbf{F} = \nabla \times \mathbf{A}$, where

$$\mathbf{A} = (y + z^2)\mathbf{i} + e^{xyz}\mathbf{j} + \cos(xz)\mathbf{k}.$$

Determine the flux of $\mathbf{F}$ through the hemisphere $x^2 + y^2 + z^2 = 1$, $z \ge 0$, in the direction away from the origin.

18. Repeat Exercise 17 for the flux of $\mathbf{F}$ across the entire unit sphere.

Stokes' Theorem for Parametrized Surfaces

In Exercises 19–24, use the surface integral in Stokes' Theorem to calculate the flux of the curl of the field $\mathbf{F}$ across the surface S.

19. $\mathbf{F} = 2z\mathbf{i} + 3x\mathbf{j} + 5y\mathbf{k}$

S: $\mathbf{r}(r,\theta) = (r\cos\theta)\mathbf{i} + (r\sin\theta)\mathbf{j} + (4 - r^2)\mathbf{k}$,
$0 \le r \le 2$, $0 \le \theta \le 2\pi$,
in the direction away from the origin.

20. $\mathbf{F} = (y - z)\mathbf{i} + (z - x)\mathbf{j} + (x + z)\mathbf{k}$

S: $\mathbf{r}(r,\theta) = (r\cos\theta)\mathbf{i} + (r\sin\theta)\mathbf{j} + (9 - r^2)\mathbf{k}$,
$0 \le r \le 3$, $0 \le \theta \le 2\pi$,
in the direction away from the origin.

21. $\mathbf{F} = x^2y\mathbf{i} + 2y^3z\mathbf{j} + 3z\mathbf{k}$

S: $\mathbf{r}(r,\theta) = (r\cos\theta)\mathbf{i} + (r\sin\theta)\mathbf{j} + r\mathbf{k}$,
$0 \le r \le 1$, $0 \le \theta \le 2\pi$,
in the direction away from the z-axis.

22. $\mathbf{F} = (x - y)\mathbf{i} + (y - z)\mathbf{j} + (z - x)\mathbf{k}$

S: $\mathbf{r}(r,\theta) = (r\cos\theta)\mathbf{i} + (r\sin\theta)\mathbf{j} + (5 - r)\mathbf{k}$,
$0 \le r \le 5$, $0 \le \theta \le 2\pi$,
in the direction away from the z-axis.

23. $\mathbf{F} = 3y\mathbf{i} + (5 - 2x)\mathbf{j} + (z^2 - 2)\mathbf{k}$

S: $\mathbf{r}(\phi,\theta) = (\sqrt{3}\sin\phi\cos\theta)\mathbf{i} + (\sqrt{3}\sin\phi\sin\theta)\mathbf{j} + (\sqrt{3}\cos\phi)\mathbf{k}$, $0 \le \phi \le \pi/2$, $0 \le \theta \le 2\pi$,
in the direction away from the origin.

24. $\mathbf{F} = y^2\mathbf{i} + z^2\mathbf{j} + x\mathbf{k}$

S: $\mathbf{r}(\phi,\theta) = (2\sin\phi\cos\theta)\mathbf{i} + (2\sin\phi\sin\theta)\mathbf{j} + (2\cos\phi)\mathbf{k}$,
$0 \le \phi \le \pi/2$, $0 \le \theta \le 2\pi$,
in the direction away from the origin.

Theory and Examples

25. Let C be the smooth curve $\mathbf{r}(t) = (2\cos t)\mathbf{i} + (2\sin t)\mathbf{j} + (3 - 2\cos^3 t)\mathbf{k}$, oriented to be traversed counterclockwise around the z-axis when viewed from above. Let S be the piecewise smooth cylindrical surface $x^2 + y^2 = 4$, below the curve for $z \ge 0$, together with the base disk in the xy-plane. Note that C lies on the cylinder S and above the xy-plane (see the accompanying figure). Verify Equation (4) in Stokes' Theorem for the vector field $\mathbf{F} = y\mathbf{i} - x\mathbf{j} + x^2\mathbf{k}$.

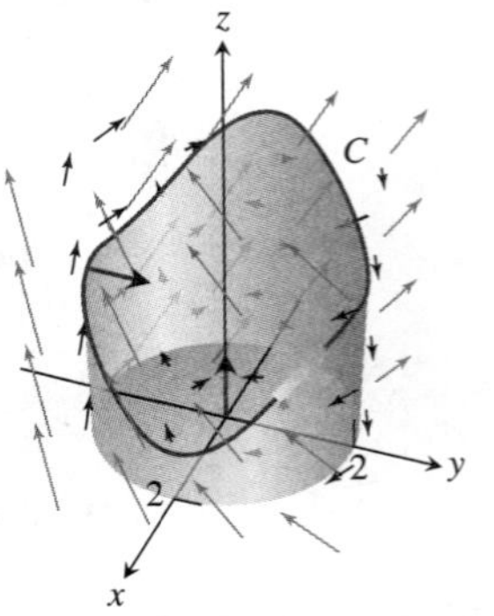

26. Verify Stokes' Theorem for the vector field $\mathbf{F} = 2xy\mathbf{i} + x\mathbf{j} + (y + z)\mathbf{k}$ and surface $z = 4 - x^2 - y^2$, $z \ge 0$, oriented with unit normal $\mathbf{n}$ pointing upward.

27. Zero circulation Use Equation (8) and Stokes' Theorem to show that the circulations of the following fields around the boundary of any smooth orientable surface in space are zero.

a. $\mathbf{F} = 2x\mathbf{i} + 2y\mathbf{j} + 2z\mathbf{k}$ **b.** $\mathbf{F} = \nabla(xy^2z^3)$

c. $\mathbf{F} = \nabla \times (x\mathbf{i} + y\mathbf{j} + z\mathbf{k})$ **d.** $\mathbf{F} = \nabla f$

28. Zero circulation Let $f(x, y, z) = (x^2 + y^2 + z^2)^{-1/2}$. Show that the clockwise circulation of the field $\mathbf{F} = \nabla f$ around the circle $x^2 + y^2 = a^2$ in the xy-plane is zero

a. by taking $\mathbf{r} = (a\cos t)\mathbf{i} + (a\sin t)\mathbf{j}$, $0 \le t \le 2\pi$, and integrating $\mathbf{F}\cdot d\mathbf{r}$ over the circle.

b. by applying Stokes' Theorem.

29. Let C be a simple closed smooth curve in the plane $2x + 2y + z = 2$, oriented as shown here. Show that

$$\oint_C 2y\,dx + 3z\,dy - x\,dz$$

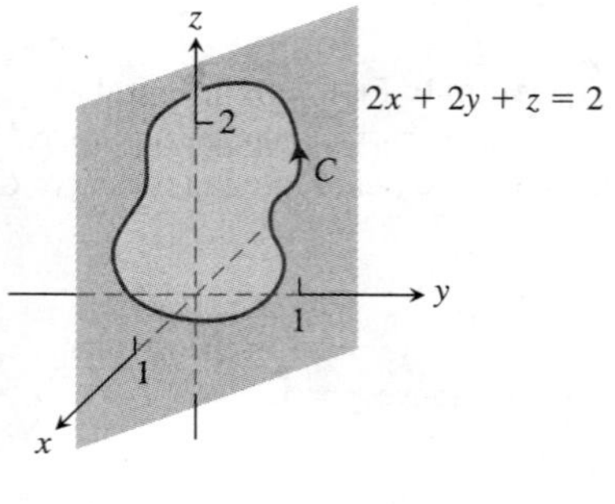

depends only on the area of the region enclosed by C and not on the position or shape of C.

30. Show that if $\mathbf{F} = x\mathbf{i} + y\mathbf{j} + z\mathbf{k}$, then $\nabla \times \mathbf{F} = \mathbf{0}$.

31. Find a vector field with twice-differentiable components whose curl is $x\mathbf{i} + y\mathbf{j} + z\mathbf{k}$, or prove that no such field exists.

32. Does Stokes' Theorem say anything special about circulation in a field whose curl is zero? Give reasons for your answer.

33. Let R be a region in the xy-plane that is bounded by a piecewise smooth simple closed curve C, and suppose that the density is $\delta = 1$ and the moments of inertia of R about the x- and y-axes are known to be I_x and I_y. Evaluate the integral

$$\oint_C \nabla(r^4) \cdot \mathbf{n}\, ds,$$

where $r = \sqrt{x^2 + y^2}$, in terms of I_x and I_y.

34. Zero curl, yet the field is not conservative Show that the curl of

$$\mathbf{F} = \frac{-y}{x^2 + y^2}\mathbf{i} + \frac{x}{x^2 + y^2}\mathbf{j} + z\mathbf{k}$$

is zero but that

$$\oint_C \mathbf{F} \cdot d\mathbf{r}$$

is not zero if C is the circle $x^2 + y^2 = 1$ in the xy-plane. (Theorem 7 does not apply here because the domain of $\mathbf{F}$ is not simply connected. The field $\mathbf{F}$ is not defined along the z-axis, so there is no way to contract C to a point without leaving the domain of $\mathbf{F}$.)

16.8 The Divergence Theorem and a Unified Theory

The divergence form of Green's Theorem in the plane states that the net outward flux of a vector field across a simple closed curve can be calculated by integrating the divergence of the field over the region enclosed by the curve. The corresponding theorem in three dimensions, called the *Divergence Theorem*, states that the net outward flux of a vector field across a closed surface in space can be calculated by integrating the divergence of the field over the solid region enclosed by the surface. In this section we prove the Divergence Theorem and show how it simplifies the calculation of flux, which is the integral of the field over the closed oriented surface. We also derive Gauss's law for flux in an electric field and the continuity equation of hydrodynamics. Finally, we summarize the chapter's vector integral theorems in a single unifying principle generalizing the Fundamental Theorem of Calculus.

Divergence in Three Dimensions

The **divergence** of a vector field $\mathbf{F} = M(x, y, z)\mathbf{i} + N(x, y, z)\mathbf{j} + P(x, y, z)\mathbf{k}$ is the scalar function

$$\text{div}\,\mathbf{F} = \nabla \cdot \mathbf{F} = \frac{\partial M}{\partial x} + \frac{\partial N}{\partial y} + \frac{\partial P}{\partial z}. \tag{1}$$

The symbol "div **F**" is read as "divergence of **F**" or "div **F**." The notation $\nabla \cdot \mathbf{F}$ is read "del dot **F**."

Div **F** has the same physical interpretation in three dimensions as it has in two. If **F** is the velocity field of a flowing gas, the value of div **F** at a point (x, y, z) is the rate at which the gas is compressing or expanding at (x, y, z). The gas is expanding if div **F** is positive and compressing if div **F** is negative. The divergence is the flux per unit volume, or *flux density*, at the point.

EXAMPLE 1 The following vector fields represent the velocity of a gas flowing in space. Find the divergence of each vector field and interpret its physical meaning. Figure 16.72 displays the vector fields.

(a) Expansion: $\mathbf{F}(x, y, z) = x\mathbf{i} + y\mathbf{j} + z\mathbf{k}$

(b) Compression: $\mathbf{F}(x, y, z) = -x\mathbf{i} - y\mathbf{j} - z\mathbf{k}$

(c) Rotation about the z-axis: $\mathbf{F}(x, y, z) = -y\mathbf{i} + x\mathbf{j}$

(d) Shearing along parallel horizontal planes: $\mathbf{F}(x, y, z) = z\mathbf{j}$

Solution

(a) $\text{div}\,\mathbf{F} = \frac{\partial}{\partial x}(x) + \frac{\partial}{\partial y}(y) + \frac{\partial}{\partial z}(z) = 3$: The gas is undergoing constant uniform expansion at all points.

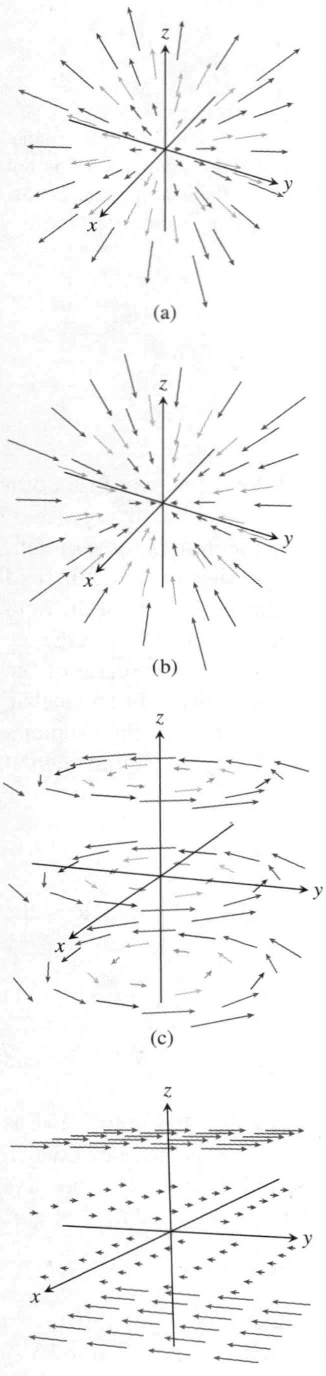

FIGURE 16.72 Velocity fields of a gas flowing in space (Example 1).

(b) $\text{div}\,\mathbf{F} = \dfrac{\partial}{\partial x}(-x) + \dfrac{\partial}{\partial y}(-y) + \dfrac{\partial}{\partial z}(-z) = -3$: The gas is undergoing constant uniform compression at all points.

(c) $\text{div}\,\mathbf{F} = \dfrac{\partial}{\partial x}(-y) + \dfrac{\partial}{\partial y}(x) = 0$: The gas is neither expanding nor compressing at any point.

(d) $\text{div}\,\mathbf{F} = \dfrac{\partial}{\partial y}(z) = 0$: Again, the divergence is zero at all points in the domain of the velocity field, so the gas is neither expanding nor compressing at any point. ■

Divergence Theorem

The Divergence Theorem says that under suitable conditions, the outward flux of a vector field across a closed surface equals the triple integral of the divergence of the field over the three-dimensional region enclosed by the surface.

THEOREM 8—Divergence Theorem
Let $\mathbf{F}$ be a vector field whose components have continuous first partial derivatives, and let S be a piecewise smooth oriented closed surface. The flux of $\mathbf{F}$ across S in the direction of the surface's outward unit normal field $\mathbf{n}$ equals the triple integral of the divergence $\nabla \cdot \mathbf{F}$ over the solid region D enclosed by the surface:

$$\underbrace{\iint_S \mathbf{F} \cdot \mathbf{n}\, d\sigma}_{\text{Outward flux}} = \underbrace{\iiint_D \nabla \cdot \mathbf{F}\, dV.}_{\text{Divergence integral}} \qquad (2)$$

EXAMPLE 2 Evaluate both sides of Equation (2) for the expanding vector field $\mathbf{F} = x\mathbf{i} + y\mathbf{j} + z\mathbf{k}$ over the sphere $x^2 + y^2 + z^2 = a^2$ (Figure 16.73).

Solution The outer unit normal to S, calculated from the gradient of $f(x, y, z) = x^2 + y^2 + z^2 - a^2$, is

$$\mathbf{n} = \frac{2(x\mathbf{i} + y\mathbf{j} + z\mathbf{k})}{\sqrt{4(x^2 + y^2 + z^2)}} = \frac{x\mathbf{i} + y\mathbf{j} + z\mathbf{k}}{a}. \qquad x^2 + y^2 + z^2 = a^2 \text{ on } S$$

It follows that

$$\mathbf{F} \cdot \mathbf{n}\, d\sigma = \frac{x^2 + y^2 + z^2}{a}\, d\sigma = \frac{a^2}{a}\, d\sigma = a\, d\sigma.$$

Therefore, the outward flux is

$$\iint_S \mathbf{F} \cdot \mathbf{n}\, d\sigma = \iint_S a\, d\sigma = a \iint_S d\sigma = a(4\pi a^2) = 4\pi a^3. \qquad \text{Area of } S \text{ is } 4\pi a^2.$$

For the right-hand side of Equation (2), the divergence of $\mathbf{F}$ is

$$\nabla \cdot \mathbf{F} = \frac{\partial}{\partial x}(x) + \frac{\partial}{\partial y}(y) + \frac{\partial}{\partial z}(z) = 3,$$

so we obtain the divergence integral,

$$\iiint_D \nabla \cdot \mathbf{F}\, dV = \iiint_D 3\, dV = 3\left(\frac{4}{3}\pi a^3\right) = 4\pi a^3. \qquad ■$$

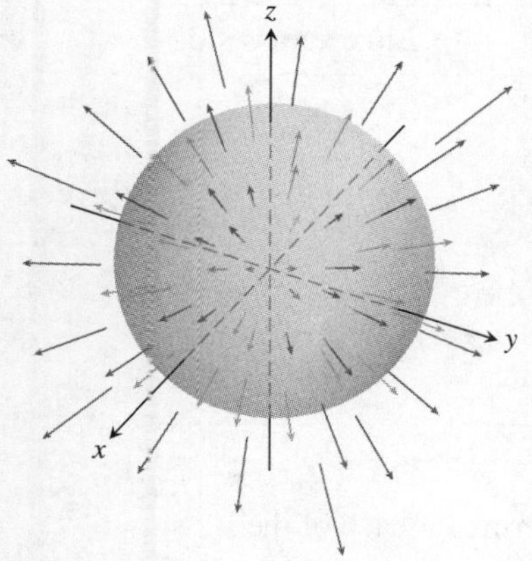

FIGURE 16.73 A uniformly expanding vector field and a sphere (Example 2).

Many vector fields of interest in applied science have zero divergence at each point. A common example is the velocity field of a circulating incompressible liquid, since it is neither expanding nor contracting. Other examples include constant vector fields $\mathbf{F} = a\mathbf{i} + b\mathbf{j} + c\mathbf{k}$, and velocity fields for shearing action along a fixed plane (see Example 1d). If $\mathbf{F}$ is a vector field whose divergence is zero at each point in the region D, then the integral on the right-hand side of Equation (2) equals 0. So if S is any closed surface for which the Divergence Theorem applies, then the outward flux of $\mathbf{F}$ across S is zero. We state this important application of the Divergence Theorem.

> **COROLLARY** The outward flux across a piecewise smooth oriented closed surface S is zero for any vector field $\mathbf{F}$ having zero divergence at every point of the region enclosed by the surface.

EXAMPLE 3 Find the flux of $\mathbf{F} = xy\mathbf{i} + yz\mathbf{j} + xz\mathbf{k}$ outward through the surface of the cube cut from the first octant by the planes $x = 1$, $y = 1$, and $z = 1$.

Solution Instead of calculating the flux as a sum of six separate integrals, one for each face of the cube, we can calculate the flux by integrating the divergence

$$\nabla \cdot \mathbf{F} = \frac{\partial}{\partial x}(xy) + \frac{\partial}{\partial y}(yz) + \frac{\partial}{\partial z}(xz) = y + z + x$$

over the cube's interior:

$$\text{Flux} = \iint\limits_{\substack{\text{Cube}\\ \text{surface}}} \mathbf{F} \cdot \mathbf{n}\, d\sigma = \iiint\limits_{\substack{\text{Cube}\\ \text{interior}}} \nabla \cdot \mathbf{F}\, dV \qquad \text{The Divergence Theorem}$$

$$= \int_0^1 \int_0^1 \int_0^1 (x + y + z)\, dx\, dy\, dz = \frac{3}{2}. \qquad \text{Routine integration}$$ ■

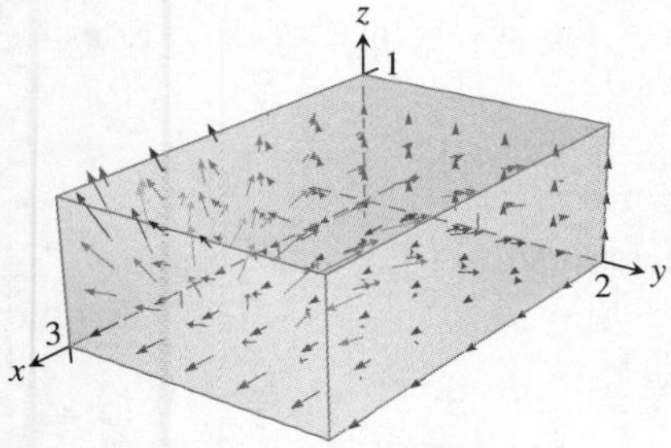

FIGURE 16.74 The integral of div **F** over this region equals the total flux across the six sides (Example 4).

EXAMPLE 4

(a) Calculate the flux of the vector field

$$\mathbf{F} = x^2\mathbf{i} + 4xyz\,\mathbf{j} + ze^x\,\mathbf{k}$$

out of the box-shaped region D: $0 \le x \le 3$, $0 \le y \le 2$, $0 \le z \le 1$. (See Figure 16.74.)

(b) Integrate div **F** over this region and show that the result is the same value as in part (a), as asserted by the Divergence Theorem.

Solution

(a) The region D has six sides. We calculate the flux across each side in turn. Consider the top side in the plane $z = 1$, having outward normal $\mathbf{n} = \mathbf{k}$. The flux across this side is given by $\mathbf{F} \cdot \mathbf{n} = ze^x$. Since $z = 1$ on this side, the flux at a point (x, y, z) on the top is e^x. The total outward flux across this side is given by the surface integral

$$\int_0^2 \int_0^3 e^x\, dx\, dy = 2e^3 - 2. \qquad \text{Routine integration}$$

The outward flux across the other sides is computed similarly, and the results are summarized in the following table.

Side	Unit normal n	$\mathbf{F}\cdot\mathbf{n}$	Flux across side
$x = 0$	$-\mathbf{i}$	$-x^2 = 0$	0
$x = 3$	$\mathbf{i}$	$x^2 = 9$	18
$y = 0$	$-\mathbf{j}$	$-4xyz = 0$	0
$y = 2$	$\mathbf{j}$	$4xyz = 8xz$	18
$z = 0$	$-\mathbf{k}$	$-ze^x = 0$	0
$z = 1$	$\mathbf{k}$	$ze^x = e^x$	$2e^3 - 2$

The total outward flux is obtained by adding the terms for each of the six sides:

$$18 + 18 + 2e^3 - 2 = 34 + 2e^3.$$

(b) We first compute the divergence of **F**, obtaining

$$\operatorname{div}\mathbf{F} = \nabla\cdot\mathbf{F} = 2x + 4xz + e^x.$$

The integral of the divergence of **F** over D is

$$\begin{aligned}\iiint_D \operatorname{div}\mathbf{F}\,dV &= \int_0^1\int_0^2\int_0^3 (2x + 4xz + e^x)\,dx\,dy\,dz\\ &= \int_0^1\int_0^2 (8 + 18z + e^3)\,dy\,dz\\ &= \int_0^1 (16 + 36z + 2e^3)\,dz\\ &= 34 + 2e^3.\end{aligned}$$

As asserted by the Divergence Theorem, the integral of the divergence over D equals the outward flux across the boundary surface of D. ■

Divergence and the Curl

If **F** is a vector field on three-dimensional space, then the curl $\nabla\times\mathbf{F}$ is also a vector field on three-dimensional space. So we can calculate the divergence of $\nabla\times\mathbf{F}$ using Equation (1). The result of this calculation is always 0.

THEOREM 9 If $\mathbf{F} = M\mathbf{i} + N\mathbf{j} + P\mathbf{k}$ is a vector field with continuous second partial derivatives, then

$$\operatorname{div}(\operatorname{curl}\mathbf{F}) = \nabla\cdot(\nabla\times\mathbf{F}) = 0.$$

Proof From the definitions of the divergence and curl, we have

$$\begin{aligned}\operatorname{div}(\operatorname{curl}\mathbf{F}) &= \nabla\cdot(\nabla\times\mathbf{F})\\ &= \frac{\partial}{\partial x}\left(\frac{\partial P}{\partial y} - \frac{\partial N}{\partial z}\right) + \frac{\partial}{\partial y}\left(\frac{\partial M}{\partial z} - \frac{\partial P}{\partial x}\right) + \frac{\partial}{\partial z}\left(\frac{\partial N}{\partial x} - \frac{\partial M}{\partial y}\right)\\ &= \frac{\partial^2 P}{\partial x\,\partial y} - \frac{\partial^2 N}{\partial x\,\partial z} + \frac{\partial^2 M}{\partial y\,\partial z} - \frac{\partial^2 P}{\partial y\,\partial x} + \frac{\partial^2 N}{\partial z\,\partial x} - \frac{\partial^2 M}{\partial z\,\partial y}\\ &= 0,\end{aligned}$$

because the mixed second partial derivatives cancel by the Mixed Derivative Theorem in Section 14.3. ■

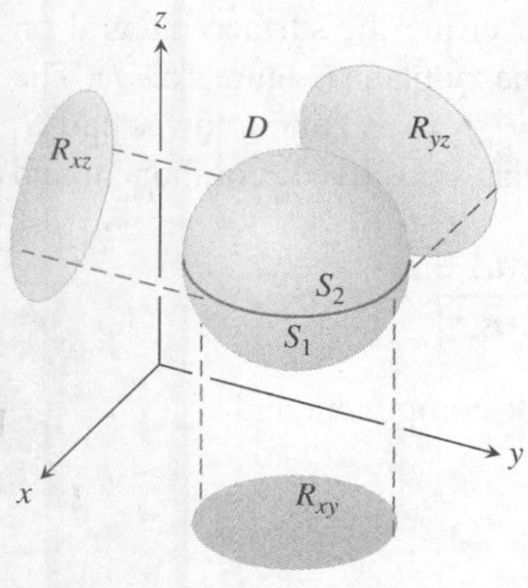

FIGURE 16.75 We prove the Divergence Theorem for the kind of three-dimensional region shown here.

Theorem 9 has some interesting applications. If a vector field $\mathbf{G} = \text{curl } \mathbf{F}$, then the field **G** must have divergence 0. Saying this another way, if $\text{div } \mathbf{G} \neq 0$, then **G** cannot be the curl of any vector field **F** having continuous second partial derivatives. Moreover, if $\mathbf{G} = \text{curl } \mathbf{F}$, then the outward flux of **G** across any closed surface *S* is zero by the corollary to the Divergence Theorem, provided the conditions of the theorem are satisfied. So if there is a closed surface for which the surface integral of the vector field **G** is nonzero, we can conclude that **G** is *not* the curl of some vector field **F**.

Proof of the Divergence Theorem for Special Regions

To prove the Divergence Theorem, we take the components of $\mathbf{F} = M\mathbf{i} + N\mathbf{j} + P\mathbf{k}$ to have continuous first partial derivatives. We first assume that *D* is a convex region with no holes or bubbles, such as a solid ball, cube, or ellipsoid, and that *S* is a piecewise smooth surface. In addition, we assume that any line perpendicular to the *xy*-plane at an interior point of the region R_{xy} that is the projection of *D* on the *xy*-plane intersects the surface *S* in exactly two points, producing surfaces

$$S_1: \quad z = f_1(x, y), \qquad (x, y) \text{ in } R_{xy}$$
$$S_2: \quad z = f_2(x, y), \qquad (x, y) \text{ in } R_{xy},$$

with $f_1 \leq f_2$. We make similar assumptions about the projection of *D* onto the other coordinate planes. See Figure 16.75, which illustrates these assumptions.

The components of the unit normal vector $\mathbf{n} = n_1\mathbf{i} + n_2\mathbf{j} + n_3\mathbf{k}$ are the cosines of the angles α, β, and γ that **n** makes with **i**, **j**, and **k** (Figure 16.76). This is true because all the vectors involved are unit vectors, giving the *direction cosines*

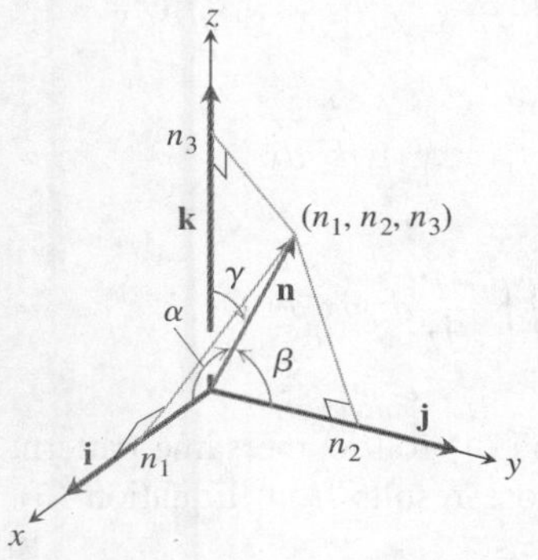

FIGURE 16.76 The components of **n** are the cosines of the angles α, β, and γ that it makes with **i**, **j**, and **k**.

$$n_1 = \mathbf{n} \cdot \mathbf{i} = |\mathbf{n}||\mathbf{i}|\cos\alpha = \cos\alpha$$
$$n_2 = \mathbf{n} \cdot \mathbf{j} = |\mathbf{n}||\mathbf{j}|\cos\beta = \cos\beta$$
$$n_3 = \mathbf{n} \cdot \mathbf{k} = |\mathbf{n}||\mathbf{k}|\cos\gamma = \cos\gamma.$$

Thus the unit normal vector is given by

$$\mathbf{n} = (\cos\alpha)\mathbf{i} + (\cos\beta)\mathbf{j} + (\cos\gamma)\mathbf{k},$$

and

$$\mathbf{F} \cdot \mathbf{n} = M\cos\alpha + N\cos\beta + P\cos\gamma.$$

In component form, the Divergence Theorem states that

$$\iint_S \underbrace{(M\cos\alpha + N\cos\beta + P\cos\gamma)}_{\mathbf{F}\cdot\mathbf{n}}\,d\sigma = \iiint_D \underbrace{\left(\frac{\partial M}{\partial x} + \frac{\partial N}{\partial y} + \frac{\partial P}{\partial z}\right)}_{\text{div } \mathbf{F}} dx\,dy\,dz.$$

We prove the theorem by establishing the following three equations:

$$\iint_S M\cos\alpha\,d\sigma = \iiint_D \frac{\partial M}{\partial x}\,dx\,dy\,dz \tag{3}$$

$$\iint_S N\cos\beta\,d\sigma = \iiint_D \frac{\partial N}{\partial y}\,dx\,dy\,dz \tag{4}$$

$$\iint_S P\cos\gamma\,d\sigma = \iiint_D \frac{\partial P}{\partial z}\,dx\,dy\,dz \tag{5}$$

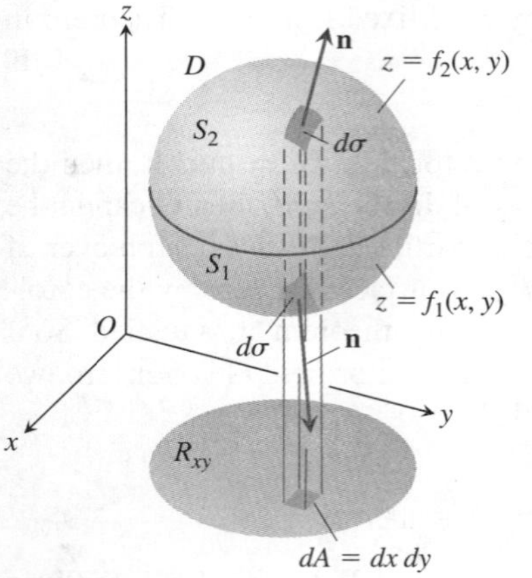

FIGURE 16.77 The region D enclosed by the surfaces S_1 and S_2 projects vertically onto R_{xy} in the xy-plane.

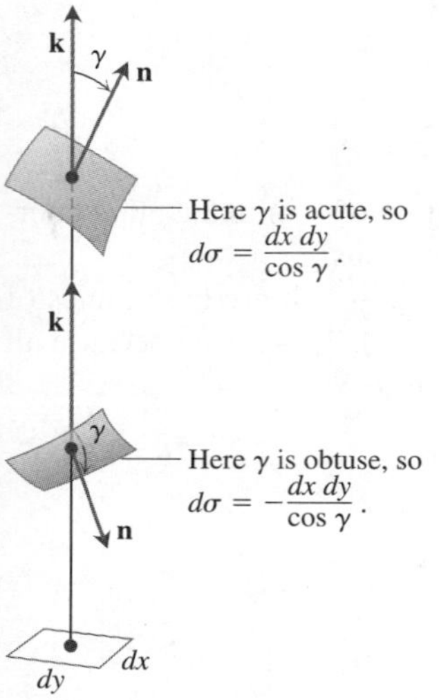

FIGURE 16.78 An enlarged view of the area patches in Figure 16.77. The relations $d\sigma = \pm dx\,dy/\cos\gamma$ come from Eq. (7) in Section 16.5 with $F = \mathbf{F}\cdot\mathbf{n}$.

Proof of Equation (5) We prove Equation (5) by converting the surface integral on the left to a double integral over the projection R_{xy} of D on the xy-plane (Figure 16.77). The surface S consists of an upper part S_2 whose equation is $z = f_2(x, y)$ and a lower part S_1 whose equation is $z = f_1(x, y)$. On S_2, the outer normal $\mathbf{n}$ has a positive $\mathbf{k}$-component and

$$\cos\gamma\, d\sigma = dx\, dy \qquad \text{because} \qquad d\sigma = \frac{dA}{|\cos\gamma|} = \frac{dx\, dy}{\cos\gamma}.$$

See Figure 16.78. On S_1, the outer normal $\mathbf{n}$ has a negative $\mathbf{k}$-component and

$$\cos\gamma\, d\sigma = -dx\, dy.$$

Therefore,

$$\begin{aligned}
\iint_S P\cos\gamma\, d\sigma &= \iint_{S_2} P\cos\gamma\, d\sigma + \iint_{S_1} P\cos\gamma\, d\sigma \\
&= \iint_{R_{xy}} P(x, y, f_2(x, y))\, dx\, dy - \iint_{R_{xy}} P(x, y, f_1(x, y))\, dx\, dy \\
&= \iint_{R_{xy}} [P(x, y, f_2(x, y)) - P(x, y, f_1(x, y))]\, dx\, dy \\
&= \iint_{R_{xy}} \left[\int_{f_1(x,y)}^{f_2(x,y)} \frac{\partial P}{\partial z} dz\right] dx\, dy = \iiint_D \frac{\partial P}{\partial z} dz\, dx\, dy.
\end{aligned}$$

This proves Equation (5). The proofs for Equations (3) and (4) follow the same pattern; just permute x, y, z; M, N, P; α, β, γ, in order, and get those results from Equation (5). This proves the Divergence Theorem for these special regions. ■

Divergence Theorem for Other Regions

The Divergence Theorem can be extended to regions that can be partitioned into a finite number of simple regions of the type just discussed and to regions that can be defined as limits of simpler regions in certain ways. For an example of one step in such a splitting process, suppose that D is the region between two concentric spheres and that $\mathbf{F}$ has continuously differentiable components throughout D and on the bounding surfaces. Split D by an equatorial plane and apply the Divergence Theorem to each half separately. The bottom half, D_1, is shown in Figure 16.79. The surface S_1 that bounds D_1 consists of an outer hemisphere, a plane washer-shaped base, and an inner hemisphere. The Divergence Theorem says that

$$\iint_{S_1} \mathbf{F}\cdot\mathbf{n}_1\, d\sigma = \iiint_{D_1} \nabla\cdot\mathbf{F}\, dV. \tag{6}$$

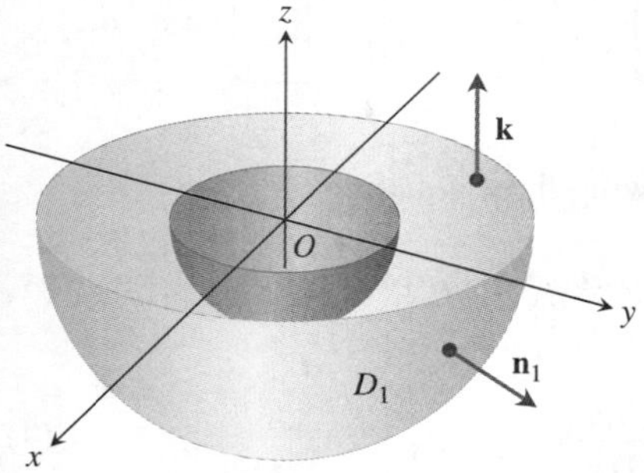

FIGURE 16.79 The lower half of the solid region between two concentric spheres.

The unit normal $\mathbf{n}_1$ that points outward from D_1 points away from the origin along the outer surface, equals $\mathbf{k}$ along the flat base, and points toward the origin along the inner surface. Next apply the Divergence Theorem to D_2 and its surface S_2 (Figure 16.80):

$$\iint_{S_2} \mathbf{F}\cdot\mathbf{n}_2\, d\sigma = \iiint_{D_2} \nabla\cdot\mathbf{F}\, dV. \tag{7}$$

As we follow $\mathbf{n}_2$ over S_2, pointing outward from D_2, we see that $\mathbf{n}_2$ equals $-\mathbf{k}$ along the washer-shaped base in the xy-plane, points away from the origin on the outer sphere, and points toward the origin on the inner sphere. When we add Equations (6) and (7), the integrals over the flat base cancel because of the opposite signs of $\mathbf{n}_1$ and $\mathbf{n}_2$. We thus arrive at the result

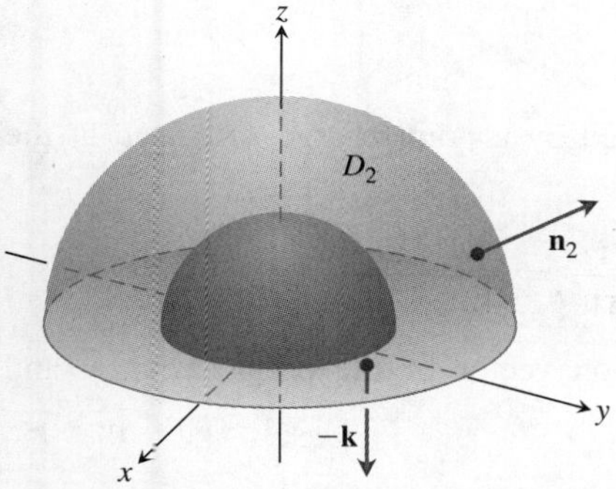

FIGURE 16.80 The upper half of the solid region between two concentric spheres.

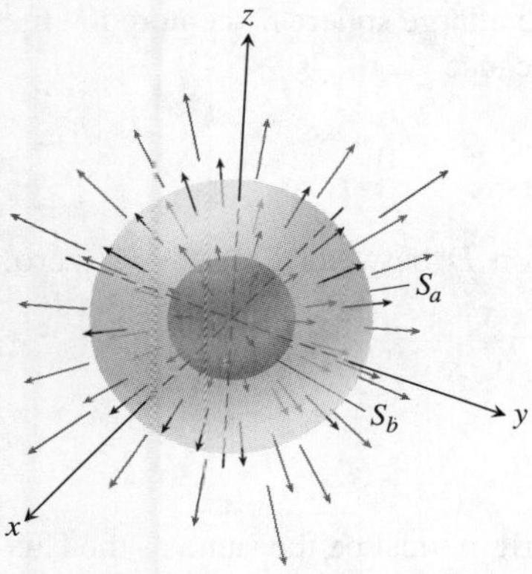

FIGURE 16.81 Two concentric spheres in an expanding vector field. The outer sphere S_a surrounds the inner sphere S_b.

$$\iint_S \mathbf{F} \cdot \mathbf{n}\, d\sigma = \iiint_D \nabla \cdot \mathbf{F}\, dV,$$

with D the region between the spheres, S the boundary of D consisting of two spheres, and $\mathbf{n}$ the unit normal to S directed outward from D.

EXAMPLE 5 Find the net outward flux of the field

$$\mathbf{F} = \frac{x\mathbf{i} + y\mathbf{j} + z\mathbf{k}}{\rho^3}, \qquad \rho = \sqrt{x^2 + y^2 + z^2} \tag{8}$$

across the boundary of the region D: $0 < b^2 \le x^2 + y^2 + z^2 \le a^2$ (Figure 16.81).

Solution The flux can be calculated by integrating $\nabla \cdot \mathbf{F}$ over D. Note that $\rho \neq 0$ in D. We have

$$\frac{\partial \rho}{\partial x} = \frac{1}{2}(x^2 + y^2 + z^2)^{-1/2}(2x) = \frac{x}{\rho}$$

and

$$\frac{\partial M}{\partial x} = \frac{\partial}{\partial x}(x\rho^{-3}) = \rho^{-3} - 3x\rho^{-4}\frac{\partial \rho}{\partial x} = \frac{1}{\rho^3} - \frac{3x^2}{\rho^5}.$$

Similarly,

$$\frac{\partial N}{\partial y} = \frac{1}{\rho^3} - \frac{3y^2}{\rho^5} \quad \text{and} \quad \frac{\partial P}{\partial z} = \frac{1}{\rho^3} - \frac{3z^2}{\rho^5}.$$

Hence,

$$\operatorname{div} \mathbf{F} = \frac{\partial M}{\partial x} + \frac{\partial N}{\partial y} + \frac{\partial P}{\partial z} = \frac{3}{\rho^3} - \frac{3}{\rho^5}(x^2 + y^2 + z^2) = \frac{3}{\rho^3} - \frac{3\rho^2}{\rho^5} = 0.$$

So the net outward flux of $\mathbf{F}$ across the boundary of D is zero by the corollary to the Divergence Theorem. There is more to learn about this vector field $\mathbf{F}$, though. The flux leaving D across the inner sphere S_b is the negative of the flux leaving D across the outer sphere S_a (because the sum of these fluxes is zero). Hence, the flux of $\mathbf{F}$ across S_b in the direction away from the origin equals the flux of $\mathbf{F}$ across S_a in the direction away from the origin. Thus, the flux of $\mathbf{F}$ across a sphere centered at the origin is independent of the radius of the sphere. What is this flux?

To find it, we evaluate the flux integral directly for an arbitrary sphere S_a. The outward unit normal on the sphere of radius a is

$$\mathbf{n} = \frac{x\mathbf{i} + y\mathbf{j} + z\mathbf{k}}{\sqrt{x^2 + y^2 + z^2}} = \frac{x\mathbf{i} + y\mathbf{j} + z\mathbf{k}}{a}.$$

Hence, on the sphere,

$$\mathbf{F} \cdot \mathbf{n} = \frac{x\mathbf{i} + y\mathbf{j} + z\mathbf{k}}{a^3} \cdot \frac{x\mathbf{i} + y\mathbf{j} + z\mathbf{k}}{a} = \frac{x^2 + y^2 + z^2}{a^4} = \frac{a^2}{a^4} = \frac{1}{a^2}$$

and

$$\iint_{S_a} \mathbf{F} \cdot \mathbf{n}\, d\sigma = \frac{1}{a^2}\iint_{S_a} d\sigma = \frac{1}{a^2}(4\pi a^2) = 4\pi.$$

The outward flux of $\mathbf{F}$ in Equation (8) across any sphere centered at the origin is 4π. This result does not contradict the Divergence Theorem because $\mathbf{F}$ is not continuous at the origin. ■

Gauss's Law: One of the Four Great Laws of Electromagnetic Theory

In electromagnetic theory, the electric field created by a point charge q located at the origin is

$$\mathbf{E}(x, y, z) = \frac{1}{4\pi\varepsilon_0}\frac{q}{|\mathbf{r}|^2}\left(\frac{\mathbf{r}}{|\mathbf{r}|}\right) = \frac{q}{4\pi\varepsilon_0}\frac{\mathbf{r}}{|\mathbf{r}|^3} = \frac{q}{4\pi\varepsilon_0}\frac{x\mathbf{i} + y\mathbf{j} + z\mathbf{k}}{\rho^3},$$

where ε_0 is a physical constant, $\mathbf{r}$ is the position vector of the point (x, y, z), and $\rho = |\mathbf{r}| = \sqrt{x^2 + y^2 + z^2}$. From Equation (8),

$$\mathbf{E} = \frac{q}{4\pi\varepsilon_0}\mathbf{F}.$$

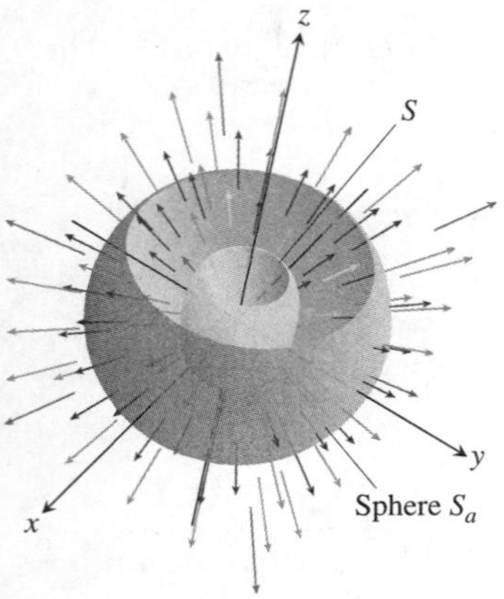

FIGURE 16.82 A sphere S_a surrounding another surface S. The tops of the surfaces are removed for visualization.

The calculations in Example 5 show that the outward flux of **E** across any sphere centered at the origin is q/ε_0, but this result is not confined to spheres. The outward flux of **E** across any closed surface S that encloses the origin (and to which the Divergence Theorem applies) is also q/ε_0. To see why, we have only to imagine a large sphere S_a centered at the origin and enclosing the surface S (see Figure 16.82). Because

$$\nabla \cdot \mathbf{E} = \nabla \cdot \frac{q}{4\pi\varepsilon_0}\mathbf{F} = \frac{q}{4\pi\varepsilon_0}\nabla \cdot \mathbf{F} = 0$$

when $\rho > 0$, the triple integral of $\nabla \cdot \mathbf{E}$ over the region D between S and S_a is zero. Hence, by the Divergence Theorem,

$$\iint_{\substack{\text{Boundary}\\ \text{of } D}} \mathbf{E} \cdot \mathbf{n}\, d\sigma = 0.$$

So the flux of **E** across S in the direction away from the origin must be the same as the flux of **E** across S_a in the direction away from the origin, which is q/ε_0. This statement, called *Gauss's law*, also applies to charge distributions that are more general than the one assumed here, as shown in most physics texts. For any closed surface that encloses the origin, we have

$$\text{Gauss's law:} \quad \iint_S \mathbf{E} \cdot \mathbf{n}\, d\sigma = \frac{q}{\varepsilon_0}.$$

Continuity Equation of Hydrodynamics

Let D be a region in space bounded by a closed oriented surface S. If $\mathbf{v}(x, y, z)$ is the velocity field of a fluid flowing smoothly through D, $\delta = \delta(t, x, y, z)$ is the fluid's density at (x, y, z) at time t, and $\mathbf{F} = \delta\mathbf{v}$, then the **continuity equation** of hydrodynamics states that

$$\nabla \cdot \mathbf{F} + \frac{\partial \delta}{\partial t} = 0.$$

If the functions involved have continuous first partial derivatives, the equation evolves naturally from the Divergence Theorem, as we now demonstrate.

First, the integral

$$\iint_S \mathbf{F} \cdot \mathbf{n}\, d\sigma$$

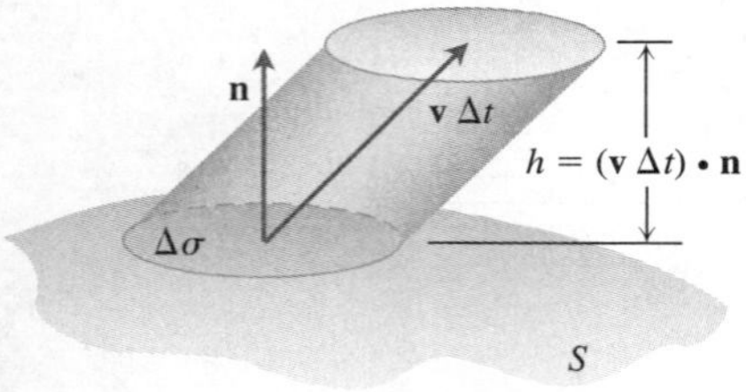

FIGURE 16.83 The fluid that flows upward through the patch $\Delta\sigma$ in a short time Δt fills a "cylinder" whose volume is approximately base × height = $\mathbf{v} \cdot \mathbf{n}\, \Delta\sigma\, \Delta t$.

is the rate at which mass leaves D across S (leaves because **n** is the outer normal). To see why, consider a patch of area $\Delta\sigma$ on the surface (Figure 16.83). In a short time interval Δt, the volume ΔV of fluid that flows across the patch is approximately equal to the volume of a cylinder with base area $\Delta\sigma$ and height $(\mathbf{v}\,\Delta t) \cdot \mathbf{n}$, where **v** is a velocity vector rooted at a point of the patch:

$$\Delta V \approx \mathbf{v} \cdot \mathbf{n}\, \Delta\sigma\, \Delta t.$$

The mass of this volume of fluid is about

$$\Delta m \approx \delta \mathbf{v} \cdot \mathbf{n}\, \Delta\sigma\, \Delta t,$$

so the rate at which mass is flowing out of D across the patch is about

$$\frac{\Delta m}{\Delta t} \approx \delta \mathbf{v} \cdot \mathbf{n}\, \Delta\sigma.$$

This leads to the approximation

$$\frac{\sum \Delta m}{\Delta t} \approx \sum \delta \mathbf{v} \cdot \mathbf{n}\, \Delta\sigma$$

as an estimate of the average rate at which mass flows across S. Finally, letting $\Delta\sigma \to 0$ and $\Delta t \to 0$ gives the instantaneous rate at which mass leaves D across S as

$$\frac{dm}{dt} = \iint_S \delta \mathbf{v} \cdot \mathbf{n}\, d\sigma,$$

which for our particular flow is

$$\frac{dm}{dt} = \iint_S \mathbf{F} \cdot \mathbf{n}\, d\sigma.$$

Now let B be a solid sphere centered at a point Q in the flow. The average value of $\nabla \cdot \mathbf{F}$ over B is

$$\frac{1}{\text{volume of } B} \iiint_B \nabla \cdot \mathbf{F}\, dV.$$

It is a consequence of the continuity of the divergence that $\nabla \cdot \mathbf{F}$ actually takes on this value at some point P in B. Thus, by the Divergence Theorem Equation (2),

$$\begin{aligned}(\nabla \cdot \mathbf{F})(P) &= \frac{1}{\text{volume of } B} \iiint_B \nabla \cdot \mathbf{F}\, dV = \frac{\iint_S \mathbf{F} \cdot \mathbf{n}\, d\sigma}{\text{volume of } B} \\ &= \frac{\text{rate at which mass leaves } B \text{ across its surface } S}{\text{volume of } B}. \qquad (9)\end{aligned}$$

The last term of the equation describes decrease in mass per unit volume.

Now let the radius of B approach zero while the center Q stays fixed. The left side of Equation (9) converges to $(\nabla \cdot \mathbf{F})_Q$, and the right side converges to $(-\partial\delta/\partial t)_Q$, since $\delta = m/V$. The equality of these two limits is the continuity equation

$$\nabla \cdot \mathbf{F} = -\frac{\partial \delta}{\partial t}.$$

The continuity equation "explains" $\nabla \cdot \mathbf{F}$: The divergence of $\mathbf{F}$ at a point is the rate at which the density of the fluid is decreasing there. The Divergence Theorem

$$\iint_S \mathbf{F} \cdot \mathbf{n}\, d\sigma = \iiint_D \nabla \cdot \mathbf{F}\, dV$$

now says that the net decrease in density of the fluid in region D (divergence integral) is accounted for by the mass transported across the surface S (outward flux integral). So, the theorem is a statement about conservation of mass (Exercise 35).

Unifying the Integral Theorems

If we think of a two-dimensional field $\mathbf{F} = M(x, y)\mathbf{i} + N(x, y)\mathbf{j}$ as a three-dimensional field whose $\mathbf{k}$-component is zero, then $\nabla \cdot \mathbf{F} = (\partial M/\partial x) + (\partial N/\partial y)$, and the normal form of Green's Theorem can be written as

$$\oint_C \mathbf{F} \cdot \mathbf{n}\, ds = \iint_R \left(\frac{\partial M}{\partial x} + \frac{\partial N}{\partial y}\right) dx\, dy = \iint_R \nabla \cdot \mathbf{F}\, dA.$$

Similarly, $(\nabla \times \mathbf{F}) \cdot \mathbf{k} = (\partial N/\partial x) - (\partial M/\partial y)$, so the tangential form of Green's Theorem can be written as

$$\oint_C \mathbf{F} \cdot \mathbf{T}\, ds = \iint_R \left(\frac{\partial N}{\partial x} - \frac{\partial M}{\partial y}\right) dx\, dy = \iint_R (\nabla \times \mathbf{F}) \cdot \mathbf{k}\, dA.$$

With the equations of Green's Theorem expressed in del notation, we can see their relationships to the equations in Stokes' Theorem and the Divergence Theorem, all summarized here.

Green's Theorem and Its Generalization to Three Dimensions

Tangential form of Green's Theorem: $\oint_C \mathbf{F} \cdot \mathbf{T}\, ds = \iint_R (\nabla \times \mathbf{F}) \cdot \mathbf{k}\, dA$

Stokes' Theorem: $\oint_C \mathbf{F} \cdot \mathbf{T}\, ds = \iint_S (\nabla \times \mathbf{F}) \cdot \mathbf{n}\, d\sigma$

Normal form of Green's Theorem: $\oint_C \mathbf{F} \cdot \mathbf{n}\, ds = \iint_R \nabla \cdot \mathbf{F}\, dA$

Divergence Theorem: $\iint_S \mathbf{F} \cdot \mathbf{n}\, d\sigma = \iiint_D \nabla \cdot \mathbf{F}\, dV$

Notice how Stokes' Theorem generalizes the tangential (curl) form of Green's Theorem from a flat surface in the plane to a surface in three-dimensional space. In each case, the surface integral of curl $\mathbf{F}$ over the interior of the oriented surface equals the circulation of $\mathbf{F}$ around the boundary.

Likewise, the Divergence Theorem generalizes the normal (flux) form of Green's Theorem from a two-dimensional region in the plane to a three-dimensional region in space. In each case, the integral of $\nabla \cdot \mathbf{F}$ over the interior of the region equals the total flux of the field across the boundary enclosing the region.

All these results can be thought of as forms of a *single fundamental theorem.* The Fundamental Theorem of Calculus in Section 5.4 says that if $f(x)$ is differentiable on (a, b) and continuous on $[a, b]$, then

$$\int_a^b \frac{df}{dx}\, dx = f(b) - f(a).$$

FIGURE 16.84 The outward unit normals at the boundary of $[a, b]$ in one-dimensional space.

If we let $\mathbf{F} = f(x)\mathbf{i}$ throughout $[a, b]$, then $df/dx = \nabla \cdot \mathbf{F}$. If we define the unit vector field $\mathbf{n}$ normal to the boundary of $[a, b]$ to be $\mathbf{i}$ at b and $-\mathbf{i}$ at a (Figure 16.84), then

$$\begin{aligned} f(b) - f(a) &= f(b)\mathbf{i} \cdot (\mathbf{i}) + f(a)\mathbf{i} \cdot (-\mathbf{i}) \\ &= \mathbf{F}(b) \cdot \mathbf{n} + \mathbf{F}(a) \cdot \mathbf{n} \\ &= \text{total outward flux of } \mathbf{F} \text{ across the boundary of } [a, b]. \end{aligned}$$

The Fundamental Theorem now says that

$$\mathbf{F}(b)\cdot\mathbf{n} + \mathbf{F}(a)\cdot\mathbf{n} = \int_{[a,b]} \nabla\cdot\mathbf{F}\,dx.$$

The Fundamental Theorem of Calculus, the normal form of Green's Theorem, and the Divergence Theorem all say that the integral of the differential operator $\nabla\,\cdot$ operating on a field $\mathbf{F}$ over a region equals the sum of the normal field components over the boundary enclosing the region. (Here we are interpreting the line integral in Green's Theorem and the surface integral in the Divergence Theorem as "sums" over the boundary.)

Stokes' Theorem and the tangential form of Green's Theorem say that, when things are properly oriented, the surface integral of the differential operator $\nabla\,\times$ operating on a field equals the sum of the tangential field components over the boundary of the surface.

The beauty of these interpretations is the observance of a single unifying principle, which we can state as follows.

A Unifying Fundamental Theorem of Vector Integral Calculus
The integral of a differential operator acting on a field over a region equals the sum of the field components appropriate to the operator over the boundary of the region.

EXERCISES 16.8

Calculating Divergence

In Exercises 1–8, find the divergence of the field.

1. $\mathbf{F} = (x - y + z)\mathbf{i} + (2x + y - z)\mathbf{j} + (3x + 2y - 2z)\mathbf{k}$
2. $\mathbf{F} = (x \ln y)\mathbf{i} + (y \ln z)\mathbf{j} + (z \ln x)\mathbf{k}$
3. $\mathbf{F} = ye^{xyz}\mathbf{i} + ze^{xyz}\mathbf{j} + xe^{xyz}\mathbf{k}$
4. $\mathbf{F} = \sin(xy)\mathbf{i} + \cos(yz)\mathbf{j} + \tan(xz)\mathbf{k}$
5. The spin field in Figure 16.13
6. The radial field in Figure 16.12
7. The gravitational field in Figure 16.9 and Exercise 38a in Section 16.3
8. The velocity field $\mathbf{v}(x, y, z) = (a^2 - x^2 - y^2)\mathbf{k}$ in Figure 16.14

Calculating Flux Using the Divergence Theorem

In Exercises 9–20, use the Divergence Theorem to find the outward flux of $\mathbf{F}$ across the boundary of the region D.

9. **Cube** $\mathbf{F} = (y - x)\mathbf{i} + (z - y)\mathbf{j} + (y - x)\mathbf{k}$
   D: The cube bounded by the planes $x = \pm 1$, $y = \pm 1$, and $z = \pm 1$
10. $\mathbf{F} = x^2\mathbf{i} + y^2\mathbf{j} + z^2\mathbf{k}$
    a. **Cube** D: The cube cut from the first octant by the planes $x = 1$, $y = 1$, and $z = 1$
    b. **Cube** D: The cube bounded by the planes $x = \pm 1$, $y = \pm 1$, and $z = \pm 1$
    c. **Cylindrical can** D: The region cut from the solid cylinder $x^2 + y^2 \le 4$ by the planes $z = 0$ and $z = 1$
11. **Cylinder and paraboloid** $\mathbf{F} = y\mathbf{i} + xy\mathbf{j} - z\mathbf{k}$
    D: The region inside the solid cylinder $x^2 + y^2 \le 4$ between the plane $z = 0$ and the paraboloid $z = x^2 + y^2$
12. **Ball** $\mathbf{F} = x^2\mathbf{i} + xz\mathbf{j} + 3z\mathbf{k}$
    D: The ball $x^2 + y^2 + z^2 \le 4$
13. **Portion of sphere** $\mathbf{F} = 4xz\mathbf{i} - y^2\mathbf{j} + yz\,\mathbf{k}$
    D: The cube bounded by $x = 0, x = 1, y = 0, y = 1, z = 0, z = 1$
14. **Cylindrical can** $\mathbf{F} = (6x^2 + 2xy)\mathbf{i} + (2y + x^2z)\mathbf{j} + 4x^2y^3\mathbf{k}$
    D: The region cut from the first octant by the cylinder $x^2 + y^2 = 4$ and the plane $z = 3$
15. **Wedge** $\mathbf{F} = 2xz\mathbf{i} - xy\mathbf{j} - z^2\mathbf{k}$
    D: The wedge cut from the first octant by the plane $y + z = 4$ and the elliptic cylinder $4x^2 + y^2 = 16$
16. **Ball** $\mathbf{F} = x^3\mathbf{i} + y^3\mathbf{j} + z^3\mathbf{k}$
    D: The ball $x^2 + y^2 + z^2 \le a^2$
17. **Thick sphere** $\mathbf{F} = \sqrt{x^2 + y^2 + z^2}\,(x\mathbf{i} + y\mathbf{j} + z\mathbf{k})$
    D: The region $1 \le x^2 + y^2 + z^2 \le 2$
18. **Thick sphere** $\mathbf{F} = 4x\mathbf{i} - 2y^2\mathbf{j} + z^2\,\mathbf{k}$
    D: The region bounded by the solid cylinder $x^2 + y^2 = 4$ and the planes $z = 0, z = 3$
19. **Thick sphere** $\mathbf{F} = (5x^3 + 12xy^2)\mathbf{i} + (y^3 + e^y \sin z)\mathbf{j} + (5z^3 + e^y \cos z)\mathbf{k}$
    D: The solid region between the spheres $x^2 + y^2 + z^2 = 1$ and $x^2 + y^2 + z^2 = 2$
20. **Thick cylinder** $\mathbf{F} = \ln(x^2 + y^2)\mathbf{i} - \left(\frac{2z}{x}\arctan\frac{y}{x}\right)\mathbf{j} + z\sqrt{x^2 + y^2}\,\mathbf{k}$
    D: The thick-walled cylinder $1 \le x^2 + y^2 \le 2$, $-1 \le z \le 2$

Theory and Examples

21. **a.** Show that the outward flux of the position vector field $\mathbf{F} = x\mathbf{i} + y\mathbf{j} + z\mathbf{k}$ through a smooth closed surface S is three times the volume of the region enclosed by the surface.

b. Let $\mathbf{n}$ be the outward unit normal vector field on S. Show that it is not possible for $\mathbf{F}$ to be orthogonal to $\mathbf{n}$ at every point of S.

22. The base of the closed cubelike surface shown here is the unit square in the xy-plane. The four sides lie in the planes $x = 0$, $x = 1$, $y = 0$, and $y = 1$. The top is an arbitrary smooth surface whose identity is unknown. Let $\mathbf{F} = x\mathbf{i} - 2y\mathbf{j} + (z + 3)\mathbf{k}$, and suppose the outward flux of $\mathbf{F}$ through Side A is 1 and through Side B is -3. Can you conclude anything about the outward flux through the top? Give reasons for your answer.

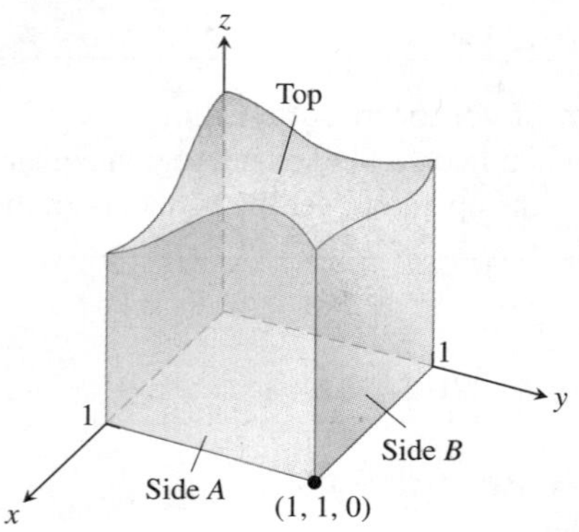

23. Let $\mathbf{F} = (y \cos 2x)\mathbf{i} + (y^2 \sin 2x)\mathbf{j} + (x^2 y + z)\mathbf{k}$. Is there a vector field $\mathbf{A}$ such that $\mathbf{F} = \nabla \times \mathbf{A}$? Explain your answer.

24. Outward flux of a gradient field Let S be the surface of the portion of the ball $x^2 + y^2 + z^2 \le a^2$ that lies in the first octant, and let $f(x, y, z) = \ln \sqrt{x^2 + y^2 + z^2}$. Calculate

$$\iint_S \nabla f \cdot \mathbf{n}\, d\sigma.$$

($\nabla f \cdot \mathbf{n}$ is the derivative of f in the direction of outward normal $\mathbf{n}$.)

25. Let $\mathbf{F}$ be a field whose components have continuous first partial derivatives throughout a portion of space containing a region D bounded by a smooth closed surface S. If $|\mathbf{F}| \le 1$, can any bound be placed on the size of

$$\iiint_D \nabla \cdot \mathbf{F}\, dV?$$

Give reasons for your answer.

26. Maximum flux Among all rectangular boxes defined by the inequalities $0 \le x \le a$, $0 \le y \le b$, $0 \le z \le 1$, find the one for which the total flux of $\mathbf{F} = (-x^2 - 4xy)\mathbf{i} - 6yz\mathbf{j} + 12z\mathbf{k}$ outward through the six sides is greatest. What *is* the greatest flux?

27. Calculate the net outward flux of the vector field

$$\mathbf{F} = xy\mathbf{i} + (\sin xz + y^2)\mathbf{j} + (e^{xy^2} + x)\mathbf{k}$$

over the surface S surrounding the region D bounded by the planes $y = 0$, $z = 0$, $z = 2 - y$ and the parabolic cylinder $z = 1 - x^2$.

28. Compute the net outward flux of the vector field $\mathbf{F} = (x\mathbf{i} + y\mathbf{j} + z\mathbf{k})/(x^2 + y^2 + z^2)^{3/2}$ across the ellipsoid $9x^2 + 4y^2 + 6z^2 = 36$.

29. Let $\mathbf{F}$ be a differentiable vector field, and let $g(x, y, z)$ be a differentiable scalar function. Verify the following identities.

a. $\nabla \cdot (g\mathbf{F}) = g\nabla \cdot \mathbf{F} + \nabla g \cdot \mathbf{F}$

b. $\nabla \times (g\mathbf{F}) = g\nabla \times \mathbf{F} + \nabla g \times \mathbf{F}$

30. Let $\mathbf{F}_1$ and $\mathbf{F}_2$ be differentiable vector fields, and let a and b be arbitrary real constants. Verify the following identities.

a. $\nabla \cdot (a\mathbf{F}_1 + b\mathbf{F}_2) = a\nabla \cdot \mathbf{F}_1 + b\nabla \cdot \mathbf{F}_2$

b. $\nabla \times (a\mathbf{F}_1 + b\mathbf{F}_2) = a\nabla \times \mathbf{F}_1 + b\nabla \times \mathbf{F}_2$

c. $\nabla \cdot (\mathbf{F}_1 \times \mathbf{F}_2) = \mathbf{F}_2 \cdot \nabla \times \mathbf{F}_1 - \mathbf{F}_1 \cdot \nabla \times \mathbf{F}_2$

31. If $\mathbf{F} = M\mathbf{i} + N\mathbf{j} + P\mathbf{k}$ is a differentiable vector field, we define the notation $\mathbf{F} \cdot \nabla$ to mean

$$M\frac{\partial}{\partial x} + N\frac{\partial}{\partial y} + P\frac{\partial}{\partial z}.$$

For differentiable vector fields $\mathbf{F}_1$ and $\mathbf{F}_2$, verify the following identities.

a. $\nabla \times (\mathbf{F}_1 \times \mathbf{F}_2) = (\mathbf{F}_2 \cdot \nabla)\mathbf{F}_1 - (\mathbf{F}_1 \cdot \nabla)\mathbf{F}_2 + (\nabla \cdot \mathbf{F}_2)\mathbf{F}_1 - (\nabla \cdot \mathbf{F}_1)\mathbf{F}_2$

b. $\nabla(\mathbf{F}_1 \cdot \mathbf{F}_2) = (\mathbf{F}_1 \cdot \nabla)\mathbf{F}_2 + (\mathbf{F}_2 \cdot \nabla)\mathbf{F}_1 + \mathbf{F}_1 \times (\nabla \times \mathbf{F}_2) + \mathbf{F}_2 \times (\nabla \times \mathbf{F}_1)$

32. Harmonic functions A function $f(x, y, z)$ is said to be *harmonic* in a region D in space if it satisfies the Laplace equation

$$\nabla^2 f = \nabla \cdot \nabla f = \frac{\partial^2 f}{\partial x^2} + \frac{\partial^2 f}{\partial y^2} + \frac{\partial^2 f}{\partial z^2} = 0$$

throughout D.

a. Suppose that f is harmonic throughout a bounded region D enclosed by a smooth surface S and that $\mathbf{n}$ is the chosen unit normal vector on S. Show that the integral over S of $\nabla f \cdot \mathbf{n}$, the derivative of f in the direction of $\mathbf{n}$, is zero.

b. Show that if f is harmonic on D, then

$$\iint_S f\nabla f \cdot \mathbf{n}\, d\sigma = \iiint_D |\nabla f|^2\, dV.$$

33. Green's first formula Suppose that f and g are scalar functions with continuous first- and second-order partial derivatives throughout a region D that is bounded by a closed piecewise smooth surface S. Show that

$$\iint_S f\nabla g \cdot \mathbf{n}\, d\sigma = \iiint_D (f\nabla^2 g + \nabla f \cdot \nabla g)\, dV. \qquad (10)$$

Equation (10) is **Green's first formula**. (*Hint:* Apply the Divergence Theorem to the field $\mathbf{F} = f\nabla g$.)

34. Green's second formula (*Continuation of Exercise 33.*) Interchange f and g in Equation (10) to obtain a similar formula. Then subtract this formula from Equation (10) to show that

$$\iint_S (f\nabla g - g\nabla f) \cdot \mathbf{n}\, d\sigma = \iiint_D (f\nabla^2 g - g\nabla^2 f)\, dV. \qquad (11)$$

This equation is **Green's second formula**.

35. Conservation of mass Let $\mathbf{v}(t, x, y, z)$ be a continuously differentiable vector field over the region D in space, and let $p(t, x, y, z)$ be a continuously differentiable scalar function. The variable t represents the time domain. The Law of Conservation of Mass asserts that

$$\frac{d}{dt}\iiint_D p(t, x, y, z)\, dV = -\iint_S p\mathbf{v} \cdot \mathbf{n}\, d\sigma,$$

where S is the surface enclosing D.

a. Give a physical interpretation of the conservation of mass law if $\mathbf{v}$ is a velocity flow field and p represents the density of the fluid at point (x, y, z) at time t.

b. Use the Divergence Theorem and Leibniz's Rule,

$$\frac{d}{dt}\iiint_D p(t, x, y, z)\, dV = \iiint_D \frac{\partial p}{\partial t}\, dV,$$

to show that the Law of Conservation of Mass is equivalent to the continuity equation,

$$\nabla \cdot p\mathbf{v} + \frac{\partial p}{\partial t} = 0.$$

(In the first term $\nabla \cdot p\mathbf{v}$, the variable t is held fixed, and in the second term $\partial p/\partial t$, it is assumed that the point (x, y, z) in D is held fixed.)

36. The heat diffusion equation Let $T(t, x, y, z)$ be a function with continuous second derivatives giving the temperature at time t at the point (x, y, z) of a solid occupying a region D in space. If the solid's heat capacity and mass density are denoted by the constants c and ρ, respectively, the quantity $c\rho T$ is called the solid's **heat energy per unit volume**.

Let $-k\nabla T$ denote the **energy flux vector**. (Here the constant k is called the **conductivity**.) Assuming the Law of Conservation of Mass with $-k\nabla T = \mathbf{v}$ and $c\rho T = p$ in Exercise 35, derive the diffusion (heat) equation

$$\frac{\partial T}{\partial t} = K\nabla^2 T,$$

where $K = k/(c\rho) > 0$ is the *diffusivity* constant. (Notice that if $T(t, x)$ represents the temperature at time t at position x in a uniform conducting rod with perfectly insulated sides, then $\nabla^2 T = \partial^2 T/\partial x^2$ and the diffusion equation reduces to the one-dimensional heat equation in Chapter 14's Additional Exercises.)

CHAPTER 16 Questions to Guide Your Review

1. What are line integrals of scalar functions? How are they evaluated? Give examples.
2. How can you use line integrals to find the centers of mass of springs or wires? Explain.
3. What is a vector field? What is the line integral of a vector field? What is a gradient field? Give examples.
4. What is the flow of a vector field along a curve? What is the work done by a vector field moving an object along a curve? How do you calculate the work done? Give examples.
5. What is the Fundamental Theorem of line integrals? Explain how it is related to the Fundamental Theorem of Calculus.
6. Specify three properties that are special about conservative fields. How can you tell when a field is conservative?
7. What is special about path independent fields?
8. What is a potential function? Show by example how to find a potential function for a conservative field.
9. What is a differential form? What does it mean for such a form to be exact? How do you test for exactness? Give examples.
10. What is Green's Theorem? Discuss how the two forms of Green's Theorem extend the Net Change Theorem in Chapter 5.
11. How do you calculate the area of a parametrized surface in space? Of an implicitly defined surface $F(x, y, z) = 0$? Of the surface that is the graph of $z = f(x, y)$? Give examples.
12. How do you integrate a scalar function over a parametrized surface? Over surfaces that are defined implicitly or in explicit form? Give examples.
13. What is an oriented surface? What is the surface integral of a vector field in three-dimensional space over an oriented surface? How is it related to the net outward flux of the field? Give examples.
14. What is the curl of a vector field? How can you interpret it?
15. What is Stokes' Theorem? Explain how it generalizes Green's Theorem to three dimensions.
16. What is the divergence of a vector field? How can you interpret it?
17. What is the Divergence Theorem? Explain how it generalizes Green's Theorem to three dimensions.
18. How are Green's Theorem, Stokes' Theorem, and the Divergence Theorem related to the Fundamental Theorem of Calculus for ordinary single integrals?

CHAPTER 16 Practice Exercises

Evaluating Line Integrals

1. The accompanying figure shows two polygonal paths in space joining the origin to the point $(1, 1, 1)$. Integrate $f(x, y, z) = 2x - 3y^2 - 2z + 3$ over each path.

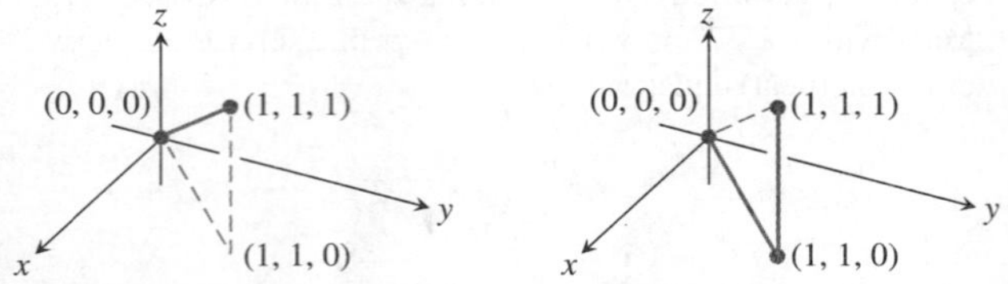

Path 1 Path 2

2. The accompanying figure shows three polygonal paths joining the origin to the point $(1, 1, 1)$. Integrate $f(x, y, z) = x^2 + y - z$ over each path.

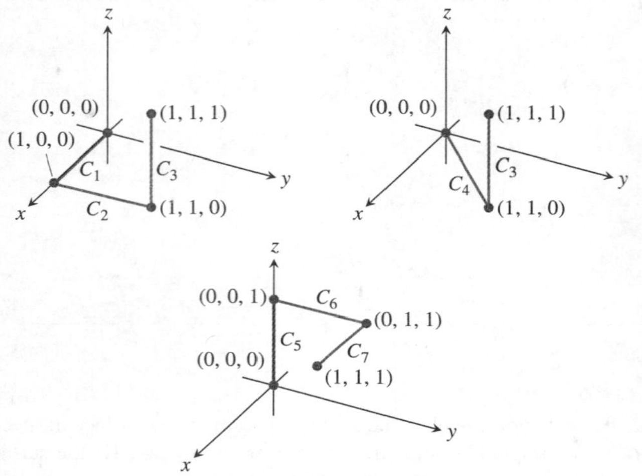

3. Integrate $f(x, y, z) = \sqrt{x^2 + z^2}$ over the circle

$$\mathbf{r}(t) = (a \cos t)\mathbf{j} + (a \sin t)\mathbf{k}, \qquad 0 \le t \le 2\pi.$$

4. Integrate $f(x, y, z) = \sqrt{x^2 + y^2}$ over the involute curve

$$\mathbf{r}(t) = (\cos t + t \sin t)\mathbf{i} + (\sin t - t \cos t)\mathbf{j}, \qquad 0 \le t \le \sqrt{3}.$$

Evaluate the integrals in Exercises 5 and 6.

5. $\displaystyle\int_{(-1,1,1)}^{(4,-3,0)} \frac{dx + dy + dz}{\sqrt{x + y + z}}$ 6. $\displaystyle\int_{(1,1,1)}^{(10,3,3)} dx - \sqrt{\frac{z}{y}}\, dy - \sqrt{\frac{y}{z}}\, dz$

7. Integrate $\mathbf{F} = -(y \sin z)\mathbf{i} + (x \sin z)\mathbf{j} + (xy \cos z)\mathbf{k}$ around the circle cut from the sphere $x^2 + y^2 + z^2 = 5$ by the plane $z = -1$, clockwise as viewed from above.

8. Integrate $\mathbf{F} = 3x^2y\mathbf{i} + (x^3 + 1)\mathbf{j} + 9z^2\mathbf{k}$ around the circle cut from the sphere $x^2 + y^2 + z^2 = 9$ by the plane $x = 2$.

Evaluate the integrals in Exercises 9 and 10.

9. $\displaystyle\int_C 8x \sin y\, dx - 8y \cos x\, dy$

   C is the square cut from the first quadrant by the lines $x = \pi/2$ and $y = \pi/2$.

10. $\displaystyle\int_C y^2\, dx + x^2\, dy$

    C is the circle $x^2 + y^2 = 4$.

Finding and Evaluating Surface Integrals

11. **Area of an elliptic region** Find the area of the elliptic region cut from the plane $x + y + z = 1$ by the cylinder $x^2 + y^2 = 1$.

12. **Area of a parabolic cap** Find the area of the cap cut from the paraboloid $y^2 + z^2 = 3x$ by the plane $x = 1$.

13. **Area of a spherical cap** Find the area of the cap cut from the top of the sphere $x^2 + y^2 + z^2 = 1$ by the plane $z = \sqrt{2}/2$.

14. a. **Hemisphere cut by cylinder** Find the area of the surface cut from the hemisphere $x^2 + y^2 + z^2 = 4$, $z \ge 0$, by the cylinder $x^2 + y^2 = 2x$.

    b. Find the area of the portion of the cylinder that lies inside the hemisphere. (*Hint:* Project onto the *xz*-plane. Or evaluate the integral $\int h\, ds$, where h is the altitude of the cylinder and ds is the element of arc length on the circle $x^2 + y^2 = 2x$ in the *xy*-plane.)

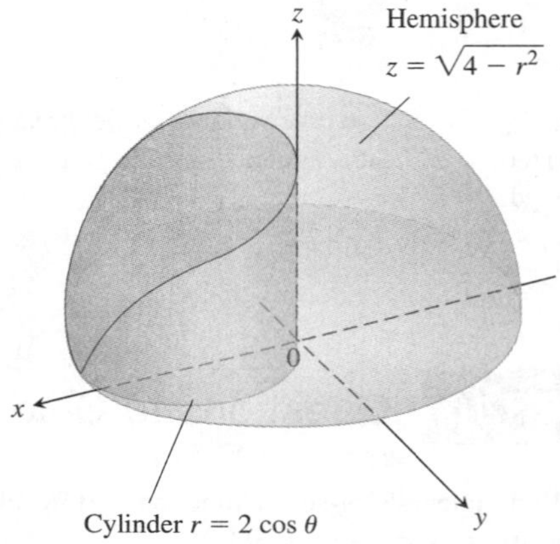

15. **Area of a triangle** Find the area of the triangle in which the plane $(x/a) + (y/b) + (z/c) = 1$ $(a, b, c > 0)$ intersects the first octant. Check your answer with an appropriate vector calculation.

16. **Parabolic cylinder cut by planes** Integrate

    a. $g(x, y, z) = \dfrac{yz}{\sqrt{4y^2 + 1}}$ b. $g(x, y, z) = \dfrac{z}{\sqrt{4y^2 + 1}}$

    over the surface cut from the parabolic cylinder $y^2 - z = 1$ by the planes $x = 0$, $x = 3$, and $z = 0$.

17. **Circular cylinder cut by planes** Integrate $g(x, y, z) = x^4y(y^2 + z^2)$ over the portion of the cylinder $y^2 + z^2 = 25$ that lies in the first octant between the planes $x = 0$ and $x = 1$ and above the plane $z = 3$.

18. **Area of Wyoming** The state of Wyoming is bounded by the meridians $111°3'$ and $104°3'$ west longitude and by the circles $41°$ and $45°$ north latitude. Assuming that Earth is a sphere of radius $R = 3959$ mi, find the area of Wyoming.

Parametrized Surfaces

Find parametrizations for the surfaces in Exercises 19–24. (There are many ways to do these, so your answers may not be the same as those in the back of the text.)

19. **Spherical band** The portion of the sphere $x^2 + y^2 + z^2 = 36$ between the planes $z = -3$ and $z = 3\sqrt{3}$

20. **Parabolic cap** The portion of the paraboloid $z = -(x^2 + y^2)/2$ above the plane $z = -2$

21. **Cone** The cone $z = 1 + \sqrt{x^2 + y^2}$, $z \le 3$

22. **Plane above square** The portion of the plane $4x + 2y + 4z = 12$ that lies above the square $0 \le x \le 2$, $0 \le y \le 2$, in the first quadrant

23. **Portion of paraboloid** The portion of the paraboloid $y = 2(x^2 + z^2)$, $y \le 2$, that lies above the *xy*-plane

24. **Portion of hemisphere** The portion of the hemisphere $x^2 + y^2 + z^2 = 10$, $y \ge 0$, in the first octant

25. **Surface area** Find the area of the surface

$$\mathbf{r}(u, v) = (u + v)\mathbf{i} + (u - v)\mathbf{j} + v\mathbf{k},$$
$$0 \le u \le 1,\ 0 \le v \le 1.$$

26. **Surface integral** Integrate $f(x, y, z) = xy - z^2$ over the surface in Exercise 25.

27. **Area of a helicoid** Find the surface area of the helicoid $\mathbf{r}(r, \theta) = (r\cos\theta)\mathbf{i} + (r\sin\theta)\mathbf{j} + \theta\mathbf{k}$, $0 \le \theta \le 2\pi$, $0 \le r \le 1$, in the accompanying figure.

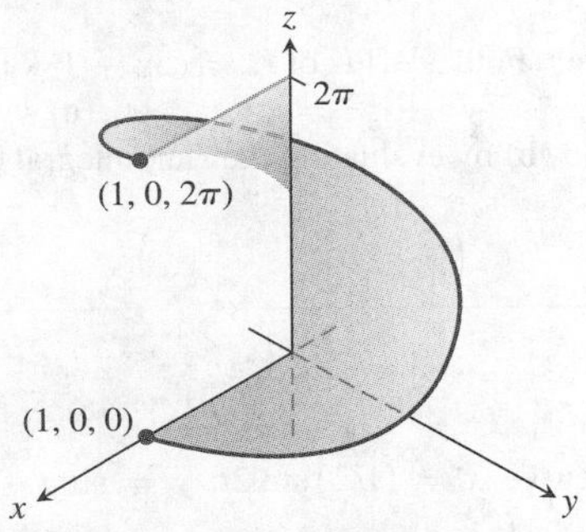

28. **Surface integral** Evaluate the integral $\iint_S \sqrt{x^2 + y^2 + 1}\, d\sigma$, where *S* is the helicoid in Exercise 27.

Conservative Fields

Which of the fields in Exercises 29–32 are conservative, and which are not?

29. $\mathbf{F} = x\mathbf{i} + y\mathbf{j} + z\mathbf{k}$

30. $\mathbf{F} = (x\mathbf{i} + y\mathbf{j} + z\mathbf{k})/(x^2 + y^2 + z^2)^{3/2}$

31. $\mathbf{F} = xe^y\mathbf{i} + ye^z\mathbf{j} + ze^x\mathbf{k}$

32. $\mathbf{F} = (\mathbf{i} + z\mathbf{j} + y\mathbf{k})/(x + yz)$

Find potential functions for the fields in Exercises 33 and 34.

33. $\mathbf{F} = 2\mathbf{i} + (2y + z)\mathbf{j} + (y + 1)\mathbf{k}$

34. $\mathbf{F} = (z\cos xz)\mathbf{i} + e^y\mathbf{j} + (x\cos xz)\mathbf{k}$

Work and Circulation

In Exercises 35 and 36, find the work done by each field along the paths from $(0, 0, 0)$ to $(1, 1, 1)$ in Exercise 1.

35. $\mathbf{F} = 2xy\mathbf{i} + \mathbf{j} + x^2\mathbf{k}$

36. $\mathbf{F} = 2xy\mathbf{i} + x^2\mathbf{j} + \mathbf{k}$

37. **Finding work in two ways** Find the work done by

$$\mathbf{F} = \frac{x\mathbf{i} + y\mathbf{j}}{(x^2 + y^2)^{3/2}}$$

over the plane curve $\mathbf{r}(t) = (e^t\cos t)\mathbf{i} + (e^t\sin t)\mathbf{j}$ from the point $(1, 0)$ to the point $(e^{2\pi}, 0)$ in two ways:

   a. By using the parametrization of the curve to evaluate the work integral.

   b. By evaluating a potential function for **F**.

38. **Flow along different paths** Find the flow of the field $\mathbf{F} = \nabla(x^2ze^y)$

   a. once around the ellipse *C* in which the plane $x + y + z = 1$ intersects the cylinder $x^2 + z^2 = 25$, clockwise as viewed from the positive *y*-axis.

   b. along the curved boundary of the helicoid in Exercise 27 from $(1, 0, 0)$ to $(1, 0, 2\pi)$.

In Exercises 39 and 40, use the curl integral in Stokes' Theorem to find the circulation of the field **F** around the curve *C* in the indicated direction.

39. **Circulation around an ellipse** $\mathbf{F} = y^2\mathbf{i} - y\mathbf{j} + 3z^2\mathbf{k}$

   C: The ellipse in which the plane $2x + 6y - 3z = 6$ meets the cylinder $x^2 + y^2 = 1$, counterclockwise as viewed from above

40. **Circulation around a circle** $\mathbf{F} = (x^2 + y)\mathbf{i} + (x + y)\mathbf{j} + (4y^2 - z)\mathbf{k}$

   C: The circle in which the plane $z = -y$ meets the sphere $x^2 + y^2 + z^2 = 4$, counterclockwise as viewed from above

Masses and Moments

41. **Wire with different densities** Find the mass of a thin wire lying along the curve $\mathbf{r}(t) = \sqrt{2}t\mathbf{i} + \sqrt{2}t\mathbf{j} + (4 - t^2)\mathbf{k}$, $0 \le t \le 1$, if the density at *t* is **(a)** $\delta = 3t$ and **(b)** $\delta = 1$.

42. **Wire with variable density** Find the center of mass of a thin wire lying along the curve $\mathbf{r}(t) = t\mathbf{i} + 2t\mathbf{j} + (2/3)t^{3/2}\mathbf{k}$, $0 \le t \le 2$, if the density at *t* is $\delta = 3\sqrt{5 + t}$.

43. **Wire with variable density** Find the center of mass and the moments of inertia about the coordinate axes of a thin wire lying along the curve

$$\mathbf{r}(t) = t\mathbf{i} + \frac{2\sqrt{2}}{3}t^{3/2}\mathbf{j} + \frac{t^2}{2}\mathbf{k}, \qquad 0 \le t \le 2,$$

   if the density at *t* is $\delta = 1/(t + 1)$.

44. **Center of mass of an arch** A slender metal arch lies along the semicircle $y = \sqrt{a^2 - x^2}$ in the *xy*-plane. The density at the point (x, y) on the arch is $\delta(x, y) = 2a - y$. Find the center of mass.

45. **Wire with constant density** A wire of constant density $\delta = 1$ lies along the curve $\mathbf{r}(t) = (e^t\cos t)\mathbf{i} + (e^t\sin t)\mathbf{j} + e^t\mathbf{k}$, $0 \le t \le \ln 2$. Find $\bar{z}$ and I_z.

46. **Helical wire with constant density** Find the mass and center of mass of a wire of constant density δ that lies along the helix $\mathbf{r}(t) = (2\sin t)\mathbf{i} + (2\cos t)\mathbf{j} + 3t\mathbf{k}$, $0 \le t \le 2\pi$.

47. **Inertia and center of mass of a shell** Find I_z and the center of mass of a thin shell of density $\delta(x, y, z) = z$ cut from the upper portion of the sphere $x^2 + y^2 + z^2 = 25$ by the plane $z = 3$.

48. **Moment of inertia of a cube** Find the moment of inertia about the *z*-axis of the surface of the cube cut from the first octant by the planes $x = 1$, $y = 1$, and $z = 1$ if the density is $\delta = 1$.

Flux Across a Plane Curve or Surface

Use Green's Theorem to find the counterclockwise circulation and outward flux for the fields and curves in Exercises 49 and 50.

49. **Square** $\mathbf{F} = (2xy + x)\mathbf{i} + (xy - y)\mathbf{j}$

   C: The square bounded by $x = 0$, $x = 1$, $y = 0$, $y = 1$

50. Triangle $\mathbf{F} = (y - 6x^2)\mathbf{i} + (x + y^2)\mathbf{j}$

C: The triangle made by the lines $y = 0$, $y = x$, and $x = 1$

51. Zero line integral Show that

$$\oint_C \ln x \sin y \, dy - \frac{\cos y}{x} dx = 0$$

for any closed curve C to which Green's Theorem applies.

52. a. Outward flux and area Show that the outward flux of the position vector field $\mathbf{F} = x\mathbf{i} + y\mathbf{j}$ across any closed curve to which Green's Theorem applies is twice the area of the region enclosed by the curve.

b. Let $\mathbf{n}$ be the outward unit normal vector to a closed curve to which Green's Theorem applies. Show that it is not possible for $\mathbf{F} = x\mathbf{i} + y\mathbf{j}$ to be orthogonal to $\mathbf{n}$ at every point of C.

In Exercises 53–56, find the outward flux of $\mathbf{F}$ across the boundary of D.

53. Cube $\mathbf{F} = 2xy\mathbf{i} + 2yz\mathbf{j} + 2xz\mathbf{k}$

D: The cube cut from the first octant by the planes $x = 1$, $y = 1$, and $z = 1$

54. Spherical cap $\mathbf{F} = xz\mathbf{i} + yz\mathbf{j} + \mathbf{k}$

D: The entire surface of the upper cap cut from the ball $x^2 + y^2 + z^2 \le 25$ by the plane $z = 3$

55. Spherical cap $\mathbf{F} = -2x\mathbf{i} - 3y\mathbf{j} + z\mathbf{k}$

D: The upper region cut from the ball $x^2 + y^2 + z^2 \le 2$ by the paraboloid $z = x^2 + y^2$

56. Cone and cylinder $\mathbf{F} = (6x + y)\mathbf{i} - (x + z)\mathbf{j} + 4yz\mathbf{k}$

D: The region in the first octant bounded by the cone $z = \sqrt{x^2 + y^2}$, the cylinder $x^2 + y^2 = 1$, and the coordinate planes

57. Hemisphere, cylinder, and plane Let S be the surface that is bounded on the left by the hemisphere $x^2 + y^2 + z^2 = a^2$, $y \le 0$, in the middle by the cylinder $x^2 + z^2 = a^2$, $0 \le y \le a$, and on the right by the plane $y = a$. Find the flux of $\mathbf{F} = y\mathbf{i} + z\mathbf{j} + x\mathbf{k}$ outward across S.

58. Cylinder and planes Find the outward flux of the field $\mathbf{F} = 3xz^2\mathbf{i} + y\mathbf{j} - z^3\mathbf{k}$ across the surface of the solid in the first octant that is bounded by the cylinder $x^2 + 4y^2 = 16$ and the planes $y = 2z$, $x = 0$, and $z = 0$.

59. Cylindrical can Use the Divergence Theorem to find the flux of $\mathbf{F} = xy^2\mathbf{i} + x^2y\mathbf{j} + y\mathbf{k}$ outward through the surface of the region enclosed by the cylinder $x^2 + y^2 = 1$ and the planes $z = 1$ and $z = -1$.

60. Hemisphere Find the flux of $\mathbf{F} = (3z + 1)\mathbf{k}$ upward across the hemisphere $x^2 + y^2 + z^2 = a^2$, $z \ge 0$, **(a)** with the Divergence Theorem and **(b)** by evaluating the flux integral directly.

CHAPTER 16 Additional and Advanced Exercises

Finding Areas with Green's Theorem

Use the Green's Theorem area formula in Exercises 16.4 to find the areas of the regions enclosed by the curves in Exercises 1–4.

1. The limaçon $x = 2\cos t - \cos 2t$, $y = 2\sin t$, $0 \le t \le 2\pi$

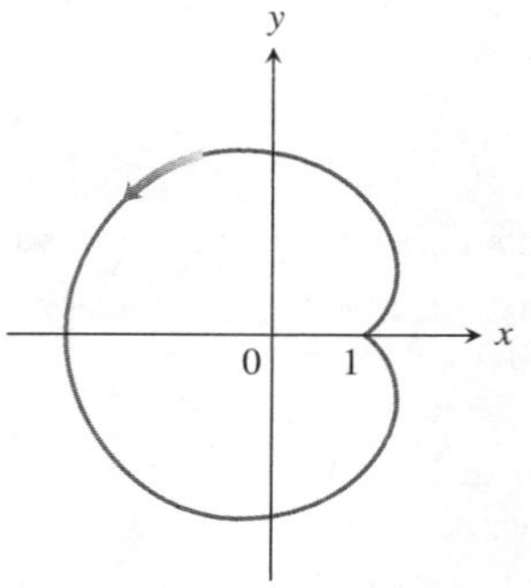

2. The deltoid $x = 2\cos t + \cos 2t$, $y = 2\sin t - \sin 2t$, $0 \le t \le 2\pi$

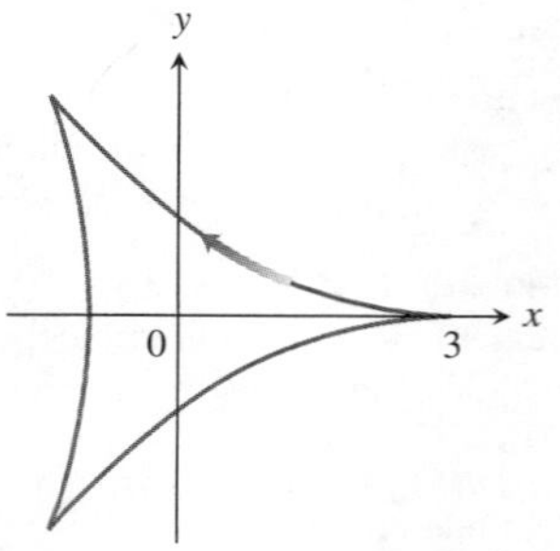

3. The eight curve $x = (1/2)\sin 2t$, $y = \sin t$, $0 \le t \le \pi$ (one loop)

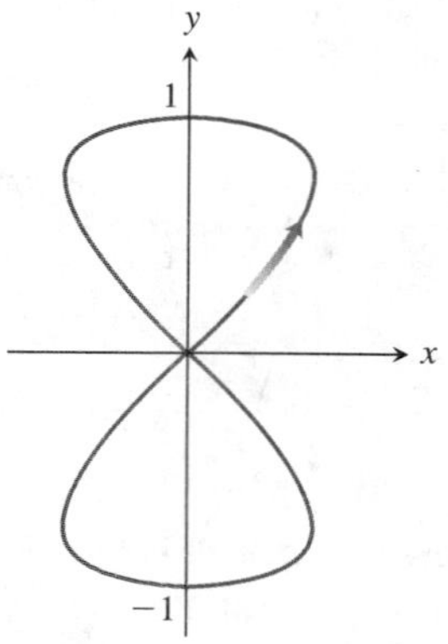

4. The teardrop $x = 2a\cos t - a\sin 2t$, $y = b\sin t$, $0 \le t \le 2\pi$

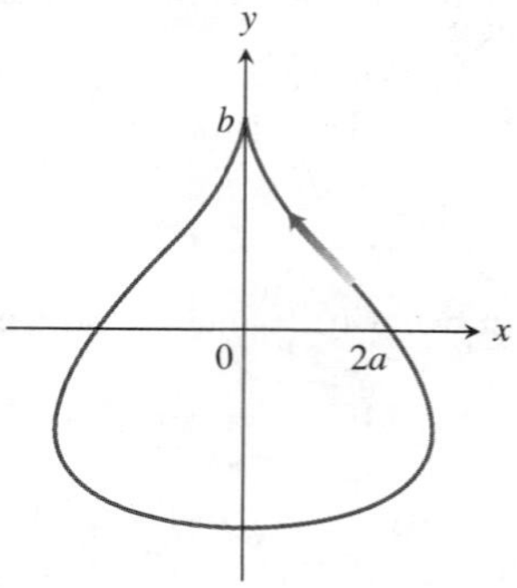

Theory and Applications

5. a. Give an example of a vector field $\mathbf{F}(x, y, z)$ that has value $\mathbf{0}$ at only one point and such that curl $\mathbf{F}$ is nonzero everywhere. Be sure to identify the point and compute the curl.

   b. Give an example of a vector field $\mathbf{F}(x, y, z)$ that has value $\mathbf{0}$ on precisely one line and such that curl $\mathbf{F}$ is nonzero everywhere. Be sure to identify the line and compute the curl.

   c. Give an example of a vector field $\mathbf{F}(x, y, z)$ that has value $\mathbf{0}$ on a surface and such that curl $\mathbf{F}$ is nonzero everywhere. Be sure to identify the surface and compute the curl.

6. Find all points (a, b, c) on the sphere $x^2 + y^2 + z^2 = R^2$ where the vector field $\mathbf{F} = yz^2\mathbf{i} + xz^2\mathbf{j} + 2xyz\mathbf{k}$ is normal to the surface and $\mathbf{F}(a, b, c) \neq \mathbf{0}$.

7. Find the mass of a spherical shell of radius R such that at each point (x, y, z) on the surface, the mass density $\delta(x, y, z)$ is its distance to some fixed point (a, b, c) of the surface.

8. Find the mass of a helicoid

$$\mathbf{r}(r, \theta) = (r\cos\theta)\mathbf{i} + (r\sin\theta)\mathbf{j} + \theta\mathbf{k},$$

$0 \le r \le 1$, $0 \le \theta \le 2\pi$, if the density function is $\delta(x, y, z) = 2\sqrt{x^2 + y^2}$. See Practice Exercise 27 for a figure.

9. Among all rectangular regions $0 \le x \le a$, $0 \le y \le b$, find the one for which the total outward flux of $\mathbf{F} = (x^2 + 4xy)\mathbf{i} - 6y\mathbf{j}$ across the four sides is least. What *is* the least flux?

10. Find an equation for the plane through the origin such that the circulation of the flow field $\mathbf{F} = z\mathbf{i} + x\mathbf{j} + y\mathbf{k}$ around the circle of intersection of the plane with the sphere $x^2 + y^2 + z^2 = 4$ is a maximum.

11. A string lies along the circle $x^2 + y^2 = 4$ from $(2, 0)$ to $(0, 2)$ in the first quadrant. The density of the string is $\rho(x, y) = xy$.

    a. Partition the string into a finite number of subarcs to show that the work done by gravity to move the string straight down to the x-axis is given by

$$\text{Work} = \lim_{n\to\infty}\sum_{k=1}^{n} g\,x_k y_k{}^2 \Delta s_k = \int_C g\,xy^2\,ds,$$

    where g is the gravitational constant.

    b. Find the total work done by evaluating the line integral in part (a).

    c. Show that the total work done equals the work required to move the string's center of mass $(\bar{x}, \bar{y})$ straight down to the x-axis.

12. A thin sheet lies along the portion of the plane $x + y + z = 1$ in the first octant. The density of the sheet is $\delta(x, y, z) = xy$.

    a. Partition the sheet into a finite number of subpieces to show that the work done by gravity to move the sheet straight down to the xy-plane is given by

$$\text{Work} = \lim_{n\to\infty}\sum_{k=1}^{n} g\,x_k y_k z_k\,\Delta\sigma_k = \iint_S g\,xyz\,d\sigma,$$

    where g is the gravitational constant.

    b. Find the total work done by evaluating the surface integral in part (a).

    c. Show that the total work done equals the work required to move the sheet's center of mass $(\bar{x}, \bar{y}, \bar{z})$ straight down to the xy-plane.

13. **Archimedes' principle** If an object such as a ball is placed in a liquid, it will either sink to the bottom, float, or sink a certain distance and remain suspended in the liquid. Suppose a fluid has constant weight density w and that the fluid's surface coincides with the plane $z = 4$. A spherical ball remains suspended in the fluid and occupies the region $x^2 + y^2 + (z - 2)^2 \le 1$.

    a. Show that the surface integral giving the magnitude of the total force on the ball due to the fluid's pressure is

$$\text{Force} = \lim_{n\to\infty}\sum_{k=1}^{n} w(4 - z_k)\,\Delta\sigma_k = \iint_S w(4 - z)\,d\sigma.$$

    b. Since the ball is not moving, it is being held up by the buoyant force of the liquid. Show that the magnitude of the buoyant force on the sphere is

$$\text{Buoyant force} = \iint_S w(z - 4)\mathbf{k}\cdot\mathbf{n}\,d\sigma,$$

    where $\mathbf{n}$ is the outer unit normal at (x, y, z). This illustrates Archimedes' principle that the magnitude of the buoyant force on a submerged solid equals the weight of the displaced fluid.

    c. Use the Divergence Theorem to find the magnitude of the buoyant force in part (b).

14. **Fluid force on a curved surface** A cone in the shape of the surface $z = \sqrt{x^2 + y^2}$, $0 \le z \le 2$, is filled with a liquid of constant weight density w. Assuming the xy-plane is "ground level," show that the total force on the portion of the cone from $z = 1$ to $z = 2$ due to liquid pressure is the surface integral

$$F = \iint_S w(2 - z)\,d\sigma.$$

Evaluate the integral.

15. **Faraday's law** If $\mathbf{E}(t, x, y, z)$ and $\mathbf{B}(t, x, y, z)$ represent the electric and magnetic fields at point (x, y, z) at time t, a basic principle of electromagnetic theory says that $\nabla \times \mathbf{E} = -\partial\mathbf{B}/\partial t$. In this expression $\nabla \times \mathbf{E}$ is computed with t held fixed and $\partial\mathbf{B}/\partial t$ is calculated with (x, y, z) fixed. Use Stokes' Theorem to derive Faraday's law,

$$\oint_C \mathbf{E}\cdot d\mathbf{r} = -\frac{\partial}{\partial t}\iint_S \mathbf{B}\cdot\mathbf{n}\,d\sigma,$$

where C represents a wire loop through which current flows counterclockwise with respect to the surface's unit normal $\mathbf{n}$, giving rise to the voltage

$$\oint_C \mathbf{E}\cdot d\mathbf{r}$$

around C. The surface integral on the right side of the equation is called the *magnetic flux*, and S is any oriented surface with boundary C.

16. Let

$$\mathbf{F} = -\frac{GmM}{|\mathbf{r}|^3}\mathbf{r}$$

be the gravitational force field defined for $\mathbf{r} \neq \mathbf{0}$. Use Gauss's law in Section 16.8 to show that there is no continuously differentiable vector field $\mathbf{H}$ satisfying $\mathbf{F} = \nabla \times \mathbf{H}$.

17. If $f(x, y, z)$ and $g(x, y, z)$ are continuously differentiable scalar functions defined over the oriented surface S with boundary curve C, prove that

$$\iint_S (\nabla f \times \nabla g) \cdot \mathbf{n}\, d\sigma = \oint_C f \nabla g \cdot d\mathbf{r}.$$

18. Suppose that $\nabla \cdot \mathbf{F}_1 = \nabla \cdot \mathbf{F}_2$ and $\nabla \times \mathbf{F}_1 = \nabla \times \mathbf{F}_2$ over a region D enclosed by the oriented surface S with outward unit normal $\mathbf{n}$ and that $\mathbf{F}_1 \cdot \mathbf{n} = \mathbf{F}_2 \cdot \mathbf{n}$ on S. Prove that $\mathbf{F}_1 = \mathbf{F}_2$ throughout D.

19. Prove or disprove that if $\nabla \cdot \mathbf{F} = 0$ and $\nabla \times \mathbf{F} = \mathbf{0}$, then $\mathbf{F} = \mathbf{0}$.

20. Let S be an oriented surface parametrized by $\mathbf{r}(u, v)$. Define the notation $d\boldsymbol{\sigma} = \mathbf{r}_u\, du \times \mathbf{r}_v\, dv$ so that $d\boldsymbol{\sigma}$ is a vector normal to the surface. Also, the magnitude $d\sigma = |d\boldsymbol{\sigma}|$ is the element of surface area (by Equation 5 in Section 16.5). Derive the identity

$$d\sigma = (EG - F^2)^{1/2}\, du\, dv,$$

where

$$E = |\mathbf{r}_u|^2, \quad F = \mathbf{r}_u \cdot \mathbf{r}_v, \quad \text{and} \quad G = |\mathbf{r}_u|^2.$$

21. Show that the volume V of a region D in space enclosed by the oriented surface S with outward normal $\mathbf{n}$ satisfies the identity

$$V = \frac{1}{3} \iint_S \mathbf{r} \cdot \mathbf{n}\, d\sigma,$$

where $\mathbf{r}$ is the position vector of the point (x, y, z) in D.

CHAPTER 16 Technology Application Projects

Mathematica/Maple Projects

Projects can be found within MyLab Math.

- ***Work in Conservative and Nonconservative Force Fields***
  Explore integration over vector fields and experiment with conservative and nonconservative force functions along different paths in the field.
- ***How Can You Visualize Green's Theorem?***
  Explore integration over vector fields and use parametrizations to compute line integrals. Both forms of Green's Theorem are explored.
- ***Visualizing and Interpreting the Divergence Theorem***
  Verify the Divergence Theorem by formulating and evaluating certain divergence and surface integrals.

Appendix A

A.1 Real Numbers and the Real Line

This section reviews real numbers, inequalities, intervals, and absolute values.

Real Numbers

Much of calculus is based on properties of the real number system. **Real numbers** are numbers that can be expressed as decimals, such as

$$-\frac{3}{4} = -0.75000\ \ldots$$
$$\frac{1}{3} = 0.33333\ \ldots$$
$$\sqrt{2} = 1.4142\ \ldots$$

The dots ... in each case indicate that the sequence of decimal digits goes on forever. Every conceivable decimal expansion represents a real number; although, some numbers have two representations. For instance, the infinite decimals .999 ... and 1.000 ... represent the same real number 1. A similar statement holds for any number with an infinite tail of 9's.

The real numbers can be represented geometrically as points on a number line called the **real line**.

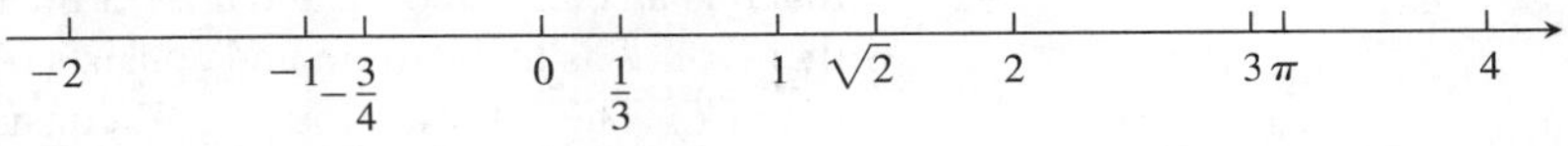

RULES FOR INEQUALITIES

If a, b, and c are real numbers, then:

1. $a < b \Rightarrow a + c < b + c$
2. $a < b \Rightarrow a - c < b - c$
3. $a < b$ and $c > 0 \Rightarrow ac < bc$
4. $a < b$ and $c < 0 \Rightarrow bc < ac$
   Special case: $a < b \Rightarrow -b < -a$
5. $a > 0 \Rightarrow \frac{1}{a} > 0$
6. If a and b are both positive or both negative, then $a < b \Rightarrow \frac{1}{b} < \frac{1}{a}$.

The symbol $\mathbb{R}$ denotes either the real number system or, equivalently, the real line.

The properties of the real number system fall into three categories: algebraic properties, order properties, and completeness. The **algebraic properties** say that the real numbers can be added, subtracted, multiplied, and divided (except by 0) to produce more real numbers under the usual rules of arithmetic. *You can never divide by* 0.

The **order properties** of real numbers are given in Appendix A.6. The useful rules at the left can be derived from them, where the symbol ⇒ means "implies."

Notice the rules for multiplying an inequality by a number. Multiplying by a positive number preserves the inequality; multiplying by a negative number reverses the inequality. Also, reciprocation reverses the inequality for numbers of the same sign. For example, $2 < 5$ but $-2 > -5$ and $1/2 > 1/5$.

The **completeness property** of the real number system is deeper and harder to define precisely. However, the property is essential to the idea of a limit (Chapter 2). Roughly speaking, it says that there are enough real numbers to "complete" the real number line, in the sense that there are no "holes" or "gaps" in it. Many theorems of calculus would fail if the real number system were not complete. Appendix A.6 introduces the ideas involved and discusses how the real numbers are constructed.

We distinguish three special subsets of real numbers.

1. The **natural numbers**, namely 1, 2, 3, 4, ...
2. The **integers**, namely 0, ± 1, ± 2, ± 3, ...
3. The **rational numbers**, namely the numbers that can be expressed in the form of a fraction m/n, where m and n are integers and $n \neq 0$. Examples are

$$\frac{1}{3}, \quad -\frac{4}{9} = \frac{-4}{9} = \frac{4}{-9}, \quad \frac{200}{13}, \qquad \text{and} \qquad 57 = \frac{57}{1}.$$

The rational numbers are precisely the real numbers with decimal expansions that are either

a. terminating (ending in an infinite string of zeros), for example,

$$\frac{3}{4} = 0.75000\ldots = 0.75 \qquad \text{or}$$

b. eventually repeating (ending with a block of digits that repeats over and over), for example,

$$\frac{23}{11} = 2.090909\ldots = 2.\overline{09}$$

The bar indicates the block of repeating digits.

A terminating decimal expansion is a special type of repeating decimal, since the ending zeros repeat.

The set of rational numbers has all the algebraic and order properties of the real numbers but lacks the completeness property. For example, there is no rational number whose square is 2; there is a "hole" in the rational line where $\sqrt{2}$ should be.

Real numbers that are not rational are called **irrational numbers**. They are characterized by having nonterminating and nonrepeating decimal expansions. Examples are π, $\sqrt{2}$, $\sqrt[3]{5}$, and $\log_{10} 3$. Since every decimal expansion represents a real number, there are infinitely many irrational numbers. Both rational and irrational numbers are found arbitrarily close to any given point on the real line.

Set notation is very useful for specifying sets of real numbers. A **set** is a collection of objects, and these objects are the **elements** of the set. If S is a set, the notation $a \in S$ means that a is an element of S, and $a \notin S$ means that a is not an element of S. If S and T are sets, then $S \cup T$ is their **union** and consists of all elements belonging to either S or T (or to both S and T). The **intersection** $S \cap T$ consists of all elements belonging to both S and T. The **empty set** $\varnothing$ is the set that contains no elements. For example, the intersection of the rational numbers and the irrational numbers is the empty set.

Some sets can be described by *listing* their elements in braces. For instance, the set A consisting of the natural numbers (or positive integers) less than 6 can be expressed as

$$A = \{1, 2, 3, 4, 5\}.$$

The entire set of integers is written as

$$\{0, \pm 1, \pm 2, \pm 3, \ldots\}.$$

Another way to describe a set is to enclose in braces a rule that generates all the elements of the set. For instance, the set

$$A = \{x \mid x \text{ is an integer and } 0 < x < 6\}$$

is the set of positive integers less than 6.

Intervals

A subset of the real line is called an **interval** if it contains at least two numbers and contains all the real numbers lying between any two of its elements. For example, the set of all real numbers x such that $x > 6$ is an interval, as is the set of all x such that $-2 \leq x \leq 5$. The set of all nonzero real numbers is not an interval; since 0 is absent, the set fails to contain every real number between -1 and 1 (for example).

Geometrically, intervals correspond to rays and line segments on the real line, along with the real line itself. Intervals of numbers corresponding to line segments are **finite intervals**; intervals corresponding to rays and the real line are **infinite intervals**.

A finite interval is said to be **closed** if it contains both of its endpoints, **half-open** if it contains one endpoint but not the other, and **open** if it contains neither endpoint. The endpoints are also called **boundary points**; they make up the interval's **boundary**. The remaining points of the interval are **interior points** and together compose the interval's **interior**. Infinite intervals are closed if they contain a finite endpoint, and open otherwise. The entire real line $\mathbb{R}$ is an infinite interval that is both open and closed. Table A.1 summarizes the various types of intervals.

TABLE A.1 Types of intervals

Notation	Set description	Type	Picture
(a, b)	$\{x \mid a < x < b\}$	Open	
$[a, b]$	$\{x \mid a \leq x \leq b\}$	Closed	
$[a, b)$	$\{x \mid a \leq x < b\}$	Half-open	
$(a, b]$	$\{x \mid a < x \leq b\}$	Half-open	
(a, ∞)	$\{x \mid x > a\}$	Open	
$[a, \infty)$	$\{x \mid x \geq a\}$	Closed	
$(-\infty, b)$	$\{x \mid x < b\}$	Open	
$(-\infty, b]$	$\{x \mid x \leq b\}$	Closed	
$(-\infty, \infty)$	$\mathbb{R}$ (set of all real numbers)	Both open and closed	

Solving Inequalities

The process of finding the interval or intervals of numbers that satisfy an inequality in x is called **solving** the inequality.

EXAMPLE 1 Solve the following inequalities and show their solution sets on the real line.

(a) $2x - 1 < x + 3$ **(b)** $\dfrac{6}{x - 1} \geq 5$

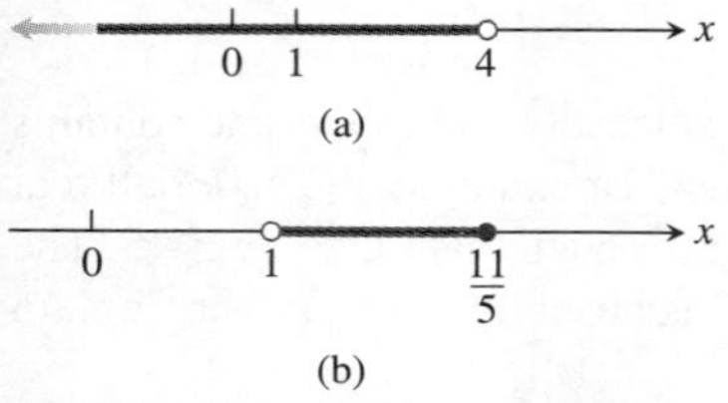

FIGURE A.1 Solution sets for the inequalities in Example 1. Hollow circles indicate endpoints that are not included in the interval, and solid dots indicate included endpoints.

Solution

(a)

$$\begin{aligned} 2x - 1 &< x + 3 & \\ 2x &< x + 4 & \text{Add 1 to both sides.} \\ x &< 4 & \text{Subtract } x \text{ from both sides.} \end{aligned}$$

The solution set is the open interval $(-\infty, 4)$ (Figure A.1a).

(b) The inequality $6/(x - 1) \geq 5$ can hold only if $x > 1$ because otherwise $6/(x - 1)$ is undefined or negative. Therefore, $(x - 1)$ is positive, and the inequality will be preserved if we multiply both sides by $(x - 1)$:

$$\begin{aligned} \frac{6}{x-1} &\geq 5 & \\ 6 &\geq 5x - 5 & \text{Multiply both sides by } (x - 1). \\ 11 &\geq 5x & \text{Add 5 to both sides.} \\ \frac{11}{5} &\geq x. & \text{Or } x \leq \tfrac{11}{5}. \end{aligned}$$

The solution set is the half-open interval $(1, 11/5]$ (Figure A.1b). ■

Absolute Value

The **absolute value** of a number x, denoted by $|x|$, is defined by the formula

$$|x| = \begin{cases} x, & x \geq 0 \\ -x, & x < 0. \end{cases}$$

EXAMPLE 2 $|3| = 3, \quad |0| = 0, \quad |-5| = -(-5) = 5, \quad |-|a|| = |a|$ ■

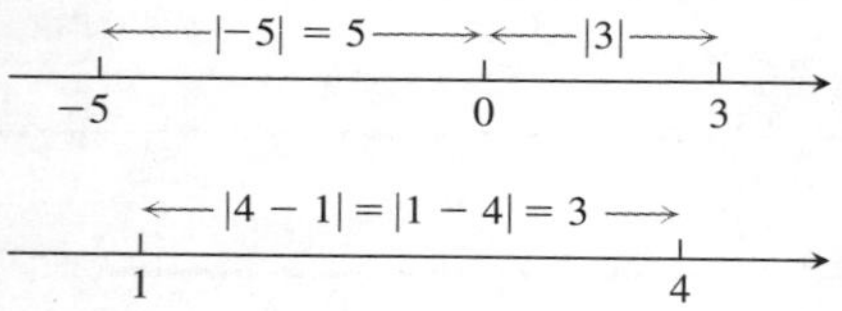

FIGURE A.2 Absolute values give distances between points on the number line.

Geometrically, the absolute value of x is the distance from x to 0 on the real number line. Since distances are always positive or 0, we see that $|x| \geq 0$ for every real number x, and $|x| = 0$ if and only if $x = 0$. Also,

$$|x - y| = \text{the distance between } x \text{ and } y$$

on the real line (Figure A.2).

Since the symbol $\sqrt{a}$ always denotes the *nonnegative* square root of a, an alternate definition of $|x|$ is

$$|x| = \sqrt{x^2}.$$

It is important to remember that $\sqrt{a^2} = |a|$. Do not write $\sqrt{a^2} = a$ unless you already know that $a \geq 0$.

The absolute value function has the following properties. (You are asked to prove these properties in the exercises.)

Absolute Value Properties

1. $|-a| = |a|$	A number and its negative have the same absolute value.
2. $|ab| = |a||b|$	The absolute value of a product is the product of the absolute values.
3. $\left|\frac{a}{b}\right| = \frac{|a|}{|b|}$	The absolute value of a quotient is the quotient of the absolute values.
4. $|a + b| \leq |a| + |b|$	The **triangle inequality**. The absolute value of the sum of two numbers is less than or equal to the sum of their absolute values.

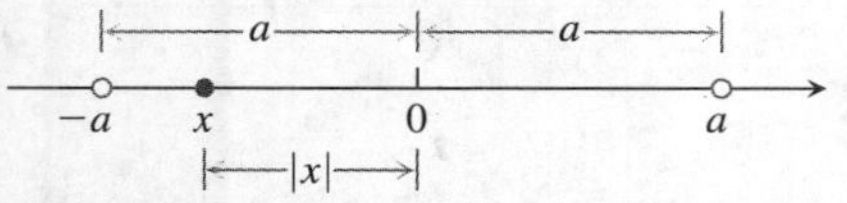

FIGURE A.3 $|x| < a$ means x lies between $-a$ and a.

Note that $|-a| \neq -|a|$. For example, $|-3| = 3$, whereas $-|3| = -3$. If a and b differ in sign, then $|a + b|$ is less than $|a| + |b|$. In all other cases, $|a + b|$ equals $|a| + |b|$. Absolute value bars in expressions like $|-3 + 5|$ work like parentheses: We do the arithmetic inside *before* taking the absolute value.

EXAMPLE 3

$$|-3 + 5| = |2| = 2 < |-3| + |5| = 8$$
$$|3 + 5| = |8| = |3| + |5|$$
$$|-3 - 5| = |-8| = 8 = |-3| + |-5|$$

■

FURTHER PROPERTIES: ABSOLUTE VALUES AND INTERVALS

If a is any positive number, then:

5. $|x| = a \quad \Leftrightarrow \quad x = \pm a$
6. $|x| < a \quad \Leftrightarrow \quad -a < x < a$
7. $|x| > a \quad \Leftrightarrow \quad x > a$ or $x < -a$
8. $|x| \leq a \quad \Leftrightarrow \quad -a \leq x \leq a$
9. $|x| \geq a \quad \Leftrightarrow \quad x \geq a$ or $x \leq -a$

The inequality $|x| < a$ says that the distance from x to 0 is less than the positive number a. This means that x must lie between $-a$ and a, as we can see from Figure A.3.

Statements 5–9 in the table at left are all consequences of the definition of absolute value and are often helpful when solving equations or inequalities involving absolute values. The symbol $\Leftrightarrow$ that appears in the table is often used by mathematicians to denote the "if and only if" logical relationship. It also means "implies and is implied by."

EXAMPLE 4 Solve the equation $|2x - 3| = 7$.

Solution By Property 5, $2x - 3 = \pm 7$, so there are two possibilities:

$2x - 3 = 7$	$2x - 3 = -7$	Equivalent equations without absolute values
$2x = 10$	$2x = -4$	Solve as usual.
$x = 5$	$x = -2$	

The solutions of $|2x - 3| = 7$ are $x = 5$ and $x = -2$. ■

EXAMPLE 5 Solve the inequality $\left|5 - \dfrac{2}{x}\right| < 1$.

Solution We have

$$\left|5 - \frac{2}{x}\right| < 1 \Leftrightarrow -1 < 5 - \frac{2}{x} < 1 \qquad \text{Property 6}$$
$$\Leftrightarrow -6 < -\frac{2}{x} < -4 \qquad \text{Subtract 5.}$$
$$\Leftrightarrow 3 > \frac{1}{x} > 2 \qquad \text{Multiply by } -\tfrac{1}{2}.$$
$$\Leftrightarrow \frac{1}{3} < x < \frac{1}{2}. \qquad \text{Take reciprocals.}$$

Notice how the various rules for inequalities were used here. Multiplying by a negative number reverses the inequality. So does taking reciprocals in an inequality in which both sides are positive. The original inequality holds if and only if $(1/3) < x < (1/2)$. The solution set is the open interval $(1/3, 1/2)$. ■

EXERCISES A.1

1. Express $1/9$ as a repeating decimal, using a bar to indicate the repeating digits. What are the decimal representations of $2/9$? $3/9$? $8/9$? $9/9$?

2. If $2 < x < 6$, which of the following statements about x are necessarily true, and which are not necessarily true?

   a. $0 < x < 4$ b. $0 < x - 2 < 4$

   c. $1 < \frac{x}{2} < 3$ d. $\frac{1}{6} < \frac{1}{x} < \frac{1}{2}$

   e. $1 < \frac{6}{x} < 3$ f. $|x - 4| < 2$

   g. $-6 < -x < 2$ h. $-6 < -x < -2$

In Exercises 3–6, solve the inequalities and show the solution sets on the real line.

3. $-2x > 4$

4. $5x - 3 \le 7 - 3x$

5. $2x - \frac{1}{2} \ge 7x + \frac{7}{6}$

6. $\frac{4}{5}(x - 2) < \frac{1}{3}(x - 6)$

Solve the equations in Exercises 7–9.

7. $|y| = 3$

8. $|2t + 5| = 4$

9. $|8 - 3s| = \frac{9}{2}$

Solve the inequalities in Exercises 10–17, expressing the solution sets as intervals or unions of intervals. Also, show each solution set on the real line.

10. $|x| < 2$

11. $|t - 1| \le 3$

12. $|3y - 7| < 4$

13. $\left|\frac{z}{5} - 1\right| \le 1$

14. $\left|3 - \frac{1}{x}\right| < \frac{1}{2}$

15. $|2s| \ge 4$

16. $|1 - x| > 1$

17. $\left|\frac{r + 1}{2}\right| \ge 1$

Solve the inequalities in Exercises 18–21. Express the solution sets as intervals or unions of intervals and show them on the real line. Use the result $\sqrt{a^2} = |a|$ as appropriate.

18. $x^2 < 2$

19. $4 < x^2 < 9$

20. $(x - 1)^2 < 4$

21. $x^2 - x < 0$

22. Do not fall into the trap of thinking $|-a| = a$. For what real numbers a is this equation true? For what real numbers is it false?

23. Solve the equation $|x - 1| = 1 - x$.

24. **A proof of the triangle inequality** Give the reason justifying each of the numbered steps in the following proof of the triangle inequality.

$$\begin{aligned}
|a + b|^2 &= (a + b)^2 && (1)\\
&= a^2 + 2ab + b^2 \\
&\le a^2 + 2|a||b| + b^2 && (2)\\
&= |a|^2 + 2|a||b| + |b|^2 && (3)\\
&= (|a| + |b|)^2 \\
|a + b| &\le |a| + |b| && (4)
\end{aligned}$$

25. Prove that $|ab| = |a||b|$ for any numbers a and b.

26. If $|x| \le 3$ and $x > -1/2$, what can you say about x?

27. Graph the inequality $|x| + |y| \le 1$.

28. For any number a, prove that $|-a| = |a|$.

29. Let a be any positive number. Prove that $|x| > a$ if and only if $x > a$ or $x < -a$.

30. a. If b is any nonzero real number, prove that $|1/b| = 1/|b|$.

    b. Prove that $\left|\frac{a}{b}\right| = \frac{|a|}{|b|}$ for any numbers a and $b \ne 0$.

A.2 Mathematical Induction

Many formulas, like

$$1 + 2 + \cdots + n = \frac{n(n + 1)}{2},$$

can be shown to hold for every positive integer n by applying an axiom called the *mathematical induction principle*. A proof that uses this axiom is called a *proof by mathematical induction* or a *proof by induction*.

The steps in proving a formula by induction are the following:

1. Check that the formula holds for $n = 1$.
2. Prove that *if* the formula holds for any positive integer $n = k$, *then* it also holds for the next integer, $n = k + 1$.

The induction axiom says that once these steps are completed, the formula holds for all positive integers n. By Step 1 it holds for $n = 1$. By Step 2 it holds for $n = 2$, and therefore by Step 2 also for $n = 3$, and by Step 2 again for $n = 4$, and so on. If the first domino falls, and if the kth domino always knocks over the $(k+1)$st when it falls, then all the dominoes fall.

EXAMPLE 1 Use mathematical induction to prove that for every positive integer n,

$$1 + 2 + \cdots + n = \frac{n(n+1)}{2}.$$

Solution We accomplish the proof by carrying out the two steps above.

1. The formula holds for $n = 1$ because

$$1 = \frac{1(1+1)}{2}.$$

2. If the formula holds for $n = k$, does it also hold for $n = k + 1$? The answer is yes, as we now show. If it is the case that

$$1 + 2 + \cdots + k = \frac{k(k+1)}{2},$$

then it follows that

$$\begin{aligned} 1 + 2 + \cdots + k + (k+1) &= \frac{k(k+1)}{2} + (k+1) = \frac{k^2 + k + 2k + 2}{2} \\ &= \frac{(k+1)(k+2)}{2} = \frac{(k+1)((k+1)+1)}{2}. \end{aligned}$$

The last expression in this string of equalities is the expression $n(n+1)/2$ for $n = (k+1)$.

The mathematical induction principle now guarantees the original formula for all positive integers n. ■

In Example 4 of Section 5.2 we gave another proof for the formula giving the sum of the first n positive integers. However, proofs by mathematical induction can also be used to find the sums of the squares and cubes of the first n positive integers (Exercises 9 and 10). Here is another example of a proof by induction.

EXAMPLE 2 Show by mathematical induction that for all positive integers n,

$$\frac{1}{2^1} + \frac{1}{2^2} + \cdots + \frac{1}{2^n} = 1 - \frac{1}{2^n}.$$

Solution We accomplish the proof by carrying out the two steps of mathematical induction.

1. The formula holds for $n = 1$ because

$$\frac{1}{2^1} = 1 - \frac{1}{2^1}.$$

2. If it is the case that

$$\frac{1}{2^1} + \frac{1}{2^2} + \cdots + \frac{1}{2^k} = 1 - \frac{1}{2^k},$$

then it follows that

$$\frac{1}{2^1}+\frac{1}{2^2}+\cdots+\frac{1}{2^k}+\frac{1}{2^{k+1}}=1-\frac{1}{2^k}+\frac{1}{2^{k+1}}=1-\frac{1\cdot 2}{2^k\cdot 2}+\frac{1}{2^{k+1}}$$
$$=1-\frac{2}{2^{k+1}}+\frac{1}{2^{k+1}}=1-\frac{1}{2^{k+1}}.$$

Thus, the original formula holds for $n=(k+1)$ whenever it holds for $n=k$.

With these steps verified, the mathematical induction principle now guarantees the formula for every positive integer n. ■

Other Starting Integers

Instead of starting at $n=1$, some induction arguments start at another integer. The steps for such an argument are as follows.

1. Check that the formula holds for $n=n_1$ (the first appropriate integer).

2. Prove that if the formula holds for any integer $n=k\geq n_1$, then it also holds for $n=(k+1)$.

Once these steps are completed, the mathematical induction principle guarantees the formula for all $n\geq n_1$.

EXAMPLE 3 Show that $n!>3^n$ if n is large enough.

Solution How large is large enough? We experiment:

n	1	2	3	4	5	6	7
$n!$	1	2	6	24	120	720	5040
3^n	3	9	27	81	243	729	2187

It looks as if $n!>3^n$ for $n\geq 7$. To be sure, we apply mathematical induction. We take $n_1=7$ in Step 1 and complete Step 2.

Suppose $k!>3^k$ for some $k\geq 7$. Then

$$(k+1)!=(k+1)(k!)>(k+1)3^k>7\cdot 3^k>3^{k+1}.$$

Thus, for $k\geq 7$,

$$k!>3^k \quad \text{implies} \quad (k+1)!>3^{k+1}.$$

The mathematical induction principle now guarantees $n!\geq 3^n$ for all $n\geq 7$. ■

Proof of the Derivative Sum Rule for Sums of Finitely Many Functions

We prove the statement

$$\frac{d}{dx}(u_1+u_2+\cdots+u_n)=\frac{du_1}{dx}+\frac{du_2}{dx}+\cdots+\frac{du_n}{dx}$$

by mathematical induction. The statement is true for $n=2$, as was proved in Section 3.3. This is Step 1 of the induction proof.

Step 2 is to show that if the statement is true for any positive integer $n=k$, where $k\geq n_0=2$, then it is also true for $n=k+1$. So suppose that

$$\frac{d}{dx}(u_1+u_2+\cdots+u_k)=\frac{du_1}{dx}+\frac{du_2}{dx}+\cdots+\frac{du_k}{dx}. \tag{1}$$

Then

$$\frac{d}{dx}(\underbrace{u_1 + u_2 + \cdots + u_k}_{\substack{\text{Call the function}\\ \text{defined by this sum } u.}} + \underbrace{u_{k+1}}_{\substack{\text{Call this}\\ \text{function } v.}})$$

$$= \frac{d}{dx}(u_1 + u_2 + \cdots + u_k) + \frac{du_{k+1}}{dx} \qquad \text{Sum Rule for } \frac{d}{dx}(u + v)$$

$$= \frac{du_1}{dx} + \frac{du_2}{dx} + \cdots + \frac{du_k}{dx} + \frac{du_{k+1}}{dx}. \qquad \text{Eq. (1)}$$

With these steps verified, the mathematical induction principle now guarantees the Sum Rule for every integer $n \geq 2$.

EXERCISES A.2

1. Assuming that the triangle inequality $|a + b| \leq |a| + |b|$ holds for any two numbers a and b, show that

$$|x_1 + x_2 + \cdots + x_n| \leq |x_1| + |x_2| + \cdots + |x_n|$$

for any n numbers.

2. Show that if $r \neq 1$, then

$$1 + r + r^2 + \cdots + r^n = \frac{1 - r^{n+1}}{1 - r}$$

for every positive integer n.

3. Use the Product Rule, $\frac{d}{dx}(uv) = u\frac{dv}{dx} + v\frac{du}{dx}$, and the fact that $\frac{d}{dx}(x) = 1$ to show that $\frac{d}{dx}(x^n) = nx^{n-1}$ for every positive integer n.

4. Suppose that a function $f(x)$ has the property that $f(x_1x_2) = f(x_1) + f(x_2)$ for any two positive numbers x_1 and x_2. Show that

$$f(x_1x_2 \cdots x_n) = f(x_1) + f(x_2) + \cdots + f(x_n)$$

for the product of any n positive numbers $x_1, x_2, \ldots, x_n$.

5. Show that

$$\frac{2}{3^1} + \frac{2}{3^2} + \cdots + \frac{2}{3^n} = 1 - \frac{1}{3^n}$$

for all positive integers n.

6. Show that $n! > n^3$ if n is large enough.

7. Show that $2^n > n^2$ if n is large enough.

8. Show that $2^n \geq 1/8$ for $n \geq -3$.

9. **Sums of squares** Show that the sum of the squares of the first n positive integers is

$$\frac{n\left(n + \frac{1}{2}\right)(n + 1)}{3}.$$

10. **Sums of cubes** Show that the sum of the cubes of the first n positive integers is $\left(n(n + 1)/2\right)^2$.

11. **Rules for finite sums** Show that the following finite sum rules hold for every positive integer n. (See Section 5.2.)

  a. $\sum_{k=1}^{n}(a_k + b_k) = \sum_{k=1}^{n} a_k + \sum_{k=1}^{n} b_k$

  b. $\sum_{k=1}^{n}(a_k - b_k) = \sum_{k=1}^{n} a_k - \sum_{k=1}^{n} b_k$

  c. $\sum_{k=1}^{n} ca_k = c \cdot \sum_{k=1}^{n} a_k$ (any number c)

  d. $\sum_{k=1}^{n} a_k = n \cdot c$ (if a_k has the constant value c)

12. Show that $|x^n| = |x|^n$ for every positive integer n and every real number x.

A.3 Lines, Circles, and Parabolas

This section reviews coordinates, lines, distance, circles, and parabolas in the plane. The notion of increment is also discussed.

Cartesian Coordinates in the Plane

In Appendix A.1 we identified the points on the line with real numbers by assigning them coordinates. Points in the plane can be identified with ordered pairs of real numbers. To begin, we draw two perpendicular coordinate lines that intersect at the 0-point of each line.

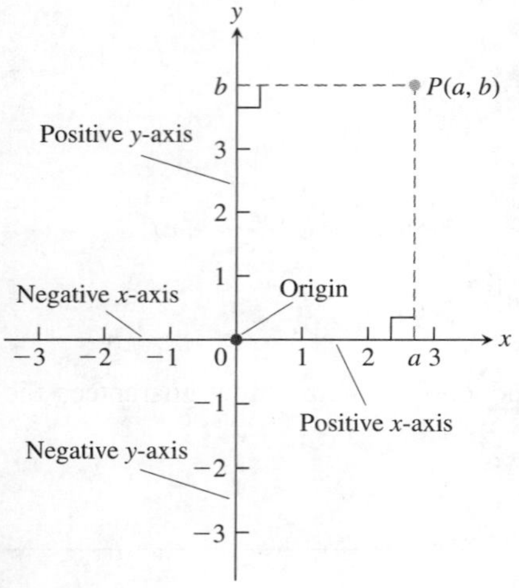

FIGURE A.4 Cartesian coordinates in the plane are based on two perpendicular axes intersecting at the origin.

These lines are called **coordinate axes**. On the horizontal x-axis, numbers are denoted by x and increase to the right. On the vertical y-axis, numbers are denoted by y and increase upward (Figure A.4). Thus "upward" and "to the right" are positive directions, whereas "downward" and "to the left" are considered negative. The **origin** O, also labeled 0, of the coordinate system is the point in the plane where x and y are both zero.

If P is any point in the plane, it can be located by exactly one ordered pair of real numbers in the following way. Draw lines through P perpendicular to the two coordinate axes. These lines intersect the axes at points with coordinates a and b (Figure A.4). The ordered pair (a, b) is assigned to the point P and is called its **coordinate pair**. The first number a is the ***x*-coordinate** (or **abscissa**) of P; the second number b is the ***y*-coordinate** (or **ordinate**) of P. The x-coordinate of every point on the y-axis is 0. The y-coordinate of every point on the x-axis is 0. The origin is the point $(0, 0)$.

Starting with an ordered pair (a, b), we can reverse the process and arrive at a corresponding point P in the plane. Often we identify P with the ordered pair and write $P(a, b)$. We sometimes also refer to "the point (a, b)" and it will be clear from the context when (a, b) refers to a point in the plane and not to an open interval on the real line. Several points labeled by their coordinates are shown in Figure A.5.

This coordinate system is called the **rectangular coordinate system** or **Cartesian coordinate system** (after the sixteenth-century French mathematician René Descartes). The coordinate axes of this coordinate or Cartesian plane divide the plane into four regions called **quadrants**, numbered counterclockwise as shown in Figure A.5.

HISTORICAL BIOGRAPHY
René Descartes
(1596–1650)
www.bit.ly/2PuKrlV

The **graph** of an equation or inequality in the variables x and y is the set of all points $P(x, y)$ in the plane whose coordinates satisfy the equation or inequality. When we plot data in the coordinate plane or graph formulas whose variables have different units of measure, we do not need to use the same scale on the two axes. If we plot time vs. thrust for a rocket motor, for example, there is no reason to place the mark that shows 1 sec on the time axis the same distance from the origin as the mark that shows 1 lb on the thrust axis.

Usually when we graph functions whose variables do not represent physical measurements and when we draw figures in the coordinate plane to study their geometry and trigonometry, we make the scales on the axes identical. A vertical unit of distance then looks the same as a horizontal unit. As on a surveyor's map or a scale drawing, line segments that are supposed to have the same length will look as if they do, and angles that are supposed to be congruent will look congruent.

Computer displays and calculator displays are another matter. The vertical and horizontal scales on machine-generated graphs usually differ, and there are corresponding distortions in distances, slopes, and angles. Circles may look like ellipses, rectangles may look like squares, right angles may appear to be acute or obtuse, and so on. We discuss these displays and distortions in greater detail in Section 1.4.

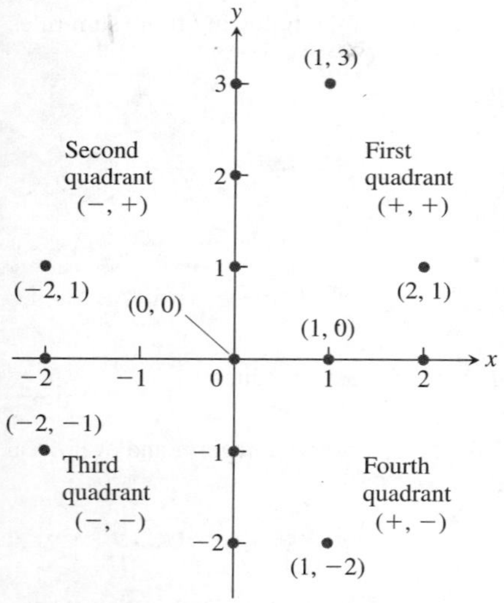

FIGURE A.5 Points labeled in the xy-coordinate or Cartesian plane. The points on the axes all have coordinate pairs but are usually labeled with single real numbers, (so $(1, 0)$ on the x-axis is labeled as 1). Notice the coordinate sign patterns of the quadrants.

Increments and Straight Lines

When a particle moves from one point in the plane to another, the net changes in its coordinates are called *increments*. They are calculated by subtracting the coordinates of the starting point from the coordinates of the ending point. If x changes from x_1 to x_2, the **increment** in x is

$$\Delta x = x_2 - x_1.$$

EXAMPLE 1 As shown in Figure A.6, in going from the point $A(4, -3)$ to the point $B(2, 5)$, the increments in the x- and y-coordinates are

$$\Delta x = 2 - 4 = -2, \qquad \Delta y = 5 - (-3) = 8.$$

From $C(5, 6)$ to $D(5, 1)$ the coordinate increments are

$$\Delta x = 5 - 5 = 0, \qquad \Delta y = 1 - 6 = -5. \quad ■$$

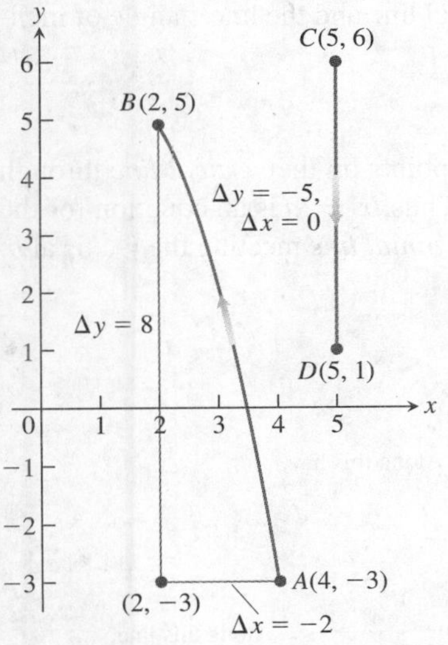

FIGURE A.6 Coordinate increments may be positive, negative, or zero (Example 1).

Given two points $P_1(x_1, y_1)$ and $P_2(x_2, y_2)$ in the plane, we call the increments $\Delta x = x_2 - x_1$ and $\Delta y = y_2 - y_1$ the **run** and the **rise**, respectively, between P_1 and P_2. Two such points always determine a unique straight line (usually called simply a line) passing through them both. We call the line P_1P_2.

Any nonvertical line in the plane has the property that the ratio

$$m = \frac{\text{rise}}{\text{run}} = \frac{\Delta y}{\Delta x} = \frac{y_2 - y_1}{x_2 - x_1}$$

has the same value for every choice of the two points $P_1(x_1, y_1)$ and $P_2(x_2, y_2)$ on the line (Figure A.7). This is because the ratios of corresponding sides for similar triangles are equal.

DEFINITION The constant ratio

$$m = \frac{\text{rise}}{\text{run}} = \frac{\Delta y}{\Delta x} = \frac{y_2 - y_1}{x_2 - x_1}$$

is the **slope** of the nonvertical line P_1P_2.

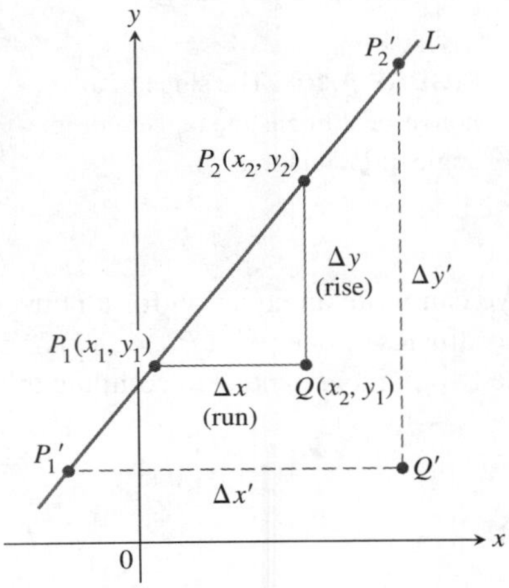

FIGURE A.7 Triangles P_1QP_2 and $P_1'Q'P_2'$ are similar, so the ratio of their sides has the same value for any two points on the line. This common value is the line's slope.

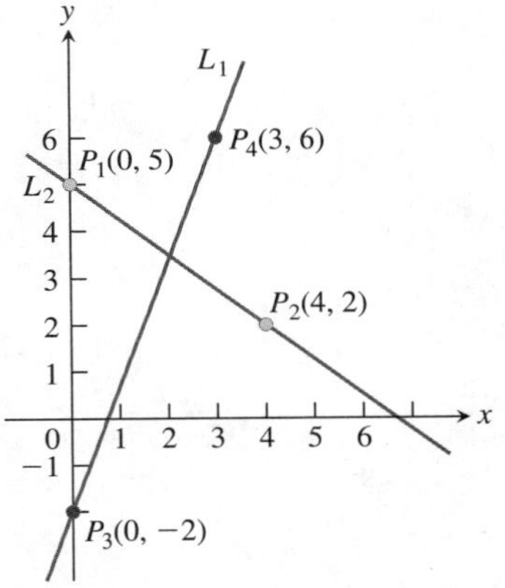

FIGURE A.8 The slope of L_1 is

$$m = \frac{\Delta y}{\Delta x} = \frac{6 - (-2)}{3 - 0} = \frac{8}{3}.$$

That is, y increases 8 units every time x increases 3 units. The slope of L_2 is

$$m = \frac{\Delta y}{\Delta x} = \frac{2 - 5}{4 - 0} = \frac{-3}{4}.$$

That is, y decreases 3 units every time x increases 4 units.

The slope tells us the direction (uphill, downhill) and steepness of a line. A line with positive slope rises uphill to the right; one with negative slope falls downhill to the right (Figure A.8). The greater the absolute value of the slope, the more rapid the rise or fall. The slope of a vertical line is *undefined*. Since the run Δx is zero for a vertical line, we cannot form the slope ratio m.

The direction and steepness of a line can also be measured with an angle. The **angle of inclination** of a line that crosses the x-axis is the smallest counterclockwise angle from the x-axis to the line (Figure A.9). The inclination of a horizontal line is $0°$. The inclination of a vertical line is $90°$. If ϕ (the Greek letter phi) is the inclination of a line, then $0 \le \phi < 180°$.

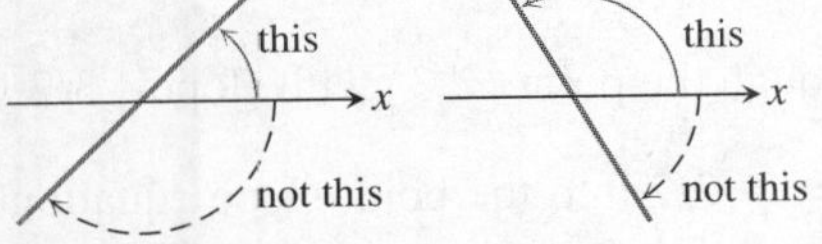

FIGURE A.9 Angles of inclination are measured counterclockwise from the x-axis.

The relationship between the slope m of a nonvertical line and the line's angle of inclination ϕ is shown in Figure A.10:

$$m = \tan \phi.$$

Straight lines have relatively simple equations. All points on the *vertical line* through the point a on the x-axis have x-coordinates equal to a. Thus, $x = a$ is an equation for the vertical line. Similarly, $y = b$ is an equation for the *horizontal line* meeting the y-axis at b. (See Figure A.11.)

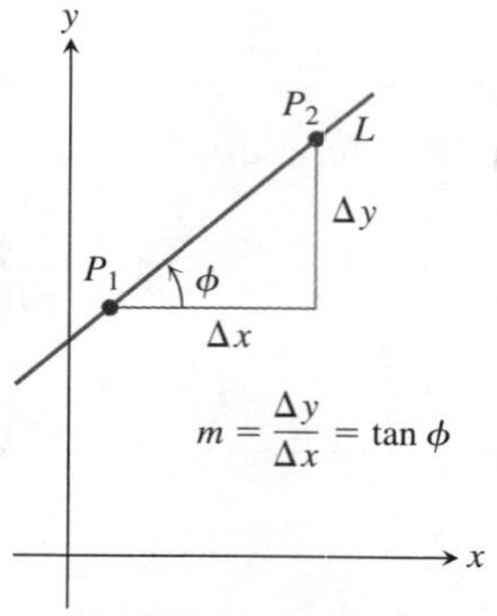

FIGURE A.10 The slope of a nonvertical line is the tangent of its angle of inclination.

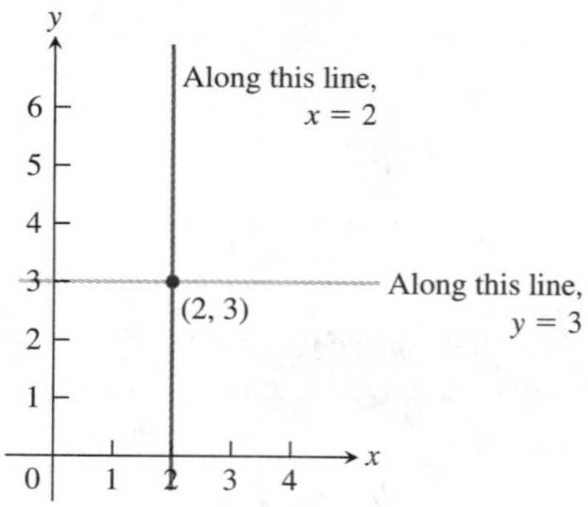

FIGURE A.11 The standard equations for the vertical and horizontal lines through $(2, 3)$ are $x = 2$ and $y = 3$.

We can write an equation for a nonvertical straight line L if we know its slope m and the coordinates of one point $P_1(x_1, y_1)$ on it. If $P(x, y)$ is *any* other point on L, then we can use the two points P_1 and P to compute the slope,

$$m = \frac{y - y_1}{x - x_1}$$

so that

$$y - y_1 = m(x - x_1), \quad \text{or} \quad y = y_1 + m(x - x_1).$$

The equation

$$y = y_1 + m(x - x_1)$$

is the **point-slope equation** of the line that passes through the point (x_1, y_1) and has slope m.

EXAMPLE 2 Write an equation for the line through the point $(2, 3)$ with slope $-3/2$.

Solution We substitute $x_1 = 2$, $y_1 = 3$, and $m = -3/2$ into the point-slope equation and obtain

$$y = 3 - \frac{3}{2}(x - 2), \quad \text{or} \quad y = -\frac{3}{2}x + 6. \qquad \blacksquare$$

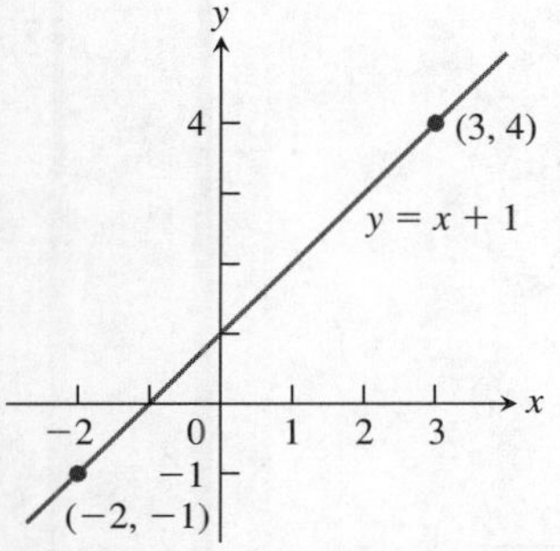

FIGURE A.12 The line in Example 3.

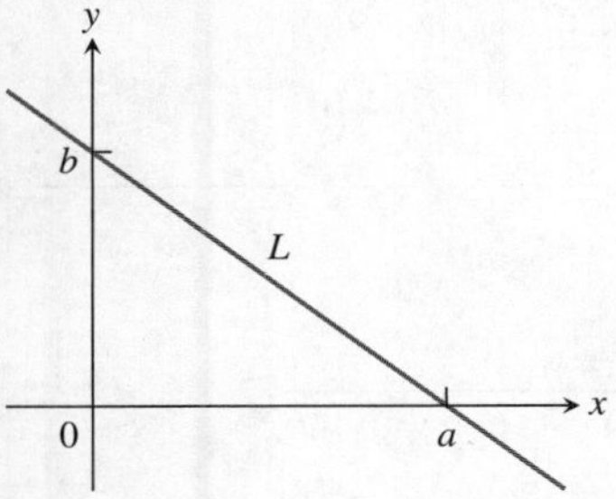

FIGURE A.13 Line L has x-intercept a and y-intercept b.

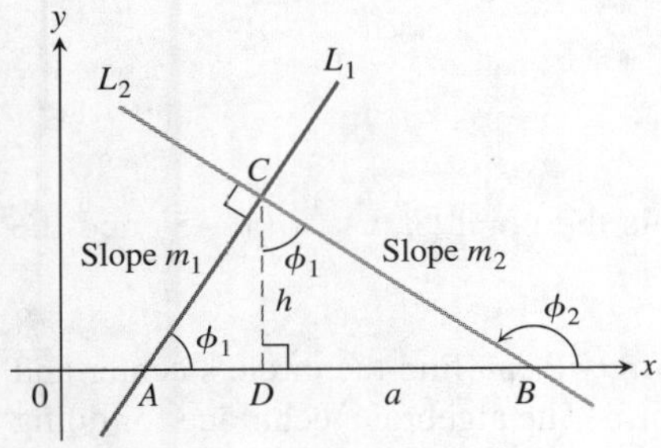

FIGURE A.14 ΔADC is similar to ΔCDB. Hence ϕ_1 is also the upper angle in ΔCDB. From the sides of ΔCDB, we read $\tan \phi_1 = a/h$.

EXAMPLE 3 Write an equation for the line through $(-2, -1)$ and $(3, 4)$.

Solution The line's slope is

$$m = \frac{-1 - 4}{-2 - 3} = \frac{-5}{-5} = 1.$$

We can use this slope with either of the two given points in the point-slope equation:

With $(x_1, y_1) = (-2, -1)$	With $(x_1, y_1) = (3, 4)$
$y = -1 + 1 \cdot (x - (-2))$	$y = 4 + 1 \cdot (x - 3)$
$y = -1 + x + 2$	$y = 4 + x - 3$
$y = x + 1$	$y = x + 1$

Same result

Either way, we see that $y = x + 1$ is an equation for the line (Figure A.12). ■

The y-coordinate of the point where a nonvertical line intersects the y-axis is called the ***y*-intercept** of the line. Similarly, the ***x*-intercept** of a nonhorizontal line is the x-coordinate of the point where it crosses the x-axis (Figure A.13). A line with slope m and y-intercept b passes through the point $(0, b)$, so it has equation

$$y = b + m(x - 0), \quad \text{or, more simply,} \quad y = mx + b.$$

> The equation
>
> $$y = mx + b$$
>
> is called the **slope-intercept equation** of the line with slope m and y-intercept b.

Lines with equations of the form $y = mx$ have y-intercept 0 and so pass through the origin. Equations of lines are called **linear** equations.

The equation

$$Ax + By = C \qquad (A \text{ and } B \text{ not both } 0)$$

is called the **general linear equation** in x and y because its graph always represents a line and every line has an equation in this form (including lines with undefined slope).

Parallel and Perpendicular Lines

Lines that are parallel have equal angles of inclination, so they have the same slope (if they are not vertical). Conversely, lines with equal slopes have equal angles of inclination and so are parallel.

If two nonvertical lines L_1 and L_2 are perpendicular, their slopes m_1 and m_2 satisfy $m_1 m_2 = -1$, so each slope is the *negative reciprocal* of the other:

$$m_1 = -\frac{1}{m_2}, \qquad m_2 = -\frac{1}{m_1}.$$

To see this, notice by inspecting similar triangles in Figure A.14 that $m_1 = a/h$, and $m_2 = -h/a$. Hence, $m_1 m_2 = (a/h)(-h/a) = -1$.

Distance and Circles in the Plane

The distance between points in the plane is calculated with a formula that comes from the Pythagorean theorem (Figure A.15).

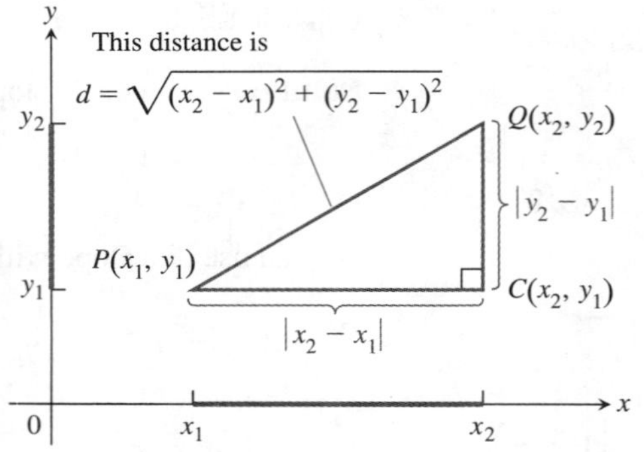

FIGURE A.15 To calculate the distance between $P(x_1, y_1)$ and $Q(x_2, y_2)$, apply the Pythagorean theorem to triangle PCQ.

Distance Formula for Points in the Plane
The distance between $P(x_1, y_1)$ and $Q(x_2, y_2)$ is

$$d = \sqrt{(\Delta x)^2 + (\Delta y)^2} = \sqrt{(x_2 - x_1)^2 + (y_2 - y_1)^2}.$$

By definition, a **circle** of radius a is the set of all points $P(x, y)$ whose distance from some center $C(h, k)$ equals a (Figure A.16). From the distance formula, P lies on the circle if and only if

$$\sqrt{(x - h)^2 + (y - k)^2} = a,$$

so

$$(x - h)^2 + (y - k)^2 = a^2. \tag{1}$$

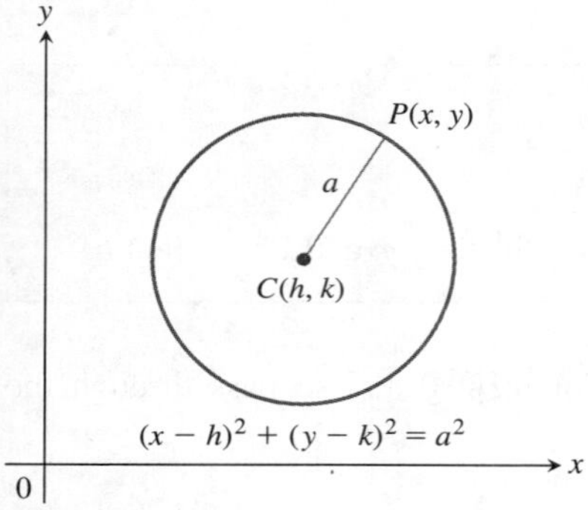

FIGURE A.16 A circle of radius a in the xy-plane, with center at (h, k).

Equation (1) is the **standard equation** of a circle with center (h, k) and radius a. The circle of radius $a = 1$ and centered at the origin is the **unit circle** with equation

$$x^2 + y^2 = 1.$$

EXAMPLE 4

(a) The standard equation for the circle of radius 2 centered at $(3, 4)$ is

$$(x - 3)^2 + (y - 4)^2 = 2^2 = 4.$$

(b) The circle

$$(x - 1)^2 + (y + 5)^2 = 3$$

has $h = 1$, $k = -5$, and $a = \sqrt{3}$. The center is the point $(h, k) = (1, -5)$ and the radius is $a = \sqrt{3}$. ■

If an equation for a circle is not in standard form, we can find the circle's center and radius by first converting the equation to standard form. The algebraic technique for doing so is *completing the square*.

EXAMPLE 5 Find the center and radius of the circle

$$x^2 + y^2 + 4x - 6y - 3 = 0.$$

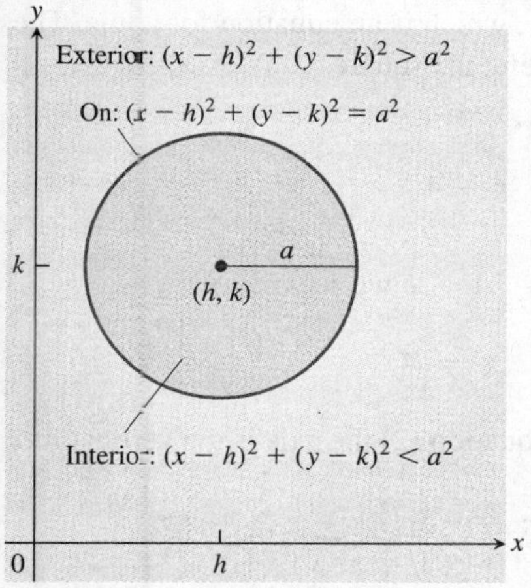

FIGURE A.17 The interior and exterior of the circle $(x-h)^2+(y-k)^2=a^2$.

Solution We convert the equation to standard form by completing the squares in x and y:

$$x^2+y^2+4x-6y-3=0$$

Start with the given equation.

$$(x^2+4x)+(y^2-6y)=3$$

Gather terms. Move the constant to the right-hand side.

$$\left(x^2+4x+\left(\frac{4}{2}\right)^2\right)+\left(y^2-6y+\left(\frac{-6}{2}\right)^2\right)=3+\left(\frac{4}{2}\right)^2+\left(\frac{-6}{2}\right)^2$$

Add the square of half the coefficient of x to each side of the equation. Do the same for y. The parenthetical expressions on the left-hand side are now perfect squares.

$$(x^2+4x+4)+(y^2-6y+9)=3+4+9$$

$$(x+2)^2+(y-3)^2=16$$

Write each quadratic as a squared linear expression.

The center is $(-2,3)$ and the radius is $a=4$. ■

The points (x, y) satisfying the inequality

$$(x-h)^2+(y-k)^2<a^2$$

make up the **interior** region of the circle with center (h, k) and radius a (Figure A.17). The circle's **exterior** consists of the points (x, y) satisfying

$$(x-h)^2+(y-k)^2>a^2.$$

Parabolas

The geometric definition and properties of general parabolas are reviewed in Chapter 11. Here we look at parabolas arising as the graphs of equations of the form $y=ax^2+bx+c$.

EXAMPLE 6 Consider the equation $y=x^2$. Some points whose coordinates satisfy this equation are $(0,0)$, $(1,1)$, $\left(\frac{3}{2},\frac{9}{4}\right)$, $(-1,1)$, $(2,4)$, and $(-2,4)$. These points (and all others satisfying the equation) make up a smooth curve called a parabola (Figure A.18). ■

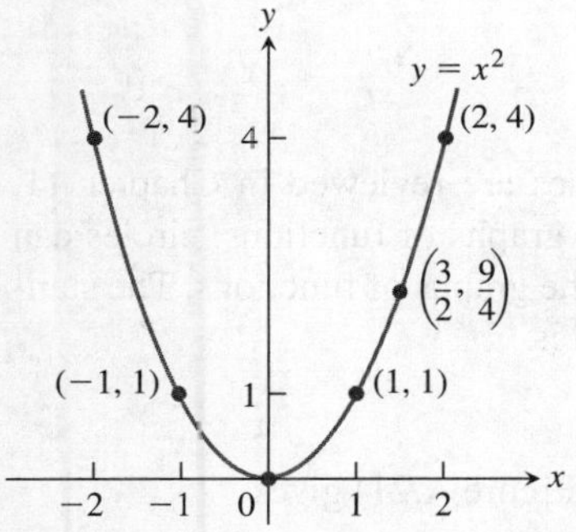

FIGURE A.18 The parabola $y=x^2$ (Example 6).

The graph of an equation of the form

$$y=ax^2$$

is a **parabola** whose **axis** (axis of symmetry) is the y-axis. The parabola's **vertex** (point where the parabola and axis cross) lies at the origin. The parabola opens upward if $a>0$ and downward if $a<0$. The larger the value of $|a|$, the narrower the parabola (Figure A.19).

Generally, the graph of $y=ax^2+bx+c$ is a shifted and scaled version of the parabola $y=x^2$. We discuss shifting and scaling of graphs in more detail in Section 1.2.

The Graph of $y=ax^2+bx+c,\quad a\neq 0$

The graph of the equation $y=ax^2+bx+c$, $a\neq 0$, is a parabola. The parabola opens upward if $a>0$ and downward if $a<0$. The **axis** is the line

$$x=-\frac{b}{2a}. \qquad (2)$$

The **vertex** of the parabola is the point where the axis and parabola intersect. Its x-coordinate is $x=-b/2a$; its y-coordinate is found by substituting $x=-b/2a$ in the parabola's equation.

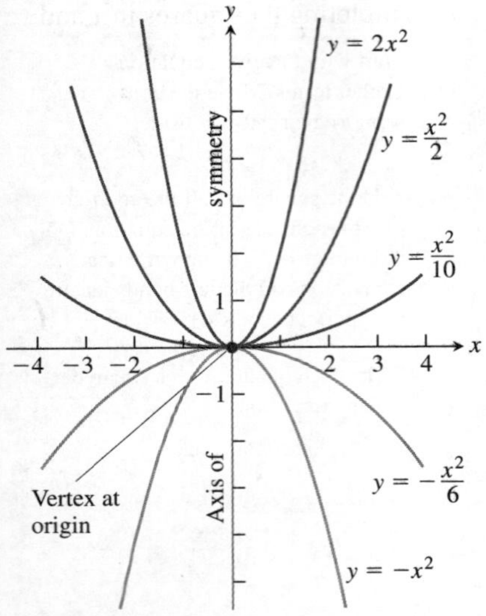

FIGURE A.19 Besides determining the direction in which the parabola $y = ax^2$ opens, the number a is a scaling factor. The parabola widens as a approaches zero and narrows as $|a|$ becomes large.

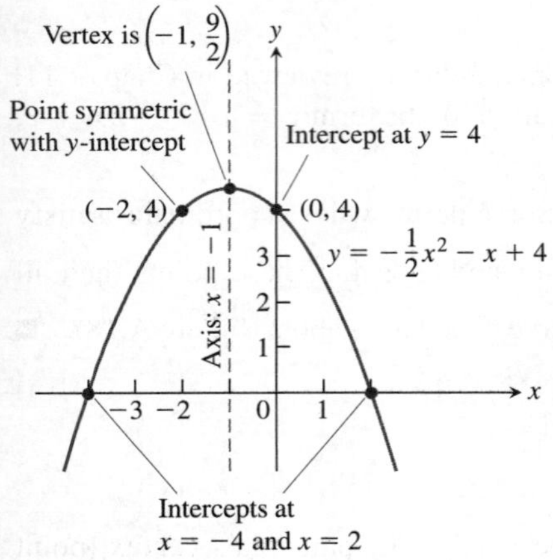

FIGURE A.20 The parabola in Example 7.

Notice that if $a = 0$, then we have $y = bx + c$, which is an equation for a line. The axis, given by Equation (2), can be found by completing the square.

EXAMPLE 7 Graph the equation $y = -\frac{1}{2}x^2 - x + 4$.

Solution Comparing the equation with $y = ax^2 + bx + c$ we see that

$$a = -\frac{1}{2}, \qquad b = -1, \qquad c = 4.$$

Since $a < 0$, the parabola opens downward. From Equation (2) the axis is the vertical line

$$x = -\frac{b}{2a} = -\frac{(-1)}{2(-1/2)} = -1.$$

When $x = -1$, we have

$$y = -\frac{1}{2}(-1)^2 - (-1) + 4 = \frac{9}{2}.$$

The vertex is $(-1, 9/2)$.

The x-intercepts are where $y = 0$:

$$-\frac{1}{2}x^2 - x + 4 = 0$$
$$x^2 + 2x - 8 = 0$$
$$(x - 2)(x + 4) = 0$$
$$x = 2, \quad x = -4$$

We plot some points, sketch the axis, and use the direction of opening to complete the graph in Figure A.20. ■

Ellipses

The geometric definition and properties of general ellipses are reviewed in Chapter 11. Here we relate them to circles. Although they are not the graphs of functions, circles can be stretched horizontally or vertically in the same way as the graphs of functions. The standard equation for a circle of radius r centered at the origin is

$$x^2 + y^2 = r^2.$$

Substituting cx for x in the standard equation for a circle (Figure A.21) gives

$$c^2x^2 + y^2 = r^2. \tag{3}$$

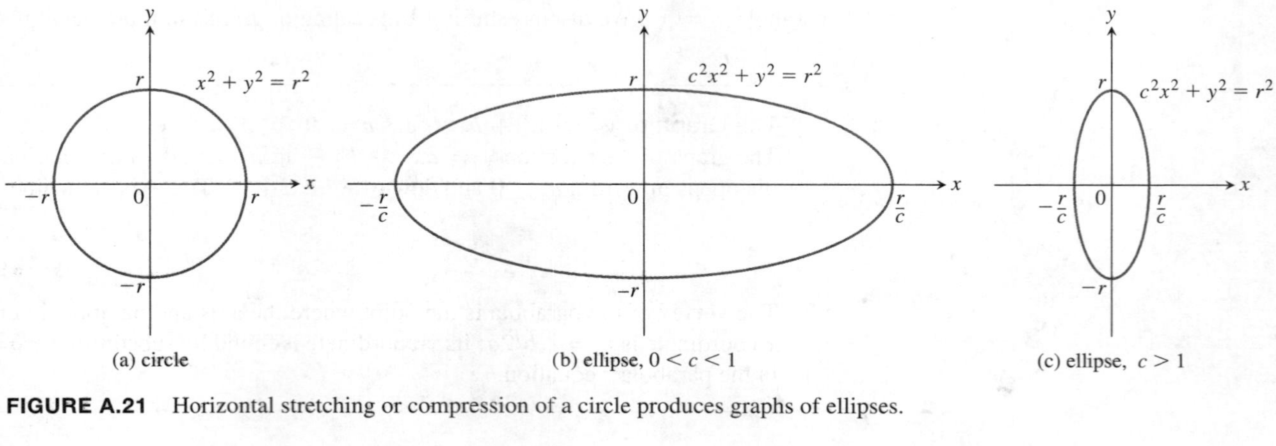

FIGURE A.21 Horizontal stretching or compression of a circle produces graphs of ellipses.

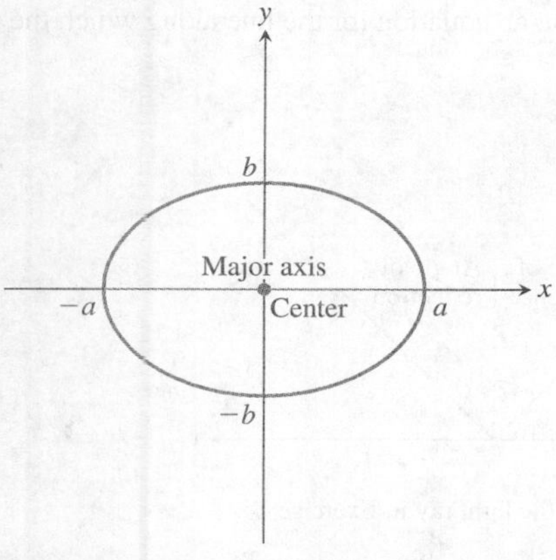

FIGURE A.22 Graph of the ellipse $\frac{x^2}{a^2} + \frac{y^2}{b^2} = 1$, $a > b$, where the major axis is horizontal.

If $0 < c < 1$, the graph of Equation (3) horizontally stretches the circle; if $c > 1$ the circle is compressed horizontally. In either case, the graph of Equation (3) is an ellipse (Figure A.21). Notice in Figure A.21 that the y-intercepts of all three graphs are always $-r$ and r. In Figure A.21b, the line segment joining the points $(\pm r/c, 0)$ is called the **major axis** of the ellipse; the **minor axis** is the line segment joining $(0, \pm r)$. The axes of the ellipse are reversed in Figure A.21c: The major axis is the line segment joining the points $(0, \pm r)$, and the minor axis is the line segment joining the points $(\pm r/c, 0)$. In both cases, the major axis is the longer line segment.

If we divide both sides of Equation (3) by r^2, we obtain

$$\frac{x^2}{a^2} + \frac{y^2}{b^2} = 1 \tag{4}$$

where $a = r/c$ and $b = r$. If $a > b$, the major axis is horizontal; if $a < b$, the major axis is vertical. The **center** of the ellipse given by Equation (4) is the origin (Figure A.22).

Substituting $x - h$ for x, and $y - k$ for y, in Equation (4) results in

$$\frac{(x-h)^2}{a^2} + \frac{(y-k)^2}{b^2} = 1. \tag{5}$$

Equation (5) is the **standard equation of an ellipse** with center at (h, k).

EXERCISES A.3

Distance, Slopes, and Lines

In Exercises 1 and 2, a particle moves from A to B in the coordinate plane. Find the increments Δx and Δy in the particle's coordinates. Also find the distance from A to B.

1. $A(-3, 2)$, $B(-1, -2)$

2. $A(-3.2, -2)$, $B(-8.1, -2)$

Describe the graphs of the equations in Exercises 3 and 4.

3. $x^2 + y^2 = 1$

4. $x^2 + y^2 \le 3$

Plot the points in Exercises 5 and 6 and find the slope (if any) of the line they determine. Also find the common slope (if any) of the lines perpendicular to line AB.

5. $A(-1, 2)$, $B(-2, -1)$

6. $A(2, 3)$, $B(-1, 3)$

In Exercises 7 and 8, find an equation for **(a)** the vertical line and **(b)** the horizontal line through the given point.

7. $(-1, 4/3)$

8. $(0, -\sqrt{2})$

In Exercises 9–15, write an equation for each line described.

9. Passes through $(-1, 1)$ with slope -1

10. Passes through $(3, 4)$ and $(-2, 5)$

11. Has slope $-5/4$ and y-intercept 6

12. Passes through $(-12, -9)$ and has slope 0

13. Has y-intercept 4 and x-intercept -1

14. Passes through $(5, -1)$ and is parallel to the line $2x + 5y = 15$

15. Passes through $(4, 10)$ and is perpendicular to the line $6x - 3y = 5$

In Exercises 16 and 17, find the line's x- and y-intercepts and use this information to graph the line.

16. $3x + 4y = 12$

17. $\sqrt{2}x - \sqrt{3}y = \sqrt{6}$

18. Is there anything special about the relationship between the lines $Ax + By = C_1$ and $Bx - Ay = C_2$ $(A \ne 0, B \ne 0)$? Give reasons for your answer.

19. A particle starts at $A(-2, 3)$ and its coordinates change by increments $\Delta x = 5$, $\Delta y = -6$. Find its new position.

20. The coordinates of a particle change by $\Delta x = 5$ and $\Delta y = 6$ as it moves from $A(x, y)$ to $B(3, -3)$. Find x and y.

Circles

In Exercises 21–23, find an equation for the circle with the given center $C(h, k)$ and radius a. Then sketch the circle in the xy-plane. Include the circle's center in your sketch. Also, label the circle's x- and y-intercepts, if any, with their coordinate pairs.

21. $C(0, 2)$, $a = 2$

22. $C(-1, 5)$, $a = \sqrt{10}$

23. $C(-\sqrt{3}, -2)$, $a = 2$

Graph the circles whose equations are given in Exercises 24–26. Label each circle's center and intercepts (if any) with their coordinate pairs.

24. $x^2 + y^2 + 4x - 4y + 4 = 0$

25. $x^2 + y^2 - 3y - 4 = 0$

26. $x^2 + y^2 - 4x + 4y = 0$

Parabolas

Graph the parabolas in Exercises 27–30. Label the vertex, axis, and intercepts in each case.

27. $y = x^2 - 2x - 3$

28. $y = -x^2 + 4x$

29. $y = -x^2 - 6x - 5$

30. $y = \frac{1}{2}x^2 + x + 4$

Inequalities

Describe the regions defined by the inequalities and pairs of inequalities in Exercises 31–34.

31. $x^2 + y^2 > 7$ **32.** $(x - 1)^2 + y^2 \leq 4$

33. $x^2 + y^2 > 1, \quad x^2 + y^2 < 4$

34. $x^2 + y^2 + 6y < 0, \quad y > -3$

35. Write an inequality that describes the points that lie inside the circle with center $(-2, 1)$ and radius $\sqrt{6}$.

36. Write a pair of inequalities that describe the points that lie inside or on the circle with center $(0, 0)$ and radius $\sqrt{2}$, and on or to the right of the vertical line through $(1, 0)$.

Theory and Examples

In Exercises 37–40, graph the two equations and find the points at which the graphs intersect.

37. $y = 2x, \quad x^2 + y^2 = 1$ **38.** $y - x = 1, \quad y = x^2$

39. $y = -x^2, \quad y = 2x^2 - 1$

40. $x^2 + y^2 = 1, \quad (x - 1)^2 + y^2 = 1$

41. Insulation By measuring slopes in the figure, estimate the temperature change in degrees per inch for **(a)** the gypsum wallboard; **(b)** the fiberglass insulation; **(c)** the wood sheathing.

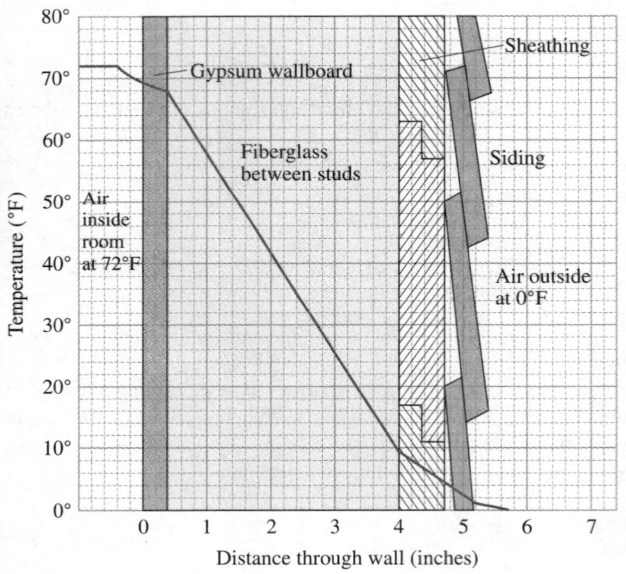

The temperature changes in the wall in Exercises 41 and 42.

42. Insulation According to the figure in Exercise 41, which of the materials is the best insulator? The poorest? Explain.

43. Pressure under water The pressure p experienced by a diver under water is related to the diver's depth d by an equation of the form $p = kd + 1$ (k a constant). At the surface, the pressure is 1 atmosphere. The pressure at 100 meters is about 10.94 atmospheres. Find the pressure at 50 meters.

44. Reflected light A ray of light comes in along the line $x + y = 1$ from the second quadrant and reflects off the x-axis (see the accompanying figure). The angle of incidence is equal to the angle of reflection. Write an equation for the line along which the departing light travels.

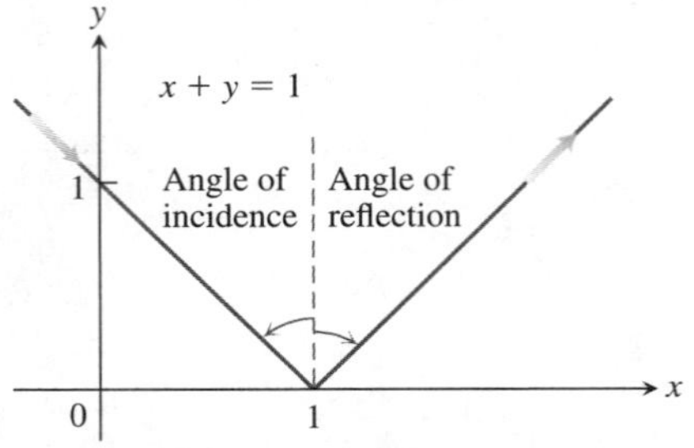

The path of the light ray in Exercise 44.

Angles of incidence and reflection are measured from the perpendicular.

45. Fahrenheit vs. Celsius In the FC-plane, sketch the graph of the equation

$$C = \frac{5}{9}(F - 32)$$

linking Fahrenheit and Celsius temperatures. On the same graph sketch the line $C = F$. Is there a temperature at which a Celsius thermometer gives the same numerical reading as a Fahrenheit thermometer? If so, find it.

46. The Mt. Washington Cog Railway Civil engineers calculate the slope of roadbed as the ratio of the distance it rises or falls to the distance it runs horizontally. They call this ratio the **grade** of the roadbed, usually written as a percentage. Along the coast, commercial railroad grades are usually less than 2%. In the mountains, they may go as high as 4%. Highway grades are usually less than 5%.

The steepest part of the Mt. Washington Cog Railway in New Hampshire has an exceptional 37.1% grade. Along this part of the track, the seats in the front of the car are 14 ft above those in the rear. About how far apart are the front and rear rows of seats?

47. By calculating the lengths of its sides, show that the triangle with vertices at the points $A(1, 2)$, $B(5, 5)$, and $C(4, -2)$ is isosceles but not equilateral.

48. Show that the triangle with vertices $A(0, 0)$, $B(1, \sqrt{3})$, and $C(2, 0)$ is equilateral.

49. Show that the points $A(2, -1)$, $B(1, 3)$, and $C(-3, 2)$ are vertices of a square, and find the fourth vertex.

50. Three different parallelograms have vertices at $(-1, 1)$, $(2, 0)$, and $(2, 3)$. Sketch them and find the coordinates of the fourth vertex of each.

51. For what value of k is the line $2x + ky = 3$ perpendicular to the line $4x + y = 1$? For what value of k are the lines parallel?

52. Midpoint of a line segment Show that the point with coordinates

$$\left(\frac{x_1 + x_2}{2}, \frac{y_1 + y_2}{2}\right)$$

is the midpoint of the line segment joining $P(x_1, y_1)$ to $Q(x_2, y_2)$.

A.4 Proofs of Limit Theorems

This appendix proves Theorem 1, Parts 2–5, and Theorem 4 from Section 2.2.

THEOREM 1—Limit Laws
If L, M, c, and k are real numbers and

$$\lim_{x\to c} f(x) = L \quad \text{and} \quad \lim_{x\to c} g(x) = M, \quad \text{then}$$

1. *Sum Rule*: $\lim_{x\to c}(f(x) + g(x)) = L + M$
2. *Difference Rule*: $\lim_{x\to c}(f(x) - g(x)) = L - M$
3. *Constant Multiple Rule*: $\lim_{x\to c}(kf(x)) = kL$
4. *Product Rule*: $\lim_{x\to c}(f(x)g(x)) = LM$
5. *Quotient Rule*: $\lim_{x\to c}\dfrac{f(x)}{g(x)} = \dfrac{L}{M}, \quad M \neq 0$
6. *Power Rule*: $\lim_{x\to c}[f(x)]^n = L^n$, n a positive integer
7. *Root Rule*: $\lim_{x\to c}\sqrt[n]{f(x)} = \sqrt[n]{L} = L^{1/n}$, n a positive integer

(If n is even, we assume that $\lim_{x\to c} f(x) = L > 0$.)

We proved the Sum Rule in Section 2.3, and the Power and Root Rules are proved in more advanced texts. We obtain the Difference Rule by replacing $g(x)$ by $-g(x)$ and M by $-M$ in the Sum Rule. The Constant Multiple Rule is the special case $g(x) = k$ of the Product Rule. This leaves only the Product and Quotient Rules.

Proof of the Limit Product Rule We show that for any $\varepsilon > 0$ there exists a $\delta > 0$ such that for all x in the intersection D of the domains of f and g,

$$|f(x)g(x) - LM| < \varepsilon \quad \text{whenever} \quad 0 < |x - c| < \delta.$$

Suppose then that ε is a positive number, and write $f(x)$ and $g(x)$ as

$$f(x) = L + (f(x) - L), \qquad g(x) = M + (g(x) - M).$$

Multiply these expressions together and subtract LM:

$$\begin{aligned}
f(x)g(x) - LM &= (L + (f(x) - L))(M + (g(x) - M)) - LM \\
&= LM + L(g(x) - M) + M(f(x) - L) \\
&\quad + (f(x) - L)(g(x) - M) - LM \\
&= L(g(x) - M) + M(f(x) - L) + (f(x) - L)(g(x) - M). \quad (1)
\end{aligned}$$

Since f and g have limits L and M as $x \to c$, there exist positive numbers δ_1, δ_2, δ_3, and δ_4 such that

$$\begin{aligned}
&|f(x) - L| < \sqrt{\varepsilon/3} && \text{whenever} \quad 0 < |x - c| < \delta_1 \\
&|g(x) - M| < \sqrt{\varepsilon/3} && \text{whenever} \quad 0 < |x - c| < \delta_2 \\
&|f(x) - L| < \varepsilon/(3(1 + |M|)) && \text{whenever} \quad 0 < |x - c| < \delta_3 \\
&|g(x) - M| < \varepsilon/(3(1 + |L|)) && \text{whenever} \quad 0 < |x - c| < \delta_4.
\end{aligned} \quad (2)$$

If we take δ to be the smallest of the numbers δ_1 through δ_4, the inequalities on the right-hand side of the Implications (2) will hold simultaneously for $0 < |x - c| < \delta$. Therefore, for all x in D, if $0 < |x - c| < \delta$ then

$$\begin{aligned}
&|f(x)g(x) - LM| && \text{Triangle inequality applied to Eq. (1)}\\
&\quad \le |L||g(x) - M| + |M||f(x) - L| + |f(x) - L||g(x) - M|\\
&\quad \le (1 + |L|)|g(x) - M| + (1 + |M|)|f(x) - L| + |f(x) - L||g(x) - M|\\
&\quad < \frac{\varepsilon}{3} + \frac{\varepsilon}{3} + \sqrt{\frac{\varepsilon}{3}}\sqrt{\frac{\varepsilon}{3}} = \varepsilon. && \text{Values from (2)}
\end{aligned}$$

This completes the proof of the Limit Product Rule. ■

Proof of the Limit Quotient Rule We show that $\lim_{x \to c} (1/g(x)) = 1/M$. We can then conclude that

$$\lim_{x \to c} \frac{f(x)}{g(x)} = \lim_{x \to c} \left(f(x) \cdot \frac{1}{g(x)} \right) = \lim_{x \to c} f(x) \cdot \lim_{x \to c} \frac{1}{g(x)} = L \cdot \frac{1}{M} = \frac{L}{M}$$

by the Limit Product Rule.

Let $\varepsilon > 0$ be given. To show that $\lim_{x \to c} (1/g(x)) = 1/M$, we need to show that there exists a $\delta > 0$ such that

$$\left| \frac{1}{g(x)} - \frac{1}{M} \right| < \varepsilon \quad \text{whenever} \quad 0 < |x - c| < \delta.$$

Since g has the limit M as $x \to c$ and since $|M| > 0$, there exists a positive number δ_1 such that

$$|g(x) - M| < \frac{M}{2} \quad \text{whenever} \quad 0 < |x - c| < \delta_1. \tag{3}$$

For any numbers A and B, the triangle inequality implies that $|A| - |B| \le |A - B|$ and $|B| - |A| \le |A - B|$, from which it follows that $\big||A| - |B|\big| \le |A - B|$. With $A = g(x)$ and $B = M$, this becomes

$$\big||g(x)| - |M|\big| \le |g(x) - M|,$$

which can be combined with the inequality on the right in Implication (3) to get, in turn,

$$\begin{aligned}
\big||g(x)| - |M|\big| &< \frac{|M|}{2}\\
-\frac{|M|}{2} < |g(x)| - |M| &< \frac{|M|}{2}\\
\frac{|M|}{2} < |g(x)| &< \frac{3|M|}{2}\\
|M| < 2|g(x)| &< 3|M|\\
\frac{1}{|g(x)|} < \frac{2}{|M|} &< \frac{3}{|g(x)|}.
\end{aligned} \tag{4}$$

Therefore, $0 < |x - c| < \delta_1$ implies that

$$\begin{aligned}
\left| \frac{1}{g(x)} - \frac{1}{M} \right| &= \left| \frac{M - g(x)}{Mg(x)} \right| \le \frac{1}{|M|} \cdot \frac{1}{|g(x)|} \cdot |M - g(x)|\\
&< \frac{1}{|M|} \cdot \frac{2}{|M|} \cdot |M - g(x)|. \qquad \text{Inequality (4)}
\end{aligned} \tag{5}$$

Since $(1/2)|M|^2\varepsilon > 0$, there exists a number $\delta_2 > 0$ such that

$$|M - g(x)| < \frac{\varepsilon}{2}|M|^2 \qquad \text{whenever} \qquad 0 < |x - c| < \delta_2. \tag{6}$$

If we take δ to be the smaller of δ_1 and δ_2, the conclusions in (5) and (6) both hold whenever $0 < |x - c| < \delta$. Combining these conclusions gives

$$\left|\frac{1}{g(x)} - \frac{1}{M}\right| < \varepsilon \qquad \text{whenever} \qquad 0 < |x - c| < \delta.$$

This concludes the proof of the Limit Quotient Rule. ■

THEOREM 4—The Sandwich Theorem
Suppose that $g(x) \le f(x) \le h(x)$ for all x in some open interval I containing c, except possibly at $x = c$ itself. Suppose also that $\lim_{x\to c} g(x) = \lim_{x\to c} h(x) = L$. Then $\lim_{x\to c} f(x) = L$.

Proof for Right-Hand Limits Suppose $\lim_{x\to c^+} g(x) = \lim_{x\to c^+} h(x) = L$. Then for any $\varepsilon > 0$ there exists a $\delta > 0$ such that the interval $(c, c + \delta)$ is contained in I, and

$$L - \varepsilon < g(x) < L + \varepsilon \qquad \text{and} \qquad L - \varepsilon < h(x) < L + \varepsilon$$

whenever $c < x < c + \delta$. Since we always have $g(x) \le f(x) \le h(x)$ it follows that if $c < x < c + \delta$, then

$$L - \varepsilon < g(x) \le f(x) \le h(x) < L + \varepsilon,$$
$$L - \varepsilon < f(x) < L + \varepsilon,$$
$$-\varepsilon < f(x) - L < \varepsilon.$$

Therefore, $|f(x) - L| < \varepsilon$ whenever $c < x < c + \delta$.

Proof for Left-Hand Limits Suppose $\lim_{x\to c^-} g(x) = \lim_{x\to c^-} h(x) = L$. Then for any $\varepsilon > 0$ there exists a $\delta > 0$ such that the interval $(c - \delta, c)$ is contained in I, and

$$L - \varepsilon < g(x) < L + \varepsilon \qquad \text{and} \qquad L - \varepsilon < h(x) < L + \varepsilon$$

whenever $c - \delta < x < c$. We conclude as before that $|f(x) - L| < \varepsilon$ whenever $c - \delta < x < c$.

Proof for Two-Sided Limits If $\lim_{x\to c} g(x) = \lim_{x\to c} h(x) = L$, then $g(x)$ and $h(x)$ both approach L as $x \to c^+$ and as $x \to c^-$; so $\lim_{x\to c^+} f(x) = L$ and $\lim_{x\to c^-} f(x) = L$. Hence $\lim_{x\to c} f(x)$ exists and equals L. ■

EXERCISES A.4

1. Suppose that functions $f_1(x)$, $f_2(x)$, and $f_3(x)$ have limits L_1, L_2, and L_3, respectively, as $x \to c$. Show that their sum has limit $L_1 + L_2 + L_3$. Use mathematical induction (Appendix A.2) to generalize this result to the sum of any finite number of functions.

2. Use mathematical induction and the Limit Product Rule in Theorem 1 to show that if functions $f_1(x), f_2(x), \ldots, f_n(x)$ have limits $L_1, L_2, \ldots, L_n$ as $x \to c$, then

$$\lim_{x\to c} f_1(x) \cdot f_2(x) \cdot \cdots \cdot f_n(x) = L_1 \cdot L_2 \cdot \cdots \cdot L_n.$$

3. Use the fact that $\lim_{x \to c} x = c$ and the result of Exercise 2 to show that $\lim_{x \to c} x^n = c^n$ for any integer $n > 1$.

4. Limits of polynomials Use the fact that $\lim_{x \to c} (k) = k$ for any number k, together with the results of Exercises 1 and 3, to show that $\lim_{x \to c} f(x) = f(c)$ for any polynomial function

$$f(x) = a_n x^n + a_{n-1} x^{n-1} + \cdots + a_1 x + a_0.$$

5. Limits of rational functions Use Theorem 1 and the result of Exercise 4 to show that if $f(x)$ and $g(x)$ are polynomial functions and $g(c) \neq 0$, then

$$\lim_{x \to c} \frac{f(x)}{g(x)} = \frac{f(c)}{g(c)}.$$

6. Composites of continuous functions Figure A.23 gives the diagram for a proof that the composite of two continuous functions is continuous. Reconstruct the proof from the diagram. The statement to be proved is this: If f is continuous at $x = c$ and g is continuous at $f(c)$, then $g \circ f$ is continuous at c.

Assume that c is an interior point of the domain of f and that $f(c)$ is an interior point of the domain of g. This will make the limits involved two-sided. (The arguments for the cases that involve one-sided limits are similar.)

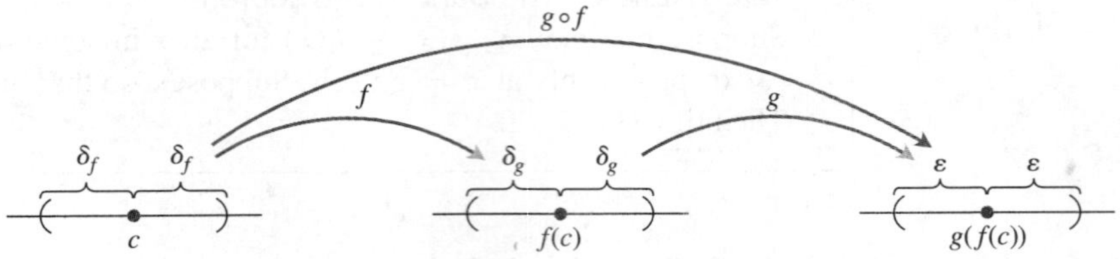

FIGURE A.23 The diagram for a proof that the composite of two continuous functions is continuous.

A.5 Commonly Occurring Limits

This appendix verifies limits (4)–(6) in Theorem 5 of Section 10.1.

Limit 4: If $|x| < 1$, $\lim_{n \to \infty} x^n = 0$ We need to show that to each $\varepsilon > 0$ there corresponds an integer N so large that $|x^n| < \varepsilon$ for all n greater than N. Since $\varepsilon^{1/n} \to 1$, while $|x| < 1$, there exists an integer N for which $\varepsilon^{1/N} > |x|$. In other words,

$$|x^N| = |x|^N < \varepsilon. \tag{1}$$

This is the integer we seek because, if $|x| < 1$, then

$$|x^n| < |x^N| \quad \text{for all } n > N. \tag{2}$$

Combining (1) and (2) produces $|x^n| < \varepsilon$ for all $n > N$, concluding the proof. ■

Limit 5: For any number x, $\lim_{n \to \infty} \left(1 + \frac{x}{n}\right)^n = e^x$ Let

$$a_n = \left(1 + \frac{x}{n}\right)^n.$$

Then

$$\ln a_n = \ln\left(1 + \frac{x}{n}\right)^n = n \ln\left(1 + \frac{x}{n}\right) \to x,$$

as we can see by the following application of L'Hôpital's Rule, in which we differentiate with respect to n:

$$\lim_{n \to \infty} n \ln\left(1 + \frac{x}{n}\right) = \lim_{n \to \infty} \frac{\ln(1 + x/n)}{1/n}$$

$$= \lim_{n \to \infty} \frac{\left(\frac{1}{1 + x/n}\right) \cdot \left(-\frac{x}{n^2}\right)}{-1/n^2} = \lim_{n \to \infty} \frac{x}{1 + x/n} = x.$$

Apply Theorem 3, Section 10.1, with $f(x) = e^x$ to conclude that

$$\left(1 + \frac{x}{n}\right)^n = a_n = e^{\ln a_n} \to e^x. \qquad \blacksquare$$

Limit 6: For any number x, $\lim_{n\to\infty} \frac{x^n}{n!} = 0$ Since

$$-\frac{|x|^n}{n!} \le \frac{x^n}{n!} \le \frac{|x|^n}{n!},$$

all we need to show is that $|x|^n/n! \to 0$. We can then apply the Sandwich Theorem for Sequences (Section 10.1, Theorem 2) to conclude that $x^n/n! \to 0$.

The first step in showing that $|x|^n/n! \to 0$ is to choose an integer $M > |x|$, so that $(|x|/M) < 1$. By Limit 4, just proved, we then have $(|x|/M)^n \to 0$. We then restrict our attention to values of $n > M$. For these values of n, we can write

$$\frac{|x|^n}{n!} = \frac{|x|^n}{1 \cdot 2 \cdot \cdots \cdot M \cdot \underbrace{(M+1)\cdot(M+2)\cdot \cdots \cdot n}_{(n-M)\text{ factors}}}$$

$$\le \frac{|x|^n}{M!M^{n-M}} = \frac{|x|^n M^M}{M!M^n} = \frac{M^M}{M!}\left(\frac{|x|}{M}\right)^n.$$

Thus,

$$0 \le \frac{|x|^n}{n!} \le \frac{M^M}{M!}\left(\frac{|x|}{M}\right)^n.$$

Now, the constant $M^M/M!$ does not change as n increases. Thus the Sandwich Theorem tells us that $|x|^n/n! \to 0$ because $(|x|/M)^n \to 0$. $\blacksquare$

A.6 Theory of the Real Numbers

A rigorous development of calculus is based on properties of the real numbers. Many results about functions, derivatives, and integrals would be false if stated for functions defined only on the rational numbers. In this appendix we briefly examine some basic concepts of the theory of the reals that hint at what might be learned in a deeper, more theoretical study of calculus.

Three types of properties make the real numbers what they are. These are the **algebraic**, **order**, and **completeness** properties. The algebraic properties involve addition and multiplication, subtraction and division. They apply to rational or complex numbers (discussed in Chapter 18) as well as to the reals.

The structure of numbers is built around a set with addition and multiplication operations. The following properties are required of addition and multiplication.

A1 $a + (b + c) = (a + b) + c$ for all a, b, c.

A2 $a + b = b + a$ for all a, b.

A3 There is a number called "0" such that $a + 0 = a$ for all a.

A4 For each number a, there is a number b such that $a + b = 0$.

M1 $a(bc) = (ab)c$ for all a, b, c.

M2 $ab = ba$ for all a, b.

M3 There is a number called "1" such that $a \cdot 1 = a$ for all a.

M4 For each nonzero number a, there is a number b such that $ab = 1$.

D $a(b + c) = ab + ac$ for all a, b, c.

A1 and M1 are *associative laws*, A2 and M2 are *commutativity laws*, A3 and M3 are *identity laws*, and D is the *distributive law*. Sets that have these algebraic properties are examples of **fields**; they are studied in depth in the area of theoretical mathematics called abstract algebra.

The **order** properties enable us to compare the sizes of any two numbers. The order properties are

O1 For any a and b, either $a \leq b$ or $b \leq a$ or both.

O2 If $a \leq b$ and $b \leq a$ then $a = b$.

O3 If $a \leq b$ and $b \leq c$ then $a \leq c$.

O4 If $a \leq b$ then $a + c \leq b + c$.

O5 If $a \leq b$ and $0 \leq c$ then $ac \leq bc$.

O3 is the *transitivity law*, and O4 and O5 relate ordering to addition and multiplication.

We can order the reals, the integers, and the rational numbers, but we cannot order the complex numbers (there is no reasonable way to decide whether a number like $i = \sqrt{-1}$ is bigger or smaller than zero). A field in which the size of any two elements can be compared as above is called an **ordered field**. Both the rational numbers and the real numbers are ordered fields, and there are many others.

We can think of real numbers geometrically, lining them up as points on a line. The **completeness property** says that the real numbers correspond to all points on the line, with no "holes" or "gaps." The rationals, in contrast, omit points such as $\sqrt{2}$ and π, and the integers even leave out fractions like 1/2. The reals, having the completeness property, omit no points.

What exactly do we mean by this vague idea of missing holes? To answer this we must give a more precise description of completeness. A number M is an **upper bound** for a set of numbers if all numbers in the set are smaller than or equal to M. M is a **least upper bound** if it is the smallest upper bound. For example, $M = 2$ is an upper bound for the negative numbers. So is $M = 1$, showing that 2 is not a least upper bound. The least upper bound for the set of negative numbers is $M = 0$. We define a **complete** ordered field to be one in which every nonempty set bounded above has a least upper bound.

If we work with just the rational numbers, the set of numbers less than $\sqrt{2}$ is bounded, but it does not have a rational least upper bound, since any rational upper bound M must be larger than $\sqrt{2}$ and can be replaced by a slightly smaller rational number that is still larger than $\sqrt{2}$. So the rationals are not complete. In the real numbers, a set that is bounded above always has a least upper bound. The reals are a complete ordered field.

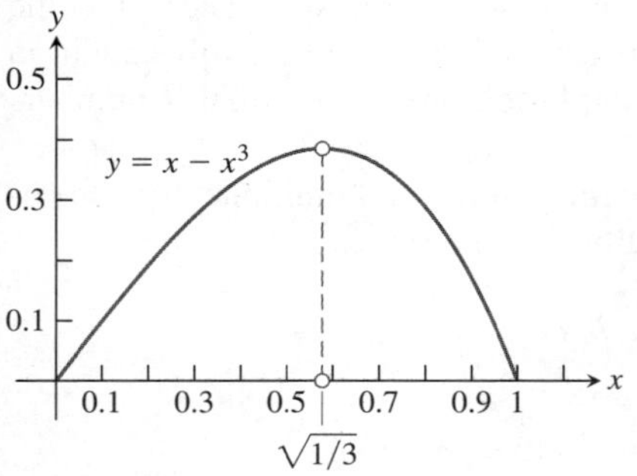

FIGURE A.24 The maximum value of $y = x - x^3$ on $[0, 1]$ occurs at the irrational number $x = \sqrt{1/3}$.

The completeness property is at the heart of many results in calculus. One example occurs when we search for a maximum value for a function on a closed interval $[a, b]$, as in Section 4.1. The function $y = x - x^3$ has a maximum value on $[0, 1]$ at the point x satisfying $1 - 3x^2 = 0$, or $x = \sqrt{1/3}$. If we limited our consideration to functions defined only on rational numbers, we would have to conclude that the function has no maximum, since $\sqrt{1/3}$ is irrational (Figure A.24). The Extreme Value Theorem (Section 4.1), which implies that continuous functions on closed intervals $[a, b]$ have a maximum value, is not true for functions defined only on the rationals.

The Intermediate Value Theorem implies that a continuous function f on an interval $[a, b]$ with $f(a) < 0$ and $f(b) > 0$ must be zero somewhere in $[a, b]$. The function values cannot jump from negative to positive without there being some point x in $[a, b]$ where $f(x) = 0$. The Intermediate Value Theorem also relies on the completeness of the real numbers and is false for continuous functions defined only on the rationals. The function $f(x) = 3x^2 - 1$ has $f(0) = -1$ and $f(1) = 2$, but if we consider f only on the rational numbers, it never equals zero. The only value of x for which $f(x) = 0$ is $x = \sqrt{1/3}$, an irrational number.

We have captured the desired properties of the reals by saying that the real numbers are a complete ordered field. But we're not quite finished. Greek mathematicians in the school of Pythagoras tried to impose another property on the numbers of the real line, the condition that all numbers are ratios of integers. They learned that their effort was doomed when they discovered irrational numbers such as $\sqrt{2}$. How do we know that our efforts to specify the real numbers are not also flawed, for some unseen reason? The artist Escher drew optical illusions of spiral staircases that went up and up until they rejoined themselves at the bottom. An engineer trying to build such a staircase would find that no structure realized the plans the architect had drawn. Could it be that our design for the reals contains some subtle contradiction, and that no construction of such a number system can be made?

We resolve this issue by giving a specific description of the real numbers and verifying that the algebraic, order, and completeness properties are satisfied in this model. This is called a **construction** of the reals, and just as stairs can be built with wood, stone, or steel, there are several approaches to constructing the reals. One construction treats the reals as all the infinite decimals,

$$a.d_1d_2d_3d_4 \cdots.$$

In this approach a real number is an integer a followed by a sequence of decimal digits $d_1, d_2, d_3, \ldots$, each between 0 and 9. This sequence may stop, or repeat in a periodic pattern, or keep going forever with no pattern. In this form, 2.00, 0.3333333 ... and 3.1415926535898 ... represent three familiar real numbers. The real meaning of the dots "..." following these digits requires development of the theory of sequences and series, as in Chapter 10. Each real number is constructed as the limit of a sequence of rational numbers given by its finite decimal approximations. An infinite decimal is then the same as a series

$$a + \frac{d_1}{10} + \frac{d_2}{100} + \cdots.$$

This decimal construction of the real numbers is not entirely straightforward. It's easy enough to check that it gives numbers that satisfy the completeness and order properties, but verifying the algebraic properties is rather involved. Even adding or multiplying two numbers requires an infinite number of operations. Making sense of division requires a careful argument involving limits of rational approximations to infinite decimals.

A different approach was taken by Richard Dedekind (1831–1916), a German mathematician, who gave the first rigorous construction of the real numbers in 1872. Given any real number x, we can divide the rational numbers into two sets: those less than or equal to x and those greater. Dedekind cleverly reversed this reasoning and defined a real number to be a division of the rational numbers into two such sets. This seems like a strange approach, but such indirect methods of constructing new structures from old are powerful tools in theoretical mathematics.

These and other approaches can be used to construct a system of numbers having the desired algebraic, order, and completeness properties. A final issue that arises is whether all the constructions give the same thing. Is it possible that different constructions result in different number systems satisfying all the required properties? If yes, which of these comprises the real numbers? Fortunately, the answer turns out to be no. The reals are the only number system satisfying the algebraic, order, and completeness properties.

Confusion about the nature of the numbers and about limits caused considerable controversy in the early development of calculus. Calculus pioneers such as Newton, Leibniz, and their successors, when looking at what happens to the difference quotient

$$\frac{\Delta y}{\Delta x} = \frac{f(x + \Delta x) - f(x)}{\Delta x}$$

as each of Δy and Δx approach zero, talked about the resulting derivative being a quotient of two infinitely small quantities. These "infinitesimals," written dx and dy, were thought to be some new kind of number, smaller than any fixed number but not zero. Similarly, a definite integral was thought of as a sum of an infinite number of infinitesimals

$$f(x) \cdot dx$$

as x varied over a closed interval. The approximating difference quotients $\Delta y/\Delta x$ were understood much as today, but it was the quotient of infinitesimal quantities, rather than a limit, that was thought to encapsulate the meaning of the derivative. This way of thinking led to logical difficulties, as attempted definitions and manipulations of infinitesimals ran into contradictions and inconsistencies. The more concrete and computable difference quotients did not cause such trouble, but they were thought of merely as useful calculation tools. Difference quotients were used to work out the numerical value of the derivative and to derive general formulas for calculation, but they were not considered to be at the heart of what the derivative actually was. Today we realize that the logical problems associated with infinitesimals can be avoided by *defining* the derivative to be the limit of its approximating difference quotients. The ambiguities of the old approach are no longer present, and in the standard theory of calculus, infinitesimals are neither needed nor used.

A.7 The Distributive Law for Vector Cross Products

In this appendix we prove the Distributive Law

$$\mathbf{u} \times (\mathbf{v} + \mathbf{w}) = \mathbf{u} \times \mathbf{v} + \mathbf{u} \times \mathbf{w},$$

which is Property 2 in Section 12.4.

Proof To derive the Distributive Law, we construct $\mathbf{u} \times \mathbf{v}$ a new way. We draw $\mathbf{u}$ and $\mathbf{v}$ from the common point O and construct a plane M perpendicular to $\mathbf{u}$ at O (Figure A.25). We then project $\mathbf{v}$ orthogonally onto M, yielding a vector $\mathbf{v}_1$ with length $|\mathbf{v}| \sin \theta$. We rotate $\mathbf{v}_1$ by 90° about $\mathbf{u}$ in the positive sense to produce a vector $\mathbf{v}_2$. Finally, we multiply $\mathbf{v}_2$ by the length of $\mathbf{u}$. The resulting vector $|\mathbf{u}|\mathbf{v}_2$ is equal to $\mathbf{u} \times \mathbf{v}$ since $\mathbf{v}_2$ has the same direction as $\mathbf{u} \times \mathbf{v}$ by its construction (Figure A.25) and

$$|\mathbf{u}||\mathbf{v}_2| = |\mathbf{u}||\mathbf{v}_1| = |\mathbf{u}||\mathbf{v}|\sin \theta = |\mathbf{u} \times \mathbf{v}|.$$

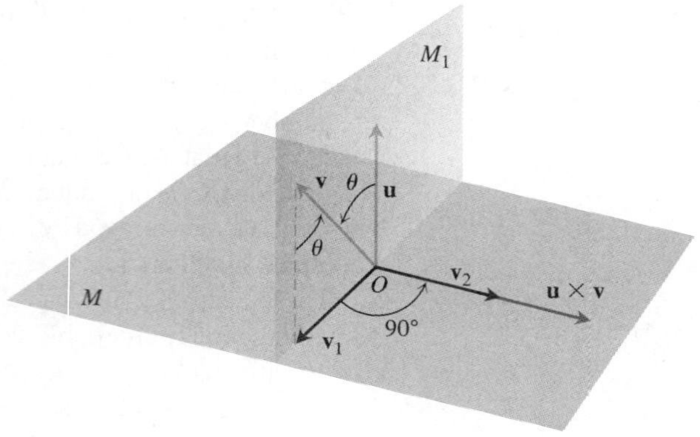

FIGURE A.25 As explained in the text, $\mathbf{u} \times \mathbf{v} = |\mathbf{u}|\mathbf{v}_2$.

Now each of these three operations, namely,

1. projection onto M
2. rotation about $\mathbf{u}$ through 90°
3. multiplication by the scalar $|\mathbf{u}|$,

when applied to a triangle whose plane is not parallel to $\mathbf{u}$, will produce another triangle. If we start with the triangle whose sides are $\mathbf{v}$, $\mathbf{w}$, and $\mathbf{s} = \mathbf{v} + \mathbf{w}$ (Figure A.26) and apply these three steps, we successively obtain the following:

1. A triangle whose sides are $\mathbf{v}_1$, $\mathbf{w}_1$, and $\mathbf{s}_1$ satisfying the vector equation

$$\mathbf{v}_1 + \mathbf{w}_1 = \mathbf{s}_1$$

2. A triangle whose sides are $\mathbf{v}_2$, $\mathbf{w}_2$, and $\mathbf{s}_2$ satisfying the vector equation

$$\mathbf{v}_2 + \mathbf{w}_2 = \mathbf{s}_2$$

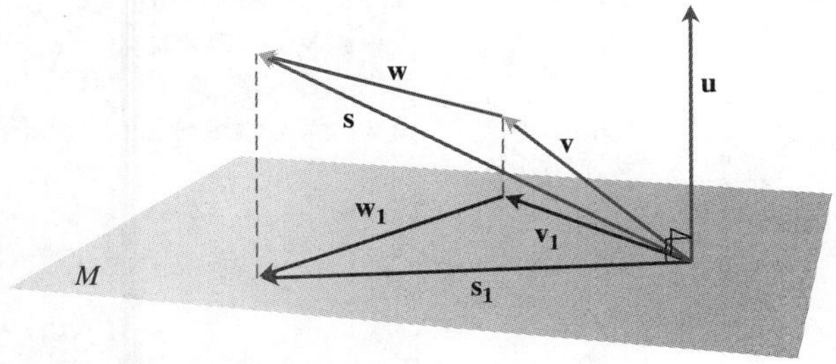

FIGURE A.26 The vectors $\mathbf{v}$, $\mathbf{w}$, and $\mathbf{v} + \mathbf{w}$ and their projections onto a plane perpendicular to $\mathbf{u}$.

3. A triangle whose sides are $|\mathbf{u}|\mathbf{v}_2$, $|\mathbf{u}|\mathbf{w}_2$, and $|\mathbf{u}|\mathbf{s}_2$ satisfying the vector equation

$$|\mathbf{u}|\mathbf{v}_2 + |\mathbf{u}|\mathbf{w}_2 = |\mathbf{u}|\mathbf{s}_2.$$

Substituting $|\mathbf{u}|\mathbf{v}_2 = \mathbf{u} \times \mathbf{v}$, $|\mathbf{u}|\mathbf{w}_2 = \mathbf{u} \times \mathbf{w}$, and $|\mathbf{u}|\mathbf{s}_2 = \mathbf{u} \times (\mathbf{v} + \mathbf{w})$ from our discussion above into this last equation gives

$$\mathbf{u} \times \mathbf{v} + \mathbf{u} \times \mathbf{w} = \mathbf{u} \times (\mathbf{v} + \mathbf{w}),$$

which is the law we wanted to establish. ■

A.8 The Mixed Derivative Theorem and the Increment Theorem

This appendix derives the Mixed Derivative Theorem (Theorem 2, Section 14.3) and the Increment Theorem for Functions of Two Variables (Theorem 3, Section 14.3). Euler first published the Mixed Derivative Theorem in 1734, in a series of papers he wrote on hydrodynamics.

THEOREM 2—The Mixed Derivative Theorem
If $f(x, y)$ and its partial derivatives f_x, f_y, f_{xy}, and f_{yx} are defined throughout an open region containing a point (a, b) and are all continuous at (a, b), then

$$f_{xy}(a, b) = f_{yx}(a, b).$$

Proof The equality of $f_{xy}(a, b)$ and $f_{yx}(a, b)$ can be established by four applications of the Mean Value Theorem (Theorem 4, Section 4.2). By hypothesis, the point (a, b) lies in the interior of a rectangle R in the xy-plane on which f, f_x, f_y, f_{xy}, and f_{yx} are all defined. We let h and k be the numbers such that the point $(a + h, b + k)$ also lies in R, and we consider the difference

$$\Delta = F(a + h) - F(a), \tag{1}$$

where

$$F(x) = f(x, b + k) - f(x, b). \tag{2}$$

We apply the Mean Value Theorem to F, which is continuous because it is differentiable. Then Equation (1) becomes

$$\Delta = hF'(c_1), \tag{3}$$

where c_1 lies between a and $a + h$. From Equation (2),

$$F'(x) = f_x(x, b + k) - f_x(x, b),$$

so Equation (3) becomes

$$\Delta = h[f_x(c_1, b + k) - f_x(c_1, b)]. \tag{4}$$

Now we apply the Mean Value Theorem to the function $g(y) = f_x(c_1, y)$ and have

$$g(b + k) - g(b) = kg'(d_1),$$

or

$$f_x(c_1, b + k) - f_x(c_1, b) = kf_{xy}(c_1, d_1)$$

for some d_1 between b and $b + k$. By substituting this into Equation (4), we get

$$\Delta = hkf_{xy}(c_1, d_1) \tag{5}$$

for some point (c_1, d_1) in the rectangle R' whose vertices are the four points (a, b), $(a + h, b)$, $(a + h, b + k)$, and $(a, b + k)$. (See Figure A.27.)

By substituting from Equation (2) into Equation (1), we may also write

$$\begin{aligned}\Delta &= f(a + h, b + k) - f(a + h, b) - f(a, b + k) + f(a, b)\\ &= [f(a + h, b + k) - f(a, b + k)] - [f(a + h, b) - f(a, b)]\\ &= \phi(b + k) - \phi(b),\end{aligned} \tag{6}$$

where

$$\phi(y) = f(a + h, y) - f(a, y). \tag{7}$$

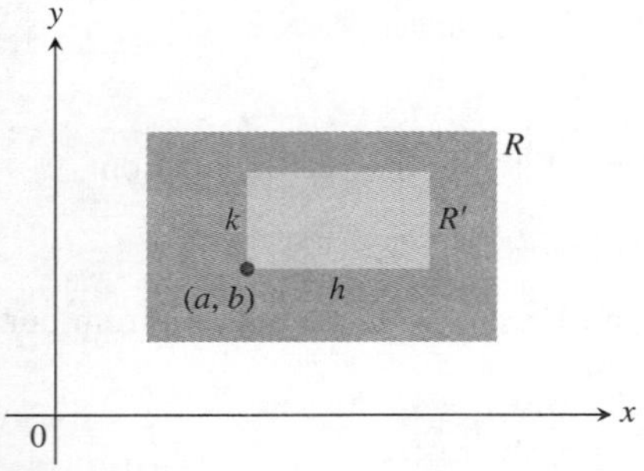

FIGURE A.27 The key to proving $f_{xy}(a, b) = f_{yx}(a, b)$ is that no matter how small R' is, f_{xy} and f_{yx} take on equal values somewhere inside R' (though not necessarily at the same point).

The Mean Value Theorem applied to Equation (6) now gives

$$\Delta = k\phi'(d_2) \tag{8}$$

for some d_2 between b and $b + k$. By Equation (7),

$$\phi'(y) = f_y(a + h, y) - f_y(a, y). \tag{9}$$

Substituting from Equation (9) into Equation (8) gives

$$\Delta = k[f_y(a + h, d_2) - f_y(a, d_2)].$$

Finally, we apply the Mean Value Theorem to the expression in brackets and get

$$\Delta = khf_{yx}(c_2, d_2) \tag{10}$$

for some c_2 between a and $a + h$.

Together, Equations (5) and (10) show that

$$f_{xy}(c_1, d_1) = f_{yx}(c_2, d_2), \tag{11}$$

where (c_1, d_1) and (c_2, d_2) both lie in the rectangle R' (Figure A.27). Equation (11) is not quite the result we want, since it says only that f_{xy} has the same value at (c_1, d_1) that f_{yx} has at (c_2, d_2). The numbers h and k in our discussion, however, may be made as small as we wish. The hypothesis that f_{xy} and f_{yx} are both continuous at (a, b) means that $f_{xy}(c_1, d_1) = f_{xy}(a, b) + \varepsilon_1$ and $f_{yx}(c_2, d_2) = f_{yx}(a, b) + \varepsilon_2$, where each of $\varepsilon_1, \varepsilon_2 \to 0$ as both $h, k \to 0$. Hence, if we let h and $k \to 0$, we have $f_{xy}(a, b) = f_{yx}(a, b)$. ■

The equality of $f_{xy}(a, b)$ and $f_{yx}(a, b)$ can be proved with hypotheses weaker than the ones we assumed. For example, it is enough for f, f_x, and f_y to exist in R and for f_{xy} to be continuous at (a, b). Then f_{yx} will exist at (a, b) and will equal f_{xy} at that point.

THEOREM 3—The Increment Theorem for Functions of Two Variables
Suppose that the first partial derivatives of $f(x, y)$ are defined throughout an open region R containing the point (x_0, y_0) and that f_x and f_y are continuous at (x_0, y_0). Then the change

$$\Delta z = f(x_0 + \Delta x, y_0 + \Delta y) - f(x_0, y_0)$$

in the value of f that results from moving from (x_0, y_0) to another point $(x_0 + \Delta x, y_0 + \Delta y)$ in R satisfies an equation of the form

$$\Delta z = f_x(x_0, y_0)\Delta x + f_y(x_0, y_0)\Delta y + \varepsilon_1 \Delta x + \varepsilon_2 \Delta y$$

in which each of $\varepsilon_1, \varepsilon_2 \to 0$ as both $\Delta x, \Delta y \to 0$.

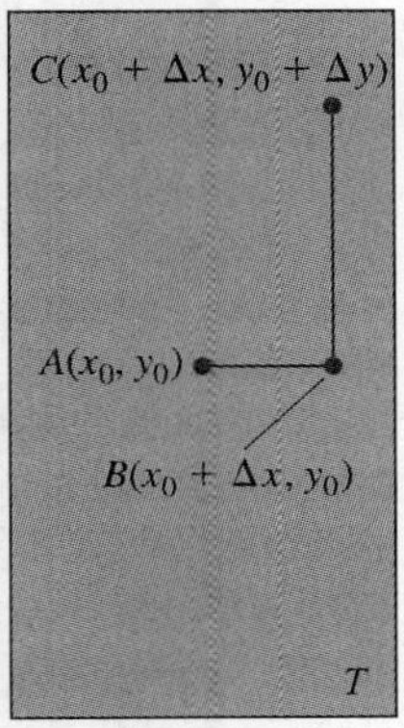

FIGURE A.28 The rectangular region T in the proof of the Increment Theorem. The figure is drawn for Δx and Δy positive, but either increment might be zero or negative.

Proof We work within a rectangle T centered at $A(x_0, y_0)$ and lying within R, and we assume that Δx and Δy are already so small that the line segment joining A to $B(x_0 + \Delta x, y_0)$ and the line segment joining B to $C(x_0 + \Delta x, y_0 + \Delta y)$ lie in the interior of T (Figure A.28).

We may think of Δz as the sum $\Delta z = \Delta z_1 + \Delta z_2$ of two increments, where

$$\Delta z_1 = f(x_0 + \Delta x, y_0) - f(x_0, y_0)$$

is the change in the value of f from A to B and

$$\Delta z_2 = f(x_0 + \Delta x, y_0 + \Delta y) - f(x_0 + \Delta x, y_0)$$

is the change in the value of f from B to C (Figure A.29).

On the closed interval of x-values joining x_0 to $x_0 + \Delta x$, the function $F(x) = f(x, y_0)$ is a differentiable (and hence continuous) function of x, with derivative

$$F'(x) = f_x(x, y_0).$$

By the Mean Value Theorem (Theorem 4, Section 4.2), there is an x-value c between x_0 and $x_0 + \Delta x$ at which

$$F(x_0 + \Delta x) - F(x_0) = F'(c)\,\Delta x$$

or

$$f(x_0 + \Delta x, y_0) - f(x_0, y_0) = f_x(c, y_0)\,\Delta x$$

or

$$\Delta z_1 = f_x(c, y_0)\,\Delta x. \tag{12}$$

Similarly, $G(y) = f(x_0 + \Delta x, y)$ is a differentiable (and hence continuous) function of y on the closed y-interval joining y_0 and $y_0 + \Delta y$, with derivative

$$G'(y) = f_y(x_0 + \Delta x, y).$$

Hence, there is a y-value d between y_0 and $y_0 + \Delta y$ at which

$$G(y_0 + \Delta y) - G(y_0) = G'(d)\,\Delta y$$

or

$$f(x_0 + \Delta x, y_0 + \Delta y) - f(x_0 + \Delta x, y) = f_y(x_0 + \Delta x, d)\,\Delta y$$

or

$$\Delta z_2 = f_y(x_0 + \Delta x, d)\,\Delta y. \tag{13}$$

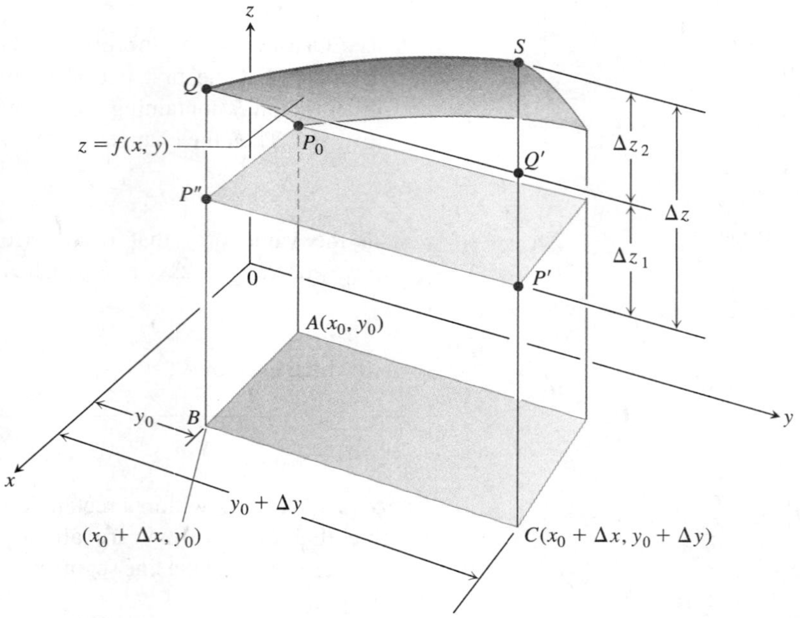

FIGURE A.29 Part of the surface $z = f(x, y)$ near $P_0(x_0, y_0, f(x_0, y_0))$. The points P_0, P', and P'' have the same height $z_0 = f(x_0, y_0)$ above the xy-plane. The change in z is $\Delta z = P'S$. The change

$$\Delta z_1 = f(x_0 + \Delta x, y_0) - f(x_0, y_0),$$

shown as $P''Q = P'Q'$, is caused by changing x from x_0 to $x_0 + \Delta x$ while holding y equal to y_0. Then, with x held equal to $x_0 + \Delta x$,

$$\Delta z_2 = f(x_0 + \Delta x, y_0 + \Delta y) - f(x_0 + \Delta x, y_0)$$

is the change in z caused by changing y_0 from $y_0 + \Delta y$, which is represented by $Q'S$. The total change in z is the sum of Δz_1 and Δz_2.

Now, as both Δx and $\Delta y \to 0$, we know that $c \to x_0$ and $d \to y_0$. Therefore, since f_x and f_y are continuous at (x_0, y_0), the quantities

$$\begin{aligned} \varepsilon_1 &= f_x(c, y_0) - f_x(x_0, y_0), \\ \varepsilon_2 &= f_y(x_0 + \Delta x, d) - f_y(x_0, y_0) \end{aligned} \tag{14}$$

both approach zero as both Δx and $\Delta y \to 0$.

Finally,

$$\begin{aligned} \Delta z &= \Delta z_1 + \Delta z_2 \\ &= f_x(c, y_0)\Delta x + f_y(x_0 + \Delta x, d)\Delta y && \text{From Eqs. (12) and (13)} \\ &= [f_x(x_0, y_0) + \varepsilon_1]\Delta x + [f_y(x_0, y_0) + \varepsilon_2]\Delta y && \text{From Eq. (14)} \\ &= f_x(x_0, y_0)\Delta x + f_y(x_0, y_0)\Delta y + \varepsilon_1 \Delta x + \varepsilon_2 \Delta y, \end{aligned}$$

where both ε_1 and $\varepsilon_2 \to 0$ as both Δx and $\Delta y \to 0$, which is what we set out to prove. ■

Analogous results hold for functions of any finite number of independent variables. Suppose that the first partial derivatives of $w = f(x, y, z)$ are defined throughout an open region containing the point (x_0, y_0, z_0) and that f_x, f_y, and f_z are continuous at (x_0, y_0, z_0). Then

$$\begin{aligned}\Delta w &= f(x_0 + \Delta x, y_0 + \Delta y, z_0 + \Delta z) - f(x_0, y_0, z_0) \\ &= f_x \Delta x + f_y \Delta y + f_z \Delta z + \varepsilon_1 \Delta x + \varepsilon_2 \Delta y + \varepsilon_3 \Delta z, \qquad (15)\end{aligned}$$

where $\varepsilon_1, \varepsilon_2$, and $\varepsilon_3 \to 0$ as $\Delta x, \Delta y$, and $\Delta z \to 0$.

The partial derivatives f_x, f_y, f_z in Equation (15) are to be evaluated at the point (x_0, y_0, z_0).

Equation (15) can be proved by treating Δw as the sum of three increments,

$$\Delta w_1 = f(x_0 + \Delta x, y_0, z_0) - f(x_0, y_0, z_0) \qquad (16)$$

$$\Delta w_2 = f(x_0 + \Delta x, y_0 + \Delta y, z_0) - f(x_0 + \Delta x, y_0, z_0) \qquad (17)$$

$$\Delta w_3 = f(x_0 + \Delta x, y_0 + \Delta y, z_0 + \Delta z) - f(x_0 + \Delta x, y_0 + \Delta y, z_0), \qquad (18)$$

and applying the Mean Value Theorem to each of these separately. Two coordinates remain constant and only one varies in each of these partial increments $\Delta w_1, \Delta w_2, \Delta w_3$. In Equation (17), for example, only y varies, since x is held equal to $x_0 + \Delta x$ and z is held equal to z_0. Since $f(x_0 + \Delta x, y, z_0)$ is a continuous function of y with a derivative f_y, it is subject to the Mean Value Theorem, and we have

$$\Delta w_2 = f_y(x_0 + \Delta x, y_1, z_0)\,\Delta y$$

for some y_1 between y_0 and $y_0 + \Delta y$.

ANSWERS TO ODD-NUMBERED EXERCISES

Chapter 1

SECTION 1.1, pp. 11–13

1. $D:(-\infty,\infty)$, $R:[1,\infty)$

3. $D:[-2,\infty)$, $R:[0,\infty)$

5. Domain: $(-\infty,-1)\cup(-1,\infty)$; Range: $(-\infty,-5)\cup(-5,\infty)$

7. (a) Not a function of x because some values of x have two values of y

(b) A function of x because for every x there is only one possible y

9. $A=\dfrac{\sqrt{3}}{4}x^2$, $p=3x$

11. $x=\dfrac{d}{\sqrt{3}}$, $A=2d^2$, $V=\dfrac{d^3}{3\sqrt{3}}$

13. $\sqrt{2x^2+2x+4}$

15. $(-\infty,\infty)$

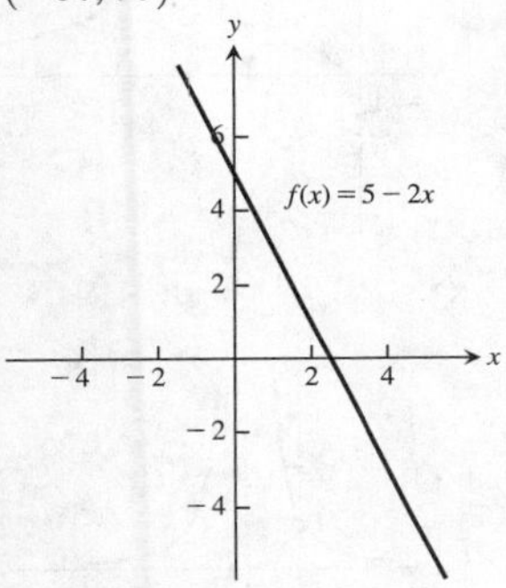

17. $(-\infty,\infty)$

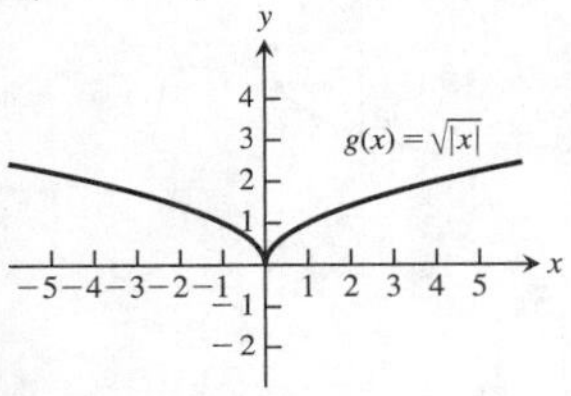

19. $(-\infty,0)\cup(0,\infty)$

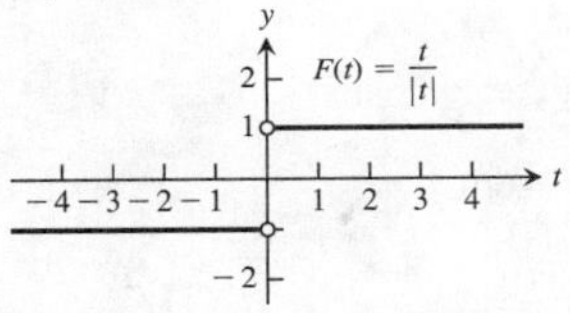

21. $(-\infty,-5)\cup(-5,-3]\cup[3,5)\cup(5,\infty)$

23. (a) For each positive value of x, there are two values of y.

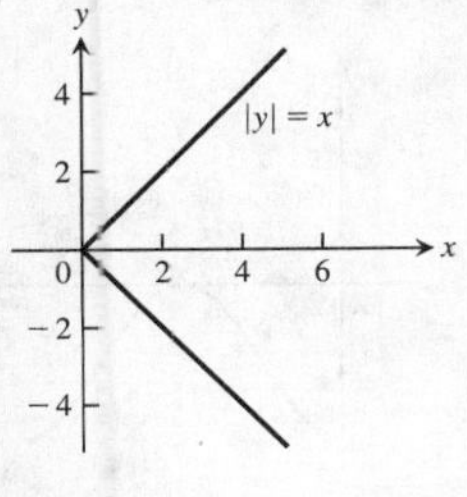

(b) For each value of $x\neq 0$, there are two values of y.

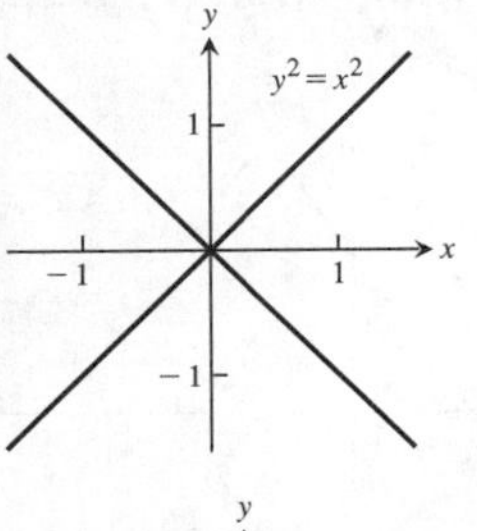

25.

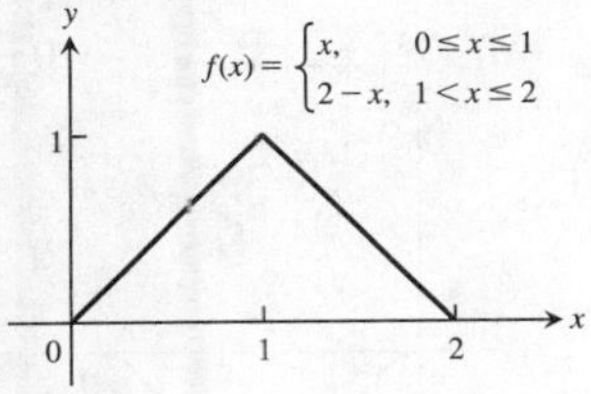

27.

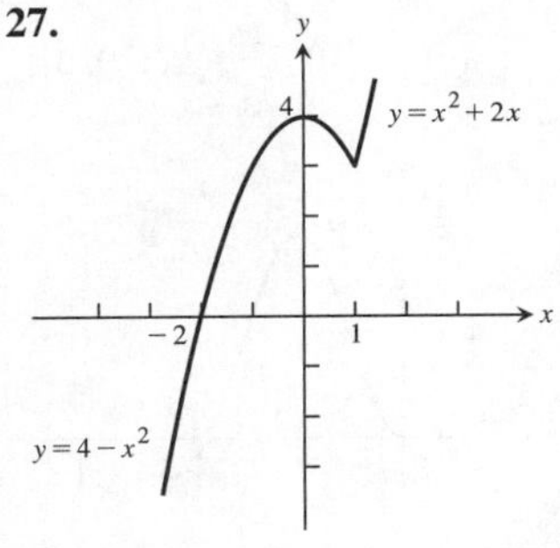

29. (a) $f(x)=\begin{cases} x, & 0\le x\le 1\\ -x+2, & 1<x\le 2\end{cases}$

(b) $f(x)=\begin{cases} 2, & 0\le x<1\\ 0, & 1\le x<2\\ 2, & 2\le x<3\\ 0, & 3\le x\le 4\end{cases}$

31. (a) $f(x)=\begin{cases} -x, & -1\le x<0\\ 1, & 0<x\le 1\\ -\frac{1}{2}x+\frac{3}{2}, & 1<x<3\end{cases}$

(b) $f(x)=\begin{cases} \frac{1}{2}x, & -2\le x\le 0\\ -2x+2, & 0<x\le 1\\ -1, & 1<x\le 3\end{cases}$

33. (a) $5\le x<6$ **(b)** $4<x\le 5$ **35.** Yes

37. Symmetric about the origin

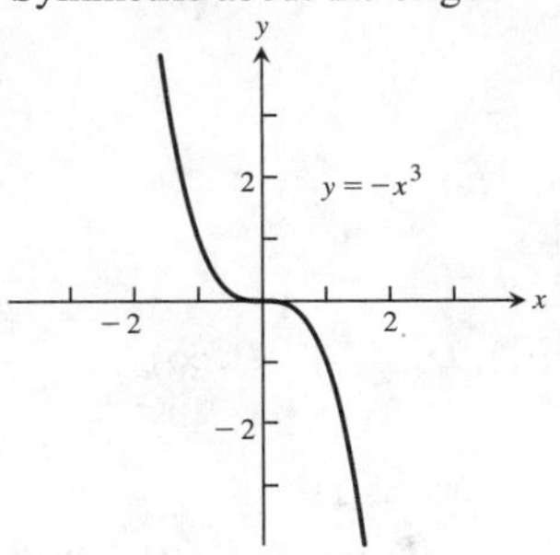

Dec. $-\infty<x<\infty$

39. Symmetric about the origin

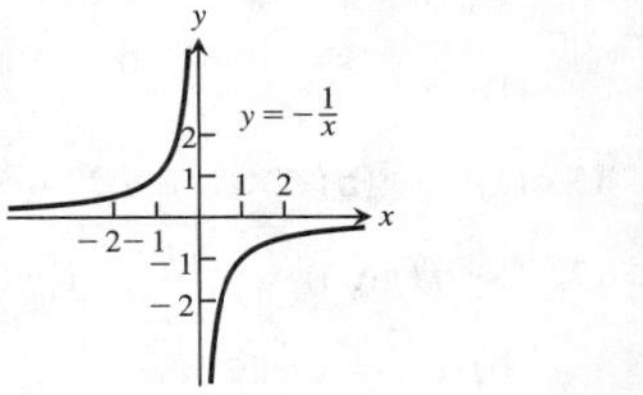

Inc. $-\infty<x<0$ and $0<x<\infty$

41. Symmetric about the y-axis

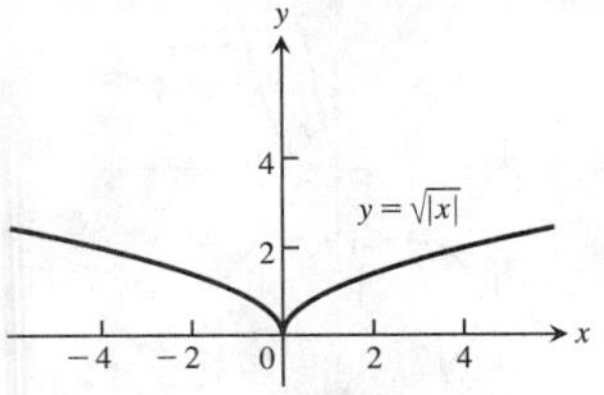

Dec. $-\infty<x\le 0$;
Inc. $0\le x<\infty$

43. Symmetric about the origin

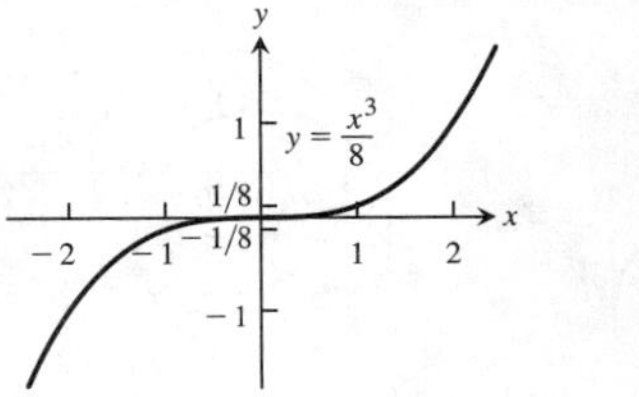

Inc. $-\infty<x<\infty$

45. No symmetry

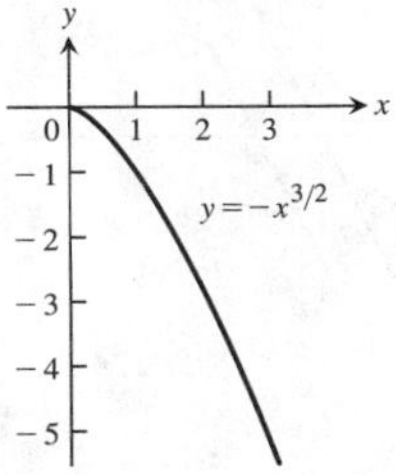

Dec. $0\le x<\infty$

47. Even **49.** Even **51.** Odd **53.** Neither odd nor even **55.** Neither **57.** Neither odd nor even **59.** Odd **61.** Even

63. $y=220$ when $x=55$ **65.** $s=2.4$

67. $V=x(14-2x)(22-2x)$

69. (a) h **(b)** f **(c)** g **71. (a)** $(-2,0)\cup(4,\infty)$

75. $C=4h\left(2\sqrt{2}+3\right)$

SECTION 1.2, pp. 18–21

1. $D_f: -\infty < x < \infty, D_g: x \geq 1, D_{f+g} = D_{f \cdot g} = D_g$

3. $D_f: 0 \leq x < \infty$; $D_g: -\infty < x \leq \infty$; $D_{f/g}: 0 < x < \infty$; $D_{g>f}: 0 < x < \infty$; $R_f: 0 \leq x < \infty$; $R_g: 0 \leq x < \infty$; $R_{f/g}: 0 \leq x < \infty$; $R_{g/f}: 0 \leq x < \infty$

5. **(a)** 2 **(b)** 3 **(c)** $x^2 + 2$ **(d)** $x^2 + 10x + 22$ **(e)** 8 **(f)** 166 **(g)** $x + 10$ **(h)** $x^4 - 6x^2 + 6$

7. $13 - 3x$ **9.** $\sqrt{\dfrac{3-5x}{1-2x}}$

11. **(a)** $f(g(x))$ **(b)** $j(g(x))$ **(c)** $g(g(x))$ **(d)** $j(j(x))$ **(e)** $f(g(h(x)))$ **(f)** $h(j(f(x)))$

13.

	$g(x)$	$f(x)$	$(f \circ g)(x)$
(a)	$x - 7$	$\sqrt{x}$	$\sqrt{x - 7}$
(b)	$x + 2$	$3x$	$3x + 6$
(c)	x^2	$\sqrt{x^2 - 6}$	$\sqrt{x^4 - 6}$
(d)	$\dfrac{x}{x-1}$	$\dfrac{x}{x-1}$	x
(e)	$\dfrac{1}{x-1}$	$1 + \dfrac{1}{x}$	x
(f)	$\dfrac{5}{x}$	$\dfrac{10}{x}$	$2x$

15. **(a)** 1 **(b)** 2 **(c)** -2 **(d)** 0 **(e)** -1 **(f)** 0

17. **(a)** $f(g(x)) = \sqrt{\dfrac{1}{x} + 1}$, $g(f(x)) = \dfrac{1}{\sqrt{x+1}}$
(b) $D_{f \circ g} = (-\infty, -1] \cup (0, \infty)$, $D_{g \circ f} = (-1, \infty)$

19. $g(x) = 4x + 8$ **21.** $V = 18t - 17$

23. **(a)** $y = -(x + 7)^2$ **(b)** $y = -(x - 4)^2$

25. **(a)** Position 4 **(b)** Position 1 **(c)** Position 2 **(d)** Position 3

27. $(x + 2)^2 + (y + 3)^2 = 49$

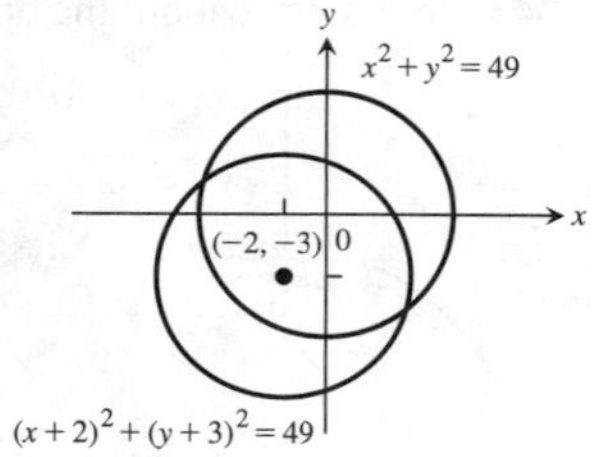

29. $y + 1 = (x + 1)^3$

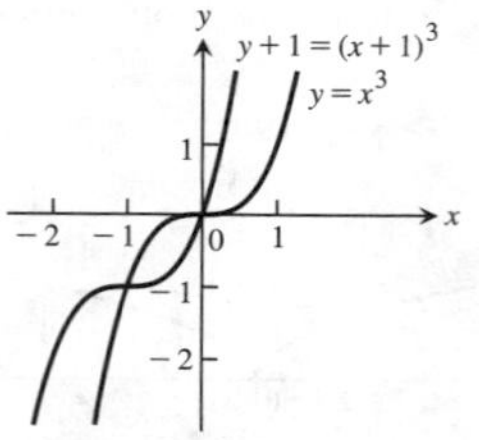

31. $y = \sqrt{x + 0.81}$

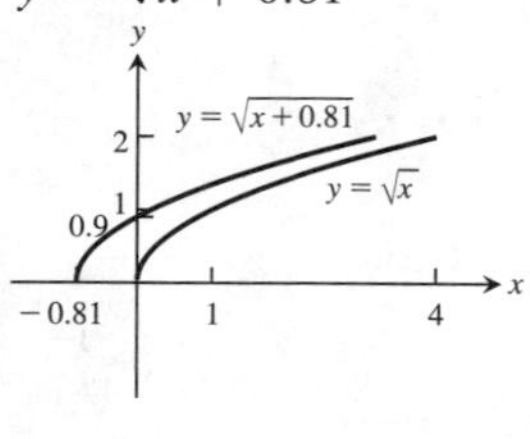

33. $y = 2x$

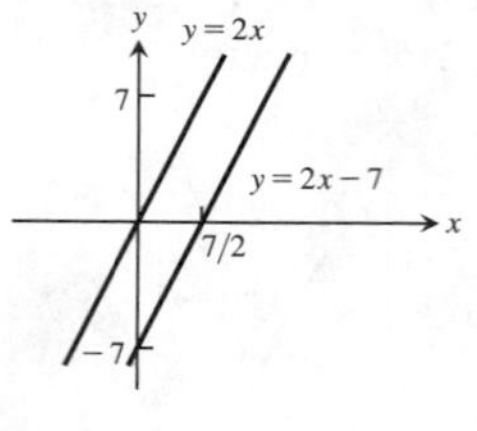

35. $y - 1 = \dfrac{1}{x - 1}$

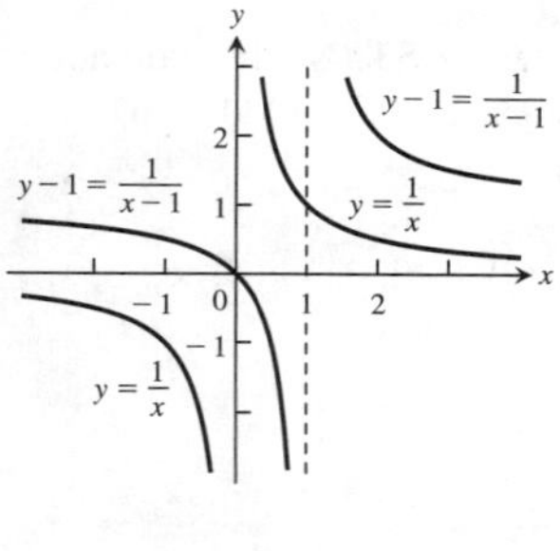

37.

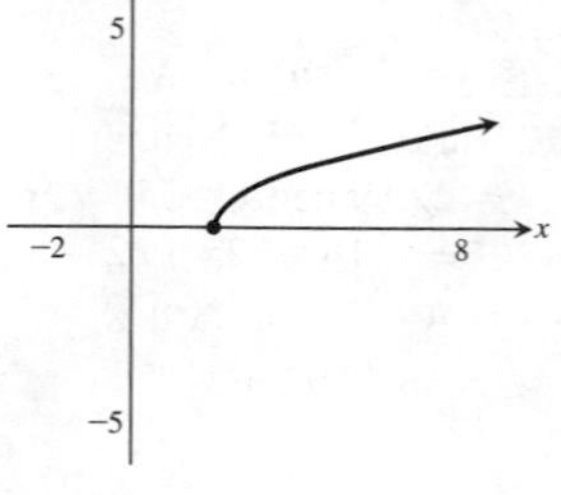

39.

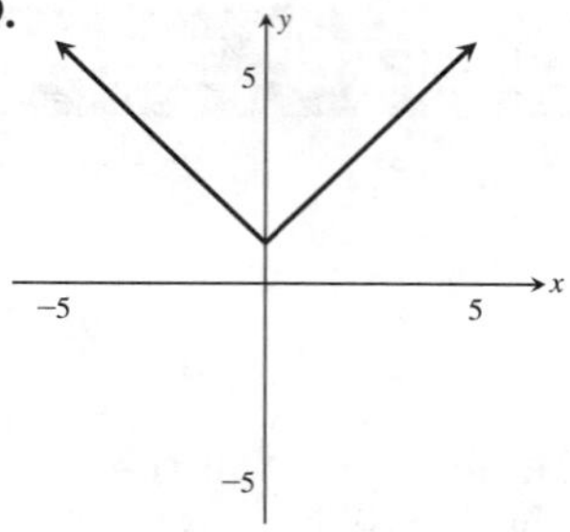

41.

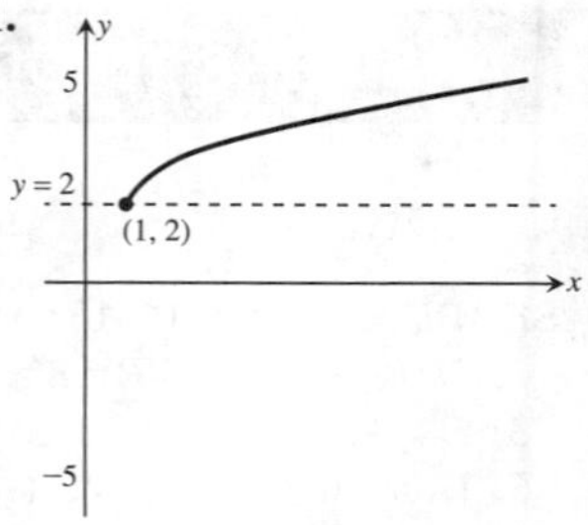

43.

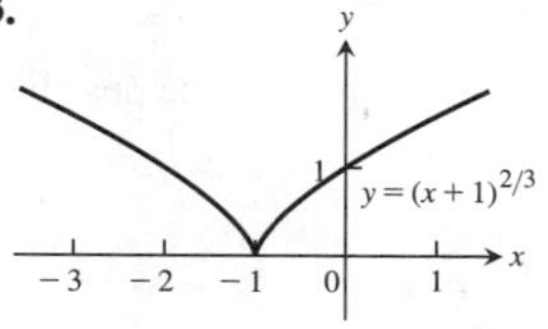

45.

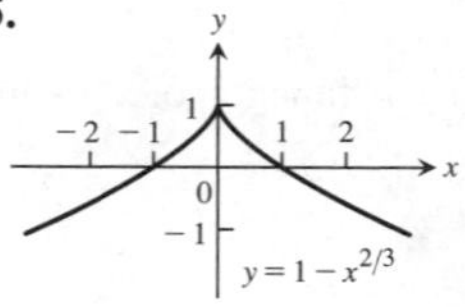

47.

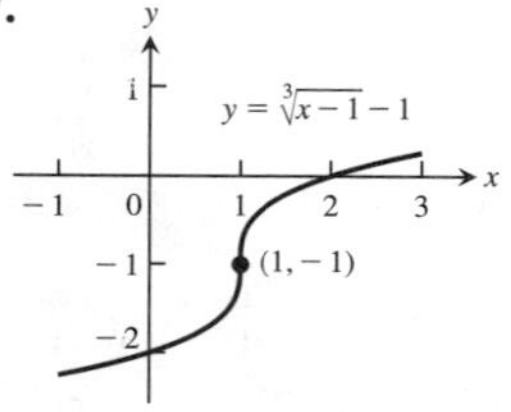

49.

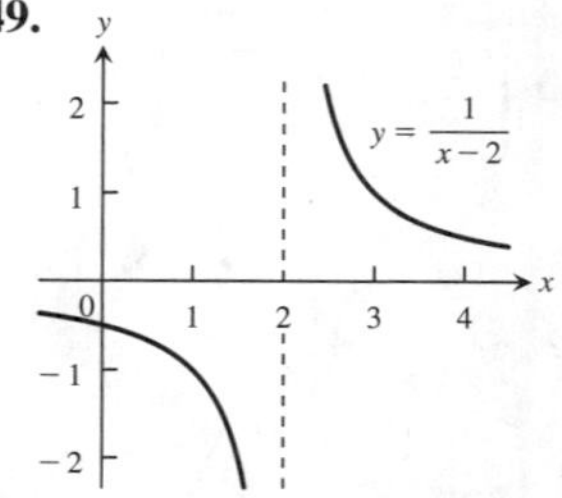

51.

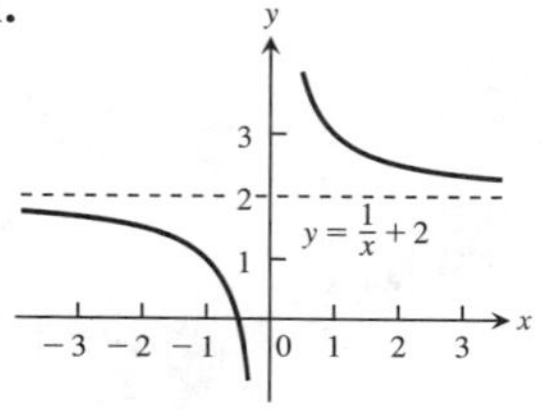

53.

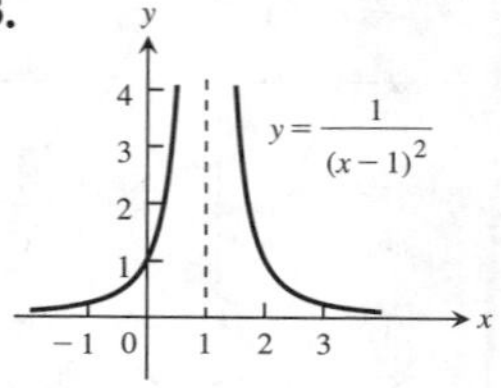

55.

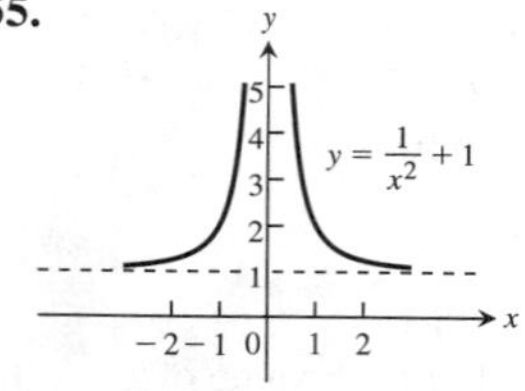

57. **(a)** $D: [0, 2]$, $R: [2, 3]$

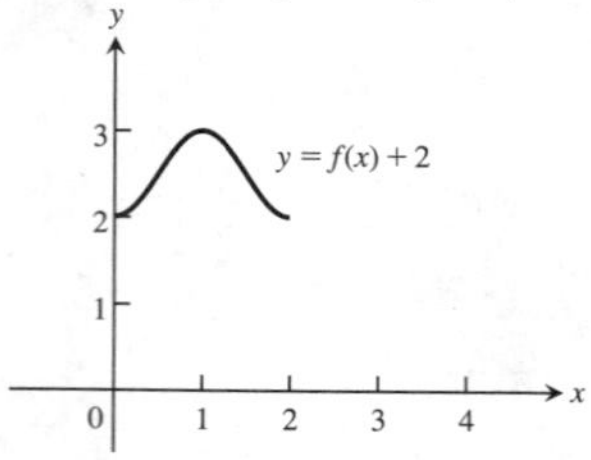

(b) $D: [0, 2]$, $R: [-1, 0]$

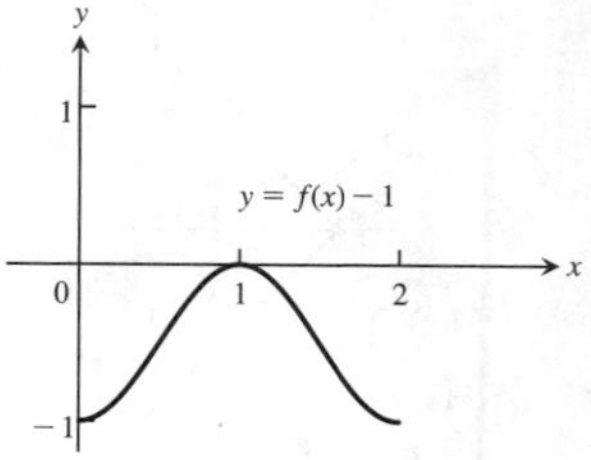

(c) $D: [0, 2]$, $R: [0, 2]$

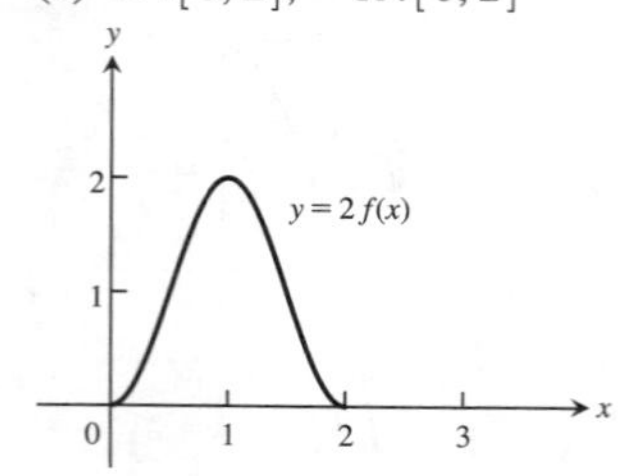

(d) $D: [0, 2]$, $R: [-1, 0]$

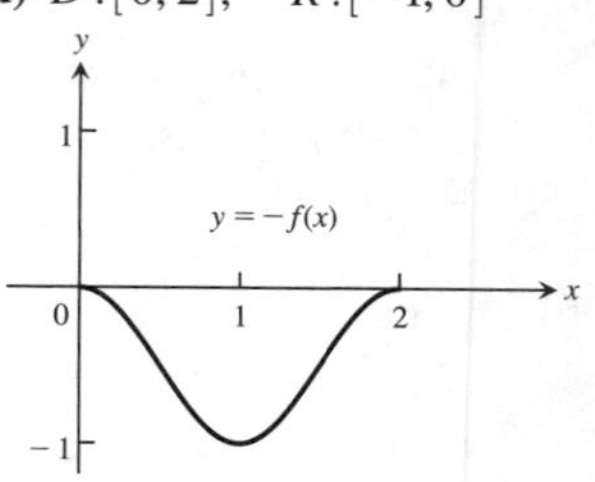

(e) $D:[-2,0]$, $R:[0,1]$

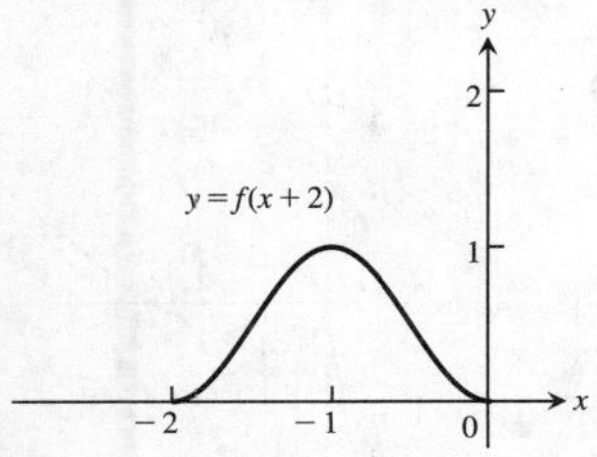

(f) $D:[1,3]$, $R:[0,1]$

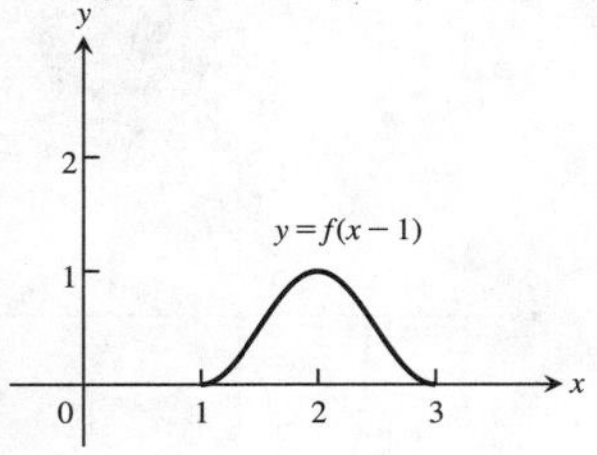

(g) $D:[-2,0]$, $R:[0,1]$

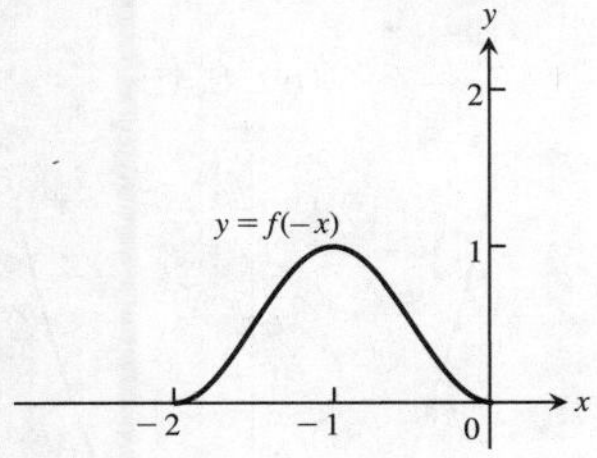

(h) $D:[-1,1]$, $R:[0,1]$

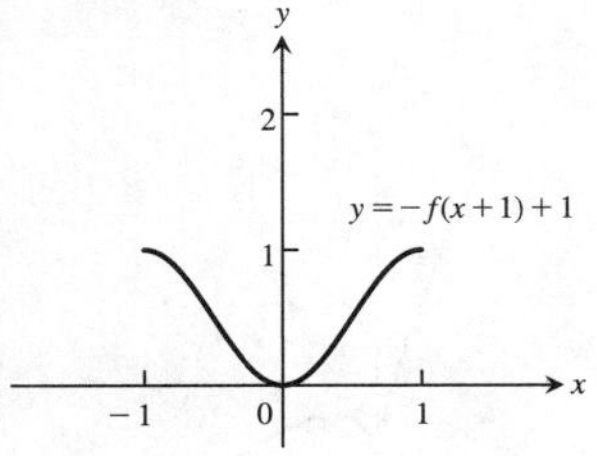

59. $y=3x^2$ **61.** $\frac{x}{2}+\frac{1}{2x}$ **63.** $\sqrt{2x+4}$

65. $y=\sqrt{4-\frac{x^2}{4}}$ **67.** $y=1-27x^3$

69.

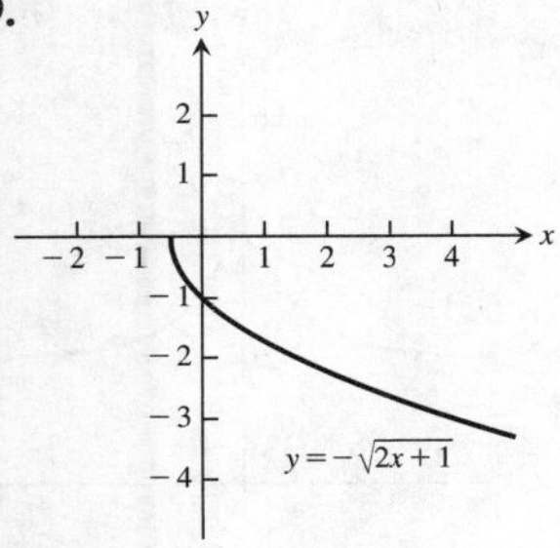

71.

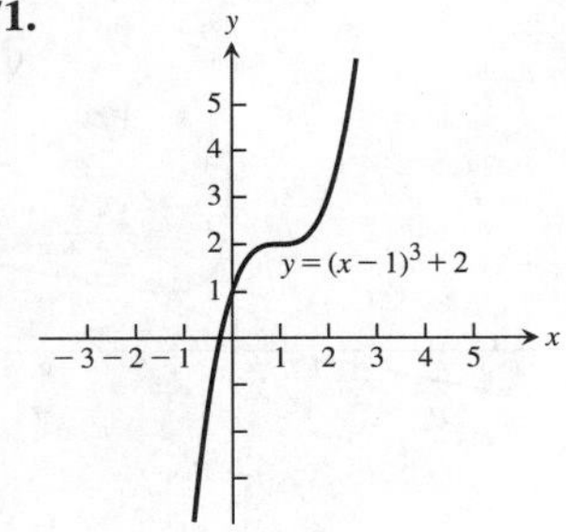

73.

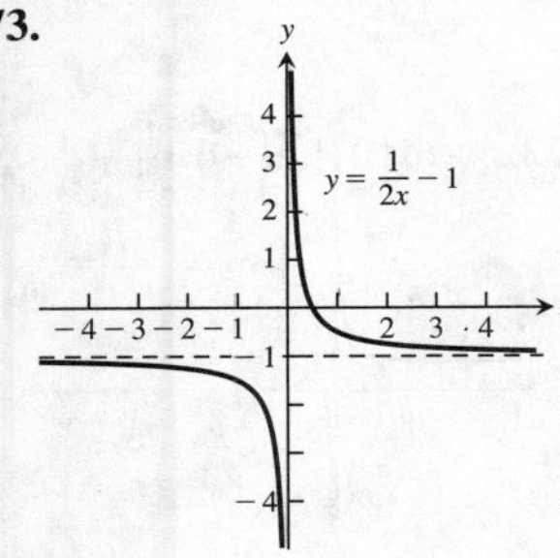

75.

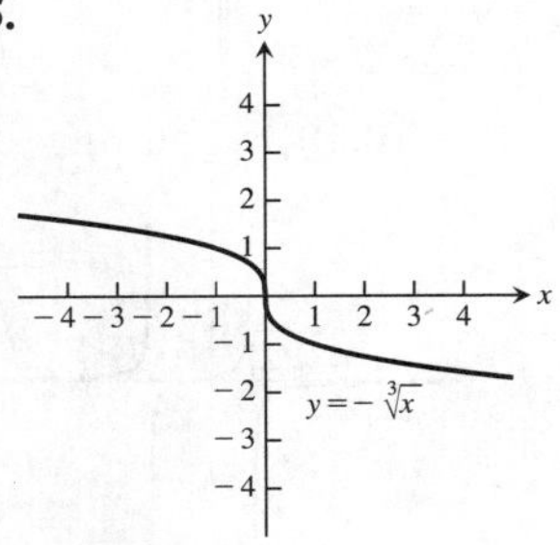

77.

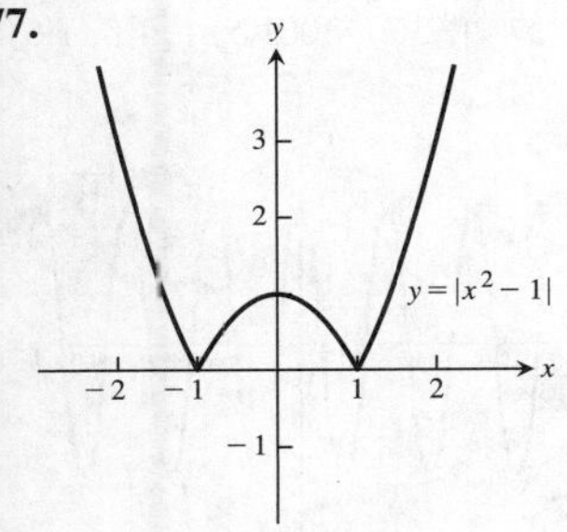

79. (a) Odd **(b)** Odd **(c)** Odd **(d)** Even **(e)** Even **(f)** Even **(g)** Even **(h)** Even **(i)** Odd

SECTION 1.3, pp. 27–29

1. (a) 9π **(b)** 8π **3.** 15.7 (approx.)

5.

θ	$-\pi$	$-2\pi/3$	0	$\pi/2$	$3\pi/4$
$\sin\theta$	0	$-\frac{\sqrt{3}}{2}$	0	1	$\frac{1}{\sqrt{2}}$
$\cos\theta$	-1	$-\frac{1}{2}$	1	0	$-\frac{1}{\sqrt{2}}$
$\tan\theta$	0	$\sqrt{3}$	0	UND	-1
$\cot\theta$	UND	$\frac{1}{\sqrt{3}}$	UND	0	-1
$\sec\theta$	-1	-2	1	UND	$-\sqrt{2}$
$\csc\theta$	UND	$-\frac{2}{\sqrt{3}}$	UND	1	$\sqrt{2}$

7. $\cos x=-\frac{12}{13}$, $\tan x=-\frac{5}{12}$

9. $\sin x=-\frac{\sqrt{8}}{3}$, $\tan x=-\sqrt{8}$

11. $\sin x=-\frac{3}{5}$, $\cos x=-\frac{4}{5}$

13. Period $=\frac{2}{3}\pi$

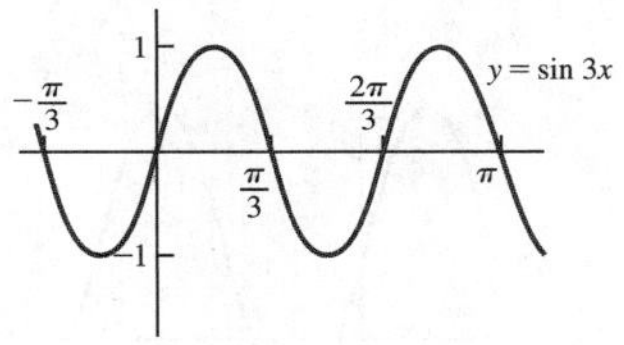

15. Period $=\pi$

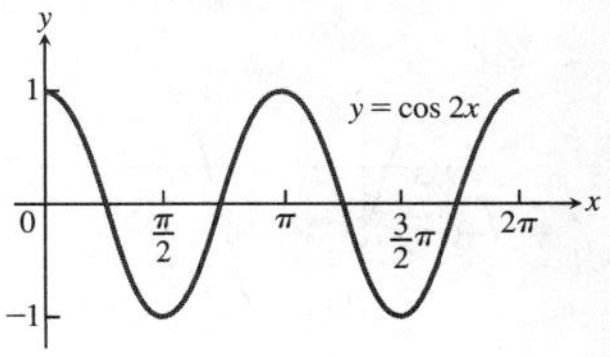

17. Period 6

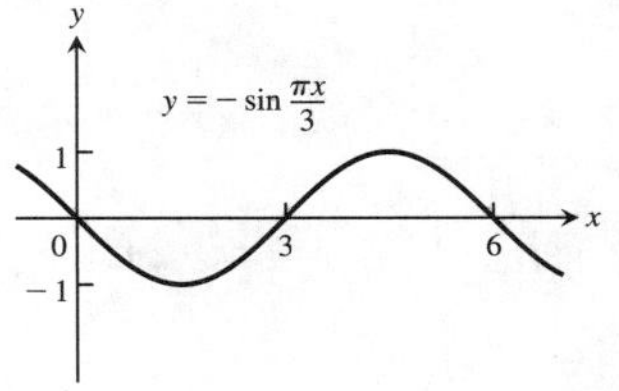

19. Period 2π

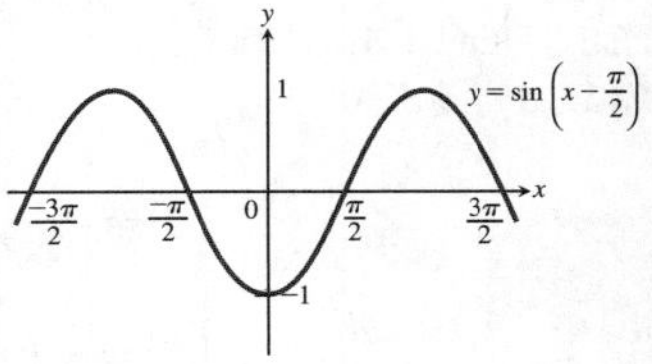

21. Period 2π

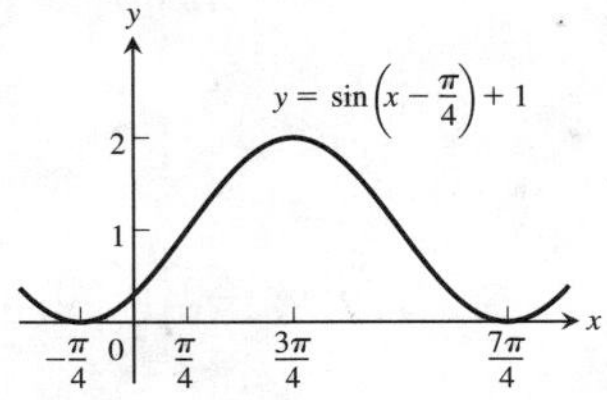

23. Period $\pi/2$, symmetric about the origin

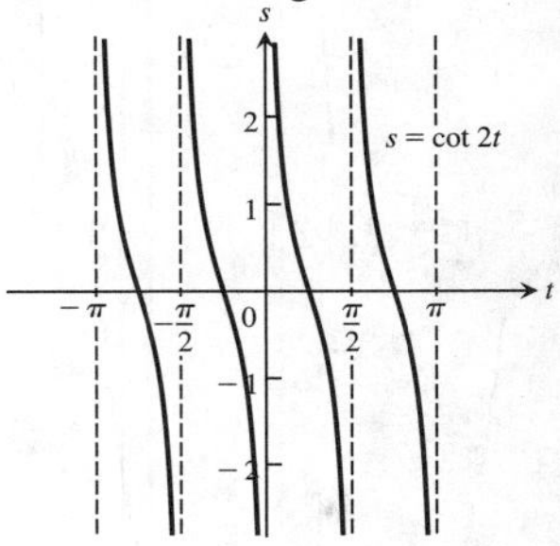

25. Period 4, symmetric about the y-axis

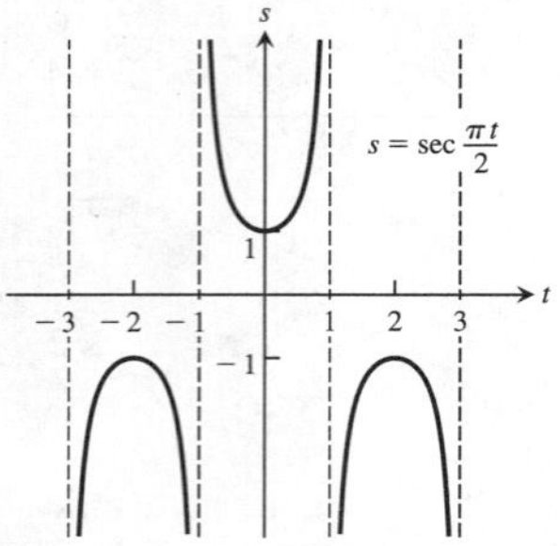

29. $D: (-\infty, \infty)$, $R: y = -1, 0, 1$

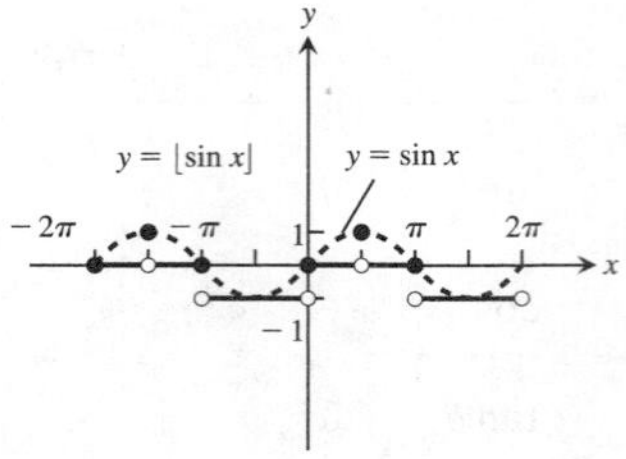

39. $-\cos x$ **41.** $-\cos x$ **43.** $\dfrac{\sqrt{6}-\sqrt{2}}{4}$ **45.** $\dfrac{\sqrt{2}+\sqrt{6}}{4}$

47. $\dfrac{2+\sqrt{2}}{4}$ **49.** $\dfrac{2+\sqrt{3}}{4}$ **51.** $\dfrac{\pi}{3}, \dfrac{2\pi}{3}, \dfrac{4\pi}{3}, \dfrac{5\pi}{3}$

53. $\dfrac{\pi}{6}, \dfrac{\pi}{2}, \dfrac{5\pi}{6}, \dfrac{3\pi}{2}$ **59.** $c = 5$ **63.** $a \approx 1.464$

65. $r = \dfrac{\alpha \sin(\theta)}{1 - \sin(\theta)}$

67. $A = 2, B = 2\pi, C = -\pi, D = -1$

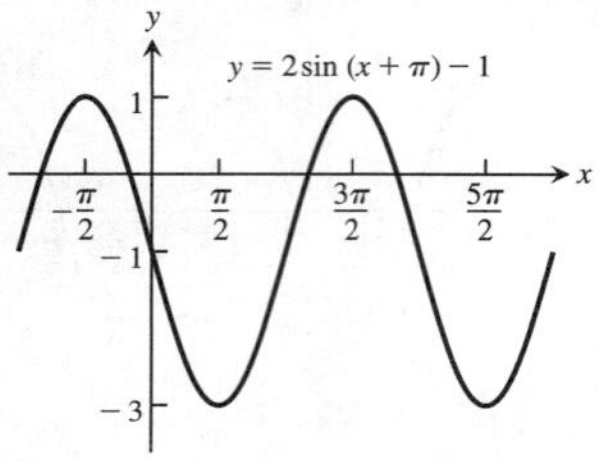

69. $A = -\dfrac{2}{\pi}, B = 4, C = 0, D = \dfrac{1}{\pi}$

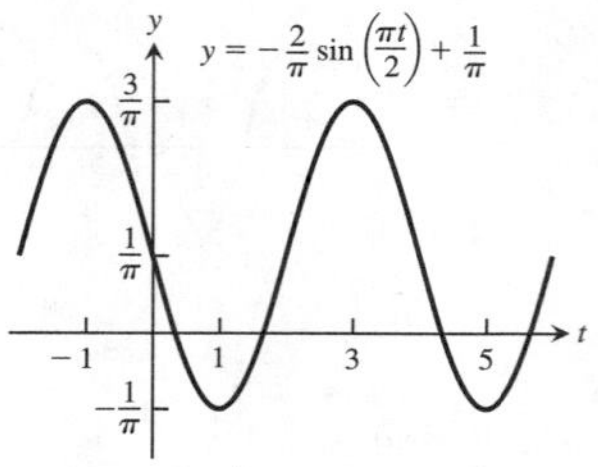

SECTION 1.4, p. 33

1. d **3.** d

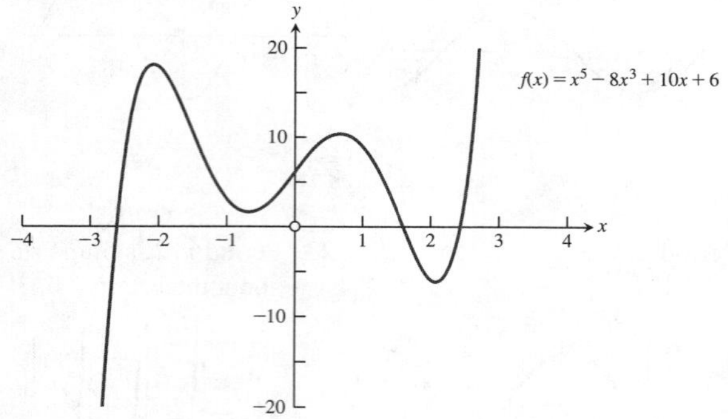

5. [−4, 4] by [−20, 100]

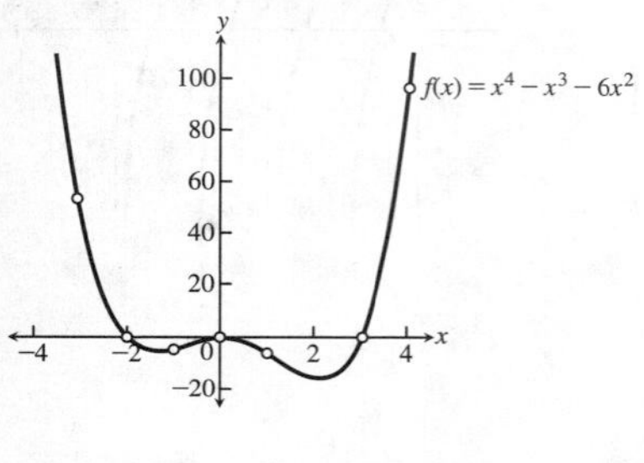

7. $[-3, 6]$ by $[-250, 50]$

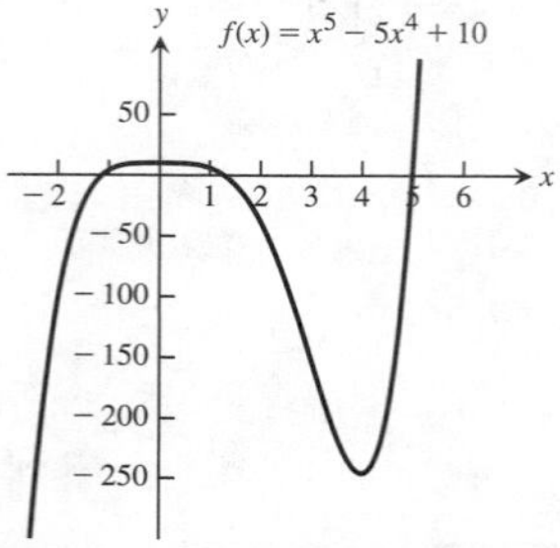

9. $[-5, 5]$ by $[-6, 6]$

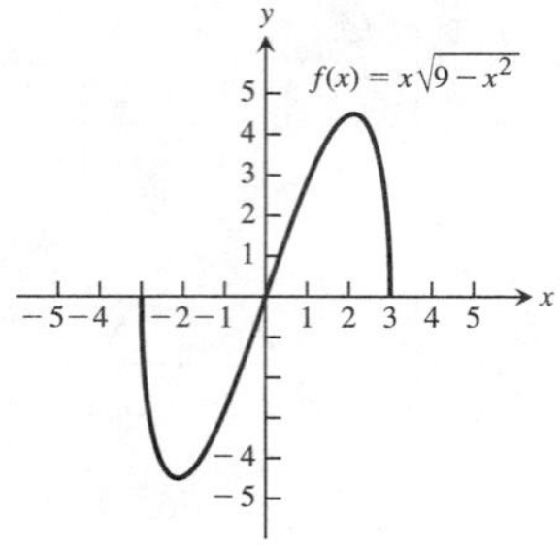

11. $[-2, 6]$ by $[-5, 4]$

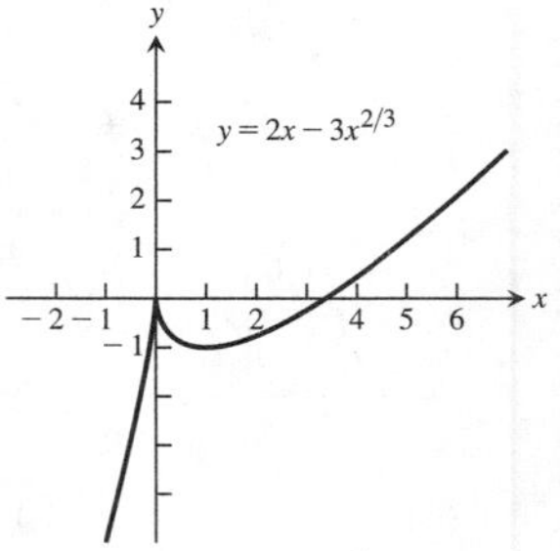

13. $[-2, 8]$ by $[-5, 10]$

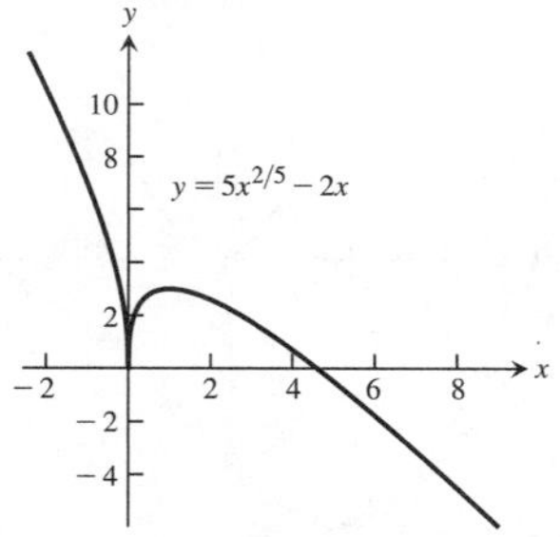

15. $[-3, 3]$ by $[0, 10]$

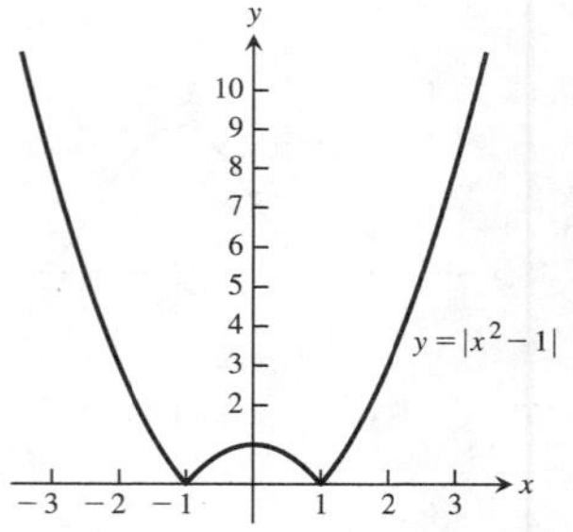

17. [−8, 8] by [−32, 32]

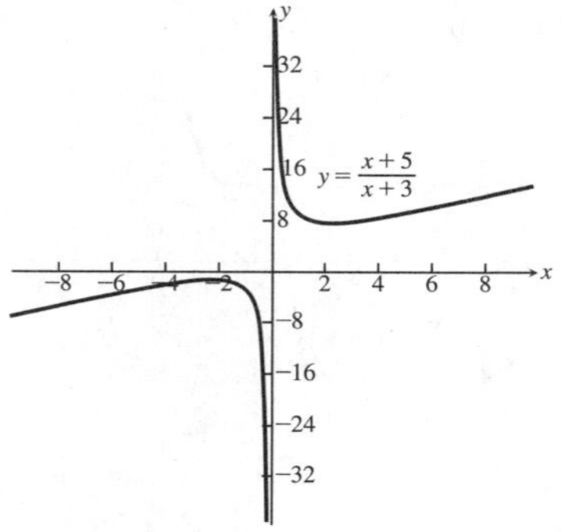

19. $[-4, 4]$ by $[0, 3]$

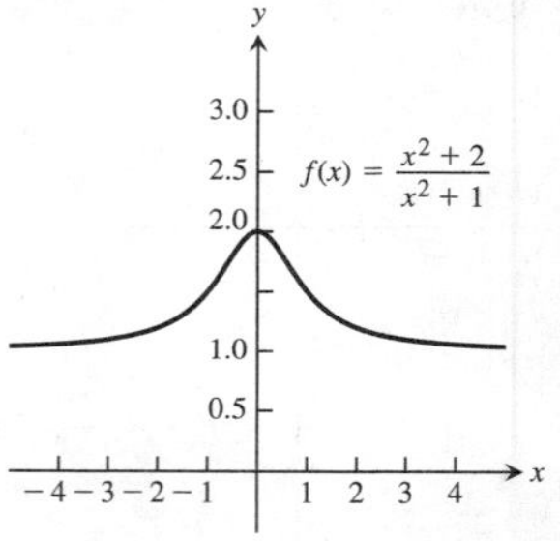

21. $[-10, 10]$ by $[-6, 6]$

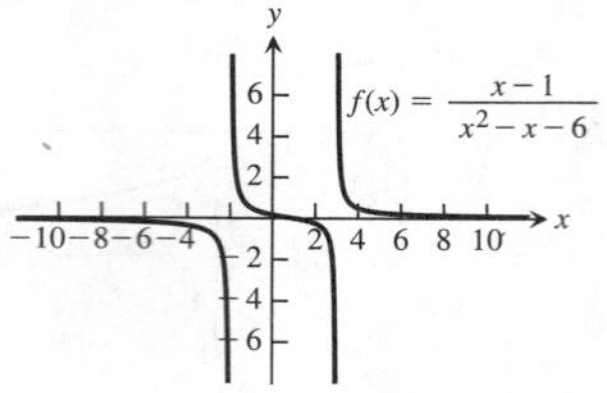

23. $[-6, 10]$ by $[-6, 6]$

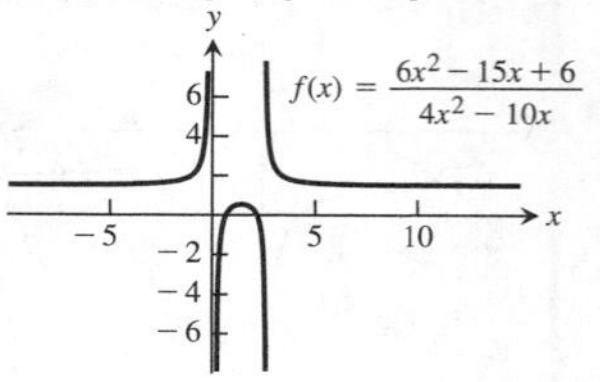

25. [−0.02, 0.02] by [−1, 1]

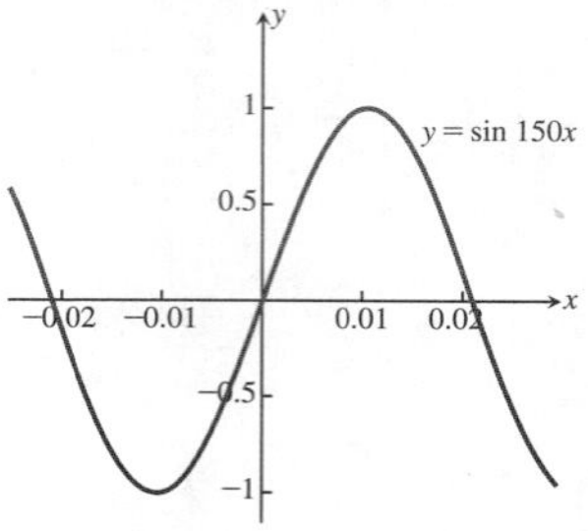

27. [−300, 300] by [−1.5, 1.5]

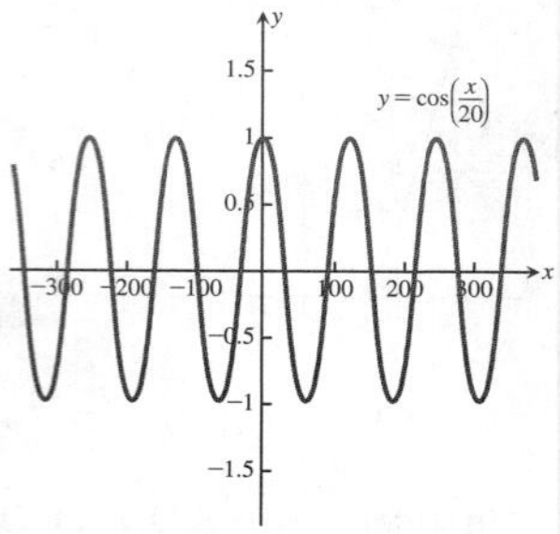

29. $\left[-\frac{\pi}{15}, \frac{\pi}{15}\right]$ by $[-0.25, 0.25]$

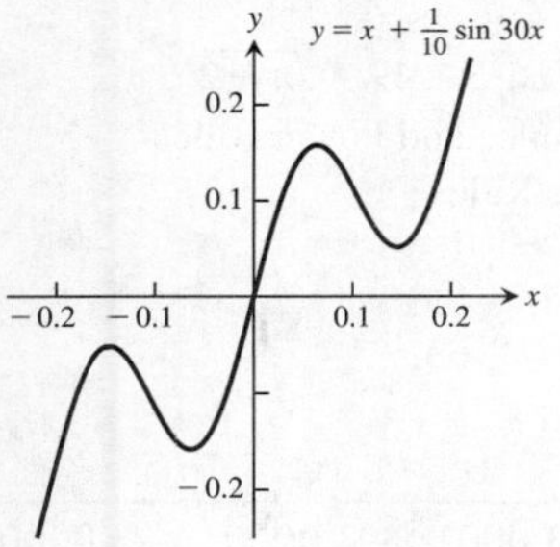

31.

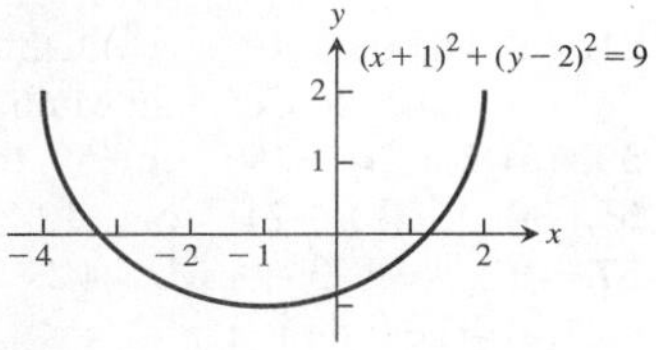

33.

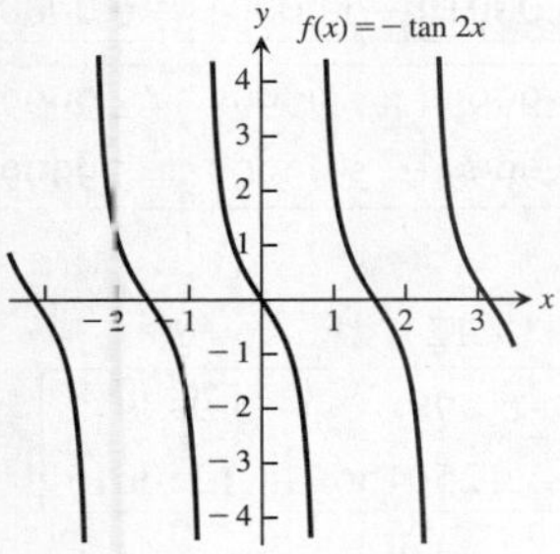

35. $[-6, 6]$ by $[-2, 2]$

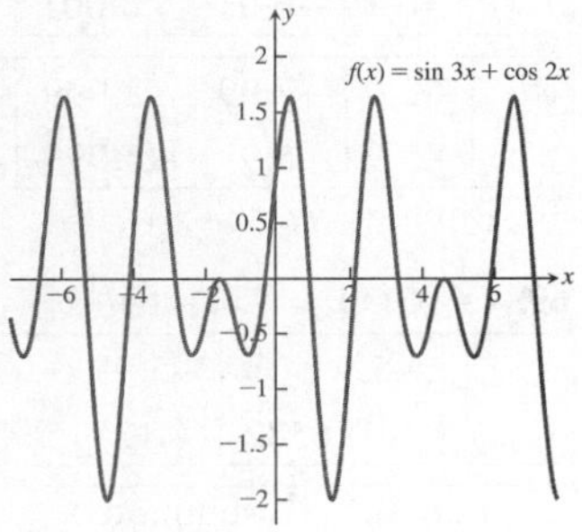

PRACTICE EXERCISES, pp. 34–35

1. $A = \pi r^2$, $C = 2\pi r$, $A = \dfrac{C^2}{4\pi}$ **3.** $x = \tan\theta$, $y = \tan^2\theta$

5. Origin **7.** Neither **9.** Even **11.** Even

13. Odd **15.** Neither

17. (a) Even **(b)** Odd **(c)** Odd **(d)** Even **(e)** Even

19. (a) Domain: all reals **(b)** Range: $[-2, \infty)$

21. (a) Domain: $[-4, 4]$ **(b)** Range: $[0, 4]$

23. (a) Domain: all reals **(b)** Range: $(-3, \infty)$

25. (a) Domain: all reals **(b)** Range: $[-3, 1]$

27. (a) Domain: all reals **(b)** Range: $[0, 2]$

29. (a) Domain: $(-\infty, -1]$ and $[3, \infty)$ **(b)** Range: $(-\infty, 5]$

31. (a) Domain: $(-\infty, 0)$ and $(0, \infty)$ **(b)** Range: $[-4, 4]$

33. (a) Increasing **(b)** Neither **(c)** Decreasing **(d)** Increasing

35. (a) Domain: $[-4, 4]$ **(b)** Range: $[0, 2]$

37. $f(x) = \begin{cases} 1 - x, & 0 \le x < 1 \\ 2 - x, & 1 \le x \le 2 \end{cases}$

39. (a) 1 **(b)** $\dfrac{1}{\sqrt{2.5}} = \sqrt{\dfrac{2}{5}}$ **(c)** $x, x \ne 0$

(d) $\dfrac{1}{\sqrt{1/\sqrt{x+2}+2}}$

41. (a) $(f \circ g)(x) = -x, x \ge -2$, $(g \circ f)(x) = \sqrt{4 - x^2}$

(b) Domain $(f \circ g)$: $[-2, \infty)$, domain $(g \circ f)$: $[-2, 2]$

(c) Range $(f \circ g)$: $(-\infty, 2]$, range $(g \circ f)$: $[0, 2]$

43.

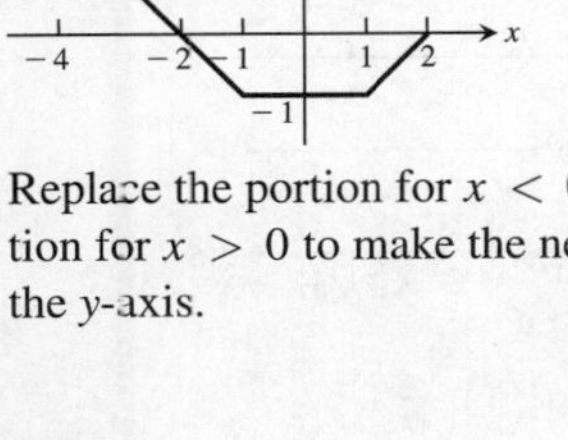

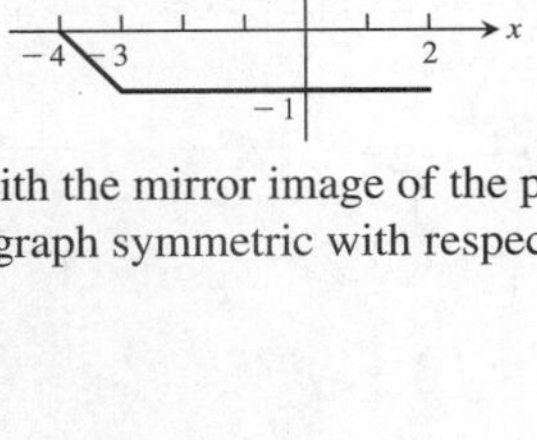

45. Replace the portion for $x < 0$ with the mirror image of the portion for $x > 0$ to make the new graph symmetric with respect to the y-axis.

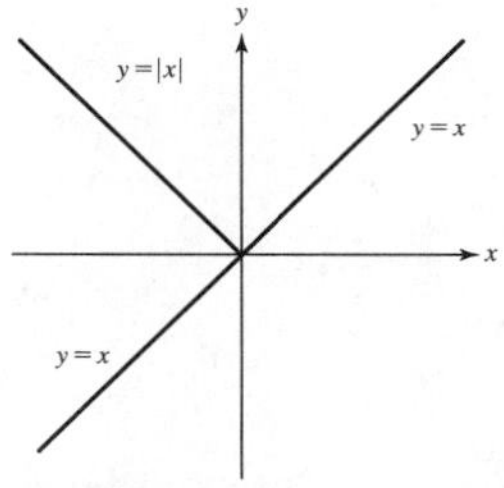

47. Reflects the portion for $y < 0$ across the x-axis

49. Reflects the portion for $y < 0$ across the x-axis

51. Adds the mirror image of the portion for $x > 0$ to make the new graph symmetric with respect to the y-axis

53. (a) $y = g(x-3) + \frac{1}{2}$ **(b)** $y = g\left(x + \frac{2}{3}\right) - 2$

(c) $y = g(-x)$ **(d)** $y = -g(x)$ **(e)** $y = 5g(x)$ **(f)** $y = g(5x)$

55.

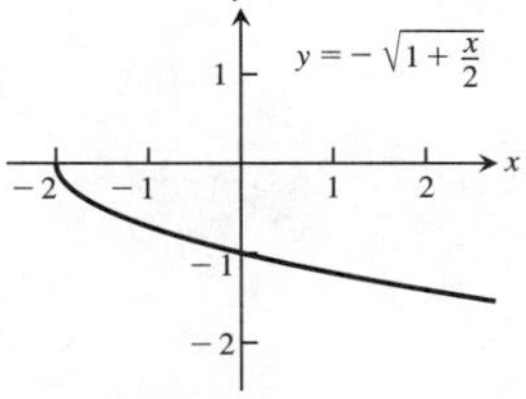

57.

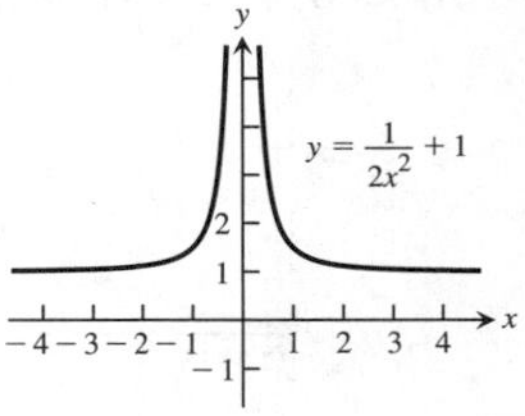

59. Period π

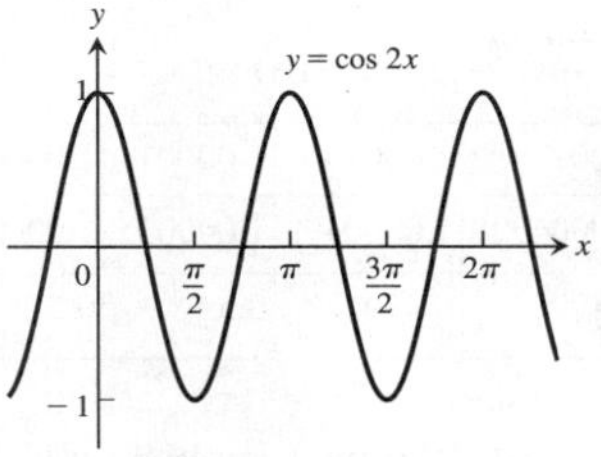

61. Period 2

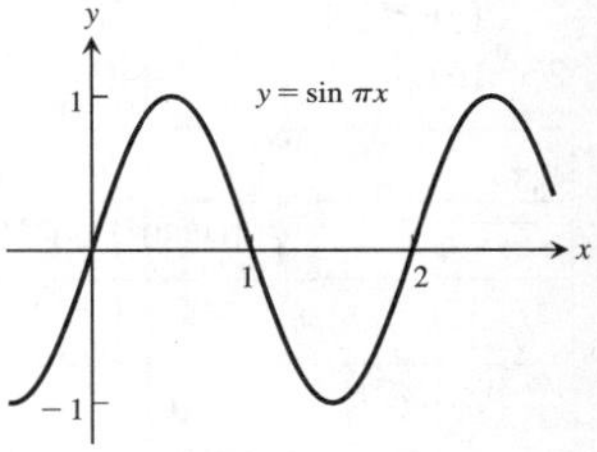

63.

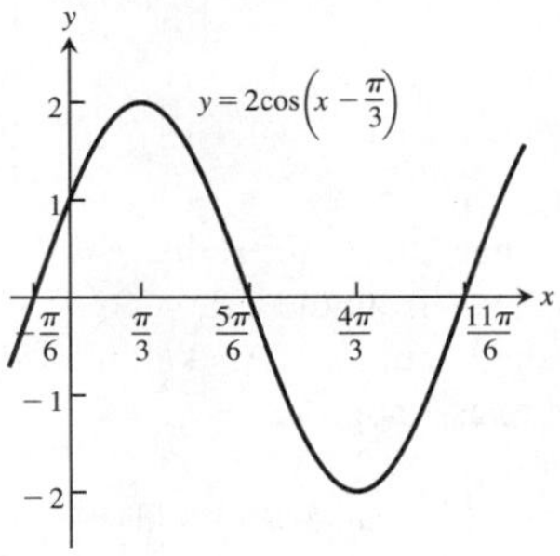

65. (a) $a = 1$ $b = \sqrt{3}$ **(b)** $a = 2\sqrt{3}/3$ $c = 4\sqrt{3}/3$

67. (a) $a = \dfrac{b}{\tan B}$ **(b)** $c = \dfrac{a}{\sin A}$ **69.** ≈ 16.98 m **71. (b)** 4π

ADDITIONAL AND ADVANCED EXERCISES, pp. 35–36

1. Yes. For instance: $f(x) = 1/x$ and $g(x) = 1/x$, or $f(x) = 2x$ and $g(x) = x/2$, or $f(x) = e^x$ and $g(x) = \ln x$.

3. If $f(x)$ is odd, then $g(x) = f(x) - 2$ is not odd. Nor is $g(x)$ even, unless $f(x) = 0$ for all x. If f is even, then $g(x) = f(x) - 2$ is also even.

5.

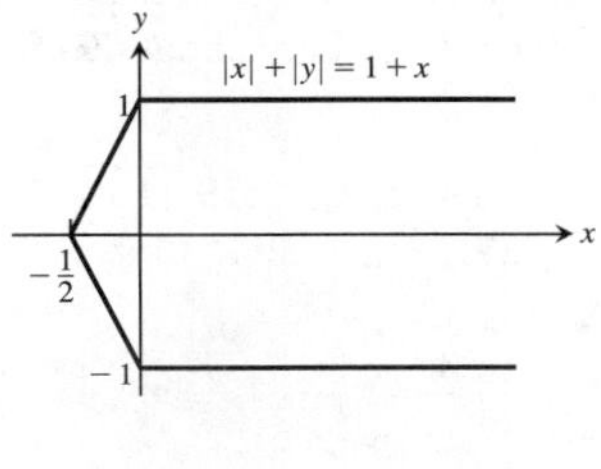

Chapter 2

SECTION 2.1, pp. 43–45

1. **(a)** 19 **(b)** 1

3. **(a)** $-\frac{4}{\pi}$ **(b)** $-\frac{3\sqrt{3}}{\pi}$

5. 1

7. **(a)** 4 **(b)** $y = 4x - 9$

9. **(a)** 2 **(b)** $y = 2x - 7$

11. **(a)** 12 **(b)** $y = 12x - 16$

13. **(a)** -9 **(b)** $y = -9x - 2$

15. **(a)** $-1/4$ **(b)** $y = -x/4 - 1$

17. **(a)** $1/4$ **(b)** $y = x/4 + 1$

19. Your estimates may not completely agree with these.

(a)

PQ_1	PQ_2	PQ_3	PQ_4
43	46	49	50

The appropriate units are m/sec.

(b) ≈50 m/sec or 180 km/h

21. **(a)**

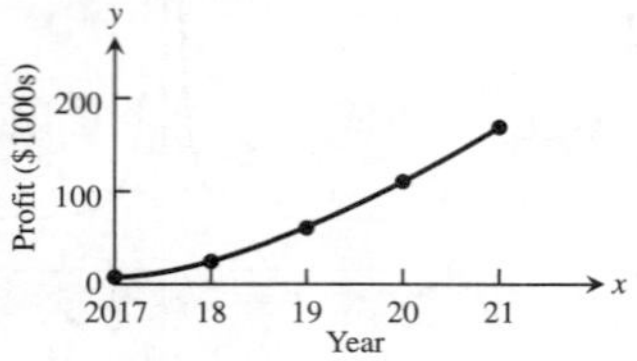

(b) ≈$56,000/year

(c) ≈$42,000/year

23. **(a)** 0.414213, 0.449489, $(\sqrt{1+h} - 1)/h$ **(b)** $g(x) = \sqrt{x}$

$1 + h$	1.1	1.01	1.001	1.0001	1.00001	1.000001
$\sqrt{1+h}$	1.04880	1.004987	1.0004998	1.0000499	1.000005	1.0000005
$(\sqrt{1+h} - 1)/h$	0.4880	0.4987	0.4998	0.499	0.5	0.5

(c) 0.5

25. **(a)** 15 mph, 3.3 mph, 10 mph **(b)** 10 mph, 0 mph, 4 mph

(c) 20 mph when $t = 3.5$ hr

SECTION 2.2, pp. 53–56

1. **(a)** 0

(b) Does not exist. As x approaches 1 from the right, $g(x)$ approaches 2. As x approaches 1 from the left, $g(x)$ approaches 1. There is no single number L that all the values of $g(x)$ get arbitrarily close to as x approaches 1.

(c) 1 **(d)** 0

3. **(a)** True **(b)** True **(c)** False **(d)** False **(e)** False **(f)** True **(g)** True **(h)** False **(i)** True **(j)** True **(k)** False

5. $\lim_{x\to 0} \frac{2x}{|2x|}$ does not exist because $\frac{2x}{|2x|} = \frac{2x}{2x} = 1$ if x approaches 0 from right, i.e., $x > 0$ and $\frac{2x}{|2x|} = \frac{-2x}{2x} = -1$ if x approaches 0 from left, i.e., $x < 0$. So, there is no single number L that all the function values get arbitrarily close to as x approaches 0.

7. Nothing can be said.

9. No, the definition does not require that f be defined at $x = 2$ in order for a limiting value to exist there. If $f(2)$ is defined, it can be any real number, so we can conclude nothing about $f(2)$ from $\lim_{x\to 2} f(x) = 10$.

11. -23 **13.** 210 **15.** 3 **17.** $-25/2$ **19.** 16

21. $3/2$ **23.** $1/10$ **25.** -7 **27.** -3 **29.** $-1/2$

31. -1 **33.** $\frac{20}{7}$ **35.** $1/6$ **37.** 4 **39.** 0

41. $3/2$ **43.** -1 **45.** 1 **47.** $1/3$ **49.** $\sqrt{\pi + 2}$

51. **(a)** Quotient Rule **(b)** Difference and Power Rules **(c)** Sum and Constant Multiple Rules

53. **(a)** 5 **(b)** -250 **(c)** 5 **(d)** -1

55. **(a)** 4 **(b)** -21 **(c)** -12 **(d)** $-7/3$

57. 10 **59.** 3 **61.** $1/(2\sqrt{7})$ **63.** 3

65. **(a)** The limit is 1.

67. **(a)** $f(x) = (x^2 - 9)/(x + 3)$

x	-3.1	-3.01	-3.001	-3.0001	-3.00001	-3.000001
$f(x)$	-6.1	-6.01	-6.001	-6.0001	-6.00001	-6.000001

x	-2.9	-2.99	-2.999	-2.9999	-2.99999	-2.999999
$f(x)$	-5.9	-5.99	-5.999	-5.9999	-5.99999	-5.999999

(c) $\lim_{x\to -3} f(x) = -6$

69. **(a)** $G(x) = (x + 6)/(x^2 + 4x - 12)$

x	-5.9	-5.99	-5.999	-5.9999	-5.99999	-5.999999
$G(x)$	$-.126582$	$-.1251564$	$-.1250156$	$-.1250015$	$-.1250001$	$-.1250000$

x	-6.1	-6.01	-6.001	-6.0001	-6.00001	-6.000001
$G(x)$	$-.123456$	$-.124843$	$-.124984$	$-.124998$	$-.124999$	$-.124999$

(c) $\lim_{x\to -6} G(x) = -1/8 = -0.125$

71. **(a)** $f(x) = (x^2 - 1)/(|x| - 1)$

x	-1.1	-1.01	-1.001	-1.0001	-1.00001	-1.000001
$f(x)$	2.1	2.01	2.001	2.0001	2.00001	2.000001

x	$-.9$	$-.99$	$-.999$	$-.9999$	$-.99999$	$-.999999$
$f(x)$	1.9	1.99	1.999	1.9999	1.99999	1.999999

(c) $\lim_{x\to -1} f(x) = 2$

73. **(a)** $g(\theta) = (\sin\theta)/\theta$

θ	.1	.01	.001	.0001	.00001	.000001
$g(\theta)$	.998334	.999983	.999999	.999999	.999999	.999999

θ	$-.1$	$-.01$	$-.001$	$-.0001$	$-.00001$	$-.000001$
$g(\theta)$	.998334	.999983	.999999	.999999	.999999	.999999

$\lim_{\theta\to 0} g(\theta) = 1$

75. $c = 0, 1, -1$; the limit is 0 at $c = 0$, and 1 at $c = 1, -1$.

77. 60 **79.** **(a)** 5 **(b)** 5 **81.** **(a)** 0 **(b)** 0

SECTION 2.3, pp. 61–64

1. $\delta = 2$

3. $\delta = 1/2$

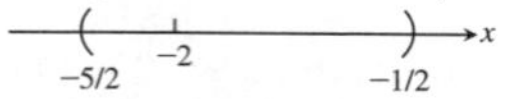

5. $\delta = 3/14$

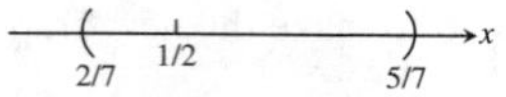

7. $\delta = 0.1$, or a smaller positive value
9. $\delta = 7/16$, or a smaller positive value
11. $\delta = \sqrt{5} - 2$, or a smaller positive value
13. $\delta = 0.36$, or a smaller positive value
15. 0.01
17. $(-0.19, 0.21)$, $\delta = 0.19$, or a smaller positive value
19. $(3, 15)$, $\delta = 5$, or a smaller positive value
21. $(10/3, 5)$, $\delta = 2/3$, or a smaller positive value
23. $(-\sqrt{4.5}, -\sqrt{3.5})$, $\delta = \sqrt{4.5} - 2 \approx 0.12$, or a smaller positive value
25. 0.0009
27. $\left(2 - \frac{0.03}{m}, 2 + \frac{0.03}{m}\right)$, $\delta = \frac{0.03}{m}$, or a smaller positive value
29. $\left(\frac{1}{2} - \frac{c}{m}, \frac{c}{m} + \frac{1}{2}\right)$, $\delta = \frac{c}{m}$, or a smaller positive value
31. $L = -7$; $\delta \approx 3.33$ **33.** $L = 0$; $\delta = 0.002496$
35. $L = 5$; $\delta = 0.99$
55. $[3.384, 3.387]$. To be safe, the left endpoint was rounded up and the right endpoint rounded down.
59. The limit does not exist as x approaches 4.

SECTION 2.4, pp. 70–72

1. (a) True **(b)** False **(c)** False **(d)** True **(e)** True **(f)** True **(g)** True **(h)** True **(i)** True **(j)** False **(k)** True
3. (a) 2; 1 **(b)** No **(c)** 2.5; 2.5 **(d)** 2.5
5. (a) No **(b)** Yes, 1 **(c)** No
7. (a)

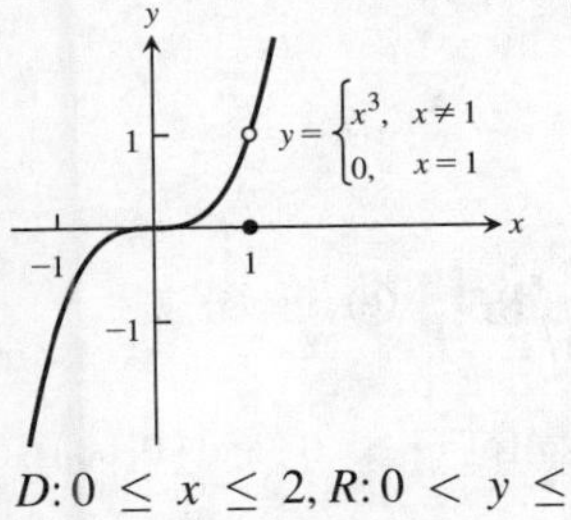

(b) 1, 1 **(c)** Yes, 1

9. (a) $D: 0 \le x \le 2, R: 0 < y \le 1$ and $y = 2$
(b) $[0, 1) \cup (1, 2]$ **(c)** $x = 2$ **(d)** $x = 0$

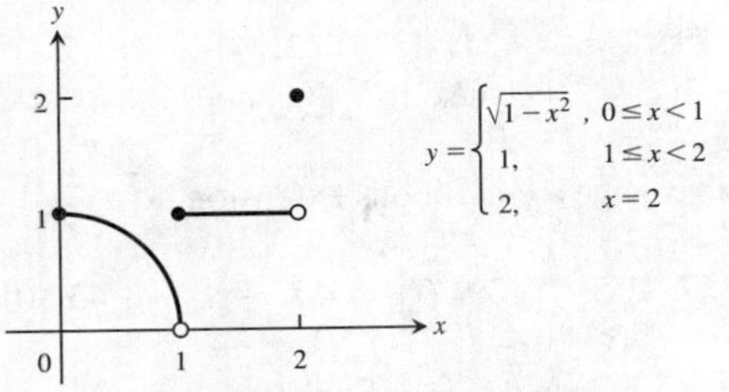

11. 3/5 **13.** −1 **15.** 3 **17. (a)** 1 **(b)** −1
19. (a) 1 **(b)** −1 **21. (a)** 1 **(b)** 2/3 **23.** 1 **25.** 5/9
27. 2 **29.** 1/2 **31.** 4 **33.** 0 **35.** 1 **37.** $\frac{1}{3}$
39. 0 **41.** 3/8 **43.** 3 **45.** 0
51. $\delta = \varepsilon^2$, or a smaller positive value, $\lim_{x \to 5^+} \sqrt{x - 5} = 0$
55. (a) 400 **(b)** 399 **(c)** The limit does not exist.

SECTION 2.5, pp. 81–83

1. No, discontinuous at $x = 1$, not defined at $x = 1$.
3. Continuous **5. (a)** Yes **(b)** Yes **(c)** Yes **(d)** Yes
7. (a) No **(b)** No **9.** 0
11. 1, nonremovable; 0, removable **13.** All x except $x = 5$
15. All x except $x = 2$ and $x = 5$ **17.** Continuous everywhere
19. All x except $x = 0$ **21.** All x except $n\pi/2$, n any integer
23. All x except $n\pi/2$, n an odd integer
25. Continuous on the interval $[-2, \infty)$
27. Continuous everywhere **29.** Continuous everywhere
31. 0, and the function is continuous at $x = 0$
33. 0, and the function is continuous at $x = 0$
35. $\frac{1}{\sqrt{2}}$ **37.** −1 **39.** 0 **41.** 8 **43.** $\frac{9}{2}$ **45.** $\frac{16}{5}$
47. $a = -2, 3$ **49.** $a = 5/2, b = -1/2$
69. $x \approx 1.8794, -1.5321, -0.3473$ **71.** $x \approx 1.7549$
73. $x \approx 0.7391$

SECTION 2.6, pp. 93–96

1. (a) 0 **(b)** 0 **(c)** 2 **(d)** Does not exist **(e)** −1 **(f)** ∞ **(g)** Does not exist **(h)** 1 **(i)** 0
3. (a) −3 **(b)** −3 **5. (a)** 1/2 **(b)** 1/2 **7. (a)** −5/3 **(b)** −5/3 **9.** 0 **11.** −1 **13. (a)** $\frac{7}{8}$ **(b)** $\frac{7}{8}$
15. (a) 0 **(b)** 0 **17. (a)** 7 **(b)** 7 **19. (a)** 0 **(b)** 0
21. (a) ∞ **(b)** ∞ **23.** $\sqrt{3}$ **25.** ∞ **27.** 0 **29.** 1
31. ∞ **33.** 1 **35.** 1/2 **37.** ∞ **39.** $-\infty$
41. $-\infty$ **43.** ∞ **45. (a)** ∞ **(b)** $-\infty$ **47.** ∞
49. 0 **51.** 1 **53. (a)** ∞ **(b)** $-\infty$ **(c)** $-\infty$ **(d)** ∞
55. (a) $-\infty$ **(b)** ∞ **(c)** 0 **(d)** 3/2
57. (a) $-\infty$ **(b)** 1/4 **(c)** 1/4 **(d)** 1/4 **(e)** It will be $-\infty$.
59. (a) $-\infty$ **(b)** ∞ **61. (a)** ∞ **(b)** ∞ **(c)** ∞ **(d)** ∞
63.

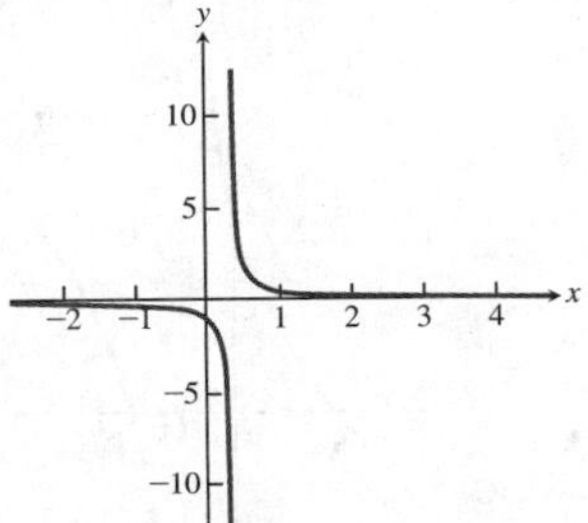

65.

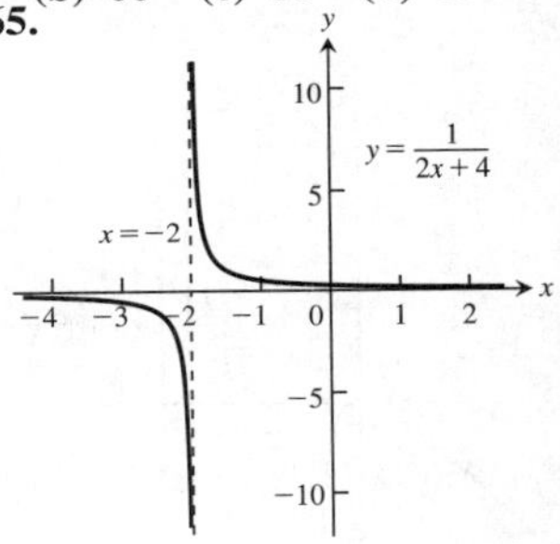

67.

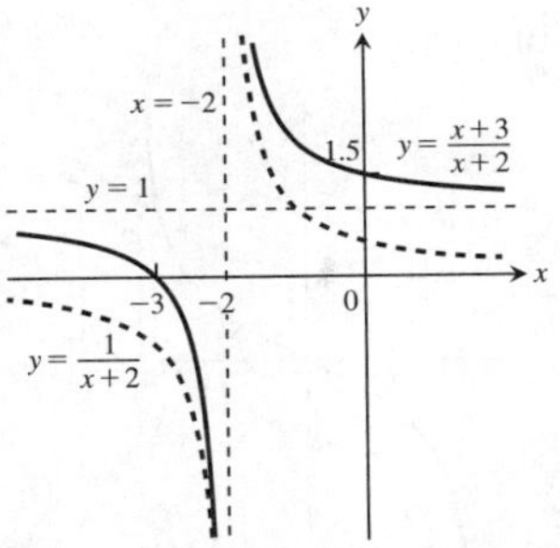

69. Domain: $(-\infty, \infty)$; horizontal asymptote: $y = 7$
71. Domain: $(-\infty, 0)$ and $(0, \infty)$; horizontal asymptotes: $y = -1, y = 1$; vertical asymptote: $x = 0$
73. Here is one possibility.

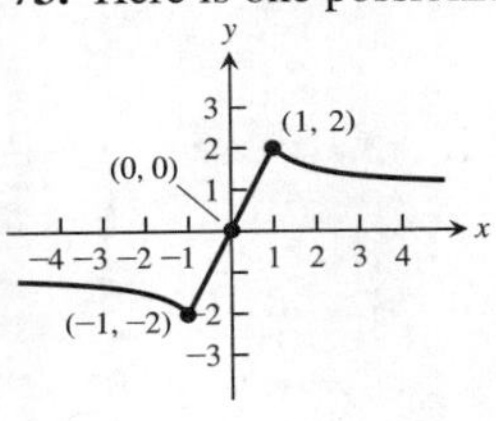

75. Here is one possibility.

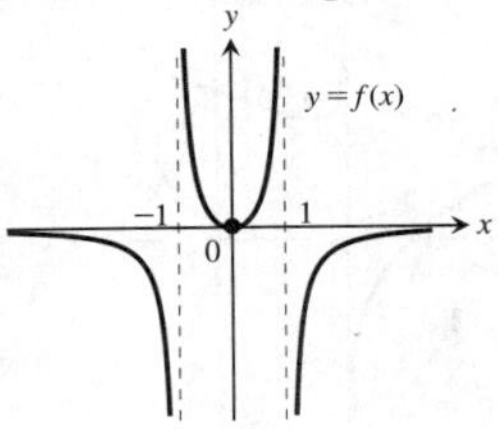

77. Here is one possibility.

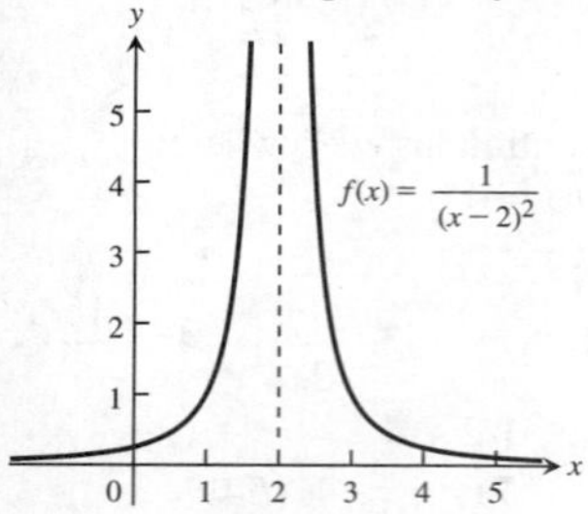

79. Here is one possibility.

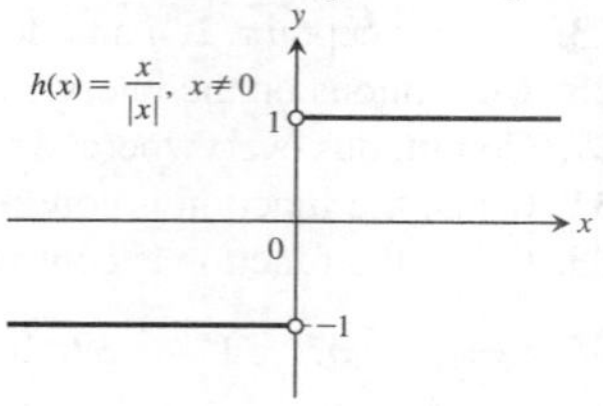

81. Yes, if $\lim_{x\to\infty}\left(\frac{f(x)}{g(x)}\right) = 5$, then $\lim_{x\to-\infty}\left(\frac{f(x)}{g(x)}\right)$ is also equal to 5.

83. At most one **85.** 0 **87.** ∞ **89.** $5/2$

97. (a) For every positive real number B there exists a corresponding number $\delta > 0$ such that for all x,

$$c - \delta < x < c \Rightarrow f(x) > B.$$

(b) For every negative real number $-B$ there exists a corresponding number $\delta > 0$ such that for all x,

$$c < x < c + \delta \Rightarrow f(x) < -B.$$

(c) For every negative real number $-B$ there exists a corresponding number $\delta > 0$ such that for all x,

$$c - \delta < x < c \Rightarrow f(x) < -B.$$

103.

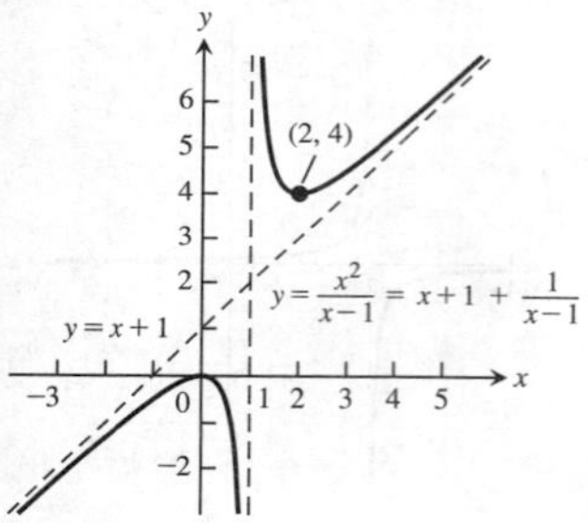

105.

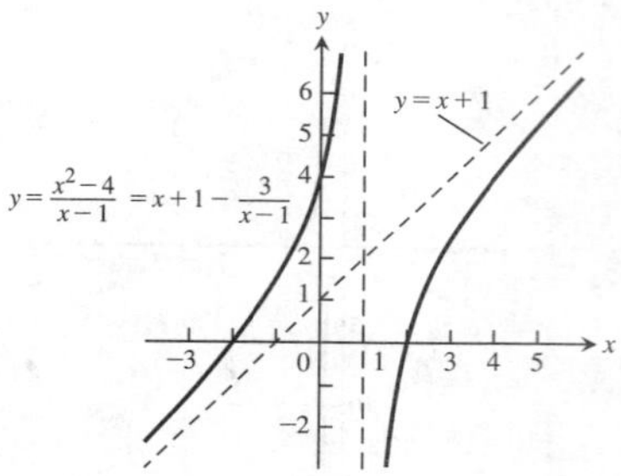

107.

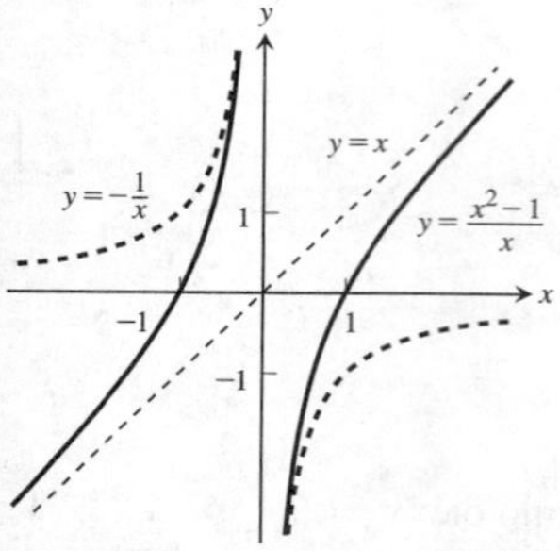

109.

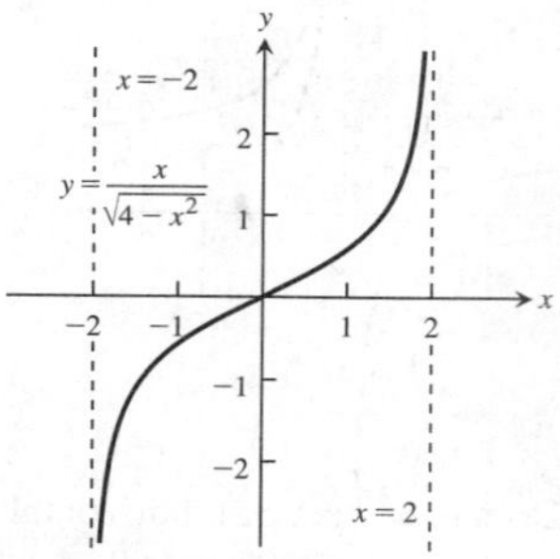

111.

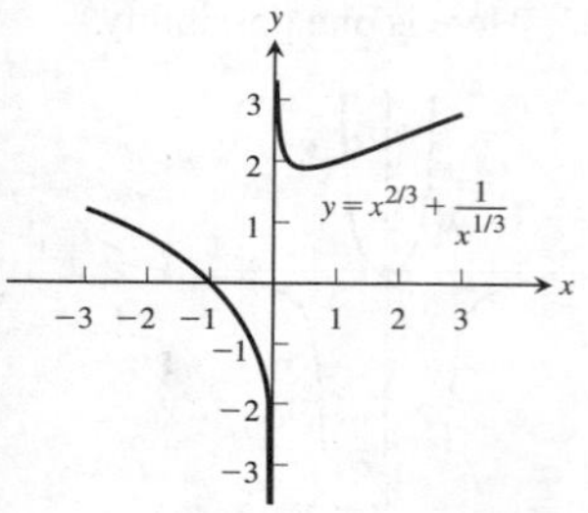

113. (a) $y \to \infty$ **(b)** $y \to \infty$ **(c)** cusps at ± 1

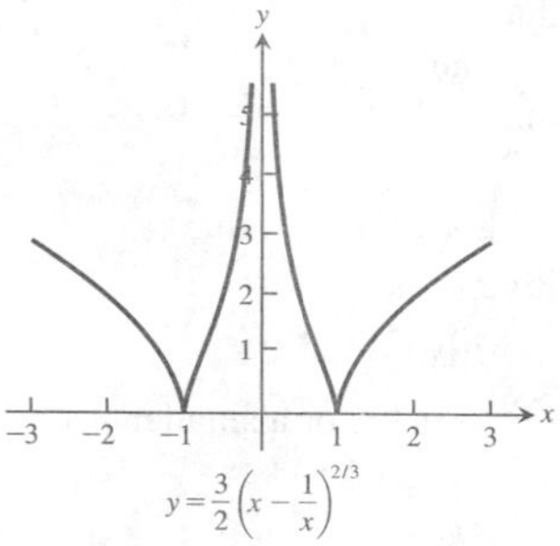

PRACTICE EXERCISES, pp. 97–98

1. At $x = -1$: $\lim_{x\to-1^-} f(x) = \lim_{x\to-1^+} f(x) = 1$, so $\lim_{x\to-1} f(x) = 1 = f(-1)$; continuous at $x = -1$

At $x = 0$: $\lim_{x\to0^-} f(x) = \lim_{x\to0^+} f(x) = 0$, so $\lim_{x\to0} f(x) = 0$. However, $f(0) \neq 0$, so f is discontinuous at $x = 0$. The discontinuity can be removed by redefining $f(0)$ to be 0.

At $x = 1$: $\lim_{x\to1^-} f(x) = -1$ and $\lim_{x\to1^+} f(x) = 1$, so $\lim_{x\to1} f(x)$ does not exist. The function is discontinuous at $x = 1$, and the discontinuity is not removable.

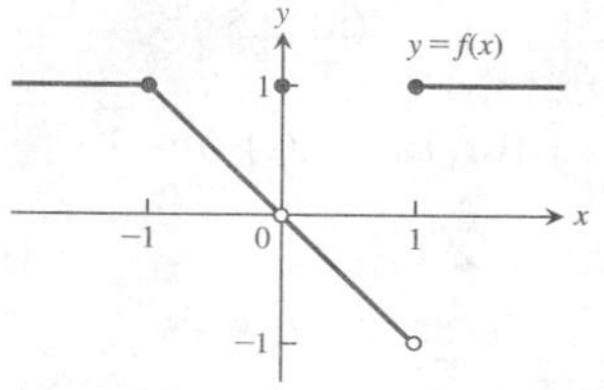

3. (a) -21 **(b)** 49 **(c)** 0 **(d)** 1 **(e)** 1 **(f)** 7 **(g)** -7 **(h)** $-\frac{1}{7}$ **5.** 4

7. (a) $(-\infty, \infty)$ **(b)** $[0, \infty)$ **(c)** $(-\infty, 0)$ and $(0, \infty)$ **(d)** $[0, \infty)$

9. (a) Does not exist **(b)** 0 **11.** $\frac{1}{2}$ **13.** $2x$ **15.** $-\frac{1}{4}$

17. $2/3$ **19.** $2/\pi$ **21.** 1 **23.** 4 **25.** 2 **27.** 0

31. No in both cases, because $\lim_{x\to1} f(x)$ does not exist, and $\lim_{x\to-1} f(x)$ does not exist.

33. Yes, f does have a continuous extension, to $a = 1$ with $f(1) = 4/3$.

35. No **37.** $2/5$ **39.** 0 **41.** $-\infty$ **43.** 0 **45.** 1

47. (a) $x = 3$ **(b)** $x = 1$ **(c)** $x = -4$

49. Domain: $[-4, 2)$ and $(2, 4]$, Range: $(-\infty, \infty)$

ADDITIONAL AND ADVANCED EXERCISES, pp. 99–101

1. 0; the left-hand limit was taken because the function is undefined for $v > c$.

3. $65 < t < 75$; within 5°F **11. (a)** B **(b)** A **(c)** A **(d)** A

19. (a) $\lim_{a\to0} r_+(a) = 0.5$, $\lim_{a\to-1^+} r_+(a) = 1$

(b) $\lim_{a\to0} r_-(a)$ does not exist, $\lim_{a\to-1^+} r_-(a) = 1$

23. 0 **25.** 1 **27.** 4 **29.** -1 **31.** $1/2$ **33.** 4

35. 1 **37.** To make g continuous at the origin, define $g(0) = 1$.

39. $y = 2x$ **41.** $y = x, y = -x$ **45.** $-4/3$

47. (a) Domain: $\{0, 1, 1/2, 1/3, 1/4, \ldots\}$

(b) The domain intersects any interval (a, b) containing 0.

(c) 0

49. **(a)** Domain: $(-\infty, -1/\pi] \cup [-1/(2\pi), -1/(3\pi)] \cup$ $[-1/(4\pi), -1/(5\pi)] \cup \cdots \cup [1/(5\pi), 1/(4\pi)] \cup$ $[1/(3\pi), 1/(2\pi)] \cup [1/\pi, \infty)$
(b) The domain intersects any interval (a, b) containing 0.
(c) 0

Chapter 3

SECTION 3.1, pp. 104–106

1. P_1: $m_1 = 1$, P_2: $m_2 = 5$
3. P_1: $m_1 = 5/2$, P_2: $m_2 = -1/2$
5. $y = -2x + 8$ **7.** $y = x + 4$

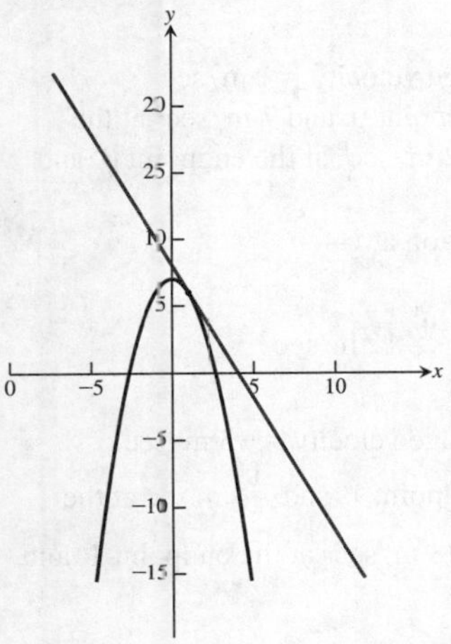

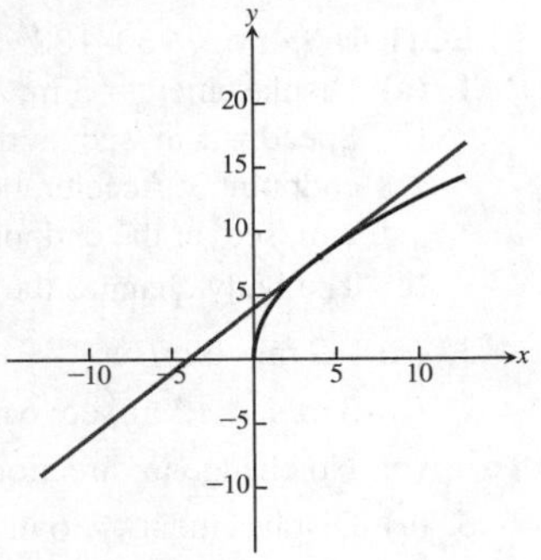

9. $y = 6x + 4$

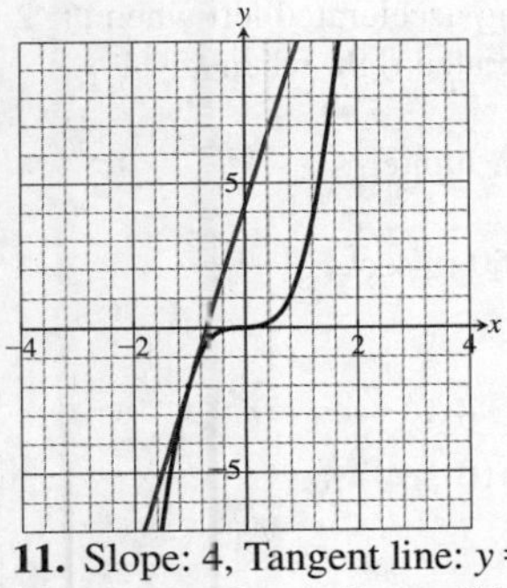

11. Slope: 4, Tangent line: $y = 4x + 3$
13. $m = -2$, $y - 3 = -2(x - 3)$
15. $m = 12$, $y - 8 = 12(t - 2)$
17. Slope: $\frac{1}{2}$, Tangent line: $y = \frac{1}{2}x + 1$
19. Slope: -1, Tangent line: $y = -x + 4$
21. Slope: -4, Tangent line: $y = -4x + 14$
23. **(a)** It is the rate of change of the number of cells when $t = 5$. The units are the number of cells per hour.
(b) $P'(3)$ because the slope of the curve is greater there.
(c) $51.72 \approx 52$ cells/h
25. $\left(\frac{-1}{4}, \frac{-21}{4}\right)$ **27.** $y = -(x + 1)$, $y = -(x - 3)$
29. The object is falling downward at a speed of 50 m/sec.
31. 2π
35. Yes, the given function does have tangent at the origin with slope 0.
37. Yes, the given function has the vertical tangent at the origin.
39. **(a)** The graph appears to have a cusp at $x = 0$.
(b) The given function does not have the vertical tangent line at $x = 0$.
41. **(a)** At $x = 0$
43. **(a)** The graph appears to have a cusp at $x = 0$.
(b) The given function does not have the vertical tangent line at $x = 0$.
45. **(a)** At $x = 1$ **47.** **(a)** At $x = 0$

SECTION 3.2, pp. 111–115

1. $-4x$; 8; 0; 4 **3.** $-\frac{1}{t^3}$; $-\frac{1}{8}$; 1; $-\frac{1}{2\sqrt{2}}$
5. $\frac{1}{\sqrt{2\theta}}$; $\frac{1}{\sqrt{2}}$; $\frac{1}{2}$; $\frac{3}{4}$ **7.** $10x$ **9.** $\frac{10}{(3t+5)^2}$
11. $\frac{2}{3q^{2/3}}$ **13.** $3 - \frac{5}{2x^2}$, and slope, $m = \frac{19}{8}$ **15.** $3t^2 - 2t$, 5
17. $\frac{-4}{(x-2)\sqrt{x-2}}$, $y - 4 = -\frac{1}{2}(x - 6)$ **19.** 8
21. $\frac{5}{54}$ **23.** $-\frac{10}{(2x+3)^2}$ **25.** $\frac{-1}{(x-1)^2}$ **27.** (b) **29.** (d)
31. **(a)** $x = 0, 1, 4$
(b)

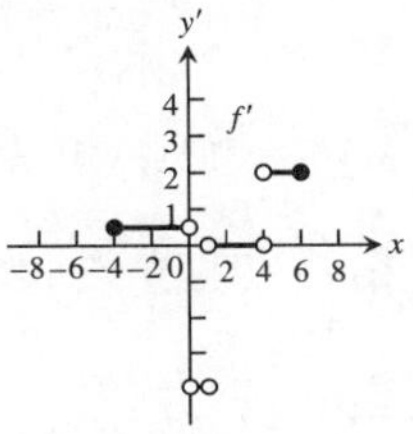

33.

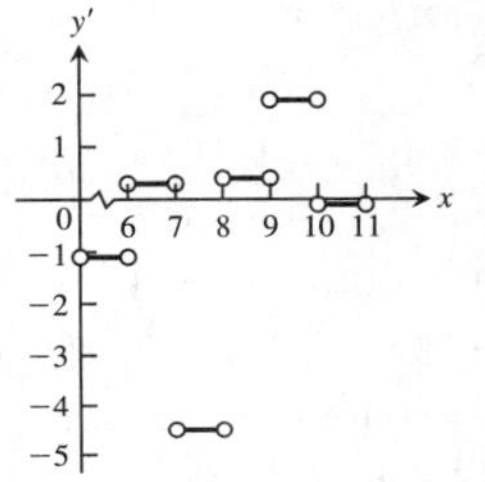

35. **(a)** **i)** ≈ 1.54 **ii)** ≈ 2.86 **iii)** ≈ 0 **iv)** ≈ -3.75
(b) 11 A.M.; 5 P.M.
(c)

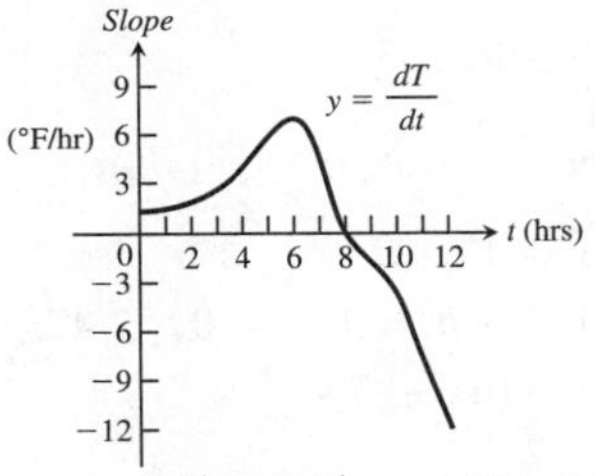

37. Since $\lim_{h\to0^+} \frac{f(0+h) - f(0)}{h} = 1$ while $\lim_{h\to0^-} \frac{f(0+h) - f(0)}{h} = 0$, $f'(0) = \lim_{h\to0} \frac{f(0+h) - f(0)}{h}$ does not exist and $f(x)$ is not differentiable at $x = 0$.

39. Since $\lim_{h\to0^+} \frac{f(1+h) - f(1)}{h} = 2$ while $\lim_{h\to0^-} \frac{f(1+h) - f(1)}{h} = \frac{1}{2}$, $f'(1) = \lim_{h\to0} \frac{f(1+h) - f(1)}{h}$ does not exist and $f(x)$ is not differentiable at $x = 1$.
41. Since $f(x)$ is not continuous at $x = 0$, $f(x)$ is not differentiable at $x = 0$.
43. Since $\lim_{h\to0^+} \frac{f(0+h) - f(0)}{h} = 3$ while $\lim_{h\to0^-} \frac{f(0+h) - f(0)}{h} = 0$, f is not differentiable at $x = 0$.
45. **(a)** $-3 \le x \le 2$ **(b)** None **(c)** None
47. **(a)** $-3 \le x < 0, 0 < x \le 3$ **(b)** None **(c)** $x = 0$

49. **(a)** $-1 \le x < 0, 0 < x \le 2$ **(b)** $x = 0$ **(c)** None

55. Equation of the tangent line: $y = 4x - 13$; and the point of tangency is $(2, -5)$

SECTION 3.3, pp. 121–124

1. $-4x; -4$ **3.** $21t^2 - 10t^4; 42t - 40t^3$

5. $\dfrac{dy}{dx} = 4x^2 - 1, \dfrac{d^2y}{dx^2} = 8x$ **7.** $\dfrac{dw}{dz} = -\dfrac{6}{z^3} + \dfrac{1}{z^2}, \dfrac{d^2w}{dz^2} = \dfrac{18}{z^4} - \dfrac{2}{z^3}$

9. $15x^2 + 14x - 10 + \dfrac{6}{x^3} - \dfrac{18}{x^4}; 30x - \dfrac{18}{x^4} + \dfrac{72}{x^5} + 14$

11. $\dfrac{dr}{ds} = \dfrac{-2}{3s^3} + \dfrac{5}{2s^2}, \dfrac{d^2r}{ds^2} = \dfrac{2}{s^4} - \dfrac{5}{s^3}$

13. **(a)** $-20x^4 + 48x^2 - 16x - 15$ **(b)** $-20x^4 + 48x^2 - 16x - 15$

15. **(a)** $3x^2 + 12x - \dfrac{12}{x^2} + 3$ **(b)** $3x^2 + 12x - \dfrac{2}{x^2} + 3$

17. $-\dfrac{4}{(x-2)^2}$ **19.** $\dfrac{x^2 + 10x + 9}{(x+5)^2}$ **21.** $\dfrac{6t^2 - 16t - 7}{(2t^2 + 7)^3}$

23. $\dfrac{18}{\sqrt{s}(2\sqrt{s} + 9)^2}$ **25.** $v' = -\dfrac{1}{x^2} + 2x^{-3/2}$

27. $y' = \dfrac{-4x^3 - 3x^2 + 1}{(x^2 - 1)^2(x^2 + x + 1)^2}$

29. $y' = 2x^3 - 3x - 1, y'' = 6x^2 - 3, y''' = 12x, y^{(4)} = 12,$ $y^{(n)} = 0$ for $n \ge 5$

31. $y' = 3x^2 + 22x - 17, y'' = 6x + 22, y''' = 6$ $y^{(n)} = 0$ for $n = 4, 5, 6, \dots$

33. $\dfrac{3x^4 - 9}{2x^2}; \dfrac{3x^4 + 9}{x^3}$

35. $\dfrac{dr}{d\theta} = 3\theta^{-4}, \dfrac{d^2r}{d\theta^2} = -12\theta^{-5}$ **37.** $\dfrac{dw}{dz} = -z^{-2} - 1, \dfrac{d^2w}{dz^2} = 2z^{-3}$

39. **(a)** 38 **(b)** $\dfrac{-9}{2}$ **(c)** $\dfrac{18}{49}$ **(d)** 35

41. **(a)** $y = -\dfrac{x}{8} + \dfrac{5}{4}$ **(b)** $m = -4$ at $(0, 1)$ **(c)** $y = 8x - 15, y = 8x + 17$

43. $y = 4x, y = 2$ **45.** $a = 1, b = 1, c = 0$ **47.** $(1, 3)$

49. $(0, 0), (4, 2)$ **51.** $y = -16x + 24$

53. **(a)** $y = 5x + 4$

(b)

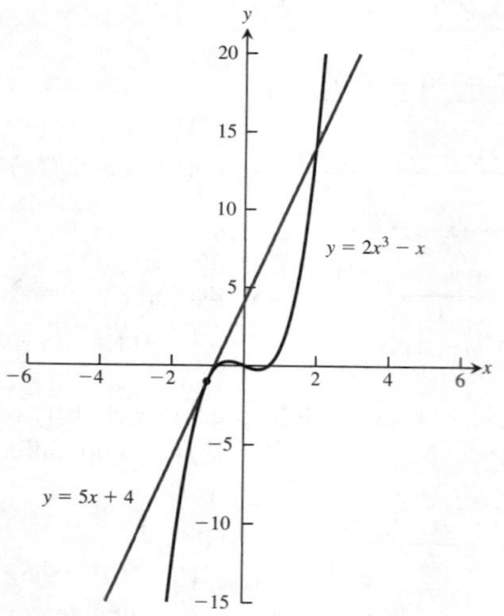

(c) $(2, 14)$

55. 50 **57.** -1

59. $P'(x) = na_nx^{n-1} + (n-1)a_{n-1}x^{n-2} + \cdots + 2a_2x + a_1$

61. The Product Rule is then the Constant Multiple Rule, so the latter is a special case of the Product Rule.

63. **(a)** $\dfrac{d}{dx}(uvw) = uvw' + uv'w + u'vw$

(b) $\dfrac{d}{dx}(u_1u_2u_3u_4) = u_1u_2u_3u_4' + u_1u_2u_3'u_4 + u_1u_2'u_3u_4 + u_1'u_2u_3u_4$

(c) $\dfrac{d}{dx}(u_1 \cdots u_n) = u_1u_2 \cdots u_{n-1}u_n' + u_1u_2 \cdots u_{n-2}u_{n-1}'u_n + \cdots + u_1'u_2 \cdots u_n$

65. $\dfrac{dP}{dV} = -\dfrac{nRT}{(V - nb)^2} + \dfrac{2an^2}{V^3}$

SECTION 3.4, pp. 130–133

1. **(a)** Displacement = 3 m; Average velocity = 1 m/sec

(b) Speed = 5 m/sec² at the endpoint 0, and 7 m/sec² at the endpoint 3; Acceleration = 2 m/sec² at the endpoint 0, and 2 m/sec² at the endpoint 3

(c) The body changes the direction at $t = \dfrac{5}{4}$.

3. **(a)** -9 m, -3 m/sec

(b) 3 m/sec, 12 m/sec; 6 m/sec², -12 m/sec²

(c) No change in direction

5. **(a)** Displacement = -6 m; Average velocity = -3 m/sec

(b) Speed = 15 m/sec at the endpoint 1, and $\dfrac{1}{3}$ m/sec at the endpoint 3; Acceleration = 48 m/sec² at the endpoint 1, and $\dfrac{4}{9}$ m/sec² at the endpoint 3

(c) The body does not change the direction in the given interval.

7. **(a)** The body is motionless but being accelerated left when $t = 2$, and motionless but being accelerated right when $t = 8$.

(b) 27 m/sec **(c)** 54 m

9. 5 sec on Mars; 1.02 sec (approx.) on Jupiter

11. 1

13. **(a)** $v = -50t$, $|v| = 50t$ ft/sec; $a = -50$ ft/sec²

(b) 2.8 sec

(c) -140 ft/sec

15. **(a)** $t = 2, t = 7$ **(b)** $3 \le t \le 6$

(c)

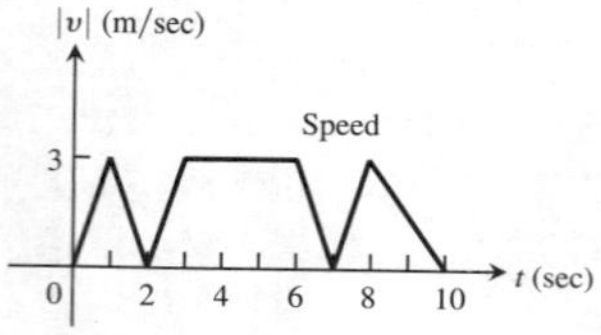

(d)

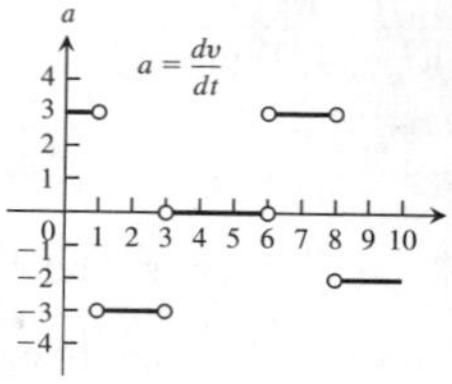

17. **(a)** 380 ft/sec **(b)** 5 sec **(c)** at 20 sec, 0 ft

(d) 27.2 sec, 180 ft/sec **(e)** 7.2 sec

(f) 5 sec after launch

(g) From $t = 5$ to $t = 27.2$ sec; 25 ft/sec² (approx.)

19. C = position, A = velocity, B = acceleration

21. **(a)** \$60 **(b)** \$30 **(c)** \$29.8

23. **(a)** $b'(0) = 10^4$ bacteria/h

(b) $b'(5) = 0$ bacteria/h

(c) $b'(10) = -10^4$ bacteria/h

25. **(a)** $-\dfrac{2}{3}\left(2 - \dfrac{t}{15}\right)$

(b) Fastest at $t = 0$, dy/dt at that time $= -\dfrac{4}{3}$ m/h; Slowest at $t = 30$ sec, dy/dt at that time = 0 m/h

(c)

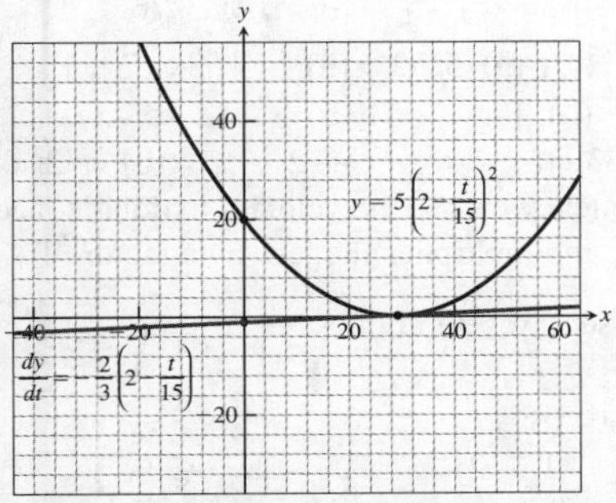

27. 4.88 ft, 8.66 ft, additional ft to stop car for 1 mph speed increase

29. $t = 25$ sec, $D = \dfrac{6250}{9}$ m

31.

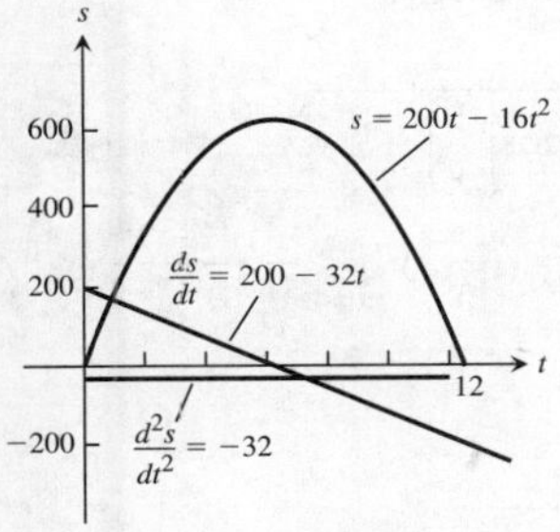

(a) $v = 0$ when $t = 6.25$ sec

(b) $v > 0$ when $0 \le t < 6.25 \Rightarrow$ the object moves up; $v < 0$ when $6.25 < t \le 12.5 \Rightarrow$ the object moves down.

(c) The object changes direction at $t = 6.25$ sec.

(d) The object speeds up on $(6.25, 12.5]$ and slows down on $[0, 6.25)$.

(e) The object is moving fastest at the endpoints $t = 0$ and $t = 12.5$ when it is traveling 200 ft/sec. It's moving slowest at $t = 6.25$ when the speed is 0.

(f) When $t = 6.25$ the object is $s = 625$ m from the origin and farthest away.

33.

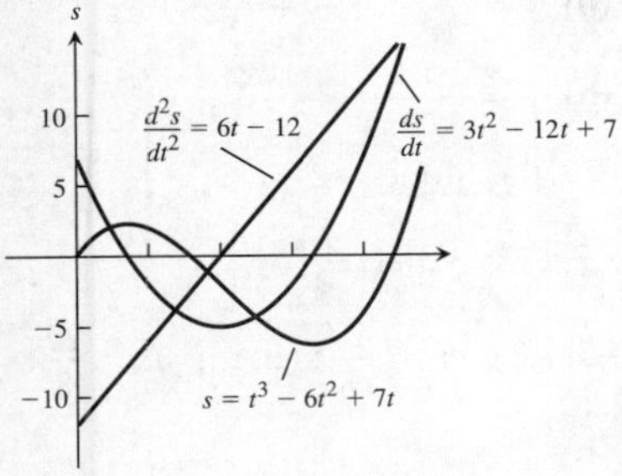

(a) $v = 0$ when $t = \dfrac{6 \pm \sqrt{15}}{3}$ sec

(b) $v < 0$ when $\dfrac{6 - \sqrt{15}}{3} < t < \dfrac{6 + \sqrt{15}}{3} \Rightarrow$ the object moves left; $v > 0$ when $0 \le t < \dfrac{6 - \sqrt{15}}{3}$ or $\dfrac{6 + \sqrt{15}}{3} < t \le 4 \Rightarrow$ the object moves right.

(c) The object changes direction at $t = \dfrac{6 \pm \sqrt{15}}{3}$ sec.

(d) The object speeds up on $\left(\dfrac{6 - \sqrt{15}}{3}, 2\right) \cup \left(\dfrac{6 + \sqrt{15}}{3}, 4\right]$ and slows down on $\left[0, \dfrac{6 - \sqrt{15}}{3}\right) \cup \left(2, \dfrac{6 + \sqrt{15}}{3}\right)$.

(e) The object is moving fastest at $t = 0$ and $t = 4$ when it is moving 7 units/sec and slowest at $t = \dfrac{6 \pm \sqrt{15}}{3}$ sec.

(f) When $t = \dfrac{6 + \sqrt{15}}{3}$ the object is at position $s \approx -6.303$ units and farthest from the origin.

SECTION 3.5, pp. 137–139

1. $-10 - 3 \sin x$ **3.** $2x \cos x - x^2 \sin x$

5. $-\csc x \cot x - \dfrac{2}{\sqrt{x}}$ **7.** $\sin x \sec^2 x + \sin x$

9. $\sec x + x \sec x \tan x - \dfrac{1}{x^2}$ **11.** $\dfrac{-\csc^2 x}{(1 + \cot x)^2}$

13. $4 \tan x \sec x - \csc^2 x$ **15.** 0

17. $\dfrac{\cos^3 x - \sec^3 x}{(\cos x + \sin x)^2}$ **19.** $\sec^2 t - 1$ **21.** $\dfrac{-2 \csc t \cot t}{(1 - \csc t)^2}$

23. $-\theta(\theta \cos \theta + 2 \sin \theta)$

25. $\sec \theta \csc \theta (\tan \theta - \cot \theta) = \sec^2 \theta - \csc^2 \theta$

27. $\sec^2 q$ **29.** $\sec^2 q$

31. $\dfrac{q^3 \cos q - q^2 \sin q - q \cos q - \sin q}{(q^2 - 1)^2}$

33. **(a)** $2 \csc^3 x - \csc x$ **(b)** $2 \sec^3 x - \sec x$

35.

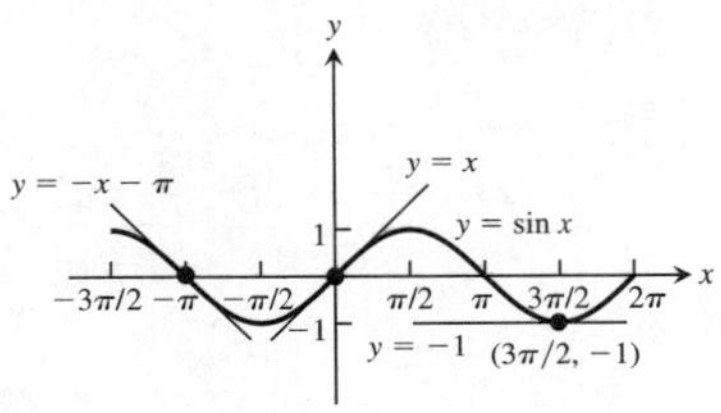

37.

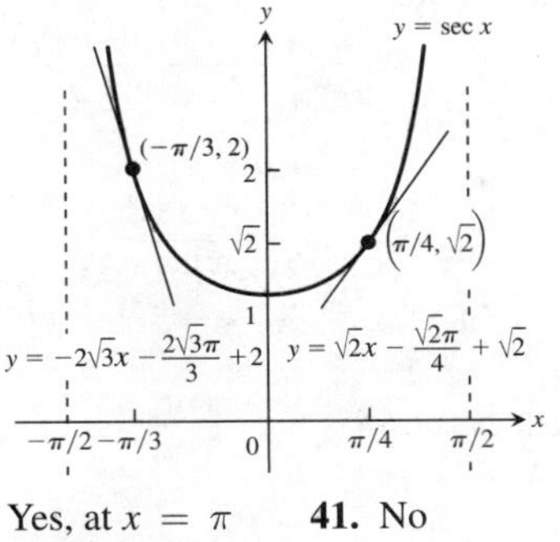

39. Yes, at $x = \pi$ **41.** No

43. Yes, at $x = 0, \pi$, and 2π

45. $\left(-\dfrac{\pi}{4}, -1\right); \left(\dfrac{\pi}{4}, 1\right)$

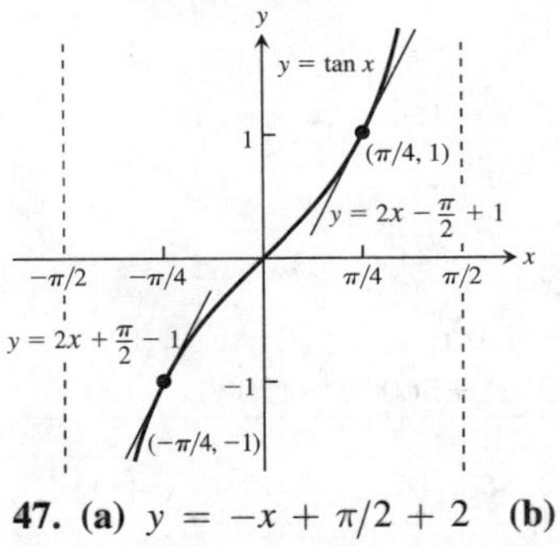

47. **(a)** $y = -x + \pi/2 + 2$ **(b)** $y = 4 - \sqrt{3}$

49. $-\sqrt{2}$ m/sec, $\sqrt{2}$ m/sec, $\sqrt{2}$ m/sec^2, $\sqrt{2}$ m/sec^3 **51.** $c = 9$

53. **(a)** $\sin x$ **(b)** $3\cos x - \sin x$ **(c)** $73\sin x + x\cos x$

55. **(a)** **i)** 10 cm **ii)** 5 cm **iii)** $-5\sqrt{2} \approx -7.1$ cm
(b) **i)** 0 cm/sec **ii)** $-5\sqrt{3} \approx -8.7$ cm/sec
iii) $-5\sqrt{2} \approx -7.1$ cm/sec

SECTION 3.6, pp. 144–146

1. $4x^2$ **3.** $2\sin(4x+10)$ **5.** $\dfrac{-\sin x}{\sqrt{\cos x}}$

7. $2\pi x \sec^2(\pi x^2)$

9. With $u = (2x+1)$, $y = u^5$: $\dfrac{dy}{dx} = \dfrac{dy}{du}\dfrac{du}{dx} = 5u^4 \cdot 2 = 10(2x+1)^4$

11. With $u = (1-(x/7))$, $y = u^{-7}$: $\dfrac{dy}{dx} = \dfrac{dy}{du}\dfrac{du}{dx} = -7u^{-8}\cdot\left(-\frac{1}{7}\right) = \left(1-\frac{x}{7}\right)^{-8}$

13. With $u = ((x^2/8) + x - (1/x))$, $y = u^4$: $\dfrac{dy}{dx} = \dfrac{dy}{du}\dfrac{du}{dx} = 4u^3\cdot\left(\frac{x}{4}+1+\frac{1}{x^2}\right) = 4\left(\frac{x^2}{8}+x-\frac{1}{x}\right)^3\left(\frac{x}{4}+1+\frac{1}{x^2}\right)$

15. With $u = \tan x$, $y = \sec u$: $\dfrac{dy}{dx} = \dfrac{dy}{du}\dfrac{du}{dx} = (\sec u \tan u)(\sec^2 x) = \sec(\tan x)\tan(\tan x)\sec^2 x$

17. With $u = \tan x$, $y = u^3$: $\dfrac{dy}{dx} = \dfrac{dy}{du}\dfrac{du}{dx} = 3u^2\sec^2 x = 3\tan^2 x(\sec^2 x)$

19. $\dfrac{-2}{2\sqrt{5-2t}}$ **21.** $\dfrac{4}{5\pi}\cos 2t - \dfrac{3}{4\pi}\sin 3t$

23. $-\dfrac{1}{1+\sin\theta}$

25. $2x\sin^4 x + 4x^2\sin^3 x\cos x + \cos^{-2} x + 2x\cos^{-3} x\sin x$

27. $(3x-2)^5 - \dfrac{1}{x^3\left(4-\frac{1}{2x^2}\right)^2}$ **29.** $\dfrac{(4x+3)^3(4x+7)}{(x+1)^4}$

31. $\sqrt{x}\sec^2(2\sqrt{x}) + \tan(2\sqrt{x})$ **33.** $\dfrac{x\sec^2 x + \tan x}{\sqrt{5+2x\tan x}}$

35. $-\dfrac{1}{1+\sin\theta}$

37. $-2\sin(\theta^2)\sin 2\theta + 2\theta\cos(2\theta)\cos(\theta^2)$

39. $\left(\dfrac{t+2}{2(t+1)^{3/2}}\right)\cos\left(\dfrac{t}{\sqrt{t+1}}\right)$

41. $2\pi\sin(\pi t-2)\cos(\pi t-2)$ **43.** $\dfrac{8\sin(2t)}{(1+\cos 2t)^5}$

45. $10t^{10}\tan^9 t\sec^2 t + 10t^9\tan^{10} t$

47. $\dfrac{-3t^6(t^2+4)}{(t^3-4t)^4}$ **49.** $-\sin\{\sin(t-4)\}\cos(t-4)$

51. $\left(1+\tan^4\left(\frac{t}{12}\right)\right)^2\left(\tan^3\left(\frac{t}{12}\right)\sec^2\left(\frac{t}{12}\right)\right)$

53. $-3\cot^2(\sin^2 x)\text{cosec}^2(\sin^2 x)\sin 2x$

55. $6\tan(\sin^3 t)\sec^2(\sin^3 t)\sin^2 t\cos t$

57. $3(2t^2-5)^3(18t^2-5)$ **59.** $\dfrac{320x^3+3600x^2+12000x+12500}{x^6}$

61. $2\csc^2(3x-1)\cot(3x-1)$ **63.** $16(2x+1)^2(5x+1)$

65. $f'(x) = 0$ for $x = 1, 4$; $f''(x) = 0$ for $x = 2, 4$ **67.** 64

69. $-\pi/4$ **71.** 0 **73.** -6

75. **(a)** $2/3$ **(b)** $2\pi+5$ **(c)** $15-8\pi$ **(d)** $37/6$
(e) -1 **(f)** $\sqrt{2}/24$ **(g)** $5/32$ **(h)** $-5/(3\sqrt{17})$

77. 3 **79.** **(a)** 1 **(b)** 1

81. $y = -8x+1$ **83.** **(a)** $y = \pi x + 2 - \pi$ **(b)** $\pi/2$

85. Doubling the frequency causes the velocity to double; acceleration to quadruple; and multiplies the jerk by a factor of 8.

87. Velocity $= \dfrac{1}{4}$ m/sec; Acceleration $= \dfrac{-1}{64}$ m/sec^2

SECTION 3.7, pp. 150–152

1. $\dfrac{-2xy-y^2}{x^2+2xy}$ **3.** $\dfrac{1-2y}{2x+2y-1}$ **5.** $\dfrac{-2x^3+3x^2y-xy^2+x}{x^2y-x^3+y}$

7. $\dfrac{1}{y(x+1)^2}$ **9.** $\dfrac{dy}{dx} = \dfrac{\cot^2 y}{2x\cot y\,\text{cosec}^2 y - 1}$

11. $\dfrac{dy}{dx} = \dfrac{y}{x}$ **13.** $\dfrac{-y^2}{y\sin\left(\frac{1}{y}\right) - \cos\left(\frac{1}{y}\right) + xy}$

15. $-\dfrac{\sqrt{r}}{\sqrt{\theta}}$ **17.** $\dfrac{-r}{\theta}$ **19.** $y' = -\dfrac{x}{y}$, $y'' = \dfrac{-y^2-x^2}{y^3}$

21. $\dfrac{dy}{dx} = \dfrac{x+1}{y}$, $\dfrac{d^2y}{dx^2} = \dfrac{-1}{y^3}$

23. $y' = \dfrac{\sqrt{y}}{\sqrt{y}+1}$, $y'' = \dfrac{1}{2(\sqrt{y}+1)^3}$

25. $y' = \dfrac{3x^2}{1-\cos y}$, $y'' = \dfrac{6x(1-\cos y)^2 - 9x^4\sin y}{(1-\cos y)^3}$

27. -2 **29.** $(-2, 1)$: $m = -1$, $(-2, -1)$: $m = 1$

31. **(a)** Equation of the tangent is $5x-4y+3=0$
(b) Equation of the normal is $4x+5y=14$

33. **(a)** $y = 3x+6$ **(b)** $y = -\dfrac{1}{3}x + \dfrac{8}{3}$

35. **(a)** $y = \dfrac{6}{7}x + \dfrac{6}{7}$ **(b)** $y = -\dfrac{7}{6}x - \dfrac{7}{6}$

37. **(a)** $y = -\dfrac{\pi}{2}x + \pi$ **(b)** $y = \dfrac{2}{\pi}x - \dfrac{2}{\pi} + \dfrac{\pi}{2}$

39. **(a)** $y = 2\pi x - 2\pi$ **(b)** $y = -\dfrac{x}{2\pi} + \dfrac{1}{2\pi}$

41. Points: $(-\sqrt{7}, 0)$ and $(\sqrt{7}, 0)$, Slope: -2

43. $m = -1$ at $\left(\dfrac{\sqrt{3}}{4}, \dfrac{\sqrt{3}}{2}\right)$, $m = \sqrt{3}$ at $\left(\dfrac{\sqrt{3}}{4}, \dfrac{1}{2}\right)$

45. $(-3, 2)$: $m = -\dfrac{27}{8}$; $(-3, -2)$: $m = \dfrac{27}{8}$; $(3, 2)$: $m = \dfrac{27}{8}$; $(3, -2)$: $m = -\dfrac{27}{8}$

47. $(3, -1)$

53. $\dfrac{dy}{dx} = -\dfrac{y^3+2xy}{x^2+3xy^2}$, $\dfrac{dx}{dy} = -\dfrac{x^2+3xy^2}{y^3+2xy}$, $\dfrac{dx}{dy} = \dfrac{1}{dy/dx}$

SECTION 3.8, pp. 157–160

1. $\dfrac{dA}{dt} = 2\pi r\dfrac{dr}{dt}$ **3.** 50 **5.** -32 **7.** -5

9. $\dfrac{14}{5}$ **11.** **(a)** -120 m^2/min **(b)** -150 m^3/min

13. **(a)** $\dfrac{dV}{dt} = \pi r^2\dfrac{dh}{dt}$ **(b)** $\dfrac{dV}{dt} = 2\pi hr\dfrac{dr}{dt}$
(c) $\dfrac{dV}{dt} = \pi r^2\dfrac{dh}{dt} + 2\pi hr\dfrac{dr}{dt}$

15. **(a)** 1 volt/sec **(b)** $-\frac{1}{3}$ amp/sec **(c)** $\frac{dR}{dt} = \frac{1}{I}\left(\frac{dV}{dt} - \frac{V}{I}\frac{dI}{dt}\right)$
(d) 3/2 ohms/sec, R is increasing.

17. **(a)** $\frac{ds}{dt} = \frac{x}{\sqrt{x^2+y^2}}\frac{dx}{dt}$
(b) $\frac{ds}{dt} = \frac{x}{\sqrt{x^2+y^2}}\frac{dx}{dt} + \frac{y}{\sqrt{x^2+y^2}}\frac{dy}{dt}$
(c) $\frac{dx}{dt} = -\frac{y}{x}\frac{dy}{dt}$

19. **(a)** $\frac{dA}{dt} = \frac{1}{2}ab\cos\theta\frac{d\theta}{dt}$
(b) $\frac{dA}{dt} = \frac{1}{2}ab\cos\theta\frac{d\theta}{dt} + \frac{1}{2}b\sin\theta\frac{da}{dt}$
(c) $\frac{dA}{dt} = \frac{1}{2}ab\cos\theta\frac{d\theta}{dt} + \frac{1}{2}b\sin\theta\frac{da}{dt} + \frac{1}{2}a\sin\theta\frac{db}{dt}$

21. **(a)** 14 cm^2/sec, increasing **(b)** 0 cm/sec, constant
(c) −14/13 cm/sec, decreasing

23. **(a)** −12 ft/sec **(b)** −59.5 ft^2/sec **(c)** −1 rad/sec

25. 20 ft/sec

27. **(a)** 10.74 cm/sec (approx.) **(b)** 14.32 cm/sec

29. **(a)** $\frac{-1}{24\pi}$ m/min **(b)** $r = \sqrt{26y - y^2}$ m
(c) $\frac{dr}{dt} = -\frac{5}{288\pi}$ m/min

31. 32π ft^2/min **33.** 11 ft/sec

35. Increasing at 466/1681 L/min^2 **37.** −5 m/sec

39. −1500 ft/sec

41. $\frac{8}{5}$ in^2/min

43. **(a)** $-32/\sqrt{13} \approx -8.875$ ft/sec
(b) $d\theta_1/dt = 8/65$ rad/sec, $d\theta_2/dt = -8/65$ rad/sec
(c) $d\theta_1/dt = 1/6$ rad/sec, $d\theta_2/dt = -1/6$ rad/sec

45. −5.5 deg/min

47. 12π km/min

SECTION 3.9, pp. 170–172

1. $19x - 30$ **3.** $-3x + 10$ **5.** $L(x) = x - \pi$

7. $2x$ **9.** $10x - 9$ **11.** $3 + \frac{1}{6}(x - 9)$

13. $f(0) = 1$. Also, $f'(x) = k(1+x)^{k-1}$, so $f'(0) = k$. This means the linearization at $x = 0$ is $L(x) = 1 + kx$.

15. **(a)** 1.01 **(b)** 1.003

17. $\left(3x^2 - \frac{3}{2\sqrt{x}}\right)dx$ **19.** $\frac{2 - 2x^2}{(1+x^2)^2}\,dx$

21. $\frac{1-y}{3\sqrt{y} + x}\,dx$ **23.** $\frac{5}{2\sqrt{x}}\cos(5\sqrt{x})\,dx$

25. $(4x^2)\sec^2\left(\frac{x^3}{3}\right)dx$

27. $\frac{3}{\sqrt{x}}(\csc(1 - 2\sqrt{x})\cot(1 - 2\sqrt{x}))\,dx$

29. **(a)** 0.41 **(b)** 0.4 **(c)** 0.01

31. **(a)** 0.231 **(b)** 0.2 **(c)** 0.031

33. **(a)** −1/3 **(b)** −2/5 **(c)** 1/15

35. $dV = 4\pi r_0^2\,dr$ **37.** $dS = 12x_0\,dx$ **39.** $dV = 2\pi r_0 h\,dr$

41. **(a)** $0.08\pi\ \text{m}^2$ **(b)** 2%

43. $dV \approx 565.5\ \text{in}^3$ **45.** **(a)** 2% **(b)** 4%

47. $\frac{1}{3}$% **49.** 3%

51. The ratio equals 37.87, so a change in the acceleration of gravity on the moon has about 38 times the effect that a change of the same magnitude has on Earth.

53. Increase $V \approx 40\%$

55. **(a)** **i)** $b_0 = f(a)$ **ii)** $b_1 = f'(a)$ **iii)** $b_2 = \frac{f''(a)}{2}$
(b) $Q(x) = 1 + x + x^2$
(d) $Q(x) = 1 - (x-1) + (x-1)^2$
(e) $Q(x) = 1 + \frac{x}{2} - \frac{x^2}{8}$
(f) The linearization of any differentiable function $u(x)$ at $x = a$ is $L(x) = u(a) + u'(a)(x - a) = b_0 + b_1(x - a)$, where b_0 and b_1 are the coefficients of the constant and linear terms of the quadratic approximation. Thus, the linearization for $f(x)$ at $x = 0$ is $1 + x$; the linearization for $g(x)$ at $x = 1$ is $1 - (x - 1)$ or $2 - x$; and the linearization for $h(x)$ at $x = 0$ is $1 + \frac{x}{2}$.

PRACTICE EXERCISES, pp. 173–176

1. $5x^4 - 0.25x + 0.25$ **3.** $3x(x-2)$

5. $2(x+1)(2x^2 + 4x + 1)$

7. $3(\theta^2 + \sec\theta + 1)^2(2\theta + \sec\theta\tan\theta)$

9. $\frac{1}{2\sqrt{t}(1+\sqrt{t})^2}$ **11.** $2\sec^2 x\tan x$

13. $8\cos^3(1 - 2t)\sin(1 - 2t)$ **15.** $5(\sec t)(\sec t + \tan t)^5$

17. $\frac{\theta\cos\theta + \sin\theta}{\sqrt{2\theta\sin\theta}}$ **19.** $\frac{\cos\sqrt{2\theta}}{\sqrt{2\theta}}$

21. $x\csc\left(\frac{2}{x}\right) + \csc\left(\frac{2}{x}\right)\cot\left(\frac{2}{x}\right)$

23. $\frac{1}{2}x^{1/2}\sec(2x)^2[16\tan(2x)^2 - x^{-2}]$ **25.** $-10x\csc^2(x^2)$

27. $8x^3\sin(2x^2)\cos(2x^2) + 2x\sin^2(2x^2)$ **29.** $\frac{-(t+1)}{8t^3}$

31. $\frac{1-x}{(x+1)^3}$ **33.** $\frac{-1}{2x^2\left(1 + \frac{1}{x}\right)^{1/2}}$ **35.** $\frac{-2\sin\theta}{(\cos\theta - 1)^2}$

37. $3\sqrt{2x+1}$ **39.** $-9\left[\frac{5x + \cos 2x}{(5x^2 + \sin 2x)^{5/2}}\right]$

41. $18x^{2.6}$ **43.** $-\frac{y+2}{x+3}$ **45.** $\frac{-3x^2 - 4y + 2}{4x - 4y^{1/3}}$

47. $-\frac{y}{x}$ **49.** $\frac{1}{2y(x+1)^2}$ **51.** $\frac{dp}{dq} = \frac{6q - 4p}{3p^2 + 4q}$

53. $\frac{dr}{ds} = (2r - 1)(\tan 2s)$

55. **(a)** $\frac{d^2y}{dx^2} = \frac{-2xy^3 - 2x^4}{y^5}$ **(b)** $\frac{d^2y}{dx^2} = \frac{-2xy^2 - 1}{x^4y^3}$

57. **(a)** 7 **(b)** −2 **(c)** 5/12 **(d)** 1/4 **(e)** 12 **(f)** 9/12 **(g)** 3/4

59. 0 **61.** $\sqrt{3}$ **63.** $-\frac{1}{2}$ **65.** $\frac{-2}{(2t+1)^2}$

67. **(a)**

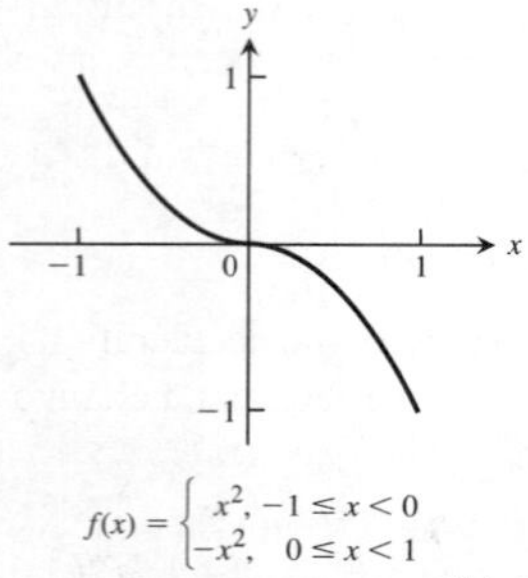

(b) Yes **(c)** Yes

69. **(a)**

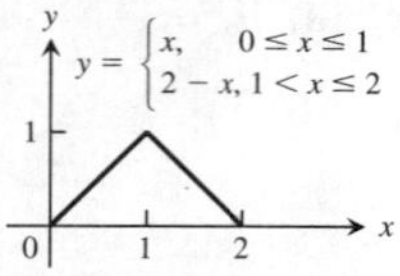

(b) Yes **(c)** No

71. $\left(\frac{5}{2}, \frac{9}{4}\right)$ and $\left(\frac{3}{2}, -\frac{1}{4}\right)$ **73.** $(-1, 27)$ and $(2, 0)$

75. **(a)** $(-2, 16), (3, 11)$ **(b)** $(0, 20), (1, 7)$

77.

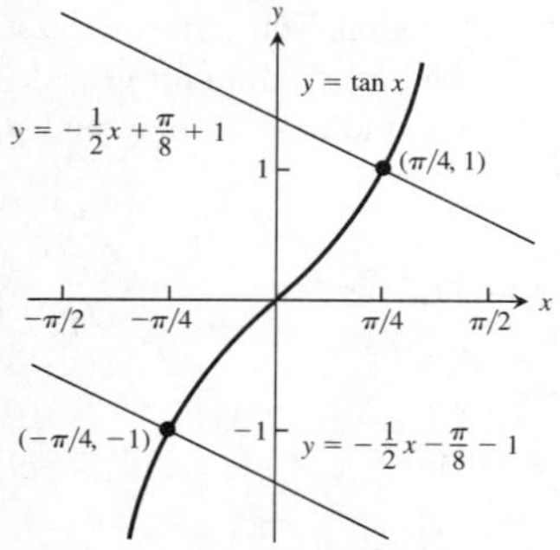

79. $\frac{1}{4}$ **81.** 4

83. Tangent: $y = -\frac{1}{4}x + \frac{9}{4}$, normal: $y = 4x - 2$

85. Tangent: $y = 2x - 4$, normal: $y = -\frac{1}{2}x + \frac{7}{2}$

87. Tangent: $y = -\frac{5}{4}x + 6$, normal: $y = \frac{4}{5}x - \frac{11}{5}$

89. $(1, 1)$: $m = -\frac{1}{2}$; $(1, -1)$: m not defined

91. B = graph of f, A = graph of f'

93.

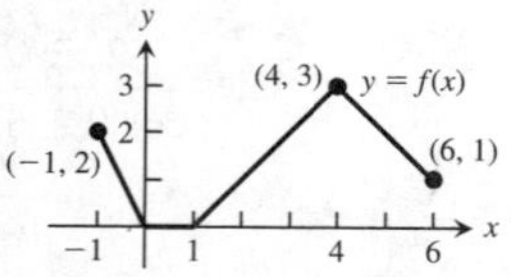

95. **(a)** $\frac{dS}{dt} = (4\pi r + 2\pi h)\frac{dr}{dt}$ **(b)** $\frac{dS}{dt} = 2\pi r\frac{dh}{dt}$

(c) $\frac{dS}{dt} = (4\pi r + 2\pi h)\frac{dr}{dt} + 2\pi r\frac{dh}{dt}$ **(d)** $\frac{dr}{dt} = -\frac{r}{2r + h}\frac{dh}{dt}$

97. $-40\ \text{m}^2/\text{sec}$ **99.** 0.02 ohm/sec **101.** 2 m/sec

103. **(a)** $r = \frac{2}{5}h$ **(b)** $-\frac{125}{144\pi}$ ft/min

105. **(a)** $\frac{3}{5}$ km/sec or 600 m/sec **(b)** $\frac{18}{\pi}$ rpm

107. **(a)** $L(x) = 2x + \frac{\pi - 2}{2}$

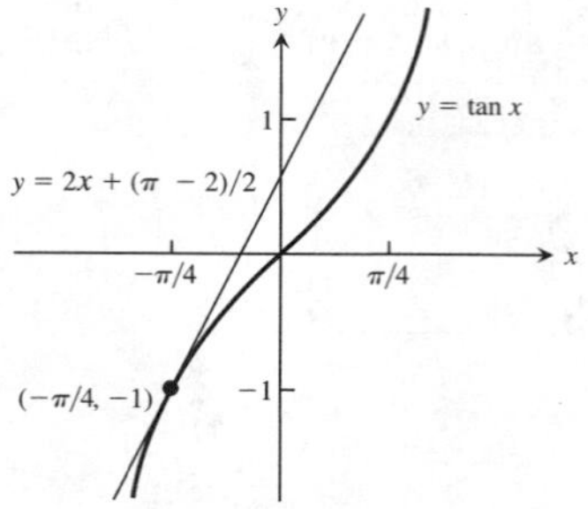

(b) $L(x) = -\sqrt{2}x + \frac{\sqrt{2}(4 - \pi)}{4}$

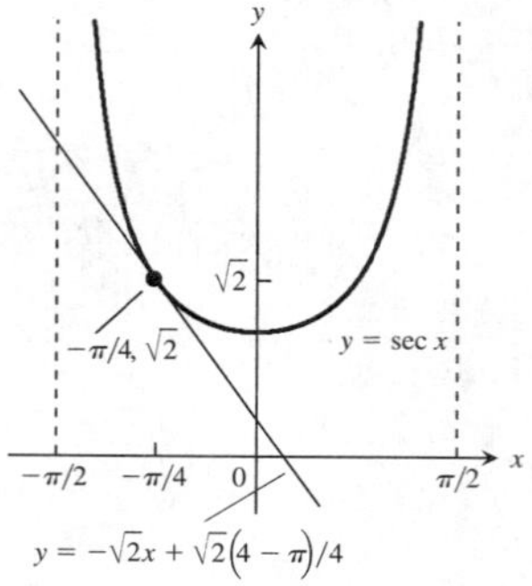

109. $L(x) = 1.5x + 0.5$ **111.** $dS = \frac{\pi r h_0}{\sqrt{r^2 + h_0^2}}dh$

113. **(a)** 4% **(b)** 8% **(c)** 12%

ADDITIONAL AND ADVANCED EXERCISES, pp. 176–179

1. **(a)** $\sin 2\theta = 2 \sin\theta\cos\theta$; $2\cos 2\theta = 2\sin\theta(-\sin\theta) + \cos\theta(2\cos\theta)$; $2\cos 2\theta = -2\sin^2\theta + 2\cos^2\theta$; $\cos 2\theta = \cos^2\theta - \sin^2\theta$
(b) $\cos 2\theta = \cos^2\theta - \sin^2\theta$; $-2\sin 2\theta = 2\cos\theta(-\sin\theta) - 2\sin\theta(\cos\theta)$; $\sin 2\theta = \cos\theta\sin\theta + \sin\theta\cos\theta$; $\sin 2\theta = 2\sin\theta\cos\theta$

3. **(a)** $a = 1, b = 0, c = -\frac{1}{2}$ **(b)** $b = \cos a, c = \sin a$

5. $h = -4, k = \frac{9}{2}, a = \frac{5\sqrt{5}}{2}$

7. **(a)** 0.09y **(b)** Increasing at 1% per year

9. Answers will vary. Here is one possibility.

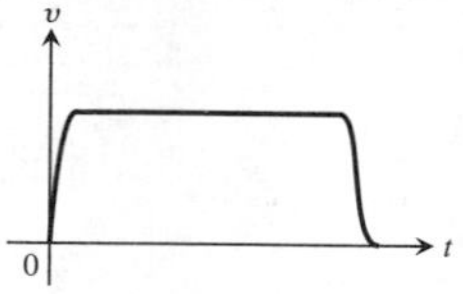

11. **(a)** 2 sec, 64 ft/sec **(b)** 12.31 sec, 393.85 ft

15. **(a)** $m = -\frac{b}{\pi}$ **(b)** $m = -1, b = \pi$

17. **(a)** $a = \frac{3}{4}, b = \frac{9}{4}$ **19.** f odd $\Rightarrow$ f' is even

23. h' is defined but not continuous at $x = 0$; k' is defined *and* continuous at $x = 0$.

25. $\frac{-7}{75}$ rad/sec

29. **(a)** 0.8156 ft **(b)** 0.00613 sec
(c) It will lose about 8.83 min/day.

Chapter 4

SECTION 4.1, pp. 186–188

1. Absolute minimum at $x = c_2$; absolute maximum at $x = b$
3. Absolute maximum at $x = c$; no absolute minimum
5. Absolute minimum at $x = a$; absolute maximum at $x = c$
7. No absolute minimum; no absolute maximum
9. Absolute maximum at $(0, 5)$ **11.** (c) **13.** (d)
15. Absolute minimum at $x = 0$; no absolute maximum

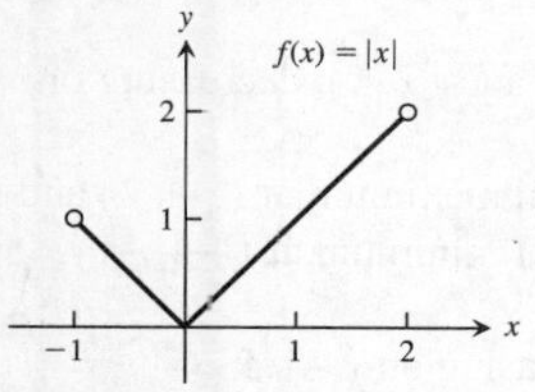

17. Absolute maximum at $x = 2$; no absolute minimum

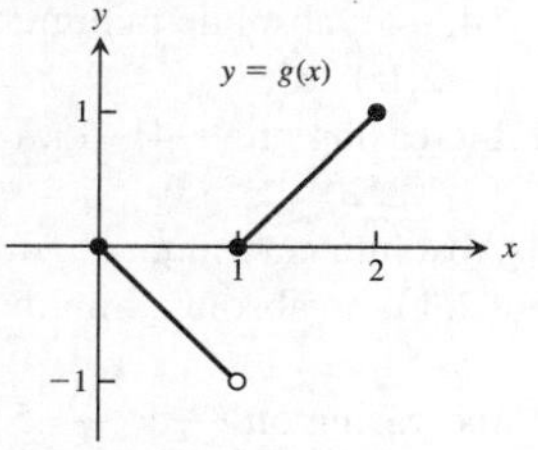

19. Absolute maximum at $x = \pi/2$; absolute minimum at $x = 3\pi/2$

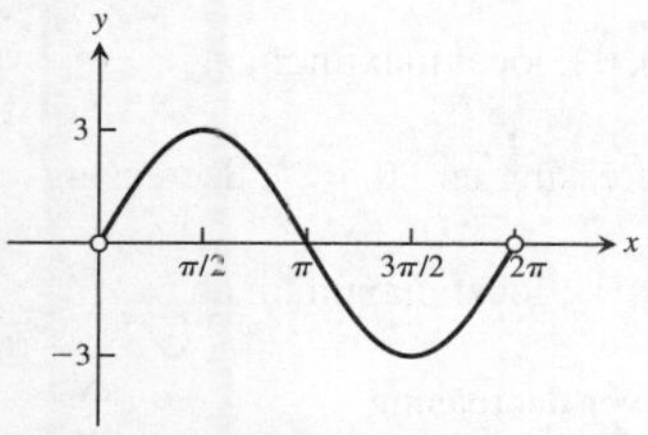

21. Absolute maximum: -3; absolute minimum: $-19/3$

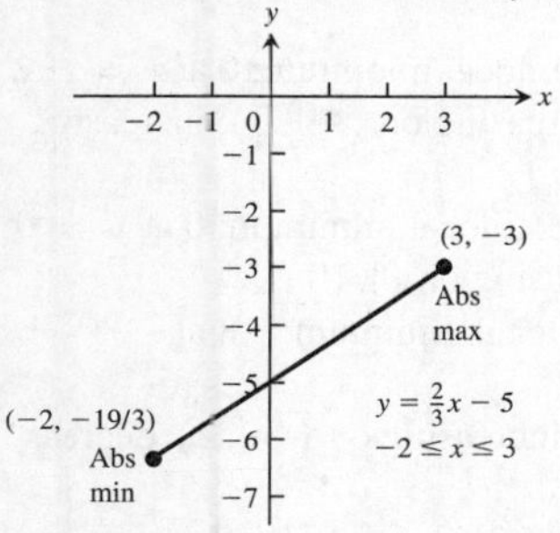

23. Absolute maximum: 3; absolute minimum: -1

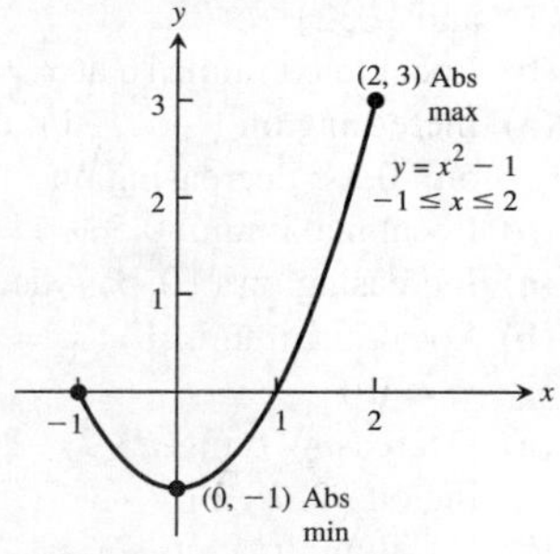

25. Absolute maximum: -0.25; absolute minimum: -4

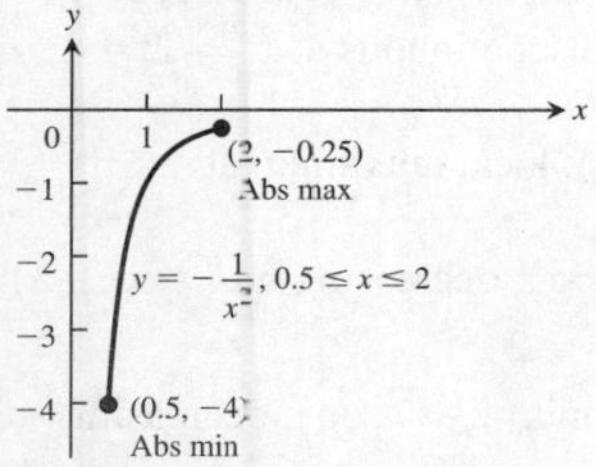

27. Absolute maximum: 2; absolute minimum: -1

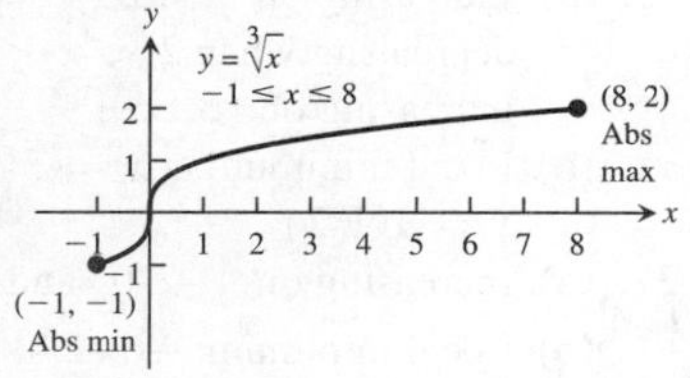

29. Absolute maximum: 2; absolute minimum: 0

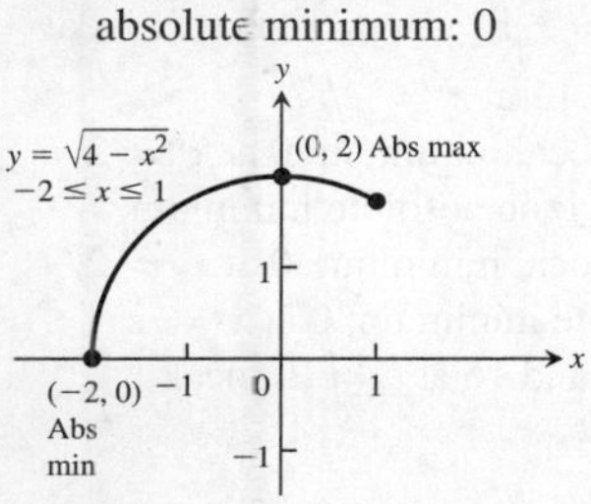

31. Absolute maximum: 1; absolute minimum: -1

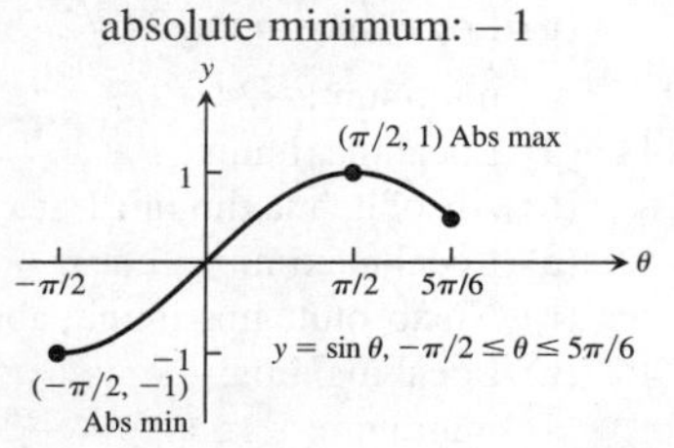

33. Absolute maximum: $2/\sqrt{3}$; absolute minimum: 1

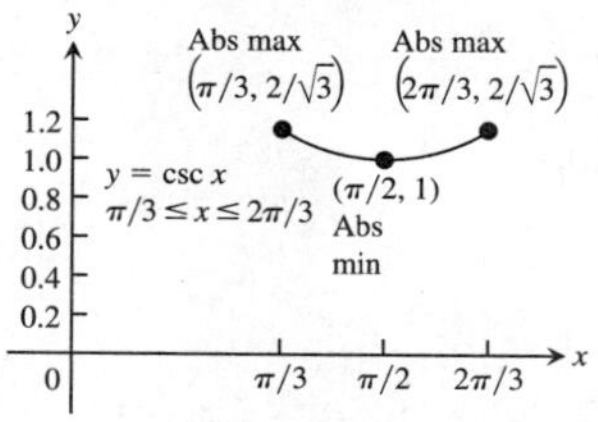

35. Absolute maximum: 2; absolute minimum: -1

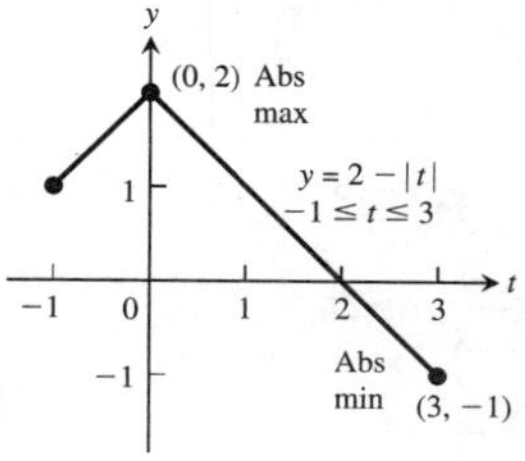

37. Increasing on $(0, 8)$, decreasing on $(-1, 0)$; absolute maximum: 16 at $x = 8$; absolute minimum: 0 at $x = 0$
39. Increasing on $(-32, 1)$; absolute maximum: 1 at $\theta = 1$; absolute minimum: -8 at $\theta = -32$
41. Maximum value is $\dfrac{3\sqrt{3}}{4}$ at $x = \dfrac{\pi}{3}$.

Minimum value is $-\dfrac{3\sqrt{3}}{4}$ at $x = 5\dfrac{\pi}{3}$.
43. $x = 1$, $x = 4$
45. $x = 1$
47. $x = 0$, $x = 4$
49. $x = 2$, $x = -4$
51. **(a)** No
(b) The derivative is defined and nonzero for $x \neq 2$. Also, $f(2) = 0$ and $f(x) > 0$ for all $x \neq 2$.
(c) No, because $(-\infty, \infty)$ is not a closed interval.
(d) The answers are the same as parts (a) and (b), with 2 replaced by a.
53. y is increasing on $(-\infty, \infty)$ so has no extrema.
55. Yes
57. g assumes a local maximum at $-c$.
59. **(a)** Maximum value is 144 at $x = 2$.
(b) The largest volume of the box is 144 cubic units, and it occurs when $x = 2$.
61. $\dfrac{{v_0}^2}{2g} + s_0$
63. Maximum value is 11 at $x = 5$; minimum value is 5 on the interval $[-3, 2]$; local maximum at $(-5, 9)$.
65. Maximum value is 5 on the interval $[3, \infty)$; minimum value is -5 on the interval $(-\infty, -2]$.

SECTION 4.2, pp. 193–194

1. 1/2 **3.** 1
5. $\frac{1}{3}(1 + \sqrt{7}) \approx 1.22$, $\frac{1}{3}(1 - \sqrt{7}) \approx -0.549$
7. Does not; f is not differentiable at the interior domain point $x = 0$.
9. Does
11. Does not; f is not differentiable at $x = -1$.
15. **(a)**

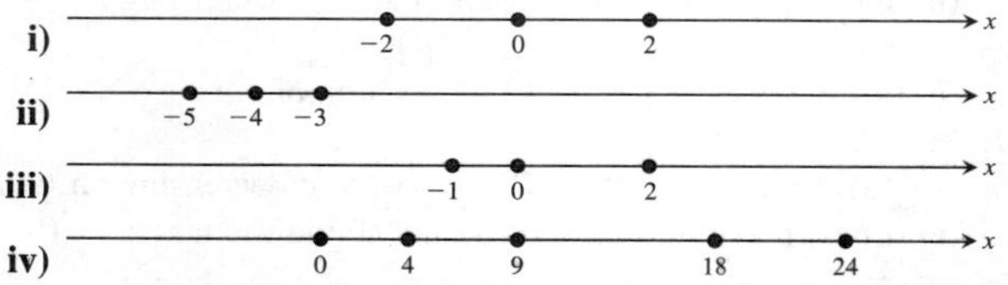

27. Yes **29.** **(a)** 4 **(b)** 3 **(c)** 3

31. **(a)** $\frac{x^2}{2} + C$ **(b)** $\frac{x^3}{3} + C$ **(c)** $\frac{x^4}{4} + C$

33. **(a)** $\frac{1}{x} + C$ **(b)** $x + \frac{1}{x} + C$ **(c)** $5x - \frac{1}{x} + C$

35. **(a)** $-\frac{1}{2}\cos 2t + C$ **(b)** $2\sin\frac{t}{2} + C$

(c) $-\frac{1}{2}\cos 2t + 2\sin\frac{t}{2} + C$

37. $f(x) = x^2 - x$ **39.** $r(\theta) = 8\theta + \cot\theta - 2\pi - 1$

41. $s = 4.9t^2 + 5t + 10$ **43.** $s = \dfrac{1 - \cos(\pi t)}{\pi}$

45. $s = 16t^2 + 20t + 5$ **47.** $s = \sin(2t) - 3$

49. If $T(t)$ is the temperature of the thermometer at time t, then $T(0) = -19\,°\text{C}$ and $T(14) = 100\,°\text{C}$. From the Mean Value Theorem, there exists a $0 < t_0 < 14$ such that $\frac{T(14) - T(0)}{14 - 0} = 8.5\,°\text{C/sec} = T'(t_0)$, the rate at which the temperature was changing at $t = t_0$ as measured by the rising mercury on the thermometer.

51. Because its average speed was approximately 7.667 knots, and by the Mean Value Theorem, it must have been going that speed at least once during the trip.

55. The conclusion of the Mean Value Theorem yields

$$\frac{\frac{1}{b} - \frac{1}{a}}{b - a} = -\frac{1}{c^2} \Rightarrow c^2\left(\frac{a-b}{ab}\right) = a - b \Rightarrow c = \sqrt{ab}.$$

59. $f(x)$ must be zero at least once between a and b by the Intermediate Value Theorem. Now suppose that $f(x)$ is zero twice between a and b. Then, by the Mean Value Theorem, $f'(x)$ would have to be zero at least once between the two zeros of $f(x)$, but this can't be true since we are given that $f'(x) \neq 0$ on this interval. Therefore, $f(x)$ is zero once and only once between a and b.

69. $1.09999 \le f(0.1) \le 1.1$

SECTION 4.3, pp. 198–200

1. **(a)** $0, 1$
(b) Increasing on $(-\infty, 0)$ and $(1, \infty)$; decreasing on $(0, 1)$
(c) Local maximum at $x = 0$; local minimum at $x = 1$

3. **(a)** $-2, 1$
(b) Increasing on $(-2, 1)$ and $(1, \infty)$; decreasing on $(-\infty, -2)$
(c) No local maximum; local minimum at $x = -2$

5. **(a)** $-2, 1, 3$
(b) Increasing on $(-2, 1)$ and $(3, \infty)$; decreasing on $(-\infty, -2)$ and $(1, 3)$
(c) Local maximum at $x = 1$; local minimum at $x = -2, 3$

7. **(a)** $0, 1$
(b) Increasing on $(-\infty, -2)$ and $(1, \infty)$; decreasing on $(-2, 0)$ and $(0, 1)$
(c) Local minimum at $x = 1$

9. **(a)** $-2, 2$
(b) Increasing on $(-\infty, -2)$ and $(2, \infty)$; decreasing on $(-2, 0)$ and $(0, 2)$
(c) Local maximum at $x = -2$; local minimum at $x = 2$

11. **(a)** $-2, 0$
(b) Increasing on $(-\infty, -2)$ and $(0, \infty)$; decreasing on $(-2, 0)$
(c) Local maximum at $x = -2$; local minimum at $x = 0$

13. **(a)** $\frac{\pi}{2}, \frac{2\pi}{3}, \frac{4\pi}{3}$
(b) Increasing on $\left(\frac{2\pi}{3}, \frac{4\pi}{3}\right)$; decreasing on $\left(0, \frac{\pi}{2}\right)$, $\left(\frac{\pi}{2}, \frac{2\pi}{3}\right)$, and $\left(\frac{4\pi}{3}, 2\pi\right)$
(c) Local maximum at $x = 0$ and $x = \frac{4\pi}{3}$; local minimum at $x = \frac{2\pi}{3}$ and $x = 2\pi$

15. **(a)** Increasing on $(-2, 0)$ and $(2, 4)$; decreasing on $(-4, -2)$ and $(0, 2)$
(b) Absolute maximum at $(-4, 2)$; local maximum at $(0, 1)$ and $(4, -1)$; absolute minimum at $(2, -3)$; local minimum at $(-2, 0)$

17. **(a)** Increasing on $(-4, -1)$, $(1/2, 2)$, and $(2, 4)$; decreasing on $(-1, 1/2)$
(b) Absolute maximum at $(4, 3)$; local maximum at $(-1, 2)$ and $(2, 1)$; no absolute minimum; local minimum at $(-4, -1)$ and $(1/2, -1)$

19. **(a)** Increasing on $(-\infty, -1.5)$; decreasing on $(-1.5, \infty)$
(b) Local maximum: 5.25 at $t = -1.5$

21. **(a)** Decreasing on $(-\infty, 0)$; increasing on $(0, 4/3)$; decreasing on $(4/3, \infty)$
(b) Local minimum at $x = 0(0, 0)$; local maximum at $x = 4/3(4/3, 32/27)$

23. **(a)** Decreasing on $(-\infty, 0)$; increasing on $(0, 1/2)$; decreasing on $(1/2, \infty)$
(b) Local minimum at $\theta = 0(0, 0)$; local maximum at $\theta = 1/2(1/2, 1/4)$

25. **(a)** Increasing on $(-\infty, \infty)$; never decreasing
(b) No local extrema

27. **(a)** Increasing on $(-2, 0)$ and $(2, \infty)$; decreasing on $(-\infty, -2)$ and $(0, 2)$
(b) Local maximum: 16 at $x = 0$; local minimum: 0 at $x = \pm 2$

29. **(a)** Increasing on $(-\infty, -1)$; decreasing on $(-1, 0)$; increasing on $(0, 1)$; decreasing on $(1, \infty)$
(b) Local maximum: 0.5 at $x = \pm 1$; local minimum: 0 at $x = 0$

31. **(a)** Increasing on $(10, \infty)$; decreasing on $(1, 10)$
(b) Local maximum: 1 at $x = 1$; local minimum: -8 at $x = 10$

33. **(a)** Decreasing on $(-2\sqrt{2}, -2)$; increasing on $(-2, 2)$; decreasing on $(2, 2\sqrt{2})$
(b) Local minima: $g(-2) = -4$, $g(2\sqrt{2}) = 0$; local maxima: $g(-2\sqrt{2}) = 0$, $g(2) = 4$

35. **(a)** Increasing on $(-\infty, 1)$; decreasing when $1 < x < 2$; decreasing when $2 < x < 3$; discontinuous at $x = 2$; increasing on $(3, \infty)$
(b) Local minimum at $x = 3(3, 6)$; local maximum at $x = 1(1, 2)$

37. **(a)** Increasing on $(-2, 0)$ and $(0, \infty)$; decreasing on $(-\infty, -2)$
(b) Local minimum: $-6\sqrt[3]{2}$ at $x = -2$

39. **(a)** Increasing on $(-\infty, -2/\sqrt{7})$ and $(2/\sqrt{7}, \infty)$; decreasing on $(-2/\sqrt{7}, 0)$ and $(0, 2/\sqrt{7})$
(b) Local maximum: $24\sqrt[3]{2}/7^{7/6} \approx 3.12$ at $x = -2/\sqrt{7}$; local minimum: $-24\sqrt[3]{2}/7^{7/6} \approx -3.12$ at $x = 2/\sqrt{7}$

41. **(a)** Local maximum: 1 at $x = 1$; local minimum: 0 at $x = 2$
(b) Absolute maximum: 1 at $x = 1$; no absolute minimum

43. **(a)** Local maximum: 1 at $x = 1$; local minimum: 0 at $x = 2$
(b) No absolute maximum; absolute minimum: 0 at $x = 2$

45. **(a)** Local maxima: -9 at $t = -3$ and 16 at $t = 2$; local minimum: -16 at $t = -2$

(b) Absolute maximum: 16 at $t = 2$; no absolute minimum

47. (a) Local minimum: 0 at $x = 0$
(b) No absolute maximum; absolute minimum: 0 at $x = 0$

49. (a) Local maximum: 5 at $x = 0$; local minimum: 0 at $x = -5$ and $x = 5$
(b) Absolute maximum: 5 at $x = 0$; absolute minimum: 0 at $x = -5$ and $x = 5$

51. (a) Local maximum: 2 at $x = 0$;
local minimum: $\dfrac{\sqrt{3}}{4\sqrt{3}-6}$ at $x = 2 - \sqrt{3}$
(b) No absolute maximum; absolute minimum at $x = 2 - \sqrt{3}$

53. (a) Local maximum: 1 at $x = \pi/4$;
local maximum: 0 at $x = \pi$;
local minimum: 0 at $x = 0$;
local minimum: -1 at $x = 3\pi/4$

55. (a) Local maximum: 2 at $x = \pi/6$;
local maximum: $\sqrt{3}$ at $x = 2\pi$;
local minimum: -2 at $x = 7\pi/6$;
local minimum: $\sqrt{3}$ at $x = 0$

57. (a) Local minimum: $(\pi/3) - \sqrt{3}$ at $x = 2\pi/3$;
local maximum: 0 at $x = 0$;
local maximum: π at $x = 2\pi$

59. (a) Local minimum: 0 at $x = \pi/4$

61. Local minimum at $x = 1$; no local maximum

63. (a) Increasing on $(-1, 2)$;
decreasing on $(-2, -1)$
(b) The sign of f cannot be determined solely from the graph of f'.

65. Local maximum: 3 at $\theta = 0$;
local minimum: -3 at $\theta = 2\pi$

67.

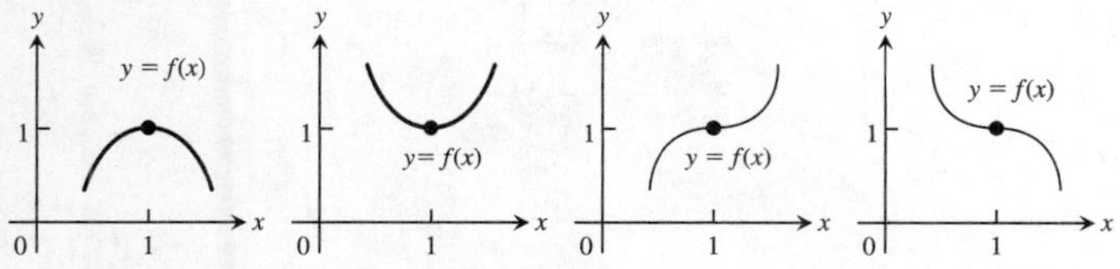

(a) (b) (c) (d)

69. (a)

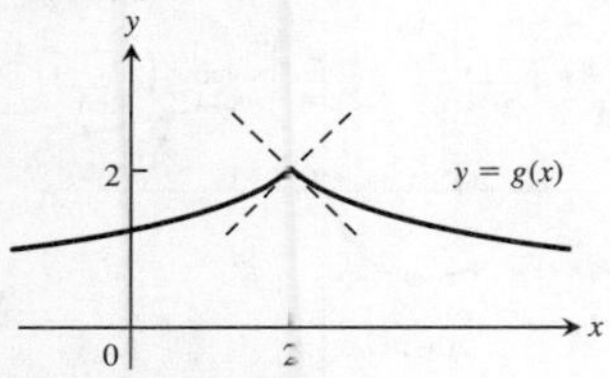

(b)

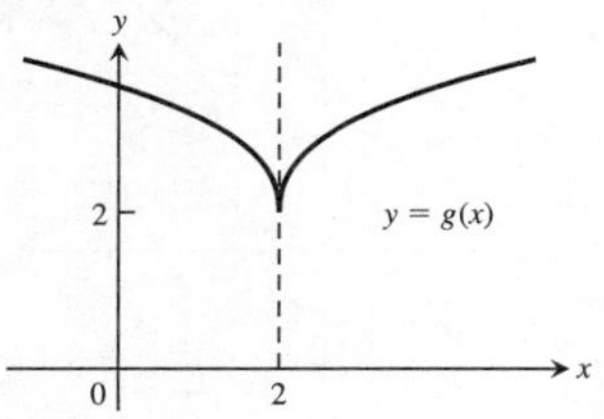

73. $a = -2, b = 4$

SECTION 4.4, pp. 209–212

1. Local maximum: $3/2$ at $x = -1$; local minimum: -3 at $x = 2$; point of inflection at $(1/2, -3/4)$; rising on $(-\infty, -1)$ and $(2, \infty)$; falling on $(-1, 2)$; concave up on $(1/2, \infty)$; concave down on $(-\infty, 1/2)$

3. Local maximum: $3/4$ at $x = 0$; local minimum: 0 at $x = \pm 1$; points of inflection at $\left(-\sqrt{3}, \dfrac{3\sqrt[3]{4}}{4}\right)$ and $\left(\sqrt{3}, \dfrac{3\sqrt[3]{4}}{4}\right)$; rising on $(-1, 0)$ and $(1, \infty)$; falling on $(-\infty, -1)$ and $(0, 1)$; concave up on $(-\infty, -\sqrt{3})$ and $(\sqrt{3}, \infty)$; concave down on $(-\sqrt{3}, -1), (-1, 1)$, and $(1, \sqrt{3})$

5. Local maxima: $\dfrac{-2\pi}{3} + \dfrac{\sqrt{3}}{2}$ at $x = -2\pi/3$, $\dfrac{\pi}{3} + \dfrac{\sqrt{3}}{2}$ at $x = \pi/3$; local minima: $-\dfrac{\pi}{3} - \dfrac{\sqrt{3}}{2}$ at $x = -\pi/3$, $\dfrac{2\pi}{3} - \dfrac{\sqrt{3}}{2}$ at $x = 2\pi/3$; points of inflection at $(-\pi/2, -\pi/2)$, $(0, 0)$, and $(\pi/2, \pi/2)$; rising on $(-\pi/3, \pi/3)$; falling on $(-2\pi/3, -\pi/3)$ and $(\pi/3, 2\pi/3)$; concave up on $(-\pi/2, 0)$ and $(\pi/2, 2\pi/3)$; concave down on $(-2\pi/3, -\pi/2)$ and $(0, \pi/2)$

7. Local maxima: 1 at $x = -\pi/2$ and $x = \pi/2$, 0 at $x = -2\pi$ and $x = 2\pi$; local minima: -1 at $x = -3\pi/2$ and $x = 3\pi/2$, 0 at $x = 0$; points of inflection at $(-\pi, 0)$ and $(\pi, 0)$; rising on $(-3\pi/2, -\pi/2), (0, \pi/2)$, and $(3\pi/2, 2\pi)$; falling on $(-2\pi, -3\pi/2), (-\pi/2, 0)$, and $(\pi/2, 3\pi/2)$; concave up on $(-2\pi, -\pi)$ and $(\pi, 2\pi)$; concave down on $(-\pi, 0)$ and $(0, \pi)$

9.

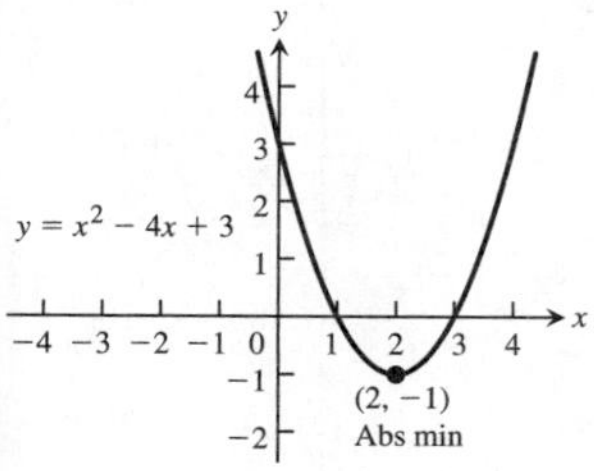

11.

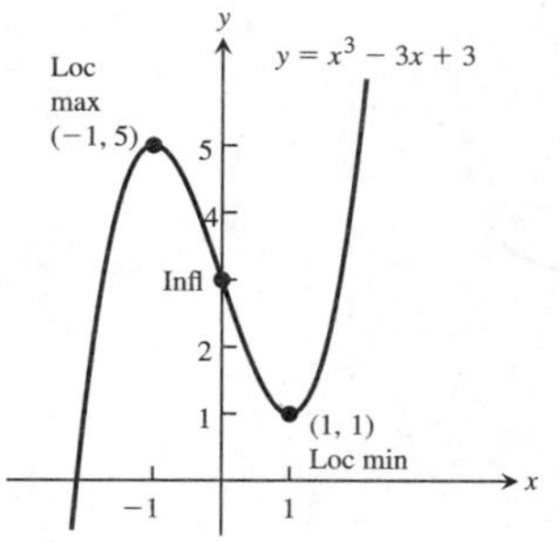

13.

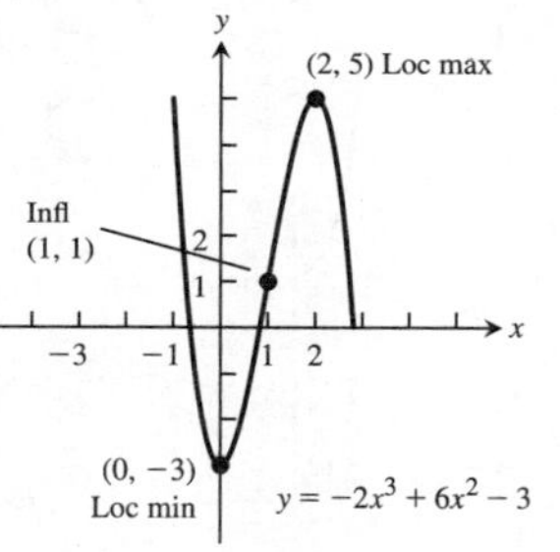

15.

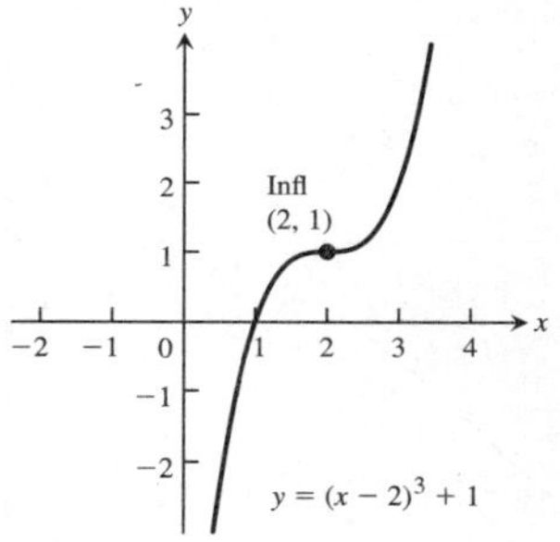

17.

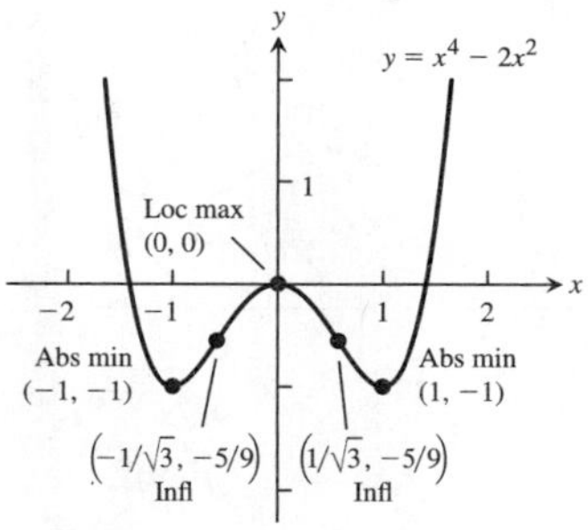

19.

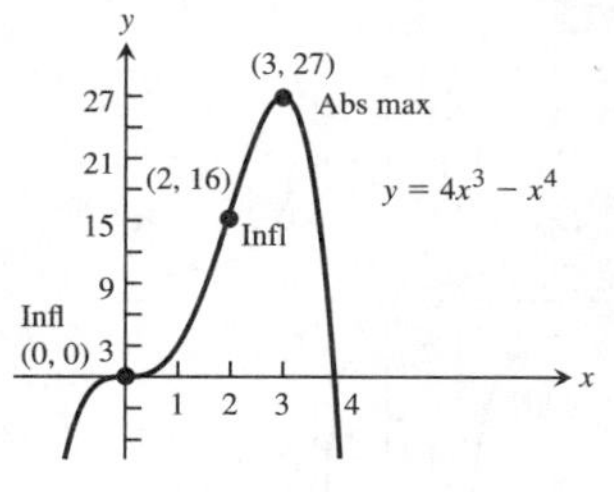

21.

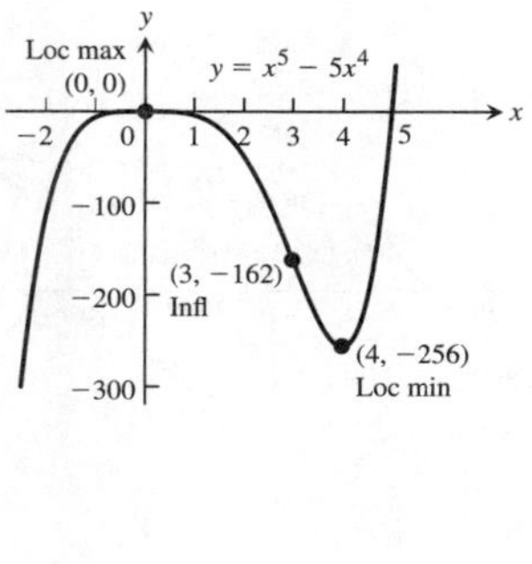

23.

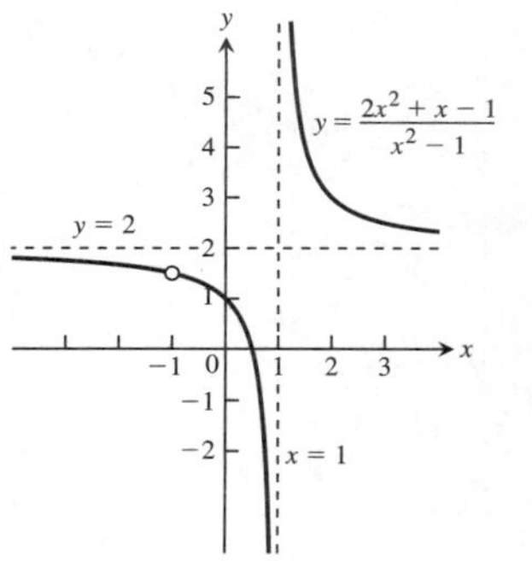

25.

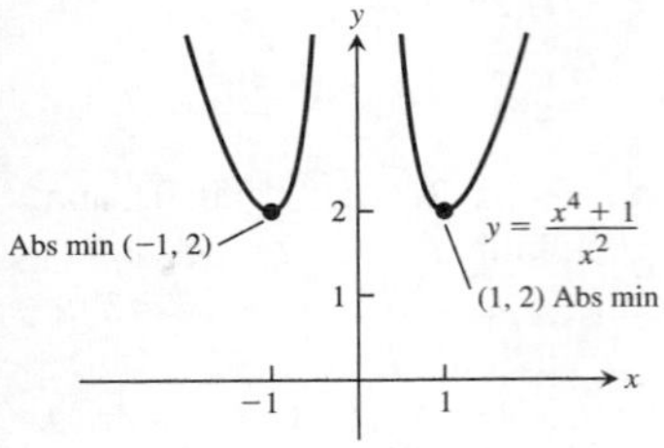

27.

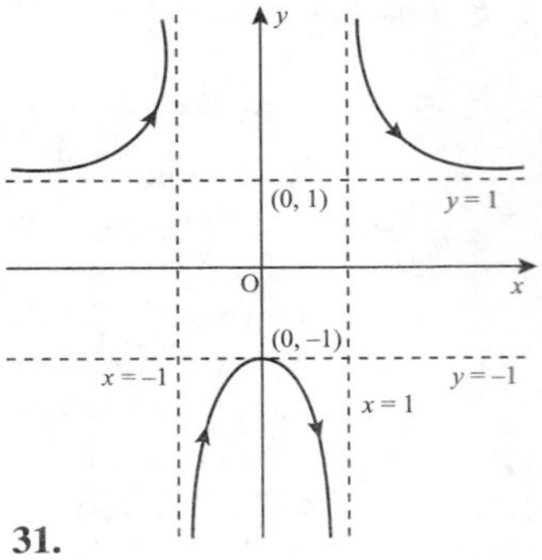

45.

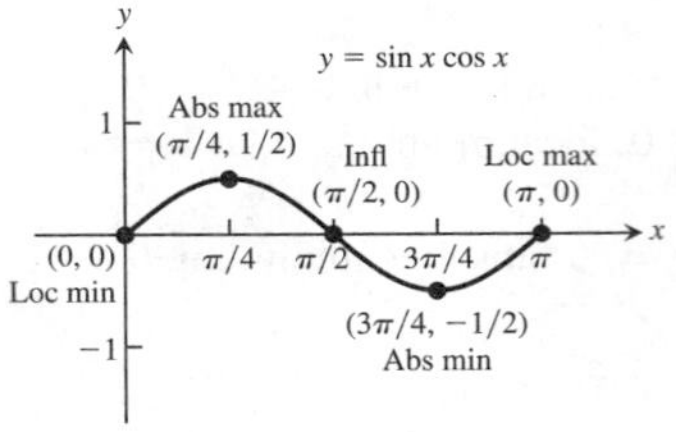

47.

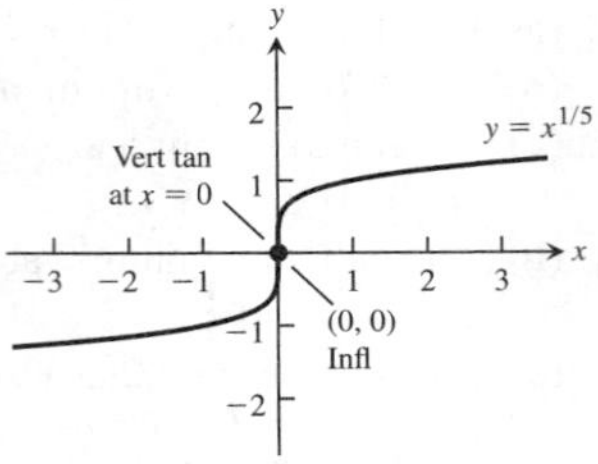

29.

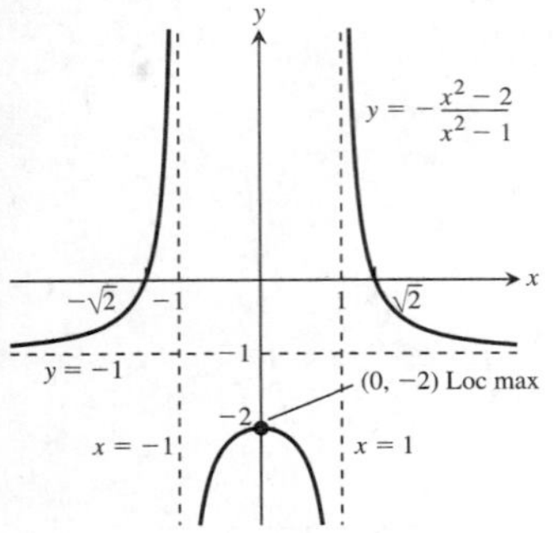

31.

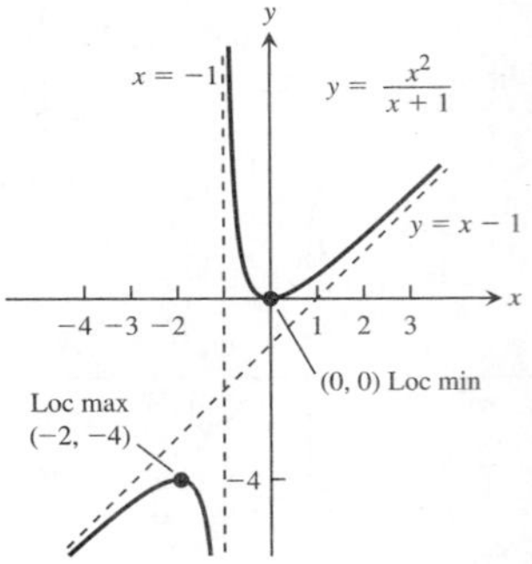

49.

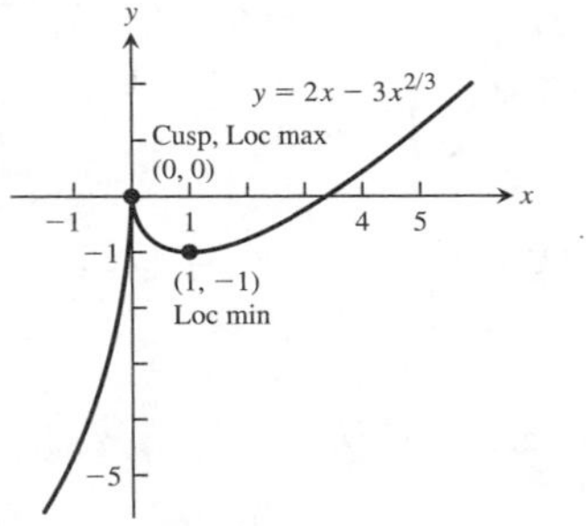

51.

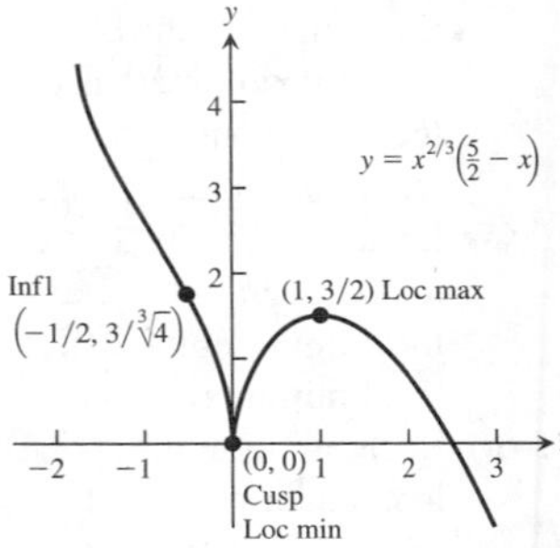

33.

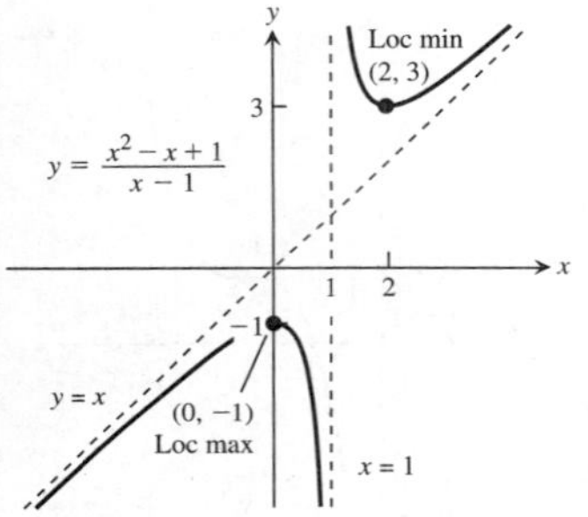

35.

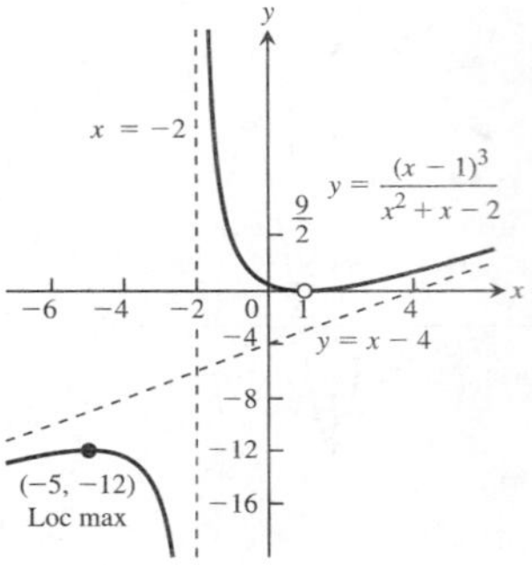

53.

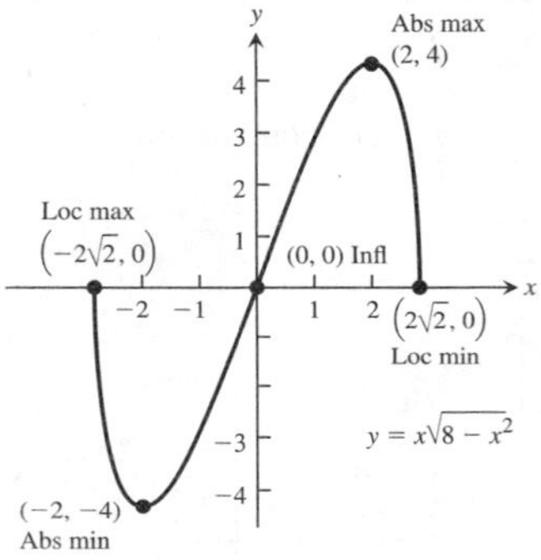

55.

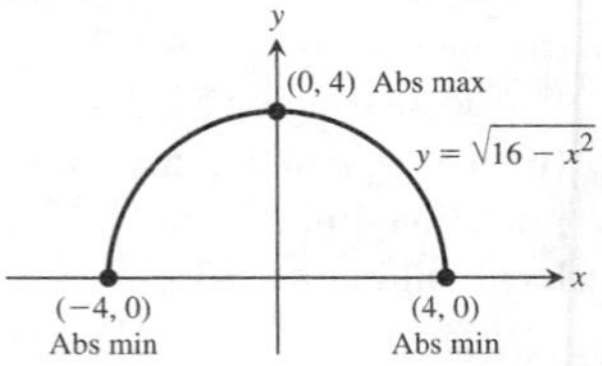

37.

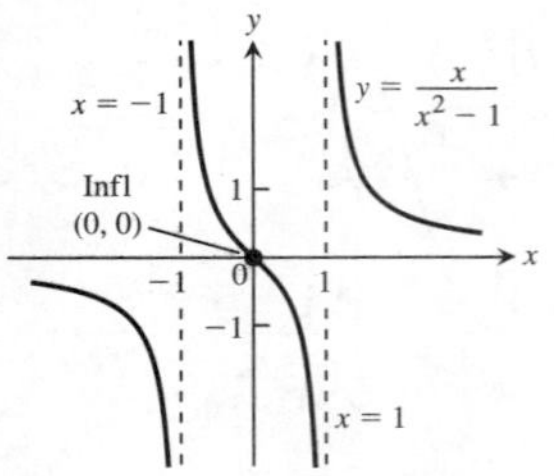

39.

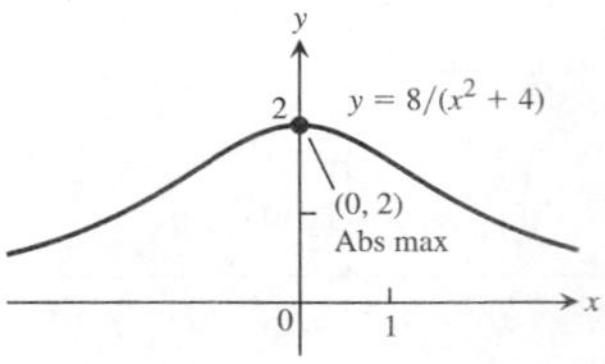

57.

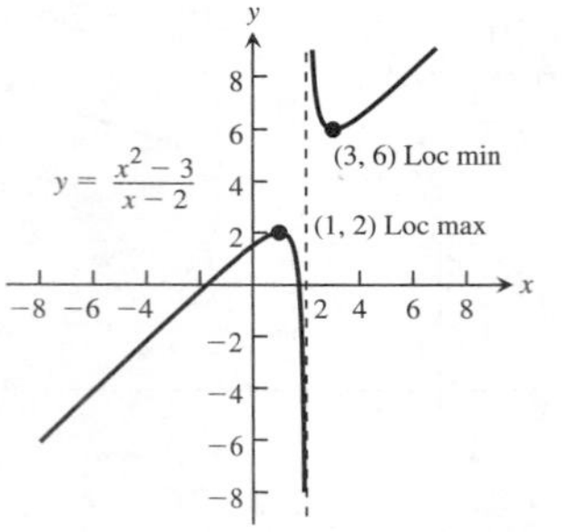

59.

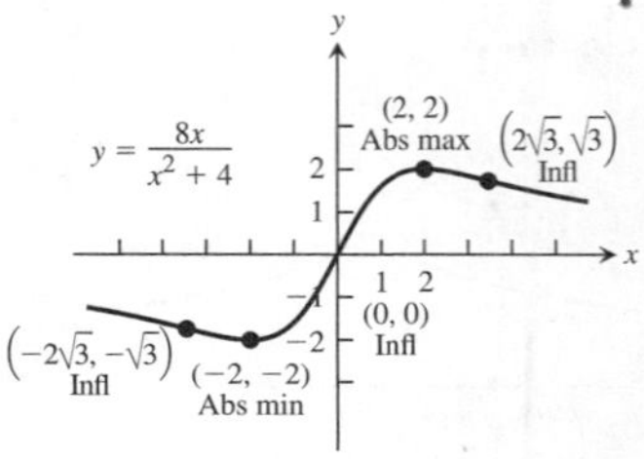

41.

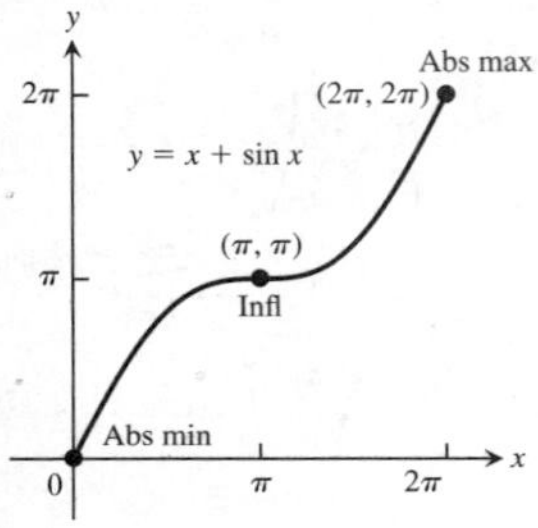

43.

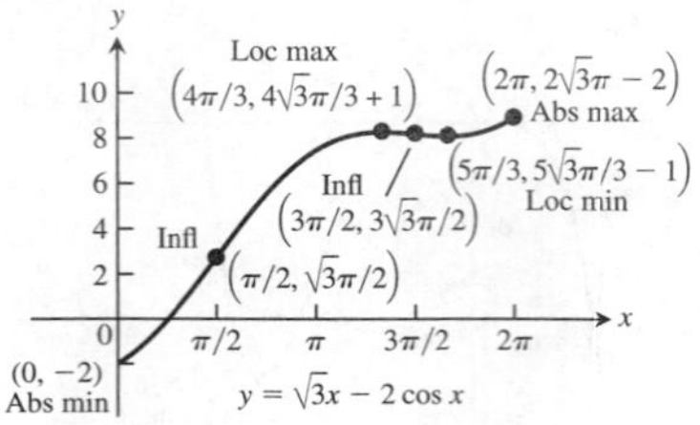

61.

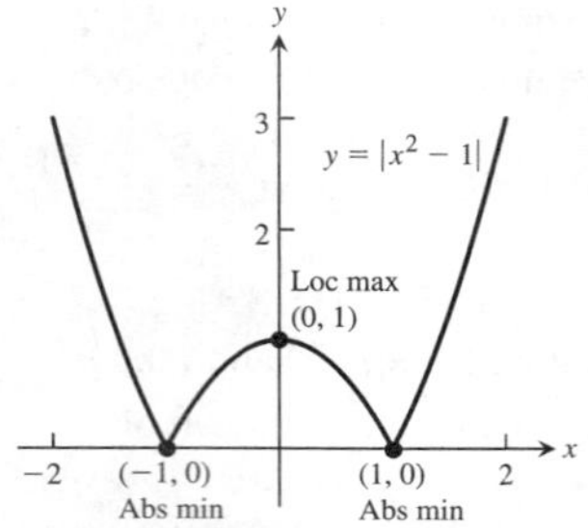

63.

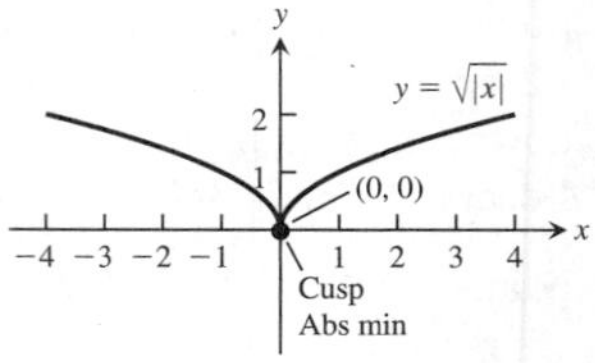

65.

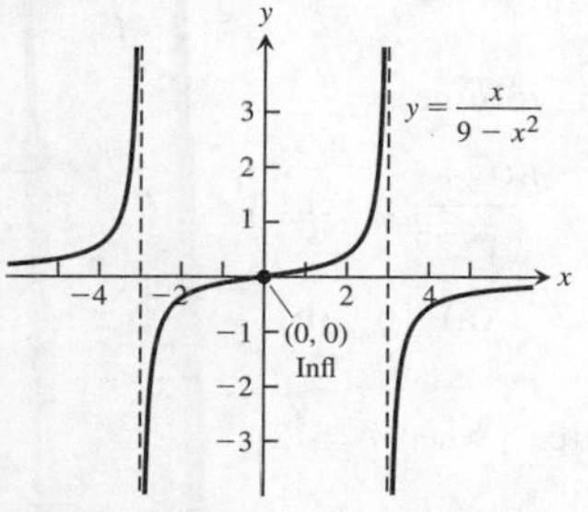

67. $y'' = 1 - 2x$

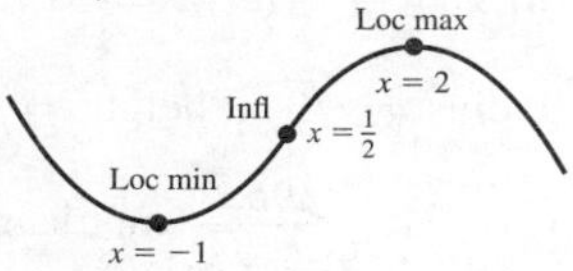

69. $y'' = 3(x - 3)(x - 1)$

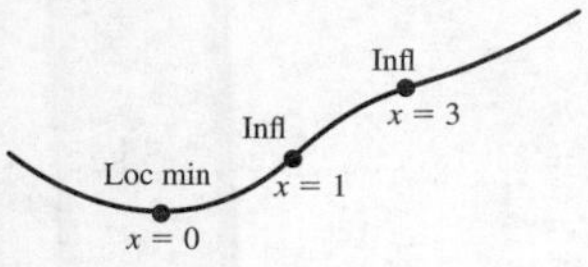

71. $y'' = 3(x - 2)(x + 2)$

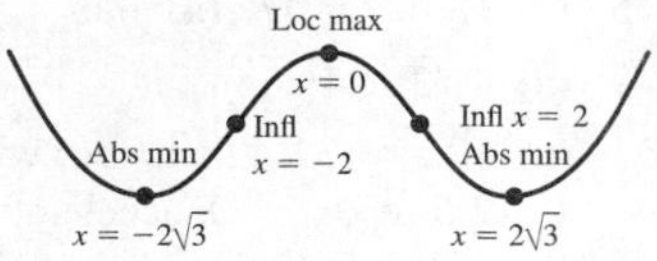

73. $y'' = 4(4 - x)(5x^2 - 16x + 8)$

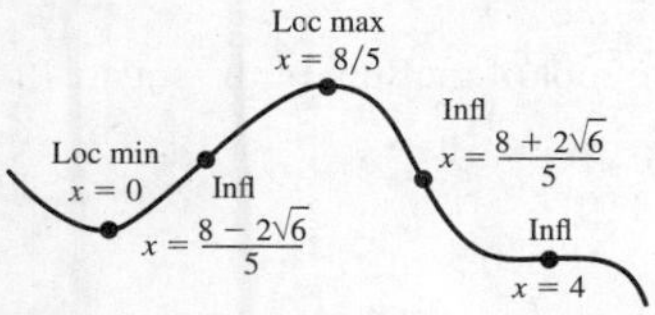

75. $y'' = 2\sec^2 x \tan x$

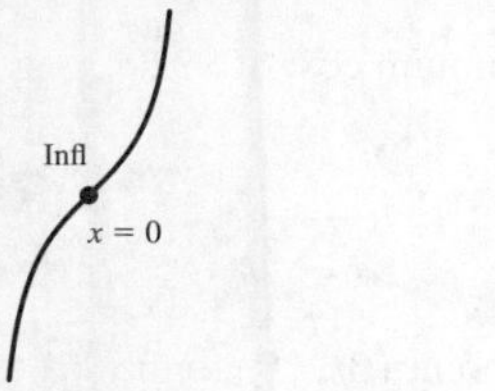

77. $y'' = -\frac{1}{2}\csc^2\frac{\theta}{2}$, $0 < \theta < 2\pi$

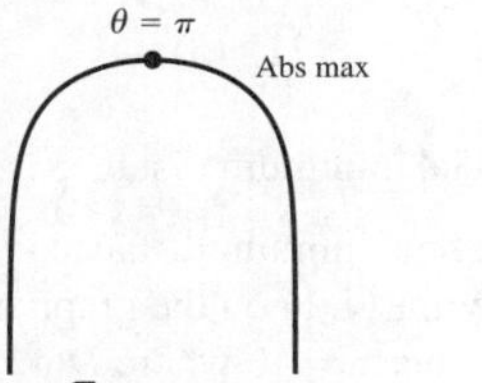

79. $y'' = 2\tan\theta\sec^2\theta$, $-\frac{\pi}{2} < \theta < \frac{\pi}{2}$

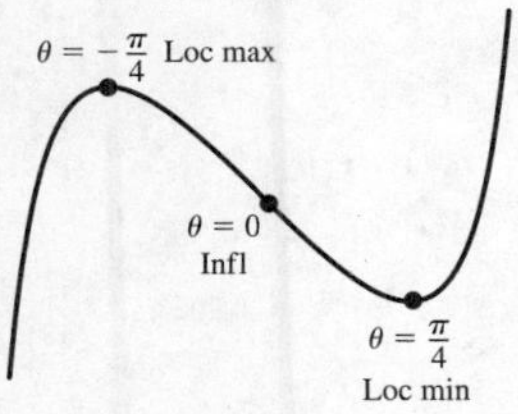

81. $y'' = -\sin t, 0 \le t \le 2\pi$

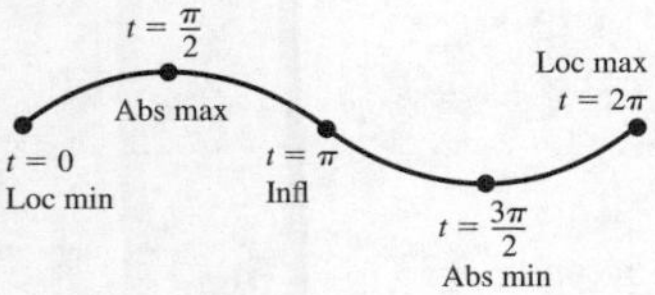

83. $y'' = -\frac{2}{3}(x + 1)^{-5/3}$

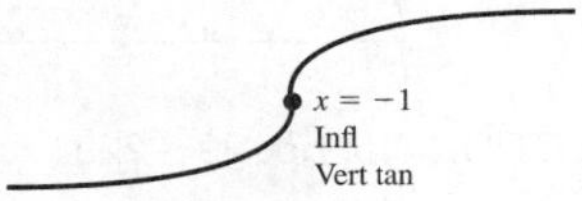

85. $y'' = \frac{1}{3}x^{-2/3} + \frac{2}{3}x^{-5/3}$

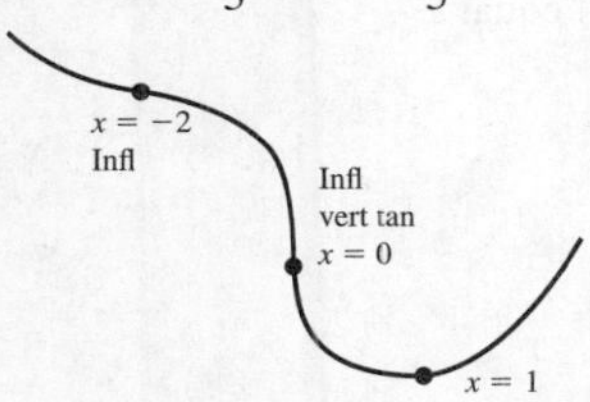

87. $y'' = \begin{cases} -2, & x < 0 \\ 2, & x > 0 \end{cases}$

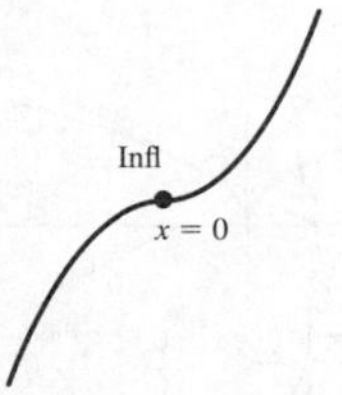

89.

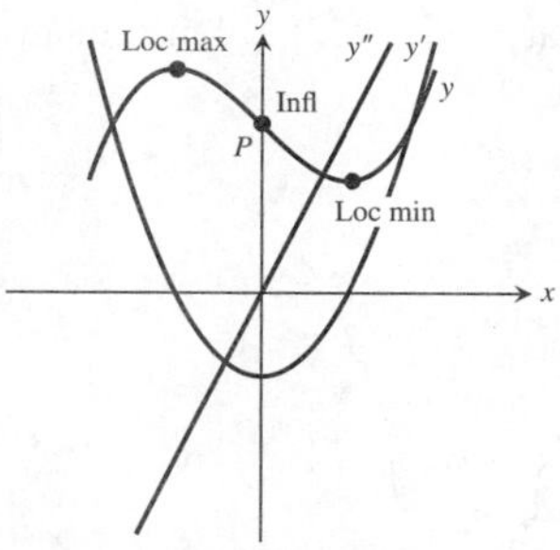

91.

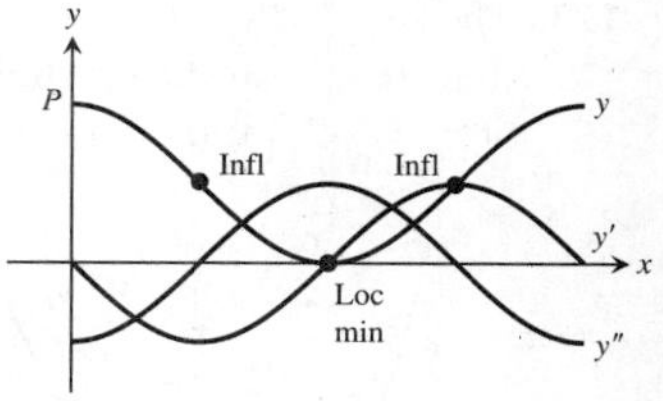

93.

Point	y'	y''
P	−	+
Q	+	0
R	+	−
S	0	−
T	−	−

95.

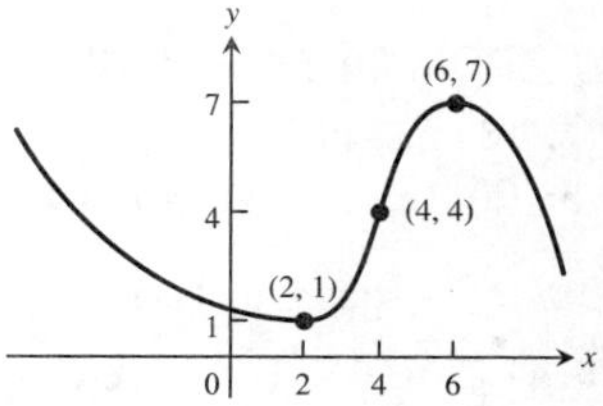

97.

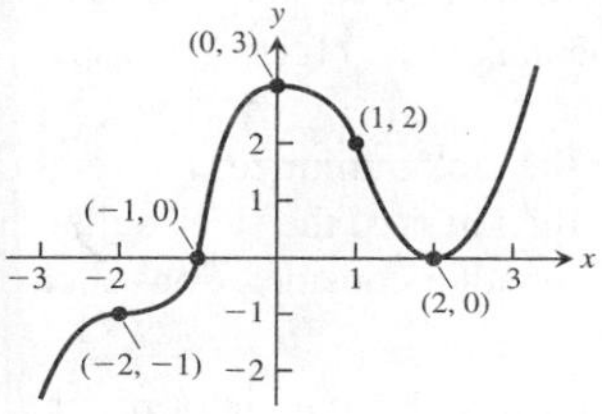

99.

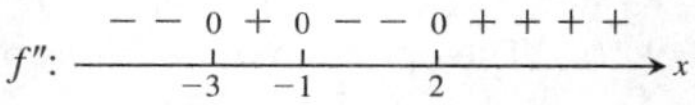

There are points of inflection at $x = -3$, $x = -1$, and $x = 2$.

101.

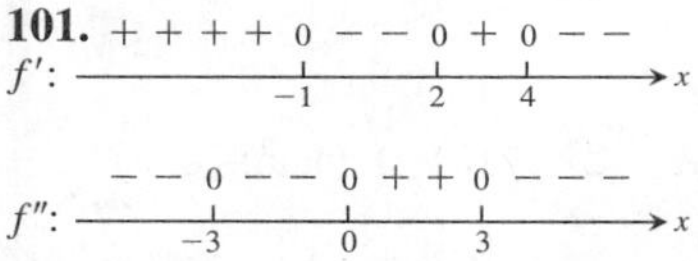

There are local maxima at $x = -1$ and $x = 4$. There is a local minimum at $x = 2$. There are points of inflection at $x = 0$ and $x = 3$.

103. Concave up on $(-2, 0)$ and $(1, 2)$; concave down on $(0, 1)$

105. **(a)** Toward origin: $0 \le t < 2$ and $6 \le t \le 10$; away from origin: $2 \le t \le 6$ and $10 \le t \le 15$
(b) $t = 2$, $t = 6$, $t = 10$
(c) $t = 5$, $t = 7$, $t = 13$
(d) Positive: $5 \le t \le 7$, $13 \le t \le 15$; negative: $0 \le t \le 5$, $7 \le t \le 13$

107. ≈ 60 thousand units

109.

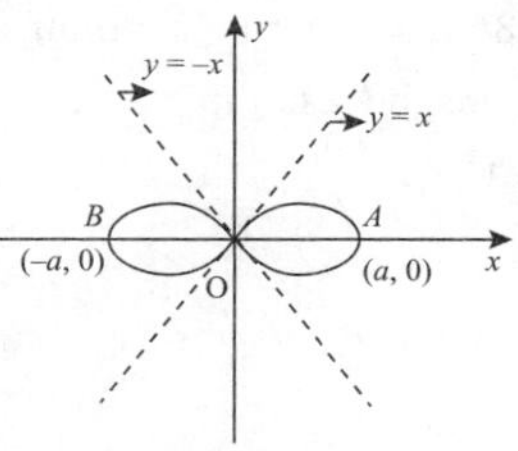

113. $b = -3$

117. $x = -1$, $x = 2$

119. Concave down in (1, 2), concave up in $(-\infty, 1)$ and $(2, \infty)$ point of inflexions are (1, 21) and (2, 34).

121. The zeros of $y' = 0$ and $y'' = 0$ are extrema and points of inflection, respectively. Inflection at $x = 3$; local maximum at $x = 0$; local minimum at $x = 4$

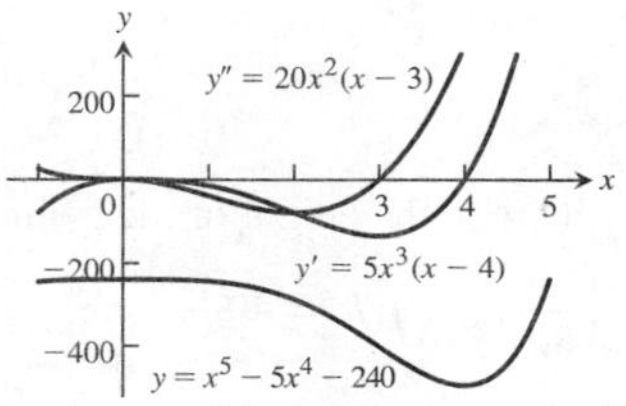

123. The zeros of $y' = 0$ and $y'' = 0$ are extrema and points of inflection, respectively. Inflection at $x = -\sqrt[3]{2}$; local maximum at $x = -2$; local minimum at $x = 0$.

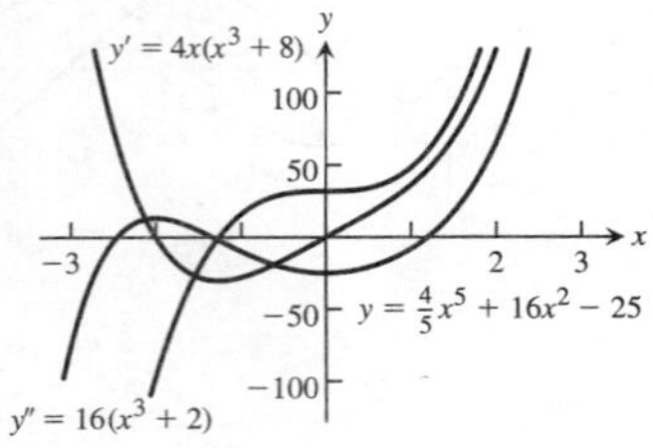

SECTION 4.5, pp. 218–224

1. 18 in.; 3 in. by 6 in.

3. (a) $(x, 1 - x)$ **(b)** $A(x) = 2x(1 - x)$

(b) $\frac{1}{2}$ square units, 1 by $\frac{1}{2}$

5. $\frac{14}{3} \times \frac{35}{3} \times \frac{5}{3}$ in., $\frac{2450}{27}$ in^3

7. 45000 sq m is the largest area. The dimensions are 300 m and 150 m.

9. (a) The optimum dimensions of the tank are 10 ft on the base edges and 5 ft deep.

(b) Minimizing the surface area of the tank minimizes its weight for a given wall thickness. The thickness of the steel walls would probably be determined by other considerations such as structural requirements.

11. The dimensions 20 by 10 inches minimize the amount of paper.

13. $\frac{\pi}{2}$ **15.** $8:\pi$

17. (a) $V(x) = 2x(24 - 2x)(18 - 2x)$ **(b)** Domain: $(0, 9)$

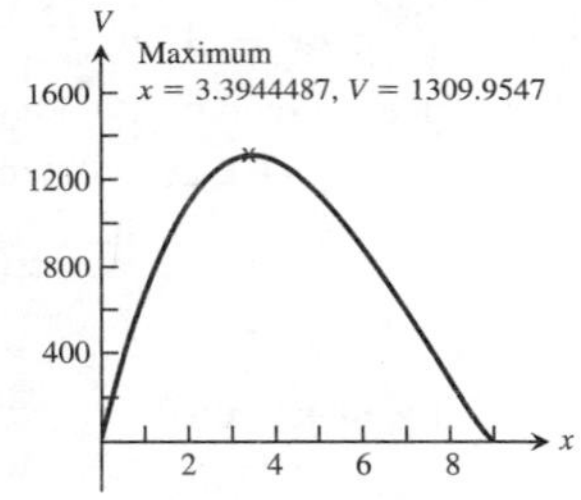

(c) Maximum volume ≈ 1309.95 in^3 when $x \approx 3.39$ in.

(d) $V'(x) = 24x^2 - 336x + 864$, so the critical point is at $x = 7 - \sqrt{13}$, which confirms the result in part (c).

(e) $x = 2$ in. or $x = 5$ in.

19. Radius $= 10\sqrt{2/3}$ cm, height $= 20/\sqrt{3}$ cm, volume $= 4000\pi/(3\sqrt{3})$ cm^3

21. (a) $h = 24, w = 18$

(b)

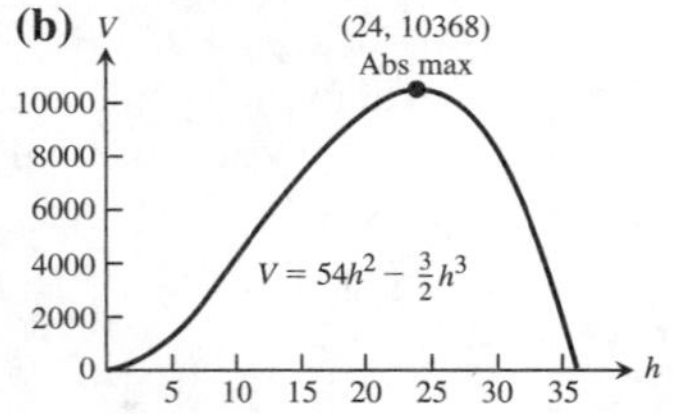

23. If r is the radius of the hemisphere, h the height of the cylinder, and V the volume, then $r = \left(\frac{3V}{8\pi}\right)^{1/3}$ and $h = \left(\frac{3V}{\pi}\right)^{1/3}$.

25. (b) $x = \frac{51}{8}$ **(c)** $L = 51\sqrt{3}/8$ in.

27. Radius $= \sqrt{2}$ m, height $= 1$ m, volume $= \frac{2\pi}{3}$ m^3

29. 1 **31.** $\frac{9b}{9 + \sqrt{3}\pi}$ m, triangle; $\frac{b\sqrt{3}\pi}{9 + \sqrt{3}\pi}$ m, circle

33. $10 + 4\sqrt{3}$ ft **35.** $\frac{3}{2} \times 2$ **37. (a)** 16 **(b)** -1

39. $r = \frac{2\sqrt{2}}{3}$ $h = \frac{4}{3}$ **41.** Area 8 when $a = 2$

43. (a) 56 ft/sec **(b)** 126 ft; it occurs at $t = 2$ sec **(c)** 28 ft/sec

45. $13\sqrt{13}$ ft **47. (a)** 6 m

49. (a) $4\sqrt{3} \times 4\sqrt{6}$ in.

51. (a) $10\pi \approx 31.42$ cm/sec; when $t = 0.5$ sec, 1.5 sec, 2.5 sec, 3.5 sec; $s = 0$, acceleration is 0.

(b) 10 cm from rest position; speed is 0.

53. (a) $s = \left((12 - 12t)^2 + 64t^2\right)^{1/2}$

(b) -12 knots; 8 knots **(c)** No

(e) $4\sqrt{13}$; this limit is the square root of the sum of the squares of the individual speeds.

55. $x = \frac{a}{2}, v = \frac{ka^2}{4}$ **57.** $\frac{c}{2} + 100$

59. (a) $\sqrt{\frac{2km}{h}}$ **(b)** $\sqrt{\frac{2km}{h}}$

63. The box is 2 ft × 2 ft × $\frac{3}{2}$ ft, with a minimum cost of \$72.

65. $M = \frac{C}{2}$ **71. (a)** $y = -1$

73. (a) The minimum distance is $\frac{\sqrt{5}}{2}$.

(b) The minimum distance is from the point $(3/2, 0)$ to the point $(1, 1)$ on the graph of $y = \sqrt{x}$, and this occurs at the value $x = 1$, where $D(x)$, the distance squared, has its minimum value.

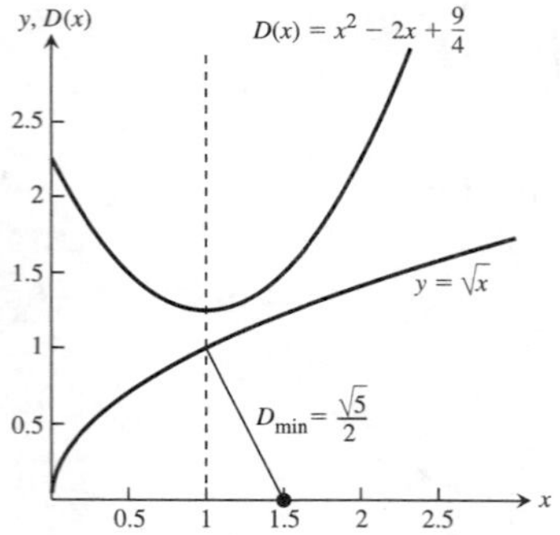

SECTION 4.6, pp. 228–229

1. 0.375 and -1.375 **3.** ≈ 0.8856 and -1.335

5. $x_2 = \frac{2387}{2000}$ **7.** ≈ 1.3310

9. x_1, and all later approximations will equal x_0.

11.

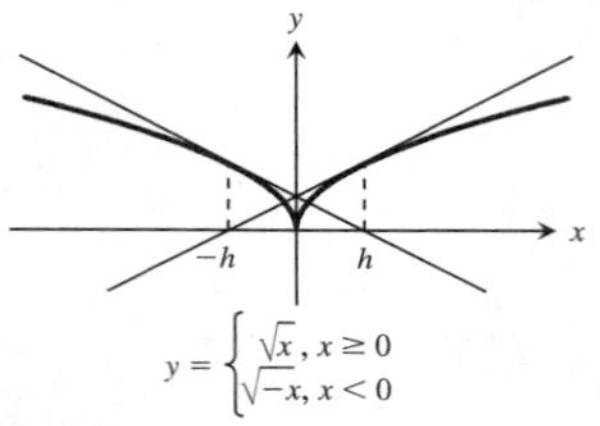

13. All four statements are equivalent to solving $f(x) = x^3 - 2x - 5$. Thus, all for ask for the same information.

15. $x \approx 1.165561185, y \approx 2.33112237$

17. (a) Two **(b)** 0.35003501505249 and -1.0261731615301

19. $\pm 1.3065629648764, \pm 0.5411961001462$ **21.** $x \approx 0.45$

23. 0.8192 **25.** The root is 1.17951.

27. (a) For $x_0 = -2$ or $x_0 = -0.8$, $x_i \to -1$ as i gets large.
(b) For $x_0 = -0.5$ or $x_0 = 0.25$, $x_i \to 0$ as i gets large.
(c) For $x_0 = 0.8$ or $x_0 = 2$, $x_i \to 1$ as i gets large.
(d) For $x_0 = -\sqrt{21}/7$ or $x_0 = \sqrt{21}/7$, Newton's method does not converge. The values of x_i alternate between $-\sqrt{21}/7$ and $\sqrt{21}/7$ as i increases.

29. Answers will vary with machine speed.

SECTION 4.7, pp. 236–239

1. (a) $\frac{3x^2}{2}$ **(b)** $\frac{x^4}{4}$ **(c)** $\frac{x^4}{4} - x^2 + x$

3. (a) $\frac{3}{x}$ **(b)** $-\frac{1}{x}$ **(c)** $-\frac{1}{3x^3} + \frac{5x^2}{2} + 3x$

5. (a) $-\frac{1}{2x}$ **(b)** $-\frac{5}{2x}$ **(c)** $2x + \frac{5}{2x}$

7. (a) $\frac{x}{3}\sqrt{x}$ **(b)** $3\sqrt{x}$ **(c)** $\frac{4x\sqrt{x}}{3} + \sqrt{x}$

9. (a) $\frac{3}{5}x^{2/3}$ **(b)** $2x^{1/3}$ **(c)** $-2x^{-1/3}$

11. (a) $\cos(\pi x)$ **(b)** $-3\cos x$ **(c)** $-\frac{1}{\pi}\cos(\pi x) + \cos(3x)$

13. (a) $\tan x$ **(b)** $2\tan\left(\frac{x}{3}\right)$ **(c)** $-\frac{2}{3}\tan\left(\frac{3x}{2}\right)$

15. (a) $-\csc x$ **(b)** $\frac{1}{5}\csc(5x)$ **(c)** $2\csc\left(\frac{\pi x}{2}\right)$

17. $\frac{x^2}{2} + x + C$ **19.** $t^3 + \frac{t^2}{4} + C$ **21.** $\frac{x^4}{2} - \frac{5x^2}{2} + 7x + C$

23. $-\frac{1}{x} - \frac{x^3}{3} - \frac{x}{3} + C$ **25.** $\frac{3}{2}x^{2/3} + C$

27. $\frac{2}{3}x^{3/2} + \frac{3}{4}x^{4/3} + C$ **29.** $4y^2 - \frac{8}{3}y^{3/4} + C$

31. $x^2 + \frac{2}{x} + C$ **33.** $2\sqrt{t} - \frac{2}{\sqrt{t}} + C$

35. $-2\sin t + C$ **37.** $-21\cos\frac{\theta}{3} + C$

39. $3\cot x + C$ **41.** $-\frac{1}{2}\csc\theta + C$

43. $4\sec x - 2\tan x + C$ **45.** $-\frac{1}{2}\cos 2x + \cot x + C$

47. $\frac{t}{2} + \frac{\sin 4t}{8} + C$ **49.** $\frac{3x^{(\sqrt{3}+1)}}{\sqrt{3}+1} + C$

51. $\tan\theta + C$ **53.** $-\cot x - x + C$ **55.** $-\cos\theta + \theta + C$

63. (a) Wrong: $\frac{d}{dx}\left(\frac{x^2}{2}\sin x + C\right) = \frac{2x}{2}\sin x + \frac{x^2}{2}\cos x = x\sin x + \frac{x^2}{2}\cos x$

(b) Wrong: $\frac{d}{dx}(-x\cos x + C) = -\cos x + x\sin x$

(c) Right: $\frac{d}{dx}(-x\cos x + \sin x + C) = -\cos x + x\sin x + \cos x = x\sin x$

65. (a) Wrong: $\frac{d}{dx}\left(\frac{(2x+1)^3}{3} + C\right) = \frac{3(2x+1)^2(2)}{3} = 2(2x+1)^2$

(b) Wrong since $\frac{d}{dx}\left(\frac{(2x+1)^3}{2} + C\right) = \frac{6(2x+1)^2}{5}$

(c) Right: $\frac{d}{dx}((2x+1)^3 + C) = 6(2x+1)^2$

67. Right since $\frac{d}{dx}\left(\frac{x+2}{x-5}\right)^3 = \frac{-21(x+2)^2}{(x-5)^4}$

69. (b) **71.** $y = 2x^2 - 12x + 18$

73. $y = -\frac{1}{x} + \frac{x^2}{2} - \frac{1}{2}$ **75.** $y = 9x^{1/3} + 4$

77. $s = t - \cos t + 6$ **79.** $r = \cos(\pi\theta) - 1$

81. $v = \frac{1}{2}\sec t + \frac{1}{2}$ **83.** $y = \frac{5x^2}{2} - \frac{x^3}{3} + 2x + 3$

85. $r = \frac{1}{t} + 2t - 2$ **87.** $y = 2x^3 - \frac{5x^2}{2} + x + 4$

89. $y = -\sin t + \cos t + t^3 - 1$ **91.** $y = 2x^{3/2} - 50$

93.

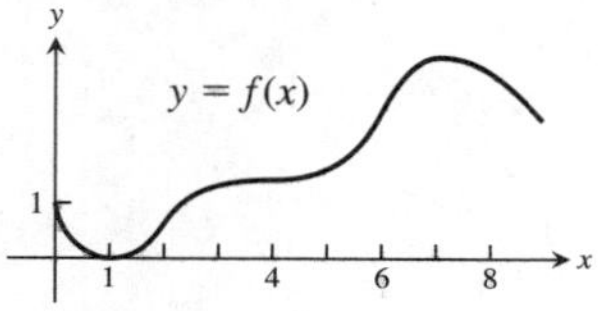

95.

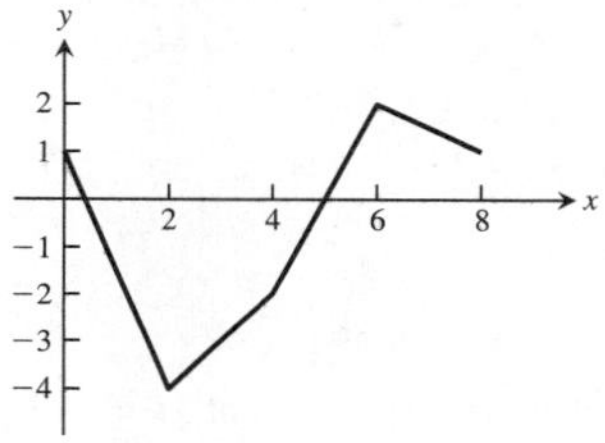

97. $y = x - x^{4/3} + \frac{1}{2}$

99. $y = -\sin x - \cos x - 2$

101. (a) (i) 33.2 units, **(ii)** 33.2 units, **(iii)** 33.2 units
(b) True

103. $t = 88/k, k = 16$

105. (a) $v = 10t^{3/2} - 6t^{1/2}$ **(b)** $s = 4t^{5/2} - 4t^{3/2}$

109. (a) $-\sqrt{x} + C$ **(b)** $x + C$ **(c)** $\sqrt{x} + C$
(d) $-x + C$
(e) $x - \sqrt{x} + C$
(f) $-x - \sqrt{x} + C$

PRACTICE EXERCISES, pp. 240–242

1. Minimum value is 1 at $x = 2$.

3. Local maximum at $(-2, 17)$; local minimum at $\left(\frac{4}{3}, -\frac{41}{27}\right)$

5. Minimum value is 0 at $x = -1$ and $x = 1$.

7. There is a local minimum at $(0, 1)$.

9. Maximum value is $\frac{1}{2}$ at $x = 1$; minimum value is $-\frac{1}{2}$ at $x = -1$.

11. No **13.** No minimum; absolute maximum: $f(1) = 16$; critical points: $x = 1$ and $11/13$

15. Yes, except at $x = 0$ **17.** No **21. (b)** one

23. (b) 0.8555 99677 2

29. Global minimum value of $\frac{1}{2}$ at $x = 2$

31. (a) $t = 0, 6, 12$ **(b)** $t = 3, 9$ **(c)** $6 < t < 12$ **(d)** $0 < t < 6, 12 < t < 14$

33.

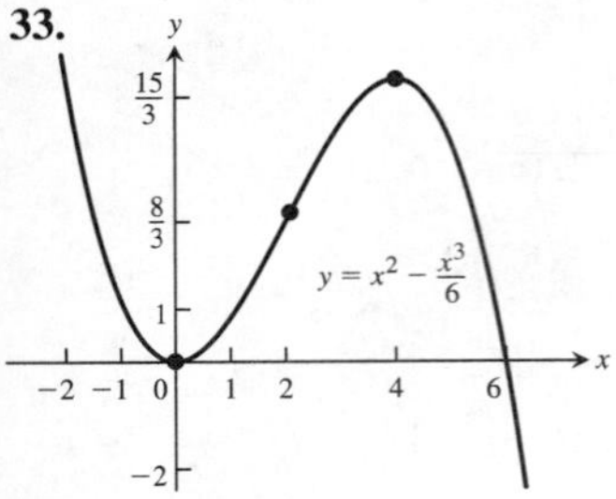

35.

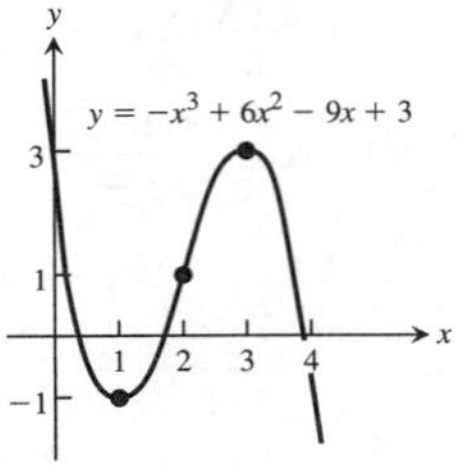

37.

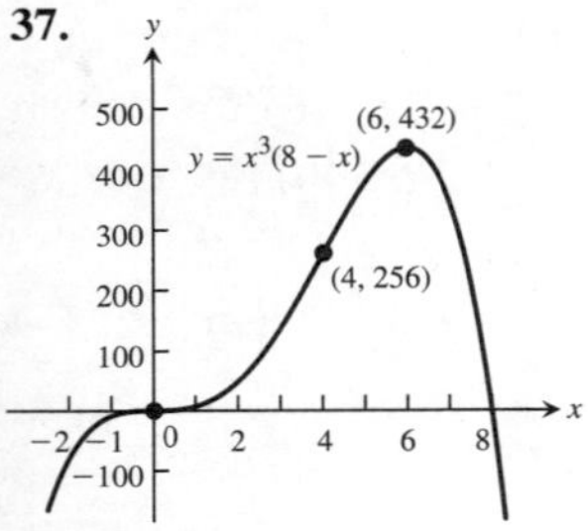

39.

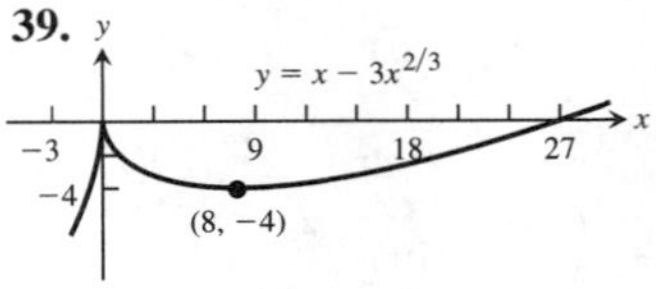

41.

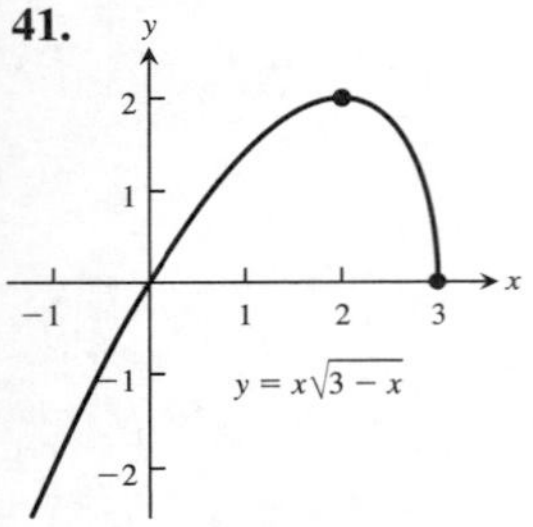

43. (a) Local maximum at $x = 4$, local minimum at $x = -4$, inflection point at $x = 0$

(b)

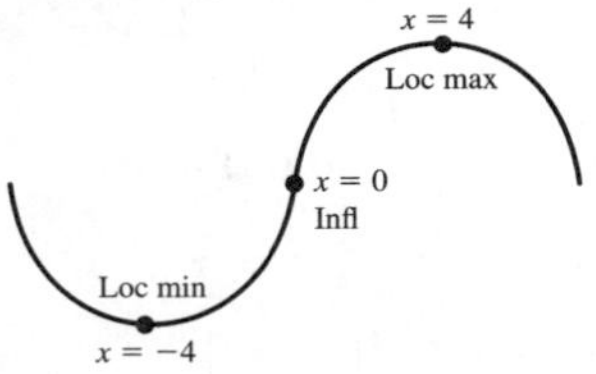

45. (a) Local maximum at $x = 0$, local minima at $x = -1$ and $x = 2$, inflection points at $x = (1 \pm \sqrt{7})/3$

(b)

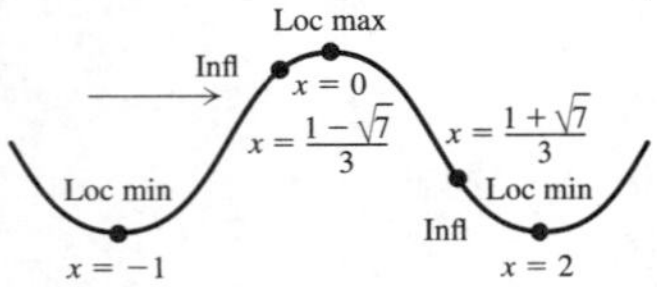

47. (a) Local maximum at $x = -\sqrt{2}$, local minimum at $x = \sqrt{2}$, inflection points at $x = \pm 1$ and 0

(b)

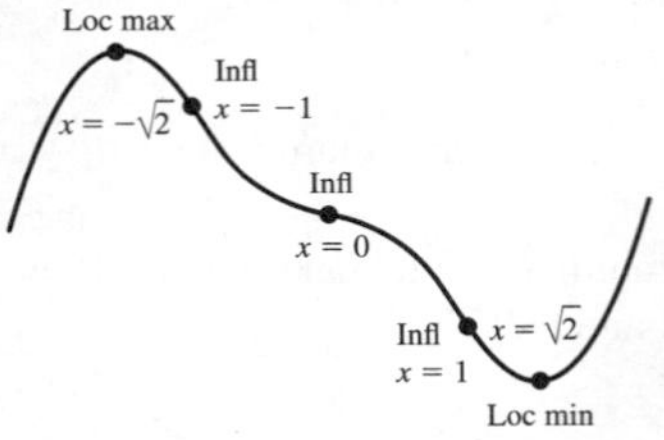

53.

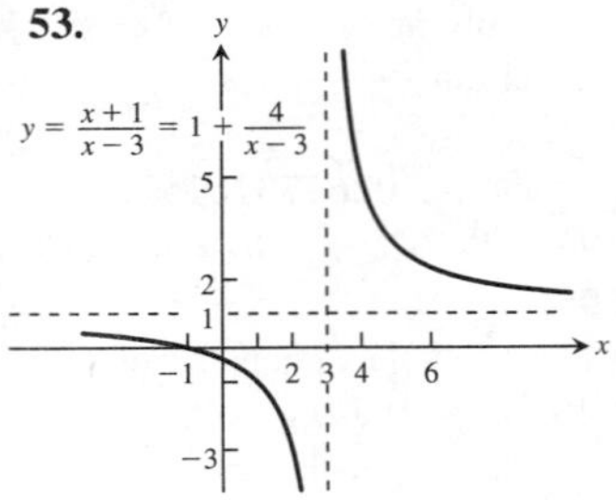

55.

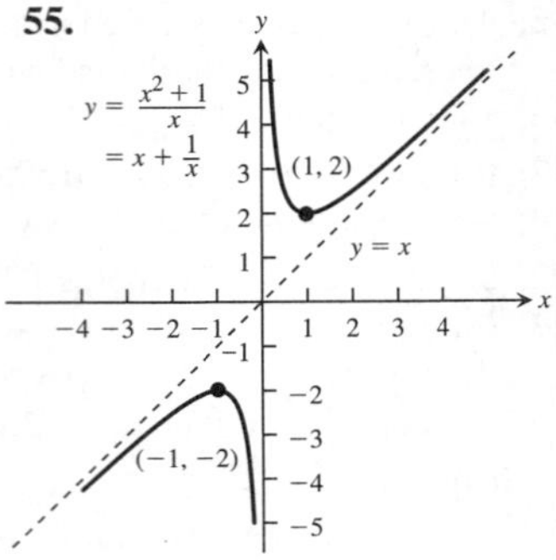

57.

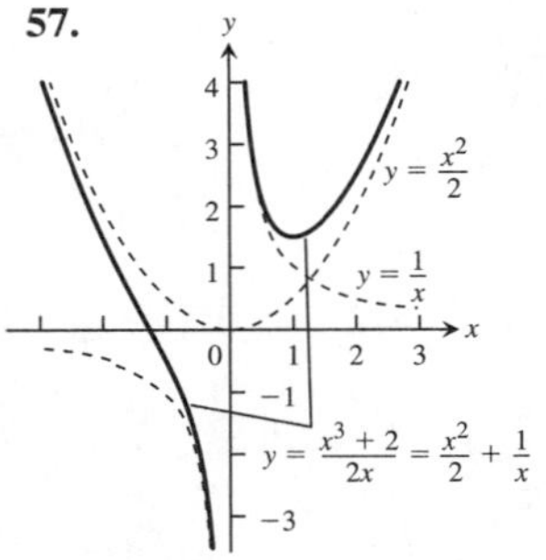

59.

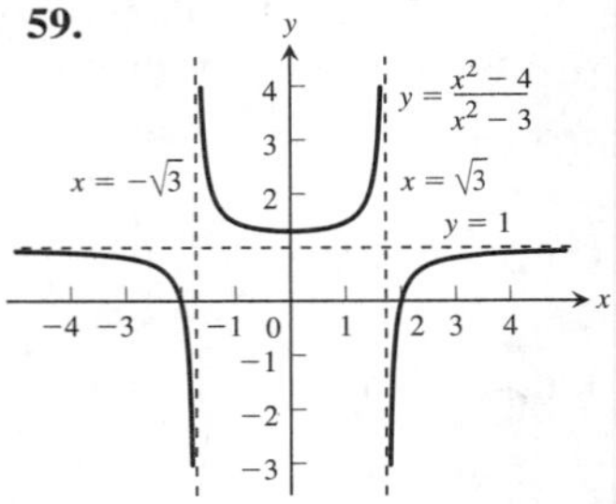

61. (a) 0, 36 **(b)** 18, 18 **63.** 54 square units

65. height $= 2$, radius $= \sqrt{2}$

67. $x = 5 - \sqrt{5}$ hundred ≈ 276 tires, $y = 2(5 - \sqrt{5})$ hundred ≈ 553 tires

69. Dimensions: base is 6 in. by 12 in., height $= 2$ in.; maximum volume $= 144$ in^3

71. $x_5 = 2.1958\ 23345$ **73.** $\frac{x^4}{4} + \frac{5}{2}x^2 - 7x + C$

75. $2t^{3/2} - \frac{4}{t} + C$ **77.** $-\frac{1}{r+5} + C$

79. $(\theta^2 + 1)^{3/2} + C$ **81.** $\frac{1}{3}(1 + x^4)^{3/4} + C$

83. $10 \tan \frac{s}{10} + C$ **85.** $-\frac{1}{\sqrt{2}} \csc \sqrt{2}\,\theta + C$

87. $\frac{1}{2}x - \sin \frac{x}{2} + C$ **89.** $y = x - \frac{1}{x} - 1$

91. $r = 4t^{5/2} + 4t^{3/2} - 8t$

ADDITIONAL AND ADVANCED EXERCISES, pp. 242–244

1. The function is constant on the interval.

3. The extreme points will not be at the end of an open interval.

5. (a) A local minimum at $x = -1$, points of inflection at $x = 0$ and $x = 2$

(b) A local maximum at $x = 0$ and local minima at $x = -1$ and $x = 2$, points of inflection at $x = \frac{1 \pm \sqrt{7}}{3}$

9. No **11.** $a = 1, b = 0, c = 1$ **13.** Yes

15. Drill the hole at $y = h/2$.

17. $r = \frac{RH}{2(H - R)}$ for $H > 2R$, $r = R$ if $H \le 2R$

19. $\frac{12}{5}$ and 5

21. (a) $\frac{c - b}{2e}$ **(b)** $\frac{c + b}{2}$ **(c)** $\frac{b^2 - 2bc + c^2 + 4ae}{4e}$

(d) $\frac{c + b + t}{2}$

23. $m_0 = 1 - \frac{1}{q}, m_1 = \frac{1}{q}$ **25. (a)** $k = -38.72$ **(b)** 25 ft

27. Yes, $y = x + C$ **29.** $v_0 = \frac{2\sqrt{2}}{3} b^{3/4}$ **33.** 3

Chapter 5

SECTION 5.1, pp. 254–256

1. (a) 0.125 **(b)** 0.21875 **(c)** 0.625 **(d)** 0.46875

3. (a) 1.066667 **(b)** 1.283333 **(c)** 2.666667 **(d)** 2.083333

5. 0.3125, 0.328125 **7.** 1.5, 1.574603

9. (a) 245 cm **(b)** 245 cm **11. (a)** 3490 ft **(b)** 3840 ft

13. (a) Because the acceleration is decreasing, an upper estimate is obtained using left end-points in summing acceleration $\cdot \Delta t$. Thus, $\Delta t = 1$ and speed = 49.96 ft/sec

(b) Using right end-points we obtain a lower estimate: speed = 30.28 ft/sec

(c) Upper estimates for the speed at each second are:

t	0	1	2	3	4
v	0	22.0	37.3	46.51	49.96

Thus, the distance fallen when $t = 2$ seconds is $s = 59.3$ ft.

15. $\frac{31}{16}$ **17.** 1

19. Since the leakage is increasing, an upper estimate uses right endpoints and a lower estimate uses left endpoints:

(a) upper estimate = 247 gal, lower estimate = 177 gal.

(b) upper estimate = 1112 gal, lower estimate = 702 gal.

(c) worst case: $1112 + 450t = 20{,}000 \Rightarrow t \approx 41.97$ hrs; best case: $702 + 450t = 20{,}000 \Rightarrow t \approx 42.88$ hrs.

21. (a) 2 **(b)** $2\sqrt{2} \approx 2.828$ **(c)** $8\sin\left(\frac{\pi}{8}\right) \approx 3.061$

(d) Each area is less than the area inside the circle, π. As n increases, the polygon area approaches π.

SECTION 5.2, pp. 262–263

1. $\frac{5}{3} + \frac{10}{4} + \frac{15}{5} = \frac{43}{6}$ **3.** $\tan\pi + \tan 2\pi + \tan 3\pi = 0$

5. $\sin\pi - \sin\frac{\pi}{2} + \sin\frac{\pi}{3} = \frac{\sqrt{3} - 2}{2}$

7. All of them represent $1 + 3 + 9 + 27 + 81 + 243$ **9.** b

11. $\sum_{k=1}^{6} k$ **13.** $\sum_{k=1}^{4} \frac{1}{2^k}$ **15.** $\sum_{k=1}^{5} (-1)^{k+1}\frac{1}{k}$

17. (a) −15 **(b)** 1 **(c)** 1 **(d)** −11 **(e)** 16

19. (a) 42 **(b)** 280 **(c)** 2016 **21.** −56 **23.** −73

25. 90 **27.** 3376 **29. (a)** 35 **(b)** 2000 **(c)** 2790

31. (a) $3n$ **(b)** $10(n-1)$ **(c)** $\frac{n(n+3)}{2}$ **33.** 2600 **35.** $-2\sqrt{3}$

37. (a)

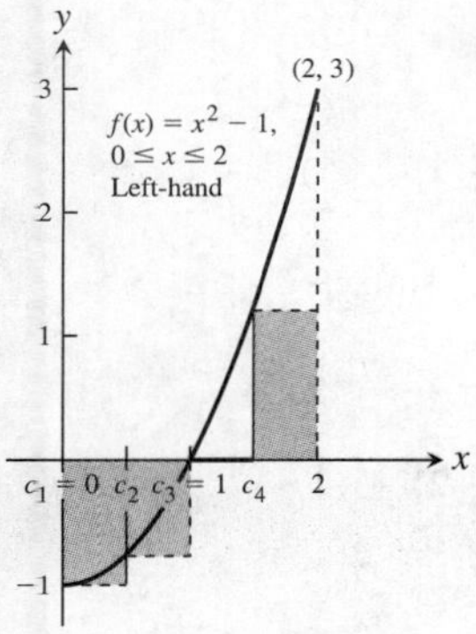

(b)

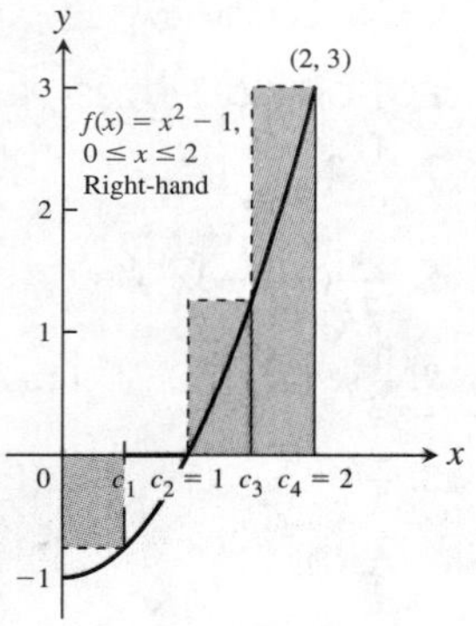

(c)

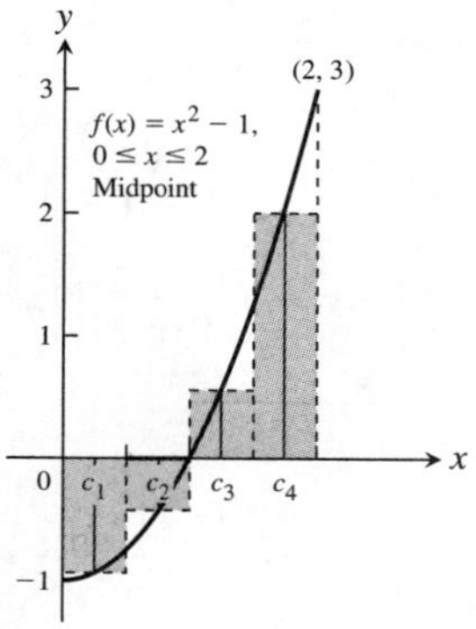

39. (a)

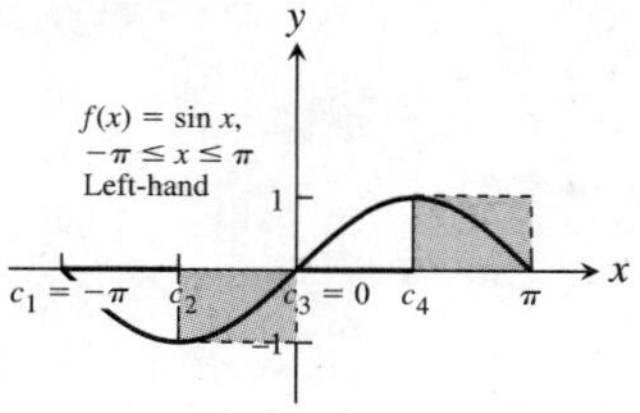

(b)

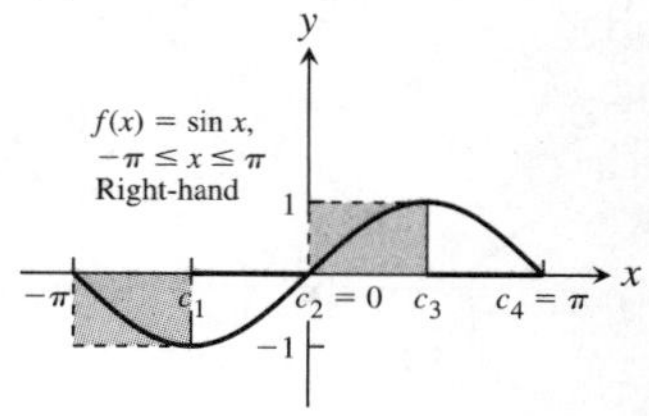

(c)

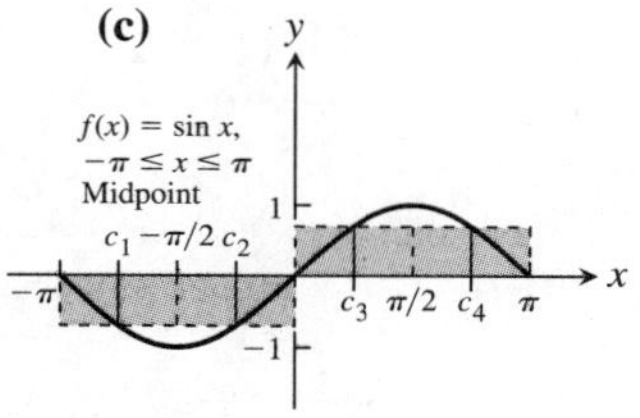

41. 1.2

43. $\frac{2}{3} - \frac{1}{2n} - \frac{1}{6n^2}, \frac{2}{3}$

45. $12 + \frac{27n + 9}{2n^2}$, 12

47. $\frac{5}{6} + \frac{6n + 1}{6n^2}, \frac{5}{6}$

49. $\frac{1}{2} + \frac{1}{n} + \frac{1}{2n^2}, \frac{1}{2}$

SECTION 5.3, pp. 272–275

1. $\int_0^3 x^3 dx$ **3.** $\int_{-5}^{2} (2x^2 - x)dx$ **5.** $\int_3^5 \frac{1}{x-1}dx$

7. $\int_{-\pi/4}^{\pi/4} \tan dx$

9. (a) 0 **(b)** −8 **(c)** −12 **(d)** 10 **(e)** 2 **(f)** 16

11. (a) 5 **(b)** $-5\sqrt{5}$ **(c)** −5 **(d)** −10

13. (a) 8 **(b)** −4 **15.** Area = 21 square units

17. Area of semicircle is: $A = \frac{1}{2}\pi r^2 = \frac{1}{2}\pi(4)^2 = 8\pi$

$\Rightarrow \int_{-4}^{4} \sqrt{16 - x^2}\, dx = 8\pi$ square units

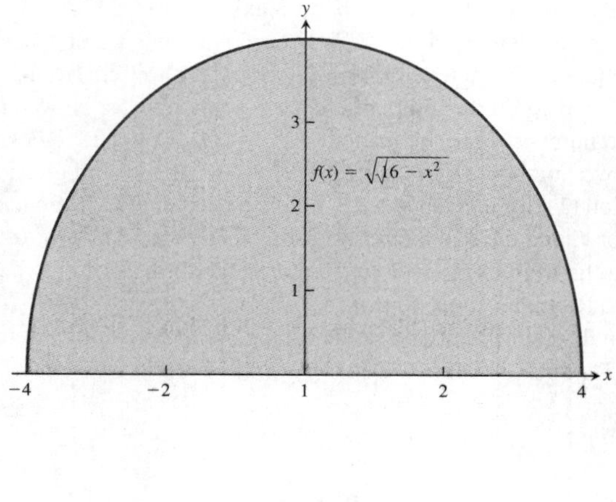

19. Area of Triangle on right is: $A = \frac{1}{2}bh = \frac{9}{2}$

$\Rightarrow \int_{-1}^{2} |x+1|\,dx = \frac{9}{2}$ square units

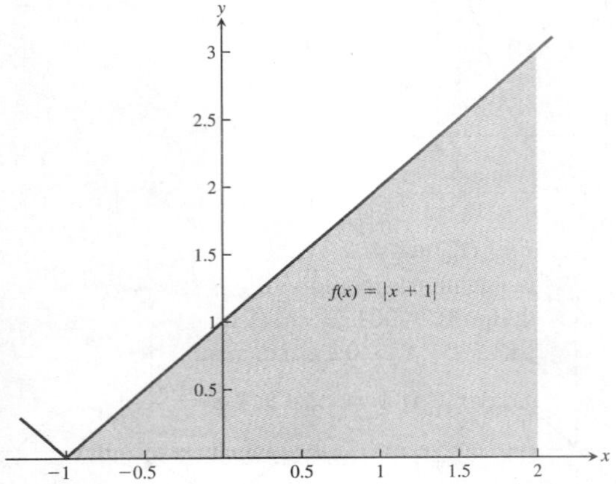

21. Area $= 3$ square units **23.** $b^2/4$ **25.** $b^2 - a^2$

27. **(a)** $\int_{-4}^{4} \left(\sqrt{16-x^2}\right)dx = 8\pi$ **(b)** $\int_{0}^{4} \left(\sqrt{16-x^2}\right)dx = 4\pi$

29. $1/2$ **31.** $3\pi^2/2$ **33.** $\frac{10\sqrt{5}}{3}$

35. $1/24$ **37.** $3a^2/2$ **39.** $b/3$ **41.** -14

43. $-3/2$ **45.** $-7/4$ **47.** $130/3$ **49.** 0

51. Using n subintervals of length $\Delta x = b/n$ and right-endpoint values:

$$\text{Area} = \int_0^b 3x^2\,dx = b^3$$

53. Using n subintervals of length $\Delta x = b/n$ and right-endpoint values:

$$\text{Area} = \int_0^b 2x\,dx = b^2$$

55. $\text{av}(f) = 0$ **57.** $\text{av}(f) = -2$ **59.** $\text{av}(f) = 1$

61. **(a)** $\text{av}(g) = -1/2$ **(b)** $\text{av}(g) = 1$ **(c)** $\text{av}(g) = 1/4$

63. $c(b - a)$ **65.** $b^3/3 - a^3/3$ **67.** 9

69. $b^4/4 - a^4/4$ **71.** $a = 0$ and $b = 1$ maximize the integral.

73. Upper bound $= 1$, lower bound $= 1/2$

75. For example, $\int_0^1 \sin(x^2)\,dx \le \int_0^1 dx = 1$

77. $\int_a^b f(x)\,dx \ge \int_a^b 0\,dx = 0$ **79.** Upper bound $= 1/2$

89. Since f is continuous, the Extreme Value Theorem implies that the continuous function f has a maximum value M and a minimum value m (Chapter 4, Theorem 1). The values of f and g are equal except at a single point c where $g(c)$ differs from $f(c)$. The magnitude of the difference is some number between the two numbers $|g(c) - m|$ and $|g(c) - M|$. Call the larger of these two numbers D. In the formula for a Riemann sum given in Equation (1), for a given n, the difference between the Riemann sum for f and the Riemann sum for g is at most $D(b - a)/n$. Taking the limit as $n \to \infty$, we see that the difference between the two Riemann sums approaches 0, so that the two limits are equal. A similar bound holds for a general partition, where $D\|P\| \to 0$.

SECTION 5.4, pp. 285–287

1. $-10/3$ **3.** $124/3$ **5.** $753/16$ **7.** 1 **9.** $2\sqrt{3}$

11. 0 **13.** $-\pi/4$ **15.** $1 - \frac{\pi}{4}$ **17.** $\frac{2-\sqrt{2}}{4}$ **19.** $-8/3$

21. $-3/4$ **23.** $\sqrt{2} - \sqrt[4]{8} + 1$ **25.** $1/2$ **27.** 16

29. $1/2$ **31.** $\sqrt{26} - \sqrt{5}$ **33.** $\left(\cos\sqrt{x}\right)\left(\frac{1}{2\sqrt{x}}\right)$ **35.** $4t^5$

37. $\frac{2x}{\sqrt{x^2+1}}$ **39.** $\sqrt{1+x^2}$ **41.** $-\frac{1}{2}x^{-1/2}\sin x$ **43.** 0

45. 1 **47.** $28/3$ **49.** $1/2$ **51.** π **53.** $\frac{\sqrt{2}\pi}{2}$

55. d, since $y' = \frac{1}{x}$ and $y(\pi) = \int_\pi^\pi \frac{1}{t}\,dt - 3 = -3$

57. b, since $y' = \sec x$ and $y(0) = \int_0^0 \sec t\,dt + 4 = 4$

59. $y = \int_2^x \sec t\,dt + 3$ **61.** $\frac{2}{3}bh$ **63.** $\frac{10}{3}$

65. **(a)** $T(0) = 70\,°\text{F}$, $T(16) = 76\,°\text{F}$, $T(25) = 85\,°\text{F}$
(b) $\text{av}(T) = 75\,°\text{F}$

67. $2x - 2$ **69.** $\frac{1}{2}$

71. **(a)** True. Since f is continuous, g is differentiable by Part 1 of the Fundamental Theorem of Calculus.
(b) True: g is continuous because it is differentiable.
(c) True, since $g'(1) = f(1) = 0$.
(d) False, since $g''(1) = f'(1) > 0$.
(e) True, since $g'(1) = 0$ and $g''(1) = f'(1) > 0$.
(f) False: $g''(x) = f'(x) > 0$, so g'' never changes sign.
(g) True, since $g'(1) = f(1) = 0$ and $g'(x) = f(x)$ is an increasing function of x (because $f'(x) > 0$).

73. **(a)** $v = \frac{ds}{dt} = \frac{d}{dt}\int_0^t f(x)\,dx = f(t) \Rightarrow v(5) = f(5) = 2$ m/sec
(b) $a = df/dt$ is negative, since the slope of the tangent line at $t = 5$ is negative.
(c) $s = \int_0^3 f(x)\,dx = \frac{1}{2}(3)(3) = \frac{9}{2}$m, since the integral is the area of the triangle formed by $y = f(x)$, the x-axis, and $x = 3$.
(d) $t = 6$, since after $t = 6$ to $t = 9$, the region lies below the x-axis.
(e) At $t = 4$ and $t = 7$, since there are horizontal tangents there.
(f) Toward the origin between $t = 6$ and $t = 9$, since the velocity is negative on this interval. Away from the origin between $t = 0$ and $t = 6$, since the velocity is positive there.
(g) Right or positive side, because the integral of f from 0 to 9 is positive, there being more area above the x-axis than below.

SECTION 5.5, pp. 293–294

1. $\frac{1}{6}(2x+4)^6 + C$ **3.** $-\frac{1}{3}(x^2+5)^{-3} + C$

5. $\frac{1}{72}(4x^3+3x^4)^6 + C$ **7.** $\frac{1}{6}\sin 6x + C$

9. $\frac{1}{2}\sec 2t + C$ **11.** $-6(1-r^3)^{1/2} + C$

13. $\frac{1}{3}(x^{3/2} - 1) - \frac{1}{6}\sin(2x^{3/2} - 2) + C$

15. **(a)** $-\frac{1}{4}(\cot^2 2\theta) + C$ **(b)** $-\frac{1}{4}(\csc^2 2\theta) + C$

17. $2\left[\frac{1}{5}(x+5)^{5/2} - \frac{10}{3}(x+5)^{3/2} + 25(x+5)^{1/2}\right] + C$

19. $-\frac{2}{5}(1-\theta^2)^{5/4} + C$

21. $(-2/(1+\sqrt{x})) + C$ **23.** $\frac{1}{3}\tan(3x+2) + C$

25. $\frac{1}{2}\sin^6\left(\frac{x}{3}\right) + C$ **27.** $\left(\frac{r^3}{18} - 1\right)^6 + C$

29. $-\frac{2}{3}\cos(x^{3/2}+1) + C$ **31.** $\frac{1}{2\cos(2t+1)} + C$

33. $-\sin\left(\frac{1}{t} - 1\right) + C$ **35.** $-\frac{\sin^2(1/\theta)}{2} + C$

37. $\frac{2}{3}(1+x)^{3/2} - 2(1+x)^{1/2} + C$ **39.** $\frac{2}{3}\left(2 - \frac{1}{x}\right)^{3/2} + C$

41. $\frac{2}{27}\left(1 - \frac{3}{x^3}\right)^{3/2} + C$ **43.** $\frac{1}{12}(x-1)^{12} + \frac{1}{11}(x-1)^{11} + C$

45. $-\frac{1}{8}(1-x)^8 + \frac{4}{7}(1-x)^7 - \frac{2}{3}(1-x)^6 + C$

47. $\frac{1}{5}(x^2+1)^{5/2} - \frac{1}{3}(x^2+1)^{3/2} + C$ **49.** $\frac{-1}{4(x^2-4)^2} + C$

51. **(a)** $-\frac{6}{2+\tan^3 x} + C$ **(b)** $-\frac{6}{2+\tan^3 x} + C$
(c) $-\frac{6}{2+\tan^3 x} + C$

53. $\frac{1}{6}\sin\sqrt{3(2r-1)^2+6} + C$ **55.** $s = \frac{1}{2}(3t^2-1)^4 - 5$

57. $s = 4t - 2\sin\left(2t + \frac{\pi}{6}\right) + 9$

59. $s = \sin\left(2t - \frac{\pi}{2}\right) + 100t + 1$ **61.** 6 m

SECTION 5.6, pp. 301–304

1. **(a)** 14/3 **(b)** 2/3 **3.** **(a)** 1/2 **(b)** −1/2
5. **(a)** 15/16 **(b)** 0 **7.** **(a)** 0 **(b)** 1/8 **9.** **(a)** 4 **(b)** 0
11. **(a)** 506/375 **(b)** 86,744/375 **13.** **(a)** 0 **(b)** 0
15. $2\sqrt{3}$ **17.** 3/4 **19.** $3^{5/2} - 1$ **21.** 3 **23.** $\pi/3$
25. 16/3 **27.** $2^{5/2}$ **29.** $\pi/2$ **31.** 128/15 **33.** 4/3
35. 5/6 **37.** 38/3 **39.** 49/6 **41.** 32/3 **43.** 48/5
45. 8/3 **47.** 8 **49.** 5/3 (There are three intersection points.)
51. 18 **53.** 243/8 **55.** 8/3 **57.** 2 **59.** 104/15
61. 56/15 **63.** 4 **65.** $\frac{4}{3} - \frac{4}{\pi}$ **67.** $\pi/2$ **69.** 2
71. 1/2 **73.** t_1 **75.** **(a)** $(\pm\sqrt{c}, c)$ **(b)** $c = 4^{2/3}$ **(c)** $c = 4^{2/3}$
77. 9/8 **79.** 3/4 **81.** Neither
83. The dark gray region has area 2, which is larger than the light gray region, which has an area $(0.6)\pi \approx 1.88$.
85. $F(6) - F(2)$ **87.** **(a)** −3 **(b)** 3 **89.** $I = a/2$

PRACTICE EXERCISES, pp. 305–307

1. **(a)** About 680 ft **(b)** h (feet)

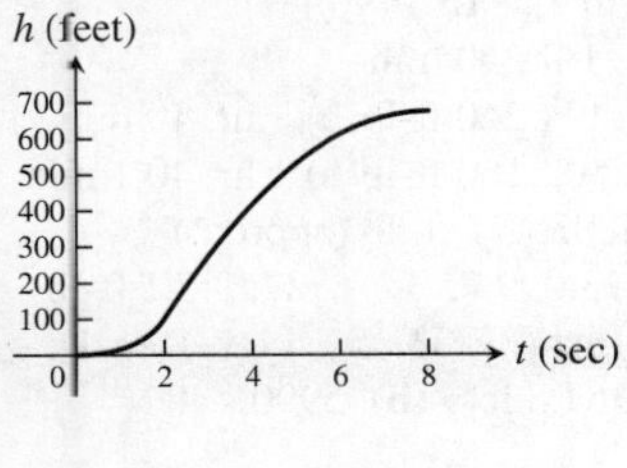

3. **(a)** −1/2 **(b)** 31 **(c)** 13 **(d)** 0

5. $\int_1^5 (2x-1)^{-1/2}\,dx = 2$ **7.** $\int_{-\pi}^0 \cos\frac{x}{2}\,dx = 2$

9. **(a)** 4 **(b)** 2 **(c)** −2 **(d)** -2π **(e)** 8/5
11. 8/3 **13.** 62 **15.** 1 **17.** 1/6 **19.** 18

21. 9/8 **23.** $\frac{\pi^2}{32} + \frac{\sqrt{2}}{2} - 1$ **25.** 4 **27.** $\frac{8\sqrt{2}-7}{6}$

29. Min: −4; max: 0; area: 27/4 **31.** $\frac{3}{8}\pi a^2$

35. $y = \int_5^x \left(\frac{\sin t}{t}\right) dt - 3$ **37.** $-4(\cos x)^{1/2} + C$

39. $\theta^2 + \theta + \sin(2\theta+1) + C$ **41.** $\frac{t^3}{3} + \frac{4}{t} + C$

43. $-\frac{1}{3}\cos(2t^{3/2}) + C$ **45.** $\frac{1}{4(\sin 2\theta + \cos 2\theta)^2} + C$

47. 16 **49.** 2 **51.** 1 **53.** 8 **55.** $27\sqrt{3}/160$
57. $\pi/2$ **59.** $\sqrt{3}$ **61.** $6\sqrt{3} - 2\pi$ **63.** −1 **65.** 2
67. 1 **69.** **(a)** b **(b)** b **73.** 25 °F **75.** $\sqrt{2+\cos^3 x}$

77. $\frac{-6}{3+x^4}$ **79.** Yes **81.** $-\sqrt{1+x^2}$

83. Cost ≈ \$10,899 using a lower sum estimate. No it cannot be done.

ADDITIONAL AND ADVANCED EXERCISES, pp. 307–310

1. **(a)** Yes **(b)** No **5.** **(a)** 1/4 **(b)** $\sqrt[3]{12}$

7. $f(x) = \frac{x}{\sqrt{x^2+1}}$ **9.** $y = x^3 + 2x - 4$

11. 36/5

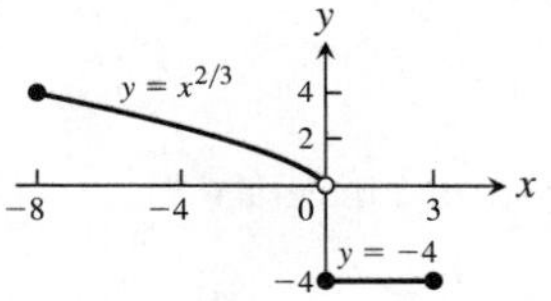

13. $\frac{1}{2} - \frac{2}{\pi}$

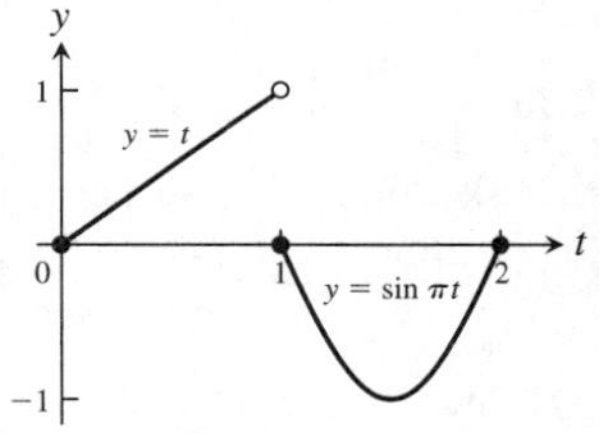

15. 13/3

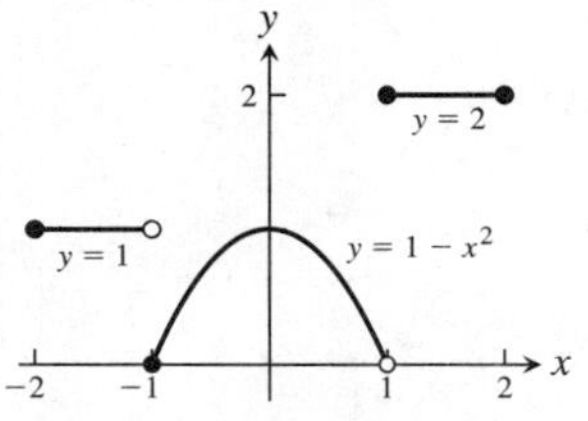

17. 19/3
19. **(a)** 0 **(b)** −1
(c) $-\pi$ **(d)** $x = 1$
(e) $y = 2x + 2 - \pi$
(f) $x = -1, x = 2$
(g) $[-2\pi, 0]$

21. $\int_0^1 f(x)\,dx$ **23.** $\frac{\sin 4y}{\sqrt{y}} - \frac{\sin y}{2\sqrt{y}}$

25. 1/6 **27.** $\frac{2x}{(1+x^4)^3} - \frac{3x^2}{(1+x^6)^3}$ **29.** **(b)** πr^2

Chapter 6

SECTION 6.1, pp. 318–322

1. 16 **3.** 16/3 **5.** **(a)** $2\sqrt{3}$ **(b)** 8 **7.** **(a)** 80 **(b)** $\frac{16}{3}$
9. 8π **11.** 10 **13.** **(a)** s^2h **(b)** s^2h **15.** 8/3
17. $\frac{2\pi}{3}$ **19.** $4 - \pi$ **21.** $\frac{32\pi}{5}$ **23.** 36π **25.** π

27. $\frac{29\pi}{30}$ **29.** $\frac{6\pi}{5}$ **31.** 2π **33.** 30π **35.** $\pi^2 - 2\pi$

37. $\frac{2\pi}{3}$ **39.** $\frac{117\pi}{5}$ **41.** $\pi(\pi - 2)$ **43.** $\frac{4\pi}{3}$ **45.** $\frac{8\pi}{3}$

47. $\frac{96\pi}{5}$ **49.** **(a)** 8π **(b)** $\frac{32\pi}{5}$ **(c)** $\frac{8\pi}{3}$ **(d)** $\frac{224\pi}{15}$

51. **(a)** $\frac{16\pi}{15}$ **(b)** $\frac{56\pi}{15}$ **(c)** $\frac{64\pi}{15}$ **53.** $V = 2a^2b\pi^2$

55. **(a)** $\frac{176\pi}{3}$ **(b)** $\frac{1}{120\pi}$ m/sec

59. $V = 3308\text{ cm}^3$ **61.** $(4 - b + a)/2$

SECTION 6.2, pp. 327–329

1. 6π **3.** 2π **5.** $14\pi/3$ **7.** 8π **9.** $5\pi/6$

11. $\frac{7\pi}{15}$ **13.** **(b)** 4π **15.** $\frac{16\pi}{15}(3\sqrt{2} + 5)$

17. $\frac{8\pi}{3}$ **19.** $\frac{4\pi}{3}$ **21.** $\frac{16\pi}{3}$

23. **(a)** 16π **(b)** 32π **(c)** 28π
(d) 24π **(e)** 60π **(f)** 48π

25. **(a)** $\frac{27\pi}{2}$ **(b)** $\frac{27\pi}{2}$ **(c)** $\frac{72\pi}{5}$ **(d)** $\frac{108\pi}{5}$

27. **(a)** $\frac{6\pi}{5}$ **(b)** $\frac{4\pi}{5}$ **(c)** 2π **(d)** 2π

29. **(a)** About the x-axis: $V = \frac{2\pi}{15}$; about the y-axis: $V = \frac{\pi}{6}$

(b) About the x-axis: $V = \frac{2\pi}{15}$; about the y-axis: $V = \frac{\pi}{6}$

31. **(a)** $\frac{5\pi}{3}$ **(b)** $\frac{4\pi}{3}$ **(c)** 2π **(d)** $\frac{2\pi}{3}$

33. **(a)** $\frac{4\pi}{15}$ **(b)** $\frac{7\pi}{30}$ **35.** **(a)** $\frac{24\pi}{5}$ **(b)** $\frac{48\pi}{5}$

37. **(a)** $\frac{9\pi}{16}$ **(b)** $\frac{9\pi}{16}$

39. Disk: 2 integrals; washer: 2 integrals; shell: 1 integral

41. **(a)** $\frac{256\pi}{3}$ **(b)** $\frac{244\pi}{3}$ **45.** 2

SECTION 6.3, pp. 333–335

1. 12 **3.** $\frac{53}{6}$ **5.** $\frac{123}{32}$

7. $\frac{99}{8}$ **9.** $\frac{53}{6}$ **11.** $\log_e(2+\sqrt{3})$

13. **(a)** $\int_{-1}^{2} \sqrt{1 + 4x^2}\, dx$ **(c)** ≈ 6.13

15. **(a)** $\int_{0}^{\pi} \sqrt{1 + \cos^2 y}\, dy$ **(c)** ≈ 3.82

17. **(a)** $\int_{-1}^{3} \sqrt{1 + (y + 1)^2}\, dy$ **(c)** ≈ 9.29

19. **(a)** $\int_{0}^{\pi/6} \sec x\, dx$ **(c)** ≈ 0.55

21. **(a)** $y = \sqrt{x}$ from (1, 1) to (4, 2)
(b) Only one. We know the derivative of the function and the value of the function at one value of x.

23. 1 **25.** $2\sqrt{5}$ **33.** $\int_{0}^{x} \sqrt{1 + 9t}\, dt,\ \frac{2}{27}(10^{3/2} - 1)$

SECTION 6.4, pp. 338–340

1. **(a)** $2\pi \int_{0}^{\pi/4} (\tan x)\sqrt{1 + \sec^4 x}\, dx$ **(c)** $S \approx 3.84$

(b)

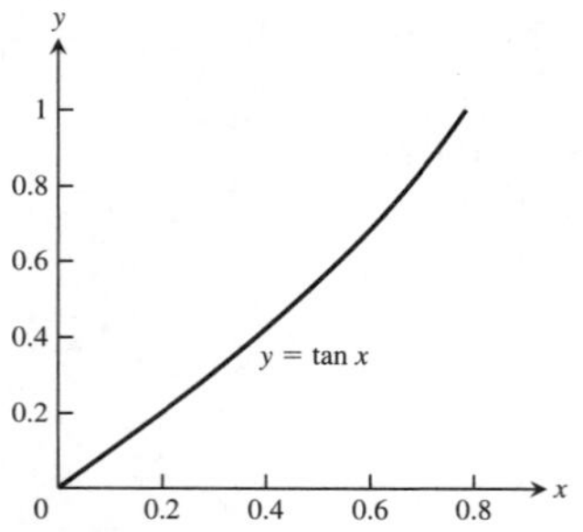

3. **(a)** $2\pi \int_{1}^{2} \frac{1}{y}\sqrt{1 + y^{-4}}\, dy$ **(c)** $S \approx 5.02$

(b)

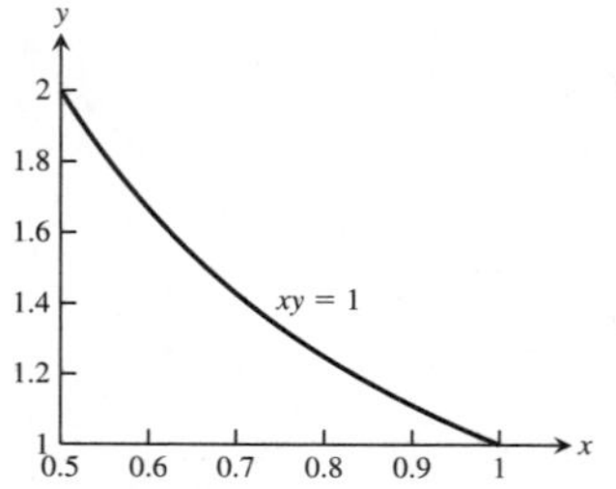

5. **(a)** $2\pi \int_{1}^{4} (3 - x^{1/2})^2 \sqrt{1 + (1 - 3x^{-1/2})^2}\, dx$ **(c)** $S \approx 63.37$

(b)

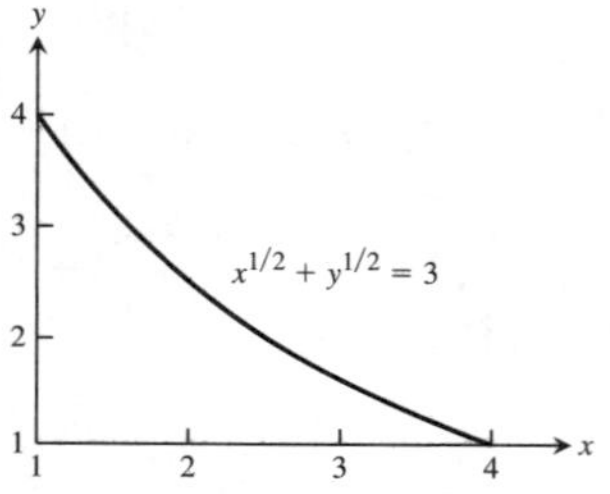

7. **(a)** $2\pi \int_{0}^{\pi/3} \left(\int_{0}^{y} \tan t\, dt\right) \sec y\, dy$ **(c)** $s \approx 2.08$

(b)

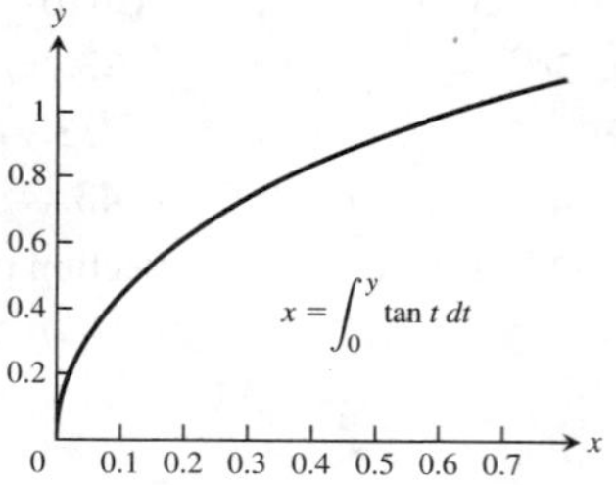

9. $4\pi\sqrt{5}$ **11.** $3\pi\sqrt{5}$ **13.** $98\pi/81$ **15.** 2π

17. $\pi(\sqrt{8} - 1)/9$ **19.** $35\pi\sqrt{5}/3$ **21.** $(2\pi/3)(2\sqrt{2} - 1)$

23. $253\pi/20$ **27.** Order 226.2 liters of each color.

SECTION 6.5, pp. 345–349

1. 116 J **3.** 600 N/m **5.** 4 cm, 0.08 J

7. **(a)** 7238 lb/in. **(b)** 905 in.-lb, 2714 in.-lb

9. 780 J **11.** 110000 ft.lb **13.** 490 J

15. **(a)** 1,497,600 ft-lb **(b)** 1 hr, 40 min
(d) At 62.26 lb/ft^3: a) 1,494,240 ft-lb b) 1 hr, 40 min
At 62.59 lb/ft^3: a) 1,502,160 ft-lb b) 1 hr, 40.1 min

17. 37,306 ft-lb **19.** 5058035.71 ft. lb (approx.)

21. 800 π ft.lb **23.** 15,073,099.75 J **27.** 85.1 ft-lb

29. 151.3 J **31.** 91.32 in.-oz **33.** 5.144×10^{10} J

35. 1684.8 lb **37.** **(a)** 6364.8 lb **(b)** 5990.4 lb

39. (a) 1164.8 lb **(b)** 1194.7 lb **41.** 1309 lb
43. (a) 12,480 lb **(b)** 8580 lb **(c)** 9722.3 lb
45. (a) 93.33 lb **(b)** 3 ft **47.** $\dfrac{wb}{2}$
49. No. The tank will overflow because the movable end will have moved only $3\frac{1}{3}$ ft by the time the tank is full.

SECTION 6.6, pp. 359–361

1. $M = \dfrac{16}{3}$, $\bar{x} = \dfrac{12}{5}$ **3.** $M = 15/2$, $\bar{x} = 9/5$
5. $M = 13$, $\bar{x} = 41/26$ **7.** $\bar{x} = 0$, $\bar{y} = \dfrac{36}{5}$
9. $\bar{x} = 1$, $\bar{y} = -3/5$ **11.** $\bar{x} = 16/105$, $\bar{y} = 8/15$
13. $\bar{x} = 0$, $\bar{y} = \pi/8$ **15. (a)** $(4/\pi, 4/\pi)$ **(b)** $(0, 4/\pi)$
17. $\bar{x} = 341/35$, $\bar{y} = 85/56$
19. $\bar{x} = 5/7$, $\bar{y} = 10/33$. $(\bar{x})^4 < \bar{y}$, so the center of mass is outside the region.
21. $\bar{x} = 3/2$, $\bar{y} = 1/2$
23. (a) $\dfrac{224\pi}{3}$ **(b)** $\bar{x} = 2$, $\bar{y} = 0$
(c)

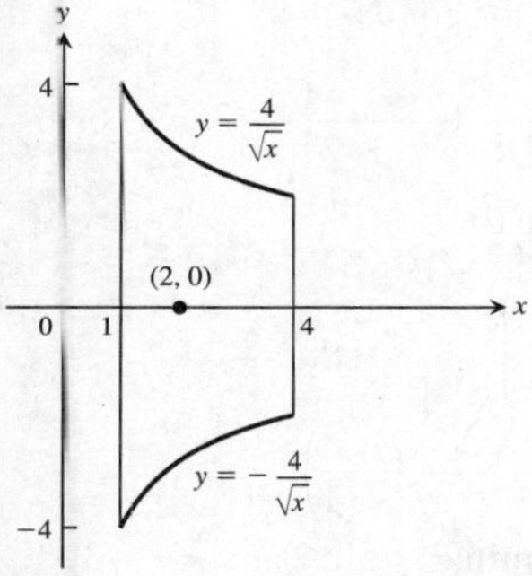

27. $\bar{x} = \bar{y} = 1/3$ **29.** $\bar{x} = a/3$, $\bar{y} = b/3$ **31.** $13\delta/6$
33. $\bar{x} = 0$, $\bar{y} = \dfrac{a\pi}{4}$ **35.** $\bar{x} = \dfrac{1}{2}$, $\bar{y} = \dfrac{8}{5}$
37. $\bar{x} = 6/5$, $\bar{y} = 8/7$ **41.** Volume $= 24\pi$; Surface area $= 48\pi$
43. $24\pi^2$ **45.** $\bar{x} = 0$, $\bar{y} = \dfrac{2a}{\pi}$ **47.** $\bar{x} = 0$, $\bar{y} = \dfrac{4b}{3\pi}$
49. $\sqrt{2}\pi a^3(4 + 3\pi)/6$ **51.** $\bar{x} = \dfrac{a}{3}$, $\bar{y} = \dfrac{b}{3}$

PRACTICE EXERCISES, pp. 362–364

1. $\dfrac{9\pi}{280}$ **3.** π^2 **5.** $\dfrac{72\pi}{35}$
7. (a) 2π **(b)** π **(c)** $12\pi/5$ **(d)** $26\pi/5$
9. (a) 8π **(b)** $1088\pi/15$ **(c)** $512\pi/15$ **11.** $\pi(3\sqrt{3} - \pi)/3$
13. (a) $16\pi/15$ **(b)** $8\pi/5$ **(c)** $8\pi/3$ **(d)** $32\pi/5$
15. $\dfrac{28\pi}{3}$ ft^3 **17.** $(\pi/3)(a^2 + ab + b^2)h$ **19.** $\dfrac{10}{3}$
21. $285/8$ **23.** $28\pi\sqrt{2}/3$ **25.** 4π
27. 4640 J **29.** $\dfrac{w}{2}(2ar - a^2)$ **31.** 418,208.81 ft-lb
33. $22{,}500\pi$ ft-lb, 257 sec **35. (a)** 128 ft-lb **(b)** 219.6 ft-lb
37. $\bar{x} = 0$, $\bar{y} = 8/5$ **39.** $\bar{x} = 3/2$, $\bar{y} = 12/5$
41. $\bar{x} = 9/5$, $\bar{y} = 11/10$ **43.** 332.8 lb **45.** 2196.48 lb

ADDITIONAL AND ADVANCED EXERCISES, pp. 364–365

1. $f(x) = \sqrt{\dfrac{2x - a}{\pi}}$ **3.** $f(x) = \sqrt{C^2 - 1}\,x + a$, where $C \geq 1$

5. $\dfrac{\pi}{30\sqrt{2}}$ **7.** $28/3$ **9.** $\dfrac{4h\sqrt{3mh}}{3}$
11. $\bar{x} = 0$, $\bar{y} = \dfrac{n}{2n + 1}$, $(0, 1/2)$
15. (a) $\bar{x} = \bar{y} = 4(a^2 + ab + b^2)/(3\pi(a + b))$
(b) $(2a/\pi, 2a/\pi)$ **17.** ≈ 2329.6 lb

Chapter 7

SECTION 7.1, pp. 373–377

1. One-to-one **3.** Not one-to-one **5.** One-to-one
7. Not one-to-one **9.** One-to-one
11. $D: (0, 1]$ $R: [0, \infty)$

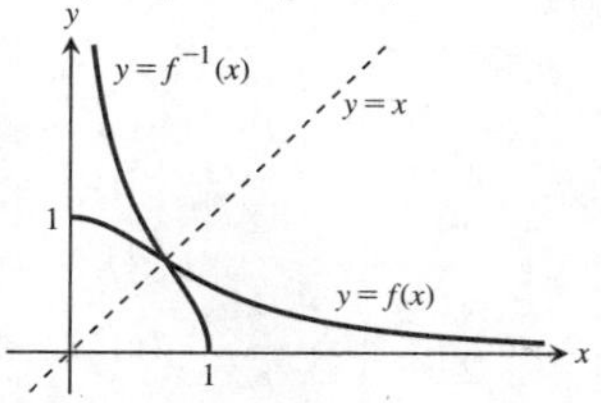

13. $D: [-1, 1]$ $R: [-\pi/2, \pi/2]$

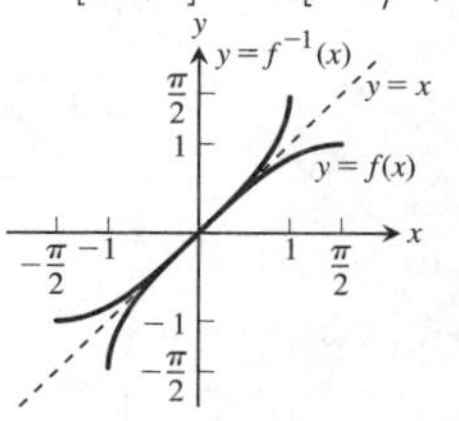

15. $D: [0, 6]$ $R: [0, 3]$

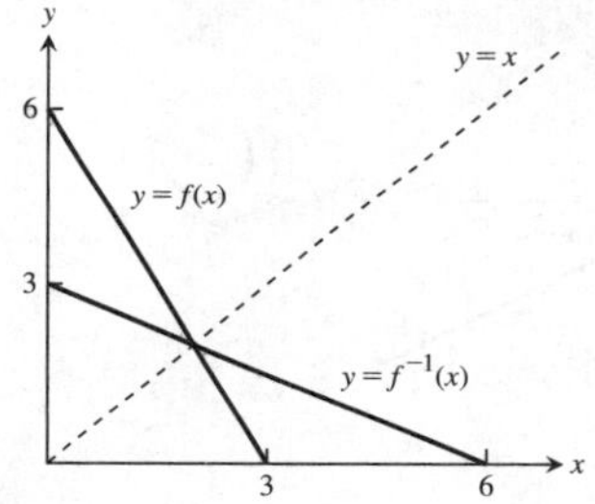

17. Symmetric about the line $y = x$

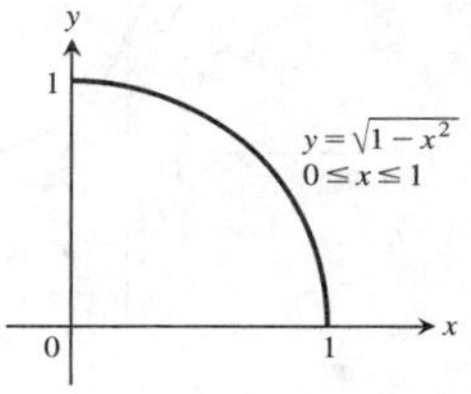

19. $f^{-1}(x) = \sqrt{x - 1}$ **21.** $f^{-1}(x) = \sqrt[3]{x + 1}$
23. $f^{-1}(x) = \sqrt{x} - 1$
25. $f^{-1}(x) = \dfrac{x}{5}$, Domain and Range of f^{-1}: all reals
27. $f^{-1}(x) = \sqrt[3]{\dfrac{x - 5}{3}}$, Domain and Range of f^{-1}: all reals
29. $f^{-1}(x) = \dfrac{1}{\sqrt{x}}$; $D: x > 0$; $R: y > 0$
31. $f^{-1}(x) = \dfrac{3}{1 - 3x}$, Domain of f^{-1}: $x \neq \dfrac{1}{3}$, Range of f^{-1}: $y \neq 0$
33. $f^{-1}(x) = 1 - \sqrt{x + 1}$; $D: -1 \leq x < \infty$; $R: -\infty < y \leq 1$
35. $f^{-1}(x) = \dfrac{2x + b}{x - 1}$;
$D: -\infty < x < \infty, x \neq 1$, $R: -\infty < y < \infty, y \neq 2$
37. (a) $f^{-1}(x) = \dfrac{1}{m}x$
(b) The graph of f^{-1} is the line through the origin with slope $1/m$.
39. (a) $f^{-1}(x) = x - 1$

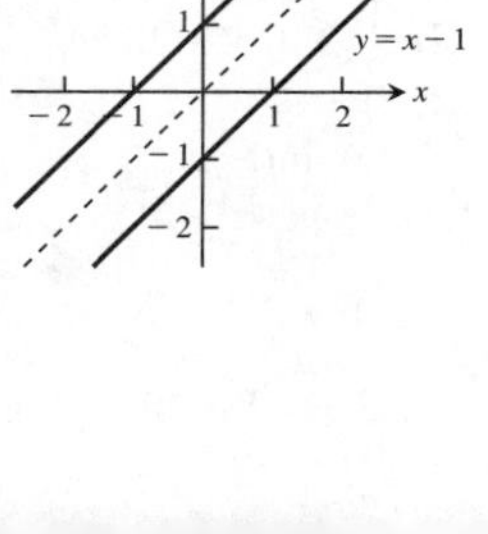

(b) $f^{-1}(x) = x - b$. The graph of f^{-1} is a line parallel to the graph of f. The graphs of f and f^{-1} lie on opposite sides of the line $y = x$ and are equidistant from that line.
(c) Their graphs will be parallel to one another and lie on opposite sides of the line $y = x$ equidistant from that line.

41. **(a)** $f^{-1}(x) = x - 2$

(b)

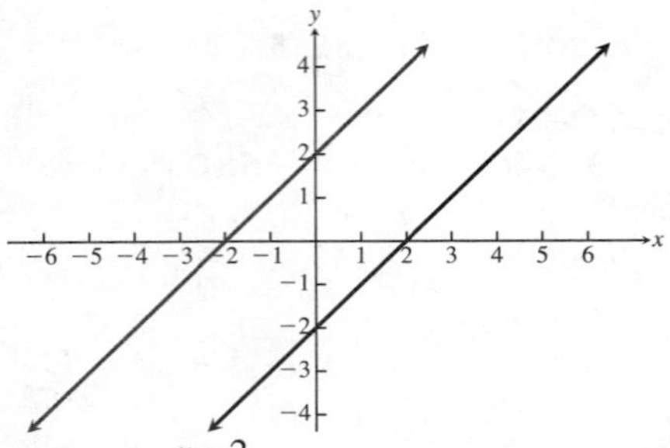

(c) 1, 1

43. **(a)** $f^{-1}(x) = -\dfrac{x-2}{2}$

(b)

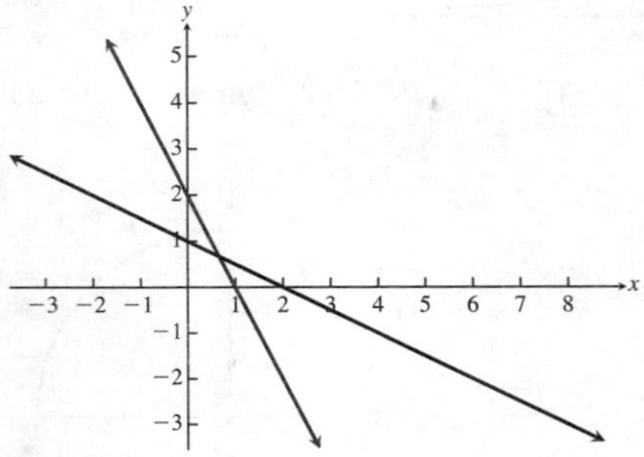

(c) 2, −1/2

45. **(a)** 2 **(b)** 1/3 **(c)** 3

47. **(a)** 2 **(b)** 7 **(c)** 3/4

49. **(b)**

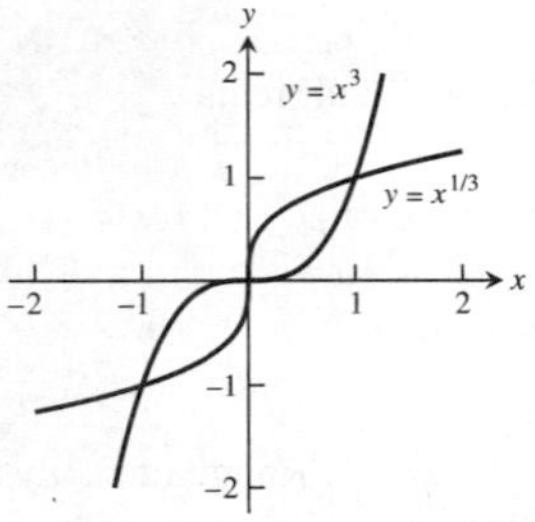

(c) Slope of f at (1, 1): 3; slope of g at (1, 1): 1/3; slope of f at (−1, −1): 3; slope of g at (−1, −1): 1/3
(d) $y = 0$ is tangent to $y = x^3$ at $x = 0$; $x = 0$ is tangent to $y = \sqrt[3]{x}$ at $x = 0$.

51. 1/9 **53.** 3

57. Increasing, therefore one-to-one; $df^{-1}/dx = \frac{1}{9}x^{-2/3}$

59. Decreasing, therefore one-to-one; $df^{-1}/dx = -\frac{1}{3}x^{-2/3}$

SECTION 7.2, pp. 384–385

1. **(a)** $\ln 3 - \ln 2$ **(b)** $2(\ln 2 - \ln 3)$ **(c)** $-\ln 2$ **(d)** $\frac{2}{3}\ln 3$ **(e)** $\ln 3 + \frac{1}{2}\ln 2$ **(f)** $\frac{1}{2}(3\ln 3 - \ln 2)$

3. **(a)** $\ln 5$ **(b)** $\frac{1}{2}\ln(3x-5) - \ln 7$ **(c)** $\ln(t^2)$

5. $e^{8/3}$ **7.** $\frac{1}{x}$ **9.** $\frac{2}{t}$ **11.** $-1/x$

13. $\dfrac{1}{\theta + 1}$ **15.** $\dfrac{3}{x}$ **17.** $2(\ln t) + (\ln t)^2$ **19.** $x^3 \ln x$

21. $\dfrac{1 - \ln t}{t^2}$ **23.** $\dfrac{1}{x(1 + \ln x)^2}$ **25.** $\dfrac{2}{x \ln 2x}$

27. $2\cos(\ln\theta)$ **29.** $-\dfrac{3x + 2}{2x(x + 1)}$ **31.** $\dfrac{10}{t(5 - \ln t)^2}$

33. $\dfrac{\tan(\ln\theta)}{\theta}$ **35.** $\dfrac{-3x^2 + 12x + 3}{2(x^2 + 3)(3 - x)}$

37. $2x\ln|x| - x\ln\dfrac{|x|}{\sqrt{2}}$ **39.** $\dfrac{1}{2}\ln\left(\dfrac{2}{3}\right)$ **41.** $\ln|y^2 - 25| + C$

43. $\ln 3$ **45.** $\dfrac{(\ln 2)^2}{2}$ **47.** $\dfrac{1}{\ln 4}$ **49.** $\ln|6 + 3\tan t| + C$

51. $\ln 8 - \dfrac{3\ln 3}{2}$ **53.** $\ln 27$ **55.** $\dfrac{2}{5}\ln|\sqrt{x} + 1| + C$

57. $-\dfrac{x}{\sqrt{25 - x^2}}$ **59.** $\left(\dfrac{1}{2}\right)\sqrt{\dfrac{t}{t+1}}\left(\dfrac{1}{t} - \dfrac{1}{t+1}\right) = \dfrac{1}{2\sqrt{t}(t+1)^{3/2}}$

61. $\sqrt{\theta + 3}(\sin\theta)\left(\dfrac{1}{2(\theta + 3)} + \cot\theta\right)$

63. $t(t+1)(t+2)\left[\dfrac{1}{t} + \dfrac{1}{t+1} + \dfrac{1}{t+2}\right] = 3t^2 + 6t + 2$

65. $\dfrac{\theta + 5}{\theta\cos\theta}\left[\dfrac{1}{\theta + 5} - \dfrac{1}{\theta} + \tan\theta\right]$

67. $\sqrt{\dfrac{x-1}{x^4+1}}\left[\dfrac{1}{2(x-1)} - \dfrac{2x^3}{x^4+1}\right]$

69. $\dfrac{1}{3}\sqrt[3]{\dfrac{x(x-2)}{x^2+1}}\left(\dfrac{1}{x} + \dfrac{1}{x-2} - \dfrac{2x}{x^2+1}\right)$

71. **(a)** Max = 0 at $x = 0$, min = $-\ln 2$ at $x = \pi/3$
(b) Max = 1 at $x = 1$, min = $\cos(\ln 2)$ at $x = 1/2$ and $x = 2$

73. $\ln 27$

75. **(a)** Increasing on $(0, e^{-2})$ and $(1, \infty)$; decreasing on $(e^{-2}, 1)$
(b) local maximum is $4/e^2$ at $x = e^{-2}$; absolute minimum is 0 at $x = 1$; no absolute maximum

77. $\pi\ln 81$ **79.** $\pi\ln 16$ **81.** **(a)** $6 + \ln 2$ **(b)** $8 + \ln 9$

83. **(a)** $\bar{x} \approx 1.44$, $\bar{y} \approx 0.36$

(b)

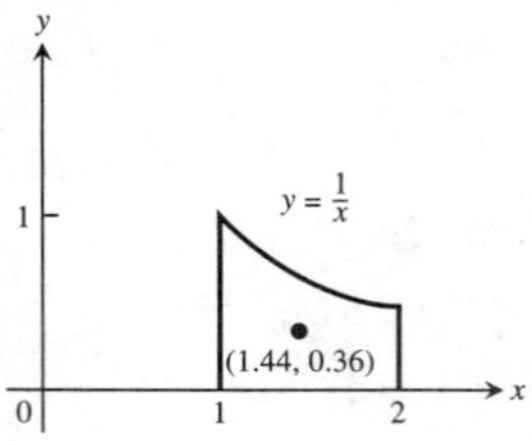

87. $y = 2x + \ln|x| + 1$

89. **(b)** 0.00469

SECTION 7.3, pp. 396–399

1. **(a)** $-10\ln 4$ **(b)** $t = -\dfrac{\ln 2}{k}$ **(c)** $t = \dfrac{\ln .4}{\ln .2}$

3. $4(\ln x)^2$ **5.** $\ln 3$ **7.** $-2e^{-2x}$ **9.** $-3e^{1-3x}$

11. $16xe^{4x}$ **13.** x^2e^x **15.** $2e^\theta\cos\theta$ **17.** $-2\theta e^{-\theta^2}\cos(e^{-\theta^2})$

19. $\dfrac{1-t}{t}$ **21.** $1/(1 + e^\theta)$ **23.** $e^{\sin t}(t\cos t + 1)$

25. $(\sin x)/x$ **27.** $\dfrac{ye^y\cos x}{1 - ye^y\sin x}$ **29.** $\dfrac{2e^{2x} - \cos(x + 3y)}{3\cos(x + 3y)}$

31. $y' = \dfrac{3x^2}{1 - \cos y}$, $y'' = \dfrac{6x(1 - \cos y)^2 - 9x^4\sin y}{(1 - \cos y)^3}$

33. $\frac{e^{2x}}{2} - 4e^{-x} + C$ **35.** 1 **37.** $\frac{5}{2}e^{2x+1} + C$ **39.** 2

41. $2e^{\sqrt{r}} + C$ **43.** $-e^{-t^2} + C$ **45.** $-e^{1/x} + C$ **47.** e

49. $\frac{1}{\pi}e^{\sec \pi t} + C$ **51.** 1 **53.** $2 \ln |5 + e^r| + C$

55. $y = 1 - \cos(e^t - 2)$ **57.** $y = -3e^{-x} + 3x - 2$

59. $\ln 2\,(2^{2x+1})$ **61.** $\left(\frac{\ln 5}{2\sqrt{s}}\right)5^{\sqrt{s}}$ **63.** $2\pi x^{2\pi - 1}$

65. $-\sqrt{2}\cos \theta^{(\sqrt{2}-1)} \sin \theta$ **67.** $7^{\sec \theta}(\ln 7)^2(\sec \theta \tan \theta)$

69. $5 \ln 3\,(3^{\sin 5t}) \cos 5t$ **71.** $\frac{1}{\theta \ln 2}$ **73.** $\frac{3}{x \ln 4}$

75. $\frac{x + 2x \ln 2x}{\ln 10}$ **77.** $\frac{-2}{(x+1)(x-1)}$

79. $2\left(\cos(\log_7 2\theta) - \frac{\sin(\log_7 2\theta)}{\ln 7}\right)$ **81.** $\frac{1}{\ln 10}$

83. $\frac{\ln 3(3^{\log_2 x})}{x \ln 2}$ **85.** $\frac{\ln 2}{t \ln 3}$ **87.** $\frac{2^x}{\ln 2} + C$ **89.** $\frac{1}{2 \ln 2}$

91. $\frac{1}{\ln 2}$ **93.** $\frac{6}{\ln 7}$ **95.** 32760 **97.** $\frac{3x^{(\sqrt{3}+1)}}{\sqrt{3}+1} + C$

99. $3^{\sqrt{2}+1}$ **101.** $\frac{\ln x}{\ln 10} + \frac{(1 - \ln x)^2}{2 \ln 10} + C$ **103.** $2(\ln 2)^2$

105. $\frac{(\ln 7)^2 - (\ln 5)^2}{\ln 4}$ **107.** $\ln 10$ **109.** $(\ln 10)\ln|\ln x| + C$

111. $\ln(\ln x),\ x > 1$ **113.** $-\ln x$

115. $(2x+1)x^2\left(2x \ln(2x+1) + \frac{2x^2}{2x+1}\right)$ **117.** $(\sqrt{t})^t\left(\frac{\ln t}{2} + \frac{1}{2}\right)$

119. $(\cos x)^x(\ln(\cos x) - x \tan x)$ **121.** $-x^x \sin x^x (1 + \ln x)$

123. $\frac{3y - xy \ln y}{x^2 - x}$ **125.** $\frac{1 - xy \ln y}{x^2(1 + \ln y)}$ **127.** $(1 + \ln t)^2$

129. Maximum: 1 at $x = 0$, minimum: $2 - 2 \ln 2$ at $x = \ln 2$

131. **(a)** Abs max: $\frac{1}{e}$ at $x = 1$ **(b)** $\left(2, \frac{2}{e^2}\right)$

133. Abs max of $1/(2e)$ assumed at $x = 1/\sqrt{e}$

135. $\frac{4}{3}$ **137.** $y = e^{x/2} - 1$ **139.** $\frac{e^2 - 1}{2e}$

141. $\ln(\sqrt{2} + 1)$

143. **(a)** $\frac{d}{dx}(x \ln x - x + C) = x \cdot \frac{1}{x} + \ln x - 1 + 0 = \ln x$

(b) $\frac{1}{e - 1}$

145. **(b)** $|\text{error}| \approx 0.02140$

(c) $L(x) = x + 1$ never overestimates e^x.

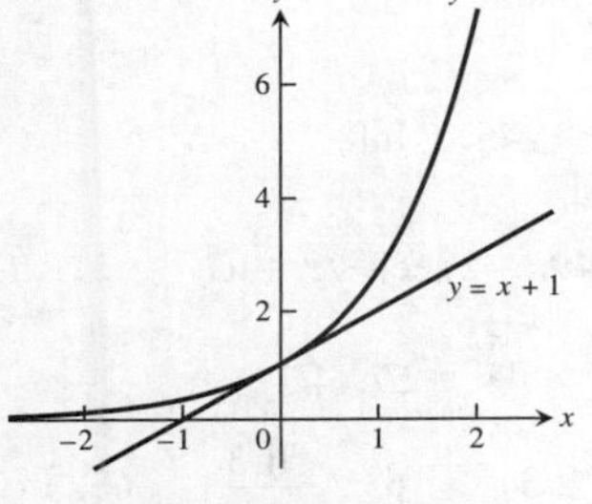

147. $2 \ln 5$ **149.** **(a)** $4 + \ln 2$ **(b)** $\left(\frac{1}{2}\right)(4 + \ln 2)$

151. $x \approx -0.76666$

153. **(a)** $L(x) = 1 + (\ln 2)x \approx 0.69x + 1$

SECTION 7.4, pp. 408–411

9. $\frac{2}{3}y^{3/2} - x^{1/2} = C$ **11.** $e^y - e^x = C$

13. $-x + 2 \tan \sqrt{y} = C$ **15.** $e^{-y} + 2e^{\sqrt{x}} = C$

17. $-(2\sqrt{1-y} - x) = C$ **19.** $\frac{1}{3}\ln|y^3 - 2| = x^3 + C$

21. $4 \ln(\sqrt{y} + 2) = e^{x^2} + C$

23. **(a)** -0.00001 **(b)** 10,536 years **(c)** 82%

25. 81 rabbits **27.** 59.8 ft **29.** 2.8147498×10^{14}

31. **(a)** 8 years **(b)** 32.02 years **33.** Yes, $y(20) < 1$

35. 15.28 years **37.** 56,562 years

41. **(a)** 17.5 minutes longer to reach 40°C

(b) 13.26 minutes (approx.)

43. -3°C **45.** About 6693 years **47.** 54.62%

49. $\approx$ 15,683 years

SECTION 7.5, pp. 418–420

1. $-1/4$ **3.** $5/7$ **5.** $1/2$ **7.** $1/4$ **9.** $-23/7$

11. $5/7$ **13.** 0 **15.** -16 **17.** -2 **19.** $\sqrt{3}/2$

21. $1/4$ **23.** 2 **25.** 3 **27.** -1 **29.** $\ln 3$ **31.** $\frac{1}{\ln 2}$

33. $\ln 2$ **35.** 1 **37.** $1/2$ **39.** $\ln 2$ **41.** $-\infty$

43. $-1/2$ **45.** -1 **47.** 1 **49.** 0 **51.** 2 **53.** $1/e$

55. $e^{1/3}$ **57.** $1/e$ **59.** $e^{1/2}$ **61.** 1 **63.** $(abcd)^{1/4}$

65. 0 **67.** 1 **69.** 3 **71.** 1 **73.** 0 **75.** ∞

77. (b) is correct. **79.** (d) is correct. **81.** $c = \frac{27}{10}$

83. **(b)** $\frac{-1}{2}$ **85.** -1 **89.** **(a)** $y = 1$ **(b)** $y = 0,\ y = \frac{3}{2}$

91. **(a)** We should assign the value 1 to $f(x) = (\sin x)^x$ to make it continuous at $x = 0$.

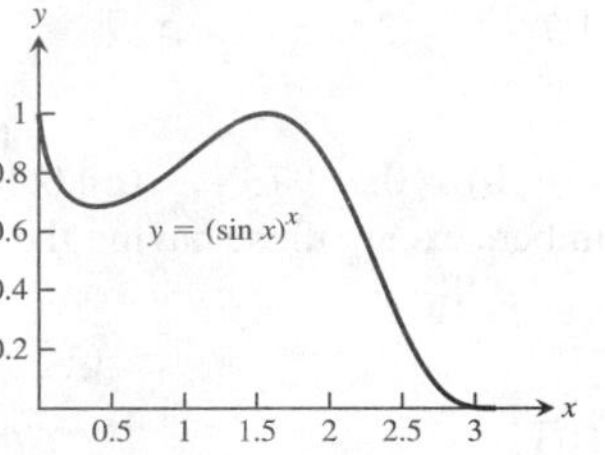

(c) The maximum value of $f(x)$ is close to 1 near the point $x \approx 1.55$ (see the graph in part (a)).

SECTION 7.6, pp. 432–435

1. **(a)** $\pi/4$ **(b)** $-\pi/3$ **(c)** $\pi/6$

3. **(a)** $-\pi/6$ **(b)** $\frac{\pi}{3}$ **(c)** $-\pi/3$

5. **(a)** $\pi/3$ **(b)** $3\pi/4$ **(c)** $\frac{5\pi}{6}$

7. **(a)** $3\pi/4$ **(b)** $\pi/6$ **(c)** $2\pi/3$

9. $1/\sqrt{2}$ **11.** $-\sqrt{3}$ **13.** 0 **15.** $\pi/2$ **17.** $\pi/2$

19. 0 **21.** $-\frac{1}{\sqrt{1-x^2}}$ **23.** $\frac{\sqrt{2}}{\sqrt{1 - 2t^2}}$ **25.** $\frac{1}{|2s+1|\sqrt{s^2 + s}}$

27. $\frac{-2x}{(x^2+1)\sqrt{x^4 + 2x^2}}$ **29.** $\frac{-1}{\sqrt{1 - t^2}}$ **31.** $-\frac{1}{\sqrt{2t}(2t+1)}$

33. $\frac{1}{(\arctan x)(1 + x^2)}$ **35.** $\frac{1}{\sqrt{e^{2t} - 1}}$ **37.** $\frac{-2s^2}{\sqrt{1 - s^2}}$

39. 0 **41.** $\arcsin x$ **43.** $\dfrac{1}{2\sqrt{\arcsin x}\,\sqrt{1-x^2}}$

45. $-\sin(x - \arccos x)\left(1 + \dfrac{1}{\sqrt{1-x^2}}\right)$

47. $\dfrac{-9x^2}{x^6+1}(\operatorname{arccot}(x^3))^2$ **49.** 0 **51.** $\dfrac{8\sqrt{2}}{4+3\pi}$

53. $\sin^{-1}\left(\dfrac{1}{2}x\right)+C$ **55.** $\dfrac{1}{\sqrt{17}}\arctan\dfrac{x}{\sqrt{17}}+C$

57. $\dfrac{1}{\sqrt{2}}\operatorname{arcsec}\left|\dfrac{5x}{\sqrt{2}}\right|+C$ **59.** $2\pi/3$ **61.** $\dfrac{\pi}{8}$ **63.** $-\pi/12$

65. $\sin^{-1}\left(\dfrac{1}{\sqrt{2}}(r-1)\right)+C$ **67.** $\dfrac{1}{\sqrt{5}}\tan^{-1}\left(\dfrac{x-2}{\sqrt{5}}\right)+C$

69. $\dfrac{1}{4}\operatorname{arcsec}\left|\dfrac{2x-1}{2}\right|+C$ **71.** π **73.** $\pi/12$

75. $\dfrac{1}{2}\sin^{-1}\left(\dfrac{1}{2}y^2\right)+C$ **77.** $\sin^{-1}(x-3)+C$ **79.** π

81. $\dfrac{1}{2}\arctan\left(\dfrac{y-1}{2}\right)+C$ **83.** $\dfrac{5\pi}{4}$

85. $\dfrac{1}{2}\ln(x^2+4)+2\arctan\dfrac{x}{2}+C$

87. $x+\ln(x^2+9)-\dfrac{10}{3}\arctan\dfrac{x}{3}+C$

89. $\operatorname{arcsec}|x+1|+C$ **91.** $e^{\arcsin x}+C$

93. $-\dfrac{1}{3}(\cos^{-1}x)^3+C$ **95.** $\ln|\arctan y|+C$ **97.** $\sqrt{3}-1$

99. $-\dfrac{1}{2}(\cos^{-1}e^x)^2+C$ **101.** $\pi^2/32$ **103.** $\dfrac{\pi}{10}$

105. 3 **107.** 1 **109.** 1 **115.** $y = \arcsin x$

117. $y = \operatorname{arcsec} x + \dfrac{2\pi}{3},\ x > 1$ **119.** **(b)** $x = 3\sqrt{5}$

121. $\theta = \arccos\left(\dfrac{1}{\sqrt{3}}\right) \approx 54.7°$ **133.** $\pi^2/2$

135. **(a)** $\pi^2/2$ **(b)** 2π

137. **(a)** 1.15928 (approx.) **(b)** -0.72973 **(c)** 0.46365

139. **(a)** Domain: all real numbers except those having the form $\dfrac{\pi}{2}+k\pi$, where k is an integer

Range: $-\dfrac{\pi}{2} < y < \dfrac{\pi}{2}$

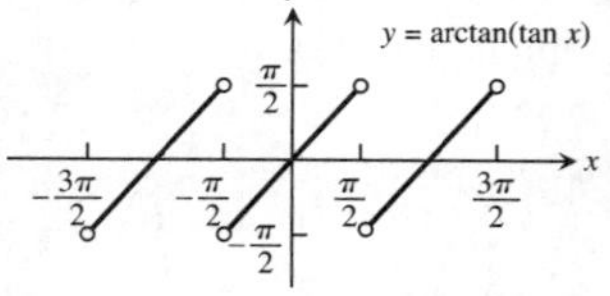

(b) Domain: $-\infty < x < \infty$; Range: $-\infty < y < \infty$

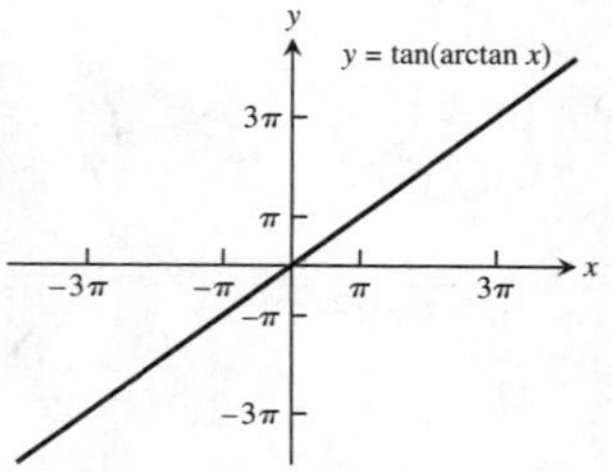

141. **(a)** Domain: $-\infty < x < \infty$;
Range: $0 \le y \le \pi$

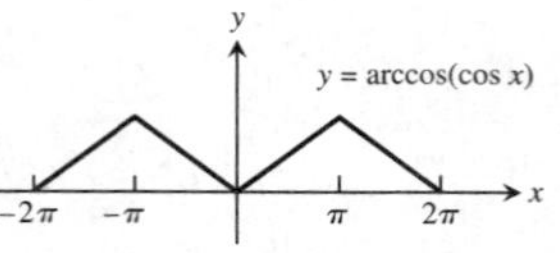

(b) Domain: $-1 \le x \le 1$;
Range: $-1 \le y \le 1$

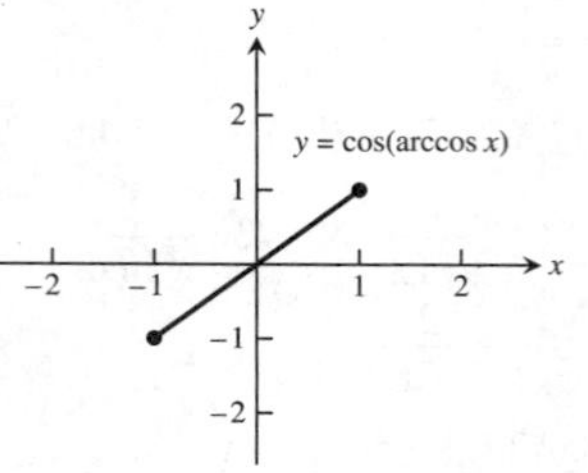

143. The graphs are identical. **145.**

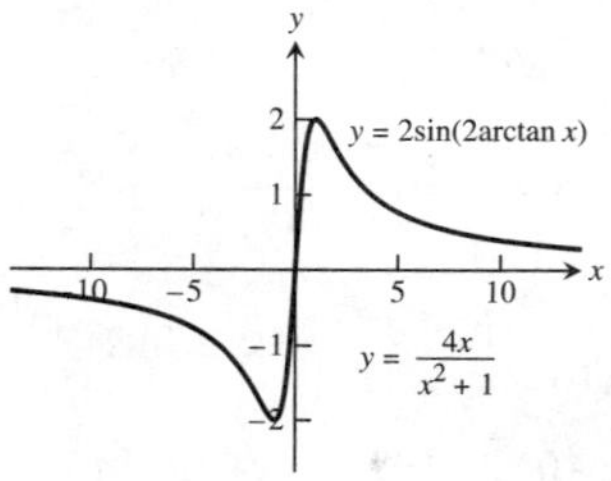

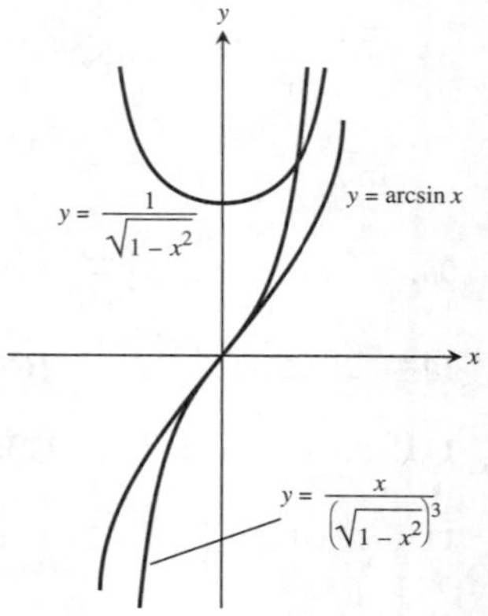

SECTION 7.7, pp. 442–445

1. $\cosh x = 5/4$, $\tanh x = -3/5$, $\coth x = -5/3$, $\operatorname{sech} x = 4/5$, $\operatorname{csch} x = -4/3$

3. $\sinh x = 8/15$, $\tanh x = 8/17$, $\coth x = 17/8$, $\operatorname{sech} x = 15/17$, $\operatorname{csch} x = 15/8$

5. $3/2\,(x+1/x)$ **7.** e^{7x} **9.** e^{8x} **13.** $2\cosh x/2$

15. $\dfrac{\operatorname{sech}^2\sqrt{t}}{2}+\dfrac{\tanh\sqrt{t}}{2\sqrt{t}}$ **17.** $-\coth z$

19. $-\operatorname{sech}^2\theta\,(\ln\tanh\theta)$ **21.** $\tanh^3 v$ **23.** 2

25. $\dfrac{1}{\sqrt{2x(2x+1)}}$ **27.** $\dfrac{2}{2\theta+1}-2\tanh^{-1}2\theta$

29. $\dfrac{1}{\sqrt{2t}}-2\coth^{-1}\sqrt{2t}$ **31.** $\dfrac{2}{\sqrt{1-x^2}}-\operatorname{sech}^{-1}x$

33. $\dfrac{\ln 2}{\sqrt{1+\left(\frac{1}{2}\right)^{2\theta}}}$ **35.** $\dfrac{2\sec^2 x}{\sqrt{4\tan^2 x+1}}$ **41.** $\dfrac{\cosh 3x}{3}+C$

43. $6\sinh\left(\dfrac{x}{2}-2\ln 2\right)+C$ **45.** $7\ln|e^{x/7}+e^{-x/7}|+C$

47. $\tanh\left(x-\dfrac{1}{3}\right)+C$ **49.** $-2\operatorname{sech}\sqrt{t}+C$ **51.** $\ln(10/3)$

53. $\dfrac{3}{32}+\ln 2$ **55.** $e-e^{-1}$ **57.** $4/3$

59. $\dfrac{3}{8}+\ln\sqrt{2}$ **61.** $\ln(2/3)$ **63.** $\dfrac{-\ln 3}{2}$ **65.** $\ln 2$

67. **(a)** $\sinh^{-1}(\sqrt{3})$ **(b)** $\ln(\sqrt{3}+2)$

69. **(a)** $\coth^{-1}(2) - \coth^{-1}(5/4)$ **(b)** $\left(\frac{1}{2}\right)\ln\left(\frac{1}{3}\right)$

71. **(a)** $-\text{sech}^{-1}\left(\frac{12}{13}\right) + \text{sech}^{-1}\left(\frac{4}{5}\right)$

(b) $-\ln\left(\frac{1 + \sqrt{1 - (12/13)^2}}{(12/13)}\right) + \ln\left(\frac{1 + \sqrt{1 - (4/5)^2}}{(4/5)}\right)$

$= -\ln\left(\frac{3}{2}\right) + \ln(2) = \ln(4/3)$

73. **(a)** 0 **(b)** 0

77. **(b)** $\sqrt{\frac{mg}{k}}$ **(c)** $80\sqrt{5} \approx 178.89$ ft/sec

79. 4π **81.** $\frac{6}{5}$

SECTION 7.8, pp. 449–450

1. **(a)** Slower **(b)** Slower **(c)** Slower **(d)** Faster
(e) Slower **(f)** Slower **(g)** Same **(h)** Slower

3. **(a)** Same **(b)** Faster **(c)** Same **(d)** Same
(e) Slower **(f)** Faster **(g)** Slower **(h)** Same

5. **(a)** Same **(b)** Same **(c)** Same **(d)** Faster
(e) Faster **(f)** Same **(g)** Slower **(h)** Faster

7. d, a, c, b

9. **(a)** False **(b)** False **(c)** True **(d)** True
(e) True **(f)** True **(g)** False **(h)** True

13. When the degree of f is less than or equal to the degree of g.

15. 1, 1

21. **(b)** $\ln(e^{17000000}) = 17{,}000{,}000 < (e^{17\times10^6})^{1/10^6}$
$= e^{17} \approx 24{,}154{,}952.75$

(c) $x \approx 3.4306311 \times 10^{15}$

(d) They cross at $x \approx 3.4306311 \times 10^{15}$.

23. **(a)** The algorithm that takes $O(n \log_2 n)$ steps

(b)

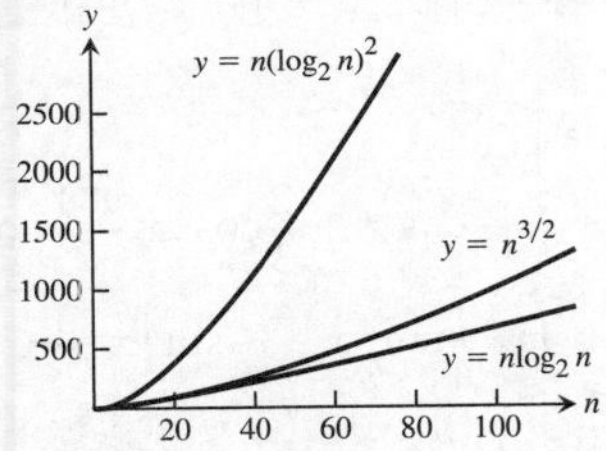

25. It could take one million for a sequential search; at most 20 steps for a binary search.

PRACTICE EXERCISES, pp. 451–454

1. $-2e^{-x/5}$ **3.** xe^{4x} **5.** $\frac{2\sin\theta\cos\theta}{\sin^2\theta} = 2\cot\theta$

7. $\frac{2}{(\ln 2)x}$ **9.** $-8^{-t}(\ln 8)$ **11.** $18x^{2.6}$

13. $(x + 2)^{x+2}(\ln(x + 2) + 1)$ **15.** $-\frac{1}{\sqrt{1 - u^2}}$

17. $\frac{-1}{\sqrt{1 - x^2}\arccos x}$ **19.** $\arctan(t) + \frac{t}{1 + t^2} - \frac{1}{2t}$

21. $\frac{1 - z}{\sqrt{z^2 - 1}} + \operatorname{arcsec} z$ **23.** -1

25. $\frac{2(x^2 + 1)}{\sqrt{\cos 2x}}\left[\frac{2x}{x^2 + 1} + \tan 2x\right]$

27. $5\left[\frac{(t + 1)(t - 1)}{(t - 2)(t + 3)}\right]^5\left[\frac{1}{t + 1} + \frac{1}{t - 1} - \frac{1}{t - 2} - \frac{1}{t + 3}\right]$

29. $\frac{1}{\sqrt{\theta}}(\sin\theta)^{\sqrt{\theta}}\left(\frac{\ln\sqrt{\sin\theta}}{2} + \theta\cot\theta\right)$ **31.** $-\cos e^x + C$

33. $\tan(e^x - 7) + C$ **35.** $e^{\tan x} + C$ **37.** $\frac{-\ln 7}{3}$

39. $\ln 8$ **41.** $\ln(9/25)$ **43.** $-[\ln|\cos(\ln v)|] + C$

45. $-\frac{1}{2}(\ln x)^{-2} + C$ **47.** $-\cot(1 + \ln r) + C$

49. $\frac{1}{2\ln 3}(3^{x^2}) + C$ **51.** $3\ln 7$ **53.** $15/16 + \ln 2$

55. $e - 1$ **57.** $1/6$ **59.** $9/14$

61. $\frac{1}{3}[(\ln 4)^3 - (\ln 2)^3]$ or $\frac{7}{3}(\ln 2)^3$ **63.** $\frac{9\ln 2}{4}$ **65.** π

67. $\pi/\sqrt{3}$ **69.** $\sec^{-1}|2y| + C$ **71.** $\pi/12$

73. $\sin^{-1}(x + 1) + C$ **75.** $\pi/2$ **77.** $\frac{1}{3}\sec^{-1}\left(\frac{t + 1}{3}\right) + C$

79. $y = \frac{\ln 2}{\ln(3/2)}$ **81.** $y = \ln x - \ln 3$ **83.** $y = \frac{1}{1 - e^x}$

85. 5 **87.** 0 **89.** 1 **91.** 3/7 **93.** 0 **95.** 1

97. ln 10 **99.** ln 2 **101.** 5 **103.** $-\infty$ **105.** 1

107. 1 **109.** **(a)** Same rate **(b)** Same rate **(c)** Faster
(d) Faster **(e)** Same rate **(f)** Same rate

111. **(a)** True **(b)** False **(c)** False **(d)** True
(e) True **(f)** True

113. 1/3

115. Absolute maximum $= 0$ at $x = e/2$, absolute minimum $= -0.5$ at $x = 0.5$

117. 1

119. $1/e$ m/sec

121. $1/\sqrt{2}$ units long by $1/\sqrt{e}$ units high, $A = 1/\sqrt{2e} \approx 0.43$ units^2

123. **(a)** Absolute maximum of $2/e$ at $x = e^2$; inflection point $(e^{8/3}, (8/3)e^{-4/3})$; concave up on $(e^{8/3}, \infty)$; concave down on $(0, e^{8/3})$

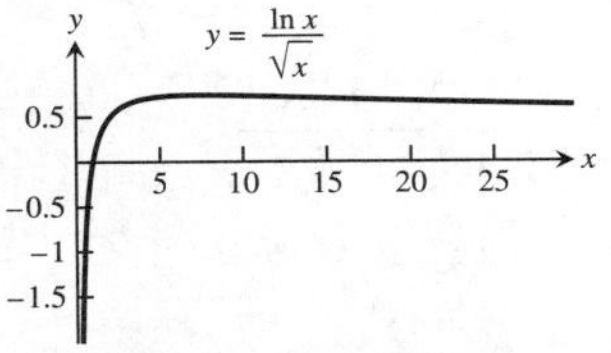

(b) Absolute maximum of 1 at $x = 0$; inflection points $(\pm 1/\sqrt{2}, 1/\sqrt{e})$; concave up on $(-\infty, -1/\sqrt{2}) \cup (1/\sqrt{2}, \infty)$; concave down on $(-1/\sqrt{2}, 1/\sqrt{2})$

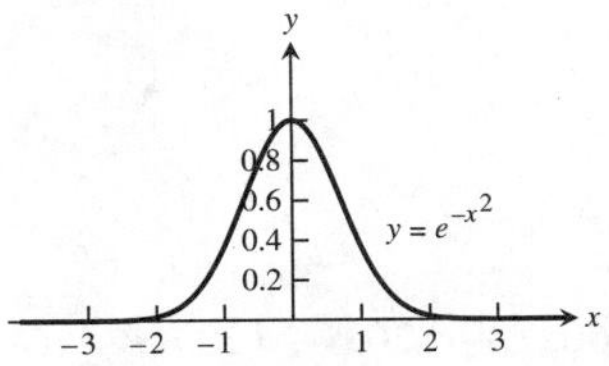

(c) Absolute maximum of 1 at $x = 0$; inflection point $(1, 2/e)$; concave up on $(1, \infty)$; concave down on $(-\infty, 1)$

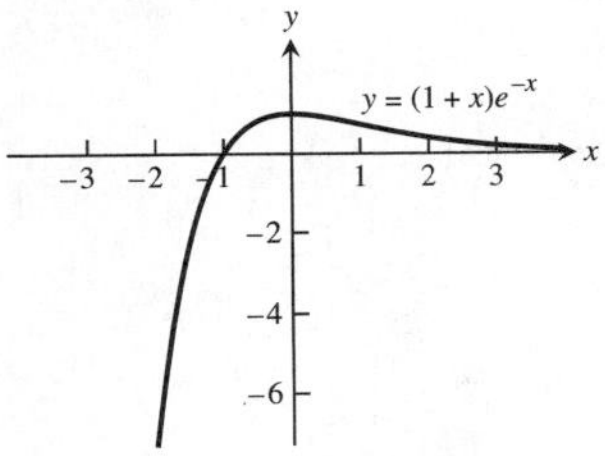

125. $y = \left(\tan^{-1}\left(\frac{x + C}{2}\right)\right)^2$ **127.** $y^2 = \sin^{-1}(2\tan x + C)$

129. $y = -2 + \ln(2 - e^{-x})$ **131.** $y = 4x - 4\sqrt{x} + 1$
133. 18,935 years **135.** $20(5 - \sqrt{17})$ m

ADDITIONAL AND ADVANCED EXERCISES, pp. 455–456

1. $\pi/2$ **3.** $1/\sqrt{e}$ **5.** $\ln 2$
7. (a) 1 **(b)** $\pi/2$ **(c)** π
9. $y' = \dfrac{xy^{-1}y^2 - ye^x(x^y + 1)\ln y}{e^x(x^y + 1) - x^y y \ln x}$ **11.** $\dfrac{1}{\ln 2}, \dfrac{1}{2\ln 2}, 2{:}1$
13. $x = 2$ **15.** $2/17$ **19.** $\bar{x} = \dfrac{\ln 4}{\pi}, \bar{y} = 0$
21. (b) $61°$

Chapter 8

SECTION 8.1, pp. 461–462

1. $\ln 5$ **3.** $2\tan x - 2\sec x - x + C$
5. $\arcsin x + \sqrt{1 - x^2} + C$ **7.** $e^{-\cot z} + C$
9. $\arctan(e^z) + C$ **11.** π **13.** $t + \cot t + \csc t + C$
15. $\sqrt{2}$ **17.** $\frac{1}{8}\ln(1 + 4\ln^2 y) + C$
19. $\ln|1 + \sin\theta| + C$ **21.** $2t^2 - t + 2\arctan\left(\frac{t}{2}\right) + C$
23. $2(\sqrt{2} - 1) \approx 0.82843$ **25.** $\operatorname{arcsec}(e^y) + C$
27. $\arcsin(2\ln x) + C$ **29.** $\ln|\sin x| + \ln|\cos x| + C$
31. $7 + \ln 8$ **33.** $\left[\arcsin y - \sqrt{1 - y^2}\right]_{-1}^{0} = \frac{\pi}{2} - 1$
35. $\operatorname{arcsec}\left|\frac{x - 1}{7}\right| + C$ **37.** $\frac{\theta^3}{3} - \frac{\theta^2}{2} + \theta + \frac{5}{4}\ln|4\theta - 5| + C$
39. $x + \ln(1 - e^{-x}) + C$ **41.** $(1/2)e^{2x} - e^x + \ln(1 + e^x) + C$
43. $2\arctan(\sqrt{x}) + C$ **45.** $2\sqrt{2} - \ln(3 + 2\sqrt{2})$
47. $\frac{1}{2}\ln 3$ **49.** $\bar{x} = 0, \bar{y} = \dfrac{1}{\ln(3 + 2\sqrt{2})}$
51. $xe^{x^3} + C$ **53.** $\frac{1}{30}(x^4 + 1)^{3/2}(3x^4 - 2) + C$

SECTION 8.2, pp. 467–469

1. $-2x\cos(x/2) + 4\sin(x/2) + C$
3. $t^2\sin t + 2t\cos t - 2\sin t + C$
5. $\ln 4 - \frac{3}{4}$ **7.** $xe^x - e^x + C$
9. $-(x^2 + 2x + 2)e^{-x} + C$
11. $y\tan^{-1}(y) - \ln\sqrt{1 + y^2} + C$
13. $x\tan x + \ln|\cos x| + C$
15. $(x^3 - 3x^2 + 6x - 6)e^x + C$
17. $(x^2 - 7x + 7)e^x + C$
19. $(x^5 - 5x^4 + 20x^3 - 60x^2 + 120x - 120)e^x + C$
21. $\frac{1}{2}(-e^\theta\cos\theta + e^\theta\sin\theta) + C$
23. $\frac{e^{2x}}{13}(3\sin 3x + 2\cos 3x) + C$
25. $\frac{2}{3}\left(\sqrt{3s + 9}\,e^{\sqrt{3s+9}} - e^{\sqrt{3s+9}}\right) + C$
27. $\frac{\pi\sqrt{3}}{3} - \ln(2) - \frac{\pi^2}{18}$
29. $\frac{1}{2}[-x\cos(\ln x) + x\sin(\ln x)] + C$
31. $\frac{1}{2}\ln|\sec x^2 + \tan x^2| + C$
33. $\frac{1}{2}x^2(\ln x)^2 - \frac{1}{2}x^2\ln x + \frac{1}{4}x^2 + C$
35. $-\frac{1}{x}\ln x - \frac{1}{x} + C$ **37.** $\frac{1}{4}e^{x^4} + C$
39. $\frac{1}{3}x^2(x^2 + 1)^{3/2} - \frac{2}{15}(x^2 + 1)^{5/2} + C$
41. $-\frac{2}{5}\sin 3x\sin 2x - \frac{3}{5}\cos 3x\cos 2x + C$
43. $\frac{2}{9}x^{3/2}(3\ln x - 2) + C$
45. $2\sqrt{x}\sin\sqrt{x} + 2\cos\sqrt{x} + C$
47. $\frac{\pi^2 - 4}{8}$ **49.** $\frac{5\pi - 3\sqrt{3}}{9}$
51. $\frac{1}{2}(x^2 + 1)\arctan x - \frac{x}{2} + C$ **53.** $xe^{x^2} + C$
55. $(2/3)x^{3/2}\arcsin(\sqrt{x}) + (2/9)\,x\sqrt{1 - x} + (4/9)\sqrt{1 - x} + C$
57. (a) π **(b)** 3π **(c)** 5π **(d)** $(2n + 1)\pi$
59. $2\pi(1 - \ln 2)$ **61. (a)** $\pi(\pi - 2)$ **(b)** 2π
63. (a) 1 **(b)** $(e - 2)\pi$ **(c)** $\frac{\pi}{2}(e^2 + 9)$
(d) $\bar{x} = \frac{1}{4}(e^2 + 1), \bar{y} = \frac{1}{2}(e - 2)$
65. $\frac{1}{2\pi}(1 - e^{-2\pi})$ **67.** $u = x^n, dv = \cos x\,dx$
69. $u = x^n, dv = e^{ax}\,dx$ **73.** $u = x^n, dv = (x + 1)^{-(1/2)}\,dx$
77. $x\arcsin x + \cos(\arcsin x) + C$
79. $x\sec^{-1}x - \ln\left|x + \sqrt{x^2 - 1}\right| + C$ **81.** Yes
83. (a) $x\sinh^{-1}x - \cosh(\sinh^{-1}x) + C$
(b) $x\sinh^{-1}x - (1 + x^2)^{1/2} + C$

SECTION 8.3, pp. 474–475

1. $\frac{1}{2}\sin 2x + C$ **3.** $-\frac{1}{4}\cos^4 x + C$
5. $\frac{1}{3}\cos^3 x - \cos x + C$ **7.** $-\cos x + \frac{2}{3}\cos^3 x - \frac{1}{5}\cos^5 x + C$
9. $\sin x - \frac{1}{3}\sin^3 x + C$ **11.** $\frac{1}{4}\sin^4 x - \frac{1}{6}\sin^6 x + C$
13. $\frac{1}{2}x + \frac{1}{4}\sin 2x + C$ **15.** $16/35$ **17.** 3π
19. $-4\sin x\cos^3 x + 2\cos x\sin x + 2x + C$
21. $-\cos^4 2\theta + C$ **23.** 4 **25.** 2
27. $\sqrt{\frac{3}{2}} - \frac{2}{3}$ **29.** $\frac{4}{5}\left(\frac{3}{2}\right)^{5/2} - \frac{18}{35} - \frac{2}{7}\left(\frac{3}{2}\right)^{7/2}$ **31.** $\sqrt{2}$
33. $\frac{1}{2}\tan^2 x + C$ **35.** $\frac{1}{3}\sec^3 x + C$ **37.** $\frac{1}{3}\tan^3 x + C$
39. $2\sqrt{3} + \ln(2 + \sqrt{3})$ **41.** $\frac{2}{3}\tan\theta + \frac{1}{3}\sec^2\theta\tan\theta + C$
43. $4/3$ **45.** $2\tan^2 x - 2\ln(1 + \tan^2 x) + C$
47. $\frac{1}{4}\tan^4 x - \frac{1}{2}\tan^2 x + \ln|\sec x| + C$ **49.** $\frac{4}{3} - \ln\sqrt{3}$
51. $\frac{43}{3}$ **53.** $-\frac{1}{10}\cos 5x - \frac{1}{2}\cos x + C$ **55.** π
57. $\frac{1}{2}\sin x + \frac{1}{14}\sin 7x + C$
59. $\frac{1}{6}\sin 3\theta - \frac{1}{4}\sin\theta - \frac{1}{20}\sin 5\theta + C$
61. $-\frac{2}{5}\cos^5\theta + C$ **63.** 1 **65.** $1/2$

67. $\cos x + \sec x + C$

69. $\frac{1}{4}x^2 - \frac{1}{4}x\sin 2x - \frac{1}{8}\cos 2x + C$ **71.** $\ln(2 + \sqrt{3})$

73. $\pi^2/2$ **75.** $\bar{x} = \frac{4\pi}{3}, \bar{y} = \frac{8\pi^2 + 3}{12\pi}$ **77.** $(\pi/4)(4 - \pi)$

SECTION 8.4, pp. 479–480

1. $\ln\left|\sqrt{9 + x^2} + x\right| + C$ **3.** $\pi/4$ **5.** $\pi/6$

7. $\frac{25}{2}\sin^{-1}\left(\frac{t}{5}\right) + \frac{t\sqrt{25 - t^2}}{2} + C$

9. $\frac{1}{2}\ln\left|\frac{2x}{7} + \frac{\sqrt{4x^2 - 49}}{7}\right| + C$

11. $7\left[\frac{\sqrt{y^2 - 49}}{7} - \sec^{-1}\left(\frac{y}{7}\right)\right] + C$ **13.** $\frac{\sqrt{x^2 - 1}}{x} + C$

15. $\operatorname{arcsec}|x| + C$ **17.** $\sqrt{x^2 - 1} + C$

19. $-\sqrt{9 - x^2} + C$ **21.** $\frac{1}{3}(x^2 + 4)^{3/2} - 4\sqrt{x^2 + 4} + C$

23. $\frac{-2\sqrt{4 - w^2}}{w} + C$ **25.** $\arcsin x - \sqrt{1 - x^2} + C$

27. $4\sqrt{3} - \frac{4\pi}{3}$ **29.** $-\frac{x}{\sqrt{x^2 - 1}} + C$

31. $-\frac{1}{5}\left(\frac{\sqrt{1 - x^2}}{x}\right)^5 + C$ **33.** $-\left(1 + \frac{1}{x^4}\right)^{1/4} + C$

35. $\frac{1}{2}x^2 + \frac{1}{2}\ln|x^2 - 1| + C$ **37.** $\frac{1}{3}\left(\frac{v}{\sqrt{1 - v^2}}\right)^3 + C$

39. $\frac{1}{\sqrt{2}}\tan^{-1}\left[\frac{x - 1/x}{\sqrt{2}}\right] + C$ **41.** $\pi/6$

43. $\frac{1}{2}\ln\left|\sqrt{1 + x^4} + x^2\right| + C$

45. $4\arcsin\frac{\sqrt{x}}{2} + \sqrt{x}\sqrt{4 - x} + C$

47. $\frac{1}{4}\arcsin\sqrt{x} - \frac{1}{4}\sqrt{x}\sqrt{1 - x}(1 - 2x) + C$

49. $(9/2)\arcsin\left(\frac{x + 1}{3}\right) + (1/2)(x + 1)\sqrt{8 - 2x - x^2} + C$

51. $\sqrt{x^2 + 4x + 3} - \operatorname{arcsec}(x + 2) + C$

53. $y = 2\left[\frac{\sqrt{x^2 - 4}}{2} - \sec^{-1}\left(\frac{x}{2}\right)\right]$

55. $y = \frac{3}{2}\arctan\left(\frac{x}{2}\right) - \frac{3\pi}{8}$ **57.** $3\pi/4$

59. (a) $\frac{1}{12}(\pi + 6\sqrt{3} - 12)$

(b) $\bar{x} = \frac{3\sqrt{3} - \pi}{4(\pi + 6\sqrt{3} - 12)}, \bar{y} = \frac{\pi^2 + 12\sqrt{3}\pi - 72}{12(\pi + 6\sqrt{3} - 12)}$

61. (a) $-\frac{1}{3}x^2(1 - x^2)^{3/2} - \frac{2}{15}(1 - x^2)^{5/2} + C$

(b) $-\frac{1}{3}(1 - x^2)^{3/2} + \frac{1}{5}(1 - x^2)^{5/2} + C$

(c) $\frac{1}{5}(1 - x^2)^{5/2} - \frac{1}{3}(1 - x^2)^{3/2} + C$

63. $\sqrt{3} - \frac{\sqrt{2}}{2} + \frac{1}{2}\ln\left(\frac{2 + \sqrt{3}}{\sqrt{2} + 1}\right)$

SECTION 8.5, pp. 486–488

1. $\frac{2}{x - 3} + \frac{3}{x - 2}$ **3.** $\frac{1}{x + 1} + \frac{3}{(x + 1)^2}$

5. $\frac{-2}{z} + \frac{-1}{z^2} + \frac{2}{z - 1}$ **7.** $1 + \frac{17}{t - 3} + \frac{-12}{t - 2}$

9. $\frac{1}{3}\log_e\left|\frac{x - 1}{x + 2}\right| + C$

11. $\frac{1}{7}\ln\left|(x + 6)^2(x - 1)^5\right| + C$ **13.** $(\ln 15)/2$

15. $-\frac{1}{2}\ln|t| + \frac{1}{6}\ln|t + 2| + \frac{1}{3}\ln|t - 1| + C$ **17.** $3\ln 2 - 2$

19. $\frac{1}{4}\ln\left|\frac{x + 1}{x - 1}\right| - \frac{x}{2(x^2 - 1)} + C$ **21.** $(\pi + 2\ln 2)/8$

23. $\tan^{-1}y - \frac{1}{y^2 + 1} + C$

25. $-(s - 1)^{-2} + (s - 1)^{-1} + \arctan s + C$

27. $\frac{2}{3}\ln|x - 1| + \frac{1}{6}\ln|x^2 + x + 1| - \sqrt{3}\arctan\left(\frac{2x + 1}{\sqrt{3}}\right) + C$

29. $\frac{1}{4}\ln\left|\frac{x - 1}{x + 1}\right| + \frac{1}{2}\arctan x + C$

31. $\frac{-1}{\theta^2 + 2\theta + 2} + \ln(\theta^2 + 2\theta + 2) - \arctan(\theta + 1) + C$

33. $x^2 + \ln\left|\frac{x - 1}{x}\right| + C$

35. $9x + 2\ln|x| + \frac{1}{x} + 7\ln|x - 1| + C$

37. $\frac{y^2}{2} - \ln|y| + \frac{1}{2}\ln(1 + y^2) + C$ **39.** $\ln\left(\frac{e^t + 1}{e^t + 2}\right) + C$

41. $\frac{1}{5}\ln\left|\frac{\sin y - 2}{\sin y + 3}\right| + C$

43. $\frac{(\tan^{-1}2x)^2}{4} - 3\ln|x - 2| + \frac{6}{x - 2} + C$

45. $\ln\left|\frac{\sqrt{x} - 1}{\sqrt{x} + 1}\right| + C$

47. $2\sqrt{1 + x} + \ln\left|\frac{\sqrt{x + 1} - 1}{\sqrt{x + 1} + 1}\right| + C$

49. $\frac{1}{4}\ln\left|\frac{x^4}{x^4 + 1}\right| + C$

51. $\frac{1}{\sqrt{2}}\ln\left|\frac{\sqrt{2}\cos\theta + 1}{\sqrt{2}\cos\theta - 1}\right| + \frac{1}{2}\ln\left|\frac{1 - \cos\theta}{1 + \cos\theta}\right| + C$

53. $4\sqrt{1 + \sqrt{x}} + 2\ln\left|\frac{\sqrt{1 + \sqrt{x}} - 1}{\sqrt{1 + \sqrt{x}} + 1}\right| + C$

55. $\frac{1}{3}x^3 - 2x^2 + 5x - 10\ln|x + 2| + C$

57. $\frac{1}{\ln 2}\ln(2^x + 2^{-x}) + C$ **59.** $\frac{1}{4}\ln\left|\frac{x - 1}{x + 1}\right| - \frac{1}{2}\arctan x + C$

61. $\frac{1}{2}\ln|(\ln x + 1)(\ln x + 3)| + C$

63. $\ln\left|x + \sqrt{x^2 - 1}\right| + C$

65. $\frac{2}{9}x^3(x^3 + 1)^{3/2} - \frac{4}{45}(x^3 + 1)^{5/2} + C$

67. $x = \ln|t - 2| - \ln|t - 1| + \ln 2$

69. $x = \frac{6t}{t + 2} - 1$ **71.** $3\pi\ln 25$

73. $\ln(3) - \frac{1}{2}$ **75.** 1.10

77. (a) $x = \frac{1000e^{4t}}{499 + e^{4t}}$ **(b)** 1.55 days

SECTION 8.6, pp. 492–494

1. $\frac{2}{\sqrt{5}}\tan^{-1}\sqrt{\frac{3x-5}{5}}+C$

3. $\sqrt{x-2}\left(\frac{2(x-2)}{3}+4\right)+C$ **5.** $\frac{(2x-3)^{3/2}(x+1)}{5}+C$

7. $-\frac{\sqrt{16-9x}}{x}-\frac{9}{8}\ln\left|\frac{\sqrt{16-9x}-4}{\sqrt{16-9x}+4}\right|+C$

9. $\frac{(x+2)(2x-6)\sqrt{4x-x^2}}{6}+4\arcsin\left(\frac{x-2}{2}\right)+C$

11. $-\frac{1}{\sqrt{7}}\ln\left|\frac{\sqrt{7}+\sqrt{7+x^2}}{x}\right|+C$

13. $\sqrt{4-x^2}-2\ln\left|\frac{2+\sqrt{4-x^2}}{x}\right|+C$

15. $\frac{e^{3t}}{13}(3\cos 2t+2\sin 2t)+C$

17. $\frac{x^2}{2}\arccos x+\frac{1}{4}\arcsin x-\frac{1}{4}x\sqrt{1-x^2}+C$

19. $\frac{x^3}{3}\arctan x-\frac{x^2}{6}+\frac{1}{6}\ln(1+x^2)+C$

21. $\frac{1}{2}$ **23.** $4\left[\frac{\sin\left(\frac{7t}{2}\right)}{7}+\frac{\sin\left(\frac{9t}{2}\right)}{9}\right]+C$

25. 0 **27.** $\frac{1}{2}\ln(x^2+1)+\frac{x}{2(1+x^2)}+\frac{1}{2}\tan^{-1}x+C$

29. $\left(x-\frac{1}{2}\right)\arcsin\sqrt{x}+\frac{1}{2}\sqrt{x-x^2}+C$

31. $2\sin^{-1}\sqrt{\frac{x}{2}}-\sqrt{2x-x^2}+C$

33. $\sqrt{1-\sin^2 t}-\ln\left|\frac{1+\sqrt{1-\sin^2 t}}{\sin t}\right|+C$

35. $\ln\left|\ln y+\sqrt{3+(\ln y)^2}\right|+C$

37. $\ln\left|x+1+\sqrt{x^2+2x+5}\right|+C$

39. $\frac{(x+1)}{2}\sqrt{8-2x-x^2}+\frac{9}{2}\sin^{-1}\left(\frac{x+1}{3}\right)+C$

41. $-\frac{\sin^3 3x\cos 3x}{12}+\frac{3}{8}x-\frac{1}{16}\sin 6x+C$

43. $\frac{\sin^3 2\theta\cos^2 2\theta}{10}+\frac{\sin^3 2\theta}{15}+C$

45. $\tan^2 2x-2\ln|\sec 2x|+C$

47. $\frac{(\sec\pi x)(\tan\pi x)}{\pi}+\frac{1}{\pi}\ln|\sec\pi x+\tan\pi x|+C$

49. $\frac{-\csc^3 x\cot x}{4}-\frac{3\csc x\cot x}{8}-\frac{3}{8}\ln|\csc x+\cot x|+C$

51. $\frac{1}{2}[\sec(e^t-1)\tan(e^t-1)+$
$\ln|\sec(e^t-1)+\tan(e^t-1)|]+C$

53. $\sqrt{2}+\ln(\sqrt{2}+1)$ **55.** $\pi/3$

57. $2\pi\sqrt{3}+\pi\sqrt{2}\ln(\sqrt{2}+\sqrt{3})$ **59.** $\bar{x}=4/3,\ \bar{y}=\ln\sqrt{2}$

61. 7.62 **63.** $\pi/2$ **67. (c)** $\pi/4$

SECTION 8.7, pp. 502–504

1. I: (a) 1.5, 0 **(b)** 1.5, 0 **(c)** 0%
II: (a) 1.5, 0 **(b)** 1.5, 0 **(c)** 0%
III: (a) 1.5, 0 **(b)** 1.5, 0 **(c)** 0%

3. I: (a) 2.63, 0.04 **(b)** 2.67, 0.04 **(c)** 1.6%
II: (a) 2.75, 0.08 **(b)** 2.67, 0.08 **(c)** 3.1%
III: (a) 2.67, 0 **(b)** 2.67, 0 **(c)** 0%

5. I: (a) 5.88, 0.25 **(b)** 6, 0.13 **(c)** 2.1%
II: (a) 6.25, 0.5 **(b)** 6, 0.25 **(c)** 4.2%
III: (a) 6, 0 **(b)** 6, 0 **(c)** 0%

7. I: (a) 0.496, 0.016 **(b)** 0.5, 0.004 **(c)** 0.9%
II: (a) 0.509, 0.031 **(b)** 0.5, 0.009 **(c)** 1.8%
III: (a) 0.5004, 0.0026 **(b)** 0.5, 0.0004 **(c)** 0.1%

9. I: (a) 2.052, 0.081 **(b)** 2, 0.052 **(c)** 2.6%
II: (a) 1.896, 0.161 **(b)** 2, 0.104 **(c)** 5.2%
III: (a) 2.005, 0.0066 **(b)** 2, 0.0046 **(c)** 0.2%

11. (a) 1 **(b)** 2 **13. (a)** 116 **(b)** 2
15. (a) 283 **(b)** 2 **17. (a)** 71 **(b)** 10
19. (a) 76 **(b)** 12 **21. (a)** 82 **(b)** 8
23. 15,990 ft^3 **25.** $\approx$10.63 ft
27. (a) $\approx$0.00021 **(b)** $\approx$1.37079 **(c)** $\approx$0.015%
31. (a) $\approx$5.870 **(b)** $E_T| \le 0.0032$
33. 21.07 in. **35.** 14.42 **39.** $\approx$28.7 mg

SECTION 8.8, pp. 512–515

1. $\pi/2$ **3.** 2 **5.** 10 **7.** $\pi/3$ **9.** ln 3 **11.** $\ln\left(\frac{9}{4}\right)$
13. 0 **15.** $\sqrt{3}$ **17.** $\frac{\pi}{\sqrt{2}}$ **19.** $\ln\left(1+\frac{\pi}{2}\right)$
21. -1 **23.** 1 **25.** $-1/4$ **27.** $\pi/2$ **29.** $\pi/3$
31. 6 **33.** ln 2 **35.** Diverges **37.** Diverges
39. Diverges **41.** Diverges **43.** Converges
45. 1 **47.** Diverges **49.** Converges
51. Converges **53.** Diverges **55.** Converges
57. Converges **59.** Diverges **61.** Converges
63. Diverges **65.** Converges **67.** Converges
69. Diverges **71.** Converges; 20 **73.** Converges; $\frac{1}{81}$
75. Diverges **77.** Diverges
79. Converges; $-\frac{1}{375}$
81. (a) Converges when $p<2$
(b) Converges when $p>1$
83. 1 **85.** 2π **87.** ln 2
89. (a) 1/2 **(b) (i)** $\pi/5$ **(ii)** 2π
(c) Diverges
91. (a) $\frac{2+\pi}{4}$ **(b)** $\pi/2$ **93. (b)** $\approx$0.88621
95. (a)

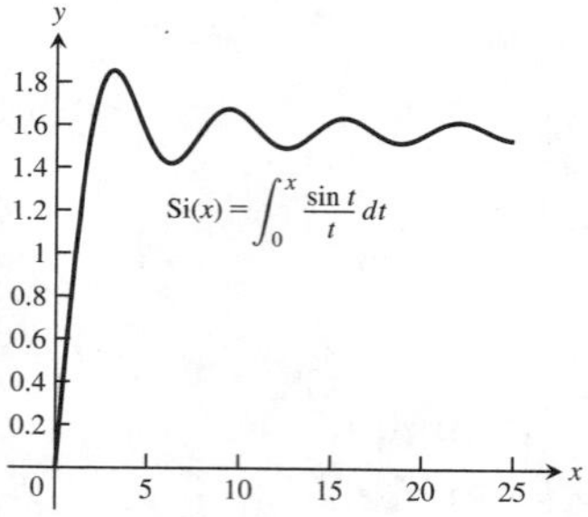

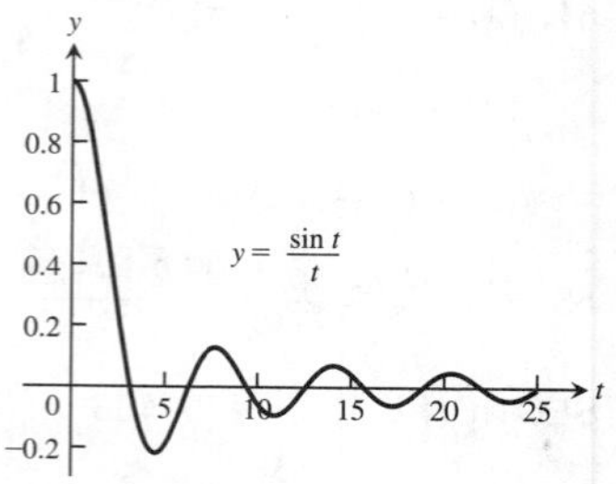

(b) $\pi/2$

97. (a)

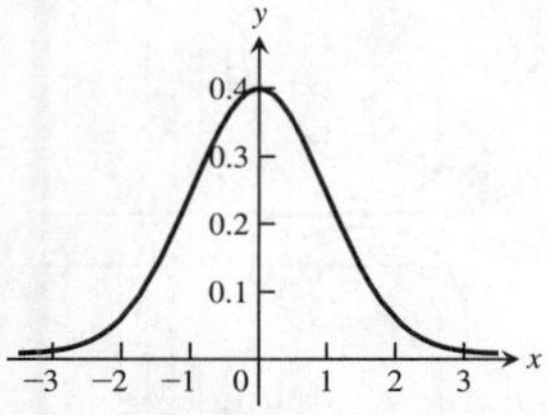

(b) $\approx 0.683, \approx 0.954, \approx 0.997$

103. ≈ 0.16462

SECTION 8.9, pp. 525–528

1. No **3.** Yes **5.** Yes **7.** No **11.** ≈ 0.497

13. ≈ 0.625 **15.** ≈ 0.0351 **17.** $2\sqrt{6}$ **19.** $\frac{1}{3}\ln 2$

21. $\frac{1}{9}, \frac{7}{27}$ **25.** mean $= \frac{8}{3} \approx 2.67$, median $= \sqrt{8} \approx 2.83$

27. mean $= 2$, median $= \sqrt{2} \approx 1.41$

29. $P(X < \frac{1}{2}) \approx 0.3935$

31. (a) ≈ 0.55, so about 55 in every 100 bulbs will fail.
(b) ≈ 693 hr

33. ≈ 60 hydra **35. (a)** ≈ 0.393 **(b)** ≈ 0.135 **(c)** 0
(d) The probability that any customer waits longer than 3 minutes is $1 - (0.997521)^{200} \approx 0.391 < 1/2$. So the most likely outcome is that all 200 would be served within 3 minutes.

37. \$15,384 **39.** $\approx 323, \approx 262$ **41.** ≈ 0.9938

43. (a) $\approx 16\%$ **(b)** ≈ 0.23832 **45.** ≈ 673 females

47. ≈ 47 adults **49.** ≈ 289 shafts

51. (a) ≈ 0.977 **(b)** ≈ 0.159 **(c)** ≈ 0.838

55. (a) {LLL, LLD, LDL, DLL, LLU, LUL, ULL, LDD, LDU, LUD, LUU, DLD, DLU, ULD, ULU, DDL, DUL, UDL, UUL, DDD, DDU, DUD, UDD, DUU, UDU, UUD, UUU}
(c) $7/27 \approx 0.26$ **(d)** $20/27 \approx 0.74$

PRACTICE EXERCISES, pp. 529–531

1. $(x+1)(\ln(x+1)) - (x+1) + C$

3. $x \arctan(3x) - \frac{1}{6}\ln(1+9x^2) + C$

5. $(x+1)^2 e^x - 2(x+1)e^x + 2e^x + C$

7. $\frac{2e^x \sin 2x}{5} + \frac{e^x \cos 2x}{5} + C$

9. $2\ln|x-2| - \ln|x-1| + C$

11. $\ln|x| - \ln|x+1| + \frac{1}{x+1} + C$

13. $-\frac{1}{3}\ln\left|\frac{\cos\theta - 1}{\cos\theta + 2}\right| + C$

15. $4\ln|x| - \frac{1}{2}\ln(x^2+1) + 4\arctan x + C$

17. $\frac{1}{16}\ln\left|\frac{(v-2)^5(v+2)}{v^6}\right| + C$

19. $\frac{1}{2}\arctan t - \frac{\sqrt{3}}{6}\tan^{-1}\frac{t}{\sqrt{3}} + C$

21. $\frac{x^2}{2} + \frac{4}{3}\ln|x+2| + \frac{2}{3}\ln|x-1| + C$

23. $\frac{x^2}{2} - \frac{9}{2}\ln|x+3| + \frac{3}{2}\ln|x+1| + C$

25. $\frac{1}{3}\ln\left|\frac{\sqrt{x+1}-1}{\sqrt{x+1}+1}\right| + C$ **27.** $\ln|1 - e^{-s}| + C$

29. $-\sqrt{16 - y^2} + C$ **31.** $-\frac{1}{2}\ln|4 - x^2| + C$

33. $\ln\frac{1}{\sqrt{9 - x^2}} + C$ **35.** $\frac{1}{6}\ln\left|\frac{x+3}{x-3}\right| + C$

37. $-\frac{\cos^5 x}{5} + \frac{\cos^7 x}{7} + C$ **39.** $\frac{\tan^5 x}{5} + C$

41. $\frac{\cos\theta}{2} - \frac{\cos 11\theta}{22} + C$ **43.** $4\sqrt{1 - \cos(t/2)} + C$

45. At least 16 **47.** All three rules yield π **49.** 25 °F

51. (a) ≈ 2.42 gal **(c)** ≈ 24.83 mi/gal

53. $\pi/2$ **55.** 6 **57.** $\ln 3$ **59.** 2 **61.** $\pi/6$

63. Diverges **65.** Diverges **67.** Converges

69. $\frac{1}{2}xe^{2x} - \frac{1}{4}e^{2x} + C$ **71.** $2\tan x - x + C$

73. $x\tan x - \ln|\sec x| + C$ **75.** $-\frac{1}{3}(\cos x)^3 + C$

77. $1 + \frac{1}{2}\ln\left(\frac{2}{1+e^2}\right)$ **79.** $2\ln\left|1 - \frac{1}{x}\right| + \frac{4x+1}{2x^2} + C$

81. $\frac{e^{2x} - 1}{e^x} + C$ **83.** $9/4$ **85.** $256/15$

87. $-\frac{1}{3}\csc^3 x + C$

89. $\frac{2x^{3/2}}{3} - x + 2\sqrt{x} - 2\ln(\sqrt{x} + 1) + C$

91. $\frac{1}{2}\sin^{-1}(x-1) + \frac{1}{2}(x-1)\sqrt{2x - x^2} + C$

93. $-2\cot x - \ln|\csc x + \cot x| + \csc x + C$

95. $\frac{1}{12}\ln\left|\frac{3+v}{3-v}\right| + \frac{1}{6}\tan^{-1}\frac{v}{3} + C$

97. $\frac{\theta\sin(2\theta+1)}{2} + \frac{\cos(2\theta+1)}{4} + C$

99. $\frac{1}{4}\sec^2\theta + C$ **101.** $2\left(\frac{\left(\sqrt{2-x}\right)^3}{3} - 2\sqrt{2-x}\right) + C$

103. $\arctan(y-1) + C$

105. $\frac{1}{4}\ln|z| - \frac{1}{4z} - \frac{1}{4}\left[\frac{1}{2}\ln(z^2+4) + \frac{1}{2}\arctan\left(\frac{z}{2}\right)\right] + C$

107. $-\frac{1}{4}\sqrt{9 - 4t^2} + C$ **109.** $\ln\left(\frac{e^t + 1}{e^t + 2}\right) + C$

111. $1/4$ **113.** $\frac{2}{3}x^{3/2} + C$ **115.** $-\frac{1}{5}\arctan(\cos 5t) + C$

117. $2\sqrt{r} - 2\ln(1 + \sqrt{r}) + C$ **119.** $\frac{1}{2}x^2 - \frac{1}{2}\ln(x^2+1) + C$

121. $\frac{2}{3}\ln|x+1| + \frac{1}{6}\ln|x^2 - x + 1| + \frac{1}{\sqrt{3}}\arctan\left(\frac{2x-1}{\sqrt{3}}\right) + C$

123. $\frac{4}{7}(1+\sqrt{x})^{7/2} - \frac{8}{5}(1+\sqrt{x})^{5/2} + \frac{4}{3}(1+\sqrt{x})^{3/2} + C$

125. $2\ln\left|\sqrt{x} + \sqrt{1+x}\right| + C$ **127.** $\ln x - \ln|1 + \ln x| + C$

129. $\frac{1}{2}x^{\ln x} + C$ **131.** $e^x\frac{x}{(1+x^2)^{3/2}} + C$

133. $x - \frac{1}{\sqrt{2}}\arctan(\sqrt{2}\tan x) + C$ **135.** $\frac{\pi}{4}$

ADDITIONAL AND ADVANCED EXERCISES, pp. 531–534

1. $x(\arcsin x)^2 + 2(\arcsin x)\sqrt{1-x^2} - 2x + C$

3. $\frac{x^2 \arcsin x}{2} + \frac{x\sqrt{1-x^2} - \arcsin x}{4} + C$

5. $\frac{1}{2}\left(\ln\left(t - \sqrt{1 - t^2}\right) - \arcsin t\right) + C$ **7.** 0

9. $\ln(4) - 1$ **11.** 1 **13.** $32\pi/35$ **15.** 2π

17. **(a)** π **(b)** $\pi(2e - 5)$

19. **(b)** $\pi\left(\frac{8(\ln 2)^2}{3} - \frac{16(\ln 2)}{9} + \frac{16}{27}\right)$ **21.** $\left(\frac{e^2 + 1}{4}, \frac{e - 2}{2}\right)$

23. $\sqrt{1 + e^2} - \ln\left(\frac{\sqrt{1 + e^2}}{e} + \frac{1}{e}\right) - \sqrt{2} + \ln\left(1 + \sqrt{2}\right)$

25. $\frac{12\pi}{5}$ **27.** $a = \frac{1}{2}, -\frac{\ln 2}{4}$ **29.** $\frac{1}{2} < p \le 1$

33. $\frac{2}{1 - \tan(x/2)} + C$ **35.** 1 **37.** $\frac{\sqrt{3}\pi}{9}$

39. $\frac{1}{\sqrt{2}}\ln\left|\frac{\tan(t/2) + 1 - \sqrt{2}}{\tan(t/2) + 1 + \sqrt{2}}\right| + C$

41. $\ln\left|\frac{1 + \tan(\theta/2)}{1 - \tan(\theta/2)}\right| + C$

Chapter 9

SECTION 9.1, pp. 541–543

1. (d) **3.** (a)

5.

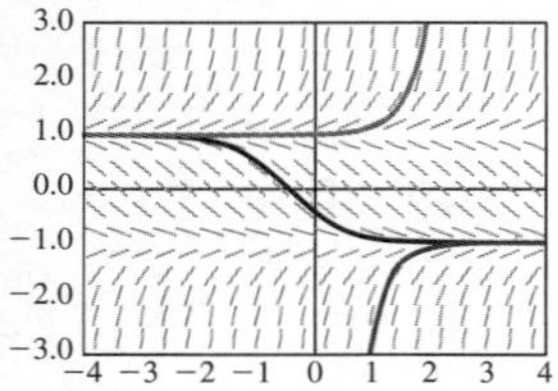

7. $y' = x - y;\ y(1) = -1$

9. $y' = -(1 + y)\sin x;\ y(0) = 2$

11. $y' = 1 + xe^y;\ y(-2) = 2$

13.

y	$f(y) = \frac{dy}{dx}$
−3	2
−2	1.5
−1	0
0	−1
1	−0.75
2	0
3	1

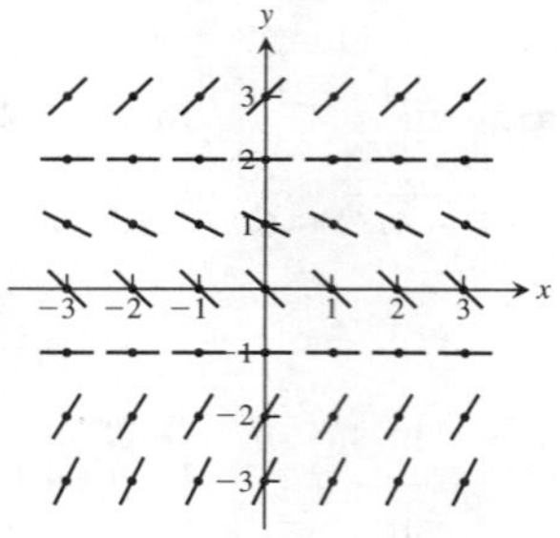

15. $y(\text{exact}) = -x^2, y_1 = -2, y_2 = -3.3333, y_3 = -5$

17. $y(\text{exact}) = \frac{1}{4}e^{2x} + \frac{1}{4}(-2x - 1)$; $y_1 = 0, y_2 = 0.0625, y_3 = 0.2186$

19. $y(\text{exact}) = e^{x^2} + 1, y_1 = 2.0, y_2 = 2.0202, y_3 = 2.0618$

21. $y_1 = -1.5$, exact value is e^{-50}

23. $y \approx -0.2272$; exact value is $1/\left(1 - 2\sqrt{5}\right) \approx -0.2880$.

27.

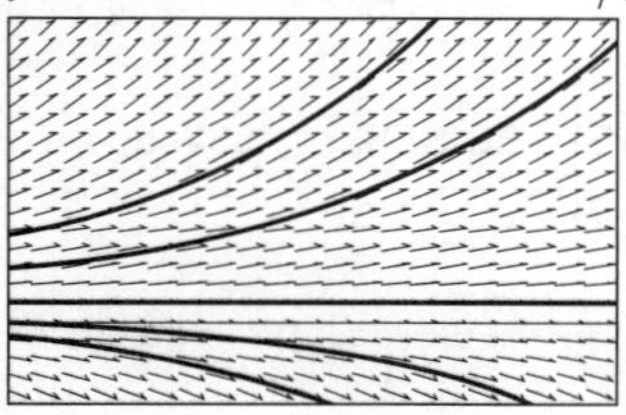

29.

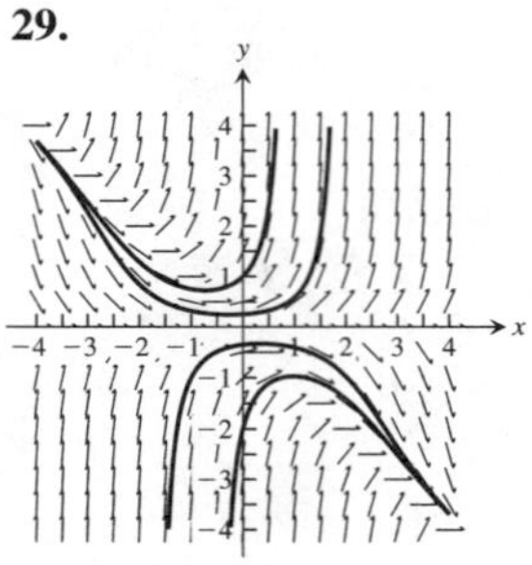

31.

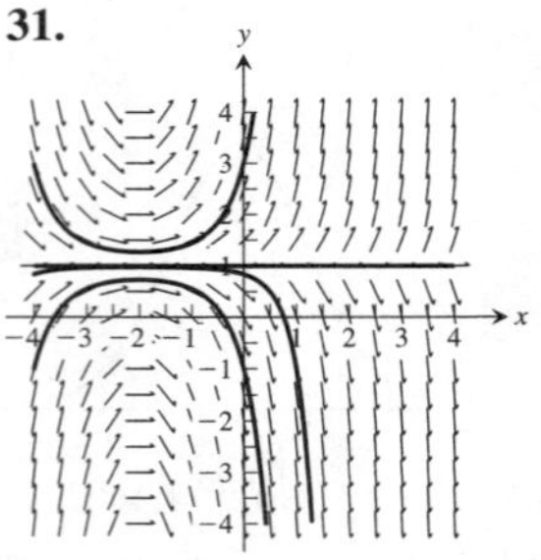

39. Euler's method gives $y \approx 3.45835$; the exact solution is $y = 1 + e \approx 3.71828$.

41. $y \approx 1.5000$; exact value is 1.5275.

SECTION 9.2, pp. 547–549

1. $y = \frac{e^x + C}{x},\quad x > 0$ **3.** $y = \frac{C - \cos x}{x^3},\quad x > 0$

5. $y = \frac{1}{2} - \frac{1}{x} + \frac{C}{x^2},\quad x > 0$ **7.** $y\cos x = e^x + C$

9. $2r\sin^2\theta + \sin^4\theta = 2C$

11. $s = \frac{t^3}{3(t-1)^4} - \frac{t}{(t-1)^4} + \frac{C}{(t-1)^4}$

13. $r = (\csc\theta)(\ln|\sec\theta| + C),\quad 0 < \theta < \pi/2$

15. $y = \frac{3}{2} - \frac{1}{2}e^{-2t}$ **17.** $y = -\frac{1}{\theta}\cos\theta + \frac{\pi}{2\theta}$

19. $y = 6e^{x^2} - \frac{e^{x^2}}{x + 1}$ **21.** $x(y - 1) = cy - 1$

23. (b) is correct, but (a) is not. **25.** $t = \frac{L}{R}\ln 2$ sec

27. **(a)** $i = \frac{V}{R} - \frac{V}{R}e^{-3} = \frac{V}{R}(1 - e^{-3}) \approx 0.95\frac{V}{R}$ amp **(b)** 86%

29. $y = \frac{1}{1 + Ce^{-x}}$ **31.** $y^2 = c\sec^2 x - \frac{2}{5}\cos^3 x$

SECTION 9.3, pp. 554–555

1. **(a)** 176.4 meters (approx.) **(b)** 45.85 sec (approx.)

3. $s(t) = 4.91\left(1 - e^{-(22.36/39.92)t}\right)$

5. $x^2 + y^2 = C$ **7.** $\ln|y| - \frac{1}{2}y^2 = \frac{1}{2}x^2 + C$

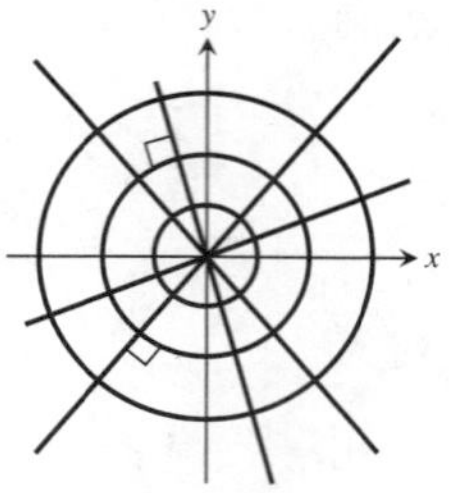

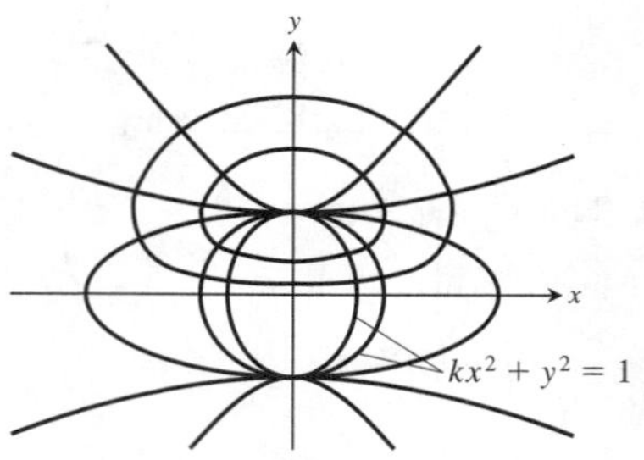

9. $y = \pm\sqrt{2x + C}$

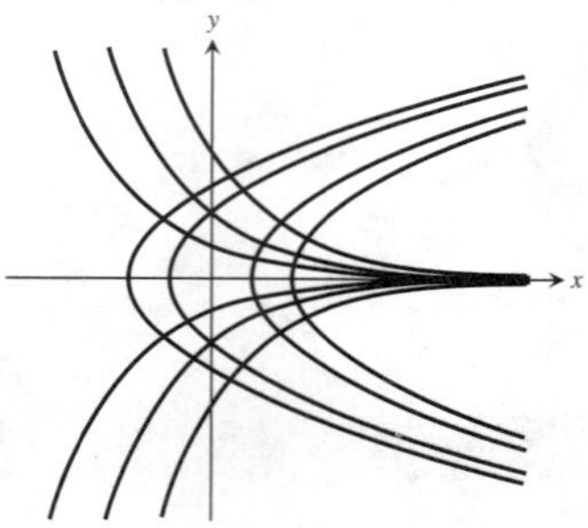

13. **(a)** 12 lb/min **(b)** $(200 + t)$ gal **(c)** $\dfrac{5y}{200+t}$

(d) $2(200+t) - \dfrac{300}{\left(1+\dfrac{t}{200}\right)^5}$

(e) 2.39 lb/gal

15. $y(27.8) \approx 14.8$ lb, $t \approx 27.8$ min

SECTION 9.4, pp. 563–564

1. $y' = (y + 2)(y - 3)$

(a) $y = -2$ is a stable equilibrium value, and $y = 3$ is an unstable equilibrium value.

(b) $y'' = 2(y + 2)\left(y - \dfrac{1}{2}\right)(y - 3)$

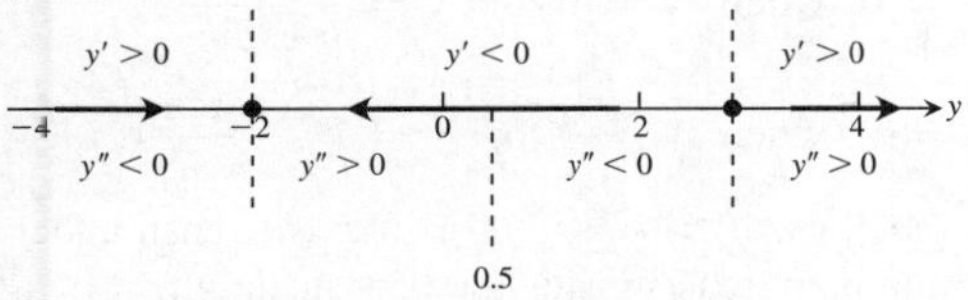

(c)

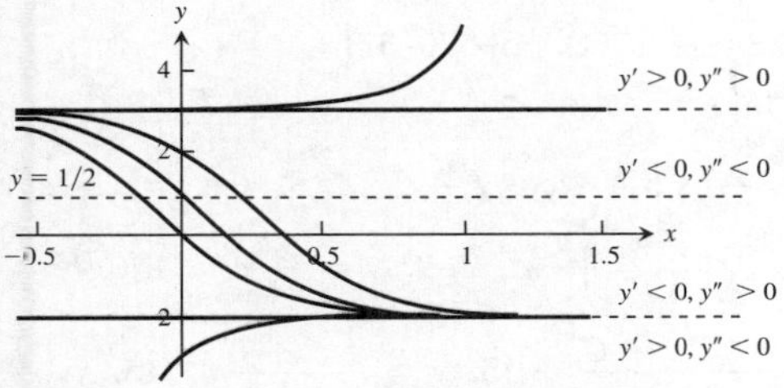

3. **(a)** $y = 2$ is an unstable equilibrium value and $y = 1$ is a semi-stable equilibrium value.

(b)

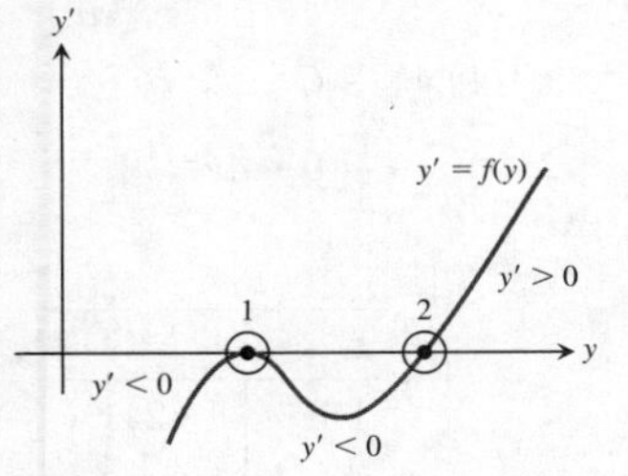

(c)

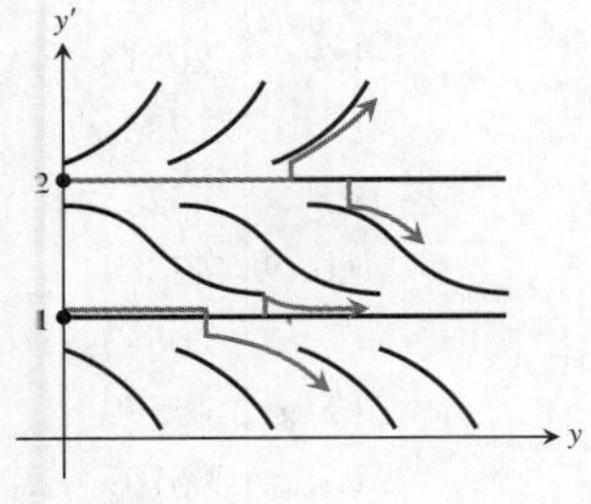

5. $y' = \sqrt{y},\ y > 0$

(a) There are no equilibrium values.

(b) $y'' = \dfrac{1}{2}$

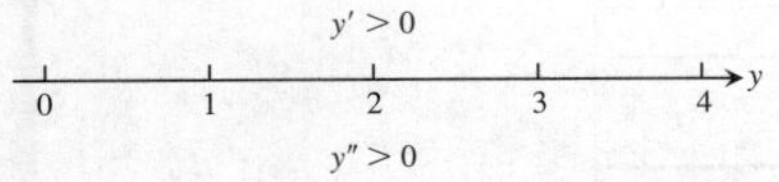

(c)

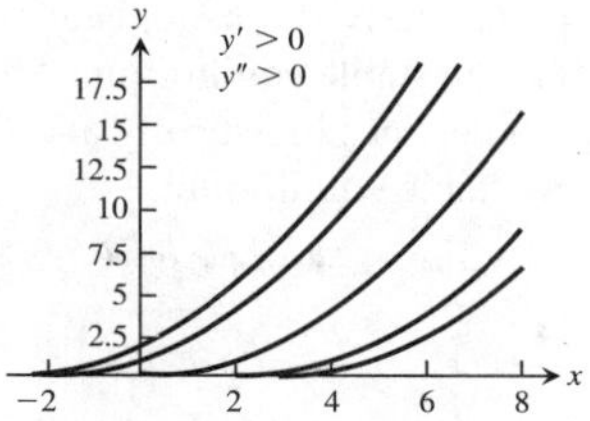

7. $y' = (y - 1)(y - 2)(y - 3)$

(a) $y = 1$ and $y = 3$ are unstable equilibria, and $y = 2$ is a stable equilibrium.

(b) $y'' = (3y^2 - 12y + 11)(y - 1)(y - 2)(y - 3) = 3(y - 1)\left(y - \dfrac{6 - \sqrt{3}}{3}\right)(y - 2)\left(y - \dfrac{6 + \sqrt{3}}{3}\right)(y - 3)$

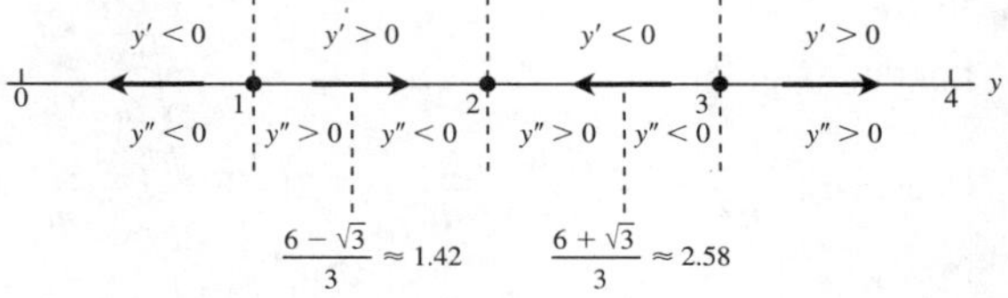

(c)

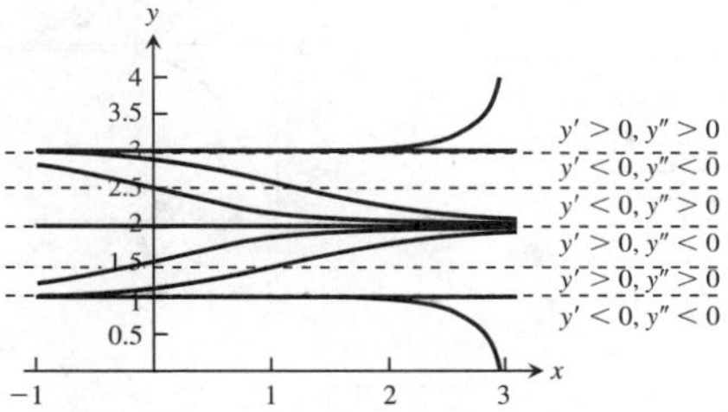

9. $\dfrac{dP}{dt} = 1 - 2P$ has a stable equilibrium at $P = \dfrac{1}{2}$; $\dfrac{d^2P}{dt^2} = -2\dfrac{dP}{dt} = -2(1 - 2P)$.

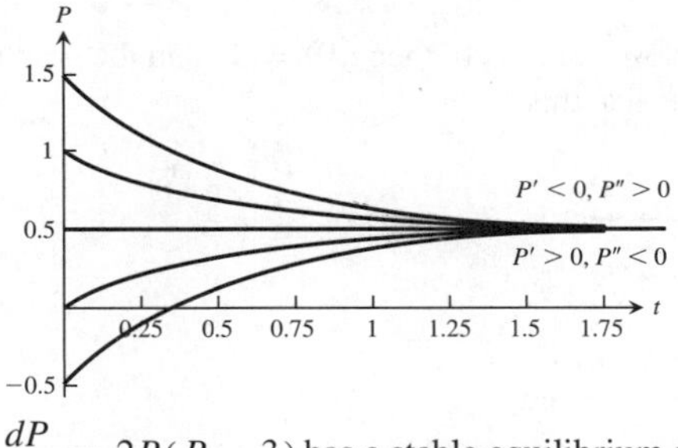

11. $\dfrac{dP}{dt} = 2P(P - 3)$ has a stable equilibrium at $P = 0$ and an unstable equilibrium at $P = 3$; $\dfrac{d^2P}{dt^2} = 2(2P - 3)\dfrac{dP}{dt} = 4P(2P - 3)(P - 3)$.

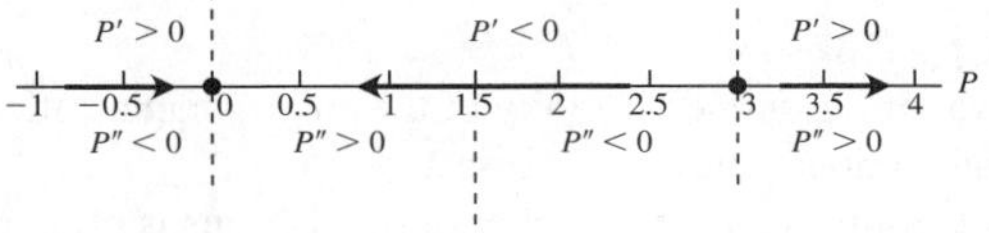

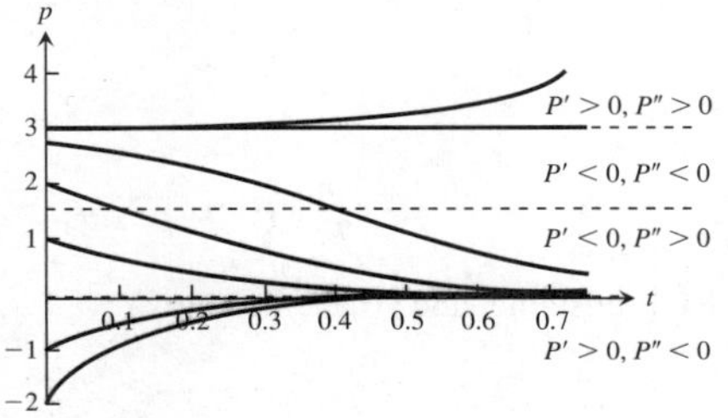

13. Before the catastrophe, the population exhibits logistic growth and $P(t)$ increases toward M_0, the stable equilibrium. After the catastrophe, the population declines logistically and $P(t)$ decreases toward M_1, the new stable equilibrium.

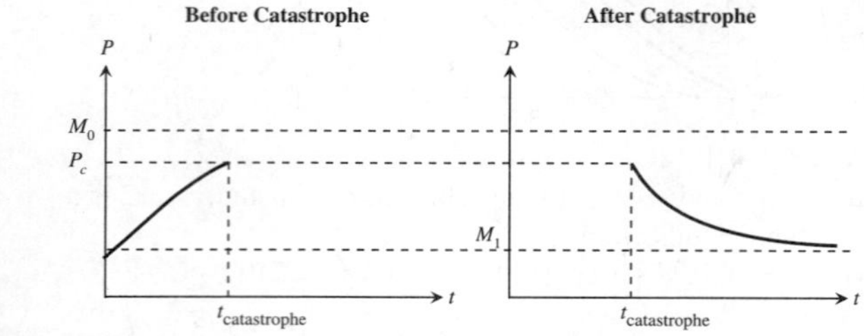

15. $\frac{dv}{dt} = g - \frac{k}{m}v^2, \quad g, k, m > 0 \text{ and } v(t) \geq 0$

Equilibrium: $\frac{dv}{dt} = g - \frac{k}{m}v^2 = 0 \Rightarrow v = \sqrt{\frac{mg}{k}}$

Concavity: $\frac{d^2v}{dt^2} = -2\left(\frac{k}{m}v\right)\frac{dv}{dt} = -2\left(\frac{k}{m}v\right)\left(g - \frac{k}{m}v^2\right)$

(a)

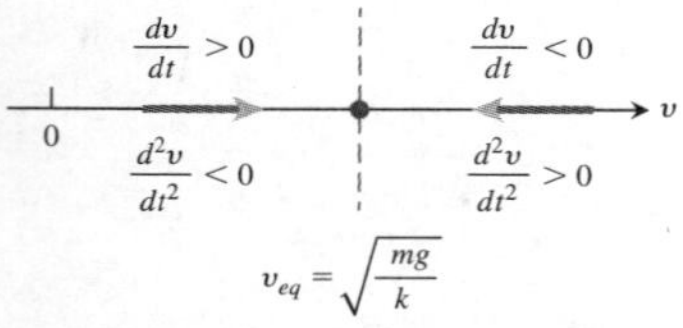

(b)

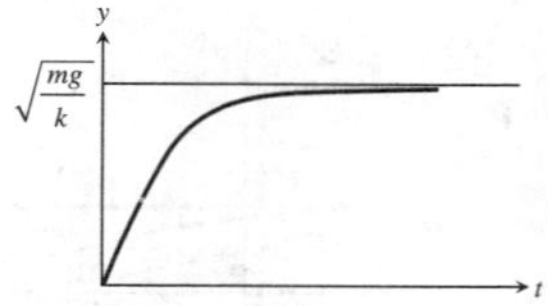

(c) 118 mph

17. 15 ft/sec

19. Phase line:

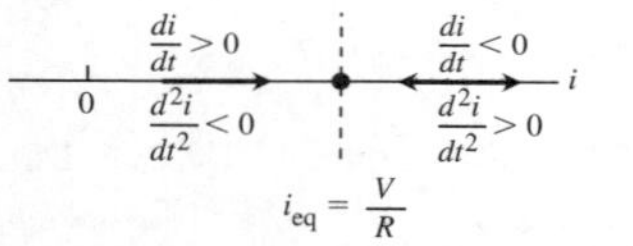

If the switch is closed at $t = 0$, then $i(0) = 0$, and the graph of the solution looks like this:

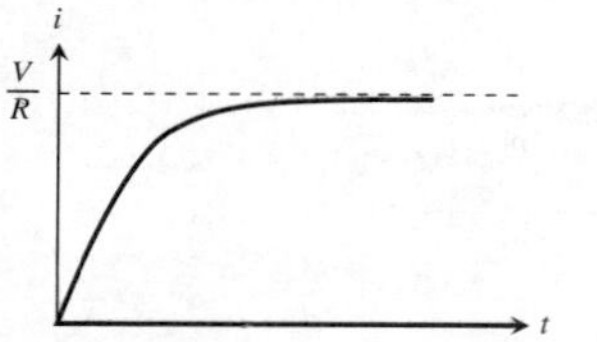

As $t \to \infty$, $i(t) \to i_{\text{steady state}} = \frac{V}{R}$.

21. $y = \frac{1}{1 + e^{-20x}}$.

SECTION 9.5, pp. 567–570

1. Seasonal variations, nonconformity of the environments, effects of other interactions, unexpected disasters, etc.

3. This model assumes that the number of interactions is proportional to the product of x and y:

$$\frac{dx}{dt} = (a - by)x, \quad a < 0,$$

$$\frac{dy}{dt} = m\left(1 - \frac{y}{M}\right)y - nxy = y\left(m - \frac{m}{M}y - nx\right).$$

Rest points are $(0, 0)$, unstable, and $(0, M)$, stable.

5. (a) Logistic growth occurs in the absence of the competitor, and involves a simple interaction between the species. Growth dominates the competition when either population is small, so it is difficult to drive either species to extinction.

(b) a: per capita growth rate for trout
m: per capita growth rate for bass
b: intensity of competition to the trout
n: intensity of competition to the bass
k_1: environmental carrying capacity for the trout
k_2: environmental carrying capacity for the bass
$\frac{a}{b}$: growth versus competition or net growth of trout
$\frac{m}{n}$: relative survival of bass

(c) $\frac{dx}{dt} = 0$ when $x = 0$ or $y = \frac{a}{b} - \frac{a}{bk_1}x,$

$\frac{dy}{dt} = 0$ when $y = 0$ or $y = k_2 - \frac{k_2 n}{m}x.$

By picking $a/b > k_2$ and $m/n > k_1$, we ensure that an equilibrium point exists inside the first quadrant.

PRACTICE EXERCISES, pp. 570–571

1. $y = -\ln\left(C - \frac{2}{5}(x-2)^{5/2} - \frac{4}{3}(x-2)^{3/2}\right)$

3. $\tan y = -x \sin x - \cos x + C$ **5.** $\tan y = c(1 - e^x)^3$

7. $y - a\tan^{-1}\left(\frac{x+y}{a}\right) = c$ **9.** $y = \frac{x^2}{4}e^{x/2} + Ce^{x/2}$

11. $y = \frac{x^2 - 2x + C}{2x^2}$ **13.** $y = \frac{e^{-x} + C}{1 + e^x}$ **15.** $xy + y^3 = C$

17. $\tan y = -\cos x + (1/3)\cos^3 x + C$

19. $e^{x^2}y = (x - 1)e^x + C$

21. $\ln|\ln y| = (1/2)x^2 \ln x - (1/4)x^2 + C$

23. $y = \frac{2x^3 + 3x^2 + 6}{6(x+1)^2}$ **25.** $y = \frac{1}{3}(1 - 4e^{-x^3})$

27. $y = e^{-x}(3x^3 - 3x^2)$

29.

x	y
0	0
0.1	0.1000
0.2	0.2095
0.3	0.3285
0.4	0.4568
0.5	0.5946
0.6	0.7418
0.7	0.8986
0.8	1.0649
0.9	1.2411
1.0	1.4273

x	y
1.1	1.6241
1.2	1.8319
1.3	2.0513
1.4	2.2832
1.5	2.5285
1.6	2.7884
1.7	3.0643
1.8	3.3579
1.9	3.6709
2.0	4.0057

31. $y(3) \approx 0.9131$

33. (a)

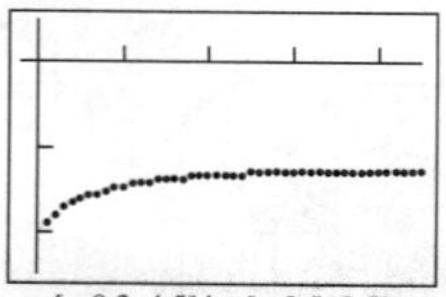

[−0.2, 4.5] by [−2.5, 0.5]

(b) Note that we choose a small interval of x values because the y values decrease very rapidly and our calculator cannot handle the calculations for $x \le -1$. (This occurs because the analytic solution is $y = -2 + \ln(2 - e^{-x})$, which has an asymptote at $x = -\ln 2 \approx -0.69$. Obviously, the Euler approximations are misleading for $x \le -0.7$.)

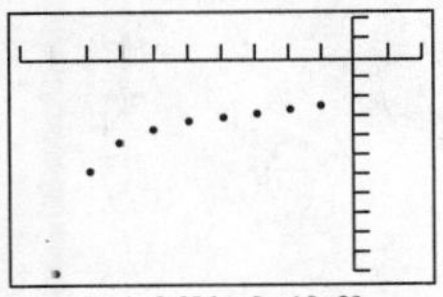

[−1, 0.2] by [−10, 2]

35. $y(\text{exact}) = \frac{1}{2}x^2 - \frac{3}{2}$; $y(2) \approx 0.4$; exact value is $\frac{1}{2}$.

37. $y(\text{exact}) = -e^{(x^2-1)/2}$; $y(2) \approx -3.4192$; exact value is $-e^{3/2} \approx -4.4817$.

39. **(a)** $y = -1$ is stable and $y = 1$ is unstable.

(b) $\frac{d^2y}{dx^2} = 2y\frac{dy}{dx} = 2y(y^2 - 1)$

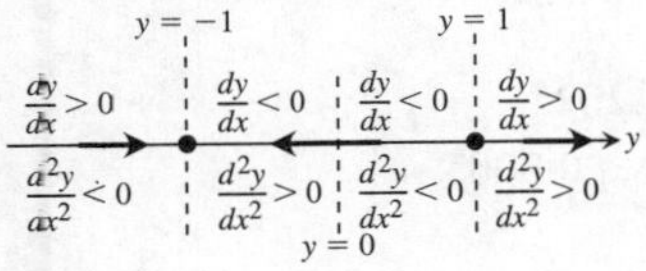

(c)

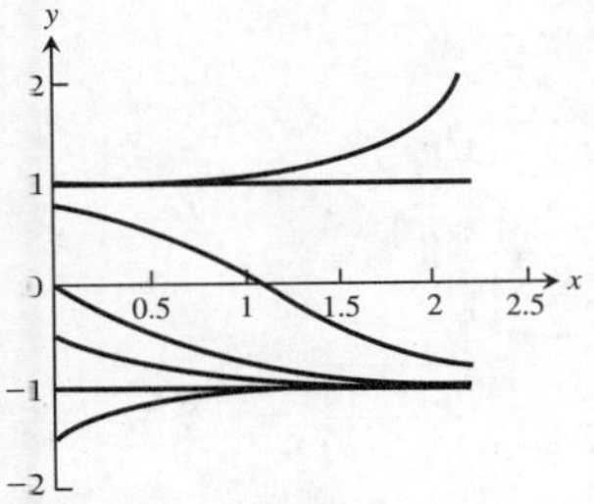

43. **(a)** 1 min: 102 gal; 10 min: 120 gal; 60 min: 220 gal

(b) $S(1) \approx 12.55$ lb, $S(10) \approx 32.22$ lb, $S(60) \approx 101.74$ lb

ADDITIONAL AND ADVANCED EXERCISES, p. 572

1. **(a)** $y = c + (y_0 - c)e^{-k(A/V)t}$

(b) Steady-state solution: $y_\infty = c$

5. $x^2(x^2 + 2y^2) = C$

7. $\ln|x| + e^{-y/x} = C$

9. $x + ye^{x/y} = c$

Chapter 10

SECTION 10.1, pp. 582–586

1. $a_1 = 1,\ a_2 = \frac{3}{4},\ a_3 = \frac{5}{9},\ a_4 = \frac{7}{16}$

3. $a_1 = 1, a_2 = -1/3, a_3 = 1/5, a_4 = -1/7$

5. $a_1 = 1/2, a_2 = 1/2, a_3 = 1/2, a_4 = 1/2$

7. $1, \frac{3}{2}, \frac{7}{4}, \frac{15}{8}, \frac{31}{16}, \frac{63}{32}, \frac{127}{64}, \frac{255}{128}, \frac{511}{256}, \frac{1023}{512}$

9. $2, 1, -\frac{1}{2}, -\frac{1}{4}, \frac{1}{8}, \frac{1}{16}, -\frac{1}{32}, -\frac{1}{64}, \frac{1}{128}, \frac{1}{256}$

11. 1, 1, 2, 3, 5, 8, 13, 21, 34, 55

13. $a_n = (-1)^{n+1}, n \ge 1$

15. $a_n = (-1)^n n^2, n = 1, 2, \ldots$

17. $a_n = \frac{3^{n-1}}{2(n+2)}, n = 1, 2, \ldots$

19. $a_n = (n^3 - 1), n = 1, 2, \ldots$

21. $a_n = 4n - 3, n \ge 1$

23. $a_n = \frac{4n+1}{n!}, n = 1, 2, \ldots$

25. $a_n = \frac{1+(-1)^n}{2}, n = 1, 2, \ldots$

27. $a_n = (-1)^{n+1}\frac{n^2}{(n+1)}, n = 1, 2, \ldots$

29. $a_n = \cos\left(\frac{\sqrt{1+n}}{\sqrt{1+n^2}}\right)$

31. Converges, 6 **33.** Converges, 1 **35.** Converges, −5
37. Diverges, ∞ **39.** Diverges **41.** Converges, 1/2
43. Converges, 0 **45.** Converges, $\sqrt{3}$ **47.** Converges, −1
49. Converges, 0 **51.** Converges, 0 **53.** Converges, 0
55. Converges, 1 **57.** Converges, e^7 **59.** Converges, 1
61. Converges, 1 **63.** Diverges, ∞ **65.** Converges, 2
67. Converges, 0 **69.** Diverges **71.** Converges, e^{-1}
73. Diverges **75.** Converges, 0 **77.** Diverges
79. Converges, $e^{2/3}$ **81.** Converges, $x (x > 0)$
83. Converges, 0 **85.** Converges, 1 **87.** Converges, 1/2
89. Converges, 1 **91.** Converges, $\pi/2$ **93.** Converges, 0
95. Converges, 0 **97.** Converges, $-\frac{1}{2}$ **99.** Converges, 0

101. 7 **103.** 4 **105.** 5 **107.** $1 + \sqrt{2}$ **109.** $x_n = 2^{n-2}$

111. **(a)** $f(x) = x^2 - 2$, $1.414213562 \approx \sqrt{2}$

(b) $f(x) = \tan(x) - 1$, $0.7853981635 \approx \pi/4$

(c) $f(x) = e^x$, diverges

113. **(b)** 1

121. Nondecreasing, bounded

123. Not nondecreasing, bounded

125. Converges, nondecreasing sequence theorem

127. Converges, nondecreasing sequence theorem

129. Not monotonic, bounded, diverges

131. Converges

133. Converges

145. **(b)** $\sqrt{3}$

SECTION 10.2, pp. 593–595

1. $s_n = \frac{2(1-(1/3)^n)}{1-(1/3)}$, 3 **3.** $s_n = \frac{1-(-1/2)^n}{1-(-1/2)}$, 2/3

5. $s_n = \frac{1}{2} - \frac{1}{n+2}, \frac{1}{2}$ **7.** $1 - \frac{1}{4} + \frac{1}{16} - \frac{1}{64} + \cdots, \frac{4}{5}$

9. $-\frac{3}{4} + \frac{9}{16} + \frac{57}{64} + \frac{249}{256} + \cdots$, diverges.

11. $(5+1) + \left(\frac{5}{2} + \frac{1}{3}\right) + \left(\frac{5}{4} + \frac{1}{9}\right) + \left(\frac{5}{8} + \frac{1}{27}\right) + \cdots, \frac{23}{2}$

13. $(1+1) + \left(\frac{1}{2} - \frac{1}{5}\right) + \left(\frac{1}{4} + \frac{1}{25}\right) + \left(\frac{1}{8} - \frac{1}{125}\right) + \cdots, \frac{17}{6}$

15. Converges, 5/3 **17.** Converges, 1/7

19. Converges, $\frac{e}{e+2}$ **21.** Diverges **23.** 37/99

25. 5/9 **27.** 1/15 **29.** 41333/33300 **31.** Diverges

33. Inconclusive **35.** Inconclusive **37.** Diverges

39. $s_n = 1 - \frac{1}{n+1}$; converges, 1

41. $s_n = \ln\sqrt{n+1}$; diverges

43. $s_n = \frac{\pi}{6} - \sin^{-1}\left(\frac{1}{n+2}\right)$, converges, $\frac{\pi}{6}$ **45.** 1/2 **47.** 5

49. -1 **51.** $-\dfrac{1}{\ln 2}$ **53.** Converges, $2 + \sqrt{2}$

55. Converges, 1 **57.** Diverges

59. Converges, $\dfrac{e^2}{e^2 - 1}$

61. Converges, $1/2$ **63.** Converges, $3/2$ **65.** Diverges

67. Converges, 4 **69.** Diverges **71.** Converges, $\dfrac{e}{e - \pi}$

73. Converges, $-5/6$ **75.** Diverges

77. $a = 1, r = -x$; converges to $1/(1 + x)$ for $|x| < 1$

79. $a = 3, r = (x - 1)/2$; converges to $6/(3 - x)$ for x in $(-1, 3)$

81. $|x| < \dfrac{1}{2}, \dfrac{1}{1 - 2x}$ **83.** $-2 < x < 0, \dfrac{1}{2 + x}$

85. $x \neq 2k\pi$, k an integer; $\dfrac{1}{1 - \cos x}$

87. **(a)** $\displaystyle\sum_{n=-2}^{\infty} \frac{1}{(n + 4)(n + 5)}$ **(b)** $\displaystyle\sum_{n=0}^{\infty} \frac{1}{(n + 2)(n + 3)}$

(c) $\displaystyle\sum_{n=5}^{\infty} \frac{1}{(n - 3)(n - 2)}$

97. **(a)** $r = \dfrac{4}{7}, 3 + 3\left(\dfrac{4}{7}\right) + 3\left(\dfrac{4}{7}\right)^2 + \cdots$

(b) $r = -\dfrac{1}{14}, \dfrac{15}{2} - \dfrac{15}{2}\left(\dfrac{1}{14}\right) + \dfrac{15}{2}\left(\dfrac{1}{14}\right)^2 - \dfrac{15}{2}\left(\dfrac{1}{14}\right)^3 + \cdots$

99. $|r| < 1, \dfrac{1 + 2r}{1 - r^2}$

101. **(a)** 16.84 mg, 17.79 mg **(b)** 17.84 mg

103. **(a)** $0, \dfrac{1}{27}, \dfrac{2}{27}, \dfrac{1}{9}, \dfrac{2}{9}, \dfrac{7}{27}, \dfrac{8}{27}, \dfrac{1}{3}, \dfrac{2}{3}, \dfrac{7}{9}, \dfrac{8}{9}, 1$

(b) $\displaystyle\sum_{n=1}^{\infty} \frac{1}{2}\left(\frac{2}{3}\right)^{n-1} = 1$ **105.** $(4/3)\pi$

SECTION 10.3, pp. 600–602

1. Converges **3.** Converges **5.** Converges **7.** Diverges

9. Converges **11.** Diverges

13. Converges; geometric series, $r = \dfrac{1}{10} < 1$

15. Diverges; $\displaystyle\lim_{n\to\infty} \frac{n}{n + 1} = 1 \neq 0$

17. Diverges; p-series, $p < 1$

19. Converges; geometric series, $r = \dfrac{1}{8} < 1$

21. Diverges; Integral Test

23. Converges; geometric series, $r = 2/3 < 1$

25. Diverges; Integral Test

27. Diverges; $\displaystyle\lim_{n\to\infty} \frac{2^n}{n + 1} \neq 0$

29. Diverges; $\displaystyle\lim_{n\to\infty} (\sqrt{n}/\ln n) \neq 0$

31. Diverges; geometric series, $r = \dfrac{1}{\ln 2} > 1$

33. Converges; Integral Test

35. Diverges; nth-Term Test

37. Converges; Integral Test

39. Diverges; nth-Term Test

41. Converges; by taking limit of partial sums

43. Converges; Integral Test

45. Converges; Integral Test **47.** $a = 1$

49. **(a)**

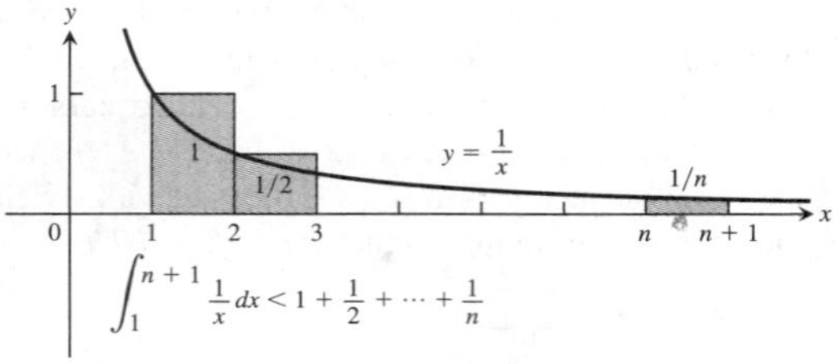

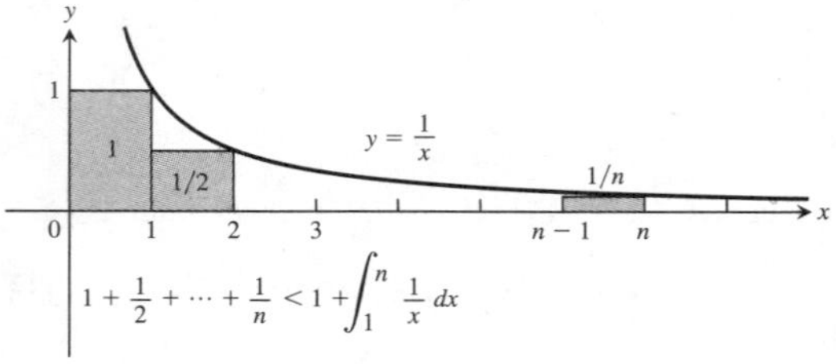

(b) ≈ 41.55

51. True **53.** $n \geq 251{,}415$

55. $s_8 = \displaystyle\sum_{n=1}^{8} \frac{1}{n^3} \approx 1.195$ **57.** 10^{60}

65. **(a)** $1.20166 \leq S \leq 1.20253$

(b) $S \approx 1.2021$, error < 0.0005

67. $\left(\dfrac{\pi^2}{6} - 1\right) \approx 0.64493$

SECTION 10.4, pp. 606–607

1. Converges; compare with $\sum (1/n^2)$

3. Diverges; compare with $\sum (1/\sqrt{n})$

5. Converges; compare with $\sum (1/n^{3/2})$

7. Converges; compare with $\displaystyle\sum \sqrt{\frac{n + 4n}{n^4 + 0}} = \sqrt{5} \sum \frac{1}{n^{3/2}}$

9. Converges

11. Diverges; limit comparison with $\sum (1/n)$

13. Diverges; limit comparison with $\sum (1/\sqrt{n})$

15. Diverges

17. Diverges; limit comparison with $\sum (1/\sqrt{n})$

19. Converges; compare with $\sum (1/2^n)$

21. Diverges; nth-Term Test

23. Converges; compare with $\sum (1/n^2)$

25. Converges; $\left(\dfrac{n}{3n + 1}\right)^n < \left(\dfrac{n}{3n}\right)^n = \left(\dfrac{1}{3}\right)^n$

27. Diverges; direct comparison with $\sum (1/n)$

29. Diverges; limit comparison with $\sum (1/n)$

31. Diverges; limit comparison with $\sum (1/n)$

33. Converges; compare with $\sum (1/n^{3/2})$

35. Converges; $\dfrac{1}{n2^n} \leq \dfrac{1}{2^n}$ **37.** Divergent **39.** Convergent

41. Diverges; comparison with $\sum (1/n)$

43. Converges; comparison with $\displaystyle\sum \frac{1}{n(n - 1)}$ or limit comparison with $\sum (1/n^2)$

45. Diverges; limit comparison with $\sum (1/n)$

47. Converges; $\dfrac{\tan^{-1} n}{n^{1.1}} < \dfrac{\pi/2}{n^{1.1}}$
49. Converges; compare with $\sum (1/n^2)$
51. Diverges; limit comparison with $\sum (1/n)$
53. Converges; limit comparison with $\sum (1/n^2)$
55. Diverges nth-Term Test
67. Converges **69.** Converges **71.** Converges

SECTION 10.5, pp. 612–613

1. Converges **3.** Diverges **5.** Converges
7. Converges **9.** Converges **11.** Diverges
13. Convergent **15.** Converges **17.** Converges; Ratio Test
19. Diverges; Ratio Test **21.** Convergent **23.** Convergent
25. Diverges; $\lim\limits_{n\to\infty}\left(1-\dfrac{3}{n}\right)^n = e^{-3} \neq 0$
27. Converges; compare with $\sum (1/n^2)$
29. Diverges; compare with $\sum (1/(2n))$ **31.** Diverges; $a_n \nrightarrow 0$
33. Converges; Ratio Test **35.** Converges; Ratio Test
37. Converges; Ratio Test **39.** Converges; Root Test
41. Converges; compare with $\sum (1/n^2)$
43. Converges; Ratio Test **45.** Diverges; Ratio Test
47. Converges; Ratio Test **49.** Diverges; Ratio Test
51. Converges; Ratio Test **53.** Converges; Ratio Test
55. Diverges; $a_n = \left(\dfrac{1}{3}\right)^{(1/n!)} \to 1$ **57.** Converges; Ratio Test
59. Diverges; Root Test **61.** Converges; Root Test
63. Converges; Ratio Test
65. **(a)** Diverges; nth-Term Test **(b)** Diverges; Root Test
(c) Converges; Root Test **(d)** Converges; Ratio Test
69. Yes **71.** Converges absolutely

SECTION 10.6, pp. 618–620

1. Converges by Alternating Series Test
3. Converges; Alternating Series Test
5. Converges; Alternating Series Test
7. Diverges; $a_n \nrightarrow 0$ **9.** Diverges; $a_n \nrightarrow 0$
11. Converges; Alternating Series Test
13. Converges by Alternating Series Test
15. Converges absolutely. Series of absolute values is a convergent geometric series.
17. Converges conditionally; $1/\sqrt{n} \to 0$ but $\sum_{n=1}^{\infty} \dfrac{1}{\sqrt{n}}$ diverges.
19. Absolutely Convergent **21.** Not Convergent
23. Diverges; $\dfrac{3+n}{5+n} \to 1$
25. Converges conditionally; $\left(\dfrac{1}{n^2}+\dfrac{1}{n}\right) \to 0$ but $(1+n)/n^2 > 1/n$
27. Converges absolutely; Ratio Test
29. Converges absolutely by Integral Test
31. Diverges; $a_n \nrightarrow 0$
33. Converges absolutely by Ratio Test
35. Converges absolutely, since $\left|\dfrac{\cos n\pi}{n\sqrt{n}}\right| = \left|\dfrac{(-1)^{n+1}}{n^{3/2}}\right| = \dfrac{1}{n^{3/2}}$ (convergent p-series)
37. Converges absolutely by Root Test
39. Diverges; $a_n \to \infty$
41. Converges conditionally;
$\sqrt{n+1} - \sqrt{n} = 1/(\sqrt{n} + \sqrt{n+1}) \to 0$, but series of absolute values diverges (compare with $\sum (1/\sqrt{n})$).
43. Diverges, $a_n \to 1/2 \neq 0$
45. Converges absolutely;
$\operatorname{sech} n = \dfrac{2}{e^n + e^{-n}} = \dfrac{2e^n}{e^{2n}+1} < \dfrac{2e^n}{e^{2n}} = \dfrac{2}{e^n}$, a term from a convergent geometric series.
47. Converges conditionally; $\sum (-1)^{n+1}\dfrac{1}{2(n+1)}$ converges by Alternating Series Test; $\sum \dfrac{1}{2(n+1)}$ diverges by limit comparison with $\sum (1/n)$.
49. $|\text{Error}| < 0.2$ **51.** $|\text{Error}| < 2 \times 10^{-11}$
53. $n \geq 31$ **55.** $n \geq 4$ **57.** Converges; Root Test
59. Converges; Limit of Partial Sums
61. Converges; Ratio Test **63.** Diverges; p-series Test
65. Converges; Root Test **67.** Converges; Limit Comparison Test
69. Diverges; Limit of Partial Sums
71. Diverges; Limit Comparison Test
73. Diverges; nth-Term Test
75. Diverges; Limit of Partial Sums
77. Converges; Limit Comparison Test
79. Converges; Limit Comparison Test
81. Converges; Ratio Test
83. 0.54030 **85.** **(a)** $a_n \geq a_{n+1}$ **(b)** $-1/2$

SECTION 10.7, pp. 629–632

1. **(a)** $1, -1 < x < 1$ **(b)** $-1 < x < 1$ **(c)** none
3. **(a)** $1/4, -1/2 < x < 0$ **(b)** $-1/2 < x < 0$ **(c)** none
5. **(a)** $10, -8 < x < 12$ **(b)** $-8 < x < 12$ **(c)** none
7. **(a)** $1, -1 < x < 1$ **(b)** $-1 < x < 1$ **(c)** none
9. **(a)** $3, -3 \leq x \leq 3$ **(b)** $-3 \leq x \leq 3$ **(c)** none
11. **(a)** ∞, for all x **(b)** for all x **(c)** none
13. **(a)** $1/2, -1/2 < x < 1/2$ **(b)** $-1/2 < x < 1/2$ **(c)** none
15. **(a)** $1, -1 \leq x < 1$ **(b)** $-1 < x < 1$ **(c)** $x = -1$
17. **(a)** $5, -8 < x < 2$ **(b)** $-8 < x < 2$ **(c)** none
19. **(a)** $\dfrac{1}{4}, -\dfrac{3}{4} \leq x \leq -\dfrac{1}{4}$ **(b)** $-3 < x < 3$ **(c)** none
21. **(a)** $\infty, x < 0$ or $x \geq 2$ **(b)** none **(c)** none
23. **(a)** $1, -1 < x < 1$ **(b)** $-1 < x < 1$ **(c)** none
25. **(a)** $0, x = 0$ **(b)** $x = 0$ **(c)** none
27. **(a)** $2, -4 < x \leq 0$ **(b)** $-4 < x < 0$ **(c)** $x = 0$
29. **(a)** $1, -1 \leq x \leq 1$ **(b)** $-1 \leq x \leq 1$ **(c)** none
31. **(a)** $1/4, 1 \leq x \leq 3/2$ **(b)** $1 \leq x \leq 3/2$ **(c)** none
33. **(a)** ∞, for all x **(b)** for all x **(c)** none
35. **(a)** $1, -1 \leq x < 1$ **(b)** $-1 < x < 1$ **(c)** -1
37. 3 **39.** 8 **41.** ∞ **43.** $-1/3 < x < 1/3, 1/(1-3x)$
45. $-1 < x < 3, 4/(3 + 2x - x^2)$
47. $0 < x < 16, 2/(4 - \sqrt{x})$
49. $-\sqrt{2} < x < \sqrt{2}, 3/(2 - x^2)$
51. $\dfrac{2}{x} = \sum\limits_{n=0}^{\infty} 2(-1)^n (x-1)^n, 0 < x < 2$
53. $\sum\limits_{n=0}^{\infty} (-\frac{1}{3})^n (x-5)^n, 2 < x < 8$
55. $1 < x < 5, 2/(x-1), \sum\limits_{n=1}^{\infty} (-\frac{1}{2})^n\, n(x-3)^{n-1}$,
$1 < x < 5, -2/(x-1)^2$

57. **(a)** $\cos x = 1 - \frac{x^2}{2!} + \frac{x^4}{4!} - \frac{x^6}{6!} + \frac{x^8}{8!} - \frac{x^{10}}{10!} + \cdots;$ converges for all x

(b) Same answer as part (c)

(c) $2x - \frac{2^3x^3}{3!} + \frac{2^5x^5}{5!} - \frac{2^7x^7}{7!} + \frac{2^9x^9}{9!} - \frac{2^{11}x^{11}}{11!} + \cdots$

59. **(a)** $\frac{x^2}{2} + \frac{x^4}{12} + \frac{x^6}{45} + \frac{17x^8}{2520} + \frac{31x^{10}}{14175}, -\frac{\pi}{2} < x < \frac{\pi}{2}$

(b) $1 + x^2 + \frac{2x^4}{3} + \frac{17x^6}{45} + \frac{62x^8}{315} + \cdots, -\frac{\pi}{2} < x < \frac{\pi}{2}$

65. **(a)** T **(b)** T **(c)** F **(d)** T **(e)** N **(f)** F **(g)** N **(h)** T

SECTION 10.8, pp. 636–637

1. $P_0(x)=1,\ P_1(x)=1+3x,\ P_2(x)=1+3x+\frac{9}{2}x^2,$
$P_3(x)=1+3x+\frac{9}{2}x^2+\frac{9}{2}x^3$

3. $P_0(x)=0,\ P_1(x)=-(x-1),\ P_2(x)=-(x-1)+\frac{1}{2}(x-1)^2,$
$P_3(x)=-(x-1)+\frac{1}{2}(x-1)^2-\frac{1}{3}(x-1)^3$

5. $P_0(x) = \frac{1}{2},\ P_1(x) = \frac{1}{2} - \frac{1}{4}(x-2),$
$P_2(x) = \frac{1}{2} - \frac{1}{4}(x-2) + \frac{1}{8}(x-2)^2,$
$P_3(x) = \frac{1}{2} - \frac{1}{4}(x-2) + \frac{1}{8}(x-2)^2 - \frac{1}{16}(x-2)^3$

7. $P_0(x) = \frac{\sqrt{2}}{2},\ P_1(x) = \frac{\sqrt{2}}{2} + \frac{\sqrt{2}}{2}\left(x - \frac{\pi}{4}\right),$
$P_2(x) = \frac{\sqrt{2}}{2} + \frac{\sqrt{2}}{2}\left(x - \frac{\pi}{4}\right) - \frac{\sqrt{2}}{4}\left(x - \frac{\pi}{4}\right)^2,$
$P_3(x) = \frac{\sqrt{2}}{2} + \frac{\sqrt{2}}{2}\left(x - \frac{\pi}{4}\right) - \frac{\sqrt{2}}{4}\left(x - \frac{\pi}{4}\right)^2 - \frac{\sqrt{2}}{12}\left(x - \frac{\pi}{4}\right)^3$

9. $P_0(x) = 2,\ P_1(x) = 2 + \frac{1}{4}(x-4),$
$P_2(x) = 2 + \frac{1}{4}(x-4) - \frac{1}{64}(x-4)^2,$
$P_3(x) = 2 + \frac{1}{4}(x-4) - \frac{1}{64}(x-4)^2 + \frac{1}{512}(x-4)^3$

11. $\sum_{n=0}^{\infty} \frac{(-x)^n}{n!} = 1 - x + \frac{x^2}{2!} - \frac{x^3}{3!} + \frac{x^4}{4!} - \cdots$

13. $\sum_{n=0}^{\infty} x^n = 1 + x + x^2 + x^3 + \cdots$

15. $\sum_{n=0}^{\infty} \frac{(-1)^n 3^{2n} x^{2n}}{(2n)!} = 1 - \frac{3^2x^2}{2!} + \frac{3^4x^4}{4!} + \cdots$

17. $7\sum_{n=0}^{\infty} \frac{(-1)^n x^{2n}}{(2n)!}$ **19.** $\sum_{n=0}^{\infty} \frac{x^{2n}}{(2n)!}$

21. $5 + 2x - 3x^3 + x^4$ **23.** $\sum_{n=1}^{\infty} (-1)^{n+1} \frac{x^{2n}}{(2n-1)!}$

25. $11 + 16(x-2) + 16(x-2)^2 + (x-2)^3$

27. $21 - 36(x+2) + 25(x+2)^2 - 8(x+2)^3 + (x+2)^4$

29. $\sum_{n=0}^{\infty} (-1)^n (n+1)(x-1)^n$ **31.** $\sum_{n=0}^{\infty} \frac{e^2}{n!}(x-2)^n$

33. $\sum_{n=0}^{\infty} (-1)^{n+1} \frac{2^{2n}}{(2n)!}\left(x - \frac{\pi}{4}\right)^{2n}$

35. $-2 - x - 2x^2 - \ldots, -1 < x < 1$

37. $x^2 - \frac{1}{2}x^3 + \frac{1}{6}x^4 + \cdots, -1 < x < 1$

39. $x^4 + x^6 + \frac{x^8}{2} + \cdots, (-\infty, \infty)$

41. $L(x) = 0,\ Q(x) = -x^2/2$ **43.** $L(x) = 1,\ Q(x) = 1 + x^2/2$

45. $L(x) = 1, Q(x) = 1 - \frac{x^2}{2}$

SECTION 10.9, pp. 643–644

1. $\sum_{n=0}^{\infty} \frac{(-3x)^n}{n!} = 1 - 3x + \frac{3^2x^2}{2!} - \frac{3^3x^3}{3!} + \cdots$

3. $\sum_{n=0}^{\infty} \frac{5(-1)^n(-x)^{2n+1}}{(2n+1)!} = \sum_{n=0}^{\infty} \frac{5(-1)^{n+1}x^{2n+1}}{(2n+1)!}$
$= -5x + \frac{5x^3}{3!} - \frac{5x^5}{5!} + \frac{5x^7}{7!} + \cdots$

5. $\sum_{n=0}^{\infty} \frac{(-1)^n(5x^2)^{2n}}{(2n)!} = 1 - \frac{25x^4}{2!} + \frac{625x^8}{4!} - \cdots$

7. $\sum_{n=1}^{\infty} (-1)^{n+1} \frac{x^{2n}}{n} = x^2 - \frac{x^4}{2} + \frac{x^6}{3} - \frac{x^8}{4} + \cdots$

9. $\sum_{n=0}^{\infty} (-1)^n \left(\frac{3}{4}\right)^n x^{3n} = 1 - \frac{3}{4}x^3 + \frac{3^2}{4^2}x^6 - \frac{3^3}{4^3}x^9 + \cdots$

11. $\ln(4) + \sum_{n=1}^{\infty} \frac{(-1)^{n+1} x^n}{2^{n+1} \cdot n} = \ln(4) + \frac{1}{4}x - \frac{1}{16}x^2 + \frac{1}{48}x^3 - \cdots$

13. $\sum_{n=0}^{\infty} \frac{(-1)^n x^{n+1}}{n!} = x - x^2 + \frac{x^3}{2!} - \frac{x^4}{3!} + \frac{x^5}{4!} - \cdots$

15. $\sum_{n=2}^{\infty} \frac{x^{2n}}{(2n)!} = \frac{x^4}{4!} + \frac{x^6}{6!} + \frac{x^8}{8!} + \frac{x^{10}}{10!} + \cdots$

17. $\sum_{n=0}^{\infty} \frac{(-1)^n \pi^{2n+1} x^{2n+2}}{(2n+1)!} = \pi x^2 - \frac{\pi^3 x^4}{3!} + \frac{\pi^5 x^6}{5!} - \frac{\pi^7 x^8}{7!} + \cdots$

19. $1 + \sum_{n=1}^{\infty} \frac{(-1)^n (2x)^{2n}}{2 \cdot (2n)!} =$
$1 - \frac{(2x)^2}{2 \cdot 2!} + \frac{(2x)^4}{2 \cdot 4!} - \frac{(2x)^6}{2 \cdot 6!} + \frac{(2x)^8}{2 \cdot 8!} - \cdots$

21. $x^3 \sum_{n=0}^{\infty} (-1)^n (2x)^n = x^3 - 2x^4 + 4x^5 - \cdots$

23. $\sum_{n=1}^{\infty} nx^{n-1} = 1 + 2x + 3x^2 + 4x^3 + \cdots$

25. $\sum_{n=1}^{\infty} (-1)^{n+1} \frac{x^{4n-1}}{2n-1} = x^3 - \frac{x^7}{3} + \frac{x^{11}}{5} - \frac{x^{15}}{7} + \cdots$

27. $\sum_{n=0}^{\infty} \left(\frac{1}{n!} - (-1)^n\right) x^n = 2 - \frac{1}{2}x^2 + \frac{7}{6}x^3 - \frac{23}{24}x^4 + \cdots$

29. $\sum_{n=1}^{\infty} \frac{(-1)^{n-1} x^{3n+1}}{5n} = \frac{x^4}{5} - \frac{x^7}{10} + \frac{x^{10}}{15} - \cdots$

31. $x + x^2 + \frac{x^3}{3} - \frac{x^5}{30} + \cdots$

33. $x^2 - \frac{2}{3}x^4 + \frac{23}{45}x^6 - \frac{44}{105}x^8 + \cdots$

35. $1 + x + \frac{1}{2}x^2 - \frac{1}{8}x^4 + \cdots$

37. $-x+\frac{1}{2}x^2-\frac{5}{24}x^4+\frac{23}{120}x^5-\cdots$

39. $|\text{Error}| \le \frac{1}{10^4 \cdot 4!} < 4.2 \times 10^{-6}$

41. $|x| < (0.048)^{1/5} < 0.54481$

43. $|\text{Error}| < \frac{(10^{-4})^3}{6} < 1.67 \times 10^{-13}, -10^{-4} < x < 0$

45. $|\text{Error}| < \frac{(3)^{0.01}(0.01)^3}{6} < 1.69 \times 10^{-7}$

53. **(a)** $Q(x) = 1 + kx + \frac{k(k-1)}{2}x^2$ **(b)** $0 \le x < 100^{-1/3}$

SECTION 10.10, pp. 651–653

1. $1 + \frac{x}{2} - \frac{x^2}{8} + \frac{x^3}{16}$ **3.** $1 + 3x + 6x^2 + 10x^3$

5. $1 - x + \frac{3x^2}{4} - \frac{x^3}{2}$ **7.** $1 - \frac{x^3}{2} + \frac{3x^6}{8} - \frac{5x^9}{16}$

9. $1 + \frac{3}{4}x^2 + \frac{3}{32}x^4 - \frac{1}{128}x^6$

11. $(1 + x)^4 = 1 + 4x + 6x^2 + 4x^3 + x^4$

13. $(1 - 2x)^3 = 1 - 6x + 12x^2 - 8x^3$ **15.** 0.0713362

17. 0.4969536 **19.** 0.0999445 **21.** 0.10000

23. $\frac{1}{13 \cdot 6!} \approx 0.00011$ **25.** $\frac{x^3}{3} - \frac{x^7}{7 \cdot 3!} + \frac{x^{11}}{11 \cdot 5!}$

27. **(a)** $\frac{x^2}{2} - \frac{x^4}{12}$

(b) $\frac{x^2}{2} - \frac{x^4}{3 \cdot 4} + \frac{x^6}{5 \cdot 6} - \frac{x^8}{7 \cdot 8} + \cdots + (-1)^{15}\frac{x^{32}}{31 \cdot 32}$

29. 1/2 **31.** −1/24 **33.** 1/3 **35.** −1 **37.** 2

39. 3/2 **41.** e **43.** $\cos\frac{3}{4}$ **45.** $\frac{\sqrt{3}}{2}$ **47.** $\frac{x^3}{1-x}$

49. $\frac{x^3}{1+x^2}$ **51.** $\frac{-1}{(1+x)^2}$ **55.** 500 terms **57.** 4 terms

59. **(a)** $x + \frac{x^3}{6} + \frac{3x^5}{40} + \frac{5x^7}{112}$, radius of convergence = 1

(b) $\frac{\pi}{2} - x - \frac{x^3}{6} - \frac{3x^5}{40} - \frac{5x^7}{112}$

61. $1 - 2x + 3x^2 - 4x^3 + \cdots$

67. **(a)** −1 **(b)** $(1/\sqrt{2})(1 + i)$ **(c)** $-i$

71. $x + x^2 + \frac{1}{3}x^3 - \frac{1}{30}x^5 + \cdots$, for all x

PRACTICE EXERCISES, pp. 654–656

1. Converges to 1 **3.** Converges to −1
5. Diverges **7.** Converges to 0
9. Converges to 1 **11.** Converges to e^{-5}
13. Converges to 3 **15.** Converges to ln 2
17. Diverges **19.** 1/6
21. 3/2 **23.** $e/(e-1)$ **25.** Diverges
27. Converges conditionally **29.** Converges conditionally
31. Converges absolutely **33.** Converges absolutely
35. Converges absolutely **37.** Converges absolutely
39. Converges absolutely **41.** Converges absolutely
43. Diverges
47. **(a)** $3, -7 \le x < -1$ **(b)** $-7 < x < -1$ **(c)** $x = -7$
49. **(a)** $1/3, 0 \le x \le 2/3$ **(b)** $0 \le x \le 2/3$ **(c)** None
51. **(a)** ∞, for all x **(b)** For all x **(c)** None
53. **(a)** $\sqrt{3}, -\sqrt{3} < x < \sqrt{3}$ **(b)** $-\sqrt{3} < x < \sqrt{3}$ **(c)** None
55. **(a)** $e, -e < x < e$ **(b)** $-e < x < e$ **(c)** Empty set

57. $\frac{1}{1+x}, \frac{1}{4}, \frac{4}{5}$ **59.** $\sin x, \pi, 0$

61. $e^x, \ln 2, 2$ **63.** $\sum_{n=0}^{\infty} 2^n x^n$

65. $\sum_{n=0}^{\infty} \frac{(-1)^n \pi^{2n+1} x^{2n+1}}{(2n+1)!}$ **67.** $\sum_{n=0}^{\infty} \frac{(-1)^n x^{10n/3}}{(2n)!}$

69. $\sum_{n=0}^{\infty} \frac{((\pi x)/2)^n}{n!}$

71. $2 - \frac{(x+1)}{2 \cdot 1!} + \frac{3(x+1)^2}{2^3 \cdot 2!} + \frac{9(x+1)^3}{2^5 \cdot 3!} + \cdots$

73. $\frac{1}{4} - \frac{1}{4^2}(x-3) + \frac{1}{4^3}(x-3)^2 - \frac{1}{4^4}(x-3)^3$

75. 0.4849171431 **77.** 0.4872223583 **79.** 7/2 **81.** 1/12

83. −2 **85.** $r = -3, s = 9/2$ **87.** 2/3

89. $\ln\left(\frac{n+1}{2n}\right)$; the series converges to $\ln\left(\frac{1}{2}\right)$.

91. **(a)** ∞ **(b)** $a = 1, b = 0$ **93.** It converges.

101. **(a)** Converges; Limit Comparison Test
(b) Converges; Direct Comparison Test
(c) Diverges; nth-Term Test

103. 2

ADDITIONAL AND ADVANCED EXERCISES, pp. 656–658

1. Converges; Comparison Test
3. Diverges; nth-Term Test
5. Converges; Comparison Test
7. Diverges; nth-Term Test

9. With $a = \pi/3$, $\cos x = \frac{1}{2} - \frac{\sqrt{3}}{2}(x - \pi/3) - \frac{1}{4}(x - \pi/3)^2 + \frac{\sqrt{3}}{12}(x - \pi/3)^3 + \cdots$

11. With $a = 0$, $e^x = 1 + x + \frac{x^2}{2!} + \frac{x^3}{3!} + \cdots$

13. With $a = 22\pi$, $\cos x = 1 - \frac{1}{2}(x - 22\pi)^2 + \frac{1}{4!}(x - 22\pi)^4 - \frac{1}{6!}(x - 22\pi)^6 + \cdots$

15. Converges, limit $= b$ **17.** $\pi/2$ **21.** $b = \pm\frac{1}{5}$

23. $a = 2, L = -7/6$ **27.** **(b)** Yes

33. **(a)** $\sum_{n=1}^{\infty} nx^{n-1}$ **(b)** 6 **(c)** $\frac{1}{q}$

35. **(a)** $R_n = C_0 e^{-kt_0}(1 - e^{-nkt_0})/(1 - e^{-kt_0})$,
$R = C_0(e^{-kt_0})/(1 - e^{-kt_0}) = C_0/(e^{kt_0} - 1)$

(b) $R_1 = 1/e \approx 0.368$,
$R_{10} = R(1 - e^{-10}) \approx R(0.9999546) \approx 0.58195$;
$R \approx 0.58198$; $0 < (R - R_{10})/R < 0.0001$

(c) 7

Chapter 11

SECTION 11.1, pp. 665–668

1. $y = x^2$

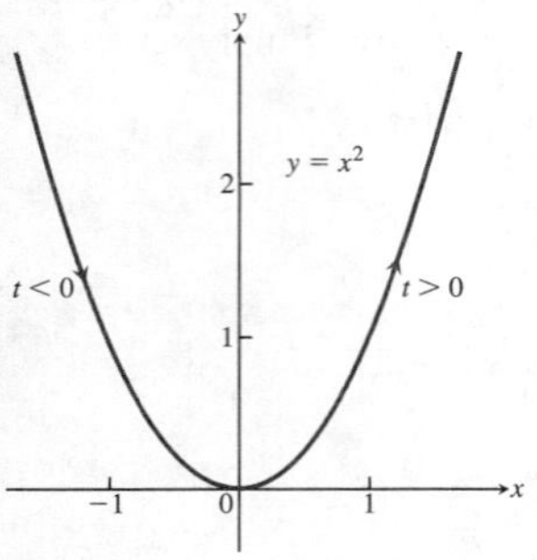

3. $y = 2x - 1$

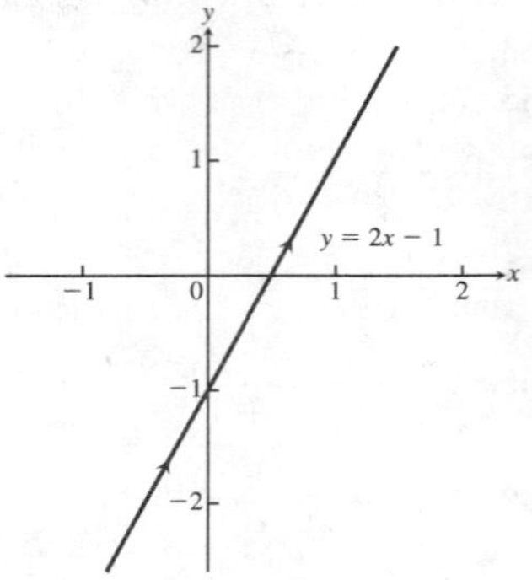

5. $x^2 + y^2 = 1$

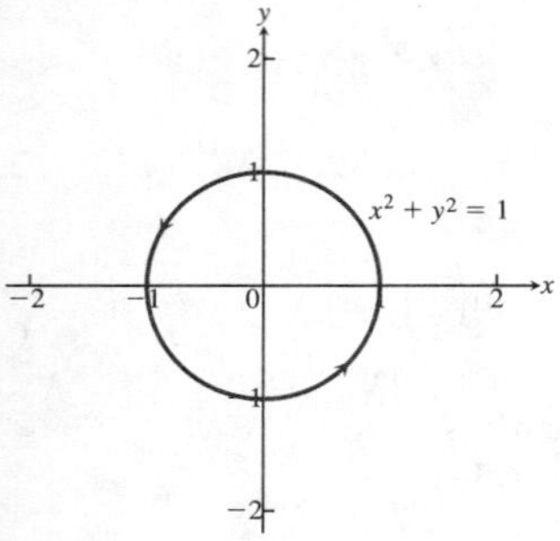

7.

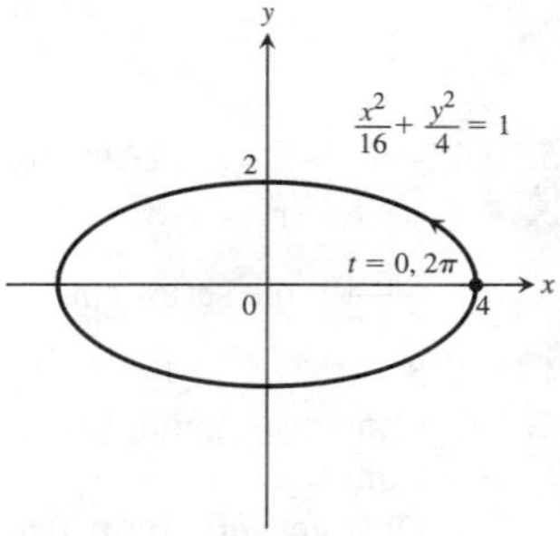

9.

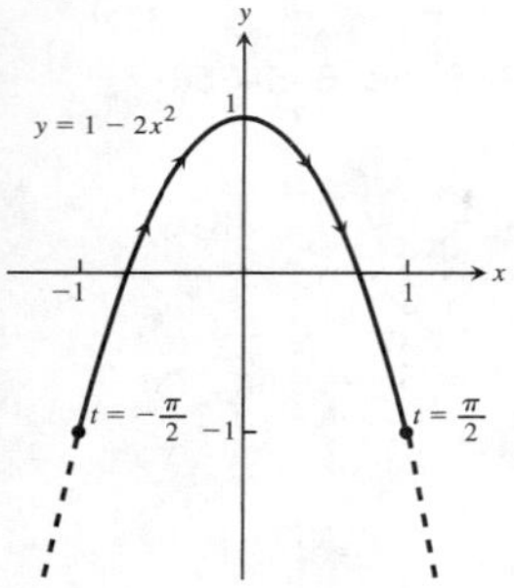

11. $y = x^3 - 2x^2$

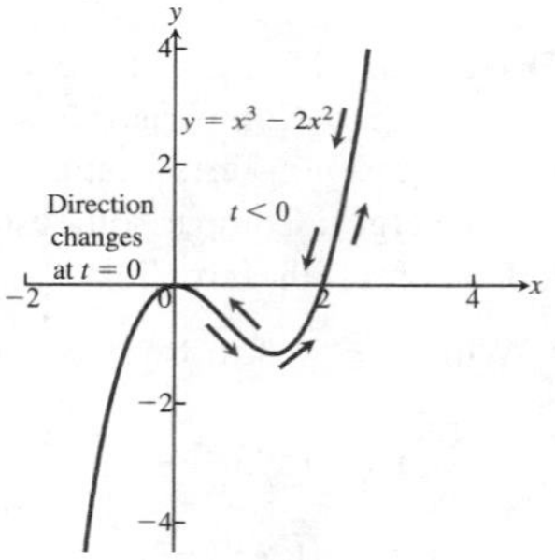

13.

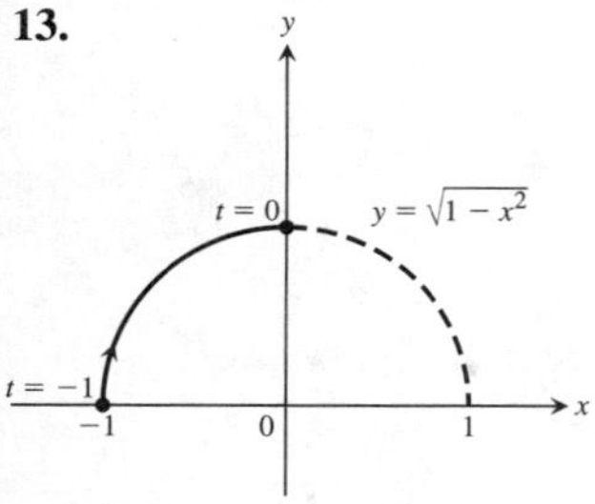

15.

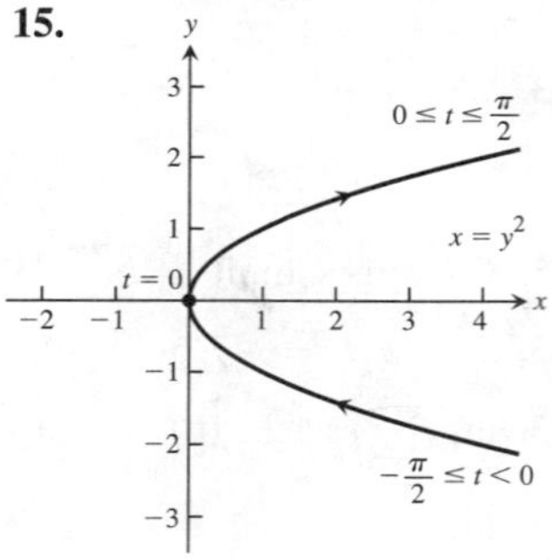

17. $x^2 - y^2 = 1$

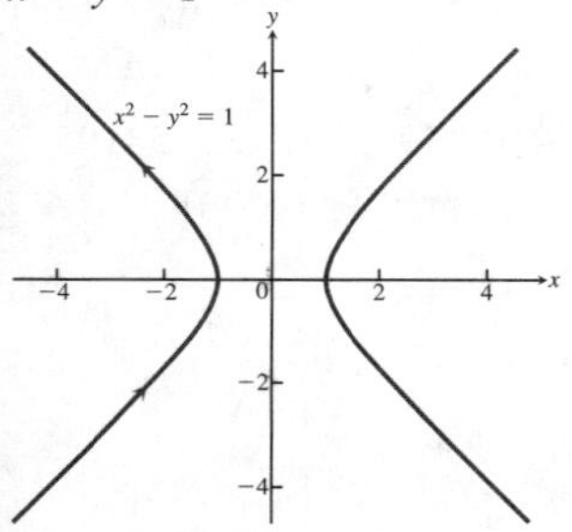

19. D **21.** E **23.** C

25.

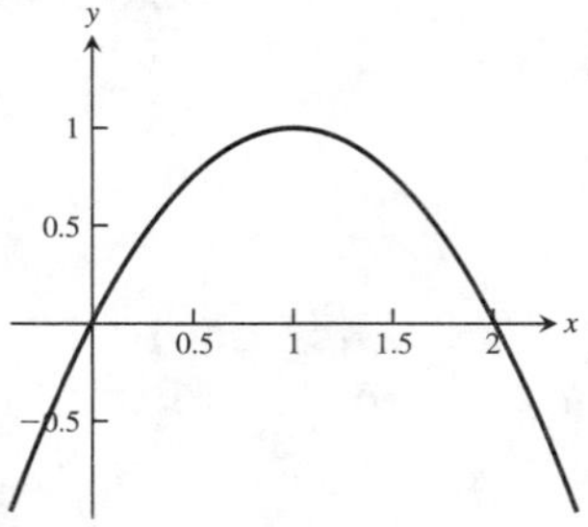

27.

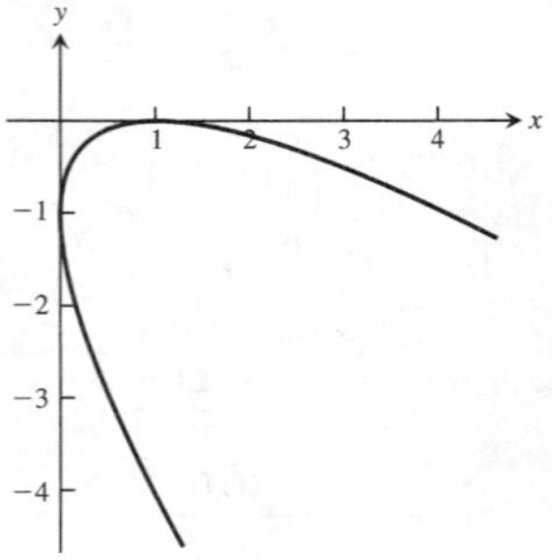

29. **(a)** $x = a\cos t,\quad y = -a\sin t,\quad 0 \le t \le 2\pi$
(b) $x = a\cos t,\quad y = a\sin t,\quad 0 \le t \le 2\pi$
(c) $x = a\cos t,\quad y = -a\sin t,\quad 0 \le t \le 4\pi$
(d) $x = a\cos t,\quad y = a\sin t,\quad 0 \le t \le 4\pi$

31. $x = -2 + 5t$ and $y = -5 + 6t,\quad 0 \le t \le 1$

33. Possible answer: $x = t^2 + 1,\quad y = t,\quad t \le 0$

35. Possible answer: $x = 2 - 3t,\quad y = 3 - 4t,\quad t \ge 0$

37. Possible answer: $x = 2\cos t,\quad y = 2|\sin t|,\quad 0 \le t \le 4\pi$

39. Possible answer: $x = \dfrac{-at}{\sqrt{2}\sqrt{1+t^2}}, y = \dfrac{a}{\sqrt{2}\sqrt{1+t^2}}, -\infty < t < \infty$

41. Possible answer: $x = \dfrac{4}{1 + 2\tan\theta},\quad y = \dfrac{4\tan\theta}{1 + 2\tan\theta},$
$0 \le \theta < \pi/2$ and $x = 0,\ y = 2$ if $\theta = \pi/2$

43. Possible answer: $x = 2 - \cos t,\quad y = \sin t,\quad 0 \le t \le 2\pi$

45. $x = 2\cot t,\quad y = 2\sin^2 t,\quad 0 < t < \pi$

47. $x = a\sin^2 t\tan t,\quad y = a\sin^2 t,\quad 0 \le t < \pi/2$ **49.** (1, 1)

53.

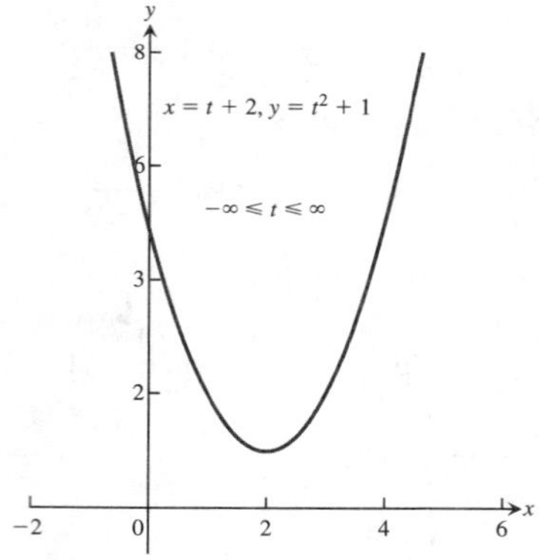

SECTION 11.2, pp. 675–677

1. $y = -x + 2\sqrt{2},\quad \dfrac{d^2y}{dx^2} = -\sqrt{2}$

3. $y = -\dfrac{1}{2}x + 2\sqrt{2},\quad \dfrac{d^2y}{dx^2} = -\dfrac{\sqrt{2}}{4}$

5. $y = x + \dfrac{1}{4},\quad \dfrac{d^2y}{dx^2} = -2$ **7.** $y = 2x - \sqrt{3},\quad \dfrac{d^2y}{dx^2} = -3\sqrt{3}$

9. $9x + y = 6,\quad \dfrac{d^2y}{dx^2} = 54$

11. $y = \sqrt{3}x - \dfrac{\pi\sqrt{3}}{3} + 2,\quad \dfrac{d^2y}{dx^2} = -4$

13. $y = 9x - 1,\quad \dfrac{d^2y}{dx^2} = 108$ **15.** $-\dfrac{3}{16}$ **17.** -6

19. 1 **21.** $3a^2\pi$ **23.** $|ab|\pi$ **25.** 4 **27.** $4\sqrt{3}$

29. π^2 **31.** $8\pi^2$ **33.** $\frac{52\pi}{3}$ **35.** $3\pi\sqrt{5}$

37. $(\bar{x}, \bar{y}) = \left(\frac{12}{\pi} - \frac{24}{\pi^2}, \frac{24}{\pi^2} - 2\right)$

39. $(\bar{x}, \bar{y}) = \left(\frac{1}{3}, \pi - \frac{4}{3}\right)$ **41. (a)** π **(b)** π

43. (a) $x = 1,\ y = 0,\ \frac{dy}{dx} = \frac{1}{2}$ **(b)** $x = 0,\ y = 3,\ \frac{dy}{dx} = 0$

(c) $x = \frac{\sqrt{3} - 1}{2},\ y = \frac{3 - \sqrt{3}}{2},\ \frac{dy}{dx} = \frac{2\sqrt{3} - 1}{\sqrt{3} - 2}$

45. $\left(\frac{\sqrt{2}}{2}, 1\right)$, $y = 2x$ at $t = 0$, $y = -2x$ at $t = \pi$

47. (a) 8a **(b)** $\frac{64\pi}{3}$ **49.** $32\pi/15$

SECTION 11.3, pp. 680–681

1. a, e; b, g; c, h; d, f **3.**

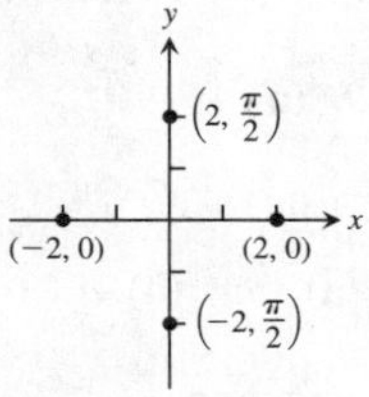

(a) $\left(2, \frac{\pi}{2} + 2n\pi\right)$ and $\left(-2, \frac{\pi}{2} + (2n + 1)\pi\right)$, n an integer

(b) $(2, 2n\pi)$ and $(-2, (2n + 1)\pi)$, n an integer

(c) $\left(2, \frac{3\pi}{2} + 2n\pi\right)$ and $\left(-2, \frac{3\pi}{2} + (2n + 1)\pi\right)$, n an integer

(d) $(2, (2n + 1)\pi)$ and $(-2, 2n\pi)$, n an integer

5. (a) $(3, 0)$ **(b)** $(-3, 0)$ **(c)** $(-1, \sqrt{3})$ **(d)** $(1, \sqrt{3})$
(e) $(3, 0)$ **(f)** $(1, \sqrt{3})$ **(g)** $(-3, 0)$ **(h)** $(-1, \sqrt{3})$

7. (a) $(2\sqrt{2}, \pi/4)$ **(b)** $(\sqrt{2}, 5\pi/4)$ **(c)** $\left(2, \frac{11\pi}{6}\right)$
(d) $\left(5, \pi - \tan^{-1}\frac{4}{3}\right)$

9. (a) $\left(-3\sqrt{2}, \frac{5\pi}{4}\right)$ **(b)** $(-1, 0)$ **(c)** $\left(-2, \frac{5\pi}{3}\right)$
(d) $\left(-5, \pi - \tan^{-1}\frac{3}{4}\right)$

11.

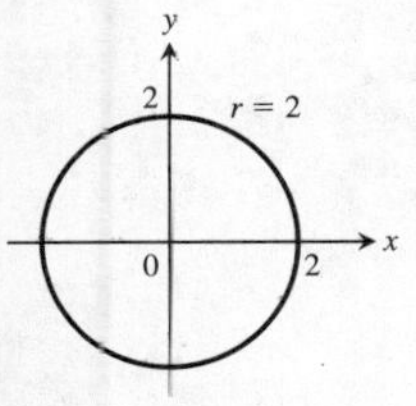

13.

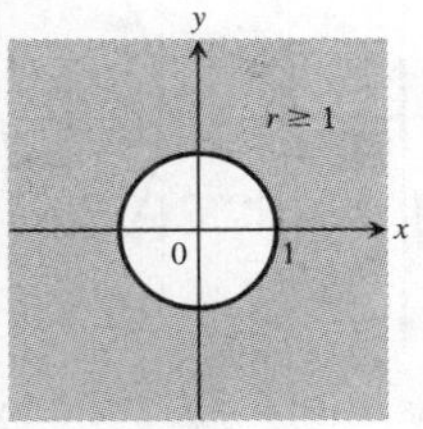

15.

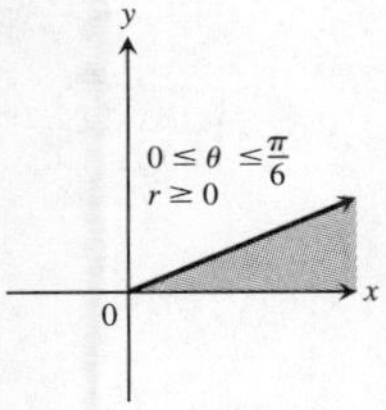

17.

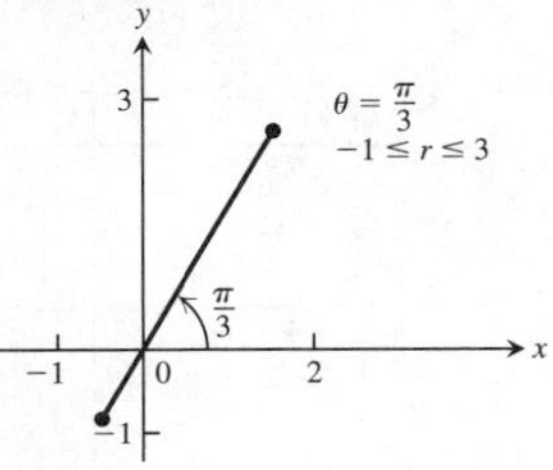

19.

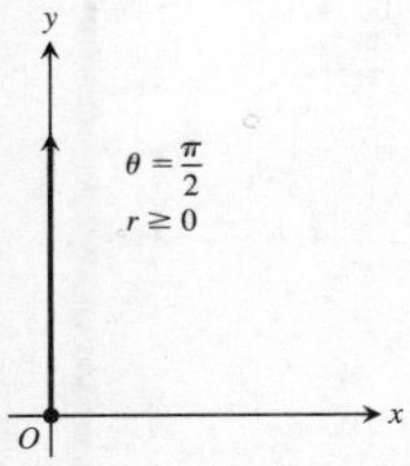

21.

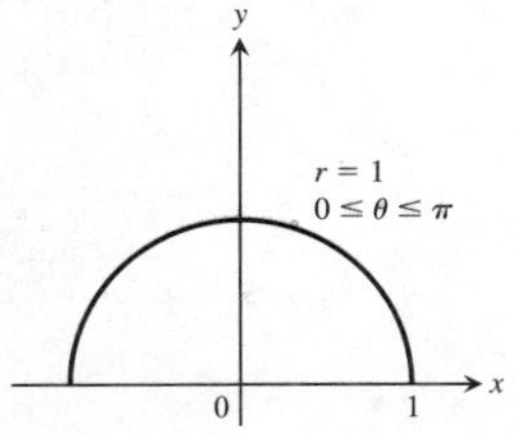

23.

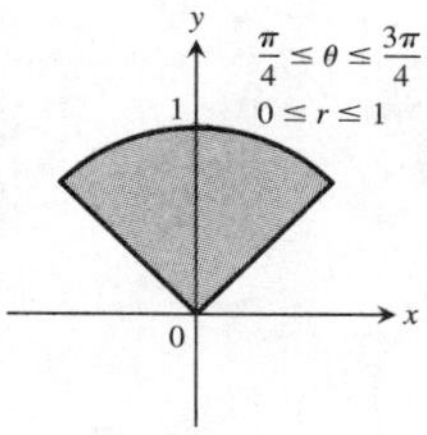

25.

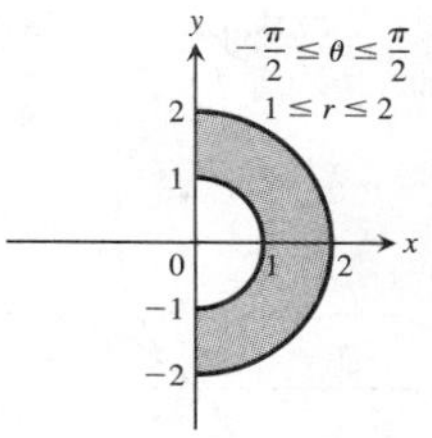

27. $x = 5$, vertical line through (5, 0) **29.** $y = 0$, the x-axis
31. $y = 3$, horizontal line through (0, 3)
33. $x + y = 1$, line, $m = -1$, $b = 1$
35. $x^2 + y^2 = 4$, circle with center (0, 0) and radius 2
37. $y - 2x = 5$, line, $m = 2$, $b = 5$
39. $y^2 = x$, parabola, vertex $(0, 0)$, opens right
41. $x = e^y$, graph of the natural exponential function
43. $x + y = \pm 1$, two straight lines of slope -1, y-intercepts $b = \pm 1$
45. $(x + 2)^2 + y^2 = 4$, circle, $C(-2, 0)$, radius 2
47. $x^2 + (y - 4)^2 = 16$, circle, $C(0, 4)$, radius 4
49. $(x - 2)^2 + (y - 2)^2 = 8$, circle with center (2, 2) and radius $2\sqrt{2}$
51. $\sqrt{3}y + x = 4$ **53.** $r\cos\theta = 5$ **55.** $\theta = \pi/6$
57. $r = 2$ or $r = -2$ **59.** $9r^2\cos^2\theta + 4r^2\sin^2\theta = 36$
61. $r\sin^2\theta = 9\cos\theta$ **63.** $r = 4\sin\theta$
65. $r^2 = 6r\cos\theta - 2r\sin\theta - 6$
67. $(0, \theta)$, where θ is any angle

SECTION 11.4, pp. 684–685

1. x-axis

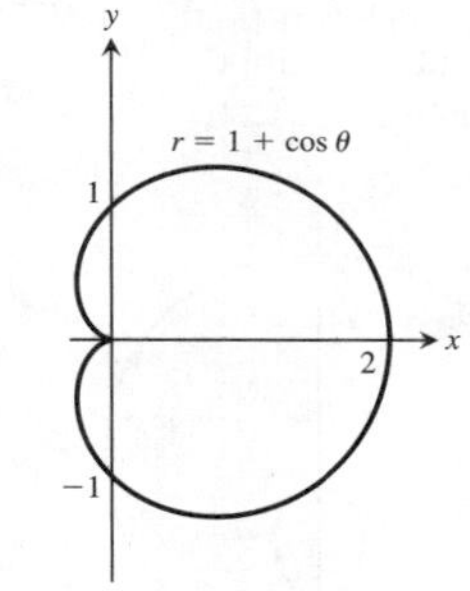

3. y-axis

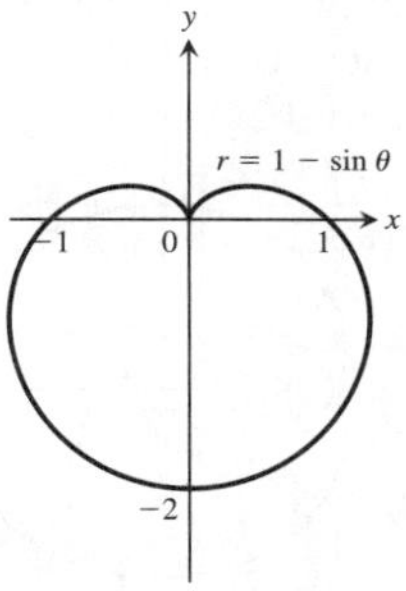

5. x-axis

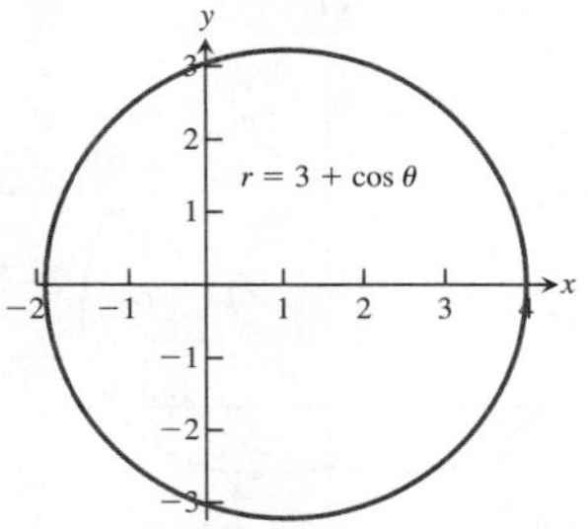

7. x-axis, y-axis, origin

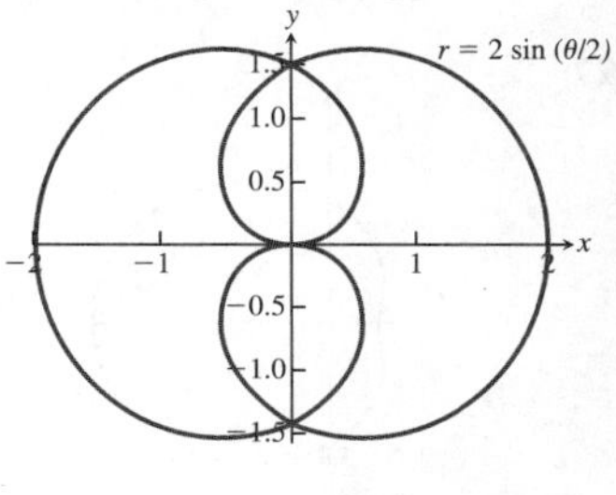

9. x-axis, y-axis, origin

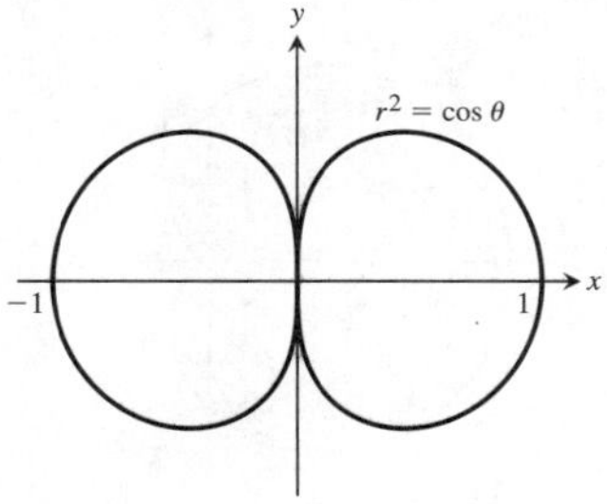

11. y-axis, x-axis, origin

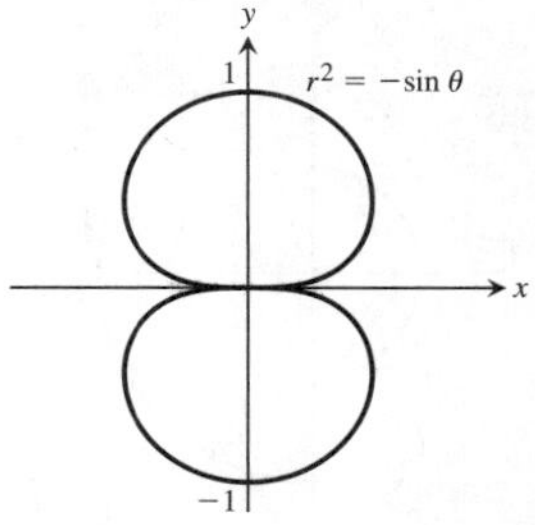

13. x-axis, y-axis, origin

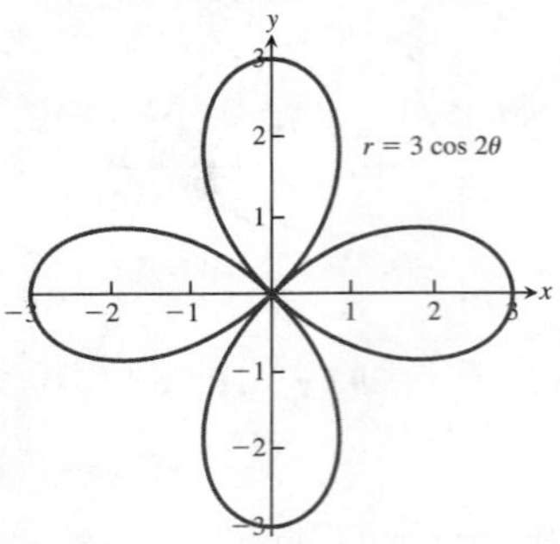

15. Origin

17. The slope at $(-1, \pi/2)$ is -1, at $(-1, -\pi/2)$ is 1.

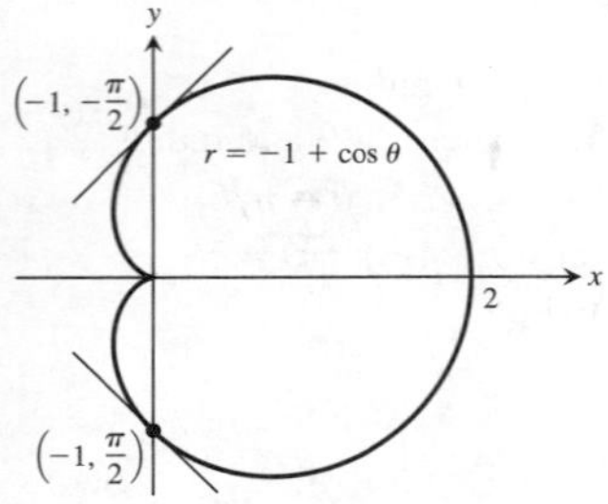

19. The slope at $(1, \pi/4)$ is -1, at $(-1, -\pi/4)$ is 1, at $(-1, 3\pi/4)$ is 1, at $(1, -3\pi/4)$ is -1.

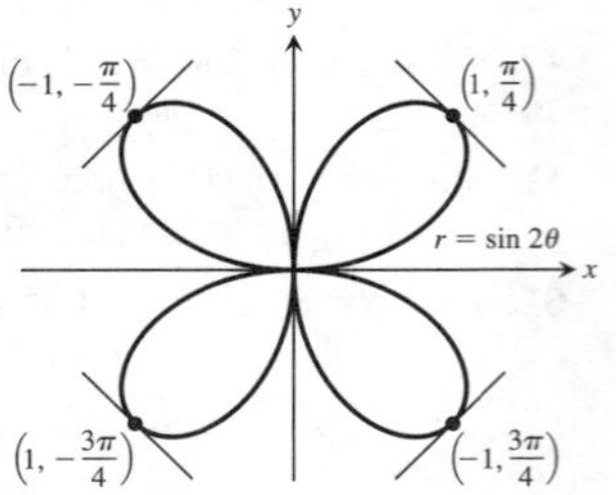

21. At $\pi/6$: slope $\sqrt{3}$, concavity 16 (concave up); at $\pi/3$: slope $-\sqrt{3}$, concavity -16 (concave down).

23. At 0: slope 0, concavity 2 (concave up); at $\pi/2$: slope $-2/\pi$, concavity $-2(8 + \pi^2)/\pi^3$ (concave down).

25. (a)

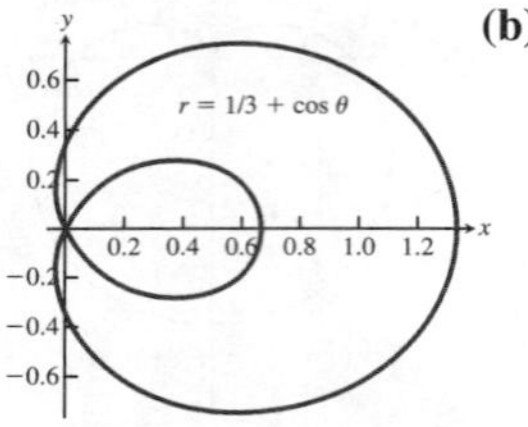

(b)

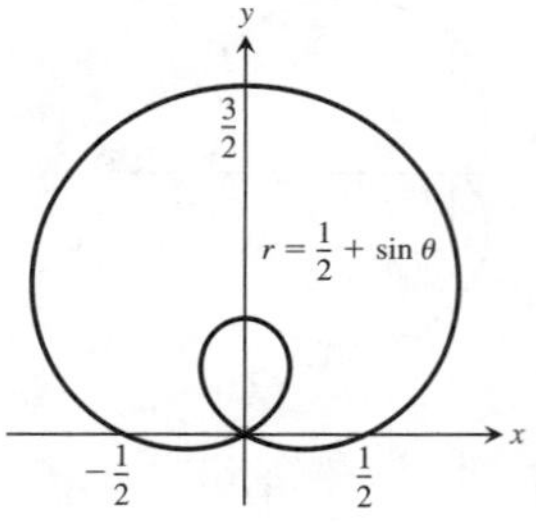

27. (a)

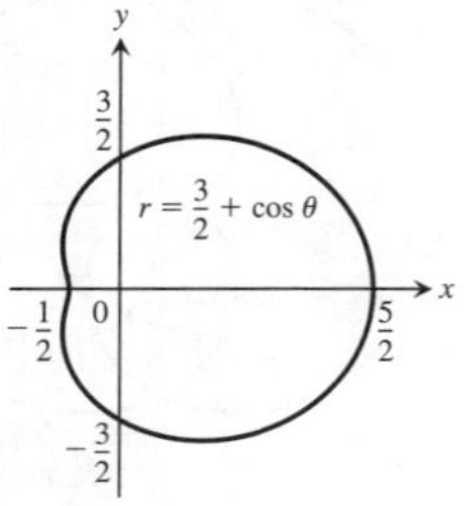

(b)

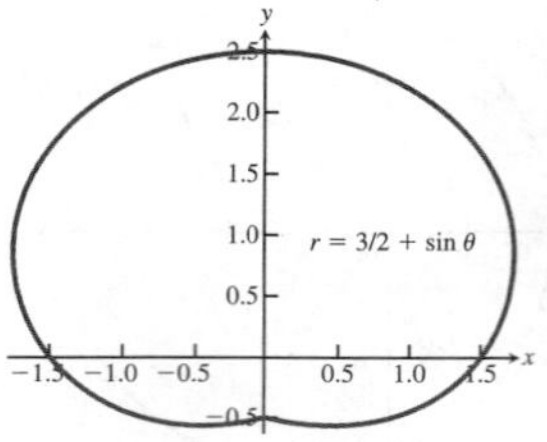

29.

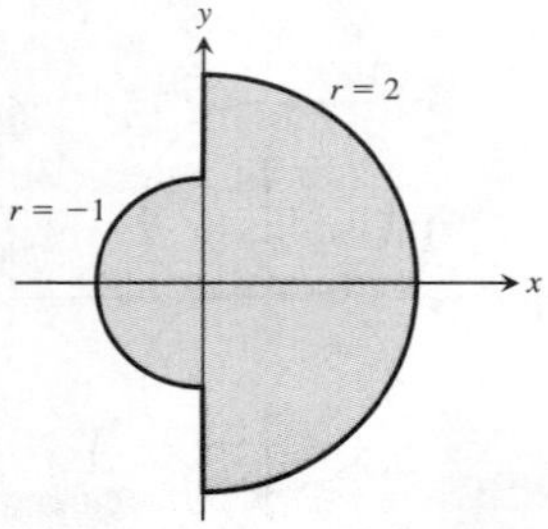

31.

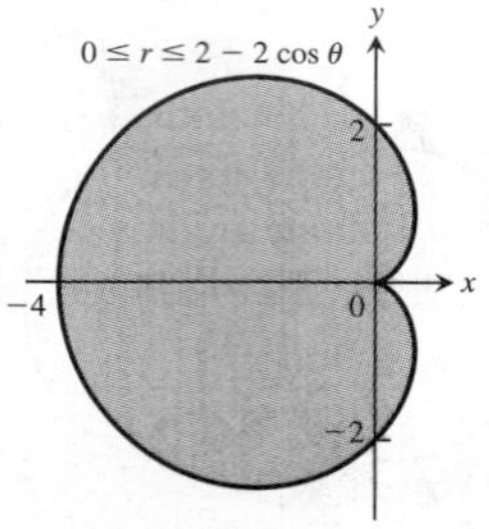

33. Equation (a)

35.

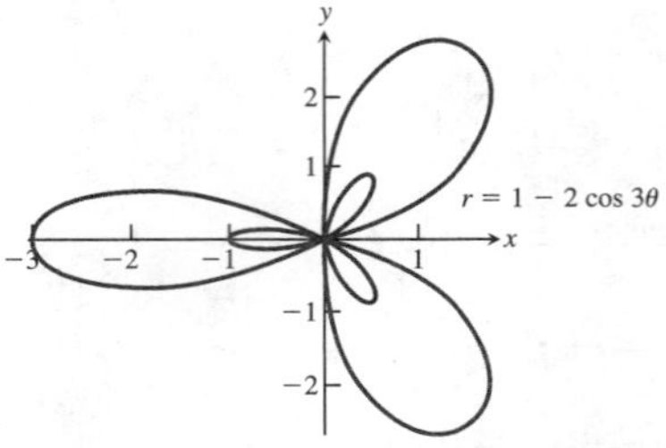

SECTION 11.5, pp. 689–690

1. $\frac{1}{6}\pi^3$ **3.** 11π **5.** $\frac{1}{24}(2+3\pi)$ **7.** $1/3$

9. $\frac{\pi}{2} - 1$ **11.** $5\pi - 8$ **13.** $3\sqrt{3} - \pi$ **15.** $\frac{\pi}{3} + \frac{\sqrt{3}}{2}$

17. $\frac{8\pi}{3} + \sqrt{3}$ **19. (a)** $\frac{3}{2} - \frac{\pi}{4}$ **21.** $19/3$

23. 8 **25.** $2(\sqrt{2} + \ln(1+\sqrt{2}))$

27. $\frac{\pi}{8} + \frac{3}{8}$ **31. (a)** a **(b)** a **(c)** $2a/\pi$

SECTION 11.6, pp. 696–698

1. $y^2 = 8x$, $F(2, 0)$, directrix: $x = -2$

3. $x^2 = -6y$, $F(0, -3/2)$, directrix: $y = 3/2$

5. $\frac{x^2}{4} - \frac{y^2}{9} = 1$, $F(\pm\sqrt{13}, 0)$, $V(\pm 2, 0)$, asymptotes: $y = \pm\frac{3}{2}x$

7. $\frac{x^2}{2} + y^2 = 1$, $F(\pm 1, 0)$, $V(\pm\sqrt{2}, 0)$

9. Focus is (4, 0), directrix $x = -4$

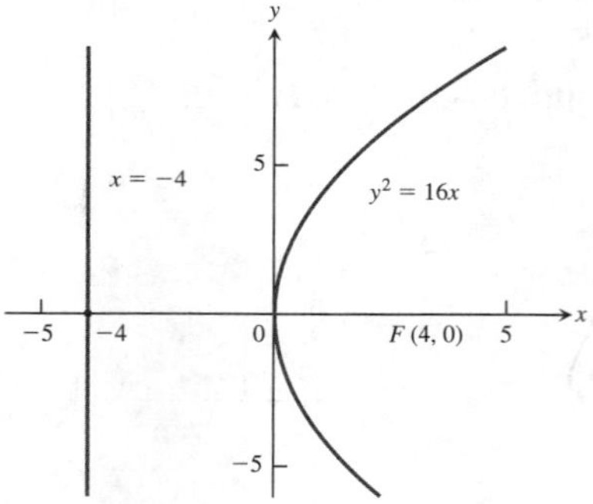

11. Focus is (0, -3), directrix $y = 3$

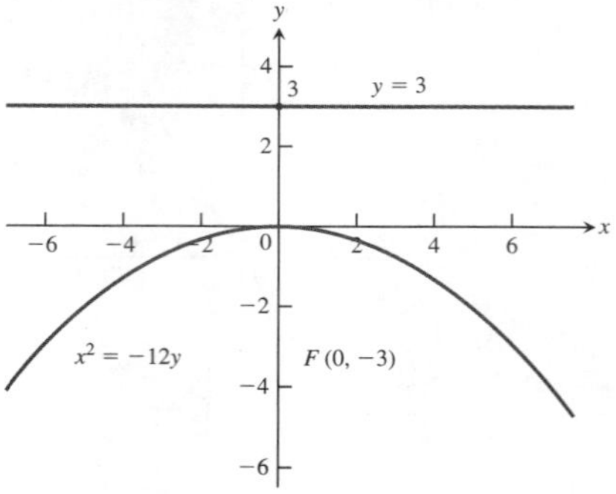

13.

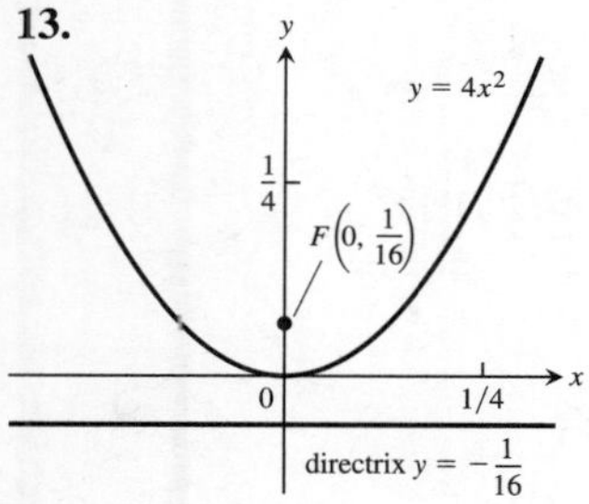

15.

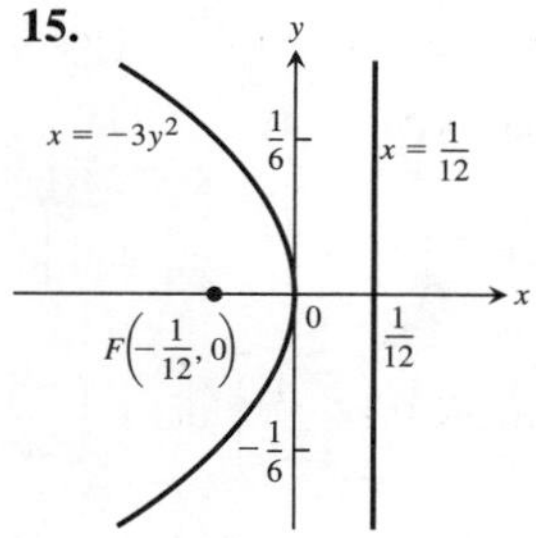

17.

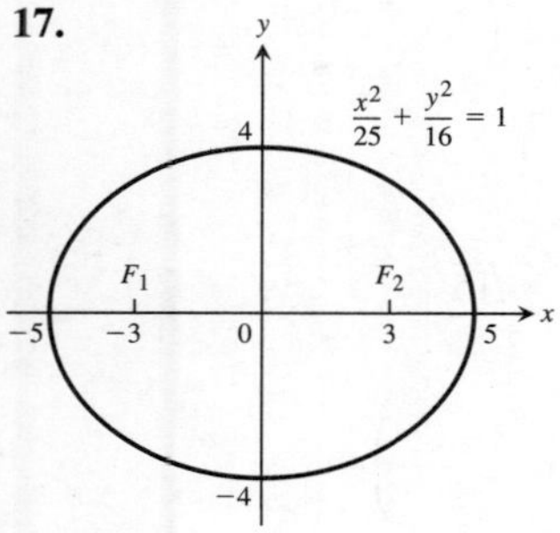

19.

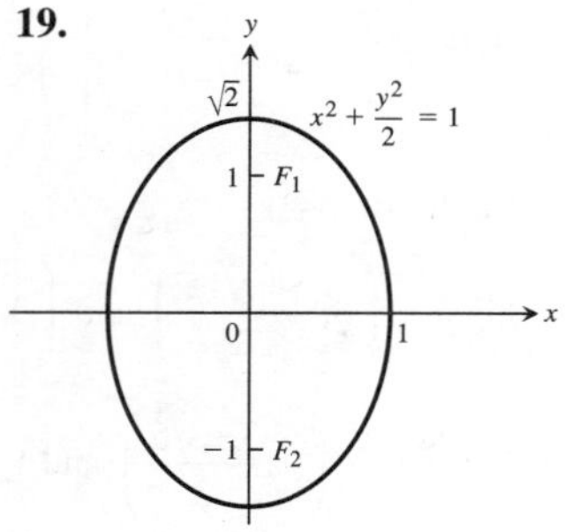

21.

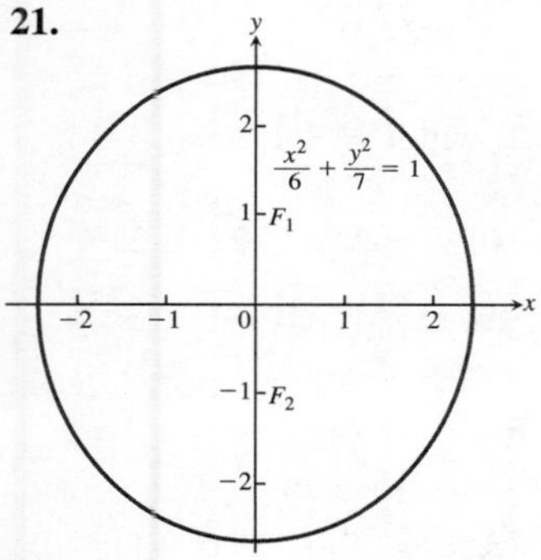

23.

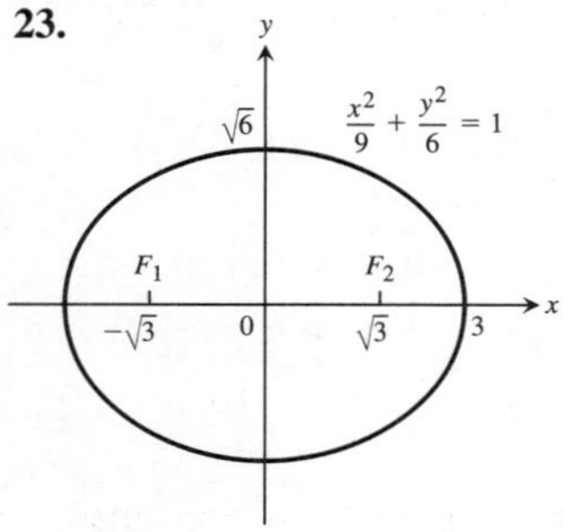

25. $\dfrac{x^2}{4} + \dfrac{y^2}{2} = 1$

27. $c = 2$, Asymptotes are $y = \pm x$

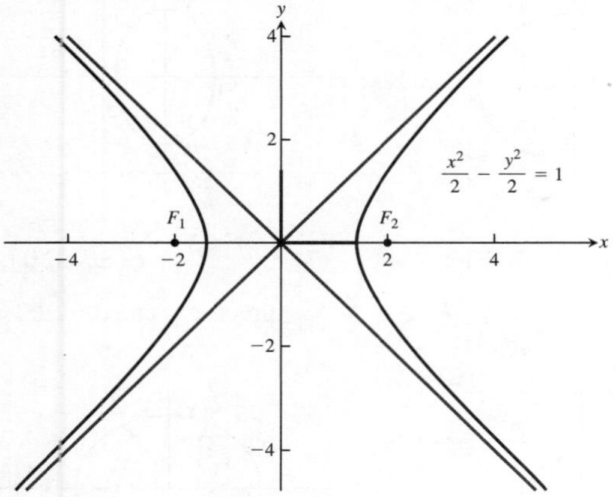

29. $c = 2\sqrt{5}$, Asymptotes are $y = \pm x$

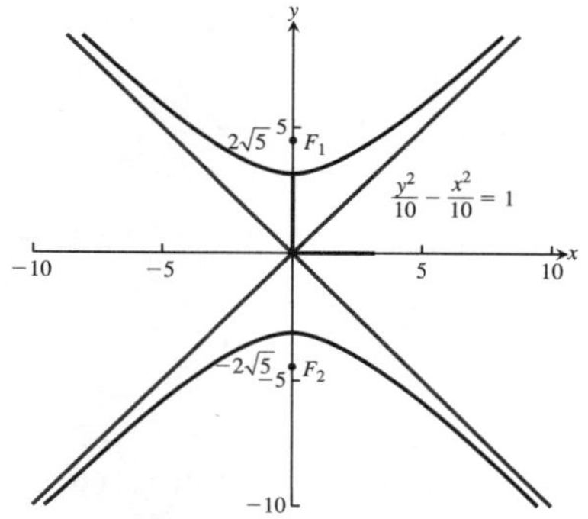

31. $c = 2\sqrt{2}$, Asymptotes are $y = \pm\sqrt{3}\,x$

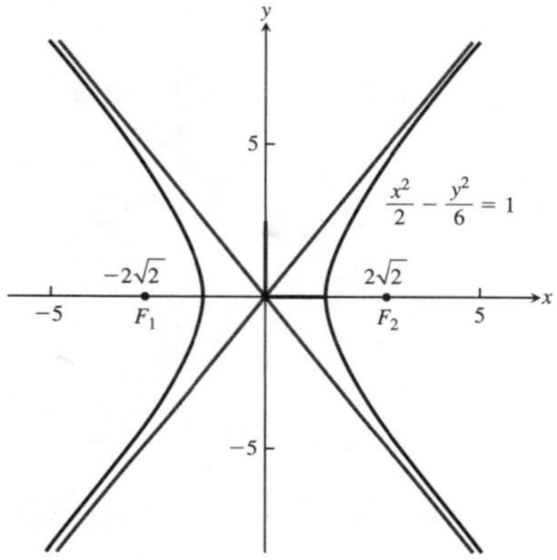

33. $c = 2\sqrt{2}$, Asymptotes are $y = \pm\dfrac{x}{\sqrt{3}}$

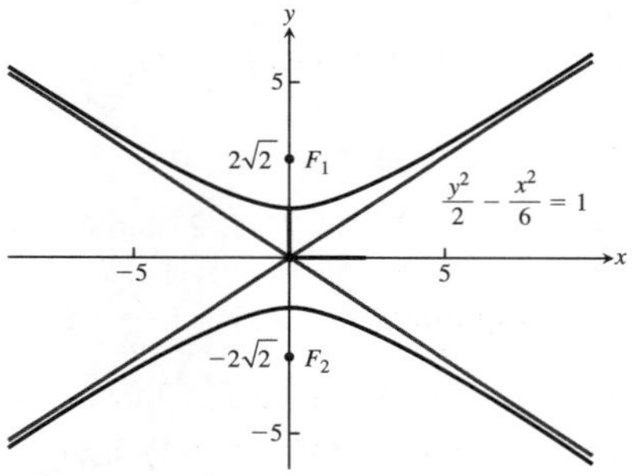

35. $y^2 - x^2 = 1$

37. $\dfrac{x^2}{9} - \dfrac{y^2}{16} = 1$

39. **(a)** Vertex: $(1, -2)$; focus: $(3, -2)$; directrix: $x = -1$

(b)

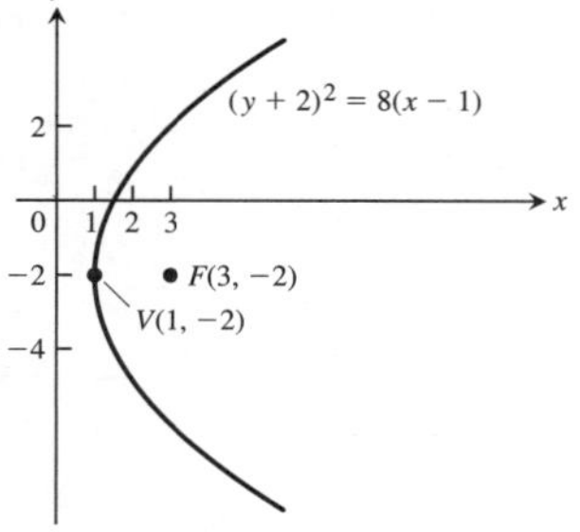

41. **(a)** Foci: $(4 \pm \sqrt{7}, 3)$; vertices: $(8, 3)$ and $(0, 3)$; center: $(4, 3)$

(b)

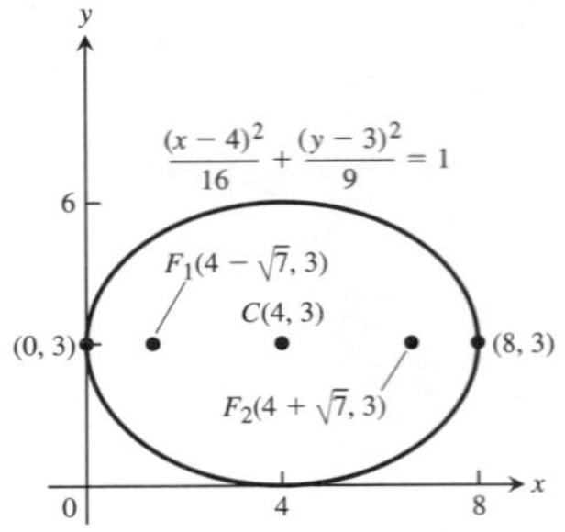

43. **(a)** Center: $(2, 0)$; foci: $(7, 0)$ and $(-3, 0)$; vertices: $(6, 0)$ and $(-2, 0)$; asymptotes: $y = \pm\frac{3}{4}(x - 2)$

(b)

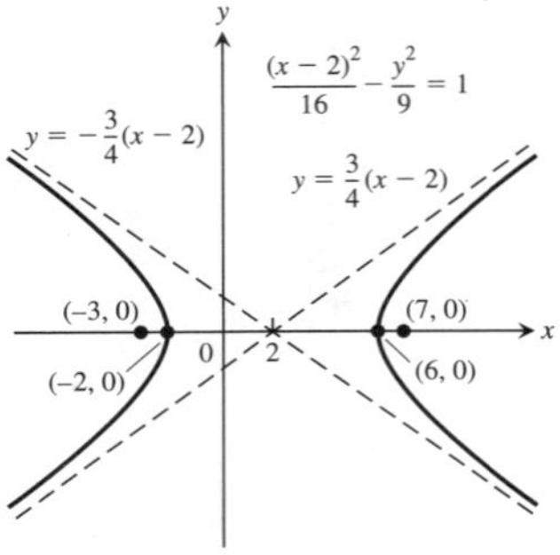

45. $(y + 3)^2 = 4(x + 2)$, $V(-2, -3)$, $F(-1, -3)$, directrix: $x = -3$

47. $(x - 1)^2 = 8(y + 7)$, $V(1, -7)$, $F(1, -5)$, directrix: $y = -9$

49. $\frac{(x + 2)^2}{6} + \frac{(y + 1)^2}{9} = 1$, $F(-2, \pm\sqrt{3} - 1)$, $V(-2, \pm 3 - 1)$, $C(-2, -1)$

51. $\frac{(x - 2)^2}{3} + \frac{(y - 3)^2}{2} = 1$, $F(3, 3)$ and $F(1, 3)$, $V(\pm\sqrt{3} + 2, 3)$, $C(2, 3)$

53. $\frac{(x - 2)^2}{4} - \frac{(y - 2)^2}{5} = 1$, $C(2, 2)$, $F(5, 2)$ and $F(-1, 2)$, $V(4, 2)$ and $V(0, 2)$; asymptotes: $(y - 2) = \pm\frac{\sqrt{5}}{2}(x - 2)$

55. $(y + 1)^2 - (x + 1)^2 = 1$, $C(-1, -1)$, $F(-1, \sqrt{2} - 1)$ and $F(-1, -\sqrt{2} - 1)$, $V(-1, 0)$ and $V(-1, -2)$; asymptotes $(y + 1) = \pm(x + 1)$

57. Circle with center $C(-1, 0)$ and radius $a = \sqrt{15}$

59. Parabola with $V(-2, 7/2)$ and $F(-2, 3)$

61. Ellipse: $\frac{(x + 2)^2}{5} + y^2 = 1$, $C(-2, 0)$, $F(0, 0)$ and $F(-4, 0)$, $V(\sqrt{5} - 2, 0)$ and $V(-\sqrt{5} - 2, 0)$

63. Ellipse: $\frac{(x-3/2)^2}{(\sqrt{57}/2)^2} + \frac{(y-1)^2}{(\sqrt{19}/2)^2} = 1$

Center $C(3/2, 1)$,

Vertices $V\left(\frac{3+\sqrt{57}}{2}, 1\right)$ and $V\left(\frac{3-\sqrt{57}}{2}, 1\right)$

Focii $F\left(\frac{3+\sqrt{38}}{2}, 1\right)$ and $\left(\frac{3-\sqrt{38}}{2}, 1\right)$

65. Hyperbola: $(x-1/4)^2 - (y-1/4)^2 = 1$

Center $(1/4, 1/4)$

Focii $F\left(\frac{1+4\sqrt{2}}{4}, \frac{1}{4}\right)$ and $\left(\frac{1-4\sqrt{2}}{4}, \frac{1}{4}\right)$

Vertices $V\left(\frac{5}{4}, \frac{1}{4}\right)$ and $V\left(\frac{-3}{4}, \frac{1}{4}\right)$

Asymptotes $y = x$ and $y = -(x-1/4) + 1/4$

67. Hyperbola: $\frac{(y-3/2)^2}{(\sqrt{5}/2)^2} - \frac{(x-0)^2}{\left(\sqrt{\frac{5}{2}}\right)^2} = 1$

Center $(0, 3/2)$

Focii $F\left(0, \frac{3+\sqrt{15}}{2}\right)$ and $\left(0, \frac{3-\sqrt{15}}{2}\right)$

Vertices $V\left(0, \frac{3+\sqrt{5}}{2}\right)$ and $V\left(0, \frac{3-\sqrt{5}}{2}\right)$

Asymptotes $y = \frac{x}{\sqrt{2}} + \frac{3}{2}$ and $y = -\frac{x}{\sqrt{2}} + \frac{3}{2}$

69. **(b)** 1:1 **73.** Length $= 2\sqrt{2}$, width $= \sqrt{2}$, area $= 4$

75. $32\pi/3$

77. $x = 0, y = 0: y = -2x$; $x = 0, y = 2: y = 2x + 2$; $x = 4, y = 0: y = 2x - 8$

79. $\bar{x} = 0$, $\bar{y} = \frac{16}{3\pi}$

SECTION 11.7, pp. 703–704

1. $e = \frac{3}{5}$, $F(\pm 3, 0)$; directrices are $x = \pm\frac{25}{3}$.

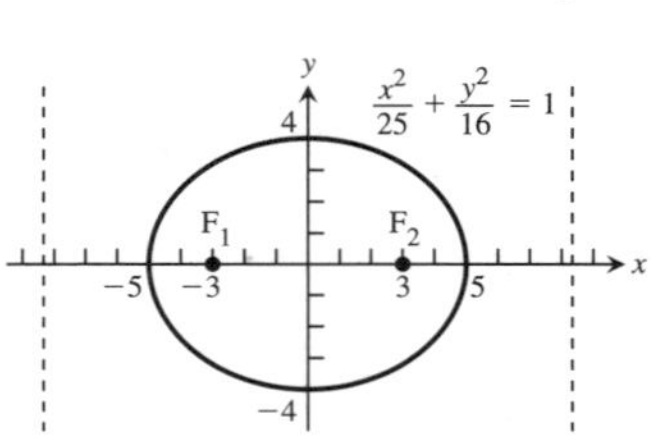

3. $e = \frac{1}{\sqrt{2}}$; $F(0, \pm 1)$; directrices are $y = \pm 2$.

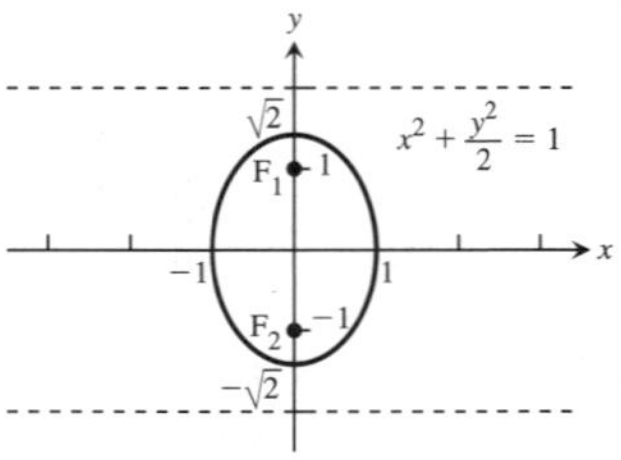

5. $e = \frac{1}{\sqrt{3}}$; $F(0, \pm 1)$; directrices are $y = \pm 3$.

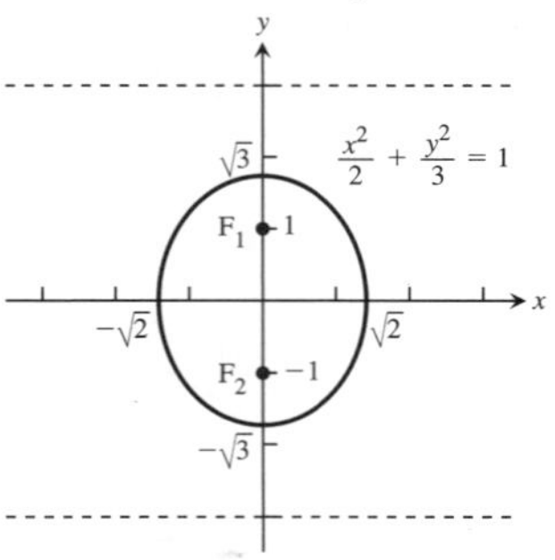

7. $e = \frac{\sqrt{3}}{3}$; $F(\pm\sqrt{3}, 0)$; directrices are $x = \pm 3\sqrt{3}$.

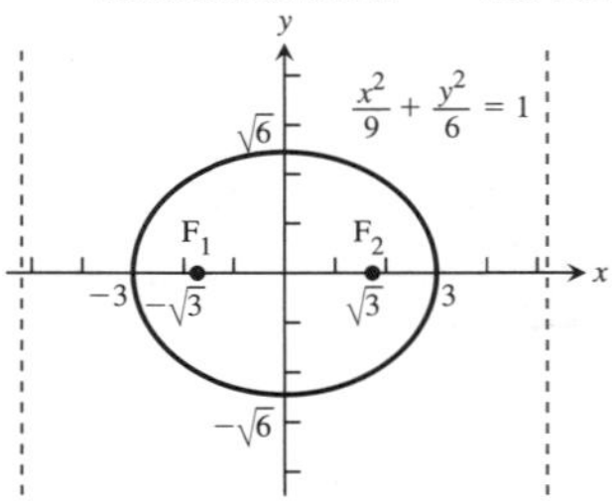

9. $\frac{x^2}{96}+\frac{y^2}{100}=1$ **11.** $\frac{x^2}{2475}+\frac{y^2}{2500}=1$

13. $\frac{x^2}{9}+\frac{y^2}{4}=1$ **15.** $\frac{x^2}{64}+\frac{y^2}{48}=1$

17. $e=\sqrt{2}$; $F(\pm\sqrt{2},0)$; directrices are $x=\pm\frac{1}{\sqrt{2}}$.

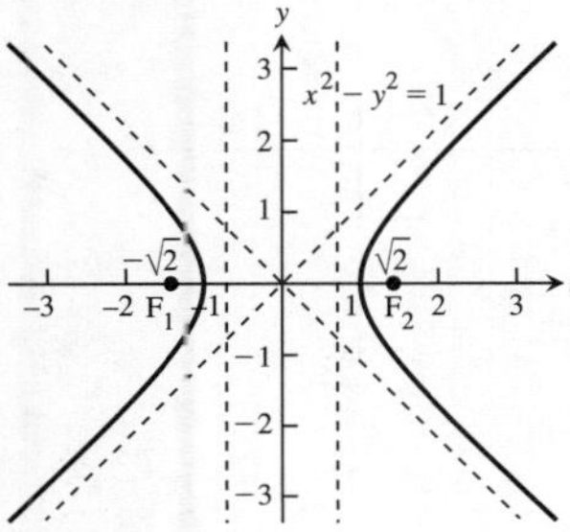

19. $e=\sqrt{2}$; $F(0,\pm 4)$; directrices are $y=\pm 2$.

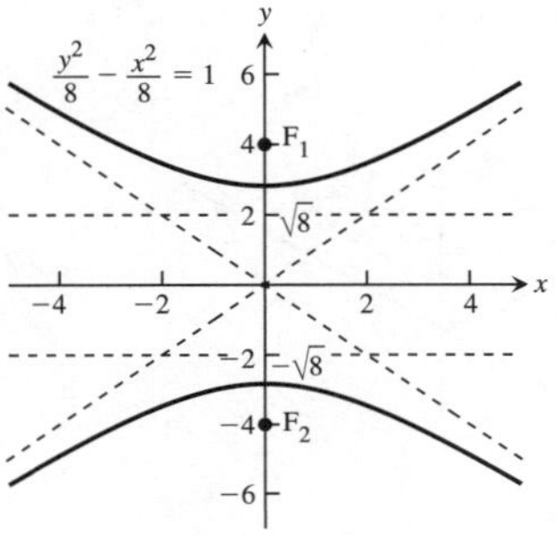

21. $e=\sqrt{5}$; $F(\pm\sqrt{10},0)$; directrices are $x=\pm\frac{2}{\sqrt{10}}$.

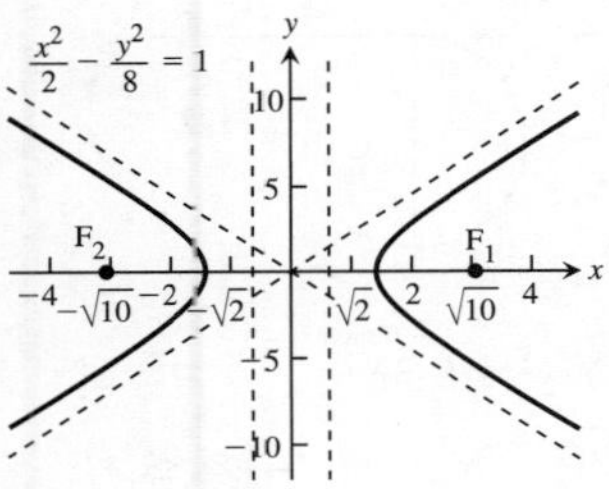

23. $e=\sqrt{5}$; $F(0,\pm\sqrt{10})$; directrices are $y=\pm\frac{2}{\sqrt{10}}$.

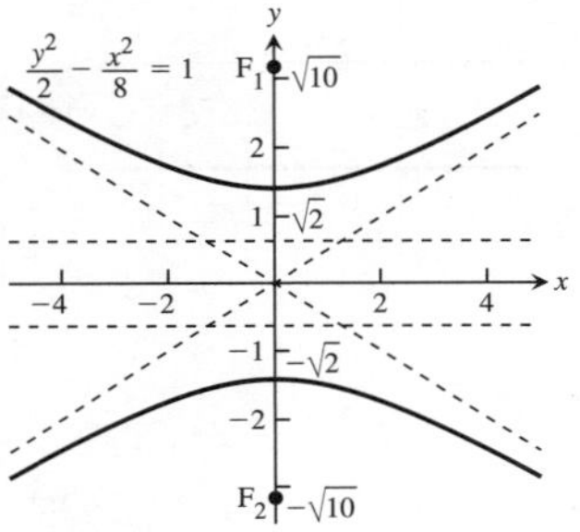

25. $\frac{y^2}{4}-\frac{x^2}{60}=1$ **27.** $x^2-\frac{y^2}{3}=1$ **29.** $r=\frac{2}{1+2\cos\theta}$

31. $r=\frac{8}{1-4\sin\theta}$ **33.** $r=\frac{1}{1+\cos\theta}$ **35.** $r=\frac{9}{3-\sin\theta}$

37.

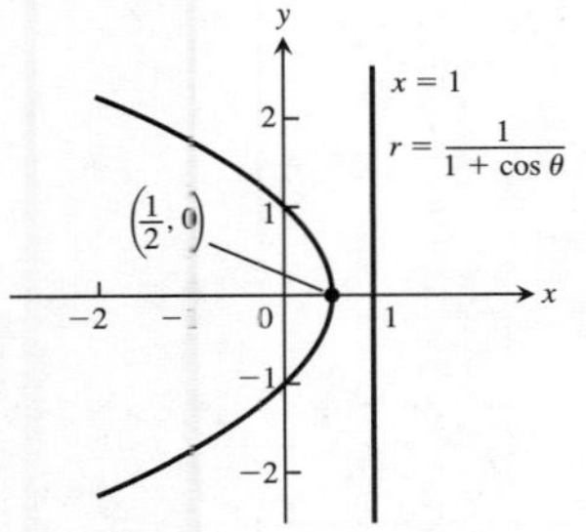

39.

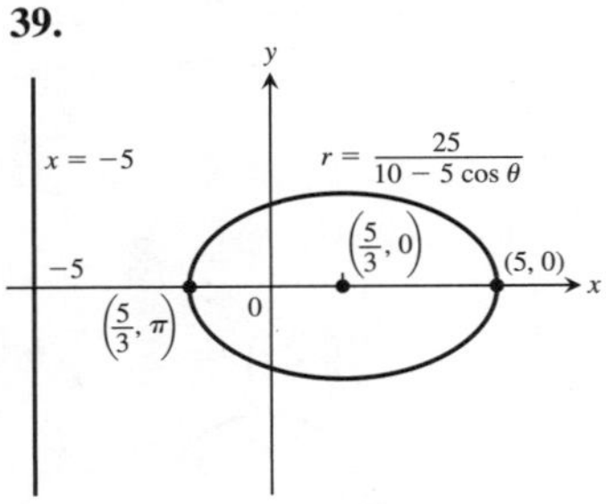

41.

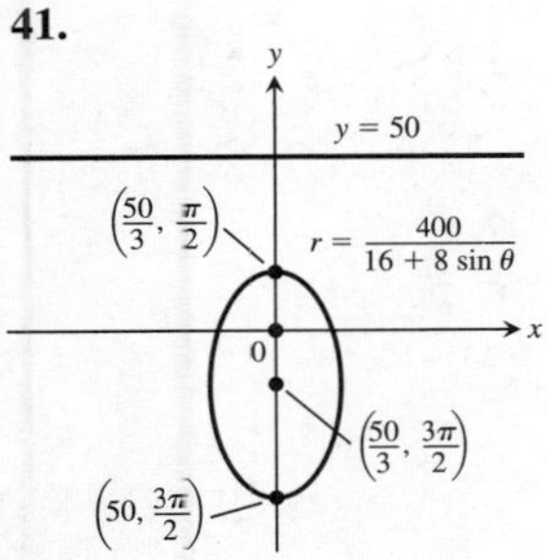

43.

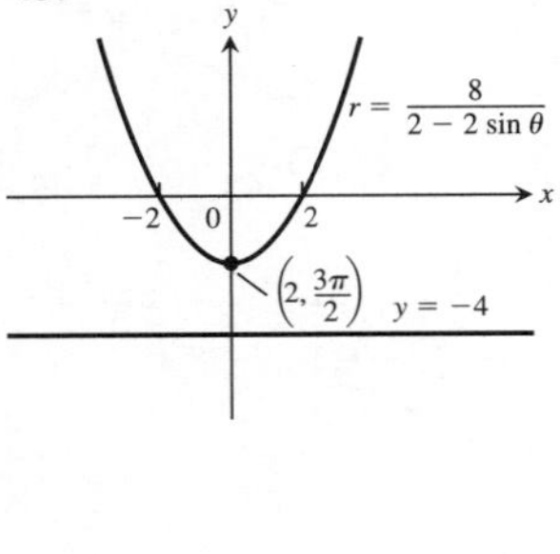

45. $y=2-x$

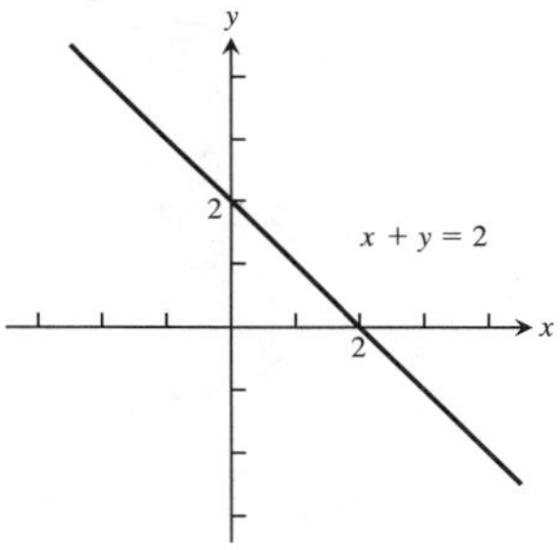

47. $y=\frac{\sqrt{3}}{3}x+2\sqrt{3}$

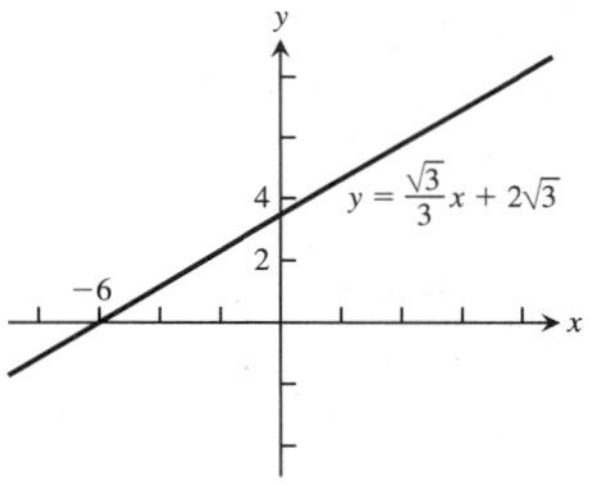

49. $r\cos\left(\theta-\frac{\pi}{4}\right)=3$ **51.** $r\cos\left(\theta+\frac{\pi}{2}\right)=3$

53. $C(6,0)$ and radius $a=6$

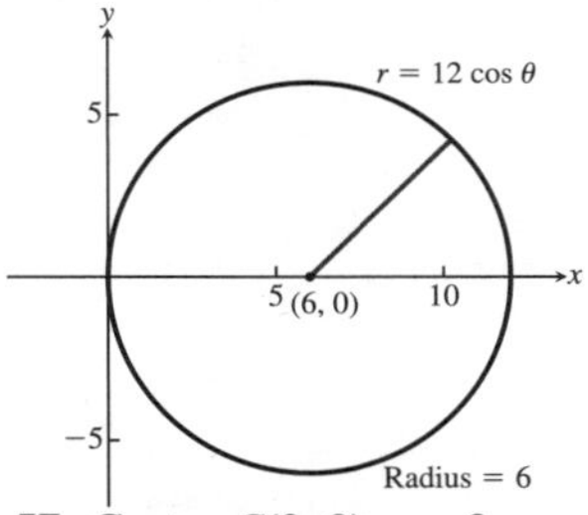

55.

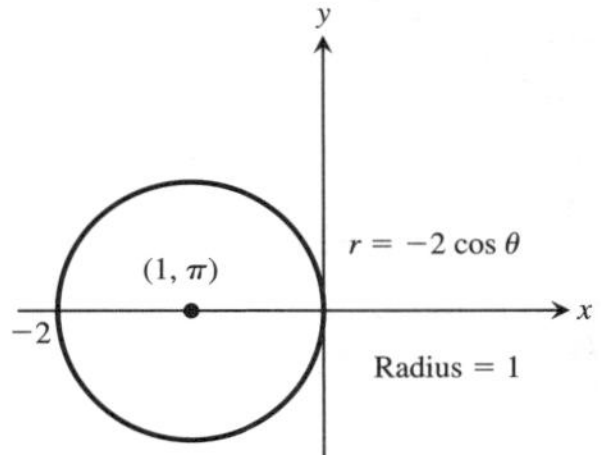

57. Center $C(3,0)$, $a=3$

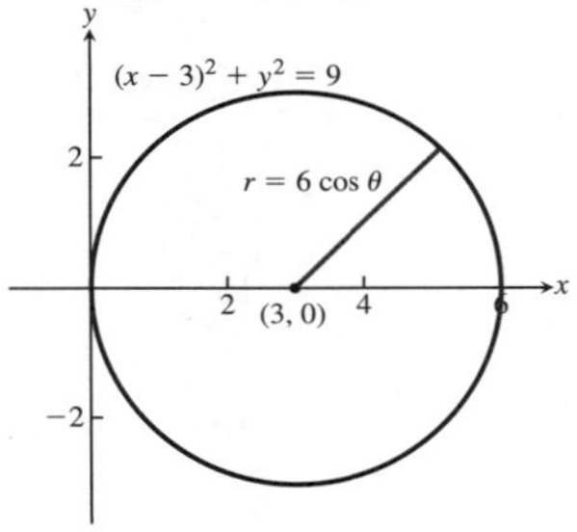

59. Center $C(0,4)$, $a=4$

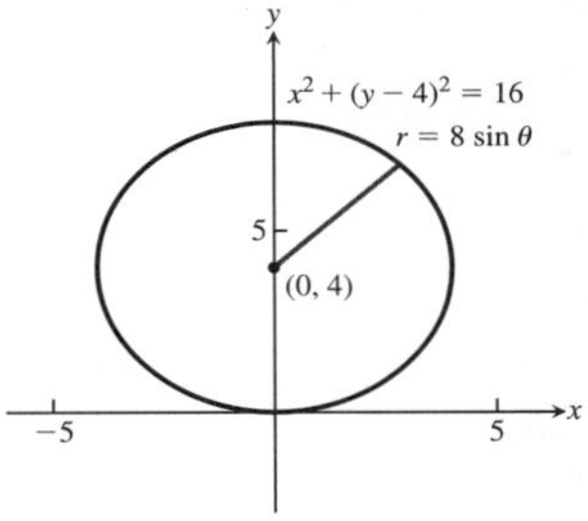

61. Center $C(0,-1)$, $a=1$

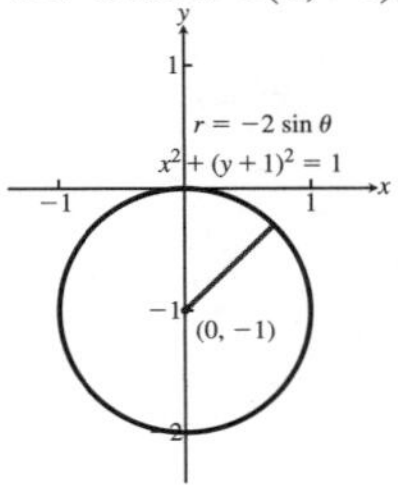

63. Center $C(-1/2,0)$, $a=1/2$

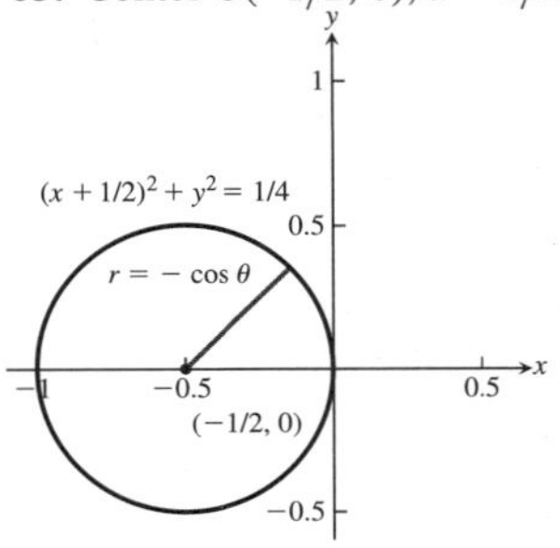

65.

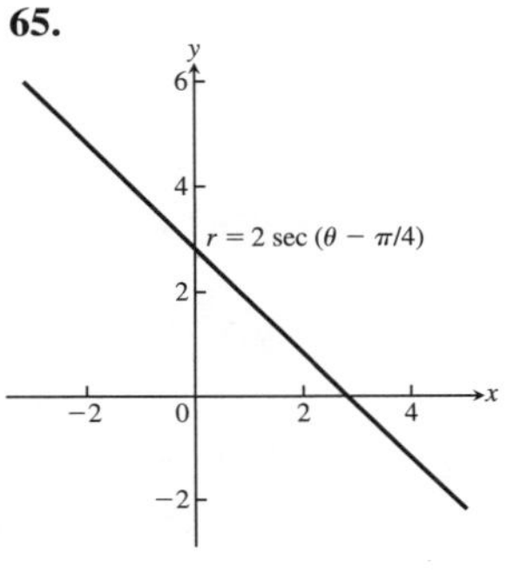

67.

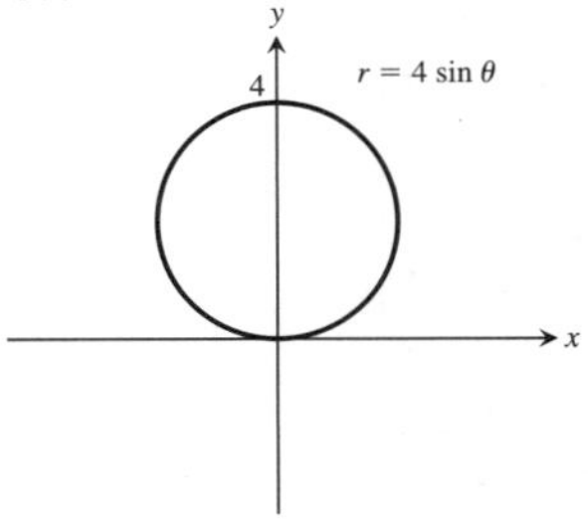

69.

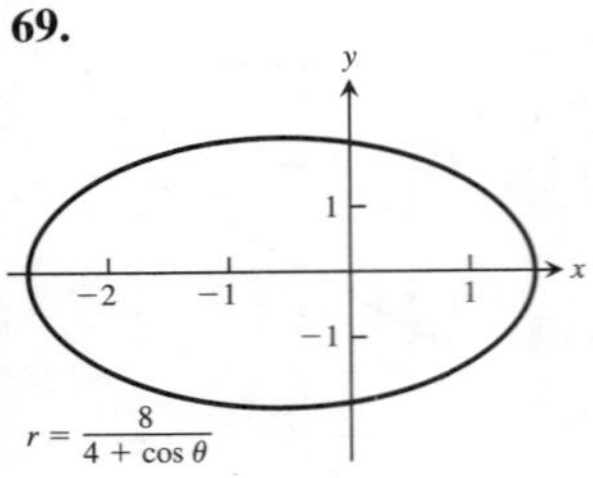

71.

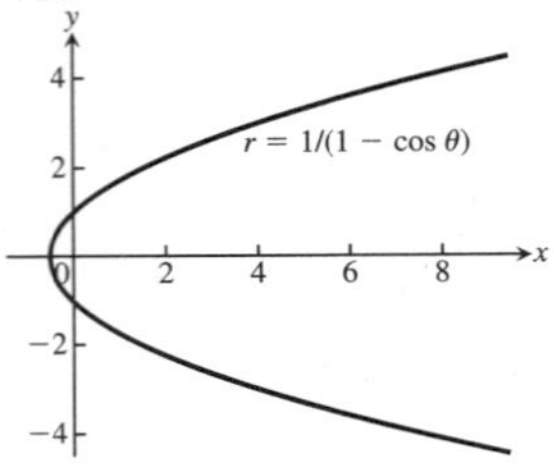

73.

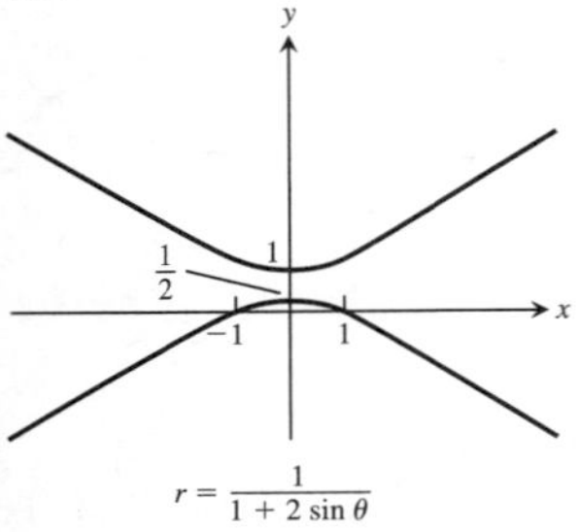

75. (b)

Planet	Perihelion	Aphelion
Mercury	0.3075 AU	0.4667 AU
Venus	0.7184 AU	0.7282 AU
Earth	0.9833 AU	1.0167 AU
Mars	1.3817 AU	1.6663 AU
Jupiter	4.9512 AU	5.4548 AU
Saturn	9.0210 AU	10.0570 AU
Uranus	18.2977 AU	20.0623 AU
Neptune	29.8135 AU	30.3065 AU

PRACTICE EXERCISES, pp. 705–707

1.

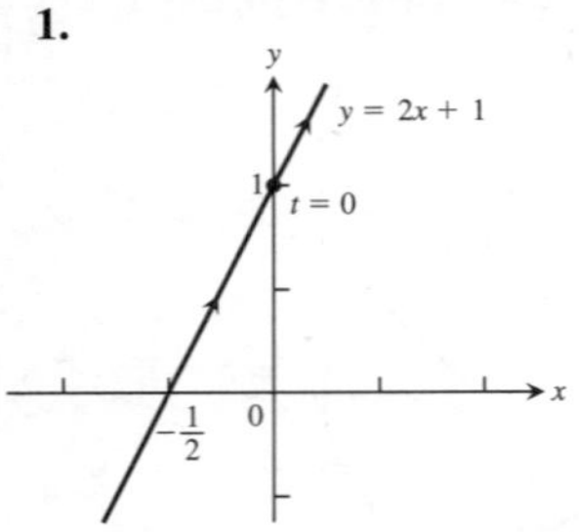

3.

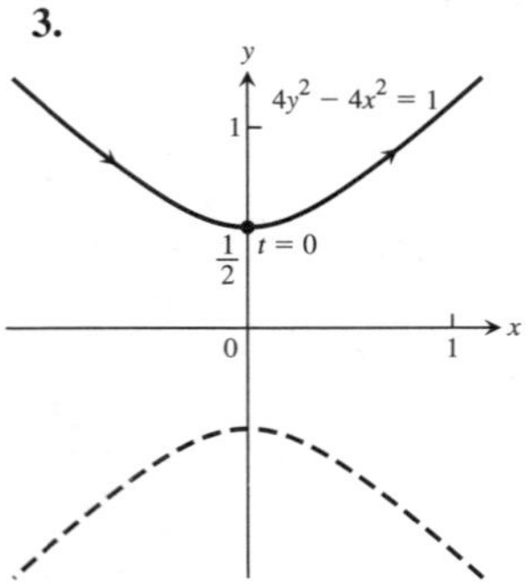

5.

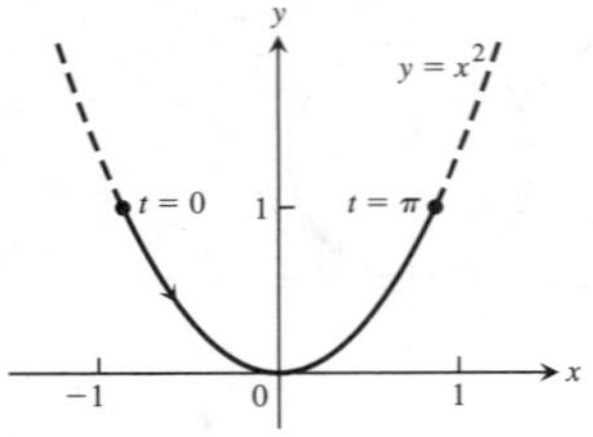

7. $x = 3\cos t,\quad y = 4\sin t,\quad 0 \le t \le 2\pi$

9. $y = \frac{\sqrt{3}}{2}x + \frac{1}{4},\ \frac{1}{4}$

11. (a) $y = \frac{\pm|x|^{3/2}}{8} - 1$ **(b)** $y = \frac{\pm\sqrt{1 - x^2}}{x}$

13. $\frac{10}{3}$ **15.** $\frac{285}{8}$ **17.** 10 **19.** $\frac{9\pi}{2}$ **21.** $\frac{76\pi}{3}$

23. $y = \frac{\sqrt{3}}{3}x - 4$

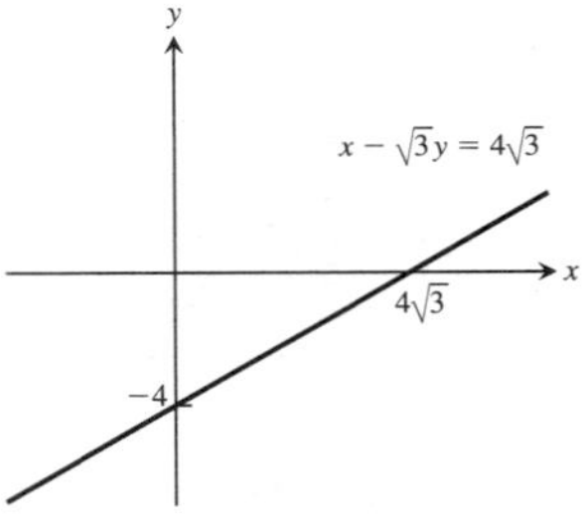

25. $x = 2$

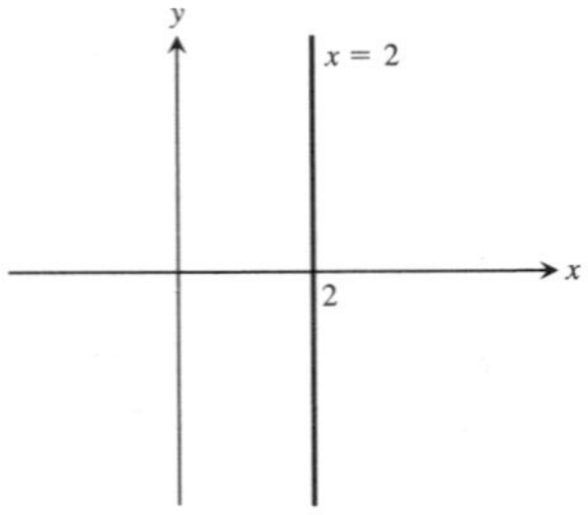

27. $y = -\frac{3}{2}$

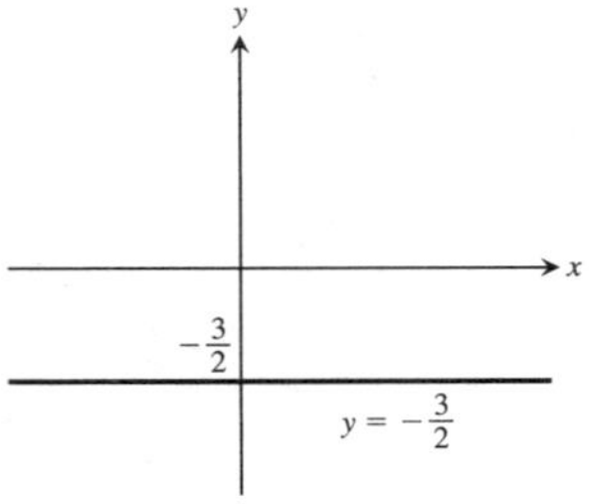

29. $x^2 + (y + 2)^2 = 4$

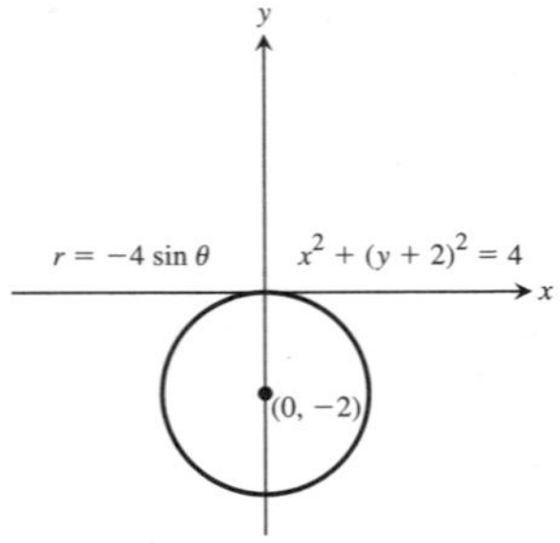

31. $\left(x - \sqrt{2}\right)^2 + y^2 = 2$

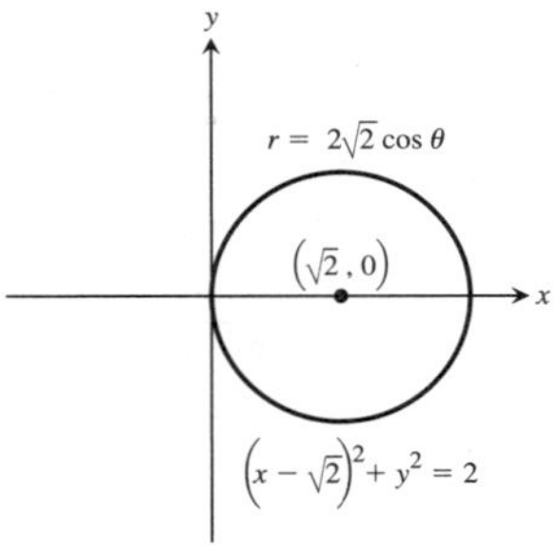

33. $r = -5\sin\theta$

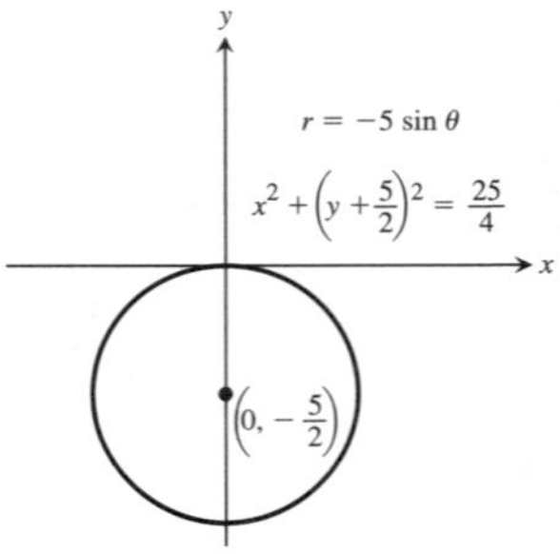

35. $r = 3\cos\theta$

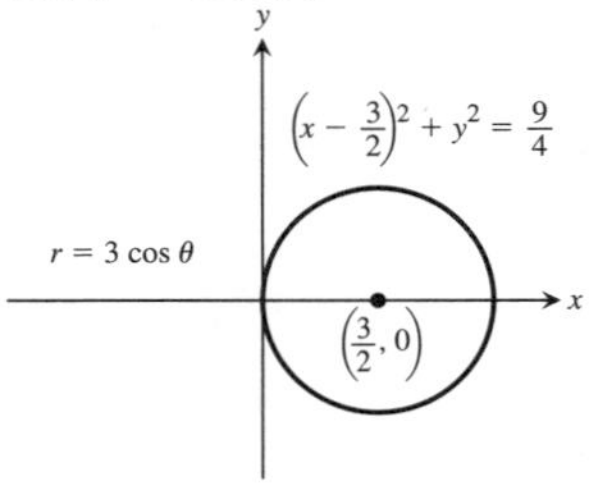

37.

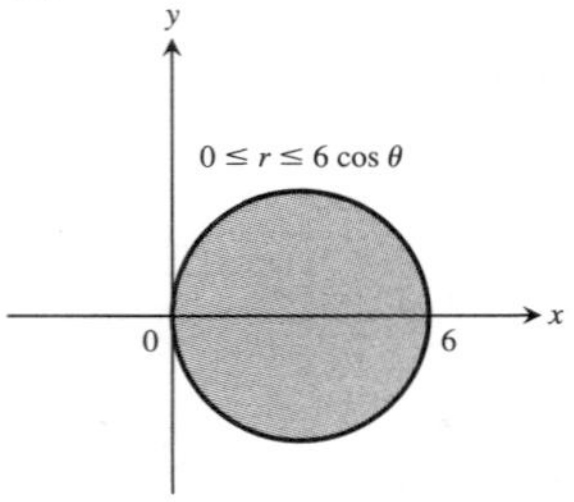

39. d **41.** l **43.** k **45.** i **47.** $\frac{9}{2}\pi$

49. $2 + \frac{\pi}{4}$ **51.** 8 **53.** $\pi - 3$

55. Focus is $(0, -1)$; directrix is $y = 1$.

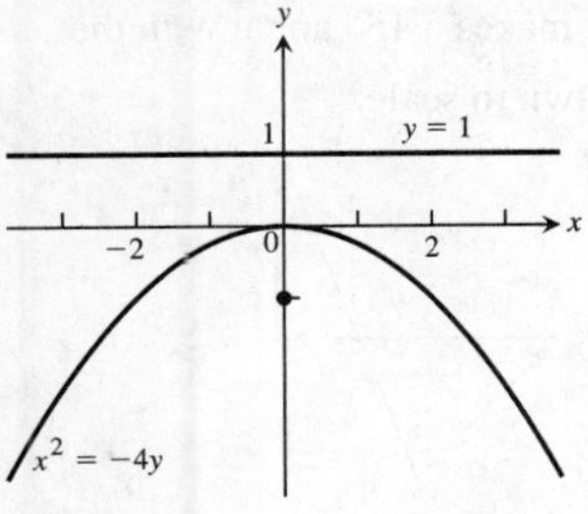

57. Focus is $\left(\frac{3}{4}, 0\right)$; directrix is $x = -\frac{3}{4}$.

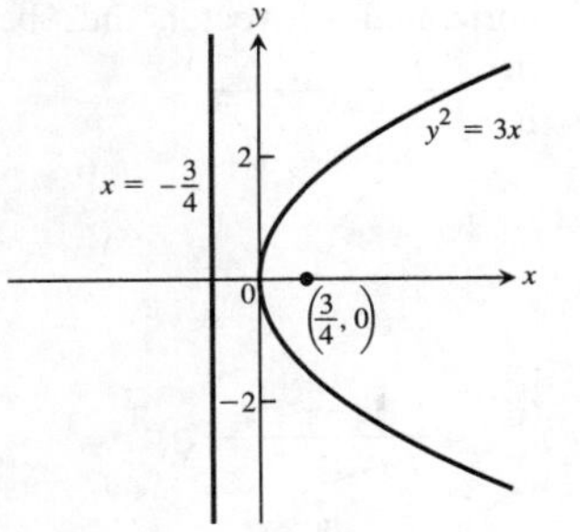

59. $e = \frac{3}{4}$

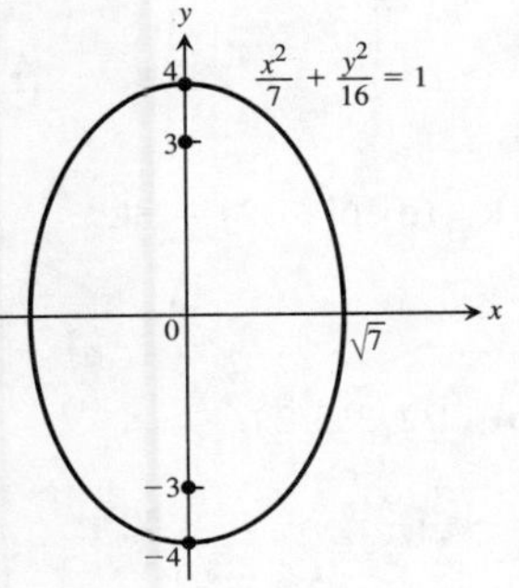

61. $e = 2$; the asymptotes are $y = \pm\sqrt{3}\,x$.

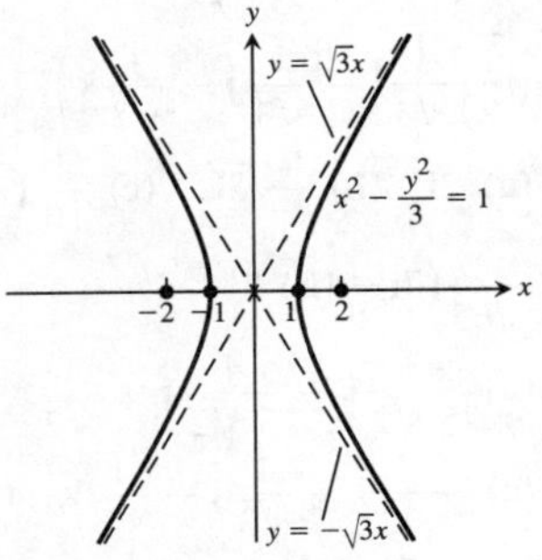

63. $(x - 2)^2 = -12(y - 3)$, $V(2, 3)$, $F(2, 0)$, directrix is $y = 6$.

65. $\frac{(x + 3)^2}{9} + \frac{(y + 5)^2}{25} = 1$, $C(-3, -5)$, $F(-3, -1)$ and $F(-3, -9)$, $V(-3, -10)$ and $V(-3, 0)$.

67. $\frac{(y - 2\sqrt{2})^2}{8} - \frac{(x - 2)^2}{2} = 1$, $C(2, 2\sqrt{2})$, $F(2, 2\sqrt{2} \pm \sqrt{10})$, $V(2, 4\sqrt{2})$ and $V(2, 0)$, the asymptotes are $y = 2x - 4 + 2\sqrt{2}$ and $y = -2x + 4 + 2\sqrt{2}$.

69. Hyperbola: $C(2, 0)$, $V(0, 0)$ and $V(4, 0)$, the foci are $F(2 \pm \sqrt{5}, 0)$, and the asymptotes are $y = \pm\frac{x - 2}{2}$.

71. Parabola: $V(-3, 1)$, $F(-7, 1)$, and the directrix is $x = 1$.

73. Ellipse: $C(-3, 2)$, $F(-3 \pm \sqrt{7}, 2)$, $V(1, 2)$ and $V(-7, 2)$

75. Circle: $C(1, 1)$ and radius $= \sqrt{2}$

77. $V(1, 0)$

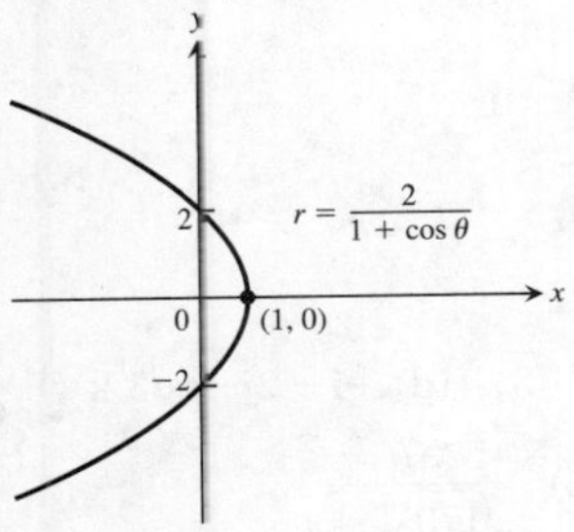

79. $V(2, \pi)$ and $V(6, \pi)$

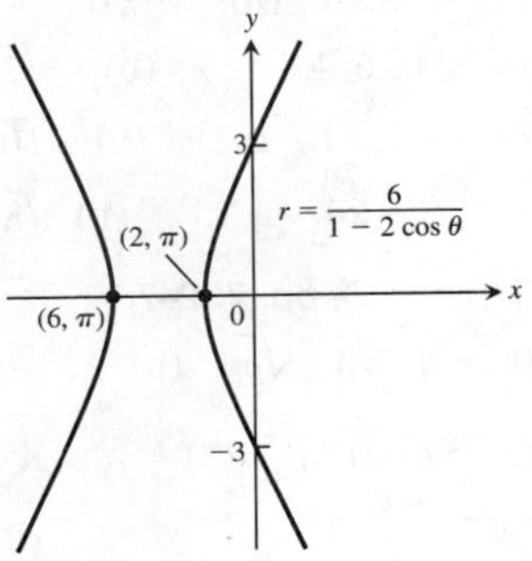

81. $r = \frac{4}{1 + 2\cos\theta}$ **83.** $r = \frac{2}{2 + \sin\theta}$

85. **(a)** 24π **(b)** 16π

ADDITIONAL AND ADVANCED EXERCISES, pp. 707–709

1. $x - \frac{7}{2} = \frac{y^2}{2}$

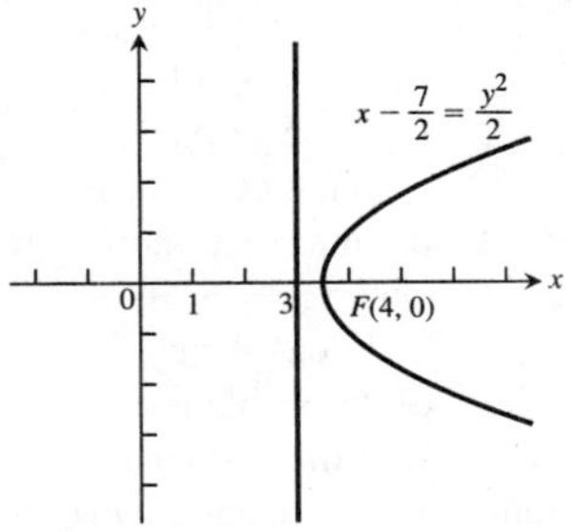

3. $3x^2 + 3y^2 - 8y + 4 = 0$ **5.** $F(0, \pm 1)$

7. **(a)** $\frac{(y - 1)^2}{16} - \frac{x^2}{48} = 1$ **(b)** $\frac{\left(y + \frac{3}{4}\right)^2}{\left(\frac{25}{16}\right)} - \frac{x^2}{\left(\frac{75}{2}\right)} = 1$

11.

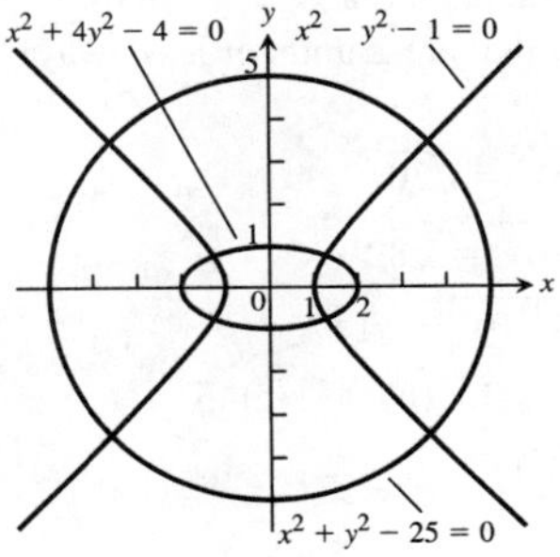

13.

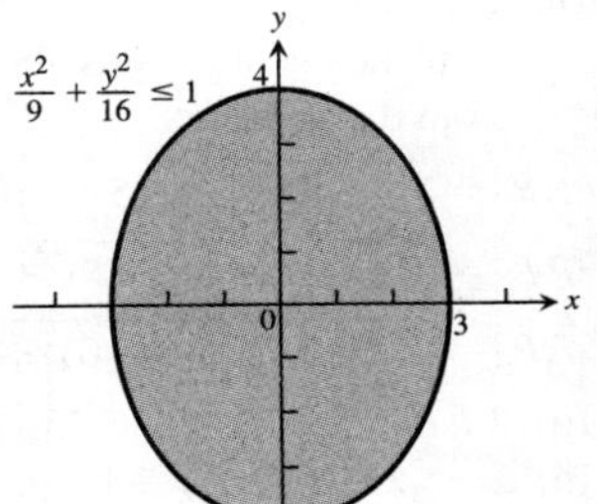

15.

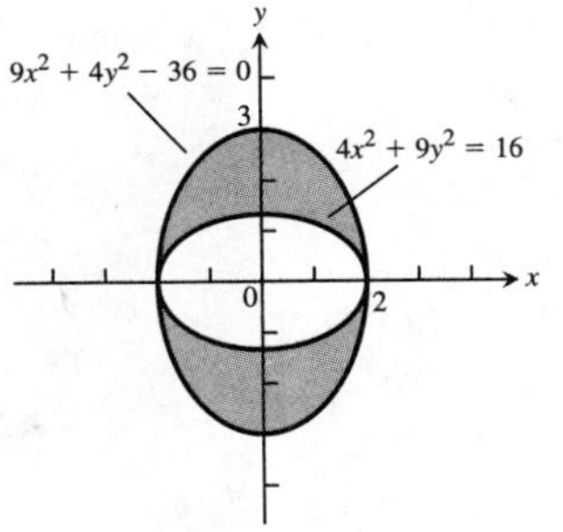

17. **(a)** $r = e^{2\theta}$ **(b)** $\frac{\sqrt{5}}{2}(e^{4\pi} - 1)$

19. $r = \frac{4}{1 + 2\cos\theta}$ **21.** $r = \frac{2}{2 + \sin\theta}$

23. $x = (a + b)\cos\theta - b\cos\left(\frac{a + b}{b}\theta\right)$,

$y = (a + b)\sin\theta - b\sin\left(\frac{a + b}{b}\theta\right)$

27. $\frac{\pi}{2}$

Chapter 12

SECTION 12.1, pp. 713–715

1. The line through the point (4, 5, 0) parallel to z-axis
3. The y-axis
5. The circle $x^2 + y^2 = 4$ in the xy-plane
7. The circle $x^2 + z^2 = 4$ in the xz-plane
9. The circle $y^2 + z^2 = 1$ in the yz-plane
11. The circle $x^2 + y^2 = 16$ in the xy-plane
13. The ellipse formed by the intersection of the cylinder $x^2 + y^2 = 4$ and the plane $z = y$
15. The parabola $y = x^2$ in the xy-plane
17. (a) The first quadrant of the xy-plane
(b) The fourth quadrant of the xy-plane
19. (a) The ball of radius 1 centered at the origin
(b) All points more than 1 unit from the origin
21. (a) The ball of radius 2 centered at the origin with the interior of the ball of radius 1 centered at the origin removed
(b) The solid upper hemisphere of radius 1 centered at the origin
23. (a) The region on or inside the parabola $y = x^2$ in the xy-plane and all points above this region
(b) The region on or to the left of the parabola $x = y^2$ in the xy-plane and all points above it that are 2 units or less away from the xy-plane
25. $|P_1P_2| = \sqrt{(7-5)^2+(4-5)^2+(0-5)^2} = \sqrt{30}$
27. $|P_1P_2| = \sqrt{(5-2)^2+(-3-5)^2+(9-6)^2} = \sqrt{82}$
29. $|P_1P_2| = \sqrt{(7-0)^2+(-7-0)^2+(-7-0)^2} = \sqrt{147}$
31. (a) 2 **(b)** 3 **(c)** 4 **33. (a)** 3 **(b)** 4 **(c)** 5
35. (a) $x = 3$ **(b)** $y = -1$ **(c)** $z = -2$
37. (a) $z = 1$ **(b)** $x = 3$ **(c)** $y = -1$
39. (a) $x^2 + (y-2)^2 = 4, z = 0$
(b) $(y-2)^2 + z^2 = 4, x = 0$ **(c)** $x^2 + z^2 = 4, y = 2$
41. (a) $y = 3, z = -1$ **(b)** $x = 1, z = -1$ **(c)** $x = 1, y = 3$
43. $x^2 + y^2 = 11$ in plane $z = 5$ **45.** $0 \le z \le 1$ **47.** $z \le 0$
49. (a) $(x-3)^2 + (y-3)^2 + (z-3)^2 < 3$
(b) $(x-3)^2 + (y-3)^2 + (z-3)^2 > 3$
51. centre (4, 0, −4), radius 4
53. centre $(\sqrt{8}, -\sqrt{8}, -\sqrt{8})$, radius 8
55. centre (−7, 0, 7), radius $\sqrt{14}$
57. centre $\left(-\frac{1}{6}, -\frac{1}{6}, -\frac{1}{6}\right)$, radius $\frac{\sqrt{171}}{6}$
59. centre (4, −5, 5), radius 77
61. $(x-2)^2 + (y-3)^2 + (z-4)^2 = 15$
63. $(x+4)^2 + \left(y - \frac{5}{2}\right)^2 + \left(z + \frac{3}{4}\right)^2 = \frac{49}{100}$
65. (a) $\sqrt{y^2 + z^2}$ **(b)** $\sqrt{x^2 + z^2}$ **(c)** $\sqrt{x^2 + y^2}$
67. Perimeter of triangle ABC is $\sqrt{45} + \sqrt{57} + \sqrt{56}$
69. $y = 1$ **71. (a)** $(0, 3, -3)$ **(b)** $(0, 5, -5)$
73. $z = x^2/4 + 1$ **75. (a)** $z^2 = x^2$ **(b)** $y^2 = x^2$

SECTION 12.2, pp. 723–725

1. (a) $\langle 9, -6\rangle$ **(b)** $3\sqrt{13}$ **3. (a)** $\langle 1, 3\rangle$ **(b)** $\sqrt{10}$
5. (a) $\langle 13, -16\rangle$ **(b)** $5\sqrt{17}$
7. (a) $\left\langle \frac{1}{5}, \frac{14}{5}\right\rangle$ **(b)** $\frac{\sqrt{197}}{5}$ **9.** $\langle 2, -3\rangle$ **11.** $\langle -3, -4\rangle$
13. $\left\langle \cos\frac{\pi}{6}, \sin\frac{\pi}{6}\right\rangle = \left\langle \frac{\sqrt{3}}{2}, \frac{1}{2}\right\rangle$ **15.** $\left\langle -\frac{\sqrt{3}}{2}, -\frac{1}{2}\right\rangle$
17. $-\mathbf{i} + 4\mathbf{j} - 2\mathbf{k}$ **19.** $-3\mathbf{i} + 16\mathbf{j}$ **21.** $3\mathbf{i} + 5\mathbf{j} - 8\mathbf{k}$
23. The vector **v** is horizontal and 1 in. long. The vectors **u** and **w** are $\frac{11}{16}$ in. long. **w** is vertical and **u** makes a 45° angle with the horizontal. All vectors must be drawn to scale.

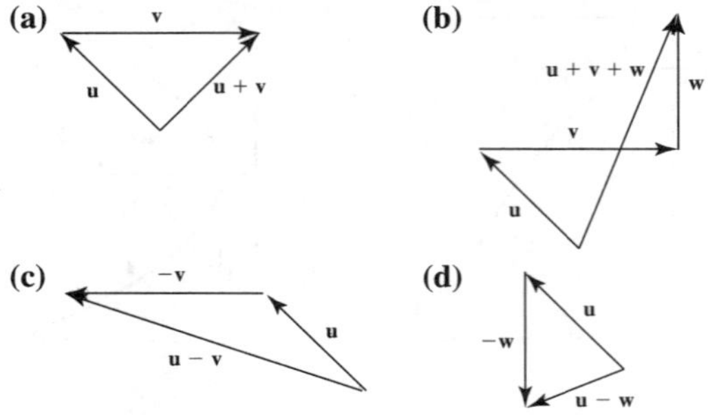

25. $3\left(\frac{1}{3}\mathbf{i} + \frac{2}{3}\mathbf{j} - \frac{2}{3}\mathbf{k}\right)$ **27.** $5(\mathbf{k})$
29. $\sqrt{\frac{1}{2}}\left(\frac{1}{\sqrt{3}}\mathbf{i} - \frac{1}{\sqrt{3}}\mathbf{j} - \frac{1}{\sqrt{3}}\mathbf{k}\right)$
31. (a) $2\mathbf{i}$ **(b)** $-\sqrt{3}\mathbf{k}$ **(c)** $\frac{3}{10}\mathbf{j} + \frac{2}{5}\mathbf{k}$ **(d)** $6\mathbf{i} - 2\mathbf{j} + 3\mathbf{k}$
33. $\frac{5}{\sqrt{65}}(7\mathbf{i} - 4\mathbf{k})$
35. (a) $\frac{3}{5\sqrt{2}}\mathbf{i} + \frac{4}{5\sqrt{2}}\mathbf{j} - \frac{1}{\sqrt{2}}\mathbf{k}$ **(b)** $(1/2, 3, 5/2)$
37. (a) $-\frac{1}{\sqrt{3}}\mathbf{i} - \frac{1}{\sqrt{3}}\mathbf{j} - \frac{1}{\sqrt{3}}\mathbf{k}$ **(b)** $\left(\frac{5}{2}, \frac{7}{2}, \frac{9}{2}\right)$
39. A is the point $(0, -3, 5)$ **41.** $a = \frac{7}{5}$ and $b = \frac{4}{5}$
43. $a = -1, b = 2, c = 1$ **45.** $\approx(-516.222, 737.235)$
47. $|\mathbf{F}_1| \approx 73.205$ N,
$|\mathbf{F}_2| \approx 89.658$ N,
$\mathbf{F}_1 = \langle -|\mathbf{F}_1|\cos 30°, |\mathbf{F}_1|\sin 30°\rangle \approx \langle -63.397, 36.603\rangle$,
$\mathbf{F}_2 = \langle |\mathbf{F}_2|\cos 45°, |\mathbf{F}_2|\sin 45°\rangle \approx \langle 63.397, 63.397\rangle$
49. $w \approx 126.093$ N,
$|\mathbf{F}_1| \approx 106.933$ N
51. (a) $(5\cos 60°, 5\sin 60°) = \left(\frac{5}{2}, \frac{5\sqrt{3}}{2}\right)$
(b) $(5\cos 60° + 10\cos 315°, 5\sin 60° + 10\sin 315°) = \left(\frac{5 + 10\sqrt{2}}{2}, \frac{5\sqrt{3} - 10\sqrt{2}}{2}\right)$
53. (a) $\frac{3}{2}\mathbf{i} + \frac{3}{2}\mathbf{j} - 3\mathbf{k}$ **(b)** $\mathbf{i} + \mathbf{j} - 2\mathbf{k}$ **(c)** $(2, 2, 1)$
59. (a) $\langle 0, 0, 0\rangle$ **(b)** $\langle 0, 0, 0\rangle$
61. (a) $\langle 0, 16, 4, -12\rangle$ **(b)** $4\sqrt{26}$
63. (a) $\langle -2, -1, -1, 2\rangle$ **(b)** $\sqrt{10}$
65. (a) $\langle 8, -24, -3, 13\rangle$ **(b)** $\sqrt{818}$

SECTION 12.3, pp. 732–734

1. (a) $-8, \sqrt{8}, \sqrt{8}$ **(b)** -1 **(c)** $-\sqrt{8}$ **(d)** $-\mathbf{i} + 2\mathbf{j} - \sqrt{3}\,\mathbf{k}$
3. (a) $57, \sqrt{170}, \sqrt{34}$ **(b)** $\frac{57}{\sqrt{170}\sqrt{34}}$
(c) $\frac{57}{\sqrt{170}}$ **(d)** $\frac{57}{170}(5\mathbf{i} + 12\mathbf{j} - \mathbf{k})$

5. **(a)** $-2, \sqrt{10}, \sqrt{3}$ **(b)** $\dfrac{-2}{\sqrt{10}\sqrt{3}}$

(c) $\dfrac{-2}{\sqrt{10}}$ **(d)** $\dfrac{-1}{5}(3\mathbf{j}-\mathbf{k})$

7. **(a)** $2\sqrt{2}+\sqrt{15}, \sqrt{5}, \sqrt{17}$ **(b)** $\dfrac{2\sqrt{2}+\sqrt{15}}{\sqrt{5}\sqrt{17}}$

(c) $\dfrac{2\sqrt{2}+\sqrt{15}}{\sqrt{5}}$ **(d)** $\dfrac{2\sqrt{2}+\sqrt{15}}{5}(2\mathbf{i}+\mathbf{j})$

9. $\theta \approx 0.75$ rad

13. Angle at $A \approx 1.02$ rad
Angle at $B \approx 0.99$ rad
Angle at $C \approx 1.124$ rad

17. $\arccos\left(\dfrac{3}{\sqrt{10}}\right) \approx 0.322$ radian or 18.43 degrees

25. Horizontal component: ≈ 1188 ft/sec, vertical component: ≈ 167 ft/sec

27. **(a)** Since $|\cos\theta| \le 1$, we have $|\mathbf{u}\cdot\mathbf{v}| = |\mathbf{u}||\mathbf{v}||\cos\theta| \le |\mathbf{u}||\mathbf{v}|(1) = |\mathbf{u}||\mathbf{v}|$.
(b) We have equality precisely when $|\cos\theta| = 1$ or when one or both of $\mathbf{u}$ and $\mathbf{v}$ are $\mathbf{0}$. In the case of nonzero vectors, we have equality when $\theta = 0$ or π, that is, when the vectors are parallel.

29. a

35. $x + 2y = 4$

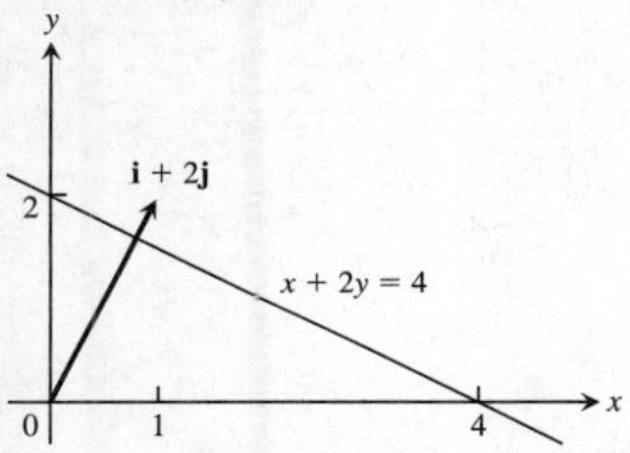

37. $-2x + y = -3$

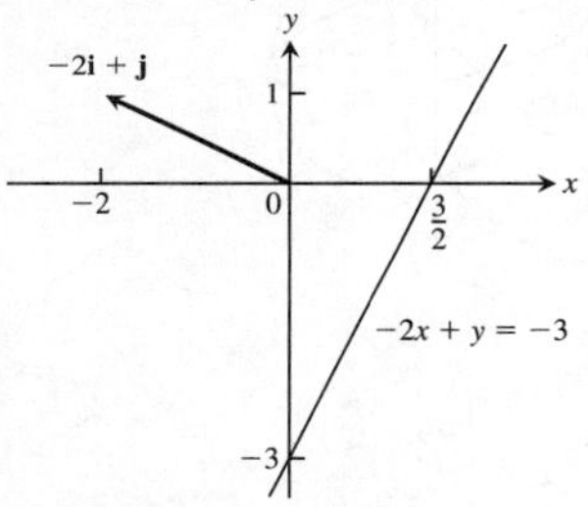

39. $x + y = -1$

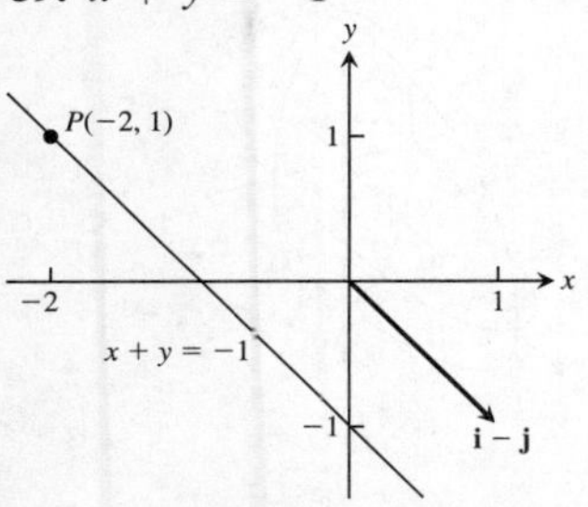

41. $2x - y = 0$

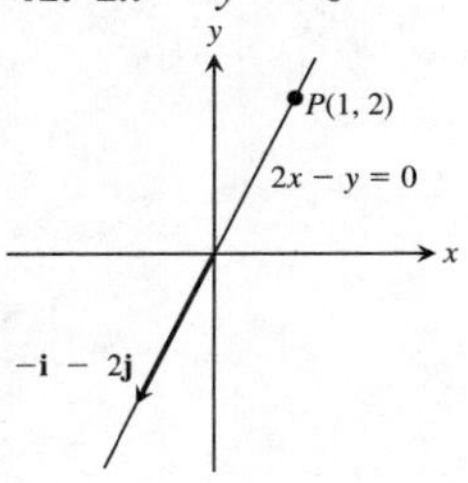

43. 7 J **45.** 600 J **47.** $\theta \approx 1.446$ rad

49. $\theta \approx \dfrac{\pi}{6}$ **51.** 0.14

53. **(a)** 3 **(b)** not orthogonal

55. **(a)** 0 **(b)** orthogonal

SECTION 12.4, pp. 739–740

1. $|\mathbf{u}\times\mathbf{v}| = 3$, direction is $\frac{2}{3}\mathbf{i} + \frac{1}{3}\mathbf{j} + \frac{2}{3}\mathbf{k}$; $|\mathbf{v}\times\mathbf{u}| = 3$, direction is $-\frac{2}{3}\mathbf{i} - \frac{1}{3}\mathbf{j} - \frac{2}{3}\mathbf{k}$

3. $|\mathbf{u}\times\mathbf{v}| = 0$, no direction; $|\mathbf{v}\times\mathbf{u}| = 0$, no direction

5. $\mathbf{u}\times\mathbf{v} = -10\mathbf{k}$
Length = 10 and has direction is $-\mathbf{k}$
$\mathbf{v}\times\mathbf{u} = 10\mathbf{k}$
Length = 10 and has direction is $\mathbf{k}$

7. $\mathbf{u}\times\mathbf{v} = 3\mathbf{i}+\mathbf{j}-12\mathbf{k}$
Length = $\sqrt{154}$ and direction is $\dfrac{3}{\sqrt{154}}\mathbf{i}+\dfrac{1}{\sqrt{154}}\mathbf{j}-\dfrac{12}{\sqrt{154}}\mathbf{k}$
$\mathbf{v}\times\mathbf{u} = -(3\mathbf{i}+\mathbf{j}-12\mathbf{k})$
Length = $\sqrt{154}$ and direction is $-\dfrac{3}{\sqrt{154}}\mathbf{i}-\dfrac{1}{\sqrt{154}}\mathbf{j}+\dfrac{12}{\sqrt{154}}\mathbf{k}$

9.

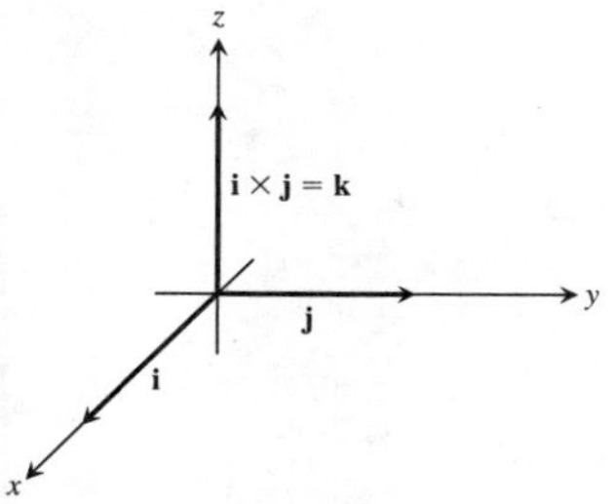

11.

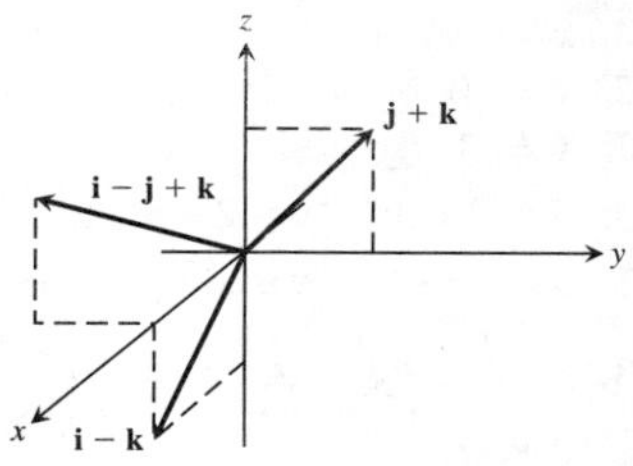

13.

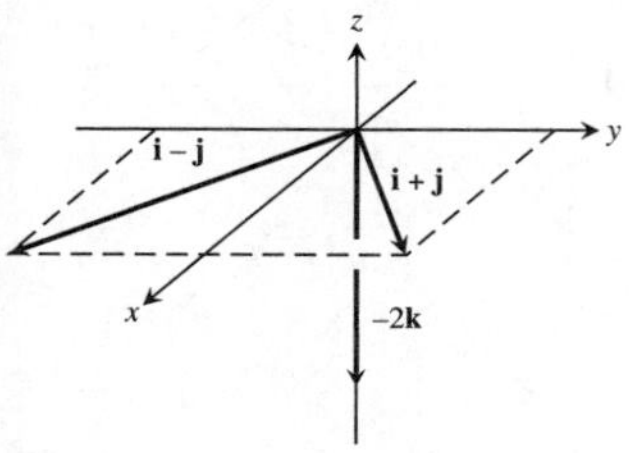

15. **(a)** Area = $\dfrac{\sqrt{3}}{2}$ **(b)** $u = \dfrac{1}{\sqrt{3}}(-\mathbf{i}+\mathbf{j}-\mathbf{k})$

17. **(a)** Area = $\dfrac{\sqrt{3}}{2}$ **(b)** $\mathbf{u} = \dfrac{1}{\sqrt{3}}(\mathbf{i}-\mathbf{j}+\mathbf{k})$

19. $|(\mathbf{u}\times\mathbf{v})\cdot\mathbf{w}| = 15$ **21.** $|(\mathbf{u}\times\mathbf{v})\cdot\mathbf{w}| = |-20| = 20$

23. **(a)** $\mathbf{u}\cdot\mathbf{v} = -4$, $\mathbf{u}\cdot\mathbf{w} = -28$, $\mathbf{v}\cdot\mathbf{w} = -16$; None are perpendicular
(b) $\mathbf{u}\times\mathbf{v} \ne 0$, $\mathbf{u}\times\mathbf{w} = 0$, $\mathbf{v}\times\mathbf{w} \ne 0$. Therefore, $\mathbf{u}$ and $\mathbf{w}$ are parallel

25. $10\sqrt{3}$ ft-lb

27. **(a)** True **(b)** Not always true **(c)** True **(d)** True
(e) Not always true **(f)** True **(g)** True **(h)** True

29. **(a)** $\text{proj}_{\mathbf{v}}\,\mathbf{u} = \dfrac{\mathbf{u}\cdot\mathbf{v}}{\mathbf{v}\cdot\mathbf{v}}\mathbf{v}$ **(b)** $k\mathbf{u}\times\mathbf{v}$ (k any constant)
(c) $k(\mathbf{u}\times\mathbf{v})\times\mathbf{w}$ (k any constant)
(d) $|(\mathbf{u}\times\mathbf{v})\cdot\mathbf{w}|$ **(e)** $k(\mathbf{u}\times\mathbf{v})\times(\mathbf{u}\times\mathbf{w})$ (k any constant)
(f) $|\mathbf{u}|\dfrac{\mathbf{v}}{|\mathbf{v}|}$

31. **(a)** Yes **(b)** No **(c)** Yes **(d)** No

33. No, $\mathbf{v}$ need not equal $\mathbf{w}$. For example, $\mathbf{i}+\mathbf{j} \ne -\mathbf{i}+\mathbf{j}$, but $\mathbf{i}\times(\mathbf{i}+\mathbf{j}) = \mathbf{i}\times\mathbf{i}+\mathbf{i}\times\mathbf{j} = \mathbf{0}+\mathbf{k} = \mathbf{k}$ and $\mathbf{i}\times(-\mathbf{i}+\mathbf{j}) = -\mathbf{i}\times\mathbf{i}+\mathbf{i}\times\mathbf{j} = \mathbf{0}+\mathbf{k} = \mathbf{k}$.

35. Area = 8 **37.** Area = 10 **39.** Area = $\sqrt{154}$

41. $\dfrac{11}{2}$ **43.** $\dfrac{25}{2}$ **45.** $\dfrac{3}{2}$ **47.** $\dfrac{\sqrt{21}}{2}$

49. If $\mathbf{A} = a_1\mathbf{i} + a_2\mathbf{j}$ and $\mathbf{B} = b_1\mathbf{i} + b_2\mathbf{j}$, then

$$\mathbf{A}\times\mathbf{B} = \begin{vmatrix} \mathbf{i} & \mathbf{j} & \mathbf{k} \\ a_1 & a_2 & 0 \\ b_1 & b_2 & 0 \end{vmatrix} = \begin{vmatrix} a_1 & a_2 \\ b_1 & b_2 \end{vmatrix}\mathbf{k},$$

and the triangle's area is

$$\frac{1}{2}\left|\mathbf{A}\times\mathbf{B}\right| = \pm\frac{1}{2}\begin{vmatrix} a_1 & a_2 \\ b_1 & b_2 \end{vmatrix}.$$

The applicable sign is $(+)$ if the acute angle from **A** to **B** runs counterclockwise in the xy-plane, and $(-)$ if it runs clockwise.

51. 4 **53.** 44/3 **55.** Coplanar **57.** Not coplanar

SECTION 12.5, pp. 747–749

1. $x = 3 + t,\quad y = -4 + t,\quad z = -1 + t$
3. $x = -2 + 5t,\quad y = 5t,\quad z = 3 - 5t$
5. $x = 0,\quad y = 2t,\quad z = t$
7. $x = 1,\quad y = 1,\quad z = 1 + t$
9. $x = 3 + 3t,\quad y = 2 - t,\quad z = -8 - 2t$
11. $x = 2 + 2t,\quad y = 1 - t,\quad z = -4t$

13. $x = t,\quad y = t,\quad z = \frac{3}{2}t,\quad 0 \le t \le 1$

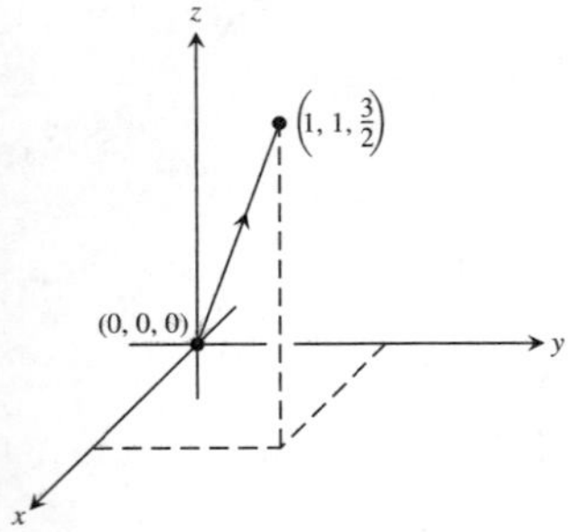

15. $x = 3 + t,\ y = 2 + 3t,\ z = 4 - 2t$

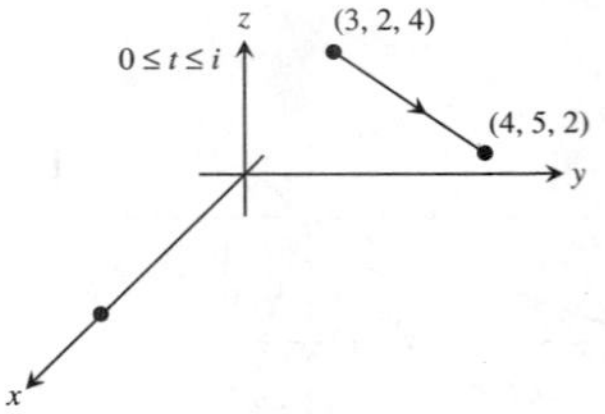

17. $x = 0,\quad y = 1 - 2t,\quad z = 1,\quad 0 \le t \le 1$

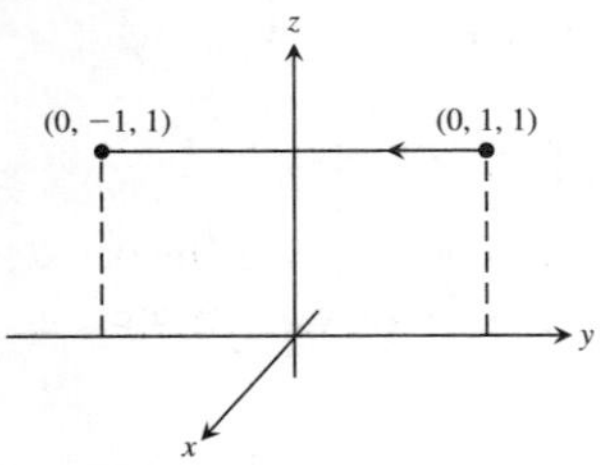

19. $x = 2 - 2t,\quad y = 2t,\quad z = 2 - 2t,\quad 0 \le t \le 1$

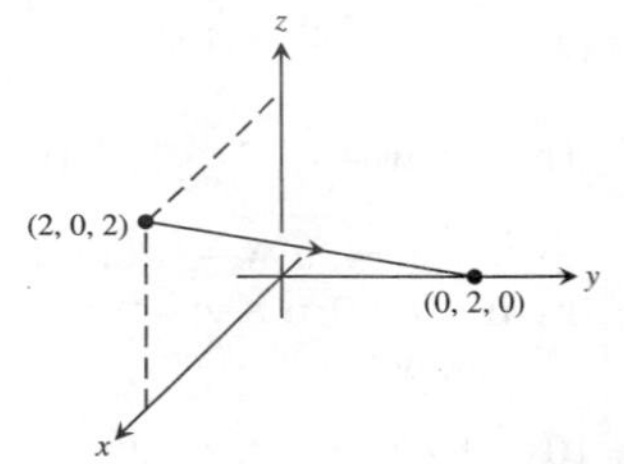

21. $3x - 2y - z = -3$ **23.** $7x - 5y - 4z = 6$
25. $x + 3y + 4z = 34$ **27.** $(1, 2, 3), -20x + 12y + z = 7$
29. $y + z = 3$ **31.** $x - y + z = 0$ **33.** $2\sqrt{30}$ **35.** $\sqrt{6}$
37. $\frac{9\sqrt{42}}{7}$ **39.** 3 **41.** $\sqrt{\frac{3}{2}}$ **43.** 5/3 **45.** $9/\sqrt{41}$
47. $\pi/4$ **49.** $\pi - \arccos(-1/6) \approx \pi - 1.738 \approx 1.403$ radians
51. 14.08 rad **53.** $\pi/2 = 1.57$ rad
55. 0.82 rad **57.** $\left(\frac{3}{2}, -\frac{3}{2}, \frac{1}{2}\right)$ **59.** $(1, 1, 0)$
61. $x = 1 - t,\quad y = 1 + t,\quad z = -1$
63. $x = 4,\quad y = 3 + 6t,\quad z = 1 + 3t$
65. $L1$ intersects $L2$; $L2$ is parallel to $L3$, $\sqrt{5}/3$; $L1$ and $L3$ are skew, $10\sqrt{2}/3$
67. $x = 2 + 2t,\quad y = -4 - t,\quad z = 7 + 3t;\quad x = -2 - t,\quad y = -2 + (1/2)t,\quad z = 1 - (3/2)t$
69. $\left(0, -\frac{1}{2}, -\frac{3}{2}\right), (-1, 0, -3), (1, -1, 0)$
73. Many possible answers. One possibility: $x + y = 3$ and $2y + z = 7$.
75. $(x/a) + (y/b) + (z/c) = 1$ describes all planes *except* those through the origin or parallel to a coordinate axis.

SECTION 12.6, pp. 753–754

1. (d), ellipsoid **3.** (a), cylinder **5.** (l), hyperbolic paraboloid
7. (b), cylinder **9.** (k), hyperbolic paraboloid **11.** (h), cone

13.

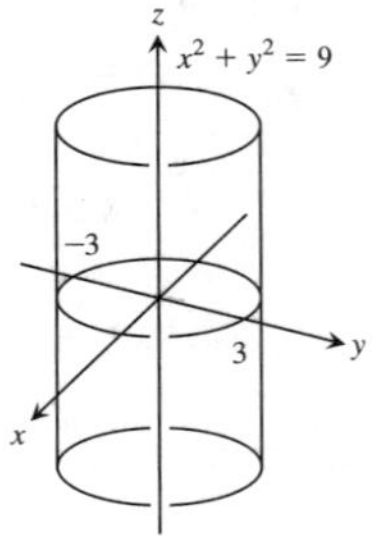

15.

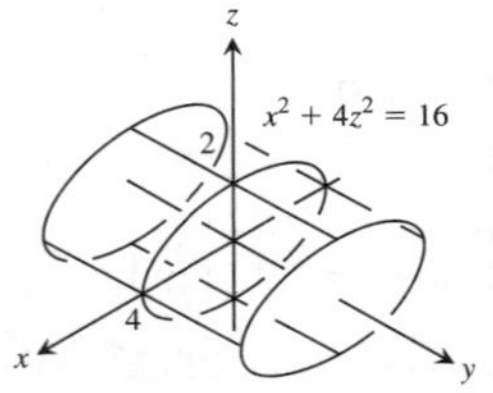

17.

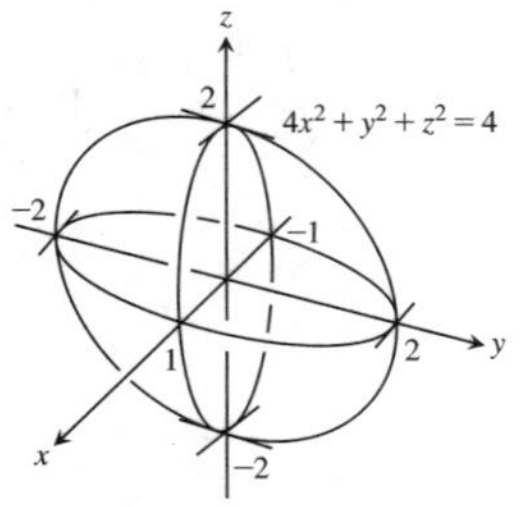

19.

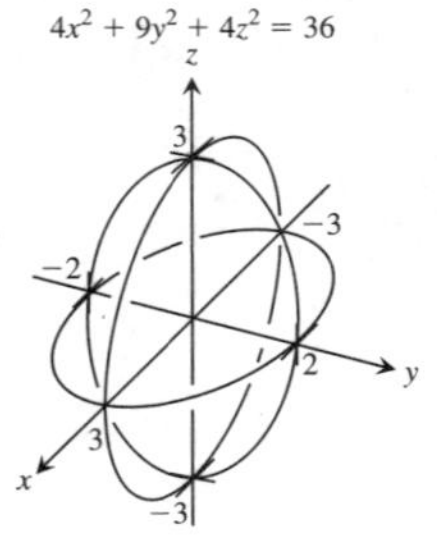

21.

23.

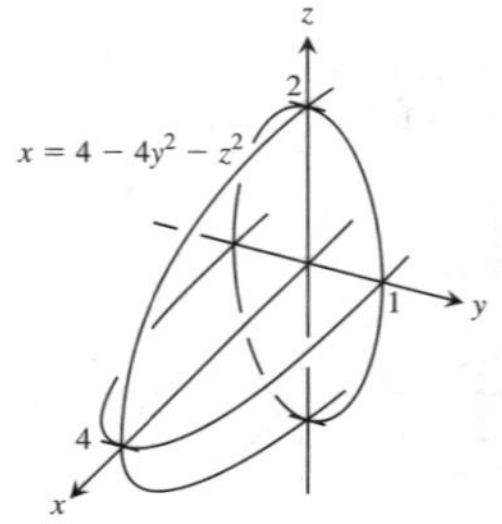

25.

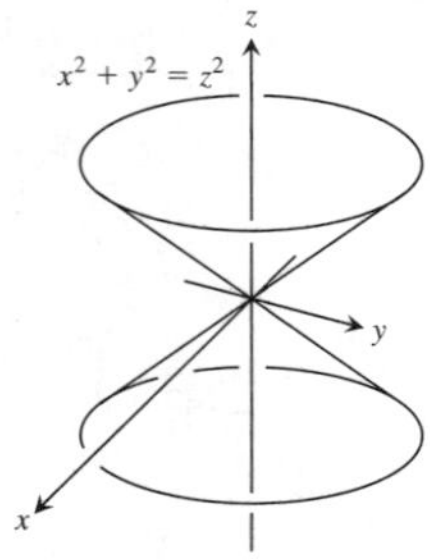

27.

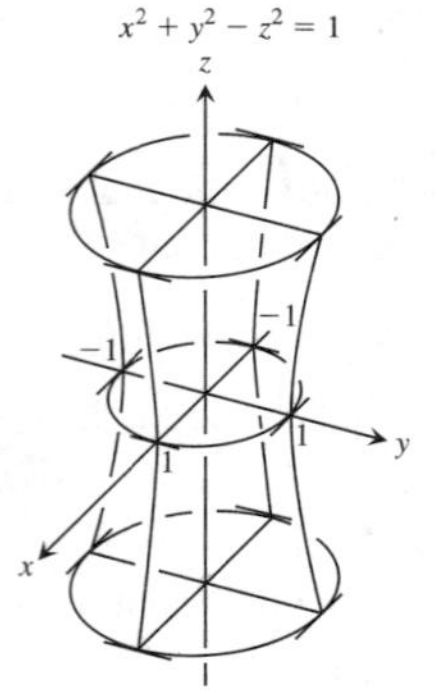

29.

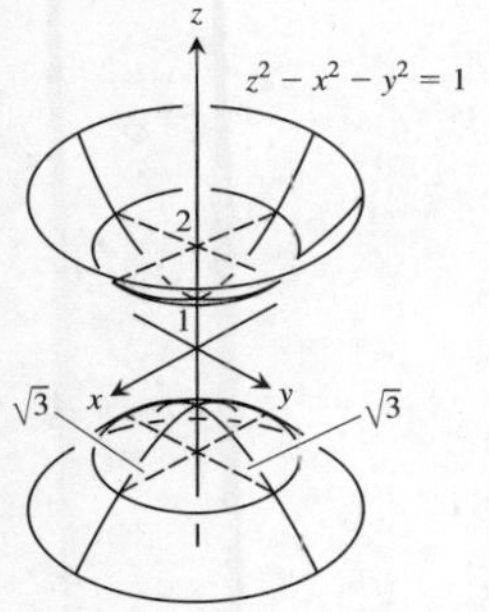

31.

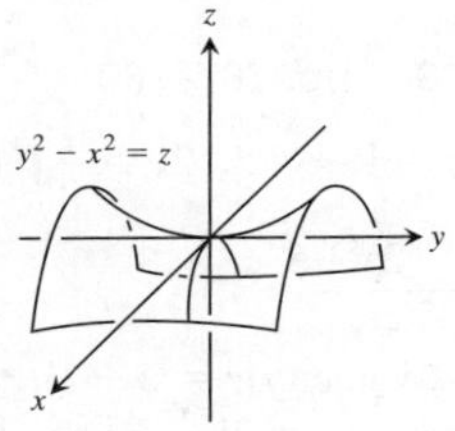

33.

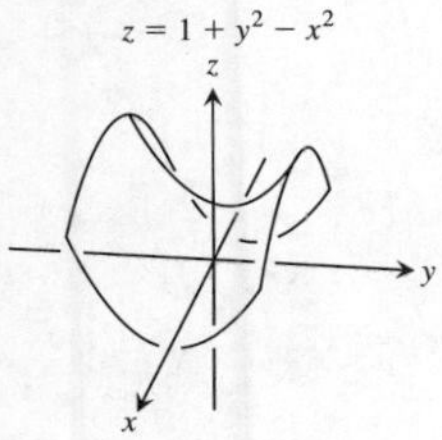

35.

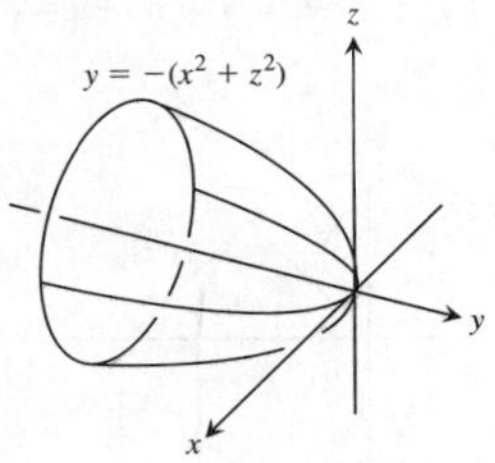

37.

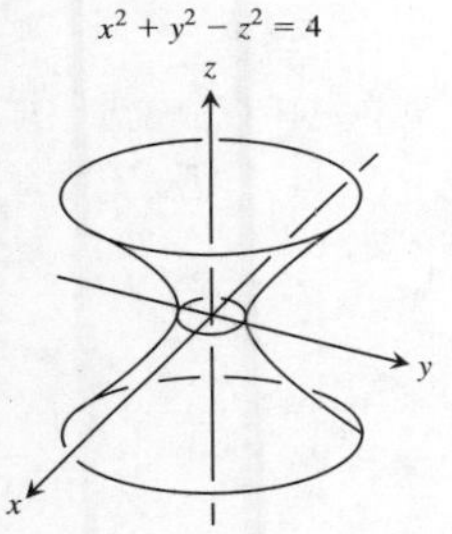

39.

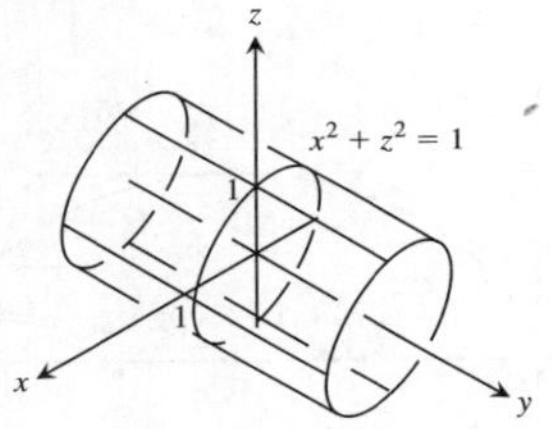

41.

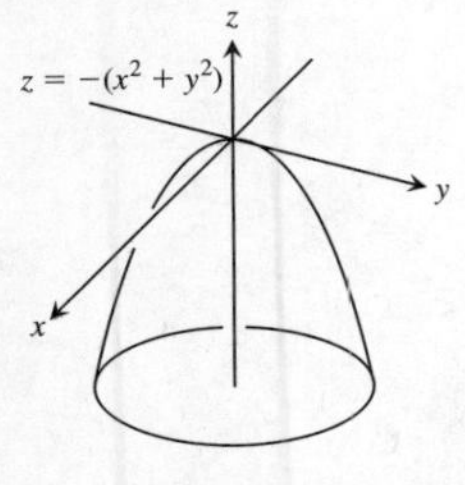

43.

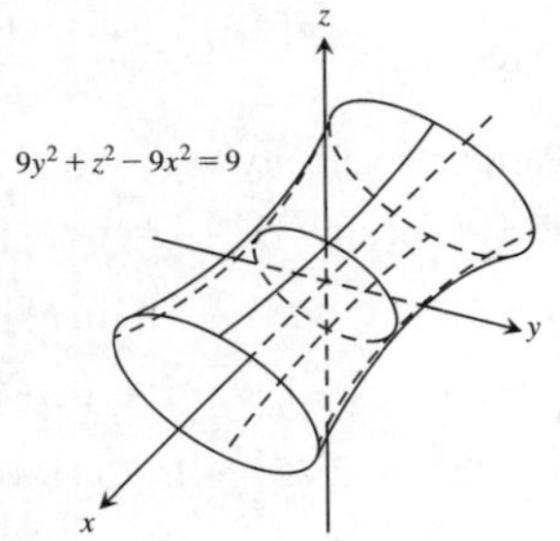

45. **(a)** $\dfrac{2\pi(9 - c^2)}{9}$ **(b)** 8π **(c)** $\dfrac{4\pi abc}{3}$

49.

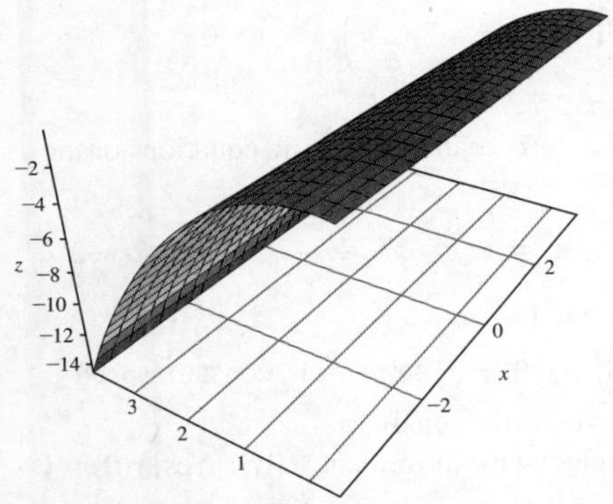

51.

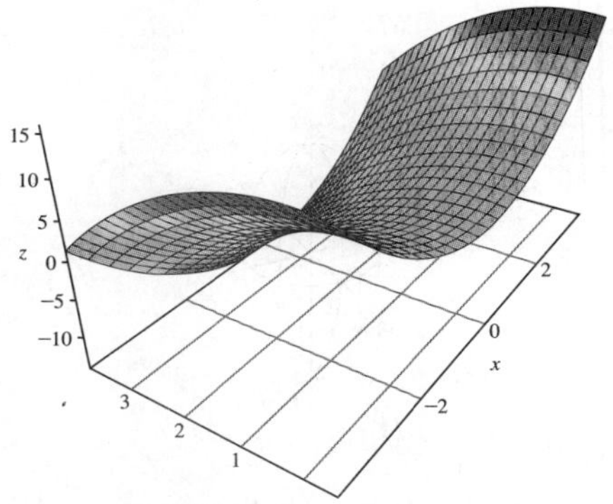

PRACTICE EXERCISES, pp. 755–757

1. **(a)** $\langle -17, 32\rangle$ **(b)** $\sqrt{1313}$ **3.** **(a)** $\langle 6, -8\rangle$ **(b)** 10

5. $\left\langle -\dfrac{\sqrt{3}}{2}, -\dfrac{1}{2}\right\rangle$ [assuming counterclockwise] **7.** $\left\langle \dfrac{8}{\sqrt{17}}, -\dfrac{2}{\sqrt{17}}\right\rangle$

9. Length $= 2$; direction is $\dfrac{1}{\sqrt{2}}\mathbf{i} + \dfrac{1}{\sqrt{2}}\mathbf{j}$.

11. Length $= 2$; direction is $-\mathbf{i}$.

13. Length $= 7$; direction is $\dfrac{2}{7}\mathbf{i} - \dfrac{3}{7}\mathbf{j} + \dfrac{6}{7}\mathbf{k}$.

15. $\dfrac{8}{\sqrt{33}}\mathbf{i} - \dfrac{2}{\sqrt{33}}\mathbf{j} + \dfrac{8}{\sqrt{33}}\mathbf{k}$

17. $|\mathbf{v}| = \sqrt{2}, |\mathbf{u}| = 3, \mathbf{v}\cdot\mathbf{u} = \mathbf{u}\cdot\mathbf{v} = 3, \mathbf{v}\times\mathbf{u} = -2\mathbf{i} + 2\mathbf{j} - \mathbf{k},$
$\mathbf{u}\times\mathbf{v} = 2\mathbf{i} - 2\mathbf{j} + \mathbf{k}, |\mathbf{v}\times\mathbf{u}| = 3, \theta = \cos^{-1}\left(\dfrac{1}{\sqrt{2}}\right) = \dfrac{\pi}{4},$
$|\mathbf{u}|\cos\theta = \dfrac{3}{\sqrt{2}}, \text{proj}_{\mathbf{v}}\mathbf{u} = \dfrac{3}{2}(\mathbf{i} + \mathbf{j})$

19. $\dfrac{4}{3}(2\mathbf{i} + \mathbf{j} - \mathbf{k})$

21. $\mathbf{u}\times\mathbf{v} = \mathbf{k}$

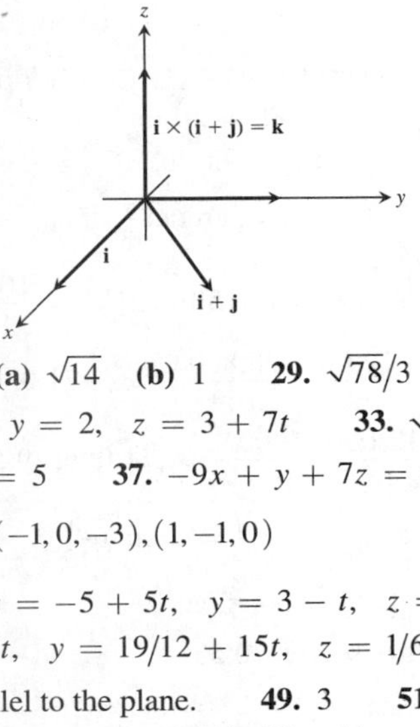

23. $2\sqrt{7}$ **25.** **(a)** $\sqrt{14}$ **(b)** 1 **29.** $\sqrt{78}/3$

31. $x = 1 - 3t,\ y = 2,\ z = 3 + 7t$ **33.** $\sqrt{2}$

35. $2x + y + z = 5$ **37.** $-9x + y + 7z = 4$

39. $\left(0, -\dfrac{1}{2}, -\dfrac{3}{2}\right), (-1, 0, -3), (1, -1, 0)$

41. $\pi/3$ **43.** $x = -5 + 5t,\ y = 3 - t,\ z = -3t$

45. **(b)** $x = -12t,\ y = 19/12 + 15t,\ z = 1/6 + 6t$

47. Yes; v is parallel to the plane. **49.** 3 **51.** $-3\mathbf{j} + 3\mathbf{k}$

53. $\dfrac{2}{\sqrt{35}}(5\mathbf{i} - \mathbf{j} - 3\mathbf{k})$ **55.** $\left(\dfrac{11}{9}, \dfrac{26}{9}, -\dfrac{7}{9}\right)$

57. $(1, -2, -1); x = 1 - 5t,\ y = -2 + 3t,\ z = -1 + 4t$

59. $2x + 7y + 2z + 10 = 0$

61. **(a)** No **(b)** No **(c)** No **(d)** No **(e)** Yes

63. $11/\sqrt{107}$

65.

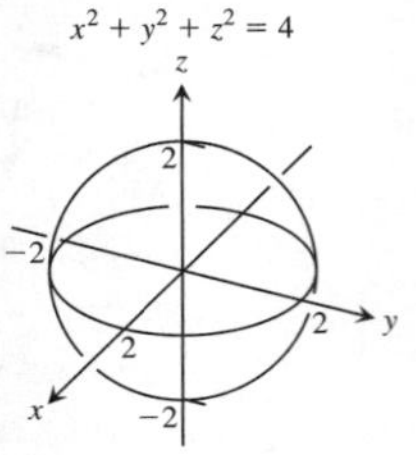

67.

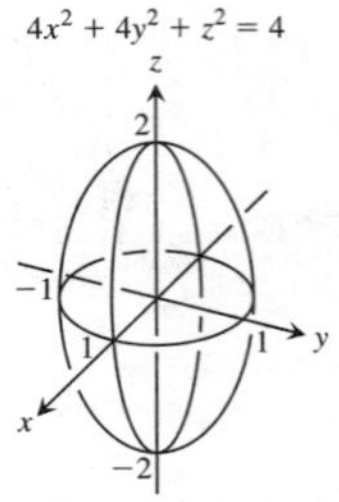

69.

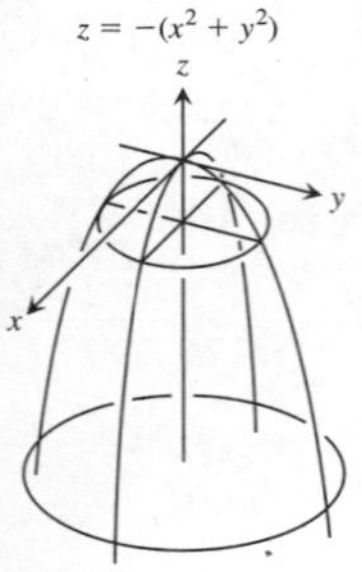

71.

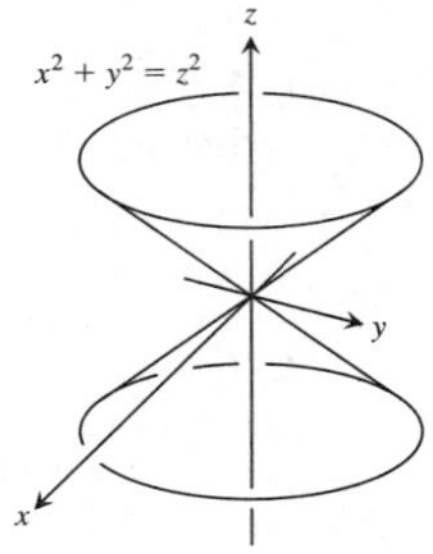

73.

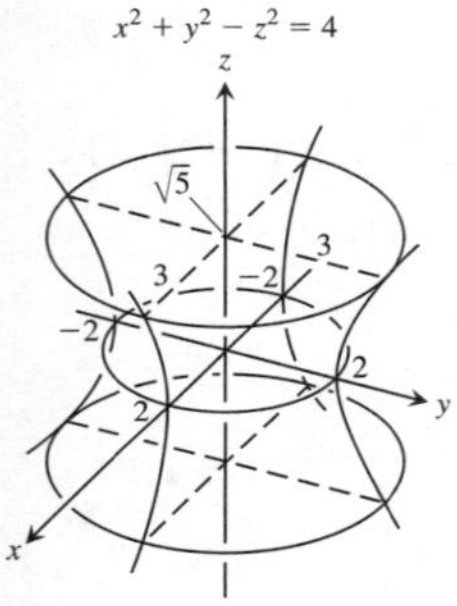

75.

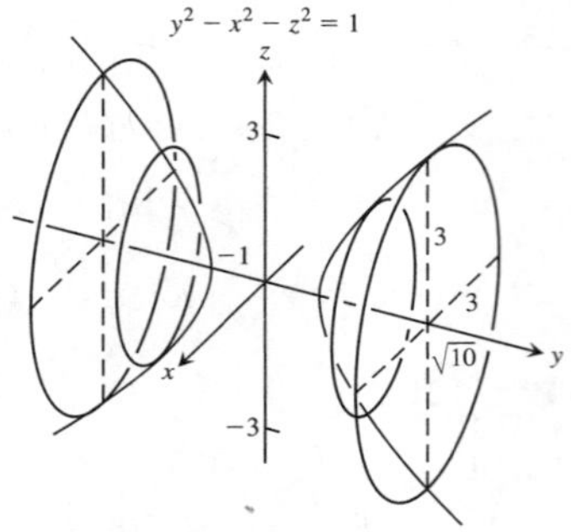

ADDITIONAL AND ADVANCED EXERCISES, pp. 757–759

1. $(26, 23, -1/3)$ **3.** $|\mathbf{F}| = 20$ lb

5. (a) $|\mathbf{F}_1| = 80$ lb, $|\mathbf{F}_2| = 60$ lb, $\mathbf{F}_1 = \langle -48, 64 \rangle$, $\mathbf{F}_2 = \langle 48, 36 \rangle$, $\alpha = \arctan \frac{4}{3}$, $\beta = \arctan \frac{3}{4}$

(b) $|\mathbf{F}_1| = \frac{2400}{13} \approx 184.615$ lb, $|\mathbf{F}_2| = \frac{1000}{13} \approx 76.923$ lb,

$\mathbf{F}_1 = \left\langle \frac{-12{,}000}{169}, \frac{28{,}800}{169} \right\rangle \approx \langle -71.006, 170.414 \rangle$,

$\mathbf{F}_2 = \left\langle \frac{12{,}000}{169}, \frac{5000}{169} \right\rangle \approx \langle 71.006, 29.586 \rangle$,

$\alpha = \arctan \frac{12}{5}$, $\beta = \arctan \frac{5}{12}$

9. (a) $\theta = \arctan \sqrt{2} \approx 54.74°$

(b) $\theta = \arctan 2\sqrt{2} \approx 70.53°$

13. (b) $\frac{6}{\sqrt{14}}$ **(c)** $2x - y + 2z = 8$

(d) $x - 2y + z = 3 + 5\sqrt{6}$ and $x - 2y + z = 3 - 5\sqrt{6}$

15. $\frac{32}{41}\mathbf{i} + \frac{23}{41}\mathbf{j} - \frac{13}{41}\mathbf{k}$

17. (a) $\mathbf{0}, \mathbf{0}$ **(b)** $-10\mathbf{i} - 2\mathbf{j} + 6\mathbf{k}, -9\mathbf{i} - 2\mathbf{j} + 7\mathbf{k}$

(c) $-4\mathbf{i} - 6\mathbf{j} + 2\mathbf{k}, \mathbf{i} - 2\mathbf{j} - 4\mathbf{k}$

(d) $-10\mathbf{i} - 10\mathbf{k}, -12\mathbf{i} - 4\mathbf{j} - 8\mathbf{k}$

19. The formula is always true.

Chapter 13

SECTION 13.1, pp. 767–769

1. $\mathbf{i} - \frac{1}{2}\mathbf{j} + \mathbf{k}$ **3.** $2\mathbf{i} + \frac{1}{2}\mathbf{j} + \frac{\pi}{4}\mathbf{k}$

5. $x = \frac{y^2 + 2y + 5}{4}$

At $t = 2$, velocity $\mathbf{v} = 4\mathbf{i} + 2\mathbf{j}$

Acceleration $\mathbf{a} = 2\mathbf{i}$

7. $y = \frac{2}{9}x^2$, $\mathbf{v} = 3\mathbf{i} + 4\mathbf{j}$, $\mathbf{a} = 3\mathbf{i} + 8\mathbf{j}$

9. $t = \frac{\pi}{4}$: $\mathbf{v} = \frac{\sqrt{2}}{2}\mathbf{i} - \frac{\sqrt{2}}{2}\mathbf{j}$, $\mathbf{a} = \frac{-\sqrt{2}}{2}\mathbf{i} - \frac{\sqrt{2}}{2}\mathbf{j}$;

$t = \pi/2$: $\mathbf{v} = -\mathbf{j}$, $\mathbf{a} = -\mathbf{i}$

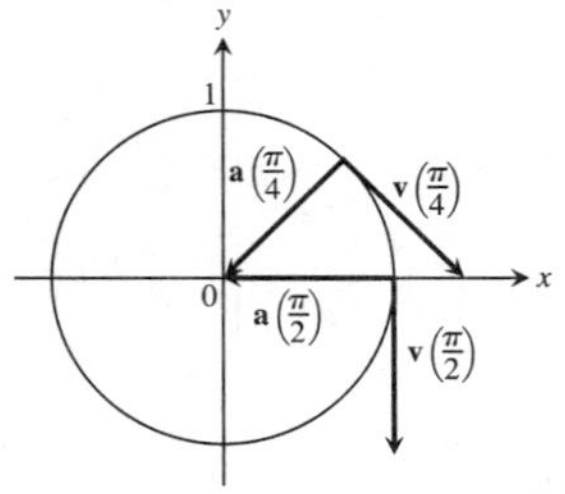

11. $t = \pi$: $\mathbf{v} = 2\mathbf{i}$, $\mathbf{a} = -\mathbf{j}$; $t = \frac{3\pi}{2}$: $\mathbf{v} = \mathbf{i} - \mathbf{j}$, $\mathbf{a} = -\mathbf{i}$

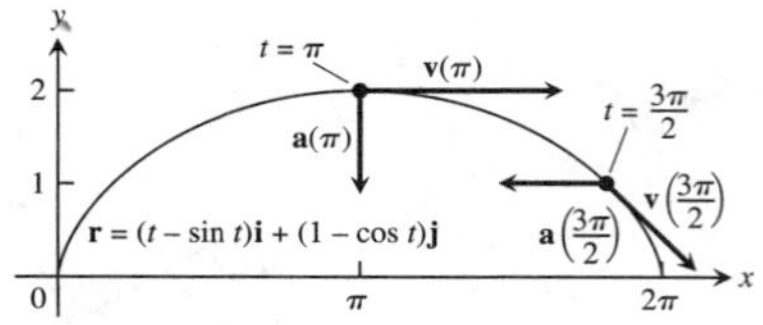

13. velocity $\mathbf{v} = 2t\mathbf{i} + 2\mathbf{j} - 2\mathbf{k}$

Acceleration $\mathbf{a} = 2\mathbf{i}$

Speed $= |\mathbf{v}(2)| = 2\sqrt{6}$

Direction $= \frac{\mathbf{v}(2)}{|\mathbf{v}(2)|} = \frac{2}{\sqrt{6}}\mathbf{i} + \frac{1}{\sqrt{6}}\mathbf{j} - \frac{1}{\sqrt{6}}\mathbf{k}$

Particle's velocity $\mathbf{v}(2) = 4\mathbf{i} + 2\mathbf{j} - 2\mathbf{k}$

15. $\mathbf{v} = (-2\sin t)\mathbf{i} + (3\cos t)\mathbf{j} + 4\mathbf{k}$;

$\mathbf{a} = (-2\cos t)\mathbf{i} - (3\sin t)\mathbf{j}$; speed: $2\sqrt{5}$;

direction: $(-1/\sqrt{5})\mathbf{i} + (2/\sqrt{5})\mathbf{k}$;

$\mathbf{v}(\pi/2) = 2\sqrt{5}[(-1/\sqrt{5})\mathbf{i} + (2/\sqrt{5})\mathbf{k}]$

17. $\mathbf{v} = \left(\frac{2}{t+1}\right)\mathbf{i} + 2t\mathbf{j} + t\mathbf{k}$; $\mathbf{a} = \left(\frac{-2}{(t+1)^2}\right)\mathbf{i} + 2\mathbf{j} + \mathbf{k}$;

speed: $\sqrt{6}$; direction: $\frac{1}{\sqrt{6}}\mathbf{i} + \frac{2}{\sqrt{6}}\mathbf{j} + \frac{1}{\sqrt{6}}\mathbf{k}$;

$\mathbf{v}(1) = \sqrt{6}\left(\frac{1}{\sqrt{6}}\mathbf{i} + \frac{2}{\sqrt{6}}\mathbf{j} + \frac{1}{\sqrt{6}}\mathbf{k}\right)$

19. $\pi/2$ **21.** $\theta = \pi/2$

23. $x = 1$, $y = -t$, and $z = -t$ are the parametric equations of the tangent line

25. $x = t$, $y = \frac{1}{3}t$, $z = t$ **27.** 4, −2 **29.** 2, −2

31. E **33.** D **35.** C

37. (a) (i): $|\mathbf{v}(t)| = \sqrt{(\cos t)^2 + (-\sin t)^2} = 1$, constant speed

(ii): $\mathbf{v} \cdot \mathbf{a} = 0$, yes, orthogonal

(iii): counterclockwise movement **(iv):** Yes, $\mathbf{r}(0) = \mathbf{j}$

(b) **(i):** It has constant speed 2. **(ii):** Yes
(iii): Counterclockwise **(iv):** Yes
(c) **(i):** It has constant speed 1. **(ii):** Yes
(iii): Counterclockwise
(iv): It starts at $(0,-1)$ instead of $(1,0)$.
(d) **(i):** $|\mathbf{v}(t)|=\sqrt{(\cos t)^2+(\sin t)^2}=1$, constant speed
(ii): $\mathbf{v}\cdot\mathbf{a}=0$, yes, orthogonal
(iii): clockwise movement **(iv):** Yes, $\mathbf{r}(0)=-\mathbf{j}$
(e) **(i):** It has variable speed. **(ii):** No
(iii): Counterclockwise **(iv):** Yes

39. Velocity vector $\mathbf{v}=\dfrac{12}{\sqrt{5}}\mathbf{i}+\dfrac{6}{\sqrt{5}}\mathbf{j}$

SECTION 13.2, pp. 774–777

1. $\dfrac{1}{2}\mathbf{i}+5\mathbf{j}-\dfrac{1}{2}\mathbf{k}$ **3.** $(2+\pi)\mathbf{j}$

5. $(\ln 4)\mathbf{i}+\left(\ln\dfrac{3}{2}\right)\mathbf{j}+\dfrac{1}{4}(\ln 2)\mathbf{k}$ **7.** $(e^4-1)\mathbf{i}+\left(2-\dfrac{2}{e^2}\right)\mathbf{j}+4\mathbf{k}$

9. $\mathbf{i}-\mathbf{j}+\dfrac{\pi}{4}\mathbf{k}$ **11.** $\mathbf{r}=(t^2+2)\mathbf{i}+(t^2+1)\mathbf{j}+(t^2+5)\mathbf{k}$

13. $\mathbf{r}(t)=((t+1)^{3/2}-1)\mathbf{i}+(-e^{-t}+1)\mathbf{j}+(\ln(t+1)+1)\mathbf{k}$

15. $\mathbf{r}(t)=(3+\ln|\sec t|)\mathbf{i}+(-2+2\sin(t/2))\mathbf{j}$
$+(1-(1/2)\ln|\sec 2t+\tan 2t|)\mathbf{k}$

17. $\mathbf{r}(t)=8t\mathbf{i}+8t\mathbf{j}+(-16t^2+100)\mathbf{k}$

19. $\mathbf{r}(t)=(e^t-2t+2)\mathbf{i}+(-e^{-t}+3t+2)\mathbf{j}+(e^{2t}-2t+1)\mathbf{k}$

21. $\mathbf{r}(t)=\left(\dfrac{3}{2}t^2+\dfrac{6}{\sqrt{11}}t+1\right)\mathbf{i}-\left(\dfrac{1}{2}t^2+\dfrac{2}{\sqrt{11}}t-2\right)\mathbf{j}$
$+\left(\dfrac{1}{2}t^2+\dfrac{2}{\sqrt{11}}t+3\right)\mathbf{k}=\left(\dfrac{1}{2}t^2+\dfrac{2t}{\sqrt{11}}\right)(3\mathbf{i}-\mathbf{j}+\mathbf{k})$
$+(\mathbf{i}+2\mathbf{j}+3\mathbf{k})$

23. 62.5 seconds

25. **(a)** $t\approx 86.58$ seconds and Range $R\approx 36734.69$ m
(b) $t\approx 14.14$ seconds and $y\approx 5020$ m **(c)** $y_{\max}\approx 9184$ m

27. **(a)** $v_0\approx 12.12$ m/s **(b)** $\alpha\approx 9.74°$ or $80.26°$

29. $\alpha\approx 39.3°$ or $50.7°$ **35.** **(b)** $\mathbf{v}_0$ would bisect $\angle AOR$.

37. **(a)** (Assuming that "x" is zero at the point of impact)
$\mathbf{r}(t)=(x(t))\mathbf{i}+(y(t))\mathbf{j}$, where $x(t)=(35\cos 27°)t$ and
$y(t)=4+(35\sin 27°)t-16t^2$.
(b) At $t\approx 0.497$ sec, it reaches its maximum height of about 7.945 ft.
(c) Range ≈ 37.45 ft; flight time ≈ 1.201 sec
(d) At $t\approx 0.254$ and $t\approx 0.740$ sec, when it is ≈ 29.532 and ≈ 14.376 ft from where it will land
(e) Yes. It changes things because the ball won't clear the net.

39. Height is about 37.2 ft.

47. **(a)** $\mathbf{r}(t)=(x(t))\mathbf{i}+(y(t))\mathbf{j}$; where

$x(t)=\left(\dfrac{1}{0.08}\right)(1-e^{-0.08t})(152\cos 20°-17.6)$ and

$y(t)=3+\left(\dfrac{152}{0.08}\right)(1-e^{-0.08t})(\sin 20°)$
$+\left(\dfrac{32}{0.08^2}\right)(1-0.08t-e^{-0.08t})$

(b) At $t\approx 1.527$ sec it reaches a maximum height of about 41.893 feet.
(c) Range ≈ 351.734 ft; flight time ≈ 3.181 sec
(d) At $t\approx 0.877$ and 2.190 sec, when it is about 106.028 and 251.530 ft from home plate
(e) No

SECTION 13.3, p. 781

1. Unit tangent vector $\mathbf{T}=\left(\dfrac{-2\sin t}{\sqrt{7}}\right)\mathbf{i}+\left(\dfrac{2\cos t}{\sqrt{7}}\right)\mathbf{j}+\left(\dfrac{\sqrt{3}}{\sqrt{7}}\right)\mathbf{k}$
Length $=\pi\sqrt{7}$

3. Unit tangent vector $\mathbf{T}=\left(\dfrac{10}{\sqrt{100+9t}}\right)\mathbf{i}+\left(\dfrac{3\sqrt{t}}{\sqrt{100+9t}}\right)\mathbf{k}$

Length $=\dfrac{2}{27}\left(29\sqrt{145}-200\right)$

5. $\mathbf{T}=-\cos t\mathbf{j}+\sin t\mathbf{k}$, $\dfrac{3}{2}$

7. $\mathbf{T}=\left(\dfrac{\cos t-t\sin t}{t+1}\right)\mathbf{i}+\left(\dfrac{\sin t+t\cos t}{t+1}\right)\mathbf{j}$
$+\left(\dfrac{\sqrt{2}t^{1/2}}{t+1}\right)\mathbf{k}$, $\dfrac{\pi^2}{2}+\pi$

9. Point $P(\pi)=(12\pi, 5\sin\pi, 5\cos\pi)=(12\pi, 0, -5)$

11. Length $=s(t)=\displaystyle\int_0^t\sqrt{2}\,dt=\sqrt{2}t=\sqrt{2}\pi$

13. $s(t)=\sqrt{3}e^t-\sqrt{3}$, $L=\dfrac{3\sqrt{3}}{4}$

15. $\sqrt{2}+\ln(1+\sqrt{2})$

17. **(a)** Cylinder is $x^2+y^2=1$; plane is $x+z=1$.
(b) and **(c)**

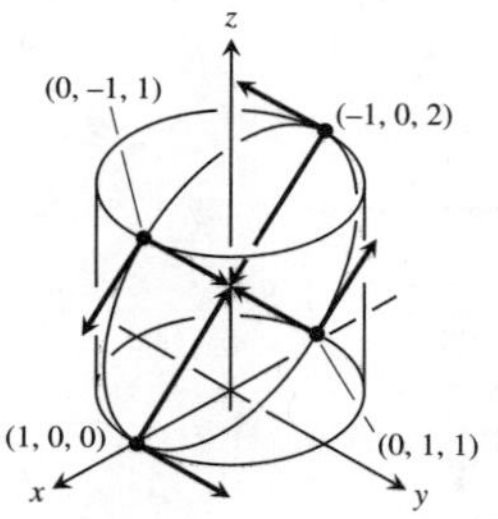

(d) $L=\displaystyle\int_0^{2\pi}\sqrt{1+\sin^2 t}\,dt$ **(e)** $L\approx 7.64$

SECTION 13.4, pp. 786–787

1. $\mathbf{T}=\sin t\mathbf{i}+\cos t\mathbf{j}$, $\mathbf{N}=\cos t\mathbf{i}-\sin t\mathbf{j}$, $\kappa=\sin t$

3. **(a)** $\mathbf{T}=\left(\dfrac{3}{\sqrt{4t^2+9}}\right)\mathbf{i}-\left(\dfrac{2t}{\sqrt{4t^2+9}}\right)\mathbf{j}$

$\mathbf{N}=\left(\dfrac{-2t}{\sqrt{4t^2+9}}\right)\mathbf{i}-\left(\dfrac{3}{\sqrt{4t^2+9}}\right)\mathbf{j}$, $\kappa=\left(\dfrac{6}{(4t^2+9)^{3/2}}\right)$

5. **(b)** $\cos x$

7. **(b)** $\mathbf{N}=\dfrac{-2e^{2t}}{\sqrt{1+4e^{4t}}}\mathbf{i}+\dfrac{1}{\sqrt{1+4e^{4t}}}\mathbf{j}$
(c) $\mathbf{N}=-\dfrac{1}{2}\left(\sqrt{4-t^2}\mathbf{i}+t\mathbf{j}\right)$

9. $\mathbf{T}=\dfrac{3\cos t}{5}\mathbf{i}-\dfrac{3\sin t}{5}\mathbf{j}+\dfrac{4}{5}\mathbf{k}$,
$\mathbf{N}=(-\sin t)\mathbf{i}-(\cos t)\mathbf{j}$, $\kappa=\dfrac{3}{25}$

11. $\mathbf{T}=\left(\dfrac{\sin t+\cos t}{\sqrt{2}}\right)\mathbf{i}+\left(\dfrac{\cos t-\sin t}{\sqrt{2}}\right)\mathbf{j}$

$\mathbf{N}=\left(\dfrac{\cos t-\sin t}{\sqrt{2}}\right)\mathbf{i}-\left(\dfrac{\sin t+\cos t}{\sqrt{2}}\right)\mathbf{j},\ \kappa=\left(\dfrac{1}{e^t 2\sqrt{2}}\right)$

13. $\mathbf{T}=\left(\dfrac{2t}{\sqrt{4t^2+9}}\right)\mathbf{i}+\left(\dfrac{3}{\sqrt{4t^2+9}}\right)\mathbf{j}$

$\mathbf{N}=\left(\dfrac{3}{\sqrt{4t^2+9}}\right)\mathbf{i}+\left(\dfrac{-2t}{\sqrt{4t^2+9}}\right)\mathbf{j},\ \kappa=\left(\dfrac{6}{t\left(4t^2+9\right)^{3/2}}\right)$

15. $\mathbf{T}=\left(\operatorname{sech}\dfrac{t}{a}\right)\mathbf{i}+\left(\tanh\dfrac{t}{a}\right)\mathbf{k},$

$\mathbf{N}=\left(-\tanh\dfrac{t}{a}\right)\mathbf{i}+\left(\operatorname{sech}\dfrac{t}{a}\right)\mathbf{k},$

$\kappa=\dfrac{1}{a}\operatorname{sech}^2\dfrac{t}{a}$

19. $1/(2b)$

21. $\left(x-\dfrac{\pi}{2}\right)^2+y^2=1$

23. $\kappa=\left(\dfrac{4}{\left(16x^2+1\right)^{3/2}}\right)$

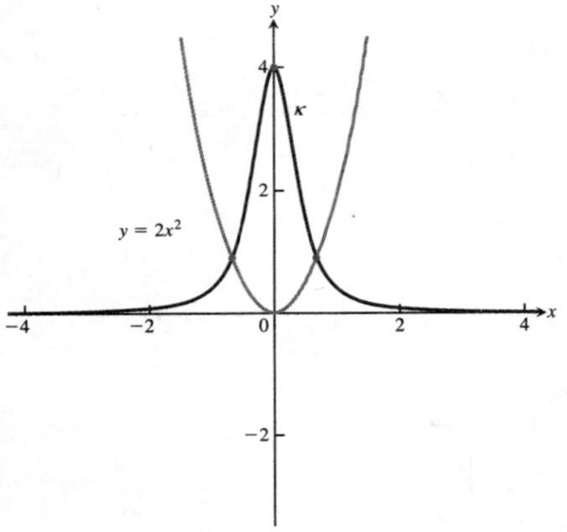

25. $\kappa=\dfrac{|2\sin x|}{\left(4\cos^2 x+1\right)^{3/2}}$

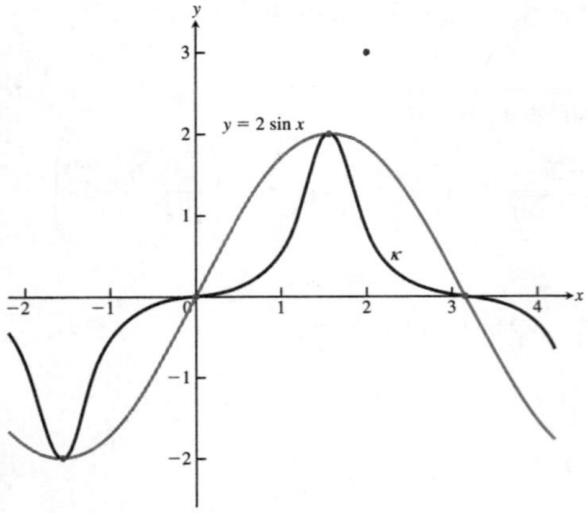

27. maximum curvature $2/(3\sqrt{3})$ at $x=1/\sqrt{2}$

SECTION 13.5, p. 793

1. $\mathbf{a}=\dfrac{2t}{\sqrt{1+t^2}}\mathbf{T}+\dfrac{2}{\sqrt{1+t^2}}\mathbf{N}$ **3.** $\mathbf{a}=|a|\,\mathbf{N}$

5. $\mathbf{a}(0)=a_{\mathrm{T}}\mathbf{T}+a_{\mathrm{N}}\mathbf{N}=\sqrt{2}\,\mathbf{N}$ **7.** $\mathbf{a}(1)=\dfrac{4}{3}\mathbf{T}+\dfrac{2\sqrt{5}}{3}\mathbf{N}$

9. $\mathbf{a}(0)=2\mathbf{N}$

17. $\mathbf{B}=-\mathbf{k},\ \tau=0$

19. $\mathbf{B}=-\mathbf{j},\ \tau=0$

21. Yes. If the car is moving on a curved path ($\kappa\neq 0$), then $a_{\mathrm{N}}=\kappa|\mathbf{v}|^2\neq 0$ and $\mathbf{a}\neq\mathbf{0}$.

27. $\kappa=\dfrac{1}{t},\ \rho=t$

31. Components of **v**: −1.8701, 0.7089, 1.0000
Components of **a**: −1.6960, −2.0307, 0
Speed: 2.2361; Components of **T**: −0.8364, 0.3170, 0.4472
Components of **N**: −0.4143, −0.8998, −0.1369
Components of **B**: 0.3590, −0.2998, 0.8839; Curvature: 0.5060
Torsion: 0.2813; Tangential component of acceleration: 0.7746
Normal component of acceleration: 2.5298

33. Components of **v**: 2.0000, 0, −0.1629
Components of **a**: 0, −1.0000, −0.0086; Speed: 2.0066
Components of **T**: 0.9967, 0, −0.0812
Components of **N**: −0.0007, −1.0000, −0.0086
Components of **B**: −0.0812, 0.0086, 0.9967;
Curvature: 0.2484
Torsion: 0.0411; Tangential component of acceleration: 0.0007
Normal component of acceleration: 1.0000

SECTION 13.6, p. 797

1. $\mathbf{v}=8\mathbf{u}_r+8\theta\,\mathbf{u}_\theta$
$\mathbf{a}=-32\theta\,\mathbf{u}_r+64\mathbf{u}_\theta$

3. $\mathbf{v}=(5a\cos\theta)\mathbf{u}_r+5a(1+\sin\theta)\,\mathbf{u}_\theta$
$\mathbf{a}=-25a(1+2\sin\theta)\,\mathbf{u}_r+50a\cos\theta\,\mathbf{u}_\theta$

5. $\mathbf{v}=(6e^\theta)\mathbf{u}_r+(6e^\theta)\mathbf{u}_\theta$
$\mathbf{a}=(36e^\theta)\mathbf{u}_\theta$

7. $\mathbf{v}=(-6\sin 6t)\mathbf{u}_r+(3\cos 6t)\mathbf{u}_\theta$
$\mathbf{a}=(-36\cos 6t-9\cos 6t)\mathbf{u}_r+(-36\sin 6t)\mathbf{u}_\theta$

13. $\approx 29.93\times 10^{10}$ m **15.** $\approx 2.25\times 10^9$ km²/sec

17. mass $\approx 1.876\times 10^{27}$ kg

PRACTICE EXERCISES, pp. 798–799

1. $\dfrac{x^2}{16}+\dfrac{y^2}{2}=1$

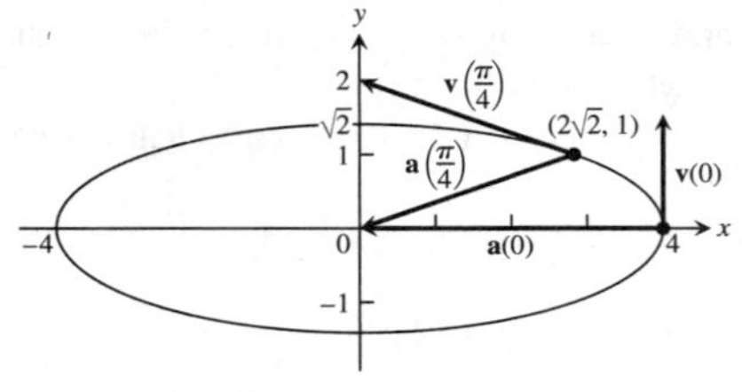

At $t=0$: $a_{\mathrm{T}}=0,\ a_{\mathrm{N}}=4,\ \kappa=2;$

At $t=\dfrac{\pi}{4}$: $a_{\mathrm{T}}=\dfrac{7}{3},\ a_{\mathrm{N}}=\dfrac{4\sqrt{2}}{3},\ \kappa=\dfrac{4\sqrt{2}}{27}$

3. $|\mathbf{v}|_{\max}=1$ **5.** $\kappa=1/5$ **7.** $dy/dt=-x$; clockwise

11. Shot put is on the ground about 66 ft 3 in. from the stopboard.

15. Length $= \frac{\pi}{4}\sqrt{1+\frac{\pi^2}{16}} + \ln\left(\frac{\pi}{4} + \sqrt{1+\frac{\pi^2}{16}}\right)$

17. $\mathbf{T}(0) = \frac{2}{3}\mathbf{i} - \frac{2}{3}\mathbf{j} + \frac{1}{3}\mathbf{k}$; $\mathbf{N}(0) = \frac{1}{\sqrt{2}}\mathbf{i} + \frac{1}{\sqrt{2}}\mathbf{j}$;

$\mathbf{B}(0) = -\frac{1}{3\sqrt{2}}\mathbf{i} + \frac{1}{3\sqrt{2}}\mathbf{j} + \frac{4}{3\sqrt{2}}\mathbf{k}$; $\kappa = \frac{\sqrt{2}}{3}$; $\tau = \frac{1}{6}$

19. $\mathbf{T}(\ln 2) = \frac{1}{\sqrt{17}}\mathbf{i} + \frac{4}{\sqrt{17}}\mathbf{j}$; $\mathbf{N}(\ln 2) = -\frac{4}{\sqrt{17}}\mathbf{i} + \frac{1}{\sqrt{17}}\mathbf{j}$;

$\mathbf{B}(\ln 2) = \mathbf{k}$; $\kappa = \frac{8}{17\sqrt{17}}$; $\tau = 0$

21. $\mathbf{a}(0) = 10\mathbf{T} + 6\mathbf{N}$

23. $\mathbf{T} = \left(\frac{1}{\sqrt{2}}\cos t\right)\mathbf{i} - (\sin t)\mathbf{j} + \left(\frac{1}{\sqrt{2}}\cos t\right)\mathbf{k}$;

$\mathbf{N} = \left(-\frac{1}{\sqrt{2}}\sin t\right)\mathbf{i} - (\cos t)\mathbf{j} - \left(\frac{1}{\sqrt{2}}\sin t\right)\mathbf{k}$;

$\mathbf{B} = \frac{1}{\sqrt{2}}\mathbf{i} - \frac{1}{\sqrt{2}}\mathbf{k}$; $\kappa = \frac{1}{\sqrt{2}}$; $\tau = 0$

25. $\frac{\pi}{3}$ **27.** $x = 1 + t,\ y = t,\ z = -t$ **31.** $\kappa = \frac{1}{a}$

ADDITIONAL AND ADVANCED EXERCISES, pp. 800–801

1. (a) $\left.\frac{d\theta}{dt}\right|_{\theta=2\pi} = 2\sqrt{\frac{\pi g b}{a^2+b^2}}$

(b) $\theta = \frac{gbt^2}{2(a^2+b^2)}$, $z = \frac{gb^2t^2}{2(a^2+b^2)}$

(c) $\mathbf{v}(t) = \frac{gbt}{\sqrt{a^2+b^2}}\mathbf{T}$;

$\frac{d^2\mathbf{r}}{dt^2} = \frac{bg}{\sqrt{a^2+b^2}}\mathbf{T} + a\left(\frac{bgt}{a^2+b^2}\right)^2\mathbf{N}$

There is no component in the direction of **B**.

5. (a) $\frac{dx}{dt} = \dot{r}\cos\theta - r\dot{\theta}\sin\theta$, $\frac{dy}{dt} = \dot{r}\sin\theta + r\dot{\theta}\cos\theta$

(b) $\frac{dr}{dt} = \dot{x}\cos\theta + \dot{y}\sin\theta$, $r\frac{d\theta}{dt} = -\dot{x}\sin\theta + \dot{y}\cos\theta$

7. (a) $\mathbf{a}(1) = -9\mathbf{u}_r - 6\mathbf{u}_\theta$, $\mathbf{v}(1) = -\mathbf{u}_r + 3\mathbf{u}_\theta$

(b) 6.5 in.

9. (c) $\mathbf{v} = \dot{r}\mathbf{u}_r + r\dot{\theta}\mathbf{u}_\theta + \dot{z}\mathbf{k}$, $\mathbf{a} = (\ddot{r} - r\dot{\theta}^2)\mathbf{u}_r + (r\ddot{\theta} + 2\dot{r}\dot{\theta})\mathbf{u}_\theta + \ddot{z}\mathbf{k}$

Chapter 14

SECTION 14.1, pp. 808–810

1. (a) 0 **(b)** 0 **(c)** 58 **(d)** 33

3. (a) 4/5 **(b)** 8/5 **(c)** 3 **(d)** 0

5. Domain: all points (x, y) on or above line $y = x + 2$

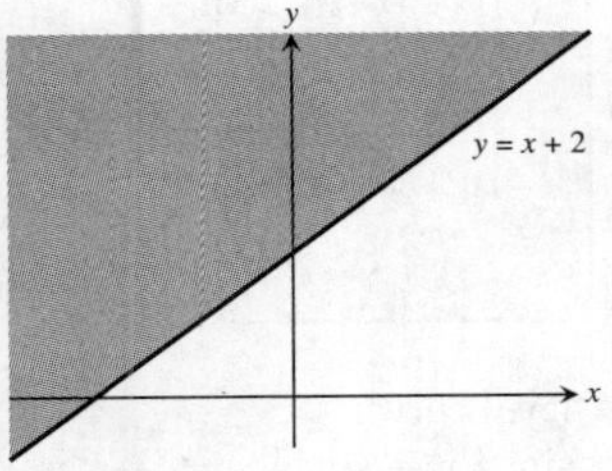

7. Domain: all points (x, y) not lying on the graph of $y = x$ or $y = x^3$

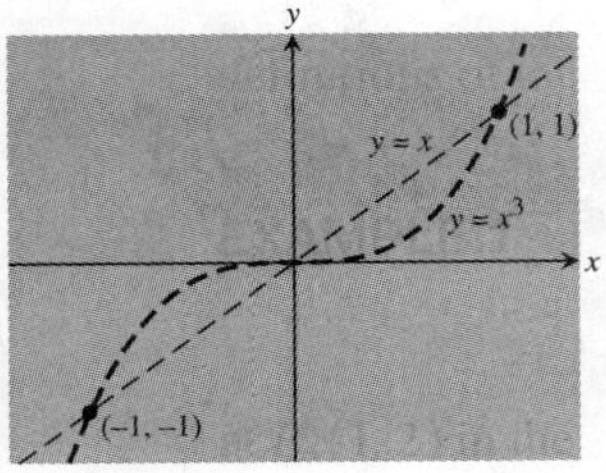

9. Domain: all points (x, y) satisfying $x^2 - 1 \le y \le x^2 + 1$

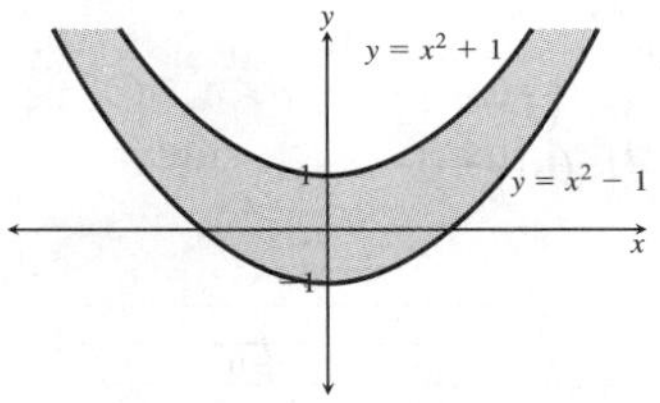

11. Domain: all points (x, y) satisfying $x^2 - y^2 \ge 0$

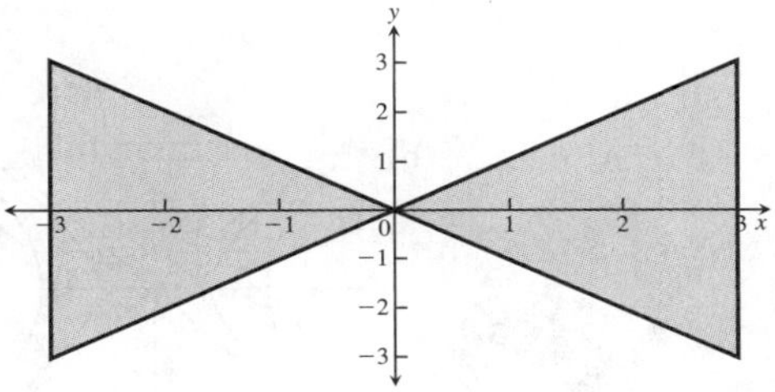

13.

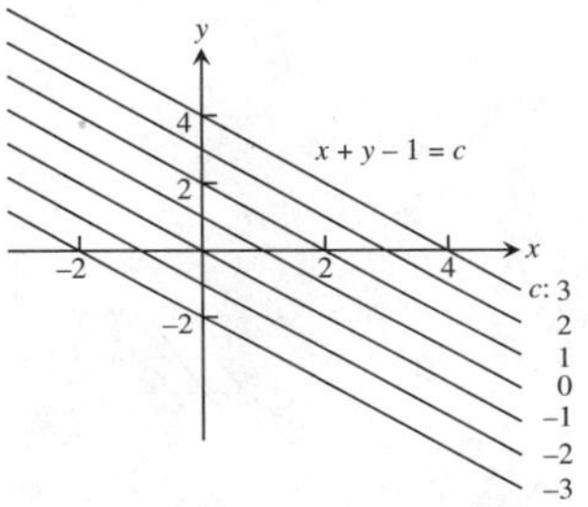

15.

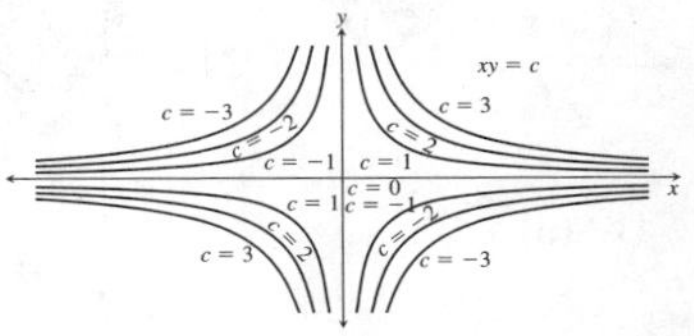

17. (a) All points in the xy-plane **(b)** All reals
(c) The lines $y - x = c$ **(d)** No boundary points
(e) Both open and closed **(f)** Unbounded

19. (a) All points in the xy-plane **(b)** $z \ge 0$
(c) For $f(x, y) = 0$, the origin; for $f(x, y) = c > 0$, ellipses with center $(0, 0)$ and minor and major axes on the x- and y-axes, respectively
(d) No boundary points **(e)** Both open and closed
(f) Unbounded

21. (a) All points in the xy-plane **(b)** All reals
(c) For $f(x, y) = 0$, the x- and y-axes; for $f(x, y) = c, c \ne 0$, hyperbolas with the x- and y-axes as asymptotes
(d) No boundary points **(e)** Both open and closed
(f) Unbounded

23. (a) All (x, y) satisfying $x^2 + y^2 < 16$ **(b)** $z \ge 1/4$
(c) Circles centered at the origin with radii $r < 4$
(d) Boundary is the circle $x^2 + y^2 = 16$
(e) Open **(f)** Bounded

25. (a) $(x, y) \ne (0, 0)$ **(b)** All reals
(c) The circles with center $(0, 0)$ and radii $r > 0$
(d) Boundary is the single point $(0, 0)$
(e) Open **(f)** Unbounded

27. (a) All (x, y) satisfying $-1 \le y - x \le 1$
(b) $0 \le z \le \pi$
(c) Straight lines of the form $y - x = c$ where $-1 \le c \le 1$
(d) Boundary is two straight lines $y = 1 + x$ and $y = -1 + x$
(e) Closed **(f)** Unbounded

29. (a) Domain: all points (x, y) outside the circle $x^2 + y^2 = 1$
(b) Range: all reals

(c) Circles centered at the origin with radii $r > 1$
(d) Boundary: $x^2 + y^2 = 1$
(e) Open **(f)** Unbounded

31. (f), (h) **33.** (a), (i) **35.** (d), (j)

37. (a)

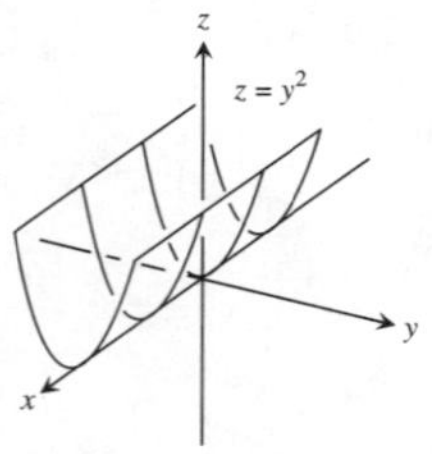

(b)

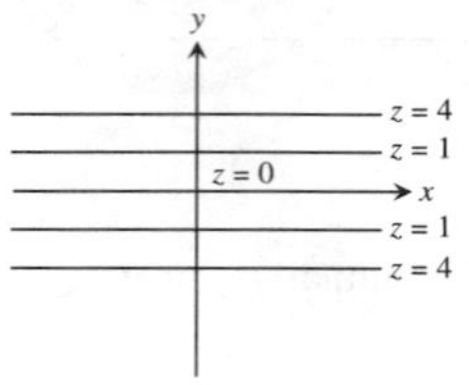

39. (a)

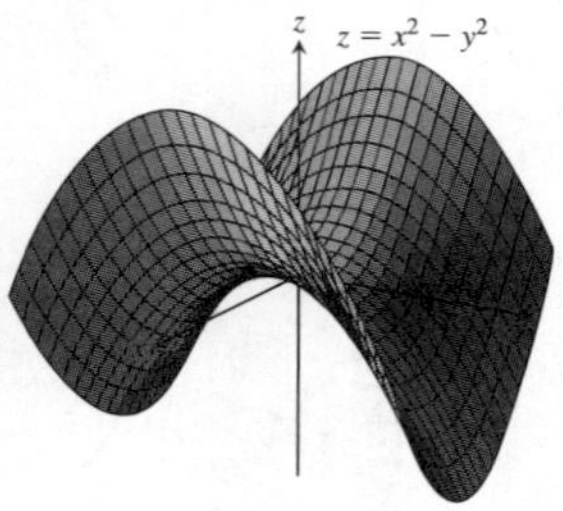

(b)

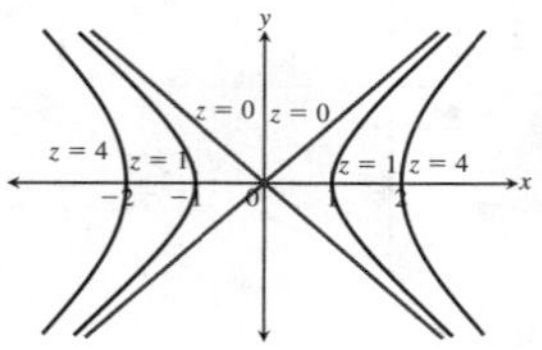

41. (a)

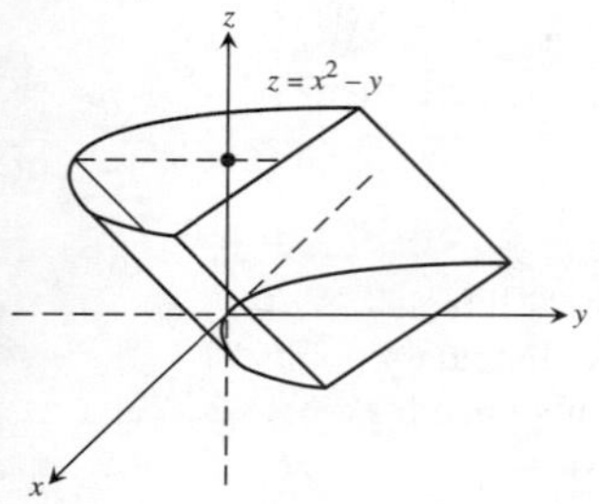

(b)

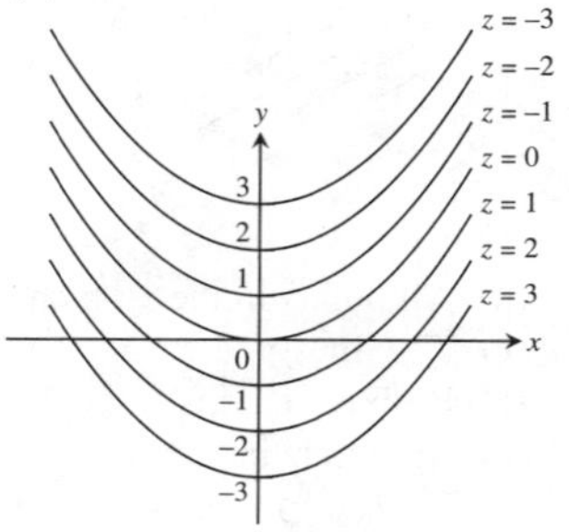

43. (a)

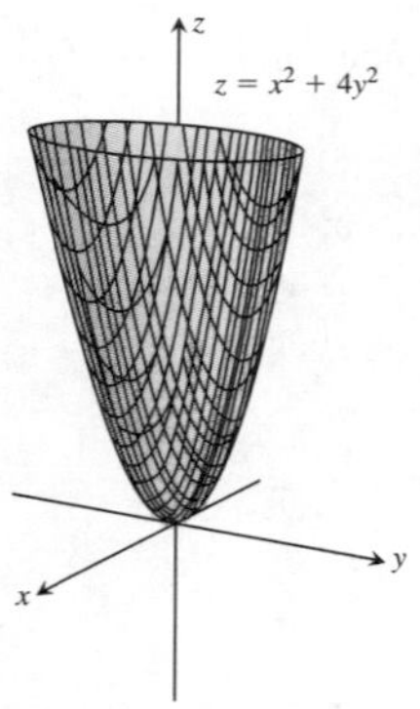

(b)

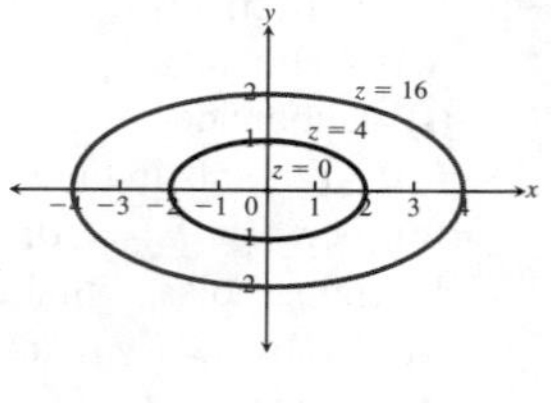

45. (a)

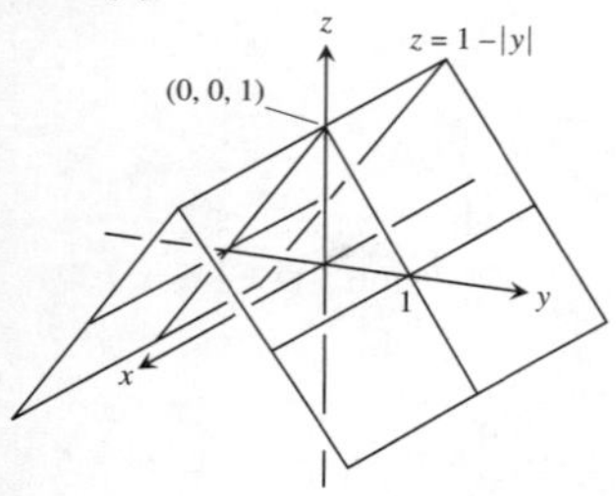

(b)

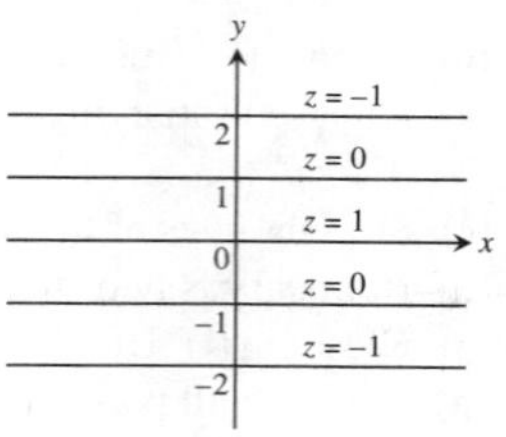

47. (a)

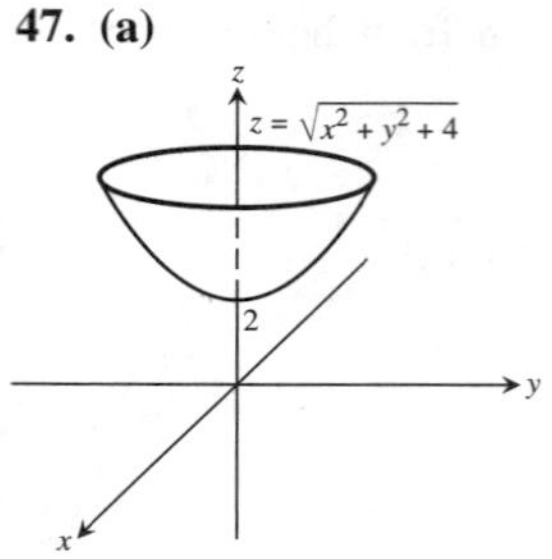

(b)

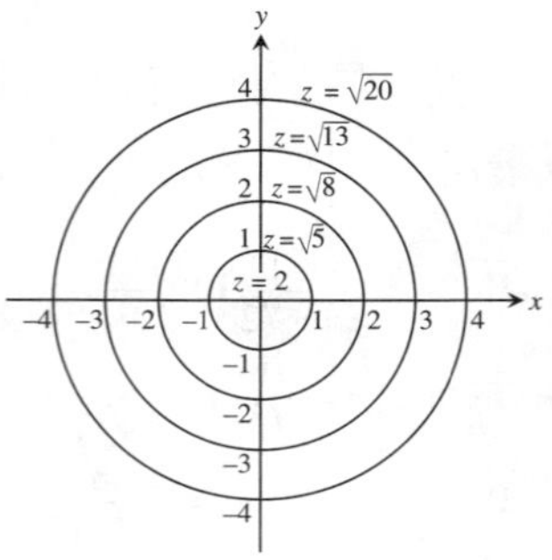

49. $x^2 + y^2 = 20$

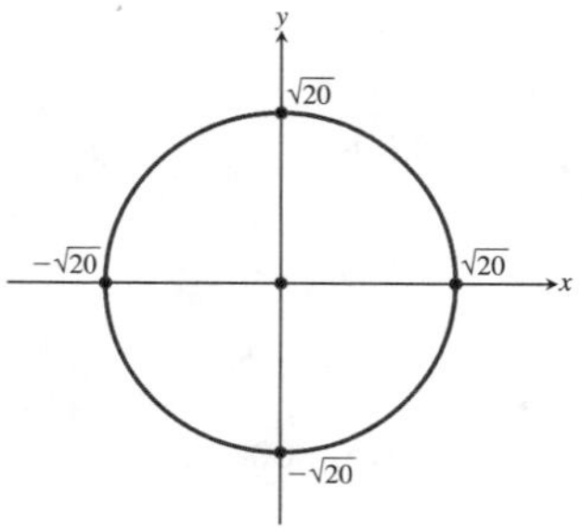

51. $x + y^2 = 4$

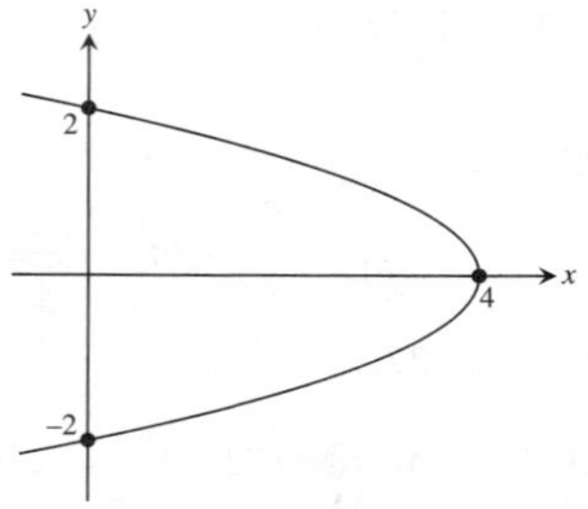

53.

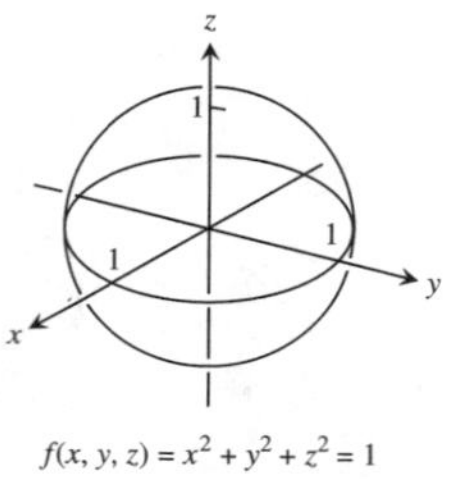

55.

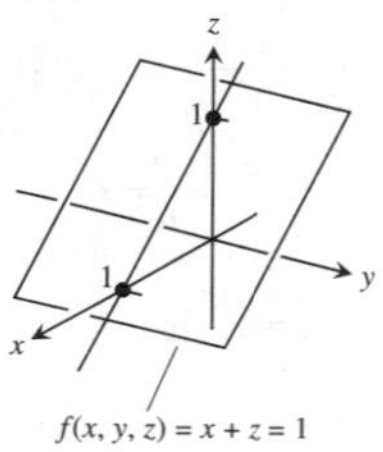

57.

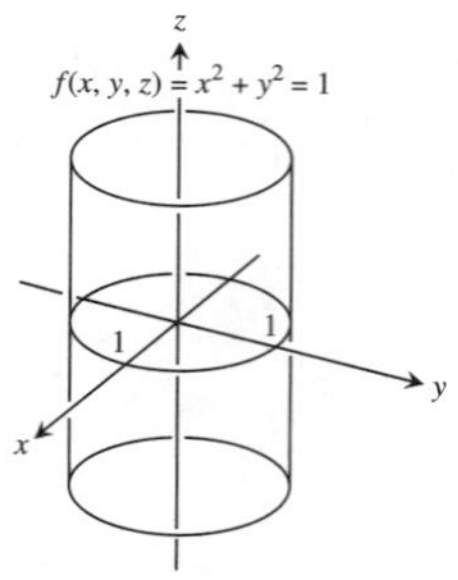

59.

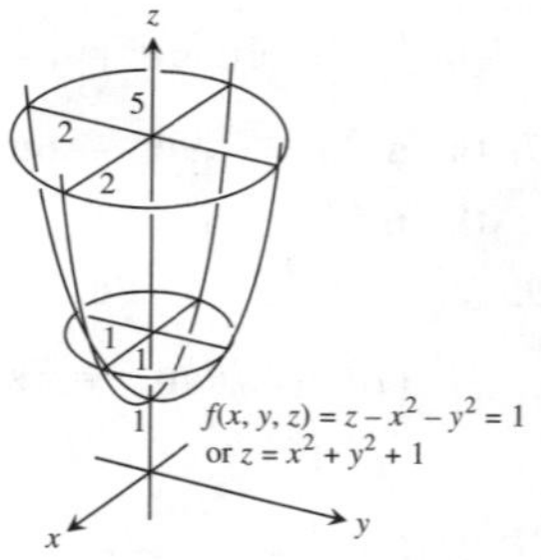

61. $\sqrt{x - y} - \ln z = 2$ **63.** $x^2 + y^2 + z^2 = 4$

65. Domain: all points (x, y) satisfying $|x| < |y|$

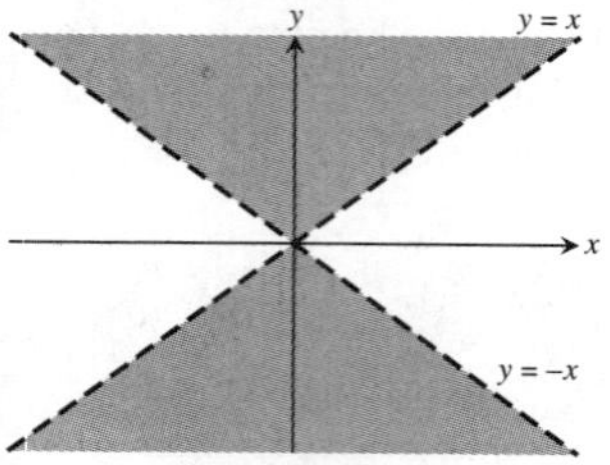

level curve: $y = 2x$

67. Domain: all points (x, y) satisfying $-1 \le x \le 1$ and $-1 \le y \le 1$

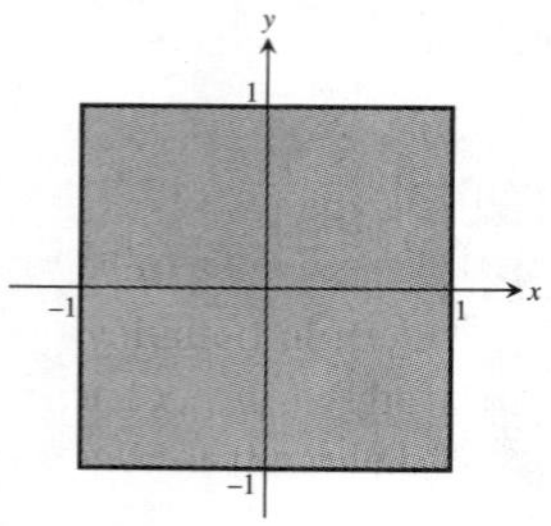

level curve: $\sin^{-1} y - \sin^{-1} x = \dfrac{\pi}{2}$

SECTION 14.2, pp. 816–819

1. 5/2 **3.** $2\sqrt{6}$ **5.** 1 **7.** 3 **9.** 1
11. 1/4 **13.** 0 **15.** −1 **17.** 1/2 **19.** 1/6
21. 1 **23.** 3 **25.** 19/12 **27.** 2 **29.** 3
31. (a) All (x, y) **(b)** All (x, y) except $(0, 0)$
33. (a) All (x, y) except where $x = 0$ or $y = 0$ **(b)** All (x, y)
35. (a) All (x, y, z)
(b) All (x, y, z) except the interior of the cylinder $x^2 + y^2 = 1$
37. (a) All (x, y, z) with $z \neq 0$ **(b)** All (x, y, z) with $y^2 + z^2 \neq 4$
39. (a) All points (x, y, z) satisfying $z > x^2 + y^2 + 1$
(b) All points (x, y, z) satisfying $z \neq \sqrt{x^2 + y^2}$
41. Consider paths along $y = x, x > 0$, and along $y = x, x < 0$.
43. Consider the paths $y = kx$, k a constant.
45. Consider the paths $y = mx$, m a constant, $m \neq -1$.
47. Consider the paths $y = kx^2$, k a constant, $k \neq 0$.
49. Consider the paths $y = 1$ and $y = x$.
51. Along $y = 1$ the limit is 0; along $y = e^x$ the limit is 1/2.
53. Along $y = 0$ the limit is 1; along $y = -\sin x$ the limit is 0.
55. (a) 1 **(b)** 0 **(c)** Does not exist
59. The limit is 1. **61.** The limit is 0.
63. (a) $f(x, y)|_{y=mx} = \sin 2\theta$, where $\tan\theta = m$ **65.** 0
67. Does not exist **69.** $\pi/2$ **71.** $f(0, 0) = \ln 5$
73. $\delta = 0.1$ **75.** $\delta = 0.005$ **77.** $\delta = 0.04$
79. $\delta = \sqrt{0.015}$ **81.** $\delta = 0.005$

SECTION 14.3, pp. 828–831

1. $\frac{\partial f}{\partial x} = 4x, \frac{\partial f}{\partial y} = -3$ **3.** $\frac{\partial f}{\partial x} = 2x(y + 2), \frac{\partial f}{\partial y} = x^2 - 1$

5. $\frac{\partial f}{\partial x} = y^x \ln y, \frac{\partial f}{\partial y} = xy^{x-1}$

7. $\frac{\partial f}{\partial x} = \frac{x}{\sqrt{x^2 + y^2}}, \frac{\partial f}{\partial y} = \frac{y}{\sqrt{x^2 + y^2}}$

9. $\frac{\partial f}{\partial x} = \frac{-1}{(x + y)^2}, \frac{\partial f}{\partial y} = \frac{-1}{(x + y)^2}$

11. $\frac{\partial f}{\partial x} = \frac{-y^2 - 1}{(xy - 1)^2}, \frac{\partial f}{\partial y} = \frac{-x^2 - 1}{(xy - 1)^2}$

13. $\frac{\partial f}{\partial x} = e^{x+y+1}, \frac{\partial f}{\partial y} = e^{x+y+1}$

15. $\frac{\partial f}{\partial x} = \frac{2x + y}{x^2 + xy + y^2}, \frac{\partial f}{\partial y} = \frac{x + 2y}{x^2 + xy + y^2}$

17. $\frac{\partial f}{\partial x} = 2\sin(x - 3y)\cos(x - 3y)$,
$\frac{\partial f}{\partial y} = -6\sin(x - 3y)\cos(x - 3y)$

19. $\frac{\partial f}{\partial x} = yx^{y-1}, \frac{\partial f}{\partial y} = x^y \ln x$

21. $\frac{\partial f}{\partial x} = -g(x), \frac{\partial f}{\partial y} = g(y)$

23. $f_x = y^2, f_y = 2xy, f_z = -4z$

25. $f_x = 1, f_y = -y(y^2 + z^2)^{-1/2}, f_z = -z(y^2 + z^2)^{-1/2}$

27. $f_x = \frac{yz}{\sqrt{1 - x^2y^2z^2}}, f_y = \frac{xz}{\sqrt{1 - x^2y^2z^2}}, f_z = \frac{xy}{\sqrt{1 - x^2y^2z^2}}$

29. $f_x = \frac{1}{x + 2y + 3z}, f_y = \frac{2}{x + 2y + 3z}, f_z = \frac{3}{x + 2y + 3z}$

31.
$f_x = -2xe^{-(x^2+y^2+z^2)}, f_y = -2ye^{-(x^2+y^2+z^2)}, f_z = -2ze^{-(x^2+y^2+z^2)}$

33. $f_x = \frac{1}{y}e^{x/y}, f_y = -\frac{x}{y^2}e^{x/y} - \frac{z}{y^2}e^{z/y}, f_z = \frac{1}{y}e^{z/y}$

35. $\frac{\partial f}{\partial t} = -2\pi\sin(2\pi t - \alpha), \frac{\partial f}{\partial \alpha} = \sin(2\pi t - \alpha)$

37. $\frac{\partial h}{\partial \rho} = \sin\phi\cos\theta, \frac{\partial h}{\partial \phi} = \rho\cos\phi\cos\theta, \frac{\partial h}{\partial \theta} = -\rho\sin\phi\sin\theta$

39. $W_P(P, V, \delta, v, g) = V, W_V(P, V, \delta, v, g) = P + \frac{\delta v^2}{2g}$,
$W_\delta(P, V, \delta, v, g) = \frac{Vv^2}{2g}, W_v(P, V, \delta, v, g) = \frac{V\delta v}{g}$,
$W_g(P, V, \delta, v, g) = -\frac{V\delta v^2}{2g^2}$

41. $\frac{\partial f}{\partial x} = 1 + y, \frac{\partial f}{\partial y} = 1 + x, \frac{\partial^2 f}{\partial x^2} = 0, \frac{\partial^2 f}{\partial y^2} = 0, \frac{\partial^2 f}{\partial y\,\partial x} = \frac{\partial^2 f}{\partial x\,\partial y} = 1$

43. $\frac{\partial g}{\partial x} = 2xy + y\cos x, \frac{\partial g}{\partial y} = x^2 - \sin y + \sin x$,
$\frac{\partial^2 g}{\partial x^2} = 2y - y\sin x, \frac{\partial^2 g}{\partial y^2} = -\cos y$,
$\frac{\partial^2 g}{\partial y\,\partial x} = \frac{\partial^2 g}{\partial x\,\partial y} = 2x + \cos x$

45.
$r_x = \frac{2x}{x^2 + y^2}, r_y = \frac{2y}{x^2 + y^2}, r_{xx} = \frac{2y^2 - 2x^2}{(x^2 + y^2)^2}$
$r_{yy} = \frac{2x^2 - 2y^2}{(x^2 + y^2)^2}, r_{xy} = r_{yx} = -\frac{4xy}{(x^2 + y^2)^2}$

47. $\frac{\partial w}{\partial x} = x^2y\sec^2(xy) + 2x\tan(xy), \frac{\partial w}{\partial y} = x^3\sec^2(xy)$,
$\frac{\partial^2 w}{\partial y\,\partial x} = \frac{\partial^2 w}{\partial x\,\partial y} = 2x^3y\sec^2(xy)\tan(xy) + 3x^2\sec^2(xy)$
$\frac{\partial^2 w}{\partial x^2} = 4xy\sec^2(xy) + 2x^2y^2\sec^2(xy)\tan(xy) + 2\tan(xy)$
$\frac{\partial^2 w}{\partial y^2} = 2x^4\sec^2(xy)\tan(xy)$

49. $\frac{\partial w}{\partial x} = \sin(x^2y) + 2x^2y\cos(x^2y), \frac{\partial w}{\partial y} = x^3\cos(x^2y)$,
$\frac{\partial^2 w}{\partial y\,\partial x} = \frac{\partial^2 w}{\partial x\,\partial y} = 3x^2\cos(x^2y) - 2x^4y\sin(x^2y)$
$\frac{\partial^2 w}{\partial x^2} = 6xy\cos(x^2y) - 4x^3y^2\sin(x^2y)$
$\frac{\partial^2 w}{\partial y^2} = -x^5\sin(x^2y)$

51. $\frac{\partial f}{\partial x} = 2xy^3 - 4x^3, \frac{\partial f}{\partial y} = 3x^2y^2 + 5y^4$,
$\frac{\partial^2 f}{\partial x^2} = 2y^3 - 12x^2, \frac{\partial^2 f}{\partial y^2} = 6x^2y + 20y^3$,
$\frac{\partial^2 f}{\partial y\,\partial x} = \frac{\partial^2 f}{\partial x\,\partial y} = 6xy^2$

53. $\dfrac{\partial z}{\partial x} = 2x\cos(2x - y^2) + \sin(2x - y^2)$,
$\dfrac{\partial z}{\partial y} = -2xy\cos(2x - y^2)$,
$\dfrac{\partial^2 z}{\partial x^2} = 4\cos(2x - y^2) - 4x\sin(2x - y^2)$,
$\dfrac{\partial^2 z}{\partial y^2} = -4xy^2\sin(2x - y^2) - 2x\cos(2x - y^2)$,
$\dfrac{\partial^2 z}{\partial x\,\partial y} = \dfrac{\partial^2 z}{\partial y\,\partial x} = 4xy\sin(2x - y^2) - 2y\cos(2x - y^2)$

55. $\dfrac{\partial w}{\partial x} = \dfrac{2}{2x+3y}, \dfrac{\partial w}{\partial y} = \dfrac{3}{2x+3y}, \dfrac{\partial^2 w}{\partial y\,\partial x} = \dfrac{\partial^2 w}{\partial x\,\partial y} = \dfrac{-6}{(2x+3y)^2}$

57. $\dfrac{\partial^2 w}{\partial x \partial y} = \dfrac{\partial^2 w}{\partial y \partial x} = 3xy(x^2+y^2)^{1/2}$

59. $\dfrac{\partial w}{\partial x} = \dfrac{2x}{y^3}, \dfrac{\partial w}{\partial y} = \dfrac{-3x^2}{y^4}$
$\dfrac{\partial^2 w}{\partial y\,\partial x} = \dfrac{-6x}{y^4}, \dfrac{\partial^2 w}{\partial x\,\partial y} = \dfrac{-6x}{y^4}$

61. **(a)** x first **(b)** y first **(c)** x first
(d) x first **(e)** y first **(f)** y first

63. $f_x(1, 2) = -13, f_y(1, 2) = -2$

65. $f_x(-2, 3) = 1/2, f_y(-2, 3) = 3/4$

67. 12 **69.** -2

71. $\dfrac{\partial A}{\partial a} = \dfrac{a}{bc\sin A}, \dfrac{\partial A}{\partial b} = \dfrac{c\cos A - b}{bc\sin A}$

73. $v_x = \dfrac{\ln v}{(\ln u)(\ln v) - 1}$

75. **(a)** 3 **(b)** 2

77. $\dfrac{\partial f}{\partial x} = 3x^2y^2 - 2x \Rightarrow$
$f(x, y) = x^3y^2 - x^2 + g(y) \Rightarrow$
$\dfrac{\partial f}{\partial y} = 2x^3y + g'(y) = 2x^3y + 64 \Rightarrow$
$g'(y) = 6y \Rightarrow g(y) = 3y^2$ works $\Rightarrow$
$f(x, y) = x^3y^2 - x^2 + 3y^2$ works

79. $\dfrac{\partial^2 f}{\partial y\,\partial x} = \dfrac{2x - 2y}{(x+y)^3} \neq \dfrac{\partial^2 f}{\partial x\,\partial y} = \dfrac{2y - 2x}{(x+y)^3}$ so impossible

81. $f_x(x, y) = 0$ for all points (x, y),

$f_y(x, y) = \begin{cases} 3y^2, & y \geq 0 \\ -2y, & y < 0 \end{cases}$,

$f_{xy}(x, y) = f_{yx}(x, y) = 0$ for all points (x, y)

99. Yes

SECTION 14.4, pp. 838–840

1. **(a)** $\dfrac{dw}{dt} = 0$, **(b)** $\dfrac{dw}{dt}(\pi) = 0$

3. **(a)** $8x + y + 8z = 24$
(b) $x = 1 + 8t, y = t, z = 2 + 8t$

5. **(a)** $\dfrac{dw}{dt} = 4t\tan^{-1}t + 1$, **(b)** $\dfrac{dw}{dt}(1) = \pi + 1$

7. **(a)** $\dfrac{\partial z}{\partial u} = 4\cos v\ln(u\sin v) + 4\cos v$,
$\dfrac{\partial z}{\partial v} = -4u\sin v\ln(u\sin v) + \dfrac{4u\cos^2 v}{\sin v}$
(b) $\dfrac{\partial z}{\partial u} = \sqrt{2}(\ln 2 + 2), \dfrac{\partial z}{\partial v} = -2\sqrt{2}(\ln 2 - 2)$

9. **(a)** $\dfrac{\partial w}{\partial u} = 2u + 4uv, \dfrac{\partial w}{\partial v} = -2v + 2u^2$
(b) $\dfrac{\partial w}{\partial u} = 3, \dfrac{\partial w}{\partial v} = -\dfrac{3}{2}$

11. **(a)** $\dfrac{\partial u}{\partial x} = 0, \dfrac{\partial u}{\partial y} = \dfrac{z}{(z-y)^2}, \dfrac{\partial u}{\partial z} = \dfrac{-y}{(z-y)^2}$
(b) $\dfrac{\partial u}{\partial x} = 0, \dfrac{\partial u}{\partial y} = 1, \dfrac{\partial u}{\partial z} = -2$

13. $\dfrac{dz}{dt} = \dfrac{\partial z}{\partial x}\dfrac{dx}{dt} + \dfrac{\partial z}{\partial y}\dfrac{dy}{dt}$

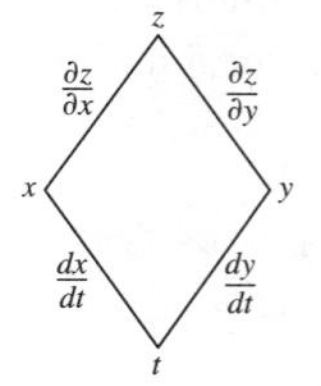

15. $x = 6 + 70t, \ y = 4 + 18t, \ z = 3 + 116t$

17. $\dfrac{\partial w}{\partial u} = \dfrac{\partial w}{\partial x}\dfrac{\partial x}{\partial u} + \dfrac{\partial w}{\partial y}\dfrac{\partial y}{\partial u}, \dfrac{\partial w}{\partial v} = \dfrac{\partial w}{\partial x}\dfrac{\partial x}{\partial v} + \dfrac{\partial w}{\partial y}\dfrac{\partial y}{\partial v}$.

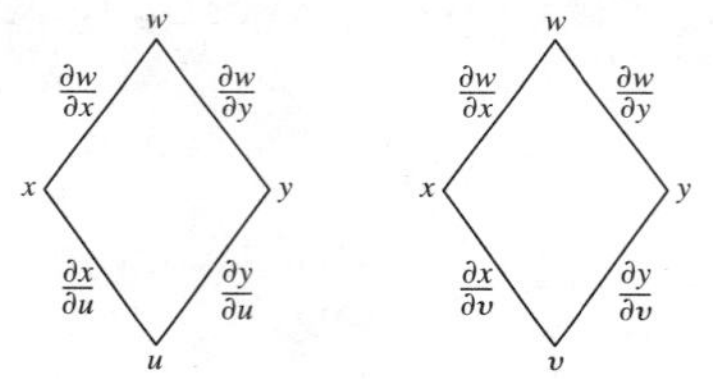

19. $\dfrac{\partial z}{\partial t} = \dfrac{\partial z}{\partial x}\dfrac{\partial x}{\partial t} + \dfrac{\partial z}{\partial y}\dfrac{\partial y}{\partial t}, \dfrac{\partial z}{\partial s} = \dfrac{\partial z}{\partial x}\dfrac{\partial x}{\partial s} + \dfrac{\partial z}{\partial y}\dfrac{\partial y}{\partial s}$

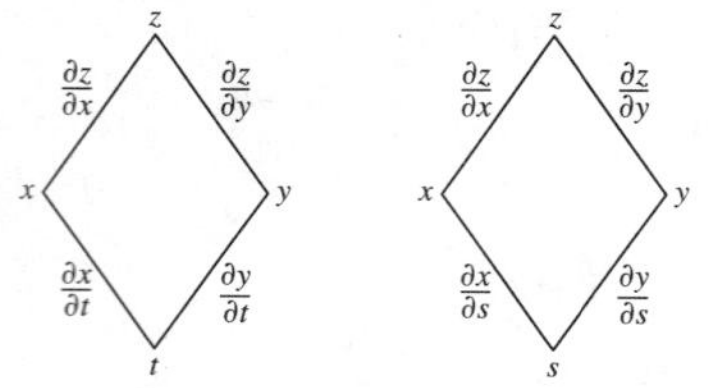

21. $\dfrac{\partial w}{\partial s} = \dfrac{dw}{du}\dfrac{\partial u}{\partial s}, \dfrac{\partial w}{\partial t} = \dfrac{dw}{du}\dfrac{\partial u}{\partial t}$

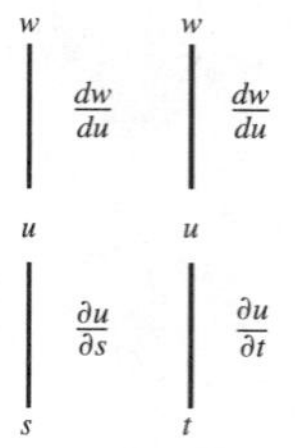

23. $\dfrac{\partial w}{\partial r} = \dfrac{\partial w}{\partial x}\dfrac{\partial x}{\partial r} + \dfrac{\partial w}{\partial y}\dfrac{\partial y}{\partial r} = \dfrac{\partial w}{\partial x}\dfrac{\partial x}{\partial r}$ since $\dfrac{\partial y}{\partial r} = 0$,
$\dfrac{\partial w}{\partial s} = \dfrac{\partial w}{\partial x}\dfrac{\partial x}{\partial s} + \dfrac{\partial w}{\partial y}\dfrac{\partial y}{\partial s} = \dfrac{\partial w}{\partial y}\dfrac{\partial y}{\partial s}$ since $\dfrac{\partial x}{\partial s} = 0$

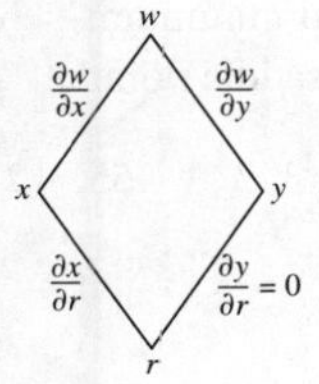

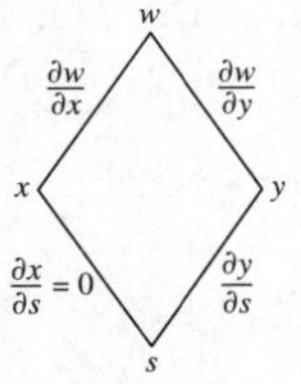

25. 4/3 **27.** $-4/5$ **29.** 20 **31.** $\frac{\partial z}{\partial x} = \frac{1}{4}, \frac{\partial z}{\partial y} = -\frac{3}{4}$

33. $\frac{\partial z}{\partial x} = -1, \frac{\partial z}{\partial y} = -1$ **35.** 12 **37.** -7

39. $\frac{\partial z}{\partial u} = 2, \frac{\partial z}{\partial v} = 1$ **41.** $\frac{\partial w}{\partial t} = 2te^{s^3+t^2}, \frac{\partial w}{\partial s} = 3s^2e^{s^3+t^2}$

43. 23 **45.** $-16, 2$ **47.** -0.00005 amp/sec

53. $(\cos 1, \sin 1, 1)$ and $(\cos(-2), \sin(-2), -2)$

55. (a) Maximum at $\left(-\frac{\sqrt{2}}{2}, \frac{\sqrt{2}}{2}\right)$ and $\left(\frac{\sqrt{2}}{2}, -\frac{\sqrt{2}}{2}\right)$; minimum at $\left(\frac{\sqrt{2}}{2}, \frac{\sqrt{2}}{2}\right)$ and $\left(-\frac{\sqrt{2}}{2}, -\frac{\sqrt{2}}{2}\right)$

(b) Max = 6; min = 2

57. 5 °C/sec **59.** $2x\sqrt{x^8 + x^3} + \int_0^{x^2} \frac{3x^2}{2\sqrt{t^4 + x^3}}\,dt$

61. 0.19 m/min **63.** $-8/61$ rad/min

SECTION 14.5, pp. 849–850

1.

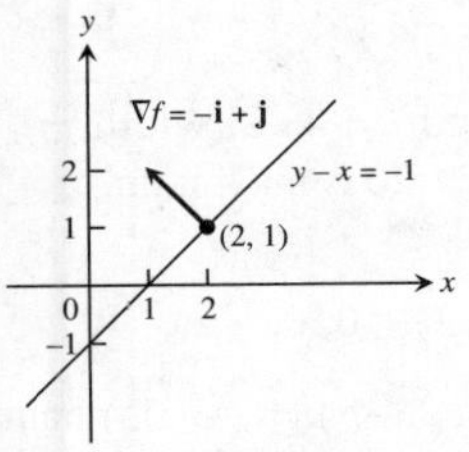

3.

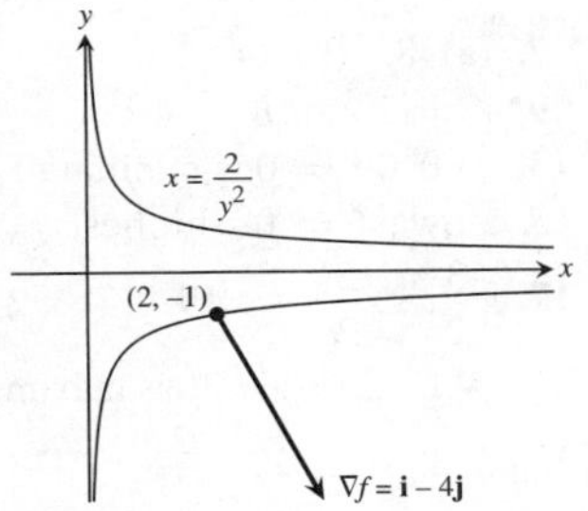

5.

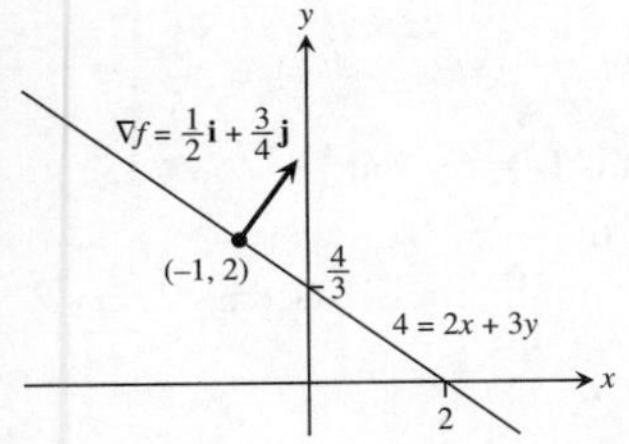

7. $\nabla f = 3\mathbf{i} + 2\mathbf{j} - 4\mathbf{k}$ **9.** $\nabla f = -\frac{26}{27}\mathbf{i} + \frac{23}{54}\mathbf{j} - \frac{23}{54}\mathbf{k}$

11. -4 **13.** 21/13 **15.** 3 **17.** 2

19. $\mathbf{u} = -\frac{1}{\sqrt{2}}\mathbf{i} + \frac{1}{\sqrt{2}}\mathbf{j}, (D_{\mathbf{u}} f)_{P_0} = \sqrt{2}; -\mathbf{u} = \frac{1}{\sqrt{2}}\mathbf{i} - \frac{1}{\sqrt{2}}\mathbf{j},$
$(D_{-\mathbf{u}} f)_{P_0} = -\sqrt{2}$

21. $\mathbf{u} = \frac{1}{3\sqrt{3}}\mathbf{i} - \frac{5}{3\sqrt{3}}\mathbf{j} - \frac{1}{3\sqrt{3}}\mathbf{k}, (D_{\mathbf{u}} f)_{P_0} = 3\sqrt{3};$
$-\mathbf{u} = -\frac{1}{3\sqrt{3}}\mathbf{i} + \frac{5}{3\sqrt{3}}\mathbf{j} + \frac{1}{3\sqrt{3}}\mathbf{k}, (D_{-\mathbf{u}} f)_{P_0} = -3\sqrt{3}$

23. $\mathbf{u} = \frac{1}{\sqrt{3}}(\mathbf{i} + \mathbf{j} + \mathbf{k}), (D_{\mathbf{u}} f)_{P_0} = 2\sqrt{3};$
$-\mathbf{u} = -\frac{1}{\sqrt{3}}(\mathbf{i} + \mathbf{j} + \mathbf{k}), (D_{-\mathbf{u}} f)_{P_0} = -2\sqrt{3}$

25.

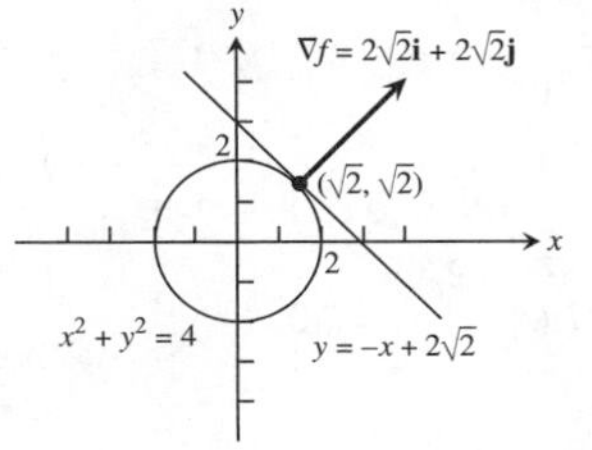

27.

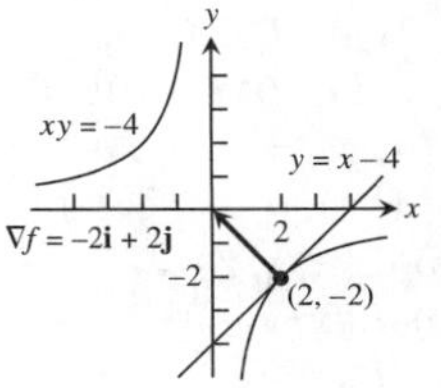

29. (a) $\mathbf{u} = \frac{3}{5}\mathbf{i} - \frac{4}{5}\mathbf{j}, D_{\mathbf{u}} f(1, -1) = 5$

(b) $\mathbf{u} = -\frac{3}{5}\mathbf{i} + \frac{4}{5}\mathbf{j}, D_{\mathbf{u}} f(1, -1) = -5$

(c) $\mathbf{u} = \frac{4}{5}\mathbf{i} + \frac{3}{5}\mathbf{j}, \mathbf{u} = -\frac{4}{5}\mathbf{i} - \frac{3}{5}\mathbf{j}$

(d) $\mathbf{u} = -\mathbf{j}, \mathbf{u} = \frac{24}{25}\mathbf{i} - \frac{7}{25}\mathbf{j}$

(e) $\mathbf{u} = -\mathbf{i}, \mathbf{u} = \frac{7}{25}\mathbf{i} + \frac{24}{25}\mathbf{j}$

31. $\mathbf{u} = \frac{7}{\sqrt{53}}\mathbf{i} - \frac{2}{\sqrt{53}}\mathbf{j}, -\mathbf{u} = -\frac{7}{\sqrt{53}}\mathbf{i} + \frac{2}{\sqrt{53}}\mathbf{j}$

33. No, the maximum rate of change is $\sqrt{185} < 14$. **35.** $-7/\sqrt{5}$

41. $r(t) = (-3 - 6t)\mathbf{i} + (4 + 8t)\mathbf{j}, -\infty < t < \infty$

43. $r(t) = (3 + 6t)\mathbf{i} + (-2 - 4t)\mathbf{j} + (1 + 2t)\mathbf{k}, -\infty < t < \infty$

45. $\left\langle \frac{14}{3}, \frac{1}{6}, -12, -\frac{4}{9} \right\rangle$ **47.** $\left\langle 30, 1, 90, \frac{1}{e}, 9 \right\rangle$

49. $\frac{3}{20e^2} - 2$ **51.** $\frac{\pi}{2} + \frac{1}{12}$

SECTION 14.6, pp. 857–860

1. (a) $x + y + z = 3$
(b) $x = 1 + 2t, y = 1 + 2t, z = 1 + 2t$

3. (a) $8x + y + 8z = 24$
(b) $x = 1 + 8t, y = t, z = 2 + 8t$

5. (a) $2x + 2y + z - 4 = 0$
(b) $x = 2t, y = 1 + 2t, z = 2 + t$

7. (a) $x + y + z - 1 = 0$
(b) $x = t, y = 1 + t, z = t$

9. (a) $-x + 3y + z/e = 2$
(b) $x = 2 - t, y = 1 + 3t, z = e + (1/e)t$

11. $2x - z - 2 = 0$ **13.** $x - y + 2z - 1 = 0$

15. $x = 6 + 70t, y = 4 + 18t, z = 3 + 116t$

17. $x = 1 - 2t, y = 1, z = \frac{1}{2} + 2t$

19. $x = 1 + 90t, y = 1 - 90t, z = 3$

21. $df = \frac{9}{11{,}830} \approx 0.0008$ **23.** $dg = 0$

25. (a) $\frac{\sqrt{3}}{2}\sin\sqrt{3} - \frac{1}{2}\cos\sqrt{3} \approx 0.935$ °C/ft

(b) $\sqrt{3}\sin\sqrt{3} - \cos\sqrt{3} \approx 1.87$ °C/sec

27. (a) $L(x, y) = 1$ **(b)** $L(x, y) = 2x + 2y - 1$

29. (a) $L(x, y) = 3x - 4y + 5$
(b) $L(x, y) = 3x - 4y + 5$

31. (a) $L(x, y) = 1 + x$
(b) $L(x, y) = -y + \frac{\pi}{2}$

33. (a)
$W(20, 25) = 11$ °F, $W(30, -10) = -39$ °F, $W(15, 15) = 0$ °F

(b) $W(10, -40) \approx -65.5$ °F, $W(50, -40) \approx -88$ °F, $W(60, 30) \approx 10.2$ °F

(c) $L(v, T) \approx -0.36\ (v - 25) + 1.337(T - 5) - 17.4088$

(d) i) $L(24, 6) \approx -15.7$ °F
ii) $L(27, 2) \approx -22.1$ °F
iii) $L(5, -10) \approx -30.2$ °F

35. $L(x, y) = 7 + x - 6y$; 0.06

37. $L(x, y) = x + y + 1$; 0.08 **39.** $L(x, y) = 1 + x$; 0.0222

41. (a) $L(x, y, z) = 2x + 2y + 2z - 3$
(b) $L(x, y, z) = y + z$ (c) $L(x, y, z) = 0$

43. (a) $L(x, y, z) = x$
(b) $L(x, y, z) = \frac{1}{\sqrt{2}}x + \frac{1}{\sqrt{2}}y$
(c) $L(x, y, z) = \frac{1}{3}x + \frac{2}{3}y + \frac{2}{3}z$

45. (a) $L(x, y, z) = 2 + x$
(b) $L(x, y, z) = x - y - z + \frac{\pi}{2} + 1$
(c) $L(x, y, z) = x - y - z + \frac{\pi}{2} + 1$

47. $L(x, y, z) = 2x - 6y - 2z + 6$, 0.0024

49. $L(x, y, z) = x + y - z - 1$, 0.00135

51. Maximum error (estimate) ≤ 0.31 in magnitude

53. Pay more attention to the smaller of the two dimensions. It will generate the larger partial derivative.

55. f is most sensitive to a change in d.

61. (a) 1.75% (b) 1.75%

SECTION 14.7, pp. 867–869

1. $f(-3, 3) = -5$, local minimum **3.** $f(-2, 1)$, saddle point

5. $f\left(3, \frac{3}{2}\right) = \frac{17}{2}$, local maximum

7. $f(2, -1) = -6$, local minimum **9.** $f(1, 2)$, saddle point

11. $f\left(\frac{16}{7}, 0\right) = -\frac{16}{7}$, local maximum

13. $f(0, 0)$, saddle point; $f\left(-\frac{2}{3}, \frac{2}{3}\right) = \frac{170}{27}$, local maximum

15. $f(0, 0) = 0$, local minimum; $f(1, -1)$, saddle point

17. $f(0, \pm\sqrt{5})$, saddle points; $f(-2, -1) = 30$, local maximum; $f(2, 1) = -30$, local minimum

19. $f(0, 0)$, saddle point; $f(1, 1) = 2$, $f(-1, -1) = 2$, local maxima

21. $f\left(\frac{\pi}{3}, \frac{\pi}{3}\right) = \frac{3\sqrt{3}}{8}$ is maximum and
$f\left(\frac{2\pi}{3}, \frac{2\pi}{3}\right) = -\frac{3\sqrt{3}}{8}$ is minimum

23. $f(n\pi, 0)$, saddle points, for every integer n

25. $f(2, 0) = e^{-4}$, local minimum

27. $f(0, 0) = 0$, local minimum; $f(0, 2)$, saddle point

29. $f\left(\frac{1}{2}, 1\right) = \ln\left(\frac{1}{4}\right) - 3$, local maximum

31. Absolute maximum: 1 at $(0, 0)$; absolute minimum: -5 at $(1, 2)$

33. Absolute maximum: 4 at $(0, 2)$; absolute minimum: 0 at $(0, 0)$

35. Absolute maximum: 11 at $(0, -3)$; absolute minimum: -10 at $(4, -2)$

37. Absolute maximum: 4 at $(2, 0)$; absolute minimum: $\frac{3\sqrt{2}}{2}$ at $\left(3, -\frac{\pi}{4}\right)$, $\left(3, \frac{\pi}{4}\right)$, $\left(1, -\frac{\pi}{4}\right)$, and $\left(1, \frac{\pi}{4}\right)$

39. $a = -3$, $b = 2$

41. Hottest is $2\frac{1}{4}^\circ$ at $\left(-\frac{1}{2}, \frac{\sqrt{3}}{2}\right)$ and $\left(-\frac{1}{2}, -\frac{\sqrt{3}}{2}\right)$; coldest is $-\frac{1}{4}^\circ$ at $\left(\frac{1}{2}, 0\right)$.

43. (a) $f(0, 0)$, saddle point (b) $f(1, 2)$, local minimum
(c) $f(1, -2)$, local minimum; $f(-1, -2)$, saddle point

49. $\left(\frac{1}{6}, \frac{1}{3}, \frac{355}{36}\right)$ **51.** $\left(\frac{9}{7}, \frac{6}{7}, \frac{3}{7}\right)$ **53.** 3, 3, 3 **55.** 12

57. $\frac{4}{\sqrt{3}} \times \frac{4}{\sqrt{3}} \times \frac{4}{\sqrt{3}}$ **59.** 2 ft × 2 ft × 1 ft

61. Points $(0, 2, 0)$ and $(0, -2, 0)$ have distance 2 from the origin.

63. (a) On the semicircle, max $f = 2\sqrt{2}$ at $t = \pi/4$, min $f = -2$ at $t = \pi$. On the quarter circle, max $f = 2\sqrt{2}$ at $t = \pi/4$, min $f = 2$ at $t = 0, \pi/2$.
(b) On the semicircle, max $g = 2$ at $t = \pi/4$, min $g = -2$ at $t = 3\pi/4$. On the quarter circle, max $g = 2$ at $t = \pi/4$, min $g = 0$ at $t = 0, \pi/2$.
(c) On the semicircle, max $h = 8$ at $t = 0, \pi$; min $h = 4$ at $t = \pi/2$. On the quarter circle, max $h = 8$ at $t = 0$, min $h = 4$ at $t = \pi/2$.

65. i) min $f = -1/2$ at $t = -1/2$; no max
ii) max $f = 0$ at $t = -1, 0$; min $f = -1/2$ at $t = -1/2$
iii) max $f = 4$ at $t = 1$; min $f = 0$ at $t = 0$

69. $y = -\frac{20}{13}x + \frac{9}{13}$, $y\Big|_{x=4} = -\frac{71}{13}$

SECTION 14.8, pp. 876–879

1. $\left(\pm\frac{1}{\sqrt{2}}, \frac{1}{2}\right)$, $\left(\pm\frac{1}{\sqrt{2}}, -\frac{1}{2}\right)$ **3.** 39 **5.** $(3, \pm 3\sqrt{2})$

7. (a) 8 (b) 64

9. $r = 2$ cm, $h = 4$ cm **11.** Length $= 4\sqrt{2}$, width $= 3\sqrt{2}$

13. $f(0, 0) = 0$ is minimum; $f(2, 4) = 20$ is maximum.

15. Lowest $= 0°$, highest $= 125°$

17. $\left(\frac{3}{2}, 2, \frac{5}{2}\right)$ **19.** 1 **21.** $(0, 0, 2)$, $(0, 0, -2)$

23. $f(1, -2, 5) = 30$ is maximum; $f(-1, 2, -5) = -30$ is minimum.

25. 3, 3, 3 **27.** $\frac{2}{\sqrt{3}}$ by $\frac{2}{\sqrt{3}}$ by $\frac{2}{\sqrt{3}}$ units

29. $\frac{8abc}{3\sqrt{3}}$ **31.** (b) ≈24,322 units

33. $U(8, 14) = \$128$ **37.** $f(2/3, 4/3, -4/3) = \frac{4}{3}$

39. $(2, 4, 4)$ **41.** Maximum is $1 + 6\sqrt{3}$ at $(\pm\sqrt{6}, \sqrt{3}, 1)$; minimum is $1 - 6\sqrt{3}$ at $(\pm\sqrt{6}, -\sqrt{3}, 1)$.

43. Maximum is 4 at $(0, 0, \pm 2)$; minimum is 2 at $(\pm\sqrt{2}, \pm\sqrt{2}, 0)$.

SECTION 14.9, p. 883

1. Quadratic: $x + xy$; cubic: $x + xy + \frac{1}{2}xy^2$

3. Quadratic: x; cubic: $x - \frac{1}{2}xy^2$

5. Quadratic: $1 - y + x^2 + \frac{1}{2}y^2$;
Cubic: $1 - y + x^2 + \frac{1}{2}y^2 - x^2y - \frac{1}{6}y^3$

7. Quadratic: $\frac{1}{2}(2x^2 + 2y^2) = x^2 + y^2$; cubic: $x^2 + y^2$

9. Quadratic: $1 + (x + y) + (x + y)^2$;
cubic: $1 + (x + y) + (x + y)^2 + (x + y)^3$

11. Quadratic: $1 - \frac{1}{2}x^2 - \frac{1}{2}y^2$; $E(x, y) \leq 0.00134$

SECTION 14.10, p. 887

1. (a) $-2y$ **(b)** $\frac{3}{2}+2z$ **(c)** $2+2z$

3. (a) $\frac{\partial U}{\partial P}+\frac{\partial U}{\partial T}\left(\frac{V}{nR}\right)$ **(b)** $\frac{\partial U}{\partial P}\left(\frac{nR}{V}\right)+\frac{\partial U}{\partial T}$

5. (a) 5 **(b)** 5 **7.** $\sin\theta, \frac{y}{\sqrt{x^2+y^2}}$

PRACTICE EXERCISES, pp. 888–891

1. Domain: all points in the xy-plane; range: $z \geq 0$. Level curves are ellipses with major axis along the y-axis and minor axis along the x-axis.

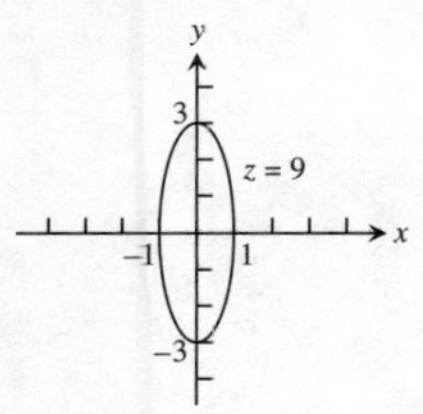

3. Domain: all (x, y) such that $x \neq 0$ and $y \neq 0$; range: $z \neq 0$. Level curves are hyperbolas with the x- and y-axes as asymptotes.

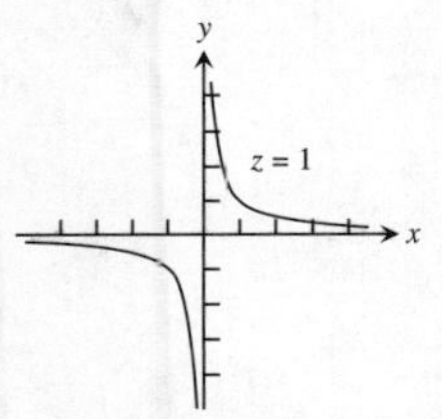

5. Domain: all points in xyz-space; range: all real numbers. Level surfaces are paraboloids of revolution with the z-axis as axis.

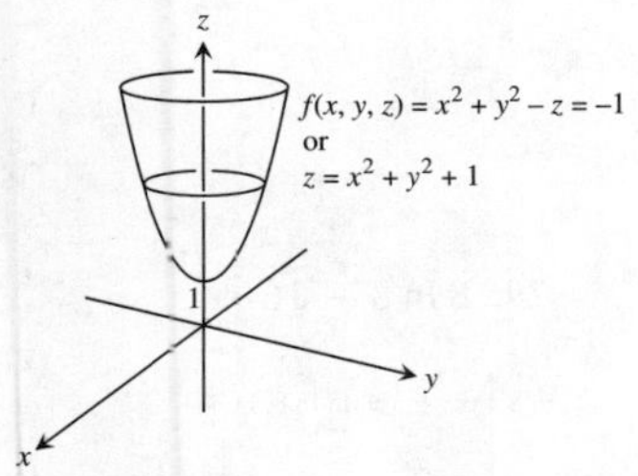

7. Domain: all (x, y, z) such that $(x, y, z) \neq (0, 0, 0)$; range: positive real numbers. Level surfaces are spheres with center $(0, 0, 0)$ and radius $r > 0$.

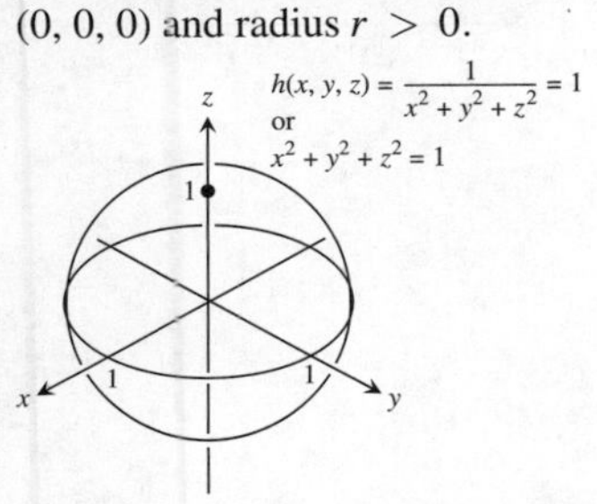

9. -2 **11.** $1/2$ **13.** 1 **15.** Let $y = kx^2, k \neq 1$

17. No; $\lim_{(x,y)\to(0,0)} f(x, y)$ does not exist.

19. $\frac{\partial g}{\partial r} = \cos\theta + \sin\theta, \frac{\partial g}{\partial \theta} = -r\sin\theta + r\cos\theta$

21. $\frac{\partial f}{\partial R_1} = -\frac{1}{R_1{}^2}, \frac{\partial f}{\partial R_2} = -\frac{1}{R_2{}^2}, \frac{\partial f}{\partial R_3} = -\frac{1}{R_3{}^2}$

23. $\frac{\partial P}{\partial n} = \frac{RT}{V}, \frac{\partial P}{\partial R} = \frac{nT}{V}, \frac{\partial P}{\partial T} = \frac{nR}{V}, \frac{\partial P}{\partial V} = -\frac{nRT}{V^2}$

25. $\frac{\partial^2 g}{\partial x^2} = 0, \frac{\partial^2 g}{\partial y^2} = \frac{2x}{y^3}, \frac{\partial^2 g}{\partial y\,\partial x} = \frac{\partial^2 g}{\partial x\,\partial y} = -\frac{1}{y^2}$

27. $\frac{\partial^2 f}{\partial x^2} = -30x + \frac{2-2x^2}{(x^2+1)^2}, \frac{\partial^2 f}{\partial y^2} = 0, \frac{\partial^2 f}{\partial y\,\partial x} = \frac{\partial^2 f}{\partial x\,\partial y} = 1$

29. $\left.\frac{dw}{dt}\right|_{t=0} = -1$

31. $\left.\frac{\partial w}{\partial r}\right|_{(r,s)=(\pi,0)} = 2, \left.\frac{\partial w}{\partial s}\right|_{(r,s)=(\pi,0)} = 2-\pi$

33. $\left.\frac{df}{dt}\right|_{t=1} = -(\sin 1 + \cos 2)(\sin 1) + (\cos 1 + \cos 2)(\cos 1)$
$-2(\sin 1 + \cos 1)(\sin 2)$

35. $\left.\frac{dy}{dx}\right|_{(x,y)=(0,1)} = -1$

37. Increases most rapidly in the direction $\mathbf{u} = -\frac{\sqrt{2}}{2}\mathbf{i} - \frac{\sqrt{2}}{2}\mathbf{j}$;
decreases most rapidly in the direction $-\mathbf{u} = \frac{\sqrt{2}}{2}\mathbf{i} + \frac{\sqrt{2}}{2}\mathbf{j}$;
$D_{\mathbf{u}} f = \frac{\sqrt{2}}{2}$; $D_{-\mathbf{u}} f = -\frac{\sqrt{2}}{2}$; $D_{\mathbf{u}_1} f = -\frac{7}{10}$ where $\mathbf{u}_1 = \frac{\mathbf{v}}{|\mathbf{v}|}$

39. Increases most rapidly in the direction $\mathbf{u} = \frac{2}{7}\mathbf{i} + \frac{3}{7}\mathbf{j} + \frac{6}{7}\mathbf{k}$;
decreases most rapidly in the direction $-\mathbf{u} = -\frac{2}{7}\mathbf{i} - \frac{3}{7}\mathbf{j} - \frac{6}{7}\mathbf{k}$;
$D_{\mathbf{u}} f = 7$; $D_{-\mathbf{u}} f = -7$; $D_{\mathbf{u}_1} f = 7$ where $\mathbf{u}_1 = \frac{\mathbf{v}}{|\mathbf{v}|}$

41. $\pi/\sqrt{2}$ **43. (a)** $f_x(1,2) = f_y(1,2) = 2$ **(b)** 14/5

45.

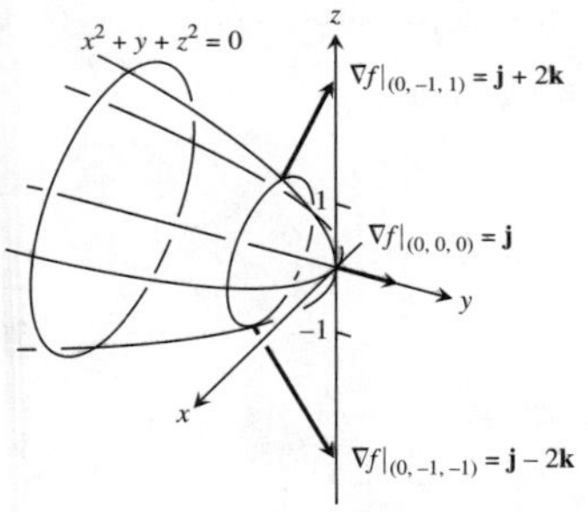

59. Be more careful with the diameter.

61. $dI = 0.038$, % change in $I = 15.83\%$, more sensitive to voltage change

63. (a) 5% **65.** Local minimum of -8 at $(-2, -2)$

67. Saddle point at $(0, 0)$, $f(0, 0) = 0$; local maximum of 1/4 at $(-1/2, -1/2)$

69. Saddle point at $(0, 0)$, $f(0, 0) = 0$; local minimum of -4 at $(0, 2)$; local maximum of 4 at $(-2, 0)$; saddle point at $(-2, 2)$, $f(-2, 2) = 0$

71. Absolute maximum: 28 at $(0, 4)$; absolute minimum: $-9/4$ at $(3/2, 0)$

73. Absolute maximum: 18 at $(2, -2)$; absolute minimum: $-17/4$ at $(-2, 1/2)$

75. Absolute maximum: 8 at $(-2, 0)$; absolute minimum: -1 at $(1, 0)$

77. Absolute maximum: 4 at $(1, 0)$; absolute minimum: -4 at $(0, -1)$

79. Absolute maximum: 1 at $(0, \pm 1)$ and $(1, 0)$; absolute minimum: -1 at $(-1, 0)$

81. Maximum: 5 at $(0, 1)$; minimum: $-1/3$ at $(0, -1/3)$

83. Maximum: $\sqrt{3}$ at $\left(\frac{1}{\sqrt{3}}, -\frac{1}{\sqrt{3}}, \frac{1}{\sqrt{3}}\right)$; minimum: $-\sqrt{3}$ at $\left(-\frac{1}{\sqrt{3}}, \frac{1}{\sqrt{3}}, -\frac{1}{\sqrt{3}}\right)$

85. Width $= \left(\frac{c^2V}{ab}\right)^{1/3}$, depth $= \left(\frac{b^2V}{ac}\right)^{1/3}$, height $= \left(\frac{a^2V}{bc}\right)^{1/3}$

87. Maximum: $\frac{3}{2}$ at $\left(\frac{1}{\sqrt{2}}, \frac{1}{\sqrt{2}}, \sqrt{2}\right)$ and $\left(-\frac{1}{\sqrt{2}}, -\frac{1}{\sqrt{2}}, -\sqrt{2}\right)$; minimum: $\frac{1}{2}$ at $\left(-\frac{1}{\sqrt{2}}, \frac{1}{\sqrt{2}}, -\sqrt{2}\right)$ and $\left(\frac{1}{\sqrt{2}}, -\frac{1}{\sqrt{2}}, \sqrt{2}\right)$

89. $\frac{\partial w}{\partial x} = \cos\theta \frac{\partial w}{\partial r} - \frac{\sin\theta}{r}\frac{\partial w}{\partial \theta}$, $\frac{\partial w}{\partial y} = \sin\theta \frac{\partial w}{\partial r} + \frac{\cos\theta}{r}\frac{\partial w}{\partial \theta}$

95. $(t, -t \pm 4, t)$, t a real number

101. **(a)** $(2y + x^2z)e^{yz}$ **(b)** $x^2e^{yz}\left(y - \frac{z}{2y}\right)$ **(c)** $(1 + x^2y)e^{yz}$

ADDITIONAL AND ADVANCED EXERCISES, pp. 891–893

1. $f_{xy}(0,0) = -1, f_{yx}(0,0) = 1$

7. **(c)** $\frac{r^2}{2} = \frac{1}{2}(x^2 + y^2 + z^2)$ **13.** $V = \frac{\sqrt{3}abc}{2}$

17. $f(x, y) = \frac{y}{2} + 4$, $g(x, y) = \frac{x}{2} + \frac{9}{2}$

19. $y = 2\ln|\sin x| + \ln 2$

21. **(a)** $\frac{1}{\sqrt{53}}(2\mathbf{i} + 7\mathbf{j})$ **(b)** $\frac{-1}{\sqrt{29{,}097}}(98\mathbf{i} - 127\mathbf{j} + 58\mathbf{k})$

23. $w = e^{-c^2\pi^2 t}\sin \pi x$

Chapter 15

SECTION 15.1, pp. 898–899

1. 24 **3.** 1 **5.** 16 **7.** $2\ln 2 - 1$ **9.** $(3/2)(5 - e)$ **11.** 3/2 **13.** $\ln 2$ **15.** $3/2, -2$ **17.** 8/3 **19.** 0 **21.** 1/2 **23.** $2\ln 2$ **25.** $(\ln 2)^2$

27.

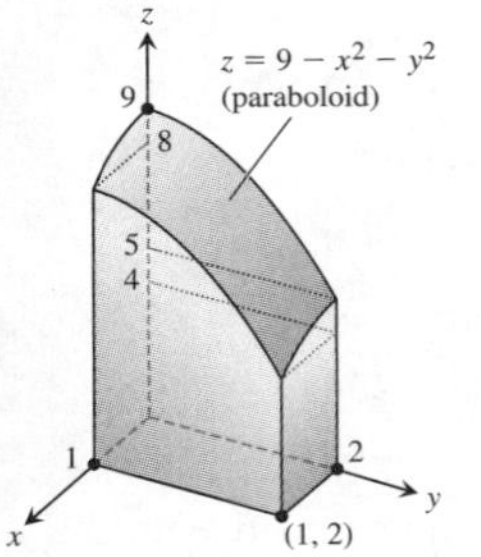

29. 8/3 **31.** 1 **33.** $\sqrt{2}$ **35.** 2/27 **37.** $\frac{3}{2}\ln 3 - 1$ **39.** **(a)** 1/3 **(b)** 2/3

SECTION 15.2, pp. 905–908

1.

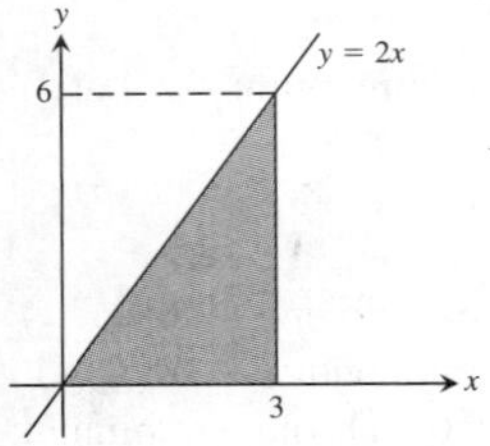

3.

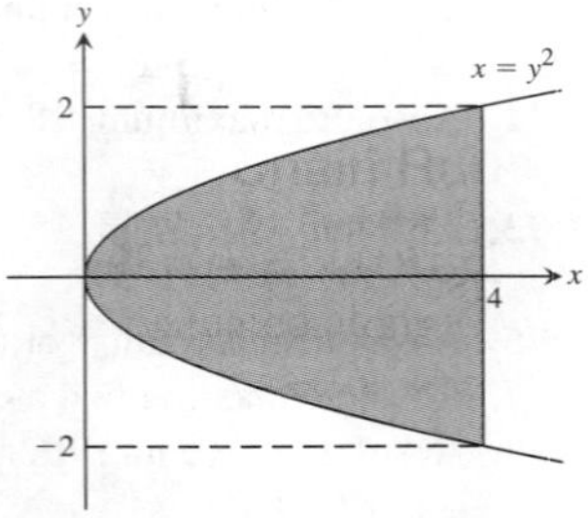

5.

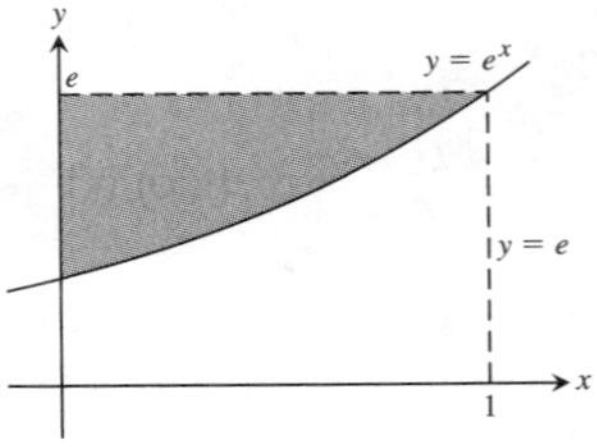

7.

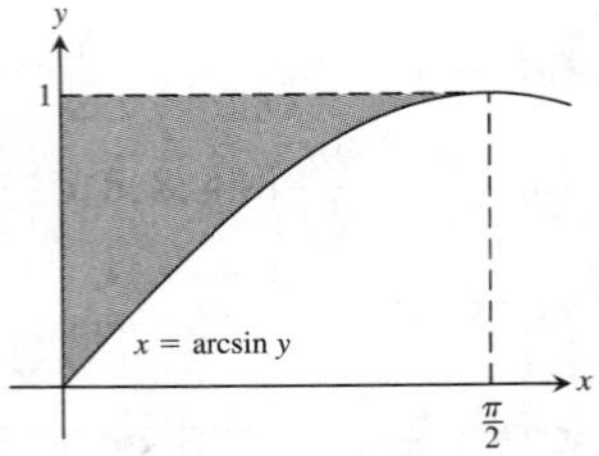

9. **(a)** $\int_0^2\int_{x^3}^8 dy\,dx$ **(b)** $\int_0^8\int_0^{y^{1/3}} dx\,dy$

11. **(a)** $\int_0^3\int_{x^2}^{3x} dy\,dx$ **(b)** $\int_0^9\int_{y/3}^{\sqrt{y}} dx\,dy$

13. **(a)** $\int_0^9\int_0^{\sqrt{x}} dy\,dx$ **(b)** $\int_0^3\int_{y^2}^9 dx\,dy$

15. **(a)** $\int_0^{\ln 3}\int_{e^{-x}}^1 dy\,dx$ **(b)** $\int_{1/3}^1\int_{-\ln y}^{\ln 3} dx\,dy$

17. **(a)** $\int_0^1\int_x^{3-2x} dy\,dx$

(b) $\int_0^1\int_0^y dx\,dy + \int_1^3\int_0^{(3-y)/2} dx\,dy$

19. 42 **21.** 21/40

23. $\frac{4}{15}$

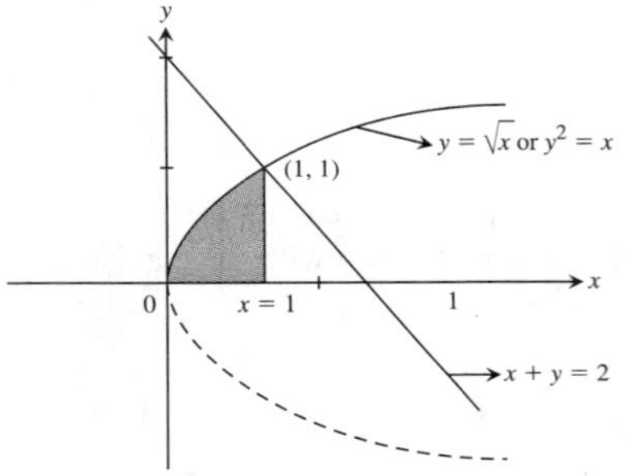

25. 49/5

27. $\frac{\pi^2}{2} + 2$

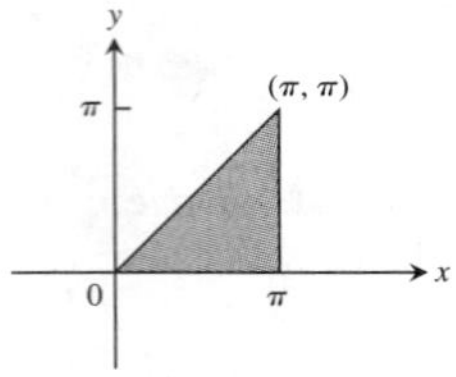

29. $8\ln 8 - 16 + e$

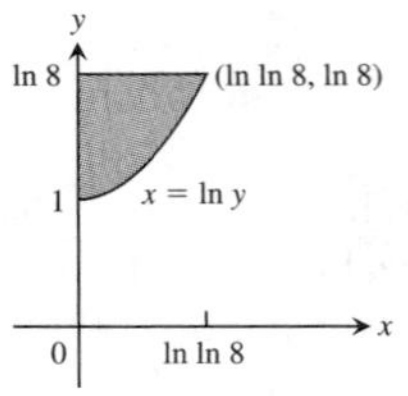

31. $e - 2$

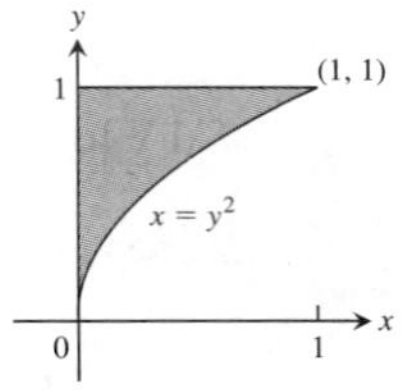

33. $\frac{a^4}{15}$ **35.** $\frac{1}{6}$

37. 8

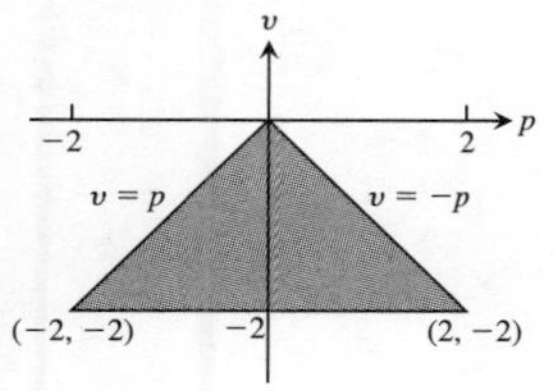

39. 2π

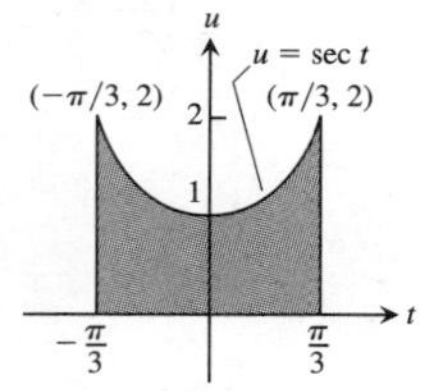

41. $\int_2^4 \int_0^{(4-y)/2} dx\, dy$

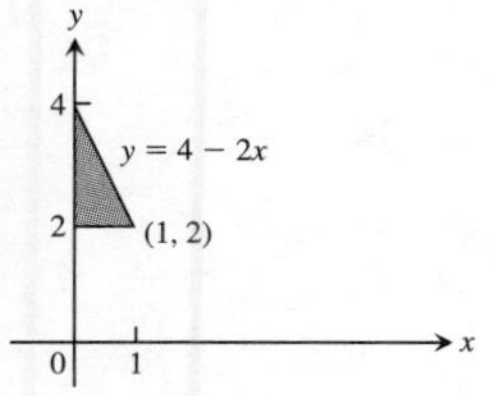

43. $\int_0^1 \int_{x^2}^{x} dy\, dx$

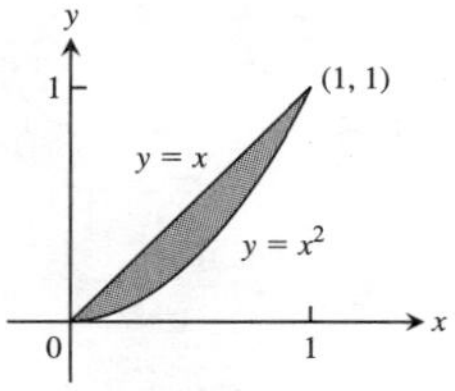

45. $\int_1^e \int_{\ln y}^{1} dx\, dy$

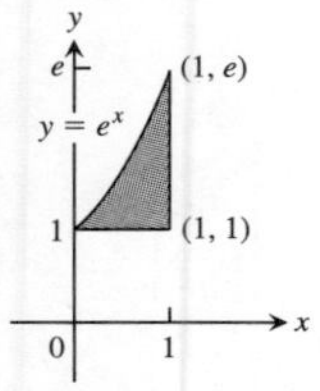

47. $\int_0^9 \int_0^{(\sqrt{9-y})/2} 16x\, dx\, dy$

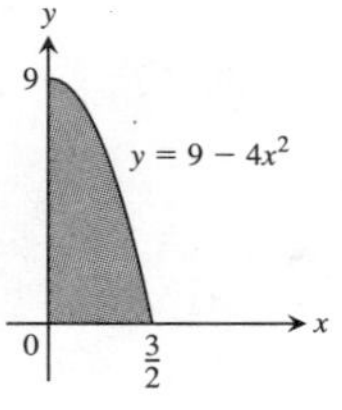

49. $\int_0^{2a} \int_0^{\sqrt{a^2-(x-a)^2}} xy\, dy\, dx$

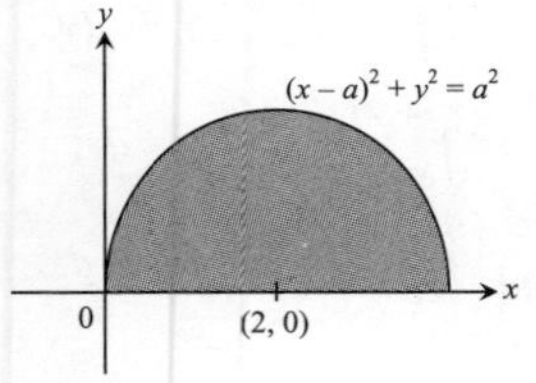

51. $\int_1^2 \int_0^{4-x^2} (x + y)\, dy\, dx$

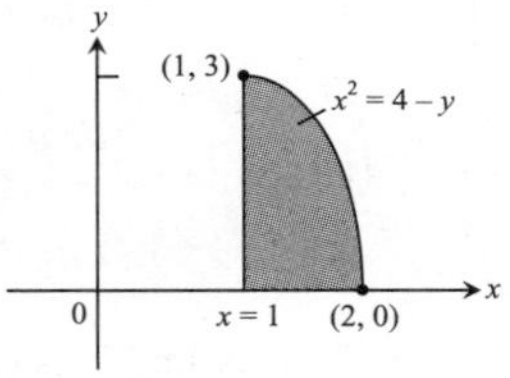

53. $\int_1^{e^3} \int_{\ln x}^{3} (x + y)\, dy\, dx$

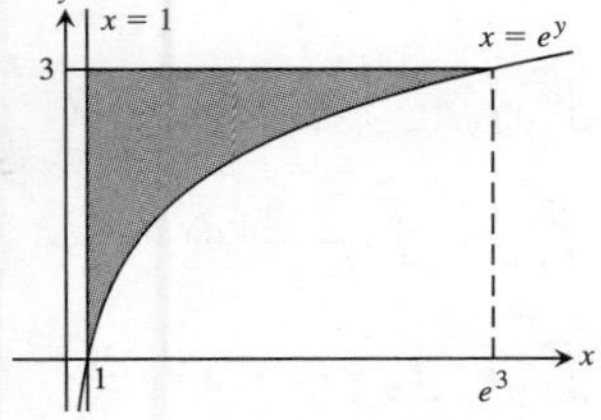

55. 2

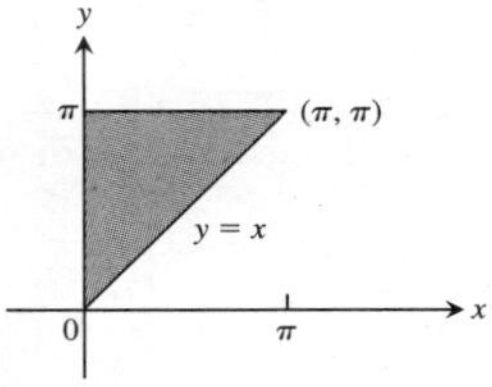

57. $\dfrac{e-2}{2}$

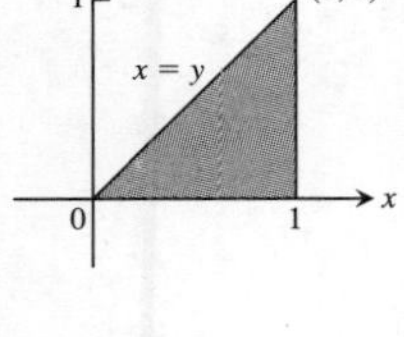

59. $1/3$

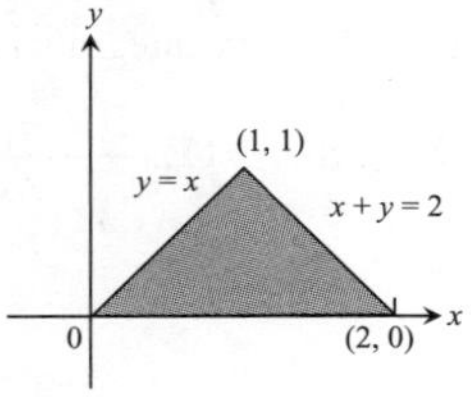

61. $1/(80\pi)$

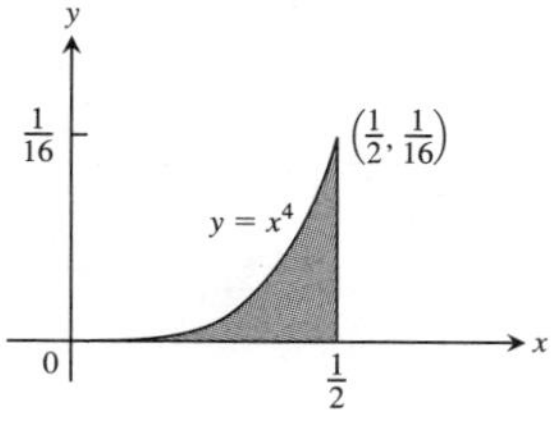

63. 6

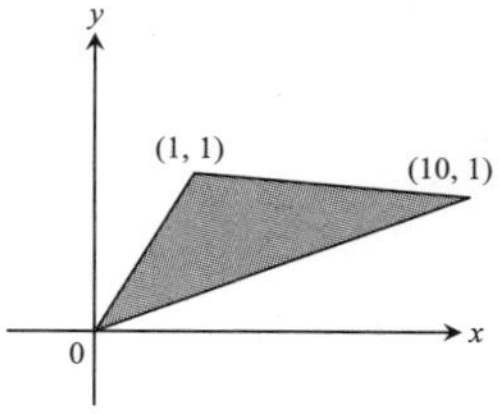

65. $4/3$ **67.** $625/12$ **69.** 16 **71.** 20 **73.** $2(1 + \ln 2)$

75.

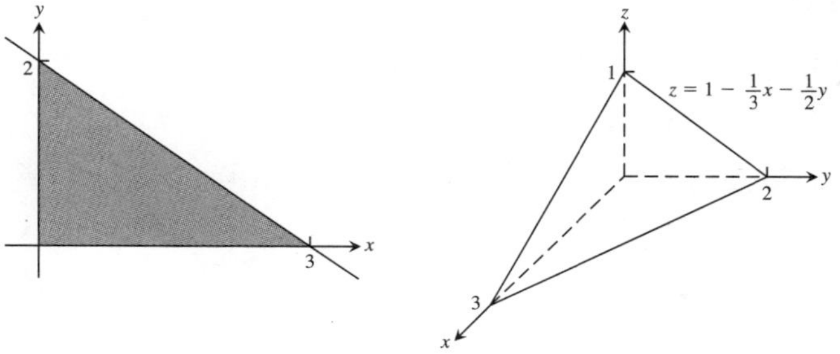

77. 1 **79.** π^2 **81.** $-\dfrac{3}{32}$ **83.** $\dfrac{20\sqrt{3}}{9}$

85. $\int_0^1 \int_x^{2-x} (x^2 + y^2)\, dy\, dx = \dfrac{4}{3}$

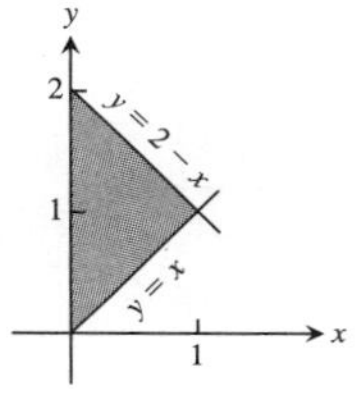

87. R is the set of points (x, y) such that $x^2 + 2y^2 \le 4$.

89. 1. **93.** 0.603 **95.** 0.233

SECTION 15.3, pp. 910–911

1. $\int_0^2 \int_0^{2-x} dy\, dx = 2$ or $\int_0^2 \int_0^{2-y} dx\, dy = 2$

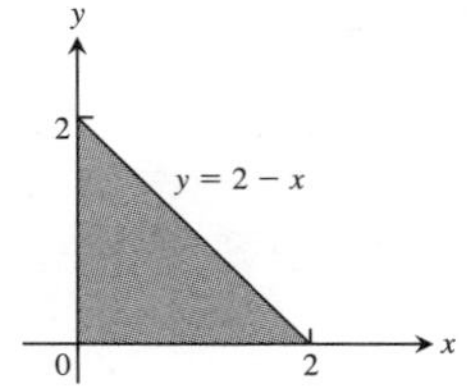

3. $\int_{-2}^{1} \int_{y-2}^{-y^2} dx\, dy = \dfrac{9}{2}$

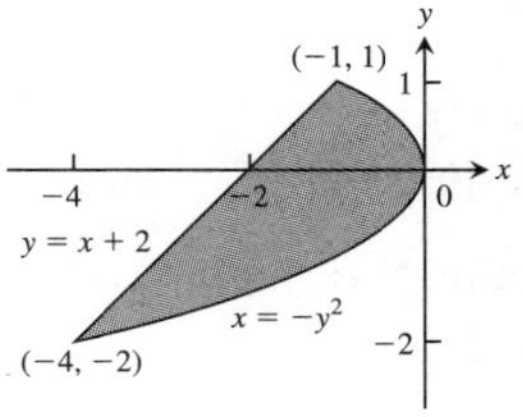

5. $\dfrac{16}{3}\sqrt{2}$

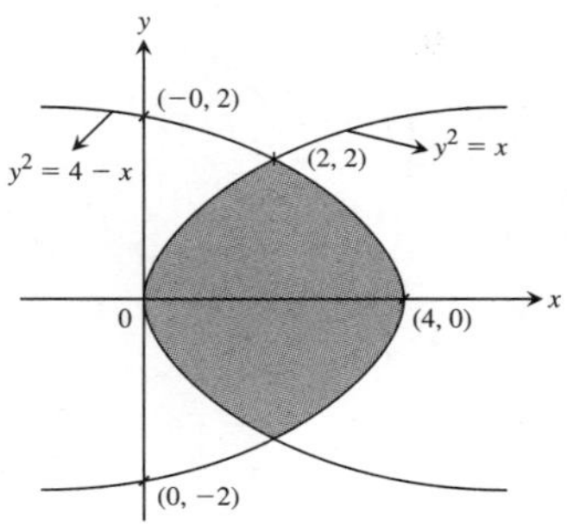

7. $\int_0^1 \int_{y^2}^{2y-y^2} dx\, dy = \dfrac{1}{3}$

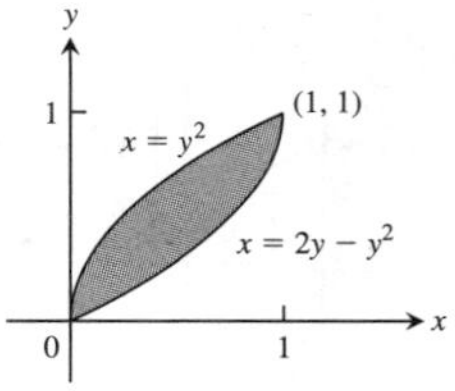

9. $\int_0^2 \int_y^{3y} 1\, dx\, dy = 4$ or

$\int_0^2 \int_{x/3}^{x} 1\, dy\, dx + \int_2^6 \int_{x/3}^{2} 1\, dy\, dx = 4$

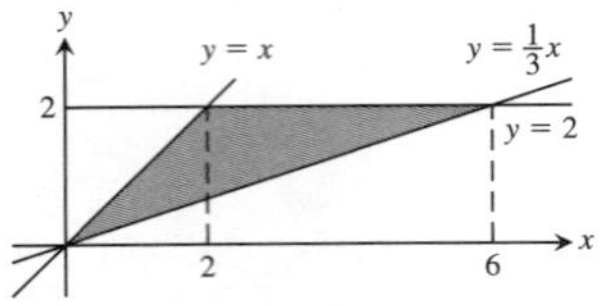

11. πab

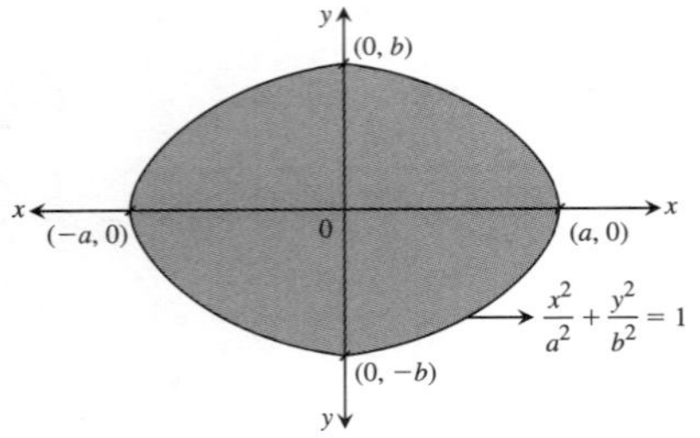

13. 12

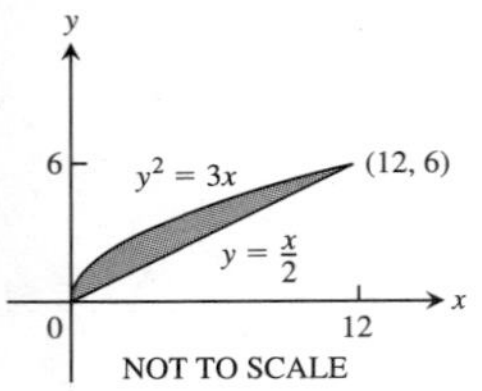

15. $\sqrt{2} - 1$

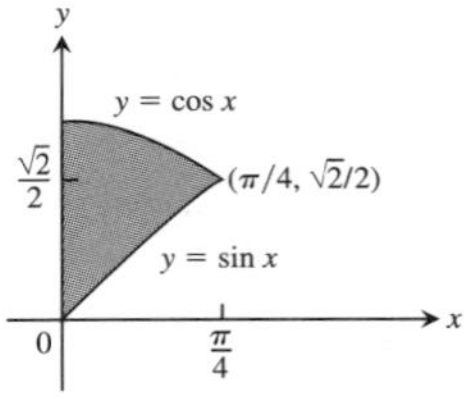

17. $\frac{3}{2}$

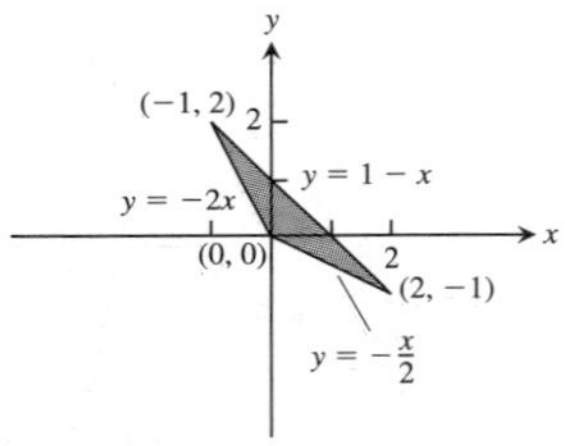

19. (a) 0 **(b)** $4/\pi^2$ **21.** $8/3$ **23.** $\pi - 2$

25. $40{,}000(1 - e^{-2})\ln(7/2) \approx 43{,}329$

SECTION 15.4, pp. 916–918

1. $\frac{\pi}{2} \le \theta \le 2\pi,\ 0 \le r \le 9$ **3.** $\frac{\pi}{4} \le \theta \le \frac{3\pi}{4},\ 0 \le r \le \csc\theta$

5. $0 \le \theta \le \frac{\pi}{6},\ 1 \le r \le 2\sqrt{3}\sec\theta;$

$\frac{\pi}{6} \le \theta \le \frac{\pi}{2},\ 1 \le r \le 2\csc\theta$

7. $-\frac{\pi}{2} \le \theta \le \frac{\pi}{2},\ 0 \le r \le 2\cos\theta$ **9.** $\frac{\pi}{2}$

11. 2π **13.** $3\pi/4$ **15.** $2 - \sqrt{3}$ **17.** $(1 - \ln 2)\pi$

19. $32/5$ **21.** $\frac{2(1 + \sqrt{2})}{3}$

23.

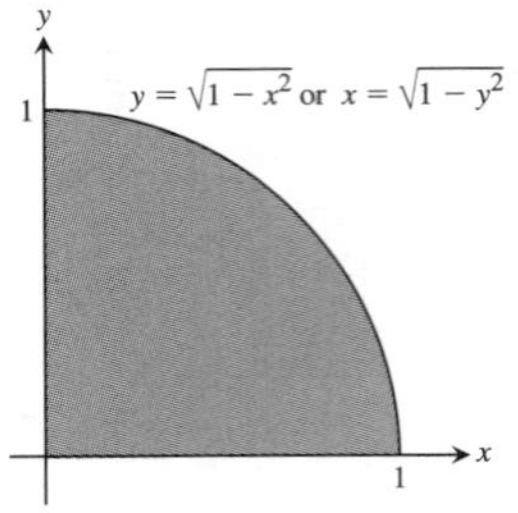

$\int_0^1 \int_0^{\sqrt{1-x^2}} xy\, dy\, dx$ or $\int_0^1 \int_0^{\sqrt{1-y^2}} xy\, dx\, dy$

25.

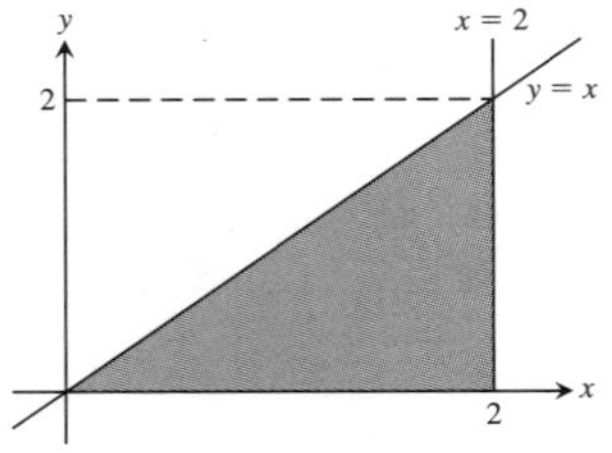

$\int_0^2 \int_0^{x} y^2(x^2 + y^2)\, dy\, dx$ or $\int_0^2 \int_y^{2} y^2(x^2 + y^2)\, dx\, dy$

27. $2(\pi - 2)$ **29.** 12π **31.** $(3\pi/8) + 1$ **33.** $\frac{2a}{3}$

35. $\frac{2a}{3}$ **37.** $2\pi(2 - \sqrt{e})$ **39.** $\frac{4}{3} + \frac{5\pi}{8}$

41. (a) $\frac{\sqrt{\pi}}{2}$ **(b)** 1 **43.** $\pi \ln 4$, no **45.** $\frac{1}{2}(a^2 + 2h^2)$

47. $\frac{8}{9}(3\pi - 4)$

SECTION 15.5, pp. 926–928

1. $1/6$

3. $\int_0^1 \int_0^{2-2x} \int_0^{3-3x-3y/2} dz\, dy\, dx,\ \int_0^2 \int_0^{1-y/2} \int_0^{3-3x-3y/2} dz\, dx\, dy,$

$\int_0^1 \int_0^{3-3x} \int_0^{2-2x-2z/3} dy\, dz\, dx,\ \int_0^3 \int_0^{1-z/3} \int_0^{2-2x-2z/3} dy\, dx\, dz,$

$\int_0^2 \int_0^{3-3y/2} \int_0^{1-y/2-z/3} dx\, dz\, dy,\ \int_0^3 \int_0^{2-2z/3} \int_0^{1-y/2-z/3} dx\, dy\, dz.$

The value of all six integrals is 1.

5.

$\int_{-2}^2 \int_{-\sqrt{4-x^2}}^{\sqrt{4-x^2}} \int_{x^2+y^2}^{8-x^2-y^2} 1\, dz\, dx\, dy,\ \int_{-2}^2 \int_{-\sqrt{4-y^2}}^{\sqrt{4-y^2}} \int_{x^2+y^2}^{8-x^2-y^2} 1\, dz\, dx\, dy,$

$\int_{-2}^2 \int_4^{8-y^2} \int_{-\sqrt{8-z-y^2}}^{\sqrt{8-z-y^2}} 1\, dx\, dz\, dy + \int_{-2}^2 \int_{y^2}^{4} \int_{-\sqrt{z-y^2}}^{\sqrt{z-y^2}} 1\, dx\, dz\, dy,$

$\int_4^8 \int_{-\sqrt{8-z}}^{\sqrt{8-z}} \int_{-\sqrt{8-z-y^2}}^{\sqrt{8-z-y^2}} 1\, dx\, dy\, dz + \int_0^4 \int_{-\sqrt{z}}^{\sqrt{z}} \int_{-\sqrt{z-y^2}}^{\sqrt{z-y^2}} 1\, dx\, dy\, dz,$

$\int_{-2}^2 \int_4^{8-x^2} \int_{-\sqrt{8-z-x^2}}^{\sqrt{8-z-x^2}} 1\, dy\, dz\, dx + \int_{-2}^2 \int_{x^2}^{4} \int_{-\sqrt{z-x^2}}^{\sqrt{z-x^2}} 1\, dy\, dz\, dx,$

$\int_4^8 \int_{-\sqrt{8-z}}^{\sqrt{8-z}} \int_{-\sqrt{8-z-x^2}}^{\sqrt{8-z-x^2}} 1\, dy\, dx\, dz + \int_0^4 \int_{-\sqrt{z}}^{\sqrt{z}} \int_{-\sqrt{z-x^2}}^{\sqrt{z-x^2}} 1\, dy\, dx\, dz.$

The value of all six integrals is 16π.

7. 1 **9.** $5/8$ **11.** $\frac{5(2 - \sqrt{3})}{4}$ **13.** 18

15. 7/6 **17.** 1/720 **19.** $\frac{1}{2} - \frac{\pi}{8}$

21. **(a)** $\int_{-1}^{1}\int_{0}^{1-x^2}\int_{x^2}^{1-z} dy\, dz\, dx$ **(b)** $\int_{0}^{1}\int_{-\sqrt{1-z}}^{\sqrt{1-z}}\int_{x^2}^{1-z} dy\, dx\, dz$

(c) $\int_{0}^{1}\int_{0}^{1-z}\int_{-\sqrt{y}}^{\sqrt{y}} dx\, dy\, dz$ **(d)** $\int_{0}^{1}\int_{0}^{1-y}\int_{-\sqrt{y}}^{\sqrt{y}} dx\, dz\, dy$

(e) $\int_{0}^{1}\int_{-\sqrt{y}}^{\sqrt{y}}\int_{0}^{1-y} dz\, dx\, dy$

23. 2/3 **25.** 20/3 **27.** 1 **29.** 16/3 **31.** $8\pi - \frac{32}{3}$

33. 2 **35.** $\frac{4\pi}{3}(a^2 - b^2)^{3/2}$ **37.** 31/3 **39.** 1

41. 2 sin 4 **43.** 4 **45.** $a = 3$ or $a = 13/3$

47. The domain is the set of all points (x, y, z) such that $4x^2 + 4y^2 + z^2 \le 4$.

SECTION 15.6, pp. 935–938

1. Center of mass is $\bar{x} = \frac{1}{2}$ and $\bar{y} = \frac{78}{55}$

3. Centroid is $\bar{x} = \frac{18}{5}$ and $\bar{y} = \frac{9}{4}$

5. $\bar{x} = \bar{y} = 4a/(3\pi)$ **7.** $I_x = I_y = \frac{81\pi}{4}$ and $I_o = \frac{81\pi}{2}$

9. $\bar{x} = -1, \bar{y} = 1/4$ **11.** $I_x = 64/105$ **13.** $\bar{x} = \frac{19}{33}$ and $\bar{y} = \frac{56}{33}$

15. $\bar{x} = 11/3, \bar{y} = 14/27, I_y = 432$

17 $\bar{x} = 0$ and $\bar{y} = \frac{724}{203}$ M.I. $= \frac{6016}{35}$

19. $\bar{x} = 0, \bar{y} = 7/10;\ I_x = 9/10\ \text{kg/m}^2,\ I_y = 3/10\ \text{kg/m}^2,\ I_0 = 6/5\ \text{kg/m}^2$

21. $I_x = \frac{M}{3}(b^2 + c^2),\ I_y = \frac{M}{3}(a^2 + c^2),\ I_z = \frac{M}{3}(a^2 + b^2)$

23. $\bar{x} = \bar{y} = 0, \bar{z} = 12/5, I_x = 7904/105 \approx 75.28, I_y = 4832/63 \approx 76.70, I_z = 256/45 \approx 5.69$

25. **(a)** $\bar{x} = \bar{y} = 0, \bar{z} = 8/3$ **(b)** $c = 2\sqrt{2}$

27. $I_L = 1386$

29. **(a)** Mass of solid $M = 81/8$

(b) Center of mass $\bar{x} = \frac{6}{5}$ and $\bar{y} = \frac{3}{5}$ and $\bar{z} = \frac{3}{5}$

31. **(a)** Mass of solid $M = 40$

(b) Center of mass $\bar{x} = \bar{y} = \bar{z} = \frac{16}{15}$, by symmetry

(c) Moment of inertia $I_x = I_y = I_z = 352/3$, by symmetry

33. 3 kg

37. **(a)** $I_{\text{c.m.}} = \frac{abc(a^2 + b^2)}{12}$

(b) $I_L = \frac{abc(a^2 + 7b^2)}{3}$

39. $\mu_x = \mu_y = 7/12$

41. $\mu_x = 3/4, \mu_y = 2/3$

43. $f(x, y) = 1/6, P(X < Y) = 2/3$

SECTION 15.7, pp. 946–950

1.

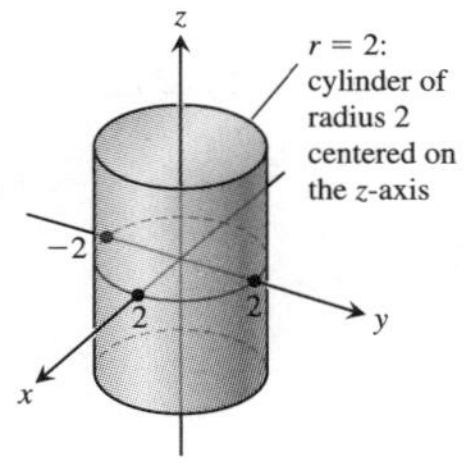

3.

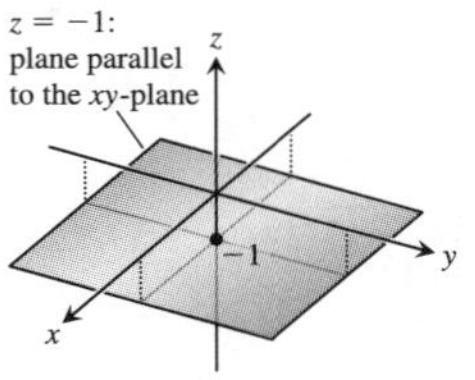

5.

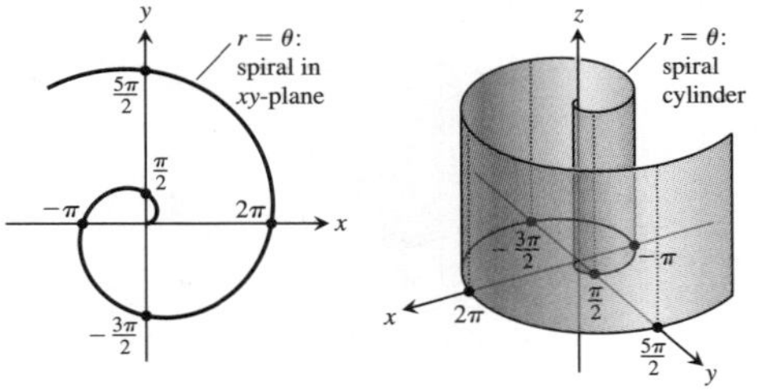

7.

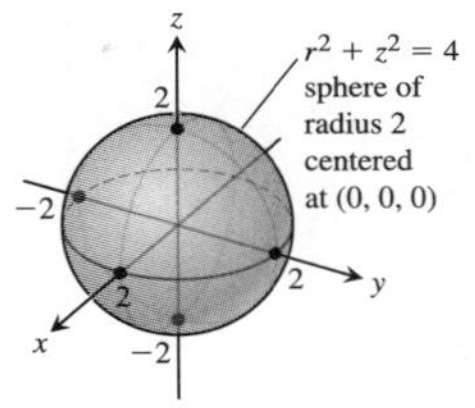

9. $r \le z \le \sqrt{9 - r^2}$: cone with vertex angle $\frac{\pi}{2}$ below a sphere of radius 3 centered at (0, 0, 0), and its interior

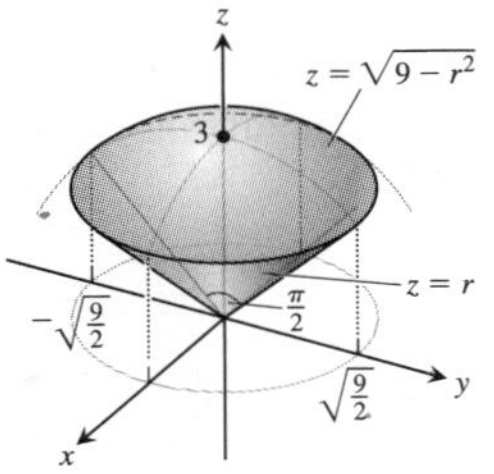

11. $0 \le r \le 4\cos\theta,\ 0 \le \theta \le \frac{\pi}{2},\ 0 \le z \le 5$: half-cylinder of height 5, radius 2, and tangent to the z-axis, and its interior

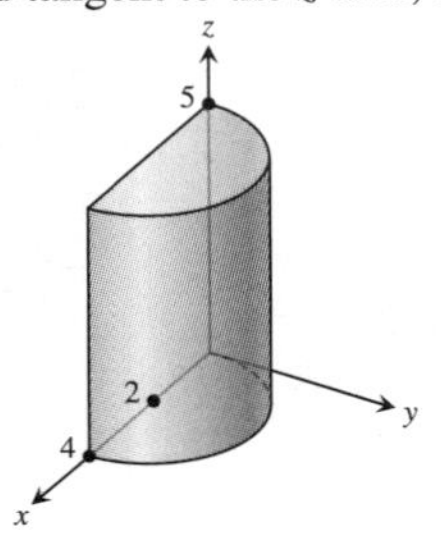

13. $\rho = 3$: sphere of radius 3 centered at $(0, 0, 0)$

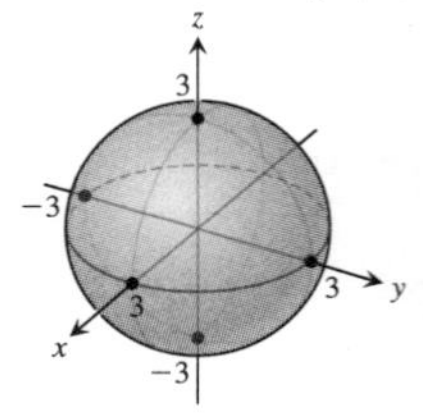

15. $\theta = \frac{2}{3}\pi$: closed half-plane along the z-axis

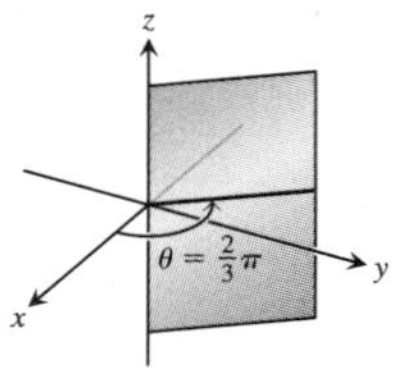

17. $\rho \cos \phi = 4$: plane with z-intercept 4 and parallel to the xy-plane

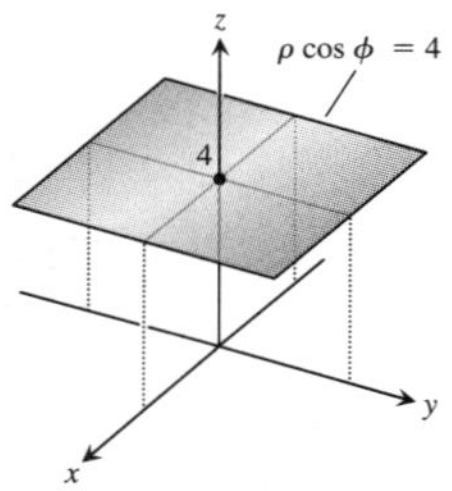

19. $0 \le \rho \le 3 \csc \phi \Rightarrow 0 \le \rho \sin \phi \le 3$: cylinder of radius 3 centered on the z-axis, and its interior

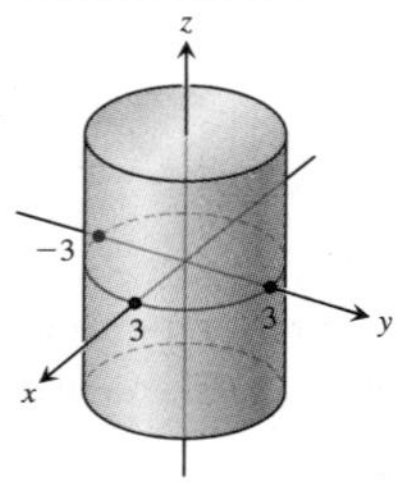

21. $0 \le \rho \cos \theta \sin \phi \le 2,\ 0 \le \rho \sin \theta \sin \phi \le 3,$ $0 \le \rho \cos \phi \le 4$: rectangular box $2 \times 3 \times 4$, and its interior

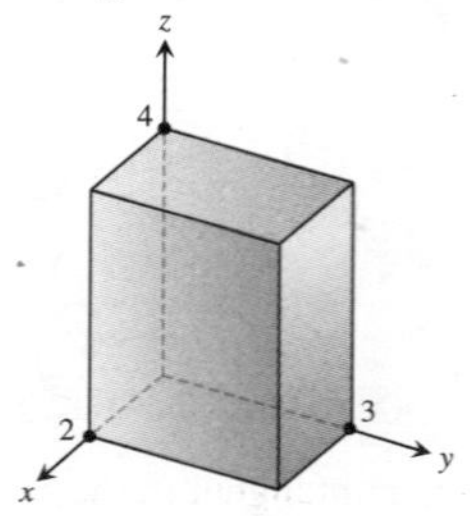

23. $\left(\sqrt{3} - \frac{2\sqrt{2}}{3}\right)\pi$ **25.** $\left(\frac{\pi^3}{3} + \frac{4\pi^5}{5}\right)$ **27.** 2π

29. $\frac{3\pi}{10}$ **31.** $\pi/3$

33. **(a)** $\int_0^{2\pi}\int_0^1\int_0^{\sqrt{4-r^2}} r\,dz\,dr\,d\theta$

(b) $\int_0^{2\pi}\int_0^{\sqrt{3}}\int_0^1 r\,dr\,dz\,d\theta + \int_0^{2\pi}\int_{\sqrt{3}}^2\int_0^{\sqrt{4-z^2}} r\,dr\,dz\,d\theta$

(c) $\int_0^1\int_0^{\sqrt{4-r^2}}\int_0^{2\pi} r\,d\theta\,dz\,dr$

35. $\int_{-\pi/2}^{\pi/2}\int_0^{\cos\theta}\int_0^{3r^2} f(r,\theta,z)\,r\,dz\,dr\,d\theta$

37. $\int_0^{\pi}\int_0^{2\sin\theta}\int_0^{4-r\sin\theta} f(r,\theta,z)\,r\,dz\,dr\,d\theta$

39. $\int_{-\pi/2}^{\pi/2}\int_1^{1+\cos\theta}\int_0^4 f(r,\theta,z)\,r\,dz\,dr\,d\theta$

41. $\int_0^{\pi/4}\int_0^{\sec\theta}\int_0^{2-r\sin\theta} f(r,\theta,z)\,r\,dz\,dr\,d\theta$ **43.** $\frac{27\pi^2}{2}$

45. $\left(\frac{1}{4} - \frac{5\pi}{64}\right)\pi$ **47.** $\frac{79\pi}{20}$ **49.** 2π **51.** $\left(\frac{8 - 5\sqrt{2}}{2}\right)\pi$

53. **(a)** $\int_0^{2\pi}\int_0^{\pi/6}\int_0^2 \rho^2 \sin\phi\,d\rho\,d\phi\,d\theta +$
$\int_0^{2\pi}\int_{\pi/6}^{\pi/2}\int_0^{\csc\phi} \rho^2 \sin\phi\,d\rho\,d\phi\,d\theta$

(b) $\int_0^{2\pi}\int_1^2\int_{\pi/6}^{\sin^{-1}(1/\rho)} \rho^2 \sin\phi\,d\phi\,d\rho\,d\theta +$
$\int_0^{2\pi}\int_0^2\int_0^{\pi/6} \rho^2 \sin\phi\,d\phi\,d\rho\,d\theta +$
$\int_0^{2\pi}\int_0^1\int_{\pi/6}^{\pi/2} \rho^2 \sin\phi\,d\phi\,d\rho\,d\theta$

55. $\int_0^{2\pi}\int_0^{\pi/2}\int_{\cos\phi}^{6} \rho^2 \sin\phi\,d\rho d\phi d\theta = \frac{863\pi}{2}$

57. $\int_0^{2\pi}\int_0^{\pi}\int_0^{1-\cos\phi} \rho^2 \sin\phi\,d\rho\,d\phi\,d\theta = \frac{8\pi}{3}$

59. $\int_0^{2\pi}\int_{\pi/4}^{\pi/2}\int_0^{2\cos\phi} \rho^2 \sin\phi\,d\rho\,d\phi\,d\theta = \frac{\pi}{3}$

61. **(a)** $8\int_0^{\pi/2}\int_0^{\pi/2}\int_0^2 \rho^2 \sin\phi\,d\rho\,d\phi\,d\theta$

(b) $8\int_0^{\pi/2}\int_0^2\int_0^{\sqrt{4-r^2}} r\,dz\,dr\,d\theta$

(c) $8\int_0^2\int_0^{\sqrt{4-x^2}}\int_0^{\sqrt{4-x^2-y^2}} dz\,dy\,dx$

63. **(a)** $\int_0^{2\pi}\int_0^{\pi/3}\int_{\sec\phi}^2 \rho^2 \sin\phi\,d\rho\,d\phi\,d\theta$

(b) $\int_0^{2\pi}\int_0^{\sqrt{3}}\int_1^{\sqrt{4-r^2}} r\,dz\,dr\,d\theta$

(c) $\int_{-\sqrt{3}}^{\sqrt{3}}\int_{-\sqrt{3-x^2}}^{\sqrt{3-x^2}}\int_1^{\sqrt{4-x^2-y^2}} dz\,dy\,dx$ **(d)** $5\pi/3$

65. $8\pi/3$ **67.** $9/4$ **69.** $\frac{3\pi - 4}{18}$ **71.** $\frac{2\pi a^3}{3}$

73. $\frac{(4\sqrt{2} - 5)\pi}{3}$ **75.** $5\pi/6$ **77.** $\frac{4(2\sqrt{2} - 1)\pi}{3}$ **79.** $\frac{16}{3}$

81. $\pi/4$ **83.** $\frac{4\pi(8 - 3\sqrt{3})}{3}$ **85.** $2/3$ **87.** $3/4$

89. $\bar{x} = 0,\ \bar{y} = 0$ and $\bar{z} = \frac{2}{5}h$ **91.** $(\bar{x}, \bar{y}, \bar{z}) = (0, 0, 3/8)$

93. $\bar{x} = \bar{y} = 0,\ \bar{z} = 5/6$ **95.** $I_x = \pi/4$ **97.** $\frac{a^4 h\pi}{10}$

99. **(a)** $(\bar{x}, \bar{y}, \bar{z}) = \left(0, 0, \frac{4}{5}\right),\ I_z = \frac{\pi}{12}$

(b) $(\bar{x}, \bar{y}, \bar{z}) = \left(0, 0, \frac{5}{6}\right),\ I_z = \frac{\pi}{14}$

101. $\frac{3M}{\pi R^3}$

103. The surface's equation $r = f(z)$ tells us that the point

$(r, \theta, z) = (f(z), \theta, z)$ will lie on the surface for all θ. In particular, $(f(z), \theta + \pi, z)$ lies on the surface whenever $(f(z), \theta, z)$ lies on the surface, so the surface is symmetric with respect to the z-axis.

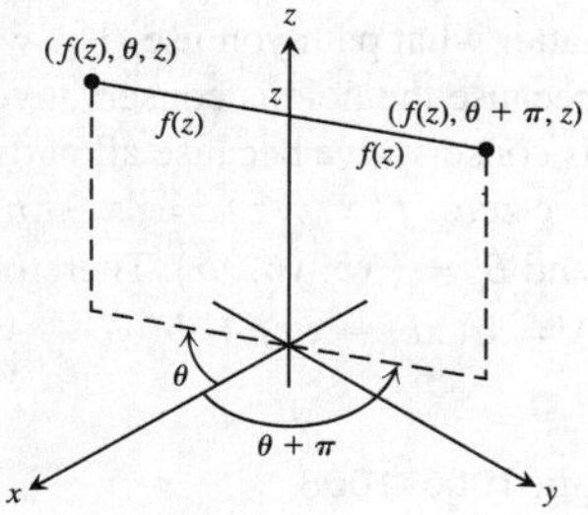

SECTION 15.8, pp. 958–959

1. (a) $x = \dfrac{u+v}{3},\ y = \dfrac{v-2u}{3};\ \dfrac{1}{3}$
(b) Triangular region with boundaries $u = 0,\ v = 0$, and $u + v = 3$

3. (a) $x = \frac{1}{5}(2u - v),\ y = \frac{1}{10}(3v - u);\ \frac{1}{10}$
(b) Triangular region with boundaries $3v = u,\ v = 2u$, and $3u + v = 10$

7. $64/5$ **9.** $6 + \dfrac{14\ln(3)}{3}$

11. $\dfrac{\pi ab(a^2+b^2)}{4}$ **13.** $\frac{1}{3}\left(1 + \frac{3}{e^2}\right) \approx 0.4687$

15. 12 **17.** 12 **19.** $\dfrac{a^2b^2c^2}{6}$

21. (a) $-v$
(b) $\begin{vmatrix} \sin v & u\cos v \\ \cos v & -u\sin v \end{vmatrix} = -u\sin^2 v - u\cos^2 v = -u$

27. $\frac{3}{2}\ln 2$

PRACTICE EXERCISES, pp. 960–962

1. $9e - 9$ **3.** $9/2$

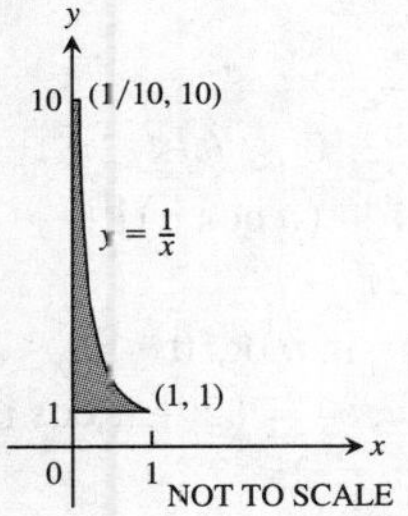

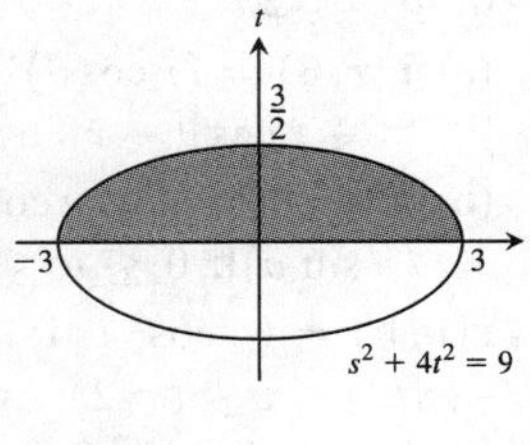

5. $\displaystyle\int_{-2}^{0}\int_{2x+4}^{4-x^2} dy\,dx = \frac{4}{3}$ **7.** $\displaystyle\int_{-3}^{3}\int_{0}^{(1/2)\sqrt{9-x^2}} y\,dy\,dx = \frac{9}{2}$

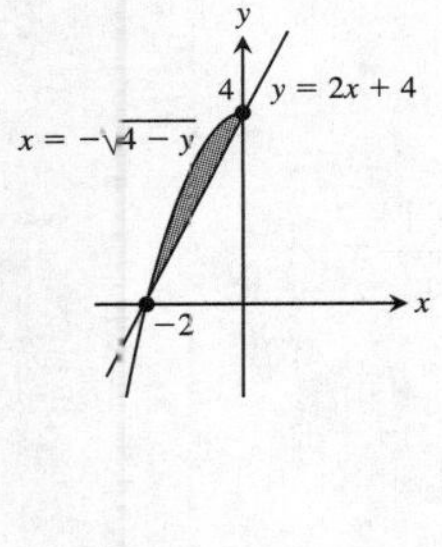

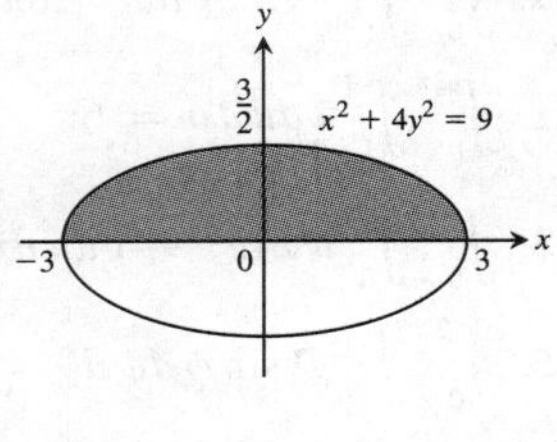

9. $\sin 4$ **11.** $\dfrac{\ln 17}{4}$ **13.** $4/3$ **15.** $4/3$ **17.** $1/4$

19. π **21.** $\dfrac{\pi - 2}{4}$ **23.** 0 **25.** $8/35$ **27.** $\pi/2$

29. $\dfrac{2(31 - 3^{5/2})}{3}$

31. (a) $\displaystyle\int_{-\sqrt{2}}^{\sqrt{2}}\int_{-\sqrt{2-y^2}}^{\sqrt{2-y^2}}\int_{\sqrt{x^2+y^2}}^{\sqrt{4-x^2-y^2}} 3\,dz\,dx\,dy$
(b) $\displaystyle\int_0^{2\pi}\int_0^{\pi/4}\int_0^{2} 3\rho^2\sin\phi\,d\rho\,d\phi\,d\theta$ **(c)** $2\pi(8 - 4\sqrt{2})$

33. $\displaystyle\int_0^{2\pi}\int_0^{\pi/4}\int_0^{\sec\phi} \rho^2\sin\phi\,d\rho\,d\phi\,d\theta = \frac{\pi}{3}$

35. $\displaystyle\int_0^1\int_{\sqrt{1-x^2}}^{\sqrt{3-x^2}}\int_1^{\sqrt{4-x^2-y^2}} z^2xy\,dz\,dy\,dx + \int_1^{\sqrt{3}}\int_0^{\sqrt{3-x^2}}\int_1^{\sqrt{4-x^2-y^2}} z^2xy\,dz\,dy\,dx$

37. (a) $\dfrac{8\pi(4\sqrt{2}-5)}{3}$ **(b)** $\dfrac{8\pi(4\sqrt{2}-5)}{3}$

39. $I_z = \dfrac{8\pi\delta(b^5 - a^5)}{15}$ **41.** $\bar{x} = \bar{y} = \dfrac{1}{2 - \ln 4}$

43. $I_0 = 104$ **45.** $I_x = 2\delta$

47. $M = 4,\ M_x = 0,\ M_y = 0$ **49.** $\bar{x} = \dfrac{3\sqrt{3}}{\pi},\ \bar{y} = 0$

51. (a) $\bar{x} = \dfrac{15\pi + 32}{6\pi + 48},\ \bar{y} = 0$
(b)

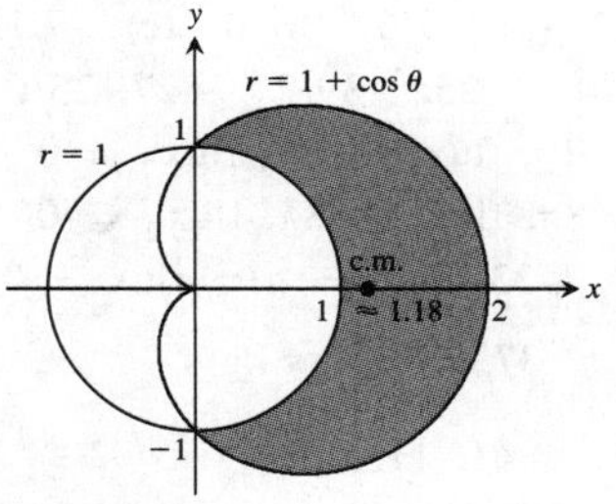

ADDITIONAL AND ADVANCED EXERCISES, pp. 962–964

1. (a) $\displaystyle\int_{-3}^{2}\int_{x}^{6-x^2} x^2\,dy\,dx$ **(b)** $\displaystyle\int_{-3}^{2}\int_{x}^{6-x^2}\int_0^{x^2} dz\,dy\,dx$
(c) $125/4$

3. 2π **5.** $3\pi/2$

7. (a) Hole radius $= 1$, sphere radius $= 2$ **(b)** $4\sqrt{3}\pi$

9. $\pi/4$ **11.** $\ln\left(\dfrac{b}{a}\right)$ **15.** $1/\sqrt[4]{3}$

17. Mass $= a^2\cos^{-1}\left(\dfrac{b}{a}\right) - b\sqrt{a^2 - b^2}$,
$I_0 = \dfrac{a^4}{2}\cos^{-1}\left(\dfrac{b}{a}\right) - \dfrac{b^3}{2}\sqrt{a^2-b^2} - \dfrac{b^3}{6}(a^2 - b^2)^{3/2}$

19. $\dfrac{1}{ab}(e^{a^2b^2} - 1)$ **21. (b)** 1 **(c)** 0

25. $h = \sqrt{20}$ in., $h = \sqrt{60}$ in.

27. $2\pi\left[\dfrac{1}{3} - \left(\dfrac{1}{3}\right)\dfrac{\sqrt{2}}{2}\right]$

Chapter 16

SECTION 16.1, pp. 970–972

1. Graph (c) **3.** Graph (g) **5.** Graph (d) **7.** Graph (f)

9. $\sqrt{2}$ **11.** -3 **13.** $3\sqrt{14}$ **15.** $\frac{1}{6}(5\sqrt{5}+9)$

17. $\frac{8}{3}$ **19. (a)** $4\sqrt{5}$ **(b)** $\frac{1}{12}(17^{3/2}-1)$

21. $\frac{15}{32}(e^{16}-e^{64})$ **23.** $\frac{1}{27}(40^{3/2}-13^{3/2})$

25. $\frac{1}{6}(5^{3/2}+7\sqrt{2}-1)$ **27.** $\frac{10\sqrt{5}-2}{3}$ **29.** 8

31. $\frac{1}{6}(17^{3/2}-1)$ **33.** $2\sqrt{2}-1$

35. (a) $4\sqrt{2}-2$ **(b)** $\sqrt{2}+\ln(1+\sqrt{2})$ **37.** $I_z = 2\pi\delta a^3$

39. (a) $I_z = 2\pi\sqrt{2}\delta$ **(b)** $I_z = 4\pi\sqrt{2}\delta$ **41.** $I_x = 2\pi - 2$

SECTION 16.2, pp. 982–985

1. $\nabla f = -(x\mathbf{i}+y\mathbf{j}+z\mathbf{k})(x^2+y^2+z^2)^{-3/2}$

3. $\nabla g = -\left(\frac{2x}{x^2+y^2}\right)\mathbf{i} - \left(\frac{2y}{x^2+y^2}\right)\mathbf{j} + e^z\mathbf{k}$

5. $\mathbf{F} = -\frac{kx}{(x^2+y^2)^{3/2}}\mathbf{i} - \frac{ky}{(x^2+y^2)^{3/2}}\mathbf{j}$, any $k > 0$

7. (a) 9/2 **(b)** 13/3 **(c)** 9/2

9. (a) 1/3 **(b)** $-1/5$ **(c)** 0

11. (a) 2 **(b)** 3/2 **(c)** 1/2

13. $-15/2$ **15.** 36 **17. (a)** $-5/6$ **(b)** 0 **(c)** $-7/12$

19. 1/2 **21.** 303 **23.** 0 **25.** $-39/2$ **27.** 25/6

29. (a) $\text{Circ}_1 = 0$, $\text{circ}_2 = 2\pi$, $\text{flux}_1 = 2\pi$, $\text{flux}_2 = 0$
(b) $\text{Circ}_1 = 0$, $\text{circ}_2 = 8\pi$, $\text{flux}_1 = 8\pi$, $\text{flux}_2 = 0$

31. Circ $= 0$, flux $= a^2\pi$ **33.** Circ $= a^2\pi$, flux $= 0$

35. (a) $-\frac{\pi}{2}$ **(b)** 0 **(c)** 1 **37.** $(.0001)\pi$ kg/s

39. (a) 32 **(b)** 32 **(c)** 32 **41.** 115.2 J

43. $5/3 - (3/2)\ln 2\ \text{m}^2/\text{s}$ **45.** 5/3 g/s

47.

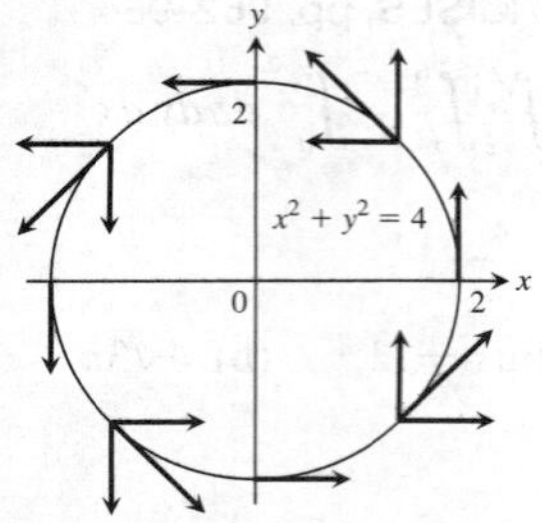

49. (a) $\mathbf{G} = -y\mathbf{i} + x\mathbf{j}$ **(b)** $\mathbf{G} = \sqrt{x^2+y^2}\mathbf{F}$

51. $\mathbf{F} = -\frac{x\mathbf{i}+y\mathbf{j}}{\sqrt{x^2+y^2}}$ **55.** 48 **57.** π **59.** 0 **61.** $\frac{1}{2}$

SECTION 16.3, pp. 994–996

1. Conservative **3.** Conservative **5.** Not conservative

7. $f(x,y,z) = x^2 + \frac{3y^2}{2} + 2z^2 + C$

9. $f(x,y,z) = xe^{y+2z} + C$

11. $f(x,y,z) = x^2y + xz^3 + c$

13. 49 **15.** 202 **17.** 1 **19.** $9\ln 2$ **21.** 0 **23.** -3

27. $\mathbf{F} = \nabla\left(\frac{x^2-1}{y}\right)$ **29. (a)** 1 **(b)** 1 **(c)** 1

31. (a) 2 **(b)** 2 **33. (a)** $c = b = 2a$ **(b)** $c = b = 2$

35. It does not matter what path you use. The work will be the same on any path because the field is conservative.

37. The force $\mathbf{F}$ is conservative because all partial derivatives of M, N, and P are zero. $f(x,y,z) = ax + by + cz + C$; $A = (xa, ya, za)$ and $B = (xb, yb, zb)$. Therefore, $\int \mathbf{F}\cdot d\mathbf{r} = f(B) - f(A) = a(xb - xa) + b(yb - ya) + c(zb - za) = \mathbf{F}\cdot\overrightarrow{AB}$.

SECTION 16.4, pp. 1006–1008

1. $2y - 1$ **3.** $ye^x - xe^y$ **5.** $\sin y - \sin x$

7. Flux $= 0$, circ $= 2\pi a^2$ **9.** Flux $= -\pi a^2$, circ $= 0$

11. Flux $= 2$, circ $= 0$ **13.** Flux $= -9$, circ $= 9$

15. Flux $= -11/60$, circ $= -7/60$

17. Flux $= 64/9$, circ $= 0$ **19.** Flux $= 1/2$, circ $= 1/2$

21. Flux $= 1/5$, circ $= -1/12$ **23.** 0 **25.** 2/33

27. 0 **29.** -16π **31.** πa^2 **33.** $3\pi/8$

35. $\frac{16}{3}$ **37.** 0

45. (a) 4π **(b)** 4π if $(0,0)$ lies inside K, 0 otherwise

SECTION 16.5, pp. 1016–1018

1. $\mathbf{r}(r,\theta) = (r\cos\theta)\mathbf{i} + (r\sin\theta)\mathbf{j} + r^2\mathbf{k}$, $0 \le r \le 2$, $0 \le \theta \le 2\pi$

3. $\mathbf{r}(r,\theta) = (r\cos\theta)\mathbf{i} + (r\sin\theta)\mathbf{j} + (r/2)\mathbf{k}$, $0 \le r \le 6$, $0 \le \theta \le \pi/2$

5. $\mathbf{r}(r,\theta) = (r\cos\theta)\mathbf{i} + (r\sin\theta)\mathbf{j} + \sqrt{9-r^2}\mathbf{k}$, $0 \le r \le 3\sqrt{2}/2$, $0 \le \theta \le 2\pi$; Also: $\mathbf{r}(\phi,\theta) = (3\sin\phi\cos\theta)\mathbf{i} + (3\sin\phi\sin\theta)\mathbf{j} + (3\cos\phi)\mathbf{k}$, $0 \le \phi \le \pi/4$, $0 \le \theta \le 2\pi$

7. $\mathbf{r}(\phi,\theta) = (\sqrt{3}\sin\phi\cos\theta)\mathbf{i} + (\sqrt{3}\sin\phi\sin\theta)\mathbf{j} + (\sqrt{3}\cos\phi)\mathbf{k}$, $\pi/3 \le \phi \le 2\pi/3$, $0 \le \theta \le 2\pi$

9. $\mathbf{r}(x,y) = x\mathbf{i} + y\mathbf{j} + (4-y^2)\mathbf{k}$, $0 \le x \le 2$, $-2 \le y \le 2$

11. $\mathbf{r}(u,v) = u\mathbf{i} + (3\cos v)\mathbf{j} + (3\sin v)\mathbf{k}$, $0 \le u \le 3$, $0 \le v \le 2\pi$

13. (a) $\mathbf{r}(r,\theta) = (r\cos\theta)\mathbf{i} + (r\sin\theta)\mathbf{j} + (1 - r\cos\theta - r\sin\theta)\mathbf{k}$, $0 \le r \le 3$, $0 \le \theta \le 2\pi$
(b) $\mathbf{r}(u,v) = (1 - u\cos v - u\sin v)\mathbf{i} + (u\cos v)\mathbf{j} + (u\sin v)\mathbf{k}$, $0 \le u \le 3$, $0 \le v \le 2\pi$

15. $\mathbf{r}(u,v) = (4\cos^2 v)\mathbf{i} + u\mathbf{j} + (4\cos v\sin v)\mathbf{k}$, $0 \le u \le 3$, $-(\pi/2) \le v \le (\pi/2)$; Another way: $\mathbf{r}(u,v) = (2 + 2\cos v)\mathbf{i} + u\mathbf{j} + (2\sin v)\mathbf{k}$, $0 \le u \le 3$, $0 \le v \le 2\pi$

17. $\int_0^{2\pi}\int_0^1 \frac{\sqrt{5}}{2} r\,dr\,d\theta = \frac{\pi\sqrt{5}}{2}$

19. $\int_0^{2\pi}\int_1^3 r\sqrt{5}\,dr\,d\theta = 8\pi\sqrt{5}$

21. $\int_0^{2\pi}\int_1^4 1\,du\,dv = 6\pi$

23. $\int_0^{2\pi}\int_0^1 u\sqrt{4u^2+1}\,du\,dv = \frac{(5\sqrt{5}-1)}{6}\pi$

25. $\int_0^{2\pi}\int_{\pi/4}^{\pi} 2\sin\phi\,d\phi\,d\theta = (4 + 2\sqrt{2})\pi$

27.

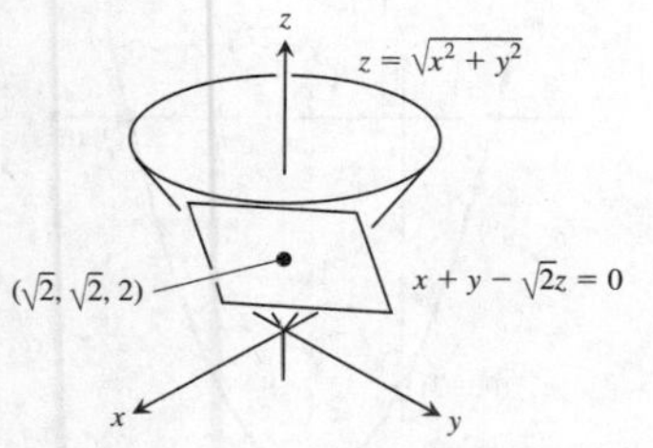

29.

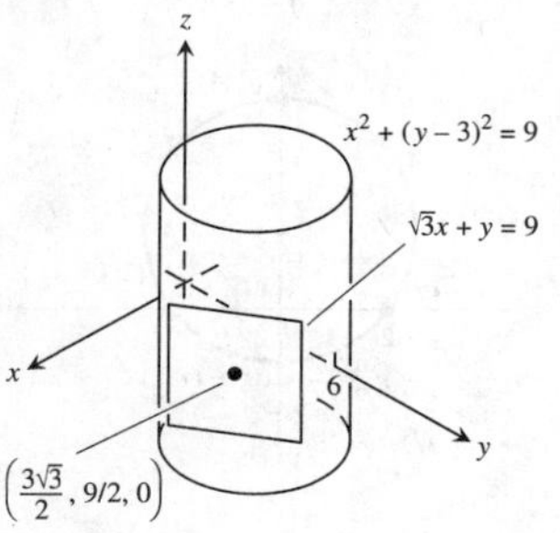

33. **(b)** $A = \int_0^{2\pi}\int_0^{\pi} [a^2b^2 \sin^2 \phi \cos^2 \phi + b^2c^2 \sin^4 \phi \cos^2 \theta + a^2c^2 \sin^4 \phi \sin^2 \theta]^{1/2}\, d\phi\, d\theta$

35. $x_0x + y_0y = 25$ **37.** $4\pi a^2$ **39.** 4

41. $6\sqrt{6} - 2\sqrt{2}$ **43.** $\pi\sqrt{c^2 + 1}$

45. $\frac{\sqrt{3}}{2}a^2$ **47.** $3 + 2\ln 2$

49. $\frac{\pi}{6}(13\sqrt{13} - 1)$ **51.** $5\pi\sqrt{2}$ **53.** $\frac{2}{3}(5\sqrt{5} - 1)$

SECTION 16.6, pp. 1026–1028

1. $\iint_S x\, d\sigma = \int_0^3\int_0^2 u\sqrt{4u^2 + 1}\, du\, dv = \frac{17\sqrt{17} - 1}{4}$

3. $\iint_S x^2\, d\sigma = \int_0^{2\pi}\int_0^{\pi} \sin^3 \phi \cos^2 \theta\, d\phi\, d\theta = \frac{4\pi}{3}$

5. $\iint_S z\, d\sigma = \int_0^1\int_0^1 (4 - u - v)\sqrt{3}\, dv\, du = 3\sqrt{3}$ (for $x = u$, $y = v$)

7. $\iint_S x^2\sqrt{5 - 4z}\, d\sigma = \int_0^1\int_0^{2\pi} u^2 \cos^2 v \cdot \sqrt{4u^2 + 1} \cdot u\sqrt{4u^2 + 1}\, dv\, du = \int_0^1\int_0^{2\pi} u^3(4u^2 + 1)\cos^2 v\, dv\, du = \frac{11\pi}{12}$

9. $9a^3$ **11.** $\frac{abc}{4}(ab + ac + bc)$ **13.** 2

15. $\frac{1}{30}(\sqrt{2} + 6\sqrt{6})$ **17.** $\sqrt{6}/30$ **19.** -32 **21.** $\frac{\pi a^3}{6}$

23. $13a^4/6$ **25.** $2\pi/3$ **27.** $-73\pi/6$ **29.** 18

31. $\frac{\pi a^3}{6}$ **33.** $\frac{\pi a^2}{4}$ **35.** $\frac{\pi a^3}{2}$ **37.** -32

39. -4 **41.** $3a^4$ **43.** $\left(\frac{a}{2}, \frac{a}{2}, \frac{a}{2}\right)$

45. $(\bar{x}, \bar{y}, \bar{z}) = \left(0, 0, \frac{14}{9}\right)$, $I_z = \frac{15\pi\sqrt{2}}{2}\delta$

47. $\frac{8\pi}{3}a^4\delta$ **49.** 70/3 mg

SECTION 16.7, pp. 1039–1041

1. $-\mathbf{i} - 4\mathbf{j} + \mathbf{k}$ **3.** $(1 - y)\mathbf{i} + (1 - z)\mathbf{j} + (1 - x)\mathbf{k}$

5. $x(z^2 - y^2)\mathbf{i} + y(x^2 - z^2)\mathbf{j} + z(y^2 - x^2)\mathbf{k}$ **7.** 4π

9. $-5/6$ **11.** 0 **13.** -6π **15.** $2\pi a^2$

17. $-\pi$ **19.** 12π **21.** $-\pi/4$ **23.** -15π

25. -8π **33.** $16I_y + 16I_x$

SECTION 16.8, pp. 1051–1053

1. 0 **3.** $(y^2z + xz^2 + x^2y)e^{xyz}$ **5.** 0 **7.** 0

9. -16 **11.** -8π **13.** $3/2$ **15.** $-40/3$

17. 12π **19.** $12\pi(4\sqrt{2} - 1)$ **23.** No

25. The integral's value never exceeds the surface area of *S*.

27. 184/35

PRACTICE EXERCISES, pp. 1054–1056

1. Path 1: $2\sqrt{3}$; path 2: $1 + 3\sqrt{2}$ **3.** $4a^2$ **5.** 0

7. $8\pi \sin(1)$ **9.** 0 **11.** $\pi\sqrt{3}$

13. $2\pi\left(1 - \frac{1}{\sqrt{2}}\right)$ **15.** $\frac{abc}{2}\sqrt{\frac{1}{a^2} + \frac{1}{b^2} + \frac{1}{c^2}}$ **17.** 50

19. $\mathbf{r}(\phi, \theta) = (6 \sin \phi \cos \theta)\mathbf{i} + (6 \sin \phi \sin \theta)\mathbf{j} + (6 \cos \phi)\mathbf{k}$, $\frac{\pi}{6} \le \phi \le \frac{2\pi}{3}$, $0 \le \theta \le 2\pi$

21. $\mathbf{r}(r, \theta) = (r \cos \theta)\mathbf{i} + (r \sin \theta)\mathbf{j} + (1 + r)\mathbf{k}$, $0 \le r \le 2$, $0 \le \theta \le 2\pi$

23. $\mathbf{r}(u, v) = (u \cos v)\mathbf{i} + 2u^2\mathbf{j} + (u \sin v)\mathbf{k}$, $0 \le u \le 1$, $0 \le v \le \pi$

25. $\sqrt{6}$ **27.** $\pi[\sqrt{2} + \ln(1 + \sqrt{2})]$ **29.** Conservative

31. Not conservative **33.** $f(x, y, z) = y^2 + yz + 2x + z$

35. Path 1: 2; path 2: 8/3 **37.** **(a)** $1 - e^{-2\pi}$ **(b)** $1 - e^{-2\pi}$

39. 0 **41.** **(a)** $4\sqrt{2} - 2$ **(b)** $\sqrt{2} + \ln(1 + \sqrt{2})$

43. $(\bar{x}, \bar{y}, \bar{z}) = \left(1, \frac{16}{15}, \frac{2}{3}\right)$; $I_x = \frac{232}{45}$, $I_y = \frac{64}{15}$, $I_z = \frac{56}{9}$

45. $\bar{z} = \frac{3}{2}$, $I_z = \frac{7\sqrt{3}}{3}$ **47.** $(\bar{x}, \bar{y}, \bar{z}) = (0, 0, 49/12)$, $I_z = 640\pi$

49. Flux: 3/2; circ: $-1/2$ **53.** 3

55. $\frac{2\pi}{3}(7 - 8\sqrt{2})$ **57.** 0 **59.** π

ADDITIONAL AND ADVANCED EXERCISES, pp. 1056–1058

1. 6π **3.** 2/3

5. **(a)** $\mathbf{F}(x, y, z) = z\mathbf{i} + x\mathbf{j} + y\mathbf{k}$
(b) $\mathbf{F}(x, y, z) = z\mathbf{i} + y\mathbf{k}$
(c) $\mathbf{F}(x, y, z) = z\mathbf{i}$

7. $\frac{16\pi R^3}{3}$ **9.** $a = 2$, $b = 1$. The minimum flux is -4.

11. **(b)** $\frac{16}{3}g$ **(c)** Work $= \left(\int_C gxy\, ds\right)\bar{y} = g\int_C xy^2\, ds = \frac{16}{3}g$

13. **(c)** $\frac{4}{3}\pi w$ **19.** False if $\mathbf{F} = y\mathbf{i} + x\mathbf{j}$

Appendices

APPENDIX A.1, p. AP-6

1. $0.\bar{1}, 0.\bar{2}, 0.\bar{3}, 0.\bar{8}, 0.\bar{9}$ or 1

3. $x < -2$

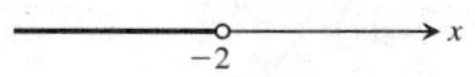

5. $x \le -\frac{1}{3}$

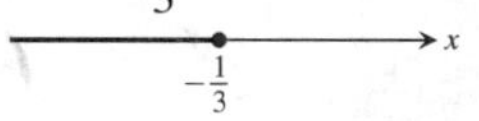

7. $3, -3$ **9.** $7/6, 25/6$

11. $[-2, 4]$

13. $[0, 10]$

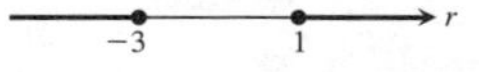

15. $(-\infty, -2] \cup [2, \infty)$

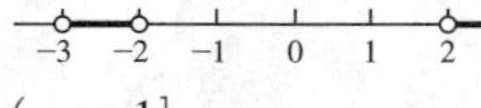

17. $(-\infty, -3] \cup [1, \infty)$

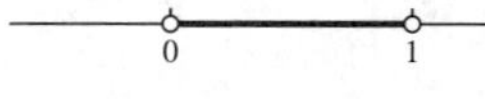

19. $(-3, -2) \cup (2, 3)$

21. $(0, 1)$

23. $(-\infty, 1]$

27. The graph of $|x| + |y| \le 1$ is the interior and boundary of the "diamond-shaped" region.

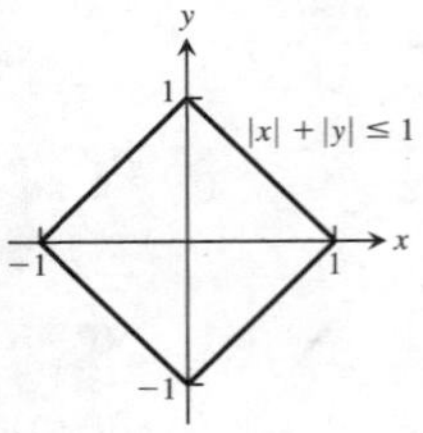

APPENDIX A.3, pp. AP-17–AP-18

1. $2, -4; 2\sqrt{5}$ **3.** Unit circle

5. $m_\perp = -\dfrac{1}{3}$

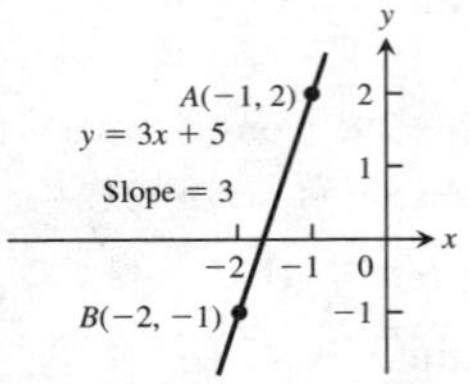

7. (a) $x = -1$ **(b)** $y = 4/3$ **9.** $y = -x$

11. $y = -\dfrac{5}{4}x + 6$ **13.** $y = 4x + 4$ **15.** $y = -\dfrac{x}{2} + 12$

17. x-intercept $= \sqrt{3}$, y-intercept $= -\sqrt{2}$

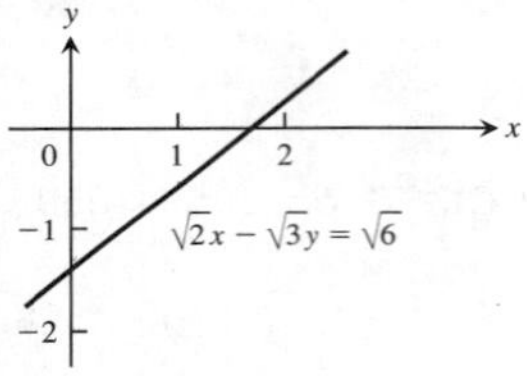

19. $(3, -3)$

21. $x^2 + (y - 2)^2 = 4$

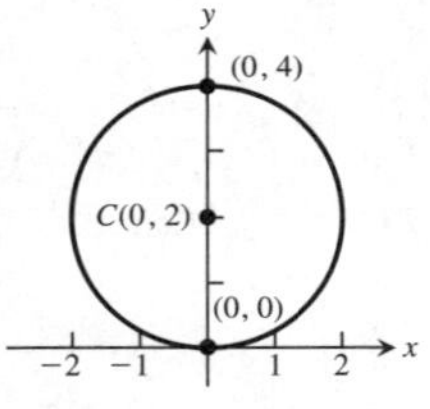

23. $\left(x + \sqrt{3}\right)^2 + (y + 2)^2 = 4$

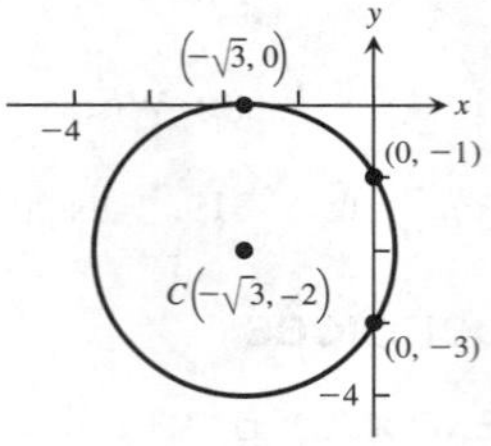

25. $x^2 + (y - 3/2)^2 = 25/4$

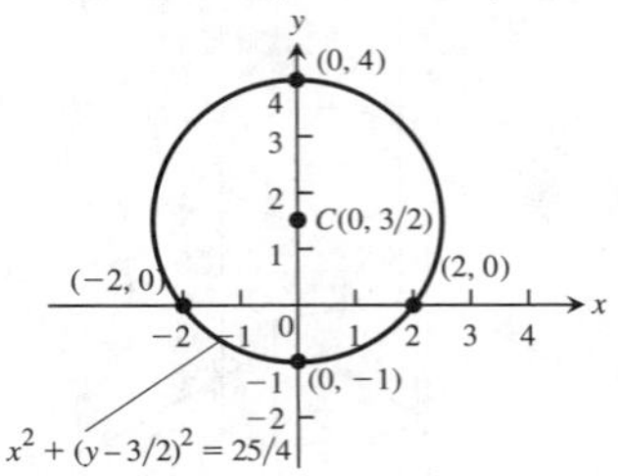

27.

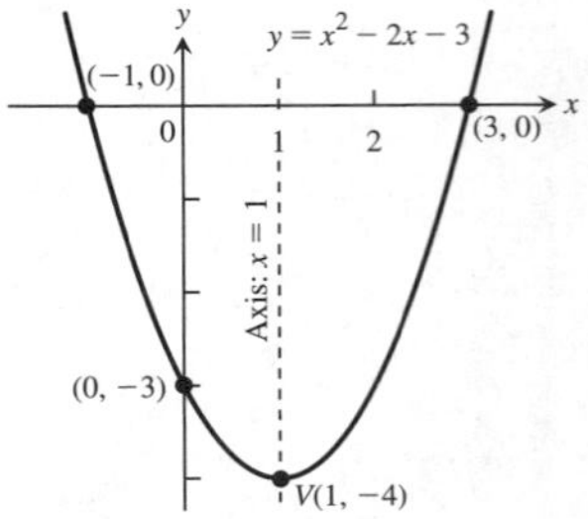

29.

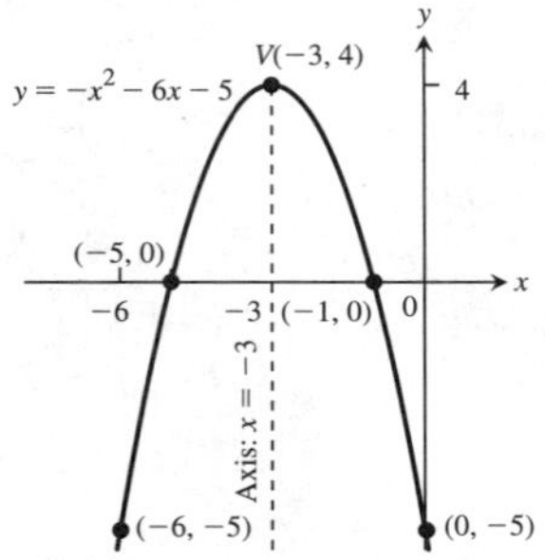

31. Exterior points of a circle of radius $\sqrt{7}$, centered at the origin

33. The washer between the circles $x^2 + y^2 = 1$ and $x^2 + y^2 = 4$ (points with distance from the origin between 1 and 2)

35. $(x + 2)^2 + (y - 1)^2 < 6$

37. $\left(\dfrac{1}{\sqrt{5}}, \dfrac{2}{\sqrt{5}}\right), \left(-\dfrac{1}{\sqrt{5}}, -\dfrac{2}{\sqrt{5}}\right)$

39. $\left(-\dfrac{1}{\sqrt{3}}, -\dfrac{1}{3}\right), \left(\dfrac{1}{\sqrt{3}}, -\dfrac{1}{3}\right)$

41. (a) ≈ -2.5 degrees/inch **(b)** ≈ -16.1 degrees/inch **(c)** ≈ -8.3 degrees/inch

43. 5.97 atm

45. Yes: $C = F = -40°$

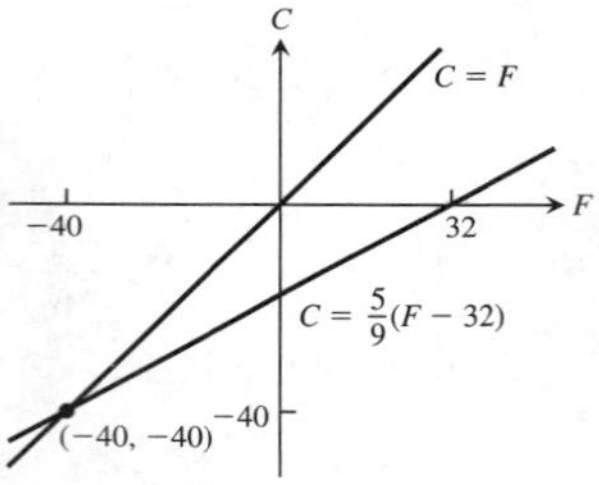

51. $k = -8$, $k = 1/2$

Applications Index

Note:

- Page numbers with the prefix "17" are in the online Chapter 17 Second-Order Differential Equations (bit.ly/3aHAZo5)
- Page numbers with the prefix "18" are in the online Chapter 18 Complex Functions (bit.ly/3aHAZo5)
- Page numbers with the prefix "19" are in the online Chapter 19 Fourier Series and Wavelets (bit.ly/3aHAZo5)
- Page numbers with the prefix "AP" are in Appendix A
- Page numbers with the prefix "B" are in the online Appendix B (bit.ly/3aHAZo5).

Automotive

Car accidents, 526
Deceleration to stop moving vehicle, 238, 244
Fuel efficiency, 530
Gasoline consumption, 45
Tire life, 527
Vehicular stopping distance, 132–133

Business and Economics

Chip lifetime, 519, 520–522
Cobb-Douglas production function, 877
Cost, 223–224
Customer service time, 526
Depreciation of 18-wheel truck, 55
Gross national product growth, 112
Industrial costs, 13
Inventory management, 124, 223
Inventory replenishment, 223
Manufacturing time, 527
Marginal costs, 129–130, 132, 216–218, 286
Marginal profit, 216–218
Marginal rates and taxes, 130
Marginal revenue, 132, 177, 216–218, 286
Maximizing utility, 878
Maximum profit
  for backpack sales, 223
  for manufacturing tires, 242
  for tour service rates, 223
Order size, 859
Production growth, 177
Production levels, 223
Profit rate of change, 44
Quality control, 527
Revenue as function of price increase, 13

Construction

Cost
  to build parking lot, 530
  for isosceles right triangle shaped fence, 13
Erecting radio telescope, 878
Industrial costs, 13
Water main, 732–733

Economics and Finance

Continuous price discounting, 410
Costs associated with rainwater catch tank, 220
Marginal cost, 129–130, 132
Marginal revenue, 132
Single-family home prices, 113

Engineering and Physical Sciences

Acceleration, 131, 145, 177, 222, 238, 244–245, 444
Aircraft flight distance, 158
Airplane flight off course, 424
Airplane ground speed and direction, 721
Airplane landing path, 223
Airplane takeoff, 133
Arc length, 444, 475, 493
Area, 157, 444, 475, 480, 917
  change with respect to diameter of circle, 124–125
  change with respect to radius of circle, 171, 175
  changing, of a disk, 105
  of circular washer, 911
  between curves, 302
  of surfaces of revolution, 363
Atmospheric pressure and altitude, 409
Average temperature, 145, 722, 911
Average values, 917
Balloon/bicycle distance increase, 158
Beam
  motion, 160
  stiffness, 222
  strength, 222
  under a load, 933
Boring a cylinder, 158
Bouncing ball, 589
Boyle's Law, 13
Building's shadow, 159
Buoyant force, 564
Carbon dating, 411
Cardiac output, 159
Center of gravity, 475
Center of mass, 929–930
  fluid forces and, 357
  of slender metal arch, 968–969
  of thin, flat plate, 353–354, 455
  for thin plate bounded by curves, 356
  of thin plate with constant density, 359–360, 363
  of thin plate with varying density, 360
  of thin shell, 1025–1026
  of thin wire, 351, 356–357, 359
Centroid, 493, 930, 941–942
  of a semicircular region of radius, 358
  of thin plates, 361, 363
  of a triangle, 357, 360
Chemical mixture problem, 552–554, 571
Chemical reaction, 409, 488
Circulation
  around bounding circle, 1031–1032
  density, 998–999
  of paddle wheel, 1035
Circumference of a circle, 334
Coffee draining from conical filter, 158–159
Conformal maps of a square, 18-33–18-34
Conical sand pile, 158
Constructing a cone, 221–222
Constructing a cylinder, 220
Cooling, 407–408, 410, 454
Cylinder pressure, 124
Damped motion, 17-17–17-18
Deceleration to stop moving vehicle, 238, 244
Density of thin wires, 360
Designing
  a box with lid, 219
  a can, 219
  a gondola for hot air balloon, 177
  a plumb bob, 321
  a suitcase, 219
  a tank, 219
  a wok, 321
Diameter of a tree, 171
Dimensions
  of fence, 218
  of silo, 220
Distance
  traveled, 502
  between two ships, 222–223
Draining
  hemispherical reservoir, 158
  a swamp, 501
  a tank, 99, 132, 158, 175
Drilling boreholes, 893
Edges of a cube, 175
Electrical resistance, 859
Electrical resistors, manufacturing, 63
Electric circuit, 17-18–17-19, 157, 175, 547, 548, 564, 890
Electric force, 348
Electricity, 157
Electromagnetic theory, 1048, 1057
Elevation angle of hot air balloon, 34, 153–154
Erecting radio telescope, 878
Error
  compounding, 176
  controlling, 176
Escape velocity, 571
Falling object, 104, 105, 131, 146, 238, 239
  resistance, 563
  skydiving, 563
Fermat's principle in optics, 223
Filling a conical tank with water, 152–153
Fluid force
  center of mass and, 357
  on a curved surface, 1057
  on end plates of a trough, 349
  out of hole in water tank, 243
  on parabolic gate, 349
  on rectangular plate, 348, 365
  of seawater, 349
  on semicircular plate, 348, 349
  on square plate in different positions, 349
  streamlines, 1008
  in a swimming pool, 345
  on triangular plate, 348, 349, 365
  on viewing portion of fish tank, 349
Flux, 981, 1004, 1023, 1024, 1043, 1047

Flying a kite, 158
Force
  of attraction, 346
  below Earth's surface, 363
  due to gravity pulling down, 17-15–17-16
  of wind passing over boat's sail, 734
Free-falling object, 44, 127, 131, 255
  in the fourteenth century, 244
  on the moon, 194
  near the surface of a planet, 239
Glider position, 770–771, 778
Gravitational fields, 987
Grinding engine cylinders, 63
Haar coefficients, 19-27–19-29
Hanging cables, 444
Harmonic motion, 17-16–17-17
Hauling in a dinghy, 158
Heat capacity of a gas, 529–530
Heat diffusion equation, 1052
Heat equation, 831, 893
Heating a plate, 157
Heat transfer (cooling), 407–408, 410
Height
  of a baseball, 774, 776
  of a building, 171
  of a bullet shot straight up, 131
  of a lamppost, 176
  of a palm tree, 286
  of a paper clip shot into the air, 177
  of a pole, 35
  of a rocket, 572
  of a weather balloon, 35
Highway patrol plane, 160
Impedance in a series circuit, 175
Inertia, moments of, 930–932
Inflating balloon, 133
Inflating hot air balloon, 153–154
Inner product, 19-15–19-16
Intensity of illumination, 222
Kinetic energy
  and potential energy, 13
  and work, 347, 365
Length
  of airplane gasoline tank, 502–503
  of an asteroid, 334
  of an object, 99
  of a beam, 878
  of curves, 330–331, 333–334, 363
  of a graph, 331
  of a ladder carried around a corner, 242
  of a line segment, 334
  of a rocket, 99
Lighthouse beam, 160
Linearization, 172, 176
Locating a planet, 228
Mass, 928–930
Maximum volume
  of a box, 214, 242
  of a can, 215, 856
Measuring cup stripes, 99
Melting ice layer, 160
Minimizing perimeter, 218
Motion
  of an object along a line, 211–212
  of a body along a line, 125
  coasting battleship, 554
  coasting bicycle, 554
  with constant acceleration, 239
  damped, 17-17–17-18
  of a helicopter, 742–743
  of a lighthouse beam, 160
  of a model rocket, 131–132
  of a particle, 130, 159, 175, 177, 222, 238, 242
  in the plane, 159
  of a projectile, 130, 222, 255, 775–777
  of a vehicle, 159
Moving shadow, 159
Norm of a function, 19-13–19-14
Ohm's law for electrical circuits, 63
Oil consumption rate, 503
Oil slick spread, 160
Orthogonal trajectory of curves, 551–552
Osculating circle, 177
Parametrization
  of a cone, 1009
  of a cylinder, 1009
  of a sphere, 1009
Particle motion, 130, 145, 159, 238, 242
Path
  of a ball, 773–774, 776
  of a baseball player, 160
  of a football, 243
  of a glider, 770–771, 778
  of a heat-seeking particle, 892
  of a javelin, 798
  of a projectile, 772–773, 775–776
  of a shot-put throw, 776
  of a water skier, 480
Periodic functions, 19-1–19-2
Planetary orbits, 704
Pressure of fluid at bottom of dam, 344
Pressure waves produced by tuning fork, 4
Projectile motion, 130, 255, 733, 775–777, 798
Pumping oil to the rim of the tank, 343
Quadratic approximations, 171–172
Quickest route, 221
Radioactive decay rate, 405–407, 410
Radius
  increase of raindrop, 158
  of inflating balloon, 158
Reservoir water, 240
Resistors connected in parallel, 824
Resultant force, 721–722
Rolle's Theorem, 194
Rotating spool speed, 175
Satellite orbit, 348
Searchlight beam motion, 176
Shortest beam, 222
Skaters stopping distances, 550, 554–555, 571
Skylab 4 view, 799
Sliding ladder, 157–158
Sonobuoy submarine location, 229
Specific heat of a gas, 307
Speed
  of aircraft approaching island, 155–156
  of a frictionless cart, 222
  of marathoner, 194
  of object dropped from tower, 105
  during parachute jump, 177
  of particle, 175, 177
  of a rocket, 105
  of rotating spool, 175
  of ships moving apart, 160
  underwater transmission cable, 242
  of a vehicle, 44, 154, 194
  velocity and, 125–126
  of warship, 194
Speed of light, 215–216
Spring's force constant, 346
Standard unit vectors, 19-7, 19-11
Stopping distance, 238
Stream flow, 845
Stretching a rubber band, 102, 346
Submarine intercept, 757
Surface area, 493
  of an asteroid, 340
  of an oil spill, 160
  of a cone, 175, 176, 339, 1011
  cut from bottom of paraboloid, 1014–1015
  of an infinite paint can or Gabriel's horn, 514
  of a melting ice cube, 179
  of a right circular cylinder, 175
  of sliced bread, 340
  of a sphere, 157, 1012
  of a torus, 359, 1017
  of U.S. National Weather Service dome, 339
  of a wok, 339
Suspension bridge cables, 697
Temperature
  annual mean air temperature, 307
  below Earth's surface, 807
  in Fairbanks, Alaska, 145, 307
  greatest, 974
  mean, 529
  and period of a pendulum, 145
  rate of change, 113, 194
  of a room, 286
Thermal expansion in laboratory equipment, 99
Tin pest, 223
Tolerance, 171
Torque, 737–738, 739, 757
Underwater transmission cable speed, 454
Vectors, 19-7, 19-11
Vehicular stopping distance, 132–133
Velocity
  after ricochet, 893
  average and instantaneous, 178
  displacement from an antiderivative of, 238
  of falling meteorite, 145
  fields, fluid, 979–980, 1001
  of moving object, 167
  of a particle, 178
  of a particle moving along a horizontal line, 125–128, 131
  in a resisting medium, 549–550
  of a rocket, 131–132
  of a skydiver, 444
  of two particles, 177
Velocity and acceleration
  of a particle, 177–178, 294
  of a piston, 145
  of weight on spring, 135–136, 138
Vertical motion, 221
Vertical spring with weight, 135–136
Vibrating springs, 17-15–17-16
Videotaping a moving car, 159
Volcanic lava fountains, 133
Voltage changes, 157, 409, 839–840
Volume, 321, 444
  of an ice cream cone, 944–945
  of a ball, 105
  of a balloon, 19
  of a bowl, 321
  of a box, 13
  of a Bundt cake, 329
  of a football, 362
  of gasoline tanks, 493–494
  of a hemisphere, 321
  of material in a cylindrical shell, 171
  of noncircular right cylinder, 917
  of a pipeline, 890
  of a pyramid, 319
  of a right circular cone, 157
  of right prism, 901
  of a solid, 319–322, 327–329
  of a solid ellipsoid, 959
  of a solid region, 915
  of a solid sphere hole, 362
  of storage tank for molasses, 856
  of a swamp, 501
  of a tetrahedron, 923
  of a torus (doughnut), 321, 358
  of a trough, 220
  using cylindrical shells, 959
  of water in swimming pool, 502
  of a wedge, 364
  of a wedgelike solid beneath surface, 904–905
Water flow, 99
Wave equations, 830
Weight on a pulley, 156
Wind chill factor, 858
Work
  along different paths, 995
  to carry water up mountain in a leaky tank, 363
  to compress a bathroom scale, 346
  done by a force field, 978–979
  done by a jack, 341
  done by a locomotive, 734

done by the conservative field, 989
done in lifting a bucket and rope, 342–343
to fill a water tower, 348
to haul up a rope, 346, 363
and kinetic energy, 347, 365
lifting a bag of sand, 346
lifting an elevator cable, 346
lifting water in a leaking bucket, 346
to move electrons, 348
to pump a reservoir, 363
to pump a tank, 363
to pump liquids from containers, 346–347
to pump oil to the rim of a tank, 343
to put a satellite in orbit, 348
required to compress a spring, 342, 346
to stretch a spring, 346, 363
to suck milkshake through a straw, 347–348

Environment

Air pollution, 256
Annual rainfall, 527
Carbon monoxide pollution, 555
Water pollution, 256

General

ACT test results, 524–525
Airport waiting time, 526–527
Box dimensions, 13, 840, 866
Cars seen by a hitchhiker, 517–518
Chip lifetime, 519, 520–522
Clock's moving hands, 160
Coin toss, 515–517, 527–528
Computer graphics
hidden lines, 749
perspective, 748
Designing a poster, 219
Dice roll, 528
Endangered species, 410
Helicopter rescue, 757
Leibniz's Rule, 178–179, 309
Linearization, 172, 176
Mixtures, 552–554, 555, 571
Oil depletion, 410
Opinion polls, 528
Paper folding, 220
Period of a clock pendulum, 171, 179
Printer lifetime, 527
Random number generator, 934–935
Rolle's Theorem, 194
Spacecraft component failures, 528
Spinning arrow direction, 523
Sugar, inversion of, 409
Swing of a clock pendulum, 663
Taxicab miles driven, 527
Urban gardening, 456
U.S. Postal Service box size, 220
Well depth, 168
Window proportions, 220
Working underwater, 409

Life Sciences

Antihistamine effects, 504
Blood flow, 171, 455
Blood pressure, 523–524, 527, 595
Blood types, 878
Body surface area, 132
Brain weights, 527
Cardiac output, 159, 890
Cattle weight, 519–520
Cholesterol levels in children, 527
Digestion time, 526
Diseases, eliminating, 410
Diseases dying out, 404–405
Drug assimilation, 504
Drug concentrations, 171, 410
Drug dosages, 658
Drug effectiveness, 105
Flow velocity of a cough, 224
Germination rate of sunflower seeds, 527
Heart, effect of flight maneuvers on, 171
Height of individuals, 527
Human evolution and tooth size, 409
Hybridization, 130
Length of pregnancy, 527
Life expectancy, 526, 527
Medicine dosages, body's reaction to, 123
Pollinating flowers, 526
Population
density of bacteria, 911
regional, 911
Population growth
of bacteria, 132, 409
of fruit flies, 112–113
of rabbits and foxes, 569–570
of trout vs bass, 565–566, 568
in U.S., 410
of world, 550–551
of yeast cells, 105
of yeast culture, 402–403
Sensitivity to medicine, 224
Serum albumin levels, 527
Transport through cell membrane, 572

Social and Behavioral Sciences

Information diffusion, 488, 563
Population, 550–551, 563
World population, 550–551

Subject Index

Note:
- Page numbers with the prefix "17" are in the online Chapter 17 Second-Order Differential Equations (bit.ly/3aHAZo5)
- Page numbers with the prefix "18" are in the online Chapter 18 Complex Functions (bit.ly/3aHAZo5)
- Page numbers with the prefix "19" are in the online Chapter 19 Fourier Series and Wavelets (bit.ly/3aHAZo5)
- Page numbers with the prefix "AP" are in Appendix A
- Page numbers with the prefix "B" are in the online Appendix B (bit.ly/3aHAZo5).

Abscissa (x-coordinate), AP-10
Absolute change, 168
Absolute convergence, 607–612
  Absolute Convergence Test, 608
  definition of, 608
  Ratio Test, 609–610
  Root Test, 611–612
Absolute maxima or minima extrema, 182–185. *See also* Extreme values
  absolute (global) maxima or minima, 180–185
  boundary points of, 865–866
  on closed bounded regions, 864–867
  of continuous functions, 864–867
  critical points and, 184
  definitions of, 860
  Extreme Value Theorem and, 181–182
  finding, 183–185
  first derivative theorem for local extreme values, 183–184
  interior points of, 865–866
  local (relative) maxima or minima, 182–185
Absolute maximum, defined, 860–861
Absolute minimum, defined, 860–861
Absolute value
  of complex numbers, 18-5
  definition of, AP-4–AP-5, 18-5
  function, derivative of, 143
  properties, AP-4
Abstract algebra, AP-24
Acceleration
  definition of, 126
  formula for calculating the normal component of, 789
  in free fall, 127
  jerk associated with, 126–127, 136
  normal component of, 788–789
  in polar coordinates, 794–797
  in simple harmonic motion, 135–136
  tangential and normal components of, 788–789
  torsion and, 789–792
  vector, 764, 788
  velocity in, 126–128
Addition
  of complex numbers, 18-3–18-4
  formulas, 25
  formulas in trigonometric functions, 25
  of functions, 14–15
  parallelogram law of, 717–718, 727
  property of real numbers, AP-23
  of terms to series, 592
  of vectors, 717–718
Additivity, line integrals and, 967–968
Additivity Rule
  for definite integrals, 268
  for double and triple integrals, 904
Agnesi, Mary, 667
Albert of Saxony, 603
Algebra
  abstract, AP-24
  functions in, 9–10
  Fundamental Theorem of Algebra, 18-9
  linear, B-4
  properties in (*See* Algebraic properties)
  rules
    for finite sums, 258
    for gradient vectors, 846
  vectors
    algebraically representing, 716
    in operations, 717–718, 722
Algebraic numbers, defined, 396
Algebraic properties
  for base a logarithms, 395
  of continuous functions, 76
  of exponential functions, 395
  of $\ln x$, 380
  of the natural logarithm, 380–381
  of real numbers, AP-1, AP-23
Alternating harmonic series, 614–618
Alternating Series Estimation Theorem, 616
Alternating Series Test, 614–615, 618
Amperes, 17-19
Amplitude, 19-4
Angle convention, 22
Angle of elevation, 772
Angle of inclination, AP-11–AP-12
Angles
  Angle Between Two Vectors, 726–728
  of elevation, 772
  firing, 772
  measured in degrees and radians, 21
  phase, 17-17
  between planes, 746–747
  in standard position, 22
  in trigonometric functions, 21–22
  between two curves, computing, 18-32
  between vectors, 746–747
Angular momentum, 801
Angular velocity, 1035
Antiderivatives, 230–239
  Constant Multiple Rules for, 232
  definition of, 230
  Difference Rule for, 232
  finding, 230–232
  indefinite integrals of, 234–235
  initial value problems and differential equations, 232–233
  integration formulas, 235
  linearity rules, 232
  motion and, 233–234
  Sum Rule for, 232
  of vector functions, 769–770
Antidifferentiation, 230, 232, 235
Aphelion, 704
Applied optimization, 213–218
  examples of, 215–218
  problems, solving, 213–214
Approximations
  convergence of, 227–228
  in converting mass to energy, 169
  differential, error in, 166–167
  finite
    for area, 249–254, 377–378
  theory of limits of (Riemann sums), 260–262, 265
  of functions, 19-16–19-19
  Haar system used in, 19-27, 19-29–19-31
  in higher dimensions vectors, 19-10–19-11
  linear, 161–163
    error formula for, 880
    standard, 854–855
  lower sum, 247–248
  Midpoint Rule for, 248, 266–267, 494
  polynomial, of two-variable functions, 882
  quadratic, 636
  at resolution level zero, 19-29–19-31
  for roots and powers, 163
  sensitivity to change in, 168–169
  of a series, error estimation in, 599–600
  by Simpson's Rule, 496–498
  tangent line in, 161
  tangent planes, 827, 854–855
  trapezoidal, 495–496
  by Trapezoidal Rule, 495–496
  by trigonometric polynomials, 19-3–19-5
  upper sum, 247–248
Arbitrary constant, 230
Arccosecant function
  definition of, 426
  derivative of, 430
  graphs/graphing, 423
  identity, 430
  inverse of, 422
Arccosine function, 422–425
  arc in, 423
  definition of, 422
  derivative of, 430
  graphs/graphing, 423–425
  identity, 425, 430
  integrals evaluated with, 431
  inverse of, 422, 425–426
  values for, 424
Archimedes' principle, 564, 1057
Archimedes spiral, 707

Arc length, 329–333
  arc length function, 332–333
  of a curve $y = f(x)$, 330–331
  differential, 673–674
  differential formula for, 332–333
  discontinuities in dy/dx, 332
Arc length in space, 778–780
  arc length formula, 778
  arc length parameter, 779
  definition of, 778
  of a function, 779
  speed on a smooth curve, 779–780
  unit tangent vector, 780
Arcsecant function
  definition of, 426
  derivative of, 429–430
  graphs/graphing, 423, 476
  integrals evaluated with, 431
  inverse of, 422, 425–426
Arcsine function, 422–425
  arc in, 423
  definition of, 422
  derivative of, 427–428
  graphs/graphing, 423–425
  identity, 425
  integrals evaluated with, 431
  inverse of, 425–426
  values for, 424
Arctangent functions
  definition of, 426
  derivative of, 428
  graphs/graphing, 423, 427, 476
  identity, 430
  integrals evaluated with, 431
  inverse of, 422, 425–426
  in Taylor series, 647–648
Area
  between curves, 297–298
  definition of, 908
  differential, change substituting, 951
  by double integrals, 908–910
    average value of, 909–910
    bounded regions in a plane, 908–909
    Riemann sums in, 908
  finite approximations for, 249–254, 377–378
  finite sums for estimating, 246–254
  Green's Theorem formula for, 1007
  infinite, 506
  lower sum approximation of, 247–248
  Midpoint Rule for approximating, 248, 266–267
  Pappus's Theorem for Surface Areas, 359
  in a plane, 685–687
  in polar coordinates, 685–688, 913–914
  of rectangles (*See* Rectangles, area of)
  of surfaces of revolution, 335–338, 674–675
  tangent, 668–669
  total, 283–284
  upper sum approximation of, 247–248
Argand Diagrams, 18-4
Argument, 18-5
Arrow diagrams, 886
Associative laws of real numbers, AP-24
ASTC rule, 23
Asymptotes
  horizontal, 85–87
  of hyperbolas, 694
  oblique, 87–88
  slant line, 87–88
  vertical, 91–92
Autonomous differential equations, 555–562
  definition of, 556
  equilibrium values or rest points in, 555–557
  falling body encountering resistance, 558–560
  graphical solutions of, 555–562
  Logistic Equation in Neural Networks and Machine Learning, 561–562
  logistic model for population growth, 560–561
  Newton's Law of Cooling, 558, 559
  phase lines, 555–557
  stable and unstable equilibria, 557–558
Auxiliary equation, 17-3
Average daily cost, 217–218
Average speed, 38–40
Average values
  for approximating area of rectangles, 253–254
  continuous, 271–272
  definition of, 272
  nonnegative continuous, 252–254
  in space, 925
Axis
  coordinate, AP-10
  of ellipses, AP-17
    focal axis, 692
    major axis, 693
    minor axis, 693
    semimajor axis, 693
    semiminor axis, 693, 701–702
  focal, of hyperbolas, 693
  imaginary, 18-4
  major, AP-17
  minor, AP-17
  moments of inertia about, 969, 1025
  of parabolas, AP-15–AP-16
  real, 18-4
  spin around, 996–999

Base a logarithms, 390, 394–396
Basis
  defined, 19-8
  orthonormal, for three-dimensional space, 19-8–19-9
Behavior
  graphical, of functions from derivatives, 208
  of indeterminate forms, 411, 414
Bell curve, 523
Bendixson's criterion, 1008
Bernoulli, Daniel, 181
Bernoulli, Jakob, 549
Bernoulli, John (Johann), 142, 411
Bernoulli differential equation, 549
"Big-oh" notation, 448
Binary search, 449
Binomial series, 644–646
Binormal vectors, 788, 792
Biology, derivatives in, 130
Birkhoff, George David, 291
Bolzano, Bernard, 126
Boundary
  of intervals, AP-3
  in a plane, 804
  in space, 806–807
  in substitution, 952–954, 957
  values, 17-6
Boundary points
  of absolute maxima or minima extrema, 865–866
  closed bounded, absolute maxima and minima on, 864–867
  of intervals, AP-3
  in a plane, 804
  in space, 806
Bounded from above/below (sequences), 581
Bounded function, 47
Bounded region
  absolute maxima and minima on closed, 864–867
  nonrectangular, 899–900
  in a plane, 804, 908–909
  in space, 806
Bounded sequences, 581–582
Bounds for the remainder in the Integral Test, 600
Bowditch curve, 677
Box function, 19-25
Boyle's Law, 13
Brachistochrones, 663–664
Branches of hyperbolas, 690
Brief Table of Integrals, 488

Calculators
  to estimate limits, 50–51
  graphs/graphing with, 29–32
Calculus of variations, 664
Calculus with parametric curves, 668–675
  arc length differential, 673–674
  areas of surfaces of revolution, 674–675
  Chain Rule in, 668
  complete elliptic integral of the second kind in, 672
  continuously differentiable functions, 670
  differentiable parametrized curves, 668–669
  length of parametrically defined curves, 670–673
  parametric formulas, 668–669
  Riemann sums in, 670
  for smooth curves, 670–671
  tangent lines and areas, 668–669
  twice-differentiable functions, 668–669
Cantor, Georg, 595
Cantor set, 595
Capacitance, 17-19
Carbon-14
  dating, 406, 407, 411
  decay, 405, 407
Cardioid curve, 683
Carrying capacity, 560
Cartesian coordinates
  equations relating spherical and cylindrical coordinates to, 942–944
  in a plane, AP-9–AP-10
  polar coordinates related to, 678, 679–680
  system of, AP-10
  in three-dimensional coordinate system, 710
Cartesian equations
  for hyperbolas, 700
  parametric, 662
Cartesian integrals, changing into polar integrals, 914–916
Cartographers, 18-32
CAS. *See* Computer algebra systems (CAS)
Catenary, 11, 444
Cauchy, Augustin-Louis, 416
Cauchy condensation test, 602
Cauchy-Riemann Equations, 18-20–18-22
Cauchy–Schwarz inequality, 244
Cauchy's Mean Value Theorem, 416–417
Cavalieri, Bonaventura, 313
Center
  of an ellipse, AP-17, 692
  of curvature for plane curves, 784
  of hyperbolas, 693
  of linear approximation, 161, 163
Centered difference quotient, 138–139
Center of mass, 929–930
  along a line, 350–351
  centroids, 356–357
  coordinates of, 969, 1025
  definition of, 351, 352
  distributed over a plane region, 352
  moments and, 350–359
  plates and, 352–356
  of plates bounded by two curves, 355–356
  system torque, 350–351
  theorems of Pappus, 357–359
  of thin, flat plates, 352–355
  thin wires and, 351
  torque, 350–351
Center-to-focus distance of ellipses, 693
Centroids, 356–357, 930
Chain curve, 444
Chain Rule, 831–838
  in calculus with parametric curves, 668
  dependency diagrams for, 835, 836

derivatives and, 139–146
in composition of functions, 139–141
in differentials, 165, 167–168
of exponential functions, 379, 388, 391, 393
in implicitly defined functions, 147–148
of inverse functions, 372
of inverse trigonometric functions, 428–430
of natural logarithm functions, 379
outside-inside rule and, 141–142
with powers of a function, 142–143
in related rates equations, 152, 154
repeated use of Chain Rule in, 142
ways to write, 141
diagram for remembering, 832
differentiable parametrized curves and, 668–669
for finding complex derivatives, 18-19
functions and
curvature, 782, 783
defined on surfaces, 834–836
Implicit Function Theorem, 837–838
of many variables, 838
of three variables, 833–834
of two variables, 831–833, 834
vector functions, 765, 766
implicit differentiation and, 836–838, 1014
motion in polar coordinates and, 794
"outside-inside" rule and, 849
for paths, 848–849
Power Chain Rule and, 142–143
proof of, 141, 167–168, 766
substitution method, 288–292
theorems, 833, 834, 836, 837–838
Chain Rule for Functions of One Independent Variable and Three Intermediate Variables, 833
Chain Rule for Functions of One Independent Variable and Two Intermediate Variables, 831
Chain Rule for Two Independent Variables and Three Intermediate Variables, 834
Formula for Implicit Differentiation, 836
Implicit Function Theorem, 837–838
Change
exponential (*See* Exponential change)
sensitivity to, 168–169
in approximations, 168–169
genetic data and, 130
Characteristic equation, 17-3
Charge, 17-18–17-19
Circles
of convergence, 18-26
of curvature for plane curves, 784–785
exterior regions of, AP-15
interior regions of, AP-15
osculating, 177, 784–785
in a plane, AP-13–AP-15
standard equation of, AP-14
unit, AP-14
Circular cylinder, 749
Circular functions, 436, 445
Circulation
density, 998–999
flux differentiated from, 980
in Green's Theorem, 1001–1003
for velocity fields, 979–980
Cissoid of Diocles, 151
Clairaut, Alexis, 826
Clairaut's Theorem, 825–826
Closed curves, 980
Closed intervals, AP-3, 74
Closed region
bounded, absolute maxima and minima on, 864–867
in a plane, 804
in space, 806–807
Coefficients
drag, 777
Fourier, 19-18–19-19
Haar, 19-27–19-29
linear second-order constant-coefficient models, 17-20
method of undetermined, 17-8–17-11
of polynomials, 9
in power-series solutions, 17-24–17-29
rational functions in, determining by differentiating, 486
undetermined, 481
Cofactor, B-2–B-5
Cofactor Expansion, B-4
Commutativity laws of real numbers, AP-24
Comparison tests, 18-24, 602–606
Direct Comparison Test, 603–604, 618
Limit Comparison Test, 604–606, 618
Competitive-hunter model, 565–566
Complementary solution, 17-8
Complete elliptic integral of the second kind, 672
Completeness property, 582
Complete ordered field, AP-24
Completing the square, AP-14–AP-15
Complex analysis, 18-1
Complex conjugate, 18-4
Complex derivatives
definition of, 18-17–18-19
differentiable, 18-18–18-19
formula for, 18-20–18-21
rules for finding, 18-19
Complex differentiable function, 18-18–18-19
Complex functions, 18-1–18-36
Cauchy-Riemann Equations, 18-20–18-22
complex numbers, 18-1–18-9
complex power series, 18-22–18-24, 18-25
conformal maps, 18-31–18-34
derivatives of, 18-17–18-19
Euler's Formula, 18-27–18-28
logarithmic, 18-28–18-29
variables of, 18-10–18-17
Complex numbers, 18-1–18-9
absolute value of, 18-5
adding, 18-3–18-4
Argand Diagrams, 18-4
complex conjugate, 18-4
complex plane and, 18-4–18-5
computations with, 18-4
computing reciprocals, 18-6
definition of, 18-1
De Moivre's Theorem, 18-7
dividing, 18-4
equality in, 18-3
Euler's Formula, 18-5
Fundamental Theorem of Algebra, 18-9
geometric representations of, 18-4–18-5
hierarchy of, 18-1–18-3
imaginary parts of, 18-3–18-4
multiplying, 18-3–18-4
powers of, 18-7
products of, 18-5–18-6
quotients in, 18-6–18-7
real parts of, 18-3–18-4
roots in, 18-7–18-9
Complex plane, 18-4–18-5
Complex power series, 18-22–18-24
complex sequences, 18-23
complex series, 18-23–18-24
Convergence Theorem for Complex Power Series, 18-25
Complex variables, 18-10–18-17
complex domains, 18-10–18-11
complex limits, 18-11–18-15
continuous, 18-15–18-17
functions of, 18-12–18-15
Component equation for a plane, 744
Component form of vectors, 715–717
Component functions, 760, 972
Component test
for conservative fields, 990–992
for exact differential form, 993
Composite functions
Chain Rule in, 139–141, 849
derivative of, 139–141
evaluating, 15–16
Composition Rule, 812
Computer algebra systems (CAS)
improper integrals and, 509–510
improper integrals in integration evaluated with, 509–510
integrate command in, 490–491
integration with, 490–491
Computer graphing, 807
Computing reciprocals, 18-6
Concavity
curve sketching and, 200–212
definition of, 201
First Derivative Test for, 204–205
graphical behavior of functions from derivatives, 208
graphing $y = f(x)$, 206–208
points of inflection and, 201–204, 206, 207, 208
Second Derivative Test for, 201, 204–208
Conditionally convergent series, 616–617
Cones
elliptical, 749–751, 752
parametrization of, 1009
surface area of, 1011
Conformal maps, 18-31–18-34
Conic sections, 690–695
ellipses, 692–693
hyperbolas, 693–695
illustrated, 690
parabolas, 691–692
Conics in polar coordinates, 698–703
directrices for, 698–700
eccentricity of, 698–702
polar equations for, 700–702
Connected domains and regions, 986
Connectedness, graphs/graphing, 78
Conservation of angular momentum, 801
Conservative fields
component test for, 990–992
line integrals in, 987–990
potentials for, finding, 990–992
properties of, 986
Stokes' Theorem and, 1038–1039
Constant-depth surface, fluid force on, 344–345
Constant force, work done by, 340–341
Constant Function Rule, 765
Constant functions, 7, 47
norm of, 19-13–19-14
zero derivatives in, 191
Constant length, vector functions of, 766
Constant Multiple Rule, 48–49, 591–592
for antiderivatives, 232
to calculate limits of sequences, 577
for definite integrals, 268
for double and triple integrals, 904
for finding complex derivatives, 18-19
for finite sums, 258
for gradients, 846
for limits, AP-19
for limits for functions of two variables, 812
for series, 591–592
Constants to describe prescribed errors, 57
Constant Value Rule, 258
Constrained maxima and minima, 869–872

Constrained variables
arrow diagrams for, 886
constrained by another equation, determining, 884–885
dependent/independent, determining, 883–884
notation for, 885–886
partial derivatives with, 883–886
Containers, lifting objects and pumping liquids from, 342–343
Continuity, 72–83
of compositions of functions, 76–78
continuous extension, 818, 888
continuously differentiable functions, 670
continuous partial derivatives, 1038
continuous random variable, 516
continuous vector fields, 972
discontinuity and, 74–75
at interior point, 73–74
at left endpoint, 73
limits and, 72–101
for multivariable functions, 814–816
one-sided, 74
partial derivatives and, 814–816, 824–825
at a point, 73–75
test, 74
two-sided, 74
of vector functions, 761–762
Continuous functions, 814–816
absolute extrema of, 864–867
algebraic properties of, 76
of compositions, 816
continuous at the point, 814–815
continuous extension of, 818, 888
definition of, 814
differentiable, 337, 338, 359, 371, 670, 673, 674
extension to a point, 80–81
Intermediate Value Theorem for, 78–79
left (or continuous from the left), 73–74
over a closed interval, 74
partial derivatives and, 824–825
right (or continuous from the right), 73–74
Contour curves of functions of two variables, 805
Convergence
absolute, 607–612
of approximations, in Newton's method, 227–228
circle of, 18-26
conditional, 616–617
of improper integrals, 507
tests for convergence, 510–512
interval of, 626
of power series, 621–626
Convergence Theorem for Power Series, 624–626
operations on, 626–629
radius of, 624–626
testing for convergence, 626
radius of, 624–626
of Riemann sums, 265
testing for
alternating series, 614–618
Alternating Series Estimation Theorem, 616
Alternating Series Test, 614–615
Direct Comparison Test, 603–604, 618
geometric series, 588–590, 618
improper integrals, 510–512
Limit Comparison Test, 604–606, 618
power series, 626
p-Series Test, 598, 618
Ratio Test, 609–610, 618
Rearrangement Theorem for Absolutely Convergent Series, 617, 618
Root Test, 611–612, 618
series with some negative terms, 608, 616, 617–618
summary of, 618
Convergence Theorem for Complex Power Series, 18-25
Convergence Theorem for Power Series, 624–626
Convex, 201
Coordinate axes, AP-10
Coordinate frame, 710
Coordinate pair, AP-10
Coordinates
Cartesian, AP-10
of center of mass, 969, 1025
cylindrical (*See* Cylindrical coordinates)
polar (*See* Polar coordinates)
rectangular (*See* Rectangular coordinates)
three-dimensional, coordinate planes in, 710–713
Coplanar vectors, 718
Corner, 110
Cosecant function, 22–23
derivatives of, 136–137
graphs/graphing, 24, 46, 423
integral of, 383
odd, 24
periodicity of, 24
in Substitution Rule, 291
Cosine function, 22–23
arcsine and arccosine functions and, 422–424, 426
derivatives of, 134–135
to eliminate a square root, 472
even, 24
graphs/graphing, 10, 24, 46, 423
inequalities and, 26
periodicity of, 24
product of, 474
products of powers of, 470–471
Cosines
direction, 732
law of, 25–26
summing, 19-3–19-5
values of, 23
Costs
fixed, 129
marginal, 128–129
of production, 128–129
variable, 129
Cotangent function, 22–23
derivatives of, 136–137
graphs/graphing, 24, 46, 423
integral of, 383
odd, 24
periodicity of, 24
in Substitution Rule, 291
Coulombs, 17-19
Courant, Richard, 115
Cramer, Gabriel, 151
Cramer's Rule, 17-12
Critical damping, 17-17–17-18
Critical points
definition of, B-8
extreme values and, 184–185
to find increasing or decreasing functions, 195–198
on graphs, 204–208
in local maxima or minima extrema, 861–864
in Max-Min Tests, 866
in optimization problems, 16, 46, 213, 215, 218
Cross Product Rule, 765–766
Cross products, 734–739
area of a parallelogram in, 735–736
calculating, as a determinant, 736–737
Cross Product Rule, proof of, 765–766
definition of, 734–735
determinant formula for, 736
distributive law for, AP-26–AP-27
pairwise, 735
parallel vectors in, 735
properties of, 735
torque vector in, 737–738
triple scalar product (box product) in, 738–739
of two vectors in space, 734–735
Cross-sections
definition of, 311
for finding limits of integration, 903
method of slicing, 311–313
solids of revolution, 313–318
disk method, 313–316
washer method, 316–318
volumes using, 311–318
Cube root function, 8
Cubic function, 9
Curl vector
divergence and, 1044–1045
field, 1028, 1029–1030
in Green's Theorem, 1001–1003
k-component of, 996–999
Current, 17-18–17-19
Curve/curvature
arc length
definition of, 331
differential formula for, 332–333
discontinuities in, 332
discontinuities in dy/dx, 332
formula for, 332
function, 333
plates bounded by two, 355–356
smooth, 330
areas between, 297–298
assumptions on, 986
Bowditch, 677
calculating, formula for, 782–784
cardioid, 683
center of, 784
chain, 444
circle of, for plane curves, 784–785
closed, 980
computing, formulas for, 791–792
contour, of functions of two variables, 805
definition of, 782
generating, 749
infinite area under, 506
level, 805, 845
normal vectors, 782–786
parametric, 565, 778–780
piecewise smooth, 763
plane, 782–785
circle of curvature for, 784–785
curvature of, 782–785
flux across, 980–981
plates bounded by two, 355–356
polar, length of a, 687–688
principal unit normal vector for smooth, 783
radius of, 784
sigmoid shape of, 561
sinusoid, 677
sketching, 200–212
slope of, 102–103
smooth, 670–671, 763–764
snowflake, 595
space
arc length and, 778–780
binormal vectors for, 788
computation formulas for, 792
curvature and normal vectors for, 785–786
definition of, 785
normal vectors for, 785–786
parametric equations for, 760, 771–773
tangents of, 760–766
work done by a force over a curve in, 977–979
synchronous, 799
tangent line to a level curve, equation for, 845
tangents to, 780
torsion along, 790
vector formula for, 791
Cusp, 110
Cycloid
brachistochrone, 663–664
calculus of variations in, 664
definition of, 663
Huygens' pendulum clock and, 663, 664
parametric equations of, 663–664
tautochrones, 663–664
Cylinders
definition of, 749
hyperbolic, 870–871
parametrization of, 1009
slicing with, 322–324
Cylindrical coordinates, 938–942
definition of, 938

equations relating rectangular coordinates to, 938–940
equations relating spherical and Cartesian coordinates to, 942–944
integration in, 938–942
limits of integration in, finding, 940–942
motion in, 794
parametrization by, 1009
Cylindrical shells, volumes using, 322–327
shell method, 324–327
slicing with cylinders, 322–324

Damped motion, 17-17–17-18
Damped vibrations, 17-18
Daubechies, Ingrid, 19-33
Daubechies scaling function, 19-33–19-34
Daubechies wavelets, 19-33–19-34
Decay
exponential, 400–401
radioactive, 405–407
rate of, 400
Decreasing functions, 5–6
Dedekind, Richard, 261, 542, 592
Definite integrals, 263–272
applications of, 311–365
arc length, 329–333
areas of surfaces of revolution, 335–338
moments and centers of mass, 350–359
volumes using cross-sections, 311–318
volumes using cylindrical shells, 322–327
work and fluid forces, 340–345
area under the graph of a non-negative function, 270–271
average value of a continuous function, 271–272
definition of, 246, 263–265
existence of, 264–265
in integrable and nonintegrable functions, 265–266
integration by parts for evaluating, 466–467
as a limit of Riemann sums with equal-width subintervals, 265
Mean Value Theorem for, 276–277, 279, 280–281
Midpoint Rule for approximating, 248, 266–267, 494
notation for, 264
properties of, 267–269
rules satisfied by, 268
shift property for, 303–304
substitution for, 294–304
of symmetric functions, 296–297
Trapezoidal Rule for the value of, 495–496
of vector functions, 770
Zero Width Interval Rule for, 377
Degrees
angles measured in, 21
of polynomials, 9
Δ, 57, 59–61
De Moivre's Theorem, 18-7, 19-5
Density
circulation, 998–999
flux, 999, 1000–1001, 1041–1042
function, exponential, 518–521
mass, of surface integrals, 1018, 1020
of plates, 352
Dependent variables
constrained, 883–884
of functions, 1
in functions of several variables, 802
Derivative Constant Multiple Rule, 116–117
Derivative Power Rule
applying, 116
general version, 116
positive integer, 115
Power Chain Rule and, 142–143
Derivative Product Rule, 134, 135, 137
Derivative Quotient Rule, 119–120, 136–137
Derivatives, 102–179
of absolute value function, 143
antiderivatives (*See* Antiderivatives)
applications of, 180–245
applied optimization, 213–225
with base a logarithms, 396
in biology, 130
Chain Rule and, 139–146
of composite functions, 139–141, 849
concavity and curve sketching, 200–212
of a constant function, 115
Cross Product Rule, 765–766
definition of, 106
calculating derivatives from, 107–108
difference quotient and, 103–104
differentials and, 164–169
in direction of vectors, 763–764
Dot Product Rule, 765
in economics, 128–130
First Derivative Test for Local Extreme Values (General Version), B-8
of functions, 103, 106–115
cosecant, 136–137
cosine, 134–135
cotangent, 136–137
definition of, 106–107
differential, inverses of, 371–373
extreme values of, on closed intervals, 180–188
f at a point x_0, 103
graphical behavior of, 208
hyperbolic, 437–438
inverses of, 427–430
monotonic, and the first derivative test, 195–200
natural exponential, 388–389
negative, 117
with no derivative at a point, 110
reciprocal, 107
recovering, from its derivative, 112
secant, 136–137
sine, 133–134
trigonometric, 133–139, 427–430
u and v for denoting, 117
vector, 762–764
General Power Rule for, 116, 391
graphs/graphing, 108–109
history of, 106
implicit differentiation and, 147–149
as instantaneous rate of change, 104
involving $\log_a x$, 394–396
left-hand, 109–110
Leibniz's Rule, 178–179, 309
linearization and, 160–163
of $\ln x$, 378–379
Mean Value Theorem in, 190–194
Mixed Derivative Theorem, AP-27–AP-28, 825–826
motion and, 762–764
Newton's dot notation for, 791
notations for, 108
one-sided, 109–110
partial (*See* Partial derivatives)
at a point, 102–106
of a positive integer power, 115
of power series, 627
as a rate of change, 124–133
rates of change and, 103–104
of reciprocal functions, 107, 371–373
of related rates, 152–160
right-hand, 109–110
Rolle's Theorem in, 188–189
Second Derivative Test for Local Extreme Values (General Version), B-9
of the sine function, 133–134
of the square root function, 108, 110
of the tangent function, 136–137
tangent lines and, 102–106
definition of, 102
finding, to graph a function, 102–103
slope of, 102–103, 108
of the tangent vector, 783
of trigonometric functions, 133–139
as velocity, 764
Derivative Sum Rule, AP-8–AP-9, 117–118, 134
Descartes, René, AP-10, 147, 149, 151
Determinants, B-1–B-6
absolute value of, 739
calculating a cross product as, 736–737
calculating the triple scalar product as, 738–739
cofactors in, B-2–B-5
definition of, B-1
evaluating, 736
formula, 736
Jacobian, 951, 955
matrix in, B-1–B-6
method of, 17-12
properties of, B-5
vector as, 738–739
Deviation standard, in probability, 522
Devil's curve, 151
Difference quotient, 103–104
definition of, 103
forms for, 107
limit of, 107
Difference Rule, 48–49
for antiderivatives, 232
to calculate limits of sequences, 577
definition of, 117
for double and triple integrals, 904
to find derivatives, 117–118, 134
for finite sums, 258
for gradients, 846
for limits, AP-19
for limits for functions of two variables, 812
for series, 591–592
for vector functions, 765
Differentiability
on an interval, 109–110
definition of, 827–828
for functions of several variables, 827–828
partial derivatives and, 827–828
Differentiable functions
absolute maxima and minima, 184–185
Chain Rule formula for
functions of many variables, 838
functions of three variables, 833–834
functions of two variables, 831–833
implicit differentiation, 837
complex, 18-18–18-19
continuous, 111, 337, 338, 359, 670, 673, 674
with first partial derivatives, 828
definition of, 107, 827–828
differentiability and, 856–857
domain of, 845
Extreme Value Theorem, 185
First Derivative Test for Local Extrema, 185, 196–198
graphical behavior, 208
increasing and decreasing functions, 195–196
Mean Value Theorem, 190–191
of more than two variables, 856–857
negative, 117
with no derivative at a point, 110
on one-sided derivatives, 109–110
optimization problems, 214, 216
parametrized curves, 668–669
plane tangent to the surface of, 850–852
points of, 862
Rolle's Theorem, 188
Second Derivative Test for, 201, 204–208
twice-differentiable, 668–669
of two variables, 821
vector functions, 763

Differentiable Implies Continuous Theorem, 828
Differential equations, 232–233
first-order (*See* First-order differential equations)
Fundamental Theorem and, 283, 309
second-order (*See* Second-order differential equations)
separable, 401–402
solution of, 401–402
for vector functions, 770–771, 775
Differentials
arc length, 673–674
formula for, 332–333
area change substituting, 951
Chain Rule in, 165, 167–168
definition of, 164, 855
error in approximation of, 166–167
estimating with, 165–166
forms of, 978, 993–994
geometric meaning of, 164
of integration by parts formula, 463–466
linearization and, 164–169
for a parametrized surface, 1011
in partial derivatives, 855–856
surface area, for a parametrized surface, 1011
tangent plane, 855–856
total, 855–857
Differentiation
Chain Rule for implicit, 1014
coefficients determined by, 486
definition of, 107
implicit, 147–152
integration and, relationship between, 282–284
in related rates equations, 155
rules, 764–766
Chain Rule, 139–146, 766
for constant function derivatives, 115
Cross Product Rule, proof of, 765–766
Derivative Constant Multiple Rule, 116–117
Derivative Product Rule, 134, 135, 137
Derivative Quotient Rule, 136
Derivative Sum Rule, 117–118, 134
Difference Rule, 117–118, 134
Dot Product Rule, proof of, 765
for vector functions, 764–766
Diocles, 151
Direct Comparison Test, 510–511, 512, 603–604, 618
Directed line segment to represent vectors, 715
Direction
along a path, 966
cosines, 732
downhill, 841
estimating change in a specific, 853
field, 537–538
forward, 974
rate of change in, 841
scalar component in, 729–730
of vectors, 719–720
Directional derivatives, 841–849
calculations of, 843–845
definition of, 841
as a dot product, 843–844
estimating change in a specific direction, 853
gradient vectors in, 843–849
interpretation of, 842–843
in a plane, 841–842
properties of, 844–845
zero, 849
Directrices
for ellipses, 699–700
for hyperbolas, 699–700
of parabolas, 691–692
Dirichlet, Lejeune, 506
Discontinuity, 110, 332
Discontinuous functions, 74–75
Discrete random variables, 516
Discriminant of local maxima or minima extrema, 862–864
Disk method, 313–316
Displacement
definition of, 251
versus distance traveled, 251–252
work by force through, 731
Distance
in a plane, AP-13–AP-15
from a point to a line, 743
from a point to a plane, AP-14, 746
spheres in space and, 712–713
squares of, 932
traveled, 249–252
displacement *versus,* 251–252
total distance traveled, 251–252, 282
velocity function, 249–252
between two functions, 19-13
between two points in the *xy*-plane, formula for, 712–714
Distributions
continuous random variable and, 516
exponentially decreasing, 518–519
memoryless, 518–519
probability, 516–519
uniform, in probability, 522–523
Distributive law for cross products, AP-26–AP-27
Divergence
curl and, 1044–1045
Divergence Theorem, 1041–1051
partial sums in, 590–591
of series, 587, 590–591
tests to determine
for alternating series, 614–618
Alternating Series Estimation Theorem, 616
Alternating Series Test, 614–615
Direct Comparison Test, 603–604, 618
geometric series, 588–590, 618
Limit Comparison Test, 604–606, 618
*n*th-Term Test, 590–591, 618
p-Series Test, 598, 618
Ratio Test, 609–610, 618
Root Test, 611–612, 618
for series with some negative terms, 608, 616, 617–618
summary of, 618
theorem (*See* Divergence Theorem)
in three dimensions, 1041–1042
of a vector field, 999–1001, 1041–1042
Divergence Theorem, 1041–1051
corollary, 1042
Green's Theorem compared to, 1041–1042, 1050
for other regions, 1046–1047
proof of, for special regions, 1045–1046
for special regions, 1045–1046
statement of, 1042
Divergent improper integrals, 507, 510–512
Division
of complex numbers, 18-4
of functions, 14–15
Domain
assumptions on, 986
connected, 986
of differentiable functions, 845
of functions, 1–2, 14
of functions of several variables, 802, 803
natural, 2
parameter, 1008
restrictions, one-to-one, 370, 421
simply connected, 986
Dominant terms, 92–93
Domination Rule
for definite integrals, 268
for double and triple integrals, 904
Dot Product Rule, 765
Dot products, 726–732
Angle Between Two Vectors, 726–728
angles and, 727–728
definition of, 726
directional derivatives as, 843–844
Dot Product Rule, 765
of gradient vectors, 843–844
orthogonal vectors and, 728
properties of, 728–731
of three-dimensional vectors, 19-6
of two *n*-dimensional vectors, 731–732
work by force through displacement, 731
Double-angle formulas, 25
Double integrals
area by, 908–910
definitions of, 895, 900
Fubini's Theorem for calculating
first form, 896–898
stronger form, 900–903
iterated or repeated, 896
over general regions, 899–905
over rectangles, 894–898
partitions formed by, 894–895
in polar form, 911–916
properties of, 904–905
Riemann sums for, 894–896, 904, 908
substitutions in, 950–954
volumes
over general regions, 900–903
over rectangles, 895–896
of a solid region in space, 918–919
Downhill direction, 841
Drag coefficient, 777
Drag force, 776–777
D-tuple of numbers, 19-9
Dummy variable, 265

Eccentricity, 698–700
definition of, 699
of ellipses, 698–702
of hyperbolas, 698–702
of parabolas, 698–702
Economics, derivatives in, 128–130
Einstein's Law in the Special Theory of Relativity, 437
Electric circuits, 17-18–17-19
Electric field, 1048
Electromagnetic theory, 1048
Elements of sets, AP-2
Elevation, angle of, 772
Ellipse
axis of, AP-17
center of, AP-17, 692
center-to-focus distance of, 693
definition of, AP-16–AP-17, 692–693
directrices for, 699–700
eccentricity of, 698–702
equations for, 693
focal axis of, 692
law (Kepler's first law), 795
major axis of, 693
minor axis of, 693
polar equations for, 700–702
semimajor axis of, 693
semiminor axis of, 693, 701–702
standard equation of, AP-17
tangent line to, 845–846, 852
vertices of, 692
Ellipsoids
graphs/graphing, 752, 753
as quadric surface, 749–751
of revolution, 750
Elliptical cones, 749–751, 752
Elliptical paraboloid, 749–751, 752
Elliptic integrals, 503
Empty set, AP-2
Endpoints
of an interval, limits at, 66–67
continuity at, 73
extreme values and, 184
Energy, approximations in converting mass to, 169
Engineering, vectors in, 721
E, 57, 59–61
Equal area law (Kepler's second law), 795–796

Equality
  in complex numbers, 18-3
  of vectors, 715, 716
Equations
  Bernoulli differential, 549
  differential (*See* Differential equations)
  for ellipses, 693
  homogeneous, 572
  for hyperbolas, 694–695
  ideal projectile motion, 772
  inverse (*See* Inverse equations)
  for parabolas, 691–692
  parametric, for lines, 741–742
  for a plane, 743–744
  predator-prey, 569–570
  solution of, 401–402
  for the sphere of radius and center, 713–714
  for the tangent line to a level curve, 845
  vector
    for lines, 741
    for planes, 744
Equilibria, stable and unstable, 557–558
Equilibrium value
  in graphical solutions of autonomous equations, 555–557
  in systems of differential equations, 565–566
Error
  analysis, in numerical integration, 498–501
  estimation, in approximating the sum of a series, 599–600
  formula for linear approximations, 880
  function, 514
  in standard linear approximation, 855
  term (or remainder of order *n*), 638–639
Euler, Leonhard, 540
Euler equations, 17-22–17-23
Euler's constant, 602
Euler's Formula, 18-5, 18-27–18-28, 19-5, 19-22
Euler's gamma function, 533
Euler's identity, 649–650
Euler's method, 538–541
Euler's solution, 540–541
Evaluation Theorem, 280–281
  Net Change Theorem, 281–282
  proof of, 280–282
Even functions, symmetry properties of, 6–7
Exact differential forms, 993–994
Expected values
  of functions, 520
  in multiple integrals, 935
Exponential change, 399–411
  definition of, 399–400
  in growth and decay, 400–401
  heat transfer (Newton's Law of Cooling) and, 407–408
  radioactive decay and, 405–407
  rate constant of, 400
  separable differential equations and, 401–402
  unlimited population growth and, 402–405
Exponential decay, 400–401
Exponential functions, 10, 386–399
  algebraic properties of, 395
  antiderivative of, 389
  with base *a,* 10, 390
  Chain Rule and, 379, 388, 391, 393
  decreasing probability density function, 518–519
  definition of, 387
  derivatives of, 388–389
  exponential decay, 400–401
  exponential density, 518–521
  exponential growth, 400–401, 403
  general, 390–391
  graphs/graphing, 10
  growth, 400–401, 403, 550–551
  integrals of, 388–389, 394, 396
  inverse of ln x and the number *e,* 386–388
  natural (*See* Natural exponential functions)
  population growth model, inaccuracy of, 550–551
Exponents
  growth in, 400–401, 403
  laws of, 389–390
  rules for, 385
Exterior regions of circle, AP-15
Extrema. *See* Extreme values
Extreme values
  absolute (*See* Absolute maxima or minima extrema)
  constrained maxima or minima extrema, 869–872
  of continuous functions on closed, bounded sets, 816
  critical points and, 184
  definition of, 180
  endpoints and, 184
  Extreme Value Theorem, 181–182
  finding extrema, 183–185
  of functions of two variables, B-6–B-11
  of functions on closed intervals, 180–188
  inflection points and, 184
  Lagrange multipliers for finding, 869–876
  local (*See* Local maxima or minima extrema)
  relative, 861
  saddle points, 860–867
  of two variables, B-6–B-11, 860–864
Extreme Value Theorem, 181–182

Falling body encountering resistance, 558–560
Faraday's law, 1057
Fermat, Pierre de, 41
Fermat's principle in optics, 215–216
Fibonacci numbers in sequences, 580
Fields
  in algebraic properties, AP-24
  conservative
    component test for, 990–992
    line integrals in, 987–990
    potentials for, finding, 990–992
    properties of, 986
    Stokes' Theorem and, 1038–1039
  curl vector, 1028, 1029–1030
  electric, 1048
  gradient, 973–974
  gravitational, 985
  ordered, AP-24
Finite approximations
  for area, 249–254, 377–378
  theory of limits of (Riemann sums), 260–262, 265
Finite intervals, AP-3
Finite limits as $x \to \pm\infty$, 83–84
Finite sums, 256–262
  algebra rules for, 258
  estimating area with, 246–254
  limits of, 259–260
  Riemann sums and, 260–262
  sigma notation and, 256–259
Firing angle, 772
First Derivative Test
  for concavity, 204–205
  for local extrema, 183–184, 196–198
  for Local Extreme Values, 861, 871
    General Version, B-8
First moments
  about coordinate planes, 929, 969, 1025
  definition of, 928–929
  formulas for, 929–930
  of a line integral, 969, 1025
  masses and, 928–930
  for three-dimensional solid, 929
  for two-dimensional plate, 929
First octant, 710
First-order differential equations, 535–572
  applications of, 549–554
  autonomous differential equations, 555–562, 564
  autonomous systems of, 564
  Bernoulli differential equation, 549
  carrying capacity, 560
  competitive-hunter model, 565–566
  curve, sigmoid shape of, 561
  definition of, 535–536
  equilibrium values, 555–557
  Euler's method, 538–541
  Euler solution, 540–541
  exponential population growth model, inaccuracy of, 550–551
  falling body encountering resistance, 558–560
  first-order initial value problem, 536–537
  first-order linear equations, 543–547
  graphical solutions of autonomous, 555–562
  homogeneous, 572
  initial value problems, 536–537
  integrating factor, 544–546
  Law of Exponential Change, 550–551
  limit cycle, 567
  limiting population, 560
  logistic equation in Neural Networks and Machine Learning, 561–562
  logistic growth, 560
  logistic model for population growth, 560–561
  mixture problems, 552–554
  motion with resistance proportional to velocity, 549–550
  Newton's Law of Cooling, 558
  Newton's second law of motion, 549, 559
  numerical method, 538–541
  numerical solution of the problem, 538–541
  orthogonal trajectories, 549, 551–552
  particular solution, 536–537
  phase-plane analysis method, limitations of, 567
  phase planes, 564–565
  predator-prey, 569–570
  resistance proportional to velocity, 549–550
  rest points, 555–557
  RL circuits, 547
  slope fields, 537–538
  solution curves, 537–538
  solutions, 535–537
  standard form of linear equations, 543–544
  steady-state value, 547
  systems of, 564–567
  terminal velocity, 560
First-order initial value problem, 536–537
First-order linear equations, 543–547
  definition of, 543–544
  integrating factor for, 544–546
  RL circuits, 547
  solving, 544–546
Fixed costs, 129
Fixed point, 83
Flight time, for ideal projectile motion, 773
Flow integrals, 979–980
Fluid flow rates, 1000
Fluid force
  centroids and, 357
  on a constant-depth surface, 344–345
  integral for, against a vertical flat plate, 345
  pressure and, 344–345
  work and (*See* Work and fluid force)
Fluid weight-density, 344

Flux
across a plane curve, 980–981
across a rectangle boundary, 1000
calculation of, 980, 1022
circulation differentiated from, 980
definition of, 980, 1022
density, 999, 1000–1001, 1041–1042
in Divergence Theorem, 1041–1045
in Green's Theorem, 1002–1003
integrals
evaluating, 980–981
surface, 1022–1023
magnetic, 1057
Focal axis
of ellipses, 692
of hyperbolas, 693
Focal length of parabolas, 691
Focus
center-to-focus distance of ellipses, 693
of parabolas, 691–692
Folium, 147, 149, 151
Force
constant, 342
work done by, 340–341
drag, 776–777
fluid (*See* Fluid force)
work and (*See* Work and fluid force)
Formula for Implicit Differentiation, 836–838
Forward direction, 974
Fourier, Jean-Baptiste Joseph, 19-1
Fourier series, 19-1
convergence of, 19-19–19-21
Fourier coefficients in, 19-18–19-19
Parseval Equality for, 19-21–19-22
Fractions, partial. *See* Partial fractions
Free fall, 38
Frenet, Jean-Frédéric, 788
Frenet frame, 788
Frequency
in periodic functions, 19-2
in probability distributions, 516
zero, 19-2
Fubini, Guido, 897
Fubini's Theorem
for calculating double integrals, 896–898, 900–903
for finding limits of integration, 896, 902
Functions, 1–37. *See also individual functions*
adding, 14–15
algebraic, 9–10
approximation of, 19-16–19-19
arc length, 332–333, 779
arcsine and arccosine, 422–425
arrow diagram of, 2
bounded, 47
box, 19-25
circular, 436, 445
combining, 14–21
common, 7–11
component, 760, 972
composition of, 15–16
constant, 7, 47
continuous/continuity (*See* Continuity; Continuous functions)
cube root, 8
cubic, 9
Daubechies scaling, 19-33–19-34
decreasing, 5–6, 195–196
definition of, 1
by formulas, 14
on surfaces, 834–836
derivatives of, 103, 106–115
cosecant, 136–137
cosine, 134–135
cotangent, 136–137
definition of, 106–107
differential, inverses of, 371–373
extreme values of, on closed intervals, 180–188
f at a point x_0, 103
graphical behavior of, 208
hyperbolic, 437–438
inverses of, 427–430
monotonic, and the first derivative test, 195–200
natural exponential, 388–389
negative, 117
with no derivative at a point, 110
reciprocal, 107
recovering, from its derivative, 112
secant, 136–137
sine, 133–134
trigonometric, 133–139, 427–430
u and *v* for denoting, 117
vector, 762–764
differentiable (*See* Differentiable functions)
discontinuous, 74–75
distance between two, 19-13
dividing, 14–15
domain of, 1–2, 14, 803
dominant terms in, 92–93
error, 514
even, symmetry properties of, 6–7
exponential (*See* Exponential functions)
finitely many, sums of, AP-8–AP-9
Fourier series, 19-19–19-22
general sine, 27
gradient of, 846–848
graphing (*See* Graphs/graphing)
greatest integer, 5, 74
growth rates of, 445–449
Hessian of, 862–864
hyperbolic (*See* Hyperbolic functions)
identity (*See* Identity function)
increasing, 5–6, 195–196
infinitely many orthonormal, 19-20
inner product of, 19-12, 19-14–19-15, 19-24
integers of, 5
integrable, 19-12–19-13, 265–266
inverse (*See* Inverse functions)
joint probability density, 934
limits of, 38–101
linear, 7
linearization of (*See* Linearization)
logarithmic, 11, 18-28–18-29
as a machine, diagram of, 2
marginal cost, 128–129, 216–217
monotonic, 195–200
multiplying, 14–15
natural domain of, 2
natural exponential (*See* Natural exponential functions)
natural logarithm (*See* Natural logarithm functions)
nonintegrable, 265–266
norm of, 19-12–19-14
numerical representation of, 4
odd, symmetry properties of, 6–7
one-to-one, 366–367, 370, 421
orthogonal, 19-15–19-16, 19-24
orthonormal, 19-12, 19-15–19-16, 19-20, 19-24, 19-27
of partial derivatives, 819–828
periodic, 19-1–19-2
piecewise-defined, 5
piecewise-smooth, 763
polynomials, 9
position, 5
potential, 986
power, 7–8
proportional relationship of, 7
quadratic, 9
range of, 1–2, 803
rational, 9
reciprocal, 107
representation as power series, 621–629
root of, 225
scalar, 760–761, 965–970, 978, 1019
scaling of, 16–18
scatterplot of, 4
shift formulas for, 16
sine-integral, 514
smooth, 330
square root, 8
subtracting, 14–15
symmetric, 6–7, 296–297
theorems for (*See* Theorems)
torsion, of a smooth curve, 790
total cost, 129
transcendental, 11, 366–456
trigonometric, 10, 21–29
unit, 19-13, 19-24
unit step, 47–48
values of (*See* Values)
variables of (*See* Variables)
velocity, 249–252, 773–774
vertical line test for, 4
Fundamental period, 19-1
Fundamental Theorem of Algebra, 18-9
Fundamental Theorem of Calculus, 276–284
for arc length differential, 673–674
for continuous vector functions, 770–771
to define natural logarithm function as an integral, 377–379
differential formula for arc length, 332–333
Divergence Theorem and, 1051
double integrals evaluated with, 897
Fundamental Theorem of Line Integrals, 987–989
Green's Theorem and, 1002
integration and differentiation in, relationship between, 282–284
for line integrals, 965, 966, 987–989
Part 1 (Mean Value Theorem), 276–279
Part 2 (Evaluation Theorem), 280–281
path independence and, 986
total area and, 283–284
Fundamental Theorem of Line Integrals, 987–989

Galileo Galilei, 38
free fall formula, 38, 127, 131
law of, 38
Gauss, Carl Friedrich, 258, 716
Gauss's Law, 1048
General exponential function, 390–391
General linear equation, AP-13
General Power Rule for derivatives, 381, 390–391
General quadric surfaces, 751, 753
General sine function, 27
General solution, 233, 401
Generating curve, 749
Genetic data and sensitivity to change, 130
Geometric series, 588–590, 618
Gibbs, Josiah Willard, 779
Gibbs phenomenon, 19-21
Gradient ascent, method of, B-14
Gradient descent, method of, B-12–B-14
Gradient fields, 973–974
Gradient vectors
algebra rules for, 846
in Chain Rule for paths, 848–849
curl of, 1038
definition of, 843
in directional derivatives, 843–849
dot product of, 843–844
in functions of more than three variables, 847–848
in functions of three variables, 846–847
to level curves, 845–846
nonzero, 845–846
Orthogonal Gradient Theorem, 872–873

Graphs/graphing, 3
arcsine and arccosine functions, 423, 424
area under the graph
of nonnegative functions, 270–271
of positive functions, 253
total, 283–284
asymptotes
horizontal, 85–87
oblique, 87–88
slant line, 87–88
vertical, 91–92
autonomous differential equations, 555–562
equilibrium values or rest points in, 555–557
falling body encountering resistance, 558–560
Logistic Equation in Neural Networks and Machine Learning, 561–562
logistic model for population growth, 560–561
Newton's Law of Cooling, 558, 559
phase lines, 555–557
stable and unstable equilibria, 557–558
terminal velocity, 560
catenary, 11
complete, software for obtaining, 32
connectedness and, 78
curve sketching, 200–212
equation or inequality, AP-10
exponential functions, 10
of functions of two variables, 805
inverse trigonometric functions, 423, 424
limit statements verified by, 57–59
logarithmic functions, 11
polar equations, 678–679, 681–684
quadric surfaces, 752
reflecting, 16–18
scaling, 16–18
secant lines, 40–43
shifting, 16
software for, 29–33, 807
complete graph obtained with, 32
graphing windows, 29–32
surface area of, formula for, 1016
symmetric, 6–7
tangent lines, 40–43
trigonometric functions, 10, 24, 26–27, 423, 424
two variables, 805
ln *x*, 381–382
Grassmann, Hermann, 719
Gravitational fields, 973, 985
Greatest integer function, 5, 74
Greatest lower bound sequence, 581
Greek letters to describe prescribed errors, 57
Green's formulas, 1052
Green's Theorem
area formula, 1007
divergence form of, 1002–1003, 1041
Divergence Theorem compared to, 1041–1042, 1050
to evaluate line integrals, 1003–1004
generalization to three dimensions, 1050
normal form of, 1050, 1051
in a plane, 996–1005
for special regions, 1004–1005
Stokes' Theorem compared to, 1030–1031, 1037–1038, 1050
tangential form of, 1001–1002, 1050
Growth
exponential, 400–401, 403, 550–551
models, 550–551
Growth rates of functions, 445–449
definition of, 446–448
order and oh-notation, 448
sequential *vs.* binary search, 449

Haar coefficients, 19-27–19-29
Haar system, 19-27, 19-29–19-31
adding details at resolution level zero, 19-30–19-31
approximations using, 19-27, 19-29–19-31
box function in, 19-25
coefficients in, 19-27–19-29
definition of, 19-27
multiresolution recursion, 19-31–19-32
scaling function in, 19-24–19-29
theorems for, 19-27, 19-29, 19-32
wavelets in, 19-24–19-26, 19-31, 19-33–19-34
Half-angle formulas, 25
Half-life, 405–407
Half-open intervals, AP-3
Harmonic analysis, 19-1
Harmonic series, 598, 614–618
alternating, 614–618
definition of, 614–618
p-Series Test and, 598
Heat equation, 831
Heat transfer, 407–408
Heaviside, Oliver, 485
Height, for ideal projectile motion, 773
Helix, 761
Hessian matrix, B-9–B-11, 863
Hessian of a function, 862–864
Higher dimension vectors, 19-9–19-11
definition of, 19-9–19-10
orthonormal bases and approximation in, 19-10–19-11
theorems, 19-10–19-11
Homogeneous equations, 17-1, 572
Hooke's Law, 17-15–17-16, 135, 341–342
Horizontal asymptotes, 85–87
Horizontal cross-sections, for finding limits of integration, 903
Horizontal line test, for one-to-one functions, 367
Horizontal scaling and reflecting formulas, 17
Horizontal shifts, 16
Huygens, Christiaan, 663
Huygens' pendulum clock, 663, 664
Hyperbolas, 693–695
asymptotes of, 694
branches of, 690
Cartesian equation for, 700
center of, 693
definition of, 693
directrices for, 699–700
eccentricity of, 698–702
equations for, 694–695
focal axis of, 693
functions of (*See* Hyperbolic functions)
polar equations for, 700–702
vertices of, 693
Hyperbolic cylinder, 870–871
Hyperbolic functions, 435–445
definition of, 435–437
derivatives of, 437–438
inverse hyperbolic functions, 440–442
identities of, 435–437, 440
integral of, 437–438
inverse, 439–442
Hyperbolic paraboloid, 750–751, 752
Hyperboloids, 749–751, 752
Hypocycloids, 667

i-component of vectors, 719–720
Identity function, 7, 46–47, 369, 394, 1038
arcsine and arccosine, 425
continuous partial derivatives with, 1038
Euler's Formula and, 18-5, 649–650
hyperbolic, 435–437, 440
inverse function–inverse cofunction, 425, 430
inverse trigonometric, 133–134
Stokes' Theorem and, 1038
trigonometric, 24–26
Image, 950
Imaginary axis, 18-4
Imaginary parts of complex numbers, 18-3–18-4
Implicit differentiation, 147–152
Chain Rule for, 147–148, 1014
definition of, 147–149
derivatives of higher order, 149
Formula for Implicit Differentiation, 836–838
lenses, tangent lines, and normal lines, 149–150
Implicit Function Theorem, 837–838, 1013
Implicit surfaces, 1013–1016
Improper integrals
with CAS, 509–510
convergent and divergent, 507
tests for, 510–512
infinite limits in, 505–507
integrands with vertical asymptotes, 508–509
in integration, 504–512
of Type I, 505–508
of Type II, 508–509
Increasing functions, 5–6
Increments, AP-10–AP-13
Increment Theorem for Functions of Two Variables, AP-29–AP-31
Indefinite integrals, 287–294
definition of, 234–235, 287
of hyperbolic function, 442
substitution for, 287–294
of vector functions, 769
Independence, path, 986
Independent constrained variables, 883–884
Independent variable, of functions, 1
Independent variables, 1
definition of, 1
for functions
of one independent variable and three intermediate variables, 833
of one independent variable and two intermediate variables, 831
of several variables, 802
for two independent variables and three intermediate variables, 834
Indeterminate forms, 411–420
$\infty/\infty, \infty \cdot 0, \infty - \infty$, 414–415
0/0, 411, 412–413
behavior of, 411, 414
indeterminate powers, 415–416
l'Hôpital's Rule and, 411–420
Indeterminate forms, in Taylor series, 648–649
Indeterminate powers, 415–416
Index
of sequences, 574–575
of summation, 257–258
Inductance, 17-19
Induction, mathematical, AP-6–AP-9
Inequality
graphs/graphing, AP-10
rules for, AP-1
solving, AP-3–AP-4
in trigonometric functions, 26
Inertia, moments of, 930–933, 969, 1025
Infinite area, 505–507
Infinite discontinuity, 75
Infinite intervals, AP-3
Infinite limits, 505–507
Infinitely many orthonormal functions, 19-20
Infinite right-hand limit, 96
Infinite sequences, 574–575
Infinite series, 573, 586–593

Infinity
divergence of sequences to, 577
limits involved in, 83–96
finite limits as $x \to \pm\infty$, 83–84
finite sums and, 259–260
horizontal asymptotes, 85–87
limits at infinity of rational functions, 85
negative, 83, 90
vertical asymptotes, 91–93
Inflection points, 184, 201–204, 206, 207, 208
Initial conditions, 17-5
Initial point of vectors, 715
Initial ray, in polar coordinates, 677
Initial speed in ideal projectile motion, 772
Initial value problems
boundary value and, 17-5–17-6
definition of, 232–233
first-order, 536–537
Inner product of functions, 19-12, 19-14–19-15, 19-24
Input variable, 1, 16, 802
Instantaneous rates of change, 42–43, 45, 68, 104
Instantaneous speed, 39
Integer ceiling, 5
Integer floor, 5
Integers, 18-1–18-2
greatest, 5, 74
least, 5
starting, AP-8
Integrable functions, 265–266
Integrals
area, 246–249
with base *a* logarithms, 396
Brief Table of Integrals, 488
complete elliptic, of the second kind, 672
definite (*See* Definite integrals)
distance traveled, 249–252
elliptic, 503
of exponential functions, 388–389, 394, 396
finite sums, 256–262
flow, 979–980
for fluid force against a vertical plate, 345
Fundamental Theorem of Calculus, 276–284
of hyperbolic functions, 437–438
improper (*See* Improper integrals)
indefinite (*See* Indefinite integrals)
inside integral (functions), 901, 920
line (*See* Line integrals)
of lower sums, 247–248
Midpoint Rule, 248, 266–267, 494
multiple (*See* Multiple integrals)
of natural logarithms, 382–383
nonelementary, 492, 646–647
nonnegative continuous function, average value of, 252–254
outside (constants), 901, 920
in polar coordinates, 911–913
Product Rule in integral form, 462–466
of a rate, 281–282
sine-integral function, 514
substitution in, 287–304
for definite integrals, 294–304
for indefinite integrals, 288–292
surface, 1018–1026, 1038
of upper sums, 247–248
of vector functions, 769–774
work, 977–979
Integral sign, 234
Integral tables, 488–489
Integral Test, 18-24, 596–600
Integrands, 234, 508–509
Integration
basic formulas in, 457–461
with CAS, 490–491
in cylindrical coordinates, 938–942
definition of, 246
differential version, 463–466
differentiation and, relationship between, 282–284
formulas, 235
improper integrals in, 504–512
with CAS, 509–510
convergent and divergent, 507
tests for, 510–512
infinite limits in integration, 505–507
integrands with vertical asymptotes, 508–509
integral tables, 488–489
integrate command, in CAS, 490–491
integrating factor, for first-order linear equations, 544–546
nonelementary integrals in, 492
numerical, 494–501
error analysis in, 498–501
Midpoint Rule for approximating integrals, 494
parabolas used for approximations, Simpson's Rule and, 496–498
trapezoidal approximations, 495–496
by parts, 462–467
evaluating definite integrals, 466–467
formula, 463–466
Product Rule in integral form, 462–466
probability in, 515–525
distributions, 516–519
expected values, means, and medians in, 519–521
normal distributions in, 523–525
random variables, 515–516
standard deviation in, 522
uniform distributions in, 522–523
variance in, 522
of rational functions by partial fractions, 481–486
coefficients, determining by differentiating, 486
method used in, general description of, 481–485
reduction formulas, 489–490
with respect to *y*, 299–300
in spherical coordinates, 942–946
techniques of, 457–534
term-by-term for power series, 627–629
by trigonometric substitutions, 476–479
variable of, 234, 264–265
in vector fields, 965–1058
of vector function, 769–774
Integration formulas, 430–432
Interior intervals, AP-3
Interior points
of absolute extrema, 865–866
closed bounded, 864–867
of intervals, AP-3
in a plane, 804
in space, 806
Interior region of a circle, AP-15
Intermediate Value Property, 78
Intermediate Value Theorem for Continuous Functions, 78–79, 276
Intermediate variables
one independent variable and three intermediate variables, 833
one independent variable and two intermediate variables, 831
two independent variables and three intermediate variables, 834
Intersection of sets, AP-2
Intervals, AP-3
closed, 74, 180–188
of convergence, 626
differentiability on, 109–110
limits at endpoints of, 66–67
Inverse function–inverse cofunction identities, 430
Inverse functions, 11, 367–370
definition of, 367
equations
for $\ln x$, 386–387
for natural exponential functions, 387
number *e*, 386–387
for a^x and $\log_a x$, 394
finding, 368–371
hyperbolic, 439–442
logarithmic, 18-29–18-29, 380–381
trigonometric (*See* Inverse trigonometric functions)
Inversely proportional, 7
Inverse trigonometric functions, 420–435
arcsine and arccosine, 422–425
defining, 421–422
derivatives of, 427–430
graphs/graphing, 423, 424
identity, 133–134
integration formulas, 430–432
inverse function–inverse cofunction identities, 430
one-to-one, 422, 429
$\tan x$, $\cot x$, $\sec x$, and $\csc x$, 425–427
Irrational numbers, AP-2
Irreducible quadratic polynomial, 482
Isolated singularity, 18-25
Iterated integrals, 919–925
definition of, 896
finding, 919–925
volume of a solid region in space, 918–919

Jacobi, Carl Gustav Jacob, 950
Jacobian determinant, 951, 955
Jacobian of the transformation, 952–954, 957
j-component of vectors, 719–720
Jerk, in acceleration, 126–127, 136
Joint probability density function, 934
Joule (J), 340–341
Joule, James Prescott, 341
Jump discontinuity, 73, 75

k (kappa), 782
k-component, 719–720, 996–999
Kepler, Johannes, 796
Kepler's laws
first (ellipse law), 795
second (equal area law), 795–796
third (time-distance law), 796–797
Kinetic energy, 13, 169
Kirchhoff, Gustav R., 17-19
Kirchhoff's law, 17-19
Koch, Helge von, 595
Korteweg-de Vries equation, 831
Kovalevsky, Sonya, 441, 860
*K*th subinterval, 261

Lagrange, Joseph-Louis, 190, 869
Lagrange multipliers, 869–876
constrained maxima and minima, 869–872
definition of, 872
geometry of the solution, 874–875
method of, 872–875
Orthogonal Gradient Theorem and, 872–873
partial derivatives and, 869–876
with two constraints, 875–876
Λ, 871
Laplace, Pierre-Simon, 825
Laplace equations, 830, 840
Law of cosines, 25–26
Laws
of cosines, 25–26
of exponents, 389–390
Law of Exponential Change, 550–551
Law of Refraction, 216
Limit Laws, 48–49
Learning rate, B-13
Least integer, 5
Least squares line, 869
Least squares solution, 19-10–19-11
Least upper bound, AP-24, 581
Left-continuous function (or continuous from the left), 73–74

Left-hand derivatives, 109–110
Left-handed coordinate frame, 710
Left-hand limits
  continuity at a point and, 73–74
  informal definition of, 65–66
  precise definition of, 67–68
  proof for, AP-21, 69–70
Legendre, Adrien-Marie, 545
Leibniz, Gottfried, 264, 276
Leibniz's formula, 648
Leibniz's notation, 108, 264
  in Chain Rule, 140
  for definite integral, 264
  in differentials, 160, 164
  in outside-inside rule, 141–142
  for writing the formula for arc length, 670
Leibniz's Rule, 178–179, 309
Leibniz's substitution method, 288–292
Length
  arc, 329–333
    length differential, 673–674
    in space, 778–780
  constant, vector functions of, 766
  focal, of parabolas, 691
  of parametrically defined curves, 670–673
  in polar coordinates, 685–688
  of a polar curve, 687–688
  of a smooth curve, 778
  of three-dimensional vectors, 19-6
  of vectors, 715, 716
Lenses, 149–150
Level curves of functions of two variables, 805
Level surfaces, 805–806
L'Hôpital, Guillaume de, 411
L'Hôpital's Rule, 412
  applied to f/g, 412
  indeterminate forms and, 411–420
  proof of, 418
  for sequences, 578–579
  using, 413
Limaçons, graphing, 685
Limit Comparison Test, 604–606, 618
Limit from the left. *See* Left-hand limits
Limit from the right. *See* Right-hand limits
Limit Product Rule, AP-19–AP-20
Limit Quotient Rule, AP-20–AP-21
Limits, 38–101
  commonly occurring, AP-22–AP-23, 580
  complex, 18-2–18-3
  continuity and, 72–83
  of continuous functions, 77–78
  for cylindrical coordinates, 940–942
  for definite integrals, 265
  definition of
    informal, 45–48
    precise, 56–61
    to prove theorems, 61
    testing, 57–59
  deltas for given epsilons, finding algebraically, 59–61
  of difference quotient, 103–104
  estimating, by using calculators and computers, 50–51
  for functions of two variables, 810–814
    definition of, 810–811
    properties of, 812–814
    two-path test for nonexistence of a limit, 815
  of function values, 45
  graphs to verify limit statements for functions, 57–59
  history of, 45
  infinite, in improper integrals, 505–507
  infinity involved in, 83–96
    finite limits as $x \to \pm\infty$, 83–84
    finite sums and, 259–260
    horizontal asymptotes, 85–87
    infinite right-hand limit, 96
    limits at infinity of rational functions, 85
    vertical asymptotes, 91–93
  laws, AP-19
  left-hand (*See* Left-hand limits)
  Limit Comparison Test, 511–512, 604–606, 618
  Limit Laws, 48–49
  number e expressed as, 391–392
  one-sided, 65–70
    approaching, 65–66
    at endpoints of an interval, 66–67
    involving sine, 68–70
    precise definitions of, 67–68
  of polynomials, evaluating, 49
  precise definitions of, 56–61
  of rational functions, evaluating, 49
  for rectangular coordinates, 919–925
  of Riemann sums, 260–262, 265, 325
  right-hand (*See* Right hand-limits)
  r-limits of integration, finding of, 913, 939, 941
  rules
    Constant Multiple Rule, 48–49
    Difference Rule, 48–49
    Limit Product Rule, AP-19–AP-20
    Limit Quotient Rule, AP-20–AP-21
    Power Rule, 48–49
    Product Rule, 48–49
    Quotient Rule, 48–49
    Root Rule, 48–49, 812
    Sum Rule, 48–49, 61
  Sandwich Theorem for calculating, 51–53, AP-21
  of sequences, 575
    calculating, 577–578
    commonly occurring, 580
    definition of, 575
  of sine, 68–70
  for spherical coordinates, 944–945
  theorems, proofs of, AP-19–AP-21
  two-sided, 65, 69, 74, 92, AP-21
  of vector-valued functions, 761–762
Limits of integration, finding
  for double integrals, 894, 896, 899, 901, 903
    in polar form, 913–914
  in Fubini's Theorem, 896, 902
  in iterated integrals, 919–925
  for iterated triple integrals, 919–925
  lemniscate for, 914
  ϕ-limits, 944, 945
  in polar coordinates, 913–914
  in rectangular coordinates, 913–914
  ρ-limits, 944, 945
  r-limits, 913, 939, 941
  by sketching, 903, 913, 919, 940, 944
  θ-limits, 913, 940, 941, 944, 945
  for triple integrals
    in cylindrical coordinates, 940–942
    in rectangular coordinates, 919–925
    in spherical coordinates, 944–946
  using horizontal cross-sections, 903
  using vertical cross-sections, 903
  x-limits, 903, 920, 921, 924
  y-limits, 903, 920–924
  z-limits, 920–924, 939, 940, 941
Linear algebra, B-4
Linear approximation, 161–163
  center of, 161, 163
  error formula for, 880
  for roots and powers, 163
  standard, 854–855
Linear combination, 17-1, 19-7
Linear drag, 777
Linear equations, AP-13, 543–544
Linear functions, 7
Linearization, 160–163
  definition of, 161, 854
  differentials and, 164–169
  of a function of two variables, 853–855
  of functions of more than two variables, 856–857
  linear approximation and, 161–163
  standard linear approximation and, 854–855
Linearly independent solutions, 17-2
Linear regression, 562
Linear second-order constant-coefficient models, 17-20
Linear transformations, 951
Line integrals
  additivity and, 967–968
  in conservative fields, 987–990
  definitions of, 966, 975
  evaluating, 966, 975
  formulas for, 976
  Fundamental Theorem of, 987–989
  mass calculations, 968–969
  moment calculations, 968–969
  in a plane, 969–970
  with respect to dx, dy, or dz, 975–976
  of scalar functions, 965–970
  of vector fields, 974–976
  work done by a force over a curve in space, 977–979
Lines
  center of mass along, 350–351
  in implicit differentiation, 149–150
  of intersection, 744–745
  least squares, 869
  line segments and, 741–743
  normal, 149–151, 850–852
  to parabolas, 151
  parallel, AP-13
  parametric equation for, 741–742
  perpendicular, AP-13, 149–150
  regression, 869
  search, B-13
  segments, 741–743
    directed, to represent vectors, 715
    midpoint of, 720
  slope of a nonvertical, AP-11
  straight, increments, AP-10–AP-13
  tangent, 149–150, 161, 668–669, 850–852
  through u_1, 19-11
  trend, 869
  vector equation for, 741
  work done by variable force along, 341
Lissajous figures, 677
"Little-oh" notation, 448
Ln x
  algebraic properties of, 380
  derivative of, 378–379
  graph and range of, 381–382
  integral of, 382
  inverse equation for, 387
  inverse of, 386–388
  number e and, 378
  two-decimal-place approximations of, 378
Local maxima or minima extrema, B-7, 860–864, 866
  critical points in, 861–864
  definitions of, 182, 860
  discriminant or Hessian of, 862–864
  finding, 183–185
  finding the absolute extrema of a continuous function f on a finite closed interval, 184–185
  First Derivative Test for Local Extrema Values, 183–184, 196–198, 861
  Second Derivative Test for, 201, 204–208
  Second Derivative Test for Local Extreme Values, 862, 864, 866, 870, 879–880
  values for functions of two variables, 860–864

Logarithmic differentiation, 383–384, 394
Logarithmic functions, 11
of complex numbers, 18-28–18-29
graphs/graphing, 11
inverses in, 18-29–18-29
principal branch of, 18-29
Logarithms
algebraic properties of, 380–381
with base *a*, 394–395
laws of exponents and, 389–390
$\log_a x$, 394–396
Logistic Equation in Neural Networks and Machine Learning, 561–562
Logistic growth, 560
Logistic model for population growth, 560–561
Loops, 980
Lotka-Volterra equations, 569
Lower bound sequence, 581
Lower sums, 247–248

Machine Learning, Logistic Equation in, 561–562
Maclaurin, Colin, 633
Maclaurin series, 633–635, 639
Magnetic flux, 1057
Magnitude, 715, 716
Main diagonal, B-4
Major axis, AP-17, 693
Marginal costs, 128–129, 216–217
Marginal profit, 216–217
Marginal revenue, 216–217
Marginals, 128–130
Mass
along a line, 350–351
calculations, line integrals and, 968–969
center of, 929–930
centroids, 356–357
converting to energy, approximations in, 169
definition of, 928
density, of surface integrals, 1018, 1020
distributed over a plane region, 352
first moments and, 928–930
formulas, 929–930
moments and center of, 350–359
in multiple integrals, 928–930
plates, 352–356
of surface integrals, 1018, 1020
theorems of Pappus, 357–359
thin shells, 1024–1026
thin wires, 351
Mathematical induction, AP-6–AP-9
Mathematics, applied optimization in, 215–218
Matrix, B-1–B-6
cofactor in, B-2–B-5
definition of, B-1
Hessian, B-9–B-11
main diagonal of, B-4
minor defined in, B-1–B-2
Maxima or minima extrema
absolute, 860–861, 864–867
constrained, 869–872
local, 860–864, 866
Max-Min Inequality Rule, 268–269
Max-Min Tests, 866
Mean
definition of, 272
expected, in probability, 519–521
in multiple integrals, 935
Mean Value Theorem, Cauchy's, 380, 416–418
Mean Value Theorems, 190–194, 276–279
arbitrary constants, 230
Cauchy's, 416–417
Corollary 1, 191, 192
Corollary 2, 191–192
Corollary 3, 195–196, 201
in defining surface area, 337
definition of, 190
for derivatives, 190–194
to find velocity and position from acceleration, 192
mathematical consequences of, 191–192
for parametrically defined curves, 670
physical interpretation of, 191
proof of, 190–191
Rolle's Theorem, 188–191
Taylor's Theorem, 637
Median, 519–521
Memoryless principle, 518–519
Mendel, Gregor Johann, 130
Mercator Projection, 18-32
Mesh size, 495
Meter (m), 340
Methods
of determinants, 17-12
of gradient ascent, B-14
of gradient descent, B-12–B-14
of Lagrange multipliers, 872–875
of partial fractions, 481–486
of slicing (*See* Slicing, methods of)
Midpoint, 720
Midpoint Rule, 248, 266–267
for approximating area, 248, 266–267
for approximating definite integrals, 494
for approximating integrals, 494
error analysis and, 498–501
Minima extrema
absolute, 860–861, 864–867
constrained, 869–872
local, 860–864, 866
Minor axis, AP-17, 693
Mixed Derivative Theorem, AP-27–AP-28, 825–826
Mixture problems, 552–554
Moments
calculations of, 968–969
center of mass and, 350–359
definition of, 350
first, 928–930
of inertia, 930–933, 969, 1025
polar, 932
of the system about the origin, 350–351
of thin shells, 1024–1026
Monotonic functions, 195–200
Monotonic sequences, 581–582
Monotonic Sequence Theorem, 582, 596
Motion
along a line, 125–128
acceleration, 126–127
simple harmonic motion, 135–136
speed, 125–126
velocity, 125–128
antiderivatives and, 233–234
in cylindrical coordinates, 794
damped, 17-17–17-18
direction of, 763–764
Kepler's laws
first (ellipse law), 795
second (equal area law), 795–796
third (time-distance law), 796–797
Newton's second law of, 549, 559, 771
of planets in planes, 795
in polar coordinates, 794
with resistance proportional to velocity, 549–550
simple harmonic, 17-16–17-17, 135–136
vibrations, 17-15–17-16
Multiple integrals, 894–964
applications, 928–935
masses and first moments, 928–930
means and expected values, 935
moments of inertia, 930–933
probability, 933–935
double integrals (*See* Double integrals)
substitutions in, 950–957
triple integrals (*See* Triple integrals)
Multiplication
of complex numbers, 18-3–18-4
of functions, 14–15
positive definite, 733
property of real numbers, AP-23
scalar, of vectors, 717–718, 729–730
Series Multiplication for Power Series, 626
M, 779

Napier, John, 380
Natural domain of functions, 2
Natural exponential functions
definition of, 387
derivative of, 388–389
integral of, 382–383
inverse equation for, 387
laws of, 389–390
Natural logarithm functions
derivatives and integrals of $\log_a x$, 396
derivatives of, 378–379, 396
general exponential function a^x in, 390
graph and range of $\ln x$ in, 381–382
integral $\int 1/x\,dx$ in, 382
inverse of $\ln x$ and the number e in, 386–387
laws of exponents and, 389–390
logarithms with base *a*, 394–395
Natural logarithms, 377–385
algebraic properties of, 380–381
definition of, 377–379
derivative of $y = \ln x$ in, 378–379
integrals of natural, 382–383
number e in, 378
Natural parametrization, 661
Navigation, vectors in, 721
N dimensional vectors, 722–723, 731–732
Negative infinity, 83, 90, 577
Negative of differentiable functions, 117
Negative reciprocal of a slope, AP-13
Net Change Theorem, 281–282
Neural Networks, Logistic Equation in, 561–562
Newton (N), 340
Newton, Sir Isaac, 276
Newton-meter (N · m), 340–341
Newton's Law of Cooling, 407–408, 558, 559
Newton's method, 225–229
applying, 226–227
convergence of the approximations, 227–228
procedure for, 225
sequences generated by, 573, 581, 584, 585
Newton's second law, 169, 549, 559, 771
Newton's serpentine, 122
Nicole Oresme's Theorem, 575, 657
Nondecreasing partial sums, 596
Nondecreasing sequences, 581–582
Nonelementary integrals, 492
Nonhomogeneous linear equations, 17-1, 17-7–17-14
form of the general solution, 17-7–17-8
method of undetermined coefficients, 17-8–17-11
method of variation of parameters, 17-12–17-14
Nonincreasing sequences, 581–582
Nonintegrable functions, 265–266
Nonlinear transformations, 953–954
Nonzero vectors
in cross product, 734–735
in dot product, 726–729
expressing, 720
gradient, 845–846
in lines and planes in space, 741, 744–745
Norm
of functions, 19-12–19-14, 19-24
of partitions, 262, 895
Normal components of acceleration, 788–789
Normal form of Green's Theorem, 1002–1003, 1050, 1051
Normalized trigonometric system, 19-16

Normal line, 149–151, 850–852
Normal probability
density function, 514
distributions, 523–525
Notation
average value of a continuous function, 271
for constrained variables, 885–886
definite integrals, 264
Evaluation Theorem, 280
for intervals, AP-3
Leibniz's, 264
for partial derivatives, 820
sigma, 19-5, 256–259
for vectors, 729
*N*th partial sums, 586–588
*N*th term of sequences, 574, 579, 586–588
*N*th-Term Test, 590–591, 618
Number *e*
definition of, 378
in exponential functions, 386–388
inverse equation for, 386–387
as a limit, 391–392
Numbers
algebraic, 396
completeness property of real, 582
counting (natural), 18-1
d-tuple of, 19-9
e (*See* Number *e*)
Fibonacci in sequences, 580
hierarchy of, 18-1–18-3
integers in, AP-2, 18-1–18-2
irrational, AP-2
rational, AP-2, 18-2
systems of, successive stages of invented, 18-1–18-2
transcendental, 386
Numerical integration, 494–501
error analysis in, 498–501
Midpoint Rule for approximating, 494
Simpson's Rule for approximating, 496–498
trapezoidal approximations, 495–496
Numerical method, 538–541
Numerical solution of the problem, 538–541

Oblique asymptote, 87–88
Octants, 710–711
Odd functions, symmetry properties of, 6–7
Oh-notation, 448
One-dimensional heat equation, 831, 893
One-dimensional wave equation, 830
One-sided derivatives, 109–110
One-sided limits, 65–70
approaching, 65–66
at endpoints of an interval, 66–67
involving sine, 68–70
precise definitions of, 67–68
One-to-one functions, 366–367
definition of, 366
domain restrictions, 370, 421
horizontal line test for, 367
inverse trigonometric, 422, 429
sine, 367, 421
Open intervals, AP-3
Open region, 804, 806–807
Optimization problem, 19-9
Order, oh-notation and, 448
Ordered field, AP-24
Order of Integration Rule, 268
Order properties of real numbers, AP-1, AP-24
Ordinate (*y*-coordinate), AP-10
Oresme, Nicole, 575, 657
Oriented surface, 1022
Origin, AP-10
graph of an odd function as symmetric about, 6–7
of planes, 710–711
in polar coordinates, 677
symmetric about, 6–7
Taylor's formula at, 882
Orthogonal functions, 19-15, 19-24
Orthogonal Gradient Theorem, 872–873
Orthogonality of the trigonometric system, 19-15–19-16
Orthogonal projection of v onto the span of $\mathbf{u}_1, \mathbf{u}_2, \ldots, \mathbf{u}_n$, 19-11
Orthogonal trajectories, 549, 551–552
Orthogonal vectors, 728
Orthonormal basis for three-dimensional space, 19-8–19-9
Orthonormal functions, 19-12, 19-15–19-16, 19-20, 19-24, 19-27
Oscillating discontinuity, 75
Osculating circle, 177, 784–785
Output variables, 802
Outside-inside rule, 141–142, 849
Overdamping, 17-18

Paddle wheel interpretation, 1034–1037
Pairs, polar coordinate, 677–678
Pairwise cross products, 735
Pappus, 357
Theorem for Surface Areas, 359
Theorem for Volumes, 357–358
Parabolas, 691–692
areas between curves, 297, 299
axis of, AP-15–AP-16
definition of, 691–692
directrix of, 691–692
eccentricity of, 698–702
equations for, 691–692
focal length of, 691
focus of, 691–692
as graphs of equations, AP-15–AP-16
normal lines to, 151
polar equations for, 700–702
reflective property of, 698
semicubical, 151–152
Simpson's Rule for approximating, 496–498
vertex of, 691
y-axis of, 691
Paraboloids
definition of, 749
elliptical, 749–751, 752
hyperbolic, 750–751, 752
region enclosed by, 924
Parallel Axis Theorem, 937
Parallel lines, AP-13
Parallelograms
area of, cross product in, 735–736
law of addition, 717
Parallel vectors, 734, 735
Parameter
definition of, 1008
domain, 1008
interval in parametric equations, 659
method of variation of, 17-7, 17-12–17-14
Parametric curves, 565, 778–780
Parametric equations, 659–663
Cartesian, 662
of cycloids, 663–664
definition of, 659
examples of, 659–663
for ideal projectile motion, 771–773
for lines, 741–742
for natural parametrization, 661
parameter interval in, 659
parametrization of the curve, 659–663
terminal point in, 659
values of, 660, 662
Parametric formulas, 668–669
Parametrization
arc length, 779
natural, 661
of plane curves, 659–664
area in a plane, 685–687
areas and lengths in polar coordinates, 685–688
conic sections, 690–695
conics in polar coordinates, 698–703
length of a polar curve, 687–688
of surfaces, 1008–1016
of cones, 1009
of cylinders, 1009
by cylindrical coordinates, 1009
of implicit surfaces, 1013–1016
of a sphere, 1009
of surface area, 1009–1013
Parseval Equality, 19-21–19-22
Partial derivatives, 819–828
calculations of, 822–823
Chain Rule for, 831–838
with constrained variables, 883–886
continuity and, 814–816, 824–825, 1038
Corollary 3, 828
Corollary of Theorem 3 and 4, 828
definitions of, 820–821
differentiability and, 827–828
Differentiable Implies Continuous Theorem, 828
differentials in, 855–856
directional derivatives and, 841–849
extreme values and, 816, 860–867
of functions of more than two variables, 823–824, 856–857
of functions of several variables, 802–807
of functions of two variables, 819–821
gradient vectors and, 841–849
identity function with continuous, 1038
Increment Theorem for Functions of Two Variables, 828
Lagrange multipliers, 869–876
limits for functions of two variables, 810–814
linearization of a function of two variables, 853–855
Mixed Derivative Theorem, 825–826
normal lines and, 850–852
notation for, 820
second-order, 825
of still higher order, 826
tangent planes and, 850–857
Partial fractions
definition of, 481
integration of rational functions by, 481–486
method of, 481–486
Partial sums
in divergent series, 590–591
nondecreasing, 596
n*th*, 586–588
of sequences, 586–587
sequences of, 586–587
Particles, 760
Particular solution, 17-8, 233, 536–537
Partitions
formed by double integrals over rectangles, 894–895
*k*th subinterval of, 261
norm of, 262, 895
for Riemann sums, 260–262
Parts, integration by, 462–467
evaluating definite integrals, 466–467
formula, 463–466
Product Rule in integral form, 462–466
Pascal, Blaise, 587
Path
Chain Rule for, 848–849
derivative along, 849
direction along, 966
independence, 986
of particles, 760
Percentage change, 168
Perihelion, 704
Periodic functions, 19-1–19-2, 24
Periodic trigonometric functions, 24

Periods
of periodic functions, 19-1
of trigonometric functions, 24
Perpendicular lines, AP-13, 149–150
Perpendicular vectors, 728
Phase angle, 17-17
Phase lines, 555–557
Phase-plane analysis method, limitations of, 567
Phase planes, 564–565
Φ, 942, 944, 945
Physics
applied optimization in, 215–218
vectors in, 721
Piecewise continuous functions, 265, 310
Piecewise-defined functions, 5
Piecewise smooth curves, 763
Plane
angles between, 746–747
area in, 685–687
Cartesian coordinates in, AP-9–AP-10
circle of curvature for plane curves, 784–785
circles in, AP-13–AP-15
coordinate, 710–711
curvature of, 782–784
curve, flux across, 980–981
determined by coordinate axes, 710
determined by or spanned by r and s, 19-9
distance in, AP-13–AP-15, 746
equation for, 743–744
flux across, 981
line integrals in, 969–970
origin of, 710–711
points in, AP-14
vector equation for, 744
Plane region
boundary of, 804
bounded/unbounded, 804
definitions of, 804
interior of, 804
mass distributed over, 352
open/closed, 804
Planetary motion, Kepler's laws of, 795–797
Plates
bounded by two curves, 355–356
density of, 352
integral for fluid force against a vertical, 345
thin, flat, 352–355
two-dimensional, formulas for, 929
Points
boundary, of a region
closed bounded, absolute maxima and minima on, 864–867
in Max-Min Tests, 866
in a plane, 804
in space, 806
continuity at, 73–75
continuous extension to, 80–81
continuous functions at, 814–815
of differentiable functions, 862
distance
formula, AP-14, 712–714
from a line to, 743
from a plane to, AP-14, 746
between two, in the *xy*-plane, 712–714
fixed, 83
functions that are continuous at, 814
of inflection, 184, 201–204, 206, 207, 208
initial, of vectors, 715
interior, of a region
closed bounded, 864–867
in a plane, 804
in space, 806
saddle, 751, 860–867
standard position for, 716
Taylor's formula at, 881
terminal, of vectors, 715
Point-slope equation, AP-12–AP-13
Poisson, Siméon-Denis, 860
Polar coordinates
acceleration in, 794–797
area in, 685–688, 913–914
of polar region, 685–687
Cartesian coordinates related to, 678, 679–680
conic sections, 690–695
conics in, 698–703
definition of, 677–678
initial ray in, 677
integrals in, 911–913
lengths in, 685–688
of lines, 702
motion in, 794
pairs in, 677–678
polar curve
length of, 687–688
slope of, 682–683
polar equations for, 678
pole in, 677
symmetry tests for, 682
velocity in, 794–797
Polar curve
length of, 687–688
slope of, 682–683
Polar equations
for circles, 702–703
for conics, 700–702
for ellipses, 700–702
graphs/graphing, 678–679, 681–684
Archimedes spiral in, 707
converting a graph from the $r\theta$-plane to the *xy*-plane, 682–683
lemniscate curves, 684
limaçons, 685
slope, 682–683
symmetry, 682
technology for, 682–683
for hyperbolas, 700–702
for lines, 702
for parabolas, 700–702
for polar coordinates, 678
relating polar and Cartesian coordinates, 679–680
Polar form, double integrals in, 911–916
Polar moment, 932
Polar rectangles, 911
Pole, in polar coordinates, 677
Polyhedral surfaces, 1037–1038
Polynomials, 9
coefficients of, 9
degrees of, 9, 19-18–19-19
irreducible quadratic, 482
limits of, evaluating, 49
Taylor, 634–636
trigonometric, 19-3–19-5, 19-18–19-19
Population growth
logistic model for, 560–561
unlimited, 402–405
Position, acceleration for finding, 192
Position vector, 760
Positive definite multiplication, 733
Potential function, 986
Potentials for conservative fields, finding, 990–992
Power Chain Rule, 142–143
Power functions, 7–8
Power Rule, 48–49
for base *a* logarithms, 395
for derivatives, 390–391
for finding complex derivatives, 18-19
General Power Rule, 381, 390–391
in law of exponents, 390
for limits, AP-19, 812
for logarithmic differentiation, 383, 394
for natural logarithms, 380
Powers
binomial series for, 644–646
Chain Rule and, 142–143
of complex numbers, 18-7
linear approximation for, 163
of sines and cosines, products of, 470–471
of $\tan x$ and $\sec x$, 472–473
Power series, 621–629
convergence of, 621–626
Convergence Theorem for Power Series, 624–626
interval of, 626
radius of, 624–626
testing for, 626
definition of, 621
method, 17-24–17-29
operations on, 626–629
convergence of power series, 627
Series Multiplication for Power Series, 626
Term-by-Term Differentiation, 627–628
Term-by-Term Integration, 628–629
solutions, 17-24–17-29
Predator-prey model, 569–570
Preimage, 950
Prescribed errors, 57
Pressure-depth equation, 344
Pressure of fluid, force and, 344–345
Principal branch of logarithmic functions, 18-29
Principal minor determinants, B-9–B-11
Principal *n*th root, 18-8
Principal unit normal vector, 783, 792
Principal value, 18-25, 18-29
Probability, 515–525
density functions
for *[a, b]*, 518
definition of, 517–518
exponentially decreasing, 518–519
normal, 514
standard normal, 525
distributions, 516–519
normal, 523–525
uniform, 522–523
expected values, means, and medians in, 519–521
joint probability density function, 934
in multiple integrals, 933–935
random variables, 515–516
standard deviation in, 522
variance in, 522
Production, costs of, 128–129
Product Rule, 48–49
for base *a* logarithms, 395
to calculate limits of sequences, 577
for finding complex derivatives, 18-19
for gradients, 846
in integral form, 462–466
in law of exponents, 390
for limits, AP-19
for limits for functions of two variables, 812
for logarithmic differentiation, 383
for natural logarithms, 380, 383
Products
of complex numbers, 18-5–18-6
of trigonometric integrals, 474
powers of, 470–471
triple scalar, 738–739
vectors of, 735
Projectile motion, 769–774
ideal
equation for, 772
height, flight time, and range for, 773
vector and parametric equations for, 771–773
integrals of vector functions, 769–771
with wind gusts, 773–774
Projections of vectors, 728–731
Proof by mathematical induction, AP-6–AP-9
Proportional relationship, of functions, 7
p-series, 598–600
p-Series Test, 598, 618
Pythagorean Theorem, 19-6–19-7, 19-15, 24, 26, 424–425
Pythagorean triples, 584

Quadrants, AP-10
Quadratic approximation, 636
Quadratic functions, 9
Quadratic polynomial, irreducible, 482
Quadric surfaces, 749–752
  definition of, 749
  ellipsoids, 749–751, 752, 753
  elliptical cones, 749–751, 752
  general, 751, 753
  graphs/graphing, 752
  hyperboloids, 749–751, 752
  saddle point of, 751
Quotient Rule, 48–49
  for base *a* logarithms, 395
  to calculate limits of sequences, 577
  for derivatives of hyperbolic functions, 438
  for finding complex derivatives, 18-19
  for gradients, 846
  for limits, AP-19
  for limits for functions of two variables, 812
  for logarithmic differentiation, 383
  for natural logarithms, 380
Quotients, in complex numbers, 18-6–18-7

Radians, 21, 22, 23
Radioactive decay, 405–407
Radioactive elements, half-life of, 405–407
Radioactivity, 405–407
Radius
  of convergence, 624–626
  of curvature, 784
Random variables, 515–516
  continuous, 516
  definition of, 515
  discrete, 516
  exponential density function for, 520–521
  probability bar graph for, 517
  standard deviation in, 522
  variance of, 522
Range
  of functions, 1–2, 802, 803
  for ideal projectile motion, 773
  of ln *x*, 381–382
Rate constant, 400
Rates
  of decay, 400
  fluid flow, 1000
  of growth, 445–450
  integrals of, 281–282
Rates of change, 38–43, 124–133
  average, 40, 42–43
  instantaneous, 42–43, 45, 68
  marginals, 128–130
  secant lines and, 40
  sensitivity to, 130
  slope of a curve and, defining, 40–42
  in speed, 38–40
  tangent lines and, 42–43
  in velocity, 125–128
Ratio, in geometric series, 588
Rational functions, 9
  coefficients in, determining by differentiating, 486
  integration of, by partial fractions, 481–486
    method used in, general description of, 481–485
  limits of
    evaluating, 49
    at infinity, 85
Rational numbers, AP-2, 18-2
Ratio Test, 18-24, 609–610, 618
Real Analysis, 19-13
Real axis, 18-4
Real line, AP-1
Real numbers
  addition property of, AP-23
  algebraic properties of, AP-1, AP-23
  associative laws of, AP-24
  commutativity laws of, AP-24
  completeness property of, AP-1–AP-2, AP-23, AP-24, 582
  construction of reals and, AP-25–AP-26
  definition of, AP-1, AP-2
  irrational, AP-2
  multiplication property of, AP-23
  order properties of, AP-1, AP-24
  real line and, AP-1
  sets of, AP-2
  subsets of, AP-2
  theory of, AP-23–AP-26
Real parts of complex numbers, 18-3–18-4
Reals, construction of, AP-25–AP-26
Real-valued function, 2, 760–761, 802–803
Rearrangement Theorem for Absolutely Convergent Series, 617, 618
Reciprocal Rule, 380, 382, 395
Reciprocals
  computing, 18-6
  derivative of, 107
  of derivatives for inverses, 371–373
  of sine function, 422
Rectangles
  average value of the function for approximating, 253–254
  between curves, 297–298
  finite approximations, 249–254, 377
  finite sums for estimating, 247–249, 259–260
  Fubini's Theorem for calculating double integrals over, 896–898
  Fundamental Theorem of Calculus, 277
  under the graph of a nonnegative function, 270
  integration with respect to *y*, 299–300
  Mean Value Theorem, 276
  polar, 911
  Riemann sums for approximating, 261–262, 283
  Riemann sums in defining, 262
  total, 283–284
Rectangles, area of
  average values for approximating, 253–254
  between curves, 297–298
  finite approximations for, 249–254, 377–378
  finite sums for estimating, 246–254
  Fundamental Theorem of Calculus, 283–284
  under the graph of a nonnegative function, 270–271
  Riemann sums for approximating, 261–262, 283
  Riemann sums in defining, 261–262
  total, 283–284
Rectangular coordinates, AP-10, 710, 918–925
  average value of a function in space, 925
  definition of, 918
  equations relating cylindrical coordinates to, 938–940
  iterated integrals, 919–925
  properties of, 925
  triple integrals in, 918–925
    average value of a function in space, 925
    in cylindrical coordinates, 938–942
    definition of, 918
    iterated integrals, 919–925
    properties of, 925
    Riemann sums for, 939, 943
    in spherical coordinates, 942–946
    volume of a solid region in space, 918–919
Recursion formula, in sequences, 580
Recursive definitions, of sequences, 580–581
Reduction formula, 466, 489–490
Reflection of a graph, 16–18
Reflective property of parabolas, 698
Refraction, Law of, 216
Regions
  of a circle
    exterior, AP-15
    interior, AP-15
  connected, 986
  Divergence Theorem
    for other regions, 1046–1047
    for special regions, 1045–1046
  Green's Theorem for special, 1004–1005
  open, 1038–1039
  simply connected, 986
  in space (*See* Space, regions of)
Regression lines, 869
Reindexing series, 592–593
Related rates, 152–156
  Chain Rule in, 152, 154
  consistent units in solving, 156
  definition of, 152
  differentiation in, 155
  strategy, 153
Relative change, 168
Relative extrema, 861
Relative maxima or minima extrema. *See* Local maxima or minima extrema
Remainder
  in Integral Test, 599–600
  of order *n* (or error term), 638–639
Remainder Estimation Theorem, 639
Remainder Formula, 599
Removable discontinuity, 73, 75
Representations, series, 632–633
Resistance
  definition of, 17-19
  falling body encountering, 558–560
  proportional to velocity, 549–550
Resolution level zero, 19-29–19-31
Rest points, 555–557
Resultant vectors, 717
Revolution
  areas of surfaces of, 335–338, 674–675
  solids of, 313–318
    disk method, 313–316
    washer method, 316–318
P-limits of integration, 944, 945
Riemann, Georg Friedrich Bernhard, 260
Riemann sums, 260–262
  for approximating area of rectangles, 261–262, 283
  for arc length, 330
  in calculus with parametric curves, 670
  convergence of, 265
  in defining area of rectangles, 262
  in defining the definite integral, 263–264
  for double integrals, 894–896
    over area, 908
    over general regions, 904
    over rectangles, 895–896
  in expected values, means, and medians, 519–520
  for integrals, 311, 353
  limits of, 260–262, 265, 325, 328
  for mass of thin, flat plates, 352
  for mass of thin wires, 351
  notion of, 260–262
  for parametrically defined curves, 670
  partitions for, 260–262
  in shell method, 325
  slicing by parallel lines, 312
  slicing with cylinders, 323–324
  for spherical coordinates, 939, 943
  total area of rectangles, 261–262
  for triple integrals, 939, 943
  for work done by a variable force, 341, 343, 345

Right-continuous function (or continuous from the right), 73–74
Right-hand derivatives, 109–110
Right-handed coordinate frame, 710
Right-hand limits
continuity at a point and, 73–74
informal definition of, 65–66
precise definition of, 67–68
proof for, AP-21, 69–70
Right-hand rule for vectors, 734
Rise, AP-11
RL circuits, 547
R-limits of integration, 913, 939, 941
Rolle, Michel, 181, 189
Rolle's Theorem, 188–189
Root finding, 78–79
Root Rule, AP-19, 48–49, 812
Roots
binomial series for, 644–646
in complex numbers, 18-7–18-9
linear approximation for, 163
*n*th, 18-7–18-8
principal *n*th, 18-8
Root Test, 611–612, 618
Run, AP-11

Saddle points, B-6–B-11, 860–867
St. Vincent, Gregory, 690
Sandwich Theorem, AP-21, 578
Scalar
components of acceleration, 788–789
definition of, 717
differential form, 978
functions, 760–761
line integrals of, 965–970
surface integrals of, formulas for, 1019
multiplication of vectors, 717–718, 729–730
Scalar Multiple Rules, 765
Scaling of a graph, 16–18
Scatterplot, of functions, 4
Secant function, 22–23
derivatives of, 136–137
even, 24
graphs/graphing, 24, 423
integrals of, 383, 472–473
periodicity of, 24
powers of, 472–473
in Substitution Rule, 291
Secant lines, 40–43
Second Derivative Test
in applied optimization, 217
for concavity, 201
for local extrema, B-9, 201, 204–208, 862, 864, 866, 870, 879–880
proving, 243
Taylor's formula to derive, 879–880
Second-order differential equations, 17-1–17-29
applications, 17-15–17-20
auxiliary equation, 17-3
boundary values, 17-6
constant-coefficient homogeneous equations, 17-2–17-5
damped motion, 17-17–17-18
damped vibrations, 17-18
definition of, 17-1
electric circuits, 17-18–17-19
Euler equations, 17-22–17-23
general solution, 17-2
homogeneous linear equations, 17-1
initial conditions, 17-5
initial value problems, 17-5–17-6
linear combination, 17-1
linearly independent solutions, 17-2
linear second-order constant-coefficient models, 17-20
nonhomogeneous linear equations, 17-1, 17-7–17-14
overdamping, 17-18
power-series solutions, 17-24–17-29
simple harmonic motion, 17-16–17-17
superposition principle, 17-1
trivial solution, 17-2
underdamping, 17-18
variation of parameters, 17-7, 17-12–17-14
vibrations, 17-15–17-16
Second-order partial derivatives, 825
Semicubical parabolas, 151–152
Semimajor axis of ellipses, 693
Semiminor axis of ellipses, 693, 701–702
Separable differential equations, 401–402, 432–434
Sequences, 573–582
bounded, 581–582
Continuous Function Theorem for, 578
convergence of, 18-23, 575–57
divergence of, 18-23, 575–577
essay on, 573
Fibonacci numbers in, 580
history of, 573
index of, 574–575
infinite, 574–575
L'Hôpital's Rule for, 578–579
limits of, 577–578
calculating, 577–578
commonly occurring, 580
definition of, 575
monotonic, 581–582
Monotonic Sequence Theorem, 582
Newton's method for generating, 573, 581, 584, 585
nondecreasing/nonincreasing, 581–582
*n*th term of, 574, 579, 586–588
of partial sums, 586–587
recursion formula in, 580
recursive definitions of, 580–581
representing, 574–575
Sandwich Theorem for, 578
terms of, 574
zipper theorem for, 585
Sequential search, 449
Series
absolute convergence, 607–612
adding or deleting terms to/from, 592
alternating, 614–618
binomial, 644–646
combining, 591–592
comparison tests for convergence of, 602–606
complex, 18-23–18-24
conditionally convergent, 616–617
Constant Multiple Rule for, 591–592
convergent, 18-23–18-24, 587
Difference Rule for, 591–592
divergent, 587, 590–591
error estimation, 599–600
essay on, 573
Fourier, 600
geometric, 588–590, 618
harmonic, 598, 614–618
history of, 573
infinite, 573, 586–593
Integral Test, 596–600
Maclaurin, 633–635, 639
Monotonic Sequence Theorem, 582, 596
*n*th partial sums in, 586–587
*n*th term of, 587
*n*th-Term Test for divergent, 590–591, 618
p-, 598–600
of partial sums, 586–588, 590–591, 596
power, 621–629
p-Series Test, 598, 618
rearranging, 617–618
reindexing, 592–593
representations, 632–633
sigma notation to write, 587
sum of, 587
Sum Rule for, 591–592
Taylor, 633–650
test to verify convergence, 18-24
Series Multiplication for Power Series, 626
Sets of real numbers, AP-2, AP-24
Shell method, 324–327
Shift formulas, for functions, 16
Shifting of a graph, 16
Shift property for definite integrals, 303
Sigma notation, 256–259
in trigonometric polynomials, 19-5
to write series, 587
Σ, 257
Sigmoid shape of curves, 561
Simple harmonic motion, 17-16–17-17, 135–136
Simply connected domains and regions, 986
Simpson, Thomas, 485
Simpson's Rule, 496–501
Sine function, 22–23
arcsine/arccosine functions and, 422–424, 426
derivatives of, 133–134
general, 27
graphs/graphing, 10, 24, 29, 423
inequalities and, 26
limits involving, 68–70
odd, 24
one-to-one, 367, 421
periodicity of, 24
powers of, 470–471
products of, 470–471, 474
reciprocal of, 422
summing, 19-3–19-5
values of, 23
Sine-integral function, 514
Sinusoid, 27, 677
SI units (*Système International* or International System), 340–341
Slant line asymptotes, 87–88
Slicing, methods of
with cylinders, 322–324
definition of, 311–312
by parallel planes, 312–313
volume and, 312–313
Slope
of the curve, definition of, 102
fields, 537–538
negative reciprocal of, AP-13
of a nonvertical line, AP-11
of polar curve, 682–683
on sine curves, 143
of tangent lines, 102–103, 108
Slope-intercept equation, AP-13
Smooth curves, 330, 763–764
curvature of, 782–784
length of, 778
principal unit normal vector for, 783
speed on, 764
torsion of a, 790
Smooth function, 330
Smooth surface, 1010–1011, 1019, 1022
Snell's Law, 216
Snell van Royen, Willebrord, 215
Snowflake curve, 595
Solids, formulas for three-dimensional, 929
Solids of revolution, 313–318
Cavalieri's principle, 313
definition of, 313
disk method, 313–316
washer method, 316–318
Solution curves, 537–538
Solutions
of differential equations, 401–402
of first-order differential equations, 535–537
of first-order linear equations, 544–546
general, 401
particular, 536–537
steady-state, 564
Solving inequalities, AP-3–AP-4
Space, geometry of, 710–759
angles between planes, 746–747
cross products, 734–739
cylinders, 749
distance
from a point to a line, 743
from a point to a plane, 746
spheres and, 712–713

dot products, 726–732
equation for a plane, 743–744
line of intersection, 744–745
lines and line segments, 741–743
quadric surfaces, 749–751
general, 751, 753
graphs/graphing, 752
three-dimensional coordinate systems, 710–713
vectors, 715–723
Space, regions of
arc length in (*See* Arc length in space)
average value of a function in, 925
boundary of, 806–807
bounded, 806
closed, 806–807
curves, 760–766, 785–786
interior point of a, 806
motion in, 769–774
open, 806–807
triple integrals in, 918–919, 925
unbounded, 806
vector fields in, 977–979
volume of solid, 918–919
Speed, 125–126
along smooth curves, 764
average, 38–40
initial, in ideal projectile motion, 772
instantaneous, 39
Spheres
parametrization of, 1009
of radius and center, equations for, 713–714
in space, 712–713
surface area of, 1012
Spherical coordinates, 942–946
algebra for finding, 943
Cartesian and cylindrical coordinates relating to, 942–944
definition of, 942
geometry for finding, 943
integration in, 942–946
limits of integration in, finding, 944–946
Riemann sums for, 939, 943
Spin around axis, 996–999
Spring constant, 342
Square
completing, AP-14–AP-15
conformal maps of, 18-33–18-34
of the distance, 932
Square root functions
definition of, 8
derivatives of, 108, 110
linearization and, 162
Square roots, eliminating, 472
Square wavelet, 19-24–19-26, 19-31, 19-33–19-34
Stable equilibria, 557–558
Standard deviation, in probability, 522
Standard equation, AP-14, AP-17, 543–544
Standard linear approximations, 161
definition of, 854
error in, 855
line, 854
tangent-plane/tangent-line approximation, 854–855
Standard normal probability density function, 525
Standard position, 22, 716
Standard unit vectors, 19-7, 719
Starting integers, AP-8
Steady-state solution, 564
Steady-state values, 547
Steepest descent, method of, B-12–B-14
Step size, 495
Still higher order, partial derivatives of, 826
Stirling, James, 533–534
Stirling's formula, 533–534
Stokes' Theorem, 1028–1039
conservative fields and, 1038–1039
definition of, 1030
Green's Theorem compared to, 1030–1031, 1037–1038, 1050
identity in, 1038
paddle wheel interpretation, 1034–1037
for polyhedral surfaces, 1037–1038
for surfaces with holes, 1038
Subsets of real numbers, AP-2
Subspace, 19-11
Substitution
boundaries in, 952–954, 957
for definite integrals, 294–304
areas between curves, 297–298
integration with respect to y, 299–300
Substitution Formula, 294–296
differential area change, 951
for double integrals, 950–954
for indefinite integrals, 288–292
Chain Rule, 288–292
Substitution Rule, 268, 289–290, 292
of tangent, cotangent, secant, and cosecant functions, 292
trying different substitutions, 292
trigonometric, 476–479
for triple integrals, 955–957
Substitution Formula, 294–296
Substitution Rule, 951
for definite integrals, 268
for indefinite integrals, 268, 289–290, 292
use of the variable u in, 290
Subtraction
derivatives of, 117–118
of functions, 14–15
of terms from series, 592
of vectors, 718
Sum Rule
for antiderivatives, 232
to calculate limits of sequences, 577
for definite integrals, 268
derivative, AP-8–AP-9
for derivatives, 117–118, 134
for double and triple integrals, 904
for finding complex derivatives, 18-19
for finite sums, 258
for gradients, 846
for limits, AP-19, 48–49, 61
for limits for functions of two variables, 812
for series, 591–592
for vector functions, 765
Sums
derivatives of, 117
of series, 587
Superposition principle, 17-1
Surface area, 1008–1016
of an implicit surface, formula for, 1014
definition of, 335–337
differential for a parametrized surface, 1011
element, 1011
of a graph, formula for, 1016
parametrization of, 1009–1013
of revolution, 335–338
for revolution about the x-axis, 337
for revolution about the y-axis, 338
of a smooth surface, 1010–1011
of a sphere, 1012
Surface integrals, 1018–1026
definitions of, 1019
for a level surface, computing, 1024
mass density of, 1018, 1020
orientation of a surface, 1022
for a parametrized surface, computing, 1023–1024
of a scalar function, formulas for, 1019
in Stokes' Theorem, 1038
thin shells, moments and masses of, 1024–1026
of vector fields, 1022
Surfaces
functions defined on, 834–836
with holes, in Stokes' Theorem, 1038
level, 805
orientation of, 1022
parametrization of, 1008–1016
of cones, 1009
of cylinders, 1009
by cylindrical coordinates, 1009
of implicit surfaces, 1013–1016
of a sphere, 1009
of surface area, 1009–1013
plane tangent to, 850–852
polyhedral, 1037–1038
saddle points of, 860–867
smooth, 1010–1011, 1019, 1022
of two-variable functions, 805
Symmetric functions, definite integrals of, 296–297
Symmetry tests for polar coordinates, 682
Synchronous curves, 799
System of first-order differential equations, 564–567
System torque, 350–351
Tangent function, 22–23
area, 668–669
components of acceleration, 788–789
of curves, 760–766, 780
definition of, 22–23
derivatives of, 136–137
graphs/graphing, 24, 423
Green's Theorem and, 1001–1002, 1050
integrals of, 383, 472–473
odd, 24
periodicity of, 24
powers of, 472–473
in Substitution Rule, 291
unit vectors, 780, 792
values of, 23
Tangent lines
to an ellipse, 845–846, 852
approximation, 854–855
functions based on, 160–161
implicit, 149–150
areas and, 668–669
to a curve, 40–41
to a level curve, 845
to a parabola, 41–42
rates of change and, 42–43
vertical, 105–106, 110
Tangent planes, 850–857
approximation, 827, 854–855
definition of, 850, 851
differentials, 855–856
estimating change in a specific direction, 853
functions of more than two variables, 856–857
linearization of a function of two variables, 853–855
normal lines and, 850–852
to a surface, 850–852
Tautochrones, 663–664
T, 779
Taylor, Brook, 633
Taylor polynomial, 19-1, 634–636
Taylor series, 633–650
applications of, 644–650
arctangents, 647–648
binomial series for powers and roots, 644–646
Euler's identity, 649–650
evaluating indeterminate forms, 648–649
evaluating nonelementary integrals, 646–647
Leibniz's formula, 648
convergence of, 18-18, 637–642
Remainder Estimation Theorem, 639
remainder of order n (or error term), 638–639
Taylor's Formula, 637–638
Taylor's Theorem, 637, 642
using Taylor series in, 640–641
definition of, 633–634
frequently used, 650
Taylor polynomials in, 634–636

Taylor's Formula, 879–882
  definition of, 637, 638
  to derive the Second Derivative Test, 879–880
  error formula for linear approximations, 880
  for functions of two variables, 881–882
  at the origin, 882
  at the point, 881
Taylor's Theorem, 498–499
  definition of, 637
  Mean Value Theorem and, 637
  proof of, 642
  Remainder Estimation Theorem and, 639
Term-by-Term Differentiation, 627–628
Term-by-Term Integration, 628–629
Terminal point, 659, 715
Terminal velocity, 560
Terms, of sequences, 574
Theorems
  absolute convergence of series, 18-24
  Absolute Convergence Test, 608
  Algebraic Properties of the Natural Logarithm, 380–381
  Alternating Series Estimation Theorem, 616
  Alternating Series Test, 614–615, 618
  Angle Between Two Vectors, 726–728
  approximations
    of functions, 19-17, 19-18
    in higher dimensions, 19-10–19-11
    using the Haar system, 19-27, 19-29–19-30
  Area of Surface of Revolution for Parametrized Curves, 675
  Chain Rule, 140
    Chain Rule for Two Independent Variables and Three Intermediate Variables, 834
    Formula for Implicit Differentiation, 836
    Functions of One Independent Variable and Three Intermediate Variables, 833
    Functions of One Independent Variable and Two Intermediate Variables, 831
    Implicit Function Theorem, 837–838
  Clairaut's Theorem, 825–826
  Cofactor Expansion, B-4
  of complex functions, 18-25
  conformal maps, 18-32–18-33
  constant-coefficient homogeneous equations, 17-2–17-5
  Constant Multiple Rule, 591–592
  Convergence Theorem for Complex Power Series, 18-25
  Convergence Theorem for Power Series, 624–626, 627
    corollary to, 625
  for convergent series, 591–592
  curl F = 0 related to the closed-loop property, 1038
  De Moivre's Theorem, 18-7
  Differentiable Implies Continuous Theorem, 111, 828
  Direct Comparison Test, 510–511, 512, 603–604, 618
  Directional Derivative Is a Dot Product, 843
  divergence and the curl, 1044–1045
  Divergence Theorem, 1041–1051
  Extreme Value Theorem, 181–182
  for finding antiderivatives, 230
  First Derivative Test for Local Extreme Values, B-8, 183–184, 196–198, 861, 871
  Formula for Implicit Differentiation, 836–838
  Fourier coefficients, 19-18
  Fourier series, convergence of, 19-19–19-21
  Fubini's Theorem, 896–898, 900–903
  Fundamental Theorem of Algebra, 18-9
  Fundamental Theorem of Calculus, 234
  Fundamental Theorem of Line Integrals, 987–989
  Green's Theorem, 996–1005, 1007
  Haar system, 19-27, 19-29, 19-32
  Implicit Function Theorem, 837–838
  Increment Theorem for Functions of Two Variables, AP-29–AP-31
  initial value and boundary value problems, 17-5
  Integral Test, 597
  interior points of a function's domain, 66
  Laws of Exponents for e_x, 389–390
  l'Hôpital's Rule, 578–579
  for limits, AP-19–AP-21
    commonly occurring, 580
    Limit Comparison Test, 511–512, 604–606, 618
    Limit Laws, 48, 84
    one-sided, 66
    of polynomials, 49
    Properties of Limits of Functions of Two Variables, 812
    to prove definition of, 61
    of rational functions, 49
    of the ratio $\sin \theta/\theta$ as $\theta \to 0$, 68
    of sequences, calculating, 577
  linearly independent solutions, 17-2
  Mean Value Theorem, 190–194, 637
  Mixed Derivative Theorem, AP-27–AP-28, 825–826
  Monotonic Sequence Theorem, 582, 596
  Nicole Oresme's, 575, 657
  nonhomogeneous linear equations, 17-8
  nth-Term Test for Divergence, 590–591
  Orthogonal Gradient Theorem, 872–873
  orthonormal bases, 19-9, 19-10–19-11
  Pappus's
    Theorem for Surface Areas, 359
    Theorem for Volumes, 357–358
  Parallel Axis Theorem, 937
  Parseval Equality, 19-21–19-22
  Properties of Determinants, B-5
  Pythagorean Theorem, 19-15, 24, 26, 424–425
  Ratio Test, 609–610, 618
  Rearrangement Theorem for Absolutely Convergent Series, 617, 618
  Remainder Estimation Theorem, 639
  Rolle's Theorem, 188–189
  Root Test, 611–612, 618
  Sandwich Theorem, 51, 578
  Second Derivative Test, B-9, 201, 204–208, 862, 864, 866, 870, 879–880
    in applied optimization, 217
    for concavity, 201
  Substitution for Double Integrals, 951
  Taylor's Formula, 637, 638
  Taylor's Theorem, 498–499, 637, 639, 642
  Unifying Fundamental Theorem of Vector Integral Calculus, 1051
  zipper theorem, 585
Θ-limits of integration, 913, 940, 941, 944, 945
Thickness variable, 325–326
Thin shells, moments and masses of, 1024–1026
Three-dimensional coordinate systems, 710–713
  Cartesian coordinates in, 710
  coordinate planes in, 710
  planes determined by coordinate axes in, 710
  rectangular coordinates in, 710
Three-dimensional Laplace equation, 830
Three-dimensional solid, formulas for, 929
Three-dimensional vectors, 19-5–19-9, 716
  definitions of, 19-6, 19-7
  distance between, 19-6
  dot product of, 19-6
  length of, 19-6
  linear combination of, 19-7
  orthogonal (or orthonormal) set of, 19-6, 19-7–19-8
  orthonormal basis for, 19-8–19-9
  Pythagorean Theorem for, 19-6–19-7
  standard unit vectors in, 19-7
  theorem, 19-9
  unit vectors, 19-6
Time constant, 548
Time-distance law (Kepler's third law), 796–797
Time measurement, fixed error in, 168
TNB frame, 788
Torque, 350–351
Torque vectors, 737–738
Torsion
  along the curve, 790
  definition of, 790
  formulas for computing, 791–792
  function of a smooth curve, 790
Total area, 283–284
Total differentials, 855–857
Total distance traveled, 251–252, 282
Training set, 562
Trajectories
  orthogonal, 549, 551–552
  systems of differential equations, 564–565
Transcendental functions, 11, 366–456
Transcendental numbers, 396
Transformations
  Jacobian of, 952–954, 957
  linear, 951
  nonlinear, 953–954
  rules applied to sine function, 27
  of trigonometric graphs, 26–27
Trapezoidal approximations, 495–496
Trapezoidal Rule, 495–496
Trend lines, 869
Trigonometric functions, 10, 21–29
  addition formulas in, 25
  angles in, 21–22
  arcsine and arccosine, 422–425
  ASTC rule for remembering, 23
  cosecant, 22–23
  cosine, 22–23
  cotangent, 22–23
  derivatives of, 133–139
    cosecant function, 136–137
    cosine function, 134–135
    cotangent function, 136–137
    secant function, 136–137
    simple harmonic motion, 135–136
    sine function, 133–134
    tangent function, 136–137
  double-angle formulas in, 25
  graphs/graphing, 24, 26–27
  half-angle formulas in, 25
  identities, 24–25
  inequalities in, 26
  inverse, 420–435
  law of cosines, 25–26
  orthogonality of, 19-15–19-16
  periodic, 24
  periods of, 24
  secant, 22–23
  sine, 22–23
  in Substitution Rule, 291
  substitutions, 476–479
  tangent, 22–23

Trigonometric integrals, 470–474
  graphed as functions, 476
  of powers of tan x and sec x, 472–473
  products of powers of sines and cosines, 470–471
  products of sines and cosines, 474
  square roots, eliminating, 472
Trigonometric polynomials, 19-3–19-5
  amplitude in, 19-4
  definition of, 19-3
  of degree *n,* 19-18–19-19
  sigma notation in, 19-5
Trigonometric system, 19-2, 19-15–19-16
Triple integrals
  iterated integrals, 919–925
  properties of, 904, 926
  in rectangular coordinates, 918–925
  in spherical coordinates, 942–946
  substitutions in, 955–957
Triple scalar product, 738–739
Trivial solution, 17-2
Trochoids, 667
Twice-differentiable functions, 668–669
Two-dimensional Laplace equation, 830
Two-dimensional plate, formulas for, 929
Two-dimensional vectors, 716, 731–732
Two-path test for nonexistence of a limit, 815
Two-sided continuity, 74
Two-sided limits, AP-21, 65, 69, 74, 78, 92, 93
Type II integrals, improper, 508–509
Type I integrals, improper, 505–508

Unbounded region
  in a plane, 804
  in space, 806
Unbounded sequence, 581
Underdamping, 17-18
Undetermined coefficients, 17-8–17-11, 481
Unified theory of vector field integrals, 1050–1051
Uniform distributions, in probability, 522–523
Uniform probability distributions, 522–523
Union of sets, AP-2
Unit circle, AP-14
Unit functions, 19-13, 19-24
Unit normal vectors, 734
Units of measurement
  for fluid's weight-density, 344
  in work, 340–341
Unit step function, 47–48
Unit tangent vector, 780, 792
Unit vectors, 19-6, 719–720
  definition of, 719
  tangent, 780, 792
  writing vectors in terms of, 719–720
Universal gravitational constant, 795
Unstable equilibria, 557–558
Upper bound, least, AP-24, 581
Upper sums, 247–248

Values
  average (*See* Average values)
  boundary, 17-6
  equilibrium, 555–557
  expected, 519–521
  extreme (*See* Extreme values)
  of functions, 2
  initial value and boundary value problems, 17-5–17-6
  limits of, 45–46
  of a line integral, 968
  of logarithmic functions, 18-29
  maximum and minimum (*See* Extreme values)
  needed by the Second Derivative Test, 864
  of parametric equations, 660, 662
  principal, 18-25, 18-29
  real-valued functions, 2, 760–761, 802–803
  steady-state, 547
  of trigonometric ratios, 23
Variable costs, 129
Variables
  dependent (*See* Dependent variables)
  dummy, 265
  force along a line, work done by, 341
  functions of many, 838
  functions of more than two
    differentiable, 856–857
    linearization of, 856–857
    many variables, 838
    in tangent planes, 856–857
  functions of several, 802–807
    definitions of, 802
    dependent/independent variables, 802
    differentiability for, 827–828
    differentiable, 827–828
    domain of, 802, 803
    input/output variables, 802
    partial derivatives of, 802–807
    range of, 802, 803
    real-valued function, 802
  functions of three, 805–807
    boundary of, 806
    boundary points of, 806
    bounded/unbounded regions of, 806
    Chain Rule for, 833–834
    definitions of, 805–807
    gradient vectors in, 846–847
    interior points of, 806
    interior regions of, 806
    level surface of, 805
    open/closed regions of, 806–807
  functions of two, 860–864
    boundary of, 804
    boundary point of, 804
    Chain Rule for, 831–833
    Composition Rule for, 812
    Constant Multiple Rule for, 812
    contours of, 805
    definitions of, 804
    extreme values of, B-6–B-11, 860–864
    graphs/graphing, 805
    Increment Theorem for Functions of Two, AP-29–AP-31, 828
    interior of the region of, 804
    interior point of, 804
    level curves of, 805
    limits for, 810–814
    linearization of, 853–855
    more than, 816, 823–824
    open/closed regions of, 804
    partial derivatives of, 819–821
    properties of limits of, 812
    saddle points of, B-6–B-11
    Sandwich Theorem for, 818
    Taylor's formula for, 881–882
  independent (*See* Independent variables)
  input, 1, 16
  of integration, 233, 234, 264–265
  output, 1, 16
  proportional relationship of, 7
  random, in probability, 515–516
    continuous random, 516
    variance of, 522
  in related rates problem, 153
  thickness, 325–326
  u, in Substitution Rule, 290
Variance, 522
Vector fields
  assumptions on, 986
  circulation for, 979–980
  continuous, 972
  curl, 1028, 1029–1030
  definition of, 972–973
  differentiable, 972
  divergence of, 999–1001, 1041–1042
  flux across a simple closed plane curve, 980–981
  gradient fields, 973–974
  gravitational, 985
  integrals
    flow, 979–980
    line, 965–970, 974–976
    surface, 1022
    unified theory of, 1050–1051
  integration in, 965–1058
  potential function for, 986
  work done by a force over a curve in space, 977–979
Vector functions, 760–801
  acceleration, 788–789, 794–797
  antiderivatives of, 769–770
  arc length in space, 778–780
  of constant length, 766
  continuity of, 761–762
  curvature
    circle of, 784–785
    formulas for calculating, 782–784, 791–792
    of normal vectors, 782–786
    of plane curve, 782–785
    in space, 760–766, 785–786
  differentiable, 763
    equations for, 770–771, 775
    rules for, 764–766
  Fundamental Theorem of Calculus for continuous, 770–771
  initial value problems for, 771–772, 775
  integrals of, 769–774
  limits of, 761–762
  motion
    derivatives and, 762–764
    path for ideal projectile, 772
    in space, 769–774
  polar coordinates, 794–797
  TNB frame, 788
  torsion, 789–792
Vectors, 715–723
  acceleration, 764, 788
  addition of, 717–718
  in algebra operations, 717–718, 722
  Angle Between Two Vectors, 726–728
  angles between, 746–747
  applications of, 721–722
  binormal vector, 788, 792
  component form of, 715–717
    i-component, 719–720
    j-component, 719–720
    k-component, 719–720, 996–999
  coplanar, 718
  cross product as area of parallelogram, 735–736
  cross product of two, 734–735
  curl, 996–999, 1001–1003, 1044–1045
  of a curve
    binormal, 788, 792
    formula for curvature, 791
    normal, 782–786
    tangent, 760–766
  definition of, 715
  derivative in direction of, 763–764
  as determinant, 738–739
  difference (subtraction) of, 718
  differential form, 978
  directed line segment to represent, 715
  direction of, 719–720
  in engineering, 721
  equality of, 715, 716
  fields, 760
  geometry in space and, 710–759
  gradient (*See* Gradient vectors)
  in gravitational field, 973
  in higher dimensions (*See* Higher dimension vectors)
  initial point of, 715
  j-component of, 719–720
  length (magnitude) of, 715, 716
  midpoint of line segments, 720
  in navigation, 721
  n dimensional, 722–723
  nonzero, 726–729, 734–735, 741, 744–745
  notation for, 729

Vectors (*continued*)
orthogonal (perpendicular), 728
parallel, 734, 735
parallelogram law of addition, 717
in physics, 721
position, 760
principal unit normal, 783, 792
product, 735
projections of, 728–731
resultant, 717
right-hand rule for, 734
scalar component in direction of, 729–730
scalar multiplication of, 717–718, 729–730
standard position, for a point, 716
standard unit, 19-7, 719
terminal point of, 715
in three dimensions (*See* Three-dimensional vectors)
torque, 737–738
triple scalar product of, 738–739
two-dimensional, 716, 731–732
unit, 19-6, 719–720
unit normal, 734, 788
unit tangent, 780, 792
velocity, 715
writing, 716
zero, 716
Velocity function
acceleration, 126–128, 192
angular, 1035
distance traveled, 249–252, 282
fields, circulation for, 979–980
motion with resistance proportional to, 549–550
in polar coordinates, 794–797
rates of change in, 125–128
resistance proportional to, 549–550
in speed, 125–126
terminal, 560
vector, 715
Vertical asymptotes, 91–92, 508–509
Vertical cross-sections, for finding limits of integration, 903
Vertical line test, for functions, 4
Vertical scaling and reflecting formulas, 17
Vertical shifts, 16
Vertical tangent lines, 105–106, 110
Vertices
of ellipses, 692
of hyperbolas, 693
of parabolas, AP-15–AP-16, 691
Vibrations, 17-15–17-16
Volts (V), 17-19
Volumes
of double integrals
over general regions, 900–903
over rectangles, 895–896
of a solid region in space, 918–919
of solids
definition of, 312
Pappus's Theorem for Volumes, 357–358
of revolution, 313–318
of solids, calculating
Cavalieri's principle, 313
by disks for rotation about the *x*-axis, 314–315
by disks for rotation about the *y*-axis, 315–316
shell formula for revolution about a vertical line, 325–326
shell method, 324–327
using cross-sections, 311–318
using cylindrical shells, 322–327
using disk method, 313–316
using washer method, 316–318
by washers for rotation about the *x*-axis, 317–318
by washers for rotation about the *y*-axis, 318

Washer method, 316–318
Wave equation, 830
Wavelets, 19-24
Daubechies, 19-33–19-34
definition of, 19-26
Haar, 19-24–19-26, 19-31, 19-33–19-34
Weierstrass, Karl, 510
Weierstrass function, 114
Weight-density, of fluid, 344
Wild oscillation, 110
Wilson lot size formula, 223, 829, 859
Windows, graphing, 29–32
Wires, thin, 351
Witch of Agnesi, 667
Work and fluid force, 340–349
along a line, 341
on a constant-depth surface, 344–345
constant force, 340–341
definition of, 340
fluid pressures, 344–345
force constant or spring constant, 342
Hooke's Law, 341–342
integrals, 345, 977–979
lifting objects and pumping liquids from containers, 342–343
over a curve in space, 977–979
pressure-depth equation, 344
through displacement, 731
units of measurement in, 340–341
Writing vectors, 716

x-axis
curve as symmetric about, 13
by disks for rotation about, 314–315
moment about, 352–355
revolution about, 337, 675
x-coordinate (or abscissa), AP-10
x-intercept, AP-13
x-limits of integration, 903, 920, 921, 924
xy-plane, 710, 712
xz-plane, 710

y, integration with respect to, 299–300
y-axis
by disks for rotation about, 315–316
graph of an even function as symmetric about, 6–7
moment about, 352–355
of the parabola, 691
revolution about, 338, 675
symmetric about, 6–7
y-coordinate (or ordinate), AP-10
y-intercept, AP-13
y-limits of integration, 903, 920–924
yz-plane, 710

z-axis, 710–711, 749, 751
Zero, frequency, 19-2
Zero denominators, eliminating common factors from, 49–50
Zero derivatives in constant functions, 191
Zero directional derivatives, 849
Zero vectors, 716
Zero Width Interval Rule, 268, 377
Zipper theorem, 585
z-limits of integration, 920–924, 939, 940, 941

A Brief Table of Integrals

Basic Forms

1. $\int k\,dx = kx + C$, k any number

2. $\int x^n\,dx = \dfrac{x^{n+1}}{n+1} + C, \quad n \neq -1$

3. $\int \dfrac{dx}{x} = \ln|x| + C$

4. $\int e^x\,dx = e^x + C$

5. $\int a^x\,dx = \dfrac{a^x}{\ln a} + C \quad (a > 0,\ a \neq 1)$

6. $\int \sin x\,dx = -\cos x + C$

7. $\int \cos x\,dx = \sin x + C$

8. $\int \sec^2 x\,dx = \tan x + C$

9. $\int \csc^2 x\,dx = -\cot x + C$

10. $\int \sec x \tan x\,dx = \sec x + C$

11. $\int \csc x \cot x\,dx = -\csc x + C$

12. $\int \tan x\,dx = \ln|\sec x| + C$

13. $\int \cot x\,dx = \ln|\sin x| + C$

14. $\int \sinh x\,dx = \cosh x + C$

15. $\int \cosh x\,dx = \sinh x + C$

16. $\int \dfrac{dx}{\sqrt{a^2 - x^2}} = \arcsin \dfrac{x}{a} + C$

17. $\int \dfrac{dx}{a^2 + x^2} = \dfrac{1}{a} \arctan \dfrac{x}{a} + C$

18. $\int \dfrac{dx}{x\sqrt{x^2 - a^2}} = \dfrac{1}{a} \operatorname{arcsec} \left|\dfrac{x}{a}\right| + C$

19. $\int \dfrac{dx}{\sqrt{a^2 + x^2}} = \sinh^{-1} \dfrac{x}{a} + C \quad (a > 0)$

20. $\int \dfrac{dx}{\sqrt{x^2 - a^2}} = \cosh^{-1} \dfrac{x}{a} + C \quad (x > a > 0)$

Forms Involving $ax + b$

21. $\int (ax + b)^n\,dx = \dfrac{(ax + b)^{n+1}}{a(n + 1)} + C, \quad n \neq -1$

22. $\int x(ax + b)^n\,dx = \dfrac{(ax + b)^{n+1}}{a^2}\left[\dfrac{ax + b}{n + 2} - \dfrac{b}{n + 1}\right] + C, \quad n \neq -1, -2$

23. $\int (ax + b)^{-1}\,dx = \dfrac{1}{a} \ln|ax + b| + C$

24. $\int x(ax + b)^{-1}\,dx = \dfrac{x}{a} - \dfrac{b}{a^2} \ln|ax + b| + C$

25. $\int x(ax + b)^{-2}\,dx = \dfrac{1}{a^2}\left[\ln|ax + b| + \dfrac{b}{ax + b}\right] + C$

26. $\int \dfrac{dx}{x(ax + b)} = \dfrac{1}{b} \ln\left|\dfrac{x}{ax + b}\right| + C$

27. $\int \left(\sqrt{ax + b}\right)^n\,dx = \dfrac{2}{a} \dfrac{\left(\sqrt{ax + b}\right)^{n+2}}{n + 2} + C, \quad n \neq -2$

28. $\int \dfrac{\sqrt{ax + b}}{x}\,dx = 2\sqrt{ax + b} + b \int \dfrac{dx}{x\sqrt{ax + b}}$

29. (a) $\int \frac{dx}{x\sqrt{ax+b}} = \frac{1}{\sqrt{b}} \ln\left|\frac{\sqrt{ax+b}-\sqrt{b}}{\sqrt{ax+b}+\sqrt{b}}\right| + C$ (b) $\int \frac{dx}{x\sqrt{ax-b}} = \frac{2}{\sqrt{b}} \arctan\sqrt{\frac{ax-b}{b}} + C$

30. $\int \frac{\sqrt{ax+b}}{x^2}\,dx = -\frac{\sqrt{ax+b}}{x} + \frac{a}{2}\int \frac{dx}{x\sqrt{ax+b}} + C$

31. $\int \frac{dx}{x^2\sqrt{ax+b}} = -\frac{\sqrt{ax+b}}{bx} - \frac{a}{2b}\int \frac{dx}{x\sqrt{ax+b}} + C$

Forms Involving $a^2 + x^2$

32. $\int \frac{dx}{a^2+x^2} = \frac{1}{a}\arctan\frac{x}{a} + C$

33. $\int \frac{dx}{(a^2+x^2)^2} = \frac{x}{2a^2(a^2+x^2)} + \frac{1}{2a^3}\arctan\frac{x}{a} + C$

34. $\int \frac{dx}{\sqrt{a^2+x^2}} = \sinh^{-1}\frac{x}{a} + C = \ln\left(x+\sqrt{a^2+x^2}\right) + C$

35. $\int \sqrt{a^2+x^2}\,dx = \frac{x}{2}\sqrt{a^2+x^2} + \frac{a^2}{2}\ln\left(x+\sqrt{a^2+x^2}\right) + C$

36. $\int x^2\sqrt{a^2+x^2}\,dx = \frac{x}{8}(a^2+2x^2)\sqrt{a^2+x^2} - \frac{a^4}{8}\ln\left(x+\sqrt{a^2+x^2}\right) + C$

37. $\int \frac{\sqrt{a^2+x^2}}{x}\,dx = \sqrt{a^2+x^2} - a\ln\left|\frac{a+\sqrt{a^2+x^2}}{x}\right| + C$

38. $\int \frac{\sqrt{a^2+x^2}}{x^2}\,dx = \ln\left(x+\sqrt{a^2+x^2}\right) - \frac{\sqrt{a^2+x^2}}{x} + C$

39. $\int \frac{x^2}{\sqrt{a^2+x^2}}\,dx = -\frac{a^2}{2}\ln\left(x+\sqrt{a^2+x^2}\right) + \frac{x\sqrt{a^2+x^2}}{2} + C$

40. $\int \frac{dx}{x\sqrt{a^2+x^2}} = -\frac{1}{a}\ln\left|\frac{a+\sqrt{a^2+x^2}}{x}\right| + C$

41. $\int \frac{dx}{x^2\sqrt{a^2+x^2}} = -\frac{\sqrt{a^2+x^2}}{a^2x} + C$

Forms Involving $a^2 - x^2$

42. $\int \frac{dx}{a^2-x^2} = \frac{1}{2a}\ln\left|\frac{x+a}{x-a}\right| + C$

43. $\int \frac{dx}{(a^2-x^2)^2} = \frac{x}{2a^2(a^2-x^2)} + \frac{1}{4a^3}\ln\left|\frac{x+a}{x-a}\right| + C$

44. $\int \frac{dx}{\sqrt{a^2-x^2}} = \arcsin\frac{x}{a} + C$

45. $\int \sqrt{a^2-x^2}\,dx = \frac{x}{2}\sqrt{a^2-x^2} + \frac{a^2}{2}\arcsin\frac{x}{a} + C$

46. $\int x^2\sqrt{a^2-x^2}\,dx = \frac{a^4}{8}\arcsin\frac{x}{a} - \frac{1}{8}x\sqrt{a^2-x^2}\,(a^2-2x^2) + C$

47. $\int \frac{\sqrt{a^2-x^2}}{x}\,dx = \sqrt{a^2-x^2} - a\ln\left|\frac{a+\sqrt{a^2-x^2}}{x}\right| + C$

48. $\int \frac{\sqrt{a^2-x^2}}{x^2}\,dx = -\arcsin\frac{x}{a} - \frac{\sqrt{a^2-x^2}}{x} + C$

49. $\int \frac{x^2}{\sqrt{a^2-x^2}}\,dx = \frac{a^2}{2}\arcsin\frac{x}{a} - \frac{1}{2}x\sqrt{a^2-x^2} + C$

50. $\int \frac{dx}{x\sqrt{a^2-x^2}} = -\frac{1}{a}\ln\left|\frac{a+\sqrt{a^2-x^2}}{x}\right| + C$

51. $\int \frac{dx}{x^2\sqrt{a^2-x^2}} = -\frac{\sqrt{a^2-x^2}}{a^2x} + C$

Forms Involving $x^2 - a^2$

52. $\int \frac{dx}{\sqrt{x^2-a^2}} = \ln\left|x+\sqrt{x^2-a^2}\right| + C$

53. $\int \sqrt{x^2-a^2}\,dx = \frac{x}{2}\sqrt{x^2-a^2} - \frac{a^2}{2}\ln\left|x+\sqrt{x^2-a^2}\right| + C$

54. $\int \left(\sqrt{x^2 - a^2}\right)^n dx = \frac{x\left(\sqrt{x^2 - a^2}\right)^n}{n+1} - \frac{na^2}{n+1}\int \left(\sqrt{x^2 - a^2}\right)^{n-2} dx, \quad n \neq -1$

55. $\int \frac{dx}{\left(\sqrt{x^2 - a^2}\right)^n} = \frac{x\left(\sqrt{x^2 - a^2}\right)^{2-n}}{(2-n)a^2} - \frac{n-3}{(n-2)a^2}\int \frac{dx}{\left(\sqrt{x^2 - a^2}\right)^{n-2}}, \quad n \neq 2$

56. $\int x\left(\sqrt{x^2 - a^2}\right)^n dx = \frac{\left(\sqrt{x^2 - a^2}\right)^{n+2}}{n+2} + C, \quad n \neq -2$

57. $\int x^2\sqrt{x^2 - a^2}\, dx = \frac{x}{8}(2x^2 - a^2)\sqrt{x^2 - a^2} - \frac{a^4}{8}\ln\left|x + \sqrt{x^2 - a^2}\right| + C$

58. $\int \frac{\sqrt{x^2 - a^2}}{x}\, dx = \sqrt{x^2 - a^2} - a \operatorname{arcsec}\left|\frac{x}{a}\right| + C$

59. $\int \frac{\sqrt{x^2 - a^2}}{x^2}\, dx = \ln\left|x + \sqrt{x^2 - a^2}\right| - \frac{\sqrt{x^2 - a^2}}{x} + C$

60. $\int \frac{x^2}{\sqrt{x^2 - a^2}}\, dx = \frac{a^2}{2}\ln\left|x + \sqrt{x^2 - a^2}\right| + \frac{x}{2}\sqrt{x^2 - a^2} + C$

61. $\int \frac{dx}{x\sqrt{x^2 - a^2}} = \frac{1}{a}\operatorname{arcsec}\left|\frac{x}{a}\right| + C = \frac{1}{a}\arccos\left|\frac{a}{x}\right| + C$

62. $\int \frac{dx}{x^2\sqrt{x^2 - a^2}} = \frac{\sqrt{x^2 - a^2}}{a^2 x} + C$

Trigonometric Forms

63. $\int \sin ax\, dx = -\frac{1}{a}\cos ax + C$

64. $\int \cos ax\, dx = \frac{1}{a}\sin ax + C$

65. $\int \sin^2 ax\, dx = \frac{x}{2} - \frac{\sin 2ax}{4a} + C$

66. $\int \cos^2 ax\, dx = \frac{x}{2} + \frac{\sin 2ax}{4a} + C$

67. $\int \sin^n ax\, dx = -\frac{\sin^{n-1} ax \cos ax}{na} + \frac{n-1}{n}\int \sin^{n-2} ax\, dx$

68. $\int \cos^n ax\, dx = \frac{\cos^{n-1} ax \sin ax}{na} + \frac{n-1}{n}\int \cos^{n-2} ax\, dx$

69. **(a)** $\int \sin ax \cos bx\, dx = -\frac{\cos(a+b)x}{2(a+b)} - \frac{\cos(a-b)x}{2(a-b)} + C, \quad a^2 \neq b^2$

(b) $\int \sin ax \sin bx\, dx = \frac{\sin(a-b)x}{2(a-b)} - \frac{\sin(a+b)x}{2(a+b)} + C, \quad a^2 \neq b^2$

(c) $\int \cos ax \cos bx\, dx = \frac{\sin(a-b)x}{2(a-b)} + \frac{\sin(a+b)x}{2(a+b)} + C, \quad a^2 \neq b^2$

70. $\int \sin ax \cos ax\, dx = -\frac{\cos 2ax}{4a} + C$

71. $\int \sin^n ax \cos ax\, dx = \frac{\sin^{n+1} ax}{(n+1)a} + C, \quad n \neq -1$

72. $\int \frac{\cos ax}{\sin ax}\, dx = \frac{1}{a}\ln|\sin ax| + C$

73. $\int \cos^n ax \sin ax\, dx = -\frac{\cos^{n+1} ax}{(n+1)a} + C, \quad n \neq -1$

74. $\int \frac{\sin ax}{\cos ax}\, dx = -\frac{1}{a}\ln|\cos ax| + C$

75. $\int \sin^n ax \cos^m ax\, dx = -\frac{\sin^{n-1} ax \cos^{m+1} ax}{a(m+n)} + \frac{n-1}{m+n}\int \sin^{n-2} ax \cos^m ax\, dx, \quad n \neq -m \quad (\text{reduces } \sin^n ax)$

76. $\int \sin^n ax \cos^m ax\, dx = \frac{\sin^{n+1} ax \cos^{m-1} ax}{a(m+n)} + \frac{m-1}{m+n}\int \sin^n ax \cos^{m-2} ax\, dx, \quad m \neq -n \quad (\text{reduces } \cos^m ax)$

77. $\displaystyle\int \frac{dx}{b + c\sin ax} = \frac{-2}{a\sqrt{b^2 - c^2}} \arctan\left[\sqrt{\frac{b-c}{b+c}} \tan\left(\frac{\pi}{4} - \frac{ax}{2}\right)\right] + C, \quad b^2 > c^2$

78. $\displaystyle\int \frac{dx}{b + c\sin ax} = \frac{-1}{a\sqrt{c^2 - b^2}} \ln\left|\frac{c + b\sin ax + \sqrt{c^2 - b^2}\cos ax}{b + c\sin ax}\right| + C, \quad b^2 < c^2$

79. $\displaystyle\int \frac{dx}{1 + \sin ax} = -\frac{1}{a}\tan\left(\frac{\pi}{4} - \frac{ax}{2}\right) + C$

80. $\displaystyle\int \frac{dx}{1 - \sin ax} = \frac{1}{a}\tan\left(\frac{\pi}{4} + \frac{ax}{2}\right) + C$

81. $\displaystyle\int \frac{dx}{b + c\cos ax} = \frac{2}{a\sqrt{b^2 - c^2}} \arctan\left[\sqrt{\frac{b-c}{b+c}} \tan\frac{ax}{2}\right] + C, \quad b^2 > c^2$

82. $\displaystyle\int \frac{dx}{b + c\cos ax} = \frac{1}{a\sqrt{c^2 - b^2}} \ln\left|\frac{c + b\cos ax + \sqrt{c^2 - b^2}\sin ax}{b + c\cos ax}\right| + C, \quad b^2 < c^2$

83. $\displaystyle\int \frac{dx}{1 + \cos ax} = \frac{1}{a}\tan\frac{ax}{2} + C$

84. $\displaystyle\int \frac{dx}{1 - \cos ax} = -\frac{1}{a}\cot\frac{ax}{2} + C$

85. $\displaystyle\int x\sin ax\,dx = \frac{1}{a^2}\sin ax - \frac{x}{a}\cos ax + C$

86. $\displaystyle\int x\cos ax\,dx = \frac{1}{a^2}\cos ax + \frac{x}{a}\sin ax + C$

87. $\displaystyle\int x^n\sin ax\,dx = -\frac{x^n}{a}\cos ax + \frac{n}{a}\int x^{n-1}\cos ax\,dx$

88. $\displaystyle\int x^n\cos ax\,dx = \frac{x^n}{a}\sin ax - \frac{n}{a}\int x^{n-1}\sin ax\,dx$

89. $\displaystyle\int \tan ax\,dx = \frac{1}{a}\ln|\sec ax| + C$

90. $\displaystyle\int \cot ax\,dx = \frac{1}{a}\ln|\sin ax| + C$

91. $\displaystyle\int \tan^2 ax\,dx = \frac{1}{a}\tan ax - x + C$

92. $\displaystyle\int \cot^2 ax\,dx = -\frac{1}{a}\cot ax - x + C$

93. $\displaystyle\int \tan^n ax\,dx = \frac{\tan^{n-1} ax}{a(n-1)} - \int \tan^{n-2} ax\,dx, \quad n \neq 1$

94. $\displaystyle\int \cot^n ax\,dx = -\frac{\cot^{n-1} ax}{a(n-1)} - \int \cot^{n-2} ax\,dx, \quad n \neq 1$

95. $\displaystyle\int \sec ax\,dx = \frac{1}{a}\ln|\sec ax + \tan ax| + C$

96. $\displaystyle\int \csc ax\,dx = -\frac{1}{a}\ln|\csc ax + \cot ax| + C$

97. $\displaystyle\int \sec^2 ax\,dx = \frac{1}{a}\tan ax + C$

98. $\displaystyle\int \csc^2 ax\,dx = -\frac{1}{a}\cot ax + C$

99. $\displaystyle\int \sec^n ax\,dx = \frac{\sec^{n-2} ax\tan ax}{a(n-1)} + \frac{n-2}{n-1}\int \sec^{n-2} ax\,dx, \quad n \neq 1$

100. $\displaystyle\int \csc^n ax\,dx = -\frac{\csc^{n-2} ax\cot ax}{a(n-1)} + \frac{n-2}{n-1}\int \csc^{n-2} ax\,dx, \quad n \neq 1$

101. $\displaystyle\int \sec^n ax\tan ax\,dx = \frac{\sec^n ax}{na} + C, \quad n \neq 0$

102. $\displaystyle\int \csc^n ax\cot ax\,dx = -\frac{\csc^n ax}{na} + C, \quad n \neq 0$

Inverse Trigonometric Forms

103. $\displaystyle\int \arcsin ax\,dx = x\arcsin ax + \frac{1}{a}\sqrt{1 - a^2x^2} + C$

104. $\displaystyle\int \arccos ax\,dx = x\arccos ax - \frac{1}{a}\sqrt{1 - a^2x^2} + C$

105. $\displaystyle\int \arctan ax\,dx = x\arctan ax - \frac{1}{2a}\ln(1 + a^2x^2) + C$

106. $\displaystyle\int x^n\arcsin ax\,dx = \frac{x^{n+1}}{n+1}\arcsin ax - \frac{a}{n+1}\int \frac{x^{n+1}\,dx}{\sqrt{1 - a^2x^2}}, \quad n \neq -1$

107. $\int x^n \arccos ax\, dx = \frac{x^{n+1}}{n+1} \arccos ax + \frac{a}{n+1} \int \frac{x^{n+1}\, dx}{\sqrt{1-a^2x^2}}, \quad n \neq -1$

108. $\int x^n \arctan ax\, dx = \frac{x^{n+1}}{n+1} \arctan ax - \frac{a}{n+1} \int \frac{x^{n+1}\, dx}{1+a^2x^2}, \quad n \neq -1$

Exponential and Logarithmic Forms

109. $\int e^{ax}\, dx = \frac{1}{a} e^{ax} + C$

110. $\int b^{ax}\, dx = \frac{1}{a} \frac{b^{ax}}{\ln b} + C, \quad b > 0, b \neq 1$

111. $\int x e^{ax}\, dx = \frac{e^{ax}}{a^2}(ax - 1) + C$

112. $\int x^n e^{ax}\, dx = \frac{1}{a} x^n e^{ax} - \frac{n}{a} \int x^{n-1} e^{ax}\, dx$

113. $\int x^n b^{ax}\, dx = \frac{x^n b^{ax}}{a \ln b} - \frac{n}{a \ln b} \int x^{n-1} b^{ax}\, dx, \quad b > 0, b \neq 1$

114. $\int e^{ax} \sin bx\, dx = \frac{e^{ax}}{a^2 + b^2}(a \sin bx - b \cos bx) + C$

115. $\int e^{ax} \cos bx\, dx = \frac{e^{ax}}{a^2 + b^2}(a \cos bx + b \sin bx) + C$

116. $\int \ln ax\, dx = x \ln ax - x + C$

117. $\int x^n (\ln ax)^m\, dx = \frac{x^{n+1}(\ln ax)^m}{n+1} - \frac{m}{n+1} \int x^n (\ln ax)^{m-1}\, dx, \quad n \neq -1$

118. $\int x^{-1} (\ln ax)^m\, dx = \frac{(\ln ax)^{m+1}}{m+1} + C, \quad m \neq -1$

119. $\int \frac{dx}{x \ln ax} = \ln |\ln ax| + C$

Forms Involving $\sqrt{2ax - x^2}$, $a > 0$

120. $\int \frac{dx}{\sqrt{2ax - x^2}} = \arcsin\left(\frac{x-a}{a}\right) + C$

121. $\int \sqrt{2ax - x^2}\, dx = \frac{x-a}{2}\sqrt{2ax - x^2} + \frac{a^2}{2} \arcsin\left(\frac{x-a}{a}\right) + C$

122. $\int \left(\sqrt{2ax - x^2}\right)^n dx = \frac{(x-a)\left(\sqrt{2ax - x^2}\right)^n}{n+1} + \frac{na^2}{n+1} \int \left(\sqrt{2ax - x^2}\right)^{n-2} dx$

123. $\int \frac{dx}{\left(\sqrt{2ax - x^2}\right)^n} = \frac{(x-a)\left(\sqrt{2ax - x^2}\right)^{2-n}}{(n-2)a^2} + \frac{n-3}{(n-2)a^2} \int \frac{dx}{\left(\sqrt{2ax - x^2}\right)^{n-2}}$

124. $\int x\sqrt{2ax - x^2}\, dx = \frac{(x+a)(2x-3a)\sqrt{2ax - x^2}}{6} + \frac{a^3}{2} \arcsin\left(\frac{x-a}{a}\right) + C$

125. $\int \frac{\sqrt{2ax - x^2}}{x}\, dx = \sqrt{2ax - x^2} + a \arcsin\left(\frac{x-a}{a}\right) + C$

126. $\int \frac{\sqrt{2ax - x^2}}{x^2}\, dx = -2\sqrt{\frac{2a - x}{x}} - \arcsin\left(\frac{x-a}{a}\right) + C$

127. $\int \frac{x\, dx}{\sqrt{2ax - x^2}} = a \arcsin\left(\frac{x-a}{a}\right) - \sqrt{2ax - x^2} + C$

128. $\int \frac{dx}{x\sqrt{2ax - x^2}} = -\frac{1}{a}\sqrt{\frac{2a - x}{x}} + C$

Hyperbolic Forms

129. $\int \sinh ax\, dx = \frac{1}{a} \cosh ax + C$

130. $\int \cosh ax\, dx = \frac{1}{a} \sinh ax + C$

131. $\int \sinh^2 ax\, dx = \frac{\sinh 2ax}{4a} - \frac{x}{2} + C$

132. $\int \cosh^2 ax\, dx = \frac{\sinh 2ax}{4a} + \frac{x}{2} + C$

133. $\int \sinh^n ax\,dx = \frac{\sinh^{n-1} ax \cosh ax}{na} - \frac{n-1}{n}\int \sinh^{n-2} ax\,dx, \quad n \neq 0$

134. $\int \cosh^n ax\,dx = \frac{\cosh^{n-1} ax \sinh ax}{na} + \frac{n-1}{n}\int \cosh^{n-2} ax\,dx, \quad n \neq 0$

135. $\int x \sinh ax\,dx = \frac{x}{a}\cosh ax - \frac{1}{a^2}\sinh ax + C$

136. $\int x \cosh ax\,dx = \frac{x}{a}\sinh ax - \frac{1}{a^2}\cosh ax + C$

137. $\int x^n \sinh ax\,dx = \frac{x^n}{a}\cosh ax - \frac{n}{a}\int x^{n-1}\cosh ax\,dx$

138. $\int x^n \cosh ax\,dx = \frac{x^n}{a}\sinh ax - \frac{n}{a}\int x^{n-1}\sinh ax\,dx$

139. $\int \tanh ax\,dx = \frac{1}{a}\ln(\cosh ax) + C$

140. $\int \coth ax\,dx = \frac{1}{a}\ln|\sinh ax| + C$

141. $\int \tanh^2 ax\,dx = x - \frac{1}{a}\tanh ax + C$

142. $\int \coth^2 ax\,dx = x - \frac{1}{a}\coth ax + C$

143. $\int \tanh^n ax\,dx = -\frac{\tanh^{n-1} ax}{(n-1)a} + \int \tanh^{n-2} ax\,dx, \quad n \neq 1$

144. $\int \coth^n ax\,dx = -\frac{\coth^{n-1} ax}{(n-1)a} + \int \coth^{n-2} ax\,dx, \quad n \neq 1$

145. $\int \operatorname{sech} ax\,dx = \frac{1}{a}\arcsin(\tanh ax) + C$

146. $\int \operatorname{csch} ax\,dx = \frac{1}{a}\ln\left|\tanh\frac{ax}{2}\right| + C$

147. $\int \operatorname{sech}^2 ax\,dx = \frac{1}{a}\tanh ax + C$

148. $\int \operatorname{csch}^2 ax\,dx = -\frac{1}{a}\coth ax + C$

149. $\int \operatorname{sech}^n ax\,dx = \frac{\operatorname{sech}^{n-2} ax \tanh ax}{(n-1)a} + \frac{n-2}{n-1}\int \operatorname{sech}^{n-2} ax\,dx, \quad n \neq 1$

150. $\int \operatorname{csch}^n ax\,dx = -\frac{\operatorname{csch}^{n-2} ax \coth ax}{(n-1)a} - \frac{n-2}{n-1}\int \operatorname{csch}^{n-2} ax\,dx, \quad n \neq 1$

151. $\int \operatorname{sech}^n ax \tanh ax\,dx = -\frac{\operatorname{sech}^n ax}{na} + C, \quad n \neq 0$

152. $\int \operatorname{csch}^n ax \coth ax\,dx = -\frac{\operatorname{csch}^n ax}{na} + C, \quad n \neq 0$

153. $\int e^{ax}\sinh bx\,dx = \frac{e^{ax}}{2}\left[\frac{e^{bx}}{a+b} - \frac{e^{-bx}}{a-b}\right] + C, \quad a^2 \neq b^2$

154. $\int e^{ax}\cosh bx\,dx = \frac{e^{ax}}{2}\left[\frac{e^{bx}}{a+b} + \frac{e^{-bx}}{a-b}\right] + C, \quad a^2 \neq b^2$

Some Definite Integrals

155. $\int_0^\infty x^{n-1}e^{-x}\,dx = \Gamma(n) = (n-1)!, \quad n > 0$

156. $\int_0^\infty e^{-ax^2}\,dx = \frac{1}{2}\sqrt{\frac{\pi}{a}}, \quad a > 0$

157. $\int_0^{\pi/2} \sin^n x\,dx = \int_0^{\pi/2} \cos^n x\,dx = \begin{cases} \frac{1\cdot 3\cdot 5\cdot \cdots \cdot (n-1)}{2\cdot 4\cdot 6\cdot \cdots \cdot n}\cdot\frac{\pi}{2}, & \text{if } n \text{ is an even integer } \geq 2 \\ \frac{2\cdot 4\cdot 6\cdot \cdots \cdot (n-1)}{3\cdot 5\cdot 7\cdot \cdots \cdot n}, & \text{if } n \text{ is an odd integer } \geq 3 \end{cases}$

Credits

Chapter 1: Page 1, Chapter 1 opening photo, Sylverarts Vectors/Shutterstock

Chapter 2: Page 56, Chapter 2 opening photo, Paper Wings/Shutterstock

Chapter 3: Page 102, Chapter 3 opening photo, Struvictory/Shutterstock

Chapter 4: Page 180, Chapter 4 opening photo, Unconventional/Shutterstock

Chapter 5: Page 246, Chapter 5 opening photo, CROCOTHERY/Shutterstock

Chapter 6: Page 311, Chapter 6 opening photo, nuchao/Shutterstock;
Page 350, Berenice Abbott/Science Source

Chapter 7: Page 366, Chapter 7 opening photo, Visible Pixels/Shutterstock

Chapter 8: Page 457, Chapter 8 opening photo, inter reality/Shutterstock

Chapter 9: Page 535, Chapter 9 opening photo, Cgicol/Shutterstock

Chapter 10: Page 573, Chapter 10 opening photo, Bill Heller/Shutterstock;
Page 589, Berenice Abbott/Science Source

Chapter 11: Page 659, Chapter 11 opening photo, Quardia/Shutterstock

Chapter 12: Page 710, Chapter 12 opening photo, korkeng/Shutterstock

Chapter 13: Page 760, Chapter 13 opening photo, Woman from Baku/Shutterstock

Chapter 14: Page 802, Chapter 14 opening photo, Theus/Shutterstock

Chapter 15: Page 894, Chapter 15 opening photo, genach_man/Shutterstock

Chapter 16: Page 965, Chapter 16 opening photo, genach_man/Shutterstock

Trigonometry Formulas

Definitions and Fundamental Identities

Sine: $\sin\theta = \dfrac{y}{r} = \dfrac{1}{\csc\theta}$

Cosine: $\cos\theta = \dfrac{x}{r} = \dfrac{1}{\sec\theta}$

Tangent: $\tan\theta = \dfrac{y}{x} = \dfrac{1}{\cot\theta}$

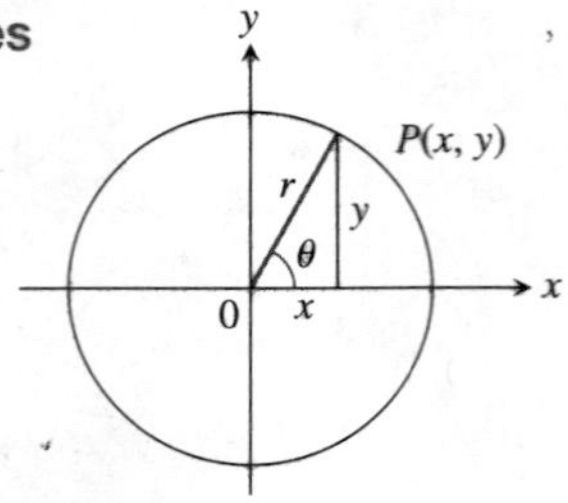

Identities

$\sin(-\theta) = -\sin\theta, \quad \cos(-\theta) = \cos\theta$

$\sin^2\theta + \cos^2\theta = 1, \quad \sec^2\theta = 1 + \tan^2\theta, \quad \csc^2\theta = 1 + \cot^2\theta$

$\sin 2\theta = 2\sin\theta\cos\theta, \quad \cos 2\theta = \cos^2\theta - \sin^2\theta$

$\cos^2\theta = \dfrac{1+\cos 2\theta}{2}, \quad \sin^2\theta = \dfrac{1-\cos 2\theta}{2}$

$\sin(A+B) = \sin A\cos B + \cos A\sin B$

$\sin(A-B) = \sin A\cos B - \cos A\sin B$

$\cos(A+B) = \cos A\cos B - \sin A\sin B$

$\cos(A-B) = \cos A\cos B + \sin A\sin B$

$\tan(A+B) = \dfrac{\tan A + \tan B}{1 - \tan A\tan B}$

$\tan(A-B) = \dfrac{\tan A - \tan B}{1 + \tan A\tan B}$

$\sin\left(A - \dfrac{\pi}{2}\right) = -\cos A, \quad \cos\left(A - \dfrac{\pi}{2}\right) = \sin A$

$\sin\left(A + \dfrac{\pi}{2}\right) = \cos A, \quad \cos\left(A + \dfrac{\pi}{2}\right) = -\sin A$

$\sin A\sin B = \dfrac{1}{2}\cos(A-B) - \dfrac{1}{2}\cos(A+B)$

$\cos A\cos B = \dfrac{1}{2}\cos(A-B) + \dfrac{1}{2}\cos(A+B)$

$\sin A\cos B = \dfrac{1}{2}\sin(A-B) + \dfrac{1}{2}\sin(A+B)$

$\sin A + \sin B = 2\sin\dfrac{1}{2}(A+B)\cos\dfrac{1}{2}(A-B)$

$\sin A - \sin B = 2\cos\dfrac{1}{2}(A+B)\sin\dfrac{1}{2}(A-B)$

$\cos A + \cos B = 2\cos\dfrac{1}{2}(A+B)\cos\dfrac{1}{2}(A-B)$

$\cos A - \cos B = -2\sin\dfrac{1}{2}(A+B)\sin\dfrac{1}{2}(A-B)$

Trigonometric Functions

Radian Measure

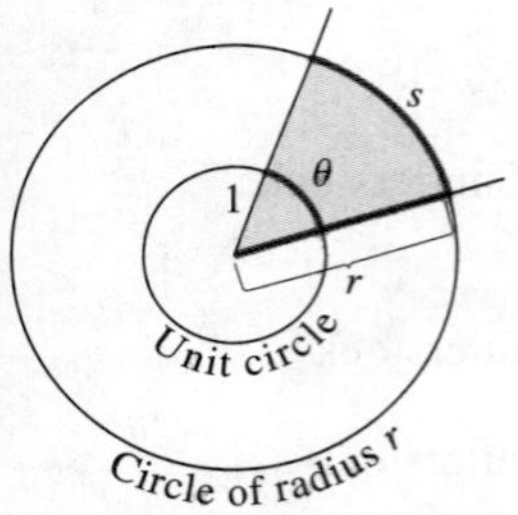

$\dfrac{s}{r} = \dfrac{\theta}{1} = \theta$ or $\theta = \dfrac{s}{r}$,

$180° = \pi$ radians.

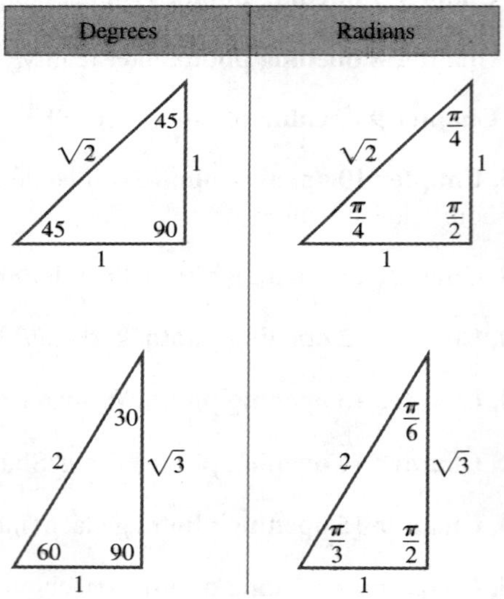

The angles of two common triangles, in degrees and radians.

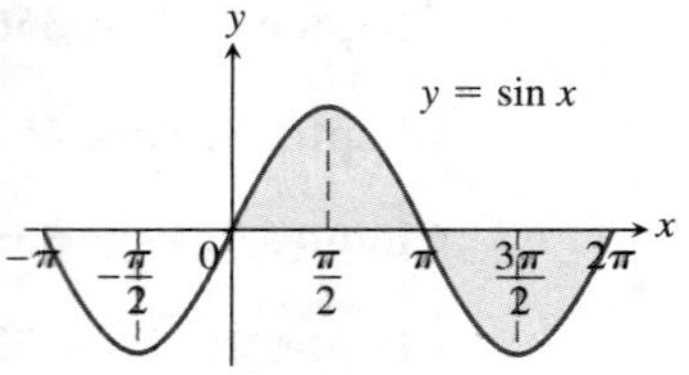

Domain: $(-\infty, \infty)$
Range: $[-1, 1]$

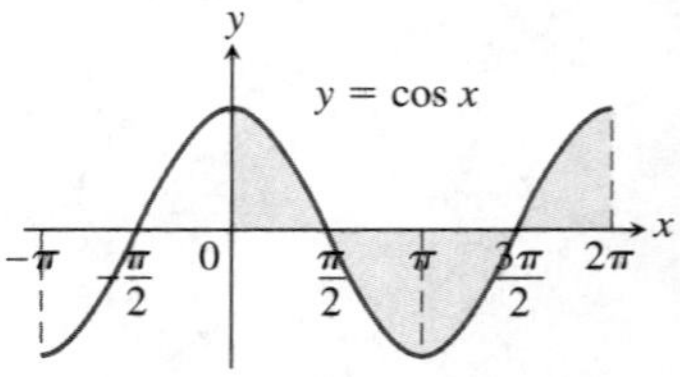

Domain: $(-\infty, \infty)$
Range: $[-1, 1]$

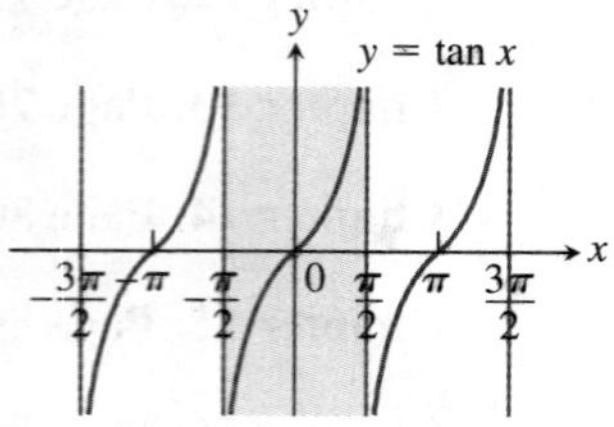

Domain: All real numbers except odd integer multiples of $\pi/2$
Range: $(-\infty, \infty)$

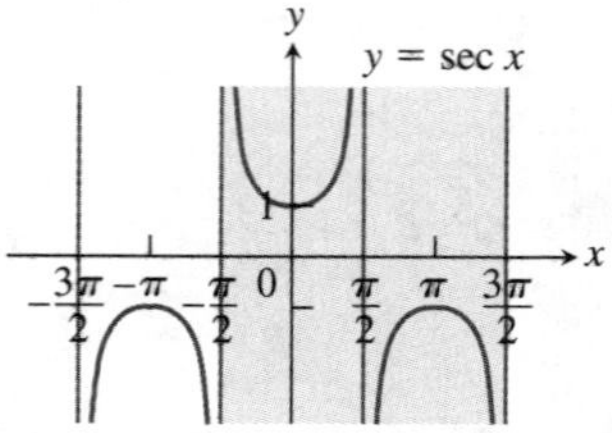

Domain: All real numbers except odd integer multiples of $\pi/2$
Range: $(-\infty, -1] \cup [1, \infty)$

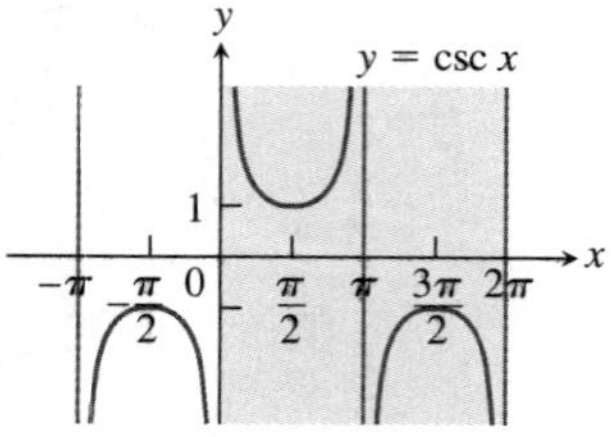

Domain: $x \neq 0, \pm\pi, \pm 2\pi, \ldots$
Range: $(-\infty, -1] \cup [1, \infty)$

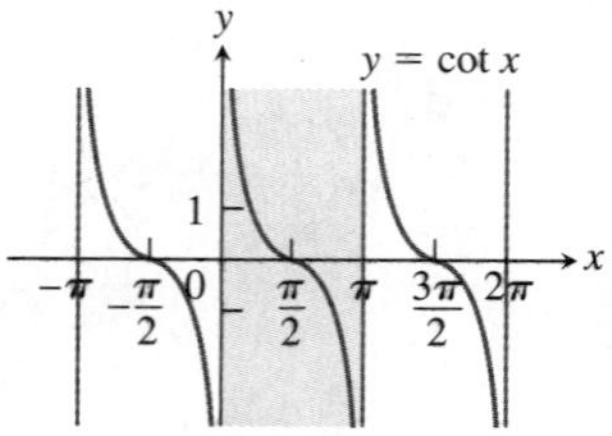

Domain: $x \neq 0, \pm\pi, \pm 2\pi, \ldots$
Range: $(-\infty, \infty)$

Series

Tests for Convergence of Infinite Series

1. **The nth-Term Test:** Unless $a_n \to 0$, the series diverges.
2. **Geometric series:** $\sum ar^n$ converges if $|r| < 1$; otherwise it diverges.
3. **p-series:** $\sum 1/n^p$ converges if $p > 1$; otherwise it diverges.
4. **Series with nonnegative terms:** Try the Integral Test, Ratio Test, or Root Test. Try comparing to a known series with the Comparison Test or the Limit Comparison Test.
5. **Series with some negative terms:** Does $\sum |a_n|$ converge? If yes, so does $\sum a_n$ because absolute convergence implies convergence.
6. **Alternating series:** $\sum a_n$ converges if the series satisfies the conditions of the Alternating Series Test.

Taylor Series

$$\frac{1}{1-x} = 1 + x + x^2 + \cdots + x^n + \cdots = \sum_{n=0}^{\infty} x^n, \qquad |x| < 1$$

$$\frac{1}{1+x} = 1 - x + x^2 - \cdots + (-x)^n + \cdots = \sum_{n=0}^{\infty} (-1)^n x^n, \qquad |x| < 1$$

$$e^x = 1 + x + \frac{x^2}{2!} + \cdots + \frac{x^n}{n!} + \cdots = \sum_{n=0}^{\infty} \frac{x^n}{n!}, \qquad |x| < \infty$$

$$\sin x = x - \frac{x^3}{3!} + \frac{x^5}{5!} - \cdots + (-1)^n \frac{x^{2n+1}}{(2n+1)!} + \cdots = \sum_{n=0}^{\infty} \frac{(-1)^n x^{2n+1}}{(2n+1)!}, \qquad |x| < \infty$$

$$\cos x = 1 - \frac{x^2}{2!} + \frac{x^4}{4!} - \cdots + (-1)^n \frac{x^{2n}}{(2n)!} + \cdots = \sum_{n=0}^{\infty} \frac{(-1)^n x^{2n}}{(2n)!}, \qquad |x| < \infty$$

$$\ln(1+x) = x - \frac{x^2}{2} + \frac{x^3}{3} - \cdots + (-1)^{n-1}\frac{x^n}{n} + \cdots = \sum_{n=1}^{\infty} \frac{(-1)^{n-1} x^n}{n}, \qquad -1 < x \le 1$$

$$\ln\frac{1+x}{1-x} = 2\tanh^{-1} x = 2\left(x + \frac{x^3}{3} + \frac{x^5}{5} + \cdots + \frac{x^{2n+1}}{2n+1} + \cdots\right) = 2\sum_{n=0}^{\infty} \frac{x^{2n+1}}{2n+1}, \qquad |x| < 1$$

$$\arctan x = x - \frac{x^3}{3} + \frac{x^5}{5} - \cdots + (-1)^n \frac{x^{2n+1}}{2n+1} + \cdots = \sum_{n=0}^{\infty} \frac{(-1)^n x^{2n+1}}{2n+1}, \qquad |x| \le 1$$

Binomial Series

$$(1+x)^m = 1 + mx + \frac{m(m-1)x^2}{2!} + \frac{m(m-1)(m-2)x^3}{3!} + \cdots + \frac{m(m-1)(m-2)\cdots(m-k+1)x^k}{k!} + \cdots$$
$$= 1 + \sum_{k=1}^{\infty} \binom{m}{k} x^k, \qquad |x| < 1,$$

where

$$\binom{m}{1} = m, \qquad \binom{m}{2} = \frac{m(m-1)}{2!}, \qquad \binom{m}{k} = \frac{m(m-1)\cdots(m-k+1)}{k!} \qquad \text{for } k \ge 3.$$

Vector Operator Formulas (Cartesian Form)

Formulas for Grad, Div, Curl, and the Laplacian

Cartesian (x, y, z) **i**, **j**, and **k** are unit vectors in the directions of increasing x, y, and z. M, N, and P are the scalar components of $\mathbf{F}(x, y, z)$ in these directions.

Gradient $$\nabla f = \frac{\partial f}{\partial x}\mathbf{i} + \frac{\partial f}{\partial y}\mathbf{j} + \frac{\partial f}{\partial z}\mathbf{k}$$

Divergence $$\nabla \cdot \mathbf{F} = \frac{\partial M}{\partial x} + \frac{\partial N}{\partial y} + \frac{\partial P}{\partial z}$$

Curl $$\nabla \times \mathbf{F} = \begin{vmatrix} \mathbf{i} & \mathbf{j} & \mathbf{k} \\ \frac{\partial}{\partial x} & \frac{\partial}{\partial y} & \frac{\partial}{\partial z} \\ M & N & P \end{vmatrix}$$

Laplacian $$\nabla^2 f = \frac{\partial^2 f}{\partial x^2} + \frac{\partial^2 f}{\partial y^2} + \frac{\partial^2 f}{\partial z^2}$$

Vector Triple Products

$$(\mathbf{u} \times \mathbf{v}) \cdot \mathbf{w} = (\mathbf{v} \times \mathbf{w}) \cdot \mathbf{u} = (\mathbf{w} \times \mathbf{u}) \cdot \mathbf{v}$$

$$\mathbf{u} \times (\mathbf{v} \times \mathbf{w}) = (\mathbf{u} \cdot \mathbf{w})\mathbf{v} - (\mathbf{u} \cdot \mathbf{v})\mathbf{w}$$

The Fundamental Theorem of Line Integrals

Part 1 Let $\mathbf{F} = M\mathbf{i} + N\mathbf{j} + P\mathbf{k}$ be a vector field whose components are continuous throughout an open connected region D in space. Then there exists a differentiable function f such that

$$\mathbf{F} = \nabla f = \frac{\partial f}{\partial x}\mathbf{i} + \frac{\partial f}{\partial y}\mathbf{j} + \frac{\partial f}{\partial z}\mathbf{k}$$

if and only if, for all points A and B in D, the value of $\int_A^B \mathbf{F} \cdot d\mathbf{r}$ is independent of the path joining A to B in D.

Part 2 If the integral is independent of the path from A to B, its value is

$$\int_A^B \mathbf{F} \cdot d\mathbf{r} = f(B) - f(A).$$

Green's Theorem and Its Generalization to Three Dimensions

Tangential form of Green's Theorem: $$\oint_C \mathbf{F} \cdot \mathbf{T}\, ds = \iint_R (\nabla \times \mathbf{F}) \cdot \mathbf{k}\, dA$$

Stokes' Theorem: $$\oint_C \mathbf{F} \cdot \mathbf{T}\, ds = \iint_S (\nabla \times \mathbf{F}) \cdot \mathbf{n}\, d\sigma$$

Normal form of Green's Theorem: $$\oint_C \mathbf{F} \cdot \mathbf{n}\, ds = \iint_R (\nabla \cdot \mathbf{F})\, dA$$

Divergence Theorem: $$\iint_S \mathbf{F} \cdot \mathbf{n}\, d\sigma = \iiint_D \nabla \cdot \mathbf{F}\, dV$$

Vector Identities

In the identities here, f and g are differentiable scalar functions; $\mathbf{F}$, $\mathbf{F}_1$, and $\mathbf{F}_2$ are differentiable vector fields; and a and b are real constants.

$$\nabla \times (\nabla f) = \mathbf{0}$$

$$\nabla(fg) = f\nabla g + g\nabla f$$

$$\nabla \cdot (g\mathbf{F}) = g\nabla \cdot \mathbf{F} + \nabla g \cdot \mathbf{F}$$

$$\nabla \times (g\mathbf{F}) = g\nabla \times \mathbf{F} + \nabla g \times \mathbf{F}$$

$$\nabla \cdot (a\mathbf{F}_1 + b\mathbf{F}_2) = a\nabla \cdot \mathbf{F}_1 + b\nabla \cdot \mathbf{F}_2$$

$$\nabla \times (a\mathbf{F}_1 + b\mathbf{F}_2) = a\nabla \times \mathbf{F}_1 + b\nabla \times \mathbf{F}_2$$

$$\nabla(\mathbf{F}_1 \cdot \mathbf{F}_2) = (\mathbf{F}_1 \cdot \nabla)\mathbf{F}_2 + (\mathbf{F}_2 \cdot \nabla)\mathbf{F}_1 + \mathbf{F}_1 \times (\nabla \times \mathbf{F}_2) + \mathbf{F}_2 \times (\nabla \times \mathbf{F}_1)$$

$$\nabla \cdot (\mathbf{F}_1 \times \mathbf{F}_2) = \mathbf{F}_2 \cdot (\nabla \times \mathbf{F}_1) - \mathbf{F}_1 \cdot (\nabla \times \mathbf{F}_2)$$

$$\nabla \times (\mathbf{F}_1 \times \mathbf{F}_2) = (\mathbf{F}_2 \cdot \nabla)\mathbf{F}_1 - (\mathbf{F}_1 \cdot \nabla)\mathbf{F}_2 + (\nabla \cdot \mathbf{F}_2)\mathbf{F}_1 - (\nabla \cdot \mathbf{F}_1)\mathbf{F}_2$$

$$\nabla \times (\nabla \times \mathbf{F}) = \nabla(\nabla \cdot \mathbf{F}) - (\nabla \cdot \nabla)\mathbf{F} = \nabla(\nabla \cdot \mathbf{F}) - \nabla^2\mathbf{F}$$

$$(\nabla \times \mathbf{F}) \times \mathbf{F} = (\mathbf{F} \cdot \nabla)\mathbf{F} - \frac{1}{2}\nabla(\mathbf{F} \cdot \mathbf{F})$$

Basic Algebra Formulas

Arithmetic Operations

$$a(b+c) = ab + ac, \qquad \frac{a}{b}\cdot\frac{c}{d} = \frac{ac}{bd}$$

$$\frac{a}{b} + \frac{c}{d} = \frac{ad+bc}{bd}, \qquad \frac{a/b}{c/d} = \frac{a}{b}\cdot\frac{d}{c}$$

Laws of Signs

$$-(-a) = a, \qquad \frac{-a}{b} = -\frac{a}{b} = \frac{a}{-b}$$

Zero Division by zero is not defined.

$$\text{If } a \neq 0\text{:} \quad \frac{0}{a} = 0, \quad a^0 = 1, \quad 0^a = 0$$

For any number a: $a \cdot 0 = 0 \cdot a = 0$

Laws of Exponents

$$a^m a^n = a^{m+n}, \qquad (ab)^m = a^m b^m, \qquad (a^m)^n = a^{mn}, \qquad a^{m/n} = \sqrt[n]{a^m} = \left(\sqrt[n]{a}\right)^m$$

If $a \neq 0$, then

$$\frac{a^m}{a^n} = a^{m-n}, \qquad a^0 = 1, \qquad a^{-m} = \frac{1}{a^m}.$$

The Binomial Theorem For any positive integer n,

$$(a+b)^n = a^n + na^{n-1}b + \frac{n(n-1)}{1\cdot 2}a^{n-2}b^2 + \frac{n(n-1)(n-2)}{1\cdot 2\cdot 3}a^{n-3}b^3 + \cdots + nab^{n-1} + b^n.$$

For instance,

$$(a+b)^2 = a^2 + 2ab + b^2, \qquad (a-b)^2 = a^2 - 2ab + b^2$$

$$(a+b)^3 = a^3 + 3a^2b + 3ab^2 + b^3, \qquad (a-b)^3 = a^3 - 3a^2b + 3ab^2 - b^3.$$

Factoring the Difference of Like Integer Powers, $n > 1$

$$a^n - b^n = (a-b)(a^{n-1} + a^{n-2}b + a^{n-3}b^2 + \cdots + ab^{n-2} + b^{n-1})$$

For instance,

$$a^2 - b^2 = (a-b)(a+b),$$
$$a^3 - b^3 = (a-b)(a^2 + ab + b^2),$$
$$a^4 - b^4 = (a-b)(a^3 + a^2b + ab^2 + b^3).$$

Completing the Square If $a \neq 0$, then

$$ax^2 + bx + c = au^2 + C \qquad \left(u = x + (b/2a),\ C = c - \frac{b^2}{4a}\right).$$

The Quadratic Formula

If $a \neq 0$ and $ax^2 + bx + c = 0$, then

$$x = \frac{-b \pm \sqrt{b^2 - 4ac}}{2a}.$$

Geometry Formulas

A = area, B = area of base, C = circumference, S = surface area, V = volume

Triangle

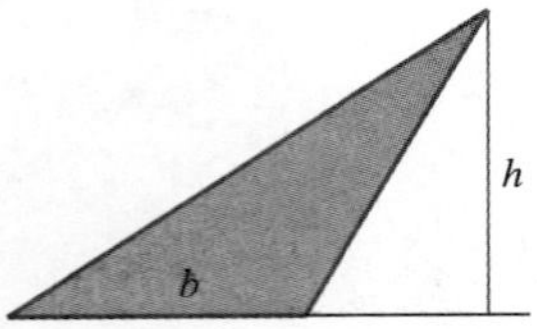

$A = \frac{1}{2}bh$

Similar Triangles

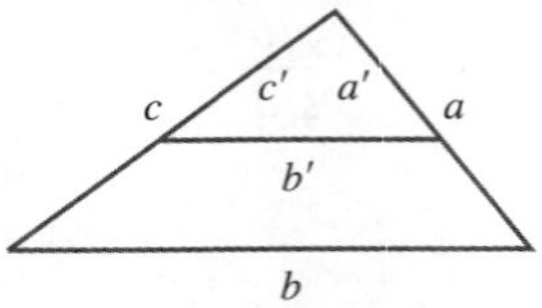

$\frac{a'}{a} = \frac{b'}{b} = \frac{c'}{c}$

Pythagorean Theorem

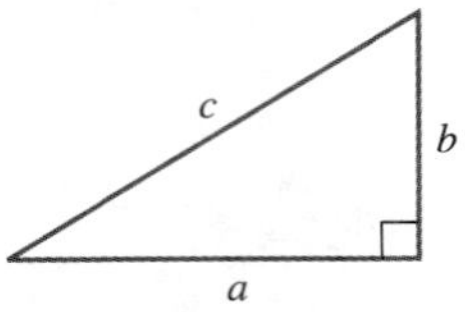

$a^2 + b^2 = c^2$

Parallelogram

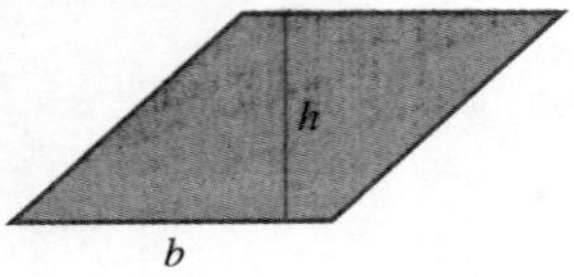

$A = bh$

Trapezoid

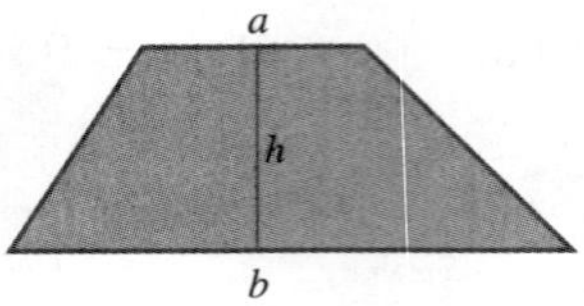

$A = \frac{1}{2}(a + b)h$

Circle

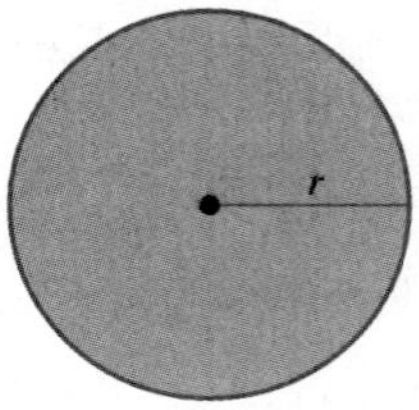

$A = \pi r^2,$
$C = 2\pi r$

Any Cylinder or Prism with Parallel Bases

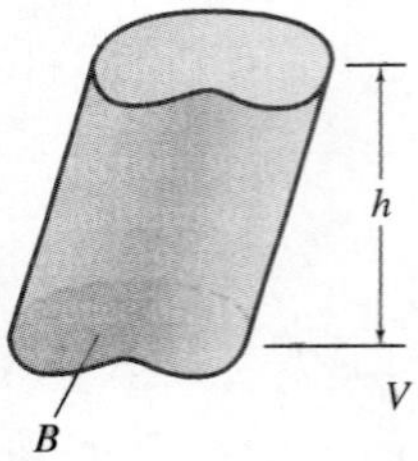

$V = Bh$

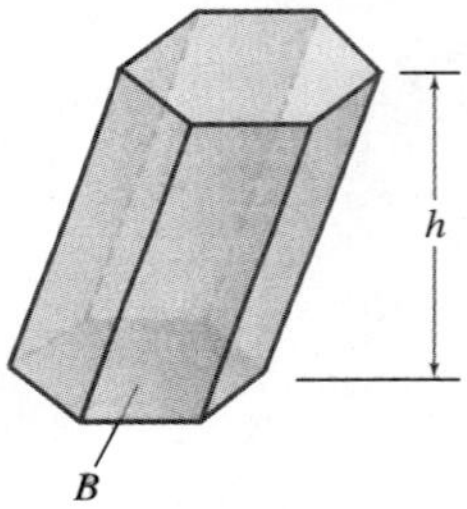

Right Circular Cylinder

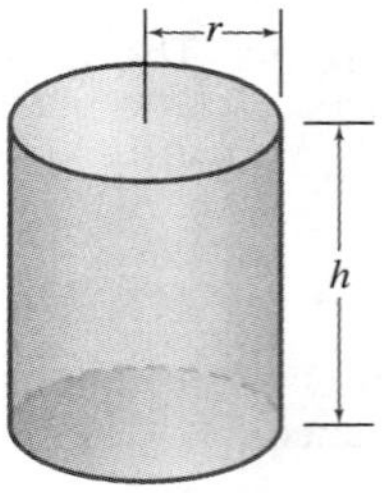

$V = \pi r^2 h$
$S = 2\pi rh$ = Area of side

Any Cone or Pyramid

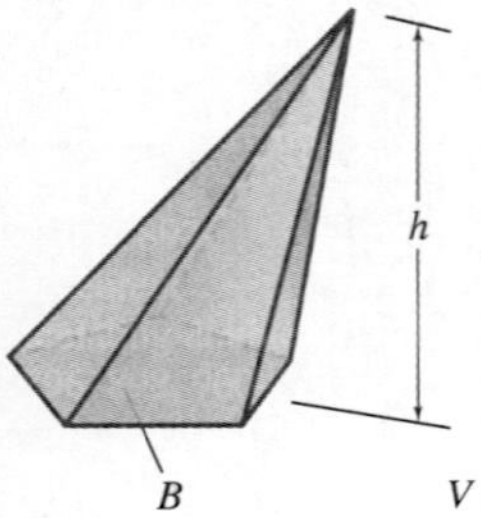

$V = \frac{1}{3}Bh$

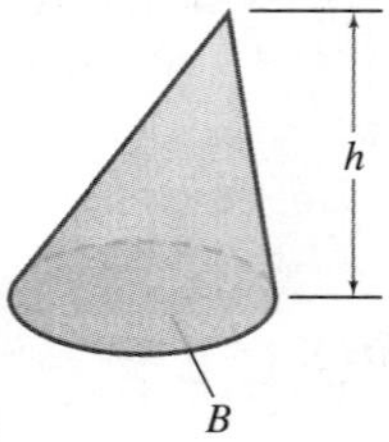

Right Circular Cone

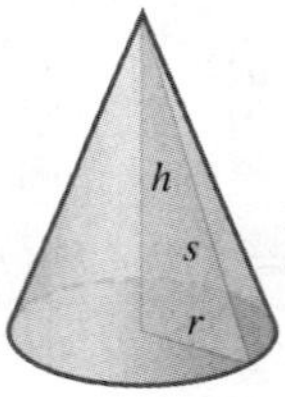

$V = \frac{1}{3}\pi r^2 h$
$S = \pi rs$ = Area of side

Sphere

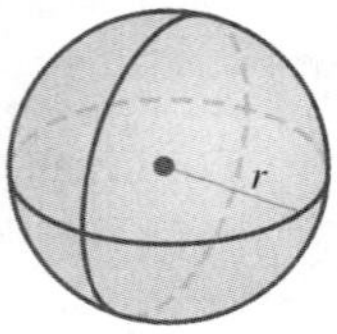

$V = \frac{4}{3}\pi r^3$, $S = 4\pi r^2$

Limits

General Laws

If L, M, c, and k are real numbers and

$$\lim_{x \to c} f(x) = L \quad \text{and} \quad \lim_{x \to c} g(x) = M, \quad \text{then}$$

Sum Rule: $\lim_{x \to c} (f(x) + g(x)) = L + M$

Difference Rule: $\lim_{x \to c} (f(x) - g(x)) = L - M$

Product Rule: $\lim_{x \to c} (f(x) \cdot g(x)) = L \cdot M$

Constant Multiple Rule: $\lim_{x \to c} (k \cdot f(x)) = k \cdot L$

Quotient Rule: $\lim_{x \to c} \dfrac{f(x)}{g(x)} = \dfrac{L}{M}, \quad M \neq 0$

The Sandwich Theorem

If $g(x) \leq f(x) \leq h(x)$ in an open interval containing c, except possibly at $x = c$, and if

$$\lim_{x \to c} g(x) = \lim_{x \to c} h(x) = L,$$

then $\lim_{x \to c} f(x) = L$.

Inequalities

If $f(x) \leq g(x)$ in an open interval containing c, except possibly at $x = c$, and both limits exist, then

$$\lim_{x \to c} f(x) \leq \lim_{x \to c} g(x).$$

Continuity

If g is continuous at L and $\lim_{x \to c} f(x) = L$, then

$$\lim_{x \to c} g(f(x)) = g(L).$$

Specific Formulas

If $P(x) = a_n x^n + a_{n-1} x^{n-1} + \cdots + a_0$, then

$$\lim_{x \to c} P(x) = P(c) = a_n c^n + a_{n-1} c^{n-1} + \cdots + a_0.$$

If $P(x)$ and $Q(x)$ are polynomials and $Q(c) \neq 0$, then

$$\lim_{x \to c} \frac{P(x)}{Q(x)} = \frac{P(c)}{Q(c)}.$$

If $f(x)$ is continuous at $x = c$, then

$$\lim_{x \to c} f(x) = f(c).$$

$$\lim_{x \to 0} \frac{\sin x}{x} = 1 \quad \text{and} \quad \lim_{x \to 0} \frac{1 - \cos x}{x} = 0$$

L'Hôpital's Rule

If $f(a) = g(a) = 0$, both f' and g' exist in an open interval I containing a, and $g'(x) \neq 0$ on I if $x \neq a$, then

$$\lim_{x \to a} \frac{f(x)}{g(x)} = \lim_{x \to a} \frac{f'(x)}{g'(x)},$$

assuming the limit on the right side exists.

Differentiation Rules

General Formulas

Assume u and v are differentiable functions of x.

Constant: $\frac{d}{dx}(c) = 0$

Sum: $\frac{d}{dx}(u + v) = \frac{du}{dx} + \frac{dv}{dx}$

Difference: $\frac{d}{dx}(u - v) = \frac{du}{dx} - \frac{dv}{dx}$

Constant Multiple: $\frac{d}{dx}(cu) = c\frac{du}{dx}$

Product: $\frac{d}{dx}(uv) = u\frac{dv}{dx} + \frac{du}{dx}v$

Quotient: $\frac{d}{dx}\left(\frac{u}{v}\right) = \frac{v\frac{du}{dx} - u\frac{dv}{dx}}{v^2}$

Power: $\frac{d}{dx}x^n = nx^{n-1}$

Chain Rule: $\frac{d}{dx}(f(g(x))) = f'(g(x)) \cdot g'(x)$

Trigonometric Functions

$\frac{d}{dx}(\sin x) = \cos x$ $\qquad$ $\frac{d}{dx}(\cos x) = -\sin x$

$\frac{d}{dx}(\tan x) = \sec^2 x$ $\qquad$ $\frac{d}{dx}(\sec x) = \sec x \tan x$

$\frac{d}{dx}(\cot x) = -\csc^2 x$ $\qquad$ $\frac{d}{dx}(\csc x) = -\csc x \cot x$

Exponential and Logarithmic Functions

$\frac{d}{dx}e^x = e^x$ $\qquad$ $\frac{d}{dx}\ln x = \frac{1}{x}$

$\frac{d}{dx}a^x = a^x \ln a$ $\qquad$ $\frac{d}{dx}(\log_a x) = \frac{1}{x \ln a}$

Inverse Trigonometric Functions

$\frac{d}{dx}(\arcsin x) = \frac{1}{\sqrt{1 - x^2}}$ $\qquad$ $\frac{d}{dx}(\arccos x) = -\frac{1}{\sqrt{1 - x^2}}$

$\frac{d}{dx}(\arctan x) = \frac{1}{1 + x^2}$ $\qquad$ $\frac{d}{dx}(\text{arcsec}\, x) = \frac{1}{|x|\sqrt{x^2 - 1}}$

$\frac{d}{dx}(\text{arccot}\, x) = -\frac{1}{1 + x^2}$ $\qquad$ $\frac{d}{dx}(\text{arccsc}\, x) = -\frac{1}{|x|\sqrt{x^2 - 1}}$

Hyperbolic Functions

$\frac{d}{dx}(\sinh x) = \cosh x$ $\qquad$ $\frac{d}{dx}(\cosh x) = \sinh x$

$\frac{d}{dx}(\tanh x) = \text{sech}^2 x$ $\qquad$ $\frac{d}{dx}(\text{sech}\, x) = -\text{sech}\, x \tanh x$

$\frac{d}{dx}(\coth x) = -\text{csch}^2 x$ $\qquad$ $\frac{d}{dx}(\text{csch}\, x) = -\text{csch}\, x \coth x$

Inverse Hyperbolic Functions

$\frac{d}{dx}(\sinh^{-1} x) = \frac{1}{\sqrt{1 + x^2}}$ $\qquad$ $\frac{d}{dx}(\cosh^{-1} x) = \frac{1}{\sqrt{x^2 - 1}}$

$\frac{d}{dx}(\tanh^{-1} x) = \frac{1}{1 - x^2}$ $\qquad$ $\frac{d}{dx}(\text{sech}^{-1} x) = -\frac{1}{x\sqrt{1 - x^2}}$

$\frac{d}{dx}(\coth^{-1} x) = \frac{1}{1 - x^2}$ $\qquad$ $\frac{d}{dx}(\text{csch}^{-1} x) = -\frac{1}{|x|\sqrt{1 + x^2}}$

Parametric Equations

If $x = f(t)$ and $y = g(t)$ are differentiable, then

$$y' = \frac{dy}{dx} = \frac{dy/dt}{dx/dt} \quad \text{and} \quad \frac{d^2y}{dx^2} = \frac{dy'/dt}{dx/dt}.$$

Integration Rules

General Formulas

Zero: $\int_a^a f(x)\,dx = 0$

Order of Integration: $\int_b^a f(x)\,dx = -\int_a^b f(x)\,dx$

Constant Multiples: $\int_a^b kf(x)\,dx = k\int_a^b f(x)\,dx, \quad k \text{ any number}$

$$\int_a^b -f(x)\,dx = -\int_a^b f(x)\,dx, \quad k = -1$$

Sums and Differences: $\int_a^b (f(x) \pm g(x))\,dx = \int_a^b f(x)\,dx \pm \int_a^b g(x)\,dx$

Additivity: $\int_a^b f(x)\,dx + \int_b^c f(x)\,dx = \int_a^c f(x)\,dx$

Max-Min Inequality: If max f and min f are the maximum and minimum values of f on $[a, b]$, then

$$\min f \cdot (b - a) \le \int_a^b f(x)\,dx \le \max f \cdot (b - a).$$

Domination: $f(x) \ge g(x)$ on $[a, b]$ implies $\int_a^b f(x)\,dx \ge \int_a^b g(x)\,dx.$

$$f(x) \ge 0 \text{ on } [a, b] \text{ implies } \int_a^b f(x)\,dx \ge 0.$$

The Fundamental Theorem of Calculus

Part 1 If f is continuous on $[a, b]$, then $F(x) = \int_a^x f(t)\,dt$ is continuous on $[a, b]$ and differentiable on (a, b), and its derivative is $f(x)$:

$$F'(x) = \frac{d}{dx}\int_a^x f(t)\,dt = f(x).$$

Part 2 If f is continuous at every point of $[a, b]$ and F is any antiderivative of f on $[a, b]$, then

$$\int_a^b f(x)\,dx = F(b) - F(a).$$

Substitution in Definite Integrals

$$\int_a^b f(g(x)) \cdot g'(x)\,dx = \int_{g(a)}^{g(b)} f(u)\,du$$

Integration by Parts

$$\int_a^b u(x)v'(x)\,dx = u(x)v(x)\Big]_a^b - \int_a^b v(x)u'(x)\,dx$$

A Brief Table of Integrals follows the Index at the back of the text.